Index of Applications

Business and Economics

General Interest

Life Sciences

(continued at back of book)

APPLIED CALCULUS

SEVENTH EDITION

Stefan Waner
Hofstra University

Steven R. Costenoble
Hofstra University

Australia • Brazil • Mexico • Singapore • United Kingdom • United States

***Applied Calculus*, Seventh Edition**
Stefan Waner, Steven R. Costenoble

Product Director: Terry Boyle
Product Manager: Rita Lombard
Content Developer: Morgan Mendoza
Product Assistant: Abby DeVeuve
Content Digitization Lead: Justin Karr
Marketing Manager: Ana Albinson
Content Project Manager: Teresa L. Trego
Art Director: Vernon Boes
Manufacturing Planner: Rebecca Cross
Production Service: Martha Emry BookCraft
Photo Researcher: Lumina Datamatics
Text Researcher: Lumina Datamatics
Text Designer: Diane Beasley
Cover Designer: Irene Morris
Cover Images: Large Image © CERN; Small Photo © Master_Andrii/ Shutterstock.com
Compositor: Graphic World, Inc.

Library of Congress Control Number: 2016952198

Student Edition:
ISBN: 978-1-337-29124-8

Loose-leaf Edition:
ISBN: 978-1-337-29140-8

Cengage
200 Pier 4 Boulevard
Boston, MA 02210
USA

Cengage is a leading provider of customized learning solutions with employees residing in nearly 40 different countries and sales in more than 125 countries around the world. Find your local representative at **www.cengage.com.**

Cengage products are represented in Canada by Nelson Education, Ltd.

To learn more about Cengage platforms and services, register or access your online learning solution, or purchase materials for your course, visit **www.cengage.com**.

About the Cover

The cover shows the liquid argon calorimeter for the ATLAS (a toroidal LHC apparatus) high-energy particle experiment at the CERN Large Hadron Collider in Geneva, Switzerland. The ongoing ATLAS experiment is one of two experiments that led to the discovery of the Higgs boson at CERN in 2012. CERN is the European Organization for Nuclear Research where physicists and engineers are probing the fundamental structure of the universe.

Printed in Mexico
Print Number: 03 Print Year: 2019

Brief Contents

Brief Contents

Contents

CHAPTER 3 Introduction to the Derivative 201

CHAPTER 4 Techniques of Differentiation with Applications 303

CHAPTER 5 Further Applications of the Derivative 385

CHAPTER 6 The Integral 469

CHAPTER 7 Further Integration Techniques and Applications of the Integral 541

CHAPTER 8 Functions of Several Variables 607

CHAPTER 9 **Trigonometric Models** 675

Preface

Applied Calculus, Seventh Edition, is intended for a one- or two-term course for students majoring in business, the social sciences, or the liberal arts. Like the earlier editions, the seventh edition of *Applied Calculus* is designed to address the challenge of generating enthusiasm and mathematical sophistication in an audience that is often underprepared and lacks motivation for traditional mathematics courses. We meet this challenge by focusing on real-life applications that students can relate to, many on topics of current interest; by presenting mathematical concepts intuitively and thoroughly; and by employing a writing style that is informal, engaging, and occasionally even humorous.

The seventh edition goes farther than earlier editions in implementing support for a wide range of instructional paradigms. On the one hand, the abundant pedagogical content available both in print and online, including comprehensive teaching videos and online tutorials, now allows us to be able to offer complete customizable courses for approaches ranging from on-campus and hybrid classes to distance learning classes. In addition, our careful integration of optional support for multiple forms of technology throughout the text makes it adaptable in classes with no technology, classes in which a single form of technology is used exclusively, and classes that incorporate several technologies.

We fully support three forms of technology in this text: TI-83/84 Plus graphing calculators, spreadsheets, and powerful online utilities we have created for the book. In particular, our comprehensive support for spreadsheet technology, both in the text and online, is highly relevant for students who are studying business and economics, in which skill with spreadsheets may be vital to their future careers.

New To This Edition

Content

- **Chapter 0:** We have added an entire new section on logarithms in the Precalculus Review, up through solving for unknowns in the exponent. Students who need additional preparation in the basics of logarithms can now be assigned this material before studying the section on logarithmic functions and models in Chapter 2.
- **Chapter 1:** In our revision of this important introductory chapter, we have downplayed the algebra sophistication somewhat so as not to present artificial barriers to the mastery of the important new concepts we discuss.
- **Chapter 2:** In view of the new section on logarithms in the Precalculus Review, the material on logarithms and logarithmic functions in Chapter 2 has been streamlined.
- **Chapter 3:** Rather than following other books that avoid discussing the important distinction between discontinuities and domain singularities (for instance, the fact that $1/x$ is continuous on its domain but singular at zero), we discuss this distinction carefully, providing lots of practice and figures.

Current Topics in the Applications

- We have added and updated numerous real data exercises and examples based on topics that are either of intense current interest or of general interest to our students, including many on social networks, on the 2009–2016 economic recovery, and on the 2014 Ebola epidemic, while retaining those of important historical interest, such as the 2008 economic crisis, the SARS outbreak of 2003, the 2010 stock market "flash crash," and many others.

Exercises

- We have added many new conceptual Communication and Reasoning exercises, including many dealing with common student errors and misconceptions.

Online Visualization and Practice Examples

- We have created a variety of web-based interactive apps available both on **www.wanermath.com** and in the new MindTap course that accompanies this edition. Instructors can use these to demonstrate important concepts such as the slopes of secant and tangent lines, the derivative function, and marginal and average cost.
- Many key examples in the text are mirrored by web-based randomizable practice examples, which allow students to test their mastery of the textbook examples and provide instructors with material for interactive presentation and class discussion.

Our Approach to Pedagogy

Real-World Orientation The diversity, breadth, and abundance of examples and exercises included in this edition continue to distinguish our book from others. A large number of these examples and exercises are based on real, referenced data from business, economics, the life sciences, and the social sciences. Our updated examples and exercises in the seventh edition are even more attuned to themes that students can identify with and relate to, from the technology used in their phones and tablets to the social networks in which they participate and many of the corporations they will instantly recognize as important in their lives. Notable events, such as the outbreaks of SARS in 2003 and Ebola in 2014, the 1990s dot-com boom, the 2005–2006 real estate bubble, the resulting 2008 economic crisis, the 2010 stock market "flash crash," and many more, are addressed in examples and exercises throughout the book.

Adapting real data for pedagogical use can be tricky; available data can be numerically complex, intimidating for students, or incomplete. We have modified and streamlined many of the real-world applications, rendering them as tractable as any "made-up" application. At the same time, we have been careful to strike a pedagogically sound balance between applications based on real data and more traditional "generic" applications. Thus, the density and selection of real data-based applications have been tailored to the pedagogical goals and appropriate difficulty level for each section.

Readability We would like students to read this book. We would like students to *enjoy* reading this book. Therefore, we have written the book in a conversational, student-oriented style and have made frequent use of question-and-answer dialogues to encourage the development of the student's mathematical curiosity and intuition. We hope that this text will give the student insight into how a mathematician develops and thinks about mathematical ideas and their applications to real life.

Pedagogical Aids We have included our favorite unique and creative approaches to solving the kinds of problems that normally cause difficulties for students and headaches for instructors. To name just a few, we discuss verbal forms of the differentiation rules in Chapter 4 to avoid a tendency to try to juggle multiple formulas in finding a derivative, "calculation thought experiments" to help the student decide which rules of differentiation to apply and the order in which to apply them, shortcut methods for common integrals involving functions of $ax + b$ in Chapter 6, and, in Chapter 7, a powerful tabular method for integration by parts that transforms what is often an agonizingly complicated topic for students into almost a triviality.

Rigor Mathematical rigor need not be antithetical to the kind of applied focus and conceptual approach that are hallmarks of this book. We have worked hard to ensure that we are always mathematically honest without being unnecessarily formal. Sometimes we do this through the question-and-answer dialogues and sometimes through the "Before we go on . . ." discussions that follow examples, but always in a manner designed to provoke the interest of the student.

Five Elements of Mathematical Pedagogy to Address Different Learning Styles The "Rule of Four" is a common theme in many texts. Implementing this approach, we discuss many of the central concepts **numerically**, **graphically**, and **algebraically** and clearly delineate these distinctions. The fourth element, **verbal communication** of mathematical concepts, is emphasized through our discussions on translating English sentences into mathematical statements and in our extensive Communication and Reasoning exercises at the end of each section. A fifth element, **interactivity**, is implemented through expanded use of question-and-answer dialogues but is seen most dramatically in the eBook in the MindTap course that accompanies this edition and at **www.wanermath.com** through our new practice and learning modules. These are small interactive apps that help a student visualize new concepts or practice examples similar to those in the text. In addition, the wanermath .com website offers interactive tutorials in the form of games, interactive chapter summaries and chapter review exercises, and online utilities that automate a variety of tasks, from graphing to regression and visual representations of Riemann sums.

Understand

Examples

Examples are a cornerstone of our approach. Many of the scenarios that we use in application examples and exercises are revisited several times throughout the book. In this way, students will find themselves analyzing the same application from a variety of different perspectives, such as graphing, the use of derivatives, and elasticity. Reusing scenarios and important functions provides unifying threads and shows students the complex texture of real-life problems. Complete solutions are provided with every example.

EXAMPLE 1 Estimating a Limit Numerically

Use a table to estimate the following limits:

a. $\lim_{x \to 2} \frac{x^3 - 8}{x - 2}$ **b.** $\lim_{x \to 0} \frac{e^{2x} - 1}{x}$

Solution

a. We cannot simply substitute $x = 2$, because the function $f(x) = \frac{x^3 - 8}{x - 2}$ is not defined at $x = 2$. (Why?)* Instead, we use a table of values as we did above, with x approaching 2 from both sides:

	x approaching 2 from the left→					←*x* approaching 2 from the right			
x	1.9	1.99	1.999	1.9999	2	2.0001	2.001	2.01	2.1
$f(x) = \frac{x^3 - 8}{x - 2}$	11.41	11.9401	11.9940	11.9994		12.0006	12.0060	12.0601	12.61

Quick Examples

Most definition boxes include quick, straightforward examples that a student can use to solidify each new concept.

Quick Example

4. $\frac{x^2 - 1}{x - 1} = x + 1$ for all x except $x = 1$. Write $\frac{x^2 - 1}{x - 1}$ as $\frac{(x + 1)(x - 1)}{x - 1}$, and cancel the $(x - 1)$.

Therefore,

$$\lim_{x \to 1} \frac{x^2 - 1}{x - 1} = \lim_{x \to 1}(x + 1) = 1 + 1 = 2.$$

Question-and-Answer Dialogues

We frequently use informal question-and-answer dialogues that anticipate the kinds of questions that may occur to the student and also guide the student through the development of new concepts.

Q: *How do we find $\lim_{x \to a} f(x)$ when $x = a$ is a singular point of the function f and we cannot simplify the given function to make a a point of the domain?*

A: In such a case it might be necessary to analyze the function by some other method, such as numerically or graphically. However, if we do not obtain the indeterminate form 0/0 upon substitution, we can often say what the limit is, as the following example shows.

Before We Go On . . .

Most examples are followed by supplementary discussions, which may include a check on the answer, a discussion of the feasibility and significance of a solution, or an in-depth look at what the solution means.

➡ **Before we go on . . .** Notice that in Example 1(b), before simplification the substitution $x = 2$ yields

$$\frac{x^3 - 8}{x - 2} = \frac{8 - 8}{2 - 2} = \frac{0}{0}.$$

Lecture Videos

Developed with Principal Lecturer, Jay Abramson, at Arizona State University, these video clips are flexible in their use as lecture starters in class or as an independent resource for students to review concepts on their own. Blending an introduction to concepts with specific examples, the videos let students quickly see the big picture of key concepts they are learning in class. Selected clips involve students and simulate a classroom-type interaction that creates a sense of the familiar and demystifies key concepts they are learning in their course. Frequently asked questions appear periodically throughout the video segments to further enhance learning. All videos are closed captioned and available in the new MindTap and Enhanced WebAssign courses that accompany the text. The topics for the lecture videos were carefully selected to accompany the subject areas that are most frequently taught and target the concepts that students struggle with most.

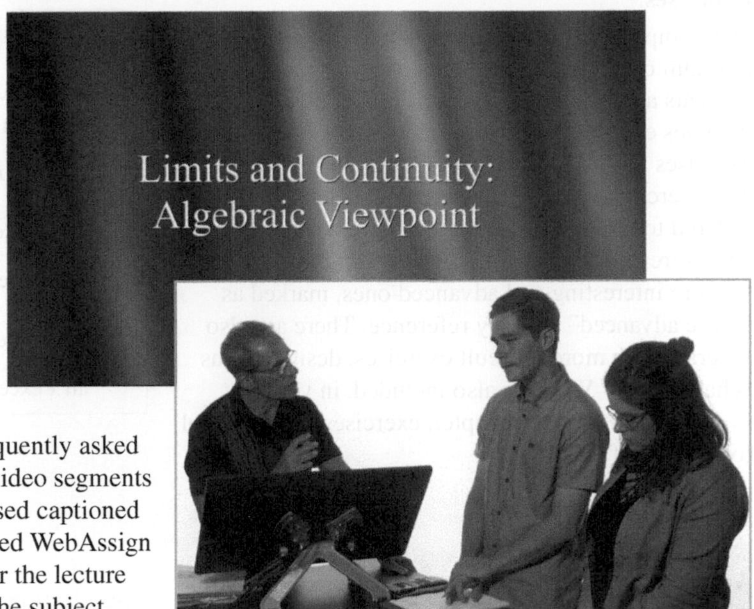

Online Visualization and Practice Examples

We have created a variety of web-based interactive apps that are available both on the wanermath.com website and in the new MindTap course accompanying this edition. Instructors can use these to demonstrate important concepts such as the slopes of secant and tangent lines, the derivative function, and marginal and average cost.

Many key examples in the text are mirrored by web-based randomizable practice examples that allow students to test their mastery of the textbook examples and provide instructors with material for interactive presentation and class discussion.

Visualize the derivative graphically

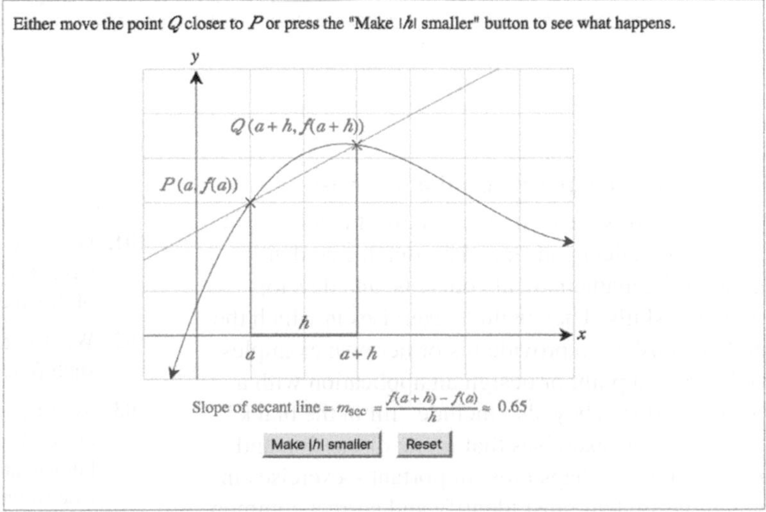

Practice and Apply

Exercises

Our comprehensive collection of exercises provides a wealth of material that can be used to challenge students at almost every level of preparation and includes everything from straightforward drill exercises to interesting and challenging applications. The exercise sets have been carefully curated and ordered to move from straightforward basic exercises and exercises that are similar to examples in the text to more interesting and advanced ones, marked as "more advanced" for easy reference. There are also several much more difficult exercises, designated as "challenging." We have also included, in virtually every section of every chapter, exercises that are ideal for the use of technology.

3.3 EXERCISES

▼ more advanced ◆ challenging

T indicates exercises that should be solved using technology

In Exercises 1–4, complete the given sentence.

1. The closed-form function $f(x) = \dfrac{1}{x-1}$ is continuous for all x except ______. [**HINT**: See Quick Example 3.]

2. The closed-form function $f(x) = \dfrac{1}{x^2-1}$ is continuous for all x except ______. [**HINT**: See Quick Example 3.]

Application Exercises

Exercises also include interesting applications based on real data to reinforce the applicability of math to real-life situations.

Applications

89. ***Processor Speeds*** The processor speeds, in megahertz (MHz), of Intel processors during the period 1996–2010 can be approximated by the following function of time t in years since the start of 1990:[17]

$$v(t) = \begin{cases} 400t - 2{,}200 & \text{if } 6 \le t < 15 \\ 3{,}800 & \text{if } 15 \le t \le 20. \end{cases}$$

a. Compute $\lim_{t \to 15^-} v(t)$ and $\lim_{t \to 15^+} v(t)$, and interpret each answer. [**HINT**: See Example 3.]

b. Is the function v continuous at $t = 15$? According to the model, was there any abrupt change in processor speeds during the period 1996–2010?

Communication and Reasoning Exercises

These exercises are designed to help students articulate mathematical concepts, broaden the student's grasp of the mathematical concepts, and develop modeling skills. They include exercises in which the student is asked to provide his or her own examples to illustrate a point or design an application with a given solution. They also include "fill in the blank" type exercises, exercises that invite discussion and debate, and—perhaps most important—exercises in which the student must identify and correct common errors. These exercises often have no single correct answer.

Communication and Reasoning Exercises

101. Describe the algebraic method of evaluating limits as discussed in this section, and give at least one disadvantage of this method.

102. What is a closed-form function? What can we say about such functions?

103. Your friends Rita and Richard are arguing. Rita claims that closed-form functions cannot have points of discontinuity, but Richard retorts, "Ever heard of $f(x) = 1/x$?" On whose side (if any) of the argument should you be? Explain.

Review

At the end of every chapter is a comprehensive list of the key concepts that were covered in each section.

Review exercises provide a great way to consolidate and check understanding and prepare for exams.

CHAPTER 3 REVIEW

KEY CONCEPTS

www.WanerMath.com Go to the Website to find a comprehensive and interactive Web-based summary of Chapter 3.

Simplifying to obtain limits [p. 226]
The indeterminate form 0/0 [p. 227]
The determinate form $k/0$ [p. 228]
Limits at infinity [p. 231]

The derivative as slope of the tangent line [p. 259]
Quick approximation of the derivative [p. 261]

REVIEW EXERCISES

T Indicates exercises that should be solved using technology

In Exercises 1–4, numerically *estimate whether the limit exists. If the limit does exist, give its approximate value.*

In Exercises 5 and 6 the graph of a function f is shown. Graphically determine whether the given limits exist. If a limit does exist, give its approximate value.

Case Studies

Each chapter ends with a section entitled "Case Study," an extended application that uses and illustrates the central ideas of the chapter, focusing on the development of mathematical models appropriate to the topics. These applications are ideal for assignment as projects.

CASE STUDY

Reducing Sulfur Emissions

Norbert Schaefer/CORBIS/Getty Images

The Environmental Protection Agency (EPA) wishes to formulate a policy that will encourage utilities to reduce sulfur emissions. Its goal is to reduce annual emissions of sulfur dioxide by a total of 10 million tons from the current level of 25 million tons by imposing a fixed charge for every ton of sulfur released into the environment per year. As a consultant to the EPA, you must determine the amount to be charged per ton of sulfur emissions.

You would like first to know the cost to the utility industry of reducing sulfur emissions. In other words, you would like to have a cost function of the form

$$C(q) = \text{Cost of removing } q \text{ tons of sulfur dioxide.}$$

Focus on Technology

Marginal Technology Notes

We give brief marginal technology notes to outline the use of graphing calculator, spreadsheet, and website technology in appropriate examples. When necessary, the reader is referred to more detailed discussion in the end-of-chapter Technology Guides.

End-of-Chapter Technology Guides

We continue to include detailed TI-83/84 Plus and Spreadsheet Guides at the end of each chapter. These Guides are referenced liberally in marginal technology notes at appropriate points in the chapter, so instructors and students can easily use this material or not, as they prefer.

Using Technology

TI-83/84 Plus

2ND CATALOG
DiagnosticOn
Then STAT CALC option #4:
LinReg(ax+b) [More details in the Technology Guide.]

Spreadsheet

Add a trendline and select the option to "Display R-squared value on chart."
[More details and other alternatives in the Technology Guide.]

Website

www.WanerMath.com

The following two utilities will show regression lines and also r^2 (link to either from Math Tools for Chapter 1):

Simple Regression Utility
Function Evaluator and Grapher

TI-83/84 Plus Technology Guide

Section 3.1

Example 1 (page 203) Use a table ... following limits.

a. $\lim_{x\to 2} \frac{x^3-8}{x-2}$ **b.** $\lim_{x\to 0} \frac{e^{2x}-1}{x}$

Solution

On the TI-83/84 Plus, use the table feat... these computations as follows:

(Y₁(9.5)-Y₁(8))/(9.5-8)

Spreadsheet Technology Guide

Section 3.1

Example 1 (page 203) Use a table to estimate the following limits.

	A	B
1	t	G(t)
2	8	=5*A2^2-85*A2+1762
3	9.5	

Instructor Resources

MindTap: Through personalized paths of dynamic assignments and applications, MindTap is a digital learning solution and representation of your course that turns cookie cutter into cutting edge, apathy into engagement, and memorizers into higher-level thinkers.

The Right Content: With MindTap's carefully curated material, you get the precise content and groundbreaking tools you need for every course you teach. This course includes a dynamic Pre-Course Assessment that tests students on their prerequisite skills, an eBook, algorithmic assignments, and new lecture videos.

Personalization: Customize every element of your course—from rearranging the learning path to inserting videos and activities.

Improved Workflow: Save time when planning lessons with all of the trusted, most current content you need in one place in MindTap.

Tracking Students' Progress in Real Time: Promote positive outcomes by tracking students in real time and tailoring your course as needed based on the analytics.

Learn more at **www.cengage.com/mindtap**.

WebAssign: Exclusively from Cengage Learning, Enhanced WebAssign combines the exceptional mathematics content that you know and love with the most powerful online homework solution, WebAssign. Enhanced WebAssign engages students with immediate feedback, rich tutorial content, and eBooks, helping students to develop a deeper conceptual understanding of their subject matter. Quick Prep and Just In Time exercises provide opportunities for students to review prerequisite skills and content, both at the start of the course and at the beginning of each section. Flexible assignment options give instructors the ability to release assignments conditionally on the basis of students' prerequisite assignment scores. Visit us at **www.cengage.com/ewa** to learn more.

Cognero: Cengage Learning Testing Powered by Cognero is a flexible, online system that allows you to author, edit, and manage test bank content; create multiple test versions in an instant; and deliver tests from your LMS, your classroom, or wherever you choose.

Instructor Companion Site: This collection of book-specific lecture and class tools is available online at **www.cengage.com/login**. Access and download PowerPoint presentations, complete solutions manual, and more.

Student Resources

Student Solutions Manual (ISBN: 978-1-337-29129-3): Go beyond the answers—see what it takes to get there and improve your grade! This manual provides worked-out, step-by-step solutions to the odd-numbered problems in the text. You'll have the information you need to truly understand how the problems are solved.

MindTap: MindTap (assigned by the instructor) is a digital representation of your course that provides you with the tools you need to better manage your limited time, stay organized, and be successful. You can complete assignments whenever and wherever you are ready to learn, with course material specially customized for you by your instructor and streamlined in one proven, easy-to-use interface. With an array of study tools, you'll get a true understanding of course concepts, achieve better grades, and lay the groundwork for your future courses. Learn more at **www.cengage.com/mindtap**.

WebAssign: Enhanced WebAssign (assigned by the instructor) provides you with instant feedback on homework assignments. This online homework system is easy to use and includes helpful links to textbook sections, video examples, and problem-specific tutorials.

CengageBrain: Visit **www.cengagebrain.com** to access additional course materials and companion resources. At the cengagebrain.com home page, search for the ISBN of your title (from the back cover of your book) using the search box at the top of the page. This will take you to the product page where free companion resources can be found.

The Author Website

The authors' website, accessible through **www.wanermath.com**, has been evolving for close to two decades with growing recognition. Students, raised in an environment in which computers suffuse both work and play, can use their web browsers to engage with the material in an active way. The following features of the authors' website are fully integrated with the text and can be used as a personalized study resource:

- **Interactive Tutorials** Highly interactive tutorials are included on major topics, with guided exercises that parallel the text and a great deal of help and feedback to assist the student.
- **Game Versions of Tutorials** More challenging tutorials with randomized questions that that work as games (complete with "health" scores, "health vials," and an assessment of one's performance at the end of the game) are offered alongside the traditional tutorials. These game tutorials, which mirror the traditional "more gentle" tutorials, randomize all the questions and do not give the student the answers but instead offer hints in exchange for "health points," so that just staying alive (not running out of health) can be quite challenging.
- **Learning and Practice Modules** These interactive demos illustrate important concepts and randomizable "practice examples" that mirror many examples and quick examples in the text.
- **Detailed Chapter Summaries** Comprehensive summaries with randomizable interactive elements review all the basic definitions and problem-solving techniques discussed in each chapter. These are a terrific pre-test study tool for students.
- **Downloadable Excel Tutorials** Detailed Excel tutorials are available for almost every section of the book. These interactive tutorials expand on the examples given in the text.
- **Online Utilities** Our collection of easy-to-use online utilities, referenced in the marginal notes of the textbook, allow students to solve many of the technology-based application exercises directly on the web. The utilities include a function grapher and evaluator that also graphs derivatives and does curve-fitting, regression tools, an interactive Riemann sum grapher with a numerical integrator, and a multifunctional line entry calculator on the main page. These utilities require nothing more than a standard web browser.
- **Chapter True-False Quizzes** Randomized quizzes that provide feedback for many incorrect answers based on the key concepts in each chapter assist the student in further mastery of the material.
- **Supplemental Topics** We include complete interactive text and exercise sets for a selection of topics that are not ordinarily included in printed texts but are often requested by instructors.
- **Spanish** A parallel Spanish version of almost the entire website is now deployed, allowing the user to switch languages on specific pages with a single mouse-click. In particular, all of the chapter summaries and most of the tutorials, game tutorials, and utilities are available in Spanish.

Acknowledgments

This project would not have been possible without the contributions and suggestions of numerous colleagues, students, and friends. We are particularly grateful to our colleagues at Hofstra and elsewhere who used and gave us useful feedback on previous editions and suggestions for this one, and to everyone at Cengage for their encouragement and guidance throughout the project. Specifically, we would like to thank Rita Lombard and Morgan Mendoza for their unflagging enthusiasm, Scott Barnett of Henry Ford Community College for his meticulous check of the mathematical accuracy, and Martha Emry and Teresa Trego for whipping the book into shape. Additionally, we would like to thank the creative force of Jay Abramson of Arizona State University for developing the new lecture videos that accompany our text, and Scott Barnett of Henry Ford Community College, Joe Rody of Arizona State University, Nada Al-Hanna of University of Texas at El Paso, and Kaat Higham of Bergen Community College for their thoughtful reviews and input into the scripts.

We would also like to thank Dario Menasce at CERN who helped us understand the fascinating new cover art, and the numerous reviewers and proofreaders who provided many helpful suggestions that have shaped the development of this book over time:

Christopher Brown, *California Lutheran University*
Melinda Camarillo, *El Paso Community College*
Nathan Carlson, *California Lutheran University*
Scott Fallstrom, *University of Oregon*
Irene Jai, *Raritan Valley Community College*
Latrice Laughlin, *University of Alaska Fairbanks*
Gabriel Mendoza, *El Paso Community College*
Charles Mundy-Castle, *Central New Mexico Community College*
Patrick Mutungi, *University of South Carolina*
Michael Price, *University of Oregon*
Christopher Quarles, *Everett Community College*
Leela Rakesh, *Central Michigan University*
Tom Rosenwinkel, *Concordia University Texas*
Bradley Stewart, *State University of New York at Oswego*
Larry Taylor, *North Dakota State University*
Daniel Wang, *Central Michigan University*

Stefan Waner
Steven R. Costenoble

0 PRECALCULUS REVIEW

DreamPictures/Taxi/Getty Images

WM www.WanerMath.com

Introduction

In this chapter we review some topics from algebra that you need to know to get the most out of this book. This chapter can be used either as a refresher course or as a reference.

There is one crucial fact you must always keep in mind: The letters used in algebraic expressions stand for numbers. All the rules of algebra are just facts about the arithmetic of numbers. If you are not sure whether some algebraic manipulation you are about to do is legitimate, try it first with numbers. If it doesn't work with numbers, it doesn't work.

0.1 Real Numbers

The **real numbers** are the numbers that can be written in decimal notation, including those that require an infinite decimal expansion. The set of real numbers includes all integers, positive, negative, and zero; all fractions; and the irrational numbers, that is, those with decimal expansions that never repeat. Examples of irrational numbers are

$$\sqrt{2} = 1.414213562373\ldots$$

and

$$\pi = 3.141592653589\ldots$$

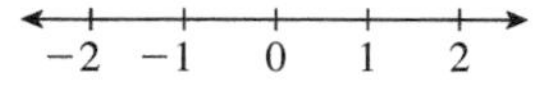

Figure 1

It is very useful to picture the real numbers as points on a line. As shown in Figure 1, larger numbers appear to the right, in the sense that if $a < b$, then the point corresponding to b is to the right of the one corresponding to a.

Intervals

Some subsets of the set of real numbers, called **intervals**, show up quite often, so we have a compact notation for them.

Interval Notation

Here is a list of types of intervals along with examples.

	Interval	Description	Picture	Example
Closed	$[a, b]$	Set of numbers x with $a \le x \le b$	 a b (includes end points)	$[0, 10]$
Open	(a, b)	Set of numbers x with $a < x < b$	 a b (excludes end points)	$(-1, 5)$
Half-Open	$(a, b]$	Set of numbers x with $a < x \le b$	 a b	$(-3, 1]$
	$[a, b)$	Set of numbers x with $a \le x < b$	 a b	$[0, 5)$

Infinite	$[a, +\infty)$	Set of numbers x with $a \le x$	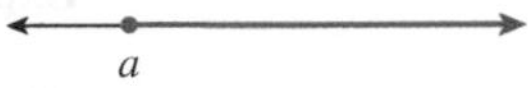	$[10, +\infty)$
	$(a, +\infty)$	Set of numbers x with $a < x$		$(-3, +\infty)$
	$(-\infty, b]$	Set of numbers x with $x \le b$		$(-\infty, -3]$
	$(-\infty, b)$	Set of numbers x with $x < b$		$(-\infty, 10)$
	$(-\infty, +\infty)$	Set of all real numbers		$(-\infty, +\infty)$

Operations

There are five important operations on real numbers: addition, subtraction, multiplication, division, and exponentiation. "Exponentiation" means raising a real number to a power; for instance, $3^2 = 3 \cdot 3 = 9$; $2^3 = 2 \cdot 2 \cdot 2 = 8$.

A note on technology: Most graphing calculators and spreadsheets use an asterisk * for multiplication and a caret ^ for exponentiation. Thus, for instance, 3×5 is entered as `3*5`, $3x$ as `3*x`, and 3^2 as `3^2`.

When we write an expression involving two or more operations, such as

$$2 \cdot 3 + 4$$

or

$$\frac{2 \cdot 3^2 - 5}{4 - (-1)},$$

we need to agree on the order in which to do the operations. Does $2 \cdot 3 + 4$ mean $(2 \cdot 3) + 4 = 10$ or $2 \cdot (3 + 4) = 14$? We all agree to use the following rules for the order in which we do the operations.

Standard Order of Operations

Parentheses and Fraction Bars First, calculate the values of all expressions inside parentheses or brackets, working from the innermost parentheses out, before using them in other operations. In a fraction, calculate the numerator and denominator separately before doing the division.

Quick Examples

1. $6(2 + [3 - 5] - 4) = 6(2 + (-2) - 4) = 6(-4) = -24$
2. $\dfrac{(4 - 2)}{3(-2 + 1)} = \dfrac{2}{3(-1)} = \dfrac{2}{-3} = -\dfrac{2}{3}$
3. $3/(2 + 4) = \dfrac{3}{2 + 4} = \dfrac{3}{6} = \dfrac{1}{2}$
4. $(x + 4x)/(y + 3y) = (5x)/(4y)$

Exponents Next, perform exponentiation.

Quick Examples

5. $2 + 4^2 = 2 + 16 = 18$
6. $(2 + 4)^2 = 6^2 = 36$
} Note the difference.

7. $2\left(\frac{3}{4-5}\right)^2 = 2\left(\frac{3}{-1}\right)^2 = 2(-3)^2 = 2 \times 9 = 18$

8. $2(1 + 1/10)^2 = 2(1.1)^2 = 2 \times 1.21 = 2.42$

Multiplication and Division Next, do all multiplications and divisions, from left to right.

Quick Examples

9. $2(3 - 5)/4 \cdot 2 = 2(-2)/4 \cdot 2$ — Parentheses first
$= -4/4 \cdot 2$ — Leftmost product
$= -1 \cdot 2 = -2$ — Multiplications and divisions, left to right

10. $2(1 + 1/10)^2 \times 2/10 = 2(1.1)^2 \times 2/10$ — Parentheses first
$= 2 \times 1.21 \times 2/10$ — Exponent
$= 4.84/10 = 0.484$ — Multiplications and divisions, left to right

11. $4\frac{2(4-2)}{3(-2 \cdot 5)} = 4\frac{2(2)}{3(-10)} = 4\frac{4}{-30} = \frac{16}{-30} = -\frac{8}{15}$

Addition and Subtraction Last, do all additions and subtractions, from left to right.

Quick Examples

12. $2(3 - 5)^2 + 6 - 1 = 2(-2)^2 + 6 - 1 = 2(4) + 6 - 1$
$= 8 + 6 - 1 = 13$

13. $\left(\frac{1}{2}\right)^2 - (-1)^2 + 4 = \frac{1}{4} - 1 + 4 = -\frac{3}{4} + 4 = \frac{13}{4}$

14. $3/2 + 4 = 1.5 + 4 = 5.5$
15. $3/(2 + 4) = 3/6 = 1/2 = 0.5$
} Note the difference.

16. $4/2^2 + (4/2)^2 = 4/2^2 + 2^2 = 4/4 + 4 = 1 + 4 = 5$

17. `-2^4 = (-1)2^4 = -16` — A negative sign before an expression means multiplication by -1.[1]

[1] Spreadsheets and some programming languages interpret `-2^4` (wrongly!) as `(-2)^4=16`. So when working with spreadsheets, write `-2^4` as `(-1)*2^4` to avoid this issue.

T indicates material discussing the use of technologies such as graphing calculators, spreadsheets, and web utilities.

T Entering Formulas

Any good calculator or spreadsheet will respect the standard order of operations. However, we must be careful with division and exponentiation and use parentheses as necessary. The following table gives some examples of simple mathematical expressions and their equivalents in the functional format used in most graphing calculators, spreadsheets, and computer programs.

Mathematical Expression	Formula	Comments
$\frac{2}{3-x}$	`2/(3-x)`	Note the use of parentheses instead of the fraction bar. If we omit the parentheses, we get the expression shown next.
$\frac{2}{3}-x$	`2/3-x`	The calculator follows the usual order of operations.
$\frac{2}{3\times 5}$	`2/(3*5)`	Putting the denominator in parentheses ensures that the multiplication is carried out first. The asterisk is usually used for multiplication in graphing calculators and computers.
$\frac{2}{x}\times 5$	`(2/x)*5`	Putting the fraction in parentheses ensures that it is calculated first. Some calculators will interpret `2/3*5` as $\frac{2}{3\times 5}$ but `2/3(5)` as $\frac{2}{3}\times 5$.
$\frac{2-3}{4+5}$	`(2-3)/(4+5)`	Note once again the use of parentheses in place of the fraction bar.
2^3	`2^3`	The caret ^ is commonly used to denote exponentiation.
2^{3-x}	`2^(3-x)`	Be careful to use parentheses to tell the calculator where the exponent ends. Enclose the *entire exponent* in parentheses.
2^3-x	`2^3-x`	Without parentheses, the calculator will follow the usual order of operations: exponentiation and then subtraction.
3×2^{-4}	`3*2^(-4)`	On some calculators, the negation key is separate from the minus key.
$2^{-4\times 3}\times 5$	`2^(-4*3)*5`	Note once again how parentheses enclose the entire exponent.
$100\left(1+\frac{0.05}{12}\right)^{60}$	`100*(1+0.05/12)^60`	This is a typical calculation for compound interest.
$PV\left(1+\frac{r}{m}\right)^{mt}$	`PV*(1+r/m)^(m*t)`	This is the compound interest formula. *PV* is understood to be a single number (present value) and not the product of *P* and *V* (or else we would have used `P*V`).
$\frac{2^{3-2}\times 5}{y-x}$	`2^(3-2)*5/(y-x)` or `(2^(3-2)*5)/(y-x)`	Notice again the use of parentheses to hold the denominator together. We could also have enclosed the numerator in parentheses, although this is optional. (Why?)
$\frac{2^y+1}{2-4^{3x}}$	`(2^y+1)/(2-4^(3*x))`	Here, it is necessary to enclose both the numerator and the denominator in parentheses.
$2^y+\frac{1}{2}-4^{3x}$	`2^y+1/2-4^(3*x)`	This is the effect of leaving out the parentheses around the numerator and denominator in the previous expression.

Accuracy and Rounding

When we use a calculator or computer, the results of our calculations are often given to far more decimal places than are useful. For example, suppose we are told that a square has an area of 2.0 square feet and we are asked how long its sides are. Each side is the square root of the area, which the calculator tells us is

$$\sqrt{2} \approx 1.414213562.$$

However, the measurement of 2.0 square feet is probably accurate to only two digits, so our estimate of the lengths of the sides can be no more accurate than that. Therefore, we round the answer to two digits:

$$\text{Length of one side} \approx 1.4 \text{ feet.}$$

The digits that follow 1.4 are meaningless. The following guide makes these ideas more precise.

Significant Digits, Decimal Places, and Rounding

The number of **significant digits** in a decimal representation of a number is the number of digits that are not leading zeros after the decimal point (as in .0005) or trailing zeros before the decimal point (as in 5,400,000). We say that a value is **accurate to *n* significant digits** if only the first n significant digits are meaningful.

When to Round

After doing a computation in which all the quantities are accurate to no more than n significant digits, round the final result to n significant digits.

Quick Examples

18. 0.00067 has two significant digits. — The 000 before 67 are leading zeros.

19. 0.000670 has three significant digits. — The 0 after 67 is significant.

20. 5,400,000 has two or more significant digits. — We can't say how many of the zeros are trailing.[2]

21. 5,400,001 has seven significant digits. — The string of zeros is not trailing.

22. Rounding 63,918 to three significant digits gives 63,900.

23. Rounding 63,958 to three significant digits gives 64,000.

24. $\pi = 3.141592653\ldots$ $\frac{22}{7} = 3.142857142\ldots$ Therefore, $\frac{22}{7}$ is an approximation of π that is accurate to only three significant digits: 3.14.

25. $4.02(1 + 0.02)^{1.4} \approx 4.13$ — We rounded to three significant digits.

[2] If we obtained 5,400,000 by rounding 5,401,011, then it has three significant digits because the zero after the 4 is significant. On the other hand, if we obtained it by rounding 5,411,234, then it has only two significant digits. The use of scientific notation avoids this ambiguity: 5.40×10^6 (or `5.40E6` on a calculator or computer) is accurate to three digits, and 5.4×10^6 is accurate to two digits.

One more point, though: If, in a long calculation, you round the intermediate results, your final answer may be even less accurate than you think. As a general rule,

When calculating, don't round intermediate results. Rather, use the most accurate results obtainable, or have your calculator or computer store them for you.

When you are done with the calculation, *then* round your answer to the appropriate number of digits of accuracy.

0.1 EXERCISES

Calculate each expression in Exercises 1–24, giving the answer as a whole number or a fraction in lowest terms.

1. $2(4+(-1))(2\cdot -4)$

2. $3+([4-2]\cdot 9)$

3. `20/(3*4)-1`

4. `2-(3*4)/10`

5. $\dfrac{3+([3+(-5)])}{3-2\times 2}$

6. $\dfrac{12-(1-4)}{2(5-1)\cdot 2-1}$

7. `(2-5*(-1))/1-2*(-1)`

8. `2-5*(-1)/(1-2*(-1))`

9. $2\cdot(-1)^2/2$

10. $2+4\cdot 3^2$

11. $2\cdot 4^2+1$

12. $1-3\cdot(-2)^2\times 2$

13. `3^2+2^2+1`

14. `2^(2^2-2)`

15. $\dfrac{3-2(-3)^2}{-6(4-1)^2}$

16. $\dfrac{1-2(1-4)^2}{2(5-1)^2\cdot 2}$

17. `10*(1+1/10)^3`

18. `121/(1+1/10)^2`

19. $3\left(\dfrac{-2\cdot 3^2}{-(4-1)^2}\right)$

20. $-\left(\dfrac{8(1-4)^2}{-9(5-1)^2}\right)$

21. $3\left(1-\left(-\frac{1}{2}\right)^2\right)^2+1$

22. $3\left(\frac{1}{9}-\left(\frac{2}{3}\right)^2\right)^2+1$

23. `(1/2)^2-1/2^2`

24. `2/(1^2)-(2/1)^2`

Convert each expression in Exercises 25–50 into its technology formula equivalent as in the table in the text.

25. $3\times(2-5)$

26. $4+\dfrac{5}{9}$

27. $\dfrac{3}{2-5}$

28. $\dfrac{4-1}{3}$

29. $\dfrac{3-1}{8+6}$

30. $3+\dfrac{3}{2-9}$

31. $3-\dfrac{4+7}{8}$

32. $\dfrac{4\times 2}{\left(\frac{2}{3}\right)}$

33. $\dfrac{2}{3+x}-xy^2$

34. $3+\dfrac{3+x}{xy}$

35. $3.1x^3-4x^{-2}-\dfrac{60}{x^2-1}$

36. $2.1x^{-3}-x^{-1}+\dfrac{x^2-3}{2}$

37. $\dfrac{\left(\frac{2}{3}\right)}{5}$

38. $\dfrac{2}{\left(\frac{3}{5}\right)}$

39. $3^{4-5}\times 6$

40. $\dfrac{2}{3+5^{7-9}}$

41. $3\left(1+\dfrac{4}{100}\right)^{-3}$

42. $3\left(\dfrac{1+4}{100}\right)^{-3}$

43. $3^{2x-1}+4^x-1$

44. $2^{x^2}-(2^{2x})^2$

45. 2^{2x^2-x+1}

46. $2^{2x^2-x}+1$

47. $\dfrac{4e^{-2x}}{2-3e^{-2x}}$

48. $\dfrac{e^{2x}+e^{-2x}}{e^{2x}-e^{-2x}}$

49. $3\left(1-\left(-\frac{1}{2}\right)^2\right)^2+1$

50. $3\left(\frac{1}{9}-\left(\frac{2}{3}\right)^2\right)^2+1$

0.2 Exponents and Radicals

In Section 0.1 we discussed exponentiation, or "raising to a power"; for example, $2^3=2\cdot 2\cdot 2$. In this section we discuss the algebra of exponentials more fully. First, we look at *integer* exponents: cases in which the powers are positive or negative whole numbers.

Integer Exponents

Positive Integer Exponents

If a is any real number and n is any positive integer, then by a^n we mean the quantity $a \cdot a \cdot \ldots \cdot a$ (n times); thus, $a^1 = a$, $a^2 = a \cdot a$, $a^5 = a \cdot a \cdot a \cdot a \cdot a$. In the expression a^n the number n is called the **exponent**, and the number a is called the **base**.

Quick Examples

$3^2 = 9$ $\quad$ $2^3 = 8$

$0^{34} = 0$ $\quad$ $(-1)^5 = -1$

$10^3 = 1{,}000$ $\quad$ $10^5 = 100{,}000$

Negative Integer Exponents

If a is any real number *other than zero* and n is any positive integer, then we define

$$a^{-n} = \frac{1}{a^n} = \frac{1}{a \cdot a \cdot \ldots \cdot a} \quad (n \text{ times}).$$

Quick Examples

$2^{-3} = \frac{1}{2^3} = \frac{1}{8}$ $\quad$ $1^{-27} = \frac{1}{1^{27}} = 1$

$x^{-1} = \frac{1}{x^1} = \frac{1}{x}$ $\quad$ $(-3)^{-2} = \frac{1}{(-3)^2} = \frac{1}{9}$

$y^7y^{-2} = y^7\frac{1}{y^2} = y^5$ $\quad$ 0^{-2} is not defined

Zero Exponent

If a is any real number other than zero, then we define

$$a^0 = 1.$$

Quick Examples

$3^0 = 1$ $\quad$ $1{,}000{,}000^0 = 1$

0^0 is not defined

When combining exponential expressions, we use the following identities.

Exponent Identity	Quick Examples
1. $a^m a^n = a^{m+n}$	$2^3 2^2 = 2^{3+2} = 2^5 = 32$
	$x^3 x^{-4} = x^{3-4} = x^{-1} = \frac{1}{x}$
	$\frac{x^3}{x^{-2}} = x^3 \frac{1}{x^{-2}} = x^3 x^2 = x^5$
2. $\frac{a^m}{a^n} = a^{m-n}$ if $a \neq 0$	$\frac{4^3}{4^2} = 4^{3-2} = 4^1 = 4$
	$\frac{x^3}{x^{-2}} = x^{3-(-2)} = x^5$
	$\frac{3^2}{3^4} = 3^{2-4} = 3^{-2} = \frac{1}{9}$
3. $(a^n)^m = a^{nm}$	$(3^2)^2 = 3^4 = 81$
	$(2^x)^2 = 2^{2x}$
4. $(ab)^n = a^n b^n$	$(4 \cdot 2)^2 = 4^2 2^2 = 64$
	$(-2y)^4 = (-2)^4 y^4 = 16y^4$
5. $\left(\frac{a}{b}\right)^n = \frac{a^n}{b^n}$ if $b \neq 0$	$\left(\frac{4}{3}\right)^2 = \frac{4^2}{3^2} = \frac{16}{9}$
	$\left(\frac{x}{-y}\right)^3 = \frac{x^3}{(-y)^3} = -\frac{x^3}{y^3}$

Caution

- In the first two identities, the bases of the expressions must be the same. For example, the first identity gives $3^2 3^4 = 3^6$ but does *not* apply to $3^2 4^2$.
- People sometimes invent their own identities, such as $a^m + a^n = a^{m+n}$, which is wrong! (Try it with $a = m = n = 1$.) If you wind up with something like $2^3 + 2^4$, you are stuck with it; there are no identities around to simplify it further. (You can factor out 2^3, but whether or not that is a simplification depends on what you are going to do with the expression next.)

EXAMPLE 1 **Combining the Identities**

$$\frac{(x^2)^3}{x^3} = \frac{x^6}{x^3} \quad \text{By identity (3)}$$

$$= x^{6-3} \quad \text{By identity (2)}$$

$$= x^3$$

$$\frac{(x^4 y)^3}{y} = \frac{(x^4)^3 y^3}{y} \quad \text{By identity (4)}$$

$$= \frac{x^{12} y^3}{y} \quad \text{By identity (3)}$$

$$= x^{12} y^{3-1} \quad \text{By identity (2)}$$

$$= x^{12} y^2$$

EXAMPLE 2 **Eliminating Negative Exponents**

Simplify the following and express the answer using no negative exponents.

a. $\dfrac{x^4y^{-3}}{x^5y^2}$ **b.** $\left(\dfrac{x^{-1}}{x^2y}\right)^5$

Solution

a. $\dfrac{x^4y^{-3}}{x^5y^2} = x^{4-5}y^{-3-2} = x^{-1}y^{-5} = \dfrac{1}{xy^5}$

b. $\left(\dfrac{x^{-1}}{x^2y}\right)^5 = \dfrac{(x^{-1})^5}{(x^2y)^5} = \dfrac{x^{-5}}{x^{10}y^5} = \dfrac{1}{x^{15}y^5}$

Radicals

If a is any nonnegative real number, then its **square root** is the nonnegative number whose square is a. For example, the square root of 16 is 4, because $4^2 = 16$. We write the square root of n as $\sqrt{n}$. (Roots are also referred to as **radicals**.) It is important to remember that $\sqrt{n}$ is never negative. Thus, for instance, $\sqrt{9}$ is 3 and not -3, even though $(-3)^2 = 9$. If we want to speak of the "negative square root" of 9, we write it as $-\sqrt{9} = -3$. If we want to write both square roots at once, we write $\pm\sqrt{9} = \pm 3$.

The **cube root** of a real number a is the number whose cube is a. The cube root of a is written as $\sqrt[3]{a}$ so that, for example, $\sqrt[3]{8} = 2$ (because $2^3 = 8$). Note that we can take the cube root of any number, positive, negative, or zero. For instance, the cube root of -8 is $\sqrt[3]{-8} = -2$ because $(-2)^3 = -8$. Unlike square roots, the cube root of a number may be negative. In fact, the cube root of a always has the same sign as a.

Higher roots are defined similarly. The **fourth root** of the *nonnegative* number a is defined as the nonnegative number whose fourth power is a and is written $\sqrt[4]{a}$. The **fifth root** of any number a is the number whose fifth power is a, and so on.

Note We cannot take an even-numbered root of a negative number, but we can take an odd-numbered root of any number. Even roots are always positive, whereas odd roots have the same sign as the number we start with. ■

EXAMPLE 3 ***n*th Roots**

$\sqrt{4} = 2$	Because $2^2 = 4$
$\sqrt{16} = 4$	Because $4^2 = 16$
$\sqrt{1} = 1$	Because $1^2 = 1$
If $x \geq 0$, then $\sqrt{x^2} = x$.	Because $x^2 = x^2$
$\sqrt{2} \approx 1.414213562$	$\sqrt{2}$ is not a whole number.
$\sqrt{1+1} = \sqrt{2} \approx 1.414213562$	First add, then take the square root.[3]
$\sqrt{9+16} = \sqrt{25} = 5$	Contrast with $\sqrt{9} + \sqrt{16} = 3 + 4 = 7$.
$\dfrac{1}{\sqrt{2}} = \dfrac{\sqrt{2}}{2}$	Multiply top and bottom by $\sqrt{2}$.

[3] In general, $\sqrt{a+b}$ means the square root of the *quantity* $(a + b)$. The radical sign acts as a pair of parentheses or a fraction bar, telling us to evaluate what is inside before taking the root. (See the Caution on the next page.)

$\sqrt[3]{27} = 3$	Because $3^3 = 27$
$\sqrt[3]{-64} = -4$	Because $(-4)^3 = -64$
$\sqrt[4]{16} = 2$	Because $2^4 = 16$
$\sqrt[4]{-16}$ is not defined.	Even-numbered root of a negative number
$\sqrt[5]{-1} = -1$, since $(-1)^5 = -1$.	Odd-numbered root of a negative number
$\sqrt[n]{-1} = -1$ if n is any odd number.	

Q: *In the example we saw that $\sqrt{x^2} = x$ if x is nonnegative. What happens if x is negative?*

A: If x is negative, then x^2 is positive, so $\sqrt{x^2}$ is still defined as the nonnegative number whose square is x^2. This number must be $|x|$, the **absolute value of x**, which is the nonnegative number with the same size as x. For instance, $|-3| = 3$, while $|3| = 3$, and $|0| = 0$. It follows that

$$\sqrt{x^2} = |x|$$

for every real number x, positive or negative. For instance,

$$\sqrt{(-3)^2} = \sqrt{9} = 3 = |-3|$$

and $\sqrt{3^2} = \sqrt{9} = 3 = |3|$.

In general, we find that

$$\sqrt[n]{x^n} = x \text{ if } n \text{ is odd and } \sqrt[n]{x^n} = |x| \text{ if } n \text{ is even.}$$

We use the following identities to evaluate radicals of products and quotients.

Radicals of Products and Quotients

If a and b are any real numbers (nonnegative in the case of even-numbered roots), then

$$\sqrt[n]{ab} = \sqrt[n]{a}\sqrt[n]{b}$$ Radical of a product = Product of radicals

$$\sqrt[n]{\frac{a}{b}} = \frac{\sqrt[n]{a}}{\sqrt[n]{b}} \quad \text{if } b \neq 0.$$ Radical of a quotient = Quotient of radicals

Notes

- The first rule is similar to the rule $(a \cdot b)^2 = a^2b^2$ for the square of a product, and the second rule is similar to the rule $\left(\frac{a}{b}\right)^2 = \frac{a^2}{b^2}$ for the square of a quotient.
- ***Caution*** There is no corresponding identity for addition. In general,

$$\sqrt{a+b} \text{ is } \textit{not} \text{ equal to } \sqrt{a} + \sqrt{b}.$$

(Consider $a = b = 1$, for example.) Equating these expressions is a common error, so be careful! ■

Quick Examples

1. $\sqrt{9 \cdot 4} = \sqrt{9}\sqrt{4} = 3 \times 2 = 6$ Alternatively, $\sqrt{9 \cdot 4} = \sqrt{36} = 6$.
2. $\sqrt{\dfrac{9}{4}} = \dfrac{\sqrt{9}}{\sqrt{4}} = \dfrac{3}{2}$
3. $\dfrac{\sqrt{2}}{\sqrt{5}} = \dfrac{\sqrt{2}\sqrt{5}}{\sqrt{5}\sqrt{5}} = \dfrac{\sqrt{10}}{5}$
4. $\sqrt{4(3 + 13)} = \sqrt{4(16)} = \sqrt{4}\sqrt{16} = 2 \times 4 = 8$
5. $\sqrt[3]{-216} = \sqrt[3]{(-27)8} = \sqrt[3]{-27}\sqrt[3]{8} = (-3)2 = -6$
6. $\sqrt{x^3} = \sqrt{x^2 \cdot x} = \sqrt{x^2}\sqrt{x} = x\sqrt{x}$ if $x \geq 0$
7. $\sqrt{\dfrac{x^2 + y^2}{z^2}} = \dfrac{\sqrt{x^2 + y^2}}{\sqrt{z^2}} = \dfrac{\sqrt{x^2 + y^2}}{|z|}$ We can't simplify the numerator any further.

Rational Exponents

We already know what we mean by expressions such as x^4 and a^{-6}. The next step is to make sense of *rational* exponents: exponents of the form p/q with p and q integers as in $a^{1/2}$ and $3^{-2/3}$.

Q: *What should we mean by $a^{1/2}$?*

A: The overriding concern here is that all the exponent identities should remain true. In this case the identity to look at is the one that says that $(a^m)^n = a^{mn}$. This identity tells us that

$$(a^{1/2})^2 = a^1 = a.$$

That is, $a^{1/2}$, when squared, gives us a. But that must mean that $a^{1/2}$ is the *square root* of a, or

$$a^{1/2} = \sqrt{a}.$$

A similar argument tells us that if q is any positive whole number, then

$$a^{1/q} = \sqrt[q]{a}, \text{ the } q\text{th root of } a.$$

Notice that if a is negative, this makes sense only for q odd. To avoid this problem, we usually stick to positive a.

Q: *If p and q are integers (q positive), what should we mean by $a^{p/q}$?*

A: By the exponent identities, $a^{p/q}$ should equal both $(a^p)^{1/q}$ and $(a^{1/q})^p$. The first is the qth root of a^p, and the second is the pth power of $a^{1/q}$.

These arguments give us the following formulas for conversion between rational exponents and radicals.

Conversion Between Rational Exponents and Radicals

If a is any nonnegative number, then

$$a^{p/q} = \sqrt[q]{a^p} = (\sqrt[q]{a})^p.$$

↑ Using exponents ↑ Using radicals ↑

In particular,

$$a^{1/q} = \sqrt[q]{a}, \text{ the } q\text{th root of } a.$$

Notes

- If a is negative, all of this makes sense only if q is odd.
- All of the exponent identities continue to work when we allow rational exponents p/q. In other words, we are free to use all the exponent identities even though the exponents are not integers. ■

Quick Examples

8. $4^{3/2} = (\sqrt{4})^3 = 2^3 = 8$

9. $8^{2/3} = (\sqrt[3]{8})^2 = 2^2 = 4$

10. $9^{-3/2} = \dfrac{1}{9^{3/2}} = \dfrac{1}{(\sqrt{9})^3} = \dfrac{1}{3^3} = \dfrac{1}{27}$

11. $\dfrac{\sqrt{3}}{\sqrt[3]{3}} = \dfrac{3^{1/2}}{3^{1/3}} = 3^{1/2-1/3} = 3^{1/6} = \sqrt[6]{3}$

12. $2^2 2^{7/2} = 2^2 2^{3+1/2} = 2^2 2^3 2^{1/2} = 2^5 2^{1/2} = 2^5\sqrt{2}$

EXAMPLE 4 **Simplifying Algebraic Expressions**

Simplify the following.

a. $\dfrac{(x^3)^{5/3}}{x^3}$ **b.** $\sqrt[4]{a^6}$ **c.** $\dfrac{(xy)^{-3}y^{-3/2}}{x^{-2}\sqrt{y}}$

Solution

a. $\dfrac{(x^3)^{5/3}}{x^3} = \dfrac{x^5}{x^3} = x^2$

b. $\sqrt[4]{a^6} = a^{6/4} = a^{3/2} = a \cdot a^{1/2} = a\sqrt{a}$

c. $\dfrac{(xy)^{-3}y^{-3/2}}{x^{-2}\sqrt{y}} = \dfrac{x^{-3}y^{-3}y^{-3/2}}{x^{-2}y^{1/2}} = \dfrac{1}{x^{-2+3}y^{1/2+3+3/2}} = \dfrac{1}{xy^5}$

Radical Form, Positive Exponent Form, and Power Form

In calculus we must often convert algebraic expressions involving powers of x, such as $\dfrac{3}{2x^2}$, into expressions in which x does not appear in the denominator, such as $\dfrac{3}{2}x^{-2}$. Also, we must often convert expressions with radicals, such as $\dfrac{1}{\sqrt{1+x^2}}$, into expressions with no radicals and all powers in the numerator, such as $(1 + x^2)^{-1/2}$. In these cases, we are converting from **positive exponent form** or **radical form** to **power form**.

Radical Form

An expression is in **radical form** if it is written with integer powers and roots only.

Quick Examples

13. $\dfrac{2}{5\sqrt[3]{x}} + \dfrac{2}{x}$ is in radical form.
14. $\dfrac{2x^{-1/3}}{5} + 2x^{-1}$ is not in radical form because $x^{-1/3}$ appears.
15. $\dfrac{1}{\sqrt{1 + x^2}}$ is in radical form, but $(1 + x^2)^{-1/2}$ is not.

Positive Exponent Form

An expression is in **positive exponent form** if it is written with positive exponents only.

Quick Examples

16. $\dfrac{2}{3x^2}$ is in positive exponent form.
17. $\dfrac{2x^{-1}}{3}$ is not in positive exponent form because the exponent of x is negative.
18. $\dfrac{x}{6} + \dfrac{6}{x}$ is in positive exponent form.

Power Form

An expression is in **power form** if there are no radicals and all powers of unknowns occur in the numerator. We write such expressions as sums or differences of terms of the form

$$\text{Constant} \times (\text{Expression with } x)^p. \qquad \text{As in } \frac{1}{3}x^{-3/2}$$

Quick Examples

19. $\dfrac{2}{3}x^4 - 3x^{-1/3}$ is in power form.
20. $\dfrac{x}{6} + \dfrac{6}{x}$ is not in power form because the second expression has x in the denominator.
21. $\sqrt[3]{x}$ is not in power form because it has a radical.
22. $(1 + x^2)^{-1/2}$ is in power form, but $\dfrac{1}{\sqrt{1 + x^2}}$ is not.

EXAMPLE 5 **Converting from One Form to Another**

Convert the following to positive exponent form:

a. $\frac{1}{2}x^{-2} + \frac{4}{3}x^{-5}$ **b.** $\frac{2}{\sqrt{x}} - \frac{2}{x^{-4}}$

Convert the following to radical form:

c. $\frac{1}{2}x^{-1/2} + \frac{4}{3}x^{-5/4}$ **d.** $\frac{(3 + x)^{-1/3}}{5}$

Convert the following to power form:

e. $\frac{3}{4x^2} - \frac{x}{6} + \frac{6}{x} + \frac{4}{3\sqrt{x}}$ **f.** $\frac{2}{(x+1)^2} - \frac{3}{4\sqrt[5]{2x-1}}$

Solution For parts (a) and (b) we eliminate negative exponents, as we did in Example 2:

a. $\frac{1}{2}x^{-2} + \frac{4}{3}x^{-5} = \frac{1}{2} \cdot \frac{1}{x^2} + \frac{4}{3} \cdot \frac{1}{x^5} = \frac{1}{2x^2} + \frac{4}{3x^5}$

b. $\frac{2}{\sqrt{x}} - \frac{2}{x^{-4}} = \frac{2}{\sqrt{x}} - 2x^4$

For parts (c) and (d) we rewrite all terms with fractional exponents as radicals:

c.
$$\frac{1}{2}x^{-1/2} + \frac{4}{3}x^{-5/4} = \frac{1}{2} \cdot \frac{1}{x^{1/2}} + \frac{4}{3} \cdot \frac{1}{x^{5/4}}$$
$$= \frac{1}{2} \cdot \frac{1}{\sqrt{x}} + \frac{4}{3} \cdot \frac{1}{\sqrt[4]{x^5}} = \frac{1}{2\sqrt{x}} + \frac{4}{3\sqrt[4]{x^5}}$$

d. $\frac{(3 + x)^{-1/3}}{5} = \frac{1}{5(3 + x)^{1/3}} = \frac{1}{5\sqrt[3]{3 + x}}$

For parts (e) and (f) we eliminate any radicals and move all expressions involving x to the numerator:

e.
$$\frac{3}{4x^2} - \frac{x}{6} + \frac{6}{x} + \frac{4}{3\sqrt{x}} = \frac{3}{4}x^{-2} - \frac{1}{6}x + 6x^{-1} + \frac{4}{3x^{1/2}}$$
$$= \frac{3}{4}x^{-2} - \frac{1}{6}x + 6x^{-1} + \frac{4}{3}x^{-1/2}$$

f.
$$\frac{2}{(x+1)^2} - \frac{3}{4\sqrt[5]{2x-1}} = 2(x+1)^{-2} - \frac{3}{4(2x-1)^{1/5}}$$
$$= 2(x+1)^{-2} - \frac{3}{4}(2x-1)^{-1/5}$$

Solving Equations with Exponents

EXAMPLE 6 **Solving Equations**

Solve the following equations:

a. $x^3 + 8 = 0$ **b.** $x^2 - \frac{1}{2} = 0$ **c.** $x^{3/2} - 64 = 0$

Solution

a. Subtracting 8 from both sides gives $x^3 = -8$. Taking the cube root of both sides gives $x = -2$.

b. Adding $\frac{1}{2}$ to both sides gives $x^2 = \frac{1}{2}$. Thus, $x = \pm\sqrt{\frac{1}{2}} = \pm\frac{1}{\sqrt{2}}$.

c. Adding 64 to both sides gives $x^{3/2} = 64$. Taking the reciprocal (2/3) power of both sides gives

$$(x^{3/2})^{2/3} = 64^{2/3}$$
$$x^1 = (\sqrt[3]{64})^2 = 4^2 = 16$$

so $$x = 16.$$

0.2 EXERCISES

Evaluate the expressions in Exercises 1–16.

1. 3^3 **2.** $(-2)^3$ **3.** $-(2 \cdot 3)^2$ **4.** $(4 \cdot 2)^2$

5. $\left(\frac{-2}{3}\right)^2$ **6.** $\left(\frac{3}{2}\right)^3$ **7.** $(-2)^{-3}$ **8.** -2^{-3}

9. $\left(\frac{1}{4}\right)^{-2}$ **10.** $\left(\frac{-2}{3}\right)^{-2}$ **11.** $2 \cdot 3^0$ **12.** $3 \cdot (-2)^0$

13. $2^3 2^2$ **14.** $3^2 3$ **15.** $2^2 2^{-1} 2^4 2^{-4}$ **16.** $5^2 5^{-3} 5^2 5^{-2}$

Simplify each expression in Exercises 17–30, expressing your answer in positive exponent form.

17. x^3x^2 **18.** x^4x^{-1} **19.** $-x^2x^{-3}y$ **20.** $-xy^{-1}x^{-1}$

21. $\frac{x^3}{x^4}$ **22.** $\frac{y^5}{y^3}$ **23.** $\frac{x^2y^2}{x^{-1}y}$ **24.** $\frac{x^{-1}y}{x^2y^2}$

25. $\frac{(xy^{-1}z^3)^2}{x^2yz^2}$ **26.** $\frac{x^2yz^2}{(xyz^{-1})^{-1}}$ **27.** $\left(\frac{xy^{-2}z}{x^{-1}z}\right)^3$

28. $\left(\frac{x^2y^{-1}z^0}{xyz}\right)^2$ **29.** $\left(\frac{x^{-1}y^{-2}z^2}{xy}\right)^{-2}$ **30.** $\left(\frac{xy^{-2}}{x^2y^{-1}z}\right)^{-3}$

Convert the expressions in Exercises 31–36 to positive exponent form.

31. $3x^{-4}$ **32.** $\frac{1}{2}x^{-4}$ **33.** $\frac{3}{4}x^{-2/3}$

34. $\frac{4}{5}y^{-3/4}$ **35.** $1 - \frac{0.3}{x^{-2}} - \frac{6}{5}x^{-1}$ **36.** $\frac{1}{3x^{-4}} + \frac{0.1x^{-2}}{3}$

Evaluate the expressions in Exercises 37–56, rounding your answer to four significant digits where necessary.

37. $\sqrt{4}$ **38.** $\sqrt{5}$ **39.** $\sqrt{\frac{1}{4}}$

40. $\sqrt{\frac{1}{9}}$ **41.** $\sqrt{\frac{16}{9}}$ **42.** $\sqrt{\frac{9}{4}}$

43. $\frac{\sqrt{4}}{5}$ **44.** $\frac{6}{\sqrt{25}}$ **45.** $\sqrt{9} + \sqrt{16}$

46. $\sqrt{25} - \sqrt{16}$ **47.** $\sqrt{9 + 16}$ **48.** $\sqrt{25 - 16}$

49. $\sqrt[3]{8 - 27}$ **50.** $\sqrt[4]{81 - 16}$ **51.** $\sqrt[3]{27/8}$

52. $\sqrt[3]{8 \times 64}$ **53.** $\sqrt{(-2)^2}$ **54.** $\sqrt{(-1)^2}$

55. $\sqrt{\frac{1}{4}(1 + 15)}$ **56.** $\sqrt{\frac{1}{9}(3 + 33)}$

Simplify the expressions in Exercises 57–64, given that x, y, z, a, b, and c are positive real numbers.

57. $\sqrt{a^2b^2}$ **58.** $\sqrt{\frac{a^2}{b^2}}$ **59.** $\sqrt{(x + 9)^2}$

60. $(\sqrt{x + 9})^2$ **61.** $\sqrt[3]{x^3(a^3 + b^3)}$ **62.** $\sqrt[4]{\frac{x^4}{a^4b^4}}$

63. $\sqrt{\frac{4xy^3}{x^2y}}$ **64.** $\sqrt{\frac{4(x^2 + y^2)}{c^2}}$

Convert the expressions in Exercises 65–84 to power form.

65. $\sqrt{3}$ **66.** $\sqrt{8}$ **67.** $\sqrt{x^3}$

68. $\sqrt[3]{x^2}$ **69.** $\sqrt[3]{xy^2}$ **70.** $\sqrt{x^2y}$

71. $\frac{x^2}{\sqrt{x}}$ **72.** $\frac{x}{\sqrt{x}}$ **73.** $\frac{3}{5x^2}$

74. $\frac{2}{5x^{-3}}$ **75.** $\frac{3x^{-1.2}}{2} - \frac{1}{3x^{2.1}}$ **76.** $\frac{2}{3x^{-1.2}} - \frac{x^{2.1}}{3}$

77. $\frac{2x}{3} - \frac{x^{0.1}}{2} + \frac{4}{3x^{1.1}}$ **78.** $\frac{4x^2}{3} + \frac{x^{3/2}}{6} - \frac{2}{3x^2}$

79. $\frac{3\sqrt{x}}{4} - \frac{5}{3\sqrt{x}} + \frac{4}{3x\sqrt{x}}$ **80.** $\frac{3}{5\sqrt{x}} - \frac{5\sqrt{x}}{8} + \frac{7}{2\sqrt[3]{x}}$

81. $\frac{3\sqrt[5]{x^2}}{4} - \frac{7}{2\sqrt{x^3}}$ **82.** $\frac{1}{8x\sqrt{x}} - \frac{2}{3\sqrt[5]{x^3}}$

83. $\frac{1}{(x^2 + 1)^3} - \frac{3}{4\sqrt[3]{(x^2 + 1)}}$ **84.** $\frac{2}{3(x^2 + 1)^{-3}} - \frac{3\sqrt[3]{(x^2 + 1)}}{4}$

Convert the expressions in Exercises 85–96 to radical form.

85. $2^{2/3}$ **86.** $3^{4/5}$ **87.** $x^{4/3}$ **88.** $y^{7/4}$

89. $(x^{1/2}y^{1/3})^{1/5}$ **90.** $x^{-1/3}y^{3/2}$ **91.** $-\frac{3}{2}x^{-1/4}$ **92.** $\frac{4}{5}x^{3/2}$

93. $0.2x^{-2/3} + \frac{3}{7x^{-1/2}}$ **94.** $\frac{3.1}{x^{-4/3}} - \frac{11}{7}x^{-1/7}$

95. $\frac{3}{4(1-x)^{5/2}}$ **96.** $\frac{9}{4(1-x)^{-7/3}}$

Simplify the expressions in Exercises 97–106.

97. $4^{-1/2}4^{7/2}$ **98.** $2^{1/a}/2^{2/a}$ **99.** $3^{2/3}3^{-1/6}$

100. $2^{1/3}2^{-1}2^{2/3}2^{-1/3}$ **101.** $\frac{x^{3/2}}{x^{5/2}}$ **102.** $\frac{y^{5/4}}{y^{3/4}}$

103. $\frac{x^{1/2}y^2}{x^{-1/2}y}$ **104.** $\frac{x^{-1/2}y}{x^2y^{3/2}}$

105. $\left(\frac{x}{y}\right)^{1/3}\left(\frac{y}{x}\right)^{2/3}$ **106.** $\left(\frac{x}{y}\right)^{-1/3}\left(\frac{y}{x}\right)^{1/3}$

Solve each equation in Exercises 107–120 for x, rounding your answer to four significant digits where necessary.

107. $x^2 - 16 = 0$ **108.** $x^2 - 1 = 0$

109. $x^2 - \frac{4}{9} = 0$ **110.** $x^2 - \frac{1}{10} = 0$

111. $x^2 - (1 + 2x)^2 = 0$ **112.** $x^2 - (2 - 3x)^2 = 0$

113. $x^5 + 32 = 0$ **114.** $x^4 - 81 = 0$

115. $x^{1/2} - 4 = 0$ **116.** $x^{1/3} - 2 = 0$

117. $1 - \frac{1}{x^2} = 0$ **118.** $\frac{2}{x^3} - \frac{6}{x^4} = 0$

119. $(x - 4)^{-1/3} = 2$ **120.** $(x - 4)^{2/3} + 1 = 5$

0.3 Multiplying and Factoring Algebraic Expressions

Multiplying Algebraic Expressions

Distributive Law

The **distributive law** for real numbers states that

$$a(b \pm c) = ab \pm ac$$
$$(a \pm b)c = ac \pm bc$$

for any real numbers a, b, and c.

Quick Examples

1. $2(x - 3)$ is *not* equal to $2x - 3$ but is equal to $2x - 2(3) = 2x - 6$.
2. $x(x + 1) = x^2 + x$
3. $2x(3x - 4) = 6x^2 - 8x$
4. $(x - 4)x^2 = x^3 - 4x^2$
5. $(x + 2)(x + 3) = (x + 2)x + (x + 2)3$
   $= (x^2 + 2x) + (3x + 6) = x^2 + 5x + 6$
6. $(x + 2)(x - 3) = (x + 2)x - (x + 2)3$
   $= (x^2 + 2x) - (3x + 6) = x^2 - x - 6$

There is a quicker way of expanding expressions like the last two, called the FOIL method (First, Outer, Inner, Last). Consider, for instance, the expression $(x + 1)(x - 2)$. The FOIL method says: Take the product of the first terms: $x \cdot x = x^2$, the product of the outer terms: $x \cdot (-2) = -2x$, the product of the inner terms: $1 \cdot x = x$, and the product of the last terms: $1 \cdot (-2) = -2$, and then add them all up, getting $x^2 - 2x + x - 2 = x^2 - x - 2$.

EXAMPLE 1 **FOIL**

a. $(x - 2)(2x + 5) = 2x^2 + 5x - 4x - 10 = 2x^2 + x - 10$

$\uparrow$ First $\uparrow$ Outer $\uparrow$ Inner $\uparrow$ Last

b. $(x^2 + 1)(x - 4) = x^3 - 4x^2 + x - 4$

c. $(a - b)(a + b) = a^2 + ab - ab - b^2 = a^2 - b^2$

d. $(a + b)^2 = (a + b)(a + b) = a^2 + ab + ab + b^2 = a^2 + 2ab + b^2$

e. $(a - b)^2 = (a - b)(a - b) = a^2 - ab - ab + b^2 = a^2 - 2ab + b^2$

The formulas in parts (c), (d), and (e) of Example 1 are particularly important and worth memorizing, so let's repeat them.

Special Formulas

$(a - b)(a + b) = a^2 - b^2$ Difference of two squares

$(a + b)^2 = a^2 + 2ab + b^2$ Square of a sum

$(a - b)^2 = a^2 - 2ab + b^2$ Square of a difference

Quick Examples

7. $(2 - x)(2 + x) = 4 - x^2$

8. $(1 + a)(1 - a) = 1 - a^2$

9. $(x + 3)^2 = x^2 + 6x + 9$

10. $(4 - x)^2 = 16 - 8x + x^2$

Here are some longer examples that require the distributive law.

EXAMPLE 2 **Multiplying Algebraic Expressions**

a.
$$\begin{aligned}(x + 1)(x^2 + 3x - 4) &= (x + 1)x^2 + (x + 1)3x - (x + 1)4 \\ &= (x^3 + x^2) + (3x^2 + 3x) - (4x + 4) \\ &= x^3 + 4x^2 - x - 4\end{aligned}$$

b.
$$\begin{aligned}\left(x^2 - \frac{1}{x} + 1\right)(2x + 5) &= \left(x^2 - \frac{1}{x} + 1\right)2x + \left(x^2 - \frac{1}{x} + 1\right)5 \\ &= (2x^3 - 2 + 2x) + \left(5x^2 - \frac{5}{x} + 5\right) \\ &= 2x^3 + 5x^2 + 2x + 3 - \frac{5}{x}\end{aligned}$$

c.
$$\begin{aligned}(x - y)(x - y)(x - y) &= (x^2 - 2xy + y^2)(x - y) \\ &= (x^2 - 2xy + y^2)x - (x^2 - 2xy + y^2)y \\ &= (x^3 - 2x^2y + xy^2) - (x^2y - 2xy^2 + y^3) \\ &= x^3 - 3x^2y + 3xy^2 - y^3\end{aligned}$$

Factoring Algebraic Expressions

We can think of factoring as applying the distributive law in reverse—for example,

$$2x^2 + x = x(2x + 1),$$

which can be checked by using the distributive law. Factoring is an art that you will learn with experience and the help of a few useful techniques.

Factoring Using a Common Factor

To use this technique, locate a **common factor**—a term that occurs as a factor in each of the expressions being added or subtracted (for example, x is a common factor in $2x^2 + x$, because it is a factor of both $2x^2$ and x). Once you have located a common factor, factor it out by applying the distributive law.

Quick Examples

11. $2x^3 - x^2 + x$ has x as a common factor, so

$$2x^3 - x^2 + x = x(2x^2 - x + 1).$$

12. $2x^2 + 4x$ has $2x$ as a common factor, so

$$2x^2 + 4x = 2x(x + 2).$$

13. $2x^2y + xy^2 - x^2y^2$ has xy as a common factor, so

$$2x^2y + xy^2 - x^2y^2 = xy(2x + y - xy).$$

14. $(x^2 + 1)(x + 2) - (x^2 + 1)(x + 3)$ has $x^2 + 1$ as a common factor, so

$$\begin{aligned}(x^2 + 1)(x + 2) - (x^2 + 1)(x + 3) &= (x^2 + 1)[(x + 2) - (x + 3)] \\ &= (x^2 + 1)(x + 2 - x - 3) \\ &= (x^2 + 1)(-1) = -(x^2 + 1).\end{aligned}$$

15. $12x(x^2 - 1)^5(x^3 + 1)^6 + 18x^2(x^2 - 1)^6(x^3 + 1)^5$ has $6x(x^2 - 1)^5(x^3 + 1)^5$ as a common factor, so

$$\begin{aligned}&12x(x^2 - 1)^5(x^3 + 1)^6 + 18x^2(x^2 - 1)^6(x^3 + 1)^5 \\ &= 6x(x^2 - 1)^5(x^3 + 1)^5[2(x^3 + 1) + 3x(x^2 - 1)] \\ &= 6x(x^2 - 1)^5(x^3 + 1)^5(2x^3 + 2 + 3x^3 - 3x) \\ &= 6x(x^2 - 1)^5(x^3 + 1)^5(5x^3 - 3x + 2).\end{aligned}$$

We would also like to be able to reverse calculations such as $(x + 2)(2x - 5) = 2x^2 - x - 10$. That is, starting with the expression $2x^2 - x - 10$, we would like to **factor** it to get the expression $(x + 2)(2x - 5)$. An expression of the form $ax^2 + bx + c$, where a, b, and c are real numbers, is called a **quadratic** expression in x. Thus, given a quadratic expression $ax^2 + bx + c$, we would like to write it in the form $(dx + e)(fx + g)$ for some real numbers d, e, f, and g. There are some quadratics, such as $x^2 + x + 1$, that cannot be factored in this form at all. Here, we consider only quadratics that do factor and do so in such a way that the numbers d, e, f, and g are integers (other cases are discussed in Section 0.5). The usual technique of factoring such quadratics is a trial and error approach.

Factoring Quadratics by Trial and Error

To factor the quadratic $ax^2 + bx + c$, factor ax^2 as $(a_1x)(a_2x)$ (with a_1 positive) and c as c_1c_2, and then check whether or not $ax^2 + bx + c = (a_1x \pm c_1)(a_2x \pm c_2)$. If not, try other factorizations of ax^2 and c.

Quick Examples

16. To factor $x^2 - 6x + 5$, first factor x^2 as $(x)(x)$, and 5 as $(5)(1)$:

$(x+5)(x+1) = x^2 + 6x + 5$ No good

$(x-5)(x-1) = x^2 - 6x + 5.$ Desired factorization

17. To factor $x^2 - 4x - 12$, first factor x^2 as $(x)(x)$, and -12 as $(1)(-12)$, $(2)(-6)$, or $(3)(-4)$. Trying them one by one gives

$(x+1)(x-12) = x^2 - 11x - 12$ No good

$(x-1)(x+12) = x^2 + 11x - 12$ No good

$(x+2)(x-6) = x^2 - 4x - 12.$ Desired factorization

18. To factor $4x^2 - 25$, we can follow the above procedure or recognize $4x^2 - 25$ as the difference of two squares:

$$4x^2 - 25 = (2x)^2 - 5^2 = (2x-5)(2x+5).$$

Note Not all quadratic expressions factor. In Section 0.5 we look at a test that tells us whether or not a given quadratic factors. ■

Here are examples that require either a little more work or a little more thought.

EXAMPLE 3 Factoring Quadratics

Factor the following: **a.** $4x^2 - 5x - 6$ **b.** $x^4 - 5x^2 + 6$

Solution

a. Possible factorizations of $4x^2$ are $(2x)(2x)$ or $(x)(4x)$. Possible factorizations of -6 are $(1)(-6)$ or $(2)(-3)$. We now systematically try out all the possibilities until we come up with the correct one:

$(2x)(2x)$ and $(1)(-6)$:	$(2x+1)(2x-6) = 4x^2 - 10x - 6$	No good
$(2x)(2x)$ and $(2)(-3)$:	$(2x+2)(2x-3) = 4x^2 - 2x - 6$	No good
$(x)(4x)$ and $(1)(-6)$:	$(x+1)(4x-6) = 4x^2 - 2x - 6$	No good
$(x)(4x)$ and $(2)(-3)$:	$(x+2)(4x-3) = 4x^2 + 5x - 6$	Almost!
Change signs:	$(x-2)(4x+3) = 4x^2 - 5x - 6.$	Correct

b. The expression $x^4 - 5x^2 + 6$ is not a quadratic, you say? Correct. It's a quartic (a fourth degree expression). However, it looks rather like a quadratic. In fact, it is quadratic *in* x^2, meaning that it is

$$(x^2)^2 - 5(x^2) + 6 = y^2 - 5y + 6,$$

where $y = x^2$. The quadratic $y^2 - 5y + 6$ factors as

$$y^2 - 5y + 6 = (y - 3)(y - 2),$$

so

$$x^4 - 5x^2 + 6 = (x^2 - 3)(x^2 - 2).$$

This is a sometimes useful technique.

Our last example is here to remind you why we should want to factor polynomials in the first place. We shall return to this in Section 0.5.

EXAMPLE 4 Solving a Quadratic Equation by Factoring

Solve the equation $3x^2 + 4x - 4 = 0$.

Solution We first factor the left-hand side to get

$$(3x - 2)(x + 2) = 0.$$

Thus, the product of the two quantities $(3x - 2)$ and $(x + 2)$ is zero. Now, if a product of two numbers is zero, one of the two must be zero. In other words, either $3x - 2 = 0$, giving $x = \frac{2}{3}$, or $x + 2 = 0$, giving $x = -2$. Thus, there are two solutions: $x = \frac{2}{3}$ and $x = -2$.

0.3 EXERCISES

Expand each expression in Exercises 1–22.

1. $x(4x + 6)$
2. $(4y - 2)y$
3. $(2x - y)y$
4. $x(3x + y)$
5. $(x + 1)(x - 3)$
6. $(y + 3)(y + 4)$
7. $(2y + 3)(y + 5)$
8. $(2x - 2)(3x - 4)$
9. $(2x - 3)^2$
10. $(3x + 1)^2$
11. $\left(x + \frac{1}{x}\right)^2$
12. $\left(y - \frac{1}{y}\right)^2$
13. $(2x - 3)(2x + 3)$
14. $(4 + 2x)(4 - 2x)$
15. $\left(y - \frac{1}{y}\right)\left(y + \frac{1}{y}\right)$
16. $(x - x^2)(x + x^2)$
17. $(x^2 + x - 1)(2x + 4)$
18. $(3x + 1)(2x^2 - x + 1)$
19. $(x^2 - 2x + 1)^2$
20. $(x + y - xy)^2$
21. $(y^3 + 2y^2 + y)(y^2 + 2y - 1)$
22. $(x^3 - 2x^2 + 4)(3x^2 - x + 2)$

In Exercises 23–30, factor each expression and simplify as much as possible.

23. $(x + 1)(x + 2) + (x + 1)(x + 3)$
24. $(x + 1)(x + 2)^2 + (x + 1)^2(x + 2)$
25. $(x^2 + 1)^5(x + 3)^4 + (x^2 + 1)^6(x + 3)^3$
26. $10x(x^2 + 1)^4(x^3 + 1)^5 + 15x^2(x^2 + 1)^5(x^3 + 1)^4$
27. $(x^3 + 1)\sqrt{x + 1} - (x^3 + 1)^2\sqrt{x + 1}$
28. $(x^2 + 1)\sqrt{x + 1} - \sqrt{(x + 1)^3}$
29. $\sqrt{(x + 1)^3} + \sqrt{(x + 1)^5}$
30. $(x^2 + 1)\sqrt[3]{(x + 1)^4} - \sqrt[3]{(x + 1)^7}$

In Exercises 31–48, ***(a)*** *factor the given expression, and* ***(b)*** *set the expression equal to zero and solve for the unknown (x in the odd-numbered exercises and y in the even-numbered exercises).*

31. $2x + 3x^2$
32. $y^2 - 4y$
33. $6x^3 - 2x^2$
34. $3y^3 - 9y^2$
35. $x^2 - 8x + 7$
36. $y^2 + 6y + 8$
37. $x^2 + x - 12$
38. $y^2 + y - 6$
39. $2x^2 - 3x - 2$
40. $3y^2 - 8y - 3$
41. $6x^2 + 13x + 6$
42. $6y^2 + 17y + 12$
43. $12x^2 + x - 6$
44. $20y^2 + 7y - 3$
45. $x^2 + 4xy + 4y^2$
46. $4y^2 - 4xy + x^2$
47. $x^4 - 5x^2 + 4$
48. $y^4 + 2y^2 - 3$

0.4 Rational Expressions

Rational Expression

A **rational expression** is an algebraic expression of the form $\frac{P}{Q}$, where P and Q are simpler expressions (usually polynomials) and the denominator Q is not zero.

Quick Examples

1. $\frac{x^2 - 3x}{x}$ $\quad P = x^2 - 3x, Q = x$
2. $\frac{x + \frac{1}{x} + 1}{2x^2y + 1}$ $\quad P = x + \frac{1}{x} + 1, Q = 2x^2y + 1$
3. $3xy - x^2$ $\quad P = 3xy - x^2, Q = 1$

Algebra of Rational Expressions

We manipulate rational expressions in the same way that we manipulate fractions, using the following rules:

	Algebraic Rule	**Quick Examples**
Product:	$\frac{P}{Q} \cdot \frac{R}{S} = \frac{PR}{QS}$	$\frac{x+1}{x} \cdot \frac{x-1}{2x+1} = \frac{(x+1)(x-1)}{x(2x+1)} = \frac{x^2-1}{2x^2+x}$
Sum:	$\frac{P}{Q} + \frac{R}{S} = \frac{PS + RQ}{QS}$	$\frac{2x-1}{3x+2} + \frac{1}{x} = \frac{(2x-1)x + 1(3x+2)}{x(3x+2)} = \frac{2x^2+2x+2}{3x^2+2x}$
Difference:	$\frac{P}{Q} - \frac{R}{S} = \frac{PS - RQ}{QS}$	$\frac{x}{3x+2} - \frac{x-4}{x} = \frac{x^2 - (x-4)(3x+2)}{x(3x+2)} = \frac{-2x^2+10x+8}{3x^2+2x}$
Reciprocal:	$\frac{1}{\left(\frac{P}{Q}\right)} = \frac{Q}{P}$	$\frac{1}{\left(\frac{2xy}{3x-1}\right)} = \frac{3x-1}{2xy}$
Quotient:	$\frac{\left(\frac{P}{Q}\right)}{\left(\frac{R}{S}\right)} = \frac{P}{Q} \cdot \frac{S}{R} = \frac{PS}{QR}$	$\frac{\left(\frac{x}{x-1}\right)}{\left(\frac{y-1}{y}\right)} = \frac{xy}{(x-1)(y-1)} = \frac{xy}{xy - x - y + 1}$
Cancellation:	$\frac{P\cancel{R}}{Q\cancel{R}} = \frac{P}{Q}$	$\frac{(x-1)(xy+4)}{(x^2y-8)(x-1)} = \frac{xy+4}{x^2y-8}$

Caution Cancellation of summands is *invalid*. For instance,

$$\frac{\cancel{x} + (2xy^2 - y)}{\cancel{x} + 4y} = \frac{(2xy^2 - y)}{4y} \quad \text{✗ WRONG! Do *not* cancel a summand.}$$

$$\frac{\cancel{x}(2xy^2 - y)}{4\cancel{x}y} = \frac{(2xy^2 - y)}{4y}. \quad \text{✔ CORRECT Do cancel a factor.}$$

Here are some examples that require several algebraic operations.

EXAMPLE 1 Simplifying Rational Expressions

a. $$\frac{\left(\frac{1}{x+y} - \frac{1}{x}\right)}{y} = \frac{\left(\frac{x-(x+y)}{x(x+y)}\right)}{y} = \frac{\left(\frac{-y}{x(x+y)}\right)}{y} = \frac{-y}{xy(x+y)} = -\frac{1}{x(x+y)}$$

b. $$\frac{(x+1)(x+2)^2 - (x+1)^2(x+2)}{(x+2)^4} = \frac{(x+1)(x+2)[(x+2)-(x+1)]}{(x+2)^4}$$

$$= \frac{(x+1)(x+2)(x+2-x-1)}{(x+2)^4} = \frac{(x+1)(x+2)}{(x+2)^4} = \frac{x+1}{(x+2)^3}$$

c. $$\frac{2x\sqrt{x+1} - \frac{x^2}{\sqrt{x+1}}}{x+1} = \frac{\left(\frac{2x(\sqrt{x+1})^2 - x^2}{\sqrt{x+1}}\right)}{x+1} = \frac{2x(x+1) - x^2}{(x+1)\sqrt{x+1}}$$

$$= \frac{2x^2 + 2x - x^2}{(x+1)\sqrt{x+1}} = \frac{x^2 + 2x}{\sqrt{(x+1)^3}} = \frac{x(x+2)}{\sqrt{(x+1)^3}}$$

0.4 EXERCISES

Rewrite each expression in Exercises 1–16 as a single rational expression, simplified as much as possible.

1. $\frac{x-4}{x+1} \cdot \frac{2x+1}{x-1}$

2. $\frac{2x-3}{x-2} \cdot \frac{x+3}{x+1}$

3. $\frac{x-4}{x+1} + \frac{2x+1}{x-1}$

4. $\frac{2x-3}{x-2} + \frac{x+3}{x+1}$

5. $\frac{x^2}{x+1} - \frac{x-1}{x+1}$

6. $\frac{x^2-1}{x-2} - \frac{1}{x-1}$

7. $\frac{1}{\left(\frac{x}{x-1}\right)} + x - 1$

8. $\frac{2}{\left(\frac{x-2}{x^2}\right)} - \frac{1}{x-2}$

9. $\frac{1}{x}\left[\frac{x-3}{xy} + \frac{1}{y}\right]$

10. $\frac{y^2}{x}\left[\frac{2x-3}{y} + \frac{x}{y}\right]$

11. $\frac{(x+1)^2(x+2)^3 - (x+1)^3(x+2)^2}{(x+2)^6}$

12. $\frac{6x(x^2+1)^2(x^3+2)^3 - 9x^2(x^2+1)^3(x^3+2)^2}{(x^3+2)^6}$

13. $\frac{(x^2-1)\sqrt{x^2+1} - \frac{x^4}{\sqrt{x^2+1}}}{x^2+1}$

14. $\frac{x\sqrt{x^3-1} - \frac{3x^4}{\sqrt{x^3-1}}}{x^3-1}$

15. $\frac{\frac{1}{(x+y)^2} - \frac{1}{x^2}}{y}$

16. $\frac{\frac{1}{(x+y)^3} - \frac{1}{x^3}}{y}$

0.5 Solving Polynomial Equations

Polynomial Equation

A **polynomial equation** in one unknown is an equation that can be written in the form

$$ax^n + bx^{n-1} + \cdots + rx + s = 0,$$

where $a, b, \ldots, r$, and s are constants.

We call the largest exponent of x appearing in a nonzero term of a polynomial equation the **degree** of that polynomial equation.

Quick Examples

1. $3x + 1 = 0$ has degree 1 because the largest power of x that occurs is $x = x^1$. Degree 1 equations are called **linear** equations.
2. $x^2 - x - 1 = 0$ has degree 2 because the largest power of x that occurs is x^2. Degree 2 equations are also called **quadratic equations**, or just **quadratics**.
3. $x^3 = 2x^2 + 1$ is a degree 3 polynomial equation (or **cubic** equation) in disguise. It can be rewritten as $x^3 - 2x^2 - 1 = 0$, which is in the standard form for a degree 3 equation.
4. $x^4 - x = 0$ has degree 4. It is called a **quartic**.

Now comes the question: How do we solve these equations for x? This question was asked by mathematicians as early as 1600 BCE. Let's look at these equations one degree at a time.

Solution of Linear Equations

By definition a linear equation can be written in the form

$$ax + b = 0. \qquad \textit{a and b are fixed numbers with } a \neq 0.$$

Solving this is a nice mental exercise: Subtract b from both sides, and then divide by a, getting $x = -b/a$. Don't bother memorizing this formula; just go ahead and solve linear equations as they arise. If you feel you need practice, see the exercises at the end of the section.

Solution of Quadratic Equations

By definition a quadratic equation has the form

$$ax^2 + bx + c = 0. \qquad \textit{a, b, and c are fixed numbers and } a \neq 0.^{4}$$

The solutions of this equation are also called the **roots** of $ax^2 + bx + c$. We're assuming that you saw quadratic equations somewhere in high school but may be a

[4] What happens if $a = 0$?

little hazy about the details of their solution. There are two ways of solving these equations—one works sometimes, and the other works every time.

Solving Quadratic Equations by Factoring (works sometimes)

If we can factor[5] the left-hand side of a quadratic equation $ax^2 + bx + c = 0$, we can solve the equation by setting each factor equal to zero.

Quick Examples

5. $x^2 + 7x + 10 = 0$

$(x + 5)(x + 2) = 0$ — Factor the left-hand side.

$x + 5 = 0$ or $x + 2 = 0$ — If a product is zero, one or both factors are zero.

Solutions: $x = -5$ and $x = -2$

6. $2x^2 - 5x - 12 = 0$

$(2x + 3)(x - 4) = 0$ — Factor the left-hand side.

$2x + 3 = 0$ or $x - 4 = 0$

Solutions: $x = -\frac{3}{2}$ and $x = 4$

Test for Factoring

The quadratic $ax^2 + bx + c$, with a, b, and c being integers (whole numbers), factors into an expression of the form $(rx + s)(tx + u)$ with r, s, t, and u integers precisely when the quantity $b^2 - 4ac$ is a perfect square. (That is, it is the square of an integer.) If this happens, we say that the quadratic **factors over the integers**.

Quick Examples

7. $x^2 + x + 1$ has $a = 1$, $b = 1$, and $c = 1$, so $b^2 - 4ac = -3$, which is not a perfect square. Therefore, this quadratic does not factor over the integers.

8. $2x^2 - 5x - 12$ has $a = 2$, $b = -5$, and $c = -12$, so $b^2 - 4ac = 121$. Because $121 = 11^2$, this quadratic does factor over the integers. (We factored it above.)

Solving Quadratic Equations with the Quadratic Formula (works every time)

The solutions of the general quadratic $ax^2 + bx + c = 0$ $(a \neq 0)$ are given by

$$x = \frac{-b \pm \sqrt{b^2 - 4ac}}{2a}.$$

[5] See Section 0.3 for a review of how to factor quadratics.

We call the quantity $\Delta = b^2 - 4ac$ the **discriminant** of the quadratic (Δ is the Greek letter delta), and we have the following general rules:

- If Δ is positive, there are two distinct real solutions.
- If Δ is zero, there is only one real solution: $x = -\frac{b}{2a}$. (Why?)
- If Δ is negative, there are no real solutions.

Quick Examples

9. $2x^2 - 5x - 12 = 0$ has $a = 2$, $b = -5$, and $c = -12$.

$$x = \frac{-b \pm \sqrt{b^2 - 4ac}}{2a} = \frac{5 \pm \sqrt{25 + 96}}{4} = \frac{5 \pm \sqrt{121}}{4} = \frac{5 \pm 11}{4}$$

$$= \frac{16}{4} \text{ or } -\frac{6}{4} = 4 \text{ or } -\frac{3}{2}$$

Δ is positive in this example.

10. $4x^2 = 12x - 9$ can be rewritten as $4x^2 - 12x + 9 = 0$, which has $a = 4$, $b = -12$, and $c = 9$.

$$x = \frac{-b \pm \sqrt{b^2 - 4ac}}{2a} = \frac{12 \pm \sqrt{144 - 144}}{8} = \frac{12 \pm 0}{8} = \frac{12}{8} = \frac{3}{2}$$

Δ is zero in this example.

11. $x^2 + 2x - 1 = 0$ has $a = 1$, $b = 2$, and $c = -1$.

$$x = \frac{-b \pm \sqrt{b^2 - 4ac}}{2a} = \frac{-2 \pm \sqrt{8}}{2} = \frac{-2 \pm 2\sqrt{2}}{2} = -1 \pm \sqrt{2}$$

The two solutions are $x = -1 + \sqrt{2} = 0.414\ldots$ and $x = -1 - \sqrt{2} = -2.414\ldots$.

Δ is positive in this example.

12. $x^2 + x + 1 = 0$ has $a = 1$, $b = 1$, and $c = 1$. Because $\Delta = -3$ is negative, there are no real solutions.

Δ is negative in this example.

Q: *This is all very useful, but where does the quadratic formula come from?*

A: To see where it comes from, we will solve a general quadratic equation using brute force. Start with the general quadratic equation,

$$ax^2 + bx + c = 0.$$

First, divide out the nonzero number a to get

$$x^2 + \frac{bx}{a} + \frac{c}{a} = 0.$$

Now we **complete the square**: Add and subtract the quantity $\frac{b^2}{4a^2}$ to get

$$x^2 + \frac{bx}{a} + \frac{b^2}{4a^2} - \frac{b^2}{4a^2} + \frac{c}{a} = 0.$$

We do this to get the first three terms to factor as a perfect square:

$$\left(x + \frac{b}{2a}\right)^2 - \frac{b^2}{4a^2} + \frac{c}{a} = 0.$$

(Check this by multiplying out.) Adding $\frac{b^2}{4a^2} - \frac{c}{a}$ to both sides gives

$$\left(x + \frac{b}{2a}\right)^2 = \frac{b^2}{4a^2} - \frac{c}{a} = \frac{b^2 - 4ac}{4a^2}.$$

Taking square roots gives

$$x + \frac{b}{2a} = \frac{\pm\sqrt{b^2 - 4ac}}{2a}.$$

Finally, adding $-\frac{b}{2a}$ to both sides yields the result

$$x = -\frac{b}{2a} + \frac{\pm\sqrt{b^2 - 4ac}}{2a}$$

or

$$x = \frac{-b \pm \sqrt{b^2 - 4ac}}{2a}.$$

Solution of Cubic Equations

By definition, a cubic equation can be written in the form

$$ax^3 + bx^2 + cx + d = 0. \qquad a, b, c, \text{ and } d \text{ are fixed numbers, and } a \neq 0.$$

Now we get into something of a bind. Although there is a perfectly respectable formula for the solutions, it is very complicated and involves the use of complex numbers rather heavily.[6] So we discuss instead a much simpler method that *sometimes* works nicely. Here is the method in a nutshell.

Solving Cubics by Finding One Factor

Start with a given cubic equation $ax^3 + bx^2 + cx + d = 0$.

Step 1 By trial and error, find one solution $x = s$. If a, b, c, and d are integers, the only possible *rational* solutions[7] are those of the form $s = \pm$ (factor of d)/(factor of a).

Step 2 It will now be possible to factor the cubic as

$$ax^3 + bx^2 + cx + d = (x - s)(ax^2 + ex + f) = 0.$$

To find $ax^2 + ex + f$, divide the cubic by $x - s$, using long division.[8]

[6] It was when this formula was discovered in the sixteenth century that complex numbers were first taken seriously. Although we would like to show you the formula, it is too large to fit in this footnote.

[7] There may be *irrational* solutions, however; for example, $x^3 - 2 = 0$ has the single solution $x = \sqrt[3]{2}$.

[8] Alternatively, use synthetic division, a shortcut that would take us too far afield to describe.

Step 3 The factored equation says that either $x - s = 0$ or $ax^2 + ex + f = 0$. We already know that s is a solution, and now we see that the other solutions are the roots of the quadratic. Note that this quadratic may or may not have any real solutions, as usual.

Quick Example

13. To solve the cubic $x^3 - x^2 + x - 1 = 0$, we first find a single solution. Here, $a = 1$ and $d = -1$. Because the only factors of ± 1 are ± 1, the only possible rational solutions are $x = \pm 1$. By substitution we see that $x = 1$ is a solution. Thus, $(x - 1)$ is a factor. Dividing by $(x - 1)$ yields the quotient $(x^2 + 1)$. Thus,

$$x^3 - x^2 + x - 1 = (x - 1)(x^2 + 1) = 0,$$

so either $x - 1 = 0$ or $x^2 + 1 = 0$.

Because the discriminant of the quadratic $x^2 + 1$ is negative, we don't get any real solutions from $x^2 + 1 = 0$, so the only real solution is $x = 1$.

Possible Outcomes When Solving a Cubic Equation

If you consider all the cases, there are three possible outcomes when solving a cubic equation:

1. One real solution (as in Quick Example 13)
2. Two real solutions (try, for example, $x^3 + x^2 - x - 1 = 0$)
3. Three real solutions (see the next example)

EXAMPLE 1 Solving a Cubic

Solve the cubic $2x^3 - 3x^2 - 17x + 30 = 0$.

Solution First, we look for a single solution. Here, $a = 2$ and $d = 30$. The factors of a are ± 1 and ± 2, and the factors of d are $\pm 1, \pm 2, \pm 3, \pm 5, \pm 6, \pm 10, \pm 15$, and ± 30. This gives us a large number of possible ratios: $\pm 1, \pm 2, \pm 3, \pm 5, \pm 6, \pm 10, \pm 15, \pm 30, \pm 1/2, \pm 3/2, \pm 5/2, \pm 15/2$. Undaunted, we first try $x = 1$ and $x = -1$, getting nowhere. So we move on to $x = 2$, and we hit the jackpot, because substituting $x = 2$ gives $16 - 12 - 34 + 30 = 0$. Thus, $(x - 2)$ is a factor. Dividing yields the quotient $2x^2 + x - 15$. Here is the calculation:

$$\begin{array}{r|l} & 2x^2 + x - 15 \\ \hline x - 2 & 2x^3 - 3x^2 - 17x + 30 \\ & \underline{2x^3 - 4x^2} \\ & \qquad x^2 - 17x \\ & \qquad \underline{x^2 - 2x} \\ & \qquad\qquad -15x + 30 \\ & \qquad\qquad \underline{-15x + 30} \\ & \qquad\qquad\qquad 0. \end{array}$$

Thus,

$$2x^3 - 3x^2 - 17x + 30 = (x - 2)(2x^2 + x - 15) = 0.$$

Setting the factors equal to zero gives either $x - 2 = 0$ or $2x^2 + x - 15 = 0$. We could solve the quadratic using the quadratic formula, but luckily, we notice that it factors as

$$2x^2 + x - 15 = (x + 3)(2x - 5).$$

Thus, the solutions are $x = 2$, $x = -3$, and $x = 5/2$.

Solution of Higher Order Polynomial Equations

Logically speaking, our next step should be a discussion of quartics, then quintics (fifth degree equations), and so on forever. Well, we've got to stop somewhere, and cubics may be as good a place as any. On the other hand, since we've gotten this far, we ought to at least tell you what is known about higher order polynomials.

Quartics Just as in the case of cubics, there is a formula to find the solutions of quartics.[9]

Quintics and Beyond All good things must come to an end, we're afraid. It turns out that there is no "quintic formula." In other words, there is no single algebraic formula or collection of algebraic formulas that gives the solutions to all quintics. This question was settled by the Norwegian mathematician Niels Henrik Abel in 1824 after almost 300 years of controversy about this question. (In fact, several notable mathematicians had previously claimed to have devised formulas for solving the quintic, but these were all shot down by other mathematicians—this being one of the favorite pastimes of practitioners of our art.) The same negative answer applies to polynomial equations of degree 6 and higher. It's not that these equations don't have solutions; it's just that the solutions can't be found by using algebraic formulas.[10] However, there are certain special classes of polynomial equations that can be solved with algebraic methods. The way of identifying such equations was discovered around 1829 by the French mathematician Évariste Galois.[11]

[9] See, for example, *First Course in the Theory of Equations* by L. E. Dickson (New York: Wiley, 1922) or *Modern Algebra* by B. L. van der Waerden (New York: Frederick Ungar, 1953).

[10] What we mean by an "algebraic formula" is a formula in the coefficients using the operations of addition, subtraction, multiplication, division, and the taking of radicals. Mathematicians call the use of such formulas in solving polynomial equations "solution by radicals." If you were a math major, you would eventually go on to study this under the heading of Galois theory.

[11] Both Abel (1802–1829) and Galois (1811–1832) died young. Abel died of tuberculosis at the age of 26. Galois was killed in a duel at the age of 20.

0.5 EXERCISES

Solve the equations in Exercises 1–12 for x (mentally, if possible).

1. $x + 1 = 0$ **2.** $x - 3 = 1$ **3.** $-x + 5 = 0$

4. $2x + 4 = 1$ **5.** $4x - 5 = 8$ **6.** $\frac{3}{4}x + 1 = 0$

7. $7x + 55 = 98$ **8.** $3x + 1 = x$ **9.** $x + 1 = 2x + 2$

10. $x + 1 = 3x + 1$ **11.** $ax + b = c \quad (a \neq 0)$

12. $x - 1 = cx + d \quad (c \neq 1)$

By any method, determine all possible real solutions of each equation in Exercises 13–30. Check your answers by substitution.

13. $2x^2 + 7x - 4 = 0$

14. $x^2 + x + 1 = 0$

15. $x^2 - x + 1 = 0$

16. $2x^2 - 4x + 3 = 0$

17. $2x^2 - 5 = 0$

18. $3x^2 - 1 = 0$

19. $-x^2 - 2x - 1 = 0$

20. $2x^2 - x - 3 = 0$

21. $\frac{1}{2}x^2 - x - \frac{3}{2} = 0$

22. $-\frac{1}{2}x^2 - \frac{1}{2}x + 1 = 0$

23. $x^2 - x = 1$

24. $16x^2 = -24x - 9$

25. $x = 2 - \frac{1}{x}$

26. $x + 4 = \frac{1}{x - 2}$

27. $x^4 - 10x^2 + 9 = 0$

28. $x^4 - 2x^2 + 1 = 0$

29. $x^4 + x^2 - 1 = 0$

30. $x^3 + 2x^2 + x = 0$

Find all possible real solutions of each equation in Exercises 31–44.

31. $x^3 + 6x^2 + 11x + 6 = 0$

32. $x^3 - 6x^2 + 12x - 8 = 0$

33. $x^3 + 4x^2 + 4x + 3 = 0$

34. $y^3 + 64 = 0$

35. $x^3 - 1 = 0$

36. $x^3 - 27 = 0$

37. $y^3 + 3y^2 + 3y + 2 = 0$

38. $y^3 - 2y^2 - 2y - 3 = 0$

39. $x^3 - x^2 - 5x + 5 = 0$

40. $x^3 - x^2 - 3x + 3 = 0$

41. $2x^6 - x^4 - 2x^2 + 1 = 0$

42. $3x^6 - x^4 - 12x^2 + 4 = 0$

43. $(x^2 + 3x + 2)(x^2 - 5x + 6) = 0$

44. $(x^2 - 4x + 4)^2(x^2 + 6x + 5)^3 = 0$

0.6 Solving Miscellaneous Equations

Equations often arise in calculus that are not polynomial equations of low degree. Many of these complicated-looking equations can be solved easily if you remember the following, which we used in the previous section.

Solving an Equation of the Form $P \cdot Q = 0$

If a product is equal to 0, then at least one of the factors must be 0. That is, if $P \cdot Q = 0$, then either $P = 0$ or $Q = 0$.

Quick Examples

1. $x^5 - 4x^3 = 0$

$x^3(x^2 - 4) = 0$ — Factor the left-hand side.

Either $x^3 = 0$ or $x^2 - 4 = 0$ — Either $P = 0$ or $Q = 0$.

$x = 0, 2$ or -2. — Solve the individual equations.

2. $(x^2 - 1)(x + 2) + (x^2 - 1)(x + 4) = 0$

$(x^2 - 1)[(x + 2) + (x + 4)] = 0$ — Factor the left-hand side.

$(x^2 - 1)(2x + 6) = 0$

Either $x^2 - 1 = 0$ or $2x + 6 = 0$ — Either $P = 0$ or $Q = 0$.

$x = -3, -1,$ or 1. — Solve the individual equations.

EXAMPLE 1 **Solving by Factoring**

Solve $12x(x^2 - 4)^5(x^2 + 2)^6 + 12x(x^2 - 4)^6(x^2 + 2)^5 = 0$.

Solution We start by factoring the left-hand side:

$$\begin{aligned}&12x(x^2 - 4)^5(x^2 + 2)^6 + 12x(x^2 - 4)^6(x^2 + 2)^5\\&= 12x(x^2 - 4)^5(x^2 + 2)^5[(x^2 + 2) + (x^2 - 4)]\\&= 12x(x^2 - 4)^5(x^2 + 2)^5(2x^2 - 2)\\&= 24x(x^2 - 4)^5(x^2 + 2)^5(x^2 - 1).\end{aligned}$$

Setting this equal to 0, we get

$$24x(x^2 - 4)^5(x^2 + 2)^5(x^2 - 1) = 0,$$

which means that at least one of the factors of this product must be zero. It certainly cannot be the 24, but it could be the x: $x = 0$ is one solution. It could also be that

$$(x^2 - 4)^5 = 0$$

or

$$x^2 - 4 = 0,$$

which has solutions $x = \pm 2$. Could it be that $(x^2 + 2)^5 = 0$? If so, then $x^2 + 2 = 0$, but this is impossible because $x^2 + 2 \geq 2$ no matter what x is. Finally, it could be that $x^2 - 1 = 0$, which has solutions $x = \pm 1$. This gives us five solutions to the original equation:

$$x = -2, -1, 0, 1, \text{ or } 2.$$

EXAMPLE 2 **Solving by Factoring**

Solve $(x^2 - 1)(x^2 - 4) = 10$.

Solution Watch out! You may be tempted to say that $x^2 - 1 = 10$ or $x^2 - 4 = 10$, but this does not follow. If two numbers multiply to give you 10, what must they be? There are lots of possibilities: 2 and 5, 1 and 10, and $-500{,}000$ and -0.00002 are just a few. The fact that the left-hand side is factored is nearly useless to us if we want to solve this equation. What we will have to do is multiply out, bring the 10 over to the left, and hope that we can factor what we get. Here goes:

$$\begin{aligned}x^4 - 5x^2 + 4 &= 10\\x^4 - 5x^2 - 6 &= 0\\(x^2 - 6)(x^2 + 1) &= 0.\end{aligned}$$

(Here, we used a sometimes useful trick that we mentioned in Section 0.3: We treated x^2 like x and x^4 like x^2, so factoring $x^4 - 5x^2 - 6$ is essentially the same as factoring $x^2 - 5x - 6$.) *Now* we are allowed to say that one of the factors must be 0: $x^2 - 6 = 0$ has solutions $x = \pm\sqrt{6} = \pm 2.449\ldots$, and $x^2 + 1 = 0$ has no real solutions. Therefore, we get exactly two solutions: $x = \pm\sqrt{6} = \pm 2.449\ldots$.

To solve equations involving rational expressions, the following rule is very useful.

Solving an Equation of the Form $P/Q = 0$

If $\dfrac{P}{Q} = 0$, then $P = 0$.

How else could a fraction equal 0? If that is not convincing, multiply both sides by Q (which cannot be 0 if the quotient is defined).

Quick Example

3. $\dfrac{(x+1)(x+2)^2 - (x+1)^2(x+2)}{(x+2)^4} = 0$

$(x+1)(x+2)^2 - (x+1)^2(x+2) = 0$ — If $\frac{P}{Q} = 0$, then $P = 0$.

$(x+1)(x+2)[(x+2) - (x+1)] = 0$ — Factor.

$(x+1)(x+2)(1) = 0$

Either $x + 1 = 0$ or $x + 2 = 0$,

$x = -1$ or $x = -2$

$x = -1$ — $x = -2$ does not make sense in the original equation: It makes the denominator 0. So it is not a solution, and $x = -1$ is the only solution.

EXAMPLE 3 Solving a Rational Equation

Solve $1 - \dfrac{1}{x^2} = 0$.

Solution Write 1 as $\frac{1}{1}$ so that we now have a difference of two rational expressions:

$$\frac{1}{1} - \frac{1}{x^2} = 0.$$

To combine these, we can put both over a common denominator of x^2, which gives

$$\frac{x^2 - 1}{x^2} = 0.$$

Now we can set the numerator, $x^2 - 1$, equal to zero. Thus,

$$x^2 - 1 = 0,$$

so

$$(x-1)(x+1) = 0,$$

giving $x = \pm 1$.

➡ **Before we go on . . .** This equation could also have been solved by writing

$$1 = \frac{1}{x^2}$$

and then multiplying both sides by x^2. ■

EXAMPLE 4 **Another Rational Equation**

Solve $\frac{2x-1}{x} + \frac{3}{x-2} = 0.$

Solution We *could* first perform the addition on the left and then set the top equal to 0, but here is another approach. Subtracting the second expression from both sides gives

$$\frac{2x-1}{x} = \frac{-3}{x-2}.$$

Cross-multiplying [multiplying both sides by both denominators—that is, by $x(x-2)$] now gives

$$(2x-1)(x-2) = -3x,$$

so

$$2x^2 - 5x + 2 = -3x.$$

Adding $3x$ to both sides gives the quadratic equation

$$2x^2 - 2x + 2 = 0.$$

The discriminant is $(-2)^2 - 4 \cdot 2 \cdot 2 = -8 < 0$, so we conclude that there is no real solution.

➡ **Before we go on . . .** Notice that when we said that $(2x-1)(x-2) = -3x$, we were *not* allowed to conclude that $2x - 1 = -3x$ or $x - 2 = -3x$. ■

EXAMPLE 5 **A Rational Equation with Radicals**

Solve $\frac{\left(2x\sqrt{x+1} - \frac{x^2}{\sqrt{x+1}}\right)}{x+1} = 0.$

Solution Setting the top equal to 0 gives

$$2x\sqrt{x+1} - \frac{x^2}{\sqrt{x+1}} = 0.$$

This still involves fractions. To get rid of the fractions, we could put everything over a common denominator $(\sqrt{x+1})$ and then set the top equal to 0, or we could multiply the whole equation by that common denominator in the first place to clear fractions. If we do the second, we get

$$2x(x+1) - x^2 = 0$$
$$2x^2 + 2x - x^2 = 0$$
$$x^2 + 2x = 0.$$

Factoring, we have

$$x(x+2) = 0,$$

so either $x = 0$ or $x + 2 = 0$, giving us $x = 0$ or $x = -2$. Again, one of these is not really a solution. The problem is that $x = -2$ cannot be substituted into $\sqrt{x+1}$, because we would then have to take the square root of -1, and we are not allowing ourselves to do that. Therefore, $x = 0$ is the only solution.

0.6 EXERCISES

Solve the equations in Exercises 1–26.

1. $x^4 - 3x^3 = 0$
2. $x^6 - 9x^4 = 0$
3. $x^4 - 4x^2 = -4$
4. $x^4 - x^2 = 6$
5. $(x+1)(x+2) + (x+1)(x+3) = 0$
6. $(x+1)(x+2)^2 + (x+1)^2(x+2) = 0$
7. $(x^2+1)^5(x+3)^4 + (x^2+1)^6(x+3)^3 = 0$
8. $10x(x^2+1)^4(x^3+1)^5 - 10x^2(x^2+1)^5(x^3+1)^4 = 0$
9. $(x^3+1)\sqrt{x+1} - (x^3+1)^2\sqrt{x+1} = 0$
10. $(x^2+1)\sqrt{x+1} - \sqrt{(x+1)^3} = 0$
11. $\sqrt{(x+1)^3} + \sqrt{(x+1)^5} = 0$
12. $(x^2+1)\sqrt[3]{(x+1)^4} - \sqrt[3]{(x+1)^7} = 0$
13. $(x+1)^2(2x+3) - (x+1)(2x+3)^2 = 0$
14. $(x^2-1)^2(x+2)^3 - (x^2-1)^3(x+2)^2 = 0$
15. $\dfrac{(x+1)^2(x+2)^3 - (x+1)^3(x+2)^2}{(x+2)^6} = 0$
16. $\dfrac{6x(x^2+1)^2(x^2+2)^4 - 8x(x^2+1)^3(x^2+2)^3}{(x^2+2)^8} = 0$
17. $\dfrac{2(x^2-1)\sqrt{x^2+1} - \frac{x^4}{\sqrt{x^2+1}}}{x^2+1} = 0$
18. $\dfrac{4x\sqrt{x^3-1} - \frac{3x^4}{\sqrt{x^3-1}}}{x^3-1} = 0$
19. $x - \dfrac{1}{x} = 0$
20. $1 - \dfrac{4}{x^2} = 0$
21. $\dfrac{1}{x} - \dfrac{9}{x^3} = 0$
22. $\dfrac{1}{x^2} - \dfrac{1}{x+1} = 0$
23. $\dfrac{x-4}{x+1} - \dfrac{x}{x-1} = 0$
24. $\dfrac{2x-3}{x-1} - \dfrac{2x+3}{x+1} = 0$
25. $\dfrac{x+4}{x+1} + \dfrac{x+4}{3x} = 0$
26. $\dfrac{2x-3}{x} - \dfrac{2x-3}{x+1} = 0$

0.7 The Coordinate Plane

Q: *Just what is the xy-plane?*

A: The *xy*-plane is an infinite flat surface with two perpendicular lines, usually labeled the ***x*-axis** and ***y*-axis**. These axes are calibrated as shown in Figure 2. (Notice also how the plane is divided into four **quadrants**.)

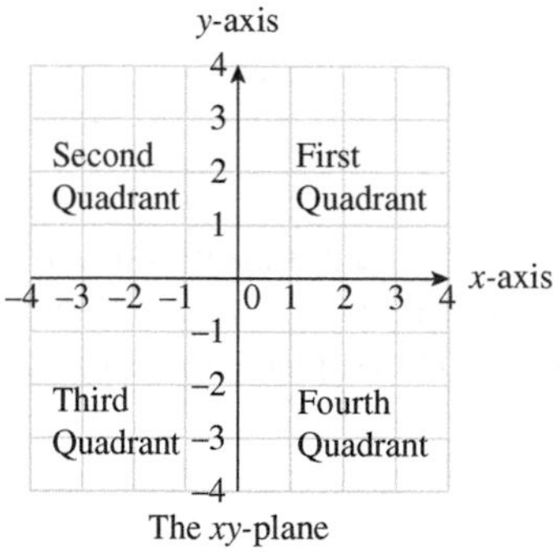

y-axis
x-axis
Second Quadrant
First Quadrant
Third Quadrant
Fourth Quadrant
The xy-plane

Figure 2

The xy-plane is nothing more than a very large—in fact, infinitely large—flat surface. The purpose of the axes is to allow us to locate specific positions, or **points**, on the plane, with the use of **coordinates**. (If Captain Picard wants to have himself beamed to a specific location, he must supply its coordinates, or he's in trouble.)

Q: *So how do we use coordinates to locate points?*

A: The rule is simple: Each point in the plane has two coordinates, an ***x*-coordinate** and a ***y*-coordinate**. These can be determined in two ways:

1. The x-coordinate measures a point's distance to the right or left of the y-axis. It is positive if the point is to the right of the axis, negative if it is to the left of the axis, and 0 if it is on the axis. The y-coordinate measures a point's distance above or below the x-axis. It is positive if the point is above the axis, negative if it is below the axis, and 0 if it is on the axis. Briefly, the x-coordinate tells us the *horizontal* position (distance left or right), and the y-coordinate tells us the *vertical* position (height).

2. Given a point P, we get its x-coordinate by drawing a vertical line from P and seeing where it intersects the x-axis. Similarly, we get the y-coordinate by extending a horizontal line from P and seeing where it intersects the y-axis.

This way of assigning coordinates to points in the plane is often called the system of **Cartesian** coordinates, in honor of the mathematician and philosopher René Descartes (1596–1650), who was the first to use them extensively.

Here are a few examples to help you review coordinates.

EXAMPLE 1 Coordinates of Points

a. Find the coordinates of the indicated points. (See Figure 3. The grid lines are placed at intervals of 1 unit.)

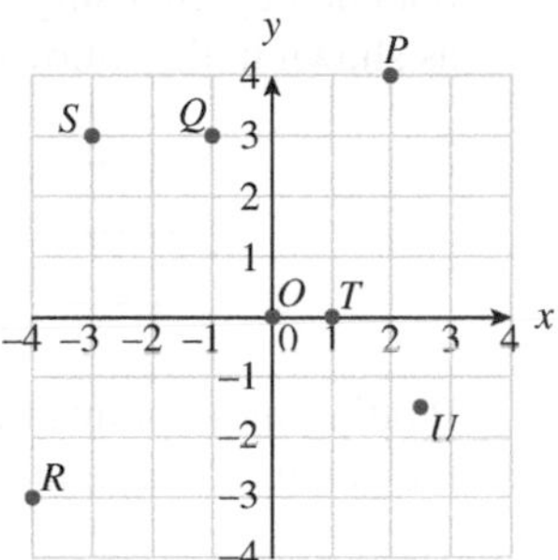

Figure 3

b. Locate the following points in the xy-plane:

$$A(2, 3), B(-4, 2), C(3, -2.5), D(0, -3), E(3.5, 0), F(-2.5, -1.5).$$

Solution

a. Taking the points in alphabetical order, we start with the origin O. This point has height zero and is also zero units to the right of the y-axis, so its coordinates are $(0, 0)$. Turning to P, dropping a vertical line gives $x = 2$ and extending a horizontal line gives $y = 4$. Thus, P has coordinates $(2, 4)$. For practice, determine the coordinates of the remaining points, and check your work against the list that follows:

$$Q(-1, 3), R(-4, -3), S(-3, 3), T(1, 0), U(2.5, -1.5).$$

b. To locate the given points, we start at the origin $(0, 0)$, and proceed as follows. (See Figure 4.)

To locate A, we move 2 units to the right and 3 up, as shown.

To locate B, we move -4 units to the right (that is, 4 to the *left*) and 2 up, as shown.

To locate C, we move 3 units right and 2.5 down.

We locate the remaining points in a similar way.

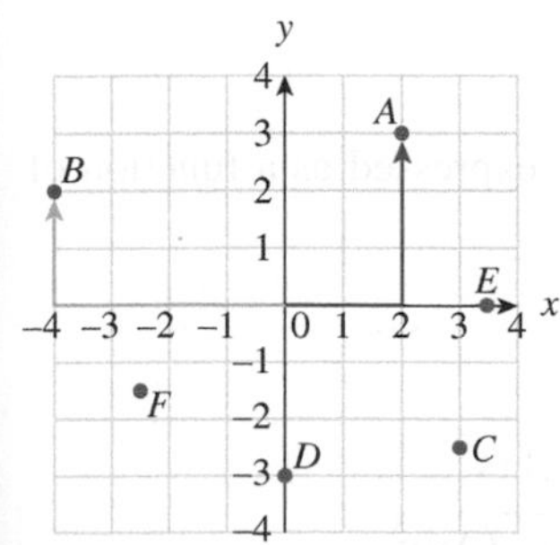

Figure 4

The Graph of an Equation

One of the more surprising developments of mathematics was the realization that equations, which are algebraic objects, can be represented by graphs, which are geometric objects. The kinds of equations that we have in mind are equations in x and y, such as

$$y = 4x - 1, \quad 2x^2 - y = 0, \quad y = 3x^2 + 1, \quad y = \sqrt{x - 1}.$$

The **graph** of an equation in the two variables x and y consists of all points (x, y) in the plane whose coordinates are solutions of the equation.

EXAMPLE 2 Graph of an Equation

Obtain the graph of the equation $y - x^2 = 0$.

Solution We can solve the equation for y to obtain $y = x^2$. Solutions can then be obtained by choosing values for x and then computing y by squaring the value of x, as shown in the following table:

x	-3	-2	-1	0	1	2	3
$y = x^2$	9	4	1	0	1	4	9

Plotting these points (x, y) gives the picture on the left side of Figure 5, suggesting the graph on the right in Figure 5.

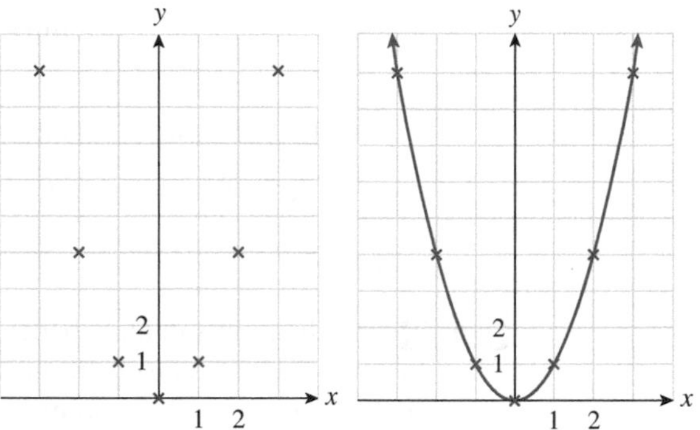

Figure 5

Distance

The distance between two points in the xy-plane can be expressed as a function of their coordinates, as follows.

Distance Formula

The distance between the points $P(x_1, y_1)$ and $Q(x_2, y_2)$ is

$$d = \sqrt{(x_2 - x_1)^2 + (y_2 - y_1)^2} = \sqrt{(\Delta x)^2 + (\Delta y)^2}.$$

Derivation

The distance d is shown in the figure below:

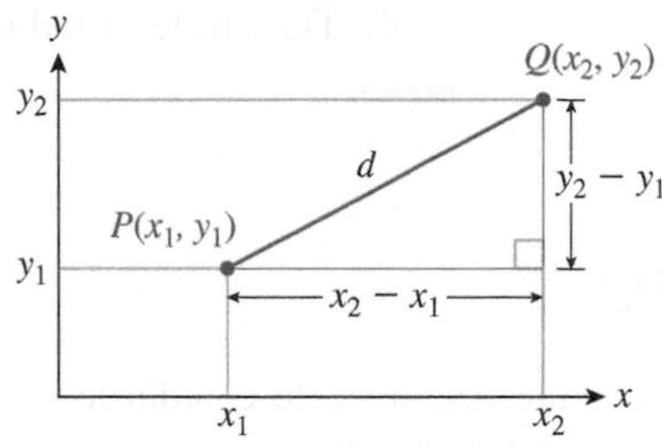

By the Pythagorean theorem applied to the right triangle shown, we get

$$d^2 = (x_2 - x_1)^2 + (y_2 - y_1)^2.$$

Taking square roots (d is a distance, so we take the positive square root), we get the distance formula. Notice that if we switch x_1 with x_2 or y_1 with y_2, we get the same result.

Quick Examples

1. The distance between the points $(3, -2)$ and $(-1, 1)$ is

$$d = \sqrt{(-1 - 3)^2 + (1 + 2)^2} = \sqrt{25} = 5.$$

2. The distance from (x, y) to the origin $(0, 0)$ is

$$d = \sqrt{(x - 0)^2 + (y - 0)^2} = \sqrt{x^2 + y^2}.$$ Distance to the origin

The set of all points (x, y) whose distance from the origin $(0, 0)$ is a fixed quantity r is a circle centered at the origin with radius r. From the second Quick Example we get the following equation for the circle centered at the origin with radius r:

$$\sqrt{x^2 + y^2} = r.$$ Distance from the origin $= r$

Squaring both sides gives the following equation.

Equation of the Circle of Radius *r* Centered at the Origin

$$x^2 + y^2 = r^2$$

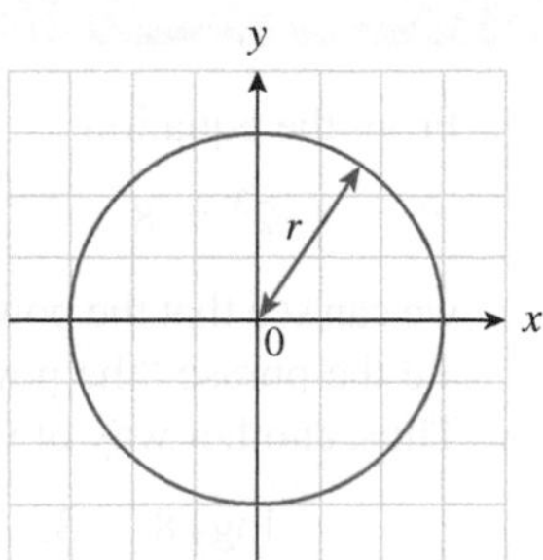

Quick Examples

3. The circle of radius 1 centered at the origin has equation $x^2 + y^2 = 1$.
4. The circle of radius 2 centered at the origin has equation $x^2 + y^2 = 4$.

0.7 EXERCISES

1. Referring to the following figure, determine the coordinates of the indicated points as accurately as you can.

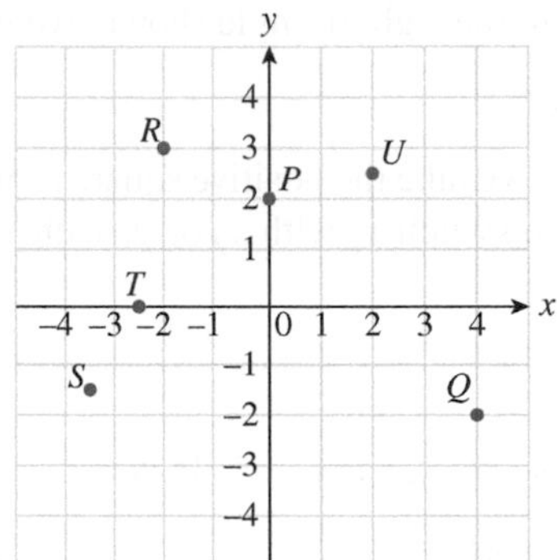

2. Referring to the following figure, determine the coordinates of the indicated points as accurately as you can.

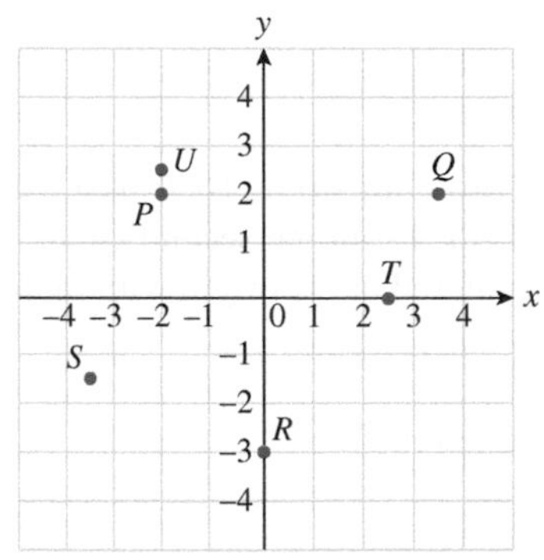

3. Graph the following points:

   $P(4, 4), Q(4, -4), R(3, 0), S(4, 0.5), T(0.5, 2.5), U(-2, 0), V(-4, 4)$

4. Graph the following points:

   $P(4, -2), Q(2, -4), R(1, -3), S(-4, 2), T(2, -1), U(-2, 0), V(-4, -4)$

Sketch the graphs of the equations in Exercises 5–12.

5. $x + y = 1$
6. $y - x = -1$
7. $2y - x^2 = 1$
8. $2y + \sqrt{x} = 1$
9. $xy = 4$
10. $x^2y = -1$
11. $xy = x^2 + 1$
12. $xy = 2x^3 + 1$

In Exercises 13–16, find the distance between the given pairs of points.

13. $(1, -1)$ and $(2, -2)$
14. $(1, 0)$ and $(6, 1)$
15. $(a, 0)$ and $(0, b)$
16. (a, a) and (b, b)
17. Find the value of k such that $(1, k)$ is equidistant from $(0, 0)$ and $(2, 1)$.
18. Find the value of k such that (k, k) is equidistant from $(-1, 0)$ and $(0, 2)$.
19. Describe the set of points (x, y) such that $x^2 + y^2 = 9$.
20. Describe the set of points (x, y) such that $x^2 + y^2 = 0$.

0.8 Logarithms

From the equation

$$2^3 = 8$$

we can see that the power to which we need to raise 2 in order to get 8 is 3. We abbreviate the phrase "the power to which we need to raise 2 in order to get 8" as "$\log_2 8$." Thus, another way of writing the equation $2^3 = 8$ is

$$\log_2 8 = 3. \quad \text{The power to which we need to raise 2 in order to get 8 is 3.}$$

This is read "The base 2 logarithm of 8 is 3" or "The log, base 2, of 8 is 3."

The general definition is as follows.

Base *b* Logarithm

The **base *b* logarithm of *x***, $\log_b x$, is the power to which we need to raise b in order to get x. Symbolically,

$$\log_b x = y \quad \text{means} \quad b^y = x.$$

Logarithmic form — Exponential form

Quick Examples

The following table lists some exponential equations and their equivalent logarithmic forms:

Exponential Form	$10^3 = 1{,}000$	$4^2 = 16$	$5^1 = 5$	$7^0 = 1$	$4^{-2} = \frac{1}{16}$	$25^{1/2} = 5$
Logarithmic Form	$\log_{10} 1{,}000 = 3$	$\log_4 16 = 2$	$\log_5 5 = 1$	$\log_7 1 = 0$	$\log_4 \frac{1}{16} = -2$	$\log_{25} 5 = \frac{1}{2}$

1. $\log_2 8$ = the power to which we need to raise 2 in order to get 8.

   Because $2^{\boxed{3}} = 8$, this power is 3, so $\log_2 8 = 3$.

2. $\log_2 16$ = the power to which we need to raise 2 in order to get 16.

   Because $2^{\boxed{4}} = 16$, this power is 4, so $\log_2 16 = 4$.

3. $\log_2 2$ = the power to which we need to raise 2 in order to get 2.

   Because $2^{\boxed{1}} = 2$, this power is 1, so $\log_2 2 = 1$.

4. $\log_3 9$ = the power to which we need to raise 3 in order to get 9.

   Because $3^{\boxed{2}} = 9$, this power is 2, so $\log_3 9 = 2$.

5. $\log_3 27$ = the power to which we need to raise 3 in order to get 27.

   Because $3^{\boxed{3}} = 27$, this power is 3, so $\log_3 27 = 3$.

6. $\log_3 3$ = the power to which we need to raise 3 in order to get 3.

   Because $3^{\boxed{1}} = 3$, this power is 1, so $\log_3 3 = 1$.

7. $\log_{10} 10{,}000$ = the power to which we need to raise 10 in order to get 10,000.

   Because $10^{\boxed{4}} = 10{,}000$, this power is 4, so $\log_{10} 10{,}000 = 4$.

8. $\log_{10} 1$ = the power to which we need to raise 10 in order to get 1.

   Because $10^{\boxed{0}} = 1$, this power is 0, so $\log_{10} 1 = 0$.

9. $\log_2 1$ = the power to which we need to raise 2 in order to get 1.

   Because $2^{\boxed{0}} = 1$, this power is 0, so $\log_2 1 = 0$.

10. $\log_3 1$ = the power to which we need to raise 3 in order to get 1.

    Because $3^{\boxed{0}} = 1$, this power is 0, so $\log_3 1 = 0$.

11. $\log_2 \frac{1}{2}$ is the power to which we need to raise 2 in order to get $\frac{1}{2}$.

Because $2^{\boxed{-1}} = \frac{1}{2}$, this power is -1, so $\log_2 \frac{1}{2} = -1$.

12. $\log_2 \frac{1}{4}$ is the power to which we need to raise 2 in order to get $\frac{1}{4}$.

Because $2^{\boxed{-2}} = \frac{1}{4}$, this power is -2, so $\log_2 \frac{1}{4} = -2$.

13. $\log_3 \frac{1}{27}$ is the power to which we need to raise 3 in order to get $\frac{1}{27}$.

Because $3^{\boxed{-3}} = \frac{1}{27}$, this power is -3, so $\log_3 \frac{1}{27} = -3$.

Note The number $\log_b x$ is defined only if b and x are both positive and $b \neq 1$. For example, it is impossible to compute $\log_3(-9)$ (because there is no power of 3 that equals -9) or $\log_1 2$ (because there is no power of 1 that equals 2). ■

The logarithm with base 10, $\log_{10}$, is called the **common logarithm** and is often written as "log" without the base.

Quick Examples

	Logarithmic Form	Exponential Form
14.	$\log 10{,}000 = \log_{10} 10{,}000 = 4$	$10^4 = 10{,}000$
15.	$\log 10 = \log_{10} 10 = 1$	$10^1 = 10$
16.	$\log \frac{1}{10{,}000} = -4$	$10^{-4} = \frac{1}{10{,}000}$

Algebraic Properties of Logarithms

The following are important properties of logarithms. The fourth property is suggested by some of the Quick Examples above.

Logarithm Identities

The following identities hold for all positive bases $a \neq 1$ and $b \neq 1$, all positive numbers x and y, and every real number r. These identities follow from the laws of exponents.

Identity	Quick Examples
1. $\log_b(xy) = \log_b x + \log_b y$	$\log_2 16 = \log_2 8 + \log_2 2$ $\log_b 5 + \log_b 6 = \log_b 30$
2. $\log_b\left(\frac{x}{y}\right) = \log_b x - \log_b y$	$\log_2\left(\frac{5}{3}\right) = \log_2 5 - \log_2 3$ $\log_b 5 - \log_b 6 = \log_b\left(\frac{5}{6}\right)$
3. $\log_b(x^r) = r \log_b x$	$\log_2(6^5) = 5 \log_2 6$ $3 \log_b x = \log_b x^3$
4. $\log_b b = 1$ and $\log_b 1 = 0$	$\log_2 2 = 1; \log_3 3 = 1$ $\log 10 = 1; \log_2 1 = 0; \log 1 = 0$
5. $\log_b\left(\frac{1}{x}\right) = -\log_b x$	$\log_2\left(\frac{1}{3}\right) = -\log_2 3$
6. $\log_b x = \frac{\log_a x}{\log_a b}$	$\log_2 5 = \frac{\log_{10} 5}{\log_{10} 2} = \frac{\log 5}{\log 2}$

EXAMPLE 1 Using the Logarithm Identities

Let $a = \log 2$, $b = \log 3$, and $c = \log 5$. Write the following in terms of a, b, and c:

a. $\log 6$ **b.** $\log 15$ **c.** $\log 30$ **d.** $\log 9$

e. $\log \frac{1}{9}$ **f.** $\log 1.5$ **g.** $\log 32$ **h.** $\log \frac{8}{81}$

Solution

a. We recognize 6 as the product of 2 and 3, so we use identity (1):

$$\log 6 = \log(2 \times 3) = \log 2 + \log 3 = a + b.$$

b. As in part (a), we recognize 15 as the product of 3 and 5, so we use again identity (1):

$$\log 15 = \log(3 \times 5) = \log 3 + \log 5 = b + c.$$

c. 30 can be written as $2 \times 15 = 2 \times 3 \times 5$, so we use identity (1) twice:

$$\log 30 = \log(2 \times 15) = \log 2 + \log 15 = \log 2 + (\log 3 + \log 5) = a + b + c.$$

More simply, we can simplify the above steps and write

$$\log 30 = \log(2 \times 3 \times 5) = \log 2 + \log 3 + \log 5 = a + b + c.$$

d. We can think of 9 either as a product 3×3 or as a power 3^2. Let us use the second interpretation, which would call for identity (3):

$$\log 9 = \log(3^2) = 2 \log 3 = 2b.$$

e. For a reciprocal we can use identity (5):

$$\log\frac{1}{9} = -\log 9 = -2b. \quad \text{Using the answer to part (d)}$$

f. We recognize 1.5 as the ratio $\frac{3}{2}$, so we use identity (3):

$$\log 1.5 = \log\frac{3}{2} = \log 3 - \log 2 = b - a.$$

g. We recognize 32 as 2^5, so we use identity (3):

$$\log 32 = \log(2^5) = 5\log 2 = 5a.$$

h. As $\frac{8}{81}$ is a ratio, we start by using identity (2):

$$\begin{aligned}\log\frac{8}{81} &= \log 8 - \log 81 && \text{By identity (2)}\\ &= \log(2^3) - \log(3^4)\\ &= 3\log(2) - 4\log(3) && \text{By identity (5)}\\ &= 3a - 4b.\end{aligned}$$

Using Logarithms to Solve for Unknowns in the Exponent

One important use of logarithms is to solve equations in which the unknown is in the exponent.

EXAMPLE 2 Solving for Unknowns in the Exponent

Solve the following equations:

a. $4^{-x} = \frac{1}{64}$ **b.** $10(1.005)^{3x} = 200$ **c.** $4\cdot 3^{4x-2} = 81$

Solution

a. As the base of the exponent is 4, we start by taking the base 4 logarithm of both sides. (We could also solve it by taking the logarithm to *any* base, as we will illustrate in part (b).)

$$\log_4(4^{-x}) = \log_4\left(\frac{1}{64}\right).$$

The left-hand side is

$$\begin{aligned}\log_4(4^{-x}) &= -x\log_4(4) && \text{By identity (3)}\\ &= -x\cdot 1 = -x. && \log_4(4) = 1 \text{ by identity (4)}\end{aligned}$$

while the right-hand side is

$$\begin{aligned}\log_4\left(\frac{1}{64}\right) &= -\log_4(64) && \text{By identity (5)}\\ &= -\log_4(4^3) = -3\log_4(4) && \log_4(4) = 1 \text{ by identity (3)}\\ &= -3\cdot 1 = -3. && \text{By identity (4)}\end{aligned}$$

Equating the left- and right-hand sides thus gives

$$-x = -3,$$

so

$$x = 3.$$

Alternatively, we can translate the given equation from exponent form into logarithmic form:

$$4^{-x} = \frac{1}{64} \qquad \text{Exponent form}$$

$$\log_4\left(\frac{1}{64}\right) = -x. \qquad \text{Logarithmic form}$$

Thus,

$$x = -\log_4\left(\frac{1}{64}\right) = -(-3) = 3. \qquad \text{We calculated this logarithm above.}$$

b. We solve this problem by taking the common logarithm of both sides (see the comment in the solution to part (a)). We *could* take the base 1.005 logarithm of both sides, but this would lead to a solution in terms of logarithms to that base. On the other hand, common logarithms are readily computable on a calculator (there is usually a button for it), so we take the common logarithm of both sides instead. But first, we simplify by dividing both sides by 10:

$$1.005^{3x} = \frac{200}{10} = 20 \qquad \text{Divide both sides by 10.}$$

$$\log 1.005^{3x} = \log 20 \qquad \text{Take the log of both sides.}$$

$$3x \log 1.005 = \log 20 \qquad \text{Apply identity (3) on the left.}$$

$$x = \frac{\log 20}{3 \log 1.005} \approx 200.21. \qquad \text{Solve for } x\text{, and approximate with a calculator.}$$

Rather than using a calculator to approximate the answer by a decimal, we could have left it in the exact form $\frac{\log 20}{3 \log 1.005}$.

c. As in part (b), we solve this problem by taking the common logarithm of both sides; taking the base 3 logarithm will result in a solution in terms of $\log_3(20.25)$. (Try it!) But first, we simplify by dividing both sides by 4:

$$3^{4x-2} = \frac{81}{4} \qquad \text{Divide both sides by 4.}$$

$$\log 3^{4x-2} = \log\left(\frac{81}{4}\right) \qquad \text{Take the log of both sides.}$$

$$(4x - 2)\log 3 = \log 81 - \log 4 \qquad \text{Apply identity (3) on the left and identity (2) on the right.}$$

$$(4x - 2) = \frac{\log 81 - \log 4}{\log 3} \qquad \text{Divide.}$$

$$4x = \frac{\log 81 - \log 4}{\log 3} + 2 \qquad \text{Add 2 to both sides.}$$

$$x = \frac{1}{4}\left(\frac{\log 81 - \log 4}{\log 3} + 2\right) \approx 1.1845.$$

0.8 EXERCISES

In Exercises 1–4, complete the given tables.

1.

Exponential Form	$10^2 = 100$	$4^3 = 64$	$4^4 = 256$	$0.45^0 = 1$	$8^{1/2} = 2\sqrt{2}$	$4^{-3} = \frac{1}{64}$
Logarithmic Form						

2.

Exponential Form	$10^1 = 10$	$5^5 = 3{,}125$	$6^3 = 216$	$0.5^3 = 0.125$	$4^{1/4} = \sqrt{2}$	$3^{-4} = \frac{1}{81}$
Logarithmic Form						

3.

Exponential Form						
Logarithmic Form	$\log_{0.3} 0.09 = 2$	$\log_{1/2} 1 = 0$	$\log_{10} 0.001 = -3$	$\log_9 \frac{1}{81} = -2$	$\log_2 1{,}024 = 10$	$\log_{64} \frac{1}{4} = -\frac{1}{3}$

4.

Exponential Form						
Logarithmic Form	$\log_2 \frac{1}{2} = -1$	$\log_{10} 100{,}000 = 5$	$\log_{10} 0.00001 = -5$	$\log_2 2{,}048 = 11$	$\log_{16} \frac{1}{2} = -\frac{1}{4}$	$\log_{0.25} 0.0625 = 2$

In Exercises 5–16, evaluate the given quantity.

5. $\log_4 16$ **6.** $\log_5 125$ **7.** $\log_5 \frac{1}{25}$

8. $\log_4 \frac{1}{64}$ **9.** $\log 100{,}000$ **10.** $\log 1{,}000$

11. $\log_{16} 16$ **12.** $\log_{1/2} \frac{1}{2}$ **13.** $\log_4 \frac{1}{16}$

14. $\log_2 \frac{1}{8}$ **15.** $\log_2 \sqrt{2}$ **16.** $\log_4 \sqrt{2}$

In Exercises 17–28, use the logarithm identities to obtain the missing quantity.

17. $\log_b 3 + \log_b 4 = \log_b \square$

18. $\log_b 3 - \log_b 4 = \log_b \square$

19. $\log_b 2 - \log_b 5 - \log_b 4 = \log_b \square$

20. $\log_b 3 + \log_b 2 - \log_b 7 = \log_b \square$

21. $\log_b 3 - 3\log_b 2 = \log_b \square$

22. $3\log_b 2 + 2\log_b 3 = \log_b \square$

23. $4\log_b x + 5\log_b y = \log_b \square$

24. $3\log_b p - 2\log_b q = \log_b \square$

25. $2\log_b x + 3\log_b y - 4\log_b z = \log_b \square$

26. $4\log_b p - 3\log_b q - 2\log_b r = \log_b \square$

27. $x\log_b 2 - 2\log_b x = \log_b \square$

28. $p\log_b q + q\log_b p = \log_b \square$

Let $a = \log 2$, $b = \log 3$, and $c = \log 7$. In Exercises 29–46, use the logarithm identities to express the given quantity in terms of a, b, and c.

29. $\log 21$ **30.** $\log 14$ **31.** $\log 42$

32. $\log 28$ **33.** $\log\left(\frac{1}{7}\right)$ **34.** $\log\left(\frac{1}{3}\right)$

35. $\log\left(\frac{2}{3}\right)$ **36.** $\log\left(\frac{7}{2}\right)$ **37.** $\log\left(\frac{4}{7}\right)$

38. $\log\left(\frac{2}{9}\right)$ **39.** $\log 16$ **40.** $\log 81$

41. $\log 0.03$ **42.** $\log 7{,}000$ **43.** $\log 5$

44. $\log 25$ **45.** $\log \sqrt{7}$ **46.** $\log\left(\frac{2}{\sqrt{3}}\right)$

In Exercises 47–56, solve the given equation for the indicated variable.

47. $4 = 2^x$ **48.** $81 = 3^x$

49. $27 = 3^{2x-1}$ **50.** $4^{2-3x} = 256$

51. $5^{-x+1} = \frac{1}{125}$ **52.** $3^{x^2-12} = \frac{1}{27}$

53. $120 = 50(2^{3t})$ (Round the answer to four decimal places.)

54. $1{,}000 = 500(1.1^{2t})$ (Round the answer to four decimal places.)

55. $1{,}000 = 300(1.3^{4t-1})$ (Round the answer to four decimal places.)

56. $10{,}000 = 700(1.04^{3t+1})$ (Round the answer to four decimal places.)

1 FUNCTIONS AND APPLICATIONS

CASE STUDY

Modeling Spending on Mobile Advertising

You are the new director of *Impact Advertising Inc.*'s mobile division, which has enjoyed a steady 0.25% of the worldwide mobile advertising market. You have drawn up an ambitious proposal to expand your division in light of your anticipation that mobile advertising will continue to skyrocket. The VP for Financial Affairs feels that current projections (based on a linear model) do not warrant the level of expansion you propose.

How can you persuade the VP that those projections do not fit the data convincingly?

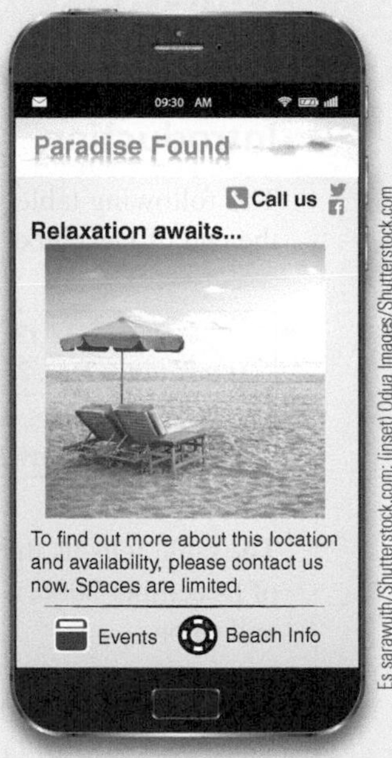

Es sarawuth/Shutterstock.com; (inset) Odua Images/Shutterstock.com

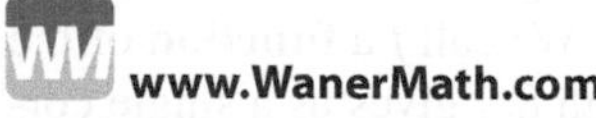

www.WanerMath.com

At the Website, in addition to the resources listed in the Preface, you will find:

The following extra topic:

- New Functions from Old: Scaled and Shifted Functions

Introduction

To analyze recent trends in spending on mobile advertising and to make reasonable projections, we need a mathematical model of this spending. Where do we start? To apply mathematics to real-world situations like this, we need a good understanding of basic mathematical concepts. Perhaps the most fundamental of these concepts is that of a function: a relationship that shows how one quantity depends on another. Functions may be described numerically and, often, algebraically. They can also be described graphically—a viewpoint that is extremely useful.

The simplest functions—the ones with the simplest formulas and the simplest graphs—are linear functions. Because of their simplicity, they are also among the most useful functions and can often be used to model real-world situations, at least over short periods of time. In discussing linear functions, we will meet the concepts of slope and rate of change, which are the starting point of the mathematics of change.

In the last section of this chapter, we discuss *simple linear regression*: construction of linear functions that best fit given collections of data. Regression is used extensively in applied mathematics, statistics, and quantitative methods in business. The inclusion of regression utilities in computer spreadsheets like Excel makes this powerful mathematical tool readily available for anyone to use.

Precalculus Review
For this chapter, you should be familiar with real numbers and intervals (see Section 0.1) and exponents and radicals (see Section 0.2).

1.1 Functions from the Numerical, Algebraic, and Graphical Viewpoints

Introduction

The following table gives accumulated sales of **Nintendo** Wii game consoles from the end of December 2006 to the end of December 2014:[1]

Time t (years since Dec. 2006)	0	1	2	3	4	5	6	7	8
Accumulated Sales f (millions of consoles)	3	9	27	53	73	88	98	105	109

The table tells us that a total of around 3 million consoles had been sold by the end of December 2006, a total of 9 million had been sold by the end of the following year, and so on.

Let's write $f(0)$ for the accumulated sales (in millions of consoles) when $t = 0$, $f(1)$ for the accumulated sales when $t = 1$, and so on. (We read $f(0)$ as "f of 0.") Thus, $f(0) = 3, f(1) = 9, f(2) = 27, \ldots, f(8) = 109$. In general, we write $f(t)$ for the accumulated sales (in millions of consoles) at time t. We call f a **function** of the variable t, meaning that for each value of t from 0 through 8, f gives us a single corresponding number $f(t)$ that tells us how many consoles were sold up to time t.

[1] Accumulated sales of Wii and WiiU consoles since the Wii was first released in November 2006. Sources: www.statista.com, Nintendo Inc.

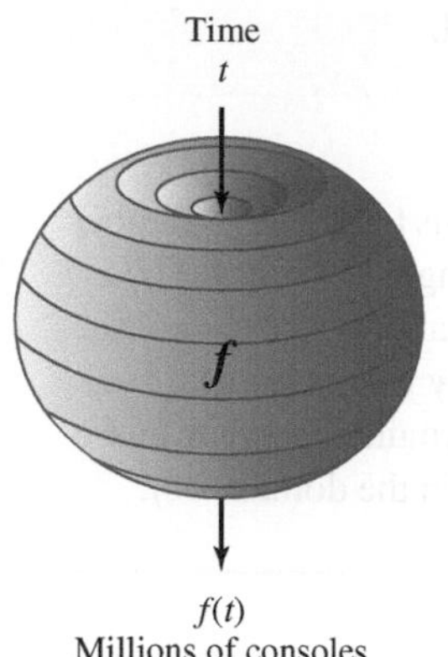

Figure 1

In general, we think of a function as a way of producing new objects from old ones. The functions we deal with in this text produce new numbers from old numbers. The numbers we have in mind are the real numbers, including not only positive and negative integers and fractions but also numbers like $\sqrt{2}$ or π. (See Chapter 0 for more on real numbers.) For this reason, the functions we use are called **real-valued functions of a real variable**. For example, the function f takes the time in years since December 2006 as input and returns the accumulated console sales at that time (Figure 1).

The letter t in $f(t)$ is called the **argument** of the function f.* A function may be specified in several different ways. Here, we have specified the function f **numerically** by giving the values of the function for a number of values of the argument, as in the preceding table.

* There is nothing special about the choice of the letter t as the argument of a function; we could have used any letter whatsoever to represent time. In general, it is customary to use t for time and x for (almost) anything else, as in $f(x)$. Likewise, there is nothing special about our choice of the letter f for the function name except for tradition.

Q: *For which values of the argument t does it make sense to ask for f(t)? In other words, for which times t is f(t) defined?*

A: Because our function f refers to the accumulated sales over the period December 2006 ($t = 0$) through December 2014 ($t = 8$), $f(t)$ should reasonably be defined when t is any number between 0 and 8, that is, when $0 \leq t \leq 8$. Using interval notation (see Chapter 0), we can say that $f(t)$ is defined when t is in the interval $[0, 8]$.

The set of values of the argument for which a function is defined is called its **domain** and is a necessary part of the definition of the function. Notice that the preceding table gives the value of $f(t)$ at only some of the infinitely many possible values in the domain $[0, 8]$. The domain of a function is not always specified explicitly; if no domain is specified for a function f, we take the domain to be the largest set of numbers x for which $f(x)$ makes sense. This "largest possible domain" is sometimes called the **natural domain**.

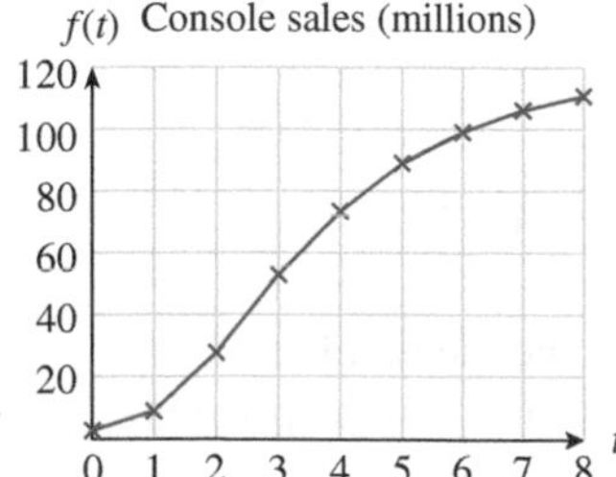

Figure 2

The above Nintendo data can also be represented on a graph by plotting the given pairs of numbers $(t, f(t))$ in the xy-plane. (See Figure 2. We have connected successive points by line segments.) In general, the graph of a function f consists of all points $(x, f(x))$ in the plane with x in the domain of f.

In Figure 2 we specified the function f **graphically** by using a graph to display its values. Suppose now that we had only the graph without the table of data. We could use the graph to find approximate values of the function. For instance, to find $f(2.5)$ from the graph, we do the following:

1. Find the desired value of t at the bottom of the graph ($t = 2.5$ in this case).
2. Estimate the height (y-coordinate) of the corresponding point on the graph (around 40 in this case).

Thus, $f(2.5) \approx 40$ million consoles sold by the end of June 2009 ($t = 2.5$).†

In some cases we may be able to use an algebraic formula to calculate the values of a function, and we say that the function is specified **algebraically**. These are not the only ways in which a function can be specified; for instance, a function can sometimes be specified *verbally*, as in "Let $f(x)$ be the number of miles my **Tesla** can go at 30 mph after being charged for x hours at the local station." Notice that any function can be represented graphically by plotting the points $(x, f(x))$ for a number of values of x in its domain.

† In a graphically defined function we can never know the y-coordinates of points exactly; no matter how accurately a graph is drawn, we can obtain only approximate values of the coordinates of points. That is why we have been using the word "estimate" rather than "calculate" and why we say $f(2.5) \approx 40$ rather than $f(2.5) = 40$.

Here is a summary of the terms we have just introduced.

Functions

A **real-valued function f of a real-valued variable x** assigns to each real number x in a specified set of numbers, called the **domain of f**, a single real number $f(x)$, read "f of x." The quantity x is called the **argument** of f, and $f(x)$ is called the **value of f at x**. A function is usually specified **numerically** using a table of values, **graphically** using a graph, or **algebraically** using a formula. The **graph of a function** consists of all points $(x, f(x))$ in the plane with x in the domain of f.

Quick Examples

1. **A function specified numerically:** Take $c(t)$ to be the world emission of carbon dioxide in year t since 2000, represented by the following table:[2]

Year t (year since 2000)	CO_2 Emissions $c(t)$ (billions of metric tons)
0	24
5	28
10	31
15	33
20	36
25	38
30	41

Graph of c
Plotting the pairs $(t, c(t))$ gives the following graph:

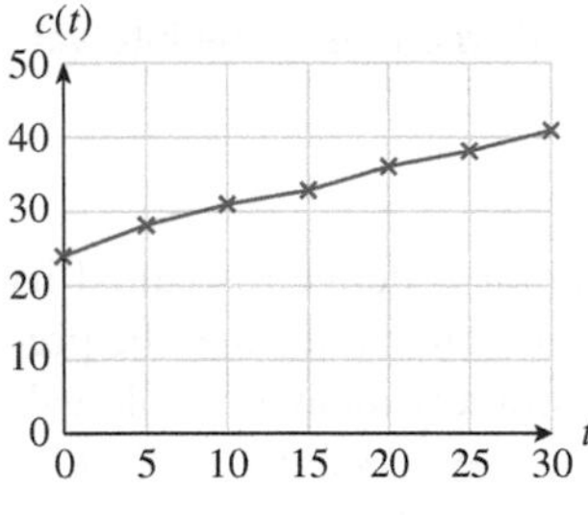

The domain of c is $[0, 30]$, the argument is t, the number of years since 2000, and the values $c(t)$ give the world production of carbon dioxide in a given year. The graph of c is shown on the left. Some values of c are

$c(0) = 24$ — 24 billion metric tons of CO_2 were produced in 2000.

$c(10) = 31$ — 31 billion metric tons of CO_2 were produced in 2010.

$c(30) = 41.$ — 41 billion metric tons of CO_2 were projected to be produced in 2030.

2. **A function specified graphically:** Take $q(p)$ to be the number of smartphones (in millions) sold worldwide in a year when the price is set at p dollars, as represented by the following graph:[3]

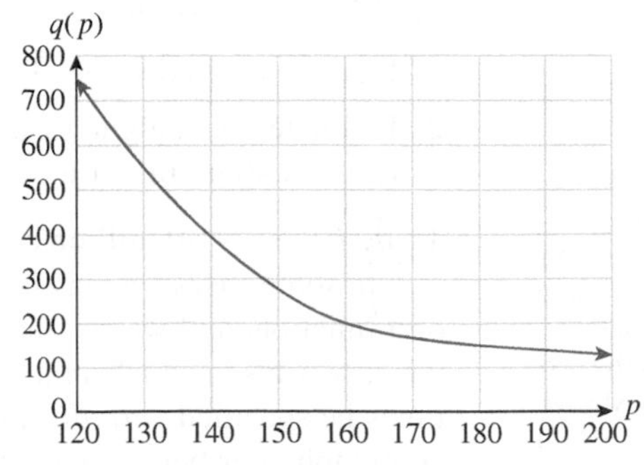

[2] Figures for 2015 and later are projections. Source: Energy Information Administration (EIA), www.eia.doe.gov.

[3] Based on prices and sales figures in www.businessweek.com and http://techcrunch.com.

The domain of q is $[120, 200]$, the argument is p, the price of a smartphone in dollars, and the values $q(p)$ give the number (in millions) of smartphones sold worldwide in a year at a given price.* Some values of q are

$q(160) \approx 200$ — When the price is \$160, about 200 million smartphones are sold worldwide in a year.

$q(130) \approx 550.$ — When the price is \$130, about 550 million smartphones are sold worldwide in a year.

* The function q is an example of a *demand function*. (See Section 1.2.)

3. **A function specified algebraically:** Let $f(x) = \frac{1}{x}$. The function f is specified algebraically. The argument is x, and the dependent variable is f. The natural domain of f consists of all real numbers except zero because $f(x)$ makes sense for all values of x other than $x = 0$. Some specific values of f are

$$f(2) = \frac{1}{2},\quad f(3) = \frac{1}{3},\quad f(-1) = \frac{1}{-1} = -1,$$

$f(0)$ is not defined because 0 is not in the domain of f.

4. **The graph of a function:** Let $f(x) = x^2$, with domain the set of all real numbers. To draw the graph of f, first choose some convenient values of x in the domain and compute the corresponding y-coordinates $f(x)$:

x	−3	−2	−1	0	1	2	3
$f(x) = x^2$	9	4	1	0	1	4	9

Plotting these points $(x, f(x))$ gives the picture on the left, suggesting the graph on the right.†

† If you plot more points, you will find that they lie on a smooth curve as shown. That is why we did not use line segments to connect the points.

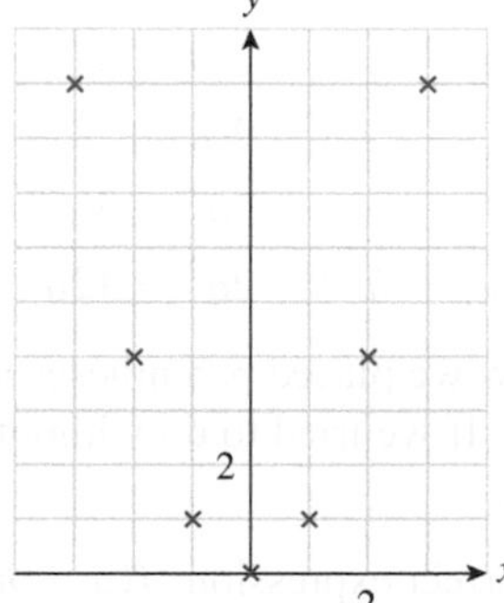

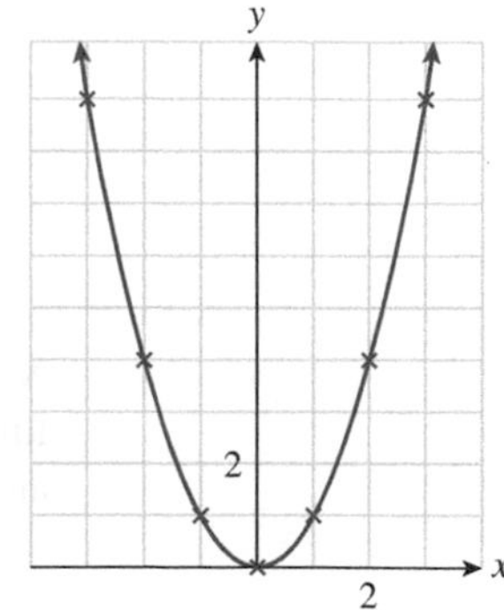

(This particular curve happens to be called a **parabola**, and its lowest point, at the origin, is called its **vertex**.)

EXAMPLE 1 iPod Sales

The total number of iPods sold by **Apple** in year x can be approximated by

$$f(x) = -2x^2 + 12x + 36 \text{ million iPods} \qquad (0 \le x \le 7),$$

where $x = 0$ represents 2005.[4]

a. What is the domain of f? Compute $f(0)$, $f(2)$, $f(4)$, and $f(6)$. What do these answers tell you about iPod sales? Is $f(-1)$ defined?

b. Compute $f(a)$, $f(-b)$, $f(a+h)$, and $f(a)+h$, assuming that the quantities a, $-b$, and $a+h$ are in the domain of f.

c. Sketch the graph of f. Does the shape of the curve suggest that iPod sales were accelerating or decelerating over the period 2005–2008?

Solution

a. The domain of f is the set of numbers x with $0 \le x \le 7$, that is, the interval $[0, 7]$. If we substitute 0 for x in the formula for $f(x)$, we get

$$f(0) = -2(0)^2 + 12(0) + 36 = 36.$$ Approximately 36 million iPods were sold in 2005.

Similarly,

$$f(2) = -2(2)^2 + 12(2) + 36 = 52$$ Approximately 52 million iPods were sold in 2007.

$$f(4) = -2(4)^2 + 12(4) + 36 = 52$$ Approximately 52 million iPods were sold in 2009.

$$f(6) = -2(6)^2 + 12(6) + 36 = 36.$$ Approximately 36 million iPods were sold in 2011.

As -1 is not in the domain of f, $f(-1)$ is not defined.

b. To find $f(a)$, we substitute a for x in the formula for $f(x)$ to get

$$f(a) = -2a^2 + 12a + 36.$$ Substitute a for x.

Similarly,

$$\begin{aligned} f(-b) &= -2(-b)^2 + 12(-b) + 36 && \text{Substitute } -b \text{ for } x. \\ &= -2b^2 - 12b + 36 && (-b)^2 = b^2 \\ f(a+h) &= -2(a+h)^2 + 12(a+h) + 36 && \text{Substitute } (a+h) \text{ for } x. \\ &= -2(a^2 + 2ah + h^2) + 12a + 12h + 36 && \text{Expand.} \\ &= -2a^2 - 4ah - 2h^2 + 12a + 12h + 36 \\ f(a) + h &= -2a^2 + 12a + 36 + h. && \text{Add } h \text{ to } f(a). \end{aligned}$$

Note how we placed parentheses around the quantities at which we evaluated the function. If we tried to do without any of these parentheses, we would likely get an error:

Correct expression: $f(a+h) = -2(a+h)^2 + 12(a+h) + 36$ ✓

NOT $-2a^2 + 12a + 36 + h$ ✗

Also notice the distinction between $f(a+h)$ and $f(a)+h$: To find $f(a+h)$, we replace x by the quantity $(a+h)$; to find $f(a)+h$, we add h to $f(a)$.

---

[4] Sales are by calendar year. Source for data: Apple quarterly earnings reports at www.apple.com/investor.

c. To draw the graph of f, we plot points of the form $(x, f(x))$ for several values of x in the domain of f. Let us use the values at the integers from 0 to 7:

x	0	1	2	3	4	5	6	7
$f(x) = -2x^2 + 12x + 36$	36	46	52	54	52	46	36	22

Graphing these points gives the graph shown on the left in Figure 3, suggesting the curve shown on the right.

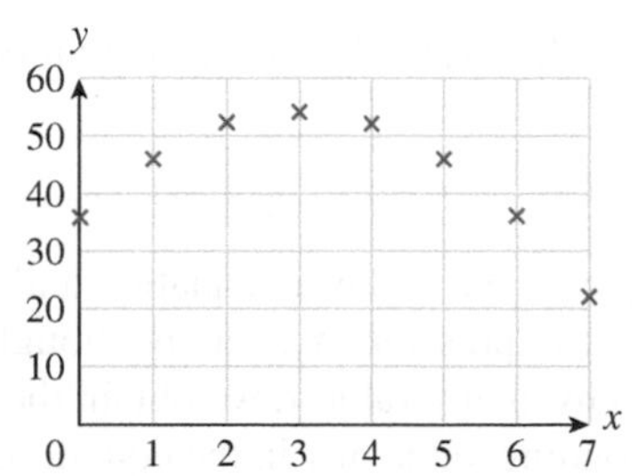

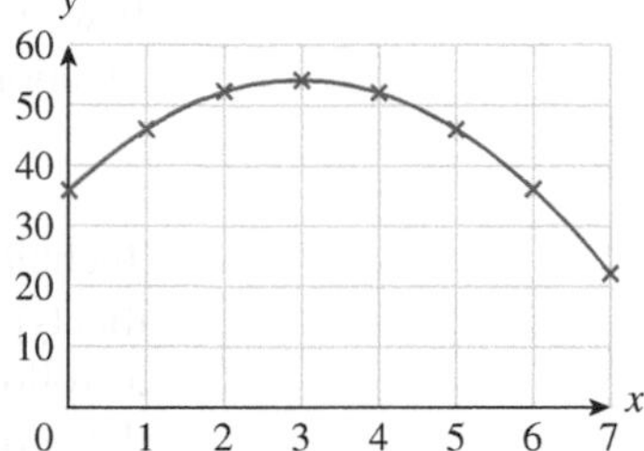

Figure 3

The period 2005–2008 is represented by the interval $[0, 3]$ on the x-axis, and the graph becomes less steep as we move from $x = 0$ to $x = 3$, suggesting that iPod sales were decelerating over the given period.

Using Technology

See the Technology Guides at the end of the chapter for detailed instructions on how to obtain the table of values and graph in Example 1 using a TI-83/84 Plus or Excel. Here is an outline:

TI-83/84 Plus

Table of values:
`Y1=-2X^2+12X+36`
`2ND` `TABLE`

Graph: `WINDOW`; Xmin = 0, Xmax = 7 `ZOOM` `0`

[More details in the Technology Guide.]

Spreadsheet

Table of values: Headings x and $f(x)$ in A1–B1; x-values 0, 2, 4, and 6 in A2–A5.
`=-2*A2^2+12*A2+36`
in B2; copy down through B5.
Graph: Highlight A1 through B5, and insert a scatter chart.
[More details in the Technology Guide.]

Website

www.WanerMath.com

Go to the Function evaluator and grapher under Online Utilities, and enter
`-2x^2+12x+36`
for y_1. To obtain a table of values, enter the x-values 0, 2, 4, 6 in the Evaluator box, and press "Evaluate" at the top of the box.
Graph: Set Xmin = 0, Xmax = 7, and press "Plot Graphs".

➡ **Before we go on...** The following table compares the value of f in Example 1 with the actual sales figures:

x	0	2	4	6
$f(x) = -2x^2 + 12x + 36$	36	52	52	36
Actual iPod Sales (millions)	32	53	52	39

The actual figures are stated here for only (some) integer values of x; for instance, $x = 4$ gives the sales in 2009. But what were, for instance, the sales in the year beginning in June 2009 ($x = 4.5$)? This is where our formula comes in handy: We can use the formula for f to **interpolate**—that is, to find sales at values of x between values that are stated:

$$f(4.5) = -2(4.5)^2 + 12(4.5) + 36 = 49.5 \text{ million iPods.}$$

We can also use the formula to **extrapolate**—that is, to predict sales at values of x *outside* the domain, say, for $x = 8$ (that is, sales in 2013):

$$f(8) = -2(8)^2 + 12(8) + 36 = 4 \text{ million iPods.}$$

As a general rule, extrapolation is far less reliable than interpolation. The extrapolated values depend heavily on the type of mathematical function used to model the data, so the same set of data can lead to vastly different extrapolated values. Further, the vagaries of the marketplace make the use of current data trends to predict the future problematic in the first place.

We call the algebraic function f an **algebraic model** of iPod sales because it uses an algebraic formula to model—or mathematically represent (approximately)—the annual sales. The particular kind of algebraic model we used is called a **quadratic model**. (See the end of this section for the names of some commonly used models.) ■

Functions and Equations

To specify a function algebraically, we need to write down a defining equation, as in, say,

$$f(x) = 3x - 2.$$

If we replace the "$f(x)$" by "y" we get an equation with no explicit mention of any function:

$$y = 3x - 2. \qquad \text{An equation in two variables: } x \text{ and } y$$

Technically, $y = 3x - 2$ is an equation and not a function. However, an equation of this type, y = expression in x, can be thought of as "specifying y as a function of x" as follows: Given any value x, we obtain the value of the function at x by calculating the corresponding value of y in the equation. So the value of the function at $x = 1$ is just

$$y = 3(1) - 2 = 1,$$

which is the same as $f(1)$ for our original function $f(x) = 3x - 2$.

Function Notation and Equation Notation

Instead of using the usual function notation to specify a function, as in, say,

$$f(x) = -2x^2 + 12x + 36, \qquad \text{Function notation}$$

we can specify it by an *equation with two variables* by replacing $f(x)$ by y or f or any letter we choose:

$$y = -2x^2 + 12x + 36 \qquad \text{Equation notation}$$

or we could choose

$$f = -2x^2 + 12x + 36.$$

In an equation of the form y = expression in x, the variable x is called the **independent variable**, and y is called the **dependent variable** (because the value of y *depends on* a choice of the value for x). So we can think of the equation $y = -2x^2 + 12x + 36$ in two ways:

1. As specifying a function y of x, as in $y(x) = -2x^2 + 12x + 36$ (or, say, $f(x) = -2x^2 + 12x + 36$)
2. As simply an equation with independent variable x and dependent variable y

Note that when we think of a function as an equation in this way, the argument of the function becomes the independent variable, and the letter we choose for the left-hand side (sometimes the same letter as we use for the name of the function) becomes the dependent variable.

Quick Example

5. If the cost to manufacture x items is given by the "cost function"* C specified by

$$C(x) = 40x + 2{,}000, \qquad \text{Cost function}$$

we could instead write

$$C = 40x + 2{,}000 \qquad \text{Cost equation}$$

and still think of C, the cost, as a function of x.

* We will discuss cost functions more fully in Section 1.2.

Function notation and equation notation, sometimes using the same letter for the function name and the dependent variable, are often used interchangeably. It is important to be able to switch back and forth between function notation and equation notation, and we shall do so when it is convenient.

Using Technology

See the Technology Guides at the end of the chapter for detailed instructions on how to obtain the table of values and graph in Example 2 using a TI-83/84 Plus or Excel. Here is an outline:

TI-83/84 Plus

Table of values:

```
Y1=(X≤3)*(5X^2+2X+2)
  +(X>3)*(-2X^2+33X-28)
```

[2ND] [TABLE]

Graph: [WINDOW]; Xmin = 0, Xmax = 8 [ZOOM] [0].

[More details in the Technology Guide.]

Spreadsheet

Table of values: Headings t and $f(t)$ in A1–B1; t-values 0, 1, . . . , 8 in A2–A10.

```
=(A2<=3)*(5*A2^2+2*A2+2)
 +(A2>3)*(-2*A2^2+33*A2-28)
```

in B2; copy down through B10.

Graph: Highlight A1 through B10, and insert a scatter chart.

[More details in the Technology Guide.]

Website

www.WanerMath.com

Go to the Function evaluator and grapher under Online Utilities, and enter

```
(x≤3)*(5x^2+2x+2)
  +(x>3)*(-2x^2+33x-28)
```

for y1. To obtain a table of values, enter the x-values 0, 1, . . . , 8 in the Evaluator box, and press "Evaluate" at the top of the box.

Graph: Set Xmin = 0, Xmax = 8, and press "Plot Graphs".

Piecewise-Defined Functions

Look again at the graph of the accumulated sales of **Nintendo** Wii game consoles in Figure 2. From year 0 through year 3, the sales seem to accelerate, but then, from year 3 to year 8, the sales seem to slow. In the following example, we model this behavior using a function with two different formulas: one that applies to the interval $[0, 3]$ and another for the interval $[3, 8]$. A function specified by two or more different formulas is called a **piecewise-defined function**.

EXAMPLE 2 A Piecewise-Defined Function: Nintendo Wii Sales

The accumulated sales of **Nintendo** Wii game consoles from December 2006 to December 2014 can be approximated by the following function of time t in years ($t = 0$ represents the end of December 2006):

$$f(t) = \begin{cases} 5t^2 + 2t + 2 & \text{if } 0 \le t \le 3 \\ -2t^2 + 33t - 28 & \text{if } 3 < t \le 8 \end{cases} \quad \text{million consoles.}$$

What were the accumulated sales of Nintendo Wii consoles at the end of December 2007, December 2009, and June 2010? Sketch the graph of f by plotting several points.

Solution We evaluate the given function at the corresponding values of t:

Dec. 2007 ($t = 1$): $f(1) = 5(1)^2 + 2(1) + 2 = 9$ — Use the first formula because $0 \le t \le 3$.

Dec. 2009 ($t = 3$): $f(3) = 5(3)^2 + 2(3) + 2 = 53$ — Use the first formula because $0 \le t \le 3$.

June 2010 ($t = 3.5$): $f(3.5) = -2(3.5)^2 + 33(3.5) - 28 = 63$ — Use the second formula because $3 < t \le 9$.

Thus, Nintendo had sold around 9 million consoles by the end of December 2007, 53 million by the end of December 2009, and 63 million by the end of June 2010.

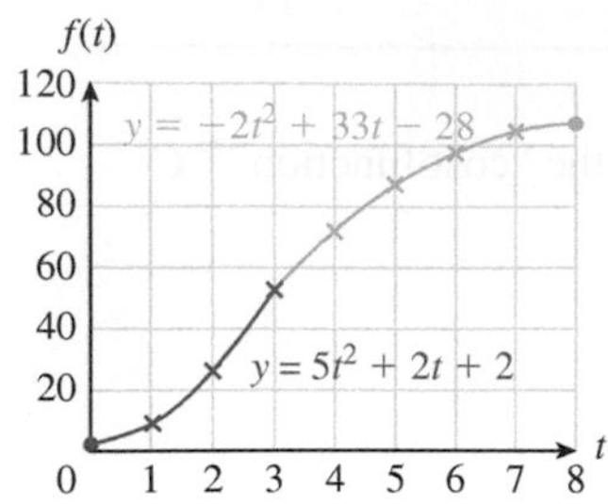

Figure 4

To sketch the graph of f, we use a table of values of $f(t)$ (two of which we have already calculated above), plot the points, and connect them to sketch the graph:

t	0	1	2	3	4	5	6	7	8
$f(t)$	2	9	26	53	72	87	98	105	108
	First formula				Second formula				

The graph (Figure 4) has the following features:

1. The first formula is used for $0 \le t \le 3$.
2. The second formula is used for $3 < t \le 8$.
3. The domain is $[0, 8]$, so the graph is cut off at $t = 0$ and $t = 8$.
4. The heavy solid dots at the ends indicate the endpoints of the domain.

EXAMPLE 3 More Complicated Piecewise-Defined Functions

Let f be the function specified by

$$f(x) = \begin{cases} -1 & \text{if } -4 \le x < -1 \\ x & \text{if } -1 \le x \le 1 \\ x^2 - 1 & \text{if } 1 < x \le 2. \end{cases}$$

Technology formula: `(X<-1)*(-1)+(-1≤X)*(X≤1)*X+(1<X)*(X^2-1)`

a. What is the domain of f? Find $f(-2)$, $f(-1)$, $f(0)$, $f(1)$, and $f(2)$.

b. Sketch the graph of f.

Solution

a. The domain of f is $[-4, 2]$, because $f(x)$ is specified only when $-4 \le x \le 2$.

$f(-2) = -1$ We used the first formula because $-4 \le x < -1$.

$f(-1) = -1$ We used the second formula because $-1 \le x \le 1$.

$f(0) = 0$ We used the second formula because $-1 \le x \le 1$.

$f(1) = 1$ We used the second formula because $-1 \le x \le 1$.

$f(2) = 2^2 - 1 = 3$ We used the third formula because $1 < x \le 2$.

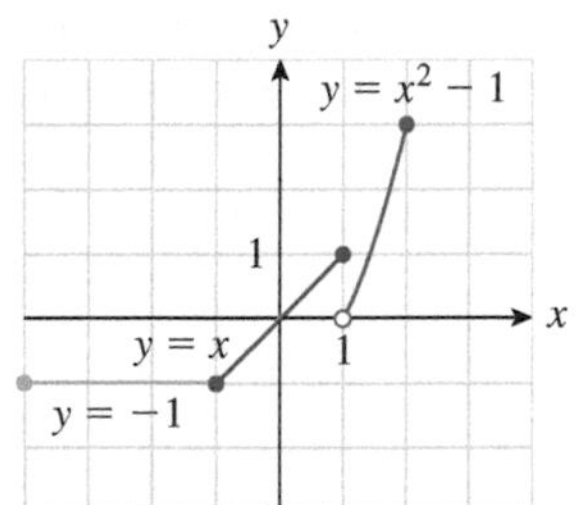

Figure 5

b. To sketch the graph by hand, we first sketch the three graphs $y = -1$, $y = x$, and $y = x^2 - 1$ and then use the appropriate portion of each (Figure 5). Note that solid dots indicate points on the graph, whereas open dots indicate points not on the graph. For example, when $x = 1$, the inequalities in the formula tell us that we are to use the middle formula (x) rather than the bottom one $(x^2 - 1)$. Thus, $f(1) = 1$, not 0, so we place a solid dot at $(1, 1)$ and an open dot at $(1, 0)$.

Vertical Line Test

Every point in the graph of a function has the form $(x, f(x))$ for some x in the domain of f. Because f assigns a *single* value $f(x)$ to each value of x in the domain, it follows that, in the graph of f, there should be only one y corresponding to any such value of x—namely, $y = f(x)$. In other words, *the graph of a function cannot contain two or more points with the same x-coordinate—that is, two or more points on the same vertical line.* On the other hand, a vertical line at a value of x not in the domain will not contain any points in the graph. This gives us the following rule.

Vertical Line Test

For a graph to be the graph of a function, every vertical line must intersect the graph in *at most* one point.

Quick Examples

6. As illustrated below, only graph B passes the vertical line test, so only graph B is the graph of a function.

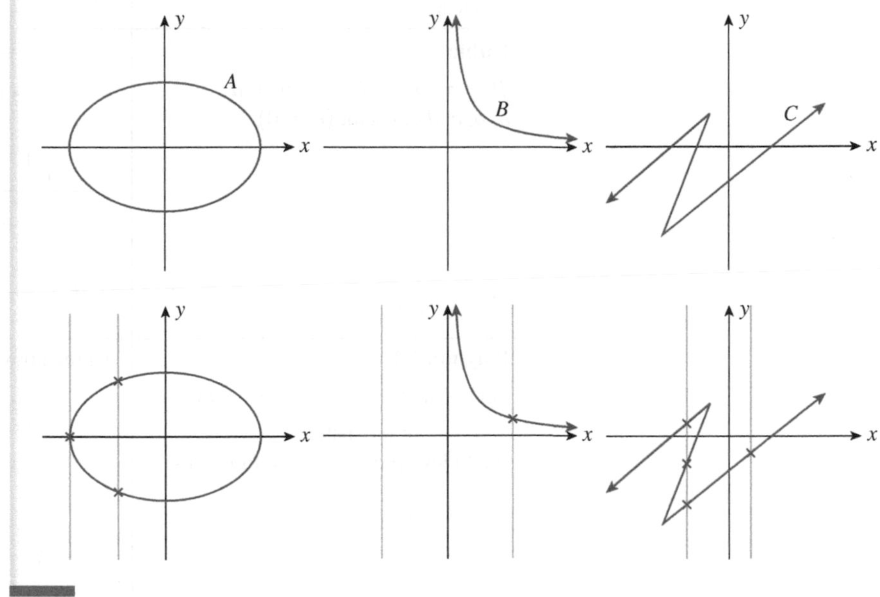

Common Functions

Table 1 lists some common types of functions that are often used to model real-world situations.

Table 1 A Compendium of Functions and Their Graphs

Type of Function	Examples	
Linear $f(x) = mx + b$ m, b constant Graphs of linear functions are straight lines. The quantity m is the **slope** of the line; the quantity b is the ***y*-intercept** of the line. (See Section 1.3.)	$y = x$ 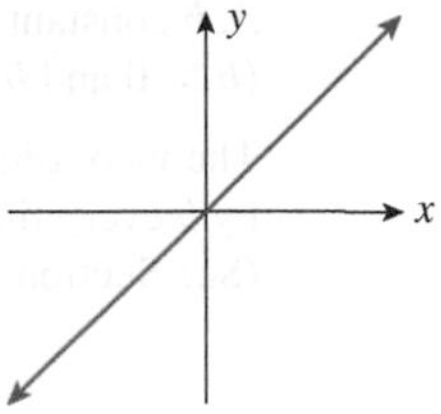	$y = -2x + 2$ 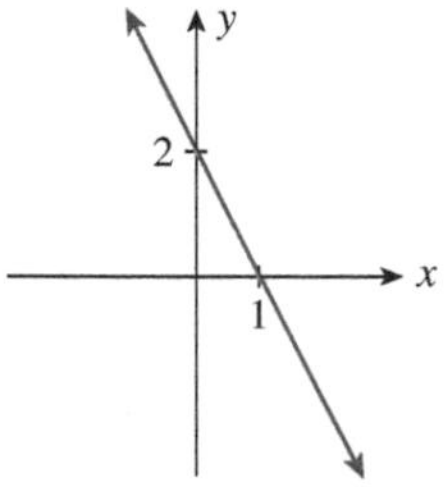
Technology formulas:	x	-2*x+2

Table 1 (*Continued*)

Type of Function	Examples	
Quadratic $f(x) = ax^2 + bx + c$ a, b, c constant $(a \neq 0)$ Graphs of quadratic functions are called **parabolas**. When a is positive, the parabola is **concave up** (example shown on the left). When a is negative, the parabola is **concave down** (example shown on the right). (See Section 2.1.)	$y = x^2$ 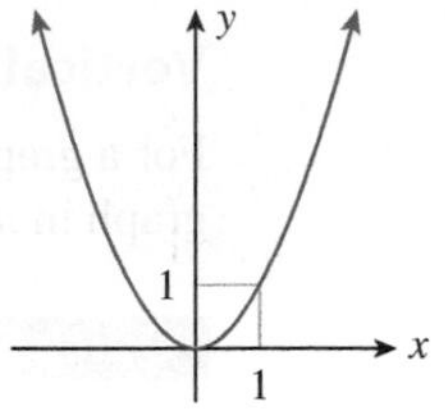	$y = -2x^2 + 2x + 4$ 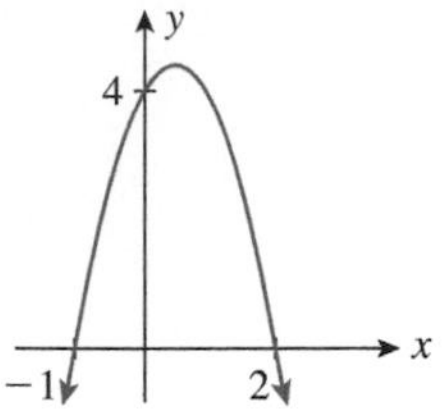
Technology formulas:	`x^2`	`-2*x^2 + 2*x + 4`
Cubic $f(x) = ax^3 + bx^2 + cx + d$ a, b, c, d constant $(a \neq 0)$	$y = x^3$ 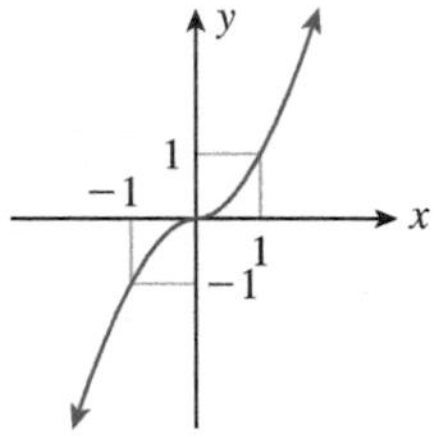	$y = -x^3 + 3x^2 + 1$ 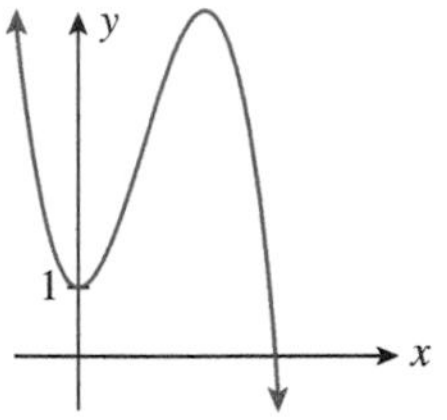
Technology formulas:	`x^3`	`-x^3 + 3*x^2 + 1`
Polynomial $f(x) = ax^n + bx^{n-1} + \cdots + rx + s$ $a, b, \ldots, r, s$ constant (includes all of the above functions)	All the above, and $f(x) = x^6 - 2x^5 - 2x^4 + 4x^2$ 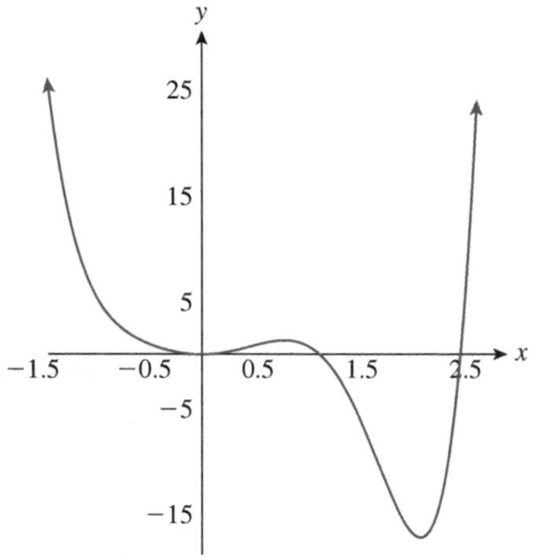	
Technology formula:	`x^6-2x^5-2x^4+4x^2`	
Exponential $f(x) = Ab^x$ A, b constant $(b > 0$ and $b \neq 1)$ The y-coordinate is multiplied by b every time x increases by 1. (See Section 2.2.)	$y = 2^x$ 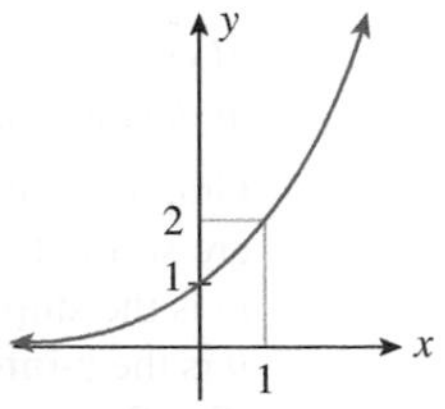 y is doubled every time x increases by 1.	$y = 4(0.5)^x$ 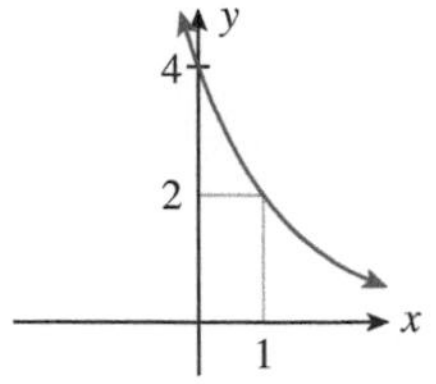 y is halved every time x increases by 1.
Technology formulas:	`2^x`	`4*0.5^x`

Table 1 *(Continued)*

Type of Function	Examples	
Logarithmic $f(x) = \log_b x + C$ b, C constant $(b > 0, b \neq 1)$ (See Section 2.3.)	$y = \log_2 x$ *x* is doubled every time *y* increases by 1.	$y = \log_{1/2} x$ *x* is doubled every time *y* decreases by 1.
Technology formulas:	`ln(x)/ln(2)`	`ln(x)/ln(1/2)`
Rational $f(x) = \frac{P(x)}{Q(x)}$; $P(x)$ and $Q(x)$ polynomials The graph of $y = 1/x$ is a **hyperbola**. The domain excludes zero because $1/0$ is not defined.	$y = \frac{1}{x}$ 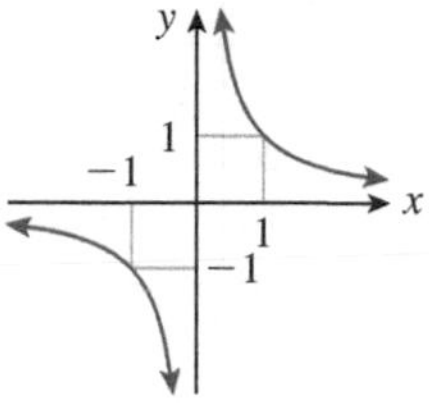	$y = \frac{x}{x-1}$ 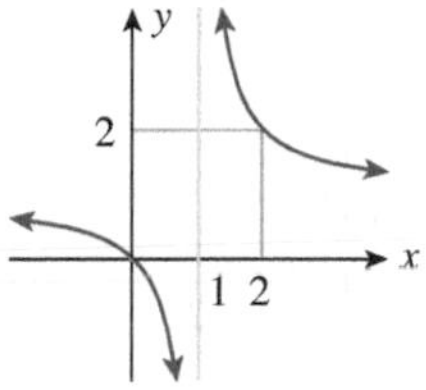
Technology formulas:	`1/x`	`x/(x-1)`
Absolute Value For *x* positive or zero the graph of $y = \|x\|$ is the same as that of $y = x$. For *x* negative or zero it is the same as that of $y = -x$.	$y = \|x\|$	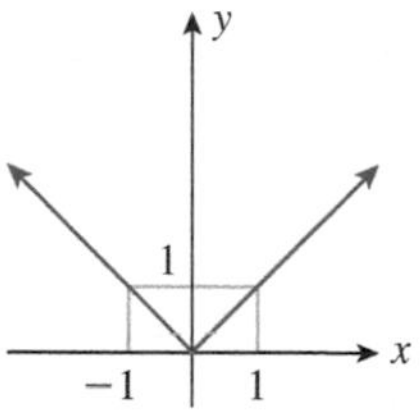 $y = \|2x + 2\|$
Technology formulas:	`abs(x)`	`abs(2*x + 2)`
Square Root The domain of $y = \sqrt{x}$ must be restricted to the nonnegative numbers because the square root of a negative number is not real. Its graph is the top half of a horizontally oriented parabola.	$y = \sqrt{x}$ 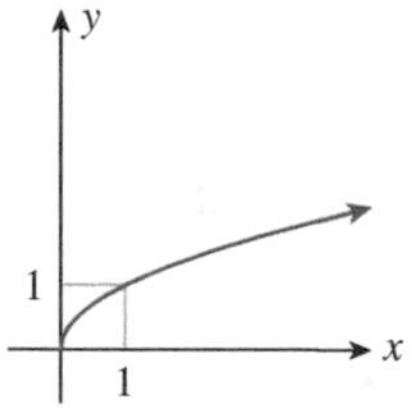	$y = \sqrt{4x - 2}$ 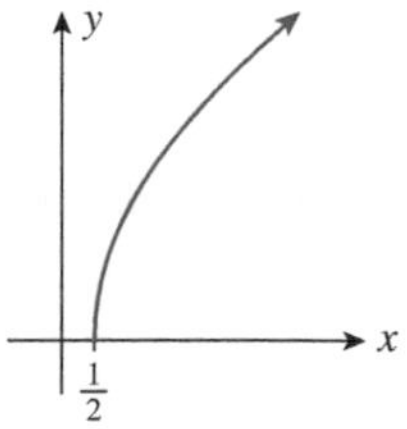
Technology formulas:	`x^0.5` or `√(x)`	`(4*x-2)^0.5` or `√(4*x-2)`

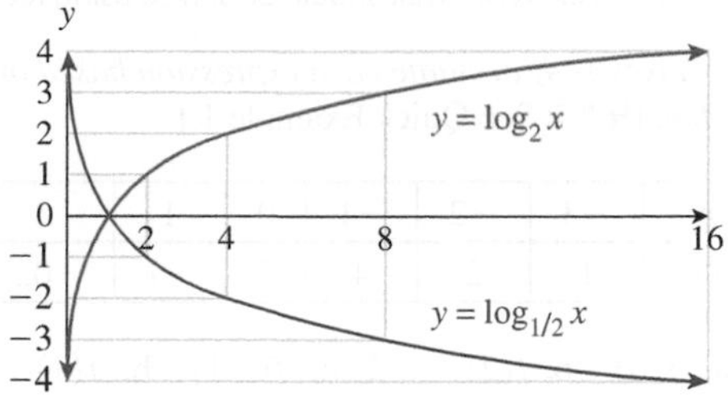

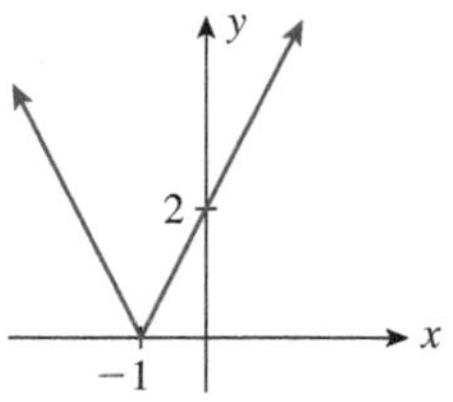

Website
www.WanerMath.com
Follow the path
Online Text →
New Functions from Old:
Scaled and Shifted Functions
where you will find complete online interactive text, examples, and exercises on scaling and translating the graph of a function by changing the formula.

Functions and models other than linear ones are called **nonlinear**.

1.1 EXERCISES

▼ more advanced ◆ challenging

T indicates exercises that should be solved using technology

In Exercises 1–4, evaluate each expression based on the following table. [**HINT**: See Quick Example 1.]

x	−3	−2	−1	0	1	2	3
$f(x)$	1	2	4	2	−1	−0.5	0.25

1. a. $f(0)$ **b.** $f(2)$ **2. a.** $f(-1)$ **b.** $f(1)$

3. a. $f(2) - f(-2)$ **b.** $f(-1)f(-2)$ **c.** $-2f(-1)$

4. a. $f(1) - f(-1)$ **b.** $f(1)f(-2)$ **c.** $3f(-2)$

In Exercises 5–8, use the graph of the function f to find approximations of the given values. [**HINT**: See Example 1.]

5.

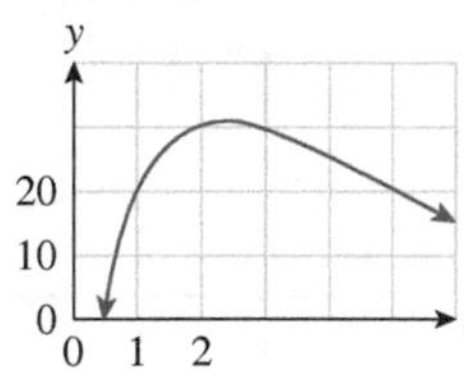

a. $f(1)$ **b.** $f(2)$
c. $f(3)$ **d.** $f(5)$
e. $f(3) - f(2)$ **f.** $f(3 - 2)$

6.

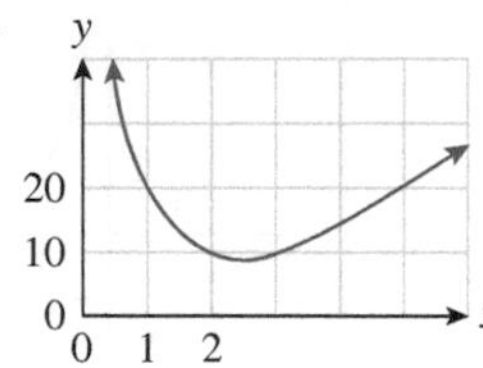

a. $f(1)$ **b.** $f(2)$
c. $f(3)$ **d.** $f(5)$
e. $f(3) - f(2)$ **f.** $f(3 - 2)$

7. a. $f(-1)$ **b.** $f(1)$ **c.** $f(3)$ **d.** $\dfrac{f(3) - f(1)}{3 - 1}$

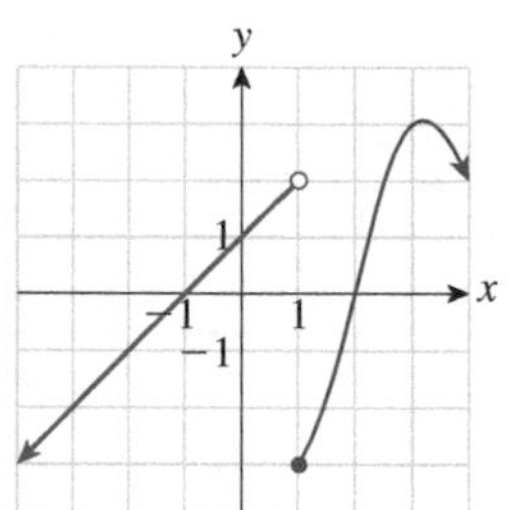

8. a. $f(-3)$ **b.** $f(-1)$ **c.** $f(1)$ **d.** $\dfrac{f(3) - f(1)}{3 - 1}$

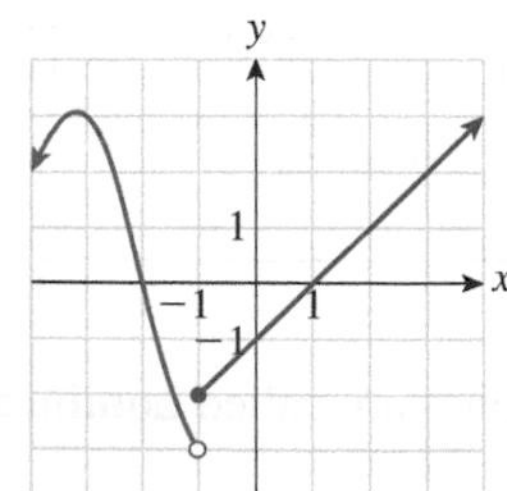

In Exercises 9–12, say whether or not $f(x)$ is defined for the given values of x. If it is defined, give its value. [**HINT**: See Quick Example 3.]

9. $f(x) = x - \dfrac{1}{x^2}$, with its natural domain
a. $x = 4$ **b.** $x = 0$ **c.** $x = -1$

10. $f(x) = \dfrac{2}{x} - x^2$, with domain $[2, +\infty)$
a. $x = 4$ **b.** $x = 0$ **c.** $x = 1$

11. $f(x) = \sqrt{x + 10}$, with domain $[-10, 0)$
a. $x = 0$ **b.** $x = 9$ **c.** $x = -10$

12. $f(x) = \sqrt{9 - x^2}$, with domain $(-3, 3)$
a. $x = 0$ **b.** $x = 3$ **c.** $x = -3$

13. Given $f(x) = 4x - 3$, find **a.** $f(-1)$ **b.** $f(0)$ **c.** $f(1)$ **d.** $f(y)$ **e.** $f(a + b)$ [**HINT**: See Example 1.]

14. Given $f(x) = -3x + 4$, find
a. $f(-1)$ **b.** $f(0)$ **c.** $f(1)$ **d.** $f(y)$ **e.** $f(a + b)$

15. Given $f(x) = x^2 + 2x + 3$, find
a. $f(0)$ **b.** $f(1)$ **c.** $f(-1)$ **d.** $f(-3)$
e. $f(a)$ **f.** $f(x + h)$ [**HINT**: See Example 1.]

16. Given $g(x) = 2x^2 - x + 1$, find
a. $g(0)$ **b.** $g(-1)$ **c.** $g(r)$ **d.** $g(x + h)$

17. Given $g(s) = s^2 + \dfrac{1}{s}$, find
a. $g(1)$ **b.** $g(-1)$ **c.** $g(4)$ **d.** $g(x)$ **e.** $g(s + h)$
f. $g(s + h) - g(s)$

18. Given $h(r) = \dfrac{1}{r + 4}$, find
a. $h(0)$ **b.** $h(-3)$ **c.** $h(-5)$ **d.** $h(x^2)$
e. $h(x^2 + 1)$ **f.** $h(x^2) + 1$

In Exercises 19–24, graph the given functions. Give the technology formula, and use technology to check your graph. We suggest that you become familiar with these graphs in addition to those in Table 1. [**HINT**: See Quick Example 4.]

19. $f(x) = -x^3$ (domain $(-\infty, +\infty)$)

20. $f(x) = x^3$ (domain $[0, +\infty)$)

21. $f(x) = x^4$ (domain $(-\infty, +\infty)$)

22. $f(x) = \sqrt[3]{x}$ (domain $(-\infty, +\infty)$)

23. $f(x) = \dfrac{1}{x^2}$ $(x \neq 0)$

24. $f(x) = x + \dfrac{1}{x}$ $(x \neq 0)$

In Exercises 25 and 26, match the functions to the graphs. (The gridlines are 1 unit apart.) Using technology to draw the graphs is suggested but not required.

25. **a.** $f(x) = x \quad (-1 \le x \le 1)$
b. $f(x) = -x \quad (-1 \le x \le 1)$
c. $f(x) = \sqrt{x} \quad (0 < x < 4)$
d. $f(x) = x + \dfrac{1}{x} - 2 \quad (0 < x < 4)$
e. $f(x) = |x| \quad (-1 \le x \le 1)$
f. $f(x) = x - 1 \quad (-1 \le x \le 1)$

(A)

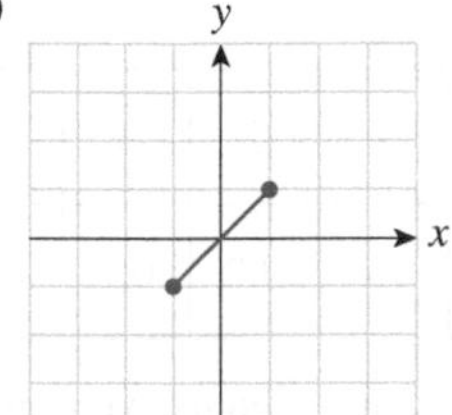

(B)

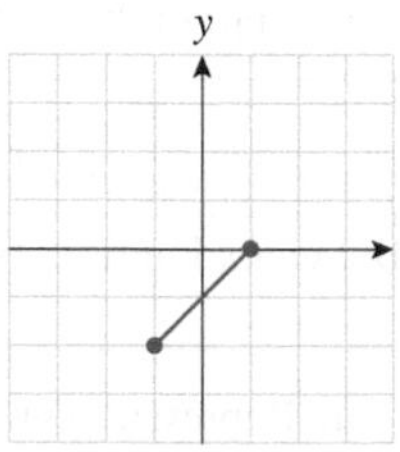

(C)

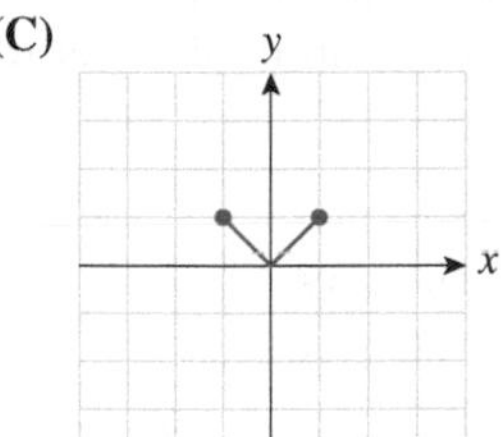

(D)

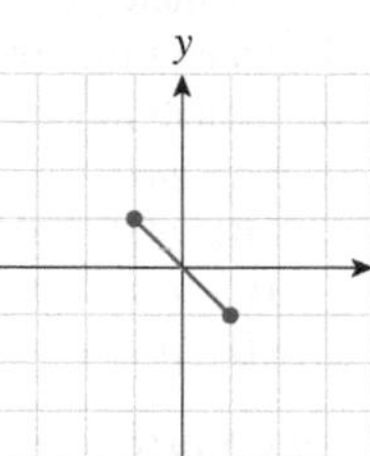

(E)

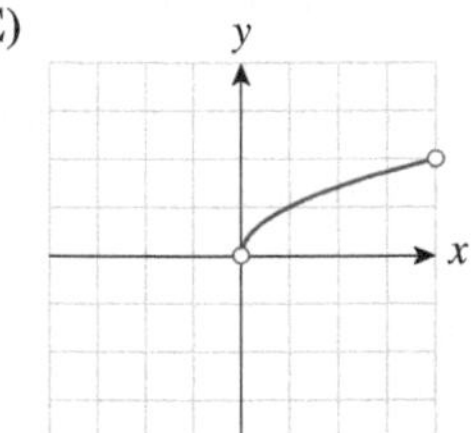

(F)

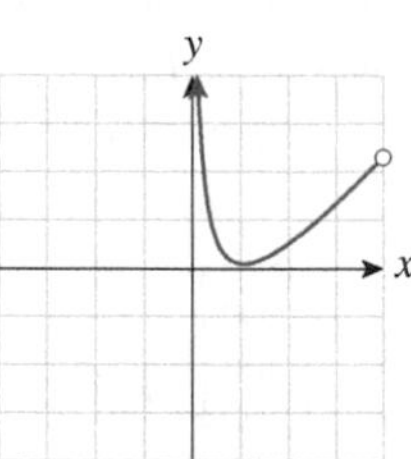

26. **a.** $f(x) = -x + 3 \quad (0 < x \le 3)$
b. $f(x) = 2 - |x| \quad (-2 < x \le 2)$
c. $f(x) = \sqrt{x + 2} \quad (-2 < x \le 2)$
d. $f(x) = -x^2 + 2 \quad (-2 < x \le 2)$
e. $f(x) = \dfrac{1}{x} - 1 \quad (0 < x \le 3)$
f. $f(x) = x^2 - 1 \quad (-2 < x \le 2)$

(A)

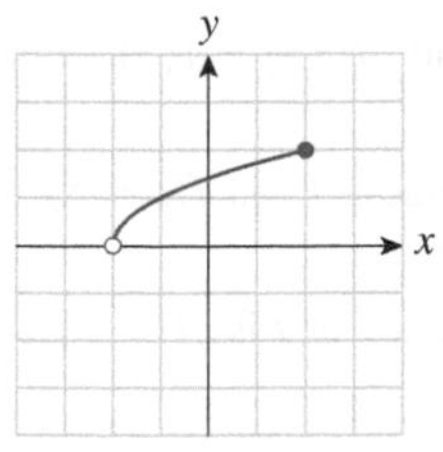

(B)

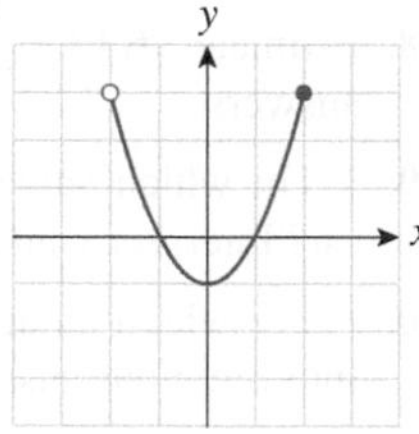

(C)

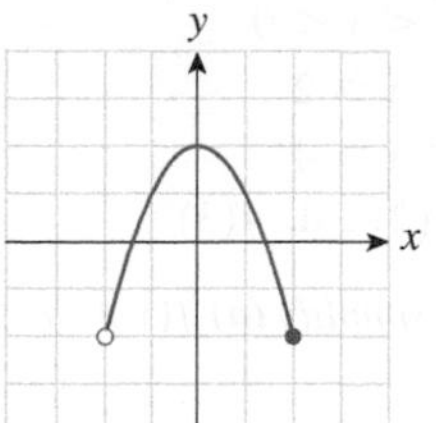

(D)

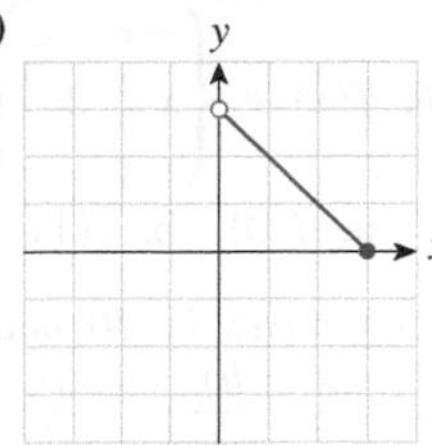

(E)

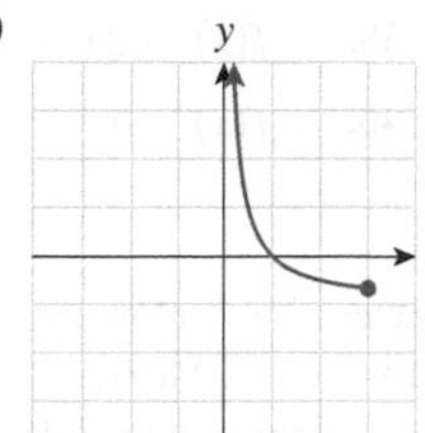

(F)

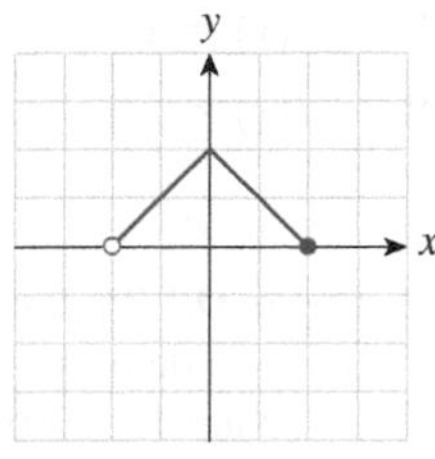

T *In Exercises 27–30, first give the technology formula for the given function, and then use technology to evaluate the function f for the given values of x (when f is defined there).*

27. T $f(x) = 0.1x^2 - 4x + 5;\ x = 0, 1, \ldots, 10$

28. T $g(x) = 0.4x^2 - 6x - 0.1;\ x = -5, -4, \ldots, 4, 5$

29. T $h(x) = \dfrac{x^2 - 1}{x^2 + 1};\ x = 0.5, 1.5, 2.5, \ldots, 10.5$ (Round all answers to four decimal places.)

30. T $r(x) = \dfrac{2x^2 + 1}{2x^2 - 1};\ x = -1, 0, 1, \ldots, 9$ (Round all answers to four decimal places.)

In Exercises 31–36, sketch the graph of the given function, evaluate the given expressions, and then use technology to duplicate the graphs. Give the technology formula.
[**HINT:** See Example 2.]

31. $f(x) = \begin{cases} x & \text{if } -4 \le x < 0 \\ 2 & \text{if } 0 \le x \le 4 \end{cases}$

a. $f(-1)$ **b.** $f(0)$ **c.** $f(1)$

32. $f(x) = \begin{cases} -1 & \text{if } -4 \le x \le 0 \\ x & \text{if } 0 < x \le 4 \end{cases}$

a. $f(-1)$ **b.** $f(0)$ **c.** $f(1)$

33. $f(x) = \begin{cases} x^2 & \text{if } -2 < x \le 0 \\ 1/x & \text{if } 0 < x \le 4 \end{cases}$

a. $f(-1)$ **b.** $f(0)$ **c.** $f(1)$

34. $f(x) = \begin{cases} -x^2 & \text{if } -2 < x \le 0 \\ \sqrt{x} & \text{if } 0 < x < 4 \end{cases}$

a. $f(-1)$ **b.** $f(0)$ **c.** $f(1)$

35. $f(x) = \begin{cases} x & \text{if } -1 < x \le 0 \\ x + 1 & \text{if } 0 < x \le 2 \\ x & \text{if } 2 < x \le 4 \end{cases}$

a. $f(0)$ **b.** $f(1)$ **c.** $f(2)$ **d.** $f(3)$
[**HINT:** See Example 3.]

36. $f(x) = \begin{cases} -x & \text{if } -1 < x < 0 \\ x - 2 & \text{if } 0 \le x \le 2 \\ -x & \text{if } 2 < x \le 4 \end{cases}$

a. $f(0)$ **b.** $f(1)$ **c.** $f(2)$ **d.** $f(3)$

In Exercises 37–40, find and simplify (a) $f(x+h) - f(x)$ *(b)* $\dfrac{f(x+h) - f(x)}{h}$.

37. ▼ $f(x) = x^2$

38. ▼ $f(x) = 3x - 1$

39. ▼ $f(x) = 2 - x^2$

40. ▼ $f(x) = x^2 + x$

Applications

41. ***Crude Oil Production: Mexico*** The following table shows daily crude oil production by **Pemex**, Mexico's national oil company, for 2008–2014 ($t = 0$ represents 2008):[5]

Year t (year since 2008)	0	1	2	3	4	5	6
Crude Oil Production $p(t)$ (million barrels/day)	3.16	2.97	2.95	2.94	2.91	2.88	2.79

a. Find $p(2)$, $p(3)$, and $p(6)$. Interpret your answers.

b. Find $p(4) - p(2)$. Interpret your answer. [**HINT**: See Quick Example 1.]

42. ***Offshore Crude Oil Production: Mexico*** The following table shows daily offshore crude oil production by **Pemex**, Mexico's national oil company, for 2008–2014 ($t = 0$ represents 2008):[6]

Year t (year since 2008)	0	1	2	3	4	5	6
Offshore Crude Oil Production $s(t)$ (million barrels/day)	2.25	2.01	1.94	1.90	1.90	1.90	1.85

a. Find $s(0)$, $s(2)$, and $s(4)$. Interpret your answers.

b. Find $s(4) - s(0)$. Interpret your answer.

43. ***Social Website Popularity: Twitter*** The following table shows the popularity of **Twitter** among social media sites as rated by StatCounter.com (t is the number of years since the start of 2008):[7]

Year t (year since start of 2008)	1	2	4	5
Twitter Popularity $p(t)$ (%)	7	6	6	7

a. Represent p graphically, and then use your graph to estimate $p(4.5)$. Interpret your answer.

b. One of the following models fits the data almost exactly. Which model is it?

(A) $p(t) = 0.4t^3 - 4t^2 + 12.5t + 15$

(B) $p(t) = -0.33t^2 + 2t - 8.7$

(C) $p(t) = -0.4t^3 + 4t^2 - 12.5t + 15$

(D) $p(t) = 0.33t^2 - 2t + 8.7$

44. ***Social Website Popularity: Delicious*** The following table shows the popularity of **Delicious** among social media sites as rated by StatCounter.com (t is the number of years since the start of 2008):[8]

Year t (year since start of 2008)	1	2	3	4	5
Delicious Popularity $p(t)$ (%)	0.4	0.2	0.1	0.05	0.02

a. Represent p graphically, and then use your graph to estimate $p(3.5)$. Interpret your answer.

b. One of the following models fits the data exactly. Which model is it?

(A) $p(t) = 0.8(2^{-t})$

(B) $p(t) = 0.8(2^t)$

(C) $p(t) = 0.02t^2 - 0.2t + 0.6$

(D) $p(t) = -0.02t^2 + 0.2t - 0.6$

Housing Starts *Exercises 45–48 refer to the following graph, which shows the number $f(t)$ of housing starts for single-family homes in the United States each year from 2000 through 2014 ($t = 0$ represents 2000, and $f(t)$ is in thousands of units):*[9]

45. Estimate $f(7)$, $f(14)$, and $f(9.5)$. Interpret your answers.

46. Estimate $f(3)$, $f(6)$, and $f(8.5)$. Interpret your answers.

47. Estimate $f(7 - 3)$ and $f(7) - f(3)$. Interpret your answers.

48. Estimate $f(13 - 3)$ and $f(13) - f(3)$. Interpret your answers.

49. ▼ For which value or values of t is $f(t + 5) - f(t)$ greatest? Interpret your answer.

50. ▼ For which value or values of t is $f(t) - f(t - 1)$ least? Interpret your answer

[5] Source: www.pemex.com (March 2015).

[6] *Ibid.*

[7] Percentages are based on worldwide page views. Source: http://gs.statcounter.com.

[8] Figures are approximate. Source: *Ibid.*

[9] Data are approximate. Source: www.census.gov.

51. ***Net Income: Casual Apparel*** In the following graph, $n(t)$ is **Abercrombie & Fitch**'s approximate net income, in millions of dollars, for the year ending at time t (t is time in years since December 2004):[10]

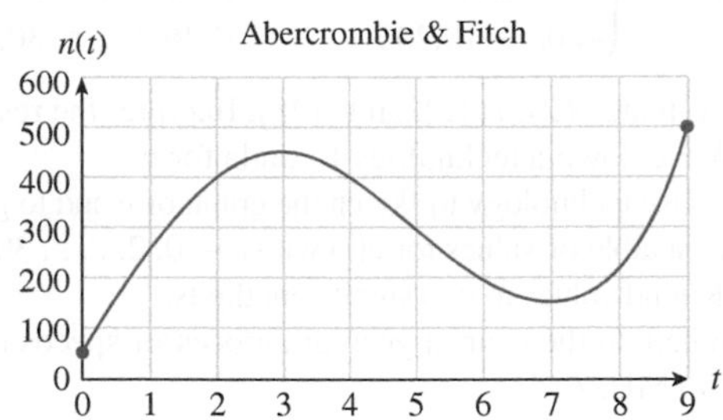

a. Estimate $n(2)$, $n(4)$, and $n(4.5)$ to the nearest 25. Interpret your answers.

b. At approximately which value of t in the interval $[3, 8]$ is $n(t)$ *increasing* most rapidly? Interpret your answer.

c. At approximately which value of t in the interval $[3, 8]$ is $n(t)$ *decreasing* most rapidly? Interpret your answer.

52. ***Net Income: Casual Apparel*** In the following graph, $n(t)$ is **Pacific Sunwear**'s approximate net income, in millions of dollars, for the year ending at time t (t is time in years since December 2004):[11]

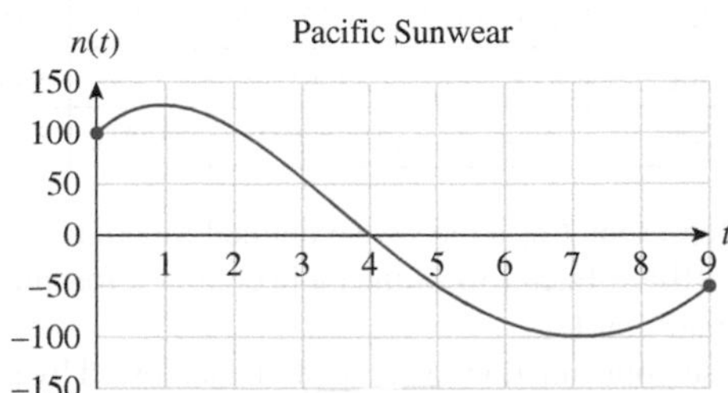

a. Estimate $n(0)$, $n(4)$, and $n(5.5)$ to the nearest 25. Interpret your answers.

b. At which of the following values of t is $n(t)$ *increasing* most rapidly: 1, 2, 4, 7, 8, or 9? Interpret your answer.

c. At which of the following values of t is $n(t)$ *decreasing* most rapidly: 1, 2, 4, 7, 8, or 9? Interpret your answer.

53. ***Funding for NASA: 1958–1966*** The percentage of the U.S. federal budget allocated to **NASA** from 1958 to 1966 can be approximated by

$$p(t) = \frac{4.5}{1.07^{(t-8)^2}} \text{ percentage points}$$

(t is time in years since 1958).[12] The following graph shows the data with the model:

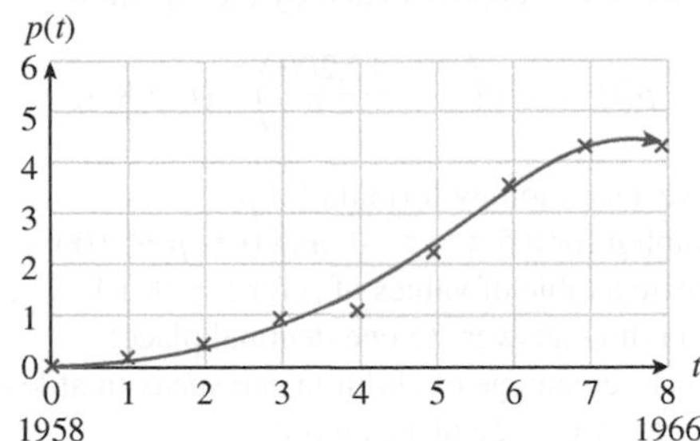

a. Find an appropriate domain of p. Is $t \geq 0$ an appropriate domain? Why or why not?

b. Compute $p(5)$ accurate to one decimal place. What does the answer say about the budget allocation to NASA?

c. At which of the following values of t is $p(t)$ increasing most rapidly: 0, 3, 5, 8? Interpret your answer.

54. ***Funding for NASA: 1966–2015*** The percentage of the U.S. federal budget allocated to **NASA** from 1966 to 2015 can be approximated by

$$p(t) = 0.03 + \frac{5}{t^{0.6}} \text{ percentage points} \quad (t \geq 1)$$

(t is time in years since 1965).[13] The following graph shows the data with the model:

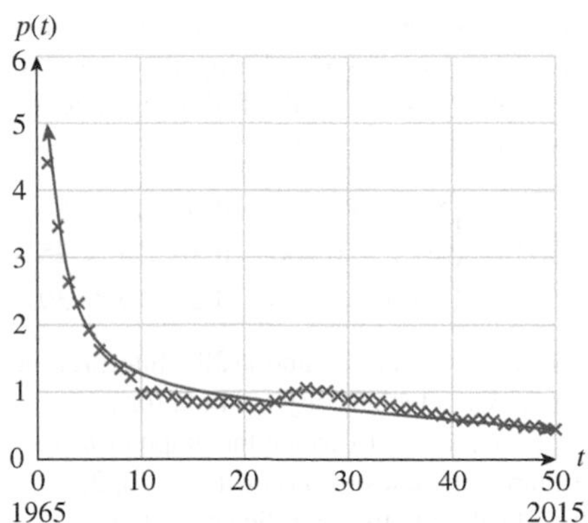

a. Find an appropriate domain of p. Is $[0, 50]$ an appropriate domain? Why or why not?

b. Compute $p(40)$ accurate to two decimal places. What does the answer say about the budget allocation to NASA?

c. If the model is extrapolated to larger and larger values of t, what does it suggest about long-term financing of NASA?

---

[10] "Net income" is an accounting term for profit (see Section 1.2). Model is the authors'. Source of data: www.wikinvest.com.

[11] *Ibid.*

[12] Model is the authors'. Source of data: U.S. Office of Management and Budget/www.wikipedia.org.

[13] *Ibid.*

55. T ***Acquisition of Language*** The percentage $p(t)$ of children who can speak in at least single words by the age of t months can be approximated by the equation[14]

$$p(t) = 100\left(1 - \frac{12{,}200}{t^{4.48}}\right) \quad (t \geq 8.5).$$

a. Give a technology formula for p.
b. Graph p for $8.5 \leq t \leq 20$ and $0 \leq p \leq 100$.
c. Create a table of values of p for $t = 9, 10, \ldots, 20$ (rounding answers to one decimal place).
d. What percentage of children can speak in at least single words by the age of 12 months?
e. By what age are 90% or more children speaking in at least single words?

56. T ***Acquisition of Language*** The percentage $p(t)$ of children who can speak in sentences of five or more words by the age of t months can be approximated by the equation[15]

$$p(t) = 100\left(1 - \frac{5.27 \times 10^{17}}{t^{12}}\right) \quad (t \geq 30).$$

a. Give a technology formula for p.
b. Graph p for $30 \leq t \leq 45$ and $0 \leq p \leq 100$.
c. Create a table of values of p for $t = 30, 31, \ldots, 40$ (rounding answers to one decimal place).
d. What percentage of children can speak in sentences of five or more words by the age of 36 months?
e. By what age are 75% or more children speaking in sentences of five or more words?

57. ▼ ***Processor Speeds*** The processor speed, in megahertz (MHz), of **Intel** processors during the period 1980–2010 could be approximated by the following function of time t in years since the start of 1980:[16]

$$v(t) = \begin{cases} 8(1.22)^t & \text{if } 0 \leq t < 16 \\ 400t - 6{,}200 & \text{if } 16 \leq t < 25 \\ 3{,}800 & \text{if } 25 \leq t \leq 30. \end{cases}$$

a. Evaluate $v(10)$, $v(16)$, and $v(28)$. Interpret the results.
b. Write down a technology formula for v.
c. T Use technology to sketch the graph of v and to generate a table of values for $v(t)$ with $t = 0, 2, \ldots, 30$. (Round values to two significant digits.)
d. When, to the nearest year, did processor speeds reach 3 gigahertz (1 gigahertz = 1,000 megahertz), according to the model?

58. ▼ ***Processor Speeds*** The processor speed, in megahertz (MHz), of **Intel** processors during the period 1970–2000 could be approximated by the following function of time t in years since the start of 1970:[17]

$$v(t) = \begin{cases} 0.12t^2 + 0.04t + 0.2 & \text{if } 0 \leq t < 12 \\ 1.1(1.22)^t & \text{if } 12 \leq t < 26 \\ 400t - 10{,}200 & \text{if } 26 \leq t \leq 30. \end{cases}$$

a. Evaluate $v(2)$, $v(12)$, and $v(28)$. Interpret the results.
b. Write down a technology formula for v.
c. T Use technology to sketch the graph of v and to generate a table of values for $v(t)$ with $t = 0, 2, \ldots, 30$. (Round values to two significant digits.)
d. When, to the nearest year, did processor speeds reach 500 MHz?

59. ▼ ***Income Taxes*** The U.S. federal income tax is a function of taxable income. Write $T(x)$ for the tax owed on a taxable income of x dollars. For tax year 2015 the function T for a single taxpayer was specified as follows:

If your taxable income was over . . .	But not over . . .	Your tax is . . .	Of the amount over . . .
\$0	9,225	10%	\$0
9,225	37,450	\$922.50 + 15%	\$9,225
37,450	90,750	5,156.25 + 25%	\$37,450
90,750	189,300	18,481.25 + 28%	\$90,750
189,300	411,500	46,075.25 + 33%	\$189,300
411,500	413,200	119,401.25 + 35%	\$411,500
413,200	—	119,996.25 + 39.6%	\$413,200

a. Represent T as a piecewise-defined function of income x. [**HINT**: Each row of the table defines a formula with a condition.]
b. Use your function to compute the tax owed by a single taxpayer on a taxable income of \$45,000.

60. ▼ ***Income Taxes*** Repeat Exercise 59 using the following information for tax year 2012:

If your taxable income was over . . .	But not over . . .	Your tax is . . .	Of the amount over . . .
\$0	8,700	10%	\$0
8,700	35,350	\$870.00 + 15%	\$8,700
35,350	85,650	4,867.50 + 25%	\$35,350
85,650	178,650	17,442.50 + 28%	\$85,650
178,650	388,350	43,482.50 + 33%	\$178,650
388,350	—	112,683.50 + 35%	\$388,350

[14] The model is the authors' and is based on data presented in the article *The Emergence of Intelligence* by William H. Calvin, *Scientific American,* October 1994, pp. 101–107.

[15] *Ibid.*

[16] Based on the fastest processors produced by Intel. Source for data: www.intel.com.

[17] *Ibid.*

Communication and Reasoning Exercises

61. Complete the following sentence: If the market price m of gold varies with time t, then the independent variable is ___, and the dependent variable is ___.

62. Complete the following sentence: If weekly profit P is specified as a function of selling price s, then the independent variable is ___, and the dependent variable is ___.

63. Complete the following: The function notation for the equation $y = 4x^2 - 2$ is ____.

64. Complete the following: The equation notation for $C(t) = -0.34t^2 + 0.1t$ is ____.

65. True or false? Every graphically specified function can also be specified numerically. Explain.

66. True or false? Every algebraically specified function can also be specified graphically. Explain.

67. True or false? Every numerically specified function with domain $[0, 10]$ can also be specified algebraically. Explain.

68. True or false? Every graphically specified function can also be specified algebraically. Explain.

69. ▼ True or false? Every function can be specified numerically. Explain.

70. ▼ Which supplies more information about a situation: a numerical model or an algebraic model?

71. ▼ Why is the following assertion false? "If $f(x) = x^2 - 1$, then $f(x + h) = x^2 + h - 1$."

72. ▼ Why is the following assertion false? "If $f(2) = 2$ and $f(4) = 4$, then $f(3) = 3$."

73. How do the graphs of two functions differ if they are specified by the same formula but have different domains?

74. How do the graphs of two functions f and g differ if $g(x) = f(x) + 10$? (Try an example.)

75. ▼ How do the graphs of two functions f and g differ if $g(x) = f(x - 5)$? (Try an example.)

76. ▼ How do the graphs of two functions f and g differ if $g(x) = f(-x)$? (Try an example.)

1.2 Functions and Models

The functions we used in Examples 1 and 2 in Section 1.1 are **mathematical models** of real-life situations because they model, or represent, situations in mathematical terms.

Mathematical Modeling

To mathematically model a situation means to represent it in mathematical terms. The particular representation used is called a **mathematical model** of the situation. Mathematical models do not always represent a situation perfectly or completely. Some (like Example 1 of Section 1.1) represent a situation only approximately; others represent only some aspects of the situation.

Quick Examples

1. The temperature is now 10°F and increasing by 20°F per hour.

   Model: $T(t) = 10 + 20t$ (t = time in hours, T = temperature)

2. I invest $1,000 at 5% interest compounded quarterly. Find the value of the investment after t years.

   Model: $A(t) = 1{,}000\left(1 + \frac{0.05}{4}\right)^{4t}$ (This is the compound interest formula we will study in Example 6.)

3. I am fencing a rectangular area whose perimeter is 100 feet. Find the area as a function of the width x.

Model: Take y to be the length, so the perimeter is

$$100 = x + y + x + y = 2(x + y).$$

This gives

$$x + y = 50.$$

Thus the length is $y = 50 - x$, and the area is

$$A = xy = x(50 - x).$$

4. You work 8 hours a day Monday through Friday and 5 hours on Saturday, and you have Sunday off. Model the number of hours you work as a function of the day of the week n, with $n = 1$ being Sunday.

Model: Take $f(n)$ to be the number of hours you work on the nth day of the week, so

$$f(n) = \begin{cases} 0 & \text{if } n = 1 \\ 8 & \text{if } 2 \le n \le 6 \\ 5 & \text{if } n = 7. \end{cases}$$

Note that the domain of f is $\{1, 2, 3, 4, 5, 6, 7\}$—a discrete set rather than a continuous interval of the real line.

5. The function

$$f(x) = -2x^2 + 12x + 36 \text{ million iPods sold } (x = \text{years since } 2005)$$

in Example 1 of Section 1.1 is a model of iPod sales.

6. The function

$$f(t) = \begin{cases} 5t^2 + 2t + 2 & \text{if } 0 \le t \le 3 \\ -2t^2 + 33t - 28 & \text{if } 3 < t \le 8 \end{cases} \quad \text{million consoles}$$

(t = years since the end of December 2006) in Example 2 of Section 1.1 is a model of accumulated Nintendo Wii game console sales.

Types of Models

Quick Examples 1–4 are **analytical models**, obtained by analyzing the situation being modeled. Quick Examples 5 and 6 are **curve-fitting models**, obtained by finding mathematical formulas that approximate observed data. All the models except for Quick Example 4 are **continuous models**, defined by functions whose domains are intervals of the real line. Quick Example 4 is a **discrete model**, as its domain is a discrete set, as was mentioned above. Discrete models are used extensively in probability and statistics.

Cost, Revenue, and Profit Models

EXAMPLE 1 Modeling Cost: Cost Function

As of March 2015, **Yellow Cab Chicago**'s rates amounted to \$3.05 on entering the cab plus \$1.80 for each mile.[18]

a. Find the cost C of an x-mile trip.

b. Use your answer to calculate the cost of a 40-mile trip.

c. What is the cost of the second mile? What is the cost of the tenth mile?

d. Graph C as a function of x.

Solution

a. We are being asked to find how the cost C depends on the length x of the trip, or to find C as a function of x. Here is the cost in a few cases:

Cost of a 1-mile trip: $C = 1.80(1) + 3.05 = 4.85$ — 1 mile at \$1.80 per mile plus \$3.05

Cost of a 2-mile trip: $C = 1.80(2) + 3.05 = 6.65$ — 2 miles at \$1.80 per mile plus \$3.05

Cost of a 3-mile trip: $C = 1.80(3) + 3.05 = 8.45.$ — 3 miles at \$1.80 per mile plus \$3.05

Do you see the pattern? The cost of an x-mile trip is given by the linear function*

$$C(x) = 1.80x + 3.05.$$

* See the table of common functions at the end of the preceding section. Linear functions are discussed in detail in Section 1.3.

Notice that the cost function is a sum of two terms: the **variable cost** $1.80x$, which depends on x, and the **fixed cost** 3.05, which is independent of x:

$$\text{Cost} = \text{Variable cost} + \text{Fixed cost}.$$

The quantity 1.80 by itself is the incremental cost per mile; you might recognize it as the *slope* of the given linear function. In this context we call 1.80 the **marginal cost**. You might recognize the fixed cost 3.05 as the *C-intercept* of the given linear function.

b. We can use the formula for the cost function to calculate the cost of a 40-mile trip as

$$C(40) = 1.80(40) + 3.05 = \$75.05.$$

c. To calculate the cost of the second mile, we *could* proceed as follows:

Find the cost of a 1-mile trip: $C(1) = 1.80(1) + 3.05 = \$4.85.$

Find the cost of a 2-mile trip: $C(2) = 1.80(2) + 3.05 = \$6.65.$

Therefore, the cost of the second mile is $\$6.65 - \$4.85 = \$1.80.$

But notice that this is just the marginal cost. In fact, the marginal cost is the cost of each additional mile, so we could have done this more simply:

$$\text{Cost of second mile} = \text{Cost of tenth mile} = \text{Marginal cost} = \$1.80.$$

[18] According to their website at www.yellowcabchicago.com.

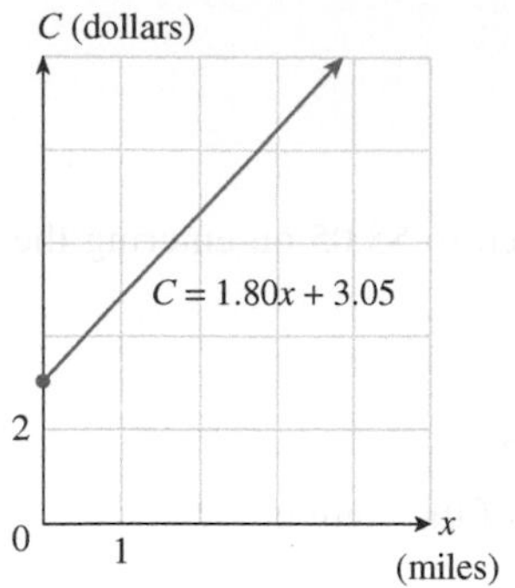

Figure 6

d. Figure 6 shows the graph of the cost function, which we can interpret as a *cost vs. miles* graph. The fixed cost is the starting height 3.05 on the left, while the marginal cost is the slope of the line: It rises 1.80 units per unit of x. (See Section 1.3 for a discussion of properties of straight lines.)

➡ **Before we go on...** The cost function in Example 1 is an example of an *analytical model:* We derived the form of the cost function from a knowledge of the cost per mile and the fixed cost.

As we discussed in Section 1.1, we can specify the cost function in Example 1 using equation notation:

$$C = 1.80x + 3.05. \quad \text{Equation notation}$$

Here, the independent variable is x, and the dependent variable is C. (This is the notation we have used in Figure 6. Remember that we will often switch between function and equation notation when it is convenient to do so.) ■

Here is a summary of some terms we used in Example 1, along with an introduction to some new terms.

Cost, Revenue, and Profit Functions

A **cost function** specifies the cost C as a function of the number of items x. Thus, $C(x)$ is the cost of x items and has the form

$$\text{Cost} = \text{Variable cost} + \text{Fixed cost}$$

where the variable cost is a function of x and the fixed cost is a constant. A cost function of the form

$$C(x) = mx + b$$

is called a **linear cost function**; the variable cost is mx, and the fixed cost is b. The slope m, the **marginal cost**, measures the incremental cost per item.

The **revenue**, or **net sales**, resulting from one or more business transactions is the total income received. If $R(x)$ is the revenue from selling x items at a price of m each, then R is the linear function $R(x) = mx$, and the selling price m can also be called the **marginal revenue**.

The **profit**, or **net income**, on the other hand, is what remains of the revenue when costs are subtracted.* If the profit depends linearly on the number of items, the slope m is called the **marginal profit**. Profit, revenue, and cost are related by the following formula.

$$\text{Profit} = \text{Revenue} - \text{Cost}$$

$$P(x) = R(x) - C(x).^{\dagger}$$

If the profit is negative, say $-\$500$, we refer to a **loss** (of \$500 in this case). To **break even** means to make neither a profit nor a loss. Thus, breakeven occurs when $P = 0$, or

$$R(x) = C(x). \quad \text{Breakeven}$$

The **break-even point** is the number of items x at which breakeven occurs.

* Note that taxes are also included in the costs. "Taxable profit" or "taxable net income" refers to profits before taxes are deducted.

† We say that the profit function P is the **difference** between the revenue and cost functions, and we express this fact as a formula about functions: $P = R - C$. (We will discuss this further when we talk about the algebra of functions at the end of this section.)

Quick Example

7. If the daily cost (including operating costs) of manufacturing x T-shirts is $C(x) = 8x + 100$ and the revenue obtained by selling x T-shirts is $R(x) = 10x$, then the daily profit resulting from the manufacture and sale of x T-shirts is

$$P(x) = R(x) - C(x) = 10x - (8x + 100) = 2x - 100.$$

Breakeven occurs when $P(x) = 0$, or $x = 50$.

EXAMPLE 2 Cost, Revenue, and Profit

The annual operating cost of *YSport Fitness* gym is estimated to be

$$C(x) = -2x^2 + 600x + 40{,}000 \text{ dollars} \qquad (0 \le x \le 150),$$

where x is the number of members. Annual revenue from membership averages \$800 per member. What is the variable cost? What is the fixed cost? What is the profit function? How many members must YSport have to make a profit? What will happen if it has fewer members? If it has more?

Solution The variable cost is the part of the cost function that depends on x:

$$\text{Variable cost} = -2x^2 + 600x.$$

The fixed cost is the constant term:

$$\text{Fixed cost} = 40{,}000.$$

The annual revenue YSport obtains from a single member is \$800. So if it has x members, it earns an annual revenue of

$$R(x) = 800x.$$

For the profit we use the formula

$$\begin{aligned} P(x) &= R(x) - C(x) && \text{Formula for profit} \\ &= 800x - (-2x^2 + 600x + 40{,}000) && \text{Substitute } R(x) \text{ and } C(x). \\ &= 2x^2 + 200x - 40{,}000. \end{aligned}$$

To make a profit, YSport needs to do better than break even, so let us find the break-even point: the value of x such that $P(x) = 0$. All we have to do is set $P(x) = 0$ and solve for x:

$$\begin{aligned} 2x^2 + 200x - 40{,}000 &= 0 \\ 2(x^2 + 100x - 20{,}000) &= 0 \\ 2(x + 200)(x - 100) &= 0 && \text{Factor the quadratic.*} \\ x = -200 \quad \text{or} \quad x &= 100. \end{aligned}$$

* Had this quadratic not factored, we would have needed to use the quadratic formula. (See Section 0.5.)

We reject the negative solution (as the domain is $[0, 150]$) and conclude that $x = 100$ members. To make a profit, should YSport have more than 100 members or fewer than 100 members? To decide, take a look at Figure 7, which shows two graphs: On the left we see the graphs of revenue and cost, and on the right we see the graph of the profit function.

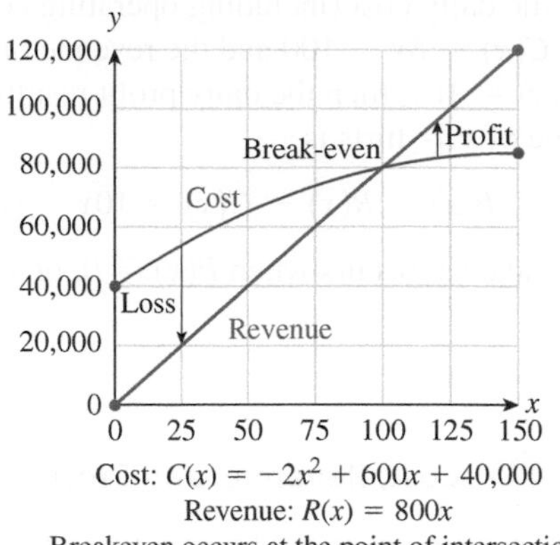

Cost: $C(x) = -2x^2 + 600x + 40{,}000$
Revenue: $R(x) = 800x$
Breakeven occurs at the point of intersection.

y
40,000
30,000
20,000
10,000
0
−10,000
−20,000
−30,000
−40,000
−50,000
Break-even
Profit
x
0 25 50 75 100 125 150

Profit: $P(x) = 2x^2 + 200x - 40{,}000$
Breakeven occurs when $P(x) = 0$.

Figure 7

For values of x less than the break-even point of 100, $P(x)$ is negative, so the company will have a loss. For values of x greater than the break-even point, $P(x)$ is positive, so the company will make a profit. Thus, YSport needs at least 101 members to make a profit.

Using Technology

Excel has a feature called "Goal Seek," which can be used to find the point of intersection of the cost and revenue graphs numerically rather than graphically. See the downloadable Excel tutorial for this section at the Website.

Demand and Supply Models

The demand for a commodity usually goes down as its price goes up. It is traditional to use the letter q for the (quantity of) demand as measured, for example, in sales. Consider the following example.

EXAMPLE 3 Demand: Private Schools

The demand, as meaured by total enrollment, for private schools in Michigan depends on the tuition cost and can be approximated by

$$q(p) = 32 + \frac{2{,}000}{p} \text{ thousand students enrolled} \qquad (200 \le p \le 2{,}200), \qquad \text{Demand function}$$

where p is the net tuition cost in dollars.[19] The graph of the demand function, shown in Figure 8, is called the associated **demand curve**.

What is the effect on demand if the tuition cost is increased from \$1,000 to \$2,000?

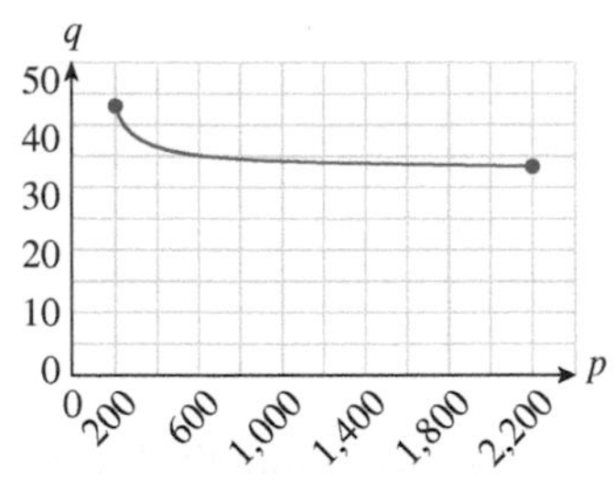

Technology formula:
`32+2000/x`

Figure 8

Solution The demand at tuition costs of \$1,000 and \$2,000 is

$$q(1{,}000) = 32 + \frac{2{,}000}{1{,}000} = 34 \text{ thousand students}$$

$$q(2{,}000) = 32 + \frac{2{,}000}{2{,}000} = 33 \text{ thousand students.}$$

[19] The tuition cost is net cost: tuition minus tax credit. The model is based on projections of enrollment under a proposed tax credit in "The Universal Tuition Tax Credit: A proposal to Advance Personal Choice in Education," Patrick L. Anderson, Richard McLellan, J.D., Joseph P. Overton, J.D., Gary Wolfram, Ph.D., Mackinac Center for Public Policy, www.mackinac.org.

The change in demand is therefore

$$q(2{,}000) - q(1{,}000) = 33 - 34 = -1 \text{ thousand students},$$

so demand decreases by around 1,000 students.

➡ **Before we go on . . .** As usual, we can represent the demand function in Example 3 as an equation:

$$q = 32 + \frac{2{,}000}{p}, \qquad \text{Demand equation}$$

where the independent variable is the tuition cost p and the dependent variable is the demand (total enrollment) q. ■

We have seen that a demand function gives the number of items consumers are willing to buy at a given price, and a higher price generally results in a lower demand. However, as the price rises, suppliers will be more inclined to produce these items (as opposed to spending their time and money on other products), so supply will generally rise. A **supply function** gives q, the number of items suppliers are willing to make available for sale,* as a function of p, the price per item.

* Although a bit confusing at first, it is traditional to use the same letter q for the quantity of supply and the quantity of demand, particularly when we want to compare them, as in the next example.

Demand, Supply, and Equilibrium Price

A **demand equation** or **demand function** expresses demand q (the number of items demanded) as a function of the unit price p (the price per item). A **supply equation** or **supply function** expresses supply q (the number of items a supplier is willing to make available) as a function of the unit price p (the price per item). It is usually the case that demand decreases and supply increases as the unit price increases.

Demand and supply are said to be in **equilibrium** when demand equals supply. The corresponding values of p and q are called the **equilibrium price** and **equilibrium demand**. To find the equilibrium price, determine the unit price p where the demand and supply curves cross (sometimes we can determine this value analytically by setting demand equal to supply and solving for p). To find the equilibrium demand, evaluate the demand (or supply) function at the equilibrium price.

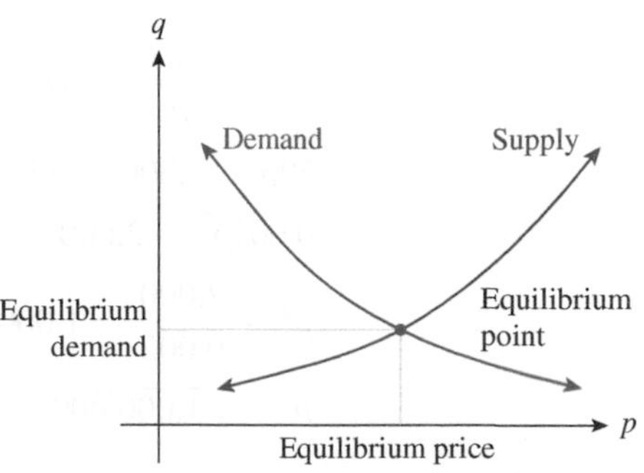

Quick Example

8. If the demand for your exclusive T-shirts is $q = -20p + 800$ shirts sold per day and the supply is $q = 10p - 100$ shirts per day, then the equilibrium point is obtained when demand = supply:

$$-20p + 800 = 10p - 100$$
$$30p = 900, \text{ giving } p = \$30.$$

The equilibrium price is therefore \$30, and the equilibrium demand is $q = -20(30) + 800 = 200$ shirts per day. What happens at prices other than the equilibrium price is discussed in Example 4.

Note In economics it is customary to plot the independent variable (price) on the vertical axis and the dependent variable (demand or supply) on the horizontal axis, but in this book we follow the usual mathematical convention for all graphs and plot the independent variable on the horizontal axis. ■

EXAMPLE 4 Demand, Supply, and Equilibrium Price

Continuing with Example 3, suppose that private school institutions are willing to create private schools to accommodate

$$q = 32 + 0.002p \text{ thousand students} \qquad (200 \le p \le 2{,}200) \qquad \text{Supply curve}$$

who pay a (net) tuition of p dollars.

a. What is the equilibrium tuition at private schools? Approximately how many students will be accommodated at that price?

b. What happens if the tuition is higher than the equilibrium tuition? What happens if it is lower?

c. Estimate the shortage or surplus of openings at private schools if the tuition is \$2,000.

Solution The equilibrium point is obtained when demand = supply:

$$32 + \frac{2{,}000}{p} = 32 + 0.002p$$

$$32p + 2{,}000 = 32p + 0.002p^2 \qquad \text{Multiply both sides by } p.$$

$$0.002p^2 = 2{,}000 \qquad \text{Cancel the } 32p.$$

$$p^2 = \frac{2{,}000}{0.002} = 1{,}000{,}000$$

$$p = \sqrt{1{,}000{,}000} = 1{,}000.$$

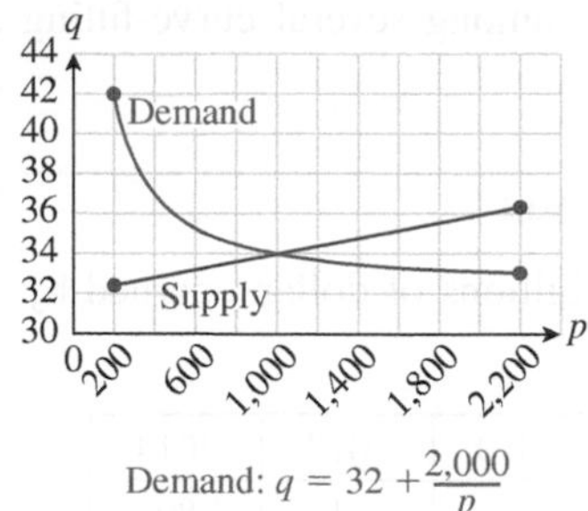

Demand: $q = 32 + \frac{2{,}000}{p}$

Supply: $q = 32 + 0.002p$

Figure 9

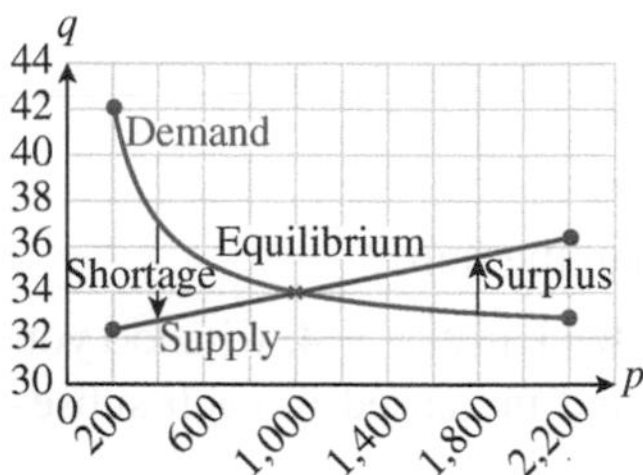

Figure 10

a. Figure 9 shows the graphs of demand $q = 32 + \dfrac{2{,}000}{p}$ and supply $q = 32 + 0.002p$. (See the margin note for a brief description of how to plot them.) The lines cross at $p = \$1{,}000$, so we conclude that demand = supply when $p = \$1{,}000$. This is the equilibrium tuition price. At that price we can calculate the demand or supply as

$$\text{Demand: } q = 32 + 2{,}000/1{,}000 = 34$$

$$\text{Supply: } q = 32 + 0.002(1{,}000) = 34, \qquad \text{Demand = Supply at equilibrium}$$

or 34,000 students.

b. Take a look at Figure 10, which shows what happens if schools charge more or less than the equilibrium price. If tuition is, say, \$2,000, then the supply will be larger than demand, and there will be a surplus of available openings at private schools. Similarly, if tuition is less—say \$400—then the supply will be less than the demand, and there will be a shortage of available openings.

c. The discussion in part (b) shows that if tuition is set at \$2,000 there will be a surplus of available openings. To estimate that number, we calculate the projected demand and supply when $p = \$2{,}000$:

$$\text{Demand: } q(2{,}000) = 32 + \frac{2{,}000}{2{,}000} = 33 \text{ thousand seats}$$

$$\text{Supply: } q(2{,}000) = 32 + 0.002(2{,}000) = 36 \text{ thousand seats}$$

$$\text{Surplus} = \text{Supply} - \text{Demand} = 36 - 33 = 3 \text{ thousand seats.}$$

So the models predict a surplus of 3,000 available seats.

➡ **Before we go on . . .** We saw in Example 4 that if tuition is less than the equilibrium price, there will be a shortage. If schools were to raise their tuition toward the equilibrium, they would create and fill more openings and increase revenue, since it is the supply equation—and not the demand equation—that determines what one can sell below the equilibrium price. On the other hand, if they were to charge more than the equilibrium price, they will be left with a possibly costly surplus of unused openings (and will want to lower tuition to reduce the surplus). Prices tend to move toward the equilibrium, so supply tends to equal demand. When supply equals demand, we say that the market **clears**. ■

Modeling Change over Time

Things around us change with time. Thus, there are many quantities, such as your income or the temperature in Honolulu, that are natural to think of as functions of time. Example 1 (on iPod sales) and Example 2 (on Nintendo Wii sales) in Section 1.1 are models of change over time. Both of those models are curve-fitting models: We used algebraic functions to approximate observed data.

Note We usually use the independent variable t to denote time (in seconds, hours, days, years, etc.). If a quantity q changes with time, then we can regard q as a function of t. ■

Using Technology

TI-83/84 Plus

Graphs:
`Y1=32+2000/X`
`Y2=32+0.002*X`
[2ND] [TABLE]
Graph: Xmin = 200, Xmax = 2200; [ZOOM] [0]

Spreadsheet

Headings *p*, Demand, Supply in A1–C1; *p*-values 200, 300, . . . , 2200 in A2–A22.
`=32+2000/A2` in B2
`=32+0.002*A2` in C2
Copy down through C22.
Highlight A1–C22; insert a scatter chart.

Website

www.WanerMath.com

Go to

Online Utilities → Function Evaluator and Grapher

and enter
`32+2000/x` for y_1 and
`32+0.002*x` for y_2.
Graph: Set Xmin = 200, Xmax = 2200, and press "Plot Graphs".

In the next example we are asked to select from among several curve-fitting models for given data.

EXAMPLE 5 Model Selection: Amazon Revenue

The following table shows some annual revenues, in billions of dollars, earned by **Amazon** from 2004 through 2014:[20]

Year	2004	2006	2008	2010	2012	2014
Revenue ($ billion)	7	11	19	34	61	89

Take t to be the number of years since 2004, and consider the following three models:

(1) $r(t) = 8t - 3$ Linear model

(2) $r(t) = 0.9t^2 - 0.6t + 7$ Quadratic model

(3) $r(t) = 7(1.3^t)$. Exponential model

a. Which two models fit the data significantly better than the third?

b. If you extrapolate the models you selected in part (a) back to $t = -4$, what do you find? (Amazon's revenue in 2000 was $2.8 billion and increased in each subsequent year.)

Using Technology

See the Technology Guide at the end of the chapter for detailed instructions on how to obtain the table and graphs in Example 5 using a TI-83/84 Plus or Excel. Here is an outline:

TI-83/84 Plus

STAT EDIT; enter the values of t in L1, and $r(t)$ in L2.
Plotting the points: ZOOM 9
Adding a curve:
Turn on Plot1 in Y= screen. Then (for the second curve)
`Y1=0.9X^2-0.6X+7`
Press GRAPH [More details in the Technology Guide.]

Spreadsheet

Table of values: Headings t, Revenue, $r(t)$ in A1–C1; t-values in A2–A7; revenue values in B2–B7.
Formula in C2:
`=0.9*A2^2-0.6*A2+7`
(second model)
Copy down through C7
Graph: Highlight A1–C7; insert a scatter chart.
[More details in the Technology Guide.]

Website
www.WanerMath.com
In the Function evaluator and grapher, enter the data and model(s) as shown below. Set xMin = 0, set xMax = 10, and press "Plot Graphs".

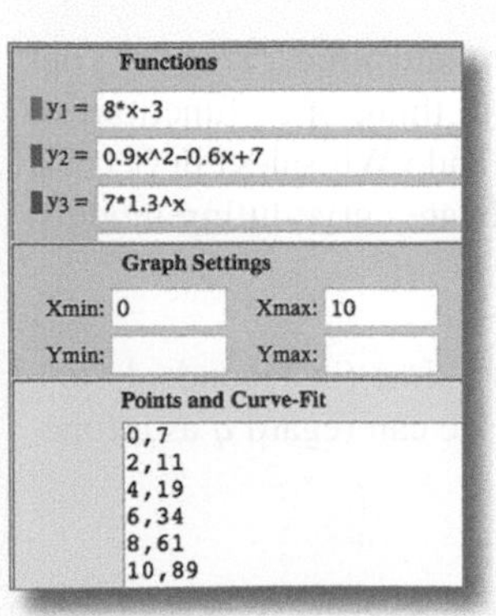

Solution

a. The following table shows the original data together with the values, rounded to the nearest 0.1, for all three models:

t	0	2	4	6	8	10
Revenue ($ billion)	7	11	19	34	61	89
Linear: $r(t) = 8t - 3$ Technology: `8*x-3`	−3	13	29	45	61	77
Quadratic: $r(t) = 0.9t^2 - 0.6t + 7$ Technology: `0.9x^2-0.6x+7`	7	9.4	19	35.8	59.8	91
Exponential: $r(t) = 7(1.3^t)$ Technology: `7*1.3^x`	7	11.8	20	33.8	57.1	96.5

Notice that all three models give values that seem reasonably close to the actual sales values. However, the graphs show the quadratic and exponential functions to be a lot more accurate than the linear one (see Figure 11).

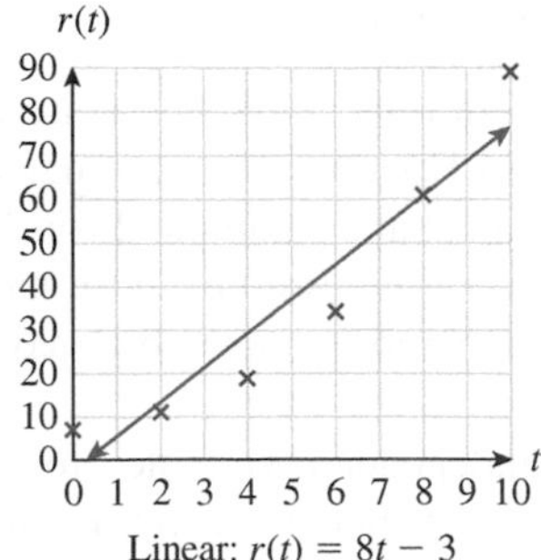

Linear: $r(t) = 8t - 3$

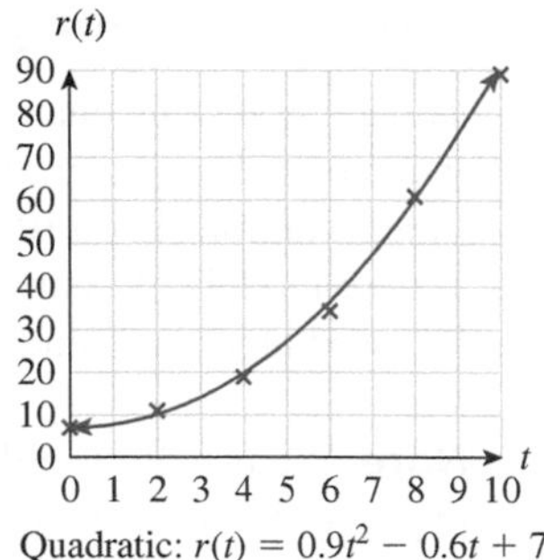

Quadratic: $r(t) = 0.9t^2 - 0.6t + 7$

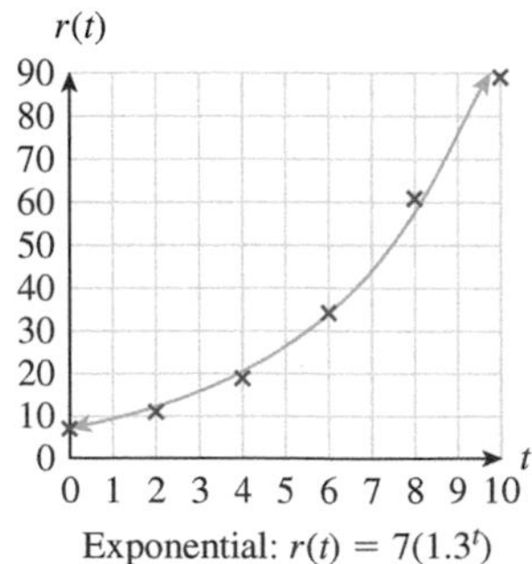

Exponential: $r(t) = 7(1.3^t)$

Figure 11

[20] Figures are rounded. Source: www.wikinvest.com.

b. Although the quadratic and exponential models both appear to fit the data well, when we extrapolate, we find

$$\text{Quadratic model: } r(-4) = 0.9(-4)^2 - 0.6(-4) + 7 = 23.8$$
$$\text{Exponential model: } r(-4) = 7(1.3^{-4}) \approx 2.5.$$

Notice that the quadratic model predicts *more revenue* in 2000 than in 2004, whereas the exponential model predicts a value closer to the actual value of \$2.8 billion. This discrepancy can be seen quite dramatically in Figure 12.

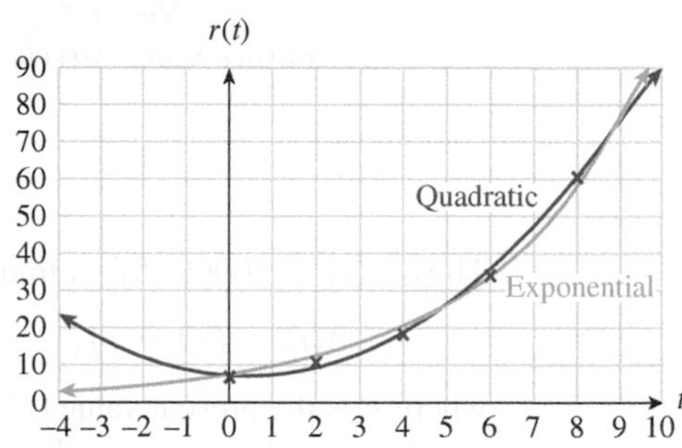

Figure 12

➡ Before we go on . . .

Q: *Does the result of Example 5(b) mean that the exponential model is more trustworthy than the quadratic model for projecting Amazon's revenue in years beyond 2013?*

A: The answer is a *qualified* "no:" No matter how well a mathematical curve fits a set of data over a specified period of time, there is nothing we can say with any certainty—either mathematically or statistically—as to whether that model applies to data beyond that period. On the other hand, the growth of many natural phenomena, from the revenues earned by a new company to population growth and the spread of an epidemic, tends to follow an exponential curve for a period of time,* so it is not surprising that extrapolating our exponential model backwards gave a good agreement with the real data. It is also quite reasonable to assume that our exponential model will also predict Amazon's revenue in the not-too-distant future, assuming that nothing out of the ordinary occurs. However, Amazon's revenue may begin to level off (as common sense predicts that it has to do eventually), experience a new spurt of growth due to the introduction of a popular new product, or take a nosedive due to the emergence of a strong competitor.

* To model their behavior over long periods of time requires a *logistic* model, which we will talk about in Section 2.4.

Q: *So what's the use of mathematical models of time-dependent data in the first place?*

A: Curve-fitting models can be used to analyze the trends of data we already have. For instance, the slope in the linear model in Example 5 gives us information about the rate of increase in Amazon's revenue, the coefficient of t^2 in the quadratic model gives us information about its *acceleration*, and the exponent in the exponential model gives us information about its *percentage* rate of increase. All this information can guide us in shaping an *analytical* model of the data, and such models *can* give meaningful predictions of what will happen in the future if they are based on reasonable assumptions about ongoing trends and other external factors. ■

We now derive an analytical model of change over time based on the idea of **compound interest**. Suppose you invest \$500 (the **present value**) in an investment account with an annual yield of 15% and the interest is reinvested at the end of every year (we say that the interest is **compounded** or **reinvested** once a year). Let t represent the number of years since you made the initial \$500 investment. Each year, the investment is worth 115% of (or 1.15 times) its value the previous year. The **future value** A of your investment changes over time t, so we think of A as a function of t. The following table illustrates how we can calculate the future value for several values of t:

Year t	0	1	2	3
Future Value $A(t)$ (\$)	500	575	661.25	760.44
	A	$500(1.15)$	$500(1.15)^2$	$500(1.15)^3$
		$\times 1.15$	$\times 1.15$	$\times 1.15$

Thus, $A(t) = 500(1.15)^t$. A traditional way to write this formula is

$$A(t) = P(1 + r)^t,$$

where P is the present value ($P = 500$) and r is the annual interest rate ($r = 0.15$).

If, instead of compounding the interest once a year, we compound it every three months (four times a year), we would earn one quarter of the interest ($r/4$ of the current investment) every three months. Because this would happen $4t$ times in t years, the formula for the future value becomes

$$A(t) = P\left(1 + \frac{r}{4}\right)^{4t}.$$

Compound Interest

If an amount (**present value**) P is invested for t years at an annual rate of r and if the interest is compounded (reinvested) m times per year, then the **future value** A is

$$A(t) = P\left(1 + \frac{r}{m}\right)^{mt}.$$

A special case is **interest compounded once a year**:

$$A(t) = P(1 + r)^t.$$

Quick Example

9. If \$2,000 is invested for two and a half years in a mutual fund with an annual yield of 12.6% and the earnings are reinvested each month, then $P = 2{,}000$, $r = 0.126$, $m = 12$, and $t = 2.5$, which gives

$$A(2.5) = 2{,}000\left(1 + \frac{0.126}{12}\right)^{12\times 2.5}$$

```
2000*(1+0.126/12)^(12*2.5)
```

$$= 2{,}000(1.0105)^{30} = \$2{,}736.02.$$

EXAMPLE 6 Compound Interest: Investments

Consider the scenario in Quick Example 9: You invest \$2,000 in a mutual fund with an annual yield of 12.6%, and the interest is reinvested each month.

a. Find the associated exponential model.

b. Compute the value of your investment after 7, 8, and 9 years. During which year does the value of your investment reach \$5,000?

Solution

a. Apply the formula

$$A(t) = P\left(1 + \frac{r}{m}\right)^{mt}$$

with $P = 2{,}000$, $r = 0.126$, and $m = 12$. We get

$$A(t) = 2{,}000\left(1 + \frac{0.126}{12}\right)^{12t}$$

$$= 2{,}000(1.0105)^{12t}.$$ `2000*(1+0.126/12)^(12*t)`

This is the exponential model. (What would happen if we left out the last set of parentheses in the technology formula?)

b. To find the value of your investment after 7, 8, and 9 years, we calculate $A(t)$ for these values of t:

$$C(7) = 2{,}000\left(1 + \frac{0.126}{12}\right)^{12(7)} \approx \$4{,}809.29$$ `2000*(1+0.126/12)^(12*7)`

$$C(8) = 2{,}000\left(1 + \frac{0.126}{12}\right)^{12(8)} \approx \$5{,}451.51$$ `2000*(1+0.126/12)^(12*8)`

$$C(9) = 2{,}000\left(1 + \frac{0.126}{12}\right)^{12(9)} \approx \$6{,}179.49.$$ `2000*(1+0.126/12)^(12*9)`

Because the balance first exceeds \$5,000 at $t = 8$ (the end of year 8), your investment reaches \$5,000 during year 8.

The compound interest examples we saw above are instances of **exponential growth**: a quantity whose magnitude is an increasing exponential function of time. The decay of unstable radioactive isotopes provides instances of **exponential decay**: a quantity whose magnitude is a *decreasing* exponential function of time. For example, carbon 14, an unstable isotope of carbon, decays exponentially to nitrogen. Because carbon 14 decay is extremely slow, it has important applications in the dating of fossils.

EXAMPLE 7 Exponential Decay: Carbon Dating

The amount of carbon 14 remaining in a sample that originally contained A grams is approximately

$$C(t) = A(0.999879)^t,$$

where t is time in years.

a. What percentage of the original amount remains after 1 year? After 2 years?

b. A fossilized plant unearthed in an archaeological dig contains 0.50 grams of carbon 14 and is known to be 50,000 years old. How much carbon 14 did the plant originally contain?

c. T Graph the function C for a sample originally containing 50 grams of carbon 14, and use your graph to estimate how long, to the nearest 1,000 years, it takes for half the original carbon 14 to decay.

Solution Notice that the given model is exponential as it has the form $f(t) = Ab^t$. (See the table at the end of Section 1.1.)

a. At the start of the first year, $t = 0$, so there are

$$C(0) = A(0.999879)^0 = A \text{ grams.}$$

At the end of the first year, $t = 1$, so there are

$$C(1) = A(0.999879)^1 = 0.999879\,A \text{ grams;}$$

that is, 99.9879% of the original amount remains. After the second year, the amount remaining is

$$C(2) = A(0.999879)^2 \approx 0.999758\,A \text{ grams,}$$

or about 99.9758% of the original sample.

b. We are given the following information: $C = 0.50$, $A =$ the unknown, and $t = 50{,}000$. Substituting gives

$$0.50 = A(0.999879)^{50{,}000}.$$

Solving for A gives

$$A = \frac{0.5}{0.999879^{50{,}000}} \approx 212 \text{ grams.}$$

Thus, the plant originally contained 212 grams of carbon 14.

c. For a sample originally containing 50 grams of carbon 14, $A = 50$, so $C(t) = 50(0.999879)^t$. Its graph is shown in Figure 13. We have also plotted the line $y = 25$ on the same graph. The graphs intersect at the point where the original sample has decayed to 25 grams: about $t = 6{,}000$ years.

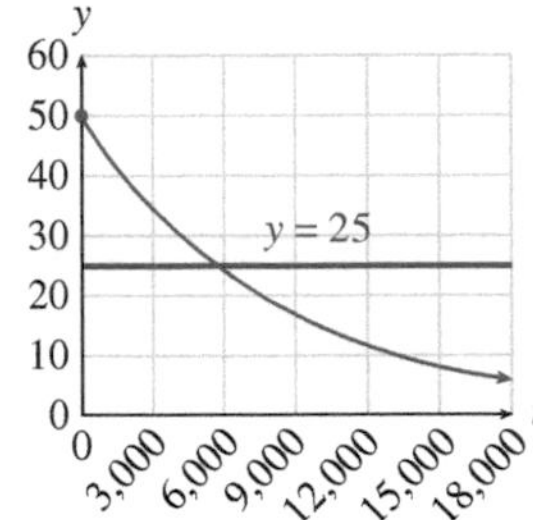

Technology formula:
`50*0.999879^x`

Figure 13

➡ **Before we go on . . .** The formula we used for A in Example 7(b) has the form

$$A(t) = \frac{C}{0.999879^t},$$

which gives the original amount of carbon 14 t years ago in terms of the amount C that is left now. A similar formula can be used in finance to find the present value, given the future value. ■

Algebra of Functions

If you look back at some of the functions considered in this section, you will notice that we frequently constructed them by combining simpler or previously constructed functions. For instance:

Quick Example 3: Area = Width × Length: $A(x) = x(50 - x)$

Example 1: Cost = Variable cost + Fixed cost: $C(x) = 1.80x + 3.05$

Quick Example 7: Profit = Revenue − Cost: $P(x) = 10x - (8x + 100)$.

Let us look a little more deeply at each of the above examples.

Area Example: $A(x) = \text{Width} \times \text{Length} = x(50 - x)$:
Think of the width and length as separate functions of x:

$$\text{Width: } W(x) = x; \qquad \text{Length: } L(x) = 50 - x$$

so

$$A(x) = W(x)L(x). \qquad \text{Area = Width × Length}$$

We say that the area function A is the **product of the functions** W and L, and we write

$$A = WL. \qquad A \text{ is the product of the functions } W \text{ and } L.$$

To calculate $A(x)$, we multiply $W(x)$ by $L(x)$.

Cost Example: $C(x) = \text{Variable cost} + \text{Fixed cost} = 1.80x + 3.05$:
Think of the variable and fixed costs as separate functions of x:

$$\text{Variable cost: } V(x) = 1.80x; \qquad \text{Fixed cost: } F(x) = 3.05^{*}$$

* F is called a **constant function** as its value, 3.05, is the same for every value of x.

so

$$C(x) = V(x) + F(x). \qquad \text{Cost = Variable cost + Fixed cost}$$

We say that the cost function C is the **sum of the functions V and F**, and we write

$$C = V + F. \qquad C \text{ is the sum of the functions } V \text{ and } F.$$

To calculate $C(x)$, we add $V(x)$ to $F(x)$.

Profit Example: $P(x) = \text{Revenue} - \text{Cost} = 10x - (8x + 100)$:
Think of the revenue and cost as separate functions of x:

$$\text{Revenue: } R(x) = 10x; \qquad \text{Cost: } C(x) = 8x + 100$$

so

$$P(x) = R(x) - C(x). \qquad \text{Profit = Revenue − Cost}$$

We say that the profit function P is the **difference between the functions R and C**, and we write

$$P = R - C. \qquad P \text{ is the difference of the functions } R \text{ and } C.$$

To calculate $P(x)$, we subtract $C(x)$ from $R(x)$.

Algebra of Functions

If f and g are real-valued functions of the real variable x, then we define their **sum s, difference d, product p,** and **quotient q** as follows:

$s = f + g$ is the function specified by $s(x) = f(x) + g(x)$.

$d = f - g$ is the function specified by $d(x) = f(x) - g(x)$.

$p = fg$ is the function specified by $p(x) = f(x)g(x)$.

$q = \dfrac{f}{g}$ is the function specified by $q(x) = \dfrac{f(x)}{g(x)}$.

Also, if f is as above and c is a constant (real number), then we define the associated **constant multiple m of f** by

$m = cf$ is the function specified by $m(x) = cf(x)$.

Note on Domains

In order for any of the expressions $f(x) + g(x)$, $f(x) - g(x)$, $f(x)g(x)$, or $f(x)/g(x)$ to make sense, x must be simultaneously in the domains of both f and g. Further, for the quotient, the denominator $g(x)$ cannot be zero. Thus, we specify the domains of these functions as follows:

Domain of $f + g$, $f - g$, and fg: All real numbers x simultaneously in the domains of f and g

Domain of f/g: All real numbers x simultaneously in the domains of f and g such that $g(x) \neq 0$

Domain of cf: Same as the domain of f

Quick Examples

10. If $f(x) = x^2 - 1$ and $g(x) = \sqrt{x}$ with domain $[0, +\infty)$, then the sum s of f and g has domain $[0, +\infty)$ and is specified by $s(x) = f(x) + g(x) = x^2 - 1 + \sqrt{x}$.

11. If $f(x) = x^2 - 1$ and $c = 3$, then the associated constant multiple m of f is specified by $m(x) = 3f(x) = 3(x^2 - 1)$.

12. If $c = -1$, then the associated constant multiple $(-1)f$ of f is often written as $-f$. Thus, if $f(x) = x^2 - 1$, then $(-f)(x) = (-1)(x^2 - 1) = -x^2 + 1$.

13. If there are $N = 1{,}000t$ Mars shuttle passengers in year t who pay a total cost of $C = 40{,}000 + 800t$ million dollars, then the cost per passenger is given by the quotient of the two functions:

$$\text{Cost per passenger} = q(t) = \frac{C(t)}{N(t)}$$
$$= \frac{40{,}000 + 800t}{1{,}000t} \text{ million dollars per passenger.}$$

The largest possible domain of C/N is $(0, +\infty)$, as the quotient is not defined if $t = 0$.

1.2 EXERCISES

▼ more advanced ◆ challenging

T indicates exercises that should be solved using technology

Exercises 1–8 are based on the following functions:

$f(x) = x^2 + 1$ *with domain* $(-\infty, +\infty)$
$g(x) = x - 1$ *with domain* $(-\infty, +\infty)$
$h(x) = x + 4$ *with domain* $[10, +\infty)$
$u(x) = \sqrt{x + 10}$ *with domain* $[-10, 0)$
$v(x) = \sqrt{10 - x}$ *with domain* $[0, 10]$

*In each exercise, **(a)** write a formula for the indicated function, **(b)** give its domain, and **(c)** specify its value at the given point a, if defined.*

1. $s = f + g; a = -3$

2. $d = g - f; a = -1$

3. $p = gu; a = -6$

4. $p = hv; a = 1$

5. $q = \frac{v}{g}; a = 1$

6. $q = \frac{g}{v}; a = 1$

7. $m = 5f; a = 1$

8. $m = 3u; a = -1$

Applications

9. ***Resources*** You now have 200 music files on your hard drive, and this number is increasing by 10 music files each day. Find a mathematical model for this situation. [**HINT:** See Quick Example 1.]

10. ***Resources*** The amount of free space left on your hard drive is now 50 gigabytes (GB) and is decreasing by 5 GB per month. Find a mathematical model for this situation.

11. ***Soccer*** My rectangular soccer field site has a length equal to twice its width. Find its area in terms of its length x. [**HINT:** See Quick Example 3.]

12. ***Cabbage*** My rectangular cabbage patch has a total area of 100 square feet. Find its perimeter in terms of the width x.

13. ***Vegetables*** I want to fence in a square vegetable patch. The fencing for the east and west sides costs $4 per foot, and the fencing for the north and south sides costs only $2 per foot. Find the total cost of the fencing as a function of the length of a side x.

14. ***Orchids*** My square orchid garden abuts my house so that the house itself forms the northern boundary. The fencing for the southern boundary costs $4 per foot, and the fencing for the east and west sides costs $2 per foot. Find the total cost of the fencing as a function of the length of a side x.

15. ***Study*** You study math 4 hours a day on Sunday through Thursday and take Friday and Saturday off. Model the number of hours h you study math as a function of the day of the week n (with $n = 1$ being Sunday). [**HINT:** See Quick Example 6.]

16. ***Recreation*** You spend 5 hours per day on Saturdays and Sundays watching movies but only 2 hours per day during the week. Model the number of hours h you watch movies as a function of the day of the week n (with $n = 1$ being Sunday).

17. ***Cost*** A piano manufacturer has a daily fixed cost of $1,000 and a marginal cost of $1,500 per piano. Find the cost $C(x)$ of manufacturing x pianos in one day. Use your function to answer the following questions:
   a. On a given day, what is the cost of manufacturing three pianos?
   b. What is the cost of manufacturing the third piano that day?
   c. What is the cost of manufacturing the 11th piano that day?
   d. What is the variable cost? What is the fixed cost? What is the marginal cost?
   e. Graph C as a function of x. [**HINT:** See Example 1.]

18. ***Cost*** The cost of renting tuxes for the Choral Society's formal is $20 down plus $88 per tux. Express the cost C as a function of x, the number of tuxedos rented. Use your function to answer the following questions:
   a. What is the cost of renting two tuxes?
   b. What is the cost of the second tux?
   c. What is the cost of the 4,098th tux?
   d. What is the variable cost? What is the fixed cost? What is the marginal cost?
   e. Graph C as a function of x.

19. ***Break-Even Analysis*** Your college newspaper, *The Collegiate Investigator*, has fixed production costs of $70 per edition and marginal printing and distribution costs of 40¢ per copy. *The Collegiate Investigator* sells for 50¢ per copy.
   a. Write down the associated cost, revenue, and profit functions. [**HINT:** See Examples 1 and 2.]
   b. What profit (or loss) results from the sale of 500 copies of *The Collegiate Investigator*?
   c. How many copies should be sold to break even?

20. ***Break-Even Analysis*** The Audubon Society at *Enormous State University* (ESU) is planning its annual fund-raising "Eat-a-thon." The society will charge students 50¢ per serving of pasta. The only expenses the society will incur are the cost of the pasta, estimated at 15¢ per serving, and the $350 cost of renting the facility for the evening.
   a. Write down the associated cost, revenue, and profit functions.
   b. How many servings of pasta must the Audubon Society sell to break even?
   c. What profit (or loss) results from the sale of 1,500 servings of pasta?

21. ***Break-Even Analysis*** *Gymnast Clothing* manufactures expensive hockey jerseys for sale to college bookstores in runs of up to 200. Its cost (in dollars) for a run of x hockey jerseys is

$$C(x) = 2{,}000 + 10x + 0.2x^2 \quad (0 \le x \le 200).$$

Gymnast Clothing sells the jerseys at $100 each. Find the revenue and profit functions. How many jerseys should Gymnast Clothing manufacture to make a profit? [**HINT:** See Example 2.]

22. ***Break-Even Analysis*** *Gymnast Clothing* also manufactures expensive soccer cleats for sale to college bookstores in runs of up to 500. Its cost (in dollars) for a run of x pairs of cleats is

$$C(x) = 3{,}000 + 8x + 0.1x^2 \quad (0 \le x \le 500).$$

Gymnast Clothing sells the cleats at $120 per pair. Find the revenue and profit functions. How many pairs of cleats should Gymnast Clothing manufacture to make a profit?

23. ***Break-Even Analysis: School Construction Costs*** The cost, in millions of dollars, of building a two-story high school in New York State was estimated to be

$$C(x) = 1.7 + 0.12x - 0.0001x^2 \quad (20 \le x \le 400),$$

where x is the number of thousands of square feet.[21] Suppose that you are contemplating building a for-profit two-story high school and estimate that your total revenue will be $0.1 million per thousand square feet. What is the profit function? What size school should you build to break even?

---

[21] The model is the authors'. Source for data: *Project Labor Agreements and Public Construction Cost in New York State,* Paul Bachman and David Tuerck, Beacon Hill Institute at Suffolk University, April 2006, www.beaconhill.org.

24. ***Break-Even Analysis: School Construction Costs*** The cost, in millions of dollars, of building a three-story high school in New York State was estimated to be

$$C(x) = 1.7 + 0.14x - 0.0001x^2 \quad (20 \le x \le 400),$$

where x is the number of thousands of square feet.[22] Suppose that you are contemplating building a for-profit three-story high school and estimate that your total revenue will be $0.2 million per thousand square feet. What is the profit function? What size school should you build to break even?

25. ▼ ***Profit Analysis: Aviation*** The hourly operating cost of a **Boeing** 747-100, which seats up to 405 passengers, is estimated to be[23] $5,132. If an airline charges each passenger a fare of $100 per hour of flight, find the hourly profit P it earns operating a 747-100 as a function of the number of passengers x. (Be sure to specify the domain.) What is the least number of passengers it must carry to make a profit? [**HINT:** The cost function is constant (Variable cost = 0).]

26. ▼ ***Profit Analysis: Aviation*** The hourly operating cost of a **McDonnell Douglas** DC 10-10, which seats up to 295 passengers, is estimated to be[24] $3,885. If an airline charges each passenger a fare of $100 per hour of flight, find the hourly profit P it earns operating a DC 10-10 as a function of the number of passengers x. (Be sure to specify the domain.) What is the least number of passengers it must carry to make a profit? [**HINT:** The cost function is constant (Variable cost = 0).]

27. ▼ ***Break-Even Analysis*** *(based on a question from a CPA exam)* The *Oliver Company* plans to market a new product. Based on its market studies, Oliver estimates that it can sell up to 5,500 units in 2005. The selling price will be $2 per unit. Variable costs are estimated to be 40% of total revenue. Fixed costs are estimated to be $6,000 for 2005. How many units should the company sell to break even?

28. ▼ ***Break-Even Analysis*** *(based on a question from a CPA exam)* The *Metropolitan Company* sells its latest product at a unit price of $5. Variable costs are estimated to be 30% of the total revenue, while fixed costs amount to $7,000 per month. How many units should the company sell per month to break even, assuming that it can sell up to 5,000 units per month at the planned price?

29. ◆ ***Break-Even Analysis*** *(from a CPA exam)* Given the following notations, write a formula for the break-even sales level:

SP = Selling price per unit
FC = Total fixed cost
VC = Variable cost per unit.

30. ◆ ***Break-Even Analysis*** *(based on a question from a CPA exam)* Given the following notation, give a formula for the total fixed cost:

SP = Selling price per unit
VC = Variable cost per unit
BE = Break-even sales level in units.

31. ◆ ***Break-Even Analysis: Organized Crime*** The organized crime boss and perfume king Butch (Stinky) Rose has daily overheads (bribes to corrupt officials, motel photographers, wages for hit men, explosives, and so on) amounting to $20,000 per day. On the other hand, he has a substantial income from his counterfeit perfume racket: He buys imitation French perfume (Chanel № 22.5) at $20 per gram, pays an additional $30 per 100 grams for transportation, and sells the perfume via his street thugs for $600 per gram. Specify Stinky's profit function, $P(x)$, where x is the quantity (in grams) of perfume he buys and sells, and use your answer to calculate how much perfume should pass through his hands per day in order that he break even.

32. ◆ ***Break-Even Analysis: Disorganized Crime*** Butch (Stinky) Rose's counterfeit Chanel № 22.5 racket has run into difficulties: It seems that the *authentic* Chanel № 22.5 perfume is selling for less than his counterfeit perfume. However, he has managed to reduce his fixed costs to zero, and his overall costs are now $400 per gram plus $30 per gram transportation costs and commission. (The perfume's smell is easily detected by specially trained Chanel Hounds, and this necessitates elaborate packaging measures.) He therefore decides to sell the perfume for $420 per gram to undercut the competition. Specify Stinky's profit function, $P(x)$, where x is the quantity (in grams) of perfume he buys and sells, and use your answer to calculate how much perfume should pass through his hands per day in order that he break even. Interpret your answer.

33. ***Demand: E-Readers*** The demand for **Amazon**'s Kindle e-reader can be approximated by

$$q(p) = \frac{760}{p} - 1 \text{ million units per year} \quad (60 \le p \le 400),$$

where p is the price charged by Amazon.[25]

a. Graph the demand function.
b. What is the result on demand if the unit price is increased from $100 to $200? [**HINT:** See Example 3.]
c. According to the graph in part (a), if the price is $200 and successively increases in $10 increments, then the demand
(A) increases at a greater and greater rate.
(B) decreases at a greater and greater rate.
(C) increases at a smaller and smaller rate.
(D) decreases at a smaller and smaller rate.
(E) increases at the same rate.
(F) decreases at the same rate.

[22] See footnote for Exercise 23.
[23] In 1992. Source: Air Transportation Association of America.
[24] *Ibid.*
[25] Based on data from 2007 to 2013. Source: www.e-reader-info.com.

34. ***Demand for Monorail Service: Mars*** The demand for monorail service on the Utarek monorail, which links the three urbynes (or districts) of Utarek on Mars, can be approximated by

$$q(p) = 7.5 + \frac{30}{p} \text{ million rides per day} \quad (3 \le p \le 8),$$

where p is the cost per ride in zonars (Ƶ).[26]

a. Graph the demand function.
b. What is the result on demand if the cost per ride is decreased from Ƶ5.00 to Ƶ3.00?
c. If the demand function is extrapolated, what does its graph suggest will be the effect of increasing the price to extremely large values?

35. ▼ ***Demand: Smartphones*** The worldwide demand for smartphones may be modeled by

$$q(p) = 0.17p^2 - 63p + 5{,}900 \text{ million units sold annually} \quad (100 \le p \le 200),$$

where p is the unit price in dollars.[27]

a. Use the demand function to estimate, to the nearest million units, worldwide sales of smartphones if the price is \$110.
b. Extrapolate the demand function to estimate, to the nearest million units, worldwide sales of smartphones if the price is \$90.
c. Model the worldwide annual revenue from the sale of smartphones as a function of unit price, and use your model to estimate, to the nearest billion dollars, worldwide annual revenue when the price is set at \$110. [**HINT:** Revenue = Price × Quantity = $p \cdot q(p)$.]
d. Graph the function you obtained in part (c). According to the graph, does worldwide revenue increase or decrease as the price decreases past \$110?

36. ▼ ***Demand: Smartphones*** (See Exercise 35.) Here is another model for worldwide demand for smartphones:

$$q(p) = 36{,}900(0.968^p) \text{ million units sold annually} \quad (100 \le p \le 200),$$

where p is the unit price in dollars.[28]

a. Use the demand function to estimate, to the nearest million units, worldwide sales of smartphones if the price is \$120.
b. Extrapolate the demand function to estimate, to the nearest million units, worldwide sales of smartphones if the price is \$210.
c. Model the worldwide annual revenue from the sale of smartphones as a function of unit price, and use your model to estimate, to the nearest billion dollars, worldwide annual revenue when the price is set at \$120. [**HINT:** Revenue = Price × Quantity = $p \cdot q(p)$.]
d. Graph the function you obtained in part (c). According to the graph, would increased worldwide revenue result from increasing or decreasing the price beyond \$120?

37. ***Equilibrium Price: Skateboards*** The demand for your hand-made skateboards, in weekly sales, is

$$q = -3p + 700$$

if the selling price is \$$p$. You are prepared to supply $q = 2p - 500$ skateboards per week at the price \$$p$. At what price should you sell your skateboards so that there is neither a shortage nor a surplus? [**HINT:** See Quick Example 8.]

38. ***Equilibrium Price: Skateboards*** The demand for your factory-made skateboards, in weekly sales, is

$$q = -5p + 50$$

if the selling price is \$$p$. If you are selling them at that price, you can obtain $q = 3p - 30$ skateboards per week from the factory. At what price should you sell your skateboards so that there is neither a shortage nor a surplus?

39. ***Equilibrium Price: Cell Phones*** Worldwide quarterly sales of **Nokia** cell phones were approximately $q = -p + 156$ million phones when the wholesale price[29] was \$$p$.

a. If Nokia was prepared to supply $q = 4p - 394$ million phones per quarter at a wholesale price of \$$p$, what would have been the equilibrium price?
b. The actual wholesale price was \$105 in the fourth quarter of 2004. Estimate the projected shortage or surplus at that price. [**HINT:** See Quick Example 8 and also Example 4.]

40. ***Equilibrium Price: Cell Phones*** Worldwide annual sales of all cell phones were approximately $-10p + 1{,}600$ million phones when the wholesale price[30] was \$$p$.

a. If manufacturers were prepared to supply $q = 14p - 800$ million phones per year at a wholesale price of \$$p$, what would have been the equilibrium price?
b. The actual wholesale price was projected to be \$80 in the fourth quarter of 2008. Estimate the projected shortage or surplus at that price.

---

[26] The zonar (Ƶ) is the official currency in the city-state of Utarek, Mars (formerly www.Marsnext.com, a now extinct virtual society).

[27] The model is the authors' based on data available in 2013. Sources for data: www.businessweek.com, http://techcrunch.com, www.wikipedia.com, www.idc.com.

[28] *Ibid.*

[29] Source for data: Embedded.com/Company reports, December 2004.

[30] Wholesale price projections are the authors'. Source for sales prediction: I-Stat/NDR, December 2004.

41. ***Demand: E-Readers*** The demand for Amazon's Kindle e-reader can be approximated by

$$q = \frac{760}{p} - 1 \text{ million units per year} \quad (60 \le p \le 400),$$

where p is the price charged by Amazon.[31] Assume that Amazon is prepared to supply

$$q = 0.019p - 1 \text{ million units per year} \quad (60 \le p \le 400)$$

at a price of $\$p$ per unit.
a. Calculate the equilibrium price and equilibrium demand.
b. T Graph the demand and supply functions to confirm your answer in part (a) graphically.
c. Estimate, to the nearest 0.1 million units, the surplus or shortage of Kindle e-readers if the price is set at $72.

42. ***Equilibrium Price: Mars Monorail Service*** The demand for monorail service on the Utarek monorail, which links the three urbynes (or districts) of Utarek on Mars, can be approximated by

$$q = 7.5 + \frac{30}{p} \text{ million rides per day} \quad (3 \le p \le 8),$$

where p is the fare the Utarek Monorail Cooperative charges in zonars. (Z̄).[32] Assume that the cooperative is prepared to provide service for

$$q = 1.2p + 7.5 \text{ million rides per day} \quad (3 \le p \le 8)$$

at a fare of Z̄p.
a. Calculate the equilibrium price and equilibrium demand.
b. T Graph the demand and supply functions to confirm your answer in part (a) graphically.
c. Estimate the shortage or surplus of monorail service at the December 2085 fare of Z̄6 per ride.

43. ▼ ***Toxic Waste Treatment*** The cost of treating waste by removing PCPs goes up rapidly as the quantity of PCPs removed goes up. Here is a possible model:

$$C(q) = 2{,}000 + 100q^2,$$

where q is the reduction in toxicity (in pounds of PCPs removed per day) and $C(q)$ is the daily cost (in dollars) of this reduction.
a. Find the cost of removing 10 pounds of PCPs per day.
b. Government subsidies for toxic waste cleanup amount to

$$S(q) = 500q,$$

where q is as above and $S(q)$ is the daily dollar subsidy. The *net cost* function is given by $N = C - S$. Give a formula for $N(q)$, and interpret your answer.
c. Find $N(20)$, and interpret your answer.

44. ▼ ***Dental Plans*** A company pays for its employees' dental coverage at an annual cost C given by

$$C(q) = 1{,}000 + 100\sqrt{q},$$

where q is the number of employees covered and $C(q)$ is the annual cost in dollars.
a. If the company has 100 employees, find its annual outlay for dental coverage.
b. Assume that the government subsidizes coverage by an annual dollar amount of

$$S(q) = 200q.$$

The *net cost* function is given by $N = C - S$. Give a formula for $N(q)$, and interpret your answer.
c. Find $N(100)$, and interpret your answer.

45. ***Spending on Corrections in the 1990s*** The following table shows the annual spending by all states in the United States on corrections:[33]

Year t (year since 1990)	0	2	4	6	7
Spending ($ billion)	16	18	22	28	30

a. Which of the following functions best fits the given data? (Warning: None of them fits exactly, but one fits more closely than the others.) [**HINT:** See Example 5.]
(A) $S(t) = -0.2t^2 + t + 16$
(B) $S(t) = 0.2t^2 + t + 16$
(C) $S(t) = t + 16$
b. Use your answer to part (a) to "predict" spending on corrections in 1998, assuming that the trend continued.

46. ***Spending on Corrections in the 1990s*** Repeat Exercise 45, this time choosing from the following functions:
(A) $S(t) = 16 + 2t$
(B) $S(t) = 16 + t + 0.5t^2$
(C) $S(t) = 16 + t - 0.5t^2$

47. ***Soccer Gear*** The *East Coast College* soccer team is planning to buy new gear for its road trip to California. The cost per shirt depends on the number of shirts the team orders as shown in the following table:

Shirts Ordered x	5	25	40	100	125
Cost/Shirt $A(x)$ ($)	22.91	21.81	21.25	21.25	22.31

a. Which of the following functions best models the data?
(A) $A(x) = 0.005x + 20.75$
(B) $A(x) = 0.01x + 20 + \frac{25}{x}$

[31] Based on data from 2007 to 2013. Source: www.e-reader-info.com.

[32] The official currency of Utarek, Mars. (See the footnote to Exercise 34.)

[33] Data are rounded. Source: National Association of State Budget Officers/*New York Times*, February 28, 1999, p. A1.

(C) $A(x) = 0.0005x^2 - 0.07x + 23.25$
(D) $A(x) = 25.5(1.08)^{(x-5)}$

b. T Graph the model you chose in part (a) for $10 \le x \le 100$. Use your graph to estimate the lowest cost per shirt and the number of shirts the team should order to obtain the lowest price per shirt.

48. *Hockey Gear* The *South Coast College* hockey team wants to purchase wool hats for its road trip to Alaska. The cost per hat depends on the number of hats the team orders as shown in the following table:

Hats Ordered x	5	25	40	100	125
Cost/Hat $A(x)$ ($)	25.50	23.50	24.63	30.25	32.70

a. Which of the following functions best models the data?
(A) $A(x) = 0.05x + 20.75$
(B) $A(x) = 0.1x + 20 + \dfrac{25}{x}$
(C) $A(x) = 0.0008x^2 - 0.07x + 23.25$
(D) $A(x) = 25.5(1.08)^{(x-5)}$

b. T Graph the model you chose in part (a) with $5 \le x \le 30$. Use your graph to estimate the lowest cost per hat and the number of hats the team should order to obtain the lowest price per hat.

Cost: Hard Drive Storage *Exercises 49 and 50 are based on the following data showing how the approximate retail cost of a gigabyte of hard drive storage has fallen since 2000:*[34]

Year t (year since 2000)	Cost/Gigabyte $c(t)$ ($)
0	7.5
2	2.5
4	1.2
6	0.6
8	0.2
10	0.1
12	0.06

49. a. T Graph each of the following models together with the data points above, and use your graph to decide which two of the models best fit the data: [**HINT:** See Example 5 and accompanying technology note.]
(A) $c(t) = 6.3(0.67)^t$
(B) $c(t) = 0.093t^2 - 1.6t + 6.7$
(C) $c(t) = 4.75 - 0.50t$
(D) $c(t) = \dfrac{12.8}{t^{1.7} + 1.7}$

b. Of the two models you chose in part (a), which predicts the lower price in 2020? What price does that model predict?

50. a. T Graph each of the following models together with the data points above, and use your graph to decide which three of the models best fit the given data:
(A) $c(t) = \dfrac{15}{1 + 2^t}$
(B) $c(t) = (7.32)0.59^t + 0.10$
(C) $c(t) = 0.00085(t - 9.6)^4$
(D) $c(t) = 7.5 - 0.82t$

b. One of the three best-fit models in part (a) gives an unreasonable prediction for the price in 2020. Which is it, what price does it predict, and why is the prediction unreasonable?

51. *Social Website Popularity: Pinterest* The following table shows the popularity of Pinterest among social media sites as rated by StatCounter.com:[35]

Year t (year since start of 2008)	3	4	5
Pinterest Popularity $p(t)$ (%)	0.0	6.5	13.0

Which of the following kinds of models would best fit the given data? Explain your choice of model. (A, a, b, c, and m are constants.)
(A) $p(t) = mt + b$
(B) $p(t) = at^2 + bt + c$
(C) $p(t) = Ab^t$

52. *Social Website Popularity: Twitter* The following table shows the popularity of Twitter among social media sites as rated by StatCounter.com:[36]

Year t (year since start of 2008)	2	4	5
Twitter Popularity $p(t)$ (%)	6	6	7

Which of the following kinds of models would best fit the given data? Explain your choice of model. (A, a, b, c, and m are constants.)
(A) $p(t) = mt + b$
(B) $p(t) = a^2 + bt + c$
(C) $p(t) = Ab^t$

[34] The 2012 price is estimated. Source for data: "Cost of Hard Drive Storage Space," http://ns1758.ca/winch/winchest.html.

[35] Percentages are approximate and based on worldwide page views. Source: http://gs.statcounter.com.

[36] *Ibid.*

53. ***Demand for Gasoline*** The following table shows the demand for gasoline in the United States in terms of the price per gallon:[37]

Price p ($/gallon)	1.50	2.50	3	3.50
Demand q (gallons sold/person/day)	1.7	1.65	1.55	1.4

Which of the following kinds of models would best fit the given data? Explain your choice of model. (A, a, b, and c are constants.)

(A) $q = Ab^p \quad (b > 1)$
(B) $q = Ab^p \quad (b < 1)$
(C) $q = ap^2 + bp + c \quad (a > 0)$
(D) $q = ap^2 + bp + c \quad (a < 0)$

54. ***Demand for Mobile Data*** The following table shows the worldwide demand for mobile smartphone data in terms of its price per megabyte (MB):[38]

Price p ($/MB)	0.01	0.03	0.06	0.10	0.20	0.46
Demand q (MB/smartphone user)	4,000	1,000	700	500	300	150

Which of the following kinds of models would best fit the given data? Explain your choice of model. (A, a, b, and c are constants.)

(A) $q = Ap^b \quad (b > 0)$
(B) $q = Ap^b \quad (b < 0)$
(C) $q = ap^2 + bp + c \quad (a > 0)$
(D) $q = ap^2 + bp + c \quad (a < 0)$

55. ***Investments*** In August 2013, **E*TRADE Financial** was offering only 0.05% interest on its online checking accounts, with interest reinvested monthly.[39] Find the associated exponential model for the value of a $5,000 deposit after t years. Assuming that this rate of return continued for 7 years, how much would a deposit of $5,000 in August 2013 be worth in August 2020? (Answer to the nearest $1.) [**HINT:** See Quick Example 8.]

56. ***Investments*** In August 2013, **Ally Bank** was offering 0.61% interest on its Online Savings Account, with interest reinvested daily.[40] Find the associated exponential model for the value of a $4,000 deposit after t years. Assuming that this rate of return continued for 8 years, how much would a deposit of $4,000 in August 2013 be worth in August 2021? (Answer to the nearest $1.)

57. ***Investments*** Refer to Exercise 55. In August of which year will an investment of $5,000 made in August 2013 first exceed $5,050? [**HINT:** See Example 6.]

58. ***Investments*** Refer to Exercise 56. In August of which year will an investment of $4,000 made in August 2013 first exceed $4,400?

59. ***Carbon Dating*** A fossil originally contained 104 grams of carbon 14. Refer to the formula for $C(t)$ in Example 7 and estimate the amount of carbon 14 left in the sample after 10,000 years, 20,000 years, and 30,000 years. [**HINT:** See Example 7.]

60. ***Carbon Dating*** A fossil contains 4.06 grams of carbon 14. Refer to the formula for $A(t)$ at the end of Example 7, and estimate the amount of carbon 14 in the sample 10,000 years, 20,000 years, and 30,000 years ago.

61. ***Carbon Dating*** A fossil contains 4.06 grams of carbon 14. It is estimated that the fossil originally contained 46 grams of carbon 14. By calculating the amount left after 5,000 years, 10,000 years, . . . , 35,000 years, estimate the age of the sample to the nearest 5,000 years. (Refer to the formula for $C(t)$ in Example 7.)

62. ***Carbon Dating*** A fossil contains 2.8 grams of carbon 14. It is estimated that the fossil originally contained 104 grams of carbon 14. By calculating the amount 5,000 years, 10,000 years, . . . , 35,000 years ago, estimate the age of the sample to the nearest 5,000 years. (Refer to the formula for $C(t)$ at the end of Example 7.)

63. ***Radium Decay*** The amount of radium 226 remaining in a sample that originally contained A grams is approximately

$$C(t) = A(0.999567)^t$$

where t is time in years.

a. Find, to the nearest whole number, the percentage of radium 226 left in an originally pure sample after 1,000 years, 2,000 years, and 3,000 years.
b. Use a graph to estimate, to the nearest 100 years, when one half of a sample of 100 grams will have decayed.

64. ***Iodine Decay*** The amount of iodine 131 remaining in a sample that originally contained A grams is approximately

$$C(t) = A(0.9175)^t$$

where t is time in days.

a. Find, to the nearest whole number, the percentage of iodine 131 left in an originally pure sample after 2 days, 4 days, and 6 days.
b. Use a graph to estimate, to the nearest day, when one half of a sample of 100 grams will have decayed.

---

[37] Source: www.advisorperspectives.com.

[38] Data are approximate. Source: http://mobithinking.com.

[39] Source: https://us.etrade.com, August 2013.

[40] Interest rate based on annual percentage yield. Source: www.ally.com, August 2013.

Communication and Reasoning Exercises

65. If the population of the lunar station at Clavius is given by $P = 200 + 30t$, where t is time in years since the station was established, then the population is increasing by ____ per year.

66. My bank balance can be modeled by $B(t) = 5{,}000 - 200t$ dollars, where t is time in days since I opened the account. The balance on my account is __________ by $200 per day.

67. Classify the following model as analytical or curve fitting, and give a reason for your choice: The price of gold was $700 on Monday, $710 on Tuesday, and $700 on Wednesday. Therefore, the price can be modeled by $p(t) = -10t^2 + 20t + 700$ where t is the day since Monday.

68. Classify the following model as analytical or curve fitting, and give a reason for your choice: The width of a small animated square on my computer screen is currently 10 mm and is growing by 2 mm per second. Therefore, its area can be modeled by $a(t) = (10 + 2t)^2$ square mm where t is time in seconds.

69. Fill in the missing information for the following *analytical model* (answers may vary): _____________. Therefore, the cost of downloading a movie can be modeled by $c(t) = 4 - 0.2t$, where t is time in months since January.

70. Repeat Exercise 69, but this time regard the given model as a *curve-fitting model.*

71. Fill in the blanks: In a linear cost function, the _______ cost is x times the _______ cost.

72. Complete the following sentence: In a linear cost function, the marginal cost is the _______________.

73. ▼ We said in the discussion of demand and supply models that the demand for a commodity generally goes down as the price goes up. Assume that the demand for a certain commodity goes up as the price goes up. Is it still possible for there to be an equilibrium price? Explain with the aid of a demand and supply graph.

74. ▼ What would happen to the price of a certain commodity if the demand was always greater than the supply? Illustrate with a demand and supply graph.

75. You have a set of data points showing the sales of videos on your website versus time that are closely approximated by two different mathematical models. Give one criterion that would lead you to choose one over the other. (Answers may vary.)

76. Would it ever be reasonable to use a quadratic model $s(t) = at^2 + bt + c$ to predict long-term sales if a is negative? Explain.

77. If f and g are functions with $f(x) \geq g(x)$ for every x, what can you say about the values of the function $f - g$?

78. If f and g are functions with $f(x) > g(x) > 0$ for every x, what can you say about the values of the function $\frac{f}{g}$?

79. If f is measured in books and g is measured in people, what are the units of measurement of the function $\frac{f}{g}$?

80. If f and g are linear functions, then what can you say about $f - g$?

1.3 Linear Functions and Models

Linear functions are among the simplest functions and are perhaps the most useful of all mathematical functions.

Linear Function

A **linear function** is one that can be written in the form

$$f(x) = mx + b \quad \text{Function form}$$

or

$$y = mx + b \quad \text{Equation form}$$

where m and b are fixed numbers. (The names m and b are traditional.*)

Quick Example

$f(x) = 3x - 1$

$y = 3x - 1$

* Actually, c is sometimes used instead of b. As for m, there has even been some research into the question of its origin, but no one knows exactly why that particular letter is used.

Linear Functions from the Numerical and Graphical Point of View

The following table shows values of $y = 3x - 1$ ($m = 3, b = -1$) for some values of x:

x	-4	-3	-2	-1	0	1	2	3	4
y	-13	-10	-7	-4	-1	2	5	8	11

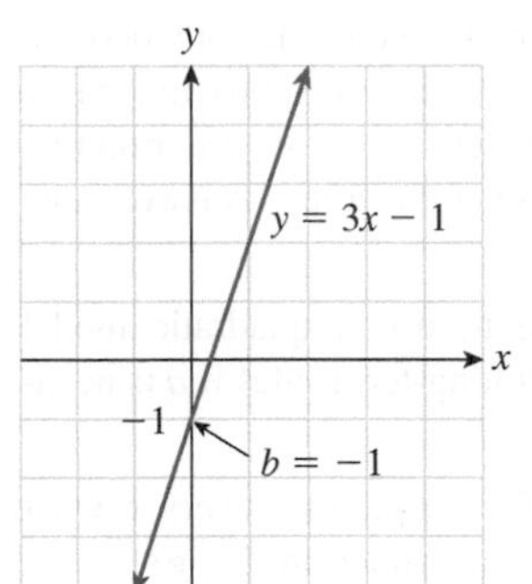

y-intercept $= b = -1$
Graphically, b is the y-intercept of the graph.

Figure 14

Notice that setting $x = 0$ gives $y = -1$, the value of b.

Numerically, b is the value of y when $x = 0$.

The graph of $f(x) = 3x - 1$ is shown in Figure 14. On the graph, the point $(0, b) = (0, -1)$ is the point where the graph crosses the y-axis, so we say that $b = -1$ is the **y-intercept** of the graph (Figure 14).

What about m? Looking once again at the table, notice that y increases by $m = 3$ units for every increase of 1 unit in x. This is caused by the term $3x$ in the formula: For every increase of 1 in x, we get an increase of $3 \times 1 = 3$ in y.

Numerically, y increases by m units for every 1-unit increase of x.

Likewise, for every increase of 2 in x, we get an increase of $3 \times 2 = 6$ in y. In general, if x increases by some amount, y will increase by three times that amount. We write

$$\text{Change in } y = 3 \times \text{Change in } x.$$

The Change in a Quantity: Delta Notation

If a quantity q changes from q_1 to q_2, the **change in q** is just the difference:

$$\begin{aligned}\text{Change in } q &= \text{Second value} - \text{First value}\\ &= q_2 - q_1.\end{aligned}$$

Mathematicians traditionally use Δ (delta, the Greek equivalent of the Roman letter D) to stand for change and write the change in q as Δq:

$$\Delta q = \text{Change in } q = q_2 - q_1.$$

Quick Examples

1. If x is changed from 1 to 3, we write

   $$\Delta x = \text{Second value} - \text{First value} = 3 - 1 = 2.$$

2. Looking at our linear equation $y = 3x - 1$, we see that, when x changes from 1 to 3, y changes from 2 to 8. So

   $$\Delta y = \text{Second value} - \text{First value} = 8 - 2 = 6.$$

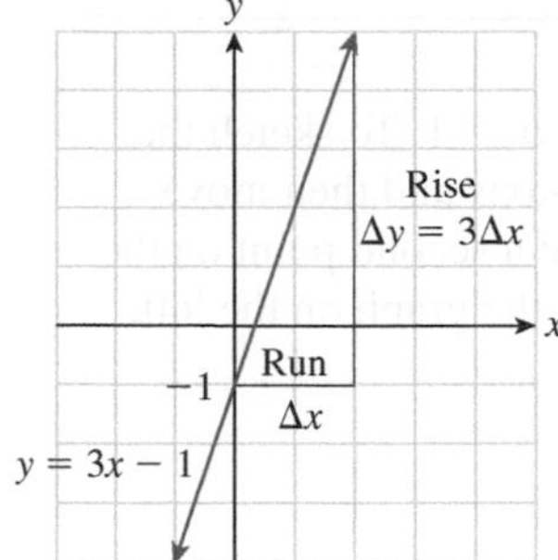

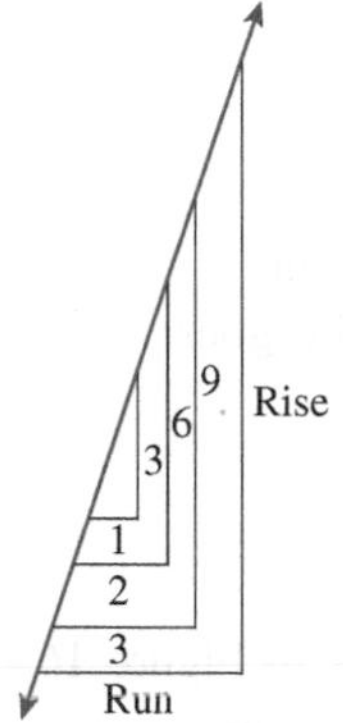

Slope $= m = 3$

Graphically, m is the slope of the graph.

Figure 15

Using delta notation, we can now write, for our linear equation $y = 3x - 1$,

$$\Delta y = 3\Delta x \qquad \text{Change in } y = 3 \times \text{Change in } x$$

or

$$\frac{\Delta y}{\Delta x} = 3.$$

Because the value of y increases by exactly 3 units for every increase of 1 unit in x, the graph is a straight line rising by 3 units for every 1 unit we go to the right. We say that we have a **rise** of 3 units for each **run** of 1 unit. Because the value of y changes by $\Delta y = 3\Delta x$ units for every change of Δx units in x, in general we have a rise of $\Delta y = 3\Delta x$ units for each run of Δx units (Figure 15). Thus, we have a rise of 6 for a run of 2, a rise of 9 for a run of 3, and so on. So $m = 3$ is a measure of the steepness of the line; we call m the **slope of the line**:

$$\text{Slope} = m = \frac{\Delta y}{\Delta x} = \frac{\text{Rise}}{\text{Run}}.$$

In general (replace the number 3 by a general number m), we can say the following.

The Roles of *m* and *b* in the Linear Function $f(x) = mx + b$

Role of *m*

Numerically If $y = mx + b$, then y changes by m units for every 1-unit change in x. A change of Δx units in x results in a change of $\Delta y = m\Delta x$ units in y. Thus,

$$m = \frac{\Delta y}{\Delta x} = \frac{\text{Change in } y}{\text{Change in } x}.$$

Graphically m is the slope of the line $y = mx + b$:

$$m = \frac{\Delta y}{\Delta x} = \frac{\text{Rise}}{\text{Run}} = \text{Slope}.$$

For positive m the graph rises m units for every 1-unit move to the right, and rises $\Delta y = m\Delta x$ units for every Δx units moved to the right. For negative m the graph drops $|m|$ units for every 1-unit move to the right, and drops $|m|\Delta x$ units for every Δx units moved to the right.

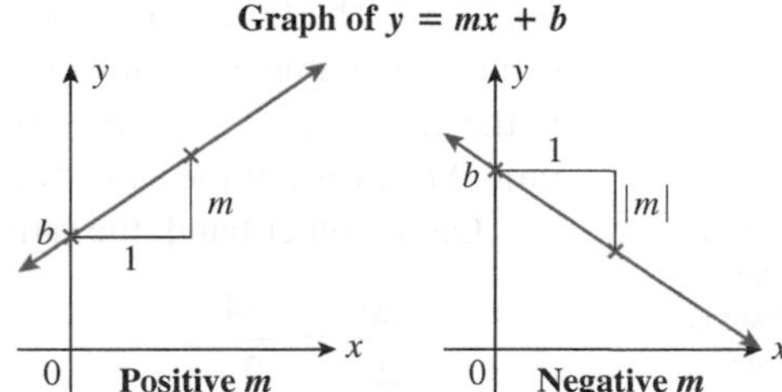

Role of *b*

Numerically When $x = 0$, $y = b$.

Graphically b is the y-intercept of the line $y = mx + b$.

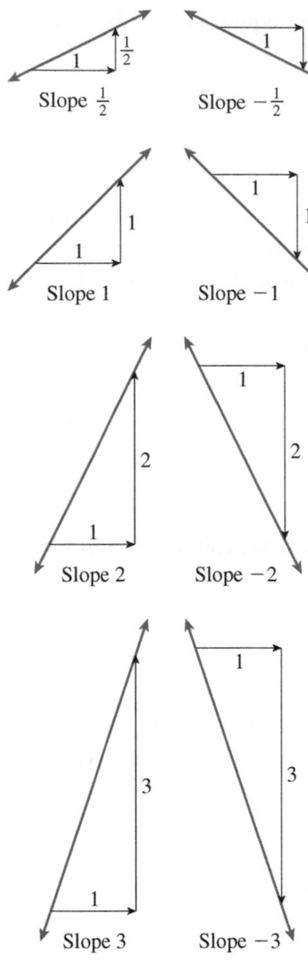

Figure 16

Quick Examples

3. $f(x) = 2x + 1$ has slope $m = 2$ and y-intercept $b = 1$. To sketch the graph, we start at the y-intercept $b = 1$ on the y-axis and then move 1 unit to the right and up $m = 2$ units to arrive at a second point on the graph. Now we connect the two points to obtain the graph on the left.

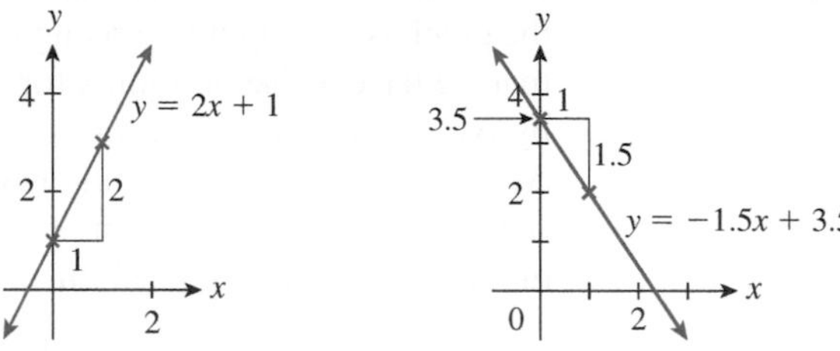

4. The line $y = -1.5x + 3.5$ has slope $m = -1.5$ and y-intercept $b = 3.5$. Because the slope is negative, the graph (above right) goes *down* 1.5 units for every 1 unit it moves to the right.

It helps to be able to picture what different slopes look like, as in Figure 16. Notice that the larger the absolute value of the slope, the steeper is the line.

EXAMPLE 1 Recognizing Linear Data Numerically and Graphically

Which of the following two tables gives the values of a linear function? What is the formula for that function?

x	0	2	4	6	8	10	12
$f(x)$	3	−1	−3	−6	−8	−13	−15

x	0	2	4	6	8	10	12
$g(x)$	3	−1	−5	−9	−13	−17	−21

Using Technology

See the Technology Guides at the end of the chapter for detailed instructions on how to obtain a table with the successive quotients $m = \Delta y/\Delta x$ for the functions *f* and *g* in Example 1 using a TI-83/84 Plus or Excel. These tables show at a glance that *f* is not linear. Here is an outline:

TI-83/84 Plus

STAT EDIT; Enter values of x and $f(x)$ in lists L_1 and L_2.
Highlight the heading L_3 and then enter the following formula (including the quotes):
"ΔList(L₂)/ΔList(L₁)"
[More details in the Technology Guide.]

Solution The function *f* cannot be linear. If it were, we would have $\Delta f = m\Delta x$ for some fixed number m. However, although the change in x between successive entries in the table is $\Delta x = 2$ each time, the change in *f* is not the same each time. Thus, the ratio $\Delta f/\Delta x$ is not the same for every successive pair of points.

On the other hand, the ratio $\Delta g/\Delta x$ is the same each time, namely,

$$\frac{\Delta g}{\Delta x} = \frac{-4}{2} = -2,$$

as we see in the following table:

Δx		$2-0=2$	$4-2=2$	$6-4=2$	$8-6=2$	$10-8=2$	$12-10=2$
x	0	2	4	6	8	10	12
$g(x)$	3	−1	−5	−9	−13	−17	−21
Δg		$-1-3 = -4$	$-5-(-1) = -4$	$-9-(-5) = -4$	$-13-(-9) = -4$	$-17-(-13) = -4$	$-21-(-17) = -4$

Spreadsheet
Enter headings x, $f(x)$, Df/Dx in cells A1–C1, and the corresponding values from one of the tables in cells A2–B8. Enter
`=(B3-B2)/(A3-A2)`
in cell C2, and copy down through C7.
[More details in the Technology Guide.]

Thus, g is linear with slope $m = -2$. By the table, $g(0) = 3$; hence, $b = 3$. Thus,

$$g(x) = -2x + 3. \quad \text{Check that this formula gives the values in the table.}$$

If you graph the points in the tables defining f and g above, it becomes easy to see that g is linear and f is not; the points of g lie on a straight line (with slope -2), whereas the points of f do not lie on a straight line (Figure 17).

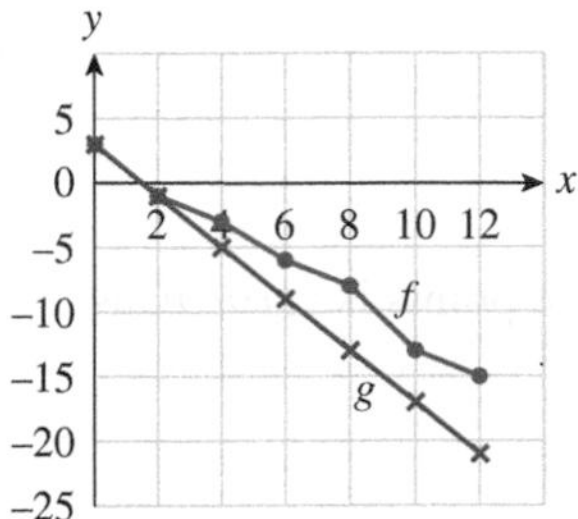

Figure 17

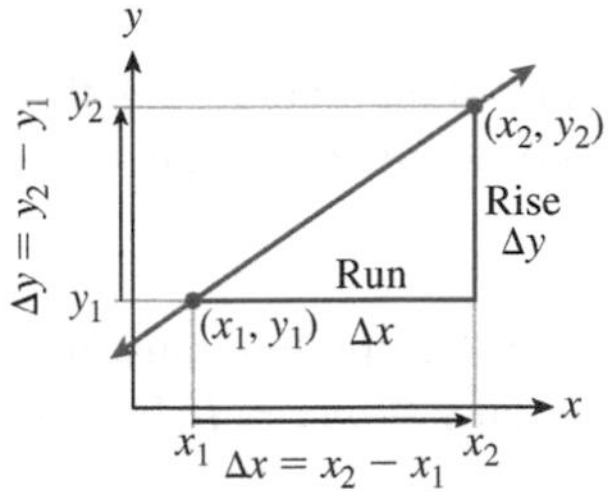

Figure 18

Finding a Linear Equation from Data

If we happen to know the slope and y-intercept of a line, writing down its equation is straightforward. For example, if we know that the slope is 3 and the y-intercept is -1, then the equation is $y = 3x - 1$. Sadly, the information we are given is seldom so convenient. For instance, we may know the slope and a point other than the y-intercept, two points on the line, or other information. We therefore need to know how to use the information we are given to obtain the slope and the intercept.

Computing the Slope

We can always determine the slope of a line if we are given two (or more) points on the line, because any two points—say, (x_1, y_1) and (x_2, y_2)—determine the line and hence its slope. To compute the slope when given two points, recall the formula

$$\text{Slope} = m = \frac{\text{Rise}}{\text{Run}} = \frac{\Delta y}{\Delta x}.$$

To find its slope, we need a run Δx and corresponding rise Δy. In Figure 18 we see that we can use $\Delta x = x_2 - x_1$, the change in the x-coordinate from the first point to the second, as our run, and $\Delta y = y_2 - y_1$, the change in the y-coordinate, as our rise. The resulting formula for computing the slope is given as follows.

Computing the Slope of a Line

We can compute the slope m of the line through the points (x_1, y_1) and (x_2, y_2) using

$$m = \frac{\Delta y}{\Delta x} = \frac{y_2 - y_1}{x_2 - x_1}.$$

Quick Examples

5. The slope of the line through $(x_1, y_1) = (1, 3)$ and $(x_2, y_2) = (5, 11)$ is

$$m = \frac{\Delta y}{\Delta x} = \frac{y_2 - y_1}{x_2 - x_1} = \frac{11 - 3}{5 - 1} = \frac{8}{4} = 2.$$

Notice that we can use the points in the reverse order: If we take $(x_1, y_1) = (5, 11)$ and $(x_2, y_2) = (1, 3)$, we obtain the same answer:

$$m = \frac{\Delta y}{\Delta x} = \frac{y_2 - y_1}{x_2 - x_1} = \frac{3 - 11}{1 - 5} = \frac{-8}{-4} = 2.$$

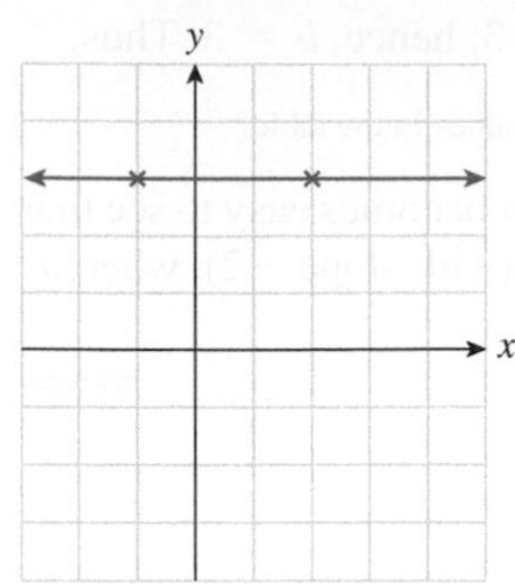

Figure 19

6. The slope of the line through $(x_1, y_1) = (1, 2)$ and $(x_2, y_2) = (2, 1)$ is

$$m = \frac{\Delta y}{\Delta x} = \frac{y_2 - y_1}{x_2 - x_1} = \frac{1 - 2}{2 - 1} = \frac{-1}{1} = -1.$$

7. The slope of the line through (2, 3) and (−1, 3) is

$$m = \frac{\Delta y}{\Delta x} = \frac{y_2 - y_1}{x_2 - x_1} = \frac{3 - 3}{-1 - 2} = \frac{0}{-3} = 0.$$

A line of slope 0 has zero rise, so it is a *horizontal* line, as shown in Figure 19.

8. The line through (3, 2) and (3, −1) has slope

$$m = \frac{\Delta y}{\Delta x} = \frac{y_2 - y_1}{x_2 - x_1} = \frac{-1 - 2}{3 - 3} = \frac{-3}{0},$$

which is undefined. The line passing through these points is *vertical*, as shown in Figure 20.

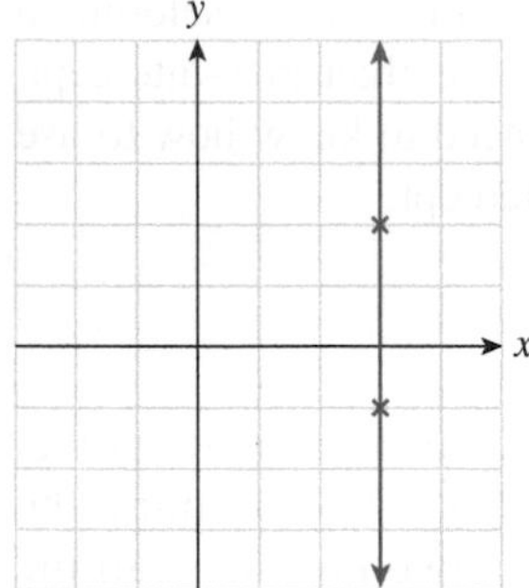

Vertical lines have undefined slope.

Figure 20

Computing the *y*-Intercept

Once we know the slope m of a line and also the coordinates of a point (x_1, y_1), then we can calculate its y-intercept b as follows: The equation of the line must be

$$y = mx + b,$$

where b is as yet unknown. To determine b, we use the fact that the line must pass through the point (x_1, y_1), and so (x_1, y_1) satisfies the equation $y = mx + b$. In other words,

$$y_1 = mx_1 + b.$$

Solving for b gives

$$b = y_1 - mx_1.$$

In summary, we have the following.

Computing the *y*-Intercept of a Line

The y-intercept of the line passing through (x_1, y_1) with slope m is

$$b = y_1 - mx_1.$$

Quick Example

9. The line through (2, 3) with slope 4 has

$$b = y_1 - mx_1 = 3 - (4)(2) = -5.$$

Its equation is therefore

$$y = mx + b = 4x - 5.$$

EXAMPLE 2 Finding Linear Equations

Find equations for the following straight lines.

a. Through the points (1, 2) and (3, −1)

b. Through (2, −2) and parallel to the line $3x + 4y = 5$

c. Horizontal and through (−9, 5)

d. Vertical and through (−9, 5)

Solution

a. To write down the equation of the line, we need the slope m and the y-intercept b.

- ***Slope*** Because we are given two points on the line, we can use the slope formula:

$$m = \frac{y_2 - y_1}{x_2 - x_1} = \frac{-1 - 2}{3 - 1} = -\frac{3}{2}.$$

- ***Intercept*** We now have the slope of the line, $m = -3/2$, and also a point—we have two to choose from, so let us choose $(x_1, y_1) = (1, 2)$. We can now use the formula for the y-intercept:

$$b = y_1 - mx_1 = 2 - \left(-\frac{3}{2}\right)(1) = \frac{7}{2}.$$

Thus, the equation of the line is

$$y = -\frac{3}{2}x + \frac{7}{2}. \qquad y = mx + b$$

b. Proceeding as before, we have the following.

- ***Slope*** We are not given two points on the line, but we are given a parallel line. We use the fact that *parallel lines have the same slope*. (Why?) We can find the slope of $3x + 4y = 5$ by solving for y and then looking at the coefficient of x:

$$y = -\frac{3}{4}x + \frac{5}{4}, \qquad \text{To find the slope, solve for } y.$$

so the slope is $-3/4$.

- ***Intercept*** We now have the slope of the line, $m = -3/4$ and also a point $(x_1, y_1) = (2, -2)$. We can now use the formula for the y-intercept:

$$b = y_1 - mx_1 = -2 - \left(-\frac{3}{4}\right)(2) = -\frac{1}{2}.$$

Thus, the equation of the line is

$$y = -\frac{3}{4}x - \frac{1}{2}. \qquad y = mx + b$$

c. We are given a point: (−9, 5). Furthermore, we are told that the line is horizontal, which tells us that the slope is $m = 0$. Therefore, all that remains is the calculation of the y-intercept:

$$b = y_1 - mx_1 = 5 - (0)(-9) = 5,$$

so the equation of the line is

$$y = 5. \qquad y = mx + b$$

Using Technology

See the Technology Guides at the end of the chapter for detailed instructions on how to obtain the slope and intercept in Example 2(a) using a TI-83/84 Plus or a spreadsheet. Here is an outline:

TI-83/84 Plus

[STAT] EDIT; Enter values of x and y in lists L_1 and L_2.

Slope: Home screen

`(L2(2)-L2(1))/(L1(2)-L1(1))→M`

Intercept: Home screen

`L2(1)-M*L1(1)`

[More details in the Technology Guide.]

Spreadsheet

Enter headings x, y, m, b, in cells A1–D1 and the values (x, y) in cells A2–B3. Enter

`=(B3-B2)/(A3-A2)`

in cell C2 and

`=B2-C2*A2`

in cell D2.

[More details in the Technology Guide.]

d. We are given a point: $(-9, 5)$. This time, we are told that the line is vertical, which means that the slope is undefined. Thus, we can't express the equation of the line in the form $y = mx + b$. (This formula makes sense only when the slope m of the line is defined.) What can we do? Well, here are some points on the desired line:

$$(-9, 1), (-9, 2), (-9, 3), \ldots$$

so $x = -9$ and $y = $ *anything*. If we simply say that $x = -9$, then these points are all solutions, so the equation is $x = -9$.

Applications: Linear Models

Cost Functions

Using linear functions to describe or approximate relationships in the real world is called **linear modeling**. Recall from Section 1.2 that a **cost function** specifies the cost C as a function of the number of items x.

EXAMPLE 3 Linear Cost Function from Data

The manager of the *FrozenAir Refrigerator* factory notices that on Monday it cost the company a total of \$25,000 to build 30 refrigerators and on Tuesday it cost \$30,000 to build 40 refrigerators. Find a linear cost function based on this information. What is the daily fixed cost, and what is the marginal cost?

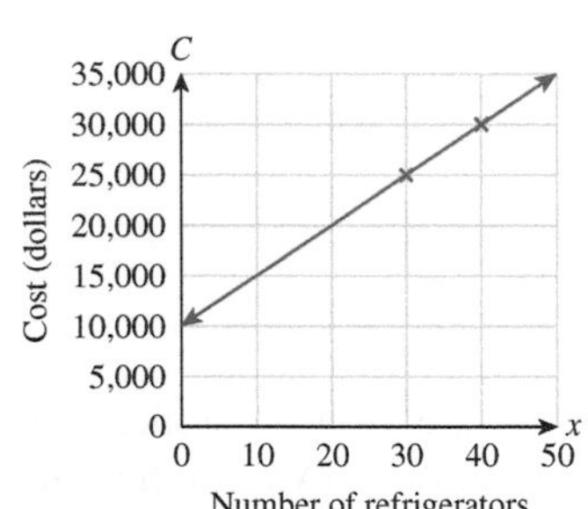

Figure 21

Solution We are seeking the cost C as a linear function of x, the number of refrigerators sold:

$$C = mx + b. \quad \text{Linear cost function (equation form)}$$

We are told that $C = 25{,}000$ when $x = 30$, and this amounts to being told that $(30, 25{,}000)$ is a point on the graph of the cost function. Similarly, $(40, 30{,}000)$ is another point on the line (Figure 21).

We can use the two points on the line to construct the linear cost equation:

- ***Slope*** $m = \dfrac{C_2 - C_1}{x_2 - x_1} = \dfrac{30{,}000 - 25{,}000}{40 - 30} = 500$ — C plays the role of y.
- ***Intercept*** $b = C_1 - mx_1 = 25{,}000 - (500)(30) = 10{,}000.$ — We used the point $(x_1, C_1) = (30, 25{,}000)$.

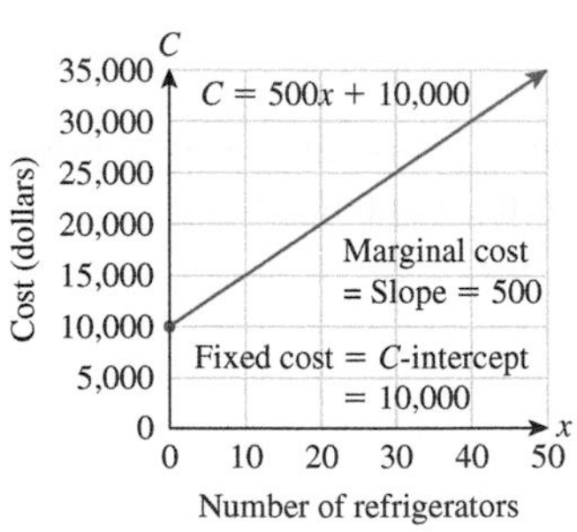

Figure 22

The linear cost function is therefore

$$C(x) = 500x + 10{,}000.$$

Because $m = 500$ and $b = 10{,}000$, the factory's fixed cost is \$10,000 each day, and its marginal cost is \$500 per refrigerator. (See the discussion of cost functions in Section 1.2.) These are illustrated in Figure 22.

➡ **Before we go on . . .** Recall that, in general, the slope m measures the number of units of change in y per 1-unit change in x, so it is measured in units of y per unit of x:

$$\text{Units of slope} = \text{Units of } y \text{ per unit of } x.$$

Using Technology

To obtain the cost equation for Example 3 with technology, apply the Technology note for Example 2(a) to the given points (30, 25,000) and (40, 30,000) on the graph of the cost equation.

In Example 3, y is the cost C, measured in dollars, and x is the number of items, measured in refrigerators. Hence,

$$\text{Units of slope} = \text{Units of } y \text{ per unit of } x = \text{Dollars per refrigerator.}$$

The y-intercept b, being a value of y, is measured in the same units as y. In Example 3, b is measured in dollars. ■

Demand Functions

In Section 1.2 we saw that a **demand function** specifies the demand q as a function of the price p per item.

EXAMPLE 4 Linear Demand Function from Data

You run a small supermarket and must determine how much to charge for *Hot'n'Spicy* brand baked beans. The following chart shows weekly sales figures (the demand) for Hot'n'Spicy at two different prices:

Price ($/can)	0.50	0.75
Demand (cans sold/week)	400	350

a. Model these data with a linear demand function. (See Example 3 in Section 1.2.)

b. How do we interpret the slope and q-intercept of the demand function?

Solution

Using Technology

To obtain the demand equation for Example 4 with technology, apply the Technology note for Example 2(a) to the given points (0.50, 400) and (0.75, 350) on the graph of the demand equation.

a. Recall that a demand equation—or demand function—expresses demand q (in this case, the number of cans of beans sold per week) as a function of the unit price p (in this case, dollars per can). We model the demand using the two points we are given: (0.50, 400) and (0.75, 350).

$$\textbf{\textit{Slope:}} \quad m = \frac{q_2 - q_1}{p_2 - p_1} = \frac{350 - 400}{0.75 - 0.50} = \frac{-50}{0.25} = -200$$

$$\textbf{\textit{Intercept:}} \quad b = q_1 - mp_1 = 400 - (-200)(0.50) = 500$$

So the demand equation is

$$q = -200p + 500. \qquad q = mp + b$$

b. The key to interpreting the slope in a demand equation is to recall (see the "Before we go on" note at the end of Example 3) that we measure the slope in *units of y per unit of x*. Here, $m = -200$, and the units of m are units of q per unit of p, or the number of cans sold per \$1 change in the price. Because m is negative, we see that the number of cans sold decreases as the price increases. We conclude that the weekly sales will drop by 200 cans per \$1 increase in the price.

To interpret the q-intercept, recall that it gives the q-coordinate when $p = 0$. Hence, it is the number of cans the supermarket can "sell" every week if it were to give them away.*

* Does this seem realistic? Demand is not always unlimited if items are given away. For instance, campus newspapers are sometimes given away, yet piles of them are often left untaken. Also see the "Before we go on" discussion at the end of this example.

➡ Before we go on...

 : *Just how reliable is the linear model used in Example 4?*

 : The *actual* demand graph could in principle be obtained by tabulating demand figures for a large number of different prices. If the resulting points were plotted on the *pq*-plane, they would probably suggest a curve and not a straight line. However, if you looked at a small enough portion of any curve, you could closely *approximate* it by a straight line. In other words, *over a small range of values of p, a linear model is accurate.* Linear models of real-world situations are generally reliable only for small ranges of the variables. (This point will come up again in some of the exercises.)

■

Time Change Models

The next example illustrates modeling change over time t with a linear function of t.

EXAMPLE 5 Growth of Sales

Worldwide sales of tablet computers were expected to increase from around 130 million units in 2012 to around 350 million units in 2017.[41]

a. Use this information to model annual worldwide sales of tablet computers as a linear function of time t in years since 2012. What is the significance of the slope?

b. According to the model, in which year will tablet computer sales surpass 280 million units?

Solution

a. Since we are interested in worldwide sales s of tablet computers as a function of time, we take time t to be the independent variable (playing the role of x) and the annual sales s, in millions of units, to be the dependent variable (in the role of y). Notice that 2012 corresponds to $t = 0$ and 2017 corresponds to $t = 5$, so we are given the coordinates of two points on the graph of s as a function of t: (0, 130) and (5, 350). We model the sales using these two points:

$$m = \frac{s_2 - s_1}{t_2 - t_1} = \frac{350 - 130}{5 - 0} = \frac{220}{5} = 44$$
$$b = s_1 - mt_1 = 130 - (44)(0) = 130.$$

So

$$s = 44t + 130 \text{ million units.} \qquad s = mt + b$$

The slope m is measured in units of s per unit of t; that is, millions of tablet computers per year, and is thus the *rate of change* of annual tablet sales. To say that $m = 44$ is to say that annual sales are increasing at a rate of 44 million tablets per year.

[41] Source: ID Press Release, March 26, 2013, www.idc.com.

b. Our model of annual sales as a function of time is

$$s = 44t + 130 \text{ million units.}$$

Annual sales of tablet computers are 280 million when $s = 280$, or

$$280 = 44t + 130$$

Solving for t, we have

$$44t = 280 - 130 = 150$$

$$t = \frac{150}{44} \approx 3.4 \text{ years.}$$

Thus, when $t = 3$ (2015), predicted sales are less than 280 million units, and when $t = 4$ (2016), predicted sales exceed 280 million units. Thus, sales will first surpass 280 million units in 2016. (Notice that, instead of rounding $t = 3.4$ to the nearest whole number, we needed to round upwards to $t = 4$.)

Using Technology

To use technology to obtain s as a function of t in Example 5, apply the Technology note for Example 2(a) to the points (0, 130) and (5, 350) on its graph.

EXAMPLE 6 Velocity

You are driving down the Ohio Turnpike, watching the mileage markers to stay awake. Measuring time in hours after you see the 20-mile marker, you see the following markers each half hour:

Time (hours)	0	0.5	1	1.5	2
Marker (miles)	20	47	74	101	128

Find your location s as a function of t, the number of hours you have been driving. (The number s is also called your **position** or **displacement**.)

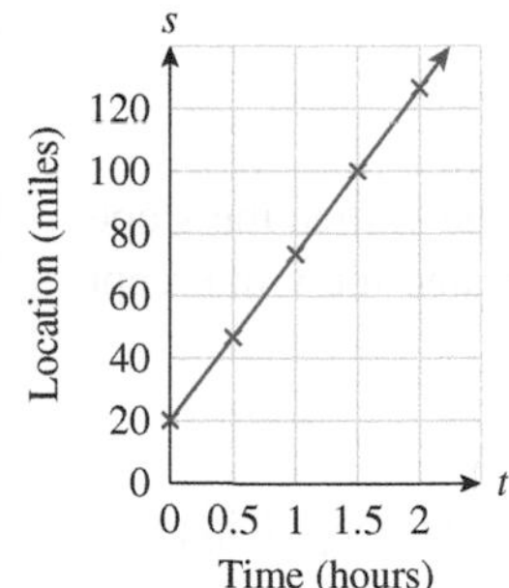

Figure 23

Solution If we plot the location s versus the time t, the five markers listed give us the graph in Figure 23. These points appear to lie along a straight line. We can verify this by calculating how far you traveled in each half hour. In the first half hour you traveled $47 - 20 = 27$ miles. In the second half hour you traveled $74 - 47 = 27$ miles also. In fact, you traveled exactly 27 miles each half hour. The points we plotted lie on a straight line that rises 27 units for every 0.5 units we go to the right, for a slope of $27/0.5 = 54$.

To get the equation of that line, notice that we have the s-intercept, which is the starting marker of 20. Thus, the equation of s as a function of time t is

$$s(t) = 54t + 20. \quad \text{We used } s \text{ in place of } y \text{ and } t \text{ in place of } x.$$

Notice the significance of the slope: For every hour you travel, you drive a distance of 54 miles. In other words, you are traveling at a constant velocity of 54 mph. We have uncovered a very important principle:

In the graph of displacement versus time, velocity is given by the slope.

Using Technology

To use technology to obtain s as a function of t in Example 6, apply the Technology note for Example 2(a) to the points (0, 20) and (1, 74) on its graph.

Linear Change over Time

If a quantity q is a linear function of time t,

$$q = mt + b,$$

then the slope m measures the **rate of change** of q, and b is the quantity at time $t = 0$, the **initial quantity**. If q represents the position of a moving object, then the rate of change is also called the **velocity**.

Units of *m* and *b*

The units of measurement of m are units of q per unit of time. For instance, if q is income in dollars and t is time in years, then the rate of change m is measured in dollars per year.

The units of b are units of q. For instance, if q is income in dollars and t is time in years, then b is measured in dollars.

Quick Example

10. If the accumulated revenue from sales of your video game software is given by $R = 2{,}000t + 500$ dollars, where t is time in years from now, then you have earned \$500 in revenue so far, and the accumulated revenue is increasing at a rate of \$2,000 per year.

Examples 5 and 6 share the following common theme.

General Linear Models

If $y = mx + b$ is a linear model of changing quantities x and y, then the slope m is the rate at which y is increasing per unit increase in x, and the y-intercept b is the value of y that corresponds to $x = 0$.

Units of *m* and *b*

The slope m is measured in units of y per unit of x, and the intercept b is measured in units of y.

Quick Example

11. If the number n of spectators at a soccer game is related to the number g of goals your team has scored so far by the equation $n = 20g + 4$, then you can expect four spectators if no goals have been scored and 20 additional spectators per additional goal scored.

FAQs

What to Use as x and y and How to Interpret a Linear Model

Q: *In a problem where I must find a linear relationship between two quantities, which quantity do I use as x, and which do I use as y?*

A: The key is to decide which of the two quantities is the independent variable and which is the dependent variable. Then use the independent variable as x and the dependent variable as y. In other words, *y depends on x*.

Here are examples of phrases that convey this information, usually of the form *Find y [dependent variable] in terms of x [independent variable]:*

- Find the cost in terms of the number of items. y = Cost, x = Number of items
- How does color depend on wavelength? y = Color, x = Wavelength

If no information is conveyed about which variable is intended to be independent, then you can use whichever is convenient.

Q: *How do I interpret a general linear model $y = mx + b$?*

A: The key to interpreting a linear model is to remember the units we use to measure m and b:

The slope m is measured in units of y per unit of x. The intercept b is measured in units of y.

For instance, if $y = 4.3x + 8.1$ and you know that x is measured in feet and y in kilograms, then you can already say, "y is 8.1 kilograms when $x = 0$ feet and increases at a rate of 4.3 kilograms per foot" without knowing anything more about the situation.

1.3 EXERCISES

▼ more advanced ◆ challenging

T indicates exercises that should be solved using technology

In Exercises 1–6, a table of values for a linear function is given. Fill in the missing value and calculate m in each case.

1.

x	−1	0	1
y	5	8	

2.

x	−1	0	1
y	−1	−3	

3.

x	2	3	5
y	−1	−2	

4.

x	2	4	5
y	−1	−2	

5.

x	−2	0	2
y	4		10

6.

x	0	3	6
y	−1		−5

In Exercises 7–10, first find f(0), if not supplied, and then find the equation of the given linear function.

7.

x	−2	0	2	4
$f(x)$	−1	−2	−3	−4

8.

x	−6	−3	0	3
$f(x)$	1	2	3	4

9.

x	−4	−3	−2	−1
$f(x)$	−1	−2	−3	−4

10.

x	1	2	3	4
$f(x)$	4	6	8	10

In Exercises 11–14, decide which of the two given functions is linear, and find its equation. [**HINT**: See Example 1.]

11.

x	0	1	2	3	4
$f(x)$	6	10	14	18	22
$g(x)$	8	10	12	16	22

12.

x	−10	0	10	20	30
$f(x)$	−1.5	0	1.5	2.5	3.5
$g(x)$	−9	−4	1	6	11

13.

x	0	3	6	10	15
$f(x)$	0	3	5	7	9
$g(x)$	−1	5	11	19	29

14.

x	0	3	5	6	9
$f(x)$	2	6	9	12	15
$g(x)$	−1	8	14	17	26

In Exercises 15–24, find the slope of the given line if it is defined.

15. $y = -\frac{3}{2}x - 4$ **16.** $y = \frac{2x}{3} + 4$

17. $y = \frac{x+1}{6}$ **18.** $y = -\frac{2x-1}{3}$

19. $3x + 1 = 0$ **20.** $8x - 2y = 1$

21. $3y + 1 = 0$ **22.** $2x + 3 = 0$

23. $4x + 3y = 7$ **24.** $2y + 3 = 0$

In Exercises 25–38, graph the given equation. [**HINT**: See Quick Examples 3 and 4.]

25. $y = 2x - 1$ **26.** $y = x - 3$

27. $y = -\frac{2}{3}x + 2$ **28.** $y = -\frac{1}{2}x + 3$

29. $y + \frac{1}{4}x = -4$ **30.** $y - \frac{1}{4}x = -2$

31. $7x - 2y = 7$ **32.** $2x - 3y = 1$

33. $3x = 8$ **34.** $2x = -7$

35. $6y = 9$ **36.** $3y = 4$

37. $2x = 3y$ **38.** $3x = -2y$

In Exercises 39–58, calculate the exact slope (rather than a decimal approximation) of the straight line through the given pair of points, if defined. Try to do as many as you can without writing anything down except the answer. [**HINT**: See Quick Example 5.]

39. $(0, 0)$ and $(1, 2)$ **40.** $(0, 0)$ and $(-1, 2)$

41. $(-1, -2)$ and $(0, 0)$ **42.** $(2, 1)$ and $(0, 0)$

43. $(4, 3)$ and $(5, 1)$ **44.** $(4, 3)$ and $(4, 1)$

45. $(1, -1)$ and $(1, -2)$ **46.** $(-2, 2)$ and $(-1, -1)$

47. $(2, 3.5)$ and $(4, 6.5)$ **48.** $(10, -3.5)$ and $(0, -1.5)$

49. $(300, 20.2)$ and $(400, 11.2)$

50. $(1, -20.2)$ and $(2, 3.2)$

51. $(0, 1)$ and $\left(-\frac{1}{2}, \frac{3}{4}\right)$

52. $\left(\frac{1}{2}, 1\right)$ and $\left(-\frac{1}{2}, \frac{3}{4}\right)$

53. (a, b) and (c, d) $(a \neq c)$

54. (a, b) and (c, b) $(a \neq c)$

55. (a, b) and (a, d) $(b \neq d)$

56. (a, b) and $(-a, -b)$ $(a \neq 0)$

57. $(-a, b)$ and $(a, -b)$ $(a \neq 0)$

58. (a, b) and (b, a) $(a \neq b)$

59. In the following figure, estimate the slopes of all line segments:

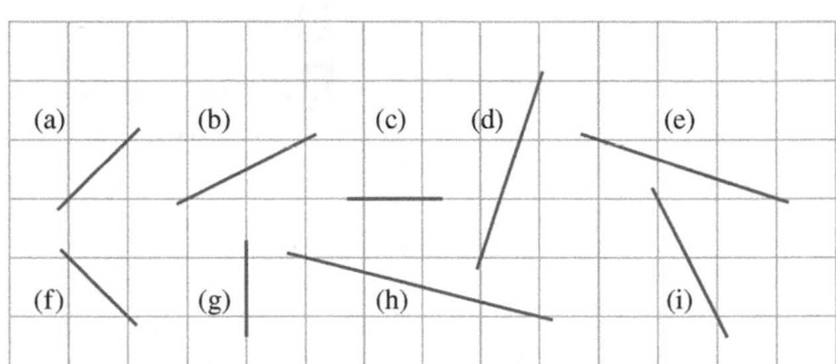

60. In the following figure, estimate the slopes of all line segments:

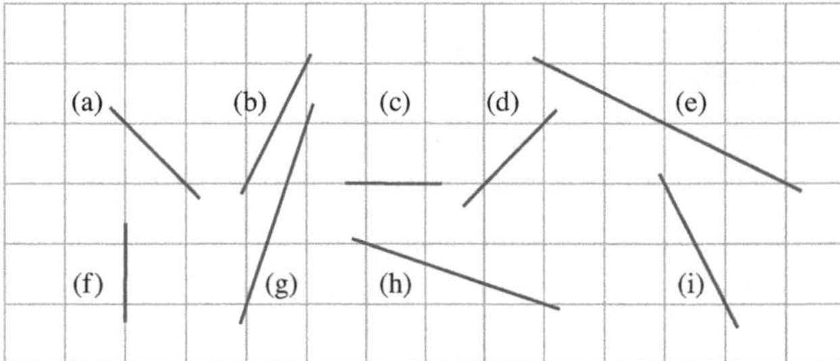

In Exercises 61–80, find a linear equation whose graph is the straight line with the given properties. [**HINT**: See Example 2.]

61. Through $(1, 3)$ with slope 3

62. Through $(2, 1)$ with slope 2

63. Through $\left(1, -\frac{3}{4}\right)$ with slope $\frac{1}{4}$

64. Through $\left(0, -\frac{1}{3}\right)$ with slope $\frac{1}{3}$

65. Through $(20, -3.5)$ and increasing at a rate of 10 units of y per unit of x

66. Through $(3.5, -10)$ and increasing at a rate of 1 unit of y per 2 units of x

67. Through $(2, -4)$ and $(1, 1)$

68. Through $(1, -4)$ and $(-1, -1)$

69. Through $(1, -0.75)$ and $(0.5, 0.75)$
70. Through $(0.5, -0.75)$ and $(1, -3.75)$
71. Through $(6, 6)$ and parallel to the line $x + y = 4$
72. Through $(\frac{1}{3}, -1)$ and parallel to the line $3x - 4y = 8$
73. Through $(0.5, 5)$ and parallel to the line $4x - 2y = 11$
74. Through $(\frac{1}{3}, 0)$ and parallel to the line $6x - 2y = 11$
75. ▼ Through $(0, 0)$ and (p, q) $(p \neq 0)$
76. ▼ Through (p, q) parallel to $y = rx + s$
77. ▼ Through (p, q) and (r, q) $(p \neq r)$
78. ▼ Through (p, q) and $(-p, -q)$ $(p \neq 0)$
79. ▼ Through $(-p, q)$ and $(p, -q)$ $(p \neq 0)$
80. ▼ Through (p, q) and (r, s) $(p \neq r)$

Applications

81. ***Cost*** The *RideEm Bicycles* factory can produce 100 bicycles in a day at a total cost of \$10,500. It can produce 120 bicycles in a day at a total cost of \$11,000. What are the company's daily fixed costs, and what is the marginal cost per bicycle? [**HINT**: See Example 3.]

82. ***Cost*** A soft-drink manufacturer can produce 1,000 cases of soda in a week at a total cost of \$6,000 and 1,500 cases of soda at a total cost of \$8,500. Find the manufacturer's weekly fixed costs and marginal cost per case of soda.

83. ***Cost: iPhone 5 (16 GB)*** If it costs **Apple** \$2,070 to manufacture 10 iPhones per hour and \$4,120 to manufacture 20 per hour at a particular plant,[42] obtain the corresponding linear cost function. What was the cost to manufacture each additional iPhone? Use the cost function to estimate the cost of manufacturing 40 iPhones in an hour.

84. ***Cost: Kinects*** If it costs **Microsoft** \$1,230 to manufacture 8 Kinects per hour for the Xbox 360 and \$2,430 to manufacture 16 per hour at a particular plant,[43] obtain the corresponding linear cost function. What was the cost to manufacture each additional Kinect? Use the cost function to estimate the cost of manufacturing 30 Kinects in an hour.

85. ***Demand*** Sales figures show that your company sold 1,960 pen sets each week when they were priced at \$1.00 per pen set and 1,800 pen sets each week when they were priced at \$5.00 per pen set. What is the linear demand function for your pen sets? [**HINT**: See Example 4.]

86. ***Demand*** A large department store is prepared to buy 3,950 of your tie-dye shower curtains per month for \$5 each but only 3,700 per month for \$10 each. What is the linear demand function for your tie-dye shower curtains?

87. ***Demand for Smartphones*** The following table shows worldwide sales of smartphones and their average selling prices in 2012 and 2013:[44]

Year	2012	2013
Selling Price (\$)	385	335
Sales (millions)	720	1,010

   a. Use the data to obtain a linear demand function for smartphones, and use your demand equation to predict sales if the price is lowered to \$265.
   b. Fill in the blanks: For every ____ increase in price, sales of smartphones decrease by ___ units.

88. ***Demand for Smartphones*** The following table shows worldwide sales of smartphones and their average selling prices in 2013 and 2017:[45]

Year	2013	2017
Selling Price (\$)	335	265
Sales (millions)	1,010	1,710

   a. Use the data to obtain a linear demand function for smartphones, and use your demand equation to predict sales if the price is raised to \$385.
   b. Fill in the blanks: For every ____ increase in price, sales of smartphones decrease by ___ units.

89. ***Demand for Monorail Service: Las Vegas*** In 2005 the Las Vegas monorail charged \$3 per ride and had an average ridership of about 28,000 per day. In December 2005 the Las Vegas Monorail Company raised the fare to \$5 per ride, and average ridership in 2006 plunged to around 19,000 per day.[46]
   a. Use the given information to find a linear demand equation.
   b. Give the units of measurement and interpretation of the slope.
   c. What would have been the effect on ridership of raising the fare to \$6 per ride?

90. ***Demand for Monorail Service: Mars*** The Utarek monorail, which links the three urbynes (or districts) of Utarek on Mars, charged Z̄5 per ride[47] and sold about 14 million rides per day. When the Utarek City Council lowered the fare to Z̄3 per ride, the number of rides increased to 18 million per day.
   a. Use the given information to find a linear demand equation.
   b. Give the units of measurement and interpretation of the slope.
   c. What would have been the effect on ridership of raising the fare to Z̄10 per ride?

---

42 Marginal costs are approximate, based on marginal cost data at www.isuppli.com and www.iphoneincanada.ca. Fixed costs are fictitious.

43 Based on marginal cost data provided by a "highly-positioned, trusted source" (www.develop-online.net).

44 Data are approximate. Source: IDC Worldwide Quarterly Mobile Phone Tracker, Nov. 26, 2013, www.zdnet.com.

45 2017 data are estimates. Source: *Ibid.*

46 Source: *New York Times*, February 10, 2007, p. A9.

47 The zonar (Z̄) is the official currency in the city-state of Utarek, Mars (formerly www.Marsnext.com, a now extinct virtual society).

91. ***Pasta Imports in the 1990s*** During the period 1990–2001, U.S. imports of pasta increased from 290 million pounds in 1990 ($t = 0$) by an average of 40 million pounds per year.[48]

a. Use this information to express y, the annual U.S. imports of pasta (in millions of pounds), as a linear function of t, the number of years since 1990.

b. Use your model to estimate U.S. pasta imports in 2005, assuming that the import trend continued.

92. ***Mercury Imports in the 2210s*** During the period 2210–2220, Martian imports of mercury (from the planet of that name) increased from 550 million kilograms in 2210 ($t = 0$) by an average of 60 million kilograms per year.

a. Use this information to express y, the annual Martian imports of mercury (in millions of kilograms), as a linear function of t, the number of years since 2210.

b. Use your model to estimate Martian mercury imports in 2230, assuming that the import trend continued.

93. ***Net Income*** The net income of **Amazon** decreased from \$0.63 billion in 2011 to $-$\$0.24 billion in 2014.[49]

a. Use this information to find a linear model for Amazon's net income N (in billions of dollars) as a function of time t in years since 2010.

b. Give the units of measurement and interpretation of the slope.

c. Use the model from part (a) to estimate the 2013 net income. (The actual 2013 net income was approximately \$0.27 billion.)

94. ***Operating Expenses*** The operating expenses of **Amazon** increased from \$3.6 billion in 2008 to \$16.3 billion in 2012.[50]

a. Use this information to find a linear model for Amazon's operating expenses E (in billions of dollars) as a function of time t in years since 2010.

b. Give the units of measurement and interpretation of the slope.

c. Use the model from part (a) to estimate the 2011 operating expenses. (The actual 2011 operating expenses were \$10.9 billion.)

95. ***Velocity*** The position of a model train, in feet along a railroad track, is given by

$$s(t) = 2.5t + 10$$

after t seconds.

a. How fast is the train moving?

b. Where is the train after 4 seconds?

c. When will the train be 25 feet along the track?

[48] Data are rounded. Sources: Department of Commerce/*New York Times*, September 5, 1995, p. D4; International Trade Administration, March 31, 2002, www.ita.doc.gov.

[49] Recall that "net income" is another term for "profit." Data are approximate. Source: www.wikinvest.com.

[50] *Ibid.*

96. ***Velocity*** The height of a falling sheet of paper, in feet from the ground, is given by

$$s(t) = -1.8t + 9$$

after t seconds.

a. What is the velocity of the sheet of paper?

b. How high is the sheet of paper after 4 seconds?

c. When will the sheet of paper reach the ground?

97. ▼ ***Fast Cars*** A police car was traveling down Ocean Parkway in a high-speed chase from Jones Beach. The police car was at Jones Beach at exactly 10 pm ($t = 10$) and was at Oak Beach, 13 miles from Jones Beach, at exactly 10:06 pm.

a. How fast was the police car traveling? [**HINT**: See Example 6.]

b. How far was the police car from Jones Beach at time t?

98. ▼ ***Fast Cars*** The car that was being pursued by the police in Exercise 97 was at Jones Beach at exactly 9:54 pm ($t = 9.9$) and passed Oak Beach (13 miles from Jones Beach) at exactly 10:06 pm, where it was overtaken by the police.

a. How fast was the car traveling? [**HINT**: See Example 6.]

b. How far was the car from Jones Beach at time t?

99. ***Textbook Sizes*** The second edition of *Applied Calculus* by Waner and Costenoble was 585 pages long. By the time we got to the sixth edition, the book had grown to 755 pages.

a. Use this information to obtain the page length L as a linear function of the edition number n.

b. What are the units of measurement of the slope? What does the slope tell you about the length of *Applied Calculus*?

c. At this rate, by which edition will the book have grown to over 1,500 pages?

100. ***Textbook Sizes*** The second edition of *Finite Mathematics* by Waner and Costenoble was 603 pages long. By the time we got to the fifth edition, the book had grown to 690 pages.

a. Use this information to obtain the page length L as a linear function of the edition number n.

b. What are the units of measurement of the slope? What does the slope tell you about the length of *Finite Mathematics*?

c. At this rate, by which edition will the book have grown to over 1,000 pages?

101. ***Fahrenheit and Celsius*** In the Fahrenheit temperature scale, water freezes at 32°F and boils at 212°F. In the Celsius scale, water freezes at 0°C and boils at 100°C. Given that the Fahrenheit temperature F and the Celsius temperature C are related by a linear equation, find F in terms of C. Use your equation to find the Fahrenheit temperatures corresponding to 30°C, 22°C, -10°C, and -14°C, to the nearest degree.

102. ***Fahrenheit and Celsius*** Use the information about Celsius and Fahrenheit given in Exercise 101 to obtain a linear equation for C in terms of F, and use your equation to find the Celsius temperatures corresponding to 104°F, 77°F, 14°F, and -40°F, to the nearest degree.

Airline Net Income *Exercises 103 and 104 are based on the following table, which compares the net incomes, in millions of dollars, of* Southwest Airlines, JetBlue Airways, *and* Alaska Air Group:[51]

Year	2010	2011	2012	2013	2014
Southwest Airlines	450	200	400	750	900
JetBlue Airways	100	90	130	170	400
Alaska Air Group	250	250	300	500	600

103. **a.** Use the 2012 and 2014 data to obtain the JetBlue net income J as a linear function of the Southwest Airlines net income S. (Use millions of dollars for all units, and round coefficients to two significant digits.)
b. How far off is your model in estimating the JetBlue net income based on the Southwest Airlines income in 2010?
c. What are the units of measurement of the slope? What does the slope of the linear function from part (a) suggest about the net incomes of these two airlines?

104. **a.** Use the 2010 and 2013 data to obtain the JetBlue net income J as a linear function of Alaska Air Group's net income A. (Use millions of dollars for all units, and round coefficients to two significant digits.)
b. In which of the remaining years does the model give the best prediction of JetBlue's net income?
c. What are the units of measurement of the slope? What does the slope of the linear function from part (a) suggest about the net incomes of these two airlines?

105. ▼ ***Income*** The well-known romance novelist Celestine A. Lafleur (a.k.a. Bertha Snodgrass) has decided to sell the screen rights to her latest book, *Henrietta's Heaving Heart*, to *Boxoffice Success Productions* for \$50,000. In addition, the contract ensures Ms. Lafleur royalties of 5% of the net profits.[52] Express her income I as a function of the net profit N, and determine the net profit necessary to bring her an income of \$100,000. What is her marginal income (share of each dollar of net profit)?

106. ▼ ***Income*** Because of the enormous success of the movie *Henrietta's Heaving Heart* based on a novel by Celestine A. Lafleur (see Exercise 105), *Boxoffice Success Productions* decides to film the sequel, *Henrietta, Oh Henrietta.* At this point, Bertha Snodgrass (whose novels now top the best-seller lists) feels that she is in a position to demand \$100,000 for the screen rights and royalties of 8% of the net profits. Express her income I as a function of the net profit N, and determine the net profit necessary to bring her an income of \$1,000,000. What is her marginal income (share of each dollar of net profit)?

107. ***Processor Speeds*** The processor speed, in megahertz (MHz), of Intel processors during the period 1996–2010 could be approximated by the following function of time t in years since the start of 1990:[53]

$$v(t) = \begin{cases} 400t - 2{,}200 & \text{if } 6 \le t < 15 \\ 3{,}800 & \text{if } 15 \le t \le 20. \end{cases}$$

How fast and in what direction was processor speed changing in 2000?

108. ***Processor Speeds*** The processor speed, in megahertz (MHz), of Intel processors during the period 1970–2000 could be approximated by the following function of time t in years since the start of 1970:[54]

$$v(t) = \begin{cases} 3t & \text{if } 0 \le t < 20 \\ 174t - 3{,}420 & \text{if } 20 \le t \le 30. \end{cases}$$

How fast and in what direction was processor speed changing in 1995?

Superbowl Advertising *Exercises 109 and 110 are based on the following graph and data showing the increasing cost of a 30-second television ad during the Super Bowl.*[55]

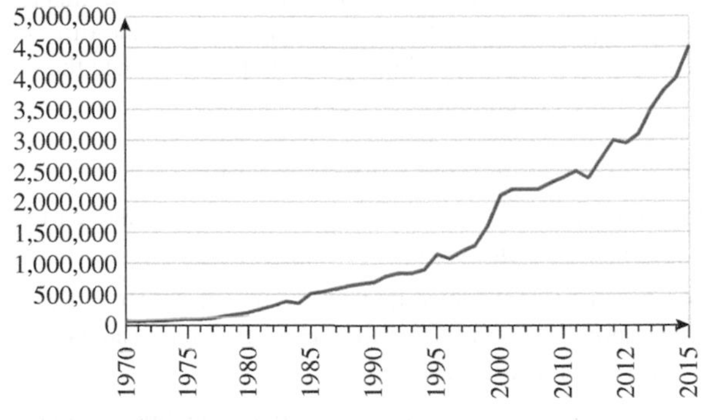

Year	1970	1980	1990	2000	2010
Cost (\$1,000)	78	222	700	2,100	2,950

109. ▼ Take t to be the number of years since 1970 and y to be the cost, in thousands of dollars, of a Super Bowl ad.
a. Model the 1970 and 1990 data with a linear equation.
b. Model the 1990 and 2010 data with a linear equation.
c. Use the results of parts (a) and (b) to obtain a piecewise-linear model of the cost of a Super Bowl ad during 1970–2010.

---

[51] Net incomes from continuing operations. Data are rounded. Source for data: www.wikinvest.com.

[52] Percentages of net profit are commonly called "monkey points." Few movies ever make a net profit on paper, and anyone with any clout in the business gets a share of the *gross*, not the net.

[53] A rough model based on the fastest processors produced by Intel. Source: www.intel.com.

[54] *Ibid.*

[55] Source: Nielsen Media Research, http://superbowl-ads.com.

d. Use your model to estimate the cost in 2004. Is your answer in rough agreement with the graph? Explain any discrepancy.

110 ▼ Take t to be the number of years since 1980 and y to be the cost, in thousands of dollars, of a Super Bowl ad.

a. Model the 1980 and 2000 data with a linear equation.

b. Model the 2000 and 2010 data with a linear equation.

c. Use the results of parts (a) and (b) to obtain a piecewise-linear model of the cost of a Super Bowl ad during 1980–2010.

d. Use your model to estimate the cost in 1992. Is your answer in rough agreement with the graph? Explain any discrepancy.

111. ▼ ***Employment in Mexico*** The number of workers employed in manufacturing jobs in Mexico was 3 million in 1995, rose to 4.1 million in 2000, and then dropped to 3.5 million in 2004.[56] Model this number N as a piecewise-linear function of the time t in years since 1995, and use your model to estimate the number of manufacturing jobs in Mexico in 2002. (Take the units of N to be millions.)

112. ▼ ***Mortgage Delinquencies*** The percentage of borrowers in the highest risk category who were delinquent on their payments decreased from 9.7% in 2001 to 4.3% in 2004 and then shot up to 10.3% in 2007.[57] Model this percentage P as a piecewise-linear function of the time t in years since 2001, and use your model to estimate the percentage of delinquent borrowers in 2006.

Communication and Reasoning Exercises

113. How would you test a table of values of x and y to see whether it comes from a linear function?

114. You have ascertained that a table of values of x and y corresponds to a linear function. How do you find an equation for that linear function?

115. To what linear function of x does the linear equation $ax + by = c$ $(b \neq 0)$ correspond? Why did we specify $b \neq 0$?

116. Complete the following. The slope of the line with equation $y = mx + b$ is the number of units that ____ increases per unit increase in ____.

117. Complete the following. If, in a straight line, y is increasing three times as fast as x, then its ____ is ____.

118. Suppose that y is decreasing at a rate of 4 units per 3-unit increase of x. What can we say about the slope of the linear relationship between x and y? What can we say about the intercept?

119. If y and x are related by the linear expression $y = mx + b$, how will y change as x changes if m is positive? negative? zero?

120. Your friend April tells you that $y = f(x)$ has the property that, whenever x is changed by Δx, the corresponding change in y is $\Delta y = -\Delta x$. What can you tell her about f?

121. Consider the following worksheet:

	A	B	C	D
1	x	y	m	b
2	1	2	=(B3-B2)/(A3-A2)	=B2-C2*A2
3	3	-1	Slope	Intercept

What is the effect on the slope of increasing the y-coordinate of the second point (the point whose coordinates are in row 3)? Explain.

122. Referring to the worksheet in Exercise 121, what is the effect on the slope of increasing the x-coordinate of the second point (the point whose coordinates are in row 3)? Explain.

123. If y is measured in bootlags,[58] x is measured in zonars,[59] and $y = mx + b$, then m is measured in _______, and b is measured in _______.

124. If the slope in a linear relationship is measured in miles per dollar, then the independent variable is measured in ______, and the dependent variable is measured in ______.

125. If a quantity is changing linearly with time and it increases by 10 units in the first day, what can you say about its behavior in the third day?

126. The quantities Q and T are related by a linear equation of the form

$$Q = mT + b$$

Q is positive when $T = 0$ but decreases to a negative quantity when T is 10. What are the signs of m and b? Explain your answers.

127. ▼ The velocity of an object is given by $v = 0.1t + 20$ m/sec, where t is time in seconds. The object is

(A) moving with fixed speed.
(B) accelerating.
(C) decelerating.
(D) impossible to say from the given information.

128. ▼ The position of an object is given by $x = 0.2t - 4$, where t is time in seconds. The object is

(A) moving with fixed speed.
(B) accelerating.
(C) decelerating.
(D) impossible to say from the given information.

---

[56] Source: *New York Times*, Februrary 18, 2007, p. WK4.

[57] The 2007 figure was projected from data through October 2006. Source: *New York Times*, Februrary 18, 2007, p. BU9.

[58] An ancient Martian unit of length; one bootlag is the mean distance from a Martian's foreleg to its rearleg.

[59] The official currency of Utarek, Mars. (See the footnote to Exercise 90.)

129. If f and g are linear functions with slope m and n, respectively, then what can you say about $f + g$?

130. If f and g are linear functions, then is $\frac{f}{g}$ linear? Explain.

131. Give examples of nonlinear functions f and g whose product is linear.

132. Give examples of nonlinear functions f and g whose quotient is linear (on a suitable domain).

133. ▼ Suppose the cost function is $C(x) = mx + b$ (with m and b positive), the revenue function is $R(x) = kx$ $(k > m)$, and the number of items is increased from the break-even quantity. Does this result in a loss or a profit, or is it impossible to say? Explain your answer.

134. ▼ You have been constructing a demand equation, and you obtained a (correct) expression of the form $p = mq + b$, whereas you would have preferred one of the form $q = mp + b$. Should you simply switch p and q in the answer, should you start again from scratch, using p in the role of x and q in the role of y, or should you solve your demand equation for q? Give reasons for your decision.

1.4 Linear Regression

Observed and Predicted Values

We have seen how to find a linear model given two data points: We find the equation of the line that passes through them. However, we often have more than two data points, and they will rarely all lie on a single straight line, but they may often come close to doing so. The problem is to find the line coming *closest* to passing through all of the points.

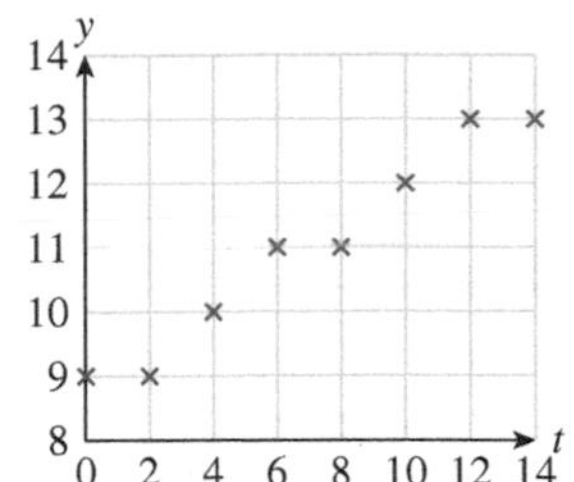

Figure 24(a)

Suppose, for example, that we are conducting research for a company that is interested in expanding into Mexico. Of interest to us would be current and projected growth in that country's economy. The following table shows past and projected per capita gross domestic product (GDP)[60] of Mexico for 2000–2014:[61]

Year t (year since 2000)	0	2	4	6	8	10	12	14
Per Capita GDP y (\$1,000)	9	9	10	11	11	12	13	13

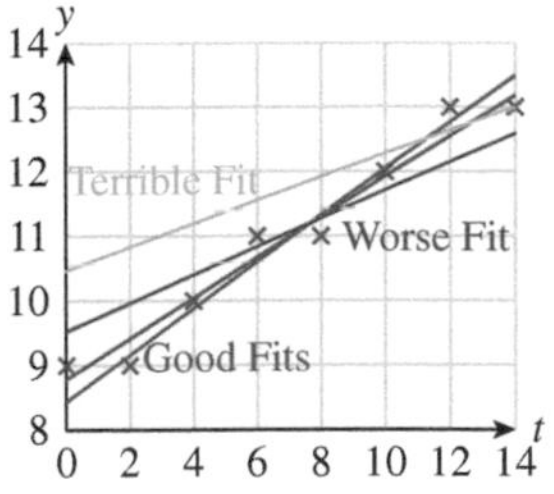

Figure 24(b)

A plot of these data suggests a roughly linear growth of the GDP (Figure 24(a)). These points suggest a roughly linear relationship between t and y, although they clearly do not all lie on a single straight line. Figure 24(b) shows the points together with several lines, some fitting better than others. Can we precisely measure which lines fit better than others? For instance, which of the two lines labeled as "good" fits in Figure 24(b) models the data more accurately? We begin by considering, for each value of t, the difference between the actual GDP (the **observed value**) and the GDP predicted by a linear equation (the **predicted value**). The difference between the predicted value and the observed value is called the **residual**.

$$\text{Residual} = \text{Observed value} - \text{Predicted value}$$

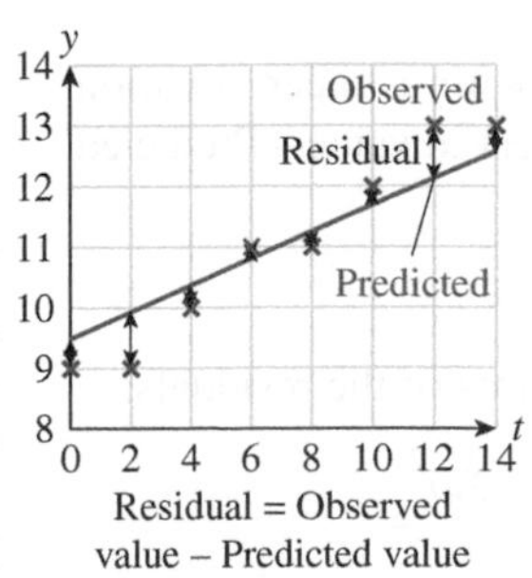

Residual = Observed value – Predicted value

Figure 25

On the graph, the residuals measure the vertical distances between the (observed) data points and the line (Figure 25), and they tell us how far the linear model is from predicting the actual GDP.

[60] The GDP is a measure of the total market value of all goods and services produced within a country.

[61] Data are approximate and/or projected. Sources: CIA World Factbook, www.indexmundi.com, www.economist.com.

The more accurate our model, the smaller the residuals should be. We can combine all the residuals into a single measure of accuracy by adding their *squares*. (We square the residuals in part to make them all positive.*) The sum of the squares of the residuals is called the **sum-of-squares error, SSE**. Smaller values of SSE indicate more accurate models.

* Why not add the absolute values of the residuals instead? Mathematically, using the squares rather than the absolute values results in a simpler and more elegant solution. Further, using the squares always results in a *single* best-fit line in cases where the x-coordinates are all different, whereas this is not the case if we use absolute values.

Here are some definitions and formulas for what we have been discussing.

Observed and Predicted Values

Suppose we are given a collection of data points $(x_1, y_1), \ldots, (x_n, y_n)$. The n quantities $y_1, y_2, \ldots, y_n$ are called the **observed y-values.** If we model these data with a linear equation

$$\hat{y} = mx + b, \qquad \hat{y} \text{ stands for "estimated } y\text{" or "predicted } y\text{."}$$

then the y-values we get by substituting the given x-values into the equation are called the **predicted y-values:**

$$\hat{y}_1 = mx_1 + b \qquad \text{Substitute } x_1 \text{ for } x.$$

$$\hat{y}_2 = mx_2 + b \qquad \text{Substitute } x_2 \text{ for } x.$$

$$\vdots$$

$$\hat{y}_n = mx_n + b. \qquad \text{Substitute } x_n \text{ for } x.$$

Quick Example

1. Consider the three data points (0, 2), (2, 5), and (3, 6). The observed y-values are $y_1 = 2$, $y_2 = 5$, and $y_3 = 6$. If we model these data with the equation $\hat{y} = x + 2.5$, then the predicted y-values are

$$\hat{y}_1 = x_1 + 2.5 = 0 + 2.5 = 2.5$$

$$\hat{y}_2 = x_2 + 2.5 = 2 + 2.5 = 4.5$$

$$\hat{y}_3 = x_3 + 2.5 = 3 + 2.5 = 5.5.$$

Residuals, Sum-of-Squares Error

Residuals and Sum-of-Squares Error (SSE)

If we model a collection of data $(x_1, y_1), \ldots, (x_n, y_n)$ with a linear equation $\hat{y} = mx + b$, then the **residuals** are the n quantities (Observed value – Predicted value):

$$(y_1 - \hat{y}_1), (y_2 - \hat{y}_2), \ldots, (y_n - \hat{y}_n).$$

The **sum-of-squares error (SSE)** is the sum of the squares of the residuals:

$$\text{SSE} = (y_1 - \hat{y}_1)^2 + (y_2 - \hat{y}_2)^2 + \cdots + (y_n - \hat{y}_n)^2.$$

Quick Example

2. For the data and linear approximation given in Quick Example 1, the residuals are

$$y_1 - \hat{y}_1 = 2 - 2.5 = -0.5$$
$$y_2 - \hat{y}_2 = 5 - 4.5 = 0.5$$
$$y_3 - \hat{y}_3 = 6 - 5.5 = 0.5,$$

and so $\text{SSE} = (-0.5)^2 + (0.5)^2 + (0.5)^2 = 0.75$.

EXAMPLE 1 Computing SSE

Using the data above on the GDP in Mexico, compute SSE for the linear models $y = 0.5t + 8$ and $y = 0.25t + 9$. Which model is the better fit?

Solution We begin by creating a table showing the values of t, the observed (given) values of y, and the values predicted by the first model:

Year t	Observed y	Predicted $\hat{y} = 0.5t + 8$
0	9	8
2	9	9
4	10	10
6	11	11
8	11	12
10	12	13
12	13	14
14	13	15

We now add two new columns for the residuals and their squares:

Year t	Observed y	Predicted $\hat{y} = 0.5t + 8$	Residual $y - \hat{y}$	Residual² $(y - \hat{y})^2$
0	9	8	$9 - 8 = 1$	$1^2 = 1$
2	9	9	$9 - 9 = 0$	$0^2 = 0$
4	10	10	$10 - 10 = 0$	$0^2 = 0$
6	11	11	$11 - 11 = 0$	$0^2 = 0$
8	11	12	$11 - 12 = -1$	$(-1)^2 = 1$
10	12	13	$12 - 13 = -1$	$(-1)^2 = 1$
12	13	14	$13 - 14 = -1$	$(-1)^2 = 1$
14	13	15	$13 - 15 = -2$	$(-2)^2 = 4$

SSE, the sum of the squares of the residuals, is then the sum of the entries in the last column,

$$SSE = 8.$$

Repeating the process using the second model, $0.25t + 9$, yields the following table:

Year t	Observed y	Predicted $\hat{y} = 0.25t + 9$	Residual $y - \hat{y}$	Residual^2 $(y - \hat{y})^2$
0	9	9	$9 - 9 = 0$	$0^2 = 0$
2	9	9.5	$9 - 9.5 = -0.5$	$(-0.5)^2 = 0.25$
4	10	10	$10 - 10 = 0$	$0^2 = 0$
6	11	10.5	$11 - 10.5 = 0.5$	$0.5^2 = 0.25$
8	11	11	$11 - 11 = 0$	$0^2 = 0$
10	12	11.5	$12 - 11.5 = 0.5$	$0.5^2 = 0.25$
12	13	12	$13 - 12 = 1$	$1^2 = 1$
14	13	12.5	$13 - 12.5 = 0.5$	$0.5^2 = 0.25$

This time, SSE = 2, so the second model is a better fit. Figure 26 shows the data points and the two linear models in question.

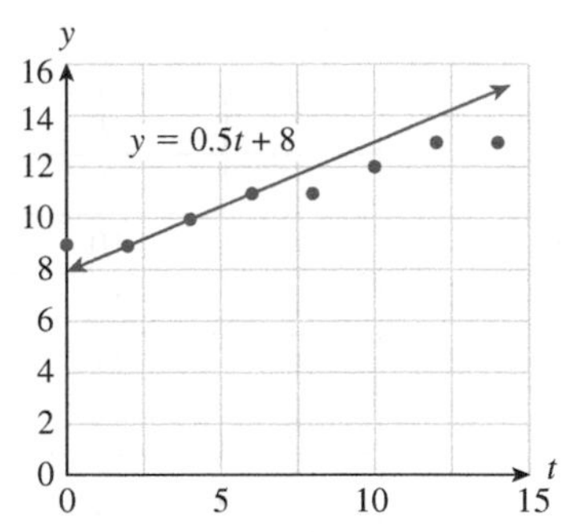

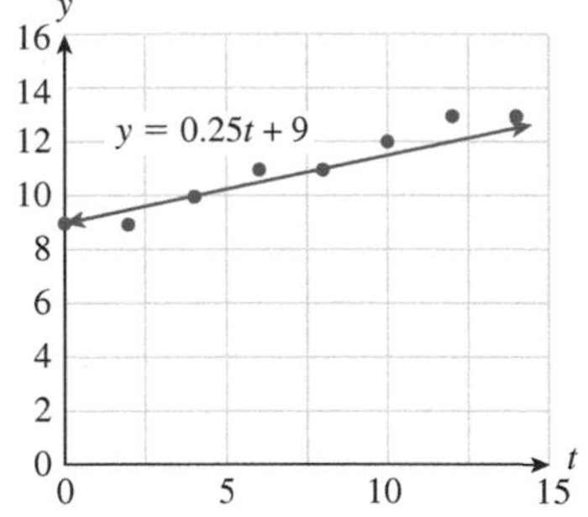

Figure 26

Using Technology

See the Technology Guides at the end of the chapter for detailed instructions on how to obtain the tables and graphs in Example 1 using a TI-83/84 Plus or a spreadsheet. Here is an outline:

TI-83/84 Plus
[STAT] EDIT
Values of t in L_1 and y in L_2. Predicted y: Highlight L_3. Enter `0.5*L1+8`
Squares of residuals: Highlight L_4. Enter the following (including the quotes):
`"(L2-L3)^2"`
SSE: Home screen `sum(L4)`
Graph: `Y1=0.5X+8`
Y = screen: Turn on Plot 1 [ZOOM] (STAT) [More details in the Technology Guide.]

Spreadsheet
Headings t, y, y-hat, Residual^2, m, b, and SSE in A1–F1.
t-values in A2–A9, y-values in B2–B9; 0.25 for m and 9 for b in E2–F2
Predicted y: `=$E$2*A2+$F$2` in C2 and copy down to C9.
Squares of residuals: `=(B2-C2)^2` in D2 and copy down to D9.
SSE: `=SUM(D2:D9)` in G2
Graph: Highlight A1–C9. Insert a scatter chart.
[More details in the Technology Guide].

➡ Before we go on . . .

Q: *It seems clear from the figure that the second model in Example 1 gives a better fit. Why bother to compute SSE to tell me this?*

A: The difference between the two models we chose is so great that it is clear from the graphs which is the better fit. However, if we used a third model with $m = 0.25$ and $b = 9.1$, then its graph would be almost indistinguishable from that of the second but would be a slightly better fit as measured by SSE = 1.68.

■

The Regression Line

Among all possible lines, there ought to be one with the least possible value of SSE—that is, the greatest possible accuracy as a model. The line (and there is only one such line) that minimizes the sum of the squares of the residuals is called the **regression line**, the **least-squares line**, or the **best-fit line**.

To find the regression line, we need a way to find values of m and b that give the smallest possible value of SSE. As an example, let us take the second linear model in Example 1. We said in the "Before we go on" discussion that increasing b from 9 to 9.1 had the desirable effect of decreasing SSE from 2 to 1.68. We could then increase m to 0.26, further reducing SSE to 1.328. Imagine this as a kind of game: Alter the values of m and b alternately by small amounts until SSE is as small as you can make it. This works but is extremely tedious and time-consuming.

Fortunately, there is an algebraic way to find the regression line. Here is the calculation. We will justify it in Chapter 8 using calculus.

Regression Line

The **regression line (least squares line, best-fit line)** associated with the points $(x_1, y_1), (x_2, y_2), \ldots, (x_n, y_n)$ is the line that gives the minimum SSE. The regression line is

$$y = mx + b,$$

where m and b are computed as follows:

$$m = \frac{n(\Sigma xy) - (\Sigma x)(\Sigma y)}{n(\Sigma x^2) - (\Sigma x)^2}$$

$$b = \frac{\Sigma y - m(\Sigma x)}{n}$$

$$n = \text{number of data points.}$$

The quantities m and b are called the **regression coefficients**.

Here, "Σ" means "the sum of." Thus, for example,

$$\Sigma x = \text{Sum of the } x\text{-values} = x_1 + x_2 + \cdots + x_n$$

$$\Sigma xy = \text{Sum of products} = x_1y_1 + x_2y_2 + \cdots + x_ny_n$$

$$\Sigma x^2 = \text{Sum of squares of the } x\text{-values} = x_1{}^2 + x_2{}^2 + \cdots + x_n{}^2.$$

On the other hand,

$$(\Sigma x)^2 = \text{Square of } \Sigma x = \text{Square of the sum of the } x\text{-values.}$$

EXAMPLE 2 Per Capita Gross Domestic Product in Mexico

In Example 1 we considered the following data on the per capita gross domestic product (GDP) of Mexico:

Year x (year since 2000)	0	2	4	6	8	10	12	14
Per Capita GDP y ($1,000)	9	9	10	11	11	12	13	13

Find the best-fit linear model for these data, and use the model to predict the per capita GDP in Mexico in 2016.

Solution Let's organize our work in the form of a table, where the original data are entered in the first two columns and the bottom row contains the column sums:

	x	y	xy	x^2
	0	9	0	0
	2	9	18	4
	4	10	40	16
	6	11	66	36
	8	11	88	64
	10	12	120	100
	12	13	156	144
	14	13	182	196
Σ (Sum)	56	88	670	560

Using Technology

See the Technology Guides at the end of the chapter for detailed instructions on how to obtain the regression line and graph in Example 2 using a TI-83/84 Plus or a spreadsheet. Here is an outline:

TI-83/84 Plus

STAT EDIT

Values of x in L_1 and y in L_2.

Regression equation: STAT CALC option #4: `LinReg(ax+b)`

Graph: Y= VARS 5 EQ 1, then ZOOM 9

[More details in the Technology Guide.]

Spreadsheet

x-values in A2–A9, y-values in B2–B9

Graph: Highlight A2–B9. Insert a scatter chart.

Regression line: Add a linear trendline. [More details in the Technology Guide.]

Website

www.WanerMath.com

The following two utilities will calculate and plot regression lines (link to either from Math Tools for Chapter 1):

Simple Regression Utility

Function Evaluator and Grapher

Because there are $n = 8$ data points, we get

$$m = \frac{n(\Sigma xy) - (\Sigma x)(\Sigma y)}{n(\Sigma x^2) - (\Sigma x)^2} = \frac{8(670) - (56)(88)}{8(560) - (56)^2} \approx 0.321$$

and

$$b = \frac{\Sigma y - m(\Sigma x)}{n} \approx \frac{88 - (0.321)(56)}{8} \approx 8.75.$$

So the regression line is

$$y = 0.321x + 8.75.$$

To predict the per capita GDP in Mexico in 2016, we substitute $x = 16$ and get $y \approx 14$, or \$14,000 per capita.

Figure 27 shows the data points and the regression line (which has SSE ≈ 0.643, a lot lower than in Example 1).

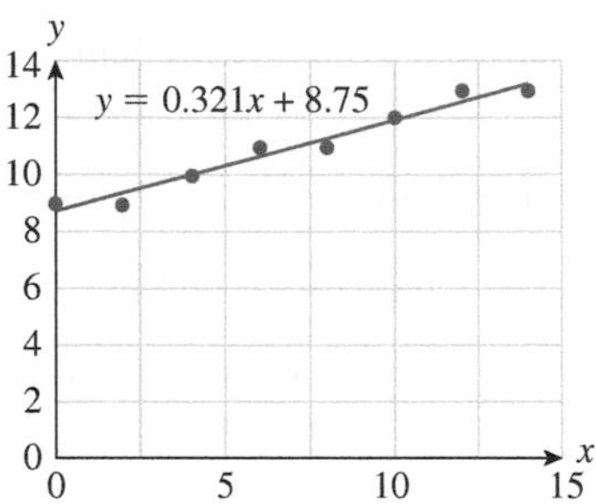

Figure 27

Coefficient of Correlation

If all the data points do not lie on one straight line, we would like to be able to measure how closely they can be approximated by a straight line. Recall that SSE measures the sum of the squares of the deviations from the regression line;

therefore, it constitutes a measurement of what is called goodness of fit. (For instance, if SSE = 0, then all the points lie on a straight line.) However, SSE depends on the units we use to measure y and also on the number of data points (the more data points we use, the larger SSE tends to be). Thus, while we can (and do) use SSE to compare the goodness of fit of two lines to the same data, we cannot use it to compare the goodness of fit of one line to one set of data with that of another to a different set of data.

To remove this dependency, statisticians have found a related quantity that can be used to compare the goodness of fit of lines to different sets of data. This quantity, called the **coefficient of correlation** or **correlation coefficient**, and usually denoted r, is between -1 and 1. The closer r is to -1 or 1, the better the fit. For an *exact* fit we would have $r = -1$ (for a line with negative slope) or $r = 1$ (for a line with positive slope). For a bad fit we would have r close to 0. Figure 28 shows several collections of data points with least-squares lines and the corresponding values of r.

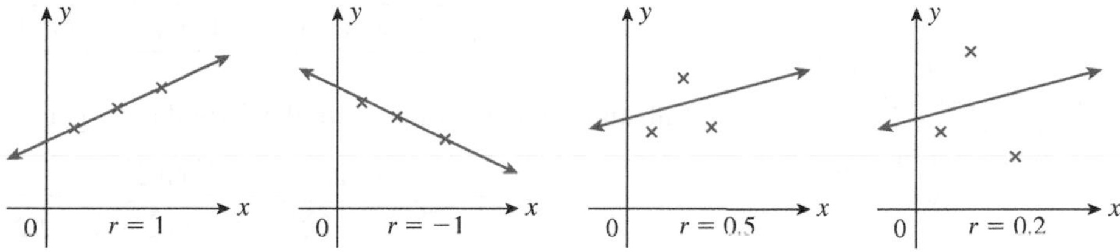

Figure 28

Correlation Coefficient

The coefficient of correlation of the n data points $(x_1, y_1), (x_2, y_2), \ldots, (x_n, y_n)$ is

$$r = \frac{n(\Sigma xy) - (\Sigma x)(\Sigma y)}{\sqrt{n(\Sigma x^2) - (\Sigma x)^2} \cdot \sqrt{n(\Sigma y^2) - (\Sigma y)^2}}.$$

It measures how closely the data points $(x_1, y_1), (x_2, y_2), \ldots, (x_n, y_n)$ fit the regression line. (The value r^2 is sometimes called the **coefficient of determination**.)

Interpretation

- If r is positive, the regression line has positive slope; if r is negative, the regression line has negative slope.
- If $r = 1$ or -1, then all the data points lie exactly on the regression line; if it is close to ± 1, then all the data points are close to the regression line.
- On the other hand, if r is not close to ± 1, then the data points are not close to the regression line, so the fit is not a good one. As a general rule of thumb, a value of $|r|$ less than around 0.8 indicates a poor fit of the data to the regression line.

EXAMPLE 3 **Computing the Coefficient of Correlation**

Find the correlation coefficient for the data in Example 2. Is the regression line a good fit?

Solution The formula for r requires Σx, Σx^2, Σxy, Σy, and Σy^2. We have all of these except for Σy^2, which we find in a new column as shown:

	x	y	xy	x^2	y^2
	0	9	0	0	81
	2	9	18	4	81
	4	10	40	16	100
	6	11	66	36	121
	8	11	88	64	121
	10	12	120	100	144
	12	13	156	144	169
	14	13	182	196	169
Σ (Sum)	56	88	670	560	986

Substituting these values into the formula, we get

$$r = \frac{n(\Sigma xy) - (\Sigma x)(\Sigma y)}{\sqrt{n(\Sigma x^2) - (\Sigma x)^2} \cdot \sqrt{n(\Sigma y^2) - (\Sigma y)^2}}$$

$$= \frac{8(670) - (56)(88)}{\sqrt{8(560) - 56^2} \cdot \sqrt{8(986) - 88^2}}$$

$$\approx 0.982.$$

As r is close to 1, the fit is a fairly good one; that is, the original points lie nearly along a straight line, as can be confirmed from the graph in Example 2.

Using Technology

See the Technology Guides at the end of the chapter for detailed instructions on how to obtain the correlation coefficient in Example 3 using a TI-83/84 Plus or a spreadsheet. Here is an outline:

TI-83/84 Plus

[2ND] [CATALOG] `DiagnosticOn`
Then [STAT] CALC option #4: `LinReg(ax+b)` [More details in the Technology Guide.]

Spreadsheet

Add a trendline and select the option to "Display R-squared value on chart."
[More details and other alternatives in the Technology Guide.]

 **Website**

www.WanerMath.com

The following two utilities will show regression lines and also r^2 (link to either from Math Tools for Chapter 1):

Simple Regression Utility

Function Evaluator and Grapher

1.4 EXERCISES

▼ more advanced ◆ challenging

T indicates exercises that should be solved using technology

In Exercises 1–4, compute the sum-of-squares error (SSE) by hand for the given set of data and linear model. [**HINT**: See Example 1.]

1. $(1, 1), (2, 2), (3, 4); y = x - 1$
2. $(0, 1), (1, 1), (2, 2); y = x + 1$
3. $(0, -1), (1, 3), (4, 6), (5, 0); y = -x + 2$
4. $(2, 4), (6, 8), (8, 12), (10, 0); y = 2x - 8$

T *In Exercises 5–8, use technology to compute the sum-of-squares error (SSE) for the given set of data and linear models. Indicate which linear model gives the better fit.*

5. $(1, 1), (2, 2), (3, 4)$; **a.** $y = 1.5x - 1$ **b.** $y = 2x - 1.5$
6. $(0, 1), (1, 1), (2, 2)$; **a.** $y = 0.4x + 1.1$ **b.** $y = 0.5x + 0.9$
7. $(0, -1), (1, 3), (4, 6), (5, 0)$; **a.** $y = 0.3x + 1.1$ **b.** $y = 0.4x + 0.9$
8. $(2, 4), (6, 8), (8, 12), (10, 0)$; **a.** $y = -0.1x + 7$ **b.** $y = -0.2x + 6$

In Exercises 9–12, find the regression line associated with the given set of points. Graph the data and the best-fit line. (Round all coefficients to four decimal places.) [**HINT**: See Example 2.]

9. $(1, 1), (2, 2), (3, 4)$
10. $(0, 1), (1, 1), (2, 2)$
11. $(0, -1), (1, 3), (3, 6), (4, 1)$
12. $(2, 4), (4, 8), (8, 12), (10, 0)$

In Exercises 13 and 14, use correlation coefficients to determine which of the given sets of data is best fit by its associated regression line and which is fit worst. Is it a perfect fit for any of the data sets? [**HINT**: See Example 3.]

13. a. (1, 3), (2, 4), (5, 6) **b.** (0, −1), (2, 1), (3, 4)
c. (4, −3), (5, 5), (0, 0)

14. a. (1, 3), (−2, 9), (2, 1) **b.** (0, 1), (1, 0), (2, 1)
c. (0, 0), (5, −5), (2, −2.1)

Applications

15. ***Mobile Broadband Subscriptions*** The following table shows the number of mobile broadband subscribers worldwide (x is the number of years since 2010):[62]

Year x	0	2	4
Subscribers y (millions)	800	1,600	2,300

Complete the following table, and obtain the associated regression line. (Round coefficients to one decimal place.) [**HINT**: See Example 2.]

	x	y	xy	x^2
	0	800		
	2	1,600		
	4	2,300		
Σ (Sum)				

Use your regression equation to project the number in 2016.

16. ***Fixed-Line Telephone Subscriptions*** The following table shows the number of fixed-line telephone subscribers in the United Kingdom (x is the number of years since 2000):[63]

Year x	0	5	14
Subscribers y (millions)	35	34	33

Complete the following table, and obtain the associated regression line. (Round coefficients to one decimal place.) [**HINT**: See Example 2.]

	x	y	xy	x^2
	0	35		
	5	34		
	14	33		
Σ (Sum)				

Use your regression equation to project the number in 2015.

17. ***Demand for Smartphones*** The following table shows worldwide sales of smartphones and their average selling prices in 2012, 2013, and 2017:[64]

Year	2012	2013	2017
Selling Price p ($100)	4	3	2
Sales q (billions)	0.7	1	2

Find the regression line (round coefficients to one decimal place), and use it to estimate the demand (in millions of units sold) when the selling price was $350.

18. ***Demand for Smartphones*** The following table shows worldwide sales of smartphones and their average selling prices in 2010, 2012, and 2013:[65]

Year	2010	2012	2013
Selling Price p ($100)	5	4	3
Sales q (billions)	0.3	0.7	1

Find the regression line (round coefficients to one decimal place), and use it to estimate the demand (in millions of units sold) when the selling price was $450.

19. ***Oil Recovery*** The Texas Bureau of Economic Geology published a study on the economic impact of using carbon dioxide enhanced oil recovery (EOR) technology to extract additional oil from fields that have reached the end of their conventional economic life. The following table gives the approximate number of jobs for the citizens of Texas that would be created at various levels of recovery:[66]

Percent Recovery (%)	20	40	80	100
Jobs Created (millions)	3	6	9	15

Find the regression line, and use it to estimate the number of jobs that would be created at a recovery level of 50%.

20. ***Oil Recovery*** (Refer to Exercise 19.) The following table gives the approximate economic value associated with various levels of oil recovery in Texas:[67]

Percent Recovery (%)	10	40	50	80
Economic Value ($ billion)	200	900	1,000	2,000

Find the regression line, and use it to estimate the economic value associated with a recovery level of 70%.

[62] Data are rounded, and 2014 figure is estimated. Source: International Telecommunication Union, www.itu.int/ITU-D/ict/statistics.

[63] *Ibid.*

[64] Data are approximate, and 2017 figures are the authors' estimates. Source: IDC Worldwide Quarterly Mobile Phone Tracker, Nov. 26, 2013, www.zdnet.com.

[65] *Ibid.*

[66] Source: "CO2–Enhanced Oil Recovery Resource Potential in Texas: Potential Positive Economic Impacts," Texas Bureau of Economic Geology, April 2004, www.rrc.state.tx.us/tepc/CO2-EOR_white_paper.pdf.

[67] *Ibid.*

21. ***Profit: Amazon*** The following table shows Amazon's approximate net sales (revenue) and net income (profit) in the period 2011–2014:[68]

Net Sales ($ billion)	50	60	70	80
Net Income ($ billion)	0.6	0.1	0.3	0.3

a. Use this information to find a linear regression model for Amazon's net income I (in millions of dollars) as a function of net sales S (in billions of dollars). Plot the data and regression line.
b. Give the units of measurement and interpretation of the slope.
c. What, according to the model, would Amazon need to earn in net sales for its net income to be $0.5 billion? (Round answer to the nearest billion dollars.)
d. Based on the graph, would you say that the linear model is reasonable? Why or why not?

22. ***Operating Expenses: Amazon*** The following table shows Amazon's approximate net sales (revenue) and operating expenses in 2011–2014:[69]

Net Sales ($ billion)	50	60	70	80
Operating Expenses ($ billion)	11	16	23	25

a. Use this information to find a linear regression model for Amazon's operating expenses E (in billions of dollars) as a function of net sales S (in billions of dollars). Plot the data and regression line.
b. Give the units of measurement and interpretation of the slope.
c. What, according to the model, would Amazon need to earn in net sales for its operating expenses to be $5 billion? (Round answer to the nearest billion dollars.)
d. Based on the graph, would you say that the linear model is reasonable? Why or why not?

23. T ***Textbook Sizes*** The following table shows the numbers of pages in previous editions of *Applied Calculus* by Waner and Costenoble:

Edition n	2	3	4	5	6
Number of Pages L	585	656	694	748	768

a. With the edition number as the independent variable, use technology to obtain a regression line and a plot of the points together with the regression line. (Round coefficients to two decimal places.)
b. Interpret the slope of the regression line.

24. T ***Textbook Sizes*** Repeat Exercise 23 using the following corresponding table for *Finite Mathematics* by Waner and Costenoble:

Edition n	2	3	4	5	6
Number of Pages L	603	608	676	692	696

25. T ***Soybean Production: Cerrados*** The following table shows soybean production, in millions of tons, in Brazil's Cerrados region as a function of the cultivated area, in millions of acres:[70]

Area (millions of acres)	25	30	32	40	52
Production (millions of tons)	15	25	30	40	60

a. Use technology to obtain the regression line and a plot of the points together with the regression line. (Round coefficients to two decimal places.)
b. Interpret the slope of the regression line.

26. T ***Soybean Production: United States*** The following table shows soybean production, in millions of tons, in the United States as a function of the cultivated area, in millions of acres:[71]

Area (millions of acres)	30	42	69	59	74	74
Production (millions of tons)	20	33	55	57	83	88

a. Use technology to obtain the regression line and a plot of the points together with the regression line. (Round coefficients to two decimal places.)
b. Interpret the slope of the regression line.

27. T ***Airline Profits and the Price of Oil*** A common perception is that airline profits are strongly correlated with the price of oil. Following are annual net incomes of Continental Airlines together with the approximate price of oil in the period 2005–2010:[72]

Year	2005	2006	2007	2008	2009	2010
Price of Oil ($/barrel)	56	63	67	92	54	71
Continental Net Income ($ million)	−70	370	430	−590	−280	150

[68] Figures are approximate, and 2014 figures are estimates. Source: www.wikinvest.com.
[69] *Ibid.*

[70] Source: Brazil Agriculture Ministry/*New York Times*, December 12, 2004, p. N32.
[71] Data are approximate. Source for data: L. David Roper, June 2010, *Crop Production in the World & the United States*, www.roperld.com/science/cropsworld&us.htm.
[72] Figures are rounded, and oil prices are inflation adjusted. Sources: www.wikinvest.com, www.inflationdata.com.

a. Use technology to obtain a regression line showing Continental's net income as a function of the price of oil and also the coefficient of correlation r.
b. What does the value of r suggest about the relationship of Continental's net income to the price of oil?
c. Support your answer to part (b) with a plot of the data and regression line.

28. T ***Airline Profits and the Price of Oil*** Repeat Exercise 27 using the following corresponding data for **American Airlines**:[73]

Year	2005	2006	2007	2008	2009	2010
Price of Oil ($/barrel)	56	63	67	92	54	71
American Net Income ($ million)	−850	250	450	−2,100	−1,450	−700

T ***Doctorates in Mexico*** *Exercises 29–32 are based on the following table showing the annual number of PhD graduates in Mexico in various fields:*[74]

	Natural Sciences	Engineering	Social Sciences	Education
1990	70	10	60	30
1995	130	40	110	40
2000	330	130	280	130
2005	490	370	460	210
2010	590	550	830	520
2012	690	590	1,000	900

29. a. With x = the number of natural science doctorates and y = the number of engineering doctorates, use technology to obtain the regression equation and graph the associated points and regression line. (Round coefficients to three significant digits.)
b. What does the slope tell you about the relationship between the number of natural science doctorates and the number of engineering doctorates?
c. Use technology to obtain the coefficient of correlation r. Does the value of r suggest a strong correlation between x and y?
d. Does the graph suggest a roughly linear relationship between x and y? Why or why not?

30. a. With x = the number of social science doctorates and y = the number of education doctorates, use technology to obtain the regression equation, and graph the associated points and regression line. (Round coefficients to three significant digits.)
b. What does the slope tell you about the relationship between the number of social science doctorates and the number of education doctorates?
c. Use technology to obtain the coefficient of correlation r. Does the value of r suggest a strong correlation between x and y?
d. Does the graph suggest a roughly linear relationship between x and y? Why or why not?

31. ▼ **a.** Use technology to obtain the regression equation and the coefficient of correlation r for the number of natural science doctorates as a function of time t in years since 1990, and graph the associated points and regression line. (Round coefficients to three significant digits.)
b. What does the slope tell you about the number of natural science doctorates?
c. Judging from the graph, would you say that the number of natural science doctorates is increasing at a faster and faster rate, a slower and slower rate, or a more-or-less constant rate? Why?
d. If r had been equal to 1, could you have drawn the same conclusion as in part (c)? Explain.

32. ▼ **a.** Use technology to obtain the regression equation and the coefficient of correlation r for the number of social science doctorates as a function of time t in years since 1990, and graph the associated points and regression line. (Round coefficients to three significant digits.)
b. What does the slope tell you about the number of social science doctorates?
c. Judging from the graph, would you say that the number of social science doctorates is increasing at a faster and faster rate, a slower and slower rate, or a more-or-less constant rate? Why?
d. If r had been equal to 1, could you have drawn the same conclusion as in part (c)? Explain.

33. ▼ **a.** Do the results of Exercises 29 and 31 suggest that the number of engineering doctorates is increasing at a faster and faster rate, a slower and slower rate, or a more-or-less constant rate? Why?
b. If the value of r in Exercise 29 had been equal to 1, would your conclusion change? Explain.
c. If the value of r in Exercise 31 had been equal to 1, would your conclusion change? Explain.

34. ▼ **a.** Do the results of Exercises 30 and 32 suggest that the number of education doctorates is increasing at a faster and faster rate, a slower and slower rate, or at a more-or-less constant rate? Why?
b. If the value of r in Exercise 30 had been equal to 1, would your conclusion change? Explain.
c. If the value of r in Exercise 32 had been equal to 1, would your conclusion change? Explain.

---

[73] See footnote for Exercise 27.

[74] Education includes humanities other than social sciences. Figures are rounded, and 2012 data are estimated. Source: Instituto Nacional de Estadística y Geografía, www.inegi.org.mx.

35. T ▼ ***New York City Housing Costs at the Turn of the Century*** The following table shows the average price of a two-bedroom apartment in downtown New York City from 1994 to 2004 ($t = 0$ represents 1994):[75]

Year t	0	2	4	6	8	10
Price p ($ million)	0.38	0.40	0.60	0.95	1.20	1.60

a. Use technology to obtain the linear regression line and correlation coefficient r, with all coefficients rounded to two decimal places, and plot the regression line and the given points.

b. Does the graph suggest that a nonlinear relationship between t and p would be more appropriate than a linear one? Why or why not?

c. Use technology to obtain the residuals. What can you say about the residuals in support of the claim in part (b)?

36. T ▼ ***Fiber-Optic Connections at the Turn of the Century*** The following table shows the number of fiber-optic cable connections to homes in the United States from 2000 to 2004 ($t = 0$ represents 2000):[76]

Year t	0	1	2	3	4
Connections c (thousands)	0	10	25	65	150

a. Use technology to obtain the linear regression line and correlation coefficient r, with all coefficients rounded to two decimal places, and plot the regression line and the given points.

b. Does the graph suggest that a nonlinear relationship between t and c would be more appropriate than a linear one? Why or why not?

c. Use technology to obtain the residuals. What can you say about the residuals in support of the claim in part (b)?

Communication and Reasoning Exercises

37. Why is the regression line associated with the two points (a, b) and (c, d) the same as the line that passes through both? (Assume that $a \neq c$.)

38. What is the smallest possible sum-of-squares error if the given points happen to lie on a straight line? Why?

39. If the points $(x_1, y_1), (x_2, y_2), \dots, (x_n, y_n)$ lie on a straight line, what can you say about the regression line associated with these points?

40. If all but one of the points $(x_1, y_1), (x_2, y_2), \dots, (x_n, y_n)$ lie on a straight line, must the regression line pass through all but one of these points?

41. ▼ Verify that the regression line for the points $(0, 0)$, $(-a, a)$, and (a, a) has slope 0. What is the value of r? (Assume that $a \neq 0$.)

42. ▼ Verify that the regression line for the points $(0, a)$, $(0, -a)$, and $(a, 0)$ has slope 0. What is the value of r? (Assume that $a \neq 0$.)

43. ▼ Must the regression line pass through at least one of the data points? Illustrate your answer with an example.

44. ▼ Why must care be taken in using mathematical models to extrapolate?

45. ▼ Your friend Imogen tells you that if r for a collection of data points is more than 0.9, then the most appropriate relationship between the variables is a linear one. Explain why she is wrong by referring to one of the exercises.

46. ▼ Your other friend Mervyn tells you that if r for a collection of data points has an absolute value less than 0.8, then the most appropriate relationship between the variables is a quadratic one and not a linear one. Explain why *he* is wrong.

---

[75] Data are rounded, and 2004 figure is an estimate. Source: Miller Samuel/*New York Times*, March 28, 2004, p. RE 11.

[76] Source: Render, Vanderslice & Associates/*New York Times*, October 11, 2004, p. C1.

CHAPTER 1 REVIEW

KEY CONCEPTS

www.WanerMath.com
Go to the Website to find a comprehensive and interactive Web-based summary of Chapter 1.

1.1 Functions from the Numerical, Algebraic, and Graphical Viewpoints

Real-valued function f of a real-valued variable x, domain [p. 48]
Numerically specified function [p. 48]
Graphically specified function [p. 48]
Algebraically defined function [p. 48]
Graph of the function f [p. 48]
Function notation and equation notation [p. 52]
Independent and dependent variables [p. 52]
Piecewise-defined function [p. 53]
Vertical line test [p. 55]
Common types of algebraic functions and their graphs [p. 55]

1.2 Functions and Models

Mathematical model [p. 63]
Analytical model [p. 64]
Curve-fitting model [p. 64]
Cost, revenue, and profit; marginal cost, revenue, and profit; break-even point [p. 66]
Demand, supply, and equilibrium price [p. 69]
Selecting a model [p. 72]
Compound interest [p. 74]
Exponential growth and decay [p. 75]
Algebra of functions (sum, difference, product, quotient) [p. 77]

1.3 Linear Functions and Models

Linear function $f(x) = mx + b$ [p. 85]
Change in q: $\Delta q = q_2 - q_1$ [p. 86]
Slope of a line:

$$m = \frac{\Delta y}{\Delta x} = \frac{\text{Change in } y}{\text{Change in } x} \quad \text{[p. 87]}$$

Interpretations of m [p. 87]
Interpretation of b: y-intercept [p. 87]
Recognizing linear data [p. 88]
Computing the slope of a line [p. 89]
Slopes of horizontal and vertical lines [p. 90]
Computing the y-intercept [p. 90]
Linear modeling [p. 92]
Linear cost [p. 92]
Linear demand [p. 93]
Linear change over time; rate of change; velocity [p. 96]
General linear models [p. 96]

1.4 Linear Regression

Observed and predicted values [p. 104]
Residuals and sum-of-squares error (SSE) [p. 104]
Regression line (least-squares line, best-fit line) [p. 107]
Correlation coefficient; coefficient of determination [p. 109]

REVIEW EXERCISES

In Exercises 1–4, use the graph of the function f to find approximations of the given values.

1.
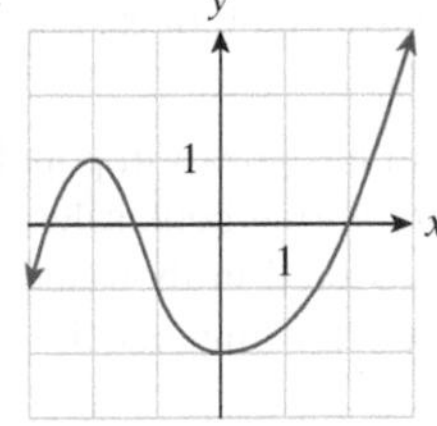

a. $f(-2)$ **b.** $f(0)$ **c.** $f(2)$ **d.** $f(2) - f(-2)$

2.
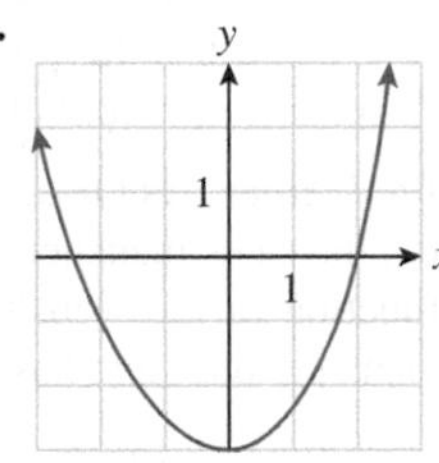

a. $f(-2)$ **b.** $f(0)$ **c.** $f(2)$ **d.** $f(2) - f(-2)$

3.
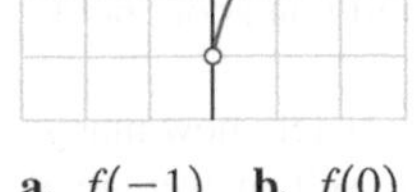

a. $f(-1)$ **b.** $f(0)$ **c.** $f(1)$ **d.** $f(1) - f(-1)$

4.
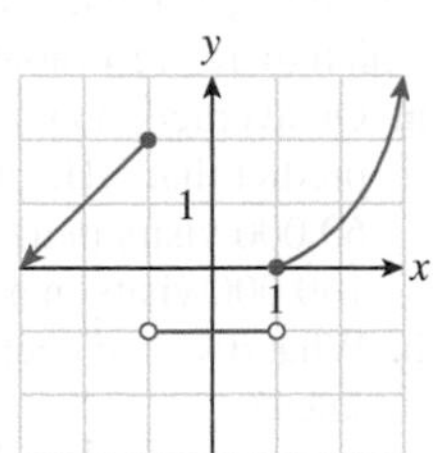
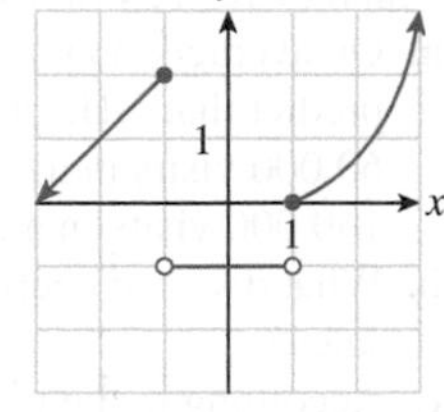

a. $f(-1)$ **b.** $f(0)$ **c.** $f(1)$ **d.** $f(1) - f(-1)$

In Exercises 5–8, graph the given function or equation.

5. $y = -2x + 5$ **6.** $2x - 3y = 12$

7. $y = \begin{cases} \frac{1}{2}x & \text{if } -1 \le x \le 1 \\ x - 1 & \text{if } 1 < x \le 3 \end{cases}$

8. $f(x) = 4x - x^2$ with domain $[0, 4]$

In Exercises 9–14, decide whether the specified values come from a linear, quadratic, exponential, or absolute value function.

9.

x	−2	0	1	2	4
$f(x)$	4	2	1	0	2

10.

x	−2	0	1	2	4
$g(x)$	−5	−3	−2	−1	1

11.

x	−2	0	1	2	4
$h(x)$	1.5	1	0.75	0.5	0

12.

x	−2	0	1	2	4
$k(x)$	0.25	1	2	4	16

13.

x	−2	0	1	2	4
$u(x)$	0	4	3	0	−12

14.

x	-2	0	1	2	4
$w(x)$	32	8	4	2	0.5

In Exercises 15–22, find the equation of the specified line.

15. Through $(3, 2)$ with slope -3
16. Through $(-2, 4)$ with slope -1
17. Through $(1, -3)$ and $(5, 2)$
18. Through $(-1, 2)$ and $(1, 0)$
19. Through $(1, 2)$ parallel to $x - 2y = 2$
20. Through $(-3, 1)$ parallel to $-2x - 4y = 5$
21. With slope 4 crossing $2x - 3y = 6$ at its x-intercept
22. With slope $1/2$ crossing $3x + y = 6$ at its x-intercept

In Exercises 23 and 24, determine which of the given lines better fits the given points.

23. $(-1, 1), (1, 2), (2, 0)$; $y = -\frac{x}{2} + 1$ or $y = -\frac{x}{4} + 1$
24. $(-2, -1), (-1, 1), (0, 1), (1, 2), (2, 4), (3, 3)$; $y = x + 1$ or $y = \frac{x}{2} + 1$

In Exercises 25 and 26, find the line that best fits the given points, and compute the correlation coefficient.

25. $(-1, 1), (1, 2), (2, 0)$
26. $(-2, -1), (-1, 1), (0, 1), (1, 2), (2, 4), (3, 3)$

Applications: OHaganBooks.com [Try the game at www.OHaganBooks.com]

27. ***Website Traffic*** John Sean O'Hagan is CEO of the online bookstore OHaganBooks.com and notices that, since the establishment of the company website 6 years ago ($t = 0$), the number of visitors to the site has grown quite dramatically, as indicated by the following table:

Year t	0	1	2	3	4	5	6
Website Traffic $V(t)$ (visits/day)	100	300	1,000	3,300	10,500	33,600	107,400

   a. Graph the function V as a function of time t. Which of the following types of function seems to fit the curve best: linear, quadratic, or exponential?
   b. Compute the ratios $\frac{V(1)}{V(0)}, \frac{V(2)}{V(1)}, \ldots,$ and $\frac{V(6)}{V(5)}$. What do you notice?
   c. Use the result of part (b) to predict website traffic next year (to the nearest 100).

28. ***Publishing Costs*** Marjory Maureen Duffin is CEO of publisher *Duffin House,* a major supplier of paperback titles to OHaganBooks.com. She notices that publishing costs over the past 5 years have varied considerably as indicated by the following table, which shows the average cost to the company of publishing a paperback novel (t is time in years, and the current year is $t = 5$):

Year t	0	1	2	3	4	5
Cost $C(t)$	\$5.42	\$5.10	\$5.00	\$5.12	\$5.40	\$5.88

   a. Graph the function C as a function of time t. Which of the following types of function seems to fit the curve best: linear, quadratic, or exponential?
   b. Compute the differences $C(1) - C(0)$, $C(2) - C(1), \ldots,$ and $C(5) - C(4)$, rounded to one decimal place. What do you notice?
   c. Use the result of part (b) to predict the cost of producing a paperback novel next year.

29. ***Website Stability*** John O'Hagan is considering upgrading the Web server equipment at OHaganBooks.com because of frequent crashes. The tech services manager has been monitoring the frequency of crashes as a function of website traffic (measured in thousands of visits per day) and has obtained the following model:

$$c(x) = \begin{cases} 0.03x + 2 & \text{if } 0 \le x \le 50 \\ 0.05x + 1 & \text{if } x > 50, \end{cases}$$

where $c(x)$ is the average number of crashes in a day in which there are x thousand visitors.

   a. On average, how many times will the website crash on a day when there are 10,000 visits? 50,000 visits? 100,000 visits?
   b. What does the coefficient 0.03 tell you about the website's stability?
   c. Last Friday, the website went down eight times. Estimate the number of visits that day.

30. ***Book Sales*** As OHaganBooks.com has grown in popularity, the sales manager has been monitoring book sales as a function of the website traffic (measured in thousands of visits per day) and has obtained the following model:

$$s(x) = \begin{cases} 1.55x & \text{if } 0 \le x \le 100 \\ 1.75x - 20 & \text{if } 100 < x \le 250, \end{cases}$$

where $s(x)$ is the average number of books sold in a day in which there are x thousand visitors.

   a. On average, how many books per day does the model predict that OHaganBooks.com will sell when it has 60,000 visits in a day? 100,000 visits in a day? 160,000 visits in a day?
   b. What does the coefficient 1.75 tell you about book sales?
   c. According to the model, approximately how many visitors per day will be needed to sell an average of 300 books per day?

31. ***New Users*** The number of registered users at OHaganBooks.com has increased substantially over the past few months.

The following table shows the number of new users registering each month for the past 6 months:

Month t	1	2	3	4	5	6
New Users (thousands)	12.5	37.5	62.5	72.0	74.5	75.0

a. Which of the following models best approximates the data?

(A) $n(t) = \dfrac{300}{4 + 100(5^{-t})}$

(B) $n(t) = 13.3t + 8.0$

(C) $n(t) = -2.3t^2 + 30.0t - 3.3$

(D) $n(t) = 7(3^{0.5t})$

b. What does each of the above models predict for the number of new users in the next few months: rising, falling, or leveling off?

32. ***Purchases*** OHaganBooks.com has been promoting a number of books published by *Duffin House.* The following table shows the number of books purchased each month from Duffin House for the past 5 months:

Month t	1	2	3	4	5
Purchases (books)	1,330	520	520	1,340	2,980

a. Which of the following models best approximates the data?

(A) $n(t) = \dfrac{3{,}000}{1 + 12(2^{-t})}$

(B) $n(t) = \dfrac{2{,}000}{4.2 - 0.7t}$

(C) $n(t) = 300(1.6^t)$

(D) $n(t) = 100(4.1t^2 - 20.4t + 29.5)$

b. What does each of the above models predict for the number of new users in the next few months: rising, falling, leveling off, or something else?

33. ***Internet Advertising*** Several months ago, John O'Hagan investigated the effect on the popularity of OHaganBooks.com of placing banner ads at well-known Internet portals. The following model was obtained from available data:

$$v(c) = -0.000005c^2 + 0.085c + 1{,}750 \text{ new visits per day},$$

where c is the monthly expenditure on banner ads.

a. John O'Hagan is considering increasing expenditure on banner ads from the current level of \$5,000 to \$6,000 per month. What will be the resulting effect on website popularity?

b. According to the model, would the website popularity continue to grow at the same rate if he continued to raise expenditure on advertising \$1,000 each month? Explain.

c. Does this model give a reasonable prediction of traffic at expenditures larger than \$8,500 per month? Why or why not?

34. ***Production Costs*** Over at *Duffin House,* Marjory Duffin is trying to decide on the size of the print runs for the best-selling new fantasy novel *Larry Plotter and the Simplex Method.* The following model shows a calculation of the total cost to produce a million copies of the novel, based on an analysis of setup and storage costs:

$$c(n) = 0.0008n^2 - 72n + 2{,}000{,}000 \text{ dollars},$$

where n is the print run size (the number of books printed in each run).

a. What would be the effect on cost if the run size was increased from 20,000 to 30,000?

b. Would increasing the run size in further steps of 10,000 result in the same changes in the total cost? Explain.

c. What approximate run size would you recommend that Marjory Duffin use for a minimum cost?

35. ***Internet Advertising*** When OHaganBooks.com actually went ahead and increased Internet advertising from \$5,000 per month to \$6,000 per month (see Exercise 33) it was noticed that the number of new visits increased from an estimated 2,050 per day to 2,100 per day. Use this information to construct a linear model giving the average number v of new visits per day as a function of the monthly advertising expenditure c.

a. What is the model?

b. Based on the model, how many new visits per day could be anticipated if OHaganBooks.com budgets \$7,000 per month for Internet advertising?

c. The goal is to eventually increase the number of new visits to 2,500 per day. Based on the model, how much should be spent on Internet advertising to accomplish this?

36. ***Production Costs*** When *Duffin House* printed a million copies of *Larry Plotter and the Simplex Method* (see Exercise 34), it used print runs of 20,000, which cost the company \$880,000. For the sequel, *Larry Plotter and the Simplex Method, Phase 2,* it used print runs of 40,000, which cost the company \$550,000. Use this information to construct a linear model giving the production cost c as a function of the run size n.

a. What is the model?

b. Based on the model, what would print runs of 25,000 have cost the company?

c. Marjory Duffin has decided to budget \$418,000 for production of the next book in the *Simplex Method* series. Based on the model, how large should the print runs be to accomplish this?

37. ***Recreation*** John O'Hagan has just returned from a sales convention in Puerto Vallarta, Mexico where, to win a bet he made with Marjory Duffin (who was also at the convention), he went bungee jumping at a nearby mountain retreat. The bungee cord he used had the property that a person weighing 70 kilograms would drop a total distance of 74.5 meters, while a 90 kg person would drop 93.5 meters. Express the distance d a jumper drops as a linear function of the jumper's weight w. John OHagan dropped 90 meters. What was his approximate weight?

38. ***Crickets*** The mountain retreat near Puerto Vallarta was so quiet at night that all one could hear was the chirping of the snowy tree crickets. These crickets behave in a rather interesting way: The rate at which they chirp depends linearly on the temperature. Early in the evening, John O'Hagan counted 140 chirps per minute and noticed that the temperature was 80°F. Later in the evening the temperature dropped to 75°F, and the chirping slowed down to 120 chirps per minute. Express the temperature T as a function of the rate of chirping r. The temperature that night dropped to a low of 65°F. At approximately what rate were the crickets chirping at that point?

39. ***Break-Even Analysis*** OHaganBooks.com has recently decided to start selling music albums online through a service it calls *o'Tunes*.[77] Users pay a fee to download an entire music album. Composer royalties and copyright fees cost an average of \$5.50 per album, and the cost of operating and maintaining *o'Tunes* amounts to \$500 per week. The company is currently charging customers \$9.50 per album.
   a. What are the associated (weekly) cost, revenue, and profit functions?
   b. How many albums must be sold per week to make a profit?
   c. If the charge is lowered to \$8.00 per album, how many albums must be sold per week to make a profit?

40. ***Break-Even Analysis*** OHaganBooks.com also generates revenue through its *o'Books* e-book service. Author royalties and copyright fees cost the company an average of \$4 per novel, and the monthly cost of operating and maintaining the service amounts to \$900 per month. The company is currently charging customers \$5.50 per novel.
   a. What are the associated cost, revenue, and profit functions?
   b. How many novels must be sold per month to break even?
   c. If the charge is lowered to \$5.00 per novel, how many books must be sold to break even?

41. ***Demand and Profit*** To generate a profit from its new *o'Tunes* service, OHaganBooks.com needs to know how the demand for music albums depends on the price it charges. During the first week of the service, it was charging \$7 per album and sold 500. Raising the price to \$9.50 had the effect of lowering demand to 300 albums per week.
   a. Use the given data to construct a linear demand equation.
   b. Use the demand equation you constructed in part (a) to estimate the demand if the price were raised to \$12 per album.
   c. Using the information on cost given in Exercise 39, determine which of the three prices (\$7, \$9.50, and \$12) would result in the largest weekly profit, and the size of that profit.

42. ***Demand and Profit*** To generate a profit from its *o'Books* e-book service, OHaganBooks.com needs to know how the demand for novels depends on the price it charges. During the first month of the service, it was charging \$10 per novel and sold 350. Lowering the price to \$5.50 per novel had the effect of increasing demand to 620 novels per month.
   a. Use the given data to construct a linear demand equation.
   b. Use the demand equation you constructed in part (a) to estimate the demand if the price were raised to \$15 per novel.
   c. Using the information on cost given in Exercise 40, determine which of the three prices (\$5.50, \$10, and \$15) would result in the largest profit, and the size of that profit.

43. ***Demand*** OHaganBooks.com has tried selling music albums on *o'Tunes* at a variety of prices, with the following results:

Price	\$8.00	\$8.50	\$10	\$11.50
Demand (weekly sales)	440	380	250	180

   a. Use the given data to obtain a linear regression model of demand.
   b. Use the demand model you constructed in part (a) to estimate the demand if the company charged \$10.50 per album. (Round the answer to the nearest album.)

44. ***Demand*** OHaganBooks.com has tried selling novels through *o'Books* at a variety of prices, with the following results:

Price	\$5.50	\$10	\$11.50	\$12
Demand (monthly sales)	620	350	350	300

   a. Use the given data to obtain a linear regression model of demand.
   b. Use the demand model you constructed in part (a) to estimate the demand if the company charged \$8 per novel. (Round the answer to the nearest novel.)

[77] The (highly original) name was suggested to John O'Hagan by Marjory Duffin over cocktails one evening.

CASE STUDY

Modeling Spending on Mobile Advertising

Es sarawuth/Shutterstock.com; (inset) Odua Images/Shutterstock.com

You are the new director of *Impact Advertising Inc.*'s mobile division, which has enjoyed a steady 0.25% of the worldwide mobile advertising market. You have drawn up an ambitious proposal to expand your division in light of your anticipation that mobile advertising will continue to skyrocket. However, upper management sees things differently and, judging by the following email, does not seem likely to approve the budget for your proposal.

TO: JCheddar@impact.com (J. R. Cheddar)
CC: CVODoylePres@impact.com (C. V. O'Doyle, CEO)
FROM: SGLombardoVP@impact.com (S. G. Lombardo, VP Financial Affairs)
SUBJECT: Your Expansion Proposal
DATE: May 30, 2017

Hi John:

Your proposal reflects exactly the kind of ambitious planning and optimism we like to see in our new upper management personnel. Your presentation last week was most impressive and obviously reflected a great deal of hard work and preparation.

I am in full agreement with you that mobile advertising is on the increase. Indeed, our Market Research department informs me that, based on a regression of the most recently available data, worldwide spending on mobile advertising will continue to grow at a rate of approximately $8.4 billion per year. This translates into $21 million in increased revenues per year for Impact, given our 0.25% market share. This rate of expansion is exactly what our planned 2018 budget anticipates. Your expansion proposal, on the other hand, is based on an increase in sales of almost five times what Market Research is projecting, even though your proposal provides no solid evidence to justify this projection.

At this stage, therefore, I am sorry to say that I am inclined not to approve the funding for your project, although I would be happy to discuss this further with you. I plan to present my final decision on the 2018 budget at next week's divisional meeting.

Regards, Sylvia

Refusing to admit defeat, you contact the Market Research department and request the details of their projections on mobile advertising. They fax you the following information:[78]

Year	2010	2011	2012	2013	2014	2015	2016	2017
Mobile Advertising Spending ($ billion)	2.34	4.02	8.80	15.82	24.91	35.55	47.16	59.67

Regression model: $y = 8.409x - 4.648$ (x = time in years since 2010)
Correlation coefficient: $r = 0.978$

Now you see where the VP got that $8.4 billion figure: The slope of the regression equation is close to 8.4, indicating a rate of increase of about $8.4 billion per year.

[78] Source for data through 2013: eMarketer. Figures from 2014 are forecasts by www.statista.com.

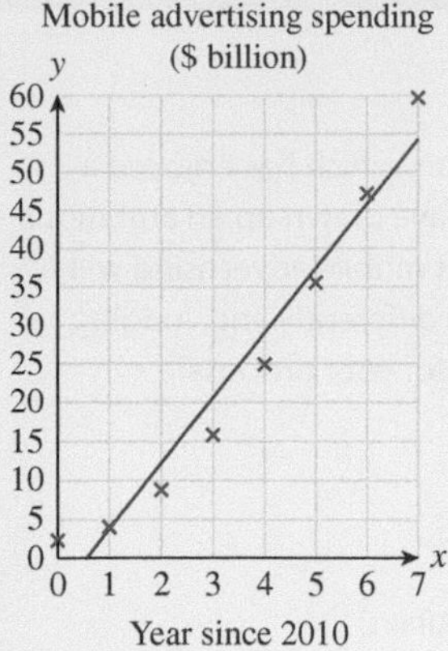

Figure 29

Also, the correlation coefficient is very high, an indication that the linear model fits the data well. In view of this strong evidence, it seems difficult to argue that revenues will increase by significantly more than the projected $8.4 billion per year.

To get a better picture of what's going on, you decide to graph the data together with the regression line in your spreadsheet. What you get is shown in Figure 29. You immediately notice that the data points in Figure 29 seem to suggest a curve, not a straight line. Then again, perhaps the suggestion of a curve is an illusion. Thus, you surmise that there are two possible interpretations of the data:

1. (Your first impression) As a function of time, mobile advertising revenue is nonlinear, and is in fact accelerating (the rate of change is increasing), so a linear model is inappropriate.
2. (Devil's advocate) Mobile advertising revenue *is* a linear function of time; the fact that the points do not lie on the regression line is simply a consequence of random factors that do not reflect a long-term trend, such as world events, mergers and acquisitions, or short-term fluctuations in the economy or the stock market.

You suspect that the VP will probably opt for the second interpretation and discount the graphical evidence of accelerating growth by claiming that it is an illusion—a "statistical fluctuation." That is, of course, a possibility, but you wonder how likely it really is. For the sake of comparison you decide to try a regression based on the simplest nonlinear model you can think of—a quadratic function:

$$y = ax^2 + bx + c.$$

Your spreadsheet allows you to fit such a function with a click of the mouse. The result is the following:

$$y = 0.8842x^2 + 2.2193x + 1.5421 \qquad (x = \text{number of years since 2010})$$
$$r = 0.9995. \qquad \text{See Note.*}$$

* Note that this r is not the linear correlation coefficient we defined in Section 1.4. What this r measures is how closely the quadratic regression model fits the data.

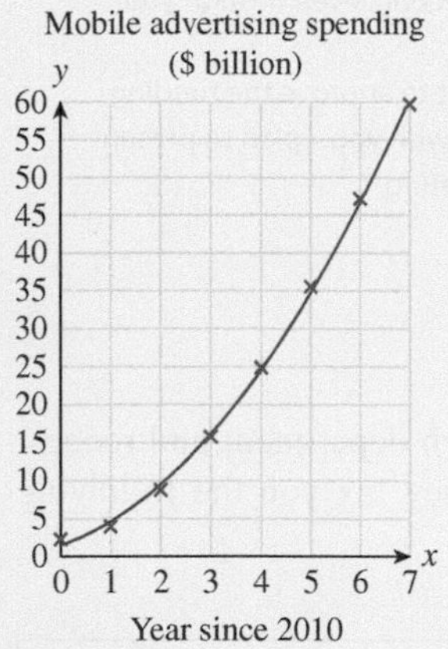

Figure 30

Figure 30 shows the graph of the regression function together with the original data.

Aha! The fit is visually far better, and the correlation coefficient is even higher! Further, the quadratic model predicts 2018 revenue as

$$y = 0.8842(8)^2 + 2.2193(8) + 1.5421 \approx \$75.9 \text{ billion},$$

which is about $16.2 billion above the 2017 spending figure in the table above. Given Impact Advertising's 0.25% market share, this translates into an increase in revenues of $40.5 million, which is about five times the increase predicted by the linear model!

You quickly draft an email to Sylvia Lombardo and are about to click "Send" when you decide, as a precaution, to check with a colleague who is knowledgeable in statistics. He tells you to be cautious: The value of r will always tend to increase if you pass from a linear model to a quadratic one because of the increase in degrees of freedom.† A good way to test whether a quadratic model is more appropriate than a linear one is to compute a statistic called the p-value associated with the coefficient of x^2. A low value of p indicates a high degree of confidence that the coefficient of x^2 cannot be zero (see below). Notice that if the coefficient of x^2 *is* zero, then you have a linear model.

You can, your colleague explains, obtain the p-value using your spreadsheet as follows. (The method we describe here works on all the popular spreadsheets, including Excel, Google Sheets, and OpenOffice Calc.)

† The number of degrees of freedom in a regression model is 1 less than the number of coefficients. For a linear model, it is 1 (there are two coefficients: the slope m and the intercept b), and for a quadratic model, it is 2. For a detailed discussion, consult a text on regression analysis.

First, set up the data in columns, with an extra column for the values of x^2:

	A	B	C
1	y	x	x^2
2	2.34	0	0
3	4.02	1	1
4	8.8	2	4
5	15.82	3	9
6	24.91	4	16
7	35.55	5	25
8	47.16	6	36
9	59.67	7	49

Then highlight a vacant 5 × 3 block (the block E1:G5, say), type the formula `=LINEST(A2:A9,B2:C9,TRUE,TRUE)`, and press Cntl+Shift+Enter (not just Enter!). You will see a table of statistics like the following:

	E	F	G
1	=LINEST(A2:A9,B2:C9,TRUE,TRUE)		
2			
3			
4			
5			

Cntl+Shift+Enter →

	E	F	G
1	0.88422619	2.21934524	1.54208333
2	0.06156915	0.44823018	0.67164014
3	0.99897427	0.7980274	#N/A
4	2434.7854	5	#N/A
5	3101.17515	3.18423869	#N/A

(Notice the coefficients of the quadratic model in the first row.) The p-value is then obtained by the formula `=TDIST(ABS(E1/E2),F4,2)`, which you can compute in any vacant cell. You should get $p \approx 0.0000295$.

Q: *What does p actually measure?*

A: *Roughly speaking,* $1 - p \approx 0.9999705$ gives the degree of confidence you can have (99.99705%) in asserting that the coefficient of x^2 is not zero. (Technically, p is the probability—allowing for random fluctuation in the data—that if the coefficient of x^2 were in fact zero, the ratio E1/E2 could be as large as it is.)

In short, you can go ahead and send your email with almost 100% confidence!

EXERCISES

Suppose you are given the following data for the spending on mobile advertising in a hypothetical country in which Impact Advertising *also has a* 0.25% *share of the market.*

Year	2010	2011	2012	2013	2014	2015	2016
Mobile Advertising Spending ($ billion)	0	0.3	1.5	2.6	3.4	4.3	5.0

1. Obtain a linear regression model and the correlation coefficient r. (Take t to be time in years since 2010.) According to the model, at what rate is spending on mobile advertising increasing in this country? How does this translate to annual revenues for Impact Advertising?
2. Use a spreadsheet or other technology to graph the data together with the best-fit line. Does the graph suggest a quadratic model (parabola)?
3. Test your impression in the preceding exercise by using technology to fit a quadratic function and graphing the resulting curve together with the data. Does the graph suggest that the quadratic model is appropriate?
4. Perform a regression analysis using the quadratic model, and find the associated p-value. What does it tell you about the appropriateness of a quadratic model?

TI-83/84 Plus Technology Guide

Section 1.1

Example 1(a) and (c) (page 50) The total number of iPods sold by **Apple** up to the end of year x can be approximated by $f(x) = -2x^2 + 12x + 36$ million iPods ($0 \le x \le 7$), where $x = 0$ represents 2005. Compute $f(0)$, $f(2)$, $f(4)$, and $f(6)$, and obtain the graph of f.

Solution

You can use the Y= screen to enter an algebraically defined function.

1. Enter the function in the Y= screen, as

```
Y1=-2X^2+12X+36
```

or

Y_1=-2X^2+12X+36

(See Chapter 0 for a discussion of technology formulas.)

2. To evaluate $f(0)$, for example, enter $Y_1(0)$ in the Home screen to evaluate the function Y_1 at 0. Alternatively, you can use the table feature: After entering the function under Y_1, press [2ND] [TBLSET], and set `Indpnt` to `Ask`. (You do this once and for all; it will permit you to specify values for x in the table screen.) Then press [2ND] [TABLE], and you will be able to evaluate the function at several values of x. Below (top) is a table showing the values requested:

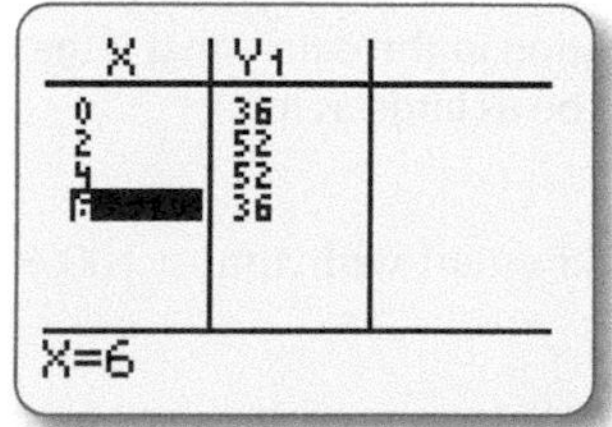

X	Y1
0	36
2	52
4	52
6	36

X=6

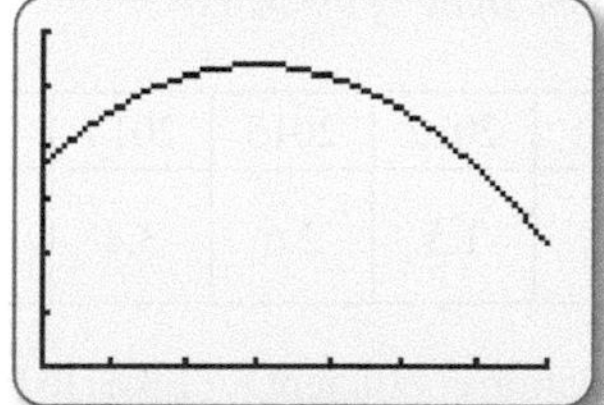

3. To obtain the graph above, press [WINDOW], set Xmin = 0, Xmax = 7 (the range of x-values we are interested in), Ymin = 0, and Ymax = 60 (we estimated Ymin and Ymax from the corresponding set of y-values in the table), and press [GRAPH] to obtain the curve. Alternatively, you can avoid having to estimate Ymin and Ymax by pressing ZoomFit ([ZOOM] [0]), which automatically sets Ymin and Ymax to the smallest and greatest values of y, respectively, in the specified range for x.

Example 2 (page 53) The accumulated sales of **Nintendo** Wii game consoles from December 2006 to December 2014 can be approximated by the following function of time t in years ($t = 0$ represents December 2006):

$$f(t) = \begin{cases} 5t^2 + 2t + 2 & \text{if } 0 \le t \le 3 \\ -2t^2 + 33t - 28 & \text{if } 3 < t \le 8 \end{cases}$$

million consoles.

Obtain a table showing the values $f(t)$ for $t = 0, \ldots, 8$, and obtain the graph of f.

Solution

You can enter a piecewise-defined function using the logical inequality operators $<$, $>$, $\le$, and $\ge$, which are found by pressing [2ND] [TEST]:

1. Enter the function f in the Y=screen as

```
Y1=(X≤3)*(5X^2+2X+2)+(X>3)*(-2X^2+33X-28)
```

When x is less than or equal to 3, the logical expression `(X≤3)` evaluates to 1 because it is true, and the expression `(X>3)` evaluates to 0 because it is false. The value of the function is therefore given by the expression `(5X^2+2X+2)`. When x is greater than 3, the expression `(X≤3)` evaluates to 0, while the expression `(X>3)` evaluates to 1, so the value of the function is given by the expression `(-2X^2+33X-28)`.

2. As in Example 1, use the Table feature to compute several values of the function at once by pressing [2ND] [TABLE]:

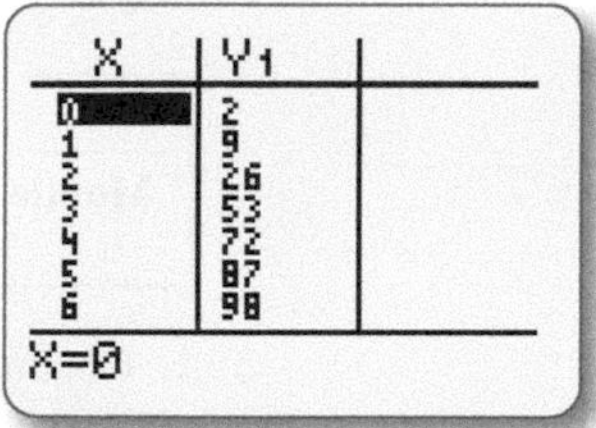

X	Y1
0	2
1	9
2	26
3	53
4	72
5	87
6	98

X=0

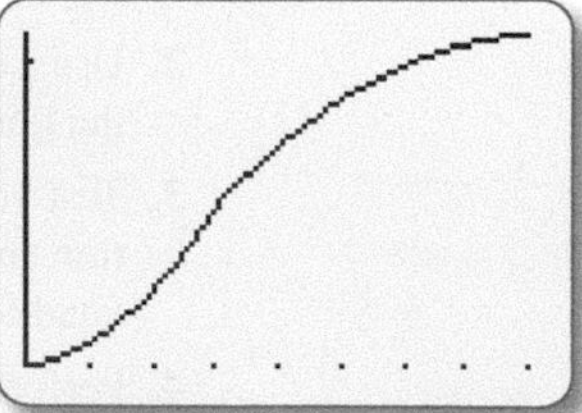

3. To obtain the graph above, we proceed as in Example 1: Press WINDOW, set Xmin = 0, Xmax = 8 (the range of x-values we are interested in), and press ZOOM 10 ("Zoomfit") to automatically compute Ymin and Ymax and plot the graph.

Section 1.2

Example 5(a) (page 72) The following table shows some annual revenues, in billions of dollars, earned by **Amazon** from 2004 through 2014:

Year	2004	2006	2008	2010	2012	2014
Revenue ($ billion)	7	11	19	34	61	89

Take t to be the number of years since 2004, and consider the following three models:

(1) $r(t) = 8t - 3$ Linear model

(2) $r(t) = 0.9t^2 - 0.6t + 7$ Quadratic model

(3) $r(t) = 7(1.3^t)$. Exponential model

a. Which two models fit the data significantly better than the third?

b. If you extrapolate the models you selected in part (a) back to $t = -4$, what do you find? (Amazon's revenue in 2000 was $2.8 billion and subsequently increased each year.).

Solution

1. First enter the actual revenue data in the stat list editor (STAT EDIT) with the values of t in L_1 and the values of $s(t)$ in L_2.

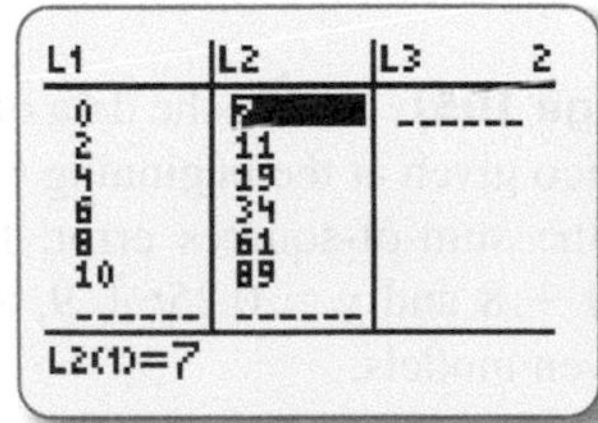

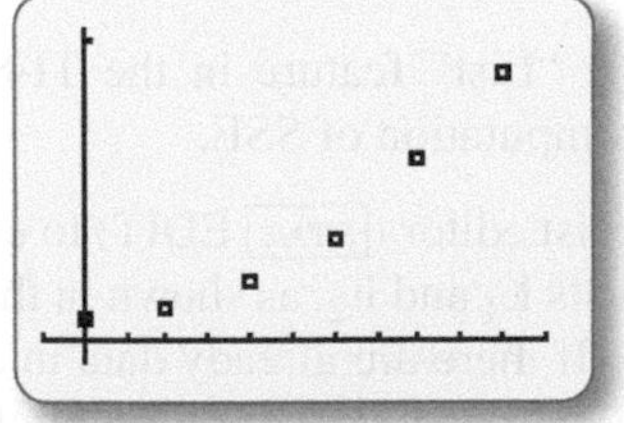

2. Now go to the Y= window, and turn `Plot1` on by selecting it and pressing ENTER. (You can also turn it on in the 2ND STAT PLOT screen.) Then press ZoomStat (ZOOM 9) to obtain a plot of the points (figure above).

3. To see any of the three curves plotted along with the points, enter its formula in the Y= screen (for instance, `Y1=0.9X²-0.6X+7` for the second model), and press GRAPH (figure on top below).

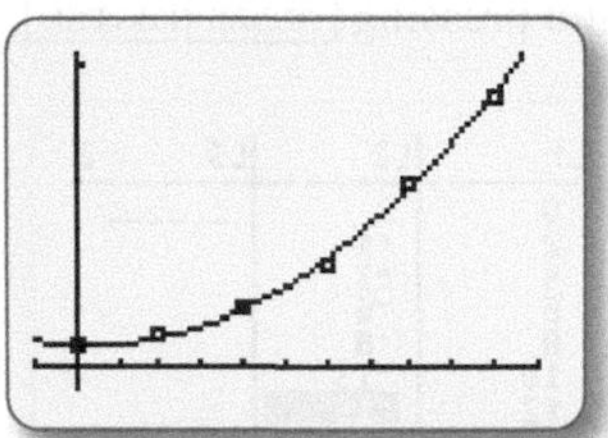

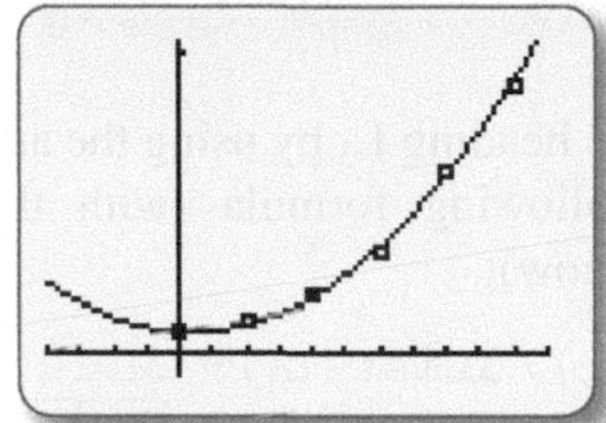

4. To see the extrapolation of the curve back to 2000, just change Xmin to −4 (in the WINDOW screen), and press GRAPH again (figure above).

5. Now change Y_1 to see similar graphs for the remaining curves.

6. When you are done, turn `Plot1` off again so that the points you entered do not show up in other graphs.

Section 1.3

Example 1 (page 88) Which of the following two tables gives the values of a linear function? What is the formula for that function?

x	0	2	4	6	8	10	12
$f(x)$	3	−1	−3	−6	−8	−13	−15

x	0	2	4	6	8	10	12
$g(x)$	3	−1	−5	−9	−13	−17	−21

Solution

We can use the "List" feature in the TI-83/84 Plus to automatically compute the successive quotients $m = \Delta y/\Delta x$ for either f or g as follows:

1. Use the stat list editor (STAT EDIT) to enter the values of x and $f(x)$ in the first two columns, called L_1 and L_2, as shown in the screenshot below. (If there are already data in a column you want to use, you can clear it by highlighting the column heading (e.g., L_1) using the arrow key, and pressing CLEAR ENTER.)

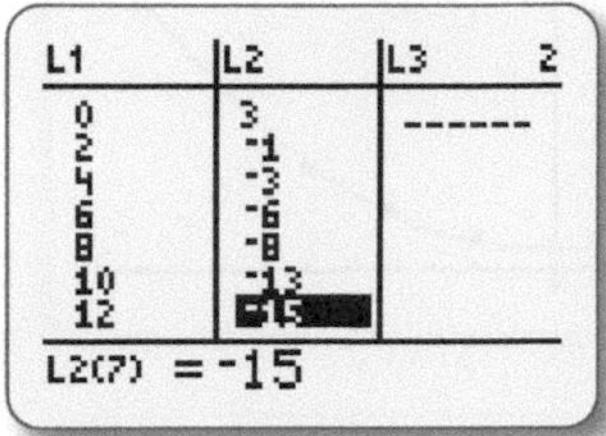

2. Highlight the heading L_3 by using the arrow keys, and enter the following formula (with the quotes, as explained below):

```
"ΔList(L2)/ΔList(L1)"
```

ΔList is found under 2ND LIST OPS. L_1 is 2ND 1

The "ΔList" function computes the differences between successive elements of a list, returning a list with one less element. The formula above then computes the quotients $\Delta y/\Delta x$ in the list L_3 as shown in the following screenshot. As you can see in the third column, $f(x)$ is not linear.

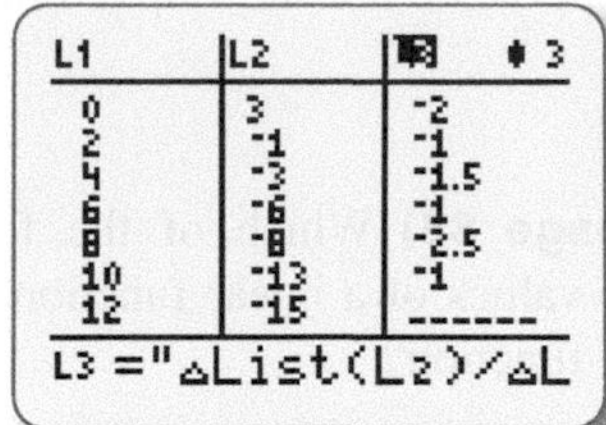

3. To redo the computation for $g(x)$, all you need to do is edit the values of L_2 in the stat list editor. By putting quotes around the formula we used for L_3, we told the calculator to remember the formula, so it automatically recalculates the values.

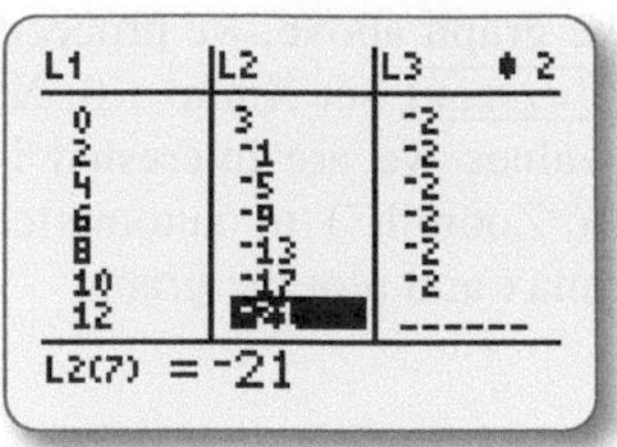

Example 2(a) (page 91) Find the equation of the line through the points $(1, 2)$ and $(3, -1)$.

Solution

1. Enter the coordinates of the given points in the stat list editor (STAT EDIT) with the values of x in L_1 and the values of y in L_2.

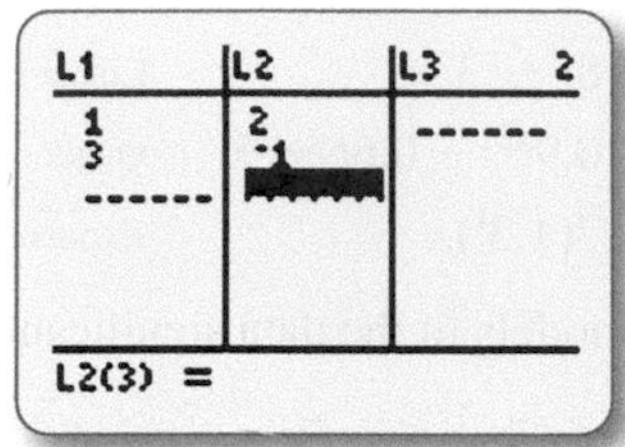

2. To compute the slope, enter the following formula in the Home screen:

```
(L2(2)-L2(1))/(L1(2)-L1(1))→M
```

L_1 and L_2 are under 2ND LIST and the arrow is STO

3. Then, to compute the y-intercept, enter

```
L2(1)-M*L1(1)
```

Section 1.4

Example 1 (page 105) Using the data on the per capita GDP in Mexico given at the beginning of Section 1.4, compute SSE, the sum-of-squares error, for the linear models $y = 0.5t + 8$ and $y = 0.25t + 9$, and graph the data with the given models.

Solution

We can use the "List" feature in the TI-83/84 Plus to automate the computation of SSE.

1. Use the stat list editor (STAT EDIT) to enter the given data in the lists L_1 and L_2, as shown in the first screenshot below. (If there are already data in a column you want to use, you can clear it by highlighting the column heading (e.g., L_1) using the arrow key, and pressing CLEAR ENTER.)

2. To compute the predicted values, highlight the heading L_3 using the arrow keys, and enter the following formula for the predicted values (figure on the top below):

`0.5*L`$_1$`+8` L_1 is [2ND] [1]

Pressing [ENTER] again will fill column 3 with the predicted values (below bottom). Note that only seven of the eight data points can be seen on the screen at one time.

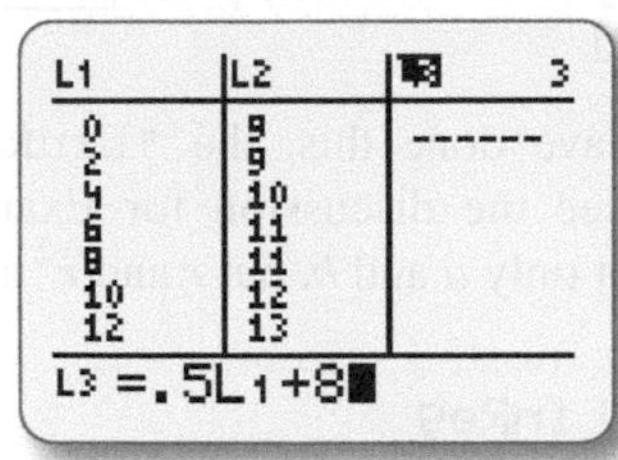

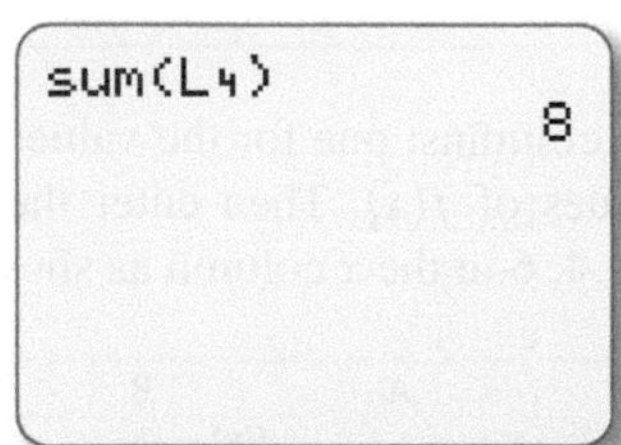

3. Highlight the heading L_4, and enter the following formula (including the quotes):

`"(L`$_2$`-L`$_3$`)^2"`   Squaring the residuals

4. Pressing [ENTER] will fill L_4 with the squares of the residuals. (Putting quotes around the formula will allow us to easily check the second model, as we shall see.)

5. To compute SSE, the sum of the entries in L_4, go to the Home screen, and enter `sum(L`$_4$`)`. (See below; "sum" is under [2ND] [LIST] [MATH].)

sum(L4)
8

6. To check the second model, go back to the List screen, highlight the heading L_3, enter the formula for the second model, `0.25*L`$_1$`+9`, and press [ENTER]. Because we put quotes around the formula for the residuals in L_4, the TI-83/84 Plus will remember the formula and automatically recalculate the values (below top). On the Home screen we can again calculate `sum(L`$_4$`)` to get SSE for the second model (below bottom).

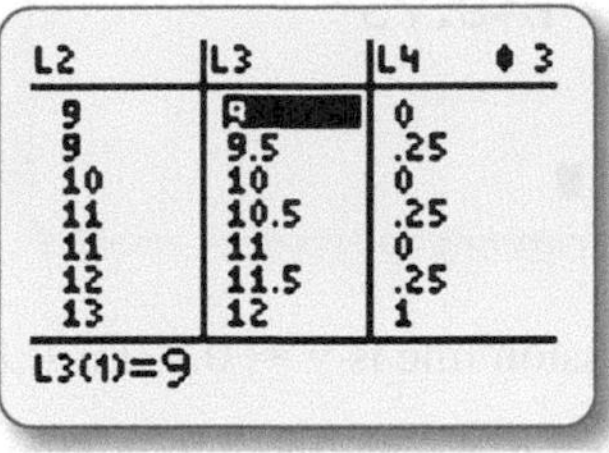

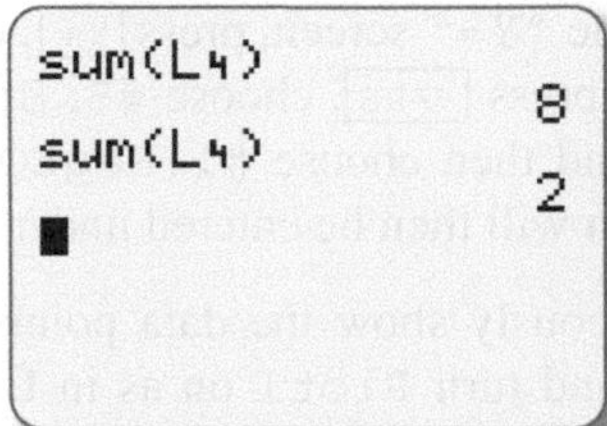

The second model gives a much smaller SSE, so it is the better fit.

7. You can also use the TI-83/84 Plus to plot both the original data points and the two lines (see below). Turn `Plot1` on in the STAT PLOT window, obtained by pressing [2ND] [STAT PLOT]. To show the lines, enter them in the "Y=" screen as usual. To obtain a convenient window showing all the points and the lines, press [ZOOM], and choose 9: `ZoomStat`.

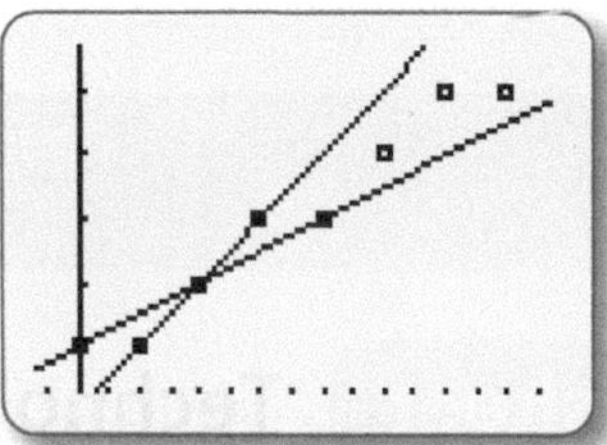

Example 2 (page 107) Use the data on the per capita GDP in Mexico to find the best-fit linear model.

Solution

1. Enter the data in the TI-83/84 Plus using the List feature, putting the x-coordinates in L_1 and the y-coordinates in L_2, just as in Example 1.

2. Press [STAT], select CALC, and choose `#4: LinReg(ax+b)`. Pressing [ENTER] will cause the

equation of the regression line to be displayed in the Home screen:

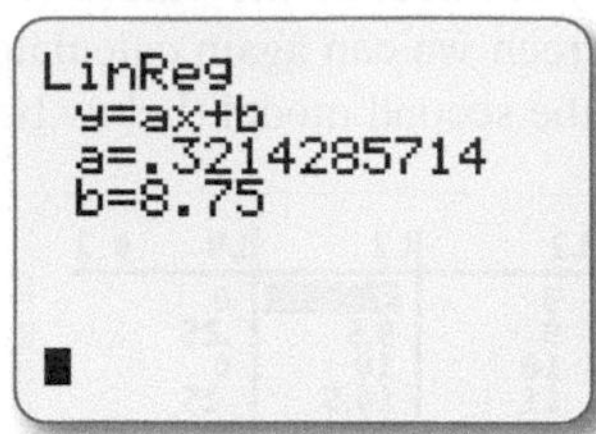

So the regression line is $y \approx 0.321x + 8.75$.

3. To graph the regression line without having to enter it by hand in the "Y=" screen, press [Y=], clear the contents of Y_1, press [VARS], choose #5: Statistics, select EQ, and then choose #1:RegEQ. The regression equation will then be entered under Y_1.

4. To simultaneously show the data points, press [2ND] [STATPLOT], and turn Plot1 on as in Example 1. To obtain a convenient window showing all the points and the line (see below), press [ZOOM] and choose #9: ZoomStat.

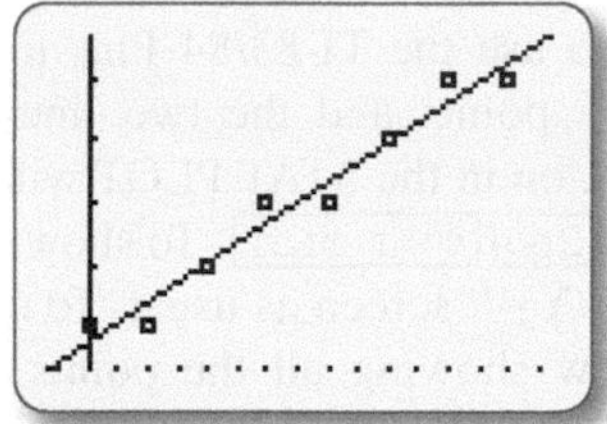

Example 3 (page 110) Find the correlation coefficient for the data in Example 2.

Solution

To find the correlation coefficient using a TI-83/84 Plus, you need to tell the calculator to show you the coefficient at the same time that it shows you the regression line.

1. Press [2ND] [CATALOG], and select DiagnosticOn from the list. The command will be pasted to the Home screen, and you should then press [ENTER] to execute the command.

2. Once you have done this, the "LinReg(ax+b)" command (see the discussion for Example 2) will show you not only a and b, but r and r^2 as well:

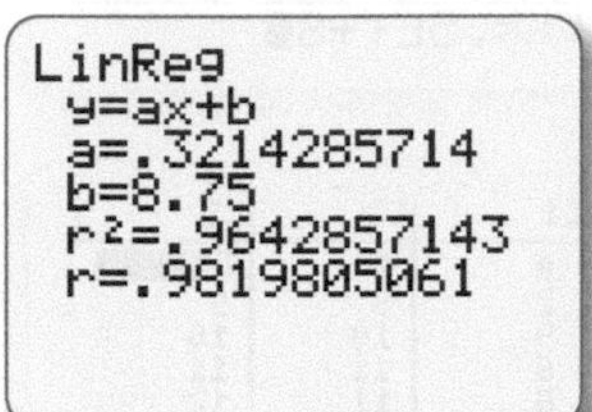

Spreadsheet Technology Guide

Section 1.1

Example 1(a) and (c) (page 50) The total number of iPods sold by **Apple** up to the end of year x can be approximated by $f(x) = -2x^2 + 12x + 36$ million iPods ($0 \le x \le 7$), where $x = 0$ represents 2005. Compute $f(0)$, $f(2)$, $f(4)$, and $f(6)$, and obtain the graph of f.

Solution

To create a table of values of f using a spreadsheet, do the following:

1. Set up two columns: one for the values of x and one for the values of $f(x)$. Then enter the sequence of values 0, 2, 4, 6 in the x column as shown below.

	A	B
1	x	f(x)
2	0	
3	2	
4	4	
5	6	

2. Now we enter a formula for $f(x)$ in cell B2 (below). The technology formula is `-2x^2+12x+36`. To use this formula in a spreadsheet, we modify it slightly:

`=-2*A2^2+12*A2+36` Spreadsheet version of tech formula

Notice that we have preceded the Excel formula by an equals sign (=) and replaced each occurrence of x by the name of the cell holding the value of x (cell A2 in this case).

	A	B	C
1	x	f(x)	
2	0	=-2*A2^2+12*A2+36	
3	2		
4	4		
5	6		

Note Instead of typing in the name of the cell "A2" each time, you can simply click on the cell A2, and "A2" will be automatically inserted. ■

3. Now highlight cell B2, and drag the **fill handle** (the little square at the lower right-hand corner of the selection) down until you reach row 5, as shown below on the top, to obtain the result shown on the bottom.

	A	B	C
1	x	f(x)	
2	0	=-2*A2^2+12*A2+36	
3	2		
4	4		
5	6		

	A	B
1	x	f(x)
2	0	36
3	2	52
4	4	52
5	6	36

4. To graph the data, highlight A1 through B5, and insert a scatter chart (the exact method of doing this depends on the specific version of the spreadsheet program). When choosing the style of the chart, choose a style that shows points connected by lines (if possible) to obtain a graph something like the following:

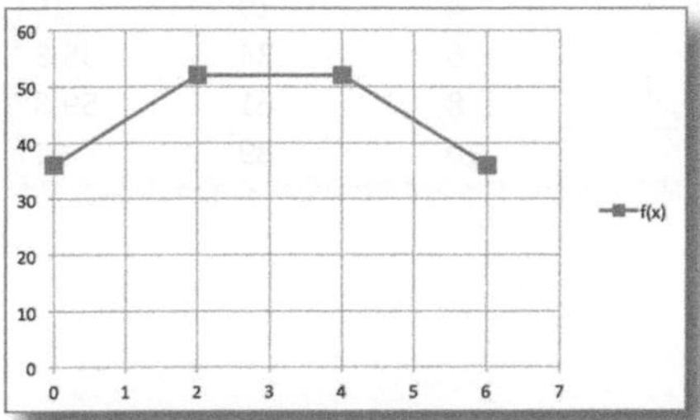

Example 2 (page 53) The accumulated sales of **Nintendo** Wii game consoles from December 2006 to December 2014 can be approximated by the following function of time t in years ($t = 0$ represents December 2006):

$$f(t) = \begin{cases} 5t^2 + 2t + 2 & \text{if } 0 \le t \le 3 \\ -2t^2 + 33t - 28 & \text{if } 3 < t \le 8 \end{cases} \quad \text{million consoles.}$$

Obtain a table showing the values $f(t)$ for $t = 0, \ldots, 8$, and also obtain the graph of f.

Solution

You can generate a table of values of $f(t)$ for $t = 0, 1, \ldots, 8$ as follows:

1. Set up two columns—one for the values of t and one for the values of $f(t)$—and enter the values $0, 1, \ldots, 8$ in the t column as shown below.
2. We must now enter the formula for f in cell B2. The following formula defines the function n in Excel:

```
=(A2<=3)*(5*A2^2+2*A2+2)+(A2>3)*
        (-2*A2^2+33*A2-28)
```

When x is less than or equal to 3, the logical expression `(x<=3)` evaluates to 1 because it is true, and the expression `(x>3)` evaluates to 0 because it is false. The value of the function is therefore given by the expression `(5*x^2+2*x+2)`. When x is greater than 3, the expression `(x<=3)` evaluates to 0 while the expression `(x>3)` evaluates to 1, so the value of the function is given by the expression `(-2*x^2+33*x-28)`. We therefore enter the formula

```
=(A2<=3)*(5*A2^2+2*A2+2)+(A2>3)*
        (-2*A2^2+33*A2-28)
```

in cell B2 and then copy down to cell B10 (first figure below) to obtain the result shown in the second figure:

	A	B	C	D	E	F
1	t	f(t)				
2	0	=(A2<=3)*(5*A2^2+2*A2+2)+(A2>3)*(-2*A2^2+33*A2-28)				
3	1					
4	2					
5	3					
6	4					
7	5					
8	6					
9	7					
10	8					

	A	B
1	t	f(t)
2	0	2
3	1	9
4	2	26
5	3	53
6	4	72
7	5	87
8	6	98
9	7	105
10	8	108

3. To graph the data, highlight A1 through B10, and insert a scatter chart as in Example 1 to obtain the result shown below:

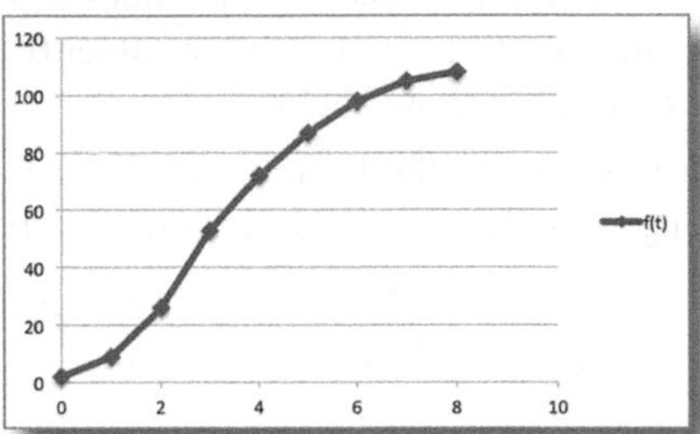

Section 1.2

Example 5(a) (page 72) The following table shows some annual revenues, in billions of dollars, earned by **Amazon** from 2004 through 2014:

Year	2004	2006	2008	2010	2012	2014
Revenue ($ billion)	7	11	19	34	61	89

Take t to be the number of years since 2004, and consider the following three models:

(1) $r(t) = 8t - 3$ Linear model

(2) $r(t) = 0.9t^2 - 0.6t + 7$ Quadratic model

(3) $r(t) = 7(1.3^t)$. Exponential model

a. Which two models fit the data significantly better than the third?

b. If you extrapolate the models you selected in part (a) back to $t = -4$, what do you find? (Amazon's revenue in 2000 was $2.8 billion and subsequently increased each year.).

Solution

1. First create a scatter plot of the given data by tabulating the data as shown below, highlighting cells A1 through B7 and inserting a scatter chart:

	A	B
1	t	Revenue
2	0	7
3	2	11
4	4	19
5	6	34
6	8	61
7	10	89

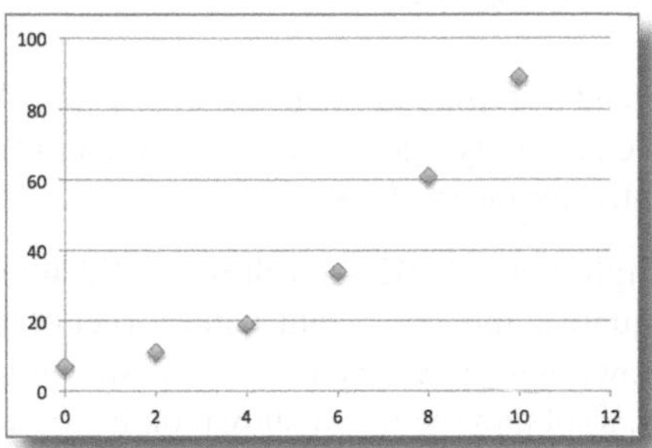

2. In column C, use the formula for the model you are interested in seeing; for example, model (2): `=0.9*A2^2-0.6*A2+7`.

	A	B	C	D
1	t	Revenue		
2	0	7	=0.9*A2^2-0.6*A2+7	
3	2	11		
4	4	19		
5	6	34		
6	8	61		
7	10	89		

	A	B	C
1	t	Revenue	
2	0	7	7
3	2	11	9.4
4	4	19	19
5	6	34	35.8
6	8	61	59.8
7	10	89	91

3. To adjust the graph to include the graph of the model you have added, you need to change the graph data from $A$1:$B$7 to $A$1:$C$7 to include column C. In Excel you can obtain this by right-clicking on the graph to select "Source Data". In OpenOffice, double-click on the graph, and then right-click it to choose "Data Ranges". In Excel you can also click once on the graph—the effect will be to outline the data you have graphed in columns A and B—and then use the fill handle at the bottom of column B to extend the selection to column C. The graph will now include markers showing the values of both the actual sales and the model you inserted in column C.
4. Right-click on any of the markers corresponding to column B in the graph (in OpenOffice you would first double-click on the graph), select "Format data series" to add lines connecting the points and remove the markers. The effect will be as shown below, with the model represented by a curve and the actual data points represented by dots (below):

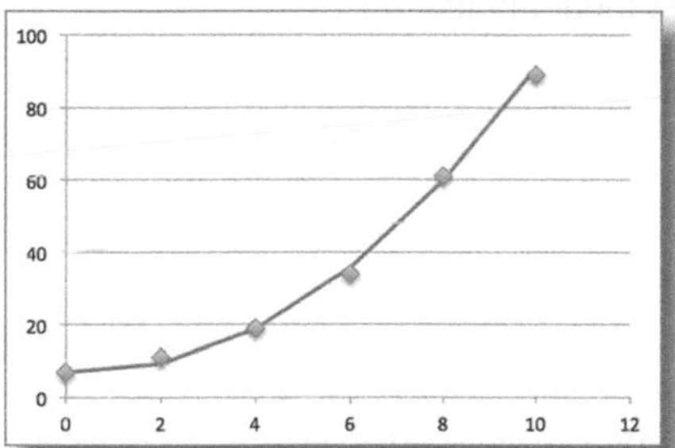

	A	B	C
1	t	Revenue	
2	-4		
3	-3		
4	-2		
5	-1		
6	0	7	7
7	2	11	9.4
8	4	19	19
9	6	34	35.8
10	8	61	59.8
11	10	89	91

5. To see the extrapolation of the curve back to 2000, first move the data down by selecting A2 through C7, clicking in the middle of the lower edge of the selected region, and dragging the data down four cells. Then add the values −1, −2, −3, and −4 in column A as shown above. The values of $r(t)$ may automatically be computed in column C as you type, depending on the spreadsheet. If not, you will need to copy the formula in cell C6 up to C2. (Do not touch column B, as that contains the observed data starting at $t = 0$ only.) The result is shown below. Now change the graph data to $A$1:$C$11 by one of the techniques shown in Step 3 to obtain the graph shown below.

	A	B	C
1	t	Revenue	
2	-4		23.8
3	-3		16.9
4	-2		11.8
5	-1		8.5
6	0	7	7
7	2	11	9.4
8	4	19	19
9	6	34	35.8
10	8	61	59.8
11	10	89	91

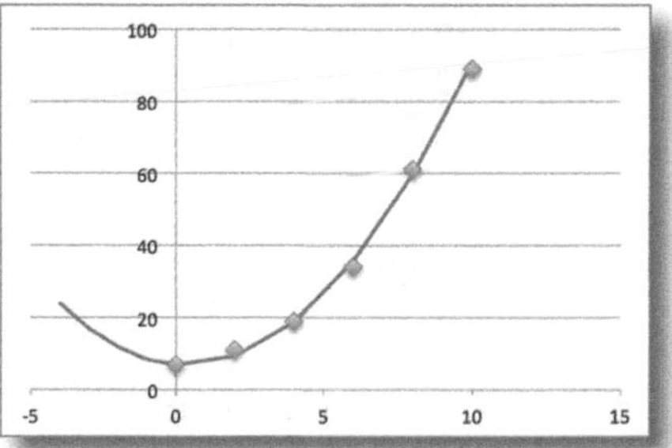

6. To see the plots for the remaining curves, change the formula in column C.

Section 1.3

Example 1 (page 88) Which of the following two tables gives the values of a linear function? What is the formula for that function?

x	0	2	4	6	8	10	12
$f(x)$	3	−1	−3	−6	−8	−13	−15

x	0	2	4	6	8	10	12
$g(x)$	3	−1	−5	−9	−13	−17	−21

Solution

1. The following worksheet shows how you can compute the successive quotients $m = \Delta y/\Delta x$ and hence check whether a given set of data shows a linear relationship, in which case all the quotients will be the same. (The

shading indicates that the formula is to be copied down only as far as cell C7. Why not cell C8?)

	A	B	C
1	x	f(x)	m
2	0	3	=(B3-B2)/(A3-A2)
3	2	-1	
4	4	-3	
5	6	-6	
6	8	-8	
7	10	-13	
8	12	-15	

2. Here are the results for both f and g:

	A	B	C
1	x	f(x)	m
2	0	3	-2
3	2	-1	-1
4	4	-3	-1.5
5	6	-6	-1
6	8	-8	-2.5
7	10	-13	-1
8	12	-15	

	A	B	C
1	x	g(x)	m
2	0	3	-2
3	2	-1	-2
4	4	-5	-2
5	6	-9	-2
6	8	-13	-2
7	10	-17	-2
8	12	-21	

Example 2(a) (page 91) Find the equation of the line through the points $(1, 2)$ and $(3, -1)$.

Solution

1. Enter the x- and y-coordinates in columns A and B, as shown below.

	A	B
1	x	y
2	1	2
3	3	-1

	A	B	C	D
1	x	y	m	b
2	1	2	=(B3-B2)/(A3-A2)	=B2-C2*A2
3	3	-1	Slope	Intercept

2. Add the headings m and b in C1 and D1, and then the formulas for the slope and intercept in C2 and D2, as shown above. The result will be as shown below:

	A	B	C	D
1	x	y	m	b
2	1	2	-1.5	3.5
3	3	-1	Slope	Intercept

Section 1.4

Example 1 (page 105) Using the data on the per capita GDP in Mexico given at the beginning of Section 1.4, compute SSE, the sum-of-squares error, for the linear models $y = 0.5t + 8$ and $y = 0.25t + 9$, and graph the data with the given models.

Solution

1. Begin by setting up your worksheet with the observed data in two columns, t and y, and the predicted data for the first model in the third.

	A	B	C	D	E	F
1	t	y (Observed)	y (Predicted)		m	b
2	0	9	=$E$2*A2+$F$2		0.5	8
3	2	9				
4	4	10				
5	6	11				
6	8	11				
7	10	12				
8	12	13				
9	14	13				

2. Notice that, instead of using the numerical equation for the first model in column C, we used absolute references to the cells containing the slope m and the intercept b. This way, we can switch from one linear model to the next by changing only m and b in cells E2 and F2. (We have deliberately left column D empty in anticipation of the next step.)

3. In column D we compute the squares of the residuals using the Excel formula `=(B2-C2)^2`.

	A	B	C	D	E	F
1	t	y (Observed)	y (Predicted)	Residual^2	m	b
2	0	9	8	=(B2-C2)^2	0.5	8
3	2	9	9			
4	4	10	10			
5	6	11	11			
6	8	11	12			
7	10	12	13			
8	12	13	14			
9	14	13	15			

4. We now compute SSE in cell F4 by summing the entries in column D.

	A	B	C	D	E	F
1	t	y (Observed)	y (Predicted)	Residual^2	m	b
2	0	9	8	1	0.5	8
3	2	9	9	0		
4	4	10	10	0	SSE:	=SUM(D2:D9)
5	6	11	11	0		
6	8	11	12	1		
7	10	12	13	1		
8	12	13	14	1		
9	14	13	15	4		

5. Here is the completed spreadsheet:

	A	B	C	D	E	F
1	t	y (Observed)	y (Predicted)	Residual^2	m	b
2	0	9	8	1	0.5	8
3	2	9	9	0		
4	4	10	10	0	SSE:	8
5	6	11	11	0		
6	8	11	12	1		
7	10	12	13	1		
8	12	13	14	1		
9	14	13	15	4		

6. Changing m to 0.25 and b to 9 gives the sum of squares error for the second model, SSE $= 2$.

	A	B	C	D	E	F
1	t	y (Observed)	y (Predicted)	Residual^2	m	b
2	0	9	9	0	0.25	9
3	2	9	9.5	0.25		
4	4	10	10	0	SSE:	2
5	6	11	10.5	0.25		
6	8	11	11	0		
7	10	12	11.5	0.25		
8	12	13	12	1		
9	14	13	12.5	0.25		

7. To plot both the original data points and each of the two lines, use a scatter plot to graph the data in columns A through C in each of the last two worksheets above.

$y = 0.5t + 8$ $\qquad\qquad$ $y = 0.25t + 9$

Example 2 (page 107) Use the data on the per capita GDP in Mexico to find the best-fit linear model.

Solution

Here are two spreadsheet shortcuts for linear regression; one is graphical, and one is based on a spreadsheet formula.

Using a Trendline

1. Start with the original data, and insert a scatter plot (below).

	A	B
1	t	y
2	0	9
3	2	9
4	4	10
5	6	11
6	8	11
7	10	12
8	12	13
9	14	13

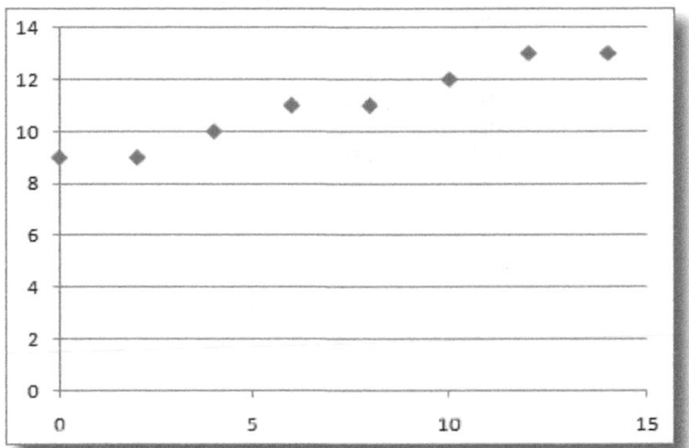

2. Insert a linear trendline, choosing the option to display the equation on the chart. The method for doing so varies from spreadsheet to spreadsheet.[79] In Excel you can right-click on one of the points in the graph and choose "Add Trendline". (In OpenOffice you would first double-click on the graph.) Then, under "Trendline Options", select "Display Equation on chart". The procedure for OpenOffice is almost identical, but you first need to double-click on the graph. The result is shown below.

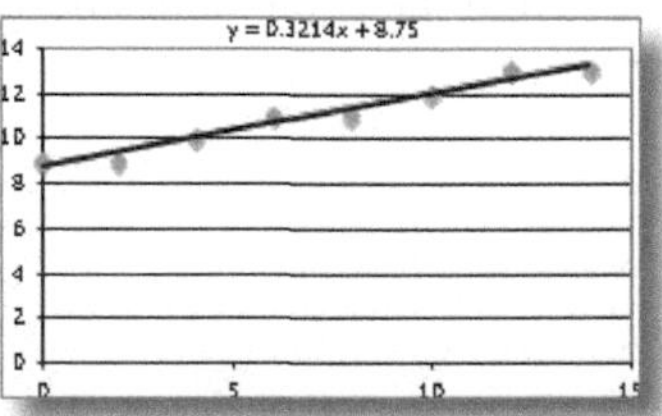

[79] At the time of this writing, Google Sheets has no trendline feature for its spreadsheet, so you would need to use the formula method.

Using a Formula

1. Enter your data as above, and select a block of unused cells two wide and one tall; for example, C2:D2. Then enter the formula

```
=LINEST(B2:B9,A2:A9)
```

as shown below. Then press Control+Shift+Enter. The result should appear as in the bottom figure below, with m and b appearing in cells C2 and D2 as shown:

	A	B	C	D
1	t	y	m	b
2	0	9	=LINEST(B2:B9,A2:A9)	
3	2	9		
4	4	10		
5	6	11		
6	8	11		
7	10	12		
8	12	13		
9	14	13		

	A	B	C	D
1	t	y	m	b
2	0	9	0.32143	8.75
3	2	9		
4	4	10		
5	6	11		
6	8	11		
7	10	12		
8	12	13		
9	14	13		

Example 3 (page 110) Find the correlation coefficient for the data in Example 2.

Solution

1. When you add a trendline to a chart, you can select the option "Display R-squared value on chart" to show the value of r^2 on the chart. (It is common to examine r^2, which takes on values between 0 and 1, instead of r.)

2. Alternatively, the LINEST function we used in Example 2 can be used to display quite a few statistics about a best-fit line, including r^2. Instead of selecting a block of cells two wide and one tall, as we did in Example 2, we select one that is two wide and *five* tall. We now enter the requisite LINEST formula with two additional arguments set to "TRUE" as shown, and press Control+Shift+Enter.

	A	B	C	D	E
1	t	y	m	b	
2	0	9	=LINEST(B2:B9,A2:A9,TRUE,TRUE)		
3	2	9			
4	4	10			
5	6	11			
6	8	11			
7	10	12			
8	12	13			
9	14	13			

	A	B	C	D
1	t	y	m	b
2	0	9	0.32143	8.75
3	2	9	0.02525	0.21129
4	4	10	0.96429	0.32733
5	6	11	162	6
6	8	11	17.3571	0.64286
7	10	12		
8	12	13		
9	14	13		

The values of m and b appear in cells C2 and D2 as before, and the value of r^2 appears in cell C4. (Among the other numbers shown is SSE in cell D6. For the meanings of the remaining numbers shown, do a web search for "LINEST"; you will see numerous articles, including many that explain all the terms. A good course in statistics wouldn't hurt either.)

2 NONLINEAR FUNCTIONS AND MODELS

www.WanerMath.com

At the Website, in addition to the resources listed in the Preface, you will find:

The following extra topics:

- Inverse Functions
- Using and Deriving Algebraic Properties of Logarithms

CASE STUDY

Checking up on Malthus

In 1798 Thomas R. Malthus (1766–1834) published an influential pamphlet, later expanded into a book, titled *An Essay on the Principle of Population as It Affects the Future Improvement of Society.* One of his main contentions was that population grows geometrically (exponentially), while the supply of resources such as food grows only arithmetically (linearly). Some 200+ years later, you have been asked to check the validity of Malthus's contention.

How do you go about doing so?

Robert Nickelsberg/Getty Images News/Getty Images

Introduction

To see whether Malthus was right, we need to see whether the data fit the models (linear and exponential) that he suggested or whether other models would be better. We saw in Chapter 1 how to fit a linear model. In this chapter we discuss how to construct models that use various *nonlinear* functions.

The nonlinear functions we consider in this chapter are the *quadratic* functions, the simplest nonlinear functions; the *exponential* functions, essential for discussing many kinds of growth and decay, including the growth (and decay) of money in finance and the initial growth of an epidemic; the *logarithmic* functions, needed to fully understand the exponential functions; and the *logistic* functions, used to model growth with an upper limit, such as the spread of an epidemic.

Precalculus Review
For this chapter, you should be familiar with the algebra reviewed in **Sections 0.2, 0.3, and 0.8**.

2.1 Quadratic Functions and Models

Quadratic Functions

In Chapter 1 we studied linear functions. Linear functions are useful, but in real-life applications they are often accurate for only a limited range of values of the variables. The relationship between two quantities is often best modeled by a curved line rather than a straight line. One of the simplest functions with a graph that is not a straight line is a quadratic function.

Quadratic Function

A **quadratic function** of the variable x is a function that can be written in the form

$$f(x) = ax^2 + bx + c \qquad \text{Function form}$$

or

$$y = ax^2 + bx + c, \qquad \text{Equation form}$$

where a, b, and c are fixed numbers (with $a \neq 0$).

Quick Examples

1.
$$\begin{aligned} f(x) &= 3x^2 - 2x + 1 \qquad a = 3, b = -2, c = 1 \\ f(-1) &= 3(-1)^2 - 2(-1) + 1 = 6 \\ f(a) &= 3a^2 - 2a + 1 \\ f(a+h) &= 3(a+h)^2 - 2(a+h) + 1 \\ &= 3(a^2 + 2ah + h^2) - 2(a+h) + 1 \\ &= 3a^2 + 6ah + 3h^2 - 2a - 2h + 1 \end{aligned}$$

2.
$$\begin{aligned} g(x) &= -x^2 \qquad a = -1, b = 0, c = 0 \\ g(0) &= -0^2 = 0 \\ g(-3) &= -(-3)^2 = -9^* \\ g(a+h) &= -(a+h)^2 \\ &= -(a^2 + 2ah + h^2) = -a^2 - 2ah - h^2 \end{aligned}$$

3.
$$\begin{aligned} R(p) &= -5{,}600p^2 + 14{,}000p \qquad a = -5{,}600, b = 14{,}000, c = 0 \\ R(10) &= -5{,}600(10)^2 + 14{,}000(10) = -420{,}000 \end{aligned}$$

* Spreadsheets actually interpret the formula $-x^2$ (wrongly!) as $(-x)^2$, so Excel would give $g(-3) = 9$ instead of -9. To get the correct answer in a spreadsheet, use the technology formula `-(x^2)`. (See Section 0.1 in the Precalculus Review for more discussion of this discrepancy.)

Quadratic Functions from the Graphical Point of View: Parabolas

Every quadratic function $f(x) = ax^2 + bx + c$ $(a \neq 0)$ has a **parabola** as its graph. Following is a summary of some features of parabolas that we can use to sketch the graph of any quadratic function.*

* We shall not fully justify the formulas for the vertex and the axis of symmetry until we have studied some calculus, although it is possible to justify them with just algebra.

Features of a Parabola

The graph of $f(x) = ax^2 + bx + c$ $(a \neq 0)$ is a **parabola**. If $a > 0$, the parabola opens upward and is called **concave up**, and if $a < 0$, it opens downward and is called **concave down**. (See the figure below.)

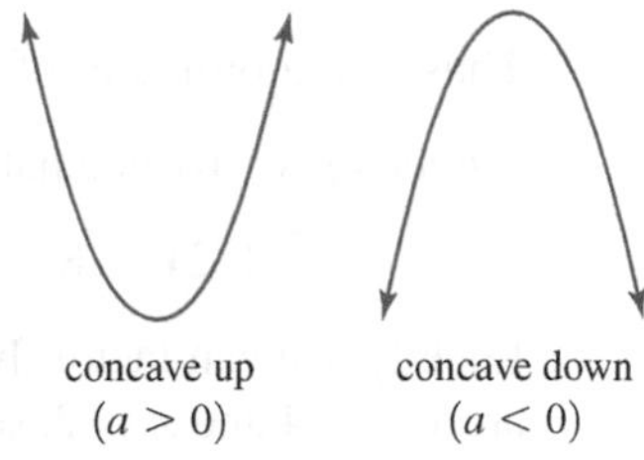

Vertex, Intercepts, and Symmetry

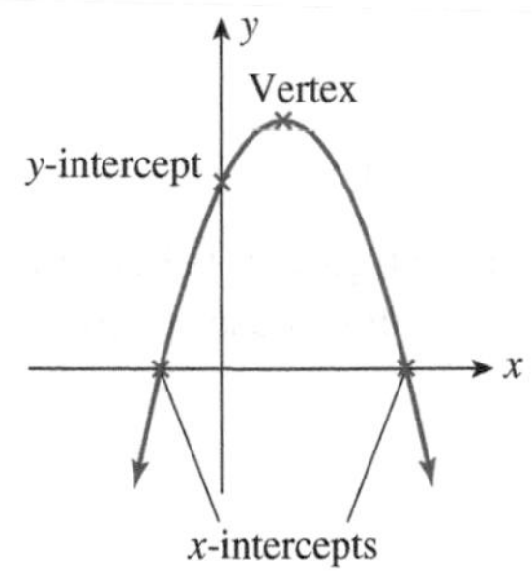

Vertex The vertex is the highest or lowest point of the parabola (see the figure in the margin). Its x-coordinate is $-\frac{b}{2a}$. Its y-coordinate is $f\left(-\frac{b}{2a}\right)$.

x-Intercepts (if any) These occur when $f(x) = 0$, that is, when

$$ax^2 + bx + c = 0.$$

Solve this equation for x by either factoring or using the quadratic formula. The x-intercepts are

$$x = \frac{-b \pm \sqrt{b^2 - 4ac}}{2a}.$$

If the **discriminant** $b^2 - 4ac$ is positive, there are two x-intercepts. If it is zero, there is a single x-intercept (at the vertex). If it is negative, there are no x-intercepts (so the parabola doesn't touch the x-axis at all).

y-Intercept This occurs when $x = 0$, so

$$y = a(0)^2 + b(0) + c = c.$$

Symmetry The parabola is symmetric with respect to the vertical line through the vertex, which is the line $x = -\frac{b}{2a}$.

Note that the x-intercepts can also be written as

$$x = -\frac{b}{2a} \pm \frac{\sqrt{b^2 - 4ac}}{2a},$$

making it clear that they are located symmetrically on either side of the line $x = -b/(2a)$. This partially justifies the claim that the whole parabola is symmetric with respect to this line.

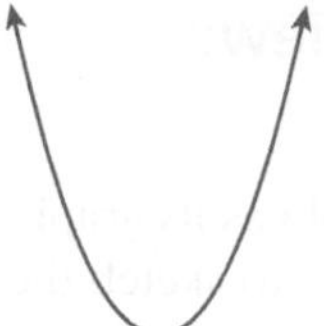

Figure 1

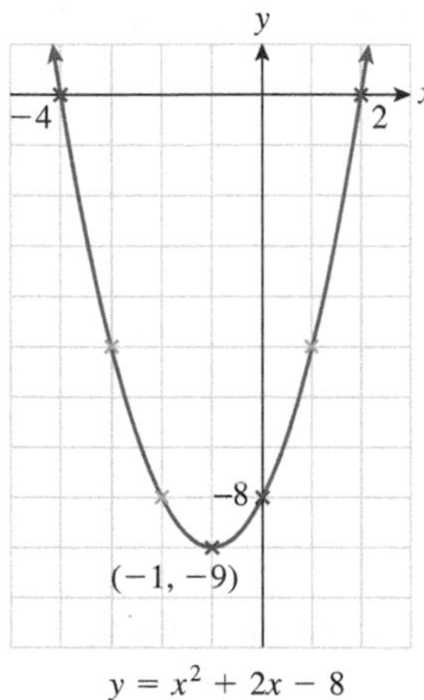

$y = x^2 + 2x - 8$

Figure 2

Using Technology

To automate the computations in Example 2 using a graphing calculator or a spreadsheet, see the Technology Guides at the end of the chapter. Outline:

TI-83/84 Plus

```
Y1=AX^2+BX+C
4→A:-12→B:9→C
[WINDOW] Xmin=0, Xmax=3
[ZOOM] [0]
```

[More details in the Technology Guide.]

Spreadsheet

Enter x-values in column A.
Compute the corresponding y-values in column B.
Graph the data in columns A and B.
[More details in the Technology Guide.]

Website
www.WanerMath.com
In the Function Evaluator and Grapher, enter `4x^2-12x+9` for y_1. For a table of values, enter the various x-values in the Evaluator box, and press "Evaluate".

EXAMPLE 1 Sketching the Graph of a Quadratic Function

Sketch the graph of $f(x) = x^2 + 2x - 8$ by hand.

Solution Here, $a = 1$, $b = 2$, and $c = -8$. Because $a > 0$, the parabola is concave up (Figure 1).

Vertex: The x-coordinate of the vertex is

$$x = -\frac{b}{2a} = -\frac{2}{2} = -1.$$

To get its y-coordinate, we substitute the value of x back into $f(x)$ to get

$$y = f(-1) = (-1)^2 + 2(-1) - 8 = 1 - 2 - 8 = -9.$$

Thus, the coordinates of the vertex are $(-1, -9)$.

x-Intercepts: To calculate the x-intercepts (if any), we solve the equation

$$x^2 + 2x - 8 = 0.$$

Luckily, we can factor the quadratic to get $(x + 4)(x - 2) = 0$. Thus, the solutions are $x = -4$ and $x = 2$, so these values are the x-intercepts. (We could also have used the quadratic formula here.)

y-Intercept: The y-intercept is given by $c = -8$.

Symmetry: The graph is symmetric around the vertical line $x = -1$.

Now we can sketch the curve as in Figure 2. (As we see in the figure, it is helpful to plot additional points by using the equation $y = x^2 + 2x - 8$ and to use symmetry to obtain others.)

EXAMPLE 2 One x-Intercept and No x-Intercepts

Sketch the graph of each quadratic function, showing the location of the vertex and intercepts.

a. $f(x) = 4x^2 - 12x + 9$

b. $g(x) = -\frac{1}{2}x^2 + 4x - 12$

Solution

a. We have $a = 4$, $b = -12$, and $c = 9$. Because $a > 0$, this parabola is concave up.

Vertex: $x = -\frac{b}{2a} = \frac{12}{8} = \frac{3}{2}$ *x*-coordinate of vertex

$y = f\left(\frac{3}{2}\right) = 4\left(\frac{3}{2}\right)^2 - 12\left(\frac{3}{2}\right) + 9 = 0$ *y*-coordinate of vertex

Thus, the vertex is at the point $(3/2, 0)$.

x-Intercepts: $4x^2 - 12x + 9 = 0$

$$(2x - 3)^2 = 0$$

The only solution is $2x - 3 = 0$, or $x = 3/2$. Note that this coincides with the vertex, which lies on the x-axis.

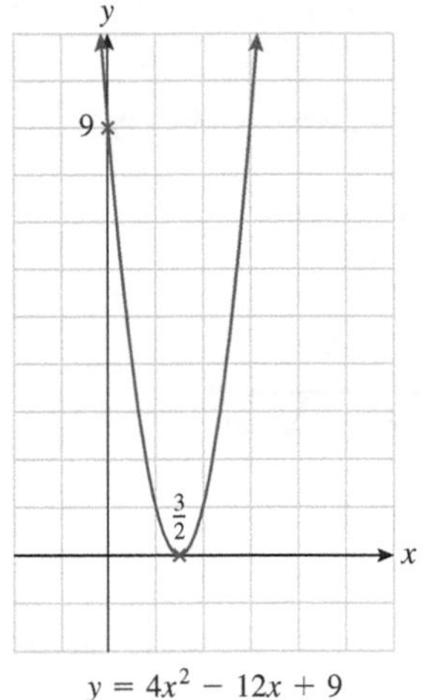

$y = 4x^2 - 12x + 9$

Figure 3

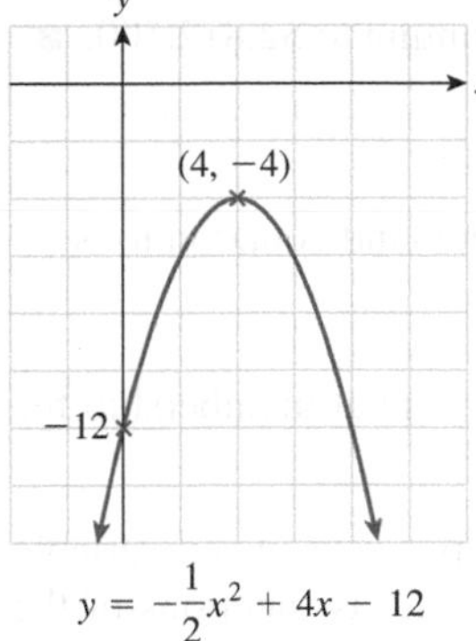

$y = -\frac{1}{2}x^2 + 4x - 12$

Figure 4

y-Intercept: $c = 9$

Symmetry: The graph is symmetric around the vertical line $x = 3/2$.

The graph is the narrow parabola shown in Figure 3. (As we remarked in Example 1, plotting additional points and using symmetry help us obtain an accurate sketch.)

b. Here, $a = -1/2$, $b = 4$, and $c = -12$. Because $a < 0$, the parabola is concave down. The vertex has x-coordinate $-b/(2a) = 4$, with corresponding y-coordinate $f(4) = -\frac{1}{2}(4)^2 + 4(4) - 12 = -4$. Thus, the vertex is at $(4, -4)$.

For the x-intercepts, we must solve $-\frac{1}{2}x^2 + 4x - 12 = 0$. If we try to use the quadratic formula, we discover that the discriminant is $b^2 - 4ac = 16 - 24 = -8$. Because the discriminant is negative, there are no solutions of the equation, so there are no x-intercepts.

The y-intercept is given by $c = -12$, and the graph is symmetric around the vertical line $x = 4$.

Because there are no x-intercepts, the graph lies entirely below the x-axis, as shown in Figure 4. (Again, you should plot additional points and use symmetry to ensure that your sketch is accurate.)

Applications

Recall that the **revenue** resulting from one or more business transactions is the total payment received. Thus, if q units of some item are sold at p dollars per unit, the revenue resulting from the sale is

$$\text{Revenue} = \text{Price} \times \text{Quantity}$$
$$R = pq.$$

EXAMPLE 3 Demand and Revenue

Alien Publications, Inc., predicts that the demand equation for the sale of its latest illustrated sci-fi novel, *Episode 93: Yoda vs. Alien,* is

$$q = -2{,}000p + 150{,}000$$

where q is the number of books it can sell each year at a price of $\$p$ per book. What price should Alien Publications, Inc., charge to obtain the maximum annual revenue?

Solution The total revenue depends on the price, as follows:

$$\begin{aligned} R &= pq && \text{Formula for revenue} \\ &= p(-2{,}000p + 150{,}000) && \text{Substitute for } q \text{ from demand equation.} \\ &= -2{,}000p^2 + 150{,}000p. && \text{Simplify.} \end{aligned}$$

We are looking for the price p that gives the maximum possible revenue. Notice that what we have is a quadratic function of the form $R(p) = ap^2 + bp + c$, where $a = -2{,}000$, $b = 150{,}000$, and $c = 0$. Because a is negative, the graph of the function is a parabola, concave down, so its vertex is its highest point (Figure 5). The p-coordinate of the vertex is

$$p = -\frac{b}{2a} = -\frac{150{,}000}{-4{,}000} = 37.5.$$

This value of p gives the highest point on the graph and thus gives the largest value of $R(p)$. We may conclude that Alien Publications, Inc., should charge \$37.50 per book to maximize its annual revenue.

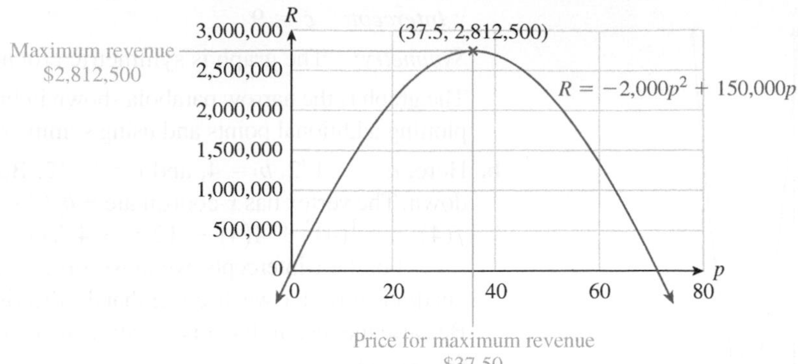

Figure 5

➡ **Before we go on . . .** You might ask what the maximum annual revenue is for the publisher in Example 3. Because $R(p)$ gives us the revenue at a price of $\$p$, the answer is $R(37.5) = -2{,}000(37.5)^2 + 150{,}000(37.5) = 2{,}812{,}500$. In other words, the company will earn total annual revenues from this book amounting to $2,812,500. ■

EXAMPLE 4 Demand, Revenue, and Profit

As the operator of *ZSport Fitness* gym, you calculate your demand equation to be

$$q = -0.06p + 84,$$

where q is the number of members in the club and p is the annual membership fee you charge.

a. Your annual operating costs are a fixed cost of $20,000 per year plus a variable cost of $20 per member. Find the annual revenue and profit as functions of the membership price p.

b. At what price should you set the annual membership fee to obtain the maximum revenue? What is the maximum possible revenue?

c. At what price should you set the annual membership fee to obtain the maximum profit? What is the maximum possible profit? What is the corresponding revenue?

Solution

a. The annual revenue is given by

$$R = pq \qquad \text{Formula for revenue}$$
$$= p(-0.06p + 84) \qquad \text{Substitute for } q \text{ from demand equation.}$$
$$= -0.06p^2 + 84p. \qquad \text{Simplify.}$$

The annual cost C is given by

$$C = 20{,}000 + 20q. \qquad \text{\$20,000 plus \$20 per member}$$

However, this is a function of q, not p. To express C as a function of p, we substitute for q, using the demand equation $q = -0.06p + 84$:

$$C = 20{,}000 + 20(-0.06p + 84)$$
$$= 20{,}000 - 1.2p + 1{,}680$$
$$= -1.2p + 21{,}680.$$

Thus, the profit function is

$$\begin{aligned} P &= R - C && \text{Formula for profit} \\ &= -0.06p^2 + 84p - (-1.2p + 21{,}680) && \text{Substitute for revenue and cost.} \\ &= -0.06p^2 + 85.2p - 21{,}680. \end{aligned}$$

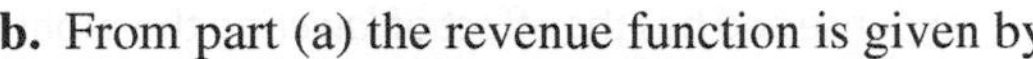

b. From part (a) the revenue function is given by

$$R = -0.06p^2 + 84p.$$

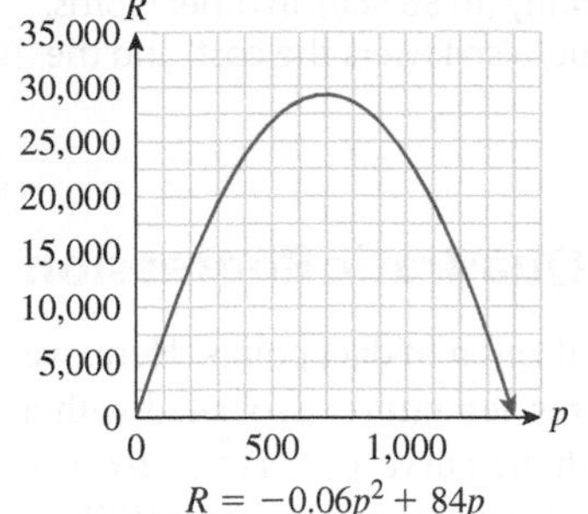

Figure 6

This is a quadratic function ($a = -0.06, b = 84, c = 0$) whose graph is a concave-down parabola (Figure 6). The maximum revenue corresponds to the highest point of the graph: the vertex, of which the p-coordinate is

$$p = -\frac{b}{2a} = -\frac{84}{2(-0.06)} = \$700.$$

This is the membership fee you should charge for the maximum revenue. The corresponding maximum revenue is given by the y-coordinate of the vertex in Figure 6:

$$R(700) = -0.06(700)^2 + 84(700) = \$29{,}400.$$

c. From part (a) the profit function is given by

$$P = -0.06p^2 + 85.2p - 21{,}680.$$

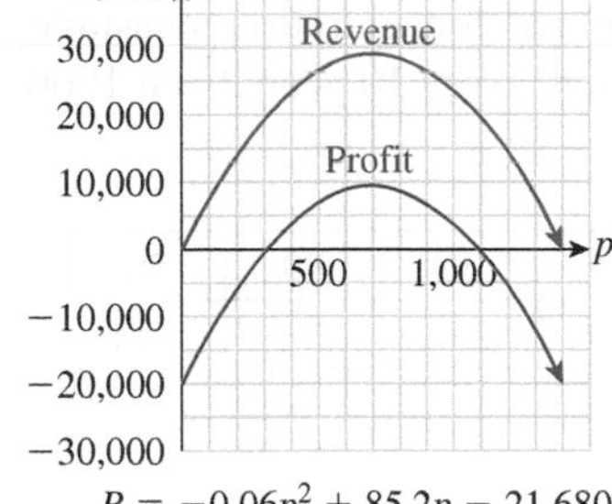

Figure 7

Like the revenue function, the profit function is quadratic ($a = -0.06, b = 85.2, c = -21{,}680$). Figure 7 shows both the revenue and profit functions. The maximum profit corresponds to the vertex, whose p-coordinate is

$$p = -\frac{b}{2a} = -\frac{85.2}{2(-0.06)} = \$710.$$

This is the membership fee you should charge for the maximum profit. The corresponding maximum profit is given by the y-coordinate of the vertex of the profit curve in Figure 7:

$$P(710) = -0.06(710)^2 + 85.2(710) - 21{,}680 = \$8{,}566.$$

The corresponding revenue is

$$R(710) = -0.06(710)^2 + 84(710) = \$29{,}394,$$

slightly less than the maximum possible revenue of \$29,400.

➡ **Before we go on . . .** The result of Example 4(c) tells us that the vertex of the profit curve in Figure 7 is slightly to the right of the vertex in the revenue curve. However, the difference is tiny in comparison to the scale of the graphs, so the graphs appear to be parallel.

In Example 4 the revenue and profit are expressed as functions of the membership price p. You could instead obtain revenue and profit as functions of the number of members q by first solving the demand equation for p as a function of q and substituting into the revenue formula, and then leaving the cost as $20{,}000 + 20q$ in the calculation of profit. The information you obtain will be analogous to that in Example 2 in Section 1.2, where revenue and cost of another gym, *YSport,* were given as functions of membership (denoted there by x rather than q.) ■

Q: *Charging \$710 membership brings in less revenue than charging \$700. So why charge \$710?*

A: A membership fee of \$700 does bring in slightly larger revenue than a fee of \$710, but it also brings in a slightly larger membership, which in turn raises the operating expense and has the effect of *lowering* the profit slightly (to \$8,560). In other words, the slightly higher fee, while bringing in less revenue, also lowers the cost, and the net result is a larger profit.

Fitting a Quadratic Function to Data: Quadratic Regression

In Section 1.4 we saw how to fit a regression line to a collection of data points. Here, we see how to use technology to obtain the **quadratic regression curve** associated with a set of points. The quadratic regression curve is the quadratic curve $y = ax^2 + bx + c$ that best fits the data points in the sense that the associated sum-of-squares error (SSE—see Section 1.4) is a minimum. Although there are algebraic methods for obtaining the quadratic regression curve, it is normal to use technology to do this.

EXAMPLE 5 Carbon Dioxide Concentration

The following table shows the annual mean carbon dioxide concentration measured at Mauna Loa Observatory in Hawaii, in parts per million, every 10 years from 1960 through 2010.[1]

Year t (years since 1960)	0	10	20	30	40	50
CO_2 Concentration C (ppm)	317	326	339	354	369	390

a. Is a linear model appropriate for these data?

b. Find the quadratic model

$$C(t) = at^2 + bt + c$$

that best fits the data.

Solution

a. To see whether a linear model is appropriate, we plot the data points and the regression line using one of the methods of Example 2 in Section 1.4 (Figure 8).

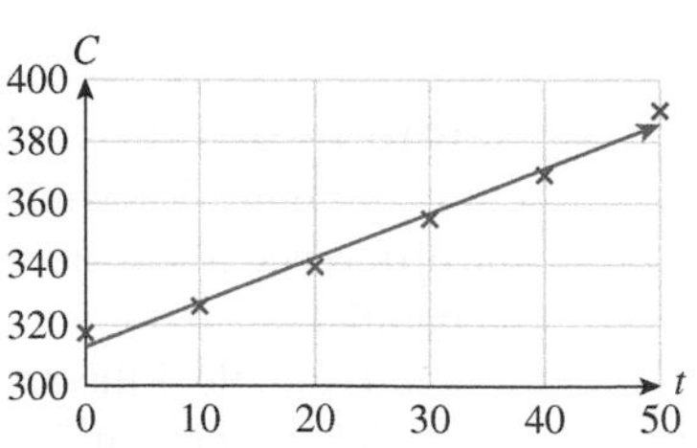

Figure 8

Using Technology

For detailed instructions on how to find and graph the regression curve in Example 5 using a graphing calculator or a spreadsheet, see the Technology Guides at the end of the chapter. Outline:

TI-83/84 Plus
[STAT] EDIT values of t in L_1 and values of C in L_2
Regression curve: [STAT] CALC option 5
QuadReg [ENTER]
Graph: [Y=] [VARS] [5] EQ [1], then [ZOOM] [9]
[More details in the Technology Guide.]

Spreadsheet
t- and C-values in columns A and B.
Graph: Highlight data, and insert a scatter chart.
Regression curve: Right-click a datapoint, and add a polynomial order 2 trendline with the option to show the equation on the chart.
[More details in the Technology Guide.]

 **Website**
www.WanerMath.com
In the Simple Regression utility, enter the data in the x- and y-columns, and press "y=ax^2+bx+c".

[1] Figures are approximate. Source: U.S. Department of Commerce/National Oceanic and Atmospheric Administration (NOAA) Earth System Research Laboratory, data downloaded from www.esrl.noaa.gov/gmd/ccgg/trends/ on March 13, 2011.

From the graph we can see that the given data suggest a curve and not a straight line: The observed points are above the regression line at the ends but below in the middle. (We would expect the data points from a linear relation to fall randomly above and below the regression line.)

b. The quadratic model that best fits the data is the quadratic regression model. As with linear regression, there are algebraic formulas to compute a, b, and c, but they are rather involved. However, we exploit the fact that these formulas are built into graphing calculators, spreadsheets, and other technology and obtain the regression curve using technology (see Figure 9):

$$C(t) = 0.012t^2 + 0.85t + 320. \quad \text{Coefficients rounded to two significant digits}$$

Notice from the graphs that the quadratic regression model appears to give a better fit than the linear regression model. This impression is supported by the values of SSE: For the linear regression model, SSE $\approx$ 58, while for the quadratic regression model, SSE is much smaller, approximately 2.6, indicating a much better fit.

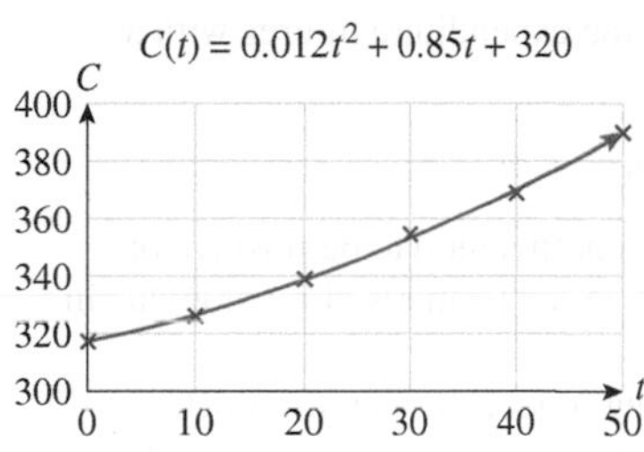

Figure 9

2.1 EXERCISES

▼ more advanced ◆ challenging

T indicates exercises that should be solved using technology

In Exercises 1–6, (a) state the values of a, b, and c in the given quadratic function $f(x) = ax^2 + bx + c$; (b) supply the missing values in the table below; (c) calculate $f(a + h)$; and (d) give a valid technology formula for $f(x)$. (Optional: Use technology to check the values in the table.) [**HINT:** See Quick Examples 1–3.]

x	−3	−2	−1	0	1	2	3
$f(x)$							

1. $f(x) = 2x^2 - x - 2$ **2.** $f(x) = -2x^2 + x + 2$

3. $f(x) = 10x^2 - 5x$ **4.** $f(x) = -x^2 - 50$

5. $f(x) = -x^2 - x - 1$ **6.** $f(x) = -3x^2 + 3x - 1$

In Exercises 7–16, sketch the graph of the quadratic function, indicating the coordinates of the vertex, the y-intercept, and the x-intercepts (if any). [**HINT:** See Example 1.]

7. $f(x) = x^2 + 3x + 2$ **8.** $f(x) = -x^2 - x$

9. $f(x) = -x^2 + 4x - 4$ **10.** $f(x) = x^2 + 2x + 1$

11. $f(x) = -x^2 - 40x + 500$ **12.** $f(x) = x^2 - 10x - 600$

13. $f(x) = x^2 + x - 1$ **14.** $f(x) = x^2 + \sqrt{2}x + 1$

15. $f(x) = x^2 + 1$ **16.** $f(x) = -x^2 + 5$

In Exercises 17–20, for each demand equation, express the total revenue R as a function of the price p per item, sketch the graph of the resulting function, and determine the price p that maximizes total revenue in each case. [**HINT:** See Example 3.]

17. $q = -4p + 100$ **18.** $q = -3p + 300$

19. $q = -2p + 400$ **20.** $q = -5p + 1{,}200$

T *In Exercises 21–24, use technology to find the quadratic regression curve through the given points. (Round all coefficients to four decimal places.)* [**HINT:** See Example 5.]

21. (1, 2), (3, 5), (4, 3), (5, 1)

22. (−1, 2), (−3, 5), (−4, 3), (−5, 1)

23. (−1, 2), (−3, 5), (−4, 3)

24. (2, 5), (3, 5), (5, 3)

Applications

25. ***World Military Expenditure*** The following chart shows total military and arms trade expenditure from 1992–2010 ($t = 0$ represents 1990).[2]

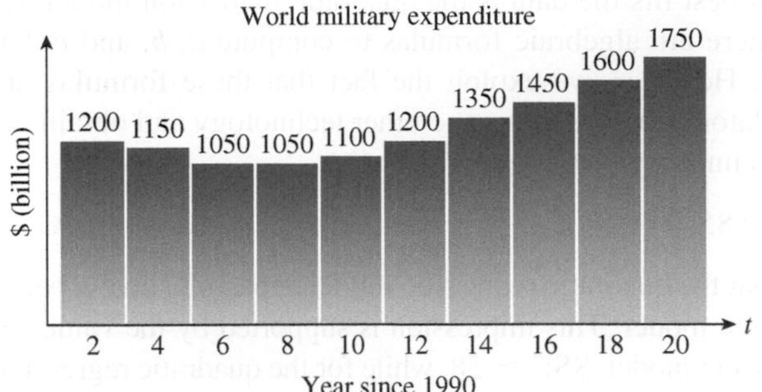

Source: www.globalissues.org/Geopolitics/ArmsTrade/Spending.asp.

a. If you want to model the expenditure figures with a function of the form

$$f(t) = at^2 + bt + c,$$

would you expect the coefficient a to be positive or negative? Why? [**HINT**: See "Features of a Parabola" in this section.]

b. Which of the following models best approximates the data given? (Try to answer this without actually computing values.)

(A) $f(t) = 4t^2 - 56t - 1{,}300$
(B) $f(t) = -4t^2 - 56t + 1{,}300$
(C) $f(t) = 4t^2 - 56t + 1{,}300$
(D) $f(t) = -4t^2 - 56t - 1{,}300$

c. What is the nearest year that would correspond to the vertex of the graph of the correct model from part (b)? What is the danger of extrapolating the data in either direction?

26. ***Education Expenditure*** The following chart shows the percentage of the U.S. discretionary budget allocated to education from 2003 to 2009. ($t = 3$ represents the start of 2003.)

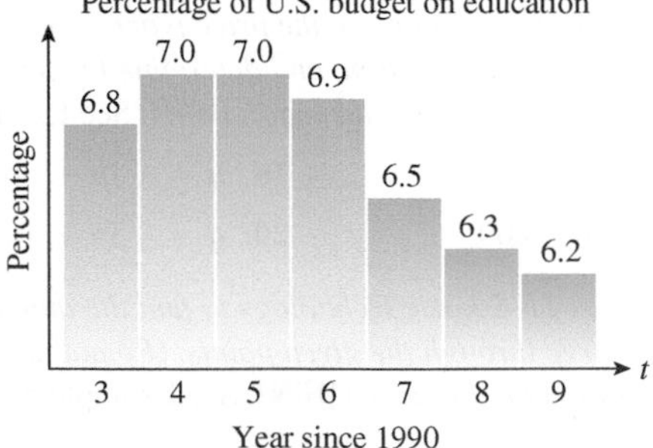

Source: www.globalissues.org/Geopolitics/ArmsTrade/Spending.asp.

a. If you want to model the percentage figures with a function of the form

$$f(t) = at^2 + bt + c,$$

would you expect the coefficient a to be positive or negative? Why? [**HINT**: See "Features of a Parabola" in this section.]

b. Which of the following models best approximates the data given? (Try to answer this without actually computing values)

(A) $f(t) = 0.04t^2 + 0.3t - 6$
(B) $f(t) = -0.04t^2 + 0.3t + 6$
(C) $f(t) = 0.04t^2 + 0.3t + 6$
(D) $f(t) = -0.04t^2 + 0.3t - 6$

c. What is the nearest year that would correspond to the vertex of the graph of the correct model from part (b)? What is the danger of extrapolating the data in either direction?

27. ***Oil Imports from Mexico*** Crude oil imports to the United States from Mexico for 2009–2013 could be approximated by

$$I(t) = -39t^2 + 800t - 3{,}000 \text{ thousand barrels per day} \quad (9 \le t \le 13)$$

where t is time in years since the start of 2000.[3] According to the model, approximately when were oil imports to the United States greatest? How many barrels per day were imported at that time? (Round the answer to two significant digits.) [**HINT**: See Example 1.]

28. ***Oil Production in Mexico*** Crude oil production by Pemex, Mexico's national oil company, for 2008–2013 could be approximated by

$$P(t) = 0.017t^2 - 0.4t + 5.23 \text{ million barrels per day} \quad (8 \le t \le 13)$$

where t is time in years since the start of 2000.[4] According to the model, approximately when was oil production by Pemex least? What was the oil production at that time? (Round the answer to two significant digits.) [**HINT**: See Example 1.]

29. ***GE Net Income 2009–2013*** The annual net income of General Electric for the period 2009–2013 could be approximated by

$$P(t) = -0.39t^2 + 5.2t - 4.1 \text{ billion dollars} \quad (4 \le t \le 8),$$

where t is time in years since the start of 2005.[5] According to the model, during what year in this period was General Electric's net income highest? What was the corresponding net income? Would you trust this model to continue to be valid long past this period? Why or why not?

[2] Approximate figures in constant 2011 dollars.

[3] Source for data: U.S. Energy Information Administration, www.eia.gov.

[4] Source for data: www.pemex.com.

[5] Source for data: www.wikinvest.com.

30. ***GE Net Income 2007–2011*** The annual net income of General Electric for the period 2007–2011 could be approximated by

$$P(t) = 1.6t^2 - 15t + 46 \text{ billion dollars} \quad (2 \le t \le 6),$$

where t is time in years since the start of 2005.[6] According to the model, during what year in this period was General Electric's net income lowest? What was the corresponding net income? Would you trust this model to continue to be valid long past this period? Why or why not?

31. ***Revenue*** The market research department of the *Better Baby Buggy Co.* predicts that the demand equation for its buggies is given by $q = -0.5p + 140$, where q is the number of buggies it can sell in a month if the price is \$$p$ per buggy. At what price should it sell the buggies to get the largest revenue? What is the largest monthly revenue? [**HINT**: See Example 3.]

32. ***Revenue*** The *Better Baby Buggy Co.* has just come out with a new model, the Turbo. The market research department predicts that the demand equation for Turbos is given by $q = -2p + 320$, where q is the number of buggies the company can sell in a month if the price is \$$p$ per buggy. At what price should it sell the buggies to get the largest revenue? What is the largest monthly revenue?

33. ***Revenue*** *Pack-Em-In Real Estate* is building a new housing development. The more houses it builds, the less people will be willing to pay, because of the crowding and smaller lot sizes. In fact, if it builds 40 houses in this particular development, it can sell them for \$200,000 each, but if it builds 60 houses, it will be able to get only \$160,000 each. Obtain a linear demand equation, and hence determine how many houses Pack-Em-In should build to get the largest revenue. What is the largest possible revenue? [**HINT**: See Example 3.]

34. ***Revenue*** *Pack-Em-In* has another development in the works. If it builds 50 houses in this development, it will be able to sell them at \$190,000 each, but if it builds 70 houses, it will get only \$170,000 each. Obtain a linear demand equation, and hence determine how many houses it should build to get the largest revenue. What is the largest possible revenue?

35. ▼ ***Revenue from Monorail Service, Las Vegas*** In 2005 the Las Vegas monorail charged \$3 per ride and had an average ridership of about 28,000 per day. In December 2005 the Las Vegas Monorail Company raised the fare to \$5 per ride, and average ridership in 2006 plunged to around 19,000 per day.[7]
    a. Use the given information to find a linear demand equation.
    b. Find the price the company should have charged to maximize revenue from ridership. What is the corresponding daily revenue?
    c. The Las Vegas Monorail Company would have needed \$44.9 million in revenues from ridership to break even in 2006. Would it have been possible to break even in 2006 by charging a suitable price?

36. ▼ ***Revenue from Monorail Service, Mars*** The Utarek monorail, which links the three urbynes (or districts) of Utarek on Mars, charged Ƶ5 per ride[8] and sold about 14 million rides per day. When the Utarek City Council lowered the fare to Ƶ3 per ride, the number of rides increased to 18 million per day.
    a. Use the given information to find a linear demand equation.
    b. Find the price the City Council should have charged to maximize revenue from ridership. What is the corresponding daily revenue?
    c. The City Council would have needed to raise Ƶ48 billion in revenues from ridership each Martian year (670 days[9]) to finance the new Mars organism research lab. Would this have been possible by charging a suitable price?

37. ***Website Profit*** You operate a gaming website, www.mudbeast.net, where users must pay a small fee to log on. When you charged \$2, the demand was 280 log-ons per month. When you lowered the price to \$1.50, the demand increased to 560 log-ons per month.
    a. Construct a linear demand function for your website and hence obtain the monthly revenue R as a function of the log-on fee x.
    b. Your Internet provider charges you a monthly fee of \$30 to maintain your site. Express your monthly profit P as a function of the log-on fee x, and hence determine the log-on fee you should charge to obtain the largest possible monthly profit. What is the largest possible monthly profit? [**HINT**: See Example 4.]

38. ***T-Shirt Profit*** Two fraternities, Sig Ep and Ep Sig, plan to raise money jointly to benefit homeless people on Long Island. They will sell Yoda vs. Alien T-shirts in the student center but are not sure how much to charge. Sig Ep treasurer Augustus recalls that they once sold 400 shirts in a week at \$8 per shirt, but Ep Sig treasurer Julius has solid research indicating that it is possible to sell 600 per week at \$4 per shirt.
    a. On the basis of this information, construct a linear demand equation for Yoda vs. Alien T-shirts. Hence, obtain the weekly revenue R as a function of the unit price x.
    b. The university administration charges the fraternities a weekly fee of \$500 for use of the student center. Write down the monthly profit P as a function of the unit

---

[6] See footnote for Exercise 29.

[7] Source: *New York Times*, Februrary 10, 2007, p. A9.

[8] The zonar (Ƶ) is the official currency in the city-state of Utarek, Mars (formerly www.Marsnext.com, a now extinct virtual society).

[9] As measured in Mars days. The actual length of a Mars year is about 670.55 Mars days, so frequent leap years are designated by the Mars Planetary Authority to adjust.

price x. Hence, determine how much the fraternities should charge to obtain the largest possible weekly profit. What is the largest possible weekly profit? [**HINT**: See Example 4.]

39. ***Website Profit*** The latest demand equation for your gaming website, www.mudbeast.net, is given by

$$q = -400x + 1{,}200$$

where q is the number of users who log on per month and x is the log-on fee you charge. Your Internet provider bills you as follows:

Site maintenance fee: $20 per month

High-volume access fee: 50¢ per log-on

Find the monthly cost as a function of the log-on fee x. Hence, find the monthly profit as a function of x, and determine the log-on fee you should charge to obtain the largest possible monthly profit. What is the largest possible monthly profit?

40. ***T-Shirt Profit*** The latest demand equation for your Yoda vs. Alien T-shirts is given by

$$q = -40x + 600,$$

where q is the number of shirts you can sell in one week if you charge \$$x$ per shirt. The Student Council charges you $400 per week for use of their facilities, and the T-shirts cost you $5 each. Find the weekly cost as a function of the unit price x. Hence, find the weekly profit as a function of x, and determine the unit price you should charge to obtain the largest possible weekly profit. What is the largest possible weekly profit?

41. ▼ ***Nightclub Management*** You have just opened a new nightclub, *Russ' Techno Pitstop,* but are unsure how high to set the cover charge (entrance fee). One week you charged $10 per guest and averaged 300 guests per night. The next week you charged $15 per guest and averaged 250 guests per night.

a. Find a linear demand equation showing the number of guests q per night as a function of the cover charge p.

b. Find the nightly revenue R as a function of the cover charge p.

c. The club will provide two free nonalcoholic drinks for each guest, costing the club $3 per head. In addition, the nightly overheads (rent, salaries, dancers, DJ, etc.) amount to $3,000. Find the cost C as a function of the cover charge p.

d. Now find the profit in terms of the cover charge p. Hence, determine the entrance fee you should charge for a maximum profit.

42. ▼ ***Television Advertising*** As sales manager for *Montevideo Productions, Inc.,* you are planning to review the prices you charge clients for television advertisement development. You currently charge each client an hourly development fee of $2,500. With this pricing structure, the demand, measured by the number of contracts Montevideo signs per month, is 15 contracts. This is down 5 contracts from the figure last year, when your company charged only $2,000.

a. Construct a linear demand equation giving the number of contracts q as a function of the hourly fee p Montevideo charges for development.

b. On average, Montevideo bills for 50 hours of production time on each contract. Give a formula for the total revenue obtained by charging \$$p$ per hour.

c. The costs to Montevideo Productions are estimated as follows:

Fixed costs: $120,000 per month

Variable costs: $80,000 per contract

Express Montevideo Productions' monthly cost **(i)** as a function of the number q of contracts and **(ii)** as a function of the hourly production charge p.

d. Express Montevideo Productions' monthly profit as a function of the hourly development fee p. Hence, find the price it should charge to maximize the profit.

43. T ***World Military Expenditure*** The following table shows total military and arms trade expenditure in 1994, 1998, and 2006. (See Exercise 25; $t = 4$ represents 1994.)[10]

Year t	4	8	16
Military Expenditure ($ billion)	1,150	1,050	1,450

Find a quadratic model for these data, and use your model to estimate world military expenditure in 2008. Compare your answer with the actual figure shown in Exercise 25. [**HINT**: See Example 5.]

44. T ***Education Expenditure*** The following table shows the percentage of the U.S. discretionary budget allocated to education in 2003, 2005, and 2009. (See Exercise 26; $t = 3$ represents the start of 2003.)[11]

Year t	3	5	9
Percentage	6.8	7	6.2

Find a quadratic model for these data, and use your model to estimate the percentage of the U.S. discretionary budget allocated to education in 2008. Compare your answer with the actual figure shown in Exercise 26.

[10] Approximate figures in constant 2011 dollars. Source: www.globalissues.org/Geopolitics/ArmsTrade/Spending.asp.

[11] Source: www.globalissues.org/Geopolitics/ArmsTrade/Spending.asp.

45. **T** ***iPod Sales*** The following table shows **Apple** iPod sales during the financial years 2010–2014. (t is time in years since 2010.)[12]

Year t	0	1	2	3	4
iPod Sales (millions)	50.4	42.6	35.2	26.4	14.4

a. Find a quadratic regression model for these data. (Round coefficients to two significant digits.) Graph the model together with the data.

b. What does the model predict for iPod sales in 2015 and 2016, to the nearest million? Comment on the answers.

46. **T** ***iPod Sales*** The following table shows **Apple** iPod sales during the financial years 2005–2009. (t is time in years since 2005.)[13]

Year t	0	1	2	3	4
iPod Sales (millions)	22.5	39.4	51.6	54.8	54.1

a. Find a quadratic regression model for these data. (Round coefficients to two significant digits.) Graph the model together with the data.

b. What does the model predict for iPod sales in 2010 and 2011, to the nearest million? Comment on the answers.

Communication and Reasoning Exercises

47. What can you say about the graph of $f(x) = ax^2 + bx + c$ if $a = 0$?

48. What can you say about the graph of $f(x) = ax^2 + bx + c$ if $c = 0$?

49. Multiple choice: Following is the graph of $f(x) = ax^2 + bx + c$:

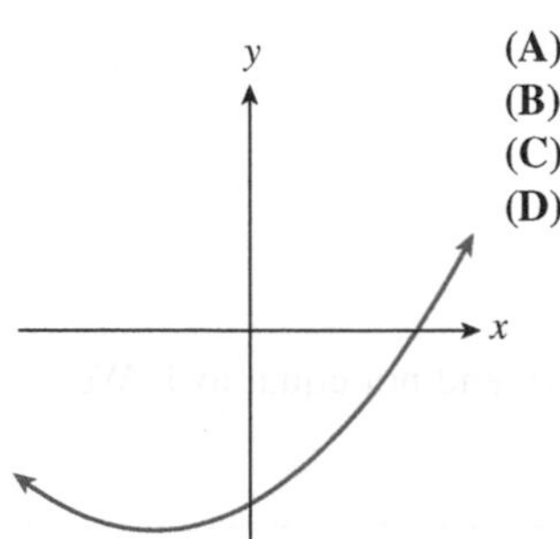

(A) a is positive and c is positive.
(B) a is negative and c is positive.
(C) a is positive and c is negative.
(D) a is negative and c is negative.

[12] Apple's financial year begins around September 25 of the preceding calendar year. Apple stopped reporting iPod sales after financial year 2014. Source: Apple quarterly press releases, http://investor.apple.com.

[13] *Ibid.*

50. Multiple choice: Following is the graph of $f(x) = ax^2 + bx + c$:

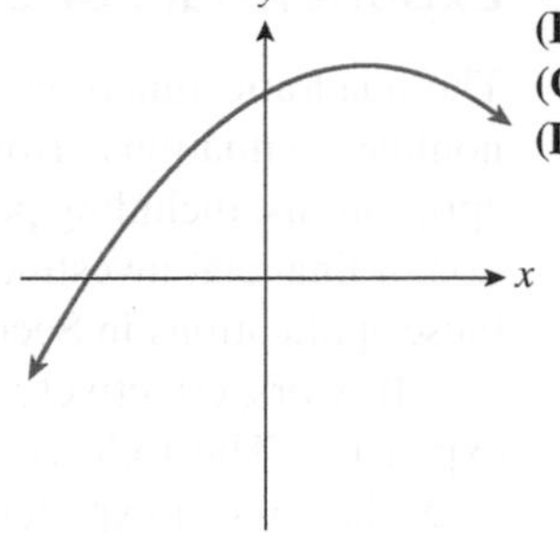

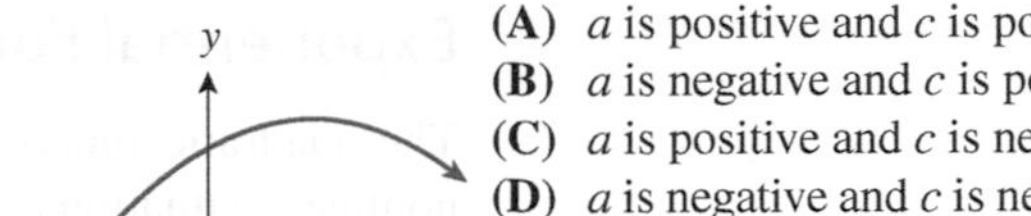

(A) a is positive and c is positive.
(B) a is negative and c is positive.
(C) a is positive and c is negative.
(D) a is negative and c is negative.

51. ▼ Refer to the graph of $f(x) = ax^2 + bx + c$ in Exercise 49. Is b positive or negative? Why?

52. ▼ Refer to the graph of $f(x) = ax^2 + bx + c$ in Exercise 50. Is b positive or negative? Why?

53. Suppose the graph of revenue as a function of unit price is a parabola that is concave down. What is the significance of the coordinates of the vertex, the x-intercepts, and the y-intercept?

54. Suppose the height of a stone thrown vertically upward is given by a quadratic function of time. What is the significance of the coordinates of the vertex, the (possible) x-intercepts, and the y-intercept?

55. How might you tell, roughly, whether a set of data should be modeled by a quadratic rather than by a linear equation?

56. A member of your study group tells you that, because the following set of data does not suggest a straight line, the data are best modeled by a quadratic.

x	0	2	4	6	8
y	1	2	1	0	1

Comment on her suggestion.

57. Is a quadratic model useful for long-term prediction of sales of an item? Why or why not?

58. Of what use is a quadratic model if not for long-term prediction?

59. ▼ Explain why, if demand is a linear function of unit price p (with negative slope), then there must be a *single value of p* that results in the maximum revenue.

60. ▼ Explain why, if the average cost of a commodity is given by $y = 0.1x^2 - 4x - 2$, where x is the number of units sold, there is a single choice of x that results in the lowest possible average cost.

61. ▼ If the revenue function for a particular commodity is $R(p) = -50p^2 + 60p$, what is the (linear) demand function? Give a reason for your answer.

62. ▼ If the revenue function for a particular commodity is $R(p) = -50p^2 + 60p + 50$, can the demand function be linear? What is the associated demand function?

2.2 Exponential Functions and Models

Exponential Functions

The quadratic functions we discussed in Section 2.1 can be used to model many nonlinear situations. However, exponential functions give better models in some applications, including population growth, radioactive decay, the growth or depreciation of financial investments, and many other phenomena. (We already saw some of these applications in Section 1.2.)

To work effectively with exponential functions, we need to know the laws of exponents. The following list, taken from Section 0.2 in the Precalculus Review, gives the laws of exponents we will be using.

The Laws of Exponents

If b and c are positive and x and y are any real numbers, then the following laws hold:

Law	Quick Examples	
1. $b^x b^y = b^{x+y}$	$2^3 2^2 = 2^5 = 32$	$2^{3-x} = 2^3 2^{-x}$
2. $\dfrac{b^x}{b^y} = b^{x-y}$	$\dfrac{4^3}{4^2} = 4^{3-2} = 4^1 = 4$	$3^{x-2} = \dfrac{3^x}{3^2} = \dfrac{3^x}{9}$
3. $\dfrac{1}{b^x} = b^{-x}$	$9^{-0.5} = \dfrac{1}{9^{0.5}} = \dfrac{1}{3}$	$2^{-x} = \dfrac{1}{2^x}$
4. $b^0 = 1$	$(3.3)^0 = 1$	$x^0 = 1$ if $x \neq 0$
5. $(b^x)^y = b^{xy}$	$(2^3)^2 = 2^6 = 64$	$\left(\dfrac{1}{2}\right)^x = (2^{-1})^x = 2^{-x}$
6. $(bc)^x = b^x c^x$	$(4 \cdot 2)^2 = 4^2 2^2 = 64$	$10^x = 5^x 2^x$
7. $\left(\dfrac{b}{c}\right)^x = \dfrac{b^x}{c^x}$	$\left(\dfrac{4}{3}\right)^2 = \dfrac{4^2}{3^2} = \dfrac{16}{9}$	$\left(\dfrac{1}{2}\right)^x = \dfrac{1^x}{2^x} = \dfrac{1}{2^x}$

Here are the functions we will study in this section.

Exponential Function

An **exponential function** has the form

$$f(x) = Ab^x, \qquad \text{Technology: } \texttt{A*b^x}$$

where A and b are constants with $A \neq 0$ and b positive and not equal to 1. We call b the **base** of the exponential function.

Quick Examples

1. $f(x) = 2^x$ — $A = 1, b = 2$; Technology: `2^x`

$f(1) = 2^1 = 2$ — `2^1`

$f(-3) = 2^{-3} = \dfrac{1}{8}$ — `2^(-3)`

$f(0) = 2^0 = 1$ — `2^0`

2. $g(x) = 20(3^x)$ — $A = 20, b = 3$; Technology: `20*3^x`

$g(2) = 20(3^2) = 20(9) = 180$ — `20*3^2`

$g(-1) = 20(3^{-1}) = 20\left(\frac{1}{3}\right) = 6\frac{2}{3}$ — `20*3^(-1)`

3. $h(x) = 2^{-x} = \left(\frac{1}{2}\right)^x$ — $A = 1, b = \frac{1}{2}$; Technology: `2^(-x)` or `(1/2)^x`

$h(1) = 2^{-1} = \frac{1}{2}$ — `2^(-1)` or `(1/2)^1`

$h(2) = 2^{-2} = \frac{1}{4}$ — `2^(-2)` or `(1/2)^2`

4. $k(x) = 3 \cdot 2^{-4x} = 3(2^{-4})^x$ — $A = 3, b = 2^{-4}$; Technology: `3*2^(-4*x)`

$k(-2) = 3 \cdot 2^{-4(-2)}$ — `3*2^(-4*(-2))`

$= 3 \cdot 2^8 = 3 \cdot 256 = 768$

Exponential Functions from the Numerical and Graphical Points of View

The following table shows values of $f(x) = 3(2^x)$ for some values of x ($A = 3$, $b = 2$):

x	-3	-2	-1	0	1	2	3
$f(x)$	$\frac{3}{8}$	$\frac{3}{4}$	$\frac{3}{2}$	3	6	12	24

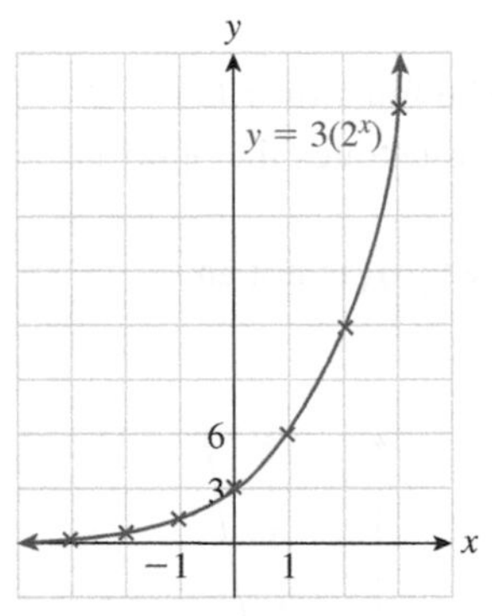

Technology formula: `3*2^x`

Figure 10

The graph of f is shown in Figure 10. Notice that the y-intercept is $A = 3$ (obtained by setting $x = 0$). In general:

In the graph of $f(x) = Ab^x$, A is the y-intercept, or the value of y when $x = 0$.

What about b? Notice from the table that the value of y is multiplied by $b = 2$ for every increase of 1 in x. If we decrease x by 1, the y-coordinate gets *divided* by $b = 2$.

The value of y is multiplied by b for every 1-unit increase of x:

x	-3	-2	-1	0	1	2	3
$f(x)$	$\frac{3}{8}$	$\frac{3}{4}$	$\frac{3}{2}$	3	6	12	24

Multiply by 2.

On the graph, if we move 1 unit to the right from any point on the curve, the y-coordinate doubles. Thus, the curve becomes dramatically steeper as the value of x increases. This phenomenon is called **exponential growth**. (See Section 1.2.)

Exponential Function Numerically and Graphically

For the exponential function $f(x) = Ab^x$:

Visualizing the Roles of *A* and *b*

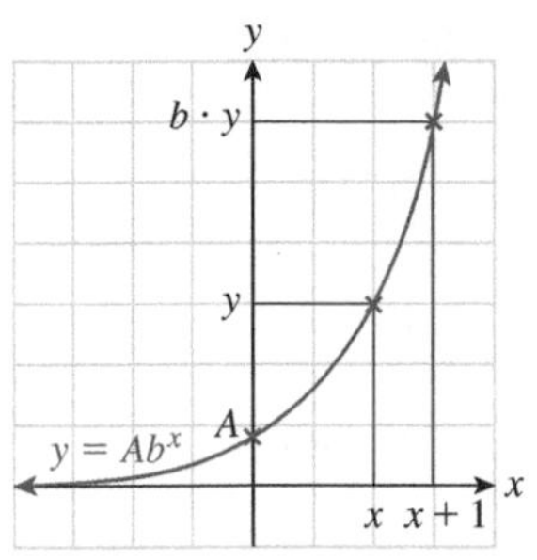

If x increases by 1, y is multiplied by b.

Role of *A*

$f(0) = A$, so A is the y-intercept of the graph of f.

Role of *b*

If x increases by 1, $f(x)$ is multiplied by b.

If x increases by 2, $f(x)$ is multiplied by b^2.

⋮

If x increases by Δx, $f(x)$ is multiplied by $b^{\Delta x}$.

Quick Examples

5. $f_1(x) = 2^x,\ f_2(x) = \left(\frac{1}{2}\right)^x = 2^{-x}$

	A	B	C
1	x	2^x	2^(-x)
2	-3	1/8	8
3	-2	1/4	4
4	-1	1/2	2
5	0	1	1
6	1	2	1/2
7	2	4	1/4
8	3	8	1/8

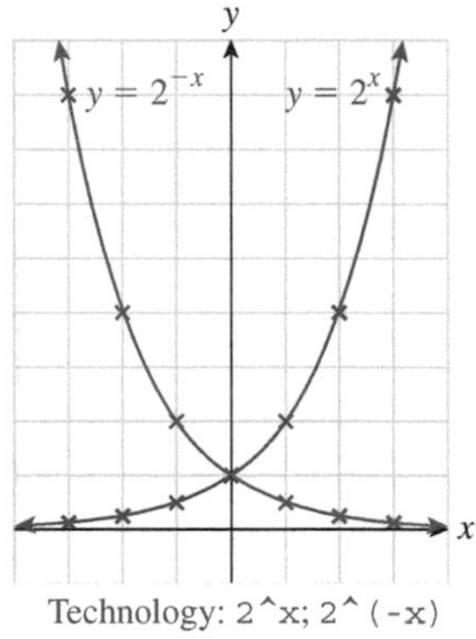

Technology: 2^x; 2^(-x)

When x increases by 1, $f_2(x)$ is multiplied by $\frac{1}{2}$. The function $f_1(x) = 2^x$ illustrates exponential growth, while $f_2(x) = \left(\frac{1}{2}\right)^x$ illustrates the opposite phenomenon: **exponential decay**.

6. $f_1(x) = 2^x,\ f_2(x) = 3^x,\ f_3(x) = 1^x$ (Can you see why f_3 is not an exponential function?)

	A	B	C	D
1	x	2^x	3^x	1^x
2	-3	1/8	1/27	1
3	-2	1/4	1/9	1
4	-1	1/2	1/3	1
5	0	1	1	1
6	1	2	3	1
7	2	4	9	1
8	3	8	27	1

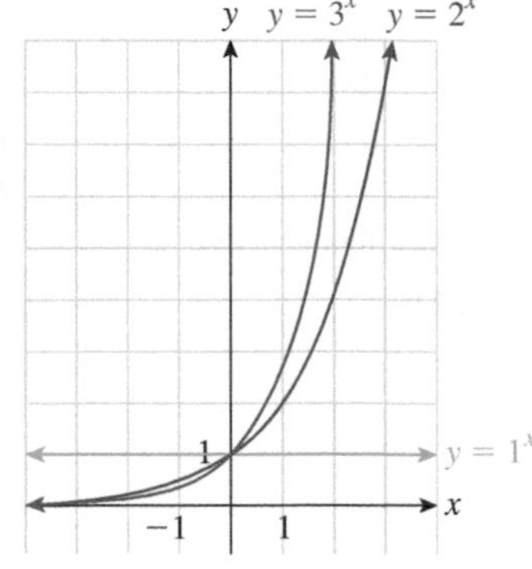

If x increases by 1, 3^x is multiplied by 3. Note also that all the graphs pass through (0, 1). (Why?)

EXAMPLE 1 Recognizing Exponential Data Numerically and Graphically

Some of the values of two functions, f and g, are given in the following table:

x	-2	-1	0	1	2
$f(x)$	-7	-3	1	5	9
$g(x)$	$\frac{2}{9}$	$\frac{2}{3}$	2	6	18

One of these functions is linear, and the other is exponential. Which is which?

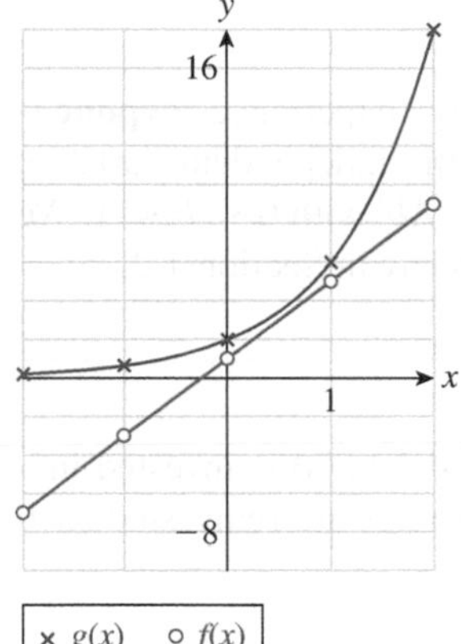

Figure 11

Solution Remember that a linear function increases (or decreases) by the same amount every time x increases by 1. The values of f behave this way: Every time x increases by 1, the value of $f(x)$ increases by 4. Therefore, f is a linear function with a *slope* of 4. Because $f(0) = 1$, we see that

$$f(x) = 4x + 1$$

is a linear formula that fits the data.

On the other hand, every time x increases by 1, the value of $g(x)$ is *multiplied* by 3. Because $g(0) = 2$, we find that

$$g(x) = 2(3^x)$$

is an exponential function fitting the data.

We can visualize the two functions f and g by plotting the data points (Figure 11). The data points for $f(x)$ clearly lie along a straight line, whereas the points for $g(x)$ lie along a curve. The y-coordinate of each point for $g(x)$ is 3 times the y-coordinate of the preceding point, demonstrating that the curve is an exponential one.

In Section 1.3 we discussed a method for calculating the equation of the line that passes through two given points. In the following example we show a method for calculating the equation of the exponential curve through two given points.

EXAMPLE 2 Finding the Exponential Curve through Two Points

Find an equation of the exponential curve through $(1, 6.3)$ and $(4, 170.1)$.

Solution We want an equation of the form

$$y = Ab^x \quad (b > 0).$$

Substituting the coordinates of the given points, we get

$$6.3 = Ab^1 \qquad \text{Substitute } (1, 6.3).$$

$$170.1 = Ab^4. \qquad \text{Substitute } (4, 170.1).$$

If we now divide the second equation by the first, we get

$$\frac{170.1}{6.3} = \frac{Ab^4}{Ab} = b^3$$

$$b^3 = 27$$

$$b = 27^{1/3} \qquad \text{Take reciprocal power of both sides.}$$

$$b = 3.$$

Now that we have b, we can substitute its value into the first equation to obtain

$$6.3 = 3A \qquad \text{Substitute } b = 3 \text{ into the equation } 6.3 = Ab^1.$$

$$A = \frac{6.3}{3} = 2.1.$$

We have both constants, $A = 2.1$ and $b = 3$, so the model is

$$y = 2.1(3^x).$$

Example 6 will show how to use technology to fit an exponential function to two or more data points.

Applications

Recall some terminology we mentioned earlier: A quantity y experiences **exponential growth** if $y = Ab^t$ with $b > 1$. (Here, we use t for the independent variable, thinking of time.) It experiences **exponential decay** if $y = Ab^t$ with $0 < b < 1$. We already saw several examples of exponential growth and decay in Section 1.2.

EXAMPLE 3 Exponential Growth and Decay

a. Compound Interest (See Example 6 of Section 1.2.) If \$2,000 is invested in a mutual fund with an annual yield of 12.6% and the earnings are reinvested each month, then the future value after t years is

$$A(t) = P\left(1 + \frac{r}{m}\right)^{mt} = 2{,}000\left(1 + \frac{0.126}{12}\right)^{12t} = 2{,}000(1.0105)^{12t},$$

which can be written as $2{,}000(1.0105^{12})^t$, so $A = 2{,}000$ and $b = 1.0105^{12}$. This is an example of exponential growth, because $b > 1$.

b. Carbon Decay (See Example 7 of Section 1.2.) The amount of carbon 14 remaining in a sample that originally contained A grams is approximately

$$C(t) = A(0.999879)^t.$$

This is an instance of exponential decay because $b < 1$.

Before we go on . . . Refer again to Example 3(a). In Example 6(b) of Section 1.2 we showed how to use technology to answer questions such as the following: "When, to the nearest year, will the value of your investment reach \$5,000?" ■

The next example shows an application to public health.

EXAMPLE 4 Exponential Growth: Epidemics

In the early stages of the AIDS epidemic during the 1980s, the number of cases in the United States was increasing by about 50% every 6 months. By the start of 1983 there were approximately 1,600 AIDS cases in the United States.[14]

---

[14] Data based on regression of the 1982–1986 figures. Source for data: Centers for Disease Control and Prevention, HIV/AIDS Surveillance Report, 2000;12 (No. 2).

a. Assuming an exponential growth model, find a function that predicts the number of people infected t years after the start of 1983.

b. Use the model to estimate the number of people infected by October 1, 1986, and also by the end of that year.

Solution

a. One way of finding the desired exponential function is to reason as follows: At time $t = 0$ (January 1, 1983) the number of people infected was 1,600, so $A = 1{,}600$. Every 6 months, the number of cases increased to 150% of the number 6 months earlier—that is, to 1.50 times that number. Each year, it therefore increased to $(1.50)^2 = 2.25$ times the number one year earlier. Hence, after t years we need to multiply the original 1,600 by 2.25^t, so the model is

$$y = 1{,}600(2.25^t) \text{ cases.}$$

Alternatively, if we wish to use the method of Example 2, we need two data points. We are given one point: (0, 1,600). Because y increased by 50% every 6 months, 6 months later it reached $1{,}600 + 800 = 2{,}400$ ($t = 0.5$). This information gives a second point: (0.5, 2,400). We can now apply the method in Example 2 to find the model above.

b. October 1, 1986, corresponds to $t = 3.75$ (because October 1 is 9 months, or $9/12 = 0.75$ of a year, after January 1). Substituting this value of t in the model gives

$$y = 1{,}600(2.25^{3.75}) \approx 33{,}481 \text{ cases.}$$

```
1600*2.25^3.75
```

By the end of 1986 the model predicts that

$$y = 1{,}600(2.25^4) \approx 41{,}006 \text{ cases.}$$

(The actual number of cases was around 41,700.)

➡ **Before we go on . . .** Increasing the number of cases by 50% every 6 months couldn't continue for very long, and this is borne out by observations. If increasing by 50% every 6 months did continue, then by January 2003 ($t = 20$) the number of infected people would have been

$$1{,}600(2.25^{20}) \approx 17{,}700{,}000{,}000$$

a number that would have been more than 50 times the size of the U.S. population! Thus, although the exponential model is fairly reliable in the early stages of an epidemic, it is unreliable for predicting long-term trends. ■

Epidemiologists use more sophisticated models to measure the spread of epidemics, and these models predict a leveling-off phenomenon as the number of cases becomes a significant part of the total population. We discuss such a model, the **logistic function**, in Section 2.4.

The Number *e* and More Applications

In nature we find examples of growth that occurs *continuously*, as though "interest" is being added more often than every second or fraction of a second. To model this, we need to see what happens to the compound interest formula of Section 1.2 as we

	A	B
1	m	(1+1/m)^m
2	1	2
3	10	2.59374246
4	100	2.704813829
5	1000	2.716923932
6	10000	2.718145927
7	100000	2.718268237
8	1000000	2.718280469
9	10000000	2.718281694
10	100000000	2.718281786
11	1000000000	2.718282031

let m (the number of times interest is added per year) become extremely large. Something very interesting does happen: We end up with a more compact and elegant formula than we began with. To see why, let's look at a very simple situation.

Suppose we invest \$1 in the bank for 1 year at 100% interest, compounded m times per year. If $m = 1$, then the 100% interest is added at the end of the year, so our money doubles. In general, the accumulated capital at the end of the year is

$$A = 1\left(1 + \frac{1}{m}\right)^m = \left(1 + \frac{1}{m}\right)^m. \qquad \texttt{(1+1/m)^m}$$

Now, we are interested in what A becomes for large values of m. On the left is a spreadsheet showing the quantity $\left(1 + \frac{1}{m}\right)^m$ for larger and larger values of m.

Something interesting *does* seem to be happening! The numbers appear to be getting closer and closer to a specific value. In mathematical terminology, we say that the numbers **converge** to a fixed number, 2.71828. . . , called the **limiting value*** of the quantities $\left(1 + \frac{1}{m}\right)^m$. This number, called e, is one of the most important in mathematics. The number e is irrational, just as the more familiar number π is, so we cannot write down its exact numerical value. To 20 decimal places,

* See Chapter 3 for more on limits.

$$e = 2.71828182845904523536\ldots$$

We now say that if \$1 is invested for 1 year at 100% interest **compounded continuously**, the accumulated money at the end of that year will amount to $\$e = \$2.72$ (to the nearest cent). But what about the following more general question?

Q: *What about a more general scenario: If we invest an amount \$P for t years at an interest rate of r, compounded continuously, what will be the accumulated amount A at the end of that period?*

A: In the special case above (P, t, and r all equal 1), we took the compound interest formula and let m get larger and larger. We do the same more generally after a little preliminary work with the algebra of exponentials.

$$\begin{aligned}
A &= P\left(1 + \frac{r}{m}\right)^{mt} \\
&= P\left(1 + \frac{1}{(m/r)}\right)^{mt} && \text{Substituting } \tfrac{r}{m} = \tfrac{1}{(m/r)} \\
&= P\left(1 + \frac{1}{(m/r)}\right)^{(m/r)rt} && \text{Substituting } m = \left(\tfrac{m}{r}\right)r \\
&= P\left[\left(1 + \frac{1}{(m/r)}\right)^{(m/r)}\right]^{rt} && \text{Using the rule } a^{bc} = (a^b)^c
\end{aligned}$$

For continuous compounding of interest we let m, and hence m/r, get very large. This affects only the term in brackets, which converges to e, and we get the formula

$$A = Pe^{rt}.$$

Q: *How do I obtain powers of e or e itself on a TI-83/84 Plus or in a spreadsheet?*

A: On the TI-83/84 Plus, enter e^x as `e^(x)`, where `e^(` can be obtained by pressing [2ND] [LN]. To obtain the number e on the TI-83/84 Plus, enter `e^(1)`. Spreadsheets have a built-in function called `EXP`; `EXP(x)` gives the value of e^x. To obtain the number e in a spreadsheet, enter `=EXP(1)`.

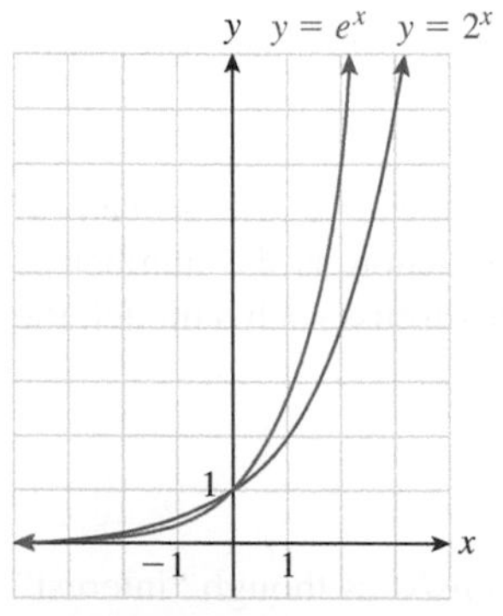

Technology formula: `e^(x)` or `EXP(x)`

Figure 12

Figure 12 shows the graph of $y = e^x$ with that of $y = 2^x$ for comparison.

The Number *e* and Continuous Compounding

The number *e* is the limiting value of the quantities $\left(1 + \frac{1}{m}\right)^m$ as *m* gets larger and larger and has the value 2.718281828459045235 36. . . .

If $P is invested at an annual interest rate *r* compounded continuously, the accumulated amount after *t* years is

$$A(t) = Pe^{rt}. \qquad \texttt{P*e\^{}(r*t)} \text{ or } \texttt{P*EXP(r*t)}$$

Quick Examples

7. If $100 is invested in an account that bears 15% interest compounded continuously, at the end of 10 years the investment will be worth

$$A(10) = 100e^{(0.15)(10)} = \$448.17. \qquad \texttt{100*e\^{}(0.15*10)} \text{ or } \texttt{100*EXP(0.15*10)}$$

8. If $1 is invested in an account that bears 100% interest compounded continuously, at the end of *x* years the investment will be worth

$$A(x) = e^x \text{ dollars.}$$

EXAMPLE 5 Continuous Compounding

a. You invest $10,000 at *Fastrack Savings & Loan,* which pays 6% compounded continuously. Express the balance in your account as a function of the number of years *t*, and calculate the amount of money you will have after 5 years.

b. Your friend has just invested $20,000 in *Constant Growth Funds,* whose stocks are continuously *declining* at a rate of 6% per year. How much will her investment be worth in 5 years?

c. T During which year will the value of your investment first exceed that of your friend?

Using Technology

TI-83/84 Plus

```
Y1=10000e^(0.06X)
Y2=20000e^(-0.06X)
```

[2ND] [TABLE]

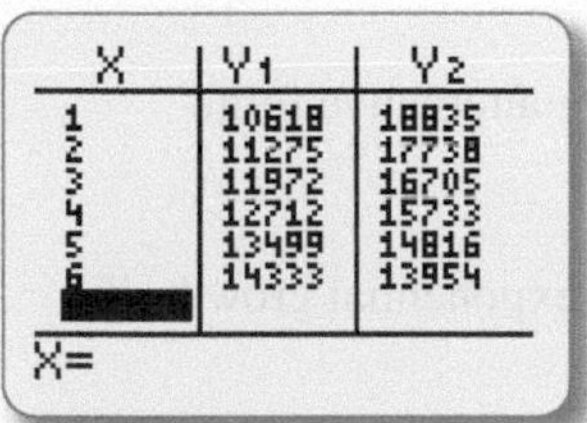

Spreadsheet

Headings t, Me, Friend in A1–C1; *t*-values 1, . . . , 6 in A2–A7.
=10000*exp(0.06*A2)
=20000*exp(-0.06*A2)
in B2 and C2, and copy down to C7.

	A	B	C
1	t	Me	Friend
2	1	10618.365	18835.291
3	2	11274.969	17738.409
4	3	11972.174	16705.404
5	4	12712.492	15732.557
6	5	13498.588	14816.364
7	6	14333.294	13953.527

Solution

a. We use the continuous growth formula with $P = 10{,}000$, $r = 0.06$, and *t* variable, getting

$$A(t) = Pe^{rt} = 10{,}000e^{0.06t}.$$

In 5 years,

$$\begin{aligned} A(5) &= 10{,}000e^{0.06(5)} \\ &= 10{,}000e^{0.3} \\ &\approx \$13{,}498.59. \end{aligned}$$

b. Because the investment is depreciating, we use a negative value for *r* and take $P = 20{,}000$, $r = -0.06$, and $t = 5$, getting

$$\begin{aligned} A(t) &= Pe^{rt} = 20{,}000e^{-0.06t} \\ A(5) &= 20{,}000e^{-0.06(5)} \\ &= 20{,}000e^{-0.3} \\ &\approx \$14{,}816.36. \end{aligned}$$

Website
www.WanerMath.com
Online Utilities
→ Function Evaluator and Grapher
10000e^(0.06X) under y_1
20000e^(-0.06X) under y_2
Enter 1, . . . , 6 as the x-values in the Evaluator, and press "Evaluate".

x-Values	y_1-Values	y_2-Values
1	10618	18835
2	11275	17738
3	11972	16705
4	12712	15733
5	13499	14816
6	14333	13954

c. We can answer the question now using a graphing calculator, a spreadsheet, or the Function Evaluator and Grapher tool at the Website. Just enter the exponential models of parts (a) and (b), and create tables to compute the values at the end of several years. (See the "Using Technology" note in the margin for details.) From the tables we see that the value of your investment overtakes that of your friend after $t = 5$ (the end of year 5) and before $t = 6$ (the end of year 6). Thus, your investment first exceeds that of your friend sometime during year 6.

➡ Before we go on . . .

Q: *How does continuous compounding compare with monthly compounding?*

A: To repeat the calculation in Example 5(a) using monthly compounding instead of continuous compounding, we use the compound interest formula with $P = 10{,}000$, $r = 0.06$, $m = 12$, and $t = 5$ and find

$$A(5) = 10{,}000(1 + 0.06/12)^{60} \approx \$13{,}488.50.$$

Thus, continuous compounding earns you approximately \$10 more than monthly compounding on a 5-year, \$10,000 investment. This is little to get excited about. ■

If we write the continuous compounding formula $A(t) = Pe^{rt}$ as $A(t) = P(e^r)^t$, we see that $A(t)$ is an exponential function of t, where the base is $b = e^r$, so we have really not introduced a new kind of function. In fact, exponential functions are often written in the following way.

Exponential Functions: Alternative Form

We can write any exponential function in the following alternative form:

$$f(x) = Ae^{rx},$$

where A and r are constants. If r is positive, f models exponential growth; if r is negative, f models exponential decay.*

Quick Examples

9. $f(x) = 100e^{0.15x}$ — Exponential growth $A = 100$, $r = 0.15$

10. $f(t) = Ae^{-0.00012101t}$ — Exponential decay of carbon 14; $r = -0.00012101$

11. $f(t) = 100e^{0.15t} = 100(e^{0.15})^t$
$= 100(1.1618)^t$ — Converting Ae^{rt} to the form Ab^t

* If r is positive, it is called the *growth constant*. If r is negative, its absolute value is called the *decay constant*. See Section 2.3 for more detailed discussion of growth and decay constants.

We will see in Chapter 4 that the exponential function with base e exhibits some interesting properties when we measure its rate of change, and this is the real mathematical importance of e.

Exponential Regression

Starting with a set of data that suggests an exponential curve, we can use technology to compute the exponential regression curve in much the same way as we did for the quadratic regression curve in Example 5 of Section 2.1.

EXAMPLE 6 T **Exponential Regression: Health Expenditures before "Obama Care"**

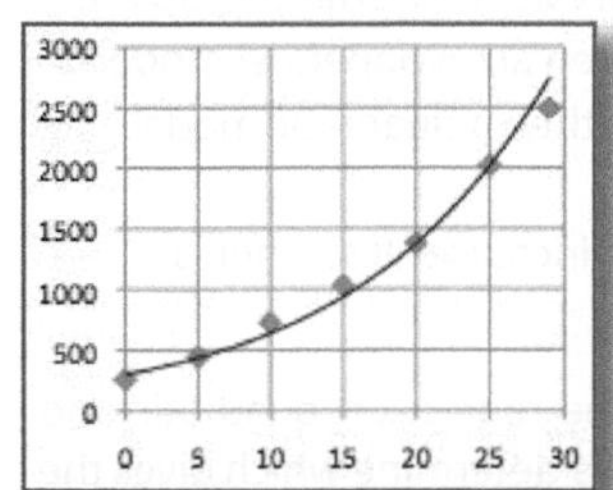

Figure 13

The following table shows annual expenditure on health in the United States from 1980 through 2009.[15]

Year t (years since 1980)	0	5	10	15	20	25	29
Expenditure ($ billion)	256	444	724	1,030	1,380	2,020	2,490

a. Find the exponential regression model

$$C(t) = Ab^t$$

for the annual expenditure.

b. Use the regression model to estimate the expenditure in 2002 ($t = 22$; the actual expenditure was approximately $1,640 billion).

Using Technology

See the Technology Guides at the end of the chapter for detailed instructions on how to obtain the regression curve and graph for Example 6 using a graphing calculator or spreadsheet. Outline:

TI-83/84 Plus
STAT EDIT values of t in L_1 and values of C in L_2.
Regression curve: STAT
CALC option #0 `ExpReg` ENTER
Graph: Y= VARS 5 EQ 1, then ZOOM 9
[More details in the Technology Guide.]

Spreadsheet
t- and C-values in columns A and B; graph these data. Regression curve: Add an exponential trendline with the option to show the equation.
[More details in the Technology Guide.]

Website
www.WanerMath.com
In the Simple Regression utility, enter the data in the x and y columns, and press "`y=a(b^x)`".

Solution

a. We use technology to obtain the exponential regression curve (see Figure 13):

$$C(t) \approx 296(1.08)^t. \quad \text{Coefficients rounded}$$

b. Using the model $C(t) \approx 296(1.08)^t$, we find that

$$C(22) \approx 296(1.08)^{22} \approx \$1,609 \text{ billion},$$

which is close to the actual number of about $1,640 billion.

➡ **Before we go on . . .** We said in the preceding section that the regression curve gives the smallest value of the sum-of-squares error, SSE (the sum of the squares of the residuals). However, exponential regression as computed via technology generally minimizes the sum of the squares of the residuals of the *logarithms* (logarithms are discussed in the next section). Using logarithms allows one easily to convert an exponential function into a linear one and then use linear regression formulas. However, in Section 2.4 we will discuss a way of using Excel's Solver to minimize SSE directly, which allows us to find the best-fit exponential curve directly without the need for devices to simplify the mathematics. If we do this, we obtain a very different equation:

$$C(t) \approx 353(1.07)^t.$$

If you plot this function, you will notice that it seems to fit the data more closely than the regression curve. ■

[15] Data are rounded. Source: U.S. Department of Health & Human Services/Centers for Medicare & Medicaid Services, National Health Expenditure Data, downloaded April 2011 from www.cms.gov.

FAQs

When to Use an Exponential Model for Data Points and When to Use *e* in Your Model

Q: *Given a set of data points that appear to be curving upward, how can I tell whether to use a quadratic model or an exponential model?*

A: Here are some things to look for:

- Do the data values appear to double at regular intervals? (For example, do the values approximately double every 5 units?) If so, then an exponential model is appropriate. If it takes longer and longer to double, then a quadratic model may be more appropriate.
- Do the values first decrease to a low point and then increase? If so, then a quadratic model is more appropriate.

It is also helpful to use technology to graph both the regression quadratic and exponential curves and to visually inspect the graphs to determine which gives the closest fit to the data.

Q: *We have two ways of writing exponential functions: $f(x) = Ab^x$ and $f(x) = Ae^{rx}$. How do we know which one to use?*

A: The two forms are equivalent, and it is always possible to convert from one form to the other.* So use whichever form seems to be convenient for a particular situation. For instance, $f(t) = A(3^t)$ conveniently models exponential growth that is tripling every unit of time, whereas $f(t) = Ae^{0.06t}$ conveniently models an investment with continuous compounding at 6%.

* Quick Example 11 shows how to convert Ae^{rx} to Ab^x. Conversion from Ab^x to Ae^{rx} involves logarithms: $r = \ln b$.

2.2 EXERCISES

▼ more advanced ◆ challenging

T indicates exercises that should be solved using technology

In Exercises 1–12, compute the missing values in the following table, and supply a valid technology formula for the given function: [**HINT:** See Quick Examples 1–4.]

x	-3	-2	-1	0	1	2	3
$f(x)$							

1. $f(x) = 4^x$

2. $f(x) = 3^x$

3. $f(x) = 3^{-x}$

4. $f(x) = 4^{-x}$

5. $g(x) = 2(2^x)$

6. $g(x) = 2(3^x)$

7. $h(x) = -3(2^{-x})$

8. $h(x) = -2(3^{-x})$

9. $r(x) = 2^x - 1$

10. $r(x) = 2^{-x} + 1$

11. $s(x) = 2^{x-1}$

12. $s(x) = 2^{1-x}$

In Exercises 13–18, graph the given function using a chart of values. (Use $-3 \le x \le 3$.)

13. $f(x) = 3^{-x}$

14. $f(x) = 4^{-x}$

15. $g(x) = 2(2^x)$

16. $g(x) = 2(3^x)$

17. $h(x) = -3(2^{-x})$

18. $h(x) = -2(3^{-x})$

In Exercises 19–24 the values of two functions, f and g, are given in a table. One, both, or neither of them may be exponential. Decide which, if any, are exponential, and give the exponential models for those that are. [**HINT:** See Example 1.]

19.

x	-2	-1	0	1	2
$f(x)$	0.5	1.5	4.5	13.5	40.5
$g(x)$	8	4	2	1	$\frac{1}{2}$

20.

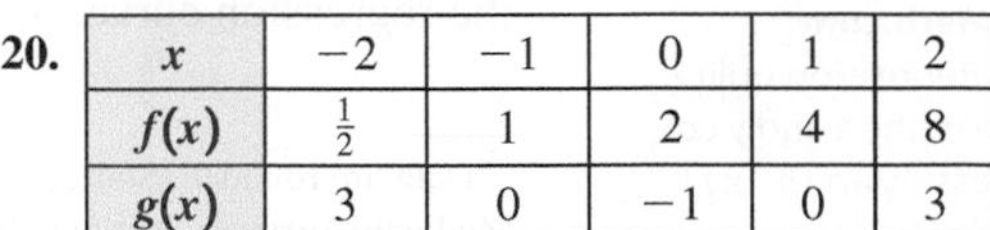

x	-2	-1	0	1	2
$f(x)$	$\frac{1}{2}$	1	2	4	8
$g(x)$	3	0	-1	0	3

21.

x	-2	-1	0	1	2
$f(x)$	22.5	7.5	2.5	7.5	22.5
$g(x)$	0.3	0.9	2.7	8.1	16.2

22.

x	-2	-1	0	1	2
$f(x)$	0.3	0.9	2.7	8.1	24.3
$g(x)$	3	1.5	0.75	0.375	0.1875

23.

x	-2	-1	0	1	2
$f(x)$	100	200	400	600	800
$g(x)$	100	20	4	0.8	0.16

24.

x	-2	-1	0	1	2
$f(x)$	0.8	0.2	0.1	0.05	0.025
$g(x)$	80	40	20	10	2

T *In Exercises 25–30, supply a valid technology formula for the given function, and then use technology to compute the missing values in the following table accurate to four decimal places.* [**HINT:** See Quick Examples 1–4.]

x	-3	-2	-1	0	1	2	3
$f(x)$							

25. $f(x) = e^{-2x}$

26. $g(x) = e^{x/5}$

27. $h(x) = 1.01(2.02^{-4x})$

28. $h(x) = 3.42(3^{-x/5})$

29. $r(x) = 50\left(1 + \frac{1}{3.2}\right)^{2x}$

30. $r(x) = 0.043\left(4.5 - \frac{5}{1.2}\right)^{-x}$

In Exercises 31–38, supply a valid technology formula for the given function.

31. 2^{x-1}

32. 2^{-4x}

33. $\frac{2}{1 - 2^{-4x}}$

34. $\frac{2^{3-x}}{1 - 2^x}$

35. $\frac{(3 + x)^{3x}}{x + 1}$

36. $\frac{20.3^{3x}}{1 + 20.3^{2x}}$

37. $2e^{(1+x)/x}$

38. $\frac{2e^{2/x}}{x}$

T *In Exercises 39–46, use technology to graph the pairs of functions on the same set of axes with $-3 \le x \le 3$. Identify which graph corresponds to which function.* [**HINT:** See Quick Examples 5 and 6.]

39. $f_1(x) = 1.6^x$, $f_2(x) = 1.8^x$

40. $f_1(x) = 2.2^x$, $f_2(x) = 2.5^x$

41. $f_1(x) = 300(1.1^x)$, $f_2(x) = 300(1.1^{2x})$

42. $f_1(x) = 100(1.01^{2x})$, $f_2(x) = 100(1.01^{3x})$

43. $f_1(x) = 2.5^{1.02x}$, $f_2(x) = e^{1.02x}$

44. $f_1(x) = 2.5^{-1.02x}$, $f_2(x) = e^{-1.02x}$

45. $f_1(x) = 1{,}000(1.045^{-3x})$, $f_2(x) = 1{,}000(1.045^{3x})$

46. $f_1(x) = 1{,}202(1.034^{-3x})$, $f_2(x) = 1{,}202(1.034^{3x})$

In Exercises 47–54, model the data using an exponential function $f(x) = Ab^x$. [**HINT:** See Example 1.]

47.

x	0	1	2
$f(x)$	500	250	125

48.

x	0	1	2
$f(x)$	500	1,000	2,000

49.

x	0	1	2
$f(x)$	10	30	90

50.

x	0	1	2
$f(x)$	90	30	10

51.

x	0	1	2
$f(x)$	500	225	101.25

52.

x	0	1	2
$f(x)$	5	3	1.8

53.

x	1	2
$f(x)$	-110	-121

54.

x	1	2
$f(x)$	-41	-42.025

In Exercises 55–62, find an equation for an exponential function that passes through the pair of points given. (Round all coefficients to four decimal places when necessary.) [**HINT:** See Example 2.]

55. Through (2, 36) and (4, 324)

56. Through (2, −4) and (4, −16)

57. Through (−2, −25) and (1, −0.2)

58. Through (1, 1.2) and (3, 0.108)

59. Through (1, 3) and (3, 6)

60. Through (1, 2) and (4, 6)

61. Through (2, 3) and (6, 2)

62. Through (−1, 2) and (3, 1)

In Exercises 63–66, obtain an exponential function in the form $f(t) = Ae^{rt}$. [**HINT:** See Example 5.]

63. $f(t)$ is the value after t years of a \$5,000 investment earning 10% interest compounded continuously.

64. $f(t)$ is the value after t years of a \$2,000 investment earning 5.3% interest compounded continuously.

65. $f(t)$ is the value after t years of a \$1,000 investment depreciating continuously at an annual rate of 6.3%.

66. $f(t)$ is the value after t years of a \$10,000 investment depreciating continuously at an annual rate of 60%.

T *In Exercises 67–70, use technology to find the exponential regression function through the given points. (Round all coefficients to four decimal places.)* [**HINT:** See Example 6.]

67. (1, 2), (3, 5), (4, 9), (5, 20)

68. (−1, 2), (−3, 5), (−4, 9), (−5, 20)

69. $(-1, 10), (-3, 5), (-4, 3)$

70. $(3, 3), (4, 5), (5, 10)$

Applications

71. ***Aspirin*** Soon after taking an aspirin, a patient has absorbed 300 milligrams of the drug. After 2 hours, only 75 milligrams remain in the bloodstream. Find an exponential model for the amount of aspirin in the bloodstream after t hours, and use your model to find the amount of aspirin in the bloodstream after 5 hours. [**HINT:** See Example 2.]

72. ***Alcohol*** After a large number of drinks, a person has a blood alcohol level of 200 mg/dL (milligrams per deciliter). If the amount of alcohol in the blood decays exponentially, and after 2 hours, 112.5 mg/dL remain, find an exponential model for the person's blood alcohol level, and use your model to estimate the person's blood alcohol level after 4 hours. [**HINT:** See Example 2.]

73. ***Freon Production*** The production of ozone-layer-damaging Freon 22 (chlorodifluoromethane) in developing countries rose from 200 tons in 2004 to a projected 590 tons in 2010.[16]

a. Use this information to find both a linear model and an exponential model for the amount F of Freon 22 (in tons) as a function of time t in years since 2000. (Round all coefficients to three significant digits.) [**HINT:** See Example 2.] Which of these models would you judge to be more appropriate to the data shown below?

Year t (years since 2000)	0	2	4	6	8	10
Amount of Freon 22 F (tons)	100	140	200	270	400	590

b. Use the better of the two models from part (a) to predict the 2008 figure, and compare it with the projected figure above.

74. ***Revenue*** The annual revenue of **Amazon** rose from approximately \$10.7 billion in 2006 to \$34.2 billion in 2010.[17]

a. Use this information to find both a linear model and an exponential model for Amazon's annual revenue I (in billions of dollars) as a function of time t in years since 2000. (Round all coefficients to three significant digits.) [**HINT:** See Example 2.] Which of these models would you judge to be more appropriate to the data shown below?

Year t (years since 2000)	6	7	8	9	10
Annual Revenue I (\$ billion)	10.7	14.8	19.2	24.5	34.2

b. Use the better of the two models from part (a) to predict the 2008 figure, and compare it with the actual figure above.

75. ▼ ***U.S. Population*** The U.S. population was 180 million in 1960 and 309 million in 2010.[18]

a. Use these data to give an exponential growth model showing the U.S. population P as a function of time t in years since 1960. Round coefficients to six significant digits. [**HINT:** See Example 2.]

b. By experimenting, determine the smallest number of significant digits to which you should round the coefficients in part (a) to obtain the correct 2010 population figure accurate to three significant digits.

c. Using the model in part (a), predict the population in 2020.

76. ▼ ***World Population*** World population was estimated at 2.56 billion people in 1950 and 6.91 billion people in 2011.[19]

a. Use these data to give an exponential growth model showing the world population P as a function of time t in years since 1950. Round coefficients to six significant digits. [**HINT:** See Example 2.]

b. By experimenting, determine the smallest number of significant digits to which you should round the coefficients in part (a) to obtain the correct 2011 population figure to three significant digits.

c. Assuming the exponential growth model from part (a), estimate the world population in the year 1000. Comment on your answer.

77. ▼ ***Frogs*** Frogs have been breeding like flies at the *Enormous State University* (ESU) campus! Each year, the pledge class of the Epsilon Delta fraternity is instructed to tag all the frogs residing on the ESU campus. Two years ago, they managed to tag all 50,000 of them (with little Epsilon Delta Fraternity tags). This year's pledge class discovered that last year's tags had all fallen off, and they wound up tagging a total of 75,000 frogs.

a. Find an exponential model for the frog population.

b. Assuming exponential population growth and that all this year's tags have fallen off, how many tags should Epsilon Delta order for next year's pledge class?

78. ▼ ***Flies*** Flies in Suffolk County have been breeding like frogs! Three years ago, the Health Commission caught 4,000 flies in a trap in 1 hour. This year, it caught 7,000 flies in 1 hour.

a. Find an exponential model for the fly population.

b. Assuming exponential population growth, how many flies should the commission expect to catch next year?

---

[16] Figures are approximate. Source: Lampert Kuijpers (Panel of the Montreal Protocol), National Bureau of Statistics in China, via CEIC Data/*New York Times*, February 23, 2007, p. C1.

[17] Source for data: www.wikinvest.com.

[18] Figures are rounded to three significant digits. Source: U.S. Census Bureau (www.census.gov).

[19] *Ibid.*

79. ***Bacteria*** A bacteria culture starts with 1,000 bacteria and doubles in size every 3 hours. Find an exponential model for the size of the culture as a function of time t in hours, and use the model to predict how many bacteria there will be after 2 days. [**HINT:** See Example 4.]

80. ***Bacteria*** A bacteria culture starts with 1,000 bacteria. Two hours later there are 1,500 bacteria. Find an exponential model for the size of the culture as a function of time t in hours, and use the model to predict how many bacteria there will be after 2 days. [**HINT:** See Example 4.]

81. ***The 2003 SARS Outbreak*** In the early stages of the deadly SARS (severe acute respiratory syndrome) epidemic in 2003, the number of cases was increasing by about 18% each day.[20] On March 17, 2003 (the first day for which statistics were reported by the World Health Organization), there were 167 cases. Find an exponential model that predicts the number of cases t days after March 17, 2003, and use it to estimate the number of cases on March 31, 2003. (The actual reported number of cases was 1,662.)

82. ***The 2003 SARS Outbreak*** A few weeks into the deadly SARS (severe acute respiratory syndrome) epidemic in 2003, the number of cases was increasing by about 4% each day.[21] On April 1, 2003 there were 1,804 cases. Find an exponential model that predicts the number of cases t days after April 1, 2003, and use it to estimate the number of cases on April 30, 2003. (The actual reported number of cases was 5,663.)

83. ***The 2014 Ebola Outbreak*** In the first six months of the 2014 Ebola outbreak, the total number of reported cases was increasing at a continuous rate of about 72% per month.[22] There were about 100 cases as of April 1, 2014. Find an exponential model in the form $C(t) = Ae^{rt}$ for the number of cases t months after April 1, 2014, and use it to predict the number of cases as of August 1, 2014. (The actual number of reported cases was 1,603.)
[**HINT:** See Example 5.]

84. ***The 2014 Ebola Outbreak*** In the first six months of the 2014 Ebola outbreak, the total number of reported deaths from the disease was increasing at a continuous rate of about 60% per month.[23] There were about 90 deaths as of April 1, 2014. Find an exponential model in the form $D(t) = Ae^{rt}$ for the number of deaths t months after April 1, 2014, and use it to predict the number of deaths as of October 1, 2014. (The actual number of reported deaths was 3,439.) [**HINT:** See Example 5.]

85. ***Investments*** In August 2013, **E*TRADE Financial** was offering only 0.05% interest on its online checking accounts, with interest reinvested monthly.[24] Find the associated exponential model for the value of a \$5,000 deposit after t years. Assuming that this rate of return continued for 7 years, how much would a deposit of \$5,000 in August 2013 be worth in August 2020? (Answer to the nearest \$1.) [**HINT:** See Example 3; you saw this exercise before in Section 1.2.]

86. ***Investments*** In August 2013, **Ally Bank** was offering 0.61% interest on its Online Savings Account, with interest reinvested daily.[25] Find the associated exponential model for the value of a \$4,000 deposit after t years. Assuming that this rate of return continued for 8 years, how much would a deposit of \$4,000 in August 2013 be worth in August 2021? (Answer to the nearest \$1.)

87. T ***Investments*** Refer to Exercise 85. In August of which year will an investment of \$5,000 made in August 2013 first exceed \$5,050? [**HINT:** See Example 5; you saw this exercise before in Section 1.2.]

88. T ***Investments*** Refer to Exercise 86. In August of which year will an investment of \$4,000 made in August 2013 first exceed \$4,400? [**HINT:** See Example 5; you saw this exercise before in Section 1.2.]

89. ***Investments*** *Rock Solid Bank & Trust* is offering a CD (certificate of deposit) that pays 4% compounded continuously. How much interest would a \$1,000 deposit earn over 10 years? [**HINT:** See Example 5.]

90. ***Savings*** *FlybynightSavings.com* is offering a savings account that pays 31% interest compounded continuously. How much interest would a deposit of \$2,000 earn over 10 years?

91. ***Home Sales*** Sales of existing homes in the United States rose continuously over the period 2011–2013 at the rate of 8.9% per year from 4.3 million in 2011.[26] Write down a formula that predicts sales of existing homes t years after 2011. Use your model to estimate, to the nearest 0.1 million, sales of existing homes in 2012 and 2014.

92. ***Home Prices*** The median selling price of an existing home in the United States increased continuously over the period 2011–2013 at the rate of 8.5% per year from approximately \$166,000 in 2011.[27] Write down a formula that predicts the median selling price of an existing home t years after 2011. Use your model to estimate, to the nearest \$1,000, the median selling price of an existing home in 2013 and 2015.

---

[20] Source: World Health Organization (www.who.int).

[21] *Ibid.*

[22] Exponential model is the authors'. Source for data: Wikipedia/Centers for Disease Control and Prevention/WHO.

[23] *Ibid.*

[24] Source: https://us.etrade.com, August 2013.

[25] Interest rate based on annual percentage yield. Source: www.ally.com, August 2013.

[26] Source: National Association of Realtors, www.realtor.org.

[27] *Ibid.*

93. ***Climate Change*** The most abundant greenhouse gas is carbon dioxide. According to figures from the Intergovernmental Panel on Climate Change (IPCC), the amount of carbon dioxide in the atmosphere (in parts of volume per million) can be approximated by

$$C(t) \approx 280e^{0.00127t} \text{ parts per million,}$$

where t is time in years since 1750.[28]

a. Use the model to estimate the amount of carbon dioxide in the atmosphere in 1950, 2000, 2050, and 2100.

b. According to the model, when, to the nearest decade, will the level surpass 400 parts per million?

94. ***Climate Change*** Another greenhouse gas is methane. According to figures from the Intergovernmental Panel on Climate Change (IPCC), the amount of methane in the atmosphere (in parts of volume per billion) can be approximated by

$$C(t) \approx 720e^{0.00351t} \text{ parts per billion,}$$

where t is time in years since 1750.[29]

a. Use the model to estimate the amount of methane in the atmosphere in 1950, 2000, 2050, and 2100. (Round your answers to the nearest 10 parts per billion.)

b. According to the model, when, to the nearest decade, will the level surpass 2,000 parts per billion?

95. T ***New York City Housing Costs: Downtown*** The following table shows the average price of a two-bedroom apartment in downtown New York City during the real estate boom from 1994 to 2004.[30]

Year t	0 (1994)	2	4	6	8	10 (2004)
Price ($ million)	0.38	0.40	0.60	0.95	1.20	1.60

a. Use exponential regression to model the price $P(t)$ as a function of time t since 1994. Include a sketch of the points and the regression curve. (Round the coefficients to three decimal places.) [**HINT**: See Example 6.]

b. Extrapolate your model to estimate the cost of a two-bedroom downtown apartment in 2005.

96. T ***New York City Housing Costs: Uptown*** The following table shows the average price of a two-bedroom apartment in uptown New York City during the real estate boom from 1994 to 2004.[31]

Year t	0 (1994)	2	4	6	8	10 (2004)
Price ($ million)	0.18	0.18	0.19	0.2	0.35	0.4

a. Use exponential regression to model the price $P(t)$ as a function of time t since 1994. Include a sketch of the points and the regression curve. (Round the coefficients to three decimal places.)

b. Extrapolate your model to estimate the cost of a two-bedroom uptown apartment in 2005.

97. T ***Facebook*** The following table gives the approximate numbers of **Facebook** members at various times early in its history.[32]

Year t (since start of 2005)	0	0.5	1	1.5	2	2.5	3	3.5
Facebook Members n (millions)	1	2	5.5	7	12	30	58	80

a. Use exponential regression to model Facebook membership as a function of time in years since the start of 2005, and graph the data points and regression curve. (Round coefficients to three decimal places.)

b. Fill in the missing quantity: According to the model, Facebook membership each year was ____ times that of the year before.

c. Use your model to estimate Facebook membership in early 2009 to the nearest million.

98. T ***Freon Production*** The following table shows Freon 22 production in developing countries in various years since 2000.[33]

Year t (years since 2000)	0	2	4	6	8	10
Amount of Freon 22 F (tons)	100	140	200	270	400	590

a. Use exponential regression to model Freon 22 production as a function of time in years since 2000, and graph the data points and regression curve. (Round coefficients to three decimal places.)

b. Fill in the missing quantity: According to the model, Freon 22 production each year was ____ times that of the year before.

c. Use your model to estimate Freon 22 production in 2009 to the nearest ton.

[28] Authors' exponential model based on the 1750 and 2011 figures. Source for data: IPCC Fifth Assessment Report: Climate Change 2013, www.ipcc.ch.

[29] *Ibid.*

[30] Data are rounded, and the 2004 figure is an estimate. Source: Miller Samuel/*New York Times*, March 28, 2004, p. RE 11.

[31] *Ibid.*

[32] Sources: www.facebook.com, www.insidehighered.com. (Some data are interpolated.)

[33] Figures are approximate. Source: Lampert Kuijpers (Panel of the Montreal Protocol), National Bureau of Statistics in China, via CEIC Data/*New York Times*, February 23, 2007, p. C1.

Communication and Reasoning Exercises

99. Which of the following three functions will be largest for large values of x?
(A) $f(x) = x^2$
(B) $r(x) = 2^x$
(C) $h(x) = x^{10}$

100. Which of the following three functions will be smallest for large values of x?
(A) $f(x) = x^{-2}$
(B) $r(x) = 2^{-x}$
(C) $h(x) = x^{-10}$

101. What limitations apply to using an exponential function to model growth in real-life situations? Illustrate your answer with an example.

102. Explain in words why 5% per year compounded continuously yields more interest than 5% per year compounded monthly.

103. ▼ The following commentary and graph appeared in politicalcalculations.blogspot.com on August 30, 2005:[34]

> One of the neater blogs I've recently encountered is The Real Returns, which offers a wealth of investing, market and economic data. Earlier this month, The Real Returns posted data related to the recent history of U.S. median house prices over the period from 1963 to 2004. The original source of the housing data is the U.S. Census Bureau.
>
> Well, that kind of data deserves some curve-fitting and a calculator to estimate what the future U.S. median house price might be, so Political Calculations has extracted the data from 1973 onward to create the following chart:

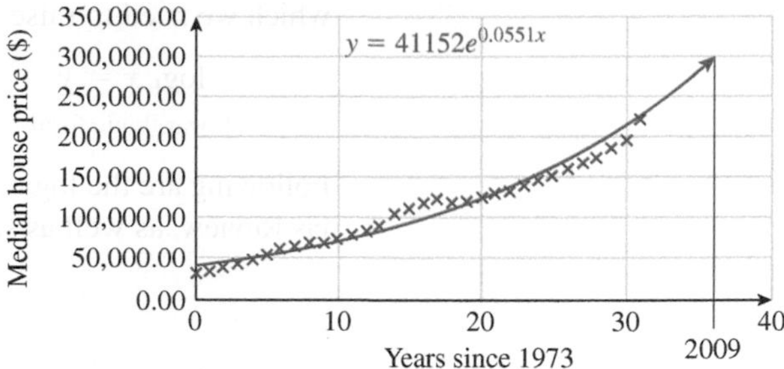

Comment on the article and graph. [**HINT:** See Exercise 92.]

104. ▼ Refer to Exercise 103. Of what possible predictive use, then, is the kind of exponential model given by the blogger in the article?

105. ▼ Describe two real-life situations in which a linear model would be more appropriate than an exponential model, and two situations in which an exponential model would be more appropriate than a linear model.

106. ▼ Describe a real-life situation in which a quadratic model would be more appropriate than an exponential model and one in which an exponential model would be more appropriate than a quadratic model.

107. How would you check whether data points of the form $(1, y_1)$, $(2, y_2)$, $(3, y_3)$ lie on an exponential curve?

108. ▼ You are told that the points $(1, y_1)$, $(2, y_2)$, $(3, y_3)$ lie on an exponential curve. Express y_3 in terms of y_1 and y_2.

109. ▼ Your local banker tells you that the reason his bank doesn't compound interest continuously is that it would be too demanding of computer resources because the computer would need to spend a great deal of time keeping all accounts updated. Comment on his reasoning.

110. ▼ Your other local banker tells you that the reason *her* bank doesn't offer continuously compounded interest is that it is equivalent to offering a fractionally higher interest rate compounded daily. Comment on her reasoning.

[34] The graph was re-created by the authors using the blog author's data source.
Source for article: politicalcalculations.blogspot.com/2005/08/projecting-us-median-housing-prices.html.
Source for data: therealreturns.blogspot.com/2005_08_01_archive.html.

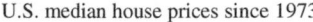

2.3 Logarithmic Functions and Models

Logarithms and Logarithmic Functions

Logarithms were invented by John Napier (1550–1617) in the late sixteenth century as a means of aiding calculation. His invention made possible the prodigious hand calculations of astronomer Johannes Kepler (1571–1630), who was the first to describe accurately the orbits and the motions of the planets. Today, computers and calculators have done away with that use of logarithms, but many other uses remain. In particular, the logarithm is used to model real-world phenomena in numerous fields, including physics, finance, and economics.

Recall from Section 0.8 that the **base b logarithm of x**, $\log_b x$, is the power to which we need to raise b in order to get x. Symbolically,

$$\log_b x = y \quad \text{means} \quad b^y = x.$$

Logarithmic form — Exponential form

Following are the logarithm identities we will need from Section 0.8 in the Precalculus Review, as well as some further properties.

Website
www.WanerMath.com
Follow the path
Chapter 2
→ Using and Deriving Algebraic Properties of Logarithms
to find a list of logarithmic identities and a discussion of where they come from.
Follow the path
Chapter 2
→ Inverse Functions
for a general discussion of inverse functions, including further discussion of the relationship between logarithmic and exponential functions.

Logarithm Identities

The following identities hold for all positive bases $a \neq 1$ and $b \neq 1$, all positive numbers x and y, and every real number r. These identities follow from the laws of exponents.

Identity	Quick Examples
1. $\log_b(xy) = \log_b x + \log_b y$	$\log_2 16 = \log_2 8 + \log_2 2$
2. $\log_b\left(\frac{x}{y}\right) = \log_b x - \log_b y$	$\log_2\left(\frac{5}{3}\right) = \log_2 5 - \log_2 3$
3. $\log_b(x^r) = r \log_b x$	$\log_2(6^5) = 5 \log_2 6$
4. $\log_b b = 1$; $\log_b 1 = 0$	$\log_2 2 = 1$; $\ln e = 1$; $\log_{11} 1 = 0$
5. $\log_b\left(\frac{1}{x}\right) = -\log_b x$	$\log_2\left(\frac{1}{3}\right) = -\log_2 3$
6. $\log_b x = \frac{\log_a x}{\log_a b}$	$\log_2 5 = \frac{\log_{10} 5}{\log_{10} 2} = \frac{\log 5}{\log 2}$

Common Logarithm, Natural Logarithm

The following are standard abbreviations.

		TI-83/84 Plus and Spreadsheet Formula
Base 10: $\log_{10} x = \log x$	Common logarithm	`log(x)`
Base e: $\log_e x = \ln x$	Natural logarithm	`ln(x)`

Quick Examples

1. $\ln e = 1$ — $e^1 = e$
2. $\ln 1 = 0$ — $e^0 = 1$
3. $\ln 2 = 0.69314718\ldots$ — $e^{0.69314718\ldots} = 2$

Log Base b and b to a Power

The following two identities, which follow directly from the definition of the base b logarithm, demonstrate that the operations of taking the base b logarithm and raising b to a power are *inverse* to each other.

Identity

1. $\log_b(b^x) = x$

In words: The power to which you raise b in order to get b^x is x. (!)

2. $b^{\log_b x} = x$

In words: Raising b to the power to which it must be raised to get x yields x. (!)

Quick Examples

$\log_2(2^7) = 7$

$5^{\log_5 8} = 8$

Some technologies (such as calculators) do not permit direct calculation of logarithms other than common and natural logarithms. To compute logarithms with other bases with these technologies, we can use identity (6) from the list above to compute them using either common or natural logarithms.

Change-of-Base Formula

$$\log_b a = \frac{\log a}{\log b} = \frac{\ln a}{\ln b}$$ Change-of-base formula*

*Here is a quick explanation of why this formula works: To calculate $\log_b a$, we ask, "to what power must we raise b to get a?" To check the formula, we try using $\log a/\log b$ as the exponent.

$$b^{\frac{\log a}{\log b}} = (10^{\log b})^{\frac{\log a}{\log b}}$$

(because $b = 10^{\log b}$)

$$= 10^{\log a} = a$$

so this exponent works!

Quick Examples

4. $\log_{11} 9 = \dfrac{\log 9}{\log 11} \approx 0.91631$ `log(9)/log(11)`

5. $\log_{11} 9 = \dfrac{\ln 9}{\ln 11} \approx 0.91631$ `ln(9)/ln(11)`

6. $\log_{3.2}\left(\dfrac{1.42}{3.4}\right) \approx -0.75065$ `log(1.42/3.4)/log(3.2)`

Using Technology to Compute Logarithms

To compute $\log_b x$ using technology, use the following formulas:

TI-83/84 Plus	`log(x)/log(b)`	Example: $\log_2(16)$ is `log(16)/log(2)`
Spreadsheet:	`=LOG(x,b)`	Example: $\log_2(16)$ is `=LOG(16,2)`

The functions that we shall study in the remainder of this section have the following form.

Logarithmic Function

A **logarithmic function** has the form

$$f(x) = \log_b x + C$$ (b and C are constants with $b > 0, b \neq 1$)

or, alternatively,

$$f(x) = A \ln x + C.$$ (A and C are constants with $A \neq 0$)

Quick Examples

7. $f(x) = \log x$
8. $g(x) = \ln x - 5$
9. $h(x) = \log_2 x + 1$
10. $k(x) = 3.2 \ln x + 7.2$

Q: *What is the difference between the two forms of the logarithmic function?*

A: None, really—they're equivalent: We can start with an equation in the first form and use the change-of-base formula to rewrite it:

$$\begin{aligned} f(x) &= \log_b x + C \\ &= \frac{\ln x}{\ln b} + C \qquad \text{Change-of-base formula} \\ &= \left(\frac{1}{\ln b}\right)\ln x + C. \end{aligned}$$

Our function now has the form $f(x) = A \ln x + C$, where $A = 1/\ln b$. We could go the other way as well, to rewrite $A \ln x + C$ in the form $\log_b x + C$.

EXAMPLE 1 Graphs of Logarithmic Functions

a. Sketch the graph of $f(x) = \log_2 x$ by hand.

b. Use technology to compare the graph in part (a) with the graphs of $\log_b x$ for $b = 1/4$, $1/2$, and 4.

Solution

a. To sketch the graph of $f(x) = \log_2 x$ by hand, we begin with a table of values. Because $\log_2 x$ is not defined when $x = 0$, we choose several values of x close to zero and also some larger values, all chosen so that their logarithms are easy to compute:

x	$\frac{1}{8}$	$\frac{1}{4}$	$\frac{1}{2}$	1	2	4	8
$f(x) = \log_2 x$	-3	-2	-1	0	1	2	3

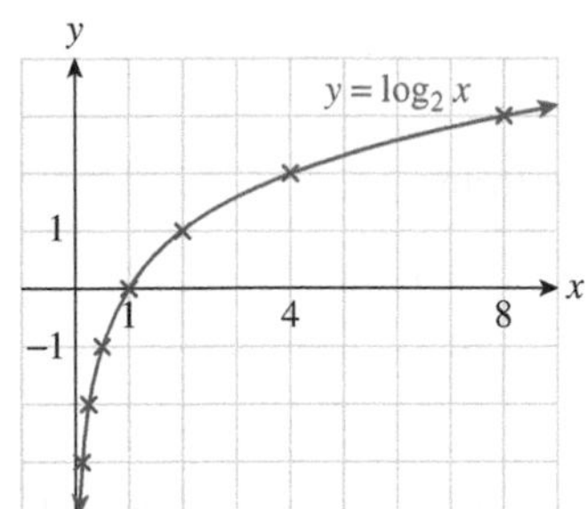

Figure 14

Graphing these points and joining them by a smooth curve, we get Figure 14.

b. We enter the logarithmic functions in graphing utilities as follows (note the use of the change-of-base formula in the TI-83/84 Plus version):

TI-83/84 Plus	Spreadsheet
`Y1=log(X)/log(0.25)`	`=LOG(x,0.25)`
`Y2=log(X)/log(0.5)`	`=LOG(x,0.5)`
`Y3=log(X)/log(2)`	`=LOG(x,2)`
`Y4=log(X)/log(4)`	`=LOG(x,4)`

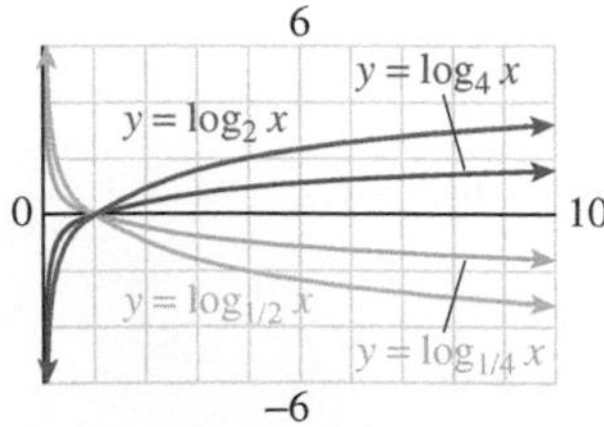

Figure 15

Figure 15 shows the resulting graphs.

➡ **Before we go on . . .** Notice that the graphs of the logarithmic functions in Example 1 all pass through the point (1, 0). (Why?) Notice further that the graphs of the logarithmic functions with bases less than 1 are upside-down versions of the others. Finally, how are these graphs related to the graphs of exponential functions? ■

Applications

EXAMPLE 2 Investments: How Long?

Ten-year government bonds sold by Mexico are yielding 3.50% per year.[35] At that interest rate, how long will it take a \$1,000 investment to be worth \$1,200 if the interest is compounded monthly?

Solution Substituting $A = 1{,}200$, $P = 1{,}000$, $r = 0.0350$, and $m = 12$ in the compound interest equation gives

$$A(t) = P\left(1 + \frac{r}{m}\right)^{mt}$$

$$1{,}200 = 1{,}000\left(1 + \frac{0.0350}{12}\right)^{12t}$$

$$\approx 1{,}000(1.002917)^{12t},$$

and we must solve for t. We first divide both sides by 1,000, getting an equation in exponential form:

$$1.2 = 1.002917^{12t}.$$

In logarithmic form, this becomes

$$12t = \log_{1.002917}(1.2).$$

We can now solve for t:

$$t = \frac{\log_{1.002917}(1.2)}{12} \approx 5.2 \text{ years.}$$ `log(1.2)/(log(1.002917)*12)`

Thus, it will take approximately 5.2 years for a \$1,000 investment to be worth \$1,200.

➡ **Before we go on . . .** We can use the logarithm identities to solve the equation

$$1.2 = 1.002917^{12t}$$

that arose in Example 2 (and also more general equations with unknowns in the exponent) by taking the natural logarithm of both sides:

$$\ln 1.2 = \ln(1.002917^{12t})$$

$$= 12t \ln 1.002917. \quad \text{By identity (3)}$$

We can now solve this for t to get

$$t = \frac{\ln 1.2}{12 \ln 1.002917},$$

which, by the change-of-base formula, is equivalent to the answer we got in Example 2. ■

[35] As of September 2014. Source: www.bloomberg.com.

EXAMPLE 3 Half-Life

a. The weight of carbon 14 that remains in a sample that originally contained A grams is given by

$$C(t) = A(0.999879)^t,$$

where t is time in years. Find the **half-life**, the time it takes half of the carbon 14 in a sample to decay.

b. Repeat part (a) using the following alternative form of the exponential model in part (a):

$$C(t) = Ae^{-0.00012101t}.$$ See Quick Example 10 in Section 2.2.

c. Another radioactive material has a half-life of 7,000 years. Find an exponential decay model in the form

$$R(t) = Ae^{-kt}$$

for the amount of undecayed material remaining. (The constant k is called the **decay constant**.)

d. How long, to the nearest 10 years, will it take for 99.95% of the substance in a sample of the material in part (c) to decay?

Solution

a. We want to find the value of t for which $C(t)$ = the weight of undecayed carbon 14 left = half the original weight = $0.5A$. Substituting, we get

$$0.5A = A(0.999879)^t.$$

Dividing both sides by A gives

$$0.5 = 0.999879^t$$

$$t = \log_{0.999879} 0.5 \approx 5{,}728 \text{ years.}$$ Logarithmic form

b. This is similar to part (a): We want to solve the equation

$$0.5A = Ae^{-0.00012101t}$$

for t. Dividing both sides by A gives

$$0.5 = e^{-0.00012101t}.$$

Taking the natural logarithm of both sides gives

$$\ln(0.5) = \ln(e^{-0.00012101t}) = -0.00012101t$$ By identity (3): $\ln(e^a) = a \ln e = a$

$$t = \frac{\ln(0.5)}{-0.00012101} \approx 5{,}728 \text{ years,}$$

as we obtained in part (a).

c. This time, we are given the half-life, which we can use to find the exponential model $R(t) = Ae^{-kt}$. At time $t = 0$ the amount of radioactive material is

$$R(0) = Ae^0 = A.$$

Because half of the sample decays in 7,000 years, this sample will decay to $0.5A$ grams in 7,000 years ($t = 7{,}000$). Substituting this information gives

$$0.5A = Ae^{-k(7{,}000)}.$$

Canceling A and taking natural logarithms (again using identity (3)), we get

$$\ln(0.5) = -7{,}000k,$$

so the decay constant k is

$$k = -\frac{\ln(0.5)}{7{,}000} \approx 0.000099021.$$

Therefore, the model is

$$R(t) = Ae^{-0.000099021t}.$$

d. If 99.95% of the substance in a sample has decayed, then the amount of undecayed material left is 0.05% of the original amount, or $0.0005A$. We have

$$0.0005A = Ae^{-0.000099021t}$$
$$0.0005 = e^{-0.000099021t}$$
$$\ln(0.0005) = -0.000099021t$$
$$t = \frac{\ln(0.0005)}{-0.000099021} \approx 76{,}760 \text{ years.}$$

Before we go on . . .

Q: *In parts (a) and (b) of Example 4 we were given two different forms of the model for carbon 14 decay. How do we convert an exponential function in one form to the other?*

A: We have already seen (Quick Example 11 in Section 2.2) how to convert from the form $f(t) = Ae^{rt}$ in Example 3(b) to the form $f(t) = Ab^t$ in Example 3(a). To go the other way, start with the model in Example 3(a), and equate it to the desired form:

$$C(t) = A(0.999879)^t = Ae^{rt}.$$

To solve for r, cancel the As, and take the natural logarithm of both sides:

$$t\ln(0.999879) = rt\ln e = rt$$

so $r = \ln(0.999879) \approx -0.000121007,$

giving

$$C(t) = Ae^{-0.00012101t}$$

as in Example 3(b).

We can use the work we did in parts (b) and (c) of Example 3 to obtain a formula for the decay constant in an exponential decay model for any radioactive substance when we know its half-life. Write the half-life as t_h. Then the calculation in Example 3(b) gives

$$k = -\frac{\ln(0.5)}{t_h} = \frac{\ln 2}{t_h}. \qquad -\ln(0.5) = -\ln\left(\frac{1}{2}\right) = \ln 2$$

Multiplying both sides by t_h gives us the relationship $t_h k = \ln 2$.

Exponential Decay Model and Half-Life

An **exponential decay function** has the form

$$Q(t) = Q_0 e^{-kt}. \qquad Q_0,\ k \text{ both positive}$$

Q_0 represents the value of Q at time $t = 0$, and k is the **decay constant**. The decay constant k and half-life t_h for Q are related by

$$t_h k = \ln 2.$$

Quick Examples

11. $Q(t) = Q_0e^{-0.00012101t}$ is the decay function for carbon 14 (see Example 4b).

12. If $t_h = 10$ years, then $10k = \ln 2$, so $k = \frac{\ln 2}{10} \approx 0.06931$, and the decay model is
$$Q(t) = Q_0e^{-0.06931t}.$$

13. If $k = 0.0123$, then $t_h(0.0123) = \ln 2$, so the half-life is
$$t_h = \frac{\ln 2}{0.0123} \approx 56.35 \text{ units of time.}$$

14. An investment of \$100 whose value decreases exponentially with a monthly decay constant of 4% would be modeled by $Q = 100e^{-0.04t}$, where t is time in months. The investment would be worth \$50 in
$$t_h = \frac{\ln 2}{0.04} \approx 17.3 \text{ months.}$$

We can repeat the analysis above for exponential growth models.

Exponential Growth Model and Doubling Time

An **exponential growth function** has the form
$$Q(t) = Q_0e^{kt}. \qquad Q_0,\ k \text{ both positive}$$

Q_0 represents the value of Q at time $t = 0$, and k is the **growth constant**. The growth constant k and doubling time t_d for Q are related by
$$t_dk = \ln 2.$$

Quick Examples

15. $P(t) = 1{,}000e^{0.05t}$ — \$1,000 invested at 5% annually with interest compounded continuously

16. If $t_d = 10$ years, then $10k = \ln 2$, so $k = \frac{\ln 2}{10} \approx 0.06931$, and the growth model is
$$Q(t) = Q_0e^{0.06931t}.$$

17. If $k = 0.0123$, then $t_d(0.0123) = \ln 2$, so the doubling time is
$$t_d = \frac{\ln 2}{0.0123} \approx 56.35 \text{ units of time.}$$

18. An epidemic that begins with 100 cases and grows exponentially with a monthly growth constant of 45% would be modeled by $Q = 100e^{0.45t}$, where t is time in months. The doubling time would then be $t_d = \frac{\ln 2}{0.45} \approx 1.54$ months.

Logarithmic Regression

If we start with a set of data that suggests a logarithmic curve, we can, by repeating the methods from previous sections, use technology to find the logarithmic regression curve $y = \log_b x + C$ approximating the data.

EXAMPLE 4 **T Research & Development**

The following table shows the total spent on research and development by institutions of higher education in the United States, in billions of dollars, for the period 2003–2012:[36]

Year t (years since 2000)	3	4	5	6	7	8	9	10	11	12
Spending ($ billion)	41	45	48	50	52	54	57	61	65	66

Find the best-fit logarithmic model of the form

$$S(t) = A \ln t + C,$$

and use the model to project total spending on research by universities and colleges in 2016, assuming that the trend continues.

Solution We use technology to get the following regression model:

$$S(t) = 17.9 \ln t + 19.4. \quad \text{Coefficients rounded}$$

Because 2016 is represented by $t = 16$, we have

$$S(16) = 17.9 \ln(16) + 19.4 \approx 69. \quad \text{Why did we round the result to two significant digits?}$$

So research and development spending by institutions of higher education is projected to be around $69 billion in 2016.

➡ **Before we go on . . .** The model in Example 4 seems to give reasonable estimates when we extrapolate forward, but extrapolating backward is quite another matter: The logarithm curve drops sharply to the left of the given range and becomes negative for small values of t (Figure 16). ■

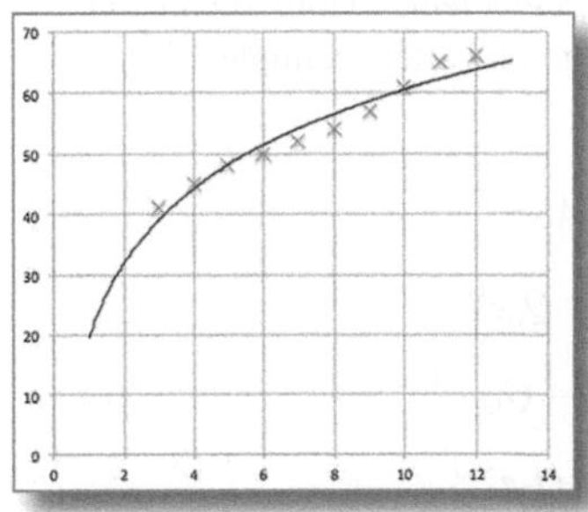

Figure 16

Using Technology

To obtain the regression curve and graph for Example 4 using a graphing calculator or a spreadsheet, see the Technology Guides at the end of the chapter. Outline:

TI-83/84 Plus

`STAT` EDIT values of t in L_1 and values of S in L_2.

Regression curve: `STAT` CALC option #9 `LnReg` `ENTER`

Graph: `Y=` `VARS` `5` EQ `1`, then `ZOOM` `9`

[More details in the Technology Guide.]

Spreadsheet

t- and S-values in columns A and B.

Graph: Highlight data and insert a scatter chart.

Regression curve: Right-click a datapoint, and add a logarithmic trendline with the option to show the equation on the chart.

[More details in the Technology Guide.]

 **Website**

www.WanerMath.com

In the Function Evaluator and Grapher, enter the data as shown, press "Examples" until the logarithmic model `$1*ln(x)+$2` shows in the first box, and press "Fit Curve".

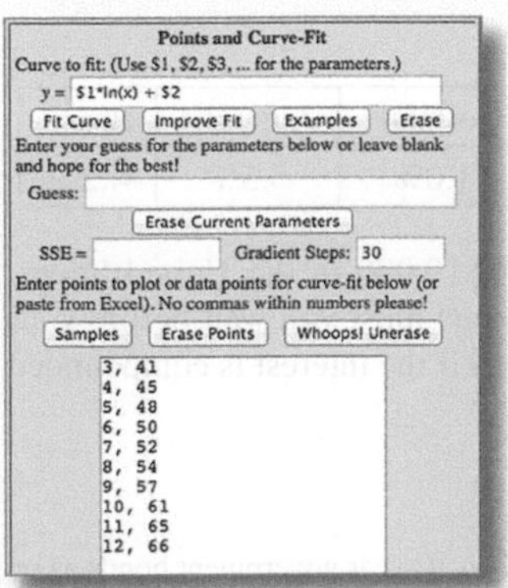

[36] Source: National Science Foundation, National Center for Science and Engineering Statistics, Higher Education Research and Development Survey, as of October 2013 (http://ncsesdata.nsf.gov/herd/2012).

2.3 EXERCISES

▼ more advanced ◆ challenging

T indicates exercises that should be solved using technology

1. If $y = 4^x$, then $x =$ ________.
2. If $y = \log_6 x$, then $x =$ ________.
3. Simplify: $2^{\log_2 8}$.
4. Simplify: $e^{\ln x}$.
5. Simplify: $\ln(e^x)$.
6. Simplify: $\ln\sqrt{a}$.

In Exercises 7–14, graph the given function. **[HINT:** See Example 1.]

7. $f(x) = \log_4 x$
8. $f(x) = \log_5 x$
9. $f(x) = \log_{1/4} x$
10. $f(x) = \log_{1/5} x$
11. $f(x) = \log_4 x + 1$
12. $f(x) = \log_5 x - 1$
13. $f(x) = \log_4(x - 1)$
14. $f(x) = \log_5(x + 1)$

In Exercises 15–20, find the associated exponential decay or growth model. **[HINT:** See Quick Examples 11–18.]

15. $Q = 1{,}000$ when $t = 0$; half-life $= 1$
16. $Q = 2{,}000$ when $t = 0$; half-life $= 5$
17. $Q = 500$ when $t = 0$; half-life $= 50$
18. $Q = 50$ when $t = 0$; half-life $= 500$
19. $Q = 1{,}000$ when $t = 0$; doubling time $= 2$
20. $Q = 2{,}000$ when $t = 0$; doubling time $= 5$

In Exercises 21–26, find the associated half-life or doubling time. **[HINT:** See Quick Examples 11–18.]

21. $Q = 1{,}000e^{0.5t}$
22. $Q = 1{,}000e^{-0.025t}$
23. $Q = 100e^{-t}$
24. $Q = 5{,}000e^{t/3}$
25. $Q = Q_0e^{-4t}$
26. $Q = Q_0e^{t}$

In Exercises 27–32, convert the given exponential function to the form indicated. Round all coefficients to four significant digits. **[HINT:** See "Before we go on" after Example 3.]

27. $f(x) = 4e^{2x}$; $f(x) = Ab^x$
28. $f(x) = 2.1e^{-0.1x}$; $f(x) = Ab^x$
29. $f(t) = 2.1(1.001)^t$; $f(t) = Q_0e^{kt}$
30. $f(t) = 23.4(0.991)^t$; $f(t) = Q_0e^{-kt}$
31. $f(t) = 10(0.987)^t$; $f(t) = Q_0e^{-kt}$
32. $f(t) = 2.3(2.2)^t$; $f(t) = Q_0e^{kt}$

Applications

33. ***Investments*** How long will it take a \$500 investment to be worth \$700 if it is continuously compounded at 10% per year? (Give the answer to two decimal places.) **[HINT:** See Example 2.]
34. ***Investments*** How long will it take a \$500 investment to be worth \$700 if it is continuously compounded at 15% per year? (Give the answer to two decimal places.) **[HINT:** See Example 2.]
35. ***Investments*** How long, to the nearest year, will it take an investment to triple if it is continuously compounded at 10% per year?
36. ***Investments*** How long, to the nearest year, will it take me to become a millionaire if I invest \$1,000 at 10% interest compounded continuously?
37. ***Investments*** I would like my investment to double in value every 3 years. At what rate of interest would I need to invest it, assuming that the interest is compounded continuously? **[HINT:** See Quick Examples 11–18.]
38. ***Depreciation*** My investment in OHaganBooks.com stocks is losing half its value every 2 years. Find and interpret the associated decay rate. **[HINT:** See Quick Examples 11–18.]
39. ***Carbon Dating*** The amount of carbon 14 remaining in a sample that originally contained A grams is given by

$$C(t) = A(0.999879)^t$$

where t is time in years. If tests on a fossilized skull reveal that 99.95% of the carbon 14 has decayed, how old, to the nearest 1,000 years, is the skull?

40. ***Carbon Dating*** Refer to Exercise 39. How old, to the nearest 1,000 years, is a fossil in which only 30% of the carbon 14 has decayed?

Long-Term Investments *Exercises 41–48 are based on the following table, which lists interest rates on long-term investments (based on 10-year government bonds) in several countries in 2014.*[37] **[HINT:** See Example 2.]

Country	U.S.	Japan	Germany	Australia	Brazil
Yield	2.5%	0.5%	1.0%	3.5%	4.2%

41. Assuming that you invest \$10,000 in the United States, how long (to the nearest year) must you wait before your investment is worth \$15,000 if the interest is compounded annually?

[37] Approximate interest rates based on 10-year government bonds as of September 27, 2014. Source: www.bloomberg.com.

42. Assuming that you invest \$10,000 in Japan, how long (to the nearest year) must you wait before your investment is worth \$15,000 if the interest is compounded annually?

43. If you invest \$10,400 in Germany and the interest is compounded monthly, when, to the nearest month, will your investment be worth \$20,000?

44. If you invest \$10,400 in the United States, and the interest is compounded monthly, when, to the nearest month, will your investment be worth \$20,000?

45. How long, to the nearest year, will it take an investment in Australia to double its value if the interest is compounded every 6 months?

46. How long, to the nearest year, will it take an investment in Brazil to double its value if the interest is compounded every 6 months?

47. If the interest on a long-term U.S. investment is compounded continuously, how long will it take the value of an investment to double? (Give the answer correct to two decimal places.)

48. If the interest on a long-term Australia investment is compounded continuously, how long will it take the value of an investment to double? (Give an answer correct to two decimal places.)

49. ***Half-Life*** The amount of radium 226 remaining in a sample that originally contained A grams is approximately

$$C(t) = A(0.999567)^t,$$

where t is time in years. Find the half-life to the nearest 100 years. [**HINT**: See Example 3a.]

50. ***Half-Life*** The amount of iodine 131 remaining in a sample that originally contained A grams is approximately

$$C(t) = A(0.9175)^t,$$

where t is time in days. Find the half-life to two decimal places. [**HINT**: See Example 3(a).]

51. ***Automobiles*** The rate of auto thefts triples every 6 months.

a. Determine, to two decimal places, the base b for an exponential model $y = Ab^t$ of the rate of auto thefts as a function of time in months.

b. Find the doubling time to the nearest tenth of a month. [**HINT**: (a) See Example 2 of Section 2.2. (b) See Quick Examples 11–18.]

52. ***Televisions*** The rate of television thefts is doubling every 4 months.

a. Determine, to two decimal places, the base b for an exponential model $y = Ab^t$ of the rate of television thefts as a function of time in months.

b. Find the tripling time to the nearest tenth of a month. [**HINT**: (a) See Example 2 of Section 2.2. (b) See Quick Examples 11–18.]

53. ***Half-Life*** The half-life of cobalt 60 is 5 years.

a. Obtain an exponential decay model for cobalt 60 in the form $Q(t) = Q_0e^{-kt}$. (Round the decay constant to three significant digits.)

b. Use your model to predict, to the nearest year, the time it takes one third of a sample of cobalt 60 to decay.

54. ***Half-Life*** The half-life of strontium 90 is 28 years.

a. Obtain an exponential decay model for strontium 90 in the form $Q(t) = Q_0e^{-kt}$. (Round the decay constant to three significant digits.)

b. Use your model to predict, to the nearest year, the time it takes three fifths of a sample of strontium 90 to decay.

55. ***Radioactive Decay*** Uranium 235 is used as fuel for some nuclear reactors. It has a half-life of 710 million years. How long will it take 10 grams of uranium 235 to decay to 1 gram? (Round your answer to three significant digits.)

56. ***Radioactive Decay*** Plutonium 239 is used as fuel for some nuclear reactors, and also as the fissionable material in atomic bombs. It has a half-life of 24,400 years. How long would it take 10 grams of plutonium 239 to decay to 1 gram? (Round your answer to three significant digits.)

57. ▼ ***Aspirin*** Soon after taking an aspirin, a patient has absorbed 300 milligrams of the drug. If the amount of aspirin in the bloodstream decays exponentially, with half being removed every 2 hours, find, to the nearest 0.1 hours, the time it will take for the amount of aspirin in the bloodstream to decrease to 100 milligrams.

58. ▼ ***Alcohol*** After a large number of drinks, a person has a blood alcohol level of 200 mg/dL (milligrams per deciliter). If the amount of alcohol in the blood decays exponentially, with one fourth being removed every hour, find the time it will take for the person's blood alcohol level to decrease to 80 mg/dL.

59. ▼ ***Radioactive Decay*** You are trying to determine the half-life of a new radioactive element you have isolated. You start with 1 gram, and 2 days later, you determine that it has decayed down to 0.7 grams. What is its half-life? (Round your answer to three significant digits.) [**HINT**: First find an exponential model, then see Example 3.]

60. ▼ ***Radioactive Decay*** You have just isolated a new radioactive element. If you can determine its half-life, you will win the Nobel Prize in physics. You purify a sample of 2 grams. One of your colleagues steals half of it, and 3 days later, you find that 0.1 grams of the radioactive material is still left. What is the half-life? (Round your answer to three significant digits.) [**HINT**: First find an exponential model, then see Example 3.]

61. **T** ***Population Aging*** The following table shows the percentage of U.S. residents over the age of 65 in 1950, 1960, . . . , 2010:[38]

Year t (years since 1900)	50	60	70	80	90	100	110
Percentage P over 65 (%)	8.2	9.2	9.9	11.3	12.6	12.6	13

a. Find the logarithmic regression model of the form $P(t) = A \ln t + C$. (Round the coefficients to four significant digits). [**HINT:** See Example 4.]

b. In 1940, 6.9% of the population was over age 65. To how many significant digits does the model reflect this figure?

c. Which of the following is correct? The model, if extrapolated into the indefinite future, predicts that

(A) the percentage of U.S. residents over the age of 65 will increase without bound.

(B) the percentage of U.S. residents over the age of 65 will level off at around 14.2%.

(C) the percentage of U.S. residents over the age of 65 will eventually decrease.

62. **T** ***Population Aging*** The following table shows the percentage of U.S. residents over the age of 85 in 1950, 1960, . . . , 2010:[39]

Year t (years since 1900)	50	60	70	80	90	100	110
Percentage P over 65 (%)	0.4	0.5	0.7	1	1.2	1.6	1.9

a. Find the logarithmic regression model of the form $P(t) = A \ln t + C$. (Round the coefficients to four significant digits). [**HINT:** See Example 4.]

b. In 2020, 2.1% of the population is projected to be over age 85. To how many significant digits does the model reflect this figure?

c. Which of the following is correct? If you increase A by 0.1 and decrease C by 0.1 in the logarithmic model, then

(A) the new model predicts eventually lower percentages.

(B) the long-term prediction is essentially the same.

(C) the new model predicts eventually higher percentages.

63. **T** ***Research & Development: Industry*** The following table shows the total spent on research and development by industry in the United States, in billions of dollars, for the period 2002–2012.[40]

Year t (years since 2000)	2	3	4	5	6	7
Spending ($ billion)	192	194	194	204	216	228
Year t (years since 2000)	8	9	10	11	12	
Spending ($ billion)	234	222	221	232	243	

Find the logarithmic regression model of the form $S(t) = A \ln t + C$ with coefficients A and C rounded to two decimal places. Also obtain a graph showing the data points and the regression curve. In which direction is it more reasonable to extrapolate the model? Why?

64. **T** ***Research & Development: Federal*** The following table shows the total spent on research and development by the federal government in the United States, in billions of dollars, for the period 2002–2012:[41]

Year t (years since 2000)	2	3	4	5	6	7
Spending ($ billion)	26	27	26	26	27	28
Year t (years since 2000)	8	9	10	11	12	
Spending ($ billion)	27	28	29	32	33	

Find the logarithmic regression model of the form $S(t) = A \ln t + C$ with coefficients A and C rounded to two decimal places. Also obtain a graph showing the data points and the regression curve. In which direction is it more reasonable to extrapolate the model? Why?

65. ▼ ***Richter Scale*** The **Richter scale** is used to measure the intensity of earthquakes. The Richter scale rating of an earthquake is given by the formula

$$R = \frac{2}{3}(\log E - 11.8),$$

where E is the energy released by the earthquake (measured in ergs[42]).

a. The San Francisco earthquake of 1906 is estimated to have registered $R = 7.9$ on the Richter scale. How many ergs of energy were released?

b. The Japan earthquake of 2011 registered 9.0 on the Richter scale. Compare the two: The energy released in the 1906 earthquake was what percentage of the energy released in the 2011 quake?

---

[38] Source: U.S. Census Bureau.

[39] *Ibid.*

[40] Constant 2005 dollars; excludes federal funding; 2012 data is preliminary. Source: National Science Foundation, National Center for Science and Engineering Statistics. December 2013. *National Patterns of R&D Resources: 2011–12 Data Update.* NSF 14-304 (www.nsf.gov/statistics/nsf14304.).

[41] Constant 2005 dollars; excludes federal funding to industry and nonprofit organizations; 2012 data is preliminary. Source: National Science Foundation, National Center for Science and Engineering Statistics. December 2013. *National Patterns of R&D Resources: 2011–12 Data Update.* NSF 14-304 (www.nsf.gov/statistics/nsf14304).

[42] An erg is a unit of energy. One erg is the amount of energy it takes to move a mass of 1 gram 1 centimeter in 1 second. The term "Richter scale" is used loosely to refer to several ways of measuring earthquake magnitudes, calibrated to agree where they overlap.

c. Solve the equation given above for E in terms of R.
d. Use the result of part (c) to show that if two earthquakes registering R_1 and R_2 on the Richter scale release E_1 and E_2 ergs of energy, respectively, then

$$\frac{E_2}{E_1} = 10^{1.5(R_2 - R_1)}.$$

e. Fill in the blank: If one earthquake registers 2 points more on the Richter scale than another, then it releases ___ times the amount of energy.

66. ▼ ***Sound Intensity*** The loudness of a sound is measured in **decibels**. The decibel level of a sound is given by the formula

$$D = 10 \log \frac{I}{I_0},$$

where D is the decibel level (dB), I is its intensity in watts per square meter (W/m^2), and $I_0 = 10^{-12}$ W/m^2 is the intensity of a barely audible "threshold" sound. A sound intensity of 90 decibels or greater causes damage to the average human ear.

a. Find the decibel levels of each of the following, rounding to the nearest decibel:

Whisper:	115×10^{-12} W/m^2
TV (average volume from 10 feet):	320×10^{-7} W/m^2
Loud music:	900×10^{-3} W/m^2
Jet aircraft (from 500 feet):	100 W/m^2

b. Which of the sounds above damages the average human ear?
c. Solve the given equation to express I in terms of D.
d. Use the answer to part (c) to show that if two sounds of intensity I_1 and I_2 register decibel levels of D_1 and D_2, respectively, then

$$\frac{I_2}{I_1} = 10^{0.1(D_2 - D_1)}.$$

e. Fill in the blank: If one sound registers 1 decibel more than another, then it is ___ times as intense.

67. ▼ ***Sound Intensity*** The decibel level of a TV set decreases with the distance from the set according to the formula

$$D = 10 \log\left(\frac{320 \times 10^7}{r^2}\right)$$

where D is the decibel level (dB) and r is the distance from the TV set in feet.

a. Find the decibel level (to the nearest decibel) at distances of 10, 20, and 50 feet.
b. Express D in the form $D = A + B \log r$ for suitable constants A and B. (Round A and B to two significant digits.)
c. How far must a listener be from a TV so that the decibel level drops to 0 decibels? (Round the answer to two significant digits.)

68. ▼ ***Acidity*** The acidity of a solution is measured by its pH, which is given by the formula

$$pH = -\log(H^+)$$

where H^+ measures the concentration of hydrogen ions in moles per liter.[43] The pH of pure water is 7. A solution is referred to as *acidic* if its pH is below 7 and as *basic* if its pH is above 7.

a. Calculate the pH of each of the following substances:

Blood:	3.9×10^{-8} moles per liter
Milk:	4.0×10^{-7} moles per liter
Soap solution:	1.0×10^{-11} moles per liter
Black coffee:	1.2×10^{-7} moles per liter

b. How many moles of hydrogen ions are contained in a liter of acid rain that has a pH of 5.0?
c. Complete the following sentence: If the pH of a solution increases by 1.0, then the concentration of hydrogen ions _________.

Communication and Reasoning Exercises

69. On the same set of axes, graph $y = \ln x$, $y = A \ln x$, and $y = A \ln x + C$ for various choices of *positive* A and C. What is the effect on the graph of $y = \ln x$ of multiplying by A? What is the effect of then adding C?

70. On the same set of axes, graph $y = -\ln x$, $y = A \ln x$, and $y = A \ln x + C$ for various choices of *negative* A and C. What is the effect on the graph of $y = \ln x$ of multiplying by A? What is the effect of then adding C?

71. Why is the logarithm of a negative number not defined?

72. Of what use are logarithms, now that they are no longer needed to perform complex calculations?

73. Your company's market share is undergoing steady growth. Explain why a logarithmic function is *not* appropriate for long-term future prediction of your market share.

74. Your company's market share is undergoing steady growth. Explain why a logarithmic function is *not* appropriate for long-term backward extrapolation of your market share.

75. ▼ If a town's population is increasing exponentially with time, how is time increasing with population? Explain.

76. ▼ If a town's population is increasing logarithmically with time, how is time increasing with population? Explain.

77. ▼ If two quantities Q_1 and Q_2 are logarithmic functions of time t, show that their sum, $Q_1 + Q_2$, is also a logarithmic function of time t.

78. T ▼ In Exercise 77 we saw that the sum of two logarithmic functions is a logarithmic function. In Exercises 63 and 64 you modeled research and development expenditure by industry and government. Now do a logarithmic regression on the sum of the two sets of figures. Does the result coincide with the sum of the two individual regression models? What does your answer tell you about the sum of logarithmic regression models?

[43] A mole corresponds to about 6.0×10^{23} hydrogen ions. (This number is known as Avogadro's number.)

2.4 Logistic Functions and Models

Logistic Functions

Figure 17 shows wired broadband penetration in the United States as a function of time t in years ($t = 0$ represents 2000).[44]

The left-hand part of the curve in Figure 17, from $t = 2$ to, say, $t = 6$, looks roughly like exponential growth: P behaves (roughly) like an exponential function, with the y-coordinates growing by a factor of around 1.5 per year. Then, as the market starts to become saturated, the growth of P slows, and its value may be approaching a "ceiling" of around 30%. **Logistic** functions have just this kind of behavior, growing exponentially at first and then leveling off. In addition to modeling the demand for a new technology or product, logistic functions are often used in epidemic and population modeling. In an epidemic the number of infected people often grows exponentially at first and then slows when a significant proportion of the entire susceptible population is infected and the epidemic has run its course. Similarly, populations may grow exponentially at first and then slow as they approach the carrying capacity of the available resources.

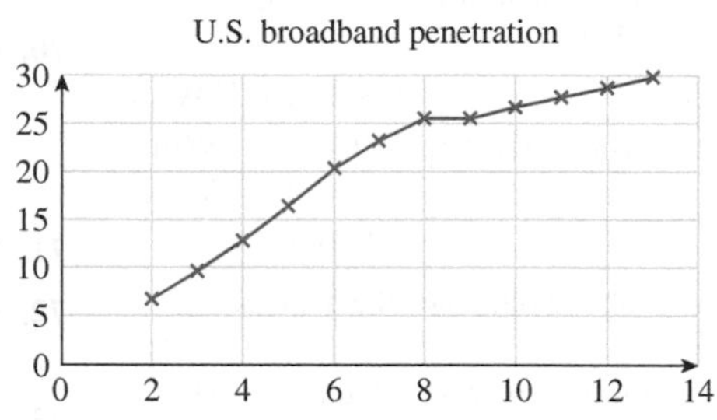

Figure 17

Graph of a Logistic Function

$$y = \frac{N}{1 + Ab^{-x}}$$

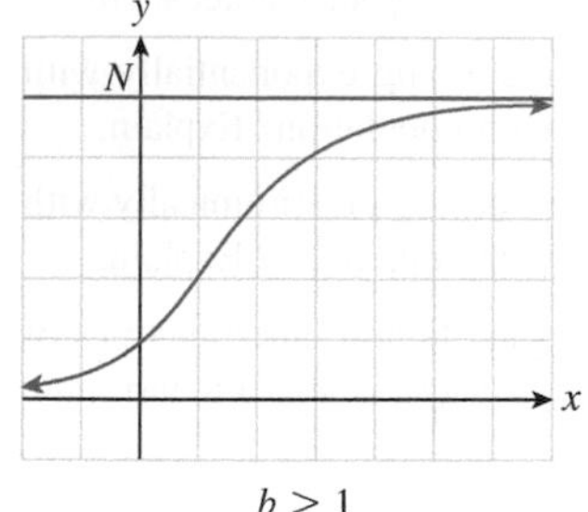

$b > 1$

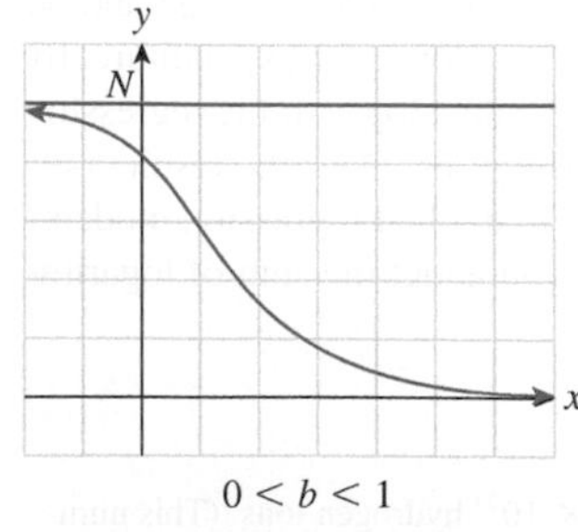

$0 < b < 1$

Logistic Function

A **logistic function** has the form

$$f(x) = \frac{N}{1 + Ab^{-x}}$$

for nonzero constants N, A, and b (A and b positive and $b \neq 1$).

Properties of the Logistic Curve $y = \dfrac{N}{1 + Ab^{-x}}$

- The graph is an S-shaped curve sandwiched between the horizontal lines $y = 0$ and $y = N$. (See the figures in the margin.) N is called the **limiting value** of the logistic curve.
- If $b > 1$, the graph rises; if $b < 1$, the graph falls.

[44] Broadband penetration is the number of broadband installations divided by the total population. Source for data: Organisation for Economic Co-operation and Development (OECD) Directorate for Science, Technology, and Industry, table of Historical Penetration Rates, December 2013, downloaded September 2014 from www.oecd.org/sti/ict/broadband.

• The y-intercept is $\dfrac{N}{1 + A}$.

• The curve is steepest when $t = \dfrac{\ln A}{\ln b}$. We will see why in Chapter 5.

Quick Example

1. $N = 6, A = 2, b = 1.1$ gives

$$f(x) = \frac{6}{1 + 2(1.1^{-x})}$$ `6/(1+2*1.1^-x)`

$$f(0) = \frac{6}{1 + 2} = 2$$ The y-intercept is $N/(1 + A)$.

$$f(1{,}000) = \frac{6}{1 + 2(1.1^{-1{,}000})} \approx \frac{6}{1 + 0} = 6 = N$$ When x is large, $f(x) \approx N$.

Because $b = 1.1 > 1$, the graph of f rises as shown on the preceding page.

Note If we write b^{-x} as e^{-kx} (so that $k = \ln b$), we get the following alternative form of the logistic function:

$$f(x) = \frac{N}{1 + Ae^{-kx}}.$$ ■

Q: *How does the constant b affect the graph?*

A: To understand the role of b, we consider only the case in which $b > 1$, so the graph rises. Rewrite the logistic function by multiplying top and bottom by b^x:

$$f(x) = \frac{N}{1 + Ab^{-x}} = \frac{Nb^x}{(1 + Ab^{-x})b^x}$$

$$= \frac{Nb^x}{b^x + A}.$$ Because $b^{-x}b^x = 1$

For values of x close to 0 the quantity b^x is close to 1, so the denominator is approximately $1 + A$, giving

$$f(x) \approx \frac{Nb^x}{1 + A} = \left(\frac{N}{1 + A}\right) b^x.$$

In other words, $f(x)$ is approximately exponential with base b for values of x close to 0. Put another way, if x represents time, then initially the logistic function behaves like an exponential function.

Note The exponential approximation above works well only when A is much larger than 1. Showing this actually works requires some calculus, but you can check it graphically by experimenting with different values of A. ■

To summarize, we have the following.

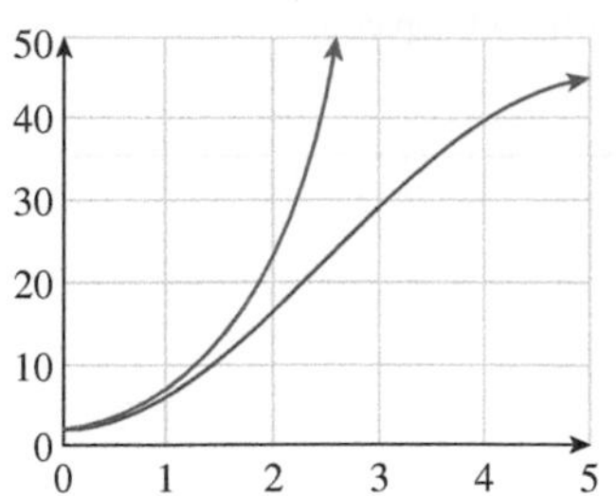

The upper curve is the exponential curve.

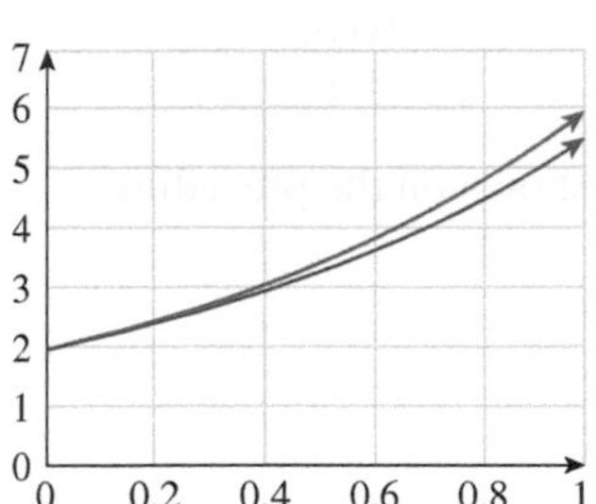

Figure 18

Logistic Function for Small x and the Role of b

If $b > 1$, then for small values of x we have

$$\frac{N}{1 + Ab^{-x}} \approx \frac{N}{1 + A}b^x.$$

Thus, for small x the logistic function grows approximately exponentially with base b.

Quick Example

2. Let

$$f(x) = \frac{50}{1 + 24(3^{-x})}. \qquad N = 50, A = 24, b = 3$$

Then

$$f(x) \approx \left(\frac{50}{1 + 24}\right)(3^x) = 2(3^x)$$

for small values of x. Figure 18 compares their graphs at different scales.

Modeling with the Logistic Function

EXAMPLE 1 Epidemics

A flu epidemic is spreading through the U.S. population. An estimated 150 million people are susceptible to this particular strain, and it is predicted that all susceptible people will eventually become infected. There are 10,000 people already infected, and the number is doubling every 2 weeks. Use a logistic function to model the number of people infected. Use your model to predict when, to the nearest week, 1 million people will be infected.

Solution Let t be time in weeks, and let $P(t)$ be the total number of people infected at time t. We want to express P as a logistic function of t, so that

$$P(t) = \frac{N}{1 + Ab^{-t}}.$$

We are told that, in the long run, 150 million people will be infected, so

$$N = 150{,}000{,}000. \qquad \text{Limiting value of } P$$

At the current time ($t = 0$), 10,000 people are infected, so

$$10{,}000 = \frac{N}{1 + A} = \frac{150{,}000{,}000}{1 + A}. \qquad \text{Value of } P \text{ when } t = 0$$

Solving for A gives

$$10{,}000(1 + A) = 150{,}000{,}000$$
$$1 + A = 15{,}000$$
$$A = 14{,}999.$$

What about b? At the beginning of the epidemic (t near 0), P is growing approximately exponentially, doubling every 2 weeks. Using the technique of Section 2.2, we find that the exponential curve passing through the points (0, 10,000) and (2, 20,000) is

$$y = 10{,}000(\sqrt{2})^t,$$

giving us $b = \sqrt{2}$. Now that we have the constants N, A, and b, we can write down the logistic model:

$$P(t) = \frac{150{,}000{,}000}{1 + 14{,}999(\sqrt{2})^{-t}}.$$

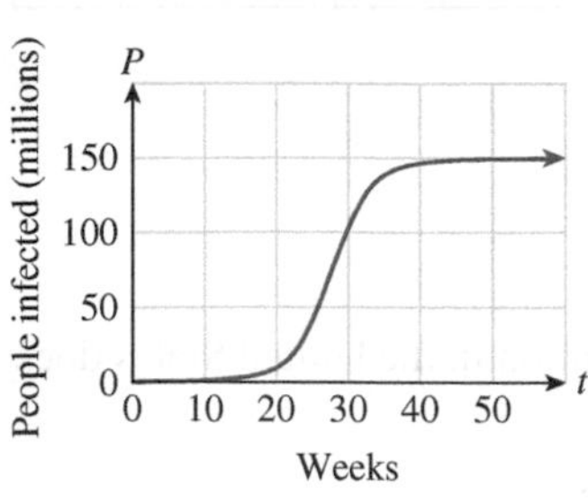

Figure 19

The graph of this function is shown in Figure 19.

Now we tackle the question of prediction: When will 1 million people be infected? In other words, when is $P(t) = 1{,}000{,}000$?

$$1{,}000{,}000 = \frac{150{,}000{,}000}{1 + 14{,}999(\sqrt{2})^{-t}}$$
$$1{,}000{,}000[1 + 14{,}999(\sqrt{2})^{-t}] = 150{,}000{,}000$$
$$1 + 14{,}999(\sqrt{2})^{-t} = 150$$
$$14{,}999(\sqrt{2})^{-t} = 149$$
$$(\sqrt{2})^{-t} = \frac{149}{14{,}999}$$
$$-t = \log_{\sqrt{2}}\left(\frac{149}{14{,}999}\right) \approx -13.31 \qquad \text{Logarithmic form}$$

Thus, 1 million people will be infected by about the thirteenth week.

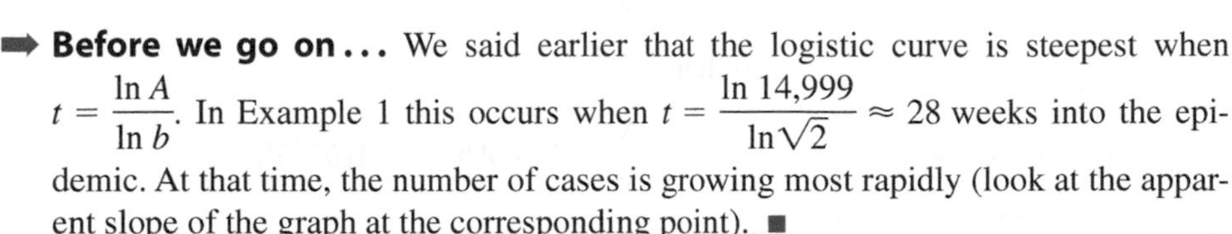

➡ **Before we go on...** We said earlier that the logistic curve is steepest when $t = \frac{\ln A}{\ln b}$. In Example 1 this occurs when $t = \frac{\ln 14{,}999}{\ln\sqrt{2}} \approx 28$ weeks into the epidemic. At that time, the number of cases is growing most rapidly (look at the apparent slope of the graph at the corresponding point). ■

Logistic Regression

Let's go back to the data on broadband penetration in the United States with which we began this section and try to determine the long-term percentage of broadband penetration. To be able to make predictions such as this, we require a model for the data, so we will need to do some form of regression.

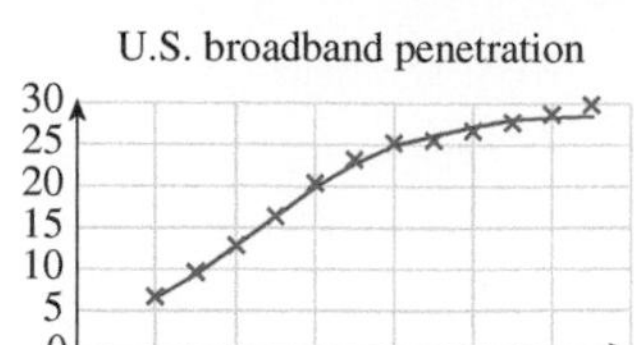

Figure 20

EXAMPLE 2 Broadband Penetration

Here are the data graphed in Figure 17:

Year t (years since 2000)	2	3	4	5	6	7	8	9	10	11	12	13
Penetration P (%)	6.7	9.6	12.8	16.4	20.3	23.2	25.5	25.5	26.7	27.7	28.7	29.8

Find a logistic regression curve of the form

$$P(t) = \frac{N}{1 + Ab^{-t}}.$$

In the long term, what percentage of broadband penetration in the United States does the model predict?

Solution We can use technology to obtain the following regression model:

$$P(t) \approx \frac{29.24}{1 + 9.00(1.64)^{-t}}. \qquad \text{Coefficients rounded to two decimal places}$$

Its graph and the original data are shown in Figure 20. Because $N = 29.24$, this model predicts that, in the long term, the percentage of broadband penetration in the United States will be 29.24%, or about 30%.

Using Technology

See the Technology Guide at the end of the chapter for detailed instructions on how to obtain the regression curve and graph for Example 2 using a graphing calculator or spreadsheet. Outline:

TI-83/84 Plus

STAT EDIT values of t in L_1 and values of P in L_2.
Regression curve: STAT
CALC option #B Logistic ENTER
Graph: Y= VARS 5 EQ 1, then ZOOM 9
[More details in the Technology Guide.]

Spreadsheet

Use the Solver Add-in to obtain the best-fit logistic curve.
[More details in the Technology Guide.]

Website

www.WanerMath.com

In the Function Evaluator and Grapher, enter the data as shown, press "Examples" until the logistic model `$1/(1+$2*$3^(-x))` shows in the first box, and press "Fit Curve".

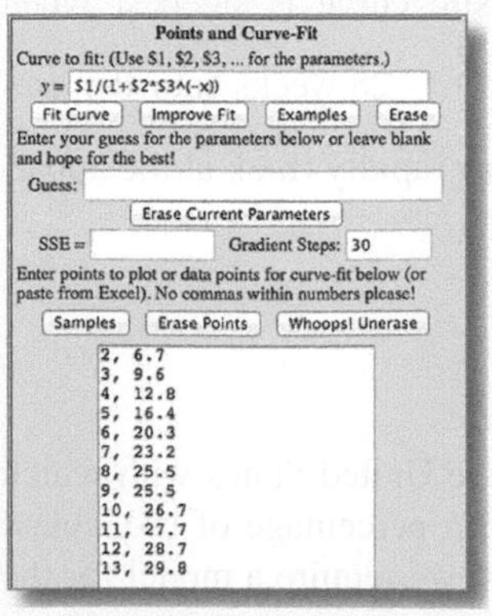

➡ **Before we go on . . .** Logistic regression programs generally estimate all three constants N, A, and b for a model $y = \frac{N}{1 + Ab^{-x}}$. However, there are times, as in Example 1, when we already know the limiting value N and require estimates of only A and b. In such cases we can use technology such as Excel Solver to find A and b for the best-fit curve with N fixed. Alternatively, we can use exponential regression to compute estimates of A and b as follows: First rewrite the logistic equation as

$$\frac{N}{y} = 1 + Ab^{-x},$$

so that

$$\frac{N}{y} - 1 = Ab^{-x} = A(b^{-1})^x.$$

This equation gives $N/y - 1$ as an exponential function of x. Thus, if we do exponential regression using the data points $(x, N/y - 1)$, we can obtain estimates for A and b^{-1} (and hence b). This is done in Exercises 35 and 36.

It is important to note that the resulting curve is not the best-fit curve (in the sense of minimizing SSE; see the "Before we go on" discussion after Example 6 in Section 2.2) and will therefore be different from that obtained by using the method in Example 2. ■

2.4 EXERCISES

▼ more advanced ◆ challenging

T indicates exercises that should be solved by using technology

In Exercises 1–6, find N, A, and b; give a technology formula for the given function; and use technology to sketch its graph for the given range of values of x. [**HINT**: See the Quick Examples on logistic functions.]

1. $f(x) = \dfrac{7}{1 + 6(2^{-x})}$; $[0, 10]$

2. $g(x) = \dfrac{4}{1 + 0.333(4^{-x})}$; $[0, 2]$

3. $f(x) = \dfrac{10}{1 + 4(0.3^{-x})}$; $[-5, 5]$

4. $g(x) = \dfrac{100}{1 + 5(0.5^{-x})}$; $[-5, 5]$

5. $h(x) = \dfrac{2}{0.5 + 3.5(1.5^{-x})}$; $[0, 15]$
(First divide top and bottom by 0.5.)

6. $k(x) = \dfrac{17}{2 + 6.5(1.05^{-x})}$; $[0, 100]$
(First divide top and bottom by 2.)

In Exercises 7–10, find the logistic function f with the given properties. [**HINT**: See Example 1.]

7. $f(0) = 10$, f has limiting value 200, and for small values of x, f is approximately exponential and doubles with every increase of 1 in x.

8. $f(0) = 1$, f has limiting value 10, and for small values of x, f is approximately exponential and grows by 50% with every increase of 1 in x.

9. f has limiting value 6 and passes through $(0, 3)$ and $(1, 4)$. [**HINT**: First find A, then substitute.]

10. f has limiting value 4 and passes through $(0, 1)$ and $(1, 2)$. [**HINT**: First find A, then substitute.]

In Exercises 11–16, choose the logistic function that best approximates the given curve.

11.

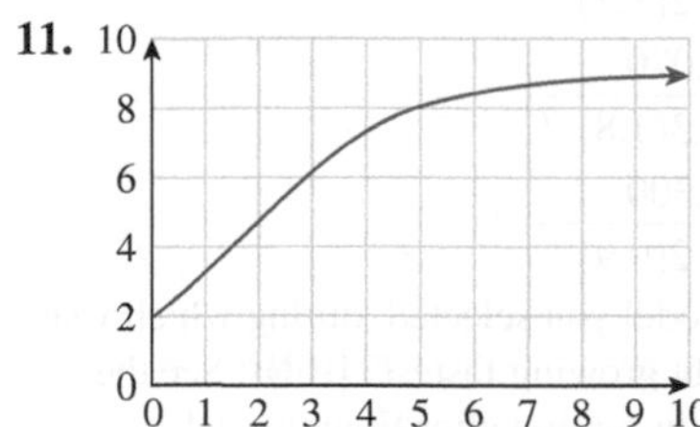

(A) $f(x) = \dfrac{6}{1 + 0.5(3^{-x})}$ (B) $f(x) = \dfrac{9}{1 + 3.5(2^{-x})}$

(C) $f(x) = \dfrac{9}{1 + 0.5(1.01)^{-x}}$

12.

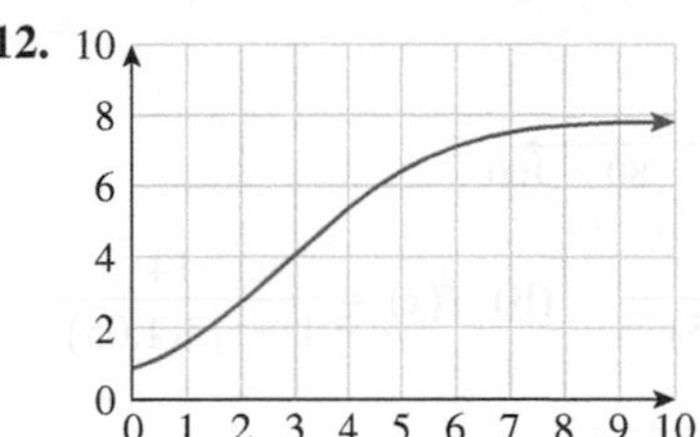

(A) $f(x) = \dfrac{8}{1 + 7(2)^{-x}}$ (B) $f(x) = \dfrac{8}{1 + 3(2)^{-x}}$

(C) $f(x) = \dfrac{6}{1 + 11(5)^{-x}}$

13.

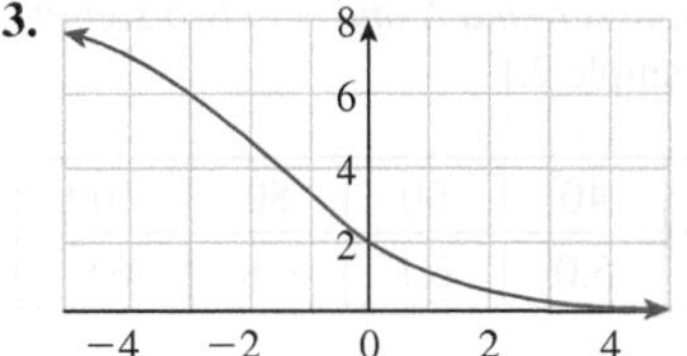

(A) $f(x) = \dfrac{8}{1 + 7(0.5)^{-x}}$ (B) $f(x) = \dfrac{8}{1 + 3(0.5)^{-x}}$

(C) $f(x) = \dfrac{8}{1 + 3(2)^{-x}}$

14.

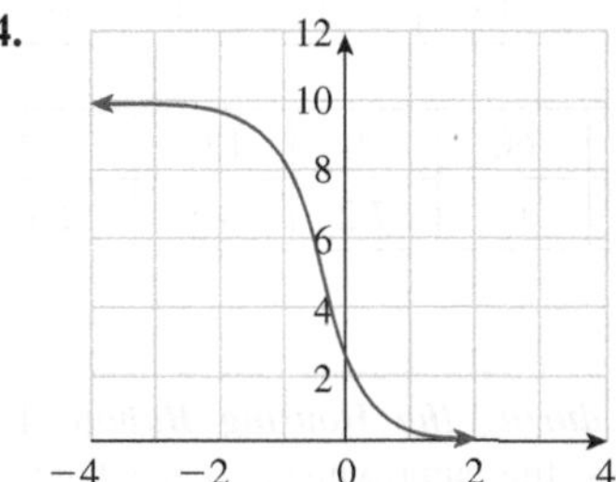

(A) $f(x) = \dfrac{10}{1 + 3(1.01)^{-x}}$ (B) $f(x) = \dfrac{8}{1 + 7(0.1)^{-x}}$

(C) $f(x) = \dfrac{10}{1 + 3(0.1)^{-x}}$

15.

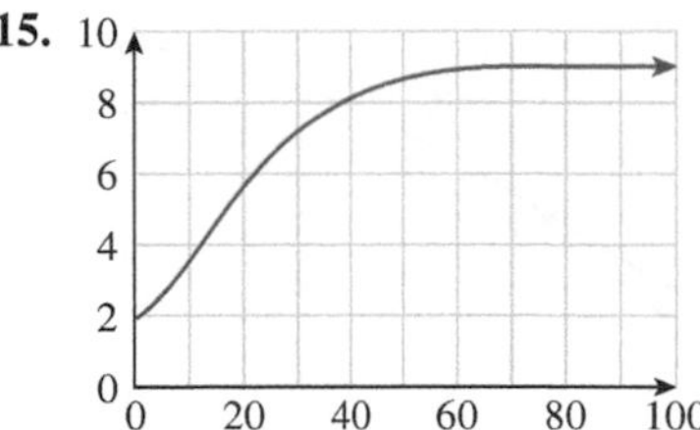

(A) $f(x) = \dfrac{18}{2 + 7(5)^{-x}}$ (B) $f(x) = \dfrac{18}{2 + 3(1.1)^{-x}}$

(C) $f(x) = \dfrac{18}{2 + 7(1.1)^{-x}}$

16. 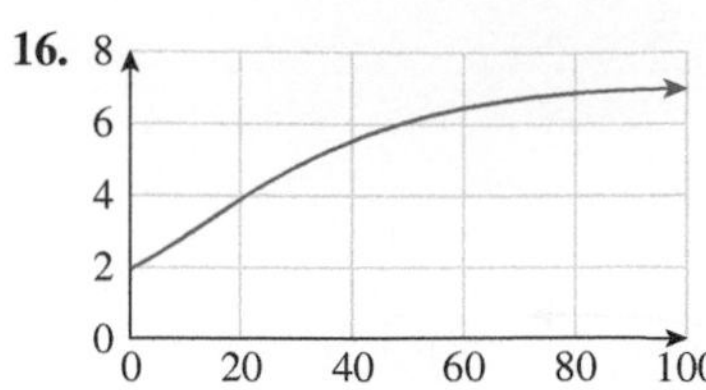

(A) $f(x) = \dfrac{14}{2 + 5(15)^{-x}}$ (B) $f(x) = \dfrac{14}{1 + 13(1.05)^{-x}}$

(C) $f(x) = \dfrac{14}{2 + 5(1.05)^{-x}}$

T *In Exercises 17–20, use technology to find a logistic regression curve* $y = \dfrac{N}{1 + Ab^{-x}}$ *approximating the given data. Draw a graph showing the data points and regression curve. (Round b to three significant digits, and round A and N to two significant digits.)* [**HINT:** See Example 2.]

17.

x	0	20	40	60	80	100
y	2.1	3.6	5.0	6.1	6.8	6.9

18.

x	0	30	60	90	120	150
y	2.8	5.8	7.9	9.4	9.7	9.9

19.

x	0	20	40	60	80	100
y	30.1	11.6	3.8	1.2	0.4	0.1

20.

x	0	30	60	90	120	150
y	30.1	20	12	7.2	3.8	2.4

Applications

21. ***Subprime Mortgages during the Housing Bubble*** The following graph shows the approximate percentage of mortgages issued in the United States during the real-estate run-up in 2000–2008 that were subprime (normally classified as risky) as well as the logistic regression curve:[45]

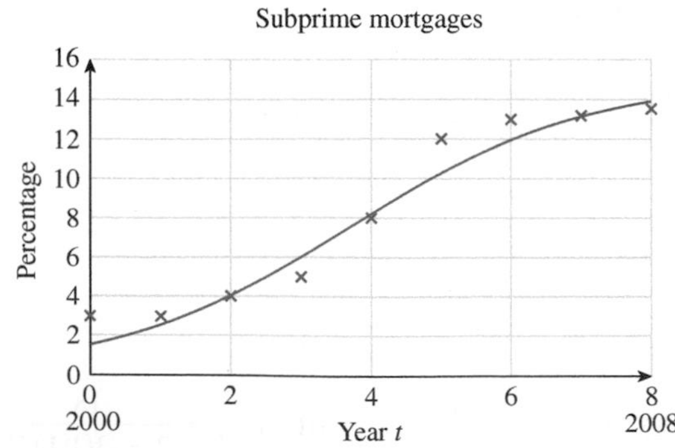

a. Which of the following logistic functions best approximates the curve? (t is the number of years since the start of 2000.) Try to determine the correct model without actually computing data points. [**HINT:** See Properties of the Logistic Curve.]

(A) $A(t) = \dfrac{15.0}{1 + 8.6(1.8)^{-t}}$

(B) $A(t) = \dfrac{2.0}{1 + 6.8(0.8)^{-t}}$

(C) $A(t) = \dfrac{2.0}{1 + 6.8(1.8)^{-t}}$

(D) $A(t) = \dfrac{15.0}{1 + 8.6(0.8)^{-t}}$

b. According to the model you selected, during which year was the percentage growing fastest? [**HINT:** See the "Before we go on" discussion after Example 1.]

22. ***Subprime Mortgage Debt during the Housing Bubble*** The following graph shows the approximate value of subprime (normally classified as risky) mortgage debt outstanding in the United States during the real-estate run-up in 2000–2008 as well as the logistic regression curve:[46]

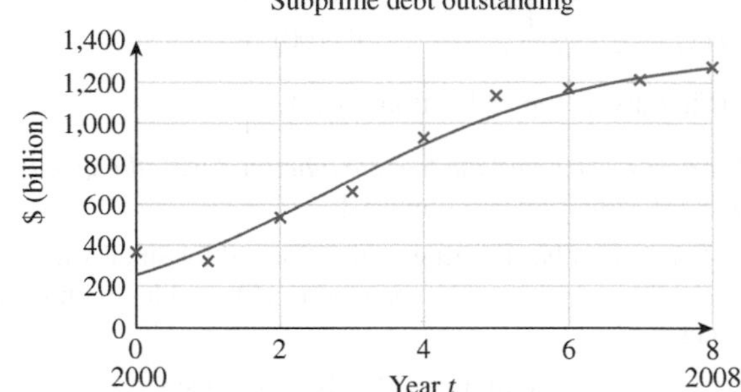

a. Which of the following logistic functions best approximates the curve? (t is the number of years since the start of 2000.) Try to determine the correct model without actually computing data points. [**HINT:** See Properties of the Logistic Curve.]

(A) $A(t) = \dfrac{1{,}850}{1 + 5.36(1.8)^{-t}}$

(B) $A(t) = \dfrac{1{,}350}{1 + 4.2(1.7)^{-t}}$

(C) $A(t) = \dfrac{1{,}020}{1 + 5.3(1.8)^{-t}}$

(D) $A(t) = \dfrac{1{,}300}{1 + 4.2(0.9)^{-t}}$

b. According to the model you selected, during which year was outstanding debt growing fastest? [**HINT:** See the "Before we go on" discussion after Example 1.]

[45] 2008 figure is an estimate. Sources: Mortgage Bankers Association, UBS.

[46] 2008–2009 figures are estimates. Source: www.data360.org/dataset.aspx?Data_Set_Id=9549.

23. ***Scientific Research*** The following graph shows the number of science research articles authored by researchers in the European Union during the period 1981–2010:[47]

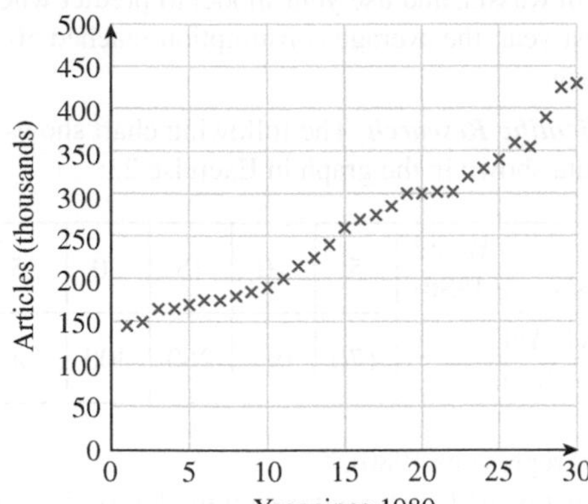

a. Which of the following logistic functions best models the data? ($t = 0$ represents 1980.) Try to determine the correct model without actually computing data points.

(A) $A(t) = \dfrac{9{,}200}{1 + 65(1.04)^{-t}}$

(B) $A(t) = \dfrac{300}{1 + 1.1(1.05)^{-t}}$

(C) $A(t) = \dfrac{9{,}200}{1 + 65(0.8)^{-t}}$

(D) $A(t) = \dfrac{9{,}200}{1 + 65(2.0)^{-t}}$

b. According to the model you selected, at what percentage was the number of articles growing around 1985?

24. ***Scientific Research*** The following graph shows the number of science research articles authored by researchers in the United States during the period 1981–2010.[48]

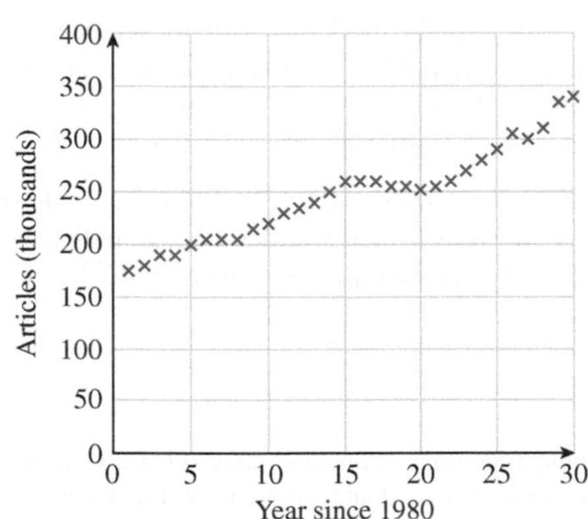

[47] Source: www.sciencewatch.com.

[48] *Ibid.*

a. Which of the following logistic functions best models the data? ($t = 0$ represents 1980.) Try to determine the correct model without actually computing data points.

(A) $A(t) = \dfrac{11{,}000}{1 + 58(1.50)^{-t}}$

(B) $A(t) = \dfrac{11{,}000}{1 + 58(0.98)^{-t}}$

(C) $A(t) = \dfrac{300}{1 + 0.70(1.04)^{-t}}$

(D) $A(t) = \dfrac{11{,}000}{1 + 58(1.02)^{-t}}$

b. According to the model you selected, at what percentage was the number of articles growing around 1985?

25. ***Internet Use*** The following graph shows the percentage of U.S. households using the Internet at home in 2010 as a function of household income (the data points) and a logistic model of these data (the curve).[49]

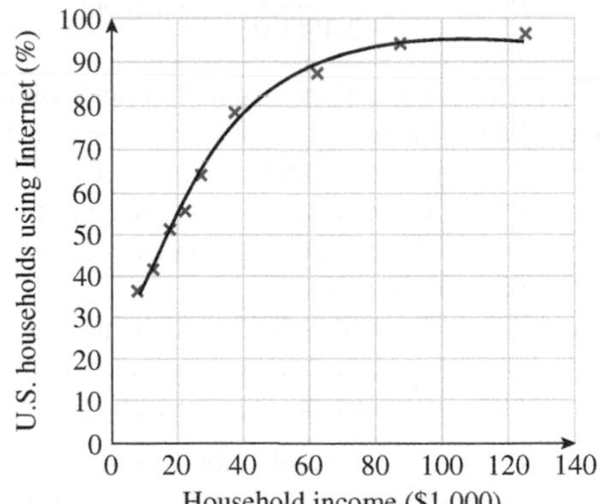

The logistic model is

$$P(x) = \frac{95.0}{1 + 2.78(1.064)^{-x}} \text{ percent,}$$

where x is the household income in thousands of dollars.

a. According to the model, what percentage of extremely wealthy households used the Internet?

b. For low incomes the logistic model is approximately exponential. Which exponential model best approximates $P(x)$ for small x?

c. According to the model, 50% of households of what income used the Internet in 2010? (Round the answer to the nearest $1,000.)

[49] Income levels are midpoints of income brackets. Source: *Current Population Survey (CPS) Internet Use 2010,* National Telecommunications and Information Administration, www.ntia.doc.gov, January 2011.

26. ***Internet Use*** The following graph shows the percentage of U.S. residents who used the Internet at home in 2010 as a function of income (the data points) and a logistic model of these data (the curve).[50]

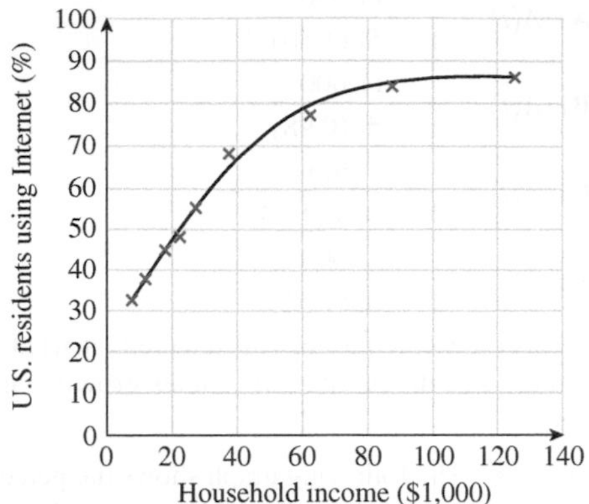

The logistic model is given by

$$P(x) = \frac{86.2}{1 + 2.49(1.054)^{-x}} \text{ percent}$$

where x is the household income in thousands of dollars.

a. According to the model, what percentage of extremely wealthy people used the Internet at home?

b. For low incomes the logistic model is approximately exponential. Which exponential model best approximates $P(x)$ for small x?

c. According to the model, 50% of individuals with what household income used the Internet at home in 2010? (Round the answer to the nearest $1,000.)

27. ***Epidemics*** There are currently 1,000 cases of Venusian flu in a total susceptible population of 10,000, and the number of cases is increasing by 25% each day. Find a logistic model for the number of cases of Venusian flu, and use your model to predict the number of flu cases a week from now. [**HINT**: See Example 1.]

28. ***Epidemics*** Last year's epidemic of Martian flu began with a single case in a total susceptible population of 10,000. The number of cases was increasing initially by 40% per day. Find a logistic model for the number of cases of Martian flu, and use your model to predict the number of flu cases 3 weeks into the epidemic. [**HINT**: See Example 1.]

29. ***Sales*** You have sold 100 "I ♥ Calculus" T-shirts, and sales appear to be doubling every 5 days. You estimate the total market for "I ♥ Calculus" T-shirts to be 3,000. Give a logistic model for your sales, and use it to predict, to the nearest day, when you will have sold 700 T-shirts.

30. ***Sales*** In Russia the average consumer drank two servings of Coca-Cola in 1993. This amount appeared to be increasing exponentially with a doubling time of 2 years.[51] Given a long-range market saturation estimate of 100 servings per year, find a logistic model for the consumption of Coca-Cola in Russia, and use your model to predict when, to the nearest year, the average consumption reached 50 servings per year.

31. T ***Scientific Research*** The following chart shows some of the data shown in the graph in Exercise 23:

Year t (years since 1980)	5	10	15	20	25	30
Research Articles A (thousands)	170	190	260	300	340	430

($t = 0$ represents 1980).[52]

a. What is the logistic regression model for the data? (Round all coefficients to three significant digits.) At what value does the model predict that the number of science research articles will level off? [**HINT**: See Example 2.]

b. According to the model, how many science research articles were published by researchers in the European Union in 2008 ($t = 28$)? (The actual number was about 390 thousand articles.)

32. T ***Scientific Research*** The following chart shows some of the data shown in the graph in Exercise 24:

Year t (years since 1980)	5	10	15	20	25	30
Research Articles A (thousands)	200	220	260	252	290	340

($t = 0$ represents 1980).[53]

a. What is the logistic regression model for the data? (Round all coefficients to three significant digits.) [**HINT**: See Example 2.]

b. According to the model, how many science research articles were published by researchers in the United States in 2008 ($t = 28$)? (The actual figure was about 310 thousand articles.)

33. T ***College Basketball: Men*** The following table shows the number of NCAA men's college basketball teams in the United States for various years since 1990:[54]

[50] See footnote for Exercise 25.

[51] The doubling time is based on retail sales of Coca-Cola products in Russia. Sales in 1993 were double those in 1991 and were expected to double again by 1995. Source: *New York Times,* September 26, 1994, p. D2.

[52] Source: www.sciencewatch.com.

[53] *Ibid.*

[54] 2010 figure is an estimate. Source: www.census.gov.

t (years since 1990)	0	5	10	11	12	13	14
Teams	767	868	932	937	936	967	981
t (years since 1990)	15	16	17	18	19	20	
Teams	983	984	982	1,017	1,017	1,011	

a. What is the logistic regression model for the data? (Round all coefficients to three significant digits.) At what value does the model predict that the number of basketball teams will level off?
b. According to the model, for what value of t is the regression curve steepest? Interpret the answer.
c. Interpret the coefficient b in the context of the number of men's basketball teams.

34. T ***College Basketball: Women*** The following table shows the number of NCAA women's college basketball teams in the United States for various years since 1990:[55]

t (years since 1990)	0	5	10	11	12	13	14
Teams	782	864	956	958	975	1,009	1,008
t (years since 1990)	15	16	17	18	19	20	
Teams	1,036	1,018	1,003	1,013	1,032	1,036	

a. What is the logistic regression model for the data? (Round all coefficients to three significant digits.) At what value does the model predict that the number of basketball teams will level off?
b. According to the model, for what value of t is the regression curve steepest? Interpret the answer.
c. Interpret the coefficient b in the context of the number of women's basketball teams.

T *Exercises 35 and 36 are based on the discussion following Example 2. If the limiting value N is known, then*

$$\frac{N}{y} - 1 = A(b^{-1})^x,$$

so $N/y - 1$ is an exponential function of x. In Exercises 35 and 36, use the given value of N and the data points $(x, N/y - 1)$ to obtain A and b and hence a logistic model.

35. T ◆ ***Population: Puerto Rico*** The following table and graph show the population of Puerto Rico in thousands from 1950 to 2025:[56]

t (years since 1950)	0	10	20	30	40	50
Population (thousands)	2,220	2,360	2,720	3,210	3,540	3,820
t (years since 1950)	55	60	65	70	75	
Population (thousands)	3,910	3,990	4,050	4,080	4,100	

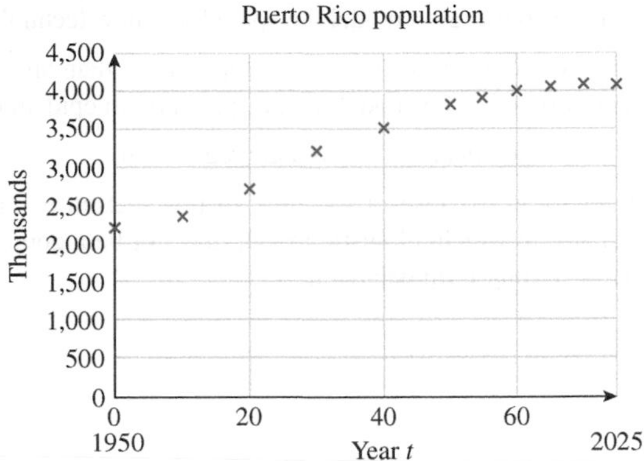

Take t to be the number of years since 1950, and find a logistic model based on the assumption that, eventually, the population of Puerto Rico will grow to 4.5 million. (Round coefficients to four decimal places.) In what year does your model predict that the population of Puerto Rico will first reach 4.0 million?

36. T ◆ ***Population: Virgin Islands*** The following table and graph show the population of the Virgin Islands in thousands from 1950 to 2025:[57]

t (years since 1950)	0	10	20	30	40	50
Population (thousands)	27	33	63	98	104	106
t (years since 1950)	55	60	65	70	75	
Population (thousands)	108	108	107	107	108	

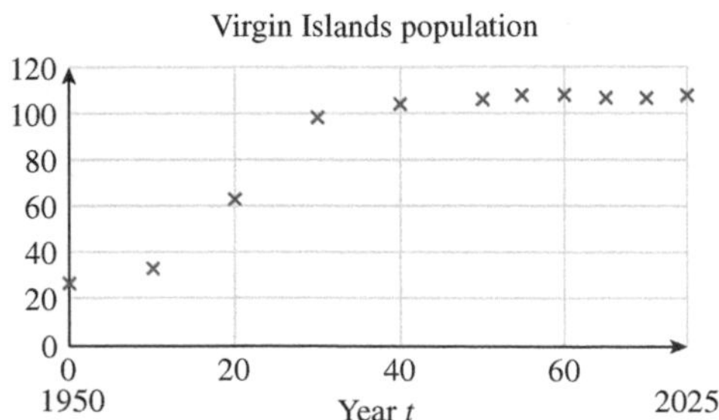

[55] See footnote for Exercise 33.

[56] Figures from 2010 on are U.S. census projections. Source: The 2008 Statistical Abstract, www.census.gov.

[57] *Ibid.*

Take t to be the number of years since 1950, and find a logistic model based on the assumption that, eventually, the population of the Virgin Islands will grow to 110,000. (Round coefficients to four decimal places.) In what year does your model predict that the population of the Virgin Islands first reached 80,000?

Communication and Reasoning Exercises

37. Logistic functions are commonly used to model the spread of epidemics. Given this fact, explain why a logistic function is also useful to model the spread of a new technology.

38. Why is a logistic function more appropriate than an exponential function for modeling the spread of an epidemic?

39. Give one practical use for logistic regression.

40. Refer to an exercise or example in this section to find a scenario in which a logistic model may not be a good predictor of long-term behavior.

41. What happens to the function $P(t) = \frac{N}{1 + Ab^{-t}}$ if we replace b^{-t} by b^t when $b > 1$? When $b < 1$?

42. ▼ What happens to the function $P(t) = \frac{N}{1 + Ab^{-t}}$ if $A = 0$? If $A < 0$?

43. ▼ We said that the logistic curve $y = \frac{N}{1 + Ab^{-t}}$ is steepest when $t = \frac{\ln A}{\ln b}$. Show that the corresponding value of y is $N/2$. [**HINT:** Use the fact that $\frac{\ln A}{\ln b} = \log_b A$.]

44. ▼ We said that the logistic curve $y = \frac{N}{1 + Ab^{-t}}$ is steepest when $t = \frac{\ln A}{\ln b}$. For which values of A and b is this value of t positive, zero, and negative?

CHAPTER 2 REVIEW

KEY CONCEPTS

www.WanerMath.com
Go to the Website to find a comprehensive and interactive Web-based summary of Chapter 2.

2.1 Quadratic Functions and Models

2.2 Exponential Functions and Models

2.3 Logarithmic Functions and Models

2.4 Logistic Functions and Models

REVIEW EXERCISES

Sketch the graph of the quadratic functions in Exercises 1 and 2, indicating the coordinates of the vertex, the y-intercept, and the x-intercepts (if any).

1. $f(x) = x^2 + 2x - 3$

2. $f(x) = -x^2 - x - 1$

In Exercises 3 and 4 the values of two functions, f and g, are given in a table. One, both, or neither of them may be exponential. Decide which, if any, are exponential, and give the exponential models for those that are.

3.

x	−2	−1	0	1	2
$f(x)$	20	10	5	2.5	1.25
$g(x)$	8	4	2	1	0

4.

x	−2	−1	0	1	2
$f(x)$	8	6	4	2	1
$g(x)$	$\frac{3}{4}$	$\frac{3}{2}$	3	6	12

In Exercises 5 and 6, graph the given pairs of functions on the same set of axes with $-3 \leq x \leq 3$.

5. $f(x) = \frac{1}{2}(3^x)$; $g(x) = \frac{1}{2}(3^{-x})$

6. $f(x) = 2(4^x)$; $g(x) = 2(4^{-x})$

T *On the same set of axes, use technology to graph the pairs of functions in Exercises 7 and 8 for the given range of x. Identify which graph corresponds to which function.*

7. $f(x) = e^x$; $g(x) = e^{0.8x}$; $-3 \leq x \leq 3$

8. $f(x) = 2(1.01)^x$; $g(x) = 2(0.99)^x$; $-100 \leq x \leq 100$

In Exercises 9–14, compute the indicated quantity.

9. A \$3,000 investment earns 3% interest, compounded monthly. Find its value after 5 years.

10. A \$10,000 investment earns 2.5% interest, compounded quarterly. Find its value after 10 years.

11. An investment earns 3% interest, compounded monthly, and is worth \$5,000 after 10 years. Find its initial value.

12. An investment earns 2.5% interest, compounded quarterly, and is worth \$10,000 after 10 years. Find its initial value.

13. A \$3,000 investment earns 3% interest, compounded continuously. Find its value after 5 years.

14. A \$10,000 investment earns 2.5% interest, compounded continuously. Find its value after 10 years.

In Exercises 15–18, find a formula of the form $f(x) = Ab^x$ using the given information.

15. $f(0) = 4.5$; the value of f triples for every half-unit increase in x.

16. $f(0) = 5$; the value of f decreases by 75% for every 1-unit increase in x.

17. $f(1) = 2$, $f(3) = 18$.

18. $f(1) = 10$, $f(3) = 5$.

On the same set of axes, graph the pairs of functions in Exercises 19 and 20.

19. $f(x) = \log_3 x$; $g(x) = \log_{(1/3)} x$

20. $f(x) = \log x$; $g(x) = \log_{(1/10)} x$

In Exercises 21–24, use the given information to find an exponential model of the form $Q = Q_0e^{-kt}$ or $Q = Q_0e^{kt}$, as appropriate. Round all coefficients to three significant digits when rounding is necessary.

21. Q is the amount of radioactive substance with a half-life of 100 years in a sample originally containing 5 grams (t is time in years).

22. Q is the number of cats on an island whose cat population was originally 10,000 but is being cut in half every 5 years (t is time in years).

23. Q is the diameter (in centimeters) of a circular patch of mold on your roommate's damp towel you have been monitoring with morbid fascination. You measured the patch at 2.5 centimeters across 4 days ago and have observed that it is doubling in diameter every 2 days (t is time in days).

24. Q is the population of cats on another island whose cat population was originally 10,000 but is doubling every 15 months. (t is time in months.)

In Exercises 25–28, find the time required, to the nearest 0.1 year, for the investment to reach the desired goal.

25. \$2,000 invested at 4%, compounded monthly; goal: \$3,000

26. \$2,000 invested at 6.75%, compounded daily; goal: \$3,000

27. \$2,000 invested at 3.75%, compounded continuously; goal: \$3,000

28. \$1,000 invested at 100%, compounded quarterly; goal: \$1,200

In Exercises 29–32, find an equation for the logistic function of x with the stated properties.

29. Through (0, 100), initially increasing by 50% per unit of x, and limiting value 900.

30. Initially exponential of the form $y = 5(1.1)^x$ with limiting value 25.

31. Passing through (0, 5) and decreasing from a limiting value of 20 to 0 at a rate of 20% per unit of x when x is near 0.

32. Initially exponential of the form $y = 2(0.8)^x$ with a value close to 10 when $x = -60$.

Applications: OHaganBooks.com
[Try the game at www.OHaganBooks.com]

33. ***Website Traffic*** The daily traffic ("hits per day") at OHaganBooks.com apparently depends on the monthly expenditure on Internet advertising. The following model is based on information collected over the past few months:

$$h = -0.000005c^2 + 0.085c + 1,750.$$

Here, h is the average number of hits per day at OHaganBooks.com, and c is the monthly advertising expenditure.

a. According to the model, what monthly advertising expenditure will result in the largest volume of traffic at OHaganBooks.com? What is that volume?

b. In addition to predicting a maximum volume of traffic, the model predicts that the traffic will eventually drop to zero if the advertising expenditure is increased too far. What expenditure (to the nearest dollar) results in no website traffic?

c. What feature of the formula for this quadratic model indicates that it will predict an eventual decline in traffic as advertising expenditure increases?

34. ***Broadband Access*** Pablo Pelogrande, a new summer intern at OHaganBooks.com in 2013, was asked by John O'Hagan to research the extent of broadband access in the United States. Pelogrande found some very old data online on broadband access from the start of 2001 to the end of 2003 and used it to construct the following quadratic model of the growth rate of broadband access:

$n(t) = 2t^2 - 6t + 12$ million new American adults with broadband per year

(t is time in years; $t = 0$ represents the start 2000.)[58]

a. What is the appropriate domain of n?

b. According to the model, when was the growth rate at a minimum?

c. Does the model predict a zero growth rate at any particular time? If so, when?

[58] Based on data for 2001–2003. Source for data: Pew Internet and American Life Project data memos dated May 18, 2003, and April 19, 2004, downloaded from www.pewinternet.org.

d. What feature of the formula for this quadratic model indicates that the growth rate eventually increases?

e. Does the fact that $n(t)$ decreases for $t \leq 1.5$ suggest that the number of broadband users actually declined before June 2001? Explain.

f. Pelogrande extrapolated the model to estimate the growth rate at the beginning of 2013 and 2014. What did he find? Comment on the answer.

35. ***Revenue and Profit*** Some time ago, a consultant formulated the following linear model of demand for online novels:

$$q = -60p + 950,$$

where q is the monthly demand for OHaganBooks.com's online novels at a price of p dollars per novel.

a. Use this model to express the monthly revenue as a function of the unit price p. Hence, determine the price you should charge for a maximum monthly revenue.

b. Author royalties and copyright fees cost the company an average of \$4 per novel, and the monthly cost of operating and maintaining the online publishing service amounts to \$900 per month. Express the monthly profit P as a function of the unit price p. Hence, determine the unit price you should charge for a maximum monthly profit. What is the resulting profit (or loss)?

36. ***Revenue and Profit*** Billy-Sean O'Hagan is John O'Hagan's son and a freshman in college. He notices that the demand for the college newspaper was 2,000 copies each week when the paper was given away free of charge but dropped to 1,000 each week when the college started charging 10¢/copy.

a. Write down the associated linear demand function.

b. Use your demand function to express the revenue as a function of the unit price p. Hence, determine the price the college should charge for a maximum revenue. At that price, what is the revenue from sales of one edition of the newspaper?

c. It costs the college 4¢ to produce each copy of the paper plus an additional fixed cost of \$200. Express the profit P as a function of the unit price p. Hence, determine the unit price the college should charge for a maximum monthly profit (or minimum loss). What is the resulting profit (or loss)?

37. ***Lobsters*** Marjory Duffin, CEO of *Duffin House,* is particularly fond of having steamed lobster at working lunches with executives from OHaganBooks.com and is therefore alarmed to learn that the yearly lobster harvest from New York's Long Island Sound has been decreasing dramatically since 1997. Indeed, the size of the annual harvest can be approximated by

$$n(t) = 9.1(0.81^t) \text{ million pounds,}$$

where t is time in years since 1997.[59]

a. The model tells us that the harvest was ____ million pounds in 1997 and decreasing by ___% each year.

b. What does the model predict for the 2013 harvest?

38. ***Stock Prices*** In the period immediately following its initial public offering (IPO), OHaganBooks.com's stock was doubling in value every 3 hours. If you bought \$10,000 worth of the stock when it was first offered, how much was your stock worth after 8 hours?

39. ***Lobsters*** (See Exercise 37.) Marjory Duffin has just left John O'Hagan, CEO of OHaganBooks.com, a frantic phone message to the effect that this year's lobster harvest from New York's Long Island Sound is predicted to dip below 200,000 pounds, making that planned lobster working lunch more urgent than ever. What year is it?

40. ***Stock Prices*** We saw in Exercise 38 that OHaganBooks.com's stock was doubling in value every 3 hours following its IPO. If you bought \$10,000 worth of the stock when it was first offered, how long from the initial offering did it take your investment to reach \$50,000?

41. ***Lobsters*** We saw in Exercise 37 that the Long Island Sound lobster harvest was given by $n(t) = 9.1(0.81^t)$ million pounds t years after 1997. However, in 2010, thanks to the efforts of *Duffin House,* it turned around and started increasing by 24% each year.[60] What, to the nearest 10,000 pounds, was the actual size of the harvest in 2013?

42. ***Stock Prices*** We saw in Exercise 38 that OHaganBooks.com's stock was doubling in value every 3 hours following its IPO. After 10 hours of trading, the stock turns around and starts losing one third of its value every 4 hours. How long (from the initial offering) will it be before your stock is once again worth \$10,000?

43. T ***Lobsters*** The following chart shows some of the data that went into the model in Exercise 37:

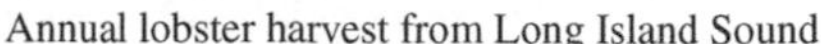
Annual lobster harvest from Long Island Sound

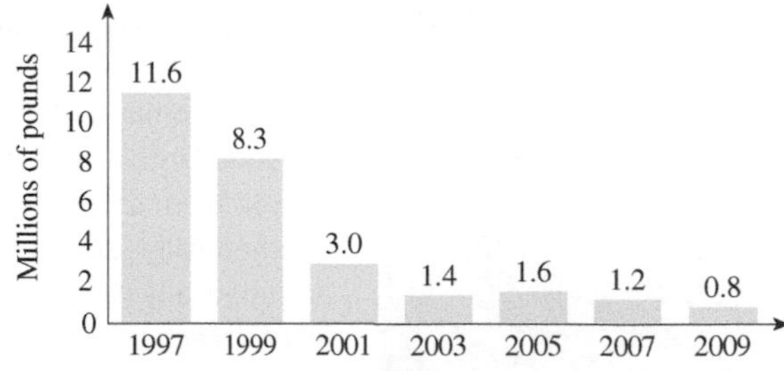

Use these data to obtain an exponential regression curve of the form $n(t) = Ab^t$, with $t = 0$ corresponding to 1997 and coefficients rounded to two significant digits.

[59] Authors' regression model. Source for data: Long Island Sound Study, data downloaded May 2011 from longislandsoundstudy.net/2010/07/lobster-landings.

[60] This claim, like Duffin House, is fiction.

44. **T** ***Stock Prices*** The actual stock price of OHaganBooks.com in the hours following its IPO is shown in the following chart:

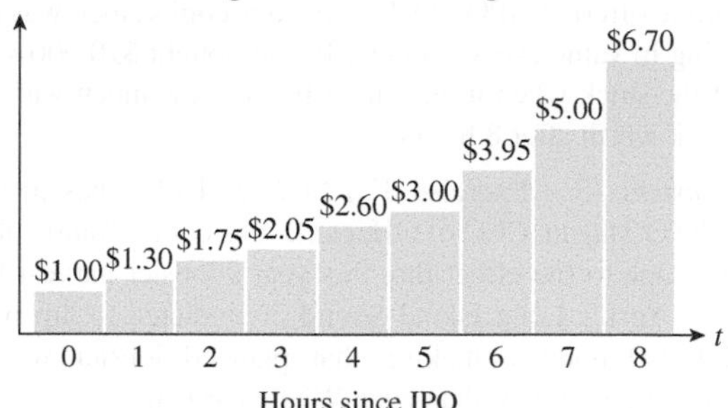

Use the data to obtain an exponential regression curve of the form $P(t) = Ab^t$, with $t = 0$ the time in hours since the IPO and coefficients rounded to three significant digits. At the end of which hour will the stock price first be above $10?

45. ***Hardware Life*** *(Based on a question from the GRE economics exam)* To estimate the rate at which new computer hard drives will have to be retired, OHaganBooks.com uses the "survivor curve":

$$L_x = L_0 e^{-x/t},$$

where

L_x = number of surviving hard drives at age x

L_0 = number of hard drives initially

t = average life in years.

All of the following are implied by the curve *except*

(A) some of the equipment is retired during the first year of service.

(B) some equipment survives three average lives.

(C) more than half the equipment survives the average life.

(D) increasing the average life of equipment by using more durable materials would increase the number surviving at every age.

(E) the number of survivors never reaches zero.

50. ***Sales*** OHaganBooks.com modeled its weekly sales over a period of time with the function

$$s(t) = 6{,}050 + \frac{4{,}470}{1 + 14(1.73^{-t})},$$

as shown in the following graph (t is measured in weeks):

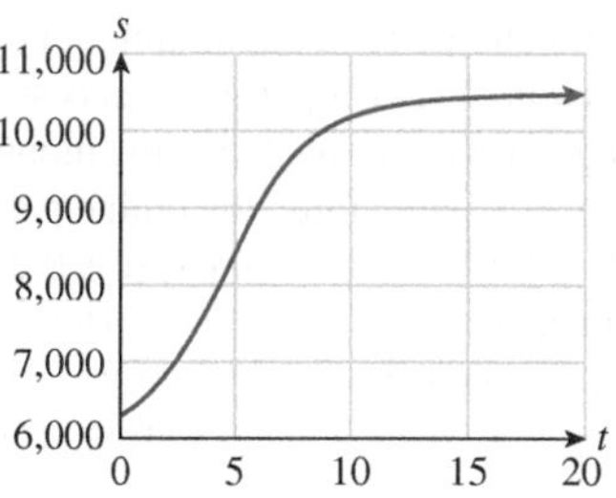

a. As time goes on, it appears that weekly sales are leveling off. At what value are they leveling off?

b. When did weekly sales rise above 10,000?

c. When, to the nearest week, were sales rising most rapidly?

CASE STUDY

Checking up on Malthus

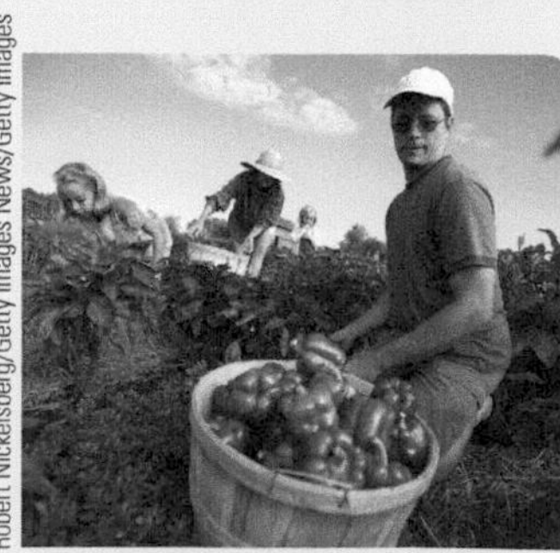
Robert Nickelsberg/Getty Images News/Getty Images

In 1798 Thomas R. Malthus (1766–1834) published an influential pamphlet, later expanded into a book, titled *An Essay on the Principle of Population as It Affects the Future Improvement of Society.* One of his main contentions was that population grows geometrically (exponentially) while the supply of resources such as food grows only arithmetically (linearly). This led him to the pessimistic conclusion that population would always reach the limits of subsistence and precipitate famine, war, and ill health unless population could be checked by other means. He advocated "moral restraint," which includes the pattern of late marriage common in Western Europe at the time and now common in most developed countries, leading to a lower reproduction rate.

Two hundred years later, you have been asked to check the validity of Malthus's contention. That population grows geometrically, at least over short periods of time, is commonly assumed. That resources grow linearly is more questionable. You decide to check the actual production of a common crop, wheat, in the United States. Agricultural statistics like these are available from the U.S. government on the Internet, through the U.S. Department of Agriculture's National Agricultural Statistics Service (NASS). As of 2016, this service was available at www.nass.usda.gov. Looking through this site, you locate data on the annual production of all wheat in the United States from 1900 through 2010.

Website
www.WanerMath.com
To download an Excel sheet with the data used in the case study, go to Everything for Applied Calc, scroll down to the case study for Chapter 2, and click on "Wheat Production Data."

Year	1900	1901	...	2009	2010
Wheat Production (thousands of bushels)	599,315	762,546	...	2,218,061	2,208,391

Graphing these data (using Excel, for example), you obtain the graph in Figure 21.

Wheat production in the United States (thousand bushels)

3,000,000
2,500,000
2,000,000
1,500,000
1,000,000
500,000
0
1900 1920 1940 1960 1980 2000 2020

Figure 21

This does not look very linear, particularly in the last half of the twentieth century, but you continue checking the mathematics. Using Excel's built-in linear regression capabilities, you find that the line that best fits these data, shown in Figure 22, has $r^2 = 0.8039$ (Recall the discussion of the correlation coefficient r in Section 1.4. A similar statistic is available for other types of regression as well.)

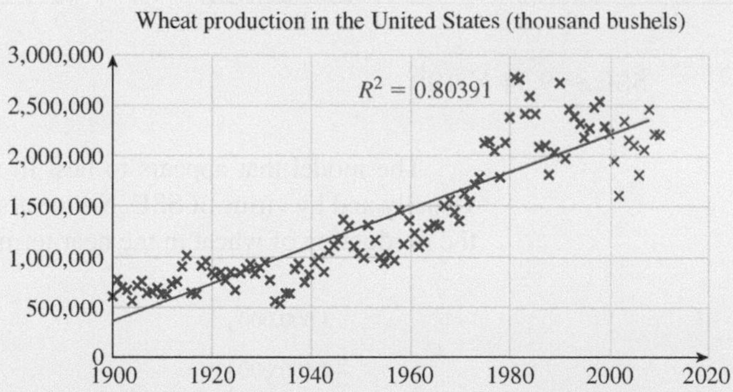

Figure 22

Although that is a fairly high correlation, you notice that the residuals* are not distributed randomly: The actual wheat production starts out higher than the line, dips below the line from about 1920 to about 1970, then rises above the line, and finally appears to dip below the line around 2002. This behavior seems to suggest a logistic curve or perhaps a cubic curve. On the other hand, it is also possible that the apparent dip at the end of the data is not statistically significant—it could be nothing

* Recall that the residuals are defined as $y_{\text{Observed}} - y_{\text{Predicted}}$ (see Section 1.4) and are the vertical distances between the observed data points and the regression line.

more than a transitory fluctuation in the wheat production industry—so perhaps we should also consider models that do not bend downward, such as exponential and quadratic models.

Following is a comparison of the four proposed models (with coefficients rounded to three significant digits). For the independent variable we used t = time in years since 1900. SSE is the sum-of-squares error.

Quadratic

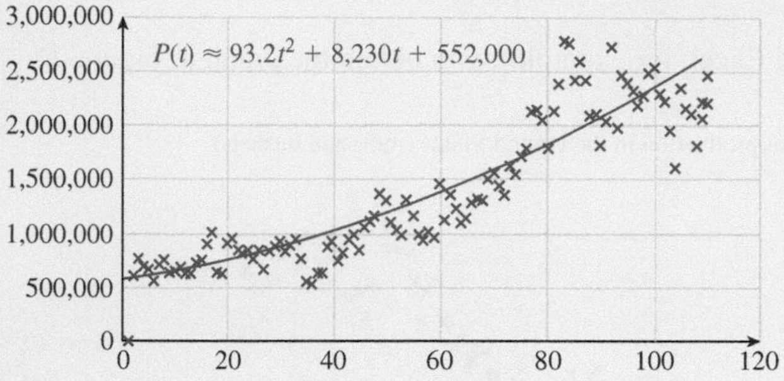

$P(t) \approx 93.2t^2 + 8{,}230t + 552{,}000$

SSE $\approx 8.68 \times 10^{12}$

Cubic

$P(t) \approx -5.30t^3 + 968t^2 - 30{,}100t + 895{,}000$

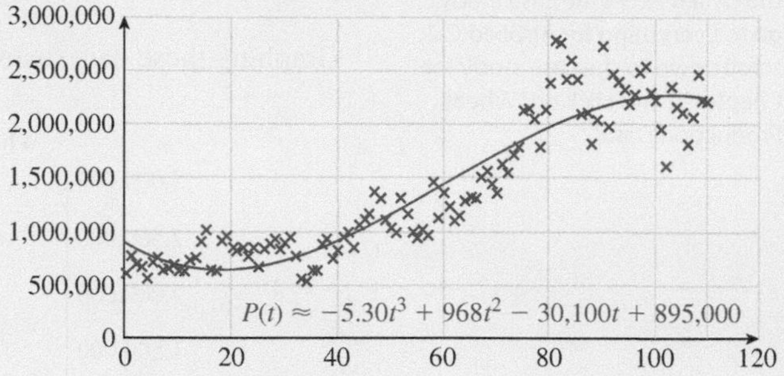

SSE $\approx 6.60 \times 10^{12}$

Exponential

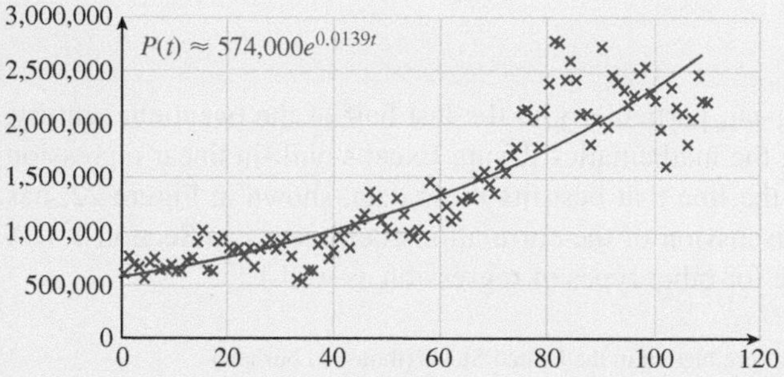

$P(t) \approx 574{,}000e^{0.0139t}$

SSE $\approx 9.34 \times 10^{12}$

Logistic

$P(t) \approx \dfrac{3{,}440{,}000}{1 + 6.33(1.026^{-t})}$

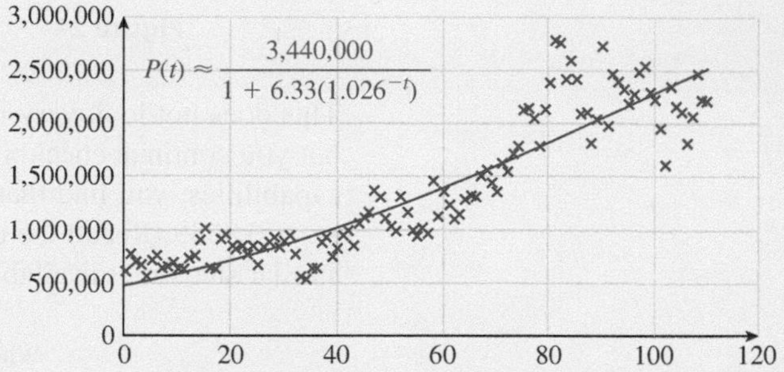

SSE $\approx 8.26 \times 10^{12}$

The model that appears to best fit the data seems to be the cubic model, both visually and by virtue of SSE. Notice also that the cubic model predicts a *decrease* in the production of wheat in the near term (see Figure 23).

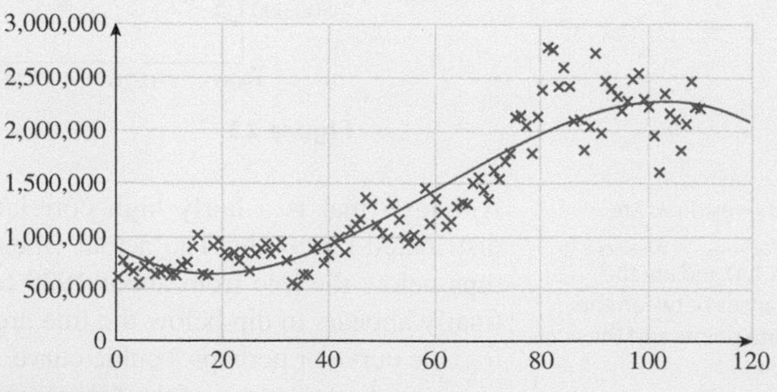

Figure 23

So you prepare a report that documents your findings and concludes that things are even worse than Malthus predicted, at least as far as wheat production in the United States is concerned: The supply is decreasing while the population is still increasing more or less exponentially. (See Exercise 75 in Section 2.2.)

You are about to hit "Send," which will dispatch copies of your report to a significant number of people on whom the success of your career depends, when you notice something strange about the pattern of data in Figure 23: The observed data points appear to hug the regression curve quite closely for small values of t but appear to become more and more scattered as the value of t increases. In the language of residuals, the residuals are small for small values of t but then tend to get larger with increasing t. Figure 24 shows a plot of the residuals that shows this trend even more clearly.

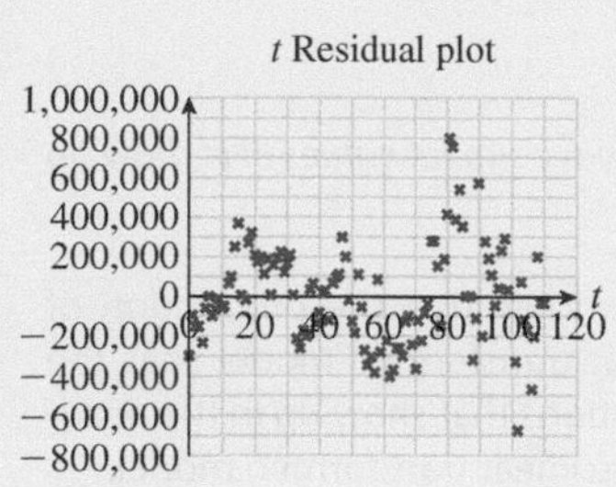

Figure 24

This reminds you vaguely of something that came up in your college business statistics course, so you consult the textbook from that class, which (fortunately) you still own, and discover that a pattern of residuals with increasing magnitude suggests that, instead of modeling y versus t directly, you should instead model $\ln y$ versus t. (The residuals for large values of t will then be scaled down by the logarithm.)

Figure 25 shows the resulting plot together with the regression line (what we call the linear transformed model).

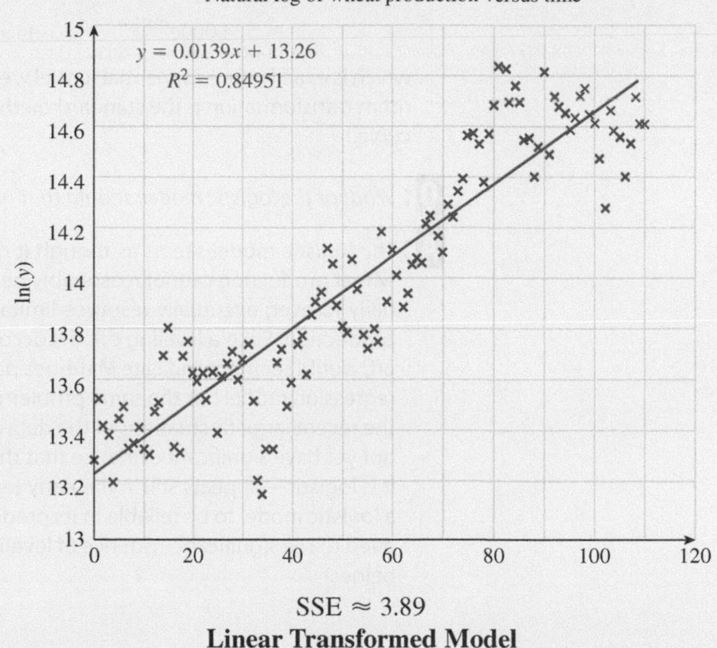

Figure 25

Notice that this time, the regression patterns no longer suggest an obvious curve. Further, they no longer appear to grow with increasing t. Although SSE is dramatically lower than the values for the earlier models, the contrast is a false one; the units of y are now different, and comparing SSE with that of the earlier models is like comparing apples and oranges. While SSE depends on the units of measurement used, the coefficient of determination r^2 discussed in Section 1.4 is independent of the units used. A similar statistic is available for other types of regression as well, as well as something called "adjusted r^2."

The value of r^2 for the transformed model is approximately 0.850, while r^2 for the cubic model* is about 0.861, which is fairly close.

* The adjusted r^2 from statistics that corrects for model size.

Q: *If the cubic model and the linear transformed model have similar values of r^2, how do I decide which is more appropriate?*

A: The cubic model, if extrapolated, predicts unrealistically that the production of wheat will plunge in the near future, but the linear transformed model sees the recent drop-off as just one of several market fluctuations that show up in the residuals. You should therefore favor the more reasonable linear transformed model.

Q: *The linear transformed model gives us ln y versus t. What does it say about y versus t?*

A: Accurately write down the equation of the transformed linear model, being careful to replace y by $\ln y$:

$$\ln y = 0.0139t + 13.26.$$

Rewriting this in exponential form gives

$$\begin{aligned} y &= e^{0.0139t+13.26} \\ &= e^{13.26}e^{0.0139t} \\ &\approx 574{,}000e^{0.0139t}, \quad \text{Coefficients rounded to three digits} \end{aligned}$$

which is exactly the exponential model we found earlier! (In fact, using the natural logarithm transformation is the standard method of computing the regression exponential curve.)

Q: *What of the logistic model; should that not be the most realistic?*

A: The logistic model seems as though it *ought* to be the most appropriate, because wheat production cannot reasonably be expected to continue increasing exponentially forever; eventually resource limitations must lead to a leveling off of wheat production. Such a leveling off, if it occurred before the population started to level off, would seem to vindicate Malthus's pessimistic predictions. However, the logistic regression model has the same problem as the cubic model: It is trying to interpret the recent large fluctuations in the data as evidence of leveling off, but we really do not yet have significant evidence that that is occurring. Wheat production—even if it is logistic—appears still in the early (exponential) stage of growth. In general, for a logistic model to be reliable in its prediction of the leveling-off value N, we would need to see significant evidence of leveling off in the data. (See, however, Exercise 2 below.)

You now conclude that wheat production for the past 100 years is better described as increasing exponentially rather than linearly, contradicting Malthus, and moreover that it shows no sign of leveling off yet.

EXERCISES

1. Use the wheat production data starting at 1950 to construct the exponential regression model in two ways: directly and using a linear transformed model as above. (Round coefficients to three digits.) Compare the growth constant k of your model with that of the exponential model based on the data from 1900 on. How would you interpret the difference?

2. Compute the least-squares logistic model for the data in the preceding exercise. (Round coefficients to three significant digits.) At what level does it predict that wheat production will level off? (Note on using Excel Solver for logistic regression: Before running Solver, press Options in the Solver window, and turn "Automatic Scaling" on. This adjusts the algorithm for the fact that the constants A, N, and b have vastly different orders of magnitude.) Give two graphs: one showing the data with the exponential regression model and the other showing the data with the logistic regression model. Which model gives a better fit visually? Justify your observation by computing SSE directly for both models. Comment on your answer in terms of Malthus's assertions.

3. Find the production figures for another common crop grown in the United States. Compare the linear, quadratic, cubic, exponential, and logistic models. What can you conclude?

4. Below are the census figures for the population of the United States (in thousands) from 1820 to 2010.[61] Compare the linear, quadratic, and exponential models. What can you conclude?

Year	1820	1830	1840	1850	1860	1870	1880	1890	1900	1910
Population (thousands)	9,638	12,861	17,063	23,192	31,443	38,558	50,189	62,980	76,212	92,228
Year	1920	1930	1940	1950	1960	1970	1980	1990	2000	2010
Population (thousands)	106,022	123,203	132,165	151,326	179,323	203,302	226,542	248,710	281,422	308,746

[61] Source: Bureau of the Census, U.S. Department of Commerce.

TI-83/84 Plus Technology Guide

Section 2.1

Example 2 (page 136) Sketch the graph of each quadratic function, showing the location of the vertex and intercepts.

a. $f(x) = 4x^2 - 12x + 9$

b. $g(x) = -\frac{1}{2}x^2 + 4x - 12$

Solution

We will do part (a).

1. Start by storing the coefficients a, b, and c using

    ```
    4→A:-12→B:9→C
    ```

    [STO>] gives the arrow [ALPHA] [.] gives the colon

2. Save your quadratic as Y_1, using the Y= screen:

    Y_1=AX^2+BX+C

3. To obtain the x-coordinate of the vertex, enter its formula as shown below on the left.

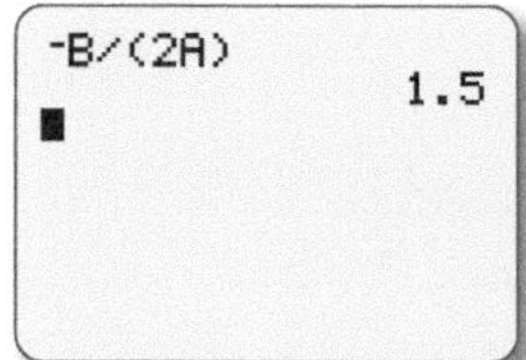

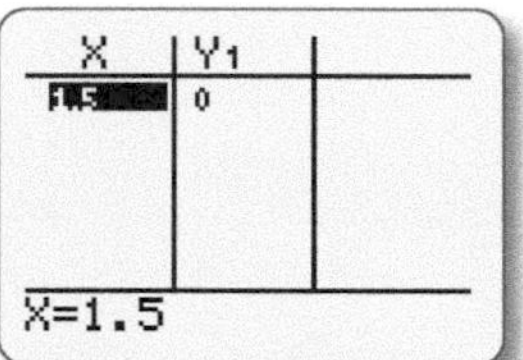

4. The y-coordinate of the vertex can be obtained from the table screen by entering $x = 1.5$ as shown above on the right. (If you can't enter values of x, press [2ND] [TBLSET], and set Indpnt to Ask.) From the table we see that the vertex is at the point (1.5, 0).

5. To obtain the x-intercepts, enter the quadratic formula on the Home screen as shown:

    ```
    (-B+√(B²-4AC))/(
    2A)
                 1.5
    (-B-√(B²-4AC))/(
    2A)
                 1.5
    ```

    Because the intercepts agree, we conclude that the graph intersects the x-axis on a single point (at the vertex).

6. To graph the function, we need to select good values for Xmin and Xmax. In general, we would like our graph to show the vertex as well as all the intercepts. To see the vertex, make sure that its x-coordinate (1.5) is between Xmin and Xmax. To see the x-intercepts, make sure that they are also between Xmin and Xmax. To see the y-intercept, make sure that $x = 0$ is between Xmin and Xmax. Thus, to see everything, choose Xmin and Xmax so that the interval [xMin, xMax] contains the x-coordinate of the vertex, the x-intercepts, and 0. For this example we can choose an interval like $[-1, 3]$.

7. Once xMin and xMax have been chosen, you can obtain convenient values of yMin and yMax by pressing [ZOOM] and selecting the option ZoomFit. (Make sure that your quadratic equation is entered in the Y= screen before doing this!)

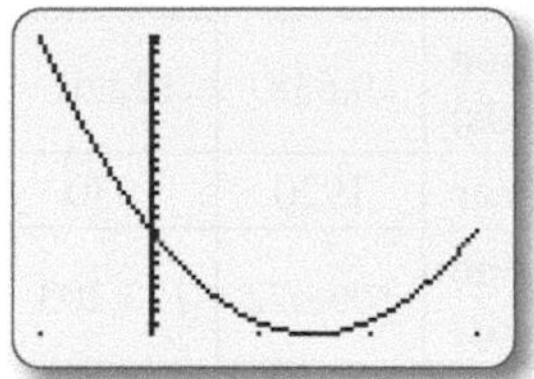

Example 5 (page 140) The following table shows the annual mean carbon dioxide concentration measured at Mauna Loa Observatory in Hawaii, in parts per million, every 10 years from 1960 through 2010:

Year t (years since 1960)	0	10	20	30	40	50
CO_2 Concentration C (ppm)	317	326	339	354	369	390

Find the quadratic model

$$C(t) = at^2 + bt + c$$

that best fits the data.

Solution

1. Using [STAT] EDIT, enter the data with the x-coordinates (values of t) in L_1 and the y-coordinates (values of C) in L_2, just as in Section 1.4:

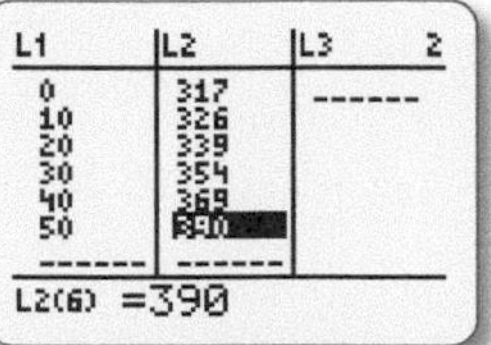

2. Press STAT, select CALC, and choose option #5 `QuadReg`. Pressing ENTER gives the quadratic regression curve in the Home screen:

$$y \approx 0.01214x^2 + 0.8471x + 316.9. \quad \text{Coefficients rounded to four decimal places}$$

3. Now go to the `Y=` window, and turn `Plot1` on by selecting it and pressing ENTER. (You can also turn it on in the 2ND STAT PLOT screen.)
4. Next, enter the regression equation in the Y= screen by pressing Y=, clearing out whatever function is there, and pressing VARS 5 and selecting `EQ` option #1: `RegEq` as shown below left.

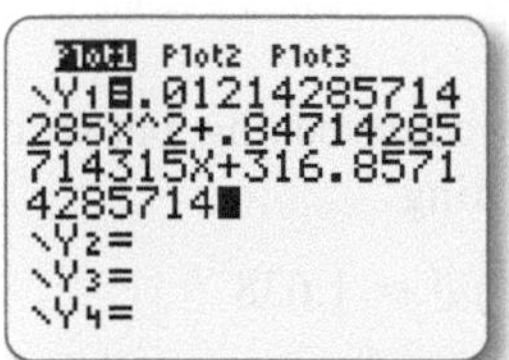

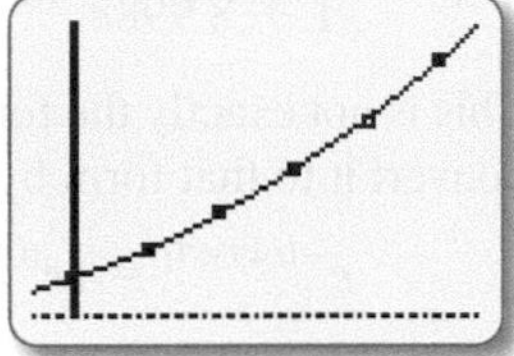

5. To obtain a convenient window showing all the points and the lines, press ZOOM, and choose option #9: `ZoomStat` as shown above on the right.

Note When you are done viewing the graph, it is a good idea to turn `Plot1` off again to avoid errors in graphing or data points showing up in your other graphs. ■

Section 2.2

Example 6(a) (page 155) The following table shows annual expenditure on health in the United States from 1980 through 2009.

Year t (years since 1980)	0	5	10	15	20	25	29
Expenditure ($ billion)	256	444	724	1,030	1,380	2,020	2,490

Find the exponential regression model

$$C(t) = Ab^t$$

for the annual expenditure.

Solution

This is very similar to Example 5 in Section 2.1 (see the Technology Guide for Section 2.1):

1. Use STAT EDIT to enter the table of values.

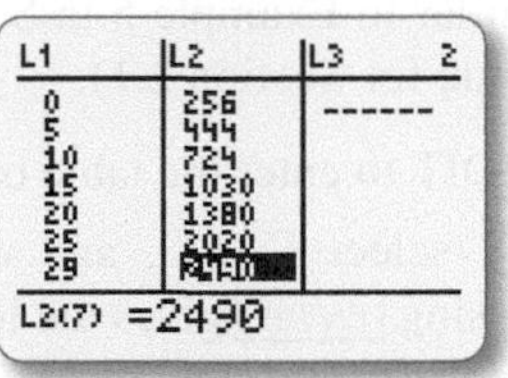

2. Press STAT, select CALC, and choose option #0 `ExpReg`. Pressing ENTER gives the exponential regression curve in the Home screen:

$$C(t) \approx 296.25(1.0798)^t \quad \text{Coefficients rounded}$$

3. To graph the points and regression line in the same window, turn `Plot1` on (see the Technology Guide for Example 5 in Section 2.1), and enter the regression equation in the `Y=` screen by pressing Y=, clearing out whatever function is there, and pressing VARS 5 and selecting `EQ` option #1: `RegEq`. Then press ZOOM and choose option #9: `ZoomStat` to see the graph.

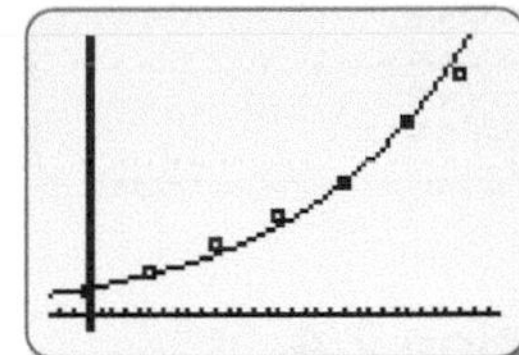

Note When you are done viewing the graph, it is a good idea to turn `Plot1` off again to avoid errors in graphing or data points showing up in your other graphs. ■

Section 2.3

Example 4 (page 169) The following table shows the total spent on research and development by institutions of higher education in the United States, in billions of dollars, for the period 2003–2012:

Year t (years since 2000)	3	4	5	6	7
Spending ($ billion)	41	45	48	50	52
Year t (years since 2000)	8	9	10	11	12
Spending ($ billion)	54	57	61	65	66

Find the best-fit logarithmic model of the form

$$S(t) = A \ln t + C,$$

and use the model to project total spending on research by universities and colleges in 2016, assuming that the trend continues.

Solution

This is very similar to Example 5 in Section 2.1 (see the Technology Guide for Section 2.1):

1. Use STAT EDIT to enter the table of values.
2. Press STAT, select CALC, and choose option #9 `LnReg`. Pressing ENTER gives the logarithmic regression curve in the Home screen:

$$S(t) \approx 17.9 \ln t + 19.4. \quad \text{Coefficients rounded}$$

3. To graph the points and regression line in the same window, turn `Plot1` on (see the Technology Guide for Example 5 in Section 2.1), and enter the regression equation in the `Y=` screen by pressing Y=, clearing out whatever function is there, and pressing VARS 5 and selecting `EQ` option #1: `RegEq`. Then press ZOOM and choose option #9: `ZoomStat` to see the graph.

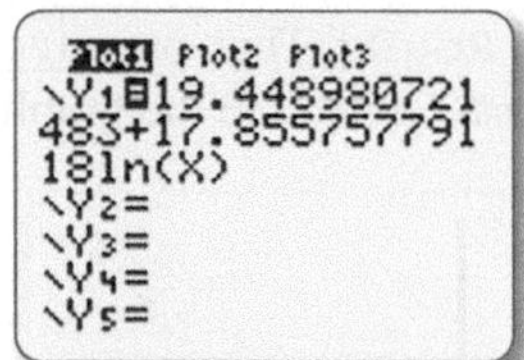

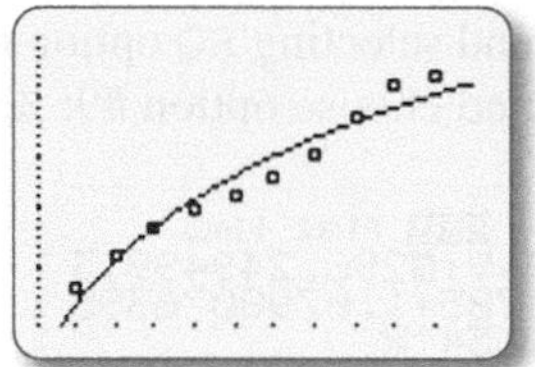

Section 2.4

Example 2 (page 178) The following table shows wired broadband penetration in the United States as a function of time t in years ($t = 0$ represents 2000).

Year t (years since 2000)	2	3	4	5	6	7
Penetration P (%)	6.7	9.6	12.8	16.4	20.3	23.2
Year t (years since 2000)	8	9	10	11	12	13
Penetration P (%)	25.5	25.5	26.7	27.7	28.7	29.8

Find a logistic regression curve of the form

$$P(t) = \frac{N}{1 + Ab^{-t}}.$$

Solution

This is very similar to Example 5 of Section 2.1 (see the Technology Guide for Section 2.1):

1. Use STAT EDIT to enter the table of values.
2. Press STAT, select CALC, and choose option #B `Logistic`. Pressing ENTER gives the logistic regression curve in the Home screen:

$$P(t) \approx \frac{29.238}{1 + 8.998e^{-0.49367t}}. \quad \text{Coefficients rounded}$$

This is not exactly the form we are seeking, but we can convert it to that form by writing

$$e^{-0.49367t} = (e^{0.49367})^{-t} \approx 1.638^{-t}$$

so

$$P(t) \approx \frac{29.238}{1 + 8.998(1.638)^{-t}}.$$

3. To graph the points and regression line in the same window, turn `Plot1` on (see the Technology Guide for Example 5 of Section 2.1), and enter the regression equation in the `Y=` screen by pressing Y=, clearing out whatever function is there, and pressing VARS 5 and selecting `EQ` option #1: `RegEq`. Then press ZOOM and choose option #9: `ZoomStat` to see the graph.

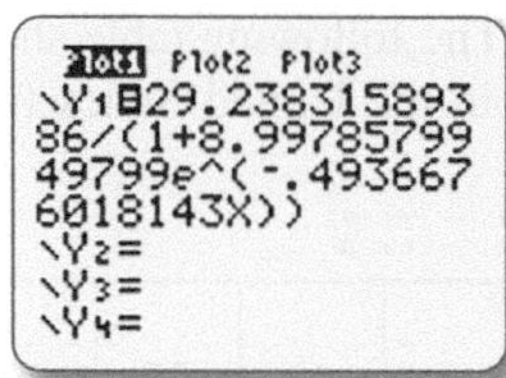

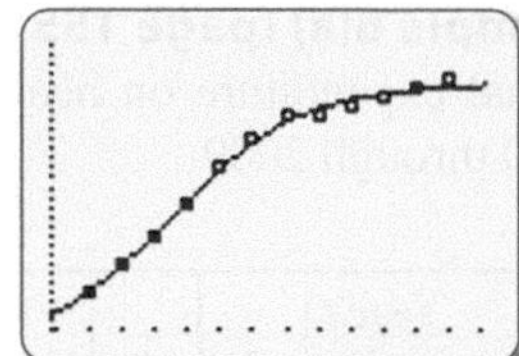

Spreadsheet Technology Guide

Section 2.1

Example 2 (page 136) Sketch the graph of each quadratic function, showing the location of the vertex and intercepts.

a. $f(x) = 4x^2 - 12x + 9$

b. $g(x) = -\frac{1}{2}x^2 + 4x - 12$

Solution

We can set up a worksheet so that all we have to enter are the coefficients a, b, and c and a range of x-values for the graph. Here is a possible layout that will plot 101 points using the coefficients for part (a):

1. First, we compute the x-coordinates.

	A	B	C	D	E
1	x	y	a	4	
2	=$D$4		b	-12	
3	=A2+$D$6		c	9	
4			Xmin	-10	
5			Xmax	10	
6			Delta X	=(D5-D4)/100	
101					
102					

2. To add the y-coordinates, we use the technology formula

```
a*x^2+b*x+c
```

replacing a, b, and c with (absolute) references to the cells containing their values.

	A	B	C	D
1	x	y	a	4
2	-10	=$D$1*A2^2+$D$2*A2+$D$3	b	-12
3	-9.8		c	9
4	-9.6		Xmin	-10
5	-9.4		Xmax	10
6	-9.2		Delta X	0.2
7				
101				
102	10			

3. Graphing the data in columns A and B gives the graph shown here:

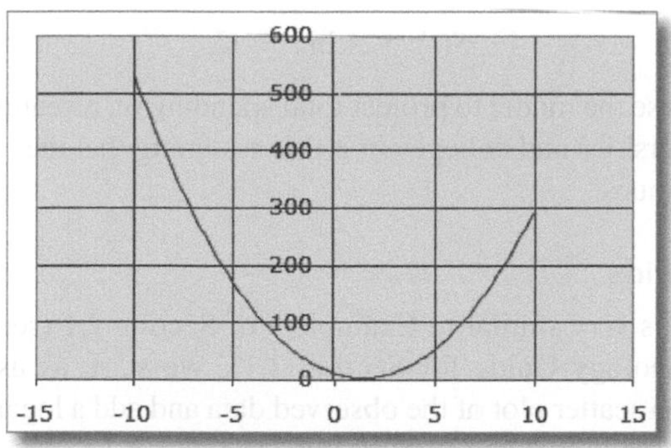

$$y = 4x^2 - 12x + 9$$

4. We can go further and compute the exact coordinates of the vertex and intercepts.

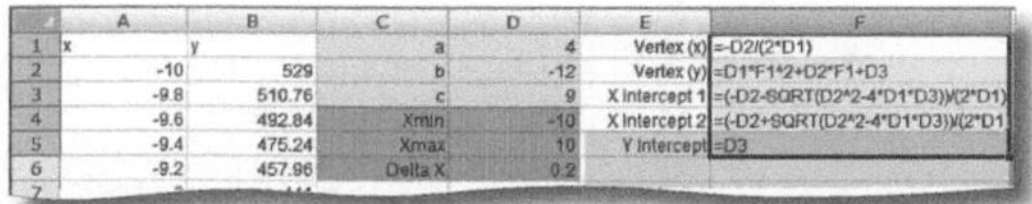

	A	B	C	D	E	F
1	x	y	a	4	Vertex (x)	=-D2/(2*D1)
2	-10	529	b	-12	Vertex (y)	=D1*F1^2+D2*F1+D3
3	-9.8	510.76	c	9	X Intercept 1	=(-D2-SQRT(D2^2-4*D1*D3))/(2*D1)
4	-9.6	492.84	Xmin	-10	X Intercept 2	=(-D2+SQRT(D2^2-4*D1*D3))/(2*D1
5	-9.4	475.24	Xmax	10	Y Intercept	=D3
6	-9.2	457.96	Delta X	0.2		

The completed sheet should look like this:

	A	B	C	D	E	F
1	x	y	a	4	Vertex (x)	1.5
2	-10	529	b	-12	Vertex (y)	0
3	-9.8	510.76	c	9	X Intercept 1	1.5
4	-9.6	492.84	Xmin	-10	X Intercept 2	1.5
5	-9.4	475.24	Xmax	10	Y Intercept	9
6	-9.2	457.96	Delta X	0.2		

We can now save this sheet as a template to handle all quadratic functions. For instance, to do part (b), we just change the values of a, b, and c in column D to $a = -1/2$, $b = 4$, and $c = -12$.

Example 5 (page 140) The following table shows the annual mean carbon dioxide concentration measured at Mauna Loa Observatory in Hawaii, in parts per million, every 10 years from 1960 through 2010:

Year t (years since 1960)	0	10	20	30	40	50
CO_2 Concentration C (ppm)	317	326	339	354	369	390

Find the quadratic model

$$C(t) = at^2 + bt + c$$

that best fits the data.

Solution

As in Example 2 of Section 1.4, we start with a scatter plot of the original data and add a trendline:

1. Start with the original data and a scatter plot (see Example 5 of Section 1.2).

	A	B
1	t	C
2	0	317
3	10	326
4	20	339
5	30	354
6	40	369
7	50	390

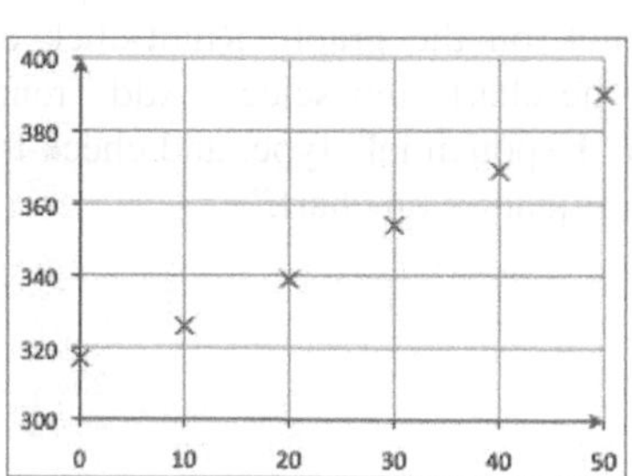

2. Add a quadratic trendline. The details vary from spreadsheet to spreadsheet. In Excel, right-click on any data point in the chart, and select "Add Trendline", then select a "Polynomial" type of order 2, and check the option to "Display Equation on chart."

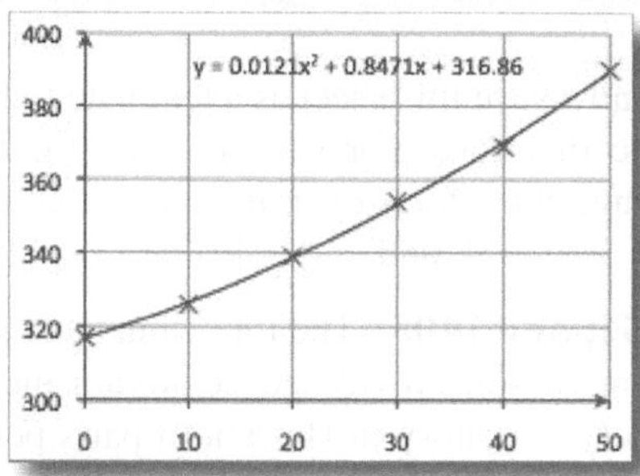

Section 2.2

Example 6(a) (page 155) The following table shows annual expenditure on health in the United States from 1980 through 2009:

Year t (years since 1980)	0	5	10	15	20	25	29
Expenditure ($ billion)	256	444	724	1,030	1,380	2,020	2,490

Find the exponential regression model

$$C(t) = Ab^t$$

for the annual expenditure.

Solution

This is very similar to Example 5 of Section 2.1 (see the Technology Guide for Section 2.1):

1. Start with a scatter plot of the observed data.
2. Add an exponential trendline. The details vary from spreadsheet to spreadsheet. In OpenOffice, first double-click on the graph. Right-click on any data point in the chart, and select "Add Trendline," then select an "Exponential" type, and check the option to "Display Equation on chart."

	A	B
1	t	C
2	0	256
3	5	444
4	10	724
5	15	1030
6	20	1380
7	25	2020
8	29	2490

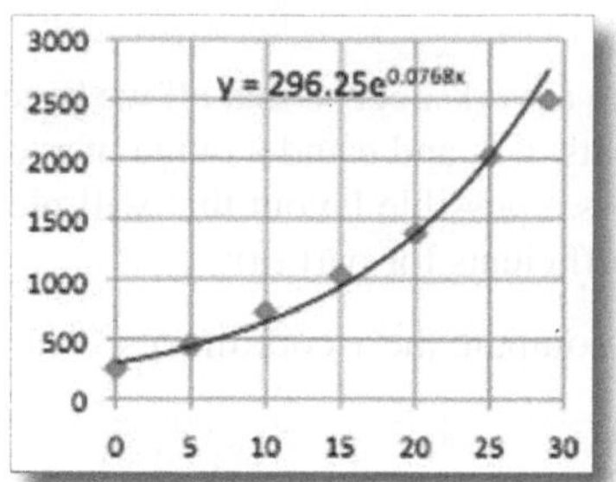

Notice that the regression curve is given in the form Ae^{kt} rather than Ab^t. To transform it, write

$$296.25e^{0.0768t} = 296.25(e^{0.0768})^t$$
$$\approx 296.25(1.0798)^t. \qquad e^{0.0768} \approx 1.0798$$

Section 2.3

Example 4 (page 169) The following table shows the total spent on research and development by institutions of higher education in the United States, in billions of dollars, for the period 2003–2012:

Year t (years since 2000)	3	4	5	6	7
Spending ($ billion)	41	45	48	50	52
Year t (years since 2000)	8	9	10	11	12
Spending ($ billion)	54	57	61	65	66

Find the best-fit logarithmic model of the form

$$S(t) = A \ln t + C,$$

and use the model to project total spending on research by universities and colleges in 2016, assuming that the trend continues.

Solution

This is very similar to Example 5 of Section 2.1 (see the Technology Guide for Section 2.1): We start, as usual, with a scatter plot of the observed data and add a logarithmic trendline. Here is the result:

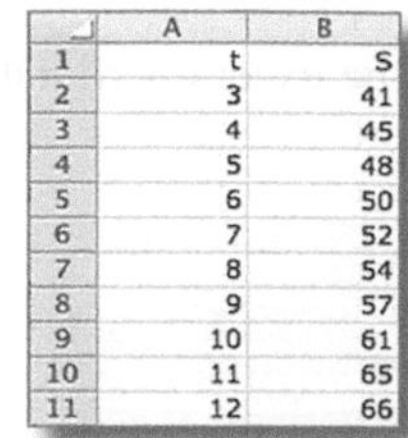

	A	B
1	t	S
2	3	41
3	4	45
4	5	48
5	6	50
6	7	52
7	8	54
8	9	57
9	10	61
10	11	65
11	12	66

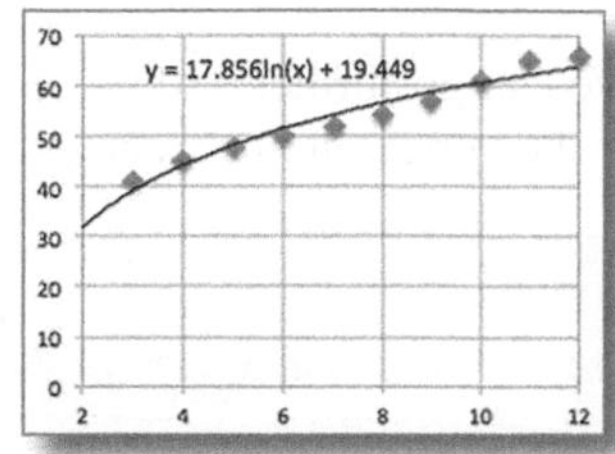

Section 2.4

Example 2 (page 178) The following table shows wired broadband penetration in the United States as a function of time t in years:

Year t (years since 2000)	2	3	4	5	6	7
Penetration P (%)	6.7	9.6	12.8	16.4	20.3	23.2
Year t (years since 2000)	8	9	10	11	12	13
Penetration P (%)	25.5	25.5	26.7	27.7	28.7	29.8

Find a logistic regression curve of the form

$$P(t) = \frac{N}{1 + Ab^{-t}}.$$

Solution

At the time of this writing, available spreadsheets did not have a built-in logistic regression calculation, so we use an alternative method that works for any type of regression curve. The Solver included with Windows versions of Excel and some Mac versions can find logistic regression curves, while the Solver included with some other spreadsheets is not yet capable of this, so the instructions here are specific to Excel.

1. First use rough estimates for N, A, and b, and compute the sum-of-squares error (SSE; see Section 1.4) directly:

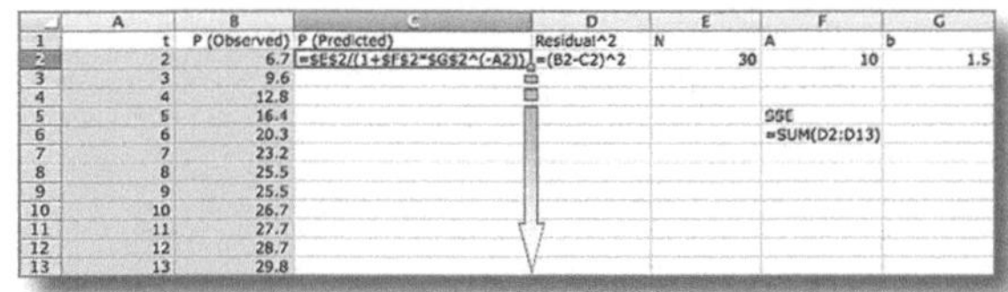

	A	B	C	D	E	F	G
1	t	P (Observed)	P (Predicted)	Residual^2	N	A	b
2	2	6.7	=$E$2/(1+$F$2*$G$2^(-A2))	=(B2-C2)^2	30	10	1.5
3	3	9.6					
4	4	12.8					
5	5	16.4				SSE	
6	6	20.3				=SUM(D2:D13)	
7	7	23.2					
8	8	25.5					
9	9	25.5					
10	10	26.7					
11	11	27.7					
12	12	28.7					
13	13	29.8					

Cells E2:G2 contain our initial rough estimates of N, A, and b. For N we used 30. (Notice that the y-coordinates do appear to level off around 30.) For A we used the fact that the y-intercept is $N/(1 + A)$ and the y-intercept appears to be approximately 3. In other words,

$$3 = \frac{30}{1 + A}.$$

Because a very rough estimate is all we are after, using $A = 10$ will do just fine. For b we chose 1.5, as the values of P appear to be increasing by around 50% per year initially. (Again, this is rough.)

2. Cell C2 contains the formula for $P(t)$, and the square of the resulting residual is computed in D2.

3. Cell F6 will contain SSE. The completed spreadsheet should look like this:

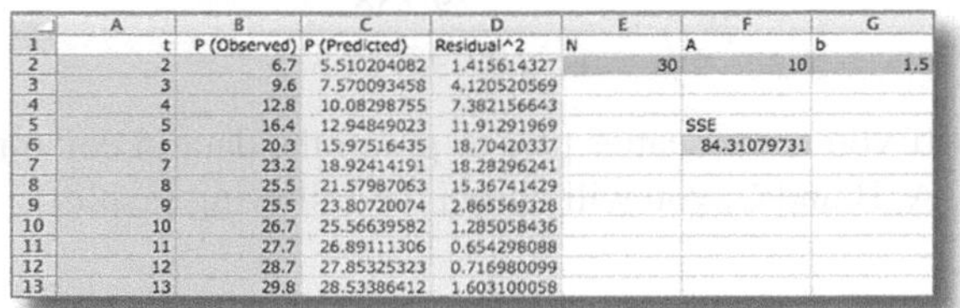

	A	B	C	D	E	F	G
1	t	P (Observed)	P (Predicted)	Residual^2	N	A	b
2	2	6.7	5.510204082	1.415614327	30	10	1.5
3	3	9.6	7.570093458	4.120520569			
4	4	12.8	10.08298755	7.382156643			
5	5	16.4	12.94849023	11.91291969		SSE	
6	6	20.3	15.97516435	18.70420337		84.31079731	
7	7	23.2	18.92414191	18.28296241			
8	8	25.5	21.57987063	15.36741429			
9	9	25.5	23.80720074	2.865569328			
10	10	26.7	25.56639582	1.285058436			
11	11	27.7	26.89111306	0.654298088			
12	12	28.7	27.85325323	0.716980099			
13	13	29.8	28.53386412	1.603100058			

The best-fit curve will result from values of N, A, and b that give a minimum value for SSE. We shall use Excel's "Solver," found in the "Analysis" group on the "Data" tab. (If "Solver" does not appear in the Analysis group, you will have to install the Solver Add-in using the Excel Options dialogue.) The Solver dialogue box with the necessary fields completed to solve the problem looks like this:

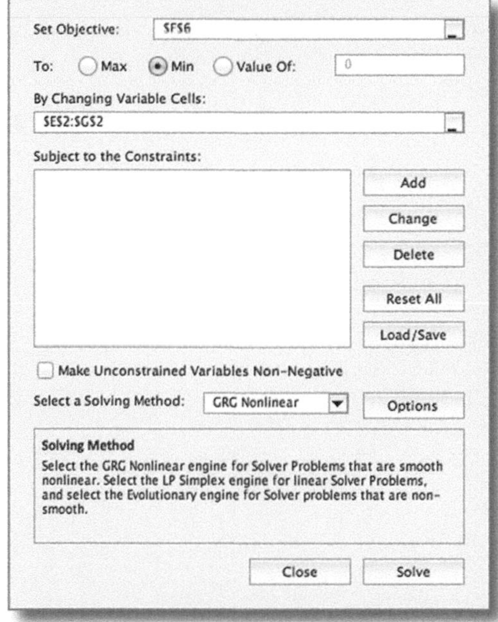

- The Objective refers to the cell that contains SSE.
- "Min" is selected because we are minimizing SSE.
- "Variable Cells" are obtained by selecting the cells that contain the current values of N, A, and b.

4. When you have filled in the values for the three items above, press "Solve" and tell Solver to Keep Solver Solution when done. You will find $N \approx 29.238$, $A \approx 8.998$, and $b \approx 1.638$, so

$$P(t) \approx \frac{29.238}{1 + 8.998(1.638)^{-t}}.$$

If you use a scatter plot to graph the data in columns A, B and C, you will obtain the following graph:

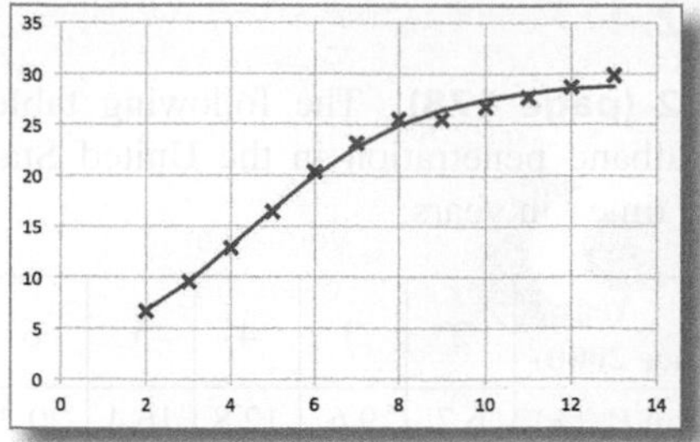

3 INTRODUCTION TO THE DERIVATIVE

CASE STUDY

Reducing Sulfur Emissions

The Environmental Protection Agency (EPA) wants to formulate a policy that will encourage utilities to reduce sulfur emissions. Its goal is to reduce annual emissions of sulfur dioxide by a total of 10 million tons from the current level of 25 million tons by imposing a fixed charge for every ton of sulfur released into the environment per year. The EPA has some data showing the marginal cost to utilities of reducing sulfur emissions. As a consultant to the EPA, you must determine the amount to be charged per ton of sulfur emissions in light of these data.

Norbert Schaefer/CORBIS/Getty Images

www.WanerMath.com

At the Website, in addition to the resources listed in the Preface, you will find:

The following extra topics:

- Sketching the Graph of the Derivative
- Continuity and Differentiability

Introduction

In the world around us, everything is changing. The mathematics of change is largely about the rate of change: how fast and in which direction the change is occurring. Is the Dow Jones average going up, and if so, how fast? If I raise my prices, how many customers will I lose? If I launch this missile, how fast will it be traveling after 2 seconds, how high will it go, and where will it come down?

We have already discussed the concept of rate of change for linear functions (straight lines), where the slope measures the rate of change. But this works only because a straight line maintains a constant rate of change along its whole length. Other functions rise faster here than there—or rise in one place and fall in another—so the rate of change varies along the graph. The first achievement of calculus is to provide a systematic and straightforward way of calculating (hence the name) these rates of change. To describe a changing world, we need a language of change, and that is what calculus is.

The history of calculus is an interesting story of personalities, intellectual movements, and controversy. Credit for its invention is given to two mathematicians: Isaac Newton (1642–1727) and Gottfried Leibniz (1646–1716). Newton, an English mathematician and scientist, developed calculus first, probably in the 1660s. We say "probably" because, for various reasons, he did not publish his ideas until much later. This allowed Leibniz, a German mathematician and philosopher, to publish his own version of calculus first, in 1684. Fifteen years later, stirred up by nationalist fervor in England and on the continent, controversy erupted over who should get the credit for the invention of calculus. The debate got so heated that the Royal Society (of which Newton and Leibniz were both members) set up a commission to investigate the question. The commission decided in favor of Newton, who happened to be president of the society at the time. The consensus today is that both mathematicians deserve credit because they came to the same conclusions working independently. This is not really surprising: Both built on well-known work of other people, and it was almost inevitable that someone would put it all together at about that time.

Precalculus Review
For this chapter you should be familiar with the algebra reviewed in **Section 0.2**.

3.1 Limits: Numerical and Graphical Viewpoints

Rates of change are calculated by derivatives, but an important part of the definition of the derivative is something called a **limit**. Arguably, much of mathematics since the eighteenth century has revolved around understanding, refining, and exploiting the idea of the limit. The basic idea is easy, but getting the technicalities right is not.

Estimating Limits Numerically

Start with a very simple example: Look at the function $f(x) = 2 + x$, and ask, "What happens to $f(x)$ as x approaches 3?" The following table shows the value of $f(x)$ for values of x close to and on either side of 3:

x approaching 3 from the left→ ←x approaching 3 from the right

x	2.9	2.99	2.999	2.9999	3	3.0001	3.001	3.01	3.1
$f(x) = 2 + x$	4.9	4.99	4.999	4.9999		5.0001	5.001	5.01	5.1

We have left the entry under 3 blank to emphasize that when calculating the limit of $f(x)$ as *x approaches* 3, we are not interested in its value when *x equals* 3.

Notice from the table that the closer x gets to 3 from either side, the closer $f(x)$ gets to 5. We write this as

$$\lim_{x\to 3} f(x) = 5. \qquad \text{The limit of } f(x)\text{, as } x \text{ approaches 3, equals 5.}$$

Q: *Why all the fuss? Can't we simply substitute $x = 3$ and avoid having to use a table?*

A: This happens to work for *some* functions but not for *all* functions. The following example illustrates this point.

* However, if you factor $x^3 - 8$, you will find that $f(x)$ can be simplified to a function that *is* defined at $x = 2$. This point will be discussed (and this example redone) in Section 3.3. The function in Example 1(b) cannot be simplified by factoring.

EXAMPLE 1 Estimating a Limit Numerically

Use a table to estimate the following limits:

a. $\lim_{x\to 2} \dfrac{x^3 - 8}{x - 2}$ **b.** $\lim_{x\to 0} \dfrac{e^{2x} - 1}{x}$

Solution

a. We cannot simply substitute $x = 2$, because the function $f(x) = \dfrac{x^3 - 8}{x - 2}$ is not defined at $x = 2$. (Why?)* Instead, we use a table of values as we did above, with x approaching 2 from both sides:

x approaching 2 from the left→ ←*x* approaching 2 from the right

x	1.9	1.99	1.999	1.9999	2	2.0001	2.001	2.01	2.1
$f(x) = \dfrac{x^3 - 8}{x - 2}$	11.41	11.9401	11.9940	11.9994		12.0006	12.0060	12.0601	12.61

We notice that as x approaches 2 from either side, $f(x)$ appears to be approaching 12. This suggests that the limit is 12, and we write

$$\lim_{x\to 2} \frac{x^3 - 8}{x - 2} = 12.$$

b. The function $g(x) = \dfrac{e^{2x} - 1}{x}$ is not defined at $x = 0$ (nor can it even be simplified to one that *is* defined at $x = 0$). In the following table we allow x to approach 0 from both sides:

x approaching 0 from the left→ ←*x* approaching 0 from the right

x	−0.1	−0.01	−0.001	−0.0001	0	0.0001	0.001	0.01	0.1
$g(x) = \dfrac{e^{2x} - 1}{x}$	1.8127	1.9801	1.9980	1.9998		2.0002	2.0020	2.0201	2.2140

The table suggests that $\lim_{x\to 0} \dfrac{e^{2x} - 1}{x} = 2.$

Using Technology

To automate the computations in Example 1 using a graphing calculator or a spreadsheet, see the Technology Guides at the end of the chapter. Outline for Example 1(a):

TI-83/84 Plus

Home screen: `Y1=(X^3-8)/(X-2)`

`2ND` `TBLSET` `Indpnt` set to `Ask`

`2ND` `TABLE` Enter some values of x from the example: 1.9, 1.99, 1.999 . . .

[More details in the Technology Guide.]

Spreadsheet

Headings x, $f(x)$ in A1–B1 and again in C1–D1.

In A2–A5 enter 1.9, 1.99, 1.999, 1.9999.

In C1–C5 enter 2.1, 2.01, 2.001, 2.0001. Enter `=(A2^3-8)/(A2-2)` in B2, and copy down to B5. Copy and paste the same formula in D2–D5.

[More details in the Technology Guide.]

Website

www.WanerMath.com

Go to

Online Utilities → Function Evaluator and Grapher

Enter `(x^3-8)/(x-2)` for y_1. For a table of values, enter the various x-values in the Evaluator box, and press "Evaluate".

➡ **Before we go on...** We will revisit the limit in Example 1(a) from the geometric point of view in Example 5. Although the table *suggests* that the limit in Example 1(b) is 2, it by no means establishes that fact conclusively. It is *conceivable* (though not in fact the case here) that putting $x = 0.000000087$ could result in, say, $g(x) = 426$. Using a table can only *suggest* a value for the limit. In the next two sections we shall discuss algebraic techniques to allow us to actually *calculate* limits. ■

Before we continue, let us make a more formal definition.

Definition of a Limit

Let f be a function such that $f(x)$ is defined for values of x arbitrarily close to, but different from, a. Then, if $f(x)$ approaches the number L as x approaches (but is not equal to) a, regardless from which side, we say that $f(x)$ **approaches** $\boldsymbol{L}$ **as** $\boldsymbol{x \to a}$ ("x approaches a") or that the **limit** of $f(x)$ as $x \to a$ is L. In other words, *we can make $f(x)$ be as close to L as we like by choosing any x in the domain of f sufficiently close to (but not equal to) a on either side.* We write

$$\lim_{x \to a} f(x) = L$$

or

$$f(x) \to L \text{ as } x \to a.$$

If $f(x)$ *fails* to approach *a single fixed number* as x approaches a, then we say that $f(x)$ **has no limit** as $x \to a$, or

$$\lim_{x \to a} f(x) \textbf{ does not exist.}$$

Quick Examples

1. $\lim_{x \to 3}(2 + x) = 5$ — See discussion before Example 1.
2. $\lim_{x \to -2}(3x) = -6$ — As x approaches -2, $3x$ approaches -6.
3. $\lim_{x \to 0}(x^2 - 2x + 1)$ exists. — In fact, the limit is 1.
4. $\lim_{x \to 5} \frac{1}{x} = \frac{1}{5}$ — As x approaches 5, $\frac{1}{x}$ approaches $\frac{1}{5}$.
5. $\lim_{x \to 2} \frac{x^3 - 8}{x - 2} = 12$ — See Example 1. (We cannot just put $x = 2$ here.)
6. $\lim_{x \to 0} \sqrt{x} = 0$. (Even though $\sqrt{x}$ is not defined to the left of 0, it still satisfies the definition: No matter what x you choose *in the domain of the function,* you can make $\sqrt{x}$ as close as you like to zero by choosing x sufficiently close to zero.)

(For examples where the limit does not exist, see Example 2.)

Notes

1. It is important that $f(x)$ approach a *single number* regardless from which side x is approaching a. For instance, if $f(x)$ approaches 5 for $x = 1.9, 1.99, 1.999, \ldots$ but approaches 4 for $x = 2.1, 2.01, 2.001, \ldots$, then the limit as $x \to 2$ does not exist. (See Example 2 for such a situation.)

2. It may happen that $f(x)$ does not approach any fixed number at all as $x \to a$ from either side. In this case we also say that the limit does not exist. ■

The next example includes instances in which a stated limit does not exist.

EXAMPLE 2 **Limits May or May Not Exist**

Do the following limits exist?

a. $\lim_{x\to 0} \frac{1}{x^2}$ **b.** $\lim_{x\to 0} \frac{|x|}{x}$ **c.** $\lim_{x\to 2} \frac{1}{x-2}$ **d.** $\lim_{x\to 1} \sqrt{x-1}$

Solution

a. Here is a table of values for $f(x) = \frac{1}{x^2}$, with x approaching 0 from both sides:

x approaching 0 from the left→ ←*x* approaching 0 from the right

x	-0.1	-0.01	-0.001	-0.0001	**0**	0.0001	0.001	0.01	0.1
$f(x) = \frac{1}{x^2}$	100	10,000	1,000,000	100,000,000		100,000,000	1,000,000	10,000	100

The table suggests that as x gets closer to zero on either side, $f(x)$ gets larger and larger **without bound**—that is, if you name any number, no matter how large, $f(x)$ will be even larger than that if x is sufficiently close to 0. Because $f(x)$ is not approaching any real number, we conclude that $\lim_{x\to 0} \frac{1}{x^2}$ does not exist. Because $f(x)$ is becoming arbitrarily large, we also say that $\lim_{x\to 0} \frac{1}{x^2}$ **diverges to** $+\infty$, or just

$$\lim_{x\to 0} \frac{1}{x^2} = +\infty.$$

Note This is not meant to imply that the limit exists; the symbol $+\infty$ does not represent any real number. We write $\lim_{x\to a} f(x) = +\infty$ to indicate two things: (1) The limit does not exist, and (2) the function gets large without bound as x approaches a. ■

b. Here is a table of values for $f(x) = \frac{|x|}{x}$, with x approaching 0 from both sides:

x approaching 0 from the left→ ←*x* approaching 0 from the right

x	-0.1	-0.01	-0.001	-0.0001	**0**	0.0001	0.001	0.01	0.1
$f(x) = \frac{\lvert x\rvert}{x}$	-1	-1	-1	-1		1	1	1	1

The table suggests that $f(x)$ does not approach the same limit as x approaches 0 from both sides. There appear to be two *different* limits: the limit as we approach 0 from the left and the limit as we approach from the right. We write

$$\lim_{x\to 0^-} f(x) = -1,$$

which is read as "the limit as x approaches 0 from the left (or from below) is -1," and

$$\lim_{x\to 0^+} f(x) = 1,$$

which is read as "the limit as x approaches 0 from the right (or from above) is 1." These are called the **one-sided limits** of $f(x)$. In order for f to have a **two-sided limit**, the two one-sided limits must be equal. Because they are not, we conclude that $\lim_{x\to 0} f(x)$ does not exist.

c. Near $x = 2$ we have the following table of values for $f(x) = \dfrac{1}{x-2}$:

x approaching 2 from the left→ ←x approaching 2 from the right

x	1.9	1.99	1.999	1.9999	2	2.0001	2.001	2.01	2.1
$f(x) = \dfrac{1}{x-2}$	−10	−100	−1,000	−10,000		10,000	1,000	100	10

Because $f(x)$ is approaching no (single) real number as $x \to 2$, we see that $\lim_{x\to 2} \dfrac{1}{x-2}$ does not exist. Notice also that $\dfrac{1}{x-2}$ diverges to $+\infty$ as $x \to 2$ from the positive side (right half of the table) and to $-\infty$ as $x \to 2$ from the left (left half of the table). In other words,

$$\lim_{x\to 2^-} \frac{1}{x-2} = -\infty, \quad \lim_{x\to 2^+} \frac{1}{x-2} = +\infty, \quad \text{and} \quad \lim_{x\to 2} \frac{1}{x-2} \text{ does not exist.}$$

d. The natural domain of $f(x) = \sqrt{x-1}$ is $[1, +\infty)$, as $f(x)$ is defined only when $x \geq 1$. Thus, we cannot evaluate $f(x)$ if x is to the left of 1. Here is a table showing values to the right of 1:

←x approaching 1 from the right

x	1	1.00001	1.0001	1.001	1.01	1.1
$f(x) = \sqrt{x-1}$		0.0032	0.0100	0.0316	0.1000	0.3162

The values suggest that

$$\lim_{x\to 1^+} \sqrt{x-1} = 0.$$

What about $\lim_{x\to 1} \sqrt{x-1}$? In the definition of a limit we need only worry about *x in the domain of f*, meaning values of x to the right of 1, and we just saw that the values of $f(x)$ are approaching 0 for such values. Thus,

$$\lim_{x\to 1} \sqrt{x-1} = 0$$

as well, even though $\lim_{x\to 1^-} \sqrt{x-1}$ does not exist, as x is not defined to the left of 1. We can also obtain this limit by substituting $x = 1$ in the formula for $f(x)$. (See the comments after the example.) To summarize,

$$\lim_{x\to 1^-} \sqrt{x-1} \text{ does not exist,} \quad \lim_{x\to 1^+} \sqrt{x-1} = 0, \quad \text{and} \quad \lim_{x\to 1} \sqrt{x-1} = 0.$$

Q: *In Example 2(d) (and in some of the Quick Examples before that) we could find a limit of an algebraically specified function by simply substituting the value of x in the formula for f(x). Does this always work?*

A: Short answer: Yes, when it makes sense. If the function is specified by a *single* algebraic formula and if $x = a$ is in the domain of f, then the limit can be obtained by substituting. We will say more about this when we discuss the algebraic approach to limits in Section 3.3. Remember, however, that, by definition the limit of a function as $x \to a$ has nothing to do with its value at $x = a$ but rather is determined by its values for *x close to, but different from, a.*

Q: *If f(x) is undefined when x = a, then the limit does not exist—right?*

A: Wrong. If $f(a)$ is not defined, then the limit may or may not exist. Example 1 shows instances in which the limit *does* exist, and Example 2 shows instances in which it does not. Again, the limit of a function as $x \to a$ has nothing to do with its value at $x = a$ but rather is determined by its values for *x close to, but different from, a.*

Limits at Infinity

In another useful kind of limit we let x approach either $+\infty$ or $-\infty$, by which we mean that we let x get arbitrarily large or let x become an arbitrarily large negative number. The next example illustrates this.

EXAMPLE 3 Limits at Infinity

Use a table to estimate the following: **a.** $\lim_{x\to+\infty} \dfrac{2x^2 - 4x}{x^2 - 1}$ and **b.** $\lim_{x\to-\infty} \dfrac{2x^2 - 4x}{x^2 - 1}$.

Solution

a. By saying that x is "approaching $+\infty$," we mean that x is getting larger and larger without bound, so we make the following table:

x approaching $+\infty$ →

x	10	100	1,000	10,000	100,000
$f(x) = \dfrac{2x^2 - 4x}{x^2 - 1}$	1.6162	1.9602	1.9960	1.9996	2.0000

(Note that we are approaching $+\infty$ only from the left because we can hardly approach it from the right!) What seems to be happening is that $f(x)$ is approaching 2. Thus we write

$$\lim_{x\to+\infty} f(x) = 2.$$

b. Here, x is approaching $-\infty$, so we make a similar table, this time with x assuming negative values of greater and greater magnitude (read this table from right to left):

← *x* approaching $-\infty$

x	−100,000	−10,000	−1,000	−100	−10
$f(x) = \dfrac{2x^2 - 4x}{x^2 - 1}$	2.0000	2.0004	2.0040	2.0402	2.4242

Once again, $f(x)$ is approaching 2. Thus, $\lim_{x\to-\infty} f(x) = 2$.

Estimating Limits Graphically

We can often estimate a limit from a graph, as the next example shows.

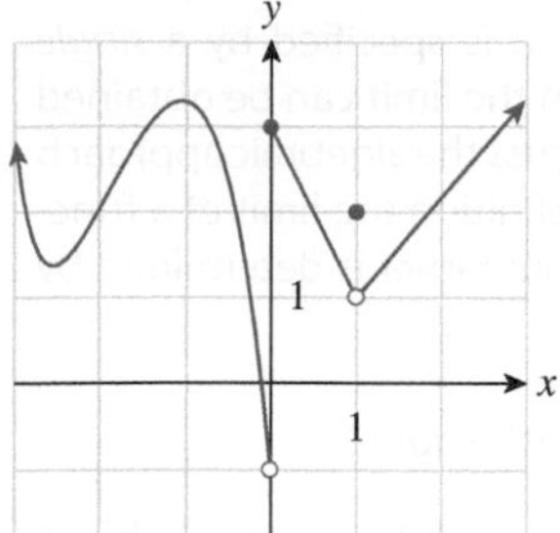

Figure 1

EXAMPLE 4 Estimating Limits Graphically

The graph of a function f is shown in Figure 1. (Recall that the solid dots indicate points on the graph and the hollow dots indicate points not on the graph.) From the graph, analyze the following limits:

a. $\lim_{x \to -2} f(x)$ **b.** $\lim_{x \to 0} f(x)$ **c.** $\lim_{x \to 1} f(x)$ **d.** $\lim_{x \to +\infty} f(x)$

Solution Since we are given only a graph of f, we must analyze these limits graphically.

a. Imagine that Figure 1 was drawn on a graphing calculator equipped with a trace feature that allows us to move a cursor along the graph and see the coordinates as we go. To simulate this, place a pencil point on the graph to the left of $x = -2$, and move it along the curve so that the x-coordinate approaches -2. (See Figure 2.) We evaluate the limit numerically by noting the behavior of the y-coordinates.*

* For a visual animation of this process, look at the online tutorial for this section at the Website.

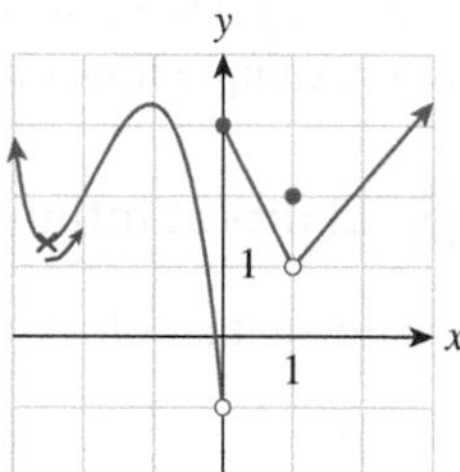

Figure 2

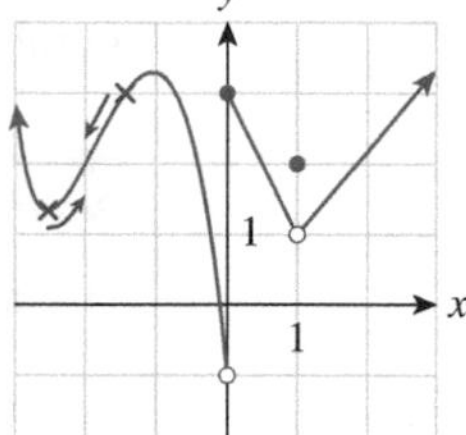

Figure 3

We can see directly from the graph that the y-coordinate approaches 2. Similarly, if we place our pencil point to the right of $x = -2$ and move it to the left, the y-coordinate will approach 2 from that side as well (Figure 3). Therefore, as x approaches -2 from either side, $f(x)$ approaches 2, so

$$\lim_{x \to -2} f(x) = 2.$$

b. This time, we move our pencil point toward $x = 0$. Referring to Figure 4, if we start from the left of $x = 0$ and approach 0 (by moving right), the y-coordinate

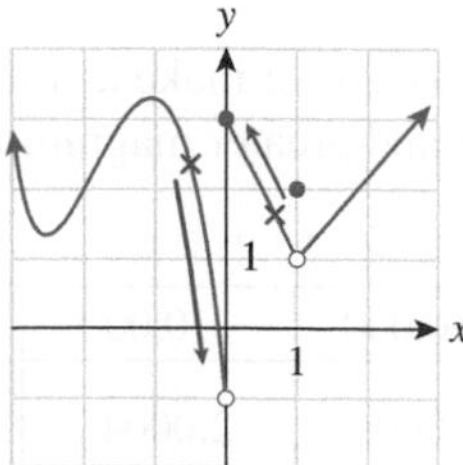

Figure 4

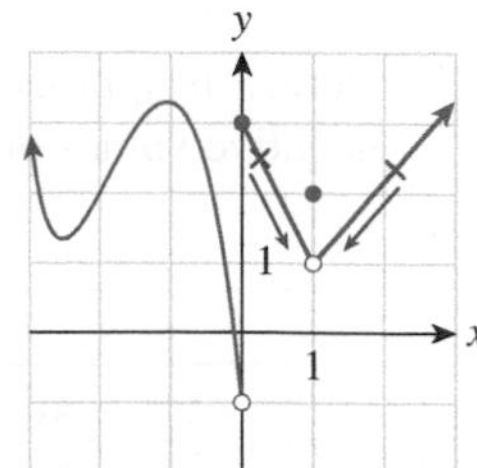

Figure 5

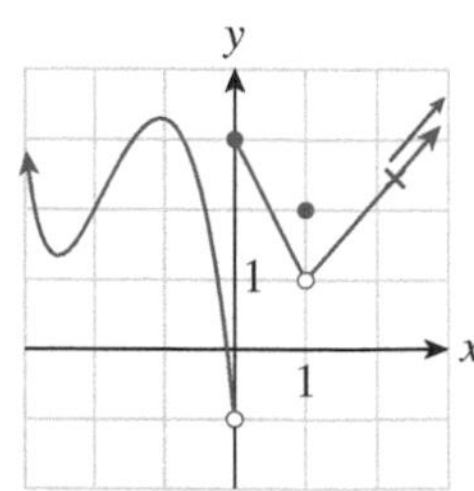

Figure 6

approaches -1. However, if we start from the right of $x = 0$ and approach 0 (by moving left), the y-coordinate approaches 3. Thus (see Example 2),

$$\lim_{x \to 0^-} f(x) = -1$$

and

$$\lim_{x \to 0^+} f(x) = 3.$$

Because these limits are not equal, we conclude that

$$\lim_{x \to 0} f(x) \text{ does not exist.}$$

In this case there is a break in the graph at $x = 0$, and we say that the function is **discontinuous** at $x = 0$ (see Section 3.2).

c. Once more, we think about a pencil point moving along the graph with the x-coordinate this time approaching $x = 1$ from the left and from the right (Figure 5). As the x-coordinate of the point approaches 1 from either side, the y-coordinate approaches 1 also. Therefore,

$$\lim_{x \to 1} f(x) = 1.$$

d. For this limit, x is supposed to approach infinity. We think about a pencil point moving along the graph farther and farther to the right, as shown in Figure 6. As the x-coordinate gets larger, the y-coordinate also gets larger and larger without bound. Thus, $f(x)$ diverges to $+\infty$:

$$\lim_{x \to +\infty} f(x) = +\infty.$$

Similarly,

$$\lim_{x \to -\infty} f(x) = +\infty.$$

➡ **Before we go on . . .** In Example 4(c), $\lim_{x \to 1} f(x) = 1$ but $f(1) = 2$ (why?). Thus, $\lim_{x \to 1} f(x) \neq f(1)$. In other words, the limit of $f(x)$ as x *approaches* 1 is not the same as the value of f *at* $x = 1$. Always keep in mind that when we evaluate a limit as $x \to a$, *we do not care about the value of the function at* $x = a$. We care only about the value of $f(x)$ as x *approaches* a. In other words, $f(a)$ may or may not equal $\lim_{x \to a} f(x)$. ■

Here is a summary of the graphical method we used in Example 4, together with some additional information.

Estimating Limits Graphically

To decide whether $\lim_{x \to a} f(x)$ exists and to estimate its value if it does:

1. Draw the graph of $f(x)$ by hand or with graphing technology.
2. Position your pencil point (or the Trace cursor) on a point of the graph to the right of $x = a$.
3. Move the point *along the graph* toward $x = a$ from the right, and read the y-coordinate as you go. The value that the y-coordinate approaches (if any) is the limit $\lim_{x \to a^+} f(x)$.

4. Repeat Steps 2 and 3, this time starting from a point on the graph to the left of $x = a$ and approaching $x = a$ along the graph from the left. The value that the y-coordinate approaches (if any) is $\lim_{x\to a^-} f(x)$.
5. If the left and right limits both exist and have the same value L, then $\lim_{x\to a} f(x) = L$. Otherwise, the limit does not exist. The value $f(a)$ has no relevance whatsoever.
6. To evaluate $\lim_{x\to+\infty} f(x)$, move the pencil point toward the far right of the graph, and estimate the value the y-coordinate approaches (if any). For $\lim_{x\to-\infty} f(x)$, move the pencil point toward the far left.
7. If $x = a$ is an endpoint of the domain of f, then only a single one-sided limit can exist there and coincides with the (overall) limit. For instance, if the domain is $(-\infty, 4]$, then $\lim_{x\to 4} f(x)$ will exist if $\lim_{x\to 4^-} f(x)$ does, and they will be equal, whereas $\lim_{x\to 4^+} f(x)$ does not exist.

In the next example we see how both the numerical and graphical approaches can be used to study the same limits.

EXAMPLE 5 **Estimating Limits Numerically and Graphically**

Determine the following limits using both the numerical and graphical approaches:

a. $\lim_{x\to 2} \frac{x^3 - 8}{x - 2}$ (Compare Example 1(a).) **b.** $\lim_{x\to 0^+} \frac{1}{x}$

Solution

a. ***Numerical Approach*** In Example 1(a) we considered the following table of values for $f(x) = \frac{x^3 - 8}{x - 2}$ near $x = 2$:

x	1.9	1.99	1.999	1.9999	2	2.0001	2.001	2.01	2.1
$f(x) = \frac{x^3 - 8}{x - 2}$	11.41	11.9401	11.9940	11.9994		12.0006	12.0060	12.0601	12.61

suggesting that $\lim_{x\to 2} \frac{x^3 - 8}{x - 2} = 12$.

Graphical Approach Figure 7 shows the graph of f in this case. (Notice the open dot at $x = 2$, indicating that $f(2)$ is not defined.) The figure also shows pencil points approaching 2 from both sides, suggesting again that $\lim_{x\to 2} \frac{x^3 - 8}{x - 2} = 12$.

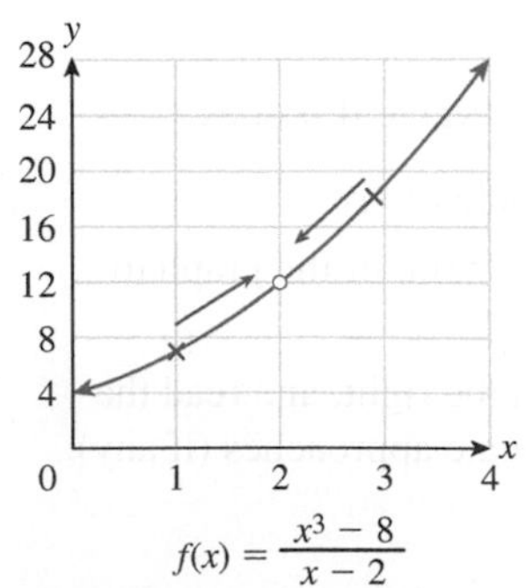

Figure 7

b. ***Numerical Approach*** Because we are asked for only the right-hand limit, we need only list values of x approaching 0 from the right:

←x approaching 0 from the right

x	0	0.0001	0.001	0.01	0.1
$f(x) = \frac{1}{x}$		10,000	1,000	100	10

What seems to be happening as x approaches 0 from the right is that $f(x)$ is increasing without bound, as in Example 4(d). That is, if you name any number, no matter how large, $f(x)$ will be even larger than that if x is sufficiently close to zero. Thus, the limit diverges to $+\infty$, so

$$\lim_{x\to 0^+} \frac{1}{x} = +\infty$$

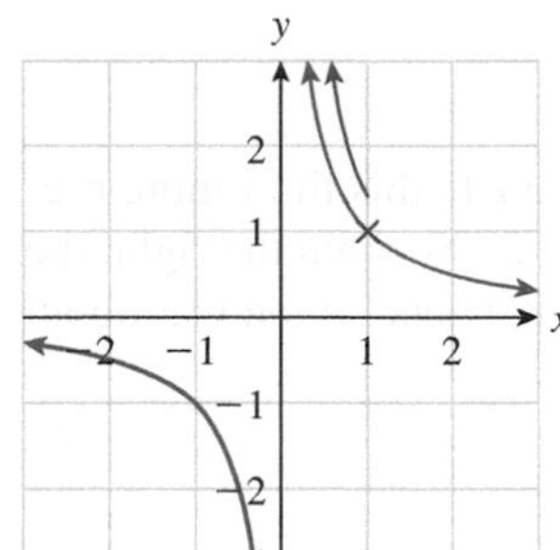

Figure 8

Graphical Approach Recall that the graph of $f(x) = \dfrac{1}{x}$ is the standard hyperbola shown in Figure 8. The figure also shows the pencil point moving so that its x-coordinate approaches 0 from the right. Because the point moves along the graph, it is forced to go higher and higher. In other words, its y-coordinate becomes larger and larger, approaching $+\infty$. Thus, we conclude that

$$\lim_{x\to 0^+} \frac{1}{x} = +\infty.$$

➡ **Before we go on . . .** In Example 5 you should check that

$$\lim_{x\to 0^-} \frac{1}{x} = -\infty. \qquad \frac{1}{x} \text{ diverges to } -\infty \text{ as } x \to 0^-.$$

Also, check that

$$\lim_{x\to +\infty} \frac{1}{x} = \lim_{x\to -\infty} \frac{1}{x} = 0. \blacksquare$$

Application

EXAMPLE 6 Broadband Penetration

Wired broadband penetration in the United States can be modeled by

$$P(t) = \frac{29.2}{1 + 9.0(1.64)^{-t}} \text{ percentage points} \qquad (t \geq 0),$$

where t is time in years since 2000.[1]

a. Estimate $\lim_{t\to +\infty} P(t)$, and interpret the answer.

b. Estimate $\lim_{t\to 0^+} P(t)$, and interpret the answer.

Solution

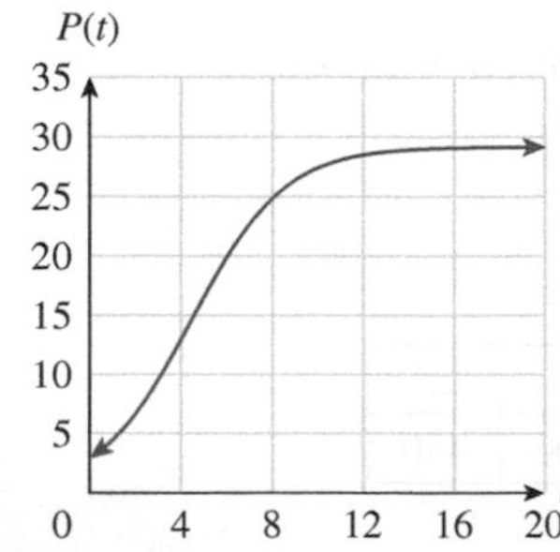

Figure 9

a. Figure 9 shows a plot of $P(t)$ for $0 \leq t \leq 20$. Using either the numerical or the graphical approach, we find

$$\lim_{t\to +\infty} P(t) = \lim_{t\to +\infty} \frac{29.2}{1 + 9.0(1.64)^{-t}} \approx 30.$$

[1] See Example 2 in Section 2.4. Broadband penetration is the number of broadband installations divided by the total population. Source for data: Organization for Economic Cooperation and Development (OECD) Directorate for Science, Technology, and Industry, table of Historical Penetration Rates, December 2013, downloaded September 2014 from www.oecd.org/sti/ict/broadband.

(The actual limit is 29.2. Why?) Thus, in the long term (as t gets larger and larger), broadband penetration in the United States is expected to approach 30%; that is, the number of installations is expected to approach 30% of the total population.

b. The limit here is

$$\lim_{t\to 0^+} P(t) = \lim_{t\to 0^+} \frac{29.2}{1 + 9.0(1.64)^{-t}} \approx 2.9.$$

(Notice that in this case we can simply put $t = 0$ to evaluate this limit more precisely as 2.92.) Thus, the closer t gets to 0 (representing 2000) from the right, the closer $P(t)$ gets to 2.9%, meaning that, in 2000, broadband penetration was about 2.9% of the population.

FAQs

Determining When a Limit Does or Does Not Exist

Q: *If I substitute $x = a$ in the formula for a function and find that the function is defined there, it means that $\lim_{x\to a} f(x)$ exists and equals $f(a)$—right?*

A: Correct, provided that the function is specified by a *single algebraic formula** and is not, say, piecewise-defined. We shall say more about this in the next two sections.

* In a sense we will make more precise in Section 3.3.

Q: *If I substitute $x = a$ in the formula for a function and find that the function is* not *defined there, it means that $\lim_{x\to a} f(x)$ does not exist—right?*

A: Wrong. The limit may still exist, as in Example 1, or may not exist, as in Example 2. In general, whether or not $\lim_{x\to a} f(x)$ exists has nothing to do with $f(a)$ but rather is determined by the values of f when x is *very close to, but not equal to a.*

Q: *Is there a quick and easy way of telling from a graph whether $\lim_{x\to a} f(x)$ exists?*

A: Yes. If you cover up the portion of the graph corresponding to $x = a$ and it appears as though the visible part of the graph could be made into a continuous curve by filling in a suitable point at $x = a$, then the limit exists. (The "suitable point" need not be $(a, f(a))$.) Otherwise, it does not. Try this method with the curves in Example 4.

3.1 EXERCISES

▼ more advanced ◆ challenging

T indicates exercises that should be solved using technology

In Exercises 1–4, use the given table of values to estimate, for the given value of a, each of the following if they exist: **(a)** $\lim_{x\to a^-} f(x)$ **(b)** $\lim_{x\to a^+} f(x)$ **(c)** $\lim_{x\to a} f(x)$ **(d)** *$f(a)$ if it is defined*

[**HINT**: See Examples 1–3.]

1. $a = 2$; table of values:

x	1.9	1.99	1.999	1.9999	2	2.0001	2.001	2.01	2.1
$f(x)$	−5.975	−5.9975	−5.99975	−5.999975	−4	440,000	44,000	4,400	440

2. $a = -2$; table of values:

x	-2.1	-2.01	-2.001	-2.0001	-2	-1.9999	-1.999	-1.99	-1.9
$f(x)$	-1.12	-11.12	-111.12	$-1,111.12$		-0.00003	-0.00031	-0.00311	-0.03111

3. $a = -5$; table of values:

x	-5.1	-5.01	-5.001	-5.0001	-5	-4.9999	-4.999	-4.99	-4.9
$f(x)$	-3.12	-31.12	-311.12	$-3,111.12$		$-4,111.12$	-411.12	-41.12	-4.12

4. $a = 0$; table of values:

x	-0.1	-0.01	-0.001	-0.0001	0	0.0001	0.001	0.01	0.1
$f(x)$	-1.0303	-1.00303	-1.0003	-1.00003	0	1.00003	1.0003	1.00303	1.0303

In Exercises 5–34, estimate the given limit numerically if it exists. [**HINT:** See Examples 1–3.]

5. $\lim_{x\to 0} \frac{x^2}{x+1}$ **6.** $\lim_{x\to 0} \frac{x-3}{x-1}$

7. $\lim_{x\to 2} \frac{x^2-4}{x-2}$ **8.** $\lim_{x\to 2} \frac{x^2-1}{x-2}$

9. $\lim_{x\to -1} \frac{x^2+1}{x+1}$ **10.** $\lim_{x\to -1} \frac{x^2+2x+1}{x+1}$

11. $\lim_{x\to 1^+} \frac{x-1}{\sqrt{x}-1}$ **12.** $\lim_{x\to 3^-} \frac{\sqrt{3-x}}{3-x}$

13. $\lim_{x\to 2^+} \frac{x^2+4x+3}{x+3}$ **14.** $\lim_{x\to 2^-} \frac{x^2-4x+4}{x-2}$

15. $\lim_{x\to 9^-} \frac{x-9}{\sqrt{x}-3}$ **16.** $\lim_{x\to 25^+} \frac{\sqrt{x}-5}{x-25}$

17. $\lim_{x\to 3^-} \frac{4}{(x-3)^2}$ **18.** $\lim_{x\to 4^-} \frac{3}{(x-4)^{1/3}}$

19. $\lim_{x\to -2^+} \frac{|x+2|}{(x+2)^{1/6}}$ **20.** $\lim_{x\to -3} \frac{(x+3)^{2/3}}{|x+3|}$

21. $\lim_{x\to +\infty} \frac{3x^2+10x-1}{2x^2-5x}$ **22.** $\lim_{x\to +\infty} \frac{6x^2+5x+100}{3x^2-9}$

23. $\lim_{x\to -\infty} \frac{x^5-1,000x^4}{2x^5+10,000}$

24. $\lim_{x\to -\infty} \frac{x^6+3,000x^3+1,000,000}{2x^6+1,000x^3}$

25. $\lim_{x\to +\infty} \frac{10x^2+300x+1}{5x+2}$ **26.** $\lim_{x\to +\infty} \frac{2x^4+20x^3}{1,000x^6+6}$

27. $\lim_{x\to +\infty} \frac{10x^2+300x+1}{5x^3+2}$ **28.** $\lim_{x\to +\infty} \frac{2x^4+20x^3}{1,000x^3+6}$

29. $\lim_{x\to 2} e^{x-2}$ **30.** $\lim_{x\to +\infty} e^{-x}$

31. $\lim_{x\to +\infty} xe^{-x}$ **32.** $\lim_{x\to -\infty} xe^{x}$

33. $\lim_{x\to -\infty} (x^{10}+2x^5+1)e^{x}$ **34.** $\lim_{x\to +\infty} (x^{50}+x^{30}+1)e^{-x}$

In Exercises 35–48 the graph of f is given. Use the graph to compute the quantities asked for. [**HINT:** See Examples 4–5.]

35. a. $\lim_{x\to 1} f(x)$ **b.** $\lim_{x\to -1} f(x)$

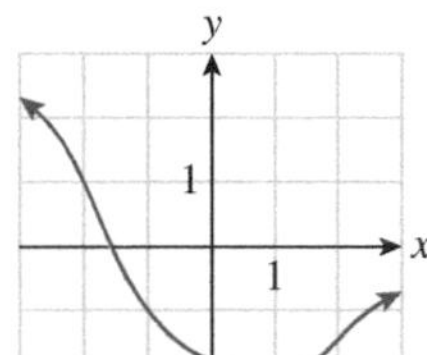

36. a. $\lim_{x\to -1} f(x)$ **b.** $\lim_{x\to 1} f(x)$

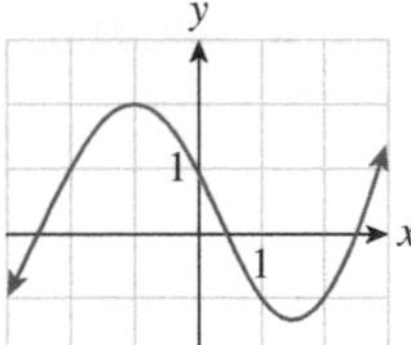

37. a. $\lim_{x\to 0} f(x)$ **b.** $\lim_{x\to 2} f(x)$

c. $\lim_{x\to -\infty} f(x)$ **d.** $\lim_{x\to +\infty} f(x)$

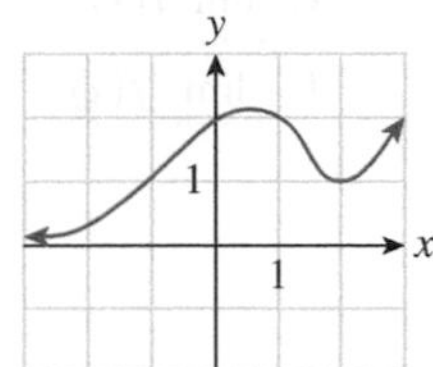

38. **a.** $\lim_{x \to -1} f(x)$ **b.** $\lim_{x \to 1} f(x)$ **c.** $\lim_{x \to +\infty} f(x)$ **d.** $\lim_{x \to -\infty} f(x)$

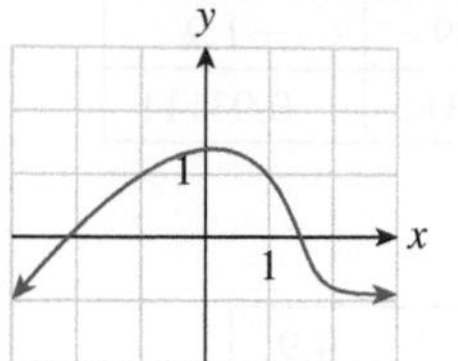

39. **a.** $\lim_{x \to 2} f(x)$ **b.** $\lim_{x \to 0^+} f(x)$ **c.** $\lim_{x \to 0^-} f(x)$
d. $\lim_{x \to 0} f(x)$ **e.** $f(0)$ **f.** $\lim_{x \to -\infty} f(x)$

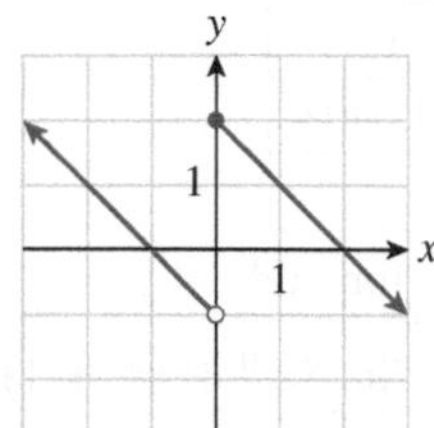

40. **a.** $\lim_{x \to 3} f(x)$ **b.** $\lim_{x \to 1^+} f(x)$ **c.** $\lim_{x \to 1^-} f(x)$
d. $\lim_{x \to 1} f(x)$ **e.** $f(1)$ **f.** $\lim_{x \to +\infty} f(x)$

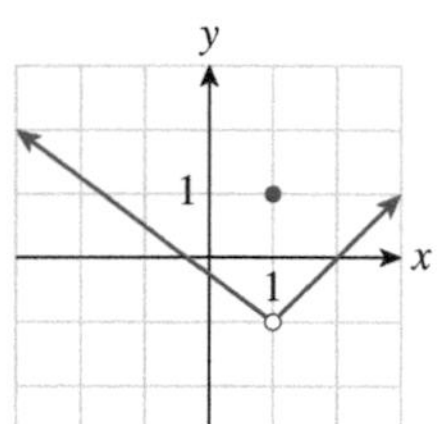

41. **a.** $\lim_{x \to -2} f(x)$ **b.** $\lim_{x \to -1^+} f(x)$ **c.** $\lim_{x \to -1^-} f(x)$
d. $\lim_{x \to -1} f(x)$ **e.** $f(-1)$ **f.** $\lim_{x \to +\infty} f(x)$

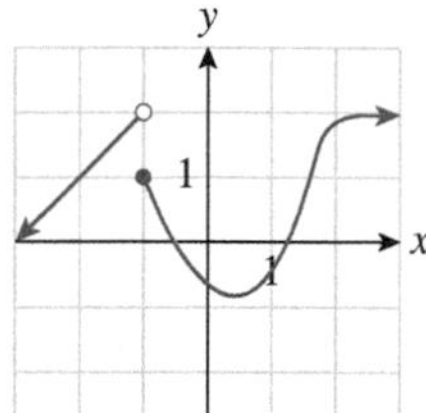

42. **a.** $\lim_{x \to -1} f(x)$ **b.** $\lim_{x \to 0^+} f(x)$ **c.** $\lim_{x \to 0^-} f(x)$
d. $\lim_{x \to 0} f(x)$ **e.** $f(0)$ **f.** $\lim_{x \to -\infty} f(x)$

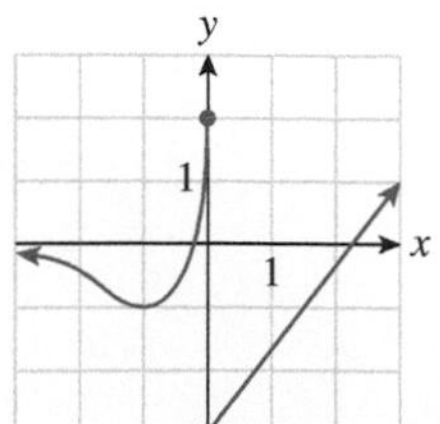

43. **a.** $\lim_{x \to -1^+} f(x)$ **b.** $\lim_{x \to -1^-} f(x)$ **c.** $\lim_{x \to -1} f(x)$ **d.** $f(-1)$

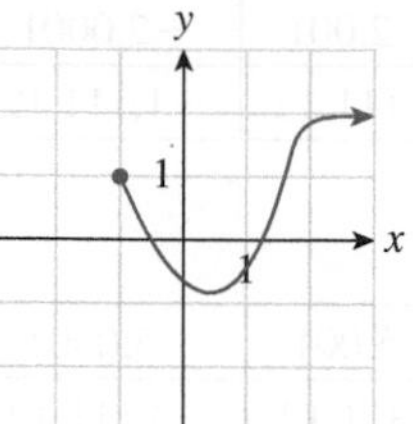

44. **a.** $\lim_{x \to 0^+} f(x)$ **b.** $\lim_{x \to 0^-} f(x)$ **c.** $\lim_{x \to 0} f(x)$ **d.** $f(0)$

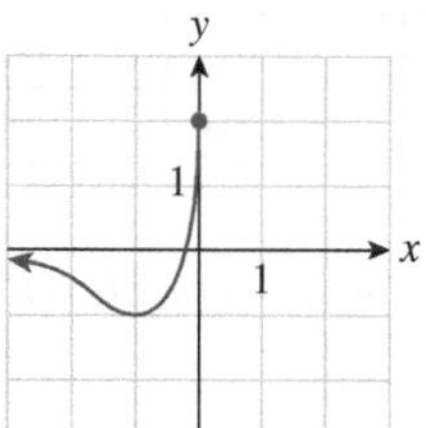

45. **a.** $\lim_{x \to -1} f(x)$ **b.** $\lim_{x \to 0^+} f(x)$ **c.** $\lim_{x \to 0^-} f(x)$
d. $\lim_{x \to 0} f(x)$ **e.** $f(0)$ **f.** $\lim_{x \to +\infty} f(x)$

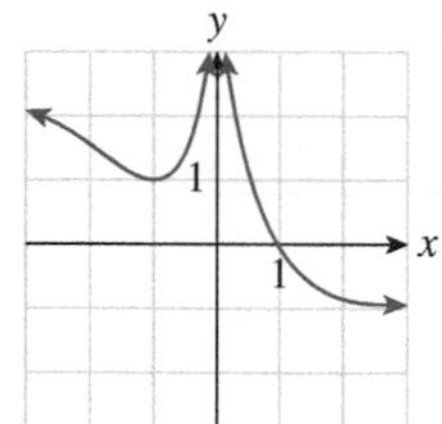

46. **a.** $\lim_{x \to 1} f(x)$ **b.** $\lim_{x \to 0^+} f(x)$ **c.** $\lim_{x \to 0^-} f(x)$
d. $\lim_{x \to 0} f(x)$ **e.** $f(0)$ **f.** $\lim_{x \to -\infty} f(x)$

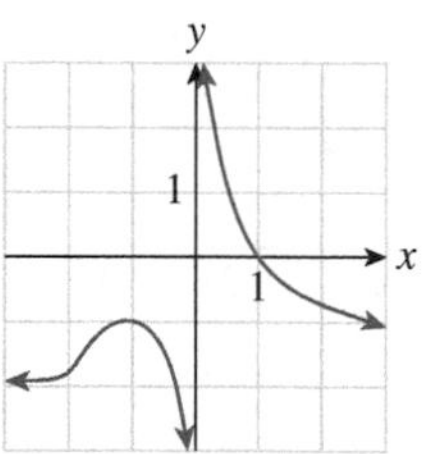

47. **a.** $\lim_{x \to -1} f(x)$ **b.** $\lim_{x \to 0^+} f(x)$ **c.** $\lim_{x \to 0^-} f(x)$
d. $\lim_{x \to 0} f(x)$ **e.** $f(0)$ **f.** $f(-1)$

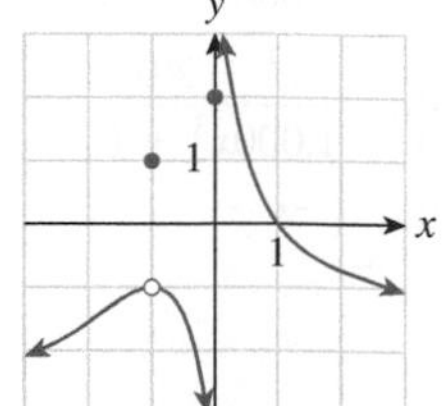

48. **a.** $\lim_{x\to 0^-} f(x)$ **b.** $\lim_{x\to 1^+} f(x)$ **c.** $\lim_{x\to 0} f(x)$
d. $\lim_{x\to 1} f(x)$ **e.** $f(0)$ **f.** $f(1)$

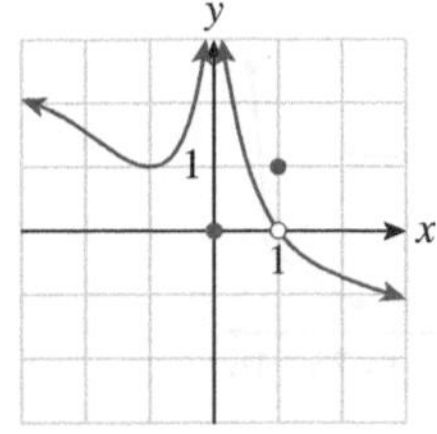

Applications

49. ***Doctorates in Mexico*** The annual number of PhD graduates in Mexico in the natural sciences for 1990–2012 can be approximated by

$$n(t) = 890(1 - e^{-0.05t}),$$

where t is time in years since 1990.[2] Numerically estimate $\lim_{t\to+\infty} n(t)$, and interpret the answer. [**HINT:** See Example 6.]

50. ***Housing Starts*** The number $s(t)$ of housing starts for single-family homes in the United States each year from 2006 through 2013 can be approximated by

$$s(t) = 500e^{0.05(x-5)^2} - 100 \text{ thousand units}$$

where t is time in years since 2006.[3] Numerically estimate $\lim_{t\to+\infty} s(t)$, and interpret the answer.
[**HINT:** See Example 6.]

51. ***Funding for NASA up to 1966*** The percentage of the U.S. federal budget allocated to NASA from 1958 to 1966 can be modeled by

$$p(t) = \frac{4.7}{1 + 139e^{-t}} \text{ percentage points}$$

(t is time in years since 1958).[4]

a. Numerically estimate $\lim_{t\to+\infty} p(t)$, and interpret the answer. [**HINT:** See Example 6.]

b. How does your answer to part (a) compare with actual current funding for NASA?

52. ***Funding for NASA up to 1966*** (Compare Exercise 51.) The percentage of the U.S. federal budget allocated to NASA from 1958 to 1966 can also be modeled by

$$p(t) = \frac{4.5}{1.07^{(t-8)^2}} \text{ percentage points}$$

(t is time in years since 1958).[5]

a. Numerically estimate $\lim_{t\to+\infty} p(t)$, and interpret the answer. [**HINT:** See Example 6.]

b. How does your answer to part (a) compare with actual current funding for NASA?

53. ***Scientific Research: 1983–2003*** The number of research articles per year, in thousands, in the prominent journal *Physical Review* written by researchers in Europe during 1983–2003 can be modeled by

$$A(t) = \frac{7.0}{1 + 5.4(1.2)^{-t}},$$

where t is time in years ($t = 0$ represents 1983).[6] Numerically estimate $\lim_{t\to+\infty} A(t)$ and interpret the answer.

54. ***Scientific Research: 1983–2003*** The percentage of research articles in the prominent journal *Physical Review* written by researchers in the United States during 1983–2003 can be modeled by

$$A(t) = 25 + \frac{36}{1 + 0.6(0.7)^{-t}},$$

where t is time in years ($t = 0$ represents 1983).[7] Numerically estimate $\lim_{t\to+\infty} A(t)$, and interpret the answer.

55. ***SAT Scores by Income*** The following bar graph shows U.S. math SAT scores as a function of household income:[8]

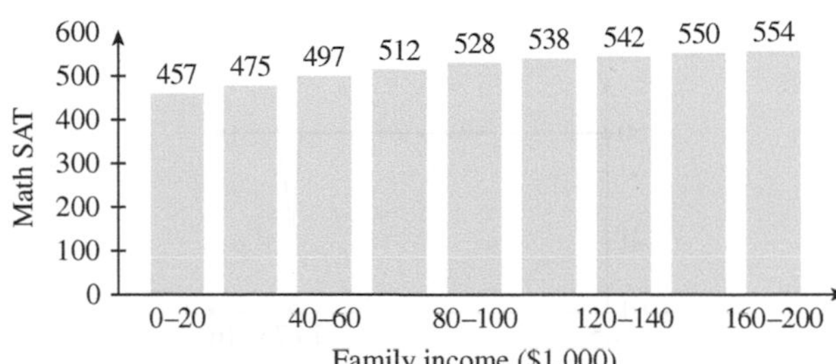

These data can be modeled by

$$S(x) = 573 - 133(0.987)^x,$$

where $S(x)$ is the average math SAT score of students whose household income is x thousand dollars per year. Numerically estimate $\lim_{x\to+\infty} S(x)$, and interpret the answer.

[2] Model is the authors'. Source for data: Instituto Nacional de Estadística y Geografía www.inegi.org.mx.

[3] Model is the authors'. Source of data: www.census.gov.

[4] Model is the authors'. Source of data: U.S. Office of Management and Budget/www.wikipedia.org.

[5] *Ibid.*

[6] Based on data from 1983 to 2003. Source: The American Physical Society/*New York Times*, May 3, 2003, p. A1.

[7] *Ibid.*

[8] 2009 data. Source: College Board/*New York Times* http://economix.blogs.nytimes.com.

56. ***SAT Scores by Income*** The following bar graph shows U.S. critical reading SAT scores as a function of household income:[9]

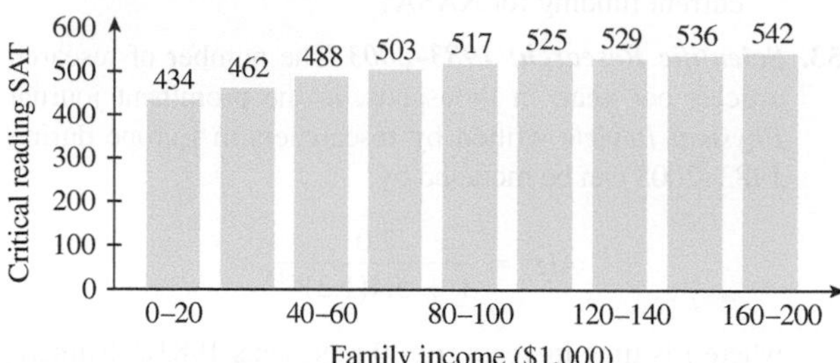

These data can be modeled by

$$S(x) = 550 - 136(0.985)^x,$$

where $S(x)$ is the average critical reading SAT score of students whose household income is x thousand dollars per year. Numerically estimate $\lim_{x\to+\infty} S(x)$, and interpret the answer.

57. ***Flash Crash*** The graph shows a rough representation of what happened to the Russell 1000 Growth Index Fund (IWF) stock price on the day of the U.S. stock market crash at 2:45 pm on May 6, 2010, the "Flash Crash" (t is the time of the day in hours, and $r(t)$ is the price of the stock in dollars).[10]

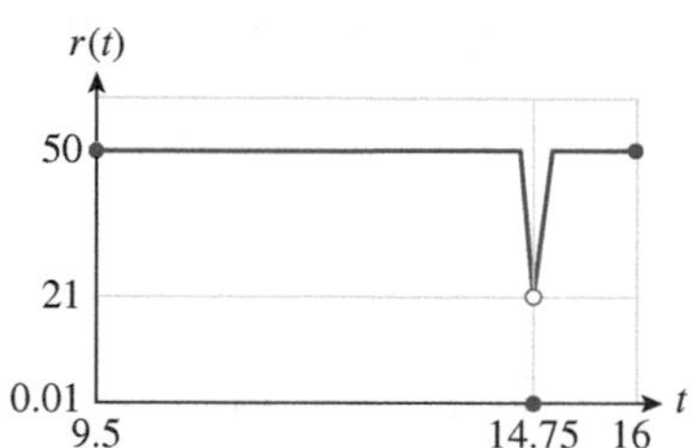

a. Compute the following (if a limit does not exist, say why):

$$\lim_{t\to 14.75^-} r(t), \quad \lim_{t\to 14.75^+} r(t), \quad \lim_{t\to 14.75} r(t), \quad r(14.75).$$

b. What do the answers to part (a) tell you about the IWF stock price?

58. ***Flash Crash*** The graph shows a rough representation of the (aggregate) market depth[11] of the stocks comprising the S&P 500 on the day of the U.S. stock market crash at 2:45 pm on

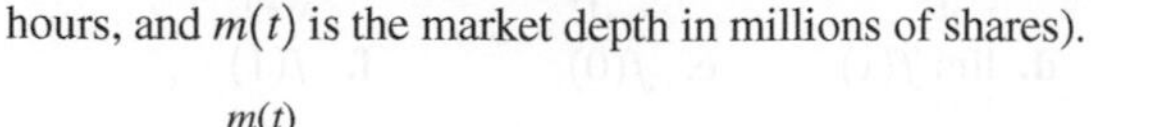

May 6, 2010, the "Flash Crash" (t is the time of the day in hours, and $m(t)$ is the market depth in millions of shares).

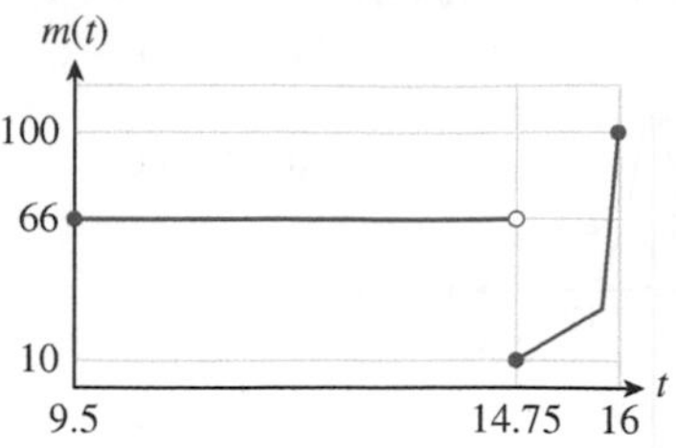

a. Compute the following (if a limit does not exist, say why):

$$\lim_{t\to 14.75^-} m(t), \quad \lim_{t\to 14.75^+} m(t), \quad \lim_{t\to 14.75} m(t), \quad m(14.75).$$

b. What do the answers to part (a) tell you about the market depth?

59. ***Home Prices*** The following graph shows the values of the home price index[12] for 2000–2014 together with a mathematical model I extrapolating the data:

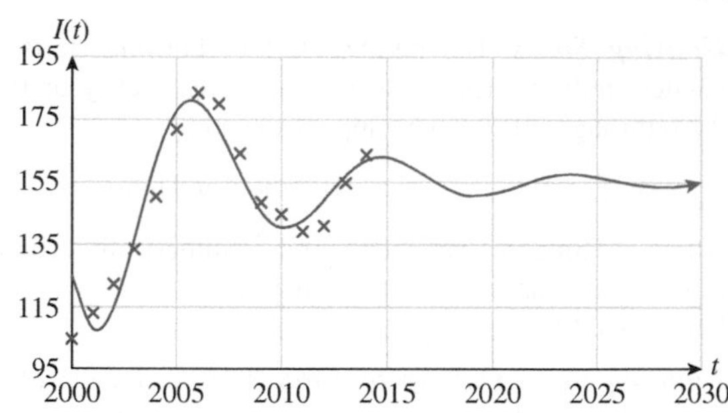

Estimate and interpret $\lim_{t\to+\infty} I(t)$.

60. ***Home Prices: Optimist Projection*** The following graph shows the values of the home price index[13] for 2000–2014 together with another mathematical model I extrapolating the data:

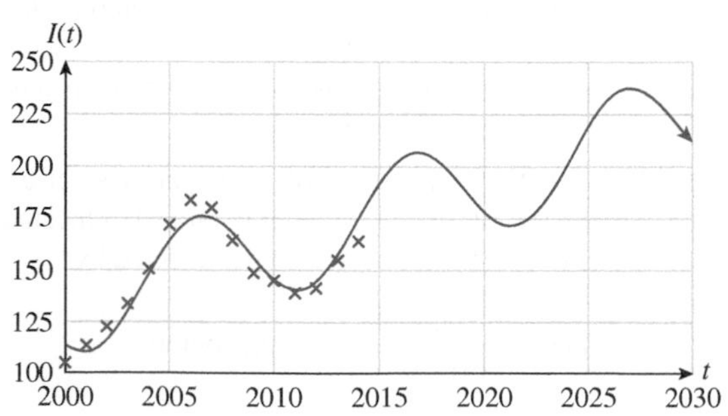

Estimate and interpret $\lim_{t\to+\infty} I(t)$.

---

[9] See footnote for Exercise 55.

[10] The actual graph can be seen at http://seekingalpha.com.

[11] The market depth of a stock is a measure of its ability to withstand relatively large market orders and is measured in orders to buy or sell the given stock. Source for data on graph: *Findings Regarding the Market Events of May 6, 2010,* U.S. Commodity Futures Trading Commission, U.S. Securities & Exchange Commission.

[12] The index is the Standard & Poor/Case-Shiller Home Price Index. Source for data: www.standardandpoors.com.

[13] *Ibid.*

61. ***Electric Rates*** The cost of electricity in Portland, Oregon, for residential customers increased suddenly on October 1, 2001, from around \$0.06 to around \$0.08 per kilowatt-hour.[14] Let $C(t)$ be this cost at time t, and take $t = 1$ to represent October 1, 2001. What does the given information tell you about $\lim_{t\to 1} C(t)$?

62. ***Airline Stocks*** Before the September 11, 2001 attacks, United Airlines stock was trading at around \$35 per share. Immediately after the attacks, the share price dropped by \$15.[15] Let $U(t)$ be this cost at time t, and take $t = 11$ to represent September 11, 2001. What does the given information tell you about $\lim_{t\to 11} U(t)$?

Foreign Trade *Annual U.S. imports from China in the years 1996–2003 can be approximated by*

$$I(t) = t^2 + 3.5t + 50 \quad (1 \le t \le 9)$$

billion dollars, where t represents time in years since 1995. Annual U.S. exports to China in the same years can be approximated by

$$E(t) = 0.4t^2 - 1.6t + 14$$

billion dollars.[16] *Exercises 63 and 64 are based on these models.*

63. ▼ Assuming that the trends shown in the above models continued indefinitely, numerically estimate

$$\lim_{t\to+\infty} I(t) \quad \text{and} \quad \lim_{t\to+\infty} \frac{I(t)}{E(t)},$$

interpret your answers, and comment on the results.

64. ▼ Repeat Exercise 63, this time calculating

$$\lim_{t\to+\infty} E(t) \quad \text{and} \quad \lim_{t\to+\infty} \frac{E(t)}{I(t)}.$$

Communication and Reasoning Exercises

65. Describe the method of evaluating limits numerically. Give at least one disadvantage of this method.

66. Describe the method of evaluating limits graphically. Give at least one disadvantage of this method.

67. Your friend Dion, a business student, claims that the study of limits that do not exist is completely unrealistic and has nothing to do with the world of business. Give two examples from the world of business that might convince him that he is wrong.

68. Your other friend Fiona claims that the study of limits is a complete farce; all you ever need to do to find the limit as x approaches a is substitute $x = a$. Give two examples that show she is wrong.

69. ▼ What is wrong with the following statement? "Because $f(a)$ is not defined, $\lim_{x\to a} f(x)$ does not exist." Illustrate your claim with an example.

70. ▼ What is wrong with the following statement? "Because $f(a)$ is defined, $\lim_{x\to a} f(x)$ exists." Illustrate your claim with an example.

71. ◆ Give an example of a function f with $\lim_{x\to 1} f(x) = f(2)$.

72. ◆ If $S(t)$ represents the size of the universe in billions of light-years at time t years since the big bang and $\lim_{t\to+\infty} s(t) = 130{,}000$, is it possible that the universe will continue to expand forever?

73. ◆ Investigate $\lim_{x\to+\infty} x^n e^{-x}$ for some large values of n. What do you find? What do you think is the value of $\lim_{x\to+\infty} p(x)e^{-x}$ if $p(x)$ is any polynomial?

74. ◆ Investigate $\lim_{x\to-\infty} x^n e^{x}$ for some large values of n. What do you find? What do you think is the value of $\lim_{x\to-\infty} p(x)e^{x}$ if $p(x)$ is any polynomial?

[14] Source: Portland General Electric/*New York Times*, February 2, 2002, p. C1.

[15] Stock prices are approximate.

[16] Based on quadratic regression using data from the U.S. Census Bureau Foreign Trade Division website www.census.gov/foreign-trade/sitc1/ as of December 2004.

3.2 Limits and Continuity

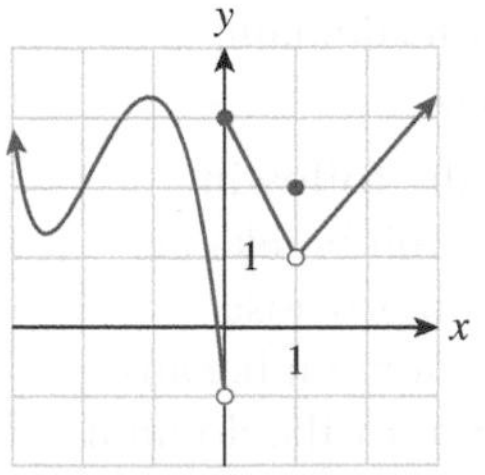

Continuous Functions

Figure 10

In Section 3.1 we saw examples of graphs that had various kinds of "breaks" or "jumps." For instance, in Example 4 we looked at the graph in Figure 10. This graph appears to have breaks, or **discontinuities**, at $x = 0$ and at $x = 1$. At $x = 0$ we saw that $\lim_{x\to 0} f(x)$ does not exist because the left- and right-hand limits are not the same. Thus, the discontinuity at $x = 0$ seems to be due to the fact that the limit does not exist there. On the other hand, at $x = 1$, $\lim_{x\to 1} f(x)$ *does* exist (it is equal to 1) but is not equal to $f(1) = 2$.

Thus, we have identified two kinds of discontinuity:

1. Points where the limit of the function does not exist — $x = 0$ in Figure 10 because $\lim_{x\to 0} f(x)$ does not exist.
2. Points where the limit exists but does not equal the value of the function — $x = 1$ in Figure 10 because $\lim_{x\to 1} f(x) = 1 \neq f(1)$.

On the other hand, there is no discontinuity at, say, $x = -2$, where we find that $\lim_{x\to -2} f(x)$ exists and equals 2 and $f(-2)$ is also equal to 2. In other words,

$$\lim_{x\to -2} f(x) = 2 = f(-2).$$

The point $x = -2$ is an example of a point where f is **continuous**. (Notice that you can draw the portion of the graph near $x = -2$ without lifting your pencil from the paper.) Similarly, f is continuous at *every* point other than $x = 0$ and $x = 1$. Here is the mathematical definition.

Continuous Function

Let f be a function, and let a be a number in the domain of f. Then f is **continuous at a** if

a. $\lim_{x\to a} f(x)$ exists and

b. $\lim_{x\to a} f(x) = f(a)$.*

The function f is said to be **continuous on its domain** if it is continuous at each point in its domain.

If f is not continuous at a particular a in its domain, we say that f is **discontinuous** at a or that f has a **discontinuity** at a. Thus, a discontinuity can occur at $x = a$ if either

a. $\lim_{x\to a} f(x)$ does not exist or

b. $\lim_{x\to a} f(x)$ exists but is not equal to $f(a)$.

* If a is an isolated point in the domain of f; that is, f is defined at a but at no other points within some distance of a (for example, $f(x) = \sqrt{-x^2}$ is defined only at $x = 0$) then, even though there can be no limit at a, we regard f as continuous at a.

Quick Examples

1. The function shown in Figure 10 is continuous at $x = -1$ and $x = 2$. It is discontinuous at $x = 0$ and $x = 1$ and so is not continuous on its domain.
2. The function $f(x) = x^2$ is continuous on its domain. (Think of its graph, which contains no breaks.)
3. The function shown in Figure 11 is continuous on its domain. In particular, it is continuous at the left endpoint $x = -1$ of its domain, because $\lim_{x\to -1} f(x) = \lim_{x\to -1^+} f(x) = 1 = f(-1)$.
4. The function f whose graph is shown on the left in the following figure is continuous on its domain. (Although the graph breaks at $x = 2$, that is not a point of its domain.) The function g whose graph is shown on the right is not continuous on its domain because it has a discontinuity at $x = 2$. (Here, $x = 2$ is a point of the domain of g.)

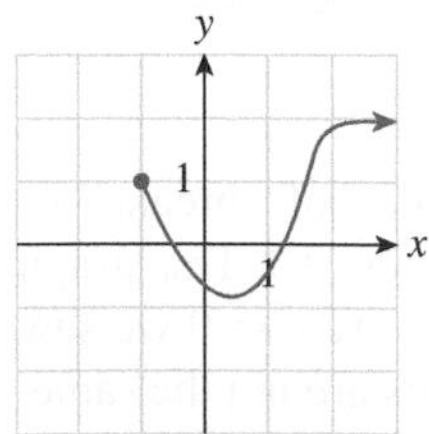

Figure 11

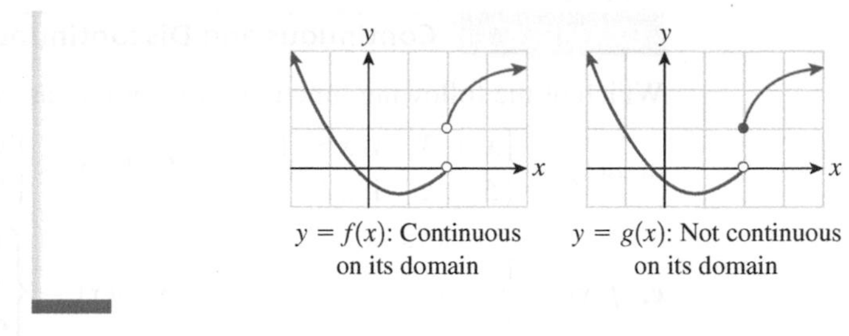

$y = f(x)$: Continuous on its domain

$y = g(x)$: Not continuous on its domain

Singularities

Continuity and discontinuity of a function are defined only for points in a function's domain; a function cannot be continuous at a point not in its domain, and it cannot be discontinuous there either. So if a is not in the domain of f—that is, if $f(a)$ is not defined—then it is meaningless to talk about whether f is continuous or discontinuous at a. For instance, for the function f in Quick Example 4, it is meaningless to talk about whether f is continuous or discontinuous at 2.

Q: *Wait a minute! The graph of the function f shown on the left in Quick Example 4 definitely breaks at $x = 2$. If that is not a point of discontinuity, then what is it?*

A: Notice that, although f is defined at values of x arbitrarily close to 2, it is not defined at $x = 2$, thus causing a break in its graph. We say that f has a *singularity* at 2.

Singularity

If $f(a)$ is not defined but $f(x)$ is defined for (at least some) values of x arbitrarily close to and on both sides of a, we will say that f has a **singularity at** a, or that a **is a singular point of** f.*

Quick Examples

5. Consider again the functions shown in Quick Example 4 above:

Singularity | **Discontinuity**

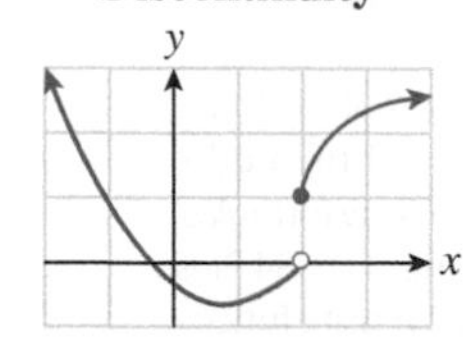

$y = f(x)$: Continuous on its domain
Singular at 2

$y = g(x)$: Not continuous on its domain
Discontinuous at 2

Although f (graph on the left) is continuous on its domain, we see that $f(2)$ is not defined, whereas $f(x)$ is defined for values of x arbitrarily close to 2. Thus, f has a singularity at 2. The function g (graph on the right) has no singularities of the type we discuss here; $f(x)$ is defined for all x.

6. $\frac{1}{x}$, $\frac{1}{x^2}$, and $\frac{1}{x^{1/3}}$ are all singular at 0.†

* In mathematics the term "singular point" or "singularity" is used quite broadly to refer to a point at which some mathematical object under consideration either is not defined or is unusual in some other manner. As a consequence, the term applies to more situations than we discuss here and has different meanings in different contexts. (See, for example, Section 5.1 for our use of the term in the context of maxima and minima.) Be aware that many people use the term "discontinuity" to apply to singular points as well, but that is contrary to the accepted definition of that term.

† What about $d(x) = \frac{1}{\sqrt{x}}$? A mathematician would also say that this function is singular at 0 because $f(x)$ approaches infinity as $x \to 0^+$ even though it is not defined on both sides of zero. Remember, what we call a "singularity" includes some—but not all—types of points a mathematician would call singular.

EXAMPLE 1 Continuous and Discontinuous Functions

Which of the following functions are continuous on their domains?

a. $h(x) = \begin{cases} x + 3 & \text{if } x \le 1 \\ 5 - x & \text{if } x > 1 \end{cases}$

b. $k(x) = \begin{cases} x + 3 & \text{if } x \le 1 \\ 1 - x & \text{if } x > 1 \end{cases}$

c. $f(x) = \dfrac{1}{x}$

d. $g(x) = \begin{cases} \dfrac{1}{x} & \text{if } x \ne 0 \\ 0 & \text{if } x = 0 \end{cases}$

Solution

a. and **b.** The graphs of h and k are shown in Figure 12.

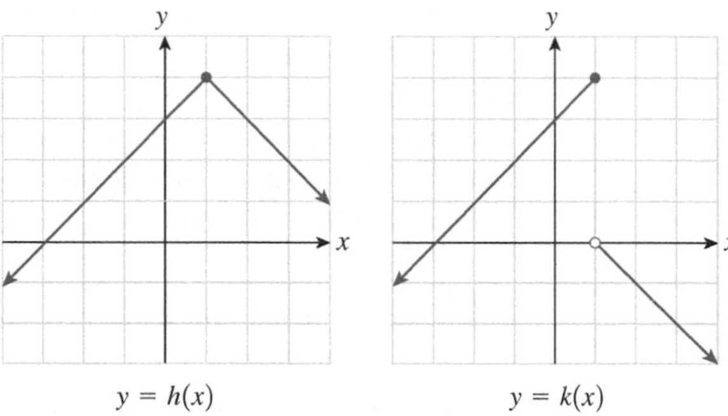

Figure 12

Even though the graph of h is made up of two different line segments, it is continuous at every point of its domain, including $x = 1$ because

$$\lim_{x \to 1} h(x) = 4 = h(1).$$

On the other hand, $x = 1$ is also in the domain of k, but $\lim_{x \to 1} k(x)$ does not exist. Thus, k has a discontinuity at $x = 1$ and is therefore not continuous on its domain.

c. and **d.** The graphs of f and g are shown in Figure 13.

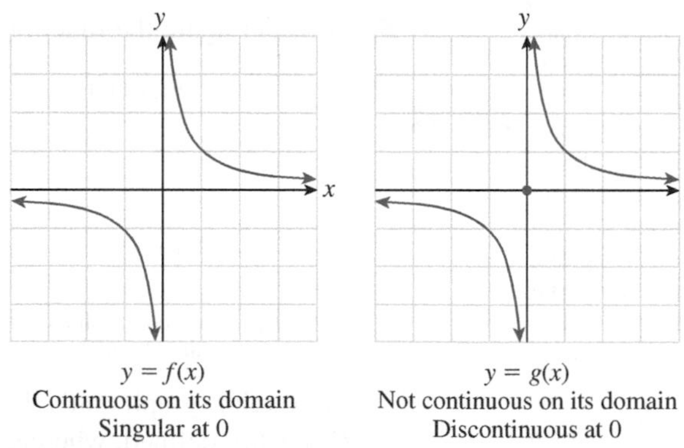

$y = f(x)$
Continuous on its domain
Singular at 0

$y = g(x)$
Not continuous on its domain
Discontinuous at 0

Figure 13

Using Technology

We can use technology to draw (approximate) graphs of the functions in Example 1(a), 1(b), and 1(c). Here are the technology formulas that will work for the TI-83/84 Plus, spreadsheets, and Website function evaluator and grapher. (In the TI-83/84 Plus, replace <= by ≤. In spreadsheets, replace x by a cell reference, and insert an equals sign in front of the formula.)

a. `(x+3)*(x<=1)+(5-x)*(x>1)`

b. `(x+3)*(x<=1)+(1-x)*(x>1)`

c. `(1/x)`

Observe in each case how technology handles the breaks in the curves.

The domain of f consists of all real numbers except 0, and f is continuous at all such numbers. (Notice that 0 is not in the domain of f, so the question of continuity at 0 does not arise.) Thus, f is continuous on its domain but singular at 0, as it is not defined there but is defined at points arbitrarily close to 0.

The function g, on the other hand, has its domain expanded to include 0, so we now need to check whether g is continuous at 0. From the graph, it is easy to see that

g is discontinuous there because $\lim_{x\to 0} g(x)$ does not exist. Thus, g is not continuous on its domain because it is discontinuous at 0.

➡ **Before we go on . . .** Remember: When we say that a function like $f(x) = \dfrac{1}{x}$ is continuous on its domain, we are claiming not that it is continuous *at every real number* but that it is continuous at every real number *in its domain,* so its graph can still break at any singular points. ■

EXAMPLE 2 Continuous Except at a Point

In each case, say what, if any, value of $f(a)$ would make f continuous at a.

a. $f(x) = \dfrac{x^3 - 8}{x - 2}$; $a = 2$ **b.** $f(x) = \dfrac{e^{2x} - 1}{x}$; $a = 0$ **c.** $f(x) = \dfrac{|x|}{x}$; $a = 0$

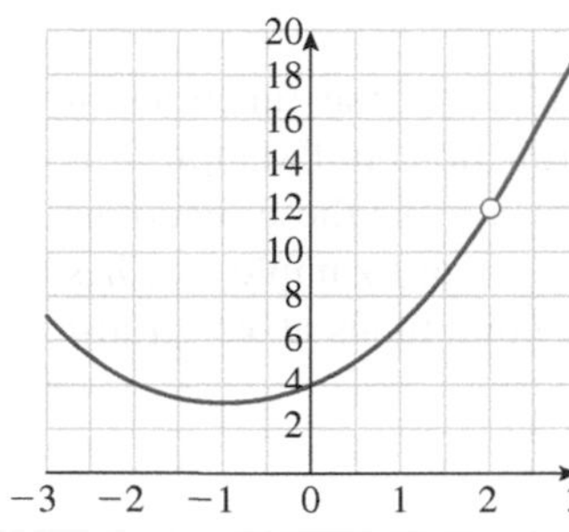

Figure 14

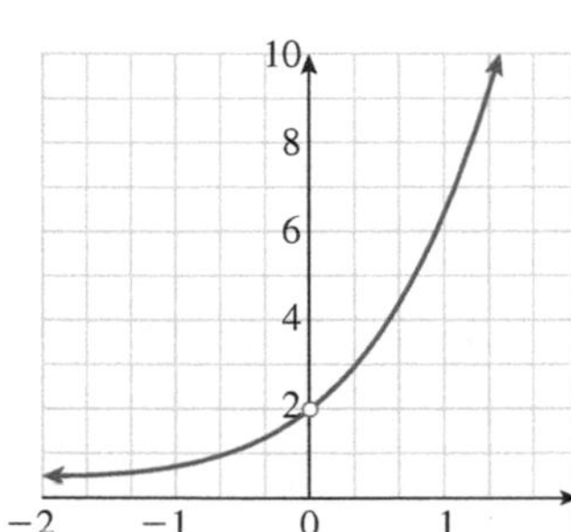

Figure 15

Solution

a. In Figure 14 we see the graph of $f(x) = \dfrac{x^3 - 8}{x - 2}$. The point corresponding to $x = 2$ is missing because f is not (yet) defined there. (Your graphing utility will probably miss this subtlety and render a continuous curve. See the technology note in the margin.) To turn f into a function that is continuous at $x = 2$, we need to "fill in the gap" so as to obtain a continuous curve. Since the graph suggests that the missing point is (2, 12), let us define $f(2) = 12$.

Does f now become continuous if we take $f(2) = 12$? From the graph or Example 1(a) of Section 3.1,

$$\lim_{x\to 2} f(x) = \lim_{x\to 2} \frac{x^3 - 8}{x - 2} = 12,$$

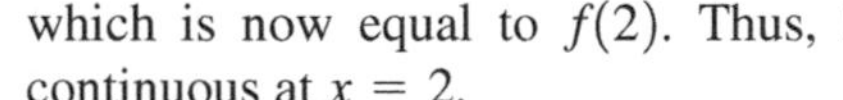

which is now equal to $f(2)$. Thus, $\lim_{x\to 2} f(x) = f(2)$, showing that f is now continuous at $x = 2$.

b. In Example 1(b) of Section 3.1, we saw that

$$\lim_{x\to 0} f(x) = \lim_{x\to 0} \frac{e^{2x} - 1}{x} = 2,$$

so, as in part (a), we must define $f(0) = 2$. This is confirmed by the graph, shown in Figure 15.

c. We considered the function $f(x) = |x|/x$ in Example 2 of Section 3.1. Its graph is shown in Figure 16.

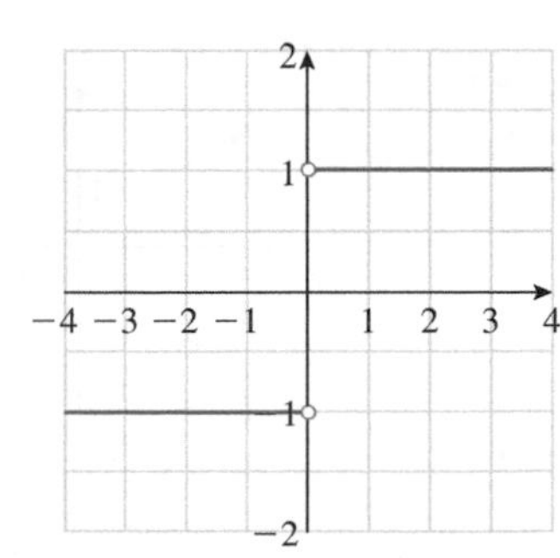

Figure 16

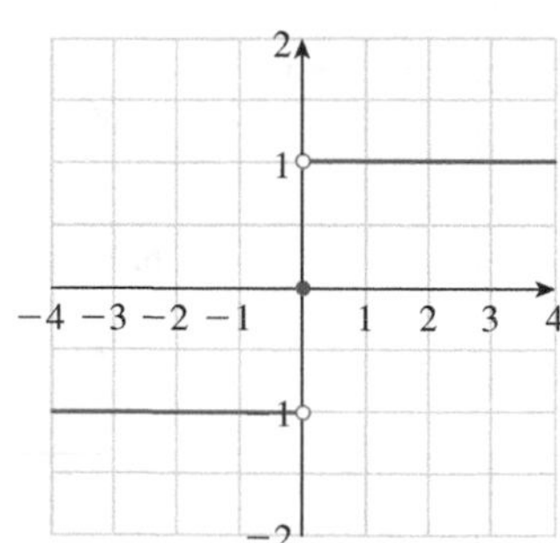

Figure 17

Using Technology

It is instructive to see how technology handles the functions in Example 2. Here are the technology formulas that will work for the TI-83/84 Plus, spreadsheets, and Website function evaluator and grapher. (In spreadsheets, replace x by a cell reference, and insert an equals sign in front of the formula.)

a. `(x^3-8)/(x-2)`

b. `(e^(2x)-1)/x`

Spreadsheet:

`=(exp(2*A2)-1)/A2`

c. `abs(x)/x`

In each case, compare the graph rendered by technology with the corresponding figure in Example 2.

Now we encounter a problem: No matter how we try to fill in the gap at the singular point at $x = 0$, the result will be a discontinuous function. For example, setting $f(0) = 0$ will result in the discontinuous function shown in Figure 17. We conclude that it is impossible to assign any value to $f(0)$ to turn f into a function that is continuous at $x = 0$.

We can also see this result algebraically: In Example 2 of Section 3.1 we saw that $\lim_{x \to 0} \frac{|x|}{x}$ does not exist. Thus, the resulting function will fail to be continuous at 0, no matter how we define $f(0)$.

Removable and Essential Singularities

The function in Example 2(a) has a singularity at 2, and the functions in Example 2(b) and 2(c) have singularities at 0. The functions in Example 2(a) and 2(b) have **removable singularities** because we can make these functions continuous at a by properly defining $f(a)$. The function in Example 2(c) has an **essential singularity** because we cannot make f continuous at $x = a$ just by defining $f(a)$ properly.

3.2 EXERCISES

▼ more advanced ◆ challenging

T indicates exercises that should be solved using technology

In Exercises 1–14 the graph of a function f is given. Determine whether f is continuous on its domain. If it is not continuous on its domain, say why. [**HINT**: See Quick Examples 1–4.]

1.

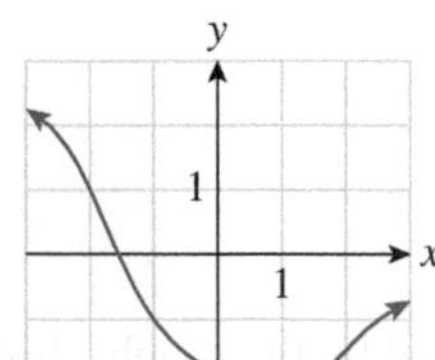

2.

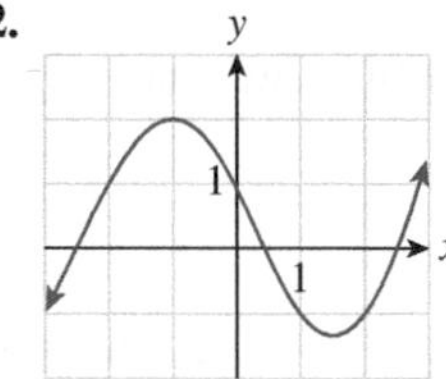

3.

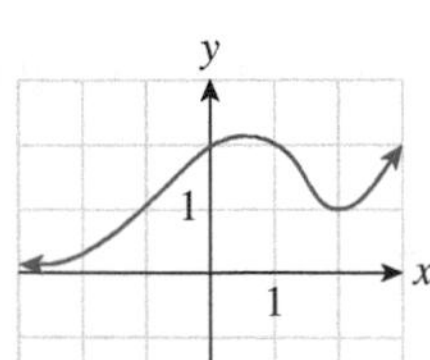

4.

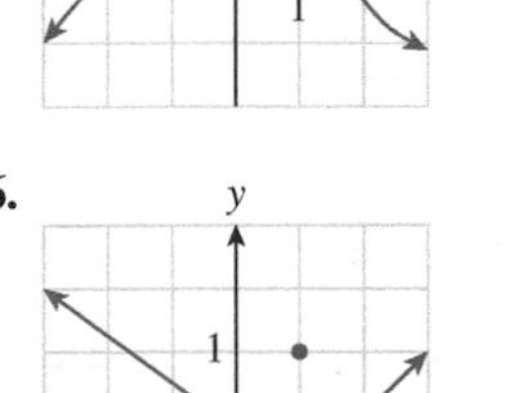

5.

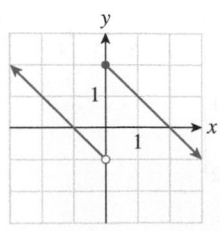

6.

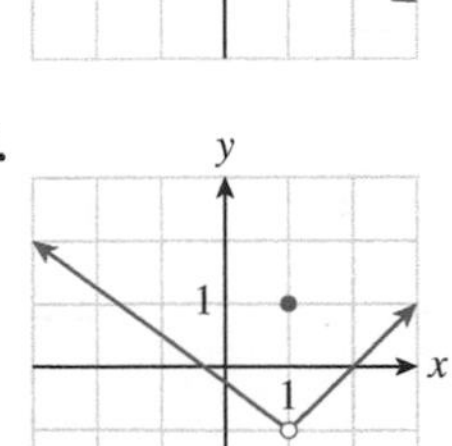

7.

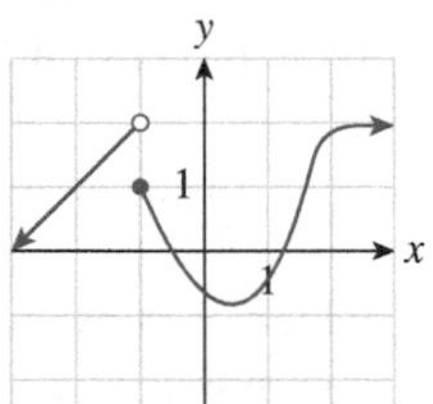

8.

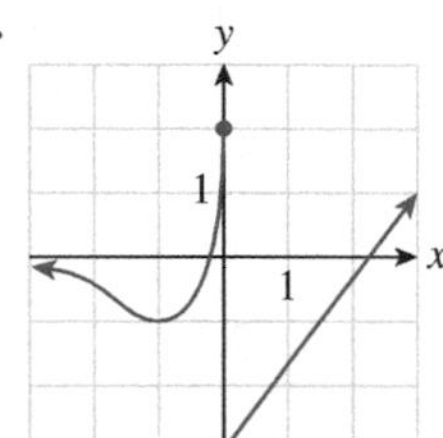

9.

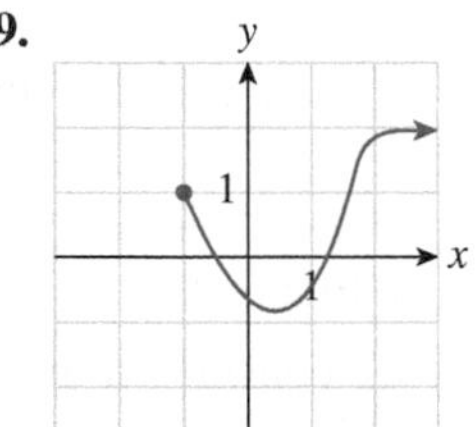

10.

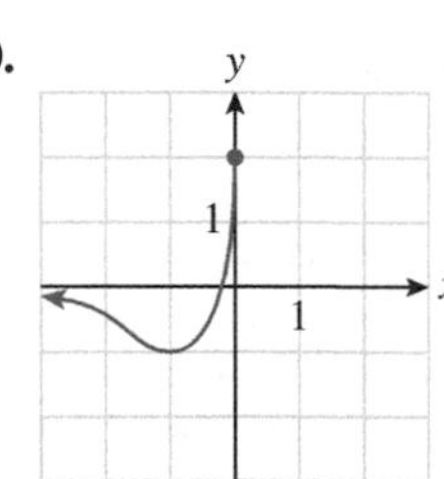

11.

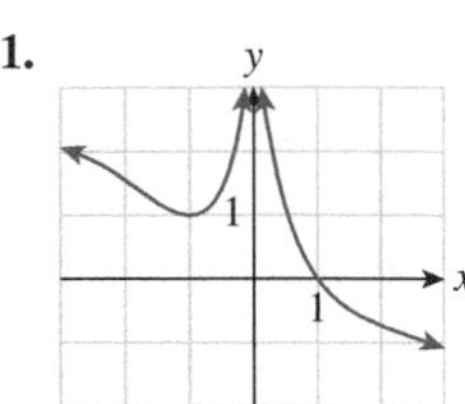

12.

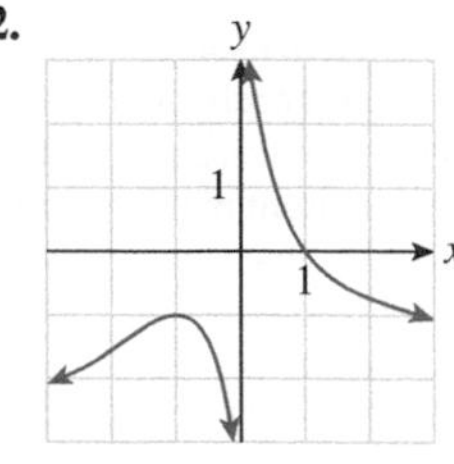

13.

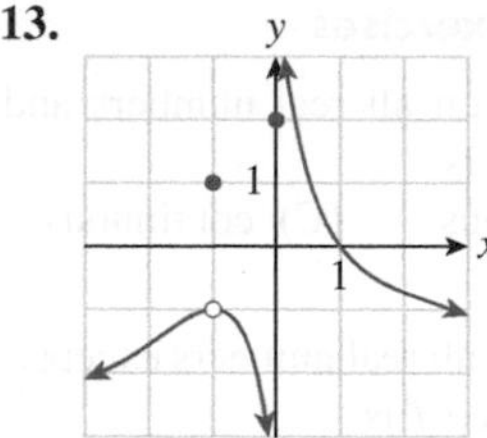

14.

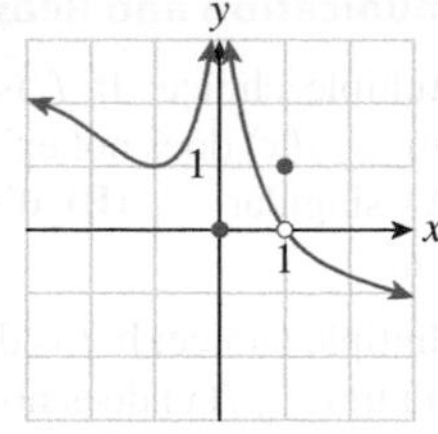

(E)

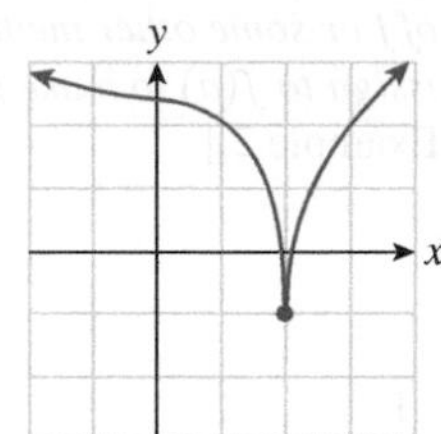

(F)

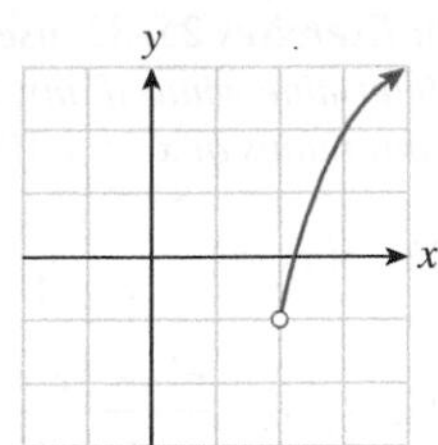

In Exercises 15 and 16, identify which (if any) of the given graphs represent functions that are continuous on their domains. [**HINT**: See Quick Examples 1–4.]

15. (A)

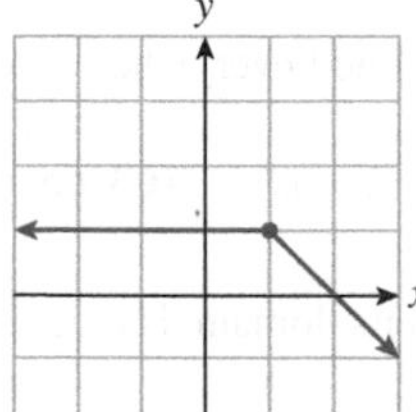

(B)

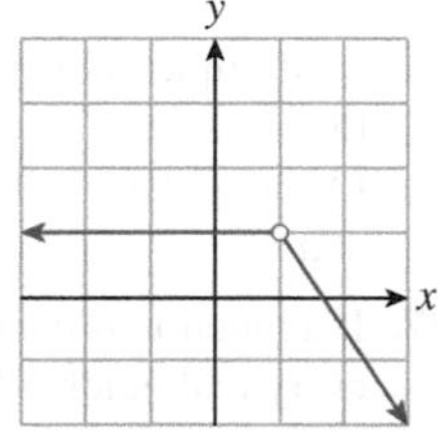

(C)

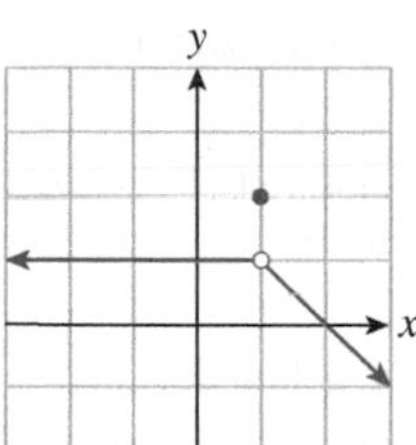

(D)

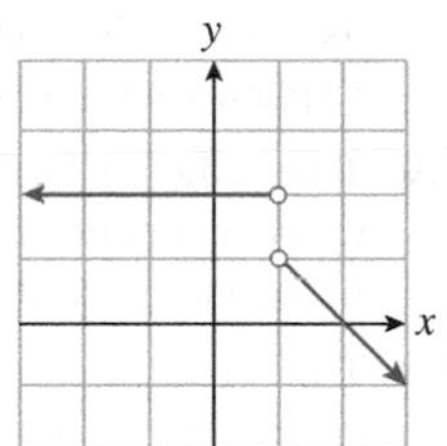

(E)

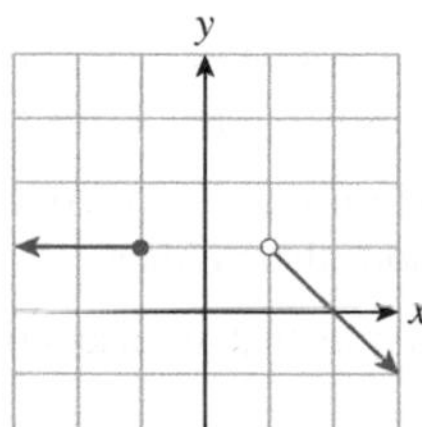

(F)

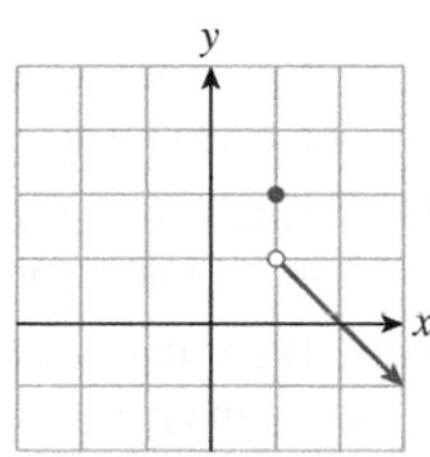

16. (A)

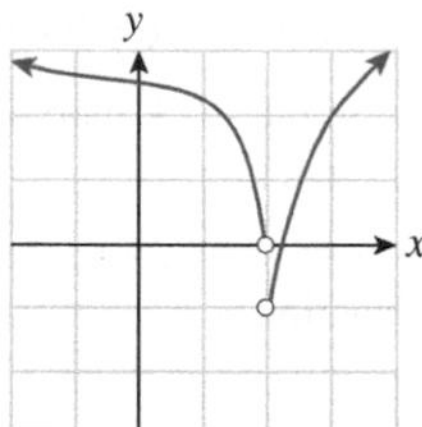

(B)

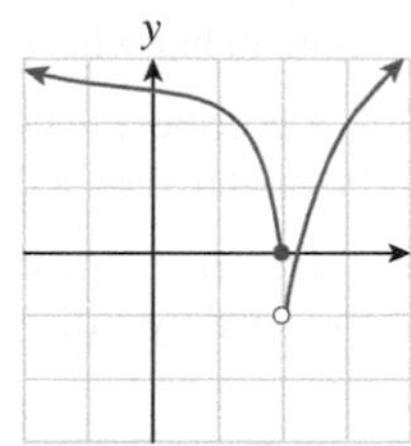

(C)

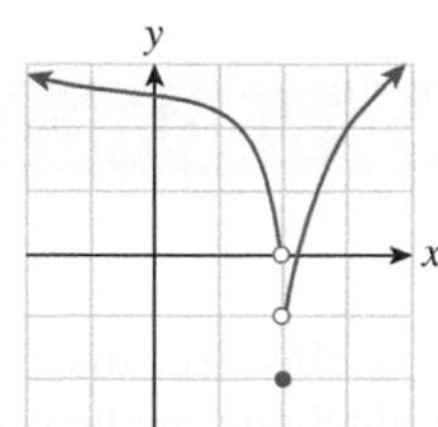

(D)

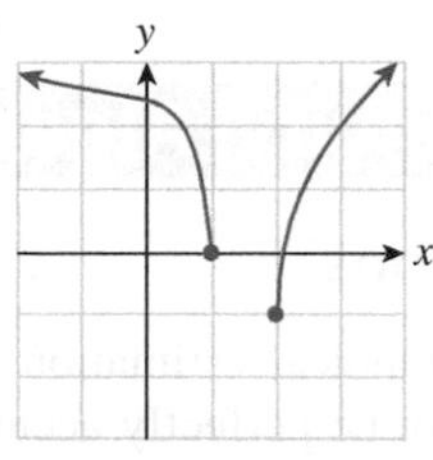

In Exercises 17–24, the graph of a function f is given. Determine whether, at the given point a, f is continuous, discontinuous, or singular. [**HINT**: See Quick Examples 5 and 6.]

17.

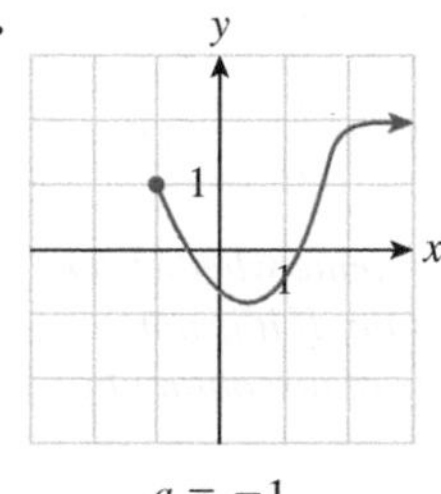

$a = -1$

18.

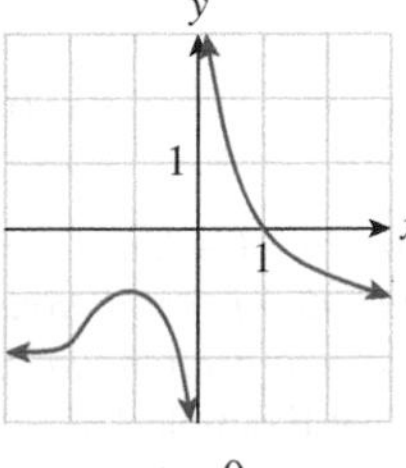

$a = 0$

19.

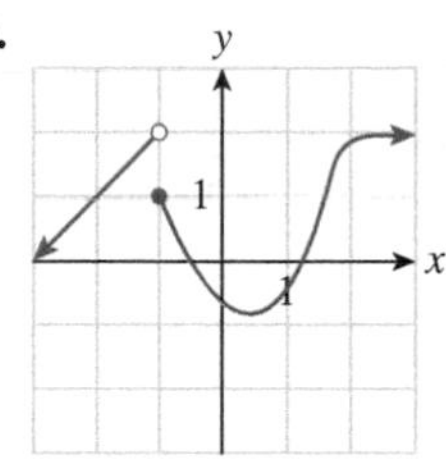

$a = -1$

20.

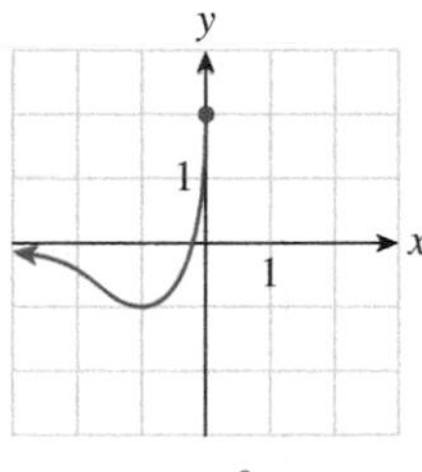

$a = 0$

21.

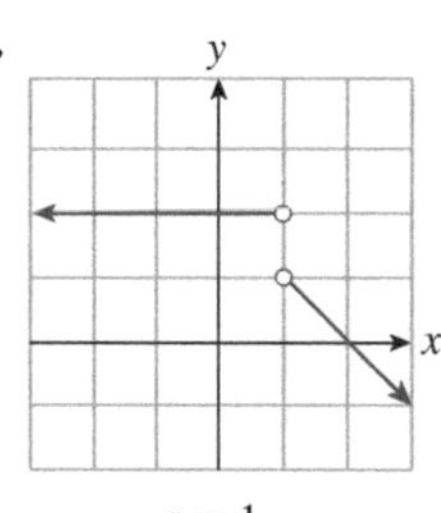

$a = 1$

22.

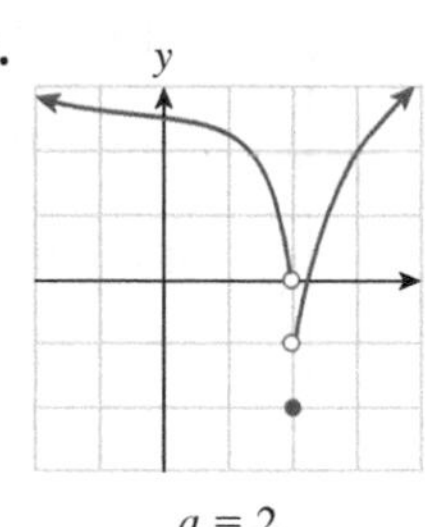

$a = 2$

23.

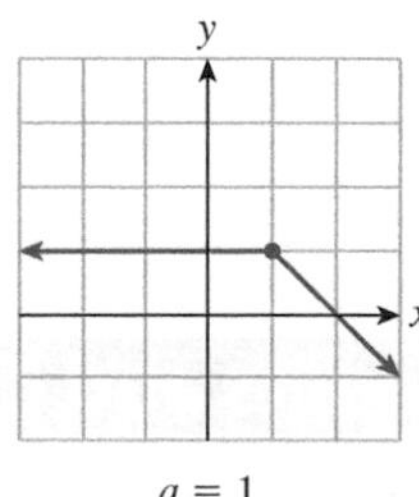

$a = 1$

24.

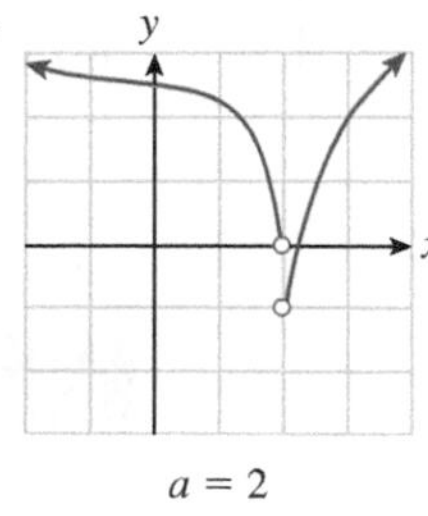

$a = 2$

In Exercises 25–32, use a graph of f or some other method to determine what, if any, value to assign to f(a) to make f continuous at x = a. [**HINT**: See Example 2.]

25. $f(x) = \dfrac{x^2 - 2x + 1}{x - 1}$; $a = 1$

26. $f(x) = \dfrac{x^2 + 3x + 2}{x + 1}$; $a = -1$

27. $f(x) = \dfrac{x}{3x^2 - x}$; $a = 0$

28. $f(x) = \dfrac{x^2 - 3x}{x + 4}$; $a = -4$

29. $f(x) = \dfrac{3}{3x^2 - x}$; $a = 0$

30. $f(x) = \dfrac{x - 1}{x^3 - 1}$; $a = 1$

31. $f(x) = \dfrac{1 - e^x}{x}$; $a = 0$

32. $f(x) = \dfrac{1 + e^x}{1 - e^x}$; $a = 0$

In Exercises 33–42, use a graph to determine whether the given function is continuous on its domain. If it is not continuous on its domain, list the points of discontinuity. [**HINT**: See Example 1.]

33. $f(x) = |x|$

34. $f(x) = \dfrac{|x|}{x}$

35. $g(x) = \dfrac{1}{x^2 - 1}$

36. $g(x) = \dfrac{x - 1}{x + 2}$

37. $f(x) = \begin{cases} x + 2 & \text{if } x < 0 \\ 2x - 1 & \text{if } x \geq 0 \end{cases}$

38. $f(x) = \begin{cases} 1 - x & \text{if } x \leq 1 \\ x - 1 & \text{if } x > 1 \end{cases}$

39. $h(x) = \begin{cases} \dfrac{|x|}{x} & \text{if } x \neq 0 \\ 0 & \text{if } x = 0 \end{cases}$

40. $h(x) = \begin{cases} \dfrac{1}{x^2} & \text{if } x \neq 0 \\ 2 & \text{if } x = 0 \end{cases}$

41. $g(x) = \begin{cases} x + 2 & \text{if } x < 0 \\ 2x + 2 & \text{if } x \geq 0 \end{cases}$

42. $g(x) = \begin{cases} 1 - x & \text{if } x \leq 1 \\ x + 1 & \text{if } x > 1 \end{cases}$

Communication and Reasoning Exercises

43. Multiple choice: If f is defined on all real numbers and $\lim_{x \to a} f(x)$ does not exist, then f is
(A) singular **(B)** discontinuous **(C)** continuous
at a.

44. Multiple choice: If f is defined on all real numbers except a and $\lim_{x \to a} f(x)$ does not exist, then f is
(A) singular **(B)** discontinuous **(C)** continuous
at a.

45. ▼ Multiple choice: If f is defined only at a, then f is
(A) singular **(B)** discontinuous **(C)** continuous
at a.

46. ▼ Multiple choice: If f is defined everywhere except at a, then f is
(A) singular **(B)** discontinuous **(C)** continuous
at a.

47. If a function is continuous on its domain, is it continuous at every real number? Explain.

48. True or false? The graph of a function that is continuous on its domain is a continuous curve with no breaks in it. Explain your answer.

49. True or false? The graph of a function that is continuous at every real number is a continuous curve with no breaks in it. Explain your answer.

50. True or false? If the graph of a function is a continuous curve with no breaks in it, then the function is continuous on its domain. Explain your answer.

51. ▼ Give a formula for a function that is continuous on its domain but whose graph consists of three distinct curves.

52. ▼ Give a formula for a function that is not continuous at $x = -1$ but is not discontinuous there either.

53. ▼ Draw the graph of a function that is discontinuous at every integer.

54. ▼ Draw the graph of a function that is continuous on its domain but whose graph has a break at every integer.

55. ▼ Describe a real-life scenario in the stock market that can be modeled by a discontinuous function.

56. ▼ Describe a real-life scenario in your room that can be modeled by a discontinuous function.

3.3 Limits and Continuity: Algebraic Viewpoint

Closed-Form Functions

Although numerical and graphical estimation of limits is effective, the estimates these methods yield may not be perfectly accurate. The algebraic method, when it can be used, will always yield an exact answer. Moreover, algebraic analysis of a function often enables us to take a function apart and see "what makes it tick."

Let's start with the function $f(x) = 2 + x$ and ask: What happens to $f(x)$ as x approaches 3? To answer this algebraically, notice that as x gets closer and closer to 3, the quantity $2 + x$ must get closer and closer to $2 + 3 = 5$. Hence,

$$\lim_{x\to 3} f(x) = \lim_{x\to 3}(2 + x) = 2 + 3 = 5.$$

Q: *Is that all there is to the algebraic method? Just substitute $x = a$?*

A: Under certain circumstances. Notice that by substituting $x = 3$, we *evaluated the function at $x = 3$*. In other words, we relied on the fact that

$$\lim_{x\to 3} f(x) = f(3).$$

In Section 3.2 we said that a function satisfying this equation is *continuous* at $x = 3$.

Thus,

If we know that the function f is continuous at a point a, we can compute $\lim_{x\to a} f(x)$ by simply substituting $x = a$ into $f(x)$.

To use this fact, we need to know how to recognize continuous functions when we see them. Geometrically, they are easy to spot: A function is continuous at $x = a$ if its graph has no break at $x = a$. Algebraically, a large class of functions are known to be continuous on their domains—those, roughly speaking, that are *specified by a single formula*.

We can be more precise: A **closed-form function** is any function that can be obtained by combining constants, powers of x, exponential functions, radicals, logarithms, absolute values, trigonometric functions (and some other functions that we do not encounter in this text) into a *single* mathematical formula by means of the usual arithmetic operations and composition of functions. (They can be as complicated as we like.)

Closed-Form Functions

A function is **written in closed form** if it is specified by combining constants, powers of x, exponential functions, radicals, logarithms, absolute values, trigonometric functions (and some other functions that we do not encounter in this text) into a *single* mathematical formula by means of the usual arithmetic operations and composition of functions. A **closed-form function** is any function that can be written in closed form.

Quick Examples

1. $3x^2 - |x| + 1$, $\dfrac{\sqrt{x^2 - 1}}{6x - 1}$, $e^{-(4x^2-1)/x}$, and $\sqrt{\log_3(x^2 - 1)}$ are written in closed form, so they are all closed-form functions.

2. $f(x) = \begin{cases} -1 & \text{if } x \le -1 \\ x^2 + x & \text{if } -1 < x \le 1 \\ 2 - x & \text{if } 1 < x \le 2 \end{cases}$ is not written in closed-form because $f(x)$ is not expressed by a *single* mathematical formula.*

* It is possible to rewrite some piecewise-defined functions in closed form (using a single formula) but not this particular function, so $f(x)$ is not a closed-form function.

What is so special about closed-form functions is the following theorem.

Theorem 3.1 Continuity of Closed-Form Functions

Every closed-form function is continuous on its domain. Thus, if f is a closed-form function and $f(a)$ is defined, then $\lim_{x\to a} f(x)$ exists, and equals $f(a)$.

Quick Example

3. $f(x) = 1/x$ is a closed-form function, and its natural domain consists of all real numbers except 0. Thus, f is continuous at every nonzero real number. That is,

$$\lim_{x\to a} \frac{1}{x} = \frac{1}{a}$$

provided that $a \neq 0$.

Mathematics majors spend a great deal of time studying the proof of this theorem. We ask you to accept it without proof.

EXAMPLE 1 Limit of a Closed-Form Function

Evaluate the following limits algebraically:

a. $\lim_{x\to 1} \frac{x^3 - 8}{x - 2}$ **b.** $\lim_{x\to 2} \frac{x^3 - 8}{x - 2}$

Solution

a. First, notice that $(x^3 - 8)/(x - 2)$ is a closed-form function because it is specified by a single algebraic formula. Also, $x = 1$ is in the domain of this function. Therefore, the theorem applies, and

$$\lim_{x\to 1} \frac{x^3 - 8}{x - 2} = \frac{1^3 - 8}{1 - 2} = 7.$$

b. Although $(x^3 - 8)/(x - 2)$ is a closed-form function, $x = 2$ is not in its domain. (It is a singular point.) Thus, the theorem does not apply, and we cannot obtain the limit by substitution. However—and this is the key to finding limits at singular points—*some preliminary algebraic simplification will allow us to obtain a closed-form function with $x = 2$ in its domain.* To do this, notice first that the numerator can be factored as

$$x^3 - 8 = (x - 2)(x^2 + 2x + 4).$$

Thus,

$$\frac{x^3 - 8}{x - 2} = \frac{(x - 2)(x^2 + 2x + 4)}{x - 2} = x^2 + 2x + 4.$$

Once we have canceled the offending $(x - 2)$ in the denominator, we are left with a closed-form function *with 2 in its domain.** Thus,

$$\lim_{x\to 2} \frac{x^3 - 8}{x - 2} = \lim_{x\to 2}(x^2 + 2x + 4)$$
$$= 2^2 + 2(2) + 4 = 12. \quad \text{Substitute } x = 2.$$

This confirms the answer we found numerically in Example 1 in Section 3.1.

* By canceling the $(x - 2)$, we have removed the singularity of $(x^3 - 8)/(x - 2)$ at $x = 2$. (See Removable and Essential Singularities in Section 3.2).

➡ **Before we go on . . .** Notice that in Example 1(b), before simplification the substitution $x = 2$ yields

$$\frac{x^3 - 8}{x - 2} = \frac{8 - 8}{2 - 2} = \frac{0}{0}.$$

Worse than the fact that $0/0$ is undefined, it also conveys absolutely no information as to what the limit might be. (The limit turned out to be 12!) We therefore call the expression $0/0$ an **indeterminate form**. Once simplified, the function became $x^2 + 2x + 4$, which, upon the substitution $x = 2$, yielded 12—no longer an indeterminate form. In general, we have the following rule of thumb:

If the substitution $x = a$ yields the indeterminate form $0/0$, try simplifying by the method in Example 1.

We will say more about indeterminate forms in Example 2. ■

Q: *There is something suspicious about Example 1(b). If 2 was not in the domain before simplifying but was in the domain after simplifying, we must have changed the function—right?*

A: Correct. In fact, when we said that

$$\underbrace{\frac{x^3 - 8}{x - 2}}_{\text{Domain excludes 2}} = \underbrace{x^2 + 2x + 4,}_{\text{Domain includes 2}}$$

we were lying a little bit. What we really meant is that these two expressions are equal *where both are defined.* The functions $(x^3 - 8)/(x - 2)$ and $x^2 + 2x + 4$ are different functions. The difference is that $x = 2$ is a singular point of $(x^3 - 8)/(x - 2)$ but is in the domain of $x^2 + 2x + 4$. Since $\lim_{x\to 2} f(x)$ explicitly *ignores* any value that f may have at 2, this does not affect the limit. From the point of view of the limit at 2, these functions *are* equal. In general, we have the following rule.

Functions with Equal Limits

If $f(x) = g(x)$ for all x except possibly $x = a$, then

$$\lim_{x\to a} f(x) = \lim_{x\to a} g(x).$$

Quick Example

4. $\dfrac{x^2 - 1}{x - 1} = x + 1$ for all x except $x = 1$. Write $\dfrac{x^2 - 1}{x - 1}$ as $\dfrac{(x + 1)(x - 1)}{x - 1}$, and cancel the $(x - 1)$.

Therefore,

$$\lim_{x \to 1} \frac{x^2 - 1}{x - 1} = \lim_{x \to 1}(x + 1) = 1 + 1 = 2.$$

Q: *How do we find $\lim_{x \to a} f(x)$ when $x = a$ is a singular point of the function f and we cannot simplify the given function to make a a point of the domain?*

A: In such a case it might be necessary to analyze the function by some other method, such as numerically or graphically. However, if we do not obtain the indeterminate form 0/0 upon substitution, we can often say what the limit is, as the following example shows.

EXAMPLE 2 Limit of a Closed-Form Function at a Singular Point: The Determinate Form $k/0$

Evaluate the following limits if they exist:

a. $\displaystyle\lim_{x \to 1^+} \frac{x^2 - 4x + 1}{x - 1}$ **b.** $\displaystyle\lim_{x \to 1} \frac{x^2 - 4x + 1}{x - 1}$ **c.** $\displaystyle\lim_{x \to 1} \frac{x^2 - 4x + 1}{x^2 - 2x + 1}$

Solution

a. Although the function $f(x) = \dfrac{x^2 - 4x + 1}{x - 1}$ is a closed-form function, $x = 1$ is a singular point. Notice that substituting $x = 1$ gives

$$\frac{x^2 - 4x + 1}{x - 1} = \frac{1^2 - 4 + 1}{1 - 1} = \frac{-2}{0}$$

The **determinate** form $\dfrac{k}{0}$

which, although not defined, conveys important information to us: As x gets closer and closer to 1, the numerator approaches -2 and the denominator gets closer and closer to 0. Now, if we divide a number close to -2 by a number close to 0, we get a number of large absolute value; for instance,

$$\frac{-2.1}{0.0001} = -21{,}000 \quad \text{and} \quad \frac{-2.1}{-0.0001} = 21{,}000$$

$$\frac{-2.01}{0.00001} = -201{,}000 \quad \text{and} \quad \frac{-2.01}{-0.00001} = 201{,}000.$$

(Compare Example 5 of Section 3.1.) In our limit for part (a), x is approaching 1 from the right, so the denominator $x - 1$ is positive (as x is to the right of 1). Thus, we have the scenario illustrated above on the left, and we can conclude that

$$\lim_{x \to 1^+} \frac{x^2 - 4x + 1}{x - 1} = -\infty.$$

Think of this as $\dfrac{-2}{0^+} = -\infty$.

b. This time, x could be approaching 1 from either side. We already have, from part (a),

$$\lim_{x\to 1^+} \frac{x^2 - 4x + 1}{x - 1} = -\infty.$$

The same reasoning we used in part (a) gives

$$\lim_{x\to 1^-} \frac{x^2 - 4x + 1}{x - 1} = +\infty \qquad \text{Think of this as } \frac{-2}{0^-} = +\infty.$$

because now the denominator is negative and still approaching zero while the numerator still approaches -2 and therefore is also negative. (See the numerical calculations above on the right.) Because the left and right limits do not agree, we conclude that

$$\lim_{x\to 1} \frac{x^2 - 4x + 1}{x - 1} \text{ does not exist.}$$

c. First notice that the denominator factors:

$$\lim_{x\to 1} \frac{x^2 - 4x + 1}{x^2 - 2x + 1} = \lim_{x\to 1} \frac{x^2 - 4x + 1}{(x - 1)^2}.$$

As x approaches 1, the numerator approaches -2 as before, and the denominator approaches 0. However, this time, the denominator $(x - 1)^2$, being a square, is ≥ 0, regardless of the side from which x is approaching 1. Thus, the entire function is negative as x approaches 1, and

$$\lim_{x\to 1} \frac{x^2 - 4x + 1}{(x - 1)^2} = -\infty. \qquad \frac{-2}{0^+} = -\infty$$

➡ **Before we go on...** In general, the determinate forms $\frac{k}{0^+}$ and $\frac{k}{0^-}$ will always yield $\pm\infty$, with the sign depending on the sign of the overall expression as $x \to a$. (When we write the form $\frac{k}{0}$, we always mean $k \neq 0$.) This and other determinate forms are discussed further after Example 4.

Figure 18 shows the graphs of $\frac{x^2 - 4x + 1}{x - 1}$ and $\frac{x^2 - 4x + 1}{(x - 1)^2}$ from Example 2. You should check that the results we obtained above agree with a geometric analysis of these graphs near $x = 1$.

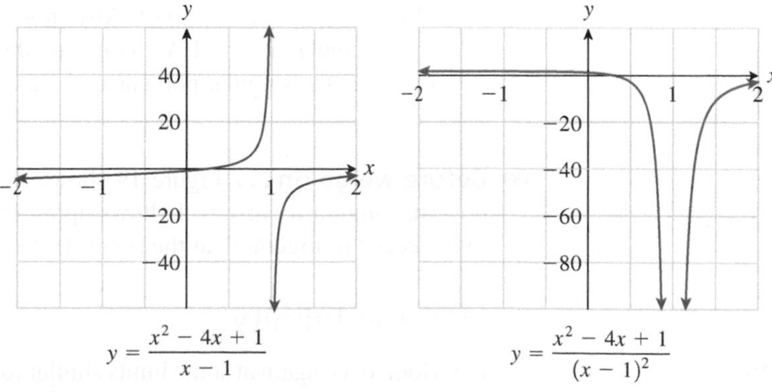

Figure 18

■

Functions Not in Closed Form

We can also use algebraic techniques to analyze functions that are not given in closed form.

EXAMPLE 3 Functions Not Written in Closed Form

For which values of x are the following piecewise-defined functions continuous?

a. $f(x) = \begin{cases} x^2 + 2 & \text{if } x < 1 \\ 2x - 1 & \text{if } x \geq 1 \end{cases}$ **b.** $g(x) = \begin{cases} x^2 - x + 1 & \text{if } x \leq 0 \\ 1 - x & \text{if } 0 < x < 1 \\ x - 3 & \text{if } x > 1 \end{cases}$

Solution

a. The function $f(x)$ is given in closed form over the intervals $(-\infty, 1)$ and $[1, +\infty)$. At $x = 1$, $f(x)$ suddenly switches from one closed-form formula to another, so $x = 1$ is the only place where there is a potential problem with continuity. To investigate the continuity of $f(x)$ at $x = 1$, let's calculate the limit there:

$$\begin{aligned} \lim_{x\to1^-} f(x) &= \lim_{x\to1^-} (x^2 + 2) && f(x) = x^2 + 2 \text{ for } x < 1. \\ &= (1)^2 + 2 = 3 && x^2 + 2 \text{ is closed-form.} \\ \lim_{x\to1^+} f(x) &= \lim_{x\to1^+} (2x - 1) && f(x) = 2x - 1 \text{ for } x > 1. \\ &= 2(1) - 1 = 1. && 2x - 1 \text{ is closed-form.} \end{aligned}$$

Because the left and right limits are different, $\lim_{x\to1} f(x)$ does not exist, so $f(x)$ is discontinuous at $x = 1$.

b. The only potential points of discontinuity for $g(x)$ occur at $x = 0$ and $x = 1$:

$$\begin{aligned} \lim_{x\to0^-} g(x) &= \lim_{x\to0^-} (x^2 - x + 1) = 1 \\ \lim_{x\to0^+} g(x) &= \lim_{x\to0^+} (1 - x) = 1. \end{aligned}$$

Thus, $\lim_{x\to0} g(x) = 1$. Further, $g(0) = 0^2 - 0 + 1 = 1$ from the formula, so

$$\lim_{x\to0} g(x) = g(0),$$

which shows that $g(x)$ is continuous at $x = 0$. At $x = 1$ we have

$$\begin{aligned} \lim_{x\to1^-} g(x) &= \lim_{x\to1^-} (1 - x) = 0 \\ \lim_{x\to1^+} g(x) &= \lim_{x\to1^+} (x - 3) = -2, \end{aligned}$$

so $\lim_{x\to1} g(x)$ does not exist. Also, notice that $x = 1$ is not in the domain of g, so $g(x)$ is singular at $x = 1$. We conclude that $g(x)$ is continuous at every real number x except at the singular point at $x = 1$ and therefore is continuous on its domain.

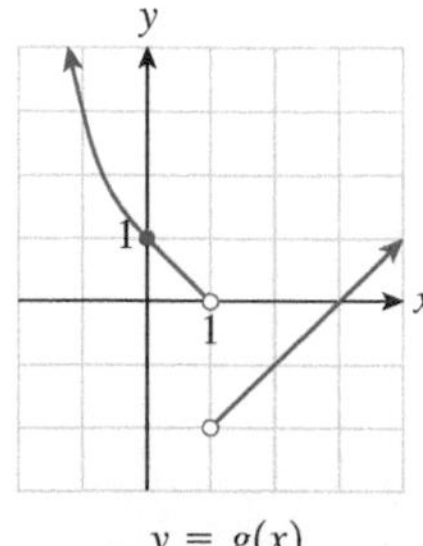

Figure 19

➡ **Before we go on . . .** Figure 19 shows the graph of g from Example 3(b). Notice how the singularity at $x = 1$ shows up as a break in the graph, whereas at $x = 0$ the two pieces "fit together" at the point $(0, 1)$. ■

Limits at Infinity

Let's look once again at some limits similar to those in Examples 3 and 6 of Section 3.1.

EXAMPLE 4 **Limits at Infinity**

Compute the following limits if they exist:

a. $\lim_{x\to+\infty} \frac{2x^2 - 4x}{x^2 - 1}$ **b.** $\lim_{x\to-\infty} \frac{2x^2 - 4x}{x^2 - 1}$

c. $\lim_{x\to+\infty} \frac{-x^3 - 4x}{2x^2 - 1}$ **d.** $\lim_{x\to+\infty} \frac{2x^2 - 4x}{5x^3 - 3x + 5}$

e. $\lim_{t\to+\infty} (e^{0.1t} - 20)$ **f.** $\lim_{x\to+\infty} \frac{80}{1 + 2.2(3.68)^{-t}}$

Solution **a.** and **b.** While calculating the values for the tables used in Example 3 in Section 3.1, you might have noticed that the highest power of x in both the numerator and denominator dominated the calculations. For instance, when $x = 100{,}000$, the term $2x^2$ in the numerator has the value of 20,000,000,000, whereas the term $4x$ has the comparatively insignificant value of 400,000. Similarly, the term x^2 in the denominator overwhelms the term -1. In other words, for large values of x (or negative values with large magnitude),

$$\frac{2x^2 - 4x}{x^2 - 1} \approx \frac{2x^2}{x^2} = 2.$$

Use only the highest powers top and bottom.

Therefore,

$$\lim_{x\to\pm\infty} \frac{2x^2 - 4x}{x^2 - 1} = \lim_{x\to\pm\infty} \frac{2x^2}{x^2} = \lim_{x\to\pm\infty} 2 = 2.$$

The procedure of using only the highest powers of x to compute the limit is stated formally and justified after this example.

c. Applying the previous technique of looking only at highest powers gives

$$\lim_{x\to+\infty} \frac{-x^3 - 4x}{2x^2 - 1} = \lim_{x\to+\infty} \frac{-x^3}{2x^2}$$

Use only the highest powers top and bottom.

$$= \lim_{x\to+\infty} \frac{-x}{2}.$$

Simplify.

As x gets large, $-x/2$ gets large in magnitude but negative, so the limit is

$$\lim_{x\to+\infty} \frac{-x}{2} = -\infty.$$

$\frac{-\infty}{2} = -\infty$ (See below.)

d. $$\lim_{x\to+\infty} \frac{2x^2 - 4x}{5x^3 - 3x + 5} = \lim_{x\to+\infty} \frac{2x^2}{5x^3}$$

Use only the highest powers top and bottom.

$$= \lim_{x\to+\infty} \frac{2}{5x}.$$

As x gets large, $2/(5x)$ gets close to zero, so the limit is

$$\lim_{x\to+\infty} \frac{2}{5x} = 0.$$

$\frac{2}{\infty} = 0$ (See below.)

e. Here, we do not have a ratio of polynomials. However, we know that, as t becomes large and positive, so does $e^{0.1t}$ and hence also $e^{0.1t} - 20$. Thus,

$$\lim_{t\to+\infty}(e^{0.1t} - 20) = +\infty. \qquad e^{+\infty} = +\infty \text{ (See below.)}$$

f. As $t \to +\infty$, the term $(3.68)^{-t} = \dfrac{1}{3.68^t}$ in the denominator, being 1 divided by a very large number, approaches zero. Hence, the denominator $1 + 2.2(3.68)^{-t}$ approaches $1 + 2.2(0) = 1$ as $t \to +\infty$. Thus,

$$\lim_{t\to+\infty}\frac{80}{1 + 2.2(3.68)^{-t}} = \frac{80}{1 + 2.2(0)} = 80. \qquad (3.68)^{-\infty} = 0 \text{ (See below.)}$$

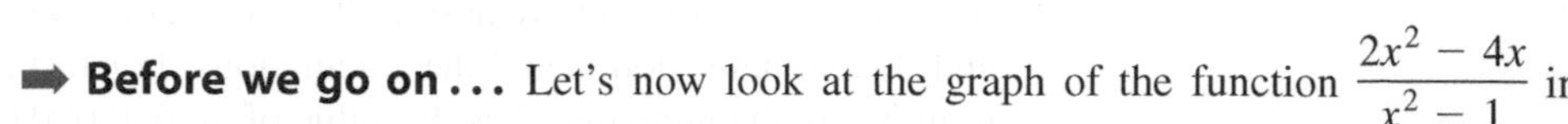

Before we go on... Let's now look at the graph of the function $\dfrac{2x^2 - 4x}{x^2 - 1}$ in Example 4(a) and 4(b). We say that the graph of f has a **horizontal asymptote** at $y = 2$ because of the limits we have just calculated. This means that the graph approaches the horizontal line $y = 2$ far to the right or left (in this case, to both the right and left). Figure 20 shows the graph of f together with the line $y = 2$. The graph reveals some additional interesting information: as $x \to 1^+$, $f(x) \to -\infty$, and as $x \to 1^-$, $f(x) \to +\infty$. Thus,

$$\lim_{x\to 1} f(x) \text{ does not exist.}$$

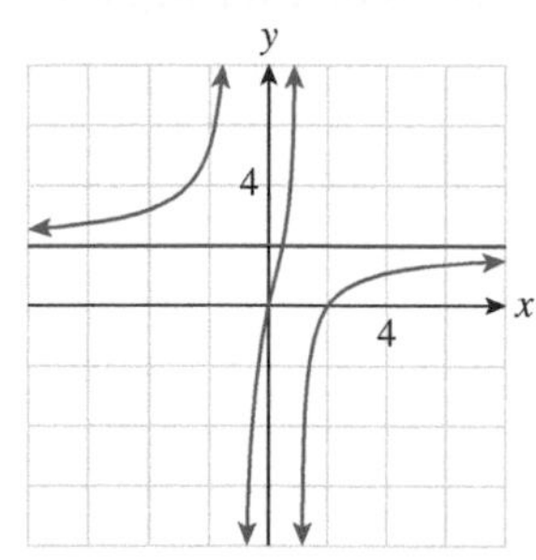

Figure 20

See whether you can determine what happens as $x \to -1$.

If you graph the functions in Example 4(d) and 4(f), you will again see a horizontal asymptote. Do the limits in Example 4(c) and 4(e) show horizontal asymptotes? ■

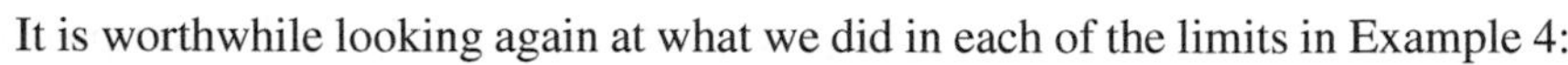

It is worthwhile looking again at what we did in each of the limits in Example 4:

a. and **b.** We saw that $\dfrac{2x^2 - 4x}{x^2 - 1} \approx \dfrac{2x^2}{x^2}$, and then we canceled the x^2. Notice that, before we cancel, letting x approach $\pm\infty$ in the numerator and denominator yields the ratio ∞/∞, which, like $0/0$, is another *indeterminate form* and indicates to us that further work is needed—in this case cancellation—before we can write down the limit.

c. We obtained $\dfrac{-x^3 - 4x}{2x^2 - 1} \approx \dfrac{-x^3}{2x^2}$, which results in another indeterminate form, $-\infty/\infty$, as $x \to +\infty$. Cancellation of the x^2 gave us $\dfrac{-x}{2}$, resulting in the *determinate* form $-\infty/2 = -\infty$. (A very large number divided by 2 is again a very large number.)

d. Here, $\dfrac{2x^2 - 4x}{5x^3 - 3x + 5} \approx \dfrac{2x^2}{5x^3} = \dfrac{2}{5x}$, and the cancellation step turns the indeterminate form ∞/∞ into the determinate form $2/\infty = 0$. (Dividing 2 by a very large number yields a very small number.)

e. We reasoned that e raised to a large positive number is large and positive. Putting $t = +\infty$ gives us the determinate form $e^{+\infty} = +\infty$.

f. Here, we reasoned that 3.68 raised to a large *negative* number is close to zero. Putting $t = -\infty$ gives us the determinate form $3.68^{-\infty} = 1/3.68^{+\infty} = 1/\infty = 0$ (see part(d)).

In Example 4(a)–(d), $f(x)$ was a **rational function**: a quotient of polynomial functions. We calculated the limit of $f(x)$ at $\pm\infty$ by ignoring all powers of x in both the numerator and denominator except for the largest. Following is a theorem that justifies this procedure.

Theorem 3.2 Evaluating the Limit of a Rational Function at $\pm\infty$

If $f(x)$ has the form

$$f(x) = \frac{c_n x^n + c_{n-1}x^{n-1} + \cdots + c_1 x + c_0}{d_m x^m + d_{m-1}x^{m-1} + \cdots + d_1 x + d_0}$$

with the c_i and d_i constants ($c_n \neq 0$ and $d_m \neq 0$), then we can calculate the limit of $f(x)$ as $x \to \pm\infty$ by ignoring all powers of x except the highest in both the numerator and denominator. Thus,

$$\lim_{x\to\pm\infty} f(x) = \lim_{x\to\pm\infty} \frac{c_n x^n}{d_m x^m}.$$

Quick Examples

(See Example 4.)

5. $\lim\limits_{x\to+\infty} \dfrac{2x^2 - 4x}{x^2 - 1} = \lim\limits_{x\to+\infty} \dfrac{2x^2}{x^2} = \lim\limits_{x\to+\infty} 2 = 2$

6. $\lim\limits_{x\to+\infty} \dfrac{-x^3 - 4x}{2x^2 - 1} = \lim\limits_{x\to+\infty} \dfrac{-x^3}{2x^2} = \lim\limits_{x\to+\infty} \dfrac{-x}{2} = -\infty$

7. $\lim\limits_{x\to+\infty} \dfrac{2x^2 - 4x}{5x^3 - 3x + 5} = \lim\limits_{x\to+\infty} \dfrac{2x^2}{5x^3} = \lim\limits_{x\to+\infty} \dfrac{2}{5x} = 0$

Proof Our function $f(x)$ is a polynomial of degree n divided by a polynomial of degree m. If n happens to be larger than m, then dividing the top and bottom by the largest power x^n of x gives

$$\begin{aligned} f(x) &= \frac{c_n x^n + c_{n-1}x^{n-1} + \cdots + c_1 x + c_0}{d_m x^m + d_{m-1}x^{m-1} + \cdots + d_1 x + d_0} \\ &= \frac{c_n x^n/x^n + c_{n-1}x^{n-1}/x^n + \cdots + c_1 x/x^n + c_0/x^n}{d_m x^m/x^n + d_{m-1}x^{m-1}/x^n + \cdots + d_1 x/x^n + d_0/x^n}. \end{aligned}$$

Canceling powers of x in each term and remembering that $n > m$ leave us with

$$f(x) = \frac{c_n + c_{n-1}/x + \cdots + c_1/x^{n-1} + c_0/x^n}{d_m/x^{n-m} + d_{m-1}/x^{n-m+1} + \cdots + d_1/x^{n-1} + d_0/x^n}.$$

As $x \to \pm\infty$, all the terms shown in blue approach 0, so we can ignore them in taking the limit. (The first term in the denominator happens to approach 0 as well, but we retain it for convenience.) Thus,

$$\lim_{x\to\pm\infty} f(x) = \lim_{x\to\pm\infty} \frac{c_n}{d_m/x^{n-m}} = \lim_{x\to\pm\infty} \frac{c_n x^n}{d_m x^m},$$

as required. The cases in which n is smaller than m and $m = n$ are proved similarly by dividing top and bottom by the largest power of x in each case. ■

Note The procedure of ignoring all but highest powers also works for arbitrary algebraic expressions involving polynomials, such as square roots of polynomials. For instance,

$$\lim_{x\to+\infty}\frac{\sqrt{25x^6-3x^2+1}}{2x^3-1}=\lim_{x\to+\infty}\frac{\sqrt{25x^6}}{2x^3}=\lim_{x\to+\infty}\frac{5x^3}{2x^3}=\lim_{x\to+\infty}\frac{5}{2}=\frac{5}{2}. \blacksquare$$

Some Determinate and Indeterminate Forms

The following summary brings these ideas together with our observations in Example 2.

Some Determinate and Indeterminate Forms

$0/0$ and $\pm\infty/\infty$ are **indeterminate**; evaluating limits in which these arise requires simplification or further analysis.* The following are **determinate** forms for any nonzero number k:

* Some other indeterminate forms are $\pm\infty\cdot 0$, $\infty-\infty$, and 1^{∞}. (These are not discussed in this text, but see the Communication and Reasoning exercises for this section.)

$\dfrac{k}{0^{\pm}}=\pm\infty$	$\dfrac{k}{\text{Small}}=\text{Big*}$ (See Example 2.)
$k(\pm\infty)=\pm\infty$	$k\times\text{Big}=\text{Big*}$
$k\pm\infty=\pm\infty$	$k\pm\text{Big}=\pm\text{Big}$
$\pm\dfrac{\infty}{k}=\pm\infty$	$\dfrac{\text{Big}}{k}=\text{Big*}$
$\pm\dfrac{k}{\infty}=0,$	$\dfrac{k}{\text{Big}}=\text{Small}$

and if $k>1$, then

$k^{+\infty}=+\infty$	$k^{\text{Big positive}}=\text{Big}$
$k^{-\infty}=0.$	$k^{\text{Big negative}}=\text{Small}$

*The sign gets switched in these forms if k is negative.

Quick Examples

8. $\displaystyle\lim_{x\to 0}\frac{60}{2x^2}=+\infty$	$\dfrac{k}{0^+}=+\infty$
9. $\displaystyle\lim_{x\to -1^-}\frac{2x-6}{x+1}=+\infty$	$\dfrac{-8}{0^-}=+\infty$
10. $\displaystyle\lim_{x\to-\infty}3x-5=-\infty$	$3(-\infty)-5=-\infty-5=-\infty$
11. $\displaystyle\lim_{x\to+\infty}\frac{2x}{60}=+\infty$	$\dfrac{2(\infty)}{60}=\infty$
12. $\displaystyle\lim_{x\to-\infty}\frac{60}{2x}=0$	$\dfrac{60}{2(-\infty)}=0$
13. $\displaystyle\lim_{x\to+\infty}\frac{60x}{2x}=30$	$\dfrac{\infty}{\infty}$ is indeterminate, but we can cancel.
14. $\displaystyle\lim_{x\to-\infty}\frac{60}{e^x-1}=\frac{60}{0-1}=-60$	$e^{-\infty}=0$

FAQs

Strategy for Evaluating Limits Algebraically

Q: *Is there a systematic way to evaluate a limit $\lim_{x\to a} f(x)$ algebraically?*

A: The following approach is often successful:

Case 1: a is a finite number (not $\pm\infty$)

1. Decide whether f is a closed-form function. If it is not, then find the left and right limits at the values of x where the function changes from one formula to another.
2. If f is a closed-form function, try substituting $x = a$ in the formula for $f(x)$. Then one of the following three things may happen:

   $f(a)$ is defined. Then $\lim_{x\to a} f(x) = f(a)$.

   $f(a)$ is not defined and has the indeterminate form $0/0$. Try to simplify the expression for f to cancel one of the terms that gives 0.

   $f(a)$ is not defined and has one of the determinate forms listed above in the above table. Use the table to determine the limit as in Quick Examples 8–14.

Case 2: $a = \pm\infty$

Remember that we can use the determinate forms $k^{+\infty} = \infty$ and $k^{-\infty} = 0$ if $k > 1$. Further, if the given function is a polynomial or ratio of polynomials, use the technique of Example 4: Focus only on the highest powers of x, and then simplify to obtain either a number L, in which case the limit exists and equals L, or one of the determinate forms $\pm\infty/k = \pm\infty$ or $\pm k/\infty = 0$.

There is another technique for evaluating certain difficult limits, called *l'Hospital's rule,* but this uses derivatives, so we'll have to wait until Section 4.1 to discuss it.

3.3 EXERCISES

▼ more advanced ◆ challenging

T indicates exercises that should be solved using technology

In Exercises 1–4, complete the given sentence.

1. The closed-form function $f(x) = \dfrac{1}{x-1}$ is continuous for all x except ______. [**HINT**: See Quick Example 3.]
2. The closed-form function $f(x) = \dfrac{1}{x^2-1}$ is continuous for all x except ______. [**HINT**: See Quick Example 3.]
3. The closed-form function $f(x) = \sqrt{x+1}$ has $x = 3$ in its domain. Therefore, $\lim_{x\to 3}\sqrt{x+1} =$ ___. [**HINT**: See Example 1.]
4. The closed-form function $f(x) = \sqrt{x-1}$ has $x = 10$ in its domain. Therefore, $\lim_{x\to 10}\sqrt{x-1} =$ ___. [**HINT**: See Example 1.]

In Exercises 5–20, determine whether the given limit leads to a determinate or indeterminate form. Evaluate the limit if it exists, or say why if not. [**HINT**: See Example 2 and Quick Examples 8–14.]

5. $\lim\limits_{x\to 0}\dfrac{60}{x^4}$
6. $\lim\limits_{x\to 0}\dfrac{2x^2}{x^2}$
7. $\lim\limits_{x\to 0}\dfrac{x^3-1}{x^3}$
8. $\lim\limits_{x\to 0}\dfrac{-2}{x^2}$
9. $\lim\limits_{x\to -\infty}(-x^2+5)$
10. $\lim\limits_{x\to 0}\dfrac{2x^2+4}{x}$
11. $\lim\limits_{x\to +\infty}4^{-x}$
12. $\lim\limits_{x\to +\infty}\dfrac{60+e^{-x}}{2-e^{-x}}$
13. $\lim\limits_{x\to 0}\dfrac{-x^3}{3x^3}$
14. $\lim\limits_{x\to -\infty}3x^2+6$
15. $\lim\limits_{x\to -\infty}\dfrac{-x^3}{3x^6}$
16. $\lim\limits_{x\to +\infty}\dfrac{-x^6}{3x^3}$

17. $\lim_{x\to-\infty} \frac{4}{-x+2}$

18. $\lim_{x\to-\infty} e^x$

19. $\lim_{x\to-\infty} \frac{60}{e^x-1}$

20. $\lim_{x\to-\infty} \frac{2}{2x^2+3}$

In Exercises 21–74, calculate the limit algebraically. If the limit does not exist, say why.

21. $\lim_{x\to0}(x+1)$ [**HINT**: See Example 1(a).]

22. $\lim_{x\to0}(2x-4)$ [**HINT**: See Example 1(a).]

23. $\lim_{x\to2} \frac{2+x}{x}$

24. $\lim_{x\to-1} \frac{4x^2+1}{x}$

25. $\lim_{x\to-1} \frac{x+1}{x}$

26. $\lim_{x\to4}(x+\sqrt{x})$

27. $\lim_{x\to8}(x-\sqrt[3]{x})$

28. $\lim_{x\to1} \frac{x-2}{x+1}$

29. $\lim_{h\to1}(h^2+2h+1)$

30. $\lim_{h\to0}(h^3-4)$

31. $\lim_{h\to3} 2$

32. $\lim_{h\to0} -5$

33. $\lim_{h\to0} \frac{h^2}{h+h^2}$ [**HINT**: See Example 1(b).]

34. $\lim_{h\to0} \frac{h^2+h}{h^2+2h}$ [**HINT**: See Example 1(b).]

35. $\lim_{x\to1} \frac{x^2-2x+1}{x^2-x}$

36. $\lim_{x\to-1} \frac{x^2+3x+2}{x^2+x}$

37. $\lim_{x\to2} \frac{x^3-8}{x-2}$

38. $\lim_{x\to-2} \frac{x^3+8}{x^2+3x+2}$

39. $\lim_{x\to0^+} \frac{1}{x^2}$ [**HINT**: See Example 2.]

40. $\lim_{x\to0^+} \frac{1}{x^2-x}$ [**HINT**: See Example 2.]

41. $\lim_{x\to-1} \frac{x^2+1}{x+1}$

42. $\lim_{x\to-1^-} \frac{x^2+1}{x+1}$

43. $\lim_{x\to-2^+} \frac{x^2+8}{x^2+3x+2}$

44. $\lim_{x\to-1} \frac{x^2+3x}{x^2+x}$

45. $\lim_{x\to-2} \frac{x^2+8}{x^2+3x+2}$

46. $\lim_{x\to-1} \frac{x^2+3x}{x^2+2x+1}$

47. $\lim_{x\to2} \frac{x^2+8}{x^2-4x+4}$

48. $\lim_{x\to-1} \frac{x^2+3x}{x^2+3x+2}$

49. ▼ $\lim_{x\to2^+} \frac{x-2}{\sqrt{x-2}}$

50. ▼ $\lim_{x\to3^-} \frac{\sqrt{3-x}}{3-x}$

51. ▼ $\lim_{x\to9} \frac{\sqrt{x}-3}{x-9}$

52. ▼ $\lim_{x\to4} \frac{x-4}{\sqrt{x}-2}$

53. $\lim_{x\to+\infty} \frac{3x^2+10x-1}{2x^2-5x}$ [**HINT**: See Example 4.]

54. $\lim_{x\to+\infty} \frac{6x^2+5x+100}{3x^2-9}$ [**HINT**: See Example 4.]

55. $\lim_{x\to+\infty} \frac{x^5-1{,}000x^4}{2x^5+10{,}000}$

56. $\lim_{x\to+\infty} \frac{x^6+3{,}000x^3+1{,}000{,}000}{2x^6+1{,}000x^3}$

57. $\lim_{x\to+\infty} \frac{10x^2+300x+1}{5x+2}$

58. $\lim_{x\to+\infty} \frac{2x^4+20x^3}{1{,}000x^3+6}$

59. $\lim_{x\to-\infty} \frac{3x^2+10x-1}{2x^2-5x}$

60. $\lim_{x\to-\infty} \frac{6x^2+5x+100}{3x^2-9}$

61. $\lim_{x\to-\infty} \frac{x^5-1{,}000x^4}{2x^5+10{,}000}$

62. $\lim_{x\to-\infty} \frac{x^6+3{,}000x^3+1{,}000{,}000}{2x^6+1{,}000x^3}$

63. $\lim_{x\to-\infty} \frac{10x^2+300x+1}{5x+2}$

64. $\lim_{x\to-\infty} \frac{2x^4+20x^3}{1{,}000x^3+6}$

65. $\lim_{x\to-\infty} \frac{10x^2+300x+1}{5x^3+2}$

66. $\lim_{x\to-\infty} \frac{2x^4+20x^3}{1{,}000x^6+6}$

67. $\lim_{x\to+\infty}(4e^{-3x}+12)$

68. $\lim_{x\to+\infty} \frac{2}{5-5.3e^{-3x}}$

69. $\lim_{x\to+\infty} \frac{2}{5-5.3(3^{3x})}$

70. $\lim_{t\to+\infty}(4.1-2e^{3t})$

71. $\lim_{t\to+\infty} \frac{2^{3t}}{1+5.3e^{-t}}$

72. $\lim_{x\to-\infty} \frac{4.2}{2-3^{2x}}$

73. $\lim_{x\to-\infty} \frac{-3^{2x}}{2+e^x}$

74. $\lim_{x\to+\infty} \frac{2^{-3x}}{1+5.3e^{-x}}$

In Exercises 75–88, identify all singular points and points of discontinuity of the given function. [**HINT**: See Example 3.]

75. $f(x)=\frac{1}{x-3}$

76. $f(x)=\begin{cases}\frac{4}{x+1} & \text{if } x\neq-1\\ 3 & \text{if } x=-1\end{cases}$

77. $f(x)=\begin{cases}\frac{4}{(x-5)^2} & \text{if } x\neq5\\ -3 & \text{if } x=5\end{cases}$

78. $f(x)=\frac{2}{x^2-9}$

79. $f(x)=\begin{cases}x+2 & \text{if } x<0\\ 2x-1 & \text{if } x\geq0\end{cases}$

80. $f(x)=\begin{cases}2x-1 & \text{if } x<1\\ -2x+3 & \text{if } x>1\end{cases}$

81. $f(x)=\begin{cases}x^2-1 & \text{if } x<0\\ x^2+1 & \text{if } x>0\end{cases}$

82. $g(x)=\begin{cases}1-x & \text{if } x\leq1\\ x-1 & \text{if } x>1\end{cases}$

83. $g(x) = \begin{cases} x + 2 & \text{if } x < 0 \\ 2x + 2 & \text{if } 0 \le x < 2 \\ x^2 + 2 & \text{if } x \ge 2 \end{cases}$

84. $f(x) = \begin{cases} 1 - x & \text{if } x \le 1 \\ x + 2 & \text{if } 1 < x < 3 \\ x^2 - 4 & \text{if } x \ge 3 \end{cases}$

85. ▼ $h(x) = \begin{cases} x + 2 & \text{if } x < 0 \\ 0 & \text{if } x = 0 \\ 2x + 2 & \text{if } x > 0 \end{cases}$

86. ▼ $h(x) = \begin{cases} 1 - x & \text{if } x < 1 \\ 1 & \text{if } x = 1 \\ x + 2 & \text{if } x > 1 \end{cases}$

87. ▼ $f(x) = \begin{cases} 1/x & \text{if } x < 0 \\ x & \text{if } 0 \le x \le 2 \\ 2^{x-1} & \text{if } x > 2 \end{cases}$

88. ▼ $f(x) = \begin{cases} x^3 + 2 & \text{if } x \le -1 \\ x^2 & \text{if } -1 < x < 0 \\ x & \text{if } x \ge 0 \end{cases}$

Applications

89. ***Processor Speeds*** The processor speeds, in megahertz (MHz), of **Intel** processors during the period 1996–2010 can be approximated by the following function of time t in years since the start of 1990:[17]

$$v(t) = \begin{cases} 400t - 2{,}200 & \text{if } 6 \le t < 15 \\ 3{,}800 & \text{if } 15 \le t \le 20. \end{cases}$$

a. Compute $\lim_{t \to 15^-} v(t)$ and $\lim_{t \to 15^+} v(t)$, and interpret each answer. [**HINT:** See Example 3.]

b. Is the function v continuous at $t = 15$? According to the model, was there any abrupt change in processor speeds during the period 1996–2010?

90. ***Processor Speeds*** The processor speeds, in megahertz (MHz), of **Intel** processors during the period 1970–2000 can be approximated by the following function of time t in years since the start of 1970:[18]

$$v(t) = \begin{cases} 3t & \text{if } 0 \le t < 20 \\ 174t - 3{,}420 & \text{if } 20 \le t \le 30. \end{cases}$$

a. Compute $\lim_{t \to 20^-} v(t)$ and $\lim_{t \to 20^+} v(t)$, and interpret each answer.

b. Is the function v continuous at $t = 20$? According to the model, was there any abrupt change in processor speeds during the period 1970–2000?

[17] A rough model based on the fastest processors produced by Intel. Source for data: www.intel.com.

[18] *Ibid.*

91. ***Movie Advertising*** Movie expenditures, in billions of dollars, on advertising in newspapers from 1995 to 2004 can be approximated by

$$f(t) = \begin{cases} 0.04t + 0.33 & \text{if } t \le 4 \\ -0.01t + 1.2 & \text{if } t > 4, \end{cases}$$

where t is time in years since 1995.[19]

a. Compute $\lim_{t \to 4^-} f(t)$ and $\lim_{t \to 4^+} f(t)$, and interpret each answer. [**HINT:** See Example 3.]

b. Is the function f continuous at $t = 4$? What does the answer tell you about movie advertising expenditures?

92. ***Movie Advertising*** The percentage of movie advertising as a share of newspapers' total advertising revenue from 1995 to 2004 can be approximated by

$$p(t) = \begin{cases} -0.07t + 6.0 & \text{if } t \le 4 \\ 0.3t + 17.0 & \text{if } t > 4, \end{cases}$$

where t is time in years since 1995.[20]

a. Compute $\lim_{t \to 4^-} p(t)$ and $\lim_{t \to 4^+} p(t)$, and interpret each answer. [**HINT:** See Example 3.]

b. Is the function p continuous at $t = 4$? What does the answer tell you about newspaper revenues?

93. ***Law Enforcement in the 1980s and 1990s*** The cost of fighting crime in the United States increased significantly during the period 1982–1999. Total spending on police and courts can be approximated by[21]

$$P(t) = 1.745t + 29.84 \text{ billion dollars} \quad (2 \le t \le 19)$$
$$C(t) = 1.097t + 10.65 \text{ billion dollars} \quad (2 \le t \le 19),$$

respectively, where t is time in years since 1980. Compute $\lim_{t \to +\infty} \frac{P(t)}{C(t)}$ to two decimal places, and interpret the result. [**HINT:** See Example 4.]

94. ***Law Enforcement in the 1980s and 1990s*** Refer to Exercise 93. Total spending on police, courts, and prisons in the period 1982–1999 could be approximated by[22]

$$P(t) = 1.745t + 29.84 \text{ billion dollars} \quad (2 \le t \le 19)$$
$$C(t) = 1.097t + 10.65 \text{ billion dollars} \quad (2 \le t \le 19)$$
$$J(t) = 1.919t + 12.36 \text{ billion dollars} \quad (2 \le t \le 19),$$

respectively, where t is time in years since 1980. Compute $\lim_{t \to +\infty} \frac{P(t)}{P(t) + C(t) + J(t)}$ to two decimal places, and interpret the result. [**HINT:** See Example 4.]

[19] Model by the authors. Source for data: Newspaper Association of America Business Analysis and Research/*New York Times*, May 16, 2005.

[20] *Ibid.*

[21] Spending is adjusted for inflation and shown in 1999 dollars. Models are based on a linear regression. Source for data: Bureau of Justice Statistics/*New York Times*, February 11, 2002, p. A14.

[22] *Ibid.*

95. ***SAT Scores by Income*** The following bar graph shows U.S. math SAT scores as a function of household income:[23]

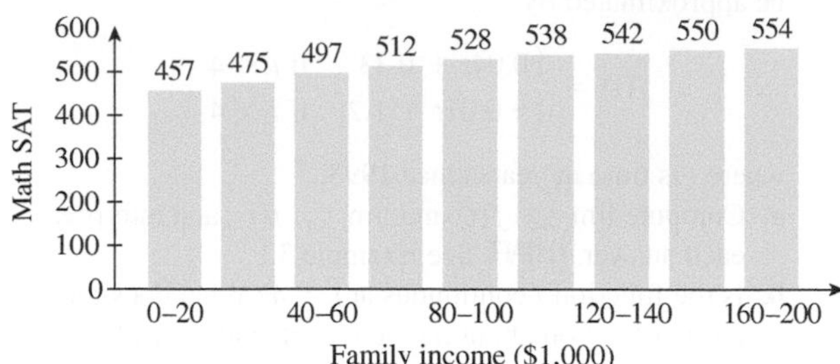

These data can be modeled by

$$S(x) = 573 - 33e^{-0.0131x},$$

where $S(x)$ is the average math SAT score of students whose household income is x thousand dollars per year. Calculate $\lim_{x \to +\infty} S(x)$, and interpret the answer.

96. ***SAT Scores by Income*** The following bar graph shows U.S. critical reading SAT scores as a function of household income:[24]

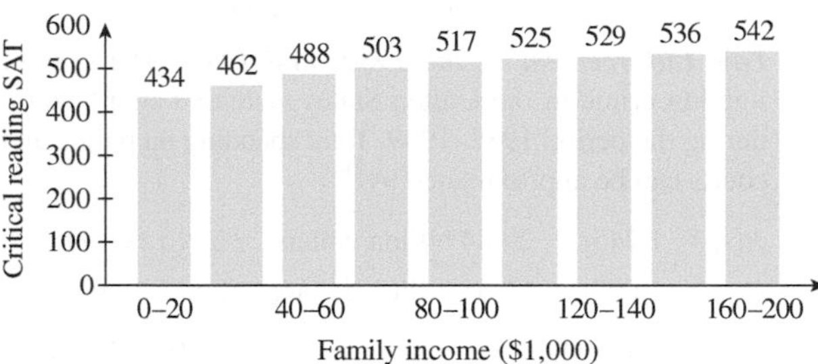

These data can be modeled by

$$S(x) = 550 - 136e^{-0.0151x},$$

where $S(x)$ is the average critical reading SAT score of students whose household income is x thousand dollars per year. Calculate $\lim_{x \to +\infty} S(x)$, and interpret the answer.

97. ***Social Website Popularity*** The following models approximate the popularity of **Twitter** and **LinkedIn** among social media sites from 2008 to 2013, as rated by **StatCounter.com**:

Twitter: $W(t) = 0.33t^2 - 2t + 8.7$ percentage points

LinkedIn: $L(t) = 0.04t^2 - 0.26t + 0.67$ percentage points.

(t is the number of years since the start of 2008.)[25] Calculate $\lim_{t \to +\infty} W(t)$ and $\lim_{t \to +\infty} \frac{W(t)}{L(t)}$ algebraically, interpret your answers, and comment on the results.

98. ***Social Website Popularity*** The following models approximate the popularity of **Facebook** and **YouTube** among social media sites from 2008 to 2013, as rated by **StatCounter.com**:

Facebook: $F(t) = -2t^2 + 16t + 35$ percentage points

YouTube: $Y(t) = -t^2 + 6.5t - 1.8$ percentage points.

(t is the number of years since the start of 2008.)[26] Calculate $\lim_{t \to +\infty} F(t)$ and $\lim_{t \to +\infty} \frac{F(t)}{Y(t)}$ algebraically, interpret your answers, and comment on the results.

99. ***Acquisition of Language*** The percentage $p(t)$ of children who can speak in at least single words by the age of t months can be approximated by the equation[27]

$$p(t) = 100\left(1 - \frac{12{,}200}{t^{4.48}}\right) \quad (t \geq 8.5).$$

Calculate $\lim_{t \to +\infty} p(t)$, and interpret the result. [**HINT**: See Example 4.]

100. ***Acquisition of Language*** The percentage $q(t)$ of children who can speak in sentences of five or more words by the age of t months can be approximated by the equation[28]

$$q(t) = 100\left(1 - \frac{5.27 \times 10^{17}}{t^{12}}\right) \quad (t \geq 30).$$

If p is the function referred to in the preceding exercise, calculate $\lim_{t \to +\infty}[p(t) - q(t)]$, and interpret the result. [**HINT**: See Example 4.]

Communication and Reasoning Exercises

101. Describe the algebraic method of evaluating limits as discussed in this section, and give at least one disadvantage of this method.

102. What is a closed-form function? What can we say about such functions?

103. Your friends Rita and Richard are arguing. Rita claims that closed-form functions cannot have points of discontinuity, but Richard retorts, "Ever heard of $f(x) = 1/x$?" On whose side (if any) of the argument should you be? Explain.

104. Your other friends, Andrew and Dorothy, are also arguing. Andrew claims that a function cannot be singular at a point of discontinuity, but Dorothy retorts, "Ever heard of $f(x) = 1/x$?" On whose side (if any) of the argument should you be? Explain.

---

[23] 2009 data. Source: College Board/*New York Times* http://economix.blogs.nytimes.com.

[24] *Ibid.*

[25] Percentages are based on worldwide page views. Source for data: http://gs.statcounter.com.

[26] *Ibid.*

[27] The model is the authors' and is based on data presented in the article *The Emergence of Intelligence* by William H. Calvin, *Scientific American*, October 1994, pp. 101–107.

[28] *Ibid.*

105. Why was the following marked wrong? What is the correct answer?

$$\lim_{x\to 3} \frac{x^3 - 27}{x - 3} = \frac{0}{0} \text{ undefined} \quad \text{✗ } \textit{WRONG!}$$

106. Why was the following marked wrong? What is the correct answer?

$$\lim_{x\to 1^-} \frac{x - 1}{x^2 - 2x + 1} = \frac{0}{0} = 0 \quad \text{✗ } \textit{WRONG!}$$

107. ▼ Your friend Karin tells you that $f(x) = 1/(x - 2)^2$ cannot be a closed-form function because it is not continuous at $x = 2$. Comment on her assertion.

108. ▼ Give an example of a function f specified by means of algebraic formulas such that the domain of f consists of all real numbers and f is not continuous at $x = 2$. Is f a closed-form function?

109. Give examples of two limits that lead to two different indeterminate forms but where both limits exist.

110. Give examples of two limits: one that leads to a determinate form and another that leads to an indeterminate form but where neither limit exists.

111. ▼ (Compare Exercise 73 in Section 3.1.) Which indeterminate form results from $\lim_{x\to +\infty} \frac{p(x)}{e^x}$ if $p(x)$ is a polynomial? Numerically or graphically estimate these limits for various polynomials $p(x)$. What does this suggest about limits that result in $\frac{p(\infty)}{e^\infty}$?

112. ▼ (Compare Exercise 74 in Section 3.1.) Which indeterminate form results from $\lim_{x\to -\infty} p(x)e^x$ if $p(x)$ is a polynomial? What does this suggest about the limits that result in $p(-\infty)e^{-\infty}$?

113. ▼ What is wrong with the following statement? If $f(x)$ is specified algebraically and $f(a)$ is defined, then $\lim_{x\to a} f(x)$ exists and equals $f(a)$. How can it be corrected?

114. ▼ What is wrong with the following statement? If $f(x)$ is specified algebraically and $f(a)$ is not defined, then $\lim_{x\to a} f(x)$ does not exist.

115. ▼ Give the formula for a function that is continuous everywhere except at two points.

116. ▼ Give the formula for a function that is continuous everywhere except at three points.

117. ◆ ***The Indeterminate Form*** $\infty - \infty$ An indeterminate form not mentioned in Section 3.3 is $\infty - \infty$. Give examples of three limits that lead to this indeterminate form and where the first limit exists and equals 5, the second limit diverges to $+\infty$, and the third limit exists and equals -5.

118. ◆ ***The Indeterminate Form*** 1^∞ An indeterminate form not mentioned in Section 3.3 is 1^∞. Give examples of three limits that lead to this indeterminate form and where the first limit exists and equals 1, the second limit exists and equals e, and the third limit diverges to $+\infty$. [**HINT:** For the third, consider modifying the second.]

3.4 Average Rate of Change

Change and Average Rate of Change

Calculus is the mathematics of change, inspired largely by observation of continuously changing quantities around us in the real world. As an example, the Consumer Price Index (CPI) C increased from 227 points in January 2012 to 234 points in January 2014.[29] As we saw in Chapter 1, the **change** in this index can be measured as the difference:

$$\Delta C = \text{Second value} - \text{First value} = 234 - 227 = 7 \text{ points.}$$

(The fact that the CPI increased is reflected in the positive sign of the change.) The kind of question we will concentrate on is *how fast* the CPI was changing. Because C increased by 7 points in 2 years, we say that it averaged a $7/2 = 3.5$ point rise each year. (It actually rose 3 points the first year and 4 points the second, giving an average rise of 3.5 points each year.)

Alternatively, we might want to measure this rate in points per month rather than points per year. Because C increased by 7 points in 24 months, it increased at an average rate of $7/24 \approx 0.292$ points per month.

[29] Figures are approximate. Source: Bureau of Labor Statistics, www.bls.gov.

In both cases we obtained the average rate of change by dividing the change by the corresponding length of time:

$$\text{Average rate of change} = \frac{\text{Change in } C}{\text{Change in time}} = \frac{7}{2} = 3.5 \text{ points per year}$$

$$\text{Average rate of change} = \frac{\text{Change in } C}{\text{Change in time}} = \frac{7}{24} \approx 0.292 \text{ points per month.}$$

EXAMPLE 1 Standard & Poor's 500

The following table lists the approximate value of Standard & Poor's 500 stock market index (S&P) during the period 2008–2014 ($t = 8$ represents 2008):[30]

Year t (year)	8	9	10	11	12	13	14
S&P Index $S(t)$ (points)	1,400	900	1,150	1,300	1,300	1,400	1,800

a. What was the average rate of change in the S&P over the 4-year period 2010–2014 (the period $10 \le t \le 14$ or $[10, 14]$ in interval notation), over the 2-year period 2008–2010 (the period $8 \le t \le 10$ or $[8, 10]$), and over the period $[8, 13]$?

b. Graph the values shown in the table. How are the rates of change reflected in the graph?

Solution

a. During the 4-year period $[10, 14]$ the S&P changed as follows:

Start of the period ($t = 10$):	$S(10) = 1{,}150$
End of the period ($t = 14$):	$S(14) = 1{,}800$
Change during the period $[10, 14]$:	$S(14) - S(10) = 650$

Thus, the S&P increased by 650 points in 4 years, giving an average rate of change of $650/4 = 162.5$ points per year. We can write the calculation this way:

$$\begin{aligned}\text{Average rate of change of } S &= \frac{\text{Change in } S}{\text{Change in } t} \\ &= \frac{\Delta S}{\Delta t} \\ &= \frac{S(14) - S(10)}{14 - 10} \\ &= \frac{1{,}800 - 1{,}150}{14 - 10} = \frac{650}{4} = 162.5 \text{ points per year.}\end{aligned}$$

Interpreting the result: During the period 2010–2014 (or $[10, 14]$ in interval notation) the S&P increased at an average rate of 162.5 points per year.

Similarly, the average rate of change during the period 2008–2010 (or $[8, 10]$) was

$$\begin{aligned}\text{Average rate of change of } S &= \frac{\Delta S}{\Delta t} = \frac{S(10) - S(8)}{10 - 8} = \frac{1{,}150 - 1{,}400}{10 - 8} \\ &= \frac{-250}{2} = -125 \text{ points per year.}\end{aligned}$$

[30] The values are approximate values at the start of the given year. Source: http://finance.google.com.

Interpreting the result: During the period 2008–2010 the S&P *decreased* at an average rate of 125 points per year.

Finally, during the period 2008–2013 (or [8, 13]) the average change was

$$\text{Average rate of change of } S = \frac{\Delta S}{\Delta t} = \frac{S(13) - S(8)}{13 - 8} = \frac{1{,}400 - 1{,}400}{13 - 8}$$

$$= \frac{0}{5} = 0 \text{ points per year.}$$

Interpreting the result: During the period 2008–2013 the average rate of change of the S&P was zero points per year (even though its value did fluctuate during that period).

b. In Chapter 1 we saw that the rate of change of a quantity that changes linearly with time is measured by the slope of its graph. However, the S&P index does not change linearly with time. Figure 21 shows the data plotted two different ways: (a) as a bar chart and (b) as a piecewise linear graph. Bar charts are more commonly used in the media, but Figure 21(b) illustrates the changing index more clearly.

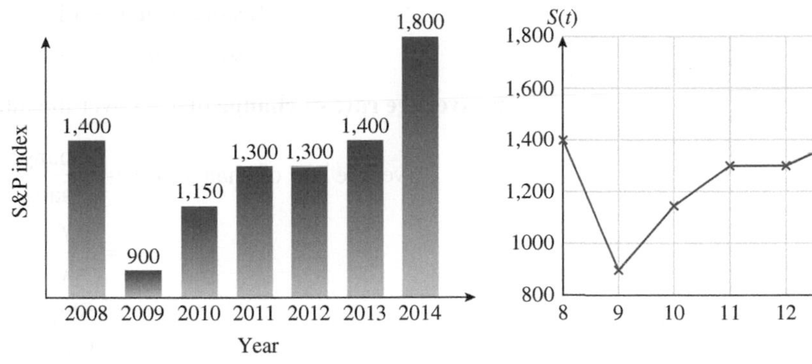

Figure 21(a) **Figure 21(b)**

We saw in part (a) that the average rate of change of S over the interval [8, 10] is the ratio

$$\text{Average rate of change of } S = \frac{\Delta S}{\Delta t} = \frac{S(10) - S(8)}{10 - 8} = -125 \text{ points per year.}$$

Notice that this rate of change is also the slope of the line through P and Q shown in Figure 22, and we can estimate this slope directly from the graph as shown.

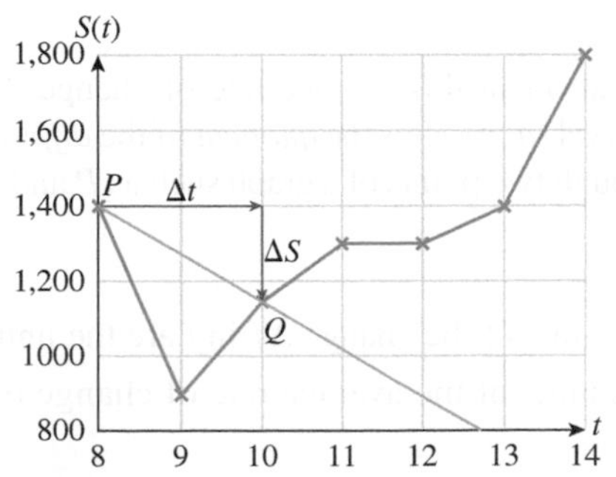

Figure 22

$$\text{Slope: } PQ = \frac{\Delta S}{\Delta t}$$

$$= \frac{1{,}150 - 1{,}400}{10 - 8}$$

$$= -125$$

Average rate of change as slope: The average rate of change of the S&P over the interval $[8, 10]$ is the slope of the line passing through the points on the graph where $t = 8$ and $t = 10$.

Similarly, the average rates of change of the S&P over the other intervals considered here are the slopes of the lines through pairs of corresponding points.

Formulas for Change and Average Rate of Change

Here is the formal definition of the average rate of change of a function over an interval.

Change and Average Rate of Change of f over $[a, b]$: Difference Quotient

The **change** in $f(x)$ over the interval $[a, b]$ is

$$\begin{aligned}\text{Change in } f &= \Delta f \\ &= \text{Second value} - \text{First value} \\ &= f(b) - f(a).\end{aligned}$$

The **average rate of change** of $f(x)$ over the interval $[a, b]$ is

$$\begin{aligned}\text{Average rate of change of } f &= \frac{\text{Change in } f}{\text{Change in } x} \\ &= \frac{\Delta f}{\Delta x} = \frac{f(b) - f(a)}{b - a} \\ &= \text{Slope of line through points } P \text{ and } Q \\ &\quad \text{(see figure).}\end{aligned}$$

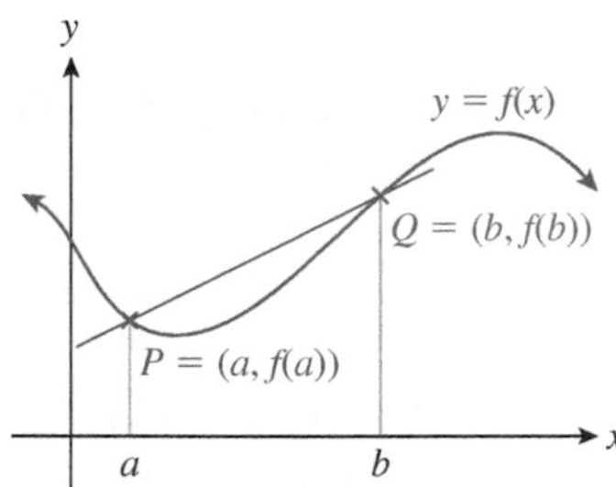

Average rate of change = Slope of PQ

We also call this average rate of change the **difference quotient** of f over the interval $[a, b]$. (It is the *quotient* of the *differences* $f(b) - f(a)$ and $b - a$.) A line through two points of a graph such as P and Q is called a **secant line** of the graph.

Units

The units of the change Δf in f are the units of $f(x)$.

The units of the average rate of change of f are units of $f(x)$ per unit of x.*

* The average rate of change is a slope, so it is measured in the same units as the slope: units of y (or $f(x)$) per unit of x.

Quick Example

1. If $f(3) = -1$ billion dollars, $f(5) = 0.5$ billion dollars, and x is measured in years, then the change and average rate of change of f over the interval $[3, 5]$ are given by

$$\text{Change in } f = f(5) - f(3) = 0.5 - (-1) = 1.5 \text{ billion dollars}$$

$$\text{Average rate of change} = \frac{f(5) - f(3)}{5 - 3} = \frac{0.5 - (-1)}{2}$$
$$= 0.75 \text{ billion dollars per year.}$$

Alternative Formula: Average Rate of Change of f over $[a, a + h]$

(Replace b in the formula for the average rate of change by $a + h$.) The average rate of change of f over the interval $[a, a + h]$ is

$$\text{Average rate of change of } f = \frac{f(a + h) - f(a)}{h}. \qquad \text{Replace } b \text{ by } a + h.$$

In Example 1 we saw that the average rate of change of a quantity can be estimated directly from a graph. Here is an example that further illustrates the graphical approach.

EXAMPLE 2 **Carbon Dioxide Concentration**

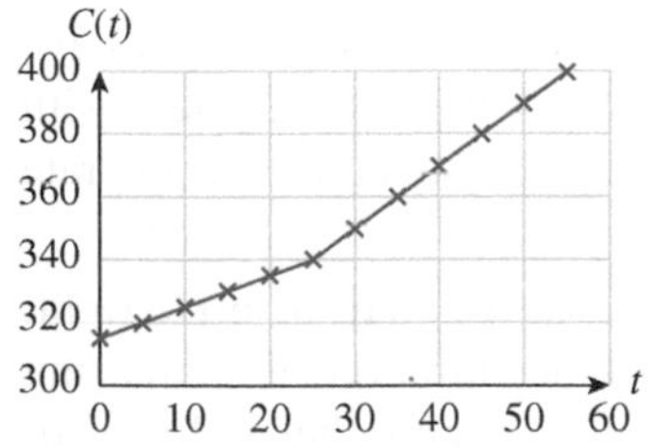

Figure 23 shows the annual mean carbon dioxide concentration measured at Mauna Loa Observatory in Hawaii, in parts per million (ppm), every 5 years from 1960 through 2015. ($t = 0$ represents 1960.)[31]

Figure 23

a. Use the graph to estimate the average rate of change of $C(t)$ with respect to t over the interval $[20, 40]$, and interpret the result.

b. Over which 10-year period(s) in the represented years was the carbon dioxide concentration increasing at an average rate of 1.5 ppm per year?

c. Multiple choice: For the period of time under consideration, carbon dioxide concentration was

(A) increasing at a constant or increasing rate.

(B) increasing at a constant or decreasing rate.

(C) decreasing at a constant or increasing rate.

(D) decreasing at a constant or decreasing rate.

[31] Figures are approximate. Source: U.S. Department of Commerce/National Oceanic and Atmospheric Administration (NOAA) Earth System Research Laboratory, data downloaded from www.esrl.noaa.gov/gmd/ccgg/trends/ on March 13, 2011.

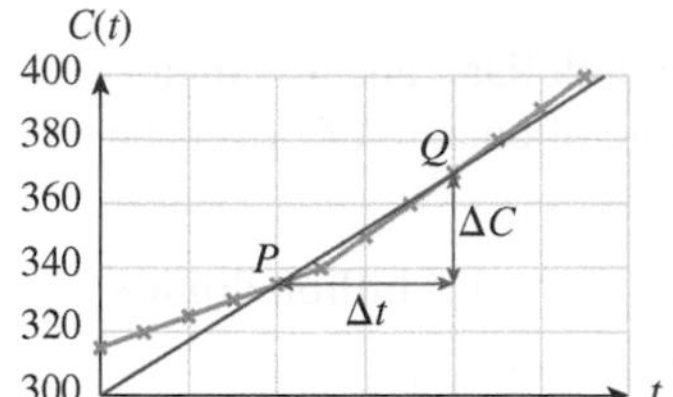

Figure 24

Solution

a. The average rate of change of C over the interval $[20, 40]$ is given by the slope of the line through the points P and Q shown in Figure 24. From the figure,

$$\text{Average rate of change of } C = \frac{\Delta C}{\Delta t} = \text{slope } PQ \approx \frac{370 - 335}{40 - 20} = \frac{35}{20} = 1.75.$$

Thus, the rate of change of C over the interval $[20, 40]$ is approximately 1.75.

Q: *How do we interpret the result?*

A: A clue is given by the units of the average rate of change: units of C per unit of t. The units of C are parts per million (ppm) of carbon dioxide, and the units of t are years. Thus, the average rate of change of C is measured in parts per million of carbon dioxide per year, and we can now interpret the result as follows.

Interpreting the average rate of change: The annual mean carbon dioxide concentration was increasing at an average rate of 1.75 ppm per year from 1980 ($t = 20$) to 2000 ($t = 40$).

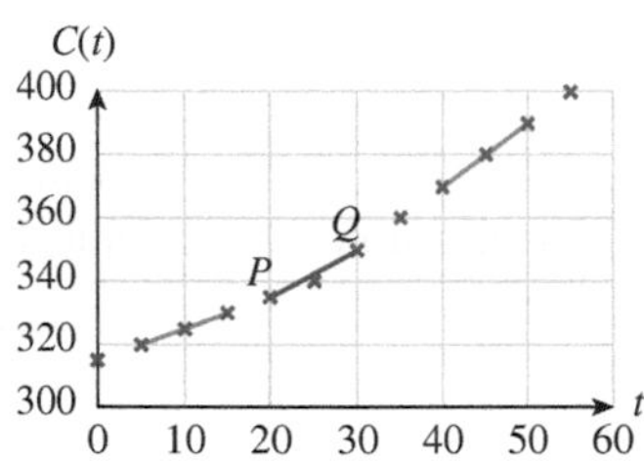

Figure 25

b. The rates of change of carbon dioxide concentration over successive 10-year periods in the represented years are given by the slopes of line segments through pairs of points 10 units apart on the t-axis in the graph in Figure 23. Figure 25 shows three such line segments. We are looking for such a segment whose slope is 1.5. Referring to Figure 25, notice that the segment PQ has the desired slope:

$$\text{Slope } PQ = \frac{350 - 335}{30 - 20} = \frac{15}{10} = 1.5.$$

Segments to the left of PQ all have slope 1, whereas segments to the right have slope 2. Thus, the segment corresponding to $[20, 30]$ is the only segment with slope 1.5, so the carbon dioxide concentration was increasing at an average rate of 1.5 ppm per year during the period 1980–1990.

c. Looking again at Figure 23, notice that the graph rises as we go from left to right; that is, the value of the function (carbon dioxide concentration) is increasing with increasing t. At the same time, the fact that the graph either is linear or bends up (is concave up) with increasing t tells us that the successive slopes are constant or increasing, so this fact applies to the average rates of change as well (choice (A)).

➡ **Before we go on . . .** Notice in Example 2 that we do not get exact answers from a graph; the best we can do is *estimate* the rates of change: Was the exact answer to part (a) closer to 1.74 or 1.76? Two people can reasonably disagree about results read from a graph, and you should bear this in mind when you check the answers to the exercises. ■

Perhaps the most sophisticated way to compute the average rate of change of a quantity is through the use of a mathematical formula or model for the quantity in question.

Average Rate of Change of a Function Specified Algebraically

EXAMPLE 3 Average Rate of Change from a Formula

You are a commodities trader, and you monitor the price of gold on the spot market very closely during an active morning. Suppose you find that the price of an ounce of gold can be approximated by the function

$$G(t) = 5t^2 - 85t + 1{,}762 \qquad (7.5 \le t \le 10.5),$$

where t is time in hours. (See Figure 26; $t = 8$ represents 8:00 am.)

Source: www.kitco.com.

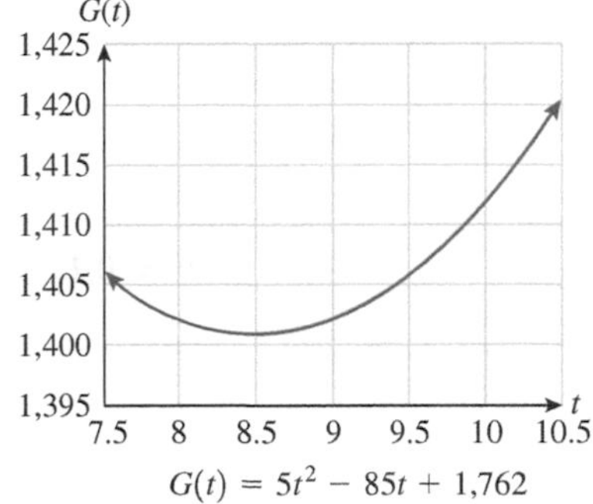

$G(t) = 5t^2 - 85t + 1{,}762$

Figure 26

Using Technology

See the Technology Guides at the end of the chapter for detailed instructions on how to calculate the average rate of change of the function in Example 3 using a TI-83/84 Plus or a spreadsheet. Here is an outline:

TI-83/84 Plus
`Y1=5X^2-85X+1762`
Home screen: `(Y1(9.5)-Y1(8))/(9.5-8)`
[More details in the Technology Guide.]

Spreadsheet
Headings t, $G(t)$, Rate of Change in A1–C1
t-values 8, 9.5 in A2–A3
`=5*A2^2-85*A2+1762` in B2, copied down to B3
`=(B3-B2)/(A3-A2)` in C2.
[More details in the Technology Guide.]

Looking at the graph on the right, we can see that the price of gold was falling at the beginning of the time period, but by $t = 8.5$ the fall had slowed to a stop, whereupon the market turned around, and the price began to rise more and more rapidly toward the end of the period. What was the average rate of change of the price of gold over the $1\frac{1}{2}$-hour period starting at 8:00 am (the interval $[8, 9.5]$ on the t-axis)?

Solution We have

$$\text{Average rate of change of } G \text{ over } [8, 9.5] = \frac{\Delta G}{\Delta t} = \frac{G(9.5) - G(8)}{9.5 - 8}.$$

From the formula for $G(t)$ we find

$$G(9.5) = 5(9.5)^2 - 85(9.5) + 1{,}762 = 1{,}405.75$$
$$G(8) = 5(8)^2 - 85(8) + 1{,}762 = 1{,}402.$$

Thus, the average rate of change of G is given by

$$\frac{G(9.5) - G(8)}{9.5 - 8} = \frac{1{,}405.75 - 1{,}402}{1.5} = \frac{3.75}{1.5} = \$2.50 \text{ per hour.}$$

In other words, the price of gold increased at an average rate of \$2.50 per hour over the $1\frac{1}{2}$-hour period.

EXAMPLE 4 T Rates of Change over Shorter Intervals

Continuing with Example 3, use technology to compute the average rate of change of

$$G(t) = 5t^2 - 85t + 1{,}762 \qquad (7.5 \le t \le 10.5)$$

over the intervals $[8, 8 + h]$, where $h = 1, 0.1, 0.01, 0.001$, and 0.0001. What do the answers tell you about the price of gold?

Solution We use the "alternative" formula

$$\text{Average rate of change of } G \text{ over } [a, a+h] = \frac{G(a+h) - G(a)}{h},$$

so

$$\text{Average rate of change of } G \text{ over } [8, 8+h] = \frac{G(8+h) - G(8)}{h}.$$

Let us calculate this average rate of change for some of the values of h listed:

$\boldsymbol{h = 1}$: $G(8+h) = G(8+1) = G(9) = 5(9)^2 - 85(9) + 1{,}762 = 1{,}402$

$$G(8) = 5(8)^2 - 85(8) + 1{,}762 = 1{,}402$$

$$\text{Average rate of change of } G = \frac{G(9) - G(8)}{1} = \frac{1{,}402 - 1{,}402}{1} = 0$$

$\boldsymbol{h = 0.1}$: $G(8+h) = G(8+0.1) = G(8.1) = 5(8.1)^2 - 85(8.1) + 1{,}762 = 1{,}401.55$

$$G(8) = 5(8)^2 - 85(8) + 1{,}762 = 1{,}402$$

$$\text{Average rate of change of } G = \frac{G(8.1) - G(8)}{0.1} = \frac{1{,}401.55 - 1{,}402}{0.1} = \frac{-0.45}{0.1} = -4.5$$

$\boldsymbol{h = 0.01}$: $G(8+h) = G(8+0.01) = G(8.01) = 5(8.01)^2 - 85(8.01) + 1{,}762 = 1{,}401.9505$

$$G(8) = 5(8)^2 - 85(8) + 1{,}762 = 1{,}402$$

$$\text{Average rate of change of } G = \frac{G(8.01) - G(8)}{0.01} = \frac{1{,}401.9505 - 1{,}402}{0.01} = \frac{-0.0495}{0.01} = -4.95$$

Continuing in this way, we get the values in the following table:

h	1	0.1	0.01	0.001	0.0001
Avg. Rate of Change $\dfrac{G(8+h) - G(8)}{h}$	0	−4.5	−4.95	−4.995	−4.9995

Each value is an average rate of change of G. For example, the value corresponding to $h = 0.01$ is -4.95, which tells us the following:

Over the interval $[8, 8.01]$ *the price of gold was decreasing at an average rate of* \$4.95 *per hour.*

In other words, during the first one hundredth of an hour (or 36 seconds) starting at $t = $ 8:00 am, the price of gold was decreasing at an average rate of \$4.95 per hour. Put another way, in those 36 seconds, the price of gold decreased at a rate that, if continued, would have produced a decrease of \$4.95 in the price of gold during the next hour. We will return to this example at the beginning of Section 3.5.

Using Technology

Example 4 is the kind of example in which the use of technology can make a huge difference. See the Technology Guides at the end of the chapter to find out how to do the above computations almost effortlessly using a TI-83/84 Plus or a spreadsheet. Here is an outline:

TI-83/84 Plus

Y_1=5X^2-85X+1762

Home screen:

(Y_1(8+1)-Y_1(8))/1

(Y_1(8+0.1)-Y_1(8))/0.1

(Y_1(8+0.01)-Y_1(8))/0.01

etc.

[More details in the Technology Guide.]

Spreadsheet

Headings *a*, *h*, *t*, *G*(*t*), Rate of Change in A1–E1

8 in A2, 1 in B2,

=A2 in C2, =A2+B2 in C3

=5*C2^2-85*C2+1762 in D2; copy down to D3

=(D3-D2)/(C3-C2) in E2.

[More details in the Technology Guide.]

FAQs

Recognizing When and How to Compute the Average Rate of Change and How to Interpret the Answer

Q: *How do I know, by looking at the wording of a problem, that it is asking for an average rate of change?*

A: If a problem does not ask for an average rate of change directly, it might do so indirectly, as in "On average, how fast is quantity q increasing?"

Q: *If I know that a problem calls for computing an average rate of change, how should I compute it? By hand or by using technology?*

A: All the computations can be done by hand, but when hand calculations are not called for, using technology might save time.

Q: *Lots of problems ask us to "interpret" the answer. How do I do that for questions involving average rates of change?*

A: The *units* of the average rate of change are often the key to interpreting the results:

The units of the average rate of change of $f(x)$ are units of $f(x)$ per unit of x.

Thus, for instance, if $f(x)$ is the cost, in dollars, of a trip of x miles in length and the average rate of change of f is calculated to be 10, then the units of the average rate of change are dollars per mile, so we can interpret the answer by saying that the cost of a trip rises an average of \$10 for each additional mile.

3.4 EXERCISES

▼ more advanced ◆ challenging

T indicates exercises that should be solved using technology

In Exercises 1–18, calculate the average rate of change of the given function over the given interval. Where appropriate, specify the units of measurement. [**HINT**: See Example 1.]

1. Interval: $[1, 3]$

x	0	1	2	3
$f(x)$	3	5	2	−1

2. Interval: $[0, 2]$

x	0	1	2	3
$f(x)$	−1	3	2	1

3. Interval: $[-3, -1]$

x	−3	−2	−1	0
$f(x)$	−2.1	0	−1.5	0

4. Interval: $[-1, 1]$

x	−2	−1	0	1
$f(x)$	−1.5	−0.5	4	6.5

5. Interval: $[2, 6]$

t (months)	2	4	6
$R(t)$ (\$ million)	20.2	24.3	20.1

6. Interval: $[1, 3]$

x (kilos)	1	2	3
$C(x)$ (£)	2.20	3.30	4.00

7. Interval: $[5, 5.5]$

p (\$)	5.00	5.50	6.00
$q(p)$ (items)	400	300	150

8. Interval: $[0.1, 0.2]$

t (hours)	0	0.1	0.2
$D(t)$ (miles)	0	3	6

9. Interval: [2, 5] [**HINT**: See Example 2.]

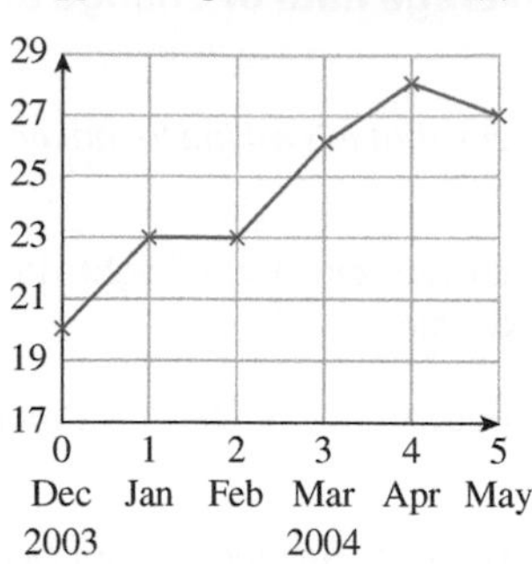

10. Interval: [1, 5] [**HINT**: See Example 2.]

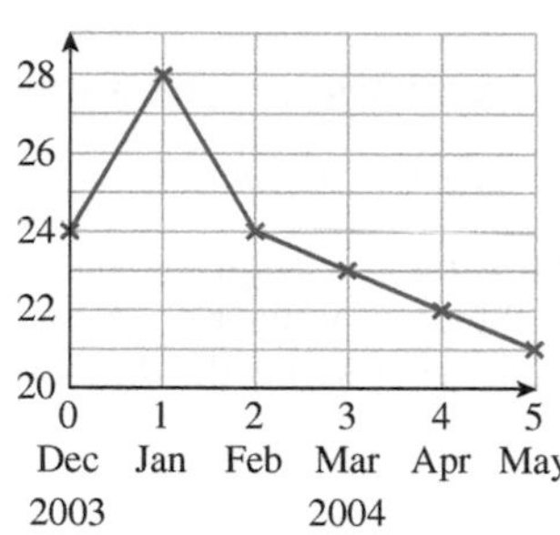

11. Interval: [0, 4]

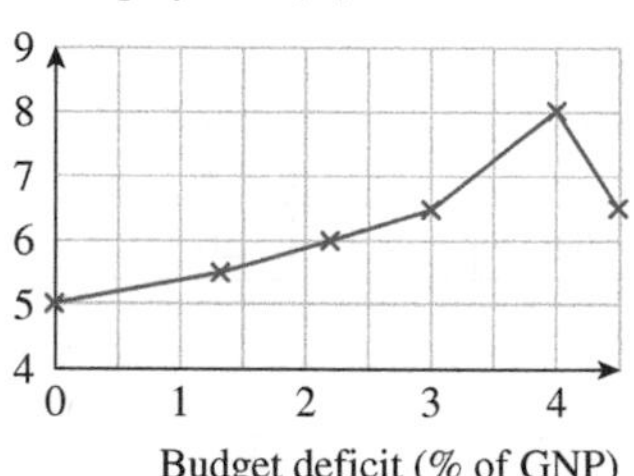

12. Interval: [0, 4]

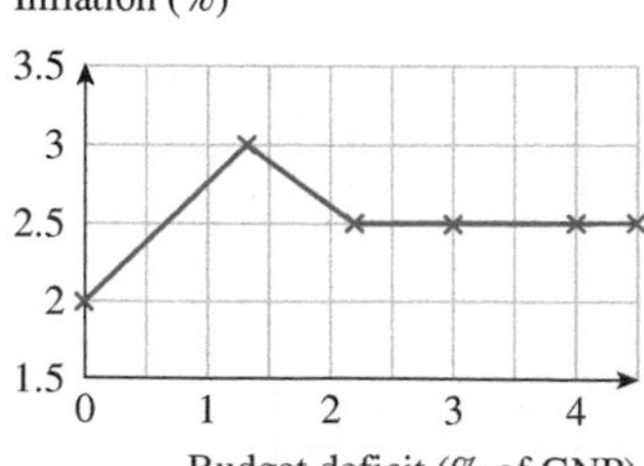

13. $f(x) = x^2 - 3; [1, 3]$ [**HINT**: See Example 3.]

14. $f(x) = 2x^2 + 4; [-1, 2]$ [**HINT**: See Example 3.]

15. $f(x) = 2x + 4; [-2, 0]$ **16.** $f(x) = \frac{1}{x}; [1, 4]$

17. $f(x) = \frac{x^2}{2} + \frac{1}{x}; [2, 3]$ **18.** $f(x) = 3x^2 - \frac{x}{2}; [3, 4]$

In Exercises 19–24, calculate the average rate of change of the given function f over the intervals $[a, a + h]$*, where* $h = 1, 0.1, 0.01, 0.001$*, and* 0.0001*. (Technology is recommended for the cases* $h = 0.01, 0.001$*, and* 0.0001*.)* [**HINT**: See Example 4.]

19. $f(x) = 2x^2; a = 0$ **20.** $f(x) = \frac{x^2}{2}; a = 1$

21. $f(x) = \frac{1}{x}; a = 2$ **22.** $f(x) = \frac{2}{x}; a = 1$

23. $f(x) = x^2 + 2x; a = 3$ **24.** $f(x) = 3x^2 - 2x; a = 0$

Applications

25. ***World Military Expenditure*** The following table shows total military and arms trade expenditure in 2000, 2006, and 2012:[32]

Year t (year since 2000)	0	6	12
Military Expenditure $C(t)$ ($ billion)	1,100	1,450	1,750

Compute and interpret the average rate of change of $C(t)$ **(a)** over the period 2006–2012 (that is, [6, 12]) and **(b)** over the period [0, 12]. Be sure to state the units of measurement. [**HINT**: See Example 1.]

26. ***Education Expenditure*** The following table shows education expenditure in the United States as a percentage of total federal spending in 2009, 2015, and 2019:[33]

Year t (year since 2000)	9	15	19
Percentage $P(t)$	25	27	26

Compute and interpret the average rate of change of $P(t)$ **(a)** over the period 2009–2019 (that is, [9, 19]) and **(b)** over the period [15, 19]. Be sure to state the units of measurement.

27. ***Crude Oil Production: Mexico*** The following table shows daily crude oil production by **Pemex**, Mexico's national oil company, for 2008–2013:[34]

Year t (year since 2008)	0	1	2	3	4	5
Daily Production $p(t)$ (million barrels)	3.16	2.97	2.95	2.94	2.91	2.92

[32] Figures are rounded. Source: www.globalissues.org/article/75/world-military-spending.

[33] Figures are rounded and figures from 2014 on are projections. Source: www.usgovernmentspending.com.

[34] 2013 figure based on data through March. Source: www.pemex.com, March 2013.

a. Compute the average rate of change of $p(t)$ over the period 2010–2013. Interpret the result. [**HINT:** See Example 1.]

b. Which of the following is true? From 2008 to 2013 the three-year average rate of change of oil production by Pemex
(A) increased in value.
(B) decreased in value.
(C) increased then decreased in value.
(D) decreased then increased in value.
[**HINT:** See Example 2.]

28. ***Offshore Crude Oil Production: Mexico*** The following table shows daily offshore crude oil production by Pemex, Mexico's national oil company, for 2008–2013:[35]

Year t (year since 2008)	0	1	2	3	4	5
Daily Offshore Production $s(t)$ (million barrels)	2.25	2.01	1.94	1.90	1.90	1.90

a. Use the data in the table to compute the average rate of change of $s(t)$ over the period 2008–2013. Interpret the result.

b. Which of the following is true? From 2008 to 2013 the two-year average rate of change of offshore crude oil production of Pemex
(A) increased in value.
(B) decreased in value.
(C) increased then decreased in value.
(D) decreased then increased in value.

29. ***Subprime Mortgages during the Housing Crisis*** The following graph shows the approximate percentage $P(t)$ of mortgages issued in the United States that were subprime (normally classified as risky):[36]

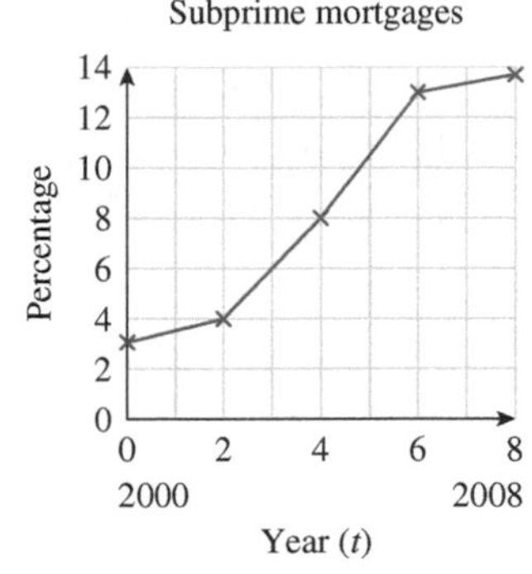

a. Use the graph to estimate, to one decimal place, the average rate of change of $P(t)$ with respect to t over the interval $[0, 6]$, and interpret the result.

b. Over which 2-year period(s) was the average rate of change of $P(t)$ the greatest? [**HINT:** See Example 2.]

30. ***Subprime Mortgage Debt during the Housing Crisis*** The following graph shows the approximate value $V(t)$ of subprime (normally classified as risky) mortgage debt outstanding in the United States:[37]

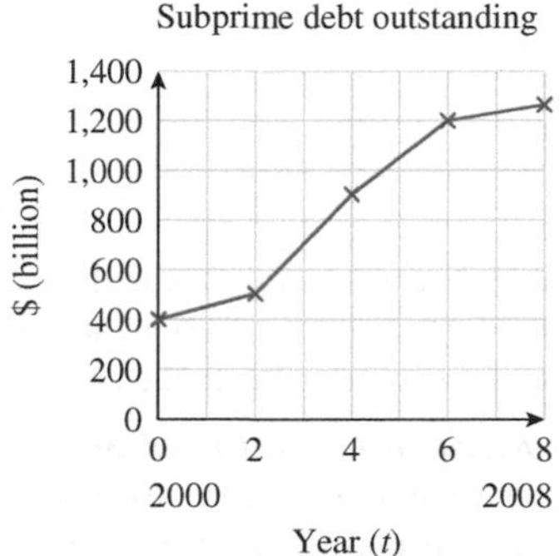

a. Use the graph to estimate, to one decimal place, the average rate of change of $V(t)$ with respect to t over the interval $[2, 6]$, and interpret the result.

b. Over which 2-year period(s) was the average rate of change of $V(t)$ the least? [**HINT:** See Example 2.]

31. ***Immigration to Ireland*** The following graph shows the approximate number (in thousands) of people who immigrated to Ireland during the period 2010–2014 (t is time in years since 2010):[38]

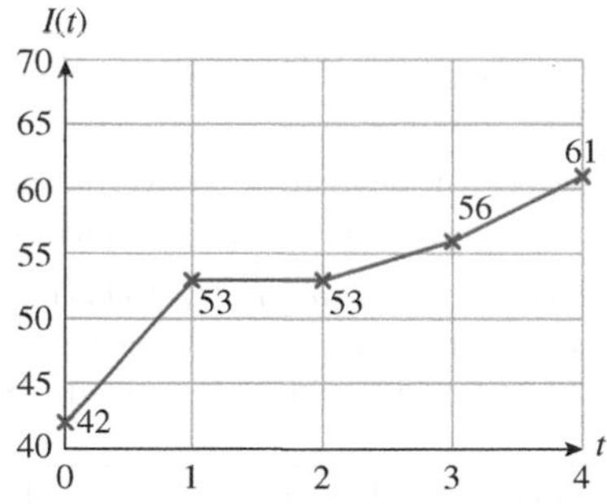

During which 2-year interval(s) was the magnitude of the average rate of change of $I(t)$ **(a)** greatest **(b)** least? Interpret your answers by referring to the rates of change.

[35] See footnote for Exercise 27.

[36] Sources: Mortgage Bankers Association, UBS.

[37] Source: Data 360 www.data360.org.

[38] Source: European Migration Network Ireland, http://emn.ie.

32. ***Emigration from Ireland*** The following graph shows the approximate number (in thousands) of people who emigrated from Ireland during the period 2010–2014 (t is time in years since 2010):[39]

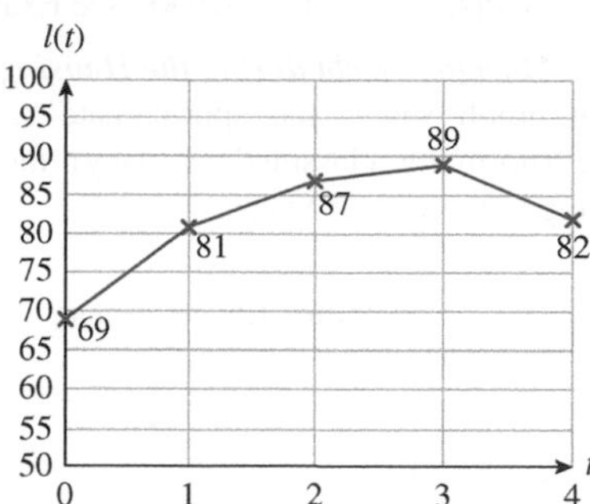

During which 2-year interval(s) was the magnitude of the average rate of change of $E(t)$ **(a)** greatest **(b)** least? Interpret your answers by referring to the rates of change.

33. ▼ ***Science Research in the United States*** The following table shows the number of science research articles authored by U.S researchers during the period 1980–2010:[40]

Year t (year since 1980)	0	5	10	15	20	25	30
Articles $N(t)$ (thousands)	170	200	220	260	252	290	340

a. Find the interval(s) over which the average rate of change of N was the greatest. What was that rate of change? Interpret your answer.

b. The **percentage change of N over the interval $[a, b]$** is defined to be

$$\text{Percentage change of } N = \frac{\text{Change in } N}{\text{First value}} = \frac{N(b) - N(a)}{N(a)}.$$

Compute the percentage change of N over the interval $[0, 30]$ and also the average rate of change. Interpret the answers.

34. ▼ ***Science Research in Europe*** The following table shows the number of science research articles authored by researchers in the European Union during the period 1980–2010:[41]

Year t (year since 1980)	0	5	10	15	20	25	30
Articles $N(t)$ (thousands)	140	170	190	260	300	340	430

a. Find the interval(s) over which the average rate of change of N was the least positive. What was that rate of change? Interpret your answer.

b. The **percentage change of N over the interval $[a, b]$** is defined to be

$$\text{Percentage change of } N = \frac{\text{Change in } N}{\text{First value}} = \frac{N(b) - N(a)}{N(a)}.$$

Compute the percentage change of N over the interval $[10, 30]$ and also the average rate of change. Interpret the answers.

35. ***College Basketball: Men*** The following chart shows the number of NCAA men's college basketball teams in the United States during the period 2000–2010:[42]

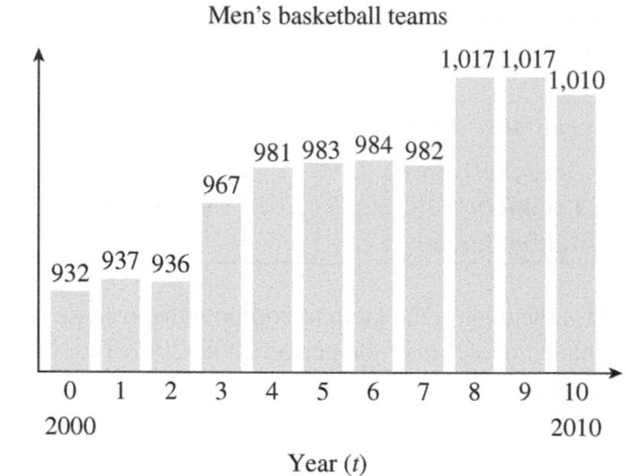

a. On average, how fast was the number of men's college basketball teams growing over the 4-year period beginning in 2002?

b. By inspecting the chart, determine whether the 3-year average rates of change increased or decreased beginning in 2005. [**HINT:** See Example 2.]

36. ***College Basketball: Women*** The following chart shows the number of NCAA women's college basketball teams in the United States during the period 2000–2010:[43]

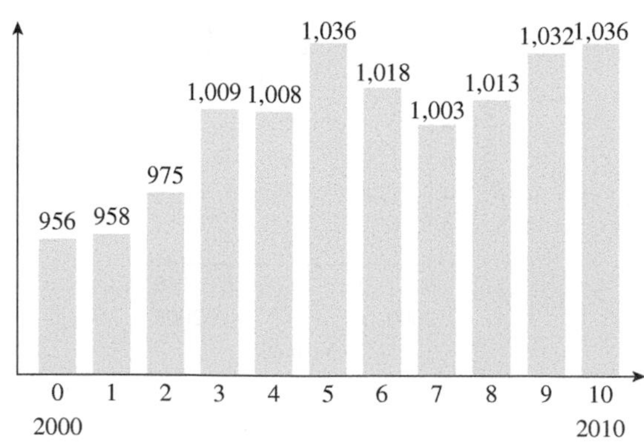

[39] See footnote for Exercise 31.

[40] 1980 data estimated. Source: www.sciencewatch.com.

[41] *Ibid.*

[42] 2010 figure is an estimate. Source: www.census.gov.

[43] *Ibid.*

a. On average, how fast was the number of women's college basketball teams growing over the 4-year period beginning in 2004?
b. By inspecting the graph, find the 3-year period over which the average rate of change was largest.

37. ***Funding for the Arts*** State governments in the United States spend between \$1 and \$2 per person on the arts and culture each year. The following chart shows the data for 2002–2010, together with the regression line:[44]

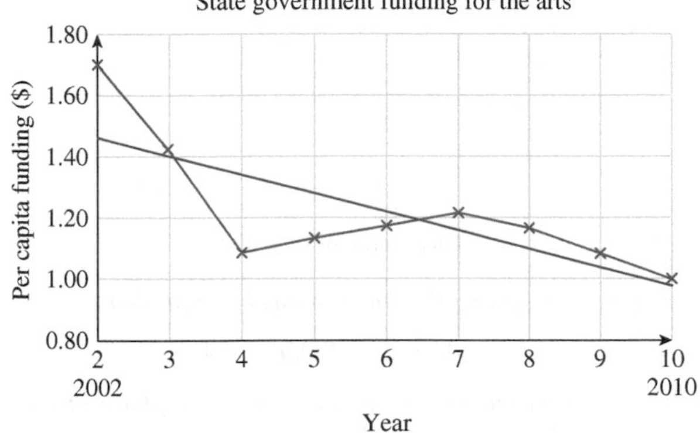

a. Over the period [2, 6] the average rate of change of state government funding for the arts was
(A) less than
(B) greater than
(C) approximately equal to
the rate predicted by the regression line.
b. Over the period [3, 10] the average rate of change of state government funding for the arts was
(A) less than
(B) greater than
(C) approximately equal to
the rate predicted by the regression line.
c. Over the period [4, 8] the average rate of change of state government funding for the arts was
(A) less than
(B) greater than
(C) approximately equal to
the rate predicted by the regression line.
d. Estimate, to two significant digits, the average rate of change of per capita state government funding for the arts over the period [2, 10]. (Be careful to state the units of measurement.) How does it compare to the slope of the regression line?

38. ***Funding for the Arts*** The U.S. federal government spends between \$6 and \$7 per person on the arts and culture each year. The following chart shows the data for 2002–2010, together with the regression line:[45]

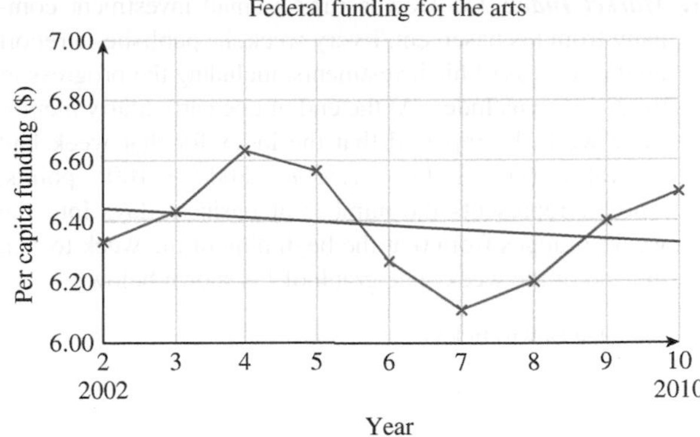

a. Over the period [4, 10] the average rate of change of federal government funding for the arts was
(A) less than (B) greater than
(C) approximately equal to
the rate predicted by the regression line.
b. Over the period [2, 7] the average rate of change of federal government funding for the arts was
(A) less than (B) greater than
(C) approximately equal to
the rate predicted by the regression line.
c. Over the period [3, 10] the average rate of change of federal government funding for the arts was
(A) less than (B) greater than
(C) approximately equal to
the rate predicted by the regression line.
d. Estimate, to one significant digit, the average rate of change of per capita federal government funding for the arts over the period [2, 10]. (Be careful to state the units of measurement.) How does it compare to the slope of the regression line?

39. ▼ ***Market Volatility during the Dot-Com Boom*** A volatility index generally measures the extent to which a market undergoes sudden changes in value. The volatility of the S&P 500 (as measured by one such index) was decreasing at an average rate of 0.2 points per year during 1991–1995 and was increasing at an average rate of about 0.3 points per year during 1995–1999. In 1995 the volatility of the S&P was 1.1.[46] Use this information to give a rough sketch of the volatility of the S&P 500 as a function of time, showing its values in 1991 and 1999.

40. ▼ ***Market Volatility during the Dot-Com Boom*** The volatility (see Exercise 39) of the NASDAQ had an average rate of change of 0 points per year during 1992–1995, and increased at an average rate of 0.2 points per year during 1995–1998. In 1995 the volatility of the NASDAQ was 1.1.[47] Use this information to give a rough sketch of the volatility of the NASDAQ as a function of time.

[44] Figures are in constant 2008 dollars, and the 2010 figure is the authors' estimate. Source: *Americans for the Arts* www.artsusa.org.
[45] *Ibid.*
[46] Source for data: Sanford C. Bernstein Company/*New York Times*, March 24, 2000, p. C1.
[47] *Ibid.*

41. ***Market Index*** Joe Downs runs a small investment company from his basement. Every week, he publishes a report on the success of his investments, including the progress of the Joe Downs Index. At the end of one particularly memorable week, he reported that the index for that week had the value $I(t) = 1{,}000 + 1{,}500t - 800t^2 + 100t^3$ points, where t represents the number of business days into the week; t ranges from 0 at the beginning of the week to 5 at the end of the week. The graph of I is shown below:

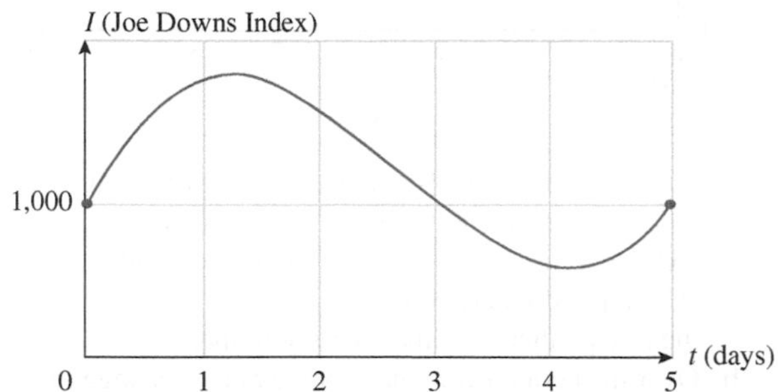

On average, how fast and in which direction was the index changing over the first two business days (the interval [0, 2])? [**HINT:** See Example 3.]

42. ***Market Index*** Refer to the Joe Downs Index in Exercise 41. On average, how fast and in which direction was the index changing over the last three business days (the interval [2, 5])? [**HINT:** See Example 3.]

43. ***Crude Oil Prices*** The price per barrel of crude oil in constant 2008 dollars can be approximated by

$$P(t) = 0.45t^2 - 12t + 105 \text{ dollars} \quad (0 \le t \le 28),$$

where t is time in years since the start of 1980.[48]

a. What, in constant 2008 dollars, was the average rate of change of the price of oil from the start of 1981 ($t = 1$) to the start of 2006 ($t = 26$)? [**HINT:** See Example 3.]

b. Your answer to part (a) is quite small. Can you conclude that the price of oil hardly changed at all over the 25-year period 1981–2006? Explain.

44. ***Median Home Prices*** The median home price in the United States over the period January 2010–January 2015 can be approximated by

$$P(t) = 4.5t^2 - 15t + 180 \text{ thousand dollars} \quad (0 \le t \le 5),$$

where t is time in years since the start of 2010.[49]

a. What was the average rate of change of the median home price from the start of 2012 to the start of 2014?

b. What, if anything, does your answer to part (a) say about the median home price in 2013? Explain.

End of the Earth *In 5 billion years the Sun will have run out of hydrogen fuel and begin to expand into a red giant, eventually engulfing the Earth and causing it to spiral into the core of the Sun 7.5 billion years from now. The following graph*[50] *shows the expanding radius of the red giant Sun (in red) and the radius of the Earth's orbit about the Sun (in green) during its final three and a half million years of existence. The radii are measured in astronomical units (AU; one AU is the current radius of the Earth's orbit around the Sun, approximately 93 million miles), and time is measured in millions of years.*

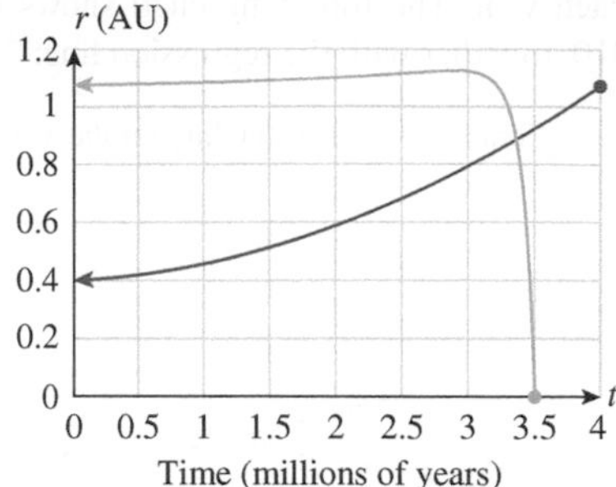

The curve representing the Sun's radius has equation

$$r = 0.037t^2 + 0.02t + 0.4.$$

($t = 4$ marks the end of the red giant expansion phase.) Exercises 45 and 46 are based on this curve.

45. **a.** Calculate the rate of change of the radius of the Sun over the successive intervals [0, 1], [1, 2], [2, 3], [3, 4].

b. The successive rates of change are a linear function of t. What is the slope of that linear function? How fast will the *rate of change* of the Sun's radius be increasing in the final 4 million years?

46. **a.** Calculate the rate of change of the radius of the Sun over the successive intervals [0, 2], [1, 3], [2, 4].

b. The successive rates of change are a linear function of t. What is the slope of that linear function? How fast will the *rate of change* of the Sun's radius be increasing in the final 4 million years?

47. ***The 2003 SARS Outbreak*** In the early stages of the SARS (severe acute respiratory syndrome) epidemic in 2003 the number of reported cases could be approximated by

$$A(t) = 167(1.18)^t \quad (0 \le t \le 20)$$

t days after March 17, 2003 (the first day for which statistics were reported by the World Health Organization).

a. What was the average rate of change of $A(t)$ from March 17 to March 23? Interpret the result.

b. Which of the following is true? For the first 20 days of the epidemic, the number of reported cases
(A) increased at a faster and faster rate.
(B) increased at a slower and slower rate.
(C) decreased at a faster and faster rate.

[48] Source for data: www.inflationdata.com.

[49] Source for data: www.zillow.com.

[50] The actual astrophysical models, of which the curves shown here are merely rough graphical representations, are described in "Distant Future of the Sun and Earth Revisited" by Klaus-Peter Schröder and Robert C. Smith, *Monthly Notices of the Royal Astronomical Society* **386** (1) (2008): 155–163.

(D) decreased at a slower and slower rate.
[HINT: See Example 2.]

48. ***The 2003 SARS Outbreak*** A few weeks into the SARS (severe acute respiratory syndrome) epidemic in 2003, the number of reported cases could be approximated by

$$A(t) = 1{,}804(1.04)^t \quad (0 \le t \le 30)$$

t days after April 1, 2003.
a. What was the average rate of change of $A(t)$ from April 19 ($t = 18$) to April 29? Interpret the result.
b. Which of the following is true? During the 30-day period beginning April 1, the number of reported cases
(A) increased at a faster and faster rate.
(B) increased at a slower and slower rate.
(C) decreased at a faster and faster rate.
(D) decreased at a slower and slower rate.
[HINT: See Example 2.]

49. ***The 2014 Ebola Outbreak*** In the first 6 months of the 2014 Ebola outbreak, the total number of reported cases could be approximated by

$$C(t) = 95.9e^{0.72t} \quad (0 \le t \le 6),$$

where t is time in months since April 1, 2014.[51]
a. Calculate the average rates of change of $C(t)$ over the successive 2-month periods $[0, 2]$, $[1, 3]$, $[2, 4]$, $[3, 5]$, and $[4, 6]$. (Round answers to two decimal places.)
b. What kind of model would best describe the successive rates of change obtained in part (a): linear, quadratic, or exponential? What does your answer tell you about the 2014 Ebola outbreak?

50. ***The 2014 Ebola Outbreak*** Repeat Exercise 49 using the following model for the total number of reported deaths from Ebola:

$$D(t) = 90.52e^{0.60t} \quad (0 \le t \le 6),$$

where t is time in months since April 1, 2014.[52]

51. ▼ ***Ecology*** Increasing numbers of manatees ("sea sirens") have been killed by boats off the Florida coast. The following graph shows the relationship between the number of boats registered in Florida and the number of manatees killed each year:

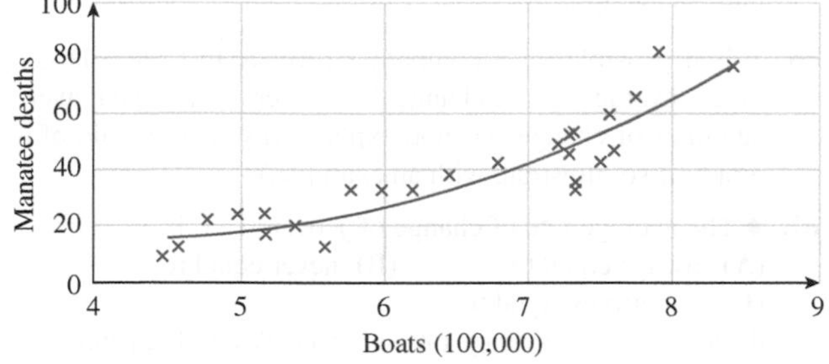

The regression curve shown is given by

$$f(x) = 3.55x^2 - 30.2x + 81 \text{ manatees} \quad (4.5 < x < 8.5),$$

where x is the number of boats (in hundreds of thousands) registered in Florida in a particular year and $f(x)$ is the number of manatees killed by boats in Florida that year.[53]
a. Compute the average rate of change of f over the intervals $[5, 6]$ and $[7, 8]$.
b. What does the answer to part (a) tell you about the manatee deaths per boat?

52. ▼ ***Ecology*** Refer to Exercise 51.
a. Compute the average rate of change of f over the intervals $[5, 7]$ and $[6, 8]$.
b. Had we used a linear model instead of a quadratic one, how would the two answers in part (a) be related to each other?

53. T ▼ ***SAT Scores by Income*** The math SAT score of a high school graduate can be approximated by

$$S(x) = 573 - 133(0.987)^x \text{ points on the math SAT test,}$$

where x is the household income of the student in thousands of dollars per year.[54]
a. Use technology to complete the following table, which shows the average rate of change of S over successive intervals of length 40. (Round all answers to two decimal places.) [HINT: See Example 4.]

Interval	[0, 40]	[40, 80]	[80, 120]	[120, 160]	[160, 200]
Avg. Rate of Change of S					

b. Interpret your answer for the interval $[40, 80]$, being sure to indicate the direction of change and the units of measurement.
c. Multiple choice: As her household income rises, a student's SAT score
(A) increases.
(B) decreases.
(C) increases, then decreases.
(D) decreases, then increases.
d. Multiple choice: As the household income increases, the effect on a student's SAT score is
(A) more pronounced.
(B) less pronounced.

[51] Exponential model is the authors'. Source for data: Wikipedia/Centers for Disease Control and Prevention/WHO.
[52] *Ibid.*
[53] Regression model is based on data from 1976 to 2000. Sources for data: Florida Department of Highway Safety & Motor Vehicles, Florida Marine Institute/*New York Times*, February 12, 2002, p. F4.
[54] The model is the authors'. Source for data: College Board/*New York Times* http://economix.blogs.nytimes.com.

54. **T** ▼ ***SAT Scores by Income*** Repeat Exercise 53 using the following model for the critical reading SAT score of a high school graduate:

$S(x) = 550 - 136(0.985)^x$ points on the critical reading SAT test,

where x is the household income of the student in thousands of dollars per year.[55]

Communication and Reasoning Exercises

55. Describe three ways we have used to determine the average rate of change of f over an interval $[a, b]$. Which of the three ways is *least* precise? Explain.

56. If f is a linear function of x with slope m, what is its average rate of change over any interval $[a, b]$?

57. Is the average rate of change of a function over $[a, b]$ affected by the values of the function between a and b? Explain.

58. If the average rate of change of a function over $[a, b]$ is zero, this means that the function is constant over that interval—right?

59. Sketch the graph of a function whose average rate of change over $[0, 3]$ is negative but whose average rate of change over $[1, 3]$ is positive.

60. Sketch the graph of a function whose average rate of change over $[0, 2]$ is positive but whose average rate of change over $[0, 1]$ is negative.

61. ▼ If the rate of change of quantity A is 2 units of quantity A per unit of quantity B, and the rate of change of quantity B is 3 units of quantity B per unit of quantity C, what is the rate of change of quantity A with respect to quantity C?

62. ▼ If the rate of change of quantity A is 2 units of quantity A per unit of quantity B, what is the rate of change of quantity B with respect to quantity A?

63. ▼ A certain function f has the property that its average rate of change over the interval $[1, 1 + h]$ (for positive h) increases as h decreases. Which of the following graphs could be the graph of f?

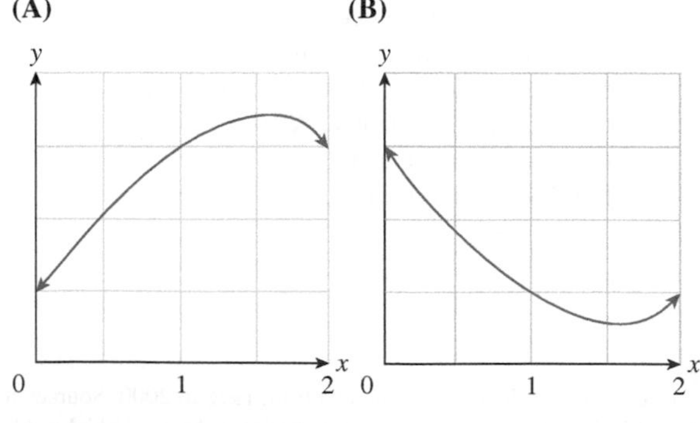

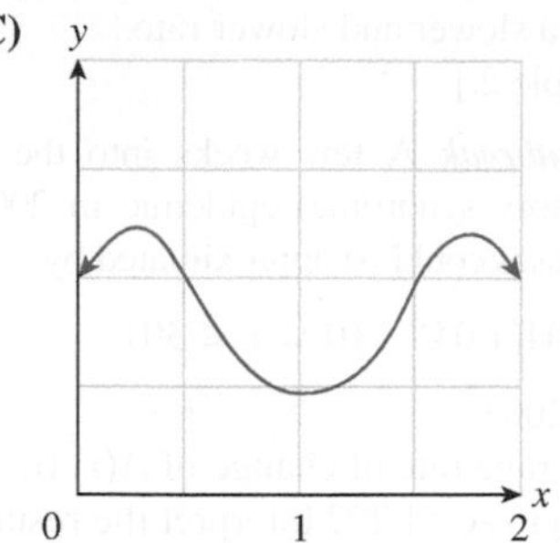

64. ▼ A certain function f has the property that its average rate of change over the interval $[1, 1 + h]$ (for positive h) decreases as h decreases. Which of the following graphs could be the graph of f?

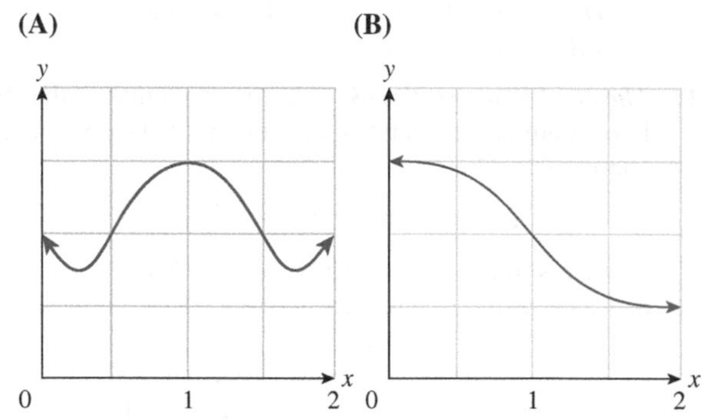

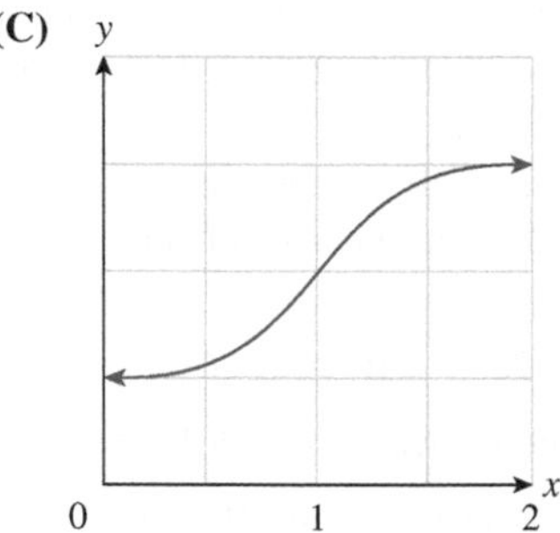

65. ▼ Is it possible for a company's revenue to have a negative 3-year average rate of growth but a positive average rate of growth in 2 of the 3 years? (If not, explain; if so, illustrate with an example.)

66. ▼ Is it possible for a company's revenue to have a larger 2-year average rate of change than either of the 1-year average rates of change? (If not, explain why with the aid of a graph; if so, illustrate with an example.)

67. ◆ The average rate of change of f over $[1, 3]$ is
(A) always equal to **(B)** never equal to
(C) sometimes equal to
the average of its average rates of change over $[1, 2]$ and $[2, 3]$.

68. ◆ The average rate of change of f over $[1, 4]$ is
(A) always equal to **(B)** never equal to
(C) sometimes equal to
the average of its average rates of change over $[1, 2]$, $[2, 3]$, and $[3, 4]$.

[55] The model is the authors'. Source for data: College Board/*New York Times* http://economix.blogs.nytimes.com.

3.5 The Derivative: Numerical and Graphical Viewpoints

Instantaneous Rate of Change of a Function

In Example 4 of Section 3.4 we looked at the average rate of change of the function $G(t) = 5t^2 - 85t + 1{,}762$ approximating the price of gold on the spot market over smaller and smaller intervals of time. We obtained the following table showing the average rates of change of G over the intervals $[8, 8 + h]$ for successively smaller values of h:

h getting smaller; interval $[8, 8 + h]$ getting smaller →

h	1	0.1	0.01	0.001	0.0001
Avg. Rate of Change over $[8, 8 + h]$	0	−4.5	−4.95	−4.995	−4.9995

Rate of change approaching −$5 per hour →

The average rates of change of the price of gold over smaller and smaller periods of time, starting at the instant $t = 8$ (8:00 am), appear to be getting closer and closer to −$5 per hour. As we look at these shrinking periods of time, we are getting closer to looking at what happens at the *instant* $t = 8$. So it seems reasonable to say that the average rates of change are approaching the **instantaneous rate of change** at $t = 8$, which the table suggests is −$5 per hour. This is how fast the price of gold was changing *exactly* at 8:00 am.

At $t = 8$ the instantaneous rate of change of $G(t)$ is -5.

We express this fact mathematically by writing $G'(8) = -5$ (which we read as "G prime of 8 equals −5"). Thus,

$G'(8) = -5$ *means that, at $t = 8$, the instantaneous rate of change of $G(t)$ is -5.*

The process of letting h get smaller and smaller is called taking the **limit** as h approaches 0 (as you recognize if you've done the sections on limits). As in the sections on limits, we write $h \to 0$ as shorthand for "h approaches 0." Thus, taking the limit of the average rates of change as $h \to 0$ gives us the instantaneous rate of change.

Q: *All these intervals $[8, 8 + h]$ are intervals to the right of 8. What about small intervals to the left of 8, such as $[7.9, 8]$?*

A: We can compute the average rate of change of our function for such intervals by choosing h to be negative ($h = -0.1, -0.01$, etc.) and using the same difference quotient formula we used for positive h:

$$\text{Average rate of change of } G \text{ over } [8 + h, 8] = \frac{G(8) - G(8 + h)}{8 - (8 + h)}.$$

Here are the results we get using negative h:

h getting closer to 0; interval $[8 + h, 8]$ getting smaller →

h	−1	−0.1	−0.01	−0.001	−0.0001
Avg. Rate of Change over $[8 + h, 8]$	−10	−5.5	−5.05	−5.005	−5.0005

Rate of change approaching −$5 per hour →

Notice that the average rates of change are again getting closer and closer to −5 as h approaches 0, suggesting once again that the instantaneous rate of change is −$5 per hour.

* For instance, a is not allowed to be an endpoint of the domain of f. A point a that satisfies the given requirement is called an **interior point** of the domain of f.

Instantaneous Rate of Change of $f(x)$ at $x = a$: Derivative

Assume that $f(x)$ is defined for all x in some open interval about $x = a$.* The **instantaneous rate of change** of $f(x)$ at $x = a$ is defined as

$$f'(a) = \lim_{h \to 0} \frac{f(a+h) - f(a)}{h},$$

f prime of a equals the limit, as h approaches 0, of the ratio $\frac{f(a+h) - f(a)}{h}$.

assuming that the limit exists. The quantity $f'(a)$ is also called the **derivative of $f(x)$ at $x = a$**. Finding the derivative of f is called **differentiating f**.

Note For $f'(a)$ to exist, two requirements need to be met:

1. a must be an interior point of the domain of f (see the side note above), and
2. the above limit must exist (and be finite).

When $f'(a)$ exists, we say that f is **differentiable at** a; otherwise, f is **not differentiable at** a. ■

Units

The units of $f'(a)$ are the same as the units of the average rate of change: units of f per unit of x.

Quick Examples

1. If $f(x) = 5x^2 - 85x + 1{,}762$, then the two tables above suggest that

$$f'(8) = \lim_{h \to 0} \frac{f(8+h) - f(8)}{h} = -5.$$

2. If $f(t)$ is the number of insects in your dorm room at time t hours, and you know that $f(3) = 5$ and $f'(3) = 8$, this means that, at time $t = 3$ hours, there are five insects in your room, and this number is growing at an instantaneous rate of eight insects per hour.

Important Notes

1. Sections 3.1–3.3 discuss limits in some detail. If you have not (yet) covered those sections, you can trust to your intuition.
2. The formula for the derivative tells us that the instantaneous rate of change is the limit of the average rates of change $[f(a+h) - f(a)]/h$ over smaller and smaller intervals. Thus, the value of $f'(a)$ can be approximated by computing the average rate of change for smaller and smaller values of h, both positive and negative.
3. As we noted above, if a happens to be an endpoint of the domain of f, then a is not an interior point of the domain of f, and so f is not differentiable at a.
4. In this section we will only *approximate* derivatives. In Section 3.6 we will begin to see how we find the *exact* values of derivatives.

5. $f'(a)$ is a number we can calculate, or at least approximate, for various values of a, as we have done in the earlier example. Since $f'(a)$ depends on the value of a, we can think of f' as *a function of a*. (We return to this idea at the end of this section.) An old name for f' is "the function *derived from f*," which has been shortened to the *derivative* of *f*.
6. It is because f' is a function that we sometimes refer to $f'(a)$ as "the derivative of *f evaluated at a*" or the "derivative of $f(x)$ evaluated at $x = a$."
7. If the average rates of change $[f(a+h) - f(a)]/h$ approach one number on the intervals using positive h and another number on the intervals using negative h, then $\lim_{h\to 0}[f(a+h) - f(a)]/h$ does not exist, so f is not differentiable at a. It is comforting to know that all polynomials and exponential functions *are* differentiable at every point, although many common functions are not; for example, neither $f(x) = |x|$ nor $f(x) = x^{1/3}$ is differentiable at 0 (see Section 4.1). ■

EXAMPLE 1 Instantaneous Rate of Change: Numerically and Graphically

The air temperature one spring morning, t hours after 7:00 am, was given by the function $f(t) = 50 + 0.1t^4$ degrees Fahrenheit $(0 \le t \le 4)$.

a. How fast was the temperature rising at 9:00 am?

b. How is the instantaneous rate of change of temperature at 9:00 am reflected in the graph of temperature vs. time?

Solution

a. We are being asked to find the instantaneous rate of change of the temperature at $t = 2$, so we need to find $f'(2)$. To do this, we examine the average rates of change

$$\frac{f(2+h) - f(2)}{h} \qquad \text{Average rate of change = Difference quotient}$$

for values of h approaching 0. Calculating the average rate of change over $[2, 2+h]$ for $h = 1$, 0.1, 0.01, 0.001, and 0.0001, we get the following values (rounded to four decimal places):*

* We can quickly compute these values using technology as in Example 4 in Section 3.4. (See the Technology Guides at the end of the chapter.)

h	1	0.1	0.01	0.001	0.0001
Avg. Rate of Change over $[2, 2+h]$	6.5	3.4481	3.2241	3.2024	3.2002

Here are the values we get using negative values of h:

h	−1	−0.1	−0.01	−0.001	−0.0001
Avg. Rate of Change over $[2+h, 2]$	1.5	2.9679	3.1761	3.1976	3.1998

The average rates of change are clearly approaching the number 3.2, so we can say that $f'(2) = 3.2$. Thus, at 9:00 in the morning, the temperature was rising at the rate of 3.2 degrees per hour.

b. We saw in Section 3.4 that the average rate of change of f over an interval is the slope of the secant line through the corresponding points on the graph of f. Figure 27 illustrates this for the intervals $[2, 2 + h]$ with $h = 1, 0.5$, and 0.1.

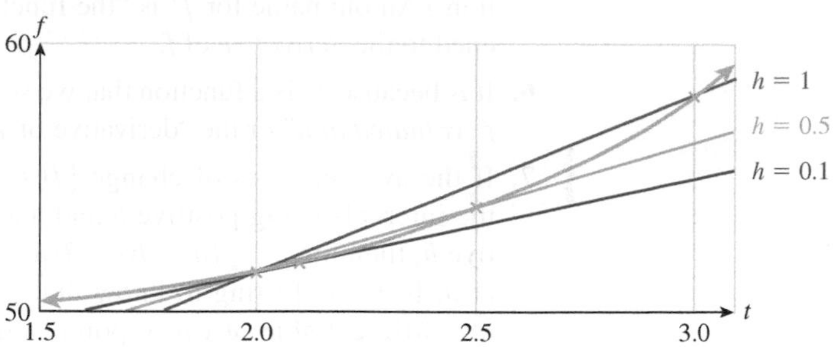

Figure 27

All three secant lines pass though the point $(2, f(2)) = (2, 51.6)$ on the graph of f. Each of them passes through a second point on the curve (the second point is different for each secant line), and this second point gets closer and closer to $(2, 51.6)$ as h gets closer to 0. What seems to be happening is that the secant lines are getting closer and closer to a line that just touches the curve at $(2, 51.6)$: the **tangent line** at $(2, 51.6)$, shown in Figure 28.

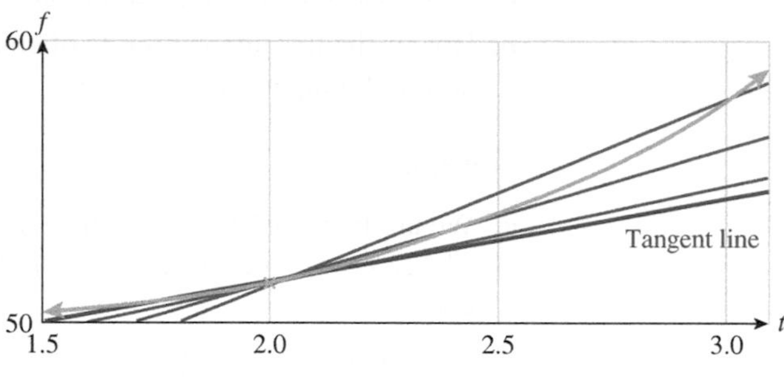

Figure 28

Q: *What is the slope of this tangent line?*

A: Because the slopes of the secant lines are getting closer and closer to 3.2, and because the secant lines are approaching the tangent line, the tangent line must have slope 3.2. In other words,

At the point on the graph where $x = 2$, the slope of the tangent line is $f'(2)$.

Q: *What is the difference between $f(2)$ and $f'(2)$?*

A: An important question. Briefly, $f(2)$ is the *value of f* when $t = 2$, while $f'(2)$ is the *rate at which f is changing* when $t = 2$. Here,

$$f(2) = 50 + 0.1(2)^4 = 51.6 \text{ degrees.}$$

Thus, at 9:00 am ($t = 2$) the temperature was 51.6 degrees. On the other hand,

$$f'(2) = 3.2 \text{ degrees per hour.}$$ Units of slope are units of f per unit of t.

This means that, at 9:00 am ($t = 2$) the temperature was increasing at a rate of 3.2 degrees per hour.

Secant and Tangent Lines

Because we have been talking about tangent lines, we should say more about what they *are.* A tangent line to a *circle* is a line that touches the circle in just one point. A tangent line gives the circle "a glancing blow," as shown in Figure 29. For a smooth curve other than a circle, a tangent line may touch the curve at more than one point or pass through it (Figure 30).

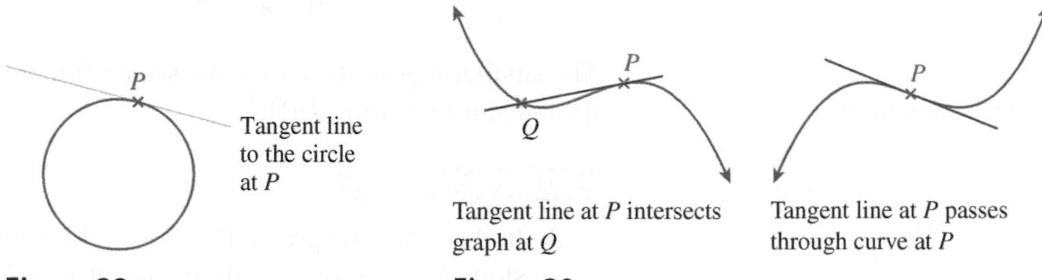

Figure 29 **Figure 30**

However, all tangent lines have the following interesting property in common: If we focus on a small portion of the curve very close to the point P—in other words, if we zoom in to the graph near the point P—the curve will appear almost straight and almost indistinguishable from the tangent line (Figure 31).*

* For a simulation of what is happening in Figure 31, go to the Website and follow
Everything for Applied Calc
→ Section 3.5 Practice/Visualize
→ Zooming In on a Curve

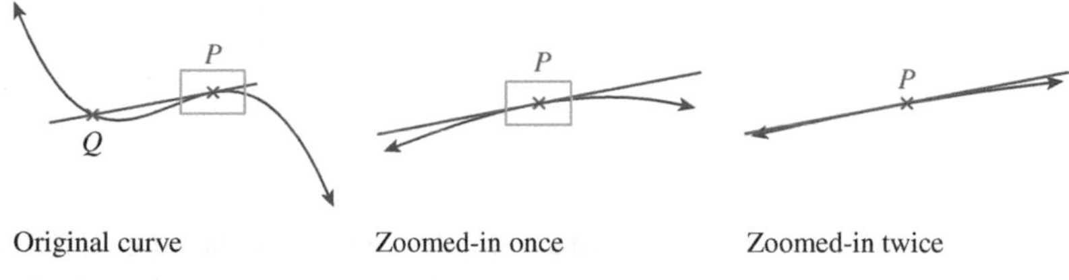

Figure 31

You can check this property by zooming in on the curve shown in Figures 27 and 28 in Example 1 near the point where $x = 2$.

Secant and Tangent Lines

The *slope of the secant line* through the points on the graph of f where $x = a$ and $x = a + h$ is given by the average rate of change, or difference quotient:

$$m_{\text{sec}} = \text{Slope of secant} = \text{Average rate of change} = \frac{f(a + h) - f(a)}{h}.$$

The *slope of the tangent line* through the point on the graph of f where $x = a$ is given by the instantaneous rate of change, or derivative:

$$m_{\tan} = \text{Slope of tangent} = \text{Instantaneous rate of change} = \text{Derivative}$$
$$= f'(a) = \lim_{h\to 0}\frac{f(a + h) - f(a)}{h},$$

assuming that the derivative exists.

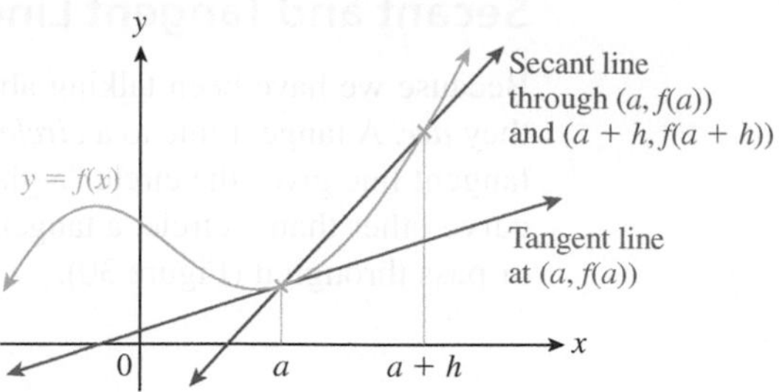

The smaller h gets, the closer the secant line in the figure above gets to being the tangent line at $(a, f(a))$.*

* On the Website, follow Everything for Applied Calc → Section 3.5 Practice/Visualize → Visualize the Derivative Graphically to see this process in an interactive graph.

Quick Example

3. In the following graph, the tangent line at the point where $x = 2$ has slope 3. Therefore, the derivative at $x = 2$ is 3. That is, $f'(2) = 3$.

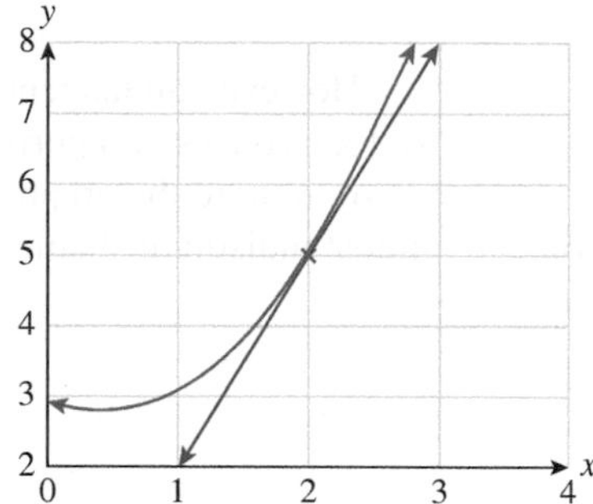

Note It might happen that the tangent line is vertical at some point or does not exist at all. These are cases in which f is not differentiable at the given point. (See Section 3.6 for examples.) ■

We can now give a more precise definition of what we mean by the tangent line to a point P on the graph of f at a given point: The **tangent line** to the graph of f at the point $P(a, f(a))$ is the straight line passing through P with slope $f'(a)$.

Quick Approximation of the Derivative

Q: *Do we always need to make tables of difference quotients as above in order to calculate an approximate value for the derivative? That seems like a large amount of work just to get an approximation.*

A: We can usually *approximate* the value of the derivative by using a single, small value of h. In the example above, the value $h = 0.0001$ would have given a pretty good approximation. The problems with using a fixed value of h are that (1) we do not get an *exact* answer, only an *approximation* of the derivative, and (2) how good an approximation it is depends on the function we're differentiating.* However, with most of the functions we'll be considering, setting $h = 0.0001$ does give us a good approximation.

* In fact, no matter how small the value we decide to use for h, it is possible to craft a function f for which the difference quotient at a is not even close to $f'(a)$.

Calculating a Quick Approximation of the Derivative

When f is differentiable at a, we can calculate an approximate value of $f'(a)$ by using the formula

$$f'(a) \approx \frac{f(a+h) - f(a)}{h} \qquad \text{Rate of change over } [a, a+h]$$

with a small value of h. The value $h = 0.0001$ works for most examples we encounter. (Students of numerical methods study the question of exactly how accurate this approximation is.)

Alternative Formula: The Balanced Difference Quotient

The following alternative formula, which measures the rate of change of f over the interval $[a - h, a + h]$, often gives a more accurate result and is the one used in many calculators:

$$f'(a) \approx \frac{f(a+h) - f(a-h)}{2h}. \qquad \text{Rate of change over } [a-h, a+h]$$

Notes

1. If f is a linear function, both approximations give the slope (the exact value of the derivative) regardless of the choice of $h \neq 0$.
2. If f is a quadratic function, the *balanced* difference quotient gives the exact value of the derivative regardless of the choice of $h \neq 0$.* ■

* To see why this is true, we would need to know some of the material that we study in the next chapter.

EXAMPLE 2 Quick Approximation of the Derivative

a. Calculate an approximate value of $f'(1.5)$ if $f(x) = x^2 - 4x$.

b. Find the equation of the tangent line at the point on the graph where $x = 1.5$.

Solution

a. We shall compute both the ordinary difference quotient and the balanced difference quotient.

Ordinary Difference Quotient: When we use $h = 0.0001$, the ordinary difference quotient is

$$\begin{aligned} f'(1.5) &\approx \frac{f(1.5 + 0.0001) - f(1.5)}{0.0001} \qquad \text{Ordinary difference quotient} \\ &= \frac{f(1.5001) - f(1.5)}{0.0001} \\ &= \frac{(1.5001^2 - 4 \times 1.5001) - (1.5^2 - 4 \times 1.5)}{0.0001} = -0.9999. \end{aligned}$$

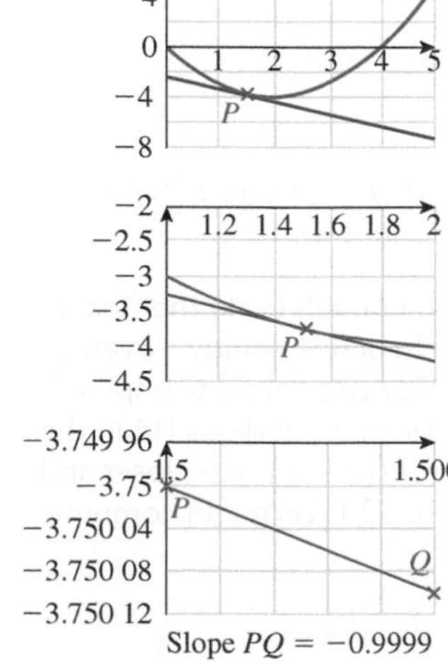

Figure 32

This answer is accurate to 0.0001; in fact, $f'(1.5) = -1$.

Graphically, we can picture this approximation as follows: Zoom in on the curve using the window $1.5 \le x \le 1.5001$, and measure the slope of the secant line joining both ends of the curve segment. Figure 32 shows close-up views of the curve and tangent line near the point P in which we are interested, the third view being the zoomed-in view used for this approximation. Notice that in the bottom window the tangent line and curve are indistinguishable. Also, the point P in which we are interested is on the left edge of the window.

Balanced Difference Quotient: For the balanced difference quotient, we get

$$f'(1.5) \approx \frac{f(1.5 + 0.0001) - f(1.5 - 0.0001)}{2(0.0001)} \qquad \text{Balanced difference quotient}$$

$$= \frac{f(1.5001) - f(1.4999)}{0.0002}$$

$$= \frac{(1.5001^2 - 4 \times 1.5001) - (1.4999^2 - 4 \times 1.4999)}{0.0002} = -1.$$

This balanced difference quotient gives the exact answer in this case. (Recall that the balanced difference quotient always gives the exact derivative for a quandratic function.) Graphically, it is as though we have zoomed in using a window that puts the point P in the *center* of the screen (Figure 33) rather than at the left edge.

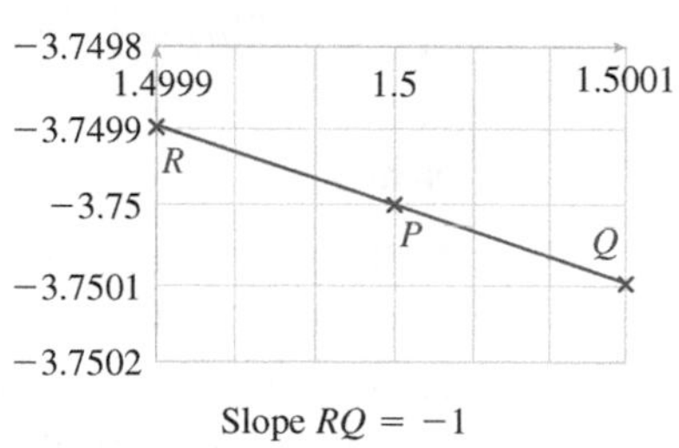

Figure 33

b. We find the equation of the tangent line from a point on the line and its slope, as we did in Chapter 1:

- ***Point*** $(1.5, f(1.5)) = (1.5, -3.75)$.
- ***Slope*** $m = f'(1.5) = -1$. Slope of the tangent line = Derivative

The equation is

$$y = mx + b,$$

where $m = -1$ and $b = y_1 - mx_1 = -3.75 - (-1)(1.5) = -2.25$. Thus, the equation of the tangent line is

$$y = -x - 2.25.$$

Using Technology

See the Technology Guides at the end of the chapter to find out how to calculate the quick approximations to the derivative in Example 2 using a TI-83/84 Plus or a spreadsheet. Here is an outline:

TI-83/84 Plus

`Y1=X^2-4*X`

Home screen:

```
(Y1(1.5001)-Y1(1.5))/
0.0001
(Y1(1.5001)-
Y1(1.4999))/0.0002
```

[More details in the Technology Guide.]

Spreadsheet

Headings a, h, x, $f(x)$, Diff Quotient, Balanced Diff Quotient in A1–F1

`1.5` in A2, `0.0001` in B2, `=A2-B2` in C2, `=A2` in C3, `=A2+B2` in C4

`=C2^2-4*C2` in D2; copy down to D4

`=(D3-D2)/(C3-C2)` in E2

`=(D4-D2)/(C4-C2)` in E3

[More details in the Technology Guide.]

Q: *Why can't we simply use h = 0.000 000 000 000 000 000 01 for an incredibly accurate approximation to the instantaneous rate of change and be done with it?*

A: This approach would certainly work if you were patient enough to do the (thankless) calculation by hand! However, doing it with the help of technology—even an ordinary calculator—will cause problems: The issue is that calculators and spreadsheets represent numbers with a maximum number of significant digits (15 in the case of Excel). As the value of h gets smaller, the value of $f(a + h)$ gets closer and closer to the value of $f(a)$. For example, if $f(x) = 50 + 0.1x^4$, Excel might compute

$$f(2 + 0.000\,000\,000\,000\,1) - f(2)$$

$$= 51.600\,000\,000\,000\,3 - 51.6 \qquad \text{Rounded to 15 digits}$$

$$= 0.000\,000\,000\,000\,3,$$

and the corresponding difference quotient would be 3, not 3.2 as it should be. If h gets even smaller, Excel will not be able to distinguish between $f(a + h)$ and $f(a)$ at all, in which case it will compute 0 for the rate of change. This loss in accuracy when subtracting two very close numbers is called **subtractive error**.

Thus, there is a trade-off in lowering the value of h: Smaller values of h yield *mathematically* more accurate approximations of the derivative, but if h gets too small, subtractive error becomes a problem and decreases the accuracy of computations that use technology.

Leibniz *d* Notation

We introduced the notation $f'(x)$ for the derivative of f at x, but there is another interesting notation. We have written the average rate of change as

$$\text{Average rate of change} = \frac{\Delta f}{\Delta x}. \qquad \frac{\text{Change in } f}{\text{Change in } x}$$

As we use smaller and smaller values for Δx, we approach the instantaneous rate of change, or derivative, for which we also have the notation df/dx, due to Leibniz:

$$\text{Instantaneous rate of change} = \lim_{\Delta x \to 0} \frac{\Delta f}{\Delta x} = \frac{df}{dx}.$$

That is, df/dx is just another notation for $f'(x)$. Do not think of df/dx as an actual quotient of two numbers: Remember that we use an actual quotient $\Delta f/\Delta x$ only to *approximate* the value of df/dx.

In Example 3 we apply the quick approximation method of estimating the derivative.

EXAMPLE 3 Velocity

My friend Eric, an enthusiastic baseball player, claims that he can "probably" throw a ball upward at a speed of 100 feet per second (ft/sec).* Our physicist friends tell us that its height s (in feet) t seconds later would be $s = 100t - 16t^2$. Find its average velocity over the interval $[2, 3]$ and its instantaneous velocity exactly 2 seconds after Eric throws it.

* Eric's claim is difficult to believe; 100 ft/sec corresponds to around 68 mph, and professional pitchers can throw *forward* at about 100 mph.

Solution The graph of the ball's height as a function of time is shown in Figure 34. Asking for the velocity is really asking for the rate of change of height with respect to time. (Why?) Consider average velocity first. To compute the **average velocity** of the ball from time 2 to time 3, we first compute the change in height:

$$\Delta s = s(3) - s(2) = 156 - 136 = 20 \text{ feet.}$$

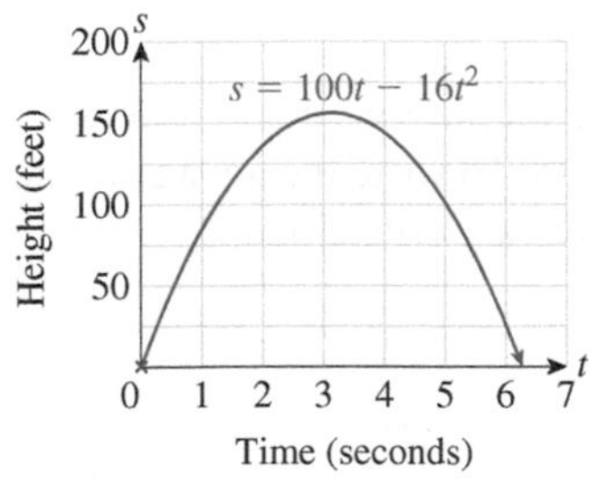

Figure 34

Since it rises 20 feet in $\Delta t = 1$ second, we use the defining formula *speed* = *distance*/*time* to get the average velocity:

$$\text{Average velocity} = \frac{\Delta s}{\Delta t} = \frac{20}{1} = 20 \text{ ft/sec}$$

from time $t = 2$ to $t = 3$. This is just the difference quotient, so we have the following:

The average velocity is the average rate of change of height.

To get the **instantaneous velocity** at $t = 2$, we find the instantaneous rate of change of height. In other words, we need to calculate the derivative ds/dt at $t = 2$.

Which approximation should we use? Because s is a quadratic function of t, we know that the balanced quick approximation will give us the exact derivative, so this is the approximation we choose:

$$\begin{aligned}\frac{ds}{dt} &\approx \frac{s(2+0.0001)-s(2-0.0001)}{2(0.0001)}\\ &= \frac{s(2.0001)-s(1.9999)}{0.0002}\\ &= \frac{100(2.0001)-16(2.0001)^2-(100(1.9999)-16(1.9999)^2)}{0.0002}\\ &= 36 \text{ ft/sec.}\end{aligned}$$

The instantaneous velocity at $t = 2$ is exactly 36 ft/sec.

➡ **Before we go on...** If we repeat the calculation in Example 3 at time $t = 5$, we get

$$\frac{ds}{dt} = -60 \text{ ft/sec.}$$

The negative sign tells us that the ball is *falling* at a rate of 60 feet per second at time $t = 5$. (How does the fact that it is falling at $t = 5$ show up on the graph?) ■

Example 3 gives another interpretation of the derivative.

Average and Instantaneous Velocity

For an object moving in a straight line with position $s(t)$ at time t, the **average velocity** from time t to time $t + h$ is the average rate of change of position with respect to time:

$$v_{\text{avg}} = \frac{s(t+h)-s(t)}{h} = \frac{\Delta s}{\Delta t}.$$

Average velocity = Average rate of change of position

The **instantaneous velocity** at time t is

$$v = \lim_{h\to 0} \frac{s(t+h)-s(t)}{h} = \frac{ds}{dt}.$$

Instantaneous velocity = Instantaneous rate of change of position

In other words, *instantaneous velocity is the derivative of position with respect to time.*

Here is one last comment on Leibniz notation. In Example 3 we could have written the velocity either as s' or as ds/dt, as we chose to do. To write the answer to the question, that the velocity at $t = 2$ sec was 36 ft/sec, we can write either

$$s'(2) = 36$$

or

$$\left.\frac{ds}{dt}\right|_{t=2} = 36.$$

The notation "$|_{t=2}$" is read "evaluated at $t = 2$." Similarly, if $y = f(x)$, we can write the instantaneous rate of change of f at $x = 5$ either in functional notation as

$$f'(5)$$ The derivative of f, evaluated at $x = 5$

or in Leibniz notation as

$$\left.\frac{dy}{dx}\right|_{x=5}. \qquad \text{The derivative of } y \text{, evaluated at } x = 5$$

The latter notation is obviously more cumbersome than the functional notation $f'(5)$, but the notation dy/dx has compensating advantages. You should practice using both notations.

The Derivative Function

The derivative $f'(x)$ is a number we can calculate, or at least approximate, for various values of x. Because $f'(x)$ depends on the value of x, we may think of f' as a function of x. This function is the **derivative function**.

Derivative Function

If f is a function, its **derivative function** f' is the function whose value $f'(x)$ is the derivative of f at x. Its domain is the set of all x at which f is differentiable. Equivalently, f' associates to each x the slope of the tangent to the graph of the function f at x, or the rate of change of f at x. The formula for the derivative function is

$$f'(x) = \lim_{h\to 0} \frac{f(x+h) - f(x)}{h}. \qquad \text{Derivative function}$$

Quick Examples

4. Let $f(x) = 3x - 1$. The graph of f is a straight line that has slope 3 everywhere. In other words, $f'(x) = 3$ for every choice of x; that is, f' is a constant function.

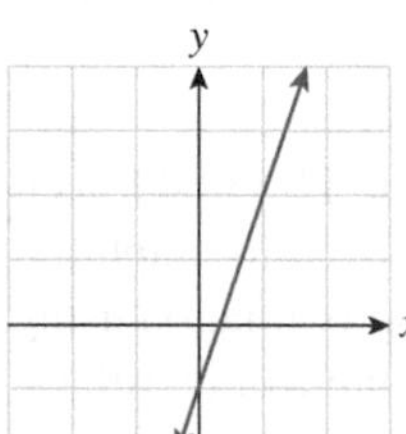

Original Function f

$f(x) = 3x - 1$

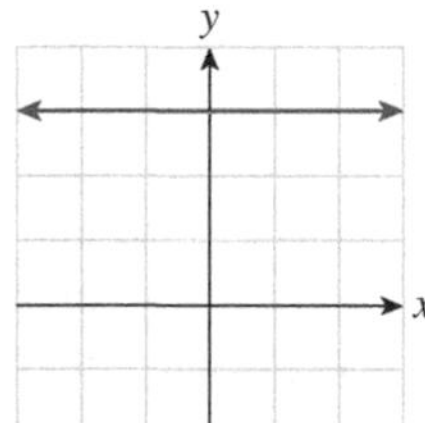

Derivative Function f'

$f'(x) = 3$

5. Given the graph of a function f, we can get a rough sketch of the graph of f' by estimating the slope of the tangent to the graph of f at several points, as illustrated below.*

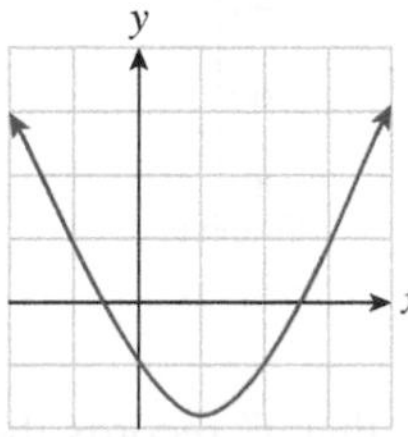

Original Function f

$y = f(x)$

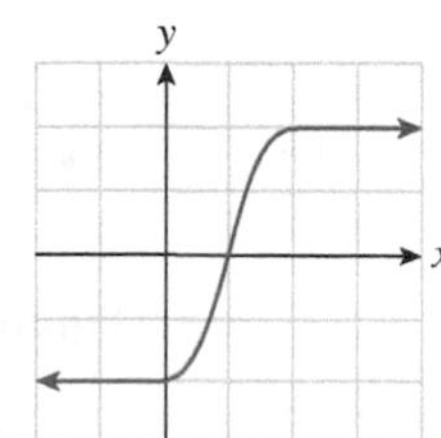

Derivative Function f'

$y = f'(x)$

* This method is discussed in detail on the Website at Online Text → Sketching the Graph of the Derivative

For x between -2 and 0 the graph of f is linear with slope -2. As x increases from 0 to 2, the slope increases from -2 to 2. For x larger than 2 the graph of f is linear with slope 2. (Notice that, when $x = 1$, the graph of f has a horizontal tangent, so $f'(1) = 0$.)

6. Look again at the graph on the left in Quick Example 5. When $x < 1$, the derivative $f'(x)$ is negative, so the graph has negative slope and f is **decreasing**; its values are going down as x increases. When $x > 1$, the derivative $f'(x)$ is positive, so the graph has positive slope and f is **increasing**; its values are going up as x increases.

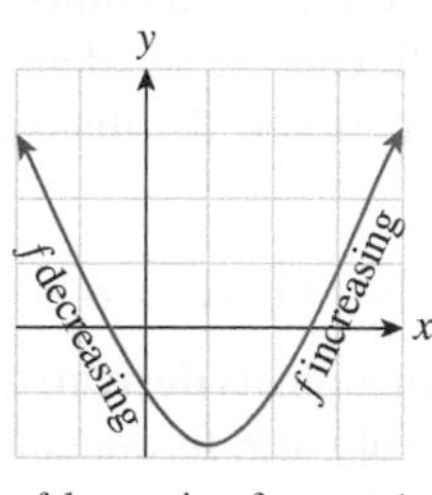

f decreasing for $x < 1$
f increasing for $x > 1$

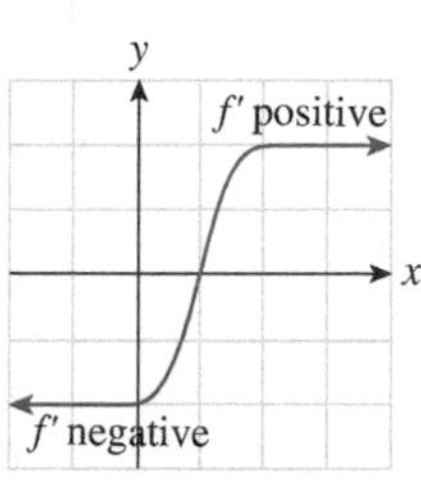

f' negative for $x < 1$
f' positive for $x > 1$

Using Technology

See the Technology Guides at the end of the chapter to find out how to obtain a table of values of and graph the derivative in Example 4 using a TI-83/84 Plus or a spreadsheet. Here is an outline:

TI-83/84 Plus
`Y1=-2X^2+6X+5`
`Y2=nDeriv(Y1,X,X)`
[More details in the Technology Guide.]

Spreadsheet
Value of h in E2
Values of x from A2 down increasing by h
`-2*A2^2+6*A2+5` from B2 down
`=(B3-B2)/$E$2` from C2 down
Insert scatter chart using columns A and C. [More details in the Technology Guide.]

 **Website**
www.WanerMath.com
Web grapher:
Online Utilities→ Function Evaluator and Grapher
Enter `deriv(-2*x^2+6*x+5)` for y_1. Alternatively, enter `-2*x^2+6*x+5` for y_1 and `deriv(y1)` for y_2.
Excel grapher:
Online Utilities→ Excel First and Second Derivative Graphing Utility Function: `-2*x^2+6*x+5`

The following example shows how we can use technology to graph the (approximate) derivative of a function, where it exists.

EXAMPLE 4 Graphing the Derivative with Technology

Use technology to graph the derivative of $f(x) = -2x^2 + 6x + 5$ for values of x starting at -5.

Solution The TI-83/84 Plus has a built-in function that approximates the derivative, and we can use it to graph the derivative of a function. In a spreadsheet we need to create the approximation using one of the quick approximation formulas, and we can then graph a table of its values. See the technology note in the margin to find out how to graph the derivative (Figure 35) using the Website graphing utility, the TI-83/84 Plus, or a spreadsheet.

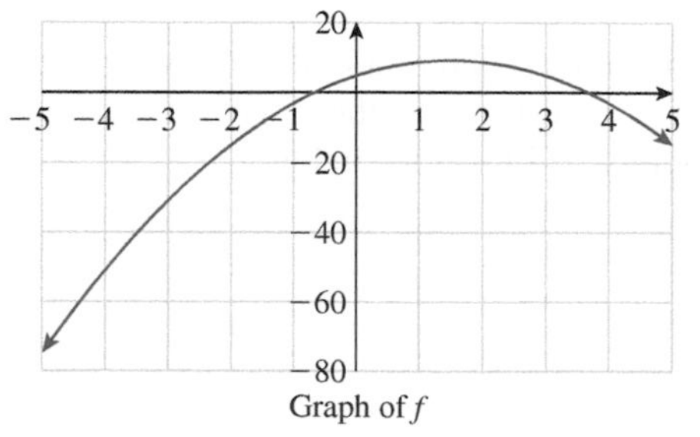

Graph of f

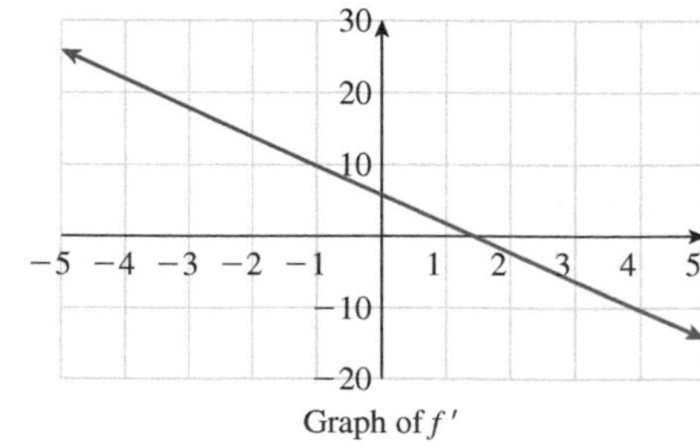

Graph of f'

Figure 35

We said that f' records the slope of (the tangent line to) the function f at each point. Notice that the graph of f' confirms that the slope of the graph of f is decreasing as x increases from -5 to 5. Note also that the graph of f reaches a high point at

$x = 1.5$ (the vertex of the parabola). At that point, the slope of the tangent is zero; that is, $f'(1.5) = 0$, as we see in the graph of f'.

EXAMPLE 5 T An Application: Broadband Penetration

Wired broadband penetration in the United States can be modeled by

$$P(t) = \frac{29.238}{1 + 8.998(1.638)^{-x}} \qquad (0 \le t \le 14),$$

where t is time in years since 2000.[56] Graph both P and its derivative, and determine when broadband penetration was growing most rapidly.

Solution Using one of the methods in Example 4, we obtain the graphs shown in Figure 36.

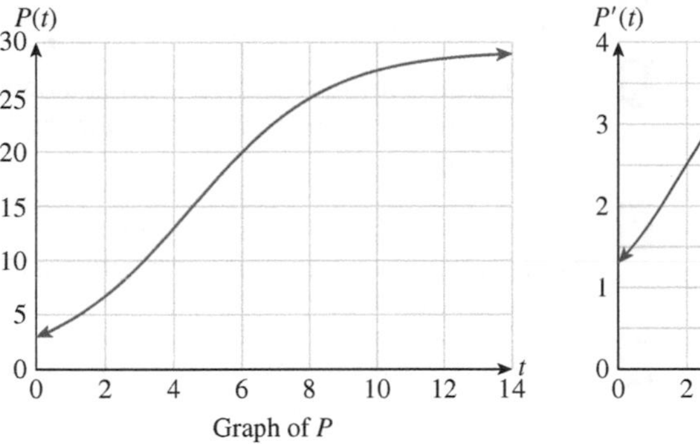

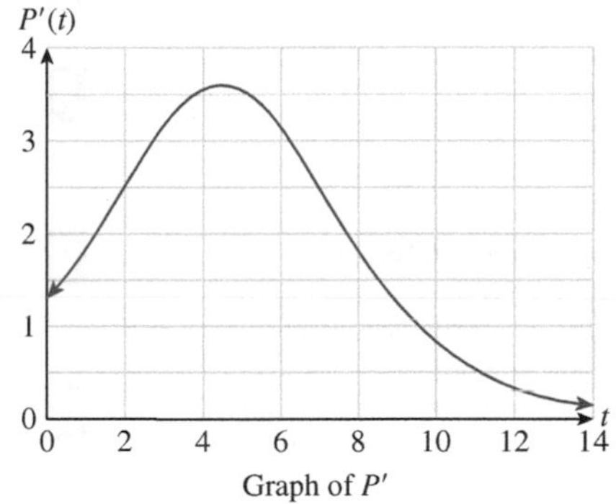

Figure 36

From the graph on the right, we see that P' reaches a peak somewhere between $t = 4$ and $t = 5$ (sometime during 2004). Recalling that P' measures the *slope* of the graph of P, we can conclude that the graph of P is steepest between $t = 4$ and $t = 5$, indicating that, according to the model, broadband penetration was growing most rapidly sometime during 2004. Notice that this is not so easy to see directly on the graph of P.

To determine the point of maximum growth more accurately, we can zoom in on the graph of P' using the range $4.0 \le t \le 5.0$ (Figure 37).

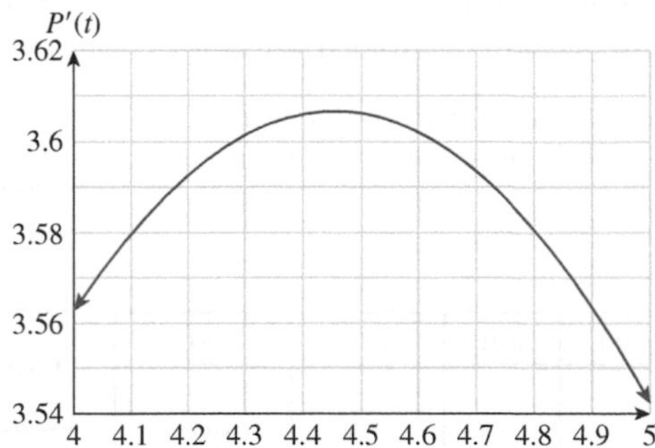

Figure 37

[56] Broadband penetration is the number of broadband installations divided by the total population. Source for data: Organisation for Economic Co-operation and Development (OECD) Directorate for Science, Technology, and Industry, table of Historical Penetration Rates, December 2013, downloaded September 2014 from www.oecd.org/sti/ict/broadband. Model is extrapolated to the range shown.

We can now see that P' reaches its highest point around $t = 4.45$, so we conclude that broadband penetration was growing most rapidly in mid-2004.

➡ **Before we go on...** Besides helping us to determine the point of maximum growth, the graph of P' in Example 5 gives us a great deal of additional information. As just one example, in Figure 37 we can see that the maximum value of P' is about 3.61, indicating that broadband penetration grew at a fastest rate of about 3.61 percentage points per year. ■

FAQs

Recognizing When and How to Compute the Instantaneous Rate of Change

Q: *How do I know, by looking at the wording of a problem, that it is asking for an instantaneous rate of change?*

A: If a problem does not ask for an instantaneous rate of change directly, it might do so indirectly, as in "How fast is quantity q increasing?" or "Find the rate of increase of q."

Q: *If I know that a problem calls for estimating an instantaneous rate of change, how should I estimate it: with a table showing smaller and smaller values of h or by using a quick approximation?*

A: For most practical purposes a quick approximation is accurate enough. Use a table showing smaller and smaller values of h when you would like to check the accuracy.

Q: *Which should I use in computing a quick approximation: the balanced difference quotient or the ordinary difference quotient?*

A: In general, the balanced difference quotient gives a more accurate answer.

Website
www.WanerMath.com
At the Website you can find the following optional interactive online sections:
- Continuity and Differentiability
- Sketching the Graph of the Derivative

You can find these sections by following
Everything for Applied Calc→ Chapter 3 (Online Sections)

3.5 EXERCISES

▼ more advanced ◆ challenging
T indicates exercises that should be solved using technology

In Exercises 1–4, estimate the derivative from the table of average rates of change. [**HINT**: See discussion at the beginning of the section.]

1. Estimate $f'(5)$.

h	1	0.1	0.01	0.001	0.0001
Avg. Rate of Change of f over $[5, 5+h]$	12	6.4	6.04	6.004	6.0004
h	−1	−0.1	−0.01	−0.001	−0.0001
Avg. Rate of Change of f over $[5+h, 5]$	3	5.6	5.96	5.996	5.9996

2. Estimate $g'(7)$.

h	1	0.1	0.01	0.001	0.0001
Avg. Rate of Change of g over $[7, 7+h]$	4	4.8	4.98	4.998	4.9998
h	−1	−0.1	−0.01	−0.001	−0.0001
Avg. Rate of Change of g over $[7+h, 7]$	5	5.3	5.03	5.003	5.0003

3. Estimate $r'(-6)$.

h	1	0.1	0.01	0.001	0.0001
Avg. Rate of Change of r over $[-6, -6+h]$	−5.4	−5.498	−5.4998	−5.499982	−5.49999822
h	−1	−0.1	−0.01	−0.001	−0.0001
Avg. Rate of Change of r over $[-6+h, -6]$	−7.52	−6.13	−5.5014	−5.5000144	−5.500001444

4. Estimate $s'(0)$.

h	1	0.1	0.01	0.001	0.0001
Avg. Rate of Change of s over $[0, h]$	−2.52	−1.13	−0.6014	−0.6000144	−0.6000001444
h	−1	−0.1	−0.01	−0.001	−0.0001
Avg. Rate of Change of s over $[h, 0]$	−0.4	−0.598	−0.5998	−0.599982	−0.59999822

Consider the functions in Exercises 5–8 as representing the value of an ounce of palladium in U.S. dollars as a function of the time t in days.[57] *Find the average rates of change of R(t) over the time intervals [t, t + h], where t is as indicated and h = 1, 0.1, and 0.01 days. Hence, estimate the instantaneous rate of change of R at time t, specifying the units of measurement. (Use smaller values of h to check your estimates.)* [**HINT**: See Example 1.]

5. $R(t) = 60 + 50t - t^2;\ t = 5$

6. $R(t) = 60t - 2t^2;\ t = 3$

7. $R(t) = 270 + 20t^3;\ t = 1$

8. $R(t) = 200 + 50t - t^3;\ t = 2$

In Exercises 9–12 the function gives the cost to manufacture x items. Find the average cost per unit of manufacturing h more items (i.e., the average rate of change of the total cost) at a production level of x, where x is as indicated and h = 10 and 1. Hence, estimate the instantaneous rate of change of the total cost at the given production level x, specifying the units of measurement. (Use smaller values of h to check your estimates.) [**HINT**: See Example 1.]

9. $C(x) = 10{,}000 + 5x - \dfrac{x^2}{10{,}000};\ x = 1{,}000$

10. $C(x) = 20{,}000 + 7x - \dfrac{x^2}{20{,}000};\ x = 10{,}000$

11. $C(x) = 15{,}000 + 100x + \dfrac{1{,}000}{x};\ x = 100$

12. $C(x) = 20{,}000 + 50x + \dfrac{10{,}000}{x};\ x = 100$

In Exercises 13–16 the graph of a function is shown together with the tangent line at a point P. Estimate the derivative of f at the corresponding x value. [**HINT**: See Quick Example 3.]

13.

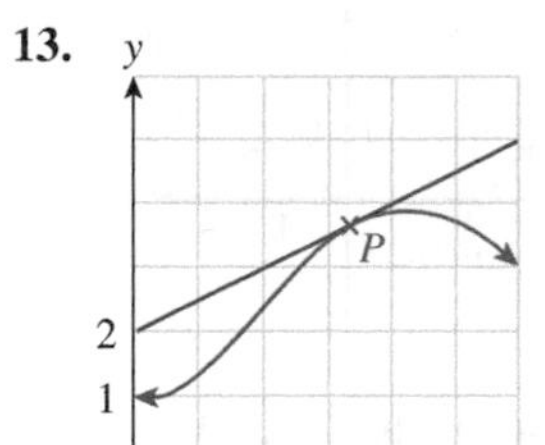

14.

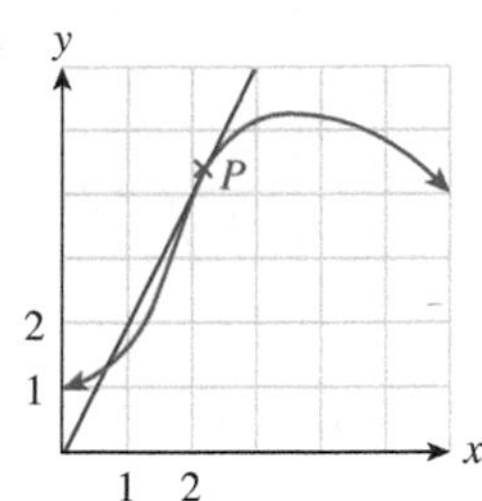

15.

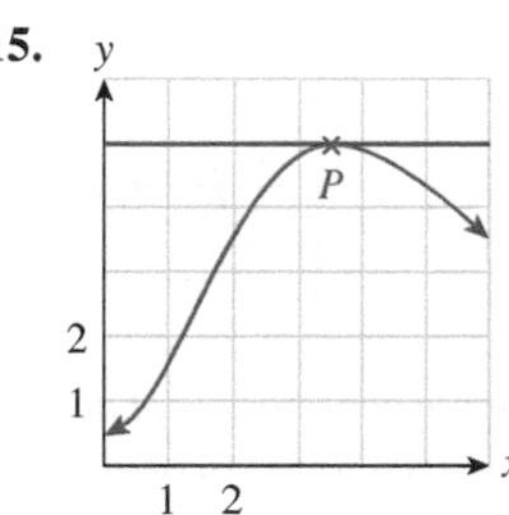

16.

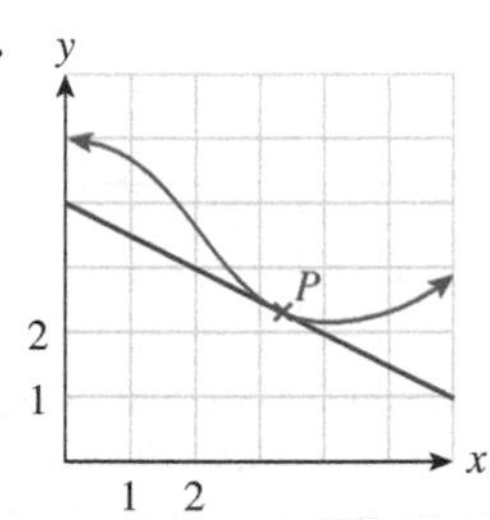

*In Exercises 17–22, say at which labeled point the slope of the tangent is **(a)** greatest and **(b)** least (in the sense that −7 is less than 1).* [**HINT**: See Quick Example 3.]

17.

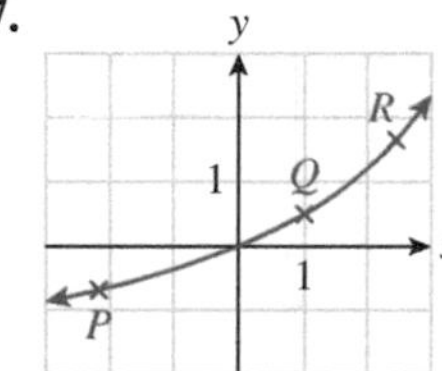

18.

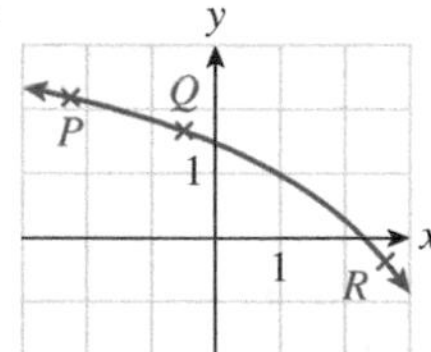

19.

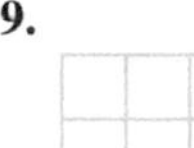
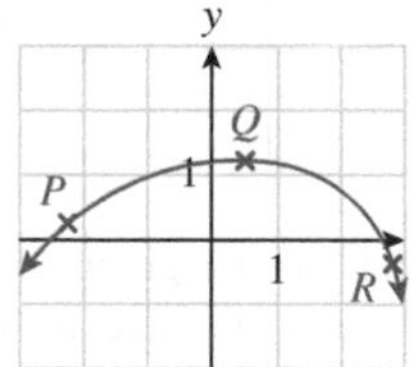

20.

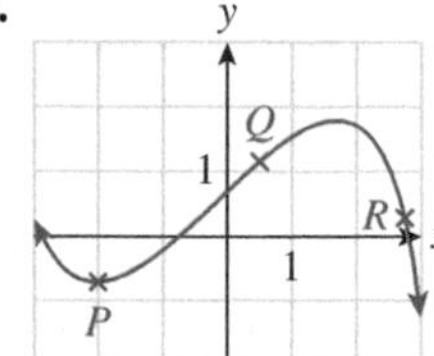

21.

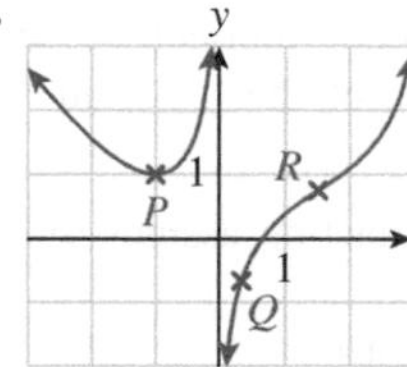

22.

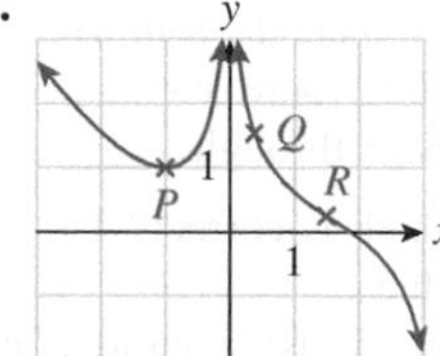

[57] Palladium was trading at around $290 in August 2008.

In each of Exercises 23–26, three slopes are given. For each slope, determine at which of the labeled points on the graph the tangent line has that slope.

23. a. 0 **b.** 4 **c.** -1

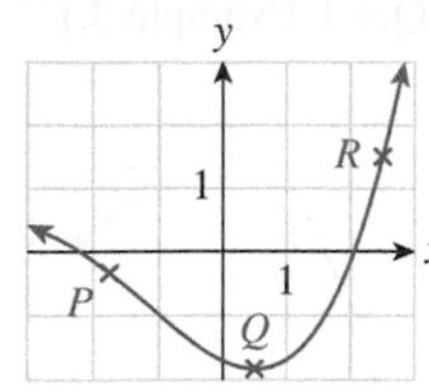

24. a. 0 **b.** 1 **c.** -1

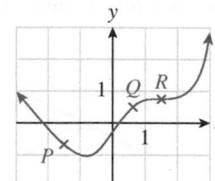

25. a. 0 **b.** 3 **c.** -3

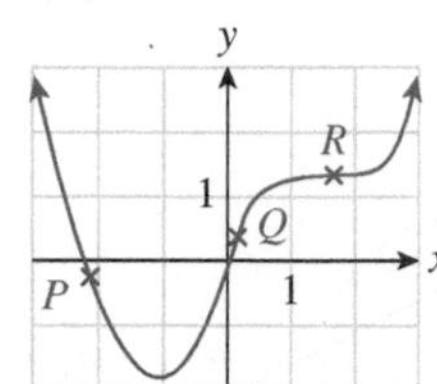

26. a. 0 **b.** 3 **c.** 1

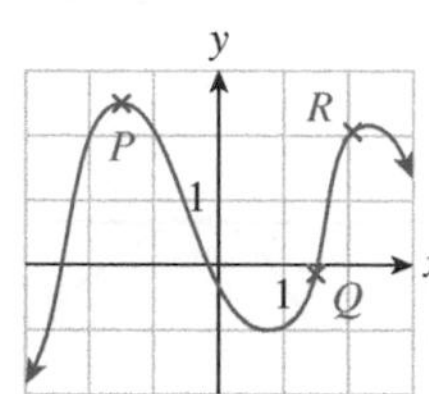

In Exercises 27–30, find the approximate coordinates of all points (if any) where the slope of the tangent is (a) 0, *(b)* 1, and *(c)* -1. [**HINT:** See Quick Example 3.]

27.

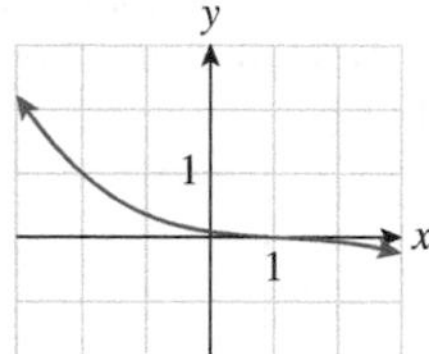

28.

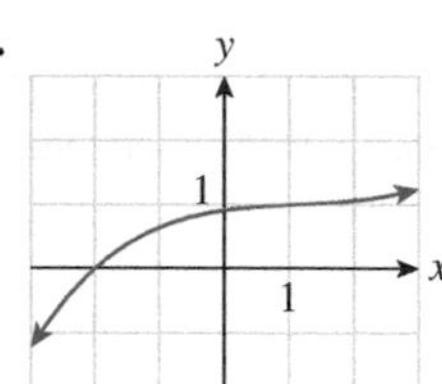

29.

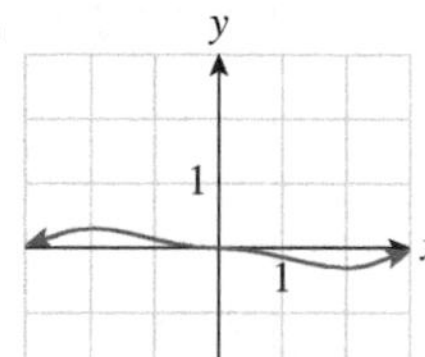

30.

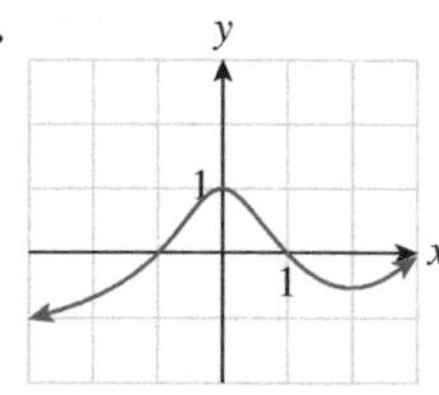

31. Complete the following: The tangent to the graph of the function f at the point where $x = a$ is the line passing through the point ____ with slope ____.

32. Complete the following: The difference quotient for f at the point where $x = a$ gives the slope of the ____ line that passes through _____.

33. Which is correct? The derivative function assigns to each value x

(A) the average rate of change of f at x.

(B) the slope of the tangent to the graph of f at $(x, f(x))$.

(C) the rate at which f is changing over the interval $[x, x + h]$ for $h = 0.0001$.

(D) the balanced difference quotient $[f(x + h) - f(x - h)]/(2h)$ for $h \approx 0.0001$.

34. Which is correct? The derivative function $f'(x)$ tells us

(A) the slope of the tangent line at each of the points $(x, f(x))$.

(B) the approximate slope of the tangent line at each of the points $(x, f(x))$.

(C) the slope of the secant line through $(x, f(x))$ and $(x + h, f(x + h))$ for $h = 0.0001$.

(D) the slope of a certain secant line through each of the points $(x, f(x))$.

35. ▼ Let f have the graph shown.

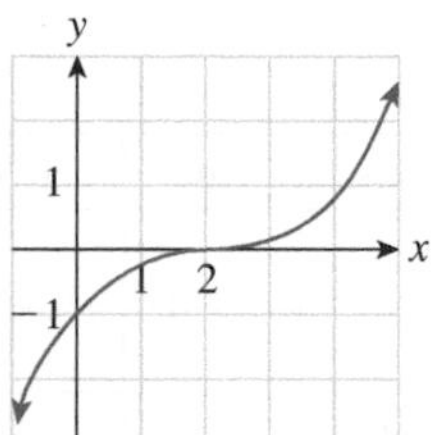

a. The average rate of change of f over the interval $[2, 4]$ is
(A) greater than $f'(2)$. **(B)** less than $f'(2)$.
(C) approximately equal to $f'(2)$.

b. The average rate of change of f over the interval $[-1, 1]$ is
(A) greater than $f'(0)$. **(B)** less than $f'(0)$.
(C) approximately equal to $f'(0)$.

c. Over the interval $[0, 2]$ the instantaneous rate of change of f is
(A) increasing. **(B)** decreasing. **(C)** neither.

d. Over the interval $[0, 4]$ the instantaneous rate of change of f is
(A) increasing, then decreasing.
(B) decreasing, then increasing.
(C) always increasing.
(D) always decreasing.

e. When $x = 4$, $f(x)$ is
(A) approximately 0 and increasing at a rate of about 0.7 units per unit of x.
(B) approximately 0 and decreasing at a rate of about 0.7 units per unit of x.
(C) approximately 0.7 and increasing at a rate of about 1 unit per unit of x.
(D) approximately 0.7 and increasing at a rate of about 3 units per unit of x.

36. ▼ A function f has the following graph.

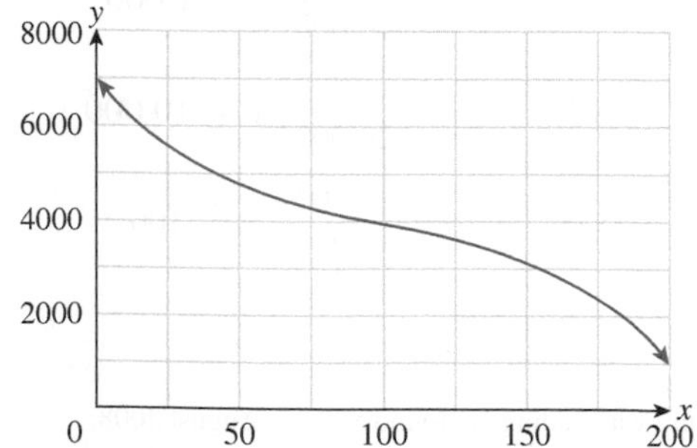

a. The average rate of change of f over $[0, 200]$ is
(A) greater than
(B) less than
(C) approximately equal to
the instantaneous rate of change at $x = 100$.

b. The average rate of change of f over $[0, 200]$ is
(A) greater than
(B) less than
(C) approximately equal to
the instantaneous rate of change at $x = 150$.

c. Over the interval $[0, 50]$ the instantaneous rate of change of f is
(A) increasing, then decreasing.
(B) decreasing, then increasing.
(C) always increasing.
(D) always decreasing.

d. Over the interval $[0, 200]$ the instantaneous rate of change of f is
(A) always positive. **(B)** always negative.
(C) negative, positive, and then negative.

e. $f'(100)$ is
(A) greater than $f'(25)$. **(B)** less than $f'(25)$.
(C) approximately equal to $f'(25)$.

In Exercises 37–40, use a quick approximation to estimate the derivative of the given function at the indicated point. [**HINT:** See Example 2(a).]

37. $f(x) = 1 - 2x; x = 2$ **38.** $f(x) = \frac{x}{3} - 1; x = -3$

39. $f(x) = \frac{x^2}{4} - \frac{x^3}{3}; x = -1$ **40.** $f(x) = \frac{x^2}{x} + \frac{x}{4}; x = 2$

In Exercises 41–48, estimate the indicated derivative by any method. [**HINT:** See Example 2.]

41. $g(t) = \frac{1}{t^5}$; estimate $g'(1)$

42. $s(t) = \frac{1}{t^3}$; estimate $s'(-2)$

43. $y = 4x^2$; estimate $\left.\frac{dy}{dx}\right|_{x=2}$

44. $y = 1 - x^2$; estimate $\left.\frac{dy}{dx}\right|_{x=-1}$

45. $s = 4t + t^2$; estimate $\left.\frac{ds}{dt}\right|_{t=-2}$

46. $s = t - t^2$; estimate $\left.\frac{ds}{dt}\right|_{t=2}$

47. $R = \frac{1}{p}$; estimate $\left.\frac{dR}{dp}\right|_{p=20}$

48. $R = \sqrt{p}$; estimate $\left.\frac{dR}{dp}\right|_{p=400}$

In Exercises 49–54, **(a)** *use any method to estimate the slope of the tangent to the graph of the given function at the point with the given x-coordinate, and* **(b)** *find an equation of the tangent line in part (a). In each case, sketch the curve together with the appropriate tangent line.* [**HINT:** See Example 2(b).]

49. $f(x) = x^3; x = -1$ **50.** $f(x) = x^2; x = 0$

51. $f(x) = x + \frac{1}{x}; x = 2$ **52.** $f(x) = \frac{1}{x^2}; x = 1$

53. $f(x) = \sqrt{x}; x = 4$ **54.** $f(x) = 2x + 4; x = -1$

In Exercises 55–58, estimate the given quantity.

55. $f(x) = e^x$; estimate $f'(0)$

56. $f(x) = 2e^x$; estimate $f'(1)$

57. $f(x) = \ln x$; estimate $f'(1)$

58. $f(x) = \ln x$; estimate $f'(2)$

In Exercises 59–64, match the graph of f to the graph of f'. (The graphs of f' are shown after Exercise 64.)

59. ▼ **60.** 

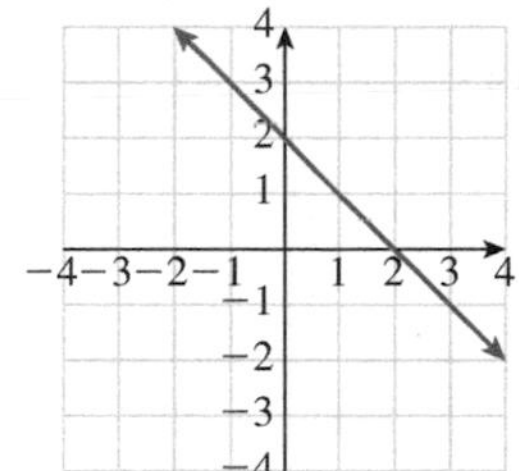

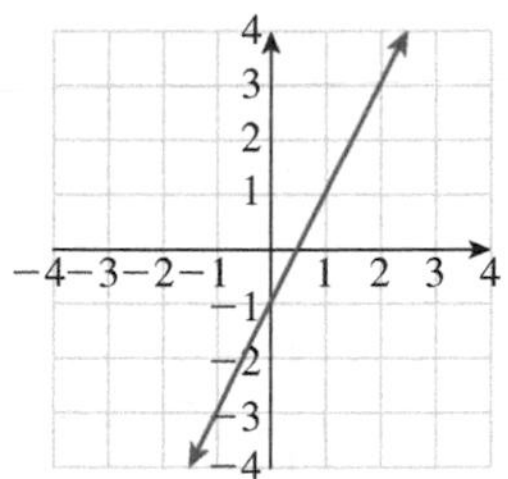

61.  **62.** 

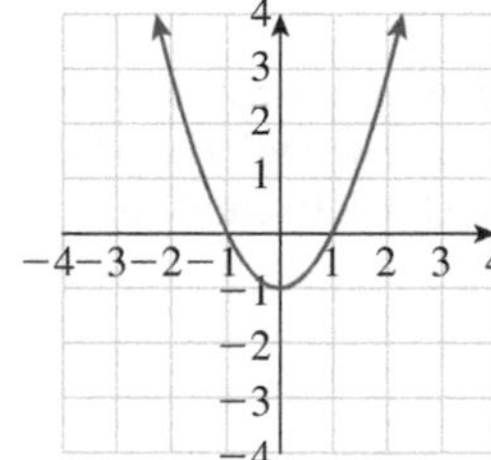

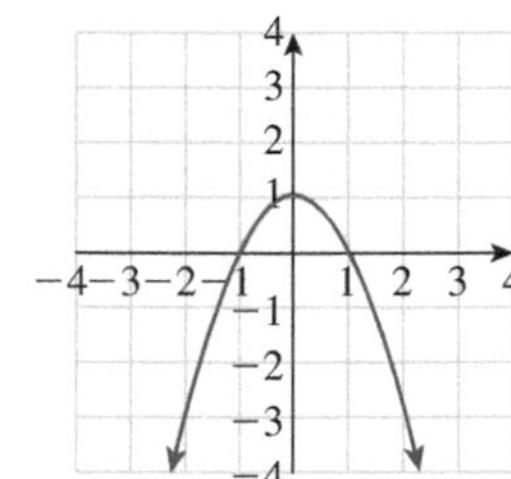

63. ▼ **64.** 

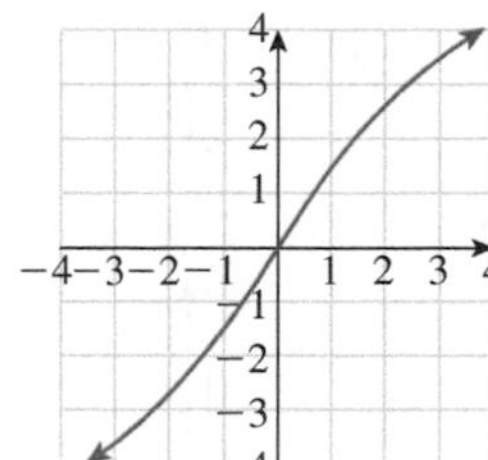

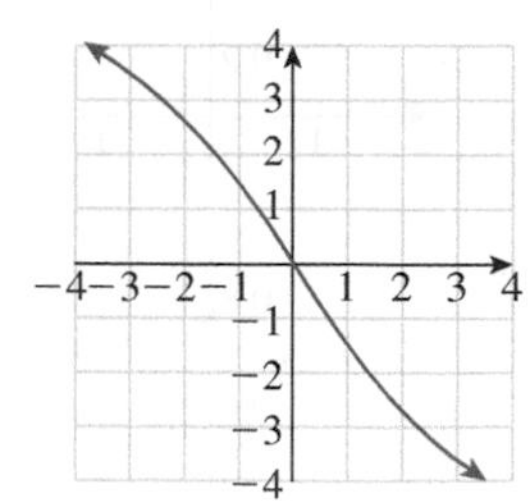

Graphs of derivatives for Exercises 59–64:

(A)

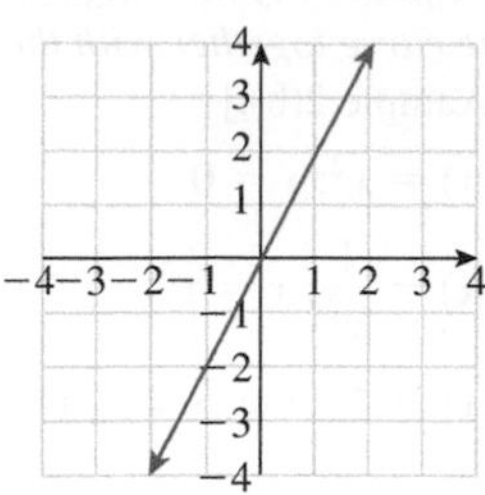

(B)

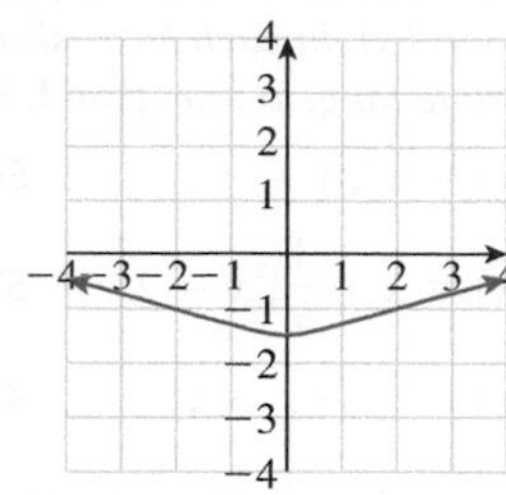

(C)

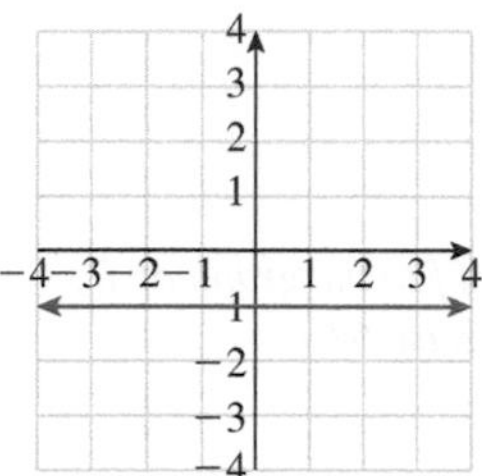

(D)

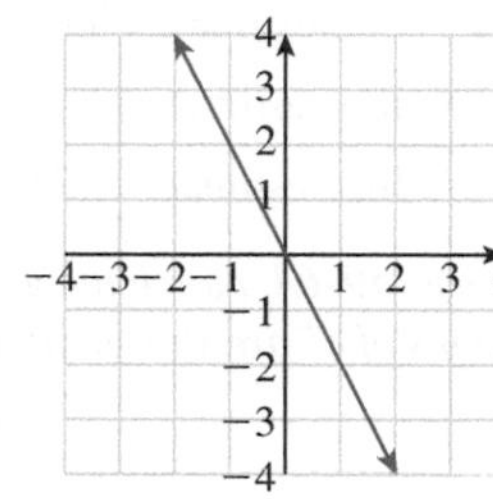

(E)

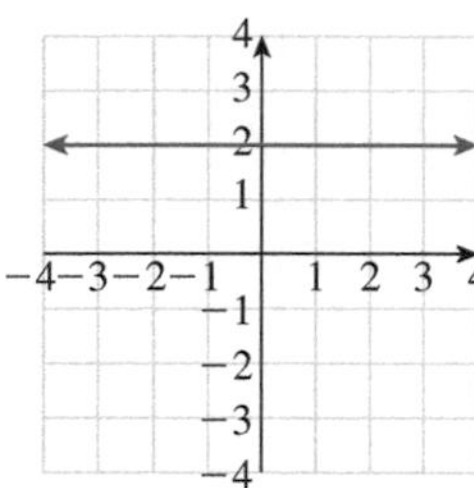

(F)

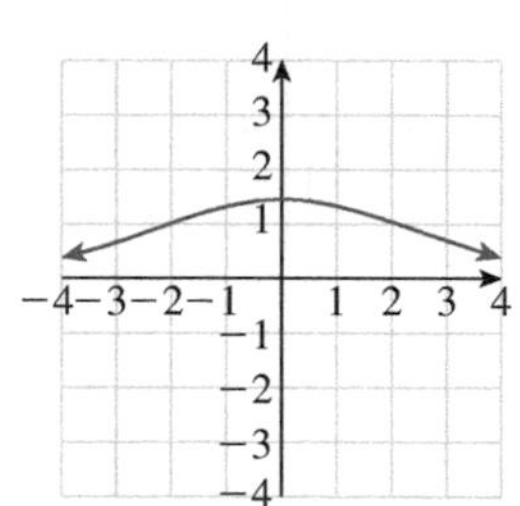

In Exercises 65–68 the graph of a function is given. For which x in the range shown is the function increasing? For which x is the function decreasing? [**HINT:** See Quick Example 6.]

65.

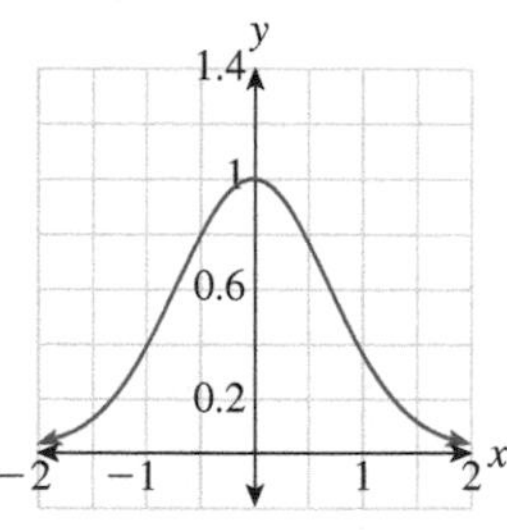

66.

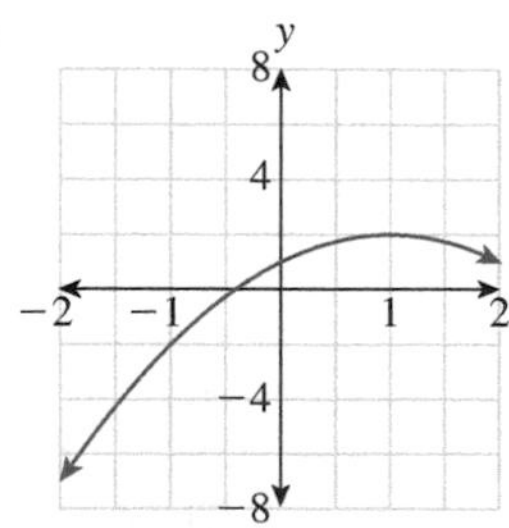

67.

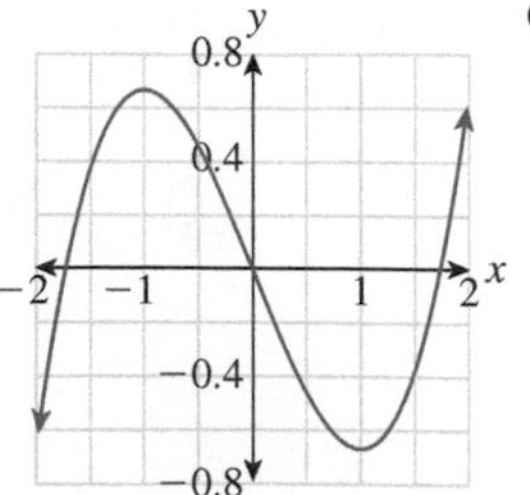

68.

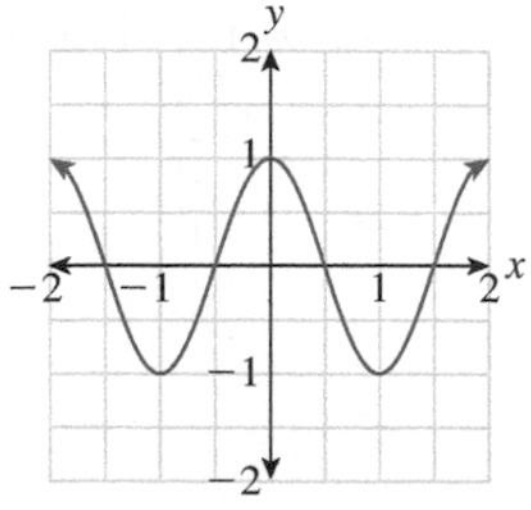

In Exercises 69–72 the graph of the derivative of a function is given. For which x is the (original) function increasing? For which x is the (original) function decreasing? [**HINT:** See Quick Example 6.]

69.

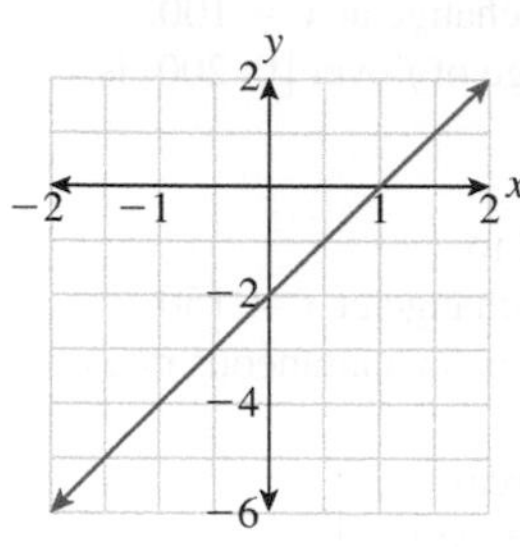

70.

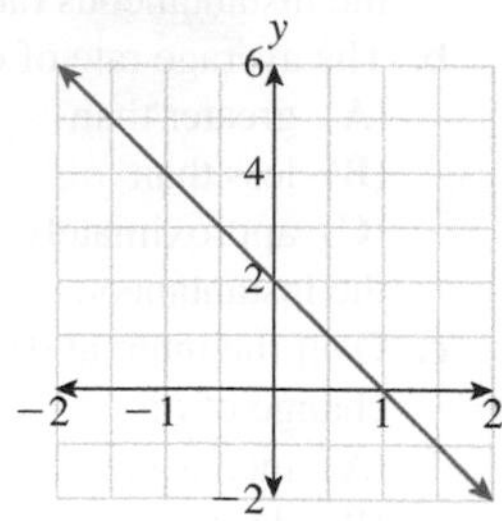

71.

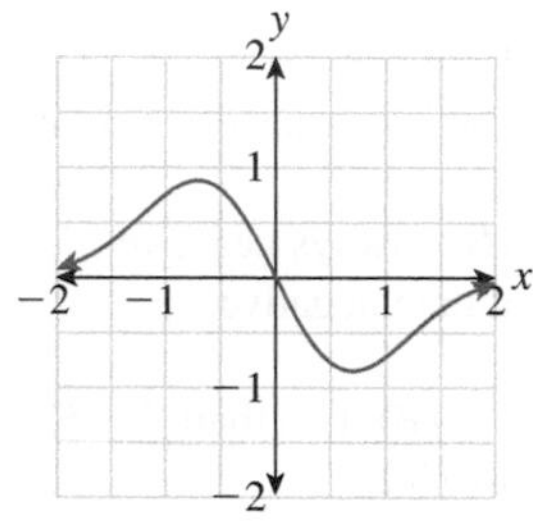

72.

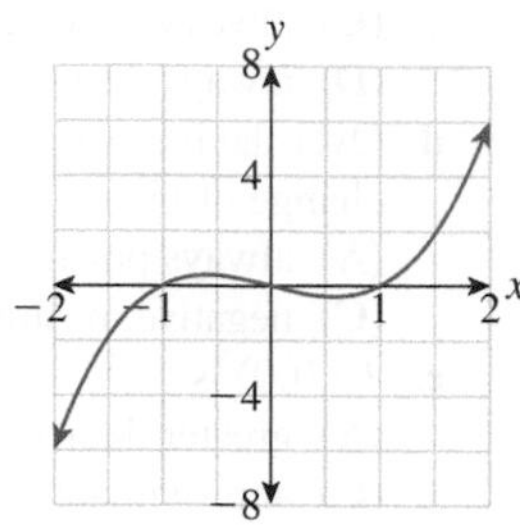

In Exercises 73 and 74, use technology to graph the derivative of the given function for the given range of values of x. Then use your graph to estimate all values of x (if any) where the tangent line to the graph of the given function is horizontal. Round answers to one decimal place.
[**HINT:** See Example 4.]

73. $f(x) = x^4 + 2x^3 - 1; \quad -2 \le x \le 1$

74. $f(x) = -x^3 - 3x^2 - 1; \quad -3 \le x \le 1$

In Exercises 75 and 76, use the method of Example 4 to list approximate values of $f'(x)$ for x in the given range. Graph $f(x)$ together with $f'(x)$ for x in the given range.

75. $f(x) = \dfrac{x+2}{x-3}; \quad 4 \le x \le 5$

76. $f(x) = \dfrac{10x}{x-2}; \quad 2.5 \le x \le 3$

Applications

77. ***Temperatures on Mars*** The air temperature one chilly spring morning at your time-share condominium at the base of Olympus Mons, t hours after 6:00 am, was given by the function $f(t) = -5t^2 + 50t - 80$ degrees Fahrenheit $(0 \le t \le 4)$.[58] What was the temperature at 7:00 am, and how fast was it rising? (Use the method of Example 1(a).)

[58] The average temperature on Mars is around −80°F but can get considerably warmer near the equator.

Stocktrek Images/Superstock

Olympus Mons

78. ***Temperatures on Venus*** The air temperature one balmy summer evening at your summer resort near Maxwell Montes, t hours after 5:00 pm, was given by the function $f(t) = 880 + 2t^2 - 20t$ degrees Fahrenheit $(0 \le t \le 4)$.[59] What was the temperature at 8:00 pm, and how fast was it dropping? (Use the method of Example 1(a).)

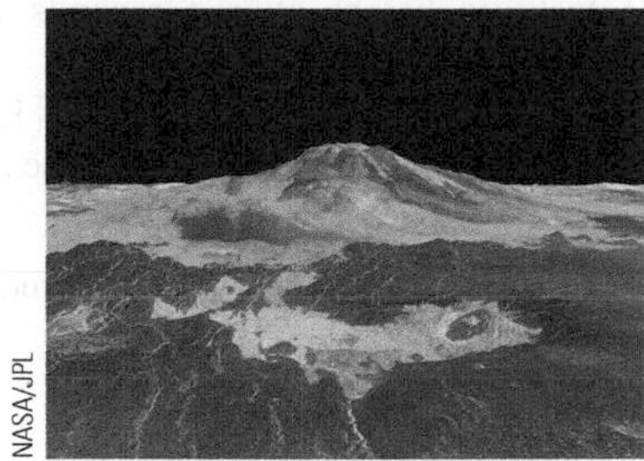
NASA/JPL

Maxwell Montes

79. ***Demand*** Suppose the demand for a new brand of sneakers is given by

$$q = \frac{5{,}000{,}000}{p},$$

where p is the price per pair of sneakers in dollars and q is the number of pairs of sneakers that can be sold at price p. Find $q(100)$, and estimate $q'(100)$. Interpret your answers. [**HINT:** See Example 1.]

80. ***Demand*** Suppose the demand for an old brand of TV is given by

$$q = \frac{100{,}000}{p + 10},$$

where p is the price per TV set in dollars and q is the number of TV sets that can be sold at price p. Find $q(190)$, and estimate $q'(190)$. Interpret your answers. [**HINT:** See Example 1.]

81. ***Oil Imports from Mexico*** The following graph shows approximate daily oil imports to the United States from Mexico.[60] Also shown is the tangent line at the point corresponding to year 2011.

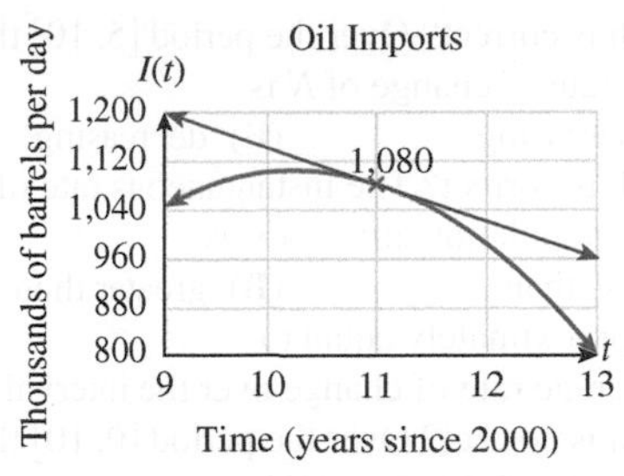

a. Estimate the slope of the tangent line shown on the graph. What does the graph tell you about oil imports from Mexico in 2011? [**HINT:** Identify two points on the tangent line. Then see Quick Example 3.]

b. According to the graph, is the rate of change of oil imports from Mexico increasing, decreasing, or increasing then decreasing? Why?

82. ***Oil Production in Mexico*** The following graph shows approximate daily oil production by **Pemex**, Mexico's national oil company.[61] Also shown is the tangent line at the point corresponding to year 2010.

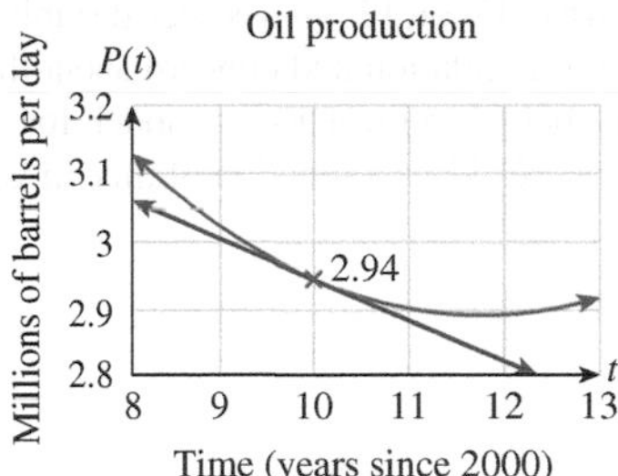

a. Estimate the slope of the tangent line shown on the graph. What does the graph tell you about oil production by Pemex in 2010? [**HINT:** Identify two points on the tangent line. Then see Quick Example 3.]

b. According to the graph, is the rate of change of oil production by Pemex increasing or decreasing over the range [8, 11]? Why?

83. ▼ ***Prison Population*** The following curve is a model of the total U.S. prison population as a function of time in years.[62]

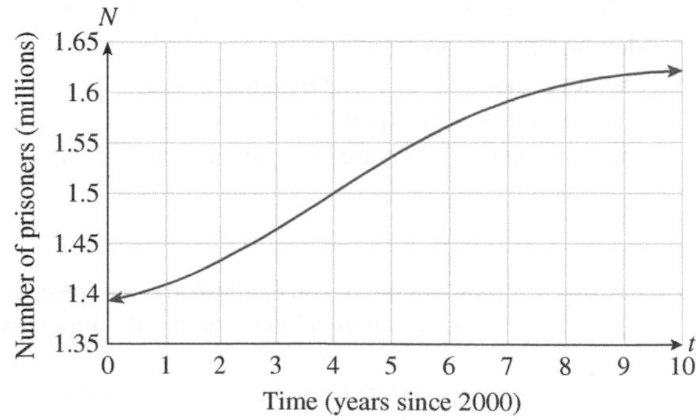

[59] The average temperature on Venus is around 860°F.

[60] Model based on data from the Department of Energy. Source for data: U.S. Energy Information Administration, www.eia.gov.

[61] Model based on company data. Source for data: www.pemex.com.

[62] Source: Bureau of Justice Statistics http://bjs.ojp.usdoj.gov.

a. Which is correct? Over the period [5, 10] the instantaneous rate of change of N is
(A) increasing. **(B)** decreasing.

b. Which is correct? The instantaneous rate of change of prison population at $t = 4$ was
(A) less than **(B)** greater than
(C) approximately equal to
the average rate of change over the interval [0, 10].

c. Which is correct? Over the period [0, 10] the instantaneous rate of change of N is
(A) increasing, then decreasing.
(B) decreasing, then increasing.
(C) always increasing.
(D) always decreasing.

d. According to the model, the U.S. prison population was increasing fastest around what year?

e. Roughly estimate the instantaneous rate of change of N at $t = 4$ by using a balanced difference quotient with $h = 1.5$. Interpret the result.

84. ▼ ***Demand for Freon 12*** The demand for chlorofluorocarbon-12 (CFC-12)—the ozone-depleting refrigerant commonly known as Freon 12[63]—has been declining significantly in response to regulation and concern about the ozone layer. The graph below represents a model for the projected demand for CFC-12 as a function of time in years.[64]

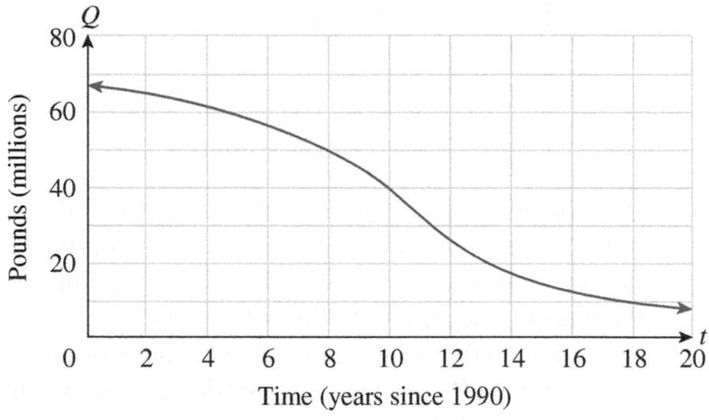

a. Which is correct? Over the period [12, 20] the instantaneous rate of change of Q is
(A) increasing. **(B)** decreasing.

b. Which is correct? The instantaneous rate of change of demand for Freon 12 at $t = 10$ was
(A) less than **(B)** greater than
(C) approximately equal to
the average rate of change over the interval [0, 20].

c. Which is correct? Over the period [0, 20] the instantaneous rate of change of Q is
(A) increasing, then decreasing.
(B) decreasing, then increasing.
(C) always increasing.
(D) always decreasing.

d. According to the model, the demand for Freon 12 was decreasing most rapidly around what year?

e. Roughly estimate the instantaneous rate of change of Q at $t = 13$ by using a balanced difference quotient with $h = 5$. Interpret the result.

85. ***Velocity*** If a stone is dropped from a height of 400 feet, its height after t seconds is given by $s = 400 - 16t^2$.

a. Find its average velocity over the period [2, 4].

b. Estimate its instantaneous velocity at time $t = 4$. [**HINT**: See Example 3.]

86. ***Velocity*** If a stone is thrown down at 120 ft/s from a height of 1,000 feet, its height after t seconds is given by $s = 1{,}000 - 120t - 16t^2$.

a. Find its average velocity over the period [1, 3].

b. Estimate its instantaneous velocity at time $t = 3$. [**HINT**: See Example 3.]

87. ***Crude Oil Prices*** The price per barrel of crude oil in constant 2008 dollars can be approximated by

$$P(t) = 0.45t^2 - 12t + 105 \text{ dollars} \quad (0 \le t \le 28),$$

where t is time in years since the start of 1980.[65]

a. Compute the average rate of change of $P(t)$ over the interval [0, 28], and interpret your answer. [**HINT**: See Example 3 of Section 3.4.]

b. Estimate the instantaneous rate of change of $P(t)$ at $t = 0$ and interpret your answer. [**HINT**: See Example 2(a).]

c. The answers to part (a) and part (b) have opposite signs. What does this indicate about the price of oil?

88. ***Median Home Prices*** The median home price in the United States over the period January 2010–January 2015 can be approximated by

$$P(t) = 4.5t^2 - 15t + 180 \text{ thousand dollars} \quad (0 \le t \le 5),$$

where t is time in years since the start of 2010.[66]

a. Compute the average rate of change of $P(t)$ over the interval [1, 5], and interpret your answer. [**HINT**: See Example 3 of Section 3.4.]

b. Estimate the instantaneous rate of change of $P(t)$ at $t = 1$, and interpret your answer. [**HINT**: See Example 2(a).]

c. The answers to parts (a) and (b) have opposite sign. What does this indicate about the median home price?

[63] The name given to it by DuPont. Freon 12 (dichlorodifluoromethane) is distinct from Freon 22 (chlorodifluoromethane, also registered by DuPont; see Exercise 73 in Section 2.2).

[64] Source for data: The Automobile Consulting Group (*New York Times*, December 26, 1993, p. F23). The exact figures were not given, and the chart is a reasonable facsimile of the chart that appeared in the *New York Times*.

[65] Source for data: www.inflationdata.com.

[66] Source for data: www.zillow.com.

89. ***The 2003 SARS Outbreak*** In the early stages of the SARS (severe acute respiratory syndrome) epidemic in 2003, the number of reported cases could be approximated by

$$A(t) = 167(1.18)^t \quad (0 \le t \le 20)$$

t days after March 17, 2003 (the first day in which statistics were reported by the World Health Organization).

a. What, approximately, was the instantaneous rate of change of $A(t)$ on March 27 ($t = 10$)? Interpret the result.

b. Which of the following is true? For the first 20 days of the epidemic, the instantaneous rate of change of the number of cases

(A) increased. **(B)** decreased.
(C) increased and then decreased.
(D) decreased and then increased.

90. ***The 2003 SARS Outbreak*** A few weeks into the SARS (severe acute respiratory syndrome) epidemic in 2003, the number of reported cases could be approximated by

$$A(t) = 1{,}804(1.04)^t \quad (0 \le t \le 30)$$

t days after April 1, 2003.

a. What, approximately, was the instantaneous rate of change of $A(t)$ on April 21 ($t = 20$)? Interpret the result.

b. Which of the following is true? During April the instantaneous rate of change of the number of cases

(A) increased. **(B)** decreased.
(C) increased and then decreased.
(D) decreased and then increased.

91. ***The 2014 Ebola Outbreak*** In the first 6 months of the 2014 Ebola outbreak, the total number of reported cases could be approximated by

$$C(t) = 95.9e^{0.72t} \quad (0 \le t \le 6),$$

where t is time in months since April 1, 2014.[67] Estimate $C(5)$ and $\left.\frac{dC}{dt}\right|_{t=5}$, and interpret your answers.

92. ***The 2014 Ebola Outbreak*** In the first 6 months of the 2014 Ebola outbreak, the total number of reported deaths from Ebola could be approximated by

$$D(t) = 90.52e^{0.60t} \quad (0 \le t \le 6),$$

where t is time in months since April 1, 2014.[68] Estimate $D(5)$ and $\left.\frac{dD}{dt}\right|_{t=5}$, and interpret your answers

93. ***Early Internet Services*** On January 1, 1996, **America Online** was the biggest online service provider, with 4.5 million subscribers, and was adding new subscribers at a rate of 60,000 per week.[69] If $A(t)$ is the number of America Online subscribers t weeks after January 1, 1996, what do the given data tell you about values of the function A and its derivative? [**HINT**: See Quick Example 2.]

94. ***Early Internet Services*** On January 1, 1996, **Prodigy** was the third-biggest online service provider, with 1.6 million subscribers, but was losing subscribers.[70] If $P(t)$ is the number of Prodigy subscribers t weeks after January 1, 1996, what do the given data tell you about values of the function P and its derivative? [**HINT**: See Quick Example 2.]

95. ▼ ***Learning to Speak*** Let $p(t)$ represent the percentage of children who are able to speak at the age of t months.

a. It is found that $p(10) = 60$ and $\left.\frac{dp}{dt}\right|_{t=10} = 18.2$. What does this mean?[71] [**HINT**: See Quick Example 2.]

b. As t increases, what happens to p and $\frac{dp}{dt}$?

96. ▼ ***Learning to Read*** Let $p(t)$ represent the percentage of children in your class who learned to read at the age of t years.

a. Assuming that everyone in your class could read by the age of 7, what does this tell you about $p(7)$ and $\left.\frac{dp}{dt}\right|_{t=7}$? [**HINT**: See Quick Example 2.]

b. Assuming that 25.0% of the people in your class could read by the age of 5 and that 25.3% of them could read by the age of 5 years and 1 month, estimate $\left.\frac{dp}{dt}\right|_{t=5}$. Remember to give its units.

97. ***Subprime Mortgages during the Housing Crisis*** (Compare Exercise 29 in Section 3.4.) The percentage of mortgages issued in the United States during the period 2000–2009 that were subprime (normally classified as risky) can be approximated by

$$A(t) = \frac{15}{1 + 8.6(1.8)^{-t}} \quad (0 \le t \le 9),$$

where t is the number of years since the start of 2000.[72]

a. Estimate $A(6)$ and $A'(6)$. (Round answers to two significant digits.) What do the answers tell you about subprime mortgages?

b. T Graph the extrapolated function and its derivative for $0 \le t \le 16$, and use your graphs to describe how the derivative behaves as t becomes large. (Express this behavior in terms of limits if you have studied the sections on limits.) What does this tell you about subprime mortgages? [**HINT**: See Example 5.]

---

[67] Exponential model is the authors'. Source for data: Wikipedia/Centers for Disease Control and Prevention/WHO.

[68] *Ibid.*

[69] Source: Information and Interactive Services Report/*New York Times*, January 2, 1996, p. C14.

[70] *Ibid.*

[71] Based on data presented in the article *The Emergence of Intelligence* by William H. Calvin, *Scientific American*, October 1994, pp. 101–107.

[72] Sources: Mortgage Bankers Association, UBS.

98. ***Subprime Mortgage Debt during the Housing Crisis*** (Compare Exercise 30 in Section 3.4.) The value of subprime (normally classified as risky) mortgage debt outstanding in the United States during the period 2000–2009 can be approximated by

$$A(t) = \frac{1,350}{1 + 4.2(1.7)^{-t}} \text{ billion dollars} \quad (0 \le t \le 9),$$

where t is the number of years since the start of 2000.[73]

a. Estimate $A(7)$ and $A'(7)$. (Round answers to three significant digits.) What do the answers tell you about subprime mortgages?

b. T Graph the function and its derivative, and use your graphs to estimate when, to the nearest year, $A'(t)$ is greatest. What does this tell you about subprime mortgages? [**HINT:** See Example 5.]

99. T ▼ ***Embryo Development*** The oxygen consumption of a turkey embryo increases from the time the egg is laid through the time the turkey chick hatches. In a brush turkey, the oxygen consumption (in milliliters per hour) can be approximated by

$$c(t) = -0.0012t^3 + 0.12t^2 - 1.83t + 3.97 \quad (20 \le t \le 50),$$

where t is the time (in days) since the egg was laid.[74] (An egg will typically hatch at around $t = 50$.) Use technology to graph $c'(t)$, and use your graph to answer the following questions. [**HINT:** See Example 5.]

a. Over the interval $[20, 32]$ the derivative c' is
(A) increasing, then decreasing.
(B) decreasing, then increasing.
(C) decreasing. (D) increasing.

b. When, to the nearest day, is the oxygen consumption increasing at the fastest rate?

c. When, to the nearest day, is the oxygen consumption increasing at the slowest rate?

100. T ▼ ***Embryo Development*** The oxygen consumption of a galliform bird embryo increases from the time the egg is laid through the time the chick hatches. In a typical galliform bird the oxygen consumption (in milliliters per hour) can be approximated by

$$c(t) = -0.0027t^3 + 0.14t^2 - 0.89t + 0.15 \quad (8 \le t \le 30),$$

where t is the time (in days) since the egg was laid.[75] (An egg will typically hatch at around $t = 28$.) Use technology to graph $c'(t)$, and use your graph to answer the following questions. [**HINT:** See Example 5.]

a. Over the interval $[8, 30]$ the derivative c' is
(A) increasing, then decreasing.
(B) decreasing, then increasing.
(C) decreasing. (D) increasing.

b. When, to the nearest day, is the oxygen consumption increasing the fastest?

c. When, to the nearest day, is the oxygen consumption increasing at the slowest rate?

The next two exercises are applications of Einstein's Special Theory of Relativity and relate to objects that are moving extremely fast. In science fiction terminology a speed of warp 1 *is the speed of light—about* 3×10^8 *meters per second. (For instance, a speed of warp* 0.8 *corresponds to* 80% *of the speed of light—about* 2.4×10^8 *meters per second.)*

101. ◆ ***Lorentz Contraction*** According to Einstein's Special Theory of Relativity, a moving object appears to get shorter to a stationary observer as its speed approaches the speed of light. If a spaceship that has a length of 100 meters at rest travels at a speed of warp p, its length in meters, as measured by a stationary observer, is given by

$$L(p) = 100\sqrt{1 - p^2}$$

with domain $[0, 1)$. Estimate $L(0.95)$ and $L'(0.95)$. What do these figures tell you?

102. ◆ ***Time Dilation*** Another prediction of Einstein's Special Theory of Relativity is that, to a stationary observer, clocks (as well as all biological processes) in a moving object appear to go more and more slowly as the speed of the object approaches that of light. If a spaceship travels at a speed of warp p, the time it takes for an onboard clock to register 1 second, as measured by a stationary observer, will be given by

$$T(p) = \frac{1}{\sqrt{1 - p^2}} \text{ seconds}$$

with domain $[0, 1)$. Estimate $T(0.95)$ and $T'(0.95)$. What do these figures tell you?

Communication and Reasoning Exercises

103. In which, if any, of the following cases is f differentiable at a? (There may be none or more than one.)

(A) $a = 2$; domain of f: all real numbers;
$$\lim_{h \to 0^+} \frac{f(2 + h) - f(2)}{h} = 3,$$
$$\lim_{h \to 0^-} \frac{f(2 + h) - f(2)}{h} = 5$$

(B) $a = 0$; domain of f: $(0, 2)$;
$$\lim_{h \to 0} \frac{f(1 + h) - f(1)}{h} = 3$$

(C) $a = 3$; domain of f: $[0, 3]$;
$$\lim_{h \to 0} \frac{f(3 + h) - f(3)}{h} = 5$$

(D) $a = 0$; domain of f: all real numbers;
$$\lim_{h \to 0} \frac{f(h) - f(0)}{h} = +\infty$$

[73] Source: Data 360 www.data360.org.

[74] The model approximates graphical data published in the article *The Brush Turkey* by Roger S. Seymour, *Scientific American*, December 1991, pp. 108–114.

[75] *Ibid.*

104. In which, if any, of the following cases is f differentiable at a? (There may be none or more than one.)

(A) $a = 4$; domain of f: all real numbers except 5;
$$\lim_{h\to 0^+} \frac{f(4+h)-f(4)}{h} = 5,$$
$$\lim_{h\to 0^-} \frac{f(4+h)-f(4)}{h} = 5$$

(B) $a = 0$; domain of f: $[0, 2)$;
$$\lim_{h\to 0} \frac{f(h)-f(0)}{h} = 3$$

(C) $a = 4$; domain of f: all real numbers except 4;
$$\lim_{h\to 0} \frac{f(4+h)-f(4)}{h} = 5$$

(D) $a = 0$; domain of f: $(-0.0001, 0.0001)$;
$$\lim_{h\to 0} \frac{f(h)-f(0)}{h} = 30$$

105. Explain why we cannot put $h = 0$ in the approximation
$$f'(x) \approx \frac{f(x+h)-f(x)}{h}$$
for the derivative of f.

106. The balanced difference quotient
$$f'(a) \approx \frac{f(a+0.0001)-f(a-0.0001)}{0.0002}$$
is the average rate of change of f on what interval?

107. Let $H(t)$ represent the number of *Handbook* members in millions t years after its inception in 2020. It is found that $H(10) = 50$ and $H'(10) = -6$. This means that, in 2030 (multiple choice),

(A) there were 6 million members and this number was decreasing at a rate of 50 million per year.

(B) there were -6 million members and this number was increasing at a rate of 50 million per year.

(C) membership had dropped by 6 million since the previous year but was now increasing at a rate of 50 million per year.

(D) there were 50 million members and this number was decreasing at a rate of 6 million per year.

(E) there were 50 million members and membership had dropped by 6 million since the previous year.

108. Let $F(t)$ represent the net earnings of *Footbook, Inc.* in millions of dollars t years after its inception in 3020. It is found that $F(100) = -10$ and $F'(100) = 60$. This means that, in 3120 (multiple choice),

(A) Footbook lost \$10 million, but its net earnings were increasing at a rate of \$60 million per year.

(B) Footbook earned \$60 million, but its earnings were decreasing at a rate of \$10 million per year.

(C) Footbook's net earnings had increased by \$60 million since the year before, but it still lost \$10 million.

(D) Footbook earned \$10 million, but its net earnings were decreasing at a rate of \$60 million per year.

(E) Footbook's net earnings had decreased by \$10 million since the year before, but it still earned \$60 million.

109. It is now 8 months since the Garden City lacrosse team won the national championship, and sales of team paraphernalia, while still increasing, have been leveling off. What does this tell you about the derivative of the sales curve?

110. Having been soundly defeated in the national lacrosse championships, Brakpan High has been faced with decreasing sales of its team paraphernalia. However, while still decreasing, sales appear to be bottoming out. What does this tell you about the derivative of the sales curve?

111. ▼ Company A's profits are given by $P(0) = \$1$ million and $P'(0) = -\$1$ million/month. Company B's profits are given by $P(0) = -\$1$ million and $P'(0) = \$1$ million per month. In which company would you rather invest? Why?

112. ▼ Company C's profits are given by $P(0) = \$1$ million and $P'(0) = \$0.5$ million/month. Company D's profits are given by $P(0) = \$0.5$ million and $P'(0) = \$1$ million per month. In which company would you rather invest? Why?

113. ▼ During the 1-month period starting last January 1, your company's profits increased at an average rate of change of \$4 million per month. On January 1, profits were increasing at an instantaneous rate of \$5 million per month. Which of the following graphs could represent your company's profits? Why?

(A)

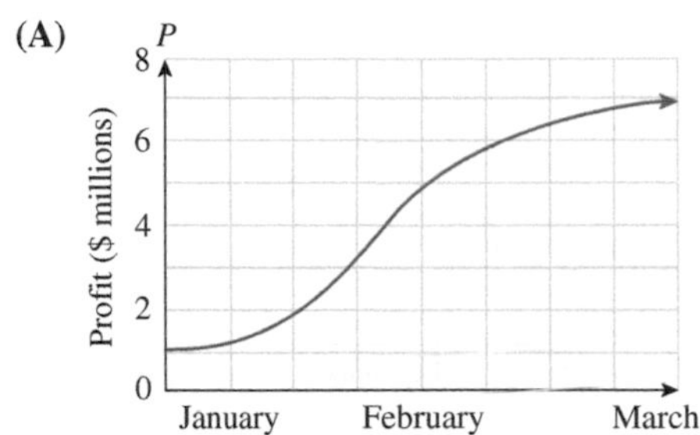

(B)

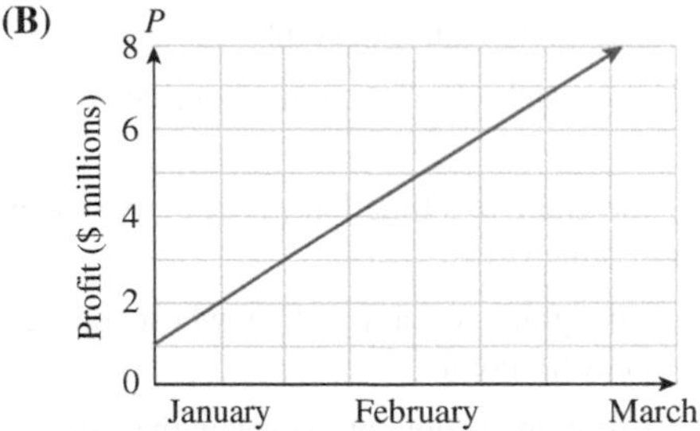

(C)

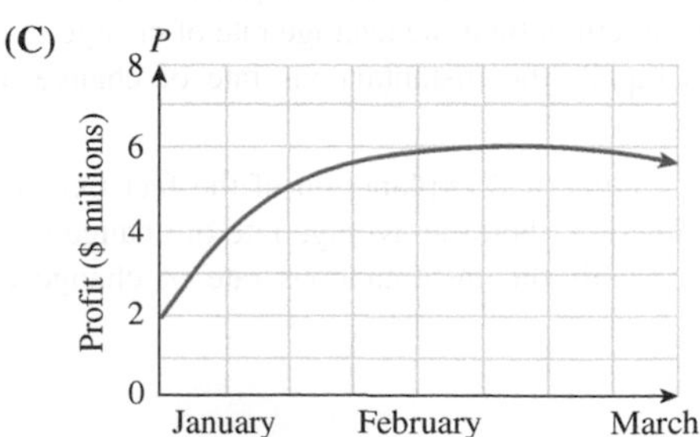

114. ▼ During the 1-month period starting last January 1, your company's sales increased at an average rate of change of \$3,000 per month. On January 1, sales were changing at an instantaneous rate of −\$1,000 per month. Which of the following graphs could represent your company's sales? Why?

(A)

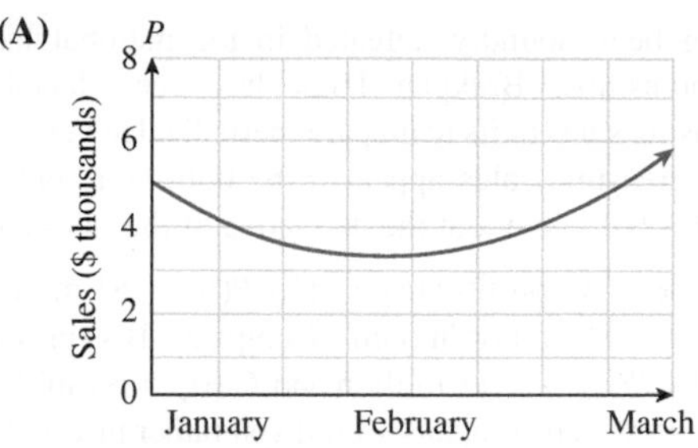

(B)

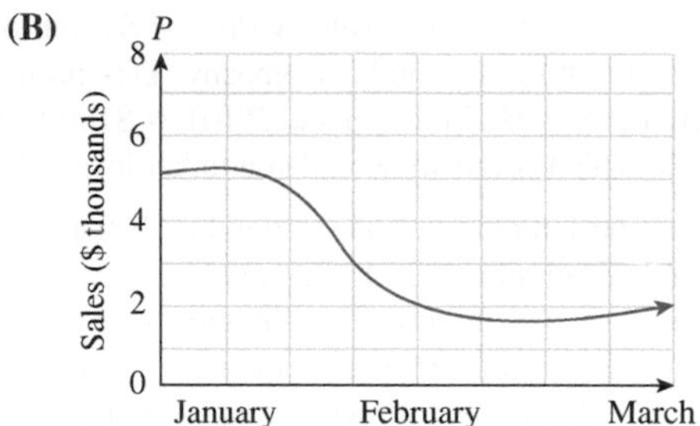

(C)

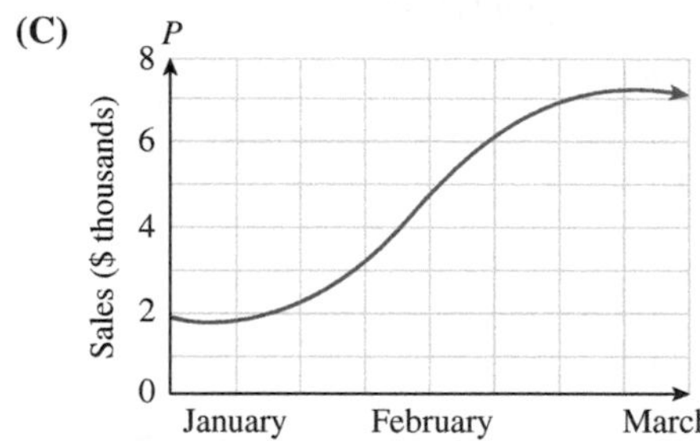

115. ▼ If the derivative of f is zero at a point, what do you know about the graph of f near that point?

116. ▼ Sketch the graph of a function whose derivative never exceeds 1.

117. ▼ Sketch the graph of a function whose derivative exceeds 1 at every point.

118. ▼ Sketch the graph of a function whose derivative is exactly 1 at every point.

119. ▼ Use the difference quotient to explain the fact that if f is a linear function, then the average rate of change over any interval equals the instantaneous rate of change at any point.

120. ▼ Give a numerical explanation of the fact that if f is a linear function, then the average rate of change over any interval equals the instantaneous rate of change at any point.

121. ◆ Consider the following values of the function f from Exercise 1:

h	0.1	0.01	0.001	0.0001
Avg. Rate of Change of f over $[5, 5+h]$	6.4	6.04	6.004	6.0004
h	−0.1	−0.01	−0.001	−0.0001
Avg. Rate of Change of f over $[5+h, 5]$	5.6	5.96	5.996	5.9996

Does the table suggest that the instantaneous rate of change of f is

(A) increasing **(B)** decreasing

as x increases toward 5?

122. ◆ Consider the following values of the function g from Exercise 2:

h	0.1	0.01	0.001	0.0001
Avg. Rate of Change of g over $[7, 7+h]$	4.8	4.98	4.998	4.9998
h	−0.1	−0.01	−0.001	−0.0001
Avg. Rate of Change of g over $[7+h, 7]$	5.3	5.03	5.003	5.0003

Does the table suggest that the instantaneous rate of change of g is

(A) increasing **(B)** decreasing

as x increases toward 7?

123. ▼ Sketch the graph of a function whose derivative is never zero but decreases as x increases.

124. ▼ Sketch the graph of a function whose derivative is never negative but is zero at exactly two points.

125. ◆ Here is the graph of the derivative f' of a function f. Give a rough sketch of the graph of f, given that $f(0) = 0$.

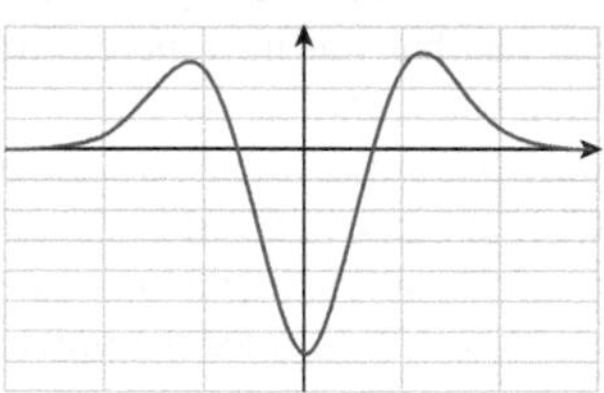

126. ◆ Here is the graph of the derivative f' of a function f. Give a rough sketch of the graph of f, given that $f(0) = 0$.

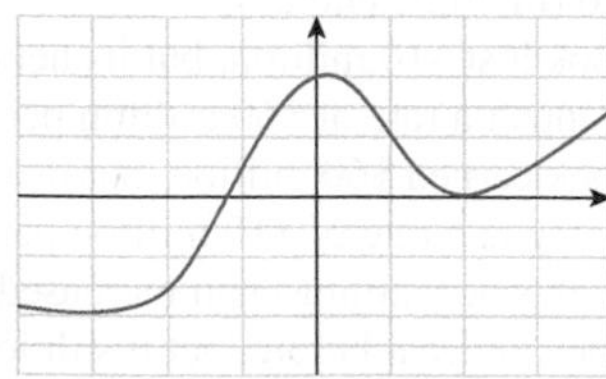

127. ◆ Professor Talker of the physics department drove a 60-mile stretch of road in exactly 1 hour. The speed limit along that stretch was 55 mph. Which of the following must be correct?
   (A) He exceeded the speed limit at no point of the journey.
   (B) He exceeded the speed limit at some point of the journey.
   (C) He exceeded the speed limit throughout the journey.
   (D) He traveled slower than the speed limit at some point of the journey.

128. ◆ Professor Silent, another physics professor, drove a 50-mile stretch of road in exactly 1 hour. The speed limit along that stretch was 55 mph. Which of the following must be correct?
   (A) She exceeded the speed limit at no point of the journey.
   (B) She exceeded the speed limit at some point of the journey.
   (C) She traveled slower than the speed limit throughout the journey.
   (D) She traveled slower than the speed limit at some point of the journey.

129. ◆ Draw the graph of a function f with the property that the balanced difference quotient gives a more accurate approximation of $f'(1)$ than the ordinary difference quotient.

130. ◆ Draw the graph of a function f with the property that the balanced difference quotient gives a less accurate approximation of $f'(1)$ than the ordinary difference quotient.

3.6 The Derivative: Algebraic Viewpoint

Calculating the Derivative Algebraically

In Section 3.5 we saw how to estimate the derivative of a function using numerical and graphical approaches. In this section we use an algebraic approach that will give us the *exact value* of the derivative rather than just an approximation, when the function is specified algebraically.

This algebraic approach is quite straightforward: Instead of subtracting numbers to estimate the average rate of change over smaller and smaller intervals, we subtract algebraic expressions. Our starting point is the definition of the derivative in terms of the difference quotient:

$$f'(a) = \lim_{h\to 0} \frac{f(a+h) - f(a)}{h}.$$

EXAMPLE 1 **Calculating the Derivative at a Point Algebraically**

Let $f(x) = x^2$. Use the definition of the derivative to compute $f'(3)$ algebraically.

Solution Substituting $a = 3$ into the definition of the derivative, we get

$$\begin{aligned}
f'(3) &= \lim_{h\to 0} \frac{f(3+h) - f(3)}{h} && \text{Formula for the derivative} \\
&= \lim_{h\to 0} \frac{\overbrace{(3+h)^2}^{f(3+h)} - \overbrace{3^2}^{f(3)}}{h} && \text{Substitute for } f(3) \text{ and } f(3+h). \\
&= \lim_{h\to 0} \frac{(9 + 6h + h^2) - 9}{h} && \text{Expand } (3+h)^2. \\
&= \lim_{h\to 0} \frac{6h + h^2}{h} && \text{Cancel the 9.} \\
&= \lim_{h\to 0} \frac{h(6+h)}{h} && \text{Factor out } h. \\
&= \lim_{h\to 0} (6 + h). && \text{Cancel the } h.
\end{aligned}$$

Now we let h approach 0. As h gets closer and closer to 0, the sum $6 + h$ clearly gets closer and closer to $6 + 0 = 6$. Thus,

$$f'(3) = \lim_{h\to 0}(6 + h) = 6. \qquad \text{As } h \to 0,\ (6 + h) \to 6.$$

(Calculations of limits like this are discussed and justified more fully in Sections 3.2 and 3.3.)

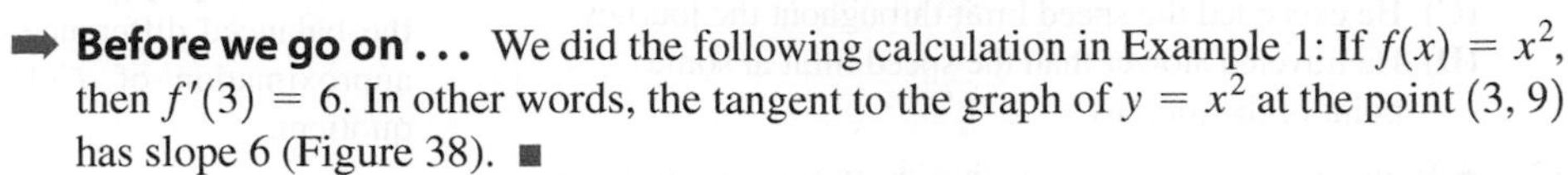

➡ **Before we go on . . .** We did the following calculation in Example 1: If $f(x) = x^2$, then $f'(3) = 6$. In other words, the tangent to the graph of $y = x^2$ at the point (3, 9) has slope 6 (Figure 38). ■

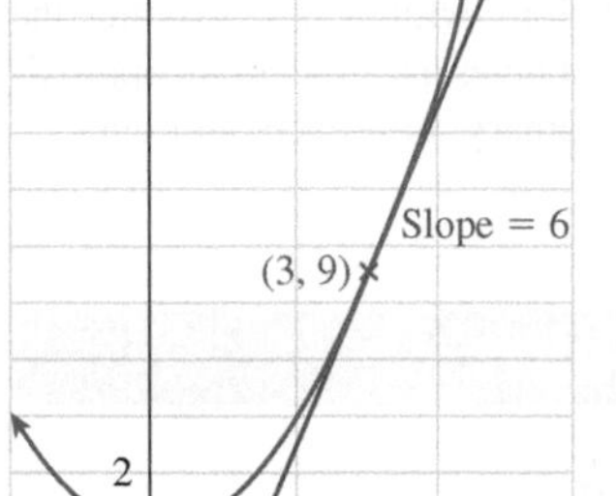

Figure 38

There is nothing very special about $a = 3$ in Example 1. Let's try to compute $f'(x)$ for general x.

EXAMPLE 2 Calculating the Derivative Function Algebraically

Let $f(x) = x^2$.

a. Use the definition of the derivative to compute $f'(x)$ algebraically.

b. Use the answer to evaluate $f'(3)$.

Solution

a. Once again, our starting point is the definition of the derivative in terms of the difference quotient:

$$\begin{aligned}
f'(x) &= \lim_{h\to 0}\frac{f(x+h) - f(x)}{h} && \text{Formula for the derivative} \\
&= \lim_{h\to 0}\frac{\overbrace{(x+h)^2}^{f(x+h)} - \overbrace{x^2}^{f(x)}}{h} && \text{Substitute for } f(x) \text{ and } f(x+h). \\
&= \lim_{h\to 0}\frac{(x^2 + 2xh + h^2) - x^2}{h} && \text{Expand } (x+h)^2. \\
&= \lim_{h\to 0}\frac{2xh + h^2}{h} && \text{Cancel the } x^2. \\
&= \lim_{h\to 0}\frac{h(2x + h)}{h} && \text{Factor out } h. \\
&= \lim_{h\to 0}(2x + h). && \text{Cancel the } h.
\end{aligned}$$

Now we let h approach 0. As h gets closer and closer to 0, the sum $2x + h$ clearly gets closer and closer to $2x + 0 = 2x$. Thus,

$$f'(x) = \lim_{h\to 0}(2x + h) = 2x.$$

This is the derivative function.

b. Now that we have a *formula* for the derivative of f, we can obtain $f'(a)$ for any value of a we choose by simply evaluating f' there. For instance,

$$f'(3) = 2(3) = 6$$

as we saw in Example 1.

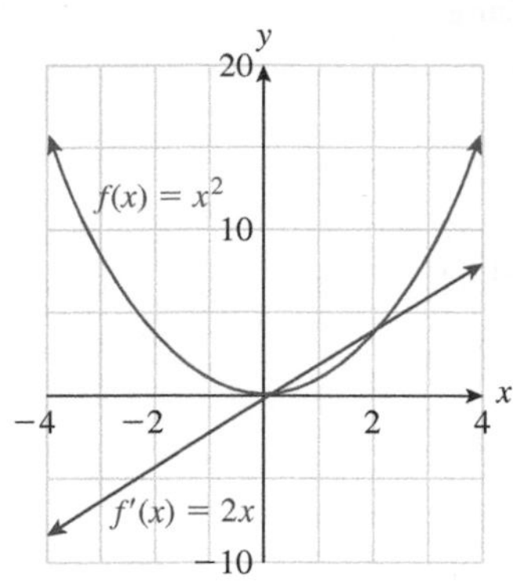

Figure 39

➡ **Before we go on . . .** The graphs of $f(x) = x^2$ and $f'(x) = 2x$ from Example 2 are familiar. Their graphs are shown in Figure 39.

When $x < 0$, the parabola slopes downward, which is reflected in the fact that the derivative $2x$ is negative there. When $x > 0$, the parabola slopes upward, which is reflected in the fact that the derivative is positive there. The parabola has a horizontal tangent line at $x = 0$, reflected in the fact that $2x = 0$ there. ■

EXAMPLE 3 More Computations of Derivative Functions

Compute the derivative $f'(x)$ for each of the following functions:

a. $f(x) = x^3$ **b.** $f(x) = 2x^2 - x$ **c.** $f(x) = \dfrac{1}{x}$

Solution

a.
$$\begin{aligned}
f'(x) &= \lim_{h\to 0} \frac{f(x+h) - f(x)}{h} && \text{Formula for the derivative} \\
&= \lim_{h\to 0} \frac{\overbrace{(x+h)^3}^{f(x+h)} - \overbrace{x^3}^{f(x)}}{h} && \text{Substitute for } f(x) \text{ and } f(x+h). \\
&= \lim_{h\to 0} \frac{(x^3 + 3x^2h + 3xh^2 + h^3) - x^3}{h} && \text{Expand } (x+h)^3. \\
&= \lim_{h\to 0} \frac{3x^2h + 3xh^2 + h^3}{h} && \text{Cancel the } x^3. \\
&= \lim_{h\to 0} \frac{h(3x^2 + 3xh + h^2)}{h} && \text{Factor out } h. \\
&= \lim_{h\to 0} (3x^2 + 3xh + h^2) && \text{Cancel the } h. \\
&= 3x^2. && \text{Let } h \text{ approach 0.}
\end{aligned}$$

b.
$$\begin{aligned}
f'(x) &= \lim_{h\to 0} \frac{f(x+h) - f(x)}{h} && \text{Formula for the derivative} \\
&= \lim_{h\to 0} \frac{\overbrace{(2(x+h)^2 - (x+h))}^{f(x+h)} - \overbrace{(2x^2 - x)}^{f(x)}}{h} && \text{Substitute for } f(x) \text{ and } f(x+h). \\
&= \lim_{h\to 0} \frac{(2x^2 + 4xh + 2h^2 - x - h) - (2x^2 - x)}{h} && \text{Expand.} \\
&= \lim_{h\to 0} \frac{4xh + 2h^2 - h}{h} && \text{Cancel the } 2x^2 \text{ and } x. \\
&= \lim_{h\to 0} \frac{h(4x + 2h - 1)}{h} && \text{Factor out } h. \\
&= \lim_{h\to 0} (4x + 2h - 1) && \text{Cancel the } h. \\
&= 4x - 1. && \text{Let } h \text{ approach 0.}
\end{aligned}$$

c. $f'(x) = \lim_{h \to 0} \frac{f(x+h) - f(x)}{h}$ Formula for the derivative

$$= \lim_{h \to 0} \frac{\left[\overbrace{\frac{1}{x+h}}^{f(x+h)} - \overbrace{\frac{1}{x}}^{f(x)}\right]}{h}$$ Substitute for $f(x)$ and $f(x + h)$.

$$= \lim_{h \to 0} \frac{\left[\frac{x - (x+h)}{(x+h)x}\right]}{h}$$ Subtract the fractions.

$$= \lim_{h \to 0} \frac{1}{h}\left[\frac{x - (x+h)}{(x+h)x}\right]$$ Division by h = Multiplication by $1/h$.

$$= \lim_{h \to 0} \left[\frac{-h}{h(x+h)x}\right]$$ Simplify.

$$= \lim_{h \to 0} \left[\frac{-1}{(x+h)x}\right]$$ Cancel the h.

$$= \frac{-1}{x^2}.$$ Let h approach 0.

In Example 4 we redo Example 3 of Section 3.5, this time getting an exact, rather than approximate, answer.

EXAMPLE 4 **Velocity**

My friend Eric, an enthusiastic baseball player, claims that he can "probably" throw a ball upward at a speed of 100 feet per second (ft/sec). Our physicist friends tell us that its height s (in feet) t seconds later would be $s(t) = 100t - 16t^2$. Find the ball's instantaneous velocity function and its velocity exactly 2 seconds after Eric throws it.

Solution The instantaneous velocity function is the derivative ds/dt, which we calculate as follows:

$$\frac{ds}{dt} = \lim_{h \to 0} \frac{s(t+h) - s(t)}{h}.$$

Let us compute $s(t + h)$ and $s(t)$ separately:

$$\begin{aligned} s(t) &= 100t - 16t^2 \\ s(t+h) &= 100(t+h) - 16(t+h)^2 \\ &= 100t + 100h - 16(t^2 + 2th + h^2) \\ &= 100t + 100h - 16t^2 - 32th - 16h^2. \end{aligned}$$

Therefore,

$$\begin{aligned}\frac{ds}{dt} &= \lim_{h\to 0}\frac{s(t+h)-s(t)}{h}\\ &= \lim_{h\to 0}\frac{100t+100h-16t^2-32th-16h^2-(100t-16t^2)}{h}\\ &= \lim_{h\to 0}\frac{100h-32th-16h^2}{h}\\ &= \lim_{h\to 0}\frac{h(100-32t-16h)}{h}\\ &= \lim_{h\to 0}(100-32t-16h)\\ &= 100-32t \text{ ft/sec.}\end{aligned}$$

Thus, the velocity exactly 2 seconds after Eric throws it is

$$\left.\frac{ds}{dt}\right|_{t=2} = 100-32(2) = 36 \text{ ft/sec.}$$

This verifies the accuracy of the approximation we made in Section 3.5.

➡ **Before we go on . . .** From the derivative function in Example 4 we can now describe the behavior of the velocity of the ball: Immediately on release ($t = 0$) the ball is traveling at 100 feet per second upward. The ball then slows down; precisely, it loses 32 feet per second of speed every second. When, exactly, does the velocity become zero, and what happens after that? ■

Q: *Do we always have to calculate the limit of the difference quotient to find a formula for the derivative function?*

A: As it turns out, no. In Section 4.1 we will start to look at shortcuts for finding derivatives that allow us to bypass the definition of the derivative in many cases.

A Function Not Differentiable at a Point

Recall from Section 3.5 that a function is **differentiable** at a point a if $f'(a)$ exists; that is, if the difference quotient $[f(a+h)-f(a)]/h$ approaches a fixed value as h approaches 0. In Section 3.5 we mentioned that the function $f(x) = |x|$ is not differentiable at $x = 0$. In Example 5 we find out why.

EXAMPLE 5 **A Function Not Differentiable at 0**

Numerically, graphically, and algebraically investigate the differentiability of the function $f(x) = |x|$ at the points **(a)** $x = 1$ and **(b)** $x = 0$.

Solution

a. We compute

$$f'(1) = \lim_{h\to 0} \frac{f(1+h) - f(1)}{h}$$
$$= \lim_{h\to 0} \frac{|1+h| - 1}{h}.$$

Numerically, we can make tables of the values of the average rate of change $(|1+h| - 1)/h$ for h positive or negative and approaching 0:

h	1	0.1	0.01	0.001	0.0001
Avg. Rate of Change over $[1, 1+h]$	1	1	1	1	1

h	−1	−0.1	−0.01	−0.001	−0.0001
Avg. Rate of Change over $[1+h, 1]$	1	1	1	1	1

From these tables it appears that $f'(1)$ is equal to 1. We can verify that algebraically: For h that is sufficiently small, $1 + h$ is positive (even if h is negative), and so

$$f'(1) = \lim_{h\to 0} \frac{1+h-1}{h}$$
$$= \lim_{h\to 0} \frac{h}{h} \qquad \text{Cancel the 1s.}$$
$$= \lim_{h\to 0} 1 \qquad \text{Cancel the } h.$$
$$= 1.$$

Graphically, we are seeing the fact that the tangent line at the point (1, 1) has slope 1 because the graph is a straight line with slope 1 near that point (Figure 40).

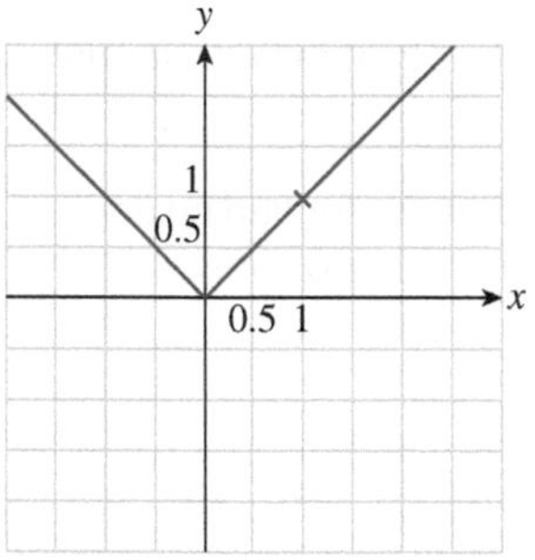

Figure 40

b. $$f'(0) = \lim_{h\to 0} \frac{f(0+h) - f(0)}{h}$$
$$= \lim_{h\to 0} \frac{|0+h| - 0}{h}$$
$$= \lim_{h\to 0} \frac{|h|}{h}$$

If we make tables of values in this case we get the following:

h	1	0.1	0.01	0.001	0.0001
Avg. Rate of Change over $[0, 0+h]$	1	1	1	1	1

h	−1	−0.1	−0.01	−0.001	−0.0001
Avg. Rate of Change over $[0+h, 0]$	−1	−1	−1	−1	−1

For the limit and hence the derivative $f'(0)$ to exist, the average rates of change should approach the same number for both positive and negative h. Because they do not, f is not differentiable at $x = 0$. We can verify this conclusion algebraically: If h is positive, then $|h| = h$, and so the ratio $|h|/h$ is 1, regardless of how small h is. Thus, according to the values of the difference quotients with $h > 0$, the limit should be 1. On the other hand, if h is negative, then $|h| = -h$ (positive), and so $|h|/h = -1$, meaning that the limit should be -1. Because the limit cannot be both -1 and 1 (it must be a single number for the derivative to exist), we conclude that $f'(0)$ does not exist.

To see what is happening graphically, take a look at Figure 41, which shows zoomed-in views of the graph of f near $x = 0$.

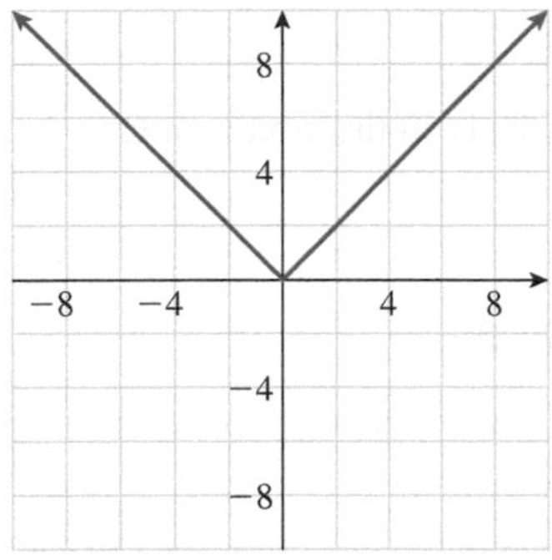

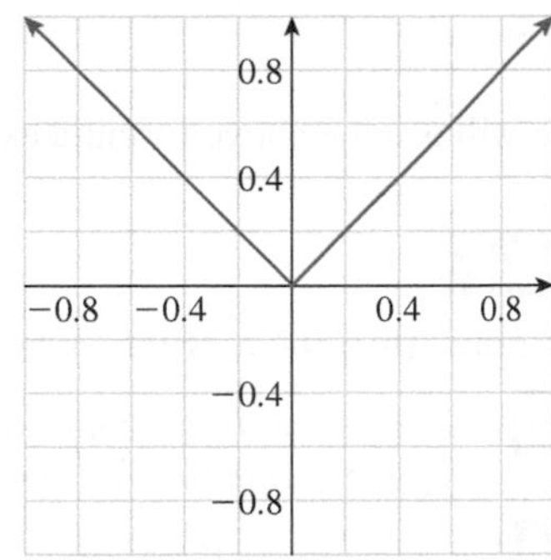

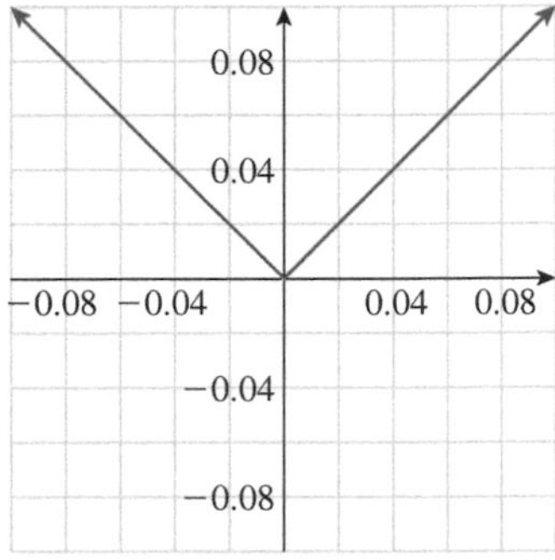

Figure 41

No matter what scale we use to view the graph, it has a sharp corner at $x = 0$ and hence has no tangent line there. Since there is no tangent line at $x = 0$, the function is not differentiable there.

➡ **Before we go on...** Notice that $|x| = \begin{cases} -x & \text{if } x < 0 \\ x & \text{if } x \geq 0 \end{cases}$ is an example of a piecewise-linear function whose graph comes to a point at $x = 0$. In general, if $f(x)$ is any piecewise-linear function whose graph comes to a point at $x = a$, it will be non-differentiable at $x = a$ for the same reason that $|x|$ fails to be differentiable at $x = 0$.

If we repeat the computation in Example 5(a) using any nonzero value for a in place of 1, we see that f is differentiable there as well. If a is positive, we find that $f'(a) = 1$, and if a is negative, $f'(a) = -1$. In other words, the derivative function is

$$f'(x) = \begin{cases} -1 & \text{if } x < 0 \\ 1 & \text{if } x > 0. \end{cases}$$

Immediately to the left of $x = 0$, we see that $f'(x) = -1$; immediately to the right, $f'(x) = 1$; and when $x = 0$, $f'(x)$ is not defined. ■

Q: *So does that mean there is no single formula for the derivative of $|x|$?*

A: Actually, there is a convenient formula. Consider the ratio $\frac{|x|}{x}$. If x is positive, then $|x| = x$, so $\frac{|x|}{x} = \frac{x}{x} = 1$. On the other hand, if x is negative, then $|x| = -x$, so $\frac{|x|}{x} = \frac{-x}{x} = -1$. In other words,

$$\frac{|x|}{x} = \begin{cases} -1 & \text{if } x < 0 \\ 1 & \text{if } x > 0, \end{cases}$$

which is exactly the formula we obtained for $f'(x)$. We have therefore obtained a convenient closed-form formula for the derivative of $|x|$!

Derivative of $|x|$

$$\text{If } f(x) = |x|, \text{ then } f'(x) = \frac{|x|}{x}.$$

Note that $|x|/x$ is not defined if $x = 0$, reflecting the fact that $f'(x)$ does not exist when $x = 0$.

We will use the above formula extensively in the next chapter.

FAQs

Computing Derivatives Algebraically

Q: *The algebraic computation of $f'(x)$ seems to require a number of steps. How do I remember what to do and when?*

A: If you examine the computations in the examples above, you will find the following pattern:

1. Write out the formula for $f'(x)$ as the limit of the difference quotient, and then substitute $f(x + h)$ and $f(x)$.
2. Expand and simplify the *numerator* of the expression but not the denominator.
3. After simplifying the numerator, factor out an h to cancel with the h in the denominator. If h does not factor out of the numerator, you might have made an error. (A frequent error is a wrong sign.)
4. After canceling the h, you should be able to see what the limit is by letting $h \to 0$.

3.6 EXERCISES

▼ more advanced ◆ challenging

T indicates exercises that should be solved using technology

In Exercises 1–14, compute $f'(a)$ algebraically for the given value of a. [**HINT**: See Example 1.]

1. $f(x) = x^2 + 1; a = 2$

2. $f(x) = x^2 - 3; a = 1$

3. $f(x) = 3x - 4; a = -1$

4. $f(x) = -2x + 4; a = -1$

5. $f(x) = 3x^2 + x; a = 1$

6. $f(x) = 2x^2 + x; a = -2$

7. $f(x) = 2x - x^2; a = -1$

8. $f(x) = -x - x^2; a = 0$

9. $f(x) = x^3 + 2x; a = 2$

10. $f(x) = x - 2x^3; a = 1$

11. $f(x) = -\frac{1}{x}; a = 1$ [**HINT**: See Example 3.]

12. $f(x) = \frac{2}{x}; a = 5$ [**HINT**: See Example 3.]

13. ▼ $f(x) = mx + b; a = 43$

14. ▼ $f(x) = \frac{x}{k} - b \quad (k \neq 0); a = 12$

In Exercises 15–28, compute the derivative function $f'(x)$ algebraically. (Notice that the functions are the same as those in Exercises 1–14.) [**HINT**: See Examples 2 and 3.]

15. $f(x) = x^2 + 1$

16. $f(x) = x^2 - 3$

17. $f(x) = 3x - 4$

18. $f(x) = -2x + 4$

19. $f(x) = 3x^2 + x$

20. $f(x) = 2x^2 + x$

21. $f(x) = 2x - x^2$

22. $f(x) = -x - x^2$

23. $f(x) = x^3 + 2x$

24. $f(x) = x - 2x^3$

25. ▼ $f(x) = -\frac{1}{x}$

26. ▼ $f(x) = \frac{2}{x}$

27. ▼ $f(x) = mx + b$

28. ▼ $f(x) = \frac{x}{k} - b \quad (k \neq 0)$

In Exercises 29–38, compute the indicated derivative.

29. $R(t) = -0.3t^2$; $R'(2)$

30. $S(t) = 1.4t^2$; $S'(-1)$

31. $U(t) = 5.1t^2 + 5.1$; $U'(3)$

32. $U(t) = -1.3t^2 + 1.1$; $U'(4)$

33. $U(t) = -1.3t^2 - 4.5t$; $U'(1)$

34. $U(t) = 5.1t^2 - 1.1t$; $U'(1)$

35. $L(r) = 4.25r - 5.01$; $L'(1.2)$

36. $L(r) = -1.02r + 5.7$; $L'(3.1)$

37. ▼ $q(p) = \frac{2.4}{p} + 3.1$; $q'(2)$

38. ▼ $q(p) = \frac{1}{0.5p} - 3.1$; $q'(2)$

In Exercises 39–44, find the equation of the tangent to the graph at the indicated point. [**HINT:** Compute the derivative algebraically; then see Example 2(b) of Section 3.5.]

39. ▼ $f(x) = x^2 - 3$; $a = 2$

40. ▼ $f(x) = x^2 + 1$; $a = 2$

41. ▼ $f(x) = -2x - 4$; $a = 3$

42. ▼ $f(x) = 3x + 1$; $a = 1$

43. ▼ $f(x) = x^2 - x$; $a = -1$

44. ▼ $f(x) = x^2 + x$; $a = -1$

Applications

45. ***Velocity*** If a stone is dropped from a height of 400 feet, its height after t seconds is given by $s = 400 - 16t^2$. Find its instantaneous velocity function and its velocity at time $t = 4$. [**HINT:** See Example 4.]

46. ***Velocity*** If a stone is thrown down at 120 feet per second from a height of 1,000 feet, its height after t seconds is given by $s = 1{,}000 - 120t - 16t^2$. Find its instantaneous velocity function and its velocity at time $t = 3$. [**HINT:** See Example 4.]

47. ***Oil Imports from Mexico*** Daily crude oil imports to the United States from Mexico for 2009–2013 could be approximated by

$$I(t) = -39t^2 + 800t - 3{,}000 \text{ thousand barrels} \quad (9 \le t \le 13),$$

where t is time in years since the start of 2000.[76] Find the derivative function $\frac{dI}{dt}$. At what rate were oil imports changing at the start of 2012 ($t = 12$)? [**HINT:** See Example 4.]

48. ***Oil Production in Mexico*** Daily crude oil production by **Pemex**, Mexico's national oil company, for 2008–2013 could be approximated by

$$P(t) = 0.017t^2 - 0.4t + 5.23 \text{ million barrels} \quad (8 \le t \le 13),$$

where t is time in years since the start of 2000.[77] Find the derivative function $\frac{dP}{dt}$. At what rate was oil production changing at the start of 2010 ($t = 10$)? [**HINT:** See Example 4.]

49. ***Bottled Water Sales*** The following chart shows the amount of bottled water sold in the United States for the period 2007–2014:[78]

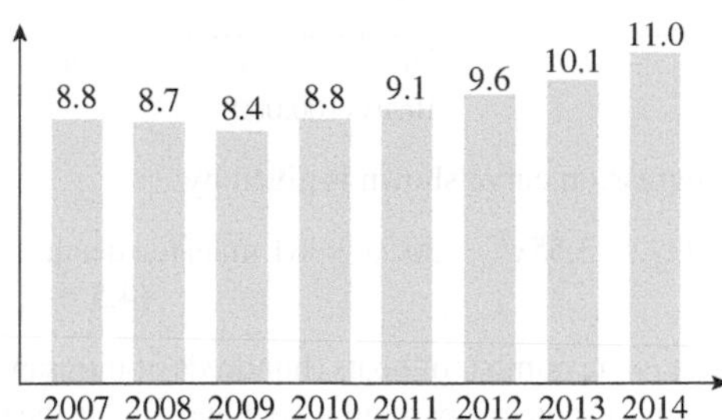

The function

$$R(t) = 0.08t^2 - 0.26t + 8.8 \text{ billion gallons} \quad (0 \le t \le 7)$$

gives a good approximation, where t is time in years since 2007. Find the derivative function $R'(t)$. According to the model, how fast were annual sales of bottled water changing in 2012?

50. ***Bottled Water Sales*** The following chart shows annual per capita sales of bottled water in the United States for the period 2007–2014:[79]

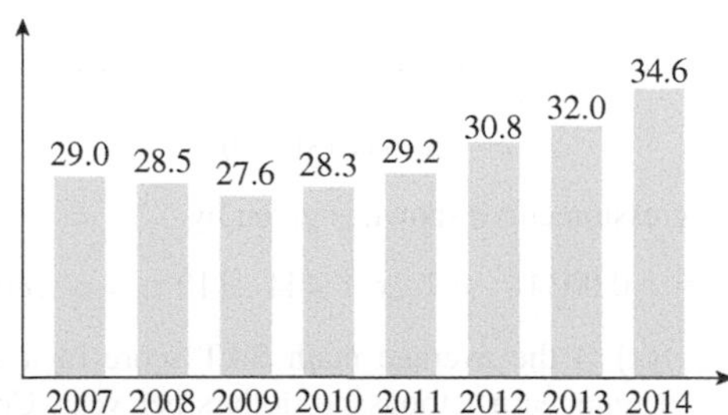

The function

$$R(t) = 0.25t^2 - t + 29 \text{ gallons} \quad (0 \le t \le 7)$$

[76] Source: U.S. Energy Information Administration, www.eia.gov.

[77] Source: www.pemex.com.

[78] The 2014 figure is a projection. Source: Beverage Marketing Corporation/www.bottledwater.org.

[79] *Ibid.*

gives a good approximation, where t is time in years since 2007. Find the derivative function $R'(t)$. According to the model, how fast were per capita sales of bottled water changing in 2011?

51. ▼ ***Ecology*** Increasing numbers of manatees have been killed by boats off the Florida coast. The following graph shows the relationship between the number of boats registered in Florida and the number of manatees killed each year.

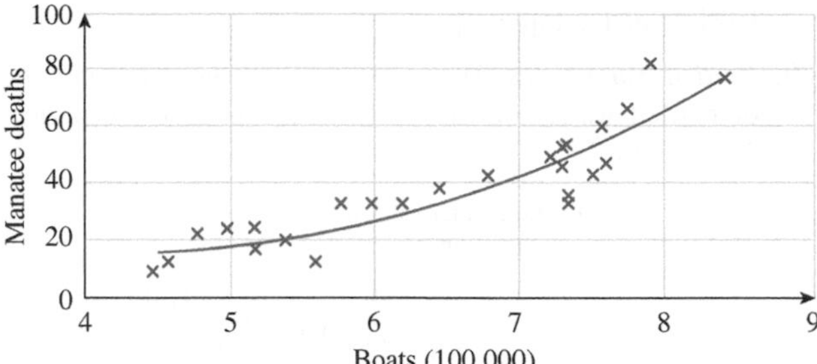

The regression curve shown is given by

$$f(x) = 3.55x^2 - 30.2x + 81 \text{ manatee deaths} \quad (4.5 \le x \le 8.5),$$

where x is the number of boats (hundreds of thousands) registered in Florida in a particular year and $f(x)$ is the number of manatees killed by boats in Florida that year.[80] Compute and interpret $f'(8)$.

52. ▼ ***SAT Scores by Income*** The following graph shows U.S. math SAT scores as a function of parents' income level.[81]

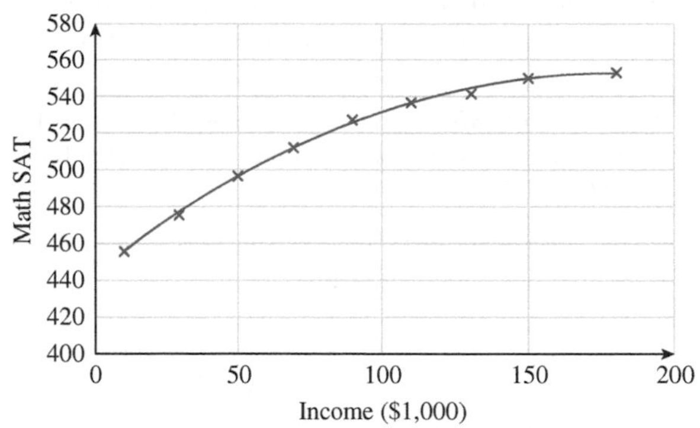

The regression curve shown is given by

$$f(x) = -0.0034x^2 + 1.2x + 444 \quad (10 \le x \le 180),$$

where $f(x)$ is the average math SAT score of a student whose parents earn x thousand dollars per year. Compute and interpret $f'(30)$.

53. ▼ ***Television Advertising*** The cost, in thousands of dollars, of a 30-second television ad during the Super Bowl in the years 1970–2010 can be approximated by the following piecewise-linear function ($t = 0$ represents 1970):[82]

$$C(t) = \begin{cases} 31.1t + 78 & \text{if } 0 \le t < 20 \\ 90t - 1{,}100 & \text{if } 20 \le t \le 40. \end{cases}$$

a. Is C a continuous function of t? Why? [**HINT**: See Example 4 of Section 3.3.]

b. Is C a differentiable function of t? Compute $\lim_{t\to 20^-} C'(t)$ and $\lim_{t\to 20^+} C'(t)$, and interpret the results. [**HINT**: See the "Before we go on" discussion after Example 5.]

54. ▼ ***Television Advertising*** (Compare Exercise 53.) The cost, in thousands of dollars, of a 30-second television ad during the Super Bowl in the years 1980–2010 can be approximated by the following piecewise-linear function ($t = 0$ represents 1980):[83]

$$C(t) = \begin{cases} 43.9t + 222 & \text{if } 0 \le t \le 20 \\ 140t - 1{,}700 & \text{if } 20 < t \le 30. \end{cases}$$

a. Is C a continuous function of t? Why? [**HINT**: See Example 4 of Section 3.3.]

b. Is C a differentiable function of t? Compute $\lim_{t\to 20^-} C'(t)$ and $\lim_{t\to 20^+} C'(t)$, and interpret the results. [**HINT**: See the "Before we go on" discussion after Example 5.]

Communication and Reasoning Exercises

55. Of the three methods (numerical, graphical, and algebraic) we can use to estimate the derivative of a function at a given value of x, which is always the most accurate? Explain.

56. Explain why we cannot put $h = 0$ in the formula

$$f'(a) = \lim_{h\to 0} \frac{f(a+h) - f(a)}{h}$$

for the derivative of f.

57. You just got your derivatives test back, and you can't understand why that teacher of yours deducted so many points for what you thought was your best work:

$$\lim_{h\to 0} \frac{f(x+h) - f(x)}{h}$$
$$= \lim_{h\to 0} \frac{f(x) + h - f(x)}{h}$$
$$= \lim_{h\to 0} \frac{h}{h} \qquad \text{Canceled the } f(x)$$
$$= 1. \qquad \text{✗ WRONG } -10$$

What was wrong with your answer? (There may be more than one error.)

[80] Regression model is based on data from 1976–2000. Sources for data: Florida Department of Highway Safety & Motor Vehicles, Florida Marine Institute/*New York Times*, February 12, 2002, p. F4.

[81] Regression model is based on 2009 data. Source: College Board/*New York Times* http://economix.blogs.nytimes.com.

[82] Source: http://en.wikipedia.org/wiki/Super_Bowl_advertising.

[83] *Ibid.*

58. Your friend just got his derivatives test back and can't understand why that teacher of his deducted so many points for the following:

$$\lim_{h\to 0} \frac{f(x+h)-f(x)}{h}$$

$$= \lim_{h\to 0} \frac{f(x)+f(h)-f(x)}{h}$$

$$= \lim_{h\to 0} \frac{f(h)}{h} \qquad \text{Canceled the } f(x)$$

$$= \lim_{h\to 0} \frac{f(\cancel{h})}{\cancel{h}} \qquad \text{Now cancel the } h.$$

$$= f. \qquad ✗ \textit{ WRONG } -50$$

What was wrong with his answer? (There may be more than one error.)

59. Your other friend just got her derivatives test back and can't understand why that teacher of hers took off so many points for the following:

$$\lim_{h\to 0} \frac{f(x+h)-f(x)}{h}$$

$$= \lim_{h\to 0} \frac{f(x+\cancel{h})-f(x)}{\cancel{h}} \qquad \text{Now cancel the } h.$$

$$= \lim_{h\to 0} f(x)-f(x) \qquad \text{Cancel the } f(x).$$

$$= 0. \qquad ✗ \textit{ WRONG } -15$$

What was wrong with her answer? (There may be more than one error.)

60. Your third friend just got her derivatives test back and can't understand why that teacher of hers took off so many points for the following:

$$\lim_{h\to 0} \frac{f(x+h)-f(x)}{h}$$

$$= \lim_{h\to 0} \frac{f(x)+h-f(x)}{h}$$

$$= \lim_{h\to 0} \frac{f(x)+\cancel{h}-f(x)}{\cancel{h}} \qquad \text{Now cancel the } h.$$

$$= \lim_{h\to 0} f(x)-f(x) \qquad \text{Cancel the } f(x).$$

$$= 0. \qquad ✗ \textit{ WRONG } -25$$

What was wrong with her answer? (There may be more than one error.)

61. Your friend Muffy claims that, because the balanced difference quotient is more accurate, it would be better to use that instead of the usual difference quotient when computing the derivative algebraically. Comment on this advice.

62. Use the balanced difference quotient formula,

$$f'(a) = \lim_{h\to 0} \frac{f(a+h)-f(a-h)}{2h},$$

to compute $f'(3)$ when $f(x) = x^2$. What do you find?

63. ▼ A certain function f has the property that $f'(a)$ does not exist. How is that reflected in the attempt to compute $f'(a)$ algebraically?

64. ▼ One cannot put $h = 0$ in the formula

$$f'(a) = \lim_{h\to 0} \frac{f(a+h)-f(a)}{h}$$

for the derivative of f. (See Exercise 56.) However, in the last step of each of the computations in the text, we are effectively setting $h = 0$ when taking the limit. What is going on here?

CHAPTER 3 REVIEW

KEY CONCEPTS

 **www.WanerMath.com**
Go to the Website to find a comprehensive and interactive Web-based summary of Chapter 3.

3.1 Limits: Numerical and Graphical Viewpoints

$\lim_{x\to a} f(x) = L$ means that $f(x)$ approaches L as x approaches a [p. 204]
What it means for a limit to exist [p. 205]
Limits at infinity [p. 207]
Estimating limits graphically [p. 208]
Interpreting limits in real-world situations [p. 211]

3.2 Limits and Continuity

f is continuous at a if $\lim_{x\to a} f(x)$ exists and $\lim_{x\to a} f(x) = f(a)$ [p. 218]
Discontinuous, continuous on domain [p. 218]
Discontinuities and singularities [p. 219]
Determining whether a given function is continuous [p. 220]

3.3 Limits and Continuity: Algebraic Viewpoint

Closed-form function [p. 225]
Limits of closed form functions [p. 226]
Simplifying to obtain limits [p. 226]
The indeterminate form $0/0$ [p. 227]
The determinate form $k/0$ [p. 228]
Limits of piecewise-defined functions [p. 230]
Limits at infinity [p. 231]
Determinate and indeterminate forms [p. 234]

3.4 Average Rate of Change

Average rate of change of $f(x)$ over $[a, b]$: $\dfrac{\Delta f}{\Delta x} = \dfrac{f(b) - f(a)}{b - a}$ [p. 242]
Average rate of change as slope of the secant line [p. 242]
Computing the average rate of change from a graph [p. 243]
Computing the average rate of change from a formula [p. 245]
Computing the average rate of change over short intervals $[a, a + h]$ [p. 245]

3.5 The Derivative: Numerical and Graphical Viewpoints

Instantaneous rate of change of $f(x)$ (derivative of f at a);

$$f'(a) = \lim_{h\to 0} \frac{f(a + h) - f(a)}{h} \quad \text{[p. 256]}$$

The derivative as slope of the tangent line [p. 259]
Quick approximation of the derivative [p. 261]
$\dfrac{d}{dx}$ Notation [p. 263]
The derivative as velocity [p. 263]
Average and instantaneous velocity [p. 264]
The derivative function [p. 265]
Graphing the derivative function with technology [p. 266]

3.6 The Derivative: Algebraic Viewpoint

Derivative at the point $x = a$:

$$f'(a) = \lim_{h\to 0} \frac{f(a + h) - f(a)}{h}$$

[p. 279]

Derivative function:

$$f'(x) = \lim_{h\to 0} \frac{f(x + h) - f(x)}{h}$$

[p. 280]

Examples of the computation of $f'(x)$ [p. 281]
$f(x) = |x|$ is not differentiable at $x = 0$ [p. 283]

REVIEW EXERCISES

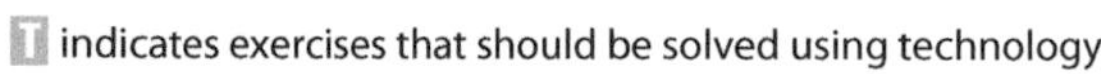

T indicates exercises that should be solved using technology

In Exercises 1–4, numerically *estimate whether the limit exists. If the limit does exist, give its approximate value.*

1. $\displaystyle\lim_{x\to 3} \frac{x^2 - x - 6}{x - 3}$

2. $\displaystyle\lim_{x\to 3} \frac{x^2 - 2x - 6}{x - 3}$

3. $\displaystyle\lim_{x\to -1} \frac{|x + 1|}{x^2 - x - 2}$

4. $\displaystyle\lim_{x\to -1} \frac{|x + 1|}{x^2 + x - 2}$

In Exercises 5 and 6 the graph of a function f is shown. Graphically determine whether the given limits exist. If a limit does exist, give its approximate value.

5.

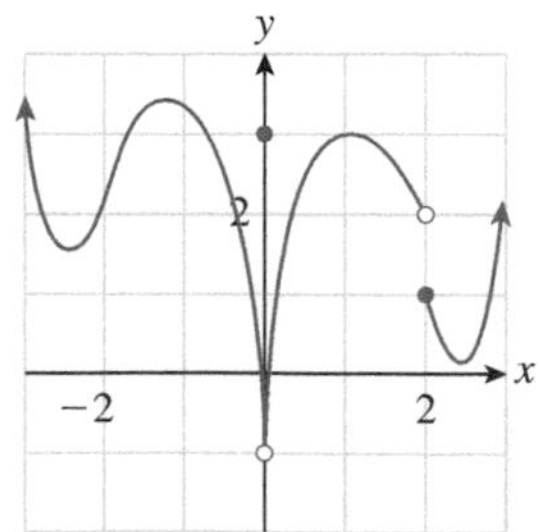

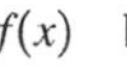

a. $\lim_{x\to 0} f(x)$ **b.** $\lim_{x\to 1} f(x)$
c. $\lim_{x\to 2} f(x)$

6.

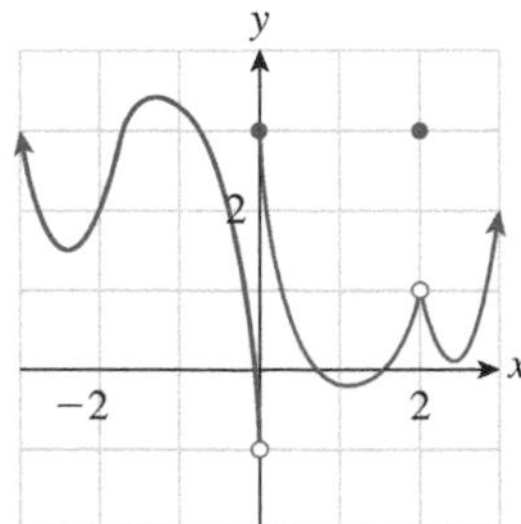

a. $\lim_{x\to 0} f(x)$ **b.** $\lim_{x\to -2} f(x)$
c. $\lim_{x\to 2} f(x)$

In Exercises 7–30, calculate the limit algebraically. If the limit does not exist, say why.

7. $\lim_{x \to -2} \frac{x^2}{x - 3}$

8. $\lim_{x \to 3} \frac{x^2 - 9}{2x - 6}$

9. $\lim_{x \to -2} \frac{x^2 - 4}{x^3 + 2x^2}$

10. $\lim_{x \to -1} \frac{x^2 - 9}{2x - 6}$

11. $\lim_{x \to 0} \frac{x}{2x^2 - x}$

12. $\lim_{x \to 1} \frac{x^2 - 9}{x - 1}$

13. $\lim_{x \to -1} \frac{x^2 + 3x}{x^2 - x - 2}$

14. $\lim_{x \to -1^+} \frac{x^2 + 1}{x^2 + 3x + 2}$

15. $\lim_{x \to 8} \frac{x^2 - 6x - 16}{x^2 - 9x + 8}$

16. $\lim_{x \to 4} \frac{x^2 + 3x}{x^2 - 8x + 16}$

17. $\lim_{x \to 4} \frac{x^2 + 8}{x^2 - 2x - 8}$

18. $\lim_{x \to 6} \frac{x^2 - 5x - 6}{x^2 - 36}$

19. $\lim_{x \to 1/2} \frac{x^2 + 8}{4x^2 - 4x + 1}$

20. $\lim_{x \to 1/2} \frac{x^2 + 3x}{2x^2 + 3x - 1}$

21. $\lim_{x \to +\infty} \frac{10x^2 + 300x + 1}{5x^3 + 2}$

22. $\lim_{x \to +\infty} \frac{2x^4 + 20x^3}{1{,}000x^6 + 6}$

23. $\lim_{x \to -\infty} \frac{x^2 - x - 6}{x - 3}$

24. $\lim_{x \to +\infty} \frac{x^2 - x - 6}{4x^2 - 3}$

25. $\lim_{t \to +\infty} \frac{-5}{5 + 5.3(3^{2t})}$

26. $\lim_{t \to +\infty} \left(3 + \frac{2}{e^{4t}}\right)$

27. $\lim_{x \to +\infty} \frac{2}{5 + 4e^{-3x}}$

28. $\lim_{x \to +\infty} (4e^{3x} + 12)$

29. $\lim_{t \to +\infty} \frac{1 + 2^{-3t}}{1 + 5.3e^{-t}}$

30. $\lim_{x \to -\infty} \frac{8 + 0.5^x}{2 - 3^{2x}}$

In Exercises 31–34, find the average rate of change of the given function over the interval $[a, a + h]$ for $h - 1$, 0.01, and 0.001. (Round answers to four decimal places.) Then estimate the slope of the tangent line to the graph of the function at a.

31. $f(x) = \frac{1}{x + 1}$; $a = 0$

32. $f(x) = x^x$; $a = 2$

33. $f(x) = e^{2x}$; $a = 0$

34. $f(x) = \ln(2x)$; $a = 1$

In Exercises 35–38 you are given the graph of a function with four points marked. Determine at which (if any) of these points the derivative of the function is (a) -1, (b) 0, (c) 1, and (d) 2.

35.

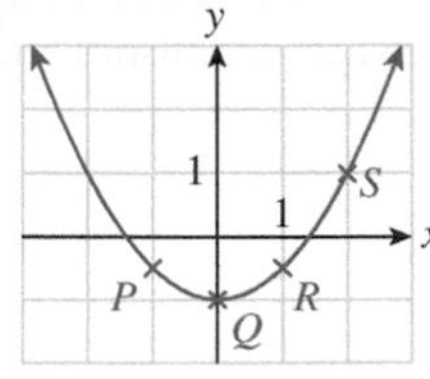

36.

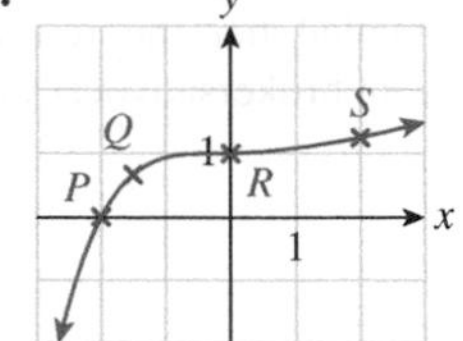

37.

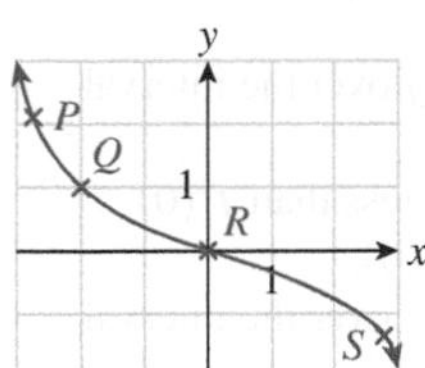

38.

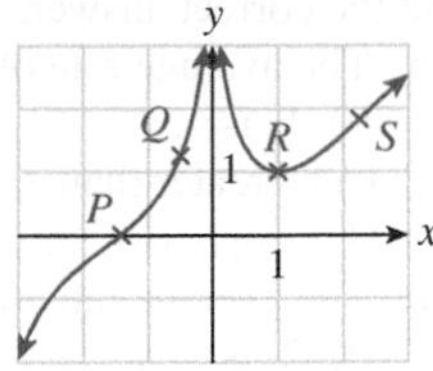

39. Let f have the graph shown.

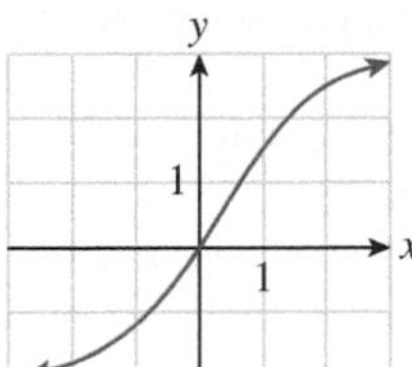

Select the correct answer.

a. The average rate of change of f over the interval $[0, 2]$ is
- **(A)** greater than $f'(0)$.
- **(B)** less than $f'(0)$.
- **(C)** approximately equal to $f'(0)$.

b. The average rate of change of f over the interval $[-1, 1]$ is
- **(A)** greater than $f'(0)$.
- **(B)** less than $f'(0)$.
- **(C)** approximately equal to $f'(0)$.

c. Over the interval $[0, 2]$ the instantaneous rate of change of f is
- **(A)** increasing.
- **(B)** decreasing.
- **(C)** neither increasing nor decreasing.

d. Over the interval $[-2, 2]$ the instantaneous rate of change of f is
- **(A)** increasing, then decreasing.
- **(B)** decreasing, then increasing.
- **(C)** approximately constant.

e. When $x = 2$, $f(x)$ is
- **(A)** approximately 1 and increasing at a rate of about 2.5 units per unit of x.
- **(B)** approximately 1.2 and increasing at a rate of about 1 unit per unit of x.
- **(C)** approximately 2.5 and increasing at a rate of about 0.5 units per unit of x.
- **(D)** approximately 2.5 and increasing at a rate of about 2.5 units per unit of x.

40. Let f have the graph shown.

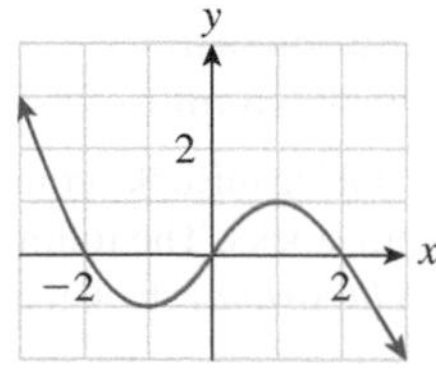

Select the correct answer.

a. The average rate of change of f over the interval $[0, 1]$ is
(A) greater than $f'(0)$. **(B)** less than $f'(0)$.
(C) approximately equal to $f'(0)$.

b. The average rate of change of f over the interval $[0, 2]$ is
(A) greater than $f'(1)$. **(B)** less than $f'(1)$.
(C) approximately equal to $f'(1)$.

c. Over the interval $[-2, 0]$ the instantaneous rate of change of f is
(A) increasing. **(B)** decreasing.
(C) neither increasing nor decreasing.

d. Over the interval $[-2, 2]$ the instantaneous rate of change of f is
(A) increasing, then decreasing.
(B) decreasing, then increasing.
(C) approximately constant.

e. When $x = 0$, $f(x)$ is
(A) approximately 0 and increasing at a rate of about 1.5 units per unit of x.
(B) approximately 0 and decreasing at a rate of about 1.5 units per unit of x.
(C) approximately 1.5 and neither increasing nor decreasing.
(D) approximately 0 and neither increasing nor decreasing.

In Exercises 41–44, use the definition of the derivative to calculate the derivative of the given function algebraically.

41. $f(x) = x^2 + x$ **42.** $f(x) = 3x^2 - x + 1$

43. $f(x) = 1 - \dfrac{2}{x}$ **44.** $f(x) = \dfrac{1}{x} + 1$

T *In Exercises 45–48, use technology to graph the derivative of the given function. In each case, choose a range of x-values and y-values that shows the interesting features of the graph.*

45. $f(x) = 10x^5 + \dfrac{1}{2}x^4 - x + 2$

46. $f(x) = \dfrac{10}{x^5} + \dfrac{1}{2x^4} - \dfrac{1}{x} + 2$

47. $f(x) = 3x^3 + 3\sqrt[3]{x}$

48. $f(x) = \dfrac{2}{x^{2.1}} - \dfrac{x^{0.1}}{2}$

Applications: OHaganBooks.com
[Try the game at www.OHaganBooks.com]

49. ***Stock Investments*** OHaganBooks.com CEO John O'Hagan has terrible luck with stocks. The following graph shows the value of *Fly-By-Night Airlines* stock that he bought acting on a "hot tip" from Marjory Duffin (CEO of *Duffin House* publishers and a close business associate):

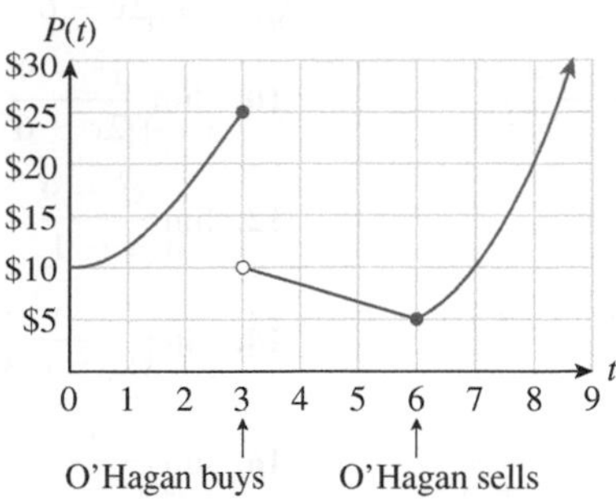

a. Compute $P(3)$, $\lim_{t\to 3^-} P(t)$ and $\lim_{t\to 3^+} P(t)$. Does $\lim_{t\to 3} P(t)$ exist? Interpret your answers in terms of Fly-By-Night stock.

b. Is P continuous at $t = 6$? Is P differentiable at $t = 6$? Interpret your answers in terms of Fly-By-Night stock.

50. ***Stock Investments*** John O'Hagan's golf partner Juan Robles seems to have had better luck with his investment in *Gapple Computer, Inc.* stocks as shown in the following graph:

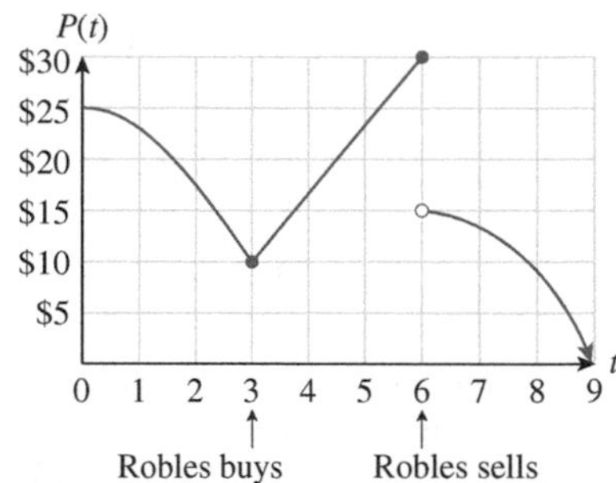

a. Compute $P(6)$, $\lim_{t\to 6^-} P(t)$ and $\lim_{t\to 6^+} P(t)$. Does $\lim_{t\to 6} P(t)$ exist? Interpret your answers in terms of Gapple stock.

b. Is P continuous at $t = 3$? Is P differentiable at $t = 3$? Interpret your answers in terms of Gapple stock.

51. ***Real Estate*** Marjory Duffin has persuaded John O'Hagan to consider investing a portion of OHaganBooks.com profits in real estate, now that the real estate market seems to have bottomed out. A real-estate broker friend of hers emailed her the following (somewhat optimistic) graph from brokersadvocacy.com:[84]

[84] Authors' note: As of March 2016, brokersadvocacy.com is unregistered.

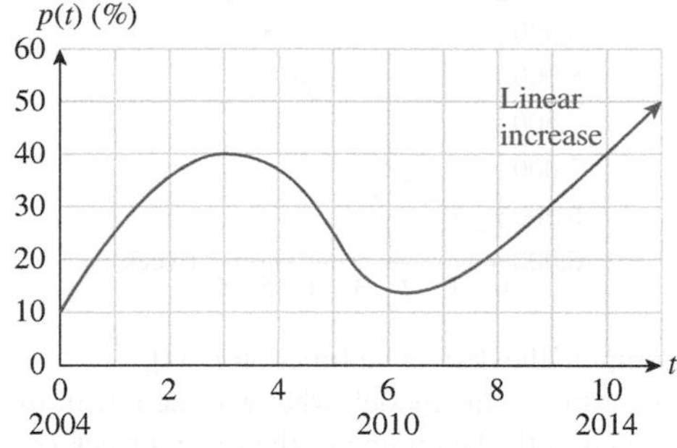

Here, $p(t)$ is the home price percentage over the 2003 level.

a. Assuming that the trend shown in the graph continues indefinitely, estimate $\lim_{t\to 3} p(t)$ and $\lim_{t\to +\infty} p(t)$, and interpret the results.

b. Estimate $\lim_{t\to +\infty} p'(t)$, and interpret the result.

52. ***Advertising Costs*** OHaganBooks.com has (on further advice from Marjory Duffin) mounted an aggressive online marketing strategy. The following graph shows the weekly cost of this campaign for the 6-week period since the start of July (t is time in weeks):

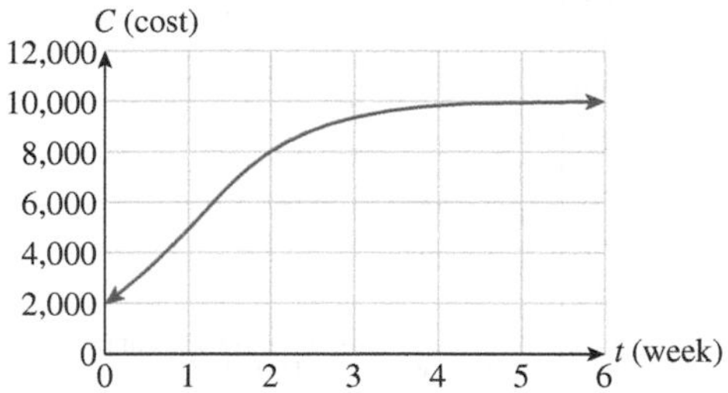

a. Assuming that the trend shown in the graph continues indefinitely, estimate $\lim_{t\to 2} C(t)$ and $\lim_{t\to +\infty} C(t)$, and interpret the results.

b. Estimate $\lim_{t\to +\infty} C'(t)$, and interpret the result.

53. ***Sales*** Since the start of July, OHaganBooks.com has seen its weekly sales increase, as shown in the following table:

Week	1	2	3	4	5	6
Sales (books)	6,500	7,000	7,200	7,800	8,500	9,000

a. What was the average rate of increase of weekly sales over this entire period?

b. During which 1-week interval(s) did the rate of increase of sales exceed the average rate?

c. During which 2-week interval(s) did the weekly sales rise at the highest average rate, and what was that average rate?

54. ***Rising Sea Level*** Marjory Duffin recently purchased a beachfront condominium in New York and is now in a panic, having just seen some disturbing figures about rising sea levels (sea levels as measured in New York relative to the 1900 level).[85]

Year since 1900	0	25	50	75	100	125
Sea Level (mm)	0	60	140	240	310	390

a. What was the average rate of increase of the sea level over this entire period?

b. During which 25-year interval(s) did the rate of increase of the sea level exceed the average rate?

c. Marjory Duffin's condominium is about 2 meters above sea level. Using the average rate of change from part (a), estimate how long she has before the sea rises to her condominium.

55. ***Real Estate*** The following graph (see Exercise 51) shows the home price index chart emailed to Marjory Duffin by a real-estate broker:

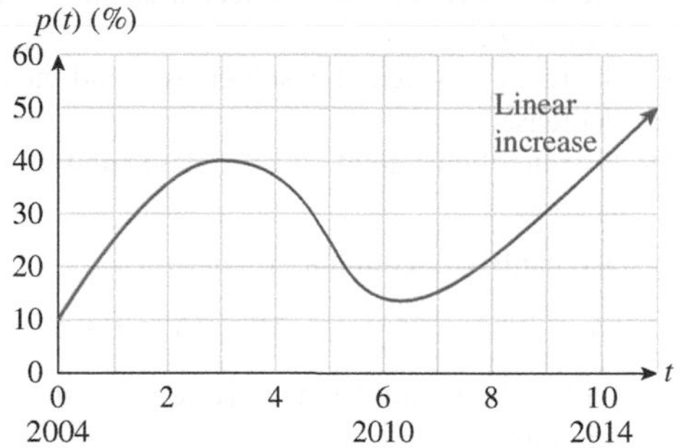

Use the graph to answer the following questions:

a. What was the average rate of change of the index over the 10-year period beginning 2004?

b. What was the average rate of change of the index over the period $[3, 10]$?

c. Which of the following is correct? Over the period $[4, 6]$

(A) the rate of change of the index increased.

(B) the rate of change of the index increased and then decreased.

(C) the rate of change of the index decreased.

(D) the rate of change of the index decreased and then increased.

[85] The 2025 level is a projection. Source: New England Integrated Science & Assessment, www.neisa.unh.edu/Climate/index.html.

56. ***Advertising Costs*** The following graph (see Exercise 52) shows the weekly cost of OHaganBooks.com's online ad campaign for the 6-week period since the start of July (t is time in weeks).

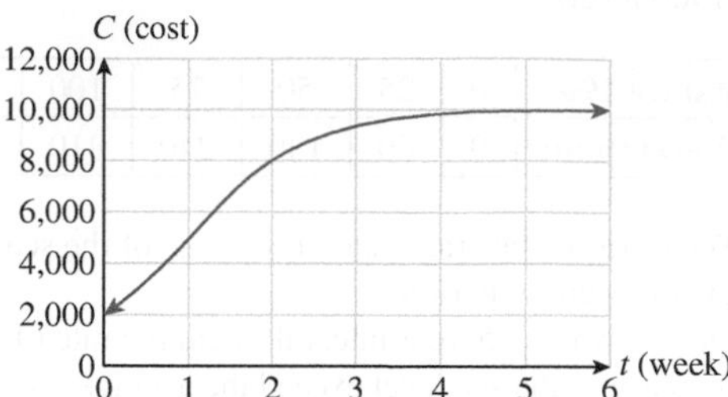

Use the graph to answer the following questions:

a. What was the average rate of change of cost over the entire 6-week period?

b. What was the average rate of change of cost over the period $[2, 6]$?

c. Which of the following is correct? Over the period $[2, 6]$

(A) the rate of change of cost increased and the cost increased.

(B) the rate of change of cost decreased and the cost increased.

(C) the rate of change of cost increased and the cost decreased.

(D) the rate of change of cost decreased and the cost decreased.

57. ***Sales*** OHaganBooks.com fits the curve

$$w(t) = 36t^2 + 250t + 6{,}240 \quad (0 \le t \le 6)$$

to its weekly sales figures from Exercise 53, as shown in the following graph:

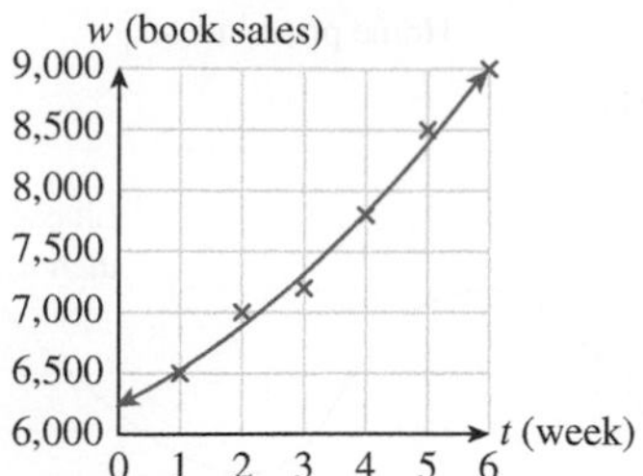

a. Compute the derivative function $w'(t)$.

b. According to the model, what was the rate of increase of sales at the beginning of the second week ($t = 1$)?

c. If we extrapolate the model, what would be the rate of increase of weekly sales at the beginning of the eighth week ($t = 7$)?

58. ***Sea Levels*** Marjory Duffin fit the curve

$$s(t) = 0.002t^2 + 3t - 6.4 \quad (0 \le t \le 125)$$

to her sea level figures from Exercise 54, as shown in the following graph:

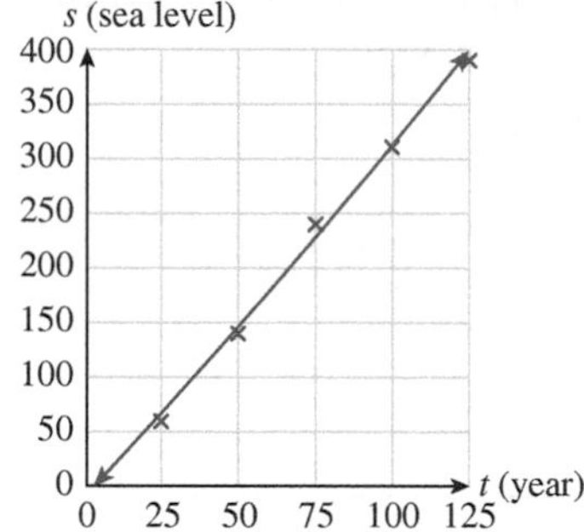

a. Compute the derivative function $s'(t)$.

b. According to the model, what was the rate of increase of the sea level in 2000 ($t = 100$)?

c. If we extrapolate the model, what would be the rate of increase of the sea level in 2100 ($t = 200$)?

CASE STUDY

Reducing Sulfur Emissions

Norbert Schaefer/CORBIS/Getty Images

The Environmental Protection Agency (EPA) wishes to formulate a policy that will encourage utilities to reduce sulfur emissions. Its goal is to reduce annual emissions of sulfur dioxide by a total of 10 million tons from the current level of 25 million tons by imposing a fixed charge for every ton of sulfur released into the environment per year. As a consultant to the EPA, you must determine the amount to be charged per ton of sulfur emissions.

You would like first to know the cost to the utility industry of reducing sulfur emissions. In other words, you would like to have a cost function of the form

$$C(q) = \text{Cost of removing } q \text{ tons of sulfur dioxide.}$$

Unfortunately, you do not have such a function handy. You do, however, have the following data, which show the *marginal* cost (that is, the *rate of change* of cost) to the utility industry of reducing sulfur emissions at several levels of reduction.[86]

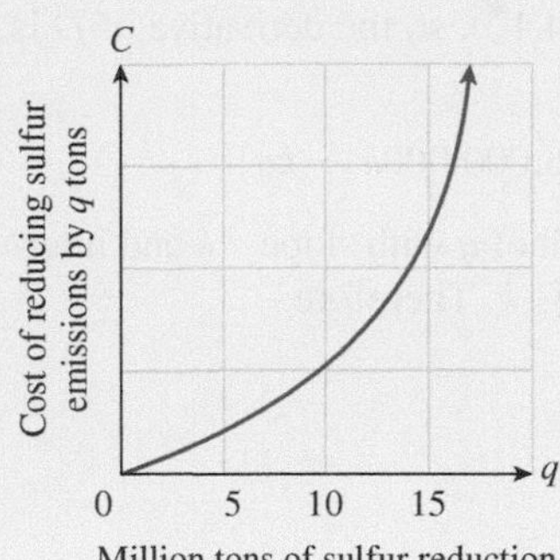

Figure 42

Reduction q (tons)	8,000,000	10,000,000	12,000,000
Marginal Cost $C'(q)$ ($/ton)	270	360	779

The table tells you that $C'(8{,}000{,}000) = \$270$ per ton, $C'(10{,}000{,}000) = \$360$ per ton, and $C'(12{,}000{,}000) = \$779$ per ton. Recalling that $C'(q)$ is the slope of the tangent to the graph of the cost function, you can see from the table that this slope is positive and increasing as q increases, so the graph of the cost function has the general shape shown in Figure 42.

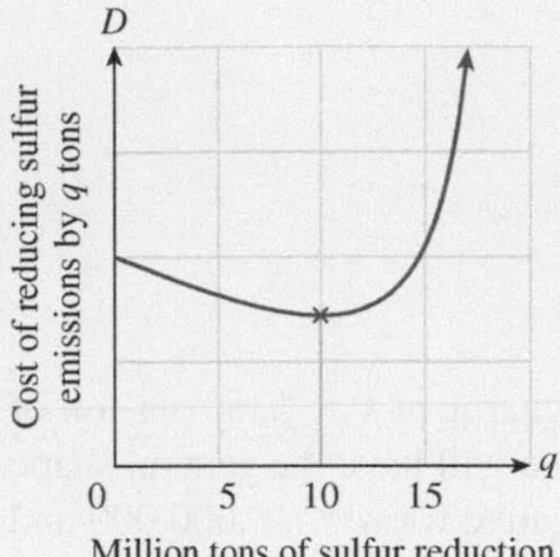

Figure 43

Notice that the slope (additional cost) is increasing as you move to the right, so the utility industry has no cost incentive to reduce emissions further, as it costs the industry significantly more per ton for each additional ton of sulfur it removes. What you would like—if the goal of reducing total emissions by 10 million tons is to be reached—is that, somehow, the imposition of a fixed charge for every ton of sulfur dioxide released will *alter* the form of the cost curve so that it has the general shape shown in Figure 43. In this ideal curve, the cost D to utilities is lowest at a reduction level of 10 million tons, so if the utilities act to minimize cost, they can be expected to reduce emissions by 10 million tons, which is exactly the EPA goal! From the graph, you can see that the tangent line to the curve at the point where $q = 10$ million tons is horizontal, and thus has zero slope: $D'(10{,}000{,}000) = \$0$ per ton. Further, the slope $D'(q)$ is negative for values of q to the left of 10 million tons and positive for values to the right.

So how much should the EPA charge per ton of sulfur released into the environment? Suppose the EPA charges $\$k$ per ton, so that

$$\text{Emission charge to utilities} = k \times \text{Sulfur emissions}.$$

It is your job to calculate k. Because you are working with q as the independent variable, you decide that it would be best to formulate the emission charge as a function of q. However, q represents the amount by which sulfur emissions have been reduced from the original 25 million tons, that is, the amount by which sulfur emissions are *lower than* the original 25 million tons:

$$q = 25{,}000{,}000 - \text{Sulfur emissions}.$$

Thus, the total annual emission charge to the utilities is

$$k \times \text{Sulfur emissions} = k(25{,}000{,}000 - q) = 25{,}000{,}000k - kq.$$

This results in a total cost to the utilities of

$$\text{Total cost} = \text{Cost of reducing emissions} + \text{Emission charge}$$
$$D(q) = C(q) + 25{,}000{,}000k - kq.$$

[86] These figures were produced in a computerized study of reducing sulfur emissions from the 1980 level by the given amounts. Source: Congress of the United States, Congressional Budget Office, *Curbing Acid Rain: Cost, Budget and Coal Market Effects* (Washington, DC: U.S. Government Printing Office, 1986): xx, xxii, 23, 80.

You now recall from calculus that the derivative of a sum of two functions is the sum of their derivatives (you will see why in Section 4.1*), so the derivative of D is given by

$$D'(q) = \text{Derivative of } C + \text{Derivative of } (25{,}000{,}000k - kq).$$

*This statement makes intuitive sense: For instance, if C is changing at a rate of 3 units per second and D is changing at a rate of 2 units per second, then their sum is changing at a rate of $3 + 2 = 5$ units per second.

The function $y = 25{,}000{,}000k - kq$ is a linear function of q with slope $-k$ and intercept $25{,}000{,}000k$. Thus, its derivative is just its slope: $-k$. Therefore,

$$D'(q) = C'(q) - k.$$

Remember that you want

$$D'(10{,}000{,}000) = 0.$$

Thus,

$$C'(10{,}000{,}000) - k = 0.$$

Referring to the table, you see that

$$360 - k = 0$$

or

$$k = \$360 \text{ per ton.}$$

In other words, all you need to do is set the emission charge at $k = \$360$ per ton of sulfur emitted. Further, to ensure that the resulting curve will have the general shape shown in Figure 43, you would like to have $D'(q)$ negative for $q < 10{,}000{,}000$ and positive for $q > 10{,}000{,}000$. To check this, write

$$\begin{aligned} D'(q) &= C'(q) - k \\ &= C'(q) - 360 \end{aligned}$$

and refer to the table to obtain

$$D'(8{,}000{,}000) = 270 - 360 = -90 < 0 \quad ✔$$

and

$$D'(12{,}000{,}000) = 779 - 360 = 419 > 0. \quad ✔$$

Thus, based on the given data, the resulting curve will have the shape you require. You therefore inform the EPA that an annual emissions charge of \$360 per ton of sulfur released into the environment will create the desired incentive: to reduce sulfur emissions by 10 million tons per year.

One week later, you are informed that this charge would be unrealistic because the utilities cannot possibly afford such a cost. You are asked whether there is an alternative plan that accomplishes the 10-million-ton reduction goal and yet is cheaper to the utilities by \$5 billion per year. You then look at your expression for the emission charge

$$25{,}000{,}000k - kq$$

and notice that, if you decrease this amount by \$5 billion, the derivative will not change at all because it will still have the same slope (only the intercept is affected). Therefore, you propose the following revised formula for the emission charge:

$$\begin{aligned} 25{,}000{,}000k - kq - 5{,}000{,}000{,}000 &= 25{,}000{,}000(360) - 360q - 5{,}000{,}000{,}000 \\ &= 4{,}000{,}000{,}000 - 360q. \end{aligned}$$

At the expected reduction level of 10 million tons, the total amount paid by the utilities will then be

$$4{,}000{,}000{,}000 - 360(10{,}000{,}000) = \$400{,}000{,}000.$$

Thus, your revised proposal is the following: Impose an annual emissions charge of \$360 per ton of sulfur released into the environment, and hand back \$5 billion in the form of subsidies. The effect of this policy will be to cause the utilities industry to reduce sulfur emissions by 10 million tons per year and will result in \$400 million in annual revenues to the government.

Notice that this policy also provides an incentive for the utilities to search for cheaper ways to reduce emissions. For instance, if they lowered costs to the point at which they could achieve a reduction level of 12 million tons, they would have a total emission charge of

$$4{,}000{,}000{,}000 - 360(12{,}000{,}000) = -\$320{,}000{,}000.$$

The fact that this is negative means that the government would be paying the utilities industry \$320 million more in annual subsidies than the industry is paying in per ton emission charges.

EXERCISES

1. Excluding subsidies, what should the annual emission charge be if the goal is to reduce sulfur emissions by 8 million tons?
2. Excluding subsidies, what should the annual emission charge be if the goal is to reduce sulfur emissions by 12 million tons?
3. What is the *marginal emission charge* (derivative of emission charge) in your revised proposal (as stated before the exercise set)? What is the relationship between the marginal cost of reducing sulfur emissions before emissions charges are implemented and the marginal emission charge, at the optimal reduction under your revised proposal?
4. We said that the revised policy provided an incentive for utilities to find cheaper ways to reduce emissions. How would $C(q)$ have to change to make 12 million tons the optimum reduction?
5. What change in $C(q)$ would make 8 million tons the optimum reduction?
6. If the scenario in Exercise 5 took place, what would the EPA have to do to make 10 million tons the optimal reduction once again?
7. Because of intense lobbying by the utility industry, you are asked to revise the proposed policy so that the utility industry will pay no charge if sulfur emissions are reduced by the desired 10 million tons. How can you accomplish this?
8. Suppose that instead of imposing a fixed charge per ton of emission, you decide to use a sliding scale, so that the total charge to the industry for annual emissions of x tons will be $\$kx^2$ for some k. What must k be to again make 10 million tons the optimum reduction? [**HINT:** The derivative of kx^2 is $2kx$.]

TI-83/84 Plus Technology Guide

Section 3.1

Example 1 (page 203) Use a table to estimate the following limits.

a. $\lim_{x \to 2} \frac{x^3 - 8}{x - 2}$ **b.** $\lim_{x \to 0} \frac{e^{2x} - 1}{x}$

Solution

On the TI-83/84 Plus, use the table feature to automate these computations as follows:

1. Define `Y1=(X^3-8)/(X-2)` for part (a) or `Y1=(e^(2X)-1)/X` for part (b).
2. Press [2ND] [TABLE] to list its values for the given values of x. (If the calculator does not allow you to enter values of x, press [2ND] [TBLSET], and set `Indpnt` to `Ask`).

   Here is the table showing some of the values for part (a):

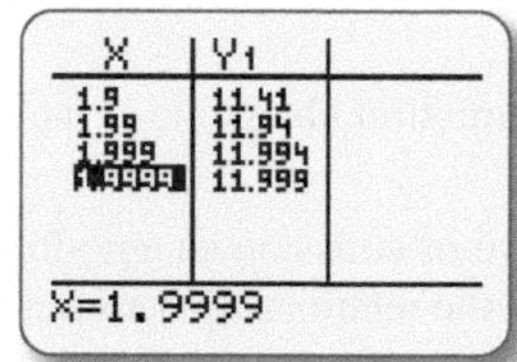

3. For part (b), use `Y1=(e^(2X)-1)/X` and values of x approaching 0 from either side.

Section 3.4

Example 3 (page 245) The price of an ounce of gold can be approximated by the function

$$G(t) = 5t^2 - 85t + 1{,}762 \quad (7.5 \le t \le 10.5)$$

where t is time in hours. ($t = 8$ represents 8:00 am.) What was the average rate of change of the price of gold over the $1\frac{1}{2}$-hour period starting at 8:00 am (the interval $[8, 9.5]$ on the t-axis)?

Solution

On the TI-83/84 Plus:

1. Enter the function G as `Y1` (using X for t):

   `Y1=5X^2-85X+1762`

2. Now find the average rate of change over $[8, 9.5]$ by evaluating the following on the Home screen:

   `(Y1(9.5)-Y1(8))/(9.5-8)`

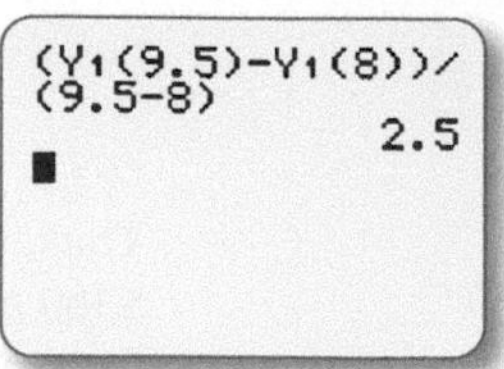

As shown on the screen, the average rate of change is 2.5.

Example 4 (page 245) Continuing with Example 3, use technology to compute the average rate of change of

$$G(t) = 5t^2 - 85t + 1{,}762 \quad (7.5 \le t \le 10.5)$$

over the intervals $[8, 8 + h]$, where $h = 1$, 0.1, 0.01, 0.001, and 0.0001.

Solution

1. As in Example 3, enter the function G as `Y1` (using X for t):

   `Y1=5X^2-85X+1762`

2. Now find the average rate of change for $h = 1$ by evaluating, on the Home screen,

   `(Y1(8+1)-Y1(8))/1`

   which gives 0.

3. To evaluate for $h = 0.1$, recall the expression using [2ND] [ENTER] and then change the 1, both places it occurs, to 0.1, getting

   `(Y1(8+0.1)-Y1(8))/0.1`

   which gives -4.95.

4. Continuing, we can evaluate the average rate of change for all the desired values of h:

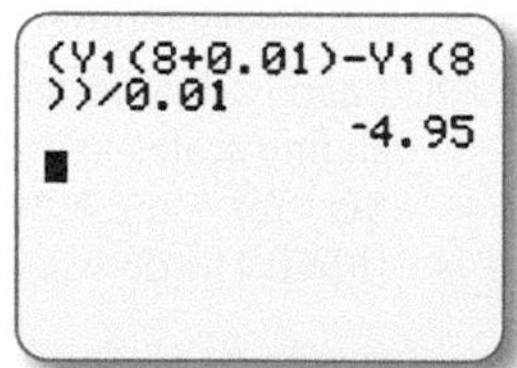

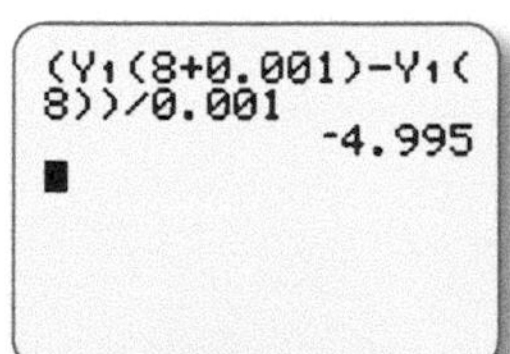

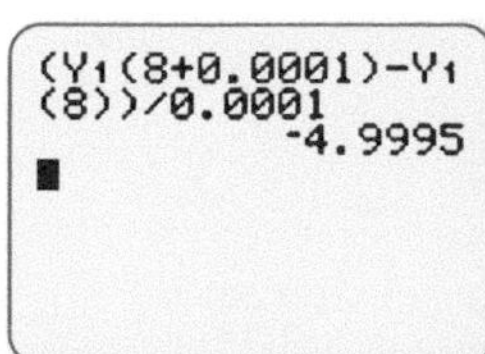

Section 3.5

Example 2 (page 261) Calculate an approximate value of $f'(1.5)$ if $f(x) = x^2 - 4x$, and then find the equation of the tangent line at the point on the graph where $x = 1.5$.

Solution

1. In the TI-83/84 Plus, enter the function f as Y_1:

$$Y_1\text{=X\^{}2-4*X}$$

2. Go to the Home screen to compute the approximations:

`(Y1(1.5001)-Y1(1.5))/0.0001` Usual difference quotient

`(Y1(1.5001)-Y1(1.4999))/0.0002` Balanced difference quotient

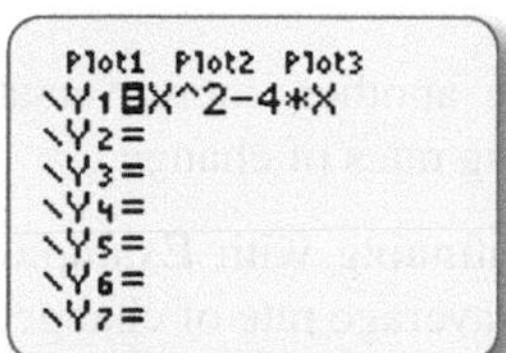

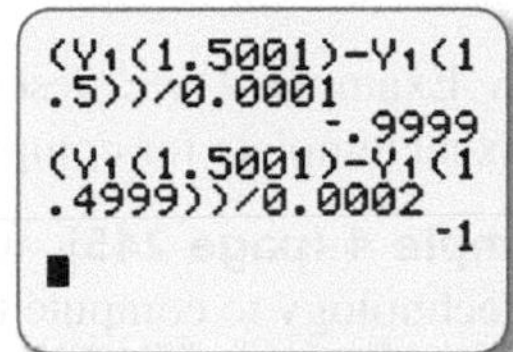

From the display on the right, we find that the difference quotient quick approximation is -0.9999 and the balanced difference quotient quick approximation is -1, which is in fact the exact value of $f'(1.5)$. See the discussion in the text for the calculation of the equation of the tangent line.

Example 4 (page 266) Use technology to graph the derivative of $f(x) = -2x^2 + 6x + 5$ for values of x in starting at -5.

Solution

On the TI-83/84 Plus, the easiest way to obtain quick approximations of the derivative of a given function is to use the built-in `nDeriv` function, which calculates balanced difference quotients.

1. On the Y= screen, first enter the function:

$$Y_1\text{=-2X\^{}2+6X+5}$$

2. Then set

`Y2=nDeriv(Y1,X,X)` For `nDeriv` press MATH 8

which is the TI-83/84 Plus's approximation of $f'(x)$ (see figure on the left below). Alternatively, we can enter the balanced difference quotient directly:

`Y2=(Y1(X+0.001)-Y1(X-0.001))/0.002`

(The TI-83/84 Plus uses $h = 0.001$ by default in the balanced difference quotient when calculating `nDeriv`, but this can be changed by giving a value of h as a fourth argument, such as `nDeriv(Y1,X,X,0.0001)`.) To see a table of approximate values of the derivative, we press 2ND TABLE and choose a collection of values for x (shown on the right below):

X	Y1	Y2
-3	-31	18
-2	-15	14
-1	-3	10
0	5	6
1	9	2
2	9	-2
3	5	-6

X=3

Here, Y_1 shows the value of f, and Y_2 shows the values of f'.

To graph the function or its derivative, we can graph Y_1 or Y_2 in a window showing the given domain $[-5, 5]$:

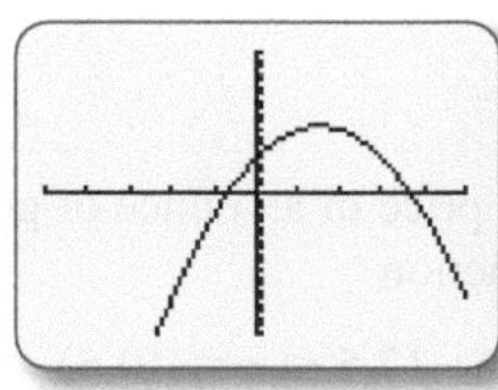

Graph of f

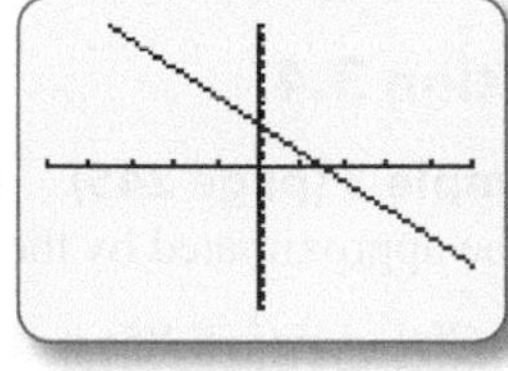

Graph of f'

Spreadsheet Technology Guide

Section 3.1

Example 1 (page 203) Use a table to estimate the following limits.

a. $\lim_{x \to 2} \frac{x^3 - 8}{x - 2}$ **b.** $\lim_{x \to 0} \frac{e^{2x} - 1}{x}$

Solution

1. Set up your spreadsheet to duplicate the table in part (a) as follows:

	A	B	C	D
1	x	f(x)	x	f(x)
2	1.9	=(A2^3-8)/(A2-2)	2.1	
3	1.99		2.01	
4	1.999		2.001	
5	1.9999		2.0001	

↓

	A	B	C	D
1	x	f(x)	x	f(x)
2	1.9	11.41	2.1	12.61
3	1.99	11.9401	2.01	12.0601
4	1.999	11.994001	2.001	12.006001
5	1.9999	11.99940001	2.0001	12.00060001

(The formula in cell B2 is copied to columns B and D, as indicated by the shading.) The values of $f(x)$ will be calculated in columns B and D.

2. For part (b), use the formula `=(EXP(2*A2)-1)/A2` in cell B2, and in columns A and C, use values of x approaching 0 from either side.

Section 3.4

Example 3 (page 245) The price of an ounce of gold can be approximated by the function

$$G(t) = 5t^2 + 85t + 1{,}762 \quad (7.5 \le t \le 10.5)$$

where t is time in hours. ($t = 8$ represents 8:00 am.) What was the average rate of change of the price of gold over the $1\frac{1}{2}$-hour period starting at 8:00 am (the interval $[8, 9.5]$ on the t-axis)?

Solution

To use a spreadsheet to compute the average rate of change of G:

1. Start with two columns, one for values of t and one for values of $G(t)$, which you enter using the formula for G:

```
=5*A2^2-85*A2+1762
```

	A	B
1	t	G(t)
2	8	=5*A2^2-85*A2+1762
3	9.5	

2. Next, calculate the average rate of change as shown here:

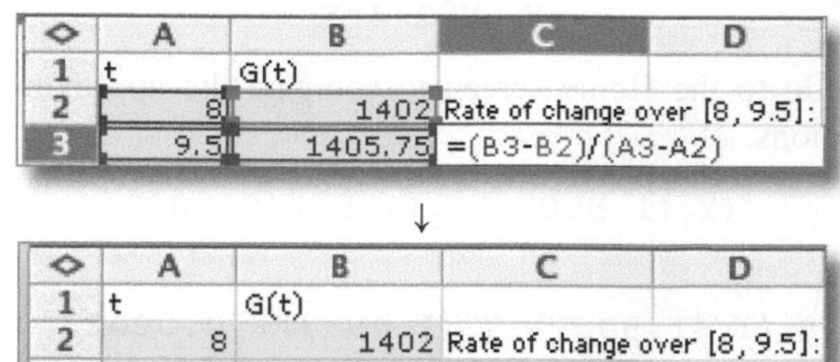

	A	B	C	D
1	t	G(t)		
2	8	1402	Rate of change over [8, 9.5]:	
3	9.5	1405.75	=(B3-B2)/(A3-A2)	

↓

	A	B	C	D
1	t	G(t)		
2	8	1402	Rate of change over [8, 9.5]:	
3	9.5	1405.75	2.5	

In Example 4 we describe another, more versatile Excel template for computing rates of change.

Example 4 (page 245) Continuing with Example 3, use technology to compute the average rate of change of

$$G(t) = 5t^2 + 85t + 1{,}762 \quad (7.5 \le t \le 10.5)$$

over the intervals $[8, 8 + h]$, where $h = 1$, 0.1, 0.01, 0.001, and 0.0001.

Solution

The template we can use to compute the rates of change is an extension of what we used in Example 3:

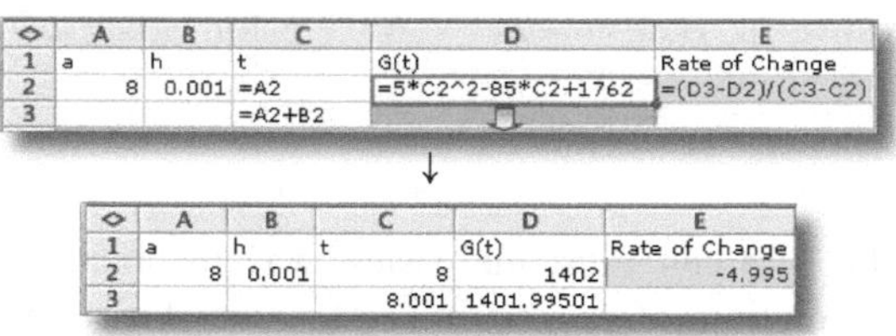

	A	B	C	D	E
1	a	h	t	G(t)	Rate of Change
2	8	0.001	=A2	=5*C2^2-85*C2+1762	=(D3-D2)/(C3-C2)
3			=A2+B2		

↓

	A	B	C	D	E
1	a	h	t	G(t)	Rate of Change
2	8	0.001	8	1402	-4.995
3			8.001	1401.99501	

1. Column C contains the values $t = a$ and $t = a + h$ we are using for the independent variable.
2. The formula in cell E2 is the average-rate-of-change formula $\Delta G/\Delta t$. Entering the different values $h = 1$, 0.1, 0.01, 0.001, and 0.0001 in cell B2 gives the results shown in Example 4.

Section 3.5

Example 2 (page 261) Calculate an approximate value of $f'(1.5)$ if $f(x) = x^2 - 4x$, and then find the equation of the tangent line at the point on the graph where $x = 1.5$.

Solution

You can compute both the difference quotient and the balanced difference quotient approximations in a spreadsheet using the following extension of the worksheet in Example 4 in Section 3.4:

	A	B	C	D	E	F
1	a	h	x	f(x)	Diff Quotients	Balanced Diff Quotient
2	1.5	0.0001	=A2-B2	=C2^2-4*C2	=(D3-D2)/(C3-C2)	=(D4-D2)/(C4-C2)
3			=A2			
4			=A2+B2			

Notice that we get two difference quotients in column E. The first uses $h = -0.0001$, while the second uses $h = 0.0001$ and is the one we use for our quick approximation. The balanced quotient is their average (column F). The results are as follows:

	A	B	C	D	E	F
1	a	h	x	f(x)	Diff Quotients	Balanced Diff Quotient
2	1.5	0.0001	1.4999	-3.7499	-1.0001	-1
3			1.5	-3.75	-0.9999	
4			1.5001	-3.7501		

From the results shown above, we find that the difference quotient quick approximation is -0.9999 and that the balanced difference quotient quick approximation is -1, which is in fact the exact value of $f'(1.5)$. See the discussion in the text for the calculation of the equation of the tangent line.

Example 4 (page 266) Use technology to graph the derivative of $f(x) = -2x^2 + 6x + 5$ for values of x starting at -5.

Solution

1. Start with a table of values for the function f:

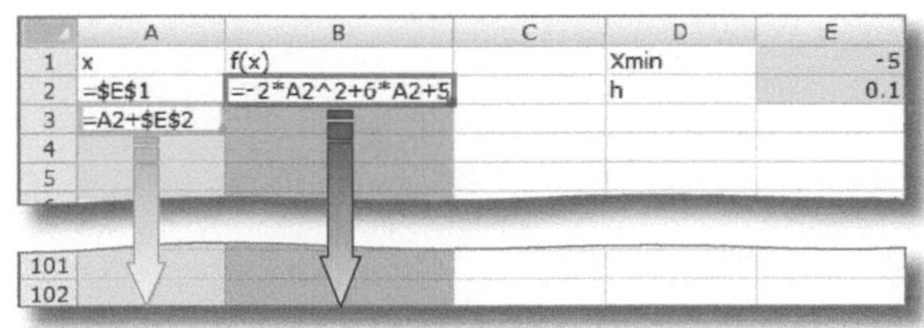

	A	B	C	D	E
1	x	f(x)		Xmin	-5
2	=$E$1	=-2*A2^2+6*A2+5		h	0.1
3	=A2+$E$2				
4					
5					
101					
102					

2. Next, compute approximate derivatives in column C:

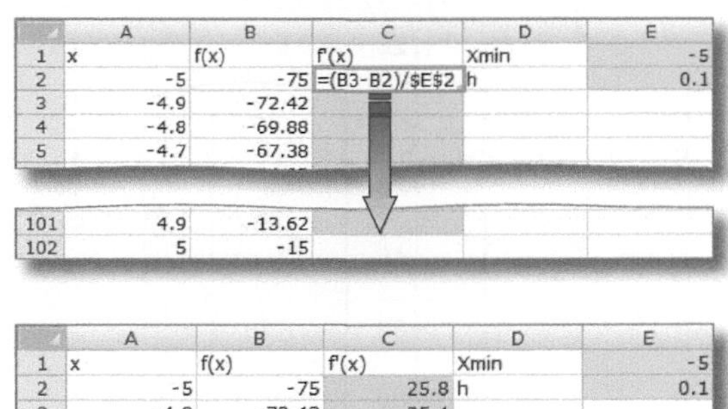

	A	B	C	D	E
1	x	f(x)	f'(x)	Xmin	-5
2	-5	-75	=(B3-B2)/$E$2	h	0.1
3	-4.9	-72.42			
4	-4.8	-69.88			
5	-4.7	-67.38			
101	4.9	-13.62			
102	5	-15			

	A	B	C	D	E
1	x	f(x)	f'(x)	Xmin	-5
2	-5	-75	25.8	h	0.1
3	-4.9	-72.42	25.4		
4	-4.8	-69.88	25		
5	-4.7	-67.38	24.6		
101	4.9	-13.62	-13.8		
102	5	-15			

You cannot paste the difference quotient formula into cell C102. (Why?) Notice that this worksheet uses the ordinary difference quotients, $[f(x + h) - f(x)]/h$. If you prefer, you can use balanced difference quotients $[f(x + h) - f(x - h)]/(2h)$, in which case cells C2 and C102 would both have to be left blank.

We now graph the function and the derivative on different graphs as follows:

1. First, graph the function f in the usual way, using columns A and B.
2. Make a copy of this graph, and click on it once. Columns A and B should be outlined, indicating that these are the columns used in the graph.
3. By dragging from the center of the bottom edge of the box, move the column B box over to column C as shown:

	A	B	C
96	4.4	-7.32	-11.8
97	4.5	-8.5	-12.2
98	4.6	-9.72	-12.6
99	4.7	-10.98	-13
100	4.8	-12.28	-13.4
101	4.9	-13.62	-13.8
102	5	-15	

↓

	A	B	C
96	4.4	-7.32	-11.8
97	4.5	-8.5	-12.2
98	4.6	-9.72	-12.6
99	4.7	-10.98	-13
100	4.8	-12.28	-13.4
101	4.9	-13.62	-13.8
102	5	-15	

The graph will then show the derivative (columns A and C):

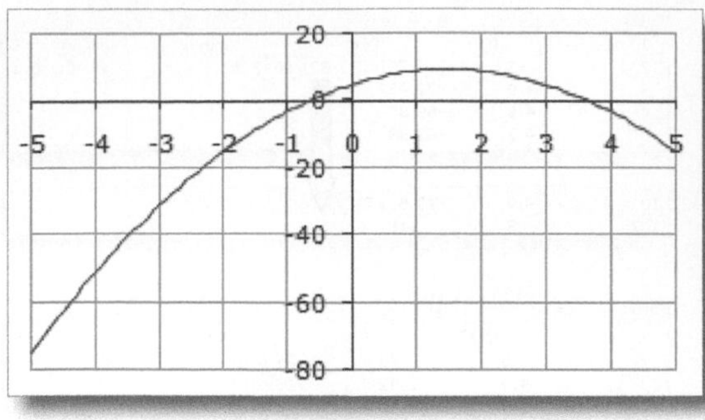

Graph of f

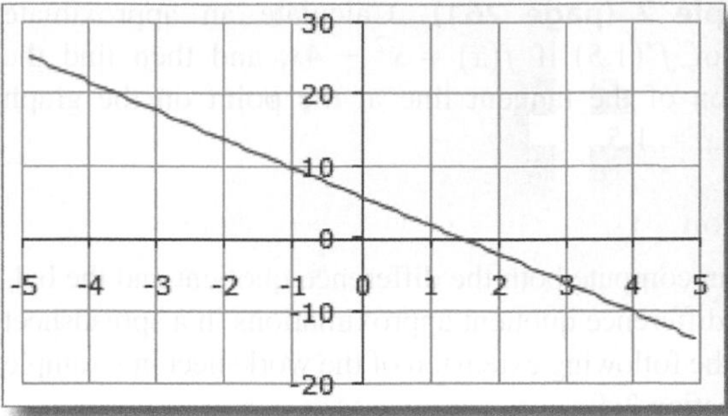

Graph of f'

4 TECHNIQUES OF DIFFERENTIATION WITH APPLICATIONS

www.WanerMath.com

At the Website, in addition to the resources listed in the Preface, you will find:

The following extra topic:

- Linear Approximation and Error Estimation

CASE STUDY

Projecting Market Growth

It is 2010, and you are on the board of directors at *Fullcourt Academic Press*. The sales director of the high school division has just burst into your office with a proposal for an expansion strategy based on the assumption that the number of graduates from private high schools in the United States will grow at a rate of at least 4,000 per year through the year 2015. Because the figures actually appear to be leveling off, you are suspicious about this estimate. You would like to devise a model that predicts this trend before tomorrow's scheduled board meeting.

How do you go about doing this?

Yuri Arcurs/Shutterstock.com

Introduction

In Chapter 3 we studied the concept of the derivative of a function, and we saw some of the applications for which derivatives are useful. However, computing the derivative of a function algebraically, from the definition, seemed to be a time-consuming process, forcing us to restrict attention to fairly simple functions.

In this chapter we develop shortcut techniques that will allow us to write down the derivative of a function directly without having to calculate any limit. These techniques will also enable us to differentiate any closed-form function—that is, any function, no matter how complicated, that can be specified by a formula involving powers, radicals, absolute values, exponents, and logarithms. (In Chapter 9 we will discuss how to add trigonometric functions to this list.) We also show how to find the derivatives of functions that are only specified *implicitly*—that is, functions for which we are not given an explicit formula for y in terms of x but only an equation relating x and y.

Precalculus Review
For this chapter you should be familiar with the algebra reviewed in **Sections 0.3 and 0.4**.

4.1 Derivatives of Powers, Sums, and Constant Multiples

Shortcut Rules

Up to this point we have approximated derivatives using difference quotients, and we have done exact calculations using the definition of the derivative as the limit of a difference quotient. While exact calculations are preferable, the calculation of a derivative as a limit is often tedious, so it would be useful to have a quicker method, or shortcut. We discuss the first of the shortcut rules in this section. By the end of this chapter we will be able to find fairly quickly the derivative of almost any function we can write.

The Power Rule

If you look at Examples 2 and 3 in Section 3.6, you may notice a pattern:

$$f(x) = x^2 \quad \Rightarrow \quad f'(x) = 2x$$
$$f(x) = x^3 \quad \Rightarrow \quad f'(x) = 3x^2.$$

This pattern generalizes to any power of x.

Theorem 4.1 The Power Rule

If n is any constant and $f(x) = x^n$, then

$$f'(x) = nx^{n-1}.$$

Quick Examples

1. If $f(x) = x^2$, then $f'(x) = 2x^1 = 2x$.
2. If $f(x) = x^3$, then $f'(x) = 3x^2$.
3. If $f(x) = x$, rewrite* as $f(x) = x^1$, so $f'(x) = 1x^0 = 1$.
4. If $f(x) = 1$, rewrite as $f(x) = x^0$, so $f'(x) = 0x^{-1} = 0$.

* To use the power rule, we rewrite expressions like this in power form: constant times x^n. See Section 0.2 in the Precalculus Review to brush up on negative and fractional exponents. Pay particular attention to radical, positive exponent, and power forms.

The proof of the power rule involves first studying the case when n is a positive integer and then studying the cases of other types of exponents (negative integer, rational number, irrational number). You can find a proof at the Website.

Website
www.WanerMath.com
At the Website you can find a proof of the power rule by following:
Everything for Applied Calc
→ Chapter 4
→ Proof of the Power Rule

EXAMPLE 1 Using the Power Rule for Negative and Fractional Exponents

Calculate the derivatives of the following:

a. $f(x) = \dfrac{1}{x}$ **b.** $f(x) = \dfrac{1}{x^2}$ **c.** $f(x) = \sqrt{x}$

Solution

a. Rewrite the function in power form as $f(x) = x^{-1}$. Then $f'(x) = (-1)x^{-2} = -\dfrac{1}{x^2}$.

b. Rewrite the function in power form as $f(x) = x^{-2}$. Then $f'(x) = (-2)x^{-3} = -\dfrac{2}{x^3}$.

c. Rewrite the function in power form as $f(x) = x^{0.5}$. Then $f'(x) = 0.5x^{-0.5} = \dfrac{0.5}{x^{0.5}}$.

Alternatively, rewrite $f(x)$ as $x^{1/2}$, so that $f'(x) = \dfrac{1}{2}x^{-1/2} = \dfrac{1}{2x^{1/2}} = \dfrac{1}{2\sqrt{x}}$.

Caution

We cannot apply the power rule to terms in the denominators or under square roots. For example:

1. The derivative of $\dfrac{1}{x^2}$ is **NOT** $\dfrac{1}{2x}$; it is $-\dfrac{2}{x^3}$. See Example 1(b).
2. The derivative of $\sqrt{x^3}$ is **NOT** $\sqrt{3x^2}$; it is $1.5x^{0.5}$. Rewrite $\sqrt{x^3}$ as $x^{3/2}$ or $x^{1.5}$, and apply the power rule.

Table 1 Table of Derivative Formulas

$f(x)$	$f'(x)$
1	0
x	1
x^2	$2x$
x^3	$3x^2$
x^n	nx^{n-1}
$\dfrac{1}{x}$	$-\dfrac{1}{x^2}$
$\dfrac{1}{x^2}$	$-\dfrac{2}{x^3}$
$\sqrt{x}$	$\dfrac{1}{2\sqrt{x}}$

Some of the derivatives in Example 1 are very useful to remember, so we summarize them in Table 1. We suggest that you add to this table as you learn more derivatives. It is *extremely* helpful to remember the derivatives of common functions such as $1/x$ and $\sqrt{x}$, even though they can be obtained by using the power rule as in the above example.

Another Notation: Differential Notation

Here is a useful notation based on the d notation we discussed in Section 3.5. **Differential notation** is based on an abbreviation for the phrase "the derivative with respect to x." For example, we learned that if $f(x) = x^3$, then $f'(x) = 3x^2$. When we say $f'(x) = 3x^2$, we mean the following:

The derivative of x^3 with respect to x equals $3x^2$.

You may wonder why we sneaked in the words "with respect to x." All this means is that the variable of the function is x and not any other variable.* Because we

* This may seem odd in the case of $f(x) = x^3$ because there are no other variables to worry about. But in expressions like st^3 that involve variables other than x, it is necessary to specify just what the variable of the function is. This is the same reason that we write $f(x) = x^3$ rather than just $f = x^3$.

use the phrase "the derivative with respect to x" often, we use the following abbreviation.

Differential Notation; Differentiation

$\frac{d}{dx}$ means "the derivative with respect to x."

Thus, $\frac{d}{dx}[f(x)]$ is the same thing as $f'(x)$, the derivative of $f(x)$ with respect to x. If y is a function of x, then the derivative of y with respect to x is

$$\frac{d}{dx}(y) \quad \text{or, more compactly,} \quad \frac{dy}{dx}.$$

To **differentiate** a function $f(x)$ with respect to x means to take its derivative with respect to x.

Quick Examples

	In Words	Formula
5.	The derivative with respect to x of x^3 is $3x^2$.	$\frac{d}{dx}(x^3) = 3x^2$
6.	The derivative with respect to t of $\frac{1}{t}$ is $-\frac{1}{t^2}$.	$\frac{d}{dt}\left(\frac{1}{t}\right) = -\frac{1}{t^2}$
7.	If $y = x^4$, then $\frac{dy}{dx} = 4x^3$.	
8.	If $u = \frac{1}{t^2}$, then $\frac{du}{dt} = -\frac{2}{t^3}$.	

Notes

1. $\frac{dy}{dx}$ is Leibniz's notation for the derivative we discussed in Section 3.5. (See the discussion before Example 3 there.)
2. Leibniz notation illustrates units nicely: Units of $\frac{dy}{dx}$ are units of y per unit of x.
3. We can (and often do!) use different kind of brackets or parentheses in Leibniz notation; for instance, $\frac{d}{dx}[x^3]$, $\frac{d}{dx}(x^3)$, and $\frac{d}{dx}\{x^3\}$ all mean the same thing (and equal $3x^2$). ■

The Rules for Sums and Constant Multiples

We can now find the derivatives of more complicated functions, such as polynomials, using the following rules. If f and g are functions and if c is a constant, we saw in Section 1.2 how to obtain the **sum**, $f + g$, **difference**, $f - g$, and **constant multiple**, cf.

Theorem 4.2 Derivatives of Sums, Differences, and Constant Multiples

If f and g are any two differentiable functions and if c is any constant, then the sum, $f + g$, the difference, $f - g$, and the constant multiple, cf, are differentiable, and

$$[f \pm g]'(x) = f'(x) \pm g'(x) \qquad \text{Sum rule}$$
$$[cf]'(x) = cf'(x). \qquad \text{Constant multiple rule}$$

In Words:

- The derivative of a sum is the sum of the derivatives, and the derivative of a difference is the difference of the derivatives.
- The derivative of c times a function is c times the derivative of the function.

Differential Notation:

$$\frac{d}{dx}[f(x) \pm g(x)] = \frac{d}{dx}f(x) \pm \frac{d}{dx}g(x)$$
$$\frac{d}{dx}[cf(x)] = c\frac{d}{dx}f(x)$$

Quick Examples

9. $\frac{d}{dx}(x^2 - x^4) = \frac{d}{dx}(x^2) - \frac{d}{dx}(x^4) = 2x - 4x^3$

10. $\frac{d}{dx}(7x^3) = 7\frac{d}{dx}(x^3) = 7(3x^2) = 21x^2$

In other words, we multiply the coefficient (7) by the exponent (3) and then decrease the exponent by 1.

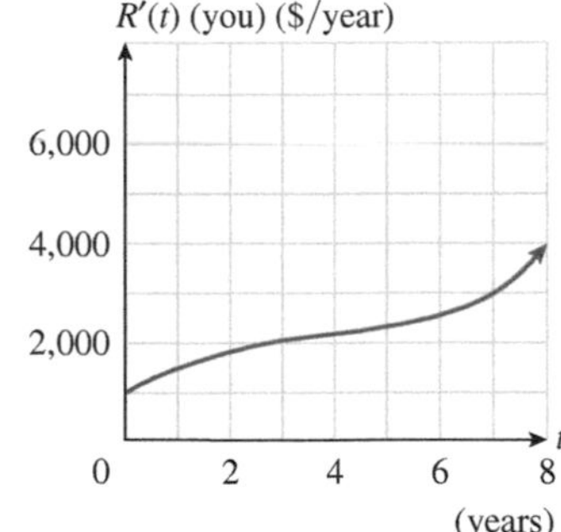

11. $\frac{d}{dx}(12x) = 12\frac{d}{dx}(x) = 12(1) = 12$

In other words, the derivative of a constant times x is that constant.

12. $\frac{d}{dx}(-x^{0.5}) = \frac{d}{dx}[(-1)x^{0.5}] = (-1)\frac{d}{dx}(x^{0.5}) = (-1)(0.5)x^{-0.5}$
$= -0.5x^{-0.5}$

13. $\frac{d}{dx}(12) = \frac{d}{dx}[12(1)] = 12\frac{d}{dx}(1) = 12(0) = 0.$

In other words, the derivative of a constant is zero.

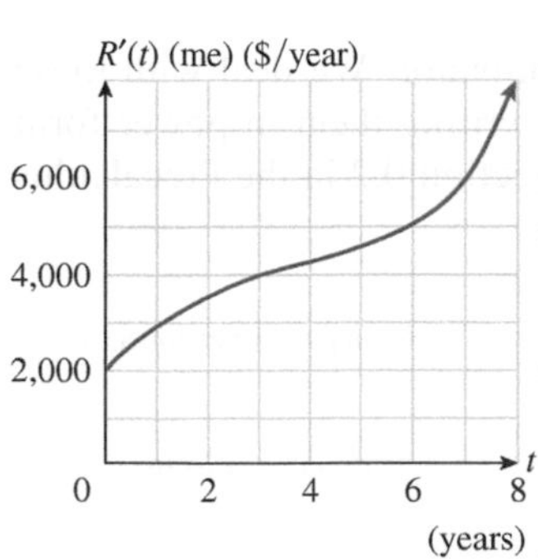

14. If my company earns twice as much (annual) revenue as yours and the derivative of your revenue function is the upper curve shown in the margin, then the derivative of my revenue function is the lower curve.

15. Suppose that a company's revenue R and cost C are changing with time. Then so is the profit, $P(t) = R(t) - C(t)$, and the rate of change of the profit is

$$P'(t) = R'(t) - C'(t).$$

In words: *The derivative of the profit is the derivative of revenue minus the derivative of cost.*

Proof of the Sum Rule

By the definition of the derivative of a function,

$$\begin{aligned}\frac{d}{dx}[f(x) + g(x)] &= \lim_{h\to 0} \frac{[f(x+h) + g(x+h)] - [f(x) + g(x)]}{h} \\ &= \lim_{h\to 0} \frac{[f(x+h) - f(x)] + [g(x+h) - g(x)]}{h} \\ &= \lim_{h\to 0} \left[\frac{f(x+h) - f(x)}{h} + \frac{g(x+h) - g(x)}{h}\right] \\ &= \lim_{h\to 0} \frac{f(x+h) - f(x)}{h} + \lim_{h\to 0} \frac{g(x+h) - g(x)}{h} \\ &= \frac{d}{dx}[f(x)] + \frac{d}{dx}[g(x)].\end{aligned}$$

The next-to-last step uses a property of limits: The limit of a sum is the sum of the limits. Think about why this should be true. The last step uses the definition of the derivative again (and the fact that the functions are differentiable).

The proofs of the rules for differences and constant multiples are similar.

EXAMPLE 2 Combining the Sum and Constant Multiple Rules and Dealing with x in the Denominator

Find the derivatives of the following:

a. $f(x) = 3x^2 + 2x - 4$ **b.** $f(x) = \frac{2x}{3} - \frac{6}{x} + \frac{2}{3x^{0.2}} - \frac{x^4}{2}$

c. $f(x) = \frac{|x|}{4} + \frac{1}{2\sqrt{x}}$

Solution

a.
$$\begin{aligned}\frac{d}{dx}(3x^2 + 2x - 4) &= \frac{d}{dx}(3x^2) + \frac{d}{dx}(2x - 4) && \text{Rule for sums} \\ &= \frac{d}{dx}(3x^2) + \frac{d}{dx}(2x) - \frac{d}{dx}(4) && \text{Rule for differences} \\ &= 3(2x) + 2(1) - 0 && \text{See Quick Example 10.} \\ &= 6x + 2\end{aligned}$$

b. Notice that f has x and powers of x in the denominator. We deal with these terms the same way we did in Example 1, by rewriting them in power form (that is, in the form constant × power of x; see Section 0.2 in the Precalculus Review):

$$\begin{aligned}f(x) &= \frac{2x}{3} - \frac{6}{x} + \frac{2}{3x^{0.2}} - \frac{x^4}{2} && \text{Given in positive exponent form} \\ &= \frac{2}{3}x - 6x^{-1} + \frac{2}{3}x^{-0.2} - \frac{1}{2}x^4. && \text{Convert to power form.}\end{aligned}$$

We are now ready to take the derivative:

$$f'(x) = \frac{2}{3}(1) - 6(-1)x^{-2} + \frac{2}{3}(-0.2)x^{-1.2} - \frac{1}{2}(4x^3)$$

$$= \frac{2}{3} + 6x^{-2} - \frac{0.4}{3}x^{-1.2} - 2x^3 \qquad \text{Answer in power form.}$$

$$= \frac{2}{3} + \frac{6}{x^2} - \frac{0.4}{3x^{1.2}} - 2x^3. \qquad \text{Answer in positive exponent form.}$$

c. Rewrite $f(x)$ using power form as follows:

$$f(x) = \frac{|x|}{4} + \frac{1}{2\sqrt{x}} \qquad \text{Given in radical form}$$

$$= \frac{1}{4}|x| + \frac{1}{2}x^{-1/2}. \qquad \text{Convert to power form.}$$

Now recall from the end of Section 3.6 that the derivative of $|x|$ is $\frac{|x|}{x}$. Thus,

$$f'(x) = \frac{1}{4}\frac{|x|}{x} + \frac{1}{2}\left(\frac{-1}{2}x^{-3/2}\right)$$

$$= \frac{|x|}{4x} - \frac{1}{4}x^{-3/2} \qquad \text{Simplify.}$$

$$= \frac{|x|}{4x} - \frac{1}{4x^{3/2}}. \qquad \text{Answer in positive exponent form.}$$

Notice that in Example 2(a) we had three terms in the expression for $f(x)$, not just two. By applying the rule for sums and differences twice, we saw that the derivative of a sum or difference of three terms is the sum or difference of the derivatives of the terms. (One of those terms had zero derivative, so the final answer had only two terms.) In fact, the derivative of a sum or difference of any number of terms is the sum or difference of the derivatives of the terms. Put another way, to take the derivative of a sum or difference of any number of terms, we take derivatives term by term.

Note Nothing forces us to use only x as the independent variable when taking derivatives (although it is traditional to give x preference). For instance, part (a) in Example 2 can be rewritten as

$$\frac{d}{dt}(3t^2 + 2t - 4) = 6t + 2 \qquad \frac{d}{dt} \text{ means "derivative with respect to } t\text{."}$$

or

$$\frac{d}{du}(3u^2 + 2u - 4) = 6u + 2. \qquad \frac{d}{du} \text{ means "derivative with respect to } u\text{."}$$ ■

In the preceding examples we saw instances of the following important facts. (Think about these graphically to see why they must be true.)

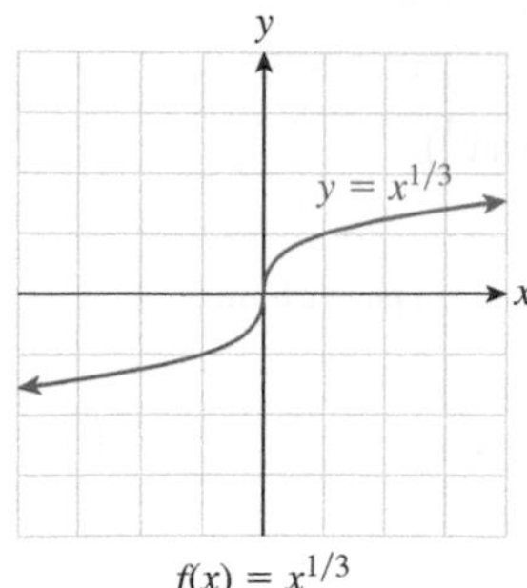

$f(x) = x^{1/3}$

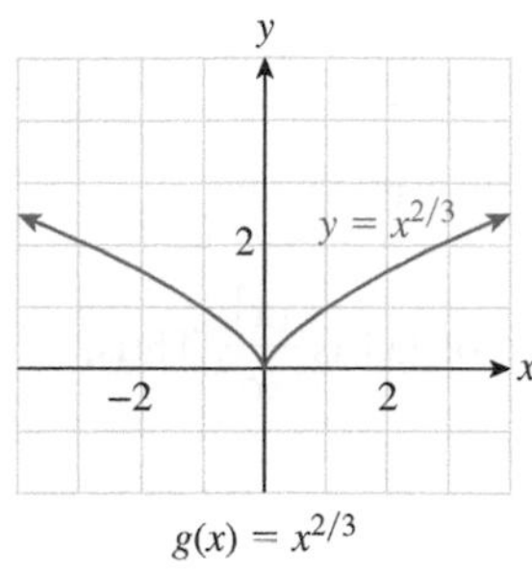

$g(x) = x^{2/3}$

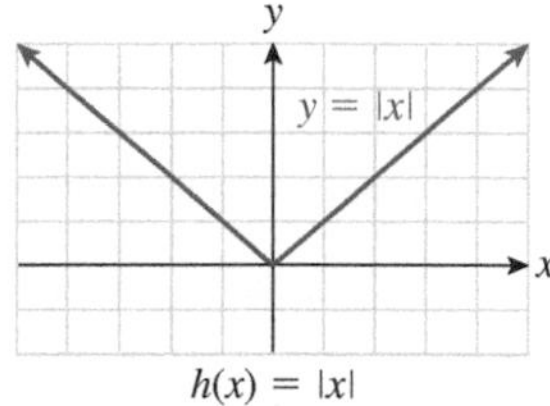

$h(x) = |x|$

Figure 1

Using Technology

If you try to graph the function $f(x) = x^{2/3}$ using the format

```
X^(2/3)
```

you may get only the right-hand portion of the graph of g in Figure 1 because graphing utilities are (often) not programmed to raise negative numbers to fractional exponents. (However, many will handle `X^(1/3)` correctly, as a special case they recognize.) To avoid this difficulty, you can take advantage of the identity

$$x^{2/3} = (x^2)^{1/3}$$

so that it is always a nonnegative number that is being raised to a fractional exponent. Thus, use the format

```
(X^2)^(1/3)
```

to obtain both portions of the graph.

The Derivative of a Constant Times x and the Derivative of a Constant

If c is any constant, then:

Rule

$$\frac{d}{dx}(cx) = c$$

$$\frac{d}{dx}(c) = 0$$

Quick Examples

$$\frac{d}{dx}(6x) = 6 \qquad \frac{d}{dx}(-x) = -1$$

$$\frac{d}{dx}(5) = 0 \qquad \frac{d}{dx}(\pi) = 0$$

In Section 3.5 we pointed out that, by definition, the derivative of a function cannot exist at an endpoint of its domain. Thus, for instance, $f(x) = \sqrt{x}$ and $g(x) = x^{1/4}$ are not differentiable at the endpoint $x = 0$ of their domains. In Example 5 of Section 3.6 we saw that $h(x) = |x|$ also fails to be differentiable at $x = 0$, even though $x = 0$ is not an endpoint of its domain (the domain of h is the set of all real numbers). In the next example we see how to spot the nondifferentiability at a point of this and other functions simply by looking at the formulas for their derivatives.

EXAMPLE 3 Functions Not Differentiable at a Point

Find the natural domains of the derivatives of $f(x) = x^{1/3}$, $g(x) = x^{2/3}$, and $h(x) = |x|$.

Solution Let's look at the derivatives of the three functions given:

$$f(x) = x^{1/3}, \quad \text{so} \quad f'(x) = \frac{1}{3}x^{-2/3} = \frac{1}{3x^{2/3}}$$

$$g(x) = x^{2/3}, \quad \text{so} \quad g'(x) = \frac{2}{3}x^{-1/3} = \frac{2}{3x^{1/3}}$$

$$h(x) = |x|, \quad \text{so} \quad h'(x) = \frac{|x|}{x}.$$

The derivatives of all three functions are defined only for nonzero values of x, and their natural domains consist of all real numbers except 0. Thus, the derivatives f', g', and h' do not exist at $x = 0$. In other words, these functions are not differentiable at $x = 0$. If we look at Figure 1, we notice why these functions fail to be differentiable at $x = 0$: The graph of f has a vertical tangent line at 0. Because a vertical line has undefined slope, the derivative is undefined at that point. The graphs of g and h come to a sharp point at 0, where it is not meaningful to speak about the slope of the tangent line; therefore, the derivatives of g and h are not defined there. (In the case of g, where the sharp point is called a *cusp,* a vertical tangent line would seem appropriate, but as in the case of f, its slope is undefined.)

You can also detect this nondifferentiability by computing some difference quotients numerically, as we did for h in Section 3.6.

Applications

EXAMPLE 4 Gold Price

You are a commodities trader, and you monitor the price of gold on the spot market very closely during an active morning. Suppose you find that the price of an ounce of gold can be approximated by the function

$$G(t) = 5t^2 - 85t + 1{,}762 \qquad (7.5 \le t \le 10.5),$$

where t is time in hours. (See Figure 2. $t = 8$ represents 8:00 am.)

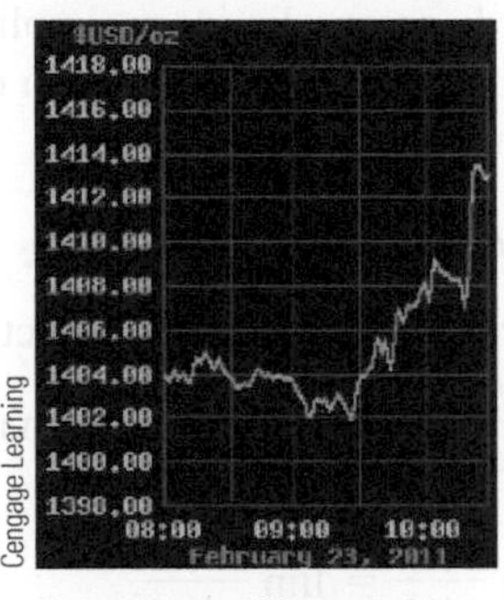

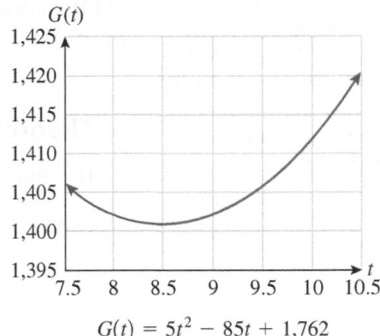

Source: www.kitco.com

$G(t)$: 1,425, 1,420, 1,415, 1,410, 1,405, 1,400, 1,395; t: 7.5, 8, 8.5, 9, 9.5, 10, 10.5

$G(t) = 5t^2 - 85t + 1{,}762$

Figure 2

a. According to the model, how fast was the price of gold changing at 8:00 am?

b. According to the model, the price of gold

(A) increased at a faster and faster rate
(B) increased at a slower and slower rate
(C) decreased at a faster and faster rate
(D) decreased at a slower and slower rate

between 7:30 and 8:30 am.

Solution

a. Differentiating the given function with respect to t gives

$$G'(t) = 10t - 85.$$

Because 8:00 am corresponds to $t = 8$, we obtain

$$G'(8) = 10(8) - 85 = -5.$$

The units of the derivative are dollars per hour, so we conclude that, at 8:00 am, the price of gold was dropping at a rate of \$5 per hour.

b. From the graph we can see that, between 7:30 and 8:30 am (the interval $[7.5, 8.5]$), the price of gold was decreasing. Also from the graph we see that the slope of the tangent becomes less and less negative as t increases, so the price of gold is decreasing at a slower and slower rate (choice (D)).

We can also see this algebraically from the derivative, $G'(t) = 10t - 85$: For values of t less than 8.5, $G'(t)$ is negative; that is, the rate of change of G is negative, so the price of gold is decreasing. Further, as t increases, $G'(t)$ becomes less and less negative, so the price of gold is decreasing at a slower and slower rate, confirming that choice (D) is the correct one.

An Application to Limits: L'Hospital's Rule

The limits that caused us some trouble in Sections 3.1–3.3 are those of the form $\lim_{x\to a} f(x)$ in which substituting $x = a$ gave us an indeterminate form, such as

$$\lim_{x\to 2} \frac{x^3 - 8}{x - 2} \qquad \text{Substituting } x = 2 \text{ yields } \tfrac{0}{0}.$$

$$\lim_{x\to +\infty} \frac{2x - 4}{x - 1}. \qquad \text{Substituting } x = +\infty \text{ yields } \tfrac{\infty}{\infty}.$$

L'Hospital's rule* gives us an alternative way of computing limits such as these without the need to do any preliminary simplification. It also allows us to compute some limits for which algebraic simplification does not work.

* Guillaume François Antoine, Marquis de l'Hospital (1661–1704) wrote the first textbook on calculus, *Analyse des infiniment petits pour l'intelligence des lignes courbes,* in 1692. The rule now known as l'Hospital's rule appeared first in this book.

Theorem 4.3 L'Hospital's Rule

If f and g are two differentiable functions such that substituting $x = a$ in the expression $\dfrac{f(x)}{g(x)}$ gives the indeterminate form $\dfrac{0}{0}$ or $\dfrac{\infty}{\infty}$, then

$$\lim_{x\to a} \frac{f(x)}{g(x)} = \lim_{x\to a} \frac{f'(x)}{g'(x)}.$$

That is, we can replace $f(x)$ and $g(x)$ with their *derivatives* and try again to take the limit.

Quick Examples

16. Substituting $x = 2$ in $\dfrac{x^3 - 8}{x - 2}$ yields $\dfrac{0}{0}$. Therefore, l'Hospital's rule applies, and

$$\lim_{x\to 2} \frac{x^3 - 8}{x - 2} = \lim_{x\to 2} \frac{3x^2}{1} = \frac{3(2)^2}{1} = 12.$$

17. Substituting $x = +\infty$ in $\dfrac{2x - 4}{x - 1}$ yields $\dfrac{\infty}{\infty}$. Therefore, l'Hospital's rule applies, and

$$\lim_{x\to +\infty} \frac{2x - 4}{x - 1} = \lim_{x\to +\infty} \frac{2}{1} = 2.$$

The proof of l'Hospital's rule is beyond the scope of this text.†

† A proof of l'Hospital's rule can be found in most advanced calculus textbooks.

EXAMPLE 5 Applying L'Hospital's Rule

Check whether l'Hospital's rule applies to each of the following limits. If it does, use it to evaluate the limit. Otherwise, use some other method to evaluate the limit.

a. $\displaystyle\lim_{x\to 1} \frac{x^2 - 2x + 1}{4x^3 - 3x^2 - 6x + 5}$ **b.** $\displaystyle\lim_{x\to +\infty} \frac{2x^2 - 4x}{5x^3 - 3x + 5}$

c. $\displaystyle\lim_{x\to 1} \frac{x - 1}{x^3 - 3x^2 + 3x - 1}$ **d.** $\displaystyle\lim_{x\to 1} \frac{x}{x^3 - 3x^2 + 3x - 1}$

Solution

a. Setting $x = 1$ yields

$$\frac{1-2+1}{4-3-6+5} = \frac{0}{0}.$$

Therefore, l'Hospital's rule applies, and

$$\lim_{x\to 1}\frac{x^2-2x+1}{4x^3-3x^2-6x+5} = \lim_{x\to 1}\frac{2x-2}{12x^2-6x-6}.$$

We are left with a closed-form function. However, we cannot substitute $x = 1$ to find the limit because the function $(2x-2)/(12x^2-6x-6)$ is still not defined at $x = 1$. In fact, if we set $x = 1$, we again get $0/0$. Thus, l'Hospital's rule applies again, and

$$\lim_{x\to 1}\frac{2x-2}{12x^2-6x-6} = \lim_{x\to 1}\frac{2}{24x-6}.$$

Once again we have a closed-form function, but this time it is defined when $x = 1$, giving

$$\frac{2}{24-6} = \frac{1}{9}.$$

Thus,

$$\lim_{x\to 1}\frac{x^2-2x+1}{4x^3-3x^2-6x+5} = \frac{1}{9}.$$

b. Setting $x = +\infty$ yields ∞/∞, so

$$\lim_{x\to +\infty}\frac{2x^2-4x}{5x^3-3x+5} = \lim_{x\to +\infty}\frac{4x-4}{15x^2-3}.$$

Setting $x = +\infty$ again yields ∞/∞, so we can apply the rule again to obtain

$$\lim_{x\to +\infty}\frac{4x-4}{15x^2-3} = \lim_{x\to +\infty}\frac{4}{30x}.$$

Note that we cannot apply l'Hospital's rule a third time because setting $x = +\infty$ yields the *determinate* form $4/\infty = 0$. (See the discussion at the end of Section 3.3.) Thus, the limit is 0.

c. Setting $x = 1$ yields $0/0$, so, by l'Hospital's rule,

$$\lim_{x\to 1}\frac{x-1}{x^3-3x^2+3x-1} = \lim_{x\to 1}\frac{1}{3x^2-6x+3}.$$

We are left with a closed-form function that is still not defined at $x = 1$. Further, l'Hospital's rule no longer applies because putting $x = 1$ yields the determinate form $1/0$. To investigate this limit, we refer to the discussion at the end of Section 3.3 and find

$$\lim_{x\to 1}\frac{1}{3x^2-6x+3} = \lim_{x\to 1}\frac{1}{3(x-1)^2} = +\infty. \qquad \frac{1}{0^+} = +\infty$$

d. Setting $x = 1$ in the expression yields the determinate form $1/0$, so l'Hospital's rule does not apply here. Using the methods of Section 3.3 again, we find that the limit does not exist.

FAQs

Using the Rules and Recognizing when a Function Is Not Differentiable

Q: *I would like to say that the derivative of $5x^2 - 8x + 4$ is just $10x - 8$ without having to go through all that stuff about derivatives of sums and constant multiples. Can I simply forget about all the rules and write down the answer?*

A: We developed the rules for sums and constant multiples precisely for that reason: so that we could simply write down a derivative without having to think about it too hard. So you are perfectly justified in simply writing down the derivative without going through the rules, but bear in mind that what you are really doing is applying the power rule, the rule for sums, and the rule for multiples over and over.

Q: *Is there a way of telling from its formula whether a function f is not differentiable at a point?*

A: Here are some indicators to look for in the formula for *f*:

- The absolute value of some expression; *f* may not be differentiable at points where that expression is zero.

  Example: $f(x) = 3x^2 - |x - 4|$ is not differentiable at $x = 4$.

- A fractional power smaller than 1 of some expression; *f* may not be differentiable at points where that expression is zero.

  Example: $f(x) = (x^2 - 16)^{2/3}$ is not differentiable at $x = \pm 4$.

4.1 EXERCISES

▼ more advanced ◆ challenging

T indicates exercises that should be solved using technology

In Exercises 1–10, use the shortcut rules to ***mentally*** *calculate the derivative of the given function.* [**HINT**: See Examples 1 and 2.]

1. $f(x) = x^5$ **2.** $f(x) = x^4$

3. $f(x) = 2x^{-2}$ **4.** $f(x) = 3x^{-1}$

5. $f(x) = -x^{0.25}$ **6.** $f(x) = -x^{-0.5}$

7. $f(x) = 2x^4 + 3x^3 - 1$ **8.** $f(x) = -x^3 - 3x^2 - 1$

9. $f(x) = -x + \frac{1}{x} + 1$ **10.** $f(x) = \frac{1}{x} + \frac{1}{x^2}$

In Exercises 11–16, obtain the derivative dy/dx, and state the rules that you use. [**HINT**: See Examples 1 and 2.]

11. $y = 10$ **12.** $y = x^3$

13. $y = x^2 + x$ **14.** $y = x - 5$

15. $y = 4x^3 + 2x - 1$ **16.** $y = 4x^{-1} - 2x - 10$

In Exercises 17–40, find the derivative of the given function. [**HINT**: See Examples 1 and 2.]

17. $f(x) = x^2 - 3x + 5$ **18.** $f(x) = 3x^3 - 2x^2 + x$

19. $f(x) = x + x^{0.5}$ **20.** $f(x) = x^{0.5} + 2x^{-0.5}$

21. $g(x) = x^{-2} - 3x^{-1} - 2$ **22.** $g(x) = 2x^{-1} + 4x^{-2}$

23. $g(x) = \frac{1}{x} - \frac{1}{x^2}$ **24.** $g(x) = \frac{1}{x^2} + \frac{1}{x^3}$

25. $h(x) = \frac{2}{x^{0.4}}$ **26.** $h(x) = -\frac{1}{2x^{0.2}}$

27. $h(x) = \frac{1}{x^2} + \frac{2}{x^3}$ **28.** $h(x) = \frac{2}{x} - \frac{2}{x^3} + \frac{1}{x^4}$

29. $r(x) = \frac{2}{3x} - \frac{1}{2x^{0.1}}$ **30.** $r(x) = \frac{4}{3x^2} + \frac{1}{x^{3.2}}$

31. $r(x) = \frac{2x}{3} - \frac{x^{0.1}}{2} + \frac{4}{3x^{1.1}} - 2$

32. $r(x) = \frac{4x^2}{3} + \frac{x^{3.2}}{6} - \frac{2}{3x^2} + 4$

33. $t(x) = |x| + \frac{1}{x}$ **34.** $t(x) = 3|x| - \sqrt{x}$

35. $s(x) = \sqrt{x} + \frac{1}{\sqrt{x}}$ **36.** $s(x) = x + \frac{7}{\sqrt{x}}$

[**HINT**: For Exercises 37–38, first expand the given function.]

37. ▼ $s(x) = x\left(x^2 - \frac{1}{x}\right)$ **38.** ▼ $s(x) = x^{-1}\left(x - \frac{2}{x}\right)$

[**HINT**: For Exercises 39–40, first rewrite the given function.]

39. ▼ $t(x) = \dfrac{x^2 - 2x^3}{x}$ **40.** ▼ $t(x) = \dfrac{2x + x^2}{x}$

In Exercises 41–46, evaluate the given expression.

41. $\dfrac{d}{dx}(2x^{1.3} - x^{-1.2})$ **42.** $\dfrac{d}{dx}(2x^{4.3} + x^{0.6})$

43. ▼ $\dfrac{d}{dx}[1.2(x - |x|)]$ **44.** ▼ $\dfrac{d}{dx}[4(x^2 + 3|x|)]$

45. ▼ $\dfrac{d}{dt}(at^3 - 4at)$ (a constant)

46. ▼ $\dfrac{d}{dt}(at^2 + bt + c)$ (a, b, c constant)

In Exercises 47–52, find the indicated derivative.

47. $y = \dfrac{x^{10.3}}{2} + 99x^{-1}; \dfrac{dy}{dx}$ **48.** $y = \dfrac{x^{1.2}}{3} - \dfrac{x^{0.9}}{2}; \dfrac{dy}{dx}$

49. $s = 2.3 + \dfrac{2.1}{t^{1.1}} - \dfrac{t^{0.6}}{2}; \dfrac{ds}{dt}$ **50.** $s = \dfrac{2}{t^{1.1}} + t^{-1.2}; \dfrac{ds}{dt}$

51. ▼ $V = \dfrac{4}{3}\pi r^3; \dfrac{dV}{dr}$ **52.** ▼ $A = 4\pi r^2; \dfrac{dA}{dr}$

In Exercises 53–58, find the slope of the tangent to the graph of the given function at the indicated point. [**HINT**: Recall that the slope of the tangent to the graph of f at $x = a$ is $f'(a)$.]

53. $f(x) = x^3; (-1, -1)$ **54.** $g(x) = x^4; (-2, 16)$

55. $f(x) = 1 - 2x; (2, -3)$ **56.** $f(x) = \dfrac{x}{3} - 1; (-3, -2)$

57. $g(t) = \dfrac{1}{t^5}; (1, 1)$ **58.** $s(t) = \dfrac{1}{t^3}; \left(-2, -\dfrac{1}{8}\right)$

In Exercises 59–64, find the equation of the tangent line to the graph of the given function at the point with the indicated x-coordinate. In each case, sketch the curve together with the appropriate tangent line.

59. ▼ $f(x) = x^3; x = -1$ **60.** ▼ $f(x) = x^2; x = 0$

61. ▼ $f(x) = x + \dfrac{1}{x}; x = 2$ **62.** ▼ $f(x) = \dfrac{1}{x^2}; x = 1$

63. ▼ $f(x) = \sqrt{x}; x = 4$ **64.** ▼ $f(x) = 2x + 4; x = -1$

In Exercises 65–70, find all values of x (if any) where the tangent line to the graph of the given equation is horizontal. [**HINT**: The tangent line is horizontal when its slope is zero.]

65. ▼ $y = 2x^2 + 3x - 1$ **66.** ▼ $y = -3x^2 - x$

67. ▼ $y = 2x + 8$ **68.** ▼ $y = -x + 1$

69. ▼ $y = x + \dfrac{1}{x}$ **70.** ▼ $y = x - \sqrt{x}$

71. ◆ Write out the proof that $\dfrac{d}{dx}(x^4) = 4x^3$.

72. ◆ Write out the proof that $\dfrac{d}{dx}(x^5) = 5x^4$.

In Exercises 73–76, determine whether f is differentiable at the given point. If $f'(a)$ exists, give its value. [**HINT**: See Example 3.]

73. $f(x) = x - x^{1/3}$ **a.** $a = 1$ **b.** $a = 0$

74. $f(x) = 2x + x^{4/3}$ **a.** $a = 8$ **b.** $a = 0$

75. $f(x) = x^{5/4} - 1$ **a.** $a = 16$ **b.** $a = 0$

76. $f(x) = x^{1/5} + 5$ **a.** $a = 1$ **b.** $a = 0$

In Exercises 77–88, say whether l'Hospital's rule applies. If it does, use it to evaluate the given limit. If not, use some other method.

77. $\lim\limits_{x\to1} \dfrac{x^2 - 2x + 1}{x^2 - x}$ **78.** $\lim\limits_{x\to-1} \dfrac{x^2 + 3x + 2}{x^2 + x}$

79. $\lim\limits_{x\to2} \dfrac{x^3 - 8}{x - 2}$ **80.** $\lim\limits_{x\to0} \dfrac{x^3 + 8}{x^2 + 3x + 2}$

81. $\lim\limits_{x\to1} \dfrac{x^2 + 3x + 2}{x^2 + x}$ **82.** $\lim\limits_{x\to-2} \dfrac{x^3 + 8}{x^2 + 3x + 2}$

83. $\lim\limits_{x\to-\infty} \dfrac{3x^2 + 10x - 1}{2x^2 - 5x}$ **84.** $\lim\limits_{x\to-\infty} \dfrac{6x^2 + 5x + 100}{3x^2 - 9}$

85. $\lim\limits_{x\to-\infty} \dfrac{10x^2 + 300x + 1}{5x + 2}$ **86.** $\lim\limits_{x\to-\infty} \dfrac{2x^4 + 20x^3}{1{,}000x^3 + 6}$

87. $\lim\limits_{x\to-\infty} \dfrac{x^3 - 100}{2x^2 + 500}$ **88.** $\lim\limits_{x\to-\infty} \dfrac{x^2 + 30x}{2x^6 + 10x}$

Applications

89. ***Crude Oil Prices*** The price per barrel of crude oil in the period 1980–2013, in constant 2014 dollars, can be approximated by

$$P(t) = 0.27t^2 - 8.6t + 93 \text{ dollars} \quad (0 \le t \le 33),$$

where t is time in years since the start of 1980.[1] Find $P'(t)$ and $P'(30)$. What does the second answer tell you about the price of crude oil? [**HINT**: See Example 2.]

90. ***Median Home Prices*** The median home price in the United States over the period January 2010–January 2015 can be approximated by

$$P(t) = 4.5t^2 - 15t + 180 \text{ thousand dollars} \quad (0 \le t \le 5),$$

where t is time in years since the start of 2010.[2] Find $P'(t)$ and $P'(1)$. What does the second answer tell you about home prices? [**HINT**: See Example 2.]

[1] Source for data: http://inflationdata.com/Inflation/Inflation_Rate/Historical_Oil_Prices_Table.asp, March 6, 2014.

[2] Source for data: www.zillow.com.

91. ***Food versus Education*** The following equation shows the approximate relationship between the percentage y of total personal consumption spent on food and the corresponding percentage x spent on education.[3]

$$y = \frac{18.8}{x^{1.05}} \text{ percentage points} \quad (0.69 \le x \le 2.42)$$

According to the model, spending on food is decreasing at a rate of _____ percentage points per 1 percentage point increase in spending on education when 2.0% of total consumption is spent on education. (Your answer should be rounded to two significant digits.) [**HINT:** See Example 2(b).]

92. ***Food versus Recreation*** The following equation shows the approximate relationship between the percentage y of total personal consumption spent on food and the corresponding percentage x spent on recreation.[4]

$$y = \frac{688}{x^{1.99}} \text{ percentage points} \quad (4.83 \le x \le 9.35)$$

According to the model, spending on food is decreasing at a rate of _____ percentage points per 1 percentage point increase in spending on recreation when 6.0% of total consumption is spent on recreation. (Your answer should be rounded to two significant digits.) [**HINT:** See Example 2(b).]

93. ***Velocity*** If a stone is dropped from a height of 400 feet, its height s after t seconds is given by $s(t) = 400 - 16t^2$, with s in feet.
    a. Compute $s'(t)$, and hence find the stone's velocity at times $t = 0, 1, 2, 3,$ and 4 seconds.
    b. When does the stone reach the ground, and how fast is it traveling when it hits the ground? [**HINT:** It reaches the ground when $s(t) = 0$.]

94. ***Velocity*** If a stone is thrown down at 120 ft/sec from a height of 1,000 feet, its height s after t seconds is given by $s(t) = 1{,}000 - 120t - 16t^2$, with s in feet.
    a. Compute $s'(t)$, and hence find the stone's velocity at times $t = 0, 1, 2, 3,$ and 4 seconds.
    b. When does the stone reach the ground, and how fast is it traveling when it hits the ground? [**HINT:** It reaches the ground when $s(t) = 0$.]

95. ***Velocity*** The height of a soccer ball kicked by Javier "Chicharito" Hernández on Mars is given by $h(t) = 76t - 1.9t^2$ meters, where t is time in seconds after he kicks the ball.[5]
    a. When the ball reaches its highest point, $h'(t)$ must equal zero. Why is this true, and at what value of t does this occur?
    b. How high does the ball go?

96. ***Velocity*** The height of a basketball thrown by Chris Paul on Neptune is given by $h(t) = v_0 t - 5.6t^2$ meters, where t is time in seconds after he throws the ball up at v_0 m/sec.[6]
    a. When the ball reaches its highest point, $h'(t)$ must equal zero (see Exercise 95). Give a formula for the value of t when this occurs.
    b. How fast would he need to throw it up in order for it to reach its highest point in half a second? How high would it get?

97. ***GE Net Income 2009–2013*** The annual net income of General Electric for the period 2009–2013 could be approximated by[7]

$$P(t) = -0.39t^2 + 5.2t - 4.1 \text{ billion dollars} \quad (4 \le t \le 8),$$

where t is time in years since 2005.

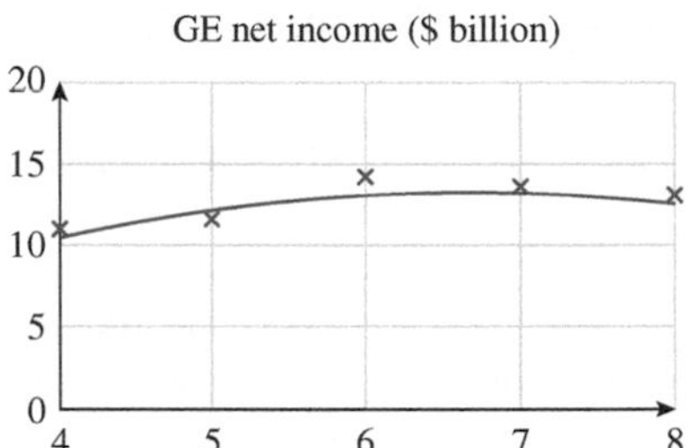

    a. Compute $P'(t)$. How fast was GE's annual net income changing in 2011? (Be careful to give correct units of measurement.)
    b. According to the model, GE's annual net income
    (A) increased at a faster and faster rate
    (B) increased at a slower and slower rate
    (C) decreased at a faster and faster rate
    (D) decreased at a slower and slower rate
    during the first 2 years shown (the interval [4, 6]). Justify your answer in two ways: geometrically, reasoning entirely from the graph, and algebraically, reasoning from the derivative of P. [**HINT:** See Example 4.]

98. ***GE Net Income 2007–2011*** The annual net income of General Electric for the period 2007–2011 could be approximated by[8]

$$P(t) = 1.6t^2 - 15t + 46 \text{ billion dollars} \quad (2 \le t \le 6),$$

where t is time in years since 2005.

[3] Model based on historical data from 1929–2013. Source for data: U.S. Bureau of Economic Analysis (www.bea.gov), August 2014.
[4] *Ibid.*
[5] The equation is quite accurate in the thin atmosphere of Mars, assuming Chicharito were to venture there and kick the ball hard enough.
[6] The equation is quite accurate if we ignore resistance due to atmospheric drag on Neptune, assuming Chris Paul were to venture there.
[7] Source for data: www.wikinvest.com.
[8] *Ibid.*

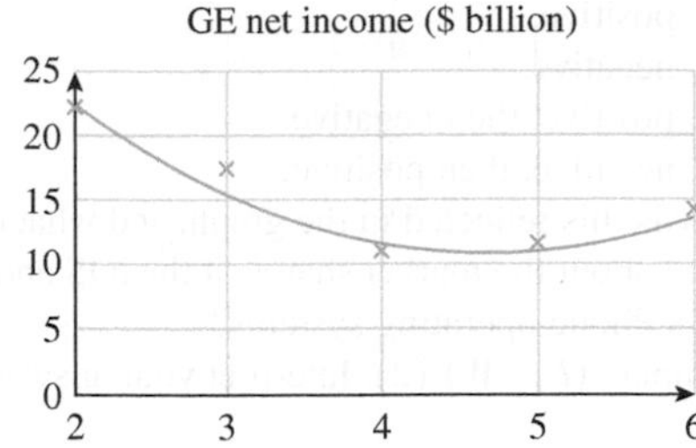

a. Compute $P'(t)$. How fast was GE's annual net income changing in 2008? (Be careful to give correct units of measurement.)

b. According to the model, GE's annual net income
 (A) increased at a faster and faster rate
 (B) increased at a slower and slower rate
 (C) decreased at a faster and faster rate
 (D) decreased at a slower and slower rate
during the first 2 years shown (the interval $[2, 4]$). Justify your answer in two ways: geometrically, reasoning entirely from the graph, and algebraically, reasoning from the derivative of P. [**HINT:** See Example 4.]

99. ***Ecology*** Increasing numbers of manatees ("sea sirens") have been killed by boats off the Florida coast. The following graph shows the relationship between the number of boats registered in Florida and the number of manatees killed each year.

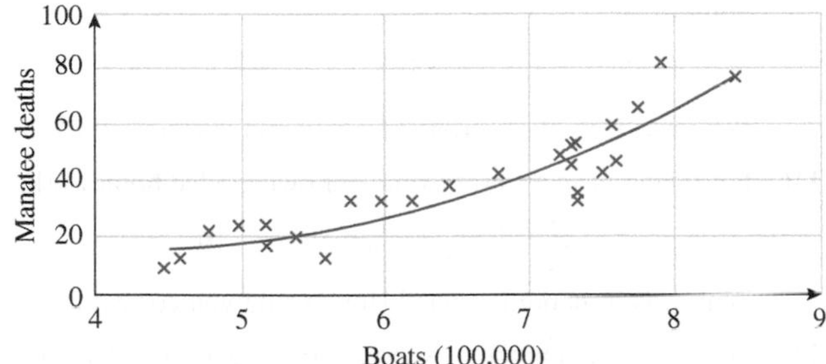

The regression curve shown is given by

$$f(x) = 3.55x^2 - 30.2x + 81 \quad (4.5 \le x \le 8.5),$$

where x is the number of boats (hundreds of thousands) registered in Florida in a particular year and $f(x)$ is the number of manatees killed by boats in Florida that year.[9]

a. Find $f'(x)$, and use your formula to compute $f'(8)$, stating its units of measurement. What does the answer say about manatee deaths?

b. Is $f'(x)$ increasing or decreasing with increasing x? Interpret the answer. [**HINT:** See Example 4.]

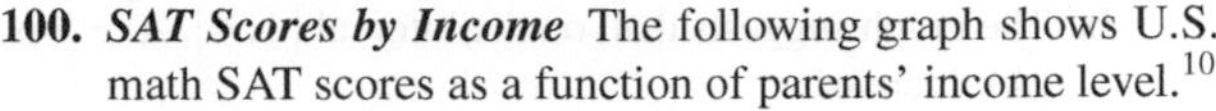
100. ***SAT Scores by Income*** The following graph shows U.S. math SAT scores as a function of parents' income level.[10]

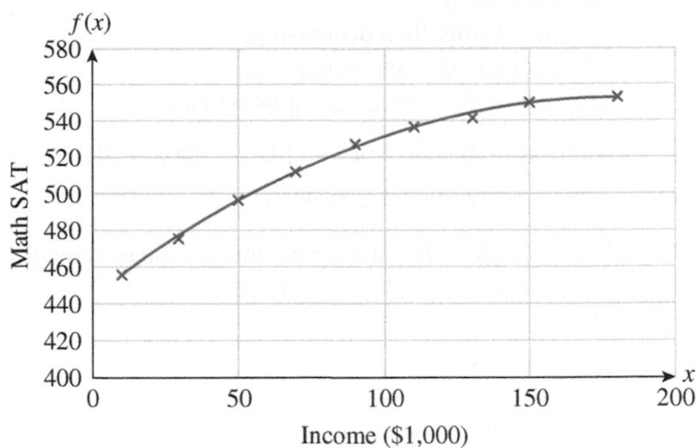

The regression curve shown is given by

$$f(x) = -0.0034x^2 + 1.2x + 444 \quad (10 \le x \le 180),$$

where $f(x)$ is the average math SAT score of a student whose parents earn x thousand dollars per year.

a. Find $f'(x)$, and use your formula to compute $f'(100)$, stating its units of measurement. What does the answer say about math SAT scores?

b. Does $f'(x)$ increase or decrease with increasing x? What does your answer say about math SAT scores? [**HINT:** See Example 4.]

101. ▼ ***Market Share: Smartphones*** The following graph shows the approximate market shares, in percentage points, of smartphones using **Google**'s Android operating system and **Apple**'s iOS operating system, from the second quarter of 2011 to 2014. (t is time in years and $t = 0$ represents the second quarter of 2010.)[11]

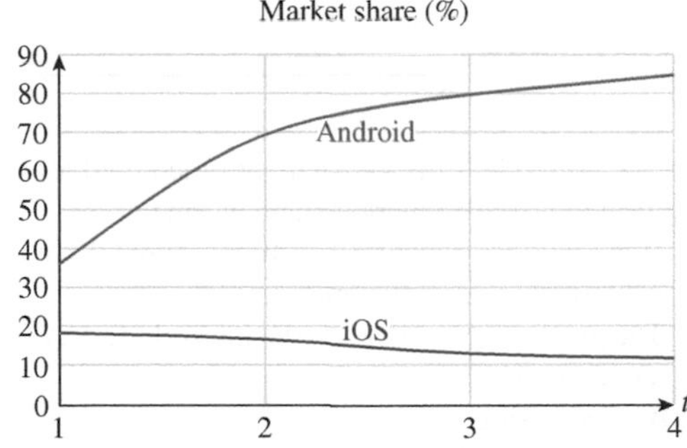

Let $A(t)$ be the Android market share at time t, and let $I(t)$ be the iOS market share at time t.

a. What does the function $A - I$ measure? What does its derivative $(A - I)'$ measure?

[9] Regression model is based on data from 1976 to 2000. Sources for data: Florida Department of Highway Safety & Motor Vehicles, Florida Marine Institute/*New York Times*, February 12, 2002, p. F4.

[10] Regression model is based on 2009 data. Source: College Board/*New York Times*, http://economix.blogs.nytimes.com.

[11] Source for data: IDC, www.idc.com.

b. The graph suggests that, on the interval $[1, 4]$, $A - I$ is
(A) increasing.
(B) decreasing.
(C) increasing, then decreasing.
(D) decreasing, then increasing.

c. The two market shares are approximated by

Android: $A(t) = 3.0t^3 - 29t^2 + 100t - 38$

iOS: $I(t) = -2.3t + 21.$

Compute $(A - I)'$, stating its units of measurement. On the interval $[1, 4]$, $(A - I)'$ is
(A) positive.
(B) negative.
(C) positive, then negative.
(D) negative, then positive.
How is this behavior reflected in the graph, and what does it mean about the market shares of the Android and iOS operating systems?

d. Compute $(A - I)'(3)$. Interpret your answer.

102. ▼ ***Market Share: Smartphones*** The following graph shows the approximate market shares, in percentage points, of smartphones using **Apple**'s iOS operating system and **Microsoft**'s Windows Phone operating system, from the second quarter of 2011 to 2014. (t is time in years and $t = 0$ represents the second quarter of 2010.)[12]

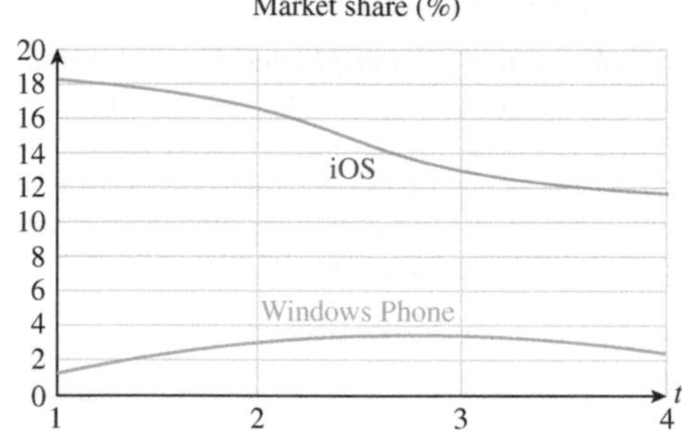

Let $I(t)$ be the iOS market share at time t, and let $W(t)$ be the Windows Phone market share at time t.

a. What does the function $I - W$ measure? What does its derivative $(I - W)'$ measure?

b. The graph suggests that, on the interval $[1, 4]$, $I - W$ is
(A) increasing.
(B) decreasing.
(C) increasing, then decreasing.
(D) decreasing, then increasing.

c. The two market shares are approximated by

iOS: $I(t) = 0.7t^3 - 5.2t^2 + 8.9t + 14$

Windows Phone: $W(t) = -0.7t^2 + 3.9t - 2.$

Compute $(I - W)'$, stating its units of measurement. On the interval $[1, 4]$, $(I - W)'$ is
(A) positive.
(B) negative.
(C) positive, then negative.
(D) negative, then positive.
How is this reflected in the graph, and what does it mean about the market shares of the iOS and Windows Phone operating systems?

d. Compute $(I - W)'(2)$. Interpret your answer.

Communication and Reasoning Exercises

103. What instructions would you give to a fellow student who wanted to accurately graph the tangent line to the curve $y = 3x^2$ at the point $(-1, 3)$?

104. What instructions would you give to a fellow student who wanted to accurately graph a line at right angles to the curve $y = 4/x$ at the point where $x = 0.5$?

105. Consider $f(x) = x^2$ and $g(x) = 2x^2$. How do the slopes of the tangent lines of f and g at the same x compare?

106. Consider $f(x) = x^3$ and $g(x) = x^3 + 3$. How do the slopes of the tangent lines of f and g compare?

107. Suppose $g(x) = -f(x)$. How do the derivatives of f and g compare?

108. Suppose $g(x) = f(x) - 50$. How do the derivatives of f and g compare?

109. Following is an excerpt from your best friend's graded homework:

$3x^4 + 11x^5 = 12x^3 + 55x^4.$ ✗ *WRONG* **−8**

Why was it marked wrong? How would you correct it?

110. Following is an excerpt from your own graded homework:

$x^n = nx^{n-1}.$ ✗ *WRONG* **−10**

Why was it marked wrong? How would you correct it?

111. Following is another excerpt from your best friend's graded homework:

$y = \dfrac{1}{2x} = 2x^{-1}$, so $\dfrac{dy}{dx} = -2x^{-2}.$ ✗ *WRONG* **−5**

Why was it marked wrong? How would you correct it?

112. Following is an excerpt from your second best friend's graded homework:

$f(x) = \dfrac{3}{4x^2}$; $f'(x) = \dfrac{3}{8x}.$ ✗ *WRONG* **−10**

Why was it marked wrong? How would you correct it?

113. Following is an excerpt from your worst enemy's graded homework:

$f(x) = 4x^2$; $f'(x) = (0)(2x) = 0.$ ✗ *WRONG* **−6**

Why was it marked wrong? How would you correct it?

[12] See footnote for Exercise 101.

114. Following is an excerpt from your second worst enemy's graded homework:

$$f(x) = \frac{3}{4x};\ f'(x) = \frac{0}{4} = 0. \qquad \text{✗ } \textit{WRONG} \quad -10$$

Why was it marked wrong? How would you correct it?

115. One of the questions in your last calculus test was "**Question 1(a)** Give the definition of the derivative of a function *f*." Following is your answer and the grade you received:

$$nx^{n-1}. \qquad \text{✗ } \textit{WRONG} \quad -10$$

Why was it marked wrong? What is the correct answer?

116. ▼ How would you respond to an acquaintance who says, "I finally understand what the derivative is: It is nx^{n-1}! Why weren't we taught that in the first place instead of the difficult way using limits?"

117. ▼ Sketch the graph of a function whose derivative is undefined at exactly two points but that has a tangent line at all but one point.

118. ▼ Sketch the graph of a function that has a tangent line at each of its points but whose derivative is undefined at exactly two points.

4.2 A First Application: Marginal Analysis

Marginal Cost, Revenue, and Profit

In Chapter 1 we considered linear *cost functions* of the form $C(x) = mx + b$, where C is the total cost, x is the number of items, and m and b are constants. The slope m is the *marginal cost.* It measures the *cost of one more item.* Notice that the derivative of $C(x) = mx + b$ is $C'(x) = m$. In other words, for a linear cost function *the marginal cost is the derivative of the cost function.*

In general, we make the following definition.

Marginal Cost

Recall from Section 1.2 that a **cost function** C specifies the total cost as a function of the number of items x, so $C(x)$ is the total cost of x items. The **marginal cost function** is the derivative C' of the cost function C. Thus, $C'(x)$ measures the rate of change of cost with respect to x.

Units

The units of marginal cost are units of cost (dollars, say) per item.

Interpretation

*We interpret $C'(x)$ as the approximate cost of one more item.**

* See Example 1 below.

Quick Example

1. If $C(x) = 400x + 1{,}000$ dollars, then the marginal cost function is $C'(x) = \$400$ per item (a constant).

EXAMPLE 1 **Marginal Cost**

Suppose that the cost in dollars to manufacture portable music players is given by

$$C(x) = 150{,}000 + 20x - 0.0001x^2,$$

where x is the number of music players manufactured.† Find the marginal cost function C', and use it to estimate the cost of manufacturing the 50,001st music player.

† The term $0.0001x^2$ may reflect a cost saving for high levels of production, such as a bulk discount in the cost of electronic components.

Solution Because

$$C(x) = 150{,}000 + 20x - 0.0001x^2,$$

the marginal cost function is

$$C'(x) = 20 - 0.0002x.$$

The units of $C'(x)$ are units of C (dollars) per unit of x (music players). Thus, $C'(x)$ is measured in dollars per music player.

The cost of the 50,001st music player is the amount by which the total cost would rise if we increased production from 50,000 music players to 50,001. Thus, we need to know the rate at which the total cost rises as we increase production. This rate of change is measured by the derivative, or marginal cost, which we just computed. At $x = 50{,}000$ we get

$$C'(50{,}000) = 20 - 0.0002(50{,}000) = \$10 \text{ per music player.}$$

In other words, we estimate that the 50,001st music player will cost approximately \$10.

➡ **Before we go on...** In Example 1 the marginal cost is really only an *approximation* to the cost of the 50,001st music player:

$$\begin{aligned} C'(50{,}000) &\approx \frac{C(50{,}001) - C(50{,}000)}{1} && \text{Set } h = 1 \text{ in the definition of the derivative.} \\ &= C(50{,}001) - C(50{,}000) \\ &= \text{cost of the 50,001st music player.} \end{aligned}$$

The exact cost of the 50,001st music player is

$$\begin{aligned} C(50{,}001) - C(50{,}000) &= [150{,}000 + 20(50{,}001) - 0.0001(50{,}001)^2] \\ &\quad -[150{,}000 + 20(50{,}000) - 0.0001(50{,}000)^2] \\ &= \$9.9999. \end{aligned}$$

So the marginal cost is a good approximation to the actual cost.

Graphically, we are using the tangent line to approximate the cost function near a production level of 50,000. Figure 3 shows the graph of the cost function together with the tangent line at $x = 50{,}000$. Notice that the tangent line is essentially indistinguishable from the graph of the function for some distance on either side of 50,000.

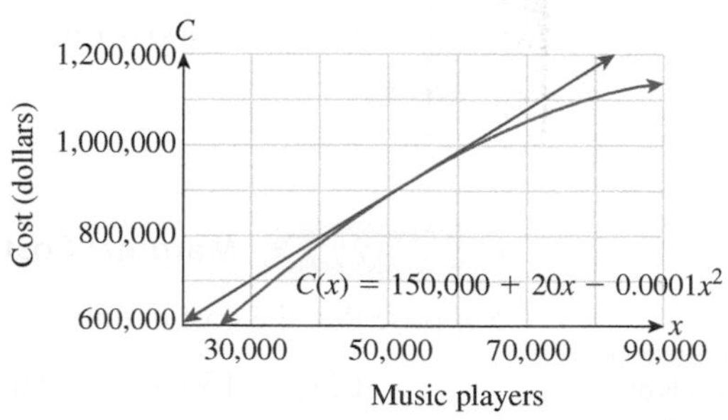

Figure 3 ■

Notes

1. In general, the difference quotient $[C(x + h) - C(x)]/h$ gives the **average cost per item** to produce h more items at a current production level of x items. (Why?)
2. Notice that $C'(x)$ is much easier to calculate than $[C(x + h) - C(x)]/h$. (Try it in this example.) ■

We can extend the idea of marginal cost to include other functions we discussed in Section 1.2, like revenue and profit.

Marginal Revenue and Profit

Recall that a **revenue** or **profit function** specifies the total revenue R or profit P as a function of the number of items x. The derivatives, R' and P', of these functions are called the **marginal revenue** and **marginal profit** functions. They measure the rate of change of revenue and profit with respect to the number of items.

Units

The units of marginal revenue and profit are the same as those of marginal cost: dollars (or euros, pesos, etc.) per item.

Interpretation

We interpret $R'(x)$ and $P'(x)$ as the approximate revenue and profit from the sale of one more item.

EXAMPLE 2 Marginal Revenue and Profit

You operate an iPad refurbishing service. (A typical refurbished iPad might have a custom color case with blinking lights and a personalized logo.) The cost to refurbish x iPads in a month is calculated to be

$$C(x) = 0.25x^2 + 40x + 1{,}000 \text{ dollars.}$$

You charge customers \$80 per iPad for the work.

a. Calculate the marginal revenue and profit functions. Interpret the results.

b. Compute the revenue and profit, and also the marginal revenue and profit, if you have refurbished 20 units this month. Interpret the results.

c. For which value of x is the marginal profit zero? Interpret your answer.

Solution

a. We first calculate the revenue and profit functions:

$$\begin{aligned} R(x) &= 80x && \text{Revenue} = \text{Price} \times \text{Quantity} \\ P(x) &= R(x) - C(x) && \text{Profit} = \text{Revenue} - \text{Cost} \\ &= 80x - (0.25x^2 + 40x + 1{,}000) \\ &= -0.25x^2 + 40x - 1{,}000. \end{aligned}$$

The marginal revenue and profit functions are then the derivatives:

$$\text{Marginal revenue} = R'(x) = 80$$
$$\text{Marginal profit} = P'(x) = -0.5x + 40.$$

Interpretation: $R'(x)$ gives the approximate revenue from the refurbishing of one more item, and $P'(x)$ gives the approximate profit from the refurbishing of one more item. Thus, if x iPads have been refurbished in a month, you will earn a revenue of \$80 and make a profit of approximately $\$(-0.5x + 40)$ if you refurbish one more that month.

Notice that the marginal revenue is a constant, so you earn the same revenue (\$80) for each iPad you refurbish. However, the marginal profit, $\$(-0.5x + 40)$, decreases as x increases, so your additional profit is about 50¢ less for each additional iPad you refurbish.

b. From part (a) the revenue, profit, marginal revenue, and marginal profit functions are

$$R(x) = 80x$$
$$P(x) = -0.25x^2 + 40x - 1{,}000$$
$$R'(x) = 80$$
$$P'(x) = -0.5x + 40.$$

Because you have refurbished 20 iPads this month, $x = 20$, so

$R(20) = 80(20) = \$1{,}600$ — Total revenue from 20 iPads

$P(20) = -0.25(20)^2 + 40(20) - 1{,}000 = -\$300$ — Total profit from 20 iPads

$R'(20) = \$80$ per unit — Approximate revenue from the 21st iPad

$P'(20) = -0.5(20) + 40 = \$30$ per unit. — Approximate profit from the 21st iPad

Interpretation: If you refurbish 20 iPads in a month, you will earn a total revenue of \$1,600 and a profit of −\$300 (indicating a loss of \$300). Refurbishing one more iPad that month will earn you an additional revenue of \$80 and an additional profit of about \$30.

c. The marginal profit is zero when $P'(x) = 0$:

$$-0.5x + 40 = 0$$
$$x = \frac{40}{0.5} = 80 \text{ iPads.}$$

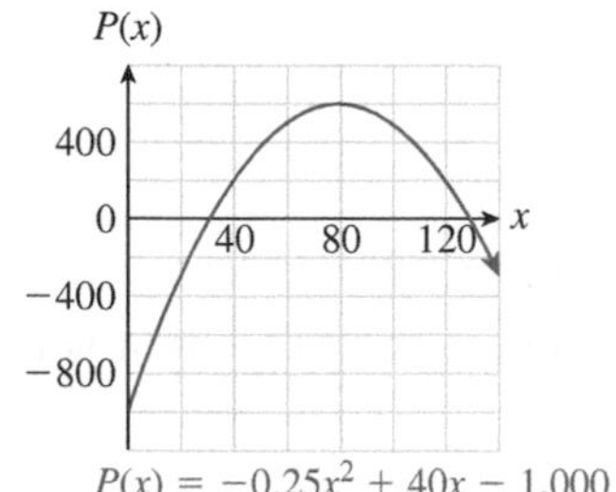

Figure 4

Thus, if you refurbish 80 iPads in a month, refurbishing one more will get you (approximately) zero additional profit. To understand this further, let us take a look at the graph of the profit function, shown in Figure 4. Notice that the graph is a parabola (the profit function is quadratic) with vertex at the point $x = 80$, where $P'(x) = 0$, so the profit is a maximum at this value of x.

➡ **Before we go on . . .** In general, setting $P'(x) = 0$ and solving for x will always give the exact values of x for which the profit peaks as in Figure 4, assuming that there is such a value. We recommend that, when finding the maximum profit in other examples, you graph the profit function to check whether the profit is indeed a maximum at such a point. ■

EXAMPLE 3 Marginal Product

A consultant determines that the number (or quantity) of precision widgets that *Precision Manufacturers* can produce annually is given by

$$Q(n) = 20{,}000n - 100n^2 - n^3 \qquad (10 \le n \le 50),$$

where n is the number of assembly-line workers it employs.

a. Compute $Q'(n)$. $Q'(n)$ is called the **marginal product of labor** at the employment level of n assembly-line workers. What are its units?

b. Calculate $Q(20)$ and $Q'(20)$, and interpret the results.

c. Precision widgets sell for $5 each, and Precision Manufacturers currently employs 20 assembly-line workers. How much more revenue will Precision Manufacturers receive annually if it hires one more assembly-line worker?

Solution

a. Taking the derivative gives

$$Q'(n) = 20{,}000 - 200n - 3n^2.$$

The units of $Q'(n)$ are widgets per worker.

b. Substituting into the formula for $Q(n)$, we get

$$Q(20) = 20{,}000(20) - 100(20)^2 - (20)^3 = 352{,}000 \text{ widgets.}$$

Thus, Precision Manufacturers will produce 352,000 precision widgets annually if it employs 20 assembly-line workers. On the other hand,

$$Q'(20) = 20{,}000 - 200(20) - 3(20)^2 = 14{,}800 \text{ widgets per worker.}$$

Thus, at an employment level of 20 assembly-line workers, annual production is increasing at a rate of 14,800 widgets per additional worker. In other words, if the company were to employ one more assembly-line worker, its annual production would increase by about 14,800 precision widgets.

c. If Precision Manufacturers hires one more worker, its annual production will rise by about 14,800 widgets. If each of these sells for $5, then its annual revenue would rise by $\$5 \times 14{,}800 = \$74{,}000$.

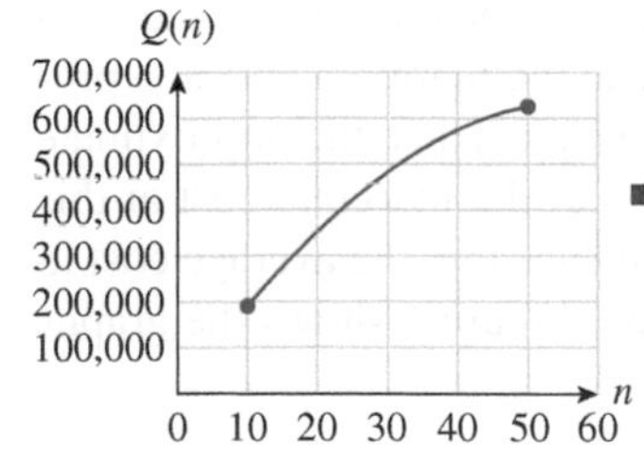

Figure 5

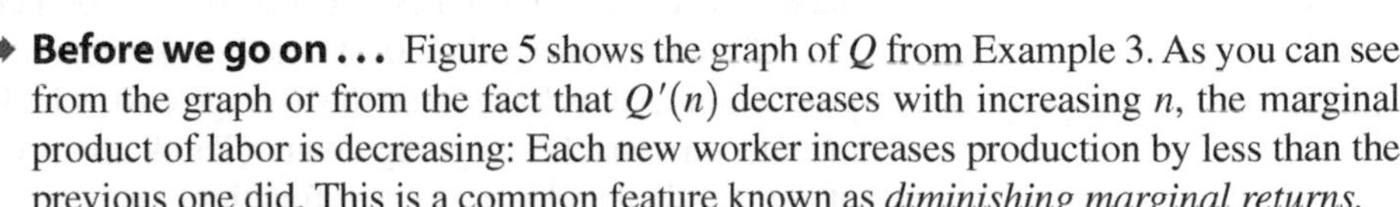

Before we go on... Figure 5 shows the graph of Q from Example 3. As you can see from the graph or from the fact that $Q'(n)$ decreases with increasing n, the marginal product of labor is decreasing: Each new worker increases production by less than the previous one did. This is a common feature known as *diminishing marginal returns.*

In Example 3(c), what would be a reasonable amount for Precision Manufacturers to pay its new worker? *Marginal productivity ethics* says that the worker should be paid his or her marginal productivity, $74,000 annually in this case, as this is the worker's worth to the company, but this argument is not widely accepted.* ■

* You might even say that it's a marginal economic theory.

Average Cost

EXAMPLE 4 **Average Cost**

Suppose the cost to manufacture watches is given by

$$C(x) = 1{,}000 + 125x - 0.6x^2 + 0.001x^3 \text{ dollars} \qquad (0 \le x \le 400)$$

when x watches are manufactured in a day.

a. Find the average cost per watch if 200 watches are manufactured in a day.

b. Find a formula for the average cost per watch if x watches are manufactured in a day. This function of x is called the **average cost function,** $\overline{C}(x)$.

Solution

a. The total cost of manufacturing 200 watches is given by

$$C(200) = 1{,}000 + 125(200) - 0.6(200)^2 + 0.001(200)^3$$
$$= \$10{,}000.$$

Because 200 watches cost a total of \$10,000 to manufacture, the average cost of manufacturing one watch is

$$\bar{C}(200) = \frac{10{,}000}{200} = \$50.00 \text{ per watch.}$$

Thus, if 200 watches are manufactured in a day, each watch costs the manufacturer an average of \$50.00 to manufacture.

b. If we replace 200 by x, we get the general formula for the average cost of manufacturing x watches:

$$\bar{C}(x) = \frac{C(x)}{x}$$
$$= \frac{1}{x}(1{,}000 + 125x - 0.6x^2 + 0.001x^3)$$
$$= \frac{1{,}000}{x} + 125 - 0.6x + 0.001x^2. \quad \text{Average cost function}$$

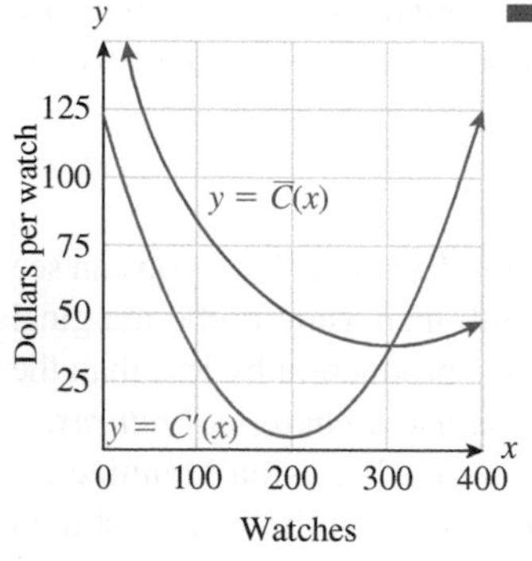

Figure 6

➡ **Before we go on . . .** Average cost and marginal cost convey different but related information. The average cost $\bar{C}(200) = \$50.00$ that we calculated in Example 4 is the cost per item of manufacturing the first 200 watches. On the other hand, the marginal cost function is $C'(x) = 125 - 1.2x + 0.003x^2$, so the marginal cost at a production level of 200 watches is $C'(200) = \$5.00$ per watch. Thus, the approximate cost of manufacturing the 201st watch is \$5.00. Note that the marginal cost at this production level is lower than the average cost, which means that the average cost to manufacture watches is going down. (Think about why.) Figure 6 shows the graphs of the marginal and average cost functions. ■

To summarize, we have the following.

Average Cost

Given a cost function C, the **average cost** of the first x items is given by

$$\bar{C}(x) = \frac{C(x)}{x}.$$

The average cost is distinct from the **marginal cost** $C'(x)$, which tells us the approximate cost of the *next* item.

Quick Example

2. For the cost function $C(x) = 20x + 100$ dollars,

$$\text{Marginal cost} = C'(x) = \$20 \text{ per additional item}$$

$$\text{Average cost} = \bar{C}(x) = \frac{C(x)}{x} = \frac{20x + 100}{x} = \$\left(20 + \frac{100}{x}\right) \text{ per item.}$$

4.2 EXERCISES

▼ more advanced ◆ challenging

T indicates exercises that should be solved using technology

In Exercises 1–4, for each cost function, find the marginal cost at the given production level x, and state the units of measurement. (All costs are in dollars.) [**HINT:** See Example 1.]

1. $C(x) = 10{,}000 + 5x - 0.0001x^2;\ x = 1{,}000$
2. $C(x) = 20{,}000 + 7x - 0.00005x^2;\ x = 10{,}000$
3. $C(x) = 15{,}000 + 100x + \dfrac{1{,}000}{x};\ x = 100$
4. $C(x) = 20{,}000 + 50x + \dfrac{10{,}000}{x};\ x = 100$

In Exercises 5 and 6, find the marginal cost, marginal revenue, and marginal profit functions, and find all values of x for which the marginal profit is zero. Interpret your answer. [**HINT:** See Example 2.]

5. $C(x) = 4x;\quad R(x) = 8x - 0.001x^2$
6. $C(x) = 5x^2;\quad R(x) = x^3 + 7x + 10$
7. ▼ A certain cost function has the following graph:

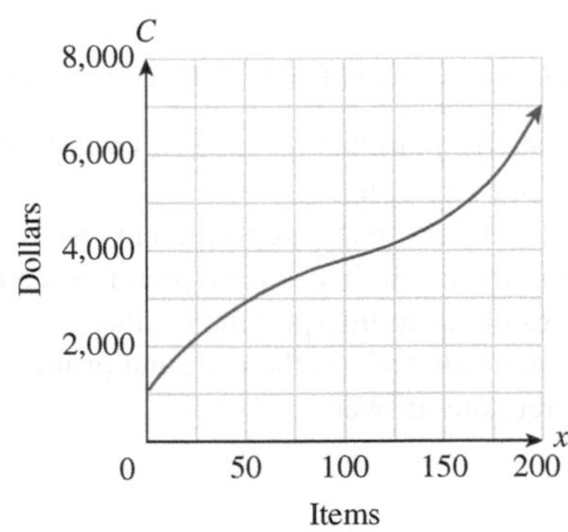

a. The associated marginal cost is
(**A**) increasing, then decreasing.
(**B**) decreasing, then increasing.
(**C**) always increasing.
(**D**) always decreasing.

b. The marginal cost is least at approximately
(**A**) $x = 0$. (**B**) $x = 50$.
(**C**) $x = 100$. (**D**) $x = 150$.

c. The cost of 50 items is
(**A**) approximately \$20 and increasing at a rate of about \$3,000 per item.
(**B**) approximately \$0.50 and increasing at a rate of about \$3,000 per item.
(**C**) approximately \$3,000 and increasing at a rate of about \$20 per item.
(**D**) approximately \$3,000 and increasing at a rate of about \$0.50 per item.

8. ▼ A certain cost function has the following graph:

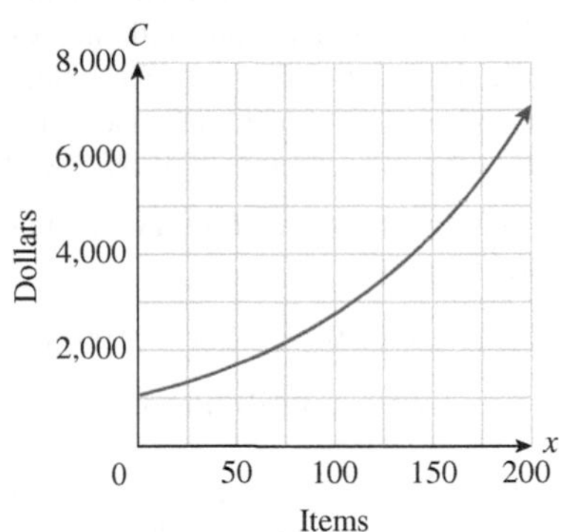

a. The associated marginal cost is
(**A**) increasing, then decreasing.
(**B**) decreasing, then increasing.
(**C**) always increasing.
(**D**) always decreasing.

b. When $x = 100$, the marginal cost is
(**A**) greater than the average cost.
(**B**) less than the average cost.
(**C**) approximately equal to the average cost.

c. The cost of 150 items is
(A) approximately \$4,400 and increasing at a rate of about \$40 per item.
(B) approximately \$40 and increasing at a rate of about \$4,400 per item.
(C) approximately \$4,400 and increasing at a rate of about \$1 per item.
(D) approximately \$1 and increasing at a rate of about \$4,400 per item.

Applications

9. ***Advertising Costs*** The cost, in thousands of dollars, of airing x television commercials during a Super Bowl game is given by[13]

$$C(x) = 20 + 4{,}000x + 0.05x^2.$$

a. Find the marginal cost function, and use it to estimate how fast the cost is increasing when $x = 4$. Compare this with the exact cost of airing the fifth commercial. [**HINT**: See Example 1.]
b. Find the average cost function $\overline{C}$, and evaluate $\overline{C}(4)$. What does the answer tell you? [**HINT**: See Example 2.]

10. ***Marginal Cost and Average Cost*** The cost of producing x teddy bears per day at the *Cuddly Companion Co.* is calculated by the company's marketing staff to be given by the formula

$$C(x) = 100 + 40x - 0.001x^2.$$

a. Find the marginal cost function, and use it to estimate how fast the cost is going up at a production level of 100 teddy bears. Compare this with the exact cost of producing the 101st teddy bear. [**HINT**: See Example 1.]
b. Find the average cost function $\overline{C}$, and evaluate $\overline{C}(100)$. What does the answer tell you? [**HINT**: See Example 4.]

11. ***Marginal and Average Cost: iPhones*** Assume that it costs **Apple** approximately

$$C(x) = 400{,}000 + 160x + 0.001x^2$$

dollars to manufacture x 32GB iPhone 6's in an hour at the **Foxconn Technology Group**.[14]
a. Find the marginal cost function, and use it to estimate how fast the cost is increasing when $x = 10{,}000$. Compare this with the exact cost of producing the 10,001st iPhone.
b. Find the average cost function $\overline{C}$ and the average cost to produce the first 10,000 iPhones.
c. Using your answers to parts (a) and (b), determine whether the average cost is rising or falling at a production level of 10,000 iPhones.

12. ***Marginal and Average Cost: PlayStation 4's*** Assume that it costs **Sony** approximately

$$C(x) = 800{,}000 + 340x + 0.0005x^2$$

dollars to manufacture x PlayStation 4's in an hour.[15]
a. Find the marginal cost function, and use it to estimate how fast the cost is increasing when $x = 60{,}000$. Compare this with the exact cost of producing the 60,001st PlayStation 4.
b. Find the average cost function $\overline{C}(x)$ and the average cost to produce the first 60,000 PlayStation 4's.
c. Using your answers to parts (a) and (b), determine whether the average cost is rising or falling at a production level of 60,000 PlayStation 4's.

13. ***Marginal Revenue and Profit*** Your college newspaper, *The Collegiate Investigator*, sells for 90¢ per copy. The cost of producing x copies of an edition is given by

$$C(x) = 70 + 0.10x + 0.001x^2 \text{ dollars.}$$

a. Calculate the marginal revenue and profit functions. [**HINT**: See Example 2.]
b. Compute the revenue and profit, and also the marginal revenue and profit, if you have produced and sold 500 copies of the latest edition. Interpret the results.
c. For which value of x is the marginal profit zero? Interpret your answer.

14. ***Marginal Revenue and Profit*** The Audubon Society at *Enormous State University* (ESU) is planning its annual fund-raising "Eatathon." The society will charge students \$1.10 per serving of pasta. The society estimates that the total cost of producing x servings of pasta at the event will be

$$C(x) = 350 + 0.10x + 0.002x^2 \text{ dollars.}$$

a. Calculate the marginal revenue and profit functions. [**HINT**: See Example 2.]
b. Compute the revenue and profit, and also the marginal revenue and profit, if you have produced and sold 200 servings of pasta. Interpret the results.
c. For which value of x is the marginal profit zero? Interpret your answer.

[13] The cost of a 30-second ad during the 2014 Super Bowl game was an estimated \$4 million. This explains the coefficient of x in the cost function. Source: "Who Bought What in Super Bowl XLVII," *Advertising Age,* Feb. 3, 2014, http://adage.com/.

[14] Not the actual cost equation; the authors do not know Apple's actual cost equation. The average costs given by this model are in rough agreement with the actual cost for one of the 2014 models. Source for cost data: http://time.com.

[15] Not the actual cost equation; the authors do not know Sony's actual cost equation. The average costs given by this model are in rough agreement with the actual cost to manufacture a PlayStation 4 in 2013. Source for estimate of marginal cost: VentureBeat (http://venturebeat.com).

15. ***Marginal Profit*** Suppose $P(x)$ represents the profit in dollars on the sale of x Blu-ray discs. If $P(1{,}000) = 3{,}000$ and $P'(1{,}000) = -3$, what do these values tell you about the profit?

16. ***Marginal Loss*** An automobile retailer calculates that its loss in dollars on the sale of *Type M* cars is given by $L(50) = 5{,}000$ and $L'(50) = -200$ where $L(x)$ represents the loss on the sale of x Type M cars. What do these values tell you about losses?

17. ***Marginal Profit*** Your monthly profit (in dollars) from selling magazines is given by

$$P = 5x + \sqrt{x}$$

where x is the number of magazines you sell in a month. If you are currently selling 50 magazines per month, find your profit and your marginal profit. Interpret your answers.

18. ***Marginal Profit*** Your monthly profit (in dollars) from your newspaper route is given by

$$P = 2n - \sqrt{n}$$

where n is the number of subscribers on your route. If you currently have 100 subscribers, find your profit and your marginal profit. Interpret your answers.

19. ▼ ***Marginal Revenue: Pricing Tuna*** Assume that the demand equation for tuna in a small coastal town is given by

$$p = \frac{20{,}000}{q^{1.5}} \quad (200 \le q \le 800),$$

where p is the price (in dollars) per pound of tuna and q is the number of pounds of tuna that can be sold at the price p in one month.[16]

a. Calculate the price that the town's fishery should charge for tuna to produce a demand of 400 pounds of tuna per month.

b. Calculate the monthly revenue R as a function of the number of pounds of tuna q.

c. Calculate the revenue and marginal revenue (derivative of the revenue with respect to q) at a demand level of 400 pounds per month, and interpret the results.

d. If the town fishery's monthly tuna catch amounted to 400 pounds of tuna and the price is at the level in part (a), would you recommend that the fishery raise or lower the price of tuna to increase its revenue?

20. ▼ ***Marginal Revenue: Pricing Tuna*** Repeat Exercise 19, assuming a demand equation of

$$p = \frac{60}{q^{0.5}} \quad (200 \le q \le 800).$$

21. ***Marginal Product*** A car wash firm calculates that its daily production (in number of cars washed) depends on the number n of workers it employs according to the formula

$$P = 40n - 0.05n^2 \text{ cars.}$$

Calculate the marginal product of labor at an employment level of 50 workers, and interpret the result. [**HINT:** See Example 3.]

22. ***Marginal Product*** Repeat Exercise 21 using the formula

$$P = -10n + 2.5n^2 - 0.0005n^4.$$

[**HINT:** See Example 3.]

23. ***Average and Marginal Cost*** The daily cost to manufacture generic trinkets for gullible tourists is given by the cost function

$$C(x) = -0.001x^2 + 0.3x + 500 \text{ dollars}$$

where x is the number of trinkets.

a. As x increases, the marginal cost
(A) increases. **(B)** decreases. **(C)** increases, then decreases. **(D)** decreases, then increases.

b. As x increases, the average cost
(A) increases. **(B)** decreases. **(C)** increases, then decreases. **(D)** decreases, then increases.

c. The marginal cost is
(A) greater than **(B)** equal to **(C)** less than
the average cost when $x = 100$. [**HINT:** See Example 4.]

24. ***Average and Marginal Cost*** Repeat Exercise 23, using the following cost function for imitation oil paintings (x is the number of "oil paintings" manufactured):

$$C(x) = 0.1x^2 - 3.5x + 500 \text{ dollars.}$$

[**HINT:** See Example 4.]

25. ***Advertising Cost*** Your company is planning to air a number of television commercials during the **ABC Television Network**'s presentation of the Academy Awards. ABC is charging your company \$1.9 million per 30-second spot.[17] Additional fixed costs (development and personnel costs) amount to \$500,000, and the network has agreed to provide a discount of $\$100{,}000\sqrt{x}$ for x television spots.

a. Write down the cost function C, marginal cost function C', and average cost function $\bar{C}$.

b. Compute $C'(3)$ and $\bar{C}(3)$. (Round all answers to three significant digits.) Use these two answers to say whether the average cost is increasing or decreasing as x increases.

---

[16] Notice that here we have specified p as a function of q and not the other way around as we did in Section 1.2. Economists frequently specify demand curves this way.

[17] ABC charged up to \$1.9 million for a 30-second spot during the 2014 Academy Awards presentation. Source: "Oscar Ad Prices Hit All-Time High as ABC Sells Out 2014 Telecast," *Variety*, October 30, 2013, http://variety.com.

26. ***Housing Costs*** The cost C of building a house is related to the number k of carpenters used and the number x of electricians used by the formula[18]

$$C = 15{,}000 + 50k^2 + 60x^2.$$

a. Assuming that 10 carpenters are currently being used, find the cost function C, marginal cost function C', and average cost function $\bar{C}$, all as functions of x.
b. Use the functions you obtained in part (a) to compute $C'(15)$ and $\bar{C}(15)$. Use these two answers to say whether the average cost is increasing or decreasing as the number of electricians increases.

27. ▼ ***Emission Control*** The cost of controlling emissions at a firm rises rapidly as the amount of emissions reduced increases. Here is a possible model:

$$C(q) = 4{,}000 + 100q^2$$

where q is the reduction in emissions (in pounds of pollutant per day) and C is the daily cost (in dollars) of this reduction.
a. If a firm is currently reducing its emissions by 10 pounds each day, what is the marginal cost of reducing emissions further?
b. Government clean-air subsidies to the firm are based on the formula

$$S(q) = 500q$$

where q is again the reduction in emissions (in pounds per day) and S is the subsidy (in dollars). At what reduction level does the marginal cost surpass the marginal subsidy?
c. Calculate the net cost function, $N(q) = C(q) - S(q)$, given the cost function and subsidy above, and find the value of q that gives the lowest net cost. What is this lowest net cost? Compare your answer to that for part (b), and comment on what you find.

28. ▼ ***Taxation Schemes*** To raise revenues during the recent recession, the governor of your state proposed the following taxation formula:

$$T(i) = 0.001i^{0.5},$$

where i represents total annual income earned by an individual in dollars and $T(i)$ is the income tax rate as a percentage of total annual income. (Thus, for example, an income of \$50,000 per year would be taxed at about 22%, while an income of double that amount would be taxed at about 32%.)[19]

a. Calculate the after-tax (net) income $N(i)$ an individual can expect to earn as a function of income i.
b. Calculate an individual's marginal after-tax income at income levels of \$100,000 and \$500,000.
c. At what income does an individual's marginal after-tax income become negative? What is the after-tax income at that level, and what happens at higher income levels?
d. What do you suspect is the most anyone can earn after taxes? (See the footnote.)

29. ▼ ***Fuel Economy*** Your Porsche's gas mileage (in miles per gallon) is given as a function $M(x)$ of speed x in miles per hour. It is found that

$$M'(x) = \frac{3{,}600x^{-2} - 1}{(3{,}600x^{-1} + x)^2}.$$

Estimate $M'(10)$, $M'(60)$, and $M'(70)$. What do the answers tell you about your car?

30. ▼ ***Marginal Revenue*** The estimated marginal revenue for sales of ESU soccer team T-shirts is given by

$$R'(p) = \frac{(8 - 2p)e^{-p^2+8p}}{10{,}000{,}000}$$

where p is the price (in dollars) that the soccer players charge for each shirt. Estimate $R'(3)$, $R'(4)$, and $R'(5)$. What do the answers tell you?

31. ◆ ***Marginal Cost*** *(from the GRE Economics Test)* In a multiplant firm in which the different plants have different and continuous cost schedules, if costs of production for a given output level are to be minimized, which of the following is essential?
(A) Marginal costs must equal marginal revenue.
(B) Average variable costs must be the same in all plants.
(C) Marginal costs must be the same in all plants.
(D) Total costs must be the same in all plants.
(E) Output per worker per hour must be the same in all plants.

32. ◆ ***Study Time*** *(from the GRE economics test)* A student has a fixed number of hours to devote to study and is certain of the relationship between hours of study and the final grade for each course. Grades are given on a numerical scale (0 to 100), and each course is counted equally in computing the grade average. To maximize his or her grade average, the student should allocate these hours to different courses so that
(A) the grade in each course is the same.
(B) the marginal product of an hour's study (in terms of final grade) in each course is zero.
(C) the marginal product of an hour's study (in terms of final grade) in each course is equal, although not necessarily equal to zero.
(D) the average product of an hour's study (in terms of final grade) in each course is equal.
(E) the number of hours spent in study for each course is equal.

---

[18] Based on an exercise in *Introduction to Mathematical Economics* by A. L. Ostrosky, Jr., and J. V. Koch (Waveland Press, Prospect Heights, Illinois, 1979).

[19] This model has the following interesting feature: An income of \$1 million per year would be taxed at 100%, leaving the individual penniless!

33. ◆ ***Marginal Product*** *(from the GRE Economics Test)* Assume that the marginal product of an additional senior professor is 50% higher than the marginal product of an additional junior professor and that junior professors are paid one half the amount that senior professors receive. With a fixed overall budget, a university that wishes to maximize its quantity of output from professors should do which of the following?
(**A**) Hire equal numbers of senior professors and junior professors.
(**B**) Hire more senior professors and junior professors.
(**C**) Hire more senior professors and discharge junior professors.
(**D**) Discharge senior professors and hire more junior professors.
(**E**) Discharge all senior professors and half of the junior professors.

34. ◆ ***Marginal Product*** *(based on a question from the GRE Economics Test)* Assume that the marginal product of an additional senior professor is twice the marginal product of an additional junior professor and that junior professors are paid two thirds the amount that senior professors receive. With a fixed overall budget, a university that wishes to maximize its quantity of output from professors should do which of the following?
(**A**) Hire equal numbers of senior professors and junior professors.
(**B**) Hire more senior professors and junior professors.
(**C**) Hire more senior professors and discharge junior professors.
(**D**) Discharge senior professors and hire more junior professors.
(**E**) Discharge all senior professors and half of the junior professors.

Communication and Reasoning Exercises

35. The marginal cost of producing the 1,001st item is
(**A**) equal to
(**B**) approximately equal to
(**C**) always slightly greater than
(**D**) always slightly less than
the actual cost of producing the 1,001st item.

36. For the cost function $C(x) = mx + b$ the marginal cost of producing the 1,001st item is
(**A**) equal to
(**B**) approximately equal to
(**C**) always slightly greater than
(**D**) always slightly less than
the actual cost of producing the 1,001st item.

37. What is a cost function? Carefully explain the difference between *average cost* and *marginal cost* in terms of (**a**) their mathematical definition, (**b**) graphs, and (**c**) interpretation.

38. The cost function for your grand piano manufacturing plant has the property that $\bar{C}(1{,}000) = \$3{,}000$ per unit and $C'(1{,}000) = \$2{,}500$ per unit. Will the average cost increase or decrease if your company manufactures a slightly larger number of pianos? Explain your reasoning.

39. Give an example of a cost function for which the marginal cost function is the same as the average cost function.

40. Give an example of a cost function for which the marginal cost function is always less than the average cost function.

41. If the average cost to manufacture one grand piano increases as the production level increases, which is greater, the marginal cost or the average cost?

42. If your analysis of a manufacturing company yielded positive marginal profit but negative profit at the company's current production levels, what would you advise the company to do?

43. ▼ If the marginal cost is decreasing, is the average cost necessarily decreasing? Explain.

44. ▼ If the average cost is decreasing, is the marginal cost necessarily decreasing? Explain.

45. ◆ If a company's marginal average cost is zero at the current production level, positive for a slightly higher production level, and negative for a slightly lower production level, what should you advise the company to do?

46. ◆ The **acceleration** of cost is defined as the derivative of the marginal cost function: that is, the derivative of the derivative—or *second derivative*—of the cost function. What are the units of acceleration of cost, and how does one interpret this measure?

4.3 The Product and Quotient Rules

Motivating the Product Rule

We know how to find the derivatives of functions that are sums of powers, such as polynomials. In general, if a function is a sum or difference of functions whose derivatives we know, then we know how to find its derivative. But what about *products and quotients* of functions whose derivatives we know? For instance, how do we calculate the derivative of something like $x^2/(x + 1)$? The derivative of $x^2/(x + 1)$ is not, as one might suspect, $2x/1 = 2x$. That calculation is based on an assumption that the derivative of a quotient is the quotient of the derivatives. But it is easy to see that this assumption is false. For instance, the derivative of $1/x$ is not $0/1 = 0$, but $-1/x^2$. Similarly, the derivative of a product is almost never the product of the derivatives. For instance, the derivative of $x = 1 \cdot x$ is not $0 \cdot 1 = 0$, but 1.

To identify the correct method of computing the derivatives of products and quotients, let's look at a simple example. We know that the daily revenue resulting from the sale of q items per day at a price of p dollars per item is given by the product, $R = pq$ dollars. Suppose you are currently selling wall posters on campus. At this time your daily sales are 50 posters, and sales are increasing at a rate of 4 per day. Furthermore, you are currently charging \$10 per poster, and you are also raising the price at a rate of \$2 per day. Let's use this information to estimate how fast your daily revenue is increasing. In other words, let us estimate the rate of change, dR/dt, of the revenue R.

There are two contributions to the rate of change of daily revenue: the increase in daily sales and the increase in the unit price. We have

$$\frac{dR}{dt} \text{ due to increasing price:} \quad \$2 \text{ per day} \times 50 \text{ posters} = \$100 \text{ per day}$$

$$\frac{dR}{dt} \text{ due to increasing sales:} \quad \$10 \text{ per poster} \times 4 \text{ posters per day} = \$40 \text{ per day.}$$

Thus, we estimate the daily revenue to be increasing at a rate of \$100 + \$40 = \$140 per day. Let us translate what we have said into symbols:

$$\frac{dR}{dt} \text{ due to increasing price:} \quad \frac{dp}{dt} \times q$$

$$\frac{dR}{dt} \text{ due to increasing sales:} \quad p \times \frac{dq}{dt}.$$

Thus, the rate of change of revenue is given by

$$\frac{dR}{dt} = \frac{dp}{dt}q + p\frac{dq}{dt}.$$

Because $R = pq$, we have discovered the following rule for differentiating a product:

$$\frac{d}{dt}(pq) = \frac{dp}{dt}q + p\frac{dq}{dt}.$$

The derivative of a product is the derivative of the first times the second, plus the first times the derivative of the second.

This rule and a similar rule for differentiating quotients are given next, and also a discussion of how these results are proved rigorously.

Product and Quotient Rules

Product Rule

If f and g are differentiable functions of x, then so is their product fg, and

$$\frac{d}{dx}[f(x)g(x)] = f'(x)g(x) + f(x)g'(x).$$

Product Rule in Words

The derivative of a product is the derivative of the first times the second, plus the first times the derivative of the second.

Quick Example

1. Let $f(x) = x^2$ and $g(x) = 3x - 1$. Because f and g are both differentiable functions of x, so is their product fg, and its derivative is

$$\frac{d}{dx}[x^2(3x-1)] = \underset{\substack{\uparrow \\ \text{Derivative of first}}}{2x} \cdot \underset{\substack{\uparrow \\ \text{Second}}}{(3x-1)} + \underset{\substack{\uparrow \\ \text{First}}}{x^2} \cdot \underset{\substack{\uparrow \\ \text{Derivative of second}}}{(3)}.$$

Quotient Rule

If f and g are differentiable functions of x, then so is their quotient f/g, and

$$\frac{d}{dx}\left(\frac{f(x)}{g(x)}\right) = \frac{f'(x)g(x) - f(x)g'(x)}{[g(x)]^2}.$$

provided that $g(x) \neq 0$.*

* If $g(x)$ is zero, then the quotient $f(x)/g(x)$ is not defined in the first place.

Quotient Rule in Words

The derivative of a quotient is the derivative of the top times the bottom, minus the top times the derivative of the bottom, all over the bottom squared.

Quick Example

2. Let $f(x) = x^3$ and $g(x) = x^2 - 1$. Because f and g are both differentiable functions of x, so is their quotient f/g, and its derivative is

$$\frac{d}{dx}\left(\frac{x^3}{x^2-1}\right) = \frac{\overset{\substack{\text{Derivative of top} \\ \downarrow}}{3x^2}\overset{\substack{\text{Bottom} \\ \downarrow}}{(x^2-1)} - \overset{\substack{\text{Top} \\ \downarrow}}{x^3} \cdot \overset{\substack{\text{Derivative of bottom} \\ \downarrow}}{2x}}{\underset{\substack{\uparrow \\ \text{Bottom squared}}}{(x^2-1)^2}},$$

provided that $x \neq 1$ or -1.

Notes

1. Don't try to remember the rules by the symbols we have used, but remember them in words. (The slogans are easy to remember, even if the terms are not precise.)
2. One more time: *The derivative of a product is* NOT *the product of the derivatives, and the derivative of a quotient is* NOT *the quotient of the derivatives.* To find the derivative of a product, you must use the product rule, and to find the derivative of a quotient, you must use the quotient rule.† ■

† Leibniz made this mistake at first, too, so you would be in good company if you forgot to use the product or quotient rule.

Q: *Wait a minute! The expression $2x^3$ is a product, and we already know that its derivative is $6x^2$. Where did we use the product rule?*

A: To differentiate functions such as $2x^3$, we have used the rule from Section 4.1:

The derivative of c times a function is c times the derivative of the function.

However, the product rule gives us the same result:

$$\frac{d}{dx}(2x^3) = \underset{\text{Derivative of first}}{(0)}\underset{\text{Second}}{(x^3)} + \underset{\text{First}}{(2)}\underset{\text{Derivative of second}}{(3x^2)} = 6x^2 \qquad \text{Product rule}$$

$$\frac{d}{dx}(2x^3) = (2)(3x^2) = 6x^2. \qquad \text{Derivative of a constant times a function}$$

We do not recommend that you use the product rule to differentiate functions such as $2x^3$. Continue to use the simpler rule when one of the factors is a constant.

Derivation of the Product Rule

Website
www.WanerMath.com
The quotient rule can be proved in a very similar way. Go to the Website and follow the path
Everything for Applied Calc
→ Chapter 4
→ Proof of Quotient Rule

Before we look at more examples of using the product and quotient rules, let's see why the product rule is true. To calculate the derivative of the product $f(x)g(x)$ of two differentiable functions, we go back to the definition of the derivative:

$$\frac{d}{dx}[f(x)g(x)] = \lim_{h\to 0}\frac{f(x+h)g(x+h) - f(x)g(x)}{h}.$$

We now rewrite this expression so that we can evaluate the limit. Notice that the numerator reflects a simultaneous change in f [from $f(x)$ to $f(x+h)$] and g [from $g(x)$ to $g(x+h)$]. To separate the two effects, we add and subtract a quantity in the numerator that reflects a change in only one of the functions:

$$\begin{aligned}
\frac{d}{dx}[f(x)g(x)] &= \lim_{h\to 0}\frac{f(x+h)g(x+h) - f(x)g(x)}{h} \\
&= \lim_{h\to 0}\frac{f(x+h)g(x+h) - f(x)g(x+h) + f(x)g(x+h) - f(x)g(x)}{h} && \text{We subtracted and added the quantity* } f(x)g(x+h). \\
&= \lim_{h\to 0}\frac{[f(x+h) - f(x)]g(x+h) + f(x)[g(x+h) - g(x)]}{h} && \text{Common factors} \\
&= \lim_{h\to 0}\left(\frac{f(x+h) - f(x)}{h}\right)g(x+h) + \lim_{h\to 0} f(x)\left(\frac{g(x+h) - g(x)}{h}\right) && \text{Limit of sum} \\
&= \lim_{h\to 0}\left(\frac{f(x+h) - f(x)}{h}\right)\lim_{h\to 0} g(x+h) + \lim_{h\to 0} f(x)\lim_{h\to 0}\left(\frac{g(x+h) - g(x)}{h}\right). && \text{Limit of product}
\end{aligned}$$

* Adding an appropriate form of zero is an age-old mathematical ploy.

Now we already know the following four limits:

$$\lim_{h\to 0}\frac{f(x+h) - f(x)}{h} = f'(x) \qquad \text{Definition of derivative of } f;\ f \text{ is differentiable.}$$

$$\lim_{h\to 0}\frac{g(x+h) - g(x)}{h} = g'(x) \qquad \text{Definition of derivative of } g;\ g \text{ is differentiable.}$$

$$\lim_{h\to 0} g(x+h) = g(x) \qquad \text{If } g \text{ is differentiable, it must be continuous.}^\dagger$$

$$\lim_{h\to 0} f(x) = f(x). \qquad \text{Limit of a constant}$$

† For a proof of the fact that, if g is differentiable, it must be continuous, go to the Website and follow the path
Everything for Applied Calc
→ Chapter 4
→ Continuity and Differentiability

Putting these limits into the one we're calculating, we get

$$\frac{d}{dx}[f(x)g(x)] = f'(x)g(x) + f(x)g'(x)$$

which is the product rule.

EXAMPLE 1 **Using the Product Rule**

Compute the following derivatives:

a. $\frac{d}{dx}[(x^{3.2}+1)(1-x)]$ Simplify the answer.

b. $\frac{d}{dx}[(x+1)(x^2+1)(x^3+1)]$ Do not expand the answer.

c. $\frac{d}{dx}\left(\frac{x|x|}{2}\right)$

Solution

a. We can do the calculation in two ways:

Using the Product Rule:

Derivative of first ↓ Second ↓ First ↓ Derivative of second ↓

$$\frac{d}{dx}[(x^{3.2}+1)(1-x)] = (3.2x^{2.2})(1-x) + (x^{3.2}+1)(-1)$$

$$= 3.2x^{2.2} - 3.2x^{3.2} - x^{3.2} - 1 \quad \text{Expand the answer.}$$

$$= -4.2x^{3.2} + 3.2x^{2.2} - 1$$

Not Using the Product Rule: First, expand the given expression:

$$(x^{3.2}+1)(1-x) = -x^{4.2} + x^{3.2} - x + 1.$$

Thus,

$$\frac{d}{dx}[(x^{3.2}+1)(1-x)] = \frac{d}{dx}(-x^{4.2} + x^{3.2} - x + 1)$$

$$= -4.2x^{3.2} + 3.2x^{2.2} - 1.$$

In this example the product rule saves us little or no work, but in later sections we shall see examples that can be done in no other way. Learn how to use the product rule now!

b. Here we have a product of *three* functions, not just two. We can find the derivative by using the product rule twice:

$$\frac{d}{dx}[(x+1)(x^2+1)(x^3+1)]$$

$$= \frac{d}{dx}(x+1)\cdot[(x^2+1)(x^3+1)] + (x+1)\cdot\frac{d}{dx}[(x^2+1)(x^3+1)]$$

$$= (1)(x^2+1)(x^3+1) + (x+1)[(2x)(x^3+1) + (x^2+1)(3x^2)]$$

$$= (1)(x^2+1)(x^3+1) + (x+1)(2x)(x^3+1) + (x+1)(x^2+1)(3x^2).$$

We can see here a more general product rule:

$$(fgh)' = f'gh + fg'h + fgh'.$$

Notice that every factor has a chance to contribute to the rate of change of the product. There are similar formulas for products of four or more functions.

c. First write $\frac{x|x|}{2}$ as $\frac{1}{2}x|x|$.

$$\frac{d}{dx}\left(\frac{1}{2}x|x|\right) = \frac{1}{2}\frac{d}{dx}(x|x|) \qquad \text{Constant multiple rule}$$

$$= \frac{1}{2}\left((1)\cdot|x| + x\cdot\frac{|x|}{x}\right) \qquad \text{Recall that } \frac{d}{dx}|x| = \frac{|x|}{x}.$$

$$= \frac{1}{2}(|x| + |x|) \qquad \text{Cancel the } x.$$

$$= \frac{1}{2}(2|x|) = |x| \qquad \text{See the note.*}$$

* Notice that we have found a function whose derivative is $|x|$, namely, $x|x|/2$. Notice also that the derivation we gave assumes that $x \neq 0$ because we divided by x in the third step. However, one can verify, using the definition of the derivative as a limit, that $x|x|/2$ is differentiable at $x = 0$ as well and that its derivative at $x = 0$ is 0, implying that the formula $\frac{d}{dx}(x|x|/2) = |x|$ is valid for all values of x, including 0.

EXAMPLE 2 Using the Quotient Rule

Compute the derivatives:

a. $\frac{d}{dx}\left(\frac{1 - 3.2x^{-0.1}}{x + 1}\right)$ **b.** $\frac{d}{dx}\left[\frac{(x + 1)(x + 2)}{x - 1}\right]$

Solution

Derivative of top ↓ Bottom ↓ Top ↓ Derivative of bottom ↓

a. $$\frac{d}{dx}\left(\frac{1 - 3.2x^{-0.1}}{x + 1}\right) = \frac{(0.32x^{-1.1})(x + 1) - (1 - 3.2x^{-0.1})(1)}{(x + 1)^2}$$

↑ Bottom squared

$$= \frac{0.32x^{-0.1} + 0.32x^{-1.1} - 1 + 3.2x^{-0.1}}{(x + 1)^2} \qquad \text{Expand the numerator.}$$

$$= \frac{3.52x^{-0.1} + 0.32x^{-1.1} - 1}{(x + 1)^2}$$

b. Here we have both a product and a quotient. Which rule do we use: the product or the quotient rule? Here is a way to decide. Think about how we would calculate, step by step, the value of $(x + 1)(x + 2)/(x - 1)$ for a specific value of x—say $x = 11$. Here is how we would probably do it:

1. Calculate $(x + 1)(x + 2) = (11 + 1)(11 + 2) = 156$.
2. Calculate $x - 1 = 11 - 1 = 10$.
3. Divide 156 by 10 to get 15.6.

Now ask: *What was the last operation we performed?* The last operation we performed was division, so we can regard the whole expression as a *quotient*—that is, as $(x + 1)(x + 2)$ *divided by* $(x - 1)$. Therefore, we should use the quotient rule.

The first thing the quotient rule tells us to do is to take the derivative of the numerator. Now, the numerator is a product, so we must use the product rule to take its derivative. Here is the calculation:

$$\frac{d}{dx}\left[\frac{(x+1)(x+2)}{x-1}\right] = \frac{\overbrace{[(1)(x+2)+(x+1)(1)]}^{\text{Derivative of top}}\overbrace{(x-1)}^{\text{Bottom}} - \overbrace{[(x+1)(x+2)]}^{\text{Top}}\overbrace{(1)}^{\text{Derivative of bottom}}}{\underbrace{(x-1)^2}_{\text{Bottom squared}}}$$

$$= \frac{(2x+3)(x-1)-(x+1)(x+2)}{(x-1)^2}$$

$$= \frac{x^2-2x-5}{(x-1)^2}.$$

What is important is to determine the *order of operations* and, in particular, to determine the last operation to be performed. Pretending to do an actual calculation reminds us of the order of operations; we call this technique the **calculation thought experiment**.

➡ **Before we go on . . .** We used the quotient rule in Example 2 because the function was a quotient; we used the product rule to calculate the derivative of the numerator because the numerator was a product. Get used to this: Differentiation rules usually must be used in combination.

Here is another way we could have done this problem. Our calculation thought experiment could have taken the following form:

1. Calculate $(x+1)/(x-1) = (11+1)/(11-1) = 1.2$.
2. Calculate $x+2 = 11+2 = 13$.
3. Multiply 1.2 by 13 to get 15.6.

We would have then regarded the expression as a *product*—the product of the factors $(x+1)/(x-1)$ and $(x+2)$—and used the product rule instead. We can't escape the quotient rule, however: We need to use it to take the derivative of the first factor, $(x+1)/(x-1)$. Try this approach for practice, and check that you get the same answer. ■

Calculation Thought Experiment

The **calculation thought experiment** is a technique to determine whether to treat an algebraic expression as a product, quotient, sum, or difference. Given an expression, consider the steps you would use in computing its value. If the last operation is multiplication, treat the expression as a product; if the last operation is division, treat the expression as a quotient; and so on.

Quick Examples

3. $(3x^2-4)(2x+1)$ can be computed by first calculating the expressions in parentheses and then multiplying. Because the last step is multiplication, we can treat the expression as a product.

4. $\dfrac{2x - 1}{x}$ can be computed by first calculating the numerator and denominator and then dividing one by the other. Because the last step is division, we can treat the expression as a quotient.
5. $x^2 + (4x - 1)(x + 2)$ can be computed by first calculating x^2, then calculating the product $(4x - 1)(x + 2)$, and finally adding the two answers. Thus, we can treat the expression as a sum.
6. $(3x^2 - 1)^5$ can be computed by first calculating the expression in parentheses and then raising the answer to the fifth power. Thus, we can treat the expression as a power. (We shall see how to differentiate powers of expressions in Section 4.4.)
7. The expression $(x + 1)(x + 2)/(x - 1)$ can be treated as either a quotient or a product: We can write it as a quotient: $\dfrac{(x + 1)(x + 2)}{x - 1}$ or as a product: $(x + 1)\left(\dfrac{x + 2}{x - 1}\right)$. (See Example 2(b).)

EXAMPLE 3 **Using the Calculation Thought Experiment**

Find $\dfrac{d}{dx}\left[6x^2 + 5\left(\dfrac{x}{x - 1}\right)\right]$.

Solution The calculation thought experiment tells us that the expression we are asked to differentiate can be treated as a *sum*. Because the derivative of a sum is the sum of the derivatives, we get

$$\frac{d}{dx}\left[6x^2 + 5\left(\frac{x}{x - 1}\right)\right] = \frac{d}{dx}(6x^2) + \frac{d}{dx}\left[5\left(\frac{x}{x - 1}\right)\right].$$

In other words, we must take the derivatives of $6x^2$ and $5\left(\dfrac{x}{x - 1}\right)$ separately and then add the answers. The derivative of $6x^2$ is $12x$. There are two ways of taking the derivative of $5\left(\dfrac{x}{x - 1}\right)$: We could first multiply the expression $\left(\dfrac{x}{x - 1}\right)$ by 5 to get $\left(\dfrac{5x}{x - 1}\right)$ and then take its derivative using the quotient rule, or we could pull the 5 out, as we do next:

$$\frac{d}{dx}\left[6x^2 + 5\left(\frac{x}{x - 1}\right)\right] = \frac{d}{dx}(6x^2) + \frac{d}{dx}\left[5\left(\frac{x}{x - 1}\right)\right] \quad \text{Derivative of sum}$$

$$= 12x + 5\frac{d}{dx}\left(\frac{x}{x - 1}\right) \quad \text{Constant} \times \text{Function}$$

$$= 12x + 5\left(\frac{(1)(x - 1) - (x)(1)}{(x - 1)^2}\right) \quad \text{Quotient rule}$$

$$= 12x + 5\left(\frac{-1}{(x - 1)^2}\right)$$

$$= 12x - \frac{5}{(x - 1)^2}.$$

Application

In the next example we return to a scenario similar to the one discussed at the start of this section.

EXAMPLE 4 Applying the Product and Quotient Rules: Revenue and Average Cost

Sales of your newly launched miniature wall posters for college dorms, *iMiniPosters,* are really taking off. (Those old-fashioned large wall posters no longer fit in today's downsized college dorm rooms.) Monthly sales to students at the start of this year were 1,500 iMiniPosters, and since that time, sales have been increasing by 300 posters each month, even though the price you charge has also been going up.

a. The price you charge for iMiniPosters is given by

$$p(t) = 10 + 0.05t^2 \text{ dollars per poster,}$$

where t is time in months since the start of January of this year. Find a formula for the monthly revenue, and then compute its rate of change at the beginning of March.

b. The number of students who purchase iMiniPosters in a month is given by

$$n(t) = 800 + 0.2t,$$

where t is as in part (a). Find a formula for the average number of posters each student buys, and hence estimate the rate at which this number was growing at the beginning of March.

Solution

a. To compute monthly revenue as a function of time t, we use

$$R(t) = p(t)q(t). \qquad \text{Revenue} = \text{Price} \times \text{Quantity}$$

We already have a formula for $p(t)$. The function $q(t)$ measures sales, which were 1,500 posters per month at time $t = 0$ and were rising by 300 per month:

$$q(t) = 1{,}500 + 300t.$$

Therefore, the formula for revenue is

$$R(t) = p(t)q(t)$$
$$R(t) = (10 + 0.05t^2)(1{,}500 + 300t).$$

Rather than expanding this expression, we shall leave it as a product so that we can use the product rule in computing its rate of change:

$$R'(t) = p'(t)q(t) + p(t)q'(t)$$
$$= [0.10t][1{,}500 + 300t] + [10 + 0.05t^2][300].$$

Because the beginning of March corresponds to $t = 2$, we have

$$R'(2) = [0.10(2)][1{,}500 + 300(2)] + [10 + 0.05(2)^2][300]$$
$$= (0.2)(2{,}100) + (10.2)(300) = \$3{,}480 \text{ per month.}$$

Therefore, your monthly revenue was increasing at a rate of \$3,480 per month at the beginning of March.

b. The average number of posters sold to each student is

$$k(t) = \frac{\text{Number of posters}}{\text{Number of students}} = \frac{q(t)}{n(t)} = \frac{1{,}500 + 300t}{800 + 0.2t}.$$

The rate of change of $k(t)$ is computed with the quotient rule:

$$k'(t) = \frac{q'(t)n(t) - q(t)n'(t)}{n(t)^2}$$

$$= \frac{(300)(800 + 0.2t) - (1{,}500 + 300t)(0.2)}{(800 + 0.2t)^2}$$

so

$$k'(2) = \frac{(300)[800 + 0.2(2)] - [1{,}500 + 300(2)](0.2)}{[800 + 0.2(2)]^2}$$

$$= \frac{(300)(800.4) - (2{,}100)(0.2)}{800.4^2} \approx 0.37 \text{ posters per student per month.}$$

Therefore, the average number of posters sold to each student was increasing at a rate of about 0.37 posters per student per month.

4.3 EXERCISES

▼ more advanced ◆ challenging

T indicates exercises that should be solved using technology

In Exercises 1–12:
(a) Calculate the derivative of the given function without using either the product rule or the quotient rule.
(b) Use the product rule or the quotient rule to find the derivative. Check that you obtain the same answer.
[**HINT:** See Quick Examples 1 and 2.]

1. $f(x) = 3x$ **2.** $f(x) = 2x^2$

3. $g(x) = x \cdot x^2$ **4.** $g(x) = x \cdot x$

5. $h(x) = x(x + 3)$ **6.** $h(x) = x(1 + 2x)$

7. $r(x) = 100x^{2.1}$ **8.** $r(x) = 0.2x^{-1}$ **9.** $s(x) = \frac{2}{x}$

10. $t(x) = \frac{x}{3}$ **11.** $u(x) = \frac{x^2}{3}$ **12.** $s(x) = \frac{3}{x^2}$

In Exercises 13–28, calculate $\frac{dy}{dx}$. *Simplify your answer.*
[**HINT:** See Examples 1 and 2.]

13. $y = 3x(4x^2 - 1)$ **14.** $y = 3x^2(2x + 1)$

15. $y = x^3(1 - x^2)$ **16.** $y = x^5(1 - x)$

17. $y = (2x + 3)^2$ **18.** $y = (4x - 1)^2$

19. $y = \frac{4x}{5x - 2}$ **20.** $y = \frac{3x}{-3x + 2}$

21. $y = \frac{2x + 4}{3x - 1}$ **22.** $y = \frac{3x - 9}{2x + 4}$

23. $y = \frac{|x|}{x}$ **24.** $y = \frac{x}{|x|}$

25. $y = \frac{|x|}{x^2}$ **26.** $y = \frac{x^2}{|x|}$

27. $y = x\sqrt{x}$ **28.** $y = x^2\sqrt{x}$

In Exercises 29–56, calculate $\frac{dy}{dx}$. *You need not expand your answers.* [**HINT:** See Examples 1 and 2.]

29. $y = (x + 1)(x^2 - 1)$

30. $y = (4x^2 + x)(x - x^2)$

31. $y = (2x^{0.5} + 4x - 5)(x - x^{-1})$

32. $y = (x^{0.7} - 4x - 5)(x^{-1} + x^{-2})$

33. $y = (2x^2 - 4x + 1)^2$

34. $y = (2x^{0.5} - x^2)^2$

35. $y = \left(\frac{x}{3.2} + \frac{3.2}{x}\right)(x^2 + 1)$

36. $y = \left(\frac{x^{2.1}}{7} + \frac{2}{x^{2.1}}\right)(7x - 1)$

37. $y = x^2(2x + 3)(7x + 2)$ [**HINT:** See Example 1(b).]

38. $y = x(x^2 - 3)(2x^2 + 1)$ [**HINT:** See Example 1(b).]

39. $y = (5.3x - 1)(1 - x^{2.1})(x^{-2.3} - 3.4)$

40. $y = (1.1x + 4)(x^{2.1} - x)(3.4 - x^{-2.1})$

41. ▼ $y = (\sqrt{x} + 1)\left(\sqrt{x} + \frac{1}{x^2}\right)$

42. ▼ $y = (4x^2 - \sqrt{x})\left(\sqrt{x} - \frac{2}{x^2}\right)$

43. $y = \frac{2x^2 + 4x + 1}{3x - 1}$ 44. $y = \frac{3x^2 - 9x + 11}{2x + 4}$

45. $y = \frac{x^2 - 4x + 1}{x^2 + x + 1}$ 46. $y = \frac{x^2 + 9x - 1}{x^2 + 2x - 1}$

47. $y = \frac{x^{0.23} - 5.7x}{1 - x^{-2.9}}$ 48. $y = \frac{8.43x^{-0.1} - 0.5x^{-1}}{3.2 + x^{2.9}}$

49. ▼ $y = \frac{\sqrt{x} + 1}{\sqrt{x} - 1}$ 50. ▼ $y = \frac{\sqrt{x} - 1}{\sqrt{x} + 1}$

51. ▼ $y = \frac{\left(\frac{1}{x} + \frac{1}{x^2}\right)}{x + x^2}$ 52. ▼ $y = \frac{\left(1 - \frac{1}{x^2}\right)}{x^2 - 1}$

53. $y = \frac{(x + 3)(x + 1)}{3x - 1}$ [**HINT**: See Example 2(b).]

54. $y = \frac{x}{(x - 5)(x - 4)}$ [**HINT**: See Example 2(b).]

55. $y = \frac{(x + 3)(x + 1)(x + 2)}{3x - 1}$

56. $y = \frac{3x - 1}{(x - 5)(x - 4)(x - 1)}$

In Exercises 57–62, compute the indicated derivatives.

57. $\frac{d}{dx}[(x^2 + x)(x^2 - x)]$ 58. $\frac{d}{dx}[(x^2 + x^3)(x + 1)]$

59. $\frac{d}{dx}[(x^3 + 2x)(x^2 - x)]\Big|_{x=2}$

60. $\frac{d}{dx}[(x^2 + x)(x^2 - x)]\Big|_{x=1}$

61. $\frac{d}{dt}[(t^2 - t^{0.5})(t^{0.5} + t^{-0.5})]\Big|_{t=1}$

62. $\frac{d}{dt}[(t^2 + t^{0.5})(t^{0.5} - t^{-0.5})]\Big|_{t=1}$

In Exercises 63–70, use the calculation thought experiment to say whether the expression is written as a sum, difference, scalar multiple, product, or quotient. Then use the appropriate rules to find its derivative. [**HINT**: See Quick Examples 3–7 and Example 3.]

63. $y = x^4 - (x^2 + 120)(4x - 1)$

64. $y = x^4 - \frac{x^2 + 120}{4x - 1}$ 65. $y = x + 1 + 2\left(\frac{x}{x + 1}\right)$

66. $y = (x + 2) - 4(x^2 - x)\left(x + \frac{1}{x}\right)$

(Do not simplify the answer.)

67. $y = (x + 2)\left(\frac{x}{x + 1}\right)$ (Do not simplify the answer.)

68. $y = \frac{(x + 2)x}{x + 1}$ (Do not simplify the answer.)

69. $y = (x + 1)(x - 2) - 2\left(\frac{x}{x + 1}\right)$

70. $y = \frac{x + 2}{x + 1} + (x + 1)(x - 2)$

In Exercises 71–76, find the equation of the line tangent to the graph of the given function at the point with the indicated x-coordinate.

71. $f(x) = (x^2 + 1)(x^3 + x); x = 1$

72. $f(x) = (x^{0.5} + 1)(x^2 + x); x = 1$

73. $f(x) = \frac{x + 1}{x + 2}; x = 0$ 74. $f(x) = \frac{\sqrt{x} + 1}{\sqrt{x} + 2}; x = 4$

75. $f(x) = \frac{x^2 + 1}{x}; x = -1$ 76. $f(x) = \frac{x}{x^2 + 1}; x = 1$

Applications

77. ***Revenue*** The monthly sales of *Sunny Electronics'* new sound system are given by $q(t) = 2{,}000t - 100t^2$ units per month, t months after its introduction. The price Sunny charges is $p(t) = 1{,}000 - t^2$ dollars per sound system, t months after introduction. Find the rate of change of monthly sales, the rate of change of the price, and the rate of change of monthly revenue 5 months after the introduction of the sound system. Interpret your answers. [**HINT**: See Example 4(a).]

78. ***Revenue*** The monthly sales of *Sunny Electronics'* new *iSun* media player is given by $q(t) = 2{,}000t - 100t^2$ units per month, t months after its introduction. The price Sunny charges is $p(t) = 100 - t^2$ dollars per iSun, t months after introduction. Find the rate of change of monthly sales, the rate of change of the price, and the rate of change of monthly revenue 6 months after the introduction of the iSun. Interpret your answers. [**HINT**: See Example 4(a).]

79. ***Saudi Oil Revenues*** The price of crude oil during the period 2000–2010 can be approximated by

$$P(t) = 6t + 18 \text{ dollars per barrel} \quad (0 \le t \le 10)$$

in year t, where $t = 0$ represents 2000. Saudi Arabia's crude oil production over the same period can be approximated by[20]

$$Q(t) = -0.036t^2 + 0.62t + 8 \text{ million barrels per day} \quad (0 \le t \le 10).$$

Use these models to estimate Saudi Arabia's daily oil revenue and also its rate of change in 2008. (Round your answers to the nearest $1 million.)

[20] Sources for data: Oil price: InflationData.com www.inflationdata.com, Production: Energy Bulletin www.energybulletin.net.

80. ***Russian Oil Revenues*** The price of crude oil during the period 2000–2010 can be approximated by

$$P(t) = 6t + 18 \text{ dollars per barrels} \quad (0 \le t \le 10)$$

in year t, where $t = 0$ represents 2000. Russia's crude oil production over the same period can be approximated by[21]

$$Q(t) = -0.08t^2 + 1.2t + 5.5 \text{ million barrels per day} \quad (0 \le t \le 10).$$

Use these models to estimate Russia's daily oil revenue and also its rate of change in 2005. (Round your answers to the nearest $1 million.)

81. ***Revenue*** Dorothy Wagner is currently selling 20 "I ♥ Calculus" T-shirts per day, but sales are dropping at a rate of 3 per day. She is currently charging $7 per T-shirt, but to compensate for dwindling sales, she is increasing the unit price by $1 per day. How fast, and in what direction, is her daily revenue currently changing?

82. ***Pricing Policy*** Let us turn Exercise 81 around a little: Dorothy Wagner is currently selling 20 "I ♥ Calculus" T-shirts per day, but sales are dropping at a rate of 3 per day. She is currently charging $7 per T-shirt, and she wishes to increase her daily revenue by $10 per day. At what rate should she increase the unit price to accomplish this (assuming that the price increase does not affect sales)?

83. ***Bus Travel*** *Thoroughbred Bus Company* finds that its monthly costs for one particular year were given by $C(t) = 10{,}000 + t^2$ dollars after t months. After t months the company had $P(t) = 1{,}000 + t^2$ passengers per month. How fast is its cost per passenger changing after 6 months? [**HINT**: See Example 4(b).]

84. ***Bus Travel*** *Thoroughbred Bus Company* finds that its monthly costs for one particular year were given by $C(t) = 100 + t^2$ dollars after t months. After t months the company had $P(t) = 1{,}000 + t^2$ passengers per month. How fast is its cost per passenger changing after 6 months? [**HINT**: See Example 4(b).]

85. ***Fuel Economy*** Your muscle car's gas mileage (in miles per gallon) is given as a function $M(x)$ of speed x in miles per hour, where

$$M(x) = \frac{3{,}000}{x + 3{,}600x^{-1}}.$$

Calculate $M'(x)$ and then $M'(10)$, $M'(60)$, and $M'(70)$. What do the answers tell you about your car?

86. ***Fuel Economy*** Your used Chevy's gas mileage (in miles per gallon) is given as a function $M(x)$ of speed x in miles per hour, where

$$M(x) = \frac{4{,}000}{x + 3{,}025x^{-1}}.$$

Calculate $M'(x)$, and hence determine *the sign* of each of the following: $M'(40)$, $M'(55)$, and $M'(60)$. Interpret your results.

87. ▼ ***Oil Imports from Mexico*** Daily oil production in Mexico and daily U.S. oil imports from Mexico during 2009–2013 could be approximated by

$$P(t) = 3.1 - 0.014t \text{ million barrels} \quad (9 \le t \le 13)$$
$$I(t) = 1.7 - 0.063t \text{ million barrels} \quad (9 \le t \le 13),$$

where t is time in years since the start of 2000.[22]

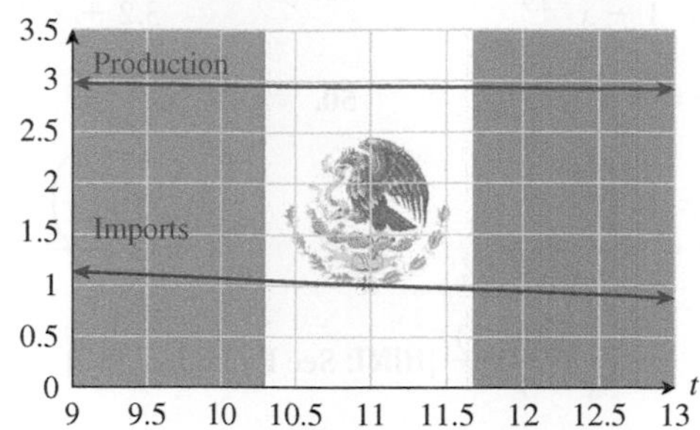

a. What are represented by the functions $P(t) - I(t)$ and $I(t)/P(t)$?

b. Compute $\frac{d}{dt}\left[\frac{I(t)}{P(t)}\right]\Big|_{t=11}$ to two significant digits. What does the answer tell you about oil imports from Mexico?

88. ▼ ***Oil Imports from Mexico*** Daily oil production in Mexico and daily U.S. oil imports from Mexico during 2000–2004 could be approximated by

$$P(t) = 3.0 + 0.13t \text{ million barrels} \quad (0 \le t \le 4)$$
$$I(t) = 1.4 + 0.06t \text{ million barrels} \quad (0 \le t \le 4),$$

where t is time in years since the start of 2000.[23]

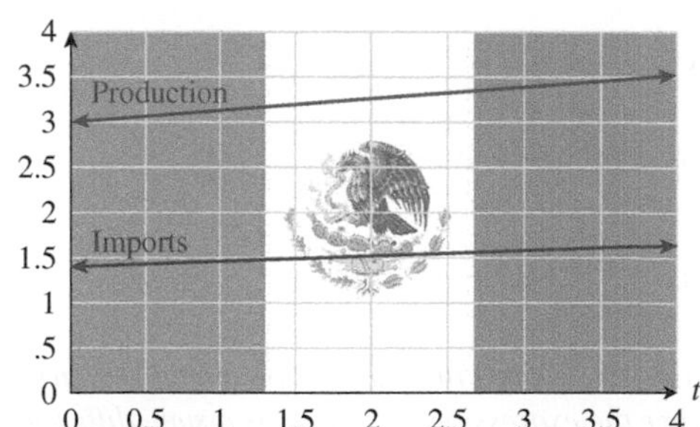

a. What are represented by the functions $P(t) - I(t)$ and $I(t)/P(t)$?

b. Compute $\frac{d}{dt}\left[\frac{I(t)}{P(t)}\right]\Big|_{t=3}$ to two significant digits. What does the answer tell you about oil imports from Mexico?

[21] See footnote for Exercise 79.

[22] Source for data: Energy Information Administration (http://tonto.eia.doe.gov)/Pemex.

[23] *Ibid.*

89. ▼ ***Military Spending*** The annual cost per active-duty armed service member in the United States was projected to increase from \$160,000 in 2014 to \$165,000 in 2018. In 2014 there were 1.36 million armed service personnel, and this number was projected to decrease to 1.32 million in 2018.[24] Use linear models for annual cost and personnel to estimate, to the nearest \$10 million, the projected rate of change of total military personnel costs in 2016.

90. ▼ ***Military Spending in the 1990s*** The annual cost per active-duty armed service member in the United States increased from \$80,000 in 1995 to \$90,000 in 2000. In 1990 there were 2 million armed service personnel and this number decreased to 1.5 million in 2000.[25] Use linear models for annual cost and personnel to estimate, to the nearest \$10 million, the rate of change of total military personnel costs in 1995.

91. ▼ ***Biology—Reproduction*** The Verhulst model for population growth specifies the reproductive rate of an organism as a function of the total population according to the following formula:

$$R(p) = \frac{r}{1 + kp},$$

where p is the total population in thousands of organisms, r and k are constants that depend on the particular circumstances and the organism being studied, and $R(p)$ is the reproduction rate in thousands of organisms per hour.[26] If $k = 0.125$ and $r = 45$, find $R'(p)$ and then $R'(4)$. Interpret the result.

92. ▼ ***Biology—Reproduction*** Another model, the predator satiation model for population growth, specifies that the reproductive rate of an organism as a function of the total population varies according to the following formula:

$$R(p) = \frac{rp}{1 + kp},$$

where p is the total population in thousands of organisms, r and k are constants that depend on the particular circumstances and the organism being studied, and $R(p)$ is the reproduction rate in new organisms per hour.[27] Given that $k = 0.2$ and $r = 0.08$, find $R'(p)$ and $R'(2)$. Interpret the result.

[24] Annual costs in constant 2014 dollars. Source: *Long-Term Implications of the 2014 Future Years Defense Program,* Congressional Budget Office, November 2013, www.cbo.gov/sites/default/files/cbofiles/attachments/44683-FYDP.pdf.

[25] Annual costs are adjusted for inflation. Sources: Department of Defense, Stephen Daggett, military analyst, Congressional Research Service/*New York Times*, April 19, 2002, p. A21.

[26] Source: *Mathematics in Medicine and the Life Sciences* by F. C. Hoppensteadt and C. S. Peskin (Springer-Verlag, New York, 1992) pp. 20–22.

[27] *Ibid.*

93. ▼ ***Embryo Development*** Bird embryos consume oxygen from the time the egg is laid through the time the chick hatches. For a typical galliform bird egg, the oxygen consumption (in milliliters) t days after the egg was laid can be approximated by[28]

$$C(t) = -0.016t^4 + 1.1t^3 - 11t^2 + 3.6t \quad (15 \le t \le 30).$$

(An egg will usually hatch at around $t = 28$.) Suppose that at time $t = 0$ you have a collection of 30 newly laid eggs and that the number of eggs decreases linearly to zero at time $t = 30$ days. How fast is the total oxygen consumption of your collection of embryos changing after 25 days? (Round your answers to two significant digits.) Comment on the result. [**HINT:** Total oxygen consumption = Oxygen consumption per egg × Number of eggs.]

94. ▼ ***Embryo Developmen*** Turkey embryos consume oxygen from the time the egg is laid through the time the chick hatches. For a brush turkey the oxygen consumption (in milliliters) t days after the egg was laid can be approximated by[29]

$$C(t) = -0.0071t^4 + 0.95t^3 - 22t^2 + 95t \quad (25 \le t \le 50).$$

(An egg will typically hatch at around $t = 50$.) Suppose that at time $t = 0$ you have a collection of 100 newly laid eggs and that the number of eggs decreases linearly to zero at time $t = 50$ days. How fast is the total oxygen consumption of your collection of embryos changing after 40 days? (Round your answer to two significant digits.) Interpret the result. [**HINT:** Total oxygen consumption = Oxygen consumption per egg × Number of eggs.]

Communication and Reasoning Exercises

95. If f and g are functions of time, and at time $t = 3$, f equals 5 and is rising at a rate of 2 units per second, and g equals 4 and is rising at a rate of 5 units per second, then the product fg equals ____ and is rising at a rate of ____ units per second.

96. If f and g are functions of time, and at time $t = 2$, f equals 3 and is rising at a rate of 4 units per second, and g equals 5 and is rising at a rate of 6 units per second, then fg equals ____ and is rising at a rate of ____ units per second.

97. If f and g are functions of time, and at time $t = 3$, f equals 5 and is rising at a rate of 2 units per second, and g equals 4 and is rising at a rate of 5 units per second, then f/g equals ____ and is changing at a rate of ____ units per second.

98. If f and g are functions of time, and at time $t = 2$, f equals 3 and is rising at a rate of 4 units per second, and g equals 5 and is rising at a rate of 6 units per second, then f/g equals ____ and is changing at a rate of ____ units per second.

[28] The model is derived from graphical data published in the article "The Brush Turkey" by Roger S. Seymour, *Scientific American,* December, 1991, pp. 108–114.

[29] *Ibid.*

99. You have come across the following in a newspaper article: "Revenues of HAL Home Heating Oil Inc. are rising by \$4.2 million per year. This is due to an annual increase of 70¢ per gallon in the price HAL charges for heating oil and an increase in sales of 6 million gallons of oil per year." Comment on this analysis.

100. Your friend says that because average cost is obtained by dividing the cost function by the number of units x, it follows that the derivative of average cost is the same as marginal cost because the derivative of x is 1. Comment on this analysis.

101. ▼ Find a demand function $q(p)$ such that, at a price per item of $p = \$100$ revenue will rise if the price per item is increased.

102. ▼ What must be true about a demand function $q(p)$ so that, at a price per item of $p = \$100$, revenue will decrease if the price per item is increased?

103. ▼ You and I are both selling a steady 20 T-shirts per day. The price I am getting for my T-shirts is increasing twice as fast as yours, but your T-shirts are currently selling for twice the price of mine. Whose revenue is increasing faster: yours, mine, or neither? Explain.

104. ▼ You and I are both selling T-shirts for a steady \$20 per shirt. Sales of my T-shirts are increasing at twice the rate of yours, but you are currently selling twice as many as I am. Whose revenue is increasing faster: yours, mine, or neither? Explain.

105. ◆ ***Marginal Product*** *(from the GRE Economics Test)* Which of the following statements about average product and marginal product is correct?

(A) If average product is decreasing, marginal product must be less than average product.

(B) If average product is increasing, marginal product must be increasing.

(C) If marginal product is decreasing, average product must be less than marginal product.

(D) If marginal product is increasing, average product must be decreasing.

(E) If marginal product is constant over some range, average product must be constant over that range.

106. ◆ ***Marginal Cost*** *(based on a question from the GRE Economics Test)* Which of the following statements about average cost and marginal cost is correct?

(A) If average cost is increasing, marginal cost must be increasing.

(B) If average cost is increasing, marginal cost must be decreasing.

(C) If average cost is increasing, marginal cost must be more than average cost.

(D) If marginal cost is increasing, average cost must be increasing.

(E) If marginal cost is increasing, average cost must be larger than marginal cost.

4.4 The Chain Rule

Introducing the Chain Rule

We can now find the derivatives of expressions involving powers of x combined using addition, subtraction, multiplication, and division, but we still cannot take the derivative of an expression like $(3x + 1)^{0.5}$. For this we need one more rule. The function $h(x) = (3x + 1)^{0.5}$ is not a sum, difference, product, or quotient. To find out what it is, we can use the calculation thought experiment and think about the last operation we would perform in calculating $h(x)$.

1. Calculate $3x + 1$.
2. Take the 0.5 power (square root) of the answer.

The last operation is "take the 0.5 power." We do not yet have a rule for finding the derivative of the 0.5 power of a quantity other than x.

There is a way to build $h(x) = (3x + 1)^{0.5}$ out of two simpler functions: $u(x) = 3x + 1$ (the function that corresponds to the first step in the calculation above) and $f(x) = x^{0.5}$ (the function that corresponds to the second step):

$$\begin{aligned} h(x) &= (3x + 1)^{0.5} \\ &= [u(x)]^{0.5} && u(x) = 3x + 1 \\ &= f(u(x)). && f(x) = x^{0.5} \end{aligned}$$

We say that h is the **composite** of f and u. We read $f(u(x))$ as "f of u of x."

To compute $h(1)$, say, we first compute $3 \cdot 1 + 1 = 4$ and then take the square root of 4, giving $h(1) = 2$. To compute $f(u(1))$, we follow exactly the same steps: First compute $u(1) = 4$, and then compute $f(u(1)) = f(4) = 2$. We always compute $f(u(x))$ from the inside out: Given x, first compute $u(x)$ and then compute $f(u(x))$.

Now, f and u are functions *whose derivatives we know.* The *chain rule* allows us to use our knowledge of the derivatives of f and u to find the derivative of $f(u(x))$. For the purposes of stating the rule, let us avoid some of the nested parentheses by abbreviating $u(x)$ as u. Thus, we write $f(u)$ instead of $f(u(x))$ and remember that u is a function of x.

Chain Rule

If f is a differentiable function of u and u is a differentiable function of x, then the composite $f(u)$ is a differentiable function of x, and

$$\frac{d}{dx}[f(u)] = f'(u)\frac{du}{dx}. \qquad \text{Chain rule}$$

In words: *The derivative of f(quantity) is the derivative of f, evaluated at that quantity, times the derivative of the quantity.*

Quick Examples

In the Quick Examples that follow, u, "the quantity," is some (unspecified) differentiable function of x.

1. Take $f(u) = u^2$. Then

$$\frac{d}{dx}(u^2) = 2u\frac{du}{dx}. \qquad \text{Because } f'(u) = 2u$$

The derivative of a quantity squared is two times the quantity, times the derivative of the quantity.

2. Take $f(u) = u^{0.5}$. Then

$$\frac{d}{dx}(u^{0.5}) = 0.5u^{-0.5}\frac{du}{dx}. \qquad \text{Because } f'(u) = 0.5u^{-0.5}$$

The derivative of a quantity raised to the 0.5 *is* 0.5 *times the quantity raised to the* −0.5, *times the derivative of the quantity.*

To motivate the chain rule, let us see why it is true in the special case when $f(u) = u^3$, where the chain rule tells us that

$$\frac{d}{dx}(u^3) = 3u^2\frac{du}{dx}. \qquad \text{Chain rule with } f(u) = u^3$$

But we could have done this using the product rule instead:

$$\frac{d}{dx}(u^3) = \frac{d}{dx}(u \cdot u \cdot u) = \frac{du}{dx} \cdot u \cdot u + u \cdot \frac{du}{dx} \cdot u + u \cdot u \cdot \frac{du}{dx} = 3u^2\frac{du}{dx},$$

which gives us the same result.*

* A similar argument works for $f(u) = u^n$, where $n = 2, 3, 4, \ldots$. For the case of a general differentiable function f the proof of the chain rule is beyond the scope of this book, but you can find one on the Website by following the path
Everything for Applied Calc
→ Chapter 4
→ Proof of Chain Rule

Using the Chain Rule

As Quick Examples 1 and 2 illustrate, for every power of a function u whose derivative we know, we now get a generalized differentiation rule. The following table gives more examples.

Original Rule	Generalized Rule	In Words
$\frac{d}{dx}(x^2) = 2x$	$\frac{d}{dx}(u^2) = 2u\frac{du}{dx}$	*The derivative of a quantity squared is twice the quantity, times the derivative of the quantity.*
$\frac{d}{dx}(x^3) = 3x^2$	$\frac{d}{dx}(u^3) = 3u^2\frac{du}{dx}$	*The derivative of a quantity cubed is 3 times the quantity squared, times the derivative of the quantity.*
$\frac{d}{dx}\left(\frac{1}{x}\right) = -\frac{1}{x^2}$	$\frac{d}{dx}\left(\frac{1}{u}\right) = -\frac{1}{u^2}\frac{du}{dx}$	*The derivative of 1 over a quantity is negative 1 over the quantity squared, times the derivative of the quantity.*
Power Rule	**Generalized Power Rule**	**In Words**
$\frac{d}{dx}(x^n) = nx^{n-1}$	$\frac{d}{dx}(u^n) = nu^{n-1}\frac{du}{dx}$	*The derivative of a quantity raised to the n is n times the quantity raised to the n − 1, times the derivative of the quantity.*
$\frac{d}{dx}\lvert x\rvert = \frac{\lvert x\rvert}{x}$	$\frac{d}{dx}\lvert u\rvert = \frac{\lvert u\rvert}{u}\frac{du}{dx}$	*The derivative of the absolute value of a quantity is the absolute value of the quantity divided by the quantity, times the derivative of the quantity.*

EXAMPLE 1 Using the Chain Rule

Compute the following derivatives:

a. $\frac{d}{dx}[(2x^2 + x)^3]$ **b.** $\frac{d}{dx}[(x^3 + x)^{100}]$ **c.** $\frac{d}{dx}\sqrt{3x + 1}$ **d.** $\frac{d}{dx}\lvert 4x^2 - x\rvert$

Solution

a. Using the calculation thought experiment, we see that the last operation we would perform in calculating $(2x^2 + x)^3$ is that of *cubing*. Thus, we think of $(2x^2 + x)^3$ as *a quantity cubed.* There are two similar methods we can use to calculate its derivative.

Method 1: Using the formula We think of $(2x^2 + x)^3$ as u^3, where $u = 2x^2 + x$. By the formula,

$$\frac{d}{dx}(u^3) = 3u^2\frac{du}{dx}. \qquad \text{Generalized power rule}$$

Now substitute for u:

$$\begin{aligned}\frac{d}{dx}[(2x^2 + x)^3] &= 3(2x^2 + x)^2\frac{d}{dx}(2x^2 + x)\\ &= 3(2x^2 + x)^2(4x + 1).\end{aligned}$$

Method 2: Using the verbal form If we prefer to use the verbal form, we get the following:

The derivative of $(2x^2 + x)$ cubed is three times $(2x^2 + x)$ squared, times the derivative of $(2x^2 + x)$.

In symbols,

$$\frac{d}{dx}[(2x^2 + x)^3] = 3(2x^2 + x)^2(4x + 1),$$

as we obtained above.

b. First, the calculation thought experiment: If we were computing $(x^3 + x)^{100}$, the last operation we would perform would be *raising a quantity to the power* 100. Thus, we are dealing with *a quantity raised to the power* 100, so we must again use the generalized power rule. According to the verbal form of the generalized power rule, the derivative of a quantity raised to the power 100 is 100 times that quantity to the power 99, times the derivative of that quantity. In symbols,

$$\frac{d}{dx}[(x^3 + x)^{100}] = 100(x^3 + x)^{99}(3x^2 + 1).$$

c. We first rewrite the expression $\sqrt{3x + 1}$ as $(3x + 1)^{0.5}$ and then use the generalized power rule as in parts (a) and (b):

The derivative of a quantity raised to the 0.5 *is* 0.5 *times the quantity raised to the* −0.5*, times the derivative of the quantity.*

Thus,

$$\begin{aligned}\frac{d}{dx}\sqrt{3x + 1} &= \frac{d}{dx}[(3x + 1)^{0.5}]\\ &= 0.5(3x + 1)^{-0.5} \cdot 3\\ &= 1.5(3x + 1)^{-0.5}\\ &= \frac{1.5}{\sqrt{3x + 1}}.\end{aligned}$$

d. The calculation thought experiment tells us that $|4x^2 - x|$ is the absolute value of a quantity, so we use the generalized rule for absolute values (above):

$$\frac{d}{dx}|u| = \frac{|u|}{u}\frac{du}{dx}, \quad \text{or, in words,}$$

The derivative of the absolute value of a quantity is the absolute value of the quantity divided by the quantity times the derivative of the quantity.

Thus,

$$\frac{d}{dx}|4x^2 - x| = \frac{|4x^2 - x|}{4x^2 - x} \cdot (8x - 1). \qquad \frac{d}{dx}|u| = \frac{|u|}{u}\frac{du}{dx}$$

➡ **Before we go on . . .** The following are examples of common errors in solving Example 1(b):

$$\text{“}\frac{d}{dx}[(x^3 + x)^{100}] = 100(3x^2 + 1)^{99}\text{”} \qquad ✗ \textit{ WRONG!}$$

$$\text{“}\frac{d}{dx}[(x^3 + x)^{100}] = 100(x^3 + x)^{99}.\text{”} \qquad ✗ \textit{ WRONG!}$$

Remember that the generalized power rule says that the derivative of a quantity to the power 100 is 100 times *that same quantity* raised to the power 99, *times the derivative of that quantity.* ■

Q: *It seems that there are now two formulas for the derivative of an nth power:*

1. $\frac{d}{dx}(x^n) = nx^{n-1}$ **2.** $\frac{d}{dx}(u^n) = nu^{n-1}\frac{du}{dx}$.

Which one do I use?

A: Formula 1 is actually a special case of Formula 2: Formula 1 is the original power rule, which applies only to a power of x. For instance, it applies to x^{10}, but it does not apply to $(2x + 1)^{10}$ because the quantity that is being raised to a power is not x. Formula 2 applies to a power of any *function of* x, such as $(2x + 1)^{10}$. It can even be used in place of the original power rule. For example, if we take $u = x$ in Formula 2, we obtain

$$\frac{d}{dx}(x^n) = nx^{n-1}\frac{dx}{dx}$$

$$= nx^{n-1}. \qquad \text{The derivative of } x \text{ with respect to } x \text{ is 1.}$$

Thus, the generalized power rule really *is* a generalization of the original power rule, as its name suggests.

EXAMPLE 2 More Examples Using the Chain Rule

Find: **a.** $\frac{d}{dx}[(2x^5 + x^2 - 20)^{-2/3}]$ **b.** $\frac{d}{dx}\left(\frac{1}{\sqrt{x+2}}\right)$ **c.** $\frac{d}{dx}\left(\frac{1}{x^2+x}\right)$

Solution Each of the given functions is, or can be rewritten as, a power of a function whose derivative we know. Thus, we can use the method of Example 1.

a. $\frac{d}{dx}[(2x^5 + x^2 - 20)^{-2/3}] = -\frac{2}{3}(2x^5 + x^2 - 20)^{-5/3}(10x^4 + 2x)$

b. $\frac{d}{dx}\left(\frac{1}{\sqrt{x+2}}\right) = \frac{d}{dx}(x+2)^{-1/2} = -\frac{1}{2}(x+2)^{-3/2} \cdot 1 = -\frac{1}{2(x+2)^{3/2}}$

c. $\frac{d}{dx}\left(\frac{1}{x^2+x}\right) = \frac{d}{dx}(x^2+x)^{-1} = -(x^2+x)^{-2}(2x+1) = -\frac{2x+1}{(x^2+x)^2}$

➡ **Before we go on . . .** In Example 2(c) we could have used the quotient rule instead of the generalized power rule. We can think of the quantity $1/(x^2 + x)$ in two different ways using the calculation thought experiment:

1. As 1 divided by something—in other words, as a quotient
2. As something raised to the −1 power

Of course, we get the same derivative using either approach. ■

We now look at some more complicated examples.

EXAMPLE 3 Harder Examples Using the Chain Rule

Find $\frac{dy}{dx}$ in each case:

a. $y = [(x+1)^{-2.5} + 3x]^{-3}$ **b.** $y = (x+10)^3\sqrt{1-x^2}$

Solution

a. The calculation thought experiment tells us that the last operation we would perform in calculating y is raising the quantity $[(x+1)^{-2.5}+3x]$ to the power -3. Thus, we use the generalized power rule.

$$\frac{dy}{dx} = -3[(x+1)^{-2.5}+3x]^{-4}\frac{d}{dx}[(x+1)^{-2.5}+3x]$$

We are not yet done; we must still find the derivative of $(x+1)^{-2.5}+3x$. Finding the derivative of a complicated function in several steps helps to keep the problem manageable. Continuing, we have

$$\frac{dy}{dx} = -3[(x+1)^{-2.5}+3x]^{-4}\frac{d}{dx}[(x+1)^{-2.5}+3x]$$

$$= -3[(x+1)^{-2.5}+3x]^{-4}\left(\frac{d}{dx}[(x+1)^{-2.5}]+\frac{d}{dx}(3x)\right). \qquad \text{Derivative of a sum}$$

Now we have two derivatives left to calculate. The second of these we know to be 3, and the first is the derivative of a quantity raised to the -2.5 power. Thus,

$$\frac{dy}{dx} = -3[(x+1)^{-2.5}+3x]^{-4}[-2.5(x+1)^{-3.5}\cdot 1+3].$$

b. The expression $(x+10)^3\sqrt{1-x^2}$ is a product, so we use the product rule:

$$\frac{d}{dx}[(x+10)^3\sqrt{1-x^2}] = \left(\frac{d}{dx}[(x+10)^3]\right)\sqrt{1-x^2}+(x+10)^3\left(\frac{d}{dx}\sqrt{1-x^2}\right)$$

$$= 3(x+10)^2\sqrt{1-x^2}+(x+10)^3\frac{1}{2\sqrt{1-x^2}}(-2x)$$

$$= 3(x+10)^2\sqrt{1-x^2}-\frac{x(x+10)^3}{\sqrt{1-x^2}}.$$

Application

The next example builds on Example 3 from Section 4.2.

EXAMPLE 4 Marginal Product and Profit

A consultant determines that *Precision Manufacturers*' annual profit (in dollars) is given by

$$P = 1{,}000Q^{0.5}+5Q,$$

where Q is the number of precision widgets it sells each year. The consultant also informs Precision's management that the number of precision widgets the company can manufacture each year depends on the number n of assembly-line workers it employs according to the equation

$$Q = 20{,}000n - 100n^2 - n^3.$$

Use the chain rule to find the marginal profit $\frac{dP}{dn}$ at an employment level of 20 workers, and interpret the answer.

Solution We could calculate the marginal profit by substituting the expression for Q in the expression for P to obtain P as a function of n and then finding dP/dn.* Alternatively—and this will simplify the calculation—we can apply the chain rule directly to the given situation. To see how the chain rule applies, notice that P is a function of Q, where Q in turn is given as a function of n. By the chain rule,

* Try it yourself before going on.

$$\frac{dP}{dn} = P'(Q)\frac{dQ}{dn} \qquad \text{Chain rule}$$

$$= \frac{dP}{dQ}\frac{dQ}{dn}. \qquad \text{Notice how the "quantities" } dQ \text{ appear to cancel.}$$

We need to calculate the two derivatives:

$$\frac{dP}{dQ} = 500Q^{-0.5} + 5$$

and

$$\frac{dQ}{dn} = 20{,}000 - 200n - 3n^2.$$

The derivative we want, dP/dn, is the product of the two derivatives above, each evaluated at an employment level of 20 workers ($n = 20$). For the second we can just substitute $n = 20$ to obtain

$$\left.\frac{dQ}{dn}\right|_{n=20} = 20{,}000 - 200(20) - 3(20)^2 = 14{,}800.^{\dagger}$$

† We also calculated this in Example 3 in Section 4.2.

For the first, we need the value of Q that corresponds to $n = 20$:

$$Q(20) = 20{,}000(20) - 100(20)^2 - (20)^3 = 352{,}000$$

so

$$\left.\frac{dP}{dQ}\right|_{n=20} = \left.\frac{dP}{dQ}\right|_{Q=352{,}000} = 500(352{,}000^{-0.5}) + 5 \approx 5.8427.$$

Multiplying the two answers gives

$$\left.\frac{dP}{dn}\right|_{n=20} = \left.\frac{dP}{dQ}\right|_{n=20} \cdot \left.\frac{dQ}{dn}\right|_{n=20} \approx (5.8427)\cdot(14{,}800)$$

$$\approx \$86{,}000 \text{ per worker.} \qquad \text{Rounded to two significant digits}$$

This means that, at a production level of 20 workers, adding more workers will increase Precision Manufacturers' annual profit by approximately \$86,000 per worker.

Recall that dP/dQ is the marginal profit per widget and that dQ/dn is the marginal product of labor. Their product, dP/dn, is the marginal profit *per worker.*

The Chain Rule in Differential Notation

The equation

$$\frac{dP}{dn} = \frac{dP}{dQ}\frac{dQ}{dn}$$

in the example above is an appealing way of writing the chain rule because it suggests that the "quantities" dQ cancel. In general, we can write the chain rule as follows.

Chain Rule: Differential Notation

If y is a differentiable function of u, and u is a differentiable function of x, then

$$\frac{dy}{dx} = \frac{dy}{du}\frac{du}{dx}. \qquad \text{The terms } du \text{ cancel.}$$

Notice how the units of measurement also cancel:

$$\frac{\text{Units of } y}{\text{Units of } x} = \frac{\text{Units of } y}{\cancel{\text{Units of } u}}\,\frac{\cancel{\text{Units of } u}}{\text{Units of } x}.$$

Quick Examples

3. If $y = u^3$, where $u = 4x + 1$, then

$$\frac{dy}{dx} = \frac{dy}{du}\frac{du}{dx} = 3u^2 \cdot 4 = 12u^2 = 12(4x + 1)^2.$$

4. If $q = 43p^2$, where p (and hence q also) is a differentiable function of t, then

$$\frac{dq}{dt} = \frac{dq}{dp}\frac{dp}{dt}$$

$$= 86p\frac{dp}{dt}. \qquad p \text{ is not specified, so we leave } dp/dt \text{ as is.}$$

5. Suppose that a company's weekly revenue R depends on the unit price p, which in turn depends on weekly sales q (by means of a demand equation). Then, if

$$\left.\frac{dR}{dp}\right|_{q=1,000} = \$40 \text{ per } \$1 \text{ increase in price}$$

and

$$\left.\frac{dp}{dq}\right|_{q=1,000} = -\$20 \text{ per additional item sold per week,}$$

the corresponding marginal revenue is

$$\left.\frac{dR}{dq}\right|_{q=1,000} = \left.\frac{dR}{dp}\right|_{q=1,000}\left.\frac{dp}{dq}\right|_{q=1,000} = (40)(-20)$$

$$= -\$800 \text{ per additional item sold.*}$$

* Notice that the units of measurement are Revenue per item = Revenue per \$1 price increase × Price increase per additional item.

Look again at the way the terms du appeared to cancel in the differential formula $\frac{dy}{dx} = \frac{dy}{du}\frac{du}{dx}$. In fact, the chain rule tells us more.

Manipulating Derivatives in Differential Notation

1. Suppose y is a function of x. Then, thinking of x as a function of y (as, for instance, when we can solve for x)† we have

$$\frac{dx}{dy} = \frac{1}{\left(\frac{dy}{dx}\right)}, \text{ provided that } \frac{dy}{dx} \neq 0. \qquad \text{Notice again how } \frac{dy}{dx} \text{ behaves like a fraction.}$$

†The notion of "thinking of x as a function of y" will be made more precise in Section 4.6.

Quick Example

6. In the demand equation $q = -0.2p - 8$ we have $\dfrac{dq}{dp} = -0.2$. Therefore,

$$\frac{dp}{dq} = \frac{1}{\left(\dfrac{dq}{dp}\right)} = \frac{1}{-0.2} = -5.$$

2. Suppose x and y are functions of t. Then, thinking of y as a function of x (as, for instance, when we can solve for t as a function of x, and hence obtain y as a function of x), we have

$$\frac{dy}{dx} = \frac{dy/dt}{dx/dt}. \qquad \text{The terms } dt \text{ appear to cancel.}$$

Quick Example

7. If $x = 3 - 0.2t$ and $y = 6 + 6t$, then

$$\frac{dy}{dx} = \frac{dy/dt}{dx/dt} = \frac{6}{-0.2} = -30.$$

To see why the above formulas work, notice that the second formula,

$$\frac{dy}{dx} = \frac{\left(\dfrac{dy}{dt}\right)}{\left(\dfrac{dx}{dt}\right)},$$

can be written as

$$\frac{dy}{dx}\frac{dx}{dt} = \frac{dy}{dt}, \qquad \text{Multiply both sides by } \frac{dx}{dt}.$$

which is just the differential form of the chain rule. For the first formula, use the second formula with y playing the role of t:

$$\frac{dy}{dx} = \frac{dy/dy}{dx/dy} = \frac{1}{dx/dy}. \qquad \frac{dy}{dy} = \frac{d}{dy}(y) = 1$$

FAQs

Using the Chain Rule

Q: *How do I decide whether or not to use the chain rule when taking a derivative?*

A: Use the calculation thought experiment (Section 4.3): Given an expression, consider the steps you would use in computing its value.

- If the last step is *raising a quantity to a power,* as in $\left(\frac{x^2-1}{x+4}\right)^4$, then the first step to use is the chain rule (in the form of the generalized power rule):

$$\frac{d}{dx}\left(\frac{x^2-1}{x+4}\right)^4 = 4\left(\frac{x^2-1}{x+4}\right)^3 \frac{d}{dx}\left(\frac{x^2-1}{x+4}\right).$$

Then use the appropriate rules to finish the computation. You may need to again use the calculation thought experiment to decide on the next step (here, the quotient rule):

$$= 4\left(\frac{x^2-1}{x+4}\right)^3 \frac{(2x)(x+4)-(x^2-1)(1)}{(x+4)^2}.$$

- If the last step is *division,* as in $\frac{(x^2-1)}{(3x+4)^4}$, then the first step to use is the quotient rule:

$$\frac{d}{dx}\frac{(x^2-1)}{(3x+4)^4} = \frac{(2x)(3x+4)^4-(x^2-1)\frac{d}{dx}(3x+4)^4}{(3x+4)^8}.$$

Then use the appropriate rules to finish the computation (here, the chain rule):

$$= \frac{(2x)(3x+4)^4-(x^2-1)[4(3x+4)^3(3)]}{(3x+4)^8}.$$

- If the last step is *multiplication, addition, subtraction, or multiplication by a constant,* then the first rule to use is the product rule or the rule for sums, differences, or constant multiples as appropriate.

Q: *Every time I compute a derivative, I leave something out. How do I make sure I am really done when taking the derivative of a complicated-looking expression?*

A: Until you are an expert at taking derivatives, the key is to use one rule at a time and write out each step rather than trying to compute the derivative in a single step. To illustrate this, try computing the derivative of $(x+10)^3\sqrt{1-x^2}$ in Example 3(b) in two ways: First try to compute it in a single step, and then compute it by writing out each step as shown in the example. How do your results compare? For more practice, try Exercises 103 and 104 in this section.

4.4 EXERCISES

▼ more advanced ◆ challenging

T indicates exercises that should be solved using technology

In Exercises 1–50, calculate the derivative of the function. [**HINT**: See Example 1.]

1. $f(x) = (2x+1)^2$ **2.** $f(x) = (3x-1)^2$

3. $f(x) = (x-1)^{-1}$ **4.** $f(x) = (2x-1)^{-2}$

5. $f(x) = (2-x)^{-2}$ **6.** $f(x) = (1-x)^{-1}$

7. $f(x) = (2x+1)^{0.5}$ **8.** $f(x) = (-x+2)^{1.5}$

9. $f(x) = \frac{1}{3x-1}$ **10.** $f(x) = \frac{1}{(x+1)^2}$

11. $f(x) = (x^2+2x)^4$ **12.** $f(x) = (x^3-x)^3$

13. $f(x) = (2x^2-2)^{-1}$ **14.** $f(x) = (2x^3+x)^{-2}$

15. $g(x) = (x^2-3x-1)^{-5}$ **16.** $g(x) = (2x^2+x+1)^{-3}$

17. $h(x) = \frac{1}{(x^2+1)^3}$ [**HINT**: See Example 2.] **18.** $h(x) = \frac{1}{(x^2+x+1)^2}$ [**HINT**: See Example 2.]

19. $r(x) = (0.1x^2 - 4.2x + 9.5)^{1.5}$

20. $r(x) = (0.1x - 4.2x^{-1})^{0.5}$

21. $r(s) = (s^2 - s^{0.5})^4$ **22.** $r(s) = (2s + s^{0.5})^{-1}$

23. $f(x) = \sqrt{1 - x^2}$ **24.** $f(x) = \sqrt{x + x^2}$

25. $f(x) = |3x - 6|$ [**HINT**: See Example 1(d).] **26.** $f(x) = |-5x + 1|$ [**HINT**: See Example 1(d).]

27. $f(x) = |-x^3 + 5x|$ **28.** $f(x) = |x - x^4|$

29. $h(x) = 2[(x + 1)(x^2 - 1)]^{-1/2}$ [**HINT**: See Example 3.]

30. $h(x) = 3[(2x - 1)(x - 1)]^{-1/3}$ [**HINT**: See Example 3.]

31. $h(x) = (3.1x - 2)^2 - \dfrac{1}{(3.1x - 2)^2}$

32. $h(x) = \left(3.1x^2 - 2 - \dfrac{1}{3.1x - 2}\right)^2$

33. $f(x) = [(6.4x - 1)^2 + (5.4x - 2)^3]^2$

34. $f(x) = (6.4x - 3)^{-2} + (4.3x - 1)^{-2}$

35. $f(x) = (x^2 - 3x)^{-2}(1 - x^2)^{0.5}$

36. $f(x) = (3x^2 + x)(1 - x^2)^{0.5}$

37. $s(x) = \left(\dfrac{2x + 4}{3x - 1}\right)^2$ **38.** $s(x) = \left(\dfrac{3x - 9}{2x + 4}\right)^3$

39. $g(z) = \left(\dfrac{z}{1 + z^2}\right)^3$ **40.** $g(z) = \left(\dfrac{z^2}{1 + z}\right)^2$

41. $f(x) = [(1 + 2x)^4 - (1 - x)^2]^3$

42. $f(x) = [(3x - 1)^2 + (1 - x)^5]^2$

43. $f(x) = (3x - 1)|3x - 1|$

44. $f(x) = |(x - 3)^{1/3}|$

45. $f(x) = |x - (2x - 3)^{1/2}|$

46. $f(x) = (3 - |3x - 1|)^{-2}$

47. ▼ $r(x) = (\sqrt{2x + 1} - x^2)^{-1}$

48. ▼ $r(x) = (\sqrt{x + 1} + \sqrt{x})^3$

49. ▼ $f(x) = (1 + (1 + (1 + 2x)^3)^3)^3$

50. ▼ $f(x) = 2x + (2x + (2x + 1)^3)^3$

In Exercises 51–66, compute the indicated derivative using the chain rule. [**HINT**: See Quick Examples 3, 6, and 7.]

51. $y = u^2, u = x + 2; \dfrac{dy}{dx}$ **52.** $y = u^3, u = x - 1; \dfrac{dy}{dx}$

53. $y = x^2 + x, x = 2t - 1; \dfrac{dy}{dt}$

54. $y = x^3 - x, x = 1 - 4t; \dfrac{dy}{dt}$

55. $y = 3x - 2; \dfrac{dx}{dy}$ **56.** $y = 8x + 4; \dfrac{dx}{dy}$

57. $y = x^2, x \geq 0; \dfrac{dx}{dy}$ **58.** $y = \sqrt[3]{x}; \dfrac{dx}{dy}$

59. $x = 2 + 3t, y = -5t; \dfrac{dy}{dx}$

60. $x = 1 - t/2, y = 4t - 1; \dfrac{dy}{dx}$

61. $x = t^2, y = 6t + 1, t \geq 0; \dfrac{dy}{dx}$

62. $x = t^3, y = -2t + 2; \dfrac{dy}{dx}$

63. $y = 3x^2 - 2x; \left.\dfrac{dx}{dy}\right|_{x=1}$ **64.** $y = 3x - \dfrac{2}{x}; \left.\dfrac{dx}{dy}\right|_{x=2}$

65. $x = t^2 + 2t, y = t^3; \left.\dfrac{dy}{dx}\right|_{t=1}$

66. $x = 2t^3 + t, y = t^2 + 1; \left.\dfrac{dy}{dx}\right|_{t=2}$

In Exercises 67–74, find the indicated derivative. In each case the independent variable is a (unspecified) differentiable function of t. [**HINT**: See Quick Example 4.]

67. $y = x^{100} + 99x^{-1}$. Find $\dfrac{dy}{dt}$.

68. $y = x^{0.5}(1 + x)$. Find $\dfrac{dy}{dt}$.

69. $s = \dfrac{1}{r^3} + r^{0.5}$. Find $\dfrac{ds}{dt}$. **70.** $s = r + r^{-1}$. Find $\dfrac{ds}{dt}$.

71. $V = \dfrac{4}{3}\pi r^3$. Find $\dfrac{dV}{dt}$. **72.** $A = 4\pi r^2$. Find $\dfrac{dA}{dt}$.

73. ▼ $y = x^3 + \dfrac{1}{x}$, $x = 2$ when $t = 1$, $\left.\dfrac{dx}{dt}\right|_{t=1} = -1$

Find $\left.\dfrac{dy}{dt}\right|_{t=1}$.

74. ▼ $y = \sqrt{x} + \dfrac{1}{\sqrt{x}}$, $x = 9$ when $t = 1$, $\left.\dfrac{dx}{dt}\right|_{t=1} = -1$

Find $\left.\dfrac{dy}{dt}\right|_{t=1}$.

Applications

75. ***Crude Oil Prices*** The price per barrel of crude oil in the period 1980–2013, in constant 2014 dollars, can be approximated by

$$P(t) = 0.27(t - 1980)^2 - 8.6(t - 1980) + 93 \text{ dollars} \quad (1980 \leq t \leq 2013),$$

where t is the year.[30] Find $P'(t)$ and $P'(2010)$. What does the second answer tell you about the price of crude oil?

[30] Source for data: http://inflationdata.com/Inflation/Inflation_Rate/Historical_Oil_Prices_Table.asp, March 6, 2014.

76. ***Median Home Prices*** The median home price in the United States over the period January 2010–January 2015 can be approximated by

$$P(t) = 4.5(t - 2010)^2 - 15(t - 2010) + 180 \text{ thousand dollars} \quad (2010 \le t \le 2015),$$

where t is the year.[31] Find $P'(t)$ and $P'(2011)$. What does the second answer tell you about home prices?

77. ***Marginal Profit*** Your monthly profit (in dollars) from selling magazines is given by

$$P = 5x + \sqrt{2x + 10},$$

where x is the number of magazines you sell in a month. If you are currently selling 50 magazines per month, find your profit and your marginal profit. Interpret your answers.

78. ***Marginal Profit*** Your monthly profit (in dollars) from your newspaper route is given by

$$P = 2n - \sqrt{3n + 100},$$

where n is the number of subscribers on your route. If you currently have 100 subscribers, find your profit and your marginal profit. Interpret your answers.

79. ***Fuel Economy*** (You saw this exercise in Section 4.3. This time, use the chain rule to calculate the derivative.) Your muscle car's gas mileage (in miles per gallon) is given as a function $M(x)$ of speed x in miles per hour, where

$$M(x) = \frac{3{,}000}{x + 3{,}600x^{-1}}.$$

Calculate $M'(x)$ and then $M'(10)$, $M'(60)$, and $M'(70)$. What do the answers tell you about your car?

80. ***Fuel Economy*** (You saw this exercise in Section 4.3. This time, use the chain rule to calculate the derivative.) Your used Chevy's gas mileage (in miles per gallon) is given as a function $M(x)$ of speed x in miles per hour, where

$$M(x) = \frac{4{,}000}{x + 3{,}025x^{-1}}.$$

Calculate $M'(x)$, and hence determine the sign of each of the following: $M'(40)$, $M'(55)$, and $M'(60)$. Interpret your results.

81. ***Marginal Profit*** *Paramount Electronics* has an annual profit given by

$$P = -100{,}000 + 5{,}000q - 0.25q^2 \text{ dollars},$$

where q is the number of laptop computers it sells each year. The number of laptop computers it can make and sell each year depends on the number n of electrical engineers Paramount employs, according to the equation

$$q = 30n + 0.01n^2.$$

Use the chain rule to find $\left.\frac{dP}{dn}\right|_{n=10}$, and interpret the result. [**HINT:** See Example 4.]

82. ***Marginal Profit*** Refer back to Exercise 81. The average profit $\bar{P}$ per computer is given by dividing the total profit P by q:

$$\bar{P} = -\frac{100{,}000}{q} + 5{,}000 - 0.25q \text{ dollars.}$$

Determine the **marginal average profit**, $d\bar{P}/dn$, at an employee level of 10 engineers. Interpret the result. [**HINT:** See Example 4.]

83. ***Food versus Education*** The percentage y (of total personal consumption) an individual spends on food is approximately

$$y = 35x^{-0.25} \text{ percentage points} \quad (6.5 \le x \le 17.5),$$

where x is the percentage the individual spends on education.[32] An individual finds that she is spending

$$x = 7 + 0.2t$$

percent of her personal consumption on education, where t is time in months since January 1. Use direct substitution to express the percentage y as a function of time t (do not simplify the expression), and then use the chain rule to estimate how fast the percentage she spends on food is changing on November 1. Be sure to specify the units. [**HINT:** See Example 3(a).]

84. ***Food versus Recreation*** The percentage y (of total personal consumption) an individual spends on food is approximately

$$y = 33x^{-0.63} \text{ percentage points} \quad (2.5 \le x \le 4.5),$$

where x is the percentage the individual spends on recreation.[33] A college student finds that he is spending

$$x = 3.5 + 0.1t$$

percent of his personal consumption on recreation, where t is time in months since January 1. Use direct substitution to express the percentage y as a function of time t (do not simplify the expression) and then use the chain rule to estimate how fast the percentage he spends on food is changing on November 1. Be sure to specify the units. [**HINT:** See Example 3(a).]

85. ***Marginal Revenue*** The weekly revenue from the sale of rubies at *Royal Ruby Retailers* is increasing at a rate of \$40 per \$1 increase in price, and the price is decreasing at a rate of \$0.75 per additional ruby sold. What is the marginal revenue? (Be sure to state the units of measurement.) Interpret the result. [**HINT:** See Quick Example 5.]

[31] Source for data: www.zillow.com.

[32] Model based on historical and projected data from 1908–2010. Sources: Historical data, Bureau of Economic Analysis; projected data, Bureau of Labor Statistics/*New York Times*, December 1, 2003, p. C2.

[33] *Ibid.*

86. ***Marginal Revenue*** The weekly revenue from the sale of emeralds at *Eduardo's Emerald Emporium* is decreasing at a rate of €500 per €1 increase in price, and the price is decreasing at a rate of €0.45 per additional emerald sold. What is the marginal revenue? (Be sure to state the units of measurement.) Interpret the result. [**HINT**: See Quick Example 5.]

87. ***Crime Statistics*** The murder rate in large cities (over 1 million residents) can be related to that in smaller cities (500,000–1,000,000 residents) by the following linear model:[34]

$$y = 1.5x - 1.9 \quad (15 \le x \le 25),$$

where y is the murder rate (in murders per 100,000 residents each year) in large cities and x is the murder rate in smaller cities. During the period 1991–1998 the murder rate in small cities was decreasing at an average rate of 2 murders per 100,000 residents each year. Use the chain rule to estimate how fast the murder rate was changing in larger cities during that period. (Show how you used the chain rule in your answer.)

88. ***Crime Statistics*** Following is a quadratic model relating the murder rates described in Exercise 87:

$$y = 0.1x^2 - 3x + 39 \quad (15 \le x \le 25).$$

In 1996 the murder rate in smaller cities was approximately 22 murders per 100,000 residents each year and was decreasing at a rate of approximately 2.5 murders per 100,000 residents each year. Use the chain rule to estimate how fast the murder rate was changing for large cities. (Show how you used the chain rule in your answer.)

89. ***Existing Home Sales*** The following graph shows the approximate value of home prices and existing home sales in 2006–2010 as a percentage change from 2003, together with quadratic approximations:[35]

Home prices and sales of existing homes

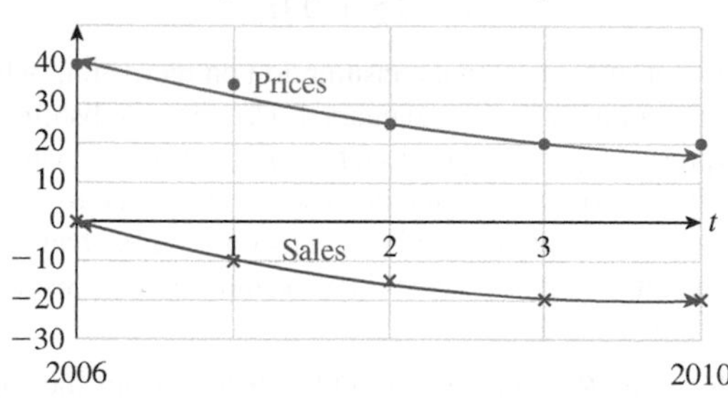

The quadratic approximations are given by

$$\text{Home prices: } P(t) = t^2 - 10t + 41 \quad (0 \le t \le 4)$$
$$\text{Existing home sales: } S(t) = 1.5t^2 - 11t \quad (0 \le t \le 4),$$

where t is time in years since the start of 2006. Use the chain rule to estimate $\left.\frac{dS}{dP}\right|_{t=2}$. What does the answer tell you about home sales and prices? [**HINT**: See Quick Examples 6 and 7.]

90. ***Existing Home Sales Leading to the Financial Crisis*** The following graph shows the approximate value of home prices and existing home sales in 2004–2007 (the 3 years prior to the 2008 economic crisis) as a percentage change from 2003, together with quadratic approximations:[36]

Home prices and sales of existing homes

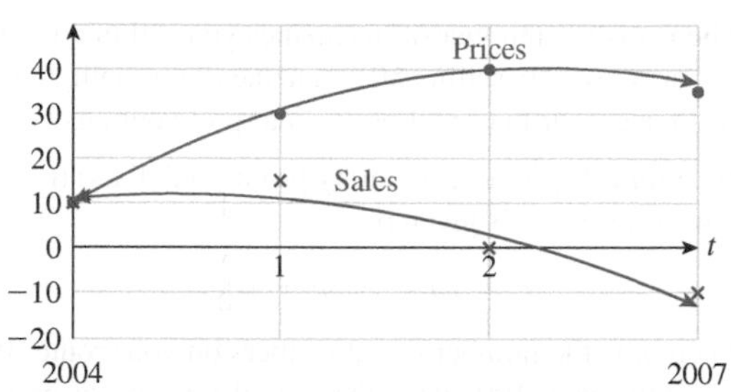

The quadratic approximations are given by

$$\text{Home prices: } P(t) = -6t^2 + 27t + 10 \quad (0 \le t \le 3)$$
$$\text{Existing home sales: } S(t) = -4t^2 + 4t + 11 \quad (0 \le t \le 3),$$

where t is time in years since the start of 2004. Use the chain rule to estimate $\left.\frac{dS}{dP}\right|_{t=2}$. What does the answer tell you about home sales and prices? [**HINT**: See Quick Examples 6 and 7.]

91. ▼ ***Pollution*** An offshore oil well is leaking oil and creating a circular oil slick. If the radius of the slick is growing at a rate of 2 miles per hour, find the rate at which the area is increasing when the radius is 3 miles. (The area of a disc of radius r is $A = \pi r^2$.) [**HINT**: See Quick Example 4.]

92. ▼ ***Mold*** A mold culture in a dorm refrigerator is circular and growing. The radius is growing at a rate of 0.3 centimeters per day. How fast is the area growing when the culture is 4 centimeters in radius? (The area of a disc of radius r is $A = \pi r^2$.) [**HINT**: See Quick Example 4.]

93. ▼ ***Budget Overruns*** The Pentagon is planning to build a new spherical satellite. As is typical in these cases, the specifications keep changing, so the size of the satellite keeps growing. In fact, the radius of the planned satellite is growing 0.5 feet per week. Its cost will be $1,000 per cubic foot. At the point when the plans call for a satellite 10 feet in radius, how fast is the cost growing? (The volume of a solid sphere of radius r is $V = \frac{4}{3}\pi r^3$.)

94. ▼ ***Soap Bubbles*** The soap bubble I am blowing has a radius that is growing at a rate of 4 centimeters per second.

[34] The model is a linear regression model. Source for data: Federal Bureau of Investigation, Supplementary Homicide Reports/*New York Times*, May 29, 2000, p. A12.

[35] Sources: Standard & Poors/Bloomberg Financial Markets/*New York Times*, September 29, 2007, p. C3. Projection is the authors'.

[36] *Ibid.*

How fast is the surface area growing when the radius is 10 centimeters? (The surface area of a sphere of radius r is $S = 4\pi r^2$.)

95. T ▼ ***Revenue Growth*** The demand for the Cyberpunk II arcade video game is modeled by the logistic curve

$$q(t) = \frac{10{,}000}{1 + 0.5e^{-0.4t}},$$

where $q(t)$ is the total number of units sold t months after its introduction.

a. Use technology to estimate $q'(4)$.

b. Assume that the manufacturers of Cyberpunk II sell each unit for \$800. What is the company's marginal revenue dR/dq?

c. Use the chain rule to estimate the rate at which revenue is growing 4 months after the introduction of the video game.

96. T ▼ ***Information Highway*** The amount of information transmitted each month in the early years of the Internet (1988–1994) can be modeled by the equation

$$q(t) = \frac{2e^{0.69t}}{3 + 1.5e^{-0.4t}} \quad (0 \le t \le 6),$$

where q is the amount of information transmitted each month in billions of data packets and t is the number of years since the start of 1988.[37]

a. Use technology to estimate $q'(2)$.

b. Assume that it costs \$5 to transmit a million packets of data. What is the marginal cost $C'(q)$?

c. How fast was the cost increasing at the start of 1990?

Money Stock *Exercises 97–100 are based on the following demand function for money (taken from a question on the GRE Economics Test):*

$$M_d = 2 \times y^{0.6} \times r^{-0.3} \times p,$$

where

M_d = *demand for nominal money balances (money stock)*
y = *real income*
r = *an index of interest rates*
p = *an index of prices.*

These exercises also use the idea of ***percentage rate of growth****:*

$$\text{Percentage rate of growth of } M = \frac{\text{Rate of growth of } M}{M} = \frac{dM/dt}{M}.$$

97. ◆ (From the GRE Economics Test) If the interest rate and price level are to remain constant while real income grows at 5% per year, the money stock must grow at what percent per year?

98. ◆ (From the GRE Economics Test) If real income and price level are to remain constant while the interest rate grows at 5% per year, the money stock must change by what percent per year?

99. ◆ (From the GRE Economics Test) If the interest rate is to remain constant while real income grows at 5% per year and the price level rises at 5% per year, the money stock must grow at what percent per year?

100. ◆ (From the GRE Economics Test) If real income grows by 5% per year, the interest rate grows by 2% per year, and the price level drops by 3% per year, the money stock must change by what percent per year?

Communication and Reasoning Exercises

101. Complete the following: The derivative of 1 over a glob is −1 over . . . .

102. Complete the following: The derivative of the square root of a glob is 1 over . . . .

103. Say why the following was marked wrong, and give the correct answer.

$$\frac{d}{dx}[(3x^3 - x)^3] = 3(9x^2 - 1)^2 \quad \text{✗ } \textit{WRONG!}$$

104. Say why the following was marked wrong, and give the correct answer.

$$\frac{d}{dx}\left[\left(\frac{3x^2 - 1}{2x - 2}\right)^3\right] = 3\left(\frac{3x^2 - 1}{2x - 2}\right)^2\left(\frac{6x}{2}\right) \quad \text{✗ } \textit{WRONG!}$$

105. Name two major errors in the following graded test question, and give the correct answer.

$$\frac{d}{dx}\left[\left(\frac{3x^2 - 1}{2x - 2}\right)^3\right] = 3\left(\frac{6x}{2}\right)^2 \quad \text{✗ } \textit{WRONG! SEE ME!}$$

106. Name two major errors in the following graded test question, and give the correct answer.

$$\frac{d}{dx}[(3x^3 - x)(2x + 1)]^4 = 4[(9x^2 - 1)(2)]^3$$

✗ *WRONG! SEE ME!*

107. ▼ Formulate a simple procedure for deciding whether to apply first the chain rule, the product rule, or the quotient rule when finding the derivative of a function.

108. ▼ Give an example of a function f with the property that calculating $f'(x)$ requires use of the following rules in the given order: (1) the chain rule, (2) the quotient rule, and (3) the chain rule.

109. ◆ Give an example of a function f with the property that calculating $f'(x)$ requires use of the chain rule five times in succession.

110. ◆ What can you say about the composite of two linear functions, and what can you say about its derivative?

---

[37] This is the authors' model, based on figures published in the *New York Times*, Nov. 3, 1993.

4.5 Derivatives of Logarithmic and Exponential Functions

Derivative of ln *x* and $\log_b x$

At this point we know how to take the derivative of any algebraic expression in x (involving powers, radicals, and so on). We now turn to the derivatives of logarithmic and exponential functions.

Derivative of the Natural Logarithm

$$\frac{d}{dx}(\ln x) = \frac{1}{x}$$

Recall that $\ln x = \log_e x$.

Quick Examples

1. $\frac{d}{dx}(3 \ln x) = 3 \cdot \frac{1}{x} = \frac{3}{x}$ Derivative of a constant times a function

2. $\frac{d}{dx}(x \ln x) = 1 \cdot \ln x + x \cdot \frac{1}{x}$ Product rule, because $x \ln x$ is a product

$= \ln x + 1.$

The above simple formula works only for the natural logarithm (the logarithm with base e). For logarithms with bases other than e we have the following:

Derivative of the Logarithm with Base *b*

$$\frac{d}{dx}(\log_b x) = \frac{1}{x \ln b}$$

Notice that, if $b = e$, we get the same formula as previously.

Quick Examples

3. $\frac{d}{dx}(\log_3 x) = \frac{1}{x \ln 3} \approx \frac{1}{1.0986x}$

4. $\frac{d}{dx}[\log_2(x^4)] = \frac{d}{dx}(4 \log_2 x)$ We used the logarithm identity $\log_b(x^r) = r \log_b x$.

$= 4 \cdot \frac{1}{x \ln 2} \approx \frac{4}{0.6931x}$

Derivation of the formulas $\frac{d}{dx}(\ln x) = \frac{1}{x}$ and $\frac{d}{dx}(\log_b x) = \frac{1}{x \ln b}$

To compute $\frac{d}{dx}(\ln x)$, we need to use the definition of the derivative. We also use properties of the logarithm to help evaluate the limit:

$$\begin{aligned}
\frac{d}{dx}(\ln x) &= \lim_{h\to 0}\frac{\ln(x+h)-\ln x}{h} && \text{Definition of the derivative}\\
&= \lim_{h\to 0}\frac{1}{h}[\ln(x+h)-\ln x] && \text{Algebra}\\
&= \lim_{h\to 0}\frac{1}{h}\ln\left(\frac{x+h}{x}\right) && \text{Properties of the logarithm}\\
&= \lim_{h\to 0}\frac{1}{h}\ln\left(1+\frac{h}{x}\right) && \text{Algebra}\\
&= \lim_{h\to 0}\ln\left(1+\frac{h}{x}\right)^{1/h} && \text{Properties of the logarithm}
\end{aligned}$$

which we rewrite as

$$\lim_{h\to 0}\ln\left[\left(1+\frac{1}{(x/h)}\right)^{x/h}\right]^{1/x}.$$

As $h \to 0^+$, the quantity x/h gets large and positive, so the quantity in brackets approaches e (see the definition of e in Section 2.2), which leaves us with

$$\ln[e]^{1/x} = \frac{1}{x}\ln e = \frac{1}{x}$$

which is the derivative we are after.* What about the limit as $h \to 0^-$? We will glide over that case and leave it for the interested reader to pursue.†

The rule for the derivative of $\log_b x$ follows from the fact that $\log_b x = \ln x/\ln b$.

* We actually used the fact that the logarithm function is continuous when we took the limit.

† Here is an outline of the argument for negative h. Because x must be positive for $\ln x$ to be defined, we find that $x/h \to -\infty$ as $h \to 0^-$, so we must consider the quantity $(1 + 1/m)^m$ for large *negative m*. It turns out the limit is still e (check it numerically!), so the computation above still works.

Derivatives of Logarithms of Functions

If we were to take the derivative of the natural logarithm of a *quantity* (a function of x) rather than just x, we would need to use the chain rule.

Derivatives of Logarithms of Functions

Original Rule	Generalized Rule	In Words
$\frac{d}{dx}\ln x = \frac{1}{x}$	$\frac{d}{dx}\ln u = \frac{1}{u}\frac{du}{dx}$	*The derivative of the natural logarithm of a quantity is* 1 *over that quantity, times the derivative of that quantity.*
$\frac{d}{dx}\log_b x = \frac{1}{x\ln b}$	$\frac{d}{dx}\log_b u = \frac{1}{u\ln b}\frac{du}{dx}$	*The derivative of the log to base b of a quantity is* 1 *over the product of* ln *b and that quantity, times the derivative of that quantity.*

Quick Examples

5. $\frac{d}{dx}\ln(x^2+1) = \frac{1}{x^2+1}\frac{d}{dx}(x^2+1)$ $u = x^2 + 1$ (See the margin note.§)

$$= \frac{1}{x^2+1}(2x) = \frac{2x}{x^2+1}$$

§ If we were to evaluate $\ln(x^2 + 1)$, the last operation we would perform would be to take the natural logarithm of a quantity. Thus, the calculation thought experiment tells us that we are dealing with ln *of a quantity,* so we need the generalized logarithm rule as stated above.

6. $$\frac{d}{dx}\log_2(x^3+x) = \frac{1}{(x^3+x)\ln 2}\frac{d}{dx}(x^3+x) \qquad u = x^3 + x$$
$$= \frac{1}{(x^3+x)\ln 2}(3x^2+1) = \frac{3x^2+1}{(x^3+x)\ln 2}$$

EXAMPLE 1 More Derivatives of Logarithms of Functions

Compute the following derivatives:

a. $\frac{d}{dx}\ln\sqrt{x+1}$ **b.** $\frac{d}{dx}\ln[(1+x)(2-x)]$ **c.** $\frac{d}{dx}\ln|x|$

Solution

a. The calculation thought experiment tells us that we have the natural logarithm of a quantity, so

$$\frac{d}{dx}\ln\sqrt{x+1} = \frac{1}{\sqrt{x+1}}\frac{d}{dx}\sqrt{x+1} \qquad \frac{d}{dx}\ln u = \frac{1}{u}\frac{du}{dx}$$
$$= \frac{1}{\sqrt{x+1}} \cdot \frac{1}{2\sqrt{x+1}} \qquad \frac{d}{dx}\sqrt{u} = \frac{1}{2\sqrt{u}}\frac{du}{dx}$$
$$= \frac{1}{2(x+1)}.$$

Q: *What happened to the square root?*

A: As with many problems involving logarithms, we could have done this one differently and much more easily if we had simplified the expression $\ln\sqrt{x+1}$ using the properties of logarithms *before* differentiating. Doing this, we get the following.

Part (a) redone by simplifying first:

$$\ln\sqrt{x+1} = \ln(x+1)^{1/2} = \frac{1}{2}\ln(x+1). \qquad \text{Simplify the logarithm first.}$$

Thus,

$$\frac{d}{dx}\ln\sqrt{x+1} = \frac{d}{dx}\left[\frac{1}{2}\ln(x+1)\right]$$
$$= \frac{1}{2}\left(\frac{1}{x+1}\right)\cdot 1 = \frac{1}{2(x+1)}.$$

A *lot* easier!

b. This time, we simplify the expression $\ln[(1+x)(2-x)]$ before taking the derivative:

$$\ln[(1+x)(2-x)] = \ln(1+x) + \ln(2-x). \qquad \text{Simplify the logarithm first.}$$

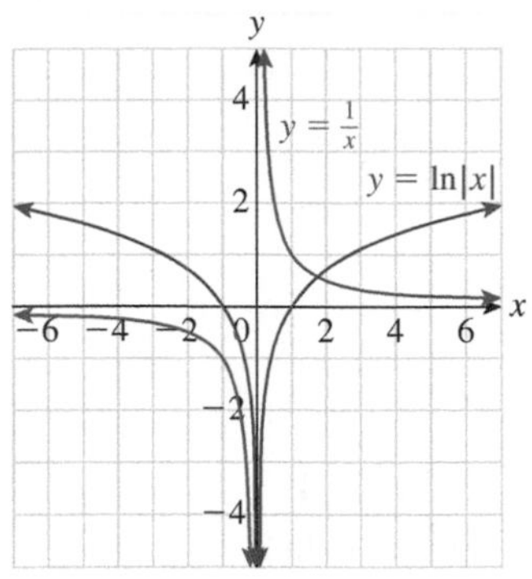

Figure 7(a)

Thus,

$$\frac{d}{dx}\ln[(1+x)(2-x)] = \frac{d}{dx}\ln(1+x) + \frac{d}{dx}\ln(2-x)$$

$$= \frac{1}{1+x} - \frac{1}{2-x}. \qquad \frac{d}{dx}\ln u = \frac{1}{u}\frac{du}{dx}$$

For practice, try doing this calculation without simplifying first. What other differentiation rule do you need to use?

c. Before we start, we note that $\ln x$ is defined only for positive values of x, so its domain is the set of positive real numbers. The domain of $\ln|x|$, on the other hand, is the set of *all* nonzero real numbers. For example, $\ln|-2| = \ln 2 \approx 0.6931$. For this reason, $\ln|x|$ often turns out to be more useful than the ordinary logarithm function.

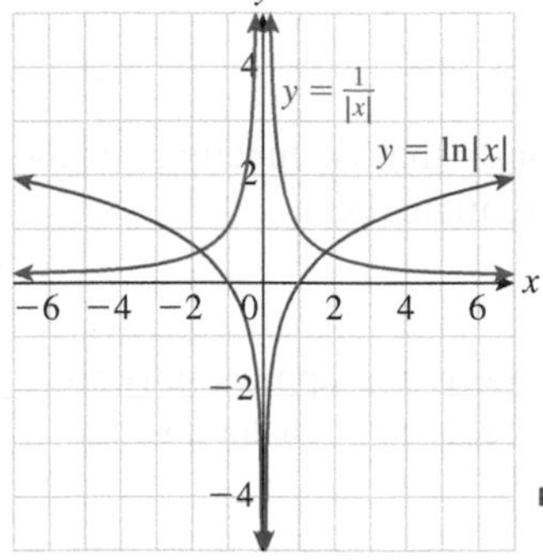

Figure 7(b)

$$\frac{d}{dx}\ln|x| = \frac{1}{|x|}\frac{d}{dx}|x| \qquad \frac{d}{dx}\ln u = \frac{1}{u}\frac{du}{dx}$$

$$= \frac{1}{|x|}\frac{|x|}{x} \qquad \text{Recall that } \frac{d}{dx}|x| = \frac{|x|}{x}.$$

$$= \frac{1}{x}$$

➡ **Before we go on . . .** Figure 7(a) shows the graphs of $y = \ln|x|$ and $y = 1/x$. Figure 7(b) shows the graphs of $y = \ln|x|$ and $y = 1/|x|$. You should be able to see from these graphs why the derivative of $\ln|x|$ is $1/x$ and not $1/|x|$. ■

This last example, in conjunction with the chain rule, gives us the following formulas.

Derivatives of Logarithms of Absolute Values

Original Rule	Generalized Rule	In Words
$\frac{d}{dx}\ln\|x\| = \frac{1}{x}$	$\frac{d}{dx}\ln\|u\| = \frac{1}{u}\frac{du}{dx}$	*The derivative of the natural logarithm of the absolute value of a quantity is* 1 *over that quantity, times the derivative of that quantity.*
$\frac{d}{dx}\log_b\|x\| = \frac{1}{x\ln b}$	$\frac{d}{dx}\log_b\|u\| = \frac{1}{u\ln b}\frac{du}{dx}$	*The derivative of the log to base b of the absolute value of a quantity is* 1 *over the product of* ln *b and that quantity, times the derivative of that quantity.*

Note Compare the above formulas with those in the box that precedes Example 1. They tell us that we can simply ignore the absolute values in $\ln|u|$ or $\log_b|u|$ when taking the derivative. ■

Quick Examples

7. $\frac{d}{dx}\ln|x^2-1| = \frac{1}{x^2-1}\frac{d}{dx}(x^2-1)$ $\quad u = x^2 - 1$

$= \frac{1}{x^2-1}(2x) = \frac{2x}{x^2-1}$

8. $\frac{d}{dx}\log_2|x^3+x| = \frac{1}{(x^3+x)\ln 2}\frac{d}{dx}(x^3+x)$ $\quad u = x^3 + x$

$= \frac{1}{(x^3+x)\ln 2}(3x^2+1) = \frac{3x^2+1}{(x^3+x)\ln 2}$

Derivatives of Exponential Functions

We now turn to the derivatives of *exponential* functions—that is, functions of the form $f(x) = b^x$. We begin by showing how *not* to differentiate them.

Caution The derivative of b^x is *not* xb^{x-1}. The power rule applies only to *constant* exponents. In this case the exponent is decidedly *not* constant, so the power rule does not apply. ■

The following shows the correct way of differentiating b^x, beginning with a special case.

Derivative of e^x

$$\frac{d}{dx}e^x = e^x$$

Quick Examples

9. $\frac{d}{dx}(3e^x) = 3\frac{d}{dx}e^x = 3e^x$ $\quad$ Constant multiple rule

10. $\frac{d}{dx}\left(\frac{e^x}{x}\right) = \frac{e^x x - e^x(1)}{x^2}$ $\quad$ Quotient rule

$= \frac{e^x(x-1)}{x^2}$

Thus, e^x has the amazing property that its derivative is itself!* For bases other than e we have the following generalization.

* There is another—very simple—function that is its own derivative. What is it?

Derivative of b^x

If b is any positive number, then

$$\frac{d}{dx}b^x = b^x \ln b.$$

Note that if $b = e$, we obtain the previous formula.

Quick Example

11. $\frac{d}{dx}3^x = 3^x \ln 3$

Derivation of the Formula $\frac{d}{dx}e^x = e^x$

To find the derivative of e^x, we use a shortcut.* Write $g(x) = e^x$. Then

$$\ln g(x) = x.$$

Take the derivative of both sides of this equation to get

$$\frac{g'(x)}{g(x)} = 1$$

or

$$g'(x) = g(x) = e^x.$$

In other words, the exponential function with base e is its own derivative. The rule for exponential functions with other bases follows from the equality $b^x = e^{x \ln b}$ (why?) and the chain rule. (Try it.)

* This shortcut is an example of a technique called *logarithmic differentiation,* which is occasionally useful. We will see it again in Section 4.6.

Derivatives of Exponentials of Functions

If we were to take the derivative of e raised to a *quantity*, not just x, we would need to use the chain rule, as follows.

Derivatives of Exponentials of Functions

Original Rule	Generalized Rule	In Words
$\frac{d}{dx}e^x = e^x$	$\frac{d}{dx}e^u = e^u \frac{du}{dx}$	*The derivative of e raised to a quantity is e raised to that quantity, times the derivative of that quantity.*
$\frac{d}{dx}b^x = b^x \ln b$	$\frac{d}{dx}b^u = b^u \ln b \frac{du}{dx}$	*The derivative of b raised to a quantity is b raised to that quantity, times* ln *b, times the derivative of that quantity.*

Quick Examples

12. $\frac{d}{dx}e^{x^2+1} = e^{x^2+1}\frac{d}{dx}(x^2+1)$ $u = x^2 + 1$ (See margin note.*)

$= e^{x^2+1}(2x) = 2x\, e^{x^2+1}$

* The calculation thought experiment tells us that we have *e* raised to a quantity.

13. $\frac{d}{dx}2^{3x} = 2^{3x}\ln 2\frac{d}{dx}(3x)$ $u = 3x$

$= 2^{3x}(\ln 2)(3) = (3\ln 2)2^{3x}$

14. $\frac{d}{dt}30e^{1.02t} = 30e^{1.02t}(1.02) = 30.6e^{1.02t}$ $u = 1.02t$

15. If \$1,000 is invested in an account earning 5% per year compounded continuously, then the rate of change of the account balance after *t* years is

$$\frac{d}{dt}(1{,}000e^{0.05t}) = 1{,}000(0.05)e^{0.05t} = 50e^{0.05t} \text{ dollars per year.}$$

Applications

EXAMPLE 2 Epidemics

In the early stages of the AIDS epidemic during the 1980s the number of cases in the United States was increasing by about 50% every 6 months. By the start of 1983 there were approximately 1,600 AIDS cases in the United States.[38] Had this trend continued, how many new cases per year would have been occurring by the start of 1993?

Solution To find the answer, we must first model this exponential growth using the methods of Chapter 2. Referring to Example 4 of Section 2.2, we find that *t* years after the start of 1983 the number of cases is

$$A = 1{,}600(2.25)^t.$$

We are asking for the number of new cases each year. In other words, we want the rate of change, dA/dt:

$$\frac{dA}{dt} = 1{,}600(2.25)^t \ln 2.25 \text{ cases per year.}$$

At the start of 1993, $t = 10$, so the number of new cases per year is

$$\left.\frac{dA}{dt}\right|_{t=10} = 1{,}600(2.25)^{10}\ln 2.25 \approx 4{,}300{,}000 \text{ cases per year.}$$

➡ **Before we go on . . .** In Example 2 the figure for the number of new cases per year is so large because we assumed that exponential growth—the 50% increase every 6 months—would continue. A more realistic model for the spread of a disease is the logistic model. (See Section 2.4 as well as the next example.) ■

[38] Data based on regression of 1982–1986 figures. Source for data: Centers for Disease Control and Prevention. HIV/AIDS Surveillance Report, 2000;12 (No. 2).

EXAMPLE 3 **Sales Growth**

The sales of the *Cyberpunk II* video game can be modeled by the logistic curve

$$q(t) = \frac{10{,}000}{1 + 0.5e^{-0.4t}}$$

where $q(t)$ is the total number of units sold t months after its introduction. How fast is the game selling 2 years after its introduction?

Solution We are asked for $q'(24)$. We can find the derivative of $q(t)$ using the quotient rule, or we can first write

$$q(t) = 10{,}000(1 + 0.5e^{-0.4t})^{-1}$$

and then use the generalized power rule:

$$\begin{aligned} q'(t) &= -10{,}000(1 + 0.5e^{-0.4t})^{-2}(0.5e^{-0.4t})(-0.4) \\ &= \frac{2{,}000e^{-0.4t}}{(1 + 0.5e^{-0.4t})^2}. \end{aligned}$$

Thus,

$$q'(24) = \frac{2{,}000e^{-0.4(24)}}{(1 + 0.5e^{-0.4(24)})^2} \approx 0.135 \text{ units per month.}$$

So after 2 years, sales are quite slow.

Application to Limits

We can now apply l'Hospital's rule and the derivatives of exponential functions to evaluate some limits of a kind we computed numerically in the exercises in Section 3.1.

EXAMPLE 4 **Exponential Functions and L'Hospital's Rule**

Evaluate the following limits using l'Hospital's rule:

a. $\lim\limits_{x \to +\infty} \dfrac{x}{e^x}$ **b.** $\lim\limits_{x \to -\infty} (x^2 + 2x)e^x$

Solution

a. This limit has the indeterminate form ∞/∞, so we can try to apply l'Hospital's rule:

$$\lim_{x \to +\infty} \frac{x}{e^x} = \lim_{x \to +\infty} \frac{1}{e^x} = 0.$$ Apply l'Hospital's rule: $\frac{d}{dx}(x) = 1, \frac{d}{dx}e^x = e^x$

b. We first rewrite the limit as a quotient:

$$\lim_{x \to -\infty} (x^2 + 2x)e^x = \lim_{x \to -\infty} \frac{x^2 + 2x}{e^{-x}}.$$

Because $x \to -\infty$, this has the indeterminate form ∞/∞, so we can apply l'Hospital's rule:

$$\lim_{x\to-\infty} \frac{x^2+2x}{e^{-x}} = \lim_{x\to-\infty} \frac{2x+2}{-e^{-x}} \quad \text{Still indeterminate}$$

$$= \lim_{x\to-\infty} \frac{2}{e^{-x}} \quad \text{Apply l'Hospital's rule again.}$$

$$= 0.$$

4.5 EXERCISES

▼ more advanced ◆ challenging

T indicates exercises that should be solved using technology

In Exercises 1–66, find the derivative of the function. [**HINT**: See Quick Examples 5–14.]

1. $f(x) = \ln(x-1)$
2. $f(x) = \ln(x+3)$
3. $g(x) = \ln|x^2+3|$
4. $g(x) = \ln|2x-4|$
5. $f(x) = \log_2 x$
6. $f(x) = \log_3 x$
7. $h(x) = \log_2(x+1)$
8. $h(x) = \log_3(x^2+x)$
9. $r(t) = \log_3(t+1/t)$
10. $r(t) = \log_3(t+\sqrt{t})$
11. $h(x) = e^{x+3}$
12. $h(x) = e^{x^2}$
13. $h(x) = e^{x^2-x+1}$
14. $h(x) = e^{2x^2-x+1/x}$
15. $r(x) = (e^{2x-1})^2$
16. $r(x) = (e^{2x^2})^3$
17. $g(x) = 4^x$
18. $g(x) = 5^x$
19. $h(x) = 2^{x^2-1}$
20. $h(x) = 3^{x^2-x}$
21. $f(x) = (x^2+1)\ln x$
22. $f(x) = (4x^2-x)\ln x$
23. $f(x) = (x^2+1)^5\ln x$
24. $f(x) = (x+1)^{0.5}\ln x$
25. $g(x) = \ln|2x^2+1|$
26. $g(x) = \ln|x^2-x|$
27. $g(x) = \ln(x^2-2.1x^{0.3})$
28. $g(x) = \ln(x-3.1x^{-1})$
29. $h(x) = \ln[(-2x+1)(x+1)]$ [**HINT**: See Example 1(b).]
30. $h(x) = \ln[(3x+1)(-x+1)]$ [**HINT**: See Example 1(b).]
31. $h(x) = \ln\left(\frac{3x+1}{4x-2}\right)$
32. $h(x) = \ln\left(\frac{9x}{4x-2}\right)$
33. $r(x) = \ln\left|\frac{(x+1)(x-3)}{-2x-9}\right|$
34. $r(x) = \ln\left|\frac{-x+1}{(3x-4)(x-9)}\right|$
35. $s(x) = \ln[(4x-2)^{1.3}]$ [**HINT**: See Example 1(a).]
36. $s(x) = \ln[(x-8)^{-2}]$ [**HINT**: See Example 1(a).]
37. $s(x) = \ln\left|\frac{(x+1)^2}{(3x-4)^3(x-9)}\right|$
38. $s(x) = \ln\left|\frac{(x+1)^2(x-3)^4}{2x+9}\right|$
39. $f(x) = (\ln|x|)^2$
40. $f(x) = \frac{1}{\ln|x|}$
41. $r(x) = \ln(x^2) - [\ln(x-1)]^2$
42. $r(x) = (\ln(x^2))^2$
43. $f(x) = xe^x$
44. $f(x) = 2e^x - x^2e^x$
45. $r(x) = \ln(x+1) + 3x^3e^x$
46. $r(x) = \ln|x+e^x|$
47. $f(x) = e^x\ln|x|$
48. $f(x) = e^x\log_2|x|$
49. $s(x) = x^2e^{2x-1}$
50. $s(x) = \frac{e^{4x-1}}{x^3-1}$
51. $v(x) = 3^{2x+1} + e^{3x+1}$
52. $v(x) = e^{2x}4^{2x}$
53. $u(x) = \frac{3^{x^2}}{x^2+1}$
54. $u(x) = (x^2+1)4^{x^2-1}$
55. $g(x) = \frac{e^x+e^{-x}}{e^x-e^{-x}}$
56. $g(x) = \frac{1}{e^x+e^{-x}}$
57. ▼ $g(x) = e^{3x-1}e^{x-2}e^x$
58. ▼ $g(x) = e^{-x+3}e^{2x-1}e^{-x+11}$
59. ▼ $f(x) = \frac{1}{x\ln x}$
60. ▼ $f(x) = \frac{e^{-x}}{xe^x}$
61. ▼ $f(x) = [\ln(e^x)]^2 - \ln[(e^x)^2]$
62. ▼ $f(x) = e^{\ln x} - e^{2\ln(x^2)}$
63. ▼ $f(x) = \ln|\ln x|$
64. ▼ $f(x) = \ln|\ln|\ln x||$
65. ▼ $s(x) = \ln\sqrt{\ln x}$
66. ▼ $s(x) = \sqrt{\ln(\ln x)}$

In Exercises 67–72, find the equation of the straight line described. Use graphing technology to check your answers by plotting the given curve together with the tangent line.

67. Tangent to $y = e^x\log_2 x$ at the point $(1, 0)$
68. Tangent to $y = e^x + e^{-x}$ at the point $(0, 2)$
69. Tangent to $y = \ln\sqrt{2x+1}$ at the point where $x = 0$

70. Tangent to $y = \ln\sqrt{2x^2 + 1}$ at the point where $x = 1$

71. At right angles to $y = e^{x^2}$ at the point where $x = 1$

72. At right angles to $y = \log_2(3x + 1)$ at the point where $x = 1$

In Exercises 73–78, use l'Hospital's rule to find the limits. [**HINT**: See Example 4.]

73. $\lim_{x\to+\infty} \dfrac{x+2}{e^x}$

74. $\lim_{x\to+\infty} \dfrac{x^2+x+1}{e^x}$

75. $\lim_{x\to-\infty} \dfrac{2x+3}{e^{-2x}}$

76. $\lim_{x\to-\infty} \dfrac{x^2-2x+1}{e^{-3x}}$

77. $\lim_{x\to0} \dfrac{e^x-1}{x}$

78. $\lim_{x\to0} \dfrac{e^x-1-x}{x^2}$

Applications

79. ***Research and Development: Industry*** The total spent on research and development by industry in the United States during 2002–2012 can be approximated by

$$S(t) = 29 \ln t + 164 \text{ billion dollars} \quad (2 \le t \le 12),$$

where t is the year since 2000.[39] What was the total spent in 2010 ($t = 10$), and how fast was it increasing? [**HINT**: See Quick Examples 1 and 2.]

80. ***Research and Development: Federal*** The total spent on research and development by the federal government in the United States during 2002–2012 can be approximated by

$$S(t) = 3.1 \ln t + 22 \text{ billion dollars} \quad (2 \le t \le 12),$$

where t is the year since 2000.[40] What was the total spent in 2005 ($t = 5$), and how fast was it increasing? [**HINT**: See Quick Examples 1 and 2.]

81. ***Research and Development: Industry*** (Refer to Exercise 79.) The function $S(t)$ in Exercise 79 can also be written (approximately) as

$$S(t) = 29 \ln(286t + 2{,}860) \text{ billion dollars} \quad (-8 \le t \le 2),$$

where this time t is the year since 2010. Use this alternative formula to estimate the amount spent in 2010 and its rate of change, and check your answers by comparing them with those in Exercise 79.

82. ***Research and Development: Federal*** (Refer to Exercise 80.) The function $S(t)$ in Exercise 80 can also be written (approximately) as

$$S(t) = 3.1 \ln(1{,}210t + 12{,}100) \text{ billion dollars} \quad (-8 \le t \le 2),$$

where this time t is the year since 2010. Use this alternative formula to estimate the amount spent in 2005 and its rate of change, and check your answers by comparing them with those in Exercise 80.

83. ▼ ***Carbon Dating*** The age in years of a specimen that originally contained 10 grams of carbon 14 is given by

$$y = \log_{0.999879}(0.1x),$$

where x is the amount of carbon 14 it currently contains. Compute $\left.\dfrac{dy}{dx}\right|_{x=5}$, and interpret your answer. [**HINT**: For the calculation, see Quick Examples 3 and 4.]

84. ▼ ***Iodine Dating*** The age in years of a specimen that originally contained 10 grams of iodine 131 is given by

$$y = \log_{0.999567}(0.1x),$$

where x is the amount of iodine 131 it currently contains. Compute $\left.\dfrac{dy}{dx}\right|_{x=8}$, and interpret your answer. [**HINT**: For the calculation, see Quick Examples 3 and 4.]

85. ***New York City Housing Costs: Downtown*** The average price of a two-bedroom apartment in downtown New York City during the real estate boom from 1994 to 2004 can be approximated by

$$p(t) = 0.33e^{0.16t} \text{ million dollars} \quad (0 \le t \le 10),$$

where t is time in years. ($t = 0$ represents 1994.)[41] What was the average price of a two-bedroom apartment in downtown New York City in 2003, and how fast was the price increasing? (Round your answers to two significant digits.) [**HINT**: See Quick Example 14.]

86. ***New York City Housing Costs: Uptown*** The average price of a two-bedroom apartment in uptown New York City during the real estate boom from 1994 to 2004 can be approximated by

$$p(t) = 0.14e^{0.10t} \text{ million dollars} \quad (0 \le t \le 10),$$

where t is time in years. ($t = 0$ represents 1994.)[42] What was the average price of a two-bedroom apartment in uptown New York City in 2002, and how fast was the price increasing? (Round your answers to two significant digits.) [**HINT**: See Quick Example 14.]

---

[39] Constant 2005 dollars; excludes federal funding; 2012 data is preliminary. Source: National Science Foundation, National Center for Science and Engineering Statistics, December 2013, *National Patterns of R&D Resources: 2011–12 Data Update.* NSF 14-304 (www.nsf.gov/statistics/nsf14304).

[40] Constant 2005 dollars; excludes federal funding to industry and nonprofit organizations; 2012 data is preliminary. Source: National Science Foundation, National Center for Science and Engineering Statistics, December 2013, *National Patterns of R&D Resources: 2011–12 Data Update.* NSF 14-304 (www.nsf.gov/statistics/nsf14304).

[41] Model is based on an exponential regression. Source for data: Miller Samuel/*New York Times*, March 28, 2004, p. RE 11.

[42] *Ibid.*

87. ***Big Brother*** The following chart shows the total number of wiretaps authorized each year by U.S. state and federal courts from 1990 to 2013. ($t = 0$ represents 1990.)[43]

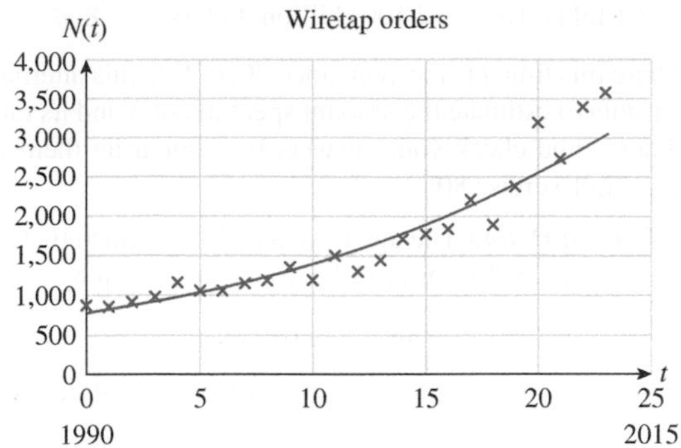

These data can be approximated by the model

$$N(t) = 770e^{0.060t} \quad (0 \le t \le 23).$$

a. Find $N(15)$ and $N'(15)$. Be sure to state the units of measurement. To how many significant digits should we round the answers? Why?

b. The number of people whose communications are intercepted averages around 100 per wiretap order. What does the answer to part (a) tell you about the number of people whose communications were intercepted?[44]

c. According to the model, the number of wiretap orders each year (choose one)
(A) increased at a linear rate
(B) decreased at a quadratic rate
(C) increased at an exponential rate
(D) increased at a logarithmic rate
over the period shown.

88. ***Big Brother*** The following chart shows the total number of wiretaps authorized each year by U.S. state courts from 1990 to 2013. ($t = 0$ represents 1990.)[45]

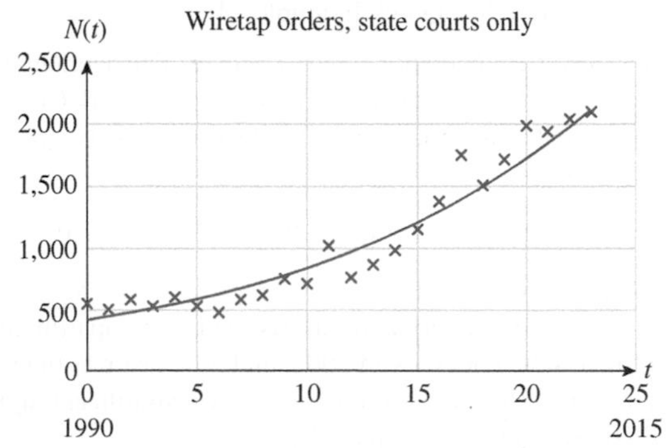

These data can be approximated by the model

$$N(t) = 410e^{0.071t} \quad (0 \le t \le 23).$$

a. Find $N(20)$ and $N'(20)$. Be sure to state the units of measurement. To how many significant digits should we round the answers? Why?

b. The number of people whose communications are intercepted averages around 100 per wiretap order. What does the answer to part (a) tell you about the number of people whose communications were intercepted?[46]

c. According to the model, the number of wiretap orders each year (choose one)
(A) increased at a linear rate
(B) decreased at a quadratic rate
(C) increased at an exponential rate
(D) increased at a logarithmic rate
over the period shown.

89. ***Investments*** If \$10,000 is invested in a savings account offering 4% per year, compounded continuously, how fast is the balance growing after 3 years? [**HINT**: See Quick Example 15.]

90. ***Investments*** If \$20,000 is invested in a savings account offering 3.5% per year, compounded continuously, how fast is the balance growing after 3 years? [**HINT**: See Quick Example 15.]

91. ***Investments*** If \$10,000 is invested in a savings account offering 4% per year, compounded semiannually, how fast is the balance growing after 3 years?

92. ***Investments*** If \$20,000 is invested in a savings account offering 3.5% per year, compounded semiannually, how fast is the balance growing after 3 years?

93. ***The 2003 SARS Outbreak*** In the early stages of the deadly SARS (severe acute respiratory syndrome) epidemic in 2003, the number of cases was increasing by about 18% each day.[47] On March 17, 2003 (the first day for which statistics were reported by the World Health Organization), there were 167 cases. Find an exponential model that predicts the number of people infected t days after March 17, 2003, and use it to estimate how fast the epidemic was spreading on March 31, 2003. (Round your answer to the nearest whole number of new cases per day.) [**HINT**: See Example 2.]

94. ***The 2003 SARS Outbreak*** A few weeks into the deadly SARS (severe acute respiratory syndrome) epidemic in 2003, the number of cases was increasing by about 4% each day.[48] On April 1, 2003, there were 1,804 cases. Find an exponential model that predicts the number $A(t)$ of people infected t days after April 1, 2003, and use it to estimate

[43] Source for data: Wiretap Reports, Administrative Office of the United States Courts, www.uscourts.gov/Statistics/WiretapReports/wiretap-report-2013.aspx.

[44] Assume that there is no significant overlap between the people whose communications are intercepted in different wiretap orders.

[45] See footnote 41.

[46] See footnote 44.

[47] Source: World Health Organization, www.who.int.

[48] *Ibid.*

how fast the epidemic was spreading on April 30, 2003. (Round your answer to the nearest whole number of new cases per day.) [**HINT**: See Example 2.]

95. ▼ ***The 2014 Ebola Outbreak*** In the first 6 months of the 2014 Ebola outbreak, the total number of reported cases was increasing exponentially with a monthly growth constant of 72%.[49] There were about 100 cases as of April 1, 2014. Find an exponential model in the form $C(t) = Ae^{rt}$ for the number of cases t months after April 1, 2014, and use it to estimate how fast the number of cases was increasing on August 1, 2014. (Round your answer to the nearest 10 new cases per month.)

96. ▼ ***The 2014 Ebola Outbreak*** In the first 6 months of the 2014 Ebola outbreak, the total number of reported deaths was increasing exponentially with a monthly growth constant of 60%.[50] There were about 90 deaths as of April 1, 2014. Find an exponential model in the form $D(t) = Ae^{rt}$ for the number of deaths t months after April 1, 2014, and use it to estimate how fast the number of deaths was increasing on October 1, 2014. (Round your answer to the nearest 10 new deaths per month.)

97. ▼ ***SAT Scores by Income*** The following bar graph shows U.S. math SAT scores as a function of household income:[51]

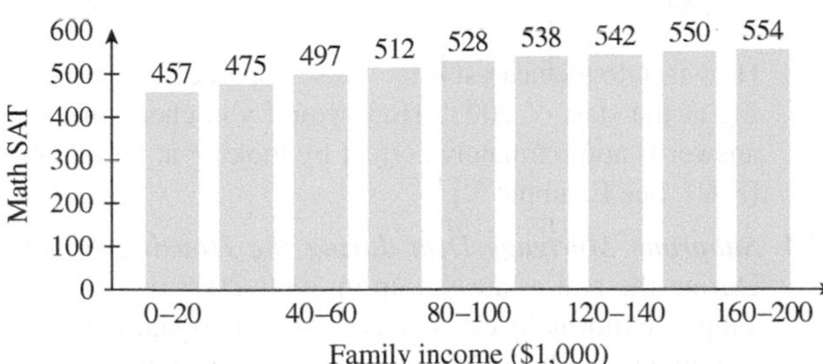

a. Which of the following best models the data (C is a constant)?

(A) $S(x) = C - 133e^{-0.0131x}$
(B) $S(x) = C + 133e^{-0.0131x}$
(C) $S(x) = C + 133e^{0.0131x}$
(D) $S(x) = C - 133e^{0.0131x}$

($S(x)$ is the average math SAT score of students whose household income is x thousand dollars per year.)

b. Use $S'(x)$ to predict how a student's math SAT score is affected by a \$1,000 increase in parents' income for a student whose parents earn \$45,000.

c. Does $S'(x)$ increase or decrease as x increases? Interpret your answer.

98. ***SAT Scores by Income*** The following bar graph shows U.S. critical reading SAT scores as a function of household income:[52]

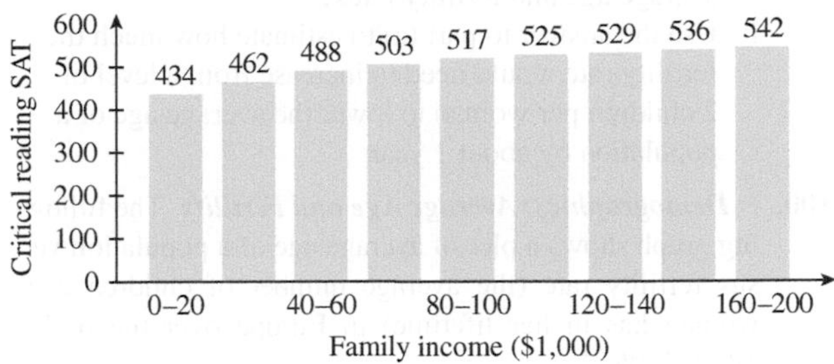

a. Which of the following best models the data (C is a constant)?

(A) $S(x) = C + \dfrac{1}{136e^{0.015x}}$

(B) $S(x) = C - 136e^{0.015x}$

(C) $S(x) = C - \dfrac{136}{e^{0.015x}}$

(D) $S(x) = C - \dfrac{e^{0.015x}}{136}$

($S(x)$ is the average critical reading SAT score of students whose household income is x thousand dollars per year.)

b. Use $S'(x)$ to predict how a student's critical reading SAT score is affected by a \$1,000 increase in parents' income for a student whose parents earn \$45,000.

c. Does $S'(x)$ increase or decrease as x increases? Interpret your answer.

99. ▼ ***Demographics: Average Age and Fertility*** The following graph shows a plot of average age of a population versus fertility rate (the average number of children each woman has in her lifetime) in the United States and Europe over the period 1950–2005:[53]

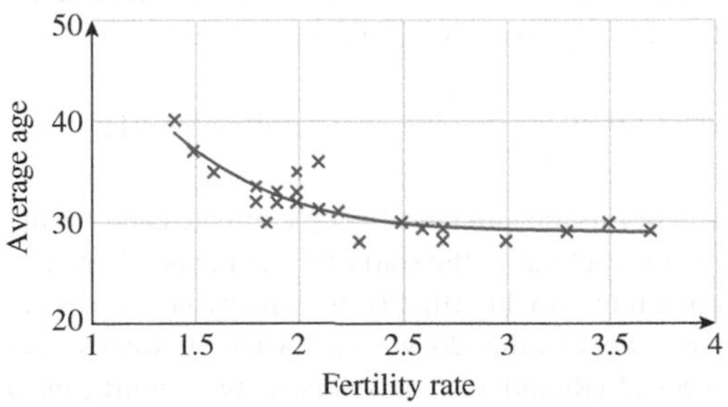

The equation of the accompanying curve is

$$a = 28.5 + 120(0.172)^x \quad (1.4 \le x \le 3.7),$$

[49] Exponential model is the authors'. Source for data: Wikipedia/Centers for Disease Control and Prevention/WHO.

[50] *Ibid.*

[51] 2009 data. Source: College Board/*New York Times*, http://economix.blogs.nytimes.com.

[52] *Ibid.*

[53] The separate data for Europe and the United States are collected in the same graph. 2005 figures are estimates. Source: United Nations World Population Division/*New York Times*, June 29, 2003, p. 3.

where a is the average age (in years) of the population and x is the fertility rate.

a. Compute $a'(2)$. What does the answer tell you about average age and fertility rates?

b. Use the answer to part (a) to estimate how much the fertility rate would need to increase from a level of 2 children per woman to lower the average age of a population by about 1 year.

100. ▼ ***Demographics: Average Age and Fertility*** The following graph shows a plot of average age of a population versus fertility rate (the average number of children each woman has in her lifetime) in Europe over the period 1950–2005:[54]

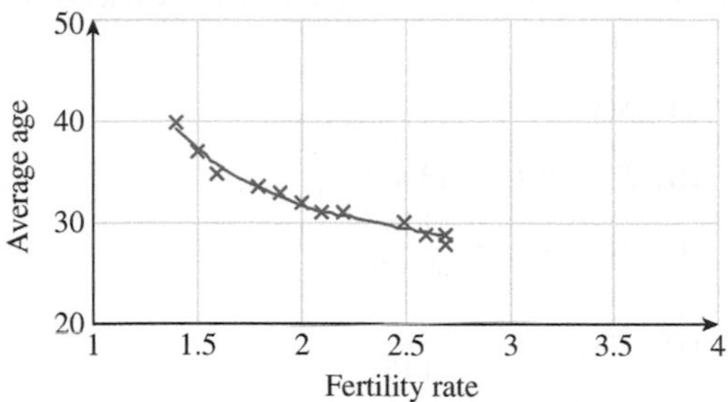

The equation of the accompanying curve is

$$g = 27.6 + 128(0.181)^x \quad (1.4 \le x \le 3.7),$$

where g is the average age (in years) of the population and x is the fertility rate.

a. Compute $g'(2.5)$. What does the answer tell you about average age and fertility rates?

b. Referring to the model that combines the data for Europe and the United States in Exercise 99, which population's average age is affected more by a changing fertility rate at the level of 2.5 children per woman?

101. ***Epidemics*** A flu epidemic described in Example 1 in Section 2.4 approximately followed the curve

$$P = \frac{150}{1 + 15{,}000e^{-0.35t}} \text{ million people,}$$

where P is the number of people infected and t is the number of weeks after the start of the epidemic. How fast is the epidemic growing (that is, how many new cases are there each week) after 20 weeks? After 30 weeks? After 40 weeks? (Round your answers to two significant digits.) [**HINT**: See Example 3.]

102. ***Epidemics*** Another epidemic follows the curve

$$P = \frac{200}{1 + 20{,}000e^{-0.549t}} \text{ million people,}$$

where P is the number of people infected and t is in years. How fast is the epidemic growing after 10 years? After 20 years? After 30 years? (Round your answers to two significant digits.) [**HINT**: See Example 3.]

103. ***Subprime Mortgages during the Housing Bubble*** During the real estate run-up in 2000–2008 the percentage of mortgages issued in the United States that were subprime (normally classified as risky) could be approximated by

$$A(t) = \frac{15.0}{1 + 8.6e^{-0.59t}} \text{ percent} \quad (0 \le t \le 8)$$

t years after the start of 2000.[55]

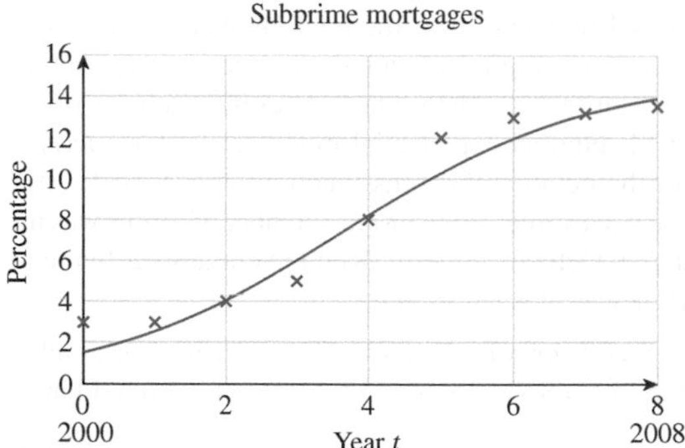

How fast, to the nearest 0.1%, was the percentage increasing at the start of 2003? How would you check that the answer is approximately correct by looking at the graph? [**HINT**: See Example 3.]

104. ***Subprime Mortgage Debt during the Housing Bubble*** During the real estate run-up in 2000–2008 the value of subprime (normally classified as risky) mortgage debt outstanding in the United States was approximately

$$A(t) = \frac{1{,}350}{1 + 4.2e^{-0.53t}} \text{ billion dollars} \quad (0 \le t \le 8)$$

t years after the start of 2000.[56]

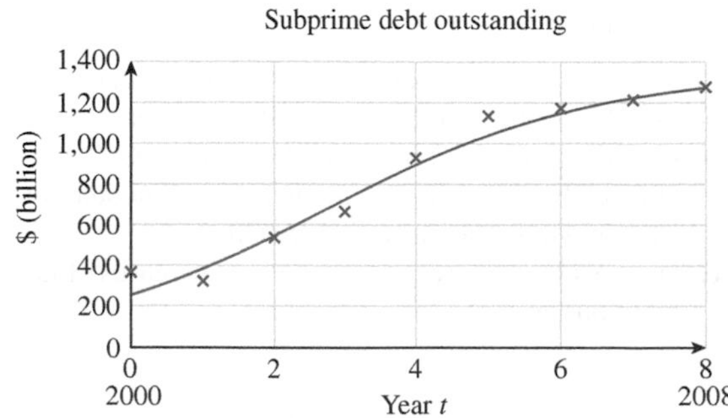

[54] All European countries including the Russian Federation. 2005 figures are estimates. Source: See footnote for Exercise 99.

[55] 2009 figure is an estimate. Sources: Mortgage Bankers Association, UBS.

[56] 2008–2009 figures are estimates. Source: www.data360.org.

How fast, to the nearest \$1 billion, was subprime mortgage debt increasing at the start of 2005? How would you check that the answer is approximately correct by looking at the graph? [**HINT:** See Example 3.]

105. ***Subprime Mortgages during the Housing Bubble*** (Compare Exercise 103.) During the real estate run-up in 2000–2008 the percentage of mortgages issued in the United States that were subprime (normally classified as risky) could be approximated by

$$A(t) = \frac{15.0}{1 + 8.6(1.8)^{-t}} \text{ percent} \quad (0 \le t \le 8)$$

t years after the start of 2000.[57]

a. How fast, to the nearest 0.1%, was the percentage increasing at the start of 2003?

b. Compute $\lim_{t \to +\infty} A(t)$ and $\lim_{t \to +\infty} A'(t)$. What do the answers tell you about subprime mortgages?

106. ***Subprime Mortgage Debt during the Housing Bubble*** (Compare Exercise 104.) During the real estate run-up in 2000–2008 the value of subprime (normally classified as risky) mortgage debt outstanding in the United States could be approximated by

$$A(t) = \frac{1{,}350}{1 + 4.2(1.7)^{-t}} \text{ billion dollars} \quad (0 \le t \le 8)$$

t years after the start of 2000.[58]

a. How fast, to the nearest \$1 billion, was subprime mortgage debt increasing at the start of 2005?

b. Compute $\lim_{t \to +\infty} A(t)$ and $\lim_{t \to +\infty} A'(t)$. What do the answers tell you about subprime mortgages?

107. ▼ ***Population Growth*** The population of Lower Anchovia was 4,000,000 at the start of 2010 and was doubling every 10 years. How fast was it growing per year at the start of 2010? (Round your answer to three significant digits.) [**HINT:** Use the method of Example 2 of Section 2.2 to obtain an exponential model for the population.]

108. ▼ ***Population Growth*** The population of Upper Anchovia was 3,000,000 at the start of 2011 and doubling every 7 years. How fast was it growing per year at the start of 2011? (Round your answer to three significant digits.) [**HINT:** Use the method of Example 2 of Section 2.2 to obtain an exponential model for the population.]

109. ▼ ***Radioactive Decay*** Plutonium 239 has a half-life of 24,400 years. How fast is a lump of 10 grams decaying after 100 years?

110. ▼ ***Radioactive Decay*** Carbon 14 has a half-life of 5,730 years. How fast is a lump of 20 grams decaying after 100 years?

111. ◆ ***Cellphone Revenues*** The number of cellphone subscribers in China for the period 2000–2005 was projected to follow the equation[59]

$$N(t) = 39t + 68 \text{ million subscribers}$$

in year t. ($t = 0$ represents 2000.) The average annual revenue per cellphone user was \$350 in 2000. Assuming that, because of competition, the revenue per cellphone user decreases exponentially with an annual decay constant of 10%, give a formula for the annual revenue in year t. Hence, project the annual revenue and its rate of change in 2002. Round all answers to the nearest billion dollars or billion dollars per year.

112. ◆ ***Cellphone Revenues*** The annual revenue for cellphone use in China for the period 2000–2005 was projected to follow the equation[60]

$$R(t) = 14t + 24 \text{ billion dollars}$$

in year t. ($t = 0$ represents 2000.) At the same time, there were approximately 68 million subscribers in 2000. Assuming that the number of subscribers increases exponentially with an annual growth constant of 10%, give a formula for the annual revenue per subscriber in year t. Hence, project to the nearest dollar the annual revenue per subscriber and its rate of change in 2002. (Be careful with units!)

Communication and Reasoning Exercises

113. Complete the following: The derivative of e raised to a glob is . . . .

114. Complete the following: The derivative of the natural logarithm of a glob is . . . .

115. Complete the following: The derivative of 2 raised to a glob is . . . .

116. Complete the following: The derivative of the base 2 logarithm of a glob is . . . .

117. What is wrong with the following?

$$\frac{d}{dx} \ln|3x + 1| = \frac{3}{|3x + 1|} \quad \text{✗ } \textit{WRONG!}$$

118. What is wrong with the following?

$$\frac{d}{dx} 2^{2x} = (2)2^{2x} \quad \text{✗ } \textit{WRONG!}$$

119. What is wrong with the following?

$$\frac{d}{dx} 3^{2x} = (2x)3^{2x-1} \quad \text{✗ } \textit{WRONG!}$$

---

[57] 2009 figure is an estimate. Sources: Mortgage Bankers Association, UBS.

[58] 2008–2009 figures are estimates. Source: www.data360.org.

[59] Based on a regression of projected figures (coefficients are rounded). Source: Intrinsic Technology/*New York Times*, Nov. 24, 2000, p. C1.

[60] Not allowing for discounting due to increased competition. Source: *Ibid.*

120. What is wrong with the following?

$$\frac{d}{dx}\ln(3x^2 - 1) = \frac{1}{6x} \quad ✗ \textit{ WRONG!}$$

121. ▼ The number N of music downloads on campus is growing exponentially with time. Can $N'(t)$ grow linearly with time? Explain.

122. ▼ The number N of graphing calculators sold on campus is decaying exponentially with time. Can $N'(t)$ grow with time? Explain.

*The **percentage rate of change** or **fractional rate of change** of a function is defined to be the ratio $f'(x)/f(x)$. (It is customary to express this as a percentage when speaking about percentage rate of change.)*

123. ◆ Show that the fractional rate of change of the exponential function e^{kx} is equal to the growth constant k, which is often called its **fractional growth rate**.

124. ◆ Show that the fractional rate of change of $f(x)$ is the rate of change of $\ln(f(x))$.

125. ◆ Let $A(t)$ represent a quantity growing exponentially. Show that the percentage rate of change, $A'(t)/A(t)$, is constant.

126. ◆ Let $A(t)$ be the amount of money in an account that pays interest that is compounded some number of times per year. Show that the percentage rate of growth, $A'(t)/A(t)$, is constant. What might this constant represent?

4.6 Implicit Differentiation

Implicit Functions

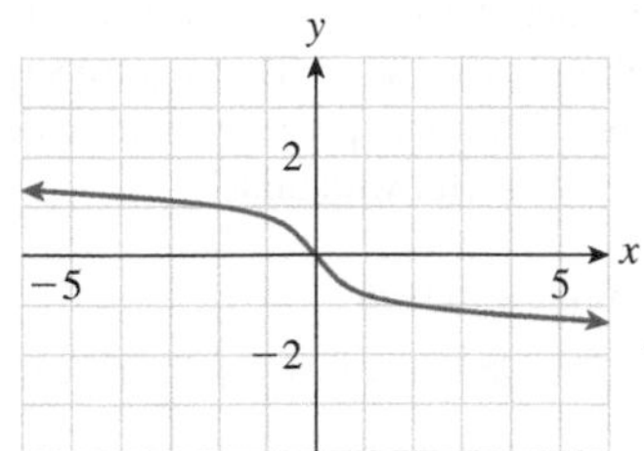

Figure 8

Consider the equation $y^5 + y + x = 0$, whose graph is shown in Figure 8. How did we obtain this graph? We did not solve for y as a function of x; that is impossible. In fact, we solved for x in terms of y to find points to plot. Nonetheless, the graph in Figure 8 is the graph of a function because it passes the vertical line test: Every vertical line crosses the graph no more than once, so for each value of x there is no more than one corresponding value of y. Because we cannot solve for y explicitly in terms of x, we say that the equation $y^5 + y + x = 0$ determines y as an **implicit function** of x.

Now, suppose we want to find the slope of the tangent line to this curve at, say, the point $(2, -1)$ (which, you should check, is a point on the curve). In the following example we find, surprisingly, that it is possible to obtain a formula for dy/dx without having to first solve the equation for y.

Implicit Differentiation

EXAMPLE 1 Implicit Differentiation

Find $\dfrac{dy}{dx}$, given that $y^5 + y + x = 0$.

Solution We use the chain rule and a little cleverness. Think of y as a function of x and take the derivative with respect to x of both sides of the equation:

$$y^5 + y + x = 0 \qquad \text{Original equation}$$

$$\frac{d}{dx}(y^5 + y + x) = \frac{d}{dx}(0) \qquad \text{Derivative with respect to } x \text{ of both sides}$$

$$\frac{d}{dx}(y^5) + \frac{d}{dx}(y) + \frac{d}{dx}(x) = 0 \qquad \text{Derivative rules}$$

Now we must be careful. The derivative *with respect to* x of y^5 is *not* $5y^4$. Rather, because y is a function of x, we must use the chain rule, which tells us that

$$\frac{d}{dx}(y^5) = 5y^4\frac{dy}{dx}.$$

Thus, we get

$$5y^4 \frac{dy}{dx} + \frac{dy}{dx} + 1 = 0.$$

We want to find dy/dx, so we *solve for it*:

$$(5y^4 + 1)\frac{dy}{dx} = -1 \qquad \text{Isolate } dy/dx \text{ on one side.}$$

$$\frac{dy}{dx} = -\frac{1}{5y^4 + 1}. \qquad \text{Divide both sides by } 5y^4 + 1.$$

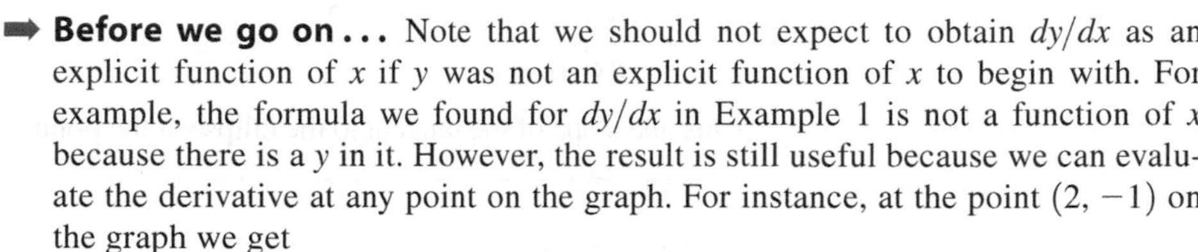

Before we go on... Note that we should not expect to obtain dy/dx as an explicit function of x if y was not an explicit function of x to begin with. For example, the formula we found for dy/dx in Example 1 is not a function of x because there is a y in it. However, the result is still useful because we can evaluate the derivative at any point on the graph. For instance, at the point $(2, -1)$ on the graph we get

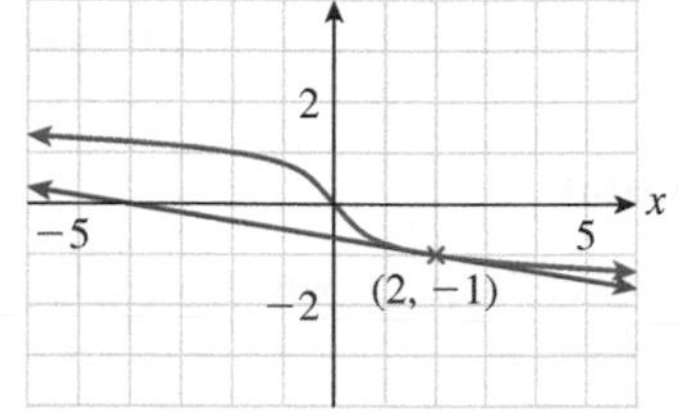

Figure 9

$$\frac{dy}{dx} = -\frac{1}{5y^4 + 1} = -\frac{1}{5(-1)^4 + 1} = -\frac{1}{6}.$$

Thus, the slope of the tangent line to the curve $y^5 + y + x = 0$ at the point $(2, -1)$ is $-1/6$. Figure 9 shows the graph and this tangent line. ■

This procedure we just used—differentiating an equation to find dy/dx without first solving the equation for y—is called **implicit differentiation**.

In Example 1 we were given an equation in x and y that determined y as an (implicit) function of x, even though we could not solve for y. But an equation in x and y need not always determine y as a function of x. Consider, for example, the equation

$$2x^2 + y^2 = 2.$$

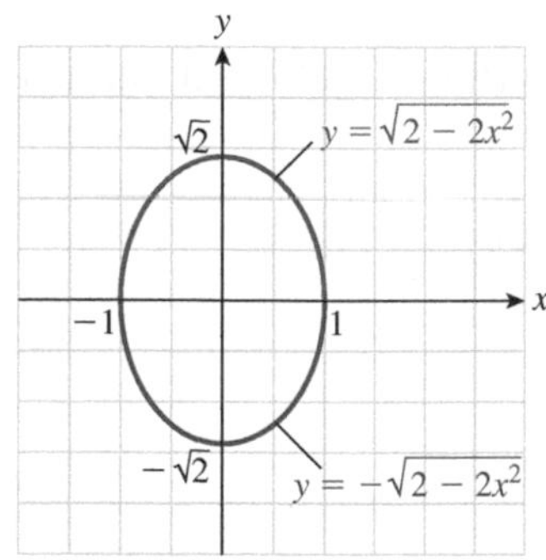

Figure 10

Solving for y yields $y = \pm\sqrt{2 - 2x^2}$. The $\pm$ sign reminds us that for some values of x there are two corresponding values for y. We can graph this equation by superimposing the graphs of

$$y = \sqrt{2 - 2x^2} \quad \text{and} \quad y = -\sqrt{2 - 2x^2}.$$

The graph, an *ellipse*, is shown in Figure 10. The graph of $y = \sqrt{2 - 2x^2}$ constitutes the top half of the ellipse, and the graph of $y = -\sqrt{2 - 2x^2}$ constitutes the bottom half.

EXAMPLE 2 **Slope of Tangent Line**

Refer to Figure 10. Find the slope of the tangent line to the ellipse $2x^2 + y^2 = 2$ at the point $(1/\sqrt{2}, 1)$.

Solution Because $(1/\sqrt{2}, 1)$ is on the top half of the ellipse in Figure 10, we *could* differentiate the function $y = \sqrt{2 - 2x^2}$ to obtain the result, but it is actually easier to apply implicit differentiation to the original equation.

$$2x^2 + y^2 = 2 \qquad \text{Original equation}$$

$$\frac{d}{dx}(2x^2 + y^2) = \frac{d}{dx}(2) \qquad \text{Derivative with respect to } x \text{ of both sides}$$

$$4x + 2y\frac{dy}{dx} = 0$$

$$2y\frac{dy}{dx} = -4x \qquad \text{Solve for } dy/dx.$$

$$\frac{dy}{dx} = -\frac{4x}{2y} = -\frac{2x}{y}$$

To find the slope at $(1/\sqrt{2}, 1)$, we now substitute for x and y:

$$\left.\frac{dy}{dx}\right|_{(1/\sqrt{2},\,1)} = -\frac{2/\sqrt{2}}{1} = -\sqrt{2}.$$

Thus, the slope of the tangent to the ellipse at the point $(1/\sqrt{2}, 1)$ is $-\sqrt{2} \approx -1.414$.

EXAMPLE 3 Tangent Line for an Implicit Function

Find the equation of the tangent line to the curve $\ln y = xy$ at the point where $y = 1$.

Solution First, we use implicit differentiation to find dy/dx:

$$\frac{d}{dx}(\ln y) = \frac{d}{dx}(xy) \qquad \text{Take } d/dx \text{ of both sides.}$$

$$\frac{1}{y}\frac{dy}{dx} = (1)y + x\frac{dy}{dx}. \qquad \text{Chain rule on left, product rule on right}$$

To solve for dy/dx, we bring all the terms containing dy/dx to the left-hand side and all terms not containing it to the right-hand side:

$$\frac{1}{y}\frac{dy}{dx} - x\frac{dy}{dx} = y \qquad \text{Bring the terms with } dy/dx \text{ to the left.}$$

$$\frac{dy}{dx}\left(\frac{1}{y} - x\right) = y \qquad \text{Factor out } dy/dx.$$

$$\frac{dy}{dx}\left(\frac{1 - xy}{y}\right) = y$$

$$\frac{dy}{dx} = y\left(\frac{y}{1 - xy}\right) = \frac{y^2}{1 - xy}. \qquad \text{Solve for } dy/dx.$$

The derivative gives the slope of the tangent line, so we want to evaluate the derivative at the point where $y = 1$. However, the formula for dy/dx requires values for both x and y. We get the value of x by substituting $y = 1$ in the original equation:

$$\ln y = xy$$

$$\ln 1 = x \cdot 1. \qquad \text{Substitute } y = 1.$$

But $\ln 1 = 0$, so $x = 0$ for this point. Thus,

$$\left.\frac{dy}{dx}\right|_{(0,\,1)} = \frac{1^2}{1 - (0)(1)} = 1.$$

Therefore, the tangent line is the line through $(x, y) = (0, 1)$ with slope 1, which is

$$y = x + 1.$$

➡ **Before we go on...** Example 3 presents an instance of an implicit function in which it is simply not possible to solve for y. Try it. ■

Logarithmic Differentiation

Sometimes, it is easiest to differentiate a complicated function of x by first taking the logarithm and then using implicit differentiation—a technique called **logarithmic differentiation.**

EXAMPLE 4 Logarithmic Differentiation

Find $\frac{d}{dx}\left[\frac{(x+1)^{10}(x^2+1)^{11}}{(x^3+1)^{12}}\right]$ without using the product or quotient rules.

Solution Write

$$y = \frac{(x+1)^{10}(x^2+1)^{11}}{(x^3+1)^{12}},$$

and then take the natural logarithm of both sides:

$$\ln y = \ln\left[\frac{(x+1)^{10}(x^2+1)^{11}}{(x^3+1)^{12}}\right].$$

We can use properties of the logarithm to simplify the right-hand side:

$$\begin{aligned}\ln y &= \ln(x+1)^{10} + \ln(x^2+1)^{11} - \ln(x^3+1)^{12}\\ &= 10\ln(x+1) + 11\ln(x^2+1) - 12\ln(x^3+1).\end{aligned}$$

Now we can find $\frac{dy}{dx}$ using implicit differentiation:

$$\frac{1}{y}\frac{dy}{dx} = \frac{10}{x+1} + \frac{22x}{x^2+1} - \frac{36x^2}{x^3+1} \qquad \text{Take } d/dx \text{ of both sides.}$$

$$\frac{dy}{dx} = y\left(\frac{10}{x+1} + \frac{22x}{x^2+1} - \frac{36x^2}{x^3+1}\right) \qquad \text{Solve for } dy/dx.$$

$$= \frac{(x+1)^{10}(x^2+1)^{11}}{(x^3+1)^{12}}\left(\frac{10}{x+1} + \frac{22x}{x^2+1} - \frac{36x^2}{x^3+1}\right). \qquad \text{Substitute for } y.$$

➡ **Before we go on...** Redo Example 4 using the product and quotient rules (and the chain rule) instead of logarithmic differentiation, and compare the answers. Compare also the amount of work involved in both methods. ■

Application

Productivity usually depends on both labor and capital. Suppose, for example, that you are managing a surfboard manufacturing company. You can measure its productivity by counting the number of surfboards the company makes each year. As a measure of labor, you can use the number of employees, and as a measure of capital you can use its operating budget. The so-called *Cobb-Douglas* model uses a function of the form

$$P = Kx^a y^{1-a}, \quad \text{Cobb-Douglas model for productivity}$$

where P stands for the number of surfboards made each year, x is the number of employees, and y is the operating budget. The numbers K and a are constants that depend on the particular situation studied, with a between 0 and 1.

EXAMPLE 5 Cobb-Douglas Production Function

The surfboard company you own has the Cobb-Douglas production function

$$P = x^{0.3}y^{0.7},$$

where P is the number of surfboards it produces per year, x is the number of employees, and y is the daily operating budget (in dollars). Assume that the production level P is constant.

a. Find $\frac{dy}{dx}$.

b. Evaluate this derivative at $x = 30$ and $y = 10{,}000$, and interpret the answer.

Solution

a. We are given the equation $P = x^{0.3}y^{0.7}$, in which P is constant. We find $\frac{dy}{dx}$ by implicit differentiation.

$$0 = \frac{d}{dx}(x^{0.3}y^{0.7}) \quad d/dx \text{ of both sides}$$

$$0 = 0.3x^{-0.7}y^{0.7} + x^{0.3}(0.7)y^{-0.3}\frac{dy}{dx} \quad \text{Product and chain rules}$$

$$-0.7x^{0.3}y^{-0.3}\frac{dy}{dx} = 0.3x^{-0.7}y^{0.7} \quad \text{Bring term with } dy/dx \text{ to left.}$$

$$\frac{dy}{dx} = -\frac{0.3x^{-0.7}y^{0.7}}{0.7x^{0.3}y^{-0.3}} \quad \text{Solve for } dy/dx.$$

$$= -\frac{3y}{7x}. \quad \text{Simplify.}$$

b. Evaluating this derivative at $x = 30$ and $y = 10{,}000$ gives

$$\left.\frac{dy}{dx}\right|_{x=30,\, y=10{,}000} = -\frac{3(10{,}000)}{7(30)} \approx -143.$$

To interpret this result, first look at the units of the derivative: We recall that the units of dy/dx are units of y per unit of x. Because y is the daily budget, its units are dollars; because x is the number of employees, its units are employees. Thus,

$$\left.\frac{dy}{dx}\right|_{x=30,\, y=10{,}000} \approx -\$143 \text{ per employee.}$$

Next, recall that dy/dx measures the rate of change of y as x changes. Because the answer is negative, the daily budget to maintain production at the fixed level is decreasing by approximately \$143 per additional employee at an employment level of 30 employees and a daily operating budget of \$10,000. In other words, increasing the workforce by one worker will result in a savings of approximately \$143 per day. Roughly speaking, *a new employee is worth \$143 per day* at the current levels of employment and production.

4.6 EXERCISES

▼ more advanced ◆ challenging

T indicates exercises that should be solved using technology

In Exercises 1–10, find dy/dx, using implicit differentiation. In each case, compare your answer with the result obtained by first solving for y as a function of x and then taking the derivative. [**HINT**: See Example 1.]

1. $2x + 3y = 7$ **2.** $4x - 5y = 9$

3. $x^2 - 2y = 6$ **4.** $3y + x^2 = 5$

5. $2x + 3y = xy$ **6.** $x - y = xy$

7. $e^x y = 1$ **8.** $e^x y - y = 2$

9. $y \ln x + y = 2$ **10.** $\dfrac{\ln x}{y} - 2 - x$

In Exercises 11–30, find the indicated derivative using implicit differentiation. [**HINT**: See Example 1.]

11. $x^2 + y^2 = 5; \dfrac{dy}{dx}$ **12.** $2x^2 - y^2 = 4; \dfrac{dy}{dx}$

13. $x^2y - y^2 = 4; \dfrac{dy}{dx}$ **14.** $xy^2 - y = x; \dfrac{dy}{dx}$

15. $3xy - \dfrac{y}{3} = \dfrac{2}{x}; \dfrac{dy}{dx}$ **16.** $\dfrac{xy}{2} - y^2 = 3; \dfrac{dy}{dx}$

17. $x^2 - 3y^2 = 8; \dfrac{dx}{dy}$ **18.** $(xy)^2 + y^2 = 8; \dfrac{dx}{dy}$

19. $p^2 - pq = 5p^2q^2; \dfrac{dp}{dq}$ **20.** $q^2 - pq = 5p^2q^2; \dfrac{dp}{dq}$

21. $xe^y - ye^x = 1; \dfrac{dy}{dx}$ **22.** $x^2e^y - y^2 = e^x; \dfrac{dy}{dx}$

23. ▼ $e^{st} = s^2; \dfrac{ds}{dt}$ **24.** ▼ $e^{s^2t} - st = 1; \dfrac{ds}{dt}$

25. ▼ $\dfrac{e^x}{y^2} = 1 + e^y; \dfrac{dy}{dx}$ **26.** ▼ $\dfrac{x}{e^y} + xy = 9y; \dfrac{dy}{dx}$

27. ▼ $\ln(y^2 - y) + x = y; \dfrac{dy}{dx}$ **28.** ▼ $\ln(xy) - x \ln y = y; \dfrac{dy}{dx}$

29. ▼ $\ln(xy + y^2) = e^y; \dfrac{dy}{dx}$ **30.** ▼ $\ln(1 + e^{xy}) = y; \dfrac{dy}{dx}$

In Exercises 31–42, use implicit differentiation to find **(a)** *the slope of the tangent line and* **(b)** *the equation of the tangent line at the indicated point on the graph. (Round answers to four decimal places as needed.) If only the x-coordinate is given, you must also find the y-coordinate.*
[**HINT**: See Examples 2 and 3.]

31. $4x^2 + 2y^2 = 12, (1, -2)$ **32.** $3x^2 - y^2 = 11, (-2, 1)$

33. $2x^2 - y^2 = xy, (-1, 2)$ **34.** $2x^2 + xy = 3y^2, (-1, -1)$

35. $x^2y - y^2 + x = 1, (1, 0)$

36. $(xy)^2 + xy - x = 8, (-8, 0)$

37. $xy - 2{,}000 = y, x = 2$ **38.** $x^2 - 10xy = 200, x = 10$

39. ▼ $\ln(x + y) - x = 3x^2, x = 0$

40. ▼ $\ln(x - y) + 1 = 3x^2, x = 0$

41. ▼ $e^{xy} - x = 4x, x = 3$ **42.** ▼ $e^{-xy} + 2x = 1, x = -1$

In Exercises 43–52, use logarithmic differentiation to find dy/dx. Do not simplify the result. [**HINT**: See Example 4.]

43. $y = \dfrac{2x + 1}{4x - 2}$ **44.** $y = (3x + 2)(8x - 5)$

45. $y = \dfrac{(3x + 1)^2}{4x(2x - 1)^3}$ **46.** $y = \dfrac{x^2(3x + 1)^2}{(2x - 1)^3}$

47. $y = (8x - 1)^{1/3}(x - 1)$ **48.** $y = \dfrac{(3x + 2)^{2/3}}{3x - 1}$

49. $y = (x^3 + x)\sqrt{x^3 + 2}$ **50.** $y = \sqrt{\dfrac{x - 1}{x^2 + 2}}$

51. ▼ $y = x^x$ **52.** ▼ $y = x^{-x}$

Applications

53. ***Productivity*** The number of CDs per hour that *Snappy Hardware* can manufacture at its plant is given by

$$P = x^{0.6}y^{0.4},$$

where x is the number of workers at the plant and y is the monthly budget (in dollars). Assume that P is constant, and compute $\dfrac{dy}{dx}$ when $x = 100$ and $y = 200{,}000$. Interpret the result. [**HINT**: See Example 5.]

54. ***Productivity*** The number of cellphone accessory kits (neon lights, matching covers, and earbuds) per day that *USA Cellular Makeover, Inc.*, can manufacture at its plant in Cambodia is given by

$$P = x^{0.5}y^{0.5},$$

where x is the number of workers at the plant and y is the monthly budget (in dollars). Assume that P is constant, and compute $\frac{dy}{dx}$ when $x = 200$ and $y = 100{,}000$. Interpret the result. [**HINT**: See Example 5.]

55. ***Demand*** The demand equation for soccer tournament T-shirts is

$$xy - 2{,}000 = y,$$

where y is the number of T-shirts the *Enormous State University* soccer team can sell at a price of \$$x$ per shirt. Find $\left.\frac{dy}{dx}\right|_{x=5}$, and interpret the result.

56. ***Cost Equations*** The cost y (in cents) of producing x gallons of *Ectoplasm* hair gel is given by the cost equation

$$y^2 - 10xy = 200.$$

Evaluate $\frac{dy}{dx}$ at $x = 1$, and interpret the result.

57. ***Housing Costs***[61] The cost C (in dollars) of building a house is related to the number k of carpenters used and the number e of electricians used by the formula

$$C = 15{,}000 + 50k^2 + 60e^2.$$

If the cost of the house is fixed at \$200,000, find $\left.\frac{dk}{de}\right|_{e=15}$, and interpret your result.

58. ***Employment*** An employment research company estimates that the value of a recent MBA graduate to an accounting company is

$$V = 3e^2 + 5g^3,$$

where V is the value of the graduate, e is the number of years of prior business experience, and g is the graduate school grade-point average. If V is fixed at 200, find $\frac{de}{dg}$ when $g = 3.0$, and interpret the result.

59. ▼ ***Grades***[62] A productivity formula for a student's performance on a difficult English examination is

$$g = 4tx - 0.2t^2 - 10x^2 \quad (t < 30),$$

where g is the score the student can expect to obtain, t is the number of hours of study for the examination, and x is the student's grade-point average.

a. For how long should a student with a 3.0 grade-point average study to score 80 on the examination?

b. Find $\frac{dt}{dx}$ for a student who earns a score of 80, evaluate it when $x = 3.0$, and interpret the result.

60. ▼ ***Grades*** Repeat Exercise 59 using the following productivity formula for a basket-weaving examination:

$$g = 10tx - 0.2t^2 - 10x^2 \quad (t < 10).$$

Comment on the result.

Exercises 61 and 62 are based on the following demand function for money (taken from a question on the GRE Economics Test):

$$M_d = (2) \times (y)^{0.6} \times (r)^{-0.3} \times (p),$$

where

M_d = *demand for nominal money balances (money stock)*
y = *real income*
r = *an index of interest rates*
p = *an index of prices.*

61. ◆ ***Money Stock*** If real income grows while the money stock and the price level remain constant, the interest rate must change at what rate? (First find dr/dy, then find dr/dt; your answers will be expressed in terms of r, y, and dy/dt.)

62. ◆ ***Money Stock*** If real income grows while the money stock and the interest rate remain constant, the price level must change at what rate?

Communication and Reasoning Exercises

63. Fill in the missing terms: The equation $x = y^3 + y - 3$ specifies ___ as a function of ___ and ___ as an implicit function of ___.

64. Fill in the missing terms: When $x \neq 0$ in the equation $xy = x^3 + 4$, it is possible to specify ___ as a function of ___. However, ___ is only an implicit function of ___.

65. ▼ Use logarithmic differentiation to give another proof of the product rule.

66. ▼ Use logarithmic differentiation to give a proof of the quotient rule.

67. ▼ If y is given explicitly as a function of x by an equation $y = f(x)$, compare finding dy/dx by implicit differentiation to finding it explicitly in the usual way.

68. ▼ Explain why one should not expect dy/dx to be a function of x if y is not a function of x.

69. ◆ If y is a function of x and $dy/dx \neq 0$ at some point, regard x as an implicit function of y and use implicit differentiation to obtain the equation

$$\frac{dx}{dy} = \frac{1}{dy/dx}.$$

70. ◆ If you are given an equation in x and y such that dy/dx is a function of x only, what can you say about the graph of the equation?

---

[61] Based on an exercise in *Introduction to Mathematical Economics* by A. L. Ostrosky Jr., and J. V. Koch (Waveland Press, Springfield, Illinois, 1979).

[62] *Ibid.*

CHAPTER 4 REVIEW

KEY CONCEPTS

www.WanerMath.com
Go to the Website to find a comprehensive and interactive Web-based summary of Chapter 4.

4.1 Derivatives of Powers, Sums, and Constant Multiples

Power Rule: If n is any constant and $f(x) = x^n$, then $f'(x) = nx^{n-1}$. [p. 304]

Using the power rule for negative and fractional exponents [p. 305]

Sums, differences, and constant multiples [p. 307]

Combining the rules [p. 308]

$\frac{d}{dx}(cx) = c, \frac{d}{dx}(c) = 0$ [p. 310]

$f(x) = x^{1/3}$, $g(x) = x^{2/3}$, and $h(x) = |x|$ are not differentiable at $x = 0$ [p. 310]

L'Hospital's rule [p. 312]

4.2 A First Application: Marginal Analysis

Marginal cost function $C'(x)$ [p. 319]

Marginal revenue and profit functions $R'(x)$ and $P'(x)$ [p. 321]

What it means when the marginal profit is zero [p. 322]

Marginal product [p. 322]

Average cost of the first x items:

$\bar{C}(x) = \frac{C(x)}{x}$ [p. 324]

4.3 The Product and Quotient Rules

Product rule: $\frac{d}{dx}[f(x)g(x)] = f'(x)g(x) + f(x)g'(x)$ [p. 330]

Quotient rule: $\frac{d}{dx}\left[\frac{f(x)}{g(x)}\right] = \frac{f'(x)g(x) - f(x)g'(x)}{[g(x)]^2}$ [p. 331]

Using the product rule [p. 333]

Using the quotient rule [p. 334]

Calculation thought experiment [p. 335]

Application to revenue and average cost [p. 337]

4.4 The Chain Rule

Chain rule: $\frac{d}{dx}[f(u)] = f'(u)\frac{du}{dx}$ [p. 343]

Generalized power rule:

$\frac{d}{dx}(u^n) = nu^{n-1}\frac{du}{dx}$ [p. 344]

Using the chain rule [p. 344]

Application to marginal product and profit [p. 347]

Chain rule in differential notation:

$\frac{dy}{dx} = \frac{dy}{du}\frac{du}{dx}$ [p. 349]

Manipulating derivatives in differential notation [p. 349]

4.5 Derivatives of Logarithmic and Exponential Functions

Derivative of the natural logarithm:

$\frac{d}{dx}\ln x = \frac{1}{x}$ [p. 356]

Derivative of logarithm with base b:

$\frac{d}{dx}\log_b x = \frac{1}{x \ln b}$ [p. 356]

Derivatives of logarithms of functions:

$\frac{d}{dx}\ln u = \frac{1}{u}\frac{du}{dx}$

$\frac{d}{dx}\log_b u = \frac{1}{u \ln b}\frac{du}{dx}$ [p. 357]

Derivatives of logarithms of absolute values:

$\frac{d}{dx}\ln|x| = \frac{1}{x}$ $\quad \frac{d}{dx}\ln|u| = \frac{1}{u}\frac{du}{dx}$

$\frac{d}{dx}\log_b|x| = \frac{1}{x \ln b}$

$\frac{d}{dx}\log_b|u| = \frac{1}{u \ln b}\frac{du}{dx}$ [p. 359]

Derivative of e^x: $\frac{d}{dx}e^x = e^x$ [p. 360]

Derivative of b^x: $\frac{d}{dx}b^x = b^x \ln b$ [p. 361]

Derivatives of exponential functions [p. 361]

Application to epidemics [p. 362]

Application to sales growth (logistic function) [p. 363]

Application to limits (l'Hospital's rule) [p. 363]

4.6 Implicit Differentiation

Implicit function of x [p. 370]

Implicit differentiation [p. 370]

Using implicit differentiation [p. 371]

Finding a tangent line [p. 372]

Logarithmic differentiation [p. 373]

Application: Cobb-Douglas production function [p. 374]

REVIEW EXERCISES

In Exercises 1–32, find the derivative of the given function.

1. $f(x) = 10x^5 + \frac{1}{2}x^4 - x + 2$

2. $f(x) = \frac{10}{x^5} + \frac{1}{2x^4} - \frac{1}{x} + 2$

3. $f(x) = 3x^3 + 3\sqrt[3]{x}$

4. $f(x) = \frac{2}{x^{2.1}} - \frac{x^{0.1}}{2}$

5. $f(x) = x + \frac{1}{x^2}$

6. $f(x) = 2x - \frac{1}{x}$

7. $f(x) = \frac{4}{3x} - \frac{2}{x^{0.1}} + \frac{x^{1.1}}{3.2} - 4$

8. $f(x) = \frac{4}{x} + \frac{x}{4} - |x|$

9. $f(x) = e^x(x^2 - 1)$

10. $f(x) = \frac{x^2 + 1}{x^2 - 1}$

11. $f(x) = \dfrac{|x| + 1}{3x^2 + 1}$

12. $f(x) = (|x| + x)(2 - 3x^2)$

13. $f(x) = (4x - 1)^{-1}$

14. $f(x) = (x + 7)^{-2}$

15. $f(x) = (x^2 - 1)^{10}$

16. $f(x) = \dfrac{1}{(x^2 - 1)^{10}}$

17. $f(x) = [2 + (x + 1)^{-0.1}]^{4.3}$

18. $f(x) = [(x + 1)^{0.1} - 4x]^{-5.1}$

19. $f(x) = e^{2x+1}$

20. $f(x) = e^{4x-5}$

21. $t(x) = 3^{2x-4}$

22. $t(x) = 4^{-x+5}$

23. $f(x) = e^x(x^2 + 1)^{10}$

24. $f(x) = \left[\dfrac{x - 1}{3x + 1}\right]^3$

25. $f(x) = \dfrac{3^x}{x - 1}$

26. $f(x) = 4^{-x}(x + 1)$

27. $f(x) = e^{x^2-1}$

28. $f(x) = (x^2 + 1)e^{x^2-1}$

29. $g(x) = \ln|3x - 1|$

30. $g(x) = \ln|5 - 9x|$

31. $f(x) = \ln(x^2 - 1)$

32. $f(x) = \dfrac{\ln(x^2 - 1)}{x^2 - 1}$

In Exercises 33–40, find all values of x (if any) where the tangent line to the graph of the given equation is horizontal.

33. $y = -3x^2 + 7x - 1$

34. $y = 5x^2 - 2x + 1$

35. $y = \dfrac{x}{2} + \dfrac{2}{x}$

36. $y = \dfrac{x^2}{2} - \dfrac{8}{x^2}$

37. $y = x - e^{2x-1}$

38. $y = e^{x^2}$

39. $y = \dfrac{x}{x + 1}$

40. $y = \sqrt{x}(x - 1)$

In Exercises 41–46, find dy/dx for the given equation.

41. $x^2 - y^2 = x$

42. $2xy + y^2 = y$

43. $e^{xy} + xy = 2$

44. $\ln\left(\dfrac{y}{x}\right) = y$

45. $y = \dfrac{(2x - 1)^4(3x + 4)}{(x + 1)(3x - 1)^3}$

46. $y = x^{x-1}3^x$

In Exercises 47–52, find the equation of the tangent line to the graph of the given equation at the specified point.

47. $y = (x^2 - 3x)^{-2}; x = 1$

48. $y = (2x^2 - 3)^{-3}; x = -1$

49. $y = x^2e^{-x}; x = -1$

50. $y = \dfrac{x}{1 + e^x}; x = 0$

51. $xy - y^2 = x^2 - 3; (-1, 1)$

52. $\ln(xy) + y^2 = 1; (-1, -1)$

In Exercises 53–56, find the limit.

53. $\lim\limits_{x\to+\infty} \dfrac{x^2}{e^{x^2}}$

54. $\lim\limits_{x\to-\infty} xe^x$

55. $\lim\limits_{x\to 0} \dfrac{x^2}{e^{x^2} - 1}$

56. $\lim\limits_{x\to 0} \dfrac{x^3}{e^x - 1 - x - x^2/2}$

Applications: OHaganBooks.com
[Try the game at www.OHaganBooks.com]

57. ***Sales*** OHaganBooks.com fits the cubic curve

$$w(t) = -3.7t^3 + 74.6t^2 + 135.5t + 6{,}300 \quad (0 \le t \le 6)$$

to its weekly sales figures (see Chapter 3 Review Exercise 57; t is time in weeks), as shown in the following graph:

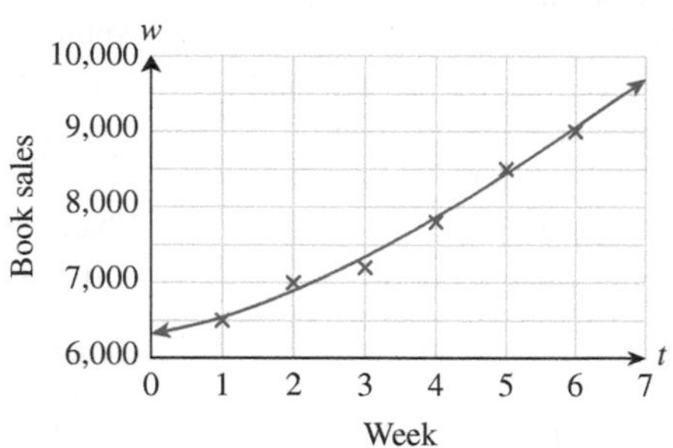

a. According to the cubic model, what was the rate of increase of sales at the beginning of the second week ($t = 1$)? (Round your answer to the nearest unit.)

b. If we extrapolate the model, what would be the rate of increase of weekly sales at the beginning of the eighth week ($t = 7$)?

c. Graph the function w for $0 \le t \le 20$. Would it be realistic to use the function to predict sales through week 20? Why?

d. By examining the graph, say why the choice of a quadratic model would result in radically different long-term predictions of sales.

58. ***Rising Sea Level*** Marjory Duffin is still toying with various models to fit to the New York sea level figures she had seen after purchasing a beachfront condominium in New York (see Chapter 3 Review Exercise 58). Following is a cubic curve she obtained using regression:

$$L(t) = -0.0001t^3 + 0.02t^2 + 2.2t \text{ mm}.$$

(t is time in years since 1900.) The curve and data are shown in the following graph:

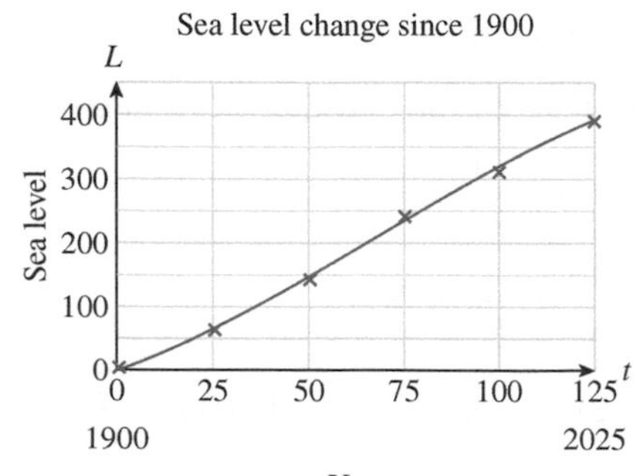

a. According to the cubic model, what was the rate at which the sea level was rising in 2000 ($t = 100$)? (Round your answer to two significant digits.)
b. If we extrapolate the model, what would be the rate at which the sea level is rising in 2025 ($t = 125$)?
c. Graph the function L for $0 \le t \le 200$. Why is it not realistic to use the function to predict the sea level through 2100?
d. James Stewart, a summer intern at *Duffin House Publishers*, differs. As he puts it, "The cubic curve came from doing regression on the actual data, and thus reflects the actual trend of the data. We can't argue against reality!" Comment on this assertion.

59. ***Cost*** As OHaganBooks.com's sales increase, so do its costs. If we take into account volume discounts from suppliers and shippers, the weekly cost of selling x books is

$$C(x) = -0.00002x^2 + 3.2x + 5{,}400 \text{ dollars.}$$

a. What is the marginal cost at a sales level of 8,000 books per week?
b. What is the average cost per book at a sales level of 8,000 books per week?
c. What is the marginal average cost ($d\bar{C}/dx$) at a sales level of 8,000 books per week?
d. Interpret the results of parts (a)–(c).

60. ***Cost*** OHaganBooks.com has been experiencing a run of bad luck with its summer college intern program in association with PCU (*Party Central University*), begun as a result of a suggestion by Marjory Duffin over dinner one evening. The frequent errors in filling orders, charges from movie download sites and dating sites, and beverages spilled on computer equipment have resulted in an estimated weekly cost to the company of

$$C(x) = 25x^2 - 5.2x + 4{,}000 \text{ dollars,}$$

where x is the number of college interns employed.
a. What is the marginal cost at a level of 10 interns?
b. What is the average cost per intern at a level of 10 interns?
c. What is the marginal average cost at a level of 10 interns?
d. Interpret the results of parts (a)–(c).

61. ***Revenue*** At the moment, OHaganBooks.com is selling 1,000 books per week, and its sales are rising at a rate of 200 books per week. Also, it is now selling all its books for \$20 each, but its price is dropping at a rate of \$1 per week.
a. At what rate is OHaganBooks.com's weekly revenue rising or falling?
b. John O'Hagan would like to see the company's weekly revenue increase at a rate of \$5,000 per week. At what rate would sales have to have been increasing to accomplish that goal, assuming that all the other information is as given above?

62. ***Revenue*** Because of ongoing problems with its large college intern program in association with PCU (see Exercise 60), OHaganBooks.com has arranged to transfer its interns to its competitor *JungleBooks.com* (whose headquarters happens to be across the road) for a small fee. At the moment, it is transferring 5 students per week, and this number is rising at a rate of 4 students per week. Also, it is now charging JungleBooks \$400 per intern, but this amount is decreasing at a rate of \$20 per week.
a. At what rate is OHaganBooks.com's weekly revenue from this transaction rising or falling?
b. Flush with success of the transfer program, John O'Hagan would like to see the company's resulting revenue increase at a rate of \$3,900 per week. At what rate would the transfer of interns have to increase to accomplish that goal, assuming all the other information is as given above?

63. ***Percentage Rate of Change of Revenue*** The percentage rate of change of a quantity Q is Q'/Q. Why is the percentage rate of change of revenue always equal to the sum of the percentage rates of change of unit price and weekly sales?

64. ***P/E Ratios*** At the beginning of last week, OHaganBooks .com stock was selling for \$100 per share, rising at a rate of \$50 per year. Its earnings amounted to \$1 per share, rising at a rate of \$0.10 per year. At what rate was its price-to-earnings (P/E) ratio, the ratio of its stock price to its earnings per share, rising or falling?

65. ***P/E Ratios*** Refer to Exercise 64. Jay Campbell, who recently invested in OHaganBooks.com stock, would have liked to see the P/E ratio increase at a rate of 100 points per year. How fast would the stock have to have been rising, assuming that all the other information is as given in Exercise 64?

66. ***Percentage Rate of Change of P/E Ratios*** Refer to Exercise 64. The percentage rate of change of a quantity Q is Q'/Q. Why is the percentage rate of change of the P/E ratio always equal to the percentage rate of change of unit price minus the percentage rate of change of earnings?

67. ***Sales*** OHaganBooks.com decided that the cubic curve in Exercise 57 was not suitable for extrapolation, so instead it tried

$$s(t) = 6{,}000 + \frac{4{,}500}{1 + e^{-0.55(t-4.8)}},$$

as shown in the following graph:

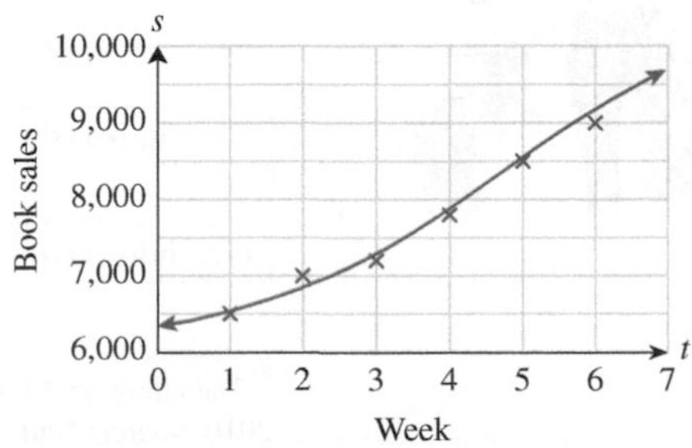

a. Compute $s'(t)$, and use the answer to estimate the rate of increase of weekly sales at the beginning of the seventh week ($t = 6$). (Round your answer to the nearest unit.)

b. Compute $\lim_{t\to+\infty} s'(t)$, and interpret the answer.

68. ***Rising Sea Level*** Upon some reflection, Marjory Duffin decided that the curve in Exercise 58 was not suitable for extrapolation, so instead she tried

$$L(t) = \frac{418}{1 + 17.2e^{-0.041t}} \quad (0 \le t \le 125)$$

(t is time in years since 1900), as shown in the following graph:

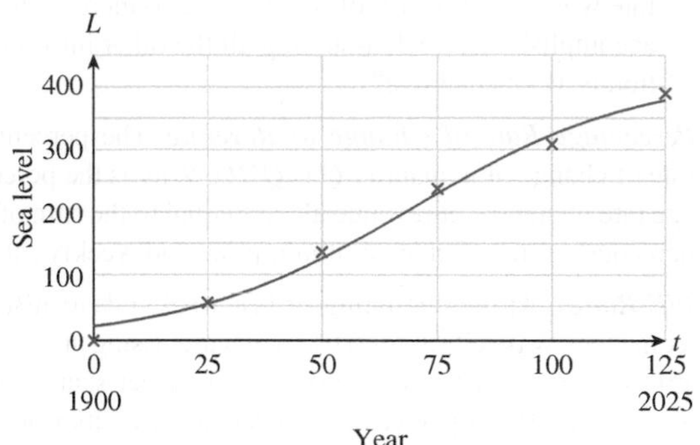

a. Compute $L'(t)$, and use the answer to estimate the rate at which the sea level was rising in 2000 ($t = 100$). (Round your answer to two decimal places.)

b. Compute $\lim_{t\to+\infty} L'(t)$, and interpret the answer.

69. ***Website Activity*** The number of hits on OHaganBooks.com's website was 1,000 per day at the beginning of the year, and was growing at a rate of 5% per week. If this growth rate continued for the whole year (52 weeks), find the rate of increase (in hits per day per week) at the end of the year.

70. ***Website Activity*** The number of hits on *ShadyDownload.net* during the summer intern program at OHaganBooks.com was 100 per day at the beginning of the intern program, and was growing at a rate of 15% per day. If this growth rate continued for the duration of the whole summer intern program (85 days), find the rate of increase (in hits per day per day) at the end of the program.

71. ***Demand and Revenue*** The price p that OHaganBooks .com charges for its latest leather-bound gift edition of *The Complete Larry Potter* is related to the demand q in weekly sales by the equation

$$250pq + q^2 = 13{,}500{,}000.$$

Suppose the price is set at \$50, which would make the demand 1,000 copies per week.

a. Using implicit differentiation, compute the rate of change of demand with respect to price, and interpret the result. (Round the answer to two decimal places.)

b. Use the result of part (a) to compute the rate of change of revenue with respect to price. Should the price be raised or lowered to increase revenue?

72. ***Demand and Revenue*** The price p that OHaganBooks .com charges for its latest leather-bound gift edition of *Lord of the Fields* is related to the demand q in weekly sales by the equation

$$100pq + q^2 = 5{,}000{,}000.$$

Suppose the price is set at \$40, which would make the demand 1,000 copies per week.

a. Using implicit differentiation, compute the rate of change of demand with respect to price, and interpret the result. (Round the answer to two decimal places.)

b. Use the result of part (a) to compute the rate of change of revenue with respect to price. Should the price be raised or lowered to increase revenue?

CASE STUDY

Projecting Market Growth

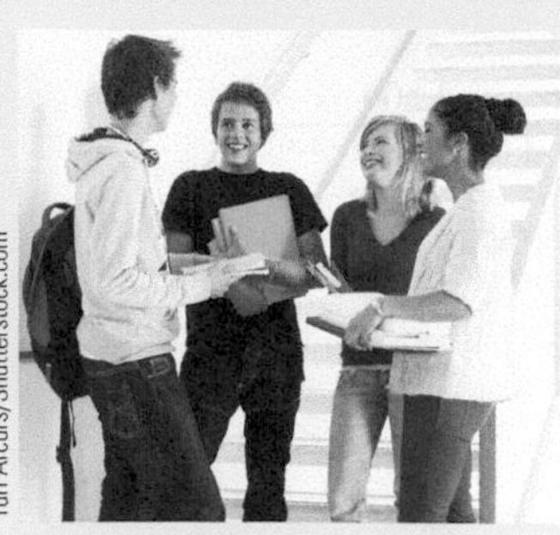

Yuri Arcurs/Shutterstock.com

It is 2010, and you are on the board of directors at *Fullcourt Academic Press*, a major textbook supplier to private schools, and various expansion strategies will be discussed at tomorrow's board meeting. TJM, the sales director of the high school division, has just burst into your office with his last-minute proposal based on data showing the number of private high school graduates in the United States each year over the past 18 years:[63]

Year t	1995	1996	1997	1998	1999	2000	2001	2002	2003
Graduates (thousands)	245	254	265	273	279	279	285	296	301
Year t	2004	2005	2006	2007	2008	2009	2010	2011	2012
Graduates (thousands)	307	307	307	314	314	315	315	316	316

[63] Data through 2011 are National Center for Educational Statistics actual and projected data as of April 2010. Source: National Center for Educational Statistics (http://nces.ed.gov).

TJM asserts that, despite the unspectacular numbers in the past few years, the long-term trend appears to support a basic premise of his proposal for an expansion strategy: that the number of high school seniors in private schools in the United States will be growing at a rate of about 4,000 per year through 2015. He points out that the rate of increase predicted by the regression line is approximately 4,080 students per year, supporting his premise.

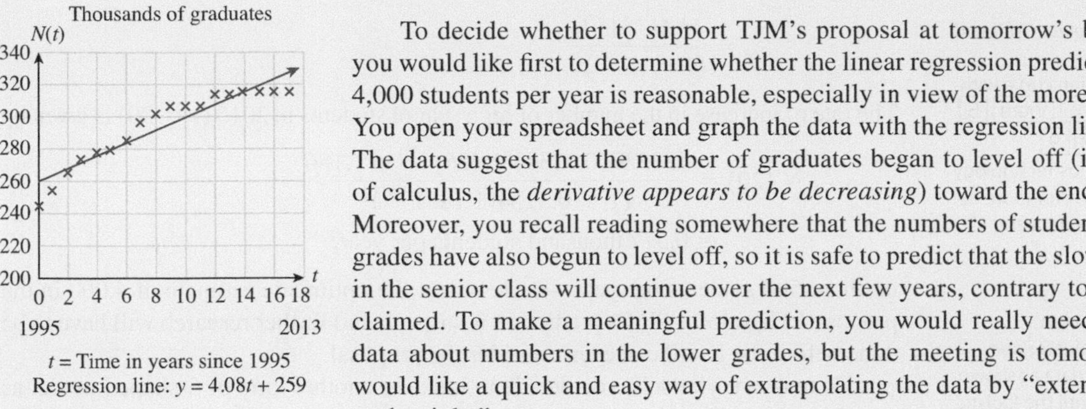

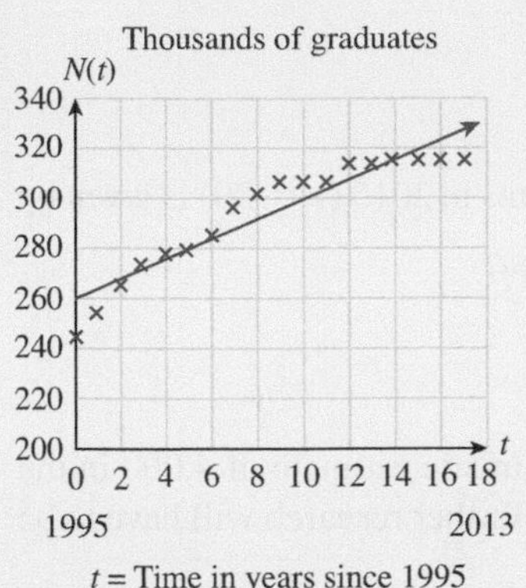

t = Time in years since 1995
Regression line: $y = 4.08t + 259$

Figure 11

To decide whether to support TJM's proposal at tomorrow's board meeting, you would like first to determine whether the linear regression prediction of around 4,000 students per year is reasonable, especially in view of the more recent figures. You open your spreadsheet and graph the data with the regression line (Figure 11). The data suggest that the number of graduates began to level off (in the language of calculus, the *derivative appears to be decreasing*) toward the end of the period. Moreover, you recall reading somewhere that the numbers of students in the lower grades have also begun to level off, so it is safe to predict that the slowing of growth in the senior class will continue over the next few years, contrary to what TJM has claimed. To make a meaningful prediction, you would really need some precise data about numbers in the lower grades, but the meeting is tomorrow, and you would like a quick and easy way of extrapolating the data by "extending the curve to the right."

It would certainly be helpful if you had a mathematical model of the data in Figure 11 that you could use to project the current trend. But what kind of model should you use? The linear model is no good because it does not show any change in the derivative (the derivative of a linear function is constant). In addition, best-fit polynomial and exponential functions do not accurately reflect the leveling off, as you realize after trying to fit a few of them (Figure 12).

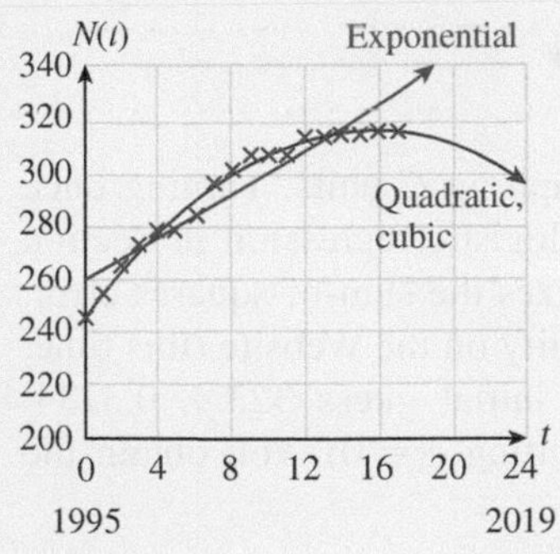

Figure 12

You then recall that a logistic curve can model the leveling-off property you desire, so you try a model of the form

$$N(t) = \frac{M}{1 + Ab^{-t}}.$$

Figure 13 shows the best-fit logistic curve, which has a sum-of-squares error (SSE) of around 109. (See Section 1.4 or any of the regression examples in Chapter 2 for a discussion of SSE.)

$$N(t) = \frac{323.9}{1 + 0.3234(1.186)^{-t}} \qquad \text{SSE} \approx 109$$

Its graph shows the leveling off and also gives more reasonable long-term predictions.

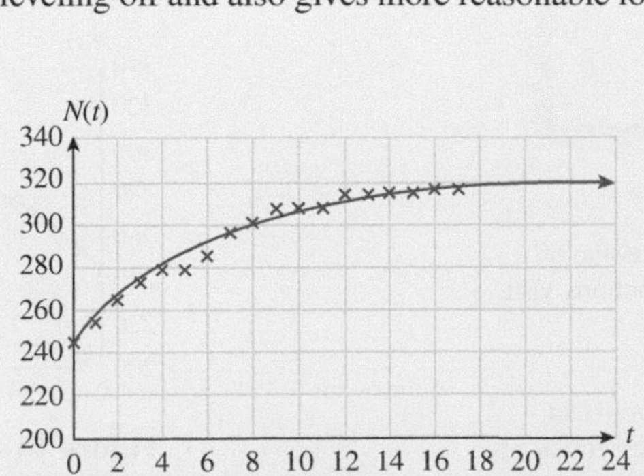

Figure 13

The rate of increase of high school students—pertinent to TJM's report—is given by the derivative, $N'(t)$:

$$N(t) = \frac{M}{1 + Ab^{-t}}$$

$$N'(t) = -\frac{M}{(1 + Ab^{-t})^2}\frac{d}{dt}(1 + Ab^{-t})$$

$$= \frac{M\,Ab^{-t}\ln b}{(1 + Ab^{-t})^2}.$$

The rate of increase in the number of high school students in 2015 ($t = 20$) is given by

$$N'(20) = \frac{(323.9)(0.3234)(1.186)^{-20}\ln 1.186}{(1 + 0.3234(1.186)^{-20})^2}$$

$$\approx 0.577 \text{ thousand students per year,}$$

or about 580 students per year—far less than the optimistic estimate of 4,000 in the proposal! Therefore, TJM's prediction is suspect, and further research will have to be done before the board can even consider the proposal.

To reassure yourself, you decide to look for another kind of S-shaped model as a backup. After flipping through a calculus book, you stumble across a function that is slightly more general than the one you have:

$$N(t) = \frac{M}{1 + Ab^{-t}} + C \qquad \text{Shifted logistic curve*}$$

The added term C has the effect of shifting the graph up C units. Turning once again to your calculus book (see the discussion of logistic regression in Section 2.4), you see that a best-fit curve is one that minimizes the sum-of-squares error, and you find the best-fit curve by again using the utility on the Website (this time, with the model `$1/(1+$2*$3^(-x))+$4` and initial guess 323.9, 0.3234, 1.186, 0; that is, keeping the current values and setting $c = 0$). You obtain the model

$$N(t) = \frac{135.5}{1 + 1.192(1.268)^{-t}} + 184.6. \qquad \text{SSE} \approx 100$$

The value of SSE has decreased only slightly, and, as seen in Figure 14, the shifted logistic curve seems almost identical to the unshifted curve but does seem to level off slightly faster. (Compare the portions of the two curves on the extreme right.)

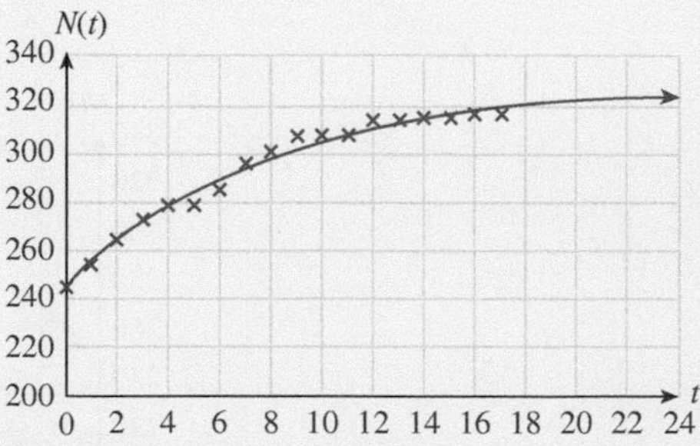

Figure 14

Using Technology

You can obtain the best-fit logistic curve using technology.

TI-83/84 Plus

[STAT] EDIT

Enter 0–17 in L_1 and the values of $N(t)$ in L_2.

[STAT] CALC and select Option #B `Logistic`

To plot the regression curve with the data: Graph: [Y=] [VARS] [5] EQ [1], then [ZOOM] [9]

[See the TI-83/84 Plus Technology Guide for Section 1.4 for more details on plotting the data and curve.]

Website

www.WanerMath.com

In the Function Evaluator and Grapher, enter the data as shown, press "Examples" until the logistic model `$1/(1+$2*$3^(-x))` shows in the first box, and press "Fit Curve". (You might need to press "Fit Curve" several times before it manages to find the curve. It also helps to read the instructions (link near the top of the Web page under the heading).)

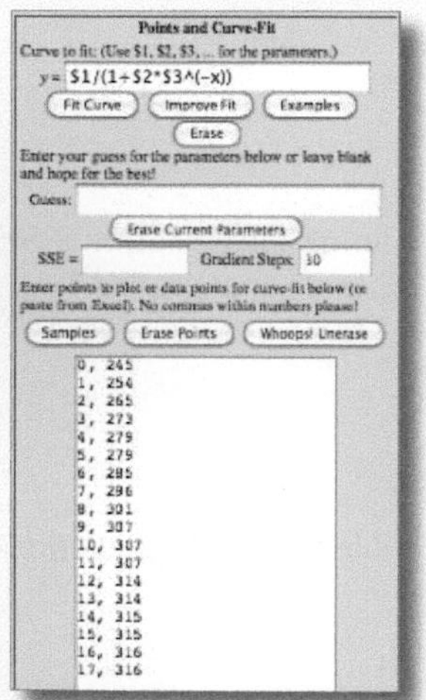

* To find a detailed discussion of scaled and shifted functions, visit the Website and follow

Chapter 1

→ New Functions from Old: Scaled and Shifted Functions.

You decide to use the shifted model to obtain another estimate of the projected rate of change in 2015. As the two models differ by a constant, their derivatives are given by the same formula, so you compute

$$N'(20) = \frac{(135.5)(1.192)(1.268)^{-20} \ln 1.268}{(1 + 1.192(1.268)^{-20})^2}$$

$$\approx 0.325 \text{ thousand students per year,}$$

or about 325 students per year, even less than the prediction of the logistic model.

Q: *Why do the two models give very different predictions of the rate of change in 2015?*

A: The long-term prediction in any logistic model is highly sensitive to small changes in the data and/or the model. This is one reason why using regression curve-fitting models to make long-term projections can be a risky undertaking.

Q: *Then what is the point of using any model to project in the first place?*

A: Projections are always tricky, as we cannot foresee the future. But a *good* model is not merely one that seems to fit the data well, but rather a model whose structure is based on the situation being modeled. For instance, a *good* model of student graduation rates should take into account such factors as the birth rate, current school populations at all levels, and the relative popularity of private schools over public schools. It is by using models of this kind that the National Center for Educational Statistics is able to make the projections shown in the data above. (And even those turned out to be overestimates.)

EXERCISES

1. In 1994 there were 246,000 private high school graduates. What do the two logistic models (unshifted and shifted) "predict" for 1994? (Round your answer to the nearest 1,000.) Which gives the better prediction?

2. What is the long-term prediction of each of the two models? (Round your answer to the nearest 1,000.)

3. Find $\lim_{t \to +\infty} N'(t)$ for both models, and interpret the results.

4. T You receive a last-minute memo from TJM to the effect that, sorry, the 2011 and 2012 figures are not accurate. Use technology to re-estimate M, A, b, and C for the shifted logistic model in the absence of this data and obtain new estimates for the 2011 and 2012 data. What does the new model predict the rate of change in the number of high school seniors will be in 2015?

5. T ***Another Model*** Using the original data, find the best-fit shifted logistic curve of the form

$$N(t) = c + b\frac{a(t - m)}{1 + a|t - m|}. \qquad (a, b, c, m \text{ constant})$$

Its graph is shown in the margin.

(Use the model `$1+$2*$3*(x-$4)/(1+$3*abs(x-$4))` and start with the following values: $a = 0.05$, $b = 160$, $c = 250$, $m = 5$; that is, input `250, 160, 0.05, 5` in the "Guess" field.) Graph the data together with the model. What is SSE? Is the model as accurate a fit as the model used in the text? What does this model predict will be the growth rate of the number of high school graduates in 2015? Comment on the answer. (Round the coefficients in the model and all answers to four decimal places.)

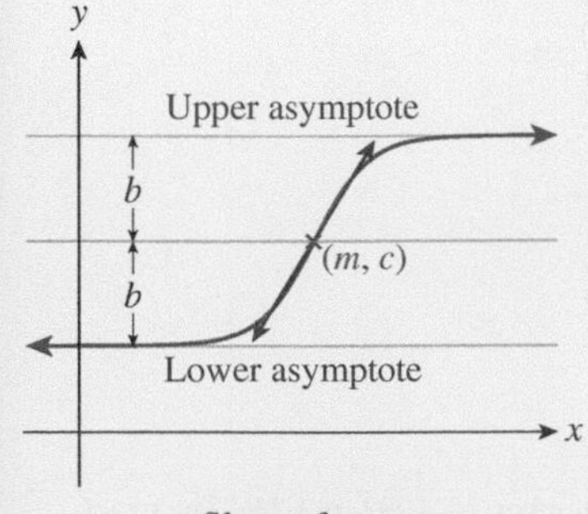

$$a = \frac{\text{Slope of tangent}}{b}$$

6. **T** ***Demand for Freon*** The demand for chlorofluorocarbon-12 (CFC-12)—the ozone-depleting refrigerant commonly known as Freon 12 (the name given to it by DuPont)—declined significantly in response to regulation and concern about the ozone layer. The chart below shows values and projections given in 1999 of the demand for CFC-12 for the period 1994–2005:[64]

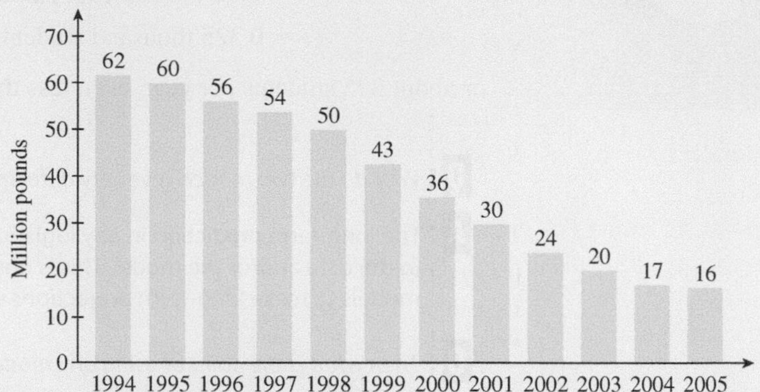

a. Use technology to obtain the best-fit equation of the form

$$N(t) = c + b\frac{a(t-m)}{1 + a|t-m|}, \qquad (a, b, c, m \text{ constant})$$

where t is the number of years since 1990. Use your function to estimate the total demand for CFC-12 from the start of the year 2000 to the start of 2010. (Start with the following values: $a = 1$, $b = -25$, $c = 35$, and $m = 10$, and round your answers to four decimal places.)

b. According to your model, how fast was the demand for Freon 12 declining in 2000?

[64] Source: The Automobile Consulting Group (*New York Times,* December 26, 1993, p. F23). The exact figures were not given, and the chart is a reasonable facsimile of the chart that appeared in the *New York Times.*

5 FURTHER APPLICATIONS OF THE DERIVATIVE

www.WanerMath.com

At the Website, in addition to the resources listed in the Preface, you will find:

The following extra topic:

- Linear Approximation and Error Estimation

CASE STUDY

Production Lot Size Management

Your publishing company is planning the production of its latest best seller, which it predicts will sell 100,000 copies each month over the coming year. The book will be printed in several batches of the same number, evenly spaced throughout the year. Each print run has a setup cost of \$5,000, a single book costs \$1 to produce, and monthly storage costs for books awaiting shipment average 1¢ per book.

To meet the anticipated demand at minimum total cost to your company, how many print runs should you plan?

SERDAR/Alamy Stock Photo

Introduction

In this chapter we begin to see the power of calculus as an optimization tool. In Chapter 2 we saw how to price an item to get the largest revenue when the demand function is linear. Using calculus, we can handle nonlinear functions, which are much more general. In Section 5.1 we show how calculus can be used to solve the problem of finding the values of a variable that lead to a maximum or minimum value of a given function. In Section 5.2 we show how this helps us in various real-world applications.

Another theme in this chapter is that calculus can help us to draw and understand the graph of a function. By the time you have completed the material in Section 5.1, you will be able to locate and sketch some of the important features of a graph, such as where it rises and where it falls. In Section 5.3 we look at the *second derivative,* the derivative of the derivative function, and what it tells us about how the graph *curves*. We also see how the second derivative is used to model the notion of *acceleration*. In Section 5.4 we put a number of ideas together that help to explain what you see in a graph (drawn, for example, using graphing technology) and to locate its most important points.

We also include sections on related rates and elasticity of demand. The first of these (Section 5.5) examines further the concept of the derivative as a rate of change. The second (Section 5.6) returns to the problem of optimizing revenue based on the demand equation, looking at it in a new way that leads to an important idea in economics: elasticity.

Precalculus Review
For this chapter you should be familiar with the algebra reviewed in **Sections 0.5 and 0.6.**

5.1 Maxima and Minima

Relative Extrema

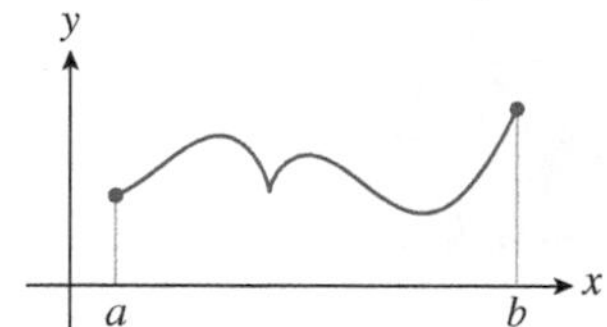

Figure 1

Figure 1 shows the graph of a function f whose domain is the closed interval $[a, b]$. A mathematician sees lots of interesting things going on here. There are hills and valleys and even a small chasm (called a *cusp*) near the center. For many purposes the important features of this curve are the highs and lows. Suppose, for example, that you know that the price of the stock of a certain company will follow this graph during the course of a week. Although you would certainly make a handsome profit if you bought at time a and sold at time b, your best strategy would be to follow the old adage to "buy low and sell high," buying at all the lows and selling at all the highs.

Figure 2 shows the graph once again with the highs and lows marked. There are names for these points: the highs (at the x-values p, r, and b) are referred to as **relative maxima**, and the lows (at the x-values a, q, and s) are referred to as **relative minima**. Collectively, these highs and lows are referred to as **relative extrema**. (A point of language: The singular forms of the plurals *minima, maxima,* and *extrema* are *minimum, maximum,* and *extremum.*)

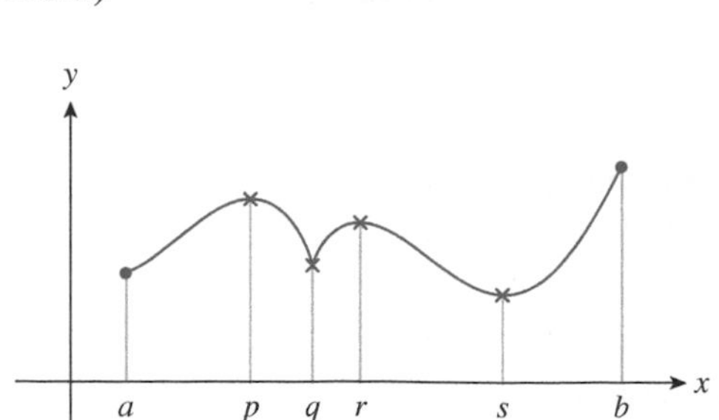

Figure 2

Why do we refer to these points as relative extrema? Take a look at the point corresponding to $x = r$. It is the highest point of the graph *compared to other points nearby*. If you were an extremely nearsighted mountaineer standing at the point where $x = r$, you would *think* that you were at the highest point of the graph, not being able to see the distant peaks at $x = p$ and $x = b$.

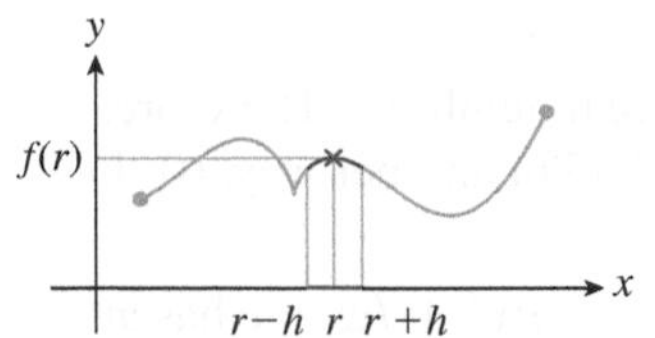

Figure 3

Let's translate into mathematical terms. We are talking about the heights of various points on the curve. The height of the curve at $x = r$ is $f(r)$, so we are saying that $f(r)$ is greater than or equal to $f(x)$ for every x near r. In other words, *$f(r)$ is the greatest value that $f(x)$ has for all choices of x between $r - h$ and $r + h$* for some (possibly small) h. (See Figure 3.)

We can phrase the formal definition as follows.

Relative Extrema: Definition

f has a **relative maximum** at r if there is some interval $(r - h, r + h)$ (even a very small one) for which $f(r) \geq f(x)$ for all x in $(r - h, r + h)$ for which $f(x)$ is defined.

f has a **relative minimum** at r if there is some interval $(r - h, r + h)$ (even a very small one) for which $f(r) \leq f(x)$ for all x in $(r - h, r + h)$ for which $f(x)$ is defined.

The relative maxima and minima are collectively referred to as **relative extrema**. If f has a relative extremum at r, then the corresponding point $(r, f(r))$ on the graph of f is also referred to as a relative maximum or relative minimum as the case may be.

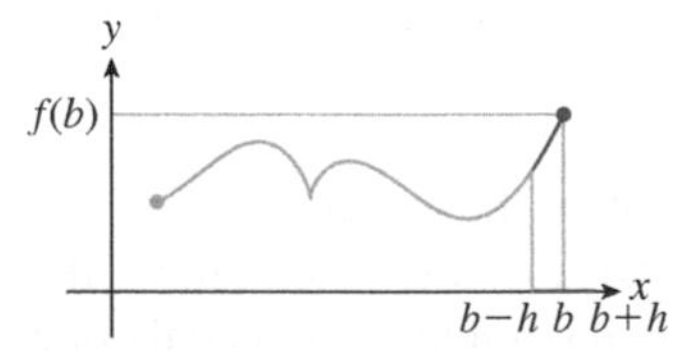

Figure 4

Quick Examples

In Figure 2, f has the following relative extrema:

1. Relative maxima at p and r.
2. A relative maximum at b. (See Figure 4.) Note that $f(x)$ is not defined for $x > b$. However, $f(b) \geq f(x)$ for every x in the interval $(b - h, b + h)$ *for which $f(x)$ is defined*—that is, for every x in $(b - h, b]$.*
3. Relative minima at a, q, and s.

* Our definition of relative extremum allows f to have a relative extremum at an endpoint of its domain; the definitions used in some books do not. In view of examples like the stock market investing strategy mentioned above, we find it more useful to allow endpoints as relative extrema.

Absolute Extrema

Looking carefully at Figure 2, we can see that the lowest point on the whole graph is where $x = s$ and the highest point is where $x = b$. This means that $f(s)$ is the least value of f on the whole domain of f (the interval $[a, b]$) and $f(b)$ is the greatest value. We call these the *absolute* minimum and maximum.

Absolute Extrema: Definition

f has an **absolute maximum** at r if $f(r) \geq f(x)$ for every x in the domain of f.

f has an **absolute minimum** at r if $f(r) \leq f(x)$ for every x in the domain of f.

As with relative extrema, if f has an *absolute* extremum at r, then the corresponding point $(r, f(r))$ on the graph of f is also referred to as an absolute maximum or absolute minimum as the case may be.

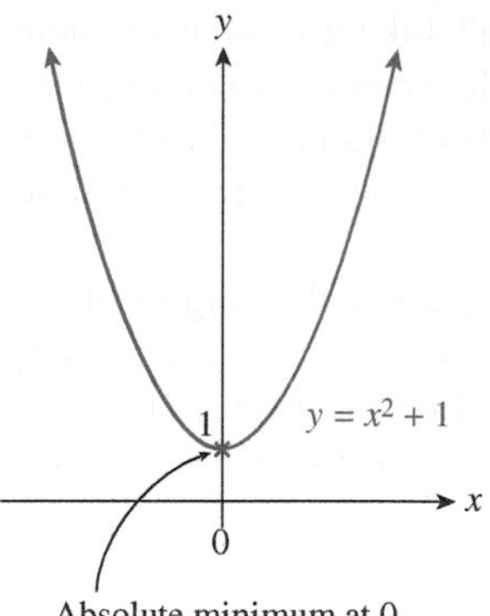

Figure 5

Quick Examples

4. In Figure 2, f has an absolute maximum at b and an absolute minimum at s.
5. If $f(x) = x^2 + 1$, then $f(x) \geq f(0)$ for every real number x. Therefore, f has an absolute minimum at 0. (See Figure 5.) Put another way, (0, 1) is an absolute minimum on the graph of f.
6. Generalizing (5), every quadratic function $f(x) = ax^2 + bx + c$ has an absolute extremum at its vertex $-b/(2a)$, an absolute minimum if $a > 0$, and an absolute maximum if $a < 0$.

A little terminology: If the point (a, b) on the graph of f represents a maximum (or minimum) of f, we will sometimes say that **f has a maximum (or minimum) value of b at a**. Thus, in Quick Example 5 we could have said any of the following:

"f has an absolute minimum value of 1 at 0."
"f has an absolute minimum at (0, 1)."
"f has an absolute minimum at 0."

Note If f has an absolute extremum at r, then it automatically satisfies the requirement for a *relative* extremum there as well; take $h = 1$ (or any other value) in the definition of relative extremum. Thus, absolute extrema are special types of relative extrema. ■

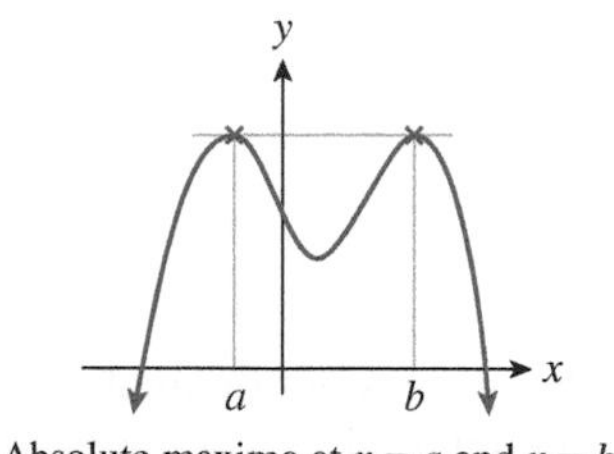

Absolute maxima at $x = a$ and $x = b$

Figure 6

Some functions have no absolute extrema at all (think of the graph of $f(x) = x$), while others might have an absolute minimum but no absolute maximum (like $f(x) = x^2$), or vice versa. When f does have an absolute maximum, there is only one absolute maximum *value* of f, but this value may occur at different values of x and similarly for absolute minima. (See Figure 6.)

Q: *At how many different values of x can f take on its absolute maximum value?*

A: An extreme case is that of a constant function; because we use ≥ in the definition of absolute maximum, a constant function has an absolute maximum (and minimum) at every point in its domain.

Locating Extrema

Now, how do we go about locating extrema? In many cases we can get a good idea by using graphing technology to zoom in on a maximum or minimum and approximate its coordinates. However, calculus gives us a way to find the exact locations of the extrema and at the same time to understand why the graph of a function behaves the way it does. In fact, it is often best to combine the powers of graphing technology with those of calculus, as we shall see.

In Figure 7 we see the graph from Figure 1 once more, but we have labeled each extreme point as one of three types. Notice that two extrema occur at endpoints of the domain and the others at interior points.* Let us look first at the extrema occurring at

* Recall from Section 3.5 that an interior point of the domain is a point a such that there is some open interval about a still in the domain. If the domain is a closed interval, an interior point is just a point other than an endpoint.

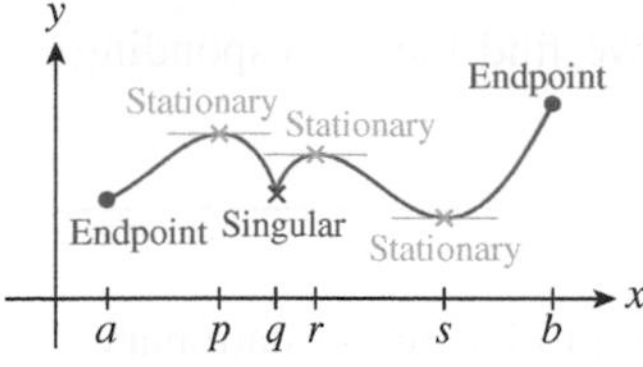

Figure 7

interior points: At the points labeled "Stationary," the tangent lines to the graph are horizontal and so have slope 0, so f' (which gives the slope) is 0. Any time $f'(x) = 0$, we say that f has a **stationary point** at x because the rate of change of f is zero there. We call an extremum that occurs at a stationary point a **stationary extremum**. In general, to find the exact location of each stationary point, we need to solve the equation $f'(x) = 0$. Note that stationary points are always interior points, as f' is defined only at interior points. (See the definition of the derivative in Section 3.5.)

There is a relative minimum in Figure 7 at $x = q$, but there is no horizontal tangent there; in fact, there is no tangent line at all. In the language of calculus, $f'(q)$ is not defined. (Recall a similar situation with the graph of $f(x) = |x|$ at $x = 0$.) We will say that the derivative f' has a **singular point** at q, and we call an extremum that occurs at a singular point a **singular extremum**. The interior points x of the domain of f that are either stationary points of f (i.e., $f'(x) = 0$) or singular points of f' (i.e., $f'(x)$ is not defined) we call collectively the **critical points** of f.

The remaining two extrema are at the **endpoints** of the domain (remember that we do allow relative extrema at endpoints). As we see in the figure, they are (almost) always either relative maxima or relative minima.

We bring all the above information together in Figure 8.

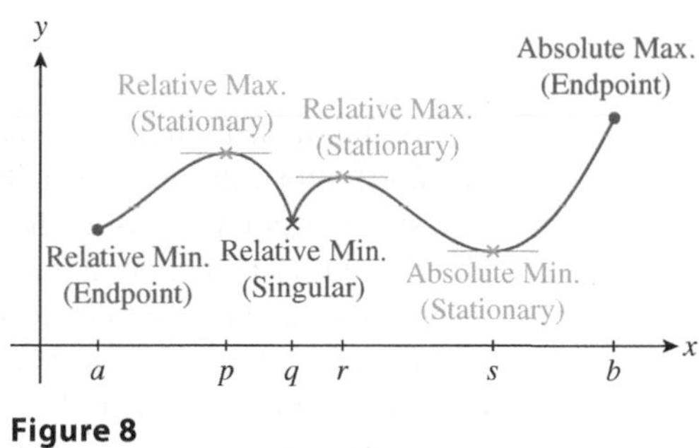

Figure 8

Q: *Are there any other types of relative extrema?*

A: No; relative extrema of a function always occur at critical points or endpoints. (A rigorous proof is beyond the scope of this book.)*

Locating Candidates for Extrema

If f is a real-valued function, then its extrema occur among the following types of points:

1. **Stationary Points:** points in the interior of the domain where the derivative is zero. To locate stationary points, set $f'(x) = 0$ and solve for x.
2. **Singular Points:**† points in the interior of the domain where the derivative is not defined. To locate singular points, find interior points x where $f'(x)$ is *not* defined but $f(x)$ *is* defined.
3. **Endpoints:** These are the endpoints, if any, of the domain. Recall that closed intervals contain endpoints but open intervals do not. If the domain of f is an open interval or the whole real line, then there are no endpoints.

* Here is an outline of the argument. Suppose f has a maximum, say, at $x = a$, at some interior point of its domain. Then either f is differentiable there, or it is not. If it is not, then we have a singular point. If f is differentiable at $x = a$, then consider the slope of the secant line through the points where $x = a$ and $x = a + h$ for small positive h. Because f has a maximum at $x = a$, it is falling (or level) to the right of $x = a$, so the slope of this secant line must be ≤ 0. Thus, we must have $f'(a) \le 0$ in the limit as $h \to 0$. On the other hand, if h is small and *negative*, then the corresponding secant line must have slope ≥ 0 because f is also falling (or level) as we move left from $x = a$, and so $f'(a) \ge 0$. Because $f'(a)$ is both ≥ 0 and ≤ 0, it must be zero, so we have a stationary point at $x = a$.

† In the context of maxima and minima, "singular point" will be understood to refer to a singular point *of the derivative* f', that is, a point in the interior of the domain where f' is undefined. ("Singular point" has another meaning in the context of limits, where it refers to a singular point of f; see the definition in Section 3.2.) When the context is not clear, we will say "singular point of f" or "singular point of f'" as the case may be.

Once we have a candidate for an extremum of f, we find the corresponding point (x, y) on the graph of f using $y = f(x)$.

Quick Examples

7. **Stationary Points:** Let $f(x) = x^3 - 12x$. Then to locate the stationary points, set $f'(x) = 0$ and solve for x. This gives $3x^2 - 12 = 0$, so f has stationary points at $x = \pm 2$. The corresponding points on the graph are $(-2, f(-2)) = (-2, 16)$ and $(2, f(2)) = (2, -16)$.
8. **Singular Points:** Let $f(x) = 3(x - 1)^{1/3}$. Then $f'(x) = (x - 1)^{-2/3} = 1/(x - 1)^{2/3}$. $f'(1)$ is not defined, although $f(1)$ *is* defined. Thus, the (only) singular point occurs at $x = 1$. The corresponding singular point on the graph is $(1, f(1)) = (1, 0)$.
9. **Endpoints:** Let $f(x) = 1/x$, with domain $(-\infty, 0) \cup [1, +\infty)$. Then the only endpoint in the domain of f occurs at $x = 1$. The corresponding endpoint on the graph is $(1, 1)$. The natural domain of $1/x$, on the other hand, has no endpoints.

Remember, though, that the three types of points we identify above are only *candidates* for extrema. It is quite possible, as we shall see, to have a stationary point or a singular point that is neither a maximum nor a minimum. (It is also possible for an endpoint to be neither a maximum nor a minimum, but this occurs only in functions whose graphs are rather bizarre—see Exercise 65.)

Now let's look at some examples of finding maxima and minima. In all of these examples we will use the following procedure: First, we find the derivative, which we examine to find the stationary points and singular points. Next, we make a table listing the x-coordinates of the critical points and endpoints, together with their y-coordinates. We use this table to make a rough sketch of the graph. From the table and rough sketch we usually have enough data to be able to say where the extreme points are and what kind they are.

EXAMPLE 1 Maxima and Minima

Find the relative and absolute maxima and minima of

$$f(x) = x^2 - 2x$$

on the interval $[0, 4]$.

Solution We first calculate $f'(x) = 2x - 2$. We use this derivative to locate the critical points (stationary and singular points).

Stationary Points To locate the stationary points, we solve the equation $f'(x) = 0$, or

$$2x - 2 = 0,$$

getting $x = 1$. The domain of the function is $[0, 4]$, so $x = 1$ is in the interior of the domain. Thus, the only candidate for a stationary relative extremum occurs when $x = 1$.

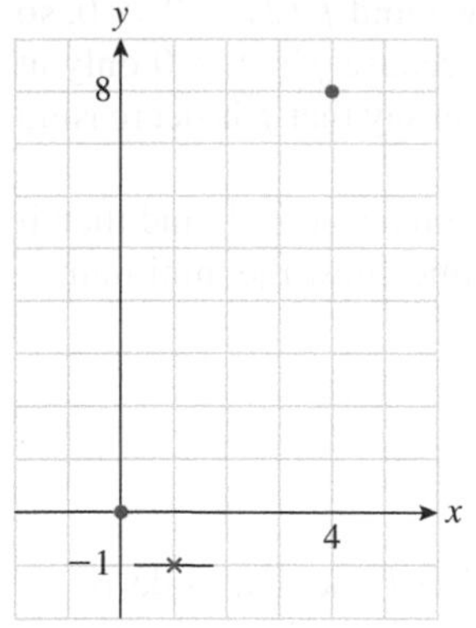

Figure 9

Singular Points We look for interior points where the derivative is not defined. However, the derivative is $2x - 2$, which is defined for every x. Thus, there are no singular points and hence no candidates for singular relative extrema.

Endpoints The domain is $[0, 4]$, so the endpoints occur when $x = 0$ and $x = 4$.

We record these values of x in a table, together with the corresponding y-coordinates (values of f):

x	0	1	4
$f(x) = x^2 - 2x$	0	−1	8

This gives us three points on the graph, $(0, 0)$, $(1, -1)$, and $(4, 8)$, which we plot in Figure 9. We remind ourselves that the point $(1, -1)$ is a stationary point of the graph by drawing in a part of the horizontal tangent line. Connecting these points must give us a graph something like that in Figure 10.

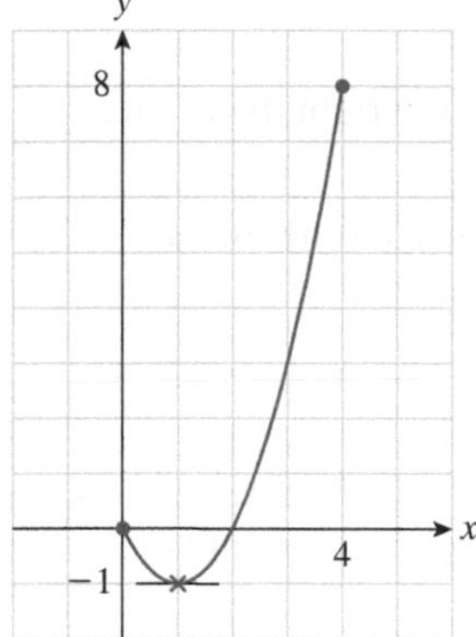

Figure 10

From Figure 10 we can see that f has the following extrema:

x	$y = x^2 - 2x$	Classification
0	0	Relative maximum (endpoint)
1	−1	Absolute minimum (stationary point)
4	8	Absolute maximum (endpoint)

➡ Before we go on . . .

Q: *How can we be sure that the graph in Example 1 doesn't look like Figure 11?*

A: If it did, there would be another critical point somewhere between $x = 1$ and $x = 4$. But we already know that there aren't any other critical points. The table we made listed all of the possible extrema; there can be no more.

■

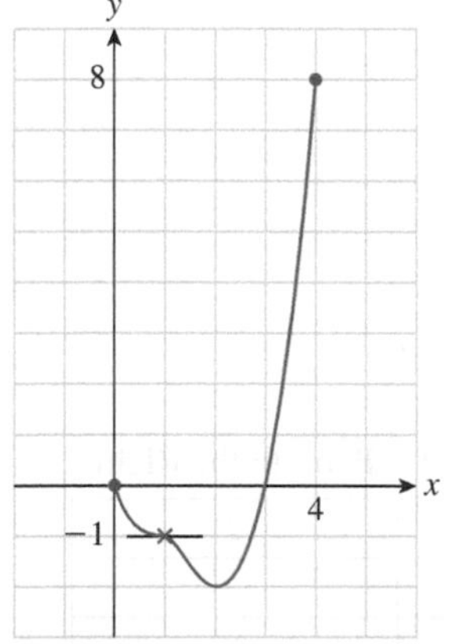

Figure 11

First Derivative Test

The **first derivative test*** gives another, very systematic, way of checking whether a critical point is a maximum or minimum. To motivate the first derivative test, consider again the critical point $x = 1$ in Example 1. If we look at some values of $f'(x)$ to the left and right of the critical point, we obtain the information shown in the following table:

	Point to the Left	Critical Point	Point to the Right
x	0.5	1	2
$f'(x) = 2x - 2$	−1	0	2
Direction of Graph	↘	→	↗

* Why "first" derivative test? To distinguish it from a test based on the **second derivative** of a function, which we shall discuss in Section 5.3.

At $x = 0.5$ (to the left of the critical point) we see that $f'(0.5) = -1 < 0$, so the graph has negative slope and f is decreasing. We note this with the downward pointing arrow. At $x = 2$ (to the right of the critical point) we find $f'(2) = 2 > 0$, so the graph has positive slope and f is increasing. In fact, because $f'(x) = 0$ only at $x = 1$, we know that $f'(x) < 0$ for all x in $(0, 1)$, and we can say that f is decreasing on the interval $(0, 1)$. Similarly, f is increasing on $(1, 4)$.

So starting at $x = 0$, the graph of f goes down until we reach $x = 1$, and then it goes back up, telling us that $x = 1$ must be a minimum. Notice how the minimum is suggested by the arrows to the left and right.

First Derivative Test for Extrema

Suppose that c is a critical point of the continuous function f and that its derivative is defined for x close to, and on both sides of, $x = c$. Then, determine the sign of the derivative to the left and right of $x = c$.

1. If $f'(x)$ is positive to the left of $x = c$ and negative to the right, then f has a maximum at $x = c$.
2. If $f'(x)$ is negative to the left of $x = c$ and positive to the right, then f has a minimum at $x = c$.
3. If $f'(x)$ has the same sign on both sides of $x = c$, then f has neither a maximum nor a minimum at $x = c$.

Quick Examples

10. In Example 1 we saw that $f(x) = x^2 - 2x$ has a critical point at $x = 1$ with $f'(x)$ negative to the left of $x = 1$ and positive to the right (see the table). Therefore, f has a minimum at $x = 1$.
11. Here is a graph showing a function f with a singular point at $x = 1$:

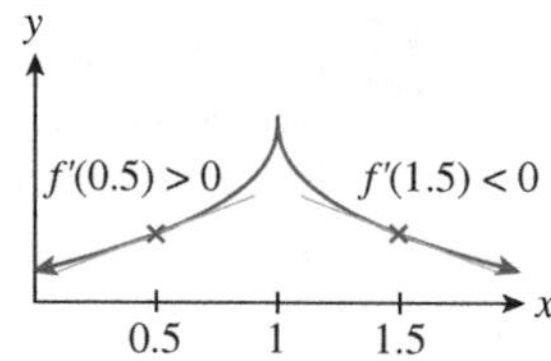

The graph gives us the information shown in the table:

	Point to the Left	Critical Point	Point to the Right
x	0.5	1	2
$f'(x)$	+	Undefined	−
Direction of Graph	↗		↘

Since $f'(x)$ is positive to the left of $x = 1$ and negative to the right, we see that f has a maximum at $x = 1$. (Notice again how this is suggested by the direction of the arrows.)

EXAMPLE 2 **Unbounded Interval**

Find all extrema of $f(x) = 3x^4 - 4x^3$ on $[-1, \infty)$.

Solution We first calculate $f'(x) = 12x^3 - 12x^2$.

Stationary Points We solve the equation $f'(x) = 0$, which is

$$12x^3 - 12x^2 = 0 \text{ or}$$
$$12x^2(x - 1) = 0.$$

There are two solutions, $x = 0$ and $x = 1$, and both are in the domain. These are our candidates for the x-coordinates of stationary extrema.

Singular Points There are no interior points where $f'(x)$ is not defined, so there are no singular points.

Endpoints The domain is $[-1, \infty)$, so there is one endpoint, at $x = -1$.

We record these points in a table with the corresponding y-coordinates:

x	-1	0	1
$f(x) = 3x^4 - 4x^3$	7	0	-1

We will illustrate three methods we can use to determine which are minima, which are maxima, and which are neither:

1. Plot these points, and sketch the graph by hand.
2. Use the first derivative test.
3. Use technology to help us.

Use the method you find most convenient.

(a) (b)

Figure 12

Using a Hand Plot: If we plot these points by hand, we obtain Figure 12(a), which suggests Figure 12(b).

We can't be sure what happens to the right of $x = 1$. Does the curve go up, or does it go down? To find out, let's plot a "test point" to the right of $x = 1$. Choosing $x = 2$, we obtain $y = 3(2)^4 - 4(2)^3 = 16$, so $(2, 16)$ is another point on the graph. Thus, it must turn upward to the right of $x = 1$, as shown in Figure 13.

From the graph, we find that f has the following extrema:

A relative (endpoint) maximum at $(-1, 7)$

An absolute (stationary) minimum at $(1, -1)$

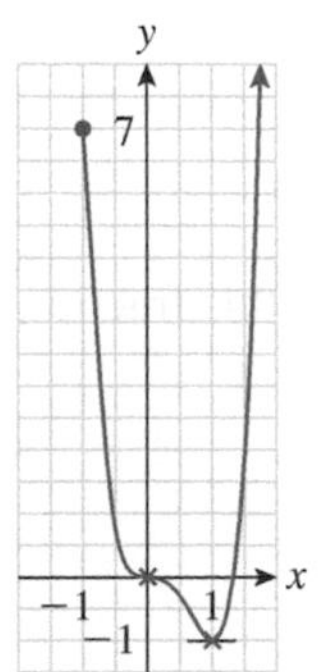

Figure 13

Using the First Derivative Test: List the critical points in a table, and add additional points as necessary so that each critical point has a noncritical point on either side. Then compute the derivative at each of these points, and draw an arrow to indicate the direction of the graph, suggesting the shape of the curve in Figure 13.

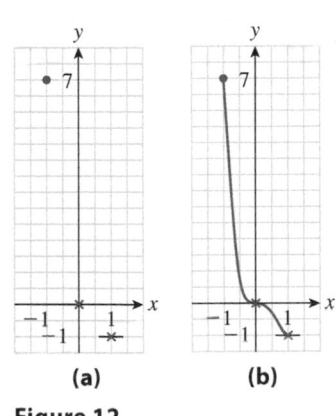

		Critical Point		**Critical Point**	
x	-0.5	0	0.5	1	2
$f'(x) = 12x^3 - 12x^2$	-1.5	0	-1.5	0	48
Direction of Graph	↘	→	↘	→	↗

The first derivative test tells us that the function has a relative maximum at $x = -1$, neither a maximum nor a minimum at $x = 0$, and a relative minimum at $x = 1$. Deciding which of these extrema are absolute and which are relative requires us to compute y-coordinates and plot the corresponding relative extrema on the graph by hand, as we did in the first method.

Using Technology

If we use technology to show the graph, we should choose the viewing window so that it contains the three interesting points we found: $x = -1$, $x = 0$, and $x = 1$. Again, we can't be sure yet what happens to the right of $x = 1$; does the graph go up or down from that point? If we set the viewing window to an interval of $[-1, 2]$ for x and $[-2, 8]$ for y, we will leave enough room to the right of $x = 1$ and below $y = -1$ to see what the graph will do. The result will be something like Figure 14.

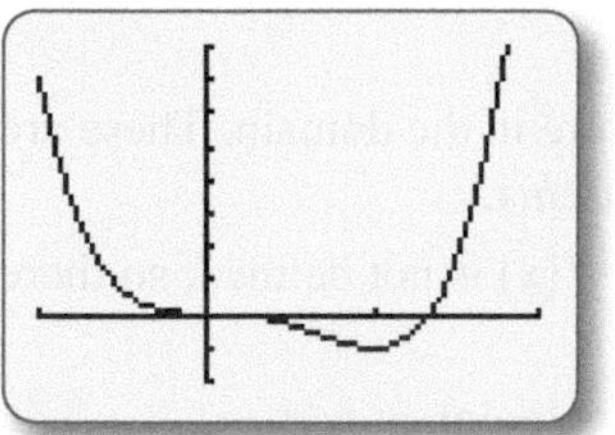

Figure 14

Now we can tell what happens to the right of $x = 1$: the function increases. We know that it cannot later decrease again because if it did, there would have to be another critical point where it turns around, and we found that there are no other critical points.

➡ **Before we go on . . .** Notice that the stationary point at $x = 0$ in Example 2 is neither a relative maximum nor a relative minimum. It is simply a place where the graph of f flattens out for a moment before it continues to fall. Notice also that f has no absolute maximum because $f(x)$ increases without bound as x gets large. ■

EXAMPLE 3 Singular Point

Find all extrema of $f(t) = t^{2/3}$ on $[-1, 1]$.

Solution First, $f'(t) = \frac{2}{3}t^{-1/3}$.

Stationary Points We need to solve

$$\frac{2}{3}t^{-1/3} = 0.$$

We can rewrite this equation without the negative exponent:

$$\frac{2}{3t^{1/3}} = 0.$$

Now, the only way a fraction can equal 0 is if the numerator is 0, so this fraction can never equal 0. Thus, there are no stationary points.

Singular Points The derivative

$$f'(t) = \frac{2}{3t^{1/3}}$$

is not defined for $t = 0$. Also, 0 is in the interior of the domain of f (although f' is not defined at $t = 0$, f itself is). Thus, f has a singular point at $t = 0$.

Endpoints There are two endpoints: -1 and 1.

We now put these three points in a table with the corresponding y-coordinates:

t	-1	0	1
$f(t)$	1	0	1

Using a Hand Plot: The derivative, $f'(t) = 2/(3t^{1/3})$, is not defined at the singular point $t = 0$. To help us sketch the graph, let's use limits to investigate what happens to the derivative as we approach 0 from either side:

$$\lim_{t\to 0^-} f'(t) = \lim_{t\to 0^-} \frac{2}{3t^{1/3}} = -\infty$$

$$\lim_{t\to 0^+} f'(t) = \lim_{t\to 0^+} \frac{2}{3t^{1/3}} = +\infty.$$

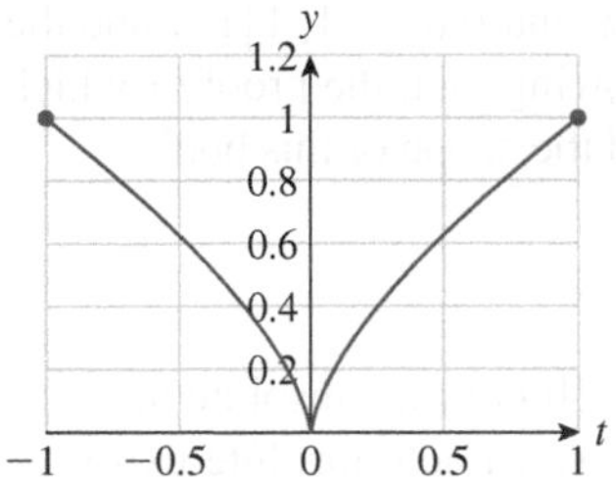

Figure 15

Thus, the graph decreases very steeply, approaching $t = 0$ from the left, and then rises very steeply as it leaves to the right. It would make sense to say that the tangent line at $x = 0$ is vertical, as seen in Figure 15.

From this graph we find the following extrema for f:

An absolute (endpoint) maximum at $(-1, 1)$
An absolute (singular) minimum at $(0, 0)$
An absolute (endpoint) maximum at $(1, 1)$.

Notice that the absolute maximum value of f is achieved at two values of t: $t = -1$ and $t = 1$.

First Derivative Test: Here is the corresponding table for the first derivative test:

t	-0.5	0	0.5
$f'(t) = \dfrac{2}{3t^{1/3}}$	$-\dfrac{2}{3(0.5)^{1/3}}$	Undefined	$\dfrac{2}{3(0.5)^{1/3}}$
Direction of Graph	↘	↕	↗

(We drew a vertical arrow at $t = 0$ to indicate a vertical tangent.) Again, notice how the arrows suggest the shape of the curve in Figure 15, and the first derivative test confirms that we have a minimum at $t = 0$.

Using Technology

Because there is only one critical point, at $t = 0$, it is clear from this table that f must decrease from $t = -1$ to $t = 0$ and then increase from $t = 0$ to $t = 1$. To graph f using technology, choose a viewing window with an interval of $[-1, 1]$ for t and $[0, 1]$ for y. The result will be something like Figure 15.*

* Many graphing calculators will give you only the right-hand half of the graph shown in Figure 15 because fractional powers of negative numbers are not, in general, real numbers. To obtain the whole curve, enter the formula as `Y=(x^2)^(1/3)`, a fractional power of the nonnegative function x^2.

In Examples 1 and 3 we could have found the absolute maxima and minima without doing any graphing. In Example 1, after finding the critical points and endpoints, we created the following table:

x	0	1	4
$f(x)$	0	-1	8

From this table we can see that f must decrease from its value of 0 at $x = 0$ to -1 at $x = 1$, and then increase to 8 at $x = 4$. The value of 8 must be the largest value it takes on, and the value of -1 must be the smallest, on the interval $[0, 4]$. Similarly, in Example 3 we created the following table:

t	-1	0	1
$f(t)$	1	0	1

From this table we can see that the largest value of f on the interval $[-1, 1]$ is 1 and the smallest value is 0. We are taking advantage of the following fact, the proof of which uses some deep and beautiful mathematics (alas, beyond the scope of this book).

Extreme Value Theorem

If f is *continuous* on a *closed interval* $[a, b]$, then it will have an absolute maximum and an absolute minimum value on that interval. Each absolute extremum must occur at either an endpoint or a critical point. Therefore, the absolute maximum is the largest value in a table of the values of f at the endpoints and critical points, and the absolute minimum is the smallest value.

Quick Example

12. The function $f(x) = 3x - x^3$ on the interval $[0, 2]$ has one critical point at $x = 1$. The values of f at the critical point and the endpoints of the interval are given in the following table:

	Endpoint	Critical Point	Endpoint
x	0	1	2
$f(x)$	0	2	-2

From this table we can say that the absolute maximum value of f on $[0, 2]$ is 2, which occurs at $x = 1$, and the absolute minimum value of f is -2, which occurs at $x = 2$.

As we can see in Example 2 and the following examples, if the domain is not a closed interval, then f may not have an absolute maximum and minimum, and a table of values as above is of little help in determining whether it does.

EXAMPLE 4 Domain Not a Closed Interval

Find all extrema of $f(x) = x + \dfrac{1}{x}$.

Solution Because no domain is specified, we take the domain to be as large as possible. The function is not defined at $x = 0$ but is defined at all other points, so we take its domain to be $(-\infty, 0) \cup (0, +\infty)$. We calculate

$$f'(x) = 1 - \frac{1}{x^2}.$$

Stationary Points Setting $f'(x) = 0$, we solve

$$1 - \frac{1}{x^2} = 0$$

to find $x = \pm 1$. Calculating the corresponding values of f, we get the two stationary points $(1, 2)$ and $(-1, -2)$.

Singular Points The only value of x for which $f'(x)$ is not defined is $x = 0$, but then f is not defined there either, so there are no singular points in the domain.

Endpoints The domain, $(-\infty, 0) \cup (0, +\infty)$, has no endpoints.

From this scant information it is hard to tell what f does. If we are sketching the graph by hand, or using the first derivative test, we will need to plot additional "test points" to the left and right of the stationary points $x = \pm 1$. Instead, we will use technology to draw the graph:

Using Technology

For the technology approach, let's choose a viewing window with an interval of $[-3, 3]$ for x and $[-4, 4]$ for y, which should leave plenty of room to see how f behaves near the stationary points. The result is something like Figure 16.

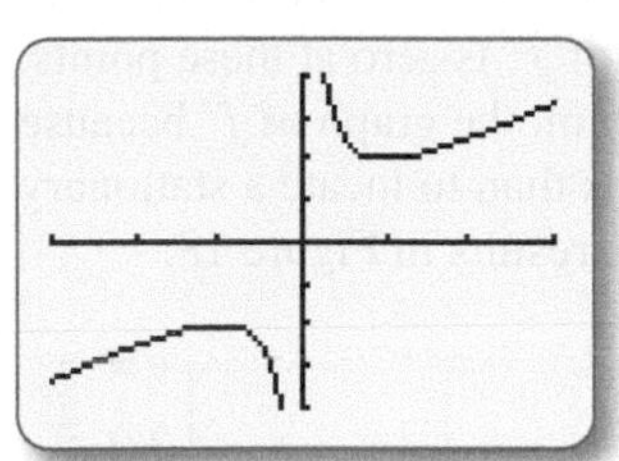

Figure 16

From this graph we can see that f has

a relative (stationary) maximum at $(-1, -2)$,

a relative (stationary) minimum at $(1, 2)$.

Curiously, the relative maximum is lower than the relative minimum! Notice also that, because of the break in the graph at $x = 0$, the graph did not need to rise to get from $(-1, -2)$ to $(1, 2)$.

So far, we have been solving the equation $f'(x) = 0$ to obtain our candidates for stationary extrema. However, it is often not easy—or even possible—to solve equations analytically. In the next example, we show a way around this problem by using graphing technology.

EXAMPLE 5 T Finding Approximate Extrema Using Technology

Graph the function $f(x) = (x - 1)^{2/3} - \frac{x^2}{2}$ with domain $[-2, +\infty)$. Also graph its derivative and hence locate and classify all extrema of f, with coordinates accurate to two decimal places.

Solution In Example 4 of Section 3.5 we saw how to draw the graphs of functions and their derivatives using technology. Note that the technology formula to use for the graph of f is

```
((x-1)^2)^(1/3)-0.5*x^2
```

instead of

```
(x-1)^(2/3)-0.5*x^2
```

(Why?)

Figure 17 shows the resulting graphs of f and f'.

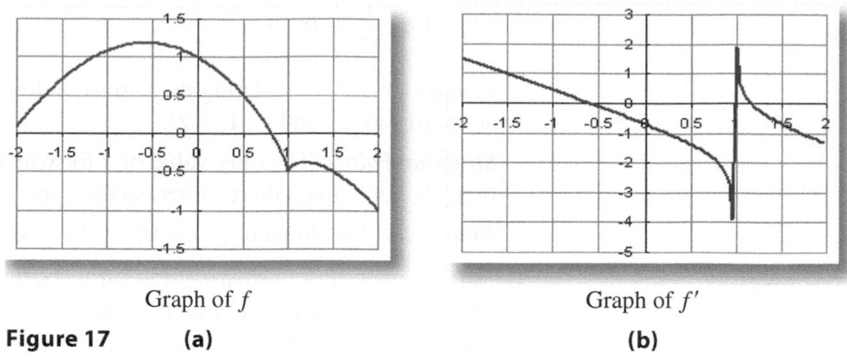

Graph of f Graph of f'

Figure 17 **(a)** **(b)**

If we extend Xmax beyond $x = 2$, we find that the graph continues downward, apparently without any further interesting behavior.

Stationary Points The graph of f shows two stationary points, both maxima, at around $x = -0.6$ and $x = 1.2$. Notice that the graph of f' is zero at these points. Moreover, it is easier to locate these values accurately on the graph of f' because it is easier to pinpoint where a graph crosses the x-axis than to locate a stationary point. Zooming in to the stationary point at $x \approx -0.6$ results in Figure 18.

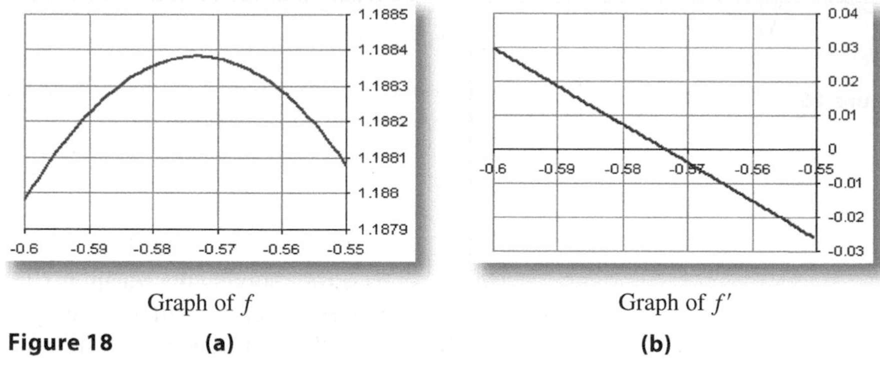

Graph of f Graph of f'

Figure 18 **(a)** **(b)**

From the graph of f we can see that the stationary point is somewhere between -0.58 and -0.57. The graph of f' shows more clearly that the zero of f', hence the stationary point of f, lies somewhat closer to -0.57 than to -0.58. Thus, the stationary point occurs at $x \approx -0.57$, rounded to two decimal places.

In a similar way we find the second stationary point at $x \approx 1.18$.

Singular Points Going back to Figure 17, we notice what appears to be a cusp (singular point) at the relative minimum around $x = 1$, and this is confirmed by a glance at the graph of f', which seems to take a sudden jump at that value. Zooming in closer suggests that the singular point occurs at exactly $x = 1$. In fact, we can calculate

$$f'(x) = \frac{2}{3(x-1)^{1/3}} - x.$$

From this formula we see clearly that $f'(x)$ is defined everywhere except at $x = 1$.

Endpoints The only endpoint in the domain is $x = -2$, which gives a relative minimum.

Thus, we have found the following approximate extrema for f:

A relative (endpoint) minimum at $(-2, 0.08)$
An absolute (stationary) maximum at $(-0.57, 1.19)$
A relative (singular) minimum at $(1, -0.5)$
A relative (stationary) maximum at $(1.18, -0.38)$

5.1 EXERCISES

▼ more advanced ◆ challenging
T indicates exercises that should be solved using technology

In Exercises 1–12, locate and classify all extrema in each graph. (By classifying the extrema, we mean listing whether each extremum is a relative or absolute maximum or minimum.) Also, locate any stationary points or singular points that are not relative extrema. [**HINT:** See the box titled "Locating Candidates for Extrema."]

1. 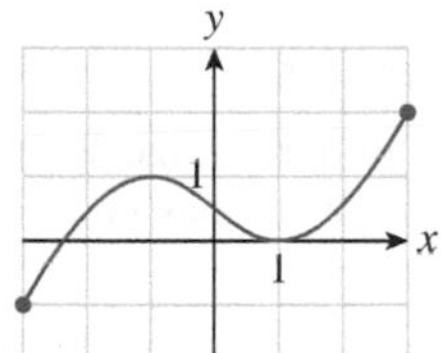

2. 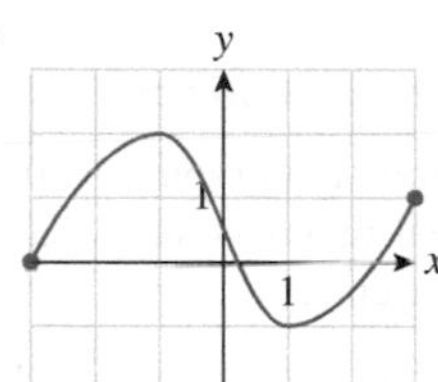

3. 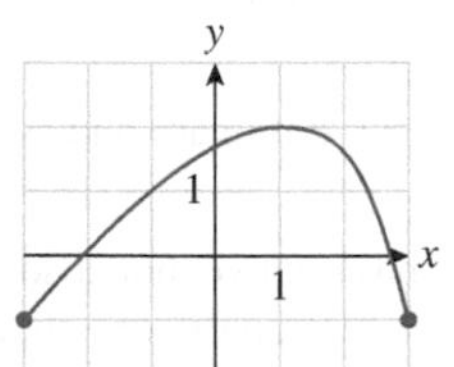

4. 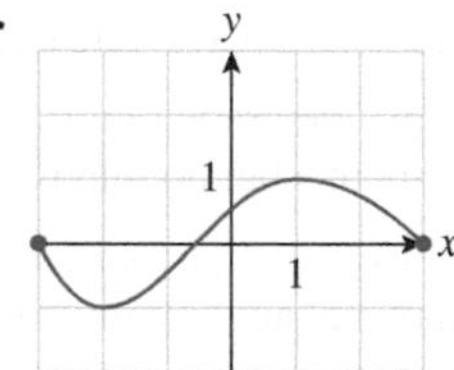

5. 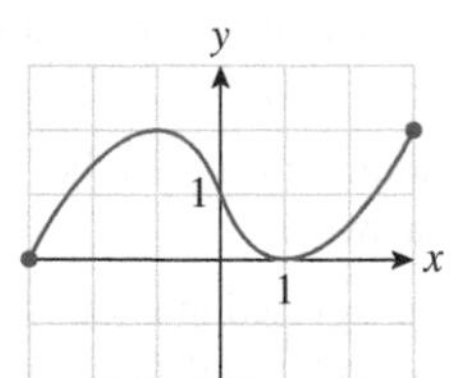

6.

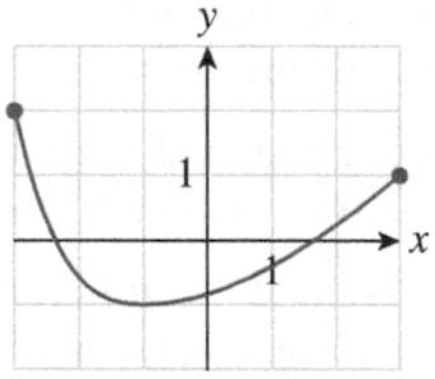

7. 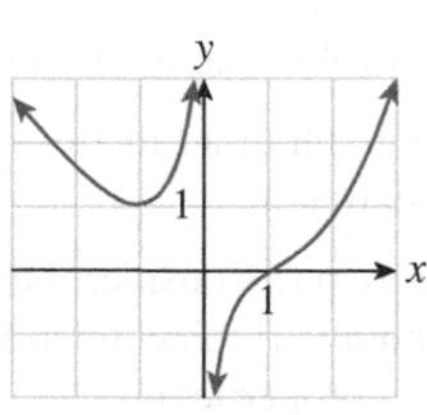

8. 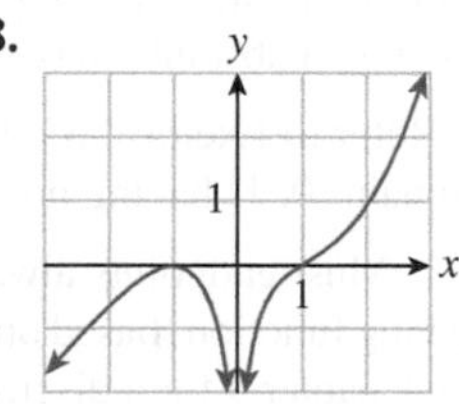

9. 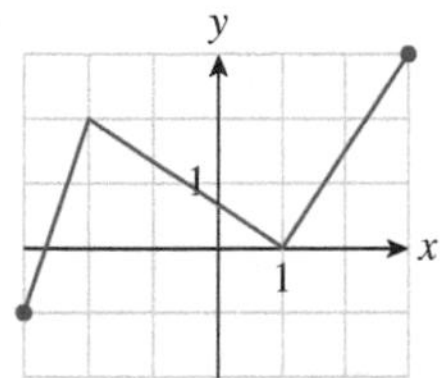

10. 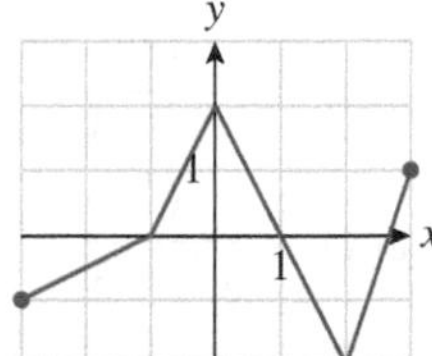

11. 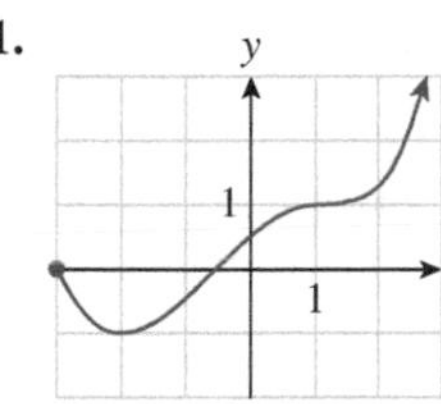

12. 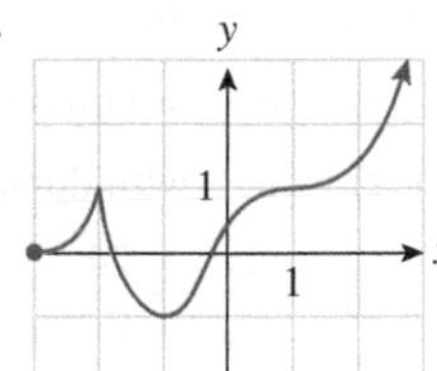

In Exercises 13–44, find the exact location of all the relative and absolute extrema of the given function. [**HINT:** See Example 1.]

13. $f(x) = x^2 - 4x + 1$ with domain $[0, 3]$
14. $f(x) = 2x^2 - 2x + 3$ with domain $[0, 3]$
15. $g(x) = x^3 - 12x$ with domain $[-4, 4]$
16. $g(x) = 2x^3 - 6x + 3$ with domain $[-2, 2]$
17. $f(t) = t^3 + t$ with domain $[-2, 2]$
18. $f(t) = -2t^3 - 3t$ with domain $[-1, 1]$
19. $h(t) = 2t^3 + 3t^2$ with domain $[-2, +\infty)$ [**HINT:** See Example 2.]
20. $h(t) = t^3 - 3t^2$ with domain $[-1, +\infty)$ [**HINT:** See Example 2.]
21. $f(x) = x^4 - 4x^3$ with domain $[-1, +\infty)$
22. $f(x) = 3x^4 - 2x^3$ with domain $[-1, +\infty)$
23. $g(t) = \frac{1}{4}t^4 - \frac{2}{3}t^3 + \frac{1}{2}t^2$ with domain $(-\infty, +\infty)$
24. $g(t) = 3t^4 - 16t^3 + 24t^2 + 1$ with domain $(-\infty, +\infty)$
25. $h(x) = (x - 1)^{2/3}$ with domain $[0, 2]$ [**HINT:** See Example 3.]
26. $h(x) = (x + 1)^{2/5}$ with domain $[-2, 0]$ [**HINT:** See Example 3.]
27. $k(x) = \frac{2x}{3} + (x + 1)^{2/3}$ with domain $(-\infty, 0]$

28. $k(x) = \dfrac{2x}{5} - (x - 1)^{2/5}$ with domain $[0, +\infty)$

29. ▼ $f(t) = \dfrac{t^2 + 1}{t^2 - 1}$; $-2 \le t \le 2, t \ne \pm 1$

30. ▼ $f(t) = \dfrac{t^2 - 1}{t^2 + 1}$ with domain $[-2, 2]$

31. ▼ $f(x) = \sqrt{x}(x - 1); x \ge 0$

32. ▼ $f(x) = \sqrt{x}(x + 1); x \ge 0$

33. ▼ $g(x) = x^2 - 4\sqrt{x}$

34. ▼ $g(x) = \dfrac{1}{x} - \dfrac{1}{x^2}$

35. ▼ $g(x) = \dfrac{x^3}{x^2 + 3}$

36. ▼ $g(x) = \dfrac{x^3}{x^2 - 3}$

37. ▼ $f(x) = x - \ln x$ with domain $(0, +\infty)$

38. ▼ $f(x) = x - \ln x^2$ with domain $(0, +\infty)$

39. ▼ $g(t) = e^t - t$ with domain $[-1, 1]$

40. ▼ $g(t) = e^{-t^2}$ with domain $(-\infty, +\infty)$

41. ▼ $f(x) = \dfrac{2x^2 - 24}{x + 4}$

42. ▼ $f(x) = \dfrac{x - 4}{x^2 + 20}$

43. ▼ $f(x) = xe^{1-x^2}$

44. ▼ $f(x) = x \ln x$ with domain $(0, +\infty)$

In Exercises 45–48, use graphing technology and the method in Example 5 to find the x-coordinates of the critical points, accurate to two decimal places. Find all relative and absolute maxima and minima. [**HINT:** See Example 5.]

45. T $y = x^2 + \dfrac{1}{x - 2}$ with domain $(-3, 2) \cup (2, 6)$

46. T $y = x^2 - 10(x - 1)^{2/3}$ with domain $(-4, 4)$

47. T $f(x) = (x - 5)^2(x + 4)(x - 2)$ with domain $[-5, 6]$

48. T $f(x) = (x + 3)^2(x - 2)^2$ with domain $[-5, 5]$

In Exercises 49–56 the graph of the derivative *of a function f is shown. Determine the x-coordinates of all stationary and singular points of f, and classify each as a relative maximum, a relative minimum, or neither. (Assume that $f(x)$ is defined and continuous everywhere in $[-3, 3]$.)* [**HINT:** See Example 5.]

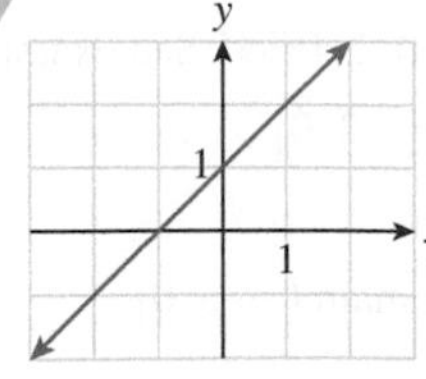

50. ▼

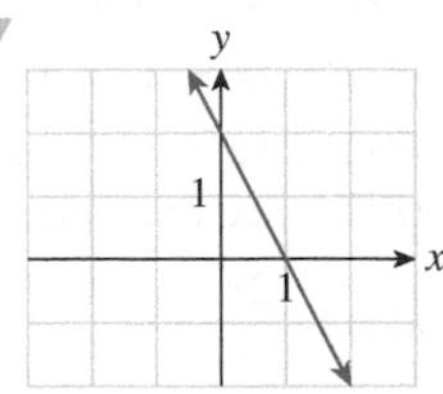

51. ▼

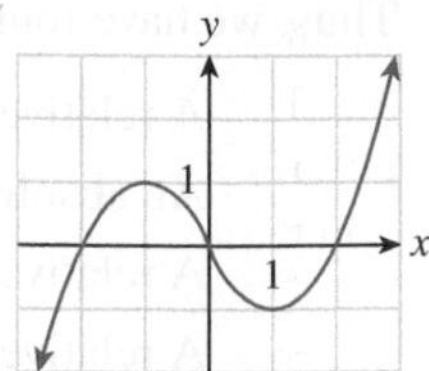

52. ▼

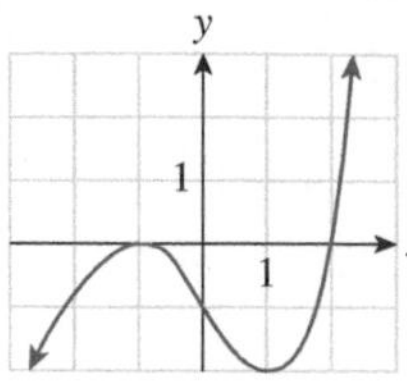

53. ▼

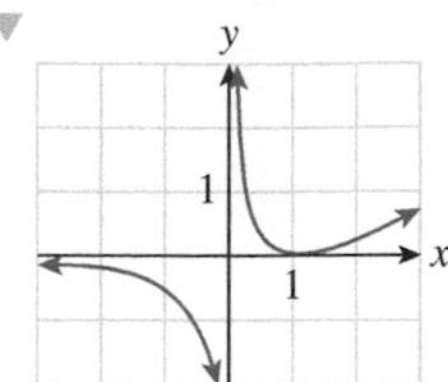

54. ▼

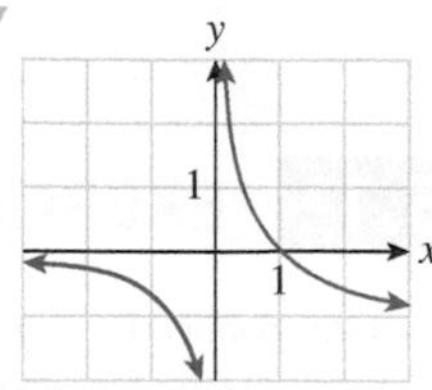

55. ▼

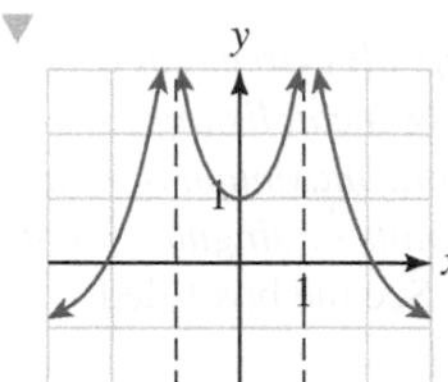

56. ▼

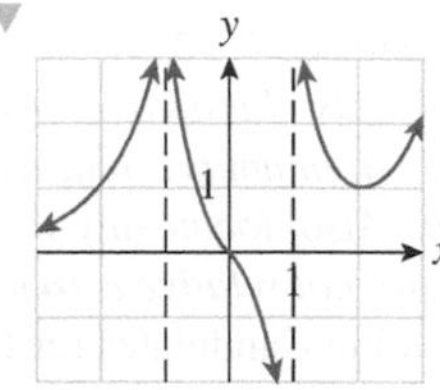

Communication and Reasoning Exercises

57. Draw the graph of a function f with domain the set of all real numbers such that f is not linear and has no relative extrema.

58. Draw the graph of a function g with domain the set of all real numbers such that g has a relative maximum and minimum but no absolute extrema.

59. Draw the graph of a function that has stationary and singular points but no relative extrema.

60. Draw the graph of a function that has relative, not absolute, maxima and minima but has no stationary or singular points.

61. If a stationary point is not a relative maximum, then must it be a relative minimum? Explain your answer.

62. If one endpoint is a relative maximum, must the other be a relative minimum? Explain your answer.

63. ▼ We said that if f is continuous on a closed interval $[a, b]$, then it will have an absolute maximum and an absolute minimum. Draw the graph of a function with domain $[0, 1]$ having an absolute maximum but no absolute minimum.

64. ▼ Refer to Exercise 63. Draw the graph of a function with domain $[0, 1]$ having no absolute extrema.

65. T ▼ Must endpoints always be extrema? Consider the following function (based on the trigonometric sine function—see Chapter 9 for a discussion of its properties):

$$f(x) = \begin{cases} x \sin\left(\dfrac{1}{x}\right) & \text{if } x > 0 \\ 0 & \text{if } x = 0. \end{cases}$$

Technology formula:
```
x*sin(1/x)
```

Graph this function using the technology formula above for $0 \le x \le h$, choosing smaller and smaller values of h, and decide whether f has either a relative maximum or relative minimum at the endpoint $x = 0$. Explain your answer. [Note: Very few graphers can draw this curve accurately; the grapher on the Website does a good job (you can increase the number of points to plot for more beautiful results), the grapher that comes with Mac computers is probably among the best, while the TI-83/84 Plus is probably among the worst.]

66. T ▼ Refer to Exercise 65, and consider the function

$$f(x) = \begin{cases} x^2 \sin\left(\frac{1}{x}\right) & \text{if } x \neq 0 \\ 0 & \text{if } x = 0. \end{cases}$$

Technology formula: `x^2*sin(1/x)`

Graph this function using the technology formula above for $-h \le x \le h$, choosing smaller and smaller values of h, and decide **(a)** whether $x = 0$ is a stationary point and **(b)** whether f has either a relative maximum or a relative minimum at $x = 0$. Explain your answers. [**HINT:** For part (a), use technology to estimate the derivative at $x = 0$.]

5.2 Applications of Maxima and Minima

In many applications we would like to find the largest or smallest possible value of some quantity—for instance, the greatest possible profit or the lowest cost. We call this the *optimal* (best) value. In this section we consider several such examples and use calculus to find the optimal value in each.

In all applications the first step is to translate a written description into a mathematical problem. In the problems we look at in this section there are *unknowns* that we are asked to find, there is an expression involving those unknowns that must be made as large or as small as possible—the **objective function**—and there may be **constraints**—equations or inequalities relating the variables.*

* If you have studied linear programming, you will notice a similarity here, but unlike the situation in linear programming, neither the objective function nor the constraints need be linear.

EXAMPLE 1 Minimizing Average Cost

Gymnast Clothing manufactures expensive hockey jerseys for sale to college bookstores in runs of up to 500. Its cost (in dollars) for a run of x hockey jerseys is

$$C(x) = 2{,}000 + 10x + 0.2x^2.$$

How many jerseys should Gymnast produce per run to minimize average cost?†

† Why don't we seek to minimize total cost? The answer would be uninteresting; to minimize total cost, we would make *no* jerseys at all. Minimizing the average cost is a more practical objective.

Solution Here is the procedure we will follow to solve problems like this:

1. ***Identify the unknown(s).*** There is one unknown: x, the number of hockey jerseys Gymnast should produce per run. (We know this because the question is "How many jerseys . . . ?")
2. ***Identify the objective function.*** The objective function is the quantity that must be made as small (in this case) as possible. In this example it is the average cost, which is given by

$$\bar{C}(x) = \frac{C(x)}{x} = \frac{2{,}000 + 10x + 0.2x^2}{x}$$
$$= \frac{2{,}000}{x} + 10 + 0.2x \text{ dollars per jersey.}$$

3. ***Identify the constraints (if any).*** At most 500 jerseys can be manufactured in a run. Also, $\bar{C}(0)$ is not defined. Thus, x is constrained by

$$0 < x \le 500.$$

Put another way, the domain of the objective function $\bar{C}(x)$ is $(0, 500]$.

4. ***State and solve the resulting optimization problem.*** Our optimization problem is:

$$\text{Minimize } \bar{C}(x) = \frac{2{,}000}{x} + 10 + 0.2x \qquad \text{Objective function}$$

$$\text{subject to } 0 < x \le 500. \qquad \text{Constraint}$$

We now solve this problem as in Section 5.1. We first calculate

$$\bar{C}'(x) = -\frac{2{,}000}{x^2} + 0.2.$$

We solve $\bar{C}'(x) = 0$ to find $x = \pm 100$. We reject $x = -100$ because -100 is not in the domain of $\bar{C}$ (and makes no sense), so we have one stationary point, at $x = 100$. There, the average cost is $\bar{C}(100) = \$50$ per jersey.

The only point at which the formula for $\bar{C}'$ is not defined is $x = 0$, but that is not in the domain of $\bar{C}$, so we have no singular points. We have one endpoint in the domain, at $x = 500$. There, the average cost is $\bar{C}(500) = \$114$.

To see the behavior of the function $\bar{C}$ at the interesting points we have found so far, we graph the function with $0 < x \le 500$ and $0 \le y \le 150$ (Figure 19). From the graph we can see that the stationary point at $x = 100$ gives the absolute minimum. We can therefore say that Gymnast Clothing should produce 100 jerseys per run, for a lowest possible average cost of \$50 per jersey.

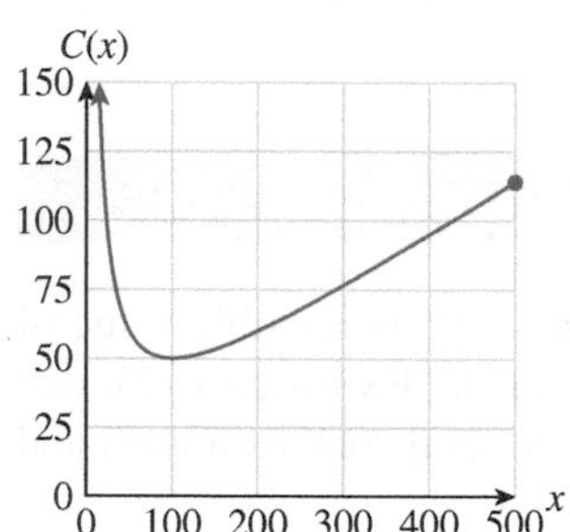

Figure 19

EXAMPLE 2 Maximizing Area

Slim wants to build a rectangular enclosure for his pet rabbit, Killer, against the side of his house, as shown in Figure 20. He has bought 100 feet of fencing. What are the dimensions of the largest area that he can enclose?

Figure 20

Solution

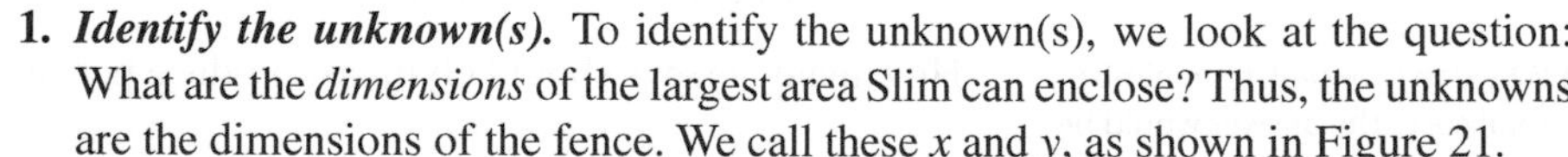

1. ***Identify the unknown(s).*** To identify the unknown(s), we look at the question: What are the *dimensions* of the largest area Slim can enclose? Thus, the unknowns are the dimensions of the fence. We call these x and y, as shown in Figure 21.

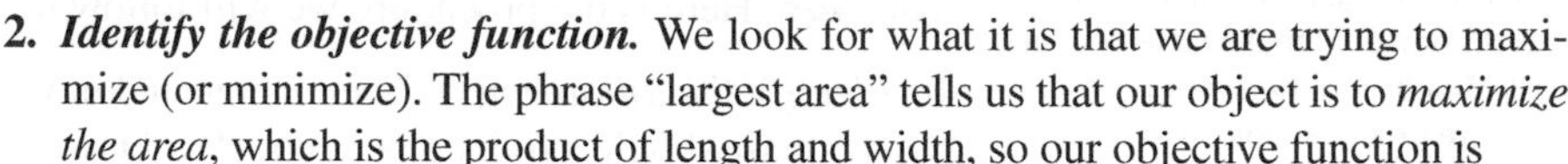

2. ***Identify the objective function.*** We look for what it is that we are trying to maximize (or minimize). The phrase "largest area" tells us that our object is to *maximize the area*, which is the product of length and width, so our objective function is

$$A = xy, \text{ where } A \text{ is the area of the enclosure.}$$

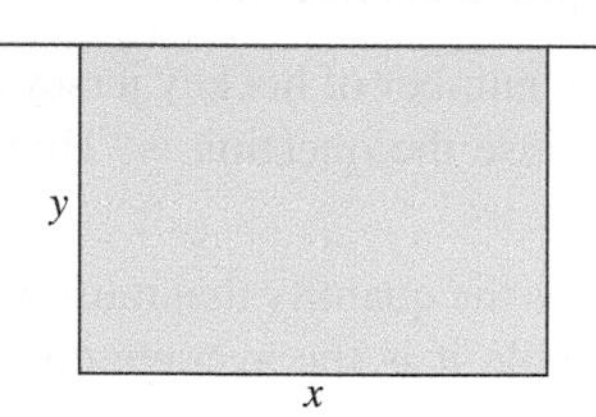

Figure 21

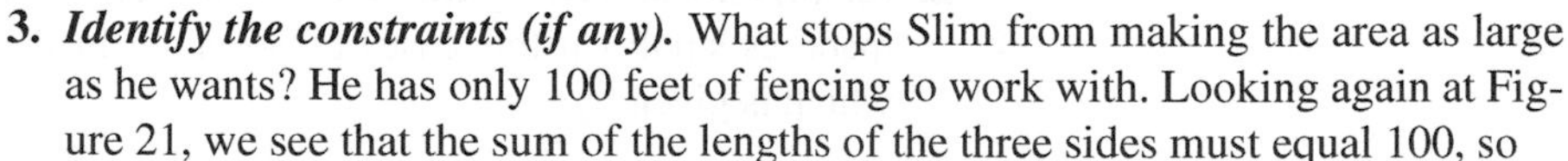

3. ***Identify the constraints (if any).*** What stops Slim from making the area as large as he wants? He has only 100 feet of fencing to work with. Looking again at Figure 21, we see that the sum of the lengths of the three sides must equal 100, so

$$x + 2y = 100.$$

One more point: Because x and y represent the lengths of the sides of the enclosure, neither can be a negative number.

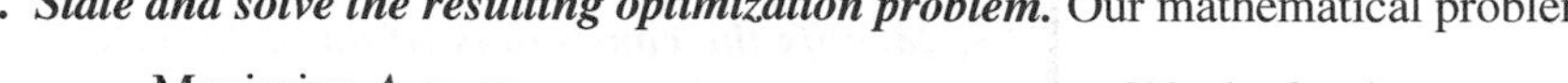

4. ***State and solve the resulting optimization problem.*** Our mathematical problem is:

$$\text{Maximize } A = xy \qquad \text{Objective function}$$

$$\text{subject to } x + 2y = 100,\ x \ge 0, \text{ and } y \ge 0. \qquad \text{Constraints}$$

We know how to find maxima and minima of a function of one variable, but A appears to depend on two variables. We can remedy this by using a constraint to express one variable in terms of the other. Let's take the constraint $x + 2y = 100$ and solve for x in terms of y:

$$x = 100 - 2y.$$

Substituting into the objective function gives

$$A = xy = (100 - 2y)y = 100y - 2y^2,$$

and we have eliminated x from the objective function. What about the inequalities? One says that $x \geq 0$, but we want to eliminate x from this as well. We substitute for x again, getting

$$100 - 2y \geq 0.$$

Solving this inequality for y gives $y \leq 50$. The second inequality says that $y \geq 0$. Now, we can restate our problem with x eliminated:

$$\text{Maximize } A(y) = 100y - 2y^2$$
$$\text{subject to } 0 \leq y \leq 50.$$

We now proceed with our usual method of solving such problems. We calculate $A'(y) = 100 - 4y$. Solving $100 - 4y = 0$, we get one stationary point at $y = 25$. There, $A(25) = 1{,}250$. There are no points at which $A'(y)$ is not defined, so there are no singular points. We have two endpoints: at $y = 0$ and $y = 50$. The corresponding areas are $A(0) = 0$ and $A(50) = 0$. We record the three points we found in a table:

y	0	25	50
$A(y)$	0	1,250	0

It's clear now how A must behave: It increases from 0 at $y = 0$ to 1,250 at $y = 25$ and then decreases back to 0 at $y = 50$. Thus, the largest possible value of A is 1,250 square feet, which occurs when $y = 25$. To completely answer the question that was asked, we need to know the corresponding value of x. We have $x = 100 - 2y$, so $x = 50$ when $y = 25$. Thus, Slim should build his enclosure 50 feet across and 25 feet deep (the "missing" 50-foot side being formed by part of the house).

➡ **Before we go on...** Notice that the problem in Example 2 came down to finding the absolute maximum value of A on the closed and bounded interval $[0, 50]$. As we noted in Section 5.1, the table of values of A at its critical points and the endpoints of the interval gives us enough information to find the absolute maximum. ■

Let's stop for a moment and summarize the steps we've taken in these two examples.

Solving an Optimization Problem

1. **Identify the unknown(s), possibly with the aid of a diagram.** These are usually the quantities asked for in the problem.
2. **Identify the objective function.** This is the quantity you are asked to maximize or minimize. You should name it explicitly, as in "Let S = surface area."
3. **Identify the constraint(s).** These can be equations relating variables or inequalities expressing limitations on the values of variables.

4. **State the optimization problem.** This will have the form "Maximize [minimize] the objective function subject to the constraint(s)."
5. **Eliminate extra variables.** If the objective function depends on several variables, solve the constraint equations to express all variables in terms of one particular variable. Substitute these expressions into the objective function to rewrite it as a function of a single variable. In short, if there is only one constraint equation:

   Solve the constraint for one of the unknowns, and substitute into the objective.

   Also substitute the expressions into any inequality constraints to help determine the domain of the objective function.
6. **Find the absolute maximum (or minimum) of the objective function.** Use the techniques of the preceding section.

Now for some further examples.

EXAMPLE 3 Maximizing Revenue

Cozy Carriage Company builds baby strollers. Using market research, the company estimates that if it sets the price of a stroller at p dollars, then it can sell $q = 300{,}000 - 10p^2$ strollers per year.* What price will bring in the greatest annual revenue?

* This equation is, of course, the demand equation for the baby strollers. However, coming up with a suitable demand equation in real life is hard, to say the least. In this regard, the very entertaining and also insightful article *Camels and Rubber Duckies* by Joel Spolsky at www.joelonsoftware.com/articles/CamelsandRubberDuckies.html is a must-read.

Solution The question we are asked identifies our main unknown: the price p. However, there is another quantity that we do not know: q, the number of strollers the company will sell per year. The question also identifies the objective function, revenue, which is

$$R = pq.$$

Including the equality constraint given to us—that $q = 300{,}000 - 10p^2$—and the "reality" inequality constraints $p \geq 0$ and $q \geq 0$, we can write our problem as follows:

$$\text{Maximize } R = pq$$
$$\text{subject to } q = 300{,}000 - 10p^2,\ p \geq 0, \text{ and } q \geq 0.$$

We are given q in terms of p, so let's substitute to eliminate q:

$$R = pq = p(300{,}000 - 10p^2) = 300{,}000p - 10p^3.$$

Substituting in the inequality $q \geq 0$, we get

$$300{,}000 - 10p^2 \geq 0.$$

Thus, $p^2 \leq 30{,}000$, which gives $-100\sqrt{3} \leq p \leq 100\sqrt{3}$. When we combine this with $p \geq 0$, we get the following restatement of our problem:

$$\text{Maximize } R(p) = 300{,}000p - 10p^3$$
$$\text{subject to } 0 \leq p \leq 100\sqrt{3}.$$

We solve this problem in much the same way we did the preceding one. We calculate $R'(p) = 300{,}000 - 30p^2$. Setting $300{,}000 - 30p^2 = 0$, we find one stationary point at $p = 100$. There are no singular points, and we have the endpoints $p = 0$ and $p = 100\sqrt{3}$. Putting these points in a table and computing the corresponding values of R, we get the following:

p	0	100	$100\sqrt{3}$
$R(p)$	0	20,000,000	0

Thus, Cozy Carriage should price its strollers at \$100 each, which will bring in the largest possible revenue of \$20,000,000.

Figure 22

EXAMPLE 4 Optimizing Resources

The Metal Can Company has an order to make cylindrical cans with a volume of 250 cubic centimeters. What should be the dimensions of the cans in order to use the least amount of metal in their production?

Solution We are asked to find the dimensions of the cans. It is traditional to take as the dimensions of a cylinder the height h and the radius of the base r, as in Figure 22.

We are also asked to minimize the amount of metal used in the can, which is the area of the surface of the cylinder. We can look up the formula or figure it out ourselves: Imagine removing the circular top and bottom and then cutting vertically and flattening out the hollow cylinder to get a rectangle, as shown in Figure 23.

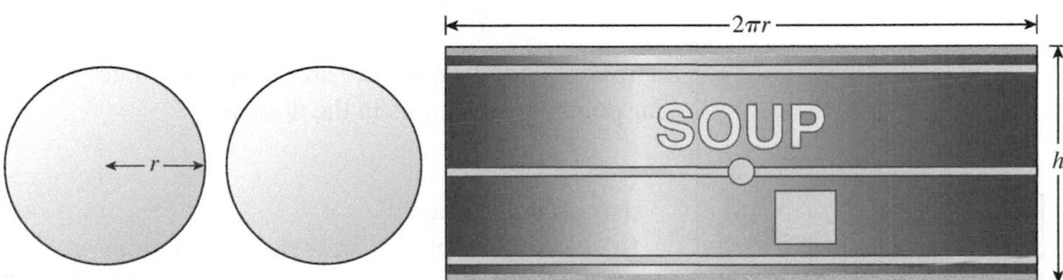

Figure 23

Our objective function is the (total) surface area S of the can. The area of each disc is πr^2, while the area of the rectangular piece is $2\pi rh$. Thus, our objective function is

$$S = 2\pi r^2 + 2\pi rh.$$

As usual, there is a constraint: The volume must be exactly 250 cubic centimeters. The formula for the volume of a cylinder is $V = \pi r^2 h$, so

$$\pi r^2 h = 250.$$

It is easiest to solve this constraint for h in terms of r:

$$h = \frac{250}{\pi r^2}.$$

Substituting in the objective function, we get

$$S = 2\pi r^2 + 2\pi r\frac{250}{\pi r^2} = 2\pi r^2 + \frac{500}{r}.$$

Now, r cannot be negative or 0, but it can become very large (a very wide but very short can could have the right volume). We therefore take the domain of $S(r)$ to be $(0, +\infty)$, so our mathematical problem is as follows:

$$\text{Minimize } S(r) = 2\pi r^2 + \frac{500}{r}$$

subject to $r > 0$.

Now we calculate

$$S'(r) = 4\pi r - \frac{500}{r^2}.$$

To find stationary points, we set this equal to 0 and solve:

$$\begin{aligned} 4\pi r - \frac{500}{r^2} &= 0 \\ 4\pi r &= \frac{500}{r^2} \\ 4\pi r^3 &= 500 \\ r^3 &= \frac{125}{\pi}. \end{aligned}$$

So

$$r = \sqrt[3]{\frac{125}{\pi}} = \frac{5}{\sqrt[3]{\pi}} \approx 3.41.$$

The corresponding surface area is approximately $S(3.41) \approx 220$. There are no singular points or endpoints in the domain.

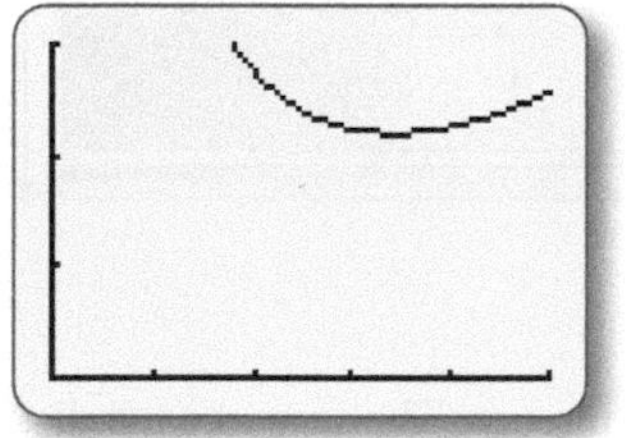

Figure 24

Using Technology

To see how S behaves near the one stationary point, let's graph it in a viewing window with interval $[0, 5]$ for r and $[0, 300]$ for S. The result is Figure 24.

From the graph we can clearly see that the smallest surface area occurs at the stationary point at $r \approx 3.41$. The height of the can will be

$$h = \frac{250}{\pi r^2} \approx 6.83.$$

Thus, the can that uses the least amount of metal has a height of approximately 6.83 centimeters and a radius of approximately 3.41 centimeters. Such a can will use approximately 220 square centimeters of metal.

➡ **Before we go on . . .** We obtained the value of r in Example 4 by solving the equation

$$4\pi r = \frac{500}{r^2}.$$

This time, let us do things differently: Divide both sides by 4π to obtain

$$r = \frac{500}{4\pi r^2} = \frac{125}{\pi r^2}$$

and compare what we got with the expression for h:

$$h = \frac{250}{\pi r^2},$$

which we see is exactly twice the expression for r. Put another way, the height is exactly equal to the diameter, so the can looks square when viewed from the side. Have you ever seen cans with that shape? Why do you think most cans do not have this shape? ■

EXAMPLE 5 Allocation of Labor

The Gym Sock Company manufactures cotton athletic socks. Production is partially automated through the use of robots. Daily operating costs amount to \$50 per laborer and \$30 per robot. The number of pairs of socks the company can manufacture in a day is given by a Cobb-Douglas* production formula

* Cobb-Douglas production formulas were discussed in Section 4.6.

$$q = 50n^{0.6}r^{0.4},$$

where q is the number of pairs of socks that can be manufactured by n laborers and r robots. Assuming that the company wishes to produce 1,000 pairs of socks per day at a minimum cost, how many laborers and how many robots should it use?

Solution The unknowns are the number of laborers n and the number of robots r. The objective is to minimize the daily cost:

$$C = 50n + 30r.$$

The constraints are given by the daily quota

$$1{,}000 = 50n^{0.6}r^{0.4}$$

and the fact that n and r are nonnegative. We solve the constraint equation for one of the variables; let's solve for n:

$$n^{0.6} = \frac{1{,}000}{50r^{0.4}} = \frac{20}{r^{0.4}}.$$

Taking the $1/0.6$ power of both sides gives

$$n = \left(\frac{20}{r^{0.4}}\right)^{1/0.6} = \frac{20^{1/0.6}}{r^{0.4/0.6}} = \frac{20^{5/3}}{r^{2/3}} \approx \frac{147.36}{r^{2/3}}.$$

Substituting in the objective equation gives us the cost as a function of r:

$$\begin{aligned} C(r) &\approx 50\left(\frac{147.36}{r^{2/3}}\right) + 30r \\ &= 7{,}368r^{-2/3} + 30r. \end{aligned}$$

The only remaining constraint on r is that $r > 0$. To find the minimum value of $C(r)$, we first take the derivative:

$$C'(r) \approx -4{,}912r^{-5/3} + 30.$$

Setting this equal to zero, we solve for r:

$$\begin{aligned} r^{-5/3} &\approx 0.006107 \\ r &\approx (0.006107)^{-3/5} \approx 21.3. \end{aligned}$$

The corresponding cost is $C(21.3) \approx \$1{,}600$. There are no singular points or endpoints in the domain of C.

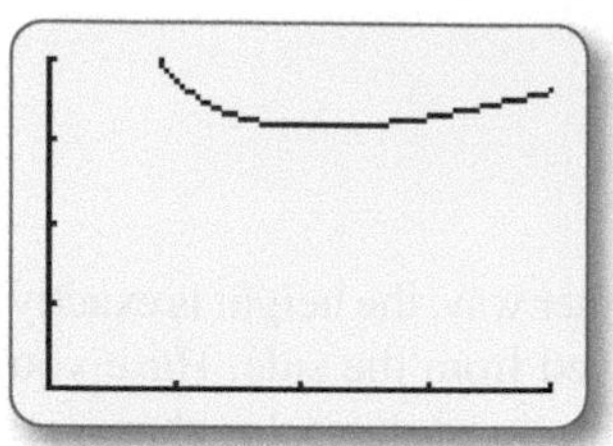

Figure 25

Using Technology

To see how C behaves near its stationary point, let's draw its graph in a viewing window with an interval of $[0, 40]$ for r and $[0, 2{,}000]$ for C. The result is Figure 25.

From the graph we can see that C does have its minimum at the stationary point. The corresponding value of n is

$$n \approx \frac{147.36}{r^{2/3}} \approx 19.2.$$

At this point, our solution appears to be this: Use (approximately) 19.2 laborers and (approximately) 21.3 robots to meet the manufacturing quota at a minimum cost. However, we are not interested in fractions of robots or people, so we need to find integer solutions for n and r. If we round these numbers, we get the solution $(n, r) = (19, 21)$. However, a quick calculation shows that

$$q = 50(19)^{0.6}(21)^{0.4} \approx 989 \text{ pairs of socks,}$$

which fails to meet the quota of 1,000. Thus, we need to round at least one of the quantities n and r *upward* to meet the quota. The three possibilities, with corresponding values of q and C, are as follows:

$$(n, r) = (20, 21), \text{ with } q \approx 1{,}020 \text{ and } C = \$1{,}630$$
$$(n, r) = (19, 22), \text{ with } q \approx 1{,}007 \text{ and } C = \$1{,}610$$
$$(n, r) = (20, 22), \text{ with } q \approx 1{,}039 \text{ and } C = \$1{,}660.$$

Of these, the solution that meets the quota at a minimum cost is $(n, r) = (19, 22)$. Thus, the Gym Sock Co. should use 19 laborers and 22 robots, at a cost of $50 \times 19 + 30 \times 22 = \$1{,}610$, to manufacture $50 \times 19^{0.6} \times 22^{0.4} \approx 1{,}007$ pairs of socks.

FAQs

Constraints and Objectives

Q: *How do I know whether or not there are constraints in an applied optimization problem?*

A: There are usually at least *inequality* constraints; the variables usually represent real quantities, such as length or number of items, and so cannot be negative, leading to constraints such as $x \geq 0$ (or $0 \leq x \leq 100$ in the event that there is an upper limit). *Equation* constraints usually arise when there is more than one unknown in the objective, and they dictate how one unknown is related to others as in, say, "the length is twice the width" or "the demand is 8 divided by the price" (a demand equation).

Q: *How do I know what to use as the objective and what to use as the constraint(s)?*

A: To identify the objective, look for a phrase such as "find the maximum (or minimum) value of." The amount you are trying to maximize or minimize is the objective. For example,

- . . . *at the least cost* . . . The objective function is the equation for cost, $C = \ldots.$
- . . . *the greatest area* . . . The objective function is the equation for area, $A = \ldots.$

To determine the constraint *inequalities,* ask yourself what limitations are placed on the unknown variables as above—are they nonnegative? are there upper limits? To identify the constraint *equations,* look for sentences that dictate restrictions in the form of relationships between the variables, as in the answer to the first question above.

5.2 EXERCISES

▼ more advanced ◆ challenging

T indicates exercises that should be solved using technology

In Exercises 1–8, solve the given optimization problems. [**HINT**: See Example 2.]

1. Maximize $P = xy$ subject to $x + y = 10$.

2. Maximize $P = xy$ subject to $x + 2y = 40$.

3. Minimize $S = x + y$ subject to $xy = 9$ and both x and $y > 0$.

4. Minimize $S = x + 2y$ subject to $xy = 2$ and both x and $y > 0$.

5. Minimize $F = x^2 + y^2$ subject to $x + 2y = 10$.

6. Minimize $F = x^2 + y^2$ subject to $xy^2 = 16$.

7. ▼ Maximize $P = xyz$ subject to $x + y = 30$, $y + z = 30$, and $x, y, z \geq 0$.

8. ▼ Maximize $P = xyz$ subject to $x + z = 12$, $y + z = 12$, and $x, y, z \geq 0$.

9. For a rectangle with perimeter 20 to have the largest area, what dimensions should it have?

10. For a rectangle with area 100 to have the smallest perimeter, what dimensions should it have?

Applications

11. ***Advertising Costs*** The cost, in thousands of dollars, of airing x 30-second television commercials during a Super Bowl game can be approximated by[1]

$$C(x) = 20 + 4{,}000x + 0.05x^2.$$

How many 30-second television commercials should your company air to minimize average costs? What is the resulting average cost of a 30-second ad? [**HINT**: See Example 1.]

12. ***Advertising Costs*** The cost, in billions of dollars, of airing x five-second hologram commercials during a Galactic Chess game can be approximated by

$$C(x) = 490 + 320x + 0.001x^2.$$

How many hologram commercials should your company air to minimize average costs? What is the resulting average cost of a five-second ad? (Round your answer to the nearest billion dollars.) [**HINT**: See Example 1.]

13. ***Average Cost: iPhones*** Assume that it costs **Apple** approximately

$$C(x) = 400{,}000 + 160x + 0.001x^2$$

dollars to manufacture x 32GB iPhone 6's in an hour at the **Foxconn Technology Group.**[2] How many iPhone 6's should be manufactured each hour to minimize average cost? What is the resulting average cost of an iPhone? How does the average cost compare with the marginal cost at the optimal production level? (Give your answer to the nearest dollar.)

14. ***Average Cost: PlayStation 4's*** Assume that it costs **Sony** approximately

$$C(x) = 800{,}000 + 340x + 0.0005x^2$$

dollars to manufacture x PlayStation 4's in an hour.[3] How many PlayStations should be manufactured each hour to minimize average cost? What is the resulting average cost of a PlayStation 4? If fewer than the optimal number are manufactured per hour, will the marginal cost be larger, smaller, or equal to the average cost at that lower production level?

15. ***Pollution Control*** The cost of controlling emissions at a firm rises rapidly as the amount of emissions reduced increases. Here is a possible model:

$$C(q) = 4{,}000 + 100q^2,$$

where q is the reduction in emissions (in pounds of pollutant per day) and C is the daily cost to the firm (in dollars) of this reduction. What level of reduction corresponds to the lowest average cost per pound of pollutant, and what would be the resulting average cost to the nearest dollar?

16. ***Pollution Control*** Repeat Exercise 15 using the following cost function:

$$C(q) = 2{,}000 + 200q^2.$$

17. ***Pollution Control*** (Compare Exercise 15.) The cost of controlling emissions at a firm is given by

$$C(q) = 4{,}000 + 100q^2,$$

where q is the reduction in emissions (in pounds of pollutant per day) and C is the daily cost to the firm (in dollars) of this reduction. Government clean-air subsidies amount to \$500 per pound of pollutant removed. How many pounds of pollutant should the firm remove each day to minimize *net* cost (cost minus subsidy)?

---

[1] The average cost of a 30-second ad during the 2014 Super Bowl game was an estimated \$4 million. Source: "Who Bought What in Super Bowl XLVIII," *Advertising Age*, Feb. 3, 2014, http://adage.com.

[2] Not the actual cost equation; the authors do not know Apple's actual cost equation. The minimum average cost in the model given is in rough agreement with the actual cost for one of the 2014 models. Source for cost data: http://time.com.

[3] Not the actual cost equation; the authors do not know Sony's actual cost equation. The minimum average cost in the model given is in rough agreement with the actual cost to manufacture a PlayStation 4 in 2013. Source for estimate of marginal cost: VentureBeat (http://venturebeat.com).

18. ***Pollution Control*** (Compare Exercise 16.) Repeat Exercise 17 using the following cost function:

$$C(q) = 2{,}000 + 200q^2$$

with government subsidies amounting to \$100 per pound of pollutant removed per day.

19. ***Fences*** I would like to create a rectangular vegetable patch. The fencing for the east and west sides costs \$4 per foot, and the fencing for the north and south sides costs only \$2 per foot. I have a budget of \$80 for the project. What are the dimensions of the vegetable patch with the largest area I can enclose? [**HINT:** See Example 2.]

20. ***Fences*** I would like to create a rectangular orchid garden that abuts my house so that the house itself forms the northern boundary. The fencing for the southern boundary costs \$4 per foot, and the fencing for the east and west sides costs \$2 per foot. If I have a budget of \$80 for the project, what are the dimensions of the garden with the largest area I can enclose? [**HINT:** See Example 2.]

21. ***Fences*** You are building a right-angled triangular flower garden along a stream as shown in the figure. (The borders can be in any direction as long as they are at right angles as shown.)

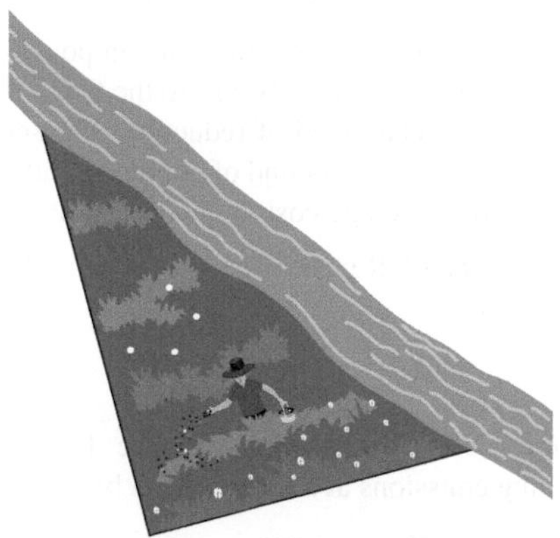

The fencing of the left border costs \$5 per foot, while the fencing of the lower border costs \$1 per foot. (No fencing is required along the river.) You want to spend \$100 and enclose as much area as possible. What are the dimensions of your garden, and what area does it enclose? [**HINT:** The area of a right-triangle is given by $A = xy/2$.]

22. ***Fences*** Repeat Exercise 21, this time assuming that the fencing of the left border costs \$8 per foot, while the fencing of the lower border costs \$2 per foot, and that you can spend \$400.

23. ▼ ***Fences*** (Compare Exercise 19.) For tax reasons I need to create a rectangular vegetable patch with an area of exactly 242 square feet. The fencing for the east and west sides costs \$4 per foot, and the fencing for the north and south sides costs only \$2 per foot. What are the dimensions of the vegetable patch with the least expensive fence? [**HINT:** Compare Exercise 3.]

24. ▼ ***Fences*** (Compare Exercise 20.) For reasons too complicated to explain, I need to create a rectangular orchid garden with an area of exactly 324 square feet abutting my house so that the house itself forms the northern boundary. The fencing for the southern boundary costs \$4 per foot, and the fencing for the east and west sides costs \$2 per foot. What are the dimensions of the orchid garden with the least expensive fence? [**HINT:** Compare Exercise 4.]

25. ***Revenue*** *Hercules Films* is deciding on the price of the video release of its film *Son of Frankenstein.* Its marketing people estimate that at a price of p dollars, it can sell a total of $q = 200{,}000 - 10{,}000p$ copies. What price will bring in the greatest revenue? [**HINT:** See Example 3.]

26. ***Profit*** *Hercules Films* is also deciding on the price of the video release of its film *Bride of the Son of Frankenstein.* Again, marketing estimates that at a price of p dollars, it can sell $q = 200{,}000 - 10{,}000p$ copies, but each copy costs \$4 to make. What price will give the greatest *profit*?

27. ***Revenue: Smartphones*** Worldwide annual sales of smartphones in 2012–2013 were approximately $q = -6p + 3{,}030$ million phones at a selling price of \$$p$ per phone.[4] What selling price would have resulted in the largest annual revenue? What, to the nearest \$10 million, would have been the resulting annual revenue? (The actual selling price in 2013 was \$335.)

28. ***Projected Revenue: Smartphones*** Worldwide annual sales of smartphones in 2013–2017 were projected to be approximately $q = -10p + 4{,}360$ million phones at a selling price of \$$p$ per phone.[5] What selling price would have resulted in the largest projected annual revenue? What would have been the resulting projected annual revenue?

29. ***Revenue: Monorail Service*** The demand for monorail service in Las Vegas in 2005 can be approximated by $q = -4{,}500p + 41{,}500$ rides per day when the fare was \$$p$. What price should have been charged to maximize total daily revenue?[6]

30. ***Revenue: Mars Monorail*** The demand for monorail service in the three urbynes (or districts) of Utarek on Mars can be approximated by $q = -2p + 24$ million riders per day when the fare is $\overline{\mathbf{Z}}p$. What price should be charged to maximize total daily revenue?[7]

31. ***Revenue*** Assume that the demand for tuna in a small coastal town is given by

$$p = \frac{500{,}000}{q^{1.5}},$$

[4] Source for data: IDC Worldwide Quarterly Mobile Phone Tracker, Nov. 26, 2013, www.zdnet.com.

[5] Data based on historical 2013 data (source: *Ibid.*) and projected 2017 data.

[6] Source for ridership data: *New York Times*, February 10, 2007, p. A9.

[7] The zonar ($\overline{\mathbf{Z}}$) is the official currency in the city-state of Utarek, Mars (formerly www.Marsnext.com, a now extinct virtual society).

where q is the number of pounds of tuna that can be sold in a month at p dollars per pound. Assume that the town's fishery wishes to sell at least 5,000 pounds of tuna per month.

a. How much should the town's fishery charge for tuna to maximize monthly revenue? [**HINT**: See Example 3, and don't neglect endpoints.]

b. How much tuna will it sell per month at that price?

c. What will be its resulting revenue?

32. ***Revenue*** In the 1930s the economist Henry Schultz devised the following demand function for corn:

$$p = \frac{6{,}570{,}000}{q^{1.3}},$$

where q is the number of bushels of corn that could be sold at p dollars per bushel in one year.[8] Assume that at least 10,000 bushels of corn per year must be sold.

a. How much should farmers charge per bushel of corn to maximize annual revenue? [**HINT**: See Example 3, and don't neglect endpoints.]

b. How much corn can farmers sell per year at that price?

c. What will be the farmers' resulting revenue?

33. ***Revenue*** During the 1950s the wholesale price for chicken in the United States fell from 25¢ per pound to 14¢ per pound, while per capita chicken consumption rose from 22 pounds per year to 27.5 pounds per year.[9] Assuming that the demand for chicken depended linearly on the price, what wholesale price for chicken would have maximized revenues for poultry farmers, and what would that revenue have amounted to?

34. ***Revenue*** Your underground used-book business is booming. Your policy is to sell all used versions of *Calculus and You* at the same price (regardless of condition). When you set the price at $10, sales amounted to 120 volumes during the first week of classes. The following semester, you set the price at $30, and sales dropped to zero. Assuming that the demand for books depends linearly on the price, what price gives you the maximum revenue, and what does that revenue amount to?

35. ***Profit: Smartphones*** (Compare Exercise 27.) Worldwide annual sales of smartphones in 2012–2013 were approximately $q = -6p + 3{,}030$ million phones at a selling price of $\$p$ per phone.[10] Assuming a manufacturing cost of $80 per phone, what selling price would have resulted in the largest annual profit? What would have been the resulting annual profit? (The actual selling price in 2013 was $335.) [**HINT**: See Example 3, and recall that Profit = Revenue − Cost.]

[8] Based on data for the period 1915–1929. Source: Henry Schultz (1938), *The Theory and Measurement of Demand,* University of Chicago Press, Chicago.

[9] Data are provided for the years 1951–1958. Source: U.S. Department of Agriculture, *Agricultural Statistics.*

[10] See footnote for Exercise 27.

36. ***Projected Profit: Smartphones*** (Compare Exercise 28.) Worldwide annual sales of smartphones in 2013–2017 were projected at approximately $q = -10p + 4{,}360$ million phones at a selling price of $\$p$ per phone.[11] Assuming a manufacturing cost of $100 per phone, what selling price would have resulted in the largest projected annual profit? What would have been the resulting annual profit? [**HINT**: See Example 3, and recall that Profit = Revenue − Cost.]

37. ▼ ***Profit*** The demand equation for your company's virtual reality video headsets is

$$p = \frac{1{,}000}{q^{0.3}},$$

where q is the total number of headsets that your company can sell in a week at a price of p dollars. The total manufacturing and shipping cost amounts to $100 per headset.

a. What is the greatest profit your company can make in a week, and how many headsets will your company sell at this level of profit? (Give answers to the nearest whole number.)

b. How much, to the nearest $1, should your company charge per headset for the maximum profit?

38. ▼ ***Profit*** Because of sales by a competing company, your company's sales of virtual reality video headsets have dropped, and your financial consultant revises the demand equation to

$$p = \frac{800}{q^{0.35}},$$

where q is the total number of headsets that your company can sell in a week at a price of p dollars. The total manufacturing and shipping cost still amounts to $100 per headset.

a. What is the greatest profit your company can make in a week, and how many headsets will your company sell at this level of profit? (Give answers to the nearest whole number.)

b. How much, to the nearest $1, should your company charge per headset for the maximum profit?

39. ***Paint Cans*** A company manufactures cylindrical paint cans with open tops with a volume of 27,000 cubic centimeters. What should be the dimensions of the cans in order to use the least amount of metal in their production? [**HINT**: See Example 4.]

40. ***Metal Drums*** A company manufactures cylindrical metal drums with open tops with a volume of 1 cubic meter. What should be the dimensions of the drums in order to use the least amount of metal in their production? [**HINT**: See Example 4.]

41. ***Tin Cans*** A company manufactures cylindrical tin cans with closed tops with a volume of 250 cubic centimeters. The metal used to manufacture the cans costs $0.01 per

[11] See footnote for Exercise 28.

square centimeter for the sides and \$0.02 per square centimeter for the (thicker) top and bottom. What should be the dimensions of the cans to minimize the cost of metal in their production? What is the ratio height/radius? [**HINT**: See Example 4.]

42. ***Metal Drums*** A company manufactures cylindrical metal drums with open tops with a volume of 2 cubic meters. The metal used to manufacture the drums costs \$2 per square meter for the sides and \$3 per square meter for the (thicker) bottom. What should be the dimensions of the drums to minimize the cost of metal in their production? What is the ratio height/radius? [**HINT**: See Example 4.]

43. ▼ ***Box Design*** *Chocolate Box Company* is going to make open-topped boxes out of 6×16-inch rectangles of cardboard by cutting squares out of the corners and folding up the sides. What is the largest volume box it can make this way?

44. ▼ ***Box Design*** *Vanilla Box Company* is going to make open-topped boxes out of 12×12-inch rectangles of cardboard by cutting squares out of the corners and folding up the sides. What is the largest volume box it can make this way?

45. ▼ ***Box Design*** A packaging company is going to make closed boxes, with square bases, that hold 125 cubic centimeters. What are the dimensions of the box that can be built with the least material?

46. ▼ ***Box Design*** A packaging company is going to make open-topped boxes, with square bases, that hold 108 cubic centimeters. What are the dimensions of the box that can be built with the least material?

47. ▼ ***Luggage Dimensions*** American Airlines requires that the total outside dimensions (length + width + height) of a checked bag not exceed 62 inches.[12] Suppose you want to check a bag whose height equals its width. What is the largest volume bag of this shape that you can check on an American flight?

48. ▼ ***Carry-on Dimensions*** American Airlines requires that the total outside dimensions (length + width + height) of a carry-on bag not exceed 45 inches.[13] Suppose you want to carry on a bag whose length is twice its height. What is the largest volume bag of this shape that you can carry on an American flight?

49. ▼ ***Luggage Dimensions*** *Fly-by-Night Airlines* has a peculiar rule about luggage: The length and width of a bag must add up to at most 45 inches, and the width and height must also add up to at most 45 inches. What are the dimensions of the bag with the largest volume that Fly-by-Night will accept?

50. ▼ ***Luggage Dimensions*** *Fair Weather Airlines* has a similar rule. It will accept only bags for which the sum of the length and width is at most 36 inches, while the sum of length, height, and twice the width is at most 72 inches. What are the dimensions of the bag with the largest volume that Fair Weather will accept?

51. ▼ ***Package Dimensions*** The U.S. Postal Service (USPS) will accept packages only if the length plus girth is no more than 108 inches.[14] (See the figure.)

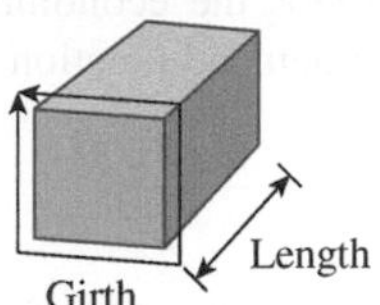

Assuming that the front face of the package (as shown in the figure) is square, what is the largest volume package that the USPS will accept?

52. ▼ ***Package Dimensions*** United Parcel Service (UPS) will accept only packages with a length of no more than 108 inches and length plus girth of no more than 165 inches.[15] (See the figure for Exercise 51.) Assuming that the front face of the package (as shown in the figure) is square, what is the largest volume package that UPS will accept?

53. ▼ ***Cellphone Revenues*** The number of cellphone subscribers in China in the years 2000–2005 was projected to follow the equation $N(t) = 39t + 68$ million subscribers in year t. ($t = 0$ represents January 2000.) The average annual revenue per cellphone user was \$350 in 2000.[16] If we assume that because of competition the revenue per cellphone user decreases continuously at an annual rate of 30%, we can model the annual revenue as

$$R(t) = 350(39t + 68)e^{-0.3t} \text{ million dollars.}$$

Determine **(a)** when to the nearest 0.1 year the revenue was projected to peak and **(b)** the revenue, to the nearest \$1 million, at that time.

54. ▼ ***Cellphone Revenues*** (Refer to Exercise 53.) If we assume instead that the revenue per cellphone user decreases continuously at an annual rate of 20%, we obtain the revenue model

$$R(t) = 350(39t + 68)e^{-0.2t} \text{ million dollars.}$$

Determine **(a)** when to the nearest 0.1 year the revenue was projected to peak and **(b)** the revenue, to the nearest \$1 million, at that time.

[12] According to information on its website (www.aa.com).

[13] *Ibid.*

[14] The requirement for packages sent other than Retail Ground, as of September 2015 (www.usps.com).

[15] The requirement as of September 2015 (www.ups.com).

[16] Based on a regression of projected figures (coefficients are rounded). Source: Intrinsic Technology/*New York Times*, Nov. 24, 2000, p. C1.

55. ▼ ***Research and Development*** Spending on research and development by drug companies in the United States t years after 1970 can be modeled by

$$S(t) = 2.5e^{0.08t} \text{ billion dollars} \quad (0 \le t \le 31).$$

The number of new drugs approved by the **Food and Drug Administration (FDA)** over the same period can be modeled by

$$D(t) = 10 + t \text{ drugs per year}^{17} \quad (0 \le t \le 31).$$

When was the function $D(t)/S(t)$ at a maximum? What is the maximum value of $D(t)/S(t)$? What does the answer tell you about the cost of developing new drugs?

56. ▼ ***Research and Development*** (Refer to Exercise 55.) If the number of new drugs approved by the FDA had been $10 + 2t$ new drugs each year, when would the function $D(t)/S(t)$ have reached a maximum? What does the answer tell you about the cost of developing new drugs?

57. ▼ ***Asset Appreciation*** As the financial consultant to a classic auto dealership, you estimate that the total value (in dollars) of its collection of 1959 Chevrolets and Fords is given by the formula

$$v = 300{,}000 + 1{,}000t^2 \quad (t \ge 5),$$

where t is the number of years from now. You anticipate a continuous inflation rate of 5% per year, so that the discounted (present) value of an item that will be worth $\$v$ in t years' time is

$$p = ve^{-0.05t}.$$

When would you advise the dealership to sell the vehicles to maximize their discounted value?

58. ▼ ***Plantation Management*** The value of a fir tree in your plantation increases with the age of the tree according to the formula

$$v = \frac{20t}{1 + 0.05t},$$

where t is the age of the tree in years. Given a continuous inflation rate of 5% per year, the discounted (present) value of a newly planted seedling is

$$p = ve^{-0.05t}.$$

At what age (to the nearest year) should you harvest your trees to ensure the greatest possible discounted value?

59. ▼ ***Marketing Strategy*** *FeatureRich Software Company* has a dilemma. Its new program, Doors-X 10.27, is almost ready to go on the market. However, the longer the company works on it, the better it can make the program and the more it can charge for it. The company's marketing analysts estimate that if it delays t days, it can set the price at $100 + 2t$ dollars. On the other hand, the longer it delays, the more market share it will lose to its main competitor (see the next exercise), so if it delays t days it will be able to sell $400{,}000 - 2{,}500t$ copies of the program. How many days should FeatureRich delay the release to get the greatest revenue?

60. ▼ ***Marketing Strategy*** *FeatureRich Software's* main competitor (see Exercise 59) is *Moon Systems*, and Moon is in a similar predicament. Its product, Walls-Y 11.4, could be sold now for \$200, but for each day Moon delays, it could increase the price by \$4. On the other hand, it could sell 300,000 copies now, but each day it waits will cut sales by 1,500. How many days should Moon delay the release to get the greatest revenue?

61. ▼ ***Average Profit*** The *FeatureRich Software Company* sells its graphing program, Dogwood, with a volume discount. If a customer buys x copies, then he or she pays[18] $\$500\sqrt{x}$. It cost the company \$10,000 to develop the program and \$2 to manufacture each copy. If a single customer were to buy all the copies of Dogwood, how many copies would the customer have to buy for FeatureRich Software's average profit per copy to be maximized? How are average profit and marginal profit related at this number of copies?

62. ▼ ***Average Profit*** Repeat Exercise 61 with the charge to the customer $\$600\sqrt{x}$ and the cost to develop the program \$9,000.

63. ***Resource Allocation*** Your company manufactures automobile alternators, and production is partially automated through the use of robots. Daily operating costs amount to \$100 per laborer and \$16 per robot. To meet production deadlines, the company calculates that the numbers of laborers and robots must satisfy the constraint

$$xy = 10{,}000,$$

where x is the number of laborers and y is the number of robots. Assuming that the company wishes to meet production deadlines at a minimum cost, how many laborers and how many robots should it use? [**HINT:** See Example 5.]

64. ***Resource Allocation*** Your company is the largest sock manufacturer in the solar system, and production is automated through the use of androids and robots. Daily operating costs amount to ₩200 per android and ₩8 per robot.[19] To meet

[17] The exponential model for R&D is based on the 1970 and 2001 spending in constant 2001 dollars, while the linear model for new drugs approved is based on the 6-year moving average from data from 1970 to 2000. Source for data: Pharmaceutical Research and Manufacturers of America, FDA/*New York Times*, April 19, 2002, p. C1.

[18] This is similar to the way site licenses have been structured for the program Maple.

[19] ₩ are Neptunian Standard Solar Units of currency.

production deadlines, the company calculates that the numbers of androids and robots must satisfy the constraint

$$xy = 1{,}000{,}000,$$

where x is the number of androids and y is the number of robots. Assuming that the company wishes to meet production deadlines at a minimum cost, how many androids and how many robots should it use? [**HINT**: See Example 5.]

65. ▼ ***Resource Allocation*** Your automobile assembly plant has a Cobb-Douglas production function given by

$$q = x^{0.4}y^{0.6},$$

where q is the number of automobiles it produces per year, x is the number of employees, and y is the daily operating budget (in dollars). Annual operating costs amount to an average of \$20,000 per employee plus the operating budget of \$365y. Assume that you wish to produce 1,000 automobiles per year at a minimum cost. How many employees should you hire? [**HINT**: See Example 5.]

66. ▼ ***Resource Allocation*** Repeat Exercise 65 using the production formula

$$q = x^{0.5}y^{0.5}.$$

[**HINT**: See Example 5.]

67. ▼ ***Incarceration Rate*** The incarceration rate (the number of persons in prison per 100,000 residents) in the United States can be approximated by

$$N(t) = 0.04t^3 - 2t^2 + 40t + 460 \quad (0 \le t \le 18).$$

(t is the year since 1990.)[20] When, to the nearest year, was the incarceration rate increasing most rapidly? When was it increasing least rapidly? [**HINT**: You are being asked to find the extreme values of the rate of change of the incarceration rate.]

68. ▼ ***Prison Population*** The prison population in the United States can be approximated by

$$N(t) = 0.02t^3 - 2t^2 + 100t + 1{,}100 \text{ thousand people} \quad (0 \le t \le 18).$$

(t is the year since 1990.)[21] When, to the nearest year, was the prison population increasing most rapidly? When was it increasing least rapidly? [**HINT**: You are being asked to find the extreme values of the rate of change of the prison population.]

69. ▼ ***Embryo Development*** The oxygen consumption of a bird embryo increases from the time the egg is laid through the time the chick hatches. In a typical galliform bird the oxygen consumption can be approximated by

$$c(t) = -0.065t^3 + 3.4t^2 - 22t + 3.6 \text{ milliliters per day} \quad (8 \le t \le 30),$$

where t is the time (in days) since the egg was laid.[22] (An egg will typically hatch at around $t = 28$.) When, to the nearest day, is $c'(t)$ a maximum? What does the answer tell you?

70. ▼ ***Embryo Development*** The oxygen consumption of a turkey embryo increases from the time the egg is laid through the time the chick hatches. In a brush turkey the oxygen consumption can be approximated by

$$c(t) = -0.028t^3 + 2.9t^2 - 44t + 95 \text{ milliliters per day} \quad (20 \le t \le 50),$$

where t is the time (in days) since the egg was laid.[23] (An egg will typically hatch at around $t = 50$.) When, to the nearest day, is $c'(t)$ a maximum? What does the answer tell you?

71. T ▼ ***Subprime Mortgages during the Housing Bubble*** During the real estate run-up in 2000–2008 the percentage of mortgages issued in the United States that were subprime (normally classified as risky) could be approximated by

$$A(t) = \frac{15.0}{1 + 8.6(1.8)^{-t}} \text{ percent} \quad (0 \le t \le 8)$$

t years after the start of 2000.[24] Graph the *derivative* of $A(t)$, and determine the year during which this derivative had an absolute maximum and also its value at that point. What does the answer tell you?

72. T ▼ ***Subprime Mortgage Debt during the Housing Bubble*** During the real estate run-up in 2000–2008 the value of subprime (normally classified as risky) mortgage debt outstanding in the United States was approximately

$$A(t) = \frac{1{,}350}{1 + 4.2(1.7)^{-t}} \text{ billion dollars} \quad (0 \le t \le 8)$$

t years after the start of 2000.[25] Graph the *derivative* of $A(t)$, and determine the year during which this derivative had an absolute maximum and also its value at that point. What does the answer tell you?

73. T ▼ ***Asset Appreciation*** You manage a small antique company that owns a collection of Louis XVI jewelry boxes. Their value v is increasing according to the formula

$$v = \frac{10{,}000}{1 + 500e^{-0.5t}},$$

[20] Source for data: Sourcebook of Criminal Justice Statistics Online (www.albany.edu/sourcebook).

[21] *Ibid.*

[22] The model approximates graphical data published in the article "The Brush Turkey" by Roger S. Seymour, *Scientific American,* December 1991, pp. 108–114.

[23] *Ibid.*

[24] Sources: Mortgage Bankers Association, UBS.

[25] Source: www.data360.org/dataset.aspx?Data_Set_Id=9549.

where t is the number of years from now. You anticipate an inflation rate of 5% per year, so the present value of an item that will be worth $\$v$ in t years' time is given by

$$p = v \cdot (1.05)^{-t}.$$

When (to the nearest year) should you sell the jewelry boxes to maximize their present value? How much (to the nearest constant dollar) will they be worth at that time?

74. T ▼ ***Harvesting Forests*** The following equation models the approximate volume in cubic feet of a typical Douglas fir tree of age t years.[26]

$$V = \frac{22{,}514}{1 + 22{,}514t^{-2.55}}$$

The lumber will be sold at \$10 per cubic foot, and you do not expect the price of lumber to appreciate in the foreseeable future. On the other hand, you anticipate a general inflation rate of 5% per year, so the present value of an item that will be worth $\$v$ in t years' time is given by

$$p = v \cdot (1.05)^{-t}.$$

At what age (to the nearest year) should you harvest a Douglas fir tree to maximize its present value? How much (to the nearest constant dollar) will a Douglas fir tree be worth at that time?

75. ◆ ***Agriculture*** The fruit yield per tree in an orchard containing 50 trees is 100 pounds per tree each year. Because of crowding, the yield decreases by 1 pound per season for every additional tree planted. How many additional trees should be planted for a maximum total annual yield?

76. ◆ ***Agriculture*** Two years ago, your orange orchard contained 50 trees, and the yield per tree was 75 bags of oranges. Last year, you removed 10 of the trees and noticed that the yield per tree increased to 80 bags. Assuming that the yield per tree depends linearly on the number of trees in the orchard, what should you do this year to maximize your total yield?

77. ◆ ***Revenue*** *(based on a question on the GRE Economics Test)*[27] If total revenue (TR) is specified by $TR = a + bQ - cQ^2$, where Q is quantity of output and a, b, and c are positive parameters, then TR is maximized for this firm when it produces Q equal to

(A) $b/2ac$. **(B)** $b/4c$.
(C) $(a + b)/c$. **(D)** $b/2c$.
(E) $c/2b$.

78. ◆ ***Revenue*** *(based on a question on the GRE Economics Test)* If total demand (Q) is specified by $Q = -aP + b$, where P is unit price and a and b are positive parameters, then total revenue is maximized for this firm when it charges P equal to

(A) $b/2a$. **(B)** $b/4a$.
(C) a/b. **(D)** $a/2b$.
(E) $-b/2a$.

Communication and Reasoning Exercises

79. You are interested in knowing the height of the tallest condominium complex that meets the city zoning requirements that the height H should not exceed eight times the distance D from the road and that the complex must provide parking for at least 50 cars. The objective function of the associated optimization problem is then

(A) H. **(B)** $H - 8D$.
(C) D. **(D)** $D - 8H$.

One of the constraints is

(A) $8H = D$. **(B)** $8D = H$.
(C) $H'(D) = 0$. **(D)** $D'(H) = 0$.

80. You are interested in building a condominium complex with a height H of at least eight times the distance D from the road and parking area of at least 1,000 square feet at the cheapest cost C. The objective function of the associated optimization problem is then

(A) H. **(B)** D.
(C) C. **(D)** $H + D - C$.

One of the constraints is

(A) $H - 8D = 0$. **(B)** $H + D - C = 0$.
(C) $C'(D) = 0$. **(D)** $8H = D$.

81. Explain why the following problem is uninteresting: A packaging company wishes to make cardboard boxes with open tops by cutting square pieces from the corners of a square sheet of cardboard and folding up the sides. What is the box with the least surface area it can make this way?

82. Explain why finding the production level that minimizes a cost function is frequently uninteresting. What would a more interesting objective be?

83. Your friend Margo claims that all you have to do to find the absolute maxima and minima in applications is set the derivative equal to zero and solve. "All that other stuff about endpoints and so on is a waste of time just to make life hard for us," according to Margo. Explain why she is wrong, and find at least one exercise in this exercise set to illustrate your point.

84. You are having a hard time persuading your other friend Marco that maximizing revenue is not the same as maximizing profit. "How on earth can you expect to obtain the largest profit if you are not taking in the largest revenue?" Explain why he is wrong, and find at least one exercise in this exercise set to illustrate your point.

85. ▼ If demand q decreases as price p increases, what does the minimum value of dq/dp measure?

86. ▼ Explain how you would solve an optimization problem of the following form. Maximize $P = f(x, y, z)$ subject to $z = g(x, y)$ and $y = h(x)$.

[26] The model is the authors' and is based on data in *Environmental and Natural Resource Economics* by Tom Tietenberg, third edition (New York: HarperCollins, 1992), p. 282.

[27] Source: GRE Economics Test, by G. Gallagher, G. E. Pollock, W. J. Simeone, G. Yohe (Piscataway, NJ: Research and Education Association, 1989).

5.3 Higher Order Derivatives: Acceleration and Concavity

The Second Derivative and Acceleration

The **second derivative** is simply the derivative of the derivative function. To explain why we would be interested in such a thing, we start by discussing one of its interpretations. Suppose a car is traveling along a straight stretch of highway. We saw in Section 3.5 that if $s(t)$ represents the position at time t of any object moving in a straight line, then its velocity is given by the derivative: $v(t) = s'(t)$. But one rarely drives a car at a constant speed; the velocity itself may be changing. The rate at which the velocity is changing is the **acceleration**. Because the derivative measures the rate of change, acceleration is the derivative of velocity: $a(t) = v'(t)$. Because v is the derivative of s, we can express the acceleration in terms of s:

$$a(t) = v'(t) = (s')'(t) = s''(t).$$

That is, a is the derivative of the derivative of s; in other words, the second derivative of s, which we write as s''. (In this context you will often hear the derivative s' referred to as the **first derivative**.)

Second Derivative, Acceleration

If a function f has a derivative that is in turn differentiable, then its **second derivative** is the derivative of the derivative of f, written as f''. If $f''(a)$ exists, we say that f is **twice differentiable at** $x = a$.

Quick Examples

1. If $f(x) = x^3 - x$, then $f'(x) = 3x^2 - 1$, so $f''(x) = 6x$ and $f''(-2) = -12$.
2. If $f(x) = 3x + 1$, then $f'(x) = 3$, so $f''(x) = 0$.
3. If $f(x) = e^x$, then $f'(x) = e^x$, so $f''(x) = e^x$ as well.

The **acceleration** of a moving object is the derivative of its velocity—that is, the second derivative of the position function.

Quick Example

4. If t is time in hours and the position of a car at time t is $s(t) = t^3 + 2t^2$ miles, then the car's velocity is $v(t) = s'(t) = 3t^2 + 4t$ miles per hour, and its acceleration is $a(t) = s''(t) = v'(t) = 6t + 4$ miles per hour per hour.

Differential Notation for the Second Derivative

We have written the second derivative of $f(x)$ as $f''(x)$. We could also use differential notation:

$$f''(x) = \frac{d^2f}{dx^2}.$$

This notation comes from writing the second derivative as the derivative of the derivative in differential notation:

$$f''(x) = \frac{d}{dx}\left[\frac{df}{dx}\right] = \frac{d^2f}{dx^2}.$$

Similarly, if $y = f(x)$, we write $f''(x)$ as $\frac{d}{dx}\left[\frac{dy}{dx}\right] = \frac{d^2y}{dx^2}$. For example, if $y = x^3$, then $\frac{d^2y}{dx^2} = 6x$.

An important example of acceleration is the acceleration due to gravity.

EXAMPLE 1 **Acceleration Due to Gravity**

According to the laws of physics, the height of an object near the surface of the Earth falling in a vacuum from an initial rest position s_0 feet above the ground under the influence of gravity is approximately

$$s(t) = s_0 - 16t^2 \text{ feet}$$

in t seconds.* Find its acceleration.

* If the object is initially moving upward with a velocity v_0, then the formula becomes

$$s(t) = s_0 + v_0t - 16t^2.$$

We will see where this formula comes from in Section 6.1.

Solution The velocity of the object is

$$v(t) = s'(t) = -32t \text{ ft/sec.}$$

Differential notation: $v = \frac{ds}{dt} = -32t$ ft/sec

The reason for the negative sign is that the height of the object is decreasing with time, so its velocity is negative. Hence, the acceleration is

$$a(t) = s''(t) = -32 \text{ ft/sec}^2.$$

Differential notation: $a = \frac{d^2s}{dt^2} = -32$ ft/sec^2

(We write ft/sec^2 as an abbreviation for feet/second/second—that is, feet per second per second. It is often read "feet per second squared.") Thus, the *downward* velocity is increasing by 32 ft/sec every second. We say that 32 ft/sec^2 is the **acceleration due to gravity**. In the absence of air resistance, all falling bodies near the surface of the Earth, no matter what their weight, will fall with this acceleration.†

† On other planets the acceleration due to gravity is different. For example, on Jupiter it is about three times as large as on Earth.

➡ **Before we go on . . .** In very careful experiments using balls rolling down inclined planes, Galileo made one of his most important discoveries: that the acceleration due to gravity is constant and does not depend on the weight or composition of the object that is falling.§ A famous, though probably apocryphal, story has him dropping cannonballs of different weights off the Leaning Tower of Pisa to prove his point.|| ■

§ An interesting aside: Galileo's experiments depended on getting extremely accurate timings. Because the timepieces of his day were very inaccurate, he used the most accurate time measurement he could: He sang and used the beat as his stopwatch.

|| Here is a true story: The point was made again during the Apollo 15 mission to the moon (July 1971) when astronaut David R. Scott dropped a feather and a hammer from the same height. The moon has no atmosphere, so the two hit the surface of the moon simultaneously.

EXAMPLE 2 **Acceleration of Sales**

For the first 15 months after the introduction of a new video game, the accumulated sales can be modeled by the curve

$$S(t) = 20e^{0.4t} \text{ units sold,}$$

where t is the time in months since the game was introduced. After about 25 months, total sales follow more closely the curve

$$S(t) = 100{,}000 - 20e^{17-0.4t}.$$

How fast are accumulated sales accelerating after 10 months? How fast are they accelerating after 30 months? What do these numbers mean?

Solution By acceleration we mean the rate of change of the rate of change, which is the second derivative. During the first 15 months the first derivative of sales is

$$\frac{dS}{dt} = 8e^{0.4t},$$

so the second derivative is

$$\frac{d^2S}{dt^2} = 3.2e^{0.4t}.$$

Thus, after 10 months the acceleration of sales is

$$\left.\frac{d^2S}{dt^2}\right|_{t=10} = 3.2e^4 \approx 175 \text{ units/month/month, or units/month}^2.$$

We can also compute total sales

$$S(10) = 20e^4 \approx 1{,}092 \text{ units}$$

and the rate of change of sales

$$\left.\frac{dS}{dt}\right|_{t=10} = 8e^4 \approx 437 \text{ units/month.}$$

What do these numbers mean? By the end of the tenth month a total of 1,092 video games have been sold. At that time the game is selling at the rate of 437 units per month. This rate of sales is increasing by 175 units per month per month. More games will be sold each month than the month before.

Analysis of the sales after 30 months is done similarly, using the formula

$$S(t) = 100{,}000 - 20e^{17-0.4t}.$$

The derivative is

$$\frac{dS}{dt} = 8e^{17-0.4t},$$

and the second derivative is

$$\frac{d^2S}{dt^2} = -3.2e^{17-0.4t}.$$

After 30 months,

$$S(30) = 100{,}000 - 20e^{17-12} \approx 97{,}032 \text{ units}$$

$$\left.\frac{dS}{dt}\right|_{t=30} = 8e^{17-12} \approx 1{,}187 \text{ units/month}$$

$$\left.\frac{d^2S}{dt^2}\right|_{t=30} = -3.2e^{17-12} \approx -475 \text{ units/month}^2.$$

By the end of the thirtieth month, 97,032 video games have been sold, the game is selling at a rate of 1,187 units per month, and the rate of sales is *decreasing* by 475 units per month. Fewer games are sold each month than the month before.

Geometric Interpretation of Second Derivative: Concavity

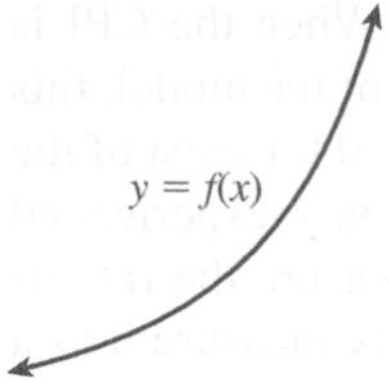

Figure 26

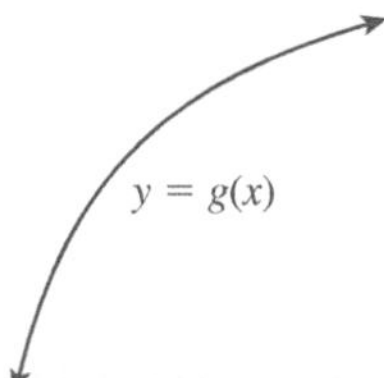

Figure 27

The first derivative of f tells us where the graph of f is rising [where $f'(x) > 0$] and where it is falling [where $f'(x) < 0$]. The second derivative tells in what direction the graph of f *curves* or *bends*. Consider the graphs in Figures 26 and 27. Think of a car driving from left to right along each of the roads shown in the two figures. A car driving along the graph of f in Figure 26 will turn to the left (upward); a car driving along the graph of g in Figure 27 will turn to the right (downward). We say that the graph of f is **concave up** and the graph of g is **concave down**. Now think about the derivatives of f and g. The derivative $f'(x)$ starts small but *increases* as the graph gets steeper. Because $f'(x)$ is increasing, its derivative $f''(x)$ must be positive. On the other hand, $g'(x)$ *decreases* as we go to the right. Because $g'(x)$ is decreasing, its derivative $g''(x)$ must be negative. Summarizing, we have the following.

Concavity and the Second Derivative

A curve is **concave up** if its slope is increasing, in which case the second derivative is positive. A curve is **concave down** if its slope is decreasing, in which case the second derivative is negative. A point in the domain of f where the graph of f changes concavity, from concave up to concave down or vice versa, is called a **point of inflection**. At a point of inflection the second derivative is either zero or undefined.

Locating Points of Inflection

To locate possible points of inflection, list points where $f''(x) = 0$ and also interior points where $f''(x)$ is not defined.

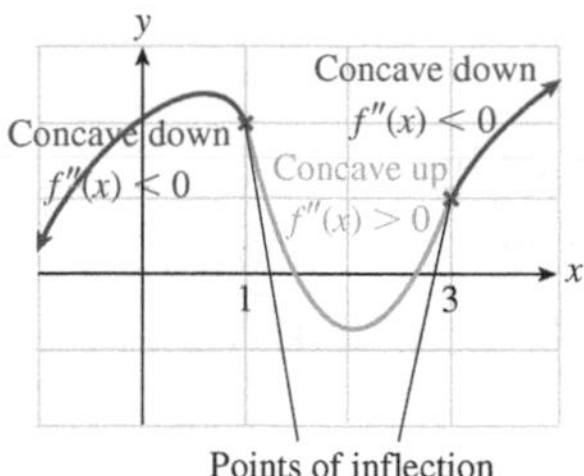

Figure 28

Quick Examples

5. The graph of the function f shown in Figure 28 is concave up when $1 < x < 3$, so $f''(x) > 0$ for $1 < x < 3$. It is concave down when $x < 1$ and $x > 3$, so $f''(x) < 0$ when $x < 1$ and $x > 3$. It has points of inflection at $x = 1$ and $x = 3$.
6. Consider $f(x) = x^3 - 3x$, whose graph is shown in Figure 29. $f''(x) = 6x$ is negative when $x < 0$ and positive when $x > 0$. The graph of f is concave down when $x < 0$ and concave up when $x > 0$. f has a point of inflection at $x = 0$, where the second derivative is 0.

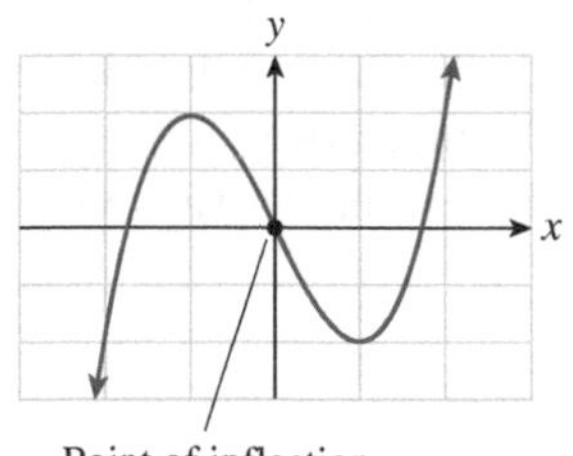

Figure 29

The following example shows one of the reasons it's useful to look at concavity.

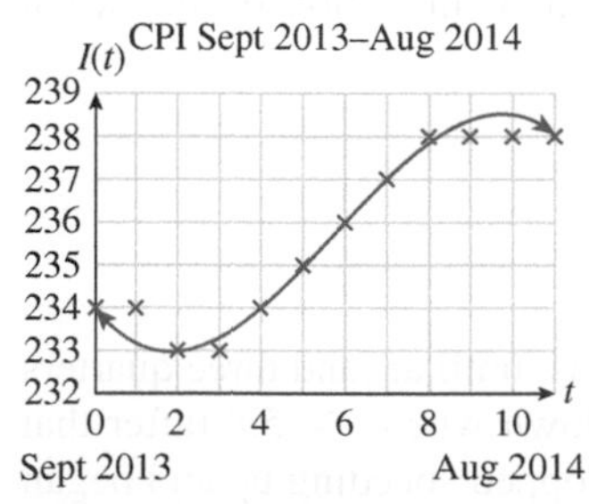

Figure 30

EXAMPLE 3 Inflation

Figure 30 shows the value of the U.S. Consumer Price Index (CPI) from September 2013 through August 2014.[28] The approximating curve shown on the figure is given by

$$I(t) = -0.0225t^3 + 0.390t^2 - 1.20t + 234 \qquad (1 \le t \le 11),$$

[28] The CPI is compiled by the Bureau of Labor Statistics and is based upon a 1982 value of 100. For instance, a CPI of 200 means the CPI has doubled since 1982. Source: InflationData.com (www.inflationdata.com).

where t is time in months. ($t = 0$ represents September 2013.) When the CPI is increasing, the U.S. economy is **experiencing inflation**. In terms of the model, this means that the derivative is positive: $I'(t) > 0$. Notice that $I'(t) > 0$ for most of the period shown (the graph is sloping upward), so the U.S. economy experienced inflation for most of the associated period of time. We *could* measure the rate of inflation by the first derivative $I'(t)$ of the CPI, but we traditionally measure it as a ratio:

$$\text{Inflation rate} = \frac{I'(t)}{I(t)}, \quad \text{Relative rate of change of the CPI}$$

expressed as a percentage per unit time (per month in this case).

a. Use the model to estimate the inflation rate in January 2014.

b. Was inflation slowing or speeding up in January 2014?

c. When was inflation slowing? When was inflation speeding up? When was inflation slowest?

Solution

a. We need to compute $I'(t)$:

$$I'(t) = -0.0675t^2 + 0.780t - 1.20.$$

Thus, the inflation rate in January 2014 ($t = 4$) was given by

$$\text{Inflation rate} = \frac{I'(4)}{I(4)} = \frac{-0.0675(4)^2 + 0.780(4) - 1.20}{-0.0225(4)^3 + 0.390(4)^2 - 1.20(4) + 234}$$

$$= \frac{0.84}{234} \approx 0.0036,$$

or 0.36% per month.*

b. We say that inflation is "slowing" when the CPI is decelerating ($I''(t) < 0$; the index rises at a slower rate or falls at a faster rate†). Similarly, inflation is "speeding up" when the CPI is accelerating ($I''(t) > 0$; the index rises at a faster rate or falls at a slower rate). From the formula for $I'(t)$ the second derivative is

$$I''(t) = -0.135t + 0.780$$

$$I''(4) = -0.135(4) + 0.780 = 0.24.$$

Since this quantity is positive, we conclude that inflation was speeding up in January 2014.

c. When inflation is speeding up, $I''(t)$ is positive, so the graph of the CPI is concave up. When inflation is slowing, it is concave down. At the point at which it switches, there is a point of inflection (Figure 31). The point of inflection occurs when $I''(t) = 0$; that is,

$$-0.135t + 0.780 = 0$$

$$t = \frac{0.780}{0.135} \approx 5.8.$$

Thus, inflation was speeding up when $t < 5.8$ (that is, until around three quarters of the way through February 2014) and slowing down when $t > 5.8$ (after that time). Inflation was "fastest" at the point when it stopped speeding up and began to slow down, $t \approx 5.8$. (Notice that the graph is steepest at that point.)

* The 0.36% monthly inflation rate corresponds to a $12 \times 0.36 = 4.32\%$ annual inflation rate. This result could be obtained directly by changing the units of the t-axis from months to years and then redoing the calculation.

† When the CPI is falling, the inflation rate is negative, and we experience *deflation*.

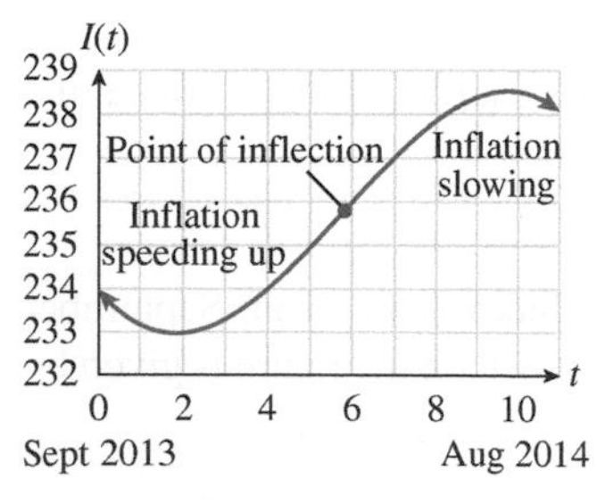

Figure 31

EXAMPLE 4 The Point of Diminishing Returns

After the introduction of a new video game, the accumulated worldwide sales are modeled by the curve

$$S(t) = \frac{1}{1 + 50e^{-0.2t}} \text{ million units sold,}$$

where t is the time in months since the game was introduced. (Compare Example 2.) The graphs of $S(t)$, $S'(t)$, and $S''(t)$ are shown in Figure 32. Where is the graph of S concave up, and where is it concave down? Where are any points of inflection? What does this all mean?

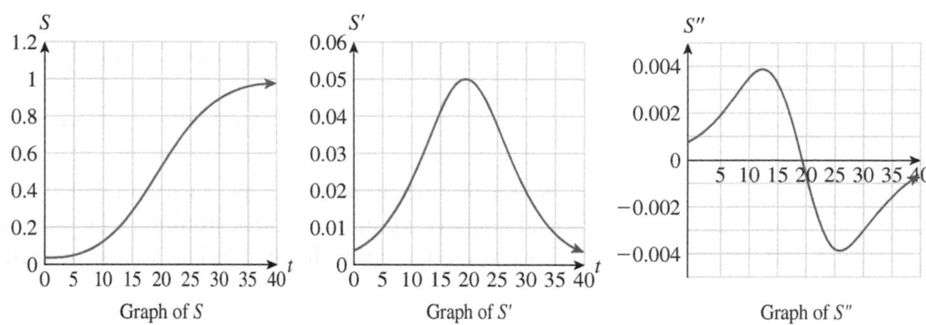

Figure 32

Solution Look at the graph of S. We see that the graph of S is concave up in the early months and then becomes concave down later. The point of inflection, where the concavity changes, is somewhere between 15 and 25 months.

Now look at the graph of S''. This graph crosses the t-axis very close to $t = 20$, is positive before that point, and negative after that point. Because positive values of S'' indicate that S is concave up and negative values indicate that it is concave down, we conclude that the graph of S is concave up for about the first 20 months, that is, for $0 < t < 20$, and concave down for $20 < t < 40$. The concavity switches at the point of inflection, which occurs at about $t = 20$ (when $S''(t) = 0$; a more accurate answer is $t \approx 19.56$).

What does this all mean? Look at the graph of S', which shows sales per unit time, or monthly sales. From this graph we see that monthly sales are increasing for $t < 20$: more units are being sold each month than the month before. Monthly sales reach a peak of 0.05 million = 50,000 games per month at the point of inflection $t = 20$ and then begin to drop off. Thus, the point of inflection occurs at the time when monthly sales stop increasing and start to fall off, that is, the time when monthly sales peak. The point of inflection is sometimes called the **point of diminishing returns**. Although the total sales figure continues to rise (see the graph of S: game units continue to be sold), the *rate* at which units are sold starts to drop. (See Figure 33.)

Figure 33

Using Technology

You can use a TI-83/84 Plus, a downloadable Excel sheet at the Website, or the Function Evaluator and Grapher at the Website to graph the second derivative of the function in Example 4:

TI-83/84 Plus

```
Y1=1/(1+50*e^(-0.2X))
Y2=nDeriv(Y1,X,X)
Y3=nDeriv(Y2,X,X)
```

Website
www.WanerMath.com
In the Function Evaluator and Grapher, enter the functions as shown.

Functions
y1 = 1/(1+50e^(-0.2x))
y2 = deriv(y1)
y3 = deriv(y2)
y4 =
y5 =

(Use Ymin = 0, yMax = 0.1 for a nice view of S', and use Ymin = −0.01 and yMax = 0.01 for S''.)
For an Excel utility, try:
Online Utilities
→ Excel First and Second Derivative Graphing Utility
(This utility needs macros, so ensure that they are enabled.)

The Second Derivative Test for Relative Extrema

The second derivative often gives us a way of knowing whether or not a stationary point is a relative extremum. Figure 34 shows a graph with two stationary points: a relative maximum at $x = a$ and a relative minimum at $x = b$.

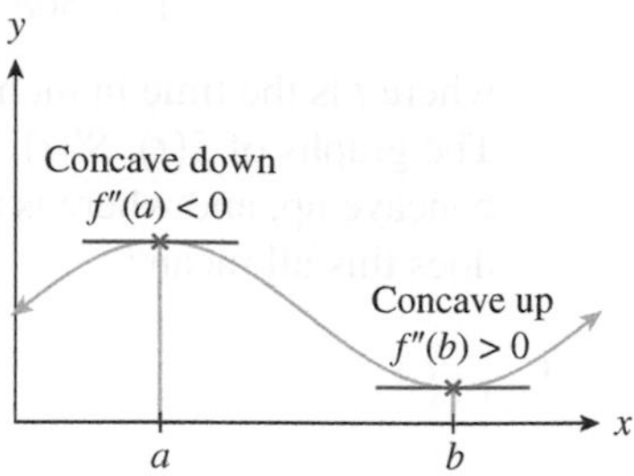

Figure 34

Notice that the curve is *concave down* at the relative maximum ($x = a$), so $f''(a) < 0$, and *concave up* at the relative minimum ($x = b$), so $f''(b) > 0$. This suggests the following. (Compare the first derivative test in Section 5.1.)

Second Derivative Test for Relative Extrema

Suppose that the function f has a stationary point at c and that $f''(c)$ exists. Determine the sign of $f''(c)$.

1. If $f''(c) > 0$, then f has a relative minimum at c.

2. If $f''(c) < 0$, then f has a relative maximum at c.

If $f''(c) = 0$, then the test is inconclusive, and you need to use one of the methods of Section 5.1 (such as the first derivative test) to determine whether or not f has a relative extremum at c.

Quick Examples

7. $f(x) = x^2 - 2x$ has $f'(x) = 2x - 2$ and hence a stationary point at $x = 1$. $f''(x) = 2$, and so $f''(1) = 2$, which is positive, so f has a relative minimum at 1.

8. Let $f(x) = x^3 - 3x^2 - 9x$. Then

$f'(x) = 3x^2 - 6x - 9 = 3(x + 1)(x - 3)$

Stationary points at $x = -1$, $x = 3$

$f''(x) = 6x - 6$

$f''(-1) = -12$, so there is a relative maximum at -1

$f''(3) = 12$, so there is a relative minimum at 3.

9. $f(x) = x^4$ has $f'(x) = 4x^3$ and hence a stationary point at $x = 0$. $f''(x) = 12x^2$, so $f''(0) = 0$, telling us that the second derivative test is inconclusive. However, we can see from the graph of f or the first derivative test that f has a minimum at $x = 0$.

Higher Order Derivatives

There is no reason to stop at the second derivative; we could once again take the derivative of the second derivative to obtain the **third derivative**, f''', and we could take the derivative once again to obtain the **fourth derivative**, written $f^{(4)}$, and then continue to obtain $f^{(5)}$, $f^{(6)}$, and so on (assuming that we get a differentiable function at each stage).

Higher Order Derivatives

We define

$$f'''(x) = \frac{d}{dx}[f''(x)]$$

$$f^{(4)}(x) = \frac{d}{dx}[f'''(x)]$$

$$f^{(5)}(x) = \frac{d}{dx}[f^{(4)}(x)],$$

and so on, assuming that all these derivatives exist.

Different Notations

$$f'(x), f''(x), f'''(x), f^{(4)}(x), \ldots, f^{(n)}(x), \ldots$$

$$\frac{df}{dx}, \frac{d^2f}{dx^2}, \frac{d^3f}{dx^3}, \frac{d^4f}{dx^4}, \ldots, \frac{d^nf}{dx^n}, \ldots$$

$$\frac{dy}{dx}, \frac{d^2y}{dx^2}, \frac{d^3y}{dx^3}, \frac{d^4y}{dx^4}, \ldots, \frac{d^ny}{dx^n}, \ldots \qquad \text{When } y = f(x)$$

$$y, y', y'', y''', y^{(4)}, \ldots, y^{(n)}, \ldots \qquad \text{When } y = f(x)$$

Quick Examples

10. If $f(x) = x^3 - x$, then $f'(x) = 3x^2 - 1$, $f''(x) = 6x$, $f'''(x) = 6$, $f^{(4)}(x) = f^{(5)}(x) = \cdots = 0$.
11. If $f(x) = e^x$, then $f'(x) = e^x$, $f''(x) = e^x$, $f'''(x) = e^x$, $f^{(4)}(x) = f^{(5)}(x) = \cdots = e^x$.

Q: *We know that the second derivative can be interpreted as acceleration. How do we interpret the third derivative and the fourth, fifth, and so on?*

A: Think of a car traveling down the road (with position $s(t)$ at time t) in such a way that its acceleration $\frac{d^2s}{dt^2}$ is changing with time (for instance, the driver may be slowly increasing pressure on the accelerator, causing the car to accelerate at a greater and greater rate). Then $\frac{d^3s}{dt^3}$ is the rate of change of acceleration.* $\frac{d^4s}{dt^4}$ would then be the *acceleration* of the acceleration, and so on.

* Sometimes called the "jerk."

Q: *How are these higher order derivatives reflected in the graph of a function f?*

A: Because the concavity is measured by f'', its derivative f''' tells us the rate of change of concavity. Similarly, $f^{(4)}$ would tell us the acceleration of concavity, and so on. These properties are very subtle and hard to discern by simply looking at the curve; the higher the order, the more subtle the property. There is a remarkable theorem by Taylor* that tells us that, for a large class of functions (including polynomial, exponential, logarithmic, and trigonometric functions) the values of all orders of derivative $f(a)$, $f'(a)$, $f''(a)$, $f'''(a)$, and so on at the single point $x = a$ are enough to describe the entire graph (even at points very far from $x = a$)! In other words, the smallest piece of a graph near any point a contains sufficient information to "clone" the entire graph!

* Brook Taylor (1685–1731) was an English mathematician.

FAQs

Interpreting Points of Inflection and Using the Second Derivative Test

Q: *It says in Example 4 that monthly sales reach a maximum at the point of inflection (second derivative is zero), but the second derivative test says that, for a maximum, the second derivative must be negative. What is going on here?*

A: What is a maximum in Example 4 is the *rate of change* of sales, which is measured in sales per unit time (monthly sales in the example). In other words, it is the *derivative* of the total sales function that is a maximum, so we located the maximum by setting its derivative (which is the *second* derivative of total sales) equal to zero. In general: To find relative (stationary) extrema of the *original* function, set $f'(x)$ equal to zero and solve for x as usual. The second derivative test can then be used to test the stationary point obtained. To find relative (stationary) extrema of the *rate of change of f*, set $f''(x) = 0$ and solve for x.

Q: *I used the second derivative test, and it was inconclusive. That means that there is neither a relative maximum nor a relative minimum at $x = a$, right?*

A: Wrong. If (as is often the case) the second derivative is zero at a stationary point, all it means is that the second derivative test itself cannot determine whether the given point is a relative maximum, minimum, or neither. For instance, $f(x) = x^4$ has a stationary minimum at $x = 0$, but the second derivative test is inconclusive. In such cases, one should use another test (such as the first derivative test) to decide if the point is a relative maximum, minimum, or neither.

5.3 EXERCISES

▼ more advanced ◆ challenging

T indicates exercises that should be solved using technology

In Exercises 1–10, calculate $\frac{d^2y}{dx^2}$. [**HINT**: See Quick Examples 1–3.]

1. $y = 3x^2 - 6$

2. $y = -x^2 + x$

3. $y = \frac{2}{x}$

4. $y = -\frac{2}{x^2}$

5. $y = 4x^{0.4} - x$

6. $y = 0.2x^{-0.1}$

7. $y = e^{-(x-1)} - x$

8. $y = e^{-x} + e^x$

9. $y = \frac{1}{x} - \ln x$

10. $y = x^{-2} + \ln x$

In Exercises 11–16 the position s of a point (in feet) is given as a function of time t (in seconds). Find **(a)** *its acceleration as a function of t and* **(b)** *its acceleration at the specified time.* [**HINT**: See Example 1.]

11. $s = 12 + 3t - 16t^2$; $t = 2$

12. $s = -12 + t - 16t^2$; $t = 2$

13. $s = \dfrac{1}{t} + \dfrac{1}{t^2}$; $t = 1$ **14.** $s = \dfrac{1}{t} - \dfrac{1}{t^2}$; $t = 2$

15. $s = \sqrt{t} + t^2$; $t = 4$ **16.** $s = 2\sqrt{t} + t^3$; $t = 1$

In Exercises 17–24 the graph of a function is given. Find the approximate coordinates of all points of inflection of each function (if any). [**HINT:** See Quick Examples 5 and 6.]

17.

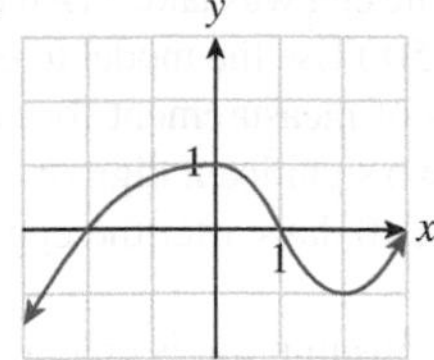

18.

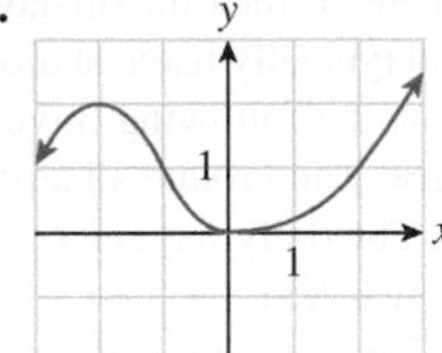

19.

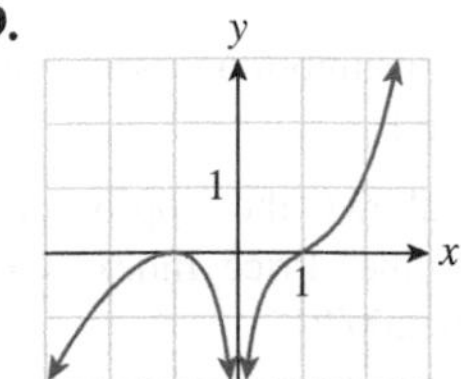

20.

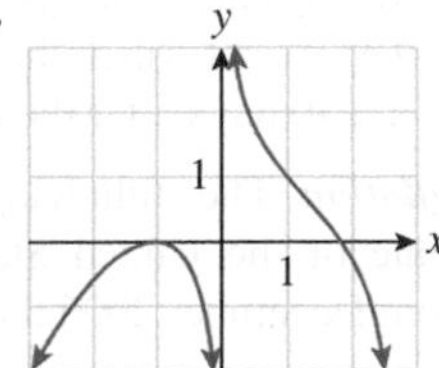

21.

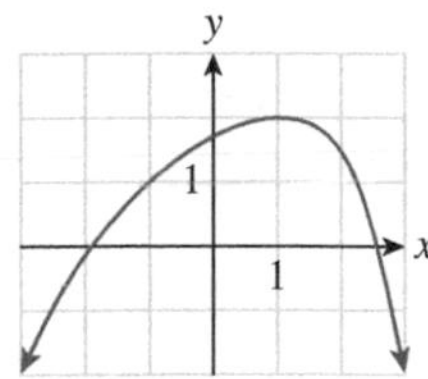

22.

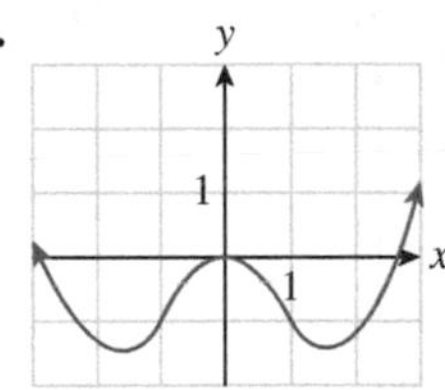

23.

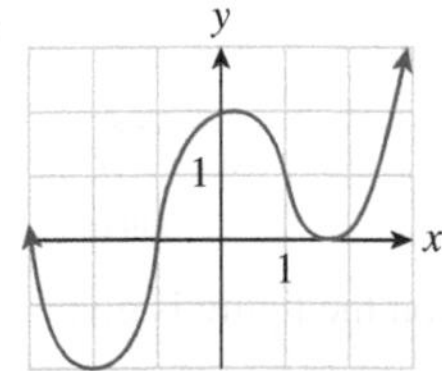

24.

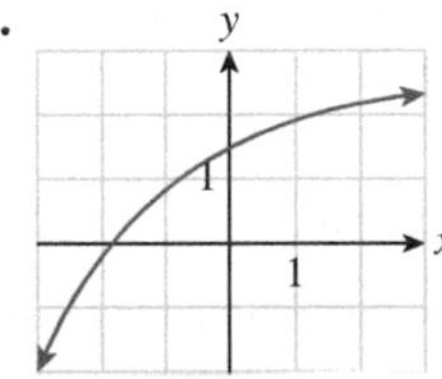

In Exercises 25–28 the graph of the derivative, $f'(x)$, *is given. Determine the x-coordinates of all points of inflection of* $f(x)$, *if any. (Assume that* $f(x)$ *is defined and continuous everywhere in* $[-3, 3]$.*)* [**HINT:** See Quick Examples 5 and 6.]

25.

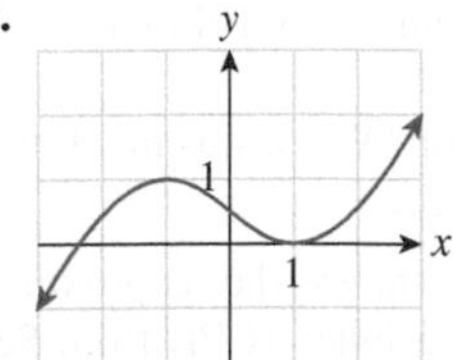

26.

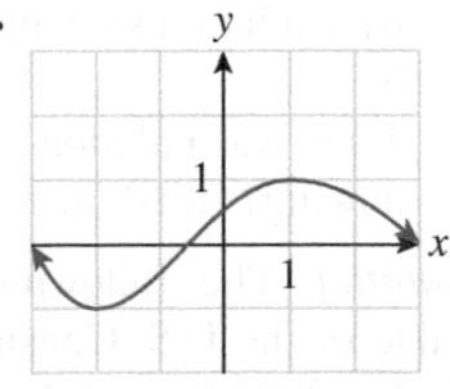

27.

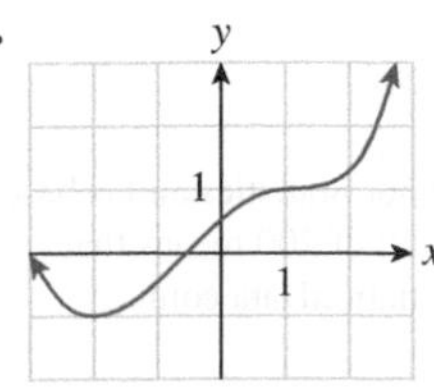

28.

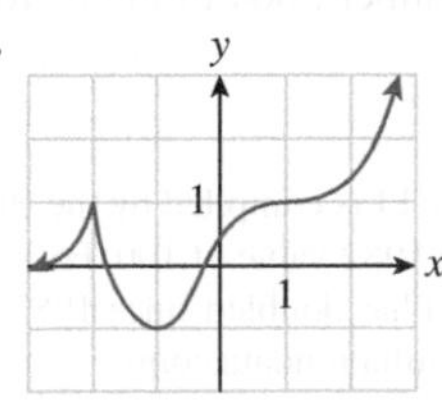

In Exercises 29–32 the graph of the second derivative, $f''(x)$, *is given. Determine the x-coordinates of all points of inflection of* $f(x)$, *if any. (Assume that* $f(x)$ *is defined and continuous everywhere in* $[-3, 3]$.*)* [**HINT:** Remember that a point of inflection of f corresponds to a point at which f'' changes sign, from positive to negative or vice versa. This could be a point where its graph crosses the x-axis or a point where its graph is broken: positive on one side of the break and negative on the other.]

29.

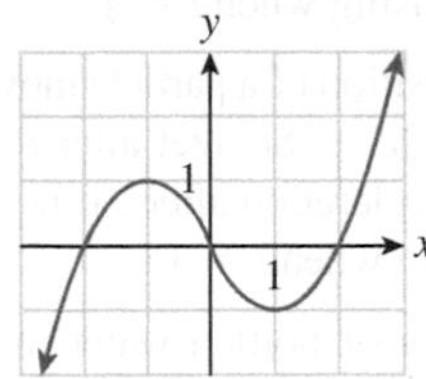

30.

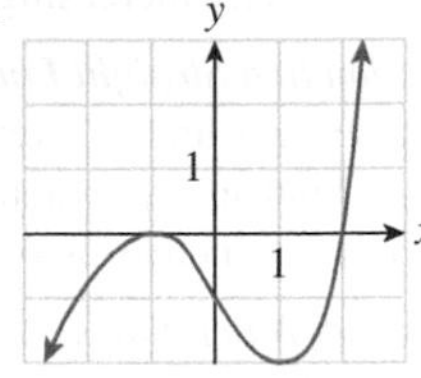

31.

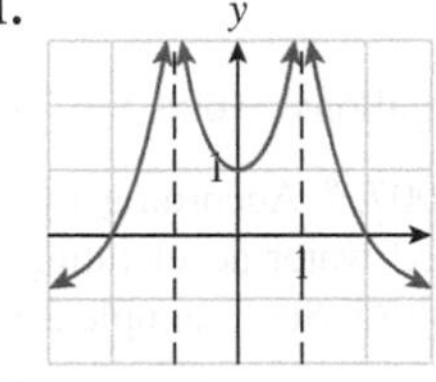

32.

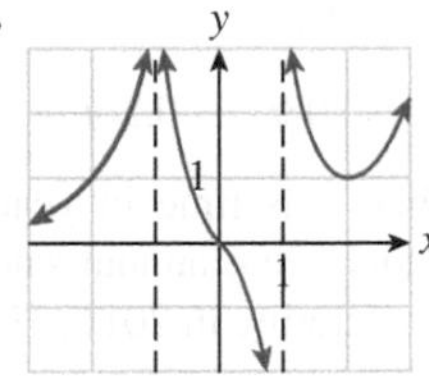

In Exercises 33–44, find the x-coordinates of all critical points of the given function. Determine whether each critical point is a relative maximum, a relative minimum, or neither, by first applying the second derivative test, and, if the test fails, by some other method. [**HINT:** See Quick Examples 7–9.]

33. $f(x) = x^2 - 4x + 1$ **34.** $f(x) = 2x^2 - 2x + 3$

35. $g(x) = x^3 - 12x$ **36.** $g(x) = 2x^3 - 6x + 3$

37. $f(t) = t^3 - t$ **38.** $f(t) = -2t^3 + 3t$

39. $f(x) = x^4 - 4x^3$ **40.** $f(x) = 3x^4 - 2x^3$

41. $f(x) = e^{-x^2}$ **42.** $f(x) = e^{2-x^2}$

43. $f(x) = xe^{1-x^2}$ **44.** $f(x) = xe^{-x^2}$

In Exercises 45–54, calculate the derivatives of all orders: $f'(x), f''(x), f'''(x), f^{(4)}(x), \ldots, f^{(n)}(x), \ldots$ [**HINT:** See Quick Examples 10 and 11.]

45. $f(x) = 4x^2 - x + 1$ **46.** $f(x) = -3x^3 + 4x$

47. $f(x) = -x^4 + 3x^2$ **48.** $f(x) = x^4 + x^3$

49. $f(x) = (2x + 1)^4$ **50.** $f(x) = (-2x + 1)^3$

51. $f(x) = e^{-x}$ **52.** $f(x) = e^{2x}$

53. $f(x) = e^{3x-1}$ **54.** $f(x) = 2e^{-x+3}$

Applications

55. ***Acceleration on Mars*** If a stone is dropped from a height of 40 meters above the Martian surface, its height in meters after t seconds is given by $s = 40 - 1.9t^2$. What is its acceleration? [**HINT:** See Example 1.]

56. ***Acceleration on the Moon*** If a stone is thrown up at 10 meters per second from a height of 100 meters above the surface of the Moon, its height in meters after t seconds is given by $s = 100 + 10t - 0.8t^2$. What is its acceleration? [**HINT**: See Example 1.]

57. ***Motion in a Straight Line*** The position of a particle moving in a straight line is given by $s = t^3 - t^2$ feet after t seconds. Find an expression for its acceleration after a time t. Is its velocity increasing or decreasing when $t = 1$?

58. ***Motion in a Straight Line*** The position of a particle moving in a straight line is given by $s = 3e^t - 8t^2$ feet after t seconds. Find an expression for its acceleration after a time t. Is its velocity increasing or decreasing when $t = 1$?

59. ***Bottled Water Sales*** Annual sales of bottled water in the United States in the period 2007–2014 could be approximated by

$$R(t) = 0.08t^2 - 0.26t + 8.8 \text{ billion gallons} \quad (0 \le t \le 7),$$

where t is time in years since 2007.[29] According to the model, were annual sales of bottled water accelerating or decelerating in 2011? How fast? [**HINT**: See Example 2.]

60. ***Bottled Water Sales*** Annual U.S. per capita sales of bottled water in the period 2007–2014 could be approximated by

$$R(t) = 0.25t^2 - t + 29 \text{ gallons} \quad (0 \le t \le 7),$$

where t is time in years since 2007.[30] According to the model, were annual U.S. per capita sales of bottled water accelerating or decelerating in 2009? How fast? [**HINT**: See Example 2.]

61. ***Embryo Development*** The daily oxygen consumption of a bird embryo increases from the time the egg is laid through the time the chick hatches. In a typical galliform bird the oxygen consumption can be approximated by

$$c(t) = -0.065t^3 + 3.4t^2 - 22t + 3.6 \text{ milliliters per day} \quad (8 \le t \le 30),$$

where t is the time (in days) since the egg was laid.[31] (An egg will typically hatch at around $t = 28$.) Use the model to estimate the following (give the units of measurement for each answer and round all answers to two significant digits):

a. The daily oxygen consumption 20 days after the egg was laid

b. The rate at which the oxygen consumption is changing 20 days after the egg was laid

c. The rate at which the oxygen consumption is accelerating 20 days after the egg was laid

62. ***Embryo Development*** The daily oxygen consumption of a turkey embryo increases from the time the egg is laid through the time the chick hatches. In a brush turkey the oxygen consumption can be approximated by

$$c(t) = -0.028t^3 + 2.9t^2 - 44t + 95 \text{ milliliters per day} \quad (20 \le t \le 50),$$

where t is the time (in days) since the egg was laid.[32] (An egg will typically hatch at around $t = 50$.) Use the model to estimate the following (give the units of measurement for each answer and round all answers to two significant digits):

a. The daily oxygen consumption 40 days after the egg was laid

b. The rate at which the oxygen consumption is changing 40 days after the egg was laid

c. The rate at which the oxygen consumption is accelerating 40 days after the egg was laid

63. ***Inflation*** The following graph shows the approximate value of the United States Consumer Price Index (CPI) from December 2006 through July 2007:[33]

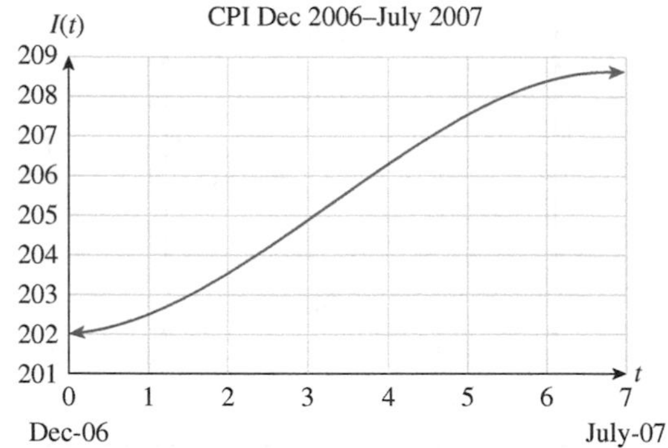

The approximating curve shown on the figure is given by

$$I(t) = -0.04t^3 + 0.4t^2 + 0.1t + 202 \quad (0 \le t \le 7),$$

where t is time in months since the start of December 2006.

a. Use the model to estimate the monthly inflation rate in February 2007 ($t = 2$). [Recall that the inflation *rate* is $I'(t)/I(t)$.]

b. Was inflation slowing or speeding up in February 2007?

c. When was inflation speeding up? When was inflation slowing? [**HINT**: See Example 3.]

64. ***Inflation*** The following graph shows the approximate value of the U.S. Consumer Price Index (CPI) from September 2004 through November 2005:[34]

---

[29] The 2014 figure is a projection. Source: Beverage Marketing Corporation (www.bottledwater.org).

[30] *Ibid.*

[31] The model approximates graphical data published in the article "The Brush Turkey" by Roger S. Seymour, *Scientific American,* December 1991, pp. 108–114.

[32] *Ibid.*

[33] The CPI is compiled by the Bureau of Labor Statistics and is based upon a 1982 value of 100. For instance, a CPI of 200 means that the CPI has doubled since 1982. Source: InflationData.com (www.inflationdata.com).

[34] *Ibid.*

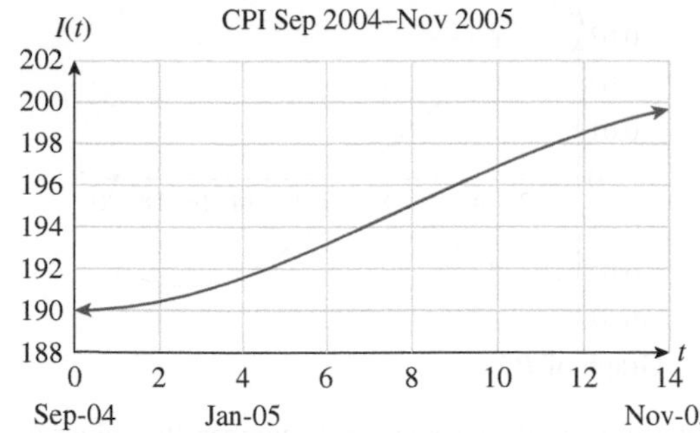

The approximating curve shown on the figure is given by

$$I(t) = -0.005t^3 + 0.12t^2 - 0.01t + 190 \quad (0 \le t \le 14),$$

where t is time in months since the start of September 2004.

a. Use the model to estimate the monthly inflation rate in July 2005 ($t = 10$). [Recall that the inflation *rate* is $I'(t)/I(t)$.]

b. Was inflation slowing or speeding up in July 2005?

c. When was inflation speeding up? When was inflation slowing? [**HINT:** See Example 3.]

65. *Inflation* The following graph shows the approximate value of the U.S. Consumer Price Index (CPI) from July 2005 through March 2006:[35]

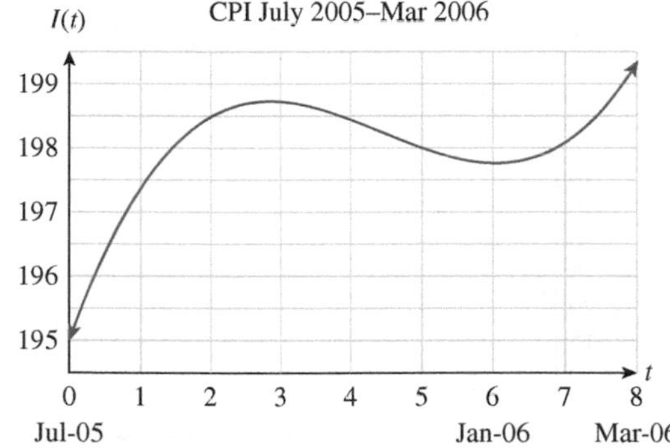

The approximating curve shown on the figure is given by

$$I(t) = 0.06t^3 - 0.8t^2 + 3.1t + 195 \quad (0 \le t \le 8),$$

where t is time in months since the start of July 2005.

a. Use the model to estimate the monthly inflation rates in December 2005 and February 2006 ($t = 5$ and $t = 7$).

b. Was inflation slowing or speeding up in February 2006?

c. When was inflation speeding up? When was inflation slowing? [**HINT:** See Example 3.]

66. *Inflation* The following graph shows the approximate value of the U.S. Consumer Price Index (CPI) from March 2006 through May 2007.[36]

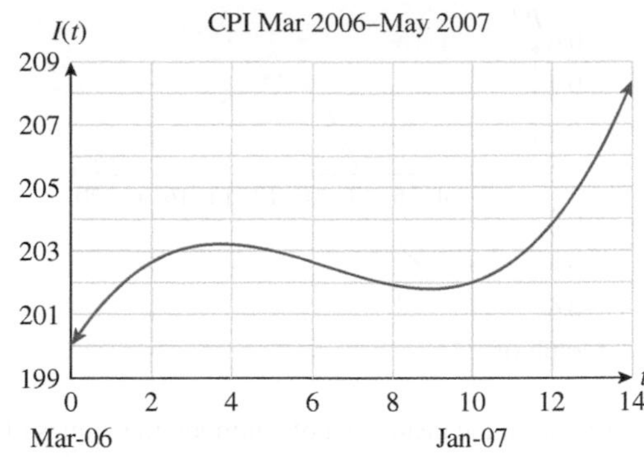

The approximating curve shown on the figure is given by

$$I(t) = 0.02t^3 - 0.38t^2 + 2t + 200 \quad (0 \le t \le 14),$$

where t is time in months since the start of March 2006.

a. Use the model to estimate the monthly inflation rates in September 2006 and January 2007 ($t = 6$ and $t = 10$).

b. Was inflation slowing or speeding up in January 2007?

c. When was inflation speeding up? When was inflation slowing? [**HINT:** See Example 3.]

67. *Scientific Research: 1983–2003* The percentage of research articles in the prominent journal *Physical Review* that were written by researchers in the United States during 1983–2003 can be modeled by

$$P(t) = 25 + \frac{36}{1 + 0.06(0.7)^{-t}},$$

where t is time in years since 1983.[37] The graphs of P, P', and P'' are shown here:

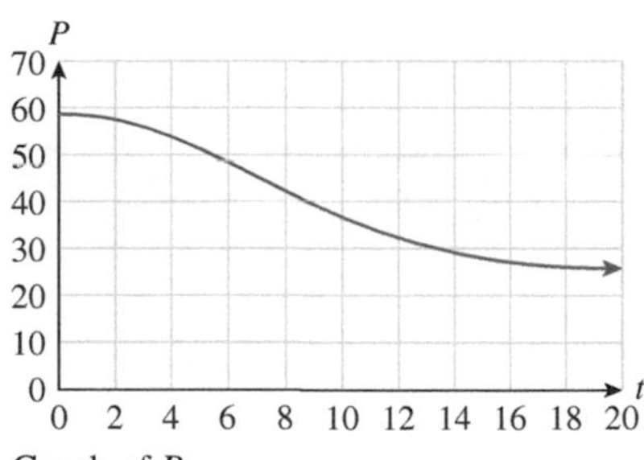

Graph of P

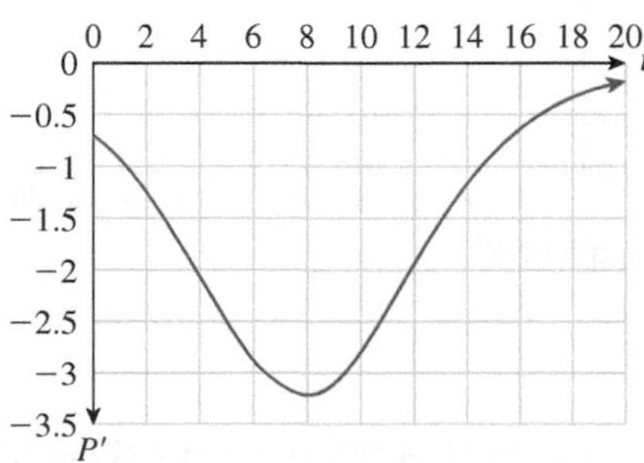

Graph of P'

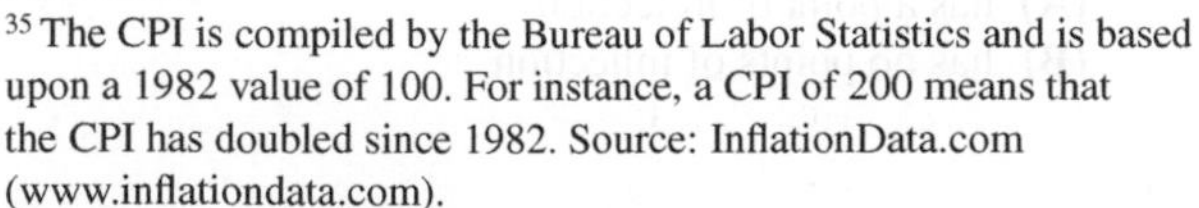
[35] The CPI is compiled by the Bureau of Labor Statistics and is based upon a 1982 value of 100. For instance, a CPI of 200 means that the CPI has doubled since 1982. Source: InflationData.com (www.inflationdata.com).

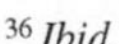
[36] *Ibid.*

[37] Source: The American Physical Society/*New York Times*, May 3, 2003, p. A1.

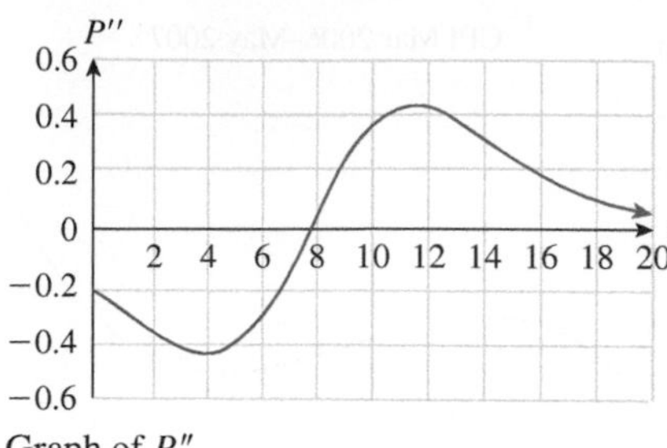

Graph of P''

Determine, to the nearest whole number, the values of t for which the graph of P is concave up and where it is concave down, and locate any points of inflection. What does the point of inflection tell you about science articles?
[**HINT**: See Example 4.]

68. ***Scientific Research: 1983–2003*** The number of research articles in the prominent journal *Physical Review* that were written by researchers in Europe during 1983–2003 can be modeled by

$$P(t) = \frac{7.0}{1 + 5.4(1.2)^{-t}},$$

where t is time in years since 1983.[38] The graphs of P, P', and P'' are shown here:

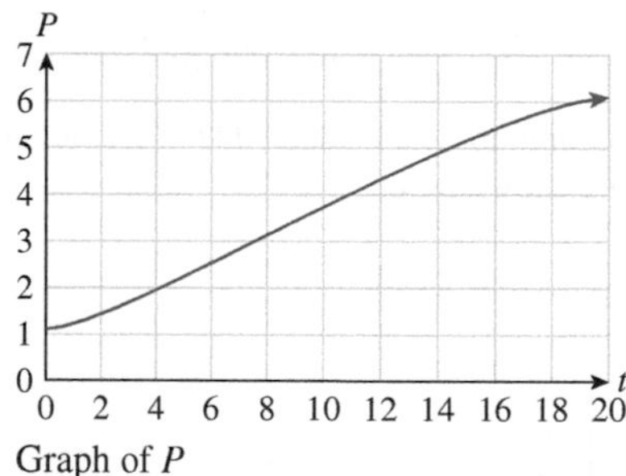

Graph of P

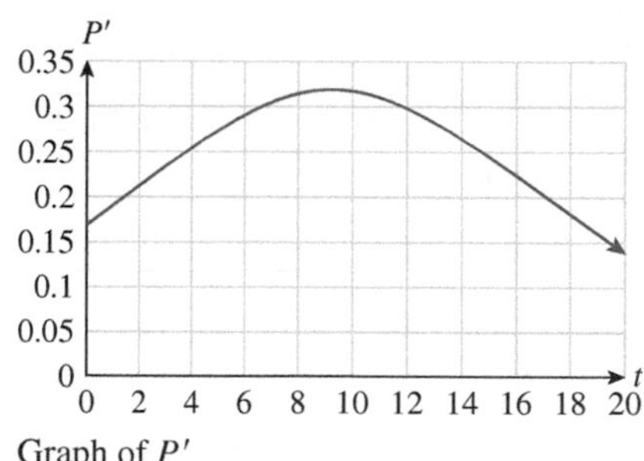

Graph of P'

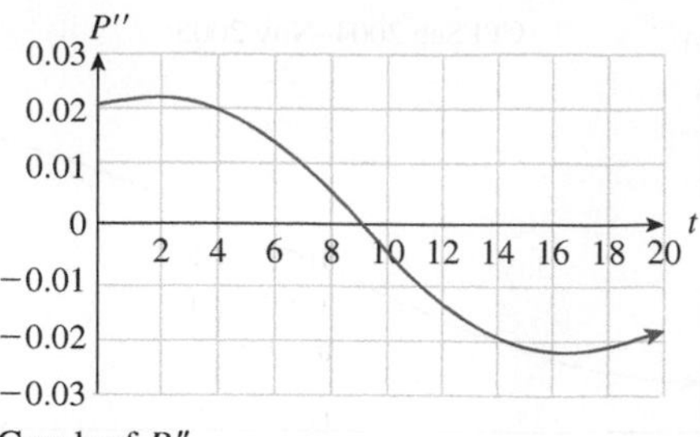

Graph of P''

Determine, to the nearest whole number, the values of t for which the graph of P is concave up and where it is concave down, and locate any points of inflection. What does the point of inflection tell you about science articles?
[**HINT**: See Example 4.]

69. ***Embryo Development*** Here are sketches of the graphs of c, c', and c'' from Exercise 61:

Graph of c

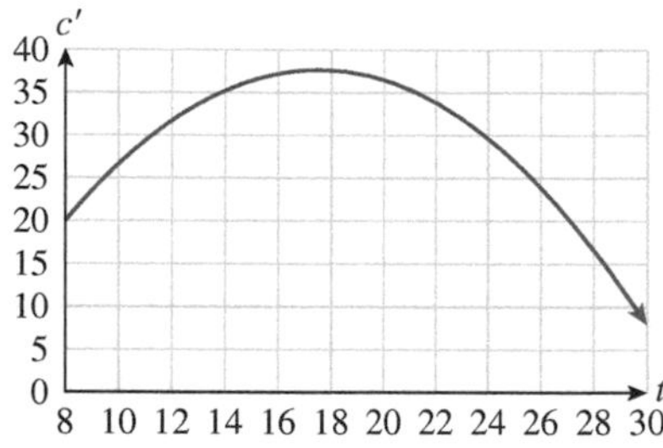

Graph of c'

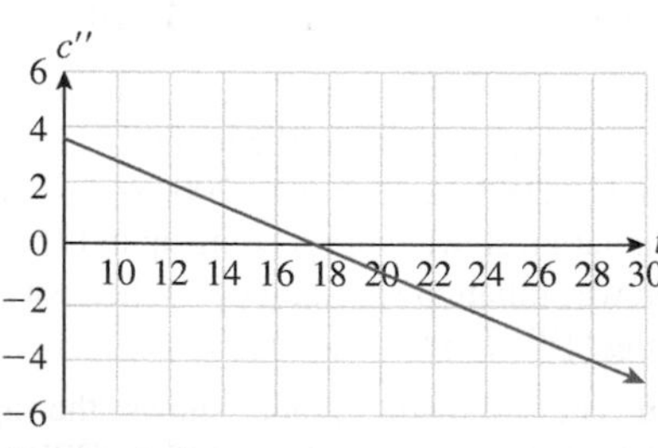

Graph of c''

a. The graph of c'
(A) has a point of inflection.
(B) has no points of inflection.

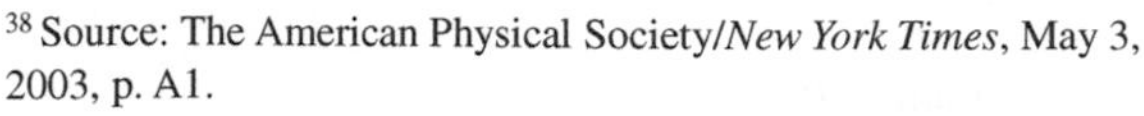
[38] Source: The American Physical Society/*New York Times*, May 3, 2003, p. A1.

b. At around 18 days after the egg is laid, daily oxygen consumption is
(**A**) at a maximum.
(**B**) increasing at a maximum rate.
(**C**) just beginning to decrease.

c. For $t > 18$ days the oxygen consumption is
(**A**) increasing at a decreasing rate.
(**B**) decreasing at an increasing rate.
(**C**) increasing at an increasing rate.

70. ***Embryo Development*** Here are sketches of the graphs of c, c', and c'' from Exercise 62:

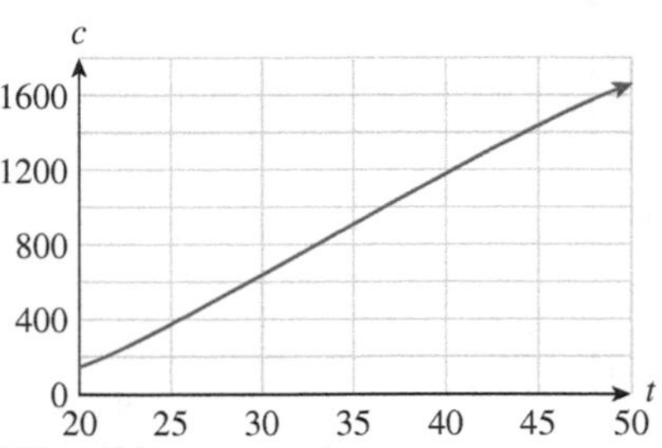

Graph of c

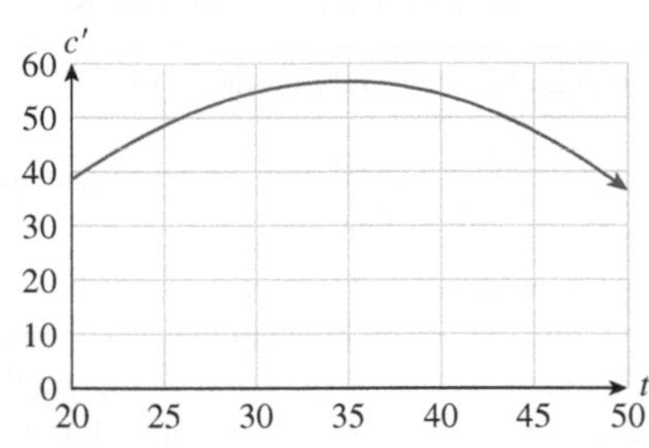

Graph of c'

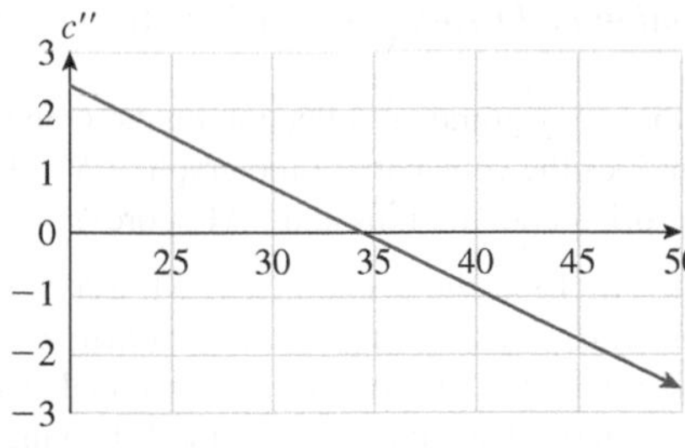

Graph of c''

a. The graph of c
(**A**) has points of inflection.
(**B**) has no points of inflection.
(**C**) may or may not have a point of inflection, but the graphs do not provide enough information.

b. At around 35 days after the egg is laid, the rate of change of daily oxygen consumption is
(**A**) at a maximum.
(**B**) increasing at a maximum rate.
(**C**) just becoming negative.

c. For $t < 35$ days the oxygen consumption is
(**A**) increasing at an increasing rate.
(**B**) increasing at a decreasing rate.
(**C**) decreasing at an increasing rate.

71. T ***Subprime Mortgages during the Housing Bubble*** During the real estate run-up in 2000–2008 the percentage of mortgages issued in the United States that were subprime (normally classified as risky) could be approximated by

$$A(t) = \frac{15.0}{1 + 8.6(1.8)^{-t}} \text{ percent} \quad (0 \le t \le 8)$$

t years after the start of 2000.[39] Graph the function as well as its first and second derivatives. Determine, to the nearest whole number, the values of t for which the graph of A is concave up and concave down and the t-coordinate of any points of inflection. What does the point of inflection tell you about subprime mortgages? [**HINT:** To graph the second derivative, see the margin note next to Example 4.]

72. T ***Subprime Mortgage Debt during the Housing Bubble*** During the real estate run-up in 2000–2008 the value of subprime (normally classified as risky) mortgage debt outstanding in the United States was approximately

$$A(t) = \frac{1{,}350}{1 + 4.2(1.7)^{-t}} \text{ billion dollars} \quad (0 \le t \le 8)$$

t years after the start of 2000.[40] Graph the function as well as its first and second derivatives. Determine, to the nearest whole number, the values of t for which the graph of A is concave up and concave down and the t-coordinate of any points of inflection. What does the point of inflection tell you about subprime mortgages? [**HINT:** To graph the second derivative, see the margin note next to Example 4.]

73. ***Epidemics*** The following graph shows the total number n of people (in millions) infected in an epidemic as a function of time t (in years):

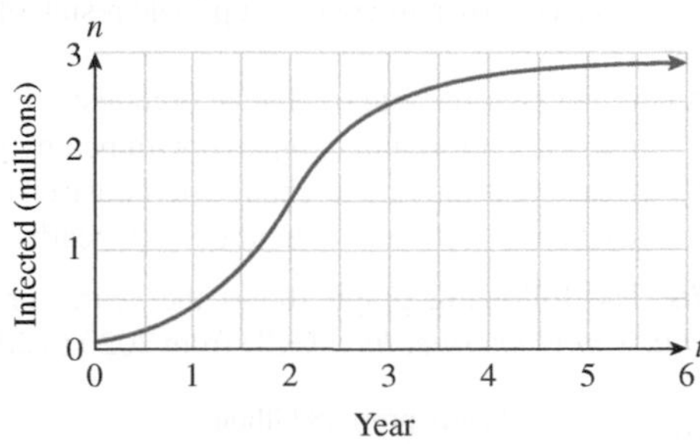

a. When, to the nearest year, was the rate of new infection largest?

b. When could the Centers for Disease Control and Prevention announce that the rate of new infection was beginning to drop? [**HINT:** See Example 4.]

[39] Sources: Mortgage Bankers Association, UBS.

[40] 2008 figure is an estimate. Source: www.data360.org.

74. ***Sales*** The following graph shows the total number of *Pomegranate Q4* computers sold since their release (t is in years):

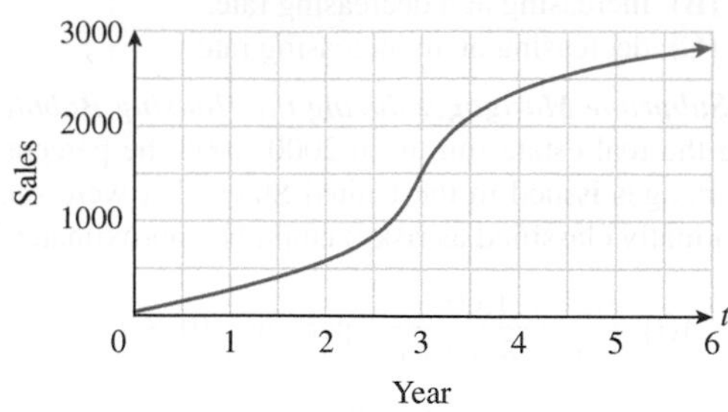

a. When were the computers selling fastest?
b. Explain why this graph might look as it does. [**HINT**: See Example 4.]

75. ***Industrial Output*** The following graph shows the yearly industrial output (measured in billions of zonars) of the city-state of Utarek on Mars over a 7-year period:

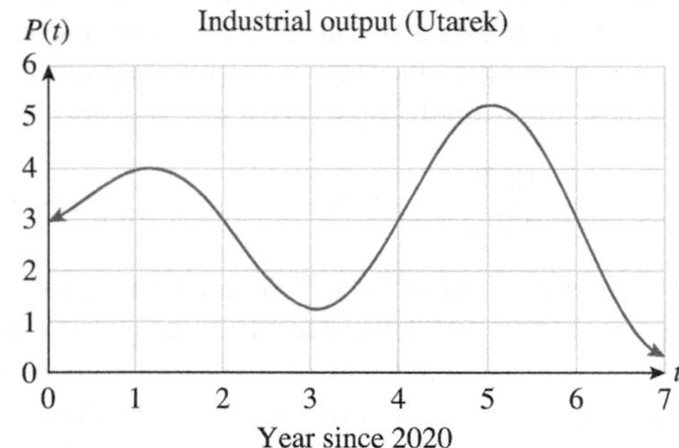

a. When, to the nearest year, did the rate of change of yearly industrial output reach a maximum?
b. When, to the nearest year, did the rate of change of yearly industrial output reach a minimum?
c. When, to the nearest year, does the graph first change from concave down to concave up? The result tells you that
   (A) in that year the rate of change of industrial output reached a minimum compared with nearby years.
   (B) in that year the rate of change of industrial output reached a maximum compared with nearby years.

76. ***Profits*** The following graph shows the yearly profits of *Gigantic Conglomerate, Inc.* (GCI) from 2020 to 2035:

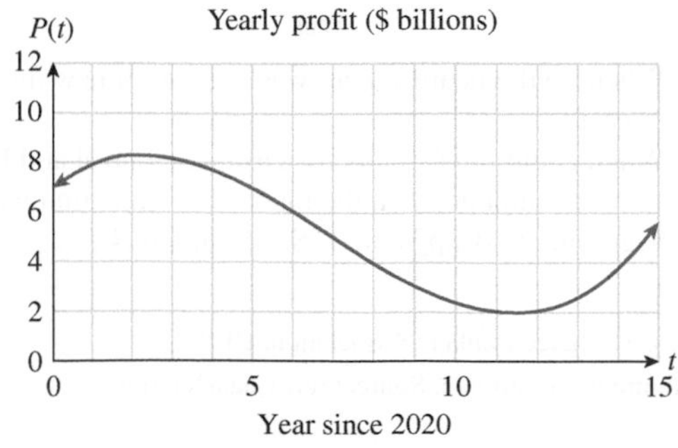

a. Approximately when were the profits rising most rapidly?
b. Approximately when were the profits falling most rapidly?
c. Approximately when could GCI's board of directors legitimately tell stockholders that they had "turned the company around"?

77. ▼ ***Education and Crime*** The following graph compares the total U.S. prison population and the average combined SAT score in the United States during the 1970s and 1980s:

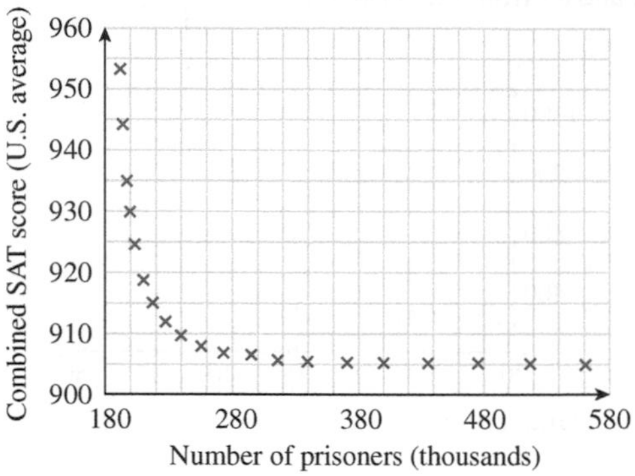

These data can be accurately modeled by

$$S(n) = 904 + \frac{1,326}{(n - 180)^{1.325}} \quad (192 \le n \le 563).$$

Here, $S(n)$ is the combined U.S. average SAT score at a time when the total U.S. prison population was n thousand.[41]

a. Are there any points of inflection on the graph of S?
b. What does the concavity of the graph of S tell you about prison populations and SAT scores?

78. ▼ ***Education and Crime*** Refer back to the model in Exercise 77.

a. Are there any points of inflection on the graph of S'?
b. What does the concavity of the graph of S' tell you about prison populations and SAT scores?

79. ▼ ***Patents*** In 1965 the economist F. M. Scherer modeled the number, n, of patents produced by a firm as a function of the size, s, of the firm (measured in annual sales in millions of dollars). He came up with the following equation based on a study of 448 large firms:[42]

$$n = -3.79 + 144.42s - 23.86s^2 + 1.457s^3.$$

a. Find $\left.\frac{d^2n}{ds^2}\right|_{s=3}$. Is the rate at which patents are produced as the size of a firm goes up increasing or decreasing

[41] Based on data for the years 1967–1989. Sources: *Sourcebook of Criminal Justice Statistics*, 1990, p. 604/Educational Testing Service.

[42] Source: F. M. Scherer, "Firm Size, Market Structure, Opportunity, and the Output of Patented Inventions," *American Economic Review* 55 (December 1965): pp. 1097–1125.

with size when $s = 3$? Comment on Scherer's words, ". . . we find diminishing returns dominating."

b. Find $\left.\dfrac{d^2n}{ds^2}\right|_{s=7}$, and interpret the answer.

c. Find the s-coordinate of any points of inflection, and interpret the result.

80. ▼ ***Returns on Investments*** A company finds that the number of new products it develops per year depends on the size of its annual R&D budget, x (in thousands of dollars), according to the formula

$$n(x) = -1 + 8x + 2x^2 - 0.4x^3.$$

a. Find $n''(1)$ and $n''(3)$, and interpret the results.

b. Find the size of the budget that gives the largest rate of return as measured in new products per dollar (again, called the point of diminishing returns).

81. T ▼ ***Oil Imports from Mexico*** Daily oil production in Mexico and daily U.S. oil imports from Mexico during 2009–2013 can be approximated by

$$P(t) = 3.1 - 0.014t \text{ million barrels} \quad (9 \le t \le 13)$$
$$I(t) = 1.7 - 0.063t \text{ million barrels} \quad (9 \le t \le 13),$$

where t is time in years since the start of 2000.[43]

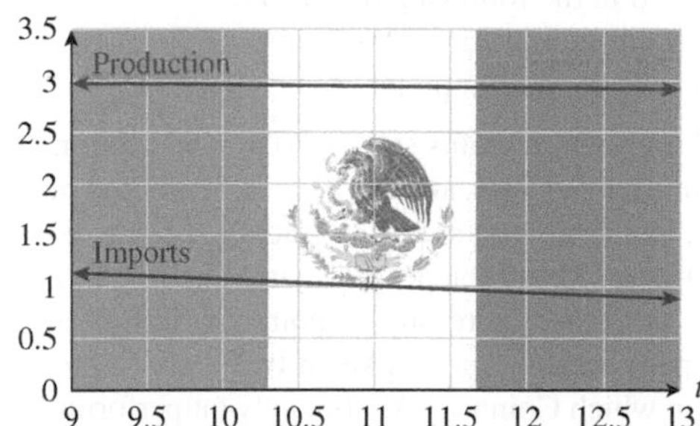

Graph the function $I(t)/P(t)$ and its derivative. Is the graph of $I(t)/P(t)$ concave up or concave down? The concavity of $I(t)/P(t)$ tells you that

(A) the percentage of oil produced in Mexico that was exported to the United States was decreasing.

(B) the percentage of oil produced in Mexico that was not exported to the United States was increasing.

(C) the percentage of oil produced in Mexico that was exported to the United States was decreasing at a slower rate.

(D) the percentage of oil produced in Mexico that was exported to the United States was decreasing at a faster rate.

[43] Source for data: Energy Information Administration (www.eia.doe.gov)/Pemex.

82. T ▼ ***Oil Imports from Mexico*** Repeat Exercise 81 using instead the models for 2000–2004 shown below:

$$P(t) = 3.0 + 0.13t \text{ million barrels} \quad (0 \le t \le 4)$$
$$I(t) = 1.4 + 0.06t \text{ million barrels} \quad (0 \le t \le 4).$$

(t is time in years since the start of 2000.)[44]

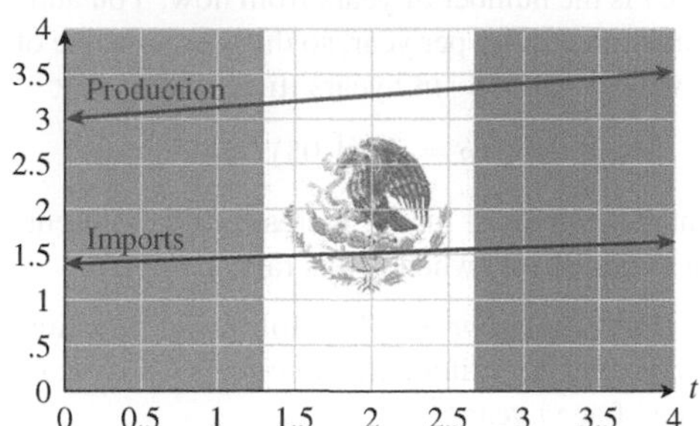

83. ◆ ***Logistic Models*** Let

$$f(x) = \frac{N}{1 + Ab^{-x}}$$

for constants N, A, and b (A and b positive and $b \ne 1$). Show that f has a single point of inflection at $x = \ln A/\ln b$.

84. ◆ ***Logistic Models*** Let

$$f(x) = \frac{N}{1 + Ae^{-kx}}$$

for constants N, A, and k (A and k positive). Show that f has a single point of inflection at $x = \ln A/k$.

85. T ***Population: Puerto Rico*** The population of Puerto Rico in 1950–2025 can be approximated by

$$P(t) = \frac{4{,}500}{1 + 1.1466(1.0357)^{-t}} \text{ thousand people} \quad (0 \le t \le 75).$$

(t is the year since 1950.)[45] Use the result of Exercise 83 to find the location of the point of inflection in the graph of P. What does the result tell you about the population of Puerto Rico?

86. T ***Population: Virgin Islands*** The population of the Virgin Islands in 1950–2025 can be approximated by

$$P(t) = \frac{110}{1 + 2.3596(1.0767)^{-t}} \text{ thousand people} \quad (0 \le t \le 75).$$

(t is the year since 1950.)[46] Use the result of Exercise 83 to find the location of the point of inflection in the graph of P. What does the result tell you about the population of the Virgin Islands?

[44] *Ibid.*

[45] Figures from 2010 on are U.S. census projections. Source for data: The 2008 Statistical Abstract (www.census.gov).

[46] *Ibid.*

87. **T** ▼ ***Asset Appreciation*** You manage a small antique store that owns a collection of Louis XVI jewelry boxes. Their value v is increasing according to the formula

$$v = \frac{10{,}000}{1 + 500e^{-0.5t}},$$

where t is the number of years from now. You anticipate an inflation rate of 5% per year, so the present value of an item that will be worth $\$v$ in t years' time is given by

$$p = v \cdot (1.05)^{-t}.$$

What is the greatest rate of increase of the present value of your antiques, and when is this rate attained?

88. **T** ▼ ***Harvesting Forests*** The following equation models the approximate volume in cubic feet of a typical Douglas fir tree of age t years[47]:

$$V = \frac{22{,}514}{1 + 22{,}514t^{-2.55}}.$$

The lumber will be sold at $10 per cubic foot, and you do not expect the price of lumber to appreciate in the foreseeable future. On the other hand, you anticipate a general inflation rate of 5% per year, so the present value of an item that will be worth $\$v$ in t years' time is given by

$$p = v \cdot (1.05)^{-t}.$$

What is the largest rate of increase of the present value of a fir tree, and when is this rate attained?

89. **T** ▼ ***Asset Appreciation*** As the financial consultant to a classic auto dealership, you estimate that the total value of its collection of 1959 Chevrolets and Fords is given by the formula

$$v = 300{,}000 + 1{,}000t^2,$$

where t is the number of years from now. You anticipate a continuous inflation rate of 5% per year, so the discounted (present) value of an item that will be worth $\$v$ in t years' time is given by

$$p = ve^{-0.05t}.$$

When is the discounted value of the collection of classic cars increasing most rapidly? When is it decreasing most rapidly?

90. **T** ▼ ***Plantation Management*** The value of a fir tree in your plantation increases with the age of the tree according to the formula

$$v = \frac{20t}{1 + 0.05t},$$

where t is the age of the tree in years. Given a continuous inflation rate of 5% per year, the discounted (present) value of a newly planted seedling is

$$p = ve^{-0.05t}.$$

When is the discounted value of a tree increasing most rapidly? Decreasing most rapidly?

Communication and Reasoning Exercises

91. Complete the following: If the graph of a function is concave up on its entire domain, then its second derivative is ______ on the domain.

92. Complete the following: If the graph of a function is concave up on its entire domain, then its first derivative is ______ on the domain.

93. Daily sales of *Kent's Tents* reached a maximum in January 2002 and declined to a minimum in January 2003 before starting to climb again. The graph of daily sales shows a point of inflection at June 2002. What is the significance of the point of inflection?

94. The graph of daily sales of *Luddington's Wellington* boots is concave down, although sales continue to increase. What properties of the graph of daily sales versus time are reflected in the following behaviors?
   a. a point of inflection next year
   b. a horizontal asymptote

95. ▼ Company A's profits satisfy $P(0) = \$1$ million, $P'(0) = \$1$ million per year, and $P''(0) = -\$1$ million per year per year. Company B's profits satisfy $P(0) = \$1$ million, $P'(0) = -\$1$ million per year, and $P''(0) = \$1$ million per year per year. There are no points of inflection in either company's profit curve. Sketch two pairs of profit curves: one in which Company A ultimately outperforms Company B and another in which Company B ultimately outperforms Company A.

96. ▼ Company C's profits satisfy $P(0) = \$1$ million, $P'(0) = \$1$ million per year, and $P''(0) = -\$1$ million per year per year. Company D's profits satisfy $P(0) = \$0$ million, $P'(0) = \$0$ million per year, and $P''(0) = \$1$ million per year per year. There are no points of inflection in either company's profit curve. Sketch two pairs of profit curves: one in which Company C ultimately outperforms Company D and another in which Company D ultimately outperforms Company C.

97. ▼ Explain geometrically why the derivative of a function has a relative extremum at a point of inflection, if it is defined there. Which points of inflection give rise to relative maxima in the derivative?

98. ▼ If we regard position, s, as a function of time, t, what is the significance of the *third* derivative, $s'''(t)$? Describe an everyday scenario in which this arises.

---

[47] The model is the authors' and is based on data in *Environmental and Natural Resource Economics* by Tom Tietenberg, third edition (New York: HarperCollins, 1992), p. 282.

5.4 Analyzing Graphs

Mathematical curves are beautiful—their subtle form can be imitated by only the best of artists—and calculus gives us the tools we need to probe their secrets. While it is easy to use graphing technology to draw a graph, we must use calculus to understand what we are seeing. Following is a list of some of the most interesting features of the graph of a function.

Features of a Graph

1. ***The x- and y-intercepts:*** If $y = f(x)$, find the x-intercept(s) by setting $y = 0$ and solving for x; find the y-intercept by setting $x = 0$ and solving for y:

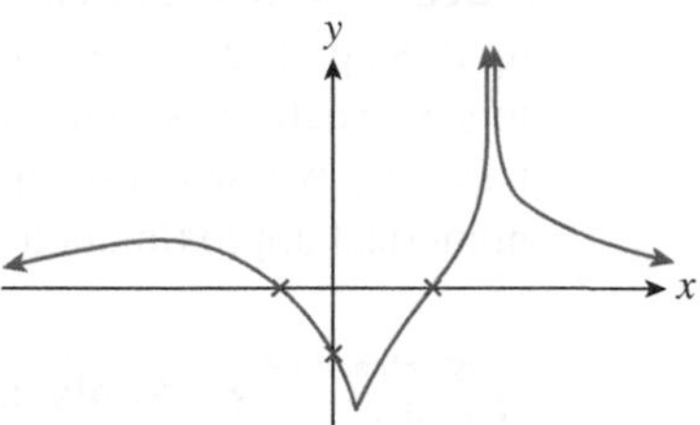

2. ***Extrema:*** Use the techniques of Section 5.1 to locate the maxima and minima:

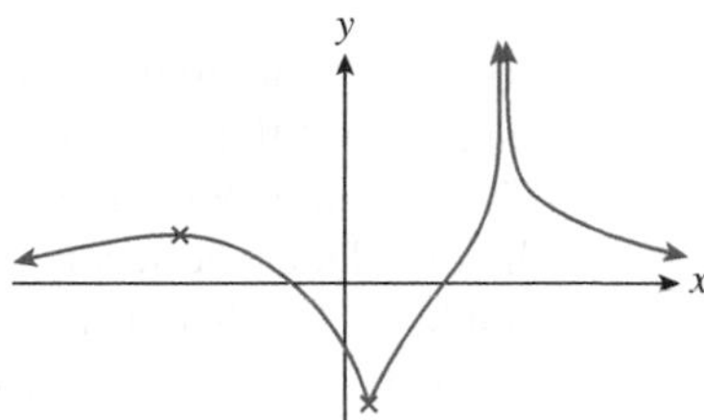

3. ***Points of inflection:*** Use the techniques of Section 5.2 to locate the points of inflection:

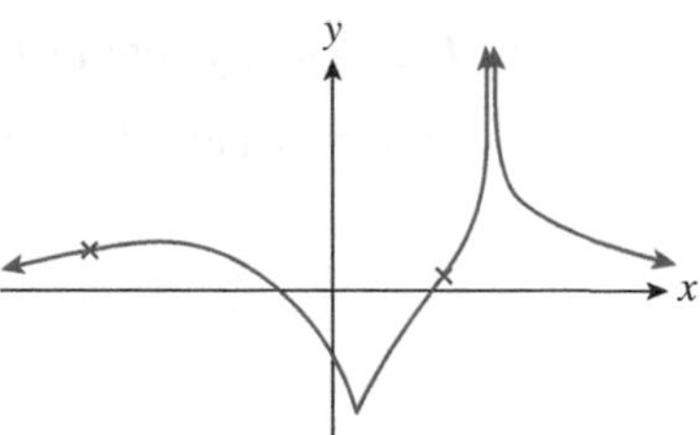

4. ***Behavior near singular points of f:**** If a is a singular point of f, consider $\lim_{x\to a^-} f(x)$ and $\lim_{x\to a^+} f(x)$ to see how the graph of f behaves as x approaches a:

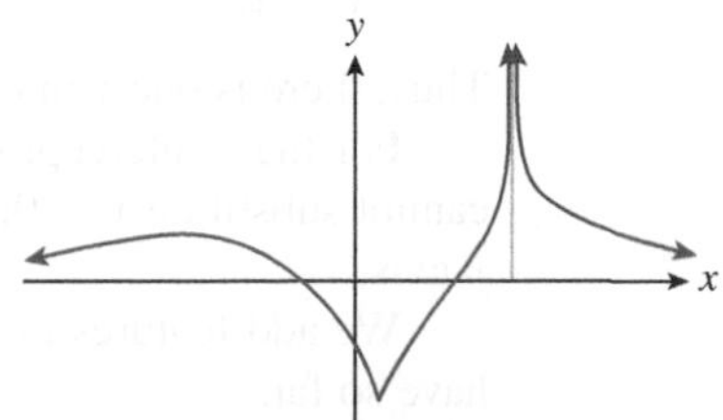

* Recall from Section 3.2 that a is a singular point of f if $f(a)$ is not defined, but $f(x)$ is defined for (at least some) points arbitrarily close to and on both sides of a.

5. ***Behavior at infinity:*** Consider $\lim_{x\to-\infty} f(x)$ and $\lim_{x\to+\infty} f(x)$ if appropriate, to see how the graph of f behaves far to the left and right:

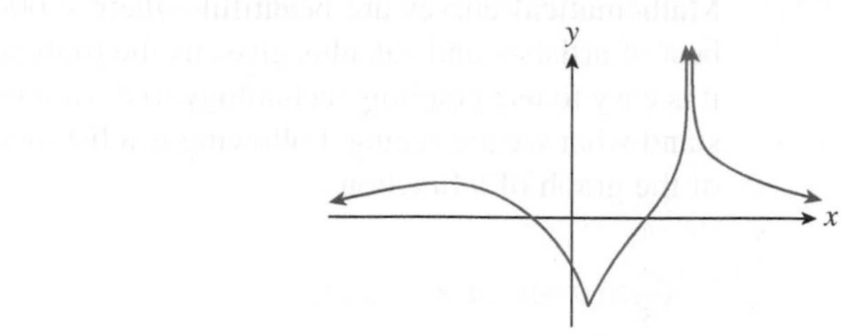

Note It is sometimes difficult or impossible to solve all of the equations that come up in Steps 1, 2, and 3 of the above analysis. As a consequence, we might not be able to say exactly where the x-intercept, extrema, or points of inflection are. When this happens, we will use graphing technology to assist us in determining accurate numerical approximations. ■

EXAMPLE 1 **Analyzing a Graph**

Analyze the graph of $f(x) = \dfrac{1}{x} - \dfrac{1}{x^2}$.

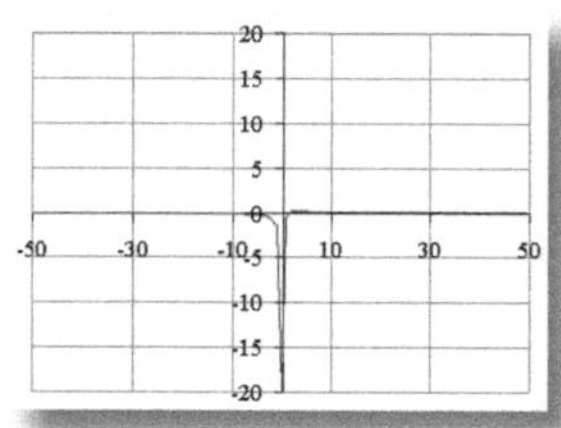

$-50 \le x \le 50, -20 \le y \le 20$

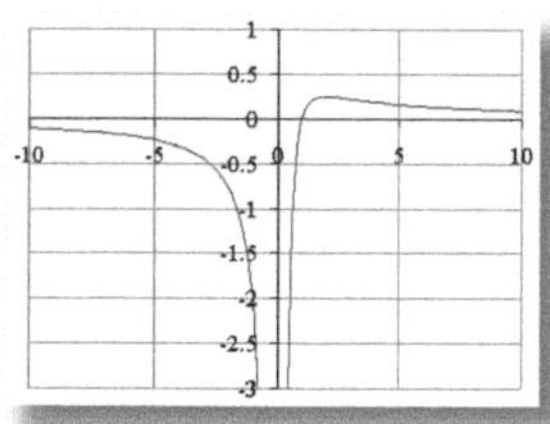

$-10 \le x \le 10, -3 \le y \le 1$

Figure 35

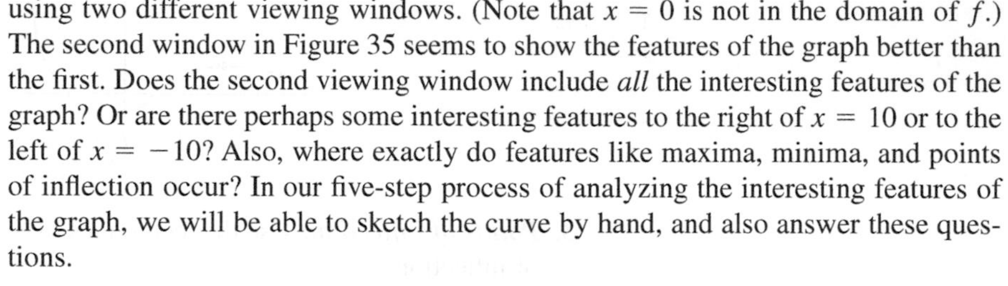

Solution The graph, as drawn using graphing technology, is shown in Figure 35, using two different viewing windows. (Note that $x = 0$ is not in the domain of f.) The second window in Figure 35 seems to show the features of the graph better than the first. Does the second viewing window include *all* the interesting features of the graph? Or are there perhaps some interesting features to the right of $x = 10$ or to the left of $x = -10$? Also, where exactly do features like maxima, minima, and points of inflection occur? In our five-step process of analyzing the interesting features of the graph, we will be able to sketch the curve by hand, and also answer these questions.

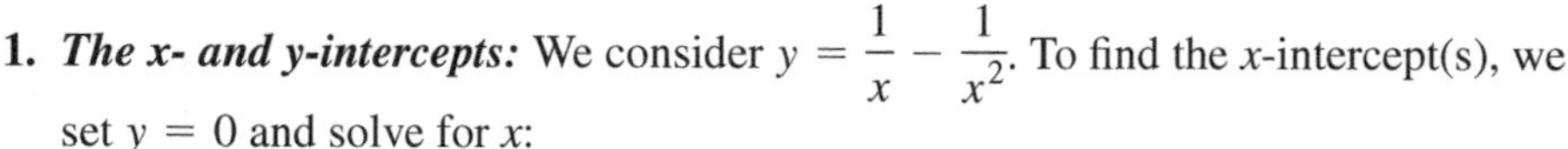

1. ***The x- and y-intercepts:*** We consider $y = \dfrac{1}{x} - \dfrac{1}{x^2}$. To find the x-intercept(s), we set $y = 0$ and solve for x:

$$0 = \frac{1}{x} - \frac{1}{x^2}$$

$$\frac{1}{x} = \frac{1}{x^2}.$$

Multiplying both sides by x^2 (we know that x cannot be zero, so we are not multiplying both sides by 0) gives

$$x = 1.$$

Thus, there is one x-intercept (which we can see in Figure 35): at $x = 1$.

For the y-intercept we would substitute $x = 0$ and solve for y. However, we cannot substitute $x = 0$; because $f(0)$ is not defined, the graph does not meet the y-axis.

We add features to our freehand sketch as we go. Figure 36 shows what we have so far.

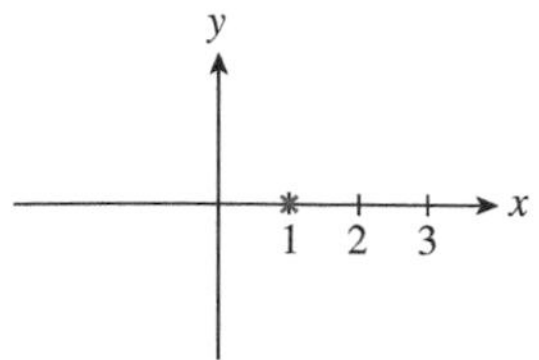

Figure 36

2. ***Relative extrema:*** We calculate $f'(x) = -\dfrac{1}{x^2} + \dfrac{2}{x^3}$. To find any stationary points, we set the derivative equal to 0 and solve for x:

$$-\frac{1}{x^2} + \frac{2}{x^3} = 0$$
$$\frac{1}{x^2} = \frac{2}{x^3}$$
$$x = 2.$$

Thus, there is one stationary point: at $x = 2$. We can use a test point to the right to determine that this stationary point is a relative maximum:

x	1 (Intercept)	2	3 (Test point)
$y = \dfrac{1}{x} - \dfrac{1}{x^2}$	0	$\dfrac{1}{4}$	$\dfrac{2}{9}$

✱ Recall from Section 5.1 that a is a singular point of f' if a is an interior point of the domain of f where the derivative $f'(a)$ is not defined. (Note the distinction from a singular point of f above.)

The only possible singular point of f'✱ is at 0 because $f'(0)$ is not defined. However, $f(0)$ is not defined either, so there are no singular points. Figure 37 shows our graph so far.

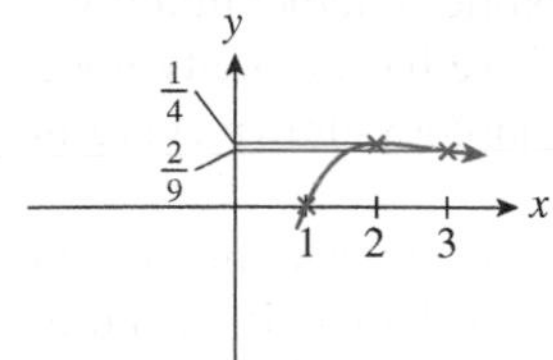

Figure 37

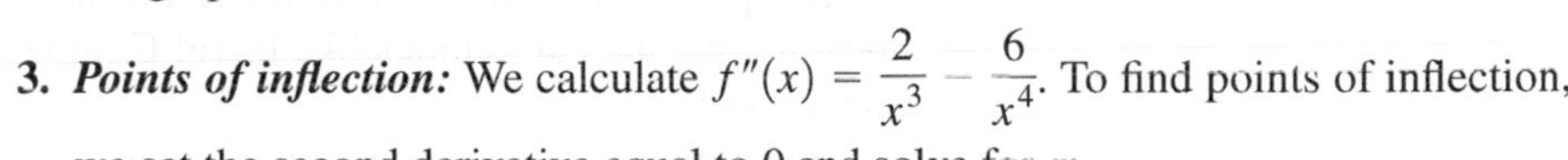

3. ***Points of inflection:*** We calculate $f''(x) = \dfrac{2}{x^3} - \dfrac{6}{x^4}$. To find points of inflection, we set the second derivative equal to 0 and solve for x:

$$\frac{2}{x^3} - \frac{6}{x^4} = 0$$
$$\frac{2}{x^3} = \frac{6}{x^4}$$
$$2x = 6$$
$$x = 3.$$

Figure 35 confirms that the graph of f changes from being concave down to being concave up at $x = 3$, so this is a point of inflection. $f''(x)$ is not defined at $x = 0$, but that is not in the domain, so there are no other points of inflection. In particular, the graph must be concave down in the whole region $(-\infty, 0)$, as we can see by calculating the second derivative at any one point in that interval: $f''(-1) = -8 < 0$. Figure 38 shows our graph so far. (We extended the curve near $x = 3$ to suggest a point of inflection at $x = 3$.)

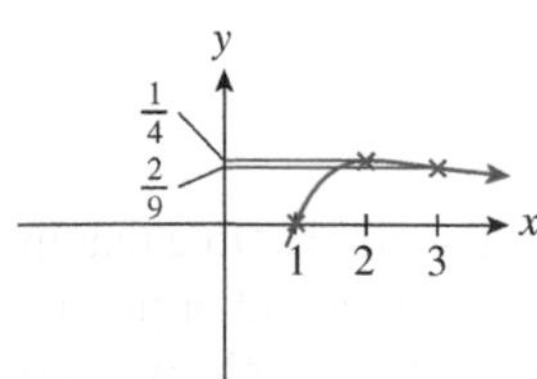

Figure 38

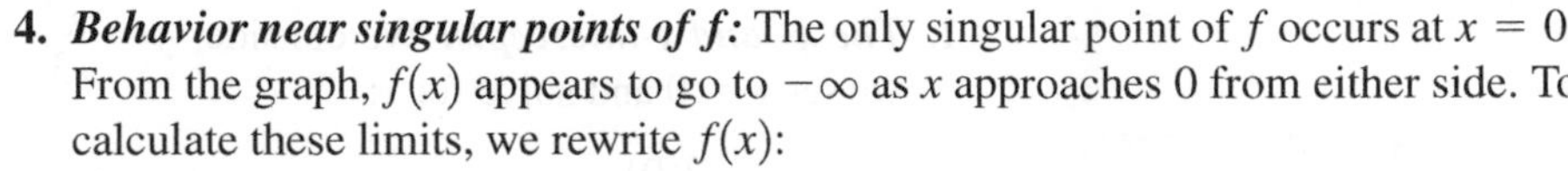

4. ***Behavior near singular points of f:*** The only singular point of f occurs at $x = 0$. From the graph, $f(x)$ appears to go to $-\infty$ as x approaches 0 from either side. To calculate these limits, we rewrite $f(x)$:

$$f(x) = \frac{1}{x} - \frac{1}{x^2} = \frac{x - 1}{x^2}.$$

Now, if x is close to 0 (on either side), the numerator $x - 1$ is close to -1, and the denominator is a very small but positive number. The quotient is therefore a negative number of very large magnitude. Therefore,

$$\lim_{x \to 0^-} f(x) = -\infty$$

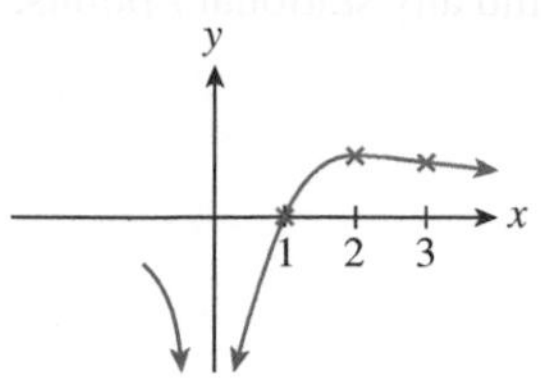

Figure 39

and

$$\lim_{x \to 0^+} f(x) = -\infty.$$

From these limits, we see the following:

(1) Immediately to the *left* of $x = 0$, the graph plunges down toward $-\infty$.

(2) Immediately to the *right* of $x = 0$, the graph also plunges down toward $-\infty$.

Figure 39 shows our graph with these features added. We say that f has a **vertical asymptote** at $x = 0$, meaning that the points on the graph of f get closer and closer to points on a vertical line (the y-axis in this case) farther and farther from the origin.

5. ***Behavior at infinity:*** Both $1/x$ and $1/x^2$ go to 0 as x goes to $-\infty$ or $+\infty$; that is,

$$\lim_{x \to -\infty} f(x) = 0$$

and

$$\lim_{x \to +\infty} f(x) = 0.$$

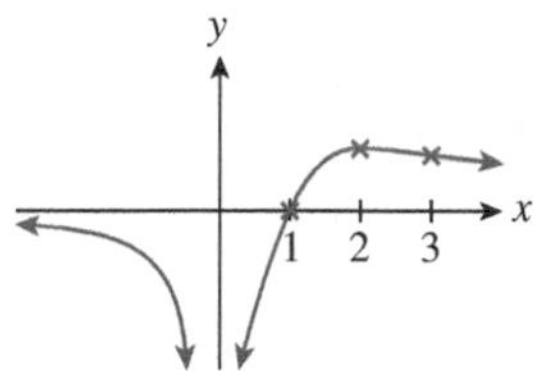

Figure 40

Thus, on the extreme left and right of our picture, the height of the curve levels off toward zero. Figure 40 shows the completed freehand sketch of the graph.

We say that f has a **horizontal asymptote** at $y = 0$. (Notice another thing: We haven't plotted a single point to the left of the y-axis, yet we have a pretty good idea of what the curve looks like there! Compare the technology-drawn curve in Figure 35.)

In summary, there is one x-intercept at $x = 1$; there is one relative maximum (which, we can now see, is also an absolute maximum) at $x = 2$; there is one point of inflection at $x = 3$, where the graph changes from being concave down to concave up. There is a vertical asymptote at $x = 0$, on both sides of which the graph goes down toward $-\infty$, and a horizontal asymptote at $y = 0$.

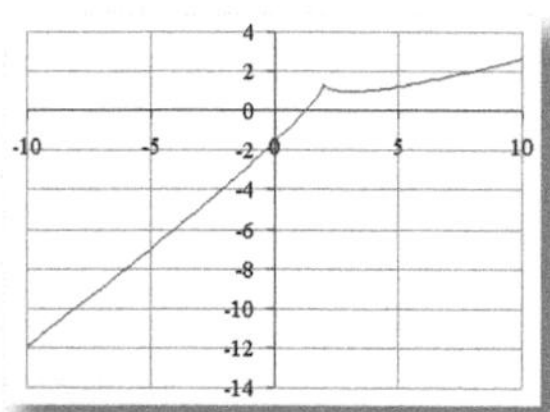

Technology:
`2*x/3-((x-2)^2)^(1/3)`
Figure 41

EXAMPLE 2 Analyzing a Graph

Analyze the graph of $f(x) = \dfrac{2x}{3} - (x - 2)^{2/3}$.

Solution Figure 41 shows a technology-generated version of the graph. Note that in the technology formulation, $(x - 2)^{2/3}$ is written as $[(x - 2)^2]^{1/3}$ to avoid problems with some graphing calculators and Excel. Let us now re-create this graph by hand and, in the process, identify the features we see in Figure 41.

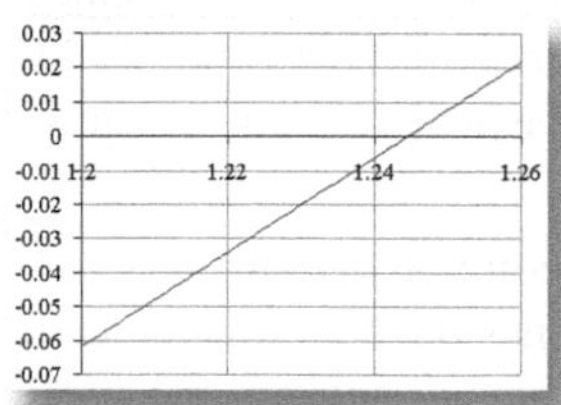

Figure 42

1. ***The x- and y-intercepts:*** We consider $y = \dfrac{2x}{3} - (x - 2)^{2/3}$. For the y-intercept we set $x = 0$ and solve for y:

$$y = \frac{2(0)}{3} - (0 - 2)^{2/3} = -2^{2/3} \approx -1.59.$$

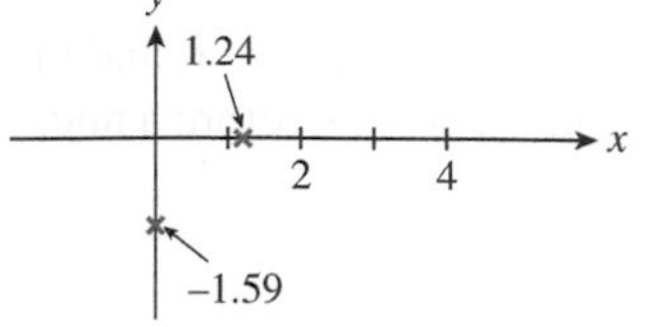

Figure 43

To find the x-intercept(s), we set $y = 0$ and solve for x. However, if we attempt this, we will find ourselves with a cubic equation that is hard to solve. (Try it!) Following the advice in the note preceding Example 1, we use graphing technology to locate the x-intercept we see in Figure 41 by zooming in (Figure 42). From Figure 42 we find $x \approx 1.24$. We shall see in the discussion to follow that there can be no other x-intercepts.

Figure 43 shows our freehand sketch so far.

2. ***Relative extrema:*** We calculate

$$f'(x) = \frac{2}{3} - \frac{2}{3}(x-2)^{-1/3}$$
$$= \frac{2}{3} - \frac{2}{3(x-2)^{1/3}}.$$

To find any stationary points, we set the derivative equal to 0 and solve for x:

$$\frac{2}{3} - \frac{2}{3(x-2)^{1/3}} = 0$$
$$(x-2)^{1/3} = 1$$
$$x - 2 = 1^3 = 1$$
$$x = 3.$$

To check for singular points, look for points where $f(x)$ is defined and $f'(x)$ is not defined. The only such point is $x = 2$: $f'(x)$ is not defined at $x = 2$, whereas $f(x)$ is defined there, so we have a singular point at $x = 2$.

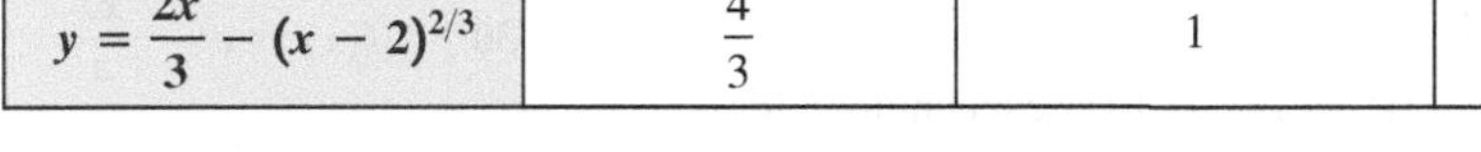

x	2 (Singular point)	3 (Stationary point)	4 (Test point)
$y = \frac{2x}{3} - (x-2)^{2/3}$	$\frac{4}{3}$	1	1.079

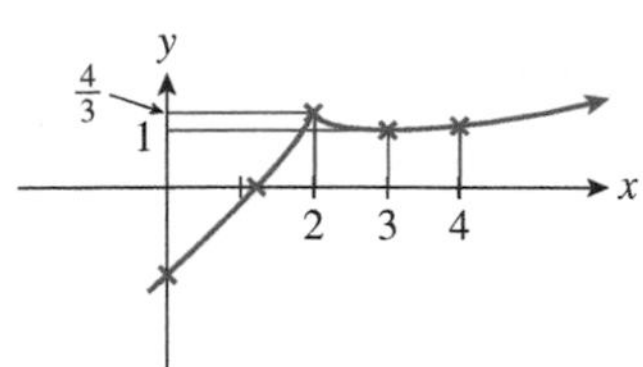

Figure 44

Figure 44 shows our graph so far.

We see that there is a singular relative maximum at $(2, 4/3)$ (we will confirm that the graph eventually gets higher on the right) and a stationary relative minimum at $x = 3$.

3. ***Points of inflection:*** We calculate

$$f''(x) = \frac{2}{9(x-2)^{4/3}}.$$

To find points of inflection, we set the second derivative equal to 0 and solve for x. But the equation

$$0 = \frac{2}{9(x-2)^{4/3}}$$

has no solution for x, so there are no points of inflection on the graph.

4. ***Behavior near points where f is not defined:*** Because $f(x)$ is defined everywhere, there are no such points to consider. In particular, there are no vertical asymptotes.

5. ***Behavior at infinity:*** We estimate the following limits numerically:

$$\lim_{x\to-\infty}\left[\frac{2x}{3} - (x-2)^{2/3}\right] = -\infty$$

and

$$\lim_{x\to+\infty}\left[\frac{2x}{3} - (x-2)^{2/3}\right] = +\infty.$$

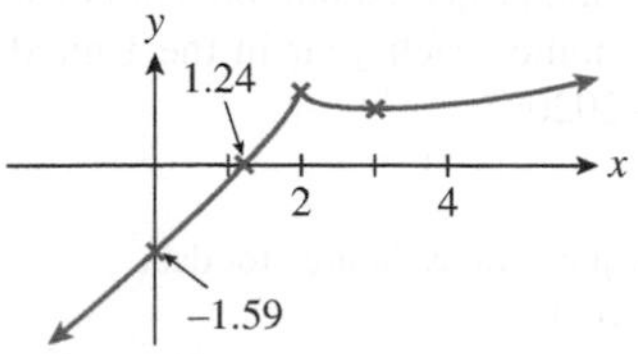

Thus, on the extreme left the curve goes down toward $-\infty$, and on the extreme right the curve rises toward $+\infty$. In particular, there are no horizontal asymptotes. (There can also be no other x-intercepts.)

Figure 45 shows the completed graph.

Figure 45

5.4 EXERCISES

▼ more advanced ◆ challenging

T indicates exercises that should be solved using technology

In Exercises 1–26, sketch the graph of the given function, indicating ***(a)*** *x- and y-intercepts,* ***(b)*** *extrema,* ***(c)*** *points of inflection,* ***(d)*** *behavior near singular points of f, and* ***(e)*** *behavior at infinity. Where indicated, technology should be used to approximate the intercepts, coordinates of extrema, and/or points of inflection to one decimal place. Check your sketch using technology.* [**HINT**: See Example 1.]

1. $f(x) = x^2 + 2x + 1$
2. $f(x) = -x^2 - 2x - 1$
3. $g(x) = x^3 - 12x$, domain $[-4, 4]$
4. $g(x) = 2x^3 - 6x$, domain $[-4, 4]$
5. $h(x) = 2x^3 - 3x^2 - 36x$ [Use technology for x-intercepts.]
6. $h(x) = -2x^3 - 3x^2 + 36x$ [Use technology for x-intercepts.]
7. $f(x) = 2x^3 + 3x^2 - 12x + 1$ [Use technology for x-intercepts.]
8. $f(x) = 4x^3 + 3x^2 + 2$ [Use technology for x-intercepts.]
9. $k(x) = -3x^4 + 4x^3 + 36x^2 + 10$ [Use technology for x-intercepts.]
10. $k(x) = 3x^4 + 4x^3 - 36x^2 - 10$ [Use technology for x-intercepts.]
11. $g(t) = \frac{1}{4}t^4 - \frac{2}{3}t^3 + \frac{1}{2}t^2$
12. $g(t) = 3t^4 - 16t^3 + 24t^2 + 1$
13. $f(x) = x + \frac{1}{x}$
14. $f(x) = x^2 + \frac{1}{x^2}$
15. $g(x) = x^3/(x^2 + 3)$
16. $g(x) = x^3/(x^2 - 3)$
17. $f(t) = \frac{t^2 + 1}{t^2 - 1}$, domain $[-2, 2]$, $t \neq \pm 1$
18. $f(t) = \frac{t^2 - 1}{t^2 + 1}$, domain $[-2, 2]$
19. $k(x) = \frac{2x}{3} + (x + 1)^{2/3}$ [Use technology for x-intercepts.] [**HINT**: See Example 2.]
20. $k(x) = \frac{2x}{5} - (x - 1)^{2/5}$ [Use technology for x-intercepts.] [**HINT**: See Example 2.]
21. $f(x) = x - \ln x$, domain $(0, +\infty)$
22. $f(x) = x - \ln x^2$, domain $(0, +\infty)$
23. $f(x) = x^2 + \ln x^2$ [Use technology for x-intercepts.]
24. $f(x) = 2x^2 + \ln x$ [Use technology for x-intercepts.]
25. $g(t) = e^t - t$, domain $[-1, 1]$
26. $g(t) = e^{-t^2}$

T *In Exercises 27–30, use technology to sketch the graph of the given function, labeling all relative and absolute extrema and points of inflection, and vertical and horizontal asymptotes. The coordinates of the extrema and points of inflection should be accurate to two decimal places.* [**HINT**: To locate extrema accurately, plot the first derivative; to locate points of inflection accurately, plot the second derivative.]

27. ▼ $f(x) = x^4 - 2x^3 + x^2 - 2x + 1$
28. ▼ $f(x) = x^4 + x^3 + x^2 + x + 1$
29. ▼ $f(x) = e^x - x^3$
30. ▼ $f(x) = e^x - \frac{x^4}{4}$

Applications

31. ***Home Prices*** The following graph shows a rough approximation of historical and projected median home prices in the United States for the period 2000–2024:[48]

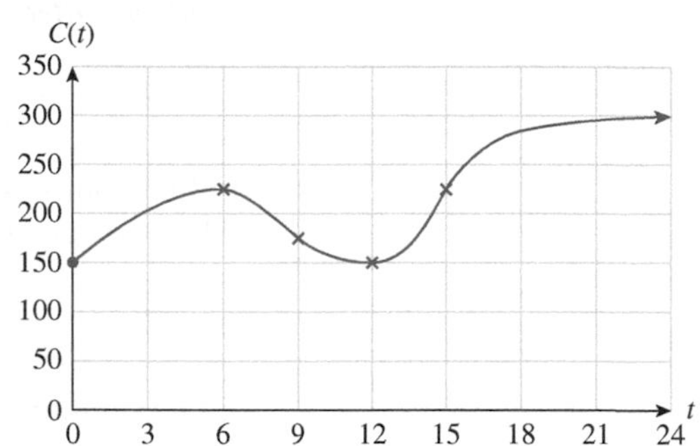

Here, t is time in years since the start of 2000, and $C(t)$ is the median home price in thousands of dollars. The locations of stationary points and points of inflection are indicated on the graph. Analyze the graph's important features, and interpret each feature in terms of the median home price.

32. ***Housing Starts*** The following graph shows a rough approximation of historical and projected numbers of housing starts of single-family homes each year in the United States for the period 2000–2020:[49]

[48] Values from 2015 on are authors' projections. Source for data through 2014: Zillow (www.zillow.com).

[49] Values from 2015 on are authors' projections. Source for data through 2014: U.S. Census Bureau (www.census.gov).

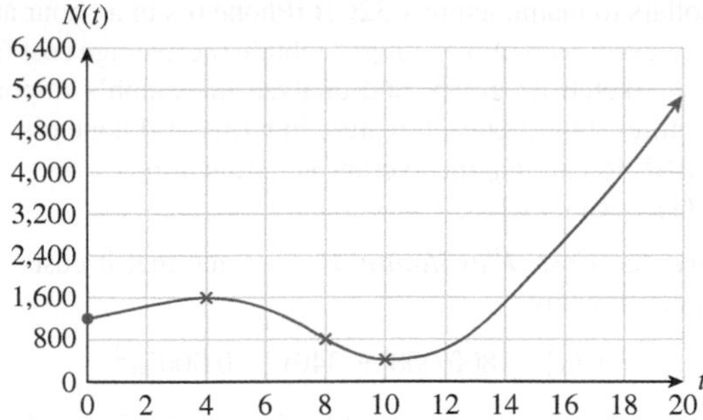

Here, t is time in years since 2000, and $N(t)$ is the number, in thousands, of housing starts per year. The locations of stationary points and points of inflection are indicated on the graph. Analyze the graph's important features, and interpret each feature in terms of the number of housing starts.

33. ***End of the Earth*** In 5 billion years the Sun will have run out of hydrogen fuel and will begin to expand into a red giant, eventually engulfing the Earth and causing it to spiral into the core of the Sun 7.5 billion years from now. The following graph shows the radius of the Earth's orbit around the Sun during its final five and a half million years. ($t = 6$ marks the end of the red giant expansion phase of the Sun.)[50]

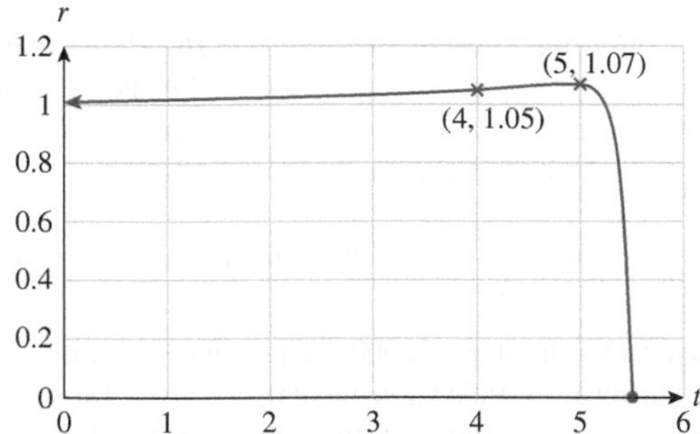

In the graph, r is in AU (astronomical units; 1 AU equals the current radius of the Earth's orbit, around 93 million miles), t is time in millions of years, and the locations of stationary points and points of inflection are indicated.

a. Analyze the graph's important features.

b. Select the correct answers: During the period $4 < t < 5$, the Earth's orbital radius will (increase/decrease/increase and then decrease), and its rate of change will (increase/decrease/increase and then decrease).

34. ***Loss of Odyssia*** The following graph shows the radius of the planet Odyssia's orbit around the star Laertes during its first five and a half million years of existence, before it was flung out of orbit by a passing planetoid and doomed to wander the galaxy forever:[51]

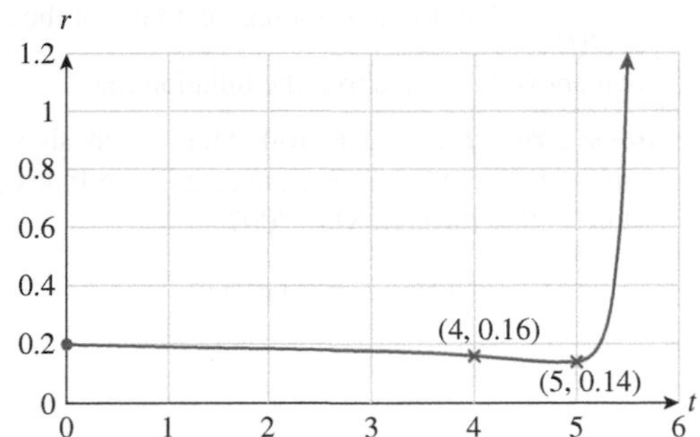

In the graph, r is in LU (Laertian units; 1 LU equals the current radius of the gas giant Pankratia, another planet orbiting Laertes, around 150 million miles), t is time in millions of years, and the locations of stationary points and points of inflection are indicated.

a. Analyze the graph's important features.

b. Select the correct answers: During the period $4 < t < 5$, Odyssia's orbital radius was (increasing/decreasing/increasing and then decreasing), and its rate of change was (increasing/decreasing/increasing and then decreasing).

35. ***Consumer Price Index*** The following graph shows the approximate value of the U.S. Consumer Price Index (CPI) from July 2005 through March 2006:[52]

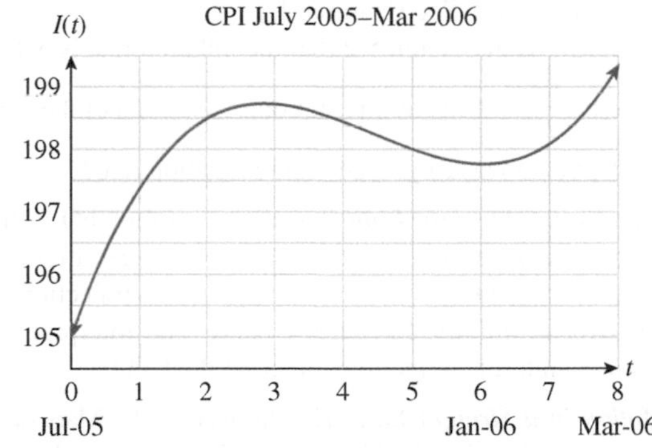

The approximating curve shown on the figure is given by

$$I(t) = 0.06t^3 - 0.8t^2 + 3.1t + 195 \quad (0 \le t \le 8),$$

where t is time in months. ($t = 0$ represents July 2005.)

[50] Based on estimates in "Distant future of the Sun and Earth revisited" by K-P Schröder and Robert Cannon Smith, *Monthly Notices of the Royal Astronomical Society* **386** (1): 155–163.

[51] Fictitious.

[52] The CPI is compiled by the Bureau of Labor Statistics and is based upon a 1982 value of 100. For instance, a CPI of 200 means the CPI has doubled since 1982. Source: InflationData.com (www.inflationdata.com).

a. Locate the intercepts, extrema, and points of inflection of the curve, and interpret each feature in terms of the CPI. (Approximate all coordinates to one decimal place.) [**HINT**: See Example 1.]

b. Recall from Section 5.2 that the inflation rate is defined to be $\frac{I'(t)}{I(t)}$. What do the stationary extrema of the curve shown above tell you about the inflation rate?

36. ***Consumer Price Index*** The following graph shows the approximate value of the U.S. Consumer Price Index (CPI) from March 2006 through May 2007:[53]

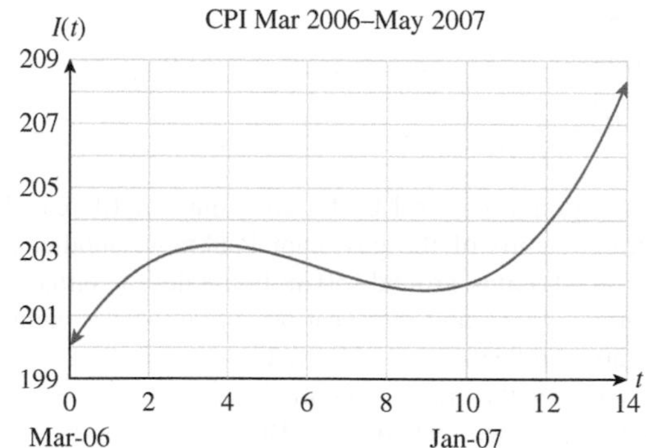

The approximating curve shown on the figure is given by

$$I(t) = 0.02t^3 - 0.38t^2 + 2t + 200 \quad (0 \le t \le 14),$$

where t is time in months. ($t = 0$ represents March 2006.)

a. Locate the intercepts, extrema, and points of inflection of the curve, and interpret each feature in terms of the CPI. (Approximate all coordinates to one decimal place.) [**HINT**: See Example 1.]

b. Recall from Section 5.2 that the inflation rate is defined to be $\frac{I'(t)}{I(t)}$. What do the stationary extrema of the curve shown above tell you about the inflation rate?

37. ***Motion in a Straight Line*** The distance of a UFO from an observer is given by $s = 2t^3 - 3t^2 + 100$ feet after t seconds ($t \ge 0$). Obtain the extrema, points of inflection, and behavior at infinity. Sketch the curve, and interpret these features in terms of the movement of the UFO.

38. ***Motion in a Straight Line*** The distance of the Mars orbiter from your location in Utarek on Mars is given by $s = 2(t-1)^3 - 3(t-1)^2 + 100$ kilometers after t seconds ($t \ge 0$). Obtain the extrema, points of inflection, and behavior at infinity. Sketch the curve, and interpret these features in terms of the movement of the Mars orbiter.

39. ***Average Cost: iPhones*** Assume that it costs **Apple** approximately

$$C(x) = 400{,}000 + 160x + 0.001x^2$$

dollars to manufacture x 32GB iPhone 6's in an hour at the **Foxconn Technology Group**.[54] Obtain the average cost function, sketch its graph, and analyze the graph's important features. Interpret each feature in terms of iPhone 6's. [**HINT**: Recall that the average cost function is $\overline{C}(x) = C(x)/x$.]

40. ***Average Cost: PlayStation 4's*** Assume that it costs **Sony** approximately

$$C(x) = 800{,}000 + 340x + 0.0005x^2$$

dollars to manufacture x PlayStation 4's in an hour.[55] Obtain the average cost function, sketch its graph, and analyze the graph's important features. Interpret each feature in terms of PlayStation 4's. [**HINT**: Recall that the average cost function is $\overline{C}(x) = C(x)/x$.]

41. T ▼ ***Subprime Mortgages during the Housing Bubble*** During the real estate run-up in 2000–2008 the percentage of mortgages issued in the United States that were subprime (normally classified as risky) could be approximated by

$$A(t) = \frac{15.0}{1 + 8.6(1.8)^{-t}} \text{ percent} \quad (0 \le t \le 8)$$

t years after the start of 2000.[56] Graph the *derivative* $A'(t)$ of $A(t)$ using an extended domain of $0 \le t \le 15$. Determine the approximate coordinates of the maximum, and determine the behavior of $A'(t)$ at infinity. What do the answers tell you?

42. T ▼ ***Subprime Mortgage Debt during the Housing Bubble*** During the real estate run-up in 2000–2008 the value of subprime (normally classified as risky) mortgage debt outstanding in the United States was approximately

$$A(t) = \frac{1{,}350}{1 + 4.2(1.7)^{-t}} \text{ billion dollars} \quad (0 \le t \le 8)$$

t years after the start of 2000.[57] Graph the *derivative* $A'(t)$ of $A(t)$ using an extended domain of $0 \le t \le 15$. Determine the approximate coordinates of the maximum, and determine the behavior of $A'(t)$ at infinity. What do the answers tell you?

Communication and Reasoning Exercises

43. A function is *bounded* if its entire graph lies between two horizontal lines. Can a bounded function have vertical

---

[53] See footnote for Exercise 35.

[54] Not the actual cost equation; the authors do not know Apple's actual cost equation. The minimum average cost in the model given is in rough agreement with the actual for one of the 2014 models. Source for cost data: http://time.com.

[55] Not the actual cost equation; the authors do not know Sony's actual cost equation. The minimum average cost in the model given is in rough agreement with the actual cost to manufacture a PlayStation 4 in 2013. Source for estimate of marginal cost: VentureBeat (http://venturebeat.com).

[56] 2009 figure is an estimate. Sources: Mortgage Bankers Association, UBS.

[57] 2008–2009 figures are estimates. Source: www.data360.org.

asymptotes? Can a bounded function have horizontal asymptotes? Explain.

44. A function is *bounded above* if its entire graph lies below some horizontal line. Can a bounded above function have vertical asymptotes? Can a bounded above function have horizontal asymptotes? Explain.

45. If the graph of a function has a vertical asymptote at $x = a$ in such a way that y increases to $+\infty$ as $x \to a$, what can you say about the graph of its derivative? Explain.

46. If the graph of a function has a horizontal asymptote at $y = a$ in such a way that y decreases to a as $x \to +\infty$, what can you say about the graph of its derivative? Explain.

47. Your friend tells you that he has found a continuous function defined on $(-\infty, +\infty)$ with exactly two critical points, each of which is a relative maximum. Can he be right?

48. Your other friend tells you that she has found a continuous function with two critical points, one a relative minimum and one a relative maximum, and no point of inflection between them. Can she be right?

49. ▼ By thinking about extrema, show that, if $f(x)$ is a polynomial, then between every pair of zeros (x-intercepts) of $f(x)$ there is a zero of $f'(x)$.

50. ▼ If $f(x)$ is a polynomial of degree 2 or higher, show that between every pair of relative extrema of $f(x)$ there is a point of inflection of $f(x)$.

5.5 Related Rates

We start by recalling some basic facts about the rate of change of a quantity.

Rate of Change of Q

If Q is a quantity changing over time t, then the derivative dQ/dt is the rate at which Q changes over time.

Quick Examples

1. If A is the area of an expanding circle, then dA/dt is the rate at which the area is increasing.
2. *Words:* The radius r of a sphere is currently 3 cm and increasing at a rate of 2 cm/sec.

   Symbols: $r = 3$ cm and $dr/dt = 2$ cm/sec.

In this section we are concerned with what are called **related rates** problems. In such a problem we have two (sometimes more) related quantities, we know the rate at which one is changing, and we wish to find the rate at which another is changing. A typical example is the following.

EXAMPLE 1 The Expanding Circle

The radius of a circle is increasing at a rate of 10 cm/sec. How fast is the area increasing at the instant when the radius has reached 5 cm?

Solution We have two related quantities: the radius of the circle, r, and its area, A. The first sentence of the problem tells us that r is increasing at a certain rate. When we see a sentence referring to speed or change, it is very helpful to rephrase the sentence using the phrase "the rate of change of." Here, we can say

The rate of change of r is 10 cm/sec.

Because the rate of change is the derivative, we can rewrite this sentence as the equation

$$\frac{dr}{dt} = 10.$$

Similarly, the second sentence of the problem asks how fast A is changing. We can rewrite that question:

What is the rate of change of A when the radius is 5 cm?

Using mathematical notation, the question is

$$\text{What is } \frac{dA}{dt} \text{ when } r = 5?$$

Thus, knowing one rate of change, dr/dt, we wish to find a related rate of change, dA/dt. To find exactly how these derivatives are related, we need the equation relating the variables, which is

$$A = \pi r^2.$$

To find the relationship between the derivatives, we take the derivative of both sides of this equation *with respect to t*. On the left we get dA/dt. On the right we need to remember that r is a function of t and use the chain rule. We get

$$\frac{dA}{dt} = 2\pi r \frac{dr}{dt}.$$

Now we substitute the given values $r = 5$ and $dr/dt = 10$. This gives

$$\left.\frac{dA}{dt}\right|_{r=5} = 2\pi(5)(10) = 100\pi \approx 314 \text{ cm}^2/\text{sec}.$$

Thus, the area is increasing at the rate of 314 cm^2/sec when the radius is 5 cm.

We can organize our work as follows.

Solving a Related Rates Problem

A. The Problem

1. List the related, changing quantities.
2. Restate the problem in terms of rates of change. Rewrite the problem using mathematical notation for the changing quantities and their derivatives.

B. The Relationship

1. Draw a diagram, if appropriate, showing the changing quantities.
2. Find an equation or equations relating the changing quantities.
3. Take the derivative with respect to time of the equation(s) relating the quantities to get the **derived equation(s)**, which relate the rates of change of the quantities.

C. The Solution

1. Substitute into the derived equation(s) the given values of the quantities and their derivatives.
2. Solve for the derivative required.

We can illustrate the procedure with the "ladder problem" that is found in almost every calculus textbook.

EXAMPLE 2 **The Falling Ladder**

Jane is at the top of a 5-foot ladder when it starts to slide down the wall at a rate of 3 feet per minute. Jack is standing on the ground behind her. How fast is the base of the ladder moving when it hits him if Jane is 4 feet from the ground at that instant?

Solution The first sentence talks about (the top of) the ladder sliding down the wall. Thus, one of the changing quantities is the height of the top of the ladder. The question asked refers to the motion of the base of the ladder, so another changing quantity is the distance of the base of the ladder from the wall. Let's record these variables and follow the outline above to obtain the solution.

A. The Problem

1. The changing quantities are

h = height of the top of the ladder

b = distance of the base of the ladder from the wall.

2. We rephrase the problem in words, using the phrase "rate of change":

The rate of change of the height of the top of the ladder is -3 *feet per minute. What is the rate of change of the distance of the base from the wall when the top of the ladder is 4 feet from the ground?*

We can now rewrite the problem mathematically:

$$\frac{dh}{dt} = -3. \text{ Find } \frac{db}{dt} \text{ when } h = 4.$$

B. The Relationship

Figure 46

1. Figure 46 shows the ladder and the variables h and b. Notice that we put in the figure the fixed length, 5, of the ladder, but any changing quantities, such as h and b, we leave as variables. We shall not use any specific values for h or b until the very end.

2. From the figure, we can see that h and b are related by the Pythagorean theorem:

$$h^2 + b^2 = 25.$$

3. Taking the derivative with respect to time of the equation above gives us the derived equation:

$$2h\frac{dh}{dt} + 2b\frac{db}{dt} = 0.$$

C. The Solution

1. We substitute the known values $dh/dt = -3$ and $h = 4$ into the derived equation:

$$2(4)(-3) + 2b\frac{db}{dt} = 0.$$

We would like to solve for db/dt, but first we need the value of b, which we can determine from the equation $h^2 + b^2 = 25$, using the value $h = 4$:

$$16 + b^2 = 25$$
$$b^2 = 9$$
$$b = 3. \quad \text{We reject the negative value because } b \text{ is a distance.}$$

Substituting into the derived equation, we get

$$-24 + 2(3)\frac{db}{dt} = 0.$$

2. Solving for db/dt gives

$$\frac{db}{dt} = \frac{24}{6} = 4.$$

Thus, the base of the ladder is sliding away from the wall at 4 feet per minute when it hits Jack.

EXAMPLE 3 Average Cost

The cost to manufacture x cellphones in a day is

$$C(x) = 10{,}000 + 20x + \frac{x^2}{10{,}000} \text{ dollars.}$$

The daily production level is currently $x = 5{,}000$ cellphones and is increasing at a rate of 100 units per day. How fast is the average cost changing?

Solution

A. The Problem

1. The changing quantities are the production level x and the average cost, $\bar{C}$.

2. We rephrase the problem as follows:

The daily production level is $x = 5{,}000$ *units, and the rate of change of* x *is* 100 *units per day. What is the rate of change of the average cost,* $\bar{C}$*?*

In mathematical notation,

$$x = 5{,}000 \text{ and } \frac{dx}{dt} = 100. \text{ Find } \frac{d\bar{C}}{dt}.$$

B. The Relationship

1. In this example the changing quantities cannot easily be depicted geometrically.

2. We are given a formula for the *total* cost. We get the *average* cost by dividing the total cost by x:

$$\bar{C} = \frac{C}{x}.$$

So

$$\bar{C} = \frac{10{,}000}{x} + 20 + \frac{x}{10{,}000}.$$

3. Taking derivatives with respect to t of both sides, we get the derived equation:

$$\frac{d\bar{C}}{dt} = \left(-\frac{10{,}000}{x^2} + \frac{1}{10{,}000}\right)\frac{dx}{dt}.$$

C. The Solution

Substituting the values from part A into the derived equation, we get

$$\frac{d\bar{C}}{dt} = \left(-\frac{10{,}000}{5{,}000^2} + \frac{1}{10{,}000}\right)100$$

$$= -0.03 \text{ dollars per day.}$$

Thus, the average cost is decreasing by 3¢ per day.

The scenario in the following example is similar to Example 5 in Section 5.2.

EXAMPLE 4 Allocation of Labor

The Gym Sock Company manufactures cotton athletic socks. Production is partially automated through the use of robots. The number of pairs of socks the company can manufacture in a day is given by a Cobb-Douglas production formula:

$$q = 50n^{0.6}r^{0.4},$$

where q is the number of pairs of socks that can be manufactured by n laborers and r robots. The company currently produces 1,000 pairs of socks each day and employs 20 laborers. It is bringing one new robot on line every month. At what rate are laborers being laid off, assuming that the number of socks produced remains constant?

Solution

A. The Problem

1. The changing quantities are the number of laborers n and the number of robots r.
2. $\dfrac{dr}{dt} = 1$. Find $\dfrac{dn}{dt}$ when $n = 20$.

B. The Relationship

1. No diagram is appropriate here.
2. The equation relating the changing quantities:

$$1{,}000 = 50n^{0.6}r^{0.4} \qquad \text{Productivity is constant at 1,000 pairs of socks each day.}$$

or

$$20 = n^{0.6}r^{0.4}.$$

3. The derived equation is

$$\begin{aligned} 0 &= 0.6n^{-0.4}\left(\frac{dn}{dt}\right)r^{0.4} + 0.4n^{0.6}r^{-0.6}\left(\frac{dr}{dt}\right) \\ &= 0.6\left(\frac{r}{n}\right)^{0.4}\left(\frac{dn}{dt}\right) + 0.4\left(\frac{n}{r}\right)^{0.6}\left(\frac{dr}{dt}\right). \end{aligned}$$

We solve this equation for dn/dt because we shall want to find dn/dt below and because the equation becomes simpler when we do this:

$$\begin{aligned} 0.6\left(\frac{r}{n}\right)^{0.4}\left(\frac{dn}{dt}\right) &= -0.4\left(\frac{n}{r}\right)^{0.6}\left(\frac{dr}{dt}\right) \\ \frac{dn}{dt} &= -\frac{0.4}{0.6}\left(\frac{n}{r}\right)^{0.6}\left(\frac{n}{r}\right)^{0.4}\left(\frac{dr}{dt}\right) \\ &= -\frac{2}{3}\left(\frac{n}{r}\right)\left(\frac{dr}{dt}\right). \end{aligned}$$

C. The Solution

Substituting the numbers in part A into the last equation in part B, we get

$$\frac{dn}{dt} = -\frac{2}{3}\left(\frac{20}{r}\right)(1).$$

We need to compute r by substituting the known value of n in the original formula:

$$20 = n^{0.6}r^{0.4}$$
$$20 = 20^{0.6}r^{0.4}$$
$$r^{0.4} = \frac{20}{20^{0.6}} = 20^{0.4}$$
$$r = 20.$$

Thus,

$$\frac{dn}{dt} = -\frac{2}{3}\left(\frac{20}{20}\right)(1) = -\frac{2}{3} \text{ laborers per month.}$$

The company is laying off laborers at a rate of 2/3 per month, or two every three months.

We can interpret this result as saying that, at the current level of production and number of laborers, one robot is as productive as 2/3 of a laborer, or 3 robots are as productive as 2 laborers.

5.5 EXERCISES

▼ more advanced ◆ challenging

T indicates exercises that should be solved using technology

Rewrite the statements and questions in Exercises 1–8 in mathematical notation. [**HINT**: See Quick Examples 1 and 2.]

1. The population P is currently 10,000 and growing at a rate of 1,000 per year.
2. There are currently 400 cases of Bangkok flu, and the number is growing by 30 new cases every month.
3. The annual revenue of your tie-dyed T-shirt operation is currently \$7,000 but is decreasing by \$700 each year. How fast are annual sales changing?
4. A ladder is sliding down a wall so that the distance between the top of the ladder and the floor is decreasing at a rate of 3 ft/sec. How fast is the base of the ladder receding from the wall?
5. The price of shoes is rising \$5 per year. How fast is the demand changing?
6. Stock prices are rising \$1,000 per year. How fast is the value of your portfolio increasing?
7. The average global temperature is 60°F and rising by 0.01°F per year. How fast are annual sales of Bermuda shorts increasing?
8. The country's population is now 260,000,000 and is increasing by 1,000,000 people per year. How fast is the annual demand for diapers increasing?

Applications

9. ***Sunspots*** The area of a circular sunspot is growing at a rate of 1,200 km^2/sec.
   a. How fast is the radius growing at the instant when it equals 10,000 kilometers? [**HINT**: See Example 1.]
   b. How fast is the radius growing at the instant when the sunspot has an area of 640,000 square kilometers? [**HINT**: Use the area formula to determine the radius at that instant.]
10. ***Puddles*** The radius of a circular puddle is growing at a rate of 5 cm/sec.
    a. How fast is its area growing at the instant when the radius is 10 centimeters? [**HINT**: See Example 1.]
    b. How fast is the area growing at the instant when it equals 36 square centimeters? [**HINT**: Use the area formula to determine the radius at that instant.]
11. ***Balloons*** A spherical party balloon is being inflated with helium pumped in at a rate of 3 ft^3/min. How fast is the radius growing at the instant when the radius has reached 1 foot? [**HINT**: See Example 1. (The volume of a sphere of radius r is $V = \frac{4}{3}\pi r^3$.)]
12. ***More Balloons*** A rather flimsy spherical balloon is designed to pop at the instant its radius has reached 10 centimeters. Assuming that the balloon is filled with helium at a rate of 10 cm^3/sec, calculate how fast the radius is growing at the instant it pops. [**HINT**: See Example 1. (The volume of a sphere of radius r is $V = \frac{4}{3}\pi r^3$.)]

13. ***End of the Earth*** In 5 billion years the Sun will have run out of hydrogen fuel and begin to expand into a red giant, eventually engulfing the Earth and causing it to spiral into the core of the Sun 7.5 billion years from now. At that point, the Sun's radius will be around 93 million miles and increasing at a rate of around 0.003 mph.[58] How fast will its volume be increasing? (Round your answer to three significant digits.) [**HINT**: See the hint for Exercise 11.]

14. ***End of Venus*** (Refer to Exercise 13.) When the Sun engulfs Venus shortly before engulfing the Earth, the Sun's radius will be around 67 million miles and increasing at a rate of around 0.002 mph.[59] How fast will its volume be increasing? (Round your answer to three significant digits.) [**HINT**: See the hint for Exercise 11.]

15. ***Sliding Ladders*** The base of a 50-foot ladder is being pulled away from a wall at a rate of 10 ft/sec. How fast is the top of the ladder sliding down the wall at the instant when the base of the ladder is 30 feet from the wall? [**HINT**: See Example 2.]

16. ***Sliding Ladders*** The top of a 5-foot ladder is sliding down a wall at a rate of 10 ft/sec. How fast is the base of the ladder sliding away from the wall at the instant when the top of the ladder is 3 feet from the ground? [**HINT**: See Example 2.]

17. ***Rising Rocket*** You are situated 600 meters from the launch pad at the Jiuquan Satellite Launch Center and are watching the launch of China's latest manned lunar vehicle. At a certain instant the vehicle is 1,000 meters away from you and rising vertically at velocity of 100 m/sec. How fast is the vehicle moving away from you at that instant?

18. ***Descending Elevator*** You are situated 300 feet from the base of Tower Glitz Plaza watching an external elevator descend down the side of the building. At a certain instant the elevator is 500 feet away from you, and its distance from you is decreasing at a rate of 16 ft/sec. How fast is the elevator descending at that instant?

19. ***Average Cost*** The average cost function for the weekly manufacture of retro portable CD players is given by

$$\bar{C}(x) = 150{,}000x^{-1} + 20 + 0.0001x \text{ dollars per player,}$$

where x is the number of CD players manufactured that week. Weekly production is currently 3,000 players and is increasing at a rate of 100 players per week. What is happening to the average cost? [**HINT**: See Example 3.]

20. ***Average Cost*** Repeat Exercise 19, using the revised average cost function

$$\bar{C}(x) = 150{,}000x^{-1} + 20 + 0.01x \text{ dollars per player.}$$

[**HINT**: See Example 3.]

21. ***Demand*** Demand for your tie-dyed T-shirts is given by the formula

$$q = 500 - 100p^{0.5},$$

where q is the number of T-shirts you can sell each month at a price of p dollars. If you currently sell T-shirts for $15 each and you raise your price by $2 per month, how fast will the demand drop? (Round your answer to the nearest whole number.)

22. ***Supply*** The number of retro portable CD players you are prepared to supply to a retail outlet every week is given by the formula

$$q = 0.1p^2 + 3p,$$

where p is the price it offers you. The retail outlet is currently offering you $40 per CD player. If the price it offers decreases at a rate of $2 per week, how will this affect the number you supply?

23. ***Revenue*** You can now sell 50 cups of lemonade per week at 30¢ per cup, but demand is dropping at a rate of 5 cups per week each week. Assuming that raising the price does not affect demand, how fast do you have to raise your price if you want to keep your weekly revenue constant? [**HINT**: Revenue = Price × Quantity.]

24. ***Revenue*** You can now sell 40 cars per month at $20,000 per car, and demand is increasing at a rate of 3 cars per month each month. What is the fastest you could drop your price before your monthly revenue starts to drop? [**HINT**: Revenue = Price × Quantity.]

25. ▼ ***Oil Revenues*** Daily oil production by **Pemex**, Mexico's national oil company, can be approximated by

$$q(t) = 0.017t^2 - 0.4t + 5.23 \text{ million barrels} \quad (8 \le t \le 13)$$

where t is time in years since the start of 2000.[60] At the start of 2010 the price of oil was $86 per barrel and decreasing at a rate of $24 per year.[61] How fast was Pemex's (daily) oil revenue changing at that time?

26. ▼ ***Oil Expenditures*** Daily oil imports to the United States from Mexico can be approximated by

$$q(t) = -39t^2 + 800t - 3{,}000 \text{ thousand barrels} \quad (9 \le t \le 13),$$

where t is time in years since the start of 2000.[62] At the start of 2012 the price of oil was $105 per barrel and increasing at a rate of $70 per year.[63] How fast was (daily) oil expenditure for imports from Mexico changing at that time?

[58] Based on estimates in "Distant future of the Sun and Earth revisited" by K-P Schröder and Robert Cannon Smith, *Monthly Notices of the Royal Astronomical Society* **386** (1): 155–163.

[59] *Ibid.*

[60] Source for data: www.pemex.com.

[61] Based on January price and January–February change. Source for data: Energy Information Administration (http://tonto.eia.doe.gov).

[62] Source for data: www.pemex.com.

[63] Based on January price and January–February change. Source for data: Energy Information Administration (http://tonto.eia.doe.gov).

27. ***Resource Allocation*** Your company manufactures automobile alternators, and production is partially automated through the use of robots. To meet production deadlines, your company calculates that the numbers of laborers and robots must satisfy the constraint

$$xy = 10{,}000,$$

where x is the number of laborers and y is the number of robots. Your company currently uses 400 robots and is increasing robot deployment at a rate of 16 per month. How fast is it laying off laborers? [**HINT**: See Example 4.]

28. ***Resource Allocation*** Your company is the largest sock manufacturer in the solar system, and production is automated through the use of androids and robots. To meet production deadlines, your company calculates that the numbers of androids and robots must satisfy the constraint

$$xy = 1{,}000{,}000,$$

where x is the number of androids and y is the number of robots. Your company currently uses 5,000 androids and is increasing android deployment at a rate of 200 per month. How fast is it scrapping robots? [**HINT**: See Example 4.]

29. ***Production*** The automobile assembly plant you manage has a Cobb-Douglas production function given by

$$P = 10x^{0.3}y^{0.7},$$

where P is the number of automobiles it produces per year, x is the number of employees, and y is the daily operating budget (in dollars). You maintain a production level of 1,000 automobiles per year. If you currently employ 150 workers and are hiring new workers at a rate of 10 per year, how fast is your daily operating budget changing? [**HINT**: See Example 4.]

30. ***Production*** Refer back to the Cobb-Douglas production formula in Exercise 29. Assume that you maintain a constant workforce of 200 workers and wish to increase production in order to meet a demand that is increasing by 100 automobiles per year. The current demand is 1,000 automobiles per year. How fast should your daily operating budget be increasing? [**HINT**: See Example 4.]

31. ***Demand*** Assume that the demand equation for tuna in a small coastal town is

$$pq^{1.5} = 50{,}000,$$

where q is the number of pounds of tuna that can be sold in one month at the price of p dollars per pound. The town's fishery finds that the demand for tuna is currently 900 pounds per month and is increasing at a rate of 100 pounds per month each month. How fast is the price changing?

32. ***Demand*** The demand equation for rubies at *Royal Ruby Retailers* is

$$q + \frac{4}{3}p = 80,$$

where q is the number of rubies RRR can sell per week at p dollars per ruby. RRR finds that the demand for its rubies is currently 20 rubies per week and is dropping at a rate of one ruby per week. How fast is the price changing?

33. ▼ ***Ships Sailing Apart*** The H.M.S. *Dreadnaught* is 40 miles south of Montauk and steaming due south at 20 mph, while the U.S.S. *Mona Lisa* is 50 miles east of Montauk and steaming due east at an even 30 mph. How fast is their distance apart increasing?

34. ▼ ***Near Miss*** My aunt and I were approaching the same intersection, she from the south and I from the west. She was traveling at a steady speed of 10 mph, while I was approaching the intersection at 60 mph. At a certain instant in time, I was one tenth of a mile from the intersection, while she was one twentieth of a mile from it. How fast were we approaching each other at that instant?

35. ▼ ***Baseball*** A baseball diamond is a square with side 90 feet.

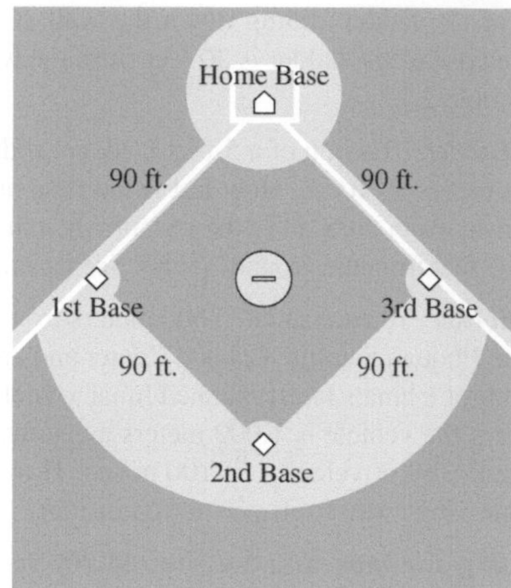

A batter at home base hits the ball and runs toward first base at a speed of 24 ft/sec. At what rate is his distance from third base increasing when he is halfway to first base?

36. ▼ ***Baseball*** Refer to Exercise 35. Another player is running from third base to home at 30 ft/sec. How fast is her distance from second base increasing when she is 60 feet from third base?

37. ▼ ***Movement along a Graph*** A point on the graph of $y = 1/x$ is moving along the curve in such a way that its x-coordinate is increasing at a rate of 4 units per second. What is happening to the y-coordinate at the instant the y-coordinate is equal to 2?

38. ▼ ***Motion around a Circle*** A point is moving along the circle $x^2 + (y - 1)^2 = 8$ in such a way that its x-coordinate is decreasing at a rate of 1 unit per second. What is happening to the y-coordinate at the instant when the point has reached $(-2, 3)$?

39. ▼ ***Education*** In 1991 the expected income of an individual depended on his or her educational level according to the following formula:

$$I(n) = 2.929n^3 - 115.9n^2 + 1{,}530n - 6{,}760 \text{ thousand dollars} \quad (12 \le n \le 15).$$

Here, n is the number of school years completed, and $I(n)$ is the individual's expected income in thousands of dollars.[64] It is 1991, and you have completed 13 years of school and are currently a part-time student. Your schedule is such that you will complete the equivalent of one year of college every three years. Assuming that your salary is linked to the above model, how fast is your income going up? (Round your answer to the nearest $1.)

40. ▼ ***Education*** Refer back to the model in Exercise 39. Assume that you have completed 14 years of school and that your income is increasing by $5,000 per year. How much schooling per year is this rate of increase equivalent to?

41. ▼ ***Employment*** An employment research company estimates that the value of a recent MBA graduate to an accounting company is

$$V = 3e^2 + 5g^3,$$

where V is the value of the graduate, e is the number of years of prior business experience, and g is the graduate school grade-point average. A company that currently employs graduates with a 3.0 average wishes to maintain a constant employee value of $V = 200$ but finds that the grade-point average of its new employees is dropping at a rate of 0.2 per year. How fast must the experience of its new employees be growing to compensate for the decline in grade point average?

42. ▼ ***Grades***[65] A production formula for a student's performance on a difficult English examination is given by

$$g = 4hx - 0.2h^2 - 10x^2,$$

where g is the grade the student can expect to obtain, h is the number of hours of study for the examination, and x is the student's grade point average. The instructor finds that students' grade point averages have remained constant at 3.0 over the years and that students currently spend an average of 15 hours studying for the examination. However, scores on the examination are dropping at a rate of 10 points per year. At what rate is the average study time decreasing?

43. ▼ ***Cones*** A right circular conical vessel is being filled with green industrial waste at a rate of 100 m³/sec. How fast is the level rising after 200π cubic meters have been poured in? The cone has a height of 50 meters and a radius of 30 meters at its brim. (The volume of a cone of height h and cross-sectional radius r at its brim is given by $V = \frac{1}{3}\pi r^2 h$.)

44. ▼ ***More Cones*** A circular conical vessel is being filled with ink at a rate of 10 cm³/sec. How fast is the level rising after 20 cubic centimeters have been poured in? The cone has height 50 centimeters and radius 20 centimeters at its brim. (The volume of a cone of height h and cross-sectional radius r at its brim is given by $V = \frac{1}{3}\pi r^2 h$.)

45. ▼ ***Cylinders*** The volume of paint in a right cylindrical can is given by $V = 4t^2 - t$, where t is time in seconds and V is the volume in cubic centimeters. How fast is the level rising when the height is 2 centimeters? The can has a height of 4 centimeters and a radius of 2 centimeters. [**HINT:** To get h as a function of t, first solve the volume $V = \pi r^2 h$ for h.]

46. ▼ ***Cylinders*** A cylindrical bucket is being filled with paint at a rate of 6 cm³/min. How fast is the level rising when the bucket starts to overflow? The bucket has a radius of 30 centimeters and a height of 60 centimeters.

47. ▼ ***Computers vs. Income*** In the 1990s the demand for personal computers in the home went up with household income. For a given community in the 1990s, the average number of computers in a home could be approximated by

$$q = 0.3454 \ln x - 3.047 \quad (10{,}000 \le x \le 125{,}000),$$

where x is mean household income.[66] A certain community had a mean income of $30,000, increasing at a rate of $2,000 per year. How many computers per household were there, and how fast was the number of computers in a home increasing? (Round your answer to four decimal places.)

48. ▼ ***Computers vs. Income*** Refer back to the model in Exercise 47. It is 1995, and the average number of computers per household in your town is 0.5 and is increasing at a rate of 0.02 computers per household per year. What is the average household income in your town, and how fast is it increasing? (Round your answers to the nearest $10).

Education and Crime *The following graph compares the total U.S. prison population and the average combined SAT score in the United States during the 1970s and 1980s:*

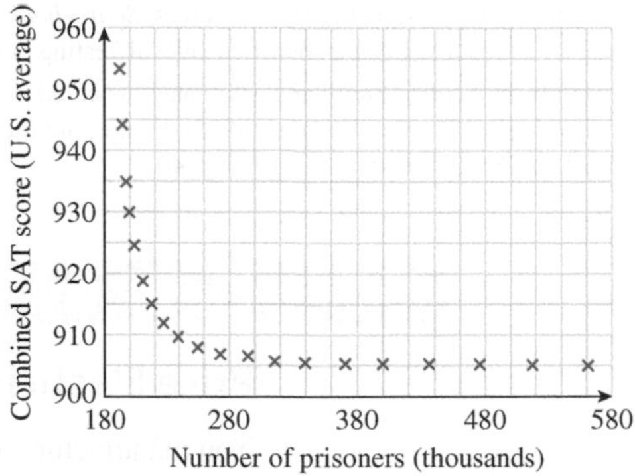

Exercises 49 and 50 are based on the following model for these data:

$$S(n) = 904 + \frac{1{,}326}{(n - 180)^{1.325}} \quad (192 \le n \le 563).$$

---

[64] The model is based on Table 358, U.S. Department of Education, *Digest of Education Statistics, 1991*, Washington, DC: Government Printing Office, 1991.

[65] Based on an exercise in *Introduction to Mathematical Economics* by A. L. Ostrosky Jr. and J. V. Koch (Waveland Press, Illinois, 1979).

[66] The model is a regression model. Source for data: Income distribution: Computer data: Forrester Research/*New York Times*, August 8, 1999, p. BU4.

Here, $S(n)$ is the combined average SAT score at a time when the total prison population is n thousand.[67]

49. ▼ In 1985 the U.S. prison population was 475,000 and increasing at a rate of 35,000 per year. What was the average SAT score, and how fast, and in what direction, was it changing? (Round your answers to two decimal places.)

50. ▼ In 1970 the U.S. combined SAT average was 940 and dropping by 10 points per year. What was the U.S. prison population, and how fast, and in what direction, was it changing? (Round your answers to the nearest 100.)

***Divorce Rates** A study found that the divorce rate d (given as a percentage) appears to depend on the ratio r of available men to available women.*[68] *This function can be approximated by*

$$d(r) = \begin{cases} -40r + 74 & \text{if } r \leq 1.3 \\ \dfrac{130r}{3} - \dfrac{103}{3} & \text{if } r > 1.3. \end{cases}$$

Exercises 51 and 52 are based on this model.

51. ◆ There are currently 1.1 available men per available woman in Littleville, and this ratio is increasing by 0.05 per year. What is happening to the divorce rate?

52. ◆ There are currently 1.5 available men per available woman in Largeville, and this ratio is decreasing by 0.03 per year. What is happening to the divorce rate?

Communication and Reasoning Exercises

53. Why is this section titled "Related Rates"?

54. If you know how fast one quantity is changing and need to compute how fast a second quantity is changing, what kind of information do you need?

55. In a related rates problem there is no limit to the number of changing quantities we can consider. Illustrate this by creating a related rates problem with four changing quantities.

56. If three quantities are related by a single equation, how would you go about computing how fast one of them is changing based on a knowledge of the other two?

57. ▼ The demand and unit price for your store's checkered T-shirts are changing with time. Show that the percentage rate of change of revenue equals the sum of the percentage rates of change of price and demand. (The percentage rate of change of a quantity Q is $Q'(t)/Q(t)$.)

58. ▼ The number N of employees and the total floor space S of your company are both changing with time. Show that the percentage rate of change of square footage per employee equals the percentage rate of change of S minus the percentage rate of change of N. (The percentage rate of change of a quantity Q is $Q'(t)/Q(t)$.)

59. ▼ In solving a related rates problem a key step is solving the derived equation for the unknown rate of change (once we have substituted the other values into the equation). Call the unknown rate of change X. The derived equation is what kind of equation in X?

60. ▼ On a recent exam you were given a related rates problem based on an algebraic equation relating two variables x and y. Your friend told you that the correct relationship between dx/dt and dy/dt was given by

$$\left(\frac{dx}{dt}\right) = \left(\frac{dy}{dt}\right)^2.$$

Could he be correct?

61. ▼ Transform the following into a mathematical statement about derivatives: If my grades are improving at twice the speed of yours, then your grades are improving at half the speed of mine.

62. ▼ If two quantities x and y are related by a linear equation, how are their rates of change related?

[67] Based on data for the years 1967–1989. Sources: *Sourcebook of Criminal Justice Statistics*, 1990, p. 604/Educational Testing Service.

[68] The cited study, by Scott J. South and associates, appeared in the *American Sociological Review* (February 1995). Figures are rounded. Source: *New York Times*, February 19, 1995, p. 40.

5.6 Elasticity

Price Elasticity of Demand

You manufacture an extremely popular brand of sneakers and want to know what will happen if you increase the selling price. Common sense tells you that demand will drop as you raise the price. But will the drop in demand be enough to cause your revenue to fall? Or will it be small enough that your revenue will rise because of the higher selling price? For example, if you raise the price by 1%, you might suffer only a 0.5% loss in sales. In this case the loss in sales will be more than offset by the increase in price, and your revenue will rise. In such a case we say that the demand is **inelastic**, because it is not very sensitive to the increase in price. On the other hand, if your 1% price increase results in a 2% drop in demand, then raising the price will cause a drop in revenues. We then say that the demand is **elastic** because it reacts strongly to a price change.

We can use calculus to measure the response of demand to price changes if we have a demand equation for the item we are selling.* We need to know the *percentage drop in demand per percentage increase in price.* This ratio is called the **elasticity of demand**, or **price elasticity of demand**, and is usually denoted by E. Let's derive a formula for E in terms of the demand equation.

* Coming up with a good demand equation is not always easy. We saw in Chapter 1 that it is possible to find a linear demand equation if we know the sales figures at two different prices. However, such an equation is only a first approximation. To come up with a more accurate demand equation, we might need to gather data corresponding to sales at several different prices and use curve-fitting techniques like regression. Another approach would be an analytic one, based on mathematical modeling techniques that an economist might use.
That said, we refer you again to *Camels and Rubber Duckies* by Joel Spolsky at www.joelonsoftware.com/articles/CamelsandRubberDuckies.html just in case you think there is nothing more to demand curves.

Assume that we have a demand equation

$$q = f(p),$$

where q stands for the number of items we would sell (per week, per month, or what have you) if we set the price per item at p. Now suppose we increase the price p by a very small amount, Δp. Then our percentage increase in price is $(\Delta p/p) \times 100\%$. This increase in p will presumably result in a decrease in the demand q. Let's denote this corresponding decrease in q by $-\Delta q$. (We use the minus sign because, by convention, Δq stands for the *increase* in demand.) Thus, the percentage decrease in demand is $(-\Delta q/q) \times 100\%$.

Now E is the ratio

$$E = \frac{\text{Percentage decrease in demand}}{\text{Percentage increase in price}},$$

so

$$E = \frac{-\dfrac{\Delta q}{q} \times 100\%}{\dfrac{\Delta p}{p} \times 100\%}.$$

Canceling the 100%s and reorganizing, we get

$$E = -\frac{\Delta q}{\Delta p} \cdot \frac{p}{q}.$$

Q: *What small change in price will we use for Δp?*

A: It should probably be pretty small. If we increased the price of sneakers to, say, \$1 million per pair, the sales would likely drop to zero. But knowing this tells us nothing about how the market would respond to a modest increase in price. In fact, we shall do the usual thing we do in calculus and let Δp approach 0.

In the expression for E, if we let Δp go to 0, then the ratio $\Delta q/\Delta p$ goes to the derivative dq/dp. This gives us our final and most useful definition of the elasticity.

Price Elasticity of Demand

The **price elasticity of demand** E is the percentage rate of decrease of demand per percentage increase in price. E is given by the formula

$$E = -\frac{dq}{dp} \cdot \frac{p}{q}.$$

We say that the demand is **elastic** if $E > 1$, is **inelastic** if $E < 1$, and has **unit elasticity** if $E = 1$.

Quick Example

1. Suppose that the demand equation is $q = 20{,}000 - 2p$, where p is the price in dollars. Then

$$E = -(-2)\frac{p}{20{,}000 - 2p} = \frac{p}{10{,}000 - p}.$$

If $p = \$2{,}000$, then $E = 1/4$, and demand is inelastic at this price.
If $p = \$8{,}000$, then $E = 4$, and demand is elastic at this price.
If $p = \$5{,}000$, then $E = 1$, and the demand has unit elasticity at this price.

We are generally interested in the price that maximizes revenue, and in ordinary cases the price that maximizes revenue must give unit elasticity. One way of seeing this is as follows:* If the demand is inelastic (which ordinarily occurs at a low unit price), then raising the price by a small percentage—1%, say—results in a smaller percentage drop in demand. For example, in Quick Example 1, if $p = \$2{,}000$, then the demand would drop by only $\frac{1}{4}\%$ for every 1% increase in price. To see the effect on revenue, we use the fact† that, for small changes in price,

* For another—more rigorous—argument, see Exercise 37.

† See, for example, Exercise 57 in Section 5.5.

$$\begin{aligned}\text{Percentage change in revenue} &\approx \text{Percentage change in price} \\ &\quad + \text{Percentage change in demand} \\ &= 1 + \left(-\frac{1}{4}\right) = \frac{3}{4}\%.\end{aligned}$$

Thus, the revenue will increase by about 3/4%. Put another way:

If the demand is inelastic, raising the price increases revenue.

On the other hand, if the price is elastic (which ordinarily occurs at a high unit price), then increasing the price slightly will lower the revenue, so:

If the demand is elastic, lowering the price increases revenue.

The price that results in the largest revenue must therefore be at unit elasticity.

EXAMPLE 1 Price Elasticity of Demand: Dolls

Suppose that the demand equation for *Bobby Dolls* is given by $q = 216 - p^2$, where p is the price per doll in dollars and q is the number of dolls sold per week.

a. Compute the price elasticity of demand when $p = \$5$ and $p = \$10$, and interpret the results.

b. Find the range of prices for which the demand is elastic and the range for which the demand is inelastic.

c. Find the price at which the weekly revenue is maximized. What is the maximum weekly revenue?

Solution

a. The price elasticity of demand is

$$E = -\frac{dq}{dp} \cdot \frac{p}{q}.$$

Taking the derivative and substituting for q gives

$$E = 2p \cdot \frac{p}{216 - p^2} = \frac{2p^2}{216 - p^2}.$$

When $p = \$5$,

$$E = \frac{2(5)^2}{216 - 5^2} = \frac{50}{191} \approx 0.26.$$

Thus, when the price is set at \$5, the demand is dropping at a rate of 0.26% per 1% increase in the price. Because $E < 1$, the demand is inelastic at this price, so raising the price will increase revenue.

When $p = \$10$,

$$E = \frac{2(10)^2}{216 - 10^2} = \frac{200}{116} \approx 1.72.$$

Thus, when the price is set at \$10, the demand is dropping at a rate of 1.72% per 1% increase in the price. Because $E > 1$, demand is elastic at this price, so raising the price will decrease revenue; lowering the price will increase revenue.

b. and **c.** We answer part (c) first. Setting $E = 1$, we get

$$\frac{2p^2}{216 - p^2} = 1$$

$$p^2 = 72.$$

Thus, we conclude that the maximum revenue occurs when $p = \sqrt{72} \approx \$8.49$. We can now answer part (b): The demand is elastic when $p > \$8.49$ (the price is too high), and the demand is inelastic when $p < \$8.49$ (the price is too low). Finally, we calculate the maximum weekly revenue, which equals the revenue corresponding to the price of \$8.49:

$$R = qp = (216 - p^2)p = (216 - 72)\sqrt{72} = 144\sqrt{72} \approx \$1{,}222.$$

Using Technology

See the Technology Guides at the end of the chapter to find out how to automate computations like those in Example 1(a) using a graphing calculator or Excel. Here is an outline:

TI-83/84 Plus

`Y1=216-X2`
`Y2=-nDeriv(Y1,X,X)*X/Y1`
2ND TABLE Enter $x = 5$
[More details in the Technology Guide.]

Spreadsheet

Enter values of p: 4.9, 4.91, . . . , 5.0, 5.01, . . . , 5.1 in A5–A25.
In B5, enter `216-A5^2` and copy down to B25.
In C5, enter `=(A6-A5)/A5` and paste the formula in C5–D24.
In E5, enter `=-D5/C5` and copy down to E24. This column contains the values of E for the values of p in column A.
[More details in the Technology Guide.]

Income Elasticity of Demand

The concept of elasticity can be applied in other situations. In the following example we consider *income* elasticity of demand—the percentage increase in demand for a particular item per percentage increase in personal income.

EXAMPLE 2 Income Elasticity of Demand: Porsches

You are the sales director at *Suburban Porsche* and have noticed that demand for Porsches depends on income according to

$$q = 0.005e^{-0.05x^2+x} \qquad (1 \le x \le 10).$$

Here, x is the income of a potential customer in hundreds of thousands of dollars and q is the probability that the person will actually purchase a Porsche.* The **income elasticity of demand** is

$$E = \frac{dq}{dx}\frac{x}{q}.$$

Compute and interpret E for $x = 2$ and 9.

* In other words, q is the fraction of visitors to your showroom having income x who actually purchase a Porsche.

Q: *Why is there no negative sign in the formula?*

A: Because we anticipate that the demand will increase as income increases, the ratio

$$\frac{\text{Percentage increase in demand}}{\text{Percentage increase in income}}$$

will be positive, so there is no need to introduce a negative sign.

Solution Turning to the calculation, since $q = 0.005e^{-0.05x^2+x}$,

$$\frac{dq}{dx} = 0.005e^{-0.05x^2+x}(-0.1x + 1),$$

so

$$\begin{aligned} E &= \frac{dq}{dx}\frac{x}{q} \\ &= 0.005e^{-0.05x^2+x}(-0.1x + 1)\frac{x}{0.005e^{-0.05x^2+x}} \\ &= x(-0.1x + 1). \end{aligned}$$

When $x = 2$, $E = 2[-0.1(2) + 1] = 1.6$. Thus, at an income level of \$200,000 the probability that a customer will purchase a Porsche increases at a rate of 1.6% per 1% increase in income.

When $x = 9$, $E = 9[-0.1(9) + 1] = 0.9$. Thus, at an income level of \$900,000 the probability that a customer will purchase a Porsche increases at a rate of 0.9% per 1% increase in income.

5.6 EXERCISES

▼ more advanced ◆ challenging

T indicates exercises that should be solved using technology

Applications

1. ***Demand for Oranges*** The weekly sales of *Honolulu Red Oranges* is given by $q = 1{,}000 - 20p$. Calculate the price elasticity of demand when the price is \$30 per orange (yes, \$30 per orange[69]). Interpret your answer. Also, calculate the price that gives a maximum weekly revenue, and find this maximum revenue. [**HINT**: See Example 1.]

2. ***Demand for Oranges*** Repeat Exercise 1 for weekly sales of $1{,}000 - 10p$. [**HINT**: See Example 1.]

3. ***Demand for Smartphones*** Worldwide annual sales of smartphones in 2012–2013 were approximately $q = -6p + 3{,}030$ million phones at a selling price of \$$p$ per phone.[70]
   a. Obtain a formula for the price elasticity of demand E.
   b. In 2013 the actual selling price was \$335 per phone. What was the corresponding price elasticity of demand? Interpret your answer.
   c. Use your formula for E to determine the selling price that would have resulted in the largest annual revenue. What, to the nearest \$10 million, would have been the resulting annual revenue?

4. ***Projected Revenue: Smartphones*** Worldwide annual sales of smartphones in 2013–2017 were projected to be approximately $q = -10p + 4{,}360$ million phones at a selling price of \$$p$ per phone.[71]
   a. Obtain a formula for the price elasticity of demand E.
   b. In 2014 the actual selling price was \$297 per phone. What was the corresponding price elasticity of demand? Interpret your answer.
   c. Use your formula for E to determine the selling price that would have resulted in the largest annual revenue. What, to the nearest \$10 million, would have been the resulting annual revenue?

[69] They are very hard to find, and their possession confers considerable social status.

[70] Source for data: IDC Worldwide Quarterly Mobile Phone Tracker, Nov. 26, 2013, www.zdnet.com.

[71] Data based on historical 2013 data (source: *Ibid.*) and projected 2017 data.

5. ***College Tuition*** An old study of about 1,800 U.S. colleges and universities resulted in the demand equation $q = 9{,}900 - 2.2p$, where q is the enrollment at a college or university and p is the average annual tuition (plus fees) it charges.[72]
   a. The study also found that the average tuition charged by universities and colleges was $2,900. What would the effect on the price elasticity of demand have been if the price had been lowered to $2,200? What does the answer suggest about the tuition price corresponding to maximum annual revenue?
   b. On the basis of the study, what would you have advised a college to charge its students to maximize total annual revenue, and what would the resulting enrollment and revenue have been?

6. ***Monorail Services*** The demand for monorail service in Las Vegas in 2005 could be approximated by $q = -4{,}500p + 41{,}500$ rides per day when the fare was $\$p$.[73]
   a. In September 2005 the Las Vegas monorail increased the price from $3 per ride to $5 per ride. What was the effect on the price elasticity of demand? What does the answer suggest about the fare corresponding to maximum daily revenue?
   b. What would you have advised the **Las Vegas Monorail Company** to charge to maximize total daily revenue, and what would the resulting daily ridership and revenue have been?

7. ***Tissues*** The consumer demand equation for tissues is given by $q = (100 - p)^2$, where p is the price per case of tissues and q is the demand in weekly sales.
   a. Determine the price elasticity of demand E when the price is set at $30, and interpret your answer.
   b. At what price should tissues be sold to maximize the revenue?
   c. Approximately how many cases of tissues would be demanded at that price?

8. ***Bodybuilding*** The consumer demand curve for *Professor Stefan Schwarzenegger* dumbbells is given by $q = (100 - 2p)^2$, where p is the price per dumbbell and q is the demand in weekly sales. Find the price Professor Schwarzenegger should charge for his dumbbells to maximize revenue.

9. ***T-Shirts*** The Physics Club sells $E = mc^2$ T-shirts at the local flea market. Unfortunately, the club's previous administration has been losing money for years, so you decide to do an analysis of the sales. A quadratic regression based on old sales data reveals the following demand equation for the T-shirts:
$$q = -2p^2 + 33p \quad (9 \le p \le 15).$$
Here, p is the price the club charges per T-shirt and q is the number it can sell each day at the flea market.
   a. Obtain a formula for the price elasticity of demand for $E = mc^2$ T-shirts.
   b. Compute the elasticity of demand if the price is set at $10 per shirt. Interpret the result.
   c. How much should the Physics Club charge for the T-shirts to obtain the maximum daily revenue? What will this revenue be?

10. ***Comics*** The demand curve for original *Iguanawoman* comics is given by
$$q = \frac{(400 - p)^2}{100} \quad (0 \le p \le 400),$$
where q is the number of copies the publisher can sell per week if it sets the price at $\$p$.
   a. Find the price elasticity of demand when the price is set at $40 per copy.
   b. Find the price at which the publisher should sell the comics to maximize weekly revenue.
   c. What, to the nearest $1, is the maximum weekly revenue the publisher can realize from sales of *Iguanawoman* comics?

11. ***E-Readers*** The demand for **Amazon**'s Kindle e-reader can be approximated by
$$q(p) = 21e^{-0.01p} \text{ million units per year} \quad (50 \le p \le 400),$$
where p is the price charged by Amazon.[74] Obtain a formula for price elasticity of demand E, and calculate its value at the two endpoints of the given range of prices. Is the price that would maximize annual revenue within the range of prices shown? How would you know this without calculating that price?

12. ***Monorail Service on Mars*** The demand for monorail service on the Utarek monorail, which links the three urbynes (or districts) of Utarek on Mars, can be approximated by
$$q(p) = 31e^{-0.7p} \text{ million rides per day} \quad (3 \le p \le 5),$$
where p is the cost per ride in zonars ($\overline{\text{Z}}$).[75] Obtain a formula for price elasticity of demand E, and calculate its value at the two endpoints of the given range of prices. Is the price that would maximize daily revenue within the range of prices shown? How would you know this without calculating that price?

13. ***Corn*** In the 1930s the economist Henry Schultz devised the following demand function for corn:
$$p = \frac{6{,}570{,}000}{q^{1.3}},$$
where q is the number of bushels of corn that could be sold at p dollars per bushel in one year.[76] Express q as a function of p, and find the price elasticity of demand if the price was set at $1.50 per bushel. Interpret the result.

---

[72] Based on a study by A. L. Ostrosky Jr. and J. V. Koch, as cited in their book *Introduction to Mathematical Economics* (Waveland Press, Illinois, 1979), p. 133.

[73] The model is the authors'. Source for data: The *New York Times,* February 10, 2007, p. A9.

[74] Model based on data from 2007 to 2013. Source: www.e-reader-info.com.

[75] The zonar ($\overline{\text{Z}}$) is the official currency in the city-state of Utarek, Mars (formerly www.Marsnext.com, a now extinct virtual society).

[76] Based on data for the period 1915–1929. Source: Henry Schultz (1938), *The Theory and Measurement of Demand,* University of Chicago Press, Chicago.

14. ***Demand for Fried Chicken*** A fried chicken franchise finds that the demand equation for its new roast chicken product, "Roasted Rooster," is given by

$$p = \frac{40}{q^{1.5}},$$

where p is the price (in dollars) per quarter-chicken serving and q is the number of quarter-chicken servings that can be sold per hour at this price. Express q as a function of p, and find the price elasticity of demand when the price is set at \$4 per serving. Interpret the result.

15. ***Paint-By-Number*** The estimated monthly sales of *Mona Lisa* paint-by-number sets is given by the formula $q = 100e^{-3p^2+p}$, where q is the demand in monthly sales and p is the retail price in hundreds of yen.
    a. Determine the price elasticity of demand E when the retail price is set at ¥300, and interpret your answer.
    b. At what price will revenue be a maximum?
    c. Approximately how many paint-by-number sets will be sold per month at the price in part (b)?

16. ***Paint-By-Number*** Repeat Exercise 15 using the demand equation $q = 100e^{p-3p^2/2}$.

17. ▼ ***Linear Demand Functions*** A general linear demand function has the form $q = mp + b$ (m and b constants, $m \neq 0$).
    a. Obtain a formula for the price elasticity of demand at a unit price of p.
    b. Obtain a formula for the price that maximizes revenue.

18. ▼ ***Exponential Demand Functions*** A general exponential demand function has the form $q = Ae^{-bp}$ (A and b nonzero constants).
    a. Obtain a formula for the price elasticity of demand at a unit price of p.
    b. Obtain a formula for the price that maximizes revenue.

19. ▼ ***Hyperbolic Demand Functions*** A general hyperbolic demand function has the form $q = \frac{k}{p^r}$ (r and k nonzero constants).
    a. Obtain a formula for the price elasticity of demand at unit price p.
    b. How does E vary with p?
    c. What does the answer to part (b) say about the model?

20. ▼ ***Quadratic Demand Functions*** A general quadratic demand function has the form $q = ap^2 + bp + c$ (a, b, and c constants with $a \neq 0$).
    a. Obtain a formula for the price elasticity of demand at a unit price p.
    b. Obtain a formula for the price or prices that could maximize revenue.

21. ▼ ***Modeling Linear Demand*** You have been hired as a marketing consultant to *Johannesburg Burger Supply, Inc.*, and you wish to come up with a unit price for its hamburgers in order to maximize its weekly revenue. To make life as simple as possible, you assume that the demand equation for Johannesburg hamburgers has the linear form $q = mp + b$, where p is the price per hamburger, q is the demand in weekly sales, and m and b are certain constants you must determine.
    a. Your market studies reveal the following sales figures: When the price is set at \$2.00 per hamburger, the sales amount to 3,000 per week, but when the price is set at \$4.00 per hamburger, the sales drop to zero. Use these data to calculate the demand equation.
    b. Now estimate the unit price that maximizes weekly revenue, and predict what the weekly revenue will be at that price.

22. ▼ ***Modeling Linear Demand*** You have been hired as a marketing consultant by *Big Book Publishing, Inc.*, and you have been approached to determine the best-selling price for the hit calculus text by Whiner and Istanbul entitled *Fun with Derivatives*. You decide to make life easy and assume that the demand equation for *Fun with Derivatives* has the linear form $q = mp + b$, where p is the price per book, q is the demand in annual sales, and m and b are certain constants you must determine.
    a. Your market studies reveal the following sales figures: When the price is set at \$50.00 per book, the sales amount to 10,000 per year; when the price is set at \$80.00 per book, the sales drop to 1,000 per year. Use these data to calculate the demand equation.
    b. Now estimate the unit price that maximizes annual revenue and predict what Big Book Publishing, Inc.'s annual revenue will be at that price.

23. ▼ ***Modeling Exponential Demand*** As the new owner of a supermarket, you have inherited a large inventory of unsold imported Limburger cheese, and you would like to set the price so that your revenue from selling it is as large as possible. Previous sales figures of the cheese are shown in the following table:

Price per Pound, p	\$3.00	\$4.00	\$5.00
Monthly Sales, q (pounds)	407	287	223

    a. Use the sales figures for the prices \$3 and \$5 per pound to construct a demand function of the form $q = Ae^{-bp}$, where A and b are constants you must determine. (Round A and b to two significant digits.)
    b. Use your demand function to find the price elasticity of demand at each of the prices listed.
    c. At what price should you sell the cheese to maximize monthly revenue?
    d. If your total inventory of cheese amounts to only 200 pounds and it will spoil 1 month from now, how should you price it to receive the greatest revenue? Is this the same answer you got in part (c)? If not, give a brief explanation.

24. ▼ ***Modeling Exponential Demand*** Repeat Exercise 23, but this time use the sales figures for \$4 and \$5 per pound to construct the demand function.

25. ***Income Elasticity of Demand: Live Drama*** The likelihood that a child will attend a live theatrical performance can be modeled by

$$q = 0.01(-0.0078x^2 + 1.5x + 4.1) \quad (15 \le x \le 100).$$

Here, q is the fraction of children with annual household income x thousand dollars who will attend a live dramatic performance at a theater during the year.[77] Compute the income elasticity of demand at an income level of $20,000 and interpret the result. (Round your answer to two significant digits.) [**HINT**: See Example 2.]

26. ***Income Elasticity of Demand: Live Concerts*** The likelihood that a child will attend a live musical performance can be modeled by

$$q = 0.01(0.0006x^2 + 0.38x + 35) \quad (15 \le x \le 100).$$

Here, q is the fraction of children with annual household income x thousand dollars who will attend a live musical performance during the year.[78] Compute the income elasticity of demand at an income level of $30,000, and interpret the result. (Round your answer to two significant digits.) [**HINT**: See Example 2.]

27. ***Income Elasticity of Demand: Broadband in 2010*** The following graph shows the percentage q of people in households with annual income x thousand dollars using broadband Internet access in 2010,[79] together with the exponential curve $q = -74e^{-0.021x} + 92$.

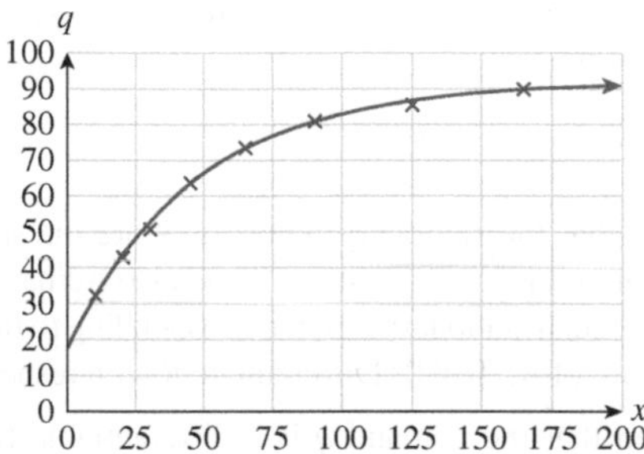

a. Find an equation for the income elasticity of demand for broadband usage, and use it to compute the elasticity for a household with annual income $100,000 to two decimal places. Interpret the result.

b. What does the model predict as the elasticity of demand for households with very large incomes?

28. ***Income Elasticity of Demand: Broadband in 2007*** The following graph shows the percentage q of people in households with annual income x thousand dollars using broadband Internet access in 2007,[80] together with the exponential curve $q = -86e^{-0.013x} + 92$.

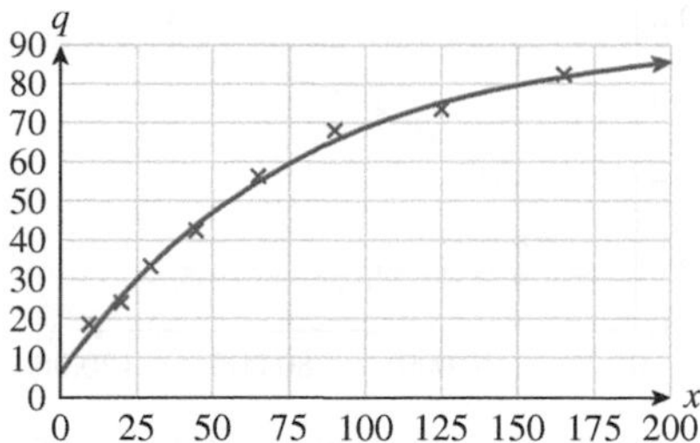

a. Find an equation for the income elasticity of demand for broadband usage, and use it to compute the elasticity for a household with annual income $60,000 to two decimal places. Interpret the result.

b. What does the model predict as the elasticity of demand for households with very large incomes?

29. ***Income Elasticity of Demand: Computer Usage in the 1990s*** The following graph shows the probability q that a household in the 1990s with annual income x dollars had a computer,[81] together with the logarithmic curve $q = 0.3454 \ln x - 3.047$.

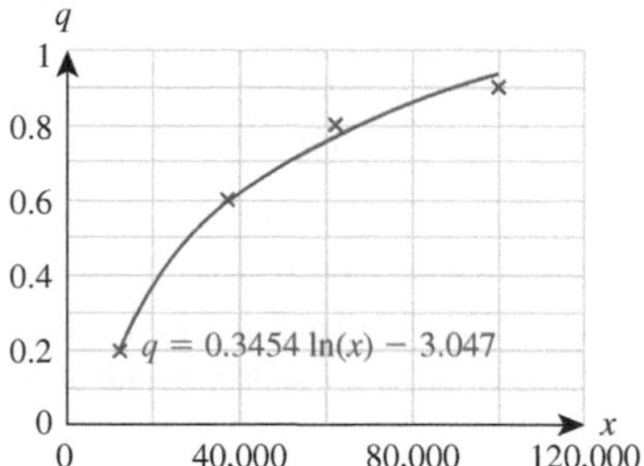

a. Compute the income elasticity of demand for computers, to two decimal places, for a household income of $60,000, and interpret the result.

b. As household income increases, how is income elasticity of demand affected?

c. How reliable is the given model of demand for incomes well above $120,000? Explain.

d. What can you say about E for incomes much larger than those shown?

30. ***Income Elasticity of Demand: Internet Usage in the 1990s*** The following graph shows the probability q that a person in the 1990s with household annual income x dollars used the Internet,[82] together with the logarithmic curve $q = 0.2802 \ln x - 2.505$.

---

[77] Based on a quadratic regression of data from a 2001 survey. Source for data: New York Foundation of the Arts (www.nyfa.org/culturalblueprint).

[78] *Ibid.*

[79] Source for data: *Digital Nation: Expanding Internet Usage,* National Telecommunications and Information Administration, U.S. Department of Commerce (http://search.ntia.doc.gov).

[80] *Ibid.*

[81] Source for data: Income distribution computer data: Forrester Research/*New York Times*, August 8, 1999, p. BU4.

[82] Sources: Luxembourg Income Study/*New York Times*, August 14, 1995, p. A9, Commerce Department, Deloitte & Touche Survey/*New York Times*, November 24, 1999, p. C1.

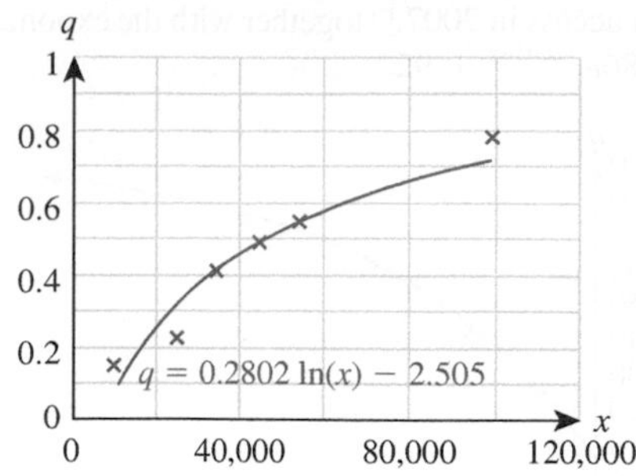

a. Compute the income elasticity of demand for Internet usage, to two decimal places, for a household income of \$60,000 and interpret the result.

b. As household income increases, how is income elasticity of demand affected?

c. The logarithmic model shown above is not appropriate for incomes well above \$100,000. Suggest a model that might be more appropriate.

d. In the model you propose, how would E behave for very large incomes?

Price Elasticity of Supply *Given a supply equation of the form* $q = f(p)$, *the associated* ***price elasticity of supply*** *is defined as the percentage rate of increase of supply per percentage increase in price:*

$$E = \frac{dq}{dp} \cdot \frac{p}{q}.$$

(Note that the formula is the same as for price elasticity of demand except for the sign.) Exercises 31 and 32 are based on this formula.

31. ***Saudi Crude Oil Supply: High Prices*** For crude oil prices of at least \$20 per barrel the supply by Saudi Arabia can be approximated by

$$q = 0.035p + 6.5 \text{ million barrels per day} \quad (20 \le p \le 105),$$

where p is the price per barrel.[83] Calculate the price elasticity of supply when the price of oil is \$60 per barrel. What does the answer tell you about Saudi oil production?

32. ***Saudi Crude Oil Supply: Low Prices*** For crude oil prices of at most \$20 per barrel the supply by Saudi Arabia can be approximated by

$$q = 0.34p + 1.2 \text{ million barrels per day} \quad (12 \le p \le 20),$$

where p is the price per barrel.[84] Calculate the price elasticity of supply when the price of oil is \$15 per barrel. What does the answer tell you about Saudi oil production?

33. ▼ ***Income Elasticity of Demand*** *(based on a question on the GRE Economics Test)* If $Q = aP^{\alpha}Y^{\beta}$ is the individual's demand function for a commodity, where P is the (fixed) price of the commodity, Y is the individual's income, and a, α, and β are parameters, explain why β can be interpreted as the income elasticity of demand.

34. ▼ ***College Tuition*** *(from the GRE Economics Test)* A time-series study of the demand for higher education, using tuition charges as a price variable, yields the following result:

$$\frac{dq}{dp} \cdot \frac{p}{q} = -0.4,$$

where p is tuition and q is the quantity of higher education. Which of the following is suggested by the result?

(A) As tuition rises, students want to buy a greater quantity of education.

(B) As a determinant of the demand for higher education, income is more important than price.

(C) If colleges lowered tuition slightly, their total tuition receipts would increase.

(D) If colleges raised tuition slightly, their total tuition receipts would increase.

(E) Colleges cannot increase enrollments by offering larger scholarships.

Communication and Reasoning Exercises

35. Complete the following: When demand is inelastic, revenue will decrease if _____.

36. Complete the following: When demand has unit elasticity, revenue will decrease if _____.

37. ▼ Given that the demand q is a differentiable function of the unit price p, show that the revenue $R = pq$ has a stationary point when

$$q + p\frac{dq}{dp} = 0.$$

Deduce that the stationary points of R are the same as the points of unit price elasticity of demand. (Ordinarily, there is only one such stationary point, corresponding to the absolute maximum of R.) [**HINT**: Differentiate R with respect to p.]

38. ▼ Given that the demand q is a differentiable function of income x, show that the quantity $R = q/x$ has a stationary point when

$$q - x\frac{dq}{dx} = 0.$$

Deduce that stationary points of R are the same as the points of unit income elasticity of demand. [**HINT**: Differentiate R with respect to x.]

39 ◆ Your calculus study group is discussing price elasticity of demand, and a member of the group asks the following question: "Since elasticity of demand measures the response of demand to change in unit price, what is the difference between elasticity of demand and the quantity $-dq/dp$?" How would you respond?

40. ◆ Another member of your study group claims that unit price elasticity of demand need not always correspond to maximum revenue. Is he correct? Explain your answer.

[83] Based on linear regression of data from 1998 through 2014. Source for data: U.S. Energy Information Administration (www.eia.gov).

[84] *Ibid.*

CHAPTER 5 REVIEW

KEY CONCEPTS

www.WanerMath.com
Go to the Website to find a comprehensive and interactive Web-based summary of Chapter 5.

5.1 Maxima and Minima

Relative maximum, relative minimum [p. 387]
Absolute maximum, absolute minimum [p. 387]
Stationary points, singular points, endpoints [p. 389]
Finding and classifying maxima and minima [p. 390]
First derivative test for relative extrema [p. 392]
Extreme value theorem [p. 396]
Using technology to locate approximate extrema [p. 397]

5.2 Applications of Maxima and Minima

Minimizing average cost [p. 401]
Maximizing area [p. 402]
Steps in solving optimization problems [p. 403]
Maximizing revenue [p. 404]
Optimizing resources [p. 405]
Allocation of labor [p. 407]

5.3 Higher Order Derivatives: Acceleration and Concavity

The second derivative of a function f is the derivative of the derivative of f, written as f'' [p. 416]
The acceleration of a moving object is the second derivative of the position function [p. 416]
Acceleration due to gravity [p. 417]
Acceleration of sales [p. 417]
Concave up, concave down, point of inflection [p. 419]
Locating points of inflection [p. 419]
Application to inflation [p. 419]
Second derivative test for relative extrema [p. 422]
Higher order derivatives [p. 423]

5.4 Analyzing Graphs

Features of a graph: x- and y-intercepts, relative extrema, points of inflection, behavior near singular points, behavior at infinity [p. 433]
Analyzing a graph [p. 434]

5.5 Related Rates

If Q is a quantity changing over time t, then the derivative dQ/dt is the rate at which Q changes over time [p. 441]
The expanding circle [p. 441]
Steps in solving related rates problems [p. 442]
The falling ladder [p. 443]
Average cost [p. 444]
Allocation of labor [p. 445]

5.6 Elasticity

Price elasticity of demand
$E = -\dfrac{dq}{dp} \cdot \dfrac{p}{q}$; demand is elastic if $E > 1$, inelastic if $E < 1$, has unit elasticity if $E = 1$ [p. 451]
Computing and interpreting elasticity, and maximizing revenue [p. 452]
Income elasticity of demand [p. 453]

REVIEW EXERCISES

In Exercises 1–8, find all the relative and absolute extrema of the given function on the given domain (if supplied) or on the largest possible domain (if no domain is supplied).

1. $f(x) = 2x^3 - 6x + 1$ on $[-2, +\infty)$
2. $f(x) = x^3 - x^2 - x - 1$ on $(-\infty, \infty)$
3. $g(x) = x^4 - 4x$ on $[-1, 1]$
4. $f(x) = \dfrac{x+1}{(x-1)^2}$ for $-2 \le x \le 2, x \ne 1$
5. $g(x) = (x-1)^{2/3}$
6. $g(x) = x^2 + \ln x$ on $(0, +\infty)$
7. $h(x) = \dfrac{1}{x} + \dfrac{1}{x^2}$
8. $h(x) = e^{x^2} + 1$

In Exercises 9–12, find the approximate x-coordinates of all relative extrema and points of inflection of f, if any. (For Exercise 12 assume f is defined on $[-4, 4]$.)

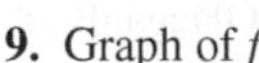

9. Graph of f:

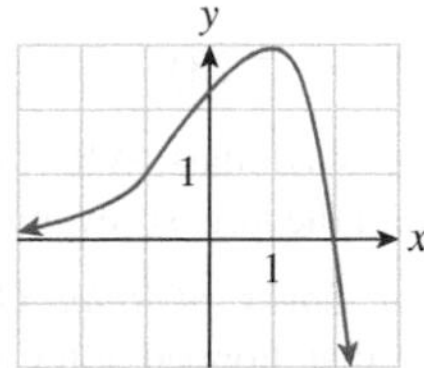

10. Graph of f:

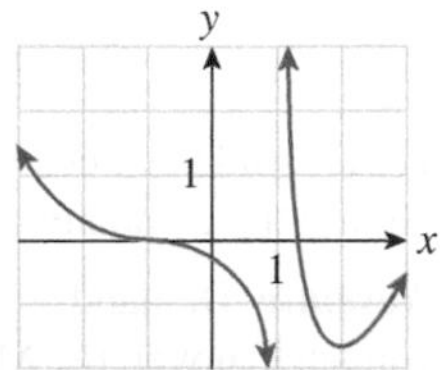

11. Graph of f':

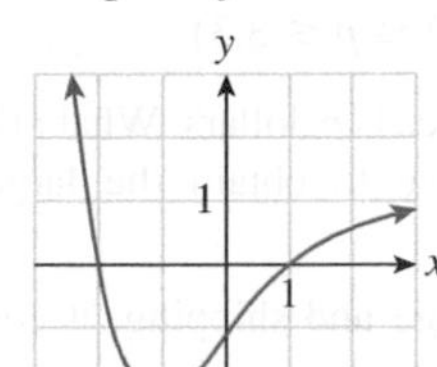

12. Graph of f':

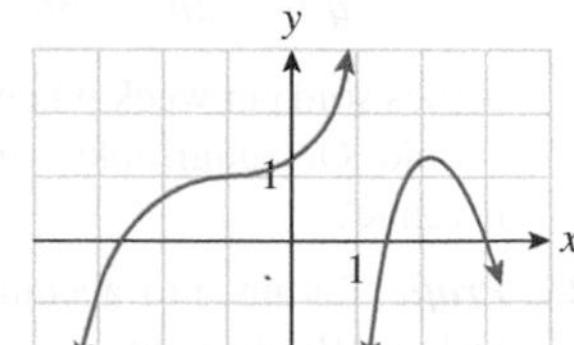

In Exercises 13 and 14, f is continuous on $[-3, 3]$, and the graph of f'' is given. Find the approximate x-coordinates of all points of inflection of the original function f (if any).

13. Graph of f''

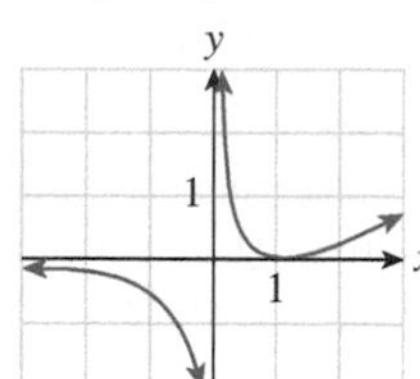

14. Graph of f''

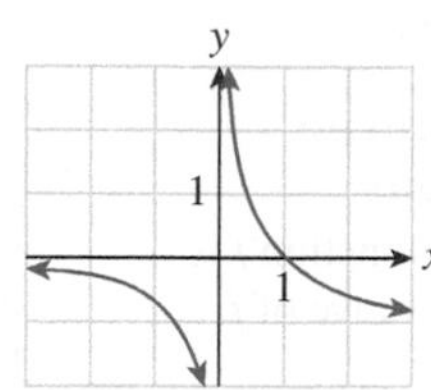

*In Exercises 15 and 16 the position s of a point (in meters) is given as a function of time t (in seconds). Find **(a)** its acceleration as a function of t and **(b)** its acceleration at the specified time.*

15. $s = \frac{2}{3t^2} - \frac{1}{t}; t = 1$

16. $s = \frac{4}{t^2} - \frac{3t}{4}; t = 2$

In Exercises 17–22, sketch the graph of the given function, indicating all relative and absolute extrema and points of inflection. Find the coordinates of these points exactly, where possible. Also indicate any horizontal and vertical asymptotes.

17. $f(x) = x^3 - 12x$ on $[-2, +\infty)$

18. $g(x) = x^4 - 4x$ on $[-1, 1]$

19. $f(x) = \frac{x^2 - 3}{x^3}$

20. $f(x) = (x - 1)^{2/3} + \frac{2x}{3}$

21. $g(x) = (x - 3)\sqrt{x}$

22. $g(x) = (x + 3)\sqrt{x}$

Applications: OHaganBooks.com [Try the game at www.OHaganBooks.com]

23. ***Revenue*** Demand for the latest best-seller at OHaganBooks.com, *A River Burns through It*, is given by

$$q = -p^2 + 33p + 9 \quad (18 \le p \le 28)$$

copies sold per week when the price is p dollars. What price should the company charge to obtain the largest revenue?

24. ***Revenue*** Demand for *The Secret Loves of John O*, a romance novel by Margó Dufón that flopped after two weeks on the market, is given by

$$q = -2p^2 + 5p + 6 \quad (0 \le p \le 3.3)$$

copies sold per week when the price is p dollars. What price should OHaganBooks.com charge to obtain the largest revenue?

25. ***Profit*** Taking into account storage and shipping, it costs OHaganBooks.com

$$C = 9q + 100$$

dollars to sell q copies of *A River Burns through It* in a week (see Exercise 23).

a. If demand is as in Exercise 23, express the weekly profit earned by OHaganBooks.com from the sale of *A River Burns through It* as a function of unit price p.

b. What price should the company charge to get the largest weekly profit? What is the maximum possible weekly profit?

c. Compare your answer in part (b) with the price the company should charge to obtain the largest revenue (Exercise 23). Explain any difference.

26. ***Profit*** Taking into account storage and shipping, it costs OHaganBooks.com

$$C = 3q$$

dollars to sell q copies of Margó Dufón's *The Secret Loves of John O* in a week (see Exercise 24).

a. If demand is as in Exercise 24, express the weekly profit earned by OHaganBooks.com from the sale of *The Secret Loves of John O* as a function of unit price p.

b. What price should the company charge to get the largest weekly profit? What is the maximum possible weekly profit?

c. Compare your answer in part (b) with the price the company should charge to obtain the largest revenue (Exercise 24). Explain any difference.

27. ***Office Space*** Although still a sophomore at college, John O'Hagan's son Billy-Sean has already created several commercial video games and is currently working on his most ambitious project to date: a game called *K* that purports to be a "simulation of the world." John O'Hagan has decided to set aside some office space for Billy-Sean against the northern wall in the headquarters penthouse. The construction of the partition will cost \$8 per foot for the south wall and \$12 per foot for the east and west walls. What are the dimensions of the office space with the largest area that can be provided for Billy-Sean with a budget of \$480, and what is its area?

28. ***Recreation Space*** As a result of complaints by the staff about noise, the coffee and recreation area for student interns at OHaganBooks.com will now be in a 384 square foot rectangular area in the headquarters basement against the southern wall. (The specified area was arrived at in complex negotiations between the student intern representative and management.) The construction of the partition will cost \$12 per foot for the north wall and \$4 per foot for the east and west walls. What are the dimensions of the cheapest recreation area that can be made, and how much will it cost?

29. ***Box Design*** The sales department at OHaganBooks.com, which has decided to send chocolate lobsters to each of its best customers, is trying to design a shipping box with a square base. It has a roll of cardboard 36 inches wide from which to make the boxes. Each box will be obtained by cutting out corners from a rectangle of cardboard as shown in the following diagram:

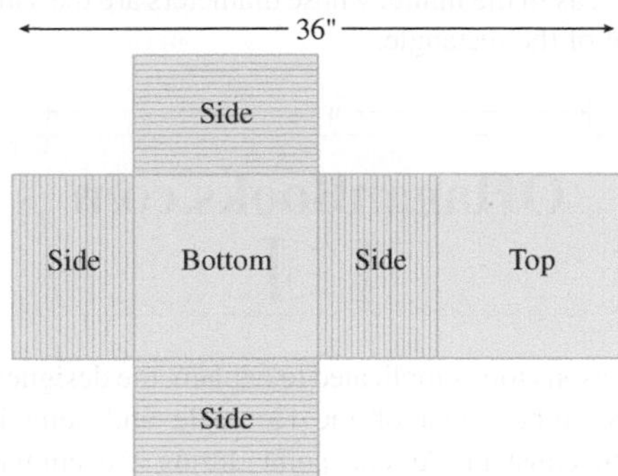

(Notice that the top and bottom of each box will be square, but the sides will not necessarily be square.) What are the dimensions of the boxes with the largest volume that can be made in this way? What is the maximum volume?

30. ***Box Redesign*** The sales department at OHaganBooks.com was not pleased with the result of the box design in Exercise 29; the resulting box was too large for the chocolate lobsters. Following a suggestion by a math major student intern, the department decided to redesign the boxes to meet the following specifications: As in Exercise 29, each box would be obtained by cutting out corners from a rectangle of cardboard, as shown in the following diagram:

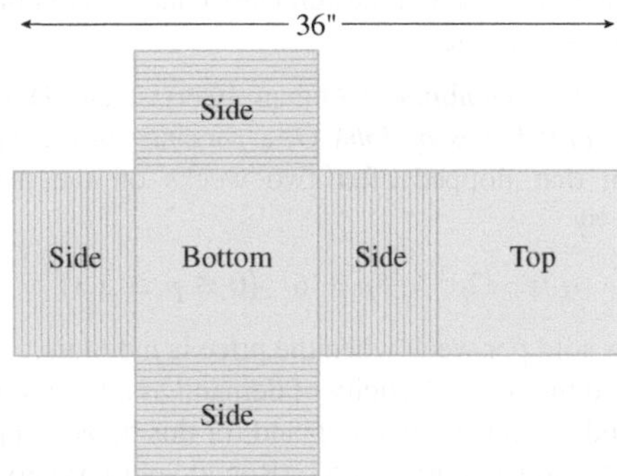

(Notice that the top and bottom of each box would be square, but not necessarily the sides.) The dimensions would be such that the total surface area of the sides plus the bottom of the box would be as large as possible. What are the dimensions of the boxes with the largest area that can be made in this way? How does this box compare with that obtained in Exercise 29?

31. ***Sales*** OHaganBooks.com modeled its weekly sales over a period of time with the function

$$s(t) = 6{,}053 + \frac{4{,}474}{1 + e^{-0.55(t-4.8)}},$$

where t is the time in weeks. Following are the graphs of s, s', and s'':

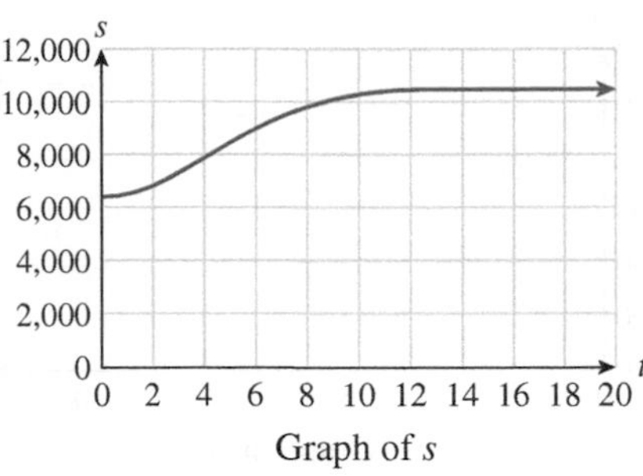

Graph of s

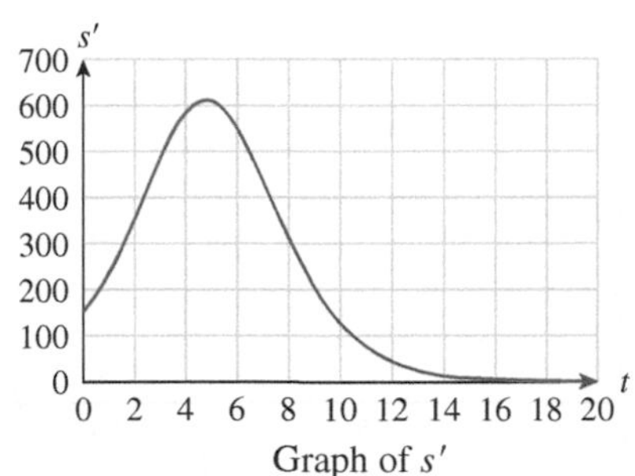

Graph of s'

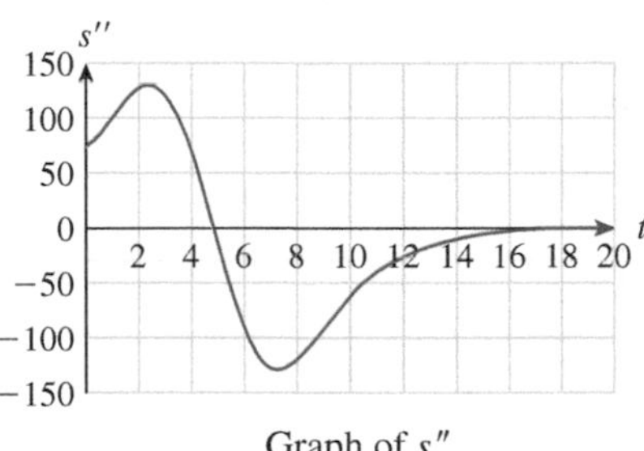

Graph of s''

a. Estimate when, to the nearest week, the weekly sales were growing fastest.

b. To what features on the graphs of s, s', and s'' does your answer to part (a) correspond?

c. The graph of s has a horizontal asymptote. What is the approximate value (s-coordinate) of this asymptote, and what is its significance in terms of weekly sales at OHaganBooks.com?

d. The graph of s' has a horizontal asymptote. What is the value (s'-coordinate) of this asymptote, and what is its significance in terms of weekly sales at OHaganBooks.com?

32. ***Sales*** The quarterly sales of OHagan *oPods* (OHaganBooks .com's answer to the iPod; a portable audio book unit with an incidental music feature) from the fourth quarter of 2009 can be roughly approximated by the function

$$N(t) = \frac{1{,}100}{1 + 9(1.8)^{-t}} \text{ oPods} \quad (t \geq 0),$$

where t is time in quarters since the fourth quarter of 2009. Following are the graphs of N, N', and N'':

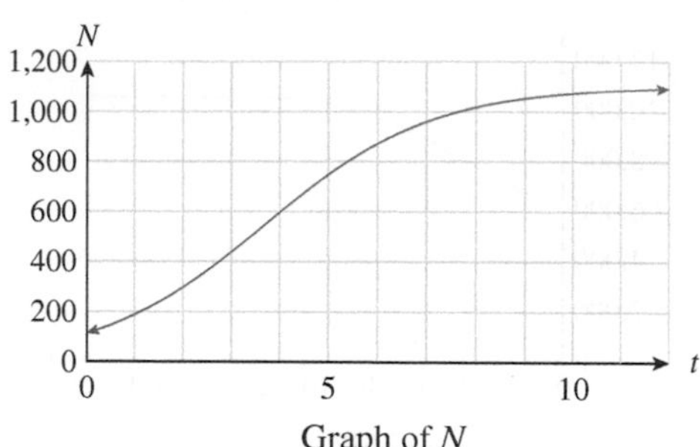
Graph of N

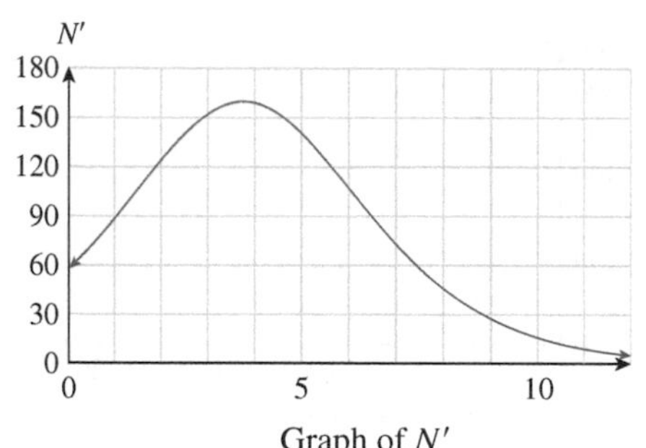
Graph of N'

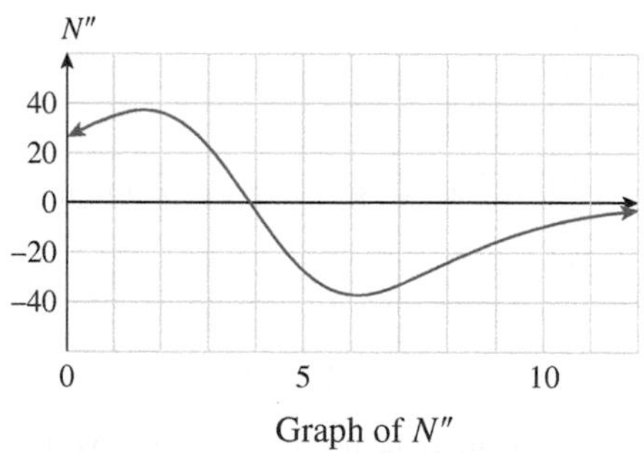
Graph of N''

a. Estimate when, to the nearest quarter, the quarterly sales were growing fastest.
b. To what features on the graphs of N, N', and N'' does your answer to part (a) correspond?
c. The graph of N has a horizontal asymptote. What is the approximate value (N-coordinate) of this asymptote, and what is its significance in terms of quarterly sales of *oPods*?
d. The graph of N' has a horizontal asymptote. What is the value (N'-coordinate) of this asymptote, and what is its significance in terms of quarterly sales of *oPods*?

33. *Chance Encounter* Marjory Duffin is walking north towards the corner entrance of OHaganBooks.com's company headquarters at 5 ft/sec, while John O'Hagan is walking west toward the same entrance, also at 5 ft/sec. How fast is their distance apart decreasing when
a. each of them is 2 feet from the corner?
b. each of them is 1 foot from the corner?
c. each of them is h feet from the corner?
d. they collide on the corner?

34. *Company Logos* OHaganBooks.com's website has an animated graphic with its name in a rectangle whose height and width change; on either side of the rectangle are semicircles, as in the figure, whose diameters are the same as the height of the rectangle.

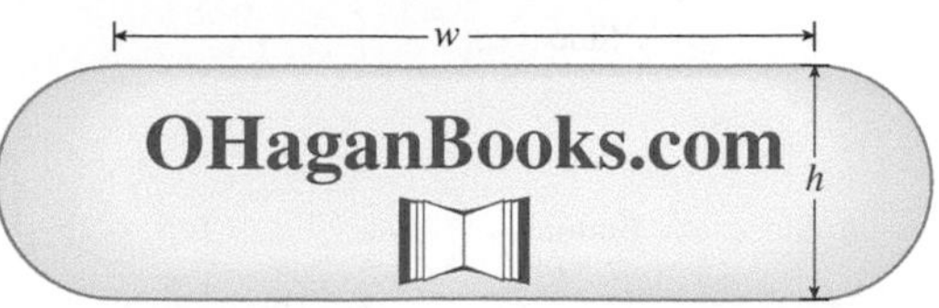

For reasons too complicated to explain, the designer wanted the combined area of the rectangle and semicircles to remain constant. At one point during the animation the width of the rectangle is 1 inch, growing at a rate of 0.5 inches per second, while the height is 3 inches. How fast is the height changing?

35. *Elasticity of Demand* (Compare Exercise 23.) Demand for the latest best-seller at OHaganBooks.com, *A River Burns through It*, is given by

$$q = -p^2 + 33p + 9 \quad (18 \le p \le 28)$$

copies sold per week when the price is p dollars.
a. Find the price elasticity of demand as a function of p.
b. Find the elasticity of demand for this book at a price of \$20 and at a price of \$25. (Round your answers to two decimal places.) Interpret the answers.
c. What price should the company charge to obtain the largest revenue?

36. *Elasticity of Demand* (Compare Exercise 24.) Demand for *The Secret Loves of John O*, a romance novel by Margó Dufón that flopped after two weeks on the market, is given by

$$q = -2p^2 + 5p + 6 \quad (0 \le p \le 3.3)$$

copies sold per week when the price is p dollars.
a. Find the price elasticity of demand as a function of p.
b. Find the elasticity of demand for this book at a price of \$2 and at a price of \$3. (Round your answers to two decimal places.) Interpret the answers.
c. What price should the company charge to obtain the largest revenue?

37. *Elasticity of Demand* Last year OHaganBooks.com experimented with an online subscriber service, Red On Line (ROL), for its e-book service. The consumer demand for ROL was modeled by the equation

$$q = 1{,}000e^{-p^2+p},$$

where p was the monthly access charge and q is the number of subscribers.
a. Obtain a formula for the price elasticity of demand, E, for ROL services.
b. Compute the elasticity of demand if the monthly access charge is set at \$2 per month. Interpret the result.

c. How much should the company have charged to obtain the maximum monthly revenue? What would this revenue have been?

38. ***Elasticity of Demand*** JungleBooks.com (one of OHaganBooks.com's main competitors) responded with its own online subscriber service, Better On Line (BOL), for its e-book service. The consumer demand for BOL was modeled by the equation

$$q = 2{,}000e^{-3p^2+2p},$$

where p was the monthly access charge and q is the number of subscribers.

a. Obtain a formula for the price elasticity of demand, E, for BOL services.

b. Compute the elasticity of demand if the monthly access charge is set at \$2 per month. Interpret the result.

c. How much should the company have charged to obtain the maximum monthly revenue? What would this revenue have been?

CASE STUDY

Production Lot Size Management

SERDAR/Alamy Stock Photo

Your publishing company, *Knockem Dead Paperbacks, Inc.*, is about to release its next best seller, *Henrietta's Heaving Heart* by Celestine A. Lafleur. The company expects to sell 100,000 books each month in the next year. You have been given the job of scheduling print runs to meet the anticipated demand and minimize total costs to the company. Each print run has a setup cost of \$5,000, each book costs \$1 to produce, and monthly storage costs for books awaiting shipment average 1¢ per book. What will you do?

If you decide to print all 1,200,000 books (the total demand for the year, 100,000 books per month for 12 months) in a single run at the start of the year and sales go as predicted, then the number of books in stock would begin at 1,200,000 and decrease to zero by the end of the year, as shown in Figure 47.

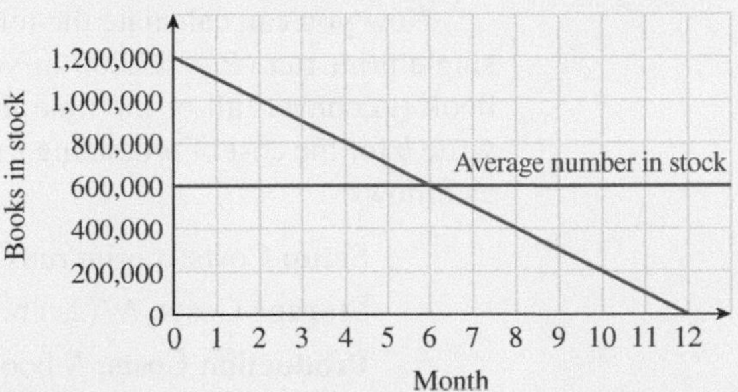

Figure 47

On average, you would be storing 600,000 books for 12 months at 1¢ per book, giving a total storage cost of $600{,}000 \times 12 \times 0.01 = \$72{,}000$. The setup cost for the single print run would be \$5,000. When you add to these the total cost of producing 1,200,000 books at \$1 per book, your total cost would be \$1,277,000.

If, on the other hand, you decide to cut down on storage costs by printing the book in two runs of 600,000 each, you would get the picture shown in Figure 48.

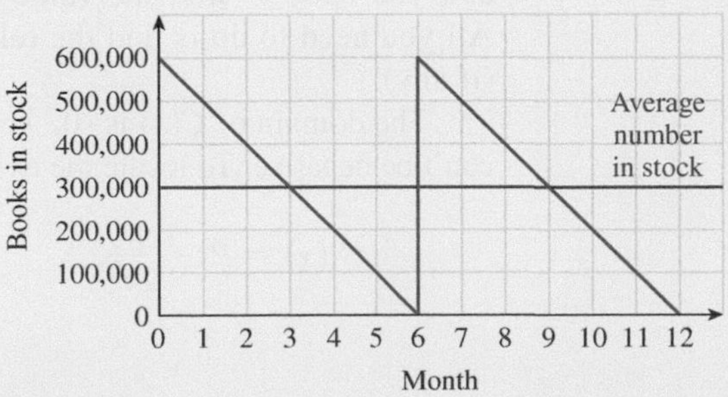

Figure 48

Now the storage cost would be cut in half because, on average, there would be only 300,000 books in stock. Thus, the total storage cost would be \$36,000, and the setup cost would double to \$10,000 (because there would now be two runs). The production costs would be the same: 1,200,000 books @ \$1 per book. The total cost would therefore be reduced to \$1,246,000, a savings of \$31,000 compared to your first scenario.

"Aha!" you say to yourself, after doing these calculations. "Why not drastically cut costs by setting up a run every month?" You calculate that the setup costs alone would be $12 \times \$5,000 = \$60,000$, which is already more than the setup plus storage costs for two runs, so a run every month will cost too much. Perhaps, then, you should investigate three runs, four runs, and so on, until you find the lowest cost. This strikes you as too laborious a process, especially considering that you will have to do it all over again when planning for Lafleur's sequel, *Lorenzo's Lost Love,* due to be released next year. Realizing that this is an optimization problem, you decide to use some calculus to help you come up with a *formula* that you can use for all future plans. So you get to work.

Instead of working with the number 1,200,000, you use the letter N so that you can be as flexible as possible. (What if *Lorenzo's Lost Love* sells more copies?) Thus, you have a total of N books to be produced for the year. You now calculate the total cost of using x print runs per year. Because you are to produce a total of N books in x print runs, you will have to produce N/x books in each print run. N/x is called the **lot size**. As you can see from the diagrams above, the average number of books in storage will be half that amount, $N/(2x)$.

Now you can calculate the total cost for a year. Write P for the setup cost of a single print run ($P = \$5,000$ in your case) and c for the *annual* cost of storing a book (to convert all of the time measurements to years; $c = \$0.12$ here). Finally, write b for the cost of producing a single book ($b = \$1$ here). The costs break down as follows:

Setup Costs: x print runs @ P dollars per run:	Px
Storage Costs: $N/(2x)$ books stored @ c dollars per year:	$cN/(2x)$
Production Costs: N books @ b dollars per book:	Nb
Total Cost:	$Px + \dfrac{cN}{2x} + Nb$

Remember that P, N, c, and b are all constants and x is the only variable. Thus, your cost function is

$$C(x) = Px + \frac{cN}{2x} + Nb,$$

and you need to find the value of x that will minimize $C(x)$. But that's easy! All you need to do is find the relative extrema and select the absolute minimum (if any).

The domain of $C(x)$ is $(0, +\infty)$ because there is an x in the denominator and x can't be negative. To locate the extrema, you start by locating the critical points:

$$C'(x) = P - \frac{cN}{2x^2}.$$

The only singular point would be at $x = 0$, but 0 is not in the domain. To find stationary points, you set $C'(x) = 0$ and solve for x:

$$P - \frac{cN}{2x^2} = 0$$

$$2x^2 = \frac{cN}{P},$$

so

$$x = \sqrt{\frac{cN}{2P}}.$$

There is only one stationary point, and there are no singular points or endpoints. To graph the function, you will need to put in numbers for the various constants. Substituting $N = 1{,}200{,}000$, $P = 5{,}000$, $c = 0.12$, and $b = 1$, you get

$$C(x) = 5{,}000x + \frac{72{,}000}{x} + 1{,}200{,}000$$

with the stationary point at

$$x = \sqrt{\frac{(0.12)(1{,}200{,}000)}{2(5000)}} \approx 3.79.$$

The total cost at the stationary point is

$$C(3.79) \approx 1{,}237{,}900.$$

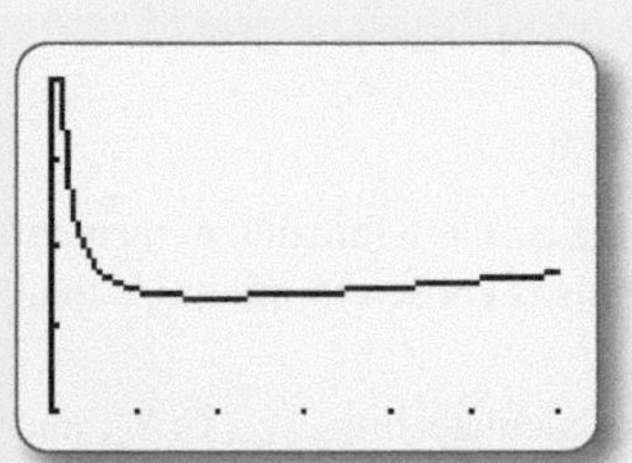

Figure 49

You now graph $C(x)$ in a window that includes the stationary point, say, $0 \le x \le 12$ and $1{,}100{,}000 \le C \le 1{,}500{,}000$, getting Figure 49.

From the graph you can see that the stationary point is an absolute minimum. In the graph it appears that the graph is always concave up, which also tells you that your stationary point is a minimum. You can check the concavity by computing the second derivative:

$$C''(x) = \frac{cN}{x^3} > 0.$$

The second derivative is always positive because c, N, and x are all positive numbers, so indeed the graph is always concave up. Now you also know that it works regardless of the particular values of the constants.

So now you are practically done! You know that the absolute minimum cost occurs when you have $x \approx 3.79$ print runs per year. Don't be disappointed that the answer is not a whole number; whole number solutions are rarely found in real scenarios. What the answer (and the graph) do indicate is that either three or four print runs per year will cost the least money. If you take $x = 3$, you get a total cost of

$$C(3) = \$1{,}239{,}000.$$

If you take $x = 4$, you get a total cost of

$$C(4) = \$1{,}238{,}000.$$

So four print runs per year will allow you to minimize your total costs.

EXERCISES

1. *Lorenzo's Lost Love* will sell 2,000,000 copies in a year. The remaining costs are the same. How many print runs should you use now?
2. In general, what happens to the number of runs that minimizes cost if both the setup cost and the total number of books are doubled?
3. In general, what happens to the number of runs that minimizes cost if the setup cost increases by a factor of 4?
4. Assuming that the total number of copies and storage costs are as originally stated, find the setup cost that would result in a single print run.
5. Assuming that the total number of copies and setup cost are as originally stated, find the storage cost that would result in a print run each month.
6. In Figure 48 we assumed that all the books in each run were manufactured in a very short time; otherwise, the figure might have looked more like the following graph, which shows the inventory, assuming a slower rate of production.

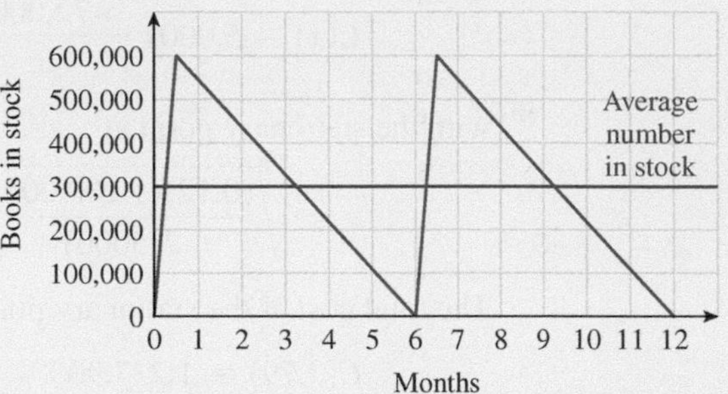

How would this affect the answer?

7. Referring to the general situation discussed in the text, find the cost as a function of the total number of books produced, assuming that the number of runs is chosen to minimize total cost. Also find the average cost per book.
8. Let $\overline{C}$ be the average cost function found in Exercise 7. Calculate $\lim_{N\to+\infty} \overline{C}(N)$, and interpret the result.

TI-83/84 Plus Technology Guide

Section 5.6

Example 1(a) (page 452) Suppose that the demand equation for *Bobby Dolls* is given by $q = 216 - p^2$, where p is the price per doll in dollars and q is the number of dolls sold per week. Compute the price elasticity of demand when $p = \$5$ and $p = \$10$, and interpret the results.

Solution

The TI-83/84 Plus function `nDeriv` can be used to compute approximations of the elasticity E at various prices.

1. Set

```
Y1=216-X^2                    Demand equation
Y2=-nDeriv(Y1,X,X)*X/Y1       Formula for E
```

2. Use the table feature to list the values of elasticity for a range of prices. For part (a) we chose values of `X` close to 5:

X	Y1	Y2
4.7	193.91	.22784
4.8	192.96	.23881
4.9	191.99	.25012
5	191	[illegible]
5.1	189.99	.2738
5.2	188.96	.2862
5.3	187.91	.29897

Y2=.261780104712

Spreadsheet Technology Guide

Section 5.6

Example 1(a) (page 452) Suppose that the demand equation for *Bobby Dolls* is given by $q = 216 - p^2$, where p is the price per doll in dollars and q is the number of dolls sold per week. Compute the price elasticity of demand when $p = \$5$ and $p = \$10$, and interpret the results.

Solution

To approximate E in a spreadsheet, we can use the following approximation of E:

$$E \approx \frac{\text{Percentage decrease in demand}}{\text{Percentage increase in price}} \approx -\frac{\left(\dfrac{\Delta q}{q}\right)}{\left(\dfrac{\Delta p}{p}\right)}.$$

The smaller Δp is, the better the approximation. Let's use $\Delta p = 1¢$, or 0.01 (which is small in comparison with the typical prices we consider—around \$5 to \$10).

1. We start by setting up our worksheet to list a range of prices, in increments of Δp, on either side of a price in which we are interested, such as $p_0 = \$5$:

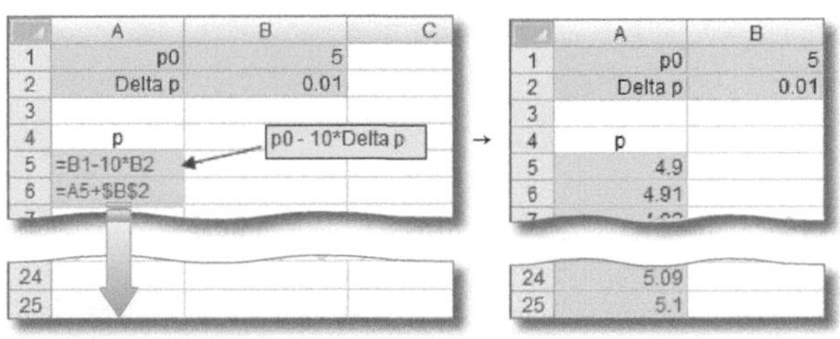

We start in cell A5 with the formula for $p_0 - 10\Delta p$ and then successively add Δp going down column A. You will find that the value $p_0 = 5$ appears midway down the list.

2. Next, we compute the corresponding values for the demand q in column B:

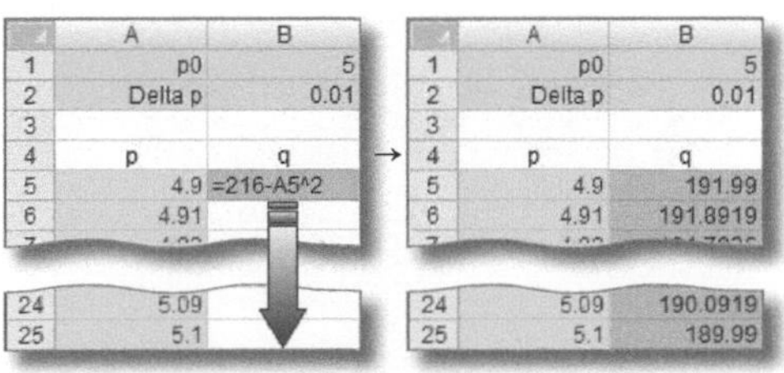

3. We add two new columns for the percentage changes in p and q. The formula shown in cell C5 is copied down columns C and D, to row 24. (Why not row 25?)

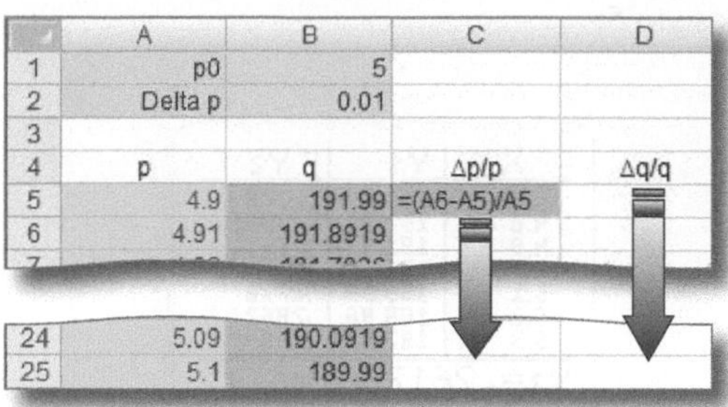

	A	B	C	D
1	p0	5		
2	Delta p	0.01		
3				
4	p	q	Δp/p	Δq/q
5	4.9	191.99	=(A6-A5)/A5	
6	4.91	191.8919		
24	5.09	190.0919		
25	5.1	189.99		

4. The elasticity can now be computed in column E as shown:

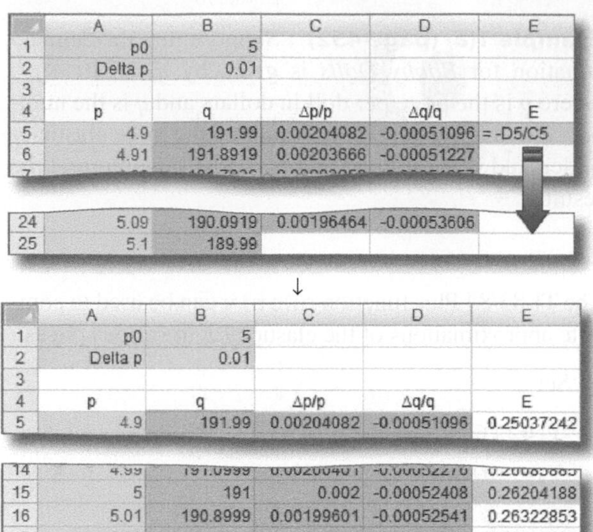

	A	B	C	D	E
1	p0	5			
2	Delta p	0.01			
3					
4	p	q	Δp/p	Δq/q	E
5	4.9	191.99	0.00204082	-0.00051096	=-D5/C5
6	4.91	191.8919	0.00203666	-0.00051227	
24	5.09	190.0919	0.00196464	-0.00053606	
25	5.1	189.99			

↓

	A	B	C	D	E
1	p0	5			
2	Delta p	0.01			
3					
4	p	q	Δp/p	Δq/q	E
5	4.9	191.99	0.00204082	-0.00051096	0.25037242
15	5	191	0.002	-0.00052408	0.26204188
16	5.01	190.8999	0.00199601	-0.00052541	0.26322853
17	5.02	190.7996	0.00199203	-0.00052673	0.26441879

6 THE INTEGRAL

www.WanerMath.com

At the Website, in addition to the resources listed in the Preface, you will find:

- A numerical integration utility with Riemann sum grapher

The following extra topic:

- Numerical Integration

CASE STUDY

Spending on Housing Construction

It is March 2007, and *Time* magazine, in its latest edition, is asking, "Will the Housing Bubble Burst in 2007?" You are a summer intern at *Schottie Construction Co.,* which is working with *Pack-Em-In Real Estate* on a major luxury condominium development to be called "Pack-Em-In/Schottie Towers." You have been asked to find formulas for monthly spending on housing construction in the United States and for the average spent per month starting 1 year ago. You have data about percentage spending changes.

How will you model the trend and estimate the total?

Rob Stothard/Stringer/Getty Images

Introduction

Roughly speaking, calculus is divided into two parts: **differential calculus** (the calculus of derivatives) and **integral calculus**, which is the subject of this chapter and the next. Integral calculus is concerned with problems that are in some sense the reverse of the problems seen in differential calculus. For example, where differential calculus shows how to compute the rate of change of a quantity, integral calculus shows how to find the quantity if we know its rate of change. This idea is made precise in the **Fundamental Theorem of Calculus**. Integral calculus and the Fundamental Theorem of Calculus allow us to solve many problems in economics, physics, and geometry, including one of the oldest problems in mathematics: computing areas of regions with curved boundaries.

6.1 The Indefinite Integral

Antiderivatives and the Indefinite Integral

Suppose that we knew the marginal cost to manufacture an item and we wanted to reconstruct the cost function. We would have to *reverse* the process of differentiation to go from the derivative (the marginal cost function) back to the original function (the total cost). We'll first discuss how to do that and then look at some applications.

Here is an example: If the derivative of $F(x)$ is $4x^3$, what was $F(x)$? We recognize $4x^3$ as the derivative of x^4. So we might have $F(x) = x^4$. However, $F(x) = x^4 + 7$ works just as well. In fact, $F(x) = x^4 + C$ works for any number C. Thus, there are *infinitely many* possible answers to this question.

In fact, we will see shortly that the formula $F(x) = x^4 + C$ covers *all* possible answers to the question. Let's give a name to what we are doing.

Antiderivative

An **antiderivative** of a function f is a function F such that $F' = f$.

Quick Examples

1. An antiderivative of $4x^3$ is x^4. Because the derivative of x^4 is $4x^3$
2. Another antiderivative of $4x^3$ is $x^4 + 7$. Because the derivative of $x^4 + 7$ is $4x^3$
3. An antiderivative of $2x$ is $x^2 + 12$. Because the derivative of $x^2 + 12$ is $2x$

Thus,

If the derivative of $A(x)$ is $B(x)$, then an antiderivative of $B(x)$ is $A(x)$.

We call the set of *all* antiderivatives of a function the **indefinite integral** of the function.

Indefinite Integral

$$\int f(x)\,dx$$

is read "the **indefinite integral** of $f(x)$ with respect to x" and stands for the set of all antiderivatives of f. Thus, $\int f(x)\,dx$ is a *collection of functions*; it is not a single function or a number. The function f that is being **integrated** is called the **integrand**, and the variable x is called the **variable of integration**.

Quick Examples

4. $\int 4x^3\,dx = x^4 + C$ Every possible antiderivative of $4x^3$ has the form $x^4 + C$.

5. $\int 2x\,dx = x^2 + C$ Every possible antiderivative of $2x$ has the form $x^2 + C$.

The **constant of integration** C reminds us that we can add any constant and get a different antiderivative.

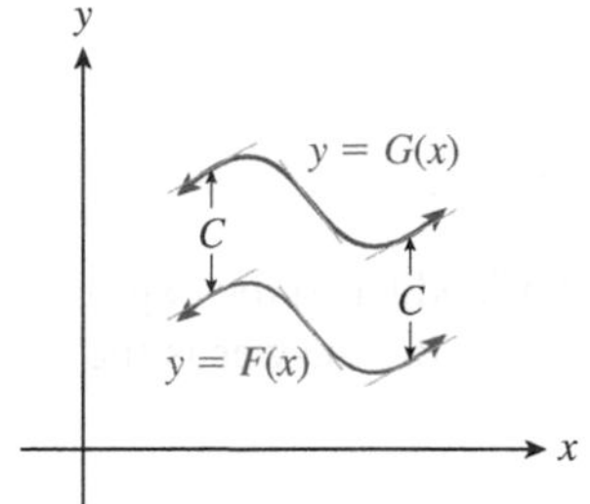

Figure 1

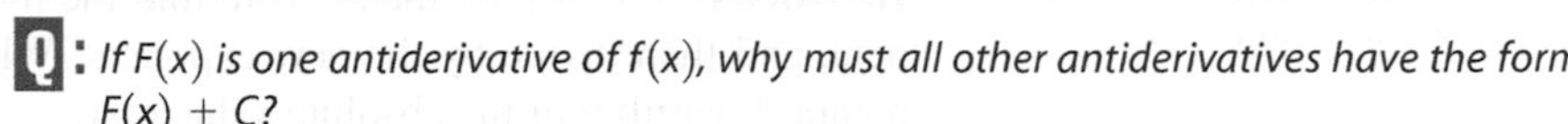

Q: *If $F(x)$ is one antiderivative of $f(x)$, why must all other antiderivatives have the form $F(x) + C$?*

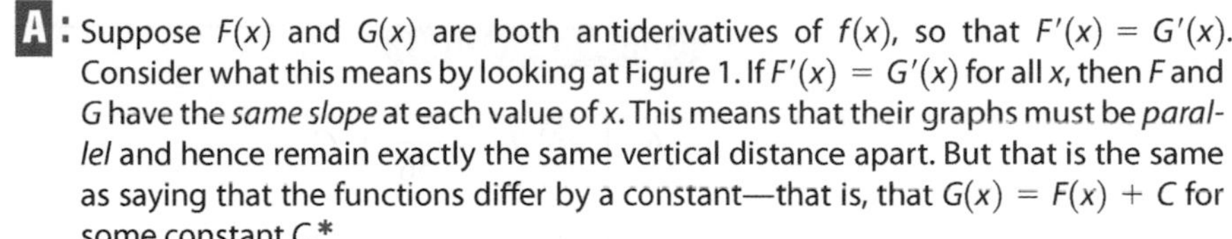

A: Suppose $F(x)$ and $G(x)$ are both antiderivatives of $f(x)$, so that $F'(x) = G'(x)$. Consider what this means by looking at Figure 1. If $F'(x) = G'(x)$ for all x, then F and G have the *same slope* at each value of x. This means that their graphs must be *parallel* and hence remain exactly the same vertical distance apart. But that is the same as saying that the functions differ by a constant—that is, that $G(x) = F(x) + C$ for some constant C.*

* This argument can be turned into a more rigorous proof—that is, a proof that does not rely on geometric concepts such as parallel graphs. We should also say that the result (and our geometric argument as well!) requires that the domain of F and G be a single (possibly infinite) open interval.

EXAMPLE 1 Indefinite Integral

Check the following:

a. $\int x\,dx = \frac{x^2}{2} + C$ **b.** $\int x^2\,dx = \frac{x^3}{3} + C$ **c.** $\int x^{-1}\,dx = \ln|x| + C$

Solution We check each equation by taking the derivative of its right-hand side and checking whether it equals the integrand on the left:

a. $\frac{d}{dx}\left(\frac{x^2}{2} + C\right) = \frac{2x}{2} + 0 = x$ ✔

b. $\frac{d}{dx}\left(\frac{x^3}{3} + C\right) = \frac{3x^2}{3} + 0 = x^2$ ✔

c. $\frac{d}{dx}(\ln|x| + C) = \frac{1}{x} + 0 = x^{-1}.$ ✔

Because the derivative of the right-hand side is the integrand in each case, we can conclude that the given statements are all valid.†

† We are glossing over a subtlety in part (c): The constant of integration C can be different for $x < 0$ and $x > 0$ because the graph breaks at $x = 0$. (See the comment at the end of the previous marginal note.) In general, our understanding will be that the constant of integration may be different on disconnected intervals of the domain.

➡ **Before we go on...** Example 1 gives us a very useful technique to check our answer every time we calculate an integral:

Take the derivative of the answer, and check that it equals the integrand. ■

Calculating Indefinite Integrals

Now, we would like to make the process of finding indefinite integrals (antiderivatives) more mechanical. For example, it would be nice to have a power rule for indefinite integrals similar to the one we already have for derivatives. Example 1 already suggests such a rule for us.

Power Rule for the Indefinite Integral

$$\int x^n\,dx = \frac{x^{n+1}}{n+1} + C$$ This holds only if $n \neq -1$.*

$$\int x^{-1}\,dx = \ln|x| + C$$ For the special case $n = -1$†

Equivalent Form of Second Formula: $\int \frac{1}{x}\,dx = \ln|x| + C$ Because $x^{-1} = \frac{1}{x}$

In Words: For n other than -1, to find the integral of x^n, add 1 to the exponent, and then divide by the new exponent. When $n = -1$, the answer is the natural logarithm of the absolute value of x.

* Note that the right-hand side of the formula makes no sense if $n = -1$ because it has $n + 1$ in the denominator.

† If x is understood to be positive, then we can drop the absolute values and write

$$\int x^{-1}\,dx = \ln x + C.$$

Quick Examples

6. $\int x^{55}\,dx = \frac{x^{56}}{56} + C$

7. $\int \frac{1}{x^{55}}\,dx = \int x^{-55}\,dx$ Power form

$= \frac{x^{-54}}{-54} + C$ When we add 1 to -55, we get -54, *not* -56.

$= -\frac{1}{54x^{54}} + C$

8. $\int 1\,dx = x + C$ Because $1 = x^0$. This is an important special case.

9. $\int \sqrt{x}\,dx = \int x^{1/2}\,dx$ Power form

$= \frac{x^{3/2}}{3/2} + C$

$= \frac{2x^{3/2}}{3} + C$

Notes

1. The integral $\int 1\,dx$ is commonly written as $\int dx$. Similarly, the integral $\int \frac{1}{x^{55}}\,dx$ may be written as $\int \frac{dx}{x^{55}}$.

2. We can easily check the power rule formula by taking the derivative of the right-hand side:

$$\frac{d}{dx}\left(\frac{x^{n+1}}{n+1}+C\right)=\frac{(n+1)x^n}{n+1}=x^n. \quad ✔$$

3. Because the derivative of $\ln x$ is also $1/x$, you might be tempted to write $\int x^{-1}\,dx = \ln x + C$. But $\ln x$, being defined only for positive x, does not have the same domain as $1/x$, whereas $\ln|x|$ does. So we must use $\ln|x| + C$ instead. ■

Following are more indefinite integrals that come from formulas for differentiation we have encountered before.

Indefinite Integral of e^x, b^x, and $|x|$

$$\int e^x\,dx = e^x + C \qquad \text{Because } \frac{d}{dx}(e^x) = e^x$$

If b is any positive number other than 1, then

$$\int b^x\,dx = \frac{b^x}{\ln b} + C \qquad \text{Because } \frac{d}{dx}\left(\frac{b^x}{\ln b}\right) = \frac{b^x \ln b}{\ln b} = b^x$$

$$\int |x|\,dx = \frac{x|x|}{2} + C. \qquad \text{Because } \frac{d}{dx}\left(\frac{x|x|}{2}\right) = |x| \quad \text{(Check this yourself!)}$$

Quick Example

10. $\int 2^x\,dx = \dfrac{2^x}{\ln 2} + C$

For more complicated functions, such as $2x^3 + 6x^5 - 1$, we need the following rules for integrating sums, differences, and constant multiples.

Sums, Differences, and Constant Multiples

Sum and Difference Rules

$$\int [f(x) \pm g(x)]\,dx = \int f(x)\,dx \pm \int g(x)\,dx$$

In Words: The integral of a sum is the sum of the integrals, and the integral of a difference is the difference of the integrals.

Constant Multiple Rule

$$\int kf(x)\,dx = k\int f(x)\,dx \quad (k \text{ constant})$$

In Words: The integral of a constant times a function is the constant times the integral of the function. (In other words, the constant "goes along for the ride.")

Quick Examples

11. Sum Rule: $\int (x^3 + 1)\,dx = \int x^3\,dx + \int 1\,dx = \frac{x^4}{4} + x + C$

$f(x) = x^3;\ g(x) = 1$

12. Constant Multiple Rule: $\int 5x^3\,dx = 5\int x^3\,dx = 5\frac{x^4}{4} + C$

$k = 5;\ f(x) = x^3$

13. Constant Multiple Rule: $\int 4\,dx = 4\int 1\,dx = 4x + C$

$k = 4;\ f(x) = 1$

14. Constant Multiple Rule: $\int 4e^x\,dx = 4\int e^x\,dx = 4e^x + C$

$k = 4;\ f(x) = e^x$

Proof of the Sum Rule

We saw above that if two functions have the same derivative, they differ by a (possibly zero) constant. Look at the rule for sums:

$$\int [f(x) + g(x)]\,dx = \int f(x)\,dx + \int g(x)\,dx.$$

If we take the derivative of the left-hand side with respect to x, we get the integrand, $f(x) + g(x)$. If we take the derivative of the right-hand side, we get

$$\frac{d}{dx}\left[\int f(x)\,dx + \int g(x)\,dx\right] = \frac{d}{dx}\left[\int f(x)\,dx\right] + \frac{d}{dx}\left[\int g(x)\,dx\right]$$

Derivative of a sum = Sum of derivatives.

$$= f(x) + g(x).$$

Because the left- and right-hand sides have the same derivative, they differ by a constant. But because both expressions are indefinite integrals, adding a constant does not affect their value, so they are the same as indefinite integrals.

Notice that a key step in the proof was the fact that the derivative of a sum is the sum of the derivatives.

A similar proof works for the difference and constant multiple rules.

EXAMPLE 2 **Using the Sum and Difference Rules**

Find the integrals:

a. $\int (x^3 + x^5 - 1)\,dx$ **b.** $\int \left(x^{2.1} + \frac{1}{x^{1.1}} + \frac{1}{x} + e^x\right)dx$ **c.** $\int (e^x + 3^x - |x|)\,dx$

Solution

a. $\int (x^3 + x^5 - 1)\,dx = \int x^3\,dx + \int x^5\,dx - \int 1\,dx$ Sum/difference rule

$= \frac{x^4}{4} + \frac{x^6}{6} - x + C$ Power rule

b. $\displaystyle\int\left(x^{2.1} + \frac{1}{x^{1.1}} + \frac{1}{x} + e^x\right) dx$

$$= \int (x^{2.1} + x^{-1.1} + x^{-1} + e^x)\, dx \qquad \text{Power form}$$

$$= \int x^{2.1}\, dx + \int x^{-1.1}\, dx + \int x^{-1}\, dx + \int e^x\, dx \qquad \text{Sum rule}$$

$$= \frac{x^{3.1}}{3.1} + \frac{x^{-0.1}}{-0.1} + \ln|x| + e^x + C \qquad \text{Power rule and exponential rule}$$

$$= \frac{x^{3.1}}{3.1} - \frac{10}{x^{0.1}} + \ln|x| + e^x + C$$

c. $\displaystyle\int (e^x + 3^x - |x|)\, dx = \int e^x\, dx + \int 3^x\, dx - \int |x|\, dx$ Sum/difference rule

$$= e^x + \frac{3^x}{\ln 3} - \frac{x|x|}{2} + C$$ Rules for powers, exponentials, and absolute value

➡ **Before we go on...** You should check each of the answers in Example 2 by differentiating.

Q: *Why is there only a single arbitrary constant C in each of the answers?*

A: We could have written the answer to part (a) as

$$\frac{x^4}{4} + D + \frac{x^6}{6} + E - x + F,$$

where D, E, and F are all arbitrary constants. Now suppose, for example, that we set $D = 1$, $E = -2$, and $F = 6$. Then the particular antiderivative we get is $x^4/4 + x^6/6 - x + 5$, which has the form $x^4/4 + x^6/6 - x + C$. Thus, we could have chosen the single constant C to be 5 and obtained the same answer. In other words, the answer $x^4/4 + x^6/6 - x + C$ is just as general as the answer $x^4/4 + D + x^6/6 + E - x + F$ but simpler.

■

In practice we do not explicitly write the integral of a sum as a sum of integrals but just "integrate term by term," much as we learned to differentiate term by term.

EXAMPLE 3 Combining the Rules

Find the integrals:

a. $\displaystyle\int (10x^4 + 2x^2 - 3e^x)\, dx$ **b.** $\displaystyle\int\left(\frac{2}{x^{0.1}} + \frac{x^{0.1}}{2} - \frac{3}{4x}\right) dx$

c. $\displaystyle\int (3e^x - 2(1.2^x) + 5|x|)\, dx$

Solution

a. We need to integrate separately each of the terms $10x^4$, $2x^2$, and $3e^x$. To integrate $10x^4$, we use the rules for constant multiples and powers:

$$\int 10x^4\,dx = 10\int x^4\,dx = 10\frac{x^5}{5} + C = 2x^5 + C.$$

The other two terms are similar. We get

$$\int (10x^4 + 2x^2 - 3e^x)\,dx = 10\frac{x^5}{5} + 2\frac{x^3}{3} - 3e^x + C = 2x^5 + \frac{2}{3}x^3 - 3e^x + C.$$

b. We first convert to power form and then integrate term by term:

$$\int\left(\frac{2}{x^{0.1}} + \frac{x^{0.1}}{2} - \frac{3}{4x}\right)dx = \int\left(2x^{-0.1} + \frac{1}{2}x^{0.1} - \frac{3}{4}x^{-1}\right)dx \qquad \text{Power form}$$

$$= 2\frac{x^{0.9}}{0.9} + \frac{1}{2}\frac{x^{1.1}}{1.1} - \frac{3}{4}\ln|x| + C \qquad \text{Integrate term by term.}$$

$$= \frac{20x^{0.9}}{9} + \frac{x^{1.1}}{2.2} - \frac{3}{4}\ln|x| + C. \qquad \text{Back to positive exponent form}$$

c. $\displaystyle\int (3e^x - 2(1.2^x) + 5|x|)\,dx = 3e^x - 2\frac{1.2^x}{\ln 1.2} + 5\frac{x|x|}{2} + C$

EXAMPLE 4 Different Variable Name

Find $\displaystyle\int\left(\frac{1}{u} + \frac{1}{u^2}\right)du.$

Solution This integral may look a little strange because we are using the letter u instead of x, but there is really nothing special about x. Using u as the variable of integration, we get

$$\int\left(\frac{1}{u} + \frac{1}{u^2}\right)du = \int (u^{-1} + u^{-2})\,du \qquad \text{Power form}$$

$$= \ln|u| + \frac{u^{-1}}{-1} + C \qquad \text{Integrate term by term.}$$

$$= \ln|u| - \frac{1}{u} + C. \qquad \text{Simplify the result.}$$

➡ **Before we go on . . .** When we compute an indefinite integral, we want the independent variable in the answer to be the same as the variable of integration. Thus, if the integral in Example 4 had been written in terms of x rather than u, we would have written

$$\int\left(\frac{1}{x} + \frac{1}{x^2}\right)dx = \ln|x| - \frac{1}{x} + C.$$

■

Applications

EXAMPLE 5 Finding Cost from Marginal Cost

The marginal cost to produce baseball caps at a production level of x caps is $4 - 0.001x$ dollars per cap, and the cost of producing 100 caps is \$500. Find the cost function.

Solution We are asked to find the cost function $C(x)$ given that the *marginal* cost function is $4 - 0.001x$. Recalling that the marginal cost function is the derivative of the cost function, we can write

$$C'(x) = 4 - 0.001x$$

and must find $C(x)$. Now $C(x)$ must be an antiderivative of $C'(x)$, so

$$\begin{aligned} C(x) &= \int (4 - 0.001x)\,dx \\ &= 4x - 0.001\frac{x^2}{2} + K \qquad \text{K is the constant of integration.*} \\ &= 4x - 0.0005x^2 + K. \end{aligned}$$

* We used K and not C for the constant of integration because we are using C for cost.

Now, unless we have a value for K, we don't really know what the cost function is. However, there is another piece of information we have ignored: The cost of producing 100 baseball caps is \$500. In symbols,

$$C(100) = 500.$$

Substituting in our formula for $C(x)$, we have

$$\begin{aligned} C(100) &= 4(100) - 0.0005(100)^2 + K \\ 500 &= 395 + K \\ K &= 105. \end{aligned}$$

Now that we know what K is, we can write down the cost function:

$$C(x) = 4x - 0.0005x^2 + 105.$$

➡ **Before we go on...** Let us consider the significance of the constant term 105 in Example 5. If we substitute $x = 0$ into the cost function, we get

$$C(0) = 4(0) - 0.0005(0)^2 + 105 = 105.$$

Thus, \$105 is the cost of producing zero items; in other words, it is the **fixed cost.** ■

EXAMPLE 6 Total Sales from Annual Sales

By the start of 2008, **Apple** had sold a total of about 3.5 million iPhones. From the start of 2008 through the end of 2014, sales of iPhones were approximately

$$s(t) = 1.5t^2 + 20t + 0.25 \text{ million iPhones per year} \qquad (0 \le t \le 6),$$

where t is time in years since the start of 2008.[1]

[1] Source for data: Apple quarterly press releases (www.apple.com/investor/).

a. Find an expression for the total sales of iPhones up to time t.

b. Use the answer to part (a) to estimate the total sales of iPhones by the end of 2014. (The actual figure was 472 million.)

Solution

a. Let $S(t)$ be the total sales of iPhones, in millions, up to time t, where t is measured in years since the start of 2008, so we know that $S(0) = 3.5$. We are also given an expression for the number of iPhones sold per year. This function is the *derivative* of $S(t)$:

$$S'(t) = 1.5t^2 + 20t + 0.25.$$

Thus, the desired total sales function must be an antiderivative of $S'(t)$:

$$\begin{aligned} S(t) &= \int (1.5t^2 + 20t + 0.25)\, dt \\ &= \frac{1.5t^3}{3} + 20\frac{t^2}{2} + 0.25t + C \\ &= 0.5t^3 + 10t^2 + 0.25t + C. \end{aligned}$$

To calculate the value of the constant C, we can, as in the preceding example, use the known value of S: $S(0) = 3.5$.

$$S(0) = 0.5(0)^3 + 10(0)^2 + 0.25(0) + C = 3.5,$$

so

$$C = 3.5.$$

We can now write down the total sales function:

$$S(t) = 0.5t^3 + 10t^2 + 0.25t + 3.5 \text{ million iPhones.}$$

b. Because the start of 2014 corresponds to $t = 6$, we calculate total sales as

$$S(3) = 0.5(6)^3 + 10(6)^2 + 0.25(6) + 3.5 = 473 \text{ million iPhones,}$$

remarkably close to the actual 472 million figure!

Motion in a Straight Line

An important application of the indefinite integral is to the study of motion. The application of calculus to problems about motion is an example of the intertwining of mathematics and physics. We begin by bringing together some facts, scattered through the last several chapters, that have to do with an object moving in a straight line. We then restate them in terms of antiderivatives.

Position, Velocity, and Acceleration: Derivative Form

If $s = s(t)$ is the **position** of an object at time t, then its **velocity** is given by the derivative

$$v = \frac{ds}{dt}.$$

In Words: Velocity is the derivative of position.

The **acceleration** of an object is given by the derivative

$$a = \frac{dv}{dt}.$$

In Words: Acceleration is the derivative of velocity.

Position, Velocity, and Acceleration: Integral Form

$$s(t) = \int v(t)\, dt \qquad \text{Because } v = \frac{ds}{dt}$$

$$v(t) = \int a(t)\, dt \qquad \text{Because } a = \frac{dv}{dt}$$

Quick Examples

15. If the velocity of a particle moving in a straight line is given by $v(t) = 4t + 1$, then its position after t seconds is given by $s(t) = \int v(t)\, dt = \int (4t + 1)\, dt = 2t^2 + t + C$.

16. If sales are accelerating at 2 golf balls per day per day, then the rate of change of sales ("velocity of sales") is $v(t) = \int a(t)\, dt = \int 2\, dt = 2t + C$ golf balls per day.

17. If the rate of change of sales is $v(t) = 2t + 5$ golf balls per day, then the total sales are $s(t) = \int v(t)\, dt = \int (2t + 5)\, dt = t^2 + 5t + C$ golf balls sold through time t.

EXAMPLE 7 **Motion in a Straight Line**

a. The velocity of a particle moving along a straight line is given by $v(t) = 4t + 1$ m/sec. Given that the particle is at position $s = 2$ meters at time $t = 1$, find an expression for s in terms of t.

b. For a freely falling body experiencing no air resistance and zero initial velocity, find an expression for the velocity v in terms of t. [Note: On Earth a freely falling body experiencing no air resistance accelerates downward at approximately 9.8 meters per second per second, or 9.8 m/sec^2 (or 32 ft/sec^2).]

Solution

a. As we saw in Quick Example 15, the position of the particle after t seconds is given by

$$\begin{aligned} s(t) &= \int v(t)\, dt \\ &= \int (4t + 1)\, dt = 2t^2 + t + C. \end{aligned}$$

But what is the value of C? Now, we are told that the particle is at position $s = 2$ at time $t = 1$. In other words, $s(1) = 2$. Substituting this into the expression for $s(t)$ gives

$$2 = 2(1)^2 + 1 + C$$

so

$$C = -1.$$

Hence the position after t seconds is given by

$$s(t) = 2t^2 + t - 1 \text{ meters.}$$

b. Let's measure heights above the ground as positive so that a rising object has positive velocity and the acceleration due to gravity is negative. (It causes the upward velocity to decrease in value.) Thus, the acceleration of the object is given by

$$a(t) = -9.8 \text{ m/sec}^2.$$

We wish to know the velocity, which is an antiderivative of acceleration, so we compute

$$v(t) = \int a(t)\,dt = \int (-9.8)\,dt = -9.8t + C.$$

To find the value of C, we use the given information that at time $t = 0$ the velocity is 0: $v(0) = 0$. Substituting this into the expression for $v(t)$ gives

$$0 = -9.8(0) + C$$

so

$$C = 0.$$

Hence, the velocity after t seconds is given by

$$v(t) = -9.8t \text{ m/sec.}$$

EXAMPLE 8 Vertical Motion under Gravity

You are standing on the edge of a cliff and toss a stone upward at a speed of $v_0 = 30$ ft/sec. (v_0 is called the *initial velocity.*)

a. Find the stone's velocity as a function of time. How fast and in what direction is it going after 5 seconds? (Neglect the effects of air resistance.)

b. Find the position of the stone as a function of time. Where will it be after 5 seconds?

c. When and where will the stone reach its zenith, its highest point?

Solution

a. This is similar to Example 7(b): Measuring height above the ground as positive, the acceleration of the stone is given by $a(t) = -32$ ft/sec^2, so

$$v(t) = \int (-32)\,dt = -32t + C.$$

To obtain C, we use the fact that you tossed the stone upward at 30 ft/sec; that is, when $t = 0$, $v = 30$, or $v(0) = 30$. Thus,

$$30 = v(0) = -32(0) + C,$$

so $C = 30$, and the formula for velocity is

$$v(t) = -32t + 30 \text{ ft/sec.} \qquad v(t) = -32t + v_0$$

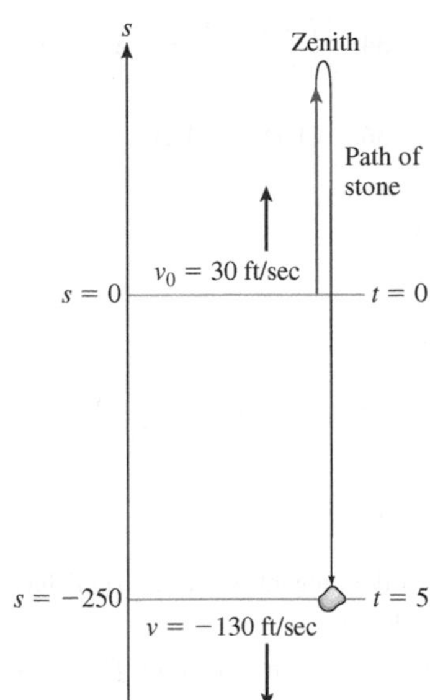

Figure 2

In particular, after 5 seconds the velocity will be

$$v(5) = -32(5) + 30 = -130 \text{ ft/sec}.$$

After 5 seconds the stone is *falling* with a speed of 130 ft/sec.

b. We wish to know the position, but position is an antiderivative of velocity. Thus,

$$s(t) = \int v(t)\,dt = \int (-32t + 30)\,dt = -16t^2 + 30t + C.$$

Now to find C, we need to know the initial position $s(0)$. We are not told this, so let's measure heights so that the initial position is zero. Then

$$0 = s(0) = C,$$

and

$$s(t) = -16t^2 + 30t \text{ feet}$$

$s(t) = -16t^2 + v_0t + s_0$
s_0 = initial position

In particular, after 5 seconds the stone has a height of

$$s(5) = -16(5)^2 + 30(5) = -250 \text{ feet}.$$

In other words, the stone is now 250 feet *below* where it was when you first threw it, as shown in Figure 2.

c. The stone reaches its zenith when its height $s(t)$ is at its maximum value, which occurs when $v(t) = s'(t)$ is zero. So we solve

$$v(t) = -32t + 30 = 0$$

getting $t = 30/32 = 15/16 = 0.9375$ seconds. This is the time when the stone reaches its zenith. The height of the stone at that time is

$$s(15/16) = -16(15/16)^2 + 30(15/16) = 14.0625 \text{ feet}.$$

➡ **Before we go on...** Here again are the formulas we obtained in Example 8, together with their metric equivalents.

Vertical Motion under Gravity: Velocity and Position

If we ignore air resistance, the vertical velocity and position of an object moving under gravity are given by

	British Units	Metric Units
Velocity:	$v(t) = -32t + v_0$ ft/sec	$v(t) = -9.8t + v_0$ m/sec
Position:	$s(t) = -16t^2 + v_0t + s_0$ feet	$s(t) = -4.9t^2 + v_0t + s_0$ meters

v_0 = initial velocity = velocity at time 0

s_0 = initial position = position at time 0

Quick Example

18. If a ball is thrown down at 2 ft/sec from a height of 200 feet, then its velocity and position after t seconds are $v(t) = -32t - 2$ ft/sec and $s(t) = -16t^2 - 2t + 200$ feet.

■

6.1 EXERCISES

▼ more advanced ◆ challenging

T indicates exercises that should be solved using technology

In Exercises 1–42, evaluate the integral.
[**HINT**: for 1–6: See Quick Examples 6–9.]

1. $\int x^5\,dx$ **2.** $\int x^7\,dx$ **3.** $\int 6\,dx$

4. $\int (-5)\,dx$ **5.** $\int x\,dx$ **6.** $\int (-x)\,dx$

[**HINT**: for 7–18: See Example 2.]

7. $\int (x^2 - x)\,dx$ **8.** $\int (x + x^3)\,dx$

9. $\int (1 + x)\,dx$ **10.** $\int (4 - x)\,dx$

11. $\int x^{-5}\,dx$ **12.** $\int x^{-7}\,dx$

13. $\int (x^{2.3} + x^{-1.3})\,dx$ **14.** $\int (x^{-0.2} - x^{0.2})\,dx$

15. $\int (u^2 - u^{-1})\,du$ [**HINT**: See Example 4.]

16. $\int (v^{-2} + 2v^{-1})\,dv$ [**HINT**: See Example 4.]

17. $\int \sqrt[4]{x}\,dx$ **18.** $\int \sqrt[3]{x}\,dx$

[**HINT**: for 19–42: See Example 3.]

19. $\int (3x^4 - 2x^{-2} + x^{-5} + 4)\,dx$

20. $\int (4x^7 - x^{-3} + 1)\,dx$

21. $\int \left(\frac{2}{u} + \frac{u}{4}\right) du$ **22.** $\int \left(\frac{2}{u^2} + \frac{u^2}{4}\right) du$

23. $\int \left(\frac{1}{x} + \frac{2}{x^2} - \frac{1}{x^3}\right) dx$ **24.** $\int \left(\frac{3}{x} - \frac{1}{x^5} + \frac{1}{x^7}\right) dx$

25. $\int (3x^{0.1} - x^{4.3} - 4.1)\,dx$ **26.** $\int \left(\frac{x^{2.1}}{2} - 2.3\right) dx$

27. $\int \left(\frac{3}{x^{0.1}} - \frac{4}{x^{1.1}}\right) dx$ **28.** $\int \left(\frac{1}{x^{1.1}} - \frac{1}{x}\right) dx$

29. $\int \left(5.1t - \frac{1.2}{t} + \frac{3}{t^{1.2}}\right) dt$ **30.** $\int \left(3.2 + \frac{1}{t^{0.9}} + \frac{t^{1.2}}{3}\right) dt$

31. $\int (2e^x + 5|x| + 1/4)\,dx$

32. $\int (-4e^x + |x|/3 - 1/8)\,dx$

33. $\int \left(\frac{6.1}{x^{0.5}} + \frac{x^{0.5}}{6} - e^x\right) dx$ **34.** $\int \left(\frac{4.2}{x^{0.4}} + \frac{x^{0.4}}{3} - 2e^x\right) dx$

35. $\int (2^x - 3^x)\,dx$ **36.** $\int (1.1^x + 2^x)\,dx$

37. $\int \left(100(1.1^x) - \frac{2|x|}{3}\right) dx$

38. $\int \left(1{,}000(0.9^x) + \frac{4|x|}{5}\right) dx$

39. ▼ $\int x^{-2}\left(x^4 - \frac{3}{2x^4}\right) dx$ **40.** ▼ $\int 3x^4\left(\frac{2}{x^4} + \frac{3}{5x^6}\right) dx$

41. ▼ $\int \frac{x+2}{x^3}\,dx$ **42.** ▼ $\int \frac{x^2-2}{x}\,dx$

43. Find $f(x)$ if $f(0) = 1$ and the tangent line at $(x, f(x))$ has slope x. [**HINT**: See Example 5.]

44. Find $f(x)$ if $f(1) = 1$ and the tangent line at $(x, f(x))$ has slope $\frac{1}{x}$. [**HINT**: See Example 5.]

45. Find $f(x)$ if $f(0) = 0$ and the tangent line at $(x, f(x))$ has slope $e^x - 1$.

46. Find $f(x)$ if $f(1) = -1$ and the tangent line at $(x, f(x))$ has slope $2e^x + 1$.

Applications

47. ***Marginal Cost*** The marginal cost of producing the xth box of light bulbs is $5 - \frac{x}{10{,}000}$, and the fixed cost is \$20,000. Find the cost function $C(x)$. [**HINT**: See Example 5.]

48. ***Marginal Cost*** The marginal cost of producing the xth box of DVDs is $10 + \frac{x^2}{100{,}000}$, and the fixed cost is \$100,000. Find the cost function $C(x)$. [**HINT**: See Example 5.]

49. ***Marginal Cost*** The marginal cost of producing the xth roll of film is $5 + 2x + \frac{1}{x}$. The total cost to produce one roll is \$1,000. Find the cost function $C(x)$.

50. ***Marginal Cost*** The marginal cost of producing the xth box of CDs is $10 + x + \frac{1}{x^2}$. The total cost to produce 100 boxes is \$10,000. Find the cost function $C(x)$.

51. ***Facebook Membership*** At the start of 2010, Facebook had 360 million members. Subsequently, new members joined at a rate of roughly

$$m(t) = -6.6t^2 + 18t + 220 \text{ million members per year} \quad (0 \le t \le 5),$$

where t is time in years since the start of 2010.[2]

a. Find an expression for total Facebook membership $M(t)$ at time t. [**HINT:** See Example 6.]

b. Use the answer to part (a) to estimate Facebook membership midway through 2014. (Round your answer to the nearest 1 million members. The actual figure was 1,317 million.)

52. ***Uploads to YouTube*** Since **YouTube** first became available to the public in mid-2005, the rate at which video has been uploaded to the site can be approximated by

$$v(t) = 1.1t^2 - 2.6t + 2.3 \text{ million hours of video per year} \quad (0 \le t \le 9),$$

where t is time in years since June 2005.[3]

a. Find an expression for the total number of hours $V(t)$ of video at time t (starting from zero hours of video at $t = 0$). [**HINT:** See Example 6.]

b. Use the answer to part (a) to estimate the total number of hours of video uploaded by the start of 2014. (Round your answer to the nearest million hours of video.)

53. ***Median Household Income*** From 2000 to 2007, median household income in the United States rose by an average of approximately \$1,200 per year.[4] Given that the median household income in 2000 was approximately \$42,000, use an indefinite integral to find a formula for the median household income I as a function of the year t ($t = 0$ represents 2000), and use your formula to estimate the median household income in 2005. (You can do this exercise without integration using the methods of Section 1.3, but here you should use an indefinite integral.)

54. ***Mean Household Income*** From 2000 to 2007 the mean household income in the United States rose by an average of approximately \$1,500 per year.[5] Given that the mean household income in 2000 was approximately \$57,000, use an indefinite integral to find a formula for the mean household income I as a function of the year t ($t = 0$ represents 2000), and use your formula to estimate the mean household income in 2006. (You can do this exercise without integration using the methods of Section 1.3, but here you should use an indefinite integral.)

55. ***Bottled Water Sales*** The rate of U.S. sales of bottled water for the period 2007–2014 can be approximated by

$$s(t) = 0.08t^2 - 0.26t + 8.8 \text{ billion gallons per year} \quad (0 \le t \le 7),$$

where t is time in years since the start of 2007.[6] Use an indefinite integral to approximate the total sales $S(t)$ of bottled water since the start of 2007. Approximately how much bottled water was sold from the start of 2007 to the start of 2014? [**HINT:** At the start of 2007, sales since that time are zero.]

56. ***Bottled Water Sales*** The rate of U.S. per capita sales of bottled water for the period 2007–2014 can be approximated by

$$s(t) = 0.25t^2 - t + 29 \text{ gallons per year} \quad (0 \le t \le 7),$$

where t is time in years since the start of 2007.[7] Use an indefinite integral to approximate the total per capita sales $S(t)$ of bottled water since the start of 2007. Approximately how much bottled water was sold, per capita, from the start of 2007 to the end of 2012? [**HINT:** At the start of 2007, sales since that time are zero.]

57. ▼ ***Health-Care Spending in the 1990s*** Write $H(t)$ for the amount spent in the United States on health care in year t, where t is measured in years since 1990. The rate of increase of $H(t)$ was approximately \$65 billion per year in 1990 and rose to \$100 billion per year in 2000.[8]

a. Find a linear model for the rate of change $H'(t)$.

b. Given that \$700 billion was spent on health care in the United States in 1990, find the function $H(t)$.

58. ▼ ***Health-Care Spending in the 2000s*** Write $H(t)$ for the amount spent in the United States on health care in year t, where t is measured in years since 2000. The rate of increase of $H(t)$ was projected to rise from \$100 billion per year in 2000 to approximately \$190 billion per year in 2010.[9]

a. Find a linear model for the rate of change $H'(t)$.

b. Given that \$1,300 billion was spent on health care in the United States in 2000, find the function $H(t)$.

59. ▼ ***Subprime Mortgage during the Housing Bubble*** At the start of 2007 the percentage of U.S. mortgages that were subprime was about 13%, was increasing at a rate of 1 percentage point per year, but was decelerating at 0.4 percentage points per year per year.[10]

a. Find an expression for the rate of change (velocity) of this percentage at time t in years since the start of 2007.

b. Use the result of part (a) to find an expression for the percentage of mortgages that were subprime at time t, and use it to estimate the percentage at the start of 2008. [**HINT:** See Quick Examples 16 and 17.]

---

[2] Sources for data: www.facebook.com, www.statista.com.

[3] Sources for data: www.YouTube.com, www.statista.com.

[4] In current dollars, unadjusted for inflation. Source for data: U.S. Census Bureau (www.census.gov).

[5] *Ibid.*

[6] Source for data: Beverage Marketing Corporation (www.bottledwater.org).

[7] *Ibid.*

[8] Source: Centers for Medicare and Medicaid Services, "National Health Expenditures," 2002 version, released January 2004 (www.cms.hhs.gov/statistics/nhe/).

[9] Source: Centers for Medicare and Medicaid Services, "National Health Expenditures 1965–2013, History and Projections" (www.cms.hhs.gov/statistics/nhe/).

[10] Sources: Mortgage Bankers Association, UBS.

60. ▼ ***Subprime Mortgage Debt during the Housing Bubble*** At the start of 2008 the value of subprime mortgage debt outstanding in the United States was about $1,300 billion, was increasing at a rate of 40 billion dollars per year, but was decelerating at 20 billion dollars per year per year.[11]
   a. Find an expression for the rate of change (velocity) of the value of subprime mortgage debt at time t in years since the start of 2008.
   b. Use the result of part (a) to find an expression for the value of subprime mortgage debt at time t, and use it to estimate the value at the start of 2009. [**HINT**: See Quick Examples 16 and 17.]

61. ***Motion in a Straight Line*** The velocity of a particle moving in a straight line is given by $v = t^2 + 1$.
   a. Find an expression for the position s after a time t.
   b. Given that $s = 1$ at time $t = 0$, find the constant of integration C, and hence find an expression for s in terms of t without any unknown constants. [**HINT**: See Example 7.]

62. ***Motion in a Straight Line*** The velocity of a particle moving in a straight line is given by $v = 3e^t + t$.
   a. Find an expression for the position s after a time t.
   b. Given that $s = 3$ at time $t = 0$, find the constant of integration C, and hence find an expression for s in terms of t without any unknown constants. [**HINT**: See Example 7.]

63. ***Vertical Motion under Gravity*** If a stone is dropped from a rest position above the ground, how fast (in feet per second) and in what direction will it be traveling after 10 seconds? (Neglect the effects of air resistance.) [**HINT**: See Example 8.]

64. ***Vertical Motion under Gravity*** If a stone is thrown upward at 10 feet per second, how fast (in feet per second) and in what direction will it be traveling after 10 seconds? (Neglect the effects of air resistance.) [**HINT**: See Example 8.]

65. ***Vertical Motion under Gravity*** Your name is Galileo Galilei, and you toss a weight upward at 16 ft/sec from the top of the Leaning Tower of Pisa (height 185 feet).
   a. Neglecting air resistance, find the weight's velocity as a function of time t in seconds.
   b. Find the height of the weight above the ground as a function of time. Where and when will it reach its zenith? [**HINT**: See Example 8 and the formulas that follow.]

66. ***Vertical Motion under Gravity*** Your name is Spaghettini Bologna (an assistant of Galileo Galilei), and, to impress your boss, you toss a weight upward at 24 ft/sec from the top of the Leaning Tower of Pisa (height 185 feet).
   a. Neglecting air resistance, find the weight's velocity as a function of time t in seconds.
   b. Find the height of the weight above the ground as a function of time. Where and when will it reach its zenith? [**HINT**: See Example 8 and the formulas that follow.]

67. ▼ ***Tailwinds*** The ground speed of an airliner is obtained by adding its air speed and the tailwind speed. On your recent trip from Mexico to the United States your plane was traveling at an air speed of 500 miles per hour and experienced tailwinds of $25 + 50t$ miles per hour, where t is the time in hours since takeoff.
   a. Obtain an expression for the distance traveled in terms of the time since takeoff. [**HINT**: Ground speed = Air speed + Tailwind speed.]
   b. Use the result of part (a) to estimate the time of your 1,800-mile trip.
   c. The equation solved in part (b) leads mathematically to two solutions. Explain the meaning of the solution you rejected.

68. ▼ ***Headwinds*** The ground speed of an airliner is obtained by subtracting its headwind speed from its air speed. On your recent trip to Mexico from the United States your plane was traveling at an air speed of 500 miles per hour and experienced headwinds of $25 + 50t$ miles per hour, where t is the time in hours since takeoff.
   a. Obtain an expression for the distance traveled in terms of the time since takeoff. [**HINT**: Ground speed = Air speed − Headwind speed.]
   b. Use the result of part (a) to estimate the time of your 1,500-mile trip.
   c. The equation solved in part (b) leads mathematically to two solutions. Explain the meaning of the solution you rejected.

69. ▼ ***Vertical Motion*** Show that if a projectile is thrown upward with a velocity of v_0 ft/sec, then (neglecting air resistance) it will reach its highest point after $v_0/32$ seconds. [**HINT**: See the formulas that follow Example 8.]

70. ▼ ***Vertical Motion*** Use the result of Exercise 69 to show that if a projectile is thrown upward with a velocity of v_0 ft/sec, its highest point will be $v_0^2/64$ feet above the starting point (if we neglect the effects of air resistance).

Exercises 71–76 use the results of Exercises 69 and 70.

71. ▼ I threw a ball up in the air to a height of 20 feet. How fast was the ball traveling when it left my hand?

72. ▼ I threw a ball up in the air to a height of 40 feet. How fast was the ball traveling when it left my hand?

73. ▼ A piece of chalk is tossed vertically upward by Prof. Schwarzenegger and hits the ceiling 100 feet above with a *BANG.*
   a. What is the minimum speed at which the piece of chalk must have been thrown to enable it to hit the ceiling?
   b. Assuming that Prof. Schwarzenegger in fact tossed the piece of chalk up at 100 ft/sec, how fast was it moving when it struck the ceiling?
   c. Assuming that Prof. Schwarzenegger tossed the chalk up at 100 ft/sec, and that it recoils from the ceiling with the same speed it had at the instant it hit, how long will it take the chalk to make the return journey and hit the ground?

[11] Source: www.data360.org/dataset.aspx?Data_Set_Id=9549.

74. ▼ A projectile is fired vertically upward from ground level at 16,000 ft/sec.
    a. How high does the projectile go?
    b. How long does it take to reach its zenith (highest point)?
    c. How fast is it traveling when it hits the ground?

75. ▼ ***Strength*** Prof. Strong can throw a 10-pound dumbbell twice as high as Prof. Weak can. How much faster can Prof. Strong throw it?

76. ▼ ***Weakness*** Prof. Weak can throw a book three times as high as Prof. Strong can. How much faster can Prof. Weak throw it?

Communication and Reasoning Exercises

77. Why is this section called "The *Indefinite* Integral"?

78. If the derivative of Julius is Augustus, then Augustus is ________ of Julius.

79. Linear functions are antiderivatives of what kind of function? Explain.

80. Constant functions are antiderivatives of what kind of function? Explain.

81. If we know the *derivative* of a function, do we know the function? Explain. If not, what further information will suffice?

82. If we know an *antiderivative* of a function, do we know the function? Explain. If not, what further information will suffice?

83. If $F(x)$ and $G(x)$ are both antiderivatives of $f(x)$, how are $F(x)$ and $G(x)$ related?

84. Your friend Marco claims that once you have one antiderivative of $f(x)$ you have all of them. Explain what he means.

85. Complete the following: The total cost function is a(n) _____ of the _____ cost function.

86. Complete the following: The distance covered is an antiderivative of the _____ function, and the velocity is an antiderivative of the _____ function.

87. If x represents the number of items manufactured and $f(x)$ represents dollars per item, what does $\int f(x)\,dx$ represent? In general, how are the units of $f(x)$ and the units of $\int f(x)\,dx$ related?

88. If t represents time in seconds since liftoff and $g(t)$ represents the volume of rocket fuel burned per second, what does $\int g(t)\,dt$ represent?

89. Why was the following marked wrong? What is the correct answer?

$$\int (3x+1)\,dx = \frac{3x^2}{2} + 0 + C = \frac{3x^2}{2} + C \quad ✗\ \textit{WRONG!}$$

90. Why was the following marked wrong? What is the correct answer?

$$\int (3x^2 - 11x)\,dx = x^3 - 11 + C \quad ✗\ \textit{WRONG!}$$

91. Why was the following marked wrong? What is the correct answer?

$$\int (12x^5 - 4x)\,dx = \int 2x^6 - 2x^2 + C \quad ✗\ \textit{WRONG!}$$

92. Why was the following marked wrong? What is the correct answer?

$$\int 5\,dt = 5x + C \quad ✗\ \textit{WRONG!}$$

93. Why was the following marked wrong? What is the correct answer?

$$\int 4(e^x - 2x)\,dx = (4x)(e^x - x^2) + C \quad ✗\ \textit{WRONG!}$$

94. Why was the following marked wrong? What is the correct answer?

$$\int (2^x - 1)\,dx = \frac{2^{x+1}}{x+1} - x + C \quad ✗\ \textit{WRONG!}$$

95. Why was the following marked wrong? How should it be corrected?

$$\frac{1}{x} = \ln|x| + C \quad ✗\ \textit{WRONG!}$$

96. Why was the following marked wrong? What is the correct answer?

$$\int \frac{1}{x^3}\,dx = \ln|x^3| + C \quad ✗\ \textit{WRONG!}$$

97. ▼ Give an argument for the rule that the integral of a sum is the sum of the integrals.

98. ▼ Give an argument for the rule that the integral of a constant multiple is the constant multiple of the integrals.

99. ▼ Give an example to show that the integral of a product is not, in general, the product of the integrals.

100. ▼ Give an example to show that the integral of a quotient is not the quotient of the integrals.

101. ▼ Complete the following: If you take the _____ of the _____ of $f(x)$, you obtain $f(x)$ back. On the other hand, if you take the _____ of the _____ of $f(x)$, you obtain $f(x) + C$.

102. ▼ If a Martian told you that the *Institute of Alien Mathematics*, after a long and difficult search, has announced the discovery of a new antiderivative of $x - 1$ called $M(x)$, completely different from $\frac{x^2}{2} - x + C$ [the formula for $M(x)$ is classified information and cannot be revealed here], how would you respond?

6.2 Substitution

The Technique of Substitution

The chain rule for derivatives gives us an extremely useful technique for finding antiderivatives. This technique is called **change of variables** or **substitution**.

Recall that to differentiate a function such as $(x^2 + 1)^6$, we first think of the function as $g(u)$, where $u = x^2 + 1$ and $g(u) = u^6$. We then compute the derivative, using the chain rule, as

$$\frac{d}{dx}g(u) = g'(u)\frac{du}{dx}.$$

Any rule for derivatives can be turned into a technique for finding antiderivatives by writing it in integral form. The integral form of the above formula is

$$\int g'(u)\frac{du}{dx}\,dx = g(u) + C.$$

But if we write $g(u) + C = \int g'(u)\,du$ we get the following interesting equation:

$$\int g'(u)\frac{du}{dx}\,dx = \int g'(u)\,du.$$

This equation is the one usually called the *change of variables formula.* We can turn it into a more useful integration technique as follows. Let $f = g'(u)(du/dx)$. We can rewrite the above change of variables formula using f:

$$\int f\,dx = \int\left(\frac{f}{du/dx}\right)du.$$

In essence, we are making the formal substitution

$$dx = \frac{1}{du/dx}du.$$

Here's the technique.

Substitution Rule

If u is a function of x, then we can use the following formula to evaluate an integral:

$$\int f\,dx = \int\left(\frac{f}{du/dx}\right)du.$$

Rather than using the formula directly, we use the following step-by-step procedure:

1. Write u as a function of x.
2. Take the derivative du/dx, and solve for the quantity dx in terms of du.
3. Use the expression you obtain in Step 2 to substitute for dx in the given integral and substitute u for its defining expression.

Now let's see how this procedure works in practice.

EXAMPLE 1 **Substitution**

Find $\int 4x(x^2 + 1)^6\,dx$.

Solution To use substitution, we need to choose an expression to be u. There is no hard and fast rule, but here is one hint that often works:

Take u to be an expression that is being raised to a power.

In this case, let's set $u = x^2 + 1$. Continuing the procedure above, we place the calculations for Step 2 in a box.

$u = x^2 + 1$	Write u as a function of x.
$\frac{du}{dx} = 2x$	Take the derivative of u with respect to x.
$dx = \frac{1}{2x}\,du$	Solve for dx: $dx = \frac{1}{du/dx}\,du$.

Now we *substitute u for its defining expression and substitute for dx* in the original integral:

$$\int 4x(x^2 + 1)^6\,dx = \int 4xu^6 \frac{1}{2x}\,du \qquad \text{Substitute* for } u \text{ and } dx.$$

$$= \int 2u^6\,du. \qquad \text{Cancel the } x\text{s and simplify.}$$

* This step is equivalent to using the formula stated in the Substitution Rule box. If it should bother you that the integral contains both x and u, note that x is now a function of u.

We have boiled the given integral down to the much simpler integral $\int 2u^6\,du$, and we can now write down the solution:

$$2\frac{u^7}{7} + C = \frac{2(x^2 + 1)^7}{7} + C. \qquad \text{Substitute } (x^2 + 1) \text{ for } u \text{ in the answer.}$$

➡ **Before we go on . . .** There are two points to notice in Example 1. First, before we can actually integrate with respect to u, *we must eliminate all xs from the integrand.* If we cannot, we may have chosen the wrong expression for u. Second, after integrating, we must substitute back to obtain an expression involving x.

It is easy to check our answer. We differentiate:

$$\frac{d}{dx}\left[\frac{2(x^2 + 1)^7}{7}\right] = \frac{2(7)(x^2 + 1)^6(2x)}{7} = 4x(x^2 + 1)^6. \quad ✔$$

Notice how we used the chain rule to check the result obtained by substitution. ■

When we use substitution, the first step is always to decide what to take as u. Again, there are no set rules, but we see some common cases in the examples.

EXAMPLE 2 **More Substitution**

Evaluate the following:

a. $\int x^2(x^3+1)^2\,dx$ **b.** $\int 3xe^{x^2}\,dx$ **c.** $\int \frac{1}{2x+5}\,dx$ **d.** $\int\left(\frac{1}{2x+5}+4x^2+1\right)dx$

Solution

a. As we said in Example 1, it often works to take u to be an expression that is being raised to a power. We usually also want to see the derivative of u as a factor in the integrand so that we can cancel terms involving x. In this case, x^3+1 is being raised to a power, so let's set $u = x^3+1$. Its derivative is $3x^2$; in the integrand we see x^2, which is missing the factor 3, but missing or incorrect constant factors are not a problem.

$$u = x^3 + 1$$ Write u as a function of x.

$$\frac{du}{dx} = 3x^2$$ Take the derivative of u with respect to x.

$$dx = \frac{1}{3x^2}\,du$$ Solve for dx: $dx = \frac{1}{du/dx}\,du$.

$$\int x^2(x^3+1)^2\,dx = \int x^2u^2\frac{1}{3x^2}\,du$$ Substitute for u and dx.

$$= \int \frac{1}{3}u^2\,du$$ Cancel the terms with x.

$$= \frac{1}{9}u^3 + C$$ Take the antiderivative.

$$= \frac{1}{9}(x^3+1)^3 + C$$ Substitute for u in the answer.

b. When we have an exponential with an expression in the exponent, it often works to substitute u for that expression. In this case, let's set $u = x^2$. (Notice again that we see a constant multiple of its derivative $2x$ as a factor in the integrand—a good sign.)

$$u = x^2$$

$$\frac{du}{dx} = 2x$$

$$dx = \frac{1}{2x}\,du$$

Substituting into the integral, we have

$$\int 3xe^{x^2}\,dx = \int 3xe^u\frac{1}{2x}\,du = \int \frac{3}{2}e^u\,du$$

$$= \frac{3}{2}e^u + C = \frac{3}{2}e^{x^2} + C.$$

c. We begin by rewriting the integrand as a power:

$$\int \frac{1}{2x+5}\,dx = \int (2x+5)^{-1}\,dx.$$

Now we take our earlier advice and set u equal to the expression that is being raised to a power:

$$u = 2x + 5$$
$$\frac{du}{dx} = 2$$
$$dx = \frac{1}{2}\,du$$

Substituting into the integral, we have

$$\int \frac{1}{2x+5}\,dx = \int \frac{1}{2}u^{-1}\,du = \frac{1}{2}\ln|u| + C$$
$$= \frac{1}{2}\ln|2x+5| + C.$$

d. Here, the substitution $u = 2x + 5$ works for the first part of the integrand, $1/(2x + 5)$ but not for the rest of it, so we break up the integral:

$$\int \left(\frac{1}{2x+5} + 4x^2 + 1\right) dx = \int \frac{1}{2x+5}\,dx + \int (4x^2 + 1)\,dx.$$

For the first integral we can use the substitution $u = 2x + 5$ (which we did in part (a)). For the second, no substitution is necessary:

$$\int \frac{1}{2x+5}\,dx + \int (4x^2 + 1)\,dx = \frac{1}{2}\ln|2x+5| + \frac{4x^3}{3} + x + C.$$

EXAMPLE 3 Choosing *u*

Evaluate $\int (x + 3)\sqrt{x^2 + 6x}\,dx$.

Solution There are two parenthetical expressions. Notice, however, that the derivative of the expression $(x^2 + 6x)$ is $2x + 6$, which is twice the term $(x + 3)$ in front of the radical. Recall that we would like the derivative of u to appear as a factor. Thus, let's take $u = x^2 + 6x$.

$$u = x^2 + 6x$$
$$\frac{du}{dx} = 2x + 6 = 2(x + 3)$$
$$dx = \frac{1}{2(x+3)}\,du$$

Substituting into the integral, we have

$$\int (x + 3)\sqrt{x^2 + 6x}\,dx$$
$$= \int (x + 3)\sqrt{u}\left(\frac{1}{2(x+3)}\right) du$$
$$= \int \frac{1}{2}\sqrt{u}\,du = \frac{1}{2}\int u^{1/2}\,du$$
$$= \frac{1}{2}\frac{2}{3}u^{3/2} + C = \frac{1}{3}(x^2 + 6x)^{3/2} + C.$$

Some cases require a little more work.

EXAMPLE 4 When the x Terms Do Not Cancel

Evaluate $\int \frac{2x}{(x-5)^2}\,dx$.

Solution We first rewrite

$$\int \frac{2x}{(x-5)^2}\,dx = \int 2x(x-5)^{-2}\,dx.$$

This suggests that we should set $u = x - 5$.

$$u = x - 5$$
$$\frac{du}{dx} = 1$$
$$dx = du$$

Substituting, we have

$$\int \frac{2x}{(x-5)^2}\,dx = \int 2xu^{-2}\,du.$$

Now, there is nothing in the integrand to cancel the x that appears. If, as here, there is still an x in the integrand after substituting, we go back to the expression for u, solve for x, and substitute the expression we obtain for x in the integrand. So we take $u = x - 5$ and solve for $x = u + 5$. Substituting, we get

$$\begin{aligned}\int 2xu^{-2}\,du &= \int 2(u+5)u^{-2}\,du \\ &= 2\int (u^{-1} + 5u^{-2})\,du \\ &= 2\ln|u| - \frac{10}{u} + C \\ &= 2\ln|x-5| - \frac{10}{x-5} + C.\end{aligned}$$

Application

EXAMPLE 5 Application: Bottled Water for Pets

Annual sales of bottled spring water for pets can be modeled by the logistic function

$$s(t) = \frac{3{,}000e^{0.5t}}{3 + e^{0.5t}} \text{ million gallons per year} \qquad (0 \le t \le 12),$$

where t is time in years since the start of 2000.[12]

a. Find an expression for the total amount of bottled spring water for pets sold since the start of 2000.

b. How much bottled spring water for pets was sold from the start of 2005 to the start of 2008?

[12] Based on data through 2008 and the recovery in 2010 of the general bottled water market to pre-recession levels. Sources: "*Liquid Assets: America's Expensive Love Affair with Bottled Water*" Daniel Gross, April 26, 2011 (finance.yahoo.com), Beverage Marketing Corporation (www.beveragemarketing.com).

Solution

a. If we write the total amount of pet spring water sold since the start of 2000 as $S(t)$, then the information we are given says that

$$S'(t) = s(t) = \frac{3{,}000e^{0.5t}}{3 + e^{0.5t}}.$$

Thus,

$$S(t) = \int \frac{3{,}000e^{0.5t}}{3 + e^{0.5t}}\, dt$$

is the function we are after. To integrate the expression, take u to be the denominator of the integrand:

$$u = 3 + e^{0.5t}$$
$$\frac{du}{dt} = 0.5e^{0.5t}$$
$$dt = \frac{1}{0.5e^{0.5t}}\, du$$

$$\begin{aligned} S(t) &= \int \frac{3{,}000e^{0.5t}}{3 + e^{0.5t}}\, dt \\ &= \int \frac{3{,}000e^{0.5t}}{u} \cdot \frac{1}{0.5e^{0.5t}}\, du \\ &= \frac{3{,}000}{0.5} \int \frac{1}{u}\, du \\ &= 6{,}000 \ln|u| + C = 6{,}000 \ln(3 + e^{0.5t}) + C. \end{aligned}$$

(Why could we drop the absolute value in the last step?)

Now what is C? Because $S(t)$ represents the total amount of bottled spring water for pets sold *since time* $t = 0$, we have $S(0) = 0$ (because that is when we started counting). Thus,

$$\begin{aligned} 0 &= 6{,}000 \ln(3 + e^{0.5(0)}) + C \\ &= 6{,}000 \ln 4 + C \\ C &= -6{,}000 \ln 4 \approx -8{,}318. \end{aligned}$$

Therefore, the total sales from the start of 2000 is approximately

$$S(t) = 6{,}000 \ln(3 + e^{0.5t}) - 8{,}318 \text{ million gallons.}$$

b. The period from the start of 2005 to the start of 2008 is represented by the interval $[5, 8]$. From part (a),

$$\begin{aligned} \text{Sales through the start of 2005} &= S(5) \\ &= 6{,}000 \ln(3 + e^{0.5(5)}) - 8{,}318 \approx 8{,}003 \text{ million gallons} \end{aligned}$$

$$\begin{aligned} \text{Sales through the start of 2008} &= S(8) \\ &= 6{,}000 \ln(3 + e^{0.5(8)}) - 8{,}318 \approx 16{,}003 \text{ million gallons.} \end{aligned}$$

Therefore, sales over the period were about $16{,}003 - 8{,}003 = 8{,}000$ million gallons.

➡ **Before we go on . . .** You might wonder why we are writing a logistic function in the form we used in Example 5 rather than in one of the standard forms $\frac{N}{1 + Ab^{-t}}$ or $\frac{N}{1 + Ae^{-kt}}$. Our only reason for doing this is to make the substitution work. To convert from the second standard form to the form we used in the example, multiply top and bottom by e^{kt}. (See Exercises 85 and 86 in Section 6.4 for further discussion.) ■

Shortcuts

The following rule allows us to simply write down the antiderivative in cases where we would otherwise need the substitution $u = ax + b$, as in Example 2. (a and b are constants with $a \neq 0$.)

Shortcut Rule: Integrals of Expressions Involving $(ax + b)$

If $\int f(x)\,dx = F(x) + C$ and a and b are constants, with $a \neq 0$, then*

$$\int f(ax+b)\,dx = \frac{1}{a}F(ax+b) + C.$$

* The shortcut rule can be justified by making the substitution $u = ax + b$ in the general case as stated.

Quick Example

(Also see the examples in the table below.)

1. Because $\int x^4\,dx = \frac{x^5}{5} + C$, it follows that

$$\int (3x-1)^4\,dx = \frac{1}{3}\frac{(3x-1)^5}{5} + C = \frac{(3x-1)^5}{15} + C.$$

Below are some instances of the shortcut rule with additional examples. (Their individual derivations using the substitution $u = ax + b$ will appear in the exercises.)

Shortcut Rule	Example
$\int (ax+b)^n\,dx = \frac{1}{a}\frac{(ax+b)^{n+1}}{n+1} + C$ (if $n \neq -1$)	$\int (3x-1)^2\,dx = \frac{(3x-1)^3}{3(3)} + C = \frac{(3x-1)^3}{9} + C$
$\int (ax+b)^{-1}\,dx = \frac{1}{a}\ln\|ax+b\| + C$	$\int (3-2x)^{-1}\,dx = \frac{1}{(-2)}\ln\|3-2x\| + C = -\frac{1}{2}\ln\|3-2x\| + C$
$\int e^{ax+b}\,dx = \frac{1}{a}e^{ax+b} + C$	$\int e^{-x+4}\,dx = \frac{1}{(-1)}e^{-x+4} + C = -e^{-x+4} + C$
$\int c^{ax+b}\,dx = \frac{1}{a\ln c}c^{ax+b} + C$	$\int 2^{-3x+4}\,dx = \frac{1}{(-3\ln 2)}2^{-3x+4} + C = -\frac{1}{3\ln 2}2^{-3x+4} + C$
$\int \|ax+b\|\,dx = \frac{1}{2a}(ax+b)\|ax+b\| + C$	$\int \|2x-1\|\,dx = \frac{1}{4}(2x-1)\|2x-1\| + C$

* Mike Fuschetto, who was a Hofstra business calculus student in spring 2010

The following more general version of the shortcut rule was suggested by a student.* We have marked it as optional so that you could skip it on a first reading, but we strongly suggest that you come back to it afterward, as the rule will allow you to easily write down the answers in most of the exercises, as well as in almost all the examples of this section.

(Optional) Mike's Shortcut Rule: Integrals of More General Expressions

If $\int f(x)\,dx = F(x) + C$ and g and u are any differentiable functions of x, then

$$\int g \cdot f(u)\,dx = \frac{g}{u'} \cdot F(u) + C \quad \text{provided that } \frac{g}{u'} \text{ is constant.}$$

Quick Example

(Also see the examples in the table below.)

2. $\int x^4\,dx = \frac{x^5}{5} + C$, so

$$\int 5x^2(x^3 - 1)^4\,dx = \frac{5x^2}{3x^2} \cdot \frac{(x^3 - 1)^5}{5} + C$$
$$= \frac{5}{3} \cdot \frac{(x^3 - 1)^5}{5} + C = \frac{(x^3 - 1)^5}{3} + C.$$

Caution

If g/u' is not constant, then Mike's rule does not apply; for instance, the following calculation is *wrong*:

$$\int x(2x - 1)^2\,dx = \frac{x}{2} \cdot \frac{(2x - 1)^3}{3} + C.$$

✗ *WRONG!* because $\frac{x}{2}$ is not constant.

Here are some instances of Mike's rule with additional examples.

Shortcut Rule	Example
$\int g \cdot u^n\,dx = \frac{g}{u'} \frac{u^{n+1}}{n + 1} + C$ (if $n \neq -1$)	$\int 3x(x^2 - 1)^3\,dx = \frac{3x}{2x} \frac{(x^2 - 1)^4}{4} + C = \frac{3(x^2 - 1)^4}{8} + C$
$\int g \cdot u^{-1}\,dx = \frac{g}{u'} \ln\lvert u \rvert + C$	$\int e^x(3 - 2e^x)^{-1}\,dx = \frac{e^x}{-2e^x} \ln\lvert 3 - 2e^x \rvert + C = -\frac{1}{2} \ln\lvert 3 - 2e^x \rvert + C$
$\int g \cdot e^u\,dx = \frac{g}{u'} e^u + C$	$\int x^2 e^{-x^3+4}\,dx = \frac{x^2}{-3x^2} e^{-x^3+4} + C = -\frac{1}{3} e^{-x^3+4} + C$

(continued)

Shortcut Rule	Example
$\displaystyle\int g \cdot c^u\,dx = \frac{g}{u' \ln c} c^u + C$	$\displaystyle\int x^3 2^{x^4-1}\,dx = \frac{x^3}{4x^3 \ln 2} 2^{x^4-1} + C = \frac{1}{4 \ln 2} 2^{x^4-1} + C$
$\displaystyle\int g \cdot \lvert u \rvert\,dx = \frac{g}{2u'} u \lvert u \rvert + C$	$\displaystyle\int x\lvert x^2 - 1\rvert\,dx = \frac{x}{4x}(x^2-1)\lvert x^2 - 1\rvert + C = \frac{1}{4}(x^2-1)\lvert x^2-1\rvert + C$

FAQs

When to Use Substitution and What to Use for *u*

Q: *If I am asked to calculate an antiderivative, how do I know when to use a substitution and when not to use one?*

A: Do *not* use substitution when integrating sums, differences, and/or constant multiples of powers of *x* and exponential functions, such as $2x^3 - \frac{4}{x^2} + \frac{1}{2x} + 3^x + \frac{2^x}{3}$.

To recognize when you should try a substitution, pretend that you are *differentiating* the given expression instead of integrating it. If differentiating the expression would require use of the chain rule, then integrating that expression may well require a substitution, as in, say, $x(3x^2 - 4)^3$ or $(x + 1)e^{x^2+2x-1}$. (In the first we have a *quantity* cubed; in the second we have *e* raised to a *quantity*.)

Q: *If an integral seems to call for a substitution, what should I use for u?*

A: There are no set rules for deciding what to use for *u*, but the preceding examples show some common patterns:

- If you see a linear expression raised to a power, try setting *u* equal to that linear expression. For example, in $(3x - 2)^{-3}$, set $u = 3x - 2$. (Alternatively, try using the shortcuts above.)
- If you see a constant raised to a linear expression, try setting *u* equal to that linear expression. For example, in $3^{(2x+1)}$, set $u = 2x + 1$. (Alternatively, try a shortcut.)
- If you see an expression raised to a power multiplied by the derivative of that expression (or a constant multiple of the derivative), try setting *u* equal to that expression. For example, in $x^2(3x^3 - 4)^{-1}$, set $u = 3x^3 - 4$.
- If you see a constant raised to an expression, multiplied by the derivative of that expression (or a constant multiple of its derivative), try setting *u* equal to that expression. For example, in $5(x + 1)e^{x^2+2x-1}$, set $u = x^2 + 2x - 1$.
- If you see an expression in the denominator and its derivative (or a constant multiple of its derivative) in the numerator, try setting *u* equal to that expression. For example, in $\frac{2^{3x}}{3 - 2^{3x}}$, set $u = 3 - 2^{3x}$.

Persistence often pays off: If a certain substitution does not work, try another approach or a different substitution.

6.2 EXERCISES

▼ more advanced ◆ challenging

T indicates exercises that should be solved using technology

In Exercises 1–10, evaluate the given integral using the substitution (or method) indicated.

1. $\int (3x - 5)^3\, dx;\ u = 3x - 5$

2. $\int (2x + 5)^{-2}\, dx;\ u = 2x + 5$

3. $\int (3x - 5)^3\, dx$; shortcut

4. $\int (2x + 5)^{-2}\, dx$; shortcut

5. $\int e^{-x}\, dx;\ u = -x$

6. $\int e^{x/2}\, dx;\ u = x/2$

7. $\int e^{-x}\, dx$; shortcut

8. $\int e^{x/2}\, dx$; shortcut

9. $\int (x + 1)e^{(x+1)^2}\, dx;\ u = (x + 1)^2$

10. $\int (x - 1)^2 e^{(x-1)^3}\, dx;\ u = (x - 1)^3$

In Exercises 11–52, decide on what substitution to use, and then evaluate the given integral using a substitution. [**HINT:** See the FAQ at the end of the section for advice on deciding on u, and the examples for the mechanics of doing the substitution.]

11. $\int (3x + 1)^5\, dx$

12. $\int (-x - 1)^7\, dx$

13. $\int 7.2\sqrt{3x - 4}\, dx$

14. $\int 4.4e^{(-3x+4)}\, dx$

15. $\int 1.2e^{(0.6x+2)}\, dx$

16. $\int 8.1\sqrt{-3x + 4}\, dx$

17. $\int x(3x^2 + 3)^3\, dx$

18. $\int x(-x^2 - 1)^3\, dx$

19. $\int 2x\sqrt{3x^2 - 1}\, dx$

20. $\int 3x\sqrt{-x^2 + 1}\, dx$

21. $\int \frac{x}{(x^2 + 1)^{1.3}}\, dx$

22. $\int \frac{x^2}{(1 + x^3)^{1.4}}\, dx$

23. $\int x|4x^2 - 1|\, dx$

24. $\int x^2|4x^3 + 1|\, dx$

25. $\int (1 + 9.3e^{3.1x-2})\, dx$

26. $\int (3.2 - 4e^{1.2x-3})\, dx$

27. $\int xe^{-x^2+1}\, dx$

28. $\int xe^{2x^2-1}\, dx$

29. $\int (x + 1)e^{-(x^2+2x)}\, dx$ [**HINT:** See Example 2(b).]

30. $\int (2x - 1)e^{2x^2-2x}\, dx$ [**HINT:** See Example 2(b).]

31. $\int \frac{-2x - 1}{(x^2 + x + 1)^3}\, dx$

32. $\int \frac{x^3 - x^2}{3x^4 - 4x^3}\, dx$

33. $\int \frac{x^2 + x^5}{\sqrt{2x^3 + x^6 - 5}}\, dx$ [**HINT:** See Example 3.]

34. $\int \frac{2(x^3 - x^4)}{(5x^4 - 4x^5)^5}\, dx$ [**HINT:** See Example 3.]

35. $\int x(x - 2)^5\, dx$ [**HINT:** See Example 4.]

36. $\int x(x - 2)^{1/3}\, dx$ [**HINT:** See Example 4.]

37. $\int 2x\sqrt{x + 1}\, dx$

38. $\int \frac{x}{\sqrt{x + 1}}\, dx$

39. $\int \frac{e^{-0.05x}}{1 - e^{-0.05x}}\, dx$ [**HINT:** See Example 5.]

40. $\int \frac{3e^{1.2x}}{2 + e^{1.2x}}\, dx$ [**HINT:** See Example 5.]

41. ▼ $\int \frac{3e^{-1/x}}{x^2}\, dx$

42. ▼ $\int \frac{2e^{2/x}}{x^2}\, dx$

43. ▼ $\int \frac{(4 + 1/x^2)^3}{x^3}\, dx$

44. ▼ $\int \frac{1}{x^2(2 - 1/x)}\, dx$

45. ▼ $\int \frac{e^x + e^{-x}}{2}\, dx$ [**HINT:** See Example 2(d).]

46. ▼ $\int (e^{x/2} + e^{-x/2})\, dx$ [**HINT:** See Example 2(d).]

47. ▼ $\int \frac{e^x - e^{-x}}{e^x + e^{-x}}\, dx$

48. ▼ $\int \frac{e^{x/2} + e^{-x/2}}{e^{x/2} - e^{-x/2}}\, dx$

49. ▼ $\int e^{3x-1}|1 - e^{3x-1}|\, dx$

50. ▼ $\int |e^{-x-1} - 1|(e^{-x-1})\, dx$

51. ▼ $\int \left((2x - 1)e^{2x^2-2x} + xe^{x^2}\right) dx$

52. ▼ $\int (xe^{-x^2+1} + e^{2x})\, dx$

In Exercises 53–56, derive the given equation, where a and b are constants with $a \neq 0$.

53. $\int (ax + b)^n\, dx = \frac{(ax + b)^{n+1}}{a(n + 1)} + C \quad (\text{if } n \neq -1)$

54. $\int (ax + b)^{-1}\, dx = \frac{1}{a}\ln|ax + b| + C$

55. $\int |ax + b|\, dx = \frac{1}{2a}(ax + b)|ax + b| + C$

56. $\int e^{ax+b}\, dx = \frac{1}{a}e^{ax+b} + C$

In Exercises 57–72, use the shortcut formulas (see the shortcuts and Exercises 53–56) to calculate the given integral.

57. $\int e^{-x}\,dx$ **58.** $\int e^{x-1}\,dx$

59. $\int e^{2x-1}\,dx$ **60.** $\int e^{-3x}\,dx$

61. $\int (2x+4)^2\,dx$ **62.** $\int (3x-2)^4\,dx$

63. $\int \frac{1}{5x-1}\,dx$ **64.** $\int (x-1)^{-1}\,dx$

65. $\int (1.5x)^3\,dx$ **66.** $\int e^{2.1x}\,dx$

67. $\int 1.5^{3x}\,dx$ **68.** $\int 4^{-2x}\,dx$

69. $\int |2x+4|\,dx$ **70.** $\int |3x-2|\,dx$

71. $\int (2^{3x+4}+2^{-3x+4})\,dx$ **72.** $\int (1.1^{-x+4}+1.1^{x+4})\,dx$

73. Find $f(x)$ if $f(0)=0$ and the tangent line at $(x, f(x))$ has slope $x(x^2+1)^3$.

74. Find $f(x)$ if $f(1)=0$ and the tangent line at $(x, f(x))$ has slope $\frac{x}{x^2+1}$.

75. Find $f(x)$ if $f(1)=1/2$ and the tangent line at $(x, f(x))$ has slope xe^{x^2-1}.

76. Find $f(x)$ if $f(2)=1$ and the tangent line at x has slope $(x-1)e^{x^2-2x}$.

In Exercises 77–84, use Mike's shortcut method (see Mike's Shortcut Rule and the examples that follow) to calculate the given integral.

77. $\int x(5x^2-3)^6\,dx$ **78.** $\int \frac{x}{(5x^2-3)^6}\,dx$

79. $\int \frac{e^x}{\sqrt{3e^x-1}}\,dx$ **80.** $\int e^x\sqrt{1+2e^x}\,dx$

81. $\int x^3 e^{(x^4-8)}\,dx$ **82.** $\int \frac{x^3}{e^{(x^4-8)}}\,dx$

83. $\int \frac{e^{3x}}{1+2e^{3x}}\,dx$ **84.** $\int \frac{e^{-2x}}{e^{-2x}-3}\,dx$

Applications

85. ***Cost*** The marginal cost of producing the xth roll of film is given by $5+1/(x+1)^2$. The total cost to produce one roll is \$1,000. Find the total cost function $C(x)$.

86. ***Cost*** The marginal cost of producing the xth box of CDs is given by $10-x/(x^2+1)^2$. The total cost to produce two boxes is \$1,000. Find the total cost function $C(x)$.

87. ***Economic Growth*** The Mexico GDP (total monetary value of all finished goods and services produced in Mexico) can be approximated by

$$g(t) = 2{,}000 - 480e^{-0.06t} \text{ billion pesos per year} \quad (0 \le t \le 5),$$

where t is time in years since January 2010.[13] Find an expression for the total GDP $G(t)$ of sold goods in Mexico from January 2010 to time t. Hence, estimate, to the nearest billion pesos, the total Mexico GDP from January 2010 through June 2014. (The actual value was 7,137 billion pesos.) [**HINT**: Use the shortcuts.]

88. ***Housing Starts: The Great Recession*** The Great Recession of 2007–2009 is largely attributed to the real estate crisis beginning in 2006, from which time the number of housing starts was approximately

$$n(t) = 2{,}400e^{-0.25t} - 200 \text{ thousand homes per year} \quad (0 \le t \le 4),$$

where t is time in years since January 2006.[14] Find an expression for the total number $N(t)$ of housing starts in the United States from January 2006 to time t. Hence estimate, to the nearest 0.1 million, the total number of housing starts from January 2006 through June 2009. (The actual number was around 4.9 million homes.) [**HINT**: Use the shortcuts.]

89. ***Revenue: Pacific Sunwear*** The annual revenue of Pacific Sunwear of California over the period January 2008–January 2015 can be approximated by

$$p(t) = (-0.075t + 0.97)^5 + 0.75 \text{ billion dollars per year} \quad (0 \le t \le 7),$$

where t is time in years since January 2008.[15] Find an expression for the total revenue $P(t)$ earned by Pacific Sunwear since the start of 2008, and hence estimate, to the nearest \$0.1 billion, the total revenue earned from the start of 2008 to the start of 2015. (The actual revenue was about \$7.1 billion.)

90. ***Revenue: Google*** The annual revenue of Google Inc. over the period January 2008–January 2015 can be approximated by

$$g(t) = (0.17t + 1.7)^4 + 10 \text{ billion dollars per year} \quad (0 \le t \le 7),$$

where t is time in years since January 2008.[16] Find an expression for the total revenue $G(t)$ earned by Google since the start of 2008, and hence estimate, to the nearest \$ billion, the total revenue earned from the start of 2008 to the start of 2014. (The actual revenue was about \$223 billion.)

[13] The GDP is in constant 2008 pesos. Source for data: Instituto Nacional de Estadística y Geografía (INEGI) (www.inegi.org.mx).

[14] Source for data: www.census.gov.

[15] Source for data: www.wikinvest.com.

[16] *Ibid.*

91. ***Scientific Research: 1983–2003*** The number of research articles in the prominent journal *Physical Review* written by researchers in Europe during 1983–2003 can be approximated by

$$E(t) = \frac{7e^{0.2t}}{5 + e^{0.2t}} \text{ thousand articles per year} \quad (0 \le t \le 20),$$

where t is time in years. ($t = 0$ represents 1983.)[17]

a. Find an (approximate) expression for the total number of articles written by researchers in Europe since 1983 ($t = 0$). [**HINT**: See Example 5.]

b. Roughly how many articles were written by researchers in Europe from 1983 to 2003? (Round your answer to the nearest 1,000 articles.)

92. ***Scientific Research: 1983–2003*** The number of research articles in the prominent journal *Physical Review* written by researchers in the United States during 1983–2003 can be approximated by

$$U(t) = \frac{4.6e^{0.6t}}{0.4 + e^{0.6t}} \text{ thousand articles per year,} \quad (0 \le t \le 20),$$

where t is time in years. ($t = 0$ represents 1983.)[18]

a. Find an (approximate) expression for the total number of articles written by researchers in the United States since 1983 ($t = 0$). [**HINT**: See Example 5.]

b. Roughly how many articles were written by researchers in the United States from 1983 to 2003?

93. ***Sales*** The rate of sales of your company's *Jackson Pollock Advanced Paint-by-Number* sets can be modeled by

$$s(t) = \frac{900e^{0.25t}}{3 + e^{0.25t}} \text{ sets per month}$$

t months after their introduction. Find an expression for the total number of paint-by-number sets $S(t)$ sold t months after their introduction, and use it to estimate the total sold in the first 12 months. [**HINT**: See Example 5.]

94. ***Sales*** The rate of sales of your company's *Jackson Pollock Beginners Paint-by-Number* sets can be modeled by

$$s(t) = \frac{1{,}800e^{0.75t}}{10 + e^{0.75t}} \text{ sets per month}$$

t months after their introduction. Find an expression for the total number of paint-by-number sets $S(t)$ sold t months after their introduction, and use it to estimate the total sold in the first 12 months. [**HINT**: See Example 5.]

95. ***Motion in a Straight Line*** The velocity of a particle moving in a straight line is given by $v = t(t^2 + 1)^4 + t$.

a. Find an expression for the position s after a time t. [**HINT**: See Example 2(d).]

b. Given that $s = 1$ at time $t = 0$, find the constant of integration C and hence an expression for s in terms of t without any unknown constants.

96. ***Motion in a Straight Line*** The velocity of a particle moving in a straight line is given by $v = 3te^{t^2} + t$.

a. Find an expression for the position s after a time t. [**HINT**: See Example 2(d).]

b. Given that $s = 3$ at time $t = 0$, find the constant of integration C and hence an expression for s in terms of t without any unknown constants.

97. ***Bottled Water Sales*** (Compare Exercise 55 in Section 6.1.) The rate of U.S. sales of bottled water for the period 2007–2014 could be approximated by

$$s(t) = 0.08(t - 2{,}007)^2 - 0.26(t - 2{,}007) + 8.8$$
$$\text{million gallons per year} \quad (2{,}007 \le t \le 2{,}014),$$

where t is the year.[19] Use an indefinite integral to approximate the total sales $S(t)$ of bottled water since 2010 ($t = 2{,}010$). Approximately how much bottled water was sold from 2010 to 2014?

98. ***Bottled Water Sales*** (Compare Exercise 56 in Section 6.1.) The rate of U.S. per capita sales of bottled water for the period 2007–2014 could be approximated by

$$s(t) = 0.25(t - 2{,}007)^2 - (t - 2{,}007) + 29$$
$$\text{gallons per year} \quad (2{,}007 \le t \le 2{,}014),$$

where t is the year.[20] Use an indefinite integral to approximate the total per capita sales $S(t)$ of bottled water since 2010 ($t = 2{,}010$). Approximately how much bottled water was sold, per capita, 2010 to 2013?

Communication and Reasoning Exercises

99. Are there any circumstances in which you should use the substitution $u = x$? Illustrate your answer by giving an example that shows the effect of this substitution.

100. You are asked to calculate $\int \frac{u}{u^2 + 1}\, du$. What is wrong with the substitution $u = u^2 + 1$?

101. If the xs do not cancel in a substitution, that means you chose the wrong expression for u, right?

102. At what stage of a calculation using a u substitution should you substitute back for u in terms of x: before or after taking the antiderivative?

---

[17] Source: The American Physical Society/*New York Times*, May 3, 2003, p. A1.

[18] *Ibid.*

[19] Source for data: Beverage Marketing Corporation/www.bottledwater.org.

[20] *Ibid.*

103. Consider $\int\left(\frac{x}{x^2-1}+\frac{3x}{x^2+1}\right)dx$. To compute it, you should use which of the following?

(A) $u = x^2 - 1$ **(B)** $u = x^2 + 1$

(C) Neither **(D)** Both

Explain your answer.

104. If the substitution $u = x^2 - 1$ works in $\int \frac{x}{x^2-1}dx$, why does it not work nearly so easily in $\int \frac{x^2-1}{x}dx$? How would you do the second integral most simply?

105. Why was the following calculation marked wrong? What is the correct answer?

$u = x^2 - 1$ $\frac{du}{dx} = 2x$ $dx = \frac{1}{2x}du$	$\int 3x(x^2-1) = \int 3xu$ $= 3x\frac{u^2}{2} + C = 3x\frac{(x^2-1)^2}{2} + C$ ✗ *WRONG!*

106. Why was the following calculation marked wrong? What is the correct answer?

$u = x^3 - 1$ $\frac{du}{dx} = 3x^2$ $dx = \frac{1}{3x^2}du$	$\int x^2(x^3-1)^2\,dx = \int x^2u^2\frac{1}{3x^2}du$ $= \frac{1}{3}u^2 + C = \frac{1}{3}(x^3-1)^2 + C$ ✗ *WRONG!*

107. Why was the following calculation marked wrong? What is the correct answer?

$u = x^3 - 1$ $\frac{du}{dx} = 3x^2$ $dx = \frac{1}{3x^2}du$	$\int x^2(x^3-1)\,dx = \int x^2u\frac{1}{3x^2}du$ $= \int \frac{1}{3}u\,du = \int \frac{1}{3}(x^3-1)\,dx$ $= \frac{1}{3}\left(\frac{x^4}{4} - x\right) + C$ ✗ *WRONG!*

108. Why was the following calculation marked wrong? What is the correct answer?

$u = x^2 - 1$ $\frac{du}{dx} = 2x$ $dx = \frac{1}{2x}du$	$\int \frac{x}{x^2-1}dx = \int \frac{x}{u}du$ $= x\int \frac{1}{u}du = x\ln\lvert u\rvert + C$ $= x\ln\lvert x^2-1\rvert + C$ ✗ *WRONG!*

109. ▼ Show that *none* of the following substitutions work for $\int e^{-x^2}dx$: $u = -x$, $u = x^2$, $u = -x^2$. (The antiderivative of e^{-x^2} involves the *error function* erf(*x*).)

110. ▼ Show that *none* of the following substitutions work for $\int \sqrt{1-x^2}\,dx$: $u = 1 - x^2$, $u = x^2$, and $u = -x^2$. (The antiderivative of $\sqrt{1-x^2}$ involves inverse trigonometric functions, discussion of which is beyond the scope of this book.)

6.3 The Definite Integral: Numerical and Graphical Viewpoints

Riemann Sums

In Sections 6.1 and 6.2 we discussed the indefinite integral. There is an older, related concept called the **definite integral**. Let's introduce this new idea with an example. (We'll drop hints now and then about how the two types of integral are related. In Section 6.4 we discuss the exact relationship, which is one of the most important results in calculus.)

In Section 6.1 we used antiderivatives to answer questions of the form "Given the marginal cost, compute the total cost." (See Example 5 in Section 6.1.) In this section we approach such questions more directly, and we will forget about antiderivatives for now.

EXAMPLE 1 Oil Spill

Your deep ocean oil rig has suffered a catastrophic failure, and oil is leaking from the ocean floor wellhead at a rate of

$$v(t) = 0.08t^2 - 4t + 60 \text{ thousand barrels per day} \qquad (0 \le t \le 20),$$

where t is time in days since the failure.[21] Use a numerical calculation to estimate the total volume of oil released during the first 20 days.

Solution The graph of $v(t)$ is shown in Figure 3.

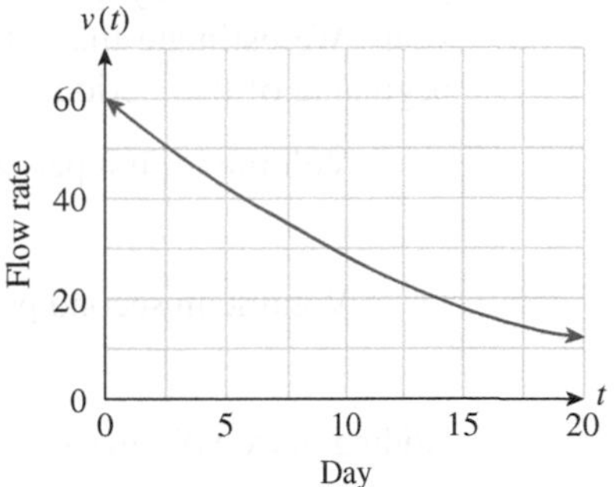

Figure 3

Let's start with a very crude estimate of the total volume of oil released, using the graph as a guide. The rate of change of this total volume at the beginning of the time period is $v(0) = 60$ thousand barrels per day. If this rate were to remain constant for the entire 20-day period, the total volume of oil released would be

$$\begin{aligned}\text{Total volume} &= \text{Volume per day} \times \text{Number of days} = 60 \times 20\\ &= 1{,}200 \text{ thousand barrels.}\end{aligned}$$

Figure 4 shows how we can represent this calculation on the graph of $v(t)$.

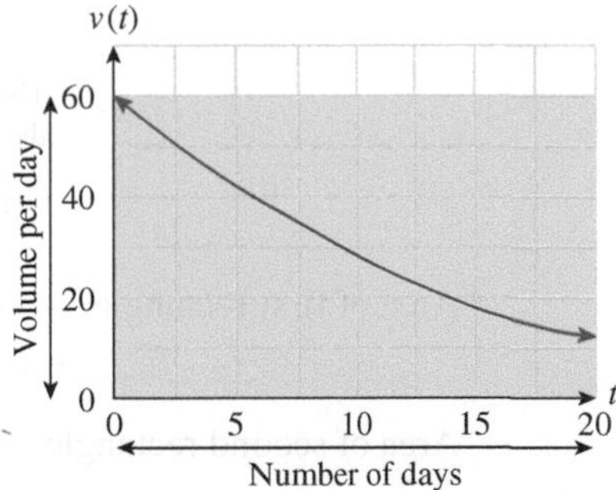

Figure 4

[21] The model is consistent with the order of magnitude of the BP Deepwater Horizon oil spill of April 20–June 15, 2010, when the rate of flow of oil was estimated by the Federal Emergency Management Agency's Flow Rate Technical Group to be between 35,000 and 60,000 barrels per day. (One barrel of oil is equivalent to about 0.16 cubic meters.) Source: www.doi.gov/deepwaterhorizon.

The volume per day based on $v(0) = 60$ is represented by the y-coordinate of the graph at its left edge, while the number of days is represented by the width of the interval $[0, 20]$ on the x-axis. Therefore, computing the area of the shaded rectangle in the figure gives the same calculation:

$$\begin{aligned}\text{Area of rectangle} &= \text{Volume per day} \times \text{Number of days}\\ &= 60 \times 20 = 1{,}200 = \text{Total volume.}\end{aligned}$$

But as we see in the graph, the flow rate does not remain constant but goes down quite significantly over the course of the 20-day interval. We can obtain a somewhat more accurate estimate of the total volume by re-estimating the volume using 10-day periods—that is, by dividing the interval $[0, 20]$ into two equal intervals, or subdivisions. We estimate the volume over each 10-day period using the flow rate at the beginning of that period:

$$\begin{aligned}\text{Volume in first period} &= \text{Volume per day} \times \text{Number of days}\\ &= v(0) \times 10 = 60 \times 10 = 600 \text{ thousand barrels}\end{aligned}$$

$$\begin{aligned}\text{Volume in second period} &= \text{Volume per day} \times \text{Number of days}\\ &= v(10) \times 10 = 28 \times 10 = 280 \text{ thousand barrels.}\end{aligned}$$

Adding these volumes gives us the more accurate estimate

$$v(0) \times 10 + v(10) \times 10 = 880 \text{ thousand barrels.}$$ Calculation using two subdivisions

In Figure 5 we see that we are computing the combined area of two rectangles, each of whose heights is determined by the height of the graph at its left edge.

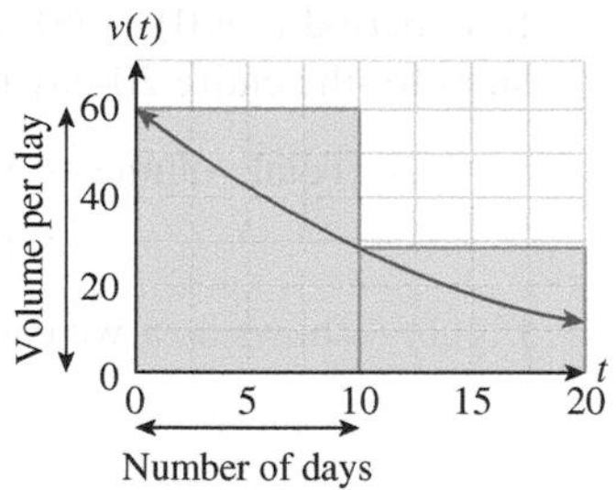

The areas of the rectangles are estimates of the volumes for successive 10-day periods.

Figure 5

$$\begin{aligned}\text{Area of first rectangle} &= \text{Volume per day} \times \text{Number of days}\\ &= v(0) \times 10 = 60 \times 10 = 600 = \text{Volume for first 10 days}\end{aligned}$$

$$\begin{aligned}\text{Area of second rectangle} &= \text{Volume per day} \times \text{Number of days} = v(10) \times 10\\ &= 28 \times 10 = 280 = \text{Volume for second 10 days.}\end{aligned}$$

We can get an even better estimate of the volume by using four divisions of $[0, 20]$ instead of two:

$$\begin{aligned}&v(0) \times 5 + v(5) \times 5 + v(10) \times 5 + v(15) \times 5\\ &\quad = 300 + 210 + 140 + 90 = 740 \text{ thousand barrels.}\end{aligned}$$ Calculation using four subdivisions

As we see in Figure 6, we have now computed the combined area of *four* rectangles, each of whose heights is again determined by the height of the graph at its left edge.

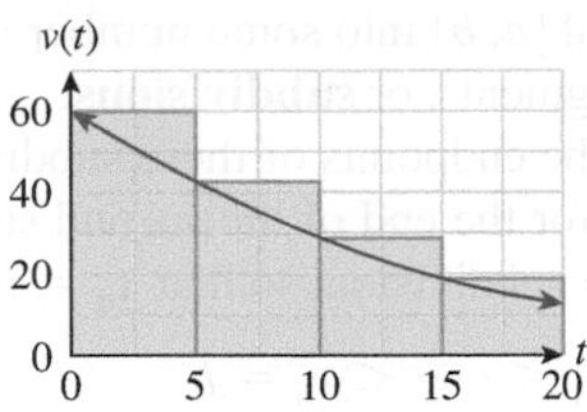

Estimated volume using four subdivisions. The areas of the rectangles are estimates of the volumes for successive 5-day periods.

Figure 6

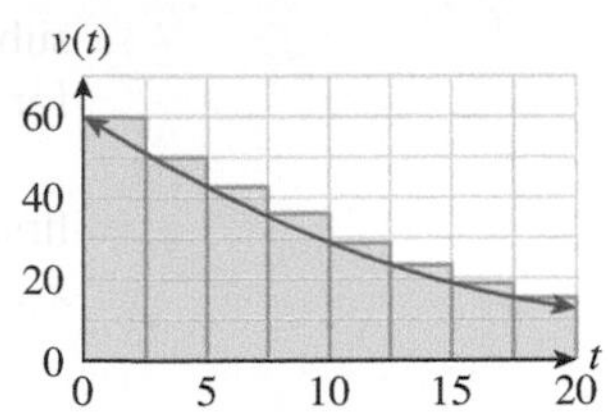

Estimated volume using eight subdivisions. The areas of the rectangles are estimates of the volumes for successive 2.5-day periods.

Figure 7

Notice how the volume seems to be decreasing as we use more subdivisions. More importantly, the total volume seems to be getting closer to the area under the graph. Figure 7 illustrates the calculation for eight equal subdivisions. The approximate total volume using eight subdivisions is the total area of the shaded region in Figure 7:

$$v(0) \times 2.5 + v(2.5) \times 2.5 + v(5) \times 2.5 + \cdots + v(17.5) \times 2.5$$
$$= 675 \text{ thousand barrels.}$$

Calculation using eight subdivisions

Looking at Figure 7, we still get the impression that we are overestimating the volume. If we want to be *really* accurate in our estimation of the volume, we should really be calculating the volume *continuously* every few hours or, better yet, minute by minute, as illustrated in Figure 8.

Using Technology

Website
www.WanerMath.com
At the Website, select the Online Utilities tab, and choose the Numerical Integration Utility and Grapher. There, enter the formula

```
0.08x^2-4x+60
```

for $f(x)$, enter 0 and 20 for the lower and upper limits, and press "Left Sum" to compute and see a Riemann sum. You can then adjust the number of subdivisions to see the effect on the graph and on the calculation and thus replicate the information in Figure 8.

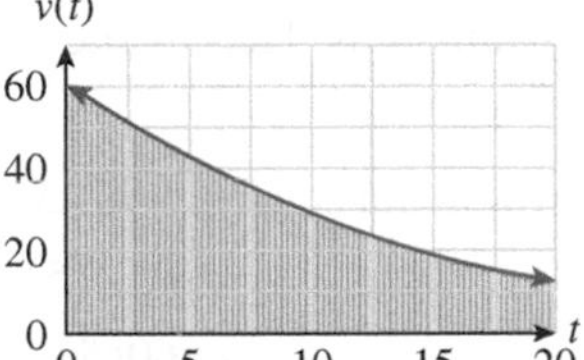

Every 6 hours (80 subdivisions)
Volume ≈ 619.35 thousand barrels

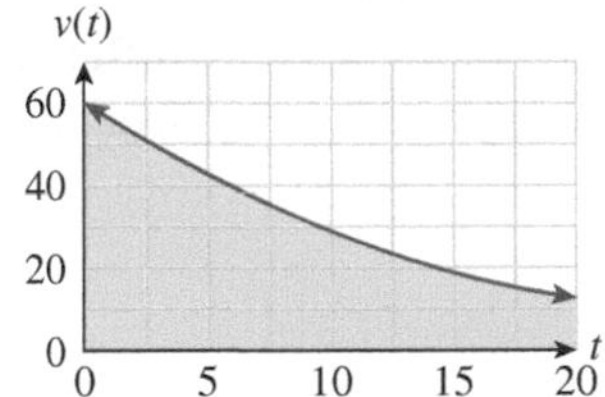

Every minute (28,800 subdivisions)
Volume ≈ 613.35 thousand barrels

Figure 8

Figure 8 strongly suggests that the more accurately we estimate the total volume, the closer the answer gets to the exact area under the portion of the graph of $v(t)$ with $0 \le t \le 20$ and leads us to the conclusion that the *exact* total volume is the exact area under the rate of change of volume curve for $0 \le t \le 20$. In other words, we have made the following remarkable discovery:

Total volume is the area under the rate of change of volume curve!

➡ **Before we go on . . .** The 80-subdivision calculation in Example 1 is tedious to do by hand, and no one in his or her right mind would even *attempt* to do the minute-by-minute calculation by hand! Below we discuss ways of doing these calculations with the aid of technology. ■

The type of calculation done in Example 1 is useful in many applications. Let's look at the general case and give the result a name.

In general, we have a function f (such as the function v in the example), and we consider an interval $[a, b]$ of possible values of the independent variable x. We subdivide the interval $[a, b]$ into some number of segments of equal length. Write n for the number of segments, or **subdivisions**.

Next, we label the endpoints of these subdivisions x_0 for a, x_1 for the end of the first subdivision, x_2 for the end of the second subdivision, and so on until we get to x_n, the end of the nth subdivision, so that $x_n = b$. Thus,

$$a = x_0 < x_1 < \cdots < x_n = b.$$

The first subdivision is the interval $[x_0, x_1]$, the second subdivision is $[x_1, x_2]$, and so on until we get to the last subdivision, which is $[x_{n-1}, x_n]$. We are dividing the interval $[a, b]$ into n subdivisions of equal length, so each segment has length $(b - a)/n$. We write Δx for $(b - a)/n$ (Figure 9).

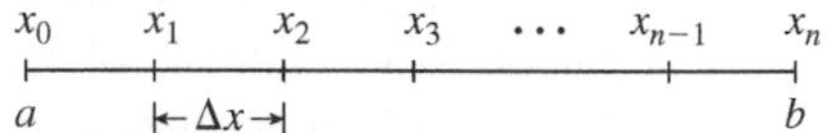

Figure 9

Having established this notation, we can write the calculation that we want to do as follows: For each subdivision $[x_{k-1}, x_k]$, compute $f(x_{k-1})$, the value of the function f at the left endpoint. Multiply this value by the length of the interval, which is Δx. Then add together all n of these products to get the number

$$f(x_0)\,\Delta x + f(x_1)\,\Delta x + \cdots + f(x_{n-1})\,\Delta x.$$

* After Georg Friedrich Bernhard Riemann (1826–1866)

This sum is called a **(left) Riemann* sum** for f. In Example 1 we computed several different Riemann sums. Here is the computation for $n = 4$ we used in the oil spill example (see Figure 10):

$$\begin{aligned}\text{Left Riemann sum} &= f(x_0)\,\Delta x + f(x_1)\,\Delta x + \cdots + f(x_{n-1})\,\Delta x \\ &= f(0)(5) + f(5)(5) + f(10)(5) + f(15)(5) \\ &= 60(5) + 42(5) + 28(5) + 18(5) = 740.\end{aligned}$$

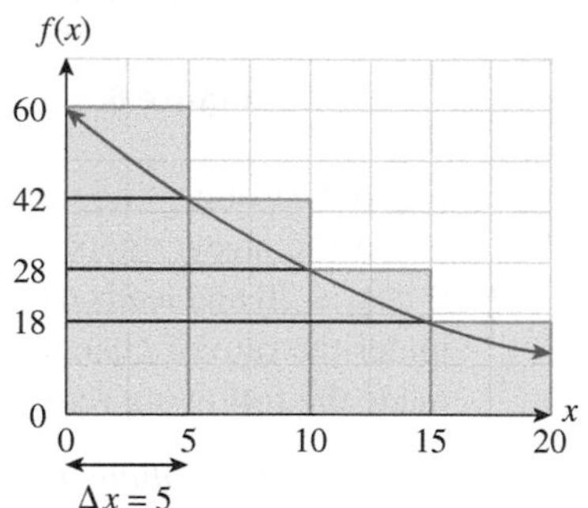

Figure 10

Because sums are often used in mathematics, mathematicians have developed a shorthand notation for them. We write

$$f(x_0)\,\Delta x + f(x_1)\,\Delta x + \cdots + f(x_{n-1})\,\Delta x \quad \text{as} \quad \sum_{k=0}^{n-1} f(x_k)\,\Delta x.$$

The symbol Σ is the Greek letter sigma and stands for **summation**. The letter k here is called the index of summation, and we can think of it as counting off the segments.

We read the notation as "the sum from $k = 0$ to $n - 1$ of the quantities $f(x_k)\,\Delta x$." Think of it as a set of instructions:

Set $k = 0$, and calculate $f(x_0)\,\Delta x$. $\quad f(0)(5)$ in the above calculation

Set $k = 1$, and calculate $f(x_1)\,\Delta x$. $\quad f(5)(5)$ in the above calculation

$\vdots$

Set $k = n - 1$, and calculate $f(x_{n-1})\,\Delta x$. $\quad f(15)(5)$ in the above calculation

Then sum all the quantities so calculated.

Riemann Sum

If f is a continuous function, the **left Riemann sum** with n equal subdivisions for f over the interval $[a, b]$ is defined to be

$$\begin{aligned}\text{Left Riemann sum} &= \sum_{k=0}^{n-1} f(x_k)\,\Delta x \\ &= f(x_0)\,\Delta x + f(x_1)\,\Delta x + \cdots + f(x_{n-1})\,\Delta x \\ &= [f(x_0) + f(x_1) + \cdots + f(x_{n-1})]\,\Delta x,\end{aligned}$$

where $a = x_0 < x_1 < \cdots < x_n = b$ are the endpoints of the subdivisions, and $\Delta x = (b - a)/n$.

Interpretation of the Riemann Sum

If f is the rate of change of a quantity F (that is, $f = F'$), then the Riemann sum of f approximates the total change of F from $x = a$ to $x = b$. The approximation improves as the number of subdivisions increases toward infinity.

Quick Examples

1. If $f(t)$ is the rate of change in the number of bats in a belfry and $[a, b] = [2, 3]$, then the Riemann sum approximates the total change in the number of bats in the belfry from time $t = 2$ to time $t = 3$.
2. If $c(x)$ is the marginal cost of producing the xth item and $[a, b] = [10, 20]$, then the Riemann sum approximates the cost of producing items 11 through 20.

Visualizing a Left Riemann Sum (Nonnegative Function)

Graphically, we can represent a left Riemann sum of a nonnegative function as an approximation of the area under a curve:

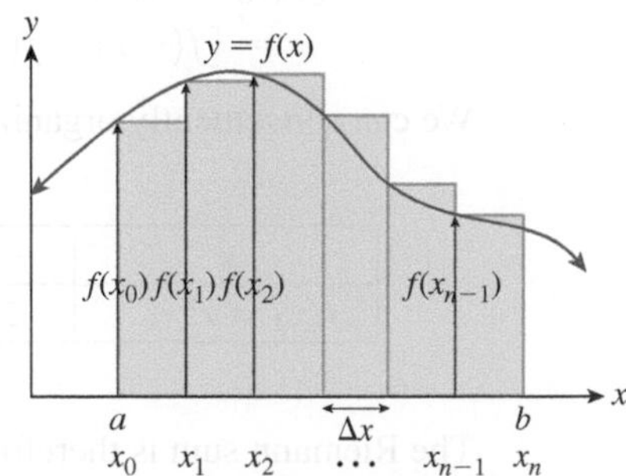

Riemann sum = Shaded area = Area of first rectangle + Area of second rectangle + $\cdots$ + Area of nth rectangle $= f(x_0)\,\Delta x + f(x_1)\,\Delta x + f(x_2)\,\Delta x + \cdots + f(x_{n-1})\,\Delta x$.

Quick Example

3. In Example 1 we computed several Riemann sums, including these:

$$\begin{aligned}
n = 1:&\quad \text{Riemann sum} = v(0)\,\Delta t = 60 \times 20 = 1{,}200\\
n = 2:&\quad \text{Riemann sum} = [v(t_0) + v(t_1)]\,\Delta t\\
&\qquad = [v(0) + v(10)](10) = 880\\
n = 4:&\quad \text{Riemann sum} = [v(t_0) + v(t_1) + v(t_2) + v(t_3)]\,\Delta t\\
&\qquad = [v(0) + v(5) + v(10) + v(15)](5) = 740\\
n = 8:&\quad \text{Riemann sum} = [v(t_0) + v(t_1) + \cdots + v(t_7)]\,\Delta t\\
&\qquad = [v(0) + v(2.5) + \cdots + v(17.5)](2.5) = 675.
\end{aligned}$$

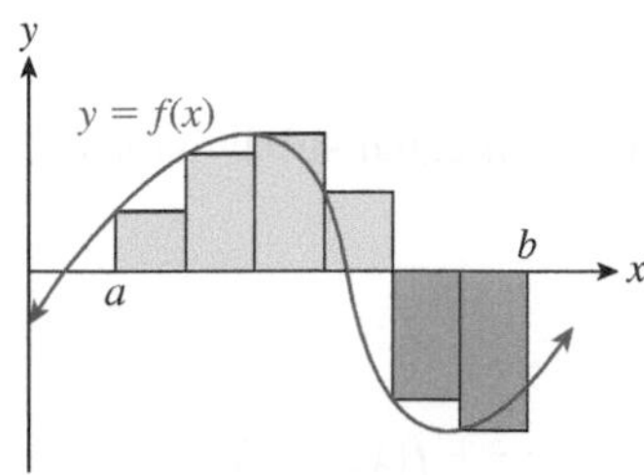

Riemann sum = Area above x-axis − Area below x-axis

Figure 11

Note To visualize the Riemann sum of a function that is negative, look again at the formula $f(x_0)\,\Delta x + f(x_1)\,\Delta x + f(x_2)\,\Delta x + \cdots + f(x_{n-1})\,\Delta x$ for the Riemann sum. Each term $f(x_k)\,\Delta x_k$ represents the area of one rectangle in the figure above. So the areas of the rectangles with negative values of $f(x_k)$ are automatically counted as negative. They appear as red rectangles in Figure 11. ■

Calculating Riemann Sums

EXAMPLE 2 Calculating a Riemann Sum from a Formula

Compute the left Riemann sum for $f(x) = x^2 + 1$ over the interval $[-1, 1]$, using $n = 5$ subdivisions.

Solution Because the interval is $[a, b] = [-1, 1]$ and $n = 5$, we have

$$\Delta x = \frac{b - a}{n} = \frac{1 - (-1)}{5} = 0.4. \qquad \text{Width of subdivisions}$$

Thus, the subdivisions of $[-1, 1]$ are determined by

$$-1 < -0.6 < -0.2 < 0.2 < 0.6 < 1. \qquad \text{Start with } -1\text{, and keep adding } \Delta x = 0.4.$$

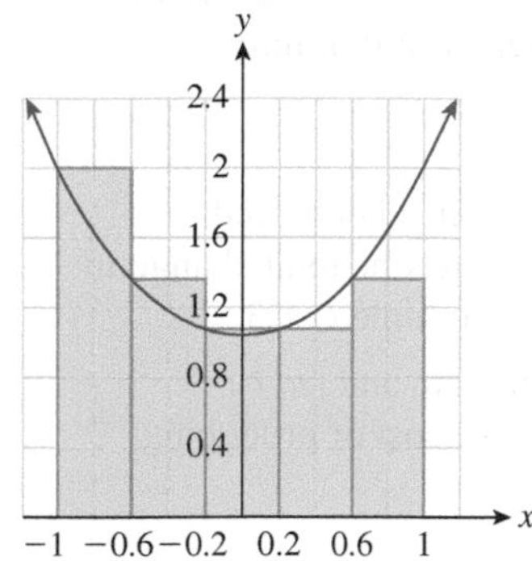

Figure 12

Figure 12 shows the graph with a representation of the Riemann sum.

The Riemann sum we want is

$$\begin{aligned}
&[f(x_0) + f(x_1) + \cdots + f(x_4)]\,\Delta x\\
&\quad = [f(-1) + f(-0.6) + f(-0.2) + f(0.2) + f(0.6)]0.4.
\end{aligned}$$

We can conveniently organize this calculation in a table as follows:

x	−1	−0.6	−0.2	0.2	0.6	**Total**
$f(x) = x^2 + 1$	2	1.36	1.04	1.04	1.36	6.8

The Riemann sum is therefore

$$6.8\,\Delta x = 6.8 \times 0.4 = 2.72.$$

EXAMPLE 3 Computing a Riemann Sum from a Graph

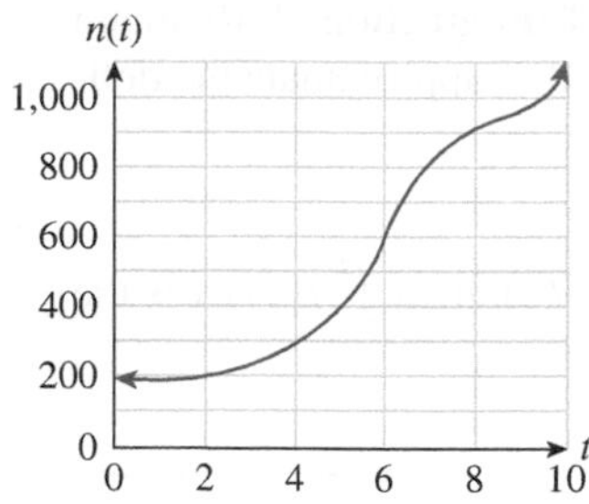

Figure 13

Figure 13 shows the approximate annual production $n(t)$ of engineering and technology PhD graduates in Mexico during the period 2000–2010.[22] Use a left Riemann sum with four subdivisions to estimate the total number of PhD graduates from 2002 to 2010.

Solution Let us represent the total number of (engineering and technology) PhD graduates up to time t (measured in years since 2000) by $N(t)$. The total number of PhD graduates from 2002 to 2010 is then the total change in $N(t)$ over the interval $[2, 10]$. In view of the above discussion we can approximate the total change in $N(t)$ using a Riemann sum of its rate of change $n(t)$. Because $n = 4$ subdivisions are specified, the width of each subdivision is

$$\Delta t = \frac{b-a}{n} = \frac{10-2}{4} = 2.$$

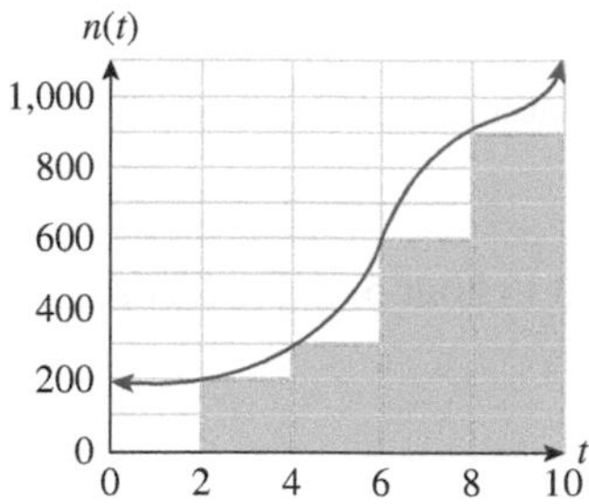

Figure 14

We can therefore represent the left Riemann sum by the shaded area shown in Figure 14.

From the graph,

$$\begin{aligned}\text{Left sum} &= n(2)\,\Delta t + n(4)\,\Delta t + n(6)\,\Delta t + n(8)\,\Delta t \\ &= (200)(2) + (300)(2) + (600)(2) + (900)(2) = 4{,}000.\end{aligned}$$

So we estimate that there was a total of about 4,000 engineering and technology PhD graduates during the given period.

➡ **Before we go on...** A glance at Figure 14 tells us that we have significantly underestimated the actual number of graduates, as the actual area under the curve is considerably larger. Figure 15 shows a **right Riemann sum** and gives us a much larger estimate. For continuous functions, the difference between these two types of Riemann sums approaches zero as the number of subdivisions approaches infinity (see below), so we will focus primarily on only one type: left Riemann sums. ■

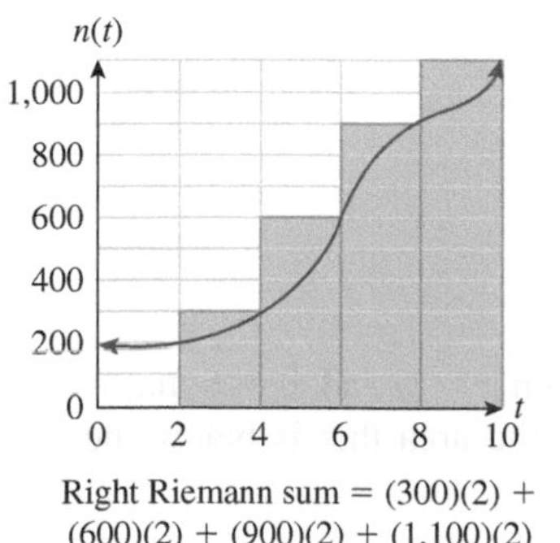

Right Riemann sum = (300)(2) + (600)(2) + (900)(2) + (1,100)(2) = 5,800

Height of each rectangle is determined by height of graph at right edge.

Figure 15

The Definite Integral

As in Example 1, we're most interested in what happens to the Riemann sum when we let n get very large. When f is continuous,* its Riemann sums will always approach a limit as n goes to infinity. (This is not meant to be obvious. Proofs may be found in advanced calculus texts.) We give the limit a name.

* This applies to some other functions as well, including "piecewise continuous" functions discussed in the next section.

The Definite Integral

If f is a continuous function, the **definite integral of f from a to b** is defined to be the limit of the Riemann sums as the number of subdivisions approaches infinity:

$$\int_a^b f(x)\,dx = \lim_{n\to\infty} \sum_{k=0}^{n-1} f(x_k)\,\Delta x.$$

In Words: The integral, from a to b, of $f(x)\,dx$ equals the limit, as $n \to \infty$, of the Riemann sum with a partition of n subdivisions.

[22] Source for data: Instituto Nacional de Estadística y Geografía (www.inegi.org.mx).

The function f is called the **integrand**, the numbers a and b are the **limits of integration**, and the variable x is the **variable of integration**. A Riemann sum with a large number of subdivisions may be used to approximate the definite integral.

Interpretation of the Definite Integral

If f is the rate of change of a quantity F (that is, $f = F'$), then $\int_a^b f(x)\,dx$ is the (exact) total change of F from $x = a$ to $x = b$.

Quick Examples

4. If $f(t)$ is the rate of change in the number of bats in a belfry and $[a, b] = [2, 3]$, then $\int_2^3 f(t)\,dt$ is the total change in the number of bats in the belfry from time $t = 2$ to time $t = 3$.
5. If, at time t hours, you are selling wall posters at a rate of $s(t)$ posters per hour, then

$$\text{Total number of posters sold from hour 3 to hour 5} = \int_3^5 s(t)\,dt.$$

Visualizing the Definite Integral

Nonnegative Functions: If $f(x) \geq 0$ for all x in $[a, b]$, then $\int_a^b f(x)\,dx$ is the area under the graph of f over the interval $[a, b]$, as shaded in the figure.

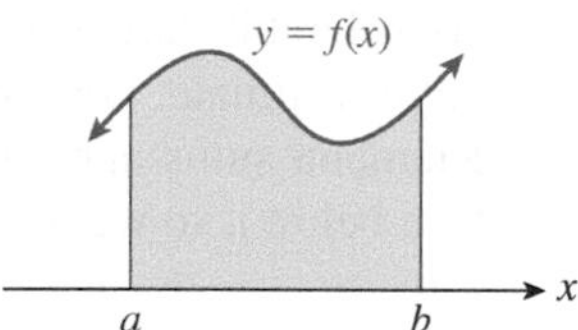

General Functions: $\int_a^b f(x)\,dx$ is the area between $x = a$ and $x = b$ that is above the x-axis and below the graph of f minus the area that is below the x-axis and above the graph of f:

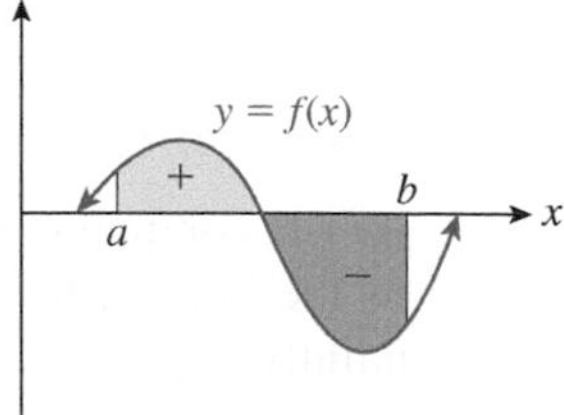

$$\int_a^b f(x)\,dx = \text{Area above } x\text{-axis} - \text{Area below } x\text{-axis}$$

Quick Examples

6.

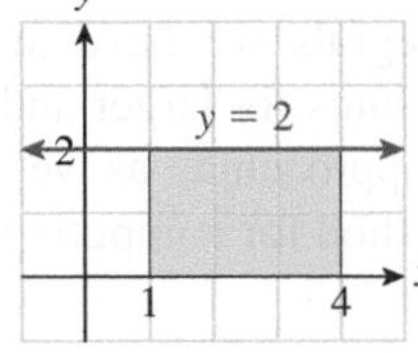

$$\int_1^4 2\,dx = \text{Area of rectangle} = 6$$

7.

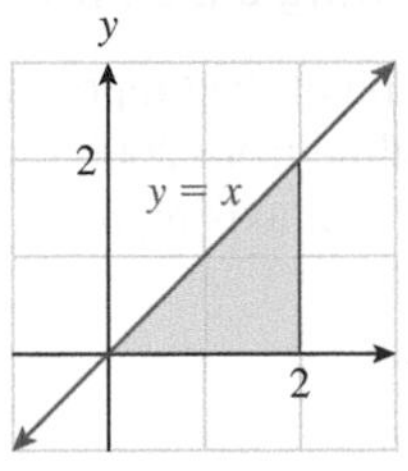

$$\int_0^2 x\,dx = \text{Area of triangle} = \frac{1}{2}\text{base} \times \text{height} = 2$$

8.

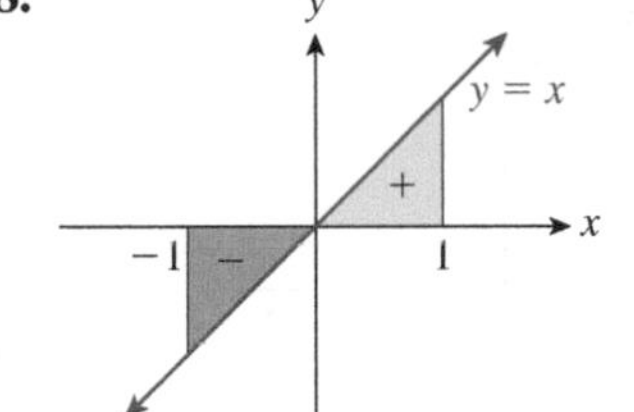

$$\int_{-1}^1 x\,dx = 0$$ The areas above and below the x-axis are equal.

Notes

1. Remember that $\int_a^b f(x)\,dx$ stands for a number that depends on f, a, and b. The variable of integration x that appears is called a **dummy variable** because it has no effect on the answer. In other words,

$$\int_a^b f(x)\,dx = \int_a^b f(t)\,dt. \qquad x \text{ or } t \text{ is just a name we give the variable.}$$

2. The notation for the definite integral (due to Leibniz) comes from the notation for the Riemann sum. The integral sign $\int$ is an elongated S, the Roman equivalent of the Greek Σ. The d in dx is the lowercase Roman equivalent of the Greek Δ.
3. The definition above is adequate for continuous functions, but more complicated definitions are needed to handle other functions. For example, we broke the interval $[a, b]$ into n subdivisions of equal length, but other definitions allow a **partition** of the interval into subdivisions of possibly unequal lengths. We have evaluated f at the left endpoint of each subdivision, but we could equally well have used the right endpoint or any other point in the subdivision. All of these variations lead to the same answer when f is continuous.
4. The similarity between the notations for the definite integral and the indefinite integral is no mistake. We will discuss the exact connection in Section 6.4. ■

Calculating Definite Integrals

In some cases we can compute the definite integral directly from the graph. (See Quick Examples 6–8 above and Example 4 below.) In general, the only method of computing definite integrals we have discussed so far is numerical estimation: Compute the Riemann sums for larger and larger values of n, and then estimate the number it seems to be approaching as we did in Example 1. (In Section 6.4 we will discuss an algebraic method for computing them.)

EXAMPLE 4 Estimating a Definite Integral from a Graph

Figure 16 shows the graph of the (approximate) rate $f'(t)$ at which the United States consumed aviation gasoline from 2000 through 2013. (t is time in years since 2000, and each unit on the vertical axis represents 100 million gallons per year.)[23]

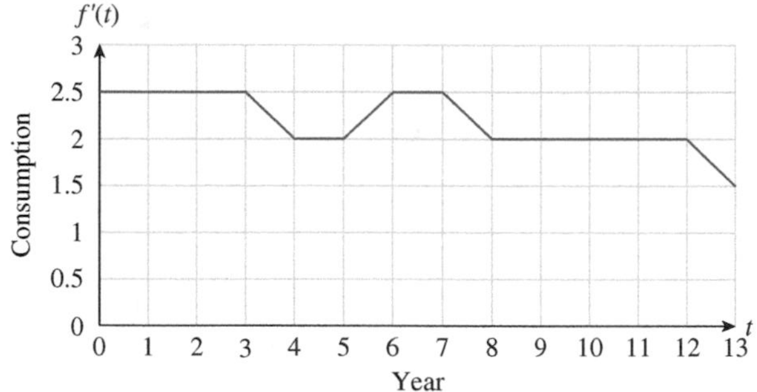

Figure 16

Use the graph to estimate the total U.S. consumption of aviation gasoline over the period 2002–2011.

Solution The derivative $f'(t)$ represents the rate of change of the total U.S. consumption of aviation gasoline, so the total U.S. consumption of aviation gasoline over the period 2002–2011 ($[2, 11]$ on the graph) is given by the definite integral

$$\text{Total U.S. consumption of aviation gasoline} = \text{Total change in } f(t) = \int_2^{11} f'(t)\,dt$$

and is given by the shaded area under the graph (Figure 17).

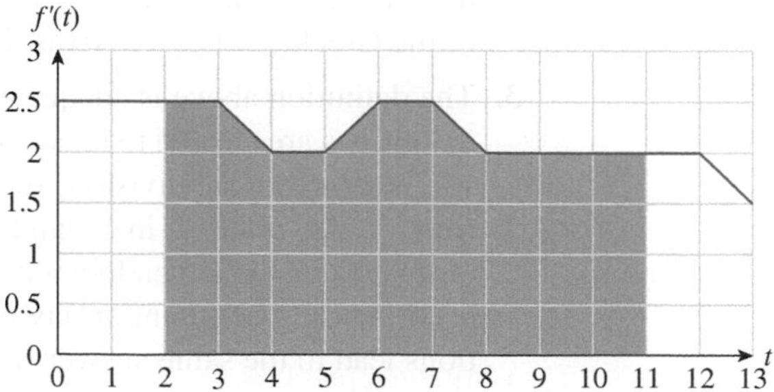

Figure 17

[23]Source: Energy Information Administration, Department of Energy (http://tonto.eia.doe.gov).

One way to determine the area is to count the number of filled rectangles as defined by the grid. Each rectangle has an area of $1 \times 0.5 = 0.5$ units (and the half-rectangles determined by diagonal portions of the graph have half that area). Counting rectangles, we find a total of 39.5 complete rectangles, so

$$\text{Total area} = 19.75.$$

Because $f'(t)$ is in 100 million gallons per year, we conclude that the total U.S. consumption of aviation gasoline over the given period was about 1,975 million gallons, or 1.975 billion gallons.

While counting rectangles might seem easy, it becomes awkward in cases involving large numbers of rectangles or partial rectangles whose area is not easy to determine. In a case like this, in which the graph consists of straight lines, rather than counting rectangles, we can get the area by averaging the left and right Riemann sums whose subdivisions are determined by the grid:

$$\text{Left sum} = (2.5 + 2.5 + 2 + 2 + 2.5 + 2.5 + 2 + 2 + 2)(1) = 20$$
$$\text{Right sum} = (2.5 + 2 + 2 + 2.5 + 2.5 + 2 + 2 + 2 + 2)(1) = 19.5$$

$$\text{Average} = \frac{20 + 19.5}{2} = 19.75.$$

To see why this works, look at the single interval $[3, 4]$. The left sum contributes $2.5 \times 1 = 2.5$, and the right sum contributes $2 \times 1 = 2$. The exact area is their average, 2.25 (Figure 18).

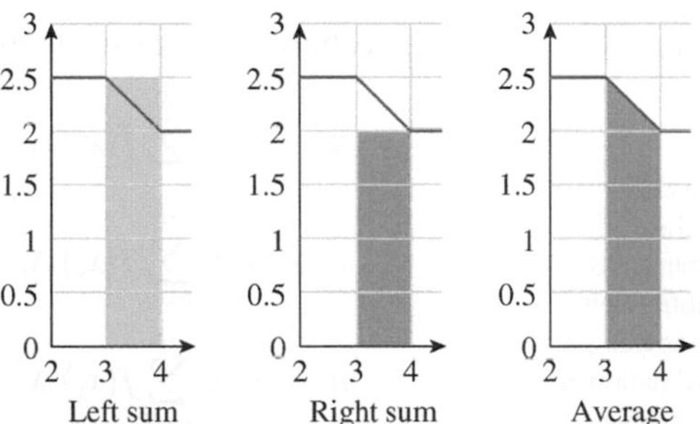

Figure 18

The average of the left and right Riemann sums is frequently a better estimate of the definite integral than either is alone.

➡ **Before we go on . . .** It is important to check that the units we are using in Example 4 match up correctly: t is given in *years*, and $f'(t)$ is given in 100 million gallons per *year*. The integral is then given in

$$\text{Years} \times \frac{\text{100 million gallons}}{\text{Year}} = \text{100 million gallons.}$$

If we had specified $f'(t)$ in, say, 100 million gallons per *day* but t in years, then we would have needed to convert either t or $f'(t)$ so that the units of time match. ■

The next example illustrates the use of technology in estimating definite integrals using Riemann sums.

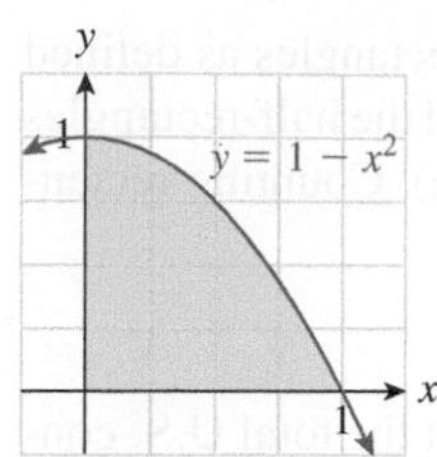

Figure 19

EXAMPLE 5 **Using Technology to Approximate the Definite Integral**

Use technology to estimate the area under the graph of $f(x) = 1 - x^2$ over the interval $[0, 1]$ using $n = 100$, $n = 200$, and $n = 500$ subdivisions.

Solution We need to estimate the area under the parabola shown in Figure 19. From the discussion above,

$$\text{Area} = \int_0^1 (1 - x^2)\, dx.$$

The Riemann sum with $n = 100$ has $\Delta x = (b - a)/n = (1 - 0)/100 = 0.01$ and is given by

$$\sum_{k=0}^{99} f(x_k)\, \Delta x = [f(0) + f(0.01) + \cdots + f(0.99)](0.01).$$

Similarly, the Riemann sum with $n = 200$ has $\Delta x = (b - a)/n = (1 - 0)/200 = 0.005$ and is given by

$$\sum_{k=0}^{199} f(x_k)\, \Delta x = [f(0) + f(0.005) + \cdots + f(0.995)](0.005).$$

For $n = 500$, $x = (b - a)/n = (1 - 0)/500 = 0.002$, and the Riemann sum is

$$\sum_{k=0}^{499} f(x_k)\, \Delta x = [f(0) + f(0.002) + \cdots + f(0.998)](0.002).$$

Using Technology

See the Technology Guides at the end of the chapter to find out how to compute these sums on a TI-83/84 Plus and a spreadsheet.

Website

www.WanerMath.com

At the Website, select the Online Utilities tab, and choose the Numerical Integration Utility and Grapher. This utility computes left and right Riemann sums as well as a good numerical approximation of the integral. There is also a downloadable Excel spreadsheet that computes and graphs Riemann sums (Riemann Sum Grapher).

Using technology to evaluate these Riemann sums, we find

$$n = 100\colon \sum_{k=0}^{99} f(x_k)\, \Delta x = 0.67165$$

$$n = 200\colon \sum_{k=0}^{199} f(x_k)\, \Delta x = 0.6691625$$

$$n = 500\colon \sum_{k=0}^{499} f(x_k)\, \Delta x = 0.667666,$$

so we estimate that the area under the curve is about 0.67. (The exact answer is 2/3, as we will be able to verify using the techniques in Section 6.4.)

EXAMPLE 6 **Motion**

A fast car has velocity $v(t) = 6t^2 + 10t$ ft/sec after t seconds (as measured by a radar gun). Use several values of n to find the distance covered by the car from time $t = 3$ seconds to time $t = 4$ seconds.

Solution Because the velocity $v(t)$ is rate of change of position, the total change in position over the interval $[3, 4]$ is

$$\text{Distance covered} = \text{Total change in position} = \int_3^4 v(t)\, dt = \int_3^4 (6t^2 + 10t)\, dt.$$

As in Examples 1 and 5, we can subdivide the 1-second interval $[3, 4]$ into smaller and smaller pieces to get more and more accurate approximations of the integral. By computing Riemann sums for various values of n, we get the following results:

$$n = 10: \sum_{k=0}^{9} v(t_k)\,\Delta t = 106.41 \qquad n = 100: \sum_{k=0}^{99} v(t_k)\,\Delta t \approx 108.740$$

$$n = 1{,}000: \sum_{k=0}^{999} v(t_k)\,\Delta t \approx 108.974 \qquad n = 10{,}000: \sum_{k=0}^{9999} v(t_k)\,\Delta t \approx 108.997$$

These calculations suggest that the total distance covered by the car, the value of the definite integral, is approximately 109 feet.

Website
www.WanerMath.com
At the Website you will find the following optional online interactive section: Internet Topic: Numerical Integration.

➡ **Before we go on . . .** Do Example 6 using antiderivatives instead of Riemann sums, as in Section 6.1. Do you notice a relationship between antiderivatives and definite integrals? This will be explored in Section 6.4. ■

6.3 EXERCISES

▼ more advanced ◆ challenging

T indicates exercises that should be solved using technology

In Exercises 1–10, calculate the left Riemann sum for the given function over the given interval, using the given value of n. (When rounding, round answers to four decimal places. If using the tabular method, values of the function in the table should be accurate to at least five decimal places.) [**HINT:** See Example 2.]

1. $f(x) = 4x - 1$ over $[0, 2]$, $n = 4$
2. $f(x) = 1 - 3x$ over $[-1, 1]$, $n = 4$
3. $f(x) = x^2$ over $[-2, 2]$, $n = 4$
4. $f(x) = x^2$ over $[1, 5]$, $n = 4$
5. $f(x) = (x - 1)^3$ over $[-1, 4]$, $n = 5$
6. $f(x) = x^3$ over $[-2, 3]$, $n = 5$
7. $f(x) = \dfrac{1}{1 + x}$ over $[0, 1]$, $n = 5$
8. $f(x) = \dfrac{x}{1 + x^2}$ over $[0, 1]$, $n = 5$
9. $f(x) = e^{-x}$ over $[0, 10]$, $n = 5$
10. $f(x) = e^{-x}$ over $[-5, 5]$, $n = 5$

In Exercises 11–18, use the given graph to estimate the left Riemann sum for the given interval with the stated number of subdivisions. [**HINT:** See Example 3.]

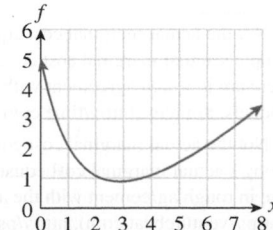

11. $[0, 5]$, $n = 5$

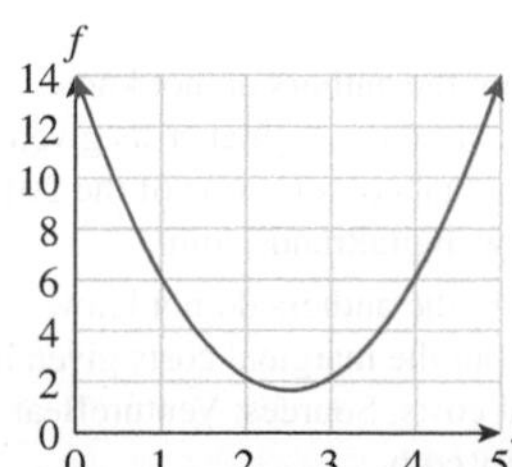

12. $[0, 8]$, $n = 4$

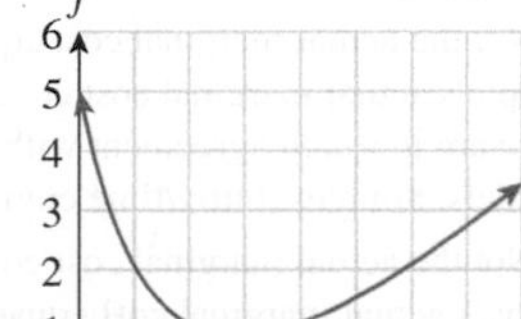

13. $[1, 9]$, $n = 4$

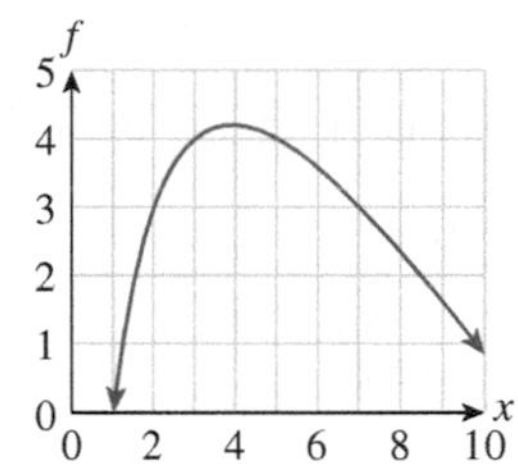

14. $[0.5, 2.5]$, $n = 4$

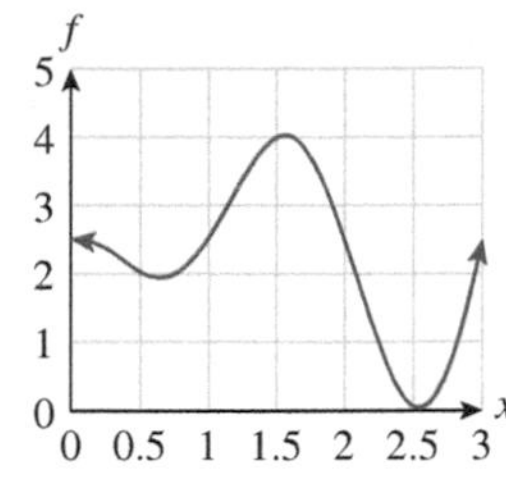

15. $[1, 3.5]$, $n = 5$

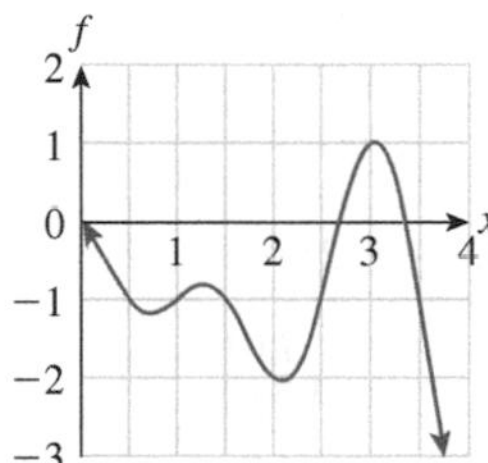

16. $[0.5, 3.5]$, $n = 3$

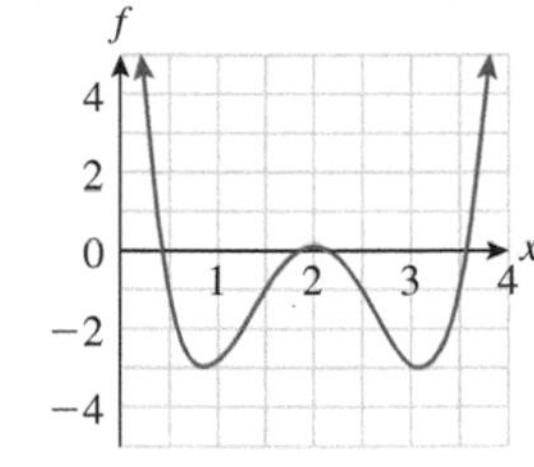

17. $[0, 3]$; $n = 3$

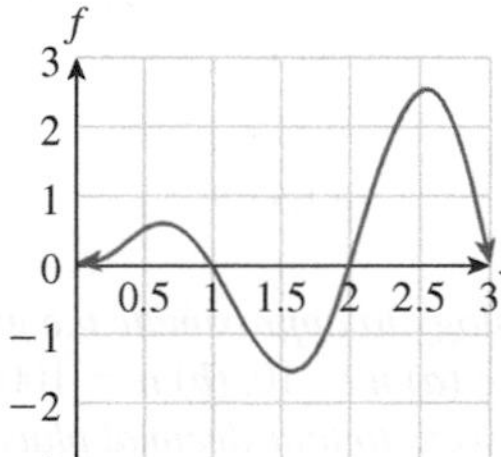

18. $[0.5, 3]$; $n = 5$

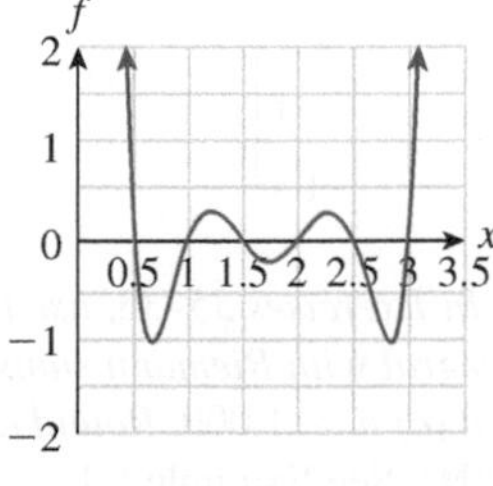

In Exercises 19–28, use geometry (not Riemann sums) to compute the integral. [**HINT:** See Quick Examples 6–8.]

19. $\displaystyle\int_0^1 1\,dx$

20. $\displaystyle\int_0^2 5\,dx$

21. $\displaystyle\int_0^1 x\,dx$ **22.** $\displaystyle\int_1^2 x\,dx$

23. $\displaystyle\int_0^1 \frac{x}{2}\,dx$ **24.** $\displaystyle\int_1^2 \frac{x}{2}\,dx$

25. $\displaystyle\int_2^4 (x-2)\,dx$ **26.** $\displaystyle\int_3^6 (x-3)\,dx$

27. $\displaystyle\int_{-1}^1 x^3\,dx$ **28.** $\displaystyle\int_{-2}^2 \frac{x}{2}\,dx$

In Exercises 29–34, the graph of the derivative $f'(t)$ of $f(t)$ is shown. Compute the total change of $f(t)$ over the given interval. [**HINT:** See Example 4.]

29. [1, 5] **30.** [2, 6]

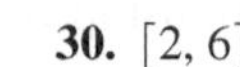

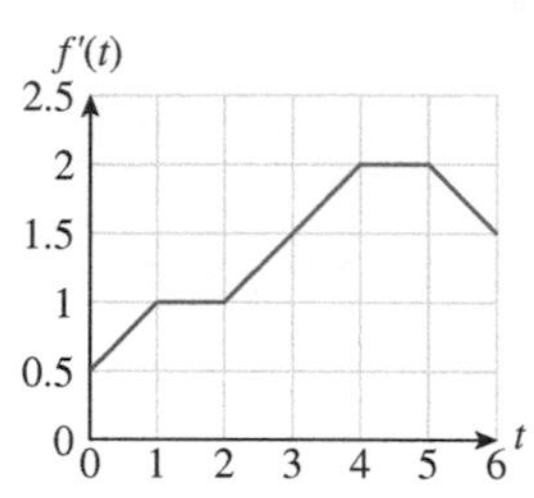

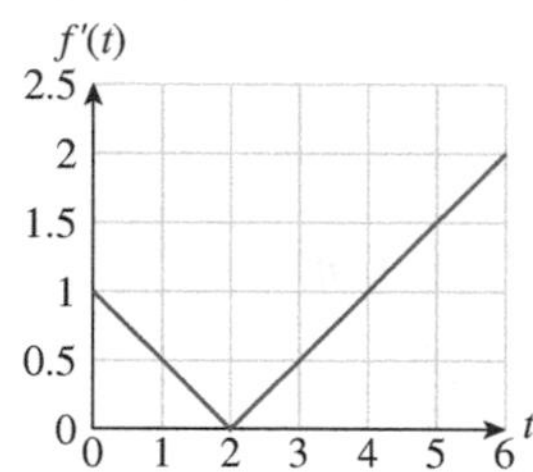

31. [2, 6] **32.** [0, 5]

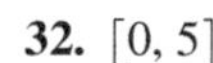

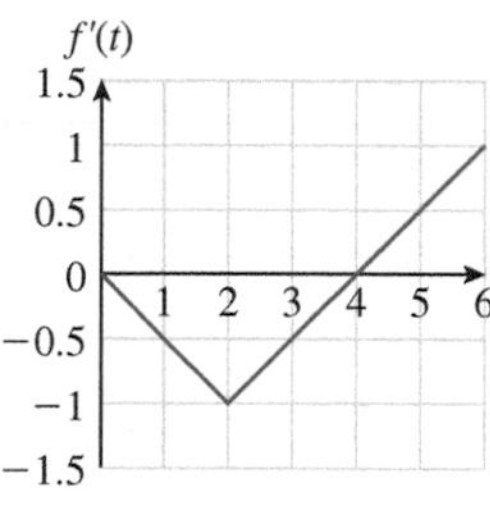

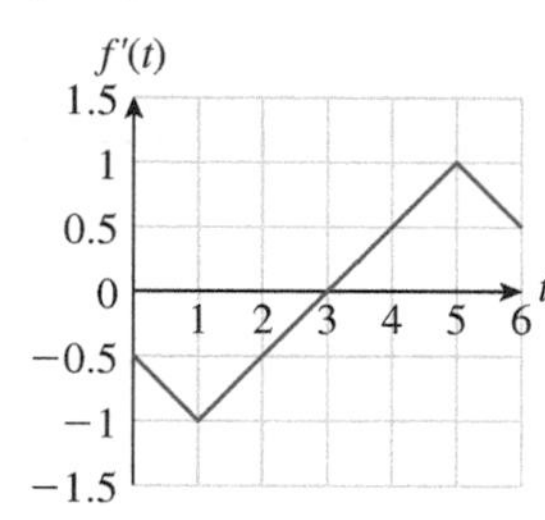

33. [−1, 2] **34.** [−1, 2]

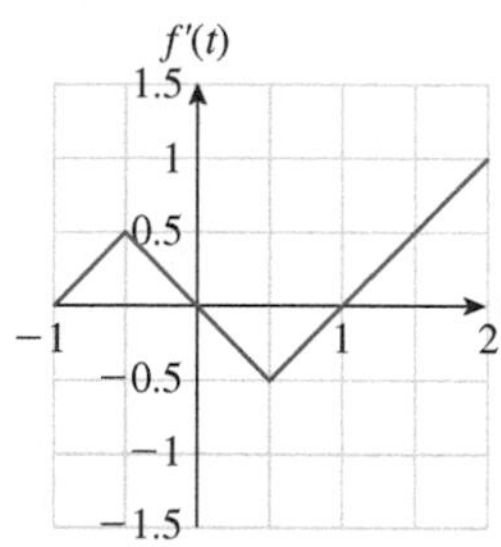

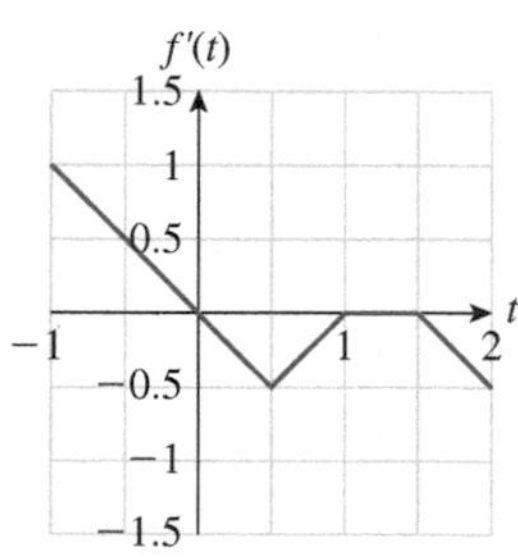

T *In Exercises 35–38, use technology to approximate the given integral with Riemann sums, using* **(a)** $n = 10$, **(b)** $n = 100$, *and* **(c)** $n = 1{,}000$. *Round all answers to four decimal places.* [**HINT:** See Example 5.]

35. $\displaystyle\int_0^1 4\sqrt{1-x^2}\,dx$ **36.** $\displaystyle\int_0^1 \frac{4}{1+x^2}\,dx$

37. $\displaystyle\int_2^3 \frac{2x^{1.2}}{1+3.5x^{4.7}}\,dx$ **38.** $\displaystyle\int_3^4 3xe^{1.3x}\,dx$

Applications

39. ***Pumps*** A pump is delivering water into a tank at a rate of

$$r(t) = 3t^2 + 5 \text{ liters per minute,}$$

where t is time in minutes since the pump is turned on. Use a Riemann sum with $n = 5$ subdivisions to estimate the total volume of water pumped in during the first 2 minutes. [**HINT:** See Examples 1 and 2. This exercise is also discussed in the tutorial on the Website.]

40. ***Pumps*** A pump is delivering water into a tank at a rate of

$$r(t) = 6t^2 + 40 \text{ liters per minute,}$$

where t is time in minutes since the pump is turned on. Use a Riemann sum with $n = 6$ subdivisions to estimate the total volume of water pumped in during the first 3 minutes. [**HINT:** See Examples 1 and 2. A similar exercise is also discussed in the tutorial on the Website.]

41. ***Cost*** The marginal cost function for the manufacture of wireless headphones is given by

$$C'(x) = 20 - \frac{x}{200},$$

where x is the number of headphones manufactured. Use a Riemann sum with $n = 5$ to estimate the cost of producing the first 5 headphones. [**HINT:** See Examples 1 and 2 and Quick Example 2.]

42. ***Cost*** Repeat Exercise 41 using the marginal cost function

$$C'(x) = 25 - \frac{x}{50}.$$

[**HINT:** See Examples 1 and 2 and Quick Example 2.]

43. ***Profit: iPhones*** Assume that **Apple**'s marginal cost function for the manufacture of x 32GB iPhone 6's per hour at the **Foxconn Technology Group** is[24]

$$c(x) = 160 - 0.002x,$$

and that Apple sells iPhone 6's for an average wholesale price of \$580. Use a Riemann sum with $n = 5$ subdivisions to estimate the total additional hourly profit corresponding to an increase in production and sales from 10,000 to 20,000 iPhone 6's per hour.

44. ***Profit: PlayStation 4's*** Assume that **Sony**'s marginal cost function for the manufacture of x PlayStation 4's per hour is[25]

$$c(x) = 340 + 0.001x,$$

[24] Not the actual marginal cost equation; the authors do not know Apple's actual marginal cost equation, but the marginal costs given here are in rough agreement with the actual costs for one of the 2014 models. Sources: http://time.com, www.digitaltrends.com.

[25] Not the actual marginal cost equation; the authors do not know Sony's actual marginal cost equation, but the marginal costs given here are in rough agreement with the actual costs. Sources: VentureBeat (http://venturebeat.com), http://ps4daily.com.

and that Sony sells PlayStation 4's for an average wholesale price of \$400. Use a Riemann sum with $n = 5$ subdivisions to estimate the total additional hourly profit corresponding to an increase in production and sales from 50,000 to 60,000 PlayStation 4's per hour.

45. ***Bottled Water Sales*** The rate of U.S. sales of bottled water for the period 2007–2014 could be approximated by

$$s(t) = 0.08t^2 - 0.26t + 8.8 \text{ billion gallons per year} \quad (0 \le t \le 7),$$

where t is time in years since the start of 2007.[26] Use a Riemann sum with $n = 5$ to estimate the total U.S. sales of bottled water from the start of 2008 to the start of 2014. (Round your answer to the nearest billion gallons.) [**HINT:** See Example 2.]

46. ***Bottled Water Sales*** The rate of U.S. per capita sales of bottled water for the period 2007–2014 could be approximated by

$$s(t) = 0.25t^2 - t + 29 \text{ gallons per year} \quad (0 \le t \le 7),$$

where t is time in years since the start of 2007.[27] Use a Riemann sum with $n = 5$ to estimate the total U.S. per capita sales of bottled water from the start of 2008 to the start of 2012. (Round your answer to the nearest gallon.) [**HINT:** See Example 2.]

47. ***Online Payments*** The following graph shows the rate of change $p(t)$ of total payments through **PayPal**, in billions of dollars per quarter, from the first quarter of 2013 through the fourth quarter of 2014. (t is time in quarters; $t = 1$ represents the first quarter of 2013.)[28]

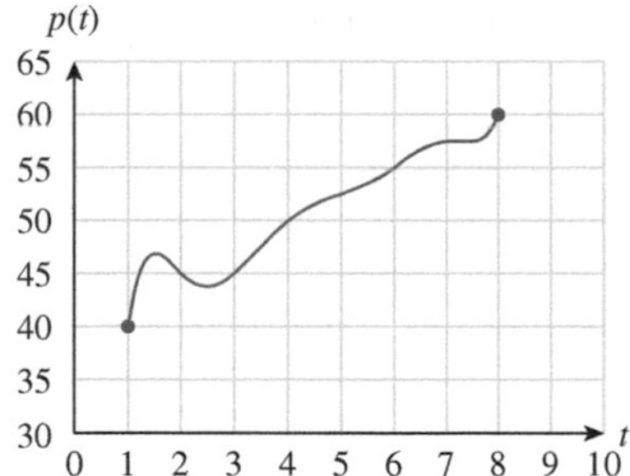

Use a left Riemann sum with three subdivisions to estimate the total value of payments through PayPal from the second quarter of 2013 to the fourth quarter of 2014 (the interval $[2, 8]$). [**HINT:** See Example 3.]

48. ***Online Auctions*** The following graph shows the rate of change $n(t)$ of the number of active **eBay** users, in millions of users per quarter, from the first quarter of 2008 through the first quarter of 2011. (t is time in quarters; $t = 1$ represents the first quarter of 2008.)[29]

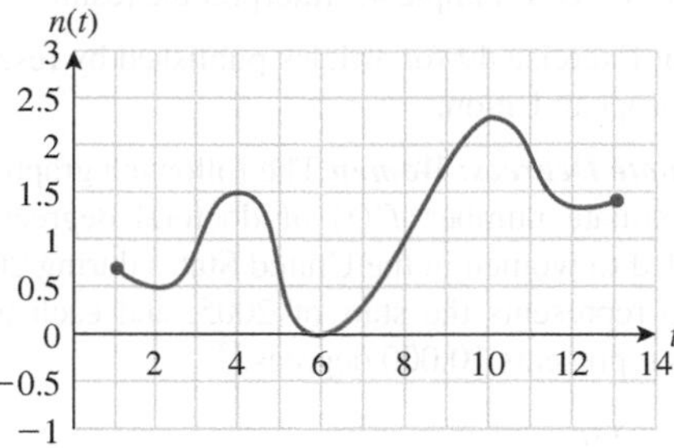

Use a left Riemann sum with four subdivisions to estimate the total change in the number of active users from the second quarter of 2008 to the second quarter of 2010 (the interval $[2, 10]$). [**HINT:** See Example 3.]

Scientific Research *Exercises 49 and 50 are based on the following figure, which shows the rate of publication of science research papers for researchers in the United States and the European Union in the years 1990–2010:*[30]

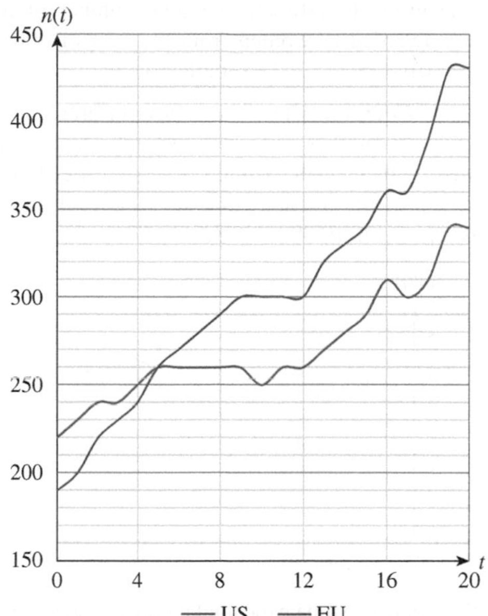

Here, t is time in years since 1990, and $n(t)$ is the publication rate in thousands of articles per year.

49. a. Use both left and right Riemann sums with five subdivisions to estimate the total number of science research papers by researchers in the United States during the 20-year period shown. [**HINT:** See Example 3.]

---

[26] Source for data: Beverage Marketing Corporation (www.bottledwater.org).

[27] *Ibid.*

[28] Source for data: Statista, www.statista.com.

[29] Source for data: eBay company reports (http://investor.ebay.com).

[30] Source for data: Jonathan Adams, David Pendlebury, Global Research Report November 2010, Thomson Reuters/www.sciencewatch.com.

b. Use the answers from part (a) to obtain an estimate of $\int_0^{20} n(t)\,dt$ for researchers in the United States. [**HINT**: See Example 4.] Interpret the result.

50. Repeat Exercise 49 for articles published by researchers in the European Union.

51. ***Graduate Degrees: Women*** The following graph shows the approximate number $f'(t)$ of doctoral degrees per year awarded to women in the United States during 2005–2013. ($t = 0$ represents the start of 2005, and each unit of the y-axis represents 10,000 degrees.)[31]

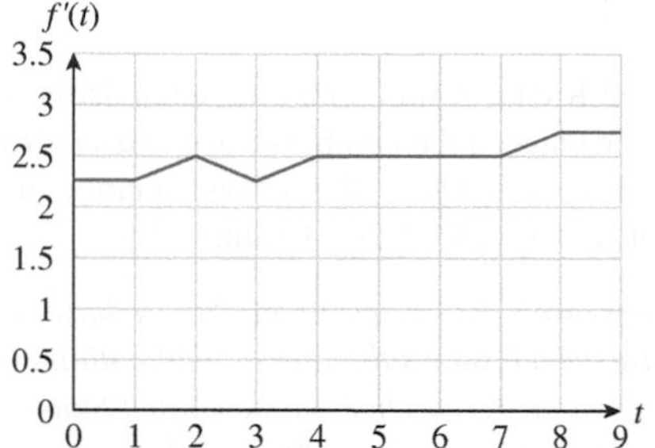

Use the graph to estimate the total number of doctoral degrees awarded to women from the start of 2005 to the start of 2010. [**HINT**: See Example 4.]

52. ***Graduate Degrees: Men*** The following graph shows the approximate number $m'(t)$ of doctoral degrees per year awarded to men in the United States during 2005–2013. ($t = 0$ represents the start of 2005, and each unit of the y-axis represents 10,000 degrees.)[32]

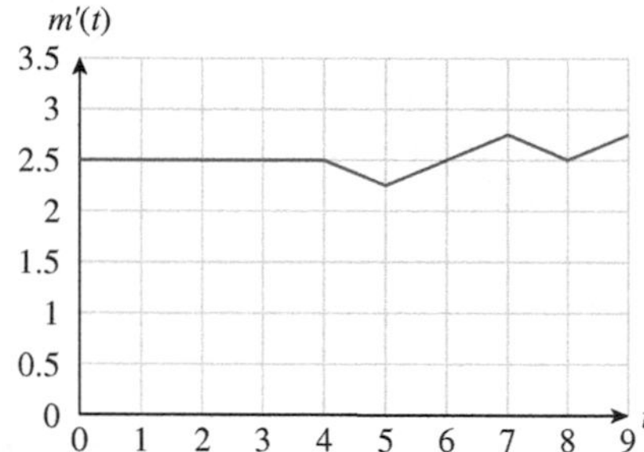

Use the graph to estimate the total number of doctoral degrees awarded to men from the start of 2009 to the end of 2013. [**HINT**: See Example 4.]

53. ***Visiting Students in the Early 2000s*** The aftermath of the September 11 attacks saw a decrease in the number of students visiting the United States. The following graph shows the approximate rate of change $c'(t)$ in the number of students from China who had taken the GRE exam required for admission to U.S. universities. (t is time in years since the start of 2000.)[33]

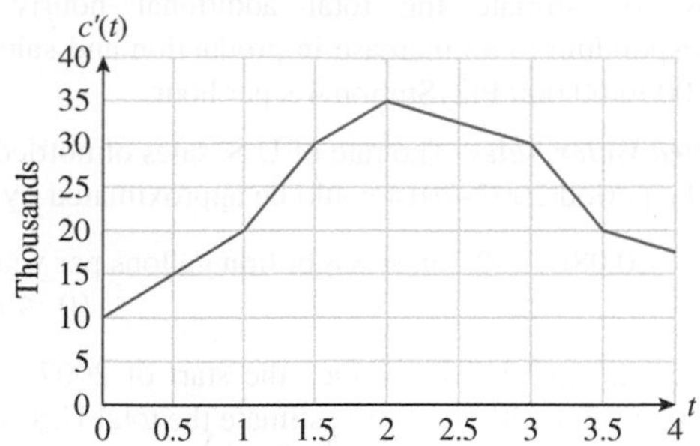

Use the graph to estimate, to the nearest 1,000, the total number of students from China who took the GRE exams from the start of 2002 to the start of 2004.

54. ***Visiting Students in the Early 2000s*** Repeat Exercise 53, using the following graph for students from India:[34]

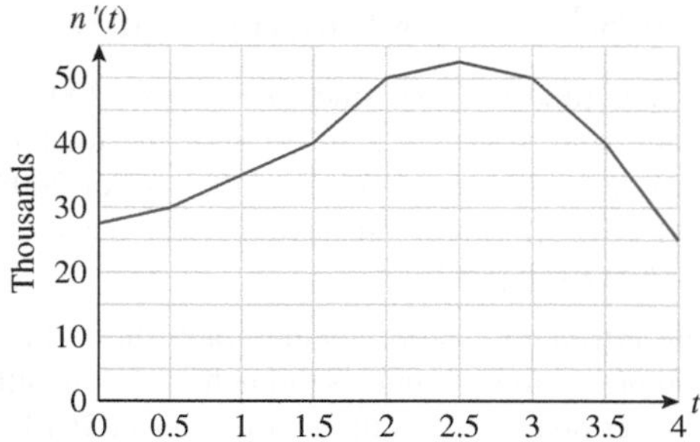

***Net Income:** Exercises 55 and 56 are based on the following graph, which shows* General Electric*'s approximate net income in billions of dollars each year from 2005 to 2011:*[35]

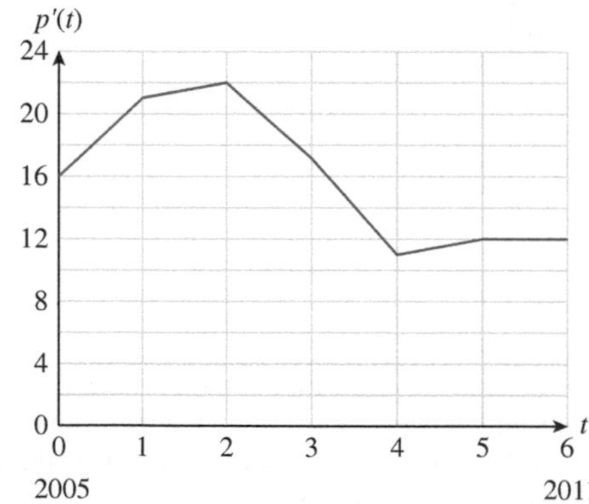

55. Compute the left and right Riemann sum estimates of $\int_0^4 p'(t)\,dt$ using $\Delta t = 1$. Which of these two sums gives

---

[31] Source for data and projections: National Center for Education Statistics (www.nces.ed.gov).

[32] *Ibid.*

[33] Source: Educational Testing Services/Shanghai and Jiao Tong University/*New York Times*, December 21, 2004, p. A25.

[34] *Ibid.*

[35] Source: Company reports (www.ge.com/investors).

the actual total net income earned by GE during the years 2005–2008? Explain.

56. Compute the left and right Riemann sum estimates of $\int_2^6 p'(t)\,dt$ using $\Delta t = 1$. Which of these two sums gives the actual total net income earned by GE during the years 2008–2011? Explain.

57. ***Crude Oil Production: Mexico*** The following table shows annual crude oil production in Mexico by **Pemex** for 2008–2014. ($t = 0$ represents 2008.)[36]

Year t (year since 2008)	0	1	2	3	4	5	6
Crude Oil Production p (billion barrels)	1.15	1.08	1.08	1.07	1.06	1.05	1.02

a. Use the table to compute the left and right Riemann sums for $p(t)$ over the interval $[0, 5]$ using five subdivisions.

b. What does the *right* Riemann sum in part (a) tell you about crude oil production by Pemex?

58. ***Offshore Crude Oil Production: Mexico*** The following table shows annual offshore crude oil production in Mexico by **Pemex** for 2008–2014. ($t = 0$ represents 2008.)[37]

Year t (year since 2008)	0	1	2	3	4	5	6
Offshore Crude Oil Production p (billion barrels)	0.82	0.73	0.71	0.69	0.69	0.69	0.68

a. Use the table to compute the left and right Riemann sums for $p(t)$ over the interval $[1, 6]$ using five subdivisions.

b. What does the *left* Riemann sum in part (a) tell you about offshore crude oil production by Pemex?

59. ***Motion under Gravity*** The velocity of a stone moving under gravity t seconds after being thrown up at 30 ft/sec is given by $v(t) = -32t + 30$ ft/sec. Use a Riemann sum with five subdivisions to estimate $\int_0^4 v(t)\,dt$. What does the answer represent? [**HINT**: See Example 6.]

60. ***Motion under Gravity*** The velocity of a stone moving under gravity t seconds after being thrown up at 4 m/sec is given by $v(t) = -9.8t + 4$ m/sec. Use a Riemann sum with five subdivisions to estimate $\int_0^1 v(t)\,dt$. What does the answer represent? [**HINT**: See Example 6.]

[36] 2014 figure is a projection based on data through Nov. 2014. Source: www.pemex.com (January 2015).

[37] *Ibid.*

61. T ***Motion*** A model rocket has upward velocity $v(t) = 40t^2$ ft/sec, t seconds after launch. Use a Riemann sum with $n = 10$ to estimate how high the rocket is 2 seconds after launch. (Use technology to compute the Riemann sum.)

62. T ***Motion*** A race car has a velocity given by $v(t) = 600(1 - e^{-0.5t})$ ft/sec, t seconds after starting. Use a Riemann sum with $n = 10$ to estimate how far the car has traveled in the first 4 seconds. (Round your answer to the nearest whole number.) (Use technology to compute the Riemann sum.)

63. T ***Facebook Membership*** From the start of 2007, new members joined **Facebook** at a rate of roughly

$$m(t) = 20t^2 + 60t + 12 \text{ million members per year} \quad (0 \le t \le 6),$$

where t is time in years since the start of 2007.[38] Estimate $\int_1^6 m(t)\,dt$ using a Riemann sum with $n = 150$. (Try the Numerical integration utility and grapher.) Round your answer to the nearest whole number, and interpret the answer. [**HINT**: See Example 5.]

64. T ***Uploads to YouTube*** Since **YouTube** first became available to the public in mid-2005, the rate at which video has been uploaded to the site can be approximated by

$$v(t) = 1.1t^2 - 2.6t + 2.3 \text{ million hours of video per year} \quad (0 \le t \le 9),$$

where t is time in years since June 2005.[39] Estimate $\int_2^9 v(t)\,dt$ using a Riemann sum with $n = 150$. (Round your answer to the nearest whole number.) Interpret the answer. [**HINT**: See Example 5.]

65. T ***Big Brother*** The total number of wiretaps authorized each year by U.S. state and federal courts from 1990 to 2015 can be approximated by

$$w(t) = 774e^{0.06t} \quad (0 \le t \le 25).$$

(t is time in years since the start of 1990.)[40] Estimate $\int_{10}^{25} w(t)\,dt$ using a (left) Riemann sum with $n = 100$. (Round your answer to the nearest 10.) Interpret the answer.

66. T ***Big Brother*** The number of wiretaps authorized each year by U.S. state courts from 1990 to 2015 can be approximated by

$$w(t) = 412e^{0.071t} \quad (0 \le t \le 25).$$

(t is time in years since the start of 1990.)[41] Estimate $\int_5^{25} w(t)\,dt$ using a (left) Riemann sum with $n = 100$. (Round your answer to the nearest 10.) Interpret the answer.

[38] Sources for data: www.facebook.com, www.insidefacebook.com.

[39] Sources for data: www.youtube.com, www.statista.com.

[40] Source for data: Wiretap Reports, Administrative Office of the United States Courts (www.uscourts.gov/Statistics/WiretapReports).

[41] *Ibid.*

67. ▼ ***Surveying*** My uncle intends to build a kidney-shaped swimming pool in his small yard, and the town zoning board will approve the project only if the total area of the pool does not exceed 500 square feet. The accompanying figure shows a diagram of the planned swimming pool, with measurements of its width at the indicated points. Will my uncle's plans be approved? Use a (left) Riemann sum to approximate the area.

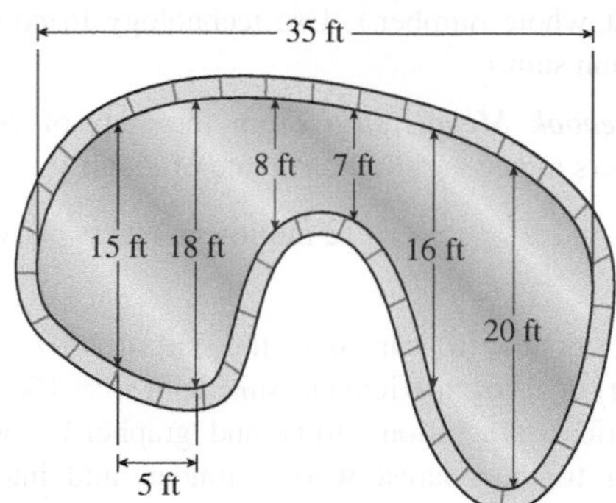

68. ▼ ***Pollution*** An aerial photograph of an ocean oil spill shows the pattern in the accompanying diagram. Assuming that the oil slick has a uniform depth of 0.01 meters, how many cubic meters of oil would you estimate to be in the spill? (Volume = Area × Thickness. Use a (left) Riemann sum to approximate the area.)

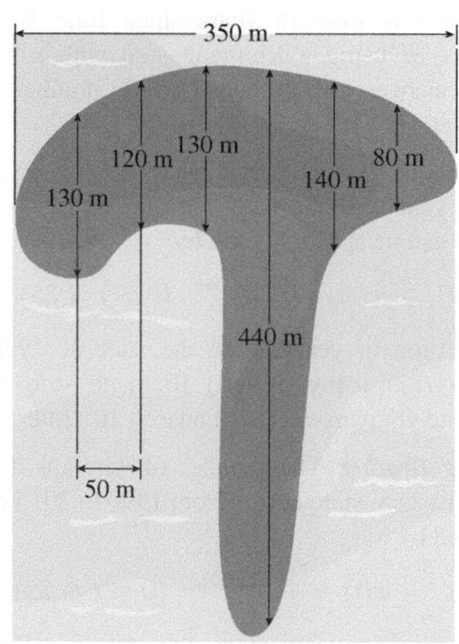

69. T ▼ ***Oil Consumption: United States*** During the period 1980–2008 the United States was consuming oil at a rate of about

$$q(t) = 76t + 5{,}540 \text{ million barrels per year} \quad (0 \le t \le 28),$$

where t is time in years since the start of 1980.[42] During the same period, the price per barrel of crude oil in constant 2008 dollars was about

$$p(t) = 0.45t^2 - 12t + 105 \text{ dollars}^{43} \quad (0 \le t \le 28).$$

a. Graph the function $r(t) = p(t)q(t)$ for $0 \le t \le 28$, indicating the area that represents $\int_{10}^{20} r(t)\,dt$. What does this area signify?

b. Estimate the area in part (a) using a Riemann sum with $n = 200$. (Round the answer to three significant digits.) Interpret the answer.

70. T ▼ ***Oil Consumption: China*** Repeat Exercise 69 using instead the rate of consumption of oil in China:

$$q(t) = 82t + 221 \text{ million barrels per year}^{44} \quad (0 \le t \le 28).$$

The Normal Curve *The normal distribution curve, which models the distributions of data in a wide range of applications, is given by the function*

$$p(x) = \frac{1}{\sqrt{2\pi}\sigma} e^{-(x-\mu)^2/(2\sigma^2)},$$

*where $\pi = 3.14159265\ldots$ and σ and μ are constants called the **standard deviation** and the **mean**, respectively. The graph of the normal distribution (when $\sigma = 1$ and $\mu = 2$) is shown in the figure. Exercises 71 and 72 illustrate its use.*

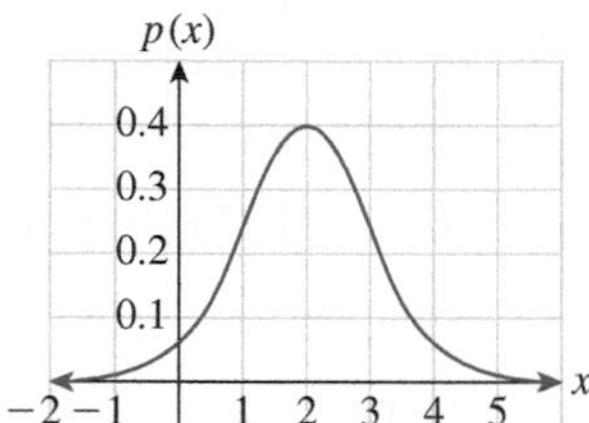

71. T ▼ ***Test Scores*** *Enormous State University*'s Calculus I test scores are modeled by a normal distribution with $\mu = 72.6$ and $\sigma = 5.2$. The percentage of students who obtained scores between a and b on the test is given by

$$\int_a^b p(x)\,dx.$$

a. Use a Riemann sum with $n = 40$ to estimate the percentage of students who obtained between 60 and 100 on the test.

b. What percentage of students scored less than 30?

---

[42] Source for data: BP Statistical Review of World Energy (www.bp.com/statisticalreview).

[43] Source for data: www.inflationdata.com.

[44] Source for data: BP Statistical Review of World Energy (www.bp.com/statisticalreview).

72. T ▼ ***Consumer Satisfaction*** In a survey, consumers were asked to rate a new toothpaste on a scale of 1–10. The resulting data are modeled by a normal distribution with $\mu = 4.5$ and $\sigma = 1.0$. The percentage of consumers who rated the toothpaste with a score between a and b on the test is given by

$$\int_a^b p(x)\,dx.$$

a. Use a Riemann sum with $n = 10$ to estimate the percentage of customers who rated the toothpaste 5 or above. (Use the range 4.5 to 10.5.)

b. What percentage of customers rated the toothpaste 0 or 1? (Use the range -0.5 to 1.5.)

Communication and Reasoning Exercises

73. If $f(x) = 6$, then the left Riemann sum _______ (increases/decreases/stays the same) as n increases.

74. If $f(x) = -1$, then the left Riemann sum _______ (increases/decreases/stays the same) as n increases.

75. If f is an increasing function of x, then the left Riemann sum _______ (increases/decreases/stays the same) as n increases.

76. If f is a decreasing function of x, then the left Riemann sum _______ (increases/decreases/stays the same) as n increases.

77. If $\int_a^b f(x)\,dx = 0$, what can you say about the graph of f?

78. Sketch the graphs of two (different) functions f and g such that $\int_a^b f(x)\,dx = \int_a^b g(x)\,dx$.

79. ▼ The definite integral counts the area under the x-axis as negative. Give an example that shows how this can be useful in applications.

80. ▼ Sketch the graph of a nonconstant function whose Riemann sum with $n = 1$ gives the exact value of the definite integral.

81. ▼ Sketch the graph of a nonconstant function whose Riemann sums with $n = 1$, 5, and 10 are all zero.

82. ▼ Besides left and right Riemann sums, another approximation of the integral is the **midpoint** approximation, in which we compute the sum

$$\sum_{k=1}^{n} f(\bar{x}_k)\,\Delta x$$

where $\bar{x}_k = (x_{k-1} + x_k)/2$ is the point midway between the left and right endpoints of the interval $[x_{k-1}, x_k]$. Why is it true that the midpoint approximation is exact if f is linear? (Draw a picture.)

83. ▼ Your cellphone company charges you $c(t) = \dfrac{20}{t + 100}$ dollars for the tth minute. You make a 60-minute phone call. What kind of (left) Riemann sum represents the total cost of the call? Explain.

84. ▼ Your friend's cellphone company charges her $c(t) = \dfrac{20}{t + 100}$ dollars for the $(t + 1)$st minute. Your friend makes a 60-minute phone call. What kind of (left) Riemann sum represents the total cost of the call? Explain.

85. ▼ Give a formula for the **right Riemann sum** with n equal subdivisions $a = x_0 < x_1 < \cdots < x_n = b$ for f over the interval $[a, b]$.

86. ▼ Refer to Exercise 85. If f is continuous, what happens to the difference between the left and right Riemann sums as $n \to \infty$? Explain.

87. ▼ Sketch the graph of a nonzero function whose left Riemann sum with n subdivisions is zero for every *even* number n.

88. ▼ When approximating a definite integral by computing Riemann sums, how might you judge whether you have chosen n large enough to get your answer accurate to, say, three decimal places?

6.4 The Definite Integral: Algebraic Viewpoint and the Fundamental Theorem of Calculus

Connection Between Definite and Indefinite Integrals

In Section 6.3 we saw that the definite integral of the marginal cost function gives the total cost. However, in Section 6.1 we used antiderivatives to recover the cost function from the marginal cost function, so we can also use antiderivatives to compute total cost. The following example, based on Example 5 in Section 6.1, compares these two approaches.

EXAMPLE 1 Finding Cost from Marginal Cost

The marginal cost of producing baseball caps at a production level of x caps is $4 - 0.001x$ dollars per cap. Find the total change of cost if production is increased from 100 to 200 caps.

Solution

Method 1: Using an Antiderivative (based on Example 5 in Section 6.1): Let $C(x)$ be the cost function. Because the marginal cost function is the derivative of the cost function, we have $C'(x) = 4 - 0.001x$, so

$$\begin{aligned} C(x) &= \int (4 - 0.001x)\,dx \\ &= 4x - 0.001\frac{x^2}{2} + K \qquad \text{K is the constant of integration.} \\ &= 4x - 0.0005x^2 + K. \end{aligned}$$

Although we do not know what to use for the value of the constant K, we can say:

$$\begin{aligned} \text{Cost at production level of 100 caps} &= C(100) \\ &= 4(100) - 0.0005(100)^2 + K \\ &= \$395 + K \\ \text{Cost at production level of 200 caps} &= C(200) \\ &= 4(200) - 0.0005(200)^2 + K \\ &= \$780 + K. \end{aligned}$$

Therefore,

$$\begin{aligned} \text{Total change in cost} &= C(200) - C(100) \\ &= (\$780 + K) - (\$395 + K) = \$385. \end{aligned}$$

Notice how the constant of integration simply canceled out! So we could choose any value for K that we wanted (such as $K = 0$) and still come out with the correct total change. Put another way, we could use *any antiderivative* of $C'(x)$, such as

$$F(x) = 4x - 0.0005x^2$$

$F(x)$ is any antiderivative of $C'(x)$, whereas $C(x)$ is the actual cost function.

or

$$F(x) = 4x - 0.0005x^2 + 4,$$

compute $F(200) - F(100)$, and obtain the total change, \$385.

Summarizing this method: To compute the total change of $C(x)$ over the interval $[100, 200]$, use any antiderivative $F(x)$ of $C'(x)$, and compute $F(200) - F(100)$.

Method 2: Using a Definite Integral (based on the interpretation of the definite integral as total change discussed in Section 6.3): Because the marginal cost $C'(x)$ is the rate of change of the total cost function $C(x)$, the total change in $C(x)$ over the interval $[100, 200]$ is given by

$$\begin{aligned} \text{Total change in cost} &= \text{Area under the marginal cost function curve} \\ &= \int_{100}^{200} C'(x)\,dx \\ &= \int_{100}^{200} (4 - 0.001x)\,dx \qquad \text{See Figure 20.} \\ &= \$385. \qquad \text{Using geometry or Riemann sums} \end{aligned}$$

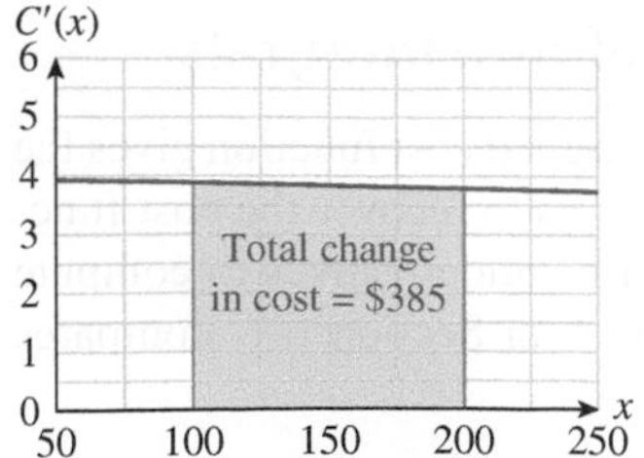

Figure 20

Putting these two methods together gives us the following surprising result:

$$\int_{100}^{200} C'(x)\,dx = F(200) - F(100),$$

where $F(x)$ is any antiderivative of $C'(x)$.

Now, there is nothing special in Example 1 about the specific function $C'(x)$ or the choice of endpoints of integration. So if we replace $C'(x)$ by a general continuous function $f(x)$, we can write

$$\int_{a}^{b} f(x)\,dx = F(b) - F(a),$$

where $F(x)$ is any antiderivative of $f(x)$. This result is known as the **Fundamental Theorem of Calculus**.

Statement of the Fundamental Theorem of Calculus

The Fundamental Theorem of Calculus (FTC)

Let f be a continuous function defined on the interval $[a, b]$, and let F be *any* antiderivative of f defined on $[a, b]$. Then

$$\int_{a}^{b} f(x)\,dx = F(b) - F(a).$$

Moreover, an antiderivative of f is guaranteed to exist.

In Words: Every continuous function has an antiderivative. To compute the definite integral of $f(x)$ over $[a, b]$, first find an antiderivative $F(x)$, then evaluate it at $x = b$, evaluate it at $x = a$, and subtract the two answers.

Q: *A technical point: What is meant by saying that F is an antiderivative of f when f is defined only on a closed interval $[a, b]$? (Remember that the derivative of F is not defined at endpoints of the domain.)*

A: It means that F is a continuous function on $[a, b]$ such that $F'(x)$ exists and equals $f(x)$ for every x in the open interval (a, b).

Quick Example

1. If $f(x) = 2x$, then an antiderivative of f is $F(x) = x^2$. Thus,

$$\int_{0}^{1} 2x\,dx = F(1) - F(0) = 1^2 - 0^2 = 1.$$

Note The Fundamental Theorem of Calculus actually applies to some other functions besides the continuous ones. The function f is **piecewise continuous** on $[a, b]$ if it is defined and continuous at all but finitely many points in the interval, and, at each point where the function is not defined or is discontinuous, the left and right limits of f exist and are finite. (See Figure 21 for some examples of piecewise continuous functions.)

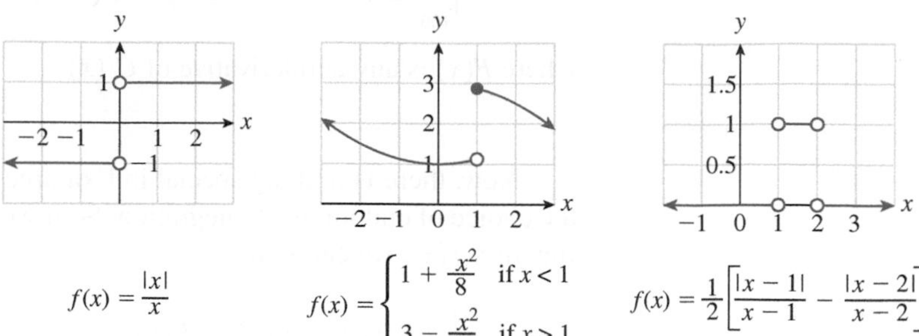

$f(x) = \frac{|x|}{x}$ $\quad f(x) = \begin{cases} 1 + \frac{x^2}{8} & \text{if } x < 1 \\ 3 - \frac{x^2}{8} & \text{if } x > 1 \end{cases}$ $\quad f(x) = \frac{1}{2}\left[\frac{|x-1|}{x-1} - \frac{|x-2|}{x-2}\right]$

Figure 21

The FTC also applies to any piecewise continuous function f, as long as we specify that the antiderivative F that we choose be continuous. To be precise, to say that F is an antiderivative of f means here that $F'(x) = f(x)$ except at the points at which f is discontinuous or not defined, where $F'(x)$ may not exist. For example, if f is the step function $f(x) = |x|/x$, shown on the left in Figure 21, then we can use $F(x) = |x|$. (Note that F is continuous, and $F'(x) = |x|/x$ except when $x = 0$.) ■

EXAMPLE 2 Using the FTC to Calculate a Definite Integral

Calculate $\int_0^1 (1 - x^2)\, dx$.

Solution To use the FTC, we need to find an antiderivative of $1 - x^2$. But we know that

$$\int (1 - x^2)\, dx = x - \frac{x^3}{3} + C.$$

We need only one antiderivative, so let's take $F(x) = x - x^3/3$. The FTC tells us that

$$\int_0^1 (1 - x^2)\, dx = F(1) - F(0) = \left(1 - \frac{1}{3}\right) - (0) = \frac{2}{3}$$

which is the value we estimated in Section 6.3.

Before we go on . . . A useful piece of notation is often used here. We write*

$$[F(x)]_a^b = F(b) - F(a).$$

* There seem to be several notations in use, actually. Another common notation is $F(x)\Big|_a^b$.

Thus, we can rewrite the computation in Example 2 more simply as follows:

$$\int_0^1 (1 - x^2)\, dx = \left[x - \frac{x^3}{3}\right]_0^1$$

Substitute $x = 1$. Substitute $x = 0$.

$$= \left(1 - \frac{1}{3}\right) - \left(0 - \frac{0}{3}\right)$$

$$= \left(1 - \frac{1}{3}\right) - (0) = \frac{2}{3}.$$

■

EXAMPLE 3 More Use of the FTC

Compute the following definite integrals:

a. $\int_0^1 (2x^3 + 10x + 1)\, dx$ **b.** $\int_1^5 \left(\frac{1}{x^2} + \frac{1}{x}\right) dx$

Solution

a.
$$\int_0^1 (2x^3 + 10x + 1)\, dx = \left[\frac{1}{2}x^4 + 5x^2 + x\right]_0^1$$

Substitute $x = 1$. Substitute $x = 0$.

$$= \left(\frac{1}{2} + 5 + 1\right) - \left(\frac{1}{2}(0) + 5(0) + 0\right)$$
$$= \left(\frac{1}{2} + 5 + 1\right) - (0) = \frac{13}{2}$$

b.
$$\int_1^5 \left(\frac{1}{x^2} + \frac{1}{x}\right) dx = \int_1^5 (x^{-2} + x^{-1})\, dx$$
$$= [-x^{-1} + \ln|x|]_1^5$$

Substitute $x = 5$. Substitute $x = 1$.

$$= \left(-\frac{1}{5} + \ln 5\right) - (-1 + \ln 1)$$
$$= \frac{4}{5} + \ln 5$$

When calculating a definite integral, we may have to use substitution to find the necessary antiderivative. We could substitute, evaluate the indefinite integral with respect to u, express the answer in terms of x, and then evaluate at the limits of integration. However, there is a shortcut, as we shall see in the next example.

EXAMPLE 4 Using the FTC with Substitution

Evaluate $\int_1^2 (2x - 1)e^{2x^2-2x}\, dx$.

Solution The shortcut we promised is to put *everything* in terms of u, including the limits of integration.

$$u = 2x^2 - 2x$$
$$\frac{du}{dx} = 4x - 2$$
$$dx = \frac{1}{4x - 2}\, du$$

When $x = 1$, $u = 0$. Substitute $x = 1$ in the formula for u.

When $x = 2$, $u = 4$. Substitute $x = 2$ in the formula for u.

We get the value $u = 0$, for example, by substituting $x = 1$ in the equation $u = 2x^2 - 2x$. We can now rewrite the integral:

$$\int_1^2 (2x-1)e^{2x^2-2x}\,dx = \int_0^4 (2x-1)e^u \frac{1}{4x-2}\,du$$
$$= \int_0^4 \frac{1}{2}e^u\,du$$
$$= \left[\frac{1}{2}e^u\right]_0^4 = \frac{1}{2}e^4 - \frac{1}{2}.$$

➡ **Before we go on . . .** The alternative, longer calculation in Example 4 is first to calculate the indefinite integral:

$$\int (2x-1)e^{2x^2-2x}\,dx = \int \frac{1}{2}e^u\,du$$
$$= \frac{1}{2}e^u + C = \frac{1}{2}e^{2x^2-2x} + C.$$

Then we can say that

$$\int_1^2 (2x-1)e^{2x^2-2x}\,dx = \left[\frac{1}{2}e^{2x^2-2x}\right]_1^2 = \frac{1}{2}e^4 - \frac{1}{2}.$$ ■

Applications

Because, as we saw in Section 6.3, the definite integral allows us to calculate total change from the rate of change, or to calculate area between a graph and the x-axis, we can now use the Fundamental Theorem of Calculus to make such calculations simpler.

EXAMPLE 5 Oil Spill

In Section 6.3 we considered the following example: Your deep ocean oil rig has suffered a catastrophic failure, and oil is leaking from the ocean floor wellhead at a rate of

$$v(t) = 0.08t^2 - 4t + 60 \text{ thousand barrels per day} \qquad (0 \le t \le 20),$$

where t is time in days since the failure. Compute the total volume of oil released during the first 20 days.

Solution We calculate

$$\text{Total volume} = \int_0^{20} (0.08t^2 - 4t + 60)\,dt = \left[0.08\frac{t^3}{3} - 2t^2 + 60t\right]_0^{20}$$
$$= \left[0.08\frac{20^3}{3} - 2(20)^2 + 60(20)\right] - [0.08(0) - 2(0) + 60(0)]$$
$$= \frac{640}{3} - 800 + 1{,}200 \approx 613.3 \text{ thousand barrels.}$$

Using Technology

TI-83/84 Plus

Home screen:

`fnInt(0.08x^2-4x+60,x,0,20)`

(`fnInt` is `MATH` → `9`)

Website

www.WanerMath.com

At the Website, select the Online Utilities tab, and choose the Numerical Integration Utility and Grapher. There, enter the formula

`0.08x^2-4x+60`

for $f(x)$, enter 0 and 20 for the lower and upper limits, and press "Integral".

EXAMPLE 6 **Computing Area**

Find the total area of the region enclosed by the graph of $y = xe^{x^2}$, the x-axis, and the vertical lines $x = -1$ and $x = 1$.

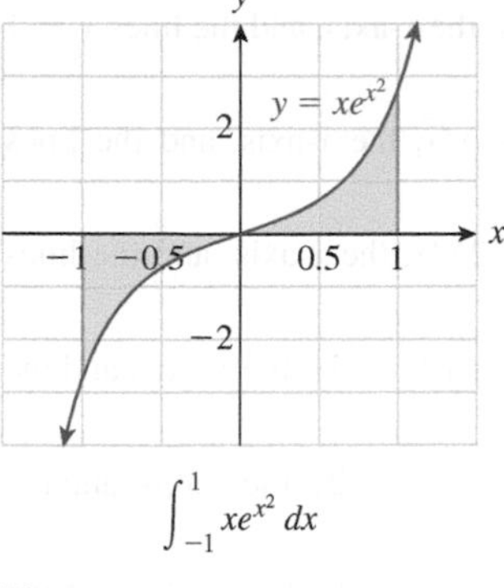

$\int_{-1}^{1} xe^{x^2}\,dx$

Figure 22

Solution The region whose area we want is shown in Figure 22. Notice the symmetry of the graph. Also, half the region we are interested in is above the x-axis, while the other half is below. If we calculated the integral $\int_{-1}^{1} xe^{x^2}\,dx$, the result would be

$$\text{Area above } x\text{-axis} - \text{Area below } x\text{-axis} = 0,$$

which does not give us the total area. To prevent the area below the x-axis from being combined with the area above the axis, we do the calculation in two parts, as illustrated in Figure 23.

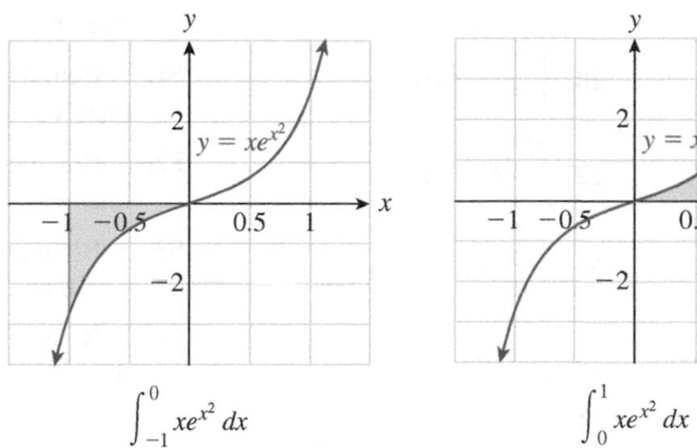

$\int_{-1}^{0} xe^{x^2}\,dx$ $\qquad$ $\int_{0}^{1} xe^{x^2}\,dx$

Figure 23

(In Figure 23 we broke the integral at $x = 0$ because that is where the graph crosses the x-axis.) These integrals can be calculated by using the substitution $u = x^2$:

$$\int_{-1}^{0} xe^{x^2}\,dx = \frac{1}{2}\Big[e^{x^2}\Big]_{-1}^{0} = \frac{1}{2}(1 - e) \approx -0.85914 \qquad \text{Why is it negative?}$$

$$\int_{0}^{1} xe^{x^2}\,dx = \frac{1}{2}\Big[e^{x^2}\Big]_{0}^{1} = \frac{1}{2}(e - 1) \approx 0.85914.$$

To obtain the total area, we should add the *absolute values* of these answers because we don't wish to count any area as negative. Thus,

$$\text{Total area} \approx 0.85914 + 0.85914 = 1.71828.$$

6.4 EXERCISES

▼ more advanced ◆ challenging

T indicates exercises that should be solved using technology

In Exercises 1–44, evaluate the integral. [**HINT:** See Example 2.]

1. $\int_{-1}^{1} (x^2 + 2)\,dx$

2. $\int_{-2}^{1} (x - 2)\,dx$

3. $\int_{0}^{1} (12x^5 + 5x^4 - 6x^2 + 4)\,dx$

4. $\int_{0}^{1} (4x^3 - 3x^2 + 4x - 1)\,dx$

5. $\int_{-2}^{2} (x^3 - 2x)\,dx$

6. $\int_{-1}^{1} (2x^3 + x)\,dx$

7. $\int_1^3 \left(\frac{2}{x^2} + 3x\right) dx$

8. $\int_2^3 \left(x + \frac{1}{x}\right) dx$

9. $\int_0^1 (2.1x - 4.3x^{1.2})\, dx$

10. $\int_{-1}^0 (4.3x^2 - 1)\, dx$

11. $\int_0^1 2e^x\, dx$

12. $\int_{-1}^0 3e^x\, dx$

13. $\int_0^1 \sqrt{x}\, dx$

14. $\int_{-1}^1 \sqrt[3]{x}\, dx$

15. $\int_0^1 2^x\, dx$

16. $\int_0^1 3^x\, dx$

[**HINT**: In Exercises 17–44, use a shortcut or see Example 4.]

17. $\int_0^1 18(3x + 1)^5\, dx$

18. $\int_0^1 8(-x + 1)^7\, dx$

19. $\int_{-1}^1 e^{2x-1}\, dx$

20. $\int_0^2 e^{-x+1}\, dx$

21. $\int_0^2 2^{-x+1}\, dx$

22. $\int_{-1}^1 3^{2x-1}\, dx$

23. $\int_0^4 |-3x + 4|\, dx$

24. $\int_{-4}^4 |-x - 2|\, dx$

25. $\int_0^1 5x(8x^2 + 1)^{-1/2}\, dx$

26. $\int_0^{\sqrt{2}} x\sqrt{2x^2 + 1}\, dx$

27. $\int_{-\sqrt{2}}^{\sqrt{2}} 3x\sqrt{2x^2 + 1}\, dx$

28. $\int_{-2}^2 xe^{-x^2+1}\, dx$

29. $\int_0^1 5xe^{x^2+2}\, dx$

30. $\int_0^2 \frac{3x}{x^2 + 2}\, dx$

31. $\int_2^3 \frac{x^2}{x^3 - 1}\, dx$

32. $\int_2^3 \frac{x}{2x^2 - 5}\, dx$

33. $\int_0^1 x(1.1)^{-x^2}\, dx$

34. $\int_0^1 x^2(2.1)^{x^3}\, dx$

35. $\int_1^2 \frac{e^{1/x}}{x^2}\, dx$

36. $\int_1^2 \frac{\sqrt{\ln x}}{x}\, dx$

37. $\int_0^2 \frac{e^{-2x}}{1 + 3e^{-2x}}\, dx$ (Round the answer to four decimal places.)

38. $\int_0^1 \frac{e^{2x}}{1 - 3e^{2x}}\, dx$ (Round the answer to four decimal places.)

39. ▼ $\int_0^2 \frac{x}{x + 1}\, dx$

40. ▼ $\int_{-1}^1 \frac{2x}{x + 2}\, dx$

41. ▼ $\int_1^2 x(x - 2)^5\, dx$

42. ▼ $\int_1^2 x(x - 2)^{1/3}\, dx$

43. ▼ $\int_0^1 x\sqrt{2x + 1}\, dx$

44. ▼ $\int_{-1}^0 2x\sqrt{x + 1}\, dx$

In Exercises 45–54, calculate the total area of the region described. Do not count area beneath the x-axis as negative. [**HINT**: See Example 6.]

45. Bounded by the line $y = x$, the x-axis, and the lines $x = 0$ and $x = 1$

46. Bounded by the line $y = 2x$, the x-axis, and the lines $x = 1$ and $x = 2$

47. Bounded by the curve $y = \sqrt{x}$, the x-axis, and the lines $x = 0$ and $x = 4$

48. Bounded by the curve $y = 2\sqrt{x}$, the x-axis, and the lines $x = 0$ and $x = 16$

49. Bounded by the graph of $y = |2x - 3|$, the x-axis, and the lines $x = 0$ and $x = 3$

50. Bounded by the graph of $y = |3x - 2|$, the x-axis, and the lines $x = 0$ and $x = 3$

51. ▼ Bounded by the curve $y = x^2 - 1$, the x-axis, and the lines $x = 0$ and $x = 4$

52. ▼ Bounded by the curve $y = 1 - x^2$, the x-axis, and the lines $x = -1$ and $x = 2$

53. ▼ Bounded by the x-axis, the curve $y = xe^{x^2}$, and the lines $x = 0$ and $x = (\ln 2)^{1/2}$

54. ▼ Bounded by the x-axis, the curve $y = xe^{x^2-1}$, and the lines $x = 0$ and $x = 1$

Applications

A number of the following exercises are similar to ones you have already seen in Section 6.3, except that this time, rather than approximating the definite integrals by Riemann sums, you are asked to calculate them exactly using the FTC.

55. ***Pumps*** (Compare Exercise 39 in Section 6.3.) A pump is delivering water into a tank at a rate of

$$r(t) = 3t^2 + 5 \text{ liters per minute,}$$

where t is time in minutes since the pump is turned on. Determine the total volume of water pumped in during the first 2 minutes. [**HINT**: See Example 5.]

56. ***Pumps*** (Compare Exercise 40 in Section 6.3.) A pump is delivering water into a tank at a rate of

$$r(t) = 6t^2 + 40 \text{ liters per minute,}$$

where t is time in minutes since the pump is turned on. Determine the total volume of water pumped in during the first 3 minutes. [**HINT**: See Example 5.]

57. ***Cost*** The marginal cost of producing the xth box of light bulbs is $5 + \frac{x^2}{1{,}000}$ dollars. Determine how much is added to the total cost by a change in production from $x = 10$ to $x = 100$ boxes. [**HINT**: See Example 5.]

58. ***Revenue*** The marginal revenue of the xth box of flash cards sold is $100e^{-0.001x}$ dollars. Find the revenue generated by selling items 101 through 1,000. [**HINT:** See Example 5.]

59. ***Profit: iPhones*** (Compare Exercise 43 in Section 6.3.) Assume that **Apple**'s marginal cost function for the manufacture of x 32GB iPhone 6's per hour at the **Foxconn Technology Group** is[45]

$$c(x) = 160 - 0.002x,$$

and that Apple sells iPhone 6's for an average wholesale price of \$580. Determine the total additional hourly profit corresponding to an increase in production and sales from 10,000 to 20,000 iPhone 6's per hour.

60. ***Profit: PlayStation 4's*** (Compare Exercise 44 in Section 6.3.) Assume that **Sony**'s marginal cost function for the manufacture of x PlayStation 4's per hour is[46]

$$c(x) = 340 + 0.001x,$$

and that Sony sells PlayStation 4's for an average wholesale price of \$400. Determine the total additional hourly profit corresponding to an increase in production and sales from 50,000 to 60,000 PlayStation 4's per hour.

61. ***Motion*** A car traveling down a road has a velocity of $v(t) = 60 - e^{-t/10}$ mph at time t hours. Find the distance it has traveled from time $t = 1$ hour to time $t = 6$ hours. (Round your answer to the nearest mile.)

62. ***Motion*** A ball thrown in the air has a velocity of $v(t) = 100 - 32t$ ft/sec at time t seconds. Find the total displacement of the ball between times $t = 1$ second and $t = 7$ seconds, and interpret your answer.

63. ***Motion*** A car slows to a stop at a stop sign, then starts up again, in such a way that its speed at time t seconds after it starts to slow is $v(t) = |-10t + 40|$ ft/sec. How far does the car travel from time $t = 0$ to time $t = 10$ seconds?

64. ***Motion*** A truck slows, doesn't quite stop at a stop sign, and then speeds up again in such a way that its speed at time t seconds is $v(t) = 10 + |-5t + 30|$ ft/sec. How far does the truck travel from time $t = 0$ to time $t = 10$?

65. ***Fuel Consumption*** The way Professor Waner drives, he burns gas at the rate of $1 - e^{-t}$ gallons each hour, t hours after a fill-up. Find the number of gallons of gas he burns in the first 10 hours after a fill-up.

66. ***Fuel Consumption*** The way Professor Costenoble drives, he burns gas at the rate of $1/(t + 1)$ gallons each hour, t hours after a fill-up. Find the number of gallons of gas he burns in the first 10 hours after a fill-up.

67. ***Bottled Water Sales*** (Compare Exercise 45 in Section 6.3.) The rate of U.S. sales of bottled water for the period 2007–2014 can be approximated by

$$s(t) = 0.08t^2 - 0.26t + 8.8 \text{ billion gallons per year} \quad (0 \le t \le 7),$$

where t is time in years since the start of 2007.[47] Use a definite integral to estimate the total U.S. sales of bottled water from the start of 2008 to the start of 2014. (Round your answer to the nearest billion gallons.)

68. ***Bottled Water Sales*** (Compare Exercise 46 in Section 6.3.) The rate of U.S. per capita sales of bottled water for the period 2007–2014 can be approximated by

$$s(t) = 0.25t^2 - t + 29 \text{ gallons per year} \quad (0 \le t \le 7),$$

where t is time in years since the start of 2007.[48] Use a definite integral to estimate the total U.S. per capita sales of bottled water from the start of 2008 to the start of 2012. (Round your answer to the nearest gallon.)

69. ***Facebook Membership*** (Compare Exercise 63 in Section 6.3.) Starting in 2007, new members joined **Facebook** at a rate of roughly

$$m(t) = 20t^2 + 60t + 12 \text{ million members per year} \quad (0 \le t \le 6),$$

where t is time in years since the start of 2007.[49] Use a definite integral to estimate, to the nearest million, the total number of new Facebook members from the start of 2008 to the start of 2012.

70. ***Uploads to YouTube*** (Compare Exercise 64 in Section 6.3.) Since **YouTube** first became available to the public in mid-2005, the rate at which video has been uploaded to the site can be approximated by

$$v(t) = 1.1t^2 - 2.6t + 2.3 \text{ million hours of video per year} \quad (0 \le t \le 9),$$

where t is time in years since June 2005.[50] Use a definite integral to estimate, to the nearest million, the total number of hours of video uploaded from June 2010 to June 2014.

---

[45] Not the actual marginal cost equation; the authors do not know Apple's actual marginal cost equation, but the marginal costs given here are in rough agreement with the actual costs for one of the 2014 models. Sources: http://time.com, www.digitaltrends.com.

[46] Not the actual marginal cost equation; the authors do not know Sony's actual marginal cost equation, but the marginal costs given here are in rough agreement with the actual costs. Sources: VentureBeat (http://venturebeat.com), http://ps4daily.com.

[47] Source for data: Beverage Marketing Corporation.

[48] *Ibid.*

[49] Sources for data: www.facebook.com, www.insidefacebook.com.

[50] Sources for data: www.youtube.com, www.statista.com.

71. ***Big Brother*** (Compare Exercise 65 in Section 6.3.) The total number of wiretaps authorized each year by U.S. state and federal courts from 1990 to 2015 can be approximated by

$$w(t) = 774e^{0.06t} \quad (0 \le t \le 25).$$

(t is time in years since the start of 1990.)[51] Compute $\int_{10}^{25} w(t)\,dt$. (Round your answer to the nearest 10.) Interpret the answer.

72. ***Big Brother*** (Compare Exercise 66 in Section 6.3.) The number of wiretaps authorized each year by U.S. state courts from 1990 to 2015 can be approximated by

$$w(t) = 412e^{0.071t} \quad (0 \le t \le 25).$$

(t is time in years since the start of 1990.)[52] Compute $\int_{5}^{25} w(t)\,dt$. (Round your answer to the nearest 10.) Interpret the answer.

73. ***Economic Growth*** (Compare Exercise 87 in Section 6.2.) The Mexico GDP (total monetary value of all finished goods and services produced in Mexico) can be approximated by

$$g(t) = 2{,}000 - 480e^{-0.06t} \text{ billion pesos per year} \quad (0 \le t \le 5),$$

where t is time in years since January 2010.[53] Use the FTC to find an expression for the total GDP $G(T)$ of sold goods in Mexico from January 2010 to time T. Hence estimate, to the nearest billion pesos, the total Mexico GDP from January 2010 through June 2014. (The actual value was 7,137 billion pesos.)

74. ***Housing Starts: The Great Recession*** (Compare Exercise 88 in Section 6.2.) The Great Recession of 2007–2009 is largely attributed to the real estate crisis beginning in 2006, from which time the number of housing starts was approximately

$$n(t) = 2{,}400e^{-0.25t} - 200 \text{ thousand homes per year} \quad (0 \le t \le 4),$$

where t is time in years since January 2006.[54] Use the FTC to find an expression for the total number $N(T)$ of housing starts in the United States from January 2006 to time T. Hence estimate, to the nearest 0.1 million, the total number of housing starts from January 2006 through June 2009. (The actual number was around 4.9 million homes.) [**HINT:** Use the shortcuts.]

75. ▼ ***Sales*** Weekly sales of your *Lord of the Rings* T-shirts have been falling by 5% per week. Assuming that you are now selling 50 T-shirts per week, how many shirts will you sell during the coming year? (Round your answer to the nearest shirt.)

76. ▼ ***Sales*** Annual sales of fountain pens in Littleville are 4,000 per year and are increasing by 10% per year. How many fountain pens will be sold over the next 5 years?

77. T ***Embryo Development*** The oxygen consumption of a bird embryo increases from the time the egg is laid through the time the chick hatches. In a typical galliform bird the oxygen consumption can be approximated by

$$c(t) = -0.065t^3 + 3.4t^2 - 22t + 3.6 \text{ milliliters per day} \quad (8 \le t \le 30),$$

where t is the time (in days) since the egg was laid.[55] (An egg will typically hatch at around $t = 28$.) Use technology to estimate the total amount of oxygen consumed during the ninth and tenth days ($t = 8$ to $t = 10$). Round your answer to the nearest milliliter. [**HINT:** See the technology note in the margin next to Example 5.]

78. T ***Embryo Development*** The oxygen consumption of a turkey embryo increases from the time the egg is laid through the time the chick hatches. In a brush turkey the oxygen consumption can be approximated by

$$c(t) = -0.028t^3 + 2.9t^2 - 44t + 95 \text{ milliliters per day} \quad (20 \le t \le 50),$$

where t is the time (in days) since the egg was laid.[56] (An egg will typically hatch at around $t = 50$.) Use technology to estimate the total amount of oxygen consumed during the 21st and 22nd days ($t = 20$ to $t = 22$). Round your answer to the nearest 10 milliliters. [**HINT:** See the technology note in the margin next to Example 5.]

79. T ***Online Payments*** The rate of change of total payments through PayPal from the first quarter of 2013 through the fourth quarter of 2014 can be approximated by

$$p(t) = -0.1t^3 + 1.18t^2 - 0.89t + 41 \text{ billion dollars per quarter} \quad (1 \le t \le 8),$$

where t is time in quarters. ($t = 1$ represents the first quarter of 2013.)[57] Use technology to estimate $\int_4^8 p(t)\,dt$. Interpret your answer.

---

[51] Source for data: Wiretap Reports, Administrative Office of the United States Courts (www.uscourts.gov/Statistics/WiretapReports).

[52] *Ibid.*

[53] The GDP is in constant 2008 pesos. Source for data: Instituto Nacional de Estadística y Geografía (INEGI) (www.inegi.org.mx).

[54] Source for data: www.census.gov.

[55] The model approximates graphical data published in the article "The Brush Turkey" by Roger S. Seymour, *Scientific American*, December 1991, pp. 108–114.

[56] *Ibid.*

[57] Source for data: Statista, www.statista.com.

80. T ***Online Auctions*** The rate of change $n(t)$ of the number of active **eBay** users could be approximated by

$$n(t) = -0.002t^4 + 0.06t^3 - 0.55t^2 + 1.9t - 1$$
million users per quarter $(1 \le t \le 13)$,

where t is time in quarters. ($t = 1$ represents the first quarter of 2008.)[58] Use technology to compute $\int_5^{13} n(t)\,dt$ correct to the nearest whole number. Interpret your answer.

81. ▼ ***Cost*** Use the Fundamental Theorem of Calculus to show that if $m(x)$ is the marginal cost at a production level of x items, then the cost function $C(x)$ is given by

$$C(x) = C(0) + \int_0^x m(t)\,dt.$$

What do we call $C(0)$?

82. ▼ ***Cost*** The total cost of producing x items is given by

$$C(x) = 246.76 + \int_0^x 5t\,dt.$$

Find the fixed cost and the marginal cost of producing the tenth item.

83. ***Scientific Research: 1983–2003*** (Compare Exercise 91 in Section 6.2.) The number of research articles in the prominent journal *Physical Review* written by researchers in Europe during 1983–2003 can be approximated by

$$E(t) = \frac{7e^{0.2t}}{5 + e^{0.2t}} \text{ thousand articles per year} \quad (0 \le t \le 20),$$

where t is time in years. ($t = 0$ represents 1983.)[59] Use a definite integral to estimate the number of articles written by researchers in Europe from 1983 to 2003. (Round your answer to the nearest 1,000 articles.) [**HINT:** See Example 5 in Section 6.2.]

84. ***Scientific Research: 1983–2003*** (Compare Exercise 92 in Section 6.2.) The number of research articles in the prominent journal *Physical Review* written by researchers in the United States during 1983–2003 can be approximated by

$$U(t) = \frac{4.6e^{0.6t}}{0.4 + e^{0.6t}} \text{ thousand articles per year} \quad (0 \le t \le 20),$$

where t is time in years. ($t = 0$ represents 1983.)[60] Use a definite integral to estimate the total number of articles written by researchers in the United States from 1983 to 2003. (Round your answer to the nearest 1,000 articles.) [**HINT:** See Example 5 in Section 6.2.]

85. ▼ ***The Logistic Function and High School Graduates***

a. Show that the logistic function $f(x) = \dfrac{N}{1 + Ab^{-x}}$ can be written in the form

$$f(x) = \frac{Nb^x}{A + b^x}.$$

[**HINT:** See the note after Example 5 in Section 6.2.]

b. Use the result of part (a) and a suitable substitution to show that

$$\int \frac{N}{1 + Ab^{-x}}\,dx = \frac{N\ln(A + b^x)}{\ln b} + C.$$

c. The rate of graduation of private high school students in the United States for the period 1994–2008 was approximately

$$r(t) = 220 + \frac{110}{1 + 3.8(1.27)^{-t}} \text{ thousand students per year} \quad (0 \le t \le 14)$$

t years since 1994.[61] Use the result of part (b) to estimate the total number of private high school graduates over the period 2000–2008.

86. ▼ ***The Logistic Function and Grant Spending***

a. Show that the logistic function $f(x) = \dfrac{N}{1 + Ae^{-kx}}$ can be written in the form

$$f(x) = \frac{Ne^{kx}}{A + e^{kx}}.$$

[**HINT:** See the note after Example 5 in Section 6.2.]

b. Use the result of part (a) and a suitable substitution to show that

$$\int \frac{N}{1 + Ae^{-kx}}\,dx = \frac{N\ln(A + e^{kx})}{k} + C.$$

c. The rate of spending on grants by U.S. foundations in the period 1993–2003 was approximately

$$s(t) = 11 + \frac{20}{1 + 1{,}800e^{-0.9t}} \text{ billion dollars per year} \quad (3 \le t \le 13),$$

where t is the number of years since 1990.[62] Use the result of part (b) to estimate, to the nearest \$10 billion, the total spending on grants from 1998 to 2003.

---

[58] Source for data: eBay company reports (http://investor.ebay.com).

[59] Based on data from 1983 to 2003. Source: The American Physical Society/*New York Times*, May 3, 2003, p. A1.

[60] *Ibid.*

[61] Based on a logistic regression. Source for data: National Center for Educational Statistics (www.nces.ed.gov).

[62] Based on a logistic regression. Source for data: The Foundation Center, *Foundation Growth and Giving Estimates*, 2004, downloaded from the Center's website (www.fdncenter.org).

87. ◆ ***Kinetic Energy*** The work done in accelerating an object from velocity v_0 to velocity v_1 is given by

$$W = \int_{v_0}^{v_1} v \frac{dp}{dv}\, dv,$$

where p is its momentum, given by $p = mv$ (m = mass). Assuming that m is a constant, show that

$$W = \frac{1}{2}mv_1^2 - \frac{1}{2}mv_0^2.$$

The quantity $\frac{1}{2}mv^2$ is referred to as the **kinetic energy** of the object, so the work required to accelerate an object is given by its change in kinetic energy.

88. ◆ ***Einstein's Energy Equation*** According to the special theory of relativity, the apparent mass of an object depends on its velocity according to the formula

$$m = \frac{m_0}{\sqrt{1 - \frac{v^2}{c^2}}},$$

where v is its velocity, m_0 is the "rest mass" of the object (that is, its mass when $v = 0$), and c is the velocity of light: approximately 3×10^8 m/sec.

a. Show that, if $p = mv$ is the momentum,

$$\frac{dp}{dv} = \frac{m_0}{\left(1 - \frac{v^2}{c^2}\right)^{3/2}}.$$

b. Use the integral formula for W in Exercise 87, together with the result in part (a), to show that the work required to accelerate an object from a velocity of v_0 to v_1 is given by

$$W = \frac{m_0c^2}{\sqrt{1 - \frac{v_1^2}{c^2}}} - \frac{m_0c^2}{\sqrt{1 - \frac{v_0^2}{c^2}}}.$$

We call the quantity $\dfrac{m_0c^2}{\sqrt{1 - \frac{v^2}{c^2}}}$ the **total relativistic energy** of an object moving at velocity v. Thus, the work to accelerate an object from one velocity to another is given by the change in its total relativistic energy.

c. Deduce (as Albert Einstein did) that the total relativistic energy E of a body at rest with rest mass m is given by the famous equation

$$E = mc^2.$$

Communication and Reasoning Exercises

89. Explain how the indefinite integral and the definite integral are related.

90. What is "definite" about the definite integral?

91. The total change of a quantity from time a to time b can be obtained from its rate of change by doing what?

92. Complete the following: The total sales from time a to time b are obtained from the marginal sales by taking its ______ ______ from ______ to ______.

93. What does the Fundamental Theorem of Calculus permit one to do?

94. If Felice and Philipe have different antiderivatives of f and each uses his or her own antiderivative to compute $\int_a^b f(x)\, dx$, they might get different answers—right?

95. ▼ Give an example of a nonzero velocity function that will produce a displacement of 0 from time $t = 0$ to time $t = 10$.

96. ▼ Give an example of a nonzero function whose definite integral over the interval $[4, 6]$ is zero.

97. ▼ Give an example of a decreasing function $f(x)$ with the property that $\int_a^b f(x)\, dx$ is positive for every choice of a and $b > a$.

98. ▼ Explain why, in computing the total change of a quantity from its rate of change, it is useful to have the definite integral subtract area below the x-axis.

99. ◆ If $f(x)$ is a continuous function defined for $x \geq a$, define a new function $F(x)$ by the formula

$$F(x) = \int_a^x f(t)\, dt.$$

Use the Fundamental Theorem of Calculus to deduce that $F'(x) = f(x)$. What, if anything, is interesting about this result?

100. T ◆ Use the result of Exercise 99 and technology to compute a table of values for $x = 1, 2, 3$ for an antiderivative $A(x)$ of e^{-x^2} with the property that $A(0) = 0$. (Round answers to two decimal places.)

CHAPTER 6 REVIEW

KEY CONCEPTS

www.WanerMath.com
Go to the Website to find a comprehensive and interactive Web-based summary of Chapter 6.

6.1 The Indefinite Integral

An antiderivative of a function *f* is a function *F* such that $F' = f$. [p. 470]

Indefinite integral $\int f(x)\,dx$ [p. 470]

Power rule for the indefinite integral:

$$\int x^n\,dx = \frac{x^{n+1}}{n+1} + C \quad (\text{if } n \neq -1)$$

$$\int x^{-1}\,dx = \ln|x| + C \quad [\text{p. } 472]$$

Indefinite integral of e^x and b^x:

$$\int e^x\,dx = e^x + C$$

$$\int b^x\,dx = \frac{b^x}{\ln b} + C \quad [\text{p. } 473]$$

Indefinite integral of $|x|$:

$$\int |x|\,dx = \frac{x|x|}{2} + C \quad [\text{p. } 473]$$

Sums, differences, and constant multiples:

$$\int [f(x) \pm g(x)]\,dx = \int f(x)\,dx \pm \int g(x)\,dx$$

$$\int kf(x)\,dx = k\int f(x)\,dx \quad (k \text{ constant}) \ [\text{p. } 473]$$

Combining the rules [p. 475]

Position, velocity, and acceleration:

$$v = \frac{ds}{dt} \qquad s(t) = \int v(t)\,dt$$

$$a = \frac{dv}{dt} \qquad v(t) = \int a(t)\,dt$$

[p. 478]

Motion in a straight line [p. 479]

Vertical motion under gravity [p. 481]

6.2 Substitution

Substitution rule:

$$\int f\,dx = \int\left(\frac{f}{du/dx}\right)du \quad [\text{p. } 486]$$

Using the substitution rule [p. 487]

Shortcuts: integrals of expressions involving $(ax + b)$:

$$\int (ax+b)^n\,dx = \frac{(ax+b)^{n+1}}{a(n+1)} + C \quad (\text{if } n \neq -1)$$

$$\int (ax+b)^{-1}\,dx = \frac{1}{a}\ln|ax+b| + C$$

$$\int e^{ax+b}\,dx = \frac{1}{a}e^{ax+b} + C$$

$$\int c^{ax+b}\,dx = \frac{1}{a\ln c}c^{ax+b} + C$$

$$\int |ax+b|\,dx = \frac{1}{2a}(ax+b)|ax+b| + C \quad [\text{p. } 492]$$

Mike's shortcut rule:

If $\int f(x)\,dx = F(x) + C$ and g and u are differentiable functions of x, then

$$\int g \cdot f(u)\,dx = \frac{g}{u'} \cdot F(u) + C$$

provided that $\frac{g}{u'}$ is constant. [p. 493]

6.3 The Definite Integral: Numerical and Graphical Viewpoints

Left Riemann sum:

$$\sum_{k=0}^{n-1} f(x_k)\,\Delta x = [f(x_0) + f(x_1) + \cdots + f(x_{n-1})]\,\Delta x$$

[p. 503]

Computing the Riemann sum from a formula [p. 504]

Computing the Riemann sum from a graph [p. 505]

Definite integral of *f* from *a* to *b*:

$$\int_a^b f(x)\,dx = \lim_{n\to\infty}\sum_{k=0}^{n-1} f(x_k)\,\Delta x$$

[p. 505]

Estimating the definite integral from a graph [p. 508]

Estimating the definite integral using technology [p. 510]

Application to motion in a straight line [p. 510]

6.4 The Definite Integral: Algebraic Viewpoint and the Fundamental Theorem of Calculus

Computing total cost from marginal cost [p. 518]

The Fundamental Theorem of Calculus (FTC) [p. 519]

Using the FTC to compute definite integrals [p. 520]

Using the FTC with substitution [p. 521]

Computing area [p. 523]

REVIEW EXERCISES

In Exercises 1–18, evaluate the indefinite integral.

1. $\int (x^2 - 10x + 2)\,dx$
2. $\int (e^x + \sqrt{x})\,dx$
3. $\int \left(\frac{4x^2}{5} - \frac{4}{5x^2}\right)dx$
4. $\int \left(\frac{3x}{5} - \frac{3}{5x}\right)dx$
5. $\int (2x)^{-1}\,dx$
6. $\int (-2x+2)^{-2}\,dx$
7. $\int e^{-2x+11}\,dx$
8. $\int \frac{dx}{(4x-3)^2}$
9. $\int x(x^2+1)^{1.3}\,dx$
10. $\int x(x^2+4)^{10}\,dx$
11. $\int \frac{4x}{(x^2-7)}\,dx$
12. $\int \frac{x}{(3x^2-1)^{0.4}}\,dx$

13. $\int (x^3 - 1)\sqrt{x^4 - 4x + 1}\, dx$

14. $\int \frac{x^2 + 1}{(x^3 + 3x + 2)^2}\, dx$

15. $\int (-xe^{x^2/2})\, dx$

16. $\int xe^{-x^2/2}\, dx$

17. $\int \frac{x + 1}{x + 2}\, dx$

18. $\int x\sqrt{x - 1}\, dx$

In Exercises 19 and 20, use the given graph to estimate the left Riemann sum for the given interval with the stated number of subdivisions.

19. $[0, 3], n = 6$

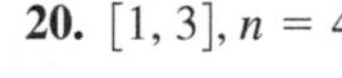

20. $[1, 3], n = 4$

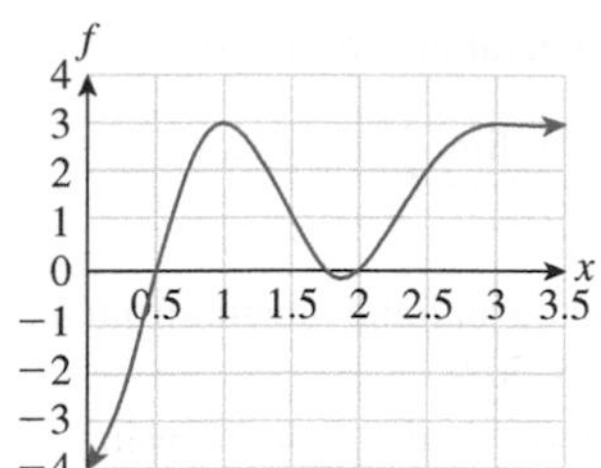

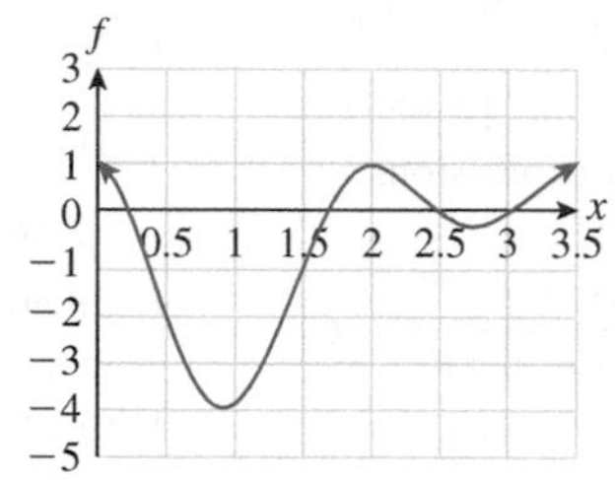

In Exercises 21–26, calculate the left Riemann sum for the given function over the given interval, using the given value of n. (When rounding, round answers to four decimal places.)

21. $f(x) = x^2 + 1$ over $[-1, 1], n = 4$

22. $f(x) = (x - 1)(x - 2) - 2$ over $[0, 4], n = 4$

23. $f(x) = x(x^2 - 1)$ over $[0, 1], n = 5$

24. $f(x) = \frac{x - 1}{x - 2}$ over $[0, 1.5], n = 3$

25. $f(x) = e^{-x^2}$ over $[0, 10], n = 4$

26. $f(x) = e^{-x^2}$ over $[0, 100], n = 4$

T *In Exercises 27 and 28, use technology to approximate the given definite integrals using left Riemann sums with n = 10, 100, and 1,000. (Round answers to four decimal places.)*

27. $\int_0^1 e^{-x^2}\, dx$

28. $\int_1^3 x^{-x}\, dx$

In Exercises 29 and 30, the graph of the derivative $f'(x)$ of $f(x)$ is shown. Compute the total change of $f(x)$ over the given interval.

29. $[-1, 2]$

30. $[0, 2]$

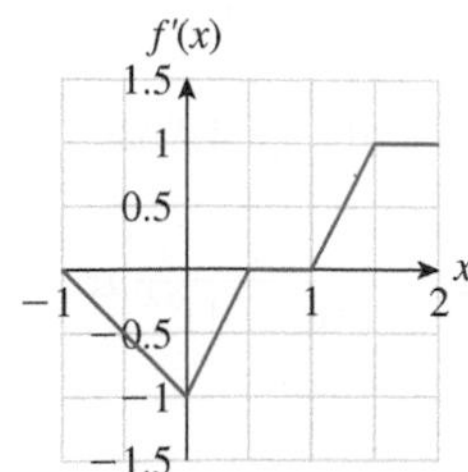

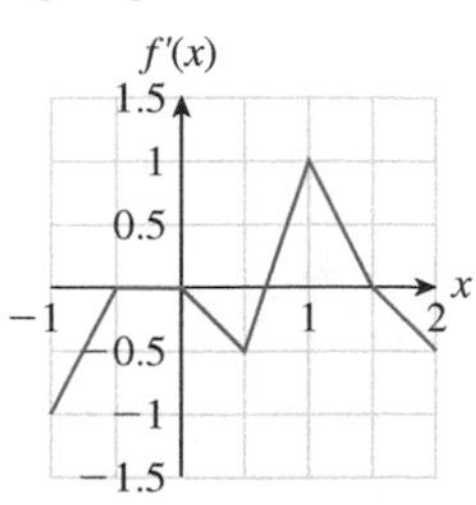

In Exercises 31–40, evaluate the definite integral using the Fundamental Theorem of Calculus.

31. $\int_{-1}^{1} (x - x^3 + |x|)\, dx$

32. $\int_0^9 (x + \sqrt{x})\, dx$

33. $\int_{-1}^{1} \frac{3}{(2x - 5)^2}\, dx$

34. $\int_0^9 \frac{1}{x + 1}\, dx$

35. $\int_0^{50} e^{-0.02x-1}\, dx$

36. $\int_{-20}^{0} 3e^{2.2x}\, dx$

37. $\int_0^2 x^2\sqrt{x^3 + 1}\, dx$

38. $\int_0^2 \frac{x^2}{\sqrt{x^3 + 1}}\, dx$

39. $\int_0^{\ln 2} \frac{e^{-2x}}{1 + 4e^{-2x}}\, dx$

40. $\int_0^{\ln 3} e^{2x}(1 - 3e^{2x})^2\, dx$

In Exercises 41–44, find the area of the specified region. (Do not count area below the x-axis as negative.)

41. The area bounded by $y = 4 - x^2$, the x-axis, and the lines $x = -2$ and $x = 2$

42. The area bounded by $y = 4 - x^2$, the x-axis, and the lines $x = 0$ and $x = 5$

43. The area bounded by $y = xe^{-x^2}$, the x-axis, and the lines $x = 0$ and $x = 5$

44. The area bounded by $y = |2x|$, the x-axis, and the lines $x = -1$ and $x = 1$

Applications: OHaganBooks.com
[Try the game at www.OHaganBooks.com]

45. ***Sales*** At OHaganBooks.com, the rate of net sales (sales minus returns) of *The Secret Loves of John O*, a romance novel by Margó Dufón, can be approximated by

$$n(t) = 196 + t^2 - 0.16t^5 \text{ copies per week}$$

t weeks since its release.

a. Find the total net sales N as a function of time t.

b. How many books are still held by customers after 6 weeks? (Round your answer to the nearest book.)

46. ***Demand*** If OHaganBooks.com were to give away its latest best seller, *A River Burns through It*, the demand q would be 100,000 books. The marginal demand (dq/dp) for the book is $-20p$ at a price of p dollars.

a. What is the demand function for this book?

b. At what price does demand drop to zero?

47. ***Motion under Gravity*** Billy-Sean O'Hagan's friend Juan says that he can throw a baseball vertically upward at 100 ft/sec. Assuming that Juan's claim is true,

a. Where would the baseball be at time t seconds?

b. How high would the ball go?

c. When would the ball return to Juan's hand?

48. ***Motion under Gravity*** An overworked employee at OHaganBooks.com goes to the top of the company's 100-foot-tall headquarters building and flings a book up into the air at a speed of 60 ft/sec.

a. When will the book hit the ground 100 feet below? (Neglect air resistance.)
b. How fast will the book be traveling when it hits the ground?
c. How high will the book go?

49. *Projected Sales* Before OHaganBooks.com launched its online site, the sales consultant contracted by John O'Hagan had conservatively projected that online sales on the website would be

$$s(t) = 6.2e^{0.25t+3} \text{ thousand books per week}$$

t weeks after going online.
a. Calculate $\int s(t)\,dt$.
b. What is the projection of total online sales beginning at the launch to time t?

50. *Bandwidth* Billy-Sean O'Hagan has been using the company servers for his (classified) student intern work at *Brain Cybernetics,* and the bandwidth he is using has been growing exponentially, following the model

$$q(t) = 5.1e^{0.1t} + 3.3 \text{ terabytes per month,}$$

where t is time in months since the beginning of the year.
a. Calculate $\int q(t)\,dt$.
b. What is the total bandwith consumed to time t given that Billy-Sean's internship started at the beginning of January?

51. *Sales* Sales at the OHaganBooks.com website of *Larry Potter and the Riemann Sum* fluctuated rather wildly in the first 5 months of last year, as the following graph shows:

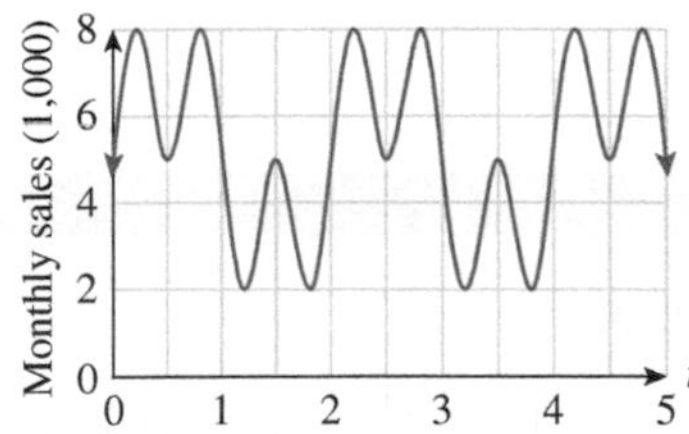

Puzzled by the graph, CEO John O'Hagan asks Jimmy Duffin[63] to estimate the total sales over the entire 5-month period shown. Jimmy decides to use a left Riemann sum with 10 partitions to estimate the total sales. What does he find?

52. *Sales* The following graph shows the approximate rate of change $s(t)$ of the total value, in thousands of dollars, of Spanish books sold online at OHaganBooks.com. (t is the number of months since January 1.)

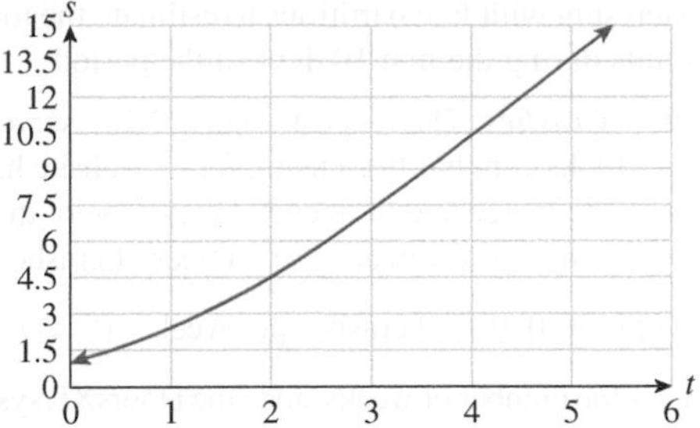

Use the graph to estimate the total value of Spanish books sold from March 1 through June 1. (Use a left Riemann sum with three subdivisions.)

53. *Promotions* Unlike sales of *Larry Potter and the Riemann Sum,* sales at OHaganBooks.com of the special leather-bound gift editions of *Calculus for Vampires* have been suffering lately, as shown in the following graph. (Negative sales indicate returns by dissatisfied customers; t is time in months since January 1 of this year.)

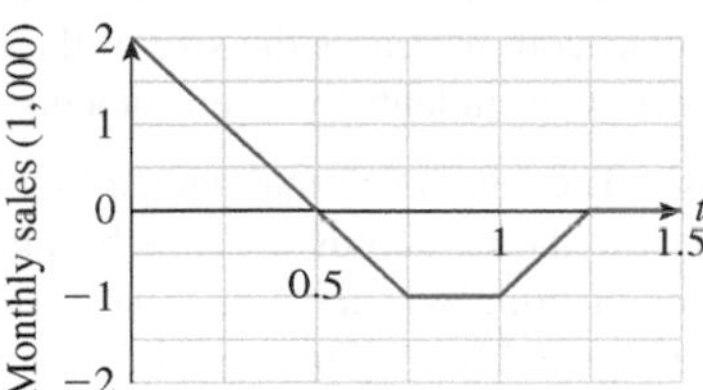

Use the graph to compute the total (net) sales over the period shown.

54. *Sales* Even worse than with the leather-bound *Calculus for Vampires*, sales of *Real Estate for Werewolves* have been dismal, as shown in the following graph. (Negative sales indicate returns by dissatisfied customers; t is time in months since January 1 of this year.)

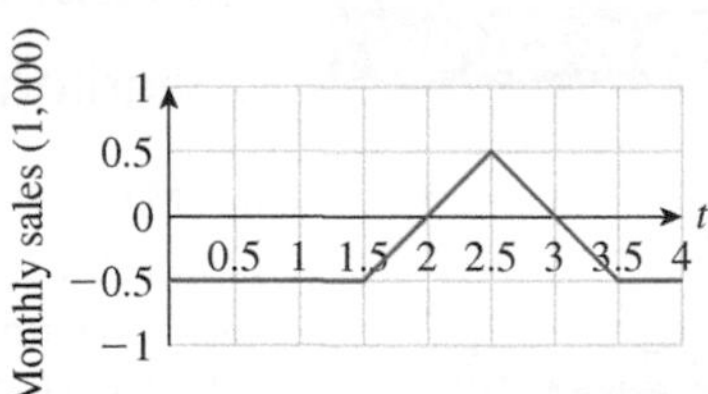

Use the graph to compute the total (net) sales over the period shown.

55. *Website Activity* The number of hits on the OHaganBooks .com website has been steadily increasing over the past month in response to recent publicity about a software glitch that caused the company to pay customers for buying books online. The activity can be modeled by

$$n(t) = 1{,}000t - 10t^2 + t^3 \text{ hits per day,}$$

where t is time in days since news about the software glitch was first publicized on GrungeReport.com. Use a left

[63] Marjory Duffin's nephew, currently at OHaganBooks.com on a summer internship.

Riemann sum with five partitions to estimate the total number of hits during the first 10 days of the period.

56. *Website Crashes* The latest DoorsXL servers that OHaganBooks.com has been using for its website have been crashing with increasing frequency lately. One of the student summer interns has estimated the number of crashes to be

$$q(t) = 0.05t^2 + 0.4t + 9 \text{ crashes per week} \quad (0 \le t \le 10),$$

where t is the number of weeks since the DoorsXL system was first installed. Use a Riemann sum with five partitions to estimate the total number of crashes from the start of week 5 to the start of week 10. (Round your answer to the nearest crash.)

57. *Student Intern Costs* The marginal monthly cost of maintaining a group of summer student interns at OHaganBooks.com is calculated to be

$$c(x) = \frac{1{,}000(x+3)^2}{(8+(x+3)^3)^{3/2}} \text{ thousand dollars per additional student.}$$

Compute, to the nearest \$100, the total monthly cost if OHaganBooks.com increases the size of the student intern program from five students to seven students.

58. *Legal Costs* The legal team maintained by OHaganBooks.com to handle the numerous lawsuits brought against the company by disgruntled clients may have to be expanded. The marginal monthly cost to maintain a team of x lawyers is estimated (by a method too complicated to explain) to be

$$c(x) = (x-2)^2[8-(x-2)^3]^{3/2} \text{ thousand dollars per additional lawyer.}$$

Compute, to the nearest \$1,000, the total monthly cost if OHaganBooks.com goes ahead with a proposal to increase the size of the legal team from two to four.

59. *Projected Sales* When OHaganBooks.com was about to go online, it estimated that its weekly sales would begin at about 6,400 books per week, with sales increasing at such a rate that weekly sales would double about every 2 weeks. If these estimates had been correct, how many books would the company have sold in the first 5 weeks? (Round your answer to the nearest 1,000 books.)

60. *Actual Sales* Once OHaganBooks.com actually went online, its weekly sales began at about 7,500 books per week, with weekly sales doubling every 3 weeks. How many books did the company actually sell in the first 5 weeks? (Round your answer to the nearest 1,000 books.)

61. *Revised Actual Sales* OHaganBooks.com modeled its revised weekly sales over a period of time after it went online with the function

$$s(t) = 6{,}053 + \frac{4{,}474e^{0.55t}}{e^{0.55t} + 14.01},$$

where t is the time in weeks after it went online. According to this model, how many books, to the nearest 100, did it actually sell in the first 5 weeks?

62. *Computer Usage* A consultant recently hired by OHaganBooks.com estimates total weekly computer usage by company employees to be

$$w(t) = 620 + \frac{900e^{0.25t}}{3 + e^{0.25t}} \text{ hours} \quad (0 \le t \le 20),$$

where t is time in weeks since January 1 of this year. Use the model to estimate the total computer usage during the first 14 weeks of the year.

CASE STUDY

Spending on Housing Construction

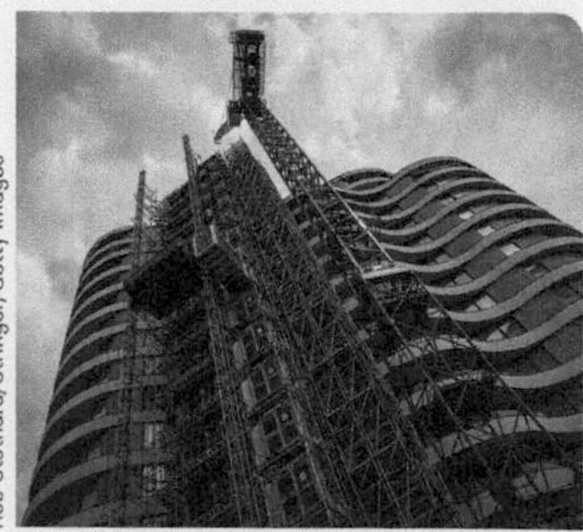

Rob Stothard/Stringer/Getty Images

It is March 2007, and *Time* magazine, in its latest edition, is asking, "Will the Housing Bubble Burst in 2007?"[64] You are a summer intern at *Schottie Construction Co.*, which is working with *Pack-Em-In Real Estate* on a major luxury condominium development to be called "Pack-Em-In/Schottie Towers."

Yesterday, you received the following memo from your supervisor:

DATE: March 15, 2007
TO: SW@EnormousStateU.edu
FROM: SC@Schottie.com (Junior VP Development)
CC: SGLombardoVP@Schottie.com (S. G. Lombardo, Senior VP Development)
SUBJECT: Residential Construction Trends. Urgent!

Help! There is a management meeting in two hours and Michelle Homestead, who, as you know, is spearheading the Pack-Em-In/Schottie Towers feasibility study, must report to

[64] *Time,* February 2007 (www.time.com/time/business/article/0,8599,1592751,00.html).

Mr. Schottie by tomorrow and has asked me to immediately produce some mathematical formulas to (1) model the trend in residential construction spending since January 2006, when it was $618.7 billion, and (2) estimate the average spent per month on residential construction over a specified period of time. All I have on hand so far is data giving the month-over-month percentage changes (attached). Do you have any ideas?

ATTACHMENT*

Month	% Change	Month	% Change
1	1.16	16	−1.59
2	1.17	17	−1.62
3	1.33	18	−1.58
4	0.67	19	−1.67
5	0.42	20	−1.77
6	−0.04	21	−1.92
7	−0.24	22	−2.06
8	−0.44	23	−2.19
9	−0.70	24	−2.45
10	−0.94	25	−2.50
11	−1.31	26	−2.65
12	−1.34	27	−2.85
13	−1.49	28	−2.65
14	−1.75	29	−2.66
15	−1.74	30	−2.66

*Based on 12-month moving average; Source for data: U.S. Census Bureau: Manufacturing, Mining and Construction Statistics, Data 360 (www.data360.org/dataset.aspx?Data_Set_Id=3627).

Getting to work, you decide that the first thing to do is fit these data to a mathematical curve that you can use to project future changes in construction spending. You graph the data to get a sense of what mathematical models might be appropriate (Figure 24).

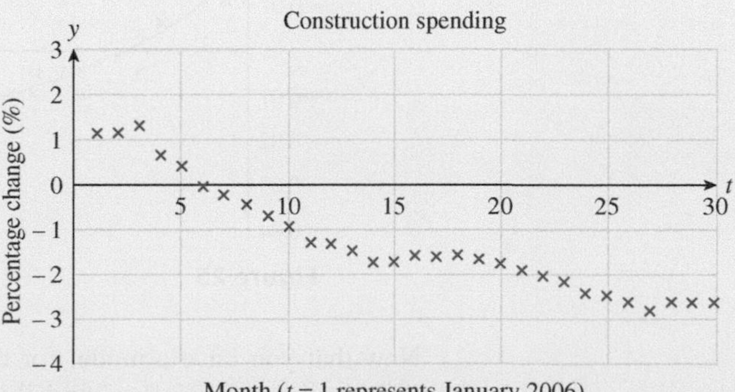

Figure 24

The graph suggests a decreasing trend, possibly concave up. You recall that there are a variety of curves that can behave this way, one of the simplest being

$$y = at^c + b \qquad (t > 0),$$

where a, b, and c are constants.

You convert all the percentages to decimals, giving the following table of data:

t	y	t	y
1	0.0116	16	−0.0159
2	0.0117	17	−0.0162
3	0.0133	18	−0.0158
4	0.0067	19	−0.0167
5	0.0042	20	−0.0177
6	−0.0004	21	−0.0192
7	−0.0024	22	−0.0206
8	−0.0044	23	−0.0219
9	−0.0070	24	−0.0245
10	−0.0094	25	−0.0250
11	−0.0131	26	−0.0265
12	−0.0134	27	−0.0285
13	−0.0149	28	−0.0265
14	−0.0175	29	−0.0266
15	−0.0174	30	−0.0266

You then find the values of a, b, and c that best fit the given data:*

$$a = -0.0200974,\ b = 0.0365316,\ c = 0.345051.$$

* To do this, you can use Excel's Solver or the Website function evaluator and grapher with model `$1*x^$2+$3`.

These values give you the following model for construction spending (with figures rounded to five significant digits):

$$y = -0.020097t^{0.34505} + 0.036532.$$

Figure 25 shows the graph of y superimposed on the data.

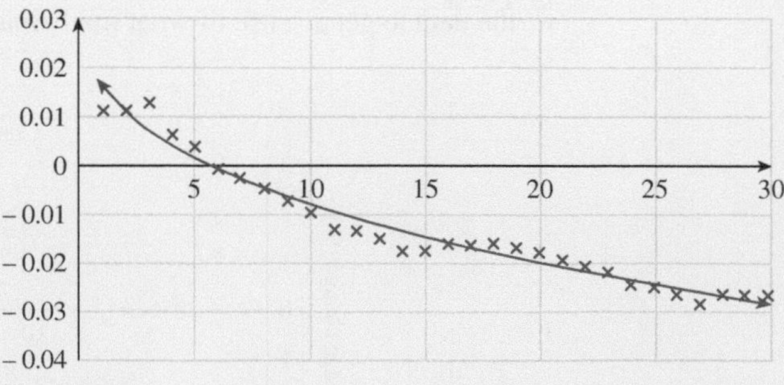

Figure 25

Now that you have a model for the month-over-month change in construction spending, you must use it to find the actual spending on construction. First, you realize that the model gives the *fractional rate of increase* of construction spending (because it is specified as a percentage, or fraction, of the total spending). In other words, if $p(t)$ represents the construction cost in month t, then

$$y = \frac{dp/dt}{p} = \frac{d}{dt}(\ln p). \quad \text{By the chain rule for derivatives}$$

You find an equation for actual monthly construction cost at time t by solving for p:

$$\ln p = \int y\, dt$$

$$= \int (at^c + b)\, dt$$

$$= \frac{at^{c+1}}{c+1} + bt + K$$

$$= dt^{c+1} + bt + K,$$

where

$$d = \frac{a}{c+1} = \frac{-0.020097}{0.34505 + 1} \approx -0.014941,$$

b and c are as above, and K is the constant of integration. So

$$p(t) = e^{dt^{c+1}+bt+K}.$$

To compute K, you substitute the initial data from the memo: $p(1) = 618.7$. Thus,

$$618.7 = e^{d+b+K} = e^{-0.014941+0.036532+K} = e^{0.021591+K}.$$

Thus,

$$\ln(618.7) = 0.021591 + K,$$

which gives

$$K = \ln(618.7) - 0.021591 \approx 6.4060 \text{ (to five significant digits).}$$

Now you can write down the following formula for the monthly spending on residential construction as a function of t, the number of months since the beginning of 2006:

$$p(t) = e^{dt^{c+1}+bt+K} = e^{-0.014941t^{0.34505+1}+0.036532t+6.4060}.$$

What remains is the calculation of the average spent per month over a specified period $[r, s]$. Since p is the rate of change of the total spent, the total spent on housing construction over this period is

$$P = \int_r^s p(t)\, dt,$$

so the average spent per month is

$$\bar{P} = \frac{1}{s-r}\int_r^s p(t)\, dt. \quad \frac{1}{\text{Number of months}} \times \text{Total spent}$$

Substituting the formula for $p(t)$ gives

$$\bar{P} = \frac{1}{s-r}\int_r^s e^{-0.014941t^{0.34505+1}+0.036532t+6.4060}\, dt.$$

You cannot find an explicit antiderivative for the integrand, so you decide that the only way to compute it is numerically. You send the following memo to SC.

DATE: March 15, 2007
TO: SC@Schottie.com (Junior VP Development)
FROM: SW@EnormousStateU.edu
CC: SGLombardoVP@Schottie.com (S. G. Lombardo, Senior VP Development)
SUBJECT: The formula you wanted

Spending in the U.S. on housing construction in the *t*th month of 2006 can be modeled by

$$p(t) = e^{-0.014941t^{0.34505+1}+0.036532t+6.4060} \text{ million dollars.}$$

Further, the average spent per month from month *r* to month *s* (since the start of January 2006) can be computed as

$$\overline{P} = \frac{1}{s-r}\int_r^s e^{-0.014941t^{0.34505+1}+0.036532t+6.4060}\,dt$$

To calculate it easily (and impress Mr. Schottie), I suggest you have a graphing calculator on hand and enter the following on your graphing calculator (watch the parentheses!):

```
Y1=1/(S-R)*fnInt(e^(-0.014941T^(0.34505+1)+0.036532T+6.4060),T,R,S)
```

Then suppose, for example, you need to estimate the average for the period March 1, 2006 ($t = 3$) to February 1, 2007 ($t = 14$). All you do is enter

```
3→R
14→S
Y1
```

and your calculator will give you the result: The average spending was $628 million per month.

Good luck with the meeting!

EXERCISES

1. Use the actual January 2006 spending figure of $618.7 million and the percentage changes in the table to compute the actual spending in February, March, and April of that year. Also use the model of monthly spending to estimate those figures, and compare the predicted values with the actual figures. Is it unacceptable that the April figures agree to only one significant digit? Explain.
2. Use the model developed above to estimate the average monthly spending on residential construction over the 12-month period beginning June 1, 2006. (Round your answer to the nearest $1 million.)
3. What (if any) advantages are there to using a model for residential construction spending when the actual residential construction spending figures are available?
4. The formulas for $p(t)$ and $\overline{P}$ were based on the January 2006 spending figure of $618.7 million. Change the models to allow for a possibly revised January 2006 spending figure of $\$p_0$ million.
5. If we had used quadratic regression to model the construction spending data, we would have obtained

$$y = 0.00005t^2 - 0.0028t + 0.0158.$$

(See the graph.)

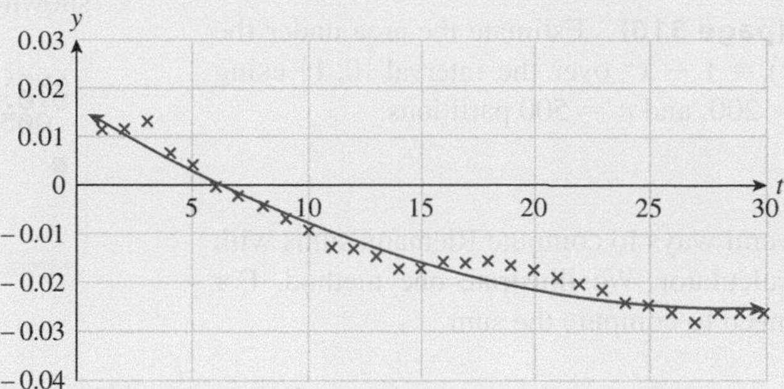

Use this formula and the given January 2006 spending figure to obtain corresponding models for $p(t)$ and $\overline{P}$.

6. Compare the model in the text with the quadratic model in Exercise 5 in terms of both short- and long-term predictions; in particular, when does the quadratic model predict construction spending will have reached its biggest monthly decrease? Are either of these models realistic in the near term? in the long term?

TI-83/84 Plus Technology Guide

Section 6.3

Example 5 (page 510) Estimate the area under the graph of $f(x) = 1 - x^2$ over the interval $[0, 1]$ using $n = 100$, $n = 200$, and $n = 500$ partitions.

Solution

There are several ways to compute Riemann sums with a graphing calculator. We illustrate one method. For $n = 100$ we need to compute the sum

$$\sum_{k=0}^{99} f(x_k)\,\Delta x = [f(0) + f(0.01) + \cdots + f(0.99)](0.01).$$

See discussion in Example 5.

Thus, we first need to calculate the numbers $f(0)$, $f(0.01)$, and so on and add them up. The TI-83/84 Plus has a built-in `sum` function (available in the LIST MATH menu), which, like the SUM function in a spreadsheet, sums the entries in a list.

1. To generate a list that contains the numbers we want to add together, use the `seq` function (available in the LIST OPS menu). If we enter

   `seq(1-X^2,X,0,0.99,0.01)`

   `seq:` [2ND] [LIST] OPS [5]

   the calculator will calculate a list by evaluating `1-X^2` for values of X from 0 to 0.99 in steps of 0.01.

2. To take the sum of all these numbers, we wrap the `seq` function in a call to `sum`:

   `sum(seq(1-X^2,X,0,0.99,0.01))`

   `sum:` [2ND] [LIST] MATH [5]

   This gives the sum

   $$f(0) + f(0.01) + \cdots + f(0.99) = 67.165.$$

3. To obtain the Riemann sum, we need to multiply this sum by $\Delta x = 0.01$, and we obtain the estimate of $67.165 \times 0.01 = 0.67165$ for the Riemann sum:

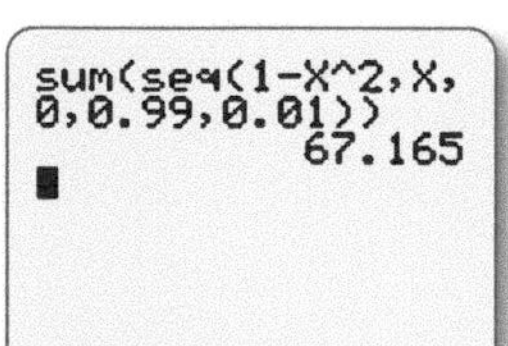

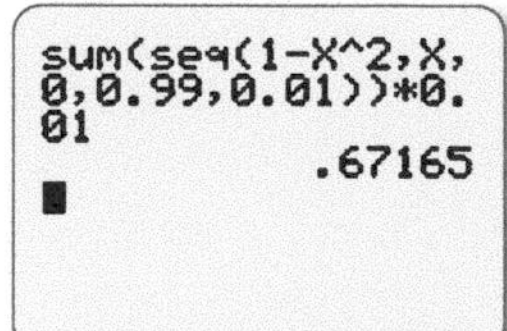

We obtain the other Riemann sums similarly, as shown here:

$n = 200$

$n = 500$

One disadvantage of this method is that the TI-83/84 Plus can generate and sum a list of at most 999 entries. The LEFTSUM program below calculates left Riemann sums for any n. The TI-83/84 Plus also has a built-in function `fnInt`, which finds a very accurate approximation of a definite integral, using a more sophisticated technique than the one we are discussing here.

The LEFTSUM program for the TI-83/84 Plus

The following program calculates (left) Riemann sums for any n. The latest version of this program (and others) is available at the Website:

PROGRAM: LEFTSUM	
`:Input "LEFT ENDPOINT? ",A`	Prompts for the left endpoint a
`:Input "RIGHT ENDPOINT? ",B`	Prompts for the right endpoint b
`:Input "N? ",N`	Prompts for the number of rectangles
`:(B-A)/N→D`	D is $\Delta x = (b - a)/n$.
`:Ø→L`	L will eventually be the left sum.
`:A→X`	X is the current x-coordinate.
`:For(I,1,N)`	Start of a loop—recall the sigma notation.
`:L+Y1→L`	Add $f(x_{i-1})$ to L.
`:A+I*D→X`	Uses formula $x_i = a + i\Delta x$
`:End`	End of loop
`:L*D→L`	Multiply by Δx.
`:Disp "LEFT SUM IS ",L`	
`:Stop`	

Section 6.3

Example 5 (page 510) Estimate the area under the graph of $f(x) = 1 - x^2$ over the interval $[0, 1]$ using $n = 100$, $n = 200$, and $n = 500$ partitions.

Solution

We need to compute various sums:

$$\sum_{k=0}^{99} f(x_k)\,\Delta x = [f(0) + f(0.01) + \cdots + f(0.99)](0.01)$$

See discussion in Example 5.

$$\sum_{k=0}^{199} f(x_k)\,\Delta x = [f(0) + f(0.005) + \cdots + f(0.995)](0.005)$$

$$\sum_{k=0}^{499} f(x_k)\,\Delta x = [f(0) + f(0.002) + \cdots + f(0.998)](0.002).$$

Here is how you can compute them all on the same spreadsheet.

1. Enter the values for the endpoints a and b, the number of subdivisions n, and the formula $\Delta x = (b - a)/n$:

	A	B	C	D
1	x	f(x)	a	0
2			b	1
3			n	100
4			Delta x	=(D2-D1)/D3

2. Next, we compute all the x-values we might need in column A. Because the largest value of n that we will be using is 500, we will need a total of 501 values of x. Note that the value in each cell below A3 is obtained from the one above by adding Δx.

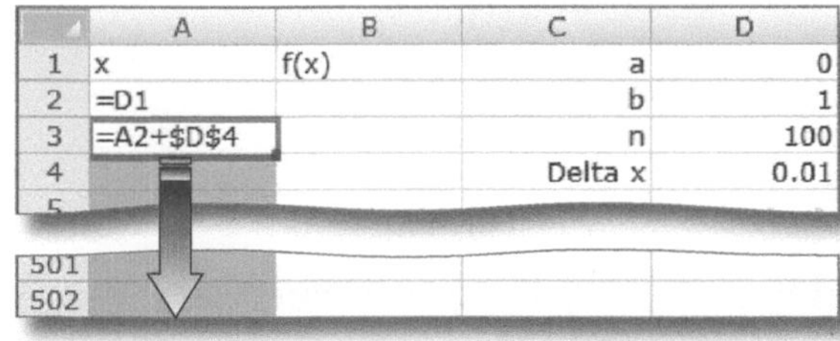

	A	B	C	D
1	x	f(x)	a	0
2	=D1		b	1
3	=A2+$D$4		n	100
4			Delta x	0.01
…				
501				
502				

	A	B	C	D
1	x	f(x)	a	0
2	0		b	1
3	0.01		n	100
4	0.02		Delta x	0.01
…				
501	4.99			
502	5			

(The fact that the values of x currently go too far will be corrected in the next step.)

3. We need to calculate the numbers $f(0)$, $f(0.01)$, and so on, but only those for which the corresponding x-value is less than b. To do this, we use a logical formula like we did with piecewise-defined functions in Chapter 1:

	A	B	C	D
1	x	f(x)	a	0
2	0	=(1-A2^2)*(A2<$D$2)	b	1
3	0.01		n	100
4	0.02		Delta x	0.01
…				
501	4.99			
502	5			

When the value of x is b or above, the function will evaluate to zero because we do not want to count it.

4. Finally, we compute the Riemann sum by adding up everything in column B and multiplying by Δx:

	A	B	C	D
1	x	f(x)	a	0
2	0	1	b	1
3	0.01	0.9999	n	100
4	0.02	0.9996	Delta x	0.01
5	0.03	0.9991	Left Sum	=SUM(B:B)*D4
6	0.04	0.9984		

↓

	A	B	C	D
1	x	f(x)	a	0
2	0	1	b	1
3	0.01	0.9999	n	100
4	0.02	0.9996	Delta x	0.01
5	0.03	0.9991	Left Sum	0.67165
6	0.04	0.9984		

Now it is easy to obtain the sums for $n = 200$ and $n = 500$: Simply change the value of n in cell D3:

	A	B	C	D
1	x	f(x)	a	0
2	0	1	b	1
3	0.005	0.999975	n	200
4	0.01	0.9999	Delta x	0.005
5	0.015	0.999775	Left Sum	0.6691625
6	0.02	0.9996		

	A	B	C	D
1	x	f(x)	a	0
2	0	1	b	1
3	0.002	0.999996	n	500
4	0.004	0.999984	Delta x	0.002
5	0.006	0.999964	Left Sum	0.667666
6	0.008	0.999936		

7

FURTHER INTEGRATION TECHNIQUES AND APPLICATIONS OF THE INTEGRAL

CASE STUDY

Estimating Tax Revenues

You have just been hired by the incoming administration to coordinate national tax policy, and the so-called experts on your staff can't seem to agree on which of several tax proposals will result in the most revenue for the government. The data you have are the two income tax proposals (graphs of tax vs. income) and the distribution of incomes in the country.

How do you use this information to decide which tax policy will result in more revenue?

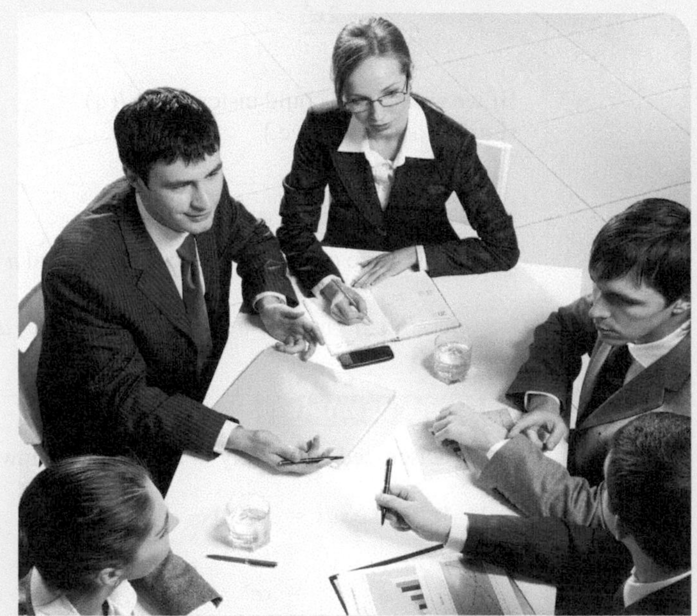

Dmitriy Shironosov/Shutterstock.com

www.WanerMath.com

At the Website, in addition to the resources listed in the Preface, you will find:

- A numerical integration utility
- Graphing calculator programs for numerical integration

as well as the following extra section:

- Linear Differitial Equations

Introduction

In Chapter 6 we learned how to compute many integrals and saw some of the applications of the integral. In this chapter we look at some further techniques for computing integrals and then at more applications of the integral. We also see how to extend the definition of the definite integral to include integrals over infinite intervals, and we show how such integrals can be used for long-term forecasting. Finally, we introduce the beautiful theory of differential equations and some of its numerous applications.

7.1 Integration by Parts

Integration-by-Parts Formula

Integration by parts is an integration technique that comes from the product rule for derivatives. The tabular method we present here has been around for some time and makes integration by parts quite simple, particularly in problems where it has to be used several times.*

We start with a little notation to simplify things while we introduce integration by parts. (We use this notation only in the next few pages.) If u is a function, denote its derivative by $D(u)$ and an antiderivative by $I(u)$. Thus, for example, if $u = 2x^2$, then

$$D(u) = 4x$$

and

$$I(u) = \frac{2x^3}{3}.$$

[If we wished, we could instead take $I(u) = \frac{2x^3}{3} + 46$, but we usually opt to take the simplest antiderivative.]

* The version of the tabular method we use was developed and taught to us by Dan Rosen at Hofstra University.

Integration by Parts

If u and v are continuous functions of x and u has a continuous derivative, then

$$\int u \cdot v\, dx = u \cdot I(v) - \int D(u)I(v)\, dx.$$

Quick Example

(Discussed more fully in Example 1 below)

1. $\displaystyle\int x \cdot e^x\, dx = xI(e^x) - \int D(x)I(e^x)\, dx$

$\displaystyle = xe^x - \int 1 \cdot e^x\, dx \qquad I(e^x) = e^x;\ D(x) = 1$

$\displaystyle = xe^x - e^x + C. \qquad \int e^x\, dx = e^x + C$

As Quick Example 1 shows, although we could not immediately integrate $u \cdot v = x \cdot e^x$, we could easily integrate $D(u)I(v) = 1 \cdot e^x = e^x$.

Derivation of Integration-by-Parts Formula

As we mentioned, the integration-by-parts formula comes from the product rule for derivatives. We apply the product rule to the function $uI(v)$:

$$\begin{aligned} D[u \cdot I(v)] &= D(u)I(v) + uD(I(v)) \\ &= D(u)I(v) + uv \end{aligned}$$

because $D(I(v))$ is the derivative of an antiderivative of v, which is v. Integrating both sides gives

$$u \cdot I(v) = \int D(u)I(v)\,dx + \int uv\,dx.$$

A simple rearrangement of the terms now gives us the integration-by-parts formula.

The Tabular Method

The integration-by-parts formula is easiest to use via the tabular method illustrated in the following example, where we repeat the calculation we did in Quick Example 1.

EXAMPLE 1 Integration by Parts: Tabular Method

Calculate $\int xe^x\,dx$.

Solution First, the reason we *need* to use integration by parts to evaluate this integral is that none of the other techniques of integration that we've talked about up to now will help us. Furthermore, we cannot simply find antiderivatives of x and e^x and multiply them together. [You should check that $(x^2/2)e^x$ is *not* an antiderivative of xe^x.] However, as we saw above, this integral can be found by integration by parts. We want to find the integral of the *product* of x and e^x. We must make a decision: Which function will play the role of u and which will play the role of v in the integration-by-parts formula? Because the derivative of x is just 1, differentiating makes it simpler, so we try letting x be u and letting e^x be v. We need to calculate $D(u)$ and $I(v)$, which we record in the following table:

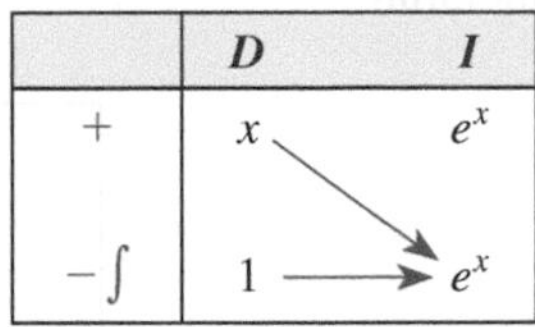

	D	I
+	x	e^x
$-\int$	1	e^x

The table is read as $+x \cdot e^x - \int 1 \cdot e^x\,dx$.

Below x in the D column, we put $D(x) = 1$; below e^x in the I column, we put $I(e^x) = e^x$. The arrow at an angle connecting x and $I(e^x)$ reminds us that the product $xI(e^x)$ will appear in the answer; the plus sign on the left of the table reminds us that it is $+x\,I(e^x)$ that appears. The integral sign and the horizontal arrow connecting $D(x)$ and $I(e^x)$ remind us that the *integral* of the product $D(x)I(e^x)$ also appears in the answer; the minus sign on the left reminds us that we need to subtract this integral. Combining these two contributions, we get

$$\int xe^x\,dx = xe^x - \int e^x\,dx.$$

The integral that appears on the right is much easier than the one we began with, so we can complete the problem:

$$\int xe^x\,dx = xe^x - \int e^x\,dx = xe^x - e^x + C.$$

➡ **Before we go on...** In Example 1, what if we had made the opposite decision and put e^x in the D column and x in the I column? Then we would have had the following table:

	D	I
$+$	e^x	x
$-\int$	e^x	$x^2/2$

This gives

$$\int xe^x\,dx = \frac{x^2}{2}e^x - \int \frac{x^2}{2}e^x\,dx.$$

The integral on the right is harder than the one we started with, not easier! How do we know beforehand which way to go? We don't. We have to be willing to do a little trial and error: We try it one way, and if it doesn't make things simpler, we try it another way. *Remember, though, that the function we put in the I column must be one that we can integrate.* ■

EXAMPLE 2 Repeated Integration by Parts

Calculate $\int x^2e^{-x}\,dx$.

Solution Again, we have a product: The integrand is the product of x^2 and e^{-x}. Because differentiating x^2 makes it simpler, we put it in the D column and get the following table:

	D	I
$+$	x^2	e^{-x}
$-\int$	$2x$	$-e^{-x}$

This table gives us

$$\int x^2e^{-x}\,dx = x^2(-e^{-x}) - \int 2x(-e^{-x})\,dx.$$

The last integral is simpler than the one we started with, but it still involves a product. It's a good candidate for another integration by parts. The table we would use would start with $2x$ in the D column and $-e^{-x}$ in the I column, which is exactly what we see in the last row of the table we've already made. Therefore, we *continue the process*, elongating the table above:

	D	I
+	x^2	e^{-x}
−	$2x$	$-e^{-x}$
$+\int$	2 →	e^{-x}

(Notice how the signs on the left alternate. Here's why: To compute $-\int 2x(-e^{-x})\,dx$, we use the negative of the following table:

	D	I
+	$2x$	$-e^{-x}$
$-\int$	2 →	e^{-x}

so we reverse all the signs.)

Now, we still have to compute an integral (the integral of the product of the functions in the bottom row) to complete the computation. But why stop here? Let's continue the process one more step:

	D	I
+	x^2	e^{-x}
−	$2x$	$-e^{-x}$
+	2	e^{-x}
$-\int$	0 →	$-e^{-x}$

In the bottom line we see that all that is left to integrate is $0(-e^{-x}) = 0$. Because the indefinite integral of 0 is C, we can read the answer from the table as

$$\int x^2e^{-x}\,dx = x^2(-e^{-x}) - 2x(e^{-x}) + 2(-e^{-x}) + C$$

$$= -x^2e^{-x} - 2xe^{-x} - 2e^{-x} + C$$

$$= -e^{-x}(x^2 + 2x + 2) + C.$$

In Example 2 we saw a technique that we can summarize as follows.

Integrating a Polynomial Times a Function

If one of the factors in the integrand is a polynomial and the other factor is a function that can be integrated repeatedly, put the polynomial in the D column and keep differentiating until you get zero. Then complete the I column to the same depth, and read off the answer.

For practice, redo Example 1 using this technique.

It is not always the case that the integrand is a polynomial times something easy to integrate, so we can't always expect to end up with a zero in the D column. In that case we hope that at some point we will be able to integrate the product of the functions in the last row. Here are some examples.

EXAMPLE 3 Polynomial Times a Logarithm

Calculate: **a.** $\int x \ln x \, dx$ **b.** $\int (x^2 - x) \ln x \, dx$ **c.** $\int \ln x \, dx$

Solution

a. This is a product and therefore a good candidate for integration by parts. Our first impulse is to differentiate x, but that would mean integrating $\ln x$, and we do not (yet) know how to do that. So we try it the other way around and hope for the best.

	D	I
$+$	$\ln x$ ↘	x
$-\int$	$\frac{1}{x}$ →	$\frac{x^2}{2}$

Why did we stop? If we continued the table, both columns would get more complicated. However, if we stop here, we get

$$\begin{aligned}\int x \ln x \, dx &= (\ln x)\left(\frac{x^2}{2}\right) - \int \left(\frac{1}{x}\right)\left(\frac{x^2}{2}\right) dx \\ &= \frac{x^2}{2} \ln x - \frac{1}{2}\int x \, dx \\ &= \frac{x^2}{2} \ln x - \frac{x^2}{4} + C.\end{aligned}$$

b. We can use the same technique we used in part (a) to integrate any polynomial times the logarithm of x:

	D	I
$+$	$\ln x$ ↘	$x^2 - x$
$-\int$	$\frac{1}{x}$ →	$\frac{x^3}{3} - \frac{x^2}{2}$

$$\begin{aligned}\int (x^2 - x) \ln x \, dx &= (\ln x)\left(\frac{x^3}{3} - \frac{x^2}{2}\right) - \int \left(\frac{1}{x}\right)\left(\frac{x^3}{3} - \frac{x^2}{2}\right) dx \\ &= \left(\frac{x^3}{3} - \frac{x^2}{2}\right) \ln x - \int \left(\frac{x^2}{3} - \frac{x}{2}\right) dx \\ &= \left(\frac{x^3}{3} - \frac{x^2}{2}\right) \ln x - \frac{x^3}{9} + \frac{x^2}{4} + C\end{aligned}$$

c. The integrand ln x is not a product. We can, however, *make* it into a product by thinking of it as $1 \cdot \ln x$. Because this is a polynomial times ln x, we proceed as in parts (a) and (b):

	D	I
$+$	$\ln x$ ↘	1
$-\int$	$1/x$ →	x

We notice that the product of $1/x$ and x is just 1, which we know how to integrate, so we can stop here:

$$\begin{aligned}\int \ln x\, dx &= x \ln x - \int \left(\frac{1}{x}\right) x\, dx \\ &= x \ln x - \int 1\, dx \\ &= x \ln x - x + C.\end{aligned}$$

FAQs

Whether to Use Integration by Parts and What Goes in the *D* and *I* Columns

Q: *Will integration by parts always work to integrate a product?*

A: No. Although integration by parts often works for products in which one factor is a polynomial, it will almost *never* work in the examples of products we saw when discussing substitution in Section 6.2. For example, although integration by parts can be used to compute $\int (x^2 - x)e^{2x-1}\, dx$ (put $x^2 - x$ in the D column and e^{2x-1} in the I column), it *cannot* be used to compute $\int (2x - 1)e^{x^2-x}\, dx$ (put $u = x^2 - x$). Recognizing when to use integration by parts is best learned by experience.

Q: *When using integration by parts, which expression goes in the D column and which in the I column?*

A: Although there is no general rule, the following guidelines are useful:

- To integrate a product in which one factor is a polynomial and the other can be integrated several times, put the polynomial in the *D* column and the other factor in the *I* column. Then differentiate the polynomial until you get zero.
- If one of the factors is a polynomial but the other factor cannot be integrated easily, put the polynomial in the *I* column and the other factor in the *D* column. Stop when the product of the functions in the bottom row can be integrated.
- If neither factor is a polynomial, put the factor that seems easier to integrate in the *I* column and the other factor in the *D* column. Again, stop the table as soon as the product of the functions in the bottom row can be integrated.
- If your method doesn't work, try switching the functions in the *D* and *I* columns or try breaking the integrand into a product in a different way. If none of this works, maybe integration by parts isn't the technique to use on this problem.

7.1 EXERCISES

▼ more advanced ◆ challenging

T indicates exercises that should be solved using technology

In Exercises 1–40, evaluate the integral using integration by parts where possible. [**HINT**: See the examples in the text.]

1. $\int 2xe^x\,dx$
2. $\int 3xe^{-x}\,dx$
3. $\int (3x-1)e^{-x}\,dx$
4. $\int (1-x)e^x\,dx$
5. $\int (x^2-1)e^{2x}\,dx$
6. $\int (x^2+1)e^{-2x}\,dx$
7. $\int (x^2+1)e^{-2x+4}\,dx$
8. $\int (x^2+1)e^{3x+1}\,dx$
9. $\int (2-x)2^x\,dx$
10. $\int (3x-2)4^x\,dx$
11. $\int (x^2-1)3^{-x}\,dx$
12. $\int (1-x^2)2^{-x}\,dx$
13. ▼ $\int \frac{x^2-x}{e^x}\,dx$
14. ▼ $\int \frac{2x+1}{e^{3x}}\,dx$
15. $\int x(x+2)^6\,dx$ (See note.[1])
16. $\int x^2(x-1)^6\,dx$ (See note.[1])
17. ▼ $\int \frac{x}{(x-2)^3}\,dx$
18. ▼ $\int \frac{x}{(x-1)^2}\,dx$
19. $\int x^3\ln x\,dx$
20. $\int x^2\ln x\,dx$
21. $\int (t^2+1)\ln(2t)\,dt$
22. $\int (t^2-1)\ln(-t)\,dt$
23. $\int t^{1/3}\ln t\,dt$
24. $\int t^{-1/2}\ln t\,dt$
25. $\int \log_3 x\,dx$
26. $\int x\log_2 x\,dx$
27. ▼ $\int (xe^{2x}-4e^{3x})\,dx$
28. ▼ $\int (x^2e^{-x}+2e^{-x+1})\,dx$
29. ▼ $\int (x^2e^x-xe^{x^2})\,dx$
30. ▼ $\int \left[(2x+1)e^{x^2+x}-x^2e^{2x+1}\right]dx$
31. ▼ $\int (3x-4)\sqrt{2x-1}\,dx$ (See note.[1])
32. ▼ $\int \frac{2x+1}{\sqrt{3x-2}}\,dx$ (See note.[1])
33. $\int_0^1 (x+1)e^x\,dx$
34. $\int_{-1}^1 (x^2+x)e^{-x}\,dx$
35. $\int_0^1 x^2(x+1)^{10}\,dx$
36. $\int_0^1 x^3(x+1)^{10}\,dx$
37. $\int_1^2 x\ln(2x)\,dx$
38. $\int_1^2 x^2\ln(3x)\,dx$
39. $\int_0^1 x\ln(x+1)\,dx$
40. $\int_0^1 x^2\ln(x+1)\,dx$
41. Find the area bounded by the curve $y = xe^{-x}$, the x-axis, and the lines $x = 0$ and $x = 10$.
42. Find the area bounded by the curve $y = x\ln x$, the x-axis, and the lines $x = 1$ and $x = e$.
43. Find the area bounded by the curve $y = (x+1)\ln x$, the x-axis, and the lines $x = 1$ and $x = 2$.
44. Find the area bounded by the curve $y = (x-1)e^x$, the x-axis, and the lines $x = 0$ and $x = 2$.

Integrals of Functions Involving Absolute Values *In Exercises 45–52, use integration by parts to evaluate the given integral using the following integral formulas where necessary. (You have seen some of these before; all can be checked by differentiating.)*

Integral Formula	Shortcut Version
$\int \frac{\lvert x\rvert}{x}\,dx = \lvert x\rvert + C$ Because $\frac{d}{dx}\lvert x\rvert = \frac{\lvert x\rvert}{x}$	$\int \frac{\lvert ax+b\rvert}{ax+b}\,dx = \frac{1}{a}\lvert ax+b\rvert + C$
$\int \lvert x\rvert\,dx = \frac{1}{2}x\lvert x\rvert + C$	$\int \lvert ax+b\rvert\,dx$ $= \frac{1}{2a}(ax+b)\lvert ax+b\rvert + C$
$\int x\lvert x\rvert\,dx = \frac{1}{3}x^2\lvert x\rvert + C$	$\int (ax+b)\lvert ax+b\rvert\,dx$ $= \frac{1}{3a}(ax+b)^2\lvert ax+b\rvert + C$
$\int x^2\lvert x\rvert\,dx = \frac{1}{4}x^3\lvert x\rvert + C$	$\int (ax+b)^2\lvert ax+b\rvert\,dx$ $= \frac{1}{4a}(ax+b)^3\lvert ax+b\rvert + C$

45. $\int x\lvert x-3\rvert\,dx$
46. $\int x\lvert x+4\rvert\,dx$
47. $\int 2x\frac{\lvert x-3\rvert}{x-3}\,dx$
48. $\int 3x\frac{\lvert x+4\rvert}{x+4}\,dx$

[1] Several exercises, including these, can also be done using substitution, although integration by parts is easier and should be used instead.

49. ▼ $\int 2x^2|-x+4|\,dx$ 50. ▼ $\int 3x^2|2x-3|\,dx$

51. ▼ $\int (x^2-2x+3)|x-4|\,dx$

52. ▼ $\int (x^2-x+1)|2x-4|\,dx$

Applications

53. ***Displacement*** A rocket rising from the ground has a velocity of $2{,}000te^{-t/120}$ ft/sec after t seconds. How far does it rise in the first 2 minutes?

54. ***Sales*** Weekly sales of graphing calculators can be modeled by the equation

$$s(t) = 10 - te^{-t/20},$$

where s is the number of calculators sold per week after t weeks. How many graphing calculators (to the nearest unit) will be sold in the first 20 weeks?

55. ***Total Cost*** The marginal cost of the xth box of light bulbs is $10 + [\ln(x+1)]/(x+1)^2$, and the fixed cost is \$5,000. Find the total cost to make x boxes of bulbs.

56. ***Total Revenue*** The marginal revenue for selling the xth box of light bulbs is $10 + 0.001x^2e^{-x/100}$. Find the total revenue generated by selling 200 boxes of bulbs.

57. ***Spending on Gasoline*** From the beginning of 2000 to the beginning of 2007 the United States consumed gasoline at a rate of about

$$q(t) = 2.0t + 131 \text{ billion gallons per year} \quad (0 \le t \le 7).$$

(t is the number of years since 2000.)[2] During the same period the price of gasoline was approximately

$$p(t) = 1.2e^{0.11t} \text{ dollars per gallon.}$$

Use an integral to estimate, to the nearest 10 billion dollars, the total spent on gasoline during the given period. [**HINT:** Rate of spending $= p(t)q(t)$.]

58. ***Spending on Gasoline*** From the beginning of 2007 to the beginning of 2014 the United States consumed gasoline at a rate of about

$$q(t) = -1.2t + 141 \text{ billion gallons per year} \quad (0 \le t \le 7).$$

(t is the number of years since 2007.)[3] During the same period the price of gasoline was approximately

$$p(t) = 3.5 - 2.38e^{-0.5t} \text{ dollars per gallon.}$$

Use an integral to estimate, to the nearest 10 billion dollars, the total spent on gasoline during the given period. [**HINT:** Rate of spending $= p(t)q(t)$.]

59. ***Housing*** The following graph shows the annual number of housing starts in the United States during 2000–2008 together with a quadratic approximating model:

$$s(t) = -30t^2 + 240t + 800 \text{ thousand homes per year} \quad (0 \le t \le 8).$$

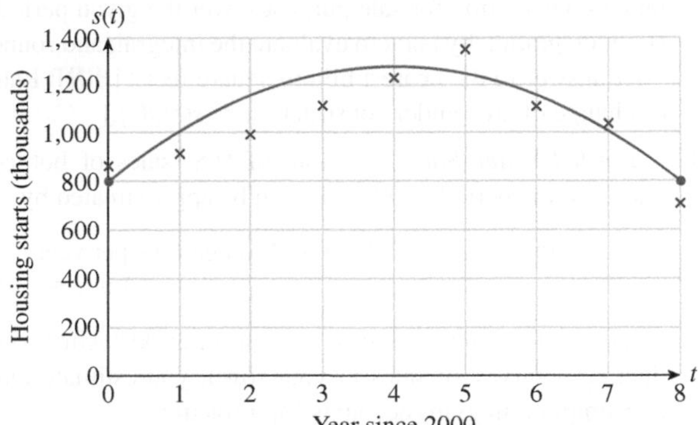

(t is the time in years since 2000.)[4] At the same time, the homes being built were getting larger: The average area per home was approximately

$$a(t) = 40t + 2{,}000 \text{ square feet.}$$

Use the given models to estimate the total housing area under construction over the given period. (Use integration by parts to evaluate the integral, and round your answer to the nearest billion square feet.) [**HINT:** Rate of change of area under construction $= s(t)a(t)$.]

60. ***Housing for Sale*** The following graph shows the number of housing starts for sale purposes in the United States during 2000–2008 together with a quadratic approximating model:

$$s(t) = -33t^2 + 240t + 700 \text{ thousand homes per year} \quad (0 \le t \le 8).$$

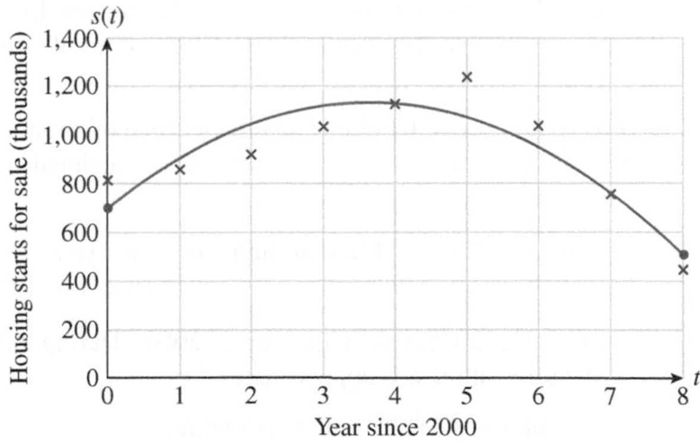

[2] Source for data: Energy Information Administration, Department of Energy (www.eia.gov).

[3] *Ibid.*

[4] Source for data: U.S. Census Bureau (www.census.gov).

(t is the time in years since 2000.)[5] At the same time, the homes being built were getting larger: The average area per home was approximately

$$a(t) = 40t + 2{,}000 \text{ square feet.}$$

Use the given models to estimate the total housing area under construction for sale purposes over the given period. (Use integration by parts to evaluate the integral, and round your answer to the nearest billion square feet.) [**HINT**: Rate of change of area under construction $= s(t)a(t)$.]

61. ▼ ***Bottled Water Sales*** The rate of U.S. sales of bottled water for the period 2000–2010 can be approximated by

$$s(t) = -45t^2 + 900t + 4{,}200 \text{ million gallons per year} \quad (0 \le t \le 10),$$

where t is time in years since the start of 2000.[6] After conducting a survey of sales in your town, you estimate that consumption in gyms accounts for a fraction

$$f(t) = \sqrt{0.1 + 0.02t}$$

of all bottled water sold in year t. Assuming that your model is correct, estimate, to the nearest hundred million gallons, the total amount of bottled water consumed in gyms from the start of 2005 to the start of 2010. [**HINT**: Rate of consumption $= s(t)f(t)$. Also see Exercise 31.]

62. ***Bottled Water Sales*** The rate of U.S. per capita sales of bottled water for the period 2000–2010 can be approximated by

$$s(t) = -0.18t^2 + 3t + 15 \text{ gallons per year} \quad (0 \le t \le 10),$$

where t is the time in years since the start of 2000.[7] After conducting a survey of sales in your state, you estimate that consumption in gyms accounts for a fraction

$$f(t) = \sqrt{0.2 + 0.04t}$$

of all bottled water consumed in year t. Assuming that your model is correct, estimate, to the nearest gallon, the total amount of bottled water consumed per capita in gyms from the start of 2005 to the start of 2010. [**HINT**: Rate of consumption $= s(t)f(t)$. Also see Exercise 31.]

63. ▼ ***Oil Production in Mexico*** The rate of crude oil production from 2008 to 2013 by **Pemex**, Mexico's national oil company, can be approximated by

$$q(t) = 6.2t^2 - 146t + 1{,}910 \text{ million barrels per year} \quad (8 \le t \le 13),$$

where t is time in years since the start of 2000.[8] During that time, the price of oil was approximately[9]

$$p(t) = 48e^{0.046t} \text{ dollars per barrel.}$$

Obtain an expression for Pemex's total oil revenue $R(x)$ from the start of 2008 to the start of year x as a function of x. (Do not simplify the answer; round all coefficients to three significant digits.) [**HINT**: Rate of revenue $= p(t)q(t)$.]

64. ▼ ***Oil Imports from Mexico*** The rate of crude oil imports to the United States from Mexico from 2009 to 2013 can be approximated by

$$r(t) = -14t^2 + 292t - 1{,}100 \text{ million barrels per year} \quad (9 \le t \le 13),$$

where t is time in years since the start of 2000.[10] During that time, the price of oil was approximately[11]

$$p(t) = 48e^{0.046t} \text{ dollars per barrel.}$$

Obtain an expression for the total oil revenue $R(x)$ Mexico earned from the United States from the start of 2009 to the start of year x as a function of x. (Do not simplify the answer; round all coefficients to three significant digits.) [**HINT**: Rate of revenue $= p(t)r(t)$.]

65. ▼ ***Revenue*** You have been raising the price of your *Lord of the Fields* T-shirts by 50¢ per week, and sales have been falling continuously at a rate of 2% per week. Assuming that you are now selling 50 T-shirts per week and charging $10 per T-shirt, how much revenue will you generate during the coming year? (Round your answer to the nearest dollar.) [**HINT**: Weekly revenue = Weekly sales × Price per T-shirt.]

66. ▼ ***Revenue*** Luckily, sales of your *Star Wars and Peace* T-shirts are now 50 T-shirts per week and increasing continuously at a rate of 5% per week. You are now charging $10 per T-shirt and are decreasing the price by 50¢ per week. How much revenue will you generate during the next 6 weeks?

Integrals of Piecewise-Linear Functions *Exercises 67 and 68 are based on the following formula that can be used to represent a piecewise-linear function as a closed-form function:*

$$\begin{cases} p(x) & \text{if } x < a \\ q(x) & \text{if } x > a \end{cases} = p(x) + \frac{1}{2}[q(x) - p(x)]\left[1 + \frac{|x-a|}{x-a}\right]$$

Such functions can then be integrated using the technique of Exercises 45–52.

67. ◆ ***Population: Mexico*** The rate of change of population in Mexico over 1990–2010 was approximately

$$r(t) = \begin{cases} -0.1t + 3 & \text{if } 0 \le t \le 10 \\ -0.05t + 2.5 & \text{if } 10 \le t \le 20 \end{cases}$$

million people per year, where t is time in years since 1990.

a. Use the formula given before the exercise to represent $r(t)$ as a closed-form function. [**HINT**: Use the formula with $a = 10$.]

[5] See footnote for Exercise 59.

[6] Source for data: Beverage Marketing Corporation (www.bottledwater.org).

[7] *Ibid.*

[8] Source for data: www.pemex.com.

[9] Source for data: www.inflationdata.com.

[10] Source for data: U.S. Energy Information Administration (www.eia.gov).

[11] Source for data: www.inflationdata.com.

b. Use the result of part (a) and a definite integral to estimate the total increase in population over the given 20-year period. [**HINT:** Break up the integral into two, and use the technique of Exercises 45–52 to evaluate one of them.]

68. ◆ ***Population: Mexico*** The rate of change of population in Mexico over 1950–1990 was approximately

$$r(t) = \begin{cases} 0.05t + 2.5 & \text{if } 0 \le t \le 20 \\ -0.075t + 5 & \text{if } 20 \le t \le 40 \end{cases}$$

million people per year, where t is time in years since 1950.

a. Use the formula given before the exercise to represent $r(t)$ as a closed-form function. [**HINT:** Use the formula with $a = 20$.]

b. Use the result of part (a) and a definite integral to estimate the total increase in population over the given 40-year period. [**HINT:** Break up the integral into two, and use the technique of Exercises 45–52 to evaluate one of them.]

Communication and Reasoning Exercises

69. Your friend Janice claims that integration by parts allows one to integrate any product of two functions. Prove her wrong by giving an example of a product of two functions that cannot be integrated using integration by parts.

70. Complete the following sentence in words: The integral of uv is the first times the integral of the second minus the integral of _____.

71. Give an example of an integral that can be computed in two ways: by substitution or integration by parts.

72. Give an example of an integral that can be computed by substitution but not by integration by parts. (You need not compute the integral.)

In Exercises 73–80, indicate whether the given integral calls for integration by parts or substitution.

73. $\int (6x - 1)e^{3x^2 - x}\,dx$

74. $\int \frac{x^2 - 3x + 1}{e^{2x-3}}\,dx$

75. $\int (3x^2 - x)e^{6x-1}\,dx$

76. $\int \frac{2x - 3}{e^{x^2 - 3x + 1}}\,dx$

77. $\int \frac{1}{(x + 1)\ln(x + 1)}\,dx$

78. $\int \frac{\ln(x + 1)}{x + 1}\,dx$

79. $\int \ln(x^2)\,dx$

80. $\int (x + 1)\ln(x + 1)\,dx$

81. ▼ If $P(x)$ is a polynomial of degree n and $f(x)$ is some function of x, how many times do we generally have to integrate $f(x)$ to compute $\int p(x)f(x)\,dx$?

82. ▼ Use integration by parts to show that $\int (\ln x)^2\,dx = x(\ln x)^2 - 2x\ln x + 2x + C$.

83. ◆ ***Hermite's Identity*** If $f(x)$ is a polynomial of degree n, show that

$$\int_0^b f(x)e^{-x}\,dx = F(0) - F(b)e^{-b},$$

where $F(x) = f(x) + f'(x) + f''(x) + \cdots + f^{(n)}(x)$. (This is the sum of f and all of its derivatives.)

84. ◆ Write down a formula similar to Hermite's identity for $\int_0^b f(x)e^x\,dx$ when $f(x)$ is a polynomial of degree n.

7.2 Area between Two Curves and Applications

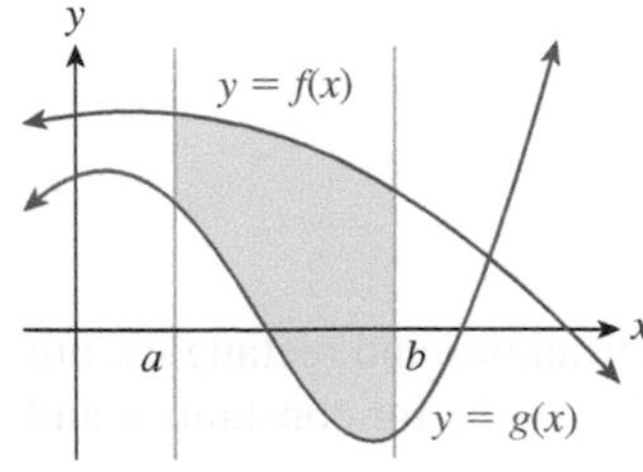

Figure 1

As we saw in Chapter 6, we can use the definite integral to calculate the area between the graph of a function and the x-axis. With only a little more work, we can use it to calculate the area between two graphs. Figure 1 shows the graphs of two functions, $f(x)$ and $g(x)$, with $f(x) \ge g(x)$ for every x in the interval $[a, b]$.

To find the shaded area between the graphs of the two functions, we use the following formula.

Area between Two Graphs

If $f(x) \ge g(x)$ for all x in $[a, b]$ (so that the graph of f does not move below that of g), then the area of the region between the graphs of f and g and between $x = a$ and $x = b$ is given by

$$A = \int_a^b [f(x) - g(x)]\,dx. \qquad \text{Integral of (Top − Bottom)}$$

Caution

If the graphs of f and g cross in the interval, the above formula does not hold. For instance, if $f(x) = x$ and $g(x) = -x$, then the total area shown in the figure is 2 square units, whereas $\int_{-1}^{1} [f(x) - g(x)]\, dx = 0$.

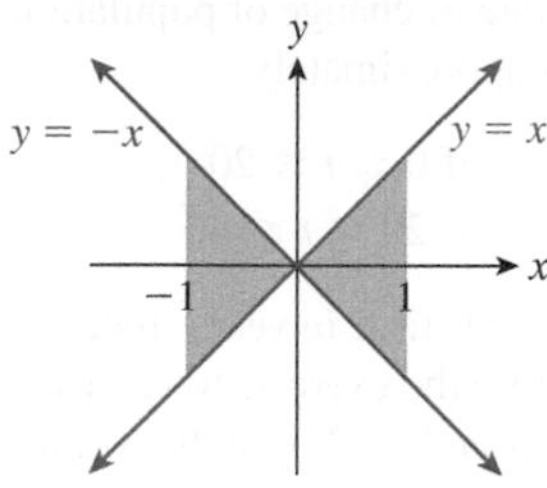

Let's look at an example and then discuss why the formula works.

EXAMPLE 1 **The Area between Two Curves**

Find the areas of the following regions:

a. Between $f(x) = -x^2 - 3x + 4$ and $g(x) = x^2 - 3x - 4$ and between $x = -1$ and $x = 1$

b. Between $f(x) = |x|$ and $g(x) = -|x - 1|$ over $[-1, 2]$

Solution

a. The area in question is shown in Figure 2. Because the graph of f lies above the graph of g in the interval $[-1, 1]$, we have $f(x) \geq g(x)$ for all x in $[-1, 1]$. Therefore, we can use the formula given above and calculate the area as follows:

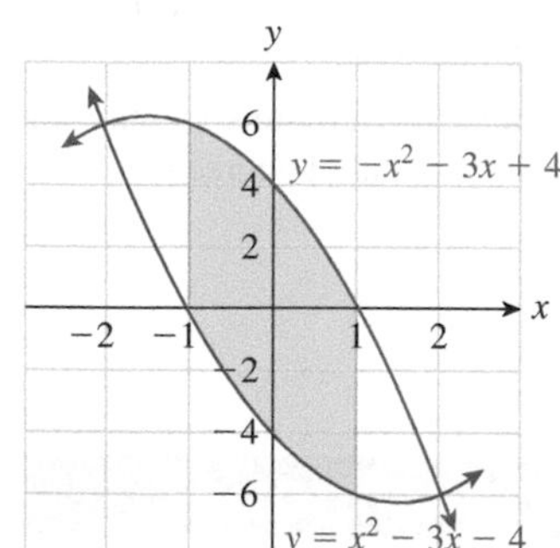

Figure 2

$$
\begin{aligned}
A &= \int_{-1}^{1} [f(x) - g(x)]\, dx \\
&= \int_{-1}^{1} [(-x^2 - 3x + 4) - (x^2 - 3x - 4)]\, dx \\
&= \int_{-1}^{1} (8 - 2x^2)\, dx \\
&= \left[8x - \frac{2}{3}x^3 \right]_{-1}^{1} \\
&= \frac{44}{3}.
\end{aligned}
$$

Figure 3

b. The given area (see Figure 3) can be broken up into triangles and rectangles, but we already know a formula for the antiderivative of $|ax + b|$ for constants a and b, so we can use calculus instead:

$$
\begin{aligned}
A &= \int_{-1}^{2} [f(x) - g(x)]\, dx \\
&= \int_{-1}^{2} [|x| - (-|x - 1|)]\, dx \\
&= \int_{-1}^{2} [|x| + |x - 1|]\, dx
\end{aligned}
$$

$$= \frac{1}{2}[x|x| + (x-1)|x-1|]_{-1}^{2} \qquad \int |ax+b|\,dx = \frac{1}{2a}(ax+b)|ax+b| + C$$

$$= \frac{1}{2}[(4+1) - (-1-4)]$$

$$= \frac{1}{2}(10) = 5.$$

Q: *Why does the formula for the area between two curves work?*

A: Let's go back once again to the general case illustrated in Figure 1, where we were given two functions f and g with $f(x) \geq g(x)$ for every x in the interval $[a, b]$. To avoid complicating the argument by the fact that the graph of g, or f, or both may dip below the x-axis in the interval $[a, b]$ (as occurs in Figure 1 and also in Example 1), we shift both graphs vertically upward by adding a big enough constant M to lift them both above the x-axis in the interval $[a, b]$, as shown in Figure 4.

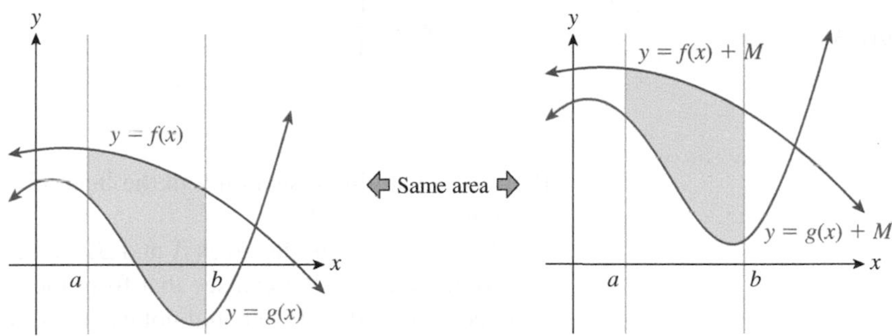

Figure 4

As the figure illustrates, the area of the region between the graphs is not affected, so we will calculate the area of the region shown on the right of Figure 4. That calculation is shown in Figure 5.

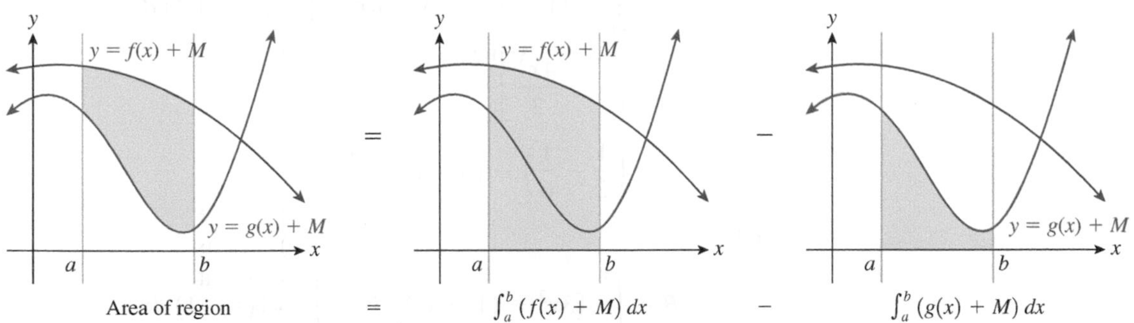

Figure 5

From the figure, the area we want is

$$\int_a^b (f(x) + M)\,dx - \int_a^b (g(x) + M)\,dx = \int_a^b [(f(x) + M) - (g(x) + M)]\,dx$$

$$= \int_a^b [f(x) - g(x)]\,dx,$$

which is the formula we gave originally.

So far, we've been assuming that $f(x) \geq g(x)$, so the graph of f never dips below the graph of g and so the graphs cannot cross (although they can touch). Example 2 shows how we compute the area between graphs that *do* cross.

EXAMPLE 2 Regions Enclosed by Crossing Curves

Find the area of the region between $y = 3x^2$ and $y = 1 - x^2$ and between $x = 0$ and $x = 1$.

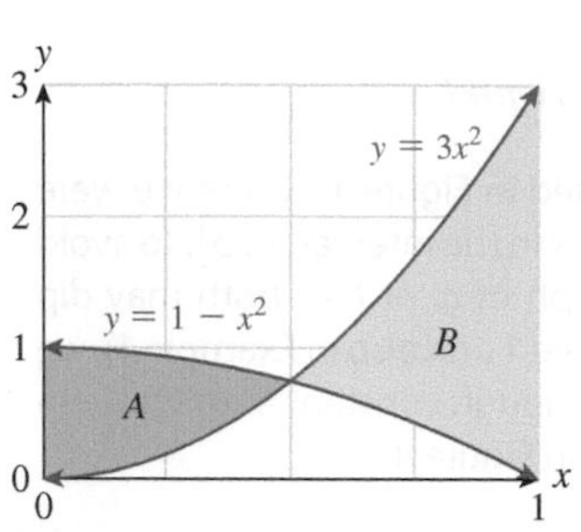

Figure 6

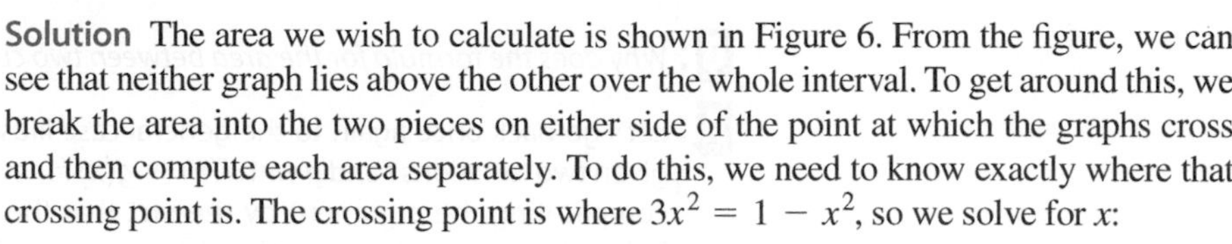

Solution The area we wish to calculate is shown in Figure 6. From the figure, we can see that neither graph lies above the other over the whole interval. To get around this, we break the area into the two pieces on either side of the point at which the graphs cross and then compute each area separately. To do this, we need to know exactly where that crossing point is. The crossing point is where $3x^2 = 1 - x^2$, so we solve for x:

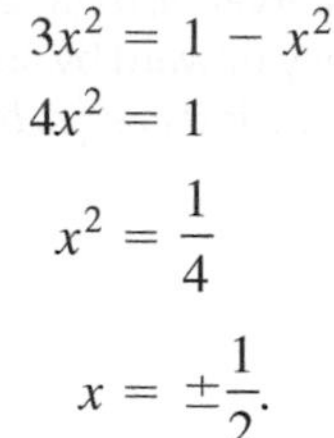

$$\begin{aligned} 3x^2 &= 1 - x^2 \\ 4x^2 &= 1 \\ x^2 &= \frac{1}{4} \\ x &= \pm\frac{1}{2}. \end{aligned}$$

Because we are interested only in the interval $[0, 1]$, the crossing point we're interested in is at $x = 1/2$.

Now, to compute the areas A and B, we need to know which graph is on top in each of these areas. We can see that from the figure, but what if the functions were more complicated and we could not easily draw the graphs? To be sure, we can test the values of the two functions at some point in each region. But we really need not worry. If we make the wrong choice for the top function, the integral will yield the negative of the area (why?), so we can simply take the absolute value of the integral to get the area of the region in question. For this example we have

$$\begin{aligned} A &= \int_0^{1/2} [(1 - x^2) - 3x^2]\,dx = \int_0^{1/2} (1 - 4x^2)\,dx \\ &= \left[x - \frac{4x^3}{3}\right]_0^{1/2} \\ &= \left(\frac{1}{2} - \frac{1}{6}\right) - (0 - 0) = \frac{1}{3} \end{aligned}$$

and

$$\begin{aligned} B &= \int_{1/2}^{1} [3x^2 - (1 - x^2)]\,dx = \int_{1/2}^{1} (4x^2 - 1)\,dx \\ &= \left[\frac{4x^3}{3} - x\right]_{1/2}^{1} \\ &= \left(\frac{4}{3} - 1\right) - \left(\frac{1}{6} - \frac{1}{2}\right) = \frac{2}{3}. \end{aligned}$$

This gives a total area of $A + B = \frac{1}{3} + \frac{2}{3} = 1$.

➡ **Before we go on . . .** What would have happened in Example 2 if we had not broken the area into two pieces but had just calculated the integral of the difference of the two functions? We would have calculated

$$\int_0^1 [(1 - x^2) - 3x^2]\, dx = \int_0^1 [1 - 4x^2]\, dx = \left[x - \frac{4x^3}{3}\right]_0^1 = -\frac{1}{3},$$

which is not even close to the right answer. What this integral calculated was actually $A - B$ rather than $A + B$. Why? ■

EXAMPLE 3 The Area Enclosed by Two Curves

Find the area enclosed by $y = x^2$ and $y = x^3$.

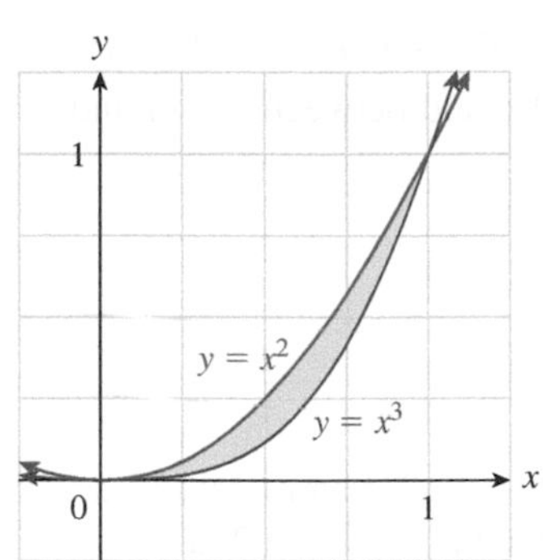

Figure 7

Solution This example has a new wrinkle: We are not told what interval to use for x. However, if we look at the graph in Figure 7, we see that the question can have only one meaning. We are being asked to find the area of the shaded sliver, which is the only region that is actually *enclosed* by the two graphs. This sliver is bounded on either side by the two points where the graphs cross, so our first task is to find those points. They are the points where $x^2 = x^3$, so we solve for x:

$$\begin{aligned} x^2 &= x^3 \\ x^3 - x^2 &= 0 \\ x^2(x - 1) &= 0 \\ x = 0 \quad &\text{or} \quad x = 1. \end{aligned}$$

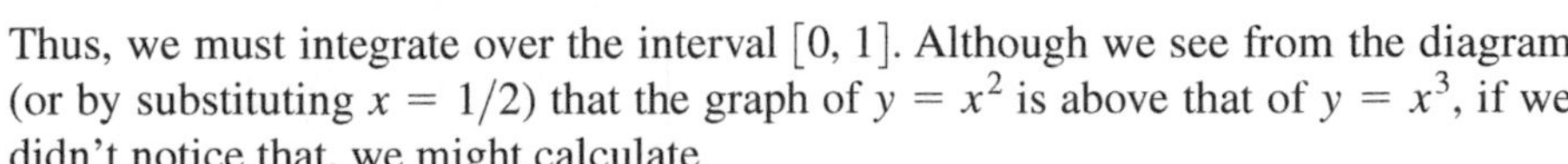

Thus, we must integrate over the interval $[0, 1]$. Although we see from the diagram (or by substituting $x = 1/2$) that the graph of $y = x^2$ is above that of $y = x^3$, if we didn't notice that, we might calculate

$$\int_0^1 (x^3 - x^2)\, dx = \left[\frac{x^4}{4} - \frac{x^3}{3}\right]_0^1 = -\frac{1}{12}.$$

This tells us that the required area is $1/12$ square units and also that we had our integral reversed. Had we calculated $\int_0^1 (x^2 - x^3)\, dx$ instead, we would have found the correct answer: $1/12$.

We can summarize the procedure we used in Examples 2 and 3.

Finding the Area between the Graphs of $f(x)$ and $g(x)$

1. Find all points of intersection by solving $f(x) = g(x)$ for x. This either determines the interval over which you will integrate or breaks up a given interval into regions between the intersection points.
2. Determine the area of each region you found by integrating the difference of the larger and the smaller function. (If you accidentally take the smaller minus the larger, the integral will give the negative of the area, so just take the absolute value.)
3. Add together the areas you found in Step 2 to get the total area.

Q: *Is there any quick and easy method to find the area between two graphs without having to find all points of intersection? What if it is hard or impossible to find out where the curves intersect?*

A: We can use technology to give the approximate area between two graphs. First recall that, if $f(x) \geq g(x)$ for all x in $[a, b]$, then the area between their graphs over $[a, b]$ is given by $\int_a^b [f(x) - g(x)]\, dx$, whereas if $g(x) \geq f(x)$, the area is given by $\int_a^b [g(x) - f(x)]\, dx$. Notice that both expressions are equal to

$$\int_a^b |f(x) - g(x)|\, dx,$$

telling us that we can use this same formula in both cases.

Area between Two Graphs: Approximation Using Technology

The area of the region between the graphs of f and g and between $x = a$ and $x = b$ is given by

$$A = \int_a^b |f(x) - g(x)|\, dx.$$

Quick Example

1. To approximate the area of the region between $y = 3x^2$ and $y = 1 - x^2$ and between $x = 0$ and $x = 1$ that we calculated in Example 2, use technology to compute

$$\int_0^1 |3x^2 - (1 - x^2)|\, dx = 1.$$

TI-83/84 Plus: Enter `fnInt(abs(3x^2-(1-x^2)),X,0,1)`

Website: Online Utilities → Numerical Integration Utility and Grapher
Enter `abs(3x^2-(1-x^2))` for $f(x)$ and 0 and 1 for the lower and upper limits, and press "Integral". (See Figure 8.)

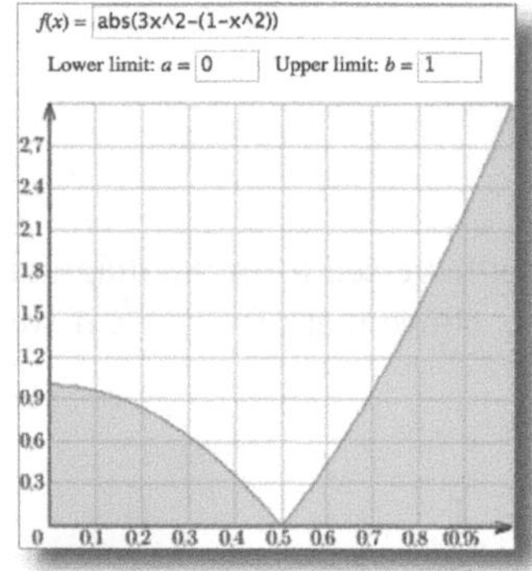

Figure 8

7.2 EXERCISES

▼ more advanced ◆ challenging

T indicates exercises that should be solved using technology

In Exercises 1–8, find the area of the shaded region. (We suggest that you use technology to check your answers.)

1.

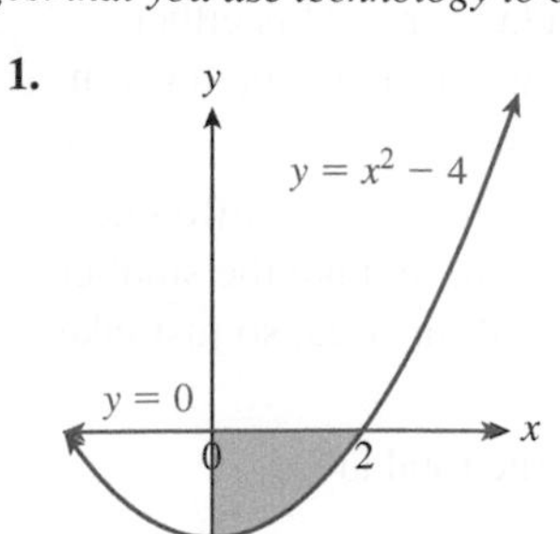

2.

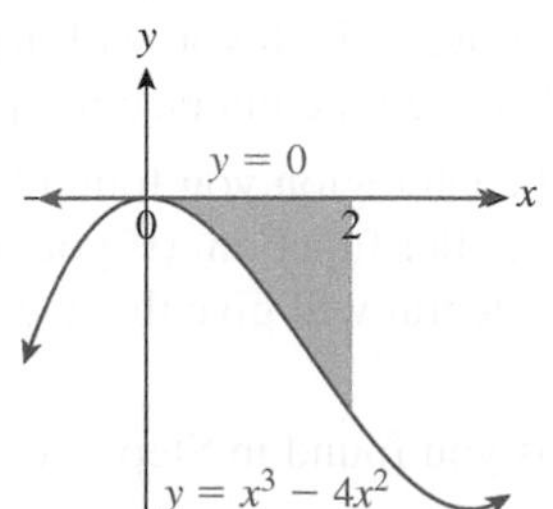

3. 4.

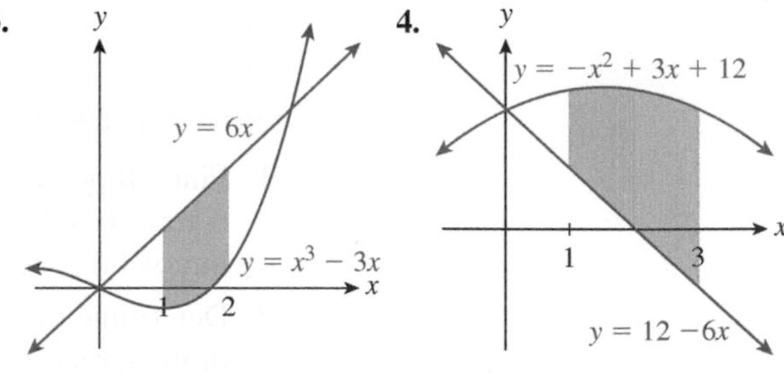

5. 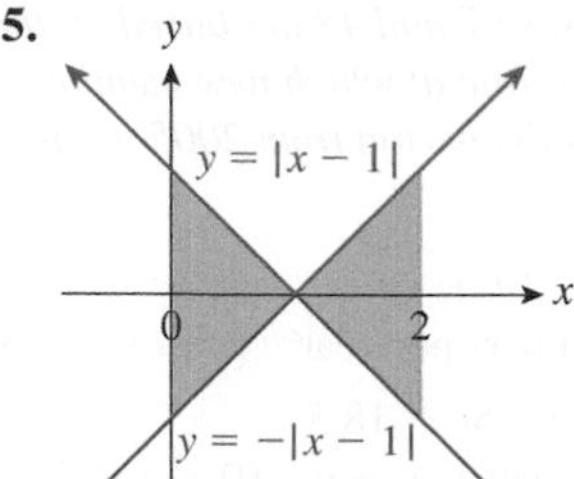

6. 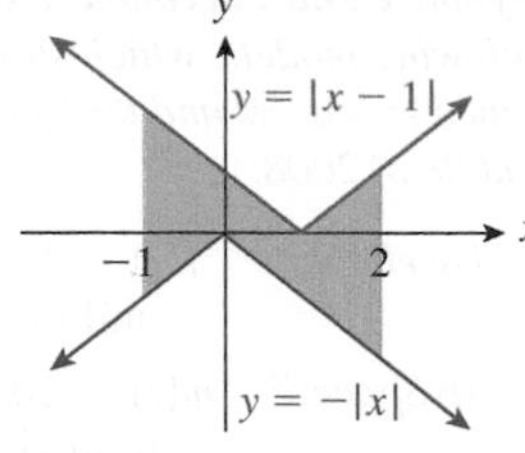

7. ▼ 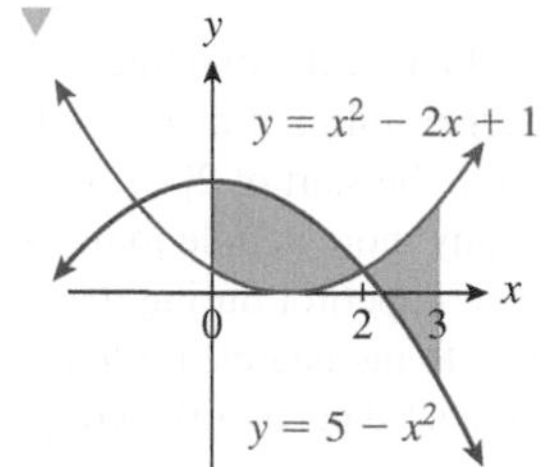

8. ▼ 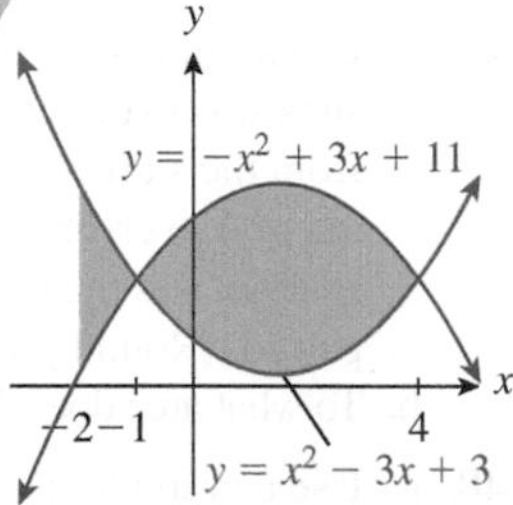

In Exercises 9–42, find the area of the indicated region. We suggest that you graph the curves to check whether one is above the other or whether they cross, and that you use technology to check your answers.

9. Between $y = x^2$ and $y = -1$ for x in $[-1, 1]$ [**HINT:** See Example 1.]
10. Between $y = x^3$ and $y = -1$ for x in $[-1, 1]$ [**HINT:** See Example 1.]
11. Between $y = -x$ and $y = x$ for x in $[0, 2]$
12. Between $y = -x$ and $y = x/2$ for x in $[0, 2]$
13. Between $y = |x|$ and $y = x^2$ for x in $[-1, 1]$
14. Between $y = -|x|$ and $y = x^2 - 2$ for x in $[-1, 1]$
15. Between $y = x$ and $y = x^2$ for x in $[-1, 1]$ [**HINT:** See Example 2.]
16. Between $y = x$ and $y = x^3$ for x in $[-1, 1]$ [**HINT:** See Example 2.]
17. Between $y = x^2 - 2x$ and $y = -x^2 + 4x - 4$ for x in $[0, 2]$
18. Between $y = x^2 - 4x + 2$ and $y = -x^2 + 4x - 4$ for x in $[0, 3]$
19. Between $y = 2x^2 + 10x - 5$ and $y = -x^2 + 4x + 4$ for x in $[-3, 2]$
20. Between $y = 2x^2 + 7x - 2$ and $y = -x^2 + 4x + 4$ for x in $[-2, 2]$
21. Between $y = e^x$ and $y = x$ for x in $[0, 1]$
22. Between $y = e^{-x}$ and $y = -x$ for x in $[0, 1]$
23. Between $y = (x - 1)^2$ and $y = -(x - 1)^2$ for x in $[0, 1]$
24. Between $y = x^2(x^3 + 1)^{10}$ and $y = -x(x^2 + 1)^{10}$ for x in $[0, 1]$
25. Enclosed by $y = x$ and $y = x^4$ [**HINT:** See Example 3.]
26. Enclosed by $y = x$ and $y = -x^4$ [**HINT:** See Example 3.]
27. Enclosed by $y = x^3$ and $y = x^4$
28. Enclosed by $y = x$ and $y = x^3$
29. Enclosed by $y = x^2$ and $y = x^4$
30. Enclosed by $y = x^4 - x^2$ and $y = x^2 - x^4$
31. Enclosed by $y = x^2 - 2x$ and $y = -x^2 + 4x - 4$
32. Enclosed by $y = x^2 - 4x + 2$ and $y = -x^2 + 4x - 4$
33. Enclosed by $y = 2x^2 + 10x - 5$ and $y = -x^2 + 4x + 4$
34. Enclosed by $y = 2x^2 + 7x - 2$ and $y = -x^2 + 4x + 4$
35. Enclosed by $y = e^x$, $y = 2$, and the y-axis
36. Enclosed by $y = e^{-x}$, $y = 3$, and the y-axis
37. Enclosed by $y = \ln x$, $y = 2 - \ln x$, and $x = 4$
38. Enclosed by $y = \ln x$, $y = 1 - \ln x$, and $x = 4$
39. T Enclosed by $y = e^x$, $y = 2x + 1$, $x = -1$, and $x = 1$ (Round answer to four significant digits.) [**HINT:** See Quick Example 1.]
40. T Enclosed by $y = 2^x$, $y = x + 2$, $x = -2$, and $x = 2$ (Round answer to four significant digits.) [**HINT:** See Quick Example 1.]
41. T Enclosed by $y = \ln x$ and $y = \frac{x}{2} - \frac{1}{2}$ (Round answer to four significant digits.) (First use technology to determine approximately where the graphs cross.)
42. T Enclosed by $y = \ln x$, and $y = x - 2$ (Round answer to four significant digits.) (First use technology to determine approximately where the graphs cross.)

Applications

43. ***Revenue and Cost*** Suppose your daily revenue from selling used DVDs is

$$R(t) = 100 + 10t \quad (0 \le t \le 5)$$

dollars per day, where t represents days from the beginning of the week, while your daily costs are

$$C(t) = 90 + 5t \quad (0 \le t \le 5)$$

dollars per day. Find the area between the graphs of $R(t)$ and $C(t)$ for $0 \le t \le 5$. What does your answer represent?

44. ***Income and Expenses*** Suppose your annual income is

$$I(t) = 50{,}000 + 2{,}000t \quad (0 \le t \le 3)$$

dollars per year, where t represents the number of years since you began your job, while your annual expenses are

$$E(t) = 45{,}000 + 1{,}500t \quad (0 \le t \le 3)$$

dollars per year. Find the area between the graphs of $I(t)$ and $E(t)$ for $0 \le t \le 3$. What does your answer represent?

45. ***Housing*** The total number of housing starts in the United States during 2000–2008 was approximately

$$h(t) = -30t^2 + 240t + 800 \text{ thousand homes per year} \quad (0 \le t \le 8),$$

where t is time in years since the start of 2000.[12] During that time, the number of housing starts for sale purposes in the United States was approximately

$$s(t) = -33t^2 + 240t + 700 \text{ thousand homes per year} \quad (0 \le t \le 8).$$

Compute the area shown in the graph, and interpret the answer.

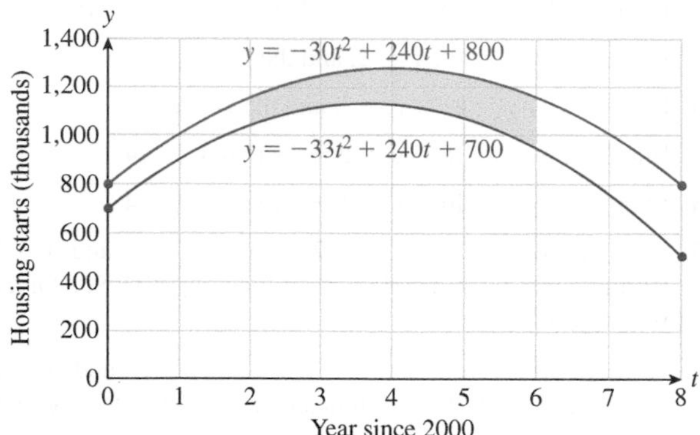

46. ***Oil Production in Mexico: Pemex*** The rate of crude oil production from 2009 to 2013 by Pemex, Mexico's national oil company, can be approximated by

$$q(t) = 6.2t^2 - 146t + 1{,}910 \text{ million barrels per year} \quad (9 \le t \le 13),$$

where t is time in years since the start of 2000.[13] During that time, Mexico exported crude oil to the United States at a rate of[14]

$$r(t) = -14t^2 + 292t - 1{,}100 \text{ million barrels per year} \quad (9 \le t \le 13).$$

Compute the area shown in the graph, and interpret the answer.

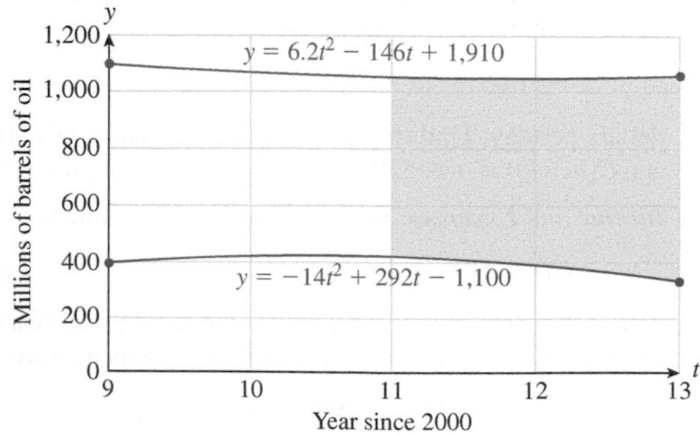

MySpace and Facebook *Exercises 47 and 48 are based on the following models, which show the rate at which new members joined* Facebook *and* MySpace *in the period from 2005 to the middle of 2008:*

Facebook:[15] $f(t) = 12t^2 - 20t + 10$ million members per year $(0 \le t \le 3.5)$

MySpace:[16] $m(t) = 10.5t^2 + 25t + 18.5$ million members per year $(0 \le t \le 3.5)$.

(t is time in years since the start of 2005.)

47. **a.** Use a graph to determine which of the two Internet sites was experiencing a larger influx of new members from the start of July 2005 to the start of 2007. Use an integral to estimate how many more people joined that Internet site than joined its competitor during that period. (Round your answer to the nearest million.)
    b. To what area does the integral used in part (a) correspond?

48. **a.** Use a graph to determine which of the two Internet sites was experiencing a larger influx of new members from the start of 2007 through June 2008. Use an integral to estimate how many more people joined that Internet site than joined its competitor during that period. (Round your answer to the nearest million.)
    b. To what area does the integral used in part (a) correspond?

Big Brother *Exercises 49 and 50 are based on the following models, which show the total number of wiretaps authorized per year by all state and federal courts in the United States and the number authorized by state courts:*

State and federal courts: $a(t) = 770e^{0.060t}$ $(0 \le t \le 23)$

State courts: $s(t) = 410e^{0.071t}$ $(0 \le t \le 23)$.

(t is time in years since the start of 1990.)[17]

49. ▼ Estimate the area between the graphs of the two functions over the interval $[0, t]$. Interpret your answer.

50. ▼ Estimate the area between the graphs of the two functions over the interval $[t, 23]$. Interpret your answer.

Communication and Reasoning Exercises

51. If f and g are continuous functions with $\int_a^b [f(x) - g(x)]\,dx = 0$, it follows that the area between the graphs of f and g is zero—right?

52. You know that f and g are continuous and their graphs do not cross on the interval $[a, b]$, so you calculate $\int_a^b [f(x) - g(x)]\,dx$ and find that the answer is -40. Why is it negative, and what is the area between the curves?

---

[12] Source for data: U.S. Census Bureau (www.census.gov).

[13] Source for data: www.pemex.com.

[14] Source for data: U.S. Energy Information Administration (www.eia.gov).

[15] Sources for data: www.facebook.com, www.insidehighered.com. (Some data are interpolated.)

[16] Source for data: www.swivel.com/data_sets.

[17] Source for data: Wiretap Reports, Administrative Office of the United States Courts (www.uscourts.gov/Statistics/WiretapReports/wiretap-report-2013.aspx).

53. The following graph shows annual U.S. exports and imports for the period 1960–2007:[18]

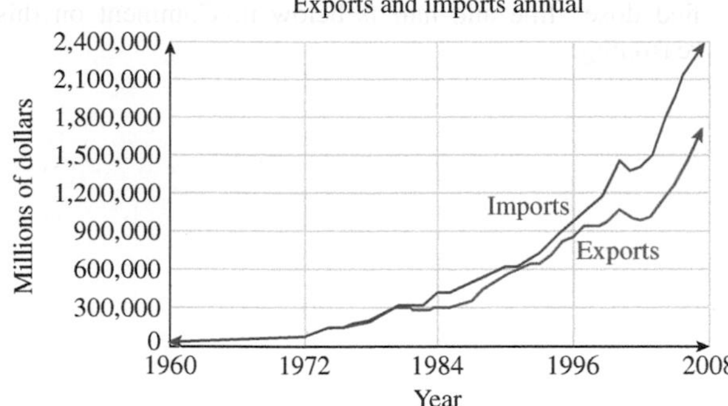

What does the area between the export and import curves represent?

54. The following graph shows a fictitious country's monthly exports and imports for the period 1997–2001:

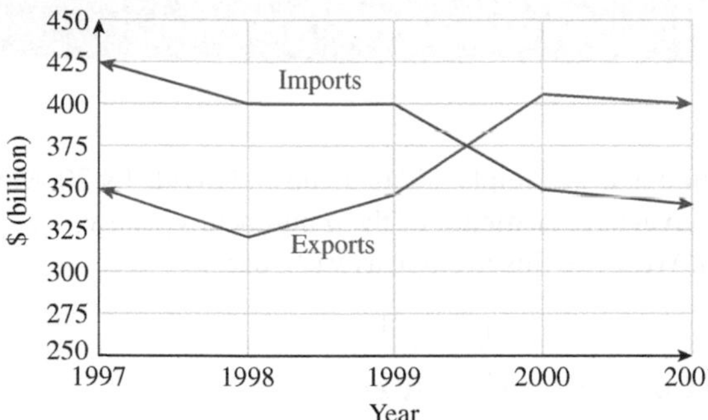

What does the total area enclosed by the export and import curves represent, and what does the definite integral of the difference, Exports − Imports, represent?

55. ▼ The following graph shows the daily revenue and cost in your *Adopt-a-Chia* operation t days from its inception:

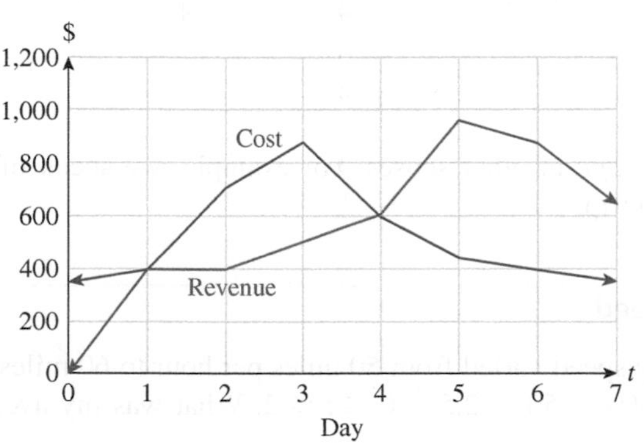

Multiple choice: The area between the cost and revenue curves represents:

(A) the accumulated loss through day 4 plus the accumulated profit for days 5 through 7.
(B) the accumulated profit for the week.
(C) the accumulated loss for the week.
(D) the accumulated cost through day 4 plus the accumulated revenue for days 5 through 7.

56. ▼ The following graph shows daily orders and inventory (stock on hand) for your *Jackson Pollock Paint-by-Number* sets t days into last week:

a. Multiple choice: Which is greatest?

(A) $\int_0^7 (\text{Orders} - \text{Inventory})\, dt$

(B) $\int_0^7 (\text{Inventory} - \text{Orders})\, dt$

(C) The area between the Orders and Inventory curves

b. Multiple choice: The answer to part (a) measures
(A) the accumulated gap between orders and inventory.
(B) the accumulated surplus through day 3 minus the accumulated shortage for days 3 through 5 plus the accumulated surplus through days 5 through 7.
(C) the total net surplus.
(D) the total net loss.

57. ▼ What is wrong with the following claim? "I purchased **Novartis AG** shares for $50 at the beginning of March 2008. My total profit per share from this investment from March through August is represented by the area between the stock price curve and the purchase price curve as shown on the following graph, where t is time in months since March 1, 2008."[19]

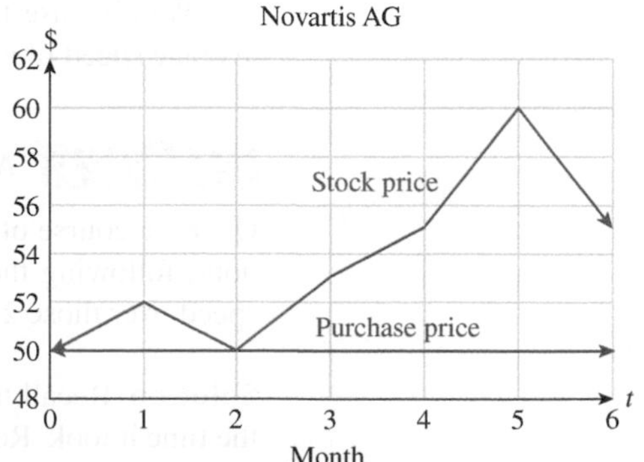

[18] Source: Data 360 (www.data360.org).

[19] Source: www.finance.google.com.

58. ▼ Your pharmaceutical company monitors the amount of medication in successive batches of 100-mg tetracycline capsules and obtains the following graph:

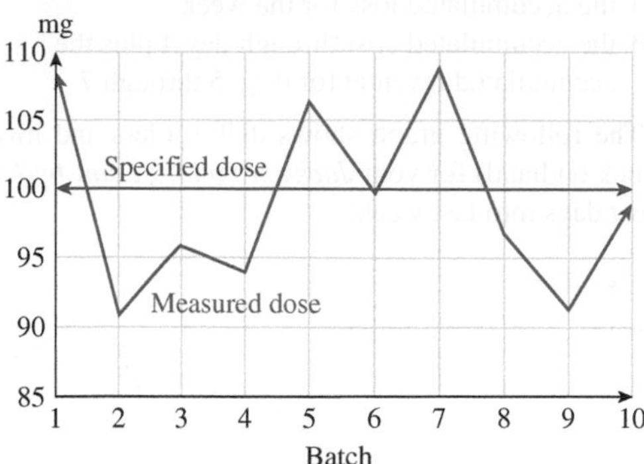

Your production manager claims that the batches of tetracycline conform to the exact dosage requirement because half of the area between the graphs is above the "Specified dose" line and half is below it. Comment on this reasoning.

7.3 Averages and Moving Averages

Averages

To find the average of, say, 20 numbers, we simply add them up and divide by 20. More generally, if we want to find the **average**, or **mean**, of the n numbers $y_1, y_2, y_3, \ldots y_n$, we add them up and divide by n. We write this average as $\bar{y}$ ("y-bar").

Average, or Mean, of a Collection of Values

$$\bar{y} = \frac{y_1 + y_2 + \cdots + y_n}{n}$$

Quick Example

1. The average of $\{0, 2, -1, 5\}$ is $\bar{y} = \dfrac{0 + 2 - 1 + 5}{4} = \dfrac{6}{4} = 1.5$.

We also use the word *average* in other senses. For example, we speak of the average speed of a car during a trip.

EXAMPLE 1 Average Speed

Over the course of 2 hours, my speed varied from 50 miles per hour to 60 miles per hour, following the function $v(t) = 50 + 2.5t^2$, $0 \le t \le 2$. What was my average speed over those 2 hours?

Solution Recall that average speed is simply the total distance traveled divided by the time it took. Recall, also, that we can find the distance traveled by integrating the speed:

$$\begin{aligned}\text{Distance traveled} &= \int_0^2 v(t)\,dt \\ &= \int_0^2 (50 + 2.5t^2)\,dt \\ &= \left[50t + \frac{2.5}{3}t^3\right]_0^2 \\ &= 100 + \frac{20}{3} \\ &\approx 106.67 \text{ miles.}\end{aligned}$$

It took 2 hours to travel this distance, so the average speed was

$$\text{Average speed} \approx \frac{106.67}{2} \approx 53.3 \text{ mph.}$$

In general, if we travel with velocity $v(t)$ from time $t = a$ to time $t = b$, we will travel a distance of $\int_a^b v(t)\,dt$ in time $b - a$, which gives an average velocity of

$$\text{Average velocity} = \frac{1}{b-a}\int_a^b v(t)\,dt.$$

Thinking of this calculation as finding the average value of the velocity function, we generalize and make the following definition.

Average Value of a Function

The **average**, or **mean**, of a function $f(x)$ on an interval $[a, b]$ is

$$\bar{f} = \frac{1}{b-a}\int_a^b f(x)\,dx.$$

Quick Example

2. The average of $f(x) = x$ on $[1, 5]$ is

$$\begin{aligned}\bar{f} &= \frac{1}{b-a}\int_a^b f(x)\,dx \\ &= \frac{1}{5-1}\int_1^5 x\,dx \\ &= \frac{1}{4}\left[\frac{x^2}{2}\right]_1^5 = \frac{1}{4}\left(\frac{25}{2} - \frac{1}{2}\right) = 3.\end{aligned}$$

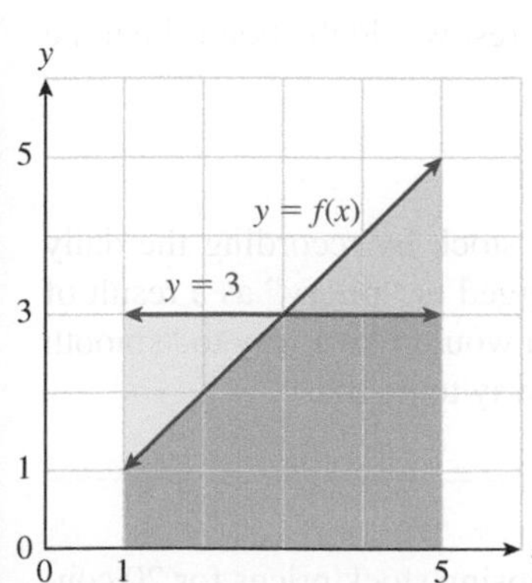

Figure 9

Interpreting the Average of a Function Geometrically

The average of a function has a geometric interpretation. Referring to Quick Example 2, we can compare the graph of $y = f(x)$ with the graph of $y = 3$, both over the interval $[1, 5]$ (Figure 9). We can find the area under the graph of $f(x) = x$ by geometry or by calculus; it is 12. The area in the rectangle under $y = 3$ is also 12.

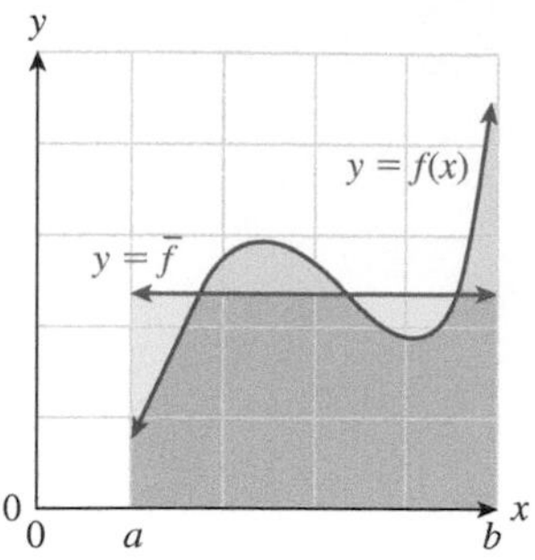

Figure 10

In general, the average $\bar{f}$ of a positive function over the interval $[a, b]$ gives the height of the rectangle over the interval $[a, b]$ that has the same area as the area under the graph of $f(x)$ as illustrated in Figure 10. The equality of these areas follows from the equation

$$(b - a)\bar{f} = \int_a^b f(x)\, dx,$$

because the left-hand side is the area of the rectangle, while the right-hand side is the area under the graph.

EXAMPLE 2 Average Balance

A savings account at the *People's Credit Union* pays 3% interest, compounded continuously, and at the end of the year you get a bonus of 1% of the average balance in the account during the year. If you deposit \$10,000 at the beginning of the year, how much interest and how large a bonus will you get?

Solution We can use the continuous compound interest formula to calculate the amount of money you have in the account at time t:

$$A(t) = 10{,}000e^{0.03t},$$

where t is measured in years. At the end of 1 year the account will have

$$A(1) = \$10{,}304.55,$$

so you will have earned \$304.55 interest. To compute the bonus, we need to find the average amount in the account, which is the average of $A(t)$ over the interval $[0, 1]$. Thus,

$$\begin{aligned}\bar{A} &= \frac{1}{b-a}\int_a^b A(t)\, dt \\ &= \frac{1}{1-0}\int_0^1 10{,}000e^{0.03t}\, dt = \frac{10{,}000}{0.03}\left[e^{0.03t}\right]_0^1 \\ &\approx \$10{,}151.51.\end{aligned}$$

The bonus is 1% of this, or \$101.52.

➡ **Before we go on . . .** The 1% bonus in Example 2 was one third of the total interest. Why did this happen? What fraction of the total interest would the bonus be if the interest rate was 4%, 5%, or 10%? ■

Moving Averages

Suppose you follow the performance of a company's stock by recording the daily closing prices. The graph of these prices may seem jagged or "jittery" as a result of random day-to-day fluctuations. To see any trends, you would like a way to "smooth out" these data. The **moving average** is one common way to do that.

EXAMPLE 3 Stock Prices

The following table shows *Colossal Conglomerate*'s closing stock prices for 20 consecutive trading days:

Day	1	2	3	4	5	6	7	8	9	10
Price	20	22	21	24	24	23	25	26	20	24
Day	11	12	13	14	15	16	17	18	19	20
Price	26	26	25	27	28	27	29	27	25	24

Plot these prices and the 5-day moving average.

Solution The 5-day moving average is the average of each day's price together with the prices of the preceding 4 days. We can compute the 5-day moving averages starting on the fifth day. We get these numbers:

Day	1	2	3	4	5	6	7	8	9	10
Moving Average					22.2	22.8	23.4	24.4	23.6	23.6
Day	11	12	13	14	15	16	17	18	19	20
Moving Average	24.2	24.4	24.2	25.6	26.4	26.6	27.2	27.6	27.2	26.4

Using Technology

See the Technology Guides at the end of the chapter to find out how to tabulate and graph moving averages using a graphing calculator or a spreadsheet. Outline:

TI-83/84 Plus

STAT EDIT; days in L_1, prices in L_2.
Home screen:

```
seq((L2(X)+L2(X-1)
+L2(X-2)+L2(X-3)
+L2(X-4))/5,X,5,20)
→L3
```

[More details in the Technology Guide.]

Spreadsheet

Day data in A2–A21
Price data in B2–B21
Enter `=AVERAGE(B2:B6)` in C6, copy down to C21.
Graph the data in columns A–C.
[More details in the Technology Guide.]

The closing stock prices and moving averages are plotted in Figure 11.

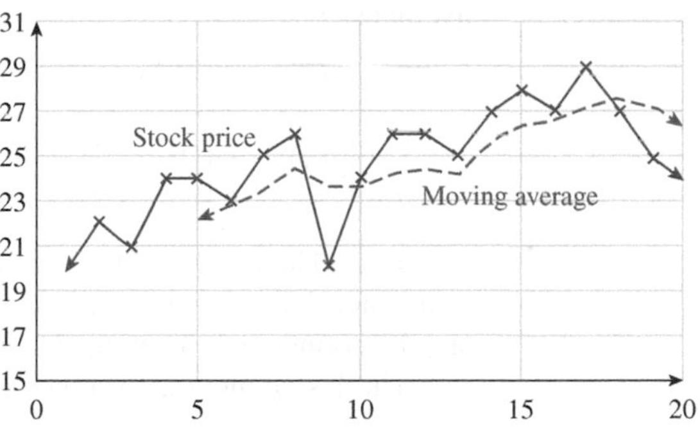

Figure 11

As you can see, the moving average is less volatile than the closing price. Because the moving average incorporates the stock's performance over 5 days at a time, a single day's fluctuation is smoothed out. Look at day 9 in particular. The moving average also tends to lag behind the actual performance because it takes past performance into account. Look at the downturns at days 6 and 18 in particular.

The period of 5 days for a moving average, as used in Example 3, is arbitrary. Using a longer period of time would smooth the data more but increase the lag. For data used as economic indicators, such as housing prices or retail sales, it is common to compute the four-quarter moving average to smooth out seasonal variations.

It is also sometimes useful to compute moving averages of continuous functions. We may want to do this if we use a mathematical model of a large collection of data. Also, some physical systems have the effect of converting an input function (an electrical signal, for example) into its moving average. By an ***n*-unit moving average** of a function $f(x)$, we mean the function $\bar{f}$ for which $\bar{f}(x)$ is the average of the value of $f(x)$ on $[x - n, x]$. Using the formula for the average of a function, we get the following formula.

n-Unit Moving Average of a Function

The n-unit moving average of a function f is

$$\bar{f}(x) = \frac{1}{n}\int_{x-n}^{x} f(t)\, dt.$$

Quick Example

3. The 2-unit moving average of $f(x) = x^2$ is

$$\bar{f}(x) = \frac{1}{2}\int_{x-2}^{x} t^2\, dt = \frac{1}{6}[t^3]_{x-2}^{x} = x^2 - 2x + \frac{4}{3}.$$

The graphs of $f(x)$ and $\bar{f}(x)$ are shown in Figure 12.

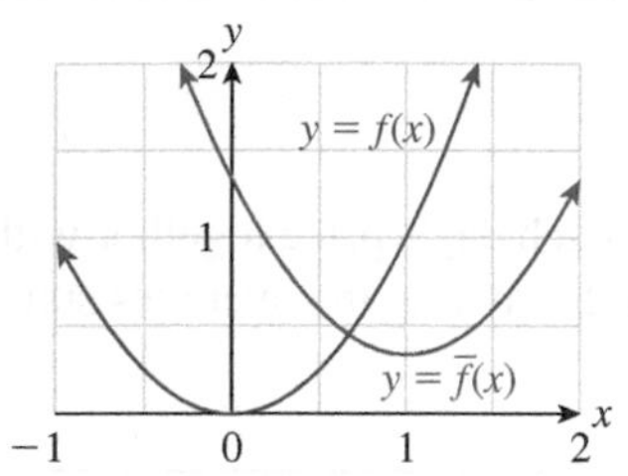

Figure 12

EXAMPLE 4 Moving Averages: Sawtooth and Step Functions

Graph the following functions, and then compute and graph their 1-unit moving averages:

$$f(x) = |x| - |x-1| + |x-2| \qquad \text{Sawtooth}$$

$$g(x) = \frac{1}{2}\left[1 + \frac{|x-1|}{x-1}\right]. \qquad \text{Unit step at } x = 1$$

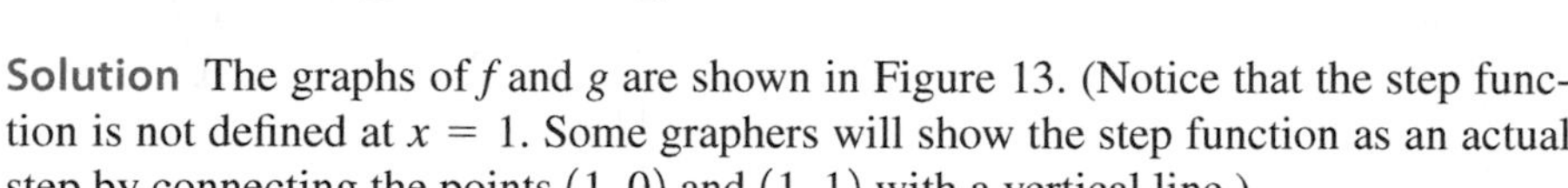

Solution The graphs of f and g are shown in Figure 13. (Notice that the step function is not defined at $x = 1$. Some graphers will show the step function as an actual step by connecting the points $(1, 0)$ and $(1, 1)$ with a vertical line.)

The 1-step moving averages are

$$\bar{f}(x) = \int_{x-1}^{x} f(t)\, dt = \int_{x-1}^{x} [|t| - |t-1| + |t-2|]\, dt$$

$$= \frac{1}{2}[t|t| - (t-1)|t-1| + (t-2)|t-2|]_{x-1}^{x}$$

$$= \frac{1}{2}([x|x| - (x-1)|x-1| + (x-2)|x-2|]$$

$$- [(x-1)|x-1| - (x-2)|x-2| + (x-3)|x-3|])$$

$$= \frac{1}{2}[x|x| - 2(x-1)|x-1| + 2(x-2)|x-2| - (x-3)|x-3|]$$

$$\bar{g}(x) = \int_{x-1}^{x} g(t)\, dt = \int_{x-1}^{x} \frac{1}{2}\left[1 + \frac{|t-1|}{t-1}\right] dt$$

$$= \frac{1}{2}[t + |t-1|]_{x-1}^{x}$$

$$= \frac{1}{2}[(x + |x-1|) - (x - 1 + |x-2|)]$$

$$= \frac{1}{2}[1 + |x-1| - |x-2|].$$

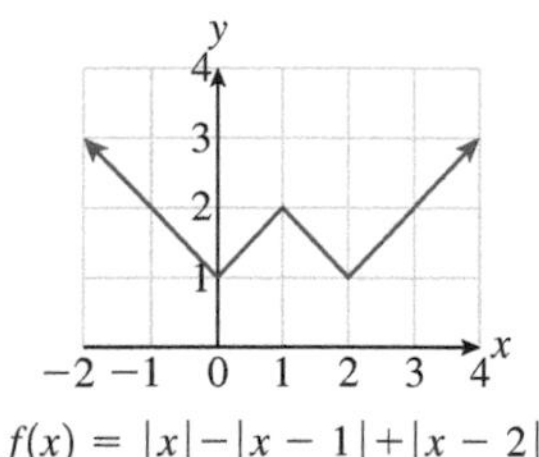

$f(x) = |x| - |x-1| + |x-2|$

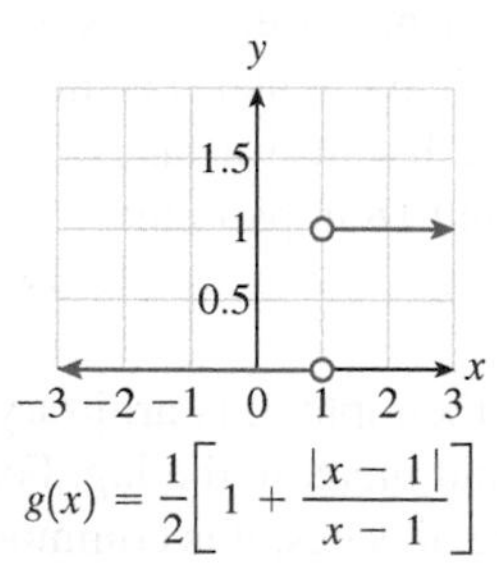

$g(x) = \frac{1}{2}\left[1 + \frac{|x-1|}{x-1}\right]$

Figure 13

The graphs of $\bar{f}$ and $\bar{g}$ are shown in Figure 14.

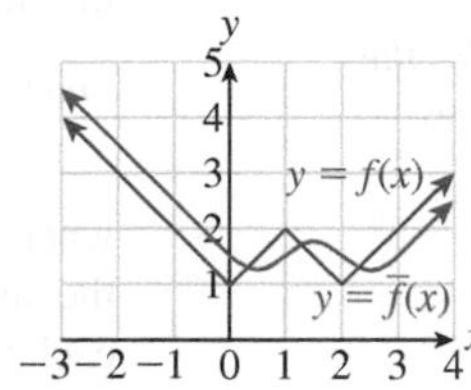

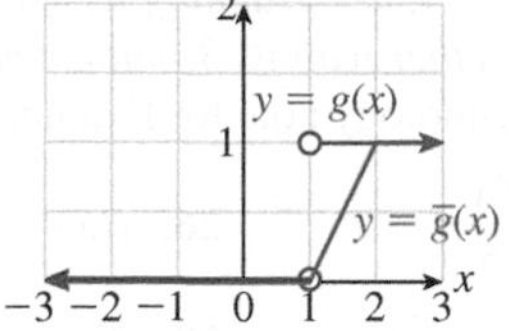

Figure 14

Notice how the graph of $\bar{f}$ smooths out the zigzags of the sawtooth function.

Using Technology

TI-83/84 Plus

We can graph the moving average of f in Example 4 on a TI-83/84 Plus as follows:

```
Y1=abs(X)-abs(X-1)
+abs(X-2)
Y2=fnInt(Y1(T),T,X-1,X)
```

[ZOOM] [0]

For g, change Y_1 to

```
Y1=.5*(1+abs(X-1)/(X-1))
```

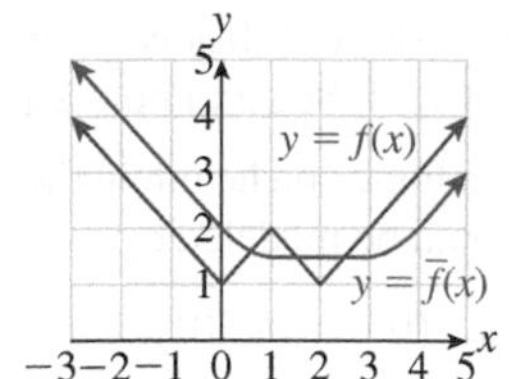

Figure 15

➡ **Before we go on . . .** Figure 15 shows the 2-unit moving average of f in Example 4,

$$\bar{f}(x) = \frac{1}{2}\int_{x-2}^{x} f(t)\,dt$$
$$= \frac{1}{4}(x|x| - (x-1)|x-1| + (x-3)|x-3| - (x-4)|x-4|).$$

Notice how the 2-point moving average has completely eliminated the zigzags, illustrating how moving averages can be used to remove seasonal fluctuations in real-life situations. ■

7.3 EXERCISES

▼ more advanced ◆ challenging

T indicates exercises that should be solved using technology

In Exercises 1–8, find the average of the function over the given interval. Plot each function and its average on the same graph (as in Figure 10). [**HINT:** See Quick Example 2.]

1. $f(x) = x^3$ over $[0, 2]$ **2.** $f(x) = x^3$ over $[-1, 1]$

3. $f(x) = x^3 - x$ over $[0, 2]$ **4.** $f(x) = x^3 - x$ over $[0, 1]$

5. $f(x) = e^{-x}$ over $[0, 2]$ **6.** $f(x) = e^x$ over $[-1, 1]$

7. $f(x) = |2x - 5|$ over $[0, 4]$

8. $f(x) = |-x + 2|$ over $[-1, 3]$

In Exercises 9 and 10, complete the given table with the values of the 3-unit moving average of the given function. [**HINT:** See Example 3.]

9.

x	0	1	2	3	4	5	6	7
$r(x)$	3	5	10	3	2	5	6	7
$\bar{r}(x)$								

10.

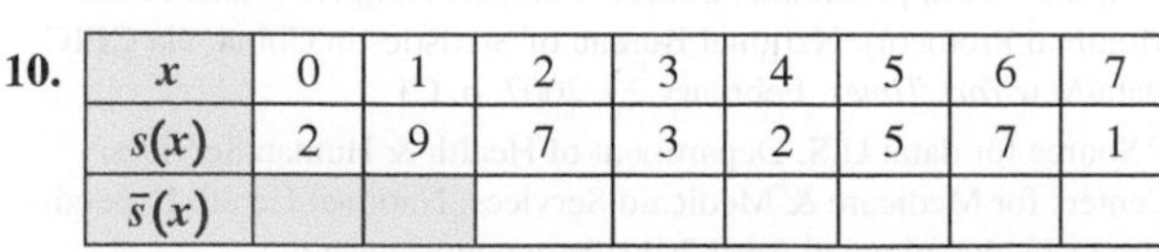

x	0	1	2	3	4	5	6	7
$s(x)$	2	9	7	3	2	5	7	1
$\bar{s}(x)$								

In Exercises 11 and 12, some values of a function and its 3-unit moving average are given. Supply the missing information.

11.

x	0	1	2	3	4	5	6	7
$r(x)$	1	2			11		10	2
$\bar{r}(x)$			3	5		11		

12.

x	0	1	2	3	4	5	6	7
$s(x)$	1	5		1				
$\bar{s}(x)$			5		5	2	3	2

In Exercises 13–24, calculate the 5-unit moving average of the given function. Plot the function and its moving average on the same graph, as in Example 4. (You may use graphing technology for these plots, but you should compute the moving averages analytically.) [**HINT:** See Quick Example 3 and Example 4.]

13. $f(x) = x^3$ **14.** $f(x) = x^3 - x$

15. $f(x) = x^{2/3}$ **16.** $f(x) = x^{2/3} + x$

17. $f(x) = e^{0.5x}$ **18.** $f(x) = e^{-0.02x}$

19. $f(x) = \sqrt{x}$ **20.** $f(x) = x^{1/3}$

21. $f(x) = 1 - \dfrac{|2x-1|}{2x-1}$ **22.** $f(x) = 2 + \dfrac{|3x+1|}{3x+1}$

23. ▼ $f(x) = 2 - |x + 1| + |x|$ [Do not simplify the answer.]

24. ▼ $f(x) = |2x + 1| - |2x| - 2$ [Do not simplify the answer.]

T *In Exercises 25–34, use graphing technology to plot the given function together with its 3-unit moving averages.* [**HINT:** See the technology note for Example 4.]

25. $f(x) = \dfrac{10x}{1 + 5|x|}$

26. $f(x) = \dfrac{1}{1 + e^x}$

27. $f(x) = \ln(1 + x^2)$

28. $f(x) = e^{1-x^2}$

29. $f(x) = |x| - |x - 1| + |x - 2| - |x - 3| + |x - 4|$

30. $f(x) = |x| - 2|x - 1| + 2|x - 2| - 2|x - 3| + |x - 4|$

31. $f(x) = \dfrac{|x|}{x} - \dfrac{|x-1|}{x-1} + \dfrac{|x-2|}{x-2} - \dfrac{|x-3|}{x-3}$

32. $f(x) = \dfrac{|x|}{x} - 2\dfrac{|x-1|}{x-1} + 2\dfrac{|x-2|}{x-2} - \dfrac{|x-3|}{x-3}$

33. $f(x) = \dfrac{|x|}{x} + \dfrac{|x-1|}{x-1} + \dfrac{|x-2|}{x-2} + \dfrac{|x-3|}{x-3}$

34. $f(x) = 4 - \dfrac{|x|}{x} - \dfrac{|x-1|}{x-1} - \dfrac{|x-2|}{x-2} - \dfrac{|x-3|}{x-3}$

Applications

35. ***Television Advertising*** The cost, in millions of dollars, of a 30-second television ad during the Super Bowl in the years 2000–2010 can be approximated by

$$C(t) = 0.14t + 1.1 \text{ million dollars} \quad (0 \le t \le 10).$$

($t = 0$ represents 2000.)[20] What was the average cost of a Super Bowl ad during the given period? [**HINT:** See Example 1.]

36. ***Television Advertising*** The cost, in millions of dollars, of a 30-second television ad during the Super Bowl in the years 1980–2000 can be approximated by

$$C(t) = 0.044t + 0.222 \text{ million dollars} \quad (0 \le t \le 20).$$

($t = 0$ represents 1980.)[21] What was the average cost of a Super Bowl ad during the given period? [**HINT:** See Example 1.]

37. ***Membership: Facebook*** The number of new members joining **Facebook** each year in the period from 2005 to the middle of 2008 can be modeled by

$$m(t) = 12t^2 - 20t + 10 \text{ million members per year} \quad (0 \le t \le 3.5),$$

where t is time in years since the start of 2005.[22] What was the average number of new members joining Facebook each year from the start of 2005 to the start of 2008?

38. ***Membership: MySpace*** The number of new members joining **MySpace** each year in the period from 2004 to the middle of 2007 can be modeled by

$$m(t) = 10.5t^2 + 14t - 6 \text{ million members per year} \quad (0 \le t \le 3.5),$$

where t is time in years since the start of 2004.[23] What was the average number of new members joining MySpace each year from the start of 2004 to the start of 2007?

39. ***Freon Production*** Annual production of ozone-layer-damaging Freon 22 (chlorodifluoromethane) in developing countries from 2000 to 2010 can be modeled by

$$F(t) = 97.2(1.20)^t \text{ million tons} \quad (0 \le t \le 10).$$

(t is the number of years since 2000.)[24] What was the average annual production over the period shown? (Round your answer to the nearest million tons.) [**HINT:** See Example 2.]

40. ***Health Expenditures*** Annual expenditures on health in the United States from 1980 to 2010 could be modeled by

$$F(t) = 296(1.08)^t \text{ billion dollars} \quad (0 \le t \le 30).$$

($t = 0$ represents 1980).[25] What was the average annual expenditure over the period shown? (Round your answer to the nearest billion dollars.) [**HINT:** See Example 2.]

41. ***Investments*** If you invest $10,000 at 8% interest compounded continuously, what is the average amount in your account over 1 year?

42. ***Investments*** If you invest $10,000 at 12% interest compounded continuously, what is the average amount in your account over 1 year?

43. ▼ ***Average Balance*** Suppose you have an account (paying no interest) into which you deposit $3,000 at the beginning of each month. You withdraw money continuously so that the amount in the account decreases linearly to 0 by the end of the month. Find the average amount in the account over a period of several months. (Assume that the account starts at $0 at $t = 0$ months.)

44. ▼ ***Average Balance*** Suppose you have an account (paying no interest) into which you deposit $4,000 at the beginning of each month. You withdraw $3,000 during the course of each month in such a way that the amount decreases linearly. Find the average amount in the account in the first 2 months. (Assume that the account starts at $0 at $t = 0$ months.)

[20] Source for data: en.wikipedia.org/wiki/Super_Bowl_advertising.

[21] *Ibid.*

[22] Sources for data: www.facebook.com, insidehighered.com. (Some data are interpolated.)

[23] Source for data: www.swivel.com/data_sets.

[24] Figures are approximate. Source: Lampert Kuijpers (Panel of the Montreal Protocol), National Bureau of Statistics in China, via CEIC Data/*New York Times*, February 23, 2007, p. C1.

[25] Source for data: U.S. Department of Health & Human Services/Centers for Medicare & Medicaid Services, National Health Expenditure Data, downloaded April 2011 from www.cms.gov.

45. **T** ***Online Payments*** The value of payments made through **PayPal** from the first quarter of 2013 through the fourth quarter of 2014 can be approximated by

$$p(t) = -0.1t^3 + 1.18t^2 - 0.89t + 41$$
billion dollars per quarter $(1 \le t \le 8)$,

where t is time in quarters. ($t = 1$ represents the first quarter of 2013.)[26] Use technology to estimate the average value of payments made each quarter through PayPal during the given period. (Round your answer to the nearest billion dollars.)

46. **T** ***Online Auctions*** The net number of new active **eBay** users each quarter from the first quarter of 2008 through the last quarter of 2010 could be approximated by

$$n(t) = -0.002x^4 + 0.06x^3 - 0.55x^2 + 1.9x - 1$$
million users per quarter $(1 \le t \le 13)$

where t is time in quarters. ($t = 1$ represents the first quarter of 2008.)[27] Use technology to estimate the average number of new active eBay users each quarter during the given period. (Round your answer to the nearest hundred thousand users.)

47. ***Stock Prices: Exxon Mobil*** The following table shows the approximate price of **Exxon Mobil** stock in December of each year from 2005 through 2014, as well as the 4-year moving averages for 2005 through 2007.[28] Complete the table by computing the remaining 4-year moving averages. Round each average to the nearest dollar.

Year t	2005	2006	2007	2008	2009	2010	2011	2012	2013	2014
Stock Price	56	77	94	80	68	73	85	87	101	92
Moving Average (rounded)	46	56	70							

The stock price spiked in 2007, dropped steeply to 2009, then recovered. What happened to the corresponding moving average? [**HINT:** See Example 3.]

48. ***Stock Prices: Nokia*** The following table shows the approximate price of **Nokia** stock in December of each year from 2005 through 2014, as well as the 4-year moving averages for 2005 through 2007.[29] Complete the table by computing the remaining 4-year moving averages. (Note the peak in 2007 and the drop in subsequent years.) Round each average to the nearest dollar.

Year t	2005	2006	2007	2008	2009	2010	2011	2012	2013	2014
Stock Price	18	20	38	16	13	10	5	4	8	8
Moving Average (rounded)	17	18	23							

How does the average size of the year-by-year change in the moving average compare with the average year-by-year change in the stock price? [**HINT:** See Example 3.]

49. ▼ ***Cancun*** The *Playa Loca Hotel* in Cancun has an advertising brochure with the following chart, showing the year-round temperature:[30]

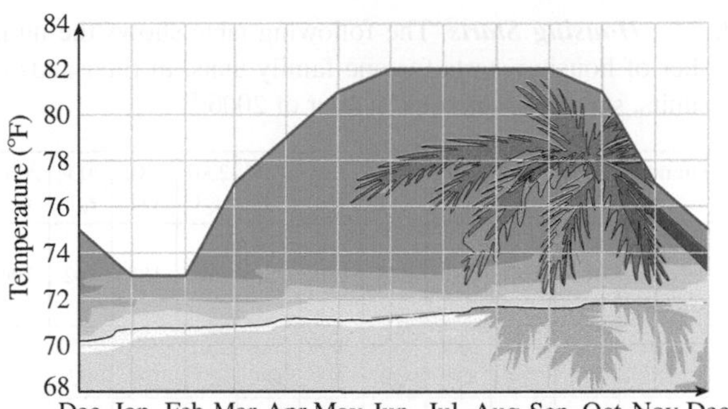

a. Estimate and plot the year-round 6-month moving average. (Use graphing technology, if available, to check your graph.)

b. What can you say about the 12-month moving average?

50. ▼ ***Reykjavik*** Repeat Exercise 49, using the following data from the brochure of the *Tough Traveler Lodge* in Reykjavik:[31]

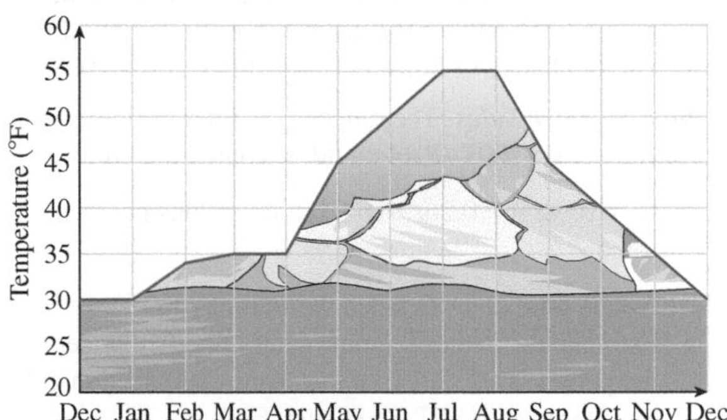

51. **T** ▼ ***Sales: Apple*** The following table shows approximate quarterly sales of **Apple** iPods in millions of units, starting in the first quarter of 2006:[32]

Quarter	2006 Q1	2006 Q2	2006 Q3	2006 Q4	2007 Q1	2007 Q2	2007 Q3	2007 Q4	2008 Q1	2008 Q2
Sales (millions)	8.5	8.1	8.7	21.1	10.5	9.8	10.2	22.1	10.6	11.0
Quarter	**2008 Q3**	**2008 Q4**	**2009 Q1**	**2009 Q2**	**2009 Q3**	**2009 Q4**	**2010 Q1**	**2010 Q2**	**2010 Q3**	**2010 Q4**
Sales (millions)	11.1	22.7	11.0	10.2	10.2	21.0	10.9	9.4	9.1	19.5

[26] Source for data: Statista, www.statista.com.

[27] Source for data: eBay company reports http://investor.ebay.com.

[28] Source: finance.yahoo.com.

[29] *Ibid.*

[30] Source: www.holiday-weather.com. (Temperatures are rounded.)

[31] *Ibid.*

[32] Source: Apple quarterly press releases, www.apple.com/investor.

a. Use technology to compute and plot the four-quarter moving average of these data.

b. The graph of the moving average for the last eight quarters will appear almost linear during 2009 and 2010. Use the 2009 Q1 and 2010 Q4 figures of the moving average to give an estimate (to the nearest 0.1 million units) of the rate of change of iPod sales during 2009–2010.

52. T ▼ ***Housing Starts*** The following table shows the number of housing starts for one-family units, in thousands of units, starting in the first quarter of 2006:[33]

Quarter	2006 Q1	2006 Q2	2006 Q3	2006 Q4	2007 Q1	2007 Q2	2007 Q3	2007 Q4	2008 Q1	2008 Q2
Housing Starts (thousands)	382	433	372	278	260	333	265	188	162	194
Quarter	**2008 Q3**	**2008 Q4**	**2009 Q1**	**2009 Q2**	**2009 Q3**	**2009 Q4**	**2010 Q1**	**2010 Q2**	**2010 Q3**	**2010 Q4**
Housing Starts (thousands)	163	103	78	124	138	105	114	142	119	96

a. Use technology to compute and plot the four-quarter moving average of these data.

b. The graph of the moving average for the eight quarters in 2007 and 2008 will appear almost linear. Use the 2007 Q1 and 2008 Q4 figures of the moving average to give an estimate (to the nearest thousand units) of the rate of change of housing starts during 2007–2008.

53. ***Bottled Water Sales*** The rate of U.S. sales of bottled water for the period 2007–2014 could be approximated by

$$s(t) = 0.08t^2 - 0.26t + 8.8 \text{ billion gallons per year} \quad (0 \le t \le 7),$$

where t is time in years since the start of 2007.[34]

a. Estimate the average annual sales of bottled water over the period 2007–2014, to the nearest 100 million gallons per year. [**HINT:** See Quick Example 2.]

b. Compute the 2-year moving average of s. (You need not simplify the answer.) [**HINT:** See Quick Example 3.]

c. Without simplifying the answer in part (b), say what kind of function the moving average is.

54. ***Bottled Water Sales*** The rate of U.S. per capita sales of bottled water for the period 2007–2014 coud be approximated by

$$s(t) = 0.25t^2 - t + 29 \text{ billion gallons per year} \quad (0 \le t \le 7),$$

where t is the time in years since the start of 2007.[35] Repeat Exercise 53 as applied to per capita sales. (Give your answer to part (a) to the nearest gallon per year.)

55. ***Medicare Spending*** Annual spending on Medicare was projected to increase from \$526 billion in 2010 to around \$977 billion in 2021.[36]

a. Use this information to express s, the annual spending on Medicare (in billions of dollars) as a linear function of t, the number of years since 2010.

b. Find the 4-year moving average of your model.

c. What can you say about the slope of the moving average?

56. ***Pasta Imports in the 1990s*** In 1990 the United States imported 290 million pounds of pasta. From 1990 to 2000, imports increased by an average of 40 million pounds per year.[37]

a. Use these data to express q, the annual U.S. imports of pasta (in millions of pounds) as a linear function of t, the number of years since 1990.

b. Find the 4-year moving average of your model.

c. What can you say about the slope of the moving average?

57. ▼ ***Moving Average of a Linear Function*** Find a formula for the a-unit moving average of a general linear function $f(x) = mx + b$.

58. ▼ ***Moving Average of an Exponential Function*** Find a formula for the a-unit moving average of a general exponential function $f(x) = Ae^{kx}$.

Communication and Reasoning Exercises

59. Explain why it is sometimes more useful to consider the moving average of a stock price rather than the stock price itself.

60. Sales this month were sharply lower than they were last month, but the 12-unit moving average this month was higher than it was last month. How can that be?

61. Your company's 6-month moving average of sales is constant. What does that say about the sales figures?

62. Your monthly salary has been increasing steadily for the past year, and your average monthly salary over the past year was x dollars. Would you have earned more money if you had been paid x dollars per month? Explain your answer.

63. ▼ What property does the graph of a (nonconstant) function have if its average value over an interval is zero? Give an example of such a function.

64. ▼ Can the average value of a function f on an interval be greater than its value at every point in that interval? Explain.

65. ▼ Criticize the following claim: The average value of a function on an interval is midway between its highest and lowest value.

[33] Source: U.S. Census Bureau, www.census.gov/const/www/newresconstindex.html.

[34] Source for data: Beverage Marketing Corporation (www.bottledwater.org).

[35] *Ibid.*

[36] Source: Congressional Budget Office, *March 2011 Medicare Baseline* (www.cbo.gov).

[37] Data are rounded. Sources: Department of Commerce/*New York Times*, September 5, 1995, p. D4; International Trade Administration (www.ita.doc.gov), March 31, 2002.

66. ▼ Your manager tells you that 12-month moving averages give at least as much information as shorter-term moving averages and very often more. How would you argue that he is wrong?

67. ▼ Which of the following most closely approximates the original function: (A) its 10-unit moving average, (B) its 1-unit moving average, or (C) its 0.8-unit moving average? Explain your answer.

68. ▼ Is an increasing function larger or smaller than its 1-unit moving average? Explain.

7.4 Applications to Business and Economics: Consumers' and Producers' Surplus and Continuous Income Streams

Consumers' Surplus

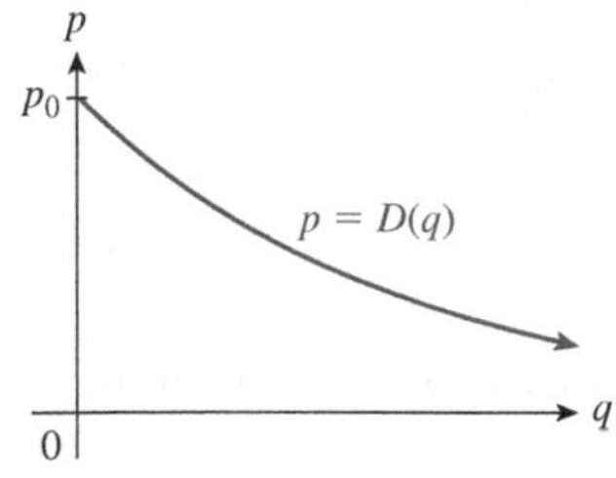

Figure 16

Consider a general demand curve presented, as is traditional in economics, as $p = D(q)$, where p is unit price and q is demand measured, say, in annual sales (Figure 16). Thus, $D(q)$ is the price at which the demand will be q units per year. The price p_0 shown on the graph is the highest price that customers are willing to pay.

Suppose, for example, that the graph in Figure 16 is the demand curve for a particular new model of computer. When the computer first comes out and supplies are low (q is small), "early adopters" will be willing to pay a high price. This is the part of the graph on the left, near the p-axis. As supplies increase and the price drops, more consumers will be willing to pay and more computers will be sold. Pick any particular number of units, $\overline{q}$. We can ask the following question: How much are consumers willing to spend for the first $\overline{q}$ units?

Consumers' Willingness to Spend

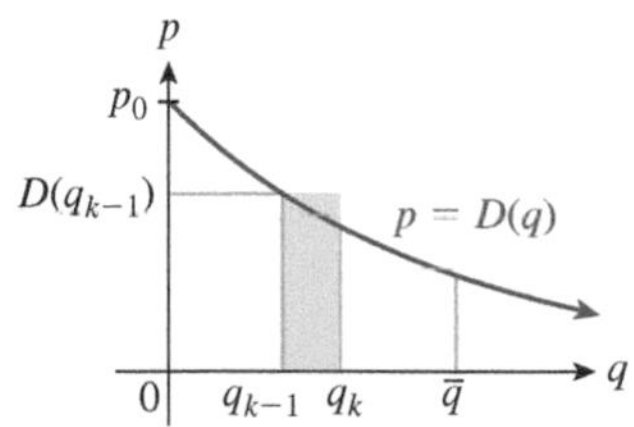

Figure 17

We can approximate consumers' willingness to spend on the first $\overline{q}$ units as follows. We partition the interval $[0, \overline{q}]$ into n subintervals of equal length, as we did when discussing Riemann sums. Figure 17 shows a typical subinterval, $[q_{k-1}, q_k]$.

The price consumers are willing to pay for each of units q_{k-1} through q_k is approximately $D(q_{k-1})$, so the total that consumers are willing to spend for these units is approximately $D(q_{k-1})(q_k - q_{k-1}) = D(q_{k-1})\,\Delta q$, the area of the shaded region in Figure 17. Thus, the total amount that consumers are willing to spend for items 0 through $\overline{q}$ is

$$W \approx D(q_0)\,\Delta q + D(q_1)\,\Delta q + \cdots + D(q_{n-1})\,\Delta q = \sum_{k=0}^{n-1} D(q_k)\,\Delta q,$$

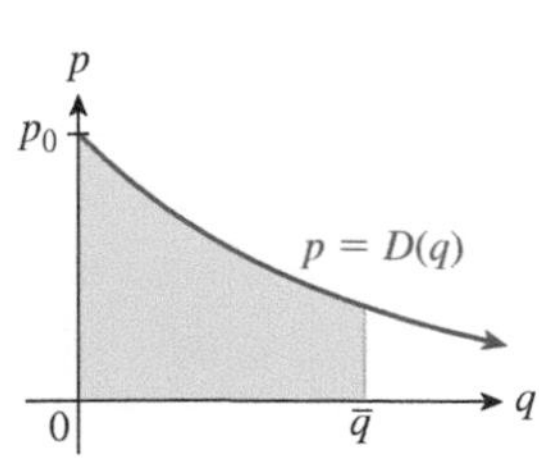

Figure 18

which is a Riemann sum. The approximation becomes better the larger n becomes, and in the limit, the Riemann sums converge to the integral

$$W = \int_0^{\overline{q}} D(q)\,dq.$$

This quantity, the area shaded in Figure 18, is the total consumers' willingness to spend to buy the first $\overline{q}$ units.

Consumers' Expenditure

Now suppose that the manufacturer simply sets the price at some value $\overline{p}$, with a corresponding demand of $\overline{q}$, so $D(\overline{q}) = \overline{p}$. Then the amount that consumers will actually spend to buy these $\overline{q}$ units is $\overline{p}\,\overline{q}$, the product of the unit price and the quantity sold.

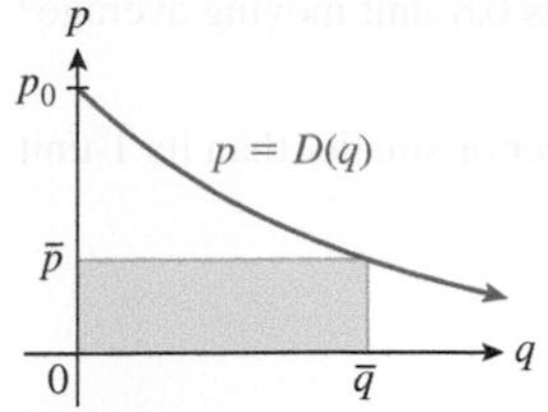

Figure 19

This is the area of the rectangle shown in Figure 19. Notice that we can write $\bar{p}\,\bar{q} = \int_0^{\bar{q}} \bar{p}\, dq$, as suggested by the figure.

The difference between what consumers are willing to pay and what they actually pay is money in their pockets and is called the **consumers' surplus.**

Consumers' Surplus

If demand for an item is given by $p = D(q)$, the selling price is $\bar{p}$, and $\bar{q}$ is the corresponding demand [so that $D(\bar{q}) = \bar{p}$], then the **consumers' surplus** is the difference between willingness to spend and actual expenditure:*

$$CS = \int_0^{\bar{q}} D(q)\, dq - \bar{p}\,\bar{q} = \int_0^{\bar{q}} (D(q) - \bar{p})\, dq.$$

Graphically, it is the area between the graphs of $p = D(q)$ and $p = \bar{p}$, as shown in the figure.

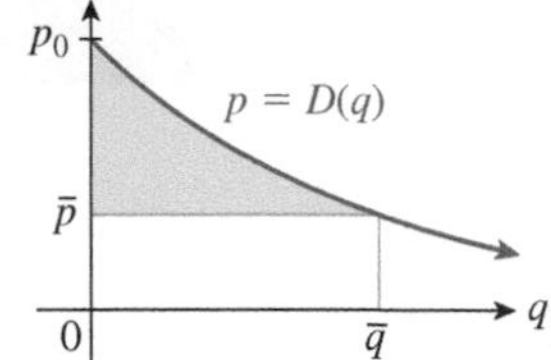

* Multiletter variables such as *CS* used here may be unusual in a math textbook but are traditional in the math of finance. In particular, the notations *PV* and *FV* used later in this section are almost universally used in finance textbooks, calculators (such as the TI-83/84 Plus), and such places as study guides for the finance portion of the Society of Actuaries exams.

EXAMPLE 1 **Consumers' Surplus**

Your video store has an exponential demand equation for used DVDs of the form

$$p = 15e^{-0.01q},$$

where q represents daily sales of used DVDs and p is the price you charge per DVD. Calculate the daily consumers' surplus if you sell your used DVDs at \$5 each.

Solution We are given $D(q) = 15e^{-0.01q}$ and $\bar{p} = 5$. We also need $\bar{q}$. By definition,

$$D(\bar{q}) = \bar{p}$$

or $\quad 15e^{-0.01\bar{q}} = 5,$

which we must solve for $\bar{q}$:

$$e^{-0.01\bar{q}} = \frac{1}{3}$$

$$-0.01\bar{q} = \ln\left(\frac{1}{3}\right) = -\ln 3$$

$$\bar{q} = \frac{\ln 3}{0.01} \approx 109.8612.$$

We now have

$$\begin{aligned} CS &= \int_0^{\bar{q}} (D(q) - \bar{p})\, dq \\ &= \int_0^{109.8612} (15e^{-0.01q} - 5)\, dq \\ &= \left[\frac{15}{-0.01}e^{-0.01q} - 5q\right]_0^{109.8612} \\ &\approx (-500 - 549.31) - (-1{,}500 - 0) \\ &= \$450.69 \text{ per day.} \end{aligned}$$

Producers' Surplus

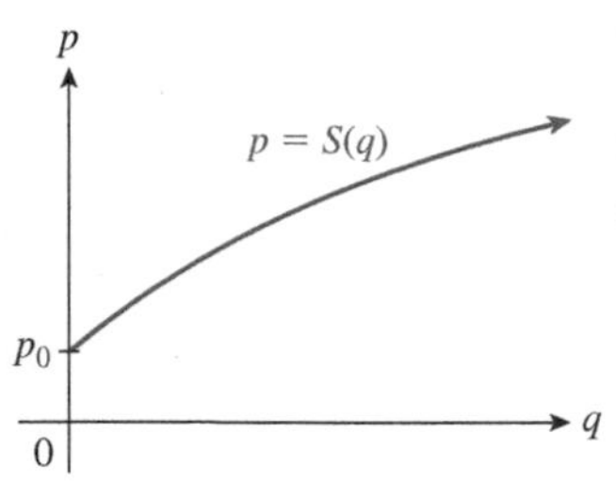

Figure 20

We can also calculate extra income earned by producers. Consider a supply equation of the form $p = S(q)$, where $S(q)$ is the price at which a supplier is willing to supply q items (per time period). Because a producer is generally willing to supply more units at a higher price per unit, a supply curve usually has a positive slope, as shown in Figure 20. The price p_0 is the lowest price that a producer is willing to charge.

Arguing as before, we see that the minimum amount of money producers are willing to receive in exchange for $\bar{q}$ items is $\int_0^{\bar{q}} S(q)\,dq$. On the other hand, if the producers charge $\bar{p}$ per item for $\bar{q}$ items, their actual revenue is $\bar{p}\,\bar{q} = \int_0^{\bar{q}} \bar{p}\,dq$.

The difference between the producers' actual revenue and the minimum they would have been willing to receive is the **producers' surplus**.

Producers' Surplus

The **producers' surplus** is the extra amount earned by producers who were willing to charge less than the selling price of $\bar{p}$ per unit and is given by

$$PS = \int_0^{\bar{q}} (\bar{p} - S(q))\,dq,$$

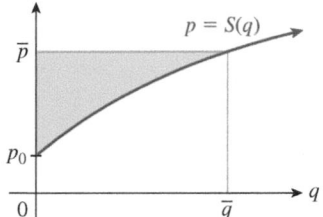

where $S(\bar{q}) = \bar{p}$. Graphically, it is the area of the region between the graphs of $p = \bar{p}$ and $p = S(q)$ for $0 \le q \le \bar{q}$, as in the figure.

EXAMPLE 2 Producers' Surplus

My tie-dyed T-shirt enterprise has grown to the extent that I am now able to produce T-shirts in bulk, and several campus groups have begun placing orders. I have informed one group that I am prepared to supply $20\sqrt{p-4}$ T-shirts at a price of p dollars per shirt. What is my total surplus if I sell T-shirts to the group at \$8 each?

Solution We need to calculate the producers' surplus when $\bar{p} = 8$. The supply equation is

$$q = 20\sqrt{p-4},$$

but to use the formula for producers' surplus, we need to express p as a function of q. First, we square both sides to remove the radical sign:

$$q^2 = 400(p-4),$$

so

$$p - 4 = \frac{q^2}{400},$$

giving

$$p = S(q) = \frac{q^2}{400} + 4.$$

We now need the value of $\bar{q}$ corresponding to $\bar{p} = 8$. Substituting $p = 8$ in the original equation gives

$$\bar{q} = 20\sqrt{8-4} = 20\sqrt{4} = 40.$$

Thus,

$$\begin{aligned}
PS &= \int_0^{\bar{q}} (\bar{p} - S(q))\, dq \\
&= \int_0^{40} \left[8 - \left(\frac{q^2}{400} + 4\right)\right] dq \\
&= \int_0^{40} \left(4 - \frac{q^2}{400}\right) dq \\
&= \left[4q - \frac{q^3}{1{,}200}\right]_0^{40} \approx \$106.67.
\end{aligned}$$

Thus, I earn a surplus of \$106.67 if I sell T-shirts to the group at \$8 each.

EXAMPLE 3 Equilibrium

To continue the preceding example: A representative informs me that the campus group is prepared to order only $\sqrt{200(16 - p)}$ T-shirts at p dollars each. I would like to produce as many T-shirts for them as possible but avoid being left with unsold T-shirts. Given the supply curve from the preceding example, what price should I charge per T-shirt, and what are the consumers' and producers' surpluses at that price?

Solution The price that guarantees neither a shortage nor a surplus of T-shirts is the **equilibrium price**, the price where supply equals demand. We have

$$\text{Supply:} \quad q = 20\sqrt{p - 4}.$$
$$\text{Demand:} \quad q = \sqrt{200(16 - p)}.$$

Equating these gives

$$\begin{aligned}
20\sqrt{p - 4} &= \sqrt{200(16 - p)} \\
400(p - 4) &= 200(16 - p) \\
400p - 1{,}600 &= 3{,}200 - 200p \\
600p &= 4{,}800 \\
p &= \$8 \text{ per T-shirt.}
\end{aligned}$$

We therefore take $\bar{p} = 8$ (which happens to be the price we used in Example 2). We get the corresponding value for q by substituting $p = 8$ into either the demand or supply equation:

$$\bar{q} = 20\sqrt{8 - 4} = 40.$$

Thus, $\bar{p} = 8$ and $\bar{q} = 40$.

We must now calculate the consumers' surplus and the producers' surplus. We calculated the producers' surplus for $\bar{p} = 8$ in the preceding example:

$$PS = \$106.67.$$

For the consumers' surplus we must first express p as a function of q for the demand equation. Thus, we solve the demand equation for p as we did for the supply equation, and we obtain

$$\text{Demand:} \quad D(q) = 16 - \frac{q^2}{200}.$$

Therefore,

$$CS = \int_0^{\bar{q}} (D(q) - \bar{p})\, dq$$

$$= \int_0^{40} \left[\left(16 - \frac{q^2}{200}\right) - 8\right] dq$$

$$= \int_0^{40} \left(8 - \frac{q^2}{200}\right) dq$$

$$= \left[8q - \frac{q^3}{600}\right]_0^{40} \approx \$213.33.$$

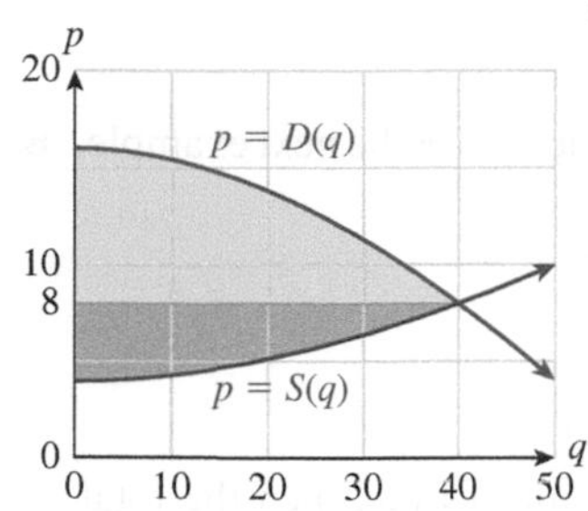

Figure 21

➡ **Before we go on...** Figure 21 shows both the consumers' surplus (top portion) and the producers' surplus (bottom portion) from Example 3. Because extra money in people's pockets is a good thing, the total of the consumers' and the producers' surpluses is called the **total social gain**. In this case it is

$$\text{Social gain} = CS + PS = 213.33 + 106.67 = \$320.00.$$

As you can see from the figure, the total social gain is also the area between two curves and equals

$$\int_0^{40} (D(q) - S(q))\, dq.$$

■

Continuous Income Streams

For purposes of calculation it is often convenient to assume that a company with a high sales volume receives money continuously. In such a case we have a function $R(t)$ that represents the rate at which money is being received by the company at time t.

EXAMPLE 4 Continuous Income

An ice cream store's business peaks in late summer. The store's summer revenue is approximated by

$$R(t) = 300 + 4.5t - 0.05t^2 \text{ dollars per day} \qquad (0 \le t \le 92),$$

where t is measured in days after June 1. What is its total revenue for the months of June, July, and August?

Solution Let's approximate the total revenue by breaking up the interval $[0, 92]$ representing the 3 months into n subintervals $[t_{k-1}, t_k]$, each with length Δt. In the interval $[t_{k-1}, t_k]$ the store receives money at a rate of approximately $R(t_{k-1})$ dollars per day for Δt days, so it will receive a total of $R(t_{k-1})\,\Delta t$ dollars. Over the whole summer, then, the store will receive approximately

$$R(t_0)\,\Delta t + R(t_1)\,\Delta t + \cdots + R(t_{n-1})\,\Delta t \text{ dollars.}$$

As we let n become large to better approximate the total revenue, this Riemann sum approaches the integral

$$\text{Total revenue} = \int_0^{92} R(t)\, dt.$$

Substituting the function we were given, we get

$$\text{Total revenue} = \int_0^{92} (300 + 4.5t - 0.05t^2)\,dt$$
$$= \left[300t + 2.25t^2 - \frac{0.05}{3}t^3\right]_0^{92}$$
$$\approx \$33{,}666.$$

Before we go on . . . We could approach the calculation in Example 4 another way: $R(t) = S'(t)$, where $S(t)$ is the total revenue earned up to day t. By the Fundamental Theorem of Calculus,

$$\text{Total revenue} = S(92) - S(0) = \int_0^{92} R(t)\,dt.$$

We did the calculation using Riemann sums mainly as practice for the next example. ■

Generalizing Example 4, we can say the following.

Total Value of a Continuous Income Stream

If the rate of receipt of income is $R(t)$ dollars per unit of time, then the total income received from time $t = a$ to $t = b$ is

$$\text{Total value} = TV = \int_a^b R(t)\,dt.$$

EXAMPLE 5 Future Value

Suppose the ice cream store in Example 4 deposits its receipts in an account paying 5% interest per year compounded continuously. How much money will it have in its account at the end of August?

Solution Now we have to take into account not only the revenue but also the interest it earns in the account. Again, we break the interval $[0, 92]$ into n subintervals. During the interval $[t_{k-1}, t_k]$, approximately $R(t_{k-1})\,\Delta t$ dollars are deposited in the account. That money will earn interest until the end of August, a period of $92 - t_{k-1}$ days, or $(92 - t_{k-1})/365$ years. The formula for continuous compounding tells us that by the end of August, those $R(t_{k-1})\,\Delta t$ dollars will have turned into

$$R(t_{k-1})\,\Delta t e^{0.05(92-t_{k-1})/365} = R(t_{k-1})e^{0.05(92-t_{k-1})/365}\,\Delta t \text{ dollars.}$$

(Recall that 5% is the *annual* interest rate.) Adding up the contributions from each subinterval, we see that the total in the account at the end of August will be approximately

$$R(t_0)e^{0.05(92-t_0)/365}\,\Delta t + R(t_1)e^{0.05(92-t_1)/365}\,\Delta t + \cdots + R(t_{n-1})e^{0.05(92-t_{n-1})/365}\,\Delta t.$$

This is a Riemann sum; as n gets large, the sum approaches the integral

$$\text{Future value} = FV = \int_0^{92} R(t)e^{0.05(92-t)/365}\,dt.$$

Substituting $R(t) = 300 + 4.5t - 0.05t^2$, we obtain

$$FV = \int_0^{92} (300 + 4.5t - 0.05t^2)e^{0.05(92-t)/365}\,dt$$
$$\approx \$33{,}880. \quad \text{Using technology or integration by parts}$$

➡ **Before we go on...** The interest earned in the account in Example 5 was fairly small. (Compare this answer to that in Example 4.) Not only was the money in the account for only 3 months, but much of it was put in the account toward the end of that period, so it had very little time to earn interest. ■

Generalizing again, we have the following.

Future Value of a Continuous Income Stream

If the rate of receipt of income from time $t = a$ to $t = b$ is $R(t)$ dollars per unit of time and the income is deposited as it is received in an account paying interest at rate r per unit of time, compounded continuously, then the amount of money in the account at time $t = b$ is

$$\text{Future value} = FV = \int_a^b R(t)e^{r(b-t)}\,dt.$$

EXAMPLE 5 **Present Value**

You are thinking of buying the ice cream store discussed in Examples 4 and 5. What is its income stream worth to you on June 1? Assume that you have access to the same account paying 5% per year compounded continuously.

Solution The value of the income stream on June 1 is the amount of money that, if deposited June 1, would give you the same future value as the income stream will. If we let PV denote this "present value," its value after 92 days will be

$$PVe^{0.05\times 92/365}.$$

We equate this with the future value of the income stream to get

$$PVe^{0.05\times 92/365} = \int_0^{92} R(t)e^{0.05(92-t)/365}\,dt,$$

so

$$PV = \int_0^{92} R(t)e^{-0.05t/365}\,dt.$$

Substituting the formula for $R(t)$ and integrating using technology or integration by parts, we get

$$PV \approx \$33{,}455.$$

The general formula is the following.

Present Value of a Continuous Income Stream

If the rate of receipt of income from time $t = a$ to $t = b$ is $R(t)$ dollars per unit of time and the income is deposited as it is received in an account paying interest at rate r per unit of time, compounded continuously, then the value of the income stream at time $t = a$ is

$$\text{Present value} = PV = \int_a^b R(t)e^{r(a-t)}\,dt.$$

We can derive this formula from the relation

$$FV = PVe^{r(b-a)}$$

because the present value is the amount that would have to be deposited at time $t = a$ to give a future value of FV at time $t = b$.

Note These formulas are more general than we've said. They still work when $R(t) < 0$ if we interpret negative values as money flowing *out* rather than in. That is, we can use these formulas for income we receive, for payments that we make, or for situations in which we sometimes receive money and sometimes pay it out. These formulas can also be used for flows of quantities other than money. For example, if we use an exponential model for population growth and we let $R(t)$ represent the rate of immigration $[R(t) > 0]$ or emigration $[R(t) < 0]$, then the future value formula gives the future population. ■

7.4 EXERCISES

▼ more advanced ◆ challenging

T indicates exercises that should be solved using technology

In Exercises 1–12, calculate the consumers' surplus at the indicated unit price $\bar{p}$ for the given demand equations. [**HINT**: See Example 1.]

1. $p = 10 - 2q; \bar{p} = 5$
2. $p = 100 - q; \bar{p} = 20$
3. $p = 100 - 3\sqrt{q}; \bar{p} = 76$
4. $p = 10 - 2q^{1/3}; \bar{p} = 6$
5. $p = 500e^{-2q}; \bar{p} = 100$
6. $p = 100 - e^{0.1q}; \bar{p} = 50$
7. $q = 100 - 2p; \bar{p} = 20$
8. $q = 50 - 3p; \bar{p} = 10$
9. $q = 100 - 0.25p^2; \bar{p} = 10$
10. $q = 20 - 0.05p^2; \bar{p} = 5$
11. $q = 500e^{-0.5p} - 50; \bar{p} = 1$
12. $q = 100 - e^{0.1p}; \bar{p} = 20$

In Exercises 13–24, calculate the producers' surplus for the given supply equations at the indicated unit price $\bar{p}$. [**HINT**: See Example 2.]

13. $p = 10 + 2q; \bar{p} = 20$
14. $p = 100 + q; \bar{p} = 200$
15. $p = 10 + 2q^{1/3}; \bar{p} = 12$
16. $p = 100 + 3\sqrt{q}; \bar{p} = 124$
17. $p = 500e^{0.5q}; \bar{p} = 1{,}000$
18. $p = 100 + e^{0.01q}; \bar{p} = 120$
19. $q = 2p - 50; \bar{p} = 40$
20. $q = 4p - 1{,}000; \bar{p} = 1{,}000$
21. $q = 0.25p^2 - 10; \bar{p} = 10$
22. $q = 0.05p^2 - 20; \bar{p} = 50$
23. $q = 500e^{0.05p} - 50; \bar{p} = 10$
24. $q = 10(e^{0.1p} - 1); \bar{p} = 5$

In Exercises 25–30, find the total value of the given income stream and also find its future value (at the end of the given interval) using the given interest rate. [**HINT**: See Examples 4 and 5.]

25. $R(t) = 30{,}000$, $0 \le t \le 10$, at 7%

26. $R(t) = 40{,}000$, $0 \le t \le 5$, at 10%

27. $R(t) = 30{,}000 + 1{,}000t$, $0 \le t \le 10$, at 7%

28. $R(t) = 40{,}000 + 2{,}000t$, $0 \le t \le 5$, at 10%

29. $R(t) = 30{,}000e^{0.05t}$, $0 \le t \le 10$, at 7%

30. $R(t) = 40{,}000e^{0.04t}$, $0 \le t \le 5$, at 10%

In Exercises 31–36, find the total value of the given income stream and also find its present value (at the beginning of the given interval) using the given interest rate. [**HINT**: See Examples 4 and 6.]

31. $R(t) = 20{,}000$, $0 \le t \le 5$, at 8%

32. $R(t) = 50{,}000$, $0 \le t \le 10$, at 5%

33. $R(t) = 20{,}000 + 1{,}000t$, $0 \le t \le 5$, at 8%

34. $R(t) = 50{,}000 + 2{,}000t$, $0 \le t \le 10$, at 5%

35. $R(t) = 20{,}000e^{0.03t}$, $0 \le t \le 5$, at 8%

36. $R(t) = 50{,}000e^{0.06t}$, $0 \le t \le 10$, at 5%

Applications

37. ***College Tuition*** A study of U.S. colleges and universities resulted in the demand equation $q = 20{,}000 - 2p$, where q is the enrollment at a public college or university and p is the average annual tuition (plus fees) it charges.[38] Officials at *Enormous State University* have developed a policy whereby the number of students it will accept per year at a tuition level of p dollars is given by $q = 7{,}500 + 0.5p$. Find the equilibrium tuition price $\bar{p}$ and the consumers' and producers' surpluses at this tuition level. What is the total social gain at the equilibrium price? [**HINT**: See Example 3.]

38. ***Fast Food*** A fast-food outlet finds that the demand equation for its new side dish, "Sweetdough Tidbit," is given by

$$p = \frac{128}{(q+1)^2},$$

where p is the price (in cents) per serving and q is the number of servings that can be sold per hour at this price. At the same time, the franchise is prepared to sell $q = 0.5p - 1$ servings per hour at a price of p cents. Find the equilibrium price $\bar{p}$ and the consumers' and producers' surpluses at this price level. What is the total social gain at the equilibrium price? [**HINT**: See Example 3.]

39. ***Revenue: Nokia*** The annual net sales (revenue) earned by **Nokia** in the years January 2004 to January 2010 can be approximated by

$$R(t) = -1.75t^2 + 12.5t + 30 \text{ billion euros per year} \quad (0 \le t \le 6),$$

where t is time in years. ($t = 0$ represents January 2004.)[39] Estimate, to the nearest €10 billion, Nokia's total revenue from January 2006 to January 2010. [**HINT**: See Example 4.]

40. ***Revenue: Nintendo*** The annual net sales (revenue) earned by **Nintendo** in the fiscal years 2000–2014, can be approximated by

$$R(t) = -3.75t^3 + 65.8t^2 - 206t + 578 \text{ billion yen per year} \quad (0 \le t \le 14),$$

where t is time in years. ($t = 0$ represents the start of fiscal year 2000.)[40] Estimate, to the nearest ¥100 billion, Nintendo's total revenue from the start of fiscal year 2006 to the start of fiscal year 2014. [**HINT**: See Example 4.]

41. ***Revenue: Walmart*** The annual revenue earned by **Walmart** in the years from January 2000 to January 2014 can be approximated by

$$R(t) = 176e^{0.079t} \text{ billion dollars per year} \quad (0 \le t \le 14),$$

where t is time in years. ($t = 0$ represents January 2000.)[41] Estimate, to the nearest \$10 billion, Walmart's total revenue from January 2004 to January 2014.

42. ***Revenue: Target*** The annual revenue earned by **Target** for fiscal years 2004 through 2010 can be approximated by

$$R(t) = 41e^{0.094t} \text{ billion dollars per year} \quad (0 \le t \le 7),$$

where t is time in years. ($t = 0$ represents the beginning of fiscal year 2004.)[42] Estimate, to the nearest \$10 billion, Target's total revenue from the beginning of fiscal year 2006 to the beginning of fiscal year 2010.

43. ▼ ***Revenue*** Refer back to Exercise 39. Suppose that, from January 2004 on, **Nokia** invested its revenue in an investment yielding 4% compounded continuously. What, to the nearest €10 billion, would the total value of Nokia's revenue from January 2006 to January 2010 have been in January 2010? [**HINT**: See Example 5.]

44. ▼ ***Revenue*** Refer back to Exercise 40. Suppose that, from the beginning of fiscal year 2006 on, **Nintendo** invested its revenue in an investment yielding 5% compounded continuously. What, to the nearest ¥100 billion, would the total value of Nintendo's revenue from the start of fiscal year

[38] Idea based on a study by A. L. Ostrosky, Jr. and J. V. Koch, as cited in their book *Introduction to Mathematical Economics* (Waveland Press, Illinois, 1979, p. 133). The data used here are fictitious, however.

[39] Source for data: Nokia financial statements (www.investors.nokia.com).

[40] Source for data: Nintendo annual reports (www.nintendo.com/corp).

[41] Source for data: Walmart annual reports (www.Walmartstores.com/Investors).

[42] Source for data: Target annual reports (investors.target.com).

2006 to the start of fiscal year 2014 have been at the start of fiscal year 2014? [**HINT**: See Example 5.]

45. ▼ ***Revenue*** Refer back to Exercise 41. Suppose that, from January 2004 on, **Walmart** invested its revenue in an investment that depreciated continuously at a rate of 5% per year. What, to the nearest \$10 billion, would the total value of Walmart's revenues from January 2004 to January 2014 have been in January 2014?

46. ▼ ***Revenue*** Refer back to Exercise 42. Suppose that, from the start of fiscal year 2004 on, **Target** invested its revenue in an investment that depreciated continuously at a rate of 3% per year. What, to the nearest \$10 billion, would the total value of Target's revenue from the beginning of fiscal year 2006 to the beginning of fiscal year 2010 have been at the beginning of fiscal year 2010?

47. ▼ ***Saving for Retirement*** You are saving for your retirement by investing \$700 per month in an annuity with a guaranteed interest rate of 6% per year. With a continuous stream of investment and continuous compounding, how much will you have accumulated in the annuity by the time you retire in 45 years?

48. ▼ ***Saving for College*** When your first child is born, you begin to save for college by depositing \$400 per month in an account paying 12% interest per year. With a continuous stream of investment and continuous compounding, how much will you have accumulated in the account by the time your child enters college 18 years later?

49. ▼ ***Saving for Retirement*** You begin saving for your retirement by investing \$700 per month in an annuity with a guaranteed interest rate of 6% per year. You increase the amount you invest at the rate of 3% per year. With continuous investment and compounding, how much will you have accumulated in the annuity by the time you retire in 45 years?

50. ▼ ***Saving for College*** When your first child is born, you begin to save for college by depositing \$400 per month in an account paying 12% interest per year. You increase the amount you save by 2% per year. With continuous investment and compounding, how much will have accumulated in the account by the time your child enters college 18 years later?

51. ▼ ***Bonds*** The U.S. Treasury issued a 30-year bond on October 15, 2014, paying 3.125% interest.[43] Thus, if you bought \$100,000 worth of these bonds, you would receive \$3,125 per year in interest for 30 years. An investor wishes to buy the rights to receive the interest on \$100,000 worth of these bonds. The amount the investor is willing to pay is the present value of the interest payments, assuming a 4% rate of return. Assuming (incorrectly but approximately) that the interest payments are made continuously, what will the investor pay? [**HINT**: See Example 6.]

52. ▼ ***Bonds*** The Megabucks Corporation is issuing a 20-year bond paying 7% interest. (See Exercise 51.) An investor wishes to buy the rights to receive the interest on \$50,000 worth of these bonds and seeks a 6% rate of return. Assuming that the interest payments are made continuously, what will the investor pay? [**HINT**: See Example 6.]

53. ▼ ***Valuing Future Income*** Inga was injured and can no longer work. As a result of a lawsuit, she is to be awarded the present value of the income she would have received over the next 20 years. Her income at the time she was injured was \$100,000 per year, increasing by \$5,000 per year. What will be the amount of her award, assuming continuous income and a 5% interest rate?

54. ▼ ***Valuing Future Income*** Max was injured and can no longer work. As a result of a lawsuit, he is to be awarded the present value of the income he would have received over the next 30 years. His income at the time he was injured was \$30,000 per year, increasing by \$1,500 per year. What will be the amount of his award, assuming continuous income and a 6% interest rate?

Communication and Reasoning Exercises

55. Complete the following: The future value of a continuous income stream earning 0% interest is the same as the ______ value.

56. Complete the following: The present value of a continuous income stream earning 0% interest is the same as the ______ value.

57. ▼ ***Linear Demand*** Given a linear demand equation $q = -mp + b\ (m > 0)$, find a formula for the consumers' surplus at a price level of $\bar{p}$ per unit.

58. ▼ ***Linear Supply*** Given a linear supply equation of the form $q = mp + b\ (m > 0)$, find a formula for the producers' surplus at a price level of $\bar{p}$ per unit.

59. ▼ Your study group friend says that the future value of a continuous stream of income is always greater than the total value, assuming a positive rate of return. Is she correct? Why?

60. ▼ Your other study group friend says that the present value of a continuous stream of income can sometimes be greater than the total value, depending on the (positive) interest rate. Is he correct? Explain.

61. ▼ Arrange from smallest to largest: total value, future value, and present value of a continuous stream of income (assuming a positive income and positive rate of return).

62. ▼ **a.** Arrange the following functions from smallest to largest $R(t)$, $R(t)e^{r(b-t)}$, $R(t)e^{r(a-t)}$, where $a \le t \le b$, and r and $R(t)$ are positive.

b. Use the result from part (a) to justify your answers in Exercises 59–61.

[43] Source: The Bureau of the Public Debt (www.publicdebt.treas.gov).

7.5 Improper Integrals and Applications

All the definite integrals we have seen so far have had the form $\int_a^b f(x)\,dx$ with a and b finite and $f(x)$ piecewise continuous on the closed interval $[a, b]$. If we relax one or both of these requirements somewhat, we obtain what are called **improper integrals**. There are various types of improper integrals.

Integrals in Which a Limit of Integration is Infinite

Integrals in which one or more limits of integration are infinite can be written as

$$\int_a^{+\infty} f(x)\,dx, \quad \int_{-\infty}^{b} f(x)\,dx, \quad \text{or} \quad \int_{-\infty}^{+\infty} f(x)\,dx.$$

Let's concentrate for a moment on the first form, $\int_a^{+\infty} f(x)\,dx$. What does the $+\infty$ mean here? As it often does, it means that we are to take a limit as something gets large. Specifically, it means the limit as the upper bound of integration gets large.

Improper Integral with an Infinite Limit of Integration

We define

$$\int_a^{+\infty} f(x)\,dx = \lim_{M\to+\infty} \int_a^M f(x)\,dx,$$

provided that the limit exists. If the limit exists, we say that $\int_a^{+\infty} f(x)\,dx$ **converges**. Otherwise, we say that $\int_a^{+\infty} f(x)\,dx$ **diverges**. Similarly, we define

$$\int_{-\infty}^{b} f(x)\,dx = \lim_{M\to-\infty} \int_M^b f(x)\,dx,$$

provided that the limit exists. Finally, we define

$$\int_{-\infty}^{+\infty} f(x)\,dx = \int_{-\infty}^{a} f(x)\,dx + \int_a^{+\infty} f(x)\,dx$$

for some convenient a, provided that *both* integrals on the right converge.

Quick Examples

1. $$\int_1^{+\infty} \frac{dx}{x^2} = \lim_{M\to+\infty} \int_1^M \frac{dx}{x^2} = \lim_{M\to+\infty} \left[-\frac{1}{x}\right]_1^M = \lim_{M\to+\infty} \left(-\frac{1}{M} + 1\right) = 1$$

Converges

2. $$\int_1^{+\infty} \frac{dx}{x} = \lim_{M\to+\infty} \int_1^M \frac{dx}{x} = \lim_{M\to+\infty} [\ln|x|]_1^M = \lim_{M\to+\infty} (\ln M - \ln 1) = +\infty$$

Diverges

3. $$\int_{-\infty}^{-1} \frac{dx}{x^2} = \lim_{M\to-\infty} \int_M^{-1} \frac{dx}{x^2} = \lim_{M\to-\infty} \left[-\frac{1}{x}\right]_M^{-1} = \lim_{M\to-\infty} \left(1 + \frac{1}{M}\right) = 1$$

Converges

4. $\displaystyle\int_{-\infty}^{+\infty} e^{-x}\,dx = \int_{-\infty}^{0} e^{-x}\,dx + \int_{0}^{+\infty} e^{-x}\,dx$

$$= \lim_{M\to-\infty}\int_M^0 e^{-x}\,dx + \lim_{M\to+\infty}\int_0^M e^{-x}\,dx$$

$$= \lim_{M\to-\infty}[-e^{-x}]_M^0 + \lim_{M\to+\infty}[-e^{-x}]_0^M$$

$$= \lim_{M\to-\infty}(e^{-M}-1) + \lim_{M\to+\infty}(1-e^{-M})$$

$$= +\infty + 1$$ Diverges

5. $\displaystyle\int_{-\infty}^{+\infty} xe^{-x^2}\,dx = \int_{-\infty}^{0} xe^{-x^2}\,dx + \int_{0}^{+\infty} xe^{-x^2}\,dx$

$$= \lim_{M\to-\infty}\int_M^0 xe^{-x^2}\,dx + \lim_{M\to+\infty}\int_0^M xe^{-x^2}\,dx$$

$$= \lim_{M\to-\infty}\left[-\frac{1}{2}e^{-x^2}\right]_M^0 + \lim_{M\to+\infty}\left[-\frac{1}{2}e^{-x^2}\right]_0^M$$

$$= \lim_{M\to-\infty}\left(-\frac{1}{2}+\frac{1}{2}e^{-M^2}\right) + \lim_{M\to+\infty}\left(-\frac{1}{2}e^{-M^2}+\frac{1}{2}\right)$$

$$= -\frac{1}{2}+\frac{1}{2} = 0$$ Converges

Q: *We learned that the integral can be interpreted as the area under the curve. Is this still true for improper integrals?*

A: Yes. Figure 22 illustrates how we can represent an improper integral as the area of an infinite region.

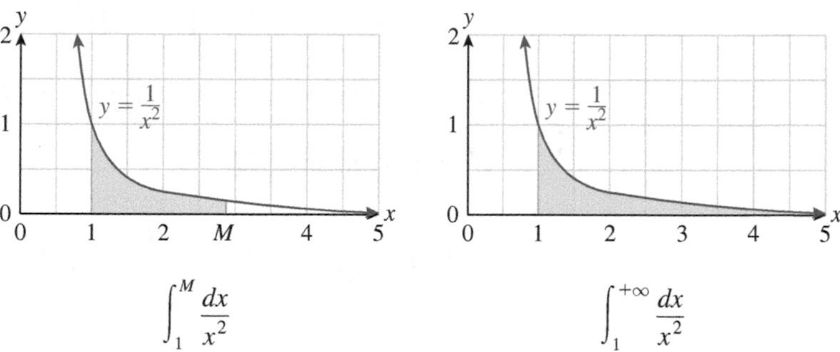

Figure 22

On the left we see the area represented by $\int_1^M dx/x^2$. As M gets larger, the integral approaches $\int_1^{+\infty} dx/x^2$. In the picture, think of M being moved farther and farther along the x-axis in the direction of increasing x, resulting in the region shown on the right.

Q: *Wait! We calculated $\int_1^{+\infty} dx/x^2 = 1$. Does this mean that the infinitely long area in Figure 22 has an area of only 1 square unit?*

A: That is exactly what it means. If you had enough paint to cover 1 square unit, you would never run out of paint while painting the region in Figure 22. This is one of the places where mathematics seems to contradict common sense. But common sense is notoriously unreliable when dealing with infinities.

Andy Crawford/Dorling Kindersley/Getty Images

EXAMPLE 1 **Future Sales of CDs**

By 2009, music downloads were making serious inroads into the sales of physical CDs. Approximately 290 million CD albums were sold in 2009, and sales declined by about 23% per year the following year.[44] Suppose that this rate of decrease were to continue indefinitely and continuously. How many CD albums, total, would be sold from 2009 on?

Solution Recall that the total sales between two dates can be computed as the definite integral of the rate of sales. So if we wanted to estimate the sales between 2009 and a time far in the future, we would compute $\int_0^M s(t)\,dt$ with a large M, where $s(t)$ is the annual sales t years after 2009. Because we want to know the *total* number of CD albums sold from 2009 on, we let $M \to +\infty$; that is, we compute $\int_0^{+\infty} s(t)\,dt$.

Because sales of CD albums are decreasing by 23% per year, we can model $s(t)$ by

$$s(t) = 290(0.77)^t \text{ million CD albums per year,}$$

where t is the number of years since 2009:

$$\begin{aligned}
\text{Total sales from 2009 on} &= \int_0^{+\infty} 290(0.77)^t\,dt \\
&= \lim_{M\to+\infty} \int_0^M 290(0.77)^t\,dt \\
&= \frac{290}{\ln 0.77} \lim_{M\to+\infty} [(0.77)^t]_0^M \\
&= \frac{290}{\ln 0.77} \lim_{M\to+\infty} (0.77^M - 0.77^0) \\
&= \frac{290}{\ln 0.77}(-1) \\
&\approx 1{,}110 \text{ million CD albums.}
\end{aligned}$$

Using Technology

You can estimate the integral in Example 1 with technology by computing $\int_0^M 290(0.77)^t\,dt$ for $M = 10, 100, 1{,}000, \ldots$. You will find that the resulting values appear to converge to about 1,110. (Stop when the effect of further increases of M has no effect at this level of accuracy.)

TI-83/84 Plus

Y_1=290*0.77^X

Home screen:

```
fnInt(Y1,X,0,10)
fnInt(Y1,X,0,100)
fnInt(Y1,X,0,1000)
```

 **Website**

www.WanerMath.com

Online Utilities

→ Numerical Integration Utility and Grapher

Enter

`290*0.77^x`

for $f(x)$. Enter 0 and 10 for the lower and upper limits, and press "Integral" for the most accurate estimate of the integral.

Repeat with the upper limit set to 100, 1,000, and higher.

Integrals in Which the Integrand Becomes Infinite

We can sometimes compute integrals $\int_a^b f(x)\,dx$ in which $f(x)$ becomes infinite. As we'll see in Example 4, the Fundamental Theorem of Calculus does not work for such integrals. The first case to consider is when $f(x)$ approaches $\pm\infty$ at either a or b.

[44] Source: *2010 Year-End Shipment Statistics*, Recording Industry Association of America (www.riaa.com).

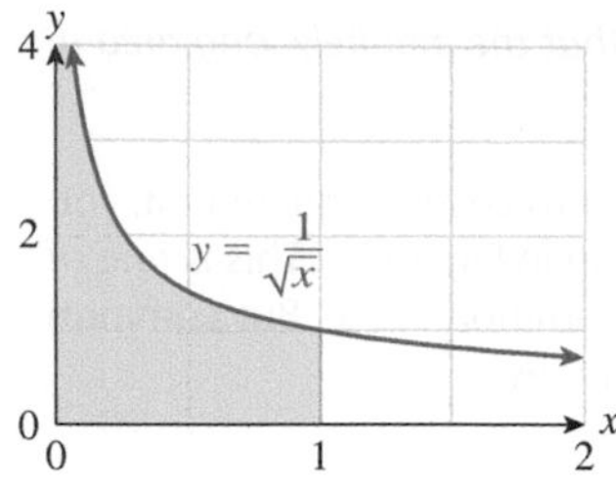

Figure 23

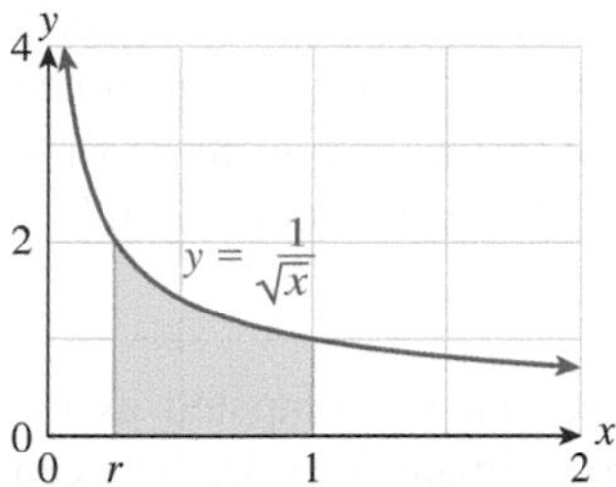

Figure 24

EXAMPLE 2 Integrand Infinite at One Endpoint

Calculate $\displaystyle\int_0^1 \frac{1}{\sqrt{x}}\,dx$.

Solution Notice that the integrand approaches $+\infty$ as x approaches 0 from the right and is not defined at 0. This makes the integral an improper integral. Figure 23 shows the region whose area we are trying to calculate; it extends infinitely vertically rather than horizontally.

Now, if $0 < r < 1$, the integral $\int_r^1 (1/\sqrt{x})\,dx$ is a proper integral because we avoid the bad behavior at 0. This integral gives the area shown in Figure 24. If we let r approach 0 from the right, the area in Figure 24 will approach the area in Figure 23. So we calculate

$$\begin{aligned}\int_0^1 \frac{1}{\sqrt{x}}\,dx &= \lim_{r\to 0^+}\int_r^1 \frac{1}{\sqrt{x}}\,dx\\ &= \lim_{r\to 0^+}\left[2\sqrt{x}\right]_r^1\\ &= \lim_{r\to 0^+}(2 - 2\sqrt{r})\\ &= 2.\end{aligned}$$

Thus, we again have an infinitely long region with finite area.

Generalizing, we make the following definition.

Improper Integral in Which the Integrand Becomes Infinite

If $f(x)$ is defined for all x with $a < x \le b$ but approaches $\pm\infty$ as x approaches a, we define

$$\int_a^b f(x)\,dx = \lim_{r\to a^+}\int_r^b f(x)\,dx$$

provided that the limit exists. Similarly, if $f(x)$ is defined for all x with $a \le x < b$ but approaches $\pm\infty$ as x approaches b, we define

$$\int_a^b f(x)\,dx = \lim_{r\to b^-}\int_a^r f(x)\,dx$$

provided that the limit exists. In either case, if the limit exists, we say that $\int_a^b f(x)\,dx$ **converges**. Otherwise, we say that $\int_a^b f(x)\,dx$ **diverges**.

Note We saw in Chapter 6 that the Fundamental Theorem of Calculus applies to piecewise continuous functions as well as continuous ones. Examples are $f(x) = |x|/x$ and $(x^2 - 1)/(x - 1)$. The integrals of such functions are not improper, and we can use the Fundamental Theorem of Calculus to evaluate such integrals in the usual way. ■

EXAMPLE 3 **Testing for Convergence**

Does $\int_{-1}^{3} \frac{x}{x^2 - 9}\, dx$ converge? If so, to what?

Solution We first check to see where, if anywhere, the integrand approaches $\pm\infty$. That will happen where the denominator becomes 0, so we solve $x^2 - 9 = 0$:

$$\begin{aligned} x^2 - 9 &= 0 \\ x^2 &= 9 \\ x &= \pm 3. \end{aligned}$$

The solution $x = -3$ is outside of the range of integration, so we ignore it. However, the solution $x = 3$ is the right endpoint of the range of integration, so the integral is improper. We need to investigate the following limit:

$$\int_{-1}^{3} \frac{x}{x^2 - 9}\, dx = \lim_{r \to 3^-} \int_{-1}^{r} \frac{x}{x^2 - 9}\, dx.$$

Now, to calculate the integral we use a substitution:

$$\begin{aligned} u &= x^2 - 9 \\ \frac{du}{dx} &= 2x \\ dx &= \frac{1}{2x}\, du \end{aligned}$$

when $x = r$, $u = r^2 - 9$

when $x = -1$, $u = (-1)^2 - 9 = -8$

Thus,

$$\begin{aligned} \int_{-1}^{r} \frac{x}{x^2 - 9}\, dx &= \int_{-8}^{r^2-9} \frac{1}{2u}\, du \\ &= \frac{1}{2}[\ln|u|]_{-8}^{r^2-9} \\ &= \frac{1}{2}(\ln|r^2 - 9| - \ln 8). \end{aligned}$$

Now we take the limit:

$$\begin{aligned} \int_{-1}^{3} \frac{x}{x^2 - 9}\, dx &= \lim_{r \to 3^-} \int_{-1}^{r} \frac{x}{x^2 - 9}\, dx \\ &= \lim_{r \to 3^-} \frac{1}{2}(\ln|r^2 - 9| - \ln 8) \\ &= -\infty \end{aligned}$$

because, as $r \to 3$, $r^2 - 9 \to 0$, so $\ln|r^2 - 9| \to -\infty$. Thus, this integral diverges.

EXAMPLE 4 Integrand Infinite between the Endpoints

Does $\displaystyle\int_{-2}^{3} \frac{1}{x^2}\,dx$ converge? If so, to what?

Solution Again we check to see whether there are any points at which the integrand approaches $\pm\infty$. There is such a point, at $x = 0$. This is between the endpoints of the range of integration. To deal with this, we break the integral into two integrals:

$$\int_{-2}^{3} \frac{1}{x^2}\,dx = \int_{-2}^{0} \frac{1}{x^2}\,dx + \int_{0}^{3} \frac{1}{x^2}\,dx.$$

Each integral on the right is an improper integral with the integrand approaching $\pm\infty$ at an endpoint. If both of the integrals on the right converge, we take the sum as the value of the integral on the left. So now we compute

$$\begin{aligned}\int_{-2}^{0} \frac{1}{x^2}\,dx &= \lim_{r\to 0^-} \int_{-2}^{r} \frac{1}{x^2}\,dx \\ &= \lim_{r\to 0^-} \left[-\frac{1}{x}\right]_{-2}^{r} \\ &= \lim_{r\to 0^-} \left(-\frac{1}{r} - \frac{1}{2}\right),\end{aligned}$$

which diverges to $+\infty$. There is no need now to check $\int_0^3 (1/x^2)\,dx$; because one of the two pieces of the integral diverges, we simply say that $\int_{-2}^{3} (1/x^2)\,dx$ diverges.

➡ **Before we go on . . .** What if we had been sloppy in Example 4 and had not checked first whether the integrand approached $\pm\infty$ somewhere? Then we probably would have applied the Fundamental Theorem of Calculus and done the following:

$$\int_{-2}^{3} \frac{1}{x^2}\,dx = \left[-\frac{1}{x}\right]_{-2}^{3} = \left(-\frac{1}{3} - \frac{1}{2}\right) = -\frac{5}{6}. \qquad \text{✗ \textit{WRONG!}}$$

Notice that the answer this "calculation" gives is patently ridiculous. Because $1/x^2 > 0$ for all x for which it is defined, the definite integral of $1/x^2$ over any interval cannot be negative. *Moral:* Always check to see whether the integrand blows up anywhere in the range of integration. If it does, the FTC does not apply, and we must use the methods of this example. ■

We end with an example of what to do if an integral is improper for more than one reason.

EXAMPLE 5 An Integral Improper in Two Ways

Does $\displaystyle\int_{0}^{+\infty} \frac{1}{\sqrt{x}}\,dx$ converge? If so, to what?

Solution This integral is improper for two reasons. First, the range of integration is infinite. Second, the integrand blows up at the endpoint 0. To separate these two problems, we break up the integral at some convenient point:

$$\int_{0}^{+\infty} \frac{1}{\sqrt{x}}\,dx = \int_{0}^{1} \frac{1}{\sqrt{x}}\,dx + \int_{1}^{+\infty} \frac{1}{\sqrt{x}}\,dx.$$

We chose to break the integral at 1. Any positive number would have sufficed, but 1 is generally easier to use in calculations.

The first piece, $\int_0^1 (1/\sqrt{x})\,dx$, we discussed in Example 2; it converges to 2. For the second piece we have

$$\begin{aligned}\int_1^{+\infty} \frac{1}{\sqrt{x}}\,dx &= \lim_{M\to+\infty} \int_1^M \frac{1}{\sqrt{x}}\,dx \\ &= \lim_{M\to+\infty} [2\sqrt{x}]_1^M \\ &= \lim_{M\to+\infty} (2\sqrt{M} - 2),\end{aligned}$$

which diverges to $+\infty$. Because the second piece of the integral diverges, we conclude that $\int_0^{+\infty} (1/\sqrt{x})\,dx$ diverges.

7.5 EXERCISES

▼ more advanced ◆ challenging

T indicates exercises that should be solved using technology

Note: For some of the exercises in this section you need to assume the fact that $\lim_{M\to+\infty} M^n e^{-M} = 0$ *for all* $n \ge 0$. *(See Exercises 73 and 74 in Section 3.1 and Exercise 111 in Section 3.3.)*

In Exercises 1–26, decide whether or not the given integral converges. If the integral converges, compute its value. [**HINT:** See Quick Examples 1–5.]

1. $\int_1^{+\infty} x\,dx$
2. $\int_0^{+\infty} e^{-x}\,dx$
3. $\int_{-2}^{+\infty} e^{-0.5x}\,dx$
4. $\int_1^{+\infty} \frac{1}{x^{1.5}}\,dx$
5. $\int_{-\infty}^{2} e^x\,dx$
6. $\int_{-\infty}^{-1} \frac{1}{x^{1/3}}\,dx$
7. $\int_{-\infty}^{-2} \frac{1}{x^2}\,dx$
8. $\int_{-\infty}^{0} e^{-x}\,dx$
9. $\int_0^{+\infty} x^2 e^{-6x}\,dx$
10. $\int_0^{+\infty} (2x - 4)e^{-x}\,dx$
11. $\int_0^5 \frac{2}{x^{1/3}}\,dx$ [**HINT:** See Example 2.]
12. $\int_0^2 \frac{1}{x^2}\,dx$
13. $\int_{-1}^2 \frac{3}{(x+1)^2}\,dx$ [**HINT:** See Example 3.]
14. $\int_{-1}^2 \frac{3}{(x+1)^{1/2}}\,dx$
15. $\int_{-1}^2 \frac{3x}{x^2 - 1}\,dx$ [**HINT:** See Example 4.]
16. $\int_{-1}^2 \frac{3}{x^{1/3}}\,dx$
17. $\int_{-2}^2 \frac{1}{(x+1)^{1/5}}\,dx$
18. $\int_{-2}^2 \frac{2x}{\sqrt{4 - x^2}}\,dx$
19. $\int_{-1}^1 \frac{2x}{x^2 - 1}\,dx$
20. $\int_{-1}^2 \frac{2x}{x^2 - 1}\,dx$
21. $\int_{-\infty}^{+\infty} xe^{-x^2}\,dx$
22. $\int_{-\infty}^{\infty} xe^{1-x^2}\,dx$
23. $\int_0^{+\infty} \frac{1}{x \ln x}\,dx$ [**HINT:** See Example 5.]
24. $\int_0^{+\infty} \ln x\,dx$
25. ▼ $\int_0^{+\infty} \frac{2x}{x^2 - 1}\,dx$
26. ▼ $\int_{-\infty}^{0} \frac{2x}{x^2 - 1}\,dx$

T *In Exercises 27–34, use technology to approximate the given integrals with* $M = 10, 100, 1{,}000, \ldots$. *Then decide whether the associated improper integral converges, and estimate its value to four significant digits if it does.* [**HINT:** See the technology note for Example 1.]

27. $\int_1^M \frac{1}{x^2}\,dx$
28. $\int_0^M e^{-x^2}\,dx$
29. $\int_0^M \frac{x}{1 + x}\,dx$
30. $\int_{1/M}^1 \frac{1}{\sqrt{x}}\,dx$
31. $\int_{1+1/M}^2 \frac{1}{\sqrt{x - 1}}\,dx$
32. $\int_1^M \frac{1}{x}\,dx$
33. $\int_0^{1-1/M} \frac{1}{(1 - x)^2}\,dx$
34. $\int_0^{2-1/M} \frac{1}{(2 - x)^3}\,dx$

Applications

35. ***New Home Sales*** Sales of new homes in the United States decreased dramatically from 2005 to 2010 as shown in the model

$$n(t) = 1.33e^{-0.299t} \text{ million homes per year} \quad (0 \le t \le 5),$$

where t is the year since 2005.[45] If this trend were to have continued into the indefinite future, estimate the total number of new homes that would have been sold in the United States from 2005 on. [**HINT**: See Example 1.]

36. ***Revenue from New Home Sales*** Revenue from the sale of new homes in the United States decreased dramatically from 2005 to 2010 as shown in the model

$$r(t) = 412e^{-0.323t} \text{ billion dollars per year} \quad (0 \le t \le 5),$$

where t is the year since 2005.[46] If this trend were to have continued into the indefinite future, estimate the total revenue from the sale of new homes in the United States from 2005 on. [**HINT**: See Example 1.]

37. ***Cigarette Sales*** According to data published by the Federal Trade Commission, the number of cigarettes sold domestically has been decreasing by about 4% per year from the 2000 total of about 415 billion.[47] Use an exponential model to forecast the total number of cigarettes sold from 2000 on. (Round your answer to the nearest 100 billion cigarettes.) [**HINT**: Use a model of the form Ab^t.]

38. ***Sales*** Sales of the text *Calculus and You* have been declining continuously at a rate of 5% per year. Assuming that *Calculus and You* currently sells 5,000 copies per year and that sales will continue this pattern of decline, calculate total future sales of the text. [**HINT**: Use a model of the form Ae^{rt}.]

39. ▼ ***Sales*** My financial adviser has predicted that annual sales of Frodo T-shirts will continue to decline by 10% each year. At the moment, I have 3,200 of the shirts in stock and am selling them at a rate of 200 per year. Will I ever sell them all?

40. ▼ ***Revenue*** Alarmed about the sales prospects for my Frodo T-shirts (see Exercise 39), I will try to make up lost revenues by increasing the price by \$1 each year. I now charge \$10 per shirt. What is the total amount of revenue I can expect to earn from sales of my T-shirts, assuming the sales levels described in the previous exercise? (Round your answer to the nearest \$1,000.)

41. ▼ ***Education*** Let $N(t)$ be the number of high school students who graduated in the United States in year t. This number has been changing at a rate of about

$$N'(t) = 0.30t^{-0.87} \text{ million graduates per year} \quad (5 \le t \le 13),$$

where t is time in years since 2000.[48] In 2005 about 2.8 million high school students graduated. By extrapolating the model, what can you say about the number of high school students who will graduate in a year far in the future?

42. ▼ ***Education: Martian*** Let $M(t)$ be the number of high school students who graduated in the Republic of Mars in year t. This number is projected to change at a rate of about

$$M'(t) = 0.321t^{-1.10} \text{ thousand graduates per year} \quad (1 \le t \le 50),$$

where t is time in years since 2150. In 2151 about 1,300 high school students graduated. By extrapolating the model, what can you say about the number of high school students who will graduate in a year far in the future?

43. ▼ ***Cellphone Revenues*** The number of cellphone subscribers in China in the early 2000s was projected to follow the equation[49]

$$N(t) = 39t + 68 \text{ million subscribers}$$

in year t. ($t = 0$ represents 2000.) The average annual revenue per cellphone user was \$350 in 2000.

a. Assuming that, because of competition, the revenue per cellphone user decreases continuously at an annual rate of 10%, give a formula for the annual revenue in year t.

b. Using the model you obtained in part (a) as an estimate of the rate of change of total revenue, estimate the total revenue from 2000 into the indefinite future.

44. ▼ ***Vidphone Revenues*** The number of vidphone subscribers in the Republic of Mars for the period 2200–2300 was projected to follow the equation

$$N(t) = 18t - 10 \text{ thousand subscribers}$$

in year t. ($t = 0$ represents 2200.) The average annual revenue per vidphone user was Z40 in 2200.[50]

a. Assuming that, because of competition, the revenue per vidphone user decreases continuously at an annual rate of 20%, give a formula for the annual revenue in year t.

b. Using the model you obtained in part (a) as an estimate of the rate of change of total revenue, estimate the total revenue from 2200 into the indefinite future.

45. T ***Development Assistance*** According to data published by the Organisation for Economic Co-operation and Development (OECD), development assistance to developing countries from 2005 through 2013 was approximately

$$q(t) = 0.13t^2 + 4.1t + 110 \text{ billion dollars per year},$$

---

[45] Based on new home sales data from the U.S. Census Bureau (www.census.gov/const/www/newressalesindex.html).

[46] *Ibid.*

[47] Source for data: Federal Trade Commission Cigarette Report for 2011, issued May 2013 (www.ftc.gov).

[48] Based on a regression model. Source for Data: U.S. Department of Education, National Center for Education Statistics, *Digest of Education Statistics: 2013* (nces.ed.gov).

[49] Based on a regression of projected figures (coefficients are rounded). Source: Intrinsic Technology/*New York Times*, Nov. 24, 2000, p. C1.

[50] The zonar (Z) is the official currency in the city-state of Utarek, Mars. Source: www.Marsnext.com, a now extinct virtual society.

where t is time in years since 2005.[51] Assuming a worldwide inflation rate of 3% per year and that the above model remains accurate into the indefinite future, find the value of all development assistance to developing countries from 2005 on in constant dollars. (The constant dollar value of $q(t)$ dollars t years from now is given by $q(t)e^{-rt}$, where r is the fractional rate of inflation. Give your answer to the nearest \$100 billion.) [**HINT:** See the technology note for Example 1.]

46. **T** ***Humanitarian Aid*** Repeat Exercise 45, using the following model for humanitarian aid:[52]

$$q(t) = 0.051t^2 + 0.34t + 9.1 \text{ billion dollars per year.}$$

47. ▼ ***Hair Mousse Sales*** The amount of extremely popular hair mousse sold online at your website can be approximated by

$$N(t) = \frac{80(7)^t}{20 + 7^t} \text{ million gallons per year.}$$

($t = 0$ represents the current year.) Investigate the integrals $\int_0^{+\infty} N(t)\,dt$ and $\int_{-\infty}^0 N(t)\,dt$, and interpret your answers.

48. ▼ ***Chocolate Mousse Sales*** The weekly demand for your company's *Lo-Cal Chocolate Mousse* is modeled by the equation

$$q(t) = \frac{50e^{2t-1}}{1 + e^{2t-1}} \text{ gallons per week,}$$

where t is time from now in weeks. Investigate the integrals $\int_0^{+\infty} q(t)\,dt$ and $\int_{-\infty}^0 q(t)\,dt$, and interpret your answers.

T ***The Normal Curve*** *Exercises 49–52 require the use of a graphing calculator or computer programmed to do numerical integration. The normal distribution curve, which models the distributions of data in a wide range of applications, is given by the function*

$$p(x) = \frac{1}{\sqrt{2\pi}\sigma} e^{-(x-\mu)^2/(2\sigma^2)},$$

where $\pi = 3.14159265\ldots$ and σ and μ are constants called the standard deviation and the mean, respectively. Its graph (for $\sigma = 1$ and $\mu = 2$) is shown in the figure.

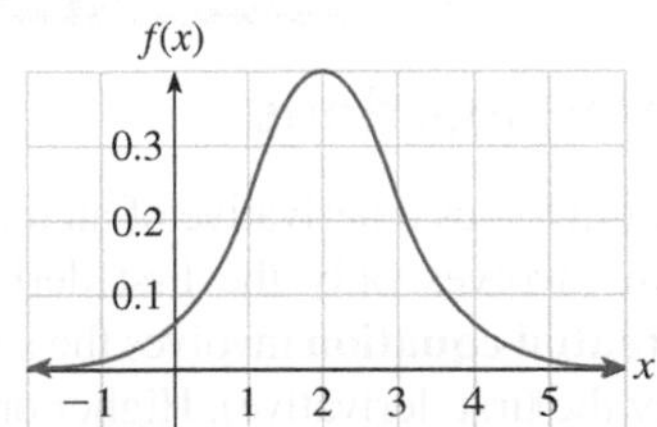

49. ▼ With $\sigma = 4$ and $\mu = 1$, approximate $\int_{-\infty}^{+\infty} p(x)\,dx$. [**HINT:** See Example 5 and the technology note for Example 1.]

50. ▼ With $\sigma = 1$ and $\mu = 0$, approximate $\int_0^{+\infty} p(x)\,dx$.

51. ▼ With $\sigma = 1$ and $\mu = 0$, approximate $\int_1^{+\infty} p(x)\,dx$.

52. ▼ With $\sigma = 1$ and $\mu = 0$, approximate $\int_{-\infty}^{1} p(x)\,dx$.

53. ◆ ***Variable Sales*** The value of your *Chateau Petit Mont Blanc* 1963 vintage burgundy is increasing continuously at an annual rate of 40%, and you have a supply of 1,000 bottles worth \$85 each at today's prices. To ensure a steady income, you have decided to sell your wine at a diminishing rate, starting at 500 bottles per year and then decreasing this figure continuously at a fractional rate of 100% per year. How much income (to the nearest dollar) can you expect to generate by this scheme? [**HINT:** Use the formula for continuously compounded interest.]

54. ◆ ***Panic Sales*** Unfortunately, your large supply of *Chateau Petit Mont Blanc* is continuously turning to vinegar at a fractional rate of 60% per year! You have thus decided to sell off your Petit Mont Blanc at \$50 per bottle, but the market is a little thin, and you can sell only 400 bottles per year. Because you have no way of knowing which bottles now contain vinegar until they are opened, you shall have to give refunds for all the bottles of vinegar. What will your net income be before all the wine turns to vinegar?

55. ◆ ***Meteor Impacts*** The frequency of meteor impacts on the Earth can be modeled by

$$n(k) = \frac{1}{5.6997k^{1.081}},$$

where $n(k) = N'(k)$, and $N(k)$ is the average number of meteors of energy less than or equal to k megatons that will hit the Earth in 1 year.[53] (A small nuclear bomb releases on the order of 1 megaton of energy.)

a. How many meteors of energy at least $k = 0.2$ hit the Earth each year?

b. Investigate and interpret the integral $\int_0^1 n(k)\,dk$.

56. ◆ ***Meteor Impacts*** (continuing Exercise 55)

a. Explain why the integral

$$\int_a^b kn(k)\,dk$$

computes the total energy released each year by meteors with energies between a and b megatons.

b. Compute and interpret

$$\int_0^1 kn(k)\,dk.$$

c. Compute and interpret

$$\int_1^{+\infty} kn(k)\,dk.$$

[51] The authors' approximation, based on data from OECD, obtained from www.oecd.org.

[52] *Ibid.*

[53] The authors' model, based on data published by NASA International Near-Earth-Object Detection Workshop (*The New York Times*, Jan. 25, 1994, p. C1).

57. ◆ ***The Gamma Function*** The gamma function is defined by the formula

$$\Gamma(x) = \int_0^{+\infty} t^{x-1} e^{-t}\, dt.$$

a. Find $\Gamma(1)$ and $\Gamma(2)$.
b. Use integration by parts to show that for every positive integer n, $\Gamma(n+1) = n\Gamma(n)$.
c. Deduce that $\Gamma(n) = (n-1)!\ [=(n-1)(n-2)\cdots 2\cdot 1]$ for every positive integer n.

58. ◆ ***Laplace Transforms*** The Laplace transform $F(x)$ of a function $f(t)$ is given by the formula

$$F(x) = \int_0^{+\infty} f(t)e^{-xt}\, dt \quad (x > 0).$$

a. Find $F(x)$ for $f(t) = 1$ and for $f(t) = t$.
b. Find a formula for $F(x)$ if $f(t) = t^n$ $(n = 1, 2, 3, \ldots)$.
c. Find a formula for $F(x)$ if $f(t) = e^{at}$ (a constant).

Communication and Reasoning Exercises

59. Why can't the Fundamental Theorem of Calculus be used to evaluate $\int_{-1}^{1} \frac{1}{x}\, dx$?

60. Why can't the Fundamental Theorem of Calculus be used to evaluate $\int_{1}^{+\infty} \frac{1}{x^2}\, dx$?

61. It sometimes happens that the Fundamental Theorem of Calculus gives the correct answer for an improper integral. Does the FTC give the correct answer for improper integrals of the form

$$\int_{-a}^{a} \frac{1}{x^{1/r}}\, dx$$

if $r = 3, 5, 7, \ldots$?

62. Does the FTC give the correct answer for improper integrals of the form

$$\int_{-a}^{a} \frac{1}{x^r}\, dx$$

if $r = 3, 5, 7, \ldots$?

63. Which of the following integrals are improper, and why? (Do not evaluate any of them.)

a. $\int_{-1}^{1} \frac{|x|}{x}\, dx$ b. $\int_{-1}^{1} x^{-1/3}\, dx$

c. $\int_0^2 \frac{x-2}{x^2-4x+4}\, dx$

64. Which of the following integrals are improper, and why? (Do not evaluate any of them.)

a. $\int_{-1}^{1} \frac{|x-1|}{x-1}\, dx$ b. $\int_0^1 \frac{1}{x^{2/3}}\, dx$

c. $\int_0^2 \frac{x^2-4x+4}{x-2}\, dx$

65. T ▼ How could you use technology to approximate improper integrals? (Your discussion should refer to each type of improper integral.)

66. T ▼ Use technology to approximate the integrals $\int_0^M e^{-(x-10)^2}\, dx$ for larger and larger values of M, using Riemann sums with 500 subdivisions. What do you find? Comment on the answer.

67. ▼ Make up an interesting application whose solution is $\int_{10}^{+\infty} 100te^{-0.2t}\, dt = \$1{,}015.01$.

68. ▼ Make up an interesting application whose solution is $\int_{100}^{+\infty} \frac{1}{r^2}\, dr = 0.01$.

7.6 Differential Equations and Applications

Differential Equations and Their Solutions

A **differential equation** is an equation that involves a derivative of an unknown function. A **first-order differential equation** involves only the first derivative of the unknown function. A **second-order differential equation** involves the second derivative of the unknown function (and possibly the first derivative). Higher order differential equations are defined similarly. In this book we will deal only with first-order differential equations.

To **solve** a differential equation means to find the unknown function. Many of the laws of science and other fields describe how things change. When expressed mathematically, these laws take the form of equations involving derivatives—that is, differential equations. The field of differential equations is a large and very active area of study in mathematics, and we shall see only a small part of it in this section.

EXAMPLE 1 **Motion**

A dragster accelerates from a stop so that its speed t seconds after starting is $40t$ ft/sec. How far will the car go in 8 seconds?*

* Notice that we can answer this question by integrating the speed from 0 to 8. However, we pretend here that we didn't notice this fact and instead approach it from the point of view of differential equations.

Solution We wish to find the car's position function $s(t)$. We are told about its speed, which is ds/dt. Precisely, we are told that

$$\frac{ds}{dt} = 40t.$$

This is the differential equation we have to solve to find $s(t)$. But we already know how to solve this kind of differential equation; we integrate:

$$s(t) = \int 40t\, dt = 20t^2 + C.$$

We now have the **general solution** to the differential equation. By letting C take on different values, we get all the possible solutions. We can specify the one **particular solution** that gives the answer to our problem by imposing the **initial condition** that $s(0) = 0$. Substituting into $s(t) = 20t^2 + C$, we get

$$0 = s(0) = 20(0)^2 + C = C,$$

so $C = 0$ and $s(t) = 20t^2$. To answer the question, the car travels $20(8)^2 = 1{,}280$ feet in 8 seconds.

Simple Differential Equations

We did not have to work hard to solve the differential equation in Example 1. In fact, any differential equation of the form $dy/dx = f(x)$ can (in theory) be solved by integrating. (Whether we can actually carry out the integration is another matter!)

Simple Differential Equations

A **simple** differential equation has the form

$$\frac{dy}{dx} = f(x).$$

Its general solution is

$$y = \int f(x)\, dx.$$

Quick Example

1. The differential equation

$$\frac{dy}{dx} = 2x^2 - 4x^3$$

is simple and has general solution

$$y = \int f(x)\, dx = \frac{2x^3}{3} - x^4 + C.$$

Separable Differential Equations

Not all differential equations are simple, as the next example shows.

EXAMPLE 2 Separable Differential Equation

Consider the differential equation $\frac{dy}{dx} = \frac{x}{y^2}$.

a. Find the general solution.

b. Find the particular solution that satisfies the initial condition $y(0) = 2$.

Solution

a. This is not a simple differential equation because the right-hand side is a function of both x and y. We cannot solve this equation by just integrating; the solution to this problem is to "separate" the variables.

Step 1: *Separate the variables algebraically.* We rewrite the equation as

$$y^2\,dy = x\,dx.$$

Step 2: *Integrate both sides.*

$$\int y^2\,dy = \int x\,dx,$$

giving

$$\frac{y^3}{3} = \frac{x^2}{2} + C.$$

Step 3: *Solve for the dependent variable.* We solve for y:

$$y^3 = \frac{3}{2}x^2 + 3C = \frac{3}{2}x^2 + D$$

(rewriting $3C$ as D, an equally arbitrary constant), so

$$y = \left(\frac{3}{2}x^2 + D\right)^{1/3}.$$

This is the general solution of the differential equation.

b. We now need to find the value for D that will give us the solution satisfying the condition $y(0) = 2$. Substituting 0 for x and 2 for y in the general solution, we get

$$2 = \left(\frac{3}{2}(0)^2 + D\right)^{1/3} = D^{1/3},$$

so

$$D = 2^3 = 8.$$

Thus, the particular solution we are looking for is

$$y = \left(\frac{3}{2}x^2 + 8\right)^{1/3}.$$

➡ Before we go on . . . We can check the general solution in Example 2 by calculating both sides of the differential equation and comparing:

$$\frac{dy}{dx} = \frac{d}{dx}\left(\frac{3}{2}x^2 + D\right)^{1/3} = x\left(\frac{3}{2}x^2 + D\right)^{-2/3}$$

$$\frac{x}{y^2} = \frac{x}{\left(\frac{3}{2}x^2 + D\right)^{2/3}} = x\left(\frac{3}{2}x^2 + D\right)^{-2/3}. \quad ✔$$

■

Q: *In Example 2 we wrote $y^2\,dy$ and $x\,dx$. What do they mean?*

A: Although it is possible to give meaning to these symbols, for us they are just a notational convenience. We could have done the following instead:

$$y^2\frac{dy}{dx} = x.$$

Now we integrate both sides with respect to x:

$$\int y^2\frac{dy}{dx}\,dx = \int x\,dx.$$

We can use substitution to rewrite the left-hand side:

$$\int y^2\frac{dy}{dx}\,dx = \int y^2\,dy,$$

which brings us back to the equation

$$\int y^2\,dy = \int x\,dx.$$

We were able to separate the variables in the preceding example because the right-hand side, x/y^2, was a *product* of a function of x and a function of y—namely,

$$\frac{x}{y^2} = x\left(\frac{1}{y^2}\right).$$

In general, we can say the following.

Separable Differential Equation

A **separable** differential equation has the form

$$\frac{dy}{dx} = f(x)g(y).$$

We solve a separable differential equation by separating the xs and the ys algebraically, writing

$$\frac{1}{g(y)}\,dy = f(x)\,dx,$$

and then integrating:

$$\int \frac{1}{g(y)}\,dy = \int f(x)\,dx.$$

Quick Examples

2. $\frac{dy}{dx} = x^2y^2$ is separable. To solve it, we write $\frac{dy}{y^2} = x^2\,dx$ and integrate:

$$\int y^{-2}\,dy = \int x^2\,dx$$

$$-y^{-1} = \frac{1}{3}x^3 + C$$

$$y = \frac{-1}{\frac{1}{3}x^3 + C} = \frac{-3}{x^3 + D}.$$

3. $\frac{dy}{dx} = x + y$ is not separable.

EXAMPLE 3 Rising Medical Costs

Spending on Medicare from 2010 to 2021 was projected to rise continuously at an instantaneous rate of 5.6% per year.[54] Find a formula for Medicare spending y as a function of time t in years since 2010.

Solution When we say that Medicare spending y was going up continuously at an instantaneous rate of 5.6% per year, we mean that

the instantaneous rate of increase of y was 5.6% of its value

or

$$\frac{dy}{dt} = 0.056y.$$

This is a separable differential equation. Separating the variables gives

$$\frac{1}{y}\,dy = 0.056\,dt.$$

Integrating both sides, we get

$$\int \frac{1}{y}\,dy = \int 0.056\,dt,$$

so

$$\ln y = 0.056t + C.$$

(We should write $\ln|y|$, but we know that the medical costs are positive.) We now solve for y:

$$y = e^{0.056t+C} = e^C e^{0.056t} = Ae^{0.056t},$$

where A is a positive constant. This is the formula we used before for continuous percentage growth.

[54] Spending is in constant 2010 dollars. Source for projected data: Congressional Budget Office, *March 2011 Medicare Baseline* (www.cbo.gov).

➡ **Before we go on...** To determine A in Example 3, we need to know, for example, Medicare spending at time $t = 0$ (the initial condition). The source cited gives Medicare spending as $525.6 billion in 2010. Substituting $t = 0$ in the equation above gives

$$525.6 = Ae^0 = A.$$

Thus, projected Medicare spending is

$$y = 525.6e^{0.056t} \text{ billion dollars}$$

t years after 2010. ■

EXAMPLE 4 Newton's Law of Cooling

Newton's Law of Cooling states that a hot object cools at a rate proportional to the difference between its temperature and the temperature of the surrounding environment (the **ambient temperature**). If a hot cup of coffee, initially at 170°F, is left to sit in a room at 70°F, how will the temperature of the coffee change over time?

Solution We let $H(t)$ denote the temperature of the coffee at time t. Newton's Law of Cooling tells us that $H(t)$ *decreases* at a rate proportional to the difference between $H(t)$ and 70°F, the ambient temperature. In other words,

$$\frac{dH}{dt} = -k(H - 70),$$

where k is some positive constant.* Note that $H \geq 70$: The coffee will never cool to less than the ambient temperature.

* When we say that a quantity Q is *proportional* to a quantity R, we mean that $Q = kR$ for some constant k. The constant k is referred to as the **constant of proportionality**.

The variables here are H and t, which we can separate as follows:

$$\frac{dH}{H - 70} = -k\,dt.$$

Integrating, we get

$$\int \frac{dH}{H - 70} = \int (-k)\,dt,$$

so

$$\ln(H - 70) = -kt + C.$$

(Note that $H - 70$ is positive, so we don't need absolute values.) We now solve for H:

$$\begin{aligned} H - 70 &= e^{-kt+C} \\ &= e^C e^{-kt} \\ &= Ae^{-kt}, \end{aligned}$$

so

$$H(t) = 70 + Ae^{-kt},$$

where A is some positive constant. We can determine the constant A using the initial condition $H(0) = 170$:

$$170 = 70 + Ae^0 = 70 + A,$$

so

$$A = 100.$$

Therefore,

$$H(t) = 70 + 100e^{-kt}.$$

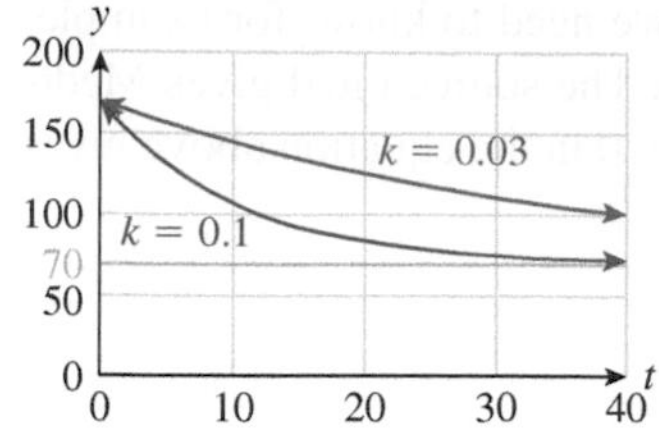

Figure 25

Q: *But what is k?*

A: The constant k determines the rate of cooling. Its value depends on the units of time we are using, on the substance cooling—in this case the coffee—and on its container. Because k depends so heavily on the particular circumstances, it's usually easiest to determine it experimentally. Figure 25 shows two possible graphs, one with $k = 0.1$ and the other with $k = 0.03$. ($k \approx 0.03$ would be reasonable for a cup of coffee in a polystyrene container with t measured in minutes.)

In any case we can see from the graph or the formula for $H(t)$ that the temperature of the coffee will approach the ambient temperature exponentially.

➡ **Before we go on...** Notice that the calculation in Example 4 shows that the temperature of an object cooling according to Newton's Law is given in general by

$$H(t) = T_a + (T_0 - T_a)e^{-kt},$$

where T_a is the ambient temperature (70° in the example) and T_0 is the initial temperature (170° in the example). The formula also holds if the ambient temperature is higher than the initial temperature ("Newton's Law of Heating"). ■

7.6 EXERCISES

▼ more advanced ◆ challenging

T indicates exercises that should be solved using technology

In Exercises 1–10, find the general solution of the given differential equation. Where possible, solve for y as a function of x.

1. $\frac{dy}{dx} = x^2 + \sqrt{x}$ [**HINT**: See Quick Example 1.]
2. $\frac{dy}{dx} = \frac{1}{x} + 3$ [**HINT**: See Quick Example 1.]
3. $\frac{dy}{dx} = \frac{x}{y}$ [**HINT**: See Example 2(a).]
4. $\frac{dy}{dx} = \frac{y}{x}$ [**HINT**: See Example 2(a).]
5. $\frac{dy}{dx} = xy$
6. $\frac{dy}{dx} = x^2y$
7. $\frac{dy}{dx} = (x + 1)y^2$
8. $\frac{dy}{dx} = \frac{1}{(x + 1)y^2}$
9. $x\frac{dy}{dx} = \frac{1}{y}\ln x$
10. $\frac{1}{x}\frac{dy}{dx} = \frac{1}{y}\ln x$

In Exercises 11–20, find the indicated particular solution of the given differential equation. [**HINT**: See Example 2(b).]

11. $\frac{dy}{dx} = x^3 - 2x$; $y = 1$ when $x = 0$
12. $\frac{dy}{dx} = 2 - e^{-x}$; $y = 0$ when $x = 0$
13. $\frac{dy}{dx} = \frac{x^2}{y^2}$; $y = 2$ when $x = 0$
14. $\frac{dy}{dx} = \frac{y^2}{x^2}$; $y = \frac{1}{2}$ when $x = 1$
15. $x\frac{dy}{dx} = y$; $y(1) = 2$
16. $x^2\frac{dy}{dx} = y$; $y(1) = 1$
17. $\frac{dy}{dx} = x(y + 1)$; $y(0) = 0$
18. $\frac{dy}{dx} = \frac{y + 1}{x}$; $y(1) = 2$
19. $\frac{dy}{dx} = \frac{xy^2}{x^2 + 1}$; $y(0) = -1$
20. $\frac{dy}{dx} = \frac{xy}{(x^2 + 1)^2}$; $y(0) = 1$

Applications

21. ***Sales*** Your monthly sales of green tea ice cream are falling at an instantaneous rate of 5% per month. If you currently sell 1,000 quarts per month, find the differential equation that describes your change in sales and then solve it to predict your monthly sales. [**HINT**: See Example 3.]

22. ***Profit*** Your monthly profit on sales of avocado ice cream is rising at an instantaneous rate of 10% per month. If you currently make a profit of \$15,000 per month, find the differential equation describing your change in profit, and solve it to predict your monthly profits. [**HINT**: See Example 3.]

23. ***Newton's Law of Cooling*** For coffee in a ceramic cup, suppose $k \approx 0.05$ with time measured in minutes. **(a)** Use Newton's Law of Cooling to predict the temperature of the coffee, initially at a temperature of 200°F, that is left to sit in a room at 75°F. **(b)** When will the coffee have cooled to 80°F? [**HINT**: See Example 4.]

24. ***Newton's Law of Cooling*** For coffee in a paper cup, suppose $k \approx 0.08$ with time measured in minutes. **(a)** Use Newton's Law of Cooling to predict the temperature of the

coffee, initially at a temperature of 210°F, that is left to sit in a room at 60°F. **(b)** When will the coffee have cooled to 70°F? [**HINT:** See Example 4.]

25. ***Cooling*** A bowl of clam chowder at 190°F is placed in a room whose air temperature is 75°F. After 10 minutes the soup has cooled to 150°F. Find the value of k in Newton's Law of Cooling, and hence find the temperature of the chowder as a function of time.

26. ***Heating*** Suppose that a pie at 20°F is put in an oven at 350°F. After 15 minutes, its temperature has risen to 80°F. Find the value of k in Newton's Law of Heating (see the note after Example 4), and hence find the temperature of the pie as a function of time.

27. ***Market Saturation*** You have just introduced a new 3D monitor to the market. You predict that you will eventually sell 100,000 monitors and that your monthly rate of sales will be 10% of the difference between the saturation value of 100,000 and the total number you have sold up to that point. Find a differential equation for your total sales (as a function of the month), and solve. (What are your total sales at the moment when you first introduce the monitor?)

28. ***Market Saturation*** Repeat Exercise 27, assuming that monthly sales will be 5% of the difference between the saturation value (of 100,000 monitors) and the total sales to that point, and assuming that you sell 5,000 monitors to corporate customers before placing the monitor on the open market.

29. ***Determining Demand*** *Nancy's Chocolates* estimates that the elasticity of demand for its dark chocolate truffles is $E = 0.05p - 1.5$, where p is the price per pound. Nancy's sells 20 pounds of truffles per week when the price is \$20 per pound. Find the formula expressing the demand q as a function of p. Recall that the elasticity of demand is given by

$$E = -\frac{dq}{dp} \times \frac{p}{q}.$$

30. ***Determining Demand*** *Nancy's Chocolates* estimates that the elasticity of demand for its chocolate strawberries is $E = 0.02p - 0.5$, where p is the price per pound. It sells 30 pounds of chocolate strawberries per week when the price is \$30 per pound. Find the formula expressing the demand q as a function of p. Recall that the elasticity of demand is given by

$$E = -\frac{dq}{dp} \times \frac{p}{q}.$$

Linear Differential Equations *Exercises 31–36 are based on **first-order linear differential equations with constant coefficients**. These have the form*

$$\frac{dy}{dt} + py = f(t) \quad (p \text{ constant})$$

and the general solution is

$$y = e^{-pt} \int f(t) e^{pt}\, dt.$$

(Check this by substituting!)

31. Solve the linear differential equation

$$\frac{dy}{dt} + y = e^{-t}; \quad y = 1 \text{ when } t = 0.$$

32. Solve the linear differential equation

$$\frac{dy}{dt} - y = e^{2t}; \quad y = 2 \text{ when } t = 0.$$

33. Solve the linear differential equation

$$2\frac{dy}{dt} - y = 2t; \quad y = 1 \text{ when } t = 0.$$

[**HINT:** First rewrite the differential equation in the form $\frac{dy}{dt} + py = f(t)$.]

34. Solve the linear differential equation

$$2\frac{dy}{dt} + y = -t; \quad y = 1 \text{ when } t = 0$$

[**HINT:** First rewrite the differential equation in the form $\frac{dy}{dt} + py = f(t)$.]

35. ▼ ***Electric Circuits*** The flow of current $i(t)$ in an electric circuit without capacitance satisfies the linear differential equation

$$L\frac{di}{dt} + Ri = V(t),$$

where L and R are constants (the *inductance* and *resistance*, respectively) and $V(t)$ is the applied voltage. (See figure.)

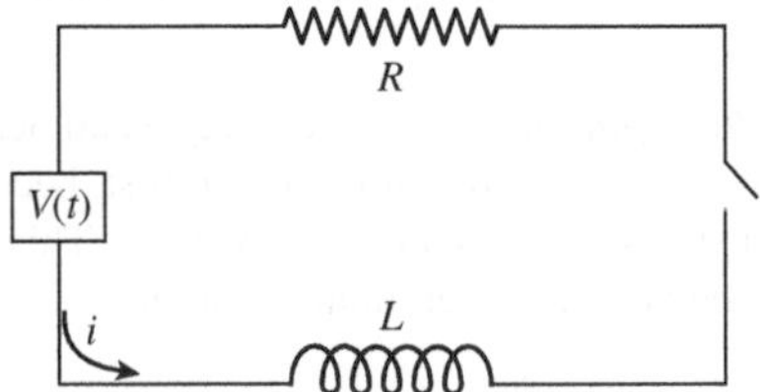

If the voltage is supplied by a 10-volt battery and the switch is turned on at time $t = 1$, then the voltage V is a step function that jumps from 0 to 10 at $t = 1$: $V(t) = 5\left[1 + \frac{|t-1|}{t-1}\right]$.

Find the current as a function of time for $L = R = 1$. Use a grapher to plot the resulting current as a function of time. (Assume that there is no current flowing at time $t = 0$.) [**HINT:** Use the following integral formula: $\int \left[1 + \frac{|t-1|}{t-1}\right] e^t\, dt = \left[1 + \frac{|t-1|}{t-1}\right](e^t - e) + C$.]

36. ▼ ***Electric Circuits*** Repeat Exercise 35 for $L = 1$, $R = 5$, and $V(t) = 5\left[1 + \frac{|t-2|}{t-2}\right]$. (The switch flipped on at time $t = 2$.) [**HINT:** Use the following integral formula: $\int \left[1 + \frac{|t-2|}{t-2}\right] e^{5t}\, dt = \left[1 + \frac{|t-2|}{t-2}\right]\left(\frac{e^{5t} - e^{10}}{5}\right) + C$.]

37. ▼ ***Approach to Equilibrium*** The *Extrasoft Toy Co.* has just released its latest creation, a plush platypus named "Eggbert." The demand function for Eggbert dolls is $D(p) = 50{,}000 - 500p$ dolls per month when the price is p dollars. The supply function is $S(p) = 30{,}000 + 500p$ dolls per month when the price is p dollars. This makes the equilibrium price \$20. The **Evans price adjustment model** assumes that if the price is set at a value other than the equilibrium price, it will change over time in such a way that its rate of change is proportional to the shortage $D(p) - S(p)$.
   a. Write the differential equation given by the Evans price adjustment model for the price p as a function of time.
   b. Find the general solution of the differential equation you wrote in part (a). (You will have two unknown constants, one being the constant of proportionality.)
   c. Find the particular solution in which Eggbert dolls are initially priced at \$10 and the price rises to \$12 after one month.

38. ▼ ***Approach to Equilibrium*** *Spacely Sprockets* has just released its latest model, the Dominator. The demand function is $D(p) = 10{,}000 - 1{,}000p$ sprockets per year when the price is p dollars. The supply function is $S(p) = 8{,}000 + 1{,}000p$ sprockets per year when the price is p dollars.
   a. Using the Evans price adjustment model described in Exercise 37, write the differential equation for the price $p(t)$ as a function of time.
   b. Find the general solution of the differential equation you wrote in part (a).
   c. Find the particular solution in which Dominator sprockets are initially priced at \$5 each but fall to \$3 each after 1 year.

39. ▼ ***Logistic Equation*** There are many examples of growth in which the rate of growth is slow at first, becomes faster, and then slows again as a limit is reached. This pattern can be described by the differential equation

$$\frac{dy}{dt} = ay(L - y),$$

where a is a constant and L is the limit of y. Show by substitution that

$$y = \frac{CL}{e^{-aLt} + C}$$

is a solution of this equation, where C is an arbitrary constant.

40. ▼ ***Logistic Equation*** Using separation of variables and integration with a table of integrals or a symbolic algebra program, solve the differential equation in Exercise 39 to derive the solution given there.

Exercises 41–44 require the use of technology.

41. ▼ ***Market Saturation*** You have just introduced a new model of Blu-ray disc player. You predict that the market will saturate at 2,000,000 Blu-ray disc players and that your total sales will be governed by the equation

$$\frac{dS}{dt} = \frac{1}{4}S(2 - S),$$

where S is the total sales in millions of Blu-ray disc players and t is measured in months. If you give away 1,000 Blu-ray disc players when you first introduce them, what will S be? Sketch the graph of S as a function of t. About how long will it take to saturate the market? (See Exercise 39.)

42. ▼ ***Epidemics*** A certain epidemic of influenza is predicted to follow the function defined by

$$\frac{dA}{dt} = \frac{1}{10}A(20 - A),$$

where A is the number of people infected in millions and t is the number of months after the epidemic starts. If 20,000 cases are reported initially, find $A(t)$ and sketch its graph. When is A growing fastest? How many people will eventually be affected? (See Exercise 39.)

43. ▼ ***Growth of Tumors*** The growth of tumors in animals can be modeled by the Gompertz equation:

$$\frac{dy}{dt} = -ay\ln\left(\frac{y}{b}\right),$$

where y is the size of a tumor, t is time, and a and b are constants that depend on the type of tumor and the units of measurement.
   a. Solve for y as a function of t.
   b. If $a = 1$, $b = 10$, and $y(0) = 5$ cubic centimeters (with t measured in days), find the specific solution and graph it.

44. ▼ ***Growth of Tumors*** Refer back to Exercise 43. Suppose that $a = 1$, $b = 10$, and $y(0) = 15$ cubic centimeters. Find the specific solution and graph it. Comparing its graph to the one obtained in Exercise 43, what can you say about tumor growth in these instances?

Communication and Reasoning Exercises

45. What is the difference between a particular solution and the general solution of a differential equation? How do we get a particular solution from the general solution?

46. Why is there always an arbitrary constant in the general solution of a differential equation? Why are there not two or more arbitrary constants in a first-order differential equation?

47. ▼ Show by example that a second-order differential equation (one involving the second derivative y'') usually has two arbitrary constants in its general solution.

48. ▼ Find a differential equation that is not separable.

49. ▼ Find a differential equation whose general solution is $y = 4e^{-x} + 3x + C$.

50. ▼ Explain how, knowing the elasticity of demand as a function of either price or demand, you may find the demand equation. (See Exercise 29.)

CHAPTER 7 REVIEW

KEY CONCEPTS

 **www.WanerMath.com**
Go to the Website to find a comprehensive and interactive Web-based summary of Chapter 7.

7.1 Integration by Parts

Integration-by-parts formula:

$$\int u \cdot v\, dx = u \cdot I(v) - \int D(u)I(v)\, dx$$

[p. 542]

Tabular method for integration by parts [p. 543]

Integrating a polynomial times a logarithm [p. 546]

7.2 Area between Two Curves and Applications

If $f(x) \geq g(x)$ for all x in $[a, b]$, then the area of the region between the graphs of f and g and between $x = a$ and $x = b$ is given by

$$A = \int_a^b [f(x) - g(x)]\, dx. \text{ [p. 551]}$$

Regions enclosed by crossing curves [p. 554]

Area enclosed by two curves [p. 555]

General instructions for finding the area between the graphs of $f(x)$ and $g(x)$ [p. 555]

Approximating the area between two curves using technology:

$$A = \int_a^b |f(x) - g(x)|\, dx \text{ [p. 556]}$$

7.3 Averages and Moving Averages

Average, or mean, of a collection of values:

$$\bar{y} = \frac{y_1 + y_2 + \cdots + y_n}{n} \text{ [p. 560]}$$

The *average*, or *mean*, of a function $f(x)$ on an interval $[a, b]$:

$$\bar{f} = \frac{1}{b - a}\int_a^b f(x)\, dx. \text{ [p. 561]}$$

Average balance [p. 562]

Computing the moving average of a set of data [p. 562]

n-Unit moving average of a function:

$$\bar{f}(x) = \frac{1}{n}\int_{x-n}^{x} f(t)\, dt \text{ [p. 564]}$$

Computing moving averages of saw-tooth and step functions [p. 564]

7.4 Applications to Business and Economics: Consumers' and Producers' Surplus and Continuous Income Streams

Consumers' surplus:

$$CS = \int_0^{\bar{q}} (D(q) - \bar{p})\, dq \text{ [p. 570]}$$

Producers' surplus:

$$PS = \int_0^{\bar{q}} (\bar{p} - S(q))\, dq \text{ [p. 571]}$$

Equilibrium price [p. 572]

Social gain $= CS + PS$ [p. 573]

Total value of a continuous income stream: $TV = \int_a^b R(t)\, dt$ [p. 574]

Future value of a continuous income stream: $FV = \int_a^b R(t)e^{r(b-t)}\, dt$ [p. 575]

Present value of a continuous income stream: $PV = \int_a^b R(t)e^{r(a-t)}\, dt$ [p. 576]

7.5 Improper Integrals and Applications

Improper integral with an infinite limit of integration:

$$\int_a^{+\infty} f(x)\, dx, \quad \int_{-\infty}^{b} f(x)\, dx,$$

$$\int_{-\infty}^{+\infty} f(x)\, dx \text{ [p. 579]}$$

Improper integral in which the integrand becomes infinite [p. 582]

Testing for convergence [p. 583]

Integrand infinite between the endpoints [p. 584]

Integral improper in two ways [p. 584]

7.6 Differential Equations and Applications

Simple differential equations:

$$\frac{dy}{dx} = f(x) \text{ [p. 589]}$$

Separable differential equations:

$$\frac{dy}{dx} = f(x)g(y) \text{ [p. 591]}$$

Newton's Law of Cooling [p. 593]

REVIEW EXERCISES

In Exercises 1–10, evaluate the given integral.

1. $\int (x^2 + 2)e^x\, dx$
2. $\int (x^2 - x)e^{-3x+1}\, dx$
3. $\int x^2 \ln(2x)\, dx$
4. $\int \log_5 x\, dx$
5. $\int 2x|2x + 1|dx$
6. $\int 3x|-x + 5|dx$
7. $\int 5x\frac{|-x + 3|}{-x + 3}dx$
8. $\int 2x\frac{|3x + 1|}{3x + 1}dx$
9. $\int_{-2}^{2} (x^3 + 1)e^{-x}\, dx$
10. $\int_1^e x^2 \ln x\, dx$

In Exercises 11–14, find the area of the given region.

11. Between $y = x^3$ and $y = 1 - x^3$ for x in $[0, 1]$
12. Between $y = e^x$ and $y = e^{-x}$ for x in $[0, 2]$
13. Enclosed by $y = 1 - x^2$ and $y = x^2$
14. Between $y = x$ and $y = xe^{-x}$ for x in $[0, 2]$

In Exercises 15–18, find the average value of the given function over the indicated interval.

15. $f(x) = x^3 - 1$ over $[-2, 2]$
16. $f(x) = \frac{x}{x^2 + 1}$ over $[0, 1]$

17. $f(x) = x^2e^x$ over $[0, 1]$

18. $f(x) = (x + 1)\ln x$ over $[1, 2e]$

In Exercises 19–22, find the 2-unit moving averages of the given function.

19. $f(x) = 3x + 1$ **20.** $f(x) = 6x^2 + 12$

21. $f(x) = x^{4/3}$ **22.** $f(x) = \ln x$

In Exercises 23 and 24, calculate the consumers' surplus at the indicated unit price $\bar{p}$ for the given demand equation.

23. $p = 50 - \frac{1}{2}q;\ \bar{p} = 10$ **24.** $p = 10 - q^{1/2};\ \bar{p} = 4$

In Exercises 25 and 26, calculate the producers' surplus at the indicated unit price $\bar{p}$ for the given supply equation.

25. $p = 50 + \frac{1}{2}q;\ \bar{p} = 100$ **26.** $p = 10 + q^{1/2};\ \bar{p} = 40$

In Exercises 27–32, decide whether the given integral converges. If the integral converges, compute its value.

27. $\int_1^{\infty} \frac{1}{x^5}\,dx$ **28.** $\int_0^1 \frac{1}{x^5}\,dx$

29. $\int_{-1}^{1} \frac{x}{(x^2 - 1)^{5/3}}\,dx$ **30.** $\int_0^2 \frac{x}{(x^2 - 1)^{1/3}}\,dx$

31. $\int_0^{+\infty} 2xe^{-x^2}\,dx$ **32.** $\int_0^{+\infty} x^2e^{-6x^3}\,dx$

In Exercises 33–36, solve the given differential equation.

33. $\frac{dy}{dx} = x^2y^2$ **34.** $\frac{dy}{dx} = xy + 2x$

35. $xy\frac{dy}{dx} = 1;\ y(1) = 1$

36. $y(x^2 + 1)\frac{dy}{dx} = xy^2;\ y(0) = 2$

Applications: OHaganBooks.com [Try the game at www.OHaganBooks.com]

37. ***Spending on Stationery*** Alarmed by the volume of pointless memos and reports being copied and circulated by management at OHaganBooks.com, John O'Hagan ordered a 5-month audit of paper usage at the company. He found that management consumed paper at a rate of

$$q(t) = 45t + 200 \text{ thousand sheets per month} \quad (0 \le t \le 5).$$

(t is the time in months since the audit began.) During the same period the price of paper was escalating; the company was charged approximately

$$p(t) = 9e^{0.09t} \text{ dollars per thousand sheets.}$$

Use an integral to estimate, to the nearest hundred dollars, the total spent on paper for management during the given period.

38. ***Spending on Shipping*** During the past 10 months, OHaganBooks.com shipped orders at a rate of about

$$q(t) = 25t + 3{,}200 \text{ packages per month} \quad (0 \le t \le 10).$$

(t is the time in months since the beginning of the year.) During the same period the cost of shipping a package averaged approximately

$$p(t) = 4e^{0.04t} \text{ dollars per package.}$$

Use an integral to estimate, to the nearest thousand dollars, the total spent on shipping orders during the given period.

39. ***Education Costs*** Billy-Sean O'Hagan, having graduated *summa cum laude* from college, has been accepted by the doctoral program in biophysics at Oxford. John O'Hagan estimates that the total cost (minus scholarships) he will need to pay is \$2,000 per month but that this cost will escalate at a continuous compounding rate of 1% per month.

a. What, to the nearest dollar, will be the average monthly cost over the course of 2 years?

b. Find the 4-month moving average of the monthly cost.

40. ***Investments*** OHaganBooks.com keeps its cash reserves in a hedge fund paying 6% compounded continuously. It starts a year with \$1 million in reserves and does not withdraw or deposit any money.

a. What is the average amount it will have in the fund over the course of 2 years?

b. Find the 1-month moving average of the amount it has in the fund.

41. ***Consumers' and Producers' Surplus*** Currently, the hottest-selling item at OHaganBooks.com is *Mensa for Dummies,*[55] with a demand curve of $q = 20{,}000(28 - p)^{1/3}$ books per week and a supply curve of $q = 40{,}000(p - 19)^{1/3}$ books per week.

a. Find the equilibrium price and demand.

b. Find the consumers' and producers' surpluses at the equilibrium price.

42. ***Consumers' and Producers' Surplus*** OHaganBooks.com is about to start selling a new coffee table book, *Computer Designs of the Late Twentieth Century.* It estimates the demand curve to be $q = 1{,}000\sqrt{200 - 2p}$, and its willingness to order books from the publisher is given by the supply curve $q = 1{,}000\sqrt{10p - 400}$.

a. Find the equilibrium price and demand.

b. Find the consumers' and producers' surpluses at the equilibrium price.

43. ***Revenue*** Sales of the bestseller *A River Burns through It* are dropping at OHaganBooks.com. To try to bolster sales, the company is decreasing the price of the book, now \$40, at a rate of \$2 per week. As a result, this week OHaganBooks.com will sell 5,000 copies, and it estimates that sales will

[55] The actual title is: *Let Us Just Have A Ball! Mensa for Dummies,* by Wendu Mekbib, Silhouette Publishing Corporation.

fall continuously at a rate of 10% per week. How much revenue will it earn on sales of this book over the next 8 weeks?

44. ***Foreign Investments*** Panicked by the performance of the U.S. stock market, Marjory Duffin is investing her 401(k) money in a Russian hedge fund at a rate of approximately

$$q(t) = 1.7t^2 - 0.5t + 8 \text{ thousand shares per month,}$$

where t is time in months since the stock market began to plummet. At the time she started making the investments, the hedge fund was selling for \$1 per share, but it subsequently declined in value at a continuous rate of 5% per month. What was the total amount of money Marjory Duffin invested after 1 year? (Answer to the nearest \$1,000.)

45. ***Investments*** OHaganBooks.com CEO John O'Hagan has started a gift account for the *Marjory Duffin Foundation.* The account pays 6% compounded continuously and is initially empty. OHaganBooks.com deposits money continuously into it, starting at the rate of \$100,000 per month and increasing by \$10,000 per month.
    a. How much money will the company have in the account at the end of 2 years?
    b. How much of the amount you found in part (a) was principal deposited and how much was interest earned? (Round answers to the nearest \$1,000.)

46. ***Savings*** John O'Hagan had been saving money for Billy-Sean's education since Billy-Sean was a wee lad. O'Hagan began depositing money at the rate of \$1,000 per month and increased his deposits by \$50 per month. If the account earned 5% compounded continuously and O'Hagan continued these deposits for 15 years,
    a. How much money did he accumulate?
    b. How much was money deposited and how much was interest?

47. ***Acquisitions*** The *Megabucks Corporation* is considering buying OHaganBooks.com. It estimates OHaganBooks.com's revenue stream at \$50 million per year, growing continuously at a 10% rate. Assuming an interest rate of 6%, how much is OHaganBooks.com's revenue for the next year worth now?

48. ***More Acquisitions*** OHaganBooks.com is thinking of buying *JungleBooks* and would like to recoup its investment after 3 years. The estimated net profit for JungleBooks is \$40 million per year, growing linearly by \$5 million per year. Assuming an interest rate of 4%, how much should OHaganBooks.com pay for JungleBooks?

49. ***Incompetence*** OHaganBooks.com is shopping around for a new bank. A junior executive at one bank offers it the following interesting deal: The bank will pay OHaganBooks.com interest continuously at a rate numerically equal to 0.01% of the square of the amount of money it has in the account at any time. By considering what would happen if \$10,000 was deposited in such an account, explain why the junior executive was fired shortly after this offer was made.

50. ***Shrewd Bankers*** The new junior officer at the bank (who replaced the one fired in Exercise 49) offers OHaganBooks.com the following deal for the \$800,000 they plan to deposit: While the amount in the account is less than \$1 million, the bank will pay interest continuously at a rate equal to 10% of the difference between \$1 million and the amount of money in the account. When it rises over \$1 million, the bank will pay interest of 20%. Why should OHaganBooks.com not take this offer?

CASE STUDY

Estimating Tax Revenues

Dmitriy Shironosov/Shutterstock.com

You have just been hired by the incoming administration of your country as chief consultant for national tax policy, and you have been getting conflicting advice from the finance experts on your staff. Several of them have come up with plausible suggestions for new tax structures, and your job is to choose the plan that results in the most revenue for the government.

Before you can evaluate their plans, you realize that it is essential to know your country's income distribution—that is, how many people earn how much money per year.* You might think that the most useful way of specifying income distribution would be to use a function that gives the exact number $f(x)$ of people who earn a given salary x. This would necessarily be a discrete function—it makes sense only if x happens to be a whole number of cents. There is, after all, no one earning a salary of exactly \$22,000.142567! Furthermore, this function would behave rather erratically because there are, for example, probably many more people making a salary of exactly \$30,000 than exactly \$30,000.01. Given these problems, it is far more convenient to start with the function defined by

$$N(x) = \text{Total number of people earning between 0 and } x \text{ dollars.}$$

* To simplify our discussion, we are assuming (1) that all tax revenues are based on earned income and (2) that everyone in the population we consider earns some income.

Actually, you would want a "smoothed" version of this function. The graph of $N(x)$ might look like the one shown in Figure 26.

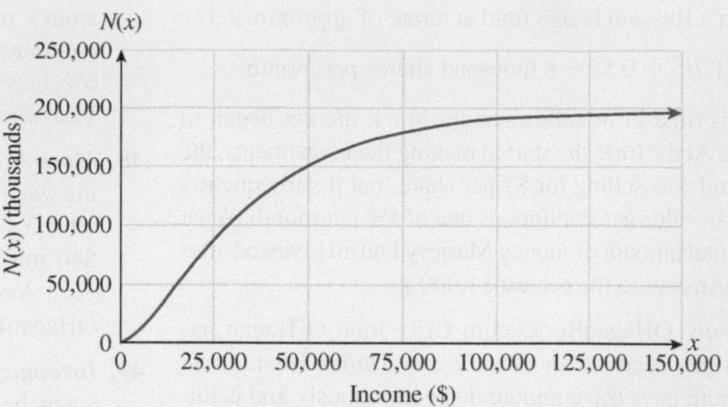

Figure 26

If we take the *derivative* of $N(x)$, we get an income distribution function. Its graph might look like the one shown in Figure 27.

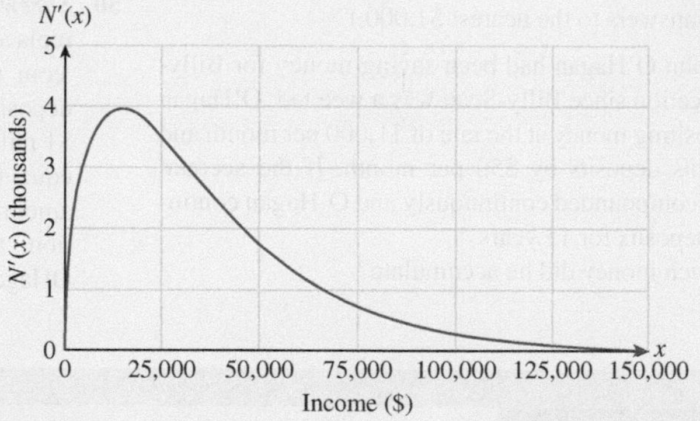

Figure 27

Because the derivative measures the rate of change, its value at x is the additional number of taxpayers per $1 increase in salary. Thus, the fact that $N'(25{,}000) \approx 3{,}500$ tells us that approximately 3,500 people are earning a salary of between $25,000 and $25,001. In other words, N' shows the distribution of incomes among the population—hence, the name "distribution function."*

You therefore send a memo to your experts requesting the income distribution function for the nation. After much collection of data, they tell you that the income distribution function is

$$N'(x) = 12x^{0.676}e^{-x/21{,}500}.$$

This is in fact the function whose graph is shown in Figure 27 and is an example of a **gamma distribution.**† (You might find it odd that you weren't given the original function N, but it will turn out that you don't need it. How would you compute it?)

Given this income distribution, your financial experts have come up with the two possible tax policies illustrated in Figures 28 and 29.

* A very similar idea is used in probability. See the optional chapter "Calculus Applied to Probability and Statistics" on the Website.

† Gamma distributions are often good models for income distributions. The one used in the text is the authors' approximation of the income distribution in the United States in 2013. Source for data: U.S. Census Bureau, Current Population Survey, 2014 Annual Social and Economic Supplement (www.census.gov).

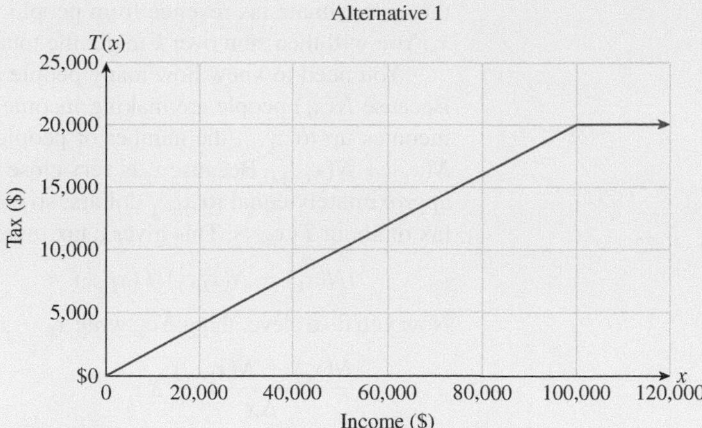

Figure 28

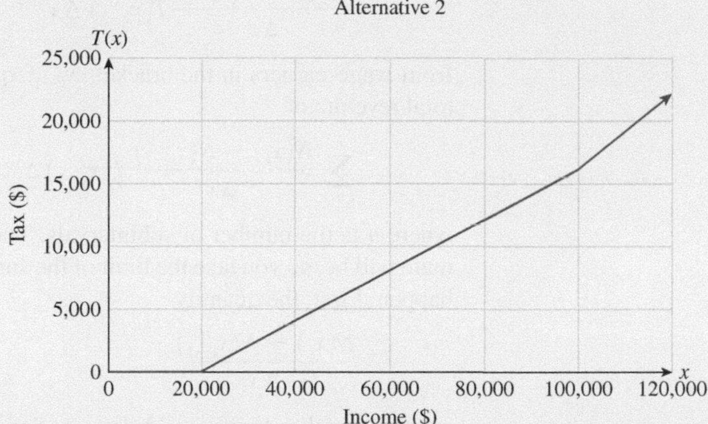

Figure 29

In the first alternative, all taxpayers pay 20% of their income in taxes, except that no one pays more than $20,000 in taxes. In the second alternative, there are three tax brackets, described by the following table:

Income	Marginal Tax Rate
$0–20,000	0%
$20,000–100,000	20%
Above $100,000	30%

Now you must determine which alternative will generate more tax revenue.

Each of Figures 28 and 29 is the graph of a function, T. Rather than using the formulas for these particular functions, you begin by working with the general situation. You have an income distribution function N' and a tax function T, both functions of annual income. You need to find a formula for total tax revenues. First you decide to use a cutoff so that you need to work only with incomes in some finite interval $[0, M]$; you might use, for example, $M = \$10$ million. (Later you will let M approach $+\infty$.) Next, you subdivide the interval $[0, M]$ into a large number of intervals of small width, Δx. If $[x_{k-1}, x_k]$ is a typical such interval, you wish to calculate the approximate tax revenue from people whose total incomes lie between x_{k-1} and x_k. You will then sum over k to get the total revenue.

You need to know how many people are making incomes between x_{k-1} and x_k. Because $N(x_k)$ people are making incomes *up to* x_k and $N(x_{k-1})$ people are making incomes up to x_{k-1}, the number of people making incomes between x_{k-1} and x_k is $N(x_k) - N(x_{k-1})$. Because x_k is very close to x_{k-1}, the incomes of these people are all approximately equal to x_{k-1} dollars, so each of these taxpayers is paying an annual tax of about $T(x_{k-1})$. This gives a tax revenue of

$$[N(x_k) - N(x_{k-1})]T(x_{k-1}).$$

Now you do a clever thing. You write $x_k - x_{k-1} = \Delta x$ and replace $N(x_k) - N(x_{k-1})$ by

$$\frac{N(x_k) - N(x_{k-1})}{\Delta x}\,\Delta x.$$

This gives you a tax revenue of about

$$\frac{N(x_k) - N(x_{k-1})}{\Delta x}\,T(x_{k-1})\,\Delta x$$

from wage-earners in the bracket $[x_{k-1}, x_k]$. Summing over k gives an approximate total revenue of

$$\sum_{k=1}^{n} \frac{N(x_k) - N(x_{k-1})}{\Delta x}\,T(x_{k-1})\,\Delta x,$$

where n is the number of subintervals. The larger n is, the more accurate your estimate will be, so you take the limit of the sum as $n \to \infty$. When you do this, two things happen. First, the quantity

$$\frac{N(x_k) - N(x_{k-1})}{\Delta x}$$

approaches the derivative, $N'(x_{k-1})$. Second, the sum, which you recognize as a Riemann sum, approaches the integral

$$\int_0^M N'(x)T(x)\,dx.$$

You now take the limit as $M \to +\infty$ to get

$$\text{Total tax revenue} = \int_0^{+\infty} N'(x)T(x)\,dx.$$

This improper integral is fine in theory, but the actual calculation will have to be done numerically, so you stick with the upper limit of \$10 million for now. You will have to check that it is reasonable at the end. (Notice that, by the graph of N', it appears that extremely few, if any, people earn that much.) Now you already have a formula for $N'(x)$, but you still need to write formulas for the tax functions $T(x)$ for both alternatives.

Alternative 1 The graph in Figure 28 rises linearly from 0 to 20,000 as x ranges from 0 to 100,000 and then stays constant at 20,000. The slope of the first part is $20{,}000/100{,}000 = 0.2$. The taxation function is therefore

$$T_1(x) = \begin{cases} 0.2x & \text{if } 0 \le x < 100{,}000 \\ 20{,}000 & \text{if } x \ge 100{,}000. \end{cases}$$

For use of technology, it's convenient to express this in closed form using absolute values:*

* To see how to obtain the formula, consult the introduction to Exercises 67–68 in Section 7.1.

$$T_1(x) = 0.2x + \frac{1}{2}\left(1 + \frac{|x - 100{,}000|}{x - 100{,}000}\right)(20{,}000 - 0.2x).$$

The total revenue generated by this tax scheme is, therefore,

$$R_1 = \int_0^{10{,}000{,}000} (12x^{0.676}e^{-x/21{,}500}) \times \left[0.2x + \frac{1}{2}\left(1 + \frac{|x - 100{,}000|}{x - 100{,}000}\right)(20{,}000 - 0.2x)\right] dx.$$

You decide not to attempt this by hand! You use numerical integration software to obtain a grand total of $R_1 = \$1{,}394{,}730{,}000{,}000$, or \$1.39473 trillion (rounded to six significant digits).†

† **Note** If you use the Numerical Integration utility on the Website,
Online Utilities
→ Numerical Integration Utility and Grapher
Enter
```
12x^(0.676)*exp(-x/21500)*
(0.2x+0.5*(20000-0.2x)*
(1+abs(x-100000)/(x-100000)))
```
for $f(x)$, and 0 and 10000000 for a and b respectively, and press "Integral".

Alternative 2 The graph in Figure 29 rises with a slope of 0.2 from 0 to 16,000 as x ranges from 20,000 to 100,000, then rises from that point on with a slope of 0.3. (This is why we say that the *marginal* tax rates are 20% and 30%, respectively.) The taxation function is therefore

$$T_2(x) = \begin{cases} 0 & \text{if } 0 \le x < 20{,}000 \\ 0.2(x - 20{,}000) & \text{if } 20{,}000 \le x < 100{,}000 \\ 16{,}000 + 0.3(x - 100{,}000) & \text{if } x \ge 100{,}000. \end{cases}$$

Again, you express this in closed form using absolute values:

$$\begin{aligned} T_2(x) &= [0.2(x - 20{,}000)]\frac{1}{2}\left(\frac{|x - 20{,}000|}{x - 20{,}000} - \frac{|x - 100{,}000|}{x - 100{,}000}\right) \\ &\quad + [16{,}000 + 0.3(x - 100{,}000)]\frac{1}{2}\left(1 + \frac{|x - 100{,}000|}{x - 100{,}000}\right) \\ &= 0.1(x - 20{,}000)\left(\frac{|x - 20{,}000|}{x - 20{,}000} - \frac{|x - 100{,}000|}{x - 100{,}000}\right) \\ &\quad + [8{,}000 + 0.15(x - 100{,}000)]\left(1 + \frac{|x - 100{,}000|}{x - 100{,}000}\right). \end{aligned}$$

Values of x between 0 and 20,000 do not contribute to the integral, so

$$R_2 = \int_{20{,}000}^{10{,}000{,}000} 12x^{0.676}e^{-x/21{,}500}\, T_2(x)\, dx$$

with $T_2(x)$ as above. Numerical integration software gives $R_2 = \$0.766843$ trillion—considerably less than Alternative 1. Thus, even though Alternative 2 taxes the wealthy more heavily, it yields less total revenue.

Now, what about the cutoff at \$10 million annual income? If you try either integral again with an upper limit of \$100 million, you will see no change in either result to six significant digits. There simply are not enough taxpayers earning an income above \$10,000,000 to make a difference. You conclude that your answers are sufficiently accurate and that the first alternative provides more tax revenue.

EXERCISES

In Exercises 1–4, calculate the total tax revenue for a country with the given income distribution and tax policies (all currency in dollars).

1. **T** $N'(x) = 100x^{0.466}e^{-x/23{,}000}$; 25% tax on all income
2. **T** $N'(x) = 100x^{0.4}e^{-x/30{,}000}$; 45% tax on all income
3. **T** $N'(x) = 100x^{0.466}e^{-x/23{,}000}$; tax brackets as in the following tax table:

Income	Marginal Tax Rate
\$0–30,000	0%
\$30,000–250,000	10%
Above \$250,000	80%

4. **T** $N'(x) = 100x^{0.4}e^{-x/30{,}000}$; no tax on any income below \$250,000, 100% marginal tax rate on any income above \$250,000
5. Let $N'(x)$ be an income distribution function.
   a. If $0 \le a < b$, what does $\int_a^b N'(x)\,dx$ represent? [**HINT:** Use the Fundamental Theorem of Calculus.]
   b. What does $\int_0^{+\infty} N'(x)\,dx$ represent?
6. Let $N'(x)$ be an income distribution function. What does $\int_0^{+\infty} xN'(x)\,dx$ represent? [**HINT:** Argue as in the text.]
7. Let $P(x)$ be the number of people earning more than x dollars.
   a. What is $N(x) + P(x)$?
   b. Show that $P'(x) = -N'(x)$.
   c. Use integration by parts to show that, if $T(0) = 0$, then the total tax revenue is

$$\int_0^{+\infty} P(x)T'(x)\,dx.$$

   [Note: You may assume that $T'(x)$ is continuous, but the result is still true if we assume only that $T(x)$ is continuous and piecewise continuously differentiable.]
8. Income tax functions T are most often described, as in the text, by tax brackets and marginal tax rates.
   a. If one tax bracket is $a < x \le b$, show that $\int_a^b P(x)\,dx$ is the total income earned in the country that falls into that bracket (P as in Exercise 7).
   b. Use part (a) to explain directly why $\int_0^{+\infty} P(x)T'(x)\,dx$ gives the total tax revenue in the case in which T is described by tax brackets and constant marginal tax rates in each bracket.

Section 7.3

Example 3 (page 562) The following table shows *Colossal Conglomerate*'s closing stock prices for 20 consecutive trading days:

Day	1	2	3	4	5	6	7	8	9	10
Price	20	22	21	24	24	23	25	26	20	24
Day	11	12	13	14	15	16	17	18	19	20
Price	26	26	25	27	28	27	29	27	25	24

Plot these prices and the 5-day moving average.

Solution

Here is how to automate this calculation on a TI-83/84 Plus.

1. Use

   `seq(X,X,1,20)→L₁`   [2ND] [STAT] → OPS → 5 [STO] [2ND] [STAT] → L1

   to enter the sequence of numbers 1 through 20 into the list L_1, representing the trading days.

2. Using the list editor accessible through the [STAT] menu, enter the daily stock prices in list L_2.

3. Calculate the list of 5-day moving averages by using the following command:

   `seq((L₂(X)+L₂(X-1)+L₂(X-2)+L₂(X-3)+L₂(X-4))/5,X,5,20)→L₃`

   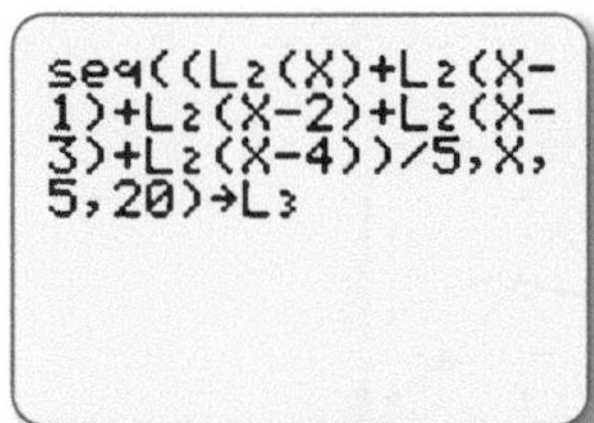

   This has the effect of putting the moving averages into elements 1 through 16 of list L_3.

4. If you wish to plot the moving average on the same graph as the daily prices, you will want the averages in L_3 to match up with the prices in L_2. One way to do this is to put four more entries at the beginning of L_3—say, copies of the first four entries of L_2. The following command accomplishes this:

   `augment(seq(L₂(X),X,1,4),L₃)→L₃`   [2ND] [STAT] → OPS → 9

   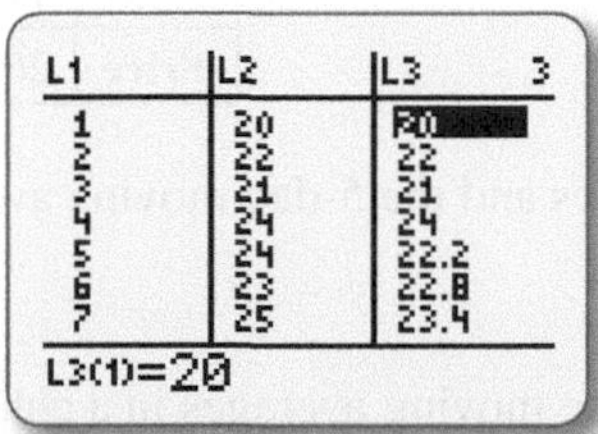

L1
L2
L3 3
1 2 3 4 5 6 7
20 22 21 24 24 23 25
20 22 21 24 22.2 22.8 23.4
L3(1)=20

5. You can now graph the prices and moving averages by creating an xyLine scatter plot through the [STAT PLOT] menu, with L_1 being the Xlist and L_2 being the Ylist for `Plot1` and L_1 being the Xlist and L_3 the Ylist for `Plot2`:

   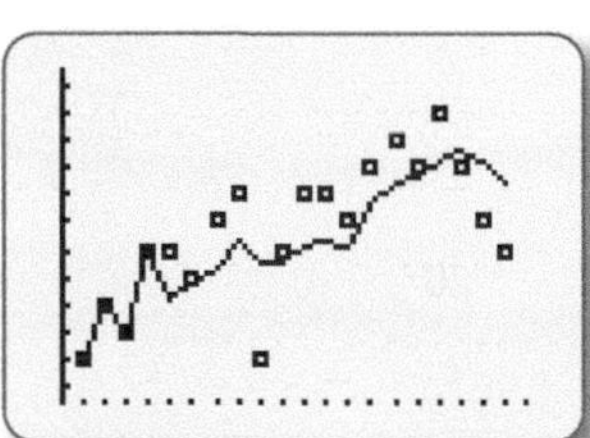

Spreadsheet Technology Guide

Section 7.3

Example 3 (page 562) The following table shows *Colossal Conglomerate*'s closing stock prices for 20 consecutive trading days:

Day	1	2	3	4	5	6	7	8	9	10
Price	20	22	21	24	24	23	25	26	20	24
Day	11	12	13	14	15	16	17	18	19	20
Price	26	26	25	27	28	27	29	27	25	24

Plot these prices and the 5-day moving average.

Solution

1. Compute the moving averages in a column next to the daily prices, as shown here:

	A	B	C
1	Day	Price	Moving Average
2	1	20	
3	2	22	
4	3	21	
5	4	24	
6	5	24	=AVERAGE(B2:B6)
7	6	23	
21	20	24	

→

	A	B	C
1	Day	Price	Moving Average
2	1	20	
3	2	22	
4	3	21	
5	4	24	
6	5	24	22.2
7	6	23	22.8
21	20	24	26.4

2. You can then graph the price and moving average using a scatter plot:

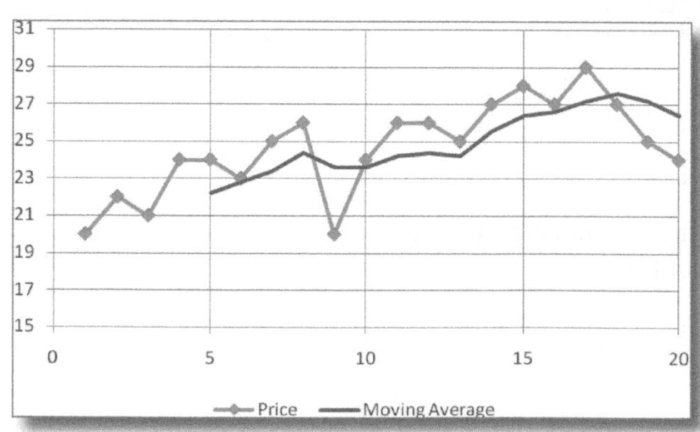

8 FUNCTIONS OF SEVERAL VARIABLES

www.WanerMath.com

At the Website, in addition to the resources listed in the Preface, you will find:

- A surface grapher
- An Excel surface grapher
- A multiple linear regression utility

The following optional extra sections:

- Maxima and Minima: Boundaries and the Extreme Value Theorem
- The Chain Rule for Functions of Several Variables

CASE STUDY

Modeling College Population

College Malls, Inc. is planning to build a national chain of shopping malls in college neighborhoods. The company is planning to lease only to stores that target the specific age demographics of the national college student population. To decide which age brackets to target, the company has asked you, a paid consultant, for an analysis of the college population by student age and of its trends over time.

How can you analyze the relevant data?

david pearson/Alamy Stock Photo

Introduction

We have studied functions of a single variable extensively. But not every useful function is a function of only one variable. In fact, most are not. For example, if you operate an online bookstore in competition with **Amazon.com**, **BN.com**, and **BooksAMillion.com**, your sales may depend on those of your competitors. Your company's daily revenue might be modeled by a function such as

$$R(x, y, z) = 10{,}000 - 0.01x - 0.02y - 0.01z + 0.00001yz,$$

where x, y, and z are the online daily revenues of Amazon.com, BN.com, and BooksAMillion.com, respectively. Here, R is a function of three variables because it *depends on* x, y, and z. As we shall see, the techniques of calculus extend readily to such functions. Among the applications we shall look at is optimization: finding, where possible, the maximum or minimum of a function of two or more variables.

8.1 Functions of Several Variables from the Numerical, Algebraic, and Graphical Viewpoints

Numerical and Algebraic Viewpoints

Recall that a function of one variable is a rule for manufacturing a new number $f(x)$ from a single independent variable x. A function of two or more variables is similar, but the new number now depends on more than one independent variable.

Function of Several Variables

A **real-valued function**, f, **of** $x, y, z, \ldots$ is a rule for manufacturing a new number, written $f(x, y, z, \ldots)$, from the values of a sequence of independent variables $(x, y, z, \ldots)$. The function f is called a **real-valued function of two variables** if there are two independent variables, a **real-valued function of three variables** if there are three independent variables, and so on.

Quick Examples

1. $f(x, y) = x - y$	Function of two variables
$f(1, 2) = 1 - 2 = -1$	Substitute 1 for x and 2 for y.
$f(2, -1) = 2 - (-1) = 3$	Substitute 2 for x and -1 for y.
$f(y, x) = y - x$	Substitute y for x and x for y.
2. $g(x, y) = x^2 + y^2$	Function of two variables
$g(-1, 3) = (-1)^2 + 3^2 = 10$	Substitute -1 for x and 3 for y.
3. $h(x, y, z) = x + y + xz$	Function of three variables
$h(2, 2, -2) = 2 + 2 + 2(-2) = 0$	Substitute 2 for x, 2 for y, and -2 for z.

Note It is often convenient to use $x_1, x_2, x_3, \ldots$ for the independent variables, so, for instance, the third example above would be $h(x_1, x_2, x_3) = x_1 + x_2 + x_1x_3$. ■

Figure 1 illustrates the concept of a function of two variables: In goes a pair of numbers, and out comes a single number.

$(x, y) \longrightarrow$ [g] $\longrightarrow x^2 + y^2$ $(2, -1) \longrightarrow$ [g] $\longrightarrow 5$

Figure 1

As with functions of one variable, functions of several variables can be represented numerically (using a table of values), algebraically (using a formula as in the above examples), and sometimes graphically (using a graph).

Let's now look at a number of examples of interesting functions of several variables.

EXAMPLE 1 Cost Function

Roy Mehta/The Image Bank/Getty Images

You own a company that makes two models of speakers: the Ultra Mini and the Big Stack. Your total monthly cost (in dollars) to make x Ultra Minis and y Big Stacks is given by

$$C(x, y) = 10{,}000 + 20x + 40y.$$

What is the significance of each term in this formula?

Solution The terms have meanings similar to those we saw for linear cost functions of a single variable. Let us look at the terms one at a time.

Constant Term Consider the monthly cost of making no speakers at all ($x = y = 0$). We find

$$C(0, 0) = 10{,}000. \quad \text{Cost of making no speakers is \$10,000.}$$

Thus, the constant term 10,000 is the **fixed cost**, the amount you have to pay each month even if you make no speakers.

Coefficients of x and y Suppose you make a certain number of Ultra Minis and Big Stacks one month and the next month you increase production by one Ultra Mini. The costs are

$$\begin{aligned} C(x, y) &= 10{,}000 + 20x + 40y && \text{First month} \\ C(x+1, y) &= 10{,}000 + 20(x+1) + 40y && \text{Second month} \\ &= 10{,}000 + 20x + 20 + 40y \\ &= C(x, y) + 20 \end{aligned}$$

Thus, each Ultra Mini adds \$20 to the total cost. We say that \$20 is the **marginal cost** of each Ultra Mini. Similarly, because of the term $40y$, each Big Stack adds \$40 to the total cost. The marginal cost of each Big Stack is \$40.

This cost function is an example of a *linear function of two variables*. The coefficients of x and y play roles similar to that of the slope of a line. In particular, they give the rates of change of the function as each variable increases while the other stays constant. (Think about it.) We shall say more about linear functions below.

Using Technology

See the Technology Guides at the end of the chapter to see how you can use a TI-83/84 Plus and a spreadsheet to display various values of $C(x, y)$ in Example 1. Here is an outline:

TI-83/84 Plus

```
Y1=10000+20X+40Y
```

To evaluate $C(10, 30)$:

```
10→X
30→Y
Y1
```

[More details in the Technology Guide.]

Spreadsheet

x-values down column A starting in A2

y-values down column B starting in B2

```
=10000+20*A2+40*B2
```

in C2; copy down column C. [More details in the Technology Guide.]

➡ **Before we go on...** In Example 1, which values of x and y may we substitute into $C(x, y)$? Certainly, we must have $x \geq 0$ and $y \geq 0$ because it makes no sense to speak of manufacturing a negative number of speakers. Also, there is certainly some

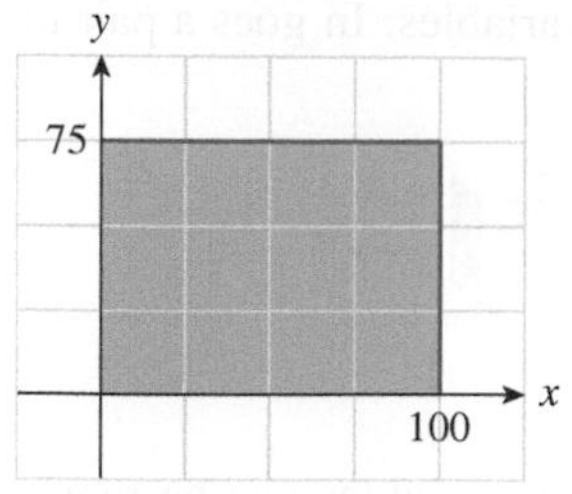

Figure 2

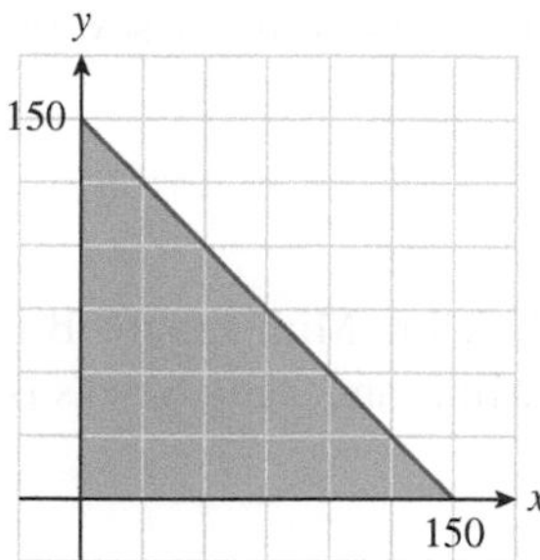

Figure 3

upper bound to the number of speakers that can be made in a month. The bound might take one of several forms. The number of each model may be bounded—say $x \le 100$ and $y \le 75$. The inequalities $0 \le x \le 100$ and $0 \le y \le 75$ describe the region in the plane shaded in Figure 2.

Another possibility is that the *total* number of speakers is bounded—say, $x + y \le 150$. This, together with $x \ge 0$ and $y \ge 0$, describes the region shaded in Figure 3.

In either case the region shown represents the pairs (x, y) for which $C(x, y)$ is defined. Just as with a function of one variable, we call this region the **domain** of the function. As before, when the domain is not given explicitly, we agree to take the largest domain possible. ■

EXAMPLE 2 Faculty Salaries

David Katz came up with the following function for the salary of a professor with 10 years of teaching experience in a large university:

$$S(x, y, z) = 13{,}005 + 230x + 18y + 102z.$$

Here, S is the salary in 1969–1970 in dollars per year, x is the number of books the professor has published, y is the number of articles published, and z is the number of "excellent" articles published.[1] What salary do you expect that a professor with 10 years' experience earned in 1969–1970 if she published 2 books, 20 articles, and 3 "excellent" articles?

Solution All we need to do is calculate

$$\begin{aligned} S(2, 20, 3) &= 13{,}005 + 230(2) + 18(20) + 102(3) \\ &= \$14{,}131. \end{aligned}$$

➡ **Before we go on . . .** In Example 1 we gave a linear function of two variables. In Example 2 we have a linear function of three variables. Katz came up with his model by surveying a large number of faculty members and then finding the linear function that "best" fit the data. Such models are called **multiple linear regression** models. In the Case Study at the end of this chapter we shall see a spreadsheet method of finding the coefficients of a multiple regression model from a set of observed data.

What does this model say about the value of a single book or a single article? If a book takes 15 times as long to write as an article, how would you recommend that a professor spend her writing time? ■

Here are two simple kinds of functions of several variables.

Linear Function

A function f of n variables is **linear** if f has the property that

$$f(x_1, x_2, \ldots, x_n) = a_0 + a_1x_1 + \cdots + a_nx_n \qquad (a_0, a_1, a_2, \ldots, a_n \text{ constants}).$$

[1] David A. Katz, "Faculty Salaries, Promotions and Productivity at a Large University," *American Economic Review*, June 1973, pp. 469–477. Prof. Katz's equation actually included other variables, such as the number of dissertations supervised; our equation assumes that all of these are zero.

Quick Examples

4. $f(x, y) = 3x - 5y$ — Linear function of x and y
5. $C(x, y) = 10{,}000 + 20x + 40y$ — Example 1
6. $S(x_1, x_2, x_3) = 13{,}005 + 230x_1 + 18x_2 + 102x_3$ — Example 2

Interaction Function

If we add to a linear function one or more terms of the form bx_ix_j (where b is a nonzero constant and $i \neq j$), we get a **second-order interaction function.**

Quick Examples

7. $C(x, y) = 10{,}000 + 20x + 40y + 0.1xy$
8. $R(x_1, x_2, x_3) = 10{,}000 - 0.01x_1 - 0.02x_2 - 0.01x_3 + 0.00001x_2x_3$

So far, we have been specifying functions of several variables **algebraically**—by using algebraic formulas. If you have ever studied statistics, you are probably familiar with statistical tables. These tables may also be viewed as representing functions **numerically**, as the next example shows.

EXAMPLE 3 Function Represented Numerically: Body Mass Index

The following table lists some values of the body mass index, which gives a measure of the massiveness of your body, taking height into account.* The variable w represents your weight in pounds, and h represents your height in inches. An individual with a body mass index of 25 or above is generally considered overweight.

* It is interesting that weight-lifting competitions are usually based on weight rather than body mass index. As a consequence, taller people are at a significant disadvantage in these competitions because they must compete with shorter, stockier people of the same weight. (An extremely thin, very tall person can weigh as much as a muscular short person, although the tall person's body mass index would be significantly lower.)

$h\downarrow$ / $w\rightarrow$	130	140	150	160	170	180	190	200	210
60	25.2	27.1	29.1	31.0	32.9	34.9	36.8	38.8	40.7
61	24.4	26.2	28.1	30.0	31.9	33.7	35.6	37.5	39.4
62	23.6	25.4	27.2	29.0	30.8	32.7	34.5	36.3	38.1
63	22.8	24.6	26.4	28.1	29.9	31.6	33.4	35.1	36.9
64	22.1	23.8	25.5	27.2	28.9	30.7	32.4	34.1	35.8
65	21.5	23.1	24.8	26.4	28.1	29.7	31.4	33.0	34.7
66	20.8	22.4	24.0	25.6	27.2	28.8	30.4	32.0	33.6
67	20.2	21.8	23.3	24.9	26.4	28.0	29.5	31.1	32.6
68	19.6	21.1	22.6	24.1	25.6	27.2	28.7	30.2	31.7
69	19.0	20.5	22.0	23.4	24.9	26.4	27.8	29.3	30.8
70	18.5	19.9	21.4	22.8	24.2	25.6	27.0	28.5	29.9
71	18.0	19.4	20.8	22.1	23.5	24.9	26.3	27.7	29.1
72	17.5	18.8	20.2	21.5	22.9	24.2	25.6	26.9	28.3
73	17.0	18.3	19.6	20.9	22.3	23.6	24.9	26.2	27.5
74	16.6	17.8	19.1	20.4	21.7	22.9	24.2	25.5	26.7
75	16.1	17.4	18.6	19.8	21.1	22.3	23.6	24.8	26.0
76	15.7	16.9	18.1	19.3	20.5	21.7	22.9	24.2	25.4

As the table shows, the value of the body mass index depends on two quantities: w and h. Let us write $M(w, h)$ for the body mass index function. What are $M(140, 62)$ and $M(210, 63)$?

Using Technology

See the Technology Guides at the end of the chapter to see how to use a spreadsheet to create the table in Example 3. Here is an outline:

Spreadsheet
w-values 130 to 210 in B1–J1
h-values 60 to 76 in A2–A18
```
=0.45*B$1/
(0.0254*$A2)^2
```
in B2; copy down and across through J18. [More details in the Technology Guide.]

Solution We can read the answers from the table:

$$M(140, 62) = 25.4 \qquad w = 140 \text{ lb}, h = 62 \text{ in}$$

and

$$M(210, 63) = 36.9. \qquad w = 210 \text{ lb}, h = 63 \text{ in}$$

The function $M(w, h)$ is actually given by the formula

$$M(w, h) = \frac{0.45w}{(0.0254h)^2}.$$

[The factor 0.45 converts the weight to kilograms, and 0.0254 converts the height to meters. If w is in kilograms and h is in meters, the formula is simpler: $M(w, h) = w/h^2$.]

Geometric Viewpoint: Three-Dimensional Space and the Graph of a Function of Two Variables

Just as functions of a single variable have graphs, so do functions of two or more variables. Recall that the graph of $f(x)$ consists of all points $(x, f(x))$ in the xy-plane. By analogy we would like to say that the graph of a function of *two* variables, $f(x, y)$, consists of all points of the form $(x, y, f(x, y))$. Thus, we need three axes: the x-, y-, and z-axes. In other words, our graph will live in **three-dimensional space**, or **3-space**.*

Just as we had two mutually perpendicular axes in two-dimensional space (the xy-plane; see Figure 4(a)), so we have three mutually perpendicular axes in three-dimensional space (Figure 4(b)).

* If we were dealing instead with a function of *three* variables, then we would need to go to *four-dimensional* space. Here, we run into visualization problems (to say the least!), so we won't discuss the graphs of functions of three or more variables in this text.

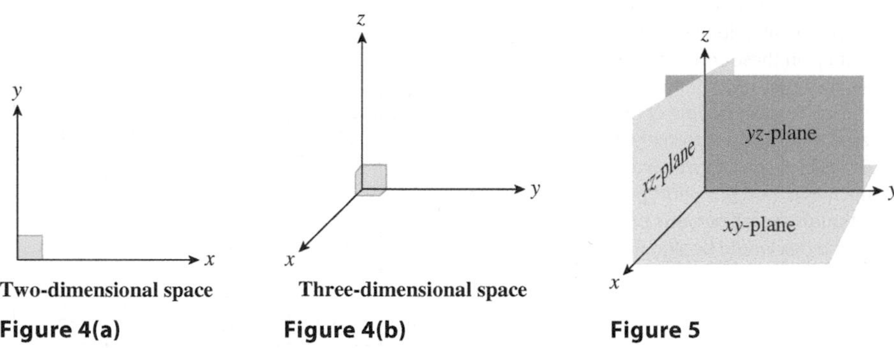

Two-dimensional space

Figure 4(a)

Three-dimensional space

Figure 4(b)

Figure 5

In both 2-space and 3-space the axis labeled with the last letter goes up. Thus, the z-direction is the "up" direction in 3-space rather than the y-direction.

Three important planes are associated with these axes: the xy-plane, the yz-plane, and the xz-plane. These planes are shown in Figure 5. Any two of these planes intersect in one of the axes (for example, the xy- and xz-planes intersect in the x-axis), and all three meet at the origin. Notice that the xy-plane consists of all points with z-coordinate zero, the xz-plane consists of all points with $y = 0$, and the yz-plane consists of all points with $x = 0$.

In 3-space, each point has *three* coordinates, as you might expect: the x-coordinate, the y-coordinate, and the z-coordinate. To see how this works, look at the following examples.

The z-coordinate of a point is its height above the xy-plane.

EXAMPLE 4 Plotting Points in Three Dimensions

Locate the points $P(1, 2, 3)$, $Q(-1, 2, 3)$, $R(1, -1, 0)$, and $S(1, 2, -2)$ in 3-space.

Solution To locate P, the procedure is similar to the one we used in 2-space: Start at the origin, proceed 1 unit in the x-direction, then proceed 2 units in the y-direction, and finally, proceed 3 units in the z-direction. We wind up at the point P shown in Figures 6(a) and 6(b).

Here is another, extremely useful way of thinking about the location of P: First, look at the x- and y-coordinates, obtaining the point $(1, 2)$ in the xy-plane. The point we want is then 3 units vertically above the point $(1, 2)$ because the z-coordinate of a point is just its height above the xy-plane. This strategy is shown in Figure 6(c).

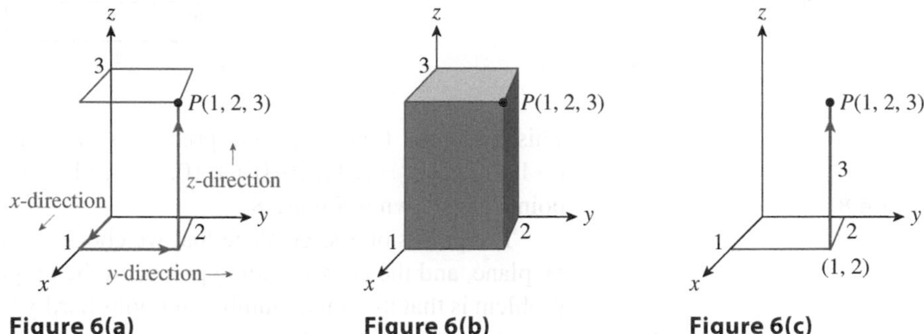

Figure 6(a) **Figure 6(b)** **Figure 6(c)**

Plotting the points Q, R, and S is similar, using the convention that negative coordinates correspond to moves back, left, or down. (See Figure 7.)

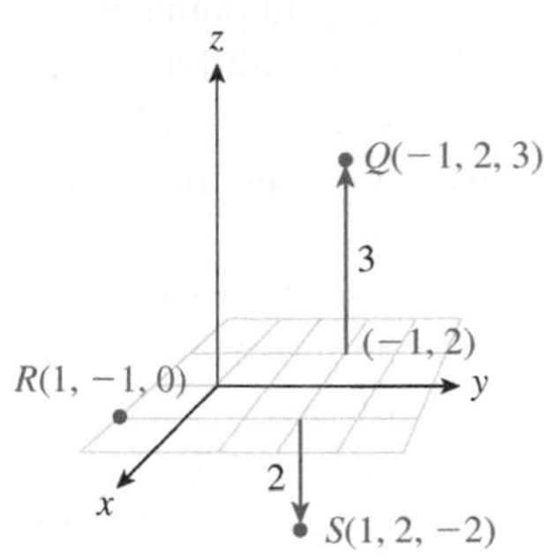

Figure 7

Our next task is to describe the graph of a function $f(x, y)$ of two variables.

Graph of a Function of Two Variables

The **graph of the function f of two variables** is the set of all points $(x, y, f(x, y))$ in three-dimensional space, where we restrict the values of (x, y) to lie in the domain of f. In other words, the graph is the set of all the points (x, y, z) with $z = f(x, y)$.

Note For *every* point (x, y) in the domain of f, the z-coordinate of the corresponding point on the graph is given by evaluating the function at (x, y). Thus, there will be a point of the graph on the vertical line through *every* point in the domain of f, so the graph is usually a *surface* of some sort (see the figure). ■

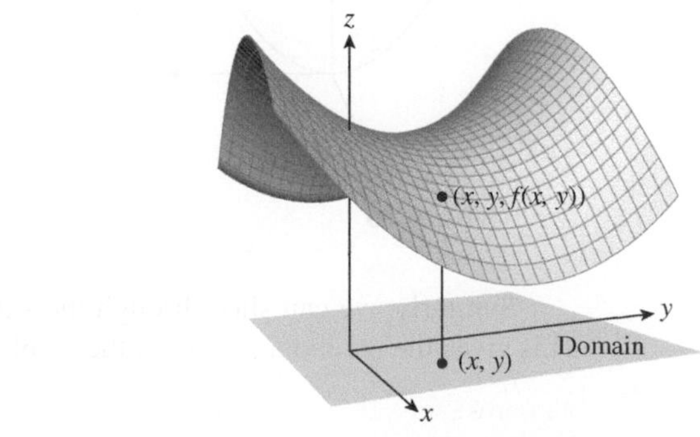

EXAMPLE 5 Graph of a Function of Two Variables

Describe the graph of $f(x, y) = x^2 + y^2$.

Solution Your first thought might be to make a table of values. You could choose some values for x and y and then, for each such pair, calculate $z = x^2 + y^2$. For example, you might get the following table:

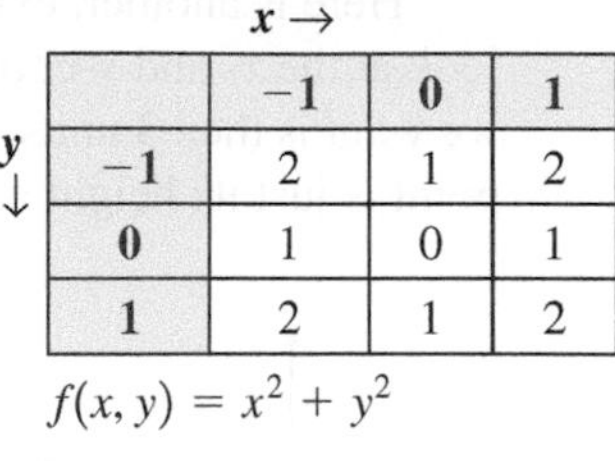
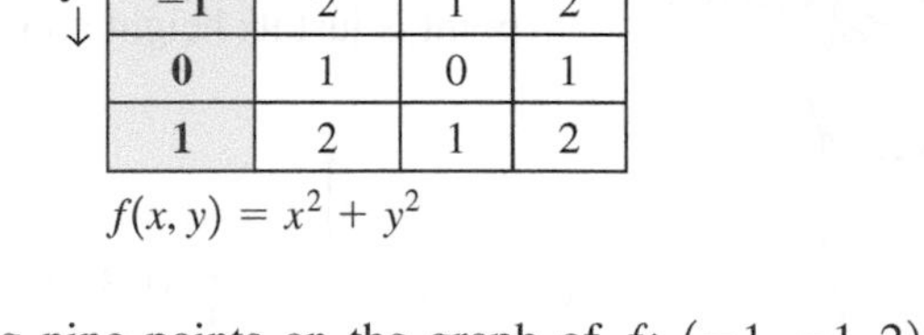

$y \downarrow$ \ $x \rightarrow$	−1	0	1
−1	2	1	2
0	1	0	1
1	2	1	2

$f(x, y) = x^2 + y^2$

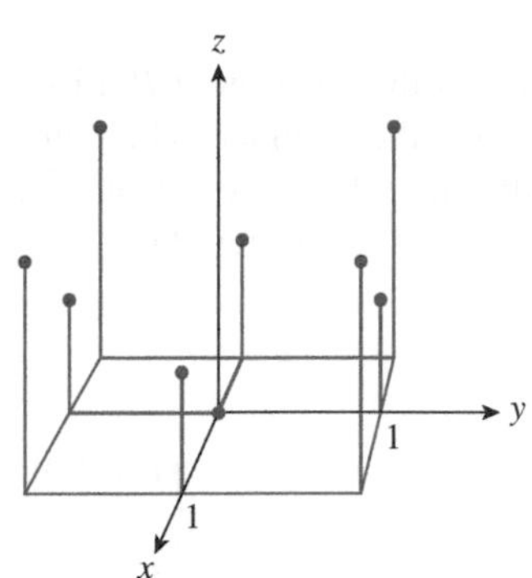

Figure 8

This gives the following nine points on the graph of f: $(-1, -1, 2)$, $(-1, 0, 1)$, $(-1, 1, 2)$, $(0, -1, 1)$, $(0, 0, 0)$, $(0, 1, 1)$, $(1, -1, 2)$, $(1, 0, 1)$, and $(1, 1, 2)$. These points are shown in Figure 8.

The points on the xy-plane that we chose for our table are the grid points in the xy-plane, and the corresponding points on the graph are marked with solid dots. The problem is that this small number of points hardly tells us what the surface looks like, and even if we plotted more points, it is not clear that we would get anything more than a mass of dots on the page.

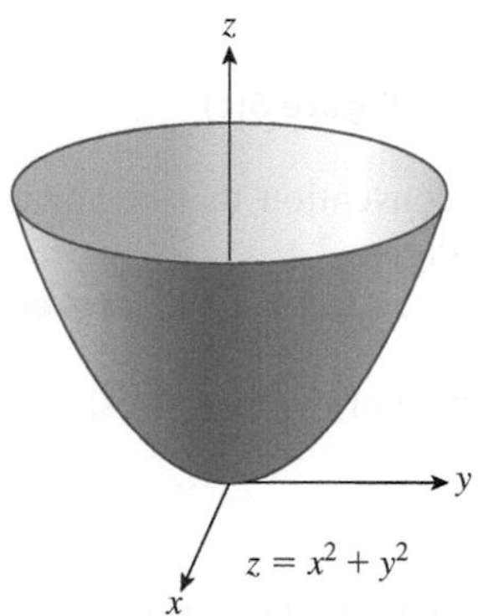

Figure 9

What can we do? There are several alternatives. One place to start is to use technology to draw the graph. (See the technology note on the next page.) We then obtain something like Figure 9. This particular surface is called a **paraboloid**.

If we slice vertically through this surface along the yz-plane, we get the picture in Figure 10. The shape of the front edge, where we cut, is a parabola. To see why, note that the yz-plane is the set of points where $x = 0$. To get the intersection of $x = 0$ and $z = x^2 + y^2$, we substitute $x = 0$ in the second equation, getting $z = y^2$. This is the equation of a parabola in the yz-plane.

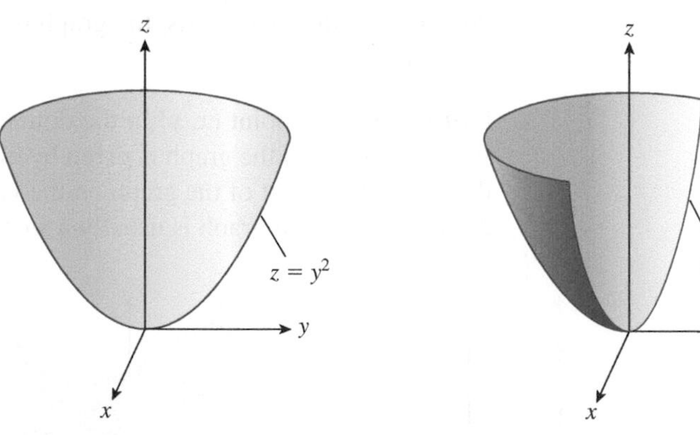

Figure 10

Figure 11

Similarly, we can slice through the surface with the xz-plane by setting $y = 0$. This gives the parabola $z = x^2$ in the xz-plane (Figure 11).

We can also look at horizontal slices through the surface, that is, slices by planes parallel to the *xy*-plane. These are given by setting $z = c$ for various numbers c. For example, if we set $z = 1$, we will see only the points with height 1. Substituting in the equation $z = x^2 + y^2$ gives the equation

$$1 = x^2 + y^2,$$

which is the equation of a circle of radius 1.* If we set $z = 4$, we get the equation of a circle of radius 2:

$$4 = x^2 + y^2.$$

In general, if we slice through the surface at height $z = c$, we get a circle (of radius $\sqrt{c}$). Figure 12 shows several of these circles.

* See Section 0.7 for a discussion of equations of circles.

Using Technology

We can use technology to obtain the graph of the function in Example 5:

Spreadsheet

Table of values:
x-values −3 to 3 in B1–H1
y-values −3 to 3 in A2–A8
`=B$1^2+$A2^2`
in B2; copy down and across through H8.
Graph: Highlight A1 through H8 and insert a surface chart. [More details in the Technology Guide.]

Website
www.WanerMath.com
Online Utilities
→ Surface Graphing Utility
Enter `x^2+y^2` for $f(x, y)$
Set xMin = −3, xMax = 3,
yMin = −3, yMax = 3
Press "Graph".

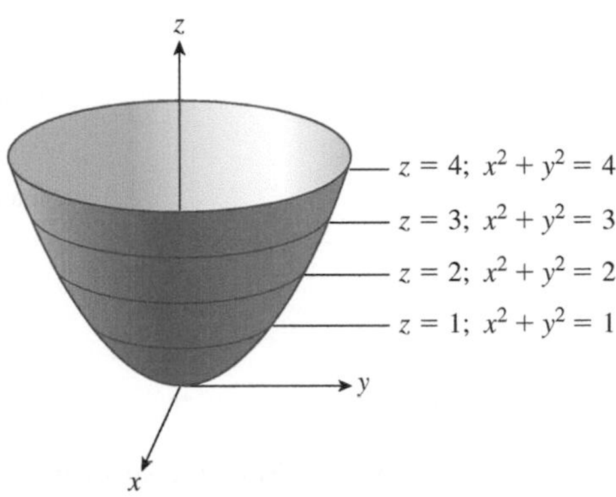

Figure 12

Looking at these circular slices, we see that this surface is the one we get by taking the parabola $z = x^2$ and spinning it around the *z*-axis. This is an example of what is known as a **surface of revolution**.

➡ **Before we go on...** The graph of any function of the form $f(x, y) = Ax^2 + By^2 + Cxy + Dx + Ey + F$ ($A, B, \dots, F$ constants), with $4AB - C^2$ positive, can be shown to be a paraboloid of the same general shape as that in Example 5 if A and B are positive or upside-down if A and B are negative. If $A \neq B$, the horizontal slices will be ellipses rather than circles.

Notice that each horizontal slice through the surface in Example 5 was obtained by putting $z = $ *constant*. This gave us an equation in x and y that described a curve. These curves are called the **level curves** of the surface $z = f(x, y)$ (see the discussion on the next page). In Example 5 the equations are of the form $x^2 + y^2 = c$ (c constant), so the level curves are circles. Figure 13 shows the level curves for $c = 0, 1, 2, 3,$ and 4.

The level curves give a contour map or topographical map of the surface. Each curve shows all of the points on the surface at a particular height c. You can use this contour map to visualize the shape of the surface. Imagine moving the contour at $c = 1$ to a height of 1 unit above the *xy*-plane, the contour at $c = 2$ to a height of 2 units above the *xy*-plane, and so on. You will end up with something like Figure 12. ■

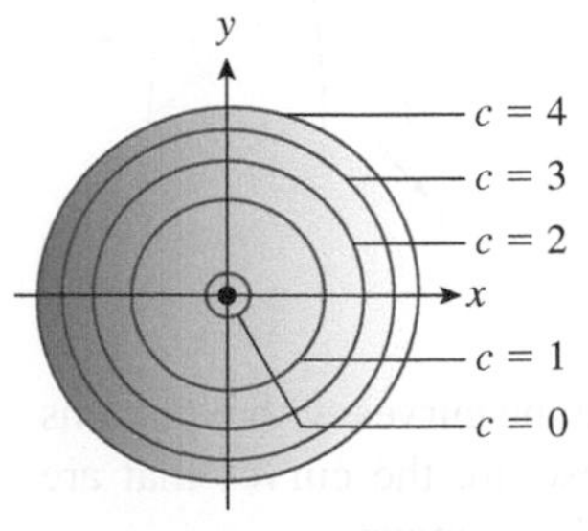

Level curves of the paraboloid
$z = x^2 + y^2$

Figure 13

The following summary includes the techniques we have just used plus some additional ones.

Analyzing the Graph of a Function of Two Variables

If possible, use technology to render the graph $z = f(x, y)$ of a given function f of two variables. You can analyze its graph as follows:

Step 1 Obtain the ***x*-, *y*-, and *z*-intercepts** (the places where the surface crosses the coordinate axes).

***x*-Intercept(s):** Set $y = 0$ and $z = 0$, and solve for x.
***y*-Intercept(s):** Set $x = 0$ and $z = 0$, and solve for y.
***z*-Intercept:** Set $x = 0$ and $y = 0$, and compute z.

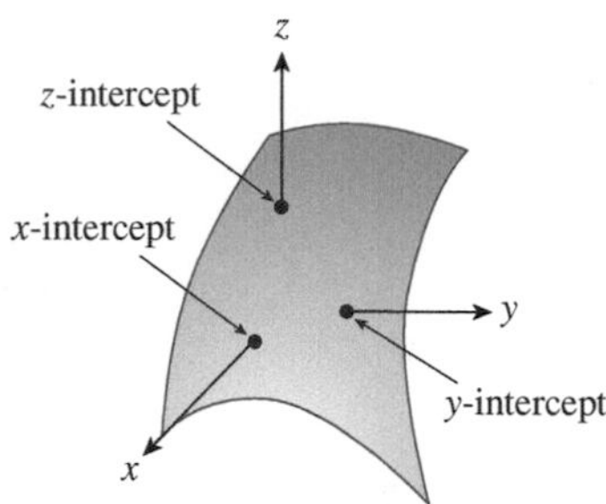

Step 2 Slice the surface along planes parallel to the *xy*-, *yz*-, and *xz*-planes.

***z* = *constant*:** Set $z = constant$, and analyze the resulting curves. These are the curves resulting from horizontal slices; they are called the **level curves** (see below).
***x* = *constant*:** Set $x = constant$, and analyze the resulting curves. These are the curves resulting from slices parallel to the *yz*-plane.
***y* = *constant*:** Set $y = constant$, and analyze the resulting curves. These are the curves resulting from slices parallel to the *xz*-plane.

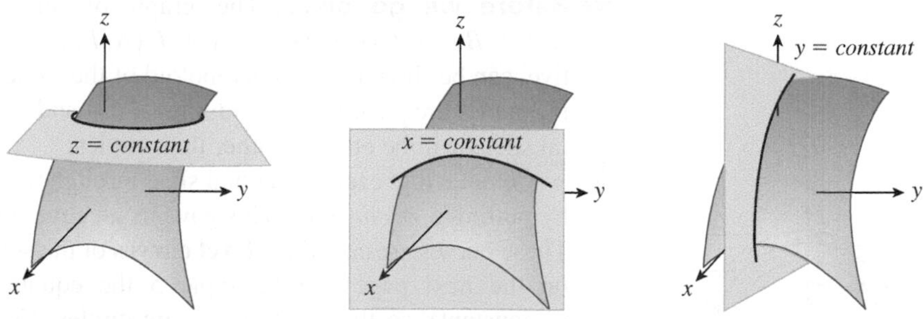

Level Curves

The **level curves** of a function f of two variables are the curves with equations of the form $f(x, y) = c$, where c is constant. These are the curves that are obtained from the graph of f by slicing it horizontally as above.

Quick Examples

9. Figure 13 shows some level curves of $f(x, y) = x^2 + y^2$. The ones shown have equations $f(x, y) = 0, 1, 2, 3,$ and 4.

10. Let $f(x, y) = y - x^2 + 4$. Its level curves have the form $y - x^2 + 4 = c$ (where c is constant). If we solve this equation for y, we see that $y = x^2 + c - 4$, the equation of a parabola with its vertex on the y-axis at the point $c - 4$. The following figure shows a portion of the graph of f and some of its level curves.

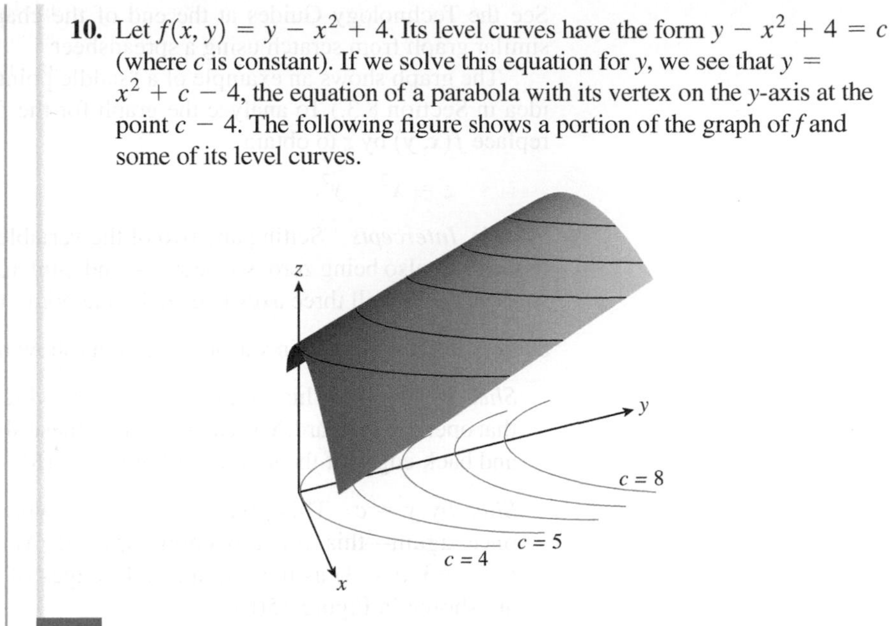

Spreadsheets often have built-in features to render surfaces such as the paraboloid in Example 5. In the following example we use Excel to graph another surface and then analyze it as above.

EXAMPLE 6 Analyzing a Surface

Describe the graph of $f(x, y) = x^2 - y^2$.

Solution First, we obtain a picture of the graph using technology. Figure 14 shows two graphs obtained using resources at the Website.

Chapter 8 → Math Tools for Chapter 8 → Surface Graphing Utility

Chapter 8 → Math Tools for Chapter 8 → Excel Surface Graphing Utility

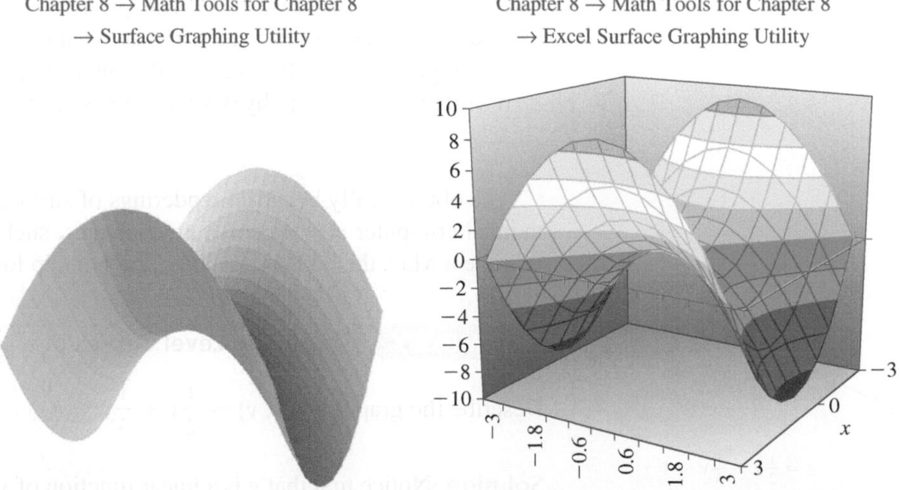

Figure 14

See the Technology Guides at the end of the chapter to find out how to obtain a similar graph from scratch using a spreadsheet.

The graph shows an example of a "saddle point" at the origin. (We return to this idea in Section 8.3.) To analyze the graph for the features shown in the box above, replace $f(x, y)$ by z to obtain

$$z = x^2 - y^2.$$

Step 1: *Intercepts* Setting any two of the variables x, y, and z equal to zero results in the third also being zero, so the x-, y-, and z-intercepts are all 0. In other words, the surface touches all three axes in exactly one point: the origin.

Step 2: *Slices* Slices in various directions show more interesting features.

Slice by $x = c$ This gives $z = c^2 - y^2$, which is the equation of a parabola that opens downward. You can see two of these slices ($c = -3$, $c = 3$) as the front and back edges of the surface in Figure 14. (More are shown in Figure 15(a).)

Slice by $y = c$ This gives $z = x^2 - c^2$, which is the equation of a parabola once again—this time, opening upward. You can see two of these slices ($c = -3$, $c = 3$) as the left and right edges of the surface in Figure 14. (More are shown in Figure 15(b).)

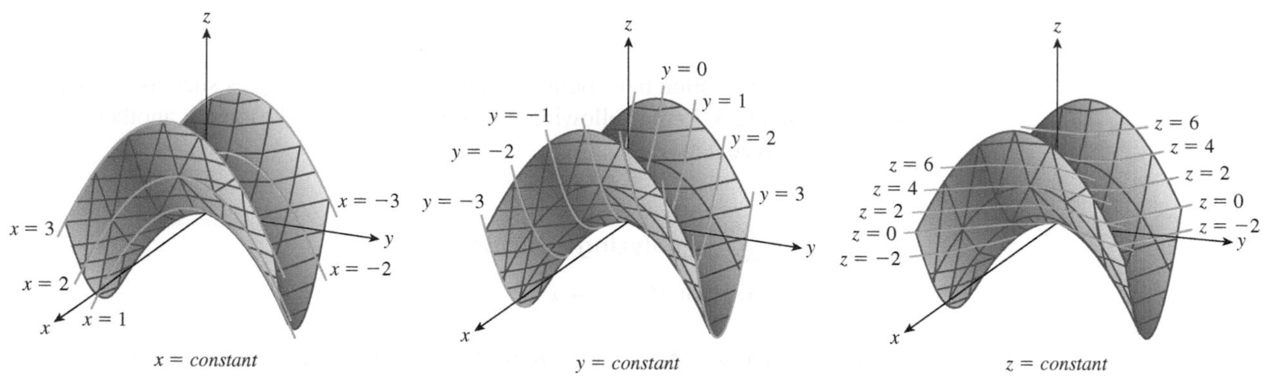

Figure 15(a) **Figure 15(b)** **Figure 15(c)**

Level Curves: Slice by $z = c$ This gives $x^2 - y^2 = c$, which is a hyperbola. The level curves for various values of c are visible in Figure 14 as the horizontal slices. (See Figure 15(c).) The case $c = 0$ is interesting: The equation $x^2 - y^2 = 0$ can be rewritten as $x = \pm y$ (why?), which represents two lines at right angles to each other.

To obtain really beautiful renderings of surfaces, you could use one of the commercial computer algebra software packages, such as Mathematica or Maple, or, if you use a Mac, the built-in grapher (grapher.app located in the Utilities folder).

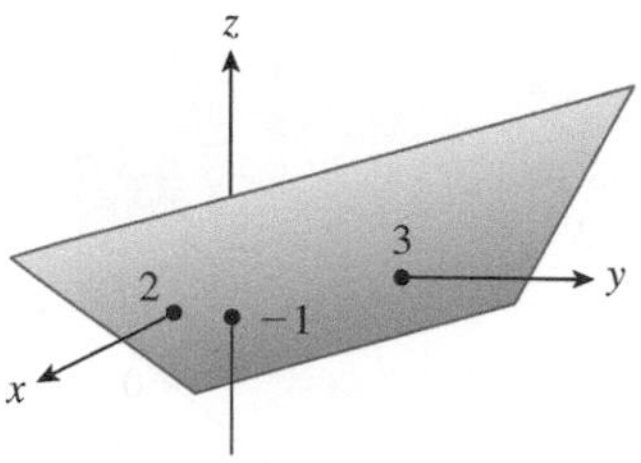

$z = \frac{1}{2}x + \frac{1}{3}y - 1$

Figure 16

EXAMPLE 7 Graph and Level Curves of a Linear Function

Describe the graph of $g(x, y) = \dfrac{1}{2}x + \dfrac{1}{3}y - 1$.

Solution Notice first that g is a linear function of x and y. Figure 16 shows a portion of the graph, which is a plane.

We can get a good idea of what plane this is by looking at the x-, y-, and z-intercepts:

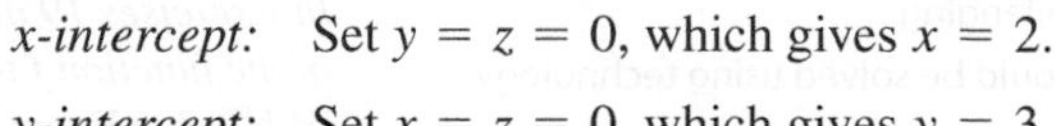

x-intercept: Set $y = z = 0$, which gives $x = 2$.

y-intercept: Set $x = z = 0$, which gives $y = 3$.

z-intercept: Set $x = y = 0$, which gives $z = -1$.

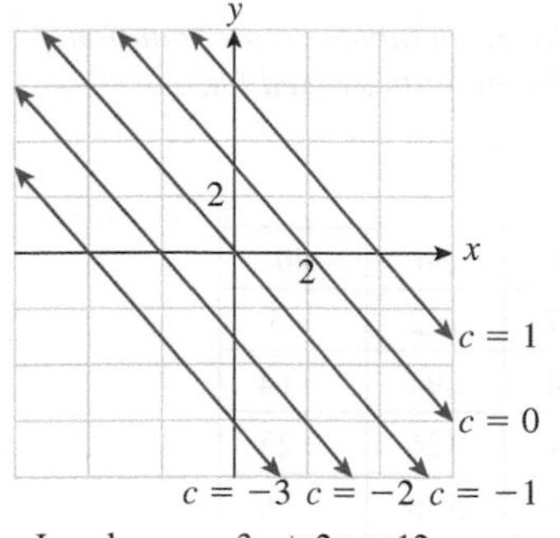

Level curves: $3x + 2y = 12$

Figure 17

Three points are enough to define a plane, so we can say that the plane is the one passing through the three points $(2, 0, 0)$, $(0, 3, 0)$, and $(0, 0, -1)$. It can be shown that the graph of every linear function of two variables is a plane.

Level curves: Set $g(x, y) = c$ to obtain $\frac{1}{2}x + \frac{1}{3}y - 1 = c$, or $\frac{1}{2}x + \frac{1}{3}y = c + 1$. We can rewrite this equation as $3x + 2y = 6(c + 1)$, which is the equation of a straight line. Choosing different values of c gives us a family of parallel lines as shown in Figure 17. (For example, the line corresponding to $c = 1$ has equation $3x + 2y = 6(1 + 1) = 12$.) In general, the set of level curves of every nonconstant linear function is a set of parallel straight lines.*

*Think about what happens when the function is constant.

EXAMPLE 8 Using Level Curves

A certain function f of two variables has level curves $f(x, y) = c$ for $c = -2, -1, 0, 1$, and 2, as shown in Figure 18. (Each grid square is 1×1.)

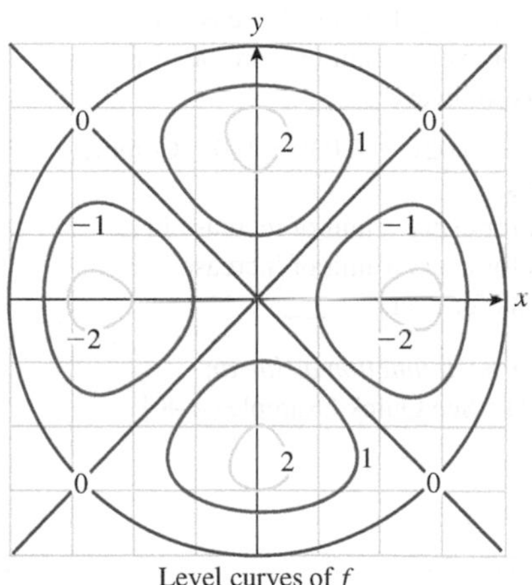

Level curves of f

Figure 18

Estimate the following: $f(1, 1)$, $f(1.5, -2)$, $f(1.5, 0)$ and $f(1, 2)$.

Solution The point $(1, 1)$ appears to lie exactly on the red level curve $c = 0$, so $f(1, 1) \approx 0$. Similarly, the point $(1.5, -2)$ appears to lie exactly on the blue level curve $c = 1$, so $f(1.5, -2) \approx 1$. The point $(1.5, 0)$ appears to lie midway between the level curves $c = -1$ and $c = -2$, so we estimate $f(1.5, 0) \approx -1.5$. Finally, the point $(1, 2)$ lies between the level curves $c = 1$ and $c = 2$ but closer to $c = 1$, so we can estimate $f(1, 2)$ at around 1.3.

8.1 EXERCISES

▼ more advanced ◆ challenging
T indicates exercises that should be solved using technology

*For each function in Exercises 1–4, evaluate **(a)** $f(0, 0)$; **(b)** $f(1, 0)$; **(c)** $f(0, -1)$; **(d)** $f(a, 2)$; **(e)** $f(y, x)$; and **(f)** $f(x + h, y + k)$.* [**HINT**: See Quick Examples 1–3.]

1. $f(x, y) = x^2 + y^2 - x + 1$

2. $f(x, y) = x^2 - y - xy + 1$

3. $f(x, y) = 0.2x + 0.1y - 0.01xy$

4. $f(x, y) = 0.4x - 0.5y - 0.05xy$

*For each function in Exercises 5–8, evaluate **(a)** $g(0, 0, 0)$; **(b)** $g(1, 0, 0)$; **(c)** $g(0, 1, 0)$; **(d)** $g(z, x, y)$; and **(e)** $g(x + h, y + k, z + l)$, provided that such a value exists.*

5. $g(x, y, z) = e^{x+y+z}$

6. $g(x, y, z) = \ln(x + y + z)$

7. $g(x, y, z) = \dfrac{xyz}{x^2 + y^2 + z^2}$

8. $g(x, y, z) = \dfrac{e^{xyz}}{x + y + z}$

9. Let $f(x, y, z) = 1.5 + 2.3x - 1.4y - 2.5z$. Complete the following sentences. [**HINT**: See Example 1.]
a. f___ by ___ units for every 1 unit of increase in x.
b. f___ by ___ units for every 1 unit of increase in y.
c. ______ by 2.5 units for every ______.

10. Let $g(x, y, z) = 0.01x + 0.02y - 0.03z - 0.05$. Complete the following sentences.
a. g ___ by ___ units for every 1 unit of increase in z.
b. g ___ by ___ units for every 1 unit of increase in x.
c. ______ by 0.02 units for every ______.

In Exercises 11–18, classify each function as linear, interaction, or neither. [**HINT**: See Quick Examples 4–8.]

11. $L(x, y) = 3x - 2y + 6xy - 4y^2$

12. $L(x, y, z) = 3x - 2y + 6xz$

13. $P(x_1, x_2, x_3) = 0.4 + 2x_1 - x_3$

14. $Q(x_1, x_2) = 4x_2 - 0.5x_1 - x_1^2$

15. $f(x, y, z) = \dfrac{x + y - z}{3}$

16. $g(x, y, z) = \dfrac{xz - 3yz + z^2}{4z}$ $(z \neq 0)$

17. $g(x, y, z) = \dfrac{xz - 3yz + z^2y}{4z}$ $(z \neq 0)$

18. $f(x, y) = x + y + xy + x^2y$

In Exercises 19 and 20, use the given tabular representation of the function f to compute the quantities asked for. [**HINT**: See Example 3.]

19.

y ↓ \ x →	10	20	30	40
10	−1	107	162	−3
20	−6	194	294	−14
30	−11	281	426	−25
40	−16	368	558	−36

a. $f(20, 10)$
b. $f(40, 20)$
c. $f(10, 20) - f(20, 10)$

20.

y ↓ \ x →	10	20	30	40
10	162	107	−5	−7
20	294	194	−22	−30
30	426	281	−39	−53
40	558	368	−56	−76

a. $f(10, 30)$
b. $f(20, 10)$
c. $f(10, 40) + f(10, 20)$

T *In Exercises 21 and 22, use a spreadsheet or some other method to complete the given tables.*

21. $P(x, y) = x - 0.3y + 0.45xy$

y ↓ \ x →	10	20	30	40
10				
20				
30				
40				

22. $Q(x, y) = 0.4x + 0.1y - 0.06xy$

y ↓ \ x →	10	20	30	40
10				
20				
30				
40				

23. T ▼ The following statistical table lists some values of the "inverse F distribution" ($\alpha = 0.5$):

$n \rightarrow$ (columns), $d \downarrow$ (rows)

	1	2	3	4	5	6	7	8	9	10
1	161.4	199.5	215.7	224.6	230.2	234.0	236.8	238.9	240.5	241.9
2	18.51	19.00	19.16	19.25	19.30	19.33	19.35	19.37	19.38	19.40
3	10.13	9.552	9.277	9.117	9.013	8.941	8.887	8.845	8.812	8.786
4	7.709	6.944	6.591	6.388	6.256	6.163	6.094	6.041	5.999	5.964
5	6.608	5.786	5.409	5.192	5.050	4.950	4.876	4.818	4.772	4.735
6	5.987	5.143	4.757	4.534	4.387	4.284	4.207	4.147	4.099	4.060
7	5.591	4.737	4.347	4.120	3.972	3.866	3.787	3.726	3.677	3.637
8	5.318	4.459	4.066	3.838	3.687	3.581	3.500	3.438	3.388	3.347
9	5.117	4.256	3.863	3.633	3.482	3.374	3.293	3.230	3.179	3.137
10	4.965	4.103	3.708	3.478	3.326	3.217	3.135	3.072	3.020	2.978

In a spreadsheet you can compute the value of this function at (n, d) by the formula

`=FINV(0.05, n, d)` The 0.05 is the value of alpha (α).

Use a spreadsheet to re-create this table.

24. T ▼ The formula for body mass index $M(w, h)$, if w is given in kilograms and h is given in meters, is

$$M(w, h) = \frac{w}{h^2}.$$ See Example 3.

Use this formula to complete the following table in a spreadsheet:

$w \rightarrow$ (columns), $h \downarrow$ (rows)

	70	80	90	100	110	120	130
1.8							
1.85							
1.9							
1.95							
2							
2.05							
2.1							
2.15							
2.2							
2.25							
2.3							

T *In Exercises 25–28, use either a graphing calculator or a spreadsheet to complete each table. Express all your answers as decimals rounded to four decimal places.*

25.

x	y	$f(x, y) = x^2\sqrt{1 + xy}$
3	1	
1	15	
0.3	0.5	
56	4	

26.

x	y	$f(x, y) = x^2e^y$
0	2	
−1	5	
1.4	2.5	
11	9	

27.

x	y	$f(x, y) = x \ln(x^2 + y^2)$
3	1	
1.4	−1	
e	0	
0	e	

28.

x	y	$f(x, y) = \dfrac{x}{x^2 - y^2}$
−1	2	
0	0.2	
0.4	2.5	
10	0	

29. ▼ Brand Z's annual sales are affected by the sales of related products X and Y as follows: Each \$1 million increase in sales of brand X causes a \$2.1 million decline in sales of brand Z, whereas each \$1 million increase in sales of brand Y results in an increase of \$0.4 million in sales of brand Z. Currently, brands X, Y, and Z are each selling \$6 million per year. Model the sales of brand Z using a linear function.

30. ▼ Brand Z's annual sales are affected by the sales of related products X and Y as follows: Each \$1 million increase in sales of brand X causes a \$2.5 million decline in sales of brand Z, whereas each \$2 million increase in sales of brand Y results in an increase of \$23 million in sales of brand Z. Currently, brands X and Y are each selling \$2 million per year, and brand Z is selling \$62 million per year. Model the sales of brand Z using a linear function.

31. Sketch the cube with vertices $(0, 0, 0)$, $(1, 0, 0)$, $(0, 1, 0)$, $(0, 0, 1)$, $(1, 1, 0)$, $(1, 0, 1)$, $(0, 1, 1)$, and $(1, 1, 1)$. [**HINT:** See Example 4.]

32. Sketch the cube with vertices $(-1, -1, -1)$, $(1, -1, -1)$, $(-1, 1, -1)$, $(-1, -1, 1)$, $(1, 1, -1)$, $(1, -1, 1)$, $(-1, 1, 1)$, and $(1, 1, 1)$. [**HINT**: See Example 4.]

33. Sketch the pyramid with vertices $(1, 1, 0)$, $(1, -1, 0)$, $(-1, 1, 0)$, $(-1, -1, 0)$, and $(0, 0, 2)$.

34. Sketch the solid with vertices $(1, 1, 0)$, $(1, -1, 0)$, $(-1, 1, 0)$, $(-1, -1, 0)$, $(0, 0, -1)$, and $(0, 0, 1)$.

In Exercises 35–40, sketch the given plane.

35. $z = -2$ **36.** $z = 4$

37. $y = 2$ **38.** $y = -3$

39. $x = -3$ **40.** $x = 2$

In Exercises 41–48, match the given equation with one of the graphs below. (If necessary, use technology to render the surfaces.) [**HINT**: See Examples 5, 6, and 7.]

41. $f(x, y) = 1 - 3x + 2y$ **42.** $f(x, y) = 1 - \sqrt{x^2 + y^2}$

43. $f(x, y) = 1 - (x^2 + y^2)$ **44.** $f(x, y) = y^2 - x^2$

45. $f(x, y) = -\sqrt{1 - (x^2 + y^2)}$

46. $f(x, y) = 1 + (x^2 + y^2)$

47. $f(x, y) = \dfrac{1}{x^2 + y^2}$ **48.** $f(x, y) = 3x - 2y + 1$

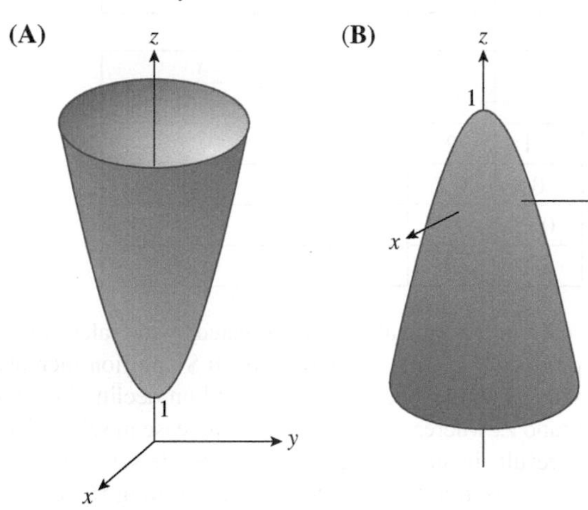

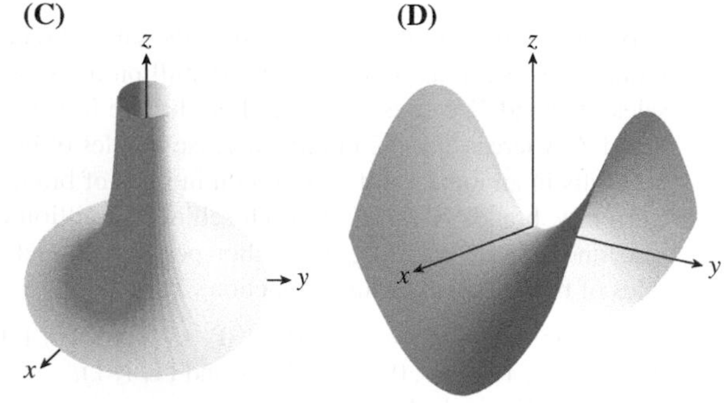

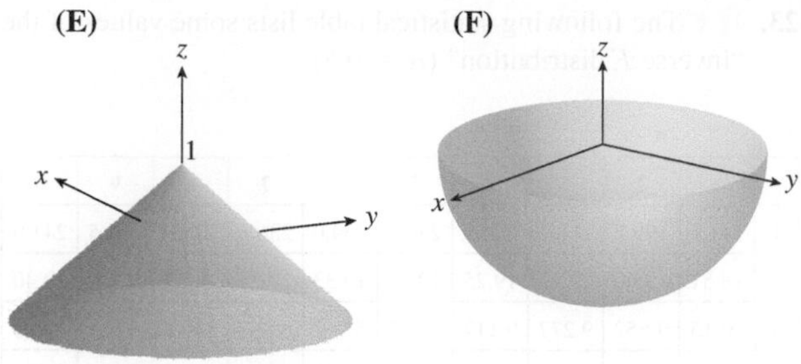

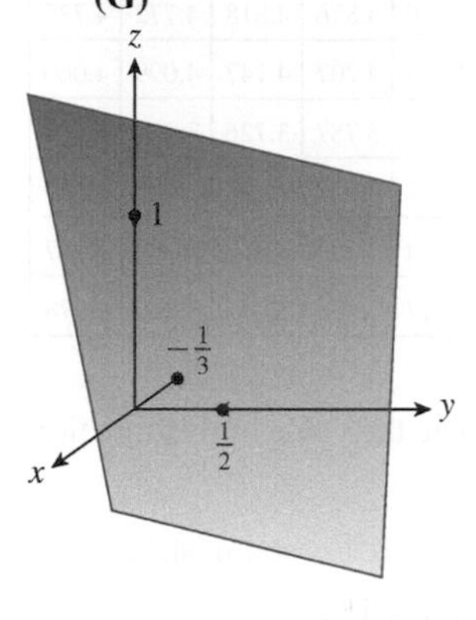

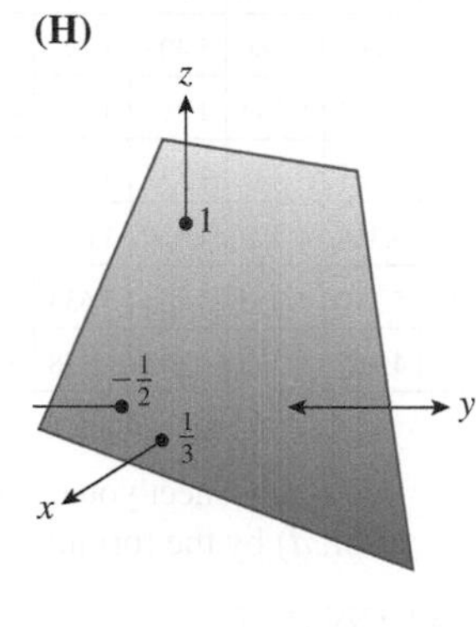

In Exercises 49–54, sketch the level curves $f(x, y) = c$ for the given function and values of c. [**HINT**: See Example 5.]

49. $f(x, y) = 2x^2 + 2y^2$; $c = 0, 2, 18$

50. $f(x, y) = 3x^2 + 3y^2$; $c = 0, 3, 27$

51. $f(x, y) = y + 2x^2$; $c = -2, 0, 2$

52. $f(x, y) = 2y - x^2$; $c = -2, 0, 2$

53. $f(x, y) = 2xy - 1$; $c = -1, 0, 1$

54. $f(x, y) = 2 + xy$; $c = -2, 0, 2$

Exercises 55–58 refer to the following plot of some level curves of $f(x, y) = c$ for $c = -2, 0, 2, 4,$ and 6. (Each grid square is 1 unit × 1 unit.) [**HINT**: See Example 8.]

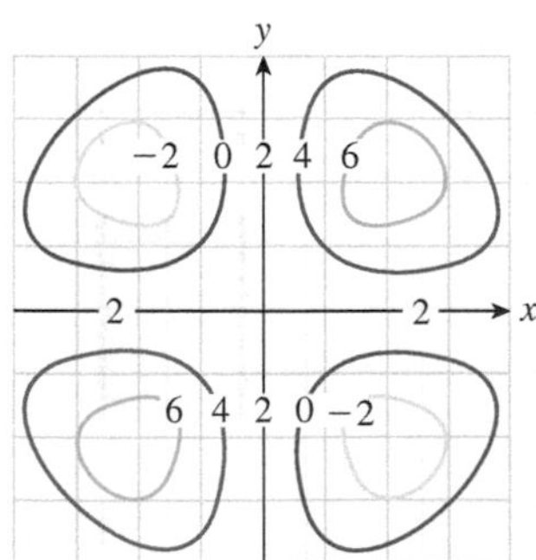

55. Estimate: **a.** $f(1, 1)$ **b.** $f(-2, -1)$ **c.** $f(3, -2.5)$

56. Estimate: **a.** $f(0, 1)$ **b.** $f(-1, -0.5)$ **c.** $f(-2, 1)$

57. At approximately which point or points does f appear to attain a maximum value?

58. At approximately which point or points does f appear to attain a minimum value?

In Exercises 59–74, sketch the graph of the function.
[**HINT:** See Example 7.]

59. $f(x, y) = 1 - x - y$
60. $f(x, y) = x + y - 2$
61. $g(x, y) = 2x + y - 2$
62. $g(x, y) = 3 - x + 2y$
63. $h(x, y) = x + 2$
64. $h(x, y) = 3 - y$

T *For Exercises 65–74, we suggest the use of technology.*
[**HINT:** See Example 6.]

65. $s(x, y) = 2x^2 + 2y^2$. Show cross sections at $z = 1$ and $z = 2$.
66. $s(x, y) = -(x^2 + y^2)$. Show cross sections at $z = -1$ and $z = -2$.
67. $f(x, y) = 2 + \sqrt{x^2 + y^2}$. Show cross sections at $z = 3$ and $y = 0$.
68. $f(x, y) = 2 - \sqrt{x^2 + y^2}$. Show cross sections at $z = 0$ and $y = 0$.
69. $f(x, y) = y^2$
70. $g(x, y) = x^2$
71. $h(x, y) = \frac{1}{y}$
72. $k(x, y) = e^y$
73. $f(x, y) = e^{-(x^2+y^2)}$
74. $g(x, y) = \frac{1}{\sqrt{x^2 + y^2}}$

Applications

75. ***Cost*** Your weekly cost (in dollars) to manufacture x cars and y trucks is

$$C(x, y) = 240{,}000 + 6{,}000x + 4{,}000y.$$

a. What is the marginal cost of a car? of a truck?
[**HINT:** See Example 1.]
b. Describe the graph of the cost function C.
[**HINT:** See Example 7.]
c. Describe the slice $x = 10$. What cost function does this slice describe?
d. Describe the level curve $z = 480{,}000$. What does this curve tell you about costs?

76. ***Cost*** Your weekly cost (in dollars) to manufacture x bicycles and y tricycles is

$$C(x, y) = 24{,}000 + 60x + 20y.$$

a. What is the marginal cost of a bicycle? of a tricycle?
[**HINT:** See Example 1.]
b. Describe the graph of the cost function C.
[**HINT:** See Example 7.]
c. Describe the slice by $y = 100$. What cost function does this slice describe?
d. Describe the level curve $z = 72{,}000$. What does this curve tell you about costs?

77. ***Cost*** Your sales of online video and audio clips are booming. Your Internet provider, Moneydrain.com, wants to get in on the action and has offered you unlimited technical assistance and consulting if you agree to pay Moneydrain 3¢ for every video clip and 4¢ for every audio clip you sell on the site. Further, Moneydrain agrees to charge you only \$10 per month to host your site. Set up a (monthly) cost function for the scenario, and describe each variable.

78. ***Cost*** Your Cabaret nightspot "Jazz on Jupiter" has become an expensive proposition: You are paying monthly costs of \$50,000 just to keep the place running. On top of that, your regular cabaret artist is charging you \$3,000 per performance, and your jazz ensemble is charging \$1,000 per hour. Set up a (monthly) cost function for the scenario, and describe each variable.

79. ***Scientific Research*** In each year from 1983 to 2003 the percentage y of research articles in *Physical Review* written by researchers in the United States can be approximated by

$$y = 82 - 0.78t - 1.02x \text{ percentage points} \quad (0 \le t \le 20),$$

where t is the year since 1983 and x is the percentage of articles written by researchers in Europe.[2]

a. In 2003, researchers in Europe wrote 38% of the articles published by the journal that year. What percentage was written by researchers in the United States?
b. In 1983, researchers in the United States wrote 61% of the articles published that year. What percentage was written by researchers in Europe?
c. What are the units of measurement of the coefficient of t?

80. ***Scientific Research*** The number z of research articles in *Physical Review* that were written by researchers in the United States from 1993 through 2003 can be approximated by

$$z = 5{,}960 - 0.71x + 0.50y \quad (3{,}000 \le x, y \le 6{,}000)$$

articles each year, where x is the number of articles written by researchers in Europe and y is the number written by researchers in other countries (excluding Europe and the United States).[3]

a. In 2000, approximately 5,500 articles were written by researchers in Europe, and 4,500 were written by researchers in other countries. How many (to the nearest 100) were written by researchers in the United States?
b. According to the model, if 5,000 articles were written in Europe and an equal number were written by researchers in the United States and other countries, what would that number be?
c. What is the significance of the fact that the coefficient of x is negative?

[2] Based on a linear regression. Source for data: The American Physical Society/*New York Times*, May 3, 2003, p. A1.

[3] *Ibid.*

81. ***Market Share in the 1990s: Chrysler, Ford, General Motors*** In the late 1990s the relationship between the domestic market shares of three major U.S. manufacturers of cars and light trucks could be modeled by

$$x_3 = 0.66 - 2.2x_1 - 0.02x_2,$$

where x_1, x_2, and x_3 are the fractions of the market held by Chrysler, Ford, and General Motors, respectively.[4] Thinking of General Motors' market share as a function of the shares of the other two manufacturers, describe the graph of the resulting function. How are the different slices by $x_1 = constant$ related to one another? What does this say about market share?

82. ***Market Share in the 1990s: Kellogg, General Mills, General Foods*** In the late 1990s the relationship among the domestic market shares of three major manufacturers of breakfast cereal was

$$x_1 = -0.4 + 1.2x_2 + 2x_3,$$

where x_1, x_2, and x_3 are the fractions of the market held by Kellogg, General Mills, and General Foods, respectively.[5] Thinking of Kellogg's market share as a function of the shares of the other two manufacturers, describe the graph of the resulting function. How are the different slices by $x_2 = constant$ related to one another? What does this say about market share?

83. ***Prison Population*** The number of prisoners in federal prisons in the United States can be approximated by

$$N(x, y) = 134 - 0.11x - 0.26y + 0.0004xy \text{ thousand inmates,}$$

where x is the number, in thousands, in state prisons and y is the number, in thousands, in local jails.[6]

a. In 2011 there were approximately 1.29 million prisoners in state prisons and 736 thousand in local jails. Estimate, to the nearest thousand, the number of prisoners in federal prisons that year.

b. Obtain N as a function of x for $y = 300$ and again for $y = 500$. Interpret the slopes of the resulting linear functions.

84. ***Prison Population*** The number of prisoners in state prisons in the United States can be approximated by

$$N(x, y) = -540 + 7.5x + 2.5y - 0.01xy \text{ thousand inmates,}$$

where x is the number, in thousands, in federal prisons and y is the number, in thousands, in local jails.[7]

a. In 2010 there were approximately 198 thousand prisoners in federal prisons and 749 thousand in local jails. Estimate, to the nearest 0.1 million, the number of prisoners in state prisons that year.

b. Obtain N as a function of y for $x = 80$ and again for $x = 100$. Interpret the slopes of the resulting linear functions.

85. ***Marginal Cost (Interaction Model)*** Your weekly cost (in dollars) to manufacture x cars and y trucks is

$$C(x, y) = 240{,}000 + 6{,}000x + 4{,}000y - 20xy.$$

(Compare with Exercise 75.)

a. Describe the slices $x = constant$ and $y = constant$.

b. Is the graph of the cost function a plane? How does your answer relate to part (a)?

c. What are the slopes of the slices $x = 10$ and $x = 20$? What does this say about cost?

86. ***Marginal Cost (Interaction Model)*** Repeat Exercise 85 using the weekly cost to manufacture x bicycles and y tricycles given by

$$C(x, y) = 24{,}000 + 60x + 20y + 0.3xy$$

(Compare with Exercise 76.)

87. ▼ ***Online Revenue*** Your major online bookstore is in direct competition with Amazon.com, BN.com, and BooksAMillion.com. Your company's daily revenue in dollars is given by

$$R(x, y, z) = 10{,}000 - 0.01x - 0.02y - 0.01z + 0.00001yz,$$

where x, y, and z are the online daily revenues of Amazon.com, BN.com, and BooksAMillion.com, respectively.

a. If, on a certain day, Amazon.com shows revenue of \$12,000, while BN.com and BooksAMillion.com each show \$5,000, what does the model predict for your company's revenue that day?

b. If Amazon.com and BN.com each show daily revenue of \$5,000, give an equation showing how your daily revenue depends on that of BooksAMillion.com.

88. ▼ ***Online Revenue*** Repeat Exercise 87 using the revised revenue function

$$R(x, y, z) = 20{,}000 - 0.02x - 0.04y - 0.01z + 0.00001yz.$$

89. ▼ ***Sales: Walmart, Target*** The following table shows the approximate net earnings, in billions of dollars, of Walmart and Target in 2008, 2010, and 2014:[8]

	2008	2010	2014
Walmart	370	420	470
Target	62	68	73

[4] Based on a linear regression. Source of data: Ward's AutoInfoBank/*New York Times*, July 29, 1998, p. D6.

[5] Based on a linear regression. Source of data: Bloomberg Financial Markets/*New York Times*, November 28, 1998, p. C1.

[6] Source for data: Sourcebook of Criminal Justice Statistics Online (www.albany.edu/sourcebook/pdf/t6132011.pdf).

[7] *Ibid.*

[8] Sources: http://walmartstores.com/Investors, http://investors.target.com, www.wikinvest.com.

Model Walmart's net earnings as a function of Target's net earnings and time, using a linear function of the form

$$f(x, t) = Ax + Bt + C \quad (A, B, C \text{ constants}),$$

where f is Walmart's net earnings (in billions of dollars), x is Target's net earnings (in billions of dollars), and t is time in years since 2008. In 2012, Target's net earnings were about \$72 billion. What, to the nearest billion dollars, does your model estimate as Walmart's net earnings that year?

90. ▼ ***Sales: Nintendo, Nokia*** The following table shows the approximate net sales of **Nintendo** (in billions of yen) and **Nokia** (in billions of euro) in 2004, 2008, and 2010:[9]

	2004	2008	2010
Nintendo	510	1,700	1,010
Nokia	30	52	42

Model Nintendo's net earnings as a function of Nokia's net earnings and time, using a linear function of the form

$$f(x, t) = Ax + Bt + C \quad (A, B, C \text{ constants}),$$

where f is Nintendo's net earnings (in billions of yen), x is Nokia's net earnings (in billions of euro), and t is time in years since 2004. In 2007, Nokia's net earnings were about €50 billion. What, to the nearest billion yen, does your model estimate as Nintendo's net earnings that year?

91. ▼ ***Utility*** Suppose your newspaper is trying to decide between two competing desktop publishing software packages: Macro Publish and Turbo Publish. You estimate that if you purchase x copies of Macro Publish and y copies of Turbo Publish, your company's daily productivity will be

$$U(x, y) = 6x^{0.8}y^{0.2} + x,$$

where $U(x, y)$ is measured in pages per day. (U is called a *utility function.*) If $x = y = 10$, calculate the effect of increasing x by 1 unit, and interpret the result.

92. ▼ ***Housing Costs***[10] The cost C (in dollars) of building a house is related to the number k of carpenters used and the number e of electricians used by

$$C(k, e) = 15{,}000 + 50k^2 + 60e^2.$$

If $k = e = 10$, compare the effects of increasing k by 1 unit and of increasing e by 1 unit. Interpret the result.

[9] Sources: www.nintendo.com/corp, http://investors.nokia.com, www.wikinvest.com.

[10] Based on an exercise in *Introduction to Mathematical Economics* by A. L. Ostrosky Jr. and J. V. Koch (Waveland Press, Illinois, 1979).

93. ▼ ***Volume*** The volume of an ellipsoid with cross-sectional radii a, b, and c is $V(a, b, c) = \frac{4}{3}\pi abc$.

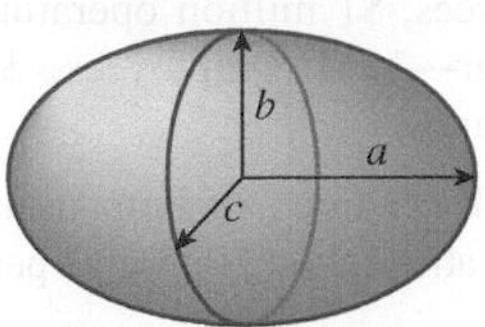

a. Find at least two sets of values for a, b, and c such that $V(a, b, c) = 1$.

b. Find the value of a such that $V(a, a, a) = 1$, and describe the resulting ellipsoid.

94. ▼ ***Volume*** The volume of a right elliptical cone with height h and radii a and b of its base is $V(a, b, h) = \frac{1}{3}\pi abh$.

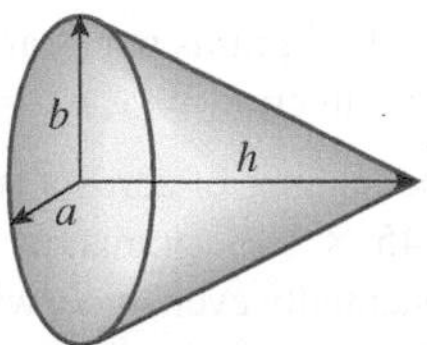

a. Find at least two sets of values for a, b, and h such that $V(a, b, h) = 1$.

b. Find the value of a such that $V(a, a, a) = 1$, and describe the resulting cone.

Exercises 95–98 involve Cobb-Douglas productivity functions. These functions have the form

$$P(x, y) = Kx^a y^{1-a},$$

where P stands for the number of items produced per year, x is the number of employees, and y is the annual operating budget. (The numbers K and a are constants that depend on the situation we are looking at, with $0 \le a \le 1$.*)*

95. ***Productivity*** How many items will be produced per year by a company with 100 employees and an annual operating budget of \$500,000 if $K = 1{,}000$ and $a = 0.5$? (Round your answer to one significant digit.)

96. ***Productivity*** How many items will be produced per year by a company with 50 employees and an annual operating budget of \$1,000,000 if $K = 1{,}000$ and $a = 0.5$? (Round your answer to one significant digit.)

97. ▼ ***Modeling Production with Cobb-Douglas*** Two years ago, my piano manufacturing plant employed 1,000 workers, had an operating budget of \$1 million, and turned out 100 pianos. Last year, I slashed the operating budget to \$10,000, and production dropped to 10 pianos.

a. Use the data for each of the two years and the Cobb-Douglas formula to obtain two equations in K and a.

b. Take logs of both sides in each equation, and obtain two linear equations in a and $\log K$.

c. Solve these equations to obtain values for a and K.

d. Use these values in the Cobb-Douglas formula to predict production if I increase the operating budget back to \$1 million but lay off half the workforce.

98. ▼ ***Modeling Production with Cobb-Douglas*** Repeat Exercise 97 using the following data: Two years ago—1,000 employees, \$1 million operating budget, 100 pianos; last year—1,000 employees, \$100,000 operating budget, 10 pianos.

99. ▼ ***Pollution*** The burden of human-made aerosol sulfate in the Earth's atmosphere, in grams per square meter, is

$$B(x, n) = \frac{xn}{A},$$

where x is the total weight of aerosol sulfate emitted into the atmosphere per year and n is the number of years it remains in the atmosphere. A is the surface area of the Earth, approximately 5.1×10^{14} square meters.[11]

a. Calculate the burden, given the 1995 estimated values of $x = 1.5 \times 10^{14}$ grams per year and $n = 5$ days.

b. What does the function $W(x, n) = xn$ measure?

100. ▼ ***Pollution*** The amount of aerosol sulfate (in grams) was approximately 45×10^{12} grams in 1940 and has been increasing exponentially ever since, with a doubling time of approximately 20 years.[12] Use the model from Exercise 99 to give a formula for the atmospheric burden of aerosol sulfate as a function of the time t in years since 1940 and the number of years n it remains in the atmosphere.

101. ▼ ***Alien Intelligence*** Frank Drake, an astronomer at the University of California at Santa Cruz, devised the following equation to estimate the number of planet-based civilizations in our Milky Way galaxy that are willing and able to communicate with Earth:[13]

$$N(R, f_p, n_e, f_l, f_i, f_c, L) = Rf_pn_ef_lf_if_cL$$

R = the number of new stars formed in our galaxy each year

f_p = the fraction of those stars that have planetary systems

n_e = the average number of planets in each such system that can support life

f_l = the fraction of such planets on which life actually evolves

f_i = the fraction of life-sustaining planets on which intelligent life evolves

f_c = the fraction of intelligent-life-bearing planets on which the intelligent beings develop the means and the will to communicate over interstellar distances

L = the average lifetime of such technological civilizations (in years).

a. What would be the effect on N if any one of the variables were doubled?

b. How would you modify the formula if you were interested only in the number of intelligent-life-bearing planets in the galaxy?

c. How could one convert this function into a linear function?

d. (For discussion) Try to come up with an estimate of N.

102. ▼ ***More Alien Intelligence*** The formula given in Exercise 101 restricts attention to planet-based civilizations in our galaxy. Give a formula that includes intelligent planet-based aliens from the galaxy Andromeda. (Assume that all the variables used in the formula for the Milky Way have the same values for Andromeda.)

Communication and Reasoning Exercises

103. Let $f(x, y) = \frac{x}{y}$. How are $f(x, y)$ and $f(y, x)$ related?

104. Let $f(x, y) = x^2y^3$. How are $f(x, y)$ and $f(-x, -y)$ related?

105. Give an example of a function of the two variables x and y with the property that interchanging x and y has no effect.

106. Give an example of a function f of the two variables x and y with the property that $f(x, y) = -f(y, x)$.

107. Give an example of a function f of the three variables x, y, and z with the property that $f(x, y, z) = f(y, x, z)$ and $f(-x, -y, -z) = -f(x, y, z)$.

108. Give an example of a function f of the three variables x, y, and z with the property that $f(x, y, z) = f(y, x, z)$ and $f(-x, -y, -z) = f(x, y, z)$.

109. Illustrate by means of an example how a real-valued function of the two variables x and y gives different real-valued functions of one variable when we restrict y to be different constants.

110. Illustrate by means of an example how a real-valued function of one variable x gives different real-valued functions of the two variables y and z when we substitute for x suitable functions of y and z.

111. ▼ If f is a linear function of x and y, show that if we restrict y to be a fixed constant, then the resulting function of x is linear. Does the slope of this linear function depend on the choice of y?

112. ▼ If f is an interaction function of x and y, show that if we restrict y to be a fixed constant, then the resulting function of x is linear. Does the slope of this linear function depend on the choice of y?

113. ▼ Suppose that $C(x, y)$ represents the cost of x CDs and y cassettes. If $C(x, y + 1) < C(x + 1, y)$ for every $x \geq 0$ and $y \geq 0$, what does this tell you about the cost of CDs and cassettes?

114. ▼ Suppose that $C(x, y)$ represents the cost of renting x DVDs and y video games. If $C(x + 2, y) < C(x, y + 1)$ for every $x \geq 0$ and $y \geq 0$, what does this tell you about the cost of renting DVDs and video games?

[11] Source: Robert J. Charlson and Tom M. L. Wigley, "Sulfate Aerosol and Climatic Change," *Scientific American*, February, 1994, pp. 48–57.

[12] *Ibid.*

[13] Source: "First Contact" (Plume Books/Penguin Group)/*New York Times*, October 6, 1992, p. C1.

115. Complete the following: The graph of a linear function of two variables is a ____.

116. Complete the following: The level curves of a linear function of two variables are ___.

117. ▼ ***Heat-Seeking Missiles*** The following diagram shows some level curves of the temperature, in degrees Fahrenheit, of a region in space, as well as the location, on the 100-degree curve, of a heat-seeking missile moving through the region. (These level curves are called **isotherms.**) In which of the three directions shown should the missile be traveling so as to experience the fastest rate of increase in temperature at the given point? Explain your answer.

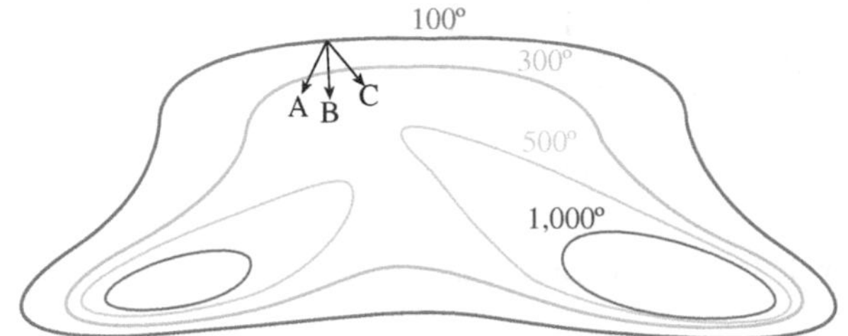

118. ▼ ***Hiking*** The following diagram shows some level curves of the altitude of a mountain valley, as well as the location, on the 2,000-ft curve, of a hiker. The hiker is currently moving at the greatest possible rate of descent. In which of the three directions shown is he moving? Explain your answer.

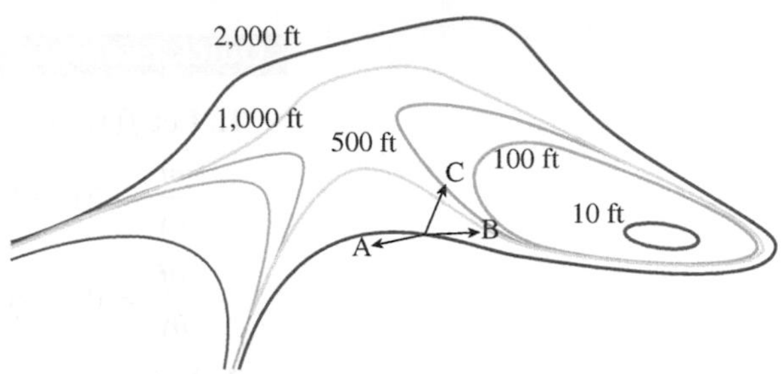

119. Your study partner Slim claims that because the surface $z = f(x, y)$ you have been studying is a plane, it follows that all the slices $x = \mathit{constant}$ and $y = \mathit{constant}$ are straight lines. Do you agree or disagree? Explain.

120. Your other study partner Shady just told you that the surface $z = xy$ you have been trying to graph must be a plane because you've already found that the slices $x = \mathit{constant}$ and $y = \mathit{constant}$ are all straight lines. Do you agree or disagree? Explain.

121. Why do we not sketch the graphs of functions of three or more variables?

122. The surface of a mountain can be thought of as the graph of what function?

123. Why is *three*-dimensional space used to represent the graph of a function of *two* variables?

124. Why is it that we can sketch the graphs of functions of two variables on the two-dimensional flat surfaces of these pages?

8.2 Partial Derivatives

Calculating and Interpreting Partial Derivatives

Recall that if f is a function of x, then the derivative df/dx measures how fast f changes as x increases. If f is a function of two or more variables, we can ask how fast f changes as each variable increases while the others remain fixed. These rates of change are called the "partial derivatives of f," and they measure how each variable contributes to the change in f. Here is a more precise definition.

Partial Derivatives

The **partial derivative of f with respect to x** is the derivative of f with respect to x, when all other variables are treated as constant. Similarly, the **partial derivative of f with respect to y** is the derivative of f with respect to y, with all other variables treated as constant, and so on for other variables. The partial derivatives are written as $\frac{\partial f}{\partial x}, \frac{\partial f}{\partial y}$, and so on. The symbol ∂ is used (instead of d) to remind us that there is more than one variable and that we are holding the other variables fixed.

Quick Examples

1. Let $f(x, y) = x^2 + y^2$.

$$\frac{\partial f}{\partial x} = 2x + 0 = 2x \qquad \text{Because } y^2 \text{ is treated as a constant}$$

$$\frac{\partial f}{\partial y} = 0 + 2y = 2y \qquad \text{Because } x^2 \text{ is treated as a constant}$$

2. Let $z = x^2 + xy$.

$$\frac{\partial z}{\partial x} = 2x + y \qquad \frac{\partial}{\partial x}(xy) = \frac{\partial}{\partial x}(x \cdot \text{constant}) = \text{constant} = y$$

$$\frac{\partial z}{\partial y} = 0 + x \qquad \frac{\partial}{\partial y}(xy) = \frac{\partial}{\partial y}(\text{constant} \cdot y) = \text{constant} = x$$

3. Let $f(x, y) = x^2y + y^2x - xy + y$.

$$\frac{\partial f}{\partial x} = 2xy + y^2 - y \qquad y \text{ is treated as a constant.}$$

$$\frac{\partial f}{\partial y} = x^2 + 2xy - x + 1 \qquad x \text{ is treated as a constant.}$$

Interpretation

$\dfrac{\partial f}{\partial x}$ is the rate at which f changes as x changes, for a fixed (constant) y.

$\dfrac{\partial f}{\partial y}$ is the rate at which f changes as y changes, for a fixed (constant) x.

EXAMPLE 1 Marginal Cost: Linear Model

We return to Example 1 from Section 8.1. Suppose that you own a company that makes two models of speakers: the Ultra Mini and the Big Stack. Your total monthly cost (in dollars) to make x Ultra Minis and y Big Stacks is given by

$$C(x, y) = 10{,}000 + 20x + 40y.$$

What is the significance of $\dfrac{\partial C}{\partial x}$ and of $\dfrac{\partial C}{\partial y}$?

Solution First, we compute these partial derivatives:

$$\frac{\partial C}{\partial x} = 20$$

$$\frac{\partial C}{\partial y} = 40.$$

We interpret the results as follows: $\dfrac{\partial C}{\partial x} = 20$ means that the cost is increasing at a rate of \$20 per additional Ultra Mini (if production of Big Stacks is held constant), and $\frac{\partial C}{\partial y} = 40$ means that the cost is increasing at a rate of \$40 per additional Big Stack (if production of Ultra Minis is held constant). In other words, these are the **marginal costs** of each model of speaker.

➡ **Before we go on . . .** How much does the cost rise if you increase x by Δx and y by Δy? In Example 1 the change in cost is given by

$$\Delta C = 20\,\Delta x + 40\,\Delta y = \frac{\partial C}{\partial x}\Delta x + \frac{\partial C}{\partial y}\Delta y.$$

This suggests the **chain rule for several variables**. Part of this rule says that if x and y are both functions of t, then C is a function of t through them, and the rate of change of C with respect to t can be calculated as

$$\frac{dC}{dt} = \frac{\partial C}{\partial x}\cdot\frac{dx}{dt} + \frac{\partial C}{\partial y}\cdot\frac{dy}{dt}.$$

See the optional section on the chain rule for several variables for further discussion and applications of this interesting result. ■

EXAMPLE 2 Marginal Cost: Interaction Model

Another possibility for the cost function in Example 1 is an interaction model:

$$C(x, y) = 10{,}000 + 20x + 40y + 0.1xy.$$

a. *Now* what are the marginal costs of the two models of speakers?

b. What is the marginal cost of manufacturing Big Stacks at a production level of 100 Ultra Minis and 50 Big Stacks per month?

Solution

a. We compute the partial derivatives:

$$\frac{\partial C}{\partial x} = 20 + 0.1y$$

$$\frac{\partial C}{\partial y} = 40 + 0.1x.$$

Thus, the marginal cost of manufacturing Ultra Minis increases by \$0.1, or 10¢, for each Big Stack that is manufactured. Similarly, the marginal cost of manufacturing Big Stacks increases by 10¢ for each Ultra Mini that is manufactured.

b. From part (a) the marginal cost of manufacturing Big Stacks is

$$\frac{\partial C}{\partial y} = 40 + 0.1x.$$

At a production level of 100 Ultra Minis and 50 Big Stacks per month, we have $x = 100$ and $y = 50$. Thus, the marginal cost of manufacturing Big Stacks at these production levels is

$$\left.\frac{\partial C}{\partial y}\right|_{(100,50)} = 40 + 0.1(100) = \$50 \text{ per Big Stack}.$$

Partial derivatives of functions of three variables are obtained in the same way as those for functions of two variables, as the following example shows.

EXAMPLE 3 Function of Three Variables

Calculate $\dfrac{\partial f}{\partial x}$, $\dfrac{\partial f}{\partial y}$, and $\dfrac{\partial f}{\partial z}$ if $f(x, y, z) = xy^2z^3 - xy$.

Solution Although we now have three variables, the calculation remains the same: $\partial f/\partial x$ is the derivative of f with respect to x, with *both* other variables, y and z, held constant:

$$\frac{\partial f}{\partial x} = y^2z^3 - y.$$

Similarly, $\partial f/\partial y$ is the derivative of f with respect to y, with both x and z held constant:

$$\frac{\partial f}{\partial y} = 2xyz^3 - x.$$

Finally, to find $\partial f/\partial z$, we hold both x and y constant and take the derivative with respect to z.

$$\frac{\partial f}{\partial z} = 3xy^2z^2.$$

Note The procedure for finding a partial derivative is the same for any number of variables: To get the partial derivative with respect to any one variable, we treat all the others as constants. ■

Geometric Interpretation of Partial Derivatives

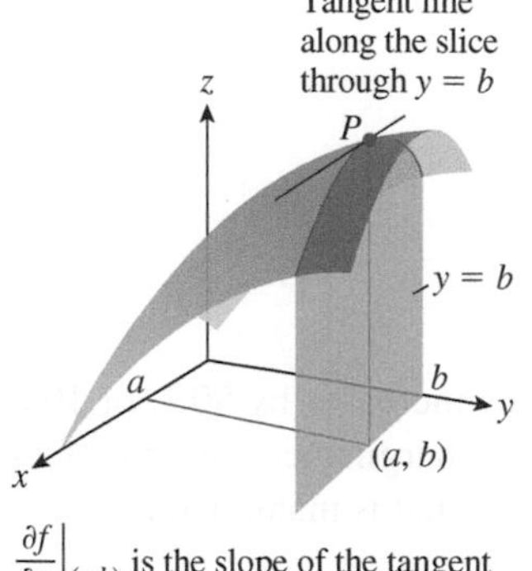

$\left.\dfrac{\partial f}{\partial x}\right|_{(a,b)}$ is the slope of the tangent line at the point $P(a, b, f(a, b))$ along the slice through $y = b$.

Figure 19

Recall that if f is a function of one variable x, then the derivative df/dx gives the slopes of the tangent lines to its graph. Now, suppose that f is a function of x and y. By definition, $\partial f/\partial x$ is the derivative of the function of x that we get by holding y fixed. If we evaluate this derivative at the point (a, b), we are holding y fixed at the value b, taking the ordinary derivative of the resulting function of x, and evaluating this at $x = a$. Now, holding y fixed at b amounts to slicing through the graph of f along the plane $y = b$, resulting in a curve. Thus, the partial derivative is the slope of the tangent line to this curve at the point where $x = a$ and $y = b$, along the plane $y = b$ (Figure 19). This fits with our interpretation of $\partial f/\partial x$ as the rate of increase of f with increasing x when y is held fixed at b.

The other partial derivative, $\partial f/\partial y|_{(a,b)}$, is, similarly, the slope of the tangent line at the same point $P(a, b, f(a, b))$ but along the slice by the plane $x = a$. You should draw the corresponding picture for this on your own.

EXAMPLE 4 Marginal Cost

Referring to the interactive cost function $C(x, y) = 10{,}000 + 20x + 40y + 0.1xy$ in Example 2, we can identify the marginal costs $\partial C/\partial x$ and $\partial C/\partial y$ of manufacturing Ultra Minis and Big Stacks at a production level of 100 Ultra Minis and 50 Big Stacks per month as the slopes of the tangent lines to the two slices by $y = 50$ and $x = 100$ at the point on the graph where $(x, y) = (100, 50)$ as seen in Figure 20.

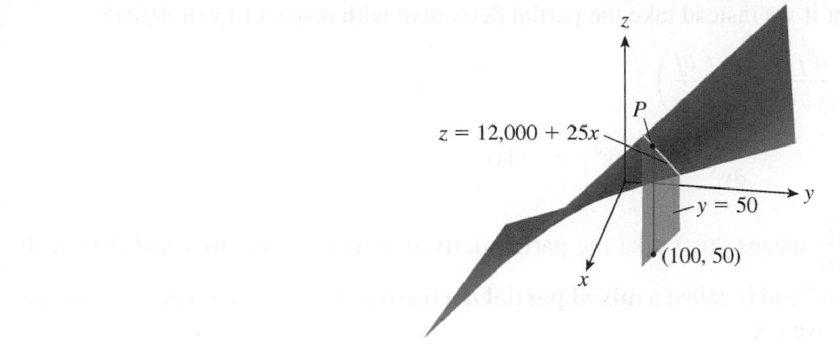

Figure 20(a)

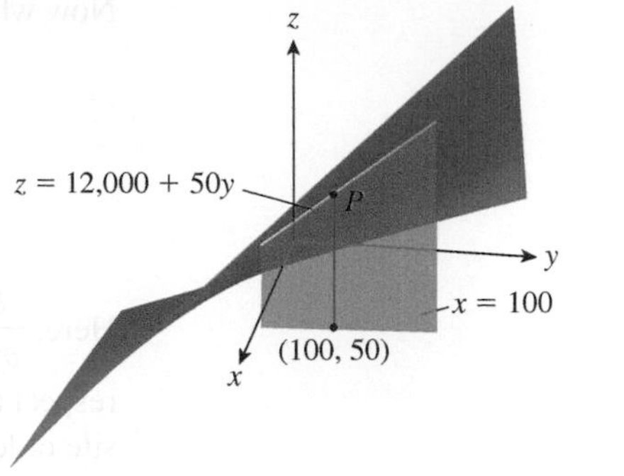

Figure 20(b)

Figure 20(a) shows the slice at $y = 50$ through the point $P = (100, 50, C(100, 50)) = (100, 50, 14{,}500)$. The equation of that slice is given by substituting $y = 50$ in the cost equation:

$$C(x, 50) = 10{,}000 + 20x + 40(50) + 0.1x(50) = 12{,}000 + 25x. \qquad \text{A line of slope 25}$$

Because the slice is already a line, it coincides with the tangent line through P as depicted in Figure 19. This slope is equal to $\partial C/\partial x|_{(100,50)}$:

$$\frac{\partial C}{\partial x} = 20 + 0.1y \qquad \text{See Example 2.}$$

so $$\left.\frac{\partial C}{\partial x}\right|_{(100,50)} = 20 + 0.1(50) = 25.$$

Similarly, Figure 20(b) shows the slice at $x = 100$ through the same point P. The equation of that slice is given by substituting $x = 100$ in the cost equation:

$$C(100, y) = 10{,}000 + 20(100) + 40y + 0.1(100)y = 12{,}000 + 50y. \qquad \text{A line of slope 50}$$

This slope is equal to $\partial C/\partial y|_{(100,50)} = 50$ as we calculated in Example 2.

Second-Order Partial Derivatives

Just as for functions of a single variable, we can calculate second derivatives. Suppose, for example, that we have a function of x and y, say, $f(x, y) = x^2 - x^2y^2$. We know that

$$\frac{\partial f}{\partial x} = 2x - 2xy^2.$$

If we take the partial derivative with respect to x once again, we obtain

$$\frac{\partial}{\partial x}\left(\frac{\partial f}{\partial x}\right) = 2 - 2y^2. \qquad \text{Take } \frac{\partial}{\partial x} \text{ of } \frac{\partial f}{\partial x}.$$

(The symbol $\partial/\partial x$ means "the partial derivative with respect to x," just as d/dx stands for "the derivative with respect to x.") This is called the **second-order partial derivative** and is written $\frac{\partial^2 f}{\partial x^2}$. We get the following derivatives similarly:

$$\frac{\partial f}{\partial y} = -2x^2y$$

$$\frac{\partial^2 f}{\partial y^2} = -2x^2. \qquad \text{Take } \frac{\partial}{\partial y} \text{ of } \frac{\partial f}{\partial y}.$$

Now what if we instead take the partial derivative with respect to y of $\partial f/\partial x$?

$$\frac{\partial^2 f}{\partial y \partial x} = \frac{\partial}{\partial y}\left(\frac{\partial f}{\partial x}\right) \qquad \text{Take } \frac{\partial}{\partial y} \text{ of } \frac{\partial f}{\partial x}.$$

$$= \frac{\partial}{\partial y}[2x - 2xy^2] = -4xy.$$

Here, $\frac{\partial^2 f}{\partial y \partial x}$ means "first take the partial derivative with respect to x and then with respect to y" and is called a **mixed partial derivative**. If we differentiate in the opposite order, we get

$$\frac{\partial^2 f}{\partial x \partial y} = \frac{\partial}{\partial x}\left(\frac{\partial f}{\partial y}\right) = \frac{\partial}{\partial x}[-2x^2y] = -4xy,$$

the same expression as $\frac{\partial^2 f}{\partial y \partial x}$. This is no coincidence: The mixed partial derivatives $\frac{\partial^2 f}{\partial x \partial y}$ and $\frac{\partial^2 f}{\partial y \partial x}$ are always the same as long as the first partial derivatives are both differentiable functions of x and y and the mixed partial derivatives are continuous. Because all the functions we shall use are of this type, we can take the derivatives in any order we like when calculating mixed derivatives.

Here is another notation for partial derivatives that is especially convenient for second-order partial derivatives:

f_x means $\frac{\partial f}{\partial x}$

f_y means $\frac{\partial f}{\partial y}$

f_{xy} means $(f_x)_y = \frac{\partial^2 f}{\partial y \partial x}$ (Note the order in which the derivatives are taken.)

f_{yx} means $(f_y)_x = \frac{\partial^2 f}{\partial x \partial y}$.

8.2 EXERCISES

▼ more advanced ◆ challenging

T indicates exercises that should be solved using technology

In Exercises 1–18, calculate $\frac{\partial f}{\partial x}, \frac{\partial f}{\partial y}, \left.\frac{\partial f}{\partial x}\right|_{(1,-1)}$, *and* $\left.\frac{\partial f}{\partial y}\right|_{(1,-1)}$ *when defined.* [**HINT**: See Quick Examples 1–3.]

1. $f(x, y) = 10{,}000 - 40x + 20y$
2. $f(x, y) = 1{,}000 + 5x - 4y$
3. $f(x, y) = 3x^2 - y^3 + x - 1$
4. $f(x, y) = x^{1/2} - 2y^4 + y + 6$
5. $f(x, y) = 10{,}000 - 40x + 20y + 10xy$
6. $f(x, y) = 1{,}000 + 5x - 4y - 3xy$
7. $f(x, y) = 3x^2y$
8. $f(x, y) = x^4y^2 - x$
9. $f(x, y) = x^2y^3 - x^3y^2 - xy$
10. $f(x, y) = x^{-1}y^2 + xy^2 + xy$
11. $f(x, y) = (2xy + 1)^3$
12. $f(x, y) = \frac{1}{(xy + 1)^2}$
13. ▼ $f(x, y) = e^{x+y}$
14. ▼ $f(x, y) = e^{2x+y}$
15. ▼ $f(x, y) = 5x^{0.6}y^{0.4}$
16. ▼ $f(x, y) = -2x^{0.1}y^{0.9}$
17. ▼ $f(x, y) = e^{0.2xy}$
18. ▼ $f(x, y) = xe^{xy}$

In Exercises 19–28, find $\frac{\partial^2 f}{\partial x^2}, \frac{\partial^2 f}{\partial y^2}, \frac{\partial^2 f}{\partial x \partial y}$, *and* $\frac{\partial^2 f}{\partial y \partial x}$, *and evaluate them all at* $(1, -1)$ *if possible.* [**HINT**: See the discussion of second-order partial derivatives.]

19. $f(x, y) = 10{,}000 - 40x + 20y$
20. $f(x, y) = 1{,}000 + 5x - 4y$
21. $f(x, y) = 10{,}000 - 40x + 20y + 10xy$

22. $f(x, y) = 1{,}000 + 5x - 4y - 3xy$

23. $f(x, y) = 3x^2y$

24. $f(x, y) = x^4y^2 - x$

25. ▼ $f(x, y) = e^{x+y}$

26. ▼ $f(x, y) = e^{2x+y}$

27. ▼ $f(x, y) = 5x^{0.6}y^{0.4}$

28. ▼ $f(x, y) = -2x^{0.1}y^{0.9}$

In Exercises 29–40, find $\frac{\partial f}{\partial x}, \frac{\partial f}{\partial y}, \frac{\partial f}{\partial z}$, *and their values at* $(0, -1, 1)$ *if possible.* [**HINT:** See Example 3.]

29. $f(x, y, z) = xyz$

30. $f(x, y, z) = xy + xz - yz$

31. ▼ $f(x, y, z) = -\frac{4}{x + y + z^2}$

32. ▼ $f(x, y, z) = \frac{6}{x^2 + y^2 + z^2}$

33. ▼ $f(x, y, z) = xe^{yz} + ye^{xz}$

34. ▼ $f(x, y, z) = xye^z + xe^{yz} + e^{xyz}$

35. ▼ $f(x, y, z) = x^{0.1}y^{0.4}z^{0.5}$

36. ▼ $f(x, y, z) = 2x^{0.2}y^{0.8} + z^2$

37. ▼ $f(x, y, z) = e^{xyz}$

38. ▼ $f(x, y, z) = \ln(x + y + z)$

39. ▼ $f(x, y, z) = \frac{2{,}000z}{1 + y^{0.3}}$

40. ▼ $f(x, y, z) = \frac{e^{0.2x}}{1 + e^{-0.1y}}$

Applications

41. ***Marginal Cost (Linear Model)*** Your weekly cost (in dollars) to manufacture x cars and y trucks is

$$C(x, y) = 240{,}000 + 6{,}000x + 4{,}000y.$$

Calculate and interpret $\frac{\partial C}{\partial x}$ and $\frac{\partial C}{\partial y}$. [**HINT:** See Example 1.]

42. ***Marginal Cost (Linear Model)*** Your weekly cost (in dollars) to manufacture x bicycles and y tricycles is

$$C(x, y) = 24{,}000 + 60x + 20y.$$

Calculate and interpret $\frac{\partial C}{\partial x}$ and $\frac{\partial C}{\partial y}$. [**HINT:** See Example 1.]

43. ***Scientific Research*** In each year from 1983 to 2003 the percentage y of research articles in *Physical Review* written by researchers in the United States can be approximated by

$$y = 82 - 0.78t - 1.02x \text{ percentage points} \quad (0 \le t \le 20),$$

where t is the year since 1983 and x is the percentage of articles written by researchers in Europe.[14] Calculate and interpret $\frac{\partial y}{\partial t}$ and $\frac{\partial y}{\partial x}$.

44. ***Scientific Research*** The number z of research articles in *Physical Review* that were written by researchers in the United States from 1993 through 2003 can be approximated by

$$z = 5{,}960 - 0.71x + 0.50y \quad (3{,}000 \le x, y \le 6{,}000)$$

articles each year, where x is the number of articles written by researchers in Europe and y is the number written by researchers in other countries (excluding Europe and the United States).[15] Calculate and interpret $\frac{\partial z}{\partial x}$ and $\frac{\partial z}{\partial y}$.

45. ***Marginal Cost (Interaction Model)*** Your weekly cost (in dollars) to manufacture x cars and y trucks is

$$C(x, y) = 240{,}000 + 6{,}000x + 4{,}000y - 20xy.$$

(Compare with Exercise 41.) Compute the marginal cost of manufacturing cars at a production level of 10 cars and 20 trucks. [**HINT:** See Example 2.]

46. ***Marginal Cost (Interaction Model)*** Your weekly cost (in dollars) to manufacture x bicycles and y tricycles is

$$C(x, y) = 24{,}000 + 60x + 20y + 0.3xy.$$

(Compare with Exercise 42.) Compute the marginal cost of manufacturing tricycles at a production level of 10 bicycles and 20 tricycles. [**HINT:** See Example 2.]

47. ***Brand Loyalty*** The fraction of **Mazda** car owners who chose another new Mazda can be modeled by the following function:[16]

$$M(c, f, g, h, t) = 1.1 - 3.8c + 2.2f + 1.9g - 1.7h - 1.3t.$$

Here, c is the fraction of **Chrysler** car owners who remained loyal to Chrysler, f is the fraction of **Ford** car owners remaining loyal to Ford, g the corresponding figure for **General Motors**, h the corresponding figure for **Honda**, and t for **Toyota**.

a. Calculate $\frac{\partial M}{\partial c}$ and $\frac{\partial M}{\partial f}$, and interpret the answers.

b. One year it was observed that $c = 0.56$, $f = 0.56$, $g = 0.72$, $h = 0.50$, and $t = 0.43$. According to the model, what percentage of Mazda owners remained loyal to Mazda? (Round your answer to the nearest percentage point.)

48. ***Brand Loyalty*** The fraction of **Mazda** car owners who chose another new Mazda can be modeled by the following function:[17]

$$M(c, f) = 9.4 + 7.8c + 3.6c^2 - 38f - 22cf + 43f^2,$$

[14] Based on a linear regression. Source for data: The American Physical Society/*New York Times*, May 3, 2003, p. A1.

[15] *Ibid.*

[16] The model is an approximation of a linear regression based on data from the period 1988–1995. Source for data: Chrysler, Maritz Market Research, Consumer Attitude Research, and Strategic Vision/*New York Times*, November 3, 1995, p. D2.

[17] The model is an approximation of a second-order regression based on data from the period 1988–1995. Source for data: Chrysler, Maritz Market Research, Consumer Attitude Research, and Strategic Vision/*New York Times*, November 3, 1995, p. D2.

where c is the fraction of **Chrysler** car owners who remained loyal to Chrysler and f is the fraction of **Ford** car owners remaining loyal to Ford.

a. Calculate $\frac{\partial M}{\partial c}$ and $\frac{\partial M}{\partial f}$ evaluated at the point(0.7, 0.7), and interpret the answers.

b. One year it was observed that $c = 0.56$, and $f = 0.56$. According to the model, what percentage of Mazda owners remained loyal to Mazda? (Round your answer to the nearest percentage point.)

49. ***Marginal Cost*** Your weekly cost (in dollars) to manufacture x cars and y trucks is

$$C(x, y) = 200{,}000 + 6{,}000x + 4{,}000y - 100{,}000e^{-0.01(x+y)}.$$

What is the marginal cost of a car? of a truck? How do these marginal costs behave as total production increases?

50. ***Marginal Cost*** Your weekly cost (in dollars) to manufacture x bicycles and y tricycles is

$$C(x, y) = 20{,}000 + 60x + 20y + 50\sqrt{xy}.$$

What is the marginal cost of a bicycle? of a tricycle? How do these marginal costs behave as x and y increase?

51. ▼ ***Income Gap*** The following model is based on data on the median incomes of Hispanic and white households in the United States for the period 2000–2013:[18]

$$z(t, x) = 44{,}200 - 330t + 17{,}500x + 40xt,$$

where

$z(t, x)$ = median household income

t = year ($t = 0$ represents 2000)

$$x = \begin{cases} 0 & \text{if the income was for a Hispanic household} \\ 1 & \text{if the income was for a white household.} \end{cases}$$

a. Use the model to estimate the median income of a Hispanic household and that of a white household in 2010.

b. According to the model, how fast was the median income for a Hispanic household increasing in 2010? How fast was the median income for a white household increasing in 2010?

c. Do the answers in part (b) suggest that the income gap between white and Hispanic households was widening or narrowing during the given period?

d. What does the coefficient of xt in the formula for $z(t, x)$ represent in terms of the income gap?

52. ▼ ***Income Gap*** The following model is based on data on the median incomes of black and white households in the United States for the period 2000–2013:[19]

$$z(t, x) = 39{,}300 - 430t + 22{,}400x + 140xt,$$

where

$z(t, x)$ = median family income

t = year ($t = 0$ represents 2000)

$$x = \begin{cases} 0 & \text{if the income was for a black household} \\ 1 & \text{if the income was for a white household.} \end{cases}$$

a. Use the model to estimate the median income of a black household and that of a white household in 2010.

b. According to the model, how fast was the median income for a black household increasing in 2010? How fast was the median income for a white household increasing in 2010?

c. Do the answers in part (b) suggest that the income gap between white and black households was widening or narrowing during the given period?

d. What does the coefficient of xt in the formula for $z(t, x)$ represent in terms of the income gap?

53. ▼ ***Average Cost*** If you average your costs over your total production, you get the **average cost**, written $\bar{C}$:

$$\bar{C}(x, y) = \frac{C(x, y)}{x + y}.$$

Find the average cost for the cost function in Exercise 49. Then find the marginal average cost of a car and the marginal average cost of a truck at a production level of 50 cars and 50 trucks. Interpret your answers.

54. ▼ ***Average Cost*** Find the average cost for the cost function in Exercise 50. (See Exercise 53.) Then find the marginal average cost of a bicycle and the marginal average cost of a tricycle at a production level of five bicycles and five tricycles. Interpret your answers.

55. ▼ ***Marginal Revenue*** As manager of an auto dealership, you offer a car rental company the following deal: You will charge \$15,000 per car and \$10,000 per truck, but you will then give the company a discount of \$5,000 times the square root of the total number of vehicles it buys from you. Looking at your marginal revenue, is this a good deal for the rental company? Why or why not?

56. ▼ ***Marginal Revenue*** As marketing director for a bicycle manufacturer, you come up with the following scheme: You will offer to sell a dealer x bicycles and y tricycles for

$$R(x, y) = 3{,}500 - 3{,}500e^{-0.02x-0.01y} \text{ dollars.}$$

Find your marginal revenue for bicycles and for tricycles. Are you likely to be fired for your suggestion? Why or why not?

57. ▼ ***Research Productivity*** Here we apply a variant of the Cobb-Douglas function to the modeling of research productivity. A mathematical model of research productivity at a particular physics laboratory is

$$P = 0.04x^{0.4}y^{0.2}z^{0.4},$$

where P is the annual number of groundbreaking research papers produced by the staff, x is the number of physicists

---

[18] Incomes are in 2013 dollars. Source for data: U.S. Census Bureau (www.census.gov).

[19] *Ibid.*

on the research team, y is the laboratory's annual research budget, and z is the annual National Science Foundation subsidy to the laboratory. Find the rate of increase of research papers per government-subsidy dollar at a subsidy level of \$1,000,000 per year and a staff level of 10 physicists if the annual budget is \$100,000.

58. ▼ ***Research Productivity*** A major drug company estimates that the annual number P of patents for new drugs developed by its research team is best modeled by the formula

$$P = 0.3x^{0.3}y^{0.4}z^{0.3},$$

where x is the number of research biochemists on the payroll, y is the annual research budget, and z is the size of the bonus awarded to discoverers of new drugs. Assuming that the company has 12 biochemists on the staff, has an annual research budget of \$500,000, and pays \$40,000 bonuses to developers of new drugs, calculate the rate of growth in the annual number of patents per new research staff member.

59. ▼ ***Utility*** Your newspaper is trying to decide between two competing desktop publishing software packages: Macro Publish and Turbo Publish. You estimate that if you purchase x copies of Macro Publish and y copies of Turbo Publish, your company's daily productivity will be

$$U(x, y) = 6x^{0.8}y^{0.2} + x \text{ pages per day.}$$

a. Calculate $U_x(10, 5)$ and $U_y(10, 5)$ to two decimal places, and interpret the results.

b. What does the ratio $\dfrac{U_x(10, 5)}{U_y(10, 5)}$ tell about the usefulness of these products?

60. ▼ ***Grades***[20] A production formula for a student's performance on a difficult English examination is given by

$$g(t, x) = 4tx - 0.2t^2 - x^2,$$

where g is the grade the student can expect to get, t is the number of hours of study for the examination, and x is the student's grade-point average.

a. Calculate $g_t(10, 3)$ and $g_x(10, 3)$, and interpret the results.

b. What does the ratio $\dfrac{g_t(10, 3)}{g_x(10, 3)}$ tell about the relative merits of study and grade-point average?

61. ▼ ***Electrostatic Repulsion*** If positive electric charges of Q and q coulombs are situated at positions (a, b, c) and (x, y, z), respectively, then the force of repulsion they experience is given by

$$F = K\frac{Qq}{(x - a)^2 + (y - b)^2 + (z - c)^2},$$

where $K \approx 9 \times 10^9$, F is given in newtons, and all positions are measured in meters. Assume that a charge of 10 coulombs is situated at the origin and that a second charge of 5 coulombs is situated at (2, 3, 3) and moving in the y-direction at 1 m/sec. How fast is the electrostatic force it experiences decreasing? (Round the answer to one significant digit.)

62. ▼ ***Electrostatic Repulsion*** Repeat Exercise 61, assuming that a charge of 10 coulombs is situated at the origin and that a second charge of 5 coulombs is situated at (2, 3, 3) and moving in the negative z-direction at 1 m/sec. (Round the answer to one significant digit.)

63. ▼ ***Investments*** Recall that the compound interest formula for annual compounding is

$$A(P, r, t) = P(1 + r)^t,$$

where A is the future value of an investment of P dollars after t years at an interest rate of r.

a. Calculate $\dfrac{\partial A}{\partial P}$, $\dfrac{\partial A}{\partial r}$, and $\dfrac{\partial A}{\partial t}$, all evaluated at (100, 0.10, 10). (Round your answers to two decimal places.) Interpret your answers.

b. What does the function $\left.\dfrac{\partial A}{\partial P}\right|_{(100,\, 0.10,\, t)}$ of t tell about your investment?

64. ▼ ***Investments*** Repeat Exercise 63, using the formula for continuous compounding:

$$A(P, r, t) = Pe^{rt}.$$

65. ▼ ***Modeling with the Cobb-Douglas Production Formula*** Assume that you are given a production formula of the form

$$P(x, y) = Kx^a y^b \quad (a + b = 1).$$

a. Obtain formulas for $\dfrac{\partial P}{\partial x}$ and $\dfrac{\partial P}{\partial y}$, and show that $\dfrac{\partial P}{\partial x} = \dfrac{\partial P}{\partial y}$ precisely when $x/y = a/b$.

b. Let x be the number of workers a firm employs, and let y be its monthly operating budget in thousands of dollars. Assume that the firm currently employs 100 workers and has a monthly operating budget of \$200,000. If each additional worker contributes as much to productivity as each additional \$1,000 per month, find values of a and b that model the firm's productivity.

66. ▼ ***Housing Costs***[21] The cost C of building a house is related to the number k of carpenters used and the number e of electricians used by

$$C(k, e) = 15{,}000 + 50k^2 + 60e^2.$$

If three electricians are currently employed in building your new house and the marginal cost per additional electrician is the same as the marginal cost per additional carpenter, how many carpenters are being used? (Round your answer to the nearest carpenter.)

---

[20] Based on an exercise in *Introduction to Mathematical Economics* by A. L. Ostrosky Jr. and J. V. Koch (Waveland Press, Illinois, 1979).

[21] *Ibid.*

67. ▼ ***Nutrient Diffusion*** Suppose that 1 cubic centimeter of nutrient is placed at the center of a circular petri dish filled with water. We might wonder how the nutrient is distributed after a time of t seconds. According to the classical theory of diffusion, the concentration of nutrient (in parts of nutrient per part of water) after a time t is given by

$$u(r, t) = \frac{1}{4\pi Dt}e^{-r^2/(4Dt)}.$$

Here, D is the *diffusivity*, which we will take to be 1, and r is the distance from the center in centimeters. How fast is the concentration increasing at a distance of 1 centimeter from the center 3 seconds after the nutrient is introduced?

68. ▼ ***Nutrient Diffusion*** Refer to Exercise 67. How fast is the concentration increasing at a distance of 4 centimeters from the center 4 seconds after the nutrient is introduced?

Communication and Reasoning Exercises

69. Given that $f(a, b) = r$, $f_x(a, b) = s$, and $f_y(a, b) = t$, complete the following: _____ is increasing at a rate of _____ units per unit of x, _____ is increasing at a rate of _____ units per unit of y, and the value of _____ is _____ when $x =$ _____ and $y =$ _____.

70. A firm's productivity depends on two variables, x and y. Currently, $x = a$ and $y = b$, and the firm's productivity is 4,000 units. Productivity is increasing at a rate of 400 units per unit *decrease* in x and is decreasing at a rate of 300 units per unit increase in y. What does all of this information tell you about the firm's productivity function $g(x, y)$?

71. Complete the following: Let $f(x, y, z)$ be the cost to build a development of x cypods (one-bedroom units) in the city-state of Utarek on Mars, y argaats (two-bedroom units), and z orbici (singular: orbicus; three-bedroom units) in **Z** (zonars, the designated currency in Utarek). Then $\frac{\partial f}{\partial z}$ measures _____ and has units _____.

72. Complete the following: Let $f(t, x, y)$ be the projected number of citizens of the Principality State of Voodice, Luna, in year t since its founding, assuming the presence of x lunar vehicle factories and y domed settlements. Then $\frac{\partial f}{\partial x}$ measures _____ and has units _____.

73. Give an example of a function $f(x, y)$ with $f_x(1, 1) = -2$ and $f_y(1, 1) = 3$.

74. Give an example of a function $f(x, y, z)$ that has all of its partial derivatives equal to nonzero constants.

75. ▼ The graph of $z = b + mx + ny$ (where b, m, and n are constants) is a plane.
    a. Explain the geometric significance of the numbers b, m, and n.
    b. Show that the equation of the plane passing through (h, k, l) with slope m in the x direction (in the sense of $\partial/\partial x$) and slope n in the y direction is

$$z = l + m(x - h) + n(y - k).$$

76. ▼ The **tangent plane** to the graph of $f(x, y)$ at $P(a, b, f(a, b))$ is the plane containing the lines tangent to the slice through the graph by $y = b$ (as in Figure 19) and the slice through the graph by $x = a$. Use the result of Exercise 75 to show that the equation of the tangent plane is

$$z = f(a, b) + f_x(a, b)(x - a) + f_y(a, b)(y - b).$$

8.3 Maxima and Minima

Relative and Absolute Maxima and Minima

In Chapter 5, on applications of the derivative, we saw how to locate relative extrema of a function of a single variable. In this section we extend our methods to functions of two variables. Similar techniques work for functions of three or more variables.

Figure 21 shows a portion of the graph of the function

$$f(x, y) = 2(x^2 + y^2) - (x^4 + y^4) + 1.$$

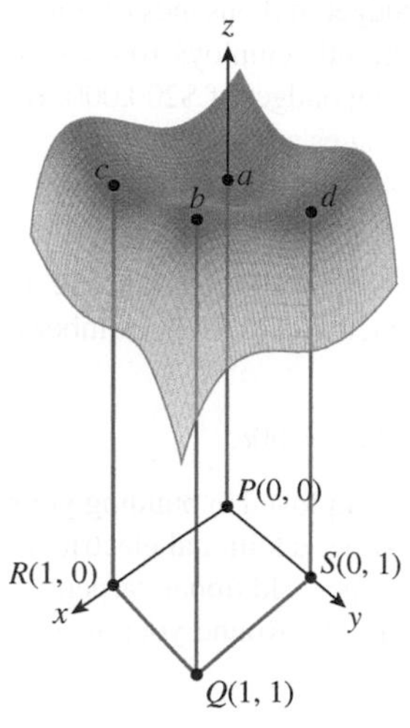

Figure 21

The graph in Figure 21 resembles a "flying carpet," and several interesting points, marked a, b, c, and d, are shown.

1. The point a has coordinates $(0, 0, f(0, 0))$, is directly above the origin $(0, 0)$, and is the lowest point in its vicinity; water would puddle there. We say that f has a **relative minimum** at $(0, 0)$ because $f(0, 0)$ is smaller than $f(x, y)$ for any (x, y) near $(0, 0)$.
2. Similarly, the point b is higher than any point in its vicinity. Thus, we say that f has a **relative maximum** at $(1, 1)$.

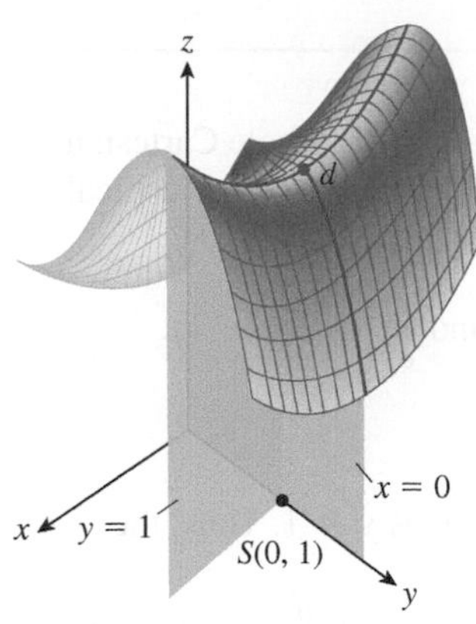

Figure 22

3. The points c and d represent a new phenomenon and are called **saddle points.** They are neither relative maxima nor relative minima but seem to be a little of both.

To see more clearly what features a saddle point has, look at Figure 22, which shows a portion of the graph near the point d.

If we slice through the graph along $y = 1$, we get a curve on which d is the *lowest* point. Thus, d looks like a relative minimum along this slice. On the other hand, if we slice through the graph along $x = 0$, we get another curve, on which d is the *highest* point, so d looks like a relative maximum along this slice. This kind of behavior characterizes a saddle point: f has a **saddle point** at (r, s) if f has a relative minimum at (r, s) along some slice through that point and a relative maximum along another slice through that point. If you look at the other saddle point, c, in Figure 21, you see the same characteristics.

While numerical information can help us to locate the approximate positions of relative extrema and saddle points, calculus permits us to locate these points accurately, as we did for functions of a single variable. Look once again at Figure 21, and notice the following:

- The points P, Q, R, and S are all in the **interior** of the domain of f; that is, none lie on the boundary of the domain. Said another way, we can move some distance in any direction from any of these points without leaving the domain of f.
- The tangent lines along the slices through these points parallel to the x- and y-axes are *horizontal*. Thus, the partial derivatives $\partial f/\partial x$ and $\partial f/\partial y$ are zero when evaluated at any of the points P, Q, R, and S. This gives us a way of locating candidates for relative extrema and saddle points.

The following summary generalizes and also expands on some of what we have just said.

Relative and Absolute Maxima and Minima

The function f of n variables has a **relative maximum** at $(r_1, r_2, \ldots, r_n)$ if $f(r_1, r_2, \ldots, r_n) \geq f(x_1, x_2, \ldots, x_n)$ for every point $(x_1, x_2, \ldots, x_n)$ near* $(r_1, r_2, \ldots, r_n)$ in the domain of f. We say that f has an **absolute maximum** at $(r_1, r_2, \ldots, r_n)$ if $f(r_1, r_2, \ldots, r_n) \geq f(x_1, x_2, \ldots, x_n)$ for every point $(x_1, x_2, \ldots, x_n)$ in the domain of f. The terms **relative minimum** and **absolute minimum** are defined in a similar way. Note that, as with functions of a single variable, absolute extrema are special kinds of relative extrema.

* For $(x_1, x_2, \ldots, x_n)$ to be near $(r_1, r_2, \ldots, r_n)$, we mean that x_1 is in some open interval centered at r_1, x_2 is in some open interval centered at r_2, and so on.

Locating Candidates for Extrema and Saddle Points in the Interior of the Domain of f

- Set $\dfrac{\partial f}{\partial x_1} = 0, \dfrac{\partial f}{\partial x_2} = 0, \ldots, \dfrac{\partial f}{\partial x_n} = 0$ simultaneously, and solve for $x_1, x_2, \ldots, x_n$.
- Check that the resulting points $(x_1, x_2, \ldots, x_n)$ are in the interior of the domain of f.

Points at which all the partial derivatives of f are zero are called **critical points.** The critical points are the only candidates for extrema and saddle points in the interior of the domain of f, assuming that its partial derivatives are defined at every point.†

† One can use the techniques of Section 8.4 to find extrema on the *boundary* of the domain of a function; for a complete discussion, see the optional extra section: *Maxima and Minima: Boundaries and the Extreme Value Theorem.* (We shall not consider the analogs of the singular points.)

Quick Examples

In each of the following Quick Examples the domain is the whole Cartesian plane, and the partial derivatives are defined at every point, so the critical points give us the only candidates for extrema and saddle points:

1. Let $f(x, y) = x^3 + (y - 1)^2$. Then $\frac{\partial f}{\partial x} = 3x^2$ and $\frac{\partial f}{\partial y} = 2(y - 1)$. Thus, we solve the system

   $$3x^2 = 0 \quad \text{and} \quad 2(y - 1) = 0.$$

   The first equation gives $x = 0$, and the second gives $y = 1$. Thus, the only critical point is $(0, 1)$.

2. Let $f(x, y) = x^2 - 4xy + 8y$. Then $\frac{\partial f}{\partial x} = 2x - 4y$ and $\frac{\partial f}{\partial y} = -4x + 8$. Thus, we solve

   $$2x - 4y = 0 \quad \text{and} \quad -4x + 8 = 0.$$

   The second equation gives $x = 2$, and the first then gives $y = 1$. Thus, the only critical point is $(2, 1)$.

3. Let $f(x, y) = e^{-(x^2+y^2)}$. Taking partial derivatives and setting them equal to zero, we get

   $$-2xe^{-(x^2+y^2)} = 0 \qquad \text{We set } \frac{\partial f}{\partial x} = 0.$$

   $$-2ye^{-(x^2+y^2)} = 0. \qquad \text{We set } \frac{\partial f}{\partial y} = 0.$$

   The first equation implies that $x = 0$,* and the second implies that $y = 0$. Thus, the only critical point is $(0, 0)$.

* Recall that if a product of two numbers is zero, then one or the other must be zero. In this case the number $e^{-(x^2+y^2)}$ can't be zero (because e^u is never zero), which gives the result claimed.

In the remainder of this section we will be interested in locating all critical points of a given function and then classifying each one as a relative maximum, relative minimum, saddle point, or none of these. Whether or not any relative extrema we find are in fact absolute is a subject that we discuss in Section 8.4.†

† In some of the applications in the exercises you will, however, need to consider whether the extrema you find are absolute.

EXAMPLE 1 Locating and Classifying Critical Points

Locate all critical points of $f(x, y) = x^2y - x^2 - 2y^2$. Graph the function to classify the critical points as relative maxima, relative minima, saddle points, or none of these.

Solution The partial derivatives are

$$f_x = 2xy - 2x = 2x(y - 1)$$
$$f_y = x^2 - 4y.$$

Setting these equal to zero gives

$$x = 0 \text{ or } y = 1$$
$$x^2 = 4y.$$

We get a solution by choosing either $x = 0$ or $y = 1$ and substituting into $x^2 = 4y$.

***Case 1: x* = 0** Substituting into $x^2 = 4y$ gives $0 = 4y$ and hence $y = 0$. Thus, the critical point for this case is $(x, y) = (0, 0)$.

Case 2: y = 1 Substituting into $x^2 = 4y$ gives $x^2 = 4$ and hence $x = \pm 2$. Thus, we get two critical points for this case: (2, 1) and (−2, 1).

We now have three critical points altogether: (0, 0), (2, 1), and (−2, 1). Because the domain of f is the whole Cartesian plane and the partial derivatives are defined at every point, these critical points are the only candidates for relative extrema and saddle points. We get the corresponding points on the graph by substituting for x and y in the equation for f to get the z-coordinates. The points are (0, 0, 0), (2, 1, −2), and (−2, 1, −2).

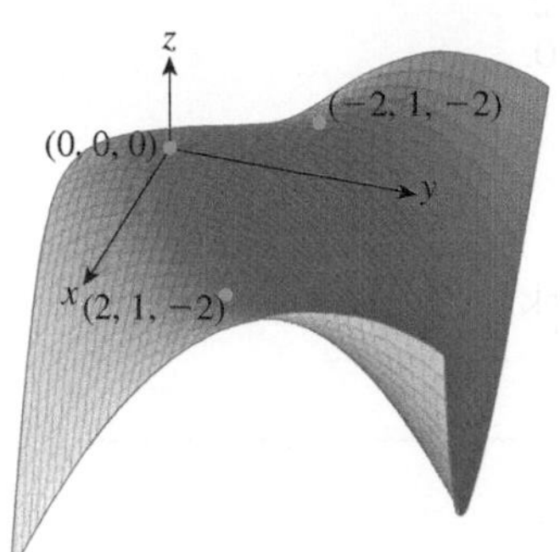

Figure 23

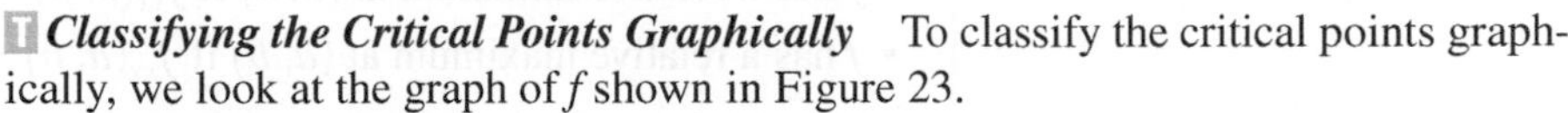

T ***Classifying the Critical Points Graphically*** To classify the critical points graphically, we look at the graph of f shown in Figure 23.

Examining the graph carefully, we see that the point (0, 0, 0) is a relative maximum. As for the other two critical points, are they saddle points or are they relative maxima? They are relative maxima along the y-direction, but they are relative minima along the lines $y = \pm x$ (see the top edge of the picture, which shows a dip at (−2, 1, −2)), so they are saddle points. If you don't believe this, we will get more evidence following and in a later example.

T ***Classifying the Critical Points Numerically*** We can use a tabular representation of the function to classify the critical points numerically. The following tabular representation of the function can be obtained by using a spreadsheet. (See the Spreadsheet Technology Guide discussion of Example 3 of Section 8.1 at the end of the chapter for information on using a spreadsheet to generate such a table.)

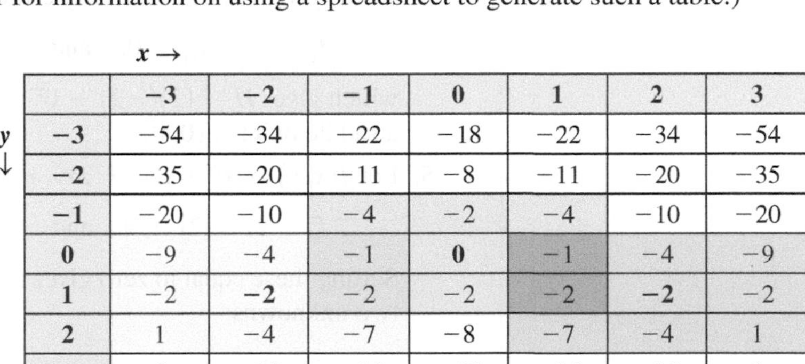

$y \downarrow$ \ $x \rightarrow$	−3	−2	−1	0	1	2	3
−3	−54	−34	−22	−18	−22	−34	−54
−2	−35	−20	−11	−8	−11	−20	−35
−1	−20	−10	−4	−2	−4	−10	−20
0	−9	−4	−1	**0**	−1	−4	−9
1	−2	**−2**	−2	−2	−2	**−2**	−2
2	1	−4	−7	−8	−7	−4	1
3	0	−10	−16	−18	−16	−10	0

The shaded and colored cells show rectangular neighborhoods of the three critical points (0, 0), (2, 1), and (−2, 1). (Notice that they overlap.) The values of f at the critical points are at the centers of these rectangles. Looking at the gray neighborhood of $(x, y) = (0, 0)$, we see that $f(0, 0) = 0$ is the largest value of f in the shaded cells, suggesting that f has a maximum at (0, 0). The shaded neighborhood of (2, 1) on the right shows $f(2, 1) = -2$ as the maximum along some slices (e.g., the vertical slice) and a minimum along the diagonal slice from top left to bottom right. This is what results in a saddle point on the graph. The point (−2, 1) is similar; thus, f also has a saddle point at (−2, 1).

Second Derivative Test

Q: *Is there an algebraic way of deciding whether a given point is a relative maximum, relative minimum, or saddle point?*

A: There is a second derivative test for functions of two variables, stated as follows.

Second Derivative Test for Functions of Two Variables

Suppose (a, b) is a critical point in the interior of the domain of the function f of two variables. Let H be the quantity

$$H = f_{xx}(a, b)f_{yy}(a, b) - [f_{xy}(a, b)]^2.$$

H is called the Hessian.

Then, if H is *positive*,

- f has a relative minimum at (a, b) if $f_{xx}(a, b) > 0$;
- f has a relative maximum at (a, b) if $f_{xx}(a, b) < 0$.

If H is *negative*,

- f has a saddle point at (a, b).

If $H = 0$ the test tells us nothing, so we need to look at the graph or a numerical table to see what is going on.

Quick Examples

4. Let $f(x, y) = x^2 - y^2$. Then

$$f_x = 2x \quad \text{and} \quad f_y = -2y,$$

which gives $(0, 0)$ as the only critical point. Also,

$$f_{xx} = 2, \quad f_{xy} = 0, \quad \text{and} \quad f_{yy} = -2,$$

Note that these are constant.

which gives $H = (2)(-2) - 0^2 = -4$. Because H is negative, we have a saddle point at $(0, 0)$.

5. Let $f(x, y) = x^2 + 2y^2 + 2xy + 4x$. Then

$$f_x = 2x + 2y + 4 \quad \text{and} \quad f_y = 2x + 4y.$$

Setting these equal to zero gives a system of two linear equations in two unknowns:

$$x + y = -2$$
$$x + 2y = 0.$$

This system has solution $(-4, 2)$, so this is our only critical point. The second partial derivatives are $f_{xx} = 2, f_{xy} = 2$, and $f_{yy} = 4$, so $H = (2)(4) - 2^2 = 4$. Because $H > 0$ and $f_{xx} > 0$, we have a relative minimum at $(-4, 2)$.

Note There is a second derivative test for functions of three or more variables, but it is considerably more complicated. We stick with functions of two variables for the most part in this book. The justification of the second derivative test is beyond the scope of this book. ■

EXAMPLE 2 Using the Second Derivative Test

Use the second derivative test to analyze the function $f(x, y) = x^2y - x^2 - 2y^2$ discussed in Example 1, and confirm the results we got there.

Solution We saw in Example 1 that the first-order derivatives are

$$f_x = 2xy - 2x = 2x(y - 1)$$
$$f_y = x^2 - 4y$$

and the critical points are $(0, 0)$, $(2, 1)$, and $(-2, 1)$. We also need the second derivatives:

$$f_{xx} = 2y - 2$$
$$f_{xy} = 2x$$
$$f_{yy} = -4.$$

The point (0, 0): $f_{xx}(0, 0) = -2$, $f_{xy}(0, 0) = 0$, and $f_{yy}(0, 0) = -4$, so $H = 8$. Because $H > 0$ and $f_{xx}(0, 0) < 0$, the second derivative test tells us that f has a relative maximum at $(0, 0)$.

The point (2, 1): $f_{xx}(2, 1) = 0$, $f_{xy}(2, 1) = 4$, and $f_{yy}(2, 1) = -4$, so $H = -16$. Because $H < 0$, we know that f has a saddle point at $(2, 1)$.

The point (−2, 1): $f_{xx}(-2, 1) = 0$, $f_{xy}(-2, 1) = -4$, and $f_{yy}(-2, 1) = -4$, so once again $H = -16$, and f has a saddle point at $(-2, 1)$.

Application: Deriving the Formulas for Linear Regression

In Section 1.4 we presented the following set of formulas for the **regression** or **best-fit** line associated with a given set of data points $(x_1, y_1), (x_2, y_2), \ldots, (x_n, y_n)$.

Regression Line

The line that best fits the n data points $(x_1, y_1), (x_2, y_2), \ldots, (x_n, y_n)$ has the form

$$y = mx + b,$$

where

$$m = \frac{n(\Sigma xy) - (\Sigma x)(\Sigma y)}{n(\Sigma x^2) - (\Sigma x)^2}$$

$$b = \frac{\Sigma y - m(\Sigma x)}{n}$$

$n =$ number of data points.

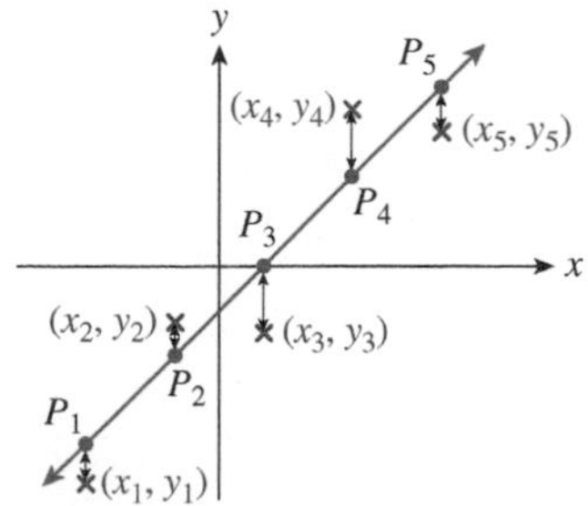

Figure 24

Derivation of the Regression Line Formulas

Recall that the regression line is defined to be the line that minimizes the sum of the squares of the **residuals**, measured by the vertical distances shown in Figure 24, which shows a regression line associated with $n = 5$ data points. In the figure, the points $P_1, \ldots, P_n$ on the regression line have coordinates $(x_1, mx_1 + b), (x_2, mx_2 + b), \ldots, (x_n, mx_n + b)$. The residuals are the quantities $y_{\text{Observed}} - y_{\text{Predicted}}$:

$$y_1 - (mx_1 + b), y_2 - (mx_2 + b), \ldots, y_n - (mx_n + b).$$

The sum of the squares of the residuals is therefore

$$S(m, b) = [y_1 - (mx_1 + b)]^2 + [y_2 - (mx_2 + b)]^2 + \cdots + [y_n - (mx_n + b)]^2,$$

and this is the quantity we must minimize by choosing m and b. Because we reason that there is a line that minimizes this quantity, there must be a relative minimum at that point. We shall see in a moment that the function S has at most one critical point, which must therefore be the desired absolute minimum. To obtain the critical points of S, we set the partial derivatives equal to zero and solve:

$$\begin{aligned} S_m = 0: &\quad -2x_1[y_1 - (mx_1 + b)] - \cdots - 2x_n[y_n - (mx_n + b)] = 0 \\ S_b = 0: &\quad -2[y_1 - (mx_1 + b)] - \cdots - 2[y_n - (mx_n + b)] = 0. \end{aligned}$$

Dividing by -2 and gathering terms allows us to rewrite the equations as

$$\begin{aligned} m(x_1^2 + \cdots + x_n^2) + b(x_1 + \cdots + x_n) &= x_1y_1 + \cdots + x_ny_n \\ m(x_1 + \cdots + x_n) + nb &= y_1 + \cdots + y_n. \end{aligned}$$

We can rewrite these equations more neatly using Σ-notation:

$$\begin{aligned} m(\Sigma x^2) + b(\Sigma x) &= \Sigma xy \\ m(\Sigma x) \ + nb &= \Sigma y. \end{aligned}$$

This is a system of two linear equations in the two unknowns m and b. It may or may not have a unique solution. When there is a unique solution, we can conclude that the best-fit line is given by solving these two equations for m and b. Alternatively, there is a general formula for the solution of any system of two equations in two unknowns, and if we apply this formula to our two equations, we get the regression formulas above.

8.3 EXERCISES

▼ more advanced ◆ challenging

T indicates exercises that should be solved using technology

In Exercises 1–4, classify each labeled point on the graph as one of the following:

Relative maximum
Relative minimum
Saddle point
Critical point but neither a relative extremum nor a saddle point
None of the above [**HINT:** See Example 1.]

1.

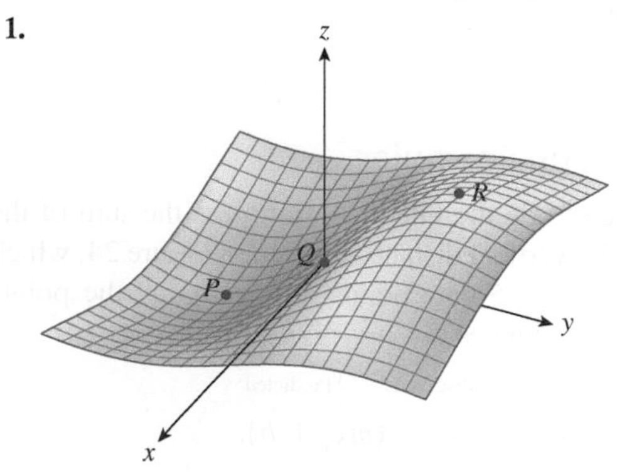

2.

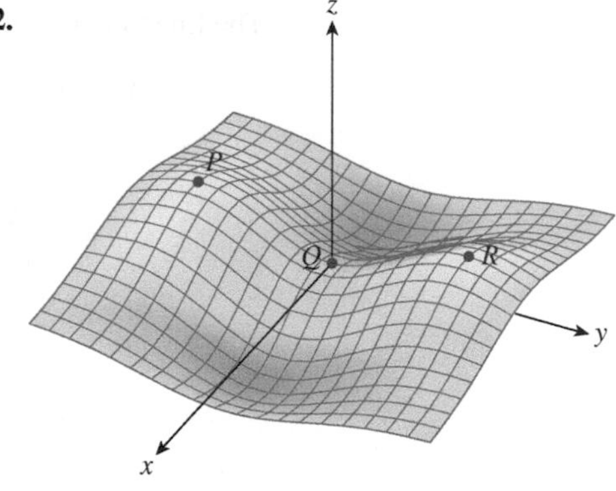

3.

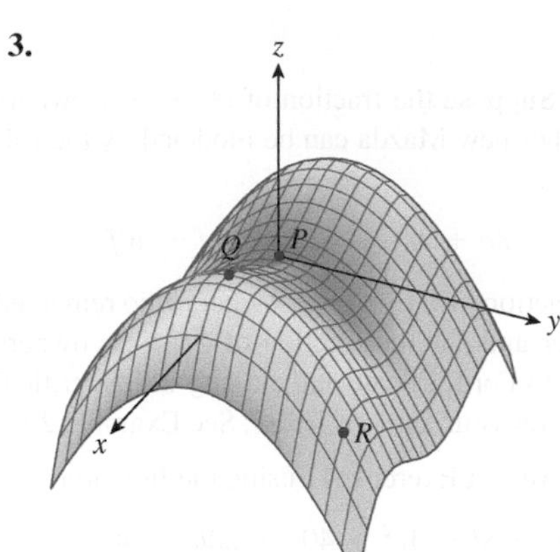

4.

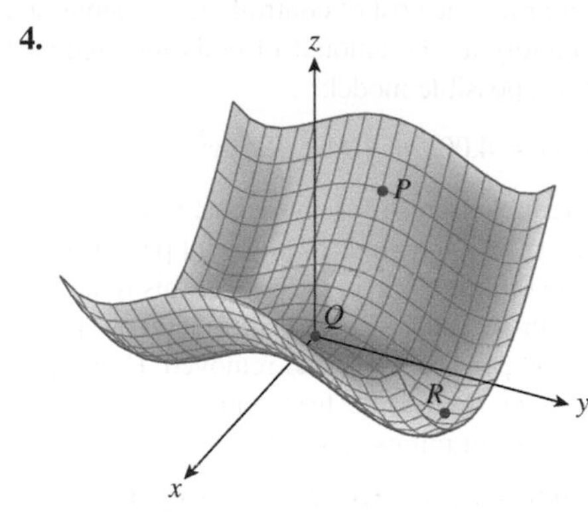

In Exercises 5–10, classify the shaded value in each table as one of the following:

Relative maximum
Relative minimum
Saddle point
Neither a relative extremum nor a saddle point

Assume that the shaded value represents a critical point.

5. $x \rightarrow$; $y \downarrow$

	−3	−2	−1	0	1	2
−3	10	5	2	1	2	5
−2	9	4	1	0	1	4
−1	10	5	2	1	2	5
0	13	8	5	4	5	8
1	18	13	10	9	10	13
2	25	20	17	16	17	20
3	34	29	26	25	26	29

6. $x \rightarrow$; $y \downarrow$

	−3	−2	−1	0	1	2
−3	5	0	−3	−4	−3	0
−2	8	3	0	−1	0	3
−1	9	4	1	0	1	4
0	8	3	0	−1	0	3
1	5	0	−3	−4	−3	0
2	0	−5	−8	−9	−8	−5
3	−7	−12	−15	−16	−15	−12

7. $x \rightarrow$; $y \downarrow$

	−3	−2	−1	0	1	2
−3	5	0	−3	−4	−3	0
−2	8	3	0	−1	0	3
−1	9	4	1	0	1	4
0	8	3	0	−1	0	3
1	5	0	−3	−4	−3	0
2	0	−5	−8	−9	−8	−5
3	−7	−12	−15	−16	−15	−12

8. $x \rightarrow$; $y \downarrow$

	−3	−2	−1	0	1	2
−3	2	3	2	−1	−6	−13
−2	3	4	3	0	−5	−12
−1	2	3	2	−1	−6	−13
0	−1	0	−1	−4	−9	−16
1	−6	−5	−6	−9	−14	−21
2	−13	−12	−13	−16	−21	−28
3	−22	−21	−22	−25	−30	−37

9. $x \rightarrow$; $y \downarrow$

	−3	−2	−1	0	1	2
−3	4	5	4	1	−4	−11
−2	3	4	3	0	−5	−12
−1	4	5	4	1	−4	−11
0	7	8	7	4	−1	−8
1	12	13	12	9	4	−3
2	19	20	19	16	11	4
3	28	29	28	25	20	13

10.

$x \rightarrow$

$y \downarrow$	-3	-2	-1	0	1	2
-3	100	101	100	97	92	85
-2	99	100	99	96	91	84
-1	98	99	98	95	90	83
0	91	92	91	88	83	76
1	72	73	72	69	64	57
2	35	36	35	32	27	20
3	-26	-25	-26	-29	-34	-41

In Exercises 11–36, locate and classify all the critical points of the given function. [**HINT:** See Example 2.]

11. $f(x, y) = x^2 + y^2 + 1$ **12.** $f(x, y) = 4 - (x^2 + y^2)$

13. $g(x, y) = 1 - x^2 - x - y^2 + y$

14. $g(x, y) = x^2 + x + y^2 - y - 1$

15. $k(x, y) = x^2 - 3xy + y^2$ **16.** $k(x, y) = x^2 - xy + 2y^2$

17. $f(x, y) = x^2 + 2xy + 2y^2 - 2x + 4y$

18. $f(x, y) = x^2 + xy - y^2 + 3x - y$

19. $g(x, y) = -x^2 - 2xy - 3y^2 - 3x - 2y$

20. $g(x, y) = -x^2 - 2xy + y^2 + x - 4y$

21. $h(x, y) = x^2y - 2x^2 - 4y^2$

22. $h(x, y) = x^2 + y^2 - y^2x - 4$

23. $f(x, y) = x^2 + 2xy^2 + 2y^2$

24. $f(x, y) = x^2 + x^2y + y^2$

25. $s(x, y) = e^{x^2+y^2}$ **26.** $s(x, y) = e^{-(x^2+y^2)}$

27. $t(x, y) = x^4 + 8xy^2 + 2y^4$

28. $t(x, y) = x^3 - 3xy + y^3$

29. $f(x, y) = x^2 + y - e^y$ **30.** $f(x, y) = xe^y$

31. $f(x, y) = e^{-(x^2+y^2+2x)}$ **32.** $f(x, y) = e^{-(x^2+y^2-2x)}$

33. ▼ $f(x, y) = xy + \dfrac{2}{x} + \dfrac{2}{y}$ **34.** ▼ $f(x, y) = xy + \dfrac{4}{x} + \dfrac{2}{y}$

35. ▼ $g(x, y) = x^2 + y^2 + \dfrac{2}{xy}$

36. ▼ $g(x, y) = x^3 + y^3 + \dfrac{3}{xy}$

37. ▼ Refer back to Exercise 11. Which (if any) of the critical points of $f(x, y) = x^2 + y^2 + 1$ are absolute extrema?

38. ▼ Refer back to Exercise 12. Which (if any) of the critical points of $f(x, y) = 4 - (x^2 + y^2)$ are absolute extrema?

39. T ▼ Refer back to Exercise 21. Which (if any) of the critical points of $h(x, y) = x^2y - 2x^2 - 4y^2$ are absolute extrema?

40. T ▼ Refer back to Exercise 22. Which (if any) of the critical points of $h(x, y) = x^2 + y^2 - y^2x - 4$ are absolute extrema?

Applications

41. ***Brand Loyalty*** Suppose the fraction of **Mazda** car owners who chose another new Mazda can be modeled by the following function:[22]

$$M(c, f) = 11 + 8c + 4c^2 - 40f - 20cf + 40f^2,$$

where c is the fraction of **Chrysler** car owners who remained loyal to Chrysler and f is the fraction of **Ford** car owners remaining loyal to Ford. Locate and classify all the critical points and interpret your answer. [**HINT:** See Example 2.]

42. ***Brand Loyalty*** Repeat Exercise 41 using the function

$$M(c, f) = -10 - 8f - 4f^2 + 40c + 20fc - 40c^2.$$

[**HINT:** See Example 2.]

43. ▼ ***Pollution Control*** The cost of controlling emissions at a firm goes up rapidly as the amount of emissions reduced goes up. Here is a possible model:

$$C(x, y) = 4{,}000 + 100x^2 + 50y^2,$$

where x is the reduction in sulfur emissions, y is the reduction in lead emissions (in pounds of pollutant per day), and C is the daily cost to the firm (in dollars) of this reduction. Government clean-air subsidies amount to \$500 per pound of sulfur and \$100 per pound of lead removed. How many pounds of pollutant should the firm remove each day to minimize *net* cost (cost minus subsidy)?

44. ▼ ***Pollution Control*** Repeat Exercise 43 using the following information:

$$C(x, y) = 2{,}000 + 200x^2 + 100y^2$$

with government subsidies amounting to \$100 per pound of sulfur and \$500 per pound of lead removed per day.

45. ▼ ***Revenue*** Your company manufactures two models of speakers, the Ultra Mini and the Big Stack. Demand for each depends partly on the price of the other. If one is expensive, then more people will buy the other. If p_1 is the price of the Ultra Mini and p_2 is the price of the Big Stack, demand for the Ultra Mini is given by

$$q_1(p_1, p_2) = 100{,}000 - 100p_1 + 10p_2,$$

where q_1 represents the number of Ultra Minis that will be sold in a year. The demand for the Big Stack is given by

$$q_2(p_1, p_2) = 150{,}000 + 10p_1 - 100p_2.$$

Find the prices for the Ultra Mini and the Big Stack that will maximize your total revenue.

[22] This model is not accurate, although it was inspired by an approximation of a second-order regression based on data from the period 1988–1995. Source for original data: Chrysler, Maritz Market Research, Consumer Attitude Research, and Strategic Vision/ *New York Times*, November 3, 1995, p. D2.

46. ▼ ***Revenue*** Repeat Exercise 45, using the following demand functions:

$$q_1(p_1, p_2) = 100{,}000 - 100p_1 + p_2$$
$$q_2(p_1, p_2) = 150{,}000 + p_1 - 100p_2.$$

47. ▼ ***Luggage Dimensions: American Airlines*** **American Airlines** requires that the total outside dimensions (length + width + height) of a checked bag not exceed 62 inches.[23] What are the dimensions of the largest-volume bag that you can check on an American flight?

48. ▼ ***Carry-on Bag Dimensions: American Airlines*** **American Airlines** requires that the total outside dimensions (length + width + height) of a carry-on bag not exceed 45 inches.[24] What are the dimensions of the largest-volume bag that you can carry on an American flight?

49. ▼ ***Package Dimensions: USPS*** The **U.S. Postal Service (USPS)** will accept only packages with a length plus girth no more than 108 inches.[25] (See the figure.)

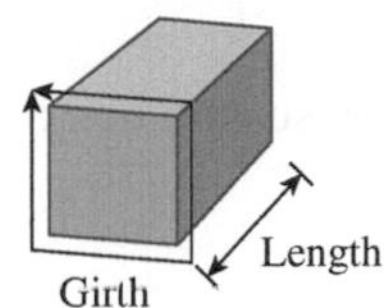

What are the dimensions of the largest-volume package that the USPS will accept? What is its volume?

50. ▼ ***Package Dimensions: UPS*** **United Parcel Service (UPS)** will accept only packages with length no more than 108 inches and length plus girth no more than 165 inches.[26] (See figure for Exercise 49.) What are the dimensions of the largest-volume package that UPS will accept? What is its volume?

[23] According to information on its website (www.aa.com).

[24] *Ibid.*

[25] The requirement for packages sent other than Retail Ground, as of September 2015 (www.usps.com).

[26] The requirement as of September 2015 (www.ups.com).

Communication and Reasoning Exercises

51. Sketch the graph of a function that has one extremum and no saddle points.

52. Sketch the graph of a function that has one saddle point and one extremum.

53. ▼ Sketch the graph of a function that has one relative extremum, no absolute extrema, and no saddle points.

54. ▼ Sketch the graph of a function that has infinitely many absolute maxima.

55. Let $H = f_{xx}(a, b)f_{yy}(a, b) - f_{xy}(a, b)^2$. What condition on H guarantees that f has a relative extremum at the point (a, b)?

56. Let H be as in Exercise 55. Give an example to show that it is possible to have $H = 0$ and a relative minimum at (a, b).

57. ▼ Suppose that when the graph of $f(x, y)$ is sliced by a vertical plane through (a, b) parallel to either the xz-plane or the yz-plane, the resulting curve has a relative maximum at (a, b). Does this mean that f has a relative maximum at (a, b)? Explain your answer.

58. ▼ Suppose that f has a relative maximum at (a, b). Does it follow that, if the graph of f is sliced by a vertical plane parallel to either the xz-plane or the yz-plane, the resulting curve has a relative maximum at (a, b)? Explain your answer.

59. ▼ ***Average Cost*** Let $C(x, y)$ be any cost function, where x and y represent the numbers of two different items manufactured. Show that when the average cost is minimized, the marginal costs C_x and C_y both equal the average cost. Explain why this is reasonable.

60. ▼ ***Average Profit*** Let $P(x, y)$ be any profit function, where x and y represent the numbers of two different items manufactured and sold. Show that when the average profit is maximized, the marginal profits P_x and P_y both equal the average profit. Explain why this is reasonable.

61. ◆ The tangent plane to a graph was introduced in Exercise 76 in Section 8.2. Use the equation of the tangent plane given there to explain why the tangent plane is parallel to the xy-plane at a relative maximum or minimum of $f(x, y)$.

62. ◆ Use the equation of the tangent plane given in Exercise 76 in Section 8.2 to explain why the tangent plane is parallel to the xy-plane at a saddle point of $f(x, y)$.

8.4 Constrained Maxima and Minima and Applications

So far, we have looked only at the relative extrema of functions with no constraints. However, in Section 5.2 we saw examples in which we needed to find the maximum or minimum of an objective function subject to one or more constraints on the independent variables. For instance, consider the following problem:

$$\text{Minimize } S = xy + 2xz + 2yz \quad \text{subject to } xyz = 4 \text{ with } x > 0, y > 0, z > 0.$$

One strategy for solving such problems is essentially the same as the strategy we used earlier: Solve the constraint equation for one of the variables, substitute into the objective function, and then optimize the resulting function using the methods of Section 8.3. We will call this the *substitution method.** An alternative method, called the *method of Lagrange multipliers*, is useful when it is difficult or impossible to solve the constraint equation for one of the variables, and even when it is possible to do so.

* Although often the method of choice, the substitution method is not infallible (see Exercises 19 and 20).

Substitution Method

EXAMPLE 1 Using Substitution

Minimize $S = xy + 2xz + 2yz$ subject to $xyz = 4$ with $x > 0$, $y > 0$, $z > 0$.

Solution As suggested in the above discussion, we proceed as follows:

Solve the constraint equation for one of the variables, and then substitute in the objective function. The constraint equation is $xyz = 4$. Solving for z gives

$$z = \frac{4}{xy}.$$

The objective function is $S = xy + 2xz + 2yz$, so substituting $z = 4/xy$ gives

$$\begin{aligned} S &= xy + 2x\frac{4}{xy} + 2y\frac{4}{xy} \\ &= xy + \frac{8}{y} + \frac{8}{x}. \end{aligned}$$

Minimize the resulting function of two variables. We use the method in Section 8.3 to find the minimum of $S = xy + \frac{8}{y} + \frac{8}{x}$ for $x > 0$ and $y > 0$. We look for critical points:

$$S_x = y - \frac{8}{x^2}, \quad S_y = x - \frac{8}{y^2}$$

$$S_{xx} = \frac{16}{x^3}, \quad S_{xy} = 1, \quad S_{yy} = \frac{16}{y^3}.$$

We now equate the first partial derivatives to zero:

$$y = \frac{8}{x^2} \quad \text{and} \quad x = \frac{8}{y^2}.$$

To solve for x and y, we substitute the first of these equations in the second, getting

$$\begin{aligned} x &= \frac{x^4}{8} \\ x^4 - 8x &= 0 \\ x(x^3 - 8) &= 0. \end{aligned}$$

The two solutions are $x = 0$, which we reject because x cannot be zero, and $x = 2$. Substituting $x = 2$ in $y = 8/x^2$ gives $y = 2$ also. Thus, the only critical point is $(2, 2)$. To apply the second derivative test, we compute

$$S_{xx}(2, 2) = 2, \quad S_{xy}(2, 2) = 1, \quad S_{yy}(2, 2) = 2$$

and find that $H = 3 > 0$ and $S_{xx}(2, 2) > 0$, so we have a relative minimum at $(2, 2)$.

The corresponding value of z is given by the constraint equation:

$$z = \frac{4}{xy} = \frac{4}{4} = 1.$$

The corresponding value of the objective function is

$$S = xy + \frac{8}{y} + \frac{8}{x} = 4 + \frac{8}{2} + \frac{8}{2} = 12.$$

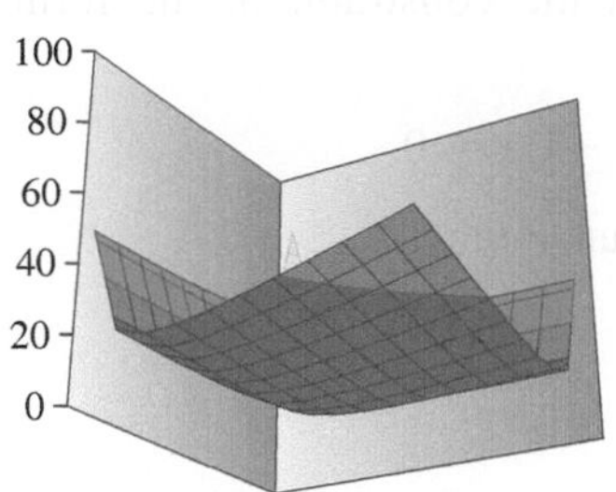

Graph of $S = xy + \frac{8}{y} + \frac{8}{x}$
$(0.2 \le x \le 5, 0.2 \le y \le 5)$

Figure 25

Figure 25 shows a portion of the graph of $S = xy + \frac{8}{y} + \frac{8}{x}$ for positive x and y (drawn by using the Excel Surface Grapher in the Chapter 8 utilities at the Website) and suggests that there is a single absolute minimum, which must be at our only candidate point: $(2, 2)$.

We conclude that the minimum of S is 12 and occurs at $(2, 2, 1)$.

The Method of Lagrange Multipliers

As we mentioned above, the method of Lagrange multipliers has the advantage that it can be used in constrained optimization problems when it is difficult or impossible to solve a constraint equation for one of the variables. We restrict attention to the case of a single constraint equation, although the method also generalizes to any number of constraint equations.

Locating Relative Extrema Using the Method of Lagrange Multipliers

To locate the candidates for relative extrema of a function $f(x, y, \ldots)$ subject to the constraint $g(x, y, \ldots) = 0$:

1. Construct the **Lagrangian function**

$$L(x, y, \ldots) = f(x, y, \ldots) - \lambda g(x, y, \ldots),$$

where λ is a new unknown called a **Lagrange multiplier**.

2. The candidates for the relative extrema occur at the critical points of $L(x, y, \ldots)$. To find them, set all the partial derivatives of $L(x, y, \ldots)$ equal to zero, and solve the resulting system, together with the constraint equation $g(x, y, \ldots) = 0$, for the unknowns $x, y, \ldots$ and λ.

The points $(x, y, \ldots)$ that occur in solutions are then the candidates for the relative extrema of f subject to $g = 0$.

Although the justification for the method of Lagrange multipliers is beyond the scope of this text (a derivation can be found in many vector calculus textbooks), we will demonstrate by example how it is used.

EXAMPLE 2 Using Lagrange Multipliers

Use the method of Lagrange multipliers to find the maximum value of $f(x, y) = 2xy$ subject to $x^2 + 4y^2 = 32$.

Solution We start by rewriting the problem with the constraint in the form $g(x, y) = 0$:

$$\text{Maximize } f(x, y) = 2xy \text{ subject to } x^2 + 4y^2 - 32 = 0.$$

Here, $g(x, y) = x^2 + 4y^2 - 32$, and the Lagrangian function is

$$\begin{aligned} L(x, y) &= f(x, y) - \lambda g(x, y) \\ &= 2xy - \lambda(x^2 + 4y^2 - 32). \end{aligned}$$

The system of equations we need to solve is thus

$$\begin{aligned} L_x = 0&: \quad 2y - 2\lambda x = 0 \\ L_y = 0&: \quad 2x - 8\lambda y = 0 \\ g = 0&: \quad x^2 + 4y^2 - 32 = 0. \end{aligned}$$

It is often convenient to solve such a system by first solving one of the equations for λ and then substituting in the remaining equations. Thus, we start by solving the first equation to obtain

$$\lambda = \frac{y}{x}.$$

(A word of caution: Because we divided by x, we made the implicit assumption that $x \neq 0$, so before continuing we should check what happens if $x = 0$. But if $x = 0$, then the first equation, $2y = 2\lambda x$, tells us that $y = 0$ as well, and this contradicts the third equation: $x^2 + 4y^2 - 32 = 0$. Thus, we can rule out the possibility that $x = 0$.) Substituting the value of λ in the second equation gives

$$2x - 8\left(\frac{y}{x}\right)y = 0 \quad \text{or} \quad x^2 = 4y^2.$$

We can now substitute $x^2 = 4y^2$ in the constraint equation, obtaining

$$\begin{aligned} 4y^2 + 4y^2 - 32 &= 0 \\ 8y^2 &= 32 \\ y &= \pm 2. \end{aligned}$$

We now substitute back to obtain

$$x^2 = 4y^2 = 16,$$

or $\qquad x = \pm 4.$

We don't need the value of λ, so we won't solve for it. Thus, the candidates for relative extrema are given by $x = \pm 4$ and $y = \pm 2$; that is, the four points $(-4, -2)$, $(-4, 2)$, $(4, -2)$, and $(4, 2)$. Recall that we are seeking the maximum value of $f(x, y) = 2xy$. Because we now have only four points to choose from, we compare the values of f at these four points and conclude that the maximum value of f occurs when $(x, y) = (-4, -2)$ or $(4, 2)$ and equals $f(-4, -2) = 2(-4)(-2) = 16$.

Something is suspicious in Example 2. We didn't check to see whether these candidates were relative extrema to begin with, let alone absolute extrema! How do we justify this omission? One of the difficulties with using the method of Lagrange multipliers is that it does not provide us with a test analogous to the second derivative test for functions of several variables. However, if you grant that the function in question does have an absolute maximum, then we require no test, because one of the candidates must give this maximum.

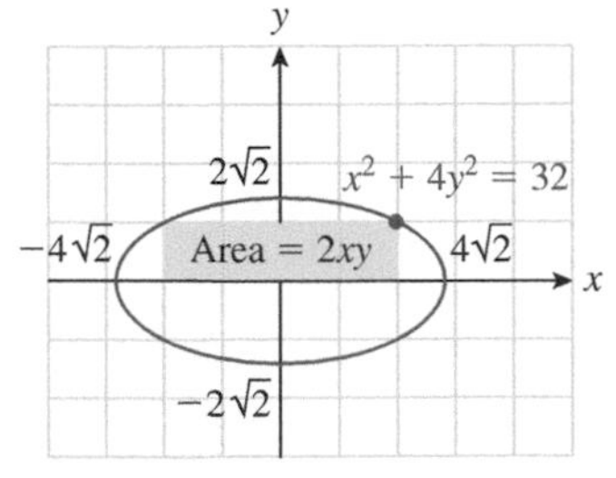

Figure 26

Q: *But how do we know that the given function has an absolute maximum?*

A: The best way to see this is by giving a geometric interpretation. The constraint $x^2 + 4y^2 = 32$ tells us that the point (x, y) must lie on the ellipse shown in Figure 26. The function $f(x, y) = 2xy$ gives the area of the rectangle shaded in the figure. There must be a *largest* such rectangle, because the area varies continuously from 0 when (x, y) is on the x-axis, to positive when (x, y) is in the first quadrant, to 0 again when (x, y) is on the y-axis, so f must have an absolute maximum for at least one pair of coordinates (x, y).

We now show how to use Lagrange multipliers to solve the minimization problem in Example 1.

EXAMPLE 3 Using Lagrange Multipliers: Function of Three Variables

Use the method of Lagrange multipliers to find the minimum value of $S = xy + 2xz + 2yz$ subject to $xyz = 4$ with $x > 0, y > 0, z > 0$.

Solution We start by rewriting the problem in standard form:

$$\text{Maximize } f(x, y, z) = xy + 2xz + 2yz$$
$$\text{subject to } xyz - 4 = 0 \text{ (with } x > 0, y > 0, z > 0).$$

Here, $g(x, y, z) = xyz - 4$, and the Lagrangian function is

$$\begin{aligned} L(x, y, z) &= f(x, y, z) - \lambda g(x, y, z) \\ &= xy + 2xz + 2yz - \lambda(xyz - 4). \end{aligned}$$

The system of equations we need to solve is thus

$$\begin{aligned} L_x &= 0: \quad y + 2z - \lambda yz = 0 \\ L_y &= 0: \quad x + 2z - \lambda xz = 0 \\ L_z &= 0: \quad 2x + 2y - \lambda xy = 0 \\ g &= 0: \quad xyz - 4 = 0. \end{aligned}$$

As in Example 2, we solve one of the equations for λ and substitute in the others. The first equation gives

$$\lambda = \frac{1}{z} + \frac{2}{y}.$$

Substituting this into the second equation gives

$$x + 2z = x + \frac{2xz}{y}$$

or $2 = \frac{2x}{y},$ Subtract x from both sides and then divide by z.

giving $y = x.$

Substituting the expression for λ into the third equation gives

$$2x + 2y = \frac{xy}{z} + 2x$$

or $2 = \frac{x}{z},$ Subtract $2x$ from both sides and then divide by y.

giving $z = \frac{x}{2}.$

Now we have both y and z in terms of x. We substitute these values in the last (constraint) equation:

$$x(x)\left(\frac{x}{2}\right) - 4 = 0$$

$$x^3 = 8$$

$$x = 2.$$

Thus, $y = x = 2$, and $z = \frac{x}{2} = 1$. Therefore, the only critical point occurs at $(2, 2, 1)$, as we found in Example 1, and the corresponding value of S is

$$S = xy + 2xz + 2yz = (2)(2) + 2(2)(1) + 2(2)(1) = 12.$$

➡ **Before we go on . . .** Again, the method of Lagrange multipliers does not tell us whether the critical point in Example 3 is a maximum, a minimum, or neither. However, if you grant that the function in question does have an absolute minimum, then the values we found must give this minimum value. ■

Applications

EXAMPLE 4 Minimizing Area

Find the dimensions of an open-topped rectangular box that has a volume of 4 cubic feet and the smallest possible surface area.

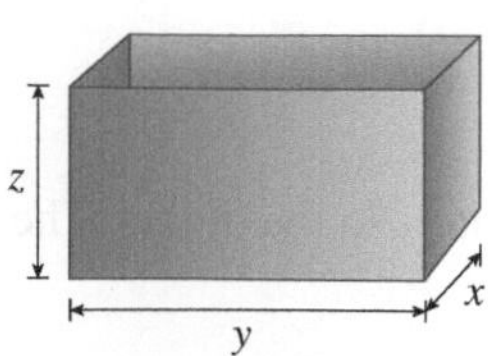

Figure 27

Solution Our first task is to rephrase this request as a mathematical optimization problem. Figure 27 shows a picture of the box with dimensions x, y, and z. We want to minimize the total surface area, which is given by

$$S = xy + 2xz + 2yz.$$ Base + Sides + Front and back

This is our objective function. We can't simply choose x, y, and z to all be zero, however, because the enclosed volume must be 4 cubic feet. So

$$xyz = 4.$$ Constraint

This is our constraint equation. Other unstated constraints are $x > 0$, $y > 0$, and $z > 0$, because the dimensions of the box must be positive. We now restate the problem as follows:

$$\text{Minimize } S = xy + 2xz + 2yz \text{ subject to } xyz = 4, x > 0, y > 0, z > 0.$$

But this is exactly the problem in Examples 1 and 3, and it has a solution $x = 2$, $y = 2$, $z = 1$, $S = 12$. Thus, the required dimensions of the box are

$$x = 2 \text{ feet}, \quad y = 2 \text{ feet}, \quad z = 1 \text{ foot},$$

requiring a total surface area of 12 square feet.

Q: *In Example 1 we checked that we had a relative minimum at $(x, y) = (2, 2)$, and we were persuaded graphically that this was probably an absolute minimum. Can we be sure that this relative minimum is an absolute minimum?*

A: Yes. There must be a least surface area among all boxes that hold 4 cubic feet. (Why?) Because this would give a relative minimum of S and because the only possible relative minimum of S occurs at (2, 2), this is the absolute minimum.

EXAMPLE 5 Maximizing Productivity

An electric motor manufacturer uses workers and robots on its assembly line and has a Cobb-Douglas productivity function* of the form

$$P(x, y) = 10x^{0.2}y^{0.8} \text{ motors manufactured per day,}$$

where x is the number of assembly-line workers and y is the number of robots. Daily operating costs amount to \$100 per worker and \$16 per robot. How many workers and robots should be used to maximize productivity if the manufacturer has a daily budget of \$4,000?

* Cobb-Douglas production formulas were discussed in Section 4.6.

Solution Our objective function is the productivity $P(x, y)$, and the constraint is

$$100x + 16y = 4{,}000.$$

So the optimization problem is

$$\text{Maximize } P(x, y) = 10x^{0.2}y^{0.8} \text{ subject to } 100x + 16y = 4{,}000 \ (x \geq 0, y \geq 0).$$

Here, $g(x, y) = 100x + 16y - 4{,}000$, and the Lagrangian function is

$$\begin{aligned} L(x, y) &= P(x, y) - \lambda g(x, y) \\ &= 10x^{0.2}y^{0.8} - \lambda(100x + 16y - 4{,}000). \end{aligned}$$

The system of equations we need to solve is thus

$$\begin{aligned} L_x = 0&: \quad 2x^{-0.8}y^{0.8} - 100\lambda = 0 \\ L_y = 0&: \quad 8x^{0.2}y^{-0.2} - 16\lambda = 0 \\ g = 0&: \quad 100x + 16y = 4{,}000. \end{aligned}$$

We can rewrite the first two equations as

$$2\left(\frac{y}{x}\right)^{0.8} = 100\lambda \quad \text{and} \quad 8\left(\frac{x}{y}\right)^{0.2} = 16\lambda.$$

Dividing the first by the second to eliminate λ gives

$$\frac{1}{4}\left(\frac{y}{x}\right)^{0.8}\left(\frac{y}{x}\right)^{0.2} = \frac{100}{16}$$

that is, $$\frac{1}{4}\frac{y}{x} = \frac{25}{4},$$

giving $$y = 25x.$$

Substituting this result into the constraint equation gives

$$100x + 16(25x) = 4{,}000$$
$$500x = 4{,}000$$

so $$x = 8 \text{ workers} \quad \text{and} \quad y = 25x = 200 \text{ robots}$$

for a productivity of

$$P(8, 200) = 10(8)^{0.2}(200)^{0.8} \approx 1{,}051 \text{ motors manufactured per day.}$$

FAQs

When to Use Lagrange Multipliers

Q: *When can I use the method of Lagrange multipliers? When should I use it?*

A: We have discussed the method only when there is a single equality constraint. There is a generalization, which we have not discussed, that works when there are more equality constraints. (We need to introduce one multiplier for each constraint.) So if you have a problem with more than one equality constraint or with any inequality constraints, you must use the substitution method. On the other hand, if you have one equality constraint and it would be difficult to solve it for one of the variables, then you should use Lagrange multipliers.

8.4 EXERCISES

▼ more advanced ◆ challenging

T indicates exercises that should be solved using technology

In Exercises 1–6, solve the given optimization problem by using substitution. [**HINT**: See Example 1.]

1. Find the maximum value of $f(x, y, z) = 1 - x^2 - y^2 - z^2$ subject to $z = 2y$. Also find the corresponding point(s) (x, y, z).
2. Find the minimum value of $f(x, y, z) = x^2 + y^2 + z^2 - 2$ subject to $x = y$. Also find the corresponding point(s) (x, y, z).
3. Find the maximum value of $f(x, y, z) = 1 - x^2 - x - y^2 + y - z^2 + z$ subject to $3x = y$. Also find the corresponding point(s) (x, y, z).
4. Find the maximum value of $f(x, y, z) = 2x^2 + 2x + y^2 - y + z^2 - z - 1$ subject to $z = 2y$. Also find the corresponding point(s) (x, y, z).
5. Minimize $S = xy + 4xz + 2yz$ subject to $xyz = 1$ with $x > 0, y > 0, z > 0$.
6. Minimize $S = xy + xz + yz$ subject to $xyz = 2$ with $x > 0, y > 0, z > 0$.

In Exercises 7–18, use Lagrange multipliers to solve the given optimization problem. [**HINT**: See Example 2.]

7. Find the maximum value of $f(x, y) = xy$ subject to $x + 2y = 40$. Also find the corresponding point(s) (x, y).
8. Find the maximum value of $f(x, y) = xy$ subject to $3x + y = 60$. Also find the corresponding point(s) (x, y).
9. Find the maximum value of $f(x, y) = 4xy$ subject to $x^2 + y^2 = 8$. Also find the corresponding point(s) (x, y).
10. Find the maximum value of $f(x, y) = xy$ subject to $y = 3 - x^2$. Also find the corresponding point(s) (x, y).
11. Find the minimum value of $f(x, y) = x^2 + y^2$ subject to $x + 2y = 10$. Also find the corresponding point(s) (x, y).

12. Find the minimum value of $f(x, y) = x^2 + y^2$ subject to $xy^2 = 16$. Also find the corresponding point(s) (x, y).

13. The problem in Exercise 1. [**HINT:** See Example 3.]

14. The problem in Exercise 2. [**HINT:** See Example 3.]

15. The problem in Exercise 3.

16. The problem in Exercise 4.

17. The problem in Exercise 5.

18. The problem in Exercise 6.

19. ◆ Consider the following constrained optimization problem:

Minimize $f(x, y, z) = (x - 3)^2 + y^2 + z^2$
subject to $x^2 + y^2 - z = 0$.

a. Explain why this minimization problem must have a solution, and solve it using the method of Lagrange multipliers.
b. Solve it again using the substitution method by solving the constraint equation for z.
c. Now try to solve it using the substitution method by solving the constraint equation for y.
d. Explain what goes wrong in part (c).

20. ◆ Consider the following constrained optimization problem:

Minimize $f(x, y, z) = x^2 + (y + 3)^2 + (z - 4)^2$
subject to $4 - x^2 - y^2 - z = 0$.

a. Explain why this minimization problem must have a solution, and solve it using the method of Lagrange multipliers.
b. Solve it again using the substitution method by solving the constraint equation for z.
c. Now try to solve it using the substitution method by solving the constraint equation for x.
d. Explain what goes wrong in part (c).

Applications

Exercises 21–24 were solved in Section 5.2. This time, use the method of Lagrange multipliers to solve them.

21. ***Fences*** I want to fence in a rectangular vegetable patch. The fencing for the east and west sides costs \$4 per foot, and the fencing for the north and south sides costs only \$2 per foot. I have a budget of \$80 for the project. What is the largest area I can enclose?

22. ***Fences*** My orchid garden abuts my house so that the house itself forms the northern boundary. The fencing for the southern boundary costs \$4 per foot, and the fencing for the east and west sides costs \$2 per foot. If I have a budget of \$80 for the project, what is the largest area I can enclose?

23. ***Revenue*** *Hercules Films* is deciding on the price of the video release of its film *Son of Frankenstein.* Its marketing people estimate that at a price of p dollars, it can sell a total of $q = 200{,}000 - 10{,}000p$ copies. What price will bring in the greatest revenue?

24. ***Profit*** *Hercules Films* is also deciding on the price of the video release of its film *Bride of the Son of Frankenstein.* Again, marketing estimates that at a price of p dollars it can sell $q = 200{,}000 - 10{,}000p$ copies, but each copy costs \$4 to make. What price will give the greatest *profit*?

25. ***Geometry*** At what points on the sphere $x^2 + y^2 + z^2 = 1$ is the product xyz a maximum? (The method of Lagrange multipliers can be used.)

26. ***Geometry*** At what point on the surface $z = (x^2 + x + y^2 + 4)^{1/2}$ is the quantity $x^2 + y^2 + z^2$ a minimum? (The method of Lagrange multipliers can be used.)

27. ▼ ***Geometry*** What point on the surface $z = x^2 + y - 1$ is closest to the origin? [**HINT:** Minimize the square of the distance from (x, y, z) to the origin.]

28. ▼ ***Geometry*** What point on the surface $z = x + y^2 - 3$ is closest to the origin? [**HINT:** Minimize the square of the distance from (x, y, z) to the origin.]

29. ▼ ***Geometry*** Find the point on the plane $-2x + 2y + z - 5 = 0$ closest to $(-1, 1, 3)$. [**HINT:** Minimize the square of the distance from the given point to a general point on the plane.]

30. ▼ ***Geometry*** Find the point on the plane $2x - 2y - z + 1 = 0$ closest to $(1, 1, 0)$. [**HINT:** Minimize the square of the distance from the given point to a general point on the plane.]

31. ***Construction Cost*** A closed rectangular box is made with two kinds of materials. The top and bottom are made with heavy-duty cardboard costing 20¢ per square foot, and the sides are made with lightweight cardboard costing 10¢ per square foot. Given that the box is to have a capacity of 2 cubic feet, what should its dimensions be if the cost is to be minimized? [**HINT:** See Example 4.]

32. ***Construction Cost*** Repeat Exercise 31 assuming that the heavy-duty cardboard costs 30¢ per square foot, the lightweight cardboard costs 5¢ per square foot, and the box is to have a capacity of 6 cubic feet. [**HINT:** See Example 4.]

33. ***Package Dimensions: USPS*** The **U.S. Postal Service (USPS)** will accept only packages with a length plus girth no more than 108 inches.[27] (See the figure.)

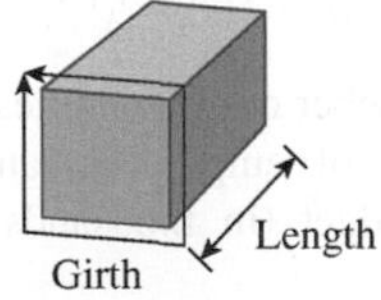

What are the dimensions of the largest-volume package that the USPS will accept? What is its volume? (This exercise is the same as Exercise 49 in Section 8.3. This time, solve it using Lagrange multipliers.)

[27] The requirement for packages sent other than Retail Ground, as of September 2015 (www.usps.com).

34. ***Package Dimensions: UPS*** **United Parcel Service (UPS)** will accept only packages with length no more than 108 inches and length plus girth no more than 165 inches.[28] (See figure for Exercise 33.) What are the dimensions of the largest-volume package that UPS will accept? What is its volume? (This exercise is the same as Exercise 50 in Section 8.3. This time, solve it using Lagrange multipliers.)

35. ▼ ***Construction Cost*** My company wishes to manufacture boxes similar to those described in Exercise 31 as cheaply as possible. Unfortunately, the company that manufactures the cardboard is unable to give me price quotes for the heavy-duty and lightweight cardboard. Find formulas for the dimensions of the box in terms of the price per square foot of heavy-duty and lightweight cardboard.

36. ▼ ***Construction Cost*** Repeat Exercise 35, assuming that only the bottoms of the boxes are to be made with heavy-duty cardboard.

37. ▼ ***Geometry*** Find the dimensions of the rectangular box with largest volume that can be inscribed above the xy-plane and under the paraboloid $z = 1 - (x^2 + y^2)$.

38. ▼ ***Geometry*** Find the dimensions of the rectangular box with largest volume that can be inscribed above the xy-plane and under the paraboloid $z = 2 - (2x^2 + y^2)$.

39. ***Productivity*** The *Gym Shirt Company* manufactures cotton socks. Production is partially automated through the use of robots. Daily operating costs amount to $150 per laborer and $60 per robot. The number of pairs of socks the company can manufacture in a day is given by a Cobb-Douglas production formula

$$q = 50n^{0.6}r^{0.4},$$

where q is the number of pairs of socks that can be manufactured by n laborers and r robots. Assuming that the company has a daily operating budget of $1,500 and wishes to maximize productivity, how many laborers and how many robots should it use? What is the productivity at these levels? [**HINT**: See Example 5.]

40. ***Productivity*** Your automobile assembly plant has a Cobb-Douglas production function given by

$$q = 100x^{0.3}y^{0.7},$$

where q is the number of automobiles it produces per year, x is the number of employees, and y is the monthly assembly-line budget (in thousands of dollars). Annual operating costs amount to an average of $60 thousand per employee plus the operating budget of $12y thousand. Your annual budget is $1,200,000. How many employees should you hire and what should your assembly-line budget be to maximize productivity? What is the productivity at these levels? [**HINT**: See Example 5.]

[28] The requirement as of September 2015 (www.ups.com).

Communication and Reasoning Exercises

41. Outline two methods of solution of the problem "*Maximize* $f(x, y, z)$ *subject to* $g(x, y, z) = 0$," and give an advantage and disadvantage of each.

42. Suppose we know that $f(x, y)$ has both partial derivatives in its domain D: $x > 0$, $y > 0$, and that (a, b) is the only point in D such that $f_x(a, b) = f_y(a, b) = 0$. Must it be the case that, if f has an absolute maximum, it occurs at (a, b)? Explain.

43. Under what circumstances would it be necessary to use the method of Lagrange multipliers?

44. Under what circumstances would the method of Lagrange multipliers not apply?

45. Restate the following problem as a maximization problem of the form "*Maximize* $f(x, y)$ *subject to* $g(x, y) = 0$":

Find the maximum value of $h(x) = 1 - 2x^2$.

46. Restate the following problem as a maximization problem of the form "*Maximize* $f(x, y, z)$ *subject to* $g(x, y, z) = 0$":

Find the maximum value of $h(x, y) = 1 - 2(x^2 + y^2)$.

47. ▼ If the partial derivatives of a function of several variables are always defined and never 0, is it possible for the function to have relative extrema when restricted to some domain? Explain your answer.

48. ▼ Give an example of a function f of three variables with an absolute maximum at $(0, 0, 0)$ but where the partial derivatives of f are never zero wherever they are defined.

49. ◆ **A linear programming problem in two variables** is a problem of the form: *Maximize (or minimize)* $f(x, y)$ *subject to constraints of the form* $C(x, y) \geq 0$ *or* $C(x, y) \leq 0$. Here, the objective function f and the constraints C are linear functions. There may be several linear constraints in one problem. Explain why the solution cannot occur in the interior of the domain of f.

50. ◆ Refer back to Exercise 49. Explain why the solution will actually be at a corner of the domain of f (where two or more of the line segments that make up the boundary meet). This result—or rather a slight generalization of it—is known as the Fundamental Theorem of Linear Programming.

8.5 Double Integrals and Applications

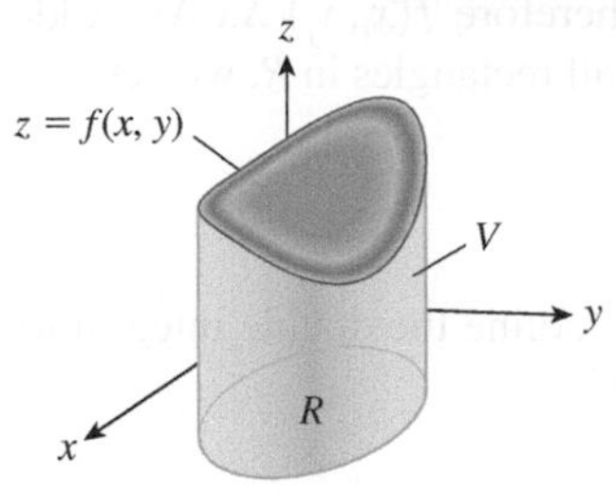

Figure 28

When discussing functions of one variable, we computed the area under a graph by integration. The analog for the graph of a function of two variables is the *volume V* under the graph, as in Figure 28. Think of the region R in the xy-plane as the "shadow" under the portion of the surface $z = f(x, y)$ shown.

By analogy with the definite integral of a function of one variable, we make the following definition.

Geometric Definition of the Double Integral

The **double integral of $f(x, y)$ over the region R in the xy-plane** is defined as

(Volume *above* the region R and under the graph of f)

−(Volume *below* the region R and above the graph of f).

We denote the double integral of $f(x, y)$ over the region R by $\iint_R f(x, y)\, dx\, dy$.

Quick Example

1. Take $f(x, y) = 2$ and take R to be the rectangle $0 \le x \le 1, 0 \le y \le 1$. Then the graph of f is a flat horizontal surface $z = 2$, and

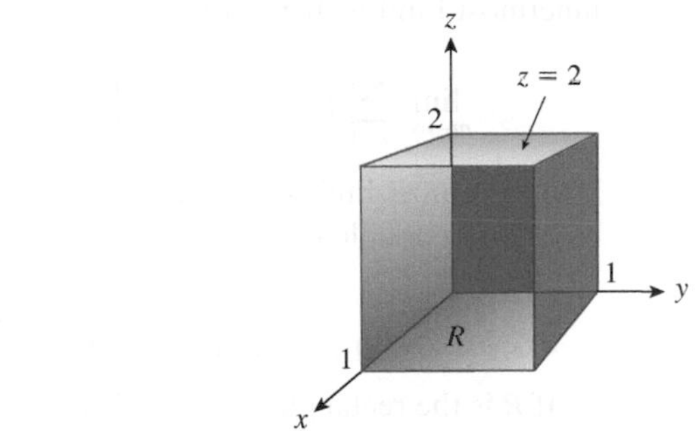

$$\iint_R f(x, y)\, dx\, dy = \text{Volume of box}$$
$$= \text{Width} \times \text{Length} \times \text{Height} = 1 \times 1 \times 2 = 2.$$

* The *Banach-Tarski paradox* is an example of the trouble we can get into if we try to rely on our intuition about volume. There are many descriptions of the paradox available on the web.

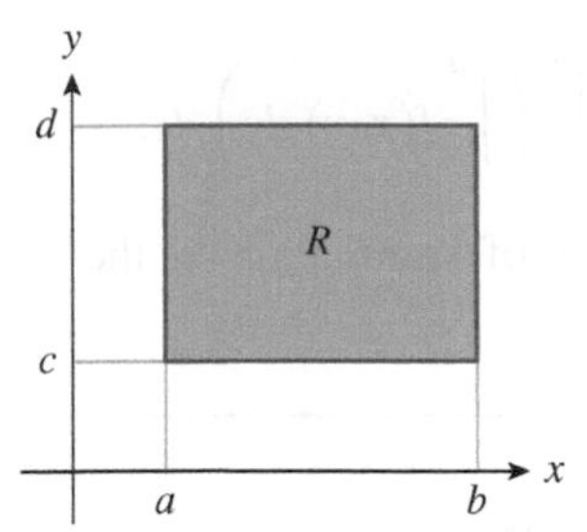

Figure 29

As we saw in the case of the definite integral of a function of one variable, we also desire *numerical* and *algebraic* definitions for two reasons: (1) to make the mathematical definition more precise, so as not to rely on any intuitive notion of "volume,"* and (2) for direct computation of the integral using technology or analytical tools.

We start with the simplest case: when the region R is a rectangle $a \le x \le b$ and $c \le y \le d$. (See Figure 29.) To compute the volume over R, we mimic what we did to find the area under the graph of a function of one variable. We break up the interval $[a, b]$ into m intervals all of width $\Delta x = (b - a)/m$, and we break up $[c, d]$ into n intervals all of width $\Delta y = (d - c)/n$. Figure 30 shows an example with $m = 4$ and $n = 5$.

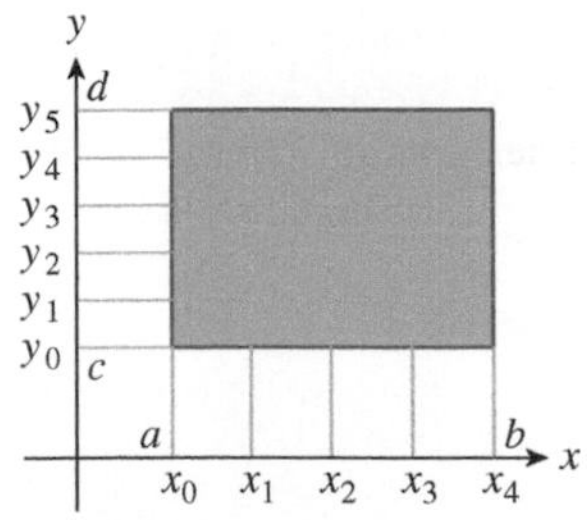

Figure 30

This gives us mn rectangles defined by $x_{i-1} \le x \le x_i$ and $y_{j-1} \le y \le y_j$. Over one of these rectangles, f is approximately equal to its value at one corner—say,

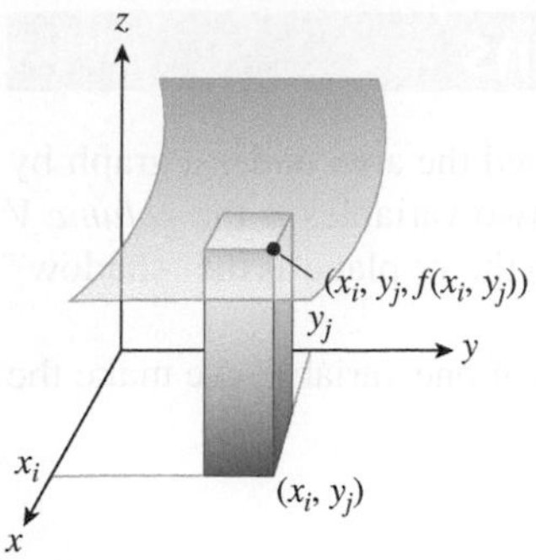

Figure 31

$f(x_i, y_j)$. The volume under f over this small rectangle is then approximately the volume of the rectangular brick (size exaggerated) shown in Figure 31. This brick has height $f(x_i, y_j)$, and its base is Δx by Δy. Its volume is therefore $f(x_i, y_j)\,\Delta x\,\Delta y$. Adding together the volumes of all of the bricks over the small rectangles in R, we get

$$\iint_R f(x, y)\,dx\,dy \approx \sum_{j=1}^{n}\sum_{i=1}^{m} f(x_i, y_j)\,\Delta x\,\Delta y.$$

This double sum is called a **double Riemann sum**. We define the double integral to be the limit of the Riemann sums as m and n go to infinity.

Algebraic Definition of the Double Integral

$$\iint_R f(x, y)\,dx\,dy = \lim_{n\to\infty}\lim_{m\to\infty}\sum_{j=1}^{n}\sum_{i=1}^{m} f(x_i, y_j)\,\Delta x\,\Delta y$$

Note This definition is adequate (the limit exists) when f is continuous. More elaborate definitions are needed for general functions. ■

This definition also gives us a clue about how to compute a double integral. The innermost sum is $\sum_{i=1}^{m} f(x_i, y_j)\,\Delta x$, which is a Riemann sum for $\int_a^b f(x, y_j)\,dx$. The innermost limit is therefore

$$\lim_{m\to\infty}\sum_{i=1}^{m} f(x_i, y_j)\,\Delta x = \int_a^b f(x, y_j)\,dx.$$

The outermost limit is then also a Riemann sum, and we get the following way of calculating double integrals.

Computing the Double Integral over a Rectangle

If R is the rectangle $a \le x \le b$ and $c \le y \le d$, then

$$\iint_R f(x, y)\,dx\,dy = \int_c^d\left(\int_a^b f(x, y)\,dx\right)dy = \int_a^b\left(\int_c^d f(x, y)\,dy\right)dx.$$

The second formula comes from switching the order of summation in the double sum.

Quick Example

2. If R is the rectangle $1 \le x \le 2$ and $1 \le y \le 3$, then

$$\iint_R 1\,dx\,dy = \int_1^3\left(\int_1^2 1\,dx\right)dy$$

$$= \int_1^3 [x]_{x=1}^{2}\,dy \qquad \text{Evaluate the inner integral.}$$

$$= \int_1^3 1\,dy \qquad [x]_{x=1}^{2} = 2 - 1 = 1.$$

$$= [y]_{y=1}^{3} = 3 - 1 = 2.$$

Quick Example 2 used a constant function for the integrand. Here is an example in which the integrand is not constant.

EXAMPLE 1 **Double Integral over a Rectangle**

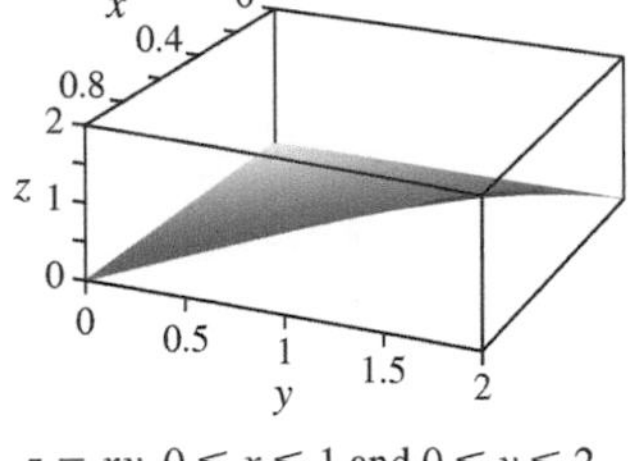

$z = xy, 0 \le x \le 1$ and $0 \le y \le 2$

Figure 32

Let R be the rectangle $0 \le x \le 1$ and $0 \le y \le 2$. Compute $\iint_R xy\,dx\,dy$. This integral gives the volume of the part of the boxed region under the surface $z = xy$ shown in Figure 32.

Solution

$$\iint_R xy\,dx\,dy = \int_0^2 \int_0^1 xy\,dx\,dy$$

(We usually drop the parentheses around the inner integral as we did here.) As in Quick Example 2, we compute this **iterated integral** from the inside out. First, we compute

$$\int_0^1 xy\,dx.$$

To do this computation, we do as we did when finding partial derivatives: We treat y as a constant. This gives

$$\int_0^1 xy\,dx = \left[\frac{x^2}{2}\cdot y\right]_{x=0}^{1} = \frac{1}{2}y - 0 = \frac{y}{2}.$$

We can now calculate the outer integral:

$$\int_0^2 \int_0^1 xy\,dx\,dy = \int_0^2 \frac{y}{2}\,dy = \left[\frac{y^2}{4}\right]_0^2 = 1.$$

➡ **Before we go on . . .** We could also reverse the order of integration in Example 1:

$$\int_0^1 \int_0^2 xy\,dy\,dx = \int_0^1 \left[x\cdot\frac{y^2}{2}\right]_{y=0}^{2} dx = \int_0^1 2x\,dx = [x^2]_0^1 = 1.$$ ■

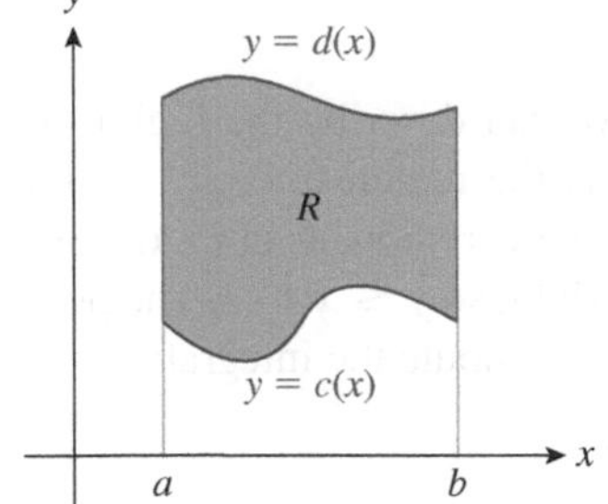

Figure 33

Often, we need to integrate over regions R that are not rectangular. There are two cases that come up. The first is a region like the one shown in Figure 33. In this region the bottom and top sides are defined by functions $y = c(x)$ and $y = d(x)$, respectively, so that the whole region can be described by the inequalities $a \le x \le b$ and $c(x) \le y \le d(x)$. To evaluate a double integral over such a region, we have the following formula.

Computing the Double Integral over a Nonrectangular Region

If R is the region $a \le x \le b$ and $c(x) \le y \le d(x)$ (Figure 33), then we integrate over R according to the following equation:

$$\iint_R f(x, y)\,dx\,dy = \int_a^b \int_{c(x)}^{d(x)} f(x, y)\,dy\,dx.$$

EXAMPLE 2 Double Integral over a Nonrectangular Region

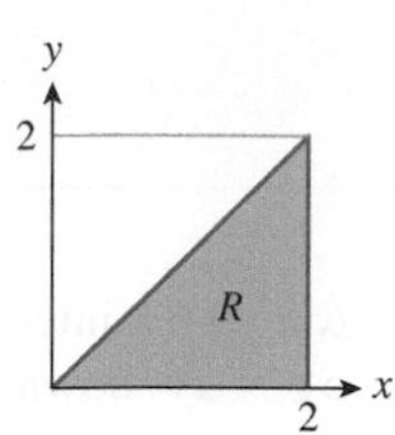

Figure 34

R is the triangle shown in Figure 34. Compute $\iint_R x\,dx\,dy$.

Solution R is the region described by $0 \le x \le 2$, $0 \le y \le x$. We have

$$\begin{aligned}\iint_R x\,dx\,dy &= \int_0^2 \int_0^x x\,dy\,dx \\ &= \int_0^2 [xy]_{y=0}^{x}\,dx \\ &= \int_0^2 x^2\,dx \\ &= \left[\frac{x^3}{3}\right]_0^2 = \frac{8}{3}.\end{aligned}$$

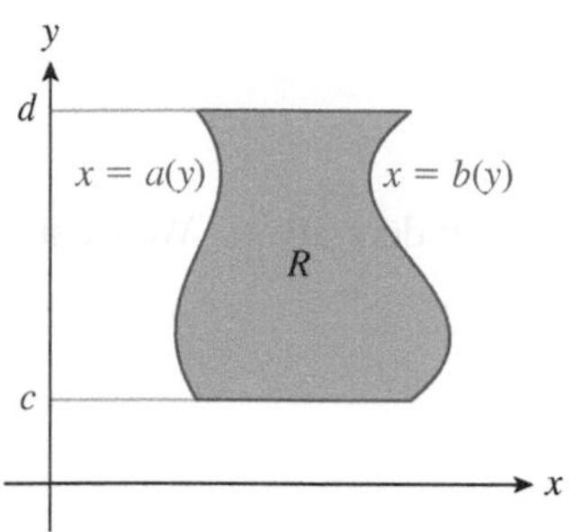

Figure 35

The second type of region is shown in Figure 35. This is the region described by $c \le y \le d$ and $a(y) \le x \le b(y)$. To evaluate a double integral over such a region, we have the following formula.

Double Integral over a Nonrectangular Region (continued)

If R is the region $c \le y \le d$ and $a(y) \le x \le b(y)$ (Figure 35), then we integrate over R according to the following equation:

$$\iint_R f(x, y)\,dx\,dy = \int_c^d \int_{a(y)}^{b(y)} f(x, y)\,dx\,dy.$$

EXAMPLE 3 Double Integral over a Nonrectangular Region

Redo Example 2, integrating in the opposite order.

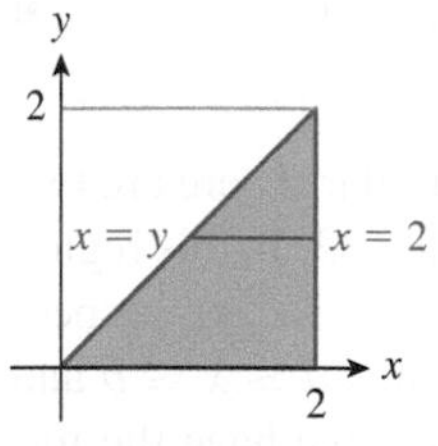

Figure 36

Solution We can integrate in the opposite order if we can describe the region in Figure 34 in the way shown in Figure 35. In fact, it is the region $0 \le y \le 2$ and $y \le x \le 2$. To see this, we draw a horizontal line through the region, as in Figure 36. The line extends from $x = y$ on the left to $x = 2$ on the right, so $y \le x \le 2$. The possible heights for such a line are $0 \le y \le 2$. We can now compute the integral:

$$\begin{aligned}\iint_R x\,dx\,dy &= \int_0^2 \int_y^2 x\,dx\,dy \\ &= \int_0^2 \left[\frac{x^2}{2}\right]_{x=y}^{2} dy \\ &= \int_0^2 \left(2 - \frac{y^2}{2}\right) dy \\ &= \left[2y - \frac{y^3}{6}\right]_0^2 = \frac{8}{3}.\end{aligned}$$

Note Many regions can be described in two different ways, as we saw in Examples 2 and 3. Sometimes one description will be much easier to work with than the other, so it pays to consider both. ■

Applications

There are many applications of double integrals besides finding volumes. For example, we can use them to find *averages*. Remember that the average of $f(x)$ on $[a, b]$ is given by $\int_a^b f(x)\,dx$ divided by $(b - a)$, the length of the interval.

Average of a Function of Two Variables

The average of $f(x, y)$ on the region R is

$$\bar{f} = \frac{1}{A}\iint_R f(x, y)\,dx\,dy.$$

Here, A is the area of R. We can compute the area A geometrically, by using the techniques from Section 7.2, or by computing

$$A = \iint_R 1\,dx\,dy.$$

Quick Example

3. The average value of $f(x, y) = xy$ on the rectangle given by $0 \le x \le 1$ and $0 \le y \le 2$ is

$$\bar{f} = \frac{1}{2}\iint_R xy\,dx\,dy \qquad \text{The area of the rectangle is 2.}$$

$$= \frac{1}{2}\int_0^2\int_0^1 xy\,dx\,dy$$

$$= \frac{1}{2}\cdot 1 = \frac{1}{2}. \qquad \text{We calculated the integral in Example 1.}$$

EXAMPLE 4 Average Revenue

Your company is planning to price its new line of subcompact cars at between \$10,000 and \$15,000. The marketing department reports that if the company prices the cars at p dollars per car, the demand will be between $q = 20{,}000 - p$ and $q = 25{,}000 - p$ cars sold in the first year. What is the average of all the possible revenues your company could expect in the first year?

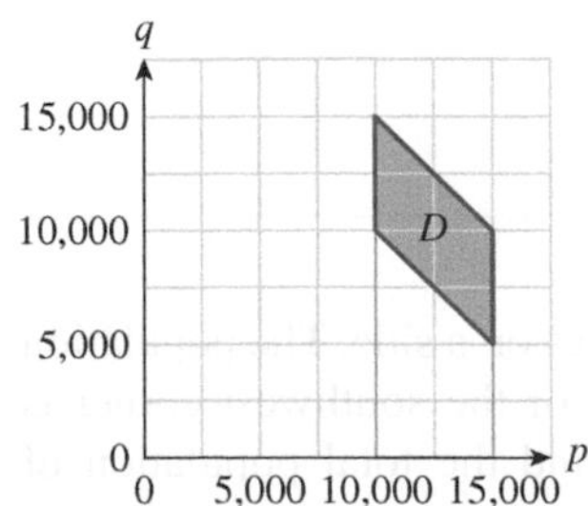

Figure 37

Solution Revenue is given by $R = pq$ as usual, and we are told that

$$10{,}000 \le p \le 15{,}000$$

and $\quad 20{,}000 - p \le q \le 25{,}000 - p.$

This domain D of prices and demands is shown in Figure 37.

To average the revenue R over the domain D, we need to compute the area A of D. Using either calculus or geometry, we get $A = 25{,}000{,}000$. We then need to integrate R over D:

$$\iint_D pq\,dp\,dq = \int_{10{,}000}^{15{,}000}\int_{20{,}000-p}^{25{,}000-p} pq\,dq\,dp$$
$$= \int_{10{,}000}^{15{,}000}\left[\frac{pq^2}{2}\right]_{q=20{,}000-p}^{25{,}000-p} dp$$
$$= \frac{1}{2}\int_{10{,}000}^{15{,}000}[p(25{,}000 - p)^2 - p(20{,}000 - p)^2]\,dp$$
$$= \frac{1}{2}\int_{10{,}000}^{15{,}000}[225{,}000{,}000p - 10{,}000p^2]\,dp$$
$$\approx 3{,}072{,}900{,}000{,}000{,}000.$$

The average of all the possible revenues your company could expect in the first year is therefore

$$\bar{R} = \frac{3{,}072{,}900{,}000{,}000{,}000}{25{,}000{,}000} \approx \$122{,}900{,}000.$$

➡ **Before we go on...** To check that the answer obtained in Example 4 is reasonable, notice that the revenues at the corners of the domain are \$100,000,000 per year, \$150,000,000 per year (at two corners), and \$75,000,000 per year. Some of these are smaller than the average and some are larger, as we would expect. ■

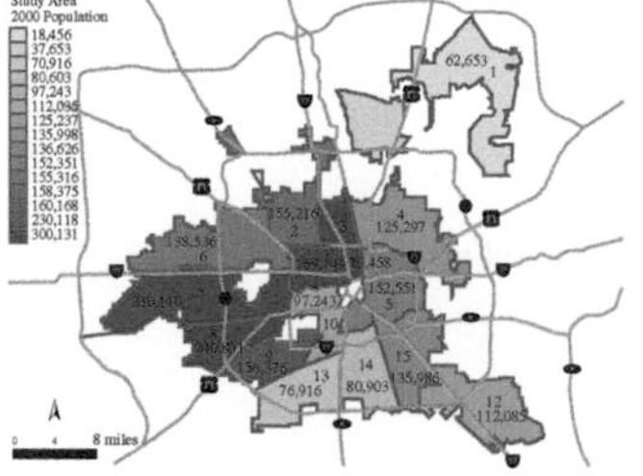

Darker regions have higher population density

Figure 38

Another useful application of the double integral comes about when we consider density. For example, suppose that $P(x, y)$ represents the population density (in people per square mile, say) in the city of Houston, shown in Figure 38.

If we break the city up into small rectangles (for example, city blocks), then the population in the small rectangle $x_{i-1} \le x \le x_i$ and $y_{j-1} \le y \le y_j$ is approximately $P(x_i, y_j)\,\Delta x\,\Delta y$. Adding up all of these population estimates, we get

$$\text{Total population} \approx \sum_{j=1}^{n}\sum_{i=1}^{m} P(x_i, y_j)\,\Delta x\,\Delta y.$$

Because this is a double Riemann sum, when we take the limit as m and n go to infinity, we get the following calculation of the population of the city:

$$\text{Total population} = \iint_{\text{City}} P(x, y)\,dx\,dy.$$

EXAMPLE 5 Population

Squaresville is a city in the shape of a square that is 5 miles on a side. The population density at a distance of x miles east and y miles north of the southwest corner is $P(x, y) = x^2 + y^2$ thousand people per square mile. Find the total population of Squaresville.

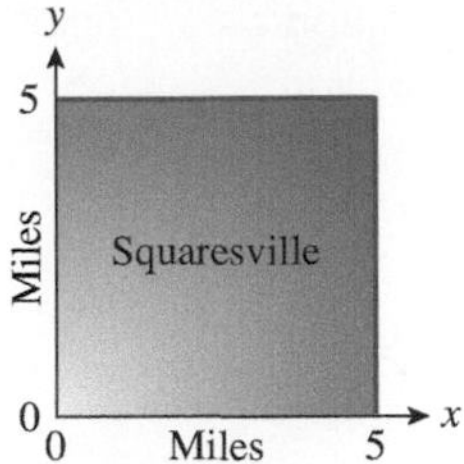

Figure 39

Solution Squaresville is pictured in Figure 39, in which we put the origin in the southwest corner of the city.

To compute the total population, we integrate the population density over the city.

$$\begin{aligned}\text{Total population} &= \iint_{\text{Squaresville}} P(x, y)\,dx\,dy \\ &= \int_0^5 \int_0^5 (x^2 + y^2)\,dx\,dy \\ &= \int_0^5 \left[\frac{x^3}{3} + xy^2\right]_{x=0}^{5} dy \\ &= \int_0^5 \left[\frac{125}{3} + 5y^2\right] dy \\ &= \frac{1{,}250}{3} \approx 417 \text{ thousand people.}\end{aligned}$$

➡ **Before we go on . . .** Note that the average population density is the total population divided by the area of the city, which is about 17,000 people per square mile in Example 5. This is the same as the calculation of the average of $P(x, y)$ over the city. (Compare to Example 4.) ■

8.5 EXERCISES

▼ more advanced ◆ challenging

T indicates exercises that should be solved using technology

In Exercises 1–16, compute the given integral. [**HINT:** See Example 1.]

1. $\int_0^1 \int_0^1 (x - 2y)\,dx\,dy$

2. $\int_{-1}^1 \int_0^2 (2x + 3y)\,dx\,dy$

3. $\int_0^1 \int_0^2 (ye^x - x - y)\,dx\,dy$

4. $\int_1^2 \int_2^3 \left(\frac{1}{x} + \frac{1}{y}\right) dx\,dy$

5. $\int_0^2 \int_0^3 e^{x+y}\,dx\,dy$

6. $\int_0^1 \int_0^1 e^{x-y}\,dx\,dy$

7. $\int_0^1 \int_0^{2-y} x\,dx\,dy$

8. $\int_0^1 \int_0^{2-y} y\,dx\,dy$

9. $\int_{-1}^1 \int_{y-1}^{y+1} e^{x+y}\,dx\,dy$
[**HINT:** See Example 2.]

10. $\int_0^1 \int_y^{y+2} \frac{1}{\sqrt{x + y}}\,dx\,dy$
[**HINT:** See Example 2.]

11. $\int_0^1 \int_{-x^2}^{x^2} x\,dy\,dx$

12. $\int_1^4 \int_{-\sqrt{x}}^{\sqrt{x}} \frac{1}{x}\,dy\,dx$

13. $\int_0^1 \int_0^x e^{x^2}\,dy\,dx$

14. $\int_0^1 \int_0^{x^2} e^{x^3+1}\,dy\,dx$

15. $\int_0^2 \int_{1-x}^{8-x} (x + y)^{1/3}\,dy\,dx$

16. $\int_1^2 \int_{1-2x}^{x^2} \frac{x + 1}{(2x + y)^3}\,dy\,dx$

In Exercises 17–24, find $\iint_R f(x, y)\,dx\,dy$*, where R is the indicated domain. (Remember that you often have a choice as to the order of integration.)* [**HINT:** See Example 2.]

17. $f(x, y) = 2$

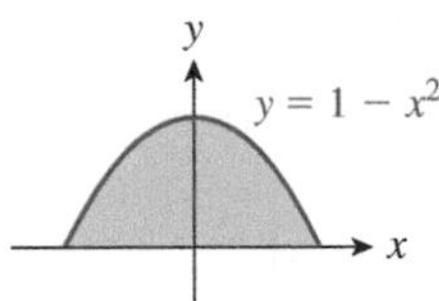

18. $f(x, y) = x$

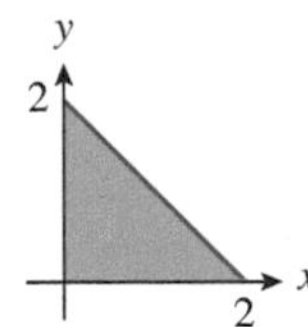

19. $f(x, y) = 1 + y$
[**HINT:** See Example 3.]

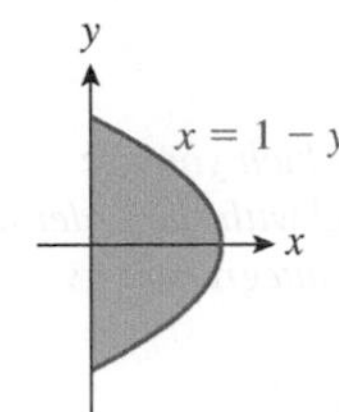

20. $f(x, y) = e^{x+y}$
[**HINT:** See Example 3.]

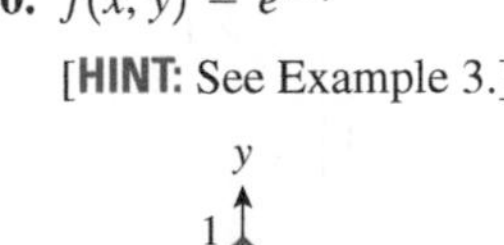
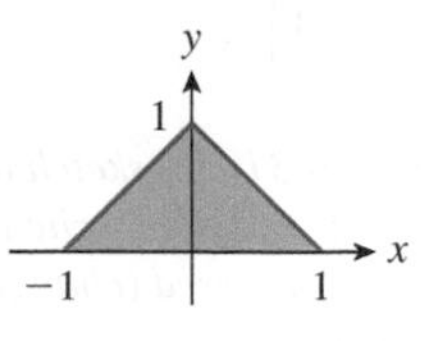

21. $f(x, y) = xy^2$ **22.** $f(x, y) = xy^2$

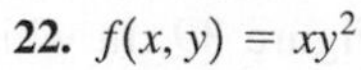

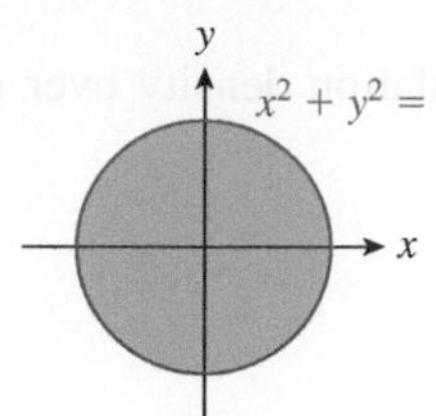

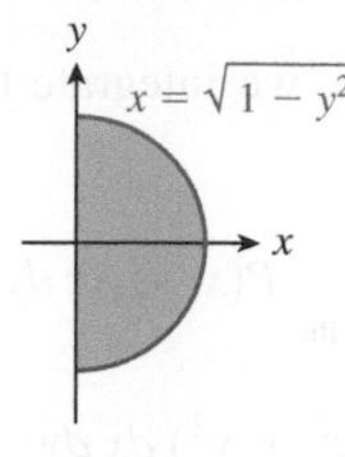

23. $f(x, y) = x^2 + y^2$ **24.** $f(x, y) = x^2$

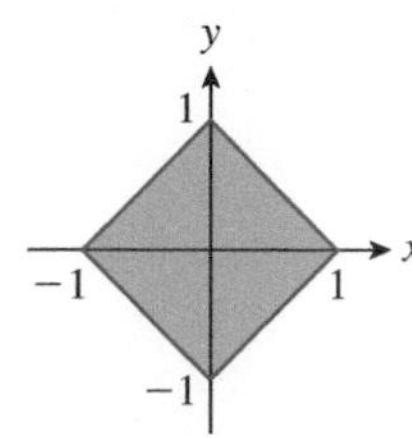

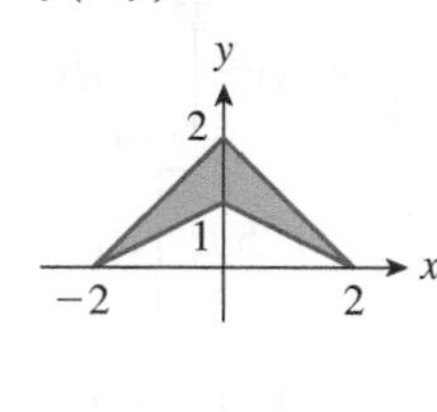

In Exercises 25–30, find the average value of the given function over the indicated domain. [**HINT**: See Quick Example 3.]

25. $f(x, y) = y$ **26.** $f(x, y) = 2 + x$

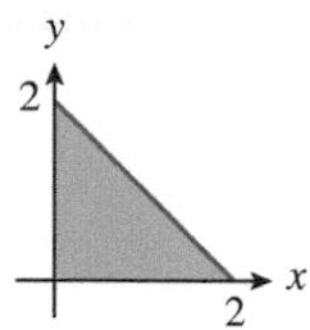

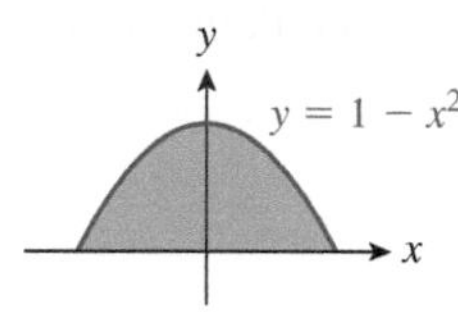

27. $f(x, y) = e^y$ **28.** $f(x, y) = y$

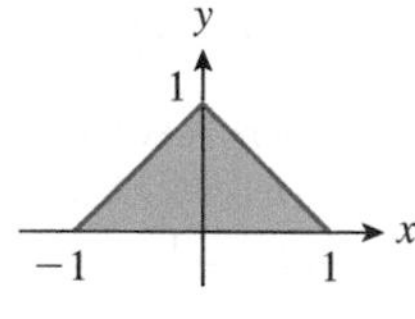

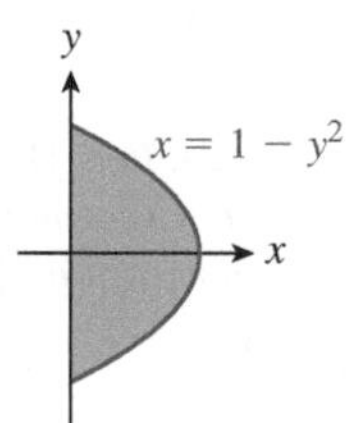

29. $f(x, y) = x^2 + y^2$ **30.** $f(x, y) = x^2$

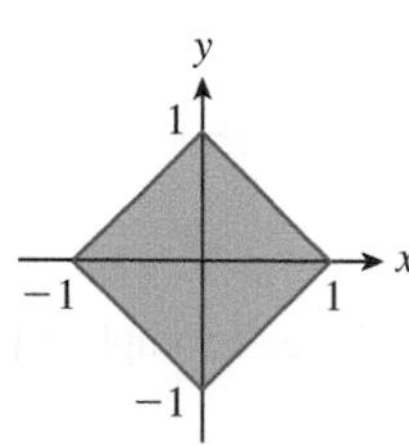

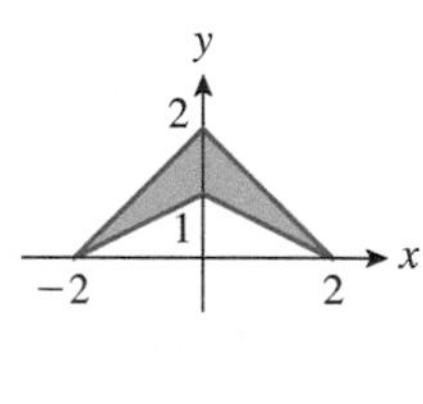

In Exercises 31–36, sketch the region over which you are integrating, and then write down the integral with the order of integration reversed (changing the limits of integration as necessary).

31. ▼ $\int_0^1 \int_0^{1-y} f(x, y)\, dx\, dy$ **32.** ▼ $\int_{-1}^1 \int_0^{1+y} f(x, y)\, dx\, dy$

33. ▼ $\int_{-1}^1 \int_0^{\sqrt{1+y}} f(x, y)\, dx\, dy$

34. ▼ $\int_{-1}^1 \int_0^{\sqrt{1-y}} f(x, y)\, dx\, dy$

35. ▼ $\int_1^2 \int_1^{4/x^2} f(x, y)\, dy\, dx$

36. ▼ $\int_1^{e^2} \int_0^{\ln x} f(x, y)\, dy\, dx$

37. Find the volume under the graph of $z = 1 - x^2$ over the region $0 \le x \le 1$ and $0 \le y \le 2$.

38. Find the volume under the graph of $z = 1 - x^2$ over the triangle $0 \le x \le 1$ and $0 \le y \le 1 - x$.

39. ▼ Find the volume of the tetrahedron shown in the figure. Its corners are $(0, 0, 0)$, $(1, 0, 0)$, $(0, 1, 0)$, and $(0, 0, 1)$.

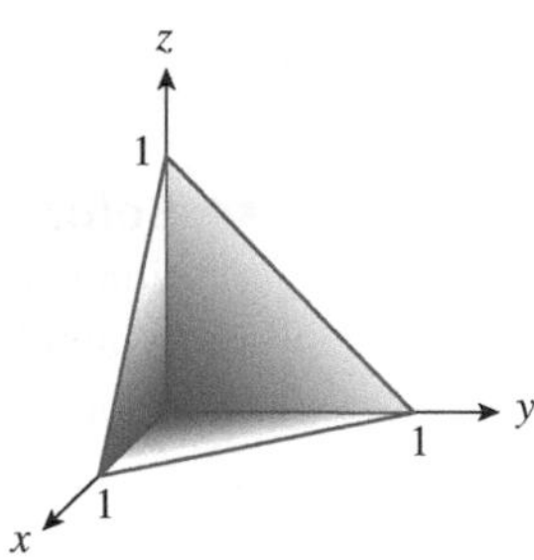

40. ▼ Find the volume of the tetrahedron with corners at $(0, 0, 0)$, $(a, 0, 0)$, $(0, b, 0)$, and $(0, 0, c)$.

Applications

41. ***Productivity*** A productivity model at the *Handy Gadget Company* is

$$P = 10{,}000x^{0.3}y^{0.7},$$

where P is the number of gadgets the company turns out per month, x is the number of employees at the company, and y is the monthly operating budget in thousands of dollars. Because the company hires part-time workers, it uses anywhere between 45 and 55 workers each month, and its operating budget varies from \$8,000 to \$12,000 per month. What is the average of the possible numbers of gadgets the company can turn out per month? (Round the answer to the nearest 1,000 gadgets.) [**HINT**: See Quick Example 3.]

42. ***Productivity*** Repeat Exercise 41 using the productivity model

$$P = 10{,}000x^{0.7}y^{0.3}.$$

43. ***Revenue*** Your latest game app is expected to sell between $q = 8{,}000 - p^2$ and $q = 10{,}000 - p^2$ copies if priced at p dollars. You plan to set the price between \$40 and \$50. What is the average of all the possible revenues you can make? [**HINT**: See Example 4.]

44. ***Revenue*** Your latest solid-state drive is expected to sell between $q = 180{,}000 - p^2$ and $q = 200{,}000 - p^2$ units if priced at p dollars. You plan to set the price between \$300 and \$400. What is the average of all the possible revenues you can make? [**HINT:** See Example 4.]

45. ***Revenue*** Your self-published novel has demand curves between $p = 15{,}000/q$ and $p = 20{,}000/q$. You expect to sell between 500 and 1,000 copies. What is the average of all the possible revenues you can make?

46. ***Revenue*** Your self-published book of poetry has demand curves between $p = 80{,}000/q^2$ and $p = 100{,}000/q^2$. You expect to sell between 50 and 100 copies. What is the average of all the possible revenues you can make?

47. ***Population Density*** The town of West Podunk is shaped like a rectangle that is 20 miles from west to east and 30 miles from north to south. (See the figure.) It has a population density of $P(x, y) = e^{-0.1(x+y)}$ hundred people per square mile x miles east and y miles north of the southwest corner of town. What is the total population of the town? [**HINT:** See Example 5.]

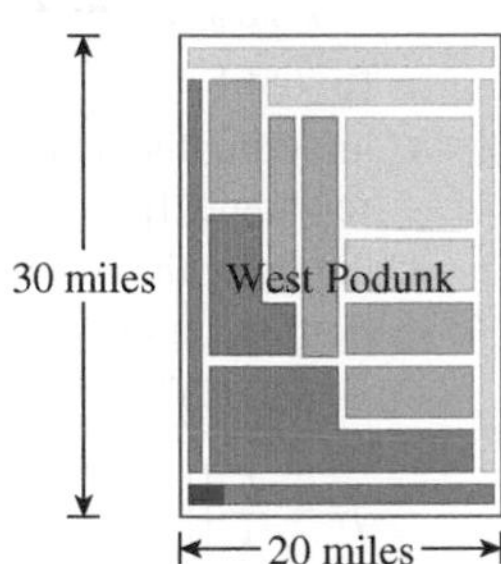

48. ***Population Density*** The town of East Podunk is shaped like a triangle with an east-west base of 20 miles and a north-south height of 30 miles. (See the figure.) It has a population density of $P(x, y) = e^{-0.1(x+y)}$ hundred people per square mile x miles east and y miles north of the southwest corner of town. What is the total population of the town? [**HINT:** See Example 5.]

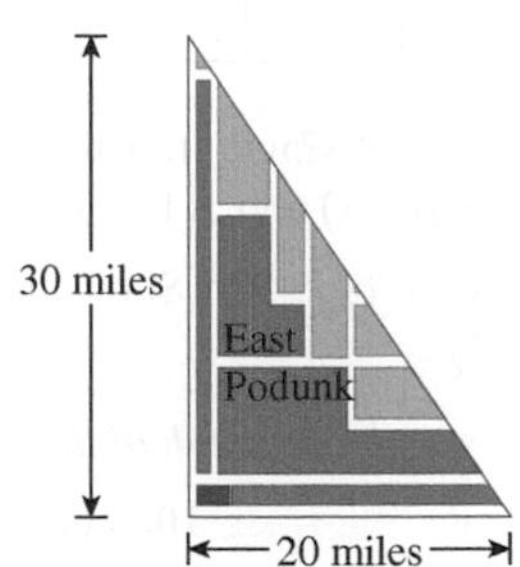

49. ***Temperature*** The temperature at the point (x, y) on the square with vertices $(0, 0)$, $(0, 1)$, $(1, 0)$, and $(1, 1)$ is given by $T(x, y) = x^2 + 2y^2$ degrees Celsius. Find the average temperature on the square.

50. ***Temperature*** The temperature at the point (x, y) on the square with vertices $(0, 0)$, $(0, 1)$, $(1, 0)$, and $(1, 1)$ is given by $T(x, y) = x^2 + 2y^2 - x$ degrees Celsius. Find the average temperature on the square.

Communication and Reasoning Exercises

51. Explain how double integrals can be used to compute the area between two curves in the xy plane.

52. Explain how double integrals can be used to compute the volume of solids in 3-space.

53. Complete the following: The first step in calculating an integral of the form

$$\int_a^b \int_{r(x)}^{s(x)} f(x, y)\, dy\, dx$$

is to evaluate the integral ______, obtained by holding ___ constant and integrating with respect to ___.

54. If the units of $f(x, y)$ are zonars per square meter and x and y are given in meters, what are the units of $\int_a^b \int_{r(x)}^{s(x)} f(x, y)\, dy\, dx$?

55. If the units of $\int_a^b \int_{r(x)}^{s(x)} f(x, y)\, dy\, dx$ are paintings, the units of x are picassos, and the units of y are dalis, what are the units of $f(x, y)$?

56. Complete the following: If the region R is bounded on the left and right by vertical lines and on the top and bottom by the graphs of functions of x, then we integrate over R by first integrating with respect to ____ and then with respect to ____.

57. ▼ Show that if a, b, c, and d are constant, then

$$\int_a^b \int_c^d f(x)g(y)\, dx\, dy = \int_c^d f(x)\, dx \int_a^b g(y)\, dy.$$

Test this result on the integral $\int_0^1 \int_1^2 ye^x\, dx\, dy$.

58. ▼ Refer to Exercise 57. If a, b, c, and d are constants, can

$$\int_a^b \int_c^d \frac{f(x)}{g(y)}\, dx\, dy$$

be expressed as a product of two integrals? Explain.

CHAPTER 8 REVIEW

KEY CONCEPTS

 **www.WanerMath.com**

Go to the Website to find a comprehensive and interactive Web-based summary of Chapter 8.

8.1 Functions of Several Variables from the Numerical, Algebraic, and Graphical Viewpoints

A real-valued function, f, of $x, y, z, \ldots$ [p. 608]

Cost functions [p. 609]

A linear function of the variables $x_1, x_2, \ldots, x_n$ is a function of the form $f(x_1, x_2, \ldots, x_n) = a_0 + a_1x_1 + \cdots + a_nx_n$ $(a_0, a_1, \ldots, a_n$ constants) [p. 610]

Representing functions of two variables numerically [p. 611]

Using a spreadsheet to represent a function of two variables [p. 612]

Plotting points in three dimensions [p. 613]

Graph of a function of two variables [p. 613]

Analyzing the graph of a function of two variables [p. 616]

Graph of a linear function [p. 618]

8.2 Partial Derivatives

Definition of partial derivatives [p. 627]

Application to marginal cost: linear cost function [p. 628]

Application to marginal cost: interaction cost function [p. 629]

Geometric interpretation of partial derivatives [p. 630]

Second-order partial derivatives [p. 631]

8.3 Maxima and Minima

Definition of relative maximum and minimum [p. 637]

Locating candidates for relative maxima and minima [p. 637]

Classifying critical points graphically [p. 639]

Classifying critical points numerically [p. 639]

Second derivative test for a function of two variables [p. 640]

Using the second derivative test [p. 640]

Formulas for linear regression:

$$m = \frac{n(\Sigma xy) - (\Sigma x)(\Sigma y)}{n(\Sigma x^2) - (\Sigma x)^2}$$

$$b = \frac{\Sigma y - m(\Sigma x)}{n}$$

n = number of data points [p. 641]

8.4 Constrained Maxima and Minima and Applications

Constrained maximum and minimum problem [p. 645]

Solving constrained maxima and minima problems using substitution [p. 646]

The method of Lagrange multipliers [p. 647]

Using Lagrange multipliers [p. 648]

8.5 Double Integrals and Applications

Geometric definition of the double integral [p. 655]

Algebraic definition of the double integral:

$$\iint_R f(x, y)\, dx\, dy = \lim_{n\to\infty} \lim_{m\to\infty} \sum_{j=1}^{n} \sum_{i=1}^{m} f(x_i, y_j)\, \Delta x\, \Delta y$$

[p. 656]

Computing the double integral over a rectangle [p. 656]

Computing the double integral over nonrectangular regions [p. 657]

Average of $f(x, y)$ on the region R:

$$\bar{f} = \frac{1}{A}\iint_R f(x, y)\, dx\, dy \quad \text{[p. 659]}$$

REVIEW EXERCISES

1. Let $f(x, y, z) = \dfrac{x}{y + xz} + x^2y$. Evaluate $f(0, 1, 1)$, $f(2, 1, 1)$, $f(-1, 1, -1)$, $f(z, z, z)$, and $f(x + h, y + k, z + l)$.

2. Let $g(x, y, z) = xy(x + y - z) + x^2$. Evaluate $g(0, 0, 0)$, $g(1, 0, 0)$, $g(0, 1, 0)$, $g(x, x, x)$, and $g(x, y + k, z)$.

3. Let $f(x, y, z) = 2.72 - 0.32x - 3.21y + 12.5z$. Complete the following: f ___ by ___ units for every 1 unit of increase in x and ___ by ___ units for every unit of increase in z.

4. Let $g(x, y, z) = 2.16x + 11y - 1.53z + 31.4$. Complete the following: g ___ by ___ units for every 1 unit of increase in y and ___ by ___ units for every unit of increase in z.

In Exercises 5 and 6, complete the given table for values for $h(x, y) = 2x^2 + xy - x$.

5.

$y \downarrow$ \ $x \to$	−1	0	1
−1			
0			
1			

6.

$y \downarrow$ \ $x \to$	−2	2	3
−2			
2			
3			

7. Give a formula for a (single) function f with the property that $f(x, y) = -f(y, x)$ and $f(1, -1) = 3$.

8. Let $f(x, y) = x^2 + (y + 1)^2$. Show that $f(y, x) = f(x + 1, y - 1)$.

In Exercises 9–14, sketch the graph of the given function.

9. $r(x, y) = x + y$

10. $r(x, y) = x - y$

11. $t(x, y) = x^2 + 2y^2$. Show cross sections at $x = 0$ and $z = 1$.

12. $t(x, y) = \dfrac{1}{2}x^2 + y^2$. Show cross sections at $x = 0$ and $z = 1$.

13. $f(x, y) = -2\sqrt{x^2 + y^2}$. Show cross sections at $z = -4$ and $y = 1$.

14. $f(x, y) = 2 + 2\sqrt{x^2 + y^2}$. Show cross sections at $z = 4$ and $y = 1$.

In Exercises 15–20, compute the partial derivatives shown for the given function.

15. $f(x, y) = x^2 + xy$; find f_x, f_y, and f_{yy}.

16. $f(x, y) = \dfrac{6}{xy} + \dfrac{xy}{6}$; find f_x, f_y, and f_{yy}.

17. $f(x, y) = 4x + 5y - 6xy$; find $f_{xx}(1, 0) - f_{xx}(3, 2)$.

18. $f(x, y) = e^{xy} + e^{3x^2-y^2}$; find $\dfrac{\partial f}{\partial x}$ and $\dfrac{\partial^2 f}{\partial x \partial y}$.

19. $f(x, y, z) = \dfrac{x}{x^2 + y^2 + z^2}$; find $\dfrac{\partial f}{\partial x}, \dfrac{\partial f}{\partial y}, \dfrac{\partial f}{\partial z}$, and $\left.\dfrac{\partial f}{\partial x}\right|_{(0, 1, 0)}$.

20. $f(x, y, z) = x^2 + y^2 + z^2 + xyz$; find $f_{xx} + f_{yy} + f_{zz}$.

In Exercises 21–26, locate and classify all critical points.

21. $f(x, y) = (x - 1)^2 + (2y - 3)^2$

22. $g(x, y) = (x - 1)^2 - 3y^2 + 9$

23. $k(x, y) = x^2y - x^2 - y^2$ **24.** $j(x, y) = xy + x^2$

25. $h(x, y) = e^{xy}$

26. $f(x, y) = \ln(x^2 + y^2) - (x^2 + y^2)$

In Exercises 27–30, solve the given constrained optimization problem by using substitution to eliminate a variable. (Do not use Lagrange multipliers.)

27. Find the largest value of xyz subject to $x + y + z = 1$ with $x > 0, y > 0, z > 0$. Also find the corresponding point(s) (x, y, z).

28. Find the minimum value of $f(x, y, z) = x^2 + y^2 + z^2 - 1$ subject to $x = y + z$. Also find the corresponding point(s) (x, y, z).

29. Find the point on the surface $z = \sqrt{x^2 + 2(y - 3)^2}$ closest to the origin.

30. Minimize $S = xy + x^2z^2 + 4yz$ subject to $xyz = 1$ with $x > 0, y > 0, z > 0$.

In Exercises 31–34, use Lagrange multipliers to solve the given optimization problem.

31. Find the minimum value of $f(x, y) = x^2 + y^2$ subject to $xy = 2$. Also find the corresponding point(s) (x, y).

32. The problem in Exercise 28.

33. The problem in Exercise 29.

34. The problem in Exercise 30.

In Exercises 35–40, compute the given quantities.

35. $\displaystyle\int_0^1 \int_0^2 2xy\, dx\, dy$

36. $\displaystyle\int_1^2 \int_0^1 xye^{x+y}\, dx\, dy$

37. $\displaystyle\int_0^2 \int_0^{2x} \frac{1}{x^2 + 1}\, dy\, dx$

38. The average value of xye^{x+y} over the rectangle $0 \le x \le 1$, $1 \le y \le 2$

39. $\iint_R (x^2 - y^2)\, dx\, dy$, where R is the region shown in the figure

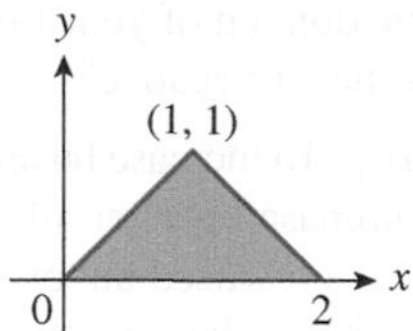

40. The volume under the graph of $z = 1 - y$ over the region in the xy-plane between the parabola $y = 1 - x^2$ and the x-axis

Applications: OHaganBooks.com [Try the game at www.OHaganBooks.com]

41. ***Website Traffic*** OHaganBooks.com has two principal competitors: *JungleBooks.com* and *FarmerBooks.com.* Current website traffic at OHaganBooks.com is estimated at 5,000 hits per day. This number is predicted to decrease by 0.8 for every new customer of JungleBooks.com and by 0.6 for every new customer of FarmerBooks.com.

a. Use this information to model the daily website traffic at OHaganBooks.com as a linear function of the new customers of its two competitors.

b. According to the model, if Junglebooks.com gets 100 new customers and OHaganBooks.com traffic drops to 4,770 hits per day, how many new customers has FarmerBooks.com obtained?

c. The model in part (a) did not take into account the growth of the total online consumer base. OHaganBooks .com expects to get approximately one additional hit per day for every 10,000 new Internet shoppers. Modify your model in part (a) to include this information using a new independent variable.

d. How many new Internet shoppers would it take to offset the effects on traffic at OHaganBooks.com of 100 new customers at each of its competitor sites?

42. ***Productivity*** Billy-Sean O'Hagan is writing his PhD thesis in biophysics but finds that his productivity is affected by the temperature and the number of text messages he receives per hour. On a brisk winter's day when the temperature is 0°C and there are no text messages, Billy-Sean can produce 15 pages of his thesis. His productivity goes down by 0.3 pages per degree Celsius increase in the temperature and by 1.2 pages for each additional text message per hour.

a. Use this information to model Billy-Sean's productivity p as a function of the temperature and the hourly rate of text messages.

b. The other day the temperature was 20°C, and Billy-Sean managed to produce only three pages of his thesis. What was the hourly rate of incoming text messages?

c. Billy-Sean finds that each cup of coffee he drinks per hour can counter the effect on his productivity of two text messages per hour. Modify the model in part (a) to take consumption of coffee into account.

d. What would the domain of your function look like to ensure that p is never negative?

43. ***Internet Advertising*** To increase business at OHaganBooks.com, you have purchased banner ads at well-known Internet portals and have advertised on television. The following interaction model shows the average number h of hits per day as a function of monthly expenditures x on banner ads and y on television advertising (x and y are in dollars):

$$h(x, y) = 1{,}800 + 0.05x + 0.08y + 0.00003xy.$$

a. Based on your model, how much traffic can you anticipate if you spend \$2,000 per month for banner ads and \$3,000 per month on television advertising?

b. Evaluate $\frac{\partial h}{\partial y}$, specify its units of measurement, and indicate whether it increases or decreases with increasing x.

c. How much should the company spend on banner ads to obtain 1 hit per day for each \$5 spent per month on television advertising?

44. ***Company Retreats*** Their companies having recently been bailed out by the government at taxpayer expense, Marjory Duffin and John O'Hagan are planning a joint winter business retreat in Cancun, but they are not sure how many sales reps to take along. The following interaction model shows the estimated cost C to their companies (in dollars) as a function of the number of sales reps x and the length of time t in days:

$$C(x, t) = 20{,}000 - 100x + 600t + 300xt.$$

a. Based on the model, how much would it cost to take five sales reps along for a 10-day retreat?

b. Evaluate $\frac{\partial C}{\partial t}$, specify its units of measurement, and indicate whether it increases or decreases with increasing x.

c. How many reps should they take along if they wish to limit the rate of increase of cost with respect to time to \$1,200 per day?

45. ***Internet Advertising*** Refer to the model in Exercise 43. One or more of the following statements is correct. Identify which one(s).

(A) If nothing is spent on television advertising, one more dollar spent per month in banner ads will buy approximately 0.05 hits per day at OHaganBooks.com.

(B) If nothing is spent on television advertising, one more hit per day at OHaganBooks.com will cost the company about 5¢ per month in banner ads.

(C) If nothing is spent on banner ads, one more hit per day at OHaganBooks.com will cost the company about 5¢ per month in banner ads.

(D) If nothing is spent on banner ads, one more dollar spent per month in banner ads will buy approximately 0.05 hits per day at OHaganBooks.com.

(E) Hits at OHaganBooks.com cost approximately 5¢ per month spent on banner ads, and this cost increases at a rate of 0.003¢ per month, per hit.

46. ***Company Retreats*** Refer to the model in Exercise 44. One or more of the following statements is correct. Identify which one(s).

(A) If the retreat lasts for 10 days, the daily cost per sales rep is \$400.

(B) If the retreat lasts for 10 days, each additional day will cost the company \$2,900.

(C) If the retreat lasts for 10 days, each additional sales rep will cost the company \$800.

(D) If the retreat lasts for 10 days, the daily cost per sales rep is \$2,900.

(E) If the retreat lasts for 10 days, each additional sales rep will cost the company \$2,900.

47. ***Productivity*** The holiday season is now at its peak, and OHaganBooks.com has been understaffed and swamped with orders. The current backlog (orders unshipped for two or more days) has grown to a staggering 50,000, and new orders are coming in at a rate of 5,000 per day. Research based on productivity data at OHaganBooks.com results in the following model:

$$P(x, y) = 1{,}000x^{0.9}y^{0.1} \text{ additional orders filled per day,}$$

where x is the number of additional personnel hired and y is the daily budget (excluding salaries) allocated to eliminating the backlog.

a. How many additional orders will be filled per day if the company hires 10 additional employees and budgets an additional \$1,000 per day? (Round the answer to the nearest 100.)

b. In addition to the daily budget, extra staffing costs the company \$150 per day for every new staff member hired. To fill at least 15,000 additional orders per day at a minimum total daily cost, how many new staff members should the company hire? (Use the method of Lagrange multipliers.)

48. ***Productivity*** The holiday season has now ended, and orders at OHaganBooks.com have plummeted, leaving staff members in the shipping department with little to do besides spend their time on Facebook, so the company is considering laying off a number of personnel and slashing the shipping budget. Research based on productivity data at OHaganBooks.com results in the following model:

$$C(x, y) = 1{,}000x^{0.8}y^{0.2} \text{ fewer orders filled per day,}$$

where x is the number of personnel laid off and y is the cut in the shipping budget (excluding salaries).

a. How many fewer orders will be filled per day if the company lays off 15 additional employees and cuts the budget by an additional \$2,000 per day? (Round the answer to the nearest 100.)

b. In addition to the cut in the shipping budget, the layoffs will save the company \$200 per day for every new staff member laid off. The company needs to meet a target of 20,000 fewer orders per day, but, for tax reasons, it must minimize the total resulting savings. How many new staff members should the company lay off? (Use the method of Lagrange multipliers.)

49. ***Profit*** If OHaganBooks.com sells x paperback books and y hardcover books per week, it will make an average weekly profit of

$$P(x, y) = 3x + 10y \text{ dollars.}$$

If it sells between 1,200 and 1,500 paperback books and between 1,800 and 2,000 hardcover books per week, what is the average of all its possible weekly profits?

50. ***Cost*** It costs *Duffin House*

$$C(x, y) = x^2 + 2y \text{ dollars}$$

to produce x coffee table art books and y paperback books per week. If it produces between 100 and 120 art books and between 800 and 1,000 paperbacks per week, what is the average of all its possible weekly costs?

CASE STUDY

Modeling College Population

david pearson/Alamy Stock Photo

College Malls, Inc. is planning to build a national chain of shopping malls in college neighborhoods. However, malls in general have been experiencing large numbers of store closings due to, among other things, misjudgments of the shopper demographics. As a result, the company is planning to lease only to stores that target the specific age demographics of the national college student population.

As a marketing consultant to College Malls, you will be providing the company with a report that addresses the following specific issues:

- A quick way of estimating the number of students of any specified age and in any particular year and the effect of increasing age on the college population
- The ages that correspond to relatively high and low college populations
- How fast the 20-year-old and 25-year-old student populations are increasing
- Some near-term projections of the student population trend

You decide that a good place to start would be with a visit to the Census Bureau's website at www.census.gov. After some time battling with search engines, all you can find is some data on college enrollment for three age brackets for the period 1980–2009, as shown in the following table:[29]

College Enrollment (thousands)

Year	1980	1985	1990	1995	2000	2001	2002	2003	2004	2005	2006	2007	2008	2009
18–24	7,229	7,537	7,964	8,541	9,451	9,629	10,033	10,365	10,611	10,834	10,587	11,161	11,466	12,072
25–34	2,703	3,063	3,161	3,349	3,207	3,422	3,401	3,494	3,690	3,600	3,658	3,838	4,013	6,141
35–44	700	963	1,344	1,548	1,454	1,557	1,678	1,526	1,615	1,657	1,548	1,520	1,672	1,848

The data are inadequate for several reasons: The data are given only for certain years and in age brackets rather than year by year and for each individual age; nor is it obvious how you would project the figures. However, you notice that the table is actually a numerical representation of a function of two variables: year and age. Since the age brackets are of different sizes, you normalize the data by dividing each figure by the number of years represented in the corresponding age bracket; for

[29] Source: Census Bureau (www.census.gov/population/www/socdemo/school.html).

instance, you divide the 1980 figure for the first age group by 7 to obtain the average enrollment for each year of age in that group. You then rewrite the resulting table representing the years by values of t and each age bracket by the (rounded) age x at its center (enrollment values are rounded):

$x \downarrow$ \ $t \rightarrow$	**0**	**5**	**10**	**15**	**20**	**21**	**22**	**23**	**24**	**25**	**26**	**27**	**28**	**29**
21	1,033	1,077	1,138	1,220	1,350	1,376	1,433	1,481	1,516	1,548	1,512	1,594	1,638	1,725
30	270	306	316	335	321	342	340	349	369	360	366	384	401	614
40	70	96	134	155	145	156	168	153	162	166	155	152	167	185

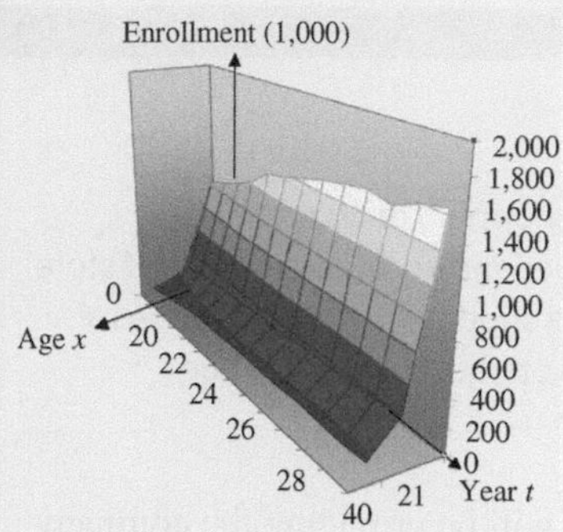

Figure 40

To see a visual representation of what the data are saying, you use Excel to graph the data as a surface (Figure 40). It is important to notice that Excel does not scale the t-axis as you would expect: It uses one subdivision for each year shown in the chart, and the result is an uneven scaling of the t-axis. Despite this drawback, you do see two trends after looking at views of the graph from various angles. First, enrollment of 21-year olds (the back edge of the graph) seems to be increasing faster than enrollment of other age groups. Second, the enrollments for all ages seem to be increasing approximately linearly with time, though at different rates for different age groups; for instance, the front and rear edges rise more or less linearly but do not seem to be parallel.

At this point you realize that a mathematical model of these data would be useful; not only would it "smooth out the bumps," but it would give you a way to estimate enrollment N at each specific age, project the enrollments, and thereby complete the project for College Malls. Although technology can give you a regression model for data such as these, it is up to you to decide on the form of the model. It is in choosing an appropriate model that your analysis of the graph comes in handy. Because N should vary linearly with time t for each value of x, you would like

$$N = mt + k$$

for each value of x. Also, because there are three values of x for every value of time, you try a quadratic model for N as a function of x:

$$N = a + bx + cx^2.$$

Putting these together, you get the following candidate model:

$$N(t, x) = a_1 + a_2t + a_3x + a_4x^2,$$

where a_1, a_2, a_3, and a_4 are constants. However, for each specific age $x = k$, you get

$$N(t, k) = a_1 + a_2t + a_3k + a_4k^2 = \text{Constant} + a_2t$$

with the same slope a_2 for every choice of the age k, contrary to your observation that enrollment for different age groups is rising at different rates, so you will need a more elaborate model. You recall from your applied calculus course that interaction functions give a way to model the effect of one variable on the rate of change of another, so, as an experiment, you try adding interaction terms to your model:

Model 1: $N(t, x) = a_1 + a_2t + a_3x + a_4x^2 + a_5xt$ Second-order model

Model 2: $N(t, x) = a_1 + a_2t + a_3x + a_4x^2 + a_5xt + a_6x^2t.$ Third-order model

(Model 1 is referred to as a second-order model because it contains no products of more than two independent variables, whereas Model 2 contains the third-order term $x^2t = x \cdot x \cdot t$.) If you study these two models for specific values k of x you get

Model 1: $N = \text{Constant} + (a_2 + a_5k)t$. Slope depends linearly on age.

Model 2: $N = \text{Constant} + (a_2 + a_5k + a_6k^2)t$. Slope depends quadratically on age.

This is encouraging: Both models show different slopes for different ages. Model 1 would predict that the slope either increases with increasing age (a_5 positive) or decreases with increasing age (a_5 negative). However, the graph suggests that the slope is larger for both younger and older students but smaller for students of intermediate age, contrary to what Model 1 predicts, so you decide to go with the more flexible Model 2, which permits the slope to decrease and then increase with increasing age, which is exactly what you observe on the graph.

You decide to use Excel to generate your model. However, the data as shown in the table are not in a form that Excel can use for regression; the data need to be organized into columns: Column A for the dependent variable N and Columns B–C for the independent variables, as shown in Figure 41.

	A	B	C
1	N	t	x
2	1033	0	21
3	1077	5	21
4	1138	10	21
5	1220	15	21
6	1350	20	21
7	1376	21	21
8	1433	22	21
9	1481	23	21
10	1516	24	21
11	1548	25	21
12	1512	26	21
13	1594	27	21
14	1638	28	21
15	1725	29	21
16	270	0	30
17	306	5	30
18	316	10	30
19	335	15	30
20	321	20	30
21	342	21	30
22	340	22	30
23	349	23	30
24	369	24	30
25	360	25	30
26	366	26	30
27	384	27	30
28	401	28	30
29	614	29	30
30	70	0	40
31	96	5	40
32	134	10	40
33	155	15	40
34	145	20	40
35	156	21	40
36	168	22	40
37	153	23	40
38	162	24	40
39	166	25	40
40	155	26	40
41	152	27	40
42	167	28	40
43	185	29	40

Figure 41

You then add columns for the higher order terms x^2, xt, and x^2t as shown below:

	A	B	C	D	E	F
1	N	t	x	x^2	x*t	x^2*t
2	1033	0	21	=C2^2	=C2*B2	=C2^2*B2
3	1077	5	21			
4	1138	10	21			
5	1220	15	21			
6	1350	20	21			
7	1376	21	21			
8	1433	22	21			
9	1481	23	21			

	A	B	C	D	E	F
1	N	t	x	x^2	x*t	x^2*t
2	1033	0	21	441	0	0
3	1077	5	21	441	105	2205
4	1138	10	21	441	210	4410
5	1220	15	21	441	315	6615
6	1350	20	21	441	420	8820
7	1376	21	21	441	441	9261
8	1433	22	21	441	462	9702
9	1481	23	21	441	483	10143

Next, highlight a vacant 5 × 6 block (the block A46:F50, say), type the formula =LINEST(A2:A43,B2:F43,TRUE,TRUE), and press Ctrl+Shift+Enter (not just Enter!). You will see a table of statistics like the following:

43	185	29	40	1600	1160	46400
44						
45						
46	=LINEST(A2:A43,B2:F43,,TRUE)					
47						
48						
49						
50						

42						
43	185	29	40	1600	1160	46400
44						
45						
46	0.088570354	-6.46546922	3.21423521	-241.273271	120.009278	4594.62978
47	0.02096866	1.28767845	0.44955302	27.6069027	18.6958148	400.824866
48	0.99337478	49.506512	#N/A	#N/A	#N/A	#N/A
49	1079.5571	36	#N/A	#N/A	#N/A	#N/A
50	13229404.1	88232.2104	#N/A	#N/A	#N/A	#N/A

The desired constants $a_1, a_2, a_3, a_4, a_5, a_6$ appear in the first row of the data but in *reverse order*. Thus, if we round to five significant digits, we have

$$a_1 = 4{,}594.6 \quad a_2 = 120.01 \quad a_3 = -241.27$$
$$a_4 = 3.2142 \quad a_5 = -6.4655 \quad a_6 = 0.088570,$$

which gives our regression model:

$$N(t, x) = 4{,}594.6 + 120.01t - 241.27x + 3.2142x^2 - 6.4655xt + 0.088570x^2t.$$

Fine, you say to yourself, now you have the model, but how good a fit is it to the data? That is where the rest of the data shown in the output comes in: In the second row are the standard errors corresponding to the corresponding coefficients. Notice that each of the standard errors is small in comparison with the magnitude of the coefficient above it; for instance, 0.021 is only around 1/4 of the magnitude of $a_6 \approx 0.088$ and indicates that the dependence of N on x^2t is statistically significant. (What we do not want to see are standard errors of magnitudes comparable to the coefficients, as those could indicate the wrong choice of independent variables.) The third figure in the left column, 0.99337478, is R^2, where R generalizes the coefficient of correlation discussed in the section on regression in Chapter 1: The closer R is to 1, the better the fit. We can interpret R^2 as indicating that approximately 99.3% of the variation in college enrollment is explained by the regression model, indicating an excellent fit. The figure 1,079.5571 beneath R^2 is called the "F-statistic." The higher the F-statistic (typically, anything above 4 or so would be considered "high"), the more confident we can be that N does depend on the independent variables we are using.*

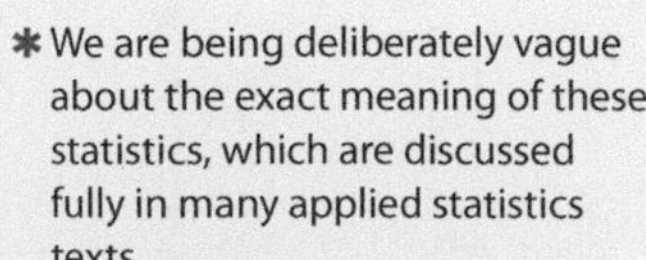
* We are being deliberately vague about the exact meaning of these statistics, which are discussed fully in many applied statistics texts.

As comforting as these statistics are, nothing can be quite as persuasive as a graph. You turn to the graphing software of your choice and notice that the graph of the model appears to be a faithful representation of the data. (See Figure 42.)

Now you get to work, using the model to address the questions posed by College Malls.

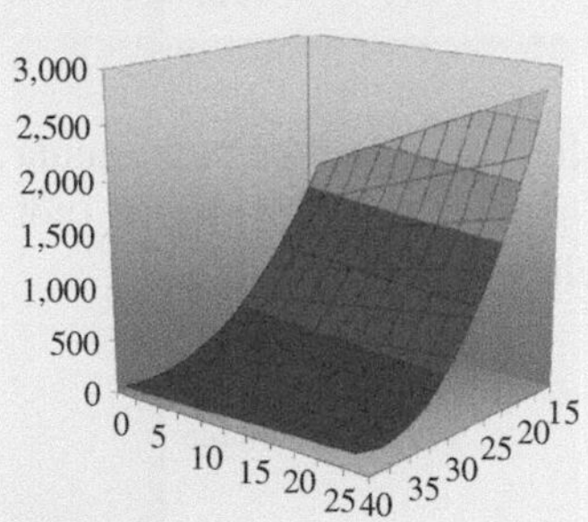

Figure 42

1. *A quick way of estimating the number of students of any specified age and in any particular year, and the effect of increasing age on the college population.* You already have a quantitative relationship in the form of the regression model. As for the second part of the question, the rate of change of college enrollment with respect to age is given by the partial derivative

$$\frac{\partial N}{\partial x} = -241.27 + 6.4284x - 6.4655t + 0.17714xt \text{ thousand students per additional year of age.}$$

Thus, for example, with $x = 20$ in 2004 ($t = 24$) we have

$$\frac{\partial N}{\partial x} = -241.27 + 6.4284(20) - 6.4655(24) + 0.17714(20)(24)$$
$$\approx -183 \text{ thousand students per additional year of age,}$$

so there were about 183,000 fewer students of age 21 than age 20 in 2004.

On the other hand, when $x = 38$ in the same year, we have

$$\frac{\partial N}{\partial x} = -241.27 + 6.4284(38) - 6.4655(24) + 0.17714(38)(24)$$

$$\approx 9.4 \text{ thousand students per additional year of age,}$$

so there were about 9,400 more students of age 39 than age 38 that year.

2. *The ages that correspond to relatively high and low college populations.* Although a glance at the graph shows you that there are no relative maxima, holding t constant (that is, on any given year) gives a parabola along the corresponding slice and hence a minimum somewhere along the slice.

$$\frac{\partial N}{\partial x} = 0$$

when $\quad -241.27 + 6.4284x - 6.4655t + 0.17714xt = 0,$

which gives $x = \dfrac{241.27 + 6.4655t}{6.4284 + 0.17714t}$ years of age.

For instance, in 2010 ($t = 30$; we are extrapolating the model slightly) the age at which there were fewest students (in the given range) is 37 years of age. The relative maxima for each slice occur at the front and back edges of the surface, meaning that there are relatively more students of the lowest and highest ages represented. The absolute maximum for each slice occurs, as expected, at the lowest age. In short, a mall catering to college students in 2010 should have focused mostly on freshman-age students, least on 37-year-olds, and somewhat more on people around age 40.

3. *How fast the 20-year-old and 25-year-old student populations are increasing.* The rate of change of student population with respect to time is

$$\frac{\partial N}{\partial t} = 120.01 - 6.4655x + 0.088570x^2 \text{ thousand students per year}$$

For the two age groups in question, we obtain

$x = 20$: $\quad 120.01 - 6.4655(20) + 0.088570(20)^2 \approx 26.1$ thousand students per year

$x = 25$: $\quad 120.01 - 6.4655(25) + 0.088570(25)^2 \approx 13.7$ thousand students per year.

(Note that these rates of change are independent of time, as we chose a model that is linear in time.)

4. *Some near-term projections of the student population trend.* As we have seen throughout the book, extrapolation can be a risky venture; however, near-term extrapolation from a good model can be reasonable. You enter the model in an Excel spreadsheet to obtain the following predicted college enrollments (in thousands) for the years 2010–2015:

$x \downarrow$ \ $t \rightarrow$	30	31	32	33	34	35
21	1,644	1,668	1,691	1,714	1,737	1,761
30	422	428	434	439	445	451
40	180	183	186	189	192	195

EXERCISES

1. Use a spreadsheet to obtain Model 1:

$$N(t, x) = a_1 + a_2 t + a_3 x + a_4 x^2 + a_5 xt.$$

Compare the fit of this model with that of the quadratic model above. Comment on the result.

2. Obtain Model 2 using only the data through 2005, and also obtain the projections for 2010–2015 using the resulting model. Compare the projections with those based on the more complete set of data in the text.

3. Compute and interpret $\left.\frac{\partial N}{\partial t}\right|_{(10,\,18)}$ and $\left.\frac{\partial^2 N}{\partial t \partial x}\right|_{(10,\,18)}$ for the model in the text. What are their units of measurement?

4. Notice that the derivatives in Exercise 3 do not depend on time. What additional polynomial term(s) would make both $\partial N/\partial t$ and $\partial^2 N/\partial t \partial x$ depend on time? (Write down the entire model.) Of what order is your model?

5. Test the model you constructed in Exercise 4 by inspecting the standard errors associated with the additional coefficients.

TI-83/84 Plus Technology Guide

Section 8.1

Example 1 (page 609) You own a company that makes two models of speakers: the Ultra Mini and the Big Stack. Your total monthly cost (in dollars) to make x Ultra Minis and y Big Stacks is given by

$$C(x, y) = 10{,}000 + 20x + 40y.$$

Compute several values of this function.

Solution

You can have a TI-83/84 Plus compute $C(x, y)$ numerically as follows:

1. In the "Y=" screen, enter

```
Y1=10000+20X+40Y
```

2. To evaluate, say, $C(10, 30)$ (the cost to make 10 Ultra Minis and 30 Big Stacks), enter

```
10→X
30→Y
Y1
```

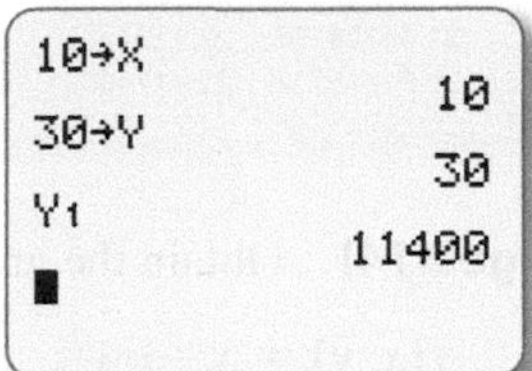

and the calculator will evaluate the function and give the answer $C(10, 30) = 11{,}400$.

This procedure is too laborious if you want to calculate $f(x, y)$ for a large number of different values of x and y.

Spreadsheet Technology Guide

Section 8.1

Example 1 (page 609) You own a company that makes two models of speakers: the Ultra Mini and the Big Stack. Your total monthly cost (in dollars) to make x Ultra Minis and y Big Stacks is given by

$$C(x, y) = 10{,}000 + 20x + 40y$$

Compute several values of this function.

Solution

Spreadsheets handle functions of several variables easily. The following setup shows how a table of values of C can be created, using values of x and y you enter:

	A	B	C
1	x	y	C(x, y)
2	10	30	=10000+20*A2+40*B2
3	20	30	
4	15	0	
5	0	30	
6	30	30	

↓

	A	B	C
1	x	y	C(x, y)
2	10	30	11400
3	20	30	11600
4	15	0	10300
5	0	30	11200
6	30	30	11800

A disadvantage of this layout is that it's not easy to enter values of x and y systematically in two columns. Can you find a way to remedy this? (See Example 3 for one method.)

Example 3 (page 611) Use technology to create a table of values of the body mass index

$$M(w, h) = \frac{0.45w}{(0.0254h)^2}.$$

Solution

We can use this formula to recreate a table in a spreadsheet, as follows:

	A	B	C	D
1		130	140	150
2	60	=0.45*B$1/(0.0254*$A2)^2		
3	61			
4	62			
5	63			
6	64			
7	65			
8	66			
9	67			

In the formula in cell B2 we have used B$1 instead of B1 for the w-coordinate because we want all references to w to use the same row (1). Similarly, we want all references to h to refer to the same column (A), so we used $A2 instead of A2.

We copy the formula in cell B2 to all of the red shaded area to obtain the desired table:

	A	B	C	D	
1		130	140	150	
2	60	25.18755038	27.12505425	29.06255813	3
3	61	24.36849808	26.24299793	28.11749778	29.9
4	62	23.58875685	25.40327661	27.21779637	29.0
5	63	22.84585068	24.60322381	26.36059694	28.1
6	64	22.13749545	23.84037971	25.54326398	27.2
7	65	21.46158138	23.11247226	24.76336314	26.4
8	66	20.81615733	22.41740021	24.01864308	25.6
9	67	20.19941665	21.75321793	23.30701921	24.8

Example 5 (page 614) Obtain the graph of

$$f(x, y) = x^2 + y^2.$$

Solution

1. Set up a table showing a range of values of x and y and the corresponding values of the function (see Example 3):

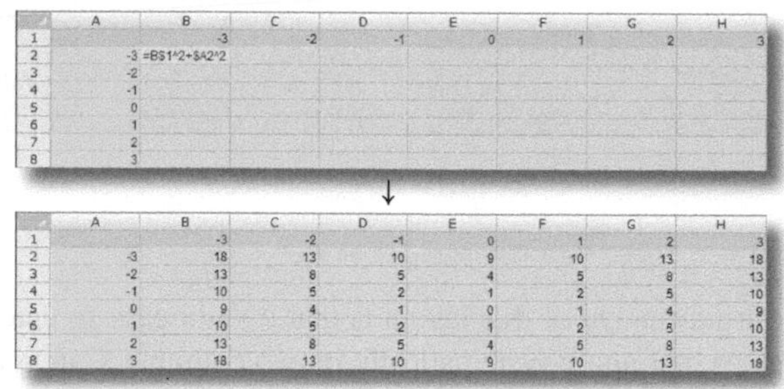

	A	B	C	D	E	F	G	H
1		-3	-2	-1	0	1	2	3
2	-3	=B1^2+A2^2						
3	-2							
4	-1							
5	0							
6	1							
7	2							
8	3							

↓

	A	B	C	D	E	F	G	H
1		-3	-2	-1	0	1	2	3
2	-3	18	13	10	9	10	13	18
3	-2	13	8	5	4	5	8	13
4	-1	10	5	2	1	2	5	10
5	0	9	4	1	0	1	4	9
6	1	10	5	2	1	2	5	10
7	2	13	8	5	4	5	8	13
8	3	18	13	10	9	10	13	18

2. Select the cells with the values (B2: H8) and insert a chart, with the "Surface" option selected and "Series in Columns" selected as the data option, to obtain a graph like the following:

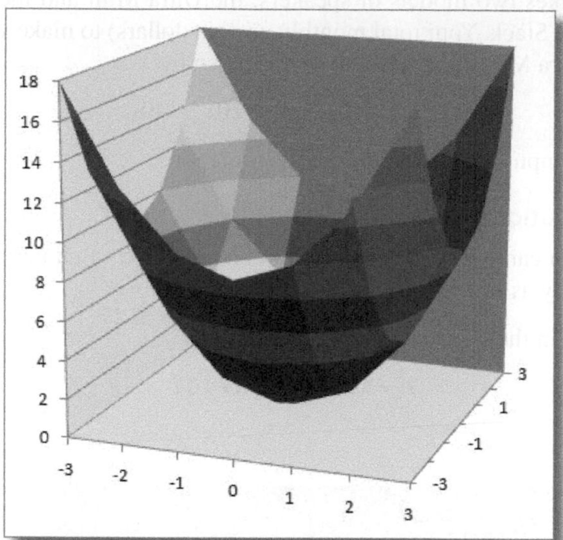

9 TRIGONOMETRIC MODELS

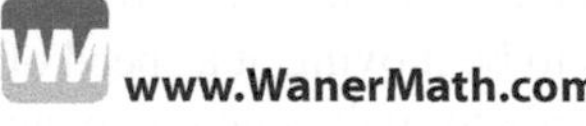

www.WanerMath.com

CASE STUDY

Predicting Airline Empty Seat Volume

You are a consultant to the Department of Transportation's Special Task Force on Air Traffic Congestion and have been asked to model the volume of empty seats on U.S. airline flights, to make short-term projections of this volume, and to give a formula that estimates the accumulated volume over a specified period of time. You have data from the Bureau of Transportation Statistics on the number of seats and passengers each month starting with January 2002.

How will you analyze these data to prepare your report?

Lawrence Manning/Corbis/Getty Images

Introduction

Cyclical behavior is common in the business world: There are seasonal fluctuations in the demand for surfing equipment, swimwear, snow shovels, and many other items. The nonlinear functions we have studied up to now cannot model this kind of behavior. To model cyclical behavior, we need the **trigonometric** functions.

In the first section we study the basic trigonometric functions—especially the **sine** and **cosine** functions from which all the trigonometric functions are built—and see how to model various kinds of periodic behavior using these functions. The rest of the chapter is devoted to the calculus of the trigonometric functions—their derivatives and integrals—and to its numerous applications.

9.1 Trigonometric Functions, Models, and Regression

The Sine Function

Figure 1 shows the approximate average daily high temperatures in New York's Central Park.[1] If we draw the graph for several years, we get the repeating pattern shown in Figure 2, where the x-coordinate represents time in years, with $x = 0$ corresponding to August 1, and where the y-coordinate represents the temperature in degrees Fahrenheit. This is an example of **cyclical** or **periodic** behavior.

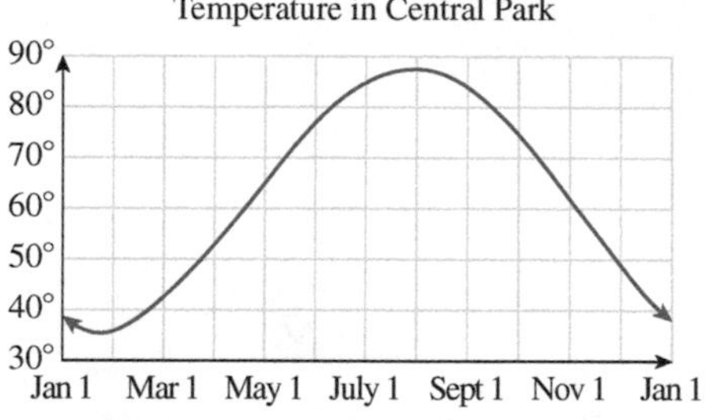

Figure 1

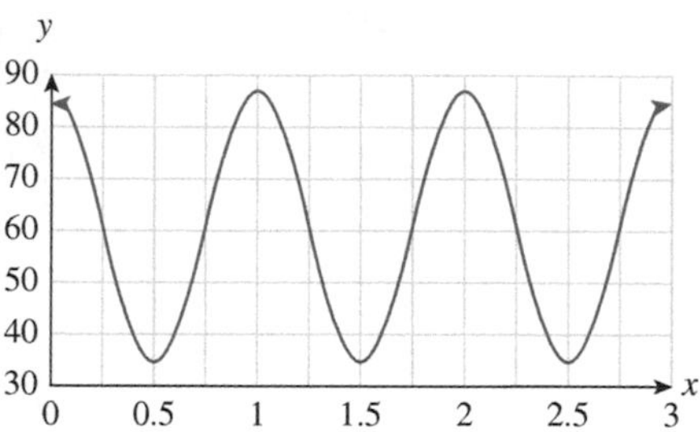

Figure 2

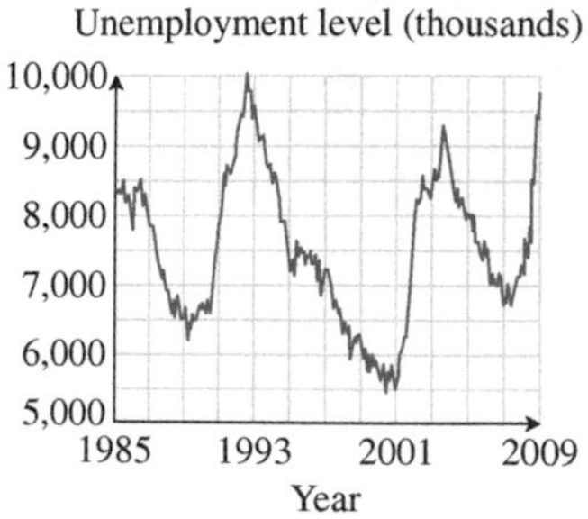

Unemployment level (thousands)

Source: Bureau of Labor Statistics, December 2008 (www.data.bls.gov).

Figure 3

Cyclical behavior is also common in the business world. The graph in Figure 3 suggests cyclical behavior in the U.S. unemployment level.

From a mathematical point of view, the simplest models of cyclical behavior are the **sine** and **cosine** functions. An easy way to describe these functions is as follows. Imagine a bicycle wheel whose radius is 1 unit, with a marker attached to the rim of the rear wheel, as shown in Figure 4 on the next page.

Now, we can measure the height $h(t)$ of the marker above the center of the wheel. As the wheel rotates, $h(t)$ fluctuates between -1 and $+1$. Suppose that, at time $t = 0$, the marker was at height zero as shown in the figure, so $h(0) = 0$. Because the wheel has a radius of 1 unit, its circumference (the distance all around) is 2π, where $\pi = 3.14159265\ldots$. If the cyclist happens to be moving at a speed of 1 unit per second, it will take the bicycle wheel 2π seconds to make one complete revolution. During the time interval $[0, 2\pi]$, the marker will first rise to a maximum height of $+1$, then drop to a low point of -1, and then return to the starting position

[1] Source: National Weather Service/*New York Times*, January 7, 1996, p. 36.

of 0 at $t = 2\pi$. This function $h(t)$ is called the **sine function**, denoted by $\sin(t)$. Figure 5 shows its graph.

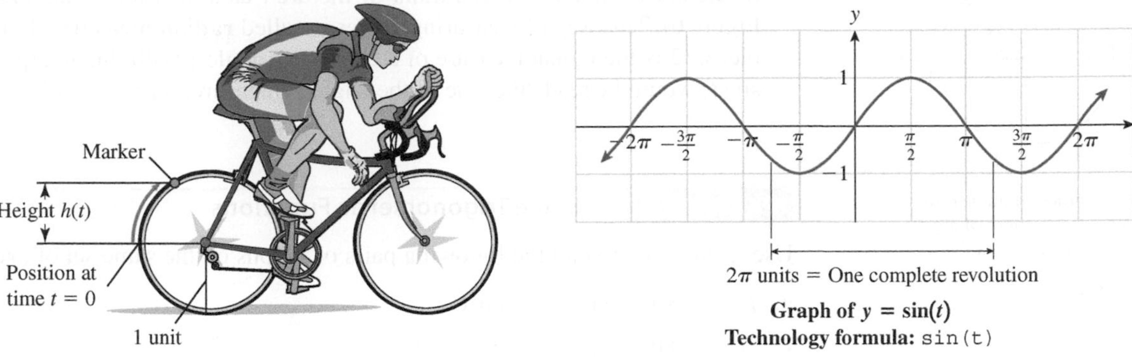

Graph of $y = \sin(t)$
Technology formula: `sin(t)`

Figure 4 **Figure 5**

Sine Function

"Bicycle Wheel" Definition

If a wheel of radius 1 unit rolls forward at a speed of 1 unit per second, then $\sin(t)$ is the height, after t seconds, of a marker on the rim of the wheel, starting in the position shown in Figure 4.

Geometric Definition

The **sine** of a real number t is the y-coordinate (height) of the point P in the following diagram, where $|t|$ is the length of the arc shown.

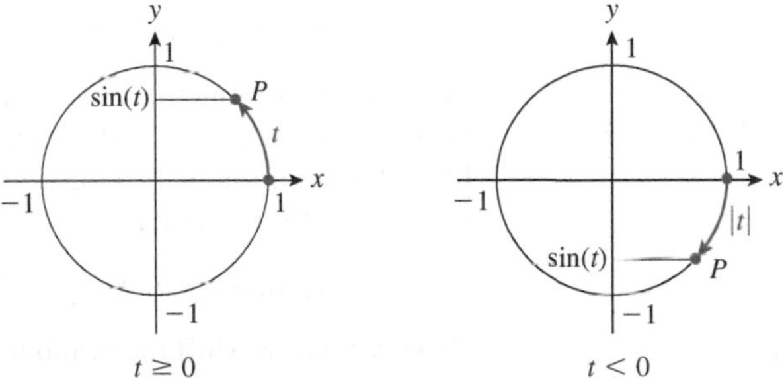

$\sin(t) = y$-coordinate of the point P

Quick Examples

From the graph we see that

1. $\sin(\pi) = 0$ — Graphing Calculator: `sin(π)` Spreadsheet: `sin(PI())`
2. $\sin\left(\frac{\pi}{2}\right) = 1$ — Graphing Calculator: `sin(π/2)` Spreadsheet: `sin(PI()/2)`
3. $\sin\left(\frac{3\pi}{2}\right) = -1.$ — Graphing Calculator: `sin(3π/2)` Spreadsheet: `sin(3*PI()/2)`

Note We often write "sin x" without the parentheses to mean $\sin(x)$; for instance, we may write $\sin(\pi)$ above as $\sin \pi$. Remember, however, that this does *not* mean that we are "multiplying" sin by π (which makes no sense). *Always read* sin x as "*the sine **of** x.*" ■

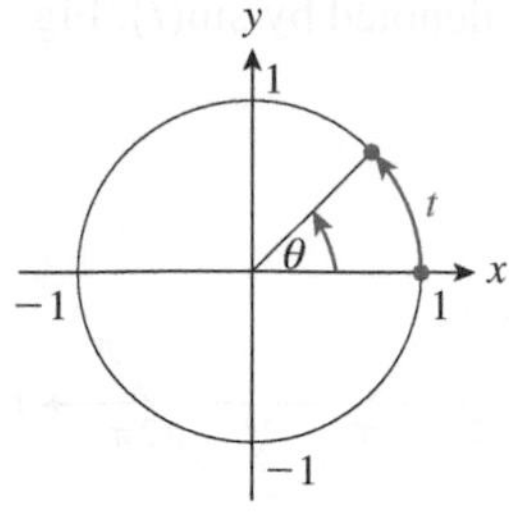

Magnitude of angle θ in radians = Length of arc t on the circle of radius 1

Figure 6

Angles and Radian Measure

In the above figure, you can think of the arc t as a measure of an *angle* (see Figure 6). This way of measuring angles is called **radian measure**. (For example, $\pi/2$ is the radian measure of a 90-degree angle.) With this interpretation, $\sin(t)$ would be read "the sine of the angle t (measured in radians)."

EXAMPLE 1 Some Trigonometric Functions

Use technology to plot the following pairs of graphs on the same set of axes:

a. $f(x) = \sin x$; $g(x) = 2 \sin x$

b. $f(x) = \sin x$; $g(x) = \sin(x + 1)$

c. $f(x) = \sin x$; $g(x) = \sin(2x)$

Solution

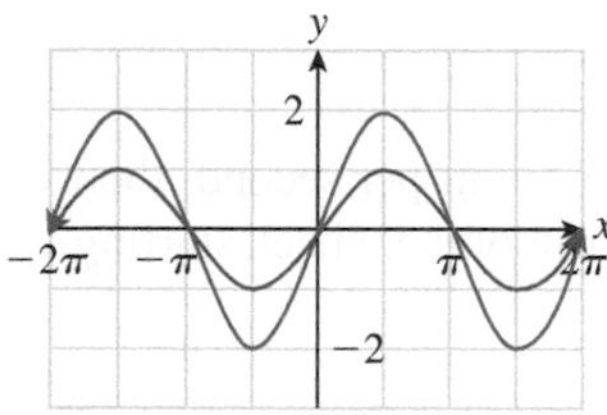

Figure 7

a. (Important note: If you are using a calculator, make sure it is set to *radian mode*, not degree mode.) We enter these functions as `sin(x)` and `2*sin(x)`, respectively. We use the range $-2\pi \le x \le 2\pi$ (approximately $-6.28 \le x \le 6.28$) for x suggested by the graph in Figure 5 but with larger range of y-coordinates (why?): $-3 \le y \le 3$. The graphs are shown in Figure 7. Here, $f(x) = \sin x$ is shown in red, and $g(x) = 2 \sin x$ in blue. Notice that multiplication by 2 has doubled the **amplitude**, or *distance it oscillates up and down*. Where the original sine curve oscillates between -1 and 1, the new curve oscillates between -2 and 2. In general:

The graph of $A \sin(x)$ *has amplitude* A.

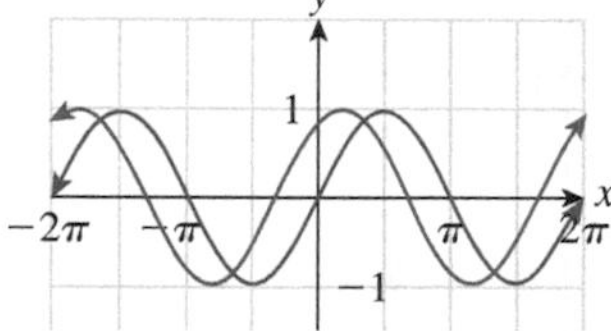

Figure 8

b. We enter these functions as `sin(x)` and `sin(x+1)`, respectively, and we get Figure 8. Once again $f(x) = \sin x$ is shown in red, and $g(x) = \sin(x + 1)$ is in blue. The addition of 1 to the argument has shifted the graph to the left by 1 unit. In general, for positive c:

Replacing x by $x + c$ shifts the graph to the left c units.

(How would we shift the graph to the *right* 1 unit?)

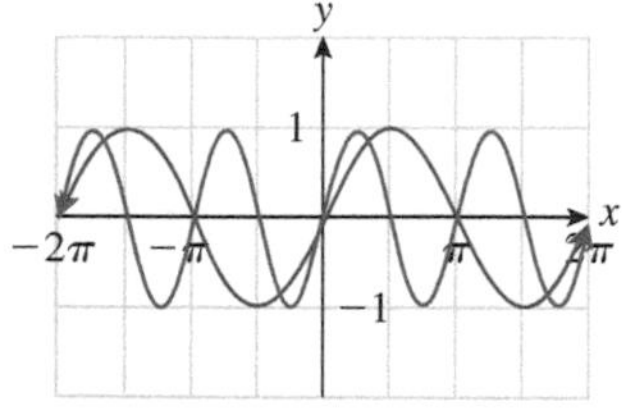

Figure 9

c. We enter these functions as `sin(x)` and `sin(2*x)`, respectively, and get the graph in Figure 9. The graph of $\sin(2x)$ oscillates twice as fast as the graph of $\sin x$. In other words, the graph of $\sin(2x)$ makes two complete cycles on the interval $[0, 2\pi]$, whereas the graph of $\sin x$ completes only one cycle. In general:

Replacing x by bx multiplies the rate of oscillation by b.

We can combine the operations in Example 1, and a vertical shift as well, to obtain the following.

Website
www.WanerMath.com
At the Website you will find the following optional online interactive section:

Internet Topic: New Functions from Old: Scaled and Shifted Functions

The General Sine Function

The general sine function is

$$f(x) = A \sin[\omega(x - \alpha)] + C.$$

Its graph is shown here:

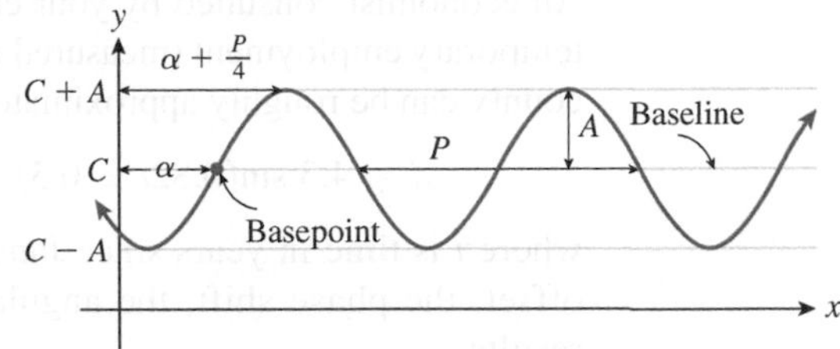

- A is the **amplitude** (the height of each peak above the baseline).
- C is the **vertical offset** (height of the baseline).
- P is the **period** or **wavelength** (the length of each cycle) and is related to ω by

$$P = \frac{2\pi}{\omega} \quad \text{or} \quad \omega = \frac{2\pi}{P}.$$

- ω is the **angular frequency** (the number of cycles in every interval of length 2π).
- α is the **phase shift**.

EXAMPLE 2 Electrical Current

The typical voltage V supplied by an electrical outlet in the United States is a sinusoidal function that oscillates between -165 volts and $+165$ volts with a frequency of 60 cycles per second. Find an equation for the voltage as a function of time t.

Solution What we are looking for is a function of the form

$$V(t) = A\sin[\omega(t - \alpha)] + C.$$

Referring to the graph of the general sine function, we can determine the constants.

Amplitude A and Vertical Offset C: Because the voltage oscillates between -165 volts and $+165$ volts, we see that $A = 165$ and $C = 0$.

Period P: The electric current completes 60 cycles in one second, so the length of time it takes to complete one cycle is $1/60$ second. Thus, the period is $P = 1/60$.

Angular Frequency ω: This is given by the formula

$$\omega = \frac{2\pi}{P} = 2\pi(60) = 120\pi.$$

Phase Shift α: The phase shift α tells us when the curve first crosses the t-axis as it ascends. As we are free to specify what time $t = 0$ represents, let us say that the curve crosses 0 when $t = 0$, so $\alpha = 0$.

Thus, the equation for the voltage at time t is

$$\begin{aligned} V(t) &= A\sin[\omega(t - \alpha)] + C \\ &= 165\sin(120\pi t), \end{aligned}$$

where t is time in seconds.

EXAMPLE 3 Cyclical Employment Patterns

An economist consulted by your employment agency indicates that the demand for temporary employment (measured in thousands of job applications per week) in your county can be roughly approximated by the function

$$d = 4.3\sin(0.82t - 0.3) + 7.3,$$

where t is time in years since January 2000. Calculate the amplitude, the vertical offset, the phase shift, the angular frequency, and the period, and interpret the results.

Solution To calculate these constants, we write

$$\begin{aligned} d = A\sin[\omega(t - \alpha)] + C &= A\sin[\omega t - \omega\alpha] + C \\ &= 4.3\sin(0.82t - 0.3) + 7.3, \end{aligned}$$

and we see right away that $A = 4.3$ (the amplitude), $C = 7.3$ (vertical offset), and $\omega = 0.82$ (angular frequency). We also have

$$\omega\alpha = 0.3,$$

so

$$\alpha = \frac{0.3}{\omega} = \frac{0.3}{0.82} \approx 0.37$$

(rounding to two significant digits; notice that all the constants are given to two digits). Finally, we get the period using the formula

$$P = \frac{2\pi}{\omega} = \frac{2\pi}{0.82} \approx 7.7.$$

We can interpret these numbers as follows: The demand for temporary employment fluctuates in cycles of 7.7 years about a baseline of 7,300 job applications per week. Every cycle, the demand peaks at 11,600 applications per week (4,300 above the baseline) and dips to a low of 3,000. In May 2000 ($t = 0.37$) the demand for employment was at the baseline level and rising.

Note The generalized sine function in Example 3 was given in the form

$$f(x) = A\sin(\omega t + d) + C$$

for some constant d. Every generalized sine function can be written in this form:

$$A\sin[\omega(t - \alpha)] + C = A\sin(\omega t - \omega\alpha) + C. \qquad d = -\omega\alpha$$

Generalized sine functions are often written in this form. ■

The Cosine Function

Closely related to the sine function is the cosine function, defined as follows. (Refer to the definition of the sine function for comparison.)

Cosine Function

Geometric Definition

The **cosine** of a real number t is the x-coordinate of the point P in the following diagram, in which $|t|$ is the length of the arc shown.

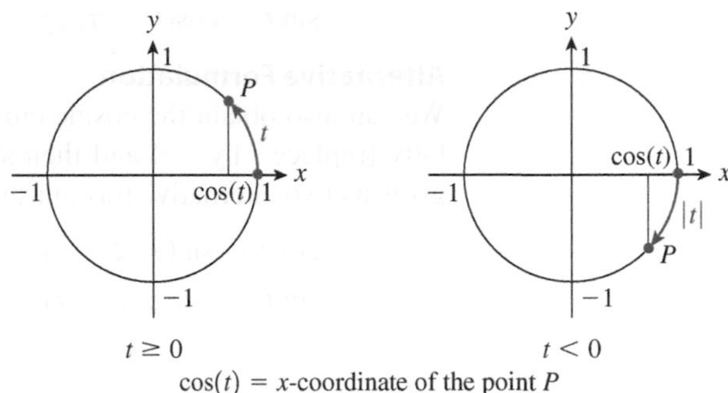

$\cos(t)$ = x-coordinate of the point P

Graph of the Cosine Function

The graph of the cosine function is identical to the graph of the sine function except that it is shifted $\pi/2$ units to the left.

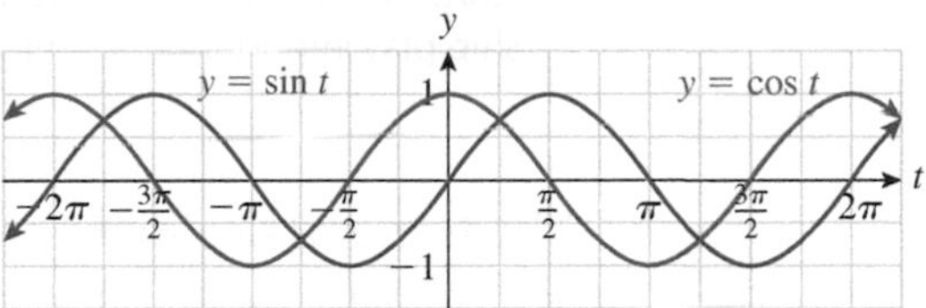

Technology formula: `cos(t)`

Notice that the coordinates of the point P in the diagram above are $(\cos t, \sin t)$ and that the distance from P to the origin is 1 unit. It follows from the Pythagorean theorem that the distance from a point (x, y) to the origin is $\sqrt{x^2 + y^2}$. Thus,

$$\text{Square of the distance from } P \text{ to } (0, 0) = 1$$
$$(\sin t)^2 + (\cos t)^2 = 1.$$

We often write $(\sin t)^2$ as $\sin^2 t$ and similarly for the cosine, so we can rewrite the equation as

$$\sin^2 t + \cos^2 t = 1.$$

This equation is one of the important relationships between the sine and cosine functions.

Fundamental Trigonometric Identities: Relationships between Sine and Cosine

The sine and cosine of a number t are related by

$$\sin^2 t + \cos^2 t = 1.$$

We can obtain the cosine curve by shifting the sine curve to the left a distance of $\pi/2$. [See Example 1(b) for a shifted sine function.] Conversely, we can obtain the sine curve from the cosine curve by shifting it $\pi/2$ units to the right. These facts can be expressed as

$$\cos t = \sin(t + \pi/2)$$
$$\sin t = \cos(t - \pi/2).$$

Alternative Formulation

We can also obtain the cosine curve by first inverting the sine curve horizontally (replace t by $-t$) and then shifting to the *right* a distance of $\pi/2$. This gives us two alternative formulas (which are easier to remember):

$$\cos t = \sin(\pi/2 - t)$$ Cosine is the sine of the complementary angle.
$$\sin t = \cos(\pi/2 - t).$$

Q: *We can rewrite the cosine function in terms of the sine function, so do we really need the cosine function?*

A: Technically, we don't need the cosine function and could get by with only the sine function. On the other hand, it is convenient to have the cosine function because it starts at its highest point rather than at zero. These two functions and their relationship play important roles throughout mathematics.

The General Cosine Function

The general cosine function is

$$f(x) = A\cos[\omega(x - \beta)] + C.$$

Its graph is as follows:

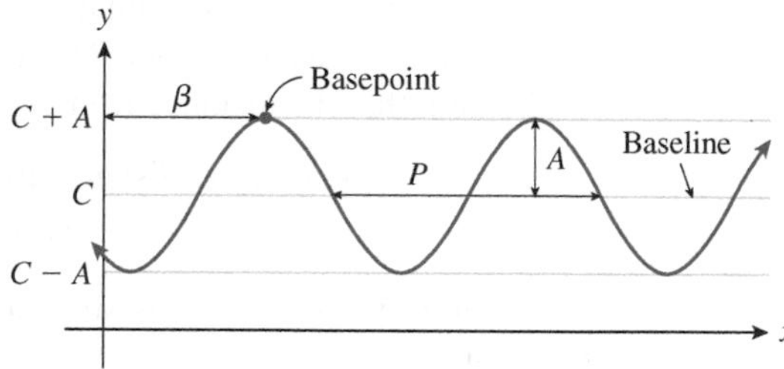

Note that the basepoint of the cosine curve is at the highest point of the curve. All the constants have the same meaning as for the general sine curve:

- A is the **amplitude** (the height of each peak above the baseline).
- C is the **vertical offset** (height of the baseline).
- P is the **period** or **wavelength** (the length of each cycle) and is related to ω by

$$P = \frac{2\pi}{\omega} \quad \text{or} \quad \omega = \frac{2\pi}{P}.$$

- ω is the **angular frequency** (the number of cycles in every interval of length 2π).
- β is the phase shift.

Notes

1. We can also describe the above curve as a generalized sine function: Observe by comparing the graph of the general cosine function to that of the general sine function that $\beta = \alpha + P/4$. Thus, $\alpha = \beta - P/4$, and the above curve is also

$$f(x) = A\sin[\omega(x - \beta + P/4)] + C.$$

2. As is the case with the generalized sine function, the cosine function above can be written in the form

$$f(x) = A\cos(\omega t + d) + C. \qquad d = -\omega\beta$$

■

EXAMPLE 4 Cash Flows into Stock Funds

The annual cash flow into stock funds (measured as a percentage of total assets) has fluctuated in cycles of approximately 40 years since 1955, when it was at a high point. The highs were roughly $+15\%$ of total assets, whereas the lows were roughly -10% of total assets.[2]

a. Model this cash flow with a cosine function of the time t in years, with $t = 0$ representing 1955.

b. Convert the answer in part (a) to a sine function model.

Solution

a. Cosine modeling is similar to sine modeling; we are seeking a function of the form

$$P(t) = A\cos[\omega(t - \beta)] + C.$$

Amplitude *A* and Vertical Offset *C*: The cash flow fluctuates between -10% and $+15\%$. We can express this as a fluctuation of $A = 12.5$ about the average $C = 2.5$.

Period *P*: This is given as $P = 40$.

Angular Frequency ω: We find ω from the formula

$$\omega = \frac{2\pi}{P} = \frac{2\pi}{40} = \frac{\pi}{20} \approx 0.157.$$

Phase Shift β: The basepoint is at the high point of the curve, and we are told that cash flow was at its high point at $t = 0$. Therefore, the basepoint occurs at $t = 0$, so $\beta = 0$.

Putting the model together gives

$$\begin{aligned} P(t) &= A\cos[\omega(t - \beta)] + C \\ &\approx 12.5\cos(0.157t) + 2.5, \end{aligned}$$

where t is time in years since 1955.

b. To convert between a sine model and a cosine model, we can use one of the relationships given earlier. Let us use the formula

$$\cos x = \sin(x + \pi/2).$$

[2] Source: Investment Company Institute/*New York Times*, February 2, 1997, p. F8.

Therefore,

$$P(t) \approx 12.5\cos(0.157t) + 2.5$$
$$= 12.5\sin(0.157t + \pi/2) + 2.5.$$

The Other Trigonometric Functions

The ratios and reciprocals of sine and cosine are given their own names.

Tangent, Cotangent, Secant, Cosecant

Tangent: $\tan x = \dfrac{\sin x}{\cos x}$

Cotangent: $\cot x = \text{cotan } x = \dfrac{\cos x}{\sin x} = \dfrac{1}{\tan x}$

Secant: $\sec x = \dfrac{1}{\cos x}$

Cosecant: $\csc x = \text{cosec } x = \dfrac{1}{\sin x}$

Trigonometric Regression

In the examples so far, we were given enough information to obtain a sine (or cosine) model directly. Often, however, we are given data that only *suggest* a sine curve. In such cases we can use regression to find the best-fit generalized sine (or cosine) curve.

EXAMPLE 5 T Spam

The authors of this book tend to get inundated with spam email. One of us systematically documented the number of spam emails arriving at his email account and noticed a curious cyclical pattern in the average number of emails arriving each week.[3] Figure 10 shows the daily spam emails for a 16-week period[4] (each point is a 1-week average):

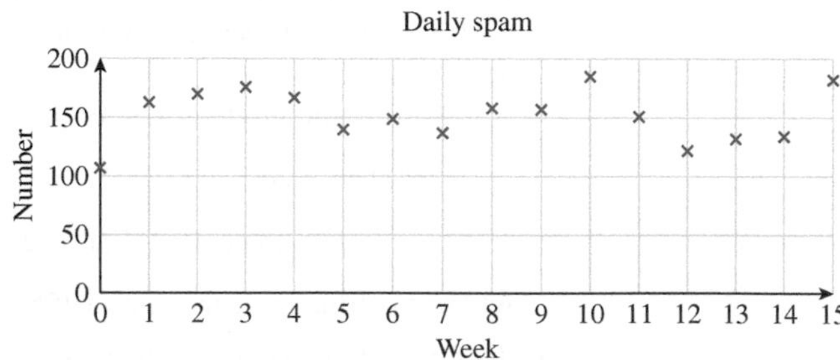

Figure 10

[3] Confirming the notion that academics have little else to do but fritter away their time in pointless pursuits.

[4] Beginning June 6, 2005.

Week	0	1	2	3	4	5	6	7	8	9	10	11	12	13	14	15
Number	107	163	170	176	167	140	149	137	158	157	185	151	122	132	134	182

a. Use technology to find the best-fit sine curve of the form

$$S(t) = A\sin[\omega(t - \alpha)] + C.$$

b. Use your model to estimate the period of the cyclical pattern in spam emails and also to predict the daily spam email average for week 23.

Solution

a. Following are the models obtained by using the TI-83/84 Plus, Excel with Solver, and the Function Evaluator and Grapher on the Website. (See the Technology Guides at the end of the chapter to find out how to obtain these models. For the Website utility we used the initial guess $A = 30$, $\omega = 1$, $\alpha = 0$, and $C = 150$.)

TI-83/84 Plus: $S(t) \approx 11.6\sin[0.910(t - 1.63)] + 155$

Excel and Website grapher: $S(t) \approx 25.8\sin[0.960(t - 1.22)] + 153$

Q: *Why do the models from the TI-83/84 Plus differ so drastically from the Solver and Website models?*

A: Not all regression algorithms are identical, and it seems that the TI-83/84 Plus's algorithm is not very efficient at finding the best-fit sine curve. Indeed, the value for the sum-of-squares error (SSE) for the TI-83/84 Plus regression curve is around 5,030, whereas it is around 2,148 for the Excel curve, indicating a far better fit.* Notice another thing: The sine curve does not appear to fit the data well in either graph. In general, we can expect better agreement between the different forms of technology for data that follow a sine curve more closely.

* This comparison is actually unfair: The method using Excel's Solver or the Website grapher starts with an initial guess of the coefficients, so the TI-83/84 Plus algorithm, which does not require an initial guess, is starting at a significant disadvantage. An initial guess that is way off can result in Solver or the Website grapher coming up with a very different result! On the other hand, the TI-83/84 Plus algorithm seems problematic and tends to fail (giving an error message) on many sets of data.

b. This model gives a period of approximately

TI-83/84 Plus: $P = \frac{2\pi}{\omega} \approx \frac{2\pi}{0.910} \approx 6.9$ weeks

Excel and Website grapher: $P = \frac{2\pi}{\omega} \approx \frac{2\pi}{0.960} \approx 6.5$ weeks.

So both models predict a very similar period.

In week 23 we obtain the following predictions:

TI-83/84 Plus: $S(23) \approx 11.6\sin[0.910(23 - 1.63)] + 155$
≈ 162 spam emails per day

Excel and Website grapher: $S(23) \approx 25.8\sin[0.960(23 - 1.22)] + 153$
≈ 176 spam emails per day.

Note The actual figure for week 23 was 213 spam emails per day. The discrepancy illustrates the danger of using regression models to extrapolate. ■

Website
www.WanerMath.com
At the Website you will find the following optional online interactive section, in which you can find further discussion of the graphs of the trigonometric functions and their relationship to right triangles:

Internet Topic: Trigonometric Functions and Calculus
→ The Six Trigonometric Functions.

9.1 EXERCISES

▼ more advanced ◆ challenging

T indicates exercises that should be solved using technology

In Exercises 1–12, graph the given functions or pairs of functions on the same set of axes.

(a) Sketch the curves without any technological help by consulting the discussion in Example 1.

(b) T Use technology to check your sketches.
[**HINT:** See Example 1.]

1. $f(t) = \sin t$; $g(t) = 3 \sin t$
2. $f(t) = \sin t$; $g(t) = 2.2 \sin t$
3. $f(t) = \sin t$; $g(t) = \sin(t - \pi/4)$
4. $f(t) = \sin t$; $g(t) = \sin(t + \pi)$
5. $f(t) = \sin t$; $g(t) = \sin(2t)$
6. $f(t) = \sin t$; $g(t) = \sin(-t)$
7. $f(t) = 2 \sin[3\pi(t - 0.5)] - 3$
8. $f(t) = 2 \sin[3\pi(t + 1.5)] + 1.5$
9. $f(t) = \cos t$; $g(t) = 5 \cos[3(t - 1.5\pi)]$
10. $f(t) = \cos t$; $g(t) = 3.1 \cos(3t)$
11. $f(t) = \cos t$; $g(t) = -2.5 \cos t$
12. $f(t) = \cos t$; $g(t) = 2 \cos(t - \pi)$

In Exercises 13–18, model each curve with a sine function. (Note that not all are drawn with the same scale on the two axes.) [**HINT:** See Example 2.]

13.

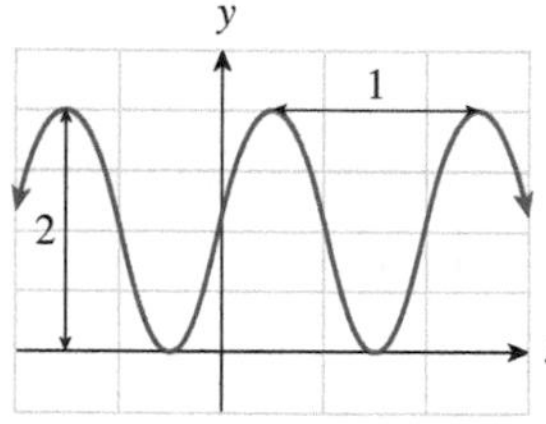

14.

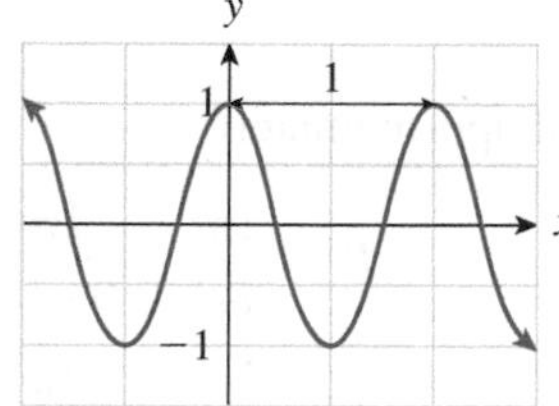

15.

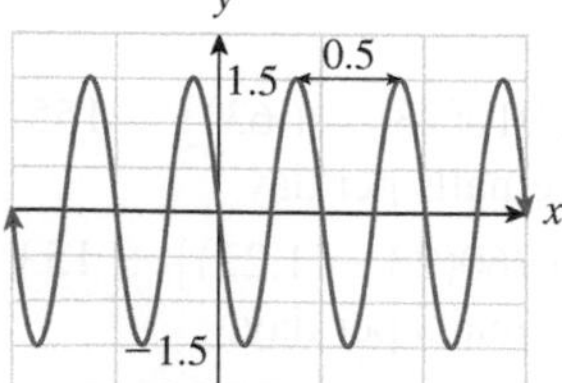

16.

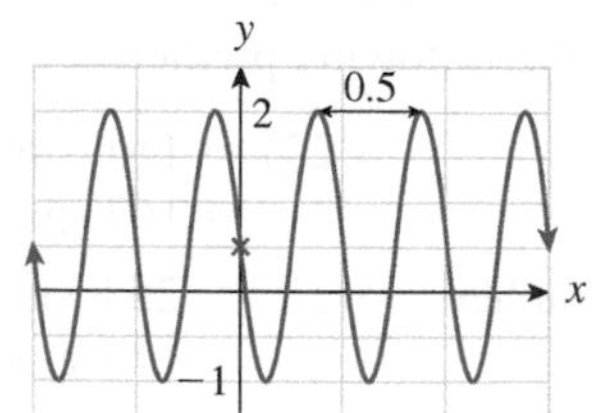

17.

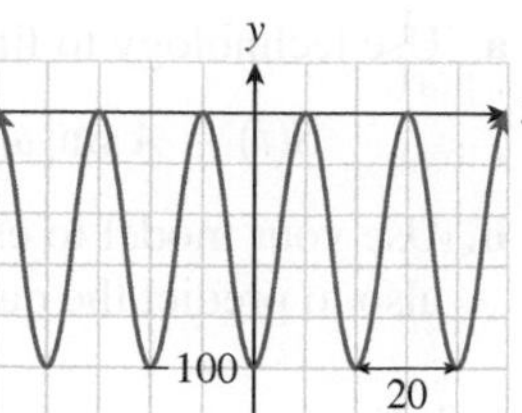

18.

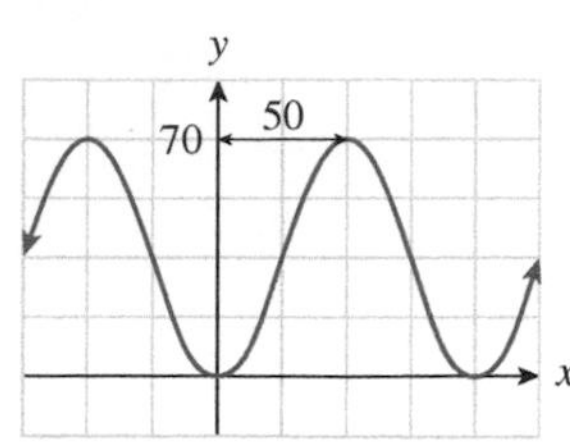

In Exercises 19–24, model each curve with a cosine function. (Note that not all are drawn with the same scale on the two axes.) [**HINT:** See Example 2.]

19.

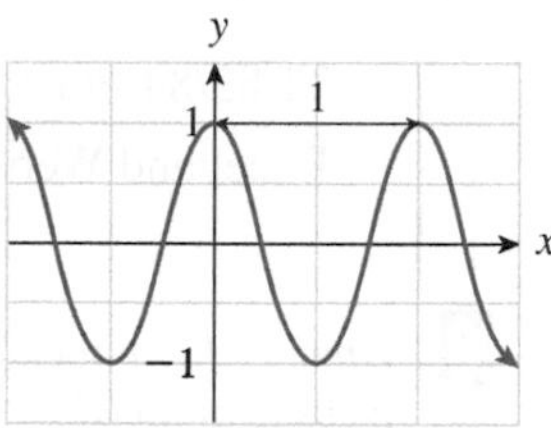

20.

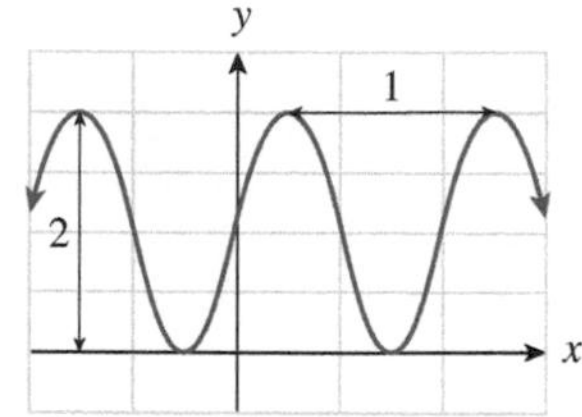

21.

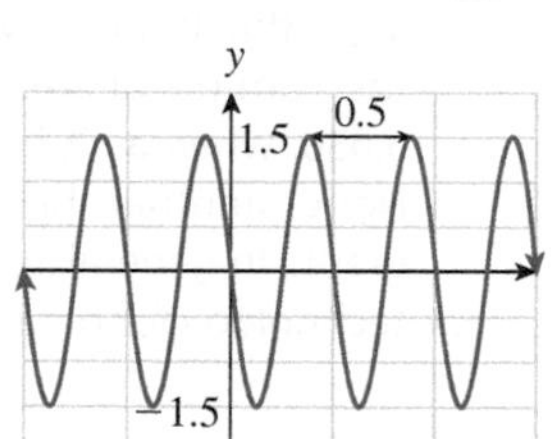

22.

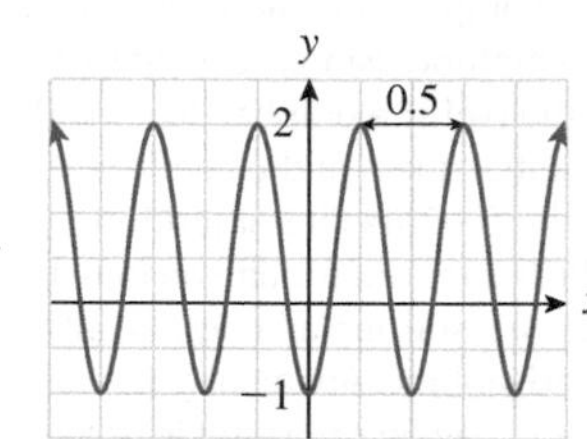

23.

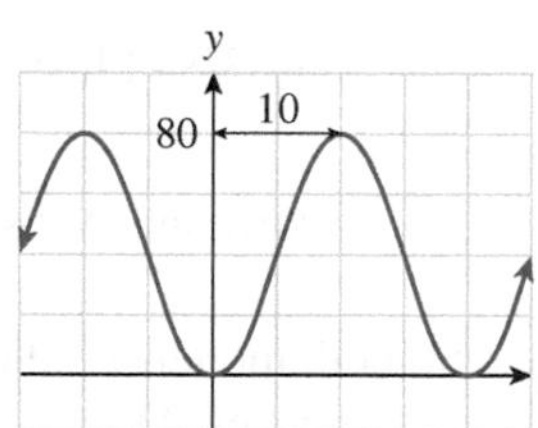

24.

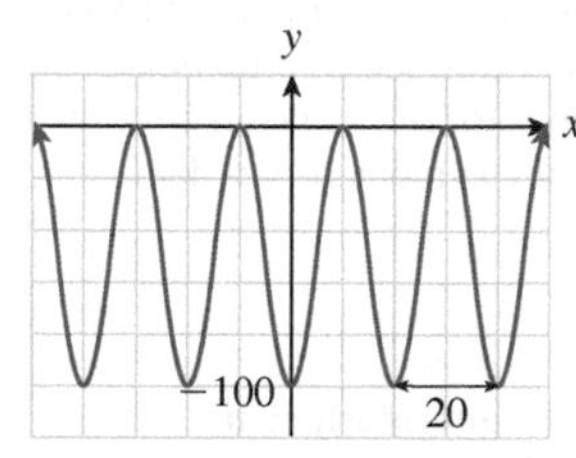

In Exercises 25–28, use the conversion formula $\cos x = \sin(\pi/2 - x)$ *to replace each expression by a sine function.*

25. ▼ $f(t) = 4.2 \cos(2\pi t) + 3$
26. ▼ $f(t) = 3 - \cos(t - 4)$
27. ▼ $g(x) = 4 - 1.3 \cos[2.3(x - 4)]$
28. ▼ $g(x) = 4.5 \cos[2\pi(3x - 1)] + 7$

Some Identities *Starting with the identity* $\sin^2 x + \cos^2 x = 1$ *and then dividing both sides of the equation by a suitable trigonometric function, derive the trigonometric identities in Exercises 29 and 30.*

29. ▼ $\sec^2 x = 1 + \tan^2 x$ **30.** ▼ $\csc^2 x = 1 + \cot^2 x$

Exercises 31–38 are based on the ***addition formulas****:*

$$\sin(x + y) = \sin x \cos y + \cos x \sin y$$
$$\sin(x - y) = \sin x \cos y - \cos x \sin y$$
$$\cos(x + y) = \cos x \cos y - \sin x \sin y$$
$$\cos(x - y) = \cos x \cos y + \sin x \sin y.$$

31. ▼ Calculate $\sin(\pi/3)$, given that $\sin(\pi/6) = 1/2$ and $\cos(\pi/6) = \sqrt{3}/2$.

32. ▼ Calculate $\cos(\pi/3)$, given that $\sin(\pi/6) = 1/2$ and $\cos(\pi/6) = \sqrt{3}/2$.

33. ▼ Use the formula for $\sin(x + y)$ to obtain the identity $\sin(t + \pi/2) = \cos t$.

34. ▼ Use the formula for $\cos(x + y)$ to obtain the identity $\cos(t - \pi/2) = \sin t$.

35. ▼ Show that $\sin(\pi - x) = \sin x$.

36. ▼ Show that $\cos(\pi - x) = -\cos x$.

37. ▼ Use the addition formulas to express $\tan(x + \pi)$ in terms of $\tan x$.

38. ▼ Use the addition formulas to express $\cot(x + \pi)$ in terms of $\cot x$.

Applications

39. ***Sunspot Activity*** The activity of the Sun (sunspots, solar flares, and coronal mass ejection) fluctuates in cycles of around 10–11 years. Sunspot activity can be modeled by the following function:[5]

$$N(t) = 57.7 \sin[0.602(t - 1.43)] + 58.8,$$

where t is the number of years since January 1, 1997, and $N(t)$ is the number of sunspots observed at time t. [**HINT:** See Example 3.]

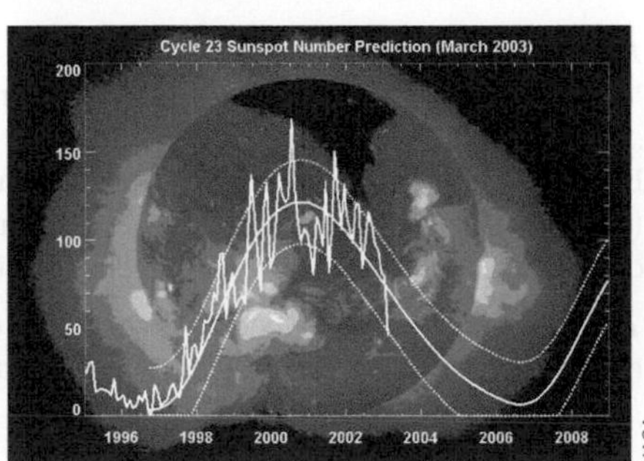

a. What is the period of sunspot activity according to this model? (Round your answer to the nearest 0.1 years.)

b. What is the maximum number of sunspots observed? What is the minimum number? (Round your answers to the nearest sunspot.)

c. When, to the nearest year, was sunspot activity expected to reach the first high point beyond 2012?

40. ***Solar Emissions*** The following model gives the flux of radio emission from the Sun:

$$F(t) = 49.6 \sin[0.602(t - 1.48)] + 111,$$

where t is the number of years since January 1, 1997, and $F(t)$ is the flux of solar emissions of a specified wavelength at time t.[6] [**HINT:** See Example 3.]

a. What is the period of radio activity according to this model? (Round your answer to the nearest 0.1 years.)

b. What is the maximum flux of radio emissions? What is the minimum flux? (Round your answers to the nearest whole number.)

c. When, to the nearest year, was radio activity expected to reach the first low point beyond 2012?

41. T ***iPod Sales*** Sales of personal electronic devices such as Apple's iPods are subject to seasonal fluctuations. Apple's sales of iPods in 2008–2010 can be approximated by the function

$$s(t) = 6.00 \sin(1.51t + 1.85) + 15 \text{ million iPods per quarter} \quad (0 \le t \le 12),$$

where t is time in quarters. ($t = 0$ represents the start of the first quarter of 2008.)[7]

a. Use technology to plot sales versus time from the beginning of 2008 through the end of 2010. Then use your graph to estimate the value of t and the quarters during which sales were lowest and highest.

b. Estimate Apple's maximum and minimum quarterly sales of iPods.

c. Indicate how the answers to part (b) can be obtained directly from the equation for $s(t)$.

[5] The model is based on a regression obtained from predicted data for 1997–2006 and the mean historical period of sunspot activity from 1755 to 1995. Source: NASA Science Directorate; Marshall Space Flight Center, August 2002 (http://science.nasa.gov/ssl/pad/solar/predict.htm).

[6] *Ibid.* Flux is measured at a wavelength of 10.7 centimeters.

[7] Authors' model. Source for data: Apple quarterly press releases (www.apple.com/investor).

42. **T** ***Housing Starts*** Housing construction is subject to seasonal fluctuations. The number of housing starts (number of new privately owned housing units started) from 2009 through mid-2011 can be approximated by the function

$$s(t) = 10.2\sin(0.537t - 1.46) + 47.6$$
thousand housing starts per month $(0 \le t \le 30)$,

where t is time in months. ($t = 0$ represents the start of January 2009.)[8]

a. Use technology to plot housing starts versus time from the beginning of 2009 through the end of June 2011. Then use your graph to estimate the value of t and the months during which housing starts were lowest and highest.

b. Estimate the maximum and minimum number of housing starts.

c. Indicate how the answers to part (b) can be obtained directly from the equation for $s(t)$.

43. ***iPod Sales*** (Based on Exercise 41, but no graphing technology required) **Apple**'s sales of iPods in 2008–2010 can be approximated by the function

$$s(t) = 6.00\sin(1.51t + 1.85) + 15$$
million iPods per quarter $(0 \le t \le 12)$,

where t is time in quarters. ($t = 0$ represents the start of the first quarter of 2008.) Calculate the amplitude, the vertical offset, the phase shift, the angular frequency, and the period, and interpret the results.

44. ***Housing Starts*** (Based on Exercise 42, but no graphing technology required) The number of housing starts (number of new privately owned housing units started) from 2009 through mid-2011 can be approximated by the function

$$s(t) = 10.2\sin(0.537t - 1.46) + 47.6$$
thousand housing starts per month $(0 \le t \le 30)$,

where t is time in months. ($t = 0$ represents the start of January 2009.) Calculate the amplitude, the vertical offset, the phase shift, the angular frequency, and the period, and interpret the results.

45. ***Biology*** Sigatoka leaf spot is a plant disease that affects bananas. In an infected plant, the percentage of leaf area affected varies from a low of around 5% at the start of each year to a high of around 20% at the middle of each year.[9] Use the sine function to model the percentage of leaf area affected by Sigatoka leaf spot t weeks since the start of a year. [**HINT**: See Example 2.]

46. ***Biology*** Apple powdery mildew is an epidemic that affects apple shoots. In a new infection the percentage of apple shoots infected varies from a low of around 10% at the start of May to a high of around 60% 6 months later.[10] Use the sine function to model the percentage of apple shoots affected by apple powdery mildew t months since the start of a year.

47. ***Cancun*** The *Playa Loca Hotel* in Cancun has an advertising brochure with a chart showing the year-round temperature.[11] The added curve is an approximate 5-month moving average.

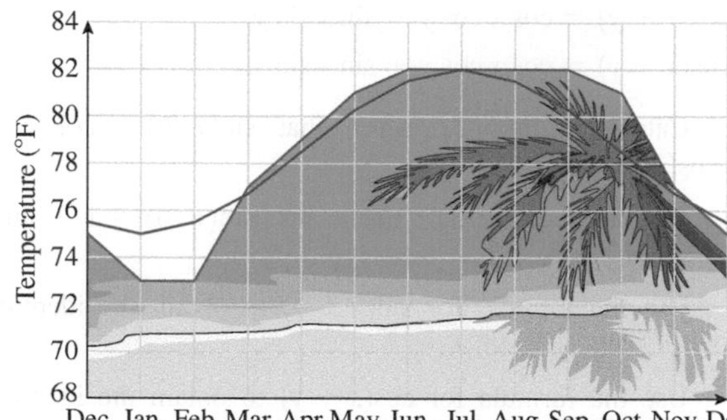

Use a cosine function to model the temperature (moving average) in Cancun as a function of time t in months since December.

48. ***Reykjavik*** Repeat Exercise 47, using the following data from the brochure of the *Tough Traveler Lodge* in Reykjavik.[12]

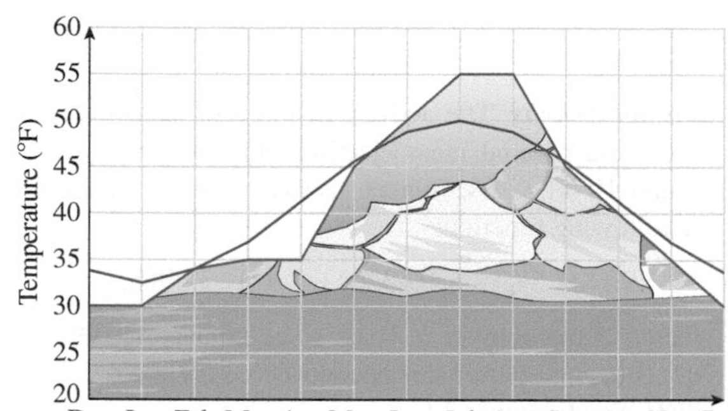

49. ***Net Income Fluctuations: General Electric*** **General Electric**'s quarterly net income $n(t)$ fluctuated from a low of \$4.5 billion in Quarter 1 of 2007 ($t = 0$) to a high of \$7 billion, and then back down to \$4.5 billion in Quarter 1 of 2008 ($t = 4$).[13] Use a sine function to model General Electric's quarterly net income $n(t)$, where t is time in quarters.

50. ***Sales Fluctuations*** Sales of cypods (one-bedroom units) in the city-state of Utarek on Mars[14] fluctuate from a low of

---

[8] Authors' regression model. Source for data: U.S. Census Bureau (www.census.gov/const/www/newresconstindex.html).

[9] Based on graphical data. Source: American Phytopathological Society, July 2002 (www.apsnet.org/education/AdvancedPlantPath/Topics/Epidemiology/CyclicalNature.htm).

[10] *Ibid.*

[11] Source: www.holiday-weather.com.

[12] *Ibid.* Temperatures are rounded.

[13] Source: Company reports (www.ge.com/investors).

[14] Based on www.Marsnext.com, a now extinct virtual society.

5 units per week each February 1 ($t = 1$) to a high of 35 units per week each August 1 ($t = 7$). Use a sine function to model the weekly sales $s(t)$ of cypods, where t is time in months.

51. ***Net Income Fluctuations*** Repeat Exercise 49, but this time use a cosine function for your model. [**HINT:** See Example 4.]

52. ***Sales Fluctuations*** Repeat Exercise 50, but this time use a cosine function for your model.

53. ***Tides*** The depth of water at my favorite surfing spot varies from 5 to 15 feet, depending on the time. Last Sunday, high tide occurred at 5:00 am, and the next high tide occurred at 6:30 pm. Use a sine function model to describe the depth of water as a function of time t in hours since midnight on Sunday morning.

54. ***Tides*** Repeat Exercise 53 using data from the depth of water at my other favorite surfing spot, where the tide last Sunday varied from a low of 6 feet at 4:00 am to a high of 10 feet at noon.

55. ▼ ***Inflation*** The uninflated cost of *Dugout* brand snow shovels currently varies from a high of $10 on January 1 ($t = 0$) to a low of $5 on July 1 ($t = 0.5$).

a. Assuming that this trend continues indefinitely, calculate the uninflated cost $u(t)$ of Dugout snow shovels as a function of time t in years. (Use a sine function.)

b. Assuming a 4% annual rate of inflation in the cost of snow shovels, the cost of a snow shovel t years from now, adjusted for inflation, will be 1.04^t times the uninflated cost. Find the cost $c(t)$ of Dugout snow shovels as a function of time t.

56. ▼ ***Deflation*** Sales of my exclusive 2010 vintage *Chateau Petit Mont Blanc* vary from a high of 10 bottles per day on April 1 ($t = 0.25$) to a low of 4 bottles per day on October 1.

a. Assuming that this trend continues indefinitely, find the undeflated sales $u(t)$ of Chateau Petit Mont Blanc as a function of time t in years. (Use a sine function.)

b. Regrettably, ever since that undercover exposé of my wine-making process, sales of Chateau Petit Mont Blanc have been declining at an annual rate of 12%. Using Exercise 55 as a guide, write down a model for the deflated sales $s(t)$ of Chateau Petit Mont Blanc t years from now.

57. T ***Air Travel: Domestic*** The following table shows total domestic air travel on U.S. air carriers in specified months from January 2006 to July 2008 ($t = 0$ represents January 2006):[15]

t	0	3	6	9	12	15	18	21	24	27	30
Revenue Passenger Miles (billions)	43	49	55	47	44	50	57	49	45	48	55

a. Plot the data and *roughly* estimate the period P and the parameters C, A, and β for a cosine model.

b. Find the best-fit cosine curve approximating the given data. [You may have to use your estimates from part (a) as initial guesses if you are using Solver.] Plot the given data together with the regression curve. (Round coefficients to three decimal places.)

c. Complete the following: Based on the regression model, domestic air travel on U.S. air carriers showed a pattern that repeats itself every ___ months, from a low of ___ to a high of ___ billion revenue passenger miles. (Round answers to the nearest whole number.) [**HINT:** See Example 5.]

58. T ***Air Travel: International*** The following table shows total international travel on U.S. air carriers in specified months from January 2006 to July 2008 ($t = 0$ represents January 2006):[16]

t	0	3	6	9	12	15	18	21	24	27	30
Revenue Passenger Miles (billions)	18	19	23	19	19	20	24	20	25	20	25

a. Plot the data, and *roughly* estimate the period P and the parameters C, A, and β for a cosine model.

b. Find the best-fit cosine curve approximating the given data. [You may have to use your estimates from part (a) as initial guesses if you are using Solver.] Plot the given data together with the regression curve. (Round coefficients to three decimal places.)

c. Complete the following: Based on the regression model, international travel on U.S. air carriers showed a pattern that repeats itself every ___ months, from a low of ___ to a high of ___ billion revenue passenger miles. (Round answers to the nearest whole number.) [**HINT:** See Example 5.]

Music *Musical sounds exhibit the same kind of periodic behavior as the trigonometric functions. High-pitched notes have short periods (less than 1/1000 second) while the lowest audible notes have periods of about 1/100 second. Some electronic synthesizers work by superimposing (adding) sinusoidal functions of different frequencies to create different textures. Exercises 59–62 show some examples of how superposition can be used to create interesting periodic functions.*

59. T ▼ ***Sawtooth Wave***

a. Graph the following functions in a window with $-7 \le x \le 7$ and $-1.5 \le y \le 1.5$:

$$y_1 = \frac{2}{\pi}\cos x$$

$$y_3 = \frac{2}{\pi}\cos x + \frac{2}{3\pi}\cos 3x$$

$$y_5 = \frac{2}{\pi}\cos x + \frac{2}{3\pi}\cos 3x + \frac{2}{5\pi}\cos 5x.$$

[15] Source: Bureau of Transportation Statistics (www.bts.gov).

[16] *Ibid.*

b. Following the pattern established above, give a formula for y_{11} and graph it.

c. How would you modify y_{11} to approximate a sawtooth wave with an amplitude of three times that in part (b) and a period of 4π?

60. T ▼ ***Square Wave*** Repeat Exercise 59 using sine functions in place of cosine functions (which results in an approximation of a square wave).

61. T ▼ ***Harmony*** If we add two sinusoidal functions with frequencies that are simple ratios of each other, the result is a pleasing sound. The following function models two notes an octave apart together with the intermediate fifth:

$$y = \cos x + \cos(1.5x) + \cos(2x).$$

Graph this function in the window $0 \le x \le 20$ and $-3 \le y \le 3$, and estimate the period of the resulting wave.

62. T ▼ ***Discord*** If we add two sinusoidal functions with similar, but unequal, frequencies, the result is a function that "pulsates," or exhibits "beats." (Piano tuners and guitar players use this phenomenon to help them tune an instrument.) Graph the function

$$y = \cos x + \cos(0.9x)$$

in the window $-50 \le x \le 50$ and $-2 \le y \le 2$, and estimate the period of the resulting wave.

Communication and Reasoning Exercises

63. What are the highs and lows for sales of a commodity modeled by a function of the form $s(t) = A\sin(2\pi t) + B$ (A, B constants)?

64. Your friend has come up with the following model for choral society Tupperware stock inventory: $r(t) = 4\sin[2\pi(t-2)/3] + 2.3$, where t is time in weeks and $r(t)$ is the number of items in stock. Why is the model not realistic?

65. Your friend is telling everybody that all six trigonometric functions can be obtained from the single function $\sin x$. Is he correct? Explain your answer.

66. Another friend claims that all six trigonometric functions can be obtained from the single function $\cos x$. Is she correct? Explain your answer.

67. If weekly sales of sodas at a movie theater are given by $s(t) = A + B\cos(\omega t)$, what is the largest B can be? Explain your answer.

68. Complete the following: If the cost of an item is given by $c(t) = A + B\cos[\omega(t-\alpha)]$, then the cost fluctuates by ____ with a period of ____ about a base of ____, peaking at time $t =$ ____.

9.2 Derivatives of Trigonometric Functions and Applications

Derivatives of Sine and Cosine

We start with the derivatives of the sine and cosine functions.

Derivatives of the Sine and Cosine Functions

The sine and cosine functions are differentiable with

$$\frac{d}{dx}\sin x = \cos x$$

$$\frac{d}{dx}\cos x = -\sin x. \quad \text{Notice the sign change.}$$

Quick Examples

1. $\frac{d}{dx}(x\cos x) = 1\cdot\cos x + x\cdot(-\sin x)$ Product rule: $x\cos x$ is a product.*

$= \cos x - x\sin x$

2. $\frac{d}{dx}\left(\frac{x^2+x}{\sin x}\right) = \frac{(2x+1)(\sin x) - (x^2+x)(\cos x)}{\sin^2 x}$ Quotient rule

* Apply the calculation thought experiment: If we were to compute $x\cos x$, the last operation we would perform is the multiplication of x and $\cos x$. Hence, $x\cos x$ is a product.

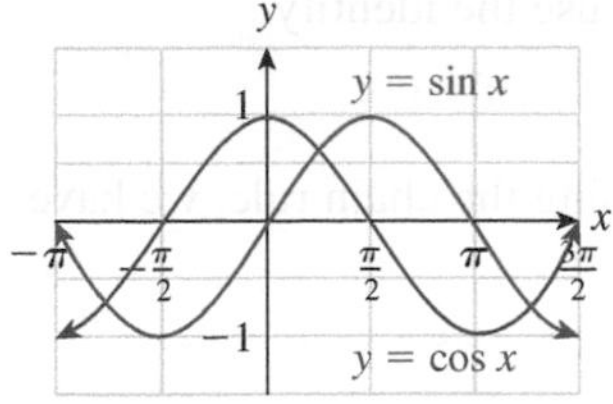

Figure 11(a)

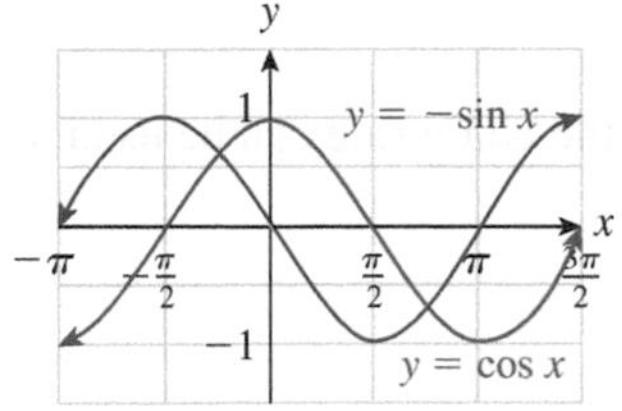

Figure 11(b)

Before deriving these formulas, we can see right away that they are plausible by examining Figures 11(a) and (b), which show the graphs of the sine and cosine functions together with their derivatives. Notice, for instance, that in Figure 11(a) the graph of $\sin x$ is rising most rapidly when $x = 0$, corresponding to the maximum value of its derivative, $\cos x$. When $x = \pi/2$, the graph of $\sin x$ levels off, so its derivative, $\cos x$, is 0. Another point to notice: Because periodic functions (such as sine and cosine) repeat their behavior, their derivatives must also be periodic.

Derivation of Formulas for Derivatives of the Sine and Cosine Functions

We first calculate the derivative of $\sin x$ from scratch, using the definition of the derivative:

$$\frac{d}{dx}f(x) = \lim_{h\to 0}\frac{f(x+h) - f(x)}{h}.$$

Thus,

$$\frac{d}{dx}\sin x = \lim_{h\to 0}\frac{\sin(x+h) - \sin x}{h}.$$

We now use the addition formula given in the exercises for Section 9.1:

$$\sin(x+h) = \sin x \cos h + \cos x \sin h.$$

Substituting this expression for $\sin(x + h)$ gives

$$\frac{d}{dx}\sin x = \lim_{h\to 0}\frac{\sin x\cos h + \cos x\sin h - \sin x}{h}.$$

Grouping the first and third terms together and factoring out the term $\sin x$, we get

$$\begin{aligned}\frac{d}{dx}\sin x &= \lim_{h\to 0}\frac{\sin x(\cos h - 1) + \cos x\sin h}{h}\\ &= \lim_{h\to 0}\frac{\sin x(\cos h - 1)}{h} + \lim_{h\to 0}\frac{\cos x\sin h}{h} \qquad \text{Limit of a sum}\\ &= \sin x\lim_{h\to 0}\frac{\cos h - 1}{h} + \cos x\lim_{h\to 0}\frac{\sin h}{h},\end{aligned}$$

and we are left with two limits to evaluate. Calculating these limits analytically requires a little trigonometry.* Alternatively, we can get a good idea of what these two limits are by estimating them numerically or graphically. Figures 12 and 13 show the graphs of $(\cos h - 1)/h$ and $(\sin h)/h$, respectively.

We find that

$$\lim_{h\to 0}\frac{\cos h - 1}{h} = 0$$

and

$$\lim_{h\to 0}\frac{\sin h}{h} = 1.$$

Therefore,

$$\frac{d}{dx}\sin x = (\sin x)(0) + (\cos x)(1) = \cos x.$$

This is the required formula for the derivative of $\sin x$.

* You can find these calculations on the Website by following the path:
Website
→ Everything for Applied Calc
→ Chapter 9
→ Proof of Some Trigonometric Limits

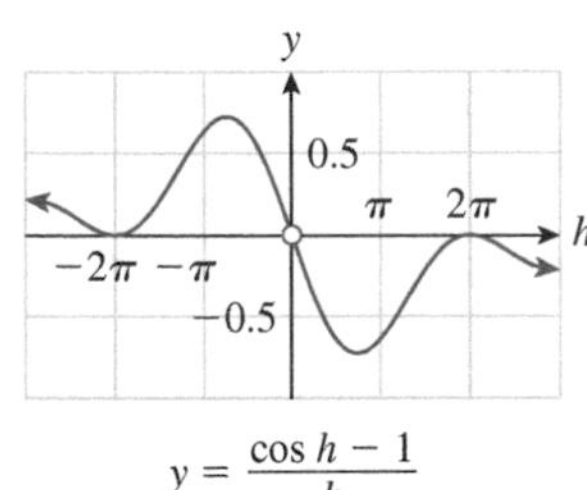

$y = \dfrac{\cos h - 1}{h}$

Figure 12

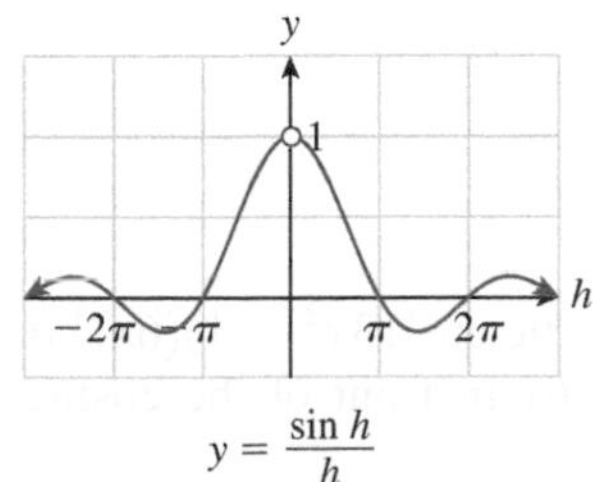

$y = \dfrac{\sin h}{h}$

Figure 13

Turning to the derivative of the cosine function, we use the identity

$$\cos x = \sin(\pi/2 - x)$$

from Section 9.1. If $y = \cos x = \sin(\pi/2 - x)$, then, using the chain rule, we have

$$\frac{dy}{dx} = \cos(\pi/2 - x)\frac{d}{dx}(\pi/2 - x)$$
$$= (-1)\cos(\pi/2 - x)$$
$$= -\sin x. \quad \text{Using the identity } \cos(\pi/2 - x) = \sin x$$

This is the required formula for the derivative of cos x.

Just as with logarithmic and exponential functions, the chain rule can be used to find more general derivatives.

Derivatives of Sines and Cosines of Functions

Original Rule	Generalized Rule	In Words
$\frac{d}{dx}\sin x = \cos x$	$\frac{d}{dx}\sin u = \cos u \frac{du}{dx}$	*The derivative of the sine of a quantity is the cosine of that quantity, times the derivative of that quantity.*
$\frac{d}{dx}\cos x = -\sin x$	$\frac{d}{dx}\cos u = -\sin u \frac{du}{dx}$	*The derivative of the cosine of a quantity is negative sine of that quantity, times the derivative of that quantity.*

Quick Examples

3. $\frac{d}{dx}\sin(3x^2 - 1) = \cos(3x^2 - 1)\frac{d}{dx}(3x^2 - 1)$ $\quad u = 3x^2 - 1$ (See margin note.*)

$= 6x\cos(3x^2 - 1)$ $\quad$ We placed the $6x$ in front. See Note below.

4. $\frac{d}{dx}\cos(x^3 + x) = -\sin(x^3 + x)\frac{d}{dx}(x^3 + x)$ $\quad u = x^3 + x$

$= -(3x^2 + 1)\sin(x^3 + x)$

* If we were to evaluate $\sin(3x^2 - 1)$, the last operation we would perform is taking the sine of a quantity. Thus, the calculation thought experiment tells us that we are dealing with the *sine of a quantity*, and we use the generalized rule.

Note Avoid writing ambiguous expressions such as $\cos(3x^2 - 1)(6x)$. Does this mean

$\cos[(3x^2 - 1)(6x)]$? $\quad$ The cosine of the quantity $(3x^2 - 1)(6x)$

Or does it mean

$[\cos(3x^2 - 1)](6x)$? $\quad$ The product of $\cos(3x^2 - 1)$ and $6x$

To avoid the ambiguity, use parentheses or brackets, and write $\cos[(3x^2 - 1)(6x)]$ if you mean the former. If you mean the latter, place the $6x$ in front of the cosine expression and write

$6x\cos(3x^2 - 1)$. $\quad$ The product of $6x$ and $\cos(3x^2 - 1)$ ■

EXAMPLE 1 **Derivatives of Trigonometric Functions**

Find the derivatives of the following functions:

a. $f(x) = \sin^2 x$ **b.** $g(x) = \sin^2(x^2)$ **c.** $h(x) = e^{-x}\cos(2x)$

Solution

a. Recall that $\sin^2 x = (\sin x)^2$. The calculation thought experiment tells us that $f(x)$ is the square of a quantity.* Therefore, we use the chain rule (or generalized power rule) for differentiating the square of a quantity:

* Notice the difference between $\sin^2 x$ and $\sin(x^2)$. The first is the square of $\sin x$, whereas the second is the sine of the quantity x^2.

$$\frac{d}{dx}(u^2) = 2u\frac{du}{dx}$$

$$\frac{d}{dx}(\sin x)^2 = 2(\sin x)\frac{d(\sin x)}{dx} \qquad u = \sin x$$

$$= 2\sin x\cos x.$$

Thus, $f'(x) = 2\sin x\cos x$.

b. We rewrite the function $g(x) = \sin^2(x^2)$ as $[\sin(x^2)]^2$. Because $g(x)$ is the square of a quantity, we have

$$\frac{d}{dx}\sin^2(x^2) = \frac{d}{dx}[\sin(x^2)]^2 \qquad \text{Rewrite } \sin^2(-) \text{ as } [\sin(-)]^2.$$

$$= 2\sin(x^2)\frac{d[\sin(x^2)]}{dx} \qquad \frac{d}{dx}[u^2] = 2u\frac{du}{dx} \text{ with } u = \sin(x^2)$$

$$= 2\sin(x^2)\cdot\cos(x^2)\cdot 2x. \qquad \frac{d}{dx}\sin u = \cos u\frac{du}{dx} \text{ with } u = x^2$$

Thus, $g'(x) = 4x\sin(x^2)\cos(x^2)$.

c. Because $h(x)$ is the product of e^{-x} and $\cos(2x)$ we use the product rule:

$$h'(x) = (-e^{-x})\cos(2x) + e^{-x}\frac{d}{dx}[\cos(2x)]$$

$$= (-e^{-x})\cos(2x) - e^{-x}\sin(2x)\frac{d}{dx}[2x] \qquad \frac{d}{dx}\cos u = -\sin u\frac{du}{dx}$$

$$= -e^{-x}\cos(2x) - 2e^{-x}\sin(2x)$$

$$= -e^{-x}[\cos(2x) + 2\sin(2x)].$$

Derivatives of Other Trigonometric Functions

Because the remaining trigonometric functions are ratios of sines and cosines, we can use the quotient rule to find their derivatives. For example, we can find the derivative of the tangent function as follows:

$$\frac{d}{dx}\tan x = \frac{d}{dx}\left(\frac{\sin x}{\cos x}\right)$$

$$= \frac{(\cos x)(\cos x) - (\sin x)(-\sin x)}{\cos^2 x}$$

$$= \frac{\cos^2 x + \sin^2 x}{\cos^2 x}$$

$$= \frac{1}{\cos^2 x}$$

$$= \sec^2 x.$$

We ask you to derive the other three derivatives in the exercises. Here is a list of the derivatives of all six trigonometric functions and their chain rule variants.

Derivatives of the Trigonometric Functions

Original Rule	Generalized Rule
$\frac{d}{dx}\sin x = \cos x$	$\frac{d}{dx}\sin u = \cos u\,\frac{du}{dx}$
$\frac{d}{dx}\cos x = -\sin x$	$\frac{d}{dx}\cos u = -\sin u\,\frac{du}{dx}$
$\frac{d}{dx}\tan x = \sec^2 x$	$\frac{d}{dx}\tan u = \sec^2 u\,\frac{du}{dx}$
$\frac{d}{dx}\cot x = -\csc^2 x$	$\frac{d}{dx}\cot u = -\csc^2 u\,\frac{du}{dx}$
$\frac{d}{dx}\sec x = \sec x \tan x$	$\frac{d}{dx}\sec u = \sec u \tan u\,\frac{du}{dx}$
$\frac{d}{dx}\csc x = -\csc x \cot x$	$\frac{d}{dx}\csc u = -\csc u \cot u\,\frac{du}{dx}$

Quick Examples

5. $\frac{d}{dx}\tan(x^2 - 1) = \sec^2(x^2 - 1)\frac{d(x^2 - 1)}{dx}$ $\quad u = x^2 - 1$

$$= 2x\sec^2(x^2 - 1)$$

6. $\frac{d}{dx}\csc(e^{3x}) = -\csc(e^{3x})\cot(e^{3x})\frac{d(e^{3x})}{dx}$ $\quad u = e^{3x}$

$$= -3e^{3x}\csc(e^{3x})\cot(e^{3x})$$ The derivative of e^{3x} is $3e^{3x}$.

EXAMPLE 2 Gas Heating Demand

In Section 9.1 we saw that seasonal fluctuations in temperature suggested a sine function. For instance, we can use the function

$$T = 60 + 25\sin\left[\frac{\pi}{6}(x - 4)\right]$$ T = temperature in °F; x = months since Jan 1

to model a temperature that fluctuates between 35°F on Feb. 1 ($x = 1$) and 85°F on Aug. 1 ($x = 7$). (See Figure 14.)

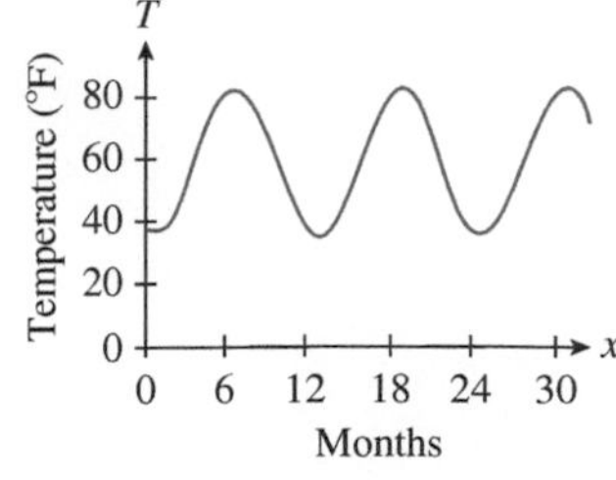

Figure 14

The demand for gas at a utility company can be expected to fluctuate in a similar way because demand grows with increased heating requirements. A reasonable model might therefore be

$$G = 400 - 100\sin\left[\frac{\pi}{6}(x - 4)\right],$$ Why did we subtract the sine term?

where G is the demand for gas in cubic yards per day. Find and interpret $G'(10)$.

Solution First, we take the derivative of G:

$$G'(x) = -100\cos\left[\frac{\pi}{6}(x-4)\right]\cdot\frac{\pi}{6}$$
$$= -\frac{50\pi}{3}\cos\left[\frac{\pi}{6}(x-4)\right] \text{ cubic yards per day per month.}$$

Thus,

$$G'(10) = -\frac{50\pi}{3}\cos\left[\frac{\pi}{6}(10-4)\right]$$
$$= -\frac{50\pi}{3}\cos(\pi) = \frac{50\pi}{3}. \qquad \text{Because } \cos\pi = -1$$

The units of $G'(10)$ are cubic yards per day per month, so we interpret the result as follows: On November 1 ($x = 10$) the daily demand for gas is increasing at a rate of $50\pi/3 \approx 52$ cubic yards per day per month. This is consistent with Figure 14, which shows the temperature decreasing on that date.

9.2 EXERCISES

▼ more advanced ◆ challenging

T indicates exercises that should be solved using technology

In Exercises 1–32, find the derivative of the given function. [**HINT**: See Quick Examples 1–6.]

1. $f(x) = \sin x - \cos x$ **2.** $f(x) = \tan x - \sin x$

3. $g(x) = (\sin x)(\tan x)$ **4.** $g(x) = (\cos x)(\cot x)$

5. $h(x) = 2\csc x - \sec x + 3x$

6. $h(x) = 2\sec x + 3\tan x + 3x$

7. $r(x) = x\cos x + x^2 + 1$ **8.** $r(x) = 2x\sin x - x^2$

9. $s(x) = (x^2 - x + 1)\tan x$ **10.** $s(x) = \dfrac{\tan x}{x^2 - 1}$

11. $t(x) = \dfrac{\cot x}{1 + \sec x}$

12. $t(x) = (1 + \sec x)(1 - \cos x)$

13. $k(x) = \cos^2 x$ **14.** $k(x) = \tan^2 x$

15. $j(x) = \sec^2 x$ **16.** $j(x) = \csc^2 x$

17. $f(x) = \sin(3x - 5)$ **18.** $f(x) = \cos(2x + 7)$

19. $f(x) = \cos(-2x + 5)$ **20.** $f(x) = \sin(-4x - 5)$

21. $p(x) = 2 + 5\sin\left[\frac{\pi}{5}(x - 4)\right]$

22. $p(x) = 10 - 3\cos\left[\frac{\pi}{6}(x + 3)\right]$

23. $u(x) = \cos(x^2 - x)$ **24.** $u(x) = \sin(3x^2 + x - 1)$

25. $v(x) = \sec(x^{2.2} + 1.2x - 1)$

26. $v(x) = \tan(x^{2.2} + 1.2x - 1)$

27. $w(x) = \sec x\tan(x^2 - 1)$ **28.** $w(x) = \cos x\sec(x^2 - 1)$

29. $y(x) = \cos(e^x) + e^x\cos x$ **30.** $y(x) = \sec(e^x)$

31. $z(x) = \ln|\sec x + \tan x|$ **32.** $z(x) = \ln|\csc x + \cot x|$

In Exercises 33–36, derive the given formula from the derivatives of sine and cosine. [**HINT**: See the discussion on derivatives of other trigonometric functions.]

33. ▼ $\dfrac{d}{dx}\sec x = \sec x\tan x$ **34.** ▼ $\dfrac{d}{dx}\cot x = -\csc^2 x$

35. ▼ $\dfrac{d}{dx}\csc x = -\csc x\cot x$ **36.** ▼ $\dfrac{d}{dx}\ln|\sec x| = \tan x$

In Exercises 37–44, calculate the derivative.

37. ▼ $\dfrac{d}{dx}[e^{-2x}\sin(3\pi x)]$ **38.** ▼ $\dfrac{d}{dx}[e^{5x}\sin(-4\pi x)]$

39. ▼ $\dfrac{d}{dx}[\sin(3x)]^{0.5}$ **40.** ▼ $\dfrac{d}{dx}\cos\left(\dfrac{x^2}{x - 1}\right)$

41. ▼ $\dfrac{d}{dx}\sec\left(\dfrac{x^3}{x^2 - 1}\right)$ **42.** ▼ $\dfrac{d}{dx}\left(\dfrac{\tan x}{2 + e^x}\right)^2$

43. ▼ $\dfrac{d}{dx}([\ln|x|][\cot(2x - 1)])$ **44.** ▼ $\dfrac{d}{dx}\ln|\sin x - 2xe^{-x}|$

In Exercises 45 and 46, investigate the differentiability of the given function at the given points. If $f'(a)$ exists, give its approximate value.

45. ▼ $f(x) = |\sin x|$ **a.** $a = 0$ **b.** $a = 1$

46. ▼ $f(x) = |\sin(1 - x)|$ **a.** $a = 0$ **b.** $a = 1$

In Exercises 47–52, evaluate the given limit (a) numerically and (b) using l'Hospital's rule.

47. ▼ $\lim_{x\to 0} \frac{\sin^2 x}{x}$

48. ▼ $\lim_{x\to 0} \frac{\sin x}{x^2}$

49. ▼ $\lim_{x\to 0} \frac{\sin(2x)}{x}$

50. ▼ $\lim_{x\to 0} \frac{\sin x}{\tan x}$

51. ▼ $\lim_{x\to 0} \frac{\cos x - 1}{x^3}$

52. ▼ $\lim_{x\to 0} \frac{\cos x - 1}{x^2}$

In Exercises 53–56, find the indicated derivative using implicit differentiation.

53. ▼ $x = \tan y$; find $\frac{dy}{dx}$

54. ▼ $x = \cos y$; find $\frac{dy}{dx}$

55. ▼ $x + y + \sin(xy) = 1$; find $\frac{dy}{dx}$

56. ▼ $xy + x\cos y = x$; find $\frac{dy}{dx}$

Applications

57. ***Cost*** The cost in dollars of *Dig-It* brand snow shovels is given by

$$c(t) = 3.5\sin[2\pi(t - 0.75)],$$

where t is time in years since January 1, 2010. How fast, in dollars per week, is the cost increasing each October 1? [**HINT**: See Example 2.]

58. ***Sales*** Daily sales of *Doggy* brand cookies can be modeled by

$$s(t) = 400\cos[2\pi(t - 2)/7]$$

cartons, where t is time in days since Monday morning. How fast are sales changing on Thursday morning? [**HINT**: See Example 2.]

59. ***Sunspot Activity*** The activity of the Sun can be approximated by the following model of sunspot activity:[17]

$$N(t) = 57.7\sin[0.602(t - 1.43)] + 58.8,$$

where t is the number of years since January 1, 1997, and $N(t)$ is the number of sunspots observed at time t. Compute and interpret $N'(6)$.

60. ***Solar Emissions*** The following model gives the flux of radio emission from the Sun:[18]

$$F(t) = 49.6\sin[0.602(t - 1.48)] + 111,$$

where t is the number of years since January 1, 1997, and $F(t)$ is the average flux of solar emissions of a specified wavelength at time t. Compute and interpret $F'(5.5)$.

61. ***Inflation*** Taking a 3.5% rate of inflation into account, the cost of *DigIn* brand snow shovels is given by

$$c(t) = 1.035^t[0.8\sin(2\pi t) + 10.2],$$

where t is time in years since January 1, 2010. How fast, in dollars per week, is the cost of DigIn shovels increasing on January 1, 2011?

62. ***Deflation*** Sales, in bottles per day, of my exclusive mass-produced 2010 vintage *Chateau Petit Mont Blanc* follow the function

$$s(t) = 4.5e^{-0.2t}\sin(2\pi t),$$

where t is time in years since January 1, 2010. How fast were sales rising or falling on January 1, 2011?

63. ***Tides*** The depth of water at my favorite surfing spot varies from 5 to 15 feet, depending on the time. Last Sunday, high tide occurred at 5:00 am, and the next high tide occurred at 6:30 pm.
    a. Obtain a cosine model describing the depth of water as a function of time t in hours since 5:00 am on Sunday morning.
    b. How fast was the tide rising (or falling) at noon on Sunday?

64. ***Tides*** Repeat Exercise 63 using data from the depth of water at my other favorite surfing spot, where the tide last Sunday varied from a low of 6 feet at 4:00 am to a high of 10 feet at noon. (As in Exercise 63, take t as time in hours since 5:00 am.)

65. ▼ ***Full-Wave Rectifier*** A *rectifier* is a circuit that converts alternating current to direct current. A *full-wave rectifier* does so by effectively converting the voltage to its absolute value as a function of time. A 110-volt 50 cycles per second AC current would be converted to a voltage as shown:

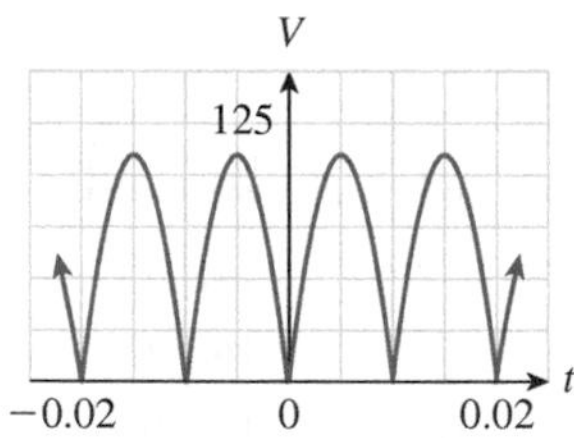

Full rectified wave
$V(t) = 110|\sin(100\pi t)|$

Compute and graph $\frac{dV}{dt}$, and explain the sudden jumps you see in the graph of the derivative.

66. ▼ ***Half-Wave Rectifier*** (See Exercise 65.) In a *half-wave rectifier* the negative voltage is "zeroed out" while the positive voltage is untouched. A 110-volt 50 cycles per second AC current would be converted to a voltage as shown:

---

[17] The model is based on a regression obtained from predicted data for 1997–2006 and the mean historical period of sunspot activity from 1755 to 1995. Source: NASA Science Directorate; Marshall Space Flight Center, August 2002 (www.science.nasa.gov/ssl/pad/solar/predict.htm).

[18] *Ibid.* Flux measured at a wavelength of 10.7 centimeters.

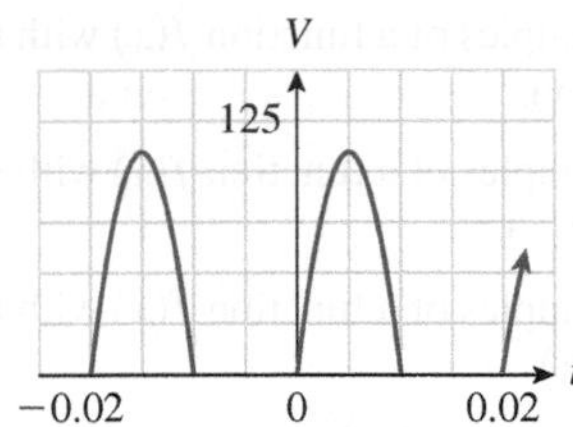

Half rectified wave

$V(t) = 55[\,|\sin(100\pi t)| + \sin(100\pi t)]$

Compute and graph $\frac{dV}{dt}$, and explain the sudden jumps you see in the graph of the derivative.

Simple Harmonic Motion and Damped Harmonic Motion *In mechanics an object whose position relative to a rest position is given by a generalized cosine (or sine) function*

$$p(t) = A\cos(\omega t + d)$$

is called a simple harmonic oscillator. *Examples of simple harmonic oscillators are a mass suspended from a spring and a pendulum swinging through a small angle, in the absence of frictional damping forces. When we take damping forces into account, we obtain a* damped harmonic oscillator:

$$p(t) = Ae^{-bt}\cos(\omega t + d)$$

(assuming that the damping forces are not so large as to prevent the system from oscillating entirely). Exercises 67–70 are based on these concepts.

67. ▼ A mass on a spring is undergoing simple harmonic motion so that its vertical position at time t seconds is given by

$p(t) = 1.2\cos(5\pi t + \pi)$ centimeters below the rest position.

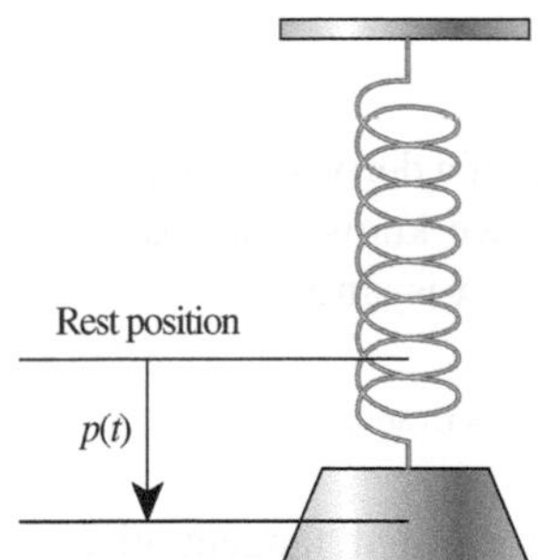

a. What is its vertical position at time $t = 0$?

b. How fast is the mass moving, and in what direction, at times $t = 0$ and $t = 0.1$?

c. The *frequency* of oscillation is defined as the reciprocal of the period. What is the frequency of oscillation of the spring?

68. ▼ A worn shock absorber on a car undergoes simple harmonic motion so that the height of the car frame after t seconds is

$p(t) = 4.2\sin(2\pi t + \pi/2)$ centimeters above the rest position.

a. What is its vertical position at time $t = 0$?

b. How fast is the height of the car changing, and in what direction, at times $t = 0$ and $t = 0.25$?

c. The *frequency* of oscillation is defined as the reciprocal of the period. What is the frequency of oscillation of the car frame?

69. ▼ A mass on a spring is undergoing damped harmonic motion so that its vertical position at time t seconds is given by

$$p(t) = 1.2e^{-0.1t}\cos(5\pi t + \pi)$$

centimeters below the rest position.

a. How fast is the mass moving, and in what direction, at times $t = 0$ and $t = 0.1$?

b. T Graph p and p' as functions of t for $0 \le t \le 10$ and also for $0 \le t \le 1$, and use your graphs and graphing technology to estimate, to the nearest tenth of a second, the time at which the (downward) velocity of the mass is greatest.

70. ▼ A worn shock absorber on a car undergoes damped harmonic motion so that the height of the car frame after t seconds is

$$p(t) = 4.2e^{-0.5t}\sin(2\pi t + \pi/2)$$

centimeters above the rest position.

a. How fast is the top of the car moving, and in what direction, at times $t = 0$ and $t = 0.25$?

b. T Graph p and p' as functions of t for $0 \le t \le 10$ and also for $0 \le t \le 2$, and use your graphs and graphing technology to estimate, to the nearest hundredth of a second, the time at which the (upward) velocity of the car is greatest.

71. ▼ ***Tilt of the Earth's Axis*** The tilt of the Earth's axis from its plane of rotation about the Sun oscillates between approximately 22.5° and 24.5° with a period of approximately 40,000 years.[19] We know that 500,000 years ago, the tilt of the Earth's axis was 24.5°.

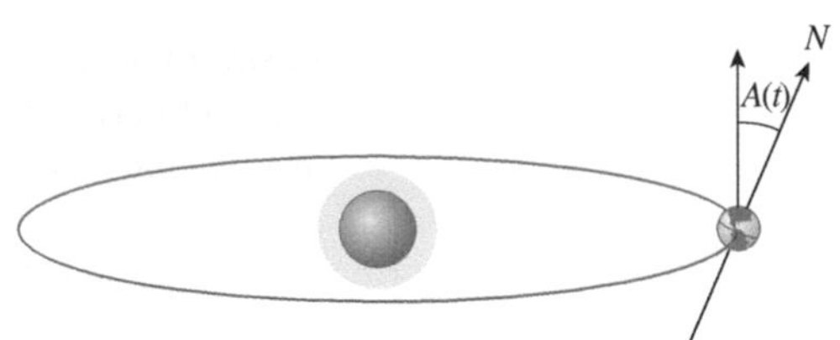

a. Which of the following functions best models the tilt of the Earth's axis?

(I) $A(t) = 23.5 + 2\sin\left(\frac{2\pi t + 500}{40}\right)$

(II) $A(t) = 23.5 + \cos\left(\frac{t + 500}{80\pi}\right)$

(III) $A(t) = 23.5 + \cos\left(\frac{2\pi(t + 500)}{40}\right)$

[19] Source: Dr. David Hodell, University of Florida/Juan Valesco/*New York Times*, February 16, 1999, p. F1.

where $A(t)$ is the tilt in degrees and t is time in thousands of years, with $t = 0$ being the present time.

b. Use the model you selected in part (a) to estimate the rate at which the tilt was changing 150,000 years ago. (Round your answer to three decimal places, and be sure to give the units of measurement.)

72. ▼ ***Eccentricity of the Earth's Orbit*** The eccentricity of the Earth's orbit (that is, the deviation of the Earth's orbit from a perfect circle) can be modeled by[20]

$$E(t) = 0.025\left[\cos\left(\frac{2\pi(t+200)}{400}\right) + \cos\left(\frac{2\pi(t+200)}{100}\right)\right],$$

where $E(t)$ is the eccentricity and t is time in thousands of years, with $t = 0$ being the present time. What was the value of the eccentricity 200,000 years ago, and how fast was it changing?

Communication and Reasoning Exercises

73. Complete the following: The rate of change of $f(x) = 3\sin(2x - 1) + 3$ oscillates between ____ and ____.

74. Complete the following: The rate of change of $g(x) = -3\cos(-x + 2) + 2x$ oscillates between ____ and ____.

[20] See footnote for Exercise 71. This is a rough model based on the actual data.

75. Give two examples of a function $f(x)$ with the property that $f''(x) = -f(x)$.

76. Give two examples of a function $f(x)$ with the property that $f''(x) = -4f(x)$.

77. Give two examples of a function $f(x)$ with the property that $f'(x) = -f(x)$.

78. Give four examples of a function $f(x)$ with the property that $f^{(4)}(x) = f(x)$.

79. ▼ By referring to the graph of $f(x) = \cos x$, explain why $f'(x) = -\sin x$, rather than $\sin x$.

80. ▼ If A and B are constants, what is the relationship between $f(x) = A\cos x + B\sin x$ and its second derivative?

81. ▼ If the value of a stock price is given by $p(t) = A\sin(\omega t + d)$ above yesterday's close for constants $A \neq 0$, $\omega \neq 0$, and d, where t is time, explain why the stock price is moving the fastest when it is at yesterday's close.

82. ▼ If the value of a stock price is given by $p(t) = A\cos(\omega t + d)$ above yesterday's close for positive constants A, ω, and d, where t is time, explain why its acceleration is greatest when its value is the lowest.

83. ▼ At what angle does the graph of $f(x) = \sin x$ depart from the origin?

84. ▼ At what angle does the graph of $f(x) = \cos x$ depart from the point $(0, 1)$?

9.3 Integrals of Trigonometric Functions and Applications

Integrals of Sine and Cosine

We saw in Section 6.1 that every calculation of a derivative also gives us a calculation of an antiderivative. For instance, because we know that $\cos x$ is the derivative of $\sin x$, we can say that an antiderivative of $\cos x$ is $\sin x$:

$$\int \cos x\, dx = \sin x + C. \qquad \text{An antiderivative of } \cos x \text{ is } \sin x.$$

The rules for the derivatives of sine, cosine, and tangent give us the following antiderivatives.

Indefinite Integrals of Some Trigonometric Functions

$$\int \cos x\, dx = \sin x + C \qquad \text{Because } \frac{d}{dx}(\sin x) = \cos x$$

$$\int \sin x\, dx = -\cos x + C \qquad \text{Because } \frac{d}{dx}(-\cos x) = \sin x$$

$$\int \sec^2 x\, dx = \tan x + C \qquad \text{Because } \frac{d}{dx}(\tan x) = \sec^2 x$$

Quick Examples

1. $\int (\sin x + \cos x)\, dx = -\cos x + \sin x + C$ — Integral of sum = Sum of integrals
2. $\int (4 \sin x - \cos x)\, dx = -4 \cos x - \sin x + C$ — Integral of constant multiple
3. $\int (e^x - \sin x + \cos x)\, dx = e^x + \cos x + \sin x + C$

EXAMPLE 1 Substitution

Evaluate $\int (x + 3) \sin(x^2 + 6x)\, dx$.

Solution There are two parenthetical expressions that we might replace with u. Notice, however, that the derivative of the expression $(x^2 + 6x)$ is $2x + 6$, which is twice the term $(x + 3)$ in front of the sine. Recall that we would like the derivative of u to appear as a factor. Thus, let us take $u = x^2 + 6x$.

$$u = x^2 + 6x$$
$$\frac{du}{dx} = 2x + 6 = 2(x + 3)$$
$$dx = \frac{1}{2(x + 3)} du$$

Substituting into the integral, we get

$$\int (x + 3) \sin(x^2 + 6x)\, dx = \int (x + 3)(\sin u) \frac{1}{2(x + 3)}\, du = \int \frac{1}{2} \sin u\, du$$
$$= -\frac{1}{2} \cos u + C = -\frac{1}{2} \cos(x^2 + 6x) + C.$$

EXAMPLE 2 Definite Integrals

Compute the following:

a. $\int_0^{\pi} \sin x\, dx$ **b.** $\int_0^{\pi} x \sin(x^2)\, dx$

Solution

a. $\int_0^{\pi} \sin x\, dx = [-\cos x]_0^{\pi} = (-\cos \pi) - (-\cos 0) = -(-1) - (-1) = 2$

Thus, the area under one "arch" of the sine curve is exactly 2 square units!

b. $\displaystyle\int_0^{\pi} x\sin(x^2)\,dx = \int_0^{\pi^2} \frac{1}{2}\sin u\,du$ After substituting $u = x^2$

$$= \left[-\frac{1}{2}\cos u\right]_0^{\pi^2}$$

$$= \left[-\frac{1}{2}\cos(\pi^2)\right] - \left[-\frac{1}{2}\cos(0)\right]$$

$$= -\frac{1}{2}\cos(\pi^2) + \frac{1}{2}$$ $\cos(0) = 1$

We can approximate $\frac{1}{2}\cos(\pi^2)$ by a decimal or leave it in the above form, depending on what we want to do with the answer.

Integrals of the Six Trigonometric Functions

The following summary gives the indefinite integrals of the six trigonometric functions. (The first two we have already seen.)

Integrals of the Trigonometric Functions

$$\int \sin x\,dx = -\cos x + C$$

$$\int \cos x\,dx = \sin x + C$$

$$\int \tan x\,dx = -\ln|\cos x| + C$$ Shown below

$$\int \cot x\,dx = \ln|\sin x| + C$$ See the exercises.

$$\int \sec x\,dx = \ln|\sec x + \tan x| + C$$ Shown below

$$\int \csc x\,dx = -\ln|\csc x + \cot x| + C$$ See the exercises.

Derivations of Formulas for Integrals of Trigonometric Functions

To show that $\int \tan x\,dx = -\ln|\cos x| + C$, we first write $\tan x$ as $\dfrac{\sin x}{\cos x}$ and put $u = \cos x$ in the integral:

$$\int \tan x\,dx = \int \frac{\sin x}{\cos x}\,dx$$

$$= -\int \frac{\sin x}{u}\frac{du}{\sin x}$$

$$= -\int \frac{du}{u}$$

$$= -\ln|u| + C$$

$$= -\ln|\cos x| + C.$$

$u = \cos x$

$\dfrac{du}{dx} = -\sin x$

$dx = -\dfrac{du}{\sin x}$

To show that $\int \sec x\, dx = \ln|\sec x + \tan x| + C$, first use a little "trick": Write $\sec x$ as $\sec x\left(\dfrac{\sec x + \tan x}{\sec x + \tan x}\right)$, and put u equal to the denominator:

$$\begin{aligned}
\int \sec x\, dx &= \int \sec x\left(\frac{\sec x + \tan x}{\sec x + \tan x}\right) dx \\
&= \int \sec x \frac{\sec x + \tan x}{u} \frac{du}{\sec x(\tan x + \sec x)} \\
&= \int \frac{du}{u} \\
&= \ln|u| + C \\
&= \ln|\sec x + \tan x| + C.
\end{aligned}$$

$$\begin{aligned}
u &= \sec x + \tan x \\
\frac{du}{dx} &= \sec x \tan x + \sec^2 x \\
&= \sec x(\tan x + \sec x) \\
dx &= \frac{du}{\sec x(\tan x + \sec x)}
\end{aligned}$$

Shortcuts

If a and b are constants with $a \neq 0$, then we have the following formulas. (All of them can be obtained by using the substitution $u = ax + b$. They will appear in the exercises.)

Shortcuts: Integrals of Expressions Involving $(ax + b)$

Rule	Quick Example
$\int \sin(ax + b)\, dx = -\frac{1}{a}\cos(ax + b) + C$	$\int \sin(-4x)\, dx = \frac{1}{4}\cos(-4x) + C$
$\int \cos(ax + b)\, dx = \frac{1}{a}\sin(ax + b) + C$	$\int \cos(x + 1)\, dx = \sin(x + 1) + C$
$\int \tan(ax + b)\, dx = -\frac{1}{a}\ln\lvert\cos(ax + b)\rvert + C$	$\int \tan(-2x)\, dx = \frac{1}{2}\ln\lvert\cos(-2x)\rvert + C$
$\int \cot(ax + b)\, dx = \frac{1}{a}\ln\lvert\sin(ax + b)\rvert + C$	$\int \cot(3x - 1)\, dx = \frac{1}{3}\ln\lvert\sin(3x - 1)\rvert + C$
$\int \sec(ax + b)\, dx = \frac{1}{a}\ln\lvert\sec(ax + b) + \tan(ax + b)\rvert + C$	$\int \sec(9x)\, dx = \frac{1}{9}\ln\lvert\sec(9x) + \tan(9x)\rvert + C$
$\int \csc(ax + b)\, dx = -\frac{1}{a}\ln\lvert\csc(ax + b) + \cot(ax + b)\rvert + C$	$\int \csc(x + 7)\, dx = -\ln\lvert\csc(x + 7) + \cot(x + 7)\rvert + C$

EXAMPLE 3 **Sales**

The rate of sales of cypods (one-bedroom living units) in the city-state of Utarek on Mars[21] can be modeled by

$$s(t) = 7.5\cos(\pi t/6) + 87.5 \text{ units per month,}$$

where t is time in months since January 1. How many cypods are sold in a calendar year?

Solution Total sales over 1 calendar year are given by

$$\int_0^{12} s(t)\,dt = \int_0^{12} [7.5\cos(\pi t/6) + 87.5]\,dt$$

$$= \left[7.5\frac{6}{\pi}\sin(\pi t/6) + 87.5t\right]_0^{12} \qquad \text{We used a shortcut on the first term.}$$

$$= \left[7.5\frac{6}{\pi}\sin(2\pi) + 87.5(12)\right] - \left[7.5\frac{6}{\pi}\sin(0) + 87.5(0)\right]$$

$$= 87.5(12) \qquad \sin(2\pi) = \sin(0) = 0$$

$$= 1{,}050 \text{ cypods.}$$

➡ **Before we go on...** Would it have made any difference in Example 3 if we had computed total sales over the period [12, 24], [6, 18], or any interval of the form $[a, a + 12]$? ■

Using Integration by Parts with Trigonometric Functions

EXAMPLE 4 **Integrating a Polynomial Times Sine or Cosine**

Calculate $\int (x^2 + 1)\sin(x + 1)\,dx$.

Solution We use the column method of integration by parts described in Section 7.1. Because differentiating $x^2 + 1$ makes it simpler, we put it in the D column and get the following table:

	D	I
$+$	$x^2 + 1$ ↘	$\sin(x + 1)$
$-$	$2x$ ↘	$-\cos(x + 1)$
$+$	2 ↘	$-\sin(x + 1)$
$-\int$	0 →	$\cos(x + 1)$

[21] Based on www.Marsnext.com, a now extinct virtual society.

[Notice that we used the shortcut formulas to repeatedly integrate $\sin(x+1)$.] We can now read the answer from the table:

$$\begin{aligned}\int (x^2+1)\sin(x+1)\,dx &= (x^2+1)[-\cos(x+1)] - 2x[-\sin(x+1)] \\ &\quad + 2[\cos(x+1)] + C \\ &= (-x^2-1+2)\cos(x+1) + 2x\sin(x+1) + C \\ &= (-x^2+1)\cos(x+1) + 2x\sin(x+1) + C.\end{aligned}$$

EXAMPLE 5 **Integrating an Exponential Times Sine or Cosine**

Calculate $\int e^x \sin x\,dx$.

Solution The integrand is the product of e^x and $\sin x$, so we put one in the D column and the other in the I column. For this example it doesn't matter much which we put where.

	D	I
$+$	$\sin x$	e^x
$-$	$\cos x$	e^x
$+\int$	$-\sin x$	e^x

It looks as though we're just spinning our wheels. Let's stop and see what we have:

$$\int e^x \sin x\,dx = e^x \sin x - e^x \cos x - \int e^x \sin x\,dx.$$

At first glance, it appears that we are back where we started, still having to evaluate $\int e^x \sin x\,dx$. However, if we add this integral to both sides of the equation above, we can solve for it:

$$2\int e^x \sin x\,dx = e^x \sin x - e^x \cos x + C.$$

(Why $+\ C$?) So

$$\int e^x \sin x\,dx = \frac{1}{2}e^x \sin x - \frac{1}{2}e^x \cos x + \frac{C}{2}.$$

Because $C/2$ is just as arbitrary as C, we write C instead of $C/2$ and obtain

$$\int e^x \sin x\,dx = \frac{1}{2}e^x \sin x - \frac{1}{2}e^x \cos x + C.$$

9.3 EXERCISES

▼ more advanced ◆ challenging

T indicates exercises that should be solved using technology

In Exercises 1–28, evaluate the given integral. [**HINT**: See Quick Examples 1–3.]

1. $\int (\sin x - 2\cos x)\,dx$
2. $\int (\cos x - \sin x)\,dx$
3. $\int (2\cos x - 4.3\sin x - 9.33)\,dx$
4. $\int (4.1\sin x + \cos x - 9.33/x)\,dx$
5. $\int \left(3.4\sec^2 x + \frac{\cos x}{1.3} - 3.2e^x\right) dx$
6. $\int \left(\frac{3\sec^2 x}{2} + 1.3\sin x - \frac{e^x}{3.2}\right) dx$
7. $\int 7.6\cos(3x - 4)\,dx$ [**HINT**: See Example 1.]
8. $\int 4.4\sin(-3x + 4)\,dx$ [**HINT**: See Example 1.]
9. $\int x\sin(3x^2 - 4)\,dx$
10. $\int x\cos(-3x^2 + 4)\,dx$
11. $\int (4x + 2)\sin(x^2 + x)\,dx$
12. $\int (x + 1)[\cos(x^2 + 2x) + (x^2 + 2x)]\,dx$
13. $\int (x + x^2)\sec^2(3x^2 + 2x^3)\,dx$
14. $\int (4x + 2)\sec^2(x^2 + x)\,dx$
15. $\int (x^2)\tan(2x^3)\,dx$
16. $\int (4x)\tan(x^2)\,dx$
17. $\int 6\sec(2x - 4)\,dx$
18. $\int 3\csc(3x)\,dx$
19. $\int e^{2x}\cos(e^{2x} + 1)\,dx$
20. $\int e^{-x}\sin(e^{-x})\,dx$
21. $\int_{-\pi}^{0} \sin x\,dx$ [**HINT**: See Example 2.]
22. $\int_{\pi/2}^{\pi} \cos x\,dx$ [**HINT**: See Example 2.]
23. $\int_{0}^{\pi/3} \tan x\,dx$
24. $\int_{\pi/6}^{\pi/2} \cot x\,dx$
25. $\int_{1}^{\sqrt{\pi+1}} x\cos(x^2 - 1)\,dx$
26. $\int_{1/2}^{(\pi+1)/2} \sin(2x - 1)\,dx$
27. ▼ $\int_{1/\pi}^{2/\pi} \frac{\sin(1/x)}{x^2}\,dx$
28. ▼ $\int_{0}^{\pi/3} \frac{\sin x}{\cos^2 x}\,dx$

In Exercises 29–32, derive the given equation, where a and b are constants with $a \neq 0$.

29. ▼ $\int \cos(ax + b)\,dx = \frac{1}{a}\sin(ax + b) + C$
30. ▼ $\int \sin(ax + b)\,dx = -\frac{1}{a}\cos(ax + b) + C$
31. ▼ $\int \cot x\,dx = \ln|\sin x| + C$
32. ▼ $\int \csc x\,dx = -\ln|\csc x + \cot x| + C$

In Exercises 33–40, use the shortcut formulas before Example 3 to calculate the given integral mentally.

33. $\int \sin(4x)\,dx$
34. $\int \cos(5x)\,dx$
35. $\int \cos(-x + 1)\,dx$
36. $\int \sin\left(\frac{1}{2}x\right) dx$
37. $\int \sin(-1.1x - 1)\,dx$
38. $\int \cos(4.2x - 1)\,dx$
39. $\int \cot(-4x)\,dx$
40. $\int \tan(6x)\,dx$

In Exercises 41–44, use geometry (not antiderivatives) to compute the given integral. [**HINT**: First draw the graph.]

41. $\int_{-\pi/2}^{\pi/2} \sin x\,dx$
42. $\int_{0}^{\pi} \cos x\,dx$
43. ▼ $\int_{0}^{2\pi} (1 + \sin x)\,dx$
44. ▼ $\int_{0}^{2\pi} (1 + \cos x)\,dx$

In Exercises 45–52, use integration by parts to evaluate the given integral. [**HINT**: See Example 4.]

45. $\int x\sin x\,dx$
46. $\int x^2\cos x\,dx$
47. $\int x^2\cos(2x)\,dx$
48. $\int (2x + 1)\sin(2x - 1)\,dx$
49. ▼ $\int e^{-x}\sin x\,dx$
50. ▼ $\int e^{2x}\cos x\,dx$
51. ▼ $\int_{0}^{\pi} x^2\sin x\,dx$
52. ▼ $\int_{0}^{\pi/2} x\cos x\,dx$

Recall from Section 7.3 that the average of a function $f(x)$ on an interval $[a, b]$ is

$$\bar{f} = \frac{1}{b-a}\int_a^b f(x)\,dx.$$

In Exercises 53 and 54, find the average of the given function over the given interval. Plot the function and its average on the same graph.

53. ▼ $f(x) = \sin x$ over $[0, \pi]$

54. ▼ $f(x) = \cos(2x)$ over $[0, \pi/4]$

In Exercises 55–58, decide whether the given integral converges. (See Section 7.5.) If the integral converges, compute its value.

55. $\int_0^{+\infty} \sin x\,dx$ **56.** $\int_0^{+\infty} \cos x\,dx$

57. ▼ $\int_0^{+\infty} e^{-x}\cos x\,dx$ **58.** ▼ $\int_0^{+\infty} e^{-x}\sin x\,dx$

Applications

59. ***Varying Cost*** The cost of producing a bottle of suntan lotion is changing at a rate of $0.04 - 0.1\sin\left[\frac{\pi}{26}(t-25)\right]$ dollars per week, t weeks after January 1. If it cost \$1.50 to produce a bottle 12 weeks into the year, find the cost $C(t)$ at time t.

60. ***Varying Cost*** The cost of producing a box of holiday tree decorations is changing at a rate of $0.05 + 0.4\cos\left[\frac{\pi}{6}(t-11)\right]$ dollars per month, t months after January 1. If it cost \$5 to produce a box on June 1, find the cost $C(t)$ at time t.

61. ▼ ***Pets*** My dog Miranda is running back and forth along a 12-foot stretch of garden in such a way that her velocity t seconds after she began is

$$v(t) = 3\pi\cos\left[\frac{\pi}{2}(t-1)\right] \text{ ft/sec.}$$

How far is she from where she began 10 seconds after starting the run? [**HINT:** See Example 3.]

62. ▼ ***Pets*** My cat, Prince Sadar, is pacing back and forth along his favorite window ledge in such a way that his velocity t seconds after he began is

$$v(t) = -\frac{\pi}{2}\sin\left[\frac{\pi}{4}(t-2)\right] \text{ ft/sec.}$$

How far is he from where he began 10 seconds after starting to pace? [**HINT:** See Example 3.]

For Exercises 63–68, recall from Section 7.3 that the average of a function $f(x)$ on an interval $[a, b]$ is

$$\bar{f} = \frac{1}{b-a}\int_a^b f(x)\,dx.$$

63. ***Sunspot Activity*** The activity of the sun (sunspots, solar flares, and coronal mass ejection) fluctuates in cycles of around 10–11 years. Sunspot activity can be modeled by the following function:[22]

$$N(t) = 57.7\sin[0.602(t-1.43)] + 58.8,$$

where t is the number of years since January 1, 1997, and $N(t)$ is the number of sunspots observed at time t. Estimate the average number of sunspots visible over the 2-year period beginning January 1, 2002. (Round your answer to the nearest whole number.)

64. ***Solar Emissions*** The following model gives the flux of radio emission from the sun:[23]

$$F(t) = 49.6\sin[0.602(t-1.48)] + 111,$$

where t is the number of years since January 1, 1997, and $F(t)$ is the flux of solar emissions of a specified wavelength at time t. Estimate the average flux of radio emission over the 5-year period beginning January 1, 2001. (Round your answer to the nearest whole number.)

65. ▼ ***Biology*** Sigatoka leaf spot is a plant disease that affects bananas. In an infected plant, the percentage of leaf area affected varies from a low of around 5% at the start of each year to a high of around 20% at the middle of each year.[24] Use a sine function model of the percentage of leaf area affected by Sigatoka leaf spot t weeks since the start of a year to estimate, to the nearest 0.1%, the average percentage of leaf area affected in the first quarter (13 weeks) of a year.

66. ▼ ***Biology*** Apple powdery mildew is an epidemic that affects apple shoots. In a new infection the percentage of apple shoots infected varies from a low of around 10% at the start of May to a high of around 60% 6 months later.[25] Use a sine function model of the percentage of apple shoots affected by apple powdery mildew t months since the start of a year to estimate, to the nearest 0.1%, the average percentage of apple shoots affected in the first 2 months of a year.

---

[22] The model is based on a regression obtained from predicted data for 1997–2006 and the mean historical period of sunspot activity from 1755 to 1995. Source: NASA Science Directorate; Marshall Space Flight Center, August 2002 (www.science.nasa.gov/ssl/pad/solar/predict.htm).

[23] *Ibid.* Flux measured at a wavelength of 10.7 centimeters.

[24] Based on graphical data. Source: American Phytopathological Society (www.apsnet.org/education/AdvancedPlantPath/Topics/Epidemiology/CyclicalNature.htm).

[25] *Ibid.*

67. T ▼ ***Electrical Current*** The typical voltage V supplied by an electrical outlet in the United States is given by

$$V(t) = 165\cos(120\pi t),$$

where t is time in seconds.

a. Find the average voltage over the interval $[0, 1/6]$. How many times does the voltage reach a maximum in 1 second? (This is referred to as the number of **cycles per second**.)
b. Plot the function $S(t) = (V(t))^2$ over the interval $[0, 1/6]$.
c. The **root mean square (RMS)** voltage is given by the formula

$$V_{\text{rms}} = \sqrt{\bar{S}},$$

where $\bar{S}$ is the average value of $S(t)$ over one cycle. Estimate V_{rms}.

68. ▼ ***Tides*** The depth of water at my favorite surfing spot varies from 5 to 15 feet, depending on the time. Last Sunday, high tide occurred at 5:00 am, and the next high tide occurred at 6:30 pm. Use the cosine function to model the depth of water as a function of time t in hours since midnight on Sunday morning. What was the average depth of the water between 10:00 am and 2:00 pm?

Income Streams *Recall from Section 7.4 that the total income received from time $t = a$ to time $t = b$ from a continuous income stream of $R(t)$ dollars per year is*

$$\text{Total value} = TV = \int_a^b R(t)\,dt.$$

In Exercises 69 and 70, find the total value of the given income stream over the given period.

69. $R(t) = 50{,}000 + 2{,}000\pi\sin(2\pi t),\ 0 \le t \le 1$

70. $R(t) = 100{,}000 - 2{,}000\pi\sin(\pi t),\ 0 \le t \le 1.5$

Communication and Reasoning Exercises

71. What can you say about the definite integral of a sine or cosine function over a whole number of periods?

72. How are the derivative and antiderivative of $\sin x$ related?

73. ▼ What is the average value of $1 + 2\cos x$ over a large interval?

74. ▼ What is the average value of $3 - \cos x$ over a large interval?

75. ▼ The acceleration of an object is given by $a = K\sin(\omega t - \alpha)$. What can you say about its displacement at time t?

76. ▼ Write down a function whose derivative is -2 times its antiderivative.

CHAPTER 9 REVIEW

KEY CONCEPTS

www.WanerMath.com
Go to the Website to find a comprehensive and interactive Web-based summary of Chapter 9.

9.1 Trigonometric Functions, Models, and Regression

The **sine** of a real number [p. 677]
Plotting the graphs of functions based on $\sin x$ [p. 678]
The general sine function:
$f(x) = A\sin[\omega(x - \alpha)] + C$
A is the **amplitude**.
C is the **vertical offset** or height of the **baseline**.
ω is the **angular frequency**.
$P = 2\pi/\omega$ is the **period** or **wavelength**.
α is the **phase shift**. [p. 678]
Modeling with the general sine function [p. 679]
The **cosine** of a real number [p. 681]
Fundamental trigonometric identities:
$\sin^2 t + \cos^2 t = 1$
$\cos t = \sin(t + \pi/2)$
$\cos t = \sin(\pi/2 - t)$
$\sin t = \cos(t - \pi/2)$
$\sin t = \cos(\pi/2 - t)$ [p. 681]
The general cosine function:
$f(x) = A\cos[\omega(x - \beta)] + C$
[p. 682]
Modeling with the general cosine function [p. 683]

Other trigonometric functions:
$\tan x = \dfrac{\sin x}{\cos x}$
$\cot x = \text{cotan } x = \dfrac{\cos x}{\sin x} = \dfrac{1}{\tan x}$
$\sec x = \dfrac{1}{\cos x}$
$\csc x = \text{cosec } x = \dfrac{1}{\sin x}$ [p. 684]

9.2 Derivatives of Trigonometric Functions and Applications

Derivatives of sine and cosine:
$\dfrac{d}{dx}\sin x = \cos x$
$\dfrac{d}{dx}\cos x = -\sin x$ [p. 690]
Some trigonometric limits:
$\lim_{h\to 0}\dfrac{\sin h}{h} = 1$
$\lim_{h\to 0}\dfrac{\cos h - 1}{h} = 0$ [p. 691]
Derivatives of sines and cosines of functions:
$\dfrac{d}{dx}\sin u = \cos u\dfrac{du}{dx}$
$\dfrac{d}{dx}\cos u = -\sin u\dfrac{du}{dx}$ [p. 692]
Derivatives of the other trigonometric functions:
$\dfrac{d}{dx}\tan x = \sec^2 x$
$\dfrac{d}{dx}\cot x = -\csc^2 x$
$\dfrac{d}{dx}\sec x = \sec x\tan x$
$\dfrac{d}{dx}\csc x = -\csc x\cot x$ [p. 694]

9.3 Integrals of Trigonometric Functions and Applications

$\int \cos x\,dx = \sin x + C$
$\int \sin x\,dx = -\cos x + C$
$\int \sec^2 x\,dx = \tan x + C$ [p. 698]

Substitution in integrals involving trigonometric functions [p. 699]
Definite integrals involving trigonometric functions [p. 699]
Integrals of the other trigonometric functions:
$\int \tan x\,dx = -\ln|\cos x| + C$
$\int \cot x\,dx = \ln|\sin x| + C$
$\int \sec x\,dx = \ln|\sec x + \tan x| + C$
$\int \csc x\,dx = -\ln|\csc x + \cot x| + C$
[p. 700]
Shortcuts: Integrals of expressions involving $(ax + b)$ [p. 701]
Using integration by parts with trigonometric functions [p. 702]

REVIEW EXERCISES

In Exercises 1–4, model the given curve with a sine function. (The scales on the two axes may not be the same.)

1.
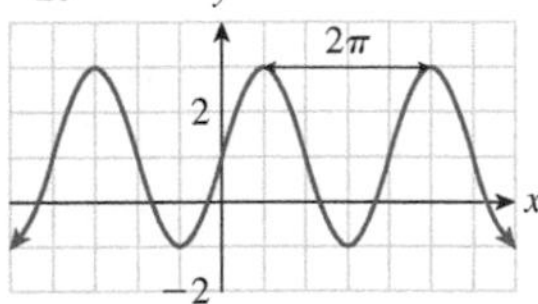

2.
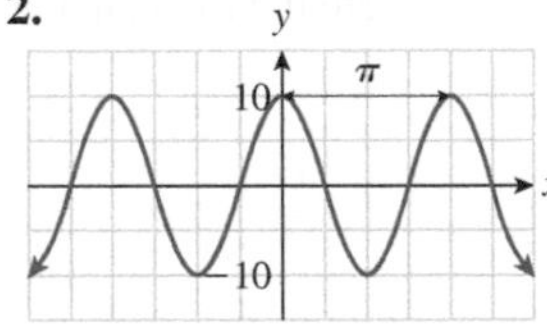

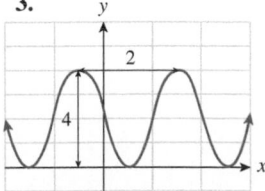

3. **4.**
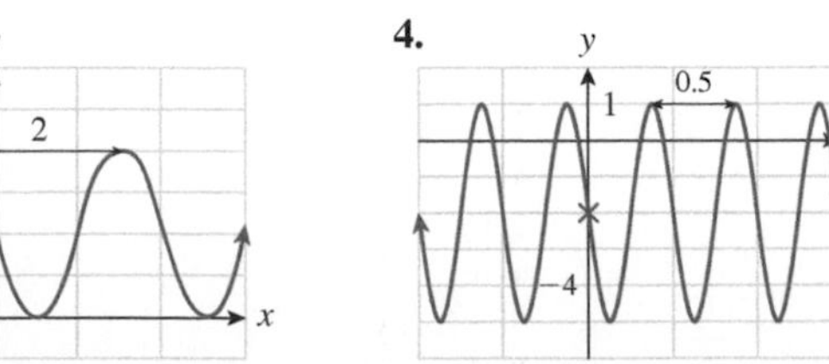

In Exercises 5–8, model the curves in Exercises 1–4 with cosine functions.

5. The curve in Exercise 1
6. The curve in Exercise 2
7. The curve in Exercise 3
8. The curve in Exercise 4

In Exercises 9–14, find the derivative of the given function.

9. $f(x) = \cos(x^2 - 1)$
10. $f(x) = \sin(x^2 + 1)\cos(x^2 - 1)$
11. $f(x) = \tan(2e^x - 1)$
12. $f(x) = \sec\sqrt{x^2 - x}$
13. $f(x) = \sin^2(x^2)$
14. $f(x) = \cos^2[1 - \sin(2x)]$

In Exercises 15–22, evaluate the given integral.

15. $\int 4\cos(2x - 1)\,dx$

16. $\int (x - 1)\sin(x^2 - 2x + 1)\,dx$

17. $\int 4x\sec^2(2x^2 - 1)\,dx$

18. $\int \frac{\cos\left(\frac{1}{x}\right)}{x^2\sin\left(\frac{1}{x}\right)}\,dx$

19. $\int x\tan(x^2 + 1)\,dx$

20. $\int_0^{\pi} \cos(x + \pi/2)\,dx$

21. $\int_{\ln(\pi/2)}^{\ln(\pi)} e^x\sin(e^x)\,dx$

22. $\int_{\pi}^{2\pi} \tan(x/6)\,dx$

In Exercises 23 and 24, use integration by parts to evaluate the integral.

23. $\int x^2\sin x\,dx$

24. $\int e^2\sin 2x\,dx$

Applications: OHaganBooks.com [Try the game at www.OHaganBooks.com]

25. ***Sales*** After several years in the business, OHaganBooks.com noticed that its sales showed seasonal fluctuations. Weekly sales oscillated in a sine wave from a low of 9,000 books per week to a high of 12,000 books per week, the high point of the year being three quarters of the way through the year, in October. Model OHaganBooks.com's weekly sales as a generalized sine function of t, the number of weeks into the year.

26. ***Mood Swings*** The shipping personnel at OHaganBooks.com are under considerable pressure to cope with the large volume of orders, and periodic emotional outbursts are commonplace. The human resources department has been logging these outbursts over the course of several years and has noticed a peak of 50 outbursts a week during the holiday season each December and a low point of 15 per week each June (probably attributable to the mild June weather). Model the weekly number of outbursts as a generalized cosine function of t, the number of months into the year. ($t = 1$ represents January.)

27. ***Precalculus for Geniuses*** The "For Geniuses" series of books has really been taking off since *Duffin House* first gained exclusive rights to the series 6 months ago. Revenues from *Precalculus for Geniuses* are expected to follow the curve

$$R(t) = 100{,}000 + 20{,}000e^{-0.05t}\sin\left[\frac{\pi}{6}(t - 2)\right] \text{ dollars} \quad (0 \le t \le 72),$$

where t is time in months from now and $R(t)$ is the monthly revenue. How fast, to the nearest dollar, will the revenue be changing 20 months from now?

28. ***Elvish for Dummies*** The sales department at OHaganBooks.com predicts that the revenue from sales of the latest blockbuster *Elvish for Dummies* will vary in accordance with annual releases of episodes of the movie series "Lord of the Rings Episodes 9–12." It has come up with the following model (which includes the effect of diminishing sales):

$$R(t) = 20{,}000 + 15{,}000e^{-0.12t}\cos\left[\frac{\pi}{6}(t - 4)\right] \text{ dollars} \quad (0 \le t \le 72),$$

where t is time in months from now and $R(t)$ is the monthly revenue. How fast, to the nearest dollar, will the revenue be changing 10 months from now?

29. ***Revenue*** Refer back to Exercise 27. Use technology or integration by parts to estimate, to the nearest \$100, the total revenue from sales of *Precalculus for Geniuses* over the next 20 months.

30. ***Revenue*** Refer back to Exercise 28. Use technology or integration by parts to estimate, to the nearest \$100, the total revenue from sales of *Elvish for Dummies* over the next 10 months.

31. ***Mars Missions*** Having completed his doctorate in biophysics, Billy-Sean O'Hagan will be accompanying the first manned mission to Mars. For reasons too complicated to explain (but having to do with the continuation of his doctoral research project and the timing of messages from his fiancée), during the voyage he will be consuming protein at a rate of

$$P(t) = 150 + 50\sin\left[\frac{\pi}{2}(t - 1)\right] \text{ grams per day}$$

t days into the voyage. Find the total amount of protein he will consume as a function of time t.

32. ***Utilities*** Expenditure for utilities at OHaganBooks.com fluctuated from a high of \$9,500 in October ($t = 0$) to a low of \$8,000 in April ($t = 6$) Construct a sinusoidal model for the monthly expenditure on utilities, and use your model to estimate the total annual cost.

CASE STUDY

Predicting Airline Empty Seat Volume

Lawrence Manning/Corbis/Getty Images

You are a consultant to the Department of Transportation's Special Task Force on Air Traffic Congestion and have been asked to model the volume of empty seats on U.S. airline flights, to make short-term projections of this volume, and to give a formula that estimates the accumulated volume over a specified period of time. You have data from the Bureau of Transportation Statistics showing, for each month starting January 2002, the number of available seat miles (the total of the number of seats times the number of miles flown) and also the number of revenue passenger miles (the total of the number of seats occupied by paying passengers times the number of miles flown), so their difference (if you ignore nonpaying passengers) measures the number of empty seat miles. (The data can be downloaded at the Website by following Everything for Applied Calc → Chapter 9 Case Study.)[26]

Month	Empty Seat Miles (billions)	Month	Empty Seat Miles (billions)	Month	Empty Seat Miles (billions)	Month	Empty Seat Miles (billions)
1	25	21	23	41	20	61	23
2	21	22	22	42	16	62	20
3	18	23	21	43	15	63	17
4	21	24	21	44	18	64	17
5	21	25	26	45	21	65	18
6	18	26	23	46	21	66	13
7	19	27	20	47	19	67	14
8	19	28	19	48	20	68	15
9	25	29	21	49	22	69	21
10	24	30	16	50	19	70	19
11	24	31	16	51	17	71	19
12	21	32	19	52	17	72	21
13	26	33	22	53	17	73	23
14	22	34	21	54	14	74	21
15	22	35	22	55	14	75	17
16	22	36	23	56	18	76	18
17	19	37	24	57	21	77	18
18	16	38	22	58	20	78	15
19	15	39	18	59	19	79	15
20	17	40	20	60	20	80	16

[26] Source: Bureau of Transportation Statistics (www.bts.gov).

On the graph (Figure 15) you notice two trends: a 12-month cyclical pattern and an overall declining trend. This overall trend is often referred to as the *secular trend* and can be seen more clearly by using the 12-month moving average (Figure 16).

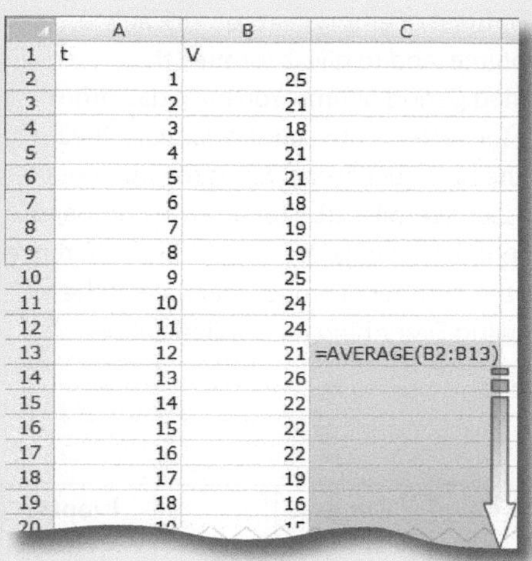

	A	B	C
1	t	V	
2	1	25	
3	2	21	
4	3	18	
5	4	21	
6	5	21	
7	6	18	
8	7	19	
9	8	19	
10	9	25	
11	10	24	
12	11	24	
13	12	21	=AVERAGE(B2:B13)
14	13	26	
15	14	22	
16	15	22	
17	16	22	
18	17	19	
19	18	16	

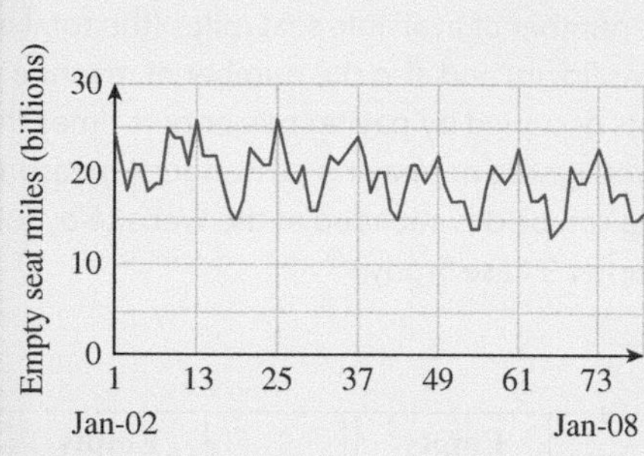

Figure 15

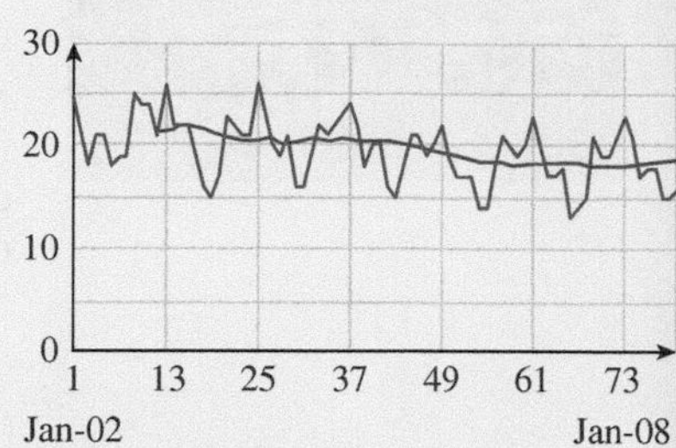

Figure 16

You notice that the secular trend appears more or less linear. The simplest model of a cyclical term with 12-month period added to a linear secular trend has the form

$$V(t) = \underbrace{A\sin(\pi t/6 + d)}_{\text{Cyclical term}} + \underbrace{Bt + C}_{\text{Secular trend}}. \qquad \omega = 2\pi/12 = \pi/6$$

You are about to use Solver to construct the model when you discover that your copy of Excel has mysteriously lost its Solver, so you wonder whether there is an alternative way to construct the model. After consulting various statistics textbooks, you discover that there is: You can use the addition formula to write

$$\begin{aligned} A\sin(\pi t/6 + d) &= A[\sin(\pi t/6)\cos d + \cos(\pi t/6)\sin d] \\ &= P\sin(\pi t/6) + Q\cos(\pi t/6) \end{aligned}$$

for constants $P = A\cos d$ and $Q = A\sin d$, so instead you could use an equivalent model of the form

$$V(t) = P\sin(\pi t/6) + Q\cos(\pi t/6) + Bt + C.$$

Note that the equations $P = A\cos d$ and $Q = A\sin d$ give

$$P^2 + Q^2 = A^2\cos^2 d + A^2\sin^2 d = A^2(\cos^2 d + \sin^2 d) = A^2,$$

so

$$A = \sqrt{P^2 + Q^2},$$

giving the amplitude in terms of P and Q.

But what has all of this to do with avoiding Solver? The point is that, now, *V is a linear function of the variables* $\sin(\pi t/6)$, $\cos(\pi t/6)$, and t, meaning that you can model y using ordinary linear regression along the lines of the Case Study in Chapter 8: First, you rearrange the data so that the V column is first, and you then add columns to calculate $\sin(\pi t/6)$ and $\cos(\pi t/6)$:

	A	B	C	D
1	V	t	sin(πt/6)	cos(πt/6)
2	25	1	=sin(PI()*B2/6)	=cos(PI()*B2/6)
3	21	2		
4	18	3		
5	21	4		
6	21	5		
7	18	6		
8	19	7		
9	19	8		

↓

	A	B	C	D
1	V	t	sin(πt/6)	cos(πt/6)
2	25	1	0.5	0.866025404
3	21	2	0.866025404	0.5
4	18	3	1	6.12574E-17
5	21	4	0.866025404	-0.5
6	21	5	0.5	-0.866025404
7	18	6	1.22515E-16	-1
8	19	7	-0.5	-0.866025404
9	19	8	-0.866025404	-0.5

Next, in the "Analysis" section of the Data tab, choose "Data analysis." (If this command is not available, you will need to load the Analysis ToolPak add-in.) Choose "Regression" from the list that appears, and in the resulting dialogue box, enter the location of the data and where you want to put the results as shown on the left in Figure 17; identify where the dependent and independent variables are (A1 through A81 for the Y range, and B1 through D81 for the X range), check "Labels", and click "OK".

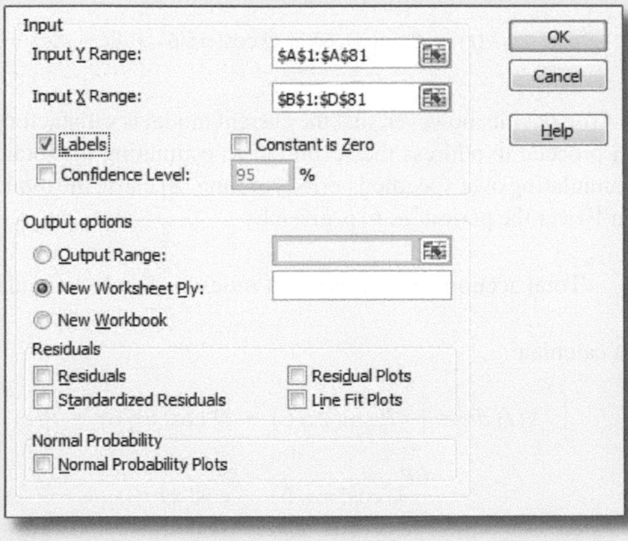

Input
Input Y Range: $A$1:$A$81
Input X Range: $B$1:$D$81
☑ Labels ☐ Constant is Zero
☐ Confidence Level: 95 %
Output options
○ Output Range:
◉ New Worksheet Ply:
○ New Workbook
Residuals
☐ Residuals ☐ Residual Plots
☐ Standardized Residuals ☐ Line Fit Plots
Normal Probability
☐ Normal Probability Plots
OK
Cancel
Help

ANOVA	
	df
Regression	3
Residual	76
Total	79
	Coefficients
Intercept	21.6833363
t	-0.051345
sin(πt/6)	0.2044244
cos(πt/6)	2.87103179

Figure 17

A portion of the output is shown above on the right, with the coefficients highlighted. You use the output to write down the regression equation (with coefficients rounded to four significant digits):

$$V(t) = 0.2044 \sin(\pi t/6) + 2.871 \cos(\pi t/6) - 0.05135t + 21.68.$$

Figure 18 shows the original data with the graph of V superimposed.

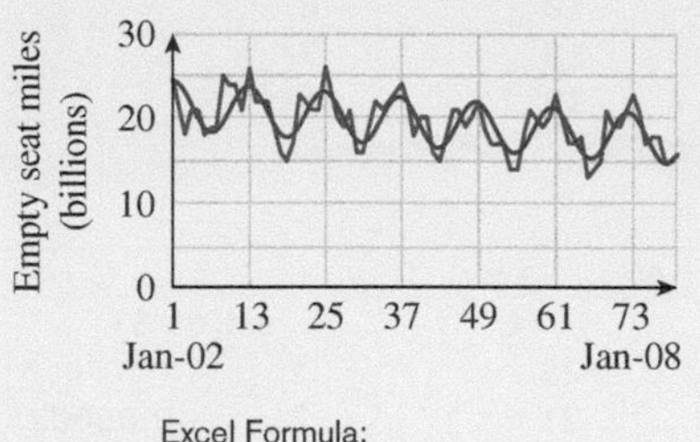

Excel Formula:
```
0.2044*sin(PI()*t/6)+2.871*
cos(PI()*t/6)-0.05135*t+21.68
```

Figure 18

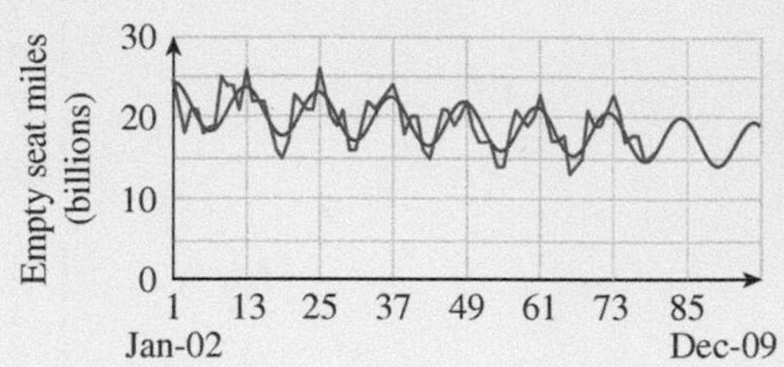

Figure 19

Although the graph does not give a perfect fit, the model captures the behavior quite accurately. Figure 19 shows the 1-year projection of the model.

Q: *How would one alter the model to capture the sharp upward spike every November and the downward spike every June-July?*

A: One could add additional terms, called seasonal variables:

$$x_1 = \begin{cases} 1 & \text{if } t = 11, 23, 35, \ldots \\ 0 & \text{if not} \end{cases} \qquad x_1 = 1 \text{ every November}$$

$$x_2 = \begin{cases} 1 & \text{if } t = 6, 7, 18, 19, \ldots \\ 0 & \text{if not} \end{cases} \qquad x_2 = 1 \text{ every June and July}$$

to obtain the following more elaborate model:

$$V(t) = P \sin(\pi t/6) + Q \cos(\pi t/6) + Bt + Cx_1 + Dx_2 + E.$$

You decide, however, that the current model is satisfactory for your purposes, and you proceed to address the second task: estimating the total volume of empty seats accumulating over specified periods of time. Because the total accumulation of a function V over the period $[a, b]$ is given by

$$\text{Total accumulated empty seat miles} = \text{Total accumulation of } V = \int_a^b V(t)\,dt,$$

you calculate

$$\begin{aligned} \int_a^b V(t)\,dt &= \int_a^b [P \sin(\pi t/6) + Q \cos(\pi t/6) + Bt + C]\,dt \\ &= \frac{6P}{\pi}[\cos(\pi a/6) - \cos(\pi b/6)] + \frac{6Q}{\pi}[\sin(\pi b/6) - \sin(\pi a/6)] \\ &\quad + \frac{B(b^2 - a^2)}{2} + C(b - a) \end{aligned}$$

billion empty seat miles.

This is a formula that, on plugging in the values of P, Q, B, and C calculated above together with the values for a and b defining the period you're interested in, gives you the total accumulated empty seat miles over that period.

EXERCISES

1. According to the regression model, the volume of empty seat miles fluctuates by ____ billion seat miles below the secular line to ____ billion miles above it.
2. Use the observed data to compute the actual accumulated empty seat miles for 2007 and compare it to the value predicted by the model.
3. T Graph the accumulated empty seat miles from January 2002 to month t as a function of t, and use your graph to project when, to the nearest month, the accumulated total will pass 1,900 billion empty seat miles.
4. Use regression on the original data to obtain a model of the form

$$V(t) = P\sin(\pi t/6) + Q\cos(\pi t/6) + Bt + Cx_1 + Dx_2 + E,$$

where

$$x_1 = \begin{cases} 1 & \text{if } t = 11, 23, 35, \dots \\ 0 & \text{if not} \end{cases} \quad \text{and} \quad x_2 = \begin{cases} 1 & \text{if } t = 6, 7, 18, 19, \dots \\ 0 & \text{if not} \end{cases}$$

as discussed in the text. Graph the resulting model together with the original data. (Round model coefficients to four significant digits.) (Use two additional columns for the independent variables: one showing the values of x_1 and the other x_2.)

5. Redo the model of the text but using instead available seat miles (in billions). For the data, go to the Website and follow Everything for Applied Calc → Chapter 9 Case Study. Note that the data are in thousands of seat miles, so you would first need to divide by 1,000,000.

TI-83/84 Plus Technology Guide

Section 9.1

Example 5(a) (page 684) The following data show the daily spam for a 16-week period (each figure is a 1-week average):

Week	0	1	2	3	4	5	6	7	8	9	10	11	12	13	14	15
Number	107	163	170	176	167	140	149	137	158	157	185	151	122	132	134	182

Use technology to find the best-fit sine curve of the form $S(t) = A \sin[\omega(t - \alpha)] + C$.

Solution The TI-83/84 Plus has a built-in sine regression utility.

1. As with the other forms of regression discussed in Chapter 2, we start by entering the coordinates of the data points in the lists L_1 and L_2, as shown below on the left:

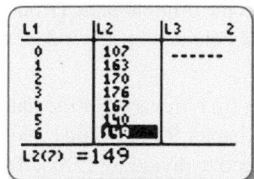

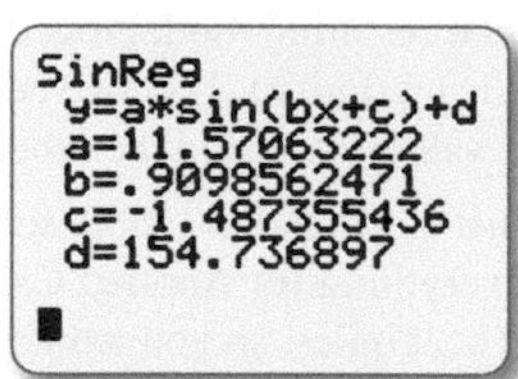

2. Press STAT, select CALC, and choose option #C: `SinReg`.
3. Pressing ENTER gives the sine regression equation in the Home screen as seen above right (we have rounded the coefficients):

$$S(t) \approx 11.57 \sin(0.9099t - 1.487) + 154.7.$$

Although this is not exactly in the form we want, we can rewrite it:

$$S(t) \approx 11.57 \sin\left[0.9099\left(t - \frac{1.487}{0.9099}\right)\right] + 154.7$$
$$\approx 11.6 \sin[0.910(t - 1.63)] + 155.$$

4. To graph the points and regression line in the same window, turn Stat Plot on by pressing 2ND STAT PLOT, selecting `1`, and turning `Plot1` on, as shown below on the left:

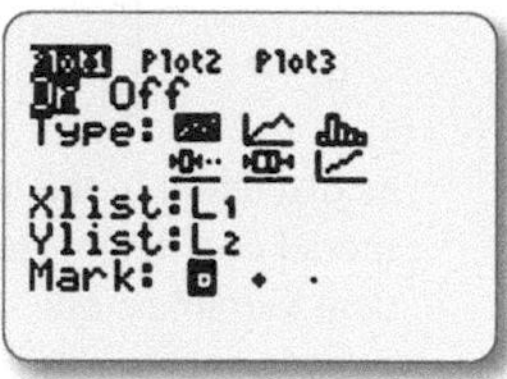

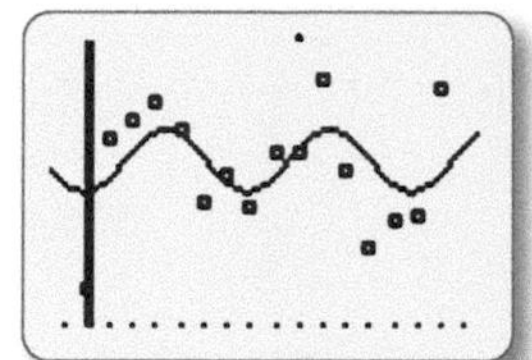

5. Enter the regression equation in the Y= screen by pressing Y=, clearing out whatever function is there, and pressing VARS 5 and selecting `EQ` option #1: `RegEq`.
6. To obtain a convenient window showing all the points and the lines, press ZOOM and choose option #9: `ZoomStat`, and you will obtain the output shown above on the right.

Spreadsheet Technology Guide

Section 9.1

Example 5(a) (page 684) The following data show the daily spam for a 16-week period (each figure is a 1-week average):

Week	0	1	2	3	4	5	6	7	8	9	10	11	12	13	14	15
Number	107	163	170	176	167	140	149	137	158	157	185	151	122	132	134	182

Use technology to find the best-fit sine curve of the form $S(t) = A\sin[\omega(t - \alpha)] + C$.

Solution We set up our worksheet, shown below, as we did for logistic regression in Section 2.4.

1. For our initial guesses, let us roughly estimate the parameters from the graph. The amplitude is around $A = 30$, and the vertical offset is roughly $C = 150$. The period seems to be around 7 weeks, so let us choose $P = 7$. This gives $\omega = 2\pi/P \approx 0.9$. Finally, let us take $\alpha = 0$ to begin with.

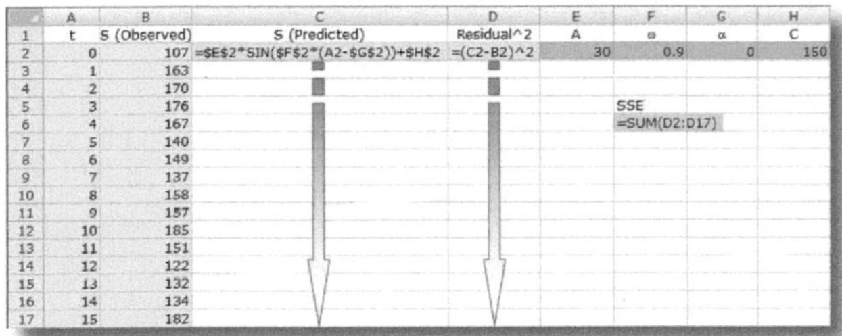

	A	B	C	D	E	F	G	H
1	t	S (Observed)	S (Predicted)	Residual^2	A	ω	α	C
2	0	107	=$E$2*SIN($F$2*(A2-$G$2))+$H$2	=(C2-B2)^2	30	0.9	0	150
3	1	163						
4	2	170						
5	3	176				SSE		
6	4	167				=SUM(D2:D17)		
7	5	140						
8	6	149						
9	7	137						
10	8	158						
11	9	157						
12	10	185						
13	11	151						
14	12	122						
15	13	132						
16	14	134						
17	15	182						

2. We then use Solver to minimize SSE by changing cells E2 through H2 as in Section 2.4 (see the setup below on the left). We obtain the following model (we have rounded the coefficients to three significant digits):

$$S(t) = 25.8\sin[0.960(t - 1.22)] + 153$$

with SSE $\approx$ 2,148. Plotting the observed and predicted values of S gives the graph shown on the right:

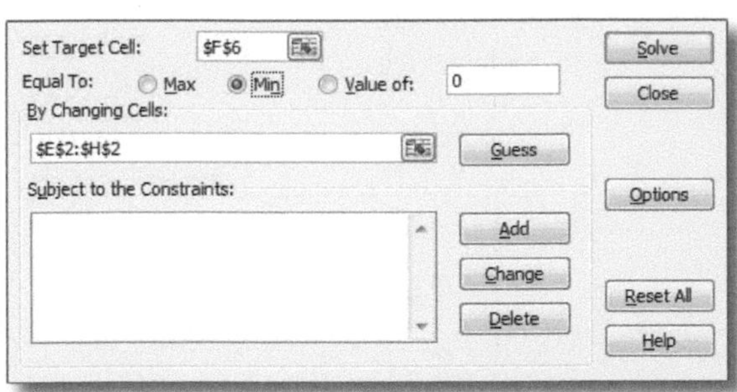

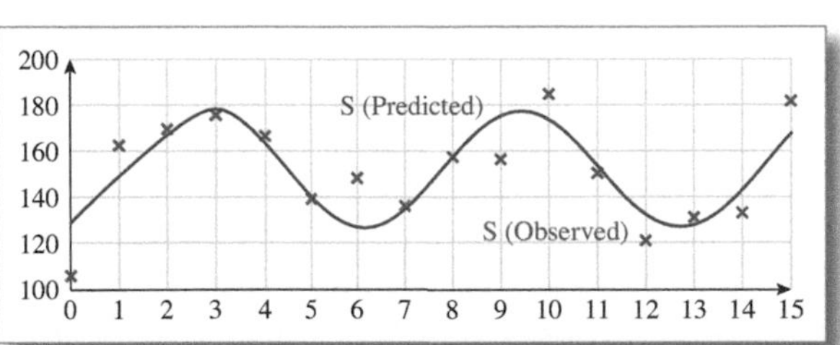

Note that Solver estimated the period for us. However, in many situations we know what to use for the period beforehand. For example, we expect sales of snow shovels to fluctuate according to an annual cycle. Thus, if we were using regression to fit a regression sine or cosine curve to snow shovel sales data, we would set $P = 1$ year, compute ω, and have Solver estimate only the remaining coefficients: A, C, and α.

Answers to Selected Exercises

Chapter 0

Section 0.1

1. −48 **3.** 2/3 **5.** −1 **7.** 9 **9.** 1 **11.** 33 **13.** 14 **15.** 5/18 **17.** 13.31 **19.** 6 **21.** 43/16 **23.** 0 **25.** `3*(2-5)` **27.** `3/(2-5)` **29.** `(3-1)/(8+6)` **31.** `3-(4+7)/8` **33.** `2/(3+x)-x*y^2` **35.** `3.1x^3-4x^(-2)-60/(x^2-1)` **37.** `(2/3)/5` **39.** `3^(4-5)*6` **41.** `3*(1+4/100)^(-3)` **43.** `3^(2*x-1)+4^x-1` **45.** `2^(2x^2-x+1)` **47.** `4*e^(-2*x)/(2-3e^(-2*x))` or `(4*e^(-2*x))/(2-3e^(-2*x))` **49.** `3(1-(-1/2)^2)^2+1`

Section 0.2

1. 27 **3.** −36 **5.** 4/9 **7.** −1/8 **9.** 16 **11.** 2 **13.** 32 **15.** 2 **17.** x^5 **19.** $-\frac{y}{x}$ **21.** $\frac{1}{x}$ **23.** x^3y **25.** $\frac{z^4}{y^3}$ **27.** $\frac{x^6}{y^6}$ **29.** $\frac{x^4y^6}{z^4}$ **31.** $\frac{3}{x^4}$ **33.** $\frac{3}{4x^{2/3}}$ **35.** $1 - 0.3x^2 - \frac{6}{5x}$ **37.** 2 **39.** 1/2 **41.** 4/3 **43.** 2/5 **45.** 7 **47.** 5 **49.** −2.668 **51.** 3/2 **53.** 2 **55.** 2 **57.** ab **59.** $x + 9$ **61.** $x\sqrt[3]{a^3 + b^3}$ **63.** $\frac{2y}{\sqrt{x}}$ **65.** $3^{1/2}$ **67.** $x^{3/2}$ **69.** $(xy^2)^{1/3}$ **71.** $x^{3/2}$ **73.** $\frac{3}{5}x^{-2}$ **75.** $\frac{3}{2}x^{-1.2} - \frac{1}{3}x^{-2.1}$ **77.** $\frac{2}{3}x - \frac{1}{2}x^{0.1} + \frac{4}{3}x^{-1.1}$ **79.** $\frac{3}{4}x^{1/2} - \frac{5}{3}x^{-1/2} + \frac{4}{3}x^{-3/2}$ **81.** $\frac{3}{4}x^{2/5} - \frac{7}{2}x^{-3/2}$ **83.** $(x^2 + 1)^{-3} - \frac{3}{4}(x^2 + 1)^{-1/3}$ **85.** $\sqrt[3]{2^2}$ **87.** $\sqrt[3]{x^4}$ **89.** $\sqrt[5]{\sqrt{x}\sqrt[3]{y}}$ **91.** $-\frac{3}{2\sqrt[4]{x}}$ **93.** $\frac{0.2}{\sqrt[3]{x^2}} + \frac{3\sqrt{x}}{7}$ **95.** $\frac{3}{4\sqrt{(1 - x)^5}}$ **97.** 64 **99.** $\sqrt{3}$ **101.** $1/x$ **103.** xy **105.** $\left(\frac{y}{x}\right)^{1/3}$ **107.** ±4 **109.** ±2/3 **111.** −1, −1/3 **113.** −2 **115.** 16 **117.** ±1 **119.** 33/8

Section 0.3

1. $4x^2 + 6x$ **3.** $2xy - y^2$ **5.** $x^2 - 2x - 3$ **7.** $2y^2 + 13y + 15$ **9.** $4x^2 - 12x + 9$ **11.** $x^2 + 2 + 1/x^2$ **13.** $4x^2 - 9$ **15.** $y^2 - 1/y^2$ **17.** $2x^3 + 6x^2 + 2x - 4$ **19.** $x^4 - 4x^3 + 6x^2 - 4x + 1$ **21.** $y^5 + 4y^4 + 4y^3 - y$ **23.** $(x + 1)(2x + 5)$ **25.** $(x^2 + 1)^5(x + 3)^3(x^2 + x + 4)$ **27.** $-x^3(x^3 + 1)\sqrt{x + 1}$ **29.** $(x + 2)\sqrt{(x + 1)^3}$ **31. a.** $x(2 + 3x)$ **b.** $x = 0, -2/3$ **33. a.** $2x^2(3x - 1)$ **b.** $x = 0, 1/3$ **35. a.** $(x - 1)(x - 7)$ **b.** $x = 1, 7$ **37. a.** $(x - 3)(x + 4)$ **b.** $x = 3, -4$ **39. a.** $(2x + 1)(x - 2)$ **b.** $x = -1/2, 2$ **41. a.** $(2x + 3)(3x + 2)$ **b.** $x = -3/2, -2/3$ **43. a.** $(3x - 2)(4x + 3)$ **b.** $x = 2/3, -3/4$ **45. a.** $(x + 2y)^2$ **b.** $x = -2y$ **47. a.** $(x^2 - 1)(x^2 - 4)$ **b.** $x = \pm 1, \pm 2$

Section 0.4

1. $\frac{2x^2 - 7x - 4}{x^2 - 1}$ **3.** $\frac{3x^2 - 2x + 5}{x^2 - 1}$ **5.** $\frac{x^2 - x + 1}{x + 1}$ **7.** $\frac{x^2 - 1}{x}$ **9.** $\frac{2x - 3}{x^2y}$ **11.** $\frac{(x + 1)^2}{(x + 2)^4}$ **13.** $\frac{-1}{\sqrt{(x^2 + 1)^3}}$ **15.** $\frac{-(2x + y)}{x^2(x + y)^2}$

Section 0.5

1. −1 **3.** 5 **5.** 13/4 **7.** 43/7 **9.** −1 **11.** $(c - b)/a$ **13.** $x = -4, 1/2$ **15.** No solutions **17.** $\pm\sqrt{\frac{5}{2}}$ **19.** −1 **21.** −1, 3 **23.** $\frac{1 \pm \sqrt{5}}{2}$ **25.** 1 **27.** ±1, ±3 **29.** $\pm\sqrt{\frac{-1 + \sqrt{5}}{2}}$ **31.** −1, −2, −3 **33.** −3 **35.** 1 **7.** −2 **39.** $1, \pm\sqrt{5}$ **41.** $\pm 1, \pm\frac{1}{\sqrt{2}}$ **43.** −2, −1, 2, 3

Section 0.6

1. 0, 3 **3.** $\pm\sqrt{2}$ **5.** −1, −5/2 **7.** −3 **9.** 0, −1 **11.** $x = -1$ ($x = -2$ is not a solution.) **13.** −2, −3/2, −1 **15.** −1 **17.** $\pm\sqrt[4]{2}$ **19.** ±1 **21.** ±3 **23.** 2/3 **25.** −4, −1/4

Section 0.7

1. $P(0, 2)$, $Q(4, -2)$, $R(-2, 3)$, $S(-3.5, -1.5)$, $T(-2.5, 0)$, $U(2, 2.5)$

3.

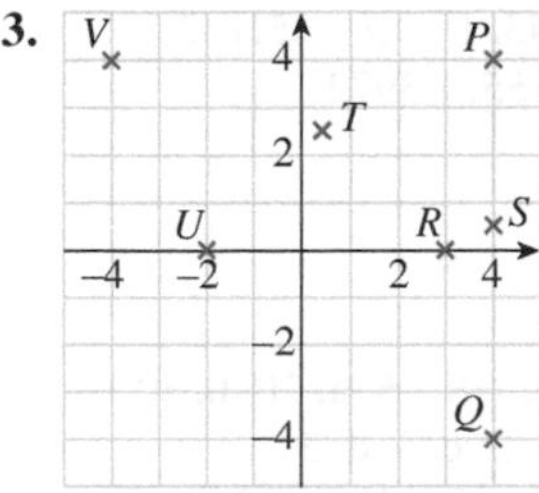

5.

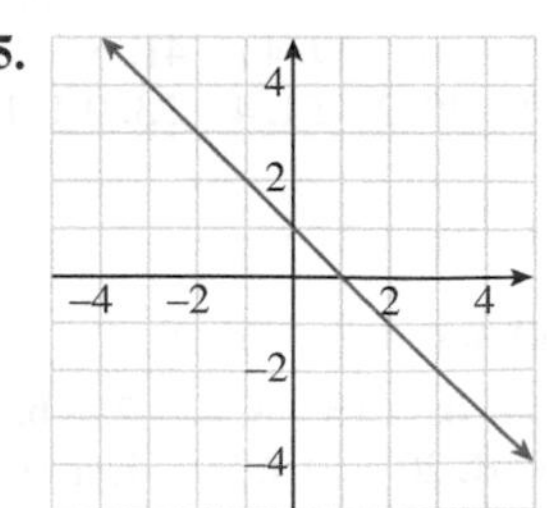

7.

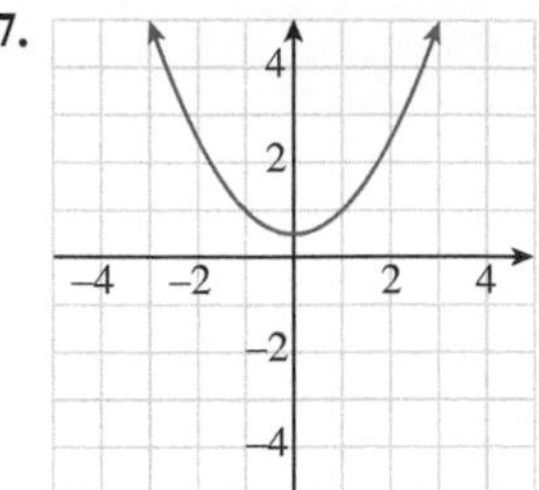

9.

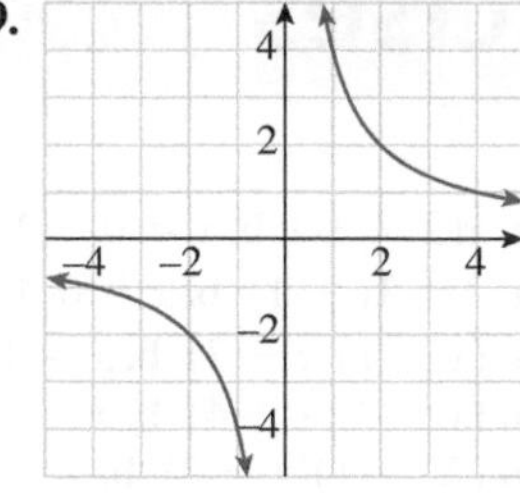

11.

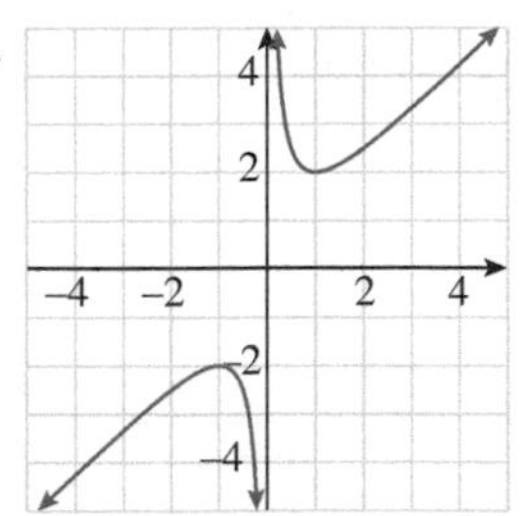

13. $\sqrt{2}$
15. $\sqrt{a^2 + b^2}$
17. 1/2
19. Circle with center (0, 0) and radius 3

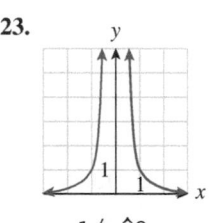

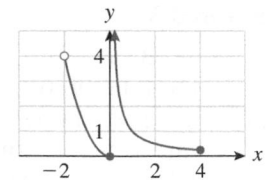

Section 0.8

1.

Exponential Form	$10^2 = 100$	$4^3 = 64$	$4^4 = 256$
Logarithmic Form	$\log_{10} 100 = 2$	$\log_4 64 = 3$	$\log_4 256 = 4$

Exponential Form	$0.45^0 = 1$	$8^{1/2} = 2\sqrt{2}$	$4^{-3} = \frac{1}{64}$
Logarithmic Form	$\log_{0.45} 1 = 0$	$\log_8 2\sqrt{2} = \frac{1}{2}$	$\log_4\left(\frac{1}{64}\right) = -3$

3.

Exponential Form	$0.3^2 = 0.09$	$\left(\frac{1}{2}\right)^0 = 1$	$10^{-3} = 0.001$
Logarithmic Form	$\log_{0.3} 0.09 = 2$	$\log_{1/2} 1 = 0$	$\log_{10} 0.001 = -3$

Exponential Form	$9^{-2} = \frac{1}{81}$	$2^{10} = 1,024$	$64^{-1/3} = \frac{1}{4}$
Logarithmic Form	$\log_9 \frac{1}{81} = -2$	$\log_2 1,024 = 10$	$\log_{64} \frac{1}{4} = -\frac{1}{3}$

5. 2 **7.** −2 **9.** 5 **11.** 1 **13.** −2 **15.** 1/2 **17.** 12 **19.** 1/10 **21.** 3/8 **23.** x^4y^5 **25.** $\frac{x^2y^3}{z^4}$ **27.** $\frac{2^x}{x^2}$ **29.** $b + c$ **31.** $a + b + c$ **33.** $-c$ **35.** $a - b$ **37.** $2a - c$ **39.** $4a$ **41.** $b - 2$ **43.** $1 - a$ **45.** $c/2$ **47.** 2 **49.** 2 **51.** 4 **53.** 0.4210 **55.** 1.3972

Chapter 1

Section 1.1

1. a. 2 **b.** −0.5 **3. a.** −2.5 **b.** 8 **c.** −8 **5. a.** 20 **b.** 30 **c.** 30 **d.** 20 **e.** 0 **f.** 20 **7. a.** 0 **b.** −3 **c.** 3 **d.** 3 **9. a.** Yes; $f(4) = 63/16$ **b.** Not defined **c.** Yes; −2 **11. a.** Not defined **b.** Not defined **c.** Yes, $f(-10) = 0$ **13. a.** −7 **b.** −3 **c.** 1 **d.** $4y - 3$ **e.** $4(a + b) - 3$ **15. a.** 3 **b.** 6 **c.** 2 **d.** 6 **e.** $a^2 + 2a + 3$ **f.** $(x + h)^2 + 2(x + h) + 3$ **17. a.** 2 **b.** 0 **c.** 65/4 **d.** $x^2 + 1/x$ **e.** $(s + h)^2 + 1/(s + h)$ **f.** $(s + h)^2 + 1/(s + h) - (s^2 + 1/s)$

19.

`-(x^3)`

21.

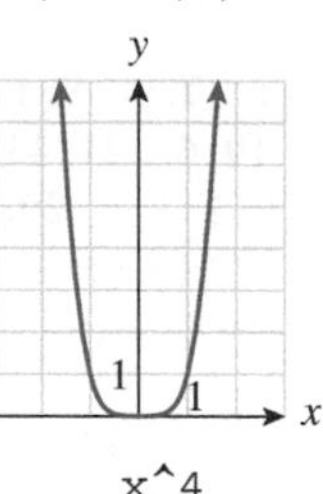

`x^4`

23.

`1/x^2`

25. a. (A) **b.** (D) **c.** (E) **d.** (F) **e.** (C) **f.** (B)

27. `0.1*x^2-4*x+5`

x	0	1	2	3
$f(x)$	5	1.1	−2.6	−6.1
x	4	5	6	7
$f(x)$	−9.4	−12.5	−15.4	−18.1
x	8	9	10	
$f(x)$	−20.6	−22.9	−25	

29. `(x^2-1)/(x^2+1)`

x	0.5	1.5	2.5	3.5
$h(x)$	−0.6000	0.3846	0.7241	0.8491
x	4.5	5.5	6.5	7.5
$h(x)$	0.9059	0.9360	0.9538	0.9651
x	8.5	9.5	10.5	
$h(x)$	0.9727	0.9781	0.9820	

31. a. −1 **b.** 2 **c.** 2

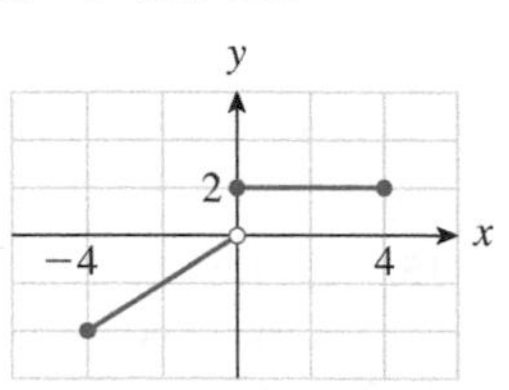

`x*(x<0)+2*(x>=0)`

33. a. 1 **b.** 0 **c.** 1

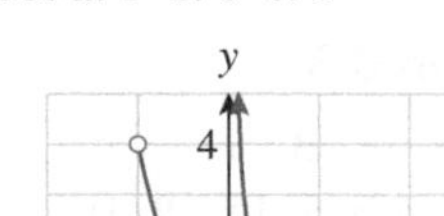

`(x^2)*(x<=0)+(1/x)*(0<x)`

35. a. 0 **b.** 2 **c.** 3 **d.** 3

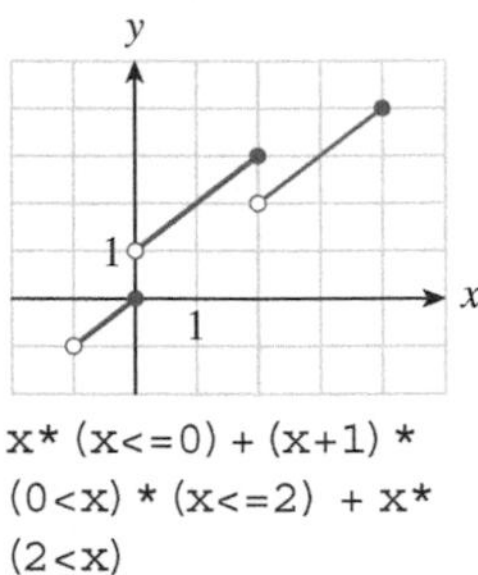


```
x*(x<=0)+(x+1)*
(0<x)*(x<=2) +x*
(2<x)
```

37. a. $h(2x + h)$ **b.** $2x + h$ **39. a.** $-h(2x + h)$ **b.** $-(2x + h)$ **41. a.** $p(2) = 2.95$; Pemex produced 2.95 million barrels of crude oil per day in 2010. $p(3) = 2.94$; Pemex produced 2.94 million barrels of crude oil per day in 2011. $p(6) = 2.79$; Pemex produced 2.79 million barrels of crude oil per day in 2014. **b.** $p(4) - p(2) = 2.91 - 2.95 = -0.04$; Crude oil production by Pemex decreased by 0.04 million barrels/day from 2010 ($t = 2$) to 2012 ($t = 4$).
43. a. Graph of p:

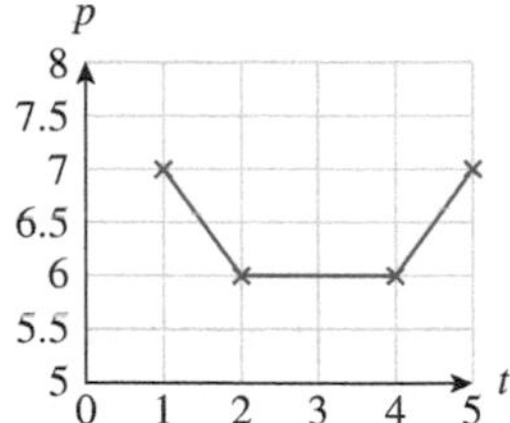

$p(4.5) \approx 6.5$. Interpretation: The popularity of Twitter midway through 2012 was about 6.5%. **b.** (D) **45.** $f(7) \approx 1{,}000$. Interpretation: Approximately 1,000,000 homes were started in 2007. $f(14) \approx 600$: Approximately 600,000 homes were started in 2014. **b.** $f(9.5) \approx 450$. Interpretation: 450,000 homes were started in the year beginning July 2009.
47. $f(7 - 3) \approx 1{,}600$, $f(7) - f(3) \approx -500$. Interpretation: 1,600,000 homes were started in 2004; there were 500,000 fewer housing starts in 2007 than in 2003. **49.** $t = 0$. Interpretation: The greatest 5-year increase in the number of housing starts occurred in 2000–2005. **51. a.** $n(2) \approx 400$, $n(4) \approx 400$, $n(4.5) \approx 350$. Interpretation: Abercrombie & Fitch's net income was $400 million in 2006, $400 million in 2008, and $350 million in the year ending June 2009.
b. $t \approx 8$. Interpretation: Between Dec. 2007 and Dec. 2012, Abercrombie & Fitch's net income was increasing most rapidly in Dec. 2012. **c.** $t \approx 5$. Interpretation: Between Dec. 2007 and Dec. 2012, Abercrombie & Fitch's net income was decreasing most rapidly in Dec. 2009. **53. a.** $[0, 8]$. $t \ge 0$ is not an appropriate domain because it would predict federal funding of NASA beyond 1966, whereas the model is based only on data up to 1966. **b.** $p(5) \approx 2.4$. In 1963, 2.4% of the U.S. federal budget was allocated to NASA. **c.** $t = 5$. The percentage of the budget allocated to NASA was increasing most rapidly in 1963. **55. a.** `100*(1-12200/t^4.48)`

b. Graph:

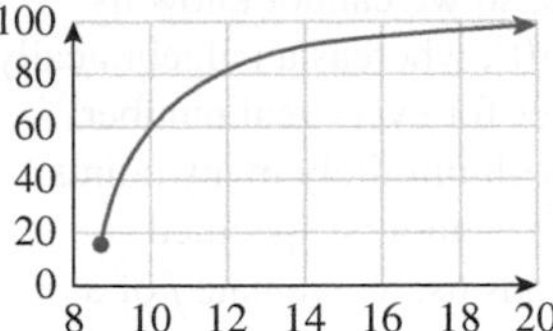

c. Table:

t	9	10	11	12	13	14
$p(t)$	35.2	59.6	73.6	82.2	87.5	91.1
t	15	16	17	18	19	20
$p(t)$	93.4	95.1	96.3	97.1	97.7	98.2

d. 82.2% **e.** 14 months **57. a.** $v(10) \approx 58$, $v(16) = 200$, $v(28) = 3{,}800$. Processor speeds were about 58 MHz in 1990, 200 MHz in 1996, and 3,800 MHz in 2008.
b. `(8*(1.22)^x)*(x<16)+(400*x-6200)*(x>=16)*(x<25)+3800*(x>=25)`
c. Graph:

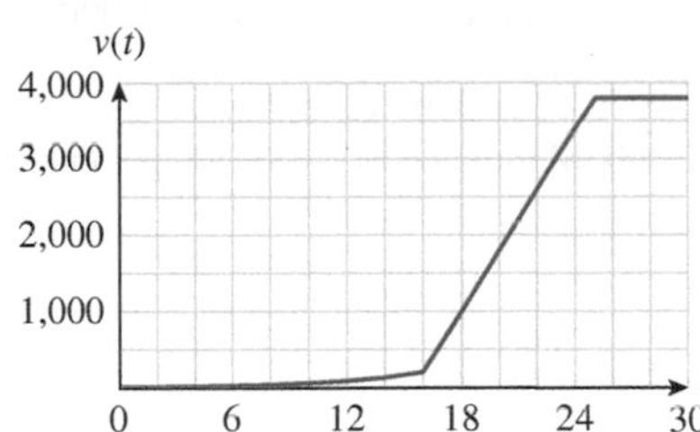

Table:

t	0	2	4	6	8	10
$v(t)$	8.0	12	18	26	39	58
t	12	14	16	18	20	22
$v(t)$	87	130	200	1,000	1,800	2,600
t	24	26	28	30		
$v(t)$	3,400	3,800	3,800	3,800		

d. 2003
59. a.

$$T(x) = \begin{cases} 0.10x & \text{if } 0 < x \le 9{,}225 \\ 922.50 + 0.15(x - 9{,}225) & \text{if } 9{,}225 < x \le 37{,}450 \\ 5{,}156.25 + 0.25(x - 37{,}450) & \text{if } 37{,}450 < x \le 90{,}750 \\ 18{,}481.25 + 0.28(x - 90{,}750) & \text{if } 90{,}750 < x \le 189{,}300 \\ 46{,}075.25 + 0.33(x - 189{,}300) & \text{if } 189{,}300 < x \le 411{,}500 \\ 119{,}401.25 + 0.35(x - 411{,}500) & \text{if } 411{,}500 < x \le 413{,}200 \\ 119{,}996.25 + 0.396(x - 413{,}200) & \text{if } 413{,}200 < x \end{cases}$$

b. $7,043.75 **61.** t; m
63. $y(x) = 4x^2 - 2$ (or $f(x) = 4x^2 - 2$)
65. False. A graph usually gives infinitely many values of the function, while a numerical table will give only a finite number of values.

67. False. In a numerically specified function, only certain values of the function are specified, so we cannot know its value on every real number in $[0, 10]$, whereas an algebraically specified function would give values for every real number in $[0, 10]$. **69.** False: Functions with infinitely many points in their domain (such as $f(x) = x^2$) cannot be specified numerically. **71.** As the text reminds us, to evaluate f of a quantity (such as $x + h$) replace x everywhere by the *whole quantity* $x + h$, getting $f(x + h) = (x + h)^2 - 1$. **73.** They are different portions of the graph of the associated equation $y = f(x)$. **75.** The graph of g is the same as the graph of f but shifted 5 units to the right.

Section 1.2

1. a. $s(x) = x^2 + x$ **b.** Domain: $(-\infty, +\infty)$ **c.** 6 **3. a.** $p(x) = (x - 1)\sqrt{x + 10}$ **b.** Domain: $[-10, 0)$ **c.** -14 **5. a.** $q(x) = \frac{\sqrt{10 - x}}{x - 1}$ **b.** Domain: $0 \le x \le 10$; $x \ne 1$ **c.** Undefined **7. a.** $m(x) = 5(x^2 + 1)$ **b.** Domain: $(-\infty, +\infty)$ **c.** 10 **9.** $N(t) = 200 + 10t$ (N = number of music files, t = time in days) **11.** $A(x) = x^2/2$ **13.** $C(x) = 12x$ **15.** $h(n) = \begin{cases} 4 & \text{if } 1 \le n \le 5 \\ 0 & \text{if } n > 5 \end{cases}$ **17.** $C(x) = 1{,}500x + 1{,}000$ per day **a.** \$5,500 **b.** \$1,500 **c.** \$1,500 **d.** Variable cost = \$1,500x; Fixed cost = \$1,000; Marginal cost = \$1,500 per piano **e.** Graph:

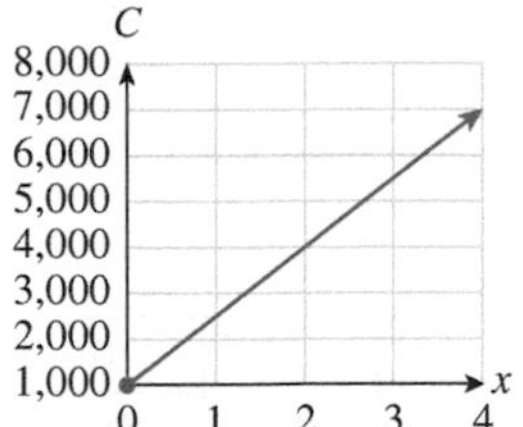

19. a. $C(x) = 0.4x + 70$, $R(x) = 0.5x$, $P(x) = 0.1x - 70$ **b.** $P(500) = -20$; a loss o \$20 **c.** 700 copies **21.** $R(x) = 100x$, $P(x) = -2{,}000 + 90x - 0.2x^2$; at least 24 jerseys **23.** $P(x) = -1.7 - 0.02x + 0.0001x^2$; approximately 264 thousand sq. ft. **25.** $P(x) = 100x - 5{,}132$, with domain $[0, 405]$. For profit, $x \ge 52$ **27.** 5,000 units **29.** $FC/(SP - VC)$ **31.** $P(x) = 579.7x - 20{,}000$, with domain $x \ge 0$; $x = 34.50$ g per day for breakeven **33. a.** Graph:

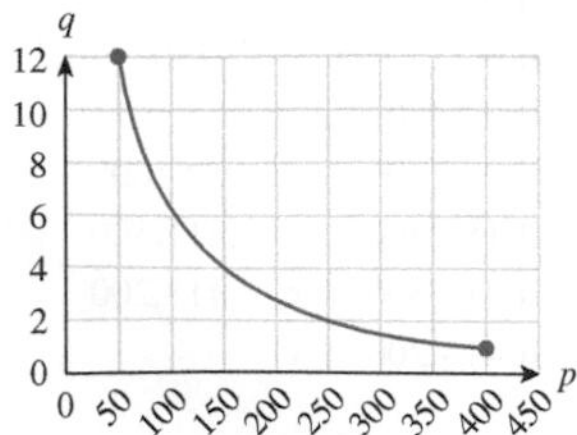

b. Demand decreases by 3.8 million units per year. **c.** (D) **35. a.** 1,027 million units **b.** 1,607 million units **c.** $R(p) = 0.17p^3 - 63p^2 + 5{,}900p$ million dollars per year; \$113 billion per year **d.** Increases; Graph:

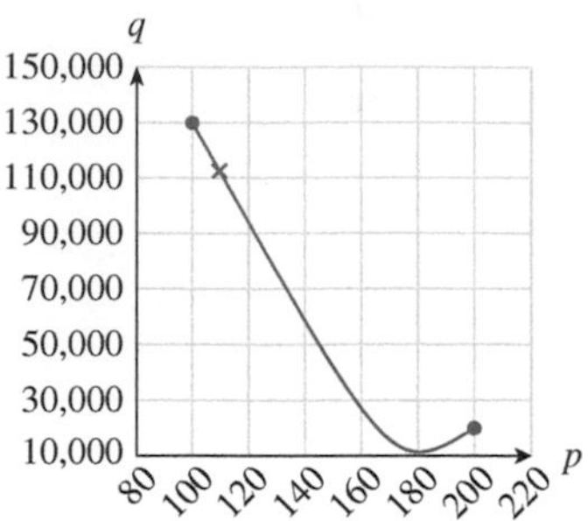

37. \$240 per skateboard **39. a.** \$110 per phone **b.** Shortage of 25 million phones **41. a.** Equilibrium price: \$200; equilibrium demand: 2.8 million units **b.** Graph:

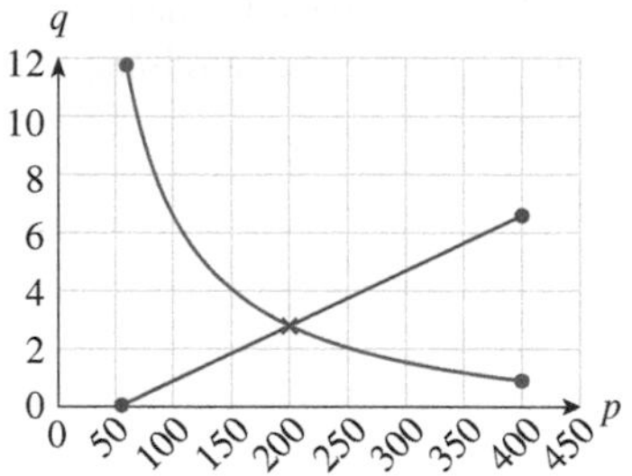

c. A shortage of around 9.2 million e-readers **43. a.** \$12,000 **b.** $N(q) = 2{,}000 + 100q^2 - 500q$; this is the cost of removing q lb of PCPs per day after the subsidy is taken into account. **c.** $N(20) = \$32{,}000$; the net cost of removing 20 lb of PCPs per day is \$32,000. **45. a.** (B) **b.** \$37 billion **47. a.** (C) **b.** \$20.80 per shirt if the team buys 70 shirts Graph:

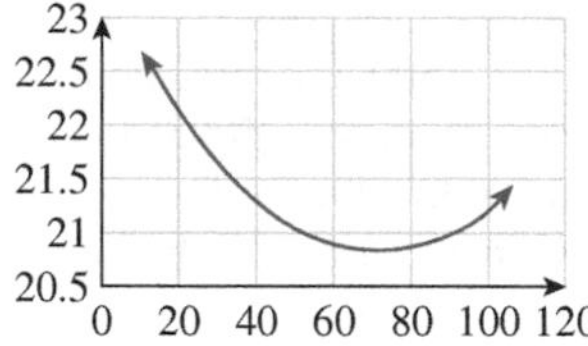

49. a. (A), (D); Graph:

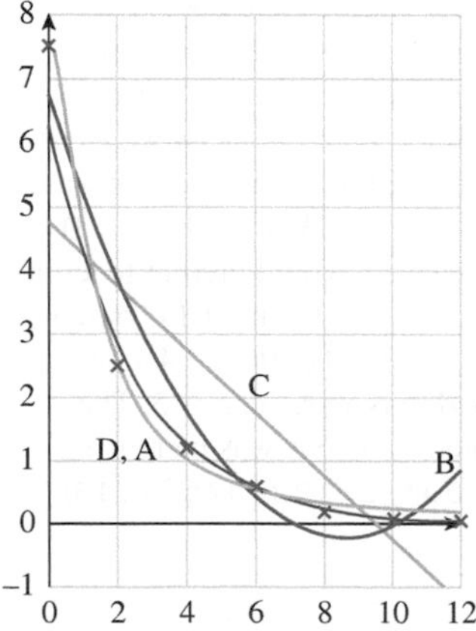

b. Model (A); Approximately \$0.0021 **51.** A linear model (A) is the best choice; a plot of the given points gives a straight line. **53.** Model (D) is the best choice; Model (A) would predict increasing demand with increasing price, Model (B) would correspond to a curve that becomes less steep as p increases, and Model (C) would give a concave-up parabola. **55.** $A(t) = 5{,}000(1 + 0.0005/12)^{12t}$; \$5,018 **57.** 2033 **59.** 31.0 g, 9.25 g, 2.76 g **61.** 20,000 years **63. a.** 1,000 years: 65%, 2,000 years: 42%, 3,000 years: 27% **b.** 1,600 years **65.** 30 **67.** Curve fitting. The model is based on fitting a curve to a given set of observed data. **69.** The cost of downloading a movie was \$4 in January and is decreasing by 20¢ per month. **71.** Variable; marginal. **73.** Yes, as long as the supply is going up at a faster rate, as illustrated by the following graph:

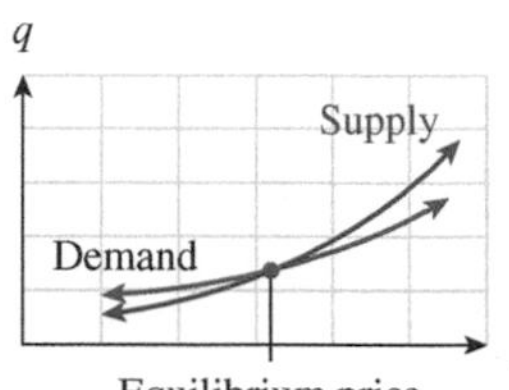

75. Extrapolate both models and choose the one that gives the most reasonable predictions. **77.** They are ≥ 0. **79.** Books per person

Section 1.3

1. 11; $m = 3$ **3.** -4; $m = -1$ **5.** 7; $m = 3/2$ **7.** $f(x) = -x/2 - 2$ **9.** $f(0) = -5$, $f(x) = -x - 5$ **11.** f is linear: $f(x) = 4x + 6$ **13.** g is linear: $g(x) = 2x - 1$ **15.** $-3/2$ **17.** 1/6 **19.** Undefined **21.** 0 **23.** $-4/3$

25.

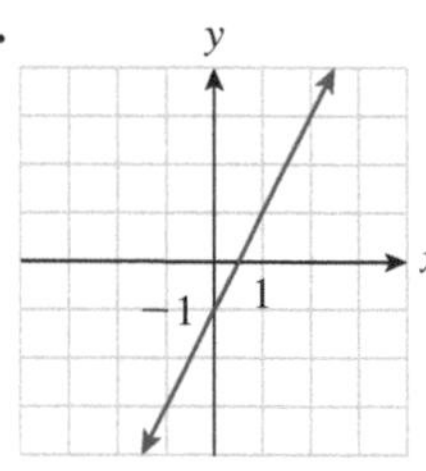

27.

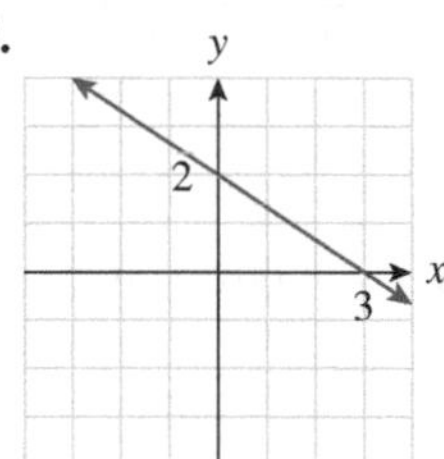

29.

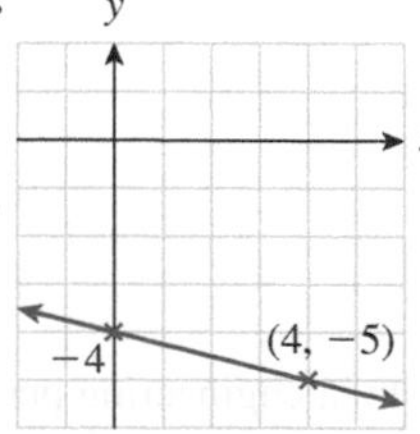

31.

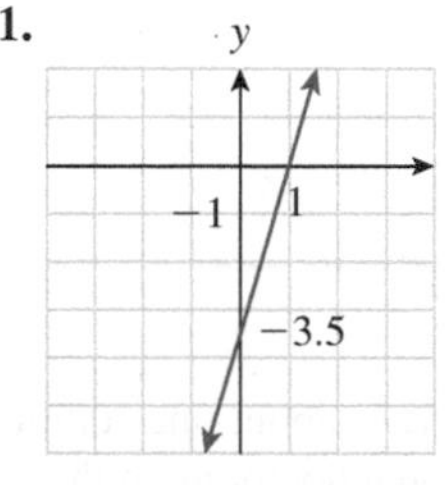

33.

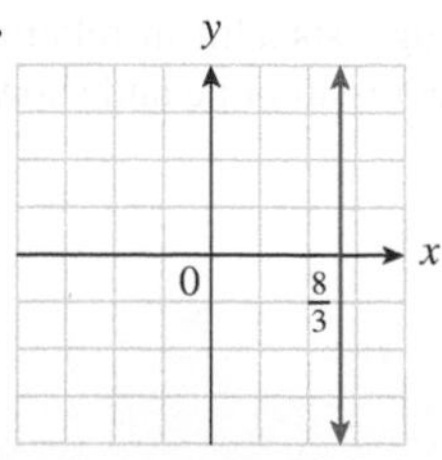

35.

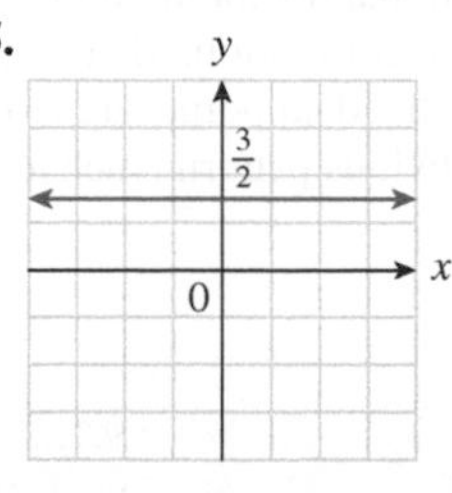

37.

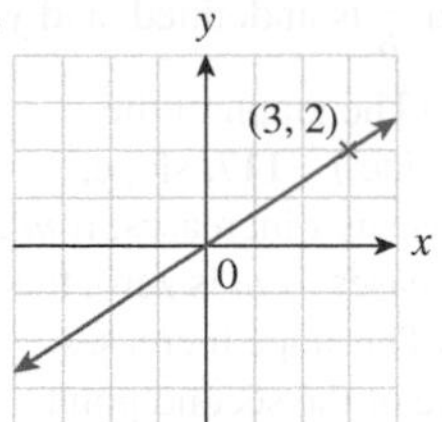

39. 2 **41.** 2 **43.** -2 **45.** Undefined **47.** 1.5 **49.** -0.09 **51.** 1/2 **53.** $(d - b)/(c - a)$ **55.** Undefined **57.** $-b/a$ **59. a.** 1 **b.** 1/2 **c.** 0 **d.** 3 **e.** $-1/3$ **f.** -1 **g.** Undefined **h.** $-1/4$ **i.** -2 **61.** $y = 3x$ **63.** $y = \frac{1}{4}x - 1$ **65.** $y = 10x - 203.5$ **67.** $y = -5x + 6$ **69.** $y = -3x + 2.25$ **71.** $y = -x + 12$ **73.** $y = 2x + 4$ **75.** $y = \frac{q}{p}x$ **77.** $y = q$ **79.** $y = -\frac{q}{p}x$ **81.** Fixed cost = \$8,000, marginal cost = \$25 per bicycle **83.** $C = 205x + 20$; \$205 per iPhone; \$8,220 **85.** $q = -40p + 2{,}000$ **87. a.** $q = -5.8p + 2{,}953$; 1,416 million phones **b.** \$1; 5.8 million **89. a.** $q = -4{,}500p + 41{,}500$ **b.** Rides per day per \$1 increase in the fare; ridership decreases by 4,500 rides per day for every \$1 increase in the fare. **c.** 14,500 rides per day **91. a.** $y = 40t + 290$ million pounds of pasta **b.** 890 million pounds **93. a.** $N = -0.29t + 0.92$ **b.** Billions of dollars per year; Amazon's net income decreased at a rate of \$0.29 billion per year. **c.** \$0.05 billion **95. a.** 2.5 ft/sec **b.** 20 ft along the track **c.** after 6 sec **97. a.** 130 mph **b.** $s = 130t - 1{,}300$ **99. a.** $L = 42.5n + 500$ **b.** Pages per edition; *Applied Calculus* is growing at a rate of 42.5 pages per edition. **c.** 24th edition **101.** $F = 1.8C + 32$; 86°F; 72°F; 14°F; 7°F **103. a.** $J = 0.54S - 86$ **b.** \$57 million **c.** Millions of dollars of JetBlue Airways net income per million dollars of Southwest Airlines net income; JetBlue Airways earned an additional net income of \$0.54 per \$1 additional net income earned by Southwest Airlines. **105.** $I(N) = 0.05N + 50{,}000$; $N = \$1{,}000{,}000$; marginal income is $m =$ 5¢ per dollar of net profit **107.** Increasing at 400 MHz per year **109. a.** $y = 31.1t + 78$ **b.** $y = 112.5t - 1{,}550$

c. $y = \begin{cases} 31.1t + 78 & \text{if } 0 \leq t < 20 \\ 112.5t - 1{,}550 & \text{if } 20 \leq t \leq 40 \end{cases}$ or

$y = \begin{cases} 31.1t + 78 & \text{if } 0 \leq t \leq 20 \\ 112.5t - 1{,}550 & \text{if } 20 < t \leq 40 \end{cases}$

d. \$2,275,000, in good agreement with the actual value shown in the graph.

111. $N = \begin{cases} 0.22t + 3 & \text{if } 0 \leq t \leq 5 \\ -0.15t + 4.85 & \text{if } 5 < t \leq 9 \end{cases}$

3.8 million jobs

113. Compute the corresponding successive changes Δx in x and Δy in y, and compute the ratios $\Delta y/\Delta x$. If the answer is always the same number, then the values in the table come from a linear function.

115. $f(x) = -\frac{a}{b}x + \frac{c}{b}$. If $b = 0$, then $\frac{a}{b}$ is undefined, and y cannot be specified as a function of x. (The graph of the resulting equation would be a vertical line.) **117.** slope, 3. **119.** If m is positive, then y will increase as x increases; if m is negative, then y will decrease as x increases; if m is zero, then y will not change as x changes. **121.** The slope increases, because an increase in the y-coordinate of the second point increases Δy while leaving Δx fixed. **123.** Bootlags per zonar; bootlags **125.** It must increase by 10 units each day, including the third. **127.** (B) **129.** It is linear with slope $m + n$. **131.** Answers may vary. For example, $f(x) = x^{1/3}$, $g(x) = x^{2/3}$ **133.** Increasing the number of items from the break-even number results in a profit: Because the slope of the revenue graph is larger than the slope of the cost graph, it is higher than the cost graph to the right of the point of intersection, and hence corresponds to a profit.

Section 1.4

1. 6 **3.** 86 **5. a.** 0.5 (better fit) **b.** 0.75
7. a. 27.42 **b.** 27.16 (better fit)
9. $y = 1.5x - 0.6667$
Graph:

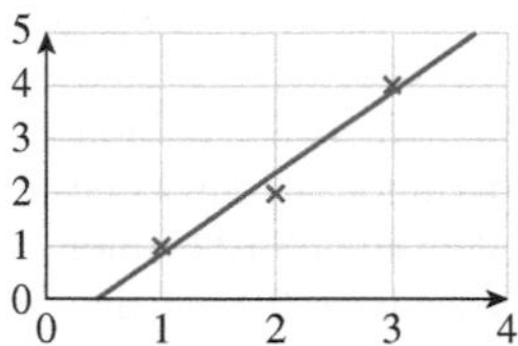

11. $y = 0.7x + 0.85$
Graph:

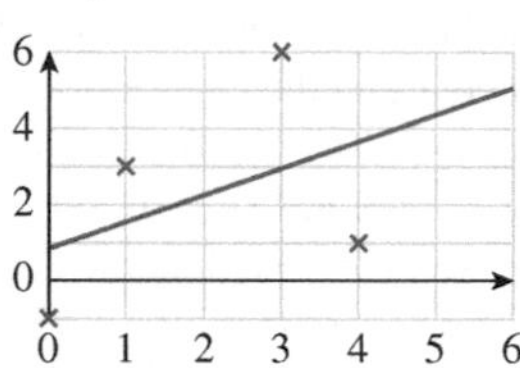

13. a. $r = 0.9959$ (best, not perfect) **b.** $r = 0.9538$
c. $r = 0.3273$ (worst)

15.

x	y	xy	x^2
0	800	0	0
2	1,600	3,200	4
4	2,300	9,200	16
6	4,700	12,400	20

$y = 375x + 816.7$; 3,066.7 million
17. $q = -0.7p + 3.2$; 750 million smartphones
19. $y = 0.135x + 0.15$; 6.9 million jobs
21. a. $I = -0.007S + 0.78$
Graph:

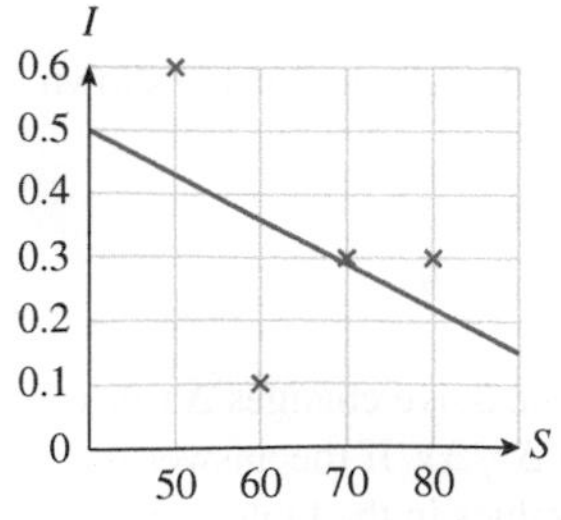

b. Amazon lost $7 million in net income per billion dollars earned in net sales. **c.** $40 billion **d.** The graph shows a poor fit, so the linear model does not seem reasonable.
23. a. $L = 45.8n + 507$
Graph:

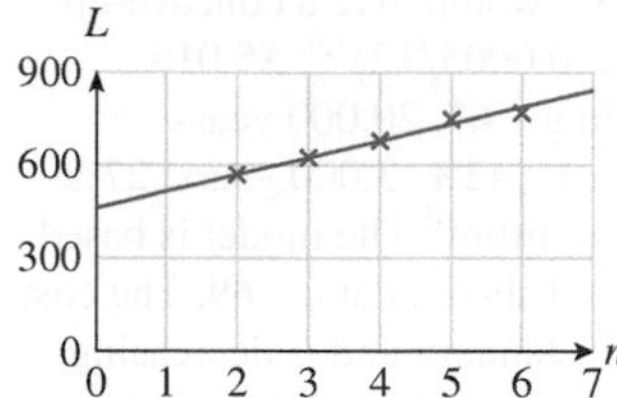

b. *Applied Calculus* is growing at a rate of 45.8 pages per edition. **25. a.** $y = 1.62x - 23.87$
Graph:

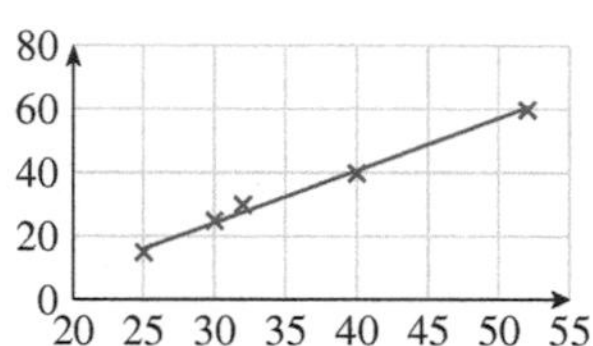

b. Each acre of cultivated land produces about 1.62 tons of soybeans. **27. a.** $y = -11.85x + 797.71$; $r \approx -0.414$
b. Continental's net income is not correlated with the price of oil. **c.** The points are nowhere near the regression line, confirming the conclusion in part (b).
Graph:

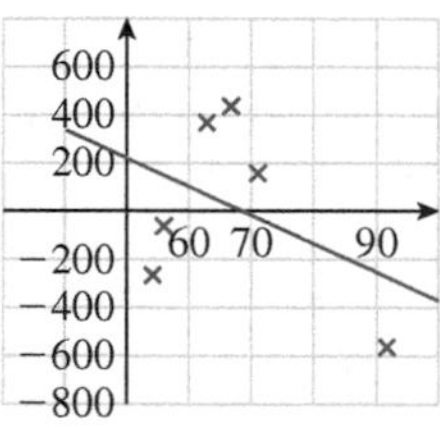

29. a. $y = 1.00x - 102$
Graph:

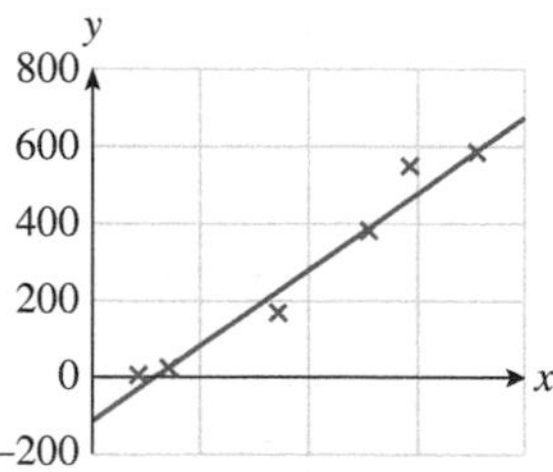

b. There is about one additional doctorate in engineering per additional doctorate in the natural sciences. **c.** $r \approx 0.976$; a strong correlation. **d.** Yes; the graph suggests a linear relationship; the data points are close to the regression line and show no obvious pattern (such as a curve).

31. a. $y = 28.9t + 37.0$
Graph:

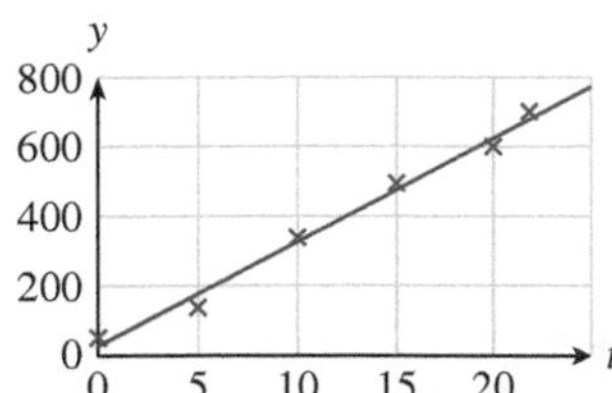

$r \approx 0.992$ **b.** The number of natural science doctorates has been increasing at a rate of about 28.9 per year. **c.** More-or-less constant rate; the slopes of successive pairs of points do not show an increasing or decreasing trend as we go from left to right. **d.** Yes; if r had been equal to 1, then the points would lie exactly on the regression line, which would indicate that the number of doctorates is growing at a constant rate. **33. a.** More-or-less constant rate; Exercise 29 suggests a roughly linear relationship between the number of natural science doctorates and the number of engineering doctorates, and Exercise 31 suggests that the number of natural science doctorates has been increasing at a more-or less constant rate. Therefore, the number of engineering doctorates is also increasing at a more-or-less constant rate. **b.** No; $r = 1$ in Exercise 29 would indicate an exactly linear relationship between the number of natural science doctorates and the number of engineering doctorates, so the conclusion would be the same. **c.** No; $r = 1$ in Exercise 31 would indicate that the number of natural science doctorates has been increasing at a constant rate, so the conclusion would be the same.
35. a. $p = 0.13t + 0.22$; $r \approx 0.97$
Graph:

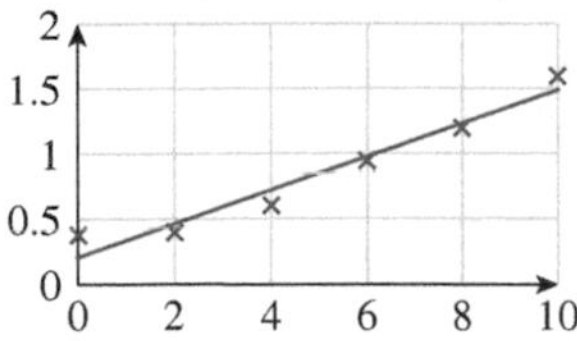

b. Yes; the first and last points lie above the regression line, while the central points lie below it, suggesting a curve.
c.

◇	A	B	C	D
1	t	p (observed)	p (predicted)	Residual
2	0	0.38	0.22	0.16
3	2	0.4	0.48	-0.08
4	4	0.6	0.74	-0.14
5	6	0.95	1	-0.05
6	8	1.2	1.26	-0.06
7	10	1.6	1.52	0.08

Notice that the residuals are positive at first, then become negative, and then become positive, confirming the impression from the graph. **37.** The line that passes through (a, b) and (c, d) gives a sum-of-squares error SSE $= 0$, which is the smallest value possible. **39.** The regression line is the line passing through the given points. **41.** 0 **43.** No. The regression line through $(-1, 1)$, $(0, 0)$, and $(1, 1)$ passes through none of these points. **45.** (Answers may vary.) The data in Exercise 35 give $r \approx 0.97$, yet the plotted points suggest a curve, not a straight line.

Chapter 1 Review

1. a. 1 **b.** -2 **c.** 0 **d.** -1 **3. a.** 1 **b.** 0 **c.** 0 **d.** -1
5.

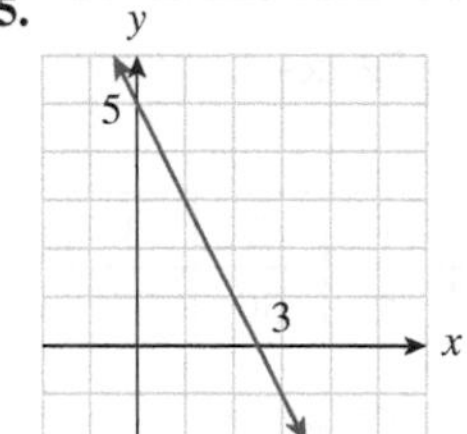

7.

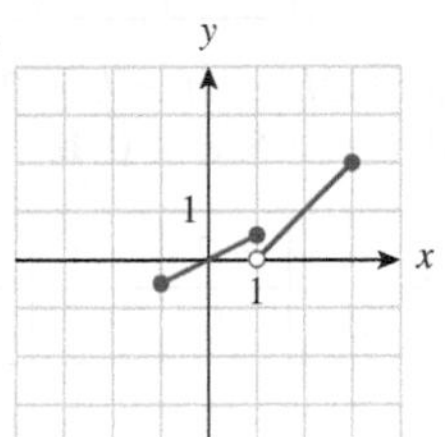

9. Absolute value **11.** Linear **13.** Quadratic
15. $y = -3x + 11$ **17.** $y = 1.25x - 4.25$
19. $y = (1/2)x + 3/2$ **21.** $y = 4x - 12$
23. $y = -\frac{x}{4} + 1$ **25.** $y = -0.214x + 1.14$, $r \approx -0.33$
27. a. Exponential
Graph:

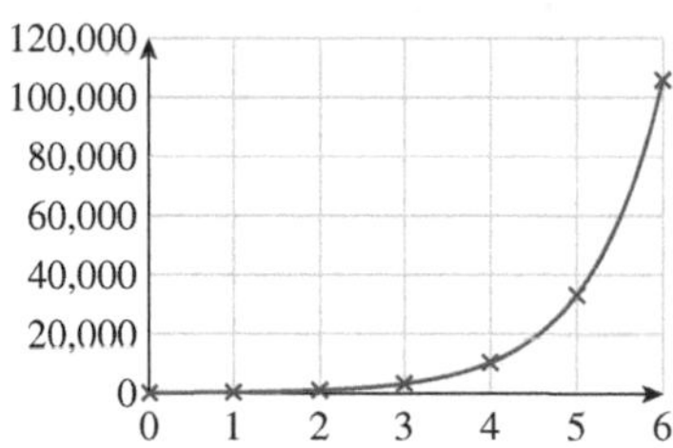

b. The ratios (rounded to 1 decimal place) are:

$V(1)/V(0)$	$V(2)/V(1)$	$V(3)/V(2)$	$V(4)/V(3)$	$V(5)/V(4)$	$V(6)/V(5)$
3	3.3	3.3	3.2	3.2	3.2

They are close to 3.2. **c.** About 343,700 visits per day
29. a. 2.3; 3.5; 6 **b.** For website traffic of up to 50,000 visits per day, the number of crashes is increasing by 0.03 per additional thousand visits. **c.** 140,000 **31. a.** (A)
b. (A) Leveling off (B) Rising (C) Rising; begins to fall after 7 months (D) Rising **33. a.** The number of visits would increase by 30 per day. **b.** No; it would increase at a slower and slower rate and then begin to decrease. **c.** Probably not. This model predicts that website popularity will start to decrease as advertising increases beyond \$8,500 per month and then drop toward zero. **35. a.** $v = 0.05c + 1{,}800$
b. 2,150 new visits per day **c.** \$14,000 per month
37. $d = 0.95w + 8$; 86 kg **39. a.** Cost: $C = 5.5x + 500$; Revenue: $R = 9.5x$; Profit $P = 4x - 500$ **b.** More than 125 albums per week **c.** More than 200 albums per week
41. a. $q = -80p + 1{,}060$ **b.** 100 albums per week
c. \$9.50, for a weekly profit of \$700
43. a. $q = -74p + 1{,}015.5$ **b.** 239 albums per week

Chapter 2

Section 2.1

1. a. $a = 2, b = -1, c = -2$ **b.**

x	-3	-2	-1	0	1	2	3
$f(x)$	19	8	1	-2	-1	4	13

c. $2a^2 + 4ah + 2h^2 - a - h - 2$ **d.** `2x^2-x-2`

3. a. $a = 10, b = -5, c = 0$ **b.**

x	-3	-2	-1	0	1	2	3
$f(x)$	105	50	15	0	5	30	75

c. $10a^2 + 20ah + 10h^2 - 5a - 5h$ **d.** `10x^2-5x`

5. a. $a = -1, b = -1, c = -1$ **b.**

x	-3	-2	-1	0	1	2	3
$f(x)$	-7	-3	-1	-1	-3	-7	-13

c. $-a^2 - 2ah - h^2 - a - h - 1$ **d.** `-(x^2)-x-1`
(See the margin note next to Quick Example 2 in the text as to the reason for the parentheses.)

7. Vertex: $(-3/2, -1/4)$, y-intercept: 2, x-intercepts: $-2, -1$

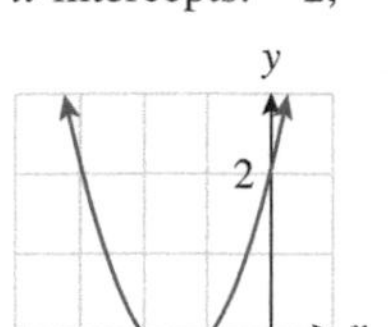

9. Vertex: $(2, 0)$, y-intercept: -4, x-intercept: 2

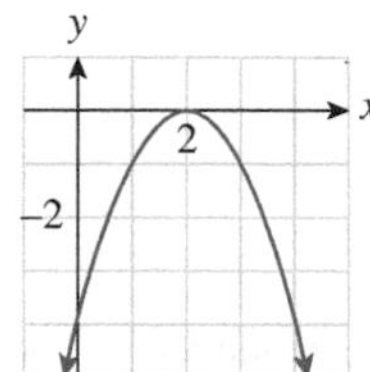

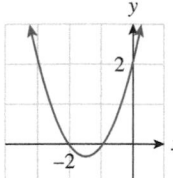

11. Vertex: $(-20, 900)$, y-intercept: 500, x-intercepts: $-50, 10$

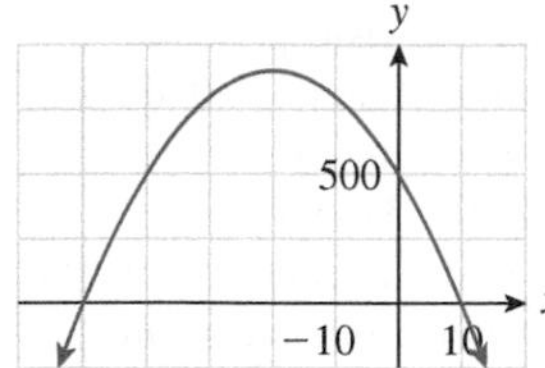

13. Vertex: $(-1/2, -5/4)$, y-intercept: -1, x-intercepts: $-1/2 \pm \sqrt{5}/2$

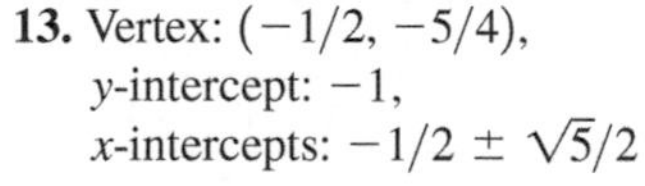

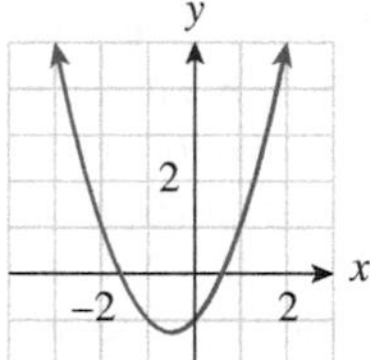

15. Vertex: $(0, 1)$, y-intercept: 1, no x-intercepts

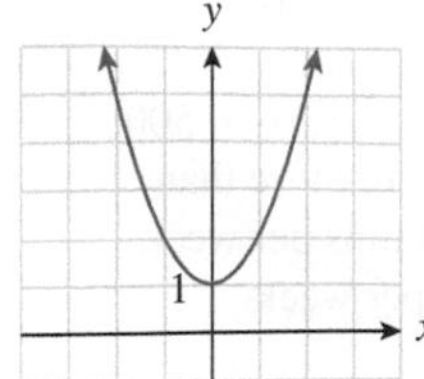

17. $R = -4p^2 + 100p$; Maximum revenue when $p = \$12.50$

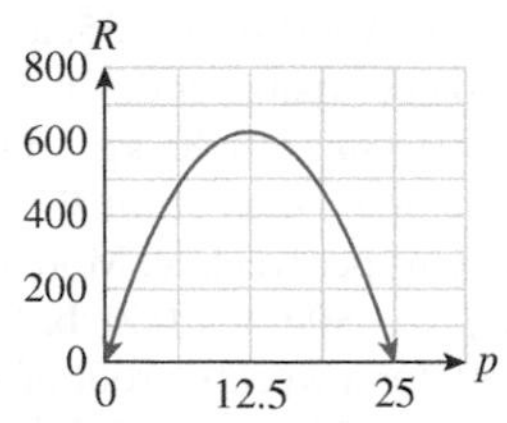

19. $R = -2p^2 + 400p$; Maximum revenue when $p = \$100$

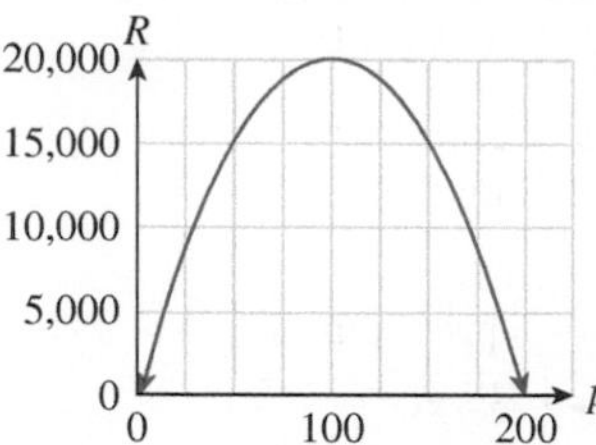

21. $y = -0.7955x^2 + 4.4591x - 1.6000$
23. $y = -1.1667x^2 - 6.1667x - 3.0000$
25. a. Positive because the data suggest a curve that is concave up. **b.** (C) **c.** 1997. Extrapolating in the positive direction leads one to predict more and more steeply rising military expenditure, which may or may not occur; extrapolating in the negative direction predicts more and more steeply increasing military expenditure as we go back in time, contradicting history. (In fact, military expenditures flattened in the years immediately after 2010.) **27.** About one quarter of the way into 2010; About 1,100 thousand barrels per day **29.** 2011; About \$13.2 billion; No, as the model predicts net income dropping without bound.
31. Maximum revenue when $p = \$140$, $R = \$9{,}800$
33. Maximum revenue with 70 houses, $R = \$9{,}800{,}000$
35. a. $q = -4{,}500p + 41{,}500$ **b.** \$4.61 for a daily revenue of \$95,680.55 **c.** No **37. a.** $q = -560x + 1{,}400$; $R = -560x^2 + 1{,}400x$ **b.** $P = -560x^2 + 1{,}400x - 30$; $x = \$1.25$; $P = \$845$ per month **39.** $C = -200x + 620$; $P = -400x^2 + 1{,}400x - 620$; $x = \$1.75$ per log-on; $P = \$605$ per month **41. a.** $q = -10p + 400$ **b.** $R = -10p^2 + 400p$ **c.** $C = -30p + 4{,}200$ **d.** $P = -10p^2 + 430p - 4{,}200$; $p = \$21.50$
43. $f(t) = 6.25t^2 - 100t + 1{,}450$; \$1,675 billion, which is \$75 billion higher than the actual value.
45. a. $S(t) = -0.70t^2 - 6.0t + 50$;

Graph:

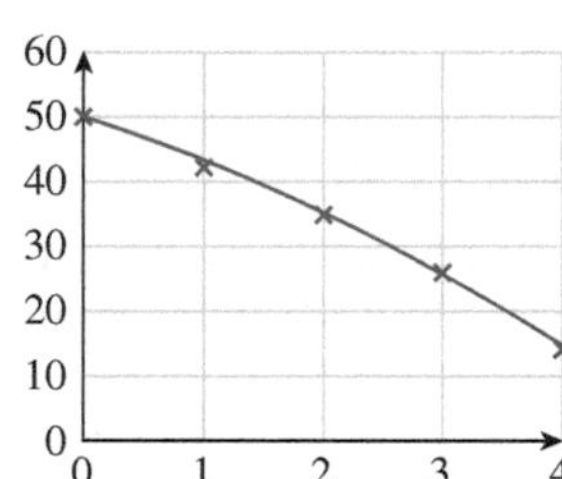

b. 3 million units, -11 million units. Even though we expected 2015 sales to be much lower than the 2014 sales, the 2016 prediction, being negative, shows the danger of exptrapolating curve-fitting models. **47.** The graph is a straight line.
49. (C) **51.** Positive; the x-coordinate of the vertex is negative, so $-b/(2a)$ must be negative. Because a is positive (the parabola is concave up), this means that b must also be positive to make $-b/(2a)$ negative. **53.** The x-coordinate of the vertex represents the unit price that leads to the maximum revenue, the y-coordinate of the vertex gives the maximum possible revenue, the x-intercepts give the unit prices that result in zero revenue, and the y-intercept gives the revenue resulting from zero unit price (which is obviously zero). **55.** Graph the data to see

whether the points suggest a curve rather than a straight line. If the curve suggested by the graph is concave up or concave down, then a quadratic model would be a likely candidate. **57.** No; the graph of a quadratic function is a parabola. In the case of a concave-up parabola, the curve would unrealistically predict sales increasing without bound in the future. In the case of a concave-down parabola, the curve would predict "negative" sales from some point on. **59.** If $q = mp + b$ (with $m < 0$), then the revenue is given by $R = pq = mp^2 + bp$. This is the equation of a parabola with $a = m < 0$ and so is concave down. Thus, the vertex is the highest point on the parabola, showing that there is a single highest value for R, namely, the y-coordinate of the vertex. **61.** Since $R = pq$, the demand must be given by $q = \dfrac{R}{p} = \dfrac{-50p^2 + 60p}{p} = -50p + 60$.

Section 2.2

1. `4^x`

x	−3	−2	−1	0	1	2	3
$f(x)$	$\frac{1}{64}$	$\frac{1}{16}$	$\frac{1}{4}$	1	4	16	64

3. `3^(-x)`

x	−3	−2	−1	0	1	2	3
$f(x)$	27	9	3	1	$\frac{1}{3}$	$\frac{1}{9}$	$\frac{1}{27}$

5. `2*2^x` or `2*(2^x)`

x	−3	−2	−1	0	1	2	3
$f(x)$	$\frac{1}{4}$	$\frac{1}{2}$	1	2	4	8	16

7. `-3*2^(-x)`

x	−3	−2	−1	0	1	2	3
$f(x)$	−24	−12	−6	−3	$-\frac{3}{2}$	$-\frac{3}{4}$	$-\frac{3}{8}$

9. `2^x-1`

x	−3	−2	−1	0	1	2	3
$f(x)$	$-\frac{7}{8}$	$-\frac{3}{4}$	$-\frac{1}{2}$	0	1	3	7

11. `2^(x-1)`

x	−3	−2	−1	0	1	2	3
$f(x)$	$\frac{1}{16}$	$\frac{1}{8}$	$\frac{1}{4}$	$\frac{1}{2}$	1	2	4

13.

$y = 3^{-x}$

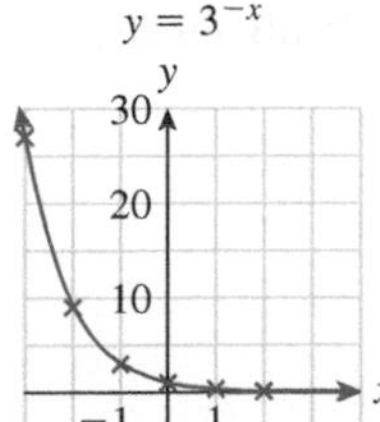

15.

$y = 2(2^x)$

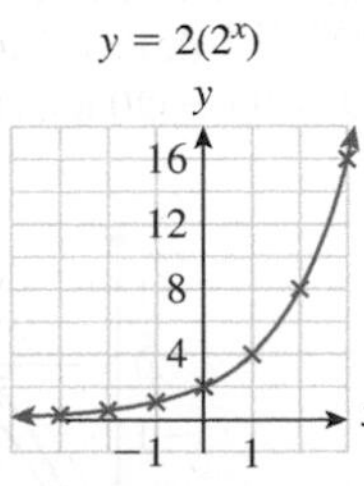

17.

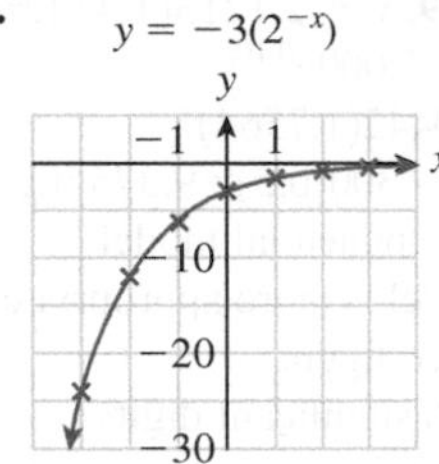

19. Both; $f(x) = 4.5(3^x)$, $g(x) = 2(1/2)^x$, or $2(2^{-x})$

21. Neither **23.** g; $g(x) = 4(0.2)^x$

25. `e^(-2*x)` or `EXP(-2*x)`

x	-3	-2	-1	0	1	2	3
$f(x)$	403.4	54.60	7.389	1	0.1353	0.01832	0.002479

27. `1.01*2.02^(-4*x)`

x	-3	-2	-1	0	1	2	3
$f(x)$	4,662	280.0	16.82	1.01	0.06066	0.003643	0.0002188

29. `50*(1+1/3.2)^(2*x)`

x	-3	-2	-1	0	1	2	3
$f(x)$	9.781	16.85	29.02	50	86.13	148.4	255.6

31. `2^(x-1)` *not* `2^x-1` **33.** `2/(1-2^(-4*x))` *not* `2/1-2^-4*x` *and not* `2/1-2^(-4*x)`

35. `(3+x)^(3*x)/(x+1)` or `((3+x)^(3*x))/(x+1)` *not* `(3+x)^(3*x)/x+1` *and not* `(3+x^(3*x))/(x+1)`

37. `2*e^((1+x)/x)` or `2*EXP((1+x)/x)` *not* `2*e^1+x/x` *and not* `2*e^(1+x)/x` *and not* `2*EXP(1+x)/x`

39.

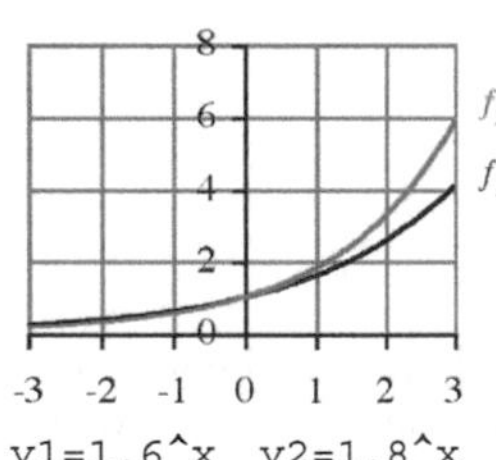

`y1=1.6^x  y2=1.8^x`

41.

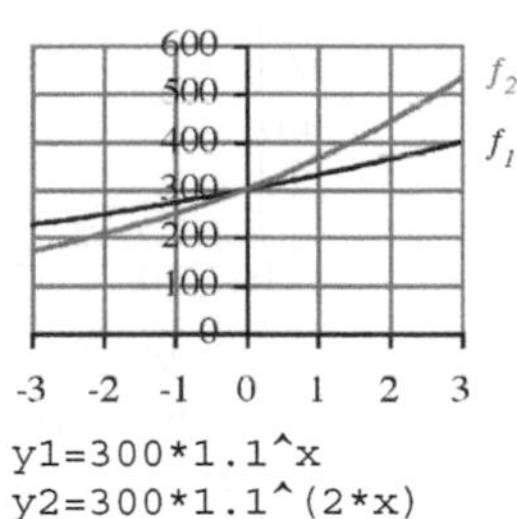

`y1=300*1.1^x`
`y2=300*1.1^(2*x)`

43.

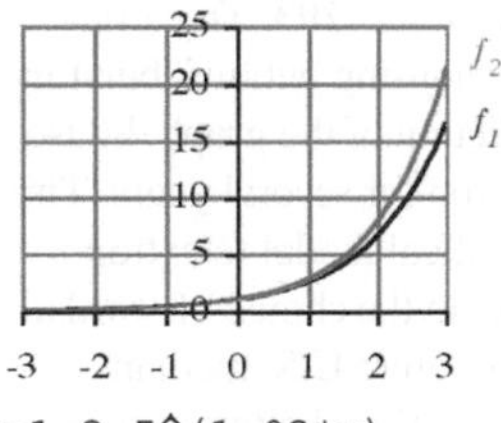

`y1=2.5^(1.02*x)`
`y2=e^(1.02*x)`
or `exp(1.02*x)`

45.

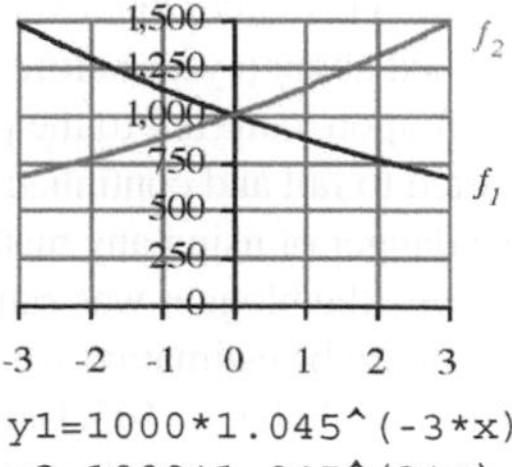

`y1=1000*1.045^(-3*x)`
`y2=1000*1.045^(3*x)`

47. $f(x) = 500(0.5)^x$ **49.** $f(x) = 10(3)^x$

51. $f(x) = 500(0.45)^x$ **53.** $f(x) = -100(1.1)^x$

55. $y = 4(3^x)$ **57.** $y = -1(0.2^x)$ **59.** $y = 2.1213(1.4142^x)$ **61.** $y = 3.6742(0.9036^x)$ **63.** $f(t) = 5{,}000e^{0.10t}$ **65.** $f(t) = 1{,}000e^{-0.063t}$ **67.** $y = 1.0442(1.7564)^x$ **69.** $y = 15.1735(1.4822)^x$ **71.** $f(t) = 300(0.5)^t$; 9.375 mg **73. a.** Linear model: $F = 65t - 60$. Exponential model: $F = 97.2(1.20)^t$. The exponential model is more appropriate. **b.** 418 tons, not too far off the projected figure **75. a.** $P = 180(1.01087)^t$ million **b.** 6 significant digits **c.** 344 million **77. a.** $y = 50{,}000(1.5^{t/2})$, t = time in years since 2 years ago **b.** 91,856 tags **79.** $y = 1{,}000(2^{t/3})$; 65,536,000 bacteria after 2 days **81.** $A(t) = 167(1.18)^t$; 1,695 cases **83.** $C(t) = 100e^{0.72t}$; 1,781 cases **85.** $A(t) = 5{,}000(1 + 0.0005/12)^{12t}$; \$5,018 **87.** 2033 **89.** \$491.82 **91.** $A(t) = 4.3e^{0.089t}$ million homes; 2012: 4.7 million homes; 2014: 5.6 million homes
93. a.

Year	1950	2000	2050	2100
$C(t)$ (ppm)	361	385	410	437

b. 2030
95. a. $P(t) = 0.339(1.169)^t$.

Graph:

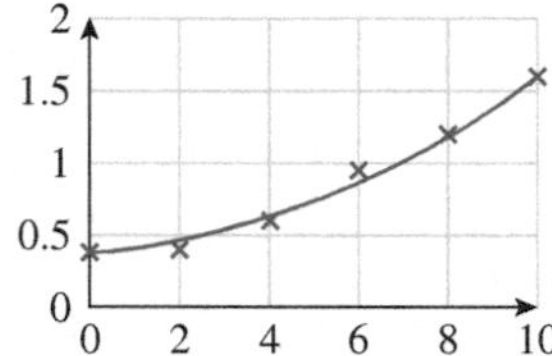

b. \$1.9 million
97. a. $n = 1.127(3.544)^t$

Graph:

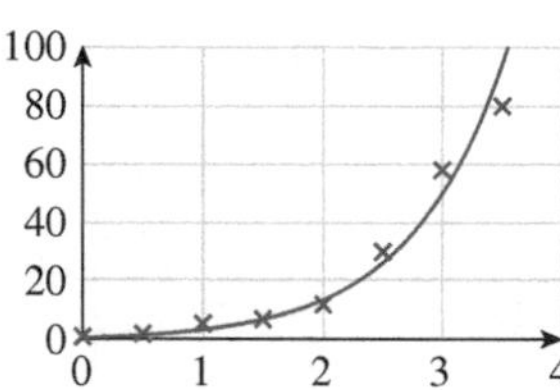

b. 3.544 **c.** 178 million **99.** (B) **101.** Exponential functions of the form $f(x) = Ab^x$ $(b > 1)$ increase rapidly for large values of x. In real-life situations, such as population growth, this model is reliable only for relatively short periods of growth. Eventually, population growth tapers off because of pressures such as limited resources and overcrowding. **103.** The article was published about a year before the "housing bubble" burst in 2006, whereupon, contrary to the prediction of the graph, house prices started to fall and continued to drop for several years. This shows the danger of using any mathematical model to extrapolate. However, the blogger was cautious in the choice of words, claiming only to be estimating what the future U.S. median house price "might be." **105.** Linear functions better: cost models where there is a fixed cost and a variable cost; simple interest, where interest is paid on the original amount invested. Exponential models better: compound interest, population growth. (In both of these, the rate of growth depends on the present number of items, rather than on some fixed quantity.)

107. Take the ratios y_2/y_1 and y_3/y_2. If they are the same, the points fit on an exponential curve. **109.** This reasoning is suspect. The bank need not use its computer resources to update all the accounts every minute but can instead use the continuous compounding formula to calculate the balance in any account at any time.

Section 2.3

1. $\log_4 y$ **3.** 8 **5.** x

7.

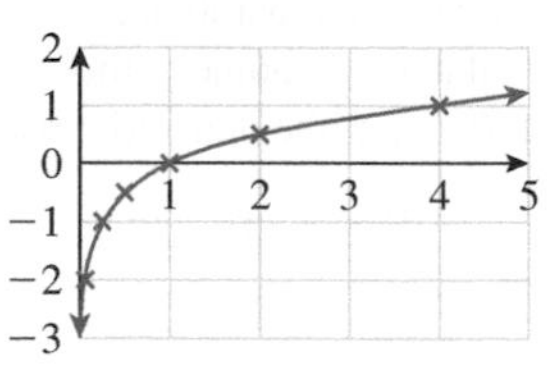

9.

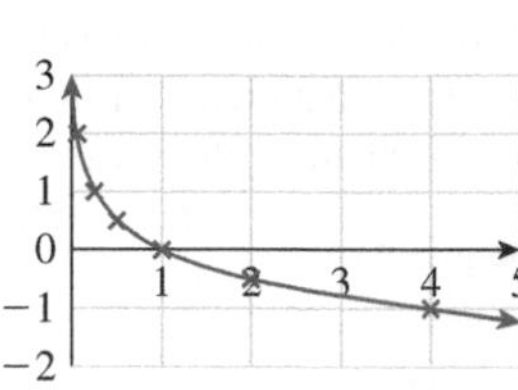

11.

13.

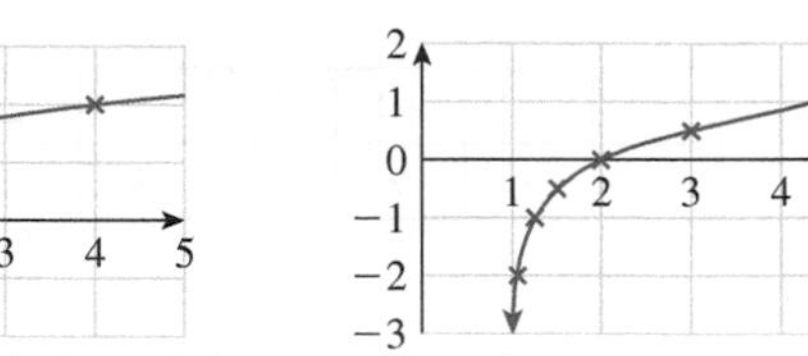

15. $Q = 1{,}000e^{-t \ln 2}$ **17.** $Q = 500e^{-t \ln 2/50}$ **19.** $Q = 1{,}000e^{t(\ln 2)/2}$ **21.** Doubling time $= 2 \ln 2$ **23.** Half-life $= \ln 2$ **25.** Half-life $= (\ln 2)/4$ **27.** $f(x) = 4(7.389)^x$ **29.** $f(t) = 2.1e^{0.0009995t}$ **31.** $f(t) = 10e^{-0.01309t}$. **33.** 3.36 years **35.** 11 years **37.** 23.1% **39.** 63,000 years old **41.** 16 years **43.** 785 months **45.** 20 years **47.** 27.73 years **49.** 1,600 years **51. a.** $b = 3^{1/6} \approx 1.20$ **b.** 3.8 months **53. a.** $Q(t) = Q_0e^{-0.139t}$ **b.** 3 years **55.** 2,360 million years **57.** 3.2 hours **59.** 3.89 days **61. a.** $P(t) = 6.591 \ln(t) - 17.69$ **b.** 1 digit **c.** (A) **63.** $S(t) = 28.56 \ln t + 164.46$

Graph:

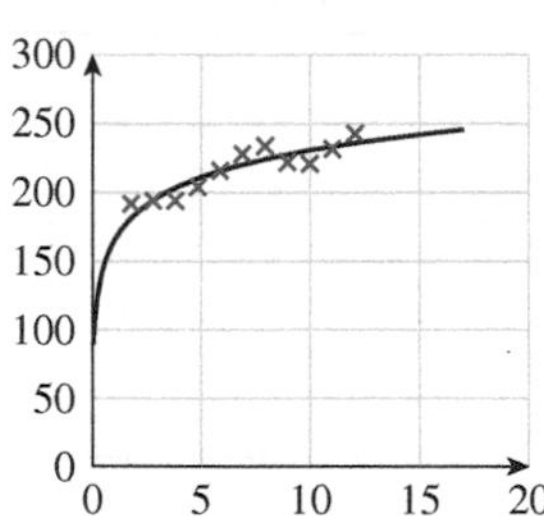

Positive direction; extrapolating in the negative direction eventually leads to negative values, which do not model reality **65. a.** About 4.467×10^{23} ergs **b.** About 2.24% **c.** $E = 10^{1.5R+11.8}$ **d.** Proof **e.** 1,000 **67. a.** 75 dB, 69 dB, 61 dB **b.** $D = 95 - 20 \log r$ **c.** 57,000 ft

Graph:

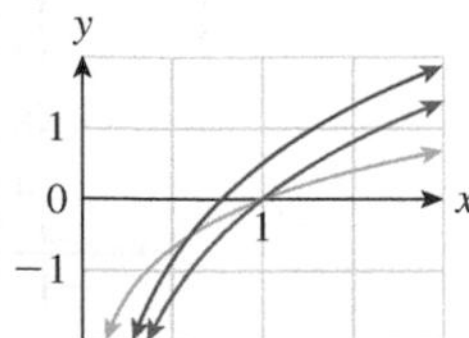

The green curve is $y = \ln x$, the blue curve is $y = 2\ln x$, and the red curve is $y = 2\ln x + 0.5$. Multiplying by A stretches the graph in the y-direction by a factor of A. Adding C moves the graph C units vertically up. **71.** The logarithm of a negative number, were it defined, would be the power to which a base must be raised to give that negative number. But raising a base to a power never results in a negative number, so there can be no such number as the logarithm of a negative number. **73.** Any logarithmic curve $y = \log_b t + C$ will eventually surpass 100% and hence will not be suitable as a long-term predictor of market share. **75.** Time is increasing logarithmically with population; solving $P = Ab^t$ for t gives $t = \log_b(P/A) = \log_b P - \log_b A$, which is of the form $t = \log_b P + C$. **77.** Proof

Section 2.4

1. $N = 7, A = 6, b = 2$;

`7/(1+6*2^-x)`

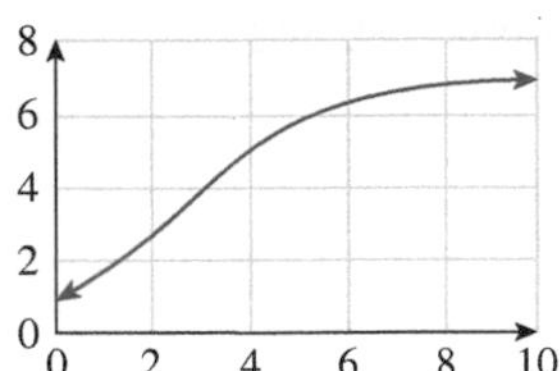

3. $N = 10, A = 4, b = 0.3$;

`10/(1+4*0.3^-x)`

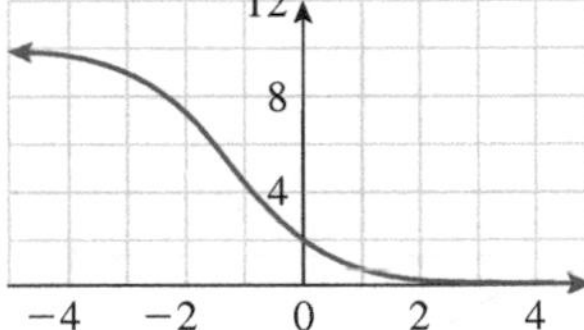

5. $N = 4, A = 7, b = 1.5$;

`4/(1+7*1.5^-x)`

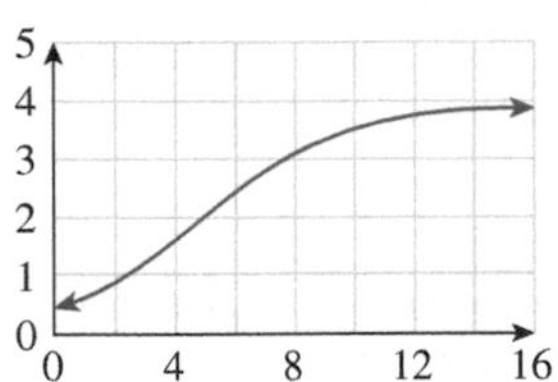

7. $f(x) = \dfrac{200}{1 + 19(2^{-x})}$ **9.** $f(x) = \dfrac{6}{1 + 2^{-x}}$

11. (B) **13.** (B) **15.** (C)

17. $y = \dfrac{7.2}{1 + 2.4(1.04)^{-x}}$

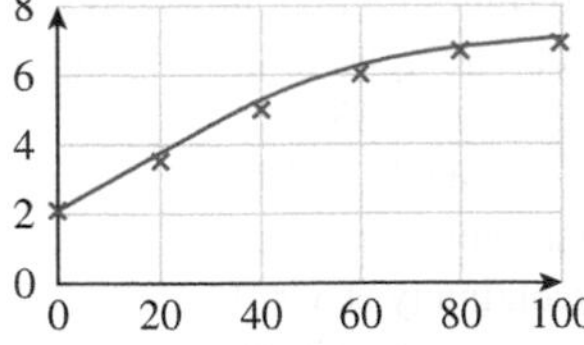

19. $y = \dfrac{97}{1 + 2.2(0.942)^{-x}}$

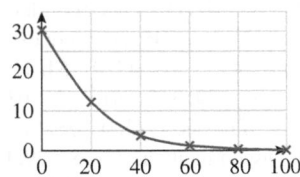

21. a. (A) **b.** 2003 **23. a.** (A) **b.** 4% per year **25. a.** 95.0% **b.** $P(x) \approx 25.13(1.064)^x$ **c.** $18,000 **27.** $N(t) = \dfrac{10{,}000}{1 + 9(1.25)^{-t}}$; $N(7) \approx 3{,}463$ cases **29.** $N(t) = \dfrac{3{,}000}{1 + 29(2^{1/5})^{-t}}$; $t = 16$ days **31. a.** $A(t) = \dfrac{9{,}790}{1 + 69.0(1.04)^{-t}}$; 9,790 thousand articles (Note: The coefficients of the model will vary depending on the initial guesses used.) **b.** 408 thousand articles **33. a.** $B(t) = \dfrac{1{,}090}{1 + 0.410(1.09)^{-t}}$; 1,090 teams **b.** $t \approx -10.3$. According to the model, the number of teams was rising fastest about 10.3 years *prior* to 1990, that is, sometime during 1979. **c.** The number of men's basketball teams was growing by about 9% per year in the past, well before 1979. **35.** $y = \dfrac{4{,}500}{1 + 1.1466(1.0357)^{-t}}$; 2013 **37.** Just as diseases are communicated via the spread of a pathogen (such as a virus), new technology is communicated via the spread of information (such as advertising and publicity). Further, just as the spread of a disease is ultimately limited by the number of susceptible individuals, so the spread of a new technology is ultimately limited by the size of the potential market. **39.** It can be used to predict where the sales of a new commodity might level off. **41.** The curve is still a logistic curve, but decreases when $b > 1$ and increases when $b < 1$. **43.** Proof

Chapter 2 Review

1.

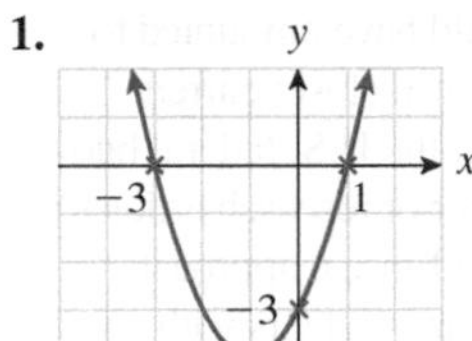

3. f: $f(x) = 5(1/2)^x$, or $5(2^{-x})$

5.

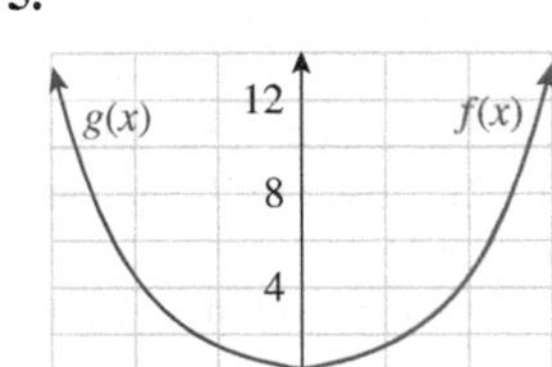

7.

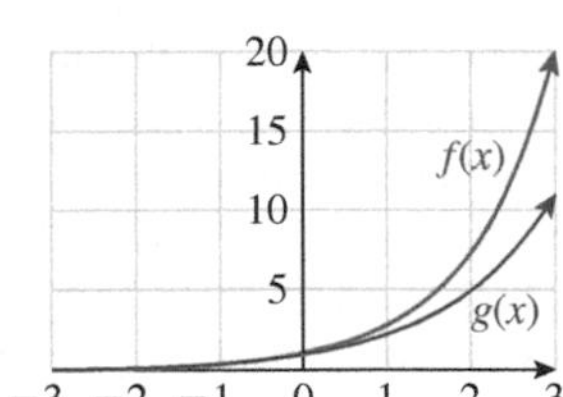

9. $3,484.85 **11.** $3,705.48 **13.** $3,485.50 **15.** $f(x) = 4.5(9^x)$ **17.** $f(x) = \dfrac{2}{3}3^x$

19.

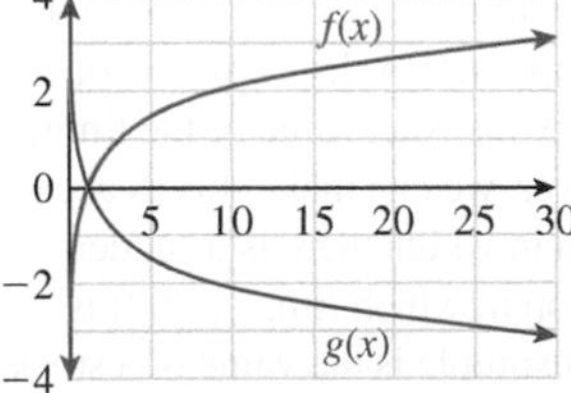

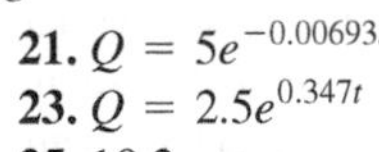

21. $Q = 5e^{-0.00693t}$ **23.** $Q = 2.5e^{0.347t}$ **25.** 10.2 years **27.** 10.8 years

29. $f(x) = \dfrac{900}{1 + 8(1.5)^{-x}}$ **31.** $f(x) = \dfrac{20}{1 + 3(0.8)^{-x}}$

33. a. $8,500 per month; an average of approximately 2,100 hits per day **b.** $29,049 per month **c.** The fact that −0.000005, the coefficient of c^2, is negative. **35. a.** $R = -60p^2 + 950p$; $p = \$7.92$ per novel, Monthly revenue = $3,760.42 **b.** $P = -60p^2 + 1{,}190p - 4{,}700$; $p = \$9.92$ per novel, Monthly profit = $1,200.42 **37. a.** 9.1, 19 **b.** About

310,000 pounds **39.** 2016 **41.** 1.12 million pounds **43.** $n(t) = 9.6(0.80^t)$ million pounds of lobster **45.** (C)

Chapter 3

Section 3.1

1. a. -6 **b.** $+\infty$ **c.** Does not exist **d.** -4 **3. a.** $-\infty$ **b.** $-\infty$ **c.** $-\infty$ **d.** Undefined **5.** 0 **7.** 4 **9.** Does not exist **11.** 0 **13.** 3 **15.** 6 **17.** Diverges to $+\infty$ **19.** 0 **21.** 1.5 **23.** 0.5 **25.** Diverges to $+\infty$ **27.** 0 **29.** 1 **31.** 0 **33.** 0 **35. a.** -2 **b.** -1 **37. a.** 2 **b.** 1 **c.** 0 **d.** $+\infty$ **39. a.** 0 **b.** 2 **c.** -1 **d.** Does not exist **e.** 2 **f.** $+\infty$ **41. a.** 1 **b.** 1 **c.** 2 **d.** Does not exist **e.** 1 **f.** 2 **43. a.** 1 **b.** Does not exist **c.** 1 **d.** 1 **45. a.** 1 **b.** $+\infty$ **c.** $+\infty$ **d.** $+\infty$ **e.** Undefined **f.** -1 **47. a.** -1 **b.** $+\infty$ **c.** $-\infty$ **d.** Does not exist **e.** 2 **f.** 1 **49.** 890 PhD graduates per year. In the long term, the model predicts that there will be 890 PhD graduates each year in the natural sciences in Mexico. **51. a.** 4.7. The model predicts that, had spending on NASA continued to follow the pattern leading up to 1966, annual spending on NASA in the long term would have amounted to 4.7% of the U.S. federal budget. **b.** Not even close; current spending (as of 2014) is less than 0.5% of the U.S. federal budget. **53.** 7.0. In the long term, the number of research articles in *Physical Review* written by researchers in Europe approaches 7,000 per year. **55.** 573. This suggests that students with an exceptionally large household income earn an average of 573 on the math SAT test. **57. a.** $\lim_{t\to 14.75^-} r(t) = 21$, $\lim_{t\to 14.75^+} r(t) = 21$, $\lim_{t\to 14.75} r(t) = 21$, $r(14.75) = 0.01$ **b.** Just before 2:45 pm, the stock was approaching \$21, but it then fell suddenly to a penny (\$0.01) at 2:45 exactly, after which time it jumped back to values close to \$21. **59.** 155; In the long term, the home price index will level off at 155 points. **61.** $\lim_{t\to 1^-} C(t) = 0.06$, $\lim_{t\to 1^+} C(t) = 0.08$, so $\lim_{t\to 1} C(t)$ does not exist. **63.** $\lim_{t\to +\infty} I(t) = +\infty$, $\lim_{t\to +\infty}(I(t)/E(t)) \approx 2.5$. In the long term, U.S. imports from China will rise without bound and be 2.5 times U.S. exports to China. In the real world, imports and exports cannot rise without bound. Thus, the given models should not be extrapolated far into the future. **65.** To approximate $\lim_{x\to a} f(x)$ numerically, choose values of x closer and closer to, and on either side of $x = a$, and evaluate $f(x)$ for each of them. The limit (if it exists) is then the number that these values of $f(x)$ approach. A disadvantage of this method is that it may never give the exact value of the limit, but only an approximation. (However, we can make this as accurate as we like.) **67.** Any situation in which there is a sudden change can be modeled by a function in which $\lim_{t\to a^+} f(t)$ is not the same as $\lim_{t\to a^-} f(t)$. One example is the value of a stock market index before and after a crash: $\lim_{t\to a^-} f(t)$ is the value immediately before the crash at time $t = a$, while $\lim_{t\to a^+} f(t)$ is the value immediately after the crash. Another example might be the price of a commodity that is suddenly increased from one level to another. **69.** It is possible for $\lim_{x\to a} f(x)$ to exist even though $f(a)$ is not defined. An example is $\lim_{x\to 1} \frac{x^2 - 3x + 2}{x - 1}$. **71.** An example is $f(x) = (x - 1)(x - 2)$. **73.** These limits are all 0.

Section 3.2

1. Continuous on its domain **3.** Continuous on its domain **5.** Discontinuous at $x = 0$ **7.** Discontinuous at $x = -1$ **9.** Continuous on its domain **11.** Continuous on its domain **13.** Discontinuous at $x = -1$ and 0 **15.** (A), (B), (D), (E) **17.** Continuous **19.** Discontinuous **21.** Singular **23.** Continuous **25.** 0 **27.** -1 **29.** No value possible **31.** -1 **33.** Continuous on its domain **35.** Continuous on its domain **37.** Discontinuity at $x = 0$ **39.** Discontinuity at $x = 0$ **41.** Continuous on its domain **43.** (B) **45.** (C) **47.** Not unless the domain of the function consists of all real numbers. (It is impossible for a function to be continuous at points not in its domain.) For example, $f(x) = 1/x$ is continuous on its domain—the set of nonzero real numbers—but not at $x = 0$. **49.** True. If the graph of a function has a break in its graph at any point a, then it cannot be continuous at the point a. **51.** Answers may vary. $f(x) = 1/[(x - 1)(x - 2)(x - 3)]$ is such a function; it is undefined at $x = 1, 2, 3$, so its graph consists of three distinct curves. **53.** Answers may vary.

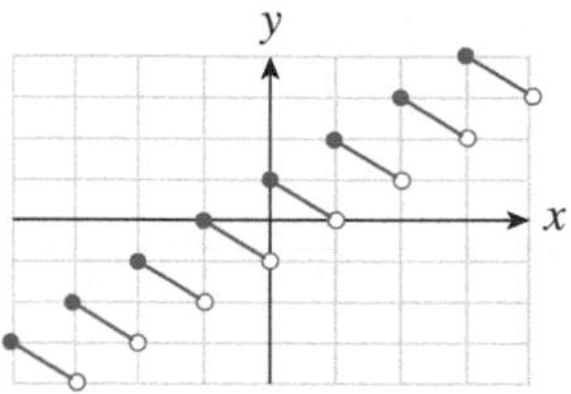

55. Answers may vary. The price of OHaganBooks.com stocks suddenly drops by \$10 as news spreads of a government investigation. Let $f(x)$ = Price of OHaganBooks.com stocks.

Section 3.3

1. $x = 1$ **3.** 2 **5.** Determinate; diverges to $+\infty$ **7.** Determinate; does not exist **9.** Determinate; diverges to $-\infty$ **11.** Determinate; 0 **13.** Indeterminate; $-1/3$ **15.** Indeterminate; 0 **17.** Determinate; 0 **19.** Determinate; -60 **21.** 1 **23.** 2 **25.** 0 **27.** 6 **29.** 4 **31.** 2 **33.** 0 **35.** 0 **37.** 12 **39.** $+\infty$ **41.** Does not exist; left and right (infinite) limits differ. **43.** $-\infty$ **45.** Does not exist: left and right (infinite) limits differ. **47.** $+\infty$ **49.** 0 **51.** 1/6 **53.** 3/2 **55.** 1/2 **57.** $+\infty$ **59.** 3/2 **61.** 1/2 **63.** $-\infty$ **65.** 0 **67.** 12 **69.** 0 **71.** $+\infty$ **73.** 0 **75.** Singular point at $x = 3$ **77.** Discontinuity at $x = 5$ **79.** Discontinuity at $x = 0$ **81.** Singularity at $x = 0$ **83.** Continuous everywhere **85.** Discontinuity at $x = 0$ **87.** Discontinuity at $x = 0$ **89. a.** $\lim_{t\to 15^-} v(t) = 3{,}800$, $\lim_{t\to 15^+} v(t) = 3{,}800$; Shortly before 2005, the speed of Intel processors was approaching 3,800 MHz. Shortly after 2005, the speed of Intel processors was close to 3,800 MHz. **b.** Continuous at $t = 15$; No abrupt change. **91. a.** 0.49, 1.16. Shortly before 1999, annual advertising expenditures were close to \$0.49 billion. Shortly after 1999, annual advertising expenditures were close to \$1.16 billion. **b.** Not continuous; movie advertising expenditures jumped suddenly in 1999. **93.** 1.59; if the trend continued indefinitely, the annual spending on police would be 1.59 times the annual spending on courts in the long run. **95.** 573. This suggests that stu-

dents with an exceptionally large household income earn an average of 573 on the math SAT test. **97.** $\lim_{t\to+\infty} W(t) = +\infty$, $\lim_{t\to+\infty} W(t)/L(t) = 8.25$. In the long term, the popularity of Twitter among social media sites will increase without bound and be 8.25 times the popularity of LinkedIn. However, a percentage cannot rise beyond 100, so extrapolating the models to obtain long-term predictions gives meaningless results. **99.** $\lim_{t\to+\infty} p(t) = 100$. The percentage of children who learn to speak approaches 100% as their age increases. **101.** To evaluate $\lim_{x\to a} f(x)$ algebraically, first check whether $f(x)$ is a closed-form function. Then check whether $x = a$ is in its domain. If so, the limit is just $f(a)$; that is, it is obtained by substituting $x = a$. If not, then try to first simplify $f(x)$ in such a way as to transform it into a new function such that $x = a$ is in its domain, and then substitute. A disadvantage of this method is that it is sometimes extremely difficult to evaluate limits algebraically, and rather sophisticated methods are often needed. **103.** Rita's side; closed-form functions are continuous at every point of their domain; a discontinuity would be a point in the domain at which the function was *not* continuous, so there cannot be any such points. Richard's example is of a function with a singular point at $x = 0$ and not a discontinuity. **105.** $x = 3$ is not in the domain of the given function f, so, yes, the *function* is undefined at $x = 3$. However, the *limit* may well be defined. In this case, it leads to the indeterminate form 0/0, telling us that we need to try to simplify, and that leads us to the correct limit of 27. **107.** She is wrong. Closed-form functions are continuous only at points in their domains, and $x = 2$ is not in the domain of the closed-form function $f(x) = 1/(x-2)^2$. (It is a singular point.) **109.** Answers may vary. (1) See Example 1(b): $\lim_{x\to2}\frac{x^3-8}{x-2}$, which leads to the indeterminate form 0/0, but the limit is 12. (2) $\lim_{x\to+\infty}\frac{60x}{2x}$, which leads to the indeterminate form ∞/∞ but where the limit exists and equals 30. **111.** $\pm\infty/\infty$; The limits are zero. This suggests that limits resulting in $\frac{p(-\infty)}{e^{\infty}}$ are zero.
113. The statement may not be true if f is not a closed-form function (for instance, a piecewise-defined function). The statement can be corrected by requiring that f be a closed-form function: "If f is a closed-form function, and $f(a)$ is defined, then $\lim_{x\to a} f(x)$ exists and equals $f(a)$."
115. Answers may vary, for example

$$f(x) = \begin{cases} 0 & \text{if } x \text{ is any number other than 1 or 2} \\ 1 & \text{if } x = 1 \text{ or } 2 \end{cases}$$

117. Answers may vary.

(1) $\lim_{x\to+\infty}[(x+5)-x] = \lim_{x\to+\infty} 5 = 5$

(2) $\lim_{x\to+\infty}[x^2-x] = \lim_{x\to+\infty} x(x-1) = +\infty$

(3) $\lim_{x\to+\infty}[(x-5)-x] = \lim_{x\to+\infty} -5 = -5$

Section 3.4

1. −3 **3.** 0.3 **5.** −\$25,000 per month **7.** −200 items per dollar **9.** \$1.33 per month **11.** 0.75 percentage point increase in unemployment per 1 percentage point increase in the deficit **13.** 4 **15.** 2 **17.** 7/3

19.

h	Avg. Rate of Change
1	2
0.1	0.2
0.01	0.02
0.001	0.002
0.0001	0.0002

21.

h	Avg. Rate of Change
1	−0.1667
0.1	−0.2381
0.01	−0.2488
0.001	−0.2499
0.0001	−0.24999

23.

h	Avg. Rate of Change
1	9
0.1	8.1
0.01	8.01
0.001	8.001
0.0001	8.0001

25. a. \$50 billion per year; World military expenditure increased at an average rate of about \$50 billion per year during 2006–2012. **b.** Approximately \$54.17 billion per year; World military expenditure increased at an average rate of about \$54.17 billion per year during 2000–2012. **27. a.** −10,000 barrels per year; During 2010–2013, daily oil production by Pemex was decreasing at an average rate of 10,000 barrels of oil per year. **b.** (A) **29. a.** 1.7; The percentage of mortgages classified as subprime was increasing at an average rate of around 1.7 percentage points per year between 2000 and 2006. **b.** 2004–2006 **31. a.** 2010–2012; During 2010–2012, immigration to Ireland was increasing at an average rate of 5,500 people per year. **b.** 2011–2013; During 2011–2013, immigration to Ireland was increasing at an average rate of 1,500 people per year. **33. a.** [25, 30]; 10 thousand articles per year. During the period 2005–2010 the number of articles authored by U.S. researchers increased at an average rate of 10,000 per year. **b.** Percentage rate 100%; Average rate ≈ 5.667 thousand articles per year. Over the period 1980–2010 the number of articles authored by U.S. researchers increased at an average rate of about 5,667 per year, representing a 100% increase over that period. **35. a.** 12 teams per year **b.** Decreased **37. a.** (A) **b.** (C) **c.** (B) **d.** Approximately −\$0.088 per year (if we round to two significant digits). This is less than the slope of the regression line, about −\$0.063 per year. **39.** Answers may vary.

Graph:

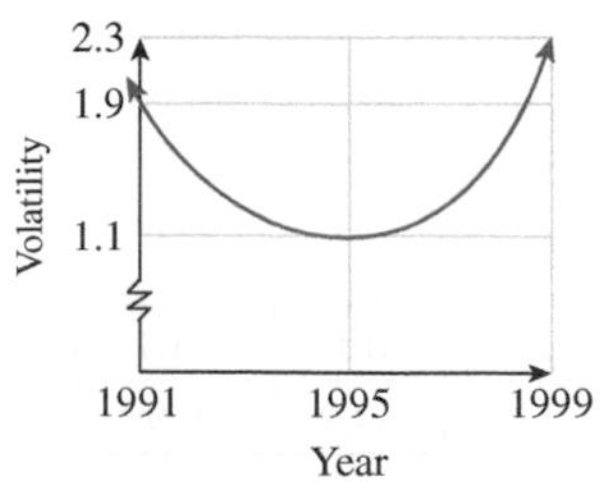

41. The index was increasing at an average rate of 300 points per day. **43. a.** \$0.15 per year **b.** No; according to the model, during that 25-year period the price of oil went down from around \$93 to a low of around \$25 in 1993 before climbing back up. **45. a.** 0.057, 0.131, 0.205, 0.279 **b.** 0.74; 0.74 AU per million years per million years. **47. a.** 47.3 new cases per day; the number of SARS cases was growing at an average rate of 47.3 new cases per day over the period March 17 to March 23. **b.** (A) **49.** Successive 2-month rates of change: 154.43, 317.27, 651.81, 1,339.10, and 2,751.10 cases per month **b.** exponential; The 2-month average rate of increase in the number of Ebola cases increased by a factor of about 2.05 each month.
51. a. 8.85 manatee deaths per 100,000 boats; 23.05 manatee deaths per 100,000 boats **b.** More boats result in more manatee deaths per additional boat. **53. a.** The average rates of change are shown in the following table:

Interval	[0, 40]	[40, 80]	[80, 120]	[120, 160]	[160, 200]
Avg. Rate of Change of S	1.35	0.80	0.48	0.28	0.17

b. For household incomes between \$40,000 and \$80,000 a student's math SAT increases at an average rate of 0.80 points per \$1,000 of additional income. **c.** (A) **d.** (B) **55.** The average rate of change of f over an interval $[a, b]$ can be determined numerically, using a table of values; graphically, by measuring the slope of the corresponding line segment through two points on the graph; or algebraically, using an algebraic formula for the function. Of these, the least precise is the graphical method, because it relies on reading coordinates of points on a graph.
57. No, the formula for the average rate of a function f over $[a, b]$ depends only on $f(a)$ and $f(b)$, and not on any values of f between a and b. **59.** Answers will vary.

Graph:

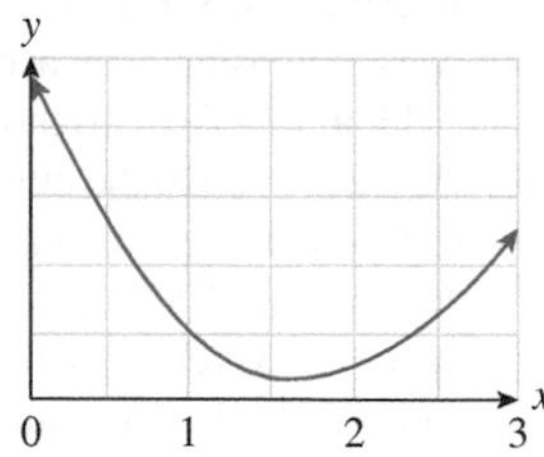

61. 6 units of quantity A per unit of quantity C **63.** (A)
65. Yes. Here is an example:

Year	2000	2001	2002	2003
Revenue (\$ billion)	10	20	30	5

67. (A)

Section 3.5

1. 6 **3.** −5.5

5.

h	1	0.1	0.01
Avg. Rate	39	39.9	39.99

Instantaneous rate = \$40 per day

7.

h	1	0.1	0.01
Avg. Rate	140	66.2	60.602

Instantaneous rate = \$60 per day

9.

h	10	1
C_{avg}	4.799	4.7999

$C'(1{,}000) = \$4.80$ per item

11.

h	10	1
C_{avg}	99.91	99.90

$C'(100) = \$99.90$ per item

13. 1/2 **15.** 0 **17. a.** R **b.** P **19. a.** P **b.** R **21. a.** Q **b.** P **23. a.** Q **b.** R **c.** P **25. a.** R **b.** Q **c.** P **27. a.** (1, 0) **b.** None **c.** (−2, 1) **29. a.** (−2, 0.3), (0, 0), (2, −0.3) **b.** None **c.** None **31.** $(a, f(a))$; $f'(a)$ **33.** (B) **35. a.** (A) **b.** (C) **c.** (B) **d.** (B) **e.** (C) **37.** −2 **39.** −1.5 **41.** −5 **43.** 16 **45.** 0 **47.** −0.0025

49. a. 3 **b.** $y = 3x + 2$

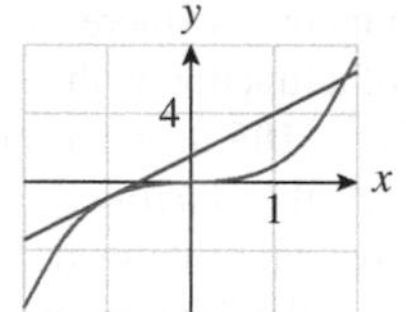

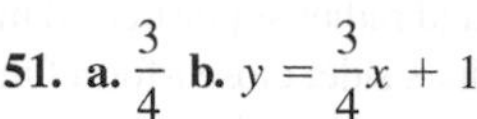

51. a. $\frac{3}{4}$ **b.** $y = \frac{3}{4}x + 1$

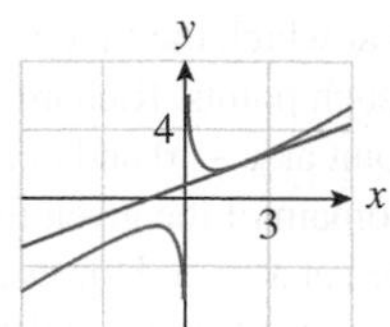

53. a. $\frac{1}{4}$ **b.** $y = \frac{1}{4}x + 1$

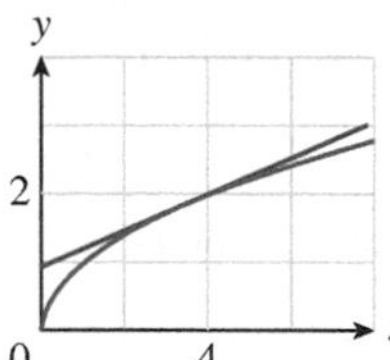

55. 1.000 **57.** 1.000
59. (C) **61.** (A) **63.** (F)
65. Increasing for $x < 0$; decreasing for $x > 0$.
67. Increasing for $x < -1$ and $x > 1$; decreasing for $-1 < x < 1$.

69. Increasing for $x > 1$; decreasing for $x < 1$.
71. Increasing for $x < 0$; decreasing for $x > 0$.
73. $x = -1.5$, $x = 0$

Graph:

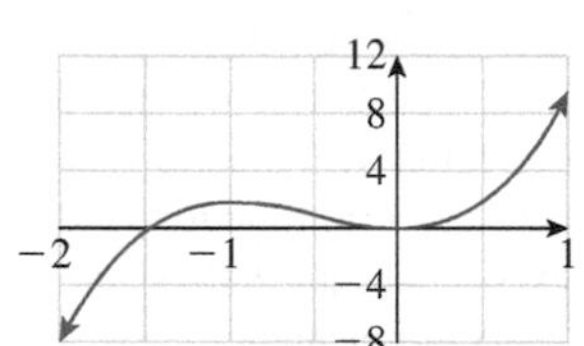

75. Note: Answers depend on the form of technology used. Excel ($h = 0.1$):

	A	B	C	D	E	F
1	x	f(x)	f'(x)	xmin	4	
2	4	6	-4.545454545	h	0.1	
3	4.1	5.545454545	-3.787878788			
4	4.2	5.166666667	-3.205128205			
5	4.3	4.846153846	-2.747252747			
6	4.4	4.571428571	-2.380952381			
7	4.5	4.333333333	-2.083333333			
8	4.6	4.125	-1.838235294			
9	4.7	3.941176471	-1.633986928			
10	4.8	3.777777778	-1.461988304			
11	4.9	3.631578947	-1.315789474			
12	5	3.5				
13						
14						

Graphs:

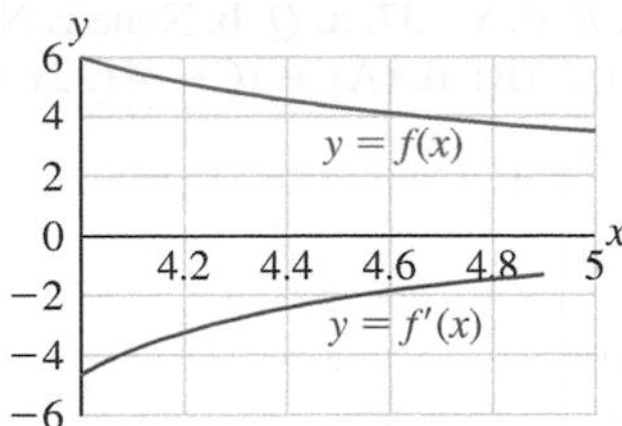

77. −35°F; 40°F per hour **79.** $q(100) = 50{,}000$, $q'(100) = -500$. A total of 50,000 pairs of sneakers can be sold at a price of $100, but the demand is decreasing at a rate of 500 pairs per $1 increase in the price. **81. a.** −60; Daily imports from Mexico in 2011 were 1.08 million barrels and declining at a rate of 0.06 million barrels (or 60,000 barrels) per year. **b.** Decreasing; the slope is decreasing. **83. a.** (B) **b.** (B) **c.** (A) **d.** 2004 **e.** 0.033; in 2004 the U.S. prison population was increasing at a rate of 0.033 million prisoners (33,000 prisoners) per year. **85. a.** −96 ft/sec **b.** −128 ft/sec **87. a.** $0.60 per year; the price per barrel of crude oil in constant 2008 dollars was growing at an average rate of about 60¢ per year over the 28-year period beginning at the start of 1980. **b.** −$12 per year; the price per barrel of crude oil in constant 2008 dollars was dropping at an instantaneous rate of about $12 per year at the start of 1980. **c.** The price of oil was decreasing in January 1980 but eventually began to increase (making the average rate of change in part (a) positive). **89. a.** 144.7 new cases per day; the number of SARS cases was growing at a rate of about 144.7 new cases per day on March 27. **b.** (A)

91. $C(5) \approx 3{,}510$, $\left.\frac{dC}{dt}\right|_{t=5} \approx 2{,}527$. Five months after the outbreak, the number of cases was around 3,510 and increasing at a rate of about 2,527 per month. **93.** $A(0) = 4.5$ million; $A'(0) = 60{,}000$ subscribers per week **95. a.** 60% of children can speak at the age of 10 months. At the age of 10 months this percentage is increasing by 18.2 percentage points per month. **b.** As t increases, p approaches 100 percentage points (all children eventually learn to speak), and dp/dt approaches zero because the percentage stops increasing. **97. a.** $A(6) \approx 12$; $A'(6) \approx 1.4$; At the start of 2006, about 12% of U.S. mortgages were subprime, and this percentage was increasing at a rate of about 1.4 percentage points per year. **b.** Graphs:

Graph of A:

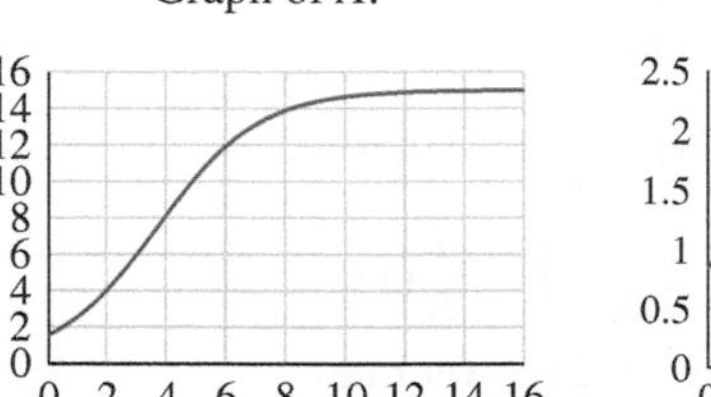

Graph of A':

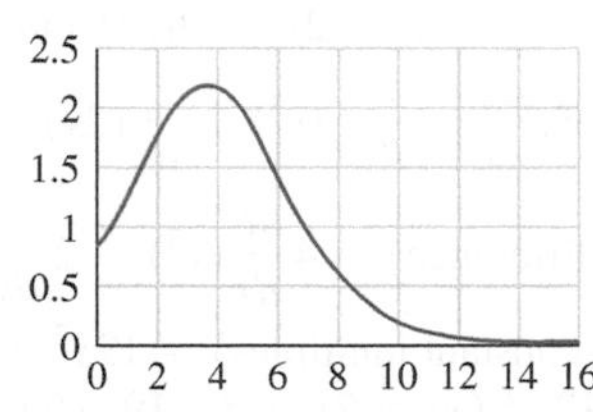

From the graphs, $A(t)$ approaches 15 as t becomes large (in terms of limits, $\lim_{x\to+\infty} A(t) = 15$) and $A'(t)$ approaches 0 as t becomes large (in terms of limits, $\lim_{x\to+\infty} A'(t) = 0$). Interpretation: If the trend modeled by the function A had continued indefinitely, in the long term 15% of U.S. mortgages would have been subprime, and this percentage would not be changing. **99. a.** (D) **b.** 33 days after the egg was laid **c.** 50 days after the egg was laid.

Graph:

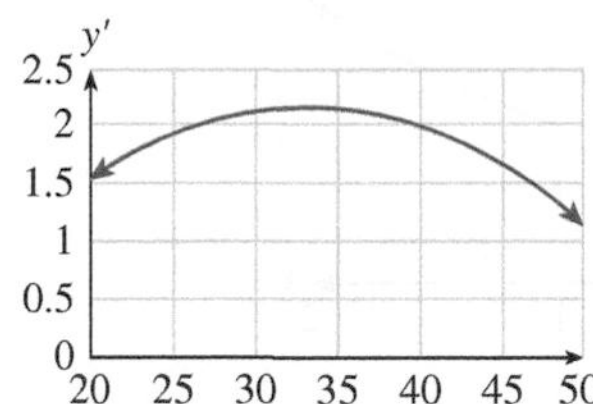

101. $L(.95) \approx 31.2$ meters and $L'(.95) \approx -304.2$ meters per warp. Thus, at a speed of warp 0.95, the spaceship has an observed length of 31.2 meters, and its length is decreasing at a rate of 304.2 meters per unit warp, or 3.042 meters per increase in speed of 0.01 warp. **103.** None **105.** The difference quotient is not defined when $h = 0$ because there is no such number as 0/0. **107.** (D) **109.** The derivative is positive and decreasing toward zero. **111.** Company B. Although the company is currently losing money, the derivative is positive, showing that the profit is increasing. Company A, on the other hand, has profits that are declining. **113.** (C) is the only graph in which the instantaneous rate of change on January 1 is greater than the 1-month average rate of change. **115.** The tangent to the graph is horizontal at that point, so the graph is almost horizontal near that point. **117.** Answers may vary.

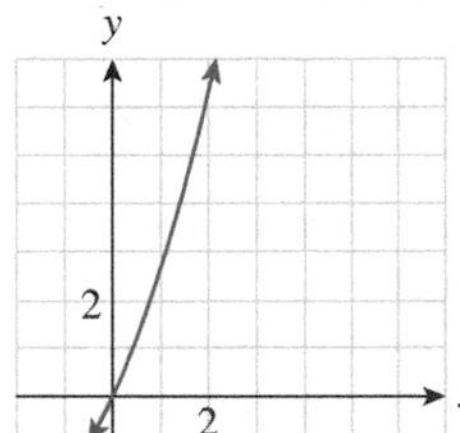

119. If $f(x) = mx + b$, then its average rate of change over any interval $[x, x + h]$ is $\frac{m(x + h) + b - (mx + b)}{h} = m$. Because this does not depend on h, the instantaneous rate is also equal to m. **121.** Increasing because the average rate of change appears to be rising as we get closer to 5 from the left. (See the bottom row.)

123. Answers may vary.

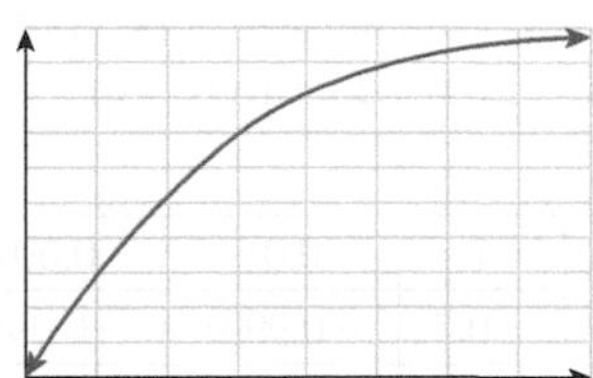

125. Answers may vary.

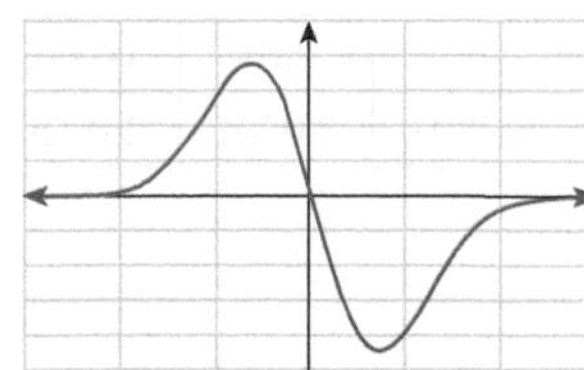

127. (B) **129.** Answers will vary.

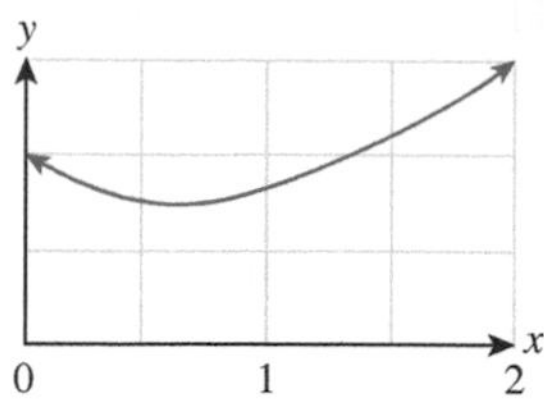

Section 3.6

1. 4 **3.** 3 **5.** 7 **7.** 4 **9.** 14 **11.** 1 **13.** m **15.** $2x$ **17.** 3 **19.** $6x + 1$ **21.** $2 - 2x$ **23.** $3x^2 + 2$ **25.** $1/x^2$ **27.** m **29.** -1.2 **31.** 30.6 **33.** -7.1 **35.** 4.25 **37.** -0.6 **39.** $y = 4x - 7$ **41.** $y = -2x - 4$ **43.** $y = -3x - 1$ **45.** $s'(t) = -32t$; $s'(4) = -128$ ft/sec **47.** $dI/dt = -78t + 800$; Daily oil imports were decreasing at a rate of 136,000 barrels per year. **49.** $R'(t) = 0.16t - 0.26$; Increasing at a rate of 540 million gallons per year **51.** $f'(8) = 26.6$ manatee deaths per 100,000 boats. At a level of 800,000 boats, the number of manatee deaths is increasing at a rate of 26.6 manatees per 100,000 additional boats. **53.** Yes; $\lim_{t\to 20^-} C(t) = \lim_{t\to 20^+} C(t) = 700 = C(20)$. **b.** No; $\lim_{t\to 20^-} C'(t) = 31.1$ while $\lim_{t\to 20^+} C'(t) = 90$. Until 1990 the cost of a Super Bowl ad was increasing at a rate of $31,100 per year. Immediately thereafter, it was increasing at a rate of $90,000 per year. **55.** The algebraic method because it gives the exact value of the derivative. The other two approaches give only approximate values (except in some special cases). **57.** The error is in the second line: $f(x + h)$ is *not* equal to $f(x) + h$. For instance, if $f(x) = x^2$, then $f(x + h) = (x + h)^2$, whereas $f(x) + h = x^2 + h$. **59.** The error is in the second line: One could cancel the h only if it were a *factor* of both the numerator and denominator; it is not a factor of the numerator. **61.** Because the algebraic computation of $f'(a)$ is exact and not an approximation, it makes no difference whether one uses the balanced difference quotient or the ordinary difference quotient in the algebraic computation. **63.** The computation results in a limit that cannot be evaluated.

Chapter 3 Review

1. 5 **3.** Does not exist **5. a.** -1 **b.** 3 **c.** Does not exist **7.** $-4/5$ **9.** -1 **11.** -1 **13.** Does not exist **15.** 10/7 **17.** Does not exist **19.** $+\infty$ **21.** 0 **23.** Diverges to $-\infty$ **25.** 0 **27.** 2/5 **29.** 1

31.

h	1	0.01	0.001
Avg. Rate of Change	-0.5	-0.9901	-0.9990

Slope ≈ -1

33.

h	1	0.01	0.001
Avg. Rate of Change	6.3891	2.0201	2.0020

Slope ≈ 2

35. a. P **b.** Q **c.** R **d.** S **37. a.** Q **b.** None **c.** None **d.** None **39. a.** (B) **b.** (B) **c.** (B) **d.** (A) **e.** (C) **41.** $2x + 1$ **43.** $2/x^2$

45.

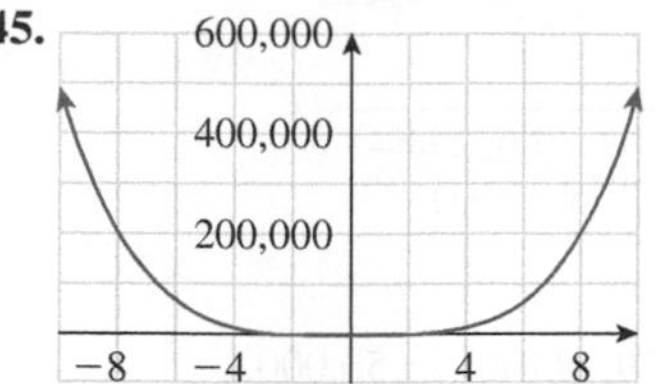

47.

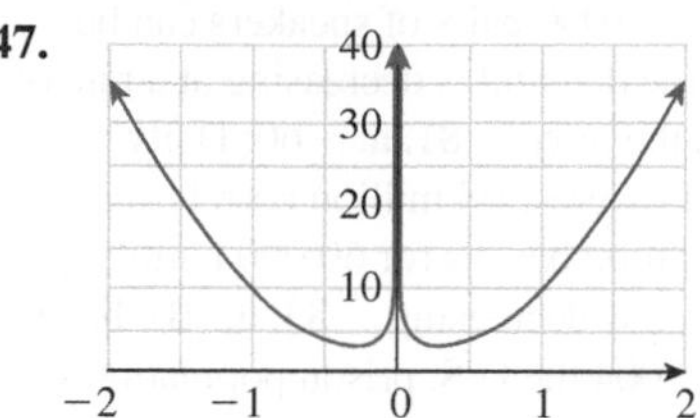

49. a. $P(3) = 25$: O'Hagan purchased the stock at $25. $\lim_{t\to 3^-} P(t) = 25$: The value of the stock had been approaching $25 up to the time he bought it. $\lim_{t\to 3^+} P(t) = 10$: The value of the stock dropped to $10 immediately after he bought it. **b.** Continuous but not differentiable. Interpretation: the stock price changed continuously but suddenly reversed direction (and started to go up) the instant O'Hagan sold it. **51. a.** $\lim_{t\to 3} p(t) \approx 40$; $\lim_{t\to +\infty} p(t) = +\infty$. Close to 2007 ($t = 3$), the home price index was about 40. In the long term, the home price index will rise without bound. **b.** 10 (The slope of the linear portion of the curve is 10.) In the long term, the home price index will rise about 10 points per year. **53. a.** 500 books per week **b.** [3, 4], [4, 5] **c.** [3, 5]; 650 books per week **55. a.** 3 percentage points per year **b.** 0 percentage points per year **c.** (D) **57. a.** $72t + 250$ **b.** 322 books per week **c.** 754 books per week.

Chapter 4

Section 4.1

1. $5x^4$ **3.** $-4x^{-3}$ **5.** $-0.25x^{-0.75}$ **7.** $8x^3 + 9x^2$ **9.** $-1 - 1/x^2$ **11.** $\frac{dy}{dx} = 10(0) = 0$ (constant multiple and power rule) **13.** $\frac{dy}{dx} = \frac{d}{dx}(x^2) + \frac{d}{dx}(x)$ (sum rule) $= 2x + 1$ (power rule) **15.** $\frac{dy}{dx} = \frac{d}{dx}(4x^3) + \frac{d}{dx}(2x) - \frac{d}{dx}(1)$ (sum and difference) $= 4\frac{d}{dx}(x^3) + 2\frac{d}{dx}(x) - \frac{d}{dx}(1)$ (constant multiples) $= 12x^2 + 2$ (power rule) **17.** $f'(x) = 2x - 3$ **19.** $f'(x) = 1 + 0.5x^{-0.5}$ **21.** $g'(x) = -2x^{-3} + 3x^{-2}$ **23.** $g'(x) = -\frac{1}{x^2} + \frac{2}{x^3}$ **25.** $h'(x) = -\frac{0.8}{x^{1.4}}$ **27.** $h'(x) = -\frac{2}{x^3} - \frac{6}{x^4}$

29. $r'(x) = -\frac{2}{3x^2} + \frac{0.1}{2x^{1.1}}$ **31.** $r'(x) = \frac{2}{3} - \frac{0.1}{2x^{0.9}} - \frac{4.4}{3x^{2.1}}$
33. $t'(x) = |x|/x - 1/x^2$ **35.** $s'(x) = \frac{1}{2\sqrt{x}} - \frac{1}{2x\sqrt{x}}$
37. $s'(x) = 3x^2\ (x \neq 0)$ **39.** $t'(x) = 1 - 4x\ (x \neq 0)$
41. $2.6x^{0.3} + 1.2x^{-2.2}$ **43.** $1.2(1 - |x|/x)$
45. $3at^2 - 4a$ **47.** $5.15x^{9.3} - 99x^{-2}$
49. $-\frac{2.31}{t^{2.1}} - \frac{0.3}{t^{0.4}}$ **51.** $4\pi r^2$ **53.** 3 **55.** −2 **57.** −5

59. $y = 3x + 2$

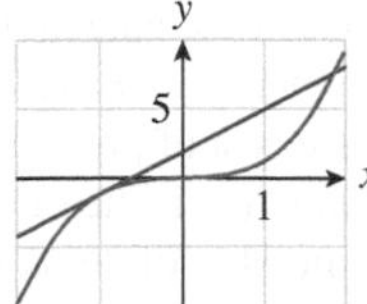

61. $y = \frac{3}{4}x + 1$

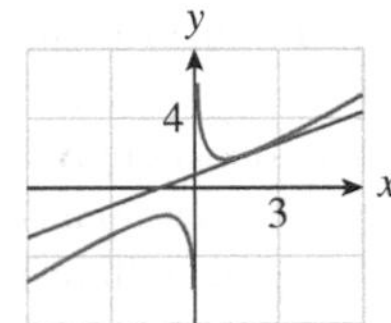

63. $y = \frac{1}{4}x + 1$

y
2
0 4 x

65. $x = -3/4$
67. No such values
69. $x = 1, -1$
71. See Solutions Manual.
73. a. 2/3 **b.** Not differentiable at 0
75. a. 5/2 **b.** Not differentiable at 0

77. Yes; 0 **79.** Yes; 12 **81.** No; 3 **83.** Yes; 3/2
85. Yes; diverges to $-\infty$ **87.** Yes; diverges to $-\infty$
89. $P'(t) = 0.54t - 8.6$; $P'(30) = 7.6$; the price of a barrel of crude oil was increasing at a rate of \$7.60 per year in 2010.
91. 4.8 **93. a.** $s'(t) = -32t$; 0, −32, −64, −96, −128 ft/sec **b.** 5 seconds; downward at 160 ft/sec **95. a.** 20 sec; at the highest point, h stops increasing and begins to decrease, so its rate of change, h', changes from positive to negative, meaning it must be zero. **b.** 760 m **97. a.** $P'(t) = -0.78t + 5.2$; increasing at a rate of \$0.52 billion per year **b.** (B)
99. a. $f'(x) = 7.1x - 30.2$; $f'(8) = 26.6$ manatees per 100,000 boats; at a level of 800,000 boats, manatee deaths are increasing at a rate of 26.6 deaths each year per 100,000 additional boats. **b.** Increasing; the number of manatees killed per additional 100,000 boats increases as the number of boats increases. **101. a.** $A - I$ measures the amount by which the Android market share exceeds the iOS market share. $(A - I)'$ measures the rate at which this difference is changing. **b.** (A) **c.** $9t^2 - 58t + 102.3$ percentage points per year; (A); The vertical distance between the graphs increases; Android increases its advantage over iOS through this range of dates. **d.** 9.3 percentage points per year; in the second quarter of 2013 ($t = 3$), Android's advantage over iOS was increasing at a rate of 9.3 percentage points per year. **103.** After graphing the curve $y = 3x^2$, draw the line passing through $(-1, 3)$ with slope −6. **105.** The slope of the tangent line of g is twice the slope of the tangent line of f. **107.** $g'(x) = -f'(x)$ **109.** The left-hand side is not equal to the right-hand side. The *derivative* of the left-hand side is equal to the right-hand side, so your friend should have written $\frac{d}{dx}(3x^4 + 11x^5) = 12x^3 + 55x^4$.
111. $\frac{1}{2x}$ is not equal to $2x^{-1}$. Your friend should have written $y = \frac{1}{2x} = \frac{1}{2}x^{-1}$, so $\frac{dy}{dx} = -\frac{1}{2}x^{-2}$. **113.** The derivative of a constant times a function is the constant times the derivative of the function, so that $f'(x) = (2)(2x) = 4x$. Your enemy mistakenly computed the *derivative* of the constant times the derivative of the function. (The derivative of a product of two functions is not the product of the derivative of the two functions. The rule for taking the derivative of a product is discussed later in the chapter.).
115. For a general function f, the derivative of f is defined to be $f'(x) = \lim_{h\to 0} \frac{f(x+h) - f(x)}{h}$. One then finds by calculation that the derivative of the specific function x^n is nx^{n-1}. In short, nx^{n-1} is the derivative of a specific function: $f(x) = x^n$, it is not the *definition* of the derivative of a general function or even the definition of the derivative of the function $f(x) = x^n$.
117. Answers may vary.

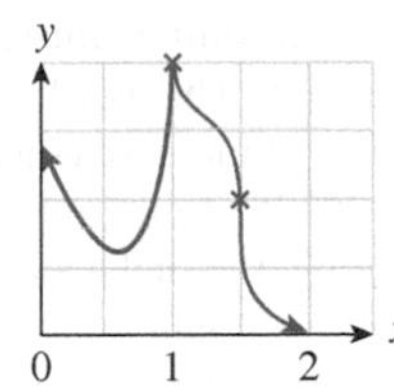

Section 4.2

1. $C'(1{,}000) = \$4.80$ per item **3.** $C'(100) = \$99.90$ per item
5. $C'(x) = 4$; $R'(x) = 8 - 0.002x$; $P'(x) = 4 - 0.002x$; $P'(x) = 0$ when $x = 2{,}000$. Thus, at a production level of 2,000, the profit is stationary (neither increasing nor decreasing) with respect to the production level. This may indicate a maximum profit at a production level of 2,000. **7. a.** (B) **b.** (C) **c.** (C) **9. a.** $C'(x) = 4{,}000 + 0.1x$; the cost is increasing at a rate of \$4,000,400 per television commercial. The exact cost of airing the fifth television commercial is \$4,000,450.
b. $\bar{C}(x) = 20/x + 4{,}000 + 0.05x$; $\bar{C}(4) = 4{,}005.2$ thousand dollars. The average cost of airing the first four television commercials is \$4,005,200. **11. a.** $C'(x) = 160 + 0.002x$; the cost is increasing at a rate of \$180 per iPhone. Actual cost of the 10,001st iPhone is \$180.001. **b.** $\bar{C}(x) = 400{,}000/x + 160 + 0.001x$; the average cost to produce the first 10,000 iPhones is \$210 per iPhone. **c.** The average cost is falling at a production level of 10,000 iPhones. **13. a.** $R'(x) = 0.90$, $P'(x) = 0.80 - 0.002x$ **b.** Revenue: \$450, profit: \$80, marginal revenue: \$0.90, marginal profit: −\$0.20. The total revenue from the sale of 500 copies is \$450. The profit from the production and sale of 500 copies is \$80. Approximate revenue from the sale of the 501st copy is 90¢. Approximate loss from the sale of the 501st copy is 20¢. **c.** $x = 400$. The profit is a maximum when you produce and sell 400 copies. **15.** The profit on the sale of 1,000 Blu-ray discs is \$3,000, and is decreasing at a rate of \$3 per additional Blu-ray disc sold.

17. Profit ≈ \$257.07; marginal profit ≈ \$5.07 per magazine. Your current profit is \$257.07 per month, and this would increase at a rate of \$5.07 per additional magazine sold. **19. a.** \$2.50 per pound **b.** $R(q) = 20{,}000/q^{0.5}$ **c.** $R(400) = \$1{,}000$. This is the monthly revenue that will result from setting the price at \$2.50 per pound. $R'(400) = -\$1.25$ per pound of tuna. Thus, at a demand level of 400 pounds per month, the revenue is decreasing at a rate of \$1.25 per pound. **d.** The fishery should raise the price (to reduce the demand). **21.** $P'(50) = 35$ cars per worker. This means that, at an employment level of 50 workers, the firm's daily production will increase at a rate of 35 cars washed per additional worker it hires. **23. a.** (B) **b.** (B) **c.** (C) **25. a.** $C(x) = 500{,}000 + 1{,}900{,}000x - 100{,}000\sqrt{x}$; $C'(x) = 1{,}900{,}000 - 50{,}000/\sqrt{x}$; $\overline{C}(x) = 500{,}000/x + 1{,}900{,}000 - 100{,}000/\sqrt{x}$ **b.** $C'(3) \approx \$1{,}870{,}000$ per spot; $\overline{C}(3) \approx \$2{,}010{,}000$ per spot. The average cost will decrease as x increases. **27. a.** \$2,000 per 1-pound reduction in emissions. **b.** 2.5 pounds per day reduction. **c.** $N(q) = 100q^2 - 500q + 4{,}000$; 2.5 pounds per day reduction. The value of q is the same as that for part (b). The net cost to the firm is minimized at the reduction level for which the cost of controlling emissions begins to increase faster than the subsidy. This is why we get the answer by setting these two rates of increase equal to each other. **29.** $M'(10) \approx 0.0002557$ mpg/mph. This means that, at a speed of 10 mph, the fuel economy is increasing at a rate of 0.0002557 miles per gallon per 1 mile per hour increase in speed. $M'(60) = 0$ mpg/mph. This means that, at a speed of 60 mph, the fuel economy is neither increasing nor decreasing with increasing speed. $M'(70) \approx -0.00001799$. This means that, at 70 mph, the fuel economy is decreasing at a rate of 0.00001799 miles per gallon per 1 mile per hour increase in speed. Thus, 60 mph is the most fuel-efficient speed for the car. **31.** (C) **33.** (D) **35.** (B) **37.** Cost is often measured as a function of the number of items x. Thus, $C(x)$ is the cost of producing (or purchasing, as the case may be) x items. **a.** The average cost function $\overline{C}(x)$ is given by $\overline{C}(x) = C(x)/x$. The marginal cost function is the derivative, $C'(x)$, of the cost function. **b.** The average cost $\overline{C}(r)$ is the slope of the line through the origin and the point on the graph where $x = r$. The marginal cost of the rth unit is the slope of the tangent to the graph of the cost function at the point where $x = r$. **c.** The average cost function $\overline{C}(x)$ gives the average cost of producing the first x items. The marginal cost function $C'(x)$ is the rate at which cost is changing with respect to the number of items x, or the incremental cost per item, and approximates the cost of producing the $(x + 1)$st item. **39.** Answers may vary. An example is $C(x) = 300x$. **41.** The marginal cost **43.** Not necessarily. For example, it may be the case that the marginal cost of the 101st item is larger than the average cost of the first 100 items (even though the marginal cost is decreasing). Thus, adding this additional item will *raise* the average cost. **45.** The circumstances described suggest that the average cost function is at a relatively low point at the current production level, so it would be appropriate to advise the company to maintain current production levels; raising or lowering the production level will result in increasing average costs.

Section 4.3

1. 3 **3.** $3x^2$ **5.** $2x + 3$ **7.** $210x^{1.1}$ **9.** $-2/x^2$ **11.** $2x/3$ **13.** $36x^2 - 3$ **15.** $3x^2 - 5x^4$ **17.** $8x + 12$ **19.** $-8/(5x - 2)^2$ **21.** $-14/(3x - 1)^2$ **23.** 0 **25.** $-|x|/x^3$ **27.** $3\sqrt{x}/2$ **29.** $(x^2 - 1) + 2x(x + 1) = (x + 1)(3x - 1)$ **31.** $(x^{-0.5} + 4)(x - x^{-1}) + (2x^{0.5} + 4x - 5)(1 + x^{-2})$ **33.** $8(2x^2 - 4x + 1)(x - 1)$ **35.** $(1/3.2 - 3.2/x^2)(x^2 + 1) + 2x(x/3.2 + 3.2/x)$ **37.** $2x(2x + 3)(7x + 2) + 2x^2(7x + 2) + 7x^2(2x + 3)$ **39.** $5.3(1 - x^{2.1})(x^{-2.3} - 3.4) - 2.1x^{1.1}(5.3x - 1)(x^{-2.3} - 3.4) - 2.3x^{-3.3}(5.3x - 1)(1 - x^{2.1})$

41. $\frac{1}{2\sqrt{x}}\left(\sqrt{x} + \frac{1}{x^2}\right) + (\sqrt{x} + 1)\left(\frac{1}{2\sqrt{x}} - \frac{2}{x^3}\right)$

43. $\frac{(4x + 4)(3x - 1) - 3(2x^2 + 4x + 1)}{(3x - 1)^2} = (6x^2 - 4x - 7)/(3x - 1)^2$

45. $\frac{(2x - 4)(x^2 + x + 1) - (x^2 - 4x + 1)(2x + 1)}{(x^2 + x + 1)^2} = (5x^2 - 5)/(x^2 + x + 1)^2$

47. $\frac{(0.23x^{-0.77} - 5.7)(1 - x^{-2.9}) - 2.9x^{-3.9}(x^{0.23} - 5.7x)}{(1 - x^{-2.9})^2}$ **49.** $\frac{\frac{1}{2}x^{-1/2}(x^{1/2} - 1) - \frac{1}{2}x^{-1/2}(x^{1/2} + 1)}{(x^{1/2} - 1)^2} = \frac{-1}{\sqrt{x}(\sqrt{x} - 1)^2}$

51. $-3/x^4$ **53.** $\frac{[(x + 1) + (x + 3)](3x - 1) - 3(x + 3)(x + 1)}{(3x - 1)^2} = (3x^2 - 2x - 13)/(3x - 1)^2$

55. $\frac{[(x + 1)(x + 2) + (x + 3)(x + 2) + (x + 3)(x + 1)](3x - 1) - 3(x + 3)(x + 1)(x + 2)}{(3x - 1)^2}$

57. $4x^3 - 2x$ **59.** 64 **61.** 3 **63.** Difference; $4x^3 - 12x^2 + 2x - 480$ **65.** Sum; $1 + 2/(x + 1)^2$

67. Product; $\left[\frac{x}{x+1}\right] + (x+2)\frac{1}{(x+1)^2}$ **69.** Difference; $2x - 1 - 2/(x+1)^2$ **71.** $y = 12x - 8$ **73.** $y = x/4 + 1/2$ **75.** $y = -2$ **77.** $q'(5) = 1{,}000$ units per month (sales are increasing at a rate of 1,000 units per month); $p'(5) = -\$10$ per month (the price of a sound system is dropping at a rate of \$10 per month); $R'(5) = 900{,}000$ (revenue is increasing at a rate of \$900,000 per month). **79.** \$703 million; increasing at a rate of \$67 million per year **81.** Decreasing at a rate of \$1 per day **83.** Decreasing at a rate of approximately \$0.10 per month **85.** $M'(x) = \frac{3{,}000(3{,}600x^{-2} - 1)}{(x + 3{,}600x^{-1})^2}$; $M'(10) \approx 0.7670$ mpg/mph. This means that, at a speed of 10 mph, the fuel economy is increasing at a rate of 0.7670 miles per gallon per 1 mile per hour increase in speed. $M'(60) = 0$ mpg/mph. This means that, at a speed of 60 mph, the fuel economy is neither increasing nor decreasing with increasing speed. $M'(70) \approx -0.0540$. This means that, at 70 mph, the fuel economy is decreasing at a rate of 0.0540 miles per gallon per 1 mile per hour increase in speed. 60 mph is the most fuel-efficient speed for the car. (In the next chapter we shall discuss how to locate largest values in general.) **87. a.** $P(t) - I(t)$ represents the daily production of oil in Mexico that was not exported to the United States. $I(t)/P(t)$ represents U.S. imports of oil from Mexico as a fraction of the total produced there. **b.** -0.020 per year; at the start of 2011 the fraction of oil produced in Mexico that was imported by the United States was decreasing at a rate of 0.020 (or 2.0 percentage points) per year.
89. Increasing at a rate of about \$50 million per year
91. $R'(p) = -\frac{5.625}{(1 + 0.125p)^2}$; $R'(4) = -2.5$ thousand organisms per hour per 1,000 organisms. This means that the reproduction rate of organisms in a culture containing 4,000 organisms is declining at a rate of 2,500 organisms per hour per 1,000 additional organisms. **93.** Oxygen consumption is decreasing at a rate of 1,600 milliliters per day. This must be due to the fact that the number of eggs is decreasing, as $C'(25)$ is positive. **95.** 20; 33 **97.** 5/4; $-17/16$ **99.** The analysis is suspect, as it seems to be asserting that the annual increase in revenue, which we can think of as dR/dt, is the product of the annual increases, dp/dt in price, and dq/dt in sales. However, because $R = pq$, the product rule implies that dR/dt is not the product of dp/dt and dq/dt but is instead $\frac{dR}{dt} = \frac{dp}{dt}\cdot q + p\cdot\frac{dq}{dt}$. **101.** Answers will vary. $q = -p + 1{,}000$ is one example. **103.** Mine; it is increasing twice as fast as yours. The rate of change of revenue is given by $R'(t) = p'(t)q(t)$ because $q'(t) = 0$. Thus, $R'(t)$ does not depend on the selling price $p(t)$. **105.** (A)

Section 4.4

1. $4(2x+1)$ **3.** $-(x-1)^{-2}$ **5.** $2(2-x)^{-3}$
7. $(2x+1)^{-0.5}$ **9.** $-3/(3x-1)^2$
11. $4(x^2+2x)^3(2x+2)$ **13.** $-4x(2x^2-2)^{-2}$
15. $-5(2x-3)(x^2-3x-1)^{-6}$ **17.** $-6x/(x^2+1)^4$
19. $1.5(0.2x - 4.2)(0.1x^2 - 4.2x + 9.5)^{0.5}$
21. $4(2s - 0.5s^{-0.5})(s^2 - s^{0.5})^3$ **23.** $-x/\sqrt{1-x^2}$
25. $\frac{3|3x-6|}{3x-6}$ **27.** $\frac{(-3x^2+5)|-x^3+5x|}{-x^3+5x}$
29. $-[(x+1)(x^2-1)]^{-3/2}(3x-1)(x+1)$
31. $6.2(3.1x-2) + 6.2/(3.1x-2)^3$
33. $2[(6.4x-1)^2 + (5.4x-2)^3] \times [12.8(6.4x-1) + 16.2(5.4x-2)^2]$
35. $-2(x^2-3x)^{-3}(2x-3)(1-x^2)^{0.5} - x(x^2-3x)^{-2}(1-x^2)^{-0.5}$
37. $-56(x+2)/(3x-1)^3$ **39.** $3z^2(1-z^2)/(1+z^2)^4$
41. $3[(1+2x)^4 - (1-x)^2]^2[8(1+2x)^3 + 2(1-x)]$
43. $6|3x-1|$ **45.** $\frac{|x-(2x-3)^{1/2}|}{x-(2x-3)^{1/2}}[1-(2x-3)^{-1/2}]$
47. $-\frac{\left(\frac{1}{\sqrt{2x+1}} - 2x\right)}{(\sqrt{2x+1} - x^2)^2}$
49. $54(1+2x)^2(1+(1+2x)^3)^2(1+(1+(1+2x)^3)^3)^2$
51. $2(x+2)$ **53.** $4(2t-1)+2$ **55.** 1/3
57. $1/(2\sqrt{y})$ **59.** $-5/3$ **61.** $3/\sqrt{x}$ **63.** 1/4
65. 3/4 **67.** $(100x^{99} - 99x^{-2})\,dx/dt$
69. $(-3r^{-4} + 0.5r^{-0.5})\,dr/dt$ **71.** $4\pi r^2\,dr/dt$ **73.** $-47/4$
75. $P'(t) = 0.54(t - 1980) - 8.6$; $P'(2010) = 7.6$; the price of a barrel of crude oil was increasing at a rate of \$7.60 per year in 2010. **77.** Profit $\approx$ \$260.49; marginal profit $\approx$ \$5.10/magazine. Your current profit is \$260.49 per month, and this would increase at a rate of \$5.10 per additional magazine sold.
79. $M'(x) = -\frac{3{,}000(1 - 3{,}600x^{-2})}{(x + 3{,}600x^{-1})^2}$; $M'(10) \approx 0.7670$ mpg/mph. This means that, at a speed of 10 mph, the fuel economy is increasing at a rate of 0.7670 miles per gallon per 1 mile per hour increase in speed. $M'(60) = 0$ mpg/mph. This means that, at a speed of 60 mph, the fuel economy is neither increasing nor decreasing with increasing speed. $M'(70) \approx -0.0540$. This means that, at 70 mph, the fuel economy is decreasing at a rate of 0.0540 miles per gallon per 1 mile per hour increase in speed. 60 mph is the most fuel-efficient speed for the car. (In the next chapter we shall discuss how to locate largest values in general.)
81. $\left.\frac{dP}{dn}\right|_{n=10} = 146{,}454.9$. At an employment level of 10 engineers, Paramount will increase its annual profit at a rate of \$146,454.90 per additional engineer hired.
83. $y = 35(7 + 0.2t)^{-0.25}$; -0.11 percentage point per month. **85.** $-\$30$ per additional ruby sold. The revenue is decreasing at a rate of \$30 per additional ruby sold.
87. $\frac{dy}{dt} = \frac{dy}{dx}\frac{dx}{dt} = (1.5)(-2) = -3$ murders per 100,000 residents per year each year. **89.** $5/6 \approx 0.833$; relative to the 2003 levels, home sales were changing at a rate of about 0.833 percentage points per percentage point change in price. (Equivalently, home sales in 2008

were dropping at a rate of about 0.833 percentage points per percentage point drop in price.) **91.** 12π square miles per hour **93.** $200,000$\pi$ per week ≈ $628,000 per week **95. a.** $q'(4) \approx 333$ units per month **b.** $dR/dq = \$800$ per unit **c.** $dR/dt \approx \$267{,}000$ per month **97.** 3% per year **99.** 8% per year **101.** The glob squared, times the derivative of the glob. **103.** The derivative of a quantity cubed is three times the *original quantity* squared times the derivative of the quantity, not three times the derivative of the quantity squared. Thus, the correct answer is $3(3x^3 - x)^2(9x^2 - 1)$. **105.** First, the derivative of a quantity cubed is three times the *original quantity* squared times the derivative of the quantity, not three times the derivative of the quantity squared. Second, the derivative of a quotient is not the quotient of the derivatives; the quotient rule needs to be used in calculating the derivative of $\frac{3x^2-1}{2x-2}$. Thus, the correct result (before simplifying) is $3\left(\frac{3x^2-1}{2x-2}\right)^2\left(\frac{6x(2x-2)-(3x^2-1)(2)}{(2x-2)^2}\right)$.
107. Following the calculation thought experiment, pretend that you were evaluating the function at a specific value of x. If the last operation you would perform is addition or subtraction, look at each summand separately. If the last operation is multiplication, use the product rule first; if it is division, use the quotient rule first; if it is any other operation (such as raising a quantity to a power or taking a radical of a quantity) then use the chain rule first. **109.** An example is

$$f(x) = \sqrt{x + \sqrt{x + \sqrt{x + \sqrt{x + \sqrt{x + 1}}}}}.$$

Section 4.5

1. $1/(x-1)$ **3.** $2x/(x^2+3)$ **5.** $1/(x \ln 2)$
7. $\frac{1}{(x+1)\ln 2}$ **9.** $\frac{1-1/t^2}{(t+1/t)\ln 3}$ **11.** e^{x+3}
13. $(2x-1)e^{x^2-x+1}$ **15.** $4(e^{2x-1})^2$ **17.** $4^x \ln 4$
19. $2^{x^2-1}2x \ln 2$ **21.** $2x \ln x + (x^2+1)/x$
23. $10x(x^2+1)^4 \ln x + (x^2+1)^5/x$
25. $4x/(2x^2+1)$ **27.** $(2x - 0.63x^{-0.7})/(x^2 - 2.1x^{0.3})$
29. $-2/(-2x+1) + 1/(x+1)$
31. $3/(3x+1) - 4/(4x-2)$
33. $1/(x+1) + 1/(x-3) - 2/(2x+9)$ **35.** $5.2/(4x-2)$
37. $2/(x+1) - 9/(3x-4) - 1/(x-9)$
39. $\frac{2\ln|x|}{x}$ **41.** $\frac{2}{x} - \frac{2\ln(x-1)}{x-1}$ **43.** $e^x(1+x)$
45. $1/(x+1) + 3e^x(x^3+3x^2)$ **47.** $e^x(\ln|x| + 1/x)$
49. $2xe^{2x-1}(1+x)$ **51.** $2 \cdot 3^{2x+1} \ln 3 + 3e^{3x+1}$
53. $\frac{2x3^{x^2}[(x^2+1)\ln 3 - 1]}{(x^2+1)^2}$ **55.** $-4/(e^x - e^{-x})^2$
57. $5e^{5x-3}$ **59.** $-\frac{\ln x + 1}{(x \ln x)^2}$ **61.** $2(x-1)$ **63.** $\frac{1}{x \ln x}$
65. $\frac{1}{2x \ln x}$ **67.** $y = (e/\ln 2)(x-1) \approx 3.92(x-1)$
69. $y = x$ **71.** $y = -[1/(2e)](x-1) + e$ **73.** 0
75. 0 **77.** 1 **79.** $231 billion and increasing at a rate of $2.9 billion per year **81.** $231 billion and increasing at a rate of $2.9 billion per year **83.** −1,653 years per gram; the age of the specimen is decreasing at a rate of about 1,653 years per additional 1 gram of carbon 14 present in the sample. (Equivalently, the age of the specimen is increasing at a rate of about 1,653 years per additional 1 gram less of carbon 14 in the sample.) **85.** Average price: $1.4 million; increasing at a rate of about $220,000 per year. **87. a.** $N(15) \approx 1{,}894 \approx 1{,}900$ (rounded to two significant digits) wiretap orders; $N'(15) \approx 113.6 \approx 110$ wiretap orders per year (rounded to two significant digits). The constants in the model are specified to two significant digits, so we cannot expect the answer to be accurate to more than two digits. **b.** In 2005 the number of people whose communications were intercepted was about 190,000 and increasing at a rate of about 11,000 people per year. **c.** (C) **89.** $451.00 per year **91.** $446.02 per year
93. $A(t) = 167(1.18)^t$; 280 new cases per day
95. $C(t) = 100e^{0.72t}$; 1,280 new cases per month
97. a. (A) **b.** The math SAT increases by approximately 0.97 points. **c.** $S'(x)$ decreases with increasing x, so as parental income increases, the effect on math SAT scores decreases.
99. a. −6.25 years per child; when the fertility rate is 2 children per woman, the average age of a population is dropping at a rate of 6.25 years per 1-child increase in the fertility rate.
b. 0.160 **101.** 3,300,000 cases per week; 11,000,000 cases per week; 640,000 cases per week **103.** 2.1 percentage points per year; the rate of change is the slope of the tangent at $t = 3$. This is also approximately the average rate of change over $[2, 4]$, which is about $4/2 = 2$, in approximate agreement with the answer. **105. a.** 2.1 percentage points per year
b. $\lim_{t\to+\infty} A(t) = 15$; had the trend continued indefinitely, the percentage of mortgages that were subprime would have approached 15% in the long term. $\lim_{t\to+\infty} A'(t) = 0$; had the trend continued indefinitely, the rate of change of the percentage of mortgages that were subprime would have approached 0 percentage points per year in the long term.
107. 277,000 people per year **109.** 0.000283 grams per year
111. $R(t) = 350e^{-0.1t}(39t + 68)$ million dollars; $R(2) \approx \$42$ billion; $R'(2) \approx \$7$ billion per year **113.** e raised to the glob, times the derivative of the glob. **115.** 2 raised to the glob, times the derivative of the glob, times the natural logarithm of 2.
117. The derivative of $\ln|u|$ is not $\frac{1}{|u|}\frac{du}{dx}$; it is $\frac{1}{u}\frac{du}{dx}$. Thus, the correct derivative is $\frac{3}{3x+1}$. **119.** The power rule does not apply when the exponent is not constant. The derivative of 3 raised to a quantity is 3 raised to the quantity, times the derivative of the quantity, times ln 3. Thus, the correct answer is $3^{2x}\, 2 \ln 3$. **121.** No. If $N(t)$ is exponential, so is its derivative. **123.** If $f(x) = e^{kx}$, then the fractional rate of change is $\frac{f'(x)}{f(x)} = \frac{ke^{kx}}{e^{kx}} = k$. **125.** If $A(t)$ is growing expo-

nentially, then $A(t) = A_0e^{kt}$ for constants A_0 and k. Its percentage rate of change is then $\dfrac{A'(t)}{A(t)} = \dfrac{kA_0e^{kt}}{A_0e^{kt}} = k$, a constant.

Section 4.6

1. $-2/3$ **3.** x **5.** $(y-2)/(3-x)$ **7.** $-y$

9. $-\dfrac{y}{x(1+\ln x)}$ **11.** $-x/y$ **13.** $-2xy/(x^2-2y)$

15. $-(6+9x^2y)/(9x^3-x^2)$ **17.** $3y/x$

19. $(p+10p^2q)/(2p-q-10pq^2)$

21. $(ye^x-e^y)/(xe^y-e^x)$ **23.** $se^{st}/(2s-te^{st})$

25. $ye^x/(2e^x+y^3e^y)$ **27.** $(y-y^2)/(-1+3y-y^2)$

29. $-y/(x+2y-xye^y-y^2e^y)$ **31. a.** 1 **b.** $y=x-3$ **33. a.** -2 **b.** $y=-2x$

35. a. -1 **b.** $y=-x+1$ **37. a.** $-2{,}000$ **b.** $y=-2{,}000x+6{,}000$ **39. a.** 0 **b.** $y=1$

41. a. -0.1898 **b.** $y=-0.1898x+1.4720$

43. $\dfrac{2x+1}{4x-2}\left[\dfrac{2}{2x+1}-\dfrac{4}{4x-2}\right]$

45. $\dfrac{(3x+1)^2}{4x(2x-1)^3}\left[\dfrac{6}{3x+1}-\dfrac{1}{x}-\dfrac{6}{2x-1}\right]$

47. $(8x-1)^{1/3}(x-1)\left[\dfrac{8}{3(8x-1)}+\dfrac{1}{x-1}\right]$

49. $(x^3+x)\sqrt{x^3+2}\left[\dfrac{3x^2+1}{x^3+x}+\dfrac{1}{2}\dfrac{3x^2}{x^3+2}\right]$

51. $x^x(1+\ln x)$ **53.** $-\$3{,}000$ per worker. The monthly budget to maintain production at the fixed level P is decreasing by approximately \$3,000 per additional worker at an employment level of 100 workers and a monthly operating budget of \$200,000. **55.** -125 T-shirts per dollar; when the price is set at \$5, the demand is dropping by 125 T-shirts per \$1 increase in price. **57.** $\left.\dfrac{dk}{de}\right|_{e=15} = -0.307$ carpenters per electrician. This means that, for a \$200,000 house whose construction employs 15 electricians, adding one more electrician would cost as much as approximately 0.307 additional carpenters. In other words, one electrician is worth approximately 0.307 carpenters. **59. a.** 22.93 hours. (The other root is rejected because it is larger than 30.) **b.** $\dfrac{dt}{dx} = \dfrac{4t-20x}{0.4t-4x}$; $\left.\dfrac{dt}{dx}\right|_{x=3.0} \approx -11.2$ hours per grade point. This means that, for a 3.0 student who scores 80 on the examination, 1 grade point is worth approximately 11.2 hours. **61.** $\dfrac{dr}{dy} = 2\dfrac{r}{y}$, so $\dfrac{dr}{dt} = 2\dfrac{r}{y}\dfrac{dy}{dt}$ by the chain rule. **63.** x, y, y, x **65.** Let $y = f(x)g(x)$. Then $\ln y = \ln f(x) + \ln g(x)$, and $\dfrac{1}{y}\dfrac{dy}{dx} = \dfrac{f'(x)}{f(x)} + \dfrac{g'(x)}{g(x)}$, so $\dfrac{dy}{dx} = y\left(\dfrac{f'(x)}{f(x)} + \dfrac{g'(x)}{g(x)}\right) = f(x)g(x)\left(\dfrac{f'(x)}{f(x)} + \dfrac{g'(x)}{g(x)}\right) = f'(x)g(x) + f(x)g'(x)$.

67. Writing $y = f(x)$ specifies y as an explicit function of x. This can be regarded as an equation giving y as an *implicit* function of x. The procedure of finding dy/dx by implicit differentiation is then the same as finding the derivative of y as an explicit function of x: We take d/dx of both sides. **69.** Differentiate both sides of the equation $y = f(x)$ with respect to y to get $1 = f'(x)\cdot\dfrac{dx}{dy}$, giving $\dfrac{dx}{dy} = \dfrac{1}{f'(x)} = \dfrac{1}{dy/dx}$.

Chapter 4 Review

1. $50x^4+2x^3-1$ **3.** $9x^2+x^{-2/3}$ **5.** $1-2/x^3$

7. $-\dfrac{4}{3x^2}+\dfrac{0.2}{x^{1.1}}+\dfrac{1.1x^{0.1}}{3.2}$ **9.** $e^x(x^2+2x-1)$

11. $(-3x|x|+|x|/x-6x)/(3x^2+1)^2$

13. $-4(4x-1)^{-2}$ **15.** $20x(x^2-1)^9$

17. $-0.43(x+1)^{-1.1}[2+(x+1)^{-0.1}]^{3.3}$

19. $2e^{2x+1}$ **21.** $2\cdot 3^{2x-4}\ln 3$

23. $e^x(x^2+1)^9(x^2+20x+1)$

25. $3^x[(x-1)\ln 3-1]/(x-1)^2$ **27.** $2xe^{x^2-1}$

29. $3/(3x-1)$ **31.** $2x/(x^2-1)$ **33.** $x=7/6$

35. $x=\pm 2$ **37.** $x=(1-\ln 2)/2$ **39.** None

41. $\dfrac{2x-1}{2y}$ **43.** $-y/x$ **45.** $\dfrac{(2x-1)^4(3x+4)}{(x+1)(3x-1)^3}\times\left[\dfrac{8}{2x-1}+\dfrac{3}{3x+4}-\dfrac{1}{x+1}-\dfrac{9}{3x-1}\right]$

47. $y=-x/4+1/2$ **49.** $y=-3ex-2e$

51. $y=x+2$ **53.** 0 **55.** 1 **57. a.** 274 books per week **b.** 636 books per week **c.** The function w begins to decrease more and more rapidly after $t = 14$. Graph:

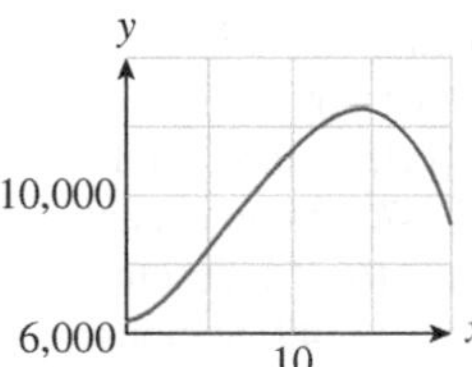

Not realistic.

d. Because the data suggest an upward curving parabola, the long-term prediction of sales for a quadratic model would be that sales will increase without bound, in sharp contrast to (c). **59. a.** \$2.88 per book **b.** \$3.715 per book **c.** Approximately $-\$0.000104$ per book per additional book sold. **d.** At a sales level of 8,000 books per week, the cost is increasing at a rate of \$2.88 per book (so that the 8,001st book costs approximately \$2.88 to sell), and it costs an average of \$3.715 per book to sell the first 8,000 books. Moreover, the average cost is decreasing at a rate of \$0.000104 per book per additional book sold. **61. a.** \$3,000 per week (rising) **b.** 300 books per week **63.** $R = pq$ gives $R' = p'q + pq'$. Thus, $R'/R = R'/(pq) = (p'q+pq')/pq = p'/p + q'/q$

65. \$110 per year **67. a.** $s'(t) = \dfrac{2{,}475e^{-0.55(t-4.8)}}{(1 + e^{-0.55(t-4.8)})^2}$; 556 books per week **b.** 0; In the long term, the rate of increase of weekly sales slows to zero. **69.** 616.8 hits per day per week. **71. a.** −17.24 copies per \$1. The demand for the gift edition of *The Complete Larry Potter* is dropping at a rate of about 17.24 copies per \$1 increase in the price. **b.** \$138 per dollar is positive, so the price should be raised.

Chapter 5

Section 5.1

1. Absolute min: (−3, −1), relative max: (−1, 1), relative min: (1, 0), absolute max: (3, 2) **3.** Absolute min: (3, −1) and (−3, −1), absolute max: (1, 2) **5.** Absolute min: (−3, 0) and (1, 0), absolute max: (−1, 2) and (3, 2) **7.** Relative min: (−1, 1) **9.** Absolute min: (−3, −1), relative max: (−2, 2), relative min: (1, 0), absolute max: (3, 3) **11.** Relative max: (−3, 0), absolute min: (−2, −1), stationary nonextreme point: (1, 1) **13.** Absolute max: (0, 1), absolute min: (2, −3), relative max: (3, −2) **15.** Absolute min: (−4, −16), absolute max: (−2, 16), absolute min: (2, −16), absolute max: (4, 16) **17.** Absolute min: (−2, −10), absolute max: (2, 10) **19.** Absolute min: (−2, −4), relative max: (−1, 1), relative min: (0, 0) **21.** Relative max: (−1, 5), absolute min: (3, −27) **23.** Absolute min: (0, 0) **25.** Absolute max: (0, 1) and (2, 1), absolute min: (1, 0) **27.** Relative max: (−2, −1/3), relative min: (−1, −2/3), absolute max: (0, 1) **29.** Relative min: (−2, 5/3), relative max: (0, −1), relative min: (2, 5/3) **31.** Relative max: (0, 0); absolute min: $(1/3, -2\sqrt{3}/9)$ **33.** Relative max: (0, 0), absolute min: (1, −3) **35.** No relative extrema **37.** Absolute min: (1, 1) **39.** Relative max: $(-1, 1 + 1/e)$, absolute min: (0, 1), absolute max: $(1, e - 1)$ **41.** Relative max: (−6, −24), relative min: (−2, −8) **43.** Absolute max: $(1/\sqrt{2}, \sqrt{e/2})$, absolute min: $(-1/\sqrt{2}, -\sqrt{e/2})$ **45.** Relative min: (0.15, −0.52) and (2.45, 8.22), relative max: (1.40, 0.29) **47.** Absolute max: (−5, 700), relative max: (3.10, 28.19) and (6, 40), absolute min: (−2.10, −392.69), relative min: (5, 0). **49.** Stationary minimum at $x = -1$ **51.** Stationary minima at $x = -2$ and $x = 2$, stationary maximum at $x = 0$ **53.** Singular minimum at $x = 0$, stationary nonextreme point at $x = 1$ **55.** Stationary minimum at $x = -2$, singular nonextreme points at $x = -1$ and $x = 1$, stationary maximum at $x = 2$

57. Answers will vary.

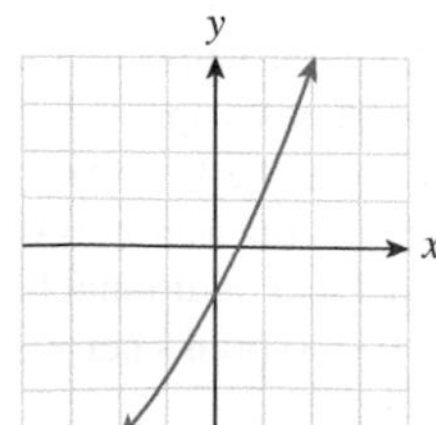

59. Answers will vary.

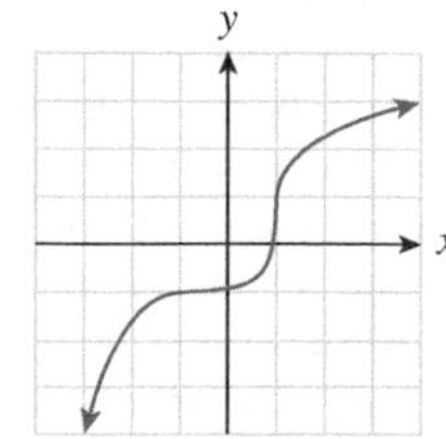

61. Not necessarily; it could be neither a relative maximum nor a relative minimum, as in the graph of $y = x^3$ at the origin. **63.** Answers will vary.

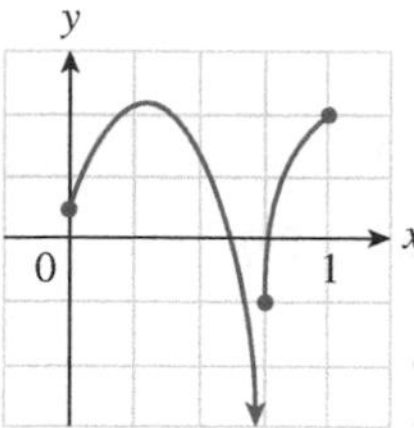

65. The graph oscillates faster and faster above and below zero as it approaches the endpoint at 0, so 0 cannot be either a relative minimum or maximum.

Section 5.2

1. $x = y = 5$; $P = 25$ **3.** $x = y = 3$; $S = 6$ **5.** $x = 2$, $y = 4$; $F = 20$ **7.** $x = 20$, $y = 10$, $z = 20$; $P = 4{,}000$ **9.** 5 × 5 **11.** 20 ads for an average cost of \$4,002,000 per ad **13.** 20,000 per hour for an average cost of \$200 per iPhone. The marginal cost is the same as the average cost at the optimal production level. **15.** $\sqrt{40} \approx 6.32$ pounds of pollutant per day, for an average cost of about \$1,265 per pound **17.** 2.5 lb **19.** 5 ft × 10 ft **21.** 50 ft × 10 ft for an area of 250 sq. ft. **23.** 11 ft × 22 ft **25.** \$10 **27.** \$252.50 for a quarterly revenue of \$382,540 million, or \$382.54 billion **29.** \$4.61 for a daily revenue of \$95,680.55 **31. a.** \$1.41 per pound **b.** 5,000 pounds **c.** \$7,071.07 per month **33.** 34.5¢ per pound, for an annual (per capita) revenue of \$5.95 **35.** \$292.50 for an annual profit of \$270,940 million, or \$270.94 billion **37. a.** 656 headsets, for a profit of \$28,120 **b.** \$143 per headset **39.** Height = Radius of base ≈ 20.48 cm. **41.** Height ≈ 10.84 cm; Radius ≈ 2.71 cm; Height/Radius = 4 **43.** $13\frac{1}{3}$ in × $3\frac{1}{3}$ in × $1\frac{1}{3}$ in for a volume of 1,600/27 ≈ 59 in^3 **45.** 5 × 5 × 5 cm **47.** $l = w = h \approx 20.67$ in, volume ≈ 8,827 in^3 **49.** $l = 30$ in, $w = 15$ in, $h = 30$ in **51.** $l = 36$ in, $w = h = 18$ in, $V = 11{,}664$ in^3 **53. a.** 1.6 years, or year 2001.6; **b.** $R_{max} = \$28{,}241$ million **55.** $t = 2.5$ or midway through 1972; $D(2.5)/S(2.5) \approx 4.09$. The number of new (approved) drugs per \$1 billion of spending on research and development reached a high of around four approved drugs per \$1 billion midway through 1972. **57.** 30 years from now **59.** 55 days **61.** 1,600 copies. At this value of x, average profit equals marginal profit; beyond this the marginal profit is smaller than the average. **63.** 40 laborers and 250 robots **65.** 71 employees **67.** Increasing most rapidly in 1990; increasing least rapidly in 2007 **69.** Maximum when $t = 17$ days. This means that the embryo's oxygen consumption is increasing most rapidly 17 days after the egg is laid.

71. Graph of derivative:

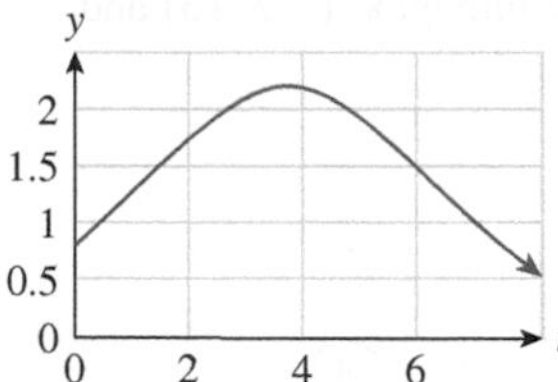

The absolute maximum occurs at approximately (3.7, 2.2) during the year 2003. The percentage of mortgages that were subprime was increasing most rapidly during 2003, when it increased at a rate of around 2.2 percentage points per year. **73.** You should sell them in 17 years' time, when they will be worth approximately \$3,960. **75.** 25 additional trees **77.** (D) **79.** (A); (B) **81.** The problem is uninteresting because the company can accomplish the objective by cutting away the entire sheet of cardboard, resulting in a box with surface area zero. **83.** Not all absolute extrema occur at stationary points; some may occur at an endpoint or singular point of the domain, as in Exercises 31 and 32. **85.** The minimum of dq/dp is the fastest that the demand is dropping in response to increasing price.

Section 5.3

1. 6 **3.** $4/x^3$ **5.** $-0.96x^{-1.6}$ **7.** $e^{-(x-1)}$ **9.** $2/x^3 + 1/x^2$ **11. a.** $a = -32$ ft/sec^2 **b.** $a = -32$ ft/sec^2 **13. a.** $a = 2/t^3 + 6/t^4$ ft/sec^2 **b.** $a = 8$ ft/sec^2 **15. a.** $a = -1/(4t^{3/2}) + 2$ ft/sec^2 **b.** $a = 63/32$ ft/sec^2 **17.** $(1, 0)$ **19.** $(1, 0)$ **21.** None **23.** $(-1, 0)$, $(1, 1)$ **25.** Points of inflection at $x = -1$ and $x = 1$ **27.** One point of inflection, at $x = -2$ **29.** Points of inflection at $x = -2$, $x = 0$, $x = 2$ **31.** Points of inflection at $x = -2$ and $x = 2$ **33.** $x = 2$; minimum **35.** Maximum at $x = -2$, minimum at $x = 2$ **37.** Maximum at $t = -1/\sqrt{3}$, minimum at $t = 1/\sqrt{3}$ **39.** Nonextreme stationary point at $x = 0$, minimum at $x = 3$ **41.** Maximum at $x = 0$ **43.** Minimum at $x = -1/\sqrt{2}$; maximum at $x = 1/\sqrt{2}$ **45.** $f'(x) = 8x - 1$; $f''(x) = 8$; $f'''(x) = f^{(4)}(x) = \cdots = f^{(n)}(x) = 0$ **47.** $f'(x) = -4x^3 + 6x$; $f''(x) = -12x^2 + 6$; $f'''(x) = -24x$; $f^{(4)}(x) = -24$; $f^{(5)}(x) = f^{(6)}(x) = \cdots = f^{(n)}(x) = 0$ **49.** $f'(x) = 8(2x + 1)^3$; $f''(x) = 48(2x + 1)^2$; $f'''(x) = 192(2x + 1)$; $f^{(4)}(x) = 384$; $f^{(5)}(x) = f^{(6)}(x) = \cdots = f^{(n)}(x) = 0$ **51.** $f'(x) = -e^{-x}$; $f''(x) = e^{-x}$; $f'''(x) = -e^{-x}$; $f^{(4)}(x) = e^{-x}$; $f^{(n)}(x) = (-1)^n e^{-x}$ **53.** $f'(x) = 3e^{3x-1}$; $f''(x) = 9e^{3x-1}$; $f'''(x) = 27e^{3x-1}$; $f^{(4)}(x) = 81e^{3x-1}$; $f^{(n)}(x) = 3^n e^{3x-1}$ **55.** -3.8 m/s^2 **57.** $6t - 2$ ft/sec^2; increasing **59.** Accelerating by 0.16 billion gal/year^2 **61. a.** 400 ml **b.** 36 ml/day **c.** -1 ml/day^2 **63. a.** 0.6% **b.** Speeding up **c.** Speeding up for $t < 3.33$ (before 1/3 of the way through March) and slowing for $t > 3.33$ (after that time) **65. a.** December 2005: -0.202% (deflation rate of 0.202%) February 2006: 0.363% **b.** Speeding up **c.** Speeding up for $t > 4.44$ (after mid-November) and decreasing for $t < 4.44$ (before that time). **67.** Concave up for $8 < t < 20$, concave down for $0 < t < 8$, point of inflection around $t = 8$. The percentage of articles written by researchers in the United States was decreasing most rapidly at around $t = 8$ (1991). **69. a.** (B) **b.** (B) **c.** (A) **71.** Graphs:

$A(t)$:

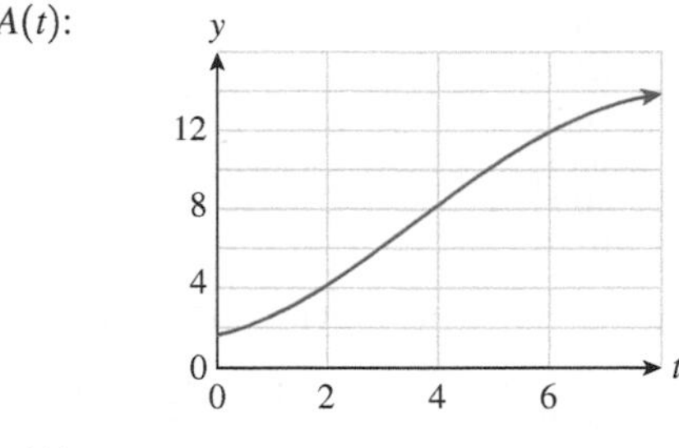

$A'(t)$:

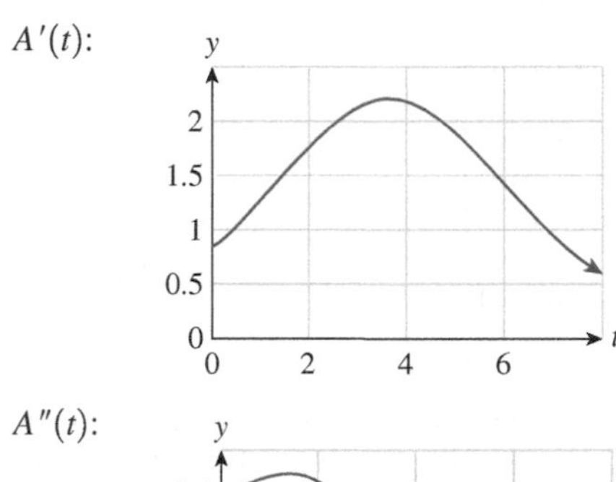

$A''(t)$:

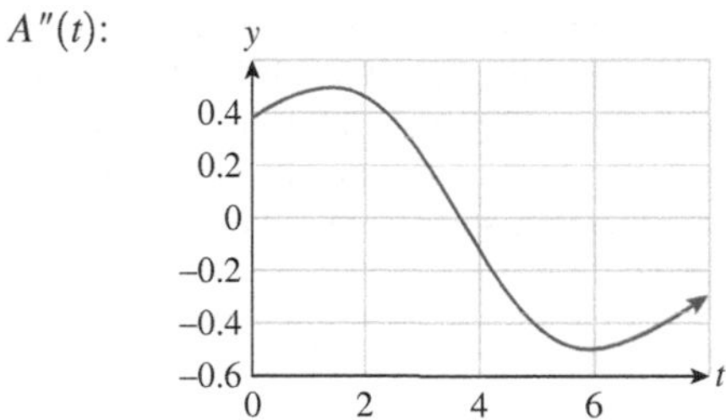

Concave up when $t < 4$; concave down when $t > 4$; point of inflection when $t \approx 4$. The percentage of U.S. mortgages that were subprime was increasing fastest at the beginning of 2004. **73. a.** 2 years into the epidemic **b.** 2 years into the epidemic **75. a.** 2024 **b.** 2026 **c.** 2022; (A) **77. a.** There are no points of inflection in the graph of *S*. **b.** Because the graph is concave up, the derivative of *S* is increasing, and so the rate of *decrease* of SAT scores with increasing numbers of prisoners was diminishing. In other words, the apparent effect of more prisoners on SAT scores was diminishing.

79. a. $\left.\dfrac{d^2n}{ds^2}\right|_{s=3} = -21.494$. Thus, for a firm with annual sales of \$3 million the rate at which new patents are produced decreases with increasing firm size. This means that the returns (as measured in the number of new patents per increase of \$1 million in sales) are diminishing as the firm size increases.

b. $\left.\dfrac{d^2n}{ds^2}\right|_{s=7} = 13.474$. Thus, for a firm with annual sales of \$7 million the rate at which new patents are produced increases with increasing firm size by 13.474 new patents per \$1 million increase in annual sales. **c.** There is a point of inflection when $s \approx 5.4587$, so in a firm with sales of \$5,458,700 per year the number of new patents produced per additional \$1 million in sales is a minimum.

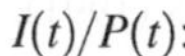

81. Graphs:

$I(t)/P(t)$:

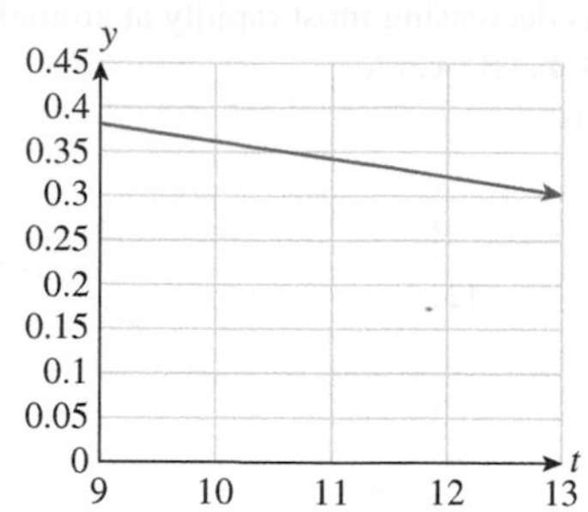

$[I(t)/P(t)]'$:

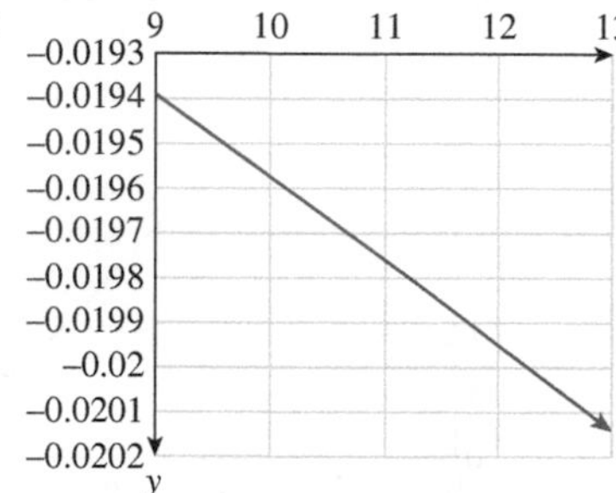

Concave down; (D) **83.** (Proof) **85.** $t \approx 4$; the population of Puerto Rico was increasing fastest in 1954. **87.** About \$570 per year, after about 12 years **89.** Increasing most rapidly in 17.64 years, decreasing most rapidly now (at $t = 0$) **91.** Nonnegative **93.** Daily sales were decreasing most rapidly in June 2002.

95.

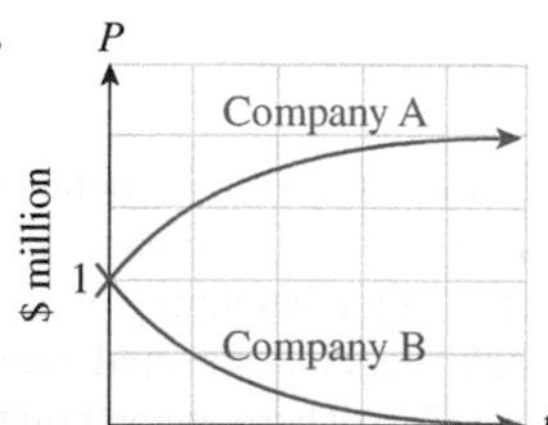

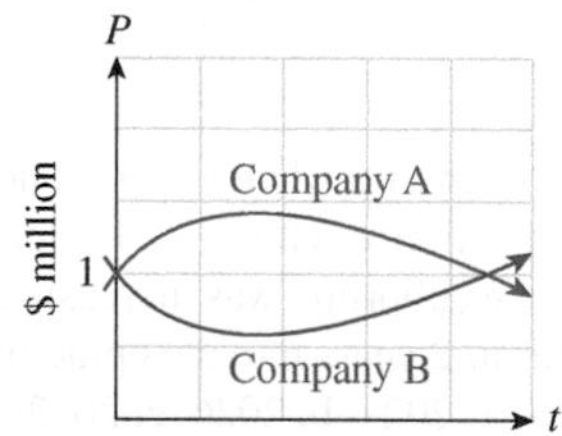

97. At a point of inflection the graph of a function changes either from concave up to concave down or vice versa. If it changes from concave up to concave down, then the derivative changes from increasing to decreasing and hence has a relative maximum. Similarly, if it changes from concave down to concave up, the derivative has a relative minimum.

Section 5.4

1. a. x-intercept: -1; y-intercept: 1 **b.** Absolute min: $(-1, 0)$ **c.** None **d.** None **e.** $y \to +\infty$ as $x \to \pm\infty$

Graph:

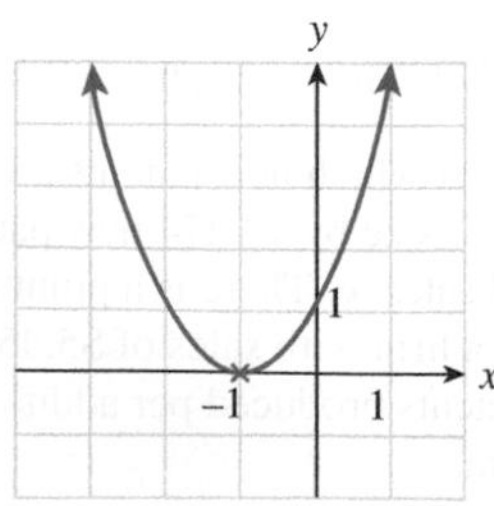

3. a. x-intercepts: $-\sqrt{12}, 0, \sqrt{12}$; y-intercept: 0 **b.** Absolute min: $(-4, -16)$ and $(2, -16)$, absolute max: $(-2, 16)$ and $(4, 16)$ **c.** $(0, 0)$ **d.** None **e.** None

Graph:

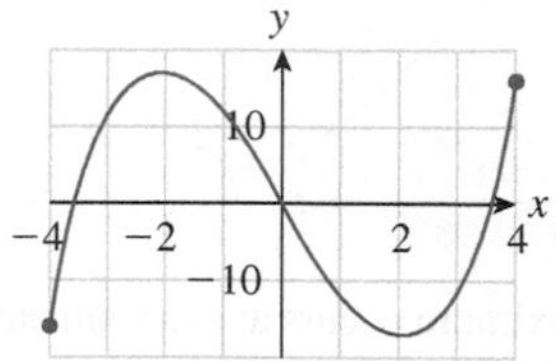

5. a. x-intercepts: $-3.6, 0, 5.1$; y-intercept: 0 **b.** Relative max: $(-2, 44)$, relative min: $(3, -81)$ **c.** $(0.5, -18.5)$ **d.** None **e.** $y \to -\infty$ as $x \to -\infty$; $y \to +\infty$ as $x \to +\infty$

Graph:

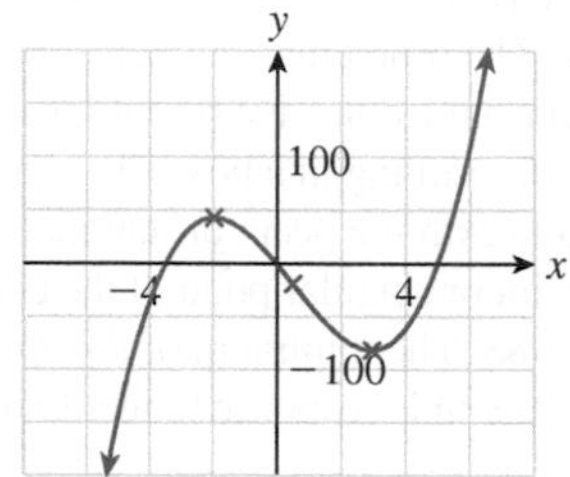

7. a. x-intercepts: $-3.3, 0.1, 1.8$; y-intercept: 1 **b.** Relative max: $(-2, 21)$, relative min: $(1, -6)$ **c.** $(-1/2, 15/2)$ **d.** None **e.** $y \to -\infty$ as $x \to -\infty$; $y \to +\infty$ as $x \to +\infty$

Graph:

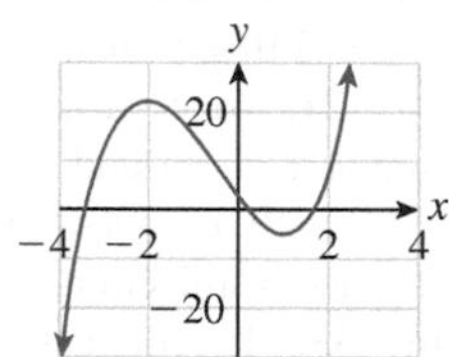

9. a. x-intercepts: $-2.9, 4.2$; y-intercept: 10 **b.** Relative max: $(-2, 74)$, relative min: $(0, 10)$, absolute max: $(3, 199)$ **c.** $(-1.12, 44.8)$, $(1.79, 117.1)$ **d.** None **e.** $y \to -\infty$ as $x \to -\infty$; $y \to -\infty$ as $x \to +\infty$

Graph:

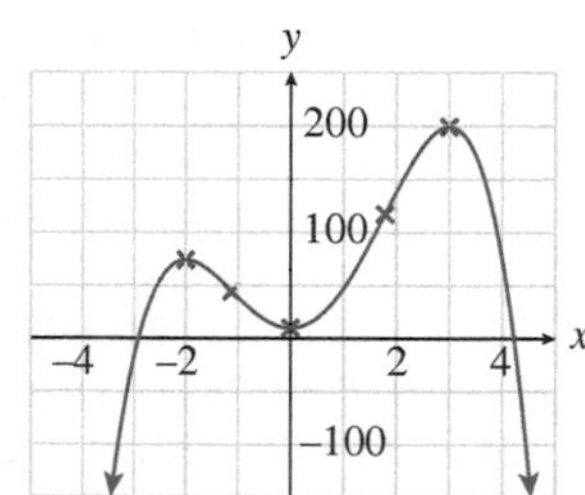

11. a. t-intercept: $t = 0$; y-intercept: 0 **b.** Absolute min: $(0, 0)$ **c.** $(1/3, 11/324)$ and $(1, 1/12)$ **d.** None **e.** $y \to +\infty$ as $t \to -\infty$; $y \to +\infty$ as $t \to +\infty$

Graph:

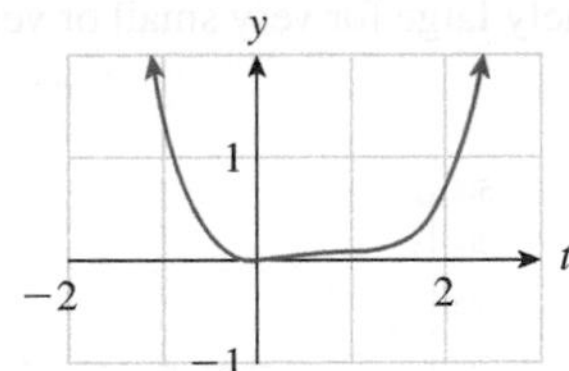

13. a. x-intercepts: None; y-intercept: None **b.** Relative min: $(1, 2)$, relative max: $(-1, -2)$ **c.** None **d.** $y \to -\infty$ as $x \to 0^-$; $y \to +\infty$ as $x \to 0^+$, so there is a vertical asymptote at $x = 0$. **e.** $y \to -\infty$ as $x \to -\infty$; $y \to +\infty$ as $x \to +\infty$

Graph:

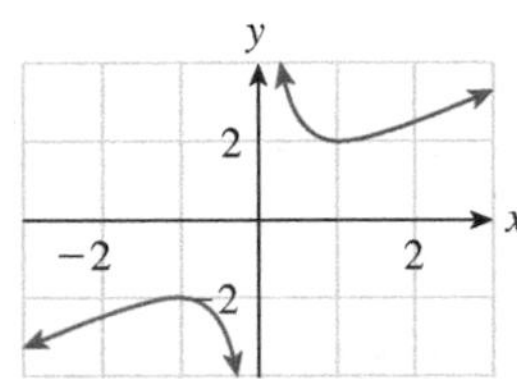

15. a. x-intercept: 0; y-intercept: 0 **b.** None **c.** $(0, 0)$, $(-3, -9/4)$, and $(3, 9/4)$ **d.** None **e.** $y \to -\infty$ as $x \to -\infty$; $y \to +\infty$ as $x \to +\infty$

Graph:

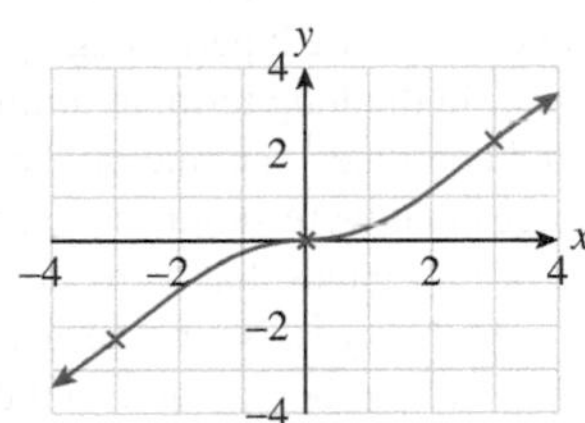

17. a. t-intercepts: None; y-intercept: -1 **b.** Relative min: $(-2, 5/3)$ and $(2, 5/3)$, relative max: $(0, -1)$ **c.** None **d.** $y \to +\infty$ as $t \to -1^-$; $y \to -\infty$ as $t \to -1^+$; $y \to -\infty$ as $t \to 1^-$; $y \to +\infty$ as $t \to 1^+$; so there are vertical asymptotes at $t = \pm 1$. **e.** None

Graph:

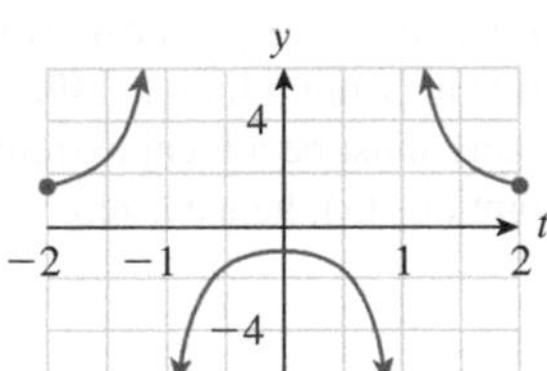

19. a. x-intercept: -0.7; y-intercept: 1 **b.** Relative max: $(-2, -1/3)$, relative min: $(-1, -2/3)$ **c.** None **d.** None. **e.** $y \to -\infty$ as $x \to -\infty$; $y \to +\infty$ as $x \to +\infty$

Graph:

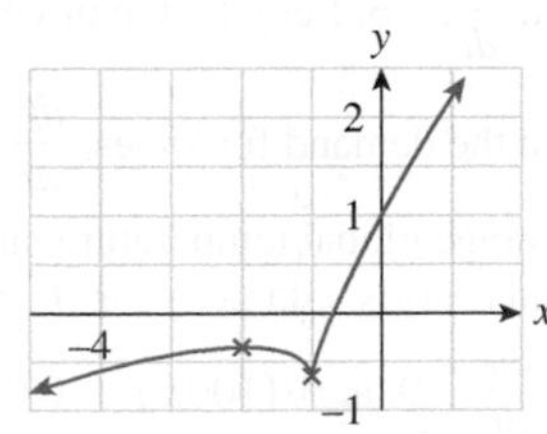

21. a. x-intercepts: None; y-intercept: None **b.** Absolute min: $(1, 1)$ **c.** None **d.** Vertical asymptote at $x = 0$ **e.** $y \to +\infty$ as $x \to +\infty$

Graph:

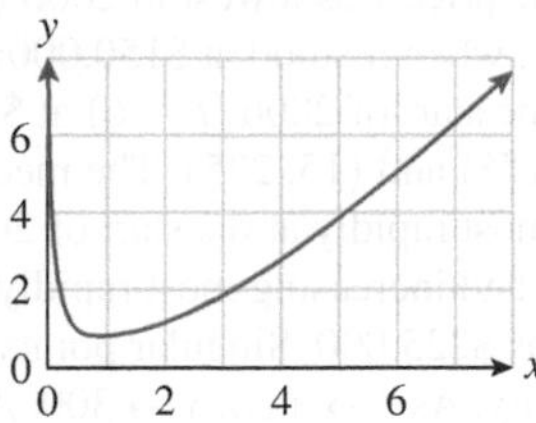

23. a. x-intercepts: ± 0.8; y-intercept: None **b.** None **c.** $(1, 1)$ and $(-1, 1)$ **d.** $y \to -\infty$ as $x \to 0$; vertical asymptote at $x = 0$ **e.** $y \to +\infty$ as $x \to \pm\infty$

Graph:

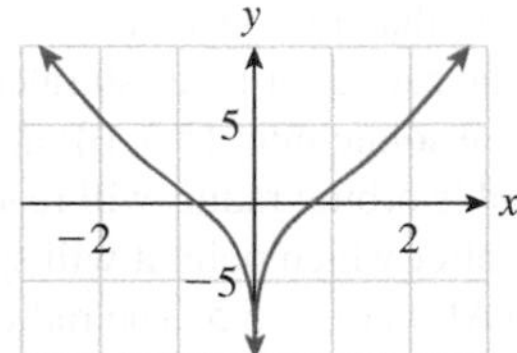

25. a. t-intercepts: None; y-intercept: 1 **b.** Absolute min: $(0, 1)$. Absolute max: $(1, e - 1)$, relative max: $(-1, e^{-1} + 1)$. **c.** None **d.** None **e.** None

Graph:

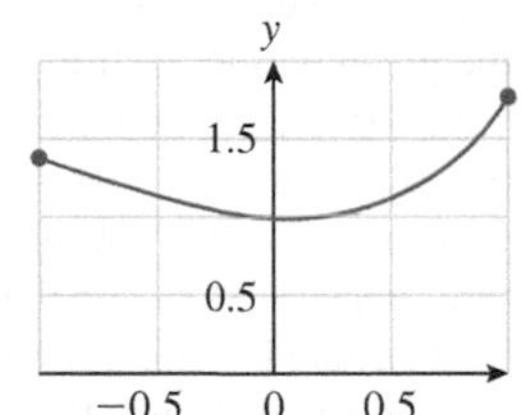

27. Absolute min: $(1.40, -1.49)$; points of inflection: $(0.21, 0.61)$, $(0.79, -0.55)$

Graph:

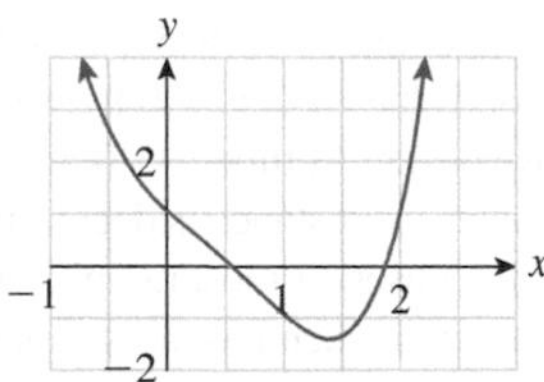

29. Relative min: $(-0.46, 0.73)$, relative max: $(0.91, 1.73)$, absolute min: $(3.73, -10.22)$; points of inflection at $(0.20, 1.22)$ and $(2.83, -5.74)$

Graph:

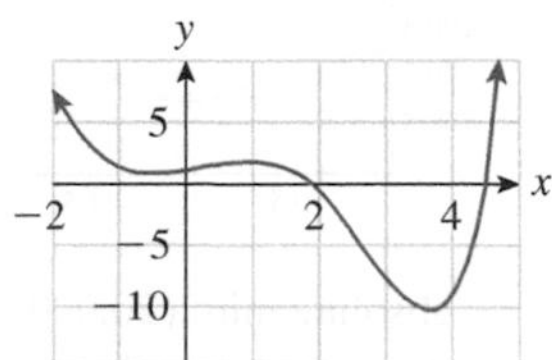

31. y-intercept: 150; t-intercepts: none. The median home price was about $150,000 at the start of 2000($t = 0$). Extrema: absolute min: (0, 150) and (12, 150); relative max: (6, 225). The median home price was lowest in 2000 ($t = 0$) and again in 2012 ($t = 12$), when it stood at $150,000; the median home price peaked at the start of 2006 ($t = 6$) at $225,000. Points of inflection at (9, 175) and (15, 225). The median home price was decreasing most rapidly at the start of 2009 ($t = 9$), when it was $175,000, and increasing most rapidly at the start of 2015, when it was $225,000. Singular points of f: none. Behavior at infinity: As $t \to +\infty$, $y \to 300$. Assuming that the trend shown in the graph continued indefinitely, the median home price would approach a value of $300,000 in the long term. **33. a.** r-intercept: slightly more than 1; t-intercept: 5.5. At the start of the period shown, the radius of the Earth's orbit will be slightly more than one AU, and five and a half million years later, it will be zero after it has spiraled into the core of the Sun. Extrema: absolute min: (5.5, 0); absolute max: (5, 1.07). The Earth's orbital radius will reach a maximum of 1.07 AU at $t = 5$, after which point it will spiral into the core of the Sun ($r = 0$ AU) at $t = 5.5$. The radius will be increasing most rapidly at $t = 4$, when it will be 1.05 AU. Singular points of f: none. Behavior at infinity: As $t \to -\infty$, $r \to 1$. At times much earlier than the period shown, the radius of the Earth was close to 1 AU. **b.** increase; decrease **35. a.** Intercepts: No t-intercept; y-intercept at $I(0) = 195$. The CPI was never zero during the given period; in July 2005 the CPI was 195. Absolute min: (0, 195), absolute max: (8, 199.3), relative max: (2.9, 198.7), relative min: (6.0, 197.8). The CPI was at a low of 195 in July 2005, rose to 198.7 around October 2005, dipped to 197.8 around January 2006, and then rose to a high of 199.3 in March 2006. There is a point of inflection at (4.4, 198.2). The rate of change of the CPI (inflation) reached a minimum around mid-November 2005 when the CPI was 198.2. **b.** The inflation rate was zero at around October 2005 and January 2006. **37.** Extrema: Relative max: (0, 100), absolute min: (1, 99); point of inflection: (0.5, 99.5); $s \to +\infty$ as $t \to +\infty$. At time $t = 0$ seconds the UFO is 100 ft away from the observer, and begins to move closer. At time $t = 0.5$ seconds, when the UFO is 99.5 feet away, its distance is decreasing most rapidly (it is moving toward the observer most rapidly). It then slows down to a stop at $t = 1$ second when it is at its closest point (99 ft away) and then begins to move farther and farther away.

Graph:

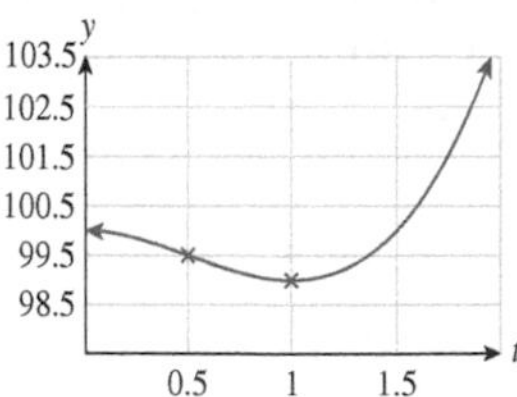

39. Intercepts: none; absolute min: (20,000, 200); no points of inflection; vertical asymptote at $x = 0$. As $x \to +\infty$, $y \to +\infty$. The average cost is never zero, nor is it defined for zero iPhone 6's. The average cost is a minimum ($200) when 20,000 iPhone 6's are manufactured per hour. The average cost becomes extremely large for very small or very large numbers of iPhone 6's.

Graph:

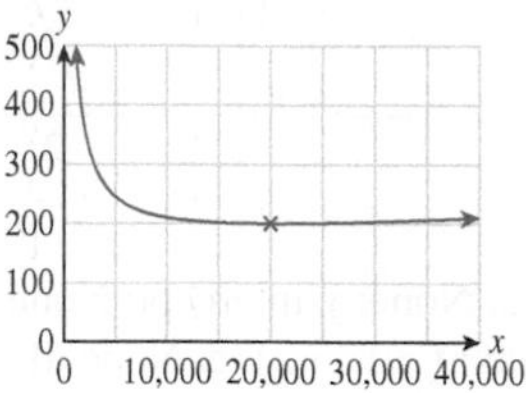

41. Graph of derivative:

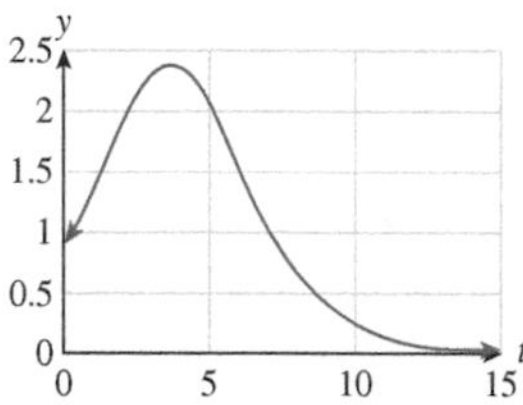

The absolute maximum occurs at approximately (3.7, 2.2); during the year 2003. The percentage of mortgages that were subprime was increasing most rapidly during 2003, when it increased at a rate of around 2.2 percentage points per year. As $t \to +\infty$, $A'(t) \to 0$; in the long term, assuming the trend shown in the model continues, the rate of change of the percentage of mortgages that were subprime approaches zero; that is, the percentage of mortgages that were subprime approaches a constant value. **43.** No; yes. Near a vertical asymptote the value of y increases without bound, and so the graph could not be included between two horizontal lines; hence, no vertical asymptotes are possible. Horizontal asymptotes are possible as, for instance, in the graph in Exercise 31. **45.** It too has a vertical asymptote at $x = a$; the magnitude of the derivative increases without bound as $x \to a$. **47.** No. If the leftmost critical point is a relative maximum, the function will decrease from there until it reaches the rightmost critical point, so it can't have a relative maximum there. **49.** Between every pair of zeros of $f(x)$ there must be a local extremum, which must be a stationary point of $f(x)$, hence a zero of $f'(x)$.

Section 5.5

1. $P = 10{,}000$; $\dfrac{dP}{dt} = 1{,}000$ **3.** Let R be the annual revenue of my company, and let q be annual sales. $R = 7{,}000$, and $\dfrac{dR}{dt} = -700$. Find $\dfrac{dq}{dt}$. **5.** Let p be the price of a pair of shoes, and let q be the demand for shoes. $\dfrac{dp}{dt} = 5$. Find $\dfrac{dq}{dt}$. **7.** Let T be the average global temperature, and let q be the number of Bermuda shorts sold per year. $T = 60$, and $\dfrac{dT}{dt} = 0.01$. Find $\dfrac{dq}{dt}$. **9. a.** $6/(100\pi) \approx 0.019$ km/sec **b.** $3/(4\sqrt{\pi}) \approx 0.4231$ km/sec **11.** $3/(4\pi) \approx 0.24$ ft/min **13.** 326×10^{12} cubic miles per hour **15.** 7.5 ft/sec

17. 80 m/sec **19.** Decreasing at a rate of \$1.66 per player per week **21.** Monthly sales will drop at a rate of 26 T-shirts per month. **23.** Raise the price by 3¢ per week. **25.** Decreasing at a rate of \$75.48 million per year **27.** 1 laborer per month **29.** Dropping at a rate of \$2.40 per year. **31.** The price is decreasing at a rate of approximately 31¢ per pound per month. **33.** $2{,}300/\sqrt{4{,}100} \approx 36$ miles per hour. **35.** About 10.7 ft/sec **37.** The y-coordinate is decreasing at a rate of 16 units per second. **39.** \$534 per year **41.** Their prior experience must increase at a rate of approximately 0.97 years every year. **43.** $\frac{2{,}500}{9\pi}\left(\frac{3}{5{,}000}\right)^{2/3} \approx 0.63$ m/sec **45.** $\frac{\sqrt{1+128\pi}}{4\pi} \approx 1.6$ cm/sec **47.** 0.5137 computers per household and increasing at a rate of 0.0230 computers per household per year. **49.** The average SAT score was 904.71 and decreasing at a rate of 0.11 per year. **51.** Decreasing by 2 percentage points per year **53.** The section is called "Related Rates" because the goal is to compute the rate of change of a quantity based on a knowledge of the rate of change of a related quantity. **55.** Answers may vary: A rectangular solid has dimensions 2 cm × 5 cm × 10 cm, and each side is expanding at a rate of 3 cm/sec. How fast is the volume increasing? **57.** Proof **59.** Linear **61.** Let $x =$ my grades and $y =$ your grades. If $dx/dt = 2\,dy/dt$, then $dy/dt = (1/2)\,dx/dt$.

Section 5.6

1. $E = 1.5$; the demand is going down 1.5% per 1% increase in price at that price level; revenue is maximized when $p = \$25$; weekly revenue at that price is \$12,500. **3. a.** $E = \frac{6p}{-6p + 3{,}030}$ **b.** 1.97; The demand was going down 1.97% per 1% increase in price at that price level. **c.** Price for maximum revenue: \$252.50 per phone for an annual revenue of \$382,540 million, or \$382.54 billion **5. a.** The elasticity would have dropped from 1.81 to 0.96, suggesting that the tuition for maximum annual revenue would have been between these two values. **b.** They should have charged an average of \$2,250 per student, and this would have resulted in an enrollment of about 4,950 students, giving an annual revenue of about \$11,137,500. **7. a.** $E = 6/7$; the demand is going down 6% per 7% increase in price at that price level; thus, a price increase is in order. **b.** Revenue is maximized when $p = 100/3 \approx \$33.33$ **c.** 4,444 cases per week **9. a.** $E = (4p - 33)/(-2p + 33)$ **b.** 0.54; the demand for $E = mc^2$ T-shirts is going down by about 0.54% per 1% increase in the price. **c.** \$11 per shirt for a daily revenue of \$1,331 **11.** $E = 0.01p$; 0.5, 4; yes; $E(50) < 1$ and $E(400) > 1$, so the price that would result in $E = 1$ lies between 50 and 400. **13.** $q = 175{,}502/p^{10/13}$; $E = 10/13$, showing an inelastic demand. Thus, increasing the price will result in increasing revenue **15. a.** $E = 51$; the demand is going down 51% per 1% increase in price at that price level; thus, a large price decrease is advised. **b.** ¥50 **c.** About 78 paint-by-number sets per month **17. a.** $E = -\frac{mp}{mp + b}$ **b.** $p = -\frac{b}{2m}$ **19. a.** $E = r$ **b.** E is independent of p. **c.** If $r = 1$, then the revenue is not affected by the price. If $r > 1$, then the revenue is always elastic, while if $r < 1$, the revenue is always inelastic. This is an unrealistic model because there should always be a price at which the revenue is a maximum. **21. a.** $q = -1{,}500p + 6{,}000$ **b.** \$2 per hamburger, giving a total weekly revenue of \$6,000 **23. a.** $q = 1{,}000e^{-0.30p}$ **b.** At $p = \$3$, $E = 0.9$; at $p = \$4$, $E = 1.2$; at $p = \$5$, $E = 1.5$ **c.** $p = \$3.33$ **d.** $p = \$5.36$. Selling at a lower price would increase demand, but you cannot sell more than 200 lb anyway. You should charge as much as you can and still be able to sell all 200 lb. **25.** $E \approx 0.77$. At a family income level of \$20,000 the fraction of children attending a live theatrical performance is increasing by 0.77% per 1% increase in household income. **27. a.** $E = \frac{1.554xe^{-0.021x}}{-74e^{-0.021x} + 92}$; $E(100) \approx 0.23$: At a household income level of \$100,000 the percentage of people using broadband in 2010 was increasing by 0.23% per 1% increase in household income. **b.** The model predicts elasticity approaching zero for households with large incomes. **29. a.** $E \approx 0.46$. The demand for computers was increasing by 0.46% per 1% increase in household income. **b.** E decreases as income increases. **c.** Unreliable; it predicts a likelihood greater than 1 at incomes of \$123,000 and above. In a more appropriate model we would expect the curve to level off at or below 1. **d.** $E \approx 0$ **31.** 0.24; When the price of oil is \$60 per barrel, Saudi production increases at a rate of 0.24% per 1% increase in the price. **33.** The income elasticity of demand is $\frac{dQ}{dY}\cdot\frac{Y}{Q} = a\beta P^{\alpha}Y^{\beta-1}\frac{Y}{aP^{\alpha}Y^{\beta}} = \beta$. **35.** The price is lowered. **37.** Start with $R = pq$, and differentiate with respect to p to obtain $\frac{dR}{dp} = q + p\frac{dq}{dp}$. For a stationary point, $dR/dp = 0$, so $q + p\frac{dq}{dp} = 0$. Rearranging this result gives $p\frac{dq}{dp} = -q$ and hence $-\frac{dq}{dp}\cdot\frac{p}{q} = 1$, or $E = 1$, showing that stationary points of R correspond to points of unit elasticity. **39.** The distinction is best illustrated by an example. Suppose that q is measured in weekly sales and p is the unit price in dollars. Then the quantity $-dq/dp$ measures the drop in weekly sales per \$1 increase in price. The elasticity of demand E, on the other hand, measures the *percentage* drop in sales per 1% increase in price. Thus, $-dq/dp$ measures absolute change, while E measures fractional, or percentage, change.

Chapter 5 Review

1. Relative max: $(-1, 5)$, absolute min: $(-2, -3)$ and $(1, -3)$ **3.** Absolute max: $(-1, 5)$, absolute min: $(1, -3)$ **5.** Absolute min: $(1, 0)$ **7.** Absolute min: $(-2, -1/4)$ **9.** Relative max: $x = 1$, point of inflection: $x = -1$

11. Relative max: $x = -2$, relative min: $x = 1$, point of inflection: $x = -1$ **13.** One point of inflection, at $x = 0$ **15. a.** $a = 4/t^4 - 2/t^3$ m/sec^2 **b.** 2 m/sec^2 **17.** Relative max: $(-2, 16)$; absolute min: $(2, -16)$; point of inflection: $(0, 0)$; no horizontal or vertical asymptotes

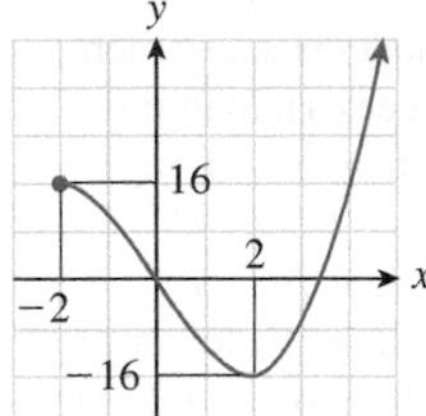

19. Relative min: $(-3, -2/9)$; relative max: $(3, 2/9)$; points of inflection: $(-3\sqrt{2}, -5\sqrt{2}/36)$, $(3\sqrt{2}, 5\sqrt{2}/36)$; vertical asymptote: $x = 0$; horizontal asymptote: $y = 0$

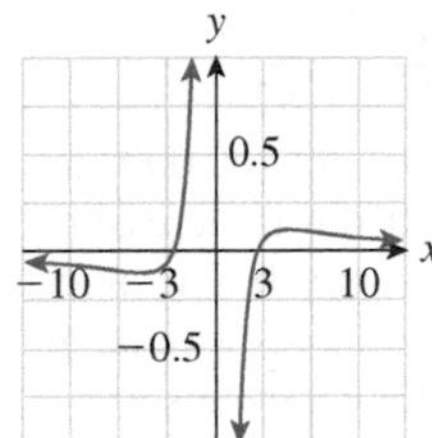

21. Relative max: $(0, 0)$, absolute min: $(1, -2)$, no asymptotes

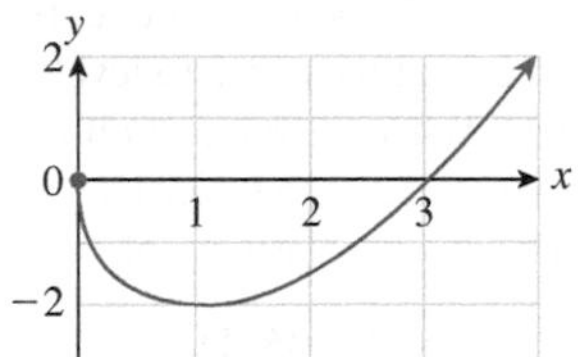

23. \$22.14 per book **25. a.** Profit $= -p^3 + 42p^2 - 288p - 181$ **b.** \$24 per copy; \$3,275 **c.** For maximum revenue the company should charge \$22.14 per copy. At this price, the cost per book is decreasing with increasing price, while the revenue is not decreasing (its derivative is zero). Thus, the profit is increasing with increasing price, suggesting that the maximum profit will occur at a higher price. **27.** 30 ft × 10 ft, for an area of 300 sq. ft. **29.** 12 in × 12 in × 6 in, for a volume of 864 in^3 **31. a.** Week 5 **b.** Point of inflection on the graph of s; maximum on the graph of s', t-intercept in the graph of s''. **c.** 10,500; if weekly sales continue as predicted by the model, they will level off at around 10,500 books per week in the long term. **d.** 0; if weekly sales continue as predicted by the model, the rate of change of sales approaches zero in the long term.

33. a.–d. $10/\sqrt{2}$ ft/sec **35. a.** $E = \dfrac{2p^2 - 33p}{-p^2 + 33p + 9}$ **b.** 0.52, 2.03; when the price is \$20, demand is dropping at a rate of 0.52% per 1% increase in the price; when the price is \$25, demand is dropping at a rate of 2.03% per 1% increase in the price. **c.** \$22.14 per book **37. a.** $E = 2p^2 - p$ **b.** 6; the demand is dropping at a rate of 6% per 1% increase in the price. **c.** \$1.00, for a monthly revenue of \$1,000

Chapter 6

Section 6.1

1. $x^6/6 + C$ **3.** $6x + C$ **5.** $x^2/2 + C$ **7.** $x^3/3 - x^2/2 + C$ **9.** $x + x^2/2 + C$ **11.** $-x^{-4}/4 + C$ **13.** $x^{3.3}/3.3 - x^{-0.3}/0.3 + C$ **15.** $u^3/3 - \ln|u| + C$ **17.** $4x^{5/4}/5 + C$ **19.** $3x^5/5 + 2x^{-1} - x^{-4}/4 + 4x + C$ **21.** $2\ln|u| + u^2/8 + C$ **23.** $\ln|x| - \dfrac{2}{x} + \dfrac{1}{2x^2} + C$ **25.** $3x^{1.1}/1.1 - x^{5.3}/5.3 - 4.1x + C$ **27.** $\dfrac{x^{0.9}}{0.3} + \dfrac{40}{x^{0.1}} + C$ **29.** $2.55t^2 - 1.2\ln|t| - \dfrac{15}{t^{0.2}} + C$ **31.** $2e^x + 5x|x|/2 + x/4 + C$ **33.** $12.2x^{0.5} + x^{1.5}/9 - e^x + C$ **35.** $\dfrac{2^x}{\ln 2} - \dfrac{3^x}{\ln 3} + C$ **37.** $\dfrac{100(1.1^x)}{\ln(1.1)} - \dfrac{x|x|}{3} + C$ **39.** $x^3/3 + 3x^{-5}/10 + C$ **41.** $-1/x - 1/x^2 + C$ **43.** $f(x) = x^2/2 + 1$ **45.** $f(x) = e^x - x - 1$ **47.** $C(x) = 5x - x^2/20{,}000 + 20{,}000$ **49.** $C(x) = 5x + x^2 + \ln x + 994$ **51. a.** $M(t) = -2.2t^3 + 9t^2 + 220t + 360$ **b.** 1,332 million members **53.** $I(t) = 42{,}000 + 1{,}200t$; \$48,000 **55.** $S(t) = 0.08t^3/3 - 0.13t^2 + 8.8t$; 64 billion gal **57. a.** $H'(t) = 3.5t + 65$ billion dollars per year **b.** $H(t) = 1.75t^2 + 65t + 700$ billion dollars **59. a.** $-0.4t + 1$ percentage points per year **b.** $-0.2t^2 + t + 13$; 13.8 **61. a.** $s = t^3/3 + t + C$ **b.** $C = 1$; $s = t^3/3 + t + 1$ **63.** 320 ft/sec downward **65. a.** $v(t) = -32t + 16$ **b.** $s(t) = -16t^2 + 16t + 185$; zenith at $t = 0.5$ sec, $s = 189$ ft, 4 ft above the top of the tower. **67. a.** $s = 525t + 25t^2$ **b.** 3 hours **c.** The negative solution indicates that, at time $t = -24$, the tailwind would have been large and negative, causing the plane to be moving backward through that position 24 hours before departure and arrive at the starting point of the flight at time 0! **69.** Proof **71.** $(1{,}280)^{1/2} \approx 35.78$ ft/sec **73. a.** 80 ft/sec **b.** 60 ft/sec **c.** 1.25 sec **75.** $\sqrt{2} \approx 1.414$ times as fast **77.** The term *indefinite* refers to the arbitrary constant term in the indefinite integral; we do not obtain a definite value for C, and hence the integral is "not definite." **79.** Constant; because the derivative of a linear function is constant, linear functions are antiderivatives of constant functions. **81.** No; there are infinitely many antiderivatives of a given function, each pair of them differing by a constant. Knowing the value of the function at a specific point suffices. **83.** They differ by a constant, $G(x) - F(x) =$ Constant **85.** Antiderivative, marginal **87.** Up to a constant, $\int f(x)\,dx$ represents the total cost of manufacturing x items. The units of $\int f(x)\,dx$ are the product of the units of $f(x)$ and the units of x. **89.** The indefinite integral of the constant 1 is not zero; it is x (+ constant). Correct answer: $3x^2/2 + x + C$ **91.** There should be no integral sign $(\int)$ in the answer. Correct answer: $2x^6 - 2x^2 + C$ **93.** The integral of a constant times a function is the constant times the integral of the function, not the *integral* of the constant times the integral of the function.

(In general, the integral of a product is *not* the product of the integrals.) Correct answer: $4(e^x - x^2) + C$ **95.** It is the *integral* of $1/x$ that equals $\ln|x| + C$, not $1/x$ itself. Correct answer: $\int \frac{1}{x}\,dx = \ln|x| + C$ **97.** $\int (f(x) + g(x))\,dx$ is, by definition, an antiderivative of $f(x) + g(x)$. Let $F(x)$ be an antiderivative of $f(x)$, and let $G(x)$ be an antiderivative of $g(x)$. Then, because the derivative of $F(x) + G(x)$ is $f(x) + g(x)$ (by the rule for sums of derivatives), this means that $F(x) + G(x)$ is an antiderivative of $f(x) + g(x)$. In symbols, $\int (f(x) + g(x))\,dx = F(x) + G(x) + C = \int f(x)\,dx + \int g(x)\,dx$, the sum of the indefinite integrals. **99.** Answers will vary. $\int x \cdot 1\,dx = \int x\,dx = x^2/2 + C$, whereas $\int x\,dx \cdot \int 1\,dx = (x^2/2 + D) \cdot (x + E)$, which is not the same as $x^2/2 + C$, no matter what values we choose for the constants C, D, and E. **101.** Derivative; indefinite integral; indefinite integral; derivative

Section 6.2

1. $(3x - 5)^4/12 + C$ **3.** $(3x - 5)^4/12 + C$ **5.** $-e^{-x} + C$ **7.** $-e^{-x} + C$ **9.** $\frac{1}{2}e^{(x+1)^2} + C$ **11.** $(3x + 1)^6/18 + C$ **13.** $1.6(3x - 4)^{3/2} + C$ **15.** $2e^{(0.6x+2)} + C$ **17.** $(3x^2 + 3)^4/24 + C$ **19.** $2(3x^2 - 1)^{3/2}/9 + C$ **21.** $-(x^2 + 1)^{-0.3}/0.6 + C$ **23.** $(4x^2 - 1)|4x^2 - 1|/16 + C$ **25.** $x + 3e^{3.1x-2} + C$ **27.** $-(1/2)e^{-x^2+1} + C$ **29.** $-(1/2)e^{-(x^2+2x)} + C$ **31.** $(x^2 + x + 1)^{-2}/2 + C$ **33.** $(2x^3 + x^6 - 5)^{1/2}/3 + C$ **35.** $(x - 2)^7/7 + (x - 2)^6/3 + C$ **37.** $4[(x + 1)^{5/2}/5 - (x + 1)^{3/2}/3] + C$ **39.** $20\ln|1 - e^{-0.05x}| + C$ **41.** $3e^{-1/x} + C$ **43.** $-\frac{(4 + 1/x^2)^4}{8} + C$ **45.** $(e^x - e^{-x})/2 + C$ **47.** $\ln(e^x + e^{-x}) + C$ **49.** $-(1 - e^{3x-1})|1 - e^{3x-1}|/6 + C$ **51.** $(e^{2x^2-2x} + e^{x^2})/2 + C$ **53.** Derivation **55.** Derivation **57.** $-e^{-x} + C$ **59.** $(1/2)e^{2x-1} + C$ **61.** $(2x + 4)^3/6 + C$ **63.** $(1/5)\ln|5x - 1| + C$ **65.** $(1.5x)^4/6 + C$ **67.** $\frac{1.5^{3x}}{3\ln(1.5)} + C$ **69.** $\frac{1}{4}(2x + 4)|2x + 4| + C$ **71.** $\frac{2^{3x+4} - 2^{-3x+4}}{3\ln 2} + C$ **73.** $f(x) = (x^2 + 1)^4/8 - 1/8$ **75.** $f(x) = (1/2)e^{x^2-1}$ **77.** $(5x^2 - 3)^7/70 + C$ **79.** $2(3e^x - 1)^{1/2}/3 + C$ **81.** $e^{(x^4-8)}/4 + C$ **83.** $\frac{1}{6}\ln|1 + 2e^{3x}| + C$ **85.** $C(x) = 5x - 1/(x + 1) + 995.5$ **87.** $G(t) = 2{,}000t + 8{,}000e^{-0.06t} - 8{,}000$; 7,107 billion pesos **89.** $P(t) = -\frac{(-0.075t + 0.97)^6}{0.45} + 0.75t + 1.9$; \$7.1 billion **91. a.** $N(t) = 35\ln(5 + e^{0.2t}) - 63$ **b.** 80,000 articles **93.** $S(t) = 3{,}600[\ln(3 + e^{0.25t}) - \ln 4]$; 6,310 sets **95. a.** $s = (t^2 + 1)^5/10 + t^2/2 + C$ **b.** $C = 9/10$; $s = (t^2 + 1)^5/10 + t^2/2 + 9/10$ **97. a.** $S(t) = 0.08(t - 2{,}007)^3/3 - 0.13(t - 2{,}007)^2 + 8.8(t - 2{,}007) - 25.95$ million gal **b.** Approximately 38.43 million gal **99.** None; the substitution $u = x$ simply replaces the letter x throughout by the letter u and thus does not change the integral at all. For instance, the integral $\int x(3x^2 + 1)\,dx$ becomes $\int u(3u^2 + 1)\,du$ if we substitute $u = x$. **101.** It may mean that, but it may not; see Example 4. **103.** (D); to compute the integral, first break it up into a sum of two integrals, $\int \frac{x}{x^2 - 1}\,dx + \int \frac{3x}{x^2 + 1}\,dx$, and then compute the first using $u = x^2 - 1$ and the second using $u = x^2 + 1$. **105.** There are several errors: First, the term "dx" is missing in the integral, and this affects all the subsequent steps (since we must substitute for dx when changing to the variable u). Second, when there is a noncanceling x in the integrand, we cannot treat it as a constant. Correct answer: $3(x^2 - 1)^2/4 + C$ **107.** In the fourth step, u was substituted back for x before the integral was taken, and du was just changed to dx; they're not equal. Correct answer: $(x^3 - 1)^2/6 + C$ **109.** Proof

Section 6.3

1. 4 **3.** 6 **5.** 0 **7.** 0.7456 **9.** 2.3129 **11.** 30 **13.** 22 **15.** −2 **17.** 0 **19.** 1 **21.** 1/2 **23.** 1/4 **25.** 2 **27.** 0 **29.** 6 **31.** 0 **33.** 0.5 **35.** 3.3045, 3.1604, 3.1436 **37.** 0.0275, 0.0258, 0.0256 **39.** 15.76 liters **41.** \$99.95 **43.** \$4,480,000 **45.** 54 billion gal **47.** \$300 billion **49. a.** Left sum: about 5,200; right sum: 5,680 **b.** 5,440; A total of about 5.44 million articles were published by researchers in the United States in the 20-year period beginning 1990. **51.** 118,750 degrees **53.** 54,000 students **55.** Left sum: 76; right sum: 71; Left sum, as it is the sum of the annual net incomes for the given years. **57. a.** Left sum = 5.44; right sum = 5.34 **b.** Pemex produced a total of 5.34 billion barrels of crude oil in the period 2009 through 2013. **59.** −84.8; after 4 sec, the stone is about 84.8 ft below where it started. **61.** 91.2 ft **63.** 2,527; A total of about 2,527 million members joined Facebook from the start of 2008 to the start of 2013. **65.** 34,150; A total of about 34,150 wiretaps were authorized by U.S. state and federal courts during the 15-year period starting January 2000. **67.** Yes. The Riemann sum gives an estimated area of 420 sq. ft. **69. a.** Graph:

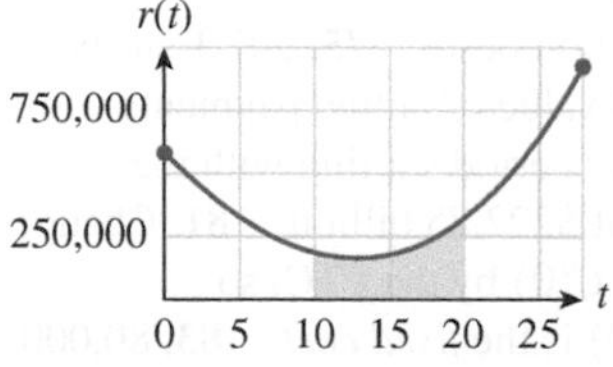

Total expenditure on oil in the United States from 1990 to 2000. **b.** 2,010,000; a total of \$2,010,000 million, or \$2.01 trillion, was spent on oil in the United States from 1990 to 2000. **71. a.** 99.4% **b.** 0 (to at least 15 decimal places) **73.** Stays the same **75.** Increases **77.** The area under the curve and above the x-axis equals the area above the curve and below the x-axis. **79.** Answers will vary. One example: Let $r(t)$ be the

rate of change of net income at time t. If $r(t)$ is negative, then the net income is decreasing, so the change in net income, represented by the definite integral of $r(t)$, is negative.
81. Answers may vary.

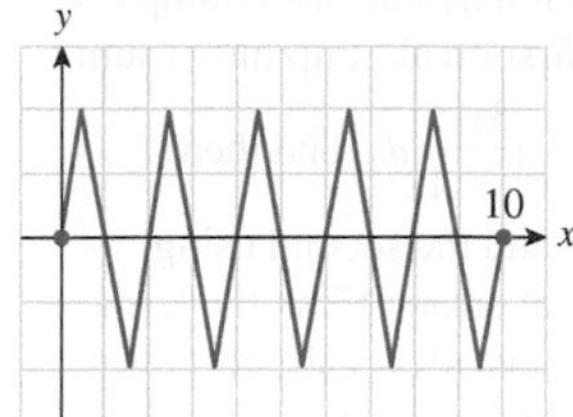

83. The total cost is $c(1) + c(2) + \cdots + c(60)$, which is represented by the Riemann sum approximation of $\int_1^{61} c(t)\,dt$ with $n = 60$.
85. $[f(x_1) + f(x_2) + \cdots + f(x_n)]\,\Delta x = \sum_{k=1}^{n} f(x_k)\,\Delta x$
87. Answers may vary.

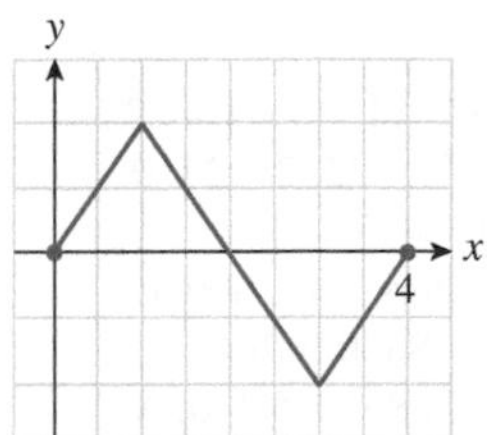

Section 6.4

1. 14/3 **3.** 5 **5.** 0 **7.** 40/3 **9.** $1.05 - 4.3/2.2 \approx -0.9045$ **11.** $2(e - 1)$ **13.** 2/3 **15.** $1/\ln 2$ **17.** 4,095 **19.** $(e^1 - e^{-3})/2$ **21.** $3/(2 \ln 2)$ **23.** 40/3 **25.** 5/4 **27.** 0 **29.** $(5/2)(e^3 - e^2)$ **31.** $(1/3)(\ln 26 - \ln 7)$ **33.** $\frac{0.1}{2.2 \ln(1.1)}$ **35.** $e - e^{1/2}$ **37.** 0.2221 **39.** $2 - \ln 3$ **41.** $-4/21$ **43.** $3^{5/2}/10 - 3^{3/2}/6 + 1/15$ **45.** 1/2 **47.** 16/3 **49.** 9/2 **51.** 56/3 **53.** 1/2 **55.** 18 liters **57.** \$783 **59.** \$4,500,000 **61.** 296 miles **63.** 260 ft **65.** 9 gal **67.** 56 billion gal **69.** 1,595 million members **71.** 34,310; A total of about 34,310 wiretaps were authorized by U.S. state and federal courts during the 15-year period starting January 2000. **73.** $G(T) = 2{,}000T + 8{,}000e^{-0.06T} - 8{,}000$; 7,107 billion pesos **75.** 907 T-shirts **77.** 68 milliliters **79.** The total value of online payments through PayPal during the 1-year period starting with the fourth quarter of 2013 was about \$222.85 billion. **81.** Change in cost $= \int_0^x m(t)\,dt = C(x) - C(0)$ by the FTC, so $C(x) = C(0) + \int_0^x m(t)\,dt$. $C(0)$ is the *fixed cost*. **83.** 80,000 articles **85. a., b.** Proofs **c.** 2,401 thousand students **87.** Proof **89.** They are related by the Fundamental Theorem of Calculus, which (briefly) states that the definite integral of a suitable function can be calculated by evaluating the indefinite integral at the two endpoints and subtracting. **91.** Computing its definite integral from a to b **93.** Calculate definite integrals by using antiderivatives. **95.** An example is $v(t) = t - 5$. **97.** An example is $f(x) = e^{-x}$. **99.** By the FTC, $\int_a^x f(t)\,dt = G(x) - G(a)$, where G is an antiderivative of f. Hence, $F(x) = G(x) - G(a)$. Taking derivatives of both sides, $F'(x) = G'(x) + 0 = f(x)$, as required. The result gives us a formula, in terms of area, for an antiderivative of any continuous function.

Chapter 6 Review

1. $x^3/3 - 5x^2 + 2x + C$ **3.** $4x^3/15 + 4/(5x) + C$ **5.** $(1/2) \ln|2x| + C$ or $(1/2) \ln|x| + C$ **7.** $-e^{-2x+11}/2 + C$ **9.** $(x^2 + 1)^{2.3}/4.6 + C$ **11.** $2 \ln|x^2 - 7| + C$ **13.** $\frac{1}{6}(x^4 - 4x + 1)^{3/2} + C$ **15.** $-e^{x^2/2} + C$ **17.** $(x + 2) - \ln|x + 2| + C$ or $x - \ln|x + 2| + C$ **19.** 1 **21.** 2.75 **23.** −0.24 **25.** 2.5048 **27.** 0.7778, 0.7500, 0.7471 **29.** 0 **31.** 1 **33.** 2/7 **35.** $50(e^{-1} - e^{-2})$ **37.** 52/9 **39.** $(\ln 5 - \ln 2)/8 = \ln(2.5)/8$ **41.** 32/3 **43.** $(1 - e^{-25})/2$ **45. a.** $N(t) = 196t + t^3/3 - 0.16t^6/6$ **b.** 4 books **47. a.** At a height of $-16t^2 + 100t$ ft **b.** 156.25 ft **c.** After 6.25 sec **49. a.** $24.8e^{0.25t+3} + C$ **b.** $24.8[e^{0.25t+3} - e^3]$ thousand books **51.** 25,000 copies **53.** 0 books **55.** 39,200 hits **57.** \$8,200 **59.** About 86,000 books **61.** About 35,800 books

Chapter 7

Section 7.1

1. $2e^x(x - 1) + C$ **3.** $-e^{-x}(2 + 3x) + C$ **5.** $e^{2x}(2x^2 - 2x - 1)/4 + C$ **7.** $-e^{-2x+4}(2x^2 + 2x + 3)/4 + C$ **9.** $2^x[(2 - x)/\ln 2 + 1/(\ln 2)^2] + C$ **11.** $-3^{-x}[(x^2 - 1)/\ln 3 + 2x/(\ln 3)^2 + 2/(\ln 3)^3] + C$ **13.** $-e^{-x}(x^2 + x + 1) + C$ **15.** $\frac{1}{7}x(x + 2)^7 - \frac{1}{56}(x + 2)^8 + C$ **17.** $-\frac{x}{2(x - 2)^2} - \frac{1}{2(x - 2)} + C$ **19.** $(x^4 \ln x)/4 - x^4/16 + C$ **21.** $(t^3/3 + t) \ln(2t) - t^3/9 - t + C$ **23.** $(3/4)t^{4/3}(\ln t - 3/4) + C$ **25.** $x \log_3 x - x/\ln 3 + C$ **27.** $e^{2x}(x/2 - 1/4) - 4e^{3x}/3 + C$ **29.** $e^x(x^2 - 2x + 2) - e^{x^2}/2 + C$ **31.** $\frac{1}{3}(3x - 4)(2x - 1)^{3/2} - \frac{1}{5}(2x - 1)^{5/2} + C$ **33.** e **35.** 38,229/286 **37.** $(7/2) \ln 2 - 3/4$ **39.** 1/4 **41.** $1 - 11e^{-10}$ **43.** $4 \ln 2 - 7/4$ **45.** $\frac{1}{2}x(x - 3)|x - 3| - \frac{1}{6}(x - 3)^2|x - 3| + C$ **47.** $2x|x - 3| - (x - 3)|x - 3| + C$ **49.** $-x^2(-x + 4)|-x + 4| - \frac{2}{3}x(-x + 4)^2|-x + 4| - \frac{1}{6}(-x + 4)^3|-x + 4| + C$

51. $\frac{1}{2}(x^2 - 2x + 3)(x - 4)|x - 4| - \frac{1}{3}(x - 1)(x - 4)^2|x - 4| + \frac{1}{12}(x - 4)^3|x - 4| + C$
53. $28{,}800{,}000(1 - 2e^{-1})$ ft
55. $5{,}001 + 10x - 1/(x + 1) - [\ln(x + 1)]/(x + 1)$
57. \$1,760 billion **59.** 19 billion square feet
61. 20,800 million gallons **63.** $e^{0.046x}[1{,}040(6.2x^2 - 146x + 1{,}910) - 22{,}600(12.4x - 146) + 491{,}000(12.4)] - 12{,}000{,}000$ **65.** 33,598
67. a. $r(t) = -0.075t + 2.75 + [0.025t - 0.25]\frac{|t - 10|}{t - 10}$
b. 42.5 million people **69.** Answers will vary. Examples are xe^{x^2} and $e^{x^2} = 1 \cdot e^{x^2}$. **71.** Answers will vary. Examples are Exercises 31 and 32, or, more simply, integrals like $\int x(x + 1)^5\,dx$. **73.** Substitution **75.** Parts **77.** Substitution
79. Parts **81.** $n + 1$ times **83.** Proof.

Section 7.2

1. 16/3 **3.** 9.75 **5.** 2 **7.** 31/3 **9.** 8/3 **11.** 4 **13.** 1/3
15. 1 **17.** 2 **19.** 39 **21.** $e - 3/2$ **23.** 2/3 **25.** 3/10
27. 1/20 **29.** 4/15 **31.** 1/3 **33.** 32 **35.** $2 \ln 2 - 1$
37. $8 \ln 4 + 2e - 16$ **39.** 0.9138 **41.** 0.3222 **43.** 112.5. This represents your total profit for the week, \$112.50.
45. 608; there were approximately 608,000 housing starts from the start of 2002 to the start of 2006 not for sale purposes.
47. a. Graph: (The upper curve is MySpace.)

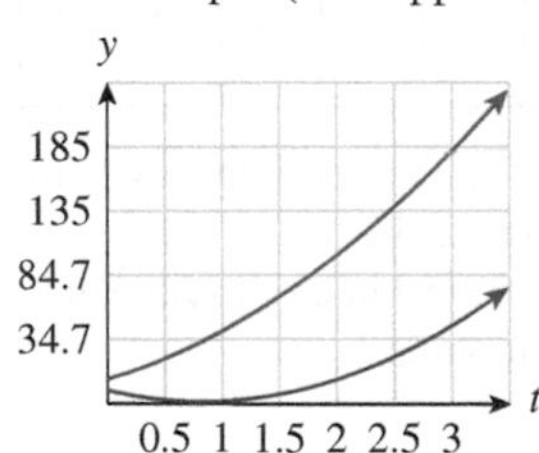

Myspace; about 93 million members

b. The area between the curves $y = f(t)$ and $y = m(t)$ over $[0.5, 2]$ **49.** $12{,}833(e^{0.060t} - 1) - 5{,}774.6(e^{0.071t} - 1)$; the total number of wiretaps authorized by federal courts from the start of 1990 up to time t was about $12{,}833(e^{0.060t} - 1) - 5{,}774.6(e^{0.071t} - 1)$. **51.** Wrong: It could mean that the graphs of f and g cross, as shown in the caution at the start of this topic in the textbook. **53.** The area between the export and import curves represents the accumulated U.S. trade deficit (that is, the total excess of imports over exports) from 1960 to 2007.
55. (A) **57.** The claim is wrong because the area under a curve can represent income only if the curve is a graph of income *per unit time.* The value of a stock price is not income per unit time—the income can be realized only when the stock is sold, and it amounts to the current market price. The total net income (per share) from the given investment would be the stock price on the date of sale minus the purchase price of \$50.

Section 7.3

1. Average = 2

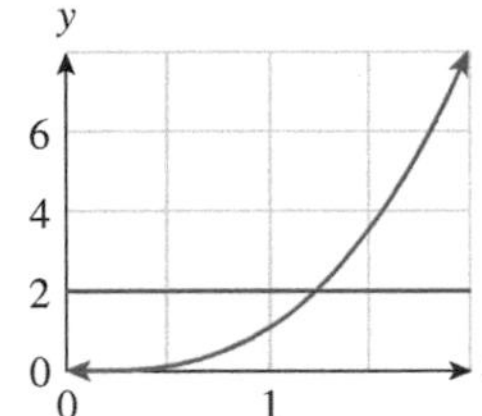

3. Average = 1

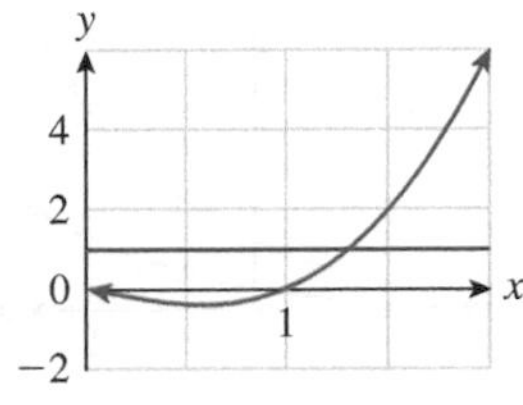

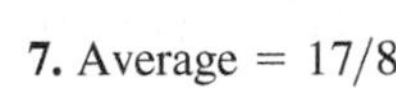

5. Average = $(1 - e^{-2})/2$

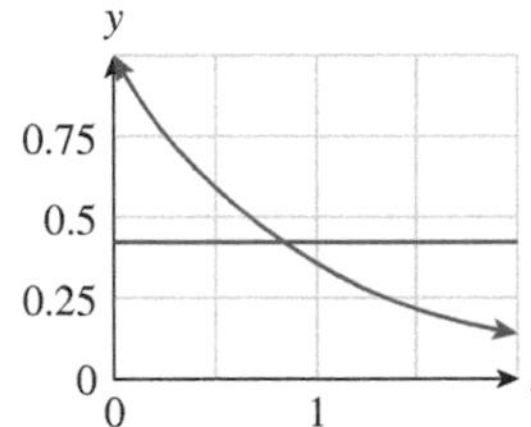

7. Average = 17/8

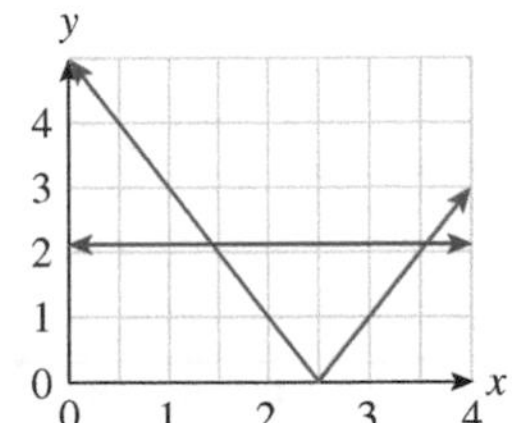

9.

x	0	1	2	3	4	5	6	7
$r(x)$	3	5	10	3	2	5	6	7
$\bar{r}(x)$			6	6	5	$\frac{10}{3}$	$\frac{13}{3}$	6

11.

x	0	1	2	3	4	5	6	7
$r(x)$	1	2	6	7	11	15	10	2
$\bar{r}(x)$			3	5	8	11	12	9

13. Moving average:
$\bar{f}(x) = x^3 - (15/2)x^2 + 25x - 125/4$

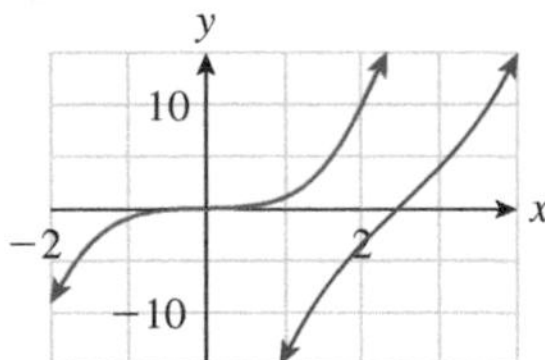

15. Moving average:
$$\bar{f}(x) = \frac{3}{25}[x^{5/3} - (x - 5)^{5/3}]$$

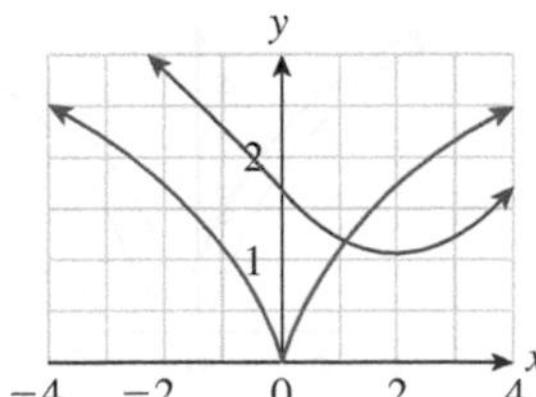

17. $\bar{f}(x) = \frac{2}{5}(e^{0.5x} - e^{0.5(x-5)})$

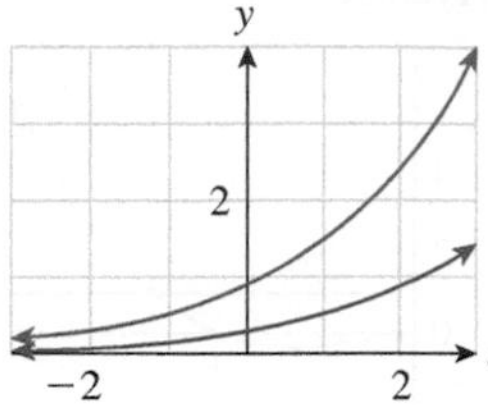

19. $\bar{f}(x) = \frac{2}{15}(x^{3/2} - (x-5)^{3/2})$

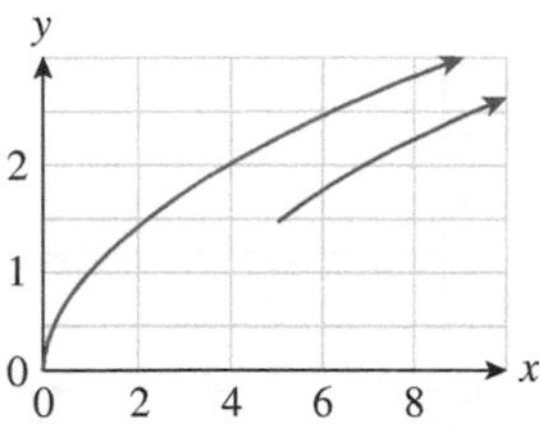

21. $\bar{f}(x) = \frac{1}{5}\left[5 - \frac{1}{2}|2x - 1| + \frac{1}{2}|2x - 11|\right]$

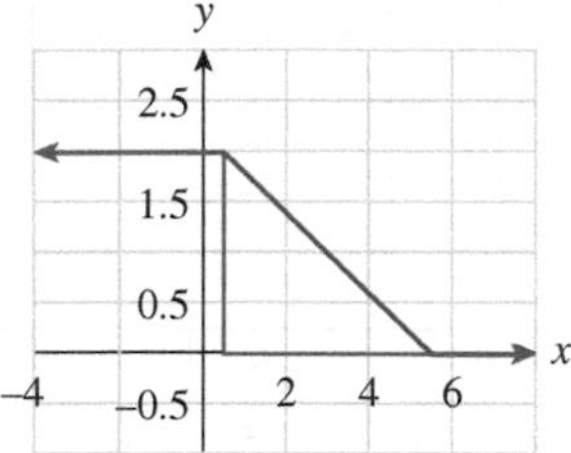

23. $\bar{f}(x) = \frac{1}{5}\left[2x - \frac{1}{2}(x+1)|x+1| + \frac{1}{2}x|x| - 2(x-5) + \frac{1}{2}(x-4)|x-4| - \frac{1}{2}(x-5)|x-5|\right]$

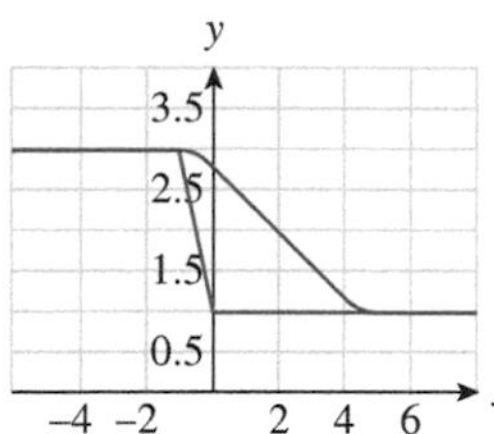

25.

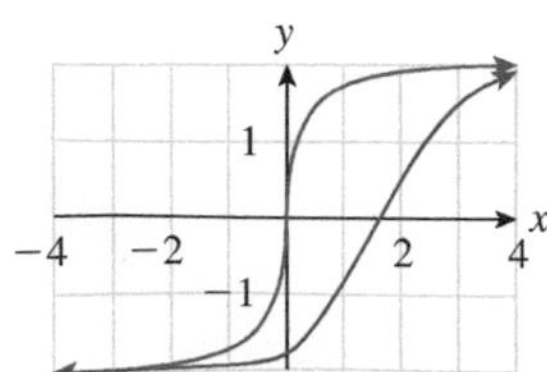

27.

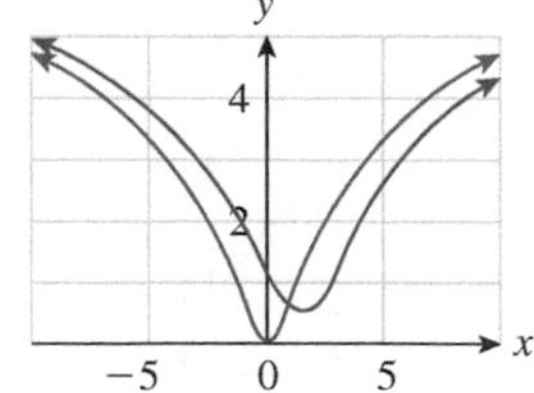

29.

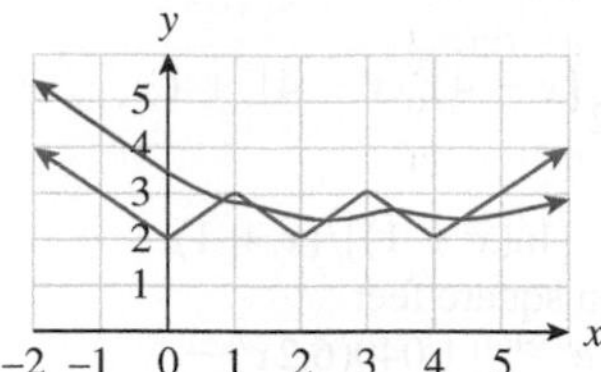

31.

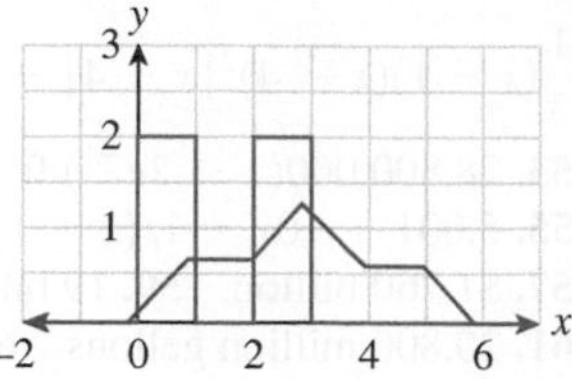

33.

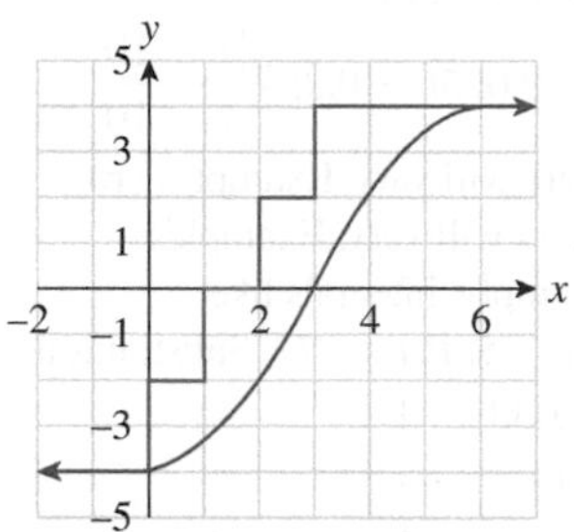

35. \$1.8 million **37.** 16 million members per year
39. 277 million tons **41.** \$10,410.88 **43.** \$1,500
45. \$51 billion per quarter
47.

Year t	2005	2006	2007	2008	2009	2010	2011	2012	2013	2014
Stock Price	56	77	94	80	68	73	85	87	101	92
Moving Avg. (rounded)	46	56	70	77	80	79	77	78	87	91

The moving average continued to rise at a lower rate, began to fall in 2010, then started to rise again in 2012.
49. a. To obtain the moving averages from January to June, use the fact that the data repeat every 12 months.

Graph:

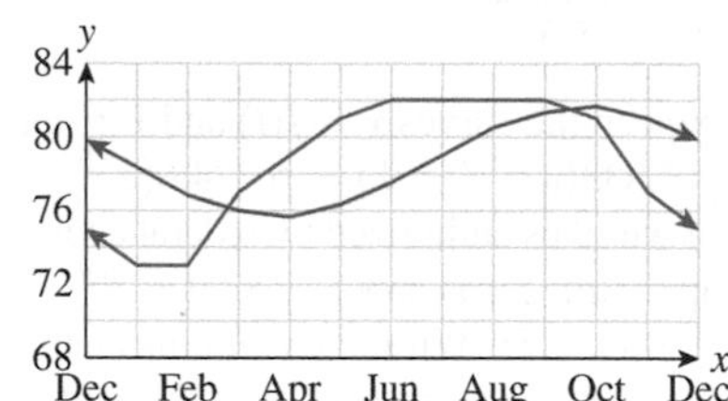

b. The 12-month moving average is constant and equal to the year-long average of approximately 79°.
51. a. Graph:

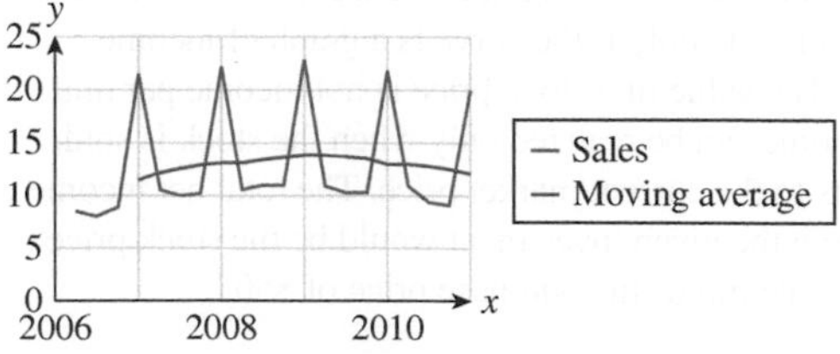

b. Approximately −0.2 million iPods per quarter

53. a. 9.2 billion gallons per year
b. $\frac{1}{2}[0.0267(t^3 - (t-2)^3) - 0.13(t^2 - (t-2)^2) + 17.6]$
c. The function is quadratic because the t^3 terms cancel.
55. a. $s = 41t + 526$ **b.** $\bar{s}(t) = 41t + 444$ **c.** The slope of the moving average is the same as the slope of the original function. **57.** $\bar{f}(x) = mx + b - \frac{ma}{2}$ **59.** The moving average "blurs" the effects of short-term oscillations in the price and shows the longer-term trend of the stock price. **61.** They repeat every 6 months. **63.** The area above the x-axis equals the area below the x-axis. Example: $y = x$ on $[-1, 1]$
65. This need not be the case; for instance, the function $f(x) = x^2$ on $[0, 1]$ has average value 1/3, whereas the value midway between the maximum and minimum is 1/2.
67. (C) A shorter-term moving average most closely approximates the original function because it averages the function over a shorter period, and continuous functions change by only a small amount over a small period.

Section 7.4

1. \$6.25 **3.** \$512 **5.** \$119.53 **7.** \$900 **9.** \$416.67
11. \$326.27 **13.** \$25 **15.** \$0.50 **17.** \$386.29 **19.** \$225
21. \$25.50 **23.** \$12,684.63 **25.** TV = \$300,000, FV = \$434,465.45 **27.** TV = \$350,000, FV = \$498,496.61
29. TV = \$389,232.76, FV = \$547,547.16
31. TV = \$100,000, PV = \$82,419.99 **33.** TV = \$112,500, PV = \$92,037.48 **35.** TV = \$107,889.50, PV = \$88,479.69
37. $\bar{p}$ = \$5,000, $\bar{q}$ = 10,000, CS = \$25 million, PS = \$100 million. The total social gain is \$125 million.
39. €200 billion **41.** \$3,680 billion **43.** €220 billion
45. \$2,990 billion **47.** \$1,943,162.44 **49.** \$3,086,245.73
51. \$54,594.20 **53.** \$1,792,723.35 **55.** total
57. $CS = \frac{1}{2m}(b - m\bar{p})^2$ **59.** She is correct, provided that there is a positive rate of return, in which case the future value (which includes interest) is greater than the total value (which does not). **61.** $PV < TV < FV$

Section 7.5

1. Diverges **3.** Converges to $2e$ **5.** Converges to e^2
7. Converges to 1/2 **9.** Converges to 1/108 **11.** Converges to $3 \times 5^{2/3}$ **13.** Diverges **15.** Diverges **17.** Converges to $\frac{5}{4}(3^{4/5} - 1)$ **19.** Diverges **21.** Converges to 0
23. Diverges **25.** Diverges **27.** 0.9, 0.99, 0.999, . . . ; converges to 1. **29.** 7.602, 95.38, 993.1, . . . ; diverges.
31. 1.368, 1.800, 1.937, 1.980, 1.994, 1.998, 1.999, 2.000, . . . ; converges to 2.000. **33.** 9.000, 99.00, 999.0, . . . ; diverges to $+\infty$. **35.** 4.45 million homes **37.** 10,200 billion cigarettes
39. No; you will not sell more than 2,000 of them. **41.** The number of graduates each year will rise without bound.
43. a. $R(t) = 350e^{-0.1t}(39t + 68)$ million dollars/year
b. \$1,603,000 million **45.** \$17,900 billion **47.** $\int_0^{+\infty} N(t)\,dt$ diverges, indicating that there is no bound to the expected future total online sales of mousse. $\int_{-\infty}^0 N(t)\,dt$ converges to approximately 2.006, indicating that total online sales of mousse prior to the current year amounted to approximately 2 million gallons. **49.** 1 **51.** 0.1587 **53.** \$70,833
55. a. 2.468 meteors on average **b.** The integral diverges. We can interpret this as saying that the number of impacts by meteors smaller than 1 megaton is very large. (This makes sense because, for example, this number includes meteors no larger than a grain of dust.) **57. a.** $\Gamma(1) = 1$; $\Gamma(2) = 1$
b. and **c.** Proofs **59.** The integrand is neither continuous nor piecewise-continuous on the interval $[-1, 1]$, so the FTC does not apply (the integral is improper). **61.** Yes; the integrals converge to 0, and the FTC also gives 0. **63. a.** Not improper. $|x|/x$ is not defined at zero, but $\lim_{x\to 0^-}|x|/x = -1$ and $\lim_{x\to 0^+}|x|/x = 1$. Because these limits are finite, the integrand is piecewise-continuous on $[-1, 1]$ and so the integral is not improper. **b.** Improper, because $x^{-1/3}$ has infinite left and right limits at 0. **c.** Improper, since $(x-2)/(x^2 - 4x + 4) = 1/(x-2)$, which has an infinite left limit at 2. **65.** In all cases you need to rewrite the improper integral as a limit and use technology to evaluate the integral of which you are taking the limit. Evaluate for several values of the endpoint approaching the limit. In the case of an integral in which one of the limits of integration is infinite, you may have to instruct the calculator or computer to use more subdivisions as you approach $+\infty$. **67.** Answers will vary.

Section 7.6

1. $y = \frac{x^3}{3} + \frac{2x^{3/2}}{3} + C$ **3.** $\frac{y^2}{2} = \frac{x^2}{2} + C$ **5.** $y = Ae^{x^2/2}$
7. $y = -\frac{2}{(x+1)^2 + C}$ **9.** $y^2 = (\ln x)^2 + C$
11. $y = \frac{x^4}{4} - x^2 + 1$ **13.** $y = (x^3 + 8)^{1/3}$ **15.** $y = 2x$
17. $y = e^{x^2/2} - 1$ **19.** $y = -\frac{2}{\ln(x^2 + 1) + 2}$ **21.** With $s(t)$ = monthly sales after t months, $\frac{ds}{dt} = -0.05s$; $s = 1{,}000$ when $t = 0$. Solution: $s = 1{,}000e^{-0.05t}$ quarts per month
23. a. $75 + 125e^{-0.05t}$ **b.** 64.4 minutes **25.** $k \approx 0.04274$; $H(t) = 75 + 115e^{-0.04274t}$ degrees Fahrenheit after t minutes
27. With $S(t)$ = total sales after t months, $\frac{dS}{dt} = 0.1(100{,}000 - S)$; $S(0) = 0$. Solution: $S = 100{,}000(1 - e^{-0.1t})$ monitors after t months. **29.** $q = 0.6078e^{-0.05p}p^{1.5}$
31. $y = e^{-t}(t + 1)$ **33.** $y = e^{t/2}[-2te^{-t/2} - 4e^{-t/2} + 5]$

35. $i = 5e^{-t}(e^t - e)\left[1 + \frac{|t-1|}{t-1}\right]$

Graph:

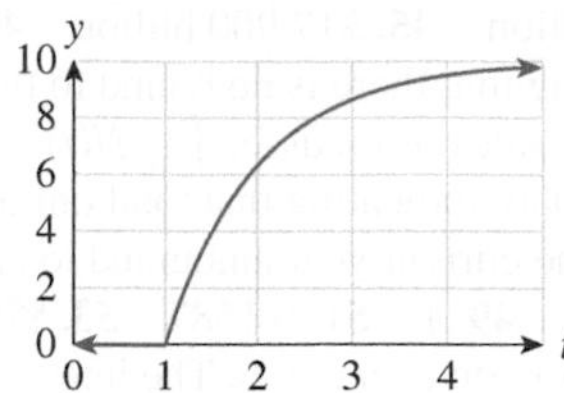

37. a. $\frac{dp}{dt} = k(D(p) - S(p)) = k(20{,}000 - 1{,}000p)$ **b.** $p = 20 - Ae^{-1{,}000kt}$ **c.** $p = 20 - 10e^{-0.2231t}$ dollars after t months **39.** Verification **41.** $S = \frac{2/1{,}999}{e^{-0.5t} + 1/1{,}999}$

Graph:

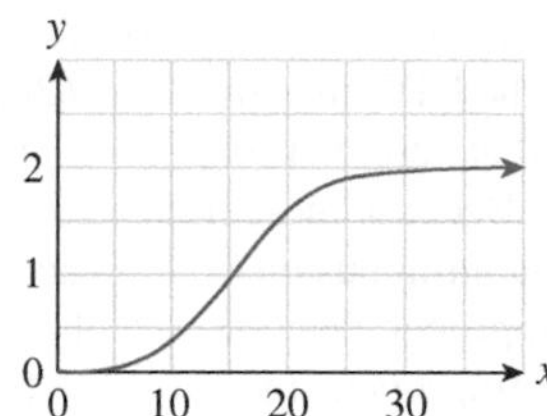

It will take about 30 months to saturate the market.
43. a. $y = be^{Ae^{-at}}$, A = constant **b.** $y = 10e^{-0.69315e^{-t}}$

Graph:

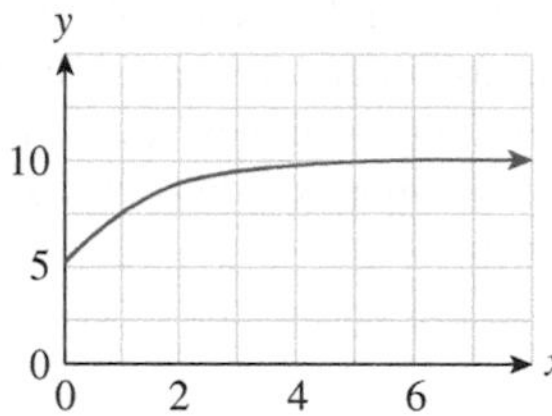

45. A general solution gives all possible solutions to the equation, using at least one arbitrary constant. A particular solution is one specific function that satisfies the equation. We obtain a particular solution by substituting specific values for any arbitrary constants in the general solution. **47.** Example: $y'' = x$ has general solution $y = \frac{1}{6}x^3 + Cx + D$ (integrate twice).
49. $y' = -4e^{-x} + 3$

Chapter 7 Review

1. $(x^2 - 2x + 4)e^x + C$ **3.** $(1/3)x^3 \ln 2x - x^3/9 + C$
5. $\frac{1}{2}x(2x+1)|2x+1| - \frac{1}{12}(2x+1)^2|2x+1| + C$
7. $-5x|-x+3| - \frac{5}{2}(-x+3)|-x+3| + C$
9. $-e^2 - 39/e^2$ **11.** $\frac{3}{2 \cdot 2^{1/3}} - \frac{1}{2}$ **13.** $\frac{2\sqrt{2}}{3}$ **15.** -1
17. $e - 2$ **19.** $3x - 2$ **21.** $\frac{3}{14}[x^{7/3} - (x-2)^{7/3}]$
23. \$1,600 **25.** \$2,500 **27.** 1/4 **29.** Diverges **31.** 1
33. $y = -\frac{3}{x^3 + C}$ **35.** $y = \sqrt{2\ln|x| + 1}$ **37.** \$18,200

39. a. \$2,260 **b.** $50{,}000e^{0.01t}(1 - e^{-0.04}) \approx 1{,}960.53e^{0.01t}$
41. a. $\bar{p} = 20$, $\bar{q} = 40{,}000$ **b.** CS = \$240,000, PS = \$30,000
43. Approximately \$910,000 **45. a.** \$5,549,000 **b.** Principal: \$5,280,000, interest: \$269,000 **47.** \$51 million **49.** The amount in the account would be given by $y = 10{,}000/(1 - t)$, where t is time in years, so would approach infinity 1 year after the deposit.

Chapter 8

Section 8.1

1. a. 1 **b.** 1 **c.** 2 **d.** $a^2 - a + 5$ **e.** $y^2 + x^2 - y + 1$
f. $(x+h)^2 + (y+k)^2 - (x+h) + 1$ **3. a.** 0 **b.** 0.2
c. -0.1 **d.** $0.18a + 0.2$ **e.** $0.1x + 0.2y - 0.01xy$
f. $0.2(x+h) + 0.1(y+k) - 0.01(x+h)(y+k)$
5. a. 1 **b.** e **c.** e **d.** e^{x+y+z} **e.** $e^{x+h+y+k+z+l}$
7. a. Does not exist **b.** 0 **c.** 0 **d.** $xyz/(x^2 + y^2 + z^2)$
e. $(x+h)(y+k)(z+l)/[(x+h)^2 + (y+k)^2 + (x+l)^2]$
9. a. Increases; 2.3 **b.** Decreases; 1.4 **c.** f decreases; 1 unit increase in z **11.** Neither **13.** Linear **15.** Linear
17. Interaction **19. a.** 107 **b.** -14 **c.** -113

21.

y ↓ \ x →	10	20	30	40
10	52	107	162	217
20	94	194	294	394
30	136	281	426	571
40	178	368	558	748

23. Spreadsheet **25.** 18, 4, 0.0965, 47,040
27. 6.9078, 1.5193, 5.4366, 0
29. Let z = annual sales of Z (in millions of dollars), x = annual sales of X, and y = annual sales of Y. The model is $z = -2.1x + 0.4y + 16.2$.

31.

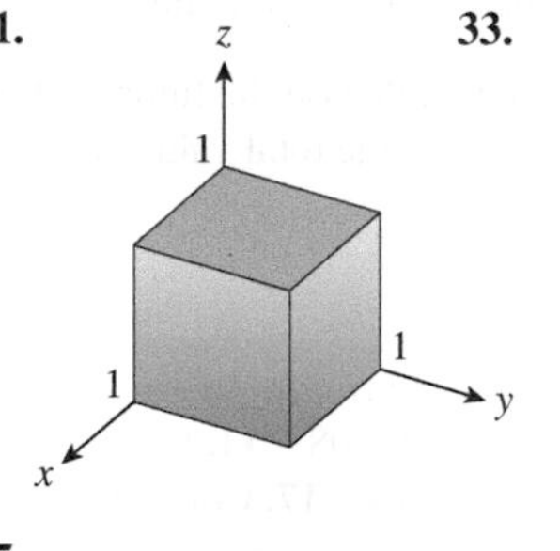

33.

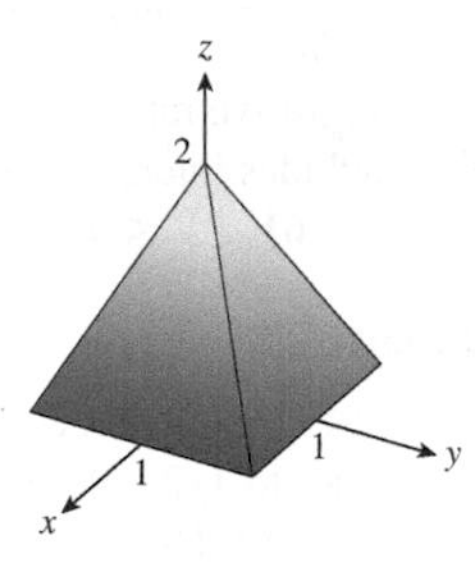

35.

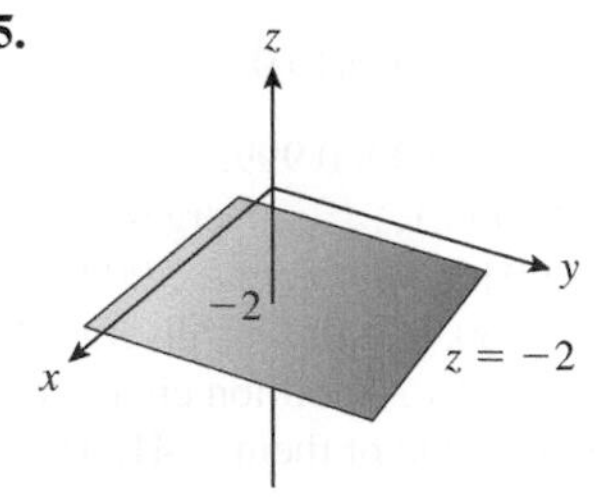

37.

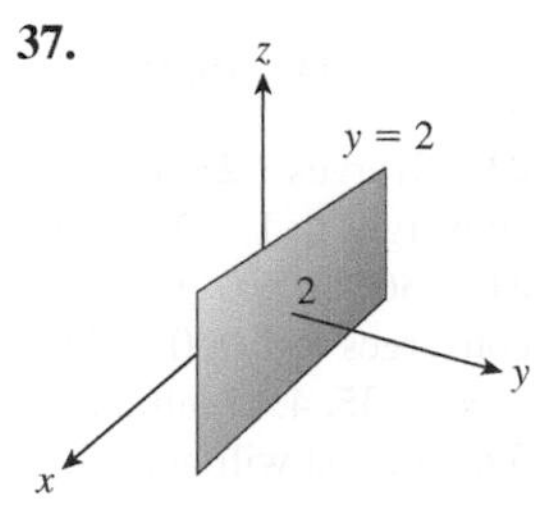

39.

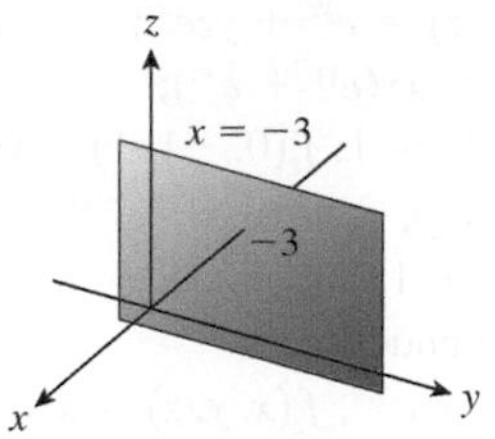

41. (H) **43.** (B) **45.** (F) **47.** (C)

49.

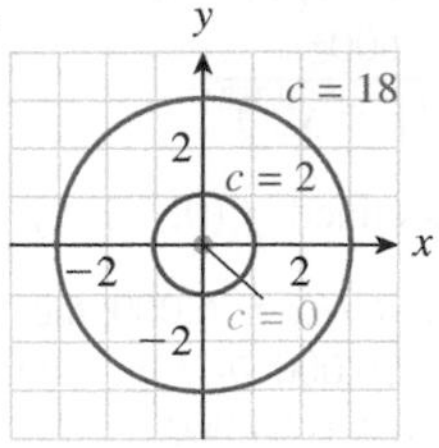

51.

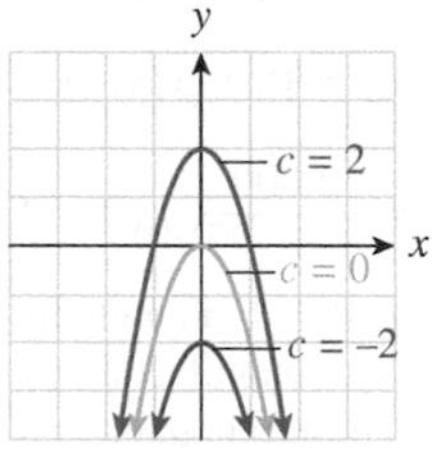

53.

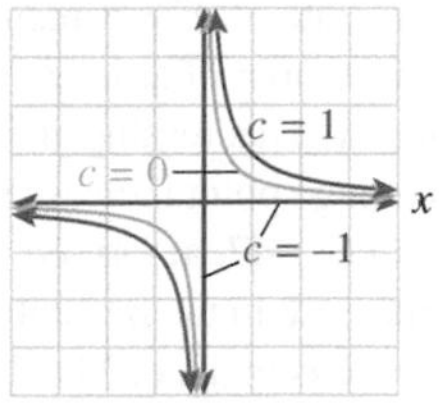

55. a. 4 **b.** 5 **c.** −1 **57.** (2, 2) and (−2, −2)

59.

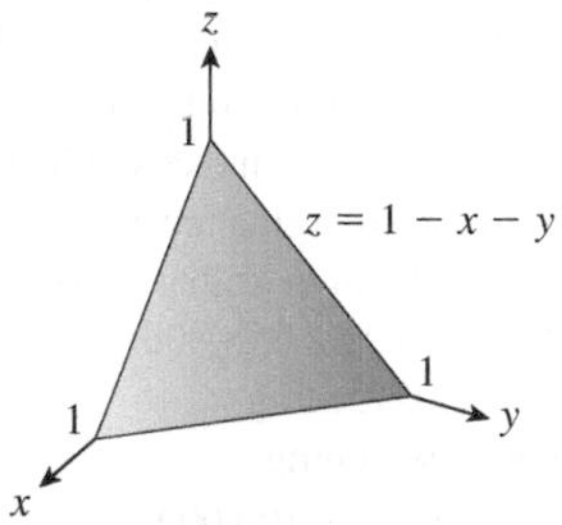

61.

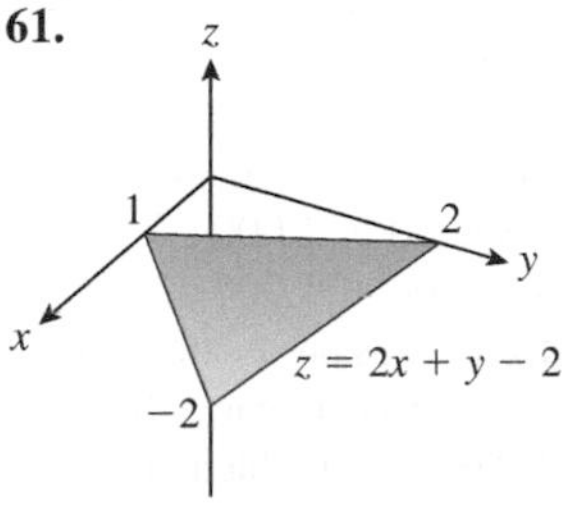

63.

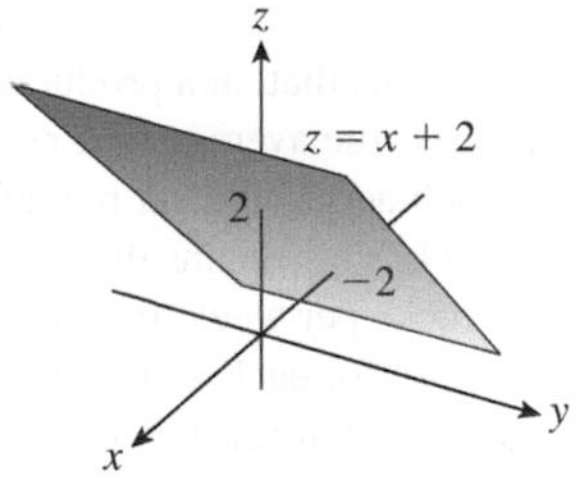

65.

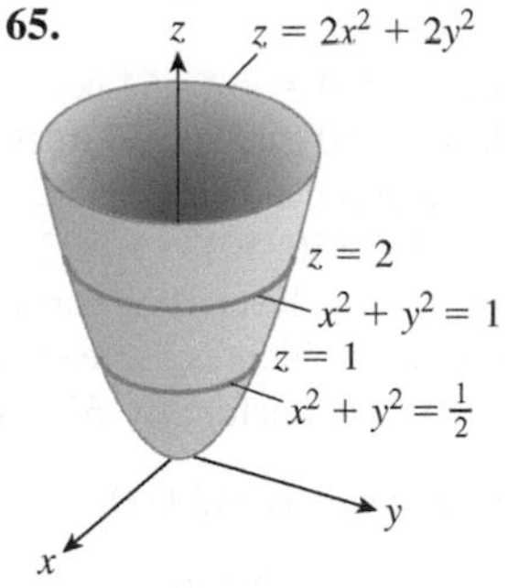

67.

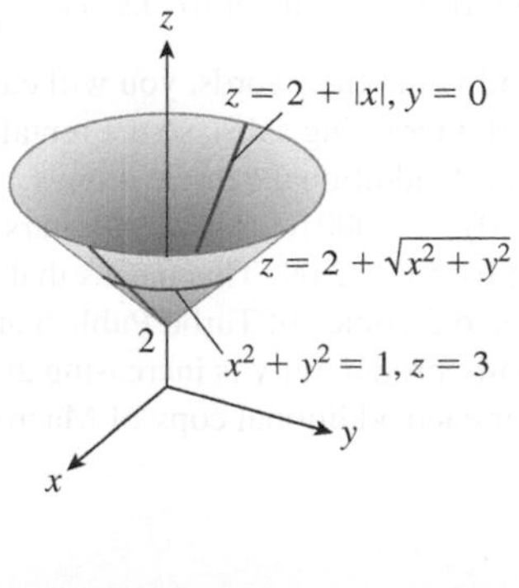

69.

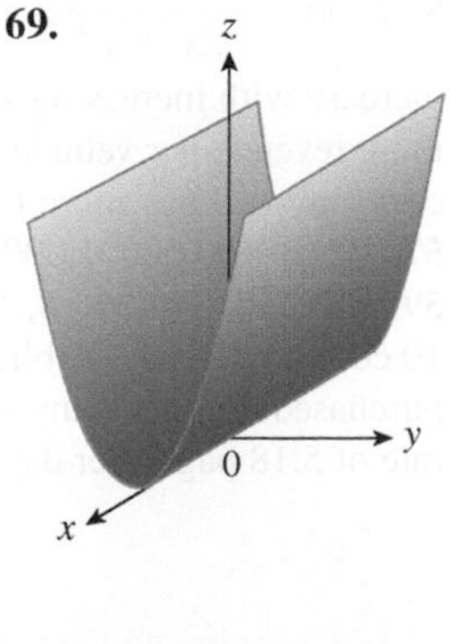

71. **73.**

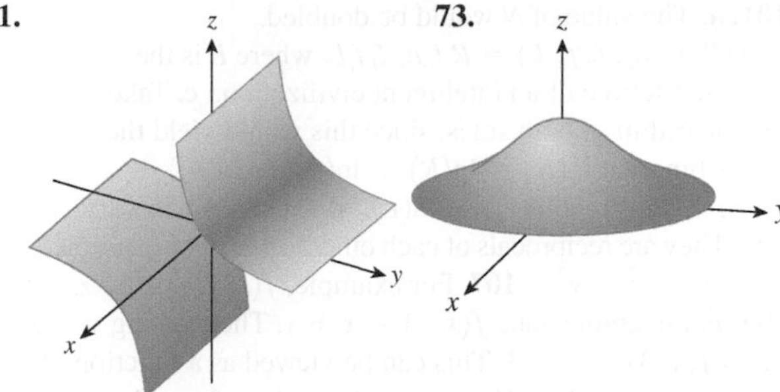

75. a. The marginal cost of cars is \$6,000 per car. The marginal cost of trucks is \$4,000 per truck. **b.** The graph is a plane with x-intercept −40, y-intercept −60, and z-intercept 240,000. **c.** The slice $x = 10$ is the straight line with equation $z = 300{,}000 + 4{,}000y$. It describes the cost function for the manufacture of trucks if car production is held fixed at 10 cars per week. **d.** The level curve $z = 480{,}000$ is the straight line $6{,}000x + 4{,}000y = 240{,}000$. It describes the number of cars and trucks you can manufacture to maintain weekly costs at \$480,000. **77.** $C(x, y) = 10 + 0.03x + 0.04y$, where C is the cost in dollars, x = # video clips sold per month, y = # audio clips sold per month **79. a.** 28% **b.** 21% **c.** Percentage points per year **81.** The graph is a plane with x_1-intercept 0.3, x_2-intercept 33, and x_3-intercept 0.66. The slices by x_1 = constant are straight lines that are parallel to each other. Thus, the rate of change of General Motors' share as a function of Ford's share does not depend on Chrysler's share. Specifically, GM's share decreases by 0.02 percentage points per 1 percentage point increase in Ford's market share, regardless of Chrysler's share. **83. a.** 181 thousand prisoners **b.** $y = 300$: $N = 0.01x + 56$; $y = 500$: $N = 0.09x + 4$. When there are 300,000 prisoners in local jails, the number in federal prisons increases by 10 per 1,000 additional prisoners in state prisons. When there are 500,000 prisoners in local jails, the number in federal prisons increases by 90 per 1,000 additional prisoners in state prisons. **85. a.** The slices x = constant and y = constant are straight lines. **b.** No. Even though the slices x = constant and y = constant are straight lines, the level curves are not, so the surface is not a plane. **c.** The slice $x = 10$ has a slope of 3,800. The slice $x = 20$ has a slope of 3,600. Manufacturing more cars lowers the marginal cost of manufacturing trucks. **87. a.** \$9,980 **b.** $R(z) = 9{,}850 + 0.04z$ **89.** $f(x, t) = 7.14x + 3.57t - 72.86$; \$456 billion **91.** $U(11, 10) - U(10, 10) \approx 5.75$. This means that, if your company now has 10 copies of Macro Publish and 10 copies of Turbo Publish, then the purchase of one additional copy of Macro Publish will result in a productivity increase of approximately 5.75 pages per day. **93. a.** Answers will vary. $(a, b, c) = (3, 1/4, 1/\pi)$; $(a, b, c) = (1/\pi, 3, 1/4)$. **b.** $a = \left(\frac{3}{4\pi}\right)^{1/3}$. The resulting ellipsoid is a sphere with radius a. **95.** 7,000,000 **97. a.** $100 = K(1{,}000)^a(1{,}000{,}000)^{1-a}$; $10 = K(1{,}000)^a(10{,}000)^{1-a}$ **b.** $\log K - 3a = -4$; $\log K - a = -3$ **c.** $a = 0.5$, $K \approx 0.003162$ **d.** $P = 71$ pianos (to the nearest piano) **99. a.** 4×10^{-3} g/m^2 **b.** The total weight of sulfates in the Earth's atmosphere

101. a. The value of N would be doubled. **b.** $N(R, f_p, n_e, f_l, f_i, L) = R f_p n_e f_l f_i L$, where L is the average lifetime of an intelligent civilization. **c.** Take the logarithm of both sides, since this would yield the linear function $\ln(N) = \ln(R) + \ln(f_p) + \ln(n_e) + \ln(f_l) + \ln(f_i) + \ln(f_c) + \ln(L)$. **d.** Answers will vary. **103.** They are reciprocals of each other. **105.** For example, $f(x, y) = x^2 + y^2$. **107.** For example, $f(x, y, z) = xyz$. **109.** For example, take $f(x, y) = x + y$. Then setting $y = 3$ gives $f(x, 3) = x + 3$. This can be viewed as a function of the single variable x. Choosing other values for y gives other functions of x. **111.** The slope is independent of the choice of $y = k$. **113.** That CDs cost more than cassettes **115.** plane **117.** (B) Traveling in the direction B results in the shortest trip to nearby isotherms and hence the fastest rate of increase in temperature. **119.** Agree: Any slice through a plane is a straight line. **121.** The graph of a function of three or more variables lives in four-dimensional (or higher) space, which makes it difficult to draw and visualize. **123.** We need one dimension for each of the variables plus one dimension for the value of the function.

Section 8.2

1. $f_x(x, y) = -40$; $f_y(x, y) = 20$; $f_x(1, -1) = -40$; $f_y(1, -1) = 20$ **3.** $f_x(x, y) = 6x + 1$; $f_y(x, y) = -3y^2$; $f_x(1, -1) = 7$; $f_y(1, -1) = -3$ **5.** $f_x(x, y) = -40 + 10y$; $f_y(x, y) = 20 + 10x$; $f_x(1, -1) = -50$; $f_y(1, -1) = 30$ **7.** $f_x(x, y) = 6xy$; $f_y(x, y) = 3x^2$; $f_x(1, -1) = -6$; $f_y(1, -1) = 3$ **9.** $f_x(x, y) = 2xy^3 - 3x^2y^2 - y$; $f_y(x, y) = 3x^2y^2 - 2x^3y - x$; $f_x(1, -1) = -4$; $f_y(1, -1) = 4$ **11.** $f_x(x, y) = 6y(2xy + 1)^2$; $f_y(x, y) = 6x(2xy + 1)^2$; $f_x(1, -1) = -6$; $f_y(1, -1) = 6$ **13.** $f_x(x, y) = e^{x+y}$; $f_y(x, y) = e^{x+y}$; $f_x(1, -1) = 1$; $f_y(1, -1) = 1$ **15.** $f_x(x, y) = 3x^{-0.4}y^{0.4}$; $f_y(x, y) = 2x^{0.6}y^{-0.6}$; $f_x(1, -1)$ undefined; $f_y(1, -1)$ undefined **17.** $f_x(x, y) = 0.2ye^{0.2xy}$; $f_y(x, y) = 0.2xe^{0.2xy}$; $f_x(1, -1) = -0.2e^{-0.2}$; $f_y(1, -1) = 0.2e^{-0.2}$ **19.** $f_{xx}(x, y) = 0$; $f_{yy}(x, y) = 0$; $f_{xy}(x, y) = f_{yx}(x, y) = 0$; $f_{xx}(1, -1) = 0$; $f_{yy}(1, -1) = 0$; $f_{xy}(1, -1) = f_{yx}(1, -1) = 0$ **21.** $f_{xx}(x, y) = 0$; $f_{yy}(x, y) = 0$; $f_{xy}(x, y) = f_{yx}(x, y) = 10$; $f_{xx}(1, -1) = 0$; $f_{yy}(1, -1) = 0$; $f_{xy}(1, -1) = f_{yx}(1, -1) = 10$ **23.** $f_{xx}(x, y) = 6y$; $f_{yy}(x, y) = 0$; $f_{xy}(x, y) = f_{yx}(x, y) = 6x$; $f_{xx}(1, -1) = -6$; $f_{yy}(1, -1) = 0$; $f_{xy}(1, -1) = f_{yx}(1, -1) = 6$ **25.** $f_{xx}(x, y) = e^{x+y}$; $f_{yy}(x, y) = e^{x+y}$; $f_{xy}(x, y) = f_{yx}(x, y) = e^{x+y}$; $f_{xx}(1, -1) = 1$; $f_{yy}(1, -1) = 1$; $f_{xy}(1, -1) = f_{yx}(1, -1) = 1$ **27.** $f_{xx}(x, y) = -1.2x^{-1.4}y^{0.4}$; $f_{yy}(x, y) = -1.2x^{0.6}y^{-1.6}$; $f_{xy}(x, y) = f_{yx}(x, y) = 1.2x^{-0.4}y^{-0.6}$; $f_{xx}(1, -1)$ undefined; $f_{yy}(1, -1)$ undefined; $f_{xy}(1, -1)$ and $f_{yx}(1, -1)$ undefined **29.** $f_x(x, y, z) = yz$; $f_y(x, y, z) = xz$; $f_z(x, y, z) = xy$; $f_x(0, -1, 1) = -1$; $f_y(0, -1, 1) = 0$; $f_z(0, -1, 1) = 0$ **31.** $f_x(x, y, z) = 4/(x + y + z^2)^2$; $f_y(x, y, z) = 4/(x + y + z^2)^2$; $f_z(x, y, z) = 8z/(x + y + z^2)^2$; $f_x(0, -1, 1)$ undefined; $f_y(0, -1, 1)$ undefined; $f_z(0, -1, 1)$ undefined **33.** $f_x(x, y, z) = e^{yz} + yze^{xz}$; $f_y(x, y, z) = xze^{yz} + e^{xz}$; $f_z(x, y, z) = xy(e^{yz} + e^{xz})$; $f_x(0, -1, 1) = e^{-1} - 1$; $f_y(0, -1, 1) = 1$; $f_z(0, -1, 1) = 0$ **35.** $f_x(x, y, z) = 0.1x^{-0.9}y^{0.4}z^{0.5}$; $f_y(x, y, z) = 0.4x^{0.1}y^{-0.6}z^{0.5}$; $f_z(x, y, z) = 0.5x^{0.1}y^{0.4}z^{-0.5}$; $f_x(0, -1, 1)$ undefined; $f_y(0, -1, 1)$ undefined, $f_z(0, -1, 1)$ undefined **37.** $f_x(x, y, z) = yze^{xyz}$, $f_y(x, y, z) = xze^{xyz}$, $f_z(x, y, z) = xye^{xyz}$; $f_x(0, -1, 1) = -1$; $f_y(0, -1, 1) = f_z(0, -1, 1) = 0$ **39.** $f_x(x, y, z) = 0$; $f_y(x, y, z) = -\dfrac{600z}{y^{0.7}(1 + y^{0.3})^2}$; $f_z(x, y, z) = \dfrac{2{,}000}{1 + y^{0.3}}$; $f_x(0, -1, 1)$ undefined; $f_y(0, -1, 1)$ undefined; $f_z(0, -1, 1)$ undefined **41.** $\partial C/\partial x = 6{,}000$; the marginal cost to manufacture each car is \$6,000. $\partial C/\partial y = 4{,}000$; the marginal cost to manufacture each truck is \$4,000. **43.** $\partial y/\partial t = -0.78$. The number of articles written by researchers in the United States was decreasing at a rate of 0.78 percentage points per year. $\partial y/\partial x = -1.02$. The number of articles written by researchers in the United States was decreasing at a rate of 1.02 percentage points per 1 percentage point increase in articles written in Europe. **45.** \$5,600 per car **47. a.** $\partial M/\partial c = -3.8$, $\partial M/\partial f = 2.2$. For every 1 point increase in the percentage of Chrysler owners who remain loyal, the percentage of Mazda owners who remain loyal decreases by 3.8 points. For every 1 point increase in the percentage of Ford owners who remain loyal, the percentage of Mazda owners who remain loyal increases by 2.2 points. **b.** 16% **49.** The marginal cost of cars is $\$6{,}000 + 1{,}000e^{-0.01(x+y)}$ per car. The marginal cost of trucks is $\$4{,}000 + 1{,}000e^{-0.01(x+y)}$ per truck. Both marginal costs decrease as production rises. **51. a.** \$40,900; \$58,800 **b.** −\$330 per year; −\$290 per year **c.** Widening **d.** The rate at which the income gap is widening **53.** $\bar{C}(x, y) = \dfrac{200{,}000 + 6{,}000x + 4{,}000y - 100{,}000e^{-0.01(x+y)}}{x + y}$; $\bar{C}_x(50, 50) = -\$2.64$ per car. This means that, at a production level of 50 cars and 50 trucks per week, the average cost per vehicle is decreasing by \$2.64 for each additional car manufactured. $\bar{C}_y(50, 50) = -\$22.64$ per truck. This means that, at a production level of 50 cars and 50 trucks per week, the average cost per vehicle is decreasing by \$22.64 for each additional truck manufactured. **55.** No. Your marginal revenue from the sale of cars is $\$15{,}000 - \dfrac{2{,}500}{\sqrt{x + y}}$ per car and $\$10{,}000 - \dfrac{2{,}500}{\sqrt{x + y}}$ per truck from the sale of trucks. These increase with increasing x and y. In other words, you will earn more revenue per vehicle with increasing sales, so the rental company will pay more for each additional vehicle it buys. **57.** $P_z(10, 100{,}000, 1{,}000{,}000) \approx 0.0001010$ papers/dollars **59. a.** $U_x(10, 5) = 5.18$, $U_y(10, 5) = 2.09$. This means that if 10 copies of Macro Publish and 5 copies of Turbo Publish are purchased, the company's daily productivity is increasing at a rate of 5.18 pages per day for each additional copy of Macro purchased and by 2.09 pages per day for each additional copy of Turbo purchased. **b.** $\frac{U_x(10, 5)}{U_y(10, 5)} \approx 2.48$ is the ratio of the usefulness of one additional copy of Macro to one of Turbo. Thus, with 10 copies of Macro and 5 copies of Turbo, the company can expect approximately 2.48 times the productivity per additional copy of Macro compared to Turbo. **61.** 6×10^9 N/sec **63. a.** $A_P(100, 0.1, 10) = 2.59$; $A_r(100, 0.1, 10) = 2{,}357.95$; $A_t(100, 0.1, 10) = 24.72$. Thus, for a \$100 investment at 10% interest, after 10 years the accumulated amount is increasing at a rate of \$2.59 per \$1 of principal, at a rate of \$2,357.95 per increase of 1 in r (note that this would correspond to an increase in the interest rate of 100%), and at a rate of \$24.72 per year. **b.** $A_P(100, 0.1, t)$ tells you the rate at which the accumulated amount in an account bearing 10% interest with a principal of \$100 is growing per \$1 increase in the principal, t years after the investment.

65. a. $P_x = Ka\left(\frac{y}{x}\right)^b$ and $P_y = Kb\left(\frac{x}{y}\right)^a$. They are equal precisely when $\frac{a}{b} = \left(\frac{x}{y}\right)^b\left(\frac{x}{y}\right)^a$. Substituting $b = 1 - a$ now gives $\frac{a}{b} = \frac{x}{y}$. **b.** The given information implies that $P_x(100, 200) = P_y(100, 200)$. By part (a) this occurs precisely when $a/b = x/y = 100/200 = 1/2$. But $b = 1 - a$, so $a/(1 - a) = 1/2$, giving $a = 1/3$ and $b = 2/3$. **67.** Decreasing at 0.0075 parts of nutrient per part of water per second **69.** $\underline{f}$ is increasing at a rate of $\underline{s}$ units per unit of x, $\underline{f}$ is increasing at a rate of $\underline{t}$ units per unit of y, and the value of $\underline{f}$ is $\underline{r}$ when $x = \underline{a}$ and $y = \underline{b}$ **71.** the marginal cost of building an additional orbicus; zonars per unit **73.** Answers will vary. One example is $f(x, y) = -2x + 3y$. Others are $f(x, y) = -2x + 3y + 9$ and $f(x, y) = xy - 3x + 2y + 10$.
75. a. b is the z-intercept of the plane. m is the slope of the intersection of the plane with the xz-plane. n is the slope of the intersection of the plane with the yz-plane. **b.** Write $z = b + rx + sy$. We are told that $\partial z/\partial x = m$, so $r = m$. Similarly, $s = n$. Thus, $z = b + mx + ny$. We are also told that the plane passes through (h, k, l). Substituting gives $l = b + mh + nk$. This gives b as $l - mh - nk$. Substituting in the equation for z therefore gives $z = l - mh - nk + mx + ny = l + m(x - h) + n(y - k)$, as required.

Section 8.3

1. P: relative minimum; Q: none of the above; R: relative maximum **3.** P: saddle point; Q: relative maximum; R: none of the above **5.** Relative minimum **7.** Neither **9.** Saddle point **11.** Relative minimum at $(0, 0, 1)$ **13.** Relative maximum at $(-1/2, 1/2, 3/2)$ **15.** Saddle point at $(0, 0, 0)$ **17.** Minimum at $(4, -3, -10)$ **19.** Maximum at $(-7/4, 1/4, 19/8)$ **21.** Relative maximum at $(0, 0, 0)$, saddle points at $(\pm 4, 2, -16)$ **23.** Relative minimum at $(0, 0, 0)$, saddle points at $(-1, \pm 1, 1)$ **25.** Relative minimum at $(0, 0, 1)$ **27.** Relative minimum at $(-2, \pm 2, -16)$, $(0, 0)$ a critical point that is not a relative extremum **29.** Saddle point at $(0, 0, -1)$ **31.** Relative maximum at $(-1, 0, e)$ **33.** Relative minimum at $(2^{1/3}, 2^{1/3}, 3(2^{2/3}))$ **35.** Relative minimum at $(1, 1, 4)$ and $(-1, -1, 4)$ **37.** Absolute minimum at $(0, 0, 1)$ **39.** None; the relative maximum at $(0, 0, 0)$ is not absolute. **41.** Minimum of 1/3 at $(c, f) = (2/3, 2/3)$. Thus, at least 1/3 of all Mazda owners would choose another new Mazda, and this lowest loyalty occurs when 2/3 of Chrysler and Ford owners remain loyal to their brands. **43.** It should remove 2.5 lb of sulfur and 1 lb of lead per day. **45.** You should charge \$580.81 for the Ultra Mini and \$808.08 for the Big Stack. **47.** $l = w = h \approx 20.67$ in, volume $\approx 8{,}827$ in³ **49.** 18 in × 18 in × 36 in, volume = 11,664 in³

51. **53.** Continues up indefinitely

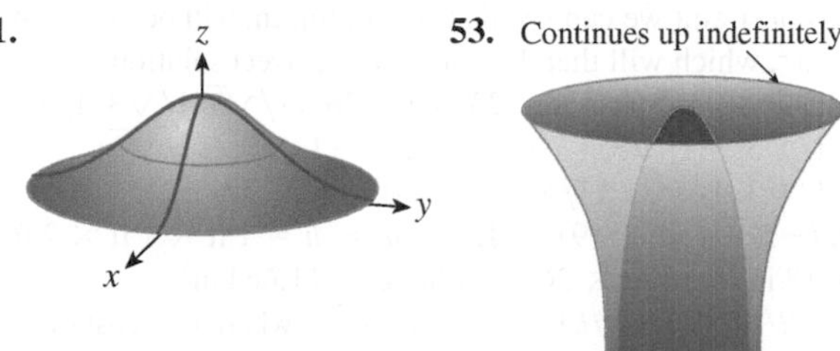

55. H must be positive. **57.** No. For there to be a relative maximum at (a, b), *all* vertical planes through (a, b) should yield a curve with a relative maximum at (a, b). It could happen that a slice by another vertical plane through (a, b) (such as $x - a = y - b$) does not yield a curve with a relative maximum at (a, b). [An example is $f(x, y) = \sqrt{xy} - x^2 - y^2$, at the point $(0, 0)$. Look at the slices through $x = 0$, $y = 0$, and $y = x$.] **59.** $\bar{C}_x = \frac{\partial}{\partial x}\left(\frac{C}{x + y}\right) = \frac{(x + y)C_x - C}{(x + y)^2}$. If this is zero, then $(x + y)C_x = C$, or $C_x = \frac{C}{x + y} = \bar{C}$. Similarly, if $\bar{C}_y = 0$, then $C_y = \bar{C}$. This is reasonable because if the average cost is decreasing with increasing x, then the average cost is greater than the marginal cost C_x. Similarly, if the average cost is increasing with increasing x, then the average cost is less than the marginal cost C_x. Thus, if the average cost is stationary with increasing x, then the average cost equals the marginal cost C_x. (The situation is similar for the case of increasing y.) **61.** The equation of the tangent plane at the point (a, b) is $z = f(a, b) + f_x(a, b)(x - a) + f_y(a, b)(y - b)$. If f has a relative extremum at (a, b), then $f_x(a, b) = 0 = f_y(a, b)$. Substituting these into the equation of the tangent plane gives $z = f(a, b)$, a constant. But the graph of $z = constant$ is a plane parallel to the xy-plane.

Section 8.4

1. 1; $(0, 0, 0)$ **3.** 1.35; $(1/10, 3/10, 1/2)$ **5.** Minimum value = 6 at $(1, 2, 1/2)$ **7.** 200; $(20, 10)$ **9.** 16; $(2, 2)$ and $(-2, -2)$ **11.** 20; $(2, 4)$ **13.** 1; $(0, 0, 0)$ **15.** 1.35; $(1/10, 3/10, 1/2)$ **17.** Minimum value = 6 at $(1, 2, 1/2)$ **19. a.** $f(x, y, z)$ is the square of the distance from the point (x, y, z) to $(3, 0, 0)$, and the constraint tells us that (x, y, z) must lie on the paraboloid $z = x^2 + y^2$. Because there must be such a point (or points) on the paraboloid closest to $(3, 0, 0)$, the given problem must have at least one solution. Solution: $(x, y, z) = (1, 0, 1)$ for a minimum value of 5. **b.** Same solution as part (a) **c.** There are no critical points using this method. **d.** The constraint equation $y^2 = z - x^2$ tells us that $z - x^2$ cannot be negative and thus restricts the domain of f to the set of points (x, y, z) with $z - x^2 \geq 0$. However, this information is lost when $z - x^2$ is substituted in the expression for f, so the substitution in part (c) results in a different optimization problem, one in which there is no requirement that $z - x^2$ be ≥ 0. If we pay attention to this constraint we can see that the minimum will occur when $z = x^2$, which will then lead us to the correct solution. **21.** $5 \times 10 = 50$ sq. ft. **23.** \$10 **25.** $(1/\sqrt{3}, 1/\sqrt{3}, 1/\sqrt{3})$, $(-1/\sqrt{3}, -1/\sqrt{3}, 1/\sqrt{3})$, $(1/\sqrt{3}, -1/\sqrt{3}, -1/\sqrt{3})$, $(-1/\sqrt{3}, 1/\sqrt{3}, -1/\sqrt{3})$ **27.** $(0, 1/2, -1/2)$ **29.** $(-5/9, 5/9, 25/9)$ **31.** $l \times w \times h = 1$ ft $\times$ 1 ft $\times$ 2 ft **33.** 18 in $\times$ 18 in $\times$ 36 in, volume $= 11{,}664$ in^3 **35.** $(2l/h)^{1/3} \times (2l/h)^{1/3} \times 2^{1/3}(h/l)^{2/3}$, where l = cost of lightweight cardboard and h = cost of heavy-duty cardboard per square foot **37.** $1 \times 1 \times 1/2$ **39.** 6 laborers, 10 robots for a productivity of 368 pairs of socks per day **41.** Method 1: Solve $g(x, y, z) = 0$ for one of the variables, and substitute in $f(x, y, z)$. Then find the maximum value of the resulting function of two variables. Advantage (answers may vary): We can use the second derivative test to check whether the resulting critical points are maxima, minima, saddle points, or none of these. Disadvantage (answers may vary): We may not be able to solve $g(x, y, z) = 0$ for one of the variables. Method 2: Use the method of Lagrange multipliers. Advantage (answers may vary): We do not need to solve the constraint equation for one of the variables. Disadvantage (answers may vary): The method does not tell us whether the critical points obtained are maxima, minima, saddle points, or none of these. **43.** If the only constraint is an equality constraint and if it is impossible to eliminate one of the variables in the objective function by substitution (solving the constraint equation for a variable or some other method). **45.** Answers may vary: Maximize $f(x, y) = 1 - x^2 - y^2$ subject to $x = y$. **47.** Yes. There may be relative extrema at points on the boundary of the domain. The partial derivatives of the function need not be 0 at such points. **49.** In a linear programming problem the objective function is linear, so the partial derivatives can never all be zero. (We are ignoring the simple case in which the objective function is constant.) It follows that the extrema cannot occur in the interior of the domain (since the partial derivatives must be zero at such points).

Section 8.5

1. $-1/2$ **3.** $e^2/2 - 7/2$ **5.** $(e^3 - 1)(e^2 - 1)$ **7.** $7/6$ **9.** $[e^3 - e - e^{-1} + e^{-3}]/2$ **11.** $1/2$ **13.** $(e - 1)/2$ **15.** $45/2$ **17.** $8/3$ **19.** $4/3$ **21.** 0 **23.** $2/3$ **25.** $2/3$ **27.** $2(e - 2)$ **29.** $1/3$

31. 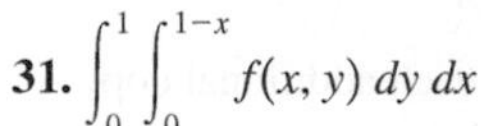 $\int_0^1 \int_0^{1-x} f(x, y)\, dy\, dx$

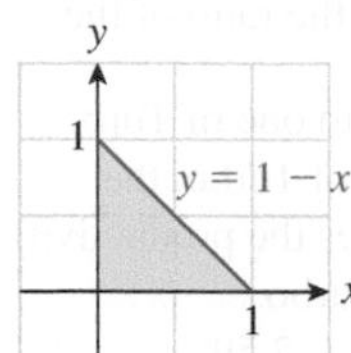

33. $\int_0^{\sqrt{2}} \int_{x^2-1}^{1} f(x, y)\, dy\, dx$

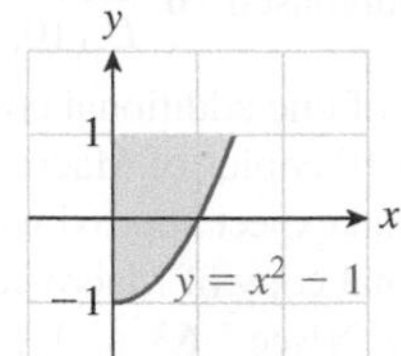

35. $\int_1^4 \int_1^{2/\sqrt{y}} f(x, y)\, dx\, dy$

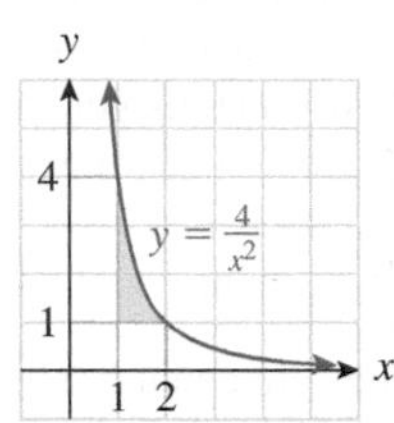

37. $4/3$ **39.** $1/6$ **41.** 162,000 gadgets **43.** \$312,750 **45.** \$17,500 **47.** 8,216 **49.** 1 degree Celsius **51.** The area between the curves $y = r(x)$ and $y = s(x)$ and the vertical lines $x = a$ and $x = b$ is given by $\int_a^b \int_{r(x)}^{s(x)} dy\, dx$, assuming that $r(x) \leq s(x)$ for $a \leq x \leq b$. **53.** The first step in calculating an integral of the form $\int_a^b \int_{r(x)}^{s(x)} f(x, y)\, dy\, dx$ is to evaluate the integral $\int_{r(x)}^{s(x)} f(x, y)\, dy$, obtained by holding x constant and integrating with respect to y. **55.** Paintings per picasso per dali **57.** The left-hand side is $\int_a^b \int_c^d f(x)g(y)\, dx\, dy = \int_a^b \left(g(y)\int_c^d f(x)\, dx\right) dy$ [because $g(y)$ is treated as a constant in the inner integral] $= \left(\int_c^d f(x)\, dx\right)\left(\int_a^b g(y)\, dy\right)$ [because $\int_c^d f(x)\, dx$ is a constant and can therefore be taken outside the integral with respect to y]. For example, $\int_0^1 \int_1^2 ye^x\, dx\, dy = \frac{1}{2}(e^2 - e)$, no matter how we compute it.

Chapter 8 Review

1. 0; 14/3; 1/2; $\dfrac{1}{1 + z} + z^3$; $\dfrac{x + h}{y + k + (x + h)(z + l)} + (x + h)^2(y + k)$ **3.** Decreases by 0.32 units; increases by 12.5 units **5.** Reading left to right, starting at the top: 4, 0, 0, 3, 0, 1, 2, 0, 2 **7.** Answers may vary; two examples are $f(x, y) = 3(x - y)/2$ and $f(x, y) = 3(x - y)^3/8$.

9. **11.**

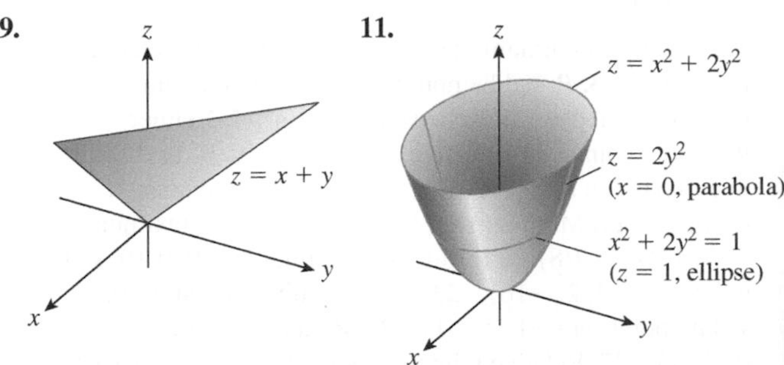

13.

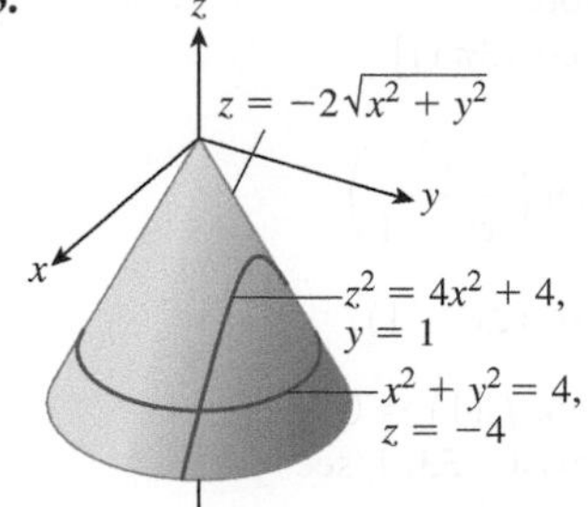

15. $f_x = 2x + y, f_y = x, f_{yy} = 0$ **17.** 0
19. $\frac{\partial f}{\partial x} = \frac{-x^2 + y^2 + z^2}{(x^2 + y^2 + z^2)^2}, \frac{\partial f}{\partial y} = -\frac{2xy}{(x^2 + y^2 + z^2)^2},$
$\frac{\partial f}{\partial z} = -\frac{2xz}{(x^2 + y^2 + z^2)^2}, \left.\frac{\partial f}{\partial x}\right|_{(0,1,0)} = 1$ **21.** Absolute minimum at $(1, 3/2)$ **23.** Maximum at $(0, 0)$, saddle points at $(\pm\sqrt{2}, 1)$ **25.** Saddle point at $(0, 0)$
27. $1/27$ at $(1/3, 1/3, 1/3)$ **29.** $(0, 2, \sqrt{2})$ **31.** 4; $(\sqrt{2}, \sqrt{2})$ and $(-\sqrt{2}, -\sqrt{2})$ **33.** $(0, 2, \sqrt{2})$ **35.** 2 **37.** ln 5
39. 1 **41. a.** $h(x, y) = 5{,}000 - 0.8x - 0.6y$ hits per day (x = number of new customers at JungleBooks.com, y = number of new customers at FarmerBooks.com) **b.** 250 **c.** $h(x, y, z) = 5{,}000 - 0.8x - 0.6y + 0.0001z$ (z = number of new Internet shoppers) **d.** 1.4 million
43. a. 2,320 hits per day **b.** $0.08 + 0.00003x$ hits (daily) per dollar spent on television advertising per month; increases with increasing x **c.** $4,000 per month **45.** (A)
47. a. About 15,800 additional orders per day **b.** 11
49. $23,050

Chapter 9

Section 9.1

1.

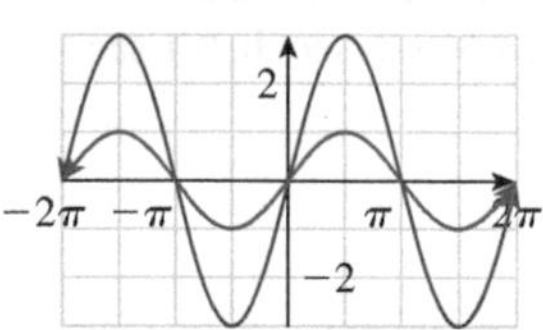

3.

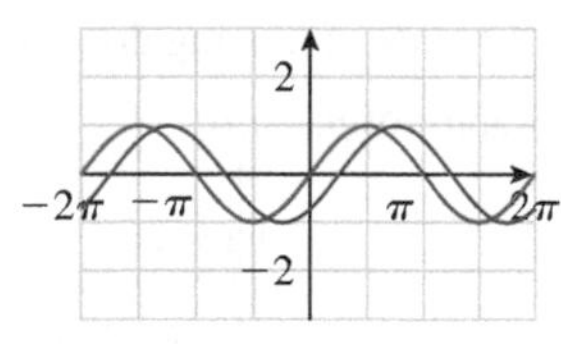

5.

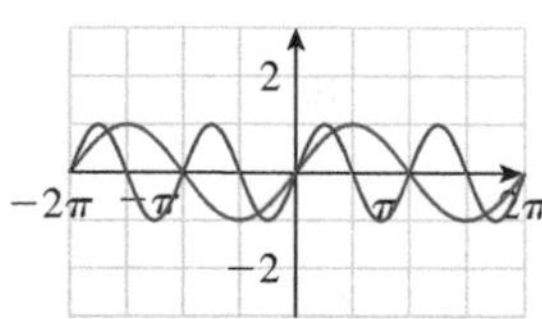

7.

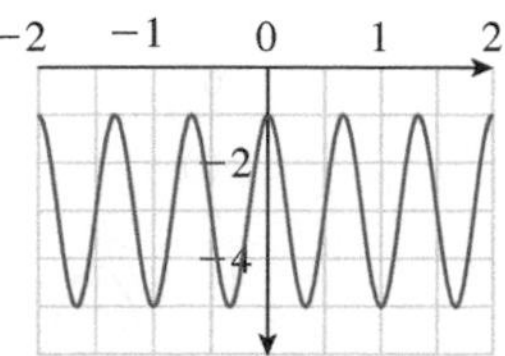

9.

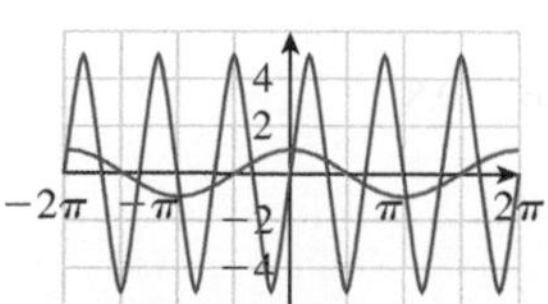

11.

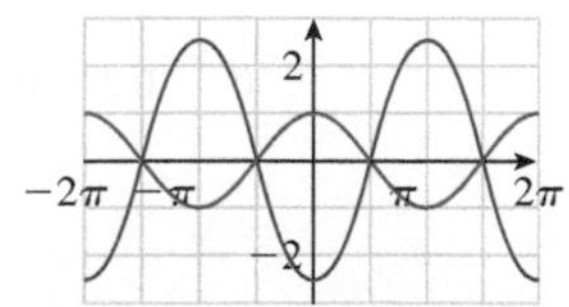

13. $f(x) = \sin(2\pi x) + 1$ **15.** $f(x) = 1.5\sin[4\pi(x - 0.25)]$
17. $f(x) = 50\sin[\pi(x - 5)/10] - 50$ **19.** $f(x) = \cos(2\pi x)$
21. $f(x) = 1.5\cos[4\pi(x - 0.375)]$
23. $f(x) = 40\cos[\pi(x - 10)/10] + 40$
25. $f(t) = 4.2\sin(\pi/2 - 2\pi t) + 3$
27. $g(x) = 4 - 1.3\sin[\pi/2 - 2.3(x - 4)]$ **29.** Proof
31. $\sqrt{3}/2$ **33.** Proof **35.** Proof **37.** $\tan(x + \pi) = \tan(x)$
39. a. $2\pi/0.602 \approx 10.4$ years. **b.** Maximum: $58.8 + 57.7 = 116.5 \approx 117$; minimum: $58.8 - 57.7 = 1.1 \approx 1$
c. $1.43 + P/4 + 2P = 1.43 + 23.48 \approx 25$ years, or about the beginning of 2022 **41. a.** Sales were highest when $t \approx 0, 4, 8$, and 12, which correspond to the end of the last quarter or beginning of the first quarter of each year. Sales were lowest when $t \approx 2, 6$, and 10, which correspond to the beginning of the third quarter of each year. **b.** The maximum quarterly sales were approximately 21 million iPods per quarter; minimum quarterly sales were approximately 9 million iPods per quarter.
c. Maximum: $15 + 6 = 21$; minimum: $15 - 6 = 9$
43. Amplitude = 6.00, vertical offset = 15, phase shift = $-1.85/1.51 \approx -1.23$, angular frequency = 1.51, period ≈ 4.16. From 2008 through 2010, Apple's sales of iPods fluctuated in cycles of 4.16 quarters about a baseline of 15 million iPods per quarter. Every cycle, sales peaked at $15 + 6 = 21$ million iPods per quarter and dipped to a low of $15 - 6 = 9$ million iPods per quarter. Sales first peaked at $t = -1.23 + (5/4) \times 4.16 = 3.97$, the end of 2008.
45. $P(t) = 7.5\sin[\pi(t - 13)/26] + 12.5$
47. $T(t) = 3.5\cos[\pi(t - 7)/6] + 78.5$
49. $n(t) = 1.25\sin[\pi(t - 1)/2] + 5.75$
51. $n(t) = 1.25\cos[\pi(t - 2)/2] + 5.75$
53. $d(t) = 5\sin[2\pi(t - 1.625)/13.5] + 10$
55. a. $u(t) = 2.5\sin(2\pi(t - 0.75)) + 7.5$
b. $c(t) = 1.04^t[2.5\sin(2\pi(t - 0.75)) + 7.5]$
57. a. $C \approx 50, A \approx 8, P \approx 12, \beta \approx 6$

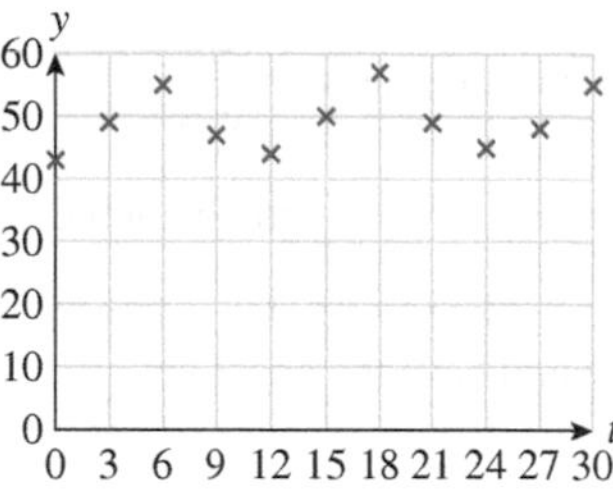

b. $f(t) = 5.882\cos[2\pi(t - 5.696)/12.263] + 49.238$

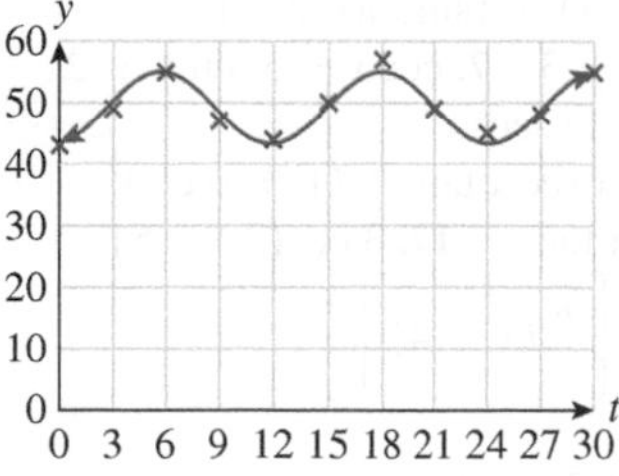

c. 12; 43; 55

59. a.

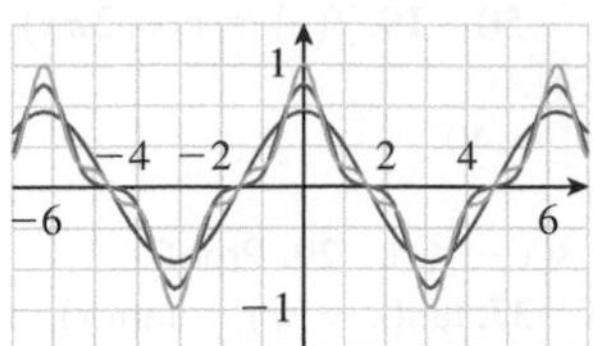

b. $y_{11} = \frac{2}{\pi}\cos x + \frac{2}{3\pi}\cos 3x + \frac{2}{5\pi}\cos 5x + \frac{2}{7\pi}\cos 7x + \frac{2}{9\pi}\cos 9x + \frac{2}{11\pi}\cos 11x$

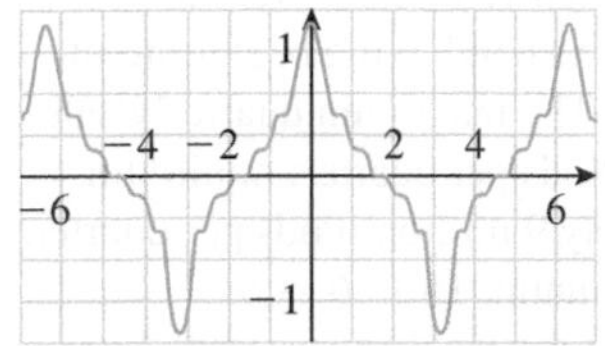

c. Multiply the amplitudes by 3 and change ω to 1/2:

$$y_{11} = \frac{6}{\pi}\cos\frac{x}{2} + \frac{6}{3\pi}\cos\frac{3x}{2} + \frac{6}{5\pi}\cos\frac{5x}{2} + \frac{6}{7\pi}\cos\frac{7x}{2} + \frac{6}{9\pi}\cos\frac{9x}{2} + \frac{6}{11\pi}\cos\frac{11x}{2}.$$

61. The period is approximately 12.6 units.

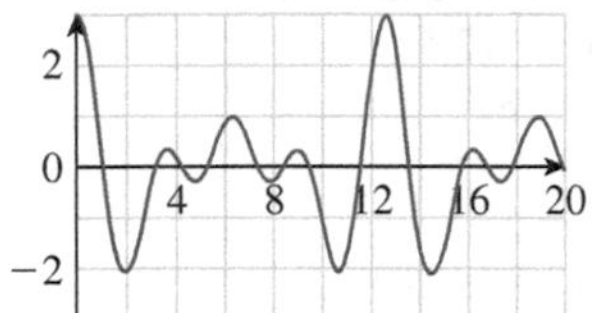

63. Lows: $B - A$; highs: $B + A$. **65.** He is correct. The other trigonometric functions can be obtained from the sine function by first using the formula $\cos x = \sin(x + \pi/2)$ to obtain cosine and then using the formulas $\tan x = \frac{\sin x}{\cos x}$, $\cot x = \frac{\cos x}{\sin x}$, $\sec x = \frac{1}{\cos x}$, and $\csc x = \frac{1}{\sin x}$ to obtain the rest. **67.** The largest B can be is A. Otherwise, if B is larger than A, the low figure for sales would have the negative value of $A - B$.

Section 9.2

1. $\cos x + \sin x$ **3.** $(\cos x)(\tan x) + (\sin x)(\sec^2 x)$ **5.** $-2\csc x \cot x - \sec x \tan x + 3$ **7.** $\cos x - x \sin x + 2x$ **9.** $(2x - 1)\tan x + (x^2 - x + 1)\sec^2 x$ **11.** $-[\csc^2 x(1 + \sec x) + \cot x \sec x \tan x]/(1 + \sec x)^2$ **13.** $-2\cos x \sin x$ **15.** $2\sec^2 x \tan x$ **17.** $3\cos(3x - 5)$ **19.** $2\sin(-2x + 5)$ **21.** $\pi \cos\left[\frac{\pi}{5}(x - 4)\right]$ **23.** $-(2x - 1)\sin(x^2 - x)$ **25.** $(2.2x^{1.2} + 1.2)\sec(x^{2.2} + 1.2x - 1) \times \tan(x^{2.2} + 1.2x - 1)$ **27.** $\sec x \tan x \tan(x^2 - 1) + 2x \sec x \sec^2(x^2 - 1)$ **29.** $e^x[-\sin(e^x) + \cos x - \sin x]$ **31.** $\sec x$ **33.** Proof **35.** Proof **37.** $e^{-2x}[-2\sin(3\pi x) + 3\pi\cos(3\pi x)]$ **39.** $1.5[\sin(3x)]^{-0.5}\cos(3x)$ **41.** $\frac{x^4 - 3x^2}{(x^2 - 1)^2}\sec\left(\frac{x^3}{x^2 - 1}\right)\tan\left(\frac{x^3}{x^2 - 1}\right)$ **43.** $\frac{\cot(2x - 1)}{x} - 2\ln|x|\csc^2(2x - 1)$

45. a. Not differentiable at 0 **b.** $f'(1) \approx 0.5403$ **47.** 0 **49.** 2 **51.** Does not exist **53.** $1/\sec^2 y$ **55.** $-[1 + y\cos(xy)]/[1 + x\cos(xy)]$ **57.** $c'(t) = 7\pi\cos[2\pi(t - 0.75)]$; $c'(0.75) \approx \$21.99$ per *year* $\approx \$0.42$ per week **59.** $N'(6) \approx -32.12$. On January 1, 2003, the number of sunspots was decreasing at a rate of 32.12 sunspots per year. **61.** $c'(t) = 1.035^t \times [\ln(1.035)(0.8\sin(2\pi t) + 10.2) + 1.6\pi\cos(2\pi t)]$; $c'(1) = 1.035[10.2\ln(1.035) + 1.6\pi] \approx \$5.57$ per year, or \$0.11 per week. **63. a.** $d(t) = 5\cos(2\pi t/13.5) + 10$ **b.** $d'(t) = -(10\pi/13.5)\sin(2\pi t/13.5)$; $d'(7) \approx 0.270$. At noon the tide was rising at a rate of 0.270 feet per hour.

65. $\frac{dV}{dt} = 11{,}000\pi\frac{|\sin(100\pi t)|}{\sin(100\pi t)}\cos(100\pi t)$

Graph:

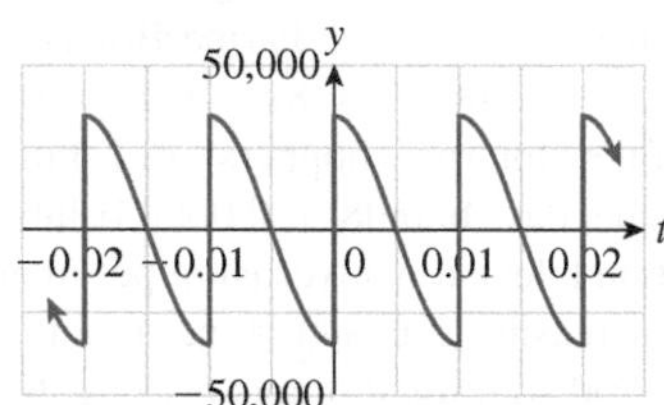

The sudden jumps in the graph are due to the nondifferentiability of V at the times 0, ± 0.01, ± 0.02, ... The derivative is negative immediately to the left and positive immediately to the right of these points. **67. a.** 1.2 cm above the rest position **b.** 0 cm/sec; not moving; moving downward at 18.85 cm/sec **c.** 2.5 cycles per second **69. a.** Moving downward at 0.12 cm/sec; moving downward at 18.66 cm/sec **b.** 0.1 sec

Graphs: of p:

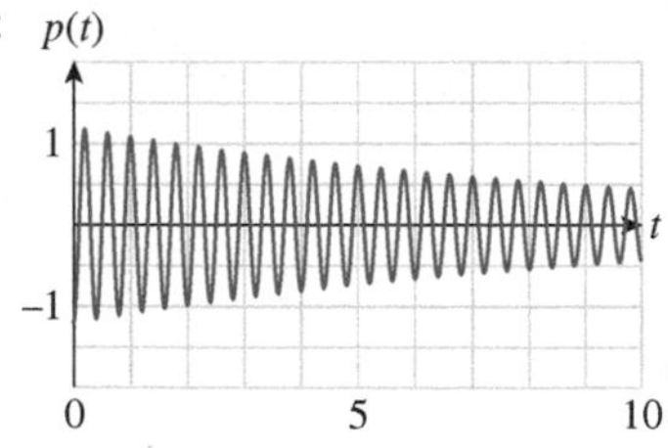

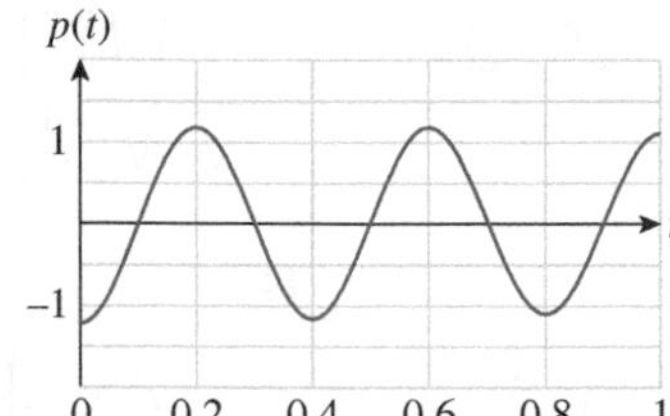

Graphs of p':

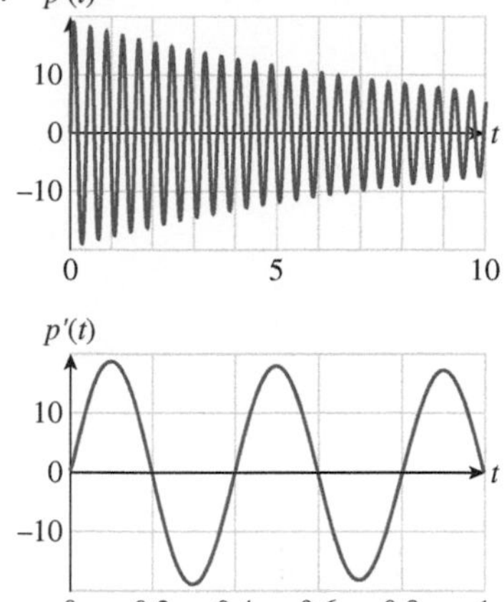

71. a. (III) **b.** Increasing at a rate of 0.157 degrees per thousand years **73.** -6; 6 **75.** Answers will vary. Examples: $f(x) = \sin x$; $f(x) = \cos x$ **77.** Answers will vary. Examples: $f(x) = e^{-x}$; $f(x) = -2e^{-x}$ **79.** The graph of $\cos x$ slopes down over the interval $(0, \pi)$, so its derivative is negative over that interval. The function $-\sin x$, and not $\sin x$, has this property. **81.** The velocity is $p'(t) = A\omega \cos(\omega t + d)$, which is a maximum when its derivative, $p''(t) = -A\omega^2 \sin(\omega t + d)$, is zero. But this occurs when $\sin(\omega t + d) = 0$, so $p(t)$ is zero as well, meaning that the stock is at yesterday's close. **83.** The derivative of $\sin x$ is $\cos x$. When $x = 0$, this is $\cos(0) = 1$. Thus, the tangent to the graph of $\sin x$ at the point $(0, 0)$ has slope 1, which means that it slopes upward at 45°.

Section 9.3

1. $-\cos x - 2 \sin x + C$ **3.** $2 \sin x + 4.3 \cos x - 9.33x + C$ **5.** $3.4 \tan x + (\sin x)/1.3 - 3.2e^x + C$ **7.** $(7.6/3) \sin(3x - 4) + C$ **9.** $-(1/6) \cos(3x^2 - 4) + C$ **11.** $-2 \cos(x^2 + x) + C$ **13.** $(1/6) \tan(3x^2 + 2x^3) + C$ **15.** $-(1/6) \ln|\cos(2x^3)| + C$ **17.** $3 \ln|\sec(2x - 4) + \tan(2x - 4)| + C$ **19.** $(1/2) \sin(e^{2x} + 1) + C$ **21.** -2 **23.** $\ln(2)$ **25.** 0 **27.** 1 **29.** Proof **31.** Proof **33.** $-\frac{1}{4}\cos(4x) + C$ **35.** $-\sin(-x + 1) + C$ **37.** $[\cos(-1.1x - 1)]/1.1 + C$ **39.** $-\frac{1}{4}\ln|\sin(-4x)| + C$ **41.** 0 **43.** 2π **45.** $-x \cos x + \sin x + C$ **47.** $\left[\frac{x^2}{2} - \frac{1}{4}\right]\sin(2x) + \frac{x}{2}\cos(2x) + C$ **49.** $-\frac{1}{2}e^{-x}\cos x - \frac{1}{2}e^{-x}\sin x + C$ **51.** $\pi^2 - 4$ **53.** Average $= 2/\pi$

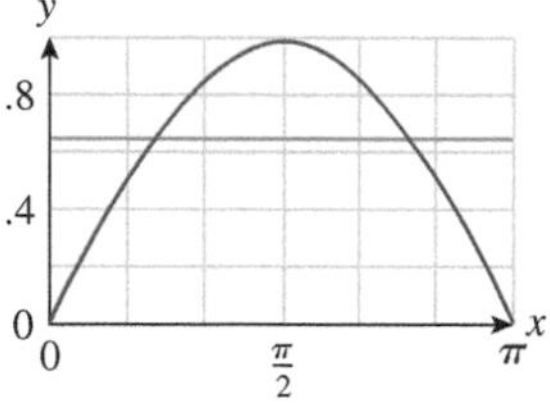

55. Diverges **57.** Converges to 1/2 **59.** $C(t) = 0.04t + \frac{2.6}{\pi}\cos\left[\frac{\pi}{26}(t - 25)\right] + 1.02$ **61.** 12 feet **63.** 79 sunspots **65.** $P(t) = 7.5 \sin[(\pi/26(t - 13)] + 12.5$; 7.7% **67. a.** Average voltage over $[0, 1/6]$ is zero; 60 cycles per second.
b.

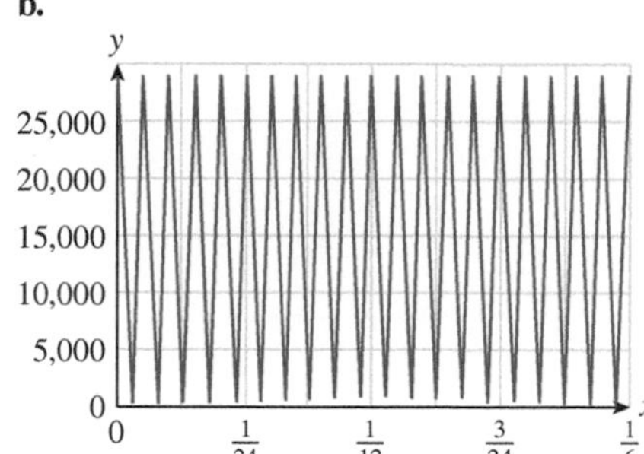

c. 116.673 volts **69.** \$50,000 **71.** It is always zero. **73.** 1 **75.** $s = -\frac{K}{\omega^2}\sin(\omega t - \alpha) + Lt + M$ for constants L and M

Chapter 9 Review

1. $f(x) = 1 + 2 \sin x$
3. $f(x) = 2 + 2 \sin[\pi(x - 1)] = 2 + 2 \sin[\pi(x + 1)]$
5. $f(x) = 1 + 2 \cos(x - \pi/2)$
7. $f(x) = 2 + 2 \cos[\pi(x + 1/2)] = 2 + 2 \cos[\pi(x - 3/2)]$
9. $-2x \sin(x^2 - 1)$ **11.** $2e^x \sec^2(2e^x - 1)$
13. $4x \sin(x^2) \cos(x^2)$ **15.** $2 \sin(2x - 1) + C$
17. $\tan(2x^2 - 1) + C$ **19.** $-\frac{1}{2}\ln|(\cos(x^2 + 1)| + C$
21. 1 **23.** $-x^2 \cos x + 2x \sin x + 2 \cos x + C$
25. $s(t) = 10{,}500 + 1{,}500 \sin[(2\pi/52)t - \pi] = 10{,}500 + 1{,}500 \sin(0.12083t - 3.14159)$ **27.** Decreasing at a rate of \$3,852 per month **29.** \$2,029,700
31. $150t - \frac{100}{\pi}\cos\left[\frac{\pi}{2}(t - 1)\right]$ grams

Index

Social Sciences

Österreich
Einwohner (2015): 8,6 Mio
0 25 50 Meilen
0 25 50 Kilometer
DEUTSCHLAND
TSCHECHIEN
SLOWAKEI
UNGARN
SLOWENIEN
ITALIEN
DIE SCHWEIZ
SÜDTIROL
OBERÖSTERREICH
NIEDERÖSTERREICH
WIEN
BURGENLAND
STEIERMARK
SALZBURG
KÄRNTEN
TIROL
VORARLBERG
Osttirol (zu Tirol)
Gmünd
Horn
Krems
Donau
Linz
Melk
Sankt Pölten
Wien
Amstetten
Baden
Neusiedler See
Eisenstadt
Wiener Neustadt
Gmunden
Salzburg
Bad Ischl
Mariazell
Hallstatt
Liezen
Enns
Bodensee
Bregenz
Reutte
Kufstein
Sankt Johann in Tirol
Wörgl
Kitzbühel
Zell am See
Bischofshofen
Bruck an der Mur
Oberwart
Feldkirch
St. Anton am Arlberg
Innsbruck
Inn
Landeck
Radstadt
Bruck
Sankt Georgen ob Judenburg
Mauterndorf
Graz
Mur
Güssing
Lienz
Spittal an der Drau
Drau
Feldkirchen
Klagenfurt
Villach
Wörther See
Meran
Bozen

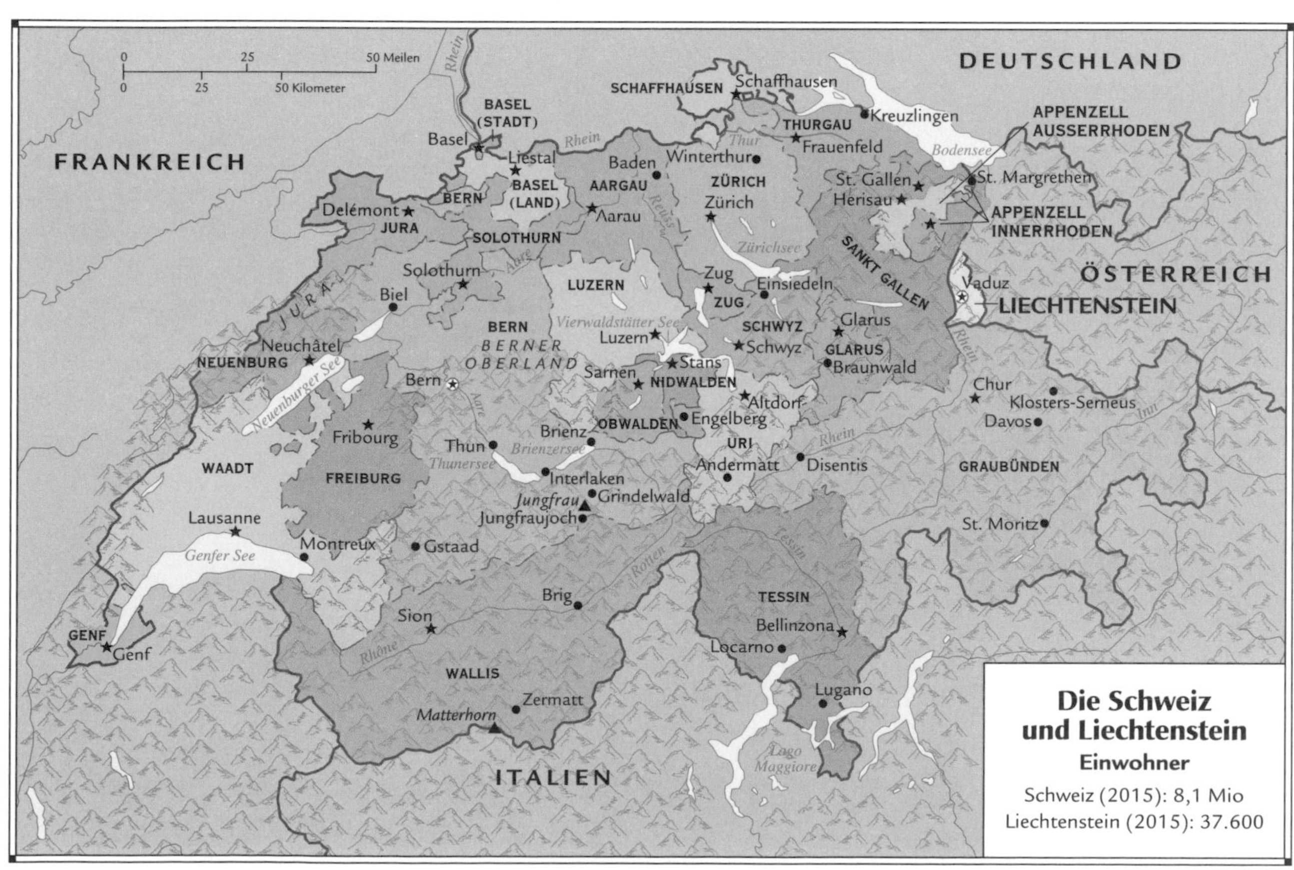
0 25 50 Meilen
0 25 50 Kilometer
Rhein
DEUTSCHLAND
FRANKREICH
ÖSTERREICH
LIECHTENSTEIN
ITALIEN
BASEL (STADT)
Basel
Liestal
BASEL (LAND)
SCHAFFHAUSEN
Schaffhausen
THURGAU
Kreuzlingen
Frauenfeld
Bodensee
APPENZELL AUSSERRHODEN
APPENZELL INNERRHODEN
St. Margrethen
St. Gallen
Herisau
Baden
Winterthur
Thur
ZÜRICH
Zürich
Zürichsee
AARGAU
Aarau
Reuss
BERN
Delémont
JURA
SOLOTHURN
Solothurn
Aare
LUZERN
Luzern
Vierwaldstätter See
Zug
ZUG
Einsiedeln
SANKT GALLEN
Vaduz
SCHWYZ
Schwyz
Glarus
GLARUS
Braunwald
Biel
Neuchâtel
NEUENBURG
Neuenburger See
BERNER OBERLAND
Bern
Stans
Sarnen
NIDWALDEN
Altdorf
OBWALDEN
Engelberg
Brienz
Brienzersee
Thun
Thunersee
Interlaken
URI
Andermatt
Disentis
Chur
Klosters-Serneus
Davos
Inn
GRAUBÜNDEN
Fribourg
FREIBURG
WAADT
Jungfrau
Grindelwald
Jungfraujoch
Lausanne
Montreux
Gstaad
Genfer See
St. Moritz
Rotten
Tessin
TESSIN
Bellinzona
Locarno
Brig
Sion
Rhône
GENF
Genf
WALLIS
Zermatt
Matterhorn
Lugano
Lago Maggiore
Die Schweiz und Liechtenstein
Einwohner
Schweiz (2015): 8,1 Mio
Liechtenstein (2015): 37.600

Kontakte

A Communicative Approach

8th Edition

Kontakte

A Communicative Approach

Erwin Tschirner
Herder-Institut, Universität Leipzig
University of Arizona

Brigitte Nikolai
Werner-von-Siemens-Gymnasium,
Bad Harzburg

KONTAKTE, A COMMUNICATIVE APPROACH, EIGHTH EDITION

Published by McGraw-Hill Education, 2 Penn Plaza, New York, NY 10121.

This book is printed on acid-free paper.

1 2 3 4 5 6 7 8 9 0 DOW/DOW 10 9 8 7 6

ISBN: 978-1-259-30742-3 (Student's Edition)
MHID: 1-259-30742-5

ISBN: 978-1-259-69261-1 (Instructor's Edition)
MHID: 1-259-69261-2

Senior Vice President, Products & Markets: *Kurt L. Strand*
Vice President, General Manager, Products & Markets: *Michael Ryan*
Vice President, Content Design & Delivery: *Kimberly Meriwether David*
Managing Director: *Katie Stevens*
Senior Brand Manager: *Katherine K. Crouch*
Executive Director of Digital Content: *Janet Banhidi*
Senior Digital Product Analyst: *Sarah Carey*
Director, Product Development: *Meghan Campbell*
Director of Marketing: *Craig Gill*
Marketing Managers: *Michael Ambrosino / Chris Brown*
Executive Market Development Manager: *Helen Greenlea*
Senior Faculty Development Manager: *Jorge Arbujas*
Senior Product Developer: *Susan Blatty*
Product Development Coordinator: *Sean Costello*
Director, Content Design & Delivery: *Terri Schiesl*
Program Manager: *Kelly Heinrichs*
Content Production Managers: *Erin Melloy / Amber Bettcher*
Buyer: *Susan C. Culbertson*
Design: *Matt Backhaus*
Content Licensing Specialists: *Carrie Burger / Beth Thole*
Cover Image: © *Private Collection / Bridgeman Images*
Compositors: *SPI Global / Lumina Datamatics*
Printer: *R.R. Donnelley*

Library of Congress Control Number: 2015958532

www.mhhe.com

Contents

Einführung A

Einführung B

Kapitel 1

Wer ich bin und was ich tue

Themen

Kapitel 2

Besitz und Vergnügen

Themen

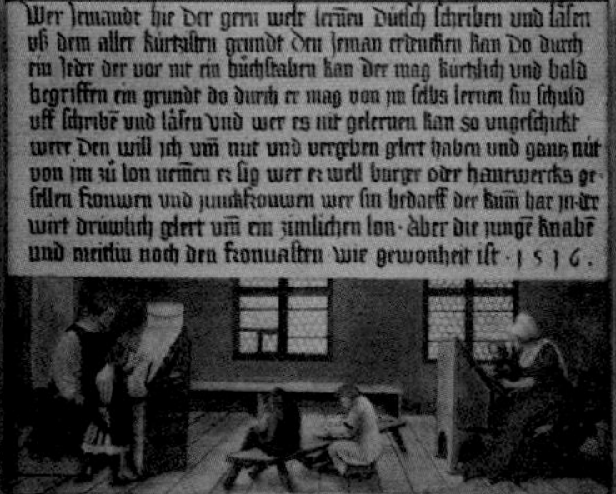

Wer jemandt hie der gern welt lernen dütsch schriben und läsen
uß dem aller kürtzisten grundt den jeman erdencken kan do durch
ein jeder der vor nit ein büchstaben kan der mag kürtzlich und bald
begriffen ein grundt do durch er mag von im selbs lernen sin schuld
uff schriben und läsen und wer es nit gelernen kan so ungeschickt
were den will ich umb nüt und vergeben gelert haben und gantz nüt
von im zü lon nemen er sig wer er well burger oder hantwercks ge-
sellen frouwen und junckfrouwen wer sin bedarff der kum har in der
wirt drüwlich gelert umb ein zimlichen lon · aber die jungen knaben
und meitlin noch den fronvasten wie gewonheit ist · 1 5 1 6 .

Kapitel 3

Talente, Pläne, Pflichten

Themen

Kapitel 4

Ereignisse und Erinnerungen

Themen

Kapitel 5

Geld und Arbeit

Themen

Kapitel 6

Wohnen

Themen

Kapitel 7

Unterwegs

Themen

Kapitel 8

Essen und Einkaufen

Themen

Kapitel 9

Kindheit und Jugend

Themen

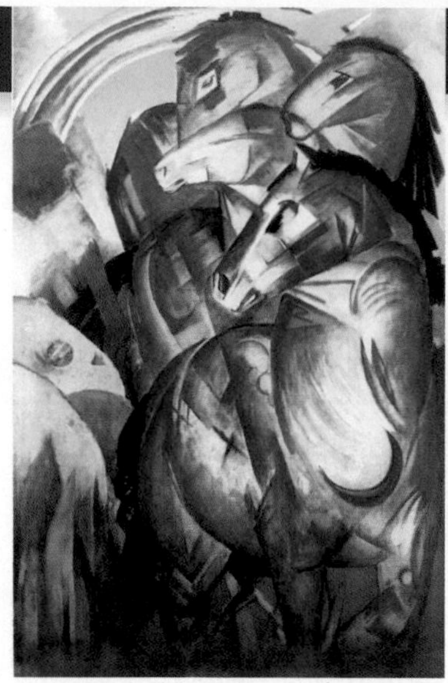

Kapitel 10

Auf Reisen

Themen

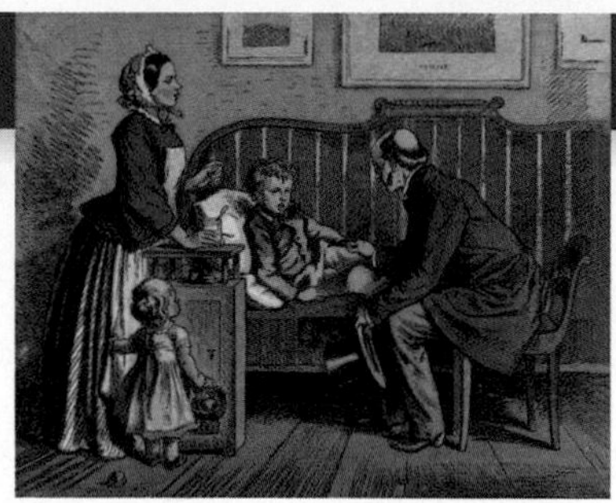

Kapitel 11

Gesundheit und Krankheit

Themen

Kapitel 12

Die moderne Gesellschaft

Themen

Preface

Kontakte continues to offer a truly communicative approach that supports functional proficiency, supported by the full suite of digital tools available in **Connect**. This proven introductory German program maintains its commitment to meaningful communicative practice as well as extensive coverage of the 5 C's and the ACTFL Proficiency Guidelines 2012. Now in its eighth edition, ***Kontakte*** has greatly expanded its digital offering: **Connect** now contains the full scope of activities originating from both the white and blue pages of the student text and the *Workbook / Laboratory Manual* **(Arbeitsbuch)**. Furthermore, the digital program now offers **LearnSmart**®, an adaptive learning program that helps students learn grammar and vocabulary more efficiently by tailoring the experience to individual student needs.

Communication in Meaningful Contexts

Throughout the ***Kontakte*** program, students have the opportunity to communicate in German in meaningful ways. Students read and listen to comprehensible German and are provided with ample opportunities to use it in interview, information-gap, role-play, autograph, writing, and other personalized activities that are theme-based, not grammar-driven. The video segments—**Perspektiven** and **Interviews**—were filmed specifically for ***Kontakte*** and feature interviews with a variety of speakers that allow students to hear authentic German in context. They provide models for talking about topics using authentic language, guiding students to communicate with one another.

In **Connect** students can also take advantage of the synchronous and asynchronous chat tools to communicate with their classmates online. For example, each chapter includes one **Rollenspiel** chat activity adapted from the role-plays in the text. After completing pre-listening tasks, students listen to a model role-play, then connect online to role-play with another student in real time. The **Interviews** and **Umfragen** activities have also been adapted to online formats, using the chat tools.

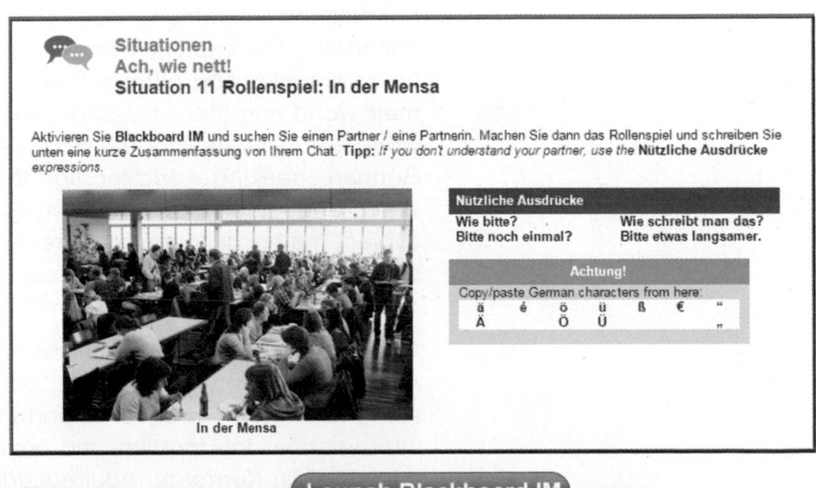

A Solid Theoretical Foundation

Firmly grounded in second-language acquisition research, ***Kontakte*** also supports the National Standards as outlined in the Standards for Foreign Language Learning in the 21st Century. As presented in the Standards, the five C's—Communication, Cultures, Connections, Comparisons, and Communities—provide a framework for what students should know and be able to do as a result of their language study.

The five C's are present in countless ways throughout the program, particularly in the wealth of communicative activities, as mentioned above, and in the cultural content. Cultural readings called **Kultur ... Landeskunde ... Informationen** develop themes such as geography, history, and society and present various perspectives on the cultures of the German-speaking world.

KULTUR ... LANDESKUNDE ... INFORMATIONEN

CHATIQUETTE: STERNCHEN, ABKÜRZUNGEN UND AKRONYME

Wenn es schnell gehen muss, verwenden[1] viele Leute im Chat, bei WhatsApp oder SMS besondere Formen der Kommunikation. Sie machen das Chatleben leichter. Viele sind lustig oder ironisch gemeint und ein fester Bestandteil[2] der Chatkultur. Sterchen[3] drücken Emotion oder Tätigkeit aus und es gibt viele Akronyme auf Englisch, aber auch auf Deutsch.

© Gerhilde Skoberne/Corbis RF

[1]*use* [2]fester ... *established part* [3]*asterisks*

Miniwörterbuch	
grinsen	to grin
knuddeln	to cuddle
doll	very (*colloquial*)
hab dich lieb	(I) love you
drücken	to hug
das **Unverständnis**	incomprehension
zeigen	to show
frech	impudently
fies	meanly

Können Sie folgende Akronyme auf Deutsch erkennen? Ordnen Sie die Akronyme den Aussagen zu.

1. *g*	a. kein Kommentar
2. *fg*	b. grinsen
3. *momtel*	c. Moment, ich telefoniere gerade
4. *knuddel*	d. liebe Grüße
5. LG	e. hab dich lieb
6. kk	f. ich knuddel/drück dich
7. N8	g. frech/fies grinsen
8. omg	h. Nacht / Gute Nacht
9. HDL	i. hab dich ganz doll lieb
10. HDGDL	j. oh mein Gott

Musikszene and **Filmclip** features highlight contributions in German-language music and film. In addition, the ***Kontakte*** video program provides a rich source of authentic language and culture that holds students' interest and draws them into interactions and discussion.

In addition to communicative practice and cultural exposure, students are encouraged to explore connections by linking their study of German with their own lives and other subjects of study, to make comparisons between their world and that of German-speaking people, and to learn about real-world German-speaking communities. They are given direct access to the German-speaking world through the post-reading **Nach dem Lesen** sections which engage students in activities where they expand the scope of the subject matter or topic to the real-world level. In several **Filmlektüren**, students complete Internet research on topics related to German cinema. All of these activities as well as the ***Kontakte*** video are available in **Connect.**

Kontakte also integrates several modes of language, as described in the ACTFL Proficiency Descriptors. The activities, exercises, and tasks offer students a wide variety of opportunities for communication and interaction in interpersonal, interpretive, and presentational modes. For example, the many interviews in ***Kontakte*** require students to negotiate meaning and therefore reinforce the interpersonal mode of communication. The diversity of

Dora Hitz: *Mädchen im Mohnfeld* (1891), Museum der Bildenden Künste, Leipzig, Deutschland

KUNST UND KÜNSTLER

Dora Hitz (1856–1924) was a German painter who studied in Munich at the "Damenmalschule der Frau Staatsrat Weber," an art school for young women, and in Paris. Later she worked in Romania as the court painter to the Romanian royal family and in Berlin where she was a member of the "Verein Berliner Künstlerinnen und Kunstfreundinnen." In 1894 she founded an art school for women. Later in life she fell into financial difficulties, became ill, and shunned social contact.

Sehen Sie das im Bild?[1]

	JA	NEIN
eine Straße[2]	☐	☐
ein Mädchen[3]	☐	☐
Blumen	☐	☐
Autos	☐	☐
die Farbe Gelb[4]	☐	☐
die Farbe Grün[5]	☐	☐
die Farbe Rot[6]	☐	☐

[1]Sehen ... *Do you see that in the picture?* [2]eine ... *a road* [3]*girl* [4]die ... *the color yellow* [5]*green* [6]*blue*

Situationen

Aufforderungen

Grammatik A.1

Situation 1 **Aufforderungen**

1. Geben Sie mir die Hausaufgabe!
2. Öffnen Sie das Buch!
3. Schließen Sie das Buch!
4. Nehmen Sie einen Stift!
5. Gehen Sie!
6. Springen Sie!
7. Laufen Sie!
8. Schauen Sie an die Tafel!

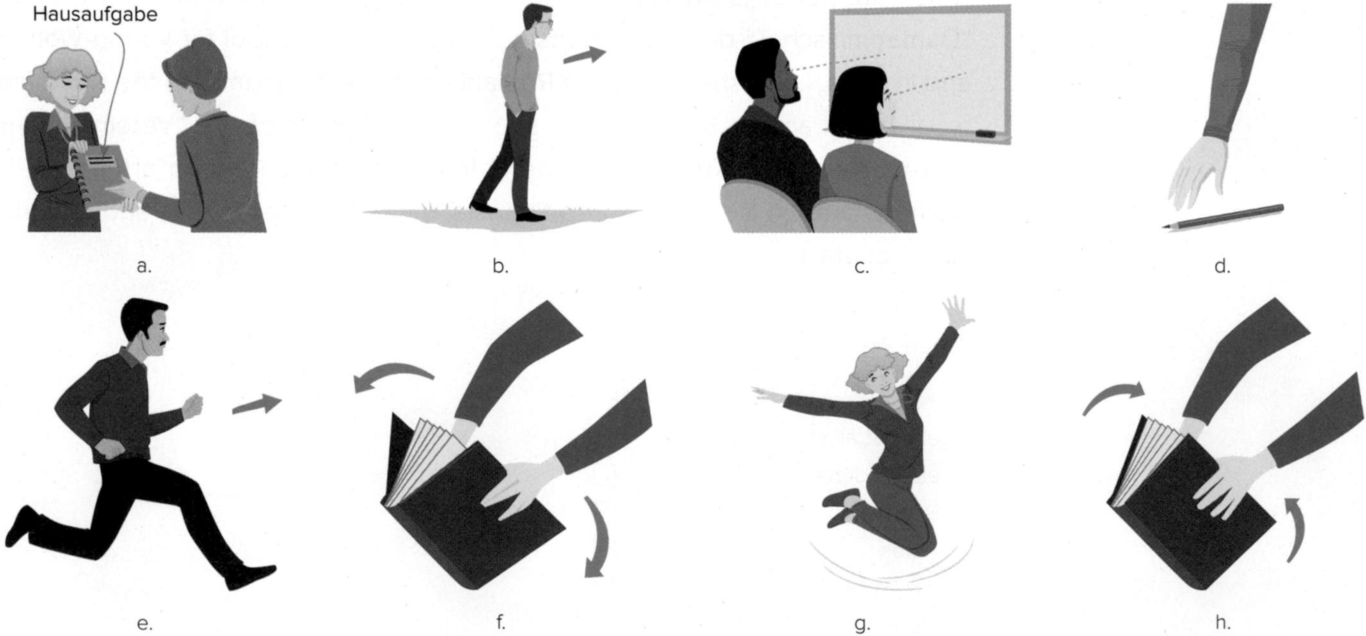

Situation 2 Wer macht das?

Hören Sie zu und schreiben Sie die Zahlen unter die Bilder.

a. ____

b. ____

c. ____

d. ____

e. ____

f. ____

g. ____

h. ____

Namen

Grammatik A.2–A.3

—Wie heißt du?
—Heidi.
—Wie schreibt man das?
—H-E-I-D-I. Und wie heißt du?

Heidi und Stefan

Buchstaben

Schreiben	*Sprechen*	*Schreiben*	*Sprechen*
A a	[aː]	***O o***	[oː]
Ä ä	[ɛː]	***Ö ö***	[øː]
B b	[beː]	***P p***	[peː]
C c	[tseː]	***Q q***	[kuː]
D d	[deː]	***R r***	[ɛr]
E e	[eː]	***S s***	[ɛs]
F f	[ɛf]	***ß***	[ɛsˈtsɛt]
G g	[geː]	***T t***	[teː]
H h	[haː]	***U u***	[uː]
I i	[iː]	***Ü ü***	[yː]
J j	[jɔt]	***V v***	[fau]
K k	[kaː]	***W w***	[veː]
L l	[ɛl]	***X x***	[ɪks]
M m	[ɛm]	***Y y***	[ˈʏpsilɔn]
N n	[ɛn]	***Z z***	[tsɛt]

KULTUR ... LANDESKUNDE ... INFORMATIONEN

VORNAMEN

- Was sind häufige[1] Vornamen in Ihrem Land für Personen über 60 Jahre? für Personen um die 40? für Personen um die 20? für Neugeborene[2]?
- Welche Vornamen gefallen Ihnen[3]?
- Welche deutschen Vornamen gibt es auch in Ihrem Kurs?
- Welche deutschen Familiennamen gibt es in Ihrem Kurs?
- Möchten Sie einen deutschen Vornamen annehmen[4]? Welchen?

DIE BELIEBTESTEN[5] VORNAMEN IN DEUTSCHLAND 2014

Mädchen	Jungen
1. Sophie/Sofie	**1.** Maximilian
2. Marie	**2.** Alexander
3. Sophia/Sofia	**3.** Paul
4. Maria	**4.** Elias
5. Mia	**5.** Luis/Louis
6. Emma	**6.** Luca/Luka
7. Hannah/Hanna	**7.** Ben
8. Emilia	**8.** Leon/Léon
9. Anna	**9.** Lukas/Lucas
10. Johanna	**10.** Noah/Noa

Source of Data: Gesellschaft für deutsche Sprache Wiesbaden.

[1]*common* [2]*newborns* [3]gefallen ... *do you like* [4]*adopt* [5]*most popular*

Situation 3 Wie heißt ...?

1. Wie heißt die Frau mit dem Buch?
2. Wie heißt der Mann mit dem Stift?
3. Wie heißt die Frau an der Tafel?
4. Wie heißt die Frau an der Tür?
5. Wie heißt der Mann mit der Brille?
6. Wie heißt der Mann mit dem Schnurrbart?
7. Wie heißt die Frau mit dem Ball?
8. Wie heißt der Mann mit dem langen Haar?

Situation 4 Interview: Wie schreibt man deinen Namen?

MODELL: ein Student / eine Studentin mit Brille →
S1: Wie heißt du?
S2 (*mit Brille*): Mark.
S1: Wie schreibt man das?
S2: M-A-R-K.

	NAME
1. ein Student / eine Studentin mit Brille	________________
2. ein Student / eine Studentin in Jeans	________________
3. ein Student / eine Studentin mit langem Haar	________________
4. ein Student / eine Studentin mit einem Buch	________________
5. ein Student / eine Studentin mit Ohrring	________________
6. ein Student / eine Studentin mit kurzem Haar	________________

Kleidung

Grammatik A.4

Michael Pusch

Jens Krüger

Maria Schneider

Josie Wagner

Situation 5 Kleidung

Wer im Deutschkurs trägt ______?

1. eine Bluse
2. einen Rock
3. eine Jacke
4. ein Kleid
5. Stiefel
6. ein Hemd
7. eine Hose
8. einen Hut
9. Sportschuhe
10. einen Pullover
11. eine Krawatte
12. einen Anzug

Situation 6* Informationsspiel: 10 Fragen

Stellen Sie zehn Fragen. Für jedes „Ja" gibt es einen Punkt.

MODELL: S1: Trägt Thomas einen Anzug?
S2: Nein. Trägt Frau Körner einen Hut?
S1: Nein.

	THOMAS JA ODER NEIN	NORA JA ODER NEIN		THOMAS JA ODER NEIN	NORA JA ODER NEIN
einen Anzug	N		einen Mantel		
eine Bluse			einen Pullover		
eine Brille			einen Rock		
ein Hemd			ein Sakko		
eine Hose			Schuhe		
einen Hut			Socken		
eine Jacke			Sportschuhe		
eine Jeans			Stiefel		
ein Kleid			ein Stirnband		
eine Krawatte			ein T-Shirt		

Herr Siebert Frau Körner

Thomas Nora

*This is the first of many information-gap activities in ***Kontakte***. Pair up with another student. One of you will work with the pictures on this page. The other will work with different pictures in Appendix A. The goal is to complete the activity while speaking only German and not looking at your partner's pictures.

Farben

Grammatik A.4

Situation 7 Meine Mitstudenten

Schauen Sie Ihre Mitstudenten und Mitstudentinnen an. Was tragen sie?

	NAME	KLEIDUNG	FARBE
1.	Heidi	Rock	blau
2.	______	______	______
3.	______	______	______
4.	______	______	______
5.	______	______	______

Situation 8 Umfrage: Was ist deine Lieblingsfarbe?

MODELL: S1: Ist deine Lieblingsfarbe blau?
S2: Ja.
S1: Unterschreib bitte hier.

	UNTERSCHRIFT
1. Ist deine Lieblingsfarbe blau?	______
2. Trägst du gern schwarz?	______
3. Hast du zu Hause braune Socken?	______
4. Ist deine Lieblingsfarbe rot?	______
5. Trägst du gern gelb?	______
6. Hast du zu Hause ein grünes T-Shirt?	______
7. Ist deine Lieblingsfarbe lila?	______
8. Hast du zu Hause ein weißes Hemd?	______

KULTUR ... LANDESKUNDE ... INFORMATIONEN

FARBEN ALS SYMBOLE

__________ ist die Liebe[1]
__________ ist die Unschuld[2]
__________ ist die Trauer[3]
__________ ist die Treue[4]
__________ ist die Hoffnung[5]
__________ ist der Neid[6]

© Julie Nicholls/Corbis

[1]*love* [2]*innocence* [3]*grief, sorrow* [4]*loyalty* [5]*hope* [6]*envy*

In der Stadt
© ullstein bild-Müller-Stauffenberg / The Image Works

Begrüßen und Verabschieden

Grammatik A.5

Guten Morgen!

Guten Tag!

Guten Abend!

—Auf Wiedersehen!
—Wiedersehen!

—Tschüss!
—Bis bald!

Situation 9 **Dialoge**

1. Jürgen Baumann spricht mit einer Studentin.

JÜRGEN: Hallo, bist du ____________ hier?
MELANIE: ____________. Du auch?
JÜRGEN: Ja. Sag mal, ____________?
MELANIE: Melanie. Und ____________?
JÜRGEN: Jürgen.

2. Frau Frisch-Okonkwo ruft Herrn Koch an.

HERR KOCH: Koch.
FRAU FRISCH-OKONKWO: Guten Tag, Herr Koch, ____________ Frisch-Okonkwo. Unser Computer ist kaputt.
HERR KOCH: ____________, ich komme morgen vorbei.
FRAU FRISCH-OKONKWO: Gut. Bis dann. ____________.

3. Jutta trifft ihren Freund Jens.

JUTTA: Servus, Jens.
JENS: Ach, ____________, Jutta.
JUTTA: Wo willst ____________ denn hin?
JENS: ____________ muss zum Fußballtraining.
JUTTA: Na, dann ____________!
JENS: ____________. Mach's gut, Jutta.

MUSIKSZENE

„A-N-N-A" (1997, Deutschland) *Freundeskreis*

Biografie *Freundeskreis* ist aus Stuttgart. Der Gründer und Lead-Sänger heißt Max Herre. „A-N-N-A" war die 1. Hitsingle der Gruppe aus dem Jahr 1997. Andere große Hits waren „Tabula rasa" und „Mit dir".

Freundeskreis

NOTE: For copyright reasons, the songs referenced in **MUSIKSZENE** have not been provided by the publisher. The song can be found online at various sites such as YouTube, Amazon, or the iTunes store.

Vor dem Hören Was ist das Besondere an dem Namen *Anna*?

☐ **1.** Er beginnt mit *A*.

☐ **2.** Er hat vier Buchstaben.

☐ **3.** Er ist von hinten und von vorne gleich.

Nach dem Hören

A. Hören Sie den Refrain! Richtig (R) oder falsch (F)?

___ **1.** Max denkt an Anna, wenn es regnet.

___ **2.** Anna war nass bis auf die Haut.

___ **3.** Max liebt Anna.

B. Wie heißt dein Freund oder deine Freundin?

Miniwörterbuch	
das Besondere an	special about
von hinten	backwards
von vorne	forwards
gleich	the same
denkt an	thinks about
regnet	rains
nass	wet
bis auf die Haut	to the skin

Situation 10* Rollenspiel: Begrüßen

S1: Begrüßen Sie einen Mitstudenten oder eine Mitstudentin. Schütteln Sie dem Mitstudenten oder der Mitstudentin die Hand. Sagen Sie Ihren Namen. Fragen Sie, wie alt er oder sie ist. Verabschieden Sie sich.

Begrüßen
© *Yavuz Arslan/ullstein bild/The Image Works*

Zahlen

0	null	10	zehn	20	zwanzig	30	dreißig
1	eins	11	elf	21	einundzwanzig	40	vierzig
2	zwei	12	zwölf	22	zweiundzwanzig	50	fünfzig
3	drei	13	dreizehn	23	dreiundzwanzig	60	sechzig
4	vier	14	vierzehn	24	vierundzwanzig	70	siebzig
5	fünf	15	fünfzehn	25	fünfundzwanzig	80	achtzig
6	sechs	16	sechzehn	26	sechsundzwanzig	90	neunzig
7	sieben	17	siebzehn	27	siebenundzwanzig	100	hundert
8	acht	18	achtzehn	28	achtundzwanzig		
9	neun	19	neunzehn	29	neunundzwanzig		

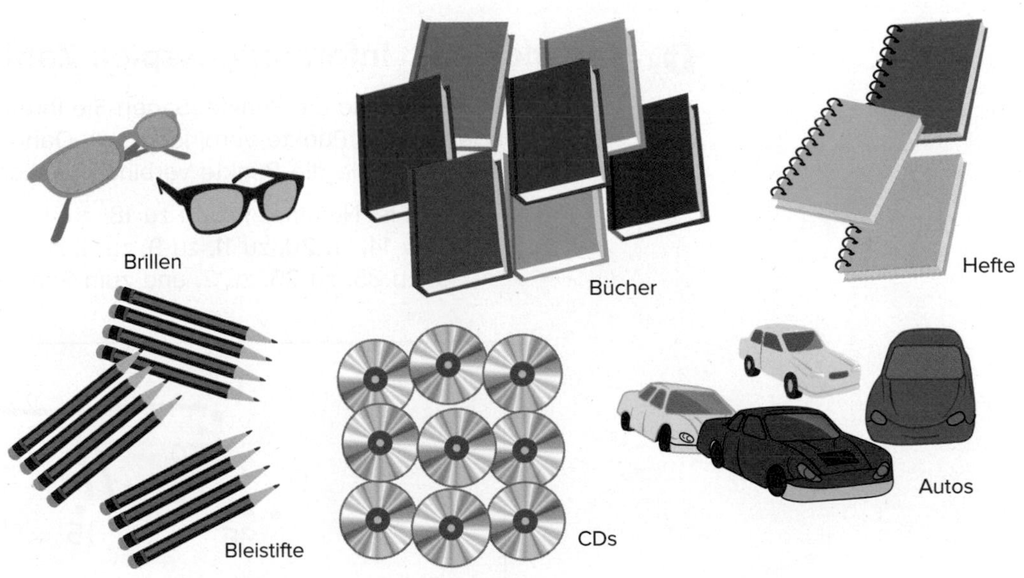

*This is the first of many role-playing activities in ***Kontakte.*** Pair up with another student. One of you takes the role of S1. The corresponding role for the other person (S2) appears in Appendix B.

Situation 11 Wie viele?

Wie viele Studenten/Studentinnen im Kurs tragen ...?

eine Hose ______
eine Brille ______
eine Armbanduhr ______
eine Bluse ______
einen Rock ______
Sportschuhe ______

KULTUR ... LANDESKUNDE ... INFORMATIONEN

SO ZÄHLT MAN ...

SO SCHREIBT MAN ...

eine Eins

eine Sieben

Situation 12 Informationsspiel: Zahlenrätsel

Verbinden Sie die Punkte. Sagen Sie Ihrem Partner oder Ihrer Partnerin, wie er oder sie die Punkte verbinden soll. Dann sagt Ihr Partner oder Ihre Partnerin Ihnen, wie Sie die Punkte verbinden sollen. Was zeigen Ihre Bilder?

S1: Start ist Nummer 1. Geh zu 18, zu 7, zu 29, zu 13, zu 60, zu 32, zu 12, zu 5, zu 14, zu 20, zu 11, zu 9, zu 3, zu 80, zu 23, zu 19, zu 4, zu 27, zu 8, zu 15, zu 35, zu 26, zu 2, und zum Schluss zu 17. Was zeigt dein Bild?

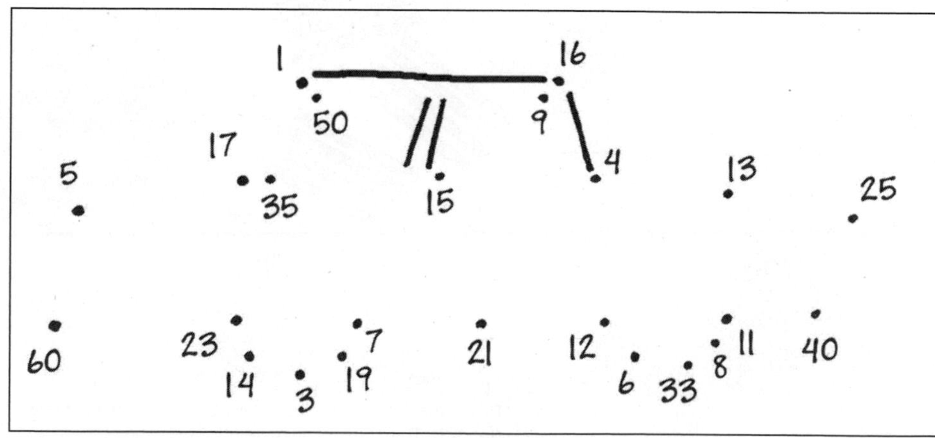

Videoecke

Perspektiven

„Hey, wie geht's?"

Aufgabe 1 Wie viele?

Miniwörterbuch	
die **Paare**	pairs
umarmen	embrace
sich	*here:* each other
küssen	kiss
zueinander	to each other

Wie viele Paare machen das?

_______ 1. Wie viele Paare schütteln sich die Hand?

_______ 2. Wie viele Paare umarmen sich?

_______ 3. Wie viele Paare küssen sich?

_______ 4. Wie viele Paare sitzen, wie viele stehen?

_______ 5. Wie viele Paare sagen: „Wie geht's?"?

_______ 6. Wie viele Paare sagen **Sie** zueinander?

Aufgabe 2 Was sagen sie?

Miniwörterbuch	
der **Zopf**	braid
die **Strickjacke**	cardigan sweater
der **Schal**	scarf
beide	both

Was sagen die folgenden Personen?

_______ 1. junger Mann mit lila Hemd

_______ 2. junge Frau mit Zopf und blondem Haar

_______ 3. junge Frau mit langem schwarzem Haar und schwarzer Strickjacke

_______ 4. Frau mit kurzem dunkelbraunem Haar und brauner Jacke

_______ 5. junge Frau mit langem blondem Haar und lila Sweatshirt

_______ 6. junger Mann mit grünkariertem Hemd

_______ 7. junge Frau mit langem dunkelbraunem Haar, lila T-Shirt und schwarzer Hose

_______ 8. junge Frau mit hellbrauner Jacke und Schal

a. „Hallo Susi."
b. „Gut, und dir?"
c. „Na, wie geht's dir?"
d. „Hey, wie geht's dir?"
e. „Hey, wie geht's?"
f. „Mir geht's gut und dir?"
g. „Ach, ganz gut und dir?"
h. „Guten Tag!"

Interviews

- Wie heißt du?
- Wie schreibt man das?
- Welche Kleidung trägst du gern?
- Welche Farben trägst du gern?
- Wie alt bist du?
- Hast du eine Glückszahl?

Nicole

Michael

Aufgabe 3 Persönliche Daten

Wer sagt das, Nicole oder Michael oder beide?

	NICOLE	MICHAEL	BEIDE
1. Ich trage gern Jeans und Pullover.	☐	☐	☐
2. Ich trage gern türkis, blau und grün.	☐	☐	☐
3. Ich trage gern rot und braun.	☐	☐	☐
4. Ich bin 45 Jahre alt.	☐	☐	☐
5. Ich bin 28 Jahre alt.	☐	☐	☐
6. Meine Glückszahl ist sieben.	☐	☐	☐
7. Meine Glückszahl ist dreizehn.	☐	☐	☐

Aufgabe 4 Interview

Interviewen Sie eine Partnerin oder einen Partner. Stellen Sie dieselben Fragen.

Wortschatz

Aufforderungen — Instructions

arbeiten Sie mit einem Partner* — work with a partner
geben Sie mir — give me
gehen Sie — go, walk
hören Sie zu — listen
laufen Sie — go, run
lesen Sie — read
nehmen Sie — take
öffnen Sie — open
sagen Sie — say
schauen Sie — look
schließen Sie — close, shut
schreiben Sie — write; spell
setzen Sie sich — sit down
springen Sie — jump
stehen Sie auf — get up, stand up

Kleidung — Clothes

er/sie hat ... — he/she has ...
hast du ...? — do you have ...?
er/sie trägt ... — he/she is wearing ...
trägst du ...? — do you wear ...? / are you wearing ...?

eine **Armbanduhr** — a watch
eine **Brille** — glasses
eine **Hose** — pants
eine **Krawatte** — a tie

einen **Anzug** — a suit
einen **Mantel** — a coat; an overcoat
einen **Ohrring** — an earring
einen **Rock** — a skirt

ein **Hemd** — a shirt
ein **Kleid** — a dress
ein **Sakko** — a sports jacket
ein **Stirnband** — a headband

Stiefel — boots

Ähnliche Wörter†
er/sie trägt ... eine **Bluse,** eine **Jacke;** einen **Hut; Schuhe, Sportschuhe**

Farben — Colors

gelb — yellow
lila — purple
rosa — pink
schwarz — black

Ähnliche Wörter
blau, braun, grau, grün, orange [oraŋʒə], rot, weiß

Zahlen (Numbers)

0	**null**	20	**zwanzig**
1	**eins**	21	**einundzwanzig**
2	**zwei**	22	**zweiundzwanzig**
3	**drei**	23	**dreiundzwanzig**
4	**vier**	24	**vierundzwanzig**
5	**fünf**	25	**fünfundzwanzig**
6	**sechs**	26	**sechsundzwanzig**
7	**sieben**	27	**siebenundzwanzig**
8	**acht**	28	**achtundzwanzig**
9	**neun**	29	**neunundzwanzig**
10	**zehn**	30	**dreißig**
11	**elf**	40	**vierzig**
12	**zwölf**	50	**fünfzig**
13	**dreizehn**	60	**sechzig**
14	**vierzehn**	70	**siebzig**
15	**fünfzehn**	80	**achtzig**
16	**sechzehn**	90	**neunzig**
17	**siebzehn**	100	**hundert**
18	**achtzehn**		
19	**neunzehn**		

Begrüßen und Verabschieden — Greeting and Leave-Taking

auf Wiedersehen — good-bye
bis bald — so long; see you soon
grüezi — hi (*Switzerland*)
grüß Gott — good afternoon; hello (*formal; southern Germany, Austria*)
guten Abend — good evening
guten Morgen — good morning
guten Tag — good afternoon; hello (*formal*)
hallo — hi (*informal*)
die **Hand schütteln** — to shake hands
mach's gut — take care (*informal*)
servus — hello; good-bye (*informal; southern Germany, Austria*)
tschüss — bye (*informal*)
viel Spaß — have fun

*The diacritic marks in the **Wortschatz** list are meant to help you learn which vowels are stressed. A dot below a single vowel indicates a short stressed vowel. An underline below a single vowel, double vowel, or diphthong (combination of two different vowels) indicates a long stressed vowel. Note that these markings are not used in written German but are provided here as an aid to pronunciation.

†**Ähnliche Wörter** (*similar words; cognates*) lists contain words that are closely related to English words in sound, form, and meaning and compound words that are composed of previously introduced vocabulary.

Personen	People
die **Frau**	woman; Mrs.; Ms.
die **Lehrerin**	female teacher, instructor
der **Herr**	gentleman; Mr.
der **Lehrer**	male teacher, instructor
die **Mitstudenten**	fellow (male) students
die **Mitstudentinnen**	fellow (female) students

Ähnliche Wörter

die **Freundin,** die **Professorin,** die **Studentin;** der **Freund,** der **Mann,** der **Professor,** der **Student**

Sonstige Substantive	Other Nouns
die **Tafel**	blackboard/whiteboard
die **Tür**	door
der **Stift**	pen
der **Bleistift**	pencil
Lieblings-	favorite
die **Lieblingsfarbe**	favorite color
der **Lieblingsname**	favorite name

Ähnliche Wörter

die **CD;** der **Ball,** der **Fußball,** der **Kurs,** der **Deutschkurs,** der **Name,** der **Familienname,** der **Vorname,** der **Teddybär;** das **Auto,** das **Buch**

Fragen	Questions
heißen	to be called, be named
wie heißen Sie?	what's your name? (*formal*)
wie heißt du?	what's your name? (*informal*)
ich heiße ...	my name is ...
was zeigen Ihre Bilder?	what do your pictures show?
welche Farbe hat ...?	what color is ...?
wer ...?	who ...?
wie schreibt man das?	how do you spell that?
wie viele ...?	how many ...?
wo willst du denn hin?	where are you going?

Wörter im Deutschkurs	Words in German Class
die **Antwort**	answer
die **Einführung**	introduction
die **Frage**	question
die **Grammatik**	grammar
die **Hausaufgabe**	homework
die **Sprechsituation**	conversational situation
die **Übung**	exercise
der **Punkt**	point
der **Wortschatz**	vocabulary
das **Kapitel**	chapter
stellen Sie Fragen	ask questions
unterschreib bitte hier	sign here, please
verbinden	to connect

Sonstige Wörter und Ausdrücke	Other Words and Expressions
aber	but
auch	also, too; as well
bitte	please
gibt es ...?	is there . . .? / are there . . .?
hübsch	pretty
kaputt	broken
mein(e)	my
mit	with
mit dem kurzen Haar	with the short hair
mit dem langen Haar	with the long hair
mit dem Ohrring	with the earring
mit dem Schnurrbart	with the mustache
nein	no
nicht	not
oder	or
schmutzig	dirty
sein	to be
sondern	but (rather/on the contrary)
trägst du gern ...?	do you like to wear ...?
viel	a lot, much
viele	many
von	of; from
zählen	to count
zu Hause	at home

Ähnliche Wörter

alt, danke, dann, hier, in, neu, oft, so, und

Strukturen und Übungen

A.1 Giving instructions: polite commands

command form = verb + **Sie**

The instructions your instructor gives you in class consist of a verb, which ends in **-en,** and the pronoun **Sie** (*you*).* Like the English *you,* the German **Sie** can be used with one person (*you*) or with more than one (*you* [*all*]). In English instructions the pronoun *you* is normally understood but not said. In German, **Sie** is a necessary part of the sentence.

Stehen Sie bitte **auf.**	*Please stand up.*
Nehmen Sie bitte das Buch.	*Please take the book.*

With certain instructions, you will also hear the word **sich** (*yourself*).†

Setzen Sie sich, bitte.	*Sit down, please.*

Übung 1 Im Seminarraum

Was sagt Frau Schulz zu den Studenten?

Nehmen Sie einen Stift!
Sagen Sie „Guten Tag"!
Schauen Sie an die Tafel!
Schließen Sie das Buch!
Schreiben Sie „Tschüss"!
Öffnen Sie das Buch!
Hören Sie zu!
Geben Sie mir die Hausaufgabe!

1. Peter

2. Heidi

3. Monika

4. Nora

5. Albert

6. Stefan

7. Thomas

8. Katrin

*The pronoun **Sie** (*you*) is capitalized to distinguish it from another pronoun, **sie** (*she; it; they*).
†**Sich** is a reflexive pronoun; its use will be explained in Kapitel 11.

A.2 What is your name? The verb *heißen*

heißen = *to be called*
Wie heißen Sie? *(formal)*
Wie heißt du? *(informal)*

Use a form of the verb **heißen** (*to be called*) to tell your name and to ask for the names of others.

Wie **heißen Sie?** / Wie **heißt du?*** — *What is your name?*
Ich heiße ... — *My name is . . .*

heißen (singular forms)		
ich	heiße	*my name is*
du	heißt	*your name is*
Sie	heißen	*your name is*
er	heißt	*his name is*
sie	heißt	*her name is*

Übung 2 Minidialoge

Ergänzen Sie[1] das Verb **heißen: heiße, heißt, heißen.**

1. ERNST: Hallo, wie _____[a] du?
   JUTTA: Ich _____[b] Jutta. Und du?
   ERNST: Ich _____[c] Ernst.
2. HERR THELEN: Guten Tag, wie _____[a] Sie bitte?
   HERR SIEBERT: Ich _____[b] Siebert, Alexander Siebert.
3. CLAIRE: Hallo, ich _____[a] Claire und wie heißt ihr?
   MELANIE: Ich _____[b] Melanie und er _____[c] Josef.

A.3 The German case system

Case shows how nouns function in a sentence.

German speakers use a *case system* (nominative for the subject, accusative for the direct object, and so on) to indicate the function of a particular noun in a sentence. The article[†] or adjective that precedes the noun shows its case. You will learn the correct endings in future lessons. For now, be aware that you will hear and read articles and adjectives with a variety of endings. These various forms will not prevent you from understanding German. Here are all the possibilities.

der, das, die, dem, den, des — *the*
ein, eine, einen, einem, einer, eines — *a, an*
blau, blaue, blauer, blaues, blauen, blauem — *blue*

*The difference between **Sie** (*formal*) and **du** (*informal*) will be explained in Section A.5.
[†]Articles are words such as the, a, and an, which precede nouns.
[1]Ergänzen ... *Supply*

In addition, definite articles may contract with some prepositions, just as *do* and *not* contract to *don't* in English. Here are some common contractions you will hear and read.

in + das = ins	*into the*
in + dem = im	*in the*
zu + der = zur	*to the*
zu + dem = zum	*to the*
an + das = ans	*to/on the*
an + dem = am	*to/at the*

A.4 Grammatical gender: nouns and pronouns

masculine = **der**
neuter = **das**
feminine = **die**
plurals (all genders) = **die**

In German, all nouns are classified grammatically as masculine, neuter, or feminine. When referring to people, grammatical gender usually matches biological sex.

MASCULINE	FEMININE
der Mann	**die** Frau
der Student	**die** Studentin

When referring to things or concepts, however, grammatical gender obviously has nothing to do with biological sex.

MASCULINE	NEUTER	FEMININE
der Rock	**das** Hemd	**die** Hose
der Hut	**das** Buch	**die** Jacke

The definite article indicates the grammatical gender of a noun. German has three nominative singular definite articles: **der** (*masculine*), **das** (*neuter*), and **die** (*feminine*). The plural article is **die** for all genders. All of these definite articles mean *the*.

	Singular	Plural
Masculine	der	die
Neuter	das	die
Feminine	die	die

der → **er** = *he, it*
das → **es** = *it*
die → **sie** = *she, it*
die (pl.) → **sie** = *they*

The personal pronouns **er, es, sie** (*he, it, she*) reflect the gender of the nouns they replace. For example, **er** (*he, it*) refers to **der Rock** because the grammatical gender is masculine; **es** (*it*) refers to **das Hemd** (*neuter*); **sie** (*she, it*) refers to **die Jacke** (*feminine*). The personal pronoun **sie** (*they*) refers to all plural nouns.

—Welche Farbe hat **der Rock?** — *What color is the skirt?*
—**Er** ist gelb. — *It is yellow.*

—Welche Farbe hat **das Hemd?** — *What color is the shirt?*
—**Es** ist weiß. — *It is white.*

—Welche Farbe hat **die Jacke?** — *What color is the jacket?*
—**Sie** ist braun. — *It is brown.*

—Welche Farbe haben **die Bleistifte?** — *What color are the pencils?*
—**Sie** sind gelb. — *They are yellow.*

> **ACHTUNG!**
> All German nouns are capitalized, whether they are common nouns (objects, concepts: **Jacke, Freund**) or proper nouns (names of people, countries, etc.: **Heidi, Deutschland**).

Sometimes gender can be determined from the ending of the noun; for example, most nouns that end in **-e,** such as **die Jacke** or **die Bluse,** are feminine. The ending **-in** indicates a female person: **die Studentin, die Professorin.**

In most cases, however, gender cannot be predicted from the form of the word. It is best, therefore, to learn the corresponding definite article along with each new noun.*

Übung 3 Kleidung

Frau Schulz spricht über die Kleidung. Ergänzen Sie **er, es, sie** oder **sie** (Plural).

Frau Schulz:

1. Hier ist die Jacke. _____ ist neu.
2. Und hier ist das Kleid. _____ ist modern.
3. Hier ist der Rock. _____ ist kurz.
4. Und hier ist die Bluse. _____ ist hübsch.
5. Hier ist das Hemd. _____ ist grün.
6. Und hier sind die Schuhe. _____ sind schmutzig.
7. Hier ist der Hut. _____ ist rot.
8. Und hier ist die Hose. _____ ist weiß.
9. Hier sind die Stiefel. _____ sind schwarz.
10. Und hier ist der Anzug. _____ ist alt.

Übung 4 Welche Farbe?

Welche Farbe haben diese Kleidungsstücke? Ergänzen Sie **er, es, sie** oder **sie** (Plural) und die richtige Farbe.

1. A: Welche Farbe hat Marias Rock?
   B: _____ ist _____.
2. A: Welche Farbe hat Michaels Hose?
   B: _____ ist _____.
3. A: Welche Farbe hat Michaels Hemd?
   B: _____ ist _____.
4. A: Welche Farbe hat Michaels Hut?
   B: _____ ist _____ und _____.
5. A: Welche Farbe haben Marias Schuhe?
   B:_____ sind _____.
6. A: Welche Farbe haben Michaels Schuhe?
   B: _____ sind _____.
7. A: Welche Farbe hat Marias Bluse?
   B: _____ ist _____.

* Some students find the following suggestion helpful. When you hear or read new nouns you consider useful, write them down in a vocabulary notebook, using different colors for the three genders; for example, use blue for masculine, black for neuter, and red for feminine. Some students also write nouns in three separate columns according to gender.

A.5 Addressing people: *Sie* versus *du* or *ihr*

Use **du** and **ihr** with friends, family, and children. Use **Sie** with almost everyone else.

German speakers use two modes of addressing others: the formal **Sie** (*singular* and *plural*) and the informal **du** (*singular*) or **ihr** (*plural*). You usually use **Sie** with someone you don't know or when you want to show respect or social distance. Children are addressed as **du.** Students generally call one another **du.**

	Singular	Plural
Informal	du	ihr
Formal	Sie	Sie

Frau Ruf, **Sie** sind 38, nicht wahr? — *Ms. Ruf, you are 38, aren't you?*
Jens und Jutta, **ihr** seid 16, nicht wahr? — *Jens and Jutta, you are 16, aren't you?*
Hans, **du** bist 13, nicht wahr? — *Hans, you are 13, aren't you?*

Übung 5 *Sie, du* oder *ihr*?

Was sagen diese Personen: **Sie, du** oder **ihr?**

1. Student → Student
2. Professor → Student
3. Freund → Freund
4. Studentin → zwei Studenten
5. Frau (40 Jahre alt) → Frau (50 Jahre alt)
6. Student → Sekretärin
7. Doktor → Patient
8. Frau → zwei Kinder

EINFÜHRUNG B

In **Einführung B,** you will continue to develop your listening skills and will begin to speak more German. You will learn to talk about your classroom, the weather, and people: their character traits, family relationships, and national origins.

Themen

Der Seminarraum

Beschreibungen

Der Körper

Die Familie

Wetter und Jahreszeiten

Herkunft und Nationalität

Kulturelles

KLI: Was ist wichtig im Leben?

KLI: Wetter und Klima

Musikszene: „36 Grad" (2raumwohnung)

KLI: Die Lage Deutschlands in Europa

Videoecke: Familie

Strukturen

B.1 Definite and indefinite articles

B.2 Who are you? The verb **sein**

B.3 What do you have? The verb **haben**

B.4 Plural forms of nouns

B.5 Personal pronouns

B.6 Origins: **Woher kommen Sie?**

B.7 Possessive determiners: **mein** and **dein/Ihr**

August Macke: *Mutter und Kind im Park (1914)*, Hamburger Kunsthalle, Hamburg

KUNST UND KÜNSTLER

August Macke (1887–1914) was one of the leading members of *Der Blaue Reiter* (The Blue Rider, 1911–1914), a group of young international artists fundamental to Expressionism who lived and worked mostly in Munich. Born in Meschede, Westphalia, Macke lived most of his short life in Berlin and produced an amazing total of 11,000 works of art. The paintings concentrate primarily on feelings and moods rather than reproducing objective reality. Macke's career was cut short by his early death in the second month of World War I, for which he, like many other young Germans of his time, had volunteered.

Was sehen Sie auf dem Gemälde[1]?

1. Welche Farben sind dominant: rot, blau, grün, grau, schwarz, braun, weiß, rosa, orange?
2. Welche Personen sehen Sie: einen Mann, eine Frau, ein Kind, einen Jungen, ein Mädchen?
3. Was ist im Vordergrund[2], was im Hintergrund[3]: die Mutter, das Kind, eine Wand, Bäume[4], ein Weg[5]?
4. Was trägt die Mutter: ein Kleid, einen Mantel, einen Hut, eine Jacke?
5. Was trägt das Kind: eine Bluse, eine Jacke, eine Mütze?
6. Welche Gefühle[6] evoziert das Gemälde: Ruhe[7], Hoffnung, Angst, Liebe, Unschuld, Glück[8], Hierarchie?

[1]*painting* [2]*foreground* [3]*background* [4]*trees* [5]*path* [6]*feelings* [7]*calm* [8]*happiness*

Situationen

Der Seminarraum

Grammatik B.1

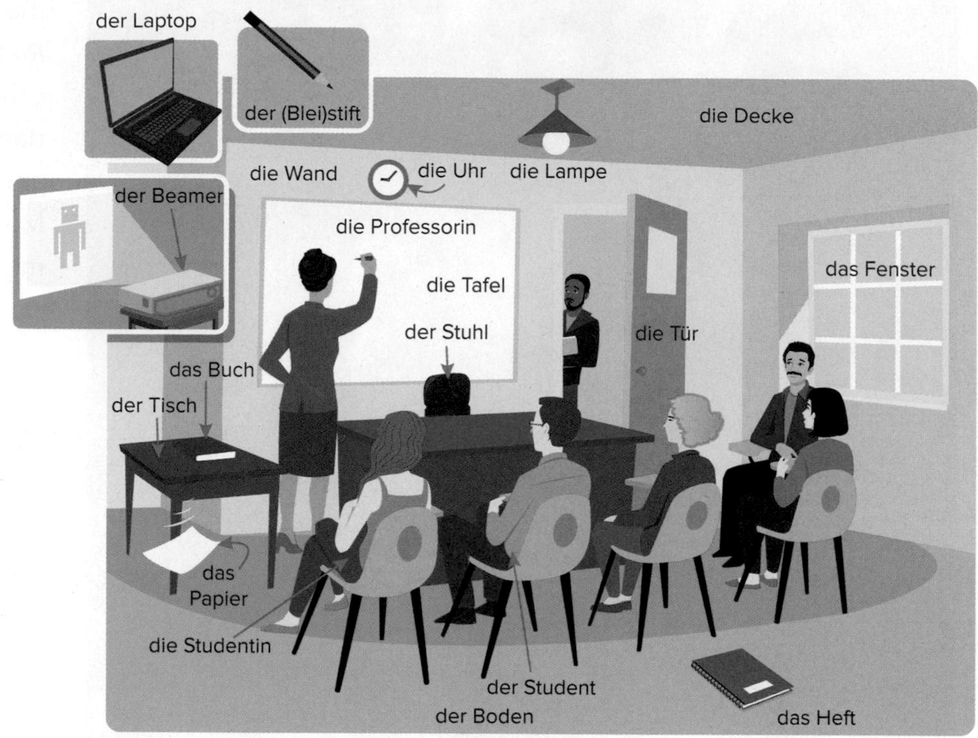

Situation 1 **Der Seminarraum**

Wie viele _______ sind im Seminarraum?

1. Studenten
2. Tische
3. Fenster
4. Lampen
5. Uhren
6. Türen
7. Bücher
8. Tafeln
9. Professoren/ Professorinnen
10. Hefte
11. Laptops

Situation 2 **Gegenstände° im Seminarraum**

°*objects*

MODELL: S1: Was ist weiß?
S2: Die Tafel (ist weiß).

1. weiß	a. der Boden
2. schmutzig	b. das Fenster
3. sauber	c. die Tafel
4. neu	d. die Uhr
5. alt	e. der Beamer
6. _______	f. _______

Beschreibungen

Grammatik B.2–B.3

Situation 3 Im Deutschkurs

1. Wer ist _______?
   a. blond
   b. groß
   c. klein
   d. schlank
   e. jung
   f. alt

2. Wer hat _______?
   a. braunes Haar
   b. graues Haar
   c. kurzes Haar
   d. langes Haar
   e. einen Bart
   f. blaue Augen
   g. braune Augen

Situation 4 Interaktion: Wie bist du?

MODELL: S1: Bist du glücklich?
S2: Ja, ich bin glücklich.
oder Nein, ich bin nicht glücklich.

	ICH	MEIN PARTNER	MEINE PARTNERIN
glücklich	☐	☐	☐
traurig	☐	☐	☐
konservativ	☐	☐	☐
schüchtern	☐	☐	☐
religiös	☐	☐	☐
ruhig	☐	☐	☐
freundlich	☐	☐	☐
verrückt	☐	☐	☐
sportlich	☐	☐	☐

Mir geht's gut.
© imageBROKER/Alamy RF

Ach, wie traurig!
© Ryan McVay/Getty Images RF

KULTUR ... LANDESKUNDE ... INFORMATIONEN

WAS IST WICHTIG IM LEBEN?

Was ist für Sie wichtig? Was ist am wichtigsten, was ist weniger wichtig? Bringen Sie die Aussagen in die Reihenfolge ihrer Wichtigkeit für Sie!

_______ Ich möchte gute Freunde haben[1].
_______ Ich möchte einen hohen Lebensstandard[2] haben.
_______ Ich möchte sozial Benachteiligten[3] helfen.
_______ Ich glaube an Gott.
_______ Ich möchte das Leben in vollen Zügen genießen[4].
_______ Ich möchte meine Bedürfnisse[5] durchsetzen[6].
_______ Ich möchte ein gutes Familienleben haben.

Schauen Sie sich die Grafik an. Welche Werte[7] haben junge Deutsche?

1. Was steht auf Platz 1?
2. Was ist wichtiger[8] für junge Deutsche: ein gutes Familienleben oder ein hoher Lebensstandard?
3. Was ist wichtiger für sie: eigene Bedürfnisse durchzusetzen oder sozial Benachteiligten zu helfen?
4. Wie viel Prozent der jungen Deutschen glauben an Gott?
5. Wie viel Prozent der jungen Deutschen wollen das Leben in vollen Zügen genießen?

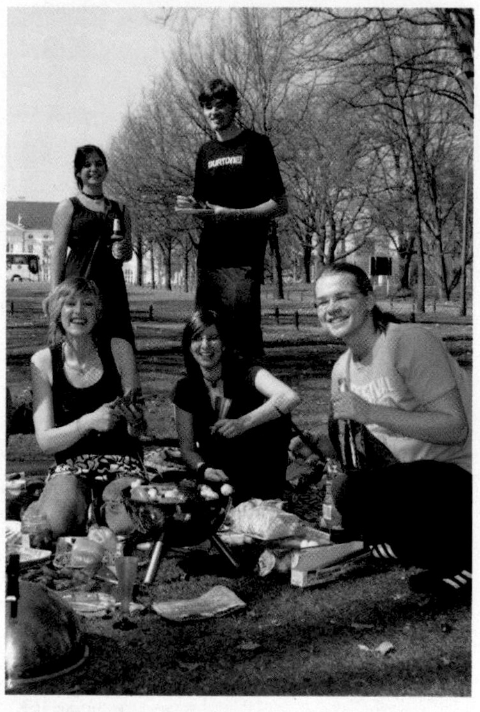

Freunde beim Grillen im Park
© ullstein bild - Volkreich/The Image Works

Wertorientierungen: Pragmatisch, aber nicht angepasst
Jugendliche im Alter von 12 bis 25 Jahren (Angaben in %)

Gute Freunde haben	97
Gutes Familienleben führen	92
Eigenverantwortlich leben und handeln	90
Fleißig und ehrgeizig sein	83
Phantasie und Kreativität entwickeln	79
Das Leben in vollen Zügen genießen	78
Hohen Lebensstandard haben	69
Sozial Benachteiligten helfen	58
Eigene Bedürfnisse durchsetzen	55
An Gott glauben	37
Das tun, was die anderen auch tun	14

Source of Data: Shell Jugendstudie - Shell Deutschland Oil, GmbH

[1]möchte haben *would like to have* [2]*standard of living* [3]*disadvantaged people* [4]das Leben ... *live life to its fullest* [5]*needs* [6]*make known* [7]*values* [8]*more important*

Der Körper

Grammatik B.4

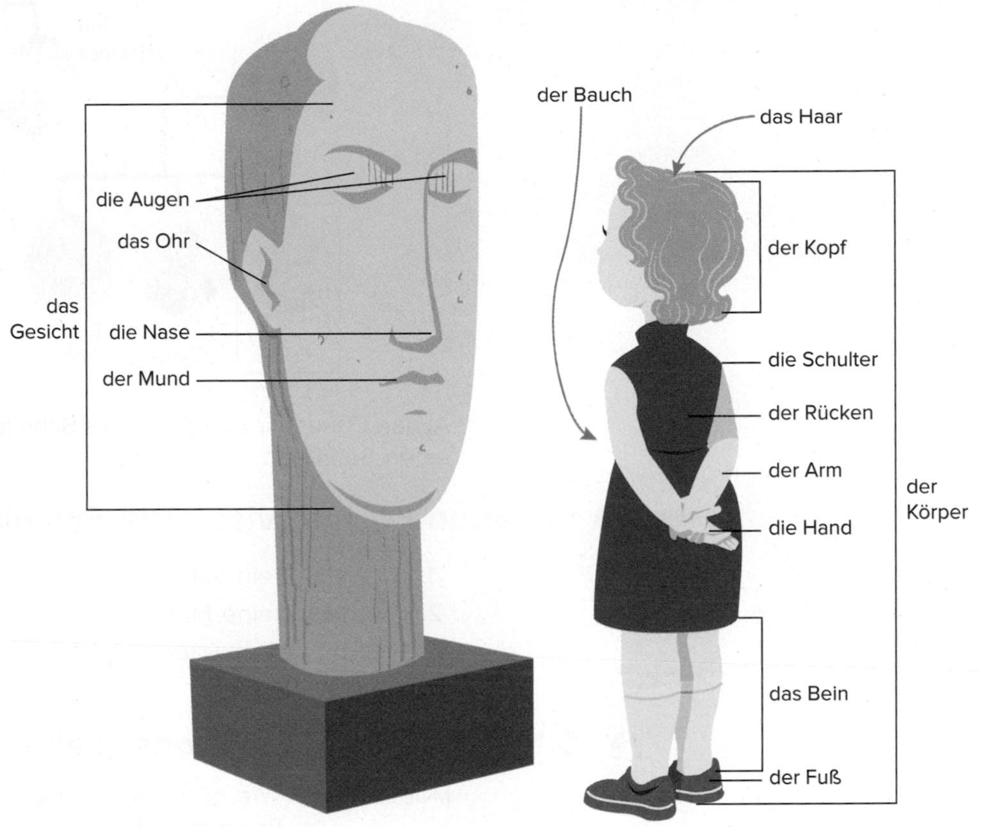

Situation 5 Welches Monster ist das?

MODELL: S1: Mein Monster hat fünf Beine und vier Arme.
S2: Das ist Momo.

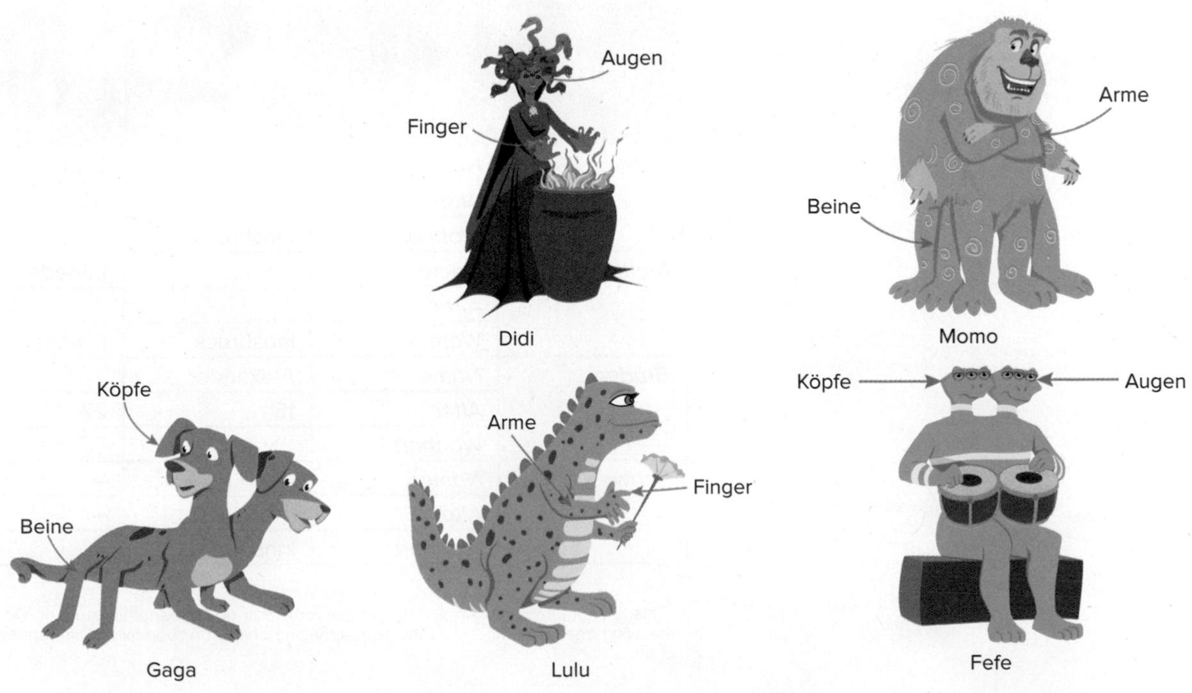

Die Familie

Grammatik B.5

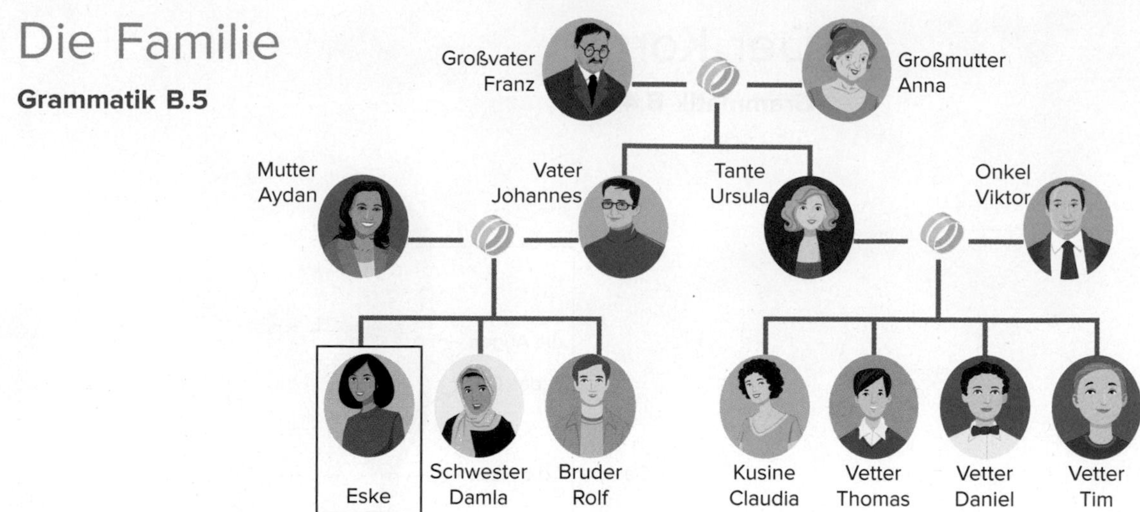

Aydan Candemir und Johannes Schmitz sind verheiratet. Sie haben drei Kinder: einen Sohn und zwei Töchter.

Situation 6 Interview: Die Familie

1. Wie heißt dein Vater/Stiefvater? Wie alt ist er? Wo wohnt er?
2. Wie heißt deine Mutter/Stiefmutter? Wie alt ist sie? Wo wohnt sie?
3. Hast du Geschwister? Wie viele? Wie heißen sie? Wie alt sind sie? Wo wohnen sie?

Situation 7* Informationsspiel: Familie

MODELL: S2: Wie heißt Richards Vater?
S1: Er heißt Werner.
S2: Wie schreibt man das?
S1: W-E-R-N-E-R. Wie alt ist er?
S2: Er ist _______ Jahre alt. Wo wohnt er?
S1: Er wohnt in Innsbruck. Wie heißt Richards Mutter?
S2: Sie heißt _______.
S1: Wie schreibt man das?
S2: _______.

		Richard	Sofie	Mehmet
Vater	*Name*	Werner	Erwin	
	Alter		50	59
	Wohnort	Innsbruck		Izmir
Mutter	*Name*		Elfriede	Sule
	Alter			
	Wohnort	Innsbruck	Dresden	
Bruder	*Name*	Alexander		Yakup
	Alter	15	27	34
	Wohnort			
Schwester	*Name*		—	
	Alter		—	
	Wohnort	Innsbruck	—	Izmir

*This is an information-gap activity in table form. Pair up with another student. One of you will work with the following chart, the other with the corresponding chart in Appendix A. Different information is missing from each chart.

Wetter und Jahreszeiten

WIE IST DAS WETTER?

1. Es ist sonnig und warm.

2. Es ist sehr heiß.

3. Es ist kalt.

4. Es regnet.

5. Es ist kühl.

6. Es schneit.

7. Es ist windig.

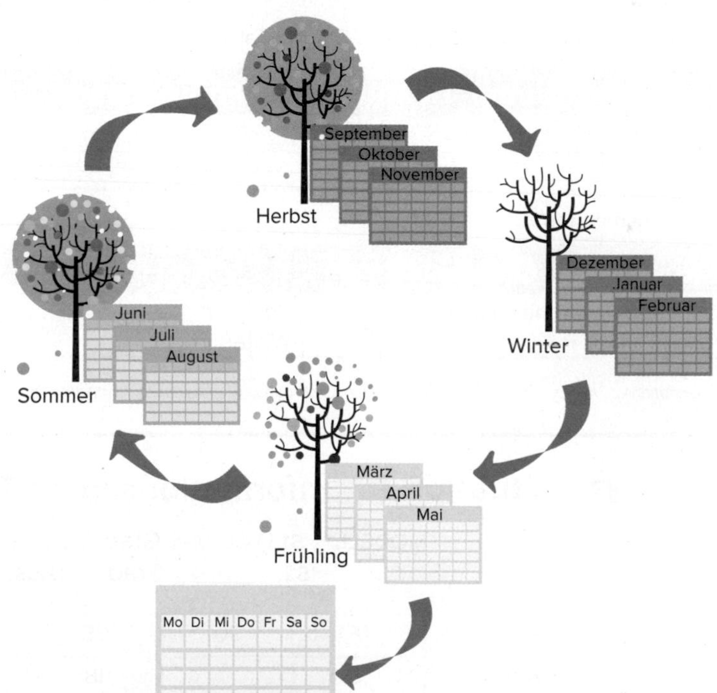

Situation 8 Dialog: Das Wetter in Regensburg

Josef trifft Claire an der Uni.

JOSEF: Schön heute, nicht?

CLAIRE: Ja, sehr __________ und __________ – wirklich schön!

JOSEF: Leider __________ es so oft hier in Bayern – auch im __________.

CLAIRE: Ist es auch oft __________ und __________ hier?

JOSEF: Ja, im __________. Und manchmal __________ es noch im April.

KULTUR ... LANDESKUNDE ... INFORMATIONEN

WETTER UND KLIMA

Wie ist das Wetter in Ihrer Stadt? Kreuzen Sie an.

	IM WINTER	IM SOMMER
sonnig	□	□
warm	□	□
(sehr) heiß	□	□
(sehr) feucht	□	□
mild	□	□
(sehr) kalt	□	□
viele Niederschläge[1] (Schnee/Regen)	□	□
windig	□	□
große Temperaturunterschiede[2]	□	□
geringe[3] Temperaturunterschiede	□	□

Winterwetter in München

Deutschland hat ein gemäßigtes[4] Klima mit Niederschlägen in allen Jahreszeiten. Im Nordwesten ist das Klima mehr ozeanisch mit warmen, aber selten heißen Sommern und relativ milden Wintern. Im Osten ist es eher[5] kontinental. Im Winter liegen die Temperaturen im Durchschnitt[6] zwischen 1,5 Grad Celsius (°C) im Tiefland[7] und minus 6°C im Gebirge[8], im Juli liegen sie zwischen 18 und 20°C.

Ausnahmen[9]: Am Rhein ist das Klima sehr mild, hier wächst[10] sogar Wein. Oberbayern hat einen warmen alpinen Südwind, den Föhn. Im Harz sind die Sommer oft kühl und im Winter gibt es viel Schnee.

Wie sind die Temperaturen in Deutschland? Benutzen Sie die Tabelle.

	Sommer	Winter Tiefland	Winter Gebirge
in °C			
in °F			

Welche Gebiete[11] bilden Ausnahmen?

wo	am Rhein	Oberbayern	im Harz
Klima	sehr _______	warmer alpiner _______	Sommer: _______ Winter: _______

[1]*precipitation* [2]*temperature variations* [3]*minor* [4]*moderate* [5]*more* [6]*im ... on average* [7]*lowlands* [8]*mountains* [9]*exceptions* [10]*grows* [11]*areas*

Temperaturen in Fahrenheit und Celsius

Fahrenheit → Celsius

32 subtrahieren und mit 5/9 multiplizieren

°F		°C
0		-17,8
32		0
50	~	10
70		21,1
90		32,2
98,6		37
212		100

Celsius → Fahrenheit

Mit 9/5 multiplizieren und 32 addieren

°C		°F
-10		14
0		32
10	~	50
20		68
30		86
37		98,6
100		212

Situation 9 Informationsspiel: Temperaturen

MODELL: S1: Wie viel Grad Celsius sind 90 Grad Fahrenheit?
S2: _______ Grad Celsius.

°F	90	65	32	0	−5	−39
°C		18		−18		−39

Sommer im Voralpenland

MUSIKSZENE

„36 Grad“ (2007, Deutschland) *2raumwohnung*

2raumwohnung im Konzert

Biografie *2raumwohnung* (Zweiraumwohnung) ist ein Duo aus Berlin. Das Duo besteht aus Inga Humpe und Tommi Eckart. „36 Grad“ war der Sommerhit des Jahres 2007 in Deutschland.

NOTE: For copyright reasons, the songs referenced in **MUSIKSZENE** have not been provided by the publisher. The song can be found online at various sites such as YouTube, Amazon, or the iTunes store.

Vor dem Hören Wie viel Grad Fahrenheit sind 36 Grad Celsius?

Nach dem Hören

A. Hören Sie den Refrain! Richtig oder falsch?

_____ **1.** Es ist heiß und wird noch heißer.

_____ **2.** Es gibt keinen Ventilator.

_____ **3.** Die Sängerin meint, das Leben ist leicht.

B. Die Sängerin zieht die Schuhe aus und den Bikini an. Was macht sie dann?

Miniwörterbuch	
der **Refrain**	chorus
keinen	no, not any
das **Leben**	life
leicht	easy
zieht ... aus	takes off
(zieht ...) an	(puts) on
regnen	to rain
tanzen	to dance
singen	to sing

Herkunft und Nationalität

Grammatik B.6–B.7

Situation 10 Dialog: Woher kommst du?

Claire trifft Melanie auf einer Party.

CLAIRE: Wie heißt du?
MELANIE: Melanie. __________?
CLAIRE: Claire.
MELANIE: Bist du __________?
CLAIRE: Ja.
MELANIE: Und __________ kommst du?
CLAIRE: __________ New York. Und du?
MELANIE: Aus Regensburg. Ich __________ von hier.

Situation 11 Herkunft

MODELL: S1: Woher kommt Silvia Mertens?
S2: Sie kommt aus _______.
S1: Wer kommt aus Dresden?
S2: _______.
S1: Kommt Kobe Okonkwo aus Innsbruck?
S2: Nein, er kommt aus _______.

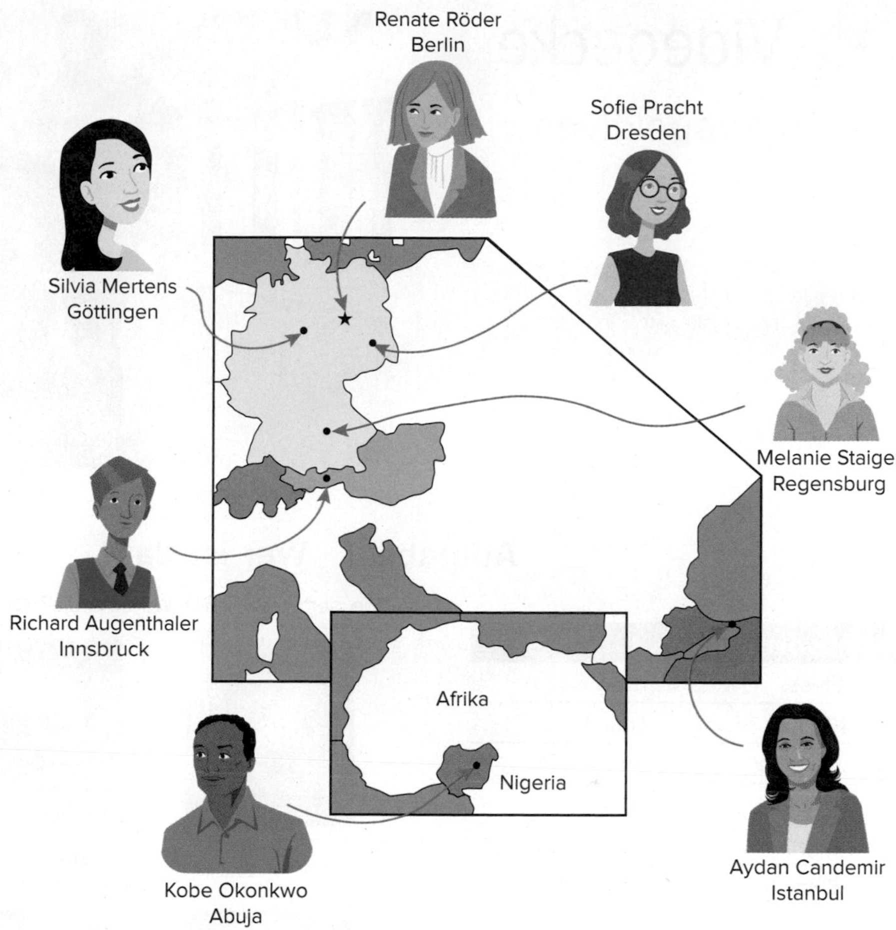

Situation 12 Rollenspiel: Herkunft

S1: Sie sind ein neuer Student / eine neue Studentin an einer Universität in Deutschland. Sie lernen einen anderen Studenten / eine andere Studentin kennen. Fragen Sie, wie er/sie heißt und woher er/sie kommt. Fragen Sie auch, ob er/sie Freunde/Freundinnen in anderen Ländern hat und welche Sprachen sie sprechen.

KULTUR ... LANDESKUNDE ... INFORMATIONEN

DIE LAGE DEUTSCHLANDS IN EUROPA

Deutschland liegt mitten in Europa. Es grenzt an[1] Dänemark, ________, Tschechien, Österreich, die ________, Frankreich, Luxemburg, ________ und die Niederlande. Die Grenzen[2] Deutschlands sind ________ Kilometer lang. Die längste Grenze ist die mit Österreich. Sie ist ________ Kilometer lang. Die Grenze zu Dänemark ist nur ________ Kilometer lang, die Grenze zu Polen ________, zu Tschechien 811, zur Schweiz ________, zu Frankreich 448, zu Luxemburg ________, zu Belgien 156 und zu den Niederlanden ________ Kilometer. Im Norden grenzt Deutschland an zwei Meere, die Nordsee und die ________.

- Deutschland gehört[3] zur Europäischen Union. Welche Länder gehören noch zur Europäischen Union? Schauen Sie auf die Karte vor[4] Situation 10.

[1]grenzt ... *has borders with* [2]*borders* [3]*belongs* [4]*before*

Videoecke

Perspektiven

Woher kommst du und woher kommen deine Eltern?

Ich komme aus Leipzig.

Aufgabe 1 Wer ist das?

Wie sehen sie aus? Wo sind sie? Ordnen Sie die Beschreibungen den Personen zu.

Miniwörterbuch	
der **Fluss**	river
die **Kirche**	church
das **Oberteil**	top
der **Tisch**	table
das **Kopftuch**	headscarf
das **Einzelkind**	only child

1. Tina ___

2. Albrecht ___

3. Simone ___

4. Sandra ___

5. Hend ___

6. Felicitas ___

7. Pascal ___

8. Sophie ___

a. Er steht an einem Fluss.
b. Er steht vor einer Kirche.
c. Sie hat langes blondes Haar und trägt ein blaues Oberteil.
d. Sie sitzt an einem Tisch.
e. Sie trägt ein graues Oberteil und eine graue Jacke.
f. Sie trägt ein Kopftuch.
g. Sie trägt ein pinkes Oberteil und eine Jacke.
h. Sie trägt einen grünen Pulli.

Aufgabe 2 Herkunft

Woher kommen sie? Woher kommen ihre Eltern? Ergänzen Sie die Tabelle mit Wörtern aus dem Kasten.

Berlin **Braunschweig** **Leipzig** **Dresden** **Prenzlau** **Grimma** **Schweiz** **Kairo**

Name	Woher?	Woher kommen die Eltern?
Tina		Leipzig
Albrecht	Dresden	
Simone		Salzgitter
Sandra		Prenzlau
Hend	Kairo	
Felicitas	Grimma	
Pascal		aus der Schweiz und aus Holland
Sophie		aus Würzburg und aus Braunschweig

Interviews

- Woher kommst du?
- Wo liegt das?
- Woher kommt deine Familie?
- Erzähl mir ein bisschen von deiner Familie!
- Welche Sprachen sprichst du?
- Wie wird morgen das Wetter?

Pascal

Nadezda

Aufgabe 3 Familie

Wer sagt das, Pascal oder Nadezda?

	PASCAL	NADEZDA
1. Ich komme aus Moskau.	☐	☐
2. Ich komme aus Zürich.	☐	☐
3. Meine Familie kommt aus Holland und aus der Schweiz.	☐	☐
4. Meine Familie kommt aus Russland und aus Europa.	☐	☐
5. Ich bin Einzelkind.	☐	☐
6. Meine Geschwister arbeiten.	☐	☐
7. Ich spreche Holländisch, Französisch, Englisch und Italienisch.	☐	☐
8. Ich spreche Deutsch, Englisch und Russisch.	☐	☐

Aufgabe 4 Interview

Interviewen Sie eine Partnerin oder einen Partner. Stellen Sie dieselben Fragen.

Wortschatz

Der Seminarraum — The Classroom

die **Decke, -n**[*]	ceiling
die **Tafel, -n** (R)[†]	blackboard/whiteboard
die **Uhr, -en**	clock
die **Wand, ¨e**	wall
der **Beamer, -** [bi:mɐ]	data projector
der **Boden, ¨**	floor
der **Laptop, -s** [lɛptɔp]	laptop (computer)
der **Stift, -e** (R)	pen
der **Bleistift, -e** (R)	pencil
der **Tisch, -e**	table
das **Fenster, -**	window
das **Heft, -e**	notebook

Ähnliche Wörter

die **Lampe, -n;** die **Professorin, -nen** (R); die **Studentin, -nen** (R); die **Uni, -s;** die **Universität, -en;** der **Professor, Professoren** (R); der **Student, -en** (R); der **Stuhl, ¨e;** das **Buch, ¨er** (R); das **Papier**

Beschreibungen — Descriptions

er/sie hat ...	he/she has ...
einen Bart	a beard
blaue Augen	blue eyes
blondes Haar	blond hair
kurzes Haar	short hair
er/sie ist ...	he/she is ...
glücklich	happy
groß	tall; big
klein	short; small
ruhig	quiet, calm
sauber	clean
schlank	slender, slim
schön	pretty, beautiful
schüchtern	shy
traurig	sad
verrückt	crazy

Ähnliche Wörter

blond, freundlich, jung, konservativ, lang, religiös, sportlich

Der Körper — The Body

der **Bauch, ¨e**	belly, stomach
der **Kopf, ¨e**	head
der **Mund, ¨er**	mouth
der **Rücken, -**	back
das **Auge, -n**	eye
das **Bein, -e**	leg
das **Gesicht, -er**	face
das **Ohr, -en**	ear

Ähnliche Wörter

die **Hand, ¨e;** die **Nase,** -n; die **Schulter, -n;** der **Arm, -e;** der **Fuß, ¨e;** das **Haar, -e**

Die Familie — The Family

die **Frau, -en** (R)	woman; wife
die **Nichte, -n**	niece
die **Schwester, -n**	sister
die **Tante, -n**	aunt
der **Mann, ¨er** (R)	man; husband
der **Vetter, -n**	male cousin
das **Kind, -er**	child
die **Eltern**	parents
die **Großeltern**	grandparents
die **Geschwister**	siblings

Ähnliche Wörter

die **Kusine, -n;** die **Mutter, ¨;** die **Großmutter, ¨;** die **Tochter, ¨;** der **Bruder, ¨;** der **Neffe, -n;** der **Onkel, -;** der **Sohn, ¨e;** der **Vater, ¨;** der **Großvater, ¨**

Wetter und Jahreszeiten — Weather and Seasons

der **Frühling**	spring
im Frühling	in the spring
der **Herbst**	fall, autumn
der **Monat, -e**	month
das **Jahr, -e**	year
es ...	it ...
ist 18 Grad Celsius	is 18 degrees Celsius
ist feucht	is humid
ist schön	is nice
regnet	is raining; rains
schneit	is snowing; snows

Ähnliche Wörter

der **Januar, im Januar,** der **Februar,** der **März,** der **April,** der **Mai,** der **Juni,** der **Juli,** der **August,** der **September,** der **Oktober,** der **November,** der **Dezember;** der **Sommer,** der **Winter; Fahrenheit, heiß, kalt, kühl, sonnig, warm, windig**

[*]Beginning with this chapter, the plural endings of nouns are indicated in the vocabulary lists. See grammar section B.4 for more explanation.

[†](R) indicates words that were listed in a previous chapter and are presented again for review.

Geografie	Geography
Deutschland	Germany
Frankreich	France
Griechenland	Greece
Österreich	Austria
Russland	Russia
Tschechien	Czech Republic
Ungarn	Hungary
die **Hauptstadt, ¨-e**	capital city
die **Ostsee**	Baltic Sea
die **Schweiz**	Switzerland
das **Mittelmeer**	Mediterranean Sea

Ähnliche Wörter

Afrika, Amerika, Asien, Australien, Belgien, Bulgarien, China, Dänemark, England, Europa, Finnland, Großbritannien, Holland, Irland, Italien, Kanada, Liechtenstein, Neuseeland, Nordamerika, Norwegen, Polen, Portugal, Rumänien, Schweden, Slowenien, Spanien, Südamerika; die **Nordsee,** die **Slowakei,** die **Türkei;** die **Niederlande** (*pl.*), die **USA** (*pl.*)

Herkunft	Origin
der/die **Deutsche, -n**	German (person)
Ich bin Deutsche/r.	I am German.
der **Franzose, -n** / die **Französin, -nen**	French (person)
der **Österreicher, -** / die **Österreicherin, -nen**	Austrian (person)
der **Schweizer, -** / die **Schweizerin, -nen**	Swiss (person)

Ähnliche Wörter

die **Amerikanerin, -nen;** die **Australierin, -nen;** die **Engländerin, -nen;** die **Kanadierin, -nen;** die **Mexikanerin, -nen;** der **Amerikaner, -;** der **Australier, -;** der **Engländer, -;** der **Kanadier, -;** der **Mexikaner, -**

Sprachen	Languages
Deutsch	German
Französisch	French

Ähnliche Wörter

Arabisch, Chinesisch, Englisch, Italienisch, Portugiesisch, Russisch, Schwedisch, Spanisch, Türkisch

Sonstige Wörter und Ausdrücke	Other Words and Expressions
das ist ...	this/that is . . .
das sind ...	these/those are . . .
dein(e)	your (*informal*)
genau	exactly
heute	today
Ihr(e)	your (*formal*)
kennen	to know
kommen (aus)	to come (*from*)
leider	unfortunately
manchmal	sometimes
noch	even, still
sehr	very
sprechen	to speak
wann	when
was	what
welch-	which
wer	who
wie	how
wirklich	really
wo	where
woher	from where
wohnen (in)	to live (in)

Strukturen und Übungen

B.1 Definite and indefinite articles

Recall that the definite article **der, das, die** (*the*) varies by gender, number, and case.* Similarly, the indefinite article **ein, eine** (*a, an*) has various forms.

WISSEN SIE NOCH?

masculine = **der**
neuter = **das**
feminine = **die**
plural (all genders) = **die**

Review grammar section A.4.

Das ist **ein** Buch.	*This is a book.*
Welche Farbe hat **das** Buch?	*What color is the book?*
Das ist **eine** Tür.	*This is a door.*
Welche Farbe hat **die** Tür?	*What color is the door?*

Here are the definite and indefinite articles for all three genders in the singular and plural, nominative case. There is only one plural definite article for all three genders: **die.** The indefinite article (*a, an*) has no plural.

der → **ein**
das → **ein**
die → **eine**
die (*pl.*) → ø

	Singular	Plural
Masculine	**der** Stift **ein** Stift	**die** Stifte Stifte
Neuter	**das** Buch **ein** Buch	**die** Bücher Bücher
Feminine	**die** Tür **eine** Tür	**die** Türen Türen

Übung 1 Im Seminarraum

Frau Schulz spricht über die Gegenstände und die Farben im Seminarraum. Ergänzen Sie den unbestimmten[1] Artikel, den bestimmten[2] Artikel und die Farbe.

MODELL: FRAU SCHULZ: Das ist eine Lampe.
Welche Farbe hat die Lampe?
STUDENT(IN): Sie ist gelb.

1. Und das ist _______[a] Stift.
   Welche Farbe hat _______[b] Stift? Er ist _______[c].

2. Und das ist _______[a] Stuhl.
   Welche Farbe hat _______[b] Stuhl? Er ist _______[c].

3. Und das ist _______[a] Tafel.
   Welche Farbe hat _______[b] Tafel? Sie ist _______[c].

*See Sections A.3 and A.4.

[1]*indefinite* [2]*definite*

4. Und das ist _______[a] Uhr.
   Welche Farbe hat _______[b] Uhr? Sie ist _______[c].

5. Und das ist _______[a] Buch.
   Welche Farbe hat _______[b] Buch? Es ist _______[c].

6. Und das ist _______[a] Brille. Welche Farbe hat _______[b] Brille? Sie ist _______[c].

Übung 2 Was ist das?

Herr Wagner spricht mit seiner kleinen Tochter.

MODELL: Ist das eine Decke? →
Nein, das ist ein Bleistift.

1. Ist das eine Tür?

2. Ist das eine Uhr?

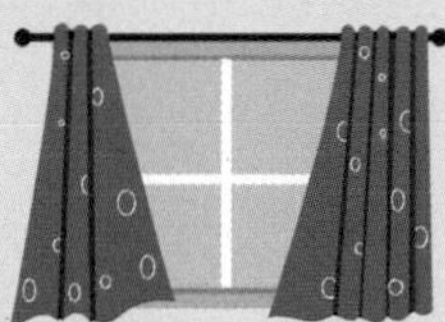

3. Ist das eine Lampe?

4. Ist das ein Tisch?

5. Ist das ein Stuhl?

6. Ist das eine Studentin?

7. Ist das ein Heft?

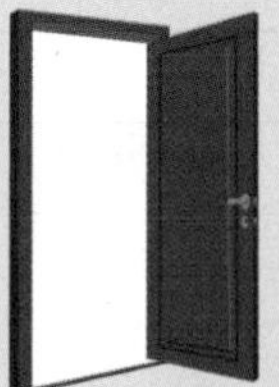

8. Ist das eine Tafel?

B.2 Who are you? The verb *sein*

sein = *to be*

ACHTUNG!

NOT = **NICHT**

—Ist Jens groß?
—Nein, er ist **nicht** groß, er ist klein.

Use a form of the verb **sein** (*to be*) to identify or describe people and things.

—**Sind Jutta und er** blond?	*Are Jutta and he blond?*
—Ja, **sie sind** blond.	*Yes, they are blond.*
Peter ist groß.	*Peter is tall.*
Das Fenster ist nicht klein.	*The window is not small.*

sein					
Singular			***Plural***		
ich	bin	*I am*	wir	sind	*we are*
du	bist	*you are*	ihr	seid	*you are*
Sie	sind	*you are*	Sie	sind	*you are*
er sie es	ist	*he* *she is* *it*	sie	sind	*they are*

Übung 3 Minidialoge

Ergänzen Sie das Verb **sein: bin, bist, ist, sind, seid.**

1. MICHAEL: Ich bin Michael. Wer _______[a] du?
   JENS: Ich _______[b] Jens. Jutta und ich, wir _______[c] gute Freunde.
2. FRAU SCHULZ: Das ist Herr Thelen. Er _______[a] alt.
   STEFAN: Herr Thelen ist alt?
   FRAU SCHULZ: Ja, Stefan. Herr Thelen ist alt, aber Maria und Michael _______[b] jung.
3. HERR THELEN: Jutta und Hans, wie alt _______[a] ihr?
   JUTTA: Ich _______[b] 16 und Hans _______[c] 13.
4. MICHAEL: Wer bist du?
   HANS: Ich _______[a] Hans.
   MICHAEL: Wie alt bist du?
   HANS: Ich _______[b] 13.

B.3 What do you have? The verb *haben*

haben = *to have*

The verb **haben** (*to have*) is often used to show possession or to describe physical characteristics.

Ich habe eine Brille. — *I have glasses.*
Hast du das Buch? — *Do you have the book?*
Nora hat braune Augen. — *Nora has brown eyes.*

haben					
Singular			***Plural***		
ich	habe	*I have*	wir	haben	*we have*
du	hast	*you have*	ihr	habt	*you have*
Sie	haben	*you have*	Sie	haben	*you have*
er sie es	hat	*he* *she has* *it*	sie	haben	*they have*

Übung 4 Minidialoge

Ergänzen Sie das Verb **haben: habe**, **hast**, **hat**, **habt**, **haben**.

1. FRAU SCHULZ: Nora, _______[a] Sie viele Freunde und Freundinnen?
   NORA: Ja, ich _______[b] viele Freunde und Freundinnen.
2. MONIKA: Stefan, _______ du einen Stift?
   STEFAN: Nein.
3. PETER: Hallo, Heidi und Katrin! _______[a] ihr das Deutschbuch?
   HEIDI: Katrin _______[b] es, aber ich nicht.
   PETER: Dann _______[c] wir zwei. Ich _______[d] es auch.

B.4 Plural forms of nouns

Just as in English, there are different ways to form plurals in German.

Albert hat ein Heft.	*Albert has one notebook.*
Peter hat zwei Hefte.	*Peter has two notebooks.*
Heidi hat eine Kusine.	*Heidi has one cousin.*
Katrin hat zwei Kusinen.	*Katrin has two cousins.*

These guidelines will help you to recognize and form the plural of German nouns.

1. Most feminine nouns add **-n** or **-en.** They add **-n** when the singular ends in **-e;** otherwise, they add **-en.** Nouns that end in **-in** add **-nen.**

   eine Lampe, zwei Lampe**n**
   eine Frau, zwei Frau**en**
   eine Tür, zwei Tür**en**
   eine Studentin, zwei Studentin**nen**

2. Masculine and neuter nouns usually add **-e** or **-er.** Those plurals that end in **-er** have an umlaut when the stem vowel is **a, o, u,** or **au.** Many masculine plural nouns ending in **-e** have an umlaut as well. Neuter plural nouns ending in **-e** do not have an umlaut.

MASCULINE **(der)**	NEUTER **(das)**
ein Rock, zwei R**ö**ck**e**	ein Heft, zwei Heft**e**
ein Mann, zwei M**ä**nn**er**	ein Buch, zwei B**ü**ch**er**

3. Masculine and neuter nouns that end in **-er** either add an umlaut or change nothing at all in the plural. Many nouns with a stem vowel of **a, o, u,** or **au** add an umlaut.

MASCULINE **(der)**	NEUTER **(das)**
ein Bruder, zwei Br**ü**der	ein Fenster, zwei Fenster
ein Computer, zwei Computer	

4. Nouns that end in a vowel other than unstressed **-e** and many nouns of English or French origin add **-s.**

ein Laptop, zwei Laptop**s**	ein Auto, zwei Auto**s**

The following chart summarizes the guidelines provided above.

Singular	Plural	Examples
ein ____er	no ending: some words add an umlaut where possible	ein Lehrer, zwei Lehrer; ein Vater, zwei V**ä**ter
ein ____	add **-e;** masculine words often add an umlaut, neuter words do not	ein Rock, zwei R**ö**ck**e;** ein Regal, zwei Regal**e**
ein ____	add **-er;** add an umlaut where possible	ein Mann, zwei M**ä**nn**er;** ein Buch, zwei B**ü**ch**er**
eine ____	add **-n, -en,** or **-nen,** depending on final letter of the word	eine Lampe, zwei Lampe**n;** eine Tür, zwei Tür**en;** eine Freundin, zwei Freundin**nen**
ein(e) ____ *(foreign words)*	add **-s**	ein Hobby, zwei Hobby**s;** eine Kamera, zwei Kamera**s**

Beginning with this chapter, the plural endings of nouns are indicated in the vocabulary lists as follows.

LISTING	PLURAL FORM
das **Fenster, -**	die **Fenster**
der **Bruder, ¨**	die **Brüder**
der **Tisch, -e**	die **Tische**
der **Stuhl, ¨e**	die **Stühle**
das **Kleid, -er**	die **Kleider**
der **Mann, ¨er**	die **Männer**
die **Tante, -n**	die **Tanten**
die **Uhr, -en**	die **Uhren**
die **Studentin, -nen**	die **Studentinnen**
das **Auto, -s**	die **Autos**

Übung 5 Der Körper

Wie viele der folgenden Körperteile hat der Mensch[1]?

MODELL: Der Mensch hat zwei Arme.

Arm
Auge
Bein
Finger
Fuß
Hand
Nase
Ohr
Schulter

Übung 6 Das Zimmer

Wie viele der folgenden Dinge sind in Ihrem[2] Zimmer? (ein[e], zwei, ..., viele, nicht viele)

das Buch
der Computer
das Fenster
die Lampe
der Stuhl
der Tisch
die Tür
die Uhr
die Wand

In meinem Zimmer ist/sind ______ Buch/Bücher, ...

[1]*person* [2]*your*

B.5 Personal pronouns

Personal pronouns refer to the speaker (first person), to the person addressed (second person), or to the person(s) or object(s) talked about (third person).

	Singular		Plural	
First person	ich	*I*	wir	*we*
Second person informal	du	*you*	ihr	*you*
Second person formal	Sie	*you*	Sie	*you*
Third person	er	*he, it*	sie	*they*
	es	*it*		
	sie	*she, it*		

WISSEN SIE NOCH?

der → **er** = *he, it*
das → **es** = *it*
die → **sie** = *she, it*
die (*pl.*) → **sie** = *they*

Review grammar section A.4.

Third-person singular pronouns reflect the grammatical gender of the nouns they replace.

—Welche Farbe hat **der Hut?** *What color is the hat?*
—**Er** ist braun. *It is brown.*

—Welche Farbe hat **das Kleid?** *What color is the dress?*
—**Es** ist grün. *It is green.*

—Welche Farbe hat **die Bluse?** *What color is the blouse?*
—**Sie** ist gelb. *It is yellow.*

The third-person plural pronoun is **sie** for all three genders.

—Welche Farbe haben **die Schuhe?** *What color are the shoes?*
—**Sie** sind schwarz. *They are black.*

Übung 7 Welche Farbe?

Frau Schulz spricht über die Farbe der Kleidung. Antworten Sie!

1. Welche Farbe hat der Hut?
2. Welche Farbe hat das Hemd?
3. Welche Farbe hat die Hose?
4. Welche Farbe hat die Bluse?
5. Welche Farbe haben die Socken?
6. Welche Farbe hat das Kleid?
7. Welche Farbe hat der Rock?
8. Welche Farbe haben die Stiefel?
9. Welche Farbe hat die Jacke?
10. Welche Farbe hat der Mantel?

B.6 Origins: *Woher kommen Sie?*

To ask about someone's origins, use the question word **woher** (*from where*) followed by the verb **kommen** (*to come*). In the answer, use the preposition **aus** (*from, out of*).

—Woher kommst du / kommen Sie? *Where do you come from?*
—Ich komme aus Berlin. *I'm from Berlin.*

kommen aus = *to come from* (a place)

kommen			
ich	komme	wir	kommen
du	kommst	ihr	kommt
Sie	kommen	Sie	kommen
er / sie / es	kommt	sie	kommen

The infinitive of German verbs, that is, the basic form of the verb, ends in **-n** or **-en.** Most verbs follow a conjugation pattern similar to that of **kommen.**

Kommen Sie heute Abend? *Are you coming this evening?*
Warten Sie! **Ich komme** mit! *Wait! I'll come along.*

Übung 8 Minidialoge

Ergänzen Sie **kommen, woher** und **aus** und die Personalpronomen.

1. MEHMET: Woher _______[a] du, Renate?
   RENATE: Ich _______[b] aus Berlin.
2. FRAU SCHULZ: Woher _______[a] Lydia?
   KATRIN: Lydia kommt _______[b] Zürich.
   FRAU SCHULZ: _______[c] kommen Josef und Melanie?
   STEFAN: Sie _______[d] aus Regensburg.
   FRAU SCHULZ: Und woher komme _______[e]?
   ALBERT: Sie, Frau Schulz, Sie kommen _______[f] Kalifornien.
3. FRAU SCHULZ: Kommt Sofie aus Regensburg?
   HEIDI: Nein, _______[a] kommt aus Dresden.
   FRAU SCHULZ: Kommen Josef und Melanie aus Innsbruck?
   STEFAN: Nein, sie _______[b] aus Regensburg.
4. ANDREAS: Silvia und Jürgen, kommt _______[a] aus Göttingen?
   SILVIA: Ja, _______[b] kommen aus Göttingen.

B.7 Possessive determiners: *mein* and *dein/Ihr*

der → **mein, dein, Ihr**
das → **mein, dein, Ihr**
die → **meine, deine, Ihre**
die (*pl.*) → **meine, deine, Ihre**

ACHTUNG!

Note that the forms of **Ihr** are capitalized, just as **Sie** is, when they mean *your.*

The possessive determiners **mein** (*my*), **dein** (*informal your*), and **Ihr** (*formal your*) have the same endings as the indefinite article **ein.** In the plural, the ending is **-e.** Here are the nominative forms of these possessive determiners.

	Onkel (*m.*)	Auto (*n.*)	Tante (*f.*)	Eltern (*pl.*)
ich	mein	mein	meine	meine
du	dein	dein	deine	deine
Sie	Ihr	Ihr	Ihre	Ihre

—Woher kommen **deine** Eltern, Albert? — *Where are your parents from, Albert?*
—**Meine** Eltern kommen aus Mexiko. — *My parents are from Mexico.*

Wie heißt **Ihr** Vater, Frau Schulz? — *What is your father's name, Ms. Schulz?*
Und **Ihre** Mutter? — *And your mother's name?*

Übung 9 Minidialoge

Ergänzen Sie die Possessivartikel.

1. FRAU SCHULZ: Wo sind _______ Hausaufgaben?
   PETER: Sie liegen leider zu Hause.
2. ONKEL: Ist das _______[a] Hund?
   NICHTE: Nein, das ist nicht _______[b] Hund. Ich habe keinen[1] Hund.
3. LYDIA: He, Yamina! Das ist _______[a] Kleid.
   YAMINA: Nein, das ist _______[b] Kleid. _______[c] Kleid ist schmutzig.
4. KATRIN: Woher kommen _______[a] Eltern, Frau Schulz?
   FRAU SCHULZ: _______[b] Mutter kommt aus Schwabing und _______[c] Vater kommt aus Germering.

ACHTUNG!

Just as in English, an ***s*** added onto someone's name in German indicates possession. In German, however, there is no apostrophe before the ***s.***

Das ist Helga. Das ist Helgas Vater.
This is Helga. That is Helga's father.

Übung 10 Woher kommen sie?

Beantworten Sie die Fragen.

1. Woher kommen Sie?
2. Woher kommt Ihre Mutter?
3. Woher kommt Ihr Vater?
4. Woher kommen Ihre Großeltern?
5. Woher kommt Ihr Professor / Ihre Professorin?
6. Wie heißt ein Student aus Ihrem Deutschkurs und woher kommt er?
7. Wie heißt eine Studentin aus Ihrem Deutschkurs und woher kommt sie?

[1]*no*

KAPITEL 1

Wer ich bin und was ich tue

In **Kapitel 1** you will learn to talk about how you spend your time: your studies, your recreational pursuits, and what you like and don't like to do.

Themen

Freizeit

Schule und Universität

Tagesablauf

Persönliche Angaben

Kulturelles

KLI: Freizeit

KLI: Schule

Filmclip: *Hilfe!* (Oliver Dommenget)

Musikszene: „Gewinner" (Clueso)

Videoecke: Tagesablauf

Lektüren

Film: *Hilfe!* (Oliver Dommenget)

Biografie: Guten Tag, ich heiße ...

Strukturen

1.1 The present tense

1.2 Expressing likes and dislikes: **gern / nicht gern**

1.3 Telling time

1.4 Word order in statements

1.5 Separable-prefix verbs

1.6 Word order in questions

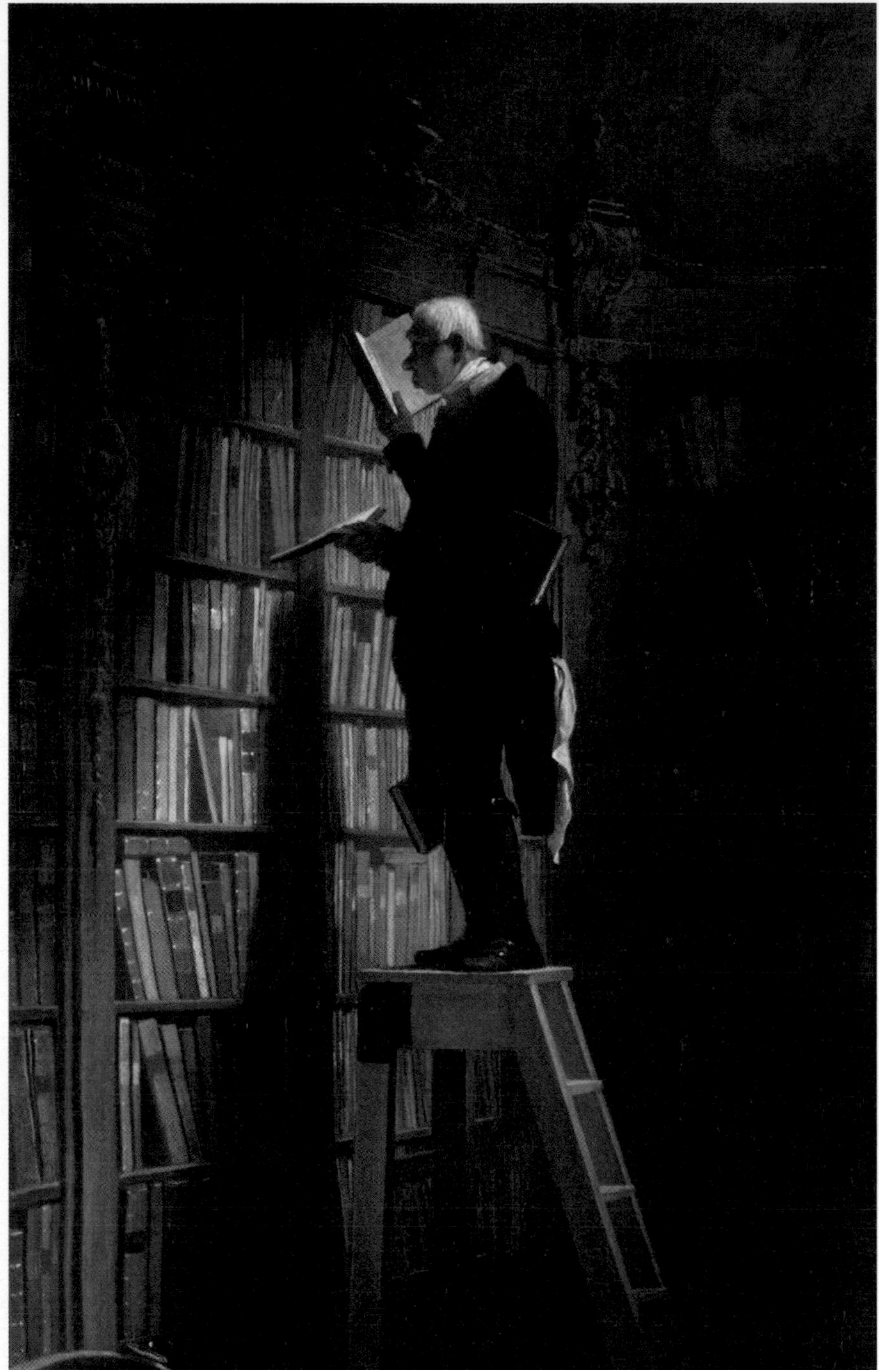

Carl Spitzweg: *Der Bücherwurm* (1850), Museum Georg Schäfer, Schweinfurt/ Deutschland

KUNST UND KÜNSTLER

Carl Spitzweg (1808–1885) ist ein deutscher Maler und Dichter[1] aus der Umgebung[2] von München. Seine Bilder sind oft ironisch. „Der Bücherwurm" ist ein gutes Beispiel[3] für Spitzwegs humorvolle Perspektive.

Schauen Sie sich das Gemälde[4] an und beantworten Sie folgende Fragen.

1. Wie ist der Mann, wie sieht er aus: jung, alt, dünn, dick, klein, groß, kurzsichtig[5], interessant, langweilig, spitze Nase, braunes Haar, in Jeans, im Anzug, trägt Stiefel, trägt Sandalen, hat ein großes Taschentuch?
2. Wo ist er: in einer alten Bibliothek, im Museum, im Kino, auf einer Leiter, auf einem Sofa?
3. Wo hat er Bücher: in den Händen, unter dem Arm, unter den Füßen, auf dem Kopf, zwischen den Knien?
4. Welche Farben dominieren: blau, braun, gelb, grau, grün, lila, orange, rosa, rot, schwarz, weiß?
5. Welche Gefühle[6] ruft das Gemälde hervor: Angst, Einfachheit[7], Elan[8], Glück, Hoffnung, Langeweile[9], Neugier[10], Ruhe, Sehnsucht[11], Selbstvergessenheit[12]?

[1]*more often* [2]*vicinity* [3]*example* [4]*painting* [5]*nearsighted* [6]*feelings* [7]*simplicity* [8]*pep* [9]*boredom* [10]*curiosity* [11]*yearning* [12]*obliviousness*

Situationen

Freizeit

Grammatik 1.1–1.2

Peter und Stefan wandern gern.

Ernst spielt gern Fußball.

Jutta und Gabi spielen gern Karten.

Melanie tanzt gern.

Michael spielt gern Gitarre.

Veronika reitet gern.

Thomas segelt gern.

Herr und Frau Ruf gehen gern spazieren.

Situation 1 Hobbys

Sagen Sie **ja** oder **nein.**

1. In den Ferien ...
   a. reise ich gern.
   b. koche ich gern.
   c. spiele ich gern Volleyball.
   d. arbeite ich gern.
2. Im Winter ...
   a. gehe ich gern ins Museum.
   b. spiele ich gern Schach.
   c. gehe ich gern Snowboard fahren.
   d. schwimme ich gern.
3. Meine Eltern ...
   a. chatten gern.
   b. spielen gern Golf.
   c. gehen gern ins Kino.
   d. trinken gern Kaffee.
4. Mein Bruder / Meine Schwester ...
   a. wandert gern in den Bergen.
   b. zeltet gern.
   c. boxt gern.
   d. singt gern.
5. Mein Deutschlehrer / Meine Deutschlehrerin ...
   a. simst gern.
   b. schreibt gern Mails.
   c. geht gern ins Konzert.
   d. spielt gern Fußball.

Situation 2 Informationsspiel: Freizeit

MODELL: S1: Wie alt ist Rolf?
S2: ______.
S1: Woher kommt Richard?
S2: Aus ______.
S1: Was macht Richard gern?
S2: Er ______.
S1: Wie alt bist du?
S2: ______.
S1: Woher kommst du?
S2: ______.
S1: Was machst du gern?
S2: ______.

	Alter	Wohnort	Hobby
Richard	18		
Rolf		Berkeley	
Jürgen	21		geht gern tanzen
Sofie	22	Dresden	
Jutta			hört gern Musik
Melanie	25		chattet gern
mein Partner / meine Partnerin			

Situation 3 Vor dem Tanzstudio

Bringen Sie die Sätze in die richtige Reihenfolge.

_____ Dann können wir ja miteinander lernen.
_____ Das macht nichts. Dann sehen wir uns wenigstens mal wieder etwas öfter. Wollen wir reingehen?
_____ Hey Nesrin!
_____ Hey Willi, lange nicht gesehen. Was machst du denn hier?
_____ Ich kann leider nur Dienstag. Freitag arbeite ich am Abend in der Studentenkneipe.
_____ Ich lerne tanzen. Und was machst du?
_____ Ja, das finde ich gut. Kommst du jeden Dienstag?
_____ Ja, ich komme jeden Dienstag und jeden Freitag. Und du?
_____ Ja, Nesrin, gehen wir rein.
_____ So ein Zufall. Ich lerne auch tanzen.

KULTUR ... LANDESKUNDE ... INFORMATIONEN

FREIZEIT

- Was machen Menschen in Ihrem Land in ihrer Freizeit?
- Was machen Sie in Ihrer Freizeit? am Wochenende? abends? in den Ferien?
- Was machen Ihre Eltern in ihrer Freizeit? am Wochenende? abends? in den Ferien?
- Wie viele Stunden Freizeit haben Sie am Tag?
- Sehen Sie sich die Grafik an. Was machen Deutsche öfter[1] als Sie? Was machen sie weniger[2] oft als Sie?
- Wie viele Stunden Freizeit haben Deutsche am Tag? Raten[3] Sie!

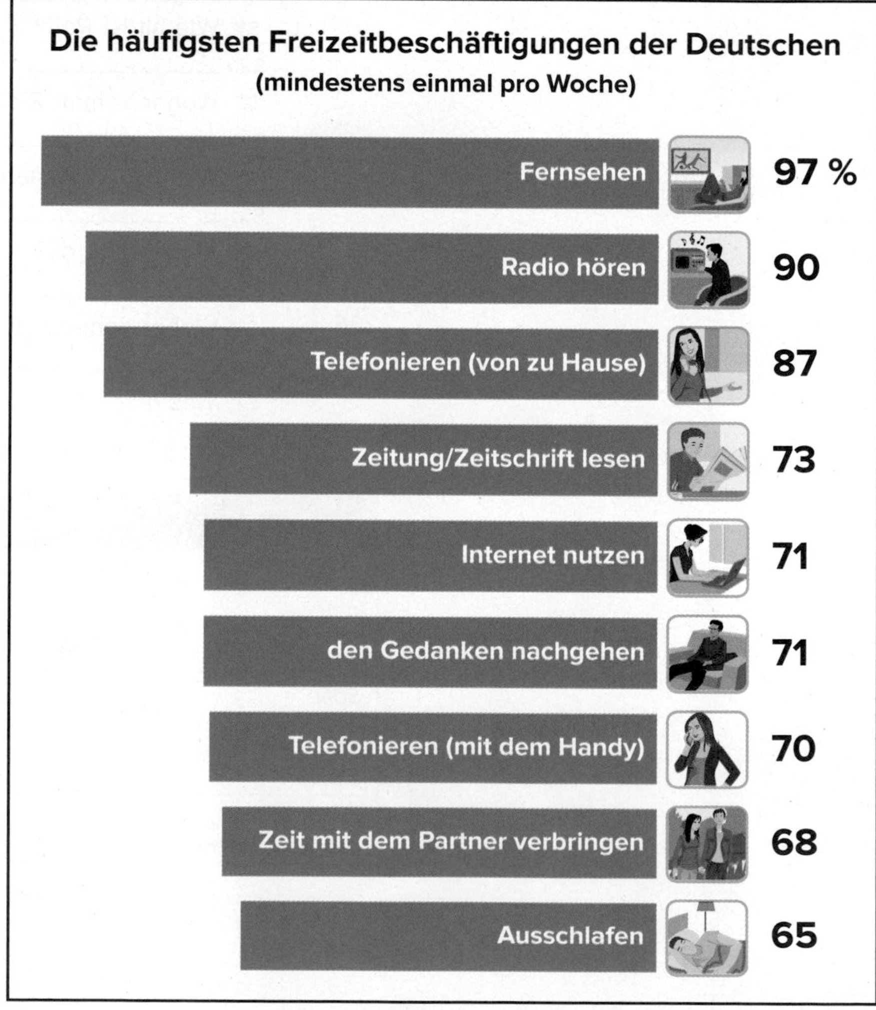

[1]*more often* [2]*less* [3]*Guess*

Situation 4 Umfrage

MODELL: S1: Schwimmst du gern im Meer?
S2: Ja.
S1: Unterschreib bitte hier.

	UNTERSCHRIFT
1. Schwimmst du gern im Schwimmbad?	______
2. Trinkst du gern Kaffee?	______
3. Spielst du gern Gitarre?	______
4. Hörst du gern Musik?	______
5. Gehst du gern zelten?	______
6. Arbeitest du gern?	______
7. Gehst du gern joggen?	______
8. Tanzt du gern?	______
9. Schreibst du gern Mails?	______
10. Machst du gern Fotos?	______

Schule und Universität

Grammatik 1.3

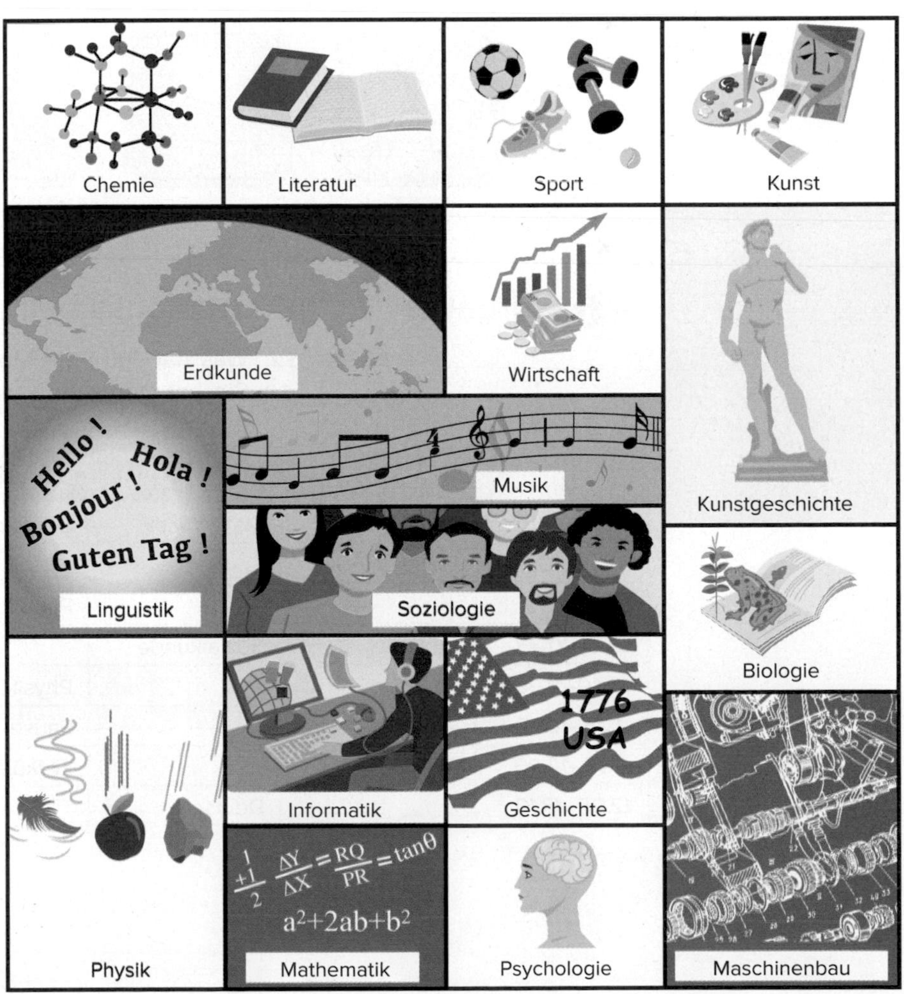

Situation 5 Dialog: Was studierst du?

Stefan trifft Rolf in der Cafeteria der Universität Berkeley.

STEFAN: Hallo, bist du ________ hier?
ROLF: Ja, ich ________ aus Deutschland.
STEFAN: Und was machst ________ hier?
ROLF: Ich ________ Psychologie. Und du?
STEFAN: ________.

Situation 6 Wie spät ist es?

MODELL: S1: Wie spät ist es?
S2: Es ist ______.

1.

2.

3.

4.

5.

6.

7.

8.

9.

10.

Situation 7 Informationsspiel: Juttas Stundenplan

MODELL: S2: Was hat Jutta am Montag um acht Uhr fünfzig?
S1: Sie hat Deutsch.

Uhr	Montag	Dienstag	Mittwoch	Donnerstag	Freitag
8.00–8.45		Mathematik	Deutsch		Französisch
8.50–9.35	Deutsch			Latein	
9.35–9.50	← Pause →				
9.50–10.35	Biologie	Sozialkunde		Geschichte	
10.40–11.25			Physik		Deutsch
11.25–11.35	← Pause →				
11.35–12.20	Sport		Erdkunde		Latein
12.25–13.10		Deutsch		Sozialkunde	frei

KULTUR ... LANDESKUNDE ... INFORMATIONEN

SCHULE

- Wann beginnt in Ihrem Land morgens die Schule?
- Wann gehen die Schüler und Schülerinnen nach Hause?
- Wann und wo machen sie Hausaufgaben?
- Wann haben sie Freizeit?
- Welche Schulfächer haben Schüler und Schülerinnen?
- Welches sind Pflichtfächer[1]?
- An welchen Tagen gehen die Schüler und Schülerinnen in die Schule?

Schauen Sie auf Juttas Stundenplan (Situation 7).

- Wann beginnt für Jutta die Schule?
- Wann geht sie nach Hause?
- Welche Fächer hat Jutta?
- Wie viele Fremdsprachen hat sie?
- An welchen Tagen geht sie in die Schule?

Was meinen Sie?

- Wann und wo macht Jutta Hausaufgaben?
- Wann hat sie Freizeit?

Große Pause an einer Schule in Berlin

[1]*required subjects*

Situation 8 Interview

1. Welche Fächer hast du in diesem Semester? Welche Fächer magst du? Welche Fächer magst du nicht?
2. Wann beginnt am Montag dein erster (1.) Kurs? Welcher Kurs ist das?
3. Wann gehst du am Montag nach Hause?
4. Wann beginnt am Dienstag dein erster Kurs? Welcher Kurs ist das?
5. Wann gehst du am Dienstag nach Hause?
6. Arbeitest du? An welchen Tagen arbeitest du? Wann beginnt deine Arbeit?
7. Wann gehst du in der Woche ins Bett? Und am Wochenende?
8. Wann machst du Hausaufgaben?
9. Wann hast du Freizeit?

Tagesablauf

Grammatik 1.4–1.5

Herr Wagner steht auf.

Er duscht.

Er frühstückt.

Er geht zur Arbeit.

Er geht einkaufen.

Er räumt die Wohnung auf.

Er geht im Park spazieren.

Er geht ins Bett.

Situation 9 Interview

1. Wann stehst du auf?
2. Wann duschst du?
3. Wann frühstückst du?
4. Wann gehst du zur Uni?
5. Wann kommst du nach Hause?
6. Wann machst du das Abendessen?
7. Wann gehst du ins Bett?

Situation 10 Am Wochenende

Was machen Sie am Wochenende sicher, wahrscheinlich, vielleicht?

S = sicher
W = wahrscheinlich
V = vielleicht

	ICH	PARTNER/PARTNERIN
1. Ich spiele Computerspiele.	______	______
2. Ich stehe spät auf.	______	______
3. Ich kaufe ein.	______	______
4. Ich surfe im Internet.	______	______
5. Ich schreibe Mails.	______	______
6. Ich chille.	______	______
7. Ich arbeite fürs Studium.	______	______
8. Ich rufe Freunde oder meine Familie an.	______	______
9. Ich räume mein Zimmer oder meine Wohnung auf.	______	______
10. Ich gehe mit Freunden aus.	______	______
11. Ich gehe ins Kino.	______	______
12. Ich jobbe.	______	______

Situation 11 Bildgeschichte: Ein Tag in Sofies Leben

Situation 12 Informationsspiel: Diese Woche

MODELL: S2: Was macht Silvia am Dienstag?
S1: Sie arbeitet am Abend in einer Bar.
S2: Was machst du am Montag?
S1: Ich ______.

	Silvia Mertens	Mehmet Sengün	mein(e) Partner(in)
Montag		Er geht um 7 Uhr zur Arbeit.	
Dienstag	Sie arbeitet am Abend in einer Bar.		
Mittwoch		Er surft im Internet.	
Donnerstag		Er geht einkaufen.	
Freitag	Sie geht tanzen.		
Samstag	Sie geht mit Freunden ins Kino.		
Sonntag	Sie besucht ihre Eltern.		

Filmlektüre

Hilfe!

Vor dem Lesen

A. Beantworten Sie die folgenden Fragen.

1. Was sehen Sie auf dem Bild? Machen Sie eine Liste, z. B. Papier, ...
2. Mickey trägt ein gelbes T-Shirt und Emma trägt eine Mütze[1], eine weiße Jacke und Jeans. Identifizieren Sie Emma und Mickey. Schreiben Sie die Namen unter das Foto.

FILMANGABEN

Titel: *Hilfe!*
Genre: Komödie
Erscheinungsjahr: 2002
Land: Deutschland
Dauer: 91 Min
Regisseur: Oliver Dommenget
Hauptrollen: Sarah Hannemann, Nick Seidensticker, Nina Petri, Dominique Horwitz, Philipp Blank, Pinkas Braun

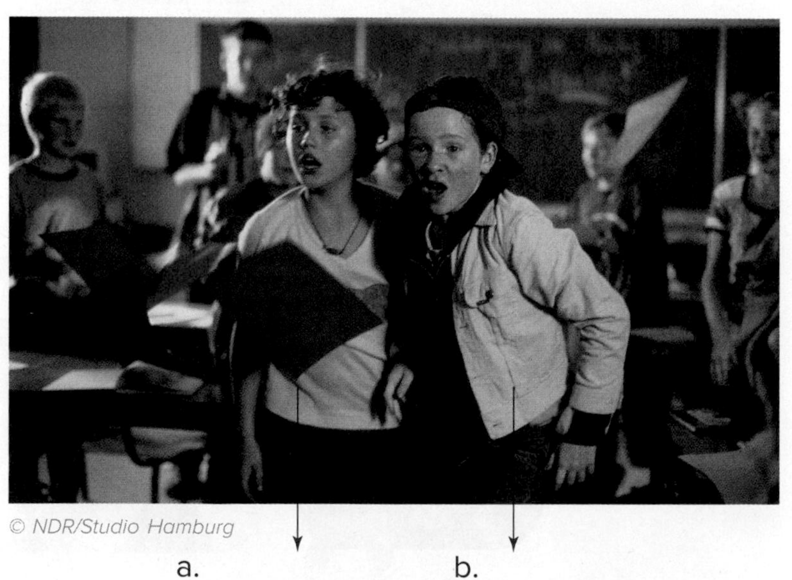

© NDR/Studio Hamburg

a. ______________ b. ______________

[1]*cap*

Miniwörterbuch	
das **Leben**	life
mögen	to like
ärgern	to tease, pick on
einmal	for once
wäre	would be
entdecken	to discover
der **Zauberspruch**	spell
der **Wunsch**	wish
in Erfüllung gehen	to come true
man	one, you
rückgängig machen	to reverse
die **Lösung**	solution
der **Mut**	courage
die **Freundschaft**	friendship
einander	one another

B. Lesen Sie die Wörter im Miniwörterbuch. Suchen Sie sie im Text und unterstreichen Sie sie.

Inhaltsangabe

Emma (Sarah Hannemann) wohnt in Hamburg. Sie ist 11 Jahre alt, Schülerin und geht in die Brecht-Schule. Sie ist sehr talentiert und in ihrer Freizeit schwimmt sie gern. Ihr Leben ist nicht leicht. Mickey (Nick Seidensticker) und seine Freunde mögen sie nicht. In der Schule ärgern sie sie oft. Emma denkt: „Einmal eine andere Person sein. Das wäre toll." Emmas Freund Freddy „Vierauge" (Philipp Blank) entdeckt in einem alten Buch einen Zauberspruch, mit dem ihr Wunsch in Erfüllung geht.

Am nächsten Morgen wacht Emma auf. Emma ist nicht mehr Emma. Sie ist jetzt Mickey, und Mickey ist nicht mehr Mickey. Er ist jetzt Emma. Das ist natürlich ein großes Problem. Wie kann man den Zauber rückgängig machen? Emma, Mickey und Vierauge suchen zusammen eine Lösung, aber sie haben nur 54 Stunden Zeit. Am Ende finden sie Mut, Freundschaft und einander.

Arbeit mit dem Text

Richtig (R) oder falsch (F)? Verbessern Sie die falschen Aussagen.

_____ 1. Emmas Hobby ist Schwimmen.
_____ 2. Emma hat viel Talent.
_____ 3. Mickey mag Emma nicht.
_____ 4. Emma und Mickey sind gute Freunde.
_____ 5. Freddy findet einen Zauberspruch.
_____ 6. Emma ist jetzt Mickey und Mickey ist jetzt Freddy.
_____ 7. Sie haben drei Tage Zeit, um eine Lösung zu finden.
_____ 8. Am Ende sind Mickey und Emma Freunde.

FILMCLIP

NOTE: For copyright reasons, the films referenced in the FILMCLIP feature have not been provided by the publisher. The film can be purchased as a DVD or found online at various sites such as YouTube, Amazon, or the iTunes store. The time codes mentioned below are for the North American DVD version of the film.

Szene: DVD Kapitel 3, „In der Schule", 17:35–19:15 Min.

Emma und Freddy („Vierauge") gehen zusammen in die Schule.

Ergänzen Sie!

1. Vierauge und Emma _____ in die Schule.
2. Die Schüler _____ heute einen Mathetest.
3. Mickey _____ Probleme mit dem Test.
4. Der Schüler mit dem Gameboy _____ eine braune Mütze.
5. Emma _____ einen Zettel[3].
6. Die Lehrerin _____, dass Mickey spickt[4].

a. glaubt[2]
b. haben
c. hat
d. kommen
e. schreibt
f. trägt

Nach dem Lesen

Kennen Sie andere Filme mit Zauberern? Wie heißen diese Filme? Mögen Sie diese Filme?

[2]*believes* [3]*note* [4]*is cheating*

Persönliche Angaben

Grammatik 1.6

Antrag auf Ausstellung eines Personalausweises

Familienname: Ruf
geborene(r): Schuler
Vornamen: Margret
Geburtstag: 13. April 1977
Geburtsort: Augsburg
Staatsangehörigkeit: deutsch
Augenfarbe: blau, grau, (grün), braun Größe 172 cm
München Sonnenstr. 11
Straße Hausnummer

München, den 30.5.2016
Margret Ruf
Unterschrift des Antragstellers

Situation 13 Dialog: Auf dem Rathaus

Melanie Staiger ist auf dem Rathaus in Regensburg. Sie braucht einen neuen Personalausweis.

BEAMTER: Grüß Gott!
MELANIE: Grüß Gott. Ich brauche einen neuen ___________.
BEAMTER: ___________ ist Ihr Name, bitte?
MELANIE: Staiger, Melanie Staiger.
BEAMTER: Und ___________ wohnen Sie?
MELANIE: In Regensburg.
BEAMTER: ___________ ist die genaue Adresse?
MELANIE: Gesandtenstraße 8.
BEAMTER: Haben Sie auch ___________?
MELANIE: Ja, die Nummer ist 24352.
BEAMTER: Was sind Sie ___________?
MELANIE: Ich bin Studentin.
BEAMTER: Sind Sie verheiratet?
MELANIE: ___________. Ich bin ledig.

Situation 14 Interview: Auf dem Rathaus

1. Wie heißen Sie?
2. Wie alt sind Sie?
3. Wo sind Sie geboren?
4. Wo wohnen Sie?
5. Was ist Ihre genaue Adresse?
6. Was ist Ihre E-Mail-Adresse?
7. Was studieren Sie?
8. Sind Sie verheiratet?
9. Welche Augenfarbe haben Sie?
10. Welche Haarfarbe?

Das Regensburger Rathaus
© Raimund Kutter/imageb/imageBROKER/Superstock

Miniwörterbuch

fühlt sich	feels
die **Beziehung**	relationship
glauben (an)	to believe (in)
dabei sein	to be close to or on the verge of
ist auch was dran	there's something to it
verlieren	to lose
gibt auf	gives up

MUSIKSZENE

„Gewinner" (2009, Deutschland) *Clueso*

Biografie Clueso kommt aus Erfurt. Sein richtiger Name ist Thomas Hübner. Er wurde 1980 geboren. Sein Künstlername kommt von Inspektor Clouseau aus dem Film *The Pink Panther.*

Clueso
© Christian Jakubaszek/Getty Images

NOTE: For copyright reasons, the songs referenced in **MUSIKSZENE** have not been provided by the publisher. The song can be found online at various sites such as YouTube, Amazon, or the iTunes store.

Vor dem Hören Wie fühlt man sich am Ende einer Beziehung?

Nach dem Hören

A. Ergänzen Sie die Sätze mit den Wörtern aus dem Kasten.

1. Ich ____________ nichts.
2. ____________ du an mich?
3. Ich ____________ dich.
4. ____________ du mich?
5. Wir ____________ dabei.

frage **glaube** **fragst** **sind** **glaubst**

B. Was sagt der Sänger zu seiner Freundin? Richtig (R) oder falsch (F)?

______ **1.** An allem, was man sagt, ist nichts dran[1].

______ **2.** Er glaubt an sie.

______ **3.** Sie verlieren einander[2].

______ **4.** Er gibt auf.

[1]An ... *There's nothing to anything we say* [2]*each other*

Situation 15 Rollenspiel: Auf dem Auslandsamt°

°*study abroad office*

S1: Sie sind Student/Studentin und möchten ein Jahr lang in Österreich studieren. Gehen Sie aufs Auslandsamt und sagen Sie, dass Sie ein Stipendium möchten. Beantworten Sie die Fragen des Beamten / der Beamtin. Sagen Sie am Ende des Gesprächs „Auf Wiedersehen".

Situation 16 Gesucht°!

°*Wanted*

Schreiben Sie die fehlenden Angaben[2] in den Steckbrief.

NÜTZLICHE WÖRTER

der Bankräuber	*bank robber*
der Spitzname	*nickname*
besonderes Kennzeichen	*distinguishing feature*
die Narbe	*scar*
das Halstuch	*bandanna*
bewaffnet	*armed*

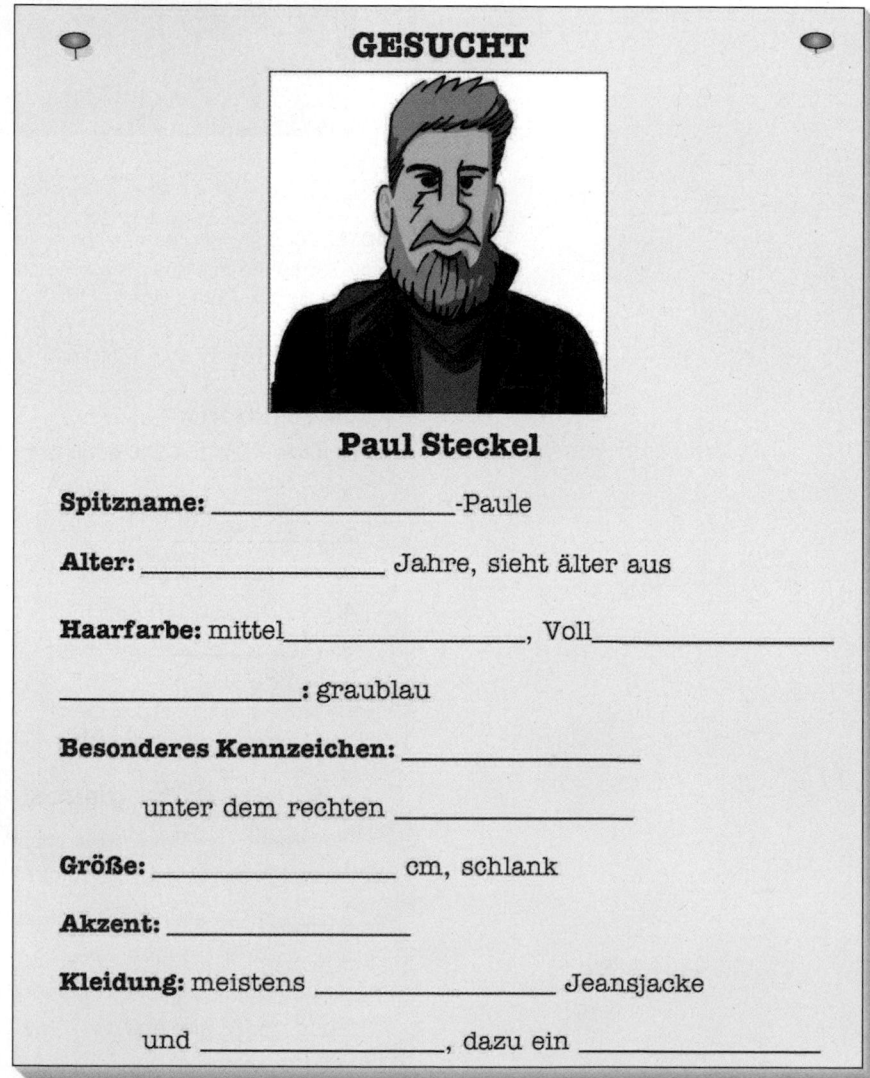

GESUCHT

Paul Steckel

Spitzname: ________________-Paule

Alter: ________________ Jahre, sieht älter aus

Haarfarbe: mittel________________, Voll________________

________________: graublau

Besonderes Kennzeichen: ________________

unter dem rechten ________________

Größe: ________________ cm, schlank

Akzent: ________________

Kleidung: meistens ________________ Jeansjacke

und ________________, dazu ein ________________

[2]*pieces of information*

Lektüre

Vor dem Lesen

LESEHILFE

In this reading, several of the characters of ***Kontakte*** are introduced. Before reading, look at each of the pictures and say as much as you can about the characters, based on the drawings. Now read the text once through. How closely do the pictures reflect what is in the text?

Welche Informationen geben Sie, wenn Sie sich vorstellen[3]? Kreuzen Sie an.

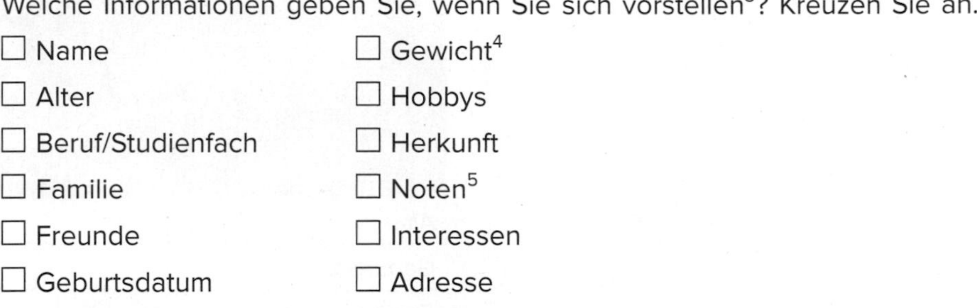

- ☐ Name
- ☐ Alter
- ☐ Beruf/Studienfach
- ☐ Familie
- ☐ Freunde
- ☐ Geburtsdatum
- ☐ Gewicht[4]
- ☐ Hobbys
- ☐ Herkunft
- ☐ Noten[5]
- ☐ Interessen
- ☐ Adresse

Miniwörterbuch	
der **Geschäftsmann**	businessman
unterrichten	to teach
die **Sozialkunde**	social studies
die **Gärtnerei**	nursery (gardening)
der **Ort**	town
das **Fahrrad**	bicycle
seit	since; *here:* for
die **Speditionsfirma**	trucking company
der **Lastwagen**	truck
unterwegs	on the road

Guten Tag, ich heiße ...

Guten Tag, ich heiße Veronika Frisch-Okonkwo. Ich bin verheiratet und habe drei Töchter. Sie heißen Sumita, Yamina und Lydia. Ich lebe mit meinem Mann Kobe Okonkwo und unseren Töchtern in der Schweiz. Wir wohnen im Kanton Zürich. Ich komme aus Zürich und mein Mann kommt aus Abuja, Nigeria. Ich bin 33 Jahre alt und Kobe ist 35. Kobe ist Geschäftsmann hier in Zürich und ich bin Lehrerin. Ich unterrichte Französisch und Sozialkunde. Meine Freizeit verbringe ich am liebsten mit meiner Familie. Außerdem reise ich gern.

Guten Tag, ich heiße Sofie Pracht, bin 22 und komme aus Dresden. Ich studiere Biologie an der Technischen Universität Dresden. Ein paar Stunden in der Woche arbeite ich in einer großen Gärtnerei. In meiner Freizeit gehe ich oft ins Kino oder ich besuche Freunde. Ich spiele Gitarre und tanze sehr gern. Mein Freund heißt Willi Schuster. Er studiert auch hier in Dresden an der Technischen Universität. Er kommt aus Radebeul. Das ist ein kleiner Ort ganz in der Nähe von Dresden. Am Wochenende fahren wir manchmal mit dem Fahrrad nach Radebeul und besuchen seine Familie.

Guten Tag, ich heiße Mehmet Sengün. Ich bin 29 und in Izmir, in der Türkei, geboren. Ich lebe jetzt seit 19 Jahren hier in Berlin. Ich wohne in Kreuzberg, einem Stadtteil von Berlin, in einer kleinen Wohnung. In Kreuzberg leben sehr viele Türken – die Berliner nennen es Klein-Istanbul – und viele meiner türkischen Freunde wohnen ganz in der Nähe. Im Moment arbeite ich für eine Speditionsfirma hier in der Stadt. Ich fahre einen Lastwagen und bin viel unterwegs. In meiner Freizeit treibe ich viel Sport. Ich spiele Fußball, Basketball und Tennis. Tennis spiele ich gern mit meiner Freundin Renate.

[3]sich ... *introduce yourself* [4]*weight* [5]*grades*

Arbeit mit dem Text

Was erfahren Sie über Veronika Frisch-Okonkwo, Sofie Pracht und Mehmet Sengün? Vervollständigen Sie die Tabelle.

Name	Veronika Frisch-Okonkwo	Sofie Pracht	Mehmet Sengün
Alter			
Geburtsort			
Familie/Freunde			
Wohnort			
Beruf			
Studienfach			
Freizeit			
Sonstiges[6]			

Nach dem Lesen

Stellen Sie sich vor.[7] Schreiben Sie einen kurzen Text. Kleben[8] Sie ein Foto auf das Papier oder zeichnen[9] Sie ein Selbstporträt. Hängen Sie Ihre Texte im Seminarraum an die Wand.

Berlin-Kreuzberg

[6]*other information* [7]Stellen ... *Introduce yourself* [8]*Glue* [9]*draw*

Videoecke

Perspektiven

Wie spät ist es?

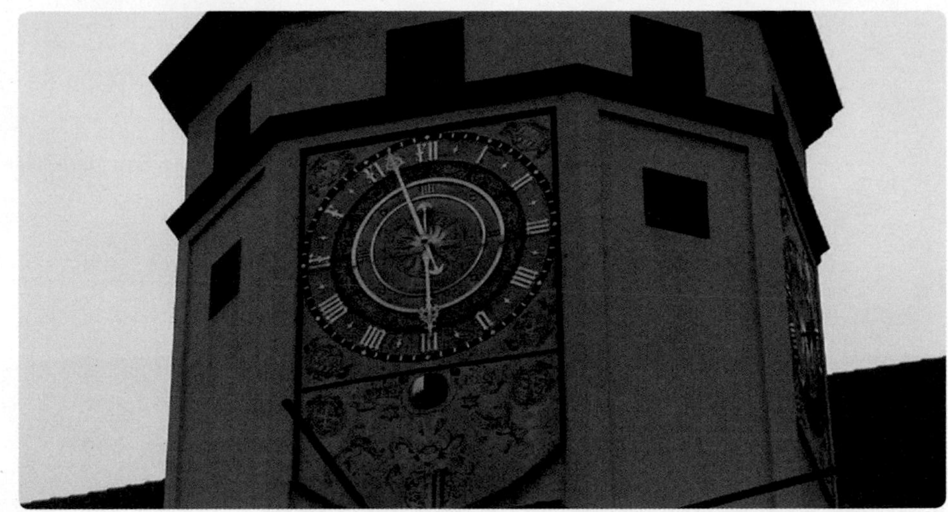

Wie spät ist es?

Aufgabe 1 Uhren

Worauf schauen sie?

1. Wie viele Leute schauen auf ihre Armbanduhr?
2. Wie viele Leute schauen auf ihr Handy?
3. Wie viele Leute schauen auf die Turmuhr?

Aufgabe 2 Wer sagt das?

Wie spät ist es? Acht Personen werden gefragt. Ordnen Sie die Aussagen den Personen zu.

1. ___

2. ___

3. ___

4. ___

5. ___

6. ___

7. ___

8. ___

a. Es ist 10 Uhr 10.
b. Es ist 12 Uhr 25.
c. Es ist 13 Uhr 28.
d. Es ist 14 Uhr 38.
e. Ich weiß es nicht. Ich habe keine Uhr.
f. Es ist drei vor sechs.
g. Es ist dreiviertel fünf.
h. Es ist neun Uhr.

Interviews

- Was studierst du?
- Welche Seminare hast du in diesem Semester?
- Wann beginnen deine Seminare?
- Wann stehst du da auf?
- Was machst du dann?
- Was machst du in deiner Freizeit?

Sandra

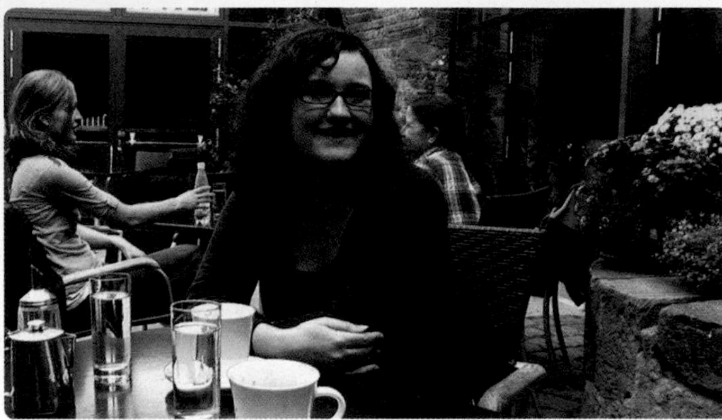
Susan

Aufgabe 3 Tagesablauf

Miniwörterbuch

die **Fremdsprache**	*foreign language*
belegen	*to take*

Wer macht das? Sandra, Susan oder beide? Kreuzen Sie an.

	Sandra	Susan	Beide
1. Sie studiert Deutsch als Fremdsprache.	☐	☐	☐
2. Sie belegt Seminare zur Phonetik, Phonologie und Grammatik.	☐	☐	☐
3. Ihre Seminare beginnen um 9 Uhr.	☐	☐	☐
4. Sie steht um halb acht auf.	☐	☐	☐
5. Sie steht um 7 Uhr auf.	☐	☐	☐
6. Sie fährt mit dem Fahrrad zur Universität.	☐	☐	☐
7. Sie geht zuerst duschen.	☐	☐	☐
8. Sie geht gern laufen und sie liest gern.	☐	☐	☐
9. Sie geht schwimmen und singt im Chor.	☐	☐	☐

Wortschatz

Freizeit	Leisure Time
chatten [tschɛt-]	to chat
chillen [tschɪl-]	to relax, to hang
lesen (R)	to read
er/sie liest	he/she reads
Zeitung lesen	to read the newspaper
liegen	to lie
in der Sonne liegen	to lie in the sun
reisen	to travel
schreiben (R)	to write
eine SMS schreiben	to write a text message
segeln	to sail
simsen	to text
spielen	to play
wandern	to hike
zelten	to camp

Ähnliche Wörter
die **E-Mail [i:meɪl], -s;** die **Mail, -s;** die **Gitarre, -n;** die **Karte, -n;** die **Musik;** die **Sonnenbrille, -n;** der **Ball, ¨-e** (R); der **Fußball, ¨-e** (R); der **Kaffee;** der **Volleyball, ¨-e;** das **Foto, -s;** das **Golf;** das **Hobby, -s;** das **Schach;** das **Snowboard, -s;** das **Tennis; boxen; hören; kochen; reiten; schwimmen gehen; singen; im Internet surfen; tanzen**

Orte	Places
die **Arbeit**	work
zur Arbeit gehen	to go to work
die **Kneipe, -n**	bar, tavern
die **Studentenkneipe, -n**	student pub
der **Berg, -e**	mountain
in die Berge gehen	to go to the mountains
in den Bergen wandern	to hike in the mountains
das **Kino, -s**	movie theater, cinema
ins Kino gehen	to go to the movies
das **Meer, -e**	sea
im Meer schwimmen	to swim in the sea
das **Rathaus, ¨-er**	town hall
auf dem Rathaus	at the town hall
das **Schwimmbad, ¨-er**	swimming pool
ins Schwimmbad fahren	to go to the swimming pool

Ähnliche Wörter
die **Party, -s; auf eine Party gehen;** die **Uni, -s** (R); **zur Uni gehen; auf der Uni sein;** der **Park, -s; im Park spazieren gehen;** das **Bett, -en; ins Bett gehen;** das **Haus, ¨-er; zu Hause sein; nach Hause gehen;** das **Konzert, -e; ins Konzert gehen;** das **Museum, Museen; ins Museum gehen**

Schule und Universität	School and University
die **Fremdsprache, -n**	foreign language
die **Geschichte**	history
die **Kunstgeschichte**	art history
die **Informatik**	computer science
die **Kunst**	art
die **Lehrerin, -nen** (R)	female teacher, instructor
die **Prüfung, -en**	test
die **Schule, -n**	school
die **Schülerin, -nen**	female pupil
die **Sozialkunde**	social studies
die **Wirtschaft**	economics
der **Lehrer, -** (R)	male teacher, instructor
der **Maschinenbau**	mechanical engineering
der **Schüler, -**	male pupil
der **Stundenplan, ¨-e**	schedule
das **Fach, ¨-er**	academic subject
das **Stipendium, Stipendien**	scholarship
das **Studium, Studien**	university studies
die **Ferien** (*pl.*)	vacation

Ähnliche Wörter
die **Biologie,** die **Chemie,** die **Geografie,** die **Linguistik,** die **Literatur;** die **Mathematik;** die **Musik;** die **Pause, -n;** die **Physik;** die **Religion;** die **Soziologie;** der **Kurs, -e** (R); der **Sport;** das **Latein;** das **Semester, -; lernen; studieren**

Persönliche Angaben	Biographical Information
die **Farbe, -n**	color
die **Größe, -n**	height
die **Staatsangehörigkeit, -en**	nationality, citizenship
die **Unterschrift, -en**	signature
der **Beruf, -e**	profession
was sind Sie von Beruf?	what's your profession?
der **Geburtstag, -e**	birthday
der **Personalausweis, -e**	(personal) ID card
der **Reisepass, ¨-e**	passport
der **Wohnort, -e**	residence
das **Alter**	age
die **Angaben** (*pl.*)	particulars
ledig	unmarried
verheiratet	married

Ähnliche Wörter
die **Adresse, -n;** die **Augenfarbe;** die **Haarfarbe;** die **Nummer, -n;** die **Hausnummer, -n;** die **Telefonnummer, -n;** die **Person, -en;** der **Name, -n** (R); der **Familienname, -n** (R); der **Vorname, -n** (R); das **Telefon, -e; geboren; wann sind Sie geboren?**

Tagesablauf	Daily Routine
die **Stunde, -n**	hour
die **Woche, -n**	week
in der Woche	during the week
der **Abend, -e**	evening
der **Tag, -e**	day
den ganzen Tag	all day long
jeden Tag	every day
der **Montag**	Monday
der **Dienstag**	Tuesday
der **Mittwoch**	Wednesday
der **Donnerstag**	Thursday
der **Freitag**	Friday
der **Samstag**	Saturday
der **Sonntag**	Sunday
das **Wochenende, -n**	weekend
am Wochenende	over the weekend
früh	early
spät(er)	late(r)
Um wie viel Uhr …?	At what time …?
Wann?	When?
um halb drei	at two thirty
um sechs (Uhr)	at six o'clock
um sieben Uhr zwanzig	at seven twenty
um Viertel vor vier	at a quarter to four
um zwanzig nach fünf	at twenty after/ past five
Welcher Tag ist heute?	What day is today?
Wie spät ist es?	What time is it?
Wie viel Uhr ist es?	What time is it?

Ähnliche Wörter
die **Sekunde, -n;** der **Moment, -e; im Moment**

Sonstige Substantive	Other Nouns
die **Lösung, -en**	solution
die **Tasche, -n**	bag; purse; pocket
die **Wohnung, -en**	apartment
der **Brief, -e**	letter
der **Mut**	courage
der **Zufall, ¨-e**	coincidence
das **Abendessen, -**	supper, evening meal
das **Motorrad, ¨-er**	motorcycle
Motorrad fahren	to ride a motorcycle
das **Zimmer, -**	room

Trennbare Verben	Separable Verbs
ab·holen	to pick (somebody) up (from a place)
an·kommen	to arrive
an·rufen	to call up
auf·geben	to give up
auf·hören (mit)	to stop (doing something)
auf·räumen	to clean (up)
auf·stehen	to get up
aus·füllen	to fill out
aus·gehen	to go out
ein·kaufen (gehen)	to (go) shop(ping), shop for
ein·packen	to pack up
fern·sehen	to watch TV
er/sie sieht fern	he/she is watching TV
hervor·rufen	to evoke, call forth
kennen·lernen	to get acquainted with
rein·gehen	to go inside

Sonstige Verben	Other Verbs
arbeiten	to work
ärgern	to tease; to annoy
besuchen	to visit
brauchen	to need; to use
duschen	to (take a) shower
fliegen	to fly
frühstücken	to eat breakfast
kaufen	to buy
können	to be able to, can
ich kann	I can
mögen	to like
ich mag	I like
du magst	you like
spazieren gehen	to go for a walk
suchen	to look for
tun	to do
unterschreiben	to sign

Ähnliche Wörter
beginnen, reparieren, trinken

Sonstige Wörter und Ausdrücke	Other Words and Expressions
Das macht nichts.	That doesn't matter.
gern	gladly, with pleasure
wir singen gern	we like to sing
ihr(e)	her
jede, jeder, jedes	each, every
jeden Tag	every day
lange nicht gesehen	haven't seen (you / each other) for a long time
miteinander	with each other, together
nervös	nervous
sein(e)	his
sicher	sure
uns	us, ourselves (*here:* each other)
wahrscheinlich	probably
wenigstens	at least

Strukturen und Übungen

1.1 The present tense

One German present-tense form expresses three different ideas in English.

Ich spiele Gitarre. { *I play the guitar.* / *I'm playing the guitar.* / *I'm going to play the guitar.* }

Most German verbs form the present tense just like **kommen (Einführung B).**

spielen			
ich	spiele	wir	spielen
du	spielst	ihr	spielt
Sie	spielen	Sie	spielen
er sie es	spielt	sie	spielen

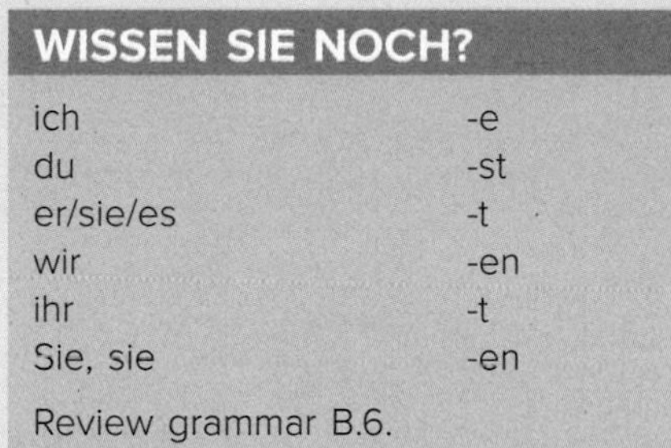

WISSEN SIE NOCH?

ich	-e
du	-st
er/sie/es	-t
wir	-en
ihr	-t
Sie, sie	-en

Review grammar B.6.

Gabi und Jutta **spielen** gern Karten. — *Gabi and Jutta like to play cards.*

Verbs whose stems end in an **s**-sound, such as **-s, -ss, -ß, -z (-ts),** or **-x (-ks),** do not add an additional **-s-** in the **du**-form: **du tanzt, du heißt, du reist.**

—Wie **heißt du?** — *What's your name?*
—**Ich heiße** Natalie. — *My name's Natalie.*

Verbs whose stems end in **-d** or **-t** (and a few other verbs such as **regnen** [*to rain*] and **öffnen** [*to open*]) insert an **-e-** between the stem and the **-st** or **-t** endings. This happens in the **du-, ihr-,** and **er/sie/es**-forms.

Reitest du jeden Tag? — *Do you go horseback riding every day?*

reiten			
ich	reite	wir	reiten
du	reitest	ihr	reitet
Sie	reiten	Sie	reiten
er sie es	reitet	sie	reiten

Übung 1 Was machen sie?

Kombinieren Sie die Wörter. Achten Sie auf die Verbendungen.

MODELL: Ich besuche Freunde.

1. ich	lernen	Freunde
2. ihr	besuche	ins Kino
3. Jutta und Jens	studiert	Spaghetti
4. du	hört	ein Buch
5. Melanie	reisen	gut Tennis
6. ich	kochen	nach Deutschland
7. wir	lese	in Regensburg
8. Richard	spielst	Spanisch
9. Jürgen und Silvia	geht	gern Musik

Übung 2 Minidialoge

Ergänzen Sie das Pronomen.

1. CLAIRE: Arbeitet Melanie?
   JOSEF: Nein, _______ arbeitet nicht.
2. MICHAEL: Schwimmen _______ gern im Meer?
   FRAU KÖRNER: Ja, sehr gern. Und Sie?
3. MEHMET: Was machst _______[a] im Sommer?
   RENATE: _______[b] fliege nach Spanien.
4. CLAIRE: Woher kommt _______[a]?
   ESKE UND DAMLA: _______[b] kommen aus Krefeld.
5. JÜRGEN: _______[a] studiere in Göttingen. Und _______[b]?
   KLAUS UND CHRISTINA: _______[c] studieren in Berlin.

Übung 3 Minidialoge

Ergänzen Sie die Verbendungen.

1. CLAIRE: Du tanz_______[a] gern, nicht?
   MELANIE: Ja, ich tanz_______[b] sehr gern, aber mein Freund tanz_______[c] nicht gern.
2. FRAU SCHULZ: Richard geh_______[a] im Sommer in den Bergen wandern.
   STEFAN: Und was mach_______[b] seine Eltern?
   FRAU SCHULZ: Seine Mutter reis_______[c] nach Frankreich und sein Vater arbeit_______[d].
3. JÜRGEN: Wir koch_______[a] heute Abend. Was mach_______[b] ihr?
   KLAUS: Wir besuch_______[c] Freunde.
4. DANIEL: Schreib_______[a] du mir eine E-Mail?
   TIM: Ja, ich schreibe dir eine E-Mail. Chatt_______[b] du auch?
   DANIEL: Ja, das mach_______[c] ich auch.

1.2 Expressing likes and dislikes: *gern / nicht gern*

verb + **gern** = *to like to do something*
verb + **nicht gern** = *to dislike doing something*

To say that you like doing something, use the word **gern** after the verb. To say that you don't like to do something, use **nicht gern.**

Ernst spielt **gern** Fußball.	*Ernst likes to play soccer.*
Josef spielt **nicht gern** Fußball.	*Josef doesn't like to play soccer.*

I	II	III	IV
Sofie	spielt	gern	Schach.
Willi	spielt	auch gern	Schach.
Ich	spiele	nicht gern	Schach.
Monika	spielt	auch nicht gern	Schach.

The position of **auch/nicht/gern** (in that order) is between the verb and its complement. The complement provides additional information and thus "completes" the meaning of the verb: **ich spiele → ich spiele Tennis; ich höre → ich höre Musik.**

Übung 4 Was machen die Studenten gern?

Bilden Sie Sätze.

MODELL: Heidi und Nora schwimmen gern.

Heidi/Nora

1. Monika/Albert

2. Heidi

3. Stefan

4. Nora

5. Peter

6. Katrin

7. Monika

8. Albert

Übung 5 Und diese Personen?

Sagen Sie, was die folgenden Personen gern machen.

MODELL: Frau Ruf liegt gern in der Sonne. Jutta liegt auch gern in der Sonne, aber Herr Ruf liegt nicht gern in der Sonne.

1. Frau Ruf

Jutta

Herr Ruf

2. Jens Ernst Jutta

3. Jens Jutta Andrea

4. Michael Maria die Rufs die Wagners

1.3 Telling time

Ask the time in German in one of two ways.

Wie spät ist es?
Wie viel Uhr ist es?

What time is it?

Es ist eins.
Es ist ein Uhr.

Es ist drei.
Es ist drei Uhr.

Es ist Viertel vor elf.
Es ist zehn Uhr fünfundvierzig.

Es ist Viertel nach elf.
Es ist elf Uhr fünfzehn.

vor = *to*
nach = *after*

Es ist zehn (Minuten) vor acht.
Es ist sieben Uhr fünfzig.

Es ist zehn (Minuten) nach acht.
Es ist acht Uhr zehn.

Es ist halb zehn.
It is nine thirty (halfway to ten).

halb = *half, thirty*
halb zehn = *half past nine, nine thirty*

The expressions **Viertel, nach, vor,** and **halb** are used in everyday speech. In German, the half hour is expressed as "half before" the following hour, not as "half after" the preceding hour, as in English.

The 24-hour clock (0.00 to 24.00) is used when giving exact or official times, as in time announcements, schedules, programs, and the like. With the 24-hour clock only the pattern [(*number*) **Uhr** (*number of minutes*)] is used.

Ankunft	km	Abfahrt	Anschlüsse	
14.22	**Potsdam Stadt**	**14.24**		
	✝	14.43	Wildpark 14.49 Werder (Havel) 14.56	(204)
	24	*E* 15.01	Wustermark 15.39 Nauen 15.57	(204.4)
			S-Bahnanschlüsse (Taktverkehr) bestehen in Richtung: Wannsee – Westkreuz – Charlottenburg – Zool Garten (Ⓢ *3*)	

Der Zug geht um vierzehn Uhr vierundzwanzig.

The train leaves at two twenty-four p.m.

Übung 6 Die Uhrzeit

Wie spät ist es?

MODELL: Es ist acht Uhr.

1.

2.

3.

4.

5.

6.

7.

8.

1.4 Word order in statements

In English, the verb usually follows the subject of a sentence.

SUBJECT	VERB	COMPLEMENT
Peter	takes	a walk.

Even when another word or phrase begins the sentence, the word order does not change.

	SUBJECT	VERB	COMPLEMENT
Every day,	Peter	takes	a walk.

In statements, verb second.

In German statements, the verb is always in second position. If the sentence begins with an element other than the subject, the subject follows the verb.

I	II	III	IV
SUBJECT	VERB		COMPLEMENT
Wir	spielen	heute	Tennis.
	VERB	SUBJECT	COMPLEMENT
Heute	spielen	wir	Tennis.

Übung 7 Rolf

Unterstreichen[1] Sie das Subjekt des Satzes. Steht das konjugierte Verb vor[2] oder nach[3] dem Subjekt?

1. Rolf kommt aus Krefeld. *nach*
2. Im Moment studiert er in Berkeley. ______
3. Seine Stiefmutter wohnt in Krefeld. ______
4. Samstags geht Rolf oft ins Kino. ______
5. Am Wochenende wandert er oft in den Bergen. ______
6. In der Woche treibt er gern Sport. ______
7. Im Sommer geht er surfen. ______
8. Er geht auch ins Schwimmbad der Uni. ______

Übung 8 Sie und Ihr Freund

Bilden Sie Sätze. Beginnen Sie die Sätze mit dem ersten Wort oder den ersten Wörtern in jeder Zeile. Beachten[4] Sie die Satzstellung[5].

MODELL: Heute (ich / sein ______) → Heute bin ich fröhlich.

1. Ich (studieren ______)
2. Im Moment (ich / wohnen in ______)
3. Heute (ich / kochen ______)
4. Manchmal (ich / trinken ______)
5. Ich (spielen gern ______)
6. Mein Freund (heißen ______)
7. Jetzt (er / wohnen in ______)
8. Manchmal (wir / spielen ______)

[1]*Underline* [2]*before* [3]*after* [4]*Pay attention to* [5]*word order*

1.5 Separable-prefix verbs

Many German verbs have prefixes that change the verb's meaning. They combine with the infinitive to form a single word.

stehen	*to stand*
gehen	*to go*
kommen	*to come*
aufstehen	*to stand up*
ausgehen	*to go out*
ankommen	*to arrive*

In statements, verb second, prefix last.

When you use a present-tense form of these verbs, put the conjugated form in second position and put the prefix at the end of the sentence. The two parts of the verb form a frame or bracket, called a **Satzklammer,** that encloses the rest of the sentence.

Claire [kommt] [an].

Claire [kommt] am Donnerstag [an].

Claire [kommt] am Donnerstag in Frankfurt [an].

Here are some common verbs with separable prefixes.

abholen	*to pick up, fetch*
ankommen	*to arrive*
anrufen	*to call up*
aufgeben	to give up
aufhören	*to stop, be over*
aufräumen	*to clean up, tidy up*
aufstehen	*to get up*
ausfüllen	*to fill out*
ausgehen	*to go out*
einkaufen	*to shop, shop for*
einpacken	*to pack up*

Übung 9 Eine Reise in die Türkei

Mehmet fliegt morgen in die Türkei. Was macht er heute? Ergänzen Sie die folgenden Wörter: **ab, an, auf, auf, auf, aus, aus, ein, ein.**

1. Er steht um 7 Uhr _______.
2. Er räumt die Wohnung _______.
3. Er packt seine Sachen[1] _______.
4. Er ruft Renate _______.
5. Er füllt ein Formular _______.
6. Er holt seinen Reisepass _______.
7. Er kauft Essen[2] _______.
8. Abends geht er _______.
9. Er geht ins Kino. Der Film hört um 22 Uhr _______.

Mehmet

[1]*things* [2]*food*

Übung 10 Was machen die Leute?

Verwenden Sie die folgenden Verben.

abholen
ankommen
anrufen
aufräumen
aufstehen
ausfüllen
ausgehen
einkaufen
einpacken

MODELL: Frau Schulz kauft Lebensmittel ein.

1. Rolf

2. Thomas

3. Heidi/Thomas

4. Albert

5. Peter/Monika

6. Peter/Monika

7. Frau Schulz

8. Stefan

1.6 Word order in questions

In **w**-questions, verb second.

When you begin a question with a question word (for example, **wie, wo, wer, was, wann, woher**), the verb follows in second position. The subject of the sentence is in third position unless the question word is the subject, e.g., **wer**. Any further elements appear in fourth position.

I	II	III	IV	
Wann	beginnt	das Spiel?		*When does the game start?*
Was	machst	du	heute Abend?	*What are you doing this evening?*
Wo	wohnst	du?		*Where do you live?*
Welches Fach	studierst	du?		*What subject are you studying?*

Here are the question words you have encountered so far.

wann	*when*
was	*what*
welcher*	*which*
wer	*who*
wie	*how*
wie viel(e)	*how much* (*many*)
wo	*where*
woher	*from where*

Questions that can be answered by *yes* or *no* begin with the verb.

Tanzt du gern?	*Do you like to dance?*
Arbeitest du hier?	*Do you work here?*
Gehst du ins Kino?	*Are you going to the movies?*

Übung 11 Ein Interview mit Nesrin Durani

Nesrin

Schreiben Sie die Fragen.

MODELL: du + heißen + wie + ? → Wie heißt du?

1. du + sein + geboren + wann + ?
2. du + kommen + woher + ?
3. du + sein + groß + wie + ?
4. du + studieren + Fächer + welch- + ?
5. du + arbeiten + Stunden + wie viele + ?
6. du + machen + gern + was + ?

Übung 12 Noch ein Interview

Sofie

Stellen Sie die Fragen.

1. —Ich heiße Sofie.
2. —Nein, ich komme nicht aus München.
3. —Ich komme aus Dresden.
4. —Ich studiere Biologie.
5. —Er heißt Willi.
6. —Er wohnt in Dresden.
7. —Nein, ich spiele nicht Tennis.
8. —Ja, ich tanze sehr gern.
9. —Ja, ich trinke gern Cola.
10. —Ja, Willi trinkt gern Bier.

*The endings of **welcher** vary according to gender, number, and case of the following noun. They are the same endings as those of the definite article. Therefore, **welcher** is called a **der**-word.

(M)	(N)	(F)	(Pl)
welch**er** Name	welch**es** Alter	welch**e** Adresse	welch**e** Studienfächer

KAPITEL 2

Besitz und Vergnügen

In **Kapitel 2** you will learn to talk more about things: your own possessions and things you give others. You will also learn how to describe what you have and don't have and to give your opinion on matters of taste or style.

Themen

Besitz

Geschenke

Kleidung und Aussehen

Vergnügen

Kulturelles

KLI: Der Euro

Musikszene: „Junge“ (Die Ärzte)

Filmclip: *Lola rennt* (Tom Tykwer)

KLI: Jugend im Netz

Videoecke: Hobbys

Lektüren

Blog Deutsch 101: Frau Schulz hat Geburtstag

Film: *Lola rennt* (Tom Tykwer)

Strukturen

2.1 The accusative case

2.2 The negative article: **kein, keine**

2.3 What would you like? **Ich möchte ...**

2.4 Possessive determiners

2.5 The present tense of stem-vowel changing verbs

2.6 Asking people to do things: the **du**-imperative

Margret Hofheinz-Döring: *Geizhals* (1926), Galerie Brigitte Mauch, Göppingen
Courtesy Galerie Brigitte Mauch

KUNST UND KÜNSTLER

Margret Hofheinz-Döring (1910–94) ist eine deutsche Malerin und Grafikerin der sogenannten Verschollenen[1] Generation, die sich während der Nazizeit und des 2. Weltkriegs nicht weiterentwickeln[2] konnte. Hofheinz-Dörings Themen sind vor allem Menschen und Bilder zur Weltliteratur. Immer wieder stellt sie typisch menschliche Situationen oder Märchenhaftes dar[3]. Dazu gehört auch ihr Bild „Der Geizhals[4]".

Schauen Sie sich das Bild an und beantworten Sie die folgenden Fragen.

1. Welche Farben dominieren: blau, braun, gelb, grau, grün, orange, rot, schwarz, weiß?
2. Welche Farbe dominiert im Zentrum des Bildes? Welche Farbe hat das Gesicht des Geizhalses? Was drücken diese Farben aus: Angst, Eifersucht[5], Habgier[6], Hoffnung, Liebe, Neid[7], Trauer, Treue, Unschuld[8]?
3. Welche Linien dominieren: runde, eckige[9], spitze[10]?
4. Wie sieht der Geizhals aus: dick und rund oder groß und dürr[11]?
5. Wo ist er: in einem Haus oder in einer Höhle[12]?
6. Was macht er: repariert er etwas, räumt er auf oder macht er ein Feuer[13]? Worauf schaut er: auf die Füße oder auf seinen Schatz?
7. Was ist im Hintergrund: Bäume[14] oder Wolken[15]?
8. Welche Gefühle ruft das Bild hervor: Angst, Freude[16], Glück, Habgier, Liebe, Neid, Neugier?

[1]*lost* [2]*continue to develop* [3]stellt dar *portrays* [4]*miser* [5]*jealousy* [6]*avarice* [7]*envy* [8]*innocence* [9]*angular* [10]*pointed* [11]*gaunt* [12]*cave* [13]*fire* [14]*trees* [15]*clouds* [16]*joy*

Situationen

Besitz

Grammatik 2.1–2.2

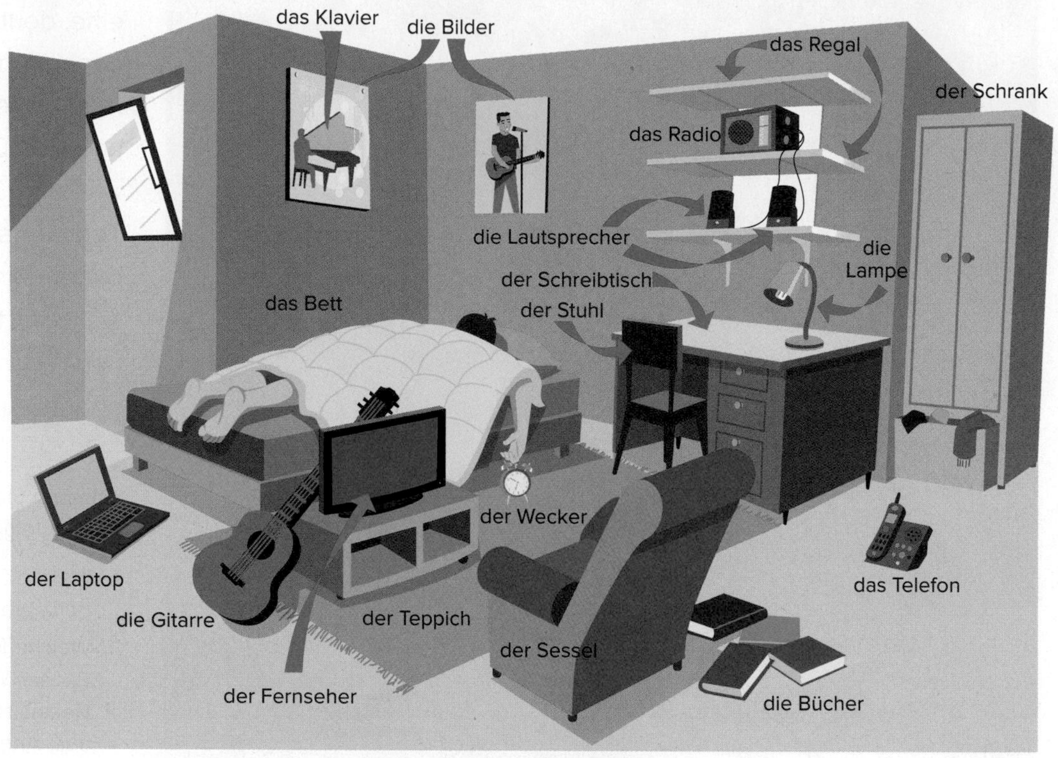

Situation 1 **Hast du einen Schlafsack?**

MODELL: S1: Hast du einen Schlafsack?
S2: Ja, ich habe einen Schlafsack.
Nein, ich habe keinen Schlafsack.

KULTUR ... LANDESKUNDE ... INFORMATIONEN

DER EURO

Fragen Sie Ihren Partner oder Ihre Partnerin.

1. Wie heißt die Währung[1] in dem Land, in dem du geboren bist?
2. Welche Münzen[2] gibt es, z. B. 1-Cent-Münzen, 2-Cent-Münzen?
3. Welche Geldscheine[3] gibt es, z. B. 1-Dollar-Scheine, 2-Dollar-Scheine?
4. Welche Farbe haben die Geldscheine?

Lesen Sie die Fragen und suchen Sie die Informationen im Text.

1. Was ist der Euro?
2. In welchen Ländern der Europäischen Union zahlt[4] man mit dem Euro? Nennen Sie fünf.
3. Welche Euroscheine gibt es? Wie sehen sie aus?
4. Welche Euromünzen gibt es? Wie sehen sie aus?

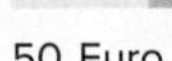

50 Euro

10 Euro

10 cent

2 Euro

10 cent (D)

2 Euro (D)

Den Euro gibt es seit[5] dem 1. Januar 2002. Der Euro ist die gemeinsame Währung[6] der Europäischen Union (EU). Doch nicht alle Länder der EU haben den Euro. Zwölf Länder haben den Euro seit 2002: Belgien, Deutschland, Finnland, Frankreich, Griechenland, Irland, Italien, Luxemburg, die Niederlande, Österreich, Portugal und Spanien. Auch Estland, Lettland, Litauen, Malta, die Slowakei, Slowenien und Zypern gehören zur Eurozone. Manche Länder akzeptieren den Euro, obwohl[7] sie nicht in der Europäischen Union sind. Dazu gehört zum Beispiel Kosovo.

Es gibt Euroscheine und Euromünzen. Euroscheine gibt es zu 5€, 10€, 20€, 50€, 100€, 200€ und 500€. Die Scheine sind in allen Ländern gleich[8]. Alle Scheine haben auf der Rückseite ein Bild von einer Brücke[9]. Euromünzen gibt es zu 1 Cent (ct), 2ct, 5ct, 10ct, 20ct und 50ct. 100 Cent sind 1 Euro. Es gibt auch 1€ und 2€ Münzen. Die Vorderseite[10] zeigt die Länder der Eurozone. Auf der Rückseite hat jedes Land ein anderes Bild.

[1]*currency* [2]*coins* [3]der Schein *bill* [4]*pays* [5]*since* [6]gemeinsame ... *common currency* [7]*although* [8]*the same* [9]*bridge* [10]*front side*

Situation 2 Dialog: Stefan zieht in sein neues Zimmer

Katrin trifft Stefan im Möbelgeschäft.

KATRIN: Hallo, Stefan. Was machst du denn hier?
STEFAN: Ach, ich brauche noch ein paar Sachen. Morgen ziehe ich in ___________.
KATRIN: Was brauchst du denn?
STEFAN: Ach, alles Mögliche.
KATRIN: Was hast du denn schon?
STEFAN: Ich habe einen ___________, eine ___________ und ... und ... und einen ___________.
KATRIN: Das ist aber nicht viel. ___________ hast du denn?
STEFAN: So 30 Dollar.
KATRIN: Ich glaube, du bist im falschen Geschäft. Der Flohmarkt ist viel besser ___________.
STEFAN: Ja, vielleicht hast du recht.

Situation 3 Informationsspiel: Was machen sie morgen?

MODELL: S2: Schreibt Jürgen morgen eine E-Mail?
S1: Nein.
S2: Schreibst du morgen eine E-Mail?
S1: Ja. (Nein.)

	Jürgen	Silvia	mein(e) Partner(in)
1. *schreibt/schreibst … eine E-Mail*	–		
2. *kauft/kaufst … ein Buch*	+		
3. *schaut/schaust … einen Film an*			
4. *ruft/rufst … eine Freundin an*	–	+	
5. *macht/machst … Hausaufgaben*	+		
6. *treibt/treibst … Sport*			
7. *besucht/besuchst … einen Freund*	+	+	
8. *räumt/räumst … das Zimmer auf*	–		

Situation 4 Interview: Besitz

1. Was hast du in deinem Zimmer? Was möchtest du haben?
2. Hast du wertvolle Sachen? Auto, Laptop, Fernseher, iPad, Smartphone? Was möchtest du haben?
3. Hast du ein Pferd, einen Hund oder eine Katze? Möchtest du ein Pferd, einen Hund oder eine Katze haben?

Geschenke

Grammatik 2.3

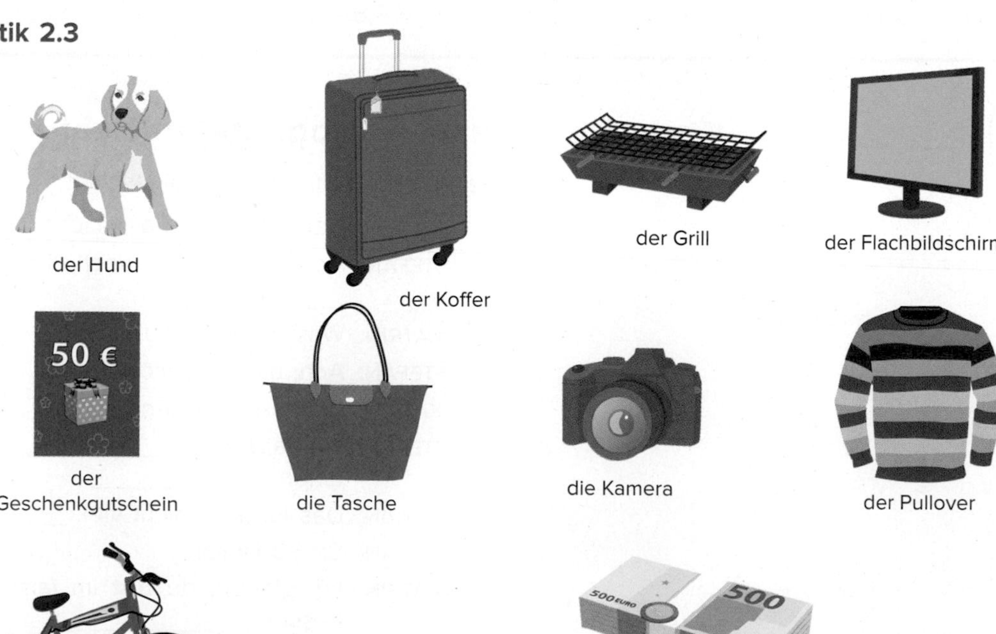

der Hund

der Koffer

der Grill

der Flachbildschirm

der Geschenkgutschein

die Tasche

die Kamera

der Pullover

das Fahrrad

das Geld

Situation 5 Welche Wörter gehören zusammen?

MODELL: S1: Welches Wort passt zu „zu Hause schlafen"?
S2: Das Bett.
S1: Ja, das passt!

1. zu Hause schlafen
2. die Haare föhnen
3. wandern
4. eine Reise machen
5. Musik hören
6. im Nationalpark schlafen
7. Würstchen braten
8. sitzen
9. Hosen aufhängen
10. aufwachen

a. der Koffer
b. das Zelt
c. der Haartrockner
d. das Bett
e. der Rucksack
f. der Schrank
g. der Sessel
h. der Wecker
i. die Lautsprecher
j. der Grill

Situation 6 Dialog: Ein Geschenk für Josef

Melanie trifft Claire in der Mensa.

MELANIE: Josef hat nächsten Donnerstag ________.
CLAIRE: Wirklich? Dann brauche ich ja noch ein ________ für ihn. Mensch, das ist schwierig. Hat er denn Hobbys?
MELANIE: Er ________ Gitarre und ________ gern Musik.
CLAIRE: Hast du schon ein Geschenk?
MELANIE: Ich ________ ein Songbuch kaufen. Aber es ist ziemlich ________. Kaufen wir es zusammen?
CLAIRE: Ja, klar. Welche Art Musik hat er denn ________?
MELANIE: Ich glaube, Soft-Rock und Oldies. Elton John, Céline Dion und so.

Situation 7 Zum Schreiben: Eine Einladung

Schreiben Sie eine Einladung zu einer Party. Benutzen Sie das Modell unten und Ihre Phantasie!

CALIGULA* PARTY
Wann: Mittwoch den 11. Juni - ab 20 Uhr
Wo: Ludwig-Thoma-Heim, Neubau, 5. Stock
Wie: Im Kostüm der Epoche, mit eigenem Kissen, um darauf zu ruhen.
B.D.E.A. (Bring Deinen Eigenen Alkohol)
* Der wahnsinnige römische Kaiser

Danke für die Einladung

Situation 8 Rollenspiel: Am Telefon

S1: Sie rufen einen Freund / eine Freundin an. Sie machen am Samstag eine Party. Laden Sie Ihren Freund / Ihre Freundin ein.

Lektüre

Vor dem Lesen

A. Frau Schulz hat bald Geburtstag. Die Studenten wollen ihr ein Geschenk machen. Im Internet diskutieren sie über ihre Ideen. Welche Geschenke kann man einer Professorin machen, wenn sie Geburtstag hat? Kreuzen Sie an.

- ☐ Blumen
- ☐ einen Rucksack
- ☐ ein Foto der Klasse
- ☐ ein Snowboard
- ☐ eine Kinokarte[1]
- ☐ Schmuck
- ☐ eine Reise nach Europa
- ☐ ein Fahrrad
- ☐ einen deutschen Film auf DVD
- ☐ einen Kuchen[2]
- ☐ einen Laptop
- ☐ ein Buch
- ☐ ein Klavier
- ☐ sie zu einer Party einladen

[1]*movie ticket* [2]*cake*

B. Suchen Sie die Wörter von A, oben, im Text. Unterstreichen Sie die Wörter, die Sie finden.

Blog Deutsch 101: Frau Schulz hat Geburtstag

Pinnwand	Etwas schreiben	Alle ansehen
Zeige 16 von 16 Einträgen		

Heidi schrieb
am Donnerstag um 14:36 Uhr

Frau Schulz hat morgen Geburtstag. Wir sollten ihr eine Freude machen. Hat jemand eine Idee?

Monika schrieb
am Donnerstag um 14:48 Uhr

Wir können einen Kuchen backen. Sie mag sehr gern Schokolade.

Thomas schrieb
am Donnerstag um 15:01 Uhr

Ich wusste gar nicht, dass Frau Schulz Geburtstag hat. Wir sollten sie überraschen!

Heidi schrieb
am Donnerstag um 15:05 Uhr

Aber womit? Sollen wir ihr etwas kaufen? Ein Buch? Oder eine Kinokarte?

Nora schrieb
am Donnerstag um 17:10 Uhr

Nein, ein Buch ist langweilig! Wir können selbst etwas machen. Ein Foto von uns allen – dann vergisst sie uns nicht.

Peter schrieb
am Donnerstag um 17:18 Uhr

Ich habe ein Bild von unserer letzten Kursfahrt. Wir sitzen alle am Tisch im Restaurant und Stefan ist ganz nass, weil die Kellnerin die Limonade verschüttet hat :)

Nora schrieb
am Donnerstag um 17:20 Uhr

Haha, daran erinnere ich mich noch sehr gut.

Katrin schrieb
am Donnerstag um 17:27 Uhr

Ich finde beide Ideen gut – den Kuchen und das Bild. Ich habe noch einen schönen Bilderrahmen.

Stefan schrieb
am Donnerstag um 17:32 Uhr

Das mit der Limonade war nicht lustig! :P Alles hat geklebt! ... Wir können alle gemeinsam etwas unternehmen. Wie alt wird Frau Schulz?

Katrin schrieb
am Donnerstag um 17:41 Uhr

Über das Alter einer Frau spricht man nicht! Wir können abends bowlen gehen. Oder wie Heidi gesagt hat: Wir schenken ihr eine Kinokarte und gehen gemeinsam mit ihr.

Nora schrieb
am Donnerstag um 17:47 Uhr

Au ja, Bowlen wäre super! Im Kino kann man nicht reden.

Thomas schrieb
am Donnerstag um 17:49 Uhr

Wir geben ihr morgen früh nur das Bild und schreiben dazu: „Treffen Sie uns heute Abend um 20 Uhr bei der Bowlingbahn!"

Heidi schrieb
am Donnerstag um 17:55 Uhr

Ja, das ist eine tolle Idee. Peter bringt das Bild mit. Katrin, bringst du bitte den Bilderrahmen mit? Monika, es wäre toll, wenn du einen Schokoladenkuchen backst und abends mitbringst.

Katrin schrieb
am Donnerstag um 17:58 Uhr

Mache ich!

Monika schrieb
am Donnerstag um 17:59 Uhr

Ja, das ist kein Problem. Der wird lecker! Hat jemand etwas von Albert gehört? Weiß er Bescheid?

Albert schrieb
am Donnerstag um 22:45 Uhr

Tut mir leid, dass ich das jetzt erst lese. Ich bin morgen auch dabei!

Arbeit mit dem Text

Beantworten Sie die Fragen.

1. Wann hat Frau Schulz Geburtstag?
2. Wer wusste nicht[3], dass Frau Schulz Geburtstag hat?
3. Welche Art[4] von Kuchen mag Frau Schulz gern?
4. Was für eine Idee hat Nora? / Was möchte Nora schenken?
5. Wo waren die Studenten mit Frau Schulz auf der letzten Kursfahrt?
6. Wer hat noch einen Bilderrahmen[5]?
7. Welche Idee hat Stefan?
8. Wie alt wird Frau Schulz?
9. Was möchten die Studierenden am Abend machen?
10. Wer backt den Kuchen?
11. Kommt Albert auch mit?

Nach dem Lesen

Fragen Sie einen Partner oder eine Partnerin. Schreiben Sie die Antworten auf.

1. Wie bleibst du mit deinen Freunden in Kontakt?
2. Wann hast du Geburtstag?
3. Was möchtest du zum Geburtstag?
4. Was machst du gern, wenn du mit Freunden weggehst? Wohin geht ihr?

[3]wusste ... *did not know* [4]*kind* [5]*picture frame*

Kleidung und Aussehen

Grammatik 2.4

Silvia

ESKE: Wie findest du ihren Haarschnitt?
DAMLA: Sieht gut aus!

Rolf

DAMLA: Wie findest du seinen Bademantel?
ESKE: Nicht schlecht!

Melanie

CLAIRE: Wie findest du ihr Nachthemd?
JOSEF: Klasse!

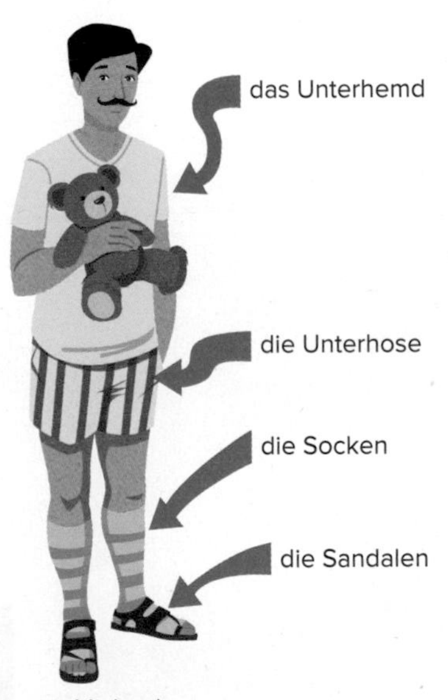

Michael

JUTTA: Na, wie findest du seine Socken?
JENS: Hässlich!

Situation 9 Veronika Frisch-Okonkwo

Welche Wörter passen in die Lücken?

Sandalen (*pl.*) **Armband** (*n.*) **Bluse** (*f.*) **Ohrringe** (*pl.*) **Haarschnitt** (*m.*) **Schal** (*m.*) **Sonnenbrille** (*f.*) **Handschuhe** (*pl.*) **Sporthose** (*f.*)

Kobe Okonkwo erzählt: „Meine Frau Veronika ist immer gut gekleidet. Meistens trägt sie einen Rock und eine __________. Sie trägt gerne Schmuck. Ihre __________ passen zu ihrer Halskette und ihr __________ ist aus Leder. Sie hat immer den neuesten __________. Im Winter trägt sie __________ aus Leder und einen __________ aus Seide. Im Sommer hat sie oft eine __________ auf der Nase, auch bei schlechtem Wetter. Selbst zu Hause ist sie immer modisch gekleidet. Wenn sie im Garten arbeitet, trägt sie eine __________ und ihre Füße stecken in schicken __________."

Situation 10 Interaktion: Wie findest du meine Sportschuhe?

1. Kreuzen Sie an, was Sie heute tragen.
2. Fragen Sie, wie Ihr Partner / Ihre Partnerin das findet.

MODELL: S1: Wie findest du meine Schuhe?
S2: Deine Schuhe? Nicht schlecht.

echt stark **klasse** **super** **schick** **hübsch** **krass** **voll süß** **grell** **Finde ich ganz toll!** **Sieht/Sehen gut aus.** **Steht/Stehen dir gut!**

	Was Sie heute tragen	Wie Ihr(e) Partner(in) das findet
meine Hose	☐	
meine Schuhe	☐	
meinen Schal	☐	
meinen Gürtel	☐	
mein Armband	☐	
meine Halskette	☐	
meinen Ohrring / meine Ohrringe	☐	

Miniwörterbuch

erfolgreich	successful
das **Schlagzeug**	drums
was gefällt Ihren Eltern	what do your parents like
an Ihnen	about you
sollen	are supposed to
die **Nachbarn**	neighbors
die **Löcher**	holes
stören	to bother
an dem Jungen	about the boy

MUSIKSZENE

„Junge" (2007, Deutschland) *Die Ärzte*

Biografie *Die Ärzte* sind eine der erfolgreichsten deutschen Punkrock-Bands. Es gibt sie seit 1982. Sie kommen aus Berlin. Farin Urlaub spielt Gitarre, Rod spielt Bass und Bela B. spielt Schlagzeug. Sie singen alle drei. Ihr Hit „Junge" stammt aus dem Jahre 2007.

Die Ärzte

NOTE: For copyright reasons, the songs referenced in **MUSIKSZENE** have not been provided by the publisher. The song can be found online at various sites such as YouTube, Amazon, or the iTunes store.

Vor dem Hören Was gefällt Ihren Eltern an Ihnen? Was gefällt Ihren Eltern nicht?

Nach dem Hören

1. Was fragt die Mutter den Jungen?
   - ☐ **a.** Junge, warum hast du nichts gelernt?
   - ☐ **b.** Warum gehst du nicht in die Stadt?
   - ☐ **c.** Was sollen die Nachbarn sagen?
   - ☐ **d.** Was sollen die Mitbewohner sagen?
2. Wie sieht der Junge aus?
   - ☐ **a.** Er hat Löcher in der Hose.
   - ☐ **b.** Er trägt kaputte Schuhe.
   - ☐ **c.** Seine Haare sind schwarz.
   - ☐ **d.** Er hat Löcher in der Nase.
3. Was stört die Eltern an dem Jungen?
   - ☐ **a.** Er hört laute Musik.
   - ☐ **b.** Er ist den ganzen Tag zu Hause.
   - ☐ **c.** Er hat lange Haare.
   - ☐ **d.** Seine Freunde nehmen Drogen.

Situation 11 **Frau Gretters neuer Mantel**

Bringen Sie die Sätze in die richtige Reihenfolge.

_______ Von Kaufland. Er ist wirklich sehr schön.
_______ Finde ich ganz toll. Woher haben Sie ihn?
_______ Guten Tag, Frau Körner.
_______ Ach, mein Mantel ist auch schon so alt. Ich brauche dringend etwas für den Winter.
_______ Guten Tag, Frau Gretter. Wie geht's denn so?
_______ Gehen Sie doch auch mal zu Kaufland. Da gibt es gute Preise.
_______ Danke, ganz gut. Wie finden Sie denn meinen neuen Mantel?

Filmlektüre

Lola rennt

Vor dem Lesen

A. Schauen Sie auf das Foto und die Filmangaben und beantworten Sie die folgenden Fragen.

1. Was macht Lola?
2. Welche Haarfarbe hat Lola? Was kann das bedeuten?
3. Wer sind der Schauspieler und die Schauspielerin in den Hauptrollen?

FILMANGABEN

Titel: *Lola rennt*
Genre: Spielfilm
Erscheinungsjahr: 1998
Land: Deutschland
Dauer: 81 min
Regisseur: Tom Tykwer
Hauptrollen: Franka Potente, Moritz Bleibtreu

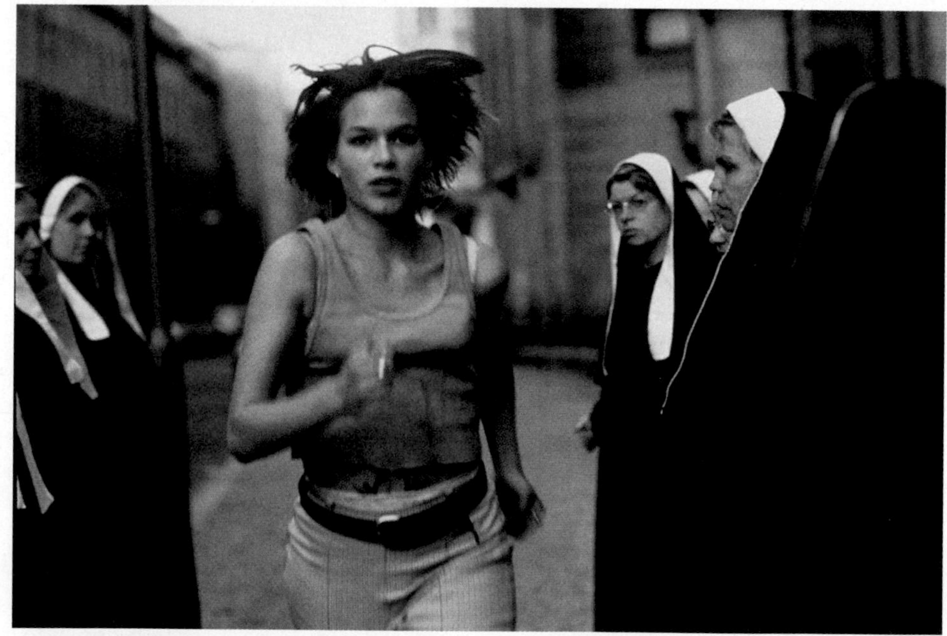

Lola hat es sehr eilig.

B. Lesen Sie die Wörter im Miniwörterbuch. Suchen Sie sie im Text und unterstreichen Sie sie.

Miniwörterbuch

der **Geldbote**	money courier
die **U-Bahn**	subway
beschaffen	to find, raise
umbringen	to kill
überfallen	to rob
verfolgen	to chase
erschießen	to kill
von vorne	from the beginning
ausrauben	to rob
überqueren	to cross
der **Krankenwagen**	ambulance
überfahren	to run over
sterben, stirbt	to die, dies

Inhaltsangabe

Manni (Moritz Bleibtreu) arbeitet als Geldbote für die Mafia. Er lässt 100.000 DM in der U-Bahn liegen. Er hat 20 Minuten Zeit, um 100.000 DM zu beschaffen. Wenn nicht, bringt ihn sein Boss, Ronnie, um. Lola (Franka Potente) ist die Freundin von Manni und versucht, ihm zu helfen. Sie rennt los.

Zuerst rennt Lola zur Bank ihres Vaters und bittet ihn um Geld. Als er ihr nicht helfen kann, überfallen Lola und Manni einen Supermarkt. Sie laufen weg. Die Polizei verfolgt sie und erschießt Lola.

Der Film beginnt von vorne. Lola rennt wieder zur Bank ihres Vaters. Diesmal raubt sie die Bank aus und bringt Manni die 100.000 DM. Manni sieht Lola auf der anderen Seite der Straße. Er überquert die Straße. Ein Krankenwagen überfährt ihn und er stirbt.

Der Film beginnt ein drittes Mal. Lola rennt zur Bank, aber ihr Vater ist nicht da. Sie rennt zu einem Casino und gewinnt beim Roulette 100.000 DM. Sie rennt zurück zu Manni. Manni hat aber die 100.000 DM wieder gefunden und sie seinem Boss gegeben. Jetzt sind sie reich.

Arbeit mit dem Text

Beantworten Sie die folgenden Fragen.

1. Für wen arbeitet Manni?
2. Wo vergisst Manni das Geld?
3. Wie viel Zeit bleibt Lola und Manni, um 100.000 DM zu finden?
4. Wen bittet Lola zuerst um[1] Geld?
5. Was passiert[2], nachdem Lola und Manni den Supermarkt überfallen?
6. Was passiert, nachdem Lola eine Bank ausraubt?
7. Wie bekommen Lola und Manni zum Schluss[3] das Geld?

FILMCLIP

NOTE: For copyright reasons, the films referenced in the **FILMCLIP** feature have not been provided by the publisher. The film can be purchased as a DVD or found online at various sites such as YouTube, Amazon, or the iTunes store. The time codes mentioned below are for the North American DVD version of the film.

Szene: DVD Kapitel 5 „Ronnie", 8:45–11:00 Min.

Manni erzählt Lola am Telefon, dass er das Geld für Ronnie in der U-Bahn vergessen hat. Er ist verzweifelt[4]. Wenn er nicht in 20 Minuten 100.000 DM beschafft, bringt Ronnie ihn um.

Schauen Sie sich die Szene an und beantworten Sie die Fragen. Es sind mehrere Antworten möglich.

1. Wo ist Manni?
   - ☐ a. in der U-Bahn
   - ☐ b. in einer Telefonzelle[5]
   - ☐ c. in der Innenstadt[6]
2. Was hat Manni bei sich?
   - ☐ a. eine Pistole
   - ☐ b. eine Tasche mit Geld
   - ☐ c. ein Handy
3. Was hat Manni vor[7]?
   - ☐ a. Er geht zurück zur U-Bahn.
   - ☐ b. Er überfällt einen Supermarkt.
   - ☐ c. Er weiß es nicht.
4. Was verspricht[8] Lola?
   - ☐ a. Sie holt Hilfe[9].
   - ☐ b. Sie ist in 20 Minuten da.
   - ☐ c. Sie kann das Geld beschaffen.
5. Wie lange wartet Manni auf Lola?
   - ☐ a. bis sie kommt
   - ☐ b. bis sie das Geld hat
   - ☐ c. bis um 12 Uhr

[1]bittet um *asks for* [2]*happens* [3]zum ... *in the end* [4]*desperate* [5]*telephone booth* [6]in ... *downtown* [7]hat vor *is planning* [8]*does promise* [9]*help*

Nach dem Lesen

Suchen Sie nach Informationen über die Schauspielerin Franka Potente im Internet.

1. Wann und wo ist Franka Potente geboren?
2. Wie heißt ihr erster Film?
3. In welchen anderen Filmen spielt sie mit?
4. Franka Potente ist nicht nur Schauspielerin. Womit ist sie noch erfolgreich?

Situation 12 **Flohmarkt**

Schreiben Sie fünf Sachen auf, die Sie verkaufen. Schreiben Sie auf, wer sie kauft und wie viel sie kosten.

MODELL: S1: Ich verkaufe meine Ohrringe. Brauchst du Ohrringe?
S2: Nein danke, ich brauche keine Ohrringe. *oder* Zeig mal. Ja, ich finde deine Ohrringe toll. Was kosten sie?
S1: 2 Euro.
S2: Gut, ich nehme sie.

	ZU VERKAUFEN	KÄUFER/KÄUFERIN	PREIS
1.	______	______	______
2.	______	______	______
3.	______	______	______
4.	______	______	______
5.	______	______	______

Vergnügen

Grammatik 2.5–2.6

Herr Wagner schläft gern.

Jens fährt gern Motorrad.

Sofie trägt gern Hosen.

Melanie lädt gern Freunde ein.

Mehmet läuft gern im Wald.

Ernst isst gern Eis.

Hans liest gern Bücher.

Sumita sieht gern fern.

Situation 13 Interview: Was machst du lieber?

MODELL: S1: Schwimmst du lieber im Meer oder lieber im Schwimmbad?
S2: Lieber im Meer.

1. Isst du lieber zu Hause oder lieber im Restaurant?
2. Spielst du lieber Volleyball oder lieber Basketball?
3. Fährst du lieber Fahrrad oder lieber Motorrad?
4. Schreibst du lieber E-Mails oder lieber Briefe?
5. Liest du lieber online oder lieber auf Papier?
6. Lädst du lieber Freunde oder lieber Verwandte ein?
7. Läufst du lieber im Wald oder lieber in der Stadt?
8. Schläfst du lieber im Hotel oder lieber im Zelt?

Situation 14 Probleme, Probleme

Peter spricht mit Heidi über seine Probleme. Heidi sagt ihm, was er machen soll.

MODELL: PETER: Ich vergesse alles.
HEIDI: Schreib es doch auf.

1. Ich vergesse alles.
2. Ich sehe den ganzen Tag fern.
3. Ich arbeite zu viel.
4. Ich bin zu dick.
5. Ich trinke zu viel Kaffee.
6. Ich esse zu viel Eis.
7. Mein Pullover ist alt.
8. Ich koche nicht gern Italienisch.
9. Das Wochenende ist langweilig.
10. Ich fahre nicht gern Auto.

a. Treib doch Sport!
b. Trink doch Cola!
c. Lies doch ein Buch!
d. Mach doch eine Pause!
e. Schreib es doch auf!
f. Fahr doch Fahrrad!
g. Iss lieber Joghurt!
h. Lade doch deine Freunde ein!
i. Kauf doch einen neuen Pullover!
j. Koch doch Chinesisch.

Situation 15 Informationsspiel: Was machen sie gern?

MODELL: S2: Was fährt Richard gern?
S1: Motorrad.
S2: Was fährst du gern?
S1: _______

	Richard	Josef und Melanie	mein(e) Partner(in)
fahren	Motorrad		
tragen		Jeans	
essen	Wiener Schnitzel		
sehen	Fußball		
vergessen		ihr Alter	
waschen	sein Auto		
treffen	seine Freundin		
einladen	seinen Bruder		
sprechen		Englisch	

KULTUR ... LANDESKUNDE ... INFORMATIONEN

JUGEND IM NETZ

Was ist im Internet für Sie am wichtigsten[1]?

Lesen Sie in der Grafik zuerst, was deutsche Teenager im Internet machen. Beantworten Sie dann die Fragen.

1. Womit verbringen[2] deutsche Jugendliche im Netz die meiste Zeit: Kommunikation, Unterhaltung, Spiele oder Informationssuche?
2. Womit verbringen mehr Jugendliche ihre Zeit: mit E-Mails oder mit Chatten?
3. Womit verbringen mehr Jugendliche ihre Zeit: mit Musik hören oder mit Videos gucken?

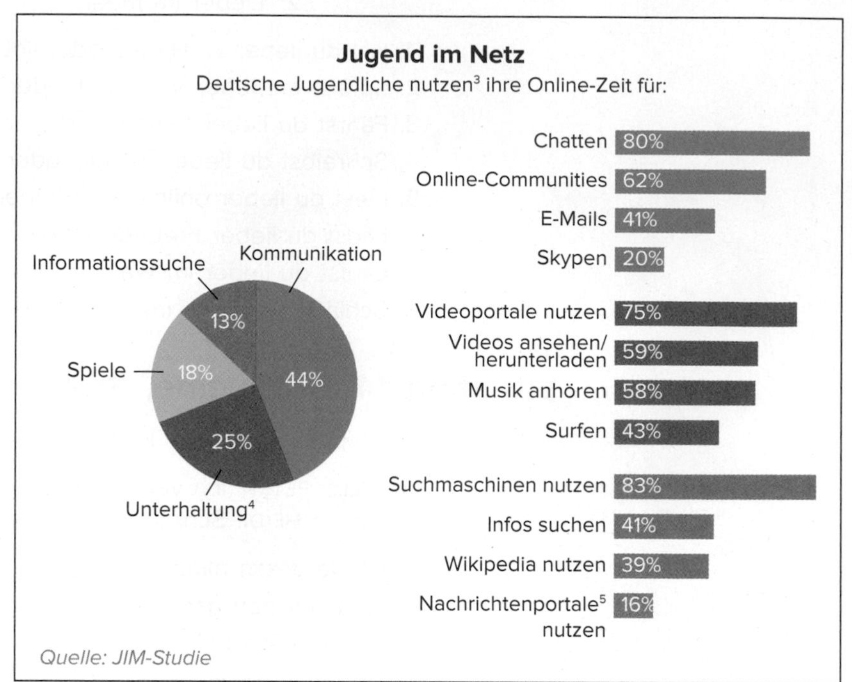

[1]*am ... most important* [2]*spend* [3]*use* [4]*entertainment* [5]*news portals*

Situation 16 Bildgeschichte: Ein Tag in Silvias Leben

Videoecke

Perspektiven

Welches elektronische Gerät[6] ist für dich am wichtigsten[7]? Warum?

Am wichtigsten ist mir mein Laptop.

Aufgabe 1 Hintergrund°

°*background*

Schauen Sie sich den Clip an. Was sehen Sie hinter[8] diesen Personen?

1. Judith ___
2. Sandra ___
3. Tina ___
4. Susan ___

5. Pascal ___
6. Martin ___
7. Felicitas ___
8. Albrecht ___

a. Ein Boot fährt vorbei.
b. Wir sehen ein schönes historisches Gebäude[9] auf einem großen Platz.
c. Zwei junge Leute schieben ihre Fahrräder vorbei[10].
d. Wir sehen schöne Blumen[11].
e. Ein Taxi fährt vorbei.
f. Ein paar Leute stehen unter einer Laterne.
g. Wir sehen ein Café.
h. Wir sehen eine Statue mit der Inschrift[12] „Johann Sebastian Bach".

[6]*device* [7]am ... *most important* [8]*behind* [9]*building* [10]ihre Fahrräder vorbeischieben *walk their bicycles past* [11]*flowers* [12]*inscription*

Aufgabe 2 Gründe°

°*reasons*

Warum sind diese elektronischen Geräte so wichtig? Ordnen Sie die Geräte und Gründe den Personen zu.

MODELL: Judith → ihr Laptop → Sie braucht ihn für Unisachen und das Internet.

PERSON	GERÄT	GRUND
1. Judith	sein Gitarrenverstärker[13]	Er kann ohne ihn keine Musik machen.
2. Sandra	sein Handy	Er schreibt seine Dissertation und schaut Filme an.
3. Tina	ihr Laptop	Er trägt es den ganzen Tag herum.
4. Susan	ihr MP3-Player	Sie braucht ihn für Unisachen und das Internet.
5. Pascal	ihr Radio	Sie hört gern Musik.
6. Martin	ihr Telefon	Sie hört gern Musik.
7. Felicitas	ihr Handy	Sie möchte immer erreichbar[14] sein.
8. Albrecht	sein Laptop	Sie ruft gern ihre Freunde und ihre Schwestern an.

Interviews

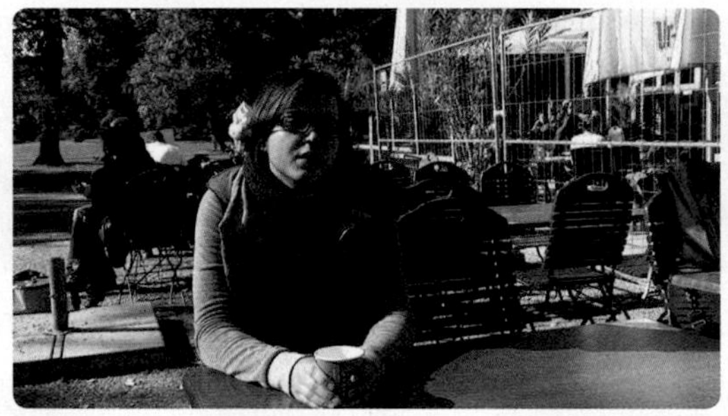

Maria

Simone

Aufgabe 3 Hobbys

- Was für Hobbys hast du?
- Welche elektronischen Geräte hast du?
- Bist du bei Facebook oder einem anderen sozialen Netzwerk?
- Was machst du damit?
- Was hältst du davon?
- Was machst du sonst? Twitterst du? Chattest du?
- Trägst du gern Schmuck?
- Hast du ein besonderes Schmuckstück?

Wer ist das, Maria, Simone oder beide?

	Maria	Simone	Beide
1. Sie geht gern schwimmen und macht Hula-Hoop.	☐	☐	☐
2. Sie macht ganz viel Sport.	☐	☐	☐
3. Sie hat ein Notebook, einen I-Pod und ein Handy.	☐	☐	☐
4. Sie hat einen Laptop und eine Stereoanlage.	☐	☐	☐
5. Sie ist bei Facebook.	☐	☐	☐
6. Sie chattet über Skype.	☐	☐	☐
7. Sie schreibt nur E-Mails.	☐	☐	☐
8. Sie trägt keinen Schmuck.	☐	☐	☐
9. Sie trägt gern Ketten und Ringe.	☐	☐	☐

Aufgabe 4 Interview

Interviewen Sie eine Partnerin oder einen Partner. Stellen Sie dieselben Fragen.

[13] guitar amplifier [14] *reachable*

Wortschatz

Besitz und Geschenke — Possessions and Gifts

der **Fernseher, -**	TV set
der **Flachbildschirm, -e**	flat-screen (monitor)
der **Geschenkgutschein, -e**	gift certificate
der **Haartrockner, -**	hair dryer
der **Lautsprecher, -**	loudspeaker
der **Rucksack, ¨-e**	backpack
der **Schallplattenspieler, -**	record player
der **Schlafsack, ¨-e**	sleeping bag
der **Wecker, -**	alarm clock
das **Bild, -er**	picture
das **Boot, -e**	boat
das **Fahrrad, ¨-er**	bicycle
das **Geld**	money
das **Geschenk, -e**	gift
das **Handy, -s [hɛndi]**	cell phone
das **Pferd, -e**	horse
das **Surfbrett, -er**	surfboard
das **Zelt, -e**	tent

Ähnliche Wörter

die **Kamera, -s;** der **CD-Spieler, -;** der **Computer, -;** der **DVD-Spieler, -;** der **Film, -e;** der **Grill, -s;** der **Laptop, -s** (R); der **MP3-Spieler, -;** das **Buch, ¨-er** (R); das **Radio, -s;** das **Smartphone, -s;** das **Songbuch, ¨-er;** das **Wörterbuch, ¨-er**

Haus und Wohnung — Home and Apartment

der **Schrank, ¨-e**	wardrobe cabinet, cupboard
der **Schreibtisch, -e**	desk
der **Sessel, -**	armchair
der **Stuhl, ¨-e** (R)	chair
der **Teppich, -e**	carpet
das **Klavier, -e**	piano
das **Regal, -e**	bookshelf, bookcase
das **Zimmer, -** (R)	room

Ähnliche Wörter

die **Katze, -n;** der **Hund, -e;** das **Haus, ¨-er** (R); das **Telefon, -e** (R)

Kleidung und Schmuck — Clothes and Jewelry

die **Halskette, -n**	necklace
die **Seide, -n**	silk
aus Seide	of/from silk
die **Sonnenbrille, -n** (R)	sunglasses
die **Sporthose, -n**	tights, sports pants
die **Unterhose, -n**	underpants
der **Bademantel, ¨-**	bathrobe
der **Gürtel, -**	belt
der **Haarschnitt, -e**	haircut
der **Handschuh, -e**	glove
der **Schmuck**	jewelry
der **Wanderschuh, -e**	hiking shoe
das **Armband, ¨-er**	bracelet
das **Leder, -**	leather
das **Nachthemd, -en**	nightshirt
das **Unterhemd, -en**	undershirt

Ähnliche Wörter

die **Bluse, -n** (R); die **Sandale, -n;** die **Socke, -n;** der **Pulli, -s;** der **Pullover, -;** der **Ring, -e; der Ohrring, -e** (R); der **Schal, -s;** das **Piercing;** das **T-Shirt, -s;** die **Jeans** (*pl.*)

Sonstige Substantive — Other Nouns

die **Art, -en**	kind, type
die **Einladung, -en**	invitation
die **Hoffnung, -en**	hope
die **Mensa, Mensen**	student cafeteria
die **Mitbewohnerin, -nen**	female roommate, housemate
die **Reihenfolge, -n**	order, sequence
die **Sache, -n**	thing
die **Stadt, ¨-e**	city
die **Stunde, -n** (R)	hour
die **Tasse, -n**	cup
die **Trauer**	sorrow
die **Zeitung, -en**	newspaper
der **Gruselfilm, -e**	horror film
der **Mensch, -en**	person
Mensch!	Man! Oh boy! (*coll.*)
der **Mitbewohner, -**	male roommate, housemate
der **Wald, ¨-er**	forest, woods
im Wald laufen	to run in the woods
das **Frühstück, -e**	breakfast
das **Geschäft, -e**	store
das **Studentenheim, -e**	dorm
das **Vergnügen**	pleasure
das **Würstchen, -**	frank(furter); hot dog
die **Verwandten** (*pl.*)	relatives

Ähnliche Wörter

die **E-Mail [i:me:l], -s;** die **Geburtstagskarte, -n;** die **Karte, -n** (R); die **Nase, -n** (R); die **Party, -s** (R); die **Pizza, Pizzen;** die **Postkarte, -n;** die **Telefonkarte, -n;** der **Basketball, ¨-e;** der **Bus, -se;** der **Flohmarkt, ¨-e;** der **Geburtstag, -e** (R); der **Kilometer, -;** der **Nationalpark, -s;** das **Bier, -e;** das **Ding, -e; das Eis;** das **Hotel, -s;** das **Restaurant, -s**

Verben	Verbs
an·schauen	to look at
auf·hängen	to hang up
auf·wachen	to wake up
aus·sehen, sieht ... aus	to look
es sieht gut aus	it looks good
braten, brät	to grill, fry
ein·laden, lädt ... ein	to invite
essen, isst	to eat
fahren, fährt	to drive, ride
föhnen	to blow dry
die Haare föhnen	to blow one's dry hair
glauben	to believe
klingeln	to ring
laufen, läuft (R)	to run
möchte	would like
passen	to match, go with
recht haben	to be right
schicken	to send
schlafen, schläft	to sleep
sitzen	to sit, be in a seated position
Sport treiben	to do sports
stehen	to stand
Das steht / Die stehen dir gut!	That looks / Those look good on you!
treffen, trifft	to meet
treffen wir uns ...	let's meet . . .
verkaufen	to sell
wissen, weiß	to know
ziehen	to move

Ähnliche Wörter

bringen; finden; kosten; sehen, sieht; vergessen, vergisst; waschen, wäscht

Adjektive und Adverbien	Adjectives and Adverbs
billig	cheap, inexpensive
dick	large, fat
dringend	urgent(ly)
echt	real(ly)
einfach	simple, simply
falsch	wrong
ganz	whole; *here:* quite
grell	gaudy, shrill; *here:* cool, neat
hässlich	ugly
hübsch (R)	pretty
langweilig	boring
modisch	fashionable
richtig	right, correct
schlecht	bad
schwierig	difficult
selbst	even; oneself
sonst	otherwise
süß	sweet
voll süß	totally sweet
teuer	expensive
toll	neat, great
wertvoll	valuable, expensive
wichtig	important
ziemlich	rather
ziemlich groß	pretty big
zu	too

Ähnliche Wörter

besser; schick

Possessivartikel	Possessive Determiners
dein, deine, deinen	your (*informal sg.*)
euer, eure, euren	your (*informal pl.*)
ihr, ihre, ihren	her, its; their
Ihr, Ihre, Ihren	your (*formal*)
mein, meine, meinen	my
sein, seine, seinen	his, its
unser, unsere, unseren	our

Präpositionen	Prepositions
an	at; on; to
am Samstag	on Saturday
am Telefon	on the phone
ans Meer	to the sea
bei	with; at
bei Monika	at Monika's
bis	until
bis acht Uhr	until eight o'clock
für	for
zu	to; for (*an occasion*)
zum Geburtstag	for someone's birthday
zur Uni	to the university

Sonstige Wörter und Ausdrücke	Other Words and Expressions
alles	everything
alles Mögliche	everything possible
da	there
dich	you (*accusative case*)
diese, diesen, dieser, dieses	this; these
ein paar	a few
etwas	something
gut gekleidet sein	to be well dressed
heute Abend	this evening
ihn	him; it (*accusative case*)
kein, keine, keinen	no; none
Klar!	Of course!
lieber	rather
ich gehe lieber ...	I'd rather go . . .
mittags	at noon
morgen	tomorrow
natürlich	naturally
nie	never
niemand	no one, nobody
pro	per
schon	already
vielleicht	perhaps
wenn	if; when
zusammen	together

Strukturen und Übungen

2.1 The accusative case

WISSEN SIE NOCH?

Case indicates the function of a noun in a sentence.

Review grammar A.3.

nominative = subject
accusative = direct object

The nominative case designates the subject of a sentence; the accusative case commonly denotes the object of the action implied by the verb, such as what is being possessed, looked at, or acted on by the subject of the sentence.

Jutta hat einen Wecker. — *Jutta has an alarm clock.*
Jens kauft eine Lampe. — *Jens buys a lamp.*

Here are the nominative and accusative forms of the definite and indefinite articles.

	Tisch (*m.*)	Bett (*n.*)	Lampe (*f.*)	Bücher (*pl.*)
Nominative	der	das	die	die
Accusative	den			
Nominative	ein	ein	eine	–
Accusative	einen			

Note that only the masculine has a different form in the accusative case.

Der Teppich ist schön. — *The rug is beautiful.*
Kaufst du **den** Teppich? — *Are you going to buy the rug?*

Übung 1 Im Kaufhaus

Was kaufen diese Leute? Was kaufen Sie?

MODELL: Jens kauft den Wecker, **das** Regal und **den** DVD-Spieler.

	Jens	Ernst	Melanie	Jutta	ich
der Pullover	–	–	–	+	
der Wecker	+	–	–	–	
die Tasche	–	+	+	–	
das Regal	+	–	+	–	
die Lampe	–	–	–	+	
die Stühle	–	+	–	–	
der DVD-Spieler	+	–	–	+	
der Schreibtisch	–	+	+	–	

Übung 2 Besitz

Was haben Sie?

MODELL: Ich habe einen/eine/ein/_______, ...

das Bett
das Bild / die Bilder
die Bücher
der CD-Spieler
der Fernseher
die Gitarre
der Grill
der Haartrockner
das Klavier
die Lampe / die Lampen
der Laptop
das Radio
das Regal / die Regale
der Schallplattenspieler
der Schrank
der Schreibtisch
der Sessel
das Smartphone
der Stuhl / die Stühle
das Telefon
der Wecker
____

2.2 The negative article: *kein, keine*

Kein and **keine** (*not a, not any, no*) are the negative forms of **ein** and **eine.**

Im Klassenzimmer sind **keine** Fenster. — *There aren't any / are no windows in the classroom.*

Stefan hat **keinen** Schreibtisch. — *Stefan doesn't have a desk.*

ein → **kein**
einen → **keinen**
eine → **keine**
[plural] → **keine**

The negative article has the same endings as the indefinite article **ein.** It also has a plural form: **keine.**

	Teppich (*m.*)	Regal (*n.*)	Uhr (*f.*)	Stühle (*pl.*)
Nominative	ein	ein	eine	–
Accusative	einen			
Nominative	kein	kein	keine	keine
Accusative	keinen			

—Hat Katrin **einen** Schrank? — *Does Katrin have a wardrobe cabinet?*

—Nein, sie hat **keinen** Schrank. — *No, she doesn't have a wardrobe cabinet.*

—Hat Katrin **Bilder** an der Wand? — *Does Katrin have pictures on the wall?*

—Nein, sie hat **keine** Bilder an der Wand. — *No, she has no pictures on the wall.*

Übung 3 Vergleiche°

°*Comparisons*

Wer hat was? Was haben Sie?

MODELL: Albert hat keinen Teppich. Er hat einen Fernseher und eine Gitarre, aber er hat kein Fahrrad. Er hat einen Computer und Bilder, aber er hat kein Smartphone.

	Albert	Heidi	Monika	ich
der Teppich	–	+	–	
der Fernseher	+	–	–	
die Gitarre	+	+	–	
das Fahrrad	–	–	+	
der Computer	+	+	+	
die Bilder	+	–	+	
das Smartphone	–	+	+	

2.3 What would you like? *Ich möchte …*

möchte = *would like*

Use **möchte** (*would like*) to express that you would like to have something. The thing you want is in the accusative case.

Ich möchte **eine Tasse Kaffee,** bitte.	*I'd like a cup of coffee, please.*
Hans möchte **einen Flachbildschirm** zum Geburtstag.	*Hans would like a flat screen for his birthday.*

Möchte is particularly common in polite exchanges, for example in shops or restaurants.

KELLNER: Was möchten Sie?	WAITER: *What would you like?*
GAST: Ich möchte ein Bier.	CUSTOMER: *I'd like a beer.*

Following are the forms of **möchte.** Note that the **er/sie/es-**form does not follow the regular pattern; it does not end in **-t.**

möchte			
ich	möchte	wir	möchten
du	möchtest	ihr	möchtet
Sie	möchten	Sie	möchten
er sie es	möchte	sie	möchten

> **WISSEN SIE NOCH?**
>
> The **Satzklammer** forms a frame or a bracket consisting of the main verb and either a separable prefix or an infinitive.
>
> Review grammar 1.5.

To say that someone would like to do something, use **möchte** with the infinitive of the verb that expresses the action. This infinitive appears at the end of the sentence. Think of the **Satzklammer** used with separable-prefix verbs, and pattern your **möchte** sentences after it. Other verbs similar to **möchte** are explained in **Kapitel 3.**

Peter **möchte** einen Mantel **kaufen**. Sofie **möchte** ein Eis **essen**.

Übung 4 Der Wunschzettel

Was, glauben Sie, möchten diese Personen?

MODELL: Meine beste Freundin möchte einen Ring.

das Auto	der Hund	der Pullover
die Digitalkamera	die Katze	das Radio
das E-Book	der Koffer	der Ring
der Fernseher	der Laptop	der Schallplattenspieler
der Haartrockner	das Motorrad	die Sonnenbrille
die Hose	die Ohrringe	die Sportschuhe

1. Ich _______
2. Mein bester Freund / Meine beste Freundin _______
3. Meine Eltern _______
4. Mein Mitbewohner / Meine Mitbewohnerin und ich _______
5. Mein Nachbar / Meine Nachbarin in der Klasse _______
6. Mein Professor / Meine Professorin _______
7. Mein Bruder / Meine Schwester _______

2.4 Possessive determiners

Use the possessive determiners **mein, dein,** and so forth to express ownership.

—Ist das **dein** Lautsprecher? *Is this your loudspeaker?*
—Nein, das ist nicht **mein** Lautsprecher. *No, that's not my loudspeaker.*

—Ist das Sofies Gitarre? *Is this Sofie's guitar?*
—Ja, das ist **ihre** Gitarre. *Yes, that's her guitar.*

Here are the nominative neuter forms of the possessive determiners.

Singular	Plural
mein Auto (*my car*)	**unser** Auto (*our car*)
dein Auto (*your car*) **Ihr** Auto (*your car*)	**euer** Auto (*your car*) **Ihr** Auto (*your car*)
sein Auto (*his/its car*) **ihr** Auto (*her/its car*)	**ihr** Auto (*their car*)

Just as the personal pronoun **sie** can mean either *she* or *they*, the possessive determiner **ihr** can mean either *her* or *their*. When it is capitalized as **Ihr,** it means *your* and corresponds to the formal **Sie** (*you*).

Note the three forms for English *your:* **dein** (*informal singular*), **euer** (*informal plural*), and **Ihr** (*formal singular* or *plural*).

Albert und Peter, wo sind **eure** Bücher? *Albert and Peter, where are your books?*
Öffnen Sie **Ihre** Bücher auf Seite 133. *Open your books to page 133.*

Possessive determiners have the same endings as the indefinite article **ein.** They agree in case (*nominative* or *accusative*), gender (*masculine, neuter,* or *feminine*), and number (*singular* or *plural*) with the noun that they precede.

Possessive determiners have the same endings as **ein** and **eine.**
ein → **mein**
eine → **meine**
einen → **meinen**
[plural] → **meine**

Mein Pulli ist warm. Möchtest du **meinen** Pulli tragen? *My sweater is warm. Would you like to wear my sweater?*
Josef verkauft **seinen** Computer. *Josef is selling his computer.*

Like **ein,** the forms of possessive determiners are the same in the nominative and accusative cases—except for the masculine singular, which has an **-en** ending in the accusative.

Possessive Determiners Nominative and Accusative Cases

	Ring (m.)	Armband (n.)	Halskette (f.)	Ohrringe (pl.)
my	mein meinen	mein	meine	meine
your	dein deinen	dein	deine	deine
your	Ihr Ihren	Ihr	Ihre	Ihre
his, its	sein seinen	sein	seine	seine
her, its	ihr ihren	ihr	ihre	ihre
our	unser unseren	unser	unsere	unsere
your	euer euren	euer	eure	eure
your	Ihr Ihren	Ihr	Ihre	Ihre
their	ihr ihren	ihr	ihre	ihre

Übung 5 Jan und Jana

Beschreiben Sie Jan und Jana.

Seine Haare sind braun.	Ihre Haare sind schwarz.
______ Augen sind grün.	______ Augen sind braun.
______ Halskette ist lang.	______ Halskette ist ...
______ Schuhe sind schmutzig.	...
______ Gitarre ist alt.	...
______ Zimmer ist groß.	...
______ Fenster ist klein.	...

Übung 6 Minidialoge

WISSEN SIE NOCH?

Use **du (dein)** and **ihr (euer)** to address people whom you know well and whom you address by their first name. Use **Sie (Ihr)** for all other people.

Review grammar A.5.

Ergänzen Sie **dein, euer** oder **Ihr.** Verwenden Sie die richtige Endung.

1. FRAU GRETTER: Wie finden Sie meinen Pullover?
   HERR WAGNER: Ich finde ______ Pullover sehr schön.
2. KOBE: Weißt du, wo meine Brille ist, Veronika?
   VERONIKA: ______ Brille ist auf dem Tisch.
3. AYDAN CANDEMIR: Eske! Damla! Räumt ______ Schuhe auf!
   ESKE UND DAMLA: Ja, gleich, Mama.
4. HERR RUF: Jutta!______ Freundin war da. Sie braucht ihr Buch zurück.
   JUTTA: Ja, gut. Ich nehme es morgen mit in die Schule.
5. HERR SIEBERT: Beißt ______ Hund?
   FRAU KÖRNER: Was glauben Sie denn! Natürlich beißt mein Hund nicht.
6. NORA: Morgen möchte ich zu meinen Eltern fahren.
   PETER: Wo wohnen ______ Eltern?
   NORA: In Santa Cruz.
7. JÜRGEN: Silvia und ich, wir verkaufen unseren Computer.
   ANDREAS: ______ Computer! Der ist so alt, den kauft doch niemand!

Übung 7 Flohmarkt

Sie und die Studenten und Studentinnen in Frau Schulz' Deutschkurs brauchen Geld und organisieren einen Flohmarkt. Schreiben Sie Sätze. Wer verkauft was?

MODELL: Monika verkauft ihre CDs.

Monika	verkaufe	ihr	Computer (der)
Thomas	verkaufen	ihre	Ohrring (der)
ich	verkaufen	ihre	Wörterbuch (das)
Katrin	verkaufen	ihren	DVD-Spieler (der)
Peter und Heidi	verkauft	ihren	CDs (*pl.*)
wir	verkauft	mein	Bücher (*pl.*)
Stefan	verkauft	seine	Gitarre (die)
Nora und Albert	verkauft	seinen	Bilder (*pl.*)
Frau Schulz	verkauft	unsere	Telefon (das)

2.5 The present tense of stem-vowel changing verbs

In some verbs, the stem vowel changes in the **du-** and the **er/sie/es-**forms.

—**Schläfst** du gern?	*Do you like to sleep?*
—Ja, ich **schlafe** sehr gern.	*Yes, I like to sleep very much.*
Ich **lese** viel, aber Ernst **liest** mehr.	*I read a lot, but Ernst reads more.*

These are the types of vowel changes you will encounter.

There are four types of stem vowel changes: **a → ä, au → äu, e → i, e → ie.**

a → ä			
braten:	du brätst*	er/sie/es brät	*to roast*
einladen:	du lädst ... ein	er/sie/es lädt ... ein	*to invite*
fahren:	du fährst	er/sie/es fährt	*to drive*
schlafen:	du schläfst	er/sie/es schläft	*to sleep*
tragen:	du trägst	er/sie/es trägt	*to wear*
waschen:	du wäschst	er/sie/es wäscht	*to wash*
au → äu†			
laufen:	du läufst	er/sie/es läuft	*to run*
e → i			
essen:	du isst‡	er/sie/es isst	*to eat*
geben:	du gibst	er/sie/es gibt	*to give*
sprechen:	du sprichst	er/sie/es spricht	*to speak*
treffen:	du triffst	er/sie/es trifft	*to meet*
vergessen:	du vergisst‡	er/sie/es vergisst	*to forget*
e → ie§			
lesen:	du liest§	er/sie/es liest	*to read*
sehen:	du siehst	er/sie/es sieht	*to see*
fernsehen:	du siehst ... fern	er/sie/es sieht ... fern	*to watch TV*

Jürgen **läuft** jeden Tag 10 Kilometer.	*Jürgen runs 10 kilometers every day.*
Ernst **isst** gern Pizza.	*Ernst likes to eat pizza.*
Michael **sieht** lieber **fern.**	*Michael prefers to watch TV.*

*Recall that verb stems ending in **-d** or **-t** insert an **-e-** before another consonant: **ich arbeite, du arbeitest.** Verb forms that contain a stem vowel change do *not* insert an **-e-**: **du brätst** but **ihr bratet.**

†Recall that **äu** is pronounced as in English *boy.*

‡Recall that verb stems that end in **-s**, **-ß**, **-z**, or **-x** do not add **-st** in the **du-**form, but only **-t**.

§Recall that **ie** is pronounced as in English *niece.*

Übung 8 Minidialoge

Ergänzen Sie das Pronomen.

ACHTUNG!

—Läufst du **gern** in der Stadt?	*Do you like to jog in the city?*
—Nein, ich laufe **lieber** im Wald.	*No, I prefer jogging in the forest.*

1. ROLF: Seht _______[a] gern fern?
   ESKE UND DAMLA: Ja, _______[b] sehen sehr gern fern.
2. FRAU GRETTER: Lesen _______[a] die Zeitung?
   MARIA: Im Moment nicht. _______[b] lese gerade ein Buch.
3. HERR SIEBERT: Isst Ihre Tochter gern Eis?
   HERR RUF: Nein, _______[a] isst lieber Joghurt. Aber da kommt mein Sohn, _______[b] isst sehr gern Eis.
4. SILVIA: Wohin[1] fährst _______[a] im Sommer?
   ANDREAS: _______[b] fahre nach Spanien. Und wohin fahrt _______[c]?
   SILVIA: _______[d] fahren nach England.

Übung 9 Jens und Jutta

Ergänzen Sie das Verb. Verwenden Sie die folgenden Wörter.

essen (3x)
fahren (2x)
lesen
machen (2x)
schlafen
sehen

MICHAEL: Was _______[a] Jutta und Jens gern?
ANDREA: Jutta _______[b] sehr gern Motorrad. Jens _______[c] lieber fern.
MICHAEL: Was essen sie gern? _______[d] Jens gern Chinesisch?
ERNST: Jens _______[e] gern Italienisch, aber nicht Chinesisch. Und Jutta _______[f] gern Fast Food.
MICHAEL: Und ihr, was _______[g] ihr gern?
ANDREA: Ich _______[h] gern Bücher und Ernst _______[i] gern. Und im Winter _______[j] wir gern Snowboard.

Übung 10 Was machen Sie gern?

Sagen Sie, was Sie gern machen, und bilden Sie Fragen.

MODELL: ich/du: Fast Food essen →
Ich esse (nicht) gern Fast Food. Isst du auch (nicht) gern Fast Food?

1. wir/ihr: Deutsch sprechen
2. ich/du: Freunde einladen
3. ich/du: im Wald laufen
4. ich/du: Pullis tragen
5. wir/ihr: fernsehen
6. ich/du: Fahrrad fahren
7. wir/ihr: die Hausaufgabe vergessen
8. ich/du: schlafen
9. wir/ihr: online lesen

[1]*Where*

2.6 Asking people to do things: the *du*-imperative

Use the **du**-imperative when addressing people you normally address with **du,** such as friends, relatives, other students, and the like. It is formed by dropping the **-(s)t** ending from the present-tense **du-**form of the verb. The pronoun **du** is not used.

Drop the **-(s)t** from the **du**-form to get the **du**-imperative.

(du) arbeitest	→	Arbeite!	*Work!*
(du) isst	→	Iss!	*Eat!*
(du) kommst	→	Komm!	*Come!*
(du) öffnest	→	Öffne!	*Open!*
(du) siehst	→	Sieh!	*See!*
(du) tanzt	→	Tanz!	*Dance!*

Verbs whose stem vowel changes from **a(u)** to **ä(u)** drop the umlaut in the **du-**imperative.

(du) fährst	→	Fahr!	*Drive!*
(du) läufst	→	Lauf!	*Run!*

WISSEN SIE NOCH?

To form commands for people you address with **Sie,** invert the subject and verb: **Sie kommen mit.** → **Kommen Sie mit!**

Review grammar A.1.

Imperative sentences always begin with the verb.

Trag mir bitte die Tasche.	*Please carry the bag for me.*
Öffne bitte das Fenster.	*Open the window please.*
Reite nicht so schnell!	*Don't ride so fast!*
Sieh nicht so viel fern!	*Don't watch so much TV!*

Übung 11 Ach, diese Geschwister!

Ihr kleiner Bruder macht alles falsch. Sagen Sie ihm, was er machen soll.

MODELL: Ihr kleiner Bruder isst zu viel. → Iss nicht so viel!

1. Ihr kleiner Bruder schläft den ganzen Tag.
2. Er liegt den ganzen Tag in der Sonne.
3. Er vergisst seine Hausaufgaben.
4. Er liest seine Bücher nicht.
5. Er sieht den ganzen Tag fern.
6. Er trinkt zu viel Cola.
7. Er sitzt den ganzen Tag am Computer.
8. Er trägt seine Brille nicht.
9. Er spielt immer Computerspiele.
10. Er treibt keinen Sport.

Übung 12 Vorschläge°

°*Suggestions*

Machen Sie Ihrem Freund / Ihrer Freundin Vorschläge.

MODELL: nicht zu spät ins Bett / gehen →
Geh nicht zu spät ins Bett!

1. heute ein T-Shirt / tragen
2. keine laute Musik / spielen
3. den Wortschatz / lernen
4. deine Freunde / anrufen
5. nicht allein im Park / laufen
6. nicht zu lange in der Sonne / liegen
7. dein Zimmer / aufräumen
8. heute Abend in einem Restaurant / essen
9. früh / aufstehen

KAPITEL 3

Talente, Pläne, Pflichten

In **Kapitel 3**, you will learn how to describe your talents and those of others. You will learn how to express your intentions and how to talk about obligation and necessity. You will also learn additional ways to describe how you or other people feel.

Themen

Talente und Pläne

Pflichten

Ach, wie nett!

Körperliche und geistige Verfassung

Kulturelles

Musikszene: „Müssen nur wollen" (Wir sind Helden)

KLI: Jugendschutz

Filmclip: *Vincent will Meer* (Ralf Huettner)

KLI: Chatiquette: Sternchen, Abkürzungen und Akronyme

Videoecke: Fähigkeiten und Pflichten

Lektüren

Zeitungsartikel: Ringe fürs Leben zu zweit

Film: *Vincent will Meer* (Ralf Huettner)

Strukturen

3.1 The modal verbs **können, wollen, mögen**

3.2 The modal verbs **müssen, sollen, dürfen**

3.3 Accusative case: personal pronouns

3.4 Word order: dependent clauses

3.5 Dependent clauses and separable-prefix verbs

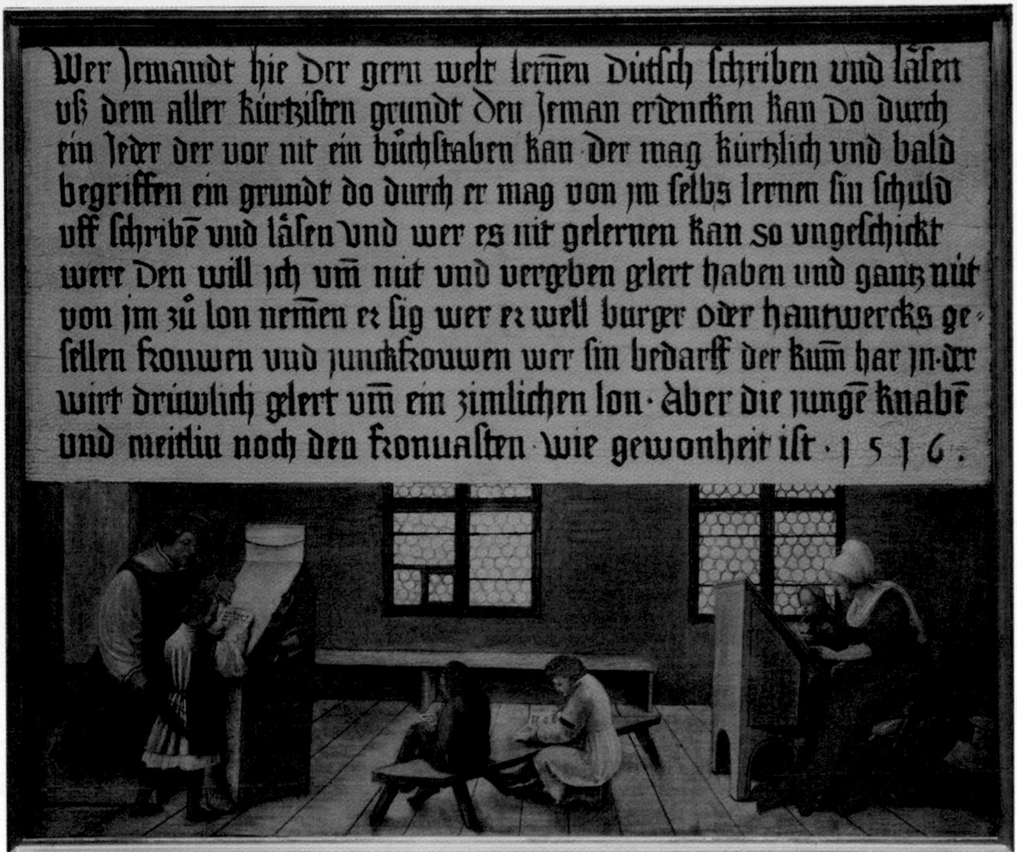

Wer Jemandt hie der gern welt lerñen dütsch schriben und läsen
uß dem aller kürtzisten grundt den Jeman erdencken kan do durch
ein Jeder der vor nit ein büchstaben kan der mag kürtzlich und bald
begriffen ein grundt do durch er mag von jm selbs lernen sin schuld
uff schribē und läsen und wer es nit gelernen kan so ungeschickt
were den will ich vm̄ nüt und vergeben glert haben und gantz nüt
von jm zů lon nem̄en er sig wer er well burger oder hantwercks ge-
sellen frouwen und junckfrouwen wer sin bedarff der kum har jn der
wirt drüwlich glert vm̄ ein zimlichen lon · aber die jungē knabē
und meitlin noch den fronuasten wie gewonheit ist · 1516 ·

Ambrosius Holbein: *Ein Schulmeister und seine Frau bringen drei Knaben und einem Mädchen das Lesen bei* (1516), Kunstmuseum, Basel

KUNST UND KÜNSTLER

Ambrosius Holbein (ca. 1494 – ca. 1519) war ein deutsch-schweizerischer Maler und Grafiker der Renaissance. Er wurde in Augsburg geboren und starb in Basel. Seine bekanntesten Werke sind das Bildnis eines Jungen mit blondem Haar und das Bildnis eines Jungen mit braunem Haar. Das Bild hier entstand als Aushängeschild[1] eines echten Schulmeisters.

Schauen Sie sich das Bild an und beantworten Sie die folgenden Fragen.

1. Wie viele Personen sehen Sie: vier, fünf, sechs?
2. Wo sind die Personen: in einer Kirche, auf der Uni, in einem Zimmer?
3. Welche Möbel sehen Sie: Bänke[2], Pulte[3], eine Tafel, Tische?
4. Was sehen Sie noch: Fenster, Uhren, Bücher, Hefte?
5. Wie sind sie gekleidet: tragen sie Hosen und Hemden, tragen sie mittelalterliche[4] Kleidung?
6. Was machen sie: sie hören zu, sie lesen, sie singen, sie sitzen, sie stehen, sie tanzen?
7. Was haben sie in der Hand: eine Rute[5], eine Blume, Bücher, Bleistifte, Karten, eine Zeitung?
8. Welche Farben dominieren: blaue Farben, braune Farben, gelbe Farben, grüne Farben, rote Farben?
9. Welche Linien dominieren: gerade[6] Linien oder krumme[7] Linien?
10. Welche Gefühle ruft das Bild hervor: Angst, Aufmerksamkeit[8], Eifersucht, Hoffnung, Neugier[9], Ruhe, Trauer?

[1]*signboard* [2]*benches* [3]*lecterns* [4]*medieval* [5]*switch* [6]*straight* [7]*curved* [8]*attentiveness* [9]*curiosity*

Situationen

Talente und Pläne

Grammatik 3.1

Peter kann ausgezeichnet kochen.

Yamina und Sumita können gut zeichnen.

Claire kann gut Deutsch.

Melanie und Josef wollen heute Abend zu Hause bleiben und lesen.

Silvia will für Jürgen einen Pullover stricken.

Sofie und Willi wollen tanzen gehen.

Situation 1 Kochen

Bringen Sie die Sätze in die richtige Reihenfolge.

_____ Spaghetti esse ich besonders gern.
_____ Dann komm doch mal vorbei.
_____ Nicht so gut. Aber ich kann sehr gut Spaghetti machen.
_____ Kannst du Chinesisch kochen?
__1__ Kochst du gern?
_____ Ja, ich koche sehr gern.
_____ Ja, gern! Vielleicht Samstag?
_____ Gut! Bis Samstag.

Situation 2 Informationsspiel: Kann Katrin kochen?

MODELL: S2: Kann Katrin kochen?
S1: Ja, ganz gut.
S2: Kannst du kochen?
S1: Ja, aber nicht so gut.

[+]
ausgezeichnet
fantastisch
sehr gut
gut

[0]
ganz gut

[−]
nicht so gut
nur ein bisschen
gar nicht
kein bisschen

	Katrin	Peter	mein(e) Partner(in)
kochen	ganz gut		
zeichnen		kein bisschen	
tippen	nur ein bisschen		
Witze erzählen	ganz gut		
tanzen		sehr gut	
stricken		kein bisschen	
Skateboard fahren	ganz gut		
Geige spielen	ausgezeichnet		
Schlittschuh laufen	gut		
ein Auto reparieren		nicht so gut	

Situation 3 Talente?

Was können die Personen? Was ist schwierig für sie? Suchen Sie die Personen im Bild und schreiben Sie ihre Namen in die Sätze.

Lydia

Eske und Damla

Sofie

Nesrin

Rolf

Melanie und Josef

Willi und Sofie

Aydan

1. __________ kann fantastisch zeichnen.
2. __________ können sehr gut Fotos machen.
3. __________ kann nicht so gut Haare schneiden.
4. __________ kann sehr gut Türkisch.
5. __________ kann sehr gut Tischtennis spielen.
6. __________ kann ihr Fahrrad reparieren.
7. __________ können sehr gut werfen.
8. __________ können nicht besonders gut Walzer tanzen.

Situation 4 Ferienpläne

Melanie und Josef wollen beide einen Teil[1] ihrer Ferien zu Hause in Regensburg verbringen, aber auch eine Reise machen. Was wollen sie wo machen? Können sie etwas zusammen machen? Hören Sie zu und ergänzen Sie die Tabelle.

NÜTZLICHE WÖRTER

die Ausstellung	*exhibition*
die Garage	*garage*
die Querflöte	*transverse flute*
sitzen	*to sit*

	Melanie	Josef	beide zusammen
in München			
zu Hause in Regensburg			
auf der Reise			

Lektüre

Vor dem Lesen

LESEHILFE

Before starting to read, it is always useful to look at the complete text, the title, and any subtitles, accompanying pictures, tables, photos, or drawings, in order to get a general idea of what the text will be about. Look at this text, its title, subtitles and its accompanying pictures. Then write down what the main topic of the text probably is and what subtopics it suggests.

German and English are closely related languages and share many words. Sometimes the words look almost identical, with minor spelling variations such as German **k** or **z** for English *c*. Sometimes you have to use a little guesswork to see the English word in the German one, as in the word **Ägypter** (*Egyptian*). In the following text, highlight the words whose meanings you think you can guess by knowing English. (In a later activity, you will work with the underlined words.)

Symbole der Liebe

[1]*part*

Ringe fürs Leben zu zweit

Symbole ewiger[2] Liebe

Der **Ehering** symbolisiert ewige Liebe: er hat keinen **Anfang** und kein Ende. So wie der Ring kein Ende hat, soll auch die Liebe nie **aufhören**. Er signalisiert aller Welt: Dieser Mann / Diese Frau ist verheiratet. Jeder Ring kann zum Ehering werden. In Deutschland ist der Ehering oft ein einfacher goldener Ring. Zum Ehering wird ein Ring durch die eingravierte Schrift. Auch auf sehr schmale Ringe kann man die Vornamen der **Eheleute** und das Hochzeitsdatum eingravieren.

Wenn[3] der Ring einmal am Finger ist, darf er nie[4] mehr **herunter** kommen. Wenn der Ring kalt wird, wird auch die Liebe kalt. Wenn der Ring **zerbricht** oder wenn er **verloren** geht, dann ist das schlecht für die Liebe.

Das Herz als Sitz der Liebe

Die alten Griechen und Ägypter trugen den Ehering am linken Ringfinger. Sie glaubten[5], dass eine Ader[6] von diesem Finger direkt zum **Herzen** führt. Sie glaubten, dass das Herz der Sitz der Liebe ist.

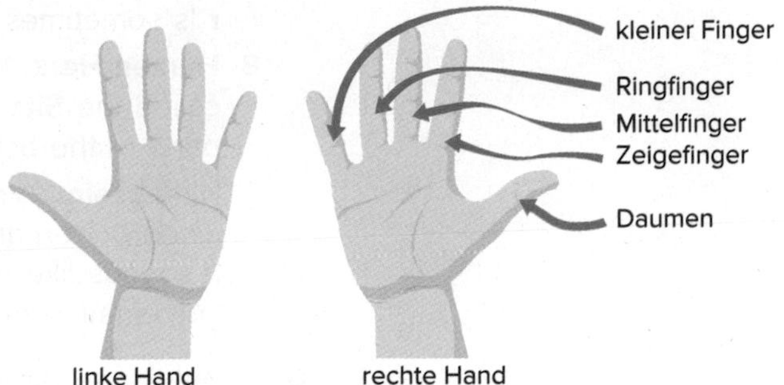

Ein bekannter **Kinderreim** lautet:

Er (oder sie) liebt mich von Herzen,
mit Schmerzen[7]*,*
oder gar nicht.

Wenn man wissen möchte, ob der Freund oder die Freundin einen[8] liebt, dann pflückt man eine Blume und reißt ihr nacheinander alle Blütenblätter ab[9]. Bei jedem Blütenblatt sagt man eine Zeile des Reims. Das, was man beim letzten Blütenblatt sagt, gilt[10].

In Italien trägt man den Ring noch heute an der linken Hand. In Deutschland trägt man nur den Verlobungsring[11] an der linken Hand. Den Ehering trägt man an der rechten Hand.

From: http://www.weddix.de

[2]*of eternal* [3]*When, If* [4]*darf ... it must never* [5]*believed* [6]*vein, artery* [7]*pain* [8]here: *you* [9]*reißt ... plucks all its petals off one by one* [10]*is valid* [11]*engagement ring*

Arbeit mit dem Text

A. Guess the meaning of the boldface underlined words in the reading by looking at the context of the sentences in which they appear. Some hints are provided.

1. **Ehering.** HINT: **Ehering** is a compound of **Ehe** and **Ring.** Look at the drawing. What kind of rings are these? What might **Ehe** mean?
2. **Anfang.** HINT: the opposite of the noun **Ende**
3. **aufhören.** HINT: a verb similar in meaning to the noun **Ende**
4. **Eheleute.** HINT: You already guessed **Ehe. Leute** means people; what might the combination of these two words mean?
5. **herunter.** HINT: Because the second clause contains the phrase *must never,* **herunter** is probably the opposite of **am Finger.**
6. **zerbricht.** HINT: What bad things can happen to a ring? The root of this word is **brich.** German **ch** is often English *k*. What English word is spelled *br*_______*k* and is something bad?
7. **verloren.** HINT: Ignore the prefix **ver**- and the -**n** for a moment. German **r** is sometimes related to English *s*. What verb is this?
8. **Herzen/Herz.** HINT: **Herzen** is the dative form of **Herz.** What might be called the **Sitz der Liebe** "seat of love" and be connected to other parts of the body by a vein?
9. **Kinderreim.** HINT: **Kinder** appears in English words such as *kindergarten* and is already familiar to you. If you pronounce **Reim,** it sounds like *rhyme,* which is in fact its meaning. What might the combination of these two words mean?

B. Beantworten Sie die folgenden Fragen.

1. Warum symbolisiert ein Ring ewige Liebe?
2. Was signalisiert ein Ehering der Welt[12]?
3. Welche Ringe trägt man in Deutschland oft als Eheringe?
4. Was ist oft in Eheringen eingraviert?
5. Was passiert, wenn der Ring vom Finger herunter kommt? Was glauben viele Leute?
6. Was macht man in Deutschland, wenn man wissen möchte, ob der Freund oder die Freundin einen liebt?
7. Was trägt man in Deutschland an der linken Hand und was an der rechten Hand?

Nach dem Lesen

A. Gibt es in Ihrer Klasse unterschiedliche Traditionen und Kulturen? Sammeln Sie in Ihrer Klasse Antworten auf die folgenden Fragen.

1. Trägt man in Ihrer Kultur Eheringe? Wenn ja, an welchem Finger welcher Hand trägt man sie? Wenn nicht, wie signalisiert man, dass Menschen verheiratet sind? Oder signalisiert man es gar nicht?
2. Was macht man in Ihrer Kultur, wenn man herausfinden möchte, ob jemand einen liebt?

B. Was halten Sie von Symbolen, die zeigen, dass zwei Menschen miteinander durchs Leben gehen wollen? Finden Sie sie wichtig? Warum (nicht)?

C. Sind Sie verheiratet? Wenn nicht, haben Sie Heiratspläne für die Zukunft?

[12]der ... *to the world*

Pflichten

Grammatik 3.2

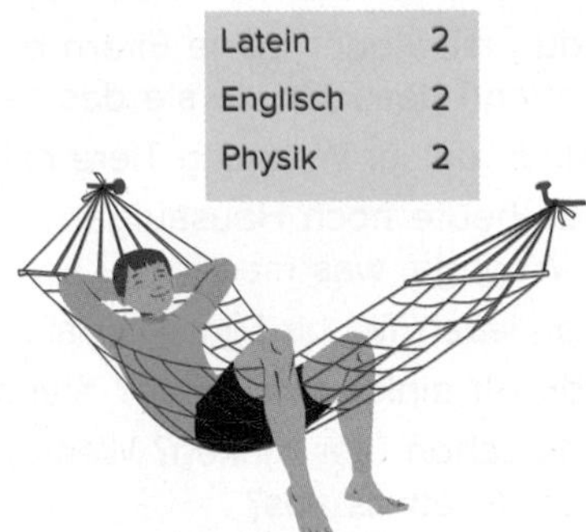

Jens hat schlechte Noten.
Er muss mehr lernen.

Er darf nicht mit seinen Freunden Skateboard fahren.

Jutta muss in der Schule besser aufpassen.

Sie darf in der Stunde nicht mit ihrer Freundin reden.

Jutta muss nach der Schule ihre Hausaufgaben machen.

Situation 5 Schlechtes Zeugnis!

Jens hat im Zeugnis in drei Fächern nur zwei Punkte.

- Was muss er machen? Was darf er nicht machen? Kreuzen Sie an.
- Schreiben Sie dann noch eine Sache dazu, die er machen muss, und eine, die er nicht machen darf.
- Entscheiden Sie schließlich, was am wichtigsten ist (1), was weniger wichtig (2) und was am unwichtigsten (3).

MUSS	DARF NICHT		WIE WICHTIG? (1–3)
☐	☐	in die Disko gehen	______
☐	☐	Latein lernen	______
☐	☐	den ganzen Tag in der Sonne liegen	______
☐	☐	seine Hausaufgaben machen	______
☐	☐	jeden Tag ins Schwimmbad gehen	______
☐	☐	eine Woche nach Italien fahren	______
☐	☐	Nachhilfe nehmen	______
☐	☐	mit seinen Lehrern sprechen	______
☒	☐	______________________________	______
☐	☒	______________________________	______

Situation 6 Interview: Studium und Alltag

1. Musst du neben dem Studium arbeiten? Wo arbeitest du? Wie viele Stunden pro Woche? Macht die Arbeit Spaß?
2. Kannst du gut Auto fahren? Hast du ein eigenes Auto? Fährst du gern Auto?
3. Musst du mal wieder deine Eltern besuchen? Wie oft besuchst du deine Eltern? Wann besuchst du sie das nächste Mal?
4. Darfst du in deiner Wohnung Tiere haben? Wenn ja, was für ein Tier hast du?
5. Musst du heute noch Hausaufgaben machen? Wenn nicht, was machst du heute? Wenn ja, was machst du sonst noch?
6. Kannst du jeden Tag bis Mittag schlafen? Wenn nicht, wann stehst du meist auf?
7. Musst du oft einkaufen gehen? Wenn nicht, wann gehst du einkaufen?
8. Darfst du schon Bier trinken? Wenn ja, trinkst du gern Bier? Wenn nicht, macht es dir etwas aus?

MUSIKSZENE

„Müssen nur wollen" (2003, Deutschland) *Wir sind Helden*

Biografie *Wir sind Helden* kommen aus Berlin. Die Lead-Sängerin heißt Judith Holofernes. Sie schreibt auch die meisten Texte und Songs der Gruppe. Die Songs sind oft sehr kritisch. Der Hit „Müssen nur wollen" stammt aus dem Jahr 2003.

Wir sind Helden
© Christian Jakubaszek/Getty Images

NOTE: For copyright reasons, the songs referenced in **MUSIKSZENE** have not been provided by the publisher. The song can be found online at various sites such as YouTube, Amazon, or the iTunes store.

Vor dem Hören Was musst du tun, was willst du tun? Willst du, was du musst?

Nach dem Hören

A. Was sagt die Sängerin? Richtig (R) oder falsch (F)?

_____ **1.** In einer Hand trägt sie die Welt, mit der anderen Hand bietet sie Getränke an.

_____ **2.** Sie kann gar nichts.

_____ **3.** Alle sollen etwas wollen.

_____ **4.** Dressierte Affen können alles schaffen.

_____ **5.** Sie kann glücklich sein und Konzerne leiten.

B. Was meinen Sie?

1. Kann man glücklich sein und Konzerne leiten?

2. Muss man alles tun, was man tun kann?

Miniwörterbuch	
tragen	to carry
anbieten	to offer
die **Getränke**	drinks
gar nichts	nothing at all
dressiert	trained (animals)
die **Affen**	monkeys
schaffen	to achieve
der **Konzern**	corporation
leiten	to lead, be head of

Situation 7 Dialog

Rolf trifft Katrin in der Cafeteria.

ROLF: Hallo, Katrin, ist hier noch ________?
KATRIN: Ja, klar.
ROLF: Ich hoffe, ich störe ________ nicht beim Lernen.
KATRIN: Nein, ich muss auch mal ________ machen.
ROLF: Was machst du denn?
KATRIN: Wir haben morgen eine ________ und ich ________ noch das Arbeitsbuch machen.
ROLF: ________ ihr viel für euren Kurs arbeiten?
KATRIN: Ja, ganz schön viel. Heute Abend ________ ich bestimmt nicht fernsehen, ________ ich so viel lernen muss.
ROLF: Ich glaube, ich störe dich nicht länger. ________ für die Prüfung.
KATRIN: Danke, tschüss.

Situation 8 Stefans Zimmer

Stefans Mutter kommt zu Besuch.

Das ist Stefans Zimmer.

So soll es sein.

Was muss Stefan machen?

den Tisch abräumen
den Papierkorb ausleeren
das Bild an die Wand hängen
die Kerzen anzünden
den Schrank zumachen
die Katze aus dem Zimmer werfen
seine Kleidung aufräumen
das Fenster zumachen
die Pflanze gießen
das Bett machen
den Fernseher ausmachen
die Bücher gerade stellen
den Boden sauber machen

KULTUR ... LANDESKUNDE ... INFORMATIONEN

JUGENDSCHUTZ

Nicht in jedem Alter darf man alles. In Deutschland regelt das Jugendschutzgesetz[1], in welchem Alter Kinder und Jugendliche etwas dürfen oder können.

Mit 13 ...

- darf man in den Ferien arbeiten.
  aber: Die Eltern müssen es erlauben[2] und die Arbeit muss leicht sein.

Mit 14 ...

- darf man im Restaurant Bier oder Wein trinken.
  aber: Die Eltern müssen dabei sein[3].

Mit 15 ...

- kann man mit der Arbeit anfangen.
  aber: Man darf nur 8 Stunden am Tag und 5 Tage in der Woche arbeiten.

Mit 16 ...

- darf man im Restaurant Bier oder Wein trinken (ohne Eltern).
- darf man von zu Hause wegziehen[4].
  aber: Die Eltern müssen es erlauben.
- darf man heiraten[5].
  aber: Die Eltern müssen es erlauben.
  und: Der Partner muss über 18 Jahre alt sein.
- darf man bis 24.00 Uhr in die Disko gehen.

Mit 17 ...

- darf man den Führerschein[6] für ein Auto machen.

Mit 18 ...

- darf man den Führerschein für ein Motorrad machen.
- darf man ohne Erlaubnis heiraten.
- darf man wählen[7].
- darf man im Kino alle Filme sehen.
- darf man im Restaurant Alkohol trinken.
- darf man so lange in die Disko gehen, wie man will.
- darf man rauchen.

In Deutschland ist man mit 18 Jahren erwachsen[8].

Mit 17 darf man Auto fahren.
© ullstein bild - Sylent Press/The Image Works

Wie ist es in Ihrem Land? Ergänzen Sie die Tabelle.

	Mit ... Jahren
Alkohol trinken	
alle Filme sehen	
arbeiten	
Auto fahren	
erwachsen sein	
heiraten	
in die Disko gehen	
rauchen	
wählen	
?	

[1]*law for the protection of minors* [2]*permit* [3]dabei ... *be present* [4]*move away* [5]*marry* [6]*driver's license* [7]*vote* [8]*grown-up*

Ach, wie nett!

Grammatik 3.3

MARIA: Der Fernseher läuft ja den ganzen Tag.
MICHAEL: Soll ich ihn ausmachen?

FRAU KÖRNER: Ich finde den Mantel einfach toll!
FRAU GRETTER: Kaufen Sie ihn doch!

AYDAN: Die Tasche ist so schwer.
ESKE: Komm, Mama, ich trage sie.

PRINZESSIN: Hier ist mein Taschentuch. Du darfst mich nie vergessen.
PRINZ: Nein, Geliebte, ich vergesse dich nie!

SILVIAS FREUNDIN: Samstag geben wir eine Party. Ich möchte euch gern einladen.

ZWEI TRAMPERINNEN: Hallo, wir wollen nach Regensburg. Nehmt ihr uns mit?

Situation 9 Minidialoge

Was passt?

1. Es ist kalt und das Fenster ist offen!
2. Der Wein ist gut.
3. Du hast nächste Woche Geburtstag?
4. Der Koffer ist so schwer.
5. Die Suppe ist wirklich gut!
6. Wie findest du *die Ärzte*?
7. Das Haus ist schmutzig.

a. Komm, ich trage ihn.
b. Machen Sie es bitte zu.
c. Darf ich ihn probieren?
d. Ich mag sie aber nicht.
e. Ja, ich gebe eine Party und ich lade euch ein.
f. Ich mache es morgen sauber.
g. Ich mag sie ganz gern.

Situation 10 Dialog

Heidi sucht einen Platz in der Cafeteria.

HEIDI: Entschuldigung, ____________?

STEFAN: Ja, sicher.

HEIDI: Danke.

STEFAN: ____________?

HEIDI: Ja, ich glaube schon. Bist du nicht auch in dem Deutschkurs um neun?

STEFAN: Na, klar. Jetzt ____________ ich's wieder. Du ____________ Stefanie, nicht wahr?

HEIDI: Nein, ich heiße Heidi.

STEFAN: Ach ja, richtig ... Heidi. Ich heiße Stefan.

HEIDI: ____________ kommst du eigentlich, Stefan?

STEFAN: ____________ Iowa City, und du?

HEIDI: Ich bin aus Berkeley.

STEFAN: Und was studierst du?

HEIDI: ____________. Vielleicht Sport, vielleicht Geschichte oder vielleicht Deutsch.

STEFAN: Ich studiere auch Deutsch, Deutsch und ____________. Ich möchte in Deutschland bei einer amerikanischen Firma arbeiten.

HEIDI: Toll! Da verdienst du sicherlich viel Geld.

STEFAN: ____________.

Situation 11 Rollenspiel: In der Mensa

S1: Sie sind Student/Studentin an der Uni in Regensburg. Sie gehen in die Mensa und setzen sich zu jemand an den Tisch. Fragen Sie, wie er/sie heißt, woher er/sie kommt und was er/sie studiert.

In der Mensa

Situation 12 Ratespiel

Was ist das?

1. _____ Man trägt sie im Sommer an den Füßen.
2. _____ Man trägt ihn nach dem Duschen.
3. _____ Man trägt es im Bett.
4. _____ Man trägt ihn im Winter um den Hals.
5. _____ Man trägt sie im Ohr.
6. _____ Man trägt sie unter der Kleidung.
7. _____ Man trägt sie im Winter an den Händen.

a. der Schal
b. die Ohrringe
c. die Handschuhe
d. die Unterhose
e. der Bademantel
f. die Sandalen
g. das Nachthemd

Filmlektüre

Vincent will Meer

Vor dem Lesen

FILMANGABEN

Titel: *Vincent will Meer*
Genre: Road Movie, Tragikomödie
Land: Deutschland
Erscheinungsjahr: 2010
Dauer: 96 Min.
Regisseur: Ralf Huettner
Hauptrollen: Florian David Fitz, Karoline Herfurth, Johannes Allmayer, Heino Ferch, Katharina Müller-Elmau

Miniwörterbuch

leiden (an)	to suffer (from)
das **Tourette-Syndrom**	Tourette's syndrome
magersüchtig	anorexic
zwangsneurotisch	obsessive-compulsive
erfüllen	to fulfill
fliehen	to flee
stehlen	to steal
die **Asche**	ashes
die **Bonbondose**	candy tin
entwickeln	to develop
der **Umweg**	detour
zusammenbrechen	to collapse
unterwegs	along the way
aussteigen	to get out
der **Herzanfall**	heart attack

Marie, Vincent und Alexander

A. Beantworten Sie die folgenden Fragen.

1. Was sehen Sie im Bild?
2. Warum sind die drei jungen Leute in dem Auto? Sind sie im Urlaub oder auf der Flucht? Sind sie Freunde oder streiten sie?
3. Wer spielt in dem Film die Hauptrollen?

B. Lesen Sie die Wörter im Miniwörterbuch. Suchen Sie sie im Text und unterstreichen Sie sie.

Inhaltsangabe

Vincent (Florian David Fitz) leidet am Tourette-Syndrom und muss nach dem Tod seiner Mutter in eine Klinik, weil sein Vater (Heino Ferch) es so will. Der Vater ist Politiker und nicht sehr nett zu seinem Sohn. In der Klinik trifft Vincent die magersüchtige Marie (Karoline Herfurth) und seinen zwangsneurotischen Zimmergenossen Alexander (Johannes Allmayer). Da Vincent den letzten Wunsch seiner Mutter erfüllen möchte, noch einmal das Meer zu sehen, fliehen die drei aus der Klinik. Sie stehlen das Auto der Therapeutin Dr. Rose (Katharina Müller-Elmau) und fahren nach Italien. Die Asche seiner Mutter hat Vincent in einer Bonbondose dabei. Vincents Vater fährt zusammen mit Dr. Rose hinterher. Zuerst will er nur seinen Sohn wieder zurück in die Klinik bringen, doch während der Zeit, die er mit der Ärztin verbringt, entwickelt er neue Gefühle für ihn. Vincent, Marie und Alex kommen über Umwege ans Meer. Marie bricht dort wegen ihrer Magersucht zusammen und kommt ins Krankenhaus. Vincent und Alex fahren zusammen mit Dr. Rose und Vincents Vater wieder zurück nach Deutschland. Unterwegs bittet Vincent seinen Vater, ihn aussteigen zu lassen, und geht zurück nach Triest, wo Marie nach ihrem Herzanfall noch immer im Krankenhaus liegt. Alex folgt ihm.

Arbeit mit dem Text

Richtig (R) oder falsch (F)? Verbessern Sie die falschen Aussagen.

______ 1. Vincent muss in eine Klinik, weil sein Vater tot ist.

______ 2. Alexander und Marie leben auch in der Klinik.

______ 3. Vincent möchte ans Meer.

______ 4. Die drei fahren mit der Therapeutin Dr. Rose nach Italien.

______ 5. Marie muss am Meer ins Krankenhaus.

______ 6. Vincent und Alexander fahren mit dem Vater nach Triest.

FILMCLIP

NOTE: For copyright reasons, the films referenced in the **FILMCLIP** feature have not been provided by the publisher. The film can be purchased as a DVD or found online at various sites such as YouTube, Amazon, or the iTunes store. The time codes mentioned below are for the North American DVD version of the film.

Szene: DVD, Kapitel 5, „Ans Meer", 18:55–22:03 Min.

Am Abend treffen sich Marie und Vincent in der Klinik. Marie hat einen Schlüssel[1] dabei. Schauen Sie sich die Szene an und beantworten Sie die Fragen.

1. Wo sind Vincent und Marie?
   - ☐ a. im Speisesaal[2]
   - ☐ b. im Waschraum
   - ☐ c. in Vincents Zimmer
2. Wohin will Vincent fahren?
   - ☐ a. nach Italien ans Meer
   - ☐ b. nach Hause zu seinem Vater
   - ☐ c. in die Berge
3. Was nimmt Vincent aus seinem Zimmer mit?
   - ☐ a. seinen Koffer
   - ☐ b. Alexanders Reisepass
   - ☐ c. Alexanders CD
4. Warum will Vincent nicht Auto fahren?
   - ☐ a. weil er Tourette-Syndrom hat
   - ☐ b. weil er müde ist
   - ☐ c. weil Marie einen Führerschein hat
5. Was will Alexander?
   - ☐ a. seine CD
   - ☐ b. seine Jacke
   - ☐ c. seinen Reisepass

[1]*key* [2]*cafeteria*

Nach dem Lesen

Suchen Sie im Internet.

1. Wer hat das Originaldrehbuch[3] zu „Vincent will Meer" geschrieben[4]?
2. Welche Preise hat der Film bekommen[5]?
3. Wie heißt die amerikanische Neuverfilmung[6]?

Körperliche und geistige Verfassung

Grammatik 3.4–3.5

Er ist glücklich.

Sie sind traurig.

Er ist wütend.

Sie ist krank.

Sie sind in Eile.

Sie ist müde.

Sie haben Hunger.

Er hat Langeweile.

Er hat Durst.

Er hat Angst.

[3]*screen play* [4]*written* [5]*received* [6]*remake*

Situation 13 Informationsspiel: Was machen sie, wenn ...?

MODELL: S2: Was macht Renate, wenn sie traurig ist?
S1: Sie ruft ihre Freundin an.
S2: Was machst du, wenn du traurig bist?
S1: Ich gehe ins Bett.

	Renate	Ernst	mein(e) Partner(in)
1. *traurig ist/bist*	ruft ihre Freundin an		
2. *müde ist/bist*		schläft	
3. *in Eile ist/bist*		ist nie in Eile	
4. *wütend ist/bist*	wirft mit Tellern		
5. *krank ist/bist*		isst Hühnersuppe	
6. *glücklich ist/bist*	lädt Freunde ein		
7. *Hunger hat/hast*	isst einen Apfel	schreit laut „Hunger!"	
8. *Langeweile hat/hast*			
9. *Durst hat/hast*	trinkt Mineralwasser		
10. *Angst hat/hast*		läuft zu Mama	

KULTUR ... LANDESKUNDE ... INFORMATIONEN

CHATIQUETTE: STERNCHEN, ABKÜRZUNGEN UND AKRONYME

Wenn es schnell gehen muss, verwenden[1] viele Leute im Chat, bei WhatsApp oder SMS besondere Formen der Kommunikation. Sie machen das Chatleben leichter. Viele sind lustig oder ironisch gemeint und ein fester Bestandteil[2] der Chatkultur. Sternchen[3] drücken Emotion oder Tätigkeit aus und es gibt viele Akronyme auf Englisch, aber auch auf Deutsch.

© *Gerhilde Skoberne/Corbis RF*

[1]*use* [2]*fester ... established part* [3]*asterisks*

Miniwörterbuch	
grinsen	to grin
knuddeln	to cuddle
doll	very (*colloquial*)
hab dich lieb	(I) love you
drücken	to hug
das **Unverständnis**	incomprehension
zeigen	to show
frech	impudently
fies	meanly

Können Sie folgende Akronyme auf Deutsch erkennen? Ordnen Sie die Akronyme den Aussagen zu.

1. *g*	a. kein Kommentar
2. *fg*	b. grinsen
3. *momtel*	c. Moment, ich telefoniere gerade
4. *knuddel*	d. liebe Grüße
5. LG	e. hab dich lieb
6. kk	f. ich knuddel/drück dich
7. N8	g. frech/fies grinsen
8. omg	h. Nacht / Gute Nacht
9. HDL	i. hab dich ganz doll lieb
10. HDGDL	j. oh mein Gott

Situation 14 Interview: Wie fühlst du dich, wenn ...?

MODELL: S1: Wie fühlst du dich, wenn du um fünf Uhr morgens aufstehst?
S2: Ausgezeichnet!

[+]	[0]	[−]
ausgezeichnet	ganz gut	nicht besonders gut
fantastisch		ziemlich schlecht
sehr gut		mies
gut		total mies

1. wenn du um fünf Uhr morgens aufstehst
2. wenn du die ganze Nacht nicht schlafen kannst
3. wenn deine Freunde dich auf eine Party einladen
4. wenn du eine Arbeit oder einen Test zurückbekommst
5. wenn du ein Referat halten musst
6. wenn das Semester zu Ende ist
7. wenn du einkaufen gehen willst, aber kein Geld hast
8. wenn alle deine T-Shirts schmutzig sind
9. wenn du eine gute Note bekommst
10. wenn du Heimweh hast
11. wenn du eifersüchtig bist

Situation 15 Warum fährt Frau Ruf mit dem Bus?

Kombinieren Sie!

MODELL: S1: Warum fährt Frau Ruf mit dem Bus?
S2: Weil ihr Auto kaputt ist.

1. Warum fährt Frau Ruf mit dem Bus?
2. Warum hat Hans Angst?
3. Warum geht Jutta nicht ins Kino?
4. Warum geht Jens nicht in die Schule?
5. Warum kauft Andrea Hans eine CD?

a. weil er Geburtstag hat
b. weil ihr Auto kaputt ist
c. weil sein Referat noch nicht fertig ist
d. weil sie für eine Klassenarbeit lernen muss
e. weil er keine Lust hat

---

6. Warum fährt Herr Wagner nach Leipzig?
7. Warum ist Ernst wütend?
8. Warum fährt Frau Gretter in die Berge?
9. Warum geht Herr Siebert um zehn Uhr ins Bett?
10. Warum ruft Maria ihre Freundin an?

f. weil er seinen Bruder besuchen will
g. weil sie wandern geht
h. weil er in Mathe so viele Hausaufgaben hat
i. weil sie sie ins Kino einladen will
j. weil er jeden Tag um sechs Uhr aufsteht

Situation 16 Zum Schreiben: Auch in Ihnen steckt ein Dichter!

Schreiben Sie ein Gedicht!

MODELL:

ein Nomen = Thema	Wasser
zwei Adjektive	kühl, nass
drei Verben	schwimmen, segeln, tauchen
vier Wörter, die ein Gefühl ausdrücken[4]	Sonne auf meiner Haut
ein Nomen = Zusammenfassung[5]	Sommer

[4]*express* [5]*summary*

Videoecke

Perspektiven

Wie viel Zeit verbringst du pro Tag am Computer? Womit verbringst du die meiste Zeit?

Ich arbeite oder ich chatte.

Aufgabe 1 Zeit am Computer

Wie viel Zeit verbringen sie am Computer? Schreiben Sie die Antworten auf.

1. Susan ____

2. Felicitas ____

3. Michael ____

4. Shaimaa ____

5. Nadezda ____

6. Pascal ____

7. Judith ____

8. Martin ____

Aufgabe 2 Tätigkeiten am Computer

Was machen sie am Computer? Ordnen Sie die Tätigkeiten den Personen unter Aufgabe 1 zu.

_____ 1. Susan	a. Die meiste Zeit verbringt sie auf Facebook.
_____ 2. Felicitas	b. Er liest E-Mails und Nachrichten.
_____ 3. Michael	c. Er macht Layout und Grafik.
_____ 4. Shaimaa	d. Er sucht potentielle Kunden[6] für seine Firma.
_____ 5. Nadezda	e. Sie arbeitet die meiste Zeit.
_____ 6. Pascal	f. Sie arbeitet oder sie chattet.
_____ 7. Judith	g. Sie checkt ihre E-Mails und ist oft bei Facebook.
_____ 8. Martin	h. Sie verwendet die meiste Zeit für ihr Studium.

[6]*customers*

Interviews

- Hast du handwerkliche oder künstlerische Fähigkeiten?
- Was kannst du besonders gut?
- Was gefällt dir daran?
- Was kannst du nicht besonders gut?
- Gibt es etwas, was du nicht gern machst?
- Wie fühlst du dich, wenn du eine Prüfung hast?
- Wie bereitest du dich darauf vor?

Carolyn

Michael

Aufgabe 3 Fähigkeiten und Pflichten

Carolyn oder Michael?

	Carolyn	Michael
1. Wer tanzt und gestaltet[7] gerne T-Shirts?	☐	☐
2. Wer hat keine handwerklichen Fähigkeiten?	☐	☐
3. Wer mag es total, sich zu den Rhythmen der Musik zu bewegen?	☐	☐
4. Wer spielt Gitarre, Akkordeon und Klavier?	☐	☐
5. Wer kann nicht gut kochen?	☐	☐
6. Wer putzt nicht gern?	☐	☐
7. Wer macht nicht gern Sport?	☐	☐
8. Wessen[8] Hände schwitzen[9], wenn er oder sie eine Prüfung hat?	☐	☐
9. Wer geht regelmäßig zu Seminaren und Vorlesungen, damit er oder sie gut vorbereitet ist?	☐	☐

Aufgabe 4 Interview

Interviewen Sie eine Partnerin oder einen Partner. Stellen Sie dieselben Fragen.

[1]*designs* [8]*Whose* [9]*sweat*

Wortschatz

Talente und Pläne	Talents and Plans
der **Besuch, -e**	visit
zu Besuch kommen	to visit
der **Schlittschuh, -e**	ice skate
Schlittschuh laufen, läuft	to go ice-skating
der **Witz, -e**	joke
Witze erzählen	to tell jokes
schneiden	to cut
Haare schneiden	to cut hair
stricken	to knit
tauchen	to dive
tippen	to type
zeichnen	to draw

Ähnliche Wörter

der **Ski, -er; Ski fahren, fährt;** der **Walzer, -;** das **Skateboard, -s; Skateboard fahren, fährt**

Pflichten	Obligations
ab·räumen	to clear
den Tisch ab·räumen	to clear the table
gerade stellen	to straighten
gießen	to water
die Blumen gießen	to water the flowers
sauber machen	to clean

Körperliche und geistige Verfassung	Physical and Mental State
die **Angst, ¨e**	fear
Angst haben	to be afraid
die **Eile**	hurry
in Eile sein	to be in a hurry
die **Langeweile**	boredom
Langeweile haben	to be bored
die **Lust**	desire
Lust haben	to feel like (doing something)
das **Glück**	luck; happiness
viel Glück!	lots of luck! good luck!
das **Heimweh**	homesickness
Heimweh haben	to be homesick
ärgern (R)	to annoy; to tease
schreien	to scream, yell
stören	to disturb
weinen	to cry
eifersüchtig	jealous
krank	sick
nett	nice
müde	tired
wütend	angry

Ähnliche Wörter

der **Durst; Durst haben;** der **Hunger; Hunger haben;** das **Gefühl, -e; fühlen; wie fühlst du dich?; ich fühle mich ...**

Schule	School
die **Nachhilfe**	tutoring
die **Sprechstunde, -n**	office hour
der **Satz, ¨e**	sentence
das **Arbeitsbuch, ¨er**	workbook
das **Beispiel, -e**	example
zum Beispiel (z. B.)	for example
das **Referat, -e**	report
das **Studium, Studien** (R)	course of studies

Sonstige Substantive	Other Nouns
die **Ärztin, -nen**	female physician
die **Blume, -n**	flower
die **Geige, -n**	violin
die **Geliebte, -n**	beloved female friend, love
die **Kerze, -n**	candle
die **Pflicht, -en**	duty; requirement
der **Arzt, ¨e**	male physician
der **Koffer, -**	suitcase
der **Papierkorb, ¨e**	wastebasket
das **Gedicht, -e**	poem
das **Krankenhaus, ¨er**	hospital
das **Mal, -e**	time
das **nächste Mal**	the next time
das **Mittagessen**	midday meal, lunch
das **Taschentuch, ¨er**	handkerchief
das **Tier, -e**	animal

Ähnliche Wörter

die **CD, -s** (R); die **Disko, -s;** die **Firma, Firmen;** die **Nacht, ¨e;** die **Pflanze, -n;** der **DVD-Spieler, -** (R); der **Mittag, -e;** der **Plan, ¨e;** der **Platz, ¨e;** das **Alphabet;** das **Licht, -er;** das **Talent, -e;** das **Taxi, -s;** das **Tischtennis**

Modalverben	Modal Verbs
dürfen, darf	to be permitted (to), may
können, kann (R)	to be able (to), can; may
mögen, mag (R)	to like, care for
möchte	would like (to)
müssen, muss	to have to, must
sollen, soll	to be supposed to
wollen, will	to want; to intend, plan (to)

Sonstige Verben	Other Verbs
an·machen	to turn on, switch on
an·sehen, sieht ... an	to look at; to watch
an·ziehen	to put on (clothes); to attract
an·zünden	to light
auf·machen	to open
auf·passen	to pay attention
aus·leeren	to empty
aus·machen	to turn off
aus·ziehen	to take off (clothes)
bekommen	to get, receive
bleiben	to remain, stay
erleben	to experience
erzählen	to tell
heiraten	to marry
lieben	to love
mit·nehmen, nimmt ... mit	to take along
probieren	to try, taste
rauchen	to smoke
stellen	to put, place (upright)
verbringen	to spend (*time*)
verreisen	to go on a trip
vorbei·kommen	to come by, visit
werfen, wirft	to throw
zu·machen	to close

Ähnliche Wörter

baden, hängen, hoffen, kämmen, kombinieren, lachen, leben, mit·bringen; das Bild an die Wand hängen

Adjektive und Adverbien	Adjectives and Adverbs
ausgezeichnet	excellent
beliebt	popular
besonders	particularly
bestimmt	definitely, certainly
eigen	own
eigentlich	actually
fertig	ready; finished
genug	enough
meist	most, mostly
nass	wet
schwer	heavy; hard, difficult
wahr	true

Sonstige Wörter und Ausdrücke	Other Words and Expressions
außerdem	besides
dreimal	three times
einander	one another, each other
hintereinander	in a row
miteinander (R)	with each other
ein bisschen	a little bit
Entschuldigung!	excuse me
die ganze Nacht	all night long
ganz schön viel	quite a bit
gar nicht	not a bit
immer	always
jede	each, every
jede Woche	every week
jemand	someone, somebody
jetzt	now
kein bisschen	not at all
mit mir	with me
na	well
nach	after; to
neben	beside, in addition to
nur	only
sicherlich	certainly
sofort	immediately
von der Arbeit	from work
warum	why
was für	what kind of
weil	because
wieder	again
schon wieder	once again
wohin	where to
zu Fuß	on foot
zum Arzt	to the doctor
zum Mittagessen	for lunch

Strukturen und Übungen

3.1 The modal verbs *können, wollen, mögen*

WISSEN SIE NOCH?

The **Satzklammer** forms a frame or a bracket consisting of a verb and either a separable prefix or an infinitive. This same structure is used with the modal verbs.

Review grammar 1.5 and 2.3.

Modal verbs, such as **können** (*can, to be able to, know how to*), **wollen** (*to want to*), and **mögen** (*to like to*) are auxiliary verbs that modify the meaning of the main verb. The main verb appears as an infinitive at the end of the clause.

The modal **können** usually indicates an ability or talent but may also be used to ask permission. The modal **wollen** expresses a desire or an intention to do something. The modal **mögen** expresses a liking; just like its English equivalent, *to like*, it is commonly used with an accusative object.

Kannst du kochen?	*Can you cook?*
Kann ich mitkommen?	*Can I come along?*
Sofie und Willi wollen tanzen gehen.	*Sofie and Willi want to go dancing.*
Ich mag aber nicht tanzen.	*I don't like to dance though.*
Magst du Sushi?	*Do you like sushi?*

Modals do not have endings in the **ich-** and **er/sie/es-**forms. Note also that these modal verbs have one stem vowel in all plural forms and in the polite **Sie**-form, and a different stem vowel in the **ich-**, **du-**, and **er/sie/es-**forms.

können = *can*
wollen = *to want to*
mögen = *to like (to)*

	können	wollen	mögen
ich	kann	will	mag
du	kannst	willst	magst
Sie	können	wollen	mögen
er/sie/es	kann	will	mag
wir	können	wollen	mögen
ihr	könnt	wollt	mögt
Sie	können	wollen	mögen
sie	können	wollen	mögen

Übung 1 Talente

A. **Wer kann das?**

MODELL: Ich kann Deutsch.
oder Wir können Deutsch.

1. Deutsch	mein Freund / meine Freundin
2. Golf spielen	meine Eltern
3. Ski fahren	ich/wir
4. Klavier spielen	mein Bruder / meine Schwester
5. gut kochen	der Professor / die Professorin
6. gut Karaoke singen	
7. Witze erzählen	
8. Snowboard fahren	

B. **Kannst du das?**

MODELL: Gedichte schreiben → Kannst du Gedichte schreiben?
oder Könnt ihr Gedichte schreiben?

1. Gedichte schreiben du
2. Auto fahren ihr
3. tippen
4. stricken
5. zeichnen

Übung 2 Pläne und Fähigkeiten

ACHTUNG!

German **will** is not the equivalent of English *will*. Instead, it means "want(s)" or "intend(s) to."

Was können oder wollen diese Personen (nicht) machen?

MODELL: am Samstag / ich / wollen →
Am Samstag **will** ich **Schlittschuh laufen**.

E-Mails lesen
Golf spielen
Haare schneiden
ins Kino gehen
nach Europa fliegen
schlafen
simsen
Ski fahren
Witze erzählen
zeichnen
_______?

1. heute Abend / ich / wollen
2. morgen / ich / nicht können
3. mein Freund (meine Freundin) / gut können
4. am Samstag / mein Freund (meine Freundin) / wollen
5. mein Freund (meine Freundin) und ich / wollen
6. im Winter / meine Eltern (meine Freunde) / wollen
7. meine Eltern (meine Freunde) / gut können

3.2 The modal verbs *müssen, sollen, dürfen*

The modal **müssen** stresses the necessity to do something. The modal **sollen** is less emphatic than **müssen** and may imply an obligation or a strong suggestion made by another person. The modal **dürfen,** used primarily to indicate permission, can also be used in polite requests.

Jens muss mehr lernen.	*Jens has to study more.*
Vati sagt, du sollst sofort nach Hause kommen.	*Dad says you're supposed to come home immediately.*
Frau Schulz sagt, du sollst morgen zu ihr kommen.	*Ms. Schulz says you should come to see her tomorrow.*
Darf ich die Kerzen anzünden?	*May I light the candles?*

müssen = *must*
sollen = *to be supposed to*
dürfen = *may*

	müssen	sollen	dürfen
ich	muss	soll	darf
du	musst	sollst	darfst
Sie	müssen	sollen	dürfen
er/sie/es	muss	soll	darf
wir	müssen	sollen	dürfen
ihr	müsst	sollt	dürft
Sie	müssen	sollen	dürfen
sie	müssen	sollen	dürfen

nicht müssen = *to not have to, to not need to*
nicht dürfen = *mustn't*

When negated, the English expressions *to have to* and *must* undergo a change in meaning. The expression *not have to* implies that there is no need to do something, while *must not* implies a strong prohibition. These two distinct meanings are expressed in German by **nicht müssen** and **nicht dürfen,** respectively.

Du musst das nicht tun. — *You don't have to do that.*
or: *You don't need to do that.*

Du darfst das nicht tun. — *You mustn't do that.*

Übung 3 Jutta hat zwei Punkte in Englisch.

Was muss sie machen? Was darf sie nicht machen?

1. mit Jens zusammen lernen
2. den ganzen Abend chatten
3. in der Klasse aufpassen und mitschreiben
4. jeden Tag tanzen gehen
5. jeden Tag ihren Wortschatz lernen
6. amerikanische Filme im Original sehen
7. ihren Englischlehrer zum Abendessen einladen
8. für eine Woche nach London fahren
9. die englische Grammatik fleißig[1] lernen

Übung 4 Minidialoge

ACHTUNG!

Remember the two characteristics of modal verbs:
1. no ending in the **ich-** and **er/sie/es-**forms;
2. one stem vowel in the **ich-, du-,** and **er/sie/es-**forms and a different one in the plural, the formal **Sie,** and the infinitive. (Note, however, that **sollen** has the same stem vowel in all forms.)

Ergänzen Sie **können, wollen, müssen, sollen, dürfen.**

1. ALBERT: Hallo, Nora. Peter und ich gehen ins Kino. _______[a] du nicht mitkommen?
   NORA: Ich _______[b] schon, aber leider _______[c] ich nicht mitkommen. Ich _______[d] arbeiten.
2. JENS: Vati, _______[a] ich mit Hans fischen gehen?
   HERR WAGNER: Nein! Du hast zwei Punkte in Physik, zwei Punkte in Latein und einen Punkt in Englisch. Du _______[b] zu Hause bleiben und deine Hausaufgaben machen.
   JENS: Aber, Vati! Meine Hausaufgaben _______[c] ich doch heute Abend machen.
   HERR WAGNER: Nein, Jens! Aber wenn du möchtest, _______[d] du zu Hans gehen und dann _______[e] ihr eure Hausaufgaben zusammen machen.
3. HEIDI: Hallo, Stefan. Frau Schulz sagt, du _______[a] morgen in ihre Sprechstunde kommen.
   STEFAN: Morgen _______[b] ich nicht, ich habe keine Zeit.
   HEIDI: Das _______[c] du Frau Schulz schon selbst sagen. Mach's gut.

[1]*diligently*

3.3 Accusative case: personal pronouns

As in English, certain German pronouns change depending on whether they are the subject or the object of a verb.

Ich möchte mitkommen.	*I would like to come along.*
Nimmst du **mich** mit?	*Will you take me with you?*
Er kommt aus Wien.	*He is from Vienna.*
Kennst du **ihn**?	*Do you know him?*

WISSEN SIE NOCH?

The accusative case is used to indicate direct objects of verbs.

Review grammar 2.1.

A. First- and second-person pronouns: nominative and accusative forms

Nominative	Accusative	
ich	mich	*me*
du	dich	*you*
Sie	Sie	*you*
wir	uns	*us*
ihr	euch	*you*
Sie	Sie	*you*

Wer bist **du**? Ich kenne **dich** nicht.	*Who are you? I don't know you.*
Wer seid **ihr?** Ich kenne **euch** nicht.	*Who are you (people)? I don't know you.*

B. Third-person pronouns: nominative and accusative forms

	Nominative	Accusative	
Masculine	er	ihn	*him, it*
Feminine	sie		*her, it*
Neuter	es		*it*
Plural	sie		*them*

der → er
den → ihn
das → es
die → sie

Recall that third-person pronouns reflect the grammatical gender of the noun they stand for: **der Film → er; die Gitarre → sie; das Foto → es.** This relationship also holds true for the accusative case: **den Film → ihn; die Gitarre → sie; das Foto → es.** Note that only the masculine singular pronoun has a different form in the accusative case.

Wo ist der Spiegel? Ich sehe **ihn** nicht.	*Where is the mirror? I don't see it.*
Das ist meine Schwester Jasmin. Du kennst **sie** noch nicht.	*This is my sister Jasmin. You don't know her yet.*
—Wann kaufst du die Bücher?	*—When will you buy the books?*
—Ich kaufe **sie** morgen.	*—I'll buy them tomorrow.*

Übung 5 Minidialoge

Ergänzen Sie **mich, dich, uns, euch, Sie.**

1. KATRIN: Holst du mich heute Abend ab, wenn wir ins Kino gehen?
   THOMAS: Natürlich hole ich _______ ab!
2. STEFAN: Hallooo! Hier bin ich, Albert! Siehst du _______[a] denn nicht?
   ALBERT: Ach, *da* bist du. Ja, jetzt sehe ich _______[b].
3. SARAH: Guten Tag, Frau Schulz. Sie kennen _______ noch nicht. Wir sind neu in Ihrer Klasse. Das ist Caleb, und ich bin Sarah.
   FRAU SCHULZ: Guten Tag, Caleb. Guten Tag, Sarah.
4. MONIKA: Hallo, Albert. Hallo, Thomas. Katrin und ich besuchen _______ heute.
   ALBERT UND THOMAS: Toll! Bringt Kuchen mit!
5. STEFAN: Heidi, ich mag _______[a]!
   HEIDI: Das ist schön, Stefan. Ich mag _______[b] auch.
6. FRAU SCHULZ: Spreche ich laut genug? Verstehen Sie _______[a]?
   KLASSE: Ja, wir verstehen _______[b] sehr gut, Frau Schulz.
7. STEFAN UND ALBERT: Auf Wiedersehen, Frau Schulz! Schöne Ferien! Und vergessen Sie uns nicht!
   FRAU SCHULZ: Natürlich nicht! Wie kann ich _______ denn je vergessen?

Übung 6 Der Deutschkurs

das → es
den → ihn
die → sie

MODELL: Machst du gern **das** Arbeitsbuch für ***Kontakte***? →
Ja, ich mache **es** gern.
oder: Nein, ich mache **es** nicht gern.

1. Machst du gern **das** Arbeitsbuch für ***Kontakte***?
2. Kannst du **das** deutsche Alphabet aufsagen?
3. Kennst du **den** beliebtesten deutschen Vornamen für Jungen?
4. Liest du gern **die** Grammatik?
5. Lernst du gern **den** Wortschatz?
6. Kennst du **die** Studenten und Studentinnen in der Klasse?
7. Vergisst du oft **die** Hausaufgaben?
8. Magst du **deinen** Lehrer oder **deine** Lehrerin?

Übung 7 Was machen diese Personen?

Beantworten Sie die Fragen negativ.

MODELL: Kauft Michael das Buch? →
Nein, er kauft es nicht, er liest es.

Verwenden Sie diese Verben.

anrufen, ruft an
anziehen, zieht an
anzünden, zündet an
ausmachen, macht aus
essen, isst
kaufen
schreiben
trinken
verkaufen
waschen, wäscht

1. Liest Maria den Brief?

2. Isst Michael die Suppe?

3. Macht Maria den Fernseher an?

4. Kauft Michael das Auto?

5. Zieht Michael die Hose aus?

6. Trägt Maria den Rock?

7. Bestellt[1] Michael das Schnitzel?

8. Besucht Michael seinen Freund?

9. Kämmt Maria ihr Haar?

10. Bläst Michael die Kerzen aus[2]?

3.4 Word order: dependent clauses

Use a conjunction such as **wenn** (*when, if*) or **weil** (*because*) to add a modifying clause to a sentence.

Mehmet hört Musik, **wenn** er traurig ist.	*Mehmet listens to music whenever he is sad.*
Renate geht nach Hause, **weil** sie müde ist.	*Renate is going home because she is tired.*

In the preceding examples, the first clause is the main clause. The clause introduced by a conjunction is called a *dependent clause*. In German, the verb in a dependent clause occurs at the end of the clause.

When **wenn** or **weil** begins a clause, the conjugated verb appears at the end of the clause.

MAIN CLAUSE	DEPENDENT CLAUSE
Ich bleibe im Bett,	wenn ich krank **bin.**
I stay in bed	*when I am sick.*

[1]bestellen *to order (in a restaurant)* [2]ausblasen *to blow out*

In sentences beginning with a dependent clause, the entire clause acts as the first element in the sentence. The verb of the main clause comes directly after the dependent clause, separated by a comma. As in all German statements, the verb is in second position. The subject of the main clause follows the verb.

I	II	III	
DEPENDENT CLAUSE	VERB	SUBJECT	
Wenn ich krank bin,	bleibe	ich	im Bett.
When I'm sick,	*I stay in bed.*		
Weil sie müde ist,	geht	Renate	nach Hause.
Because she's tired,	*Renate is going home.*		

Übung 8 Warum denn?

Beantworten Sie die Fragen.

MODELL: Warum gehst du nicht in die Schule? → Weil ich krank bin.

1. Warum gehst du nicht in die Schule?	a. Durst haben
2. Warum liegt dein Bruder im Bett?	b. krank sein
3. Warum esst ihr denn schon wieder?	c. traurig sein
4. Warum kommt Nora nicht mit ins Kino?	d. Langeweile haben
5. Warum sieht Jutta schon wieder fern?	e. Angst haben
6. Warum sitzt du allein in deinem Zimmer?	f. glücklich sein
7. Warum trinken sie Wasser?	g. lernen müssen
8. Warum machst du denn das Licht an?	h. müde sein
9. Warum singt Jens den ganzen Tag?	i. Hunger haben
10. Warum bleibst du zu Hause?	j. keine Zeit haben

Übung 9 Ist das immer so?

Sagen Sie, wie das für andere Personen ist und wie das für Sie ist.

MODELL: S1: Was macht Albert, wenn er müde ist?
S2: Wenn Albert müde ist, geht er nach Hause.
S1: Und du?
S2: Wenn ich müde bin, trinke ich einen Kaffee.

1. Albert ist müde.	a. Sie trifft Michael.
2. Maria ist glücklich.	b. Er geht nach Hause.
3. Herr Ruf hat Durst.	c. Sie fährt mit dem Taxi.
4. Frau Wagner ist in Eile.	d. Sie kauft einen Hamburger.
5. Heidi hat Hunger.	e. Er trinkt eine Cola.
6. Frau Schulz hat Ferien.	f. Er geht zum Arzt.
7. Hans hat Angst.	g. Er ruft: „Mama, Mama".
8. Stefan ist krank.	h. Sie fliegt nach Deutschland.

3.5 Dependent clauses and separable-prefix verbs

As you know, the prefix of a separable-prefix verb occurs at the end of an independent clause.

Rolf **steht** immer früh **auf.**	*Rolf always gets up early.*

In a dependent clause, the prefix is attached to the verb form, which is placed at the end of the clause.

Rolf ist immer müde, wenn er früh **aufsteht.**	*Rolf is always tired when he gets up early.*
Eske, bitte **mach** das Fenster nicht **auf!** Es wird kalt, wenn du es **aufmachst.**	*Eske, please don't open the window. It gets cold when you open it.*

When there are two verbs in a dependent clause, such as a modal verb and an infinitive, the modal verb comes last, following the infinitive.

INDEPENDENT CLAUSE	Rolf **muss** früh **aufstehen.**	*Rolf has to get up early.*
DEPENDENT CLAUSE	Er ist müde, wenn er früh **aufstehen muss.**	*He is tired when he has to get up early.*
INDEPENDENT CLAUSE	Eske hat kein Geld. Sie **kann** nichts **machen.**	*Eske doesn't have any money. She can't do anything.*
DEPENDENT CLAUSE	Sie hat Langeweile, weil sie nichts **machen kann.**	*She's bored because she can't do anything.*

Übung 10 Warum ist das so?

MODELL: Jürgen ist wütend, weil er immer so früh aufstehen muss.

1. Jürgen ist wütend.	a. Sie muss noch einkaufen.
2. Silvia ist froh.	b. Er muss immer so früh aufstehen.
3. Claire ist in Eile.	c. Seine Freundin nimmt ihn zur Uni mit.
4. Josef ist traurig.	d. Er sieht immer fern.
5. Thomas geht nicht zu Fuß.	e. Sie kann nicht schwimmen.
6. Willi hat selten Langeweile.	f. Er will seine Eltern besuchen.
7. Nesrin hat Angst vor Wasser.	g. Melanie ruft ihn nicht an.
8. Mehmet fährt in die Türkei.	h. Sie muss heute nicht arbeiten.

KAPITEL 4

Ereignisse und Erinnerungen

In **Kapitel 4,** you will begin to talk about things that happened in the past: your own experiences and those of others. You will also talk about different kinds of memories.

Themen

Der Alltag
Urlaub und Freizeit
Geburtstage und Jahrestage
Ereignisse

Kulturelles

KLI: Universität und Studium
Musikszene: „Du hast den Farbfilm vergessen" (Nina Hagen)
KLI: Feiertage und Bräuche
Filmclip: *Jenseits der Stille* (Caroline Link)
Videoecke: Feste und Feiern

Lektüren

Kurzgeschichte: Vater im Baum (Margret Steenfatt)
Film: *Jenseits der Stille* (Caroline Link)

Strukturen

4.1 Talking about the past: the perfect tense
4.2 Strong and weak past participles
4.3 Dates and ordinal numbers
4.4 Prepositions of time: **um, am, im**
4.5 Past participles with and without **ge-**

Max Liebermann: *Wannseelandschaft* (1924), Christie's Images, London

KUNST UND KÜNSTLER

Max Liebermann (1847–1935) ist ein jüdischer Maler aus Berlin und ein Pionier der modernen Malerei. Seine frühen Bilder zeigen oft naturalistische Szenen von Menschen bei der Arbeit. Später malt Liebermann sehr private impressionistische Bilder wie dieses hier. Max Liebermann hatte ein Sommerhaus direkt am Wannsee.

Schauen Sie sich das Bild an und beantworten Sie die folgenden Fragen.

1. Was sehen Sie auf dem Bild: einen Mann, eine Frau, ein Kind, Häuser, Segelboote, Bäume, Blumen, Autos, einen Garten, einen Wald, einen Hund?
2. Wo sind die Menschen: in einem Garten, in der Stadt, an einem See, auf dem Land, auf dem Rasen, in einem Zimmer?
3. Was tragen die Menschen: eine Hose, ein weißes Kleid, ein hellblaues Kleid, einen Hut, eine Sonnenbrille, einen Bademantel, Sandalen, Stiefel, Schuhe?
4. Welche Farben dominieren: blau, braun, gelb, grau, grün, lila, orange, rosa, rot, schwarz, weiß?
5. Welche Gefühle ruft das Bild hervor: Angst, Einfachheit[1], Elan[2], Glück, Hoffnung, Freiheit, Langeweile[3], Neugier[4], Ruhe, Sehnsucht[5]?
6. An was denken Sie: Arbeit, Freizeit, Schule, Urlaub, Studium?

[1]*simplicity* [2]*pep* [3]*boredom* [4]*curiosity* [5]*yearning*

Situationen

Der Alltag

Grammatik 4.1

Ich habe geduscht.

Ich habe gefrühstückt.

Ich bin in die Uni gegangen.

Ich bin in einem Kurs gewesen.

Ich habe mit meinen Freunden Kaffee getrunken.

Ich bin nach Hause gekommen.

Ich habe zu Mittag gegessen.

Ich bin nachmittags zu Hause geblieben.

Ich habe abends gelernt.

Situation 1 Umfrage: Hast du das gemacht?

MODELL: S1: Hast du zum Frühstück etwas gegessen?
S2: Ja.
S1: Unterschreib bitte hier.

	UNTERSCHRIFT
1. Hast du zum Frühstück etwas gegessen?	________
2. Hast du Kaffee getrunken?	________
3. Hast du heute die Zeitung gelesen?	________
4. Hast du gestern einen Film gesehen?	________
5. Hast du gestern mit deiner Freundin telefoniert?	________
6. Hast du in der Bibliothek gearbeitet?	________
7. Hast du deine Freunde zum Essen eingeladen?	________
8. Hast du gestern viele E-Mails geschrieben?	________
9. Bist du vor Mitternacht ins Bett gegangen?	________

Situation 2 Dialog: Das Fest

Silvia und Jürgen sitzen in der Mensa und essen zu Mittag.

SILVIA: Ich bin furchtbar ________.
JÜRGEN: Bist du wieder so spät ins Bett ________?
SILVIA: Ja. Ich bin heute früh erst um vier Uhr nach Hause ________.
JÜRGEN: Wo ________ du denn so lange?
SILVIA: Auf einem Fest.
JÜRGEN: ________?
SILVIA: Ja, ich habe ein paar alte Freunde ________ und wir haben uns sehr gut unterhalten.
JÜRGEN: Kein Wunder, ________!

Situation 3 Zum Schreiben: Ein Tagebuch

Schreiben Sie ein Tagebuch! Vielleicht haben Sie das früher schon einmal auf Englisch gemacht. Machen Sie sich zuerst ein paar Notizen. Was ist letzte Woche passiert? Was haben Sie gemacht? Was wollen Sie nicht vergessen?

MODELL: Letzte Woche habe/bin ich ...

28. Juli 2015
Habe einen total coolen Jungen kennengelernt! Er heißt Billy, eigentlich Paul, aber er sieht aus wie Billy Idol. Er ist total süß!! Habe gleich einen Brief an Geli geschrieben und ihr von Billy erzählt. Warte jetzt auf Gelis Antwort... Außerdem haben wir Zeugnisse bekommen. Das war nicht so gut...

Juttas Tagebuch

Urlaub und Freizeit

Grammatik 4.2

Jutta ist ins Schwimmbad gefahren.

Sie hat in der Sonne gelegen.

Sie ist geschwommen.

Sie hat Musik gehört.

Jens und Robert haben Postkarten geschrieben.

Sie sind in den Bergen gewandert.

Sie haben Tennis gespielt.

Sie haben viel gelesen.

Situation 4 Bildgeschichte: Familie Wagner im Urlaub am Strand

KULTUR ... LANDESKUNDE ... INFORMATIONEN

UNIVERSITÄT UND STUDIUM

- Wann haben Sie mit dem Studium am College oder an der Universität angefangen?
- Welche Voraussetzungen[1] (High-School-Abschluss, Prüfungen usw.) braucht man für ein Studium?
- An welchen Universitäten haben Sie sich beworben[2]?
- Studieren Sie an einer privaten oder einer staatlichen Hochschule[3]?
- Müssen Sie Studiengebühren[4] bezahlen?
- Wie lange dauert Ihr Studium voraussichtlich?
- Welchen Abschluss[5] haben Sie am Ende Ihres Studiums?
- Was für Kurse müssen Sie belegen?

Die meisten Universitäten in Deutschland sind öffentliche Universitäten. Es gibt nur wenige private Hochschulen. Bis vor ein paar Jahren gab es in vielen Bundesländern Studiengebühren. Sie wurden aber alle abgeschafft. Auch in Österreich gibt es keine Studiengebühren mehr. In der Schweiz bezahlt man von Uni zu Uni unterschiedliche Studiengebühren, von 425 Franken pro Semester in Neuenburg bis zu 2.000 Franken in Lugano.

Viele Studenten arbeiten während des Semesters und in den Semesterferien. Um die 30% bekommen ein Stipendium oder eine finanzielle Hilfe vom Staat, das sogenannte BAföG (Bundesausbildungsförderungsgesetz[6]). Der BAföG-Höchstsatz[7] beträgt zur Zeit 735 Euro im Monat.

Um an einer Universität zu studieren, braucht man normalerweise das Abitur[8]. Seit über 10 Jahren gibt es in Deutschland und in Europa die neuen Bachelor- und Masterstudiengänge. Diese Studiengänge wurden eingeführt, um international vergleichbare[9] Studienabschlüsse zu haben. Ein Bachelorstudium dauert meist drei Jahre, ein Masterstudium zwei weitere Jahre.

Etwa 12% der Studierenden in Deutschland kommen aus dem Ausland[10], die meisten aus der Türkei, gefolgt von China, Russland, Österreich und Italien. Zum Vergleich: Der Anteil ausländischer Studenten in den USA beträgt 4%. Allerdings studieren relativ wenige Deutsche im Ausland, nämlich nur etwas über 6%, die meisten in Österreich, gefolgt von den Niederlanden, der Schweiz, Großbritannien und den USA. US-amerikanische Studenten gehen allerdings noch seltener für ein Semester oder länger ins Ausland, nämlich nur 1%.

Studierende in der Bibliothek der FU Berlin

[1]*prerequisites* [2]sich ... *applied* [3]*college, university* [4]*fees, tuition* [5]*degree; diploma*

[6]*federal law for the promotion of higher education* [7]*maximum amount* [8]roughly: *high school diploma* [9]*comparable* [10]aus ... *from abroad*

Situation 5 Dialog: Jens' und Juttas Wochenende

Es ist Montag. Jutta und Jens treffen sich auf dem Schulhof ihrer Schule und reden über ihr Wochenende.

JENS: Hallo, Jutta!

JUTTA: Grüß dich, Jens! Was hast du am Wochenende ___________?

JENS: Ach, nichts Besonderes. Ich habe ___________ und Musik ___________. Es war langweilig. Und du?

JUTTA: Ich bin mit meinen Eltern in die Berge ___________. Wir sind viel ___________ und haben sogar ein Picknick gemacht. Das war ganz super.

JENS: Das hört sich wirklich toll an!

JUTTA: Ja, auf jeden Fall. Komm doch das nächste Mal mit.

JENS: Au ja, gern.

Am Wochenende haben wir gefeiert.

Situation 6 Am Wochenende

Schauen Sie auf die Bilder und finden Sie die passende Antwort auf jede Frage.

1. _____ Was hat Frau Ruf am Freitag gemacht?
2. _____ Was hat Jutta am Samstag gemacht?
3. _____ Was haben Jutta und Hans am Sonntag gemacht?
4. _____ Was haben die Okonkwos am Sonntag gemacht?
5. _____ Was hat Michael am Samstag gemacht?
6. _____ Was hat Jens am Sonntag gemacht?
7. _____ Was hat Herr Ruf am Freitag gemacht?
8. _____ Was hat Richard am Samstag gemacht?

a. Sie haben den Hund gebadet.
b. Er hat mit Maria zu Abend gegessen.
c. Sie sind in den Bergen gewandert.
d. Er hat stundenlang ferngesehen.
e. Sie hat Billy kennengelernt.
f. Sie ist nach Augsburg gefahren.
g. Er hat für seine Familie die Wäsche gewaschen.
h. Er ist zum Strand gefahren.

Miniwörterbuch

damalig	former
verlassen, verließ, verlassen	to leave
gelten als	to be known as
der **Strand,** die **Strände**	beach

MUSIKSZENE

„Du hast den Farbfilm vergessen" (1974, Ostdeutschland) *Nina Hagen*

Biografie Nina Hagen wurde 1955 in der damaligen DDR geboren. Sie ist die Stieftochter von Wolf Biermann, einem bekannten Liedermacher. Beide verließen 1976 die DDR. Nina Hagen gilt als die deutsche „Godmother" des Punk. Neben ihrer Karriere als Sängerin und Songwriterin ist sie auch als Schauspielerin berühmt geworden. Ihr Hit „Du hast den Farbfilm vergessen" wird von vielen gesungen.

Nina Hagen
© Sean Gallup/Getty Images

NOTE: For copyright reasons, the songs referenced in **MUSIKSZENE** have not been provided by the publisher. The song can be found online at various sites such as YouTube, Amazon, or the iTunes store.

Vor dem Hören Machen Sie gern Fotos? Haben Sie schon mal Schwarz-Weiß-Fotos gemacht?

Nach dem Hören Beantworten Sie die Fragen.

1. Wo ist die Sängerin?
2. Wie heißt ihr Freund?
3. Welche Farben nennt sie?
4. Welche Fotos hat der Freund von ihr gemacht?
5. Warum ist es schlimm, dass der Freund den Farbfilm vergessen hat?
6. Was passiert, wenn der Freund den Farbfilm noch einmal vergisst?

Situation 7 Interview: Letztes Wochenende

Was hast du am Wochenende gemacht?

1. Hast du am Samstag lang geschlafen? Wie lang?
2. Bist du tanzen gegangen?
3. Hast du mit jemandem gefrühstückt?
4. Hast du Sport getrieben? Welchen Sport?
5. Hast du Fotos gemacht?
6. Hast du viele Mails geschrieben? An wen?
7. Bist du ins Kino gegangen? Welchen Film hast du gesehen?
8. Hast du ein Buch gelesen? Welches?
9. Hast du Geld verdient? Was hast du gemacht?
10. Hast du deine Seminare vorbereitet? Welche?

Geburtstage und Jahrestage

Grammatik 4.3–4.4

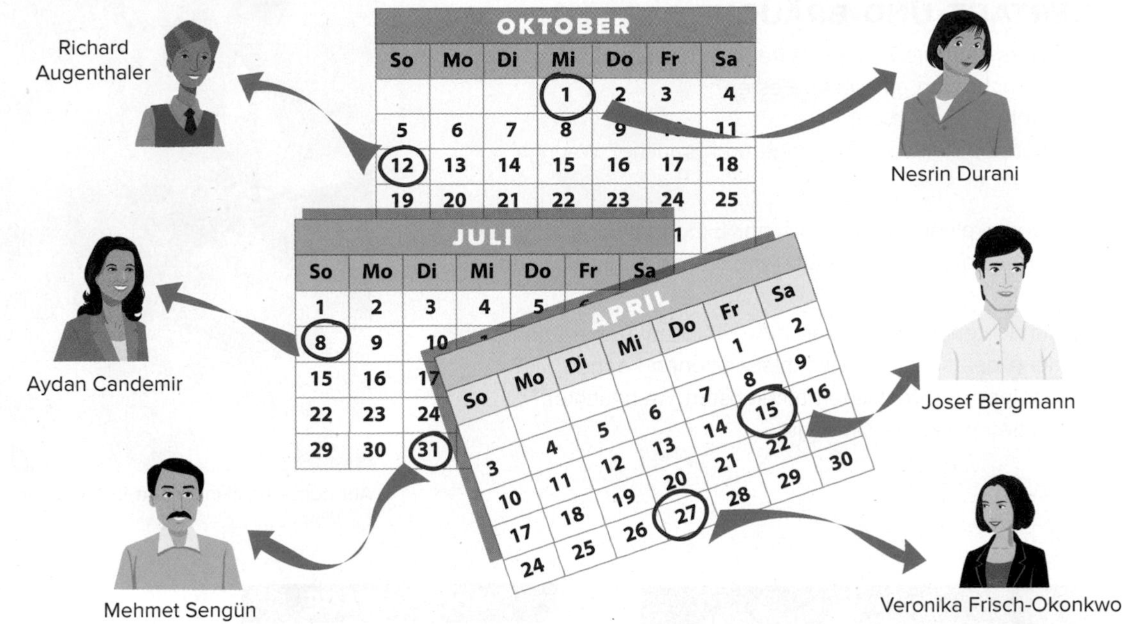

Nesrin hat am ersten Oktober Geburtstag.
Richard hat am zwölften Oktober Geburtstag.
Aydan hat am achten Juli Hochzeitstag.
Mehmet ist am einunddreißigsten Juli geboren.
Josef ist am fünfzehnten April geboren.
Veronika hat am siebenundzwanzigsten April Geburtstag.

 Situation 8 Dialog: Welcher Tag ist heute?

Bringen Sie die Sätze in die richtige Reihenfolge.

Nesrin und Sofie sitzen im Café. Sofie fragt:

_____ Nein, welches Datum?
_____ Montag.
_____ Wirklich? Ich dachte, er hat im August Geburtstag.
_____ Hast du denn schon ein Geschenk?
__1__ Welcher Tag ist heute?
_____ Ach so, der dreißigste.
_____ Der dreißigste? Mann, dann ist ja heute Willis Geburtstag!
_____ Das ist es ja! Ich hab' noch nicht einmal ein Geschenk.
_____ Nein, Christian hat im August Geburtstag, aber Willi im Mai.
_____ Na, dann viel Spaß beim Geschenke kaufen!

KULTUR ... LANDESKUNDE ... INFORMATIONEN

FEIERTAGE UND BRÄUCHE°

°customs

- Welches sind die Familienfeste in Ihrem Land?
- Was macht man an diesen Festen?
- Wer feiert[1] zusammen?
- Kennen Sie deutsche Feiertage und Bräuche? Wenn ja, welche?

Der Adventskalender: Ein deutscher Exportartikel in christlicher Tradition ist über 100 Jahre alt. Amerika ist das Importland Nummer 1.

- Weihnachten in Deutschland: An welchen Tagen feiert man?
- Welche deutschen Weihnachtstraditionen kennen Sie?
- Wie feiern die Deutschen am liebsten Weihnachten? Analysieren Sie die Umfrage.

Auf dem Christkindlmarkt in München im Jahre 1897

FOCUS-FRAGE

„Wo verbringen Sie Weihnachten?"

EIN FAMILIENFEST ZU HAUSE

von 1300 Befragten* antworteten

zu Hause	73 %
bei den Eltern/Kindern	21 %
bei Freunden	3 %
im Urlaub	3 %

83 Prozent der Deutschen verbringen Weihnachten im Kreis der Familie, 7 Prozent zusammen mit dem Partner, 6 Prozent mit Freunden, 4 Prozent feiern alleine.

* Repräsentative Umfrage des Sample-Instituts für FOCUS im Dezember

[1]celebrates

Situation 9 Informationsspiel: Geburtstage

MODELL: S1: Wann ist Willi geboren?
S2: Am dreißigsten Mai 1991.

Person	Geburtstag
Willi	30. Mai 1991
Sofie	
Claire	1. Dezember 1990
Melanie	
Nora	4. Juli 1998
Thomas	
Heidi	23. Juni 1995
mein(e) Partner(in)	
sein/ihr Vater	
seine/ihre Mutter	

Situation 10 Fest- und Feiertage

1. Wann feiert man den Valentinstag?
2. Wann feiert man den Nationalfeiertag in deinem Land?
3. Welcher Feiertag ist in deiner Familie der wichtigste? Wann feiert man ihn?
4. Was feiert man am 1. Mai?
5. Welcher Festtag ist der wichtigste für dich? Dein Geburtstag? Der Tag deiner Taufe? Der Tag deiner Religionsmündigkeit[2] (Bar-Mizwa / Bat-Mizwa)? Ein anderer Tag?
6. Was feiert man am 31. Oktober?
7. Wann feiert man Weihnachten? An welchem Tag bekommt man die Geschenke?
8. Wann feiert man das islamische Opferfest[3]?
9. Wann feiert man Jom Kippur?
10. Wann beginnt der Frühling?

Situation 11 Erfindungen und Entdeckungen

MODELL: S1: Wer hat den Bleistift erfunden?
S2: _______.
S1: Wann hat er ihn erfunden?
S2: _______.

das Toilettenpapier
der Kugelschreiber
der Bleistift
der Kaffeefilter
die Schallplatte
die Schreibmaschine
das Akkordeon

Cyril Demian
1829

Friedrich Staedtler
1662

Emil Berliner
1887

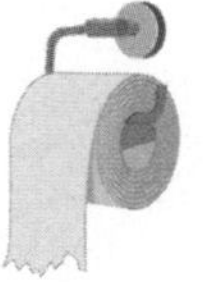

Joseph Cayetti
1857

Melitta Bentz
1908

Laszlo Biro
1938

Peter Mitterhofer
1864

[2] *religious maturity* [3] *Feast of the Sacrifice*

MODELL: S1: Wer hat das Radium entdeckt?
S2: ________.
S1: Wann hat sie es entdeckt?
S2: ________.

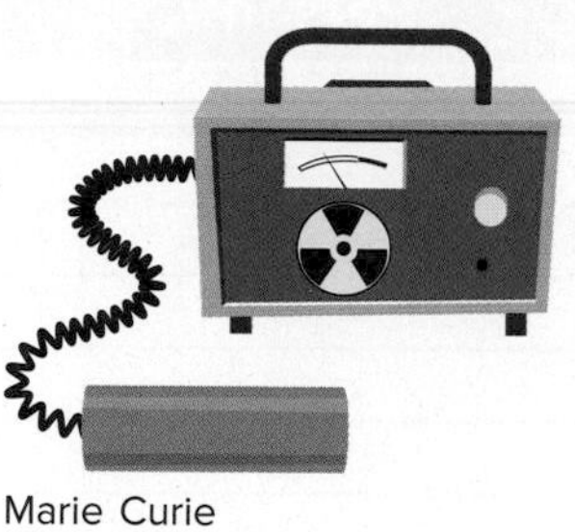

Marie Curie
1898

Friedrich Herschel
1781

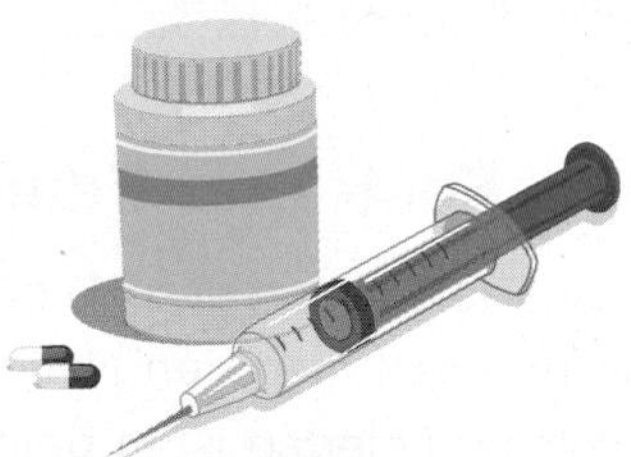

Alexander Fleming
1928

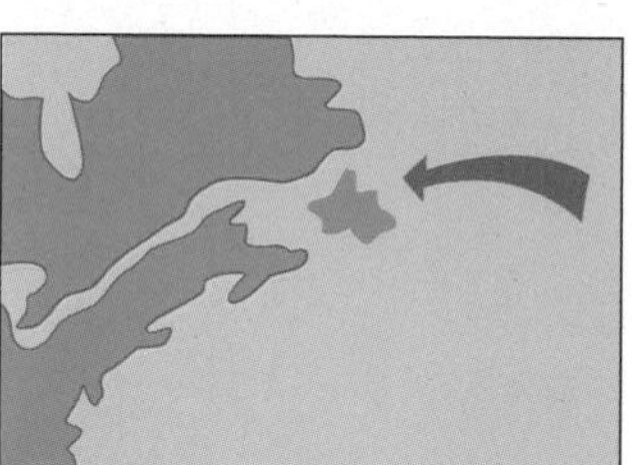

Leif Eriksson
1000

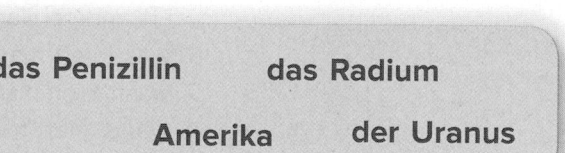

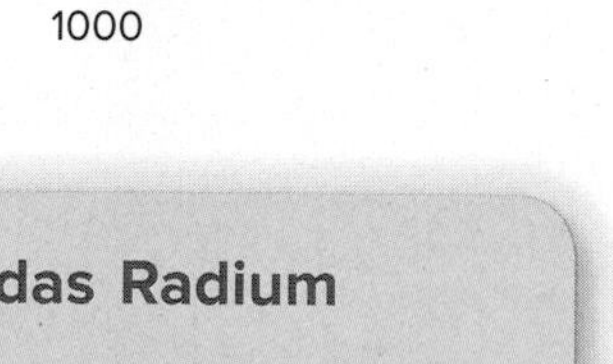

Situation 12 Interview

1. Wann bist du geboren (Tag, Monat, Jahr)? Wann ist dein Freund / deine Freundin geboren (Tag, Monat, Jahr)? Wann ist dein Vater / deine Mutter geboren (Tag, Monat, Jahr)?
2. Wann bist du in die Schule gekommen (Monat, Jahr)? Wann hast du angefangen zu studieren (Monat, Jahr)?
3. Was war der wichtigste Tag in deinem Leben? Was ist da passiert? In welchem Monat war das? In welchem Jahr?
4. In welchem Monat warst du zum ersten Mal verliebt? hast du zum ersten Mal Geld verdient? hast du einen Unfall gehabt?
5. An welchen Tagen in der Woche arbeitest du? hast du frei? gehst du ins Kino? besuchst du deine Eltern? gehst du in Vorlesungen? gehst du ins Sprachlabor? gehst du in die Bibliothek?
6. Um wie viel Uhr stehst du auf? ist dein erster Kurs? gehst du nach Hause? gehst du ins Bett?
7. Was hast du vor zwei Tagen gemacht? vor zwei Wochen?

Lektüre

Vor dem Lesen

LESEHILFE

As you learned in **Kapitel 3**, it is often useful to first read the title, glance over the complete text, and look at the images in order to get a general idea of what a text will be about. "Vater im Baum" is the title of this fictional text for young readers. Write down any questions that come to mind when you consider the title of this text. Then read the text and complete the activities.

A. **Eigenschaften von Tieren.** Vervollständigen Sie die Sätze mit den Adjektiven.

1. Füchse sind ______.
2. Hunde sind ______.
3. Vögel sind ______.
4. Schildkröten sind ______.
5. Löwen sind ______.
6. Bären sind ______.

mutig **schlau** **stark** **langsam** **treu** **frei**

B. Lesen Sie die Wörter im Miniwörterbuch. Dann suchen Sie sie im Text und unterstreichen Sie sie.

Vater im Baum

von Margret Steenfatt

Miniwörterbuch

der **Wagen**	here: *car*
der **Scherz**	*joke*
die **Wahrheit**	*truth*
glauben	*to believe*
flüstern	*to whisper*
nichts taugen	*to be no good*
das **Gerät**	*device*
der **Krimi**	*crime show*
der **Stein,** die **Steine**	*rock*
werfen	*to throw*
prügeln	*to spank*
verschwinden	*to disappear*
fassungslos	*perplexed*
hinaufsteigen	*to climb up*
klettern	*to climb*
der **Quatsch**	*nonsense*
heben	*to raise*
fort	*away*

„Mama, Vater sitzt im Baum!"

„Erzählt doch keine Märchen[4], Kinder. Papa wäscht den Wagen!"

„Nein, Mama, er sitzt im Baum!"

„Lasst mich in Ruhe mit euren Scherzen. Wir wollen gleich in die Stadt fahren. Ich habe noch zu tun."

„Aber es ist die Wahrheit, Mama. Er will nicht herunterkommen."

„Jetzt wird es mir zu bunt[5]. Geht hinaus und spielt." Die Mutter schlägt die Haustür zu.

„Sie will uns nicht glauben", sagt Christian zu Sabine. „Was tun wir jetzt?"

„Nichts."

„Und Papa?"

„Den kriegen wir schon 'runter."

„Wie denn?"

„Ich sag's dir ins Ohr." Sabine beugt sich zum Bruder und flüstert etwas. Gleich darauf stürmen beide Kinder zur Garage.

Der neue Ford steht vor der Tür. Christian und Sabine schwingen sich aufs Autodach[6]. Sie rufen laut zum Baum hinüber[7]: „Papa, schau her!" Dann trampeln sie vereint mit ungeheurem Getöse[8] auf dem Blechdach herum. Nach einer Weile beginnt der Lack[9] zu splittern. Es zeigen sich Beulen[10] im Dach. „Papa!", brüllen die Kinder aus vollem Halse. „Schau doch, Papa!"

Auf dem Baum rührt sich nichts. Ein paar Pfeifenwölkchen[11] schweben zum Himmel.

„Sabine, dein Plan taugt nichts", sagt Christian. „Ich weiß was Besseres, warte!" Er rutscht vom Autodach und läuft ins Haus. Ein paar Minuten später schleppt er den Fernseher herbei und setzt ihn unter den Baum. Er schaltet das Gerät ein und stellt es auf volle Lautstärke. „Komm endlich, Vater, 'n Krimi gibt's."

Aber noch immer regt sich droben nichts.

Die Kinder sammeln Steine, kleine zunächst, und werfen. Sie zielen nicht sehr gut. „Jetzt wird er gleich heruntersteigen, weil er uns prügeln will", sagt Sabine. „Dann müssen wir schnell verschwinden!"

Sie nehmen größere Steine und treffen hin und wieder. Doch der Vater im Baum gibt keinen Laut von sich[12] und die Kinder sehen ein[13], dass er nicht mehr herabkommen wird. Sie toben und kreischen und brüllen[14].

Da kommt die Mutter aus dem Haus, reisefertig, mit Koffer und Tasche. Sie geht zur Garage und erblickt das zerbeulte Auto. Sie sieht die Kinder mit Steinen in den Händen und im Baum den Vater, ihren Mann. „Was soll das bedeuten?", fragt sie fassungslos.

„Vater sitzt im Baum!", schreit Christian. „Er will nicht herunter!"

„Das ist unmöglich", sagt die Mutter. „Euer Vater sitzt nicht in Bäumen."

„So sieh ihn doch an, wie er dort sitzt und sich um nichts kümmert[15]!", kreischt Sabine.

„Eduard!", ruft die Mutter beschwörend, „lass diese Albernheiten[16]. Wir müssen fahren!" – „Eduard, so komm doch endlich herunter!" – „Warum antwortest du denn nicht?"

„Steigt doch mal hinauf, Kinder!", bittet die Mutter. „Ich verstehe das alles nicht."

Sabine und Christian beginnen zu klettern. Der Baum ist ziemlich hoch. Oben in der Krone sitzt der Vater. Er sagt kein Wort und rührt sich nicht[17]. Christian steigt schneller als Sabine. Er kommt dem Vater immer näher. Fast hat er ihn erreicht[18]. „Papa, was soll der Quatsch!", ruft Christian.

Mit einem Mal hebt der Vater die Arme, hebt und senkt sie, richtet sich auf und fliegt wie ein Vogel davon, fort vom Baum, fort vom Haus, fort von der Familie.

[4]Erzählt ...: *Don't talk nonsense* [5]wird ...: *this is getting to be too much for me* [6]*roof of the car* [7]rufen ...: *call loudly across to the tree* [8]*noise, racket* [9]*varnish* [10]*dents* [11]*little clouds of pipe smoke* [12]gibt ...: *makes no sound* [13]sehen ...: *realize* [14]toben ...: *romp and scream and yell* [15]sich ...: *is concerned about nothing* [16]*absurdities* [17]rührt ...: *doesn't move* [18]*reached*

Arbeit mit dem Text

A. Personen. Wer macht was?

	Vater	Mutter	Kinder
1. davonfliegen	☐	☐	☐
2. die Haustür zuschlagen	☐	☐	☐
3. flüstern	☐	☐	☐
4. im Baum sitzen	☐	☐	☐
5. in den Baum klettern	☐	☐	☐
6. keinen Laut von sich geben	☐	☐	☐
7. mit Koffer aus dem Haus kommen	☐	☐	☐
8. Pfeife[19] rauchen	☐	☐	☐
9. Steine sammeln	☐	☐	☐
10. trampeln	☐	☐	☐
11. zum Baum hinüberrufen	☐	☐	☐
12. zur Garage stürmen	☐	☐	☐

B. Handlung. Die folgenden Sätze fassen die Handlung zusammen. Bringen Sie sie in die richtige Reihenfolge.

_____ Beide Kinder stürmen zur Garage.
_____ Die Mutter schlägt die Haustür zu.
__*1*__ Der Vater sitzt im Baum.
_____ Sie trampeln auf dem Blechdach herum.
_____ Christian holt den Fernseher.
_____ Die Kinder toben und machen viel Lärm.
_____ Der Vater will nicht herunterkommen.
_____ Die Kinder werfen Steine.
_____ Sie muss in die Stadt fahren.
_____ Der Vater fliegt wie ein Vogel davon.
_____ Es gibt einen Krimi.
_____ Der Vater raucht Pfeife.
_____ Die Mutter kommt wieder aus dem Haus.
_____ Die Kinder klettern in den Baum.

C. Inhalt. Warum sitzt der Vater im Baum? Schreiben Sie drei Möglichkeiten auf.

Der Vater sitzt im Baum, ...

weil __________.
weil __________.
weil __________.

Nach dem Lesen

Beantworten Sie die folgenden Fragen.

1. Warum glaubt die Mutter nicht, dass der Vater im Baum sitzt?
2. Der Vater fliegt davon wie ein Vogel. Was symbolisiert das?
3. Wie ist das Leben des Vaters, bevor er auf den Baum klettert?
4. Wie geht das Leben weiter? Das Leben der Kinder, das Leben der Mutter, das Leben des Vaters?

[19] *pipe*

Ereignisse

Grammatik 4.5

1. Wann sind Sie aufgewacht?
2. Wann sind Sie aufgestanden?
3. Wann sind Sie von zu Hause weggegangen?
4. Wann hat Ihr Kurs angefangen?
5. Wann hat Ihr Kurs aufgehört?
6. Wann sind Sie nach Hause gekommen?
7. Wann haben Sie unsere Prüfungen korrigiert?

1. Wann hast du eingekauft?
2. Wann hast du das Geschirr gespült?
3. Wann hast du mit deiner Freundin telefoniert?
4. Wann hast du ferngesehen?
5. Wann hast du dein Fahrrad repariert?
6. Wann bist du abends ausgegangen?

Situation 13 Michaels freier Tag

Michael telefoniert mit Maria. Sie reden über Michaels freien Tag. Bringen Sie die Sätze in die richtige Reihenfolge.

_______ Tut mir leid, Maria, an dich habe ich leider nicht gedacht. Aber wenn du willst, können wir heute Abend etwas machen.

_______ Hallo Maria. Hier ist Michael. Wie geht's?

_______ Also, zuerst habe ich meinen kleinen Bruder besucht und sein Motorrad repariert.

___*13*___ Tschüss.

_______ Dann habe ich meinen Keller aufgeräumt. Und am Abend bin ich ausgegangen, in die Kneipe, mit zwei Arbeitskollegen.

_______ Nein, natürlich nicht. Mittags habe ich meinen neuen Nachbarn kennengelernt und wir haben zusammen Kaffee getrunken.

_______ Und dann?

___*1*___ Schneider, guten Tag.

_______ Und an mich hast du den ganzen Tag nicht gedacht, oder doch?

_______ Also gut. Kannst du mich um acht Uhr abholen?

_______ Ganz gut, danke. Du, sag mal, ich habe versucht, dich gestern anzurufen. Was hast du denn den ganzen Tag gemacht?

_______ Ja gern. Bis dann um acht. Tschüss.

_______ So, und das hat den ganzen Tag gedauert?

Situation 14 Interview: Gestern

1. Wann bist du aufgestanden?
2. Was hast du gefrühstückt?
3. Wie bist du zur Uni gekommen?
4. Was war dein erster Kurs?
5. Was hast du zu Mittag gegessen?
6. Was hast du getrunken?
7. Wen hast du getroffen?
8. Was hast du nachmittags gemacht?
9. Wie war das Wetter?
10. Wo bist du um sechs Uhr abends gewesen?
11. Was hast du abends gemacht?
12. Wann bist du ins Bett gegangen?
13. Ist gestern etwas Interessantes passiert? Was?

Studentenalltag
© fStop/Getty Images RF

Situation 15 Informationsspiel: Zum ersten Mal

MODELL: S2: Wann hat Frau Gretter ihren ersten Kuss bekommen?
S1: Als sie dreizehn war.

	Herr Thelen	Frau Gretter	mein(e) Partner(in)
seinen/ihren/deinen ersten Kuss bekommen		als sie 13 war	
zum ersten Mal ausgegangen	als er 14 war		
seinen/ihren/deinen Führerschein gemacht		mit 25	
sein/ihr/dein erstes Bier getrunken	mit 16		
seinen/ihren/deinen ersten Preis gewonnen		noch nie	
zum ersten Mal nachts nicht nach Hause gekommen		mit 21	

Filmlektüre

Jenseits der Stille

Vor dem Lesen

A. Beantworten Sie die folgenden Fragen.

1. Was assoziieren Sie mit Stille?
2. Schauen Sie sich das Foto an: Wie sehen die beiden jungen Leute aus?
3. Welche Charaktereigenschaften haben sie vielleicht? Finden Sie Adjektive.
4. Was machen sie mit ihren Händen?

FILMANGABEN

Titel: *Jenseits der Stille*
Genre: Drama
Erscheinungsjahr: 1996
Land: Deutschland
Dauer: 107 min
Regisseur: Caroline Link
Hauptrollen: Sylvie Testud, Tatjana Trieb, Howie Seago, Emmanuelle Laborit, Sibylle Canonica

Zeichensprache

B. Lesen Sie die Wörter im Miniwörterbuch. Suchen Sie sie dann im Text und unterstreichen Sie sie.

Miniwörterbuch	
gehörlos	deaf
die **Verständigung**	communication
die **Zeichensprache**	sign language
beherrschen	to master
übersetzen	to translate
das **Verhältnis**	relationship
zunächst	at first
der **Ärger**	trouble
sich lösen	to free oneself
(sich) vorbereiten	to prepare (oneself)
die **Aufnahmeprüfung**	entrance exam
ums Leben kommen	to die
angespannt	tense
die **Prüfungskommission**	examining board

Inhaltsangabe

Die achtjährige Lara (Tatjana Trieb) lebt mit ihren Eltern in Bayern. Sie ist die Einzige in der Familie, die sprechen und hören kann. Ihre Eltern sind beide gehörlos. Lara muss ihnen bei der Verständigung im Alltag oft helfen. Weil sie die Zeichensprache und die Sprache der Außenwelt beherrscht, übersetzt sie für ihre Eltern: bei jedem Telefonat, auf der Bank, in der Schule. Zu ihrem Vater hat Lara ein besonders gutes Verhältnis.

Eines Tages bekommt Lara eine Klarinette von ihrer Tante Clarissa (Sibylle Canonica). Lara lernt auf dem Instrument zu spielen und ist richtig gut. Sie hat Talent. Doch nicht nur das: Sie entdeckt eine große Welt außerhalb der häuslichen Stille, nämlich die Musik. Mit 18 (Sylvie Testud) will sie nach Berlin auf das Konservatorium und dort Musik studieren. Ihren Eltern sagt sie zunächst nichts davon. Als sie es dann doch erfahren, gibt es Ärger. Vor allem ihr Vater (Howie Seago) ist wütend und eifersüchtig. Er weiß, dass Lara dabei ist, sich von ihnen zu lösen und Welten zu entdecken, die ihnen verschlossen bleiben.

Lara geht trotzdem nach Berlin und bereitet sich auf die Aufnahmeprüfung vor. Auch als ihre Mutter (Emmanuelle Laborit) plötzlich bei einem Verkehrsunfall ums Leben kommt, bessert sich das angespannte Verhältnis zwischen Vater und Tochter nicht. Aber in dem Moment, in dem Lara vor die Prüfungskommission des Konservatoriums tritt, sieht sie ihren Vater im Konzertsaal. Er will sie spielen sehen.

Arbeit mit dem Text

Welche Aussagen sind falsch? Verbessern Sie die falschen Aussagen.

1. Lara kann hören und sprechen, ihre Eltern aber nicht.
2. Lara hilft ihren Eltern im Alltag, weil sie gehörlos sind.
3. Lara hat ein besonders gutes Verhältnis zu ihrer Mutter.
4. Lara bekommt von ihrer Kusine Clarissa eine Klarinette.
5. Laras Eltern möchten, dass Lara nach Berlin auf das Konservatorium geht.
6. Das Verhältnis zwischen Lara und ihrem Vater wird nach dem Tod der Mutter auch nicht besser.
7. Laras Vater akzeptiert am Ende des Films Laras Wunsch, Musik zu studieren.

FILMCLIP

NOTE: For copyright reasons, the films referenced in the **FILMCLIP** feature have not been provided by the publisher. The film can be purchased as a DVD or found online at various sites such as YouTube, Amazon, or the iTunes store. The time codes mentioned below are for the North American DVD version of the film.

Szene: DVD, Kapitel 3, „Merry Christmas", 14:20–17:36 Min.

Laras Tante, Clarissa, ist eine begeisterte Musikerin. Es ist die Weihnachtszeit. Jedes Weihnachten spielt sie auf ihrer Klarinette, begleitet von ihrem Vater auf dem Klavier, allen Gästen etwas vor.

Schauen Sie sich die Szene an. Die folgenden Aussagen beschreiben die Szene in der falschen Reihenfolge. Bringen Sie die Sätze in die richtige Reihenfolge.

_____ Lara, ihre Mutter, ihre Tante und ihre Großmutter sind in der Küche und sprechen über das neue Baby.
__1__ Lara hört ihrer Tante Clarissa und ihrem Großvater zu, wie sie Klavier und Klarinette spielen.
_____ Laras Vater erinnert sich[1], als er ein Junge war und seine Schwester spielen sah.
_____ Tante Clarissa möchte, dass Lara bei ihr bleibt, und bittet sie, ihre Mutter zu fragen, ob sie darf.
_____ Tante Clarissa schenkt Lara ihre erste Klarinette zu Weihnachten.
_____ Tante Clarissa sagt, sie ist ein Talent.
_____ Lara spielt mit der Klarinette und schafft[2] es, einen Ton zu spielen.
_____ Laras Mutter sagt widerwillig[3] ja.

Nach dem Lesen

Kreatives Schreiben. Wie geht die Geschichte weiter? Lara schreibt einen Brief an ihre beste Freundin oder ihren besten Freund und erzählt, wie es nach dem Vorspielen weitergegangen ist.

MODELL: Liebe ... (Lieber ...), wie geht es dir? Letzte Woche habe ich hier in Berlin am Konservatorium vorgespielt ...

Situation 16 Rollenspiel: Das Studentenleben

S1: Sie sind Reporter/Reporterin einer Unizeitung in Österreich und machen ein Interview zum Thema Studentenleben[4] in anderen Ländern. Fragen Sie, was Ihr Partner / Ihre Partnerin gestern alles gemacht hat: am Vormittag, am Mittag, am Nachmittag und am Abend.

[1]erinnert ... *remembers* [2]*manages* [3]*reluctantly* [4]*student life*

Videoecke

Perspektiven

Was hast du gestern Abend gemacht?

Gestern Abend habe ich ein Buch gelesen.

Aufgabe 1 Gestern Abend

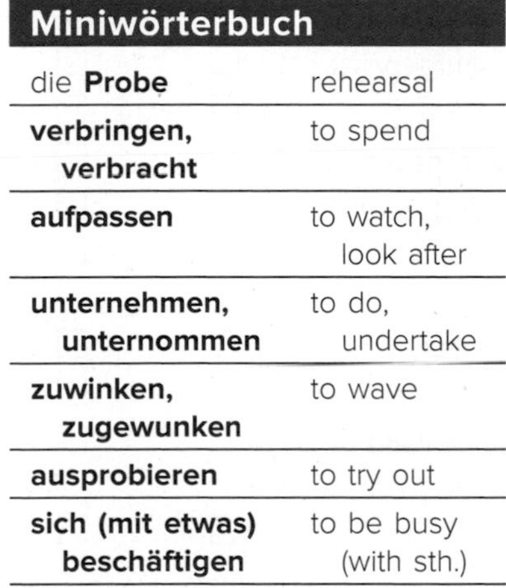

Miniwörterbuch	
die **Probe**	rehearsal
verbringen, verbracht	to spend
aufpassen	to watch, look after
unternehmen, unternommen	to do, undertake
zuwinken, zugewunken	to wave
ausprobieren	to try out
sich (mit etwas) beschäftigen	to be busy (with sth.)

Wer hat das gestern Abend gemacht? Ordnen Sie die Aussagen den Personen zu.

1. Sandra ___

2. Hend ___

3. Martin ___

4. Simone ___

5. Sophie ___

6. Jenny ___

7. Pascal ___

8. Tina ___

a. Ich habe ein ägyptisches Essen gekocht.
b. Ich habe ein Buch gelesen.
c. Ich habe etwas in einer Kneipe getrunken.
d. Ich habe mit einer Freundin etwas Schönes gekocht.
e. Ich habe mir beim Chinesen etwas zu essen geholt.
f. Ich habe mit meiner Freundin Wein getrunken.
g. Ich habe zu Hause eine DVD geguckt.
h. Ich hatte Probe mit meiner Band.

Interviews

- Wie hast du das letzte Wochenende verbracht?
- Was war das Interessanteste, was dir in den letzten Tagen passiert ist?
- Wann hast du Geburtstag?
- Wie hast du deinen letzten Geburtstag gefeiert?
- Welchen Feiertag findest du am besten? Warum?
- Was war der schönste Tag in deinem Leben?

Tanja

Felicitas

Aufgabe 2 Tanja und Felicitas

Auf wen treffen die folgenden Aussagen zu, Tanja oder Felicitas?

	Tanja	Felicitas
1. Letztes Wochenende bin ich zu meinem Freund nach Jena gefahren.	☐	☐
2. Ich habe auf die Kinder von meiner Schwester aufgepasst.	☐	☐
3. Ich habe am 2. März Geburtstag.	☐	☐
4. Ich habe ein Fußballspiel der Champions League gesehen.	☐	☐
5. An meinem letzten Geburtstag habe ich Sushi gegessen.	☐	☐
6. An meinem letzten Geburtstag habe ich abends etwas mit Freunden unternommen.	☐	☐
7. Ich finde den Tag der Deutschen Einheit am besten.	☐	☐
8. Ich finde Weihnachten am schönsten.	☐	☐
9. Ich habe ein Stipendium gewonnen, um in Deutschland zu studieren.	☐	☐
10. Justin Timberlake hat mir zugewunken.	☐	☐

Aufgabe 3 Tanjas Wochenende

Was ist passiert? Bringen Sie die Sätze in die richtige Reihenfolge.

__1__ Ich bin am Samstag zu meinem Freund nach Jena gefahren.

_____ Ich habe mir einen neuen Laptop gekauft.

_____ Wir haben Programme installiert und alles ausprobiert.

_____ Am Abend haben wir uns die ganze Zeit mit dem neuen Laptop beschäftigt.

_____ Wir sind einkaufen gegangen.

Aufgabe 4 Dies und das

Beantworten Sie die folgenden Fragen.

1. Wer hat in Tanjas Heimatstadt das Champions-League-Spiel gewonnen?
2. Wann hat Tanja Geburtstag?
3. Warum findet Tanja Weihnachten so schön?
4. Was ist am schönsten Tag in Tanjas Leben passiert?
5. Wo war Felicitas mit den Kindern ihrer Schwester?
6. Warum findet Felicitas den Tag der Deutschen Einheit so gut?
7. Was ist passiert, als Felicitas beim Justin-Timberlake-Konzert war?

Wortschatz

Unterwegs — On the Road

die **Fahrkarte, -n**	ticket
der **Bahnhof, ¨-e**	train station
der **Führerschein, -e**	driver's license
der **Unfall, ¨-e**	accident
der **Urlaub, -e**	vacation

Zeit und Reihenfolge — Time and Sequence

der **Abend, -e** (R)	evening
am Abend	in the evening
der **Alltag**	daily routine
der **Nachmittag, -e**	afternoon
der **Vormittag, -e**	late morning
das **Datum, Daten**	date
welches Datum ist heute?	what is today's date?
das **Mal, -e** (R)	time
das **letzte Mal**	the last time
zum ersten Mal	for the first time
abends	evenings, in the evening
gestern	yesterday
gestern Abend	last night
letzt-	last
letzte Woche	last week
letzten Montag	last Monday
letzten Sommer	last summer
letztes Wochenende	last weekend
nachmittags	afternoons, in the afternoon
nachts	nights, at night
an (R)	on; in
am Abend	in the evening
am ersten Oktober	on the first of October
an welchem Tag?	on what day?
bis (R)	until
bis um vier Uhr	until four o'clock
einmal	once
warst du schon einmal ...?	were you ever . . .?
erst	not until
erst um vier Uhr	not until four o'clock
früh (R)	in the morning
bis um vier Uhr früh	until four in the morning
schon (R)	already
seit	since; for
seit zwei Jahren	for two years
über	over
übers Wochenende	over the weekend
vor	ago
vor zwei Tagen	two days ago

Schule und Universität — School and University

die **Aufgabe, -n**	assignment
die **Bibliothek, -en**	library
die **Vorlesung, -en**	lecture
der **Kugelschreiber, -**	ballpoint pen
das **Abitur**	high school graduation exam
belegen	to take (a course)
halten, hält, gehalten*	to hold
ein **Referat halten**	to give a paper / oral report
vor·bereiten	to prepare

Feste und Feiertage — Holidays

die **Taufe, -n**	baptism, christening
der **Feiertag, -e**	holiday
das **Fest, -e**	celebration
(das) **Weihnachten**	Christmas

Ähnliche Wörter

die **Tradition, -en**; das **Picknick, -s**

Ordinalzahlen (Ordinal Numbers)

erst-
der erste Oktober
zweit-
dritt-
viert-
fünft-
sechst-
siebt-
acht-
neunt-
zehnt-
elft-
zwölft-
dreizehnt-
zwanzigst-
hundertst-

Sonstige Substantive — Other Nouns

die **Erinnerung, -en**	memory, remembrance
die **Kneipe, -n** (R)	bar, tavern
die **Nachbarin, -nen**	female neighbor
die **Rechnung, -en**	bill; check (in restaurant)
die **Sandburg, -en**	sandcastle
die **Umfrage, -n**	survey

*Strong and irregular verbs are listed in the **Wortschatz** with the third-person singular, if there is a stem-vowel change, and with the past participle. All verbs that use **sein** as the auxiliary in the present perfect tense are listed with **ist**.

der **Keller, -** | basement, cellar
der **Kuss, ¨e** | kiss
der **Liegestuhl, ¨e** | deck chair
der **Nachbar, -n** | male neighbor
der **Preis, -e** | prize
der **Strand, ¨e** | beach

das **Ferienhaus, ¨er** | vacation house
das **Geschirr** | dishes
Geschirr spülen | to wash the dishes
das **Jahrzehnt, -e** | decade
das **Sprachlabor, -s** | language laboratory
das **Tagebuch, ¨er** | diary

Ähnliche Wörter
die **Information, -en**; die **Reporterin, -nen**; die **Rolle, -n**; die **Wäsche**; der **Reporter, -**; der **Tee**; das **Café, -s**; das **Interview, -s**; das **Prozent, -e**; das **Thema, Themen**; das **Wunder, -**; **kein Wunder**

Sonstige Verben	Other Verbs
ab·fahren, fährt ... ab, ist abgefahren	to depart
an·fangen, fängt ... an, angefangen	to begin
antworten*	to answer
auf·wachen, ist aufgewacht (R)	to wake up
bezahlen	to pay (for)
dauern	to last
denken (**an** + *akk.*), **gedacht** (R)	to think (of)
entdecken	to discover
entscheiden, entschieden	to decide
erfinden, erfunden	to invent
ergänzen	to complete, fill in the blanks
feiern	to celebrate
los·fahren, fährt ... los, ist losgefahren	to drive off
passieren, ist passiert	to happen
spülen	to wash; to rinse
verdienen	to earn
verstehen, verstanden	to understand
versuchen	to try, attempt
war, warst, waren	was, were

Ähnliche Wörter
diskutieren; essen, isst, gegessen (R); **zu Abend essen; fotografieren; gewinnen, gewonnen; korrigieren; sitzen, gesessen** (R); **telefonieren; weg·gehen, ist weggegangen**

Adjektive und Adverbien	Adjectives and Adverbs
furchtbar	terrible
knapp	just, barely
links	left
mit dem linken Fuß auf·stehen, ist aufgestanden	to get up on the wrong side of bed
pünktlich	punctual; on time
verliebt	in love

Ähnliche Wörter
total

Sonstige Wörter und Ausdrücke	Other Words and Expressions
also	well, so, thus
auf jeden Fall	by all means
das hört sich toll an	that sounds great
deshalb	therefore; that's why
diese, dieser, dieses (R)	this, that, these, those
doch!	yes (on the contrary)!
etwas (R)	something
etwas Interessantes/Neues	something interesting/new
in (R)	in; at
im Garten	in the garden
im Café	at the cafe
ja	indeed
das ist es ja!	that's just it!
tut mir leid	I'm sorry
wen	whom (*accusative*)
zuerst	first

*Regular weak verbs are listed only with their infinitive.

Strukturen und Übungen

4.1 Talking about the past: the perfect tense

WISSEN SIE NOCH?

You've already seen how a **Satzklammer** forms a frame or a bracket consisting of a verb and either a separable prefix or an infinitive (grammar 1.5, 2.3, and 3.1). Note here how the **Satzklammer** is composed of **haben/sein** and the past participle.

In conversation, German speakers generally use the perfect tense to describe past events. The simple past tense, which you will study in **Kapitel 9,** is used more often in writing.

Ich **habe** gestern Abend ein Glas Wein **getrunken.**	*I drank a glass of wine last night.*
Nora **hat** gestern Basketball **gespielt.**	*Nora played basketball yesterday.*

German forms the perfect tense with an auxiliary (**haben** or **sein**) and a past participle (**gewaschen**). Participles usually begin with the prefix **ge-.**

	AUXILIARY		PARTICIPLE
Ich	**habe**	mein Auto	**gewaschen.**

The auxiliary is in first position in yes/no questions and in second position in statements and **w**-word questions. The past participle is at the end of the clause.

Hat Heidi gestern einen Film **gesehen?**	*Did Heidi see a movie last night?*
Ich **habe** gestern zu viel Kaffee **getrunken.**	*I drank too much coffee yesterday.*
Wann **bist** du ins Bett **gegangen?**	*When did you go to bed?*

Verbs with **sein** = no direct object; change of location or condition.

Although most verbs form the present perfect tense with **haben,** many use **sein.** To use **sein,** a verb must fulfill two conditions.

1. It cannot take a direct object.
2. It must indicate change of location or condition.

sein	**haben**
Ich **bin aufgestanden.** *I got out of bed.*	Ich **habe gefrühstückt.** *I ate breakfast.*
Stefan **ist** ins Kino **gegangen.** *Stefan went to the movies.*	Er **hat** einen Film **gesehen.** *He saw a film.*

Here is a list of common verbs that take **sein** as an auxiliary. (**Appendix E** contains a list of many other common verbs; those taking **sein** as an auxiliary are indicated.)

ankommen	*to arrive*	ich bin angekommen
aufstehen	*to get up*	ich bin aufgestanden
fahren	*to go, drive*	ich bin gefahren
gehen	*to go, walk*	ich bin gegangen
kommen	*to come*	ich bin gekommen
schwimmen	*to swim*	ich bin geschwommen
wandern	*to hike*	ich bin gewandert

In addition to these verbs, **sein** itself and the verb **bleiben** (*to stay*) take **sein** as an auxiliary.

Bist du schon in China **gewesen?**	*Have you ever been to China?*
Gestern **bin** ich zu Hause **geblieben.**	*Yesterday I stayed home.*

Übung 1 Yaminas erster Schultag

Ergänzen Sie **haben** oder **sein.** Beantworten Sie dann die Fragen.

Yamina _______[a] bis sieben Uhr geschlafen. Dann _______[b] sie aufgestanden und _______[c] mit ihren Eltern und ihren Schwestern gefrühstückt. Sie _______[d] ihre Tasche genommen und _______[e] mit ihrer Mutter zur Schule gegangen. Ihre Mutter und sie _______[f] ins Klassenzimmer gegangen und ihre Mutter _______[g] noch ein bisschen dageblieben. Die Lehrerin, Frau Dehne, _______[h] alle begrüßt. Dann _______[i] Frau Dehne „Herzlich willkommen" an die Tafel geschrieben.

1. Wann ist Yamina aufgestanden?
2. Wohin sind Yamina und ihre Mutter gegangen?
3. Was hat Frau Dehne an die Tafel geschrieben?

Übung 2 Eine Reise nach Istanbul

Ergänzen Sie **haben** oder **sein.** Beantworten Sie dann die Fragen.

JOSEF UND MELANIE:

Wir _______[a] ein Taxi genommen. Mit dem Taxi _______[b] wir zum Bahnhof gefahren. Dort _______[c] wir uns Fahrkarten gekauft. Dann _______[d] wir in den Orientexpress eingestiegen. Um 5.30 _______[e] wir abgefahren. Wir _______[f] im Speisewagen[1] gefrühstückt. Den ganzen Tag _______[g] wir Karten gespielt. Nachts _______[h] wir in den Schlafwagen gegangen. Wir _______[i] schlecht geschlafen. Aber wir _______[j] gut in Istanbul angekommen.

1. Wohin sind Josef und Melanie mit dem Taxi gefahren?
2. Wann sind sie mit dem Zug abgefahren?
3. Wo haben sie gefrühstückt?
4. Was haben sie nachts gemacht?

Übung 3 Ein ganz normaler Tag

Ergänzen Sie das Partizip.

aufgestanden **gefrühstückt** **gehört**
gearbeitet **gegangen** **getroffen**
geduscht **gegessen** **getrunken**

Heute bin ich um 7.00 Uhr _______[a]. Ich habe _______[b], _______[c] und bin an die Uni _______[d]. Ich habe einen Vortrag _______[e]. Um 10 Uhr habe ich ein paar Mitstudenten _______[f] und Kaffee _______[g]. Dann habe ich bis 12.30 Uhr in der Bibliothek _______[h] und habe in der Mensa zu Mittag _______[i].

[1]*dining car*

4.2 Strong and weak past participles

weak verbs = **ge-** + verb stem + **-(e)t**

German verbs that form the past participle with **-(e)t** are called *weak verbs.*

arbeiten	gearbeitet	*work*	*worked*
spielen	gespielt	*play*	*played*

To form the regular past participle, take the present tense **er/sie/es**-form and precede it with **ge-**.

er	spielt	→	er	hat	gespielt
sie	arbeitet	→	sie	hat	gearbeitet
es	regnet	→	es	hat	geregnet

strong verbs = **ge-** + verb stem + **-en;** the verb stem may have vowel or consonant changes.

Verbs that form the past participle with **-en** are called *strong verbs.* Many verbs have the same stem vowel in the infinitive and the past participle.

k**o**mmen → gek**o**mmen

Some verbs have a change in the stem vowel.

schw**i**mmen → geschw**o**mmen

Some also have a change in consonants.

ge**h**en → gega**ng**en

Here is a reference list of common irregular past participles.

PARTICIPLES WITH **haben**

essen, gegessen	*to eat*
halten, gehalten	*to hold*
lesen, gelesen	*to read*
liegen, gelegen	*to lie, be situated*
nehmen, genommen	*to take*
schlafen, geschlafen	*to sleep*
schreiben, geschrieben	*to write*
sehen, gesehen	*to see*
sprechen, gesprochen	*to speak*
tragen, getragen	*to wear, carry*
treffen, getroffen	*to meet*
trinken, getrunken	*to drink*
waschen, gewaschen	*to wash*

PARTICIPLES WITH **sein**

ankommen, angekommen	*to arrive*
aufstehen, aufgestanden	*to get up*
bleiben, geblieben	*to stay, remain*
fahren, gefahren	*to go (using a vehicle), drive*
gehen, gegangen	*to go (walk)*
kommen, gekommen	*to come*
schwimmen, geschwommen	*to swim*
sein, gewesen	*to be*

Übung 4 Das ungezogene° Kind

°*naughty*

Stellen Sie die Fragen!

MODELL: SIE: Hast du schon geduscht?
DAS KIND: Heute will ich nicht duschen.

1. Heute will ich nicht frühstücken.
2. Heute will ich nicht schwimmen.
3. Heute will ich keine Geschichte lesen.
4. Heute will ich nicht Klavier spielen.
5. Heute will ich nicht schlafen.
6. Heute will ich nicht essen.
7. Heute will ich nicht Geschirr spülen.
8. Heute will ich den Brief nicht schreiben.
9. Heute will ich nicht ins Bett gehen.

Übung 5 Katrins Tagesablauf

Wie war Katrins Tag gestern? Schreiben Sie zu jedem Bild einen Satz. Verwenden Sie diese Ausdrücke.

MODELL: Katrin hat bis 9 Uhr im Bett gelegen.

arbeiten
abends zu Hause bleiben
ein Referat halten
nach Hause kommen
bis neun im Bett liegen
regnen
mit Frau Schulz sprechen
einen Rock tragen
Freunde treffen
ihre Wäsche waschen

4.3 Dates and ordinal numbers

To form ordinal numbers, add **-te** to the cardinal numbers 1 through 19 and **-ste** to the numbers 20 and above. Exceptions to this pattern are **erste** (*first*), **dritte** (*third*), **siebte** (*seventh*), and **achte** (*eighth*).

Ordinals 1–19 add **-te** to the cardinal number (but note: **erste, dritte, siebte, achte**).

eins	**erste**	*first*
zwei	zweite	*second*
drei	**dritte**	*third*
vier	vierte	*fourth*
fünf	fünfte	*fifth*
sechs	sechste	*sixth*
sieben	**siebte**	*seventh*
acht	**achte**	*eighth*
neun	neunte	*ninth*
. . .		
neunzehn	neunzehnte	*nineteenth*
zwanzig	zwanzigste	*twentieth*
einundzwanzig	einundzwanzigste	*twenty-first*
zweiundzwanzig	zweiundzwanzigste	*twenty-second*
. . .		
dreißig	dreißigste	*thirtieth*
vierzig	vierzigste	*fortieth*
. . .		
hundert	hundertste	*hundredth*
. . .		

Ordinals 20 and higher add **-ste** to the cardinal number.

Ordinal numbers usually end in **-e** or **-en.** Use the construction **der** + **-e** to answer the question **Welches Datum ...?**

All dates are masculine:
***der* zwelte Mai**
***am* zweit*en* Mai**

Welches Datum ist heute?	*What is today's date?*
Heute ist **der** acht**e** Mai.	*Today is May 8th.*

Use **am** + **-en** to answer the question **Wann ...?**

Wann sind Sie geboren?	*When were you born?*
Am achtzehnt**en** Juni 1997.	*On the eighteenth of June, 1997.*

Ordinal numbers in German can be written as words or figures.

am zweiten Februar	*on the second of February*
am 2. Februar	*on the 2nd of February*

Übung 6 Wichtige Daten

Beantworten Sie die Fragen.

1. Welches Datum ist heute?
2. Welches Datum ist morgen?
3. Wann hast du Geburtstag?
4. Wann hat deine Mutter oder dein Vater Geburtstag?
5. Wann feiert man das neue Jahr?
6. Wann feiert man den Día de los Muertos?
7. Wann ist dieses Jahr Muttertag?
8. Wann ist nächstes Jahr Ostern?
9. Wann beginnt der Sommer?
10. Wann beginnt der Herbst?

4.4 Prepositions of time: *um, am, im*

Use the question word **wann** to ask for a specific time. The preposition in the answer will vary depending on whether it refers to clock time, days or parts of days, months, or seasons.

um CLOCK TIME

—Wann beginnt der Film?	*When does the film start?*
—**Um** neun Uhr.	*At nine o'clock.*

um 

am DAYS AND PARTS OF DAYS*

—Wann ist das Konzert?	*When is the concert?*
—**Am** Montag.	*On Monday.*
—Wann arbeitest du?	*When do you work?*
—**Am** Abend.	*In the evening.*

am

Mo	Di	Mi	Do	Fr	Sa	So
		1	2	3	4	5
6	7	8	9	10	11	12
13	14	15	16	17		
20	21					
27						

am 

im SEASONS AND MONTHS

—Wann ist das Wetter schön?	*When is the weather nice?*
—**Im** Sommer und besonders **im** August.	*In the summer and especially in August.*

im 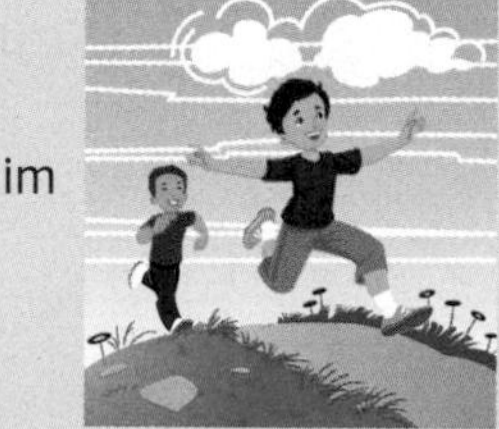

No preposition is used when stating the year in which something takes place.

—Wann bist du geboren?	*When were you born?*
—Ich bin 1995 geboren.	*I was born in 1995.*

*Note the exceptions: **in der Nacht** (*at night*) and **um Mitternacht** (*at midnight*).

Übung 7 Melanies Geburtstag

Ergänzen Sie **um, am, im** oder —.

Melanie hat _______[a] Frühling Geburtstag, _______[b] April. Sie ist _______[c] 1992 geboren, _______[d] 3. April 1992. _______[e] Dienstag kommen Claire und Josef _______[f] halb vier zum Kaffee. Melanies Mutter kommt _______[g] 16 Uhr. _______[h] Abend gehen Melanie, Claire und Josef ins Kino. Josef hat auch _______[i] April Geburtstag, aber erst _______[j] 15. April.

Übung 8 Interview

Beantworten Sie die Fragen.

1. Was machst du im Winter? im Sommer?
2. Wie ist das Wetter im Frühling? im Herbst?
3. Was machst du am Morgen? am Abend?
4. Was machst du am Freitag? am Samstag?
5. Was machst du heute um sechs Uhr abends? um zehn Uhr abends?
6. Was machst du am Sonntag um Mitternacht?

4.5 Past participles with and without *ge-*

A. Participles with **ge-**

German past participles usually begin with **ge-**. The past participles of separable-prefix verbs, however, begin with the prefix; the **ge-** goes between the prefix and the verb.

Separable-prefix verbs form their past participles with **-ge-** before the verb stem. The verb stem may have vowel or consonant changes.

WEAK VERBS

prefix + **-ge-** + stem + **-(e)t**

STRONG VERBS

prefix + **-ge-** + stem + **-en**

Separable prefixes include: **an, auf, aus, ein, mit, weg, wieder, zusammen,** and others.

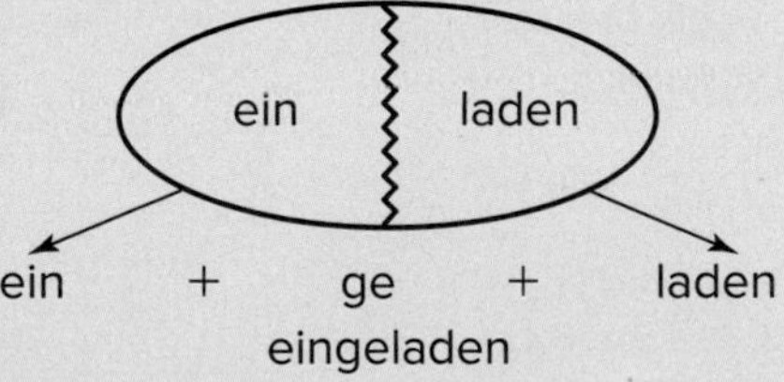

Frau Schulz **hat** Heidi und Nora zum Essen **eingeladen.** — *Frau Schulz invited Heidi and Nora to dinner.*

Here are the infinitives and past participles of some common separable-prefix verbs.

PAST PARTICIPLES WITH **haben**

anfangen	angefangen	*to start*
anrufen	angerufen	*to call up*
aufräumen	aufgeräumt	*to tidy up*
auspacken	ausgepackt	*to unpack*

PAST PARTICIPLES WITH **sein**

ankommen	angekommen	*to arrive*
aufstehen	aufgestanden	*to get up*
ausgehen	ausgegangen	*to go out*
weggehen	weggegangen	*to leave*

B. Participles without **ge-**

There are two types of verbs that do not add **ge-** to form the past participle: verbs that end in **-ieren** and verbs with inseparable prefixes.

Verbs ending in **-ieren** are weak: verb stem + **-t.**

1. Verbs ending in **-ieren** form the past participle with **-t: studieren → studiert.**

Paula **hat** Deutsch **studiert.** *Paula studied German.*

Here is a list of common verbs that end in **-ieren.**

diskutieren	diskutiert	to discuss
fotografieren	fotografiert	to take pictures
korrigieren	korrigiert	to correct
probieren	probiert	to try, taste
reparieren	repariert	to repair, fix
studieren	studiert	to study
telefonieren	telefoniert	*to telephone*

Almost all verbs ending in **-ieren** form the perfect tense with **haben.** The verb **passieren** (*to happen*) requires **sein** as an auxiliary: **Was ist passiert?** (*What happened?*)

Verbs with inseparable prefixes may be weak or strong. The verb stem may have vowel or consonant changes.

WEAK VERBS
verb stem + **-(e)t**
STRONG VERBS
verb stem + **-en**

INSEPARABLE PREFIXES:
be-, **ent-**, **er-**, **ge-**, **ver-**, **zer-**
Separable prefixes can stand alone as whole words; inseparable prefixes are always unstressed syllables.

2. The past participles of inseparable-prefix verbs do not include **ge-**: **verstehen → verstanden.**

Stefan **hat** nicht **verstanden.** *Stefan didn't understand.*

Whereas separable prefixes are words that can stand alone (**auf, aus, wieder,** and so forth), inseparable prefixes are simply syllables: **be-, ent-, er-, ge-, ver-,** and **zer-.** The past participles of most inseparable-prefix verbs require **haben** as an auxiliary. Here is a list of common inseparable-prefix verbs and their past participles.

bekommen	bekommen	*to get*
besuchen	besucht	*to visit*
bezahlen	bezahlt	*to pay*
entdecken	entdeckt	*to discover*
erfinden	erfunden	*to invent*
erzählen	erzählt	*to tell*
verdienen	verdient	*to earn*
vergessen	vergessen	*to forget*
verlieren	verloren	*to lose*
verstehen	verstanden	*to understand*

Übung 9 Ein schlechter Tag

Herr Thelen ist gestern mit dem linken Fuß aufgestanden. Zuerst hat er seinen Wecker nicht gehört und hat verschlafen. Dann ist er in die Küche gegangen und hat Kaffee gekocht. Nach dem Frühstück ist er mit seinem Auto in die Stadt zum Einkaufen gefahren. Er hat geparkt und ist erst nach zwei Stunden zurückgekommen. Herr Thelen hat einen Strafzettel[1] bekommen und 30 Euro bezahlt für falsches Parken. Er ist nach Hause gefahren, hat die Wäsche gewaschen und hat aufgeräumt. Beim Aufräumen ist eine teure Vase auf den Boden gefallen und zerbrochen[2]. Als die Wäsche fertig war, war ein Pullover eingelaufen[3]. Herr Thelen ist dann schnell ins Bett gegangen. Fünf Minuten vor Mitternacht ist das Haus abgebrannt[4].

[1]*parking ticket* [2]*broken* [3]*shrunk* [4]*burned down*

A. Richtig (R) oder falsch (F)?

1. _______ Herr Thelen hat gestern verschlafen.
2. _______ Vor dem Frühstück ist er in die Stadt gefahren.
3. _______ Herr Thelen hat falsch geparkt.
4. _______ Er hat seine Wohnung aufgeräumt.
5. _______ Herr Thelen braucht ein neues Haus.

B. Suchen Sie die Partizipien heraus, bilden Sie die Infinitive und schreiben Sie sie auf.

PARTIZIPIEN MIT **ge-**	INFINITIVE
_____________	_____________
_____________	_____________
⋮	⋮

PARTIZIPIEN OHNE **ge-**	INFINITIVE
_____________	_____________
_____________	_____________
⋮	⋮

Übung 10 In der Türkei

KASTEN FÜR a – e

gehen **schlafen** **trinken** **ankommen** **begrüßen**

KASTEN FÜR f – j

gehen **sprechen** **trinken** **fragen** **gehen**

Mehmet ist in der Türkei. Was hat er gestern gemacht? Verwenden Sie die Verben in den gelben Kästen[5].

Mehmet ist in der Türkei bei seinen Eltern. Gestern _______ er um 17 Uhr _______[a]. Er _______ seine Eltern und Geschwister _______[b] und einen Tee mit ihnen _______[c]. Dann _______ er in sein Zimmer _______[d] und _______ _______[e].

Nach einer Stunde _______ er zum Abendessen in die Küche _______[f]. Seine Eltern _______ ihn viel über sein Leben in Deutschland _______[g] und Mehmet _______ über seine Arbeit und seine Freunde _______[h]. Sie _______ noch einen Tee _______[i] und _______ um 23 Uhr ins Bett _______[j].

Übung 11 Interview

Fragen Sie Ihren Partner / Ihre Partnerin. Schreiben Sie die Antworten auf.

MODELL: mit deinen Eltern telefonieren (wie lange?) →
S1: Hast du gestern mit deinen Eltern telefoniert?
S2: Ja.
S1: Wie lange?
S2: Eine halbe Stunde.

1. früh aufstehen (wann?)
2. jemanden fotografieren (wen?)
3. jemanden besuchen (wen?)
4. ausgehen (wohin?)
5. etwas bezahlen (was?)
6. etwas reparieren (was?)
7. etwas Neues probieren (was?)
8. fernsehen (wie lange?)
9. etwas nicht verstehen (was?)
10. dein Zimmer aufräumen (wann?)

[5] *boxes*

KAPITEL 5

Geld und Arbeit

In **Kapitel 5,** you will talk about shopping, jobs and the workplace, and daily life at home. You will expand your ability to express your likes and dislikes and learn to describe your career plans.

Themen

Geschenke und Gefälligkeiten
Berufe
Arbeitsplätze
In der Küche

Kulturelles

KLI: Leipzig
Musikszene: „Millionär“ (Die Prinzen)
KLI: Ausbildung und Beruf
Filmclip: *Der Tunnel* (Roland Suso Richter)
Videoecke: Studium und Arbeit

Lektüren

Webartikel: Die coolsten Studentenjobs
Film: *Der Tunnel* (Roland Suso Richter)

Strukturen

5.1 Dative case: articles and possessive determiners
5.2 Question pronouns: **wer, wen, wem**
5.3 Expressing change: the verb **werden**
5.4 Location: **in, an, auf** + dative case
5.5 Dative case: personal pronouns

Adolph von Menzel: *Eisenwalzwerk* (1872–75), Alte Nationalgalerie, Berlin

KUNST UND KÜNSTLER

Adolph von Menzel (1815–1905) war der bedeutendste[1] Maler des Realismus. Vor allem war er ein Maler des damaligen modernen Lebens. Für sein Bild *Eisenwalzwerk* reiste[2] Menzel nach Königshütte in Schlesien, damals eine der modernsten Industrieregionen Deutschlands. Das Bild zeigt die Herstellung[3] von Eisenbahnschienen[4].

Schauen Sie sich das Bild an und beantworten Sie die folgenden Fragen.

1. Was ist im Zentrum des Bildes? Was machen die Arbeiter mit den Zangen[5]?
2. Was machen die Arbeiter vorne rechts? Wer ist die Person, die sie ansieht?
3. Was machen die Arbeiter links hinten?
4. Wer ist die Person mit dem Hut links im Hintergrund? Was macht er?
5. Welche Farben werden verwendet? Welchen Eindruck gibt das dem Bild?
6. Welche Assoziationen weckt das Bild?

[1]*most significant* [2]*traveled* [3]*production* [4]*rails for railroad tracks* [5]*tongs*

Situationen

Geschenke und Gefälligkeiten

Grammatik 5.1–5.2

1. Peter kauft seinem Freund Albert eine Konzertkarte.

2. Ernst gibt seinem Vater die Tageszeitung.

3. Michael schenkt seiner Freundin Maria einen Ausflug an die Ostsee.

4. Hans leiht seiner Schwester einen MP3-Spieler.

5. Aydan kocht ihrem Stiefsohn Rolf das Abendessen.

6. Heidi verkauft ihrem Mitstudenten Stefan ein Wörterbuch.

7. Melanie erzählt ihrer Freundin Claire ein Geheimnis.

8. Claire schreibt ihrer Mutter einen Brief.

Situation 1 Ist das normal?

Welches Bild gehört zu welchem Satz?

a.

b.

1. _______ Jens gießt seiner Tante die Blumen.

   _______ Jens gießt seine Tante.

a.

b.

2. _______ Jutta repariert ihren Bruder.

   _______ Jutta repariert ihrem Bruder das Radio.

a.

b.

3. _______ Silvia kauft das Kind.

   _______ Silvia kauft dem Kind die Schokolade.

a.

b.

4. _______ Herr Ruf kocht der Familie das Essen.

   _______ Herr Ruf kocht die Familie.

Situation 2 Sagen Sie *ja, nein* oder *vielleicht.*

1. Wem geben die Studenten ihre Hausaufgaben?
   a. dem Professor
   b. ihren Eltern
   c. dem Hausmeister
   d. dem Taxifahrer
2. Wem schreibt Rolf eine E-Mail?
   a. seiner Katze
   b. dem Präsidenten
   c. seinem Friseur
   d. seinen Eltern
3. Wem kauft Andrea das Hundefutter[1]?
   a. ihrer Mutter
   b. ihrem Freund Lukas
   c. ihrem Hund
   d. ihren Geschwistern
4. Wem repariert Herr Ruf das Fahrrad?
   a. seinem Hund
   b. seiner Mutter
   c. seinen Nachbarn
   d. seinem Sohn

Situation 3 Interaktion: Was schenkst du deiner Mutter?

Sie haben in der Lotterie 2.000 Euro gewonnen. Für 500 Euro wollen Sie Ihrer Familie und Ihren Freunden Geschenke kaufen. Was schenken Sie ihnen?

MODELL: S1: Was schenkst du deiner Mutter?
S2: Einen/Ein/Eine ______.
S1: Was schenkst du deinem Vater?
S2: Einen/Ein/Eine ______.

	ich	mein(e) Partner(in)
deiner Mutter		
deinem Vater		
deiner Schwester		
deinem Bruder		
deinem Großvater		
deiner Großmutter		
deinem Freund / deiner Freundin		
deinem Professor / deiner Professorin		
deinem Mitbewohner / deiner Mitbewohnerin		

[1] *dog food*

LEIPZIG

Beantworten Sie die folgenden Fragen.

- Wo liegt Leipzig? Suchen Sie die Stadt auf einer Landkarte.
- Was wissen Sie über Leipzig?
- Was wissen Sie über Johann Sebastian Bach und Richard Wagner?
- Was ist 1989 in Deutschland passiert?

Lesen Sie den Text und suchen Sie die Antworten auf die folgenden Fragen:

- Wann hatte Leipzig die meisten Einwohner[2]? Wie viele hat es jetzt?
- Wann wurde die Universität Leipzig gegründet? Wie viele Studierende hat sie?
- Welchen Beruf hatten die folgenden Personen: Heisenberg, Ostwald, Mommsen, Wundt, Wagner und Leibniz?
- Welche berühmte Messe[3] findet jedes Jahr in Leipzig im März statt, und welches Festival im Juni?
- Wie lange war Johann Sebastian Bach Thomaskantor in Leipzig?
- Warum nennt man Leipzig die Heldenstadt[4]?
- In welcher Straße gibt es besonders viele Cafés, Kneipen[5] und Clubs?

2009: die Leipziger Universität ist 600 Jahre alt.
Printed with permission of the German Ministry of Finance and the artist, Nadine Nill. Background: Universität Leipzig, Kustodie/Kunstsammlung (photograph: Marion Wenzel)

Leipzig ist eine der größten und bedeutendsten Städte Deutschlands. Leipzig ist die Stadt der zweitältesten Universität Deutschlands, die Stadt des Buches, die Stadt der Musik und die Stadt der friedlichen[6] Revolution von 1989.

Leipzig erhielt 1165 das Stadtrecht[7], damals mit nur 500 Einwohnern. Vor Beginn des 1. Weltkriegs[8] war sie mit 590.000 Einwohnern die viertgrößte Stadt Deutschlands. 1930 hatte sie mehr als 700.000 Einwohner. Heute hat sie ca. 530.000 Einwohner, ist immer noch ein wichtiger Verkehrsknotenpunkt[9] und eines der wichtigsten Wirtschaftszentren Ostdeutschlands.

Die Universität Leipzig wurde 1409 gegründet und ist nach Heidelberg die zweitälteste Universität Deutschlands. Im 19. Jahrhundert war sie eine der drei wichtigsten Universitäten Deutschlands. An ihr unterrichteten die Dichter[10] Gottsched und Lessing, der Physiker Werner Heisenberg (Nobelpreis 1932), der Chemiker Wilhelm Ostwald (Nobelpreis 1909), der Historiker Theodor Mommsen (Nobelpreis in Literatur 1902), Wilhelm Wundt, der Begründer der experimentellen Psychologie und der Philosoph Ernst Bloch. An ihr studierten Goethe (Jura), Nietzsche (Altphilologie) und de Saussure (Indogermanistik), die Komponisten Robert Schumann und Richard Wagner sowie Gottfried Wilhelm Leibniz (Philosophie), dessen Statue eines der Wahrzeichen[11] der Universität ist. Jetzt studieren an ihr ca. 28.000 Studenten.

Eine besonders große Rolle spielt in der Geschichte Leipzigs die Musik. Johann Sebastian Bach war Thomaskantor in Leipzig und leitete den Thomanerchor von 1723 bis 1750. Jedes Jahr im Juni erinnert das Bach Leipzig Festival an[12] diesen berühmten Musiker.

Buchmesse in Leipzig
© ullstein bild - CARO/Keunecke/The Image Works

1989 begannen in Leipzig die Montagsdemonstrationen, die zum Fall der Berliner Mauer und zur Wiedervereinigung Deutschlands 1990 führten. Seitdem wird Leipzig auch die Heldenstadt genannt.

Da es in Leipzig viele Studierende gibt, gibt es viele Möglichkeiten auszugehen oder zu feiern. Wöchentlich finden Partys in den Studentenclubs „Moritzbastei", „TV-Club" und „StuK" statt. Die Karl-Liebknecht-Straße, liebevoll „Karli" genannt, bietet jede Menge Cafés, Kneipen und Clubs zum gemütlichen Cocktailtrinken oder zum Tanzen bis in die frühen Morgenstunden.

Leipzig hat eine lange Tradition als Messestadt. Neben vielen anderen Messen findet jedes Jahr im März die Leipziger Buchmesse statt. Sie war bis 1945 die größte Buchmesse Deutschlands, heute ist sie nach Frankfurt am Main die zweitgrößte.

[2]*inhabitants* [3]*trade fair* [4]*city of heroes* [5]*bars* [6]*peaceful* [7]*town privileges* [8]*world war* [9]*transportation hub* [10]*poets* [11]*landmarks* [12]erinnert an *commemorates*

Situation 4 Fragen über Fragen

Welche Antwort passt auf welche Frage? Ordnen Sie zu.

1. Wer unterrichtet Deutsch in Berkeley?
2. Wen hat Veronika geheiratet?
3. Wem hat Melanie ein Geheimnis erzählt?
4. Wen hat Jutta letztes Wochenende kennengelernt?
5. Wem hat Kobe gezeigt, wie man Fahrrad fährt?
6. Wer zeichnet gern Hunde?
7. Wem hat Aydan ihre Kamera geliehen?
8. Wen hat Veronika gepflegt?
9. Wem hat Willi seinen Kühlschrank verkauft?
10. Wem hat Yamina eine Frage gestellt?

a. ihre kranke Tochter Yamina
b. Eske
c. Frau Schulz
d. ihrem Stiefsohn Rolf
e. ihren Mann Kobe
f. ihren neuen Freund Billy
g. ihrer Freundin Claire
h. seiner Freundin Nesrin
i. ihrer Mutter
j. seiner Tochter Sumita

Berufe

Grammatik 5.3

1. Der Arzt hilft kranken Menschen.

2. Der Verkäufer arbeitet in einem Laden.

3. Die Anwältin verteidigt den Angeklagten.

4. Der Pilot fliegt ein Flugzeug.

5. Der Richter arbeitet im Gericht.

6. Die Bauarbeiterin baut ein Parkhaus.

7. Die Architektin zeichnet ein Haus.

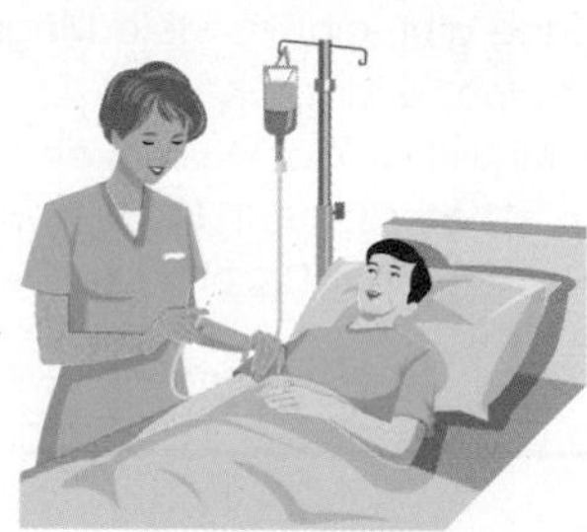

8. Die Krankenpflegerin arbeitet im Krankenhaus.

Situation 5 Definitionen

Finden Sie den richtigen Beruf.

Anwältin	Verkäufer	Pilot	Ärztin
Schriftsteller	Krankenpflegerin		
Architekt(in)			Lehrer

1. Dieser Mann unterrichtet an einer Schule. Er ist ______.
2. Diese Frau untersucht Patienten im Krankenhaus. Sie ist ______.
3. Dieser Mann fliegt ein Flugzeug. Er ist ______.
4. Dieser Mann verkauft Computer in einem Laden. Er ist ______.
5. Diese Person zeichnet Pläne für Häuser. Sie ist ______.
6. Diese Frau arbeitet auf dem Gericht. Sie ist ______.
7. Diese Frau pflegt kranke Menschen. Sie ist ______.
8. Dieser Mann schreibt Romane. Er ist ______.

Situation 6 Bildgeschichte: Was Michael Pusch schon alles gemacht hat

Situation 7 Berufe

Machen Sie Listen. Suchen Sie zu jeder Frage drei Berufe.

In welchen Berufen ...

1. verdient man sehr viel Geld?
2. verdient man nur wenig Geld?
3. gibt es mehr Männer als Frauen?
4. gibt es mehr Frauen als Männer?
5. muss man gut in Mathematik sein?
6. muss man gut in Sprachen sein?
7. muss man viel reisen?
8. muss man viel Kraft[1] haben?

Situation 8 Interview

1. Arbeitest du? Wo? Als was? Was machst du? An welchen Tagen arbeitest du? Wann fängst du an? Wann hörst du auf?
2. Was studierst du? Wie lange dauert das Studium?
3. Was möchtest du werden? Verdient man da viel Geld? Ist das ein Beruf mit viel Prestige?
4. Was ist dein Vater von Beruf? Was hat er gelernt (studiert)?
5. Was ist deine Mutter von Beruf? Was hat sie gelernt (studiert)?

Lektüre

Vor dem Lesen

A. Beantworten Sie die folgenden Fragen.

1. Arbeiten Sie neben dem Studium? Was machen Sie?
2. Wie viele Stunden pro Woche arbeiten Sie? Warum?
3. Was machen Sie mit Ihrem Geld?
4. Arbeiten Sie auch in den Semesterferien? Was machen Sie?
5. Macht Ihnen Ihr Job Spaß? Was macht Ihnen Spaß?

B. Lesen Sie die Wörter im Miniwörterbuch. Suchen Sie sie im Text und unterstreichen Sie sie.

Miniwörterbuch	
unterstützen	to support
ausreichen	to be enough, to last
kellnern	to wait tables
die **Nachhilfe**	tutoring
der **Heißluftballon**	hot air balloon
jagen	to chase
hingeweht werden	to be blown to
der **Flughafen**	airport
der **Vogel,** die **Vögel**	bird
zusammenstoßen	to smash into
der **Unfall**	accident
vertreiben	to chase away
beschäftigen	to employ
der **Straßenbahnführer**	streetcar driver
die **Trennungsagentur**	separation agency
Schluss machen	to end (*here:* end a relationship)
der **Kunde**	customer
nachsichtig	considerately
unbarmherzig	ruthlessly
zurückzahlen	to pay back

[1] *strength*

Dieser Student jobbt als Straßenbahnfahrer.
© Jörg Carstensen/dpa/Corbis

LESEHILFE

Scanning a text is one way to find details without reading word for word. How many familiar words can you identify by scanning the text?

Die coolsten Studentenjobs

Mehr als die Hälfte der Studenten in Deutschland arbeiten. Sie sind jung und brauchen das Geld. Welche Jobs sind besonders populär? Welche sind besonders interessant?

Mehr als die Hälfte der Studenten in Deutschland arbeiten neben dem Studium und in den Semesterferien. Der Staat unterstützt junge Leute mit Stipendien und BAföG und die Familie hilft auch oft. Aber das Geld reicht nicht aus.

Besonders beliebt bei den Studenten sind kellnern, Nachhilfe geben, babysitten oder im Supermarkt kassieren. Doch wer die Augen offen hält, findet auch interessantere Jobs. Man kann zum Beispiel Heißluftballons jagen: Die „Verfolger" beobachten, wo ein Ballon hingeweht wird, um die Passagiere am Landeplatz abzuholen und zurück zum Startpunkt zu fahren.

Flughäfen beschäftigen sogenannte Vogelvertreiber, denn Vögel sind auf Flughäfen nicht gern gesehen. Wenn sie mit Flugzeugen zusammenstoßen oder in die Triebwerke der Maschinen fliegen, kann es zu schweren Unfällen kommen. Die Vogelvertreiber beobachten die Start- und Landebahnen und vertreiben die Störenfriede. Die Kölner Verkehrsbetriebe beschäftigen auch Studenten als Fahrer. Sie bilden die jungen Leute in einem siebenwöchigen Intensivkurs zum Straßenbahnführer aus.

Mutige Studenten melden sich bei Trennungsagenturen. Diese beschäftigen „Schlussmacher", die einem Partner die Botschaft vom Beziehungsende telefonisch oder im direkten Gespräch übermitteln. Der Kunde kann vorher sagen, ob er möchte, dass der Schlussmacher den Partner nachsichtig oder unbarmherzig behandeln soll.

Für Studenten, die BAföG erhalten, ist es wichtig darauf zu achten, nicht mehr als rund 400 Euro im Monat zu verdienen. Wenn sie mehr verdienen, müssen sie das Geld an den Staat zurückzahlen. Es gibt viele „Minijobs", die das berücksichtigen und gerne Studenten beschäftigen.

Bearbeitung des Textes „Die verrücktesten Nebenjobs", http://home.1und1.de/themen/beruf. 19. April 2011

Arbeit mit dem Text

A. Beliebte Jobs und interessante Jobs. Welche der folgenden Jobs sind beliebt, welche sind interessant? Schreiben Sie *B* neben die Jobs, die beliebt sind, und *I* neben die Jobs, die interessant sind.

_______ als Babysitter/in arbeiten
_______ als Kassierer/in im Supermarkt arbeiten
_______ als Kellner/in arbeiten
_______ als Straßenbahnführer/in arbeiten
_______ bei einer Trennungsagentur arbeiten
_______ Heißluftballons jagen
_______ Nachhilfe geben
_______ Vögel vom Flughafen vertreiben

B. Beantworten Sie die Fragen.

1. Woher können Studierende in Deutschland Geld bekommen?
2. Was machen Heißluftballonjäger?
3. Wo arbeiten Vogelvertreiber? Warum ist ihre Arbeit wichtig?
4. In welcher Stadt kann man als Student als Straßenbahnführer arbeiten? Wie lange dauert der Vorbereitungskurs[2]?
5. Was machen Schlussmacher? Welche Optionen gibt es für die Kunden?
6. Wie viel darf man im Monat verdienen, wenn man BAföG bekommt?

Nach dem Lesen

1. Machen Sie eine Umfrage im Kurs. Stellen Sie die folgenden Fragen:
   - Welche Jobs hattest du?
   - Wie alt warst du, als du deinen ersten Job hattest?
   - Wie viel hast du gearbeitet?
   - Wie viel hast du verdient?
   - Was hast du dir von deinem Geld gekauft?

   Benutzen Sie die folgende Tabelle.

Name	Alter	Job	Stundenlohn	Geld für ...

2. Sammeln Sie die Antworten und machen Sie ein Plakat mit dem Titel: *Die Jobs unserer Kursteilnehmer.* Hängen Sie das Plakat aus.

[2]*preparatory course*

Miniwörterbuch

witzig	funny
das **Mitglied, -er**	member
das **Schlagzeug**	drums
wäre	would be
wäre gern	would like to be
das **Portemonnaie**	wallet
das **Konto**	(bank) account
weder ... noch	neither . . . nor
faul	lazy
knacken	to break into (*slang*)
ausrauben	to rob
gefährlich	dangerous
der **Knast**	slammer (*slang for jail*)
die **Witwe**	widow

MUSIKSZENE

„Millionär" (1991, Deutschland) *Die Prinzen*

Biografie Die Prinzen sind eine Musikgruppe aus Leipzig. Sie sind für ihre A-cappella-Musik und ihre witzigen Texte bekannt. Ihre Mitglieder lernten sich an der Thomanerschule in Leipzig kennen. Sebastian Krumbiegel studierte an der Hochschule für Musik und Theater Felix Mendelssohn Bartholdy Schlagzeug und Gesang. Tobias Künzel und Wolfgang Lenk waren Mitglieder des Thomanerchors. Die zwei weiteren Sänger heißen Jens Sembdner und Henri Schmidt.

Die Prinzen, eine Musikgruppe aus Leipzig

NOTE: For copyright reasons, the songs referenced in **MUSIKSZENE** have not been provided by the publisher. The song can be found online at various sites such as YouTube, Amazon, or the iTunes store.

Vor dem Hören Wie bekommt man eine Million Euro?

Nach dem Hören Welche Antworten sind richtig?

1. Als Millionär wäre das ...
   - ☐ **a.** Portemonnaie sehr schwer.
   - ☐ **b.** Konto nie leer.
2. Der Sänger ist ...
   - ☐ **a.** Professor
   - ☐ **b.** faul
3. Der Sänger möchte ...
   - ☐ **a.** eine Bank knacken.
   - ☐ **b.** Popstar werden.
4. Eine Bank auszurauben ...
   - ☐ **a.** ist nicht gefährlich.
   - ☐ **b.** bringt einen in den Knast.
5. Viele reiche Witwen wollen ...
   - ☐ **a.** seinen Körper.
   - ☐ **b.** sein Geld.

Arbeitsplätze

Grammatik 5.4

Situation 9 Der Arbeitsplatz

MODELL: S1: Wo arbeitet eine Anwältin?
S2: Auf dem Gericht.

im Krankenhaus · auf der Post · auf der Bank · in der Kirche · auf der Polizei · auf dem Gericht · auf der Universität · im Kaufhaus · im Schwimmbad · in der Schule

1. eine Anwältin
2. ein Arzt
3. eine Bademeisterin
4. ein Bankangestellter
5. ein Lehrer
6. eine Polizistin
7. ein Postbeamter
8. ein Priester
9. eine Professorin
10. eine Verkäuferin

Situation 10 Minidialoge

Sie hören neun kurze Dialoge. Wo finden sie statt?

im Hotel · an der Tankstelle · in der Gaststätte · auf der Post · in der Bäckerei · auf dem Bahnhof · an der Kinokasse · im Schwimmbad · auf der Bank

1. ______________________
2. ______________________
3. ______________________
4. ______________________
5. ______________________
6. ______________________
7. ______________________
8. ______________________
9. ______________________

KULTUR ... LANDESKUNDE ... INFORMATIONEN

AUSBILDUNG UND BERUF

Wie ist es in Ihrem Land?

- Welchen Schulabschluss[1] braucht man für eine Berufsausbildung?
- Wie bekommt man eine Berufsausbildung?
- Wo lernt man die praktische Seite des Berufs? Wie lange dauert das?
- Wo lernt man die theoretische Seite? Wie lange dauert das?
- Macht man am Ende eine Prüfung? Was ist man dann?

Max hat keine Lust auf Schule und später Studium. Wenn er die zehnte Klasse erfolgreich[2] abschließt[3], hat er den Realschulabschluss. Er möchte am liebsten eine praktische Ausbildung machen, z. B. als Tischler oder Koch. Ein Facharbeiter[4] verdient mehr als ein ungelernter Arbeiter. Die Grafik zeigt, wie die Ausbildung für Max weitergeht.

Wie ist es in Deutschland?

- Wie lange dauert eine Ausbildung oder Lehre?
- Wo bekommt man die theoretische Ausbildung?
- Wo lernt man die praktische Seite des Berufs?
- Was bekommt man am Ende der Gesellenprüfung?
- Was ist man am Schluss[5]?

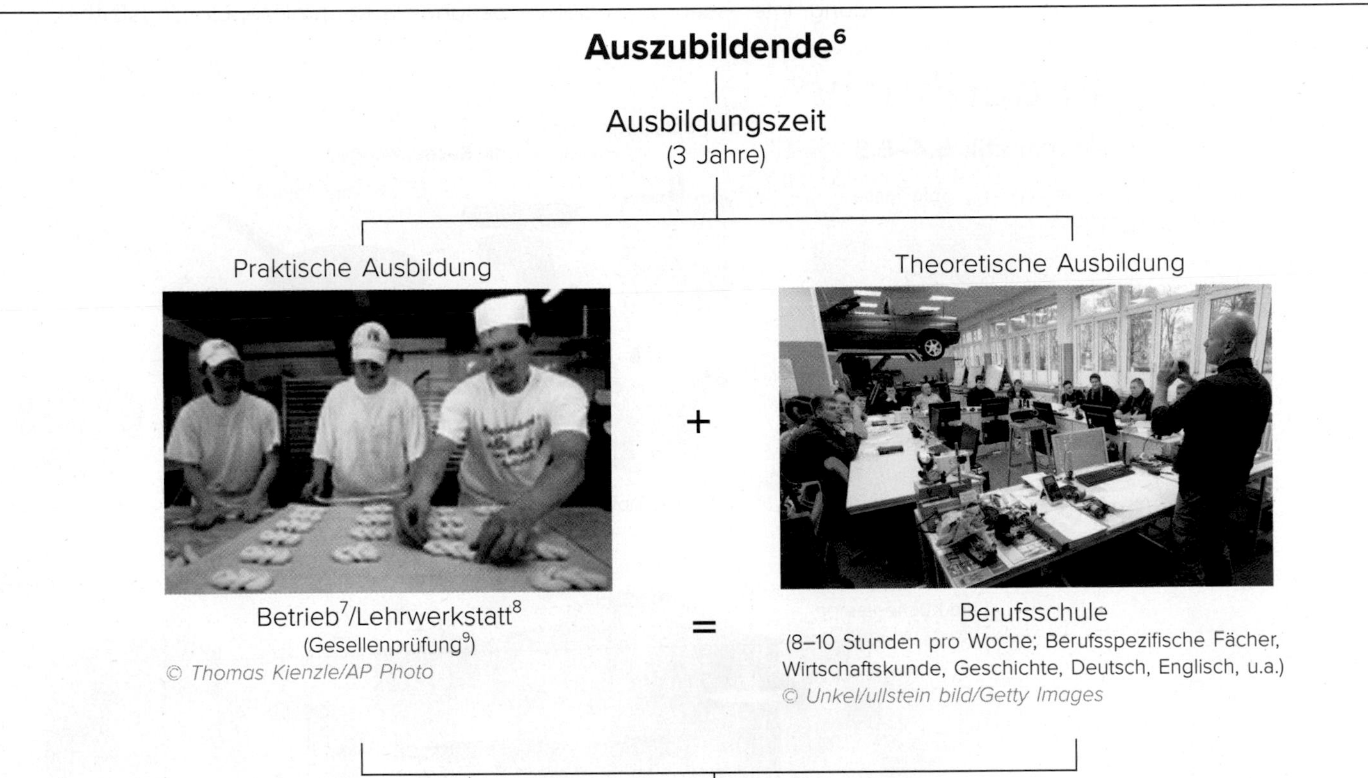

© Thomas Kienzle/AP Photo

© Unkel/ullstein bild/Getty Images

[1]*educational degree* [2]*successfully* [3]*completes* [4]*trade worker; skilled worker* [5]*am ... in the end* [6]*those receiving a specialized education; apprentices* [7]*business* [8]*apprentice shop* [9]*trade workers' examination* [10]*certificate of completed apprenticeship*

 Situation 11 Zum Schreiben: Vor der Berufsberatung

Morgen haben Sie einen Termin beim Berufsberater an Ihrer Universität. Bereiten Sie sich auf das Gespräch vor. Machen Sie sich ausführliche Notizen zu den folgenden Themen.

- Schulbildung: Nennen Sie die Schulen und Universitäten, wo Sie waren. Schreiben Sie etwas zur Art der Schule (öffentliche Gesamtschule, private Universität), zu Ihren Noten, zu Auslandsaufenthalten und zu Ihrem sozialen Engagement.

- Interessen, Hobbys: Nennen Sie Ihre Interessen und Hobbys und schreiben Sie etwas dazu, zum Beispiel seit wann Sie das Hobby haben, was Sie schon erreicht haben und wie wichtig es für Sie ist.
- Lieblingsfächer, besondere Fähigkeiten: Nennen Sie die Fächer und besondere Fähigkeiten, die Sie haben und schreiben Sie etwas dazu, zum Beispiel welches Niveau Sie erreicht haben oder was Sie damit machen können.
- Erwartungen[11] an den zukünftigen[12] Beruf: Schreiben Sie, was Ihnen wichtig oder vielleicht auch nicht so wichtig ist, zum Beispiel, wie viel Geld Sie verdienen, welche Arbeitszeiten Sie haben, wie viele Tage Urlaub Sie im Jahr haben möchten oder welche Aufstiegschancen[13] für Sie wichtig sind.

Situation 12 Rollenspiel: Bei der Berufsberatung

S1: Sie arbeiten bei der Berufsberatung. Ein Student / Eine Studentin kommt in Ihre Sprechstunde. Stellen Sie ihm/ihr Fragen zu diesen Themen: Schulbildung, Interessen und Hobbys, besondere Kenntnisse, Lieblingsfächer.

In der Küche

Grammatik 5.4–5.5

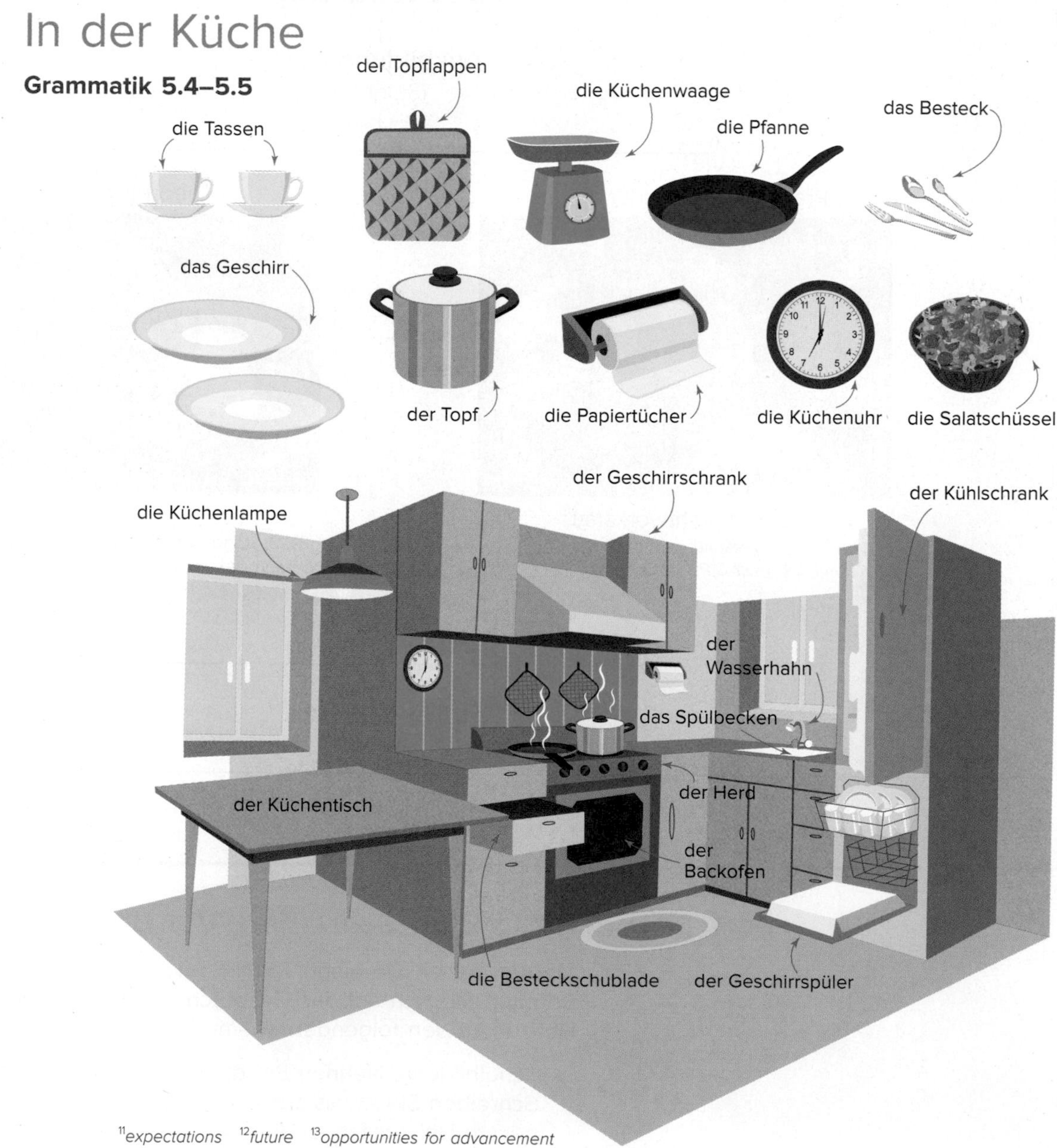

[11]*expectations* [12]*future* [13]*opportunities for advancement*

Situation 13 Wo ist ...?

MODELL: S1: Wo ist der Küchentisch?
S2: Unter der Küchenlampe.

am Fenster **unter dem Herd**
unter dem Geschirrschrank
im Geschirrschrank **auf dem Herd** **im Geschirrspüler**
an der Wand **unter dem Kühlschrank**
in der Besteckschublade

1. Wo ist der Geschirrspüler?
2. Wo ist die Küchenuhr?
3. Wo ist der Backofen?
4. Wo ist das Spülbecken?
5. Wo sind die Papiertücher?
6. Wo ist die Pfanne?
7. Wo ist das Geschirr?
8. Wo ist der Topf?
9. Wo sind die Gläser?
10. Wo ist das Besteck?

Situation 14 Interaktion: Küchenarbeit

Wie oft spülst du das Geschirr?

mehrmals am Tag
jeden Tag
fast jeden Tag
zwei- bis dreimal in der Woche
einmal in der Woche
einmal im Monat
selten
nie

Wie oft ...?	ich	mein(e) Partner(in)
gehst du einkaufen		
kochst du		
deckst du den Tisch		
spülst du das Geschirr		
stellst du das Geschirr weg		
machst du den Herd sauber		
machst du den Tisch sauber		
machst du den Kühlschrank sauber		
fegst du den Boden		
bringst du die leeren Flaschen weg		

Situation 15 Umfrage: Kochst du mir ein Abendessen?

MODELL: S1: Kochst du mir morgen ein Abendessen?
S2: Ja.
S1: Unterschreib bitte hier.

	UNTERSCHRIFT
1. Kochst du mir morgen ein Abendessen?	______________
2. Backst du mir einen Kuchen zum Geburtstag?	______________
3. Kaufst du mir ein Eis?	______________
4. Schenkst du mir deinen Kugelschreiber?	______________
5. Hilfst du mir heute bei der Hausaufgabe?	______________
6. Kannst du mir die Grammatik erklären?	______________
7. Schreibst du mir in den Ferien eine Postkarte?	______________
8. Kannst du mir mein Zimmer aufräumen?	______________
9. Kannst du mir fünf Dollar leihen?	______________

Situation 16 Dialog: Chaos in der Küche

In der Küche herrscht Chaos und Herr Ruf ist sauer.

HERR RUF: Jutta, komm mal her!
JUTTA: Ja, Papa. Warum schreist du denn so?
HERR RUF: Weil es hier aussieht wie im Schweinestall! Warum ist Marmelade ______________?
JUTTA: Ich habe mir ein Brot gemacht und das ist dann in die Schublade gefallen.
HERR RUF: Und warum ist die Kaffeemaschine ______________?
JUTTA: Hans brauchte Platz ______________ für seine Legos.
HERR RUF: Das Kochbuch liegt ______________! Unglaublich!
JUTTA: Weil es da warm ist. Es war leider nass.
HERR RUF: Und warum ist der Kuchen ______________?
JUTTA: Keine Ahnung!
HERR RUF: Ihr glaubt wohl, dass Aufräumen meine Lieblingsbeschäftigung ist!
JUTTA: Ach, Papa, das ist doch nicht so schlimm. Ich hole Hans und dann helfen wir dir.

Filmlektüre

Der Tunnel

Vor dem Lesen

FILMANGABEN

Titel: *Der Tunnel*
Genre: Drama
Erscheinungsjahr: 2001
Land: Deutschland
Dauer: 150 min
Regisseur: Roland Suso Richter
Hauptrollen: Alexandra Maria Lara, Claudia Michelsen, Heino Ferch, Nicolette Krebitz, Sebastian Koch, Felix Eitner

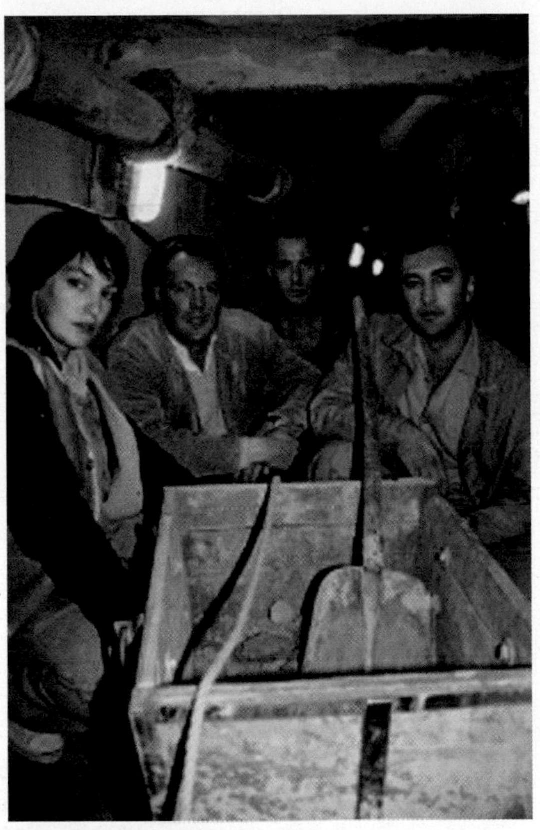

Szene aus dem Film *Der Tunnel*

A. Sehen Sie sich das Filmposter an.

1. Wie viele Leute sehen Sie?
2. Wie sehen sie aus?
3. Wie ist die Stimmung?
4. Sehen Sie Werkzeuge oder Materialien?
5. Warum bauen Menschen eigentlich Tunnel?

B. Lesen Sie die Wörter im Miniwörterbuch. Suchen Sie sie im Text und unterstreichen Sie sie.

Inhaltsangabe

13. August 1961: Die DDR-Regierung baut eine Mauer durch Berlin und grenzt den Osten der Stadt vom Westen ab. Schwimmstar Harry Melchior (Heino Ferch) hat genug von der DDR und will weg. Noch im Herbst, kurz nach dem Bau der Berliner Mauer, flieht er mit seinem Freund Matthis Hiller (Sebastian Koch) in den Westteil Berlins. Die beiden beschließen, Harrys Schwester Lotte (Alexandra Maria Lara) und Matthis' Frau Carola (Claudia Michelsen) in den Westen zu holen. Zusammen mit Fred

Miniwörterbuch	
abgrenzen	to fence off
fliehen	to flee
beschließen	to decide
graben	to dig
versperrt sein	to be blocked
sich anschließen (schließt sich an)	to join (joins)
der **Verlobte**	fiancé
vertrauen	to trust
zusammenbrechen	to collapse
das **Grundwasser**	groundwater
die **Stasi (Staatssicherheit)**	East German secret service
der **Geheimdienst**	secret service
der **Oberst**	colonel
hinter jemandem her sein	to be after someone
durchkreuzen	to thwart
gefährlich	dangerous
kriegen	to get, to catch

von Klausnitz und dem Ex-GI Vittorio „Vic" Castanza wollen sie einen Tunnel von West nach Ost graben, weil alle anderen Fluchtwege versperrt sind. Im Keller einer alten Fabrik an der Bernauer Straße finden sie den idealen Ort für den Tunnel. Die junge, attraktive Friederike „Fritzi" (Nicolette Krebitz) schließt sich der Gruppe an. Sie will für ihren Verlobten Heiner die Flucht in den Westen möglich machen.

Die Gruppe um Harry rekrutiert mehr Helfer, um die schwierige Aufgabe zu schaffen. Die Frage dabei ist immer: Wem kann man vertrauen? Einfach ist das Tunnelprojekt nicht. Einmal bricht der Tunnel beinah zusammen und ein anderes Mal läuft Grundwasser ein. Auch müssen die Fluchthelfer ihre Aktion finanzieren: Sie verkaufen die Rechte ihrer Geschichte an die NBC und werden dafür bei ihrer Arbeit gefilmt. Und dann gibt es noch die Stasi, den Geheimdienst der DDR. Vor allem Stasi-Oberst Krüger ist hinter Harry und Matthis her. Er will ihren Plan durchkreuzen. Am Ende wird es gefährlich für Harry und seine Freunde, aber Krüger kriegt sie nicht!

Arbeit mit dem Text

Was gehört zusammen?

1. Harry hat genug von der DDR, ...
2. Alle Fluchtwege sind versperrt, ...
3. Harry und seine Freunde können den Tunnel nicht allein bauen, ...
4. Harry und seine Leute haben nicht genug Geld, ...
5. Kein DDR-Bürger darf das Land verlassen, ...

a. deshalb helfen mehr Menschen beim Graben.
b. deshalb verfolgt Stasi-Oberst Krüger Fluchthelfer wie Harry und seine Freunde.
c. deshalb ist der Tunnel eine der letzten Möglichkeiten, die DDR zu verlassen.
d. deshalb verkaufen sie ihre Geschichte an das amerikanische Fernsehen.
e. deshalb will er weg.

FILMCLIP

NOTE: For copyright reasons, the films referenced in the **FILMCLIP** feature have not been provided by the publisher. The film can be purchased as a DVD or found online at various sites such as YouTube, Amazon, or the iTunes store. The time codes mentioned below are for the North American DVD version of the film.

Szene: DVD, Kapitel 1, Vorspann, 0:00:00–0:03:00 Min.

Harry Melchior, die Hauptfigur des Films, stellt sich und andere vor.

Schauen Sie sich die Szene an und beantworten Sie die Fragen.

1. Wann hat Harry sein Land verlassen?
2. Was genau ist „sein Land"?
3. Bei welchem Wettkampf[1] tritt Harry an[2]? Wie schneidet er ab[3]?
4. Wer ist Lotte?
5. Wer ist Matthis?
6. Matthis erzählt, dass man große Mengen Steine von seiner Baustelle[4] wegschafft[5]. Warum macht man das?

[1]*competition* [2]tritt an *does compete* [3]schneidet ab *does perform* [4]*construction site* [5]*carries away*

Nach dem Lesen

Kreatives Schreiben. Oberst Krüger verfolgt Harry im Tunnel. Es kommt zu einem Gespräch zwischen den beiden. Was sagen sie? Schreiben Sie einen Dialog zwischen Harry und Oberst Krüger.

Videoecke

Perspektiven

Wie finanzierst du dir dein Studium?

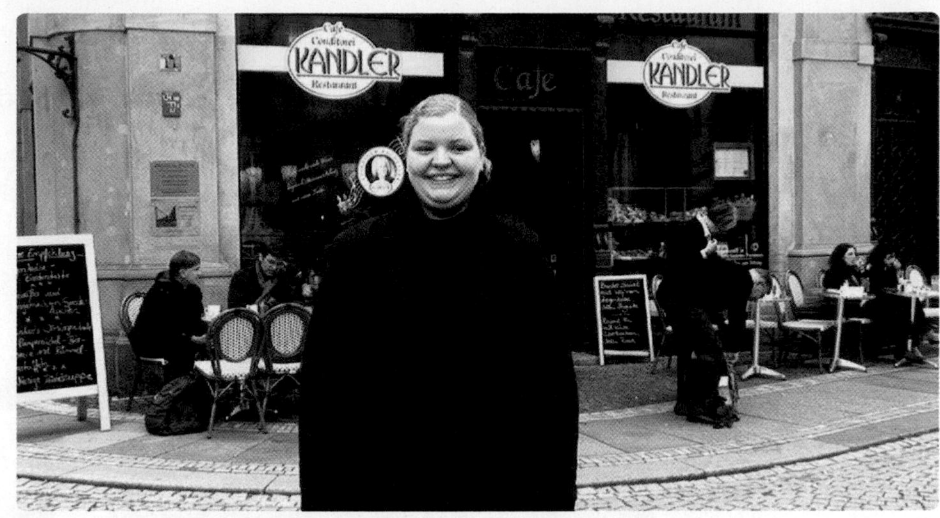

Ich bekomme BAföG.

Aufgabe 1 Das Studium

Wer bekommt BAföG? Wer hat einen Nebenjob? Schreiben Sie auf, wie sich die Studentinnen und Studenten ihr Studium finanzieren. ACHTUNG: Manche bekommen Geld aus unterschiedlichen Quellen.

Miniwörterbuch

der **Nebenjob**	side job
einen Kredit aufnehmen	to take out a loan
sparen	to save
die **Kulturwissenschaften**	cultural sciences
eigentlich	actually
die **Forschung**	research
die **Entwicklungshilfe**	developmental aid
das **Fließband**	assembly line
basteln	to do crafts

1. Judith ___

2. Susan ___

3. Shaimaa ___

4. Martin ___

5. Tina ___

6. Inna ___

7. Nadezda ___

8. Sophie ___

a. arbeiten
b. BAföG
c. einen Kredit aufnehmen
d. Eltern
e. Nebenjobs
f. Stipendium

Interviews

- Was studierst du?
- Was gefällt dir an deinem Studium?
- Was willst du damit mal machen?
- Arbeitest du neben deinem Studium?
- Was machst du da genau?
- Wie viel Geld verdienst du damit?
- Gibt es etwas, worauf du sparst?

Tabea

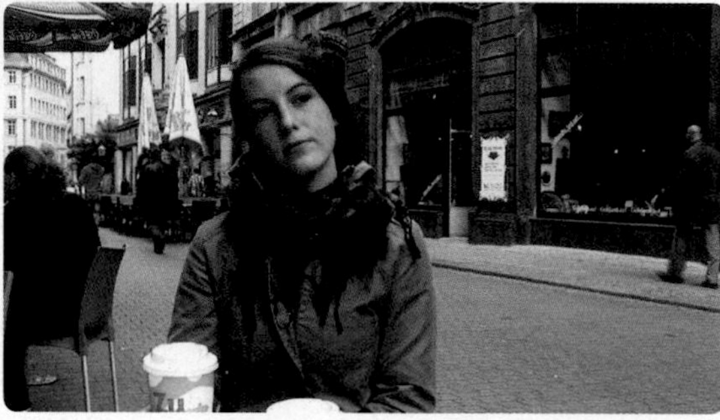

Tina

Aufgabe 2 Wer sagt was?

Wer sagt das, Tabea oder Tina?

	Tabea	Tina
1. Ich studiere Biochemie und Mathematik.	☐	☐
2. Ich studiere Kulturwissenschaften.	☐	☐
3. Eigentlich wollte[6] ich Medizin studieren.	☐	☐
4. Ich möchte gern in die Forschung gehen.	☐	☐
5. Ich möchte vielleicht in die Entwicklungshilfe gehen.	☐	☐
6. Ich arbeite in den Semesterferien am Fließband.	☐	☐
7. Ich passe auf ein dreijähriges Kind auf.	☐	☐
8. Ich verdiene 8 Euro 50 pro Stunde.	☐	☐
9. Ich möchte einmal nach Japan reisen.	☐	☐
10. Ich ziehe um und brauche eine neue Küche.	☐	☐

Aufgabe 3 Dies und das

Beantworten Sie die folgenden Fragen.

1. Warum kann Tabea nicht Medizin studieren?
2. Was möchte Tabea machen, wenn sie in die Forschung geht?
3. Was gefällt Tina an ihrem Studium?
4. Was macht Tina mit dem dreijährigen Kind?
5. Wie viel verdient Tina?

Aufgabe 4 Interview

Interviewen Sie eine Partnerin oder einen Partner. Stellen Sie die Interviewfragen.

[6]*wanted to*

Wortschatz

Berufe — Professions

der **Anwalt, ¨-e** / die **Anwältin, -nen** — lawyer
der **Arzt** (R), **¨-e** / die **Ärztin, -nen** — physician, doctor
der **Bademeister, -** / die **Bademeisterin, -nen** — swimming-pool attendant
der/die **Bankangestellte, -n** — bank employee
der **Bauarbeiter, -** / die **Bauarbeiterin, -nen** — construction worker
der **Berufsberater, -** / die **Berufsberaterin, -nen** — career counselor
der **Dirigent, -en** (*wk. masc.*) / die **Dirigentin, -nen** — (orchestra) conductor
der **Friseur, -e** / die **Friseurin, -nen** — hairdresser
der **Hausmeister, -** / die **Hausmeisterin, -nen** — custodian
der **Krankenpfleger, -** / die **Krankenpflegerin, -nen** — nurse
der **Richter, -** / die **Richterin, -nen** — judge
der **Schriftsteller, -** / die **Schriftstellerin, -nen** — writer
der **Verkäufer, -** / die **Verkäuferin, -nen** — salesperson

Ähnliche Wörter

der **Arbeiter, -** / die **Arbeiterin, -nen;** der **Architekt, -en** (*wk. masc.*) / die **Architektin, -nen;** der **Bibliothekar, -e** / die **Bibliothekarin, -nen;** der **Koch, ¨-e** / die **Köchin, -nen;** der **Pilot, -en** (*wk. masc.*) / die **Pilotin, -nen;** der **Polizist, -en** (*wk. masc.*) / die **Polizistin, -nen;** der **Präsident, -en** (*wk. masc.*) / die **Präsidentin, -nen;** der **Priester, -** / die **Priesterin, -nen;** der **Sekretär, -e** / die **Sekretärin, -nen;** der **Taxifahrer, -** / die **Taxifahrerin, -nen**

Orte — Places

die **Ecke, -n** — corner
um die Ecke — around the corner
die **Fabrik, -en** — factory
in der Fabrik — in the factory
die **Gaststätte, -n** — restaurant
in der Gaststätte — at the restaurant
die **Kasse, -n** — ticket booth
an der Kasse — at the ticket booth
die **Kirche, -n** — church
in der Kirche — at church
die **Polizei** — police station
auf der Polizei — at the police station
die **Post** — post office
auf der Post — at the post office
die **Tankstelle, -n** — gas station
an der Tankstelle — at the gas station

der **Bahnhof, ¨-e** (R) — train station
auf dem Bahnhof — at the train station
der **Schalter, -** — ticket booth
am Schalter — at the ticket booth

das **Büro, -s** — office
im Büro — at the office
das **Gericht, -e** — courthouse
auf dem Gericht — at the courthouse
das **Kaufhaus, ¨-er** — department store
im Kaufhaus — at the department store
das **Krankenhaus, ¨-er** (R) — hospital
im Krankenhaus — in the hospital
das **Schwimmbad, ¨-er** (R) — swimming pool
im Schwimmbad — at the swimming pool

Ähnliche Wörter

die **Bäckerei, -en; in der Bäckerei;** die **Bank, -en; auf der Bank;** die **Schule, -n** (R); **in der Schule;** die **Universität, -en** (R); **auf der Universität;** der **Supermarkt, ¨-e; im Supermarkt;** das **Hotel, -s** (R); **im Hotel**

In der Küche — In the Kitchen

die **Flasche, -n** — bottle
die **Küche, -n** — kitchen
die **Küchenwaage, -n** — kitchen scale
die **Salatschüssel, -n** — salad (mixing) bowl
die **Schublade, -n** — drawer
die **Tasse, -n** (R) — cup

der **Backofen, ¨-** — oven
der **Geschirrspüler, -** — dishwasher
der **Herd, -e** — stove
der **Kühlschrank, ¨-e** — refrigerator
der **Topf, ¨-e** — pot, pan
der **Topflappen, -** — potholder
der **Wasserhahn, ¨-e** — faucet

das **Besteck** — silverware, cutlery
das **Geschirr** (R) — dishes
das **Papiertuch, ¨-er** — paper towel
das **Spülbecken, -** — sink

Ähnliche Wörter

die **Kaffeemaschine, -n;** die **Küchenlampe, -n;** die **Küchenuhr, -en;** die **Pfanne, -n;** der **Küchentisch, -e;** das **Glas, ¨-er**

Einkäufe und Geschenke	Purchases and Presents
die **Badehose, -n**	swim(ming) trunks
die **Briefmarke, -n**	stamp
die **Mütze, -n**	cap
der **Regenschirm, -e**	umbrella
der **Reiseführer, -**	travel guidebook
der **Roman, -e**	novel
das **Weihnachtsgeschenk, -e**	Christmas present

Ähnliche Wörter

die **Konzertkarte, -n;** die **Tageszeitung, -en;** der **Bikini, -s;** der **Fahrradhelm, -e;** der **MP3-Spieler, -** (R); das **Parfüm, -e; das Videospiel, -e**

Schule und Beruf	School and Career
die **Ausbildung**	specialized training
praktische Ausbildung	practical (career) training
die **Bundeswehr**	German army
bei der Bundeswehr	in the German army
die **Schulbildung**	education, schooling
der **Aufenthalt, -e**	stay, sojourn

Sonstige Substantive	Other Nouns
die **Dusche, -n**	shower
die **Enkelin, -nen**	granddaughter
die **Lehre, -n**	apprenticeship
die **Möglichkeit, -en**	possibility
die **Umkleidekabine, -n**	dressing room
die **Versicherung, -en**	insurance
die **Werkstatt, ¨-en**	repair shop, garage
der **Eindruck, ¨-e**	impression
der **Enkel, -**	grandson
der **Kuchen, -**	cake
der **Rasen**	lawn
der **Rat, Ratschläge**	advice
der **Termin, -e**	appointment
der **Urlaub, -e** (R)	vacation
der **Vorschlag, ¨-e**	suggestion
das **Doppelzimmer, -**	double room
das **Geheimnis, -se**	secret
das **Interesse, -n**	interest
Interesse haben an (+ *dat.*)	to be interested in
das **Konto, Konten**	bank account
ein Konto eröffnen	to open a bank account
das **Lieblingsfach, ¨-er**	favorite subject
das **Öl**	oil
das Öl kontrollieren	to check the oil
die **Kenntnisse** (*pl.*)	skills; knowledge about a field

Ähnliche Wörter

die **Klasse, -n;** die **Liste, -n;** die **Lotterie, -n; in der Lotterie gewinnen;** die **Patientin, -nen;** der **Patient, -en** (*wk. masc.*); das **Chaos;** das **Prestige** [prɛsti:ʒ]

Verben	Verbs
aus·tragen, trägt ... aus, ausgetragen	to deliver
Zeitungen austragen	to deliver newspapers
decken	to cover; set
den Tisch decken	to set the table
ein·kaufen gehen, ist einkaufen gegangen (R)	to go shopping
erhalten, erhält, erhalten	to receive
erklären	to explain
erreichen	to reach
erzählen (R)	to tell (a story, joke)
fegen	to sweep
feiern (R)	to celebrate
heiraten (R)	to marry
interessieren	to interest
sich interessieren für	to be interested in
leid·tun, leidgetan (+ *dat.*)	to be sorry
tut mir leid (R)	I'm sorry
leihen, geliehen	to lend
mähen	to mow
pflegen	to attend to; to nurse
sagen (R)	to say, tell
schenken	to give (as a present)
sparen	to save (money)
statt·finden, stattgefunden	to take place
stellen (R)	to place, put
eine Frage stellen	to ask a question
unterrichten	to teach, instruct
untersuchen	to investigate; to examine
verkaufen (R)	to sell
voll·tanken	to fill up (with gas)
weg·stellen	to put away
werden, wird, ist geworden	to become
zahlen	to pay
zeichnen (R)	to draw

Ähnliche Wörter

backen, backt/bäckt, gebacken; heilen; weg·bringen, weggebracht

Adjektive und Adverbien	Adjectives and Adverbs
ausführlich	thorough
ausverkauft	sold out
dunkel	dark
getrennt	separately; separate checks
leer	empty
sauer	angry
unglaublich	incredible

Ähnliche Wörter

normal, praktisch

Sonstige Wörter und Ausdrücke	Other Words and Expressions
als	as; when
als was?	as what?
als ich acht Jahre alt war	when I was eight years old
etwas (R)	something, anything
sonst noch etwas?	anything else?
fast	almost
gern (R)	gladly
ich hätte gern	I would like
hin und zurück	round-trip
jede, jeder, jedes (R)	each
mehrmals	several times
nebenan	next door
von nebenan	from next door
sonst (R)	otherwise
unter	under, underneath
unter dem Fenster	under the window
wem	whom (*dative*)
zweimal	twice

Strukturen und Übungen

5.1 Dative case: articles and possessive determiners

WISSEN SIE NOCH?

The nominative case designates the subject of a sentence. The accusative case designates the object of the action of the verb.

Review grammar 2.1.

A noun or pronoun in the dative case is used to designate the person to or for whom something is done.

Ernst schenkt **seiner Mutter** ein Buch.	*Ernst gives his mother a book.*
Sofie gibt **ihrem Freund** einen Kuss.	*Sofie gives her boyfriend a kiss.*

Note that the dative case frequently appears in sentences with three nouns: a person who does something, a person who receives something, and the object that is passed from the doer to the receiver. The doer, the subject of the sentence, is in the nominative case; the recipient, or beneficiary, of the action is in the dative case; and the object is in the accusative case.

The dative case indicates to or for whom.

Doer		**Recipient**	**Object**
Nominative Case	*Verb*	*Dative Case*	*Accusative Case*
Maria	kauft	ihrem Freund	ein Hemd.

Maria is buying her boyfriend a shirt.

In German, the signal for the dative case is the ending **-m** in the masculine and neuter, **-r** in the feminine, and **-n** in the plural. Here are the dative forms of the definite, indefinite, and negative articles, and of the possessive determiners.

	Masculine and Neuter	Feminine	Plural
Definite Article	dem	der	den
Indefinite Article	einem	einer	—
Negative Article	keinem	keiner	keinen
Possessive Determiners	meinem	meiner	meinen
	deinem	deiner	deinen
	Ihrem	Ihrer	Ihren
	seinem	seiner	seinen
	ihrem	ihrer	ihren
	unserem	unserer	unseren
	eurem	eurer	euren

Jutta schreibt **einem Freund** einen Brief.	*Jutta is writing a letter to a friend.*
Jens erzählt **seinen Eltern** einen Witz.	*Jens is telling his parents a joke.*

Plural nouns add **-n** in dative.

All plural nouns add an **-n** in the dative unless they already end in **-n** or in **-s.**

Claire erzählt **ihren Freunden** von ihrer Reise nach Deutschland.	Claire is telling her friends about her trip to Germany.

Here is a short list of verbs that often take an accusative object and a dative recipient.

erklären	*to explain something to someone*
erzählen	*to tell someone (a story)*
geben	*to give someone something*
leihen	*to lend someone something*
sagen	*to tell someone something*
schenken	*to give someone something as a gift*

Certain masculine nouns, in particular a number of nouns denoting professions, add **-(e)n** in the dative and accusative singular as well as in the plural. They are often called weak masculine nouns.

	Singular	Plural
Nominative	der Student	die Studenten
Accusative	den Studenten	die Studenten
Dative	dem Studenten	den Studenten

Übung 1 Was machen Sie für diese Leute?

Schreiben Sie mit jedem Verb einen Satz.

MODELL: Ich schenke meiner Mutter eine Kamera.

backen	Bruder/Schwester	ein Abendessen
erklären	Freund/Freundin	meine Bilder
erzählen	Großvater/Großmutter	einen Brief
geben	Mitbewohner/Mitbewohnerin	ein Buch
kaufen	Onkel/Tante	eine CD
kochen	Partner/Partnerin	mein Deutschbuch
leihen	Professor/Professorin	50 Dollar
schenken	Vater/Mutter	eine E-Mail
schreiben	Vetter/Kusine	ein Geheimnis
verkaufen		eine Geschichte
		Kaffee
		eine Konzertkarte
		eine Krawatte
		einen Kuchen
		einen Kuss
		einen Laptop
		einen Witz

Übung 2 Was machen diese Leute?

Bilden Sie Sätze.

MODELL: Heidi schreibt ihren Eltern eine Karte.

Bikini (*m.*) = der Bikini
Grammatik (*f.*) = die Grammatik
Zelt (*n.*) = das Zelt

[Heidi]	erklären	***ihren*** [Eltern]	Armband (*n.*)
Peter	erzählen	Freund	Bikini (*m.*)
Thomas	geben	Freundin	Geheimnis (*n.*)
Katrin	kaufen	Mann	Grammatik (*f.*)
Stefan	kochen	Mutter	***eine*** [Karte] (*f.*)
Albert	leihen	Professor	Regenschirm (*m.*)
Monika	schenken	Schwester	Rucksack (*m.*)
Frau Schulz	[schreiben]	Tante	Suppe (*f.*)
Nora	verkaufen	Vetter	Zelt (*n.*)

5.2 Question pronouns: *wer, wen, wem*

Use the pronouns **wer, wen,** and **wem** to ask questions about people: **wer** indicates the subject, the person who performs the action; **wen** indicates the accusative object; **wem** indicates the dative object.

wer (Who is it?) = nominative
wen (Whom do you know?) = accusative
wem (Whom did you give it to?) = dative

Wer arbeitet heute Abend um acht?	*Who's working tonight at eight?*
Wen triffst du heute Abend?	*Whom are you meeting tonight?*
Wem leihst du das Zelt?	*To whom are you lending the tent?*

Übung 3 Minidialoge

Ergänzen Sie **wer, wen** oder **wem.**

1. JÜRGEN: _______ hat meinen Regenschirm?
   SILVIA: Ich habe ihn.
2. MELANIE: _______ hast du in der Stadt gesehen?
   JOSEF: Claire.
3. SOFIE: _______ willst du die DVD schenken?
   WILLI: Nesrin. Sie wünscht sie sich schon lange.
4. FRAU AUGENTHALER: Na, erzähl doch mal. _______ hast du letztes Wochenende kennengelernt?
   RICHARD: Also, sie heißt Uschi und ...
5. MEHMET: _______ wollt ihr denn euren neuen Computer verkaufen?
   RENATE: Schülern und Studenten.
6. SUMITA: Weißt du, _______ heute Abend zu uns kommt?
   LYDIA: Nein, du?
   SUMITA: Tante Christa natürlich.

5.3 Expressing change: the verb *werden*

Use a form of **werden** to talk about changing conditions.

Ich **werde** alt.	*I am getting old.*
Es **wird** dunkel.	*It is getting dark.*

werden: e → i
du wirst; er/sie/es wird

werden			
ich	werde	*wir*	werden
du	wirst	*ihr*	werdet
Sie	werden	*Sie*	werden
er *sie* *es*	wird	*sie*	werden

In German, **werden** is also used to talk about what somebody wants to be.

Was willst du **werden?**	*What do you want to be (become)?*
Sumita will Ärztin **werden.**	*Sumita wants to be (become) a physician.*

Übung 4 Was passiert?

Bilden Sie Fragen und suchen Sie dann eine logische Antwort darauf.

MODELL: Was passiert im Winter? —Es wird kalt.

1. am Abend	a. Man wird Arzt.
2. wenn man Bücher schreibt	b. Man wird bekannt[1].
3. wenn man krank wird	c. Die Blätter werden bunt[2].
4. im Frühling	d. Es wird dunkel.
5. im Herbst	e. Sie werden größer.
6. wenn Kinder älter werden	f. Es wird wärmer.
7. wenn man in der Lotterie gewinnt	g. Es wird hell[3].
8. wenn man Medizin studiert	h. Man bekommt Fieber.
9. am Morgen	i. Die Tage werden länger.
10. im Sommer	j. Man wird reich.

Übung 5 Was werden sie vielleicht?

Suchen Sie einen möglichen Beruf für jede Person.

MODELL: Jens hilft gern kranken Menschen. →
Vielleicht wird er Arzt.

1. Lydia kocht gern.
2. Damla interessiert sich für Medikamente.
3. Ernst fliegt gern.
4. Jürgen hat Interesse an Pädagogik.
5. Jutta zeichnet gern Pläne für Häuser.
6. Eske geht gern in die Bibliothek.
7. Hans möchte gern kranke Menschen heilen.
8. Andrea hört gern klassische Musik.

Apotheker/Apothekerin
Architekt/Architektin
Bibliothekar/Bibliothekarin
Dirigent/Dirigentin
Koch/Köchin
Krankenpfleger/Krankenpflegerin
Lehrer/Lehrerin
Pilot/Pilotin

5.4 Location: *in, an, auf* + dative case

To express the location of someone or something, use the following prepositions with the dative case.

For location, **in, an,** and **auf** take the dative.

in (*in, at*)
auf (*on, at*)
an (*on, at*)
\+
dem/einem ____ (*m., n.*)
der/einer ____ (*f.*)
den ____ (*pl.*)

Katrin wohnt **in der Stadt.** — *Katrin lives in the city.*
Stefan und Albert sind **auf der Bank.** — *Stefan and Albert are at the bank.*

A. Forms and Contractions

Remember the signals for dative case.

	Masculine and Neuter	Feminine	Plural
Dative	dem	der	den
	einem	einer	—

[1]*well-known* [2]*colorful* [3]*bright; light*

in + **dem** = **im**
an + **dem** = **am**

Note that the prepositions **in** + **dem** and **an** + **dem** are contracted to **im** and **am.**

Masculine and Neuter	Feminine	Plural
im Kino	**in der** Stadt	**in den** Wäldern
in einem Kino	**in einer** Stadt	**in** Wäldern
am See	**an der** Tankstelle	**an den** Wänden
an einem See	**an einer** Tankstelle	**an** Wänden
auf dem Berg	**auf der** Bank	**auf den** Bäumen
auf einem Berg	**auf einer** Bank	**auf** Bäumen

B. Uses

1. Use **in** when referring to enclosed spaces.

im Supermarkt	*in the supermarket* (*enclosed*)
in der Stadt	*in* (*within*) *the city*

2. **An,** in the sense of English *at,* denotes some kind of border or limiting area.

am Fenster	*at the window*
an der Tankstelle	*at the gas pump*
am See	*at the lake*

3. Use **auf,** in the sense of English *on,* when referring to surfaces.

auf dem Tisch	*on the table*
auf dem Herd	*on the stove*

4. **Auf** is also used to express location in public buildings such as the bank, the post office, or the police station.

auf der Bank	*at the bank*
auf der Post	*at the post office*
auf der Polizei	*at the police station*

Übung 6 Was macht man dort?

Stellen Sie einem Partner / einer Partnerin Fragen. Er/Sie soll eine Antwort darauf geben.

MODELL: S1: Was macht man am Strand?
S2: Man spielt Volleyball.

Benzin[1] tanken **tanzen** **Briefmarken kaufen** **Geld wechseln[2]** **ein Buch lesen** **beten[3]** **schwimmen** **einen Film sehen** **Volleyball spielen** **spazieren gehen** **?**

1. im Kino
2. auf der Post
3. an der Tankstelle
4. in der Disko
5. in der Kirche
6. auf der Bank
7. im Meer
8. in der Bibliothek
9. im Park

[1]*gasoline* [2]*to exchange* [3]*to pray*

Übung 7 Wo?

Wo sind die Leute? Wo sind das Poster, der Topf und der Wein?

MODELL: Stefan ist am Strand.

1.

2.

3.

4.

5.

6.

7.

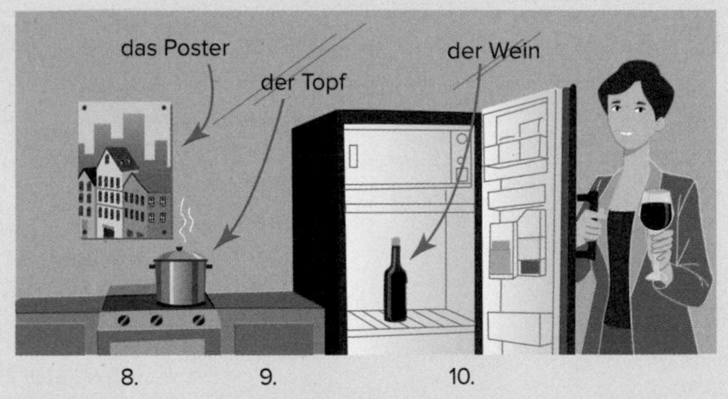

8. 9. 10.

5.5 Dative case: personal pronouns

WISSEN SIE NOCH?

The dative case designates the person to whom or for whom something is done.

Review grammar 5.1.

Personal pronouns in the dative case designate the person to or for whom something is done. (See also **Strukturen 5.1.**)

Kaufst du **mir** ein Buch?	*Are you buying me a book?*
Nein, ich schenke **dir** eine DVD.	*No, I'm giving you a DVD.*

A. First- and Second-Person Pronouns

Here are the nominative and dative forms of the first- and second-person pronouns.

Singular		**Plural**	
Nominative	*Dative*	*Nominative*	*Dative*
ich	mir	wir	uns
du	dir	ihr	euch
Sie	Ihnen	Sie	Ihnen

Note that German speakers use three different pronouns to express the recipient or beneficiary in the second person (English *you*): **dir, euch,** and **Ihnen.**

RICHARD: Leihst du mir dein Auto, Mutti? (*Will you lend me your car, Mom?*)
FRAU AUGENTHALER: Ja, ich leihe **dir** mein Auto. (*Yes, I'll lend you my car.*)

HERR THELEN: Viel Spaß in Wien! (*Have fun in Vienna!*)
HERR WAGNER: Danke! Wir schreiben **Ihnen** eine Postkarte. (*Thank you! We'll write you a postcard.*)

HANS: Ernst und Andrea! Kommt in mein Zimmer! Ich zeige **euch** meine Briefmarken. (*Ernst and Andrea! Come to my room! I'll show you my stamp collection.*)

B. Third-Person Pronouns

The third-person pronouns have the same signals as the dative articles: **-m** in the masculine and neuter, **-r** in the feminine, and **-n** in the plural.

de**m** → ih**m**
de**r** → ih**r**
de**n** → ihne**n**

	Masculine and Neuter	**Feminine**	**Plural**
Article	dem	der	den
Pronoun	**ihm**	**ihr**	**ihnen**

Was kaufst du deinem Vater?	*What are you going to buy your dad?*
Ich kaufe **ihm** ein Buch.	*I'll buy him a book.*
Was schenkst du deiner Schwester?	*What are you going to give your sister?*
Ich schenke **ihr** eine Bluse.	*I'll give her a blouse.*
Was kochen Sie Ihren Kindern?	*What are you going to cook for your kids?*
Ich koche **ihnen** Spaghetti.	*I'm making them spaghetti.*

Note that the dative-case pronoun precedes the accusative-case noun.

Ich schreibe **dir** einen Brief.	*I'll write you a letter.*

Übung 8 Minidialoge

Ergänzen Sie **mir, dir, uns, euch** oder **Ihnen.**

1. HANS: Mutti, kaufst du ______ Schokolade?
   FRAU RUF: Ja, aber du weißt, dass du vor dem Essen nichts Süßes essen sollst.
2. MARIA: Was hat denn Frau Körner gesagt?
   MICHAEL: Das erzähle ich ______ nicht.
3. ERNST: Mutti, kochst du Andrea und mir einen Pudding?
   FRAU WAGNER: Natürlich koche ich ______ einen Pudding.
4. HERR SIEBERT: Sie schulden[1] mir noch 50 Euro, Herr Pusch.
   HERR PUSCH: Was!? Wofür denn?
   HERR SIEBERT: Ich habe ______ doch für 300 Euro mein altes Motorrad verkauft, und Sie hatten nur 250 Euro dabei.
   HERR PUSCH: Ach, ja, richtig.
5. FRAU KÖRNER: Mein Mann und ich gehen heute Abend aus. Können Sie ______ vielleicht ein gutes Restaurant empfehlen, Herr Pusch?
   MICHAEL: Ja, gern ...

Übung 9 Wer? Wem? Was?

Beantworten Sie die Fragen mit Hilfe der Tabelle.

MODELL: Was hat Renate ihrem Freund geschenkt?
Sie hat ihm ein T-Shirt geschenkt.

	Renate	Mehmet
schenken	ein T-Shirt	einen Regenschirm
leihen	ihr Auto	500 Euro
erzählen	ein Geheimnis	eine Geschichte
verkaufen	ihre Sonnenbrille	seinen Fernseher
zeigen	ihr Büro	seine Wohnung
kaufen	eine neue Brille	einen Kinderwagen

1. Was hat Mehmet seiner Mutter geschenkt?
2. Was hat Renate ihrem Vater geliehen?
3. Was hat Mehmet seinem Bruder geliehen?
4. Was hat Renate ihrer Friseurin erzählt?
5. Was hat Mehmet seinen Nichten erzählt?
6. Was hat Renate ihrer Freundin verkauft?
7. Was hat Mehmet seinen Eltern verkauft?
8. Was hat Renate ihrem Schwager gezeigt?
9. Was hat Mehmet seinem Freund gezeigt?
10. Was hat Renate ihrer Großmutter gekauft?
11. Was hat Mehmet seiner Schwägerin gekauft?

[1] *owe*

KAPITEL 6

Wohnen

In **Kapitel 6,** you will learn vocabulary and expressions for describing where you live, for finding a place to live, and for talking about housework.

Themen

Haus und Wohnung

Das Stadtviertel

Auf Wohnungssuche

Hausarbeit

Kulturelles

KLI: Wohnen

KLI: Deutsch und Englisch als germanische Sprachen

Musikszene: „Haus am See" (Peter Fox)

Filmclip: *Good bye Lenin!* (Wolfgang Becker)

Videoecke: Wohnen

Lektüren

Sachtext: Städteranking 2014

Film: *Good bye Lenin!* (Wolfgang Becker)

Strukturen

6.1 Dative verbs

6.2 Location vs. destination: two-way prepositions with the dative or accusative case

6.3 Word order: time before place

6.4 Direction: **in/auf** vs. **zu/nach**

6.5 Separable-prefix verbs: the present tense and the perfect tense

6.6 The prepositions **mit** and **bei** + dative

Friedensreich Hundertwasser: *(630A) Mit der Liebe warten tut weh, wenn die Liebe woanders ist (1971)*, Galerie Koller, Zürich

KUNST UND KÜNSTLER

Friedensreich Hundertwasser (1928–2000), mit bürgerlichem Namen[1] Friedrich Stowasser, ist einer der bedeutendsten Künstler des 20. Jahrhunderts. Er war Maler und Architekt, entwarf[2] aber auch Briefmarken und Bucheinbände[3]. Er war Sohn einer Jüdin. Aufgrund seiner Erfahrungen im Dritten Reich war er politisch aktiv, vor allem gegen Diktaturen. Weiterhin engagierte er sich sehr für den Umweltschutz. Mit seinen Gebäuden schuf er Beispiele für eine natur- und menschengerechte Architektur. Seine Bilder sind von kräftigen Farben, organischen Formen und einer Ablehnung[4] von geraden[5] Linien gekennzeichnet.

Schauen Sie sich das Bild an und beantworten Sie die folgenden Fragen.

1. Welche Farben dominieren im Bild? Welche Farbe dominiert besonders? Was symbolisiert diese Farbe? Denken Sie an den Titel.
2. Was ist das Besondere an den Häusern? Laden sie ein oder schließen sie aus? Woran erkennt man das?
3. Was sehen Sie links unten im rechten Haus? Wer könnte das sein?
4. Weckt das Bild eher fröhliche oder traurige Assoziationen? Warum?

[1]mit ... *born* [2]*designed* [3]*book covers* [4]*rejection* [5]*straight*

Situationen

Haus und Wohnung

Grammatik 6.1–6.2

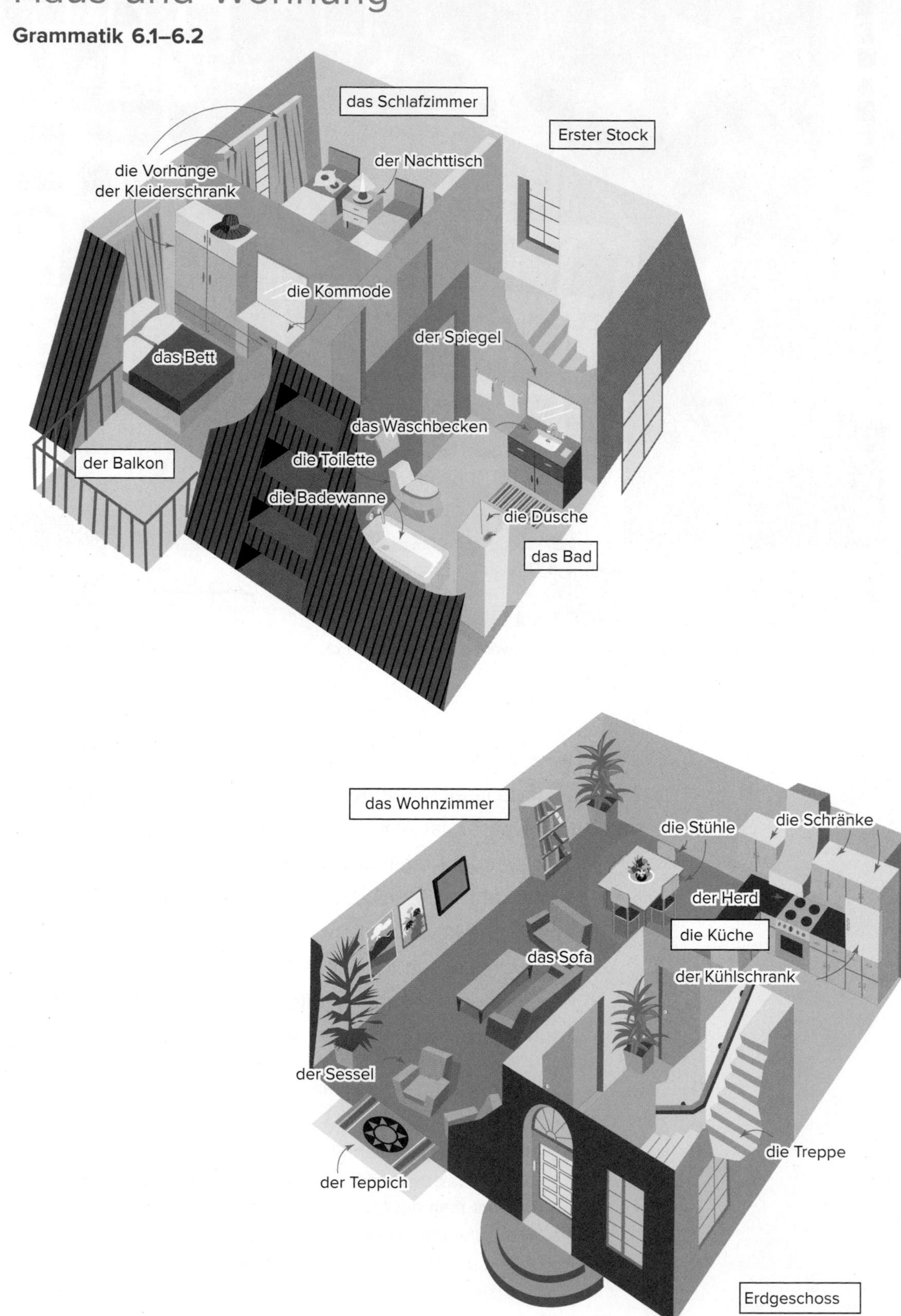

Situation 1 Das Zimmer

Wählen Sie ein Bild, aber sagen Sie die Nummer nicht. Ihr Partner oder Ihre Partnerin stellt Fragen und sagt, welches Bild Sie gewählt haben.

MODELL: S1: Ist die Katze auf dem Sofa?
S2: Ja.
S1: Ist es neun Uhr?
S2: Ja.
S1: Dann ist es Bild 1.
S2: Richtig. Jetzt bist du dran.

am Fenster **an der Wand** **auf dem Sofa**
vor dem Sofa
auf dem Tisch **über dem Schrank**
neben dem Sofa **?** **unter dem Tisch**

KULTUR ... LANDESKUNDE ... INFORMATIONEN

WOHNEN

In Ihrem Land:

- Haben moderne Häuser in Ihrem Land einen Keller[1], eine Terrasse, einen Balkon?
- Haben sie einen Garten vor oder hinter dem Haus?
- Aus welchem Material sind die Häuser normalerweise? (aus Stein, aus Holz[2], aus Beton[3])
- Gibt es einen Zaun[4] um das ganze Grundstück[5] herum oder nur um den Garten hinter dem Haus?
- Wie viele Garagen sind üblich[6]? Wie groß sind die Garagen? (Platz für ein Auto, zwei Autos, drei Autos)
- Aus welchem Material ist das Dach? (aus Asphaltschindeln[7], aus Holzschindeln[8], aus Ziegeln[9])

Einfamilienhaus in einer Neubausiedlung
© Ullstein bild - Schöning/The Image Works

Wohnblöcke im Ostteil Berlins
© Stuart Cohen

Zweifamilienhaus aus Backstein
© ullstein bild - Business Picture/The Image Works

In Deutschland:

- Schauen Sie sich die Fotos an. Welche Unterschiede[10] gibt es zu Häusern in Ihrem Land?

Hören Sie sich den Text an und beantworten Sie die folgenden Fragen.

- Wie viele Menschen leben in Deutschland?
- Wie groß ist Deutschland?
- In Deutschland leben ungefähr[11] 200 Menschen auf einem Quadratkilometer[12], das sind fast 600 auf einer Quadratmeile. In den USA z. B. sind es im Durchschnitt[13] 80 auf einer Quadratmeile. Wie viele sind es in Ihrem Bundesland?

[1]*basement* [2]*wood* [3]*concrete* [4]*fence* [5]*property* [6]*customary* [7]*asphalt shingles* [8]*wooden shingles* [9]*clay tiles* [10]*differences* [11]*approximately* [12]*square kilometer* [13]im ... *on average*

Situation 2 Interview

1. Wo wohnst du? (in einer Wohnung, in einem Studentenheim, in einem Haus, auf dem Land, in der Stadt, ______)
2. Wohnst du allein? (in einer WG [Wohngemeinschaft], bei deinen Eltern, bei einer Familie, mit einem Mitbewohner, mit einer Mitbewohnerin, ______)
3. Wie lange brauchst du zur Uni? (zehn Minuten zu Fuß, fünf Minuten mit dem Fahrrad, eine halbe Stunde mit dem Auto oder mit dem Bus, ______)
4. Was kostet dein Zimmer / deine Wohnung pro Monat?
5. Was für Möbel hast du in deinem Zimmer / in deiner Wohnung?

Situation 3 Interaktion: In der Wohnung

Beantworten Sie die Fragen für sich selbst und schreiben Sie Ihre Antworten auf. Stellen Sie dann die gleichen Fragen an Ihren Partner oder Ihre Partnerin.

	ich	mein(e) Partner(in)
Wie gefällt dir deine Wohnung oder dein Zimmer?		
Welches Möbelstück fehlt dir?		
Welches Möbelstück gehört dir nicht?		
Wie gefällt dir das Aufräumen und Putzen?		
Wer hilft dir beim Aufräumen und Putzen?		

Situation 4 Zum Schreiben: So wohne ich

Schreiben Sie einen kurzen Text darüber, wo und wie Sie wohnen. Schreiben Sie einen Absatz darüber, wo Sie wohnen und mit wem Sie wohnen, und einen Absatz darüber, wie viele Zimmer es dort gibt und was in diesen Zimmern ist. Schreiben Sie einen kurzen Schluss, in dem Sie sagen, wie es Ihnen gefällt und wie lange Sie dort noch wohnen werden.

Das Stadtviertel

Grammatik 6.3–6.4

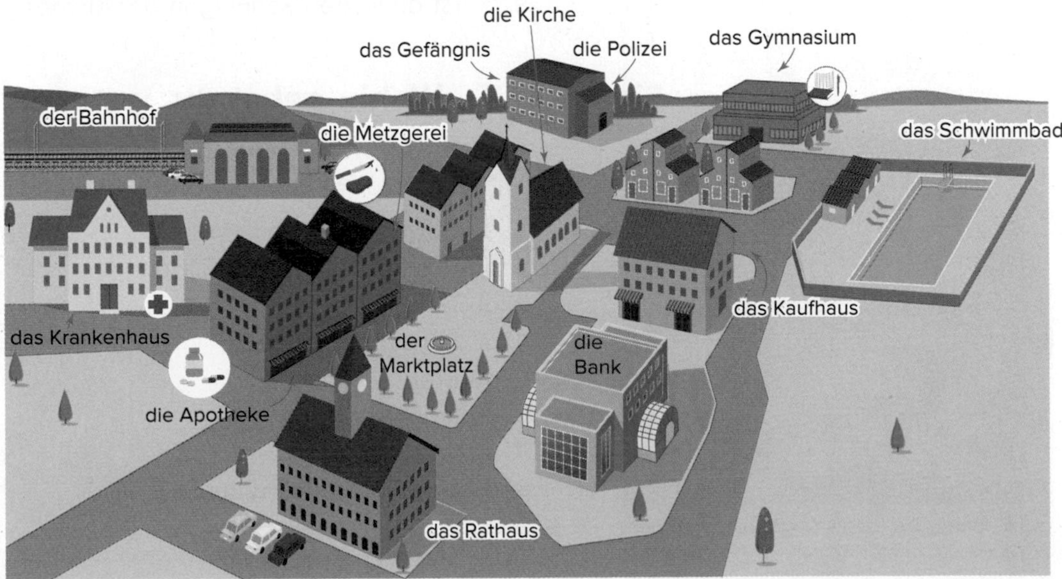

Situation 5 Wie weit weg?

MODELL: S1: Wie weit weg sollte die Apotheke von deiner Wohnung sein?
S2: ______

1. die Apotheke
2. die Universität
3. die Polizei
4. der Flughafen
5. das Kino
6. das Krankenhaus
7. das Gefängnis
8. der Kindergarten
9. der Supermarkt
10. die Kirche

so weit weg wie möglich
gleich um die Ecke
am anderen Ende der Stadt
gleich gegenüber
fünf Minuten zu Fuß
zehn Minuten mit dem Fahrrad
eine halbe Stunde mit dem Auto
mir egal
zwei Straßen weiter

Situation 6 Umfrage

MODELL: S1: Wohnst du in der Nähe der Universität?
S2: Ja.
S1: Unterschreib bitte hier.

	UNTERSCHRIFT
1. Wohnst du in der Nähe der Universität?	______
2. Übernachtest du manchmal in Hotels?	______
3. Gibt es in deiner Heimatstadt ein Schwimmbad?	______
4. Warst du letzte Woche auf der Post?	______
5. Warst du gestern im Supermarkt?	______
6. Gibt es in deiner Heimatstadt ein Rathaus?	______
7. Warst du letzten Freitag in der Disko?	______
8. Bist du oft in der Bibliothek?	______
9. Warst du letzten Sonntag in der Kirche?	______

Situation 7 Wohin gehst du, wenn ...?

Wohin fahren Sie, wenn Sie Benzin brauchen?
© Begsteiger/agefotostock

MODELL: S1: Wohin gehst du, wenn du ein Buch lesen willst?
S2: Wenn ich ein Buch lesen will? In die Bibliothek.

1. du schwimmen gehen willst?	zum Bahnnof
2. du Briefmarken kaufen willst?	in die Bäckerei
3. du Geld brauchst?	zum Flughafen
4. du Benzin brauchst?	zum Arzt
5. du Brot brauchst?	auf die Bank
6. du krank bist?	zur Tankstelle
7. du verreisen willst?	auf die Post
8. du eine Zugfahrkarte kaufen willst?	ins Schwimmbad
9. ______?	______

Situation 8 Informationsspiel: Gestern und heute

Arbeiten Sie zu zweit und stellen Sie Fragen wie im Modell.

MODELL: S2: Heute ist hier ein Schuhgeschäft. Was war früher hier?
S1: Früher war hier eine Disko.

FRÜHER

HEUTE

Lektüre

LESEHILFE

This reading, **„Städteranking 2014,“** is a **Sachtext** (informational, nonfiction text). German **Sachtexte** are usually filled with passive constructions such as **wird gefragt** (*is asked*) or **wurden untersucht** (*were studied*). When you see a form of **werden** plus a past participle, it is usually a passive construction. You will learn how to create and use passive constructions in **Kapitel 10.**

Vor dem Lesen

A. Beantworten Sie die folgenden Fragen.

1. Welche Städte sind die beliebtesten Städte in Ihrem Land? Warum sind sie beliebt?
2. Was ist Ihre Lieblingsstadt? Warum?
3. Welche Merkmale[1] einer Stadt sind für Sie wichtig? Welche sind weniger wichtig? Sortieren Sie die folgenden Merkmale von *am wichtigsten* zu *am wenigsten wichtig:* gute Schulen, viele Arbeitsplätze, große Universität, Alter und Geschichte, Größe der Stadt, Kulturangebot, Nähe zum Meer oder zu den Bergen, geringe Kriminalität, Größe des Flughafens, viele junge Leute.
4. Welche großen Städte kennen Sie in Deutschland, in Österreich oder in der Schweiz? Warum kennen Sie diese Stadt oder diese Städte? Was wissen Sie über sie?

B. Suchen Sie die Städtenamen im Text und unterstreichen Sie sie. Suchen Sie dann die Städte auf einer Landkarte.

C. Suchen Sie nun die Wörter des Miniwörterbuchs im Text und in den darauffolgenden Aktivitäten und unterstreichen Sie sie.

Miniwörterbuch

aktuell	current
blass	pale
der **Lichtblick**	ray of hope
erscheinen	to appear, be published
der **Zustand**	conditions, circumstances
der **Wohlstand**	quality of life
die **Wirtschaft**	economy
der **Arbeitsmarkt**	job market
das **Bevölkerungswachstum**	population growth
die **Anzahl**	number, amount
das **Ergebnis**	result
unterschiedlich	different
die **Behörde**	government agency
umgehen mit (etwas)	to handle (sth.)
sicher	secure
das **Einkommen**	income
die **Sicherheit**	safety
schaffen, schaffte	to manage (to do sth.), managed
punkten	to score
der **Vertreter**	representative
die **Arbeitslosenquote**	unemployment rate
die **Leistung**	performance
der **Durchschnitt**	average
bewerten	to rate

[1] *features*

Städteranking 2014

Welche Städte sind besonders grün, wirtschaftsfreundlich und wohlhabend? Auch das aktuelle Städteranking lässt die Städte im Westen, Osten und Norden Deutschlands blass aussehen. Lichtblick für die neuen Bundesländer: das Dynamikranking. Leipzig auf Platz 4, Berlin auf Platz 5, Erfurt auf Platz 9.

Die „Initiative Neue Soziale Marktwirtschaft" und die Zeitschrift *Wirtschaftswoche* haben für das Jahr 2014 ein Städteranking für Deutschland erarbeitet. Seit 2004 erscheint jährlich eine neue Studie. Das Niveauranking zeigt den aktuellen Zustand. Das Dynamikranking zeigt, wie sich die Städte in der Zeit von 2009 bis 2014 entwickelt haben. Aus dem Niveau- und dem Dynamikranking ergibt sich ein Gesamtranking. Es wird gefragt: In welcher deutschen Stadt ist der Wohlstand am größten? Und: Wo findet man die höchste wirtschaftliche Dynamik? Wichtige Faktoren sind der Arbeitsmarkt, der Wohlstand, die Kriminalitätsrate, das Bevölkerungswachstum, die Anzahl Hochqualifizierter, die Anzahl junger Leute und die Anzahl der Gästeübernachtungen. Einhundert deutsche Städte wurden untersucht.

Beim Niveauranking belegte München den ersten Platz. München ist die Landeshauptstadt von Bayern. Die Ergebnisse beruhen auf unterschiedlichen Fragen. Es wurde z. B. gefragt: Wie gehen die Behörden der jeweiligen Stadt mit den Finanzen um? Wie sicher fühlen sich die Bürger der Stadt? Wie hoch ist das Jahreseinkommen der Bürger? Bei all diesen Fragen lag München jeweils sehr weit vorn. 97,5 Prozent der Menschen sagten, dass die öffentliche Sicherheit in München sehr hoch ist. Das Jahreseinkommen liegt in München bei 23.145 Euro. In den anderen deutschen Städten beträgt es nur 18.418 Euro.

Viele andere süddeutsche Städte lagen beim Niveauranking auf den ersten 10 Plätzen. Die bayrischen Städte Erlangen, Ingolstadt und Regensburg sowie die baden-württembergischen Städte Stuttgart, Freiburg, Karlsruhe und Ulm schafften es unter die Top-Ten. Nur eine norddeutsche Stadt kam auf einen der ersten 10 Plätze: Wolfsburg in Niedersachsen. Andere große deutsche Städte schafften es nicht unter die ersten 10. So kam Hamburg auf Platz 12, Köln auf 29 und Berlin sogar nur auf Platz 43.

Beim Dynamikranking konnten allerdings einige Städte aus den neuen Bundesländern punkten. Leipzig schaffte es auf Platz 4. Leipzig liegt in Sachsen. Für die Top-Ten des Dynamikrankings ist Leipzig ein typischer Vertreter. Die Arbeitslosenquote sank in den letzten 5 Jahren um 19,5% und die Wirtschaftsleistung stieg um 17,2%. Die Zahl der Einwohner stieg zwischen den Jahren 2011 und 2013 um 5,7. Der deutschlandweite Durchschnitt war 1,8%. Leipzig erreichte bei diesem Kriterium damit Platz 1.

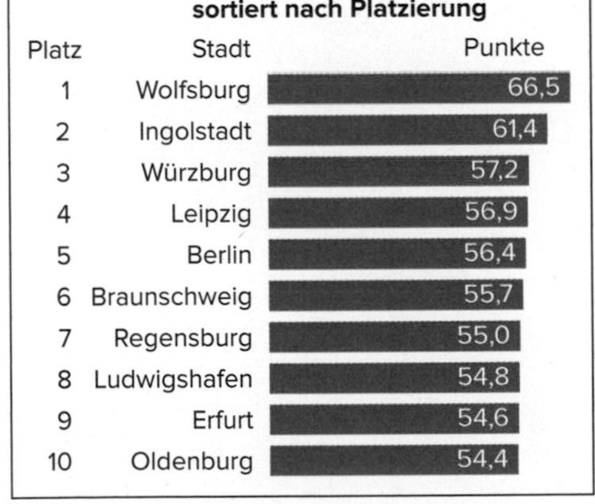

Unter den Top-Ten des Dynamikrankings lagen weitere Städte aus den neuen Bundesländern, viele Städte aus Bayern, aber auch welche aus Niedersachsen und eine Stadt aus Baden-Württemberg, nämlich Ludwigshafen. Vor allem Autostädte wie Wolfsburg (VW) und Ingolstadt (Audi) sowie Universitätsstädte wie Erfurt und Würzburg schafften es in die Top-Ten.

Die Top-Ten des Dynamikranking sieht man in der Grafik. Auf Platz 1 kam Wolfsburg. Dort lag die Arbeitslosenquote bei nur 4,4%; dazu hatten 26,4% der Einwohner einen Hochschulabschluss. Viele große deutsche Städte folgen erst auf den hinteren Plätzen: 25. Nürnberg, 30. Hannover, 31. Hamburg, 39. Bremen, 40. Köln, 50. Frankfurt am Main und 59. Kiel.

From Neue Soziale Marktwirtschaft GmbH Berlin

In Erlangen lebt man besser.
© imageBROKER/Alamy

Arbeit mit dem Text

A. Fragen zum Text. Beantworten Sie die folgenden Fragen.

1. Wie viele deutsche Städte werden im Städteranking untersucht? Wie lange gibt es das Städteranking schon?
2. Welche Faktoren sind besonders wichtig?
3. Welche Stadt liegt beim Niveauranking auf Platz 1? Warum?
4. In welchem Bundesland liegen die meisten der Top-Ten-Städte des Niveaurankings? Aus welchen anderen Bundesländern gab es weitere Top-Ten-Städte?
5. Welche Stadt liegt beim Dynamikranking auf Platz 1? Warum?
6. In welchen Bundesländern liegen die anderen Top-Ten-Städte des Dynamikrankings?
7. Was ist besonders an Leipzig?
8. Welche großen Städte werden im Dynamikranking besonders schlecht bewertet?

B. Zeitungssprache. In journalistischen Texten gibt es viele Komposita. Komposita setzen sich aus zwei oder mehr einfachen Wörtern zusammen. Bilden Sie aus den folgenden Wörtern Komposita und suchen Sie sie im Text. Was bedeuten sie?

die Arbeit
arbeitslos
die Bevölkerung
der Gast
das Jahr
die Kriminalität
das Land
das Licht
die Wirtschaft
wohnen
der Blick
das Einkommen
die Hauptstadt
der Komfort
die Leistung
der Markt
die Quote
die Rate
die Übernachtung
das Wachstum

Nach dem Lesen

Suchen Sie im Internet das aktuelle Städteranking für Deutschland, Österreich oder die Schweiz. Welche der Top-Ten-Städte eines dieser Länder finden Sie am interessantesten? Tragen Sie Informationen zu dieser Stadt zusammen und präsentieren Sie sie im Seminar.

Auf Wohnungssuche

das Reihenhaus

das Einfamilienhaus

die Villa

die Altbauwohnung

das Bauernhaus

die Skihütte

das Hochhaus

das Studentenheim

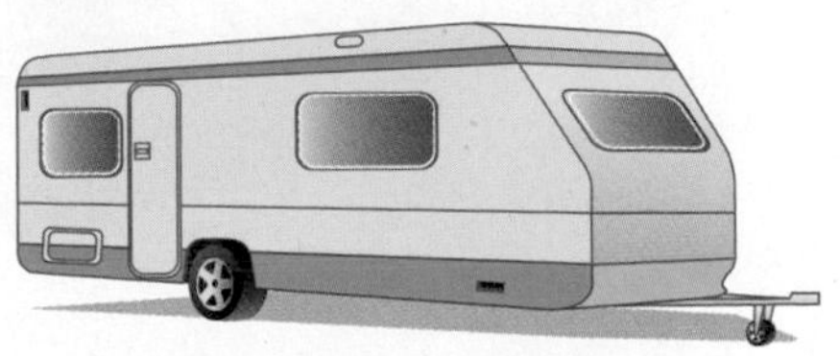
der Wohnwagen

Situation 9 Wo möchtest du gern wohnen?

Fragen Sie fünf Personen und schreiben Sie die Antworten auf.

MODELL: S1: Wo möchtest du gern wohnen?
S2: In einem Bauernhaus mit alten Möbeln.
S1: Und wo soll es stehen?
S2: Auf dem Land.

in einem Bauernhaus	mit Weinkeller	in der Innenstadt
in einem Wohnwagen	mit schönem Ausblick	am Stadtrand
in einem Hochhaus	mit Terrasse	im Ausland
in einem Einfamilienhaus	mit Balkon	auf dem Land
in einem Reihenhaus	mit alten Möbeln	in den Bergen
in einer Skihütte	mit vielen Fenstern	an einem See
in einer Villa	mit einem Garten	in der Nähe der Stadt
im Studentenheim	mit Garage	in der Nähe der Uni

Situation 10 Umfrage

MODELL: S1: Möchtest du gern in der Innenstadt wohnen?
S2: Ja.
S1: Unterschreib bitte hier.

	UNTERSCHRIFT
1. Möchtest du gern in der Innenstadt wohnen?	____________________
2. Möchtest du gern am Stadtrand wohnen?	____________________
3. Kannst du dir ein Leben auf dem Land vorstellen?	____________________
4. Möchtest du gern im Ausland wohnen?	____________________
5. Möchtest du in einer Villa wohnen?	____________________
6. Möchtest du in einem Wohnwagen wohnen?	____________________
7. Kannst du dir ein Leben auf einem Hausboot vorstellen?	____________________
8. Möchtest du gern im Studentenheim wohnen?	____________________
9. Möchtest du gern eine Woche unter Wasser wohnen?	____________________
10. Möchtest du gern im Wald wohnen?	____________________

DEUTSCH UND ENGLISCH ALS GERMANISCHE SPRACHEN

- Wie viele Sprachen spricht man in Ihrem Kurs? Welche? Welche dieser Sprachen sind verwandt mit dem Englischen?
- Gibt es Wörter in den Sprachen Ihres Kurses, die es auch im Englischen gibt? Sammeln Sie zwei bis drei Wörter pro Sprache.

Lesen Sie den Text und suchen Sie die Antworten auf die folgenden Fragen:

- Zu welcher Sprachfamilie gehören Deutsch und Englisch?
- Wie viele Wörter des Englischen (in Prozent) kommen aus dem Germanischen?
- Wo lebten die Germanen um Christi Geburt[1]?
- Zu welcher Sprachfamilie gehört das Schwedische?
- Welche Sprache hat die germanischen Konsonanten *p, t, k* behalten: das Deutsche oder das Englische?
- Wie schreibt man im Deutschen die Lautkombination **ts?**
- Wo blieb das germanische *k* im Deutschen als **k** erhalten[2]: am Wortanfang oder im Wortinneren[3]?
- Wie sprach man das westgermanische *th* aus?
- Haben formal verwandte Wörter immer dieselbe Bedeutung?

Ein altes Germanenhaus
© Bildarchiv Steffens/Bridgeman Images

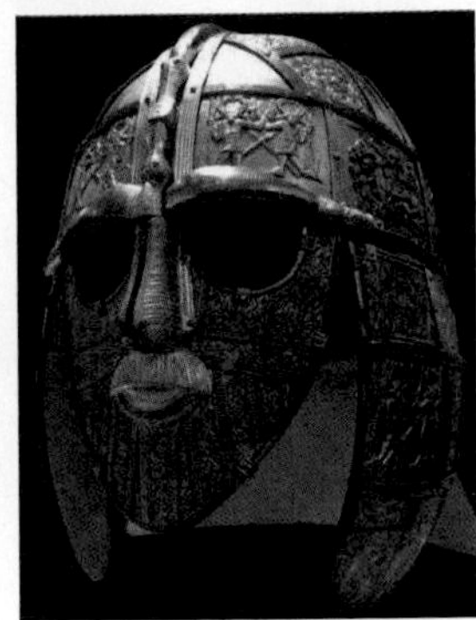

Ein Kriegerhelm aus der Germanenzeit
© World History Archive/ Newscom

Miniwörterbuch

die **Bedeutung**	meaning
gemeinsam	in common, together
allerdings	however
verändern	to change
betreffen, betrifft, betroffen	to concern

Englisch und Deutsch haben Vieles gemeinsam. Viele Wörter wie **Pfeffer** und *pepper*, **Wasser** und *water*, **brechen** und *break* zeigen, dass beide Sprachen miteinander verwandt sind. Das kommt daher, dass das Englische und Deutsche germanische Sprachen sind.

Circa 25% des englischen Wortschatzes kommt aus dem Germanischen. In der Alltagssprache verwendet man allerdings einen viel größeren Prozentsatz. So sind die häufigsten 100 Wörter des Englischen alle germanischen Ursprungs[4].

Die Germanen waren indoeuropäische Völker, die um Christi Geburt im nördlichen Mitteleuropa und südlichen Skandinavien ansässig waren. Die Angeln und Sachsen eroberten[5] im 5. Jahrhundert von der heutigen deutschen Nordseeküste aus England. Aus anderen germanischen Stämmen[6] entstanden die Deutschen. Deutsch und Englisch gehören zu den westgermanischen Sprachen, Dänisch, Schwedisch und Norwegisch zu den nordgermanischen Sprachen.

Sprachen verändern sich im Laufe der[7] Zeit. So auch die germanischen Sprachen. Sie entwickelten sich auseinander. Im Englischen blieben die germanischen Konsonanten *p, t, k* erhalten. Das deutsche **pf** oder **f** kommt aus dem germanischen *p*, **ts** oder **s** aus *t* und **ch** aus *k*. Am Wortanfang oder Wortende wurde ein germanisches *p* oder *t* zu[8] **pf** oder **ts**, zwischen zwei Vokalen wurden sie zu ***f*** und ***s***. Deshalb ist das deutsche Wort **Pfeife** verwandt mit dem englischen *pipe* und **Pfanne** mit *pan*. Die Lautkombination **ts** wird im Deutschen **tz** oder **z** geschrieben. Das deutsche **Salz** ist deshalb verwandt mit Englisch *salt* und **zu** mit *to*. Das germanische *k* blieb am Wortanfang auch im Deutschen ein **k**, zwischen Vokalen und am Wortende wurde es zu **ch**. **Kuchen** ist verwandt mit *cake* und **machen** mit *make*.

Eine weitere Veränderung betrifft die westgermanischen Konsonanten *th* (gesprochen wie im Englischen) und *d*. Das deutsche **d** war im Germanischen ein *th*, das deutsche **t** war ein *d*. Deshalb ist **danken** verwandt mit Englisch *thank*, **Ding** mit *thing*, **tot** mit *dead* und **rot** mit *red*. Ein deutsches **b** zwischen zwei Vokalen schließlich war im Germanischen oft ein *v*. So ist **Nabel** verwandt mit Englisch *navel*, **Leber** mit *liver*, **leben** mit *live* und **haben** mit *have*.

Formal verwandte Wörter haben aber nicht immer dieselbe Bedeutung. So ist **Zimmer** zwar verwandt mit *timber* und **Dach** mit *thatch*, die Bedeutungen im Deutschen und Englischen allerdings sind nicht dieselben.

	Deutsch	Englisch
pf, ff, f	**Pf**anne, o**ff**en	**p**an, o**p**en
z, tz, ss, ß, s	Sal**z**, Wa**ss**er	sal**t**, wa**t**er
k, ch	**K**u**ch**en	**c**a**k**e
d	**D**ing	**th**ing
t	ro**t**	re**d**
b	Na**b**el	na**v**el

[1]um ...: *around Christ's birth* [2]blieb erhalten: *was preserved* [3]*middle of a word* [4]*origin* [5]*conquered* [6]*tribes* [7]im Laufe der: *over the course of* [8]wurde zu *turned into*

 Situation 11 Dialog: Auf Wohnungssuche

Silvia ist auf Wohnungssuche.

FRAU SCHUSTER: __________!

SILVIA: Guten Tag. Hier Silvia Mertens. Ich rufe wegen des Zimmers an. Ist es noch __________?

FRAU SCHUSTER: Ja, das ist noch zu haben.

SILVIA: Prima, in welchem __________ ist es denn?

FRAU SCHUSTER: Frankfurt-Süd, Waldschulstraße __________.

SILVIA: Und in welchem __________ liegt das Zimmer?

FRAU SCHUSTER: Im fünften, gleich unter dem __________.

SILVIA: Gibt es einen __________?

FRAU SCHUSTER: Nein, leider nicht.

SILVIA: Schade. Was kostet denn das Zimmer?

FRAU SCHUSTER: Vierhundert Euro __________.

SILVIA: Möbliert? Was steht denn drin?

FRAU SCHUSTER: Also, ein Bett natürlich, ein Tisch mit zwei Stühlen und ein __________.

SILVIA: Ist auch ein Bad dabei?

FRAU SCHUSTER: Nein, aber baden können Sie __________. Und Sie haben natürlich Ihre __________ Toilette.

SILVIA: Wann könnte ich mir denn das Zimmer mal __________?

FRAU SCHUSTER: Wenn Sie wollen, können Sie gleich vorbeikommen.

SILVIA: Gut, dann komme ich gleich mal vorbei. Auf __________.

FRAU SCHUSTER: Auf __________.

 Situation 12 Rollenspiel: Zimmer zu vermieten

S1: Sie sind Student/Studentin und suchen ein schönes, großes Zimmer. Das Zimmer soll hell und ruhig sein. Sie haben nicht viel Geld und können nur bis zu 400 Euro Miete zahlen, inklusive Nebenkosten. Sie rauchen nicht und hören keine laute Musik. Fragen Sie den Vermieter / die Vermieterin, wie groß das Zimmer ist, was es kostet, ob es im Winter warm ist, ob Sie kochen dürfen und ob Ihre Freunde Sie besuchen dürfen. Sagen Sie dann, ob Sie das Zimmer mieten möchten.

Hausarbeit

Grammatik 6.5–6.6

Andrea putzt ihre Schuhe.

Paula wischt den Tisch ab.

Ernst mäht den Rasen.

Jens fegt den Boden.

Josie saugt Staub.

Uli bügelt sein Hemd.

Jochen macht die Toilette sauber.

Jutta wäscht die Wäsche.

Margret wischt den Boden.

Hans macht sein Bett.

Situation 13 Was macht man mit einem Besen?

MODELL: S1: Was macht man mit einem Besen?
S2: Man fegt den Boden.

Staub saugen **den Rasen mähen** **Hemden oder Blusen bügeln** **die Wäsche waschen** **den Boden fegen** **den Rasen sprengen** **die Schuhe putzen** **das Geschirr spülen** **die Blumen gießen** **den Tisch abwischen**

1. mit einem Staubsauger
2. mit einem Geschirrspüler
3. mit einer Waschmaschine
4. mit einem Besen
5. mit einem Rasenmäher
6. mit einer Gießkanne
7. mit einem Bügeleisen
8. mit einem Putzlappen
9. mit einem Gartenschlauch

Situation 14 Angenehm oder unangenehm?

Welche Hausarbeit machen Sie gern, weniger gern oder gar nicht gern? Ordnen Sie die folgenden Tätigkeiten von sehr angenehm (1) zu sehr unangenehm (10).

_______ Hosen bügeln
_______ Regale abwischen
_______ eine Einkaufsliste schreiben
_______ die Toilette putzen
_______ den Müll wegbringen
_______ die Sessel absaugen
_______ die Vorhänge waschen
_______ Töpfe und Pfannen spülen
_______ das Bett machen
_______ Fenster putzen

MUSIKSZENE

„Haus am See“ (2008, Deutschland) *Peter Fox*

Biografie Peter Fox ist ein Reggae- und Hip-Hop-Musiker aus Berlin. Sein bürgerlicher Name ist Pierre Baigorry. Seine Mutter kommt aus dem französischen Baskenland. Wegen seiner roten Haare wurde er als Kind Foxi genannt. Deshalb nennt er sich jetzt Fox. Peter Fox studierte Musik, Sonderschulpädagogik und Englisch, beendete sein Studium aber nicht. Peter Fox ist einer der Sänger der Reggae-Gruppe Seeed. Von 2007 bis 2009 hatte er eine Solokarriere. Sein Album Stadtaffe (2008), aus der die Single „Haus am See“ ausgekoppelt wurde, war in Deutschland und Österreich auf Platz 1 der Charts und in der Schweiz auf Platz 4.

Peter Fox und Auftritt der Band SEEED
© ullstein bild/Getty Images

NOTE: For copyright reasons, the songs referenced in **MUSIKSZENE** have not been provided by the publisher. The song can be found online at various sites such as YouTube, Amazon, or the iTunes store.

Vor dem Hören Wie stellen Sie sich Ihr Traumhaus vor? Wo liegt es? Wer wohnt da? Wie sieht es aus?

Nach dem Hören

1. Warum sagt der Sänger in der ersten Strophe, dass er weg muss?
   - ☐ **a.** Er kennt jedes Haus und jeden Laden und sogar jede Taube.
   - ☐ **b.** Er wartet auf eine schicke Frau mit einem schnellen Wagen.
   - ☐ **c.** Ein Frauenchor singt am Straßenrand für ihn.
2. Was gefällt dem Sänger an seinem Haus am See? (Mehrere Antworten sind möglich)
   - ☐ **a.** Er hat 20 Kinder und eine schöne Frau.
   - ☐ **b.** Auf dem Weg dorthin liegen Orangenbaumblätter.
   - ☐ **c.** Alle Menschen kommen ihn dort besuchen.
3. Wie kommt der Sänger zurück?
   - ☐ **a.** Er kommt zurück mit einer Frau.
   - ☐ **b.** Er kommt zurück mit Schnee und Sand.
   - ☐ **c.** Er kommt zurück mit beiden Taschen voll Gold.
4. Was passiert, als er wieder zu Hause ist?
   - ☐ **a.** Er fängt vor Freude an zu weinen.
   - ☐ **b.** Alle Leute laden ihn ein.
   - ☐ **c.** Er feiert eine Woche lang jede Nacht.

Miniwörterbuch

die **Strophe**	stanza
die **Taube, -n**	pigeon
der **Chor,** die **Chöre**	choir
der **Straßenrand**	side of the street
der **Orangenbaum,** die **-bäume**	orange tree
das **Blatt,** die **Blätter**	leaf, blossom
vor Freude	for joy

Situation 15 Bildgeschichte: Frühjahrsputz

Situation 16 Informationsspiel: Haus- und Gartenarbeit

MODELL: S2: Was macht Thomas am liebsten?
S1: Er mäht am liebsten den Rasen.
S2: Was hat Nora letztes Wochenende gemacht?
S1: Sie hat ihre Bluse gebügelt.
S2: Was muss Thomas diese Woche noch machen?
S1: Er muss seine Wäsche waschen.

S1: Was machst du am liebsten?
S2: Ich ______ am liebsten ______.

	Thomas	Nora	mein(e) Partner(in)
am liebsten	den Rasen mähen		
am wenigsten gern		die Fenster putzen	
jeden Tag		den Tisch abwischen	
einmal in der Woche	sein Bett machen		
letztes Wochenende		ihre Bluse bügeln	
gestern		ihr Zimmer aufräumen	
diese Woche	seine Wäsche waschen		
bald mal wieder	die Flaschen wegbringen		

Filmlektüre

Good bye Lenin!

Vor dem Lesen

FILMANGABEN

Titel: *Good bye Lenin!*
Genre: Komödie
Erscheinungsjahr: 2003
Land: Deutschland
Dauer: 150 min
Regisseur: Wolfgang Becker
Hauptrollen: Daniel Brühl, Katrin Saß, Maria Simon

„Guten Abend, meine Damen und Herren"

Miniwörterbuch

DDR (Deutsche Demokratische Republik)	GDR (German Democratic Republic)
der **Bürger** / die **Bürgerin**	citizen
überzeugt	staunch
der **Herzinfarkt**	heart attack
die **Aufregung**	excitement
schaden	to harm
die **Gesundheit**	health
die **Veränderung**	change
verheimlichen	to conceal
vorspielen	to feign
der **Sperrmüll**	bulk refuse (heap)
wohlbekannt	well-known
entschlossen	determined
der **Tod**	death
der **Zusammenbruch**	collapse
gefälscht	fake
geht es nach Alex	if you believe Alex
flüchten	to flee
der **Kosmonaut**	*East German word for astronaut*

A. Sehen Sie sich das Foto aus dem Film an.

1. Welche Art[1] Fernsehsendung ist das?
2. Der Mann im Hintergrund war 1971 bis 1989 Staatschef[2] der DDR. Wie hieß er?
3. Was wissen Sie über die ehemalige[3] DDR und die Wiedervereinigung[4]? Sammeln Sie Informationen.

B. Lesen Sie die Wörter im Miniwörterbuch. Suchen Sie sie im Text und unterstreichen Sie sie. Lesen Sie dann den Text.

Inhaltsangabe

Christiane Kerner (Katrin Saß) – eine engagierte DDR-Bürgerin und überzeugte Sozialistin – hat am 7. Oktober 1989 einen Herzinfarkt und fällt ins Koma. Während sie im Krankenhaus liegt und bewusstlos ist, fällt zwei Tage später die Berliner Mauer und die DDR wird ein Teil der Bundesrepublik Deutschland. Acht Monate später wacht sie auf und die DDR existiert nicht mehr.

Jede Art von Aufregung schadet Christiane Kerners Gesundheit, deshalb beschließt ihr Sohn Alex (Daniel Brühl) die politischen Veränderungen vor seiner Mutter zu verheimlichen. Gemeinsam mit seiner Schwester Ariane (Maria Simon) will Alex seiner Mutter den ganz normalen DDR-Alltag vorspielen. Leichter gesagt als getan: Die alten DDR-Möbel der Familie sind out und stehen im Keller oder liegen auf dem Sperrmüll; im Supermarkt gibt es jetzt westdeutsche Lebensmittel und keine aus

[1]*type* [2]*head of state* [3]*former* [4]*reunification*

DDR-Produktion; die wohlbekannten DDR-Fernsehsendungen laufen auch nicht mehr; West-Autos und Fast-Food-Restaurants überrollen den Osten; und am Haus gegenüber hängt ein großes Coca-Cola Werbeplakat. Dies alles ist für Alex und Ariane ein großes Problem. Aber Alex ist entschlossen und kreativ. Selbst Freunde und Nachbarn spielen mit.

Am Ende, kurz vor ihrem Tod, erfährt Mutter Christiane aber doch vom Zusammenbruch der DDR. Sie sagt Alex nichts davon. Alex' DDR, die er mit gefälschten DDR-Nachrichtensendungen belebt, ist ganz anders als die alte DDR: Geht es nach Alex, ist die DDR das Wunschland aller Menschen; Westdeutsche flüchten in den Osten; und Staatschef ist natürlich Alex' Idol, der DDR-Kosmonaut Sigmund Jähn.

Arbeit mit dem Text

Beantworten Sie die folgenden Fragen.

1. Warum sagt Alex seiner Mutter nicht, dass die DDR nicht mehr existiert?
2. In welchen alltäglichen Bereichen[5] ändert sich das Leben der Kerners nach dem Fall der Mauer?
3. Wie wünscht[6] sich Alex die DDR?

FILMCLIP

NOTE: For copyright reasons, the films referenced in the **FILMCLIP** feature have not been provided by the publisher. The film can be purchased as a DVD or found online at various sites such as YouTube, Amazon, or the iTunes store. The time codes mentioned below are for the North American DVD version of the film.

Szene: DVD, Kapitel 17, Geburtstagsfeier, 00:59:50–1:02:02 Min.

Christiane, die Mutter von Alex und Ariane, hat Geburtstag. Nachdem sie aus dem Koma erwacht ist, wird sie zu Hause von ihren Kindern gepflegt. Deshalb lädt Alex die Geburtstagsgäste nach Hause ein. Die Gäste dürfen aber nicht sagen, dass es die DDR nicht mehr gibt, weil sich Christiane nicht aufregen[7] soll. Nachdem die Pioniere gesungen und die Genossen[8] Ansprachen gehalten haben, beginnt Alex zu sprechen.

Schauen Sie sich die Szene an und beantworten Sie die Fragen.

1. Was sagt Alex über seine Mutter?
2. Welches Plakat wird während der Geburtstagfeier am Gebäude[9] gegenüber angebracht? Wofür steht dieses Plakat? Warum ist Christiane so entsetzt[10]?
3. Was macht Lara, als sie mit Alex im Nebenzimmer ist?
4. Worüber beschwert sich[11] Lara? Was war ihr Vater in Wirklichkeit?

Nach dem Lesen

A. Recherchieren Sie im Internet über den Schauspieler Daniel Brühl, der im Film Alex Kerner spielt. Woher kommt er? Welche Filme hat er noch gemacht? Welche Preise hat er mit *Good bye Lenin!* gewonnen? Welche Projekte hat er gerade?

B. Die Resonanz[12] auf *Good bye Lenin!* war sehr groß in Ost- und Westdeutschland. Warum ist der Film in Deutschland so beliebt? Finden Sie Antworten (auch im Internet) und präsentieren Sie Ihre Gedanken und Lösungen auf einem Poster in der Klasse.

[5]*domains, spheres* [6]*wishes* [7]sich aufregen *to get excited* [8]*comrades* [9]*building* [10]*upset*
[11]beschwert ... *complains* [12]*response*

Videoecke

Perspektiven

Wie hast du deine Wohnung gefunden?

Ich habe meine Wohnung über ein Internetportal gefunden.

Aufgabe 1 Wohnungssuche

Miniwörterbuch	
die **Annonce**	ad
das **Studentenwerk**	student services
einteilen	to arrange, divide up
unsaniert	unrenovated
der **Altbau**	old building

Wie haben die Leute ihre Wohnung gefunden?

1. Albrecht ___

2. Nadezda ___

3. Simone ___

4. Jenny ___

5. Michael ___

6. Sophie ___

7. Pascal ___

8. Sandra ___

a. über eine Annonce in der Zeitung
b. durch/über einen Freund
c. über das Internet
d. über ein Internetportal
e. durch eine ehemalige Kollegin
f. durch das Studentenwerk

Interviews

- Wo wohnst du?
- Kannst du mir deine Wohnung beschreiben?
- Was bezahlst du für deine Wohnung?
- Wohnst du gern mit Leuten zusammen?
- Wie teilt ihr euch die gemeinsame Arbeit ein?
- Gibt es da oder gab es da schon mal Probleme?

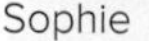
Sophie

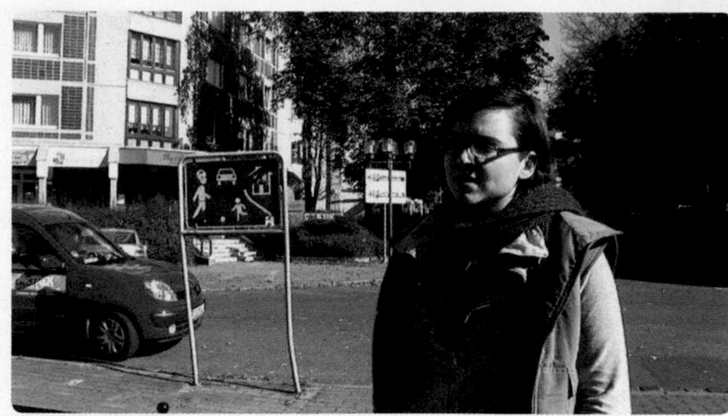

Maria

Aufgabe 2 Sophie oder Maria?

Sehen Sie sich das Video an und kreuzen Sie an.

	Sophie	Maria
1. Wer wohnt im Süden von Leipzig in einer kleinen Wohnung?	☐	☐
2. Wer wohnt in der Nähe vom Bahnhof?	☐	☐
3. Wer wohnt in einem unsanierten Altbau?	☐	☐
4. Wer bezahlt 150 Euro plus Nebenkosten?	☐	☐
5. Wer möchte nicht mit mehr Leuten zusammen wohnen?	☐	☐
6. Wer hat einen Wochenplan?	☐	☐

Aufgabe 3 Sophies Wohnung

Richtig (R) oder falsch (F)? Verbessern Sie die falschen Antworten.

_______ 1. Ihre Wohnung hat drei Zimmer, eine Küche und ein Bad.

_______ 2. Ihre Wohnung ist im 2. Stock.

_______ 3. Die Zimmer sind sehr groß und haben jeweils zwei Fenster.

_______ 4. Das Bad ist groß.

Aufgabe 4 Arbeiten im Haushalt

Welche Hausarbeiten nennt Sophie, welche nennt Maria, welche nennt keine von beiden?

	Sophie	Maria	keine von beiden
1. das Bad putzen	☐	☐	☐
2. das Geschirr spülen	☐	☐	☐
3. den Müll runterbringen	☐	☐	☐
4. die Betten machen	☐	☐	☐
5. die Küche aufräumen	☐	☐	☐
6. die Küche putzen	☐	☐	☐
7. kochen	☐	☐	☐
8. sauber machen	☐	☐	☐
9. Staub saugen	☐	☐	☐
10. die Waschmaschine füllen	☐	☐	☐

Aufgabe 5 Interview

Interviewen Sie eine Partnerin oder einen Partner. Stellen Sie dieselben Fragen.

Wortschatz

In der Stadt	In the City
die **Apotheke, -n**	pharmacy
die **Drogerie, -n**	drugstore
die **Metzgerei, -en**	butcher shop
die **Reinigung, -en**	dry cleaner's
die **Stadt, ¨-e** (R)	town, city
die **Heimatstadt, ¨-e**	hometown
die **Innenstadt, ¨-e**	downtown
die **Straße, -n**	street, road
der **Flughafen, ¨-**	airport
der **Stadtrand, ¨-er**	city limits
der **Stadtteil, -e**	district, neighborhood
das **Gefängnis, -se**	prison, jail
das **Gymnasium, Gymnasien**	high school, college preparatory school
das **Rathaus, ¨-er** (R)	town hall
das **Schreibwarengeschäft, -e**	stationery store
das **Stadtviertel, -**	district, neighborhood

Ähnliche Wörter

die **Boutique, -n;** der **Kindergarten, ¨-;** der **Marktplatz, ¨-e;** der **Supermarkt, ¨-e (R);** das **Reisebüro, -s;** das **Schuhgeschäft, -e; das Theater, -**

Haus und Wohnung	House and Apartment
die **Badewanne, -n**	bathtub
die **Treppe, -n**	stairway
die **Zentralheizung**	central heating
der **Aufzug, ¨-e**	elevator
der **Ausblick, -e**	view
der **Quadratmeter (qm), -**	square meter (m^2)
der **Stock, Stockwerke**	floor, story
im ersten Stock*	on the second floor
das **Dach, ¨-er**	roof
das **Waschbecken, -**	(wash)basin

Ähnliche Wörter

die **Garage, -n** [gara:ʒə]; die **Terrasse, -n;** die **Toilette, -n;** der **Balkon, -e;** der **Keller, -** (R); das **Bad, ¨-er;** das **Esszimmer, -;** das **Schlafzimmer, -;** das **Wohnzimmer, -**

Haus und Garten	House and Garden
die **Kommode, -n**	dresser
der **Besen, -**	broom
der **Frühjahrsputz**	spring cleaning
der **Gartenschlauch, ¨-e**	garden hose
der **Müll**	trash, garbage
der **Putzlappen, -**	cloth, rag (for cleaning)
der **Rasenmäher, -**	lawn mower
der **Schrank, ¨-e** (R)	wardrobe cabinet, cupboard
der **Kleiderschrank, ¨-e**	clothes closet, wardrobe
der **Sessel, -** (R)	armchair
der **Spiegel, -**	mirror
der **Staubsauger, -**	vacuum cleaner
der **Vorhang, ¨-e**	drapery, curtain
das **Bügeleisen, -**	iron
die **Möbel** (*pl.*)	furniture

Ähnliche Wörter

die **Pflanze, -n** (R); die **Waschmaschine, -n;** der **Nachttisch, -e; das Bett, -en** (R); das **Poster, -;** das **Sofa, -s**

Wohnmöglichkeiten	Living Arrangements
die **Skihütte, -n**	ski lodge
die **Villa, Villen**	mansion
die **WG, -s (Wohngemeinschaft, -en)**	shared housing
das **Haus, ¨-er** (R)	house
das **Bauernhaus, ¨-er**	farmhouse
das **Einfamilienhaus, ¨-er**	single-family home
das **Hochhaus, ¨-er**	high-rise building
das **Reihenhaus, ¨-er**	row house, townhouse

Ähnliche Wörter

das **Studentenheim, -e** (R)

Auf Wohnungssuche	Looking for a Room or Apartment
die **Anzeige, -n**	ad
die **Miete, -n**	rent
die **Mieterin, -nen**	female renter
die **Vermieterin, -nen**	landlady
der **Mieter, -**	male renter
der **Vermieter, -**	landlord
die **Nebenkosten** (*pl.*)	extra costs (e.g., utilities)

Sonstige Substantive	Other Nouns
die **Bedeutung, -en**	meaning
die **Nähe**	vicinity
in der Nähe	in the vicinity
die **Seite, -n**	side; page
die **Tätigkeit, -en**	activity
das **Ausland**	foreign countries
im Ausland	abroad
das **Benzin**	gasoline
das **Land, ¨-er**	country (*rural*)
auf dem Land	in the country

*The first floor is called **das Erdgeschoss**. All levels above the first floor are referred to as **Stock** or **Stockwerke**. Thus, **der erste Stock** refers to the second floor, and so on.

Verben	Verbs
ab·trocknen	to dry (dishes)
ab·wischen	to wipe clean
begegnen (+ *dat.*)	to meet
betreffen, betrifft, betroffen	to concern, deal with
bügeln	to iron
fehlen (+ *dat.*)	to be missing
geben, gibt, gegeben	to give
es gibt ...	there is/are . . .
gibt es ...? (R)	is/are there . . .?
gefallen, gefällt, gefallen (+ *dat.*)	to be to one's liking, to please
es gefällt mir	I like it
gehören (+ *dat.*)	to belong to
gratulieren (+ *dat.*)	to congratulate
helfen, hilft, geholfen (+ *dat.*)	to help
mieten	to rent
passen (+ *dat.*) (R)	to fit
putzen	to clean
schaden (+ *dat.*)	to be harmful to
schmecken (+ *dat.*)	to taste good to
Staub saugen	to vacuum
stehen, gestanden (R)	to stand
stehen, gestanden (+ *dat.*)	to suit
tippen (R)	to type
übernachten	to stay overnight
verändern	to change
vermieten	to rent out
vor·stellen	to introduce, present
sich etwas vorstellen	to imagine something
wiederholen	to repeat
wischen	to mop
zu·hören (+ *dat.*)	to listen to

Ähnliche Wörter

kosten (R); **zurück·kommen, ist zurückgekommen**

Adjektive und Adverbien	Adjectives and Adverbs
angenehm	pleasant
eigen (R)	own
gemeinsam	in common, together
hell	light
möbliert	furnished
warm	heated, heat included
weit	far
Wie weit weg?	How far away?
wunderschön	exceedingly beautiful

Ähnliche Wörter

attraktiv, dumm, leicht, modern

Sonstige Wörter und Ausdrücke	Other Words and Expressions
allerdings	however
auf Wiederhören	good-bye (*on phone*)
bei (R)	at; with
bei deinen Eltern	with/at your parents'
bei einer Bank	at a bank
Ist ein/eine ... dabei?	Does it come with a ...?
drin/darin	in it
egal	equal, same
Das ist mir egal.	It doesn't matter to me.
gegenüber	opposite; across
gleich gegenüber	right across the way
gleich	right away; right, directly
gleich um die Ecke	right around the corner
inklusive	included (*utilities*)
möglichst (+ *adverb*)	as ... as possible
ob	if, whether
prima!	great!
schade!	too bad!
unter (R)	below, beneath; among
wegen	on account of; about

Strukturen und Übungen

6.1 Dative verbs

WISSEN SIE NOCH?

The dative case is used primarily to indicate to whom or for whom something is done (or given).

Review grammar 5.1.

Dative verbs are verbs that require a dative object.

The dative object usually indicates the person to whom or for whom something is done. The dative case can be seen as the partner case. The "something" that is done (or given) is in the accusative case (it is the direct object).

Ich schenke **dir ein Bügeleisen.**	*I'll give you an iron. (I'll give an iron to you.)*
Ich kaufe **meinem Bruder ein Buch.**	*I'll buy my brother a book. (I'll buy a book for my brother.)*

Certain verbs, called "dative verbs," require only a subject and a dative object; there is no accusative object. These verbs fall into two groups. In Group 1, both the subject and the dative object are persons.

antworten	*to answer*
begegnen	*to meet*
gratulieren	*to congratulate*
helfen	*to help*
zuhören	*to listen to*

Er antwortete **mir** nicht.	*He didn't answer me.*
Wir begegneten **dem alten Vermieter.**	*We met the old landlord.*
Ich gratuliere **dir** zum Geburtstag.	*Happy Birthday! (I congratulate you on your birthday.)*
Soll **ich dir** helfen?	*Do you want me to help you?*
Ich höre **dir** genau zu.	*I'm listening to you carefully.*

In Group 2, the subject is usually a thing. The dative object is often a person who experiences or owns the thing.

gehören	*to belong to*
passen	*to fit*
schaden	*to be harmful to*
schmecken	*to taste good to*
stehen	*to suit*

Diese Poster gehören **mir.**	*These posters belong to me.*
Diese Hose passt **mir** nicht.	*These pants don't fit me.*
Rauchen schadet **der Gesundheit.**	*Smoking is bad for (damages) your health.*
Schmeckt **Ihnen** der Fisch?	*Does the fish taste good to you?*
Blau steht **dir** gut.	*Blue suits you well.*

Note that the following Group 2 verbs express ideas that are rendered very differently in English.

fehlen	*to be missing*
gefallen	*to be to one's liking, to please*

Mir fehlt ein Buch.	*I'm missing a book.*
Gefällt **Ihnen** dieser Schrank?	*Do you like this cupboard? (Does this cupboard please you?)*

Übung 1 Minidialoge

Ergänzen Sie das Verb. Nützliche Wörter:

antworten	**gefallen**	**passen**
begegnen	**gehören**	**schaden**
fehlen	**gratulieren**	**schmecken**
helfen	**stehen**	**zuhören**

1. MONIKA: Schau, ich habe mir einen neuen MP3-Spieler gekauft.
   KATRIN: Der ist aber toll! Der _______ mir!
2. NESRIN: Hallo, Willi. Ich habe gehört, du hast endlich eine Wohnung gefunden. Ich _______ dir ganz herzlich.
   WILLI: Danke. Das ist aber lieb von dir.
3. FRAU RUF: Jochen, kannst du mir bitte _______? Ich kann die Vorhänge nicht allein tragen.
   HERR RUF: Ja, ich komme.
4. FRAU GRETTER: _______ Ihnen der Salat?
   HERR SIEBERT: Ja, sehr gut, die Soße ist ausgezeichnet.
5. FRAU KÖRNER: Dieser Rock _______ mir nicht. Ich brauche doch Größe 42.
   VERKÄUFER: Ich seh mal nach, ob wir Größe 42 haben.
6. JÜRGEN: Wem _______ denn dieser neue Staubsauger?
   SILVIA: Mir. Ich habe ihn gestern gekauft.
7. FRAU SCHULZ: Was suchen Sie, Albert? _______ Ihnen etwas?
   ALBERT: Ja, ich kann mein Heft nicht finden.
8. FRAU KÖRNER: Wissen Sie, wer mir am Marktplatz _______ ist, Herr Siebert?
   HERR SIEBERT: Nein, wer denn?
   FRAU KÖRNER: Die Mutter von Maria. Und wissen Sie, was die mir erzählt hat?
   HERR SIEBERT: Nein, was denn?
   FRAU KÖRNER: Also, ...
9. ARZT: Also, Herr Ruf, Sie müssen jetzt wirklich mit dem Rauchen aufhören. Nikotin _______ Ihrer Gesundheit!
   HERR RUF: Aber, Herr Doktor, dann habe ich ja gar keine Freude mehr im Leben.
10. STEFAN: Entschuldigung, Frau Schulz, ich habe Ihnen nicht _______. Können Sie das noch mal wiederholen?
    FRAU SCHULZ: Na, gut.

Übung 2 Interview

1. Wem haben Sie neulich[1] gratuliert?
2. Wem sind Sie neulich begegnet?
3. Welches Essen schmeckt Ihnen am besten?
4. Wie steht Ihnen Ihr Lieblingshemd?
5. Wie gefällt Ihnen Ihre Wohnung oder Ihr Zimmer?
6. Welches Möbelstück fehlt Ihnen in der Wohnung oder im Zimmer?

[1]*recently*

6.2 Location vs. destination: two-way prepositions with the dative or accusative case

Wo asks about location. Questions about location are answered with a preposition + dative.

The prepositions **in** (*in*), **an** (*on, at*), **auf** (*on top of*), **vor** (*before*), **hinter** (*behind*), **über** (*above*), **unter** (*underneath*), **neben** (*next to*), and **zwischen** (*between*) are used with both the dative and accusative cases. When they refer to a fixed location, the dative case is required. In these instances, the prepositional phrase answers the question **wo** (*where* [*at*]).

WISSEN SIE NOCH?

The prepositions **in, an,** and **auf** use the dative case when they indicate location.

Review grammar 5.4.

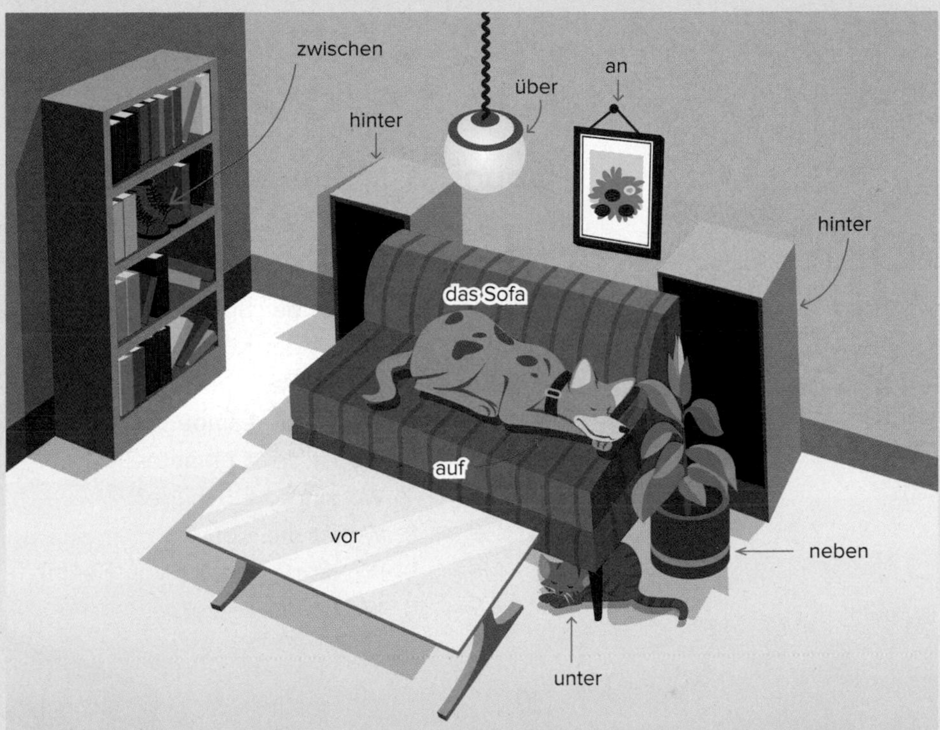

Im Wohnzimmer steht ein Sofa.
Hinter dem Sofa stehen zwei große Boxen.
An der Wand hängt ein Bild.
Auf dem Sofa liegt ein Hund.
Unter dem Sofa liegt eine Katze.
Vor dem Sofa steht ein Tisch.
Über dem Sofa hängt eine Lampe.
Neben dem Sofa steht eine große Pflanze.
Zwischen den Büchern stehen Tennisschuhe.

Wohin asks about placement or destination. Questions about placement or destination are answered with a preposition + accusative.

When these prepositions describe movement toward a place or a destination, they are used with the accusative case. In these instances, the prepositional phrase answers the question **wohin** (*where* [*to*]).

Peter hat das Sofa **ins Wohnzimmer** gestellt.
Die Boxen hat er **hinter das Sofa** gestellt.
Das Bild hat er **an die Wand** gehängt.
Der Hund hat sich gleich **auf das Sofa** gelegt.
Die Katze hat sich **unter das Sofa** gelegt.
Peter hat den Tisch **vor das Sofa** gestellt.
Die Lampe hat er **über das Sofa** gehängt.
Die große Pflanze hat er **neben das Sofa** gestellt.
Und seine Tennisschuhe hat er **zwischen die Bücher** gestellt.

ACHTUNG!

in + dem = im
an + dem = am

in + das = ins
an + das = ans

	Wo?	Wohin?
	Location *Dative*	*Placement/Destination* *Accusative*
Masculine	Es ist auf **dem** Stuhl. *It is on the table.*	Leg es auf **den** Stuhl. *Put it on the table.*
Neuter	Es ist auf **dem** Bett. *It is on the bed.*	Leg es auf **das** Bett. *Put it on the bed.*
Feminine	Es ist auf **der** Kommode. *It is on the bureau.*	Leg es auf **die** Kommode. *Put it on the bureau.*
Plural	Es steht vor **den** Boxen. *It is in front of the speakers.*	Stell es vor **die** Boxen. *Put it in front of the speakers.*

Übung 3 Alberts Zimmer

Schauen Sie sich Alberts Zimmer an.

1. Wo ist Albert?
2. Wo ist der Spiegel?
3. Wo ist der Kühlschrank?
4. Wo ist das Deutschbuch?
5. Wo ist die Lampe?
6. Wo ist der Computer?
7. Wo sind die Schuhe?
8. Wo ist die Hose?
9. Wo ist das Poster von Berlin?
10. Wo ist die Katze?

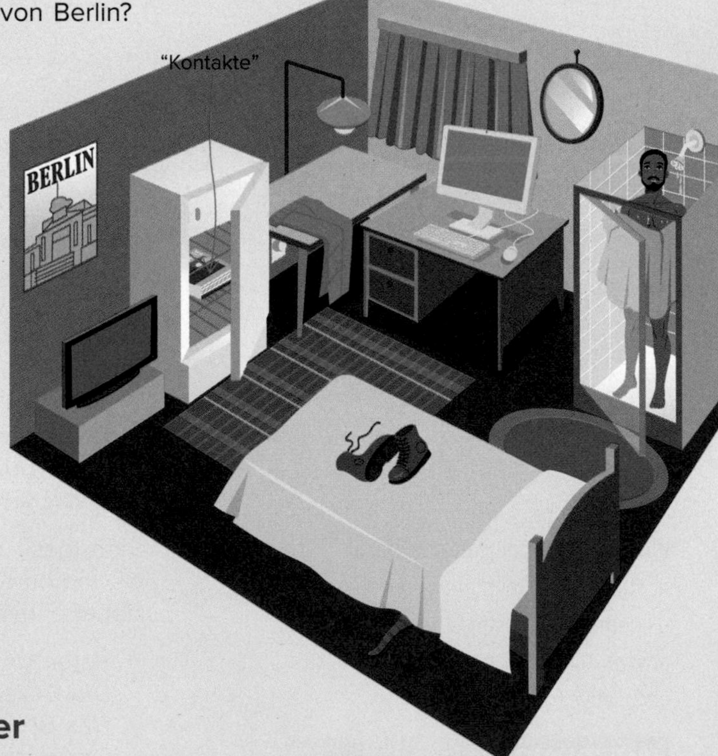

Übung 4 Mein Zimmer

Beschreiben Sie Ihr Zimmer möglichst genau.
Schreiben Sie mindestens acht Sätze mit verschiedenen Präpositionen.

MODELL: Das Bett ist unter dem Fenster. Rechts neben dem Bett steht ein Nachttisch ...

6.3 Word order: time before place

Time before place

In a German sentence, a time expression usually precedes a place expression. Note that this sequence is often reversed in English sentences.

Ich gehe heute Abend in die Bibliothek.

I'm going to the library tonight.

Übung 5 Wo sind Sie wann?

Bilden Sie Sätze aus den Satzteilen.

MODELL: heute Abend → Ich bin heute Abend im Kino.

	WANN	WO
1.	heute Abend	in der Klasse
2.	am Nachmittag	bei meinen Eltern
3.	um 16 Uhr	im Bett
4.	in der Nacht	auf einer Party
5.	am frühen Morgen	im Urlaub
6.	am Montag	am Frühstückstisch
7.	am ersten August	in der Mensa
8.	an Weihnachten	in der Bibliothek
9.	im Winter	?
10.	am Wochenende	

6.4 Direction: *in/auf* vs. *zu/nach*

Direction:
in/auf + accusative; **zu/nach** + dative

To refer to the place where you are going, use either **in** or **auf** + accusative, **zu** + dative, or **nach** + place name.

Albert geht **in die** Kirche.	*Albert goes to church.*
Katrin geht **auf die** Bank.	*Katrin goes to the bank.*
Heidi fährt **zum** Flughafen.	*Heidi drives to the airport.*
Rolf fliegt **nach** Deutschland.	*Rolf is flying to Germany.*

A. **in** + Accusative

in for most buildings and enclosed spaces

In general, use **in** when you plan to enter a building or an enclosed space.

Heute Nachmittag gehe ich **in die Bibliothek.**	*This afternoon I'll go to (into) the library.*
Abends gehe ich **ins Kino.**	*In the evening I go to (into) the movies.*
Morgen fahre ich **in die Stadt.**	*Tomorrow I'll drive to (into) the city.*

in for countries with a definite article

Also use **in** with the names of countries that have a definite article, such as **die Schweiz, die Türkei,** and **die USA.**

Herr Okonkwo fliegt oft **in die** USA.	*Mr. Okonkwo often flies to the USA.*
Claire fährt **in die** Schweiz.	*Claire is going to Switzerland.*
Mehmet fährt alle zwei Jahre **in die** Türkei.	*Mehmet goes to Turkey every two years.*

auf for public buildings

B. **auf** + Accusative

Use **auf** instead of **in** when the destination is a public building such as the post office, the bank, or the police station.

Ich brauche Briefmarken. Ich gehe **auf die** Post.	*I need stamps. I'm going to the post office.*
Ich brauche Geld. Ich gehe **auf die** Bank.	*I need money. I'm going to the bank.*

zu for specifically named buildings, places in general, open spaces, and people's places

C. **zu** + Dative

Use **zu** to refer to destinations that are specific names of buildings, places or open spaces such as a playing field, or people.

Ernst geht **zu** McDonald's.	*Ernst is going to McDonald's.*
Hans geht **zum** Sportplatz.	*Hans goes to the playing field.*
Andrea geht **zum** Arzt.	*Andrea goes to the doctor.*

zu Hause = *at home*

Note that **zu Hause** (*at home*) is an exception. It does not indicate destination but rather location.

D. **nach** + place name

Use **nach** with names of countries and cities that have no article. Note that this applies to the vast majority of countries and cities.

Renate fliegt **nach Paris.**	*Renate is flying to Paris.*
Melanie fährt **nach Österreich.**	*Melanie is driving to Austria.*

nach Hause = *(going/coming) home*

Also use **nach** in the idiomatic construction **nach Hause** (*going/coming home*).

Übung 6 Situationen

Heute ist Montag. Wohin gehen oder fahren die folgenden Personen?

MODELL: Katrin sucht ein Buch. → Sie geht in die Bibliothek.

ACHTUNG!

in + das = ins
auf + das = aufs

zu + dem = zum
zu + der = zur

zum Arzt **zum Fußballplatz** **zur Tankstelle** **zum Flughafen** **ins Hotel** **ins Theater** **auf die Post** **in die Schule** **zu ihrem Freund** **in den Supermarkt** **in den Wald**

1. Albert ist krank.
2. Hans möchte Fußball spielen.
3. Frau Schulz ist auf Reisen in einer fremden[1] Stadt. Sie braucht einen Platz zum Schlafen.
4. Herr Ruf braucht Benzin.
5. Herr Thelen braucht Lebensmittel.
6. Herr Wagner muss Briefmarken kaufen.
7. Jürgen und Silvia gehen Pilze[2] suchen.
8. Maria möchte mit ihrem Freund sprechen.
9. Mehmet möchte in die Türkei fliegen.
10. Renate möchte ein Musical sehen.
11. Jutta muss ein Referat halten.

[1]here: *unfamiliar* [2]*mushrooms*

6.5 Separable-prefix verbs: the present tense and the perfect tense

WISSEN SIE NOCH?

Separable-prefix verbs consist of a prefix plus a base verb. In the present tense, the verb and the prefix form the **Satzklammer.**

Review grammar 1.5 and 3.5.

The infinitive of a separable-prefix verb consists of a prefix such as **auf, mit,** or **zu** followed by the base verb.

aufstehen	*to get up*
mitkommen	*to come along*
zuschauen	*to watch*

Most prefixes are derived from prepositions and adverbs.

abwaschen	*to do the dishes*
vorstellen	*to introduce*

A. The Present Tense

Separable prefixes are placed at the end of the independent clause.

1. Independent clauses: In an independent clause in the present tense, the conjugated form of the base verb is in second position and the prefix is in last position.

Ich **stehe** jeden Morgen um sieben Uhr **auf.**	*I get up at seven every morning.*

Separable prefixes are "reconnected" to the base verb in dependent clauses.

2. Dependent clauses: In a dependent clause, the prefix and the base verb form a single verb. It appears at the end of the clause and is conjugated.

Rolf sagt, dass er jeden Morgen um sechs Uhr **aufsteht.**	*Rolf says that he gets up at six every morning.*
Hast du nicht gesagt, dass du heute **abwäschst?**	*Didn't you say that you would do the dishes today?*

Separable prefixes stay attached to the infinitive.

3. Modal verb constructions: In an independent clause with a modal verb (**wollen, müssen,** etc.), the infinitive of the separable-prefix verb is in last position. In a dependent clause with a modal verb, the separable-prefix verb is in the second-to-last position, and the modal verb is in the last position.

Jutta möchte ihren Freund **anrufen.**	*Jutta wants to call her boyfriend.*
Ernst hat schlechte Laune, wenn er **zuhören** muss.	*Ernst is in a bad mood when he has to pay attention.*

WISSEN SIE NOCH?

The perfect tense is formed with **haben/sein** plus the past participle.

Review grammar 4.5.

B. The Perfect Tense

Separable prefixes precede the **-ge-** marker in past participles.

The past participle of a separable-prefix verb is a single word, consisting of the past participle of the base verb + the prefix.

Infinitive	Past Participle
auf**stehen**	auf**gestanden**
um**ziehen**	um**gezogen**
weg**bringen**	weg**gebracht**

Note that the prefix does not influence the formation of the past participle of the base verb; it is simply attached to it.

Herr Wagner **hat** gestern die Garage **aufgeräumt.**	*Mr. Wagner cleaned up his garage yesterday.*
Ich **habe** vor einer Stunde **angerufen.**	*I called an hour ago.*

Übung 7 Minidialoge

Ergänzen Sie die Sätze.

aufstehen **mitkommen** **ankommen** **ausmachen** **anrufen** **einladen** **mitnehmen** **zuhören** **aufräumen** **umziehen**

1. HERR WAGNER: Ernst, aufwachen! Hast du nicht gestern gesagt, dass du heute um 7 Uhr ______?
   ERNST: Ich bin aber noch so müde!
2. FRAU WAGNER: Andrea, jetzt ______[a] mir mal ______[b]! Sonst ______[c] ich den Fernseher sofort ______[d].
   ANDREA: Aber, Mami, nur noch das Ende. Der Film ist doch gleich vorbei!
3. SILVIA: Entschuldigen Sie bitte! Wann ______[a] der Zug aus Hamburg ______[b]?
   BAHNANGESTELLTER: Um 14 Uhr 56.
4. ANDREAS: Hallo, Jürgen. Ich habe gehört, dass ihr bald eine neue Wohnung habt. Wann ______[a] ihr denn ______[b]?
   JÜRGEN: Nächstes Wochenende.
5. NESRIN: Hallo, Sofie. Ich habe morgen Geburtstag und ich möchte dich gern zu einer kleinen Feier ______.
   SOFIE: Das ist aber nett von dir. Ich komme gern.
6. CLAIRE: Hallo, Melanie. Wo ist Josef?
   MELANIE: Er ist zu Hause. Er ______[a] heute sein Zimmer ______[b] und das dauert bei ihm immer etwas länger.
7. JÜRGEN: Hallo, Silvia. Ich fahre heute mit dem Auto zur Uni. Willst du ______[a]?
   SILVIA: Ja, gern. Schön, dass du mich ______[b].
8. KATRIN: Hier ist meine Telefonnummer. Warum ______[a] du mich nicht mal ______[b]!
   HEIDI: Gut, das mach' ich mal.

Übung 8 Am Sonntag

Gestern war Sonntag. Was haben die folgenden Personen gestern gemacht?

NÜTZLICHE WÖRTER

abtrocknen	ausgehen
anrufen	ausziehen
anziehen	fernsehen
aufwachen	zurückkommen

Andrea

Katrin und Peter

Heidi / Frau Schulz

Herr Ruf

Jürgen

Jutta

Maria

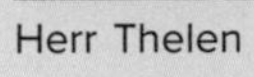
Herr Thelen

6.6 The prepositions *mit* and *bei* + dative

The prepositions **mit** (*with, by*) and **bei** (*near, with*) are followed by the dative case.

Masculine	Neuter	Feminine	Plural
mit dem Staubsauger	mit dem Bügeleisen	mit der Arbeit	mit den Eltern
beim Onkel	beim Fenster	bei der Tür	bei den Eltern

Mit corresponds to the preposition *with* in English and is used in similar ways.

Herr Wagner fegt die Terrasse **mit** seinem neuen Besen.	*Mr. Wagner sweeps the patio with his new broom.*
Ich gehe **mit** meinen Freunden ins Kino.	*I'm going to the movies with my friends.*
Ich möchte ein Haus **mit** einem offenen Kamin.	*I want a house with a fireplace.*

Use **mit** with means of transportation.

The preposition **mit** also indicates the means of transportation; in this instance it corresponds to the English preposition *by*. Note the use of the definite article in German.

Rolf fährt **mit** dem Bus zur Uni.	*Rolf goes to the university by bus.*
Renate fährt **mit** dem Auto zur Arbeit.	*Renate drives to work (goes to work by car).*

The preposition **bei** may refer to a place in the vicinity of another place; in this instance it corresponds to the English preposition *near.*

Bad Harzburg liegt **bei** Goslar.	*Bad Harzburg is near Goslar.*

The preposition **bei** also indicates placement with a person, a company, or an institution; in these instances it corresponds to the English preposition *with, at,* or *for.*

Ich wohne **bei** meinen Eltern.	*I'm living (staying) with my parents / at my parents'.*
Hans arbeitet **bei** McDonald's.	*Hans works at (for) McDonald's.*

	German	English
Instrument	mit dem Hammer	*with the hammer*
Togetherness	mit Freunden	*with friends*
Means of transportation	mit dem Flugzeug	*by airplane*
Vicinity	bei München	*near Munich*
Somebody's place	bei den Eltern	*(staying) with parents*
Place of employment	bei McDonald's	*at McDonald's*

Übung 9 Im Haus und im Garten

Womit machen Sie die folgenden Aktivitäten?

MODELL: S1: Womit mähst du den Rasen?
S2: Mit dem Rasenmäher.

der Besen **das Bügeleisen**
der Computer
der Gartenschlauch **die Gießkanne**
die Kaffeemaschine
der Putzlappen **der Staubsauger**

1. Kaffee kochen
2. Staub saugen
3. den Boden fegen
4. bügeln
5. einen Brief tippen
6. die Blumen im Garten gießen
7. den Boden wischen
8. die Blumen in der Wohnung gießen

Übung 10 Minidialoge

Ergänzen Sie die Sätze mit der Präposition **mit** oder **bei.**

1. FRAU KÖRNER: Fahren Sie ______[a] dem Bus oder ______[b] dem Fahrrad zur Arbeit?
   MICHAEL PUSCH: ______[c] dem Bus. Ich arbeite jetzt ______[d] Siemens. Das ist am anderen Ende von München.
2. PETER: Wohnst du in Krefeld ______[a] deinen Eltern?
   ROLF: Ja, sie haben ein wunderschönes Haus ______[b] einem riesigen Garten.
   PETER: Liegt Krefeld eigentlich ______[c] Dortmund?
   ROLF: Nein, nach Dortmund fährt man über eine Stunde ______[d] dem Auto.
3. JÜRGEN: Oh je, jetzt habe ich deinen Gummibaum[1] umgeworfen[2]! Soll ich die Erde[3] ______[a] dem Staubsauger aufsaugen?
   SILVIA: Mach es lieber ______[b] dem Besen. Er steht ______[c] der Kellertür.

[1]*rubber plant* [2]*knocked over* [3]*dirt*

KAPITEL 7

Unterwegs

Kapitel 7 is about geography and transportation. You will learn more about the geography of the German-speaking world and about the kinds of transportation used by people who live there.

Themen

Geografie

Transportmittel

Das Auto

Reiseerlebnisse

Kulturelles

Musikszene: „Mädchen, lach doch mal!" (Wise Guys)

Filmclip: *Im Juli* (Fatih Akin)

KLI: Volkswagen

KLI: Die Schweiz

Videoecke: Ausflüge und Verkehrsmittel

Lektüren

Kurzgeschichte: Die Motorradtour (Christine Egger)

Film: *Im Juli* (Fatih Akin)

Strukturen

7.1 Relative clauses

7.2 Making comparisons: the comparative and superlative forms of adjectives and adverbs

7.3 Referring to and asking about things and ideas: **da**-compounds and **wo**-compounds

7.4 The perfect tense (review)

7.5 The simple past tense of **haben** and **sein**

Elfriede Lohse-Wächtler: *Loschwitzer Brücke (Blaues Wunder)* (1931), Privatbesitz

KUNST UND KÜNSTLER

Elfriede Lohse-Wächtler (1899–1940) war eine Dresdner Malerin der Avantgarde. Sie studierte in Dresden an der Kunstgewerbeschule[1] und der Kunstakademie. Als sie an Schizophrenie erkrankte, wurde sie von den Nazis entmündigt[2], in ein Krankenhaus eingewiesen[3] und schließlich 1940 umgebracht[4]. Neben Stadt- und Landschaftsporträts malte sie viele Kopf- und Körperstudien von psychisch Kranken. Ihre Kunst wurde von den Nazis als *Entartete Kunst*[5] bezeichnet und viele ihrer Bilder wurden vernichtet[6]. Die Loschwitzer Brücke wurde 1883 gebaut. Sie wird das *Blaue Wunder* genannt und ist eines der Wahrzeichen von Dresden.

Schauen Sie sich das Bild an und beantworten Sie die folgenden Fragen.

1. Was sehen Sie auf dem Bild? Beschreiben Sie es.
2. Welche Farben und Linien dominieren im Bild? Welche Perspektive wird eingenommen[7]?
3. Was ist das Besondere an den Gebäuden? Wie wird die Brücke dargestellt[8]? Und der Fluss?
4. Welche Assoziationen weckt das Bild?

[1]*School of Arts and Crafts* [2]*declared legally incompetent* [3]*committed* [4]*killed* [5]Entartete ... *degenerate art* [6]*destroyed* [7]*assumed* [8]*represented*

Situationen

Geografie

Grammatik 7.1–7.2

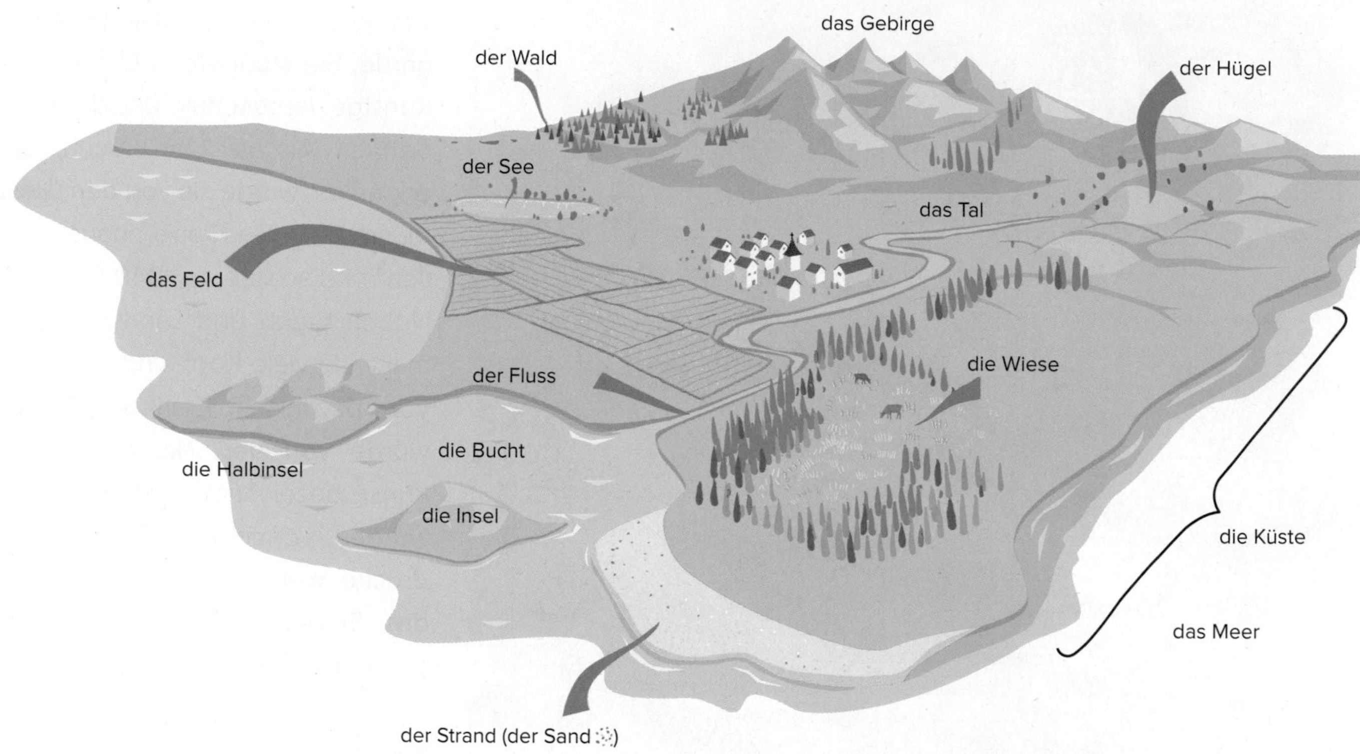

Situation 1 Erdkunde: Wer weiß – gewinnt

1. Fluss, der durch Wien fließt
2. Wald, in dem die Germanen[1] die Römer besiegt[2] haben
3. Insel in der Ostsee, auf der weiße Kreidefelsen[3] sind
4. Berg, auf dem sich die Hexen[4] treffen
5. See, der zwischen Deutschland, Österreich und der Schweiz liegt
6. Meer, das Europa von Afrika trennt
7. Gebirge in Österreich, in dem man sehr gut Ski fahren kann
8. berühmte Wüste, die in Ostasien liegt
9. Inseln, die vor der Küste von Ostfriesland liegen
10. Fluss, der durch Köln fließt

a. das Mittelmeer
b. der Brocken im Harz (1.142 Meter hoch)
c. die Kitzbühler Alpen
d. der Teutoburger Wald
e. der Bodensee
f. die Wüste Gobi
g. der Rhein
h. die Donau
i. Rügen
j. die Ostfriesischen Inseln

[1]*Germanic tribes* [2]*die ... defeated the Romans* [3]*chalk cliffs* [4]*witches*

Situation 2 Ratespiel: Stadt, Land, Fluss

1. Wie heißt der tiefste See der Schweiz?
2. Wie heißt der höchste Berg Österreichs?
3. Wie heißt der längste Fluss Deutschlands?
4. Wie heißt das salzigste Meer der Welt?
5. Wie heißt der größte Gletscher der Alpen?
6. Was ist die heißeste Wüste der Welt?
7. Wie heißt die älteste Universitätsstadt Deutschlands?
8. Wie heißt das kleinste Land, in dem man Deutsch spricht?
9. Wie heißt die berühmteste Höhle in Österreich?

a. die Dachstein-Mammuthöhle
b. das Tote Meer
c. der Genfer See
d. der Großglockner
e. die Wüste Sahara
f. der Rhein
g. Liechtenstein
h. der Große Aletschgletscher
i. Heidelberg

Situation 3 Informationsspiel: Deutschlandreise

Wo liegen die folgenden Städte? Schreiben Sie die Namen der Städte auf die Landkarte.

Aachen, Bayreuth, Dresden, Erfurt, Flensburg, Freiburg, Hannover, Heidelberg, Magdeburg, Wiesbaden

MODELL: S2: Wo liegt Braunschweig?
S1: Braunschweig liegt im Norden.
S2: Wo genau?
S1: Südlich von Hamburg.

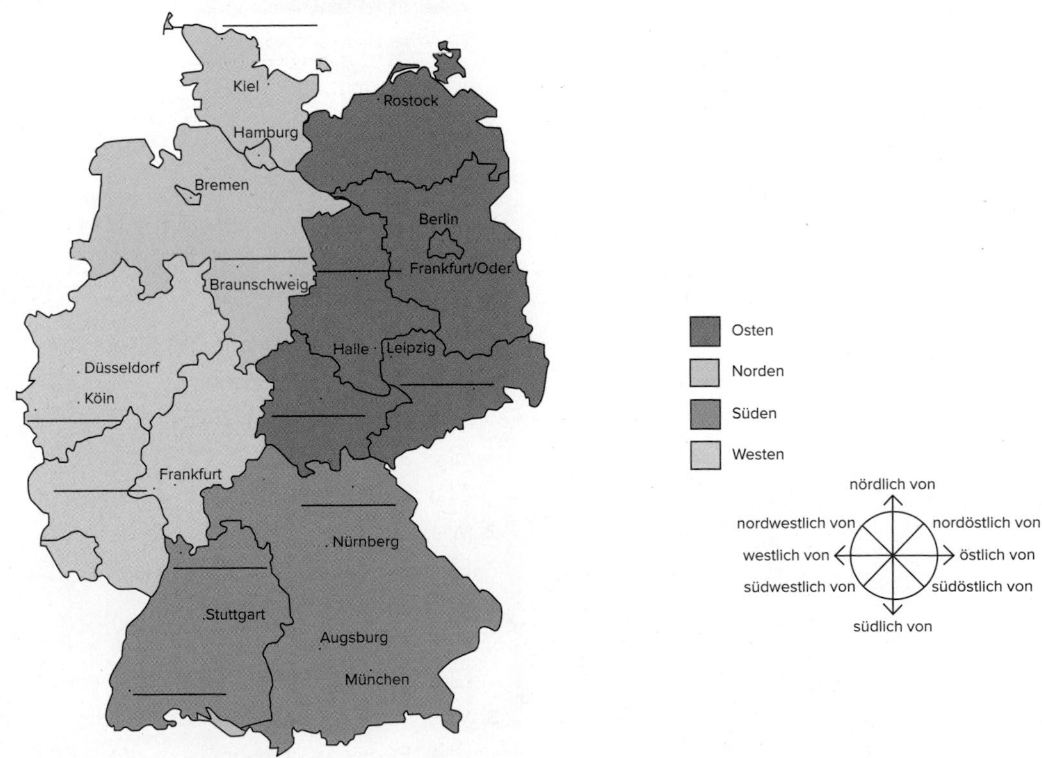

Miniwörterbuch

der **Besserwisser**	wise guy, know-it-all
auftreten	to perform
der **Kirchentag**	church congress
die **Stammkneipe**	favorite bar
der **Dom**	cathedral
der **Traum**	dream
geheimnisvoll	mysterious
der feste Freund	steady boyfriend
die **Zähne**	teeth
würde	would
sich stürzen	to jump (to one's death)
das **Grab**	grave

MUSIKSZENE

„Mädchen, lach doch mal!" (1999, Deutschland) *Wise Guys*

Biografie Die *Wise Guys* kommen aus Köln. Sie singen vor allem a cappella. Viele von ihnen sind an dasselbe Gymnasium in Köln gegangen, wo ihre Lehrer sie die Besserwisser genannt haben. Daher stammt auch ihr Name. Ihre Alben *Frei* (2008) und *Klassenfahrt* (2010) haben Platz 2 der deutschen Charts erreicht. Sie treten regelmäßig auf dem Deutschen Evangelischen Kirchentag auf ebenso wie auf anderen großen öffentlichen Veranstaltungen. Der Hit *„Mädchen, lach doch mal"* stammt aus dem Album *Skandal* von 1999.

Die Wise Guys aus Köln
© ullstein bild/Getty Images

NOTE: For copyright reasons, the songs referenced in **MUSIKSZENE** have not been provided by the publisher. The song can be found online at various sites such as YouTube, Amazon, or the iTunes store.

Vor dem Hören Wer könnte das Mädchen sein? Warum lacht sie nicht?

Nach dem Hören Kreuzen Sie die richtigen Antworten an. Nur jeweils eine Antwort pro Frage ist falsch.

1. Wo trifft der junge Mann das Mädchen?
- ☐ a. in der Straßenbahn
- ☐ b. in seiner Stammkneipe
- ☐ c. im Dom
- ☐ d. im Traum

2. Wie beschreibt er das Mädchen?
- ☐ a. Sie hat ein geheimnisvolles Gesicht.
- ☐ b. Sie hat eine Topfigur.
- ☐ c. Sie hat einen festen Freund.
- ☐ d. Sie hat strahlend weiße Zähne.

3. Was würde er tun, wenn das Mädchen lachen würde?
- ☐ a. Er würde sich vom Dom stürzen.
- ☐ b. Er würde über den Rhein schwimmen.
- ☐ c. Er würde mit dem Rad nach Rom fahren.
- ☐ d. Er würde ihren Namen auf sein Grab schreiben.

Situation 4 Interview: Landschaften

1. Warst du schon mal im Gebirge? Wo? Was hast du da gemacht? Wie heißt der höchste Berg, den du gesehen (oder bestiegen) hast?
2. Warst du schon mal am Meer? Wo und wann war das? Hast du gebadet? Was hast du sonst noch gemacht?
3. Wohnst du in der Nähe von einem großen Fluss? Wie heißt er? Wie heißt der größte Fluss, an dem du schon warst? Was hast du da gemacht?
4. Wie heißt die interessanteste Stadt, in der du schon warst?
5. Warst du schon mal in der Wüste oder im Dschungel[5]? Wie war das?

Transportmittel

Grammatik 7.1, 7.4

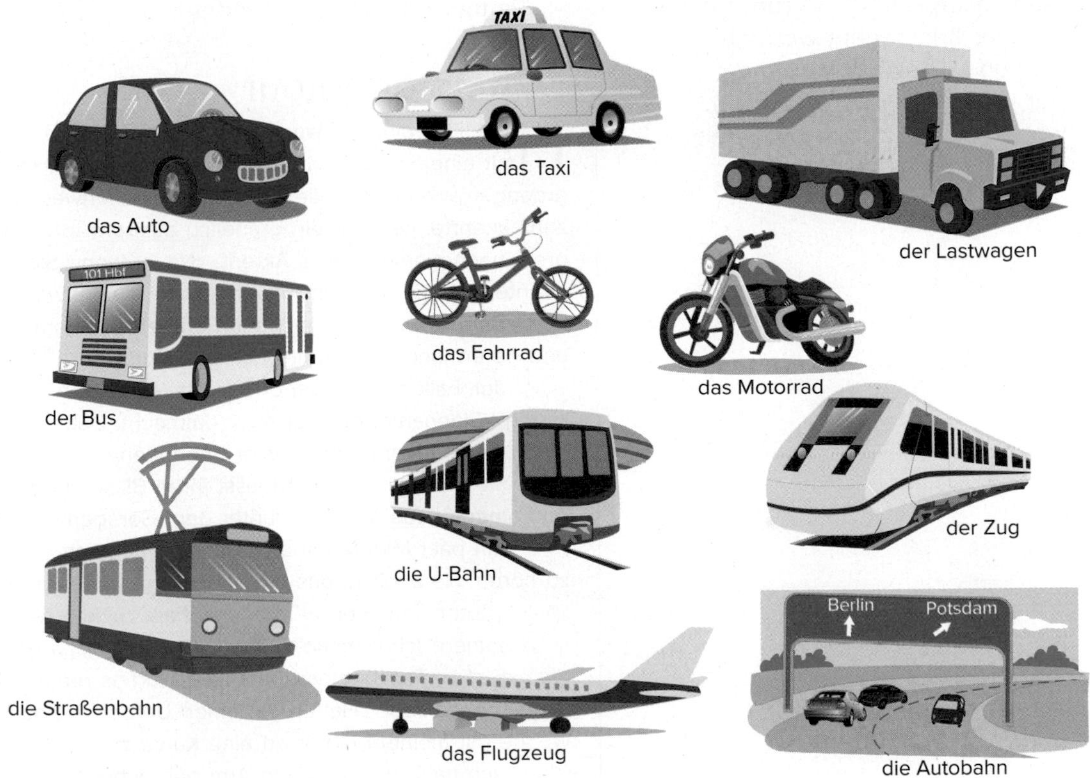

Situation 5 Definitionen: Transportmittel

1. das Flugzeug
2. die Rakete
3. das Kamel
4. das Fahrrad
5. der Kinderwagen
6. der Zeppelin
7. der Zug
8. das Taxi

a. Transportmittel, das Waggons und eine Lokomotive hat
b. Transportmittel, das fliegt
c. Tier, das viele Beduinen als Transportmittel benutzen
d. Transportmittel, mit dem man zum Mond fliegen kann
e. Auto, das in Deutschland ein gelbes Schild auf dem Dach hat
f. Transportmittel in der Luft, das wie eine Zigarre aussieht
g. Transportmittel mit zwei Rädern, das ohne Benzin fährt
h. Wagen, in dem man Babys transportiert

[5]*jungle*

Lektüre

Vor dem Lesen

So lesen Sie wie ein Detektiv ...

1. Setzen Sie sich an einen ruhigen Ort, wo Sie sich konzentrieren können.
2. Legen Sie sich Papier und Schreibzeug bereit.
3. Lesen Sie den ganzen Text durch, um zu wissen, worum es geht.
4. Lesen Sie den Text jetzt absatzweise[1] etwas genauer und machen Sie sich dabei Notizen.
5. Vergleichen Sie Ihre Notizen mit Ihrem Partner oder mit Ihrer Partnerin.

LESEHILFE

In the following story, detective Juli Falk uses her well-honed skills to investigate a crime. As you read it, you become a detective, too. Some hints from her "Handbook for a Rookie Detective" are shown. They will help you to catch the important details as you read the story. As you might expect, taking notes is part of the investigation. When you take notes during the **Vor dem Lesen** activity, be sure to include: 1) important words to look up in the dictionary, three per paragraph at most; 2) words that seem key to the plot; and 3) interesting facts.

Die Motorradtour

„Hallo, Kollegin, wie war's in den Ferien?" Oberinspektor Eichhorn begrüßt Juli Falk mit einem freundschaftlichen **Handschlag**. „Hoffentlich ist es Ihnen nicht genauso ergangen wie der Familie Andres am Blumenweg 1. Als die von ihrer Reise zurückkehrte, fand sie ein gründlich **ausgeraubtes** Haus vor." Oberinspektor Eichhorn greift nach einem Bündel Akten[2]. „Na ja, wenn Sie den Fall[3] gleich **weiterverfolgen** könnten ...? Die meisten Anwohner am Blumenweg haben wir bereits vernommen[4]. Zu befragen wären da noch ein Rentnerpaar, Familie Wächter im Haus Nummer 7, und deren junger Untermieter Heinz Hurtig."

Juli Falk drückt zum dritten Mal den Knopf[5] über dem Schildchen „Heinz Hurtig". Eigenartig, dass er nicht aufmacht. Dabei hat sie doch gerade eben noch einen jungen Mann am Fenster oben stehen sehen. Juli schüttelt **verwundert** den Kopf. Sie dreht sich um und lässt ihren Blick[6] über den verlassenen[7] Hof und das **funkelnagelneue** Motorrad unter dem Garagenvordach schweifen.

Ein paar Minuten später klingelt Juli noch ein Mal. Ein Geräusch ist von drinnen zu hören. Na endlich, das hat aber lange gedauert! Heinz Hurtig guckt durch den Türspalt. „Guten Tag, Herr Hurtig." Juli Falk zückt ihren Ausweis. „Darf ich einen Moment reinkommen? Ich ermittle[8] wegen des **Einbruchs** bei Familie Andres."

Erst im Flur bemerkt Juli, dass Hurtigs rechter Arm dick einbandagiert in einer **Armschlinge** liegt. „Hatten Sie einen Unfall?" Heinz Hurtig nickt. „Ich habe letzte Woche mit meinem Motorrad eine Kurve zu schnell genommen. Aber ich hatte noch Glück, ich habe mir bloß den Arm gebrochen."

Heinz Hurtig führt die Inspektorin in die Küche. Auf dem Küchentisch steht ein Teller mit Speck[9] und Rührei[10], daneben eine Tasse mit dampfend heißem Kaffee. „Darf ich Ihnen auch eine Tasse Kaffee anbieten? – Nein? Keinen Kaffee? Nun, was den Einbruch betrifft[11], ich bin ja erst vorgestern von meiner Motorradtour **heimgekommen**, habe nichts gesehen und gehört. Und, sorry, falls ich ein Alibi brauche – mit meinem verletzten Arm hätte ich wirklich kein Haus ausrauben können, nicht wahr?"

„Leben Sie allein hier?", fragt die Inspektorin. „Nein, mit Schnurrli, meinem Kater." Heinz Hurtig grinst und weist mit dem Kinn zum Fenstersims, wo sich eine prächtige rote Katze wohlig in der Sonne **ausstreckt**. „Tut mir leid, Herr Hurtig", meint Juli Falk sachlich. „Sie begleiten mich jetzt aufs Präsidium[12]. Mit Ihrem Alibi stimmt nämlich etwas ganz und gar nicht."

Aus: *Aufgepasst, Juli Falk!* von Christine Egger

[1]*paragraph by paragraph* [2]*files* [3]*case* [4]*questioned* [5]here: *doorbell* [6]*glance* [7]*deserted* [8]*am investigating* [9]*bacon* [10]*scrambled eggs* [11]was ... betrifft *as far as ... is concerned* [12]*police station*

Arbeit mit dem Text

A. Guess the meaning of the boldface underlined words in the reading by looking at the context of the sentences in which they appear. Some hints are provided. Then check yourself by looking up the words in the glossary at the end of the book.

1. **Handschlag.** HINT: You already know the word **Hand. Schlagen** means *to beat, strike,* or *hit.* How do people sometimes greet with their hands?
2. **ausgeraubt.** HINT: This is the past participle of the verb **ausrauben.** What English word is similar to **raub** and is related to crime and houses?
3. **weiterverfolgen.** HINT: **Weiter** is the comparative form of **weit.** The prefix **ver** adds a sense of continuation. The verb **folgen** means *to follow.*
4. **verwundert.** HINT: The verb **verwundern** means *to surprise;* verwundert is the past participle.
5. **funkelnagelneu.** HINT: The verb **funkeln** means *to sparkle* and **Nagel** means *nail.*
6. **Einbruch.** HINT: The prefix **ein** often means *in.* The word **Bruch** is a noun related to the verb **brechen,** which means *to break.*
7. **Armschlinge.** HINT: You already know the word for the body part **Arm.** What English word is like **Schlinge** and has to do with an arm injury?
8. **heimgekommen.** HINT: You know what **Heimweh** means. What English word is like **heim** and combines with *come* to indicate a destination?
9. **ausstrecken.** HINT: German **-ck-** is occasionally equivalent to English *-tch-*. What might a cat do on a sunny **Fenstersims?**

B. Was ist passiert? Bringen Sie die folgenden Sätze in die richtige Reihenfolge.

_______ Als Frau Falk bei Heinz Hurtig klingelt, macht er zuerst nicht auf.

_______ Endlich macht Hurtig auf und lässt sie in seine Wohnung.

_______ Er erzählt der Kommissarin von seinem Motorradunfall in der vergangenen Woche.

_______ Juli bemerkt, dass Hurtig seinen rechten Arm einbandagiert hat.

_______ Juli Falk schaut sich inzwischen aufmerksam im Hof um.

_______ Juli Falk zweifelt stark an Heinz Hurtigs Alibi.

___*1*___ Kommissarin Falk ist gerade aus dem Urlaub zurückgekommen.

_______ Sein Alibi ist sein verletzter Arm.

_______ Sie soll wegen des Einbruchs bei Familie Andres ermitteln.

_______ Weil er erst vor zwei Tagen von der Motorradtour zurückgekommen ist, hat er nichts gesehen und gehört.

Nach dem Lesen

1. Beschreiben Sie die Szene in Heinz Hurtigs Küche so genau wie möglich. Notieren Sie alle Einzelheiten.
2. Warum zweifelt Juli Falk am Alibi von Heinz Hurtig? Sammeln Sie alles, was nicht zusammenpasst.

 Situation 6 Pendeln, aber wie?

Viele Europäer wohnen an einem Ort und arbeiten, studieren oder gehen an einem anderen Ort zur Schule. Sie pendeln. Hören Sie den folgenden Personen zu und entscheiden Sie, womit sie zur Schule, zur Arbeit oder zum Studium kommen.

Josef

Veronika

Margret

Silvia

Volker

Miniwörterbuch	
Regenstauf	(a city in Landkreis Regensburg)
Thalwil	(a village near Zurich)
Küsnacht	(a village near Zurich)
pendeln	to commute
die **Fähre, -n**	ferry
mindestens	at least
die **Fahrt, -en**	trip
der **Fahrradweg, -e**	bicycle paths
Weende	(a district of the city of Göttingen)
das **Institut, -e**	institute
bergauf	uphill
anstrengend	strenuous
der **Betrieb, -e**	workplace, operation
Radeberg	(a small town near Dresden)
der **Dienstwagen, -**	company car

 Situation 7 Interview

1. Welche Transportmittel hast du schon benutzt?
2. Fährst du oft mit der U-Bahn oder mit dem Bus? Warum (nicht)?
3. Fährst du gern mit dem Zug (oder möchtest du gern mal mit dem Zug fahren)? Welche Vorteile/Nachteile hat das Reisen mit dem Zug?
4. Fliegst du gern? Warum (nicht)? Welche Vorteile/Nachteile hat das Reisen mit dem Flugzeug?
5. Fährst du lieber mit dem Auto oder mit öffentlichen Verkehrsmitteln? Warum? Womit fährst du am liebsten?
6. Denkst du an die Umwelt, wenn du Transportmittel benutzt?

 Situation 8 Dialog: Eine Bahnfahrt online buchen

RENATE: Okay, Mehmet, dann lass uns mal unsere Bahnfahrt nach München ______________. Bist du schon online?

MEHMET: Ja. Was ist noch mal dein ______________ bei bahn.de?

RENATE: 16. Oktober.

MEHMET: Also, wann wollen wir denn fahren? ______________ früh ______________ möglich?

RENATE: Nein, wir müssen um 17 Uhr da sein. Aber bitte ohne Umsteigen.

MEHMET: Okay, Abfahrt um 10.39 Uhr, ______________ um 16.39, mit ICE, ohne Umsteigen.

RENATE: Was kostet das?

MEHMET: Mit Sparpreis ______________, Hin- und Rückfahrt, 140 Euro pro Person.

RENATE: Müssen wir Sitzplätze reservieren?

MEHMET: Um diese Zeit ______________ nicht. Soll ich zwei Plätze für uns buchen?

RENATE: Ja.

Situation 9 Rollenspiel: Am Fahrkartenschalter

S1: Sie stehen am Fahrkartenschalter im Bahnhof von Bremen und wollen eine Fahrkarte nach München kaufen. Sie wollen billig fahren, müssen aber vor 16.30 Uhr am Bahnhof in München ankommen. Fragen Sie, wann und wo der Zug abfährt und über welche Städte der Zug fährt.

Das Auto

Grammatik 7.3

1. Damit kann man hupen.
2. Daran sieht man, woher das Auto kommt.
3. Darin kann man seine Koffer verstauen.
4. Damit wischt man die Scheiben.

Situation 10 Definitionen: Die Teile des Autos

1. die Bremsen
2. die Scheibenwischer
3. das Autoradio
4. das Lenkrad
5. die Hupe
6. das Nummernschild
7. die Sitze
8. das Benzin
9. der Tank

a. Man setzt sich darauf.
b. Man braucht sie, wenn man bei Regen fährt.
c. Damit lenkt man das Auto.
d. Damit warnt man andere Fahrer oder Fußgänger.
e. Daran sieht man, woher das Auto kommt.
f. Damit hört man Musik und Nachrichten.
g. Damit fährt das Auto.
h. Darin ist das Benzin.
i. Damit hält man den Wagen an.

Situation 11 Informationsspiel: Ein Auto kaufen

S1: Sie wollen einen älteren Gebrauchtwagen[1] kaufen und lesen deshalb Anzeigen im Internet. Die Anzeigen für einen Opel Corsa und einen Ford Fiesta sind interessant. Rufen Sie an und stellen Sie Fragen.

Sie haben auch eine Anzeige im Internet aufgegeben, weil Sie Ihren VW Golf und Ihren VW Touareg Hybrid verkaufen wollen. Antworten Sie auf die Fragen der Leute über Ihre Autos.

MODELL: Guten Tag, ich rufe wegen des Opel Corsa an.
Wie alt ist der Wagen?
Welche Farbe hat er?
Wie ist der Kilometerstand[2]?
Wie lange hat er noch TÜV?
Wie viel Benzin braucht er?
Was kostet der Wagen?

Modell	VW Golf	VW Touareg Hybrid	Opel Corsa	Ford Fiesta
Baujahr	2014	2016		
Farbe	rot	grau		
Kilometerstand	65.000 km	5.000 km		
TÜV	noch 1 Jahr	2 Jahre		
Benzinverbrauch pro 100 km	5,5 Liter	8,2 Liter		
Preis	12.500 Euro	69.000 Euro		

Situation 12 Interview: Das Auto

1. Hast du einen Führerschein? Wann hast du ihn gemacht?
2. Was für ein Auto möchtest du am liebsten haben? Warum?
3. Welche Autos findest du am schönsten?
4. Welche Autos findest du am praktischsten (unpraktischsten)? Warum?

[1]*used car* [2]*number of kilometers driven*

5. Wer von deinen Freunden hat das älteste Auto? Wie alt ist es ungefähr? Und wer hat das hässlichste (schnellste, interessanteste)?
6. Mit was für einem Auto möchtest du am liebsten in Urlaub fahren?
7. Was glaubst du: Was ist das teuerste Auto der Welt?
8. Was glaubst du: In welchem Land fährt man am schnellsten?
9. Was glaubst du: Was ist das kleinste Auto der Welt?

Situation 13 **Verkehrsschilder**

Kennen Sie diese Verkehrsschilder? Was bedeuten sie?

a. b. c. d. e. f. g. h. i. j.

1. Dieses Verkehrsschild bedeutet „Halt".
2. Hier darf man nicht halten.
3. Wer von rechts kommt, hat Vorfahrt[3].
4. Hier darf man nur in eine Richtung fahren.
5. Hier darf man nur mit dem Rad fahren.
6. Hier darf man auf dem Fußgängerweg[4] parken.
7. Hier dürfen keine Autos fahren.
8. Achtung Radfahrer!
9. Dieser Weg ist nur für Fußgänger.
10. Hier dürfen keine Motorräder fahren.

[3]*right of way* [4]*sidewalk*

KULTUR ... LANDESKUNDE ... INFORMATIONEN

VOLKSWAGEN

- Haben Sie ein Auto? Wenn ja, was für eins? Wie lange haben Sie es schon? Was gefällt Ihnen daran?
- Welche deutschen Automarken[5] kennen Sie? Was halten Sie von deutschen Autos?

Lesen Sie den Text und suchen Sie die Antworten auf die folgenden Fragen:

- Wie viel Prozent der Autos in Europa sind Volkswagen?
- Wo liegt der Firmensitz[6] des Volkswagenkonzerns?
- Wer hat den ersten Prototyp des Volkswagen Käfers gebaut?
- Wo hat Ferdinand Porsche Produktionsmethoden von Autos studiert?
- Welches Auto ist das meistverkaufte der Welt?
- In welchen Ländern hat VW große Standorte[7]?
- Wie lange gibt es den VW-Bus schon?

Jedes fünfte Auto in Europa ist ein Volkswagen. Schon seit vielen Jahren ist der VW Golf in Deutschland das meistverkaufte Auto. Auf Platz 2 steht meistens der VW Polo. Volkswagen ist der größte Automobilhersteller[8] Europas und der drittgrößte der Welt. Zum Volkswagenkonzern gehören nicht nur die Marken VW und Audi, sondern auch so schicke Autos wie Bugatti und Lamborghini, die Edelmarke Bentley, sowie seit 2011 auch Porsche. Der Firmensitz befindet sich in Deutschland, im niedersächsischen Wolfsburg.

Den ersten Prototyp des Volkswagen Käfer baute der Österreicher Ferdinand Porsche 1934. Der Volkswagen sollte[9] ein Auto für die breite Masse werden. Um ihn möglichst billig bauen zu können, ging Porsche nach Detroit und studierte die Produktionsmethoden von Ford. Das erste Volkswagenwerk wurde[10] 1938 in Niedersachsen in der Nähe von Braunschweig gebaut. Der Ausbruch des 2. Weltkriegs[11] verhinderte die Massenproduktion des VW Käfer. Erst nach dem Krieg kam es dazu. Aus dem Volkswagenwerk entstand die Stadt Wolfsburg, die heute 100.000 Einwohner hat, und der Siegeszug[12] des Käfers war nicht aufzuhalten. Bis 2002 war der Käfer mit 21,5 Millionen Exemplaren das meistverkaufte Auto der Welt. Dann wurde er vom VW Golf überholt[13], der jetzt diesen Titel besitzt.

Der VW-Konzern hat viele Standorte in der ganzen Welt. Die größten Standorte außerhalb Europas befinden sich in Mexiko (Puebla), Brasilien (Saõ Paulo) und China (Shanghai). Weitere große Standorte befinden sich in Südafrika (Uitenhage), Kenia (Nairobi) und in den USA (Chattanooga).

Beliebte Modelle sind neben dem Golf und dem Polo der Passat, die Vans Touran und Sharan sowie der Geländewagen[14] Touareg. Von 1997 bis 2010 gab es auch den New Beetle, den neuen Käfer im Retrolook, den es auch als Cabrio gab. Unschlagbar[15] unter Studenten und auf der ganzen Welt bekannt ist allerdings der VW-Bus, den es seit 1950 gibt und der immer noch hergestellt wird.

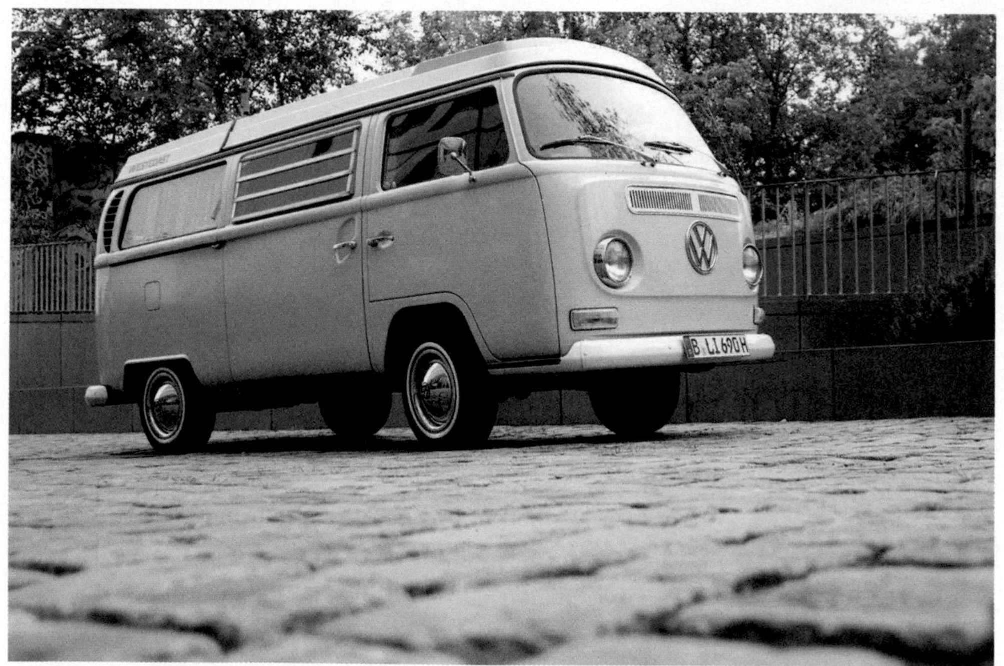

Ein VW-Bus von 1969 ist ein Oldtimer.
© ullstein bild - Lambert/The Image Works

[5]automobile makes [6]company headquarters [7]production sites [8]auto maker [9]was supposed to [10]was [11]world war [12]triumph [13]overtaken [14]SUV [15]Unbeatable

Filmlektüre

Im Juli

Vor dem Lesen

FILMANGABEN

Titel: *Im Juli*
Genre: Roadmovie
Erscheinungsjahr: 2000
Land: Deutschland
Dauer: 99 min
Regisseur: Fatih Akin
Hauptrollen: Moritz Bleibtreu, Christiane Paul, Idil Üner

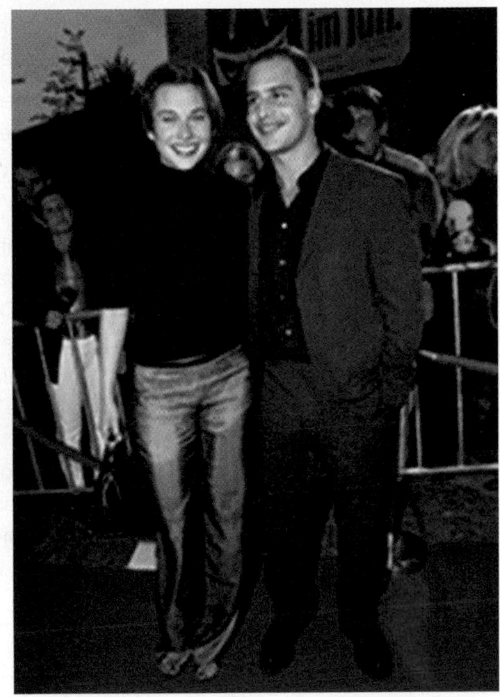

Die Hauptdarsteller Christiane Paul und Moritz Bleibtreu bei der Kinopremiere des Films *Im Juli*.

A. Beantworten Sie die folgenden Fragen.

1. Was sehen Sie auf dem Foto?
2. Warum heißt der Film wohl *Im Juli?*
3. Wer ist der Regisseur?
4. Wann startete der Film im Kino?

B. Lesen Sie die Wörter im Miniwörterbuch. Suchen Sie sie in der Inhaltsangabe und unterstreichen Sie sie.

Miniwörterbuch	
langweilig	boring
flippig	weird, funky
der **Schmuck**	jewelry
das **Pech**	bad luck
fahren	to drive
der **Liebeskummer**	love problems
schüchtern	shy

Inhaltsangabe

Der junge Lehrer Daniel (Moritz Bleibtreu) lebt in Hamburg und ist ein sehr langweiliger Typ. Nur die flippige Schmuckverkäuferin Juli (Christiane Paul) interessiert sich für ihn. In den Sommerferien trifft Daniel auf einer Party die Türkin Melek (Idil Üner). Sie ist auf dem Weg nach Istanbul. Daniel verliebt sich sofort in seine „Traumfrau". Das ist aber Pech für Juli, die auch auf die Party kommt! Als Melek am nächsten Morgen in die Türkei fliegt, fährt Daniel seiner großen Liebe mit dem Auto nach – 2.700 Kilometer bis nach Istanbul. Auf der Autobahn trifft er Juli wieder. Sie will aus Liebeskummer einfach wegtrampen. Daniel nimmt sie in seinem Auto mit. Für Juli und Daniel beginnt eine wilde Odyssee und eine Reise in ein neues Leben. Am Ende ist Daniel ein anderer Typ: nicht mehr der schüchterne und langweilige Lehrer, sondern ein cooler Lover und auch Juli ist keine traurige junge Frau mit Liebeskummer mehr.

FILMCLIP

NOTE: For copyright reasons, the films referenced in the **FILMCLIP** feature have not been provided by the publisher. The film can be purchased as a DVD or found online at various sites such as YouTube, Amazon, or the iTunes store. The time codes mentioned below are for the North American DVD version of the film.

Szene: DVD, Kapitel 2, „School's out", 7:28–11:00 Min.

Daniel ist Referendar in einer Schule in Hamburg. Er geht jeden Tag an Julis Stand auf dem Markt vorbei ... bis sie ihn eines Tages anspricht.

Schauen Sie sich die Szene an. Die folgenden Aussagen beschreiben die Szene in der falschen Reihenfolge. Bringen Sie die Sätze in die richtige Reihenfolge.

_____ Daniel fällt seine Tüte zu Boden. Juli sagt zu Daniel, „He du, komm doch mal her."
__1__ Daniel ist Lehrer in einer Schule. Es ist die letzte Stunde vor den Ferien.
_____ Die Schüler passen nicht auf und wollen Daniels Fragen nicht beantworten.
_____ Juli verkauft Daniel den Ring für 35 Euro, weil sie ihn gern hat.
_____ Juli lädt Daniel auf eine Party ein.
_____ Juli sagt, Daniel wird ein Mädchen kennenlernen, das eine Sonne trägt.
_____ Juli sagt: „Die Sonne macht Licht", und „Ein anderes Wort für Licht ist Glück".
_____ Juli (mit Dreadlocks) und ihre Freundin (mit blonden Haaren) unterhalten sich über Daniel. Juli sagt, sie ist schüchtern.
_____ Juli zeigt Daniel einen Ring. Darauf sieht man eine Sonne.
_____ Die Schülerin Kira sagt: „Wir machen Schluss." Alle Schüler stehen auf und gehen.

Arbeit mit dem Text

Richtig (R) oder falsch (F)? Verbessern Sie die falschen Aussagen.

1. Daniel kommt aus Hamburg.
2. Auf einer Reise lernt Daniel die Türkin Melek kennen.
3. Juli liebt den langweiligen Daniel.
4. Daniel fährt mit dem Motorrad nach Istanbul.
5. Juli fliegt mit Melek in die Türkei.
6. Am Ende der Reise ist Daniel nicht mehr langweilig.

Nach dem Lesen

A. Suchen Sie weitere Informationen über den Regisseur Fatih Akin im Internet.

1. Woher kommt Fatih Akin?
2. Woher kommen seine Eltern?
3. Wie alt ist er?
4. Wie heißt sein erster Film?
5. Welche Preise haben seine Filme bekommen?

B. Sehen Sie den Trailer zum Film im Internet an und machen Sie ein Poster zum Film.

Reiseerlebnisse

Grammatik 7.4–7.5

Im letzten Urlaub waren Kobe Okonkwo und Veronika Frisch-Okonkwo in Italien.

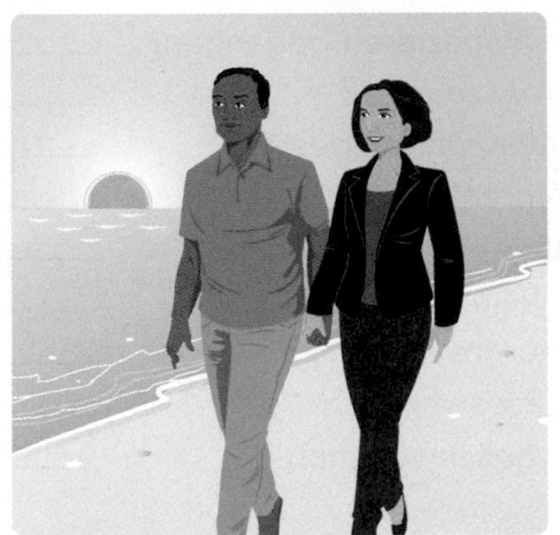

1. Am Morgen sind Kobe und Veronika am Strand spazieren gegangen.

2. Dann sind sie im Meer geschwommen.

3. Zu Mittag haben sie Spaghetti gegessen.

4. Später sind sie in die Stadt gefahren.

5. Zuerst hat Veronika dort Souvenirs gekauft.

6. Dann haben sie eine Stadtrundfahrt gemacht.

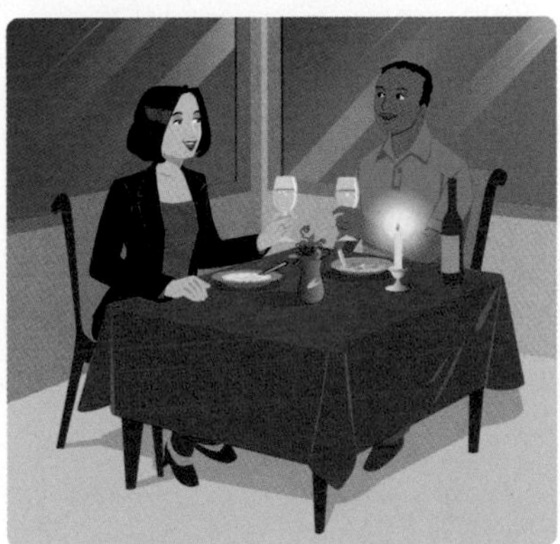

7. Am Abend haben sie Wein getrunken.

Situation 14 Umfrage: Warst du schon mal im Ausland?

MODELL: S1: Warst du schon mal im Ausland?
S2: Ja!
S1: Unterschreib bitte hier.

UNTERSCHRIFT

1. Warst du schon mal im Ausland? ____________
2. Bist du schon mal am Strand spazieren gegangen? ____________
3. Hattest du schon mal einen Autounfall? ____________
4. Warst du schon mal auf einem Oktoberfest? ____________
5. Bist du schon mal Zug gefahren? ____________
6. Hast du schon mal eine Stadtrundfahrt gemacht? ____________
7. Hattest du schon mal eine Reifenpanne? ____________
8. Warst du schon mal auf einer Insel? ____________
9. Hast du schon mal deinen Pass verloren? ____________
10. Bist du schon mal im Meer geschwommen? ____________

Situation 15 Bildgeschichte: Stefans Reise nach Österreich

Situation 16 Zum Schreiben: Ein Reiseerlebnis

Hatten Sie schon mal ein interessantes Reiseerlebnis? Schreiben Sie eine Geschichte darüber! Denken Sie an die folgenden Fragen.

1. Personen: Wer war dabei? Was muss man über diese Personen wissen, um Ihre Geschichte besser zu verstehen?
2. Ort: Wo hatten Sie das Erlebnis? Was war interessant an diesem Ort? Versuchen Sie den Ort zu visualisieren und beschreiben Sie ihn.
3. Zeit: Wann hatten Sie das Erlebnis? Vor wie vielen Jahren? Welche Tageszeit war es? War es ein besonderer Tag?

4. Handlung: Was ist zuerst passiert? Was haben Sie gefühlt und gedacht? Was ist dann passiert? Was war der Höhepunkt des Erlebnisses? Was war das Besondere? Wie ist es ausgegangen?
5. Persönliche Note: Wie denken Sie heute darüber?

KULTUR ... LANDESKUNDE ... INFORMATIONEN

DIE SCHWEIZ

- Woran denken Sie, wenn Sie *Schweiz* hören?
- Kennen Sie eine Schweizerin oder einen Schweizer? Waren Sie schon mal in der Schweiz? Was waren Ihre Eindrücke?
- Finden Sie die Schweiz auf einer Landkarte. Lokalisieren Sie die Städte Bern, Genf und Zürich.

Lesen Sie den Text und suchen Sie die Antworten auf die folgenden Fragen:

- Was für ein Land ist die Schweiz?
- Welche Sprachen spricht man dort? Wie sind sie verteilt[1]?
- Wie heißen die größten Städte?
- Wann wurde die Schweiz gegründet[2]? Wie hießen die ersten Kantone?
- Wie ist die Schweiz politisch? Welchen Organisationen gehört sie (nicht) an?
- Wofür ist die Schweiz bekannt?
- Wofür ist Genf bekannt?
- Was sagen Fußballfans, wenn sie wollen, dass die Schweiz gewinnt?

Einer für alle, alle für einen. So lautet das Motto der Schweiz, eines kleinen Alpenlandes, das eines der reichsten Länder der Welt ist. Nach Fläche[3] ist die Schweiz auf Platz 133, nach der Bevölkerungszahl auf Platz 94 der Länder der Welt. 64 Prozent der Schweizer sprechen Deutsch, 20% Französisch, knapp 7% Italienisch und nur 0,5% Rätoromanisch. Diese Sprachen sind die offiziellen Sprachen der Schweiz. Französisch spricht man im Westen; Italienisch im Südosten, im Tessin; Rätoromanisch im Osten; und Deutsch im Norden, in der Mitte und im Süden. Die größten Städte sind die deutschsprachigen Städte Zürich und Basel und das französischsprachige Genf. Allerdings[4] darf man nicht glauben, dass das Deutsch der Schweizer so einfach zu verstehen ist. Die Deutschschweizer sprechen alemannische Dialekte, die sich sehr von der deutschen Standardsprache unterscheiden. Die Schriftsprache dagegen[5] weist nur wenige Unterschiede auf.

Die Schweiz besteht aus 26 Kantonen[6]. Der Legende nach entstand die Schweiz im Jahre 1291 auf dem Rütliberg, als sich die drei Kantone Uri, Schwyz und Unterwalden zu einem Bund zusammenschlossen[7]. Seit 1815 ist die Schweiz politisch neutral. Sie ist nicht Teil der EU und auch nicht in der NATO. Seit 2002 ist sie aber Mitglied in der UNO.

Wofür ist die Schweiz am bekanntesten? Für ihre Schokolade? Für ihren Käse? Für ihre Uhren? Für ihre Berge? Für ihr Eisenbahnnetz[8]? Wer kennt nicht Johanna Spyri, die Schöpferin[9] von Heidi; den Psychologen Jean Piaget; die Reformatoren Calvin und Zwingli? Die ETH (Eidgenössische Technische Hochschule) in Zürich ist eine der besten Universitäten der Welt. Die UNO-Stadt Genf ist nicht nur der Ort, wo das Rote Kreuz gegründet wurde, sondern eine sehr schöne Stadt, Sitz von 25 großen internationalen Organisationen, unter anderen die Welthandelsorganisation (WTO) und die Weltgesundheitsorganisation (WHO). Das Matterhorn ist einer der schönsten und höchsten Berge der Schweiz (4.478 m). In der Schweiz entspringen der Rhein und die Rhône, die längsten Flüsse Deutschlands und Frankreichs. Nicht nur im Fußball sagen deshalb viele Leute: „Hopp Schwiiz!" (Los, Schweiz!)

Ein typisches Schweizer Dorf

Die Schweiz hat das dichteste[10] Eisenbahnnetz der Welt.

[1]*distributed* [2]*founded* [3]*area* [4]*However* [5]*on the other hand* [6]*cantons (roughly equivalent to a province or state)* [7]*joined* [8]*railway network* [9]*creator* [10]*densest*

Videoecke

Perspektiven

Lebst du umweltbewusst?
Was machst du?

Ich achte darauf, wo die Produkte herkommen.

Aufgabe 1 Wer sagt das?

Ordnen Sie die Aussagen den Personen zu.

Miniwörterbuch

den Müll trennen	to sort trash (for recycling)
die **Führerschein-prüfung**	driver's license test

1. Pascal ___

2. Felicitas ___

3. Nadezda ___

4. Simone ___

5. Michael ___

6. Hend ___

7. Sophie ___

8. Albrecht ___

a. Ich achte darauf, wo die Produkte herkommen.
b. Ich fahre viel Fahrrad und kaufe im Bioladen ein.
c. Ich habe kein Auto und ich trenne den Müll.
d. Ich mache das Wasser beim Zähneputzen aus.
e. Ich nutze wenig Wasser.
f. Ich recycle.
g. Ich trenne den Müll.
h. Nein, ich verbrauche sehr viel Wasser.

Interviews

- Woher kommst du?
- Wo liegt das?
- Was ist dort besonders interessant?
- Wie bist du in Leipzig unterwegs?
- Hast du einen Führerschein?
- War es schwer, ihn zu bekommen?

Albrecht

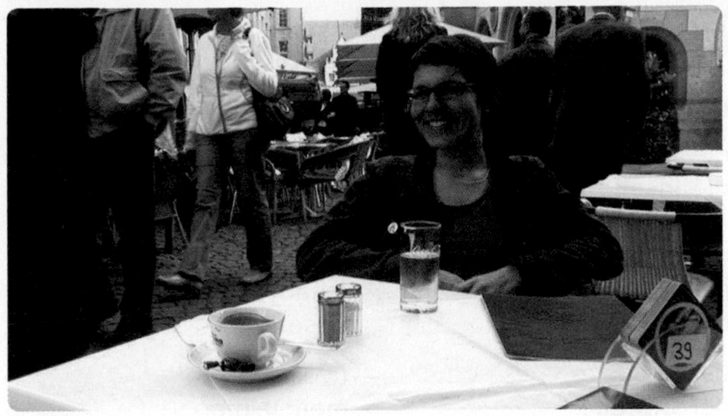

Simone

Aufgabe 2 Albrecht und Simone

Sehen Sie sich das Video an und ergänzen Sie die Tabelle.

	Albrecht	Simone
Woher kommen sie?		aus Braunschweig
Wo liegt die Stadt?	im Osten, südlich von Berlin	
Was ist dort interessant?	viele Kneipen, viel Grün	
Wie unterwegs?		zu Fuß
Führerschein?		ja
Schwer, ihn zu bekommen?	ja	

Aufgabe 3 Fragen

Beantworten Sie die folgenden Fragen.

1. Wie lange wohnt Albrecht schon in Leipzig?
2. Wie viele Kilometer sind es von Leipzig nach Berlin?
3. Wo gibt es viele Seen?
4. Wann fährt Albrecht mit dem Fahrrad?
5. Warum war es für Albrecht schwer, den Führerschein zu bekommen?
6. Was kann man vom Fluss Oker aus machen?
7. Wann bekommt Simone ihr Fahrrad?
8. Was hat Simone nicht?
9. Wovor hatte Simone große Angst?

Aufgabe 4 Interview

Interviewen Sie eine Partnerin oder einen Partner. Stellen Sie dieselben Interviewfragen.

Wortschatz

Geografie	Geography
die **Bucht, -en**	bay
die **Insel, -n**	island
die **Halbinsel, -n**	peninsula
die **Richtung, -en**	direction
die **Wiese, -n**	meadow, pasture
die **Wüste, -n**	desert
der **Fluss, ¨e**	river
der **Gipfel, -**	mountaintop
der **Gletscher, -**	glacier
der **Hügel, -**	hill
der **See, -n**	lake
der **Strand, ¨e** (R)	shore, beach
der **Wald, ¨er** (R)	forest, woods
das **Feld, -er**	field
das **Gebirge, -**	(range of) mountains
das **Meer, -e** (R)	sea
das **Tal, ¨er**	valley

Ähnliche Wörter

die **Küste, -n**; die **Landkarte, -n**; die **Alpen** (*pl.*); **nördlich (von)**; **östlich (von)**; **südlich (von)**; **westlich (von)**

Auto	Car
die **Bremse, -n**	brake
die **Hupe, -n**	horn
die **Reifenpanne, -n**	flat tire
der **Gang, ¨e**	gear
der **Kofferraum, ¨e**	trunk
der **Reifen, -**	tire
der **Scheibenwischer, -**	windshield wiper
der **Sicherheitsgurt, -e**	seat belt
der **Sitz, -e**	seat
der **Tank, -s**	(fuel) tank
das **Autoradio, -s**	car radio
das **Lenkrad, ¨er**	steering wheel
das **Nummernschild, -er**	license plate
das **Rad, ¨er**	wheel

Verkehr und Transportmittel	Traffic and Means of Transportation
die **Abfahrt, -en**	departure
die **Ankunft, ¨e**	arrival
die **Bahn, -en**	railroad
die **Autobahn, -en**	freeway
die **Straßenbahn, -en**	streetcar
die **U-Bahn, -en**	subway
die **Bahnfahrt, -en**	train trip
die **Einbahnstraße, -n**	one-way street
die **Fahrt, -en**	trip
die **Hin- und Rückfahrt**	round-trip
die **Radfahrerin, -nen**	(female) bicyclist
die **Rakete, -n**	rocket
der **Fahrkartenschalter, -**	ticket window
der **Fußgänger, -**	pedestrian
der **Unfall, ¨e** (R)	accident
der **Verkehr**	traffic
der **Wagen, -**	car
der **Kinderwagen, -**	baby carriage
der **Lastwagen, -**	truck
der **Waggon** [vagoŋ], **-s**	train car
der **Zug, ¨e**	train
das **Fahrrad, ¨er** (R)	bicycle
das **Fahrzeug, -e**	vehicle
das **Flugzeug, -e**	airplane
das **Motorrad, ¨er** (R)	motorcycle
das **Schild, -er**	sign
das **Verkehrsschild, -er**	traffic sign

Ähnliche Wörter

die **Fahrerin, -nen**; die **Kurve, -n**; die **Lokomotive, -n**; der **Bus, -se** (R); der **Fahrer, -**; das **Passwort, ¨er**; das **Taxi, -s** (R); **buchen**; **parken**; **transportieren**

Reiseerlebnisse	Travel Experiences
die **Reise, -n**	trip, journey
auf Reisen sein	to be on a trip
die **Stadtrundfahrt, -en**	tour of the city
die **Wanderung, -en**	hike
die **Welt, -en**	world
der **Höhepunkt, -e**	highlight
das **Erlebnis, -se**	experience
besichtigen	to visit, sightsee
besteigen, bestiegen	to climb

Ähnliche Wörter

der **Pass, ¨e**; der **Wein, -e**; das **Souvenir, -s**; **reservieren**

Sonstige Substantive	Other Nouns
die **Achtung**	attention
die **Angestellte, -n**	female clerk
die **Fläche, -n**	surface
die **Luft**	air
die **Scheibe, -n**	windowpane
der **Angestellte, -n**	male clerk
der **Betrieb, -e**	workplace, operation

der **Regen**	rain
bei Regen	in rainy weather
der **Teil, -e**	part
der **Nachteil, -e**	disadvantage
der **Vorteil, -e**	advantage
das **Tier, -e** (R)	animal
die **Leute** (*pl.*)	people
die **Nachrichten** (*pl.*)	news

Ähnliche Wörter

der **Euro, -**; der **Liter, -**; der **Preis, -e**; der **Sand;** das **Baby** [be:bi]**, -s**; das **Institut, -e**; das **Oktoberfest, -e**

Sonstige Verben — Other Verbs

an·halten, hält ... an, angehalten	to stop
aus·gehen, ist ausgegangen	to end, turn out
es ist gut ausgegangen	it ended well
benutzen	to use
ein·schlafen, schläft ... ein, ist eingeschlafen	to fall asleep
erlauben	to permit
fließen, ist geflossen	to flow
halten, hält, gehalten	to stop
hupen	to honk
nach·denken (über + *akk.*), **nachgedacht**	to think (about), consider
Rad fahren, fährt ... Rad, ist Rad gefahren	to ride a bicycle
rennen, ist gerannt	to run
rufen, gerufen	to call, shout
schwimmen, ist geschwommen	to swim; to float
setzen	to put, place, set
trennen	to separate
übernehmen, übernimmt, übernommen	to take on
um·steigen, ist umgestiegen	to change (trains)
vergleichen, verglichen	to compare
verlieren, verloren	to lose
versprechen, verspricht, versprochen	to promise
verstauen	to stow
warten	to wait
wischen (R)	to wipe

Ähnliche Wörter

beantworten, warnen, wecken

Sonstige Wörter und Ausdrücke — Other Words and Expressions

berühmt	famous
bitte schön?	yes please?; may I help you?
dort	there
durch	through
hoch	high
lieb	dear
am liebsten	like (*to do*) best
mindestens	at least
nah	close, nearby
öffentlich	public
rechts	to the right
schließlich	finally
schnell	quick, fast
ungefähr	approximately
zuerst (R)	first
zwischen	between

Ähnliche Wörter

exotisch, intelligent, interessant, mehr, salzig, tief, tolerant

Strukturen und Übungen

7.1 Relative clauses

WISSEN SIE NOCH?

A relative clause is a type of dependent clause. As in other dependent clauses, the conjugated verb appears at the end of the clause.

Review grammar 3.4.

Relative clauses add information about a person, place, thing, or idea already mentioned in the sentence. The relative pronoun begins the relative clause, which usually follows the noun it describes. The relative pronoun corresponds to the English words *who, whom, that,* and *which.* The conjugated verb is in the end position.

RELATIVE CLAUSE

Der Atlantik ist das Meer, das Europa und Afrika von Amerika **trennt.**

VERB IN END POSITION

The Atlantic is the ocean that separates Europe and Africa from America.

Do not omit the relative pronoun in the German sentence.

While relative pronouns may sometimes be omitted in English, they cannot be omitted from German sentences.

Das ist der Mantel, **den** ich letzte Woche gekauft habe.
That is the coat (that) I bought last week.

Relative clauses are preceded by a comma.

Likewise, the comma is not always necessary in an English sentence, but it must precede a relative clause in German. If the relative clause comes in the middle of a German sentence, it is followed by a comma as well.

Der See, **der** zwischen Deutschland und der Schweiz liegt, heißt Bodensee.
The lake that lies between Germany and Switzerland is called Lake Constance.

A. Relative Pronouns in the Nominative Case

In the nominative (subject) case, the forms of the relative pronoun are the same as the forms of the definite article **der, das, die.**

Der Fluss, **der** durch Wien fließt, heißt Donau.
Gobi heißt **die** Wüste, **die** in Innerasien liegt.

The relative pronoun and the noun it refers to have the same number and gender.

The relative pronoun has the same gender and number as the noun it refers to.

Masculine	der Mann, **der ...**	*the man who ...*
Neuter	das Auto, **das ...**	*the car that ...*
Feminine	die Frau, **die ...**	*the woman who ...*
Plural	die Leute, **die ...**	*the people who ...*

B. Relative Pronouns in the Accusative and Dative Cases

The case of a relative pronoun depends on its function within the relative clause.

When the relative pronoun functions as an accusative object or as a dative object within the relative clause, then the relative pronoun is in the accusative or dative case, respectively.

ACCUSATIVE

Nur wenige Menschen haben **den Mount Everest** bestiegen.	*Only a few people have climbed Mount Everest.*
Der Mount Everest ist ein Berg, **den** nur wenige Menschen bestiegen haben.	*Mount Everest is a mountain that only a few people have climbed.*

DATIVE

Ich habe **meinem Vater** nichts davon erzählt.	*I haven't told my father anything about it.*
Mein Vater ist der einzige Mensch, **dem** ich nichts davon erzählt habe.	*My father is the only person whom I haven't told anything about it.*

As in the nominative case, the accusative and dative relative pronouns have the same forms as the definite article, except for the dative plural, **denen.**

	Masculine	Neuter	Feminine	Plural
Accusative	den	das	die	die
Dative	dem	dem	der	denen

C. Relative Pronouns Following a Preposition

The case of the relative pronoun depends on the preposition that precedes it.

When a relative pronoun follows a preposition, the case is determined by that preposition. The gender and number of the pronoun are determined by the noun.

Ich spreche am liebsten **mit meinem** Bruder.	*Most of all I like to talk with my brother.*
Mein Bruder ist der Mensch, **mit dem** ich am liebsten spreche.	*My brother is the person (whom) I like to talk with most.*
Auf der Insel Rügen sind weiße Kreidefelsen.	*There are white chalk cliffs on the island of Rügen.*
Rügen ist eine Insel in der Ostsee, **auf der** weiße Kreidefelsen sind.	*Rügen is an island in the Baltic Sea on which there are white chalk cliffs.*

Preposition + relative pronoun = inseparable unit

The preposition and the pronoun stay together as a unit in German.

Wer war die Frau, **mit der** ich dich gestern gesehen habe?	*Who was the woman (whom) I saw you with yesterday?*

Übung 1 Das mag ich, das mag ich nicht!

Bilden Sie Sätze!

MODELL: Ich mag Leute, die spät ins Bett gehen.

1. Ich mag Leute, die ...
2. Ich mag keine Leute, die ...
3. Ich mag eine Stadt, die ...
4. Ich mag keine Stadt, die ...
5. Ich mag einen Mann, der ...
6. Ich mag keinen Mann, der ...
7. Ich mag eine Frau, die ...
8. Ich mag keine Frau, die ...
9. Ich mag einen Urlaub, der ...
10. Ich mag ein Auto, das ...

nett sein
laut lachen
Spaß machen
exotisch sein
schnell fahren
gern im Sand spielen
viel sprechen
betrunken sein
langweilig sein
gern verreisen
interessant aussehen
?

Übung 2 Risiko°

°*Jeopardy*

Hier sind die Antworten. Stellen Sie die Fragen!

MODELL: Diesen Kontinent hat Kolumbus entdeckt. →
Wie heißt der Kontinent, den Kolumbus entdeckt hat? (Amerika)

1. Europa
2. Mississippi
3. San Francisco
4. die Alpen
5. Washington
6. das Tal des Todes
7. Ellis
8. der Pazifik
9. die Sahara
10. der Große Salzsee

a. Auf diesem See in Utah kann man segeln.
b. Diese Insel sieht man von New York.
c. Diese Stadt liegt an einer Bucht.
d. Diese Wüste kennt man aus vielen Filmen.
e. Diesem Staat in den USA hat ein Präsident seinen Namen gegeben.
f. In diesem Tal ist es sehr heiß.
g. In diesen Bergen kann man sehr gut Ski fahren.
h. Dieser Kontinent ist eigentlich eine Halbinsel von Asien.
i. Über dieses Meer fliegt man nach Hawaii.
j. Von diesem Fluss erzählt Mark Twain.

7.2 Making comparisons: the comparative and superlative forms of adjectives and adverbs

A. Comparisons of Equality: **so ... wie**

so ... wie = *as . . . as*

To say that two or more persons or things are alike or equal in some way, use the phrase **so ... wie** (*as ... as*) with an adjective or adverb.

Deutschland ist ungefähr **so groß wie** Montana.	*Germany is about as big as Montana.*
Der Mount Whitney ist fast **so hoch wie** das Matterhorn.	*Mount Whitney is almost as high as the Matterhorn.*

Inequality can also be expressed with this formula and the addition of **nicht.**

Die Zugspitze ist **nicht so hoch wie** der Mount Everest.	*The Zugspitze is not as high as Mount Everest.*
Österreich ist **nicht ganz so groß wie** Maine.	*Austria is not quite as big as Maine.*

B. Comparisons of Superiority and Inferiority

All comparatives in German are formed with **-er.**

To compare two unequal persons or things, add **-er** to the adjective or adverb. Note that the comparative form of German adjectives and adverbs always ends in **-er,** whereas English sometimes uses the adjective with the word *more.*

als = *than*

Ein Fahrrad ist **billiger als** ein Motorrad.	*A bicycle is cheaper than a motorcycle.*
Sumita ist **intelligenter als** ihre Schwester.	*Sumita is more intelligent than her sister.*
Jens läuft **schneller als** Ernst.	*Jens runs faster than Ernst.*

Some adjectives that end in **-el** and **-er** drop the **-e-** in the comparative form.

teuer → teu~~e~~rer
dunkel → dunk~~e~~ler

Eine Wohnung in Regensburg ist teuer, aber eine Wohnung in München ist noch **teurer.**	*An apartment in Regensburg is expensive, but an apartment in Munich is even more expensive.*
Gestern war es dunkel, aber heute ist es **dunkler.**	*Yesterday it was dark, but today it is darker.*

C. The Superlative

Superlatives: **am** + **-sten**

To express the superlative in German, use the contraction **am** with a predicate adjective or adverb plus the ending **-sten.**

Ein Porsche ist schnell, ein Flugzeug ist schneller, und eine Rakete ist am schnellsten.	*A Porsche is fast, an airplane is faster, and a rocket is the fastest.*

Unlike the English superlative, which has two forms, all German adjectives and adverbs form the superlative in this way.

Hans ist **am jüngsten.**	*Hans is the youngest.*
Jens ist **am tolerantesten.**	*Jens is the most tolerant.*

When the adjective or adverb ends in **-d** or **-t,** or an **s**-sound such as **-s, -ß, -sch, -x,** or **-z,** an **-e-** is inserted between the stem and the ending.

frisch → am frisch**esten**
gesund → am gesünd**esten**
heiß → am heiß**esten**
intelligent → am intelligent**esten**

Um die Mittagszeit ist es oft am heißesten.	*The hottest (weather) is often around noontime.*

Groß is an exception to the rule: **am größten.**

D. Irregular Comparative and Superlative Forms

Irregular comparatives and superlatives have an umlaut whenever possible.

The following adjectives have an umlaut in the comparative and the superlative.

alt	älter	am ältesten
gesund	gesünder	am gesündesten
groß	größer	am größten
jung	jünger	am jüngsten
kalt	kälter	am kältesten
krank	kränker	am kränksten
kurz	kürzer	am kürzesten
lang	länger	am längsten
warm	wärmer	am wärmsten

Im März ist es oft **wärmer** als im Januar. Im August ist es **am wärmsten.**	*In March it's often warmer than in January. It's warmest in August.*

As in English, some comparative and superlative forms are very different from their base forms:

gern	lieber	am liebsten
gut	besser	am besten
hoch	höher	am höchsten
nah	näher	am nächsten
viel	mehr	am meisten

Ich spreche Deutsch, Englisch und Spanisch. Englisch spreche ich **am besten** und Deutsch spreche ich **am liebsten.**	*I speak German, English, and Spanish. I speak English the best, and I like to speak German the most.*

Superlatives before nouns in the nominative:
der/das/die + **-(e)ste**
die (*pl.*) + **-(e)sten**

E. Superlative Forms Preceding Nouns

When the superlative form of an adjective is used with a definite article (**der, das, die**) directly *before* a noun, it has an **-(e)ste** ending in all forms of the nominative singular and an **-(e)sten** ending in the plural. You will get used to the **-e/-en** distribution as you have more experience listening to and reading German. (A more detailed description of adjectives that precede nouns will follow in **Kapitel 8.**)

Nominative	
der längst**e**	Fluss (*m.*)
das tiefst**e**	Tal (*n.*)
die größt**e**	Wüste (*f.*)
die höchst**en**	Berge (*pl.*)

—Wie heißt der längste Fluss Europas?	*What is the name of the longest river in Europe?*
—Wolga.	*The Volga.*
—In welchem Land wohnen die meisten Menschen?	*What country has the most people?*
—In China.	*China.*

Übung 3 Vergleiche

Vergleichen Sie.

MODELL: Wien / Göttingen / klein → Göttingen ist kleiner als Wien.

1. Berlin / Zürich / groß
2. San Francisco / München / alt
3. Hamburg / Athen / warm
4. das Matterhorn / der Mount Everest / hoch
5. der Mississippi / der Rhein / lang
6. die Schweiz / Liechtenstein / klein
7. Leipzig / Kairo / kalt
8. ein Fernseher / eine Waschmaschine / billig
9. Schnaps / Bier / stark
10. ein Haus in der Stadt / ein Haus auf dem Land / schön
11. zehn Euro / zehn Cent / viel
12. eine Wohnung in einem Studentenheim / ein Appartement / teuer
13. ein Fahrrad / ein Motorrad / schnell
14. ein Sofa / ein Stuhl / schwer
15. Milch / Bier / gut

Übung 4 Biografische Daten

Vergleichen Sie. [(+) = Superlativ]

MODELL: alt / Thomas / Stefan → Thomas ist **älter** als Stefan.
alt (+) → Heidi ist **am ältesten.**

	Thomas	Heidi	Stefan	Monika
Alter	19	22	18	21
Größe	1,89 m	1,75 m	1,82 m	1,69 m
Gewicht	75 kg	65 kg	75 kg	57 kg
Haarlänge	20 cm	15 cm	5 cm	25 cm
Note in Deutsch	B	A	C	B

1. schwer / Monika / Heidi
2. schwer (+)
3. gut in Deutsch / Thomas / Stefan
4. gut in Deutsch (+)
5. klein / Heidi / Stefan
6. klein (+)
7. jung / Thomas / Stefan
8. jung (+)
9. lang / Heidis Haare / Thomas' Haare
10. lang (+)
11. kurz / Monikas Haare / Heidis Haare
12. kurz (+)
13. schlecht in Deutsch / Heidi / Monika
14. schlecht in Deutsch (+)

Übung 5 Geografie und Geschichte

MODELL: Das Tal des Todes (–86 m) liegt tiefer als das Kaspische Meer (–28 m). →
Das Tote Meer (–396 m) liegt am tiefsten.

1. In Rom (25,6°C) ist es im Sommer heißer als in München (17,2°C).
2. In Wien (–1,4°C) ist es im Winter kälter als in Paris (3,5°C).
3. Liechtenstein (157 km^2)* ist kleiner als Luxemburg (2.586 km^2).
4. Deutschland (911) ist älter als die Schweiz (1291).
5. Kanada (1840) ist jünger als die USA (1776).
6. Der Mississippi (6.021 km) ist länger als die Donau (2.850 km).
7. Philadelphia (40° nördliche Breite) liegt nördlicher als Kairo (30° nördliche Breite).
8. Der Mont Blanc (4.807 m) ist höher als der Mount Whitney (4.418 m).
9. Österreich (83.849 km^2) ist größer als die Schweiz (41.288 km^2).

a. Athen (27,6°C)
b. das Tote Meer (–396 m)
c. Deutschland (357.050 km^2)
d. Frankfurt (50° nördliche Breite)
e. Frankreich (498)
f. Monaco (1,49 km^2)
g. Moskau (–9,9°C)
h. der Mount Everest (8.848 m)
i. der Nil (6.671 km)
j. Südafrika (1884)

*km^2 = Quadratkilometer

7.3 Referring to and asking about things and ideas: *da*-compounds and *wo*-compounds

In both German and English, personal pronouns are used directly after prepositions when these pronouns refer to people or animals.

Ich werde bald **mit ihr** sprechen.	*I'll talk to her soon.*
—Bist du mit Josef gefahren?	*Did you go with Josef?*
—Ja, ich bin **mit ihm** gefahren.	*Yes, I went with him.*

da- or **dar-** + preposition

When the object of the preposition is a thing or concept, it is common in English to use the pronoun *it* or *them* with a preposition: *with it, for them,* and so on. In German, it is preferable to use compounds that begin with **da-** (or **dar-** if the preposition begins with a vowel).

dadurch	*through it/them*	daraus	*out of it/them*
dafür	*for it/them*	darin	*in it/them*
dagegen	*against it/them*	darüber	*over it/them*
dahinter	*behind it/them*	darunter	*underneath it/them*
damit	*with it/them*	davon	*from it/them*
daneben	*next to it/them*	davor	*in front of it/them*
daran	*on it/them*	dazu	*to it/them*
darauf	*on top of it/them*	dazwischen	*between it/them*

Note that the following prepositions cannot be preceded by **da(r)-: ohne, außer, seit.**

—Was macht man mit einer Hupe?	*What do you do with a horn?*
—Man warnt andere Leute **damit.**	*You warn other people with it.*
—Hast du etwas gegen das Rauchen?	*Do you have something against smoking?*
—Nein, ich habe nichts **dagegen.**	*No, I don't have anything against it.*

Some **da**-compounds are idiomatic.

Hast du Geld **dabei?**	*Do you have any money on you?*
Darum hast du auch kein Glück.	*That's why you don't have any luck.*

Use a preposition + **wem** or **wen** to ask about people.

Questions about people begin with **wer** (*who*) or **wen/wem** (*whom*). If a preposition is involved, it precedes the question word.

—Mit **wem** gehst du ins Theater?	*Who will you go to the theater with? (With whom . . .?)*
—Mit Melanie.	*With Melanie.*
—In **wen** hast du dich diesmal verliebt?	*Who did you fall in love with this time? (With whom . . .?)*

Use **wo-** + a preposition to ask about things or ideas.

Questions about things and concepts begin with **was** (*what*). If a preposition is involved, German speakers use compound words that begin with **wo-** (or **wor-** if the preposition begins with a vowel).

—**Womit** fährst du nach Berlin?	*How are you getting to Berlin?*
—Mit dem Bus.	*By bus.*
—**Worüber** sprichst du?	*What are you talking about?*
—Ich spreche über den neuen Film von Doris Dörrie.	*I'm talking about Doris Dörrie's new film.*

People	Things and Concepts
mit wem	womit
von wem	wovon
zu wem	wozu
an wen	woran
auf wen	worauf
für wen	wofür
über wen	worüber
um wen	worum

—**Von wem** ist die Oper „Parsifal"? *Who is the opera* Parzival *by?*
—Von Richard Wagner. *By Richard Wagner.*
—**Wovon** handelt diese Oper? *What is the opera about?*
—Von der Suche nach dem Gral. *About the search for the Holy Grail.*

Übung 6 Juttas Zimmer

Ergänzen Sie!

Links[1] ist eine Kommode. Eine Lampe steht *darauf* [a]. Rechts ____[b] steht der Schreibtisch. ____[c] steht Juttas Tasche. An der Wand steht ein Schrank. ____[d] hängen Juttas Sachen. Links an der Wand steht Juttas Bett. ____[e] liegt die Katze auf dem Teppich. An der Wand ____[f] hängt ein Bild. Auf dem Bild ist eine Wiese mit einem Baum. ____[g] hängen Äpfel. Mitten im Zimmer steht ein Sessel. ____[h] sieht man Juttas Schuhe und ____[i] hat sich Hans versteckt[2].

dahinter **daran** **daneben** **davor** **darauf** **darunter** **darin** **darüber** **dazwischen**

[1]*To the left* [2]hat ... *Hans has hidden himself*

Übung 7 Ein Interview mit Richard

Das folgende Interview ist nicht vollständig. Es fehlen die Fragen. Rekonstruieren Sie die Fragen aus den Antworten.

1. Ich gehe am liebsten **mit meiner Kusine** ins Theater.
2. Am meisten freue ich mich **auf die Ferien.**
3. Ich muss immer **auf meinen Freund** warten. Er kommt immer zu spät.
4. In letzter Zeit habe ich mich **über meinen Physiklehrer** geärgert.
5. Wenn ich „USA" höre, denke ich **an Hochhäuser und Disneyland, an den Grand Canyon und die Rocky Mountains und natürlich an Iowa.**
6. Zur Schule fahre ich meistens **mit dem Fahrrad, manchmal auch mit dem Bus.**
7. Ich schreibe nicht gern **über Sachen,** die mich nicht interessieren, wie zum Beispiel die Vorteile und Nachteile des Kapitalismus.
8. Meinen letzten Brief habe ich **an einen alten Freund von mir** geschrieben. Der ist vor kurzem nach Graz gezogen, um dort Jura zu studieren.
9. Ich halte nicht viel **von meinen Lehrern.** Die tun nur immer so, als wüssten sie alles; in Wirklichkeit wissen die gar nichts.

7.4 The perfect tense (review)

WISSEN SIE NOCH?

The perfect tense consists of a form of the present tense of **haben** or **sein** + the past participle.

Review grammar 4.1.

As you remember from **Kapitel 4,** it is preferable to use the perfect tense in oral communication when talking about past events.

Ich **habe** im Garten Äpfel **gepflückt.**	*I picked apples in the garden.*

To form the perfect tense, use **haben** or **sein** as an auxiliary with the past participle of the verb.

Use **haben** with most verbs.
Use **sein** if the verb:
- cannot take an accusative object
- indicates change of location or condition.

See **Appendix E** for a list of common verbs and their auxiliaries.

A. **haben** or **sein**

Haben is by far the more commonly used auxiliary. **Sein** is normally used only when both of the following conditions are met: (1) The verb cannot take an accusative object. (2) The verb implies a change of location or condition.

Bertolt Brecht **ist** 1956 in Berlin **gestorben.**	*Bertolt Brecht died in Berlin in 1956.*
Ernst **ist** mit seinem Hund **spazieren gegangen.**	*Ernst went for a walk with his dog.*

In spite of the fact that there is no change of location or condition, the following verbs also take **sein** as an auxiliary: **sein, bleiben,** and **passieren.**

Letztes Jahr **bin** ich in St. Moritz **gewesen.**	*Last year I was in St. Moritz.*
Was **ist passiert?**	*What happened?*

B. Forming the Past Participle

Past participles of strong verbs end in **-en;** past participles of weak verbs end in **-t** or **-et.**

There are basically two ways to form the past participle. Strong verbs add the prefix **ge-** and the ending **-en** to the stem. Weak verbs add the prefix **ge-** and the ending **-t** or **-et.**

rufen	hat **ge**ruf**en**	*to shout, call*
reisen	ist **ge**reis**t**	*to travel*
arbeiten	hat **ge**arbeit**et**	*to work*

In the past-participle form, most, but not all, strong verbs have a changed stem vowel or stem.

	gehen	ist geg**ang**en	*to walk*
	werfen	hat gew**o**rfen	*to throw*
but:	laufen	ist gelaufen	*to run*

Very few weak verbs have a change in the stem vowel. Here are some common weak verbs that do change.

bringen	hat gebr**ach**t	*to bring*
denken	hat ged**ach**t	*to think*
dürfen	hat ged**u**rft	*to be allowed to*
können	hat gek**o**nnt	*to be able to*
müssen	hat gem**u**sst	*to have to*
rennen	ist ger**a**nnt	*to run*
wissen	hat gew**u**sst	*to know (as a fact)*

C. Past Participles with and without **ge-**

No **ge-** with verbs ending in **-ieren** and inseparable-prefix verbs

Another group of verbs forms the past participle without **ge-.** You will recognize them because, unlike most verbs, they are not pronounced with an emphasis on the first syllable. These verbs fall into two major groups: those that end in **-ieren** and those that have inseparable prefixes.

passieren	ist passiert	*to happen*
studieren	hat studiert	*to study, go to college*

Common inseparable prefixes: **be-, ent-, er-, ge-, ver-**

The most common inseparable prefixes are **be-, ent-, er-, ge-,** and **ver-.**

besuchen	hat besucht	*to visit*
entdecken	hat entdeckt	*to discover*
erzählen	hat erzählt	*to tell*
gewinnen	hat gewonnen	*to win*
verlieren	hat verloren	*to lose*
versprechen	hat versprochen	*to promise*

The past participle of separable-prefix verbs is formed by adding the prefix to the past participle of the base verb.

anfangen	hat angefangen	*to begin*
aufstehen	ist aufgestanden	*to get up*

Übung 8 Renate

Ergänzen Sie **haben** oder **sein.**

1. In meiner Schulzeit _____ ich nie gern aufgestanden.
2. Meine Mutter _____[a] mich immer geweckt, denn ich _____[b] nie von allein aufgewacht.
3. Ich _____[a] ganz schnell etwas gegessen und _____[b] zur Schule gerannt.
4. Meistens hatte es schon zur Stunde geklingelt, wenn ich angekommen _____.
5. In der Schule war es oft langweilig; in Biologie _____ ich sogar einmal eingeschlafen.
6. Einmal in der Woche hatten wir nachmittags Sport. Am liebsten _____[a] ich Basketball gespielt und _____[b] geschwommen.
7. Auf dem Weg nach Hause _____[a] ich einmal einen Autounfall gesehen. Zum Glück _____[b] nichts passiert.
8. Aber viele Leute _____[a] herumgestanden, bis die Polizei gekommen _____[b].
9. Sie _____[a] geblieben, bis eine Autowerkstatt die kaputten Autos abgeholt _____[b].

Übung 9 Ernst

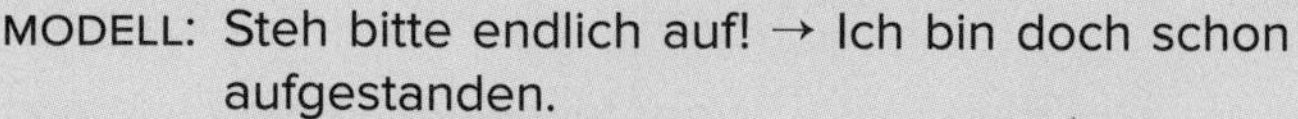

Ernst war fleißig. Er hat schon alles gemacht und spielt jetzt Fußball. Übernehmen Sie seine Rolle.

MODELL: Steh bitte endlich auf! → Ich bin doch schon aufgestanden.

1. Mach bitte Frühstück!
2. Trink bitte deine Milch!
3. Mach bitte den Tisch sauber!
4. Lauf mal schnell zum Bäcker!
5. Bring bitte Brötchen mit!
6. Nimm bitte Geld mit!
7. Füttere bitte den Hund!
8. Mach bitte die Tür zu!

7.5 The simple past tense of *haben* and *sein*

When talking about events that have already happened, people commonly use the verbs **haben** and **sein** in the simple past tense instead of the perfect tense. The conjugations appear below; notice that the **ich-** and the **er/sie/es-**forms are the same.

Warst du schon mal im Ausland? — *Have you ever been abroad?*
Letzte Woche **hatte** ich einen Autounfall. — *Last week I had a car accident.*

sein			
ich	war	*wir*	waren
du	warst	*ihr*	wart
Sie	waren	*Sie*	waren
er *sie* *es*	war	*sie*	waren

haben			
ich	hatte	*wir*	hatten
du	hattest	*ihr*	hattet
Sie	hatten	*Sie*	hatten
er *sie* *es*	hatte	*sie*	hatten

Übung 10 Minidialoge

Ergänzen Sie eine Form von **war** oder **hatte.**

1. FRAU GRETTER: Ihr Auto sieht ja so kaputt aus. _______[a] Sie einen Unfall?
   HERR THELEN: Ja, leider _______[b] ich wieder mal einen Unfall. Das ist schon der dritte in dieser Woche.
2. FRAU KÖRNER: Sie sind aber braun geworden. _______ Sie im Urlaub?
   MICHAEL PUSCH: Ja, ich war drei Wochen in der Türkei.
3. HANS: Warum _______[a] ihr gestern nicht in der Schule?
   JENS UND JUTTA: Wir _______[b] keine Zeit.
4. CLAIRE: _______[a] du schon mal in Linz, Melanie?
   MELANIE: Ja, ich _______[b] schon ein paar Mal da.
5. MARIA SCHNEIDER: Wo warst du letzte Woche, Jens?
   JENS: Ich _______ Ferien und war bei meinen Großeltern auf dem Land.
6. JUTTA: Michael, sag mal, _______ du schon mal eine Reifenpanne?
   MICHAEL PUSCH: Nein, Gott sei Dank noch nie.
7. CLAIRE: Ich habe dich gestern im Kino gesehen. _______[a] du allein?
   JOSEF: Ja, Melanie _______[b] gestern zu Hause. Sie _______[c] keine Lust, ins Kino zu gehen.

KAPITEL 8

Essen und Einkaufen

In **Kapitel 8,** you will learn to talk about shopping for food and cooking and about the kinds of foods you like. You will also talk about household appliances and about dining out.

Themen

Essen und Trinken

Haushaltsgeräte

Einkaufen und Kochen

Im Restaurant

Kulturelles

Musikszene: „Hawaii Toast Song“ (Alexander Marcus)

KLI: Österreich

Filmclip: *Bella Martha* (Sandra Nettelbeck)

KLI: Brot

Videoecke: Essen

Lektüren

Sachtext: Stichwort Fabel

Fabel: Die gebratene Ameise (Paul Scheerbart)

Film: *Bella Martha* (Sandra Nettelbeck)

Strukturen

8.1 Adjectives: an overview

8.2 Attributive adjectives in the nominative and accusative cases

8.3 Destination vs. location: **stellen/stehen, legen/liegen, setzen/sitzen, hängen/hängen**

8.4 Adjectives in the dative case

8.5 Talking about the future: the present and future tenses

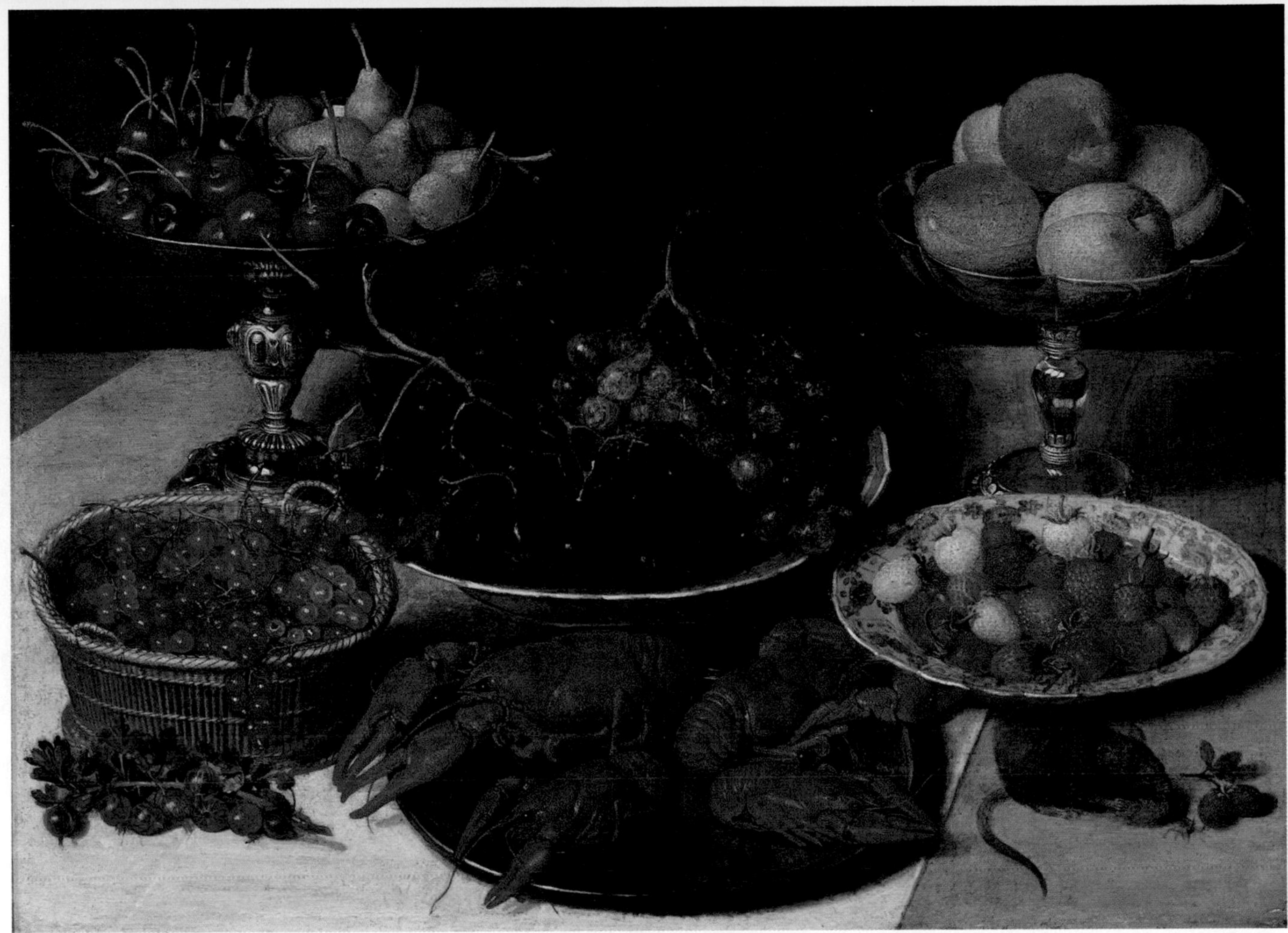

Georg Flegel: *Stillleben mit Obst und Krebsen* (ca. 1630), Nationalgalerie, Warschau

KUNST UND KÜNSTLER

Georg Flegel (1566–1638) war der erste und vielleicht wichtigste Stilllebenmaler in Deutschland. Er wurde in Olmütz in Mähren (heute Tschechische Republik) geboren, arbeitete dann im österreichischen Linz in der Werkstatt des niederländischen Malers Lucas von Valckenborch und zog mit ihm um 1592 nach Frankfurt am Main, wo Flegel bis zu seinem Tode als Maler arbeitete. Seine Bilder sind ein perfektes Abbild der Gegenstände[1], aber im Sinne[2] des Barock haben sie ein fast magisches Eigenleben[3]. Typisch für Flegels Werke ist, dass oft ein kleines Tier in Kontrast zu den leblosen Objekten des Stilllebens tritt.

Schauen Sie sich das Bild an und beantworten Sie die folgenden Fragen.

1. Was sehen Sie auf dem Bild? Identifizieren Sie die Gegenstände und Früchte.
2. Welche Farben und Linien dominieren im Bild? Wie sind die Gegenstände verteilt?
3. Welches Tier tritt in Kontrast zu den Gegenständen? Was macht es?
4. Welche Assoziationen und Gefühle weckt das Bild in Ihnen?

[1]*objects* [2]*sense* [3]*life of their own*

Situationen

Essen und Trinken

Grammatik 8.1–8.2

RENATE: Meistens esse ich ein frisches Brötchen, ein gekochtes Ei und selbst gemachte Marmelade zum Frühstück. Außerdem brauche ich einen starken Kaffee. Am Wochenende esse ich auch Schinken und Käse und trinke einen frisch gepressten Orangensaft. Als ich ein Kind war, habe ich meistens Milch mit Honig getrunken, später auch Tee.

das Frühstück
© Dirk E. Hasenpusch

HERR THELEN: Zu Mittag esse ich am liebsten einen gemischten Salat, gebratenes Fleisch oder gegrillten Fisch mit gekochten Kartoffeln. Auch Hähnchen mag ich ganz gern und Karotten mit viel Salz und Pfeffer. Meistens trinke ich eine Apfelschorle. Das ist ein Gemisch aus Apfelsaft und Mineralwasser. Am Sonntag trinke ich vielleicht auch mal ein Glas Wein, am liebsten Rotwein.

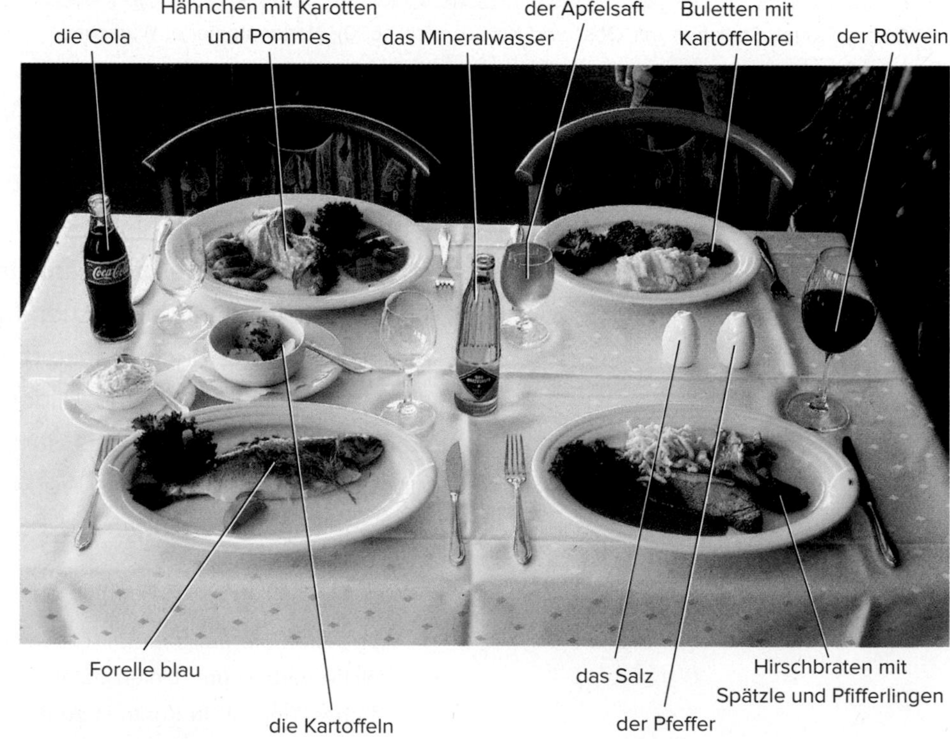

das Mittagessen
© Dirk E. Hasenpusch

das Abendessen

© Dirk E. Hasenpusch

FRAU FRISCH-OKONKWO: Am Abend esse ich gern rustikal: Brot, Butter, Schinken, Käse. Rohen Schinken esse ich gern mit Meerrettich. Manchmal mache ich mir auch ein paar warme Würstchen. Die esse ich dann mit Senf. Emmentaler esse ich gern mit sauren Essiggurken. Dazu trinke ich entweder ein Glas Milch oder Saft mit Mineralwasser.

Situation 1 Umfrage: Isst du gern fettige Hamburger?

MODELL: S1: Isst du gern fettige Hamburger?
S2: Ja!
S1: Unterschreib bitte hier!

	UNTERSCHRIFT
1. Isst du gern fettige Hamburger?	____________
2. Isst du oft Chinesisch?	____________
3. Isst du oft frisches Obst?	____________
4. Frühstückst du selten?	____________
5. Isst du zum Frühstück gern gebratene Eier mit Speck?	____________
6. Isst du meistens in der Cafeteria?	____________
7. Isst du manchmal Pizza?	____________
8. Würzt du dein Essen mit viel Pfeffer?	____________
9. Isst du selten zu Hause?	____________
10. Hast du für heute ein belegtes Brot dabei?	____________

Situation 2 Informationsspiel: Mahlzeiten und Getränke

MODELL: S1: Was isst Stefan zum Frühstück?
S2: _______

	Frau Gretter	Stefan	Andrea
zum Frühstück essen	frische Brötchen		
zum Frühstück trinken	schwarzen Kaffee		heißen Kakao
zu Mittag essen		belegte Brote und Kartoffelchips	
zu Abend essen	nichts, sie will abnehmen		Brot mit Honig
nach dem Sport trinken			Apfelsaft
auf einem Fest trinken		mexikanisches Bier	
essen, wenn er/sie groß ausgeht	etwas Gesundes		den schönsten Kinderteller

Situation 3 Ratespiel: Regionale Spezialitäten

Was glauben Sie? Wo isst oder trinkt man diese regionalen Spezialitäten? Es gibt viele richtige Antworten.

1. Wo trinkt man Berliner Weiße?
2. Wo isst man selbst gemachte Fleischchüechli?
3. Wo isst man gebratene Eier und Speck?
4. Wo isst man deftige Knödel?
5. Wo isst man frischen Fisch aus der Nordsee?
6. Wo trinkt man frisch gepressten Orangensaft?
7. Wo isst man frische Semmeln?
8. Wo trinkt man eiskalten Eistee?
9. Wo isst man Rote Grütze?
10. Wo trinkt man sächsisches Schwarzbier?

in Österreich **in Berlin**
überall **in Sachsen**
in Bayern **in den USA**
in der Schweiz **in Norddeutschland**

[1]*folk music*

MUSIKSZENE

„Hawaii Toast Song" (2009, Deutschland) *Alexander Marcus*

Biografie Alexander Marcus ist ein Berliner Musiker und Entertainer. Er singt einfache dümmliche Schlagertexte zu einer Mischung aus Volksmusik und Techno. Marcus nennt seinen Stil *Electrolore*, eine Mischung aus Elektro und Folklore[1]. Der „Hawaii Toast Song" ist von seinem zweiten Album *Mega*.

Alexander Marcus

NOTE: For copyright reasons, the songs referenced in **MUSIKSZENE** have not been provided by the publisher. The song can be found online at various sites such as YouTube, Amazon, or the iTunes store.

Vor dem Hören Kennen Sie Hawaii Toast? Welche Zutaten, glauben Sie, braucht man dafür? (Denken Sie an Hawaii Pizza.)

Nach dem Hören Beantworten Sie die folgenden Fragen.

1. Welche Zutaten braucht man für einen Hawaii Toast?
2. Wie macht man ihn?
3. Was bewirkt er?
4. Haben Sie schon mal einen Hawaii Toast gemacht oder gegessen?

Miniwörterbuch	
dümmlich	silly
der **Schlager**	German pop song of the 1950s and 60s
der **Scheiblettenkäse**	individually wrapped cheese slices
vorheizen	to preheat
die mittlere Schiene	the middle oven rack
die **Not**	time of need
Trost spenden	to provide solace
bewirken	to have an effect

KULTUR ... LANDESKUNDE ... INFORMATIONEN

BROT

- Essen Sie gern Brot?
- Welche Sorten essen Sie am liebsten?
- Gibt es Sorten, die Sie nicht mögen?
- Zu welchen Mahlzeiten essen Sie Brot?
- Haben Sie schon mal Brot gebacken? Was braucht man dazu?

Lesen Sie die Wörter im Miniwörterbuch. Suchen Sie sie im Text und unterstreichen Sie sie.

Miniwörterbuch	
die **Sorte**	sort, kind
die Hefe	yeast, starter
das **Getreide**	grain
der **Teig**	dough
der **Stein**	stone
das **(Grund)nahrungsmittel**	(basic) foodstuff
der **Brauch,** die **Bräuche**	custom
der **Wohlstand**	prosperity

Brot ist ein sehr wichtiges Nahrungsmittel. Lesen Sie den Text und suchen Sie Antworten auf die folgenden Fragen.

1. Was vermissen Deutsche, wenn sie im Ausland leben?
2. Wer will die deutsche Brotkultur zum Weltkulturerbe erklären?
3. Womit stellt man Sauerteig her?
4. Welche Erfindung war sehr wichtig für das deutsche Brot?
5. Was schenkt man jemandem, der in eine neue Wohnung zieht?
6. Welche Geste signalisiert Gastfreundschaft?
7. Was braucht man um Mischbrote aus Roggen- und Weizenmehl zu backen?
8. Nennen Sie drei weitere Brotsorten, die in Deutschland beliebt sind.

Was Deutschen und Österreichern am meisten fehlt, wenn sie im Ausland wohnen, ist ihr Brot. Vor allem in Deutschland, aber auch in Österreich, gibt es sehr viele verschiedene Brotsorten. Das ist so etwas Besonderes, dass die Vertretung[2] der Bäcker in Deutschland diese Brotkultur zum Weltkulturerbe[3] erklären will.

Die meisten Sorten in Deutschland sind Mischbrote aus Roggen- und Weizenmehl[4], die mit Sauerteig[5] gebacken werden.

Das Brotsortiment einer deutschen Bäckerei
© Nordic Photos/Superstock

Es gibt für viele Regionen verschiedene Spezialitäten, z. B. Heidebrot, Holzofenbrot, Vollkornbrot oder Schwarzwälder.

Sauerteig ist in Deutschland sehr wichtig zum Brotbacken. Man stellt ihn aus Hefe her, die man auch in anderen Ländern zum Backen verwendet.

Schon vor vielen tausend Jahren hat man Getreide gemahlen und zu Brei[6] verarbeitet. Dann hat man etwas Brei oder Teig auf einem heißen Stein flach ausgebreitet und gebacken. So konnte man das Nahrungsmittel leichter aufheben und transportieren. Aber erst die Erfindung des Backofens hat Brot, so wie wir es heute kennen, möglich gemacht. Im Backofen kann man ein rundes Brot backen, weil der Teig von allen Seiten Hitze[7] bekommt.

Aber Brot ist nicht nur ein Grundnahrungsmittel, es hat auch eine symbolische Bedeutung in nationalen Bräuchen und religiösen Traditionen. So schenkt man einem Hochzeitspaar oder auch jemandem, der in eine neue Wohnung einzieht, Brot und Salz. Das soll Glück und Wohlstand bringen. In der christlichen Tradition ist das Brot sehr wichtig, weil Jesus beim letzten Abendmahl[8] sein Brot mit den Jüngern[9] geteilt hat. Auch in anderen Religionen, wie z. B. der jüdischen, ist das Teilen des Brotes eine wichtige Geste, die Gastfreundschaft signalisiert.

Heute gibt es in Deutschland und Österreich natürlich nicht nur die traditionellen Brotsorten. Baguettes und Croissants aus Frankreich, italienisches Weißbrot und türkisches Fladenbrot sind beliebt und überall zu haben.

[2]here: *representatives* [3]*world cultural heritage* [4]Roggen- ...: *rye and wheat flour* [5]*sourdough* [6]*mush* [7]*heat* [8]letzten ...: *Last Supper* [9]*disciples*

Situation 4 Interview: Die Mahlzeiten

1. Was isst du normalerweise zum Frühstück? Was zu Mittag?
2. Isst du viel zu Abend? Was?
3. Isst du immer eine Nachspeise? Was isst du am liebsten als Nachspeise?
4. Trinkst du viel Kaffee? Energydrinks? Warum (nicht)?
5. Isst du zwischen den Mahlzeiten? Warum (nicht)?
6. Was isst du, wenn du mitten in der Nacht großen Hunger hast?
7. Was trinkst du, wenn du auf Feste gehst?
8. Was hast du heute Morgen gegessen und getrunken?
9. Was isst du heute zu Mittag?
10. Was isst du heute zu Abend?

Haushaltsgeräte

Grammatik 8.3

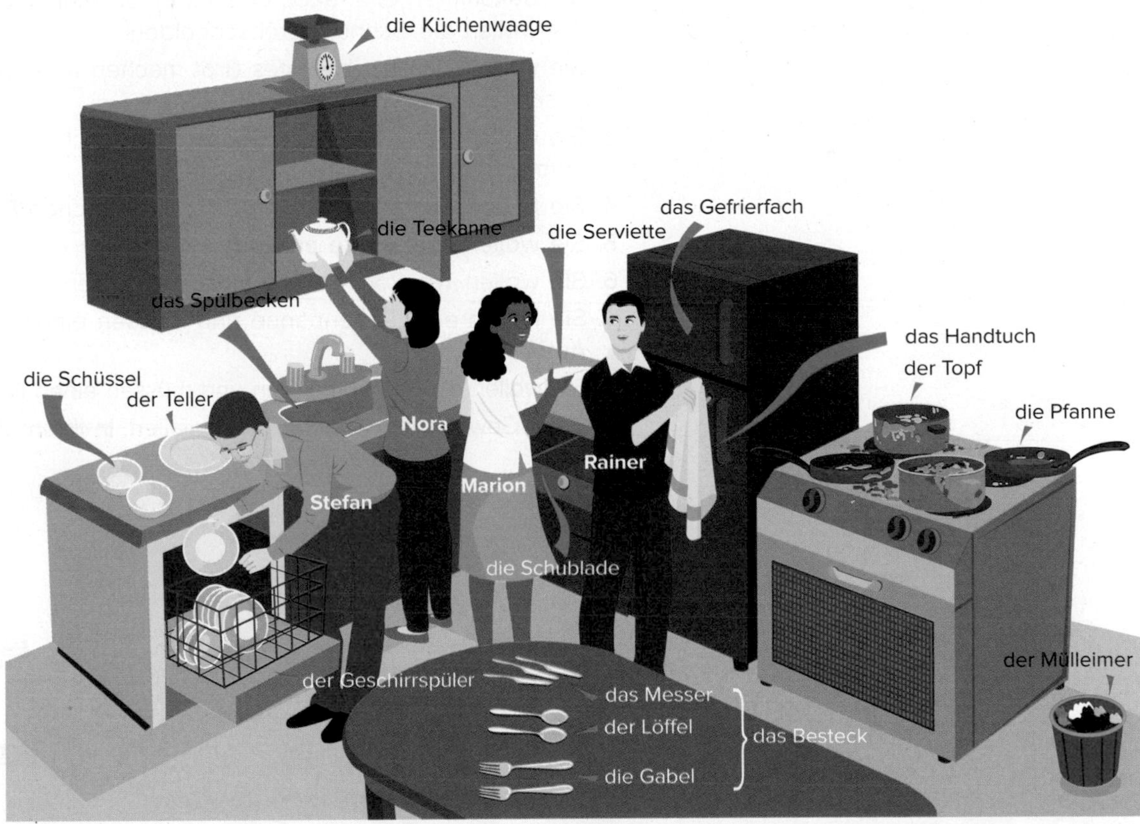

Stefan stellt die Schüsseln und Teller in den Geschirrspüler.
Nora stellt die Teekanne in den Schrank.
Marion legt die Servietten in die Schublade.
Rainer hängt das Handtuch an den Haken.
Die schmutzigen Töpfe und Pfannen stehen auf dem Herd.
Messer, Gabeln und Löffel liegen auf dem Tisch.

Situation 5 Was kosten diese Gegenstände?

Listen Sie die Gegenstände in jeder Gruppe dem Preis nach. Beginnen Sie mit dem teuersten Gegenstand. Wählen Sie dann aus jeder Gruppe die vier Gegenstände aus, auf die Sie am wenigsten verzichten[1] könnten.

GRUPPE A	GRUPPE B
1. eine Kaffeemaschine	1. eine Mikrowelle
2. ein elektrischer Dosenöffner	2. ein Kühlschrank
3. eine Küchenmaschine	3. ein Geschirrspüler
4. ein Korkenzieher	4. eine Waschmaschine
5. eine Kaffeemühle[2]	5. ein Wäschetrockner
6. ein Bügeleisen	6. ein Grill
7. eine Küchenwaage	7. ein Staubsauger
8. ein Toaster	8. eine Gefriertruhe[3]

Situation 6 Was brauchen Sie dazu?

1. Sie bekommen ein Paket, das mit einer Schnur zugebunden ist. Sie wollen die Schnur durchschneiden[4].
2. Sie wollen sich ein belegtes Brot machen und eine Scheibe Wurst abschneiden.
3. Sie wollen sich eine Dose Suppe heiß machen und müssen die Dose aufmachen.
4. Sie haben Gäste und wollen ein paar Flaschen Bier aufmachen.
5. Sie wollen eine Kerze anzünden.
6. Sie wollen Tee kochen und müssen Wasser heiß machen.
7. Sie haben eine Reifenpanne und müssen einen rostigen Nagel aus einem Autoreifen ziehen.
8. Sie wollen ein Bild aufhängen und müssen einen Nagel in die Wand schlagen.
9. Beim Gewitter ist der Strom ausgefallen. In Ihrem Zimmer ist es total dunkel.

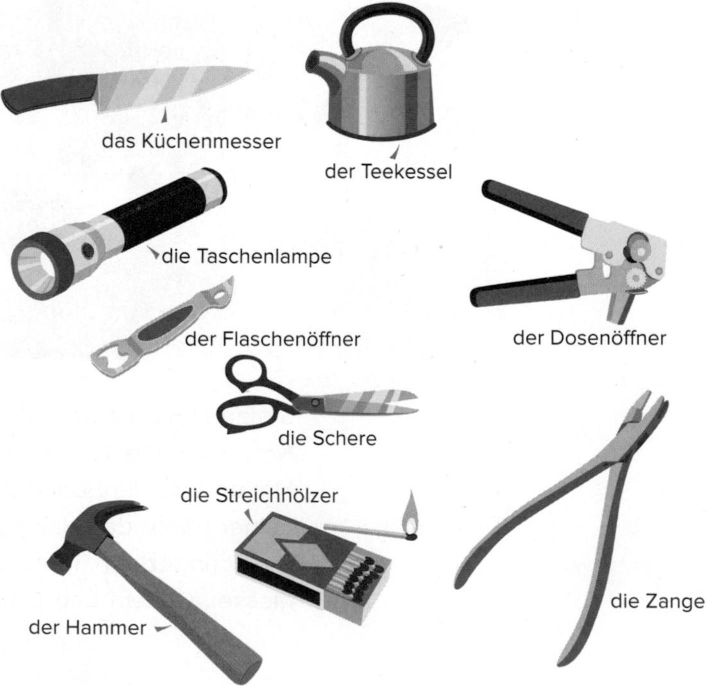

[1]do without [2]coffee mill [3]chest freezer [4]cut through

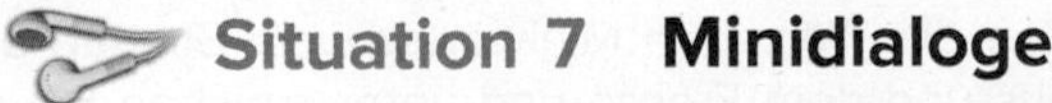

Situation 7 Minidialoge

Hören Sie zu und ergänzen Sie die fehlenden Wörter.

1. JÜRGEN: Wohin soll ich die Blumen stellen, Silvia?
   SILVIA: Stell sie doch bitte auf ________ Tisch.
2. MELANIE: Warum setzt du dich nicht zu uns an ________ Tisch?
   JOSEF: Ich sitze lieber auf ________ Sofa.
3. DAMLA: Eske, deine Bücher liegen schon wieder auf ________ Tisch. Könntest du sie bitte ________ Regal stellen?
   ESKE: Ja, klar, Damla, reg dich nicht so künstlich auf.
4. FRAU FRISCH-OKONKWO: Sumita, wo ist der Papa?
   SUMITA: Der sitzt draußen ________ Garten, auf ________ Lieblingsbank.
5. MONIKA: Hast du die Weinflaschen in ________ Schrank gestellt?
   HEIDI: Ja, sie stehen neben ________ Weingläsern.
6. SOFIE (*am Telefon*): Hi Nesrin! Was machst du denn heute Nachmittag?
   NESRIN: Nichts, Sofie. Ich bin so müde. Ich lege mich gleich ________ Bett.
7. KATRIN: Hi Stefan! Kann ich mich neben ________ setzen?
   STEFAN: Ja, klar, gerne. Hier ist noch frei.
8. FRAU RUF: Hast du die Suppe schon auf ________ Herd gestellt, Schatz?
   HERR RUF: Na klar! Sie steht schon eine halbe Stunde auf ________ Herd.
9. HERR RUF: Schatz, ich kann den Stadtplan nicht finden.
   FRAU RUF: Ich glaube, er liegt unter ________ Zeitung.

Situation 8 Diskussion: Haushaltsgeräte

1. Welche elektrischen Haushaltsgeräte haben Sie, Ihre Eltern oder Freunde? Welches Gerät finden Sie am wichtigsten?
2. Stellen Sie sich vor, Sie dürfen nur ein Gerät im Hause haben. Welches wählen Sie und warum?
3. Welche Werkzeuge sollte es in jedem Haushalt geben?
4. Sie wollen übers Wochenende zum Zelten. Machen Sie eine Liste, welche Geräte Sie zum Essen und Kochen brauchen.
5. Sie planen ein elegantes Picknick. Was packen Sie alles ein?

Lektüre

Vor dem Lesen 1

LESEHILFE

A fable is a very short story that usually deals with animals. The animals have human characteristics and represent types or characters. At the end of the fable is a moral or message, often containing a criticism of society or of human weaknesses such as lust, greed, vanity, or stupidity.

A. Kennen Sie Fabeln? Welche Tiere kommen in den Fabeln vor, die Sie kennen? Was symbolisieren sie?

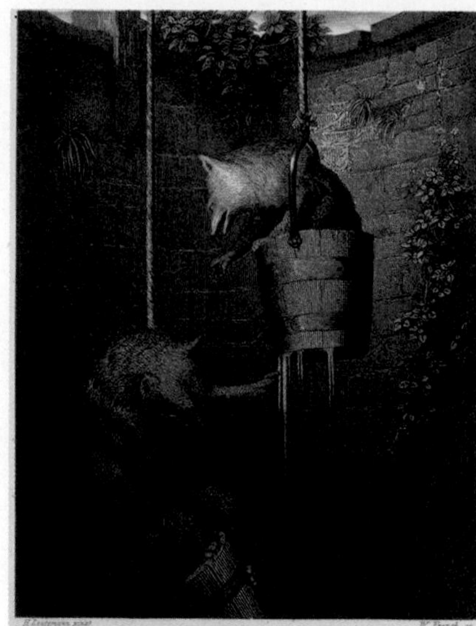

Der Wolf und der Fuchs im Brunnen.

B. Lesen Sie die Wörter im Miniwörterbuch. Suchen Sie sie in der Fabel von der Maus und dem Frosch und unterstreichen Sie sie.

Stichwort° Fabel

°Keyword

Die Idee der Fabel, die die Menschen belehrt, ist schon sehr alt. Von dem griechischen Dichter Äsop, der ungefähr 600 Jahre vor Christus gelebt hat, wird berichtet[5], dass er den einfachen Leuten Fabeln erzählt hat. Äsop soll ein Sklave[6] gewesen sein.

In Europa werden die Fabeln der Antiken Welt[7] in den Klosterschulen[8] des Mittelalters gelesen und dann im 17. und 18. Jahrhundert vor allem in Frankreich und Deutschland wiederentdeckt. Berühmte Fabeln aus der Epoche der Aufklärung[9] gibt es z. B. von Jean de La Fontaine (1621–1695) und Gotthold Ephraim Lessing (1729–1781). Aber auch der deutsche Theologe und Reformator Martin Luther (1483–1546) hat Fabeln geschrieben.

Von der Maus und dem Frosch

Eine Maus wollte gerne über ein Wasser kommen und konnte doch nicht, da bat sie einen Frosch um[10] Rat. Der Frosch war gemein und sagte: »Binde deinen Fuß an meinen, so will ich schwimmen und dich hinüberziehen.«

Als sie aber auf das Wasser kamen, tauchte der Frosch hinunter und wollte die Maus ertränken. Aber die Maus wehrte sich und kämpfte[11]. Da kam eine Weihe[12] und fing[13] die Maus, zog[14] den Frosch auch mit heraus und fraß[15] sie alle beide.

LUTHERS LEHRE: Sei vorsichtig, mit wem du dich einlässt[16]. Die Welt ist falsch und man kann niemandem trauen.

(Adaptiert nach Martin Luther)

Miniwörterbuch

der **Frosch**	frog
gemein	mean
ertränken	to drown
die **Lehre**	moral
vorsichtig	cautious
trauen	to trust

Arbeit mit dem Text 1

Beantworten Sie die folgenden Fragen.

1. Was kann die Maus nicht?
2. Was macht der Frosch, um die Maus zu töten?
3. Wer frisst am Ende beide auf?
4. Um welche schlechten Eigenschaften geht es in dieser Fabel?

Vor dem Lesen 2

LESEHILFE

The following short text is a more modern fable by the German author of fantasy and science fiction Paul Scheerbart (1853–1915). The small insects which are depicted in this text symbolize characteristics that are usually considered mostly positive. Ants live in highly organized societies working collectively together to support the colony. Most are workers or "soldiers", some are drones, and very few are queens. The parallels to human society such as division of labor, the ability to communicate, and solving problems make them almost perfect protagonists for fables.

A. Beantworten Sie die folgenden Fragen.

1. Welche Charakteristiken verbinden Sie mit Ameisen?
2. Gibt es unterschiedliche Charakteristiken für Ameisen in unterschiedlichen Kulturen?

B. Suchen Sie die Wörter des Miniwörterbuchs im Text und in den darauf folgenden Aktivitäten und unterstreichen Sie sie.

[5]*reported* [6]*slave* [7]Antiken ...: *Antiquity* [8]*monastery schools* [9]*Enlightenment* [10]bat um: *asked for*
[11]wehrte ...: *defended itself and fought* [12]*kite* (type of bird) [13]*caught* [14]*pulled* [15]*ate* [16]mit ...: *who you deal with*

Miniwörterbuch	
die **Ameise, -n**	ant
herrschen	to reign, dominate
sonderbar	peculiar
die **Sitte, -n**	custom
feierlich	ceremoniously
der **Stamm, ¨-e**	tribe
der **Geist**	spirit, mind
übergehen	to transfer
die **Ehre**	honor

Die gebratene Ameise

von Paul Scheerbart

Bei den fleißigen Ameisen herrscht eine sonderbare Sitte: Die Ameise, die in acht Tagen am meisten gearbeitet hat, wird am neunten Tage feierlich gebraten und von den Ameisen ihres Stammes gemeinschaftlich verspeist[17].

Die Ameisen glauben, dass durch dieses Gericht der Arbeitsgeist der Fleißigsten auf die Essenden[18] übergehe.

Und es ist für eine Ameise eine ganz außerordentliche Ehre, feierlich am neunten Tage gebraten und verspeist zu werden. Aber trotzdem ist es einmal vorgekommen[19], dass eine der fleißigsten Ameisen kurz vor dem Gebratenwerden noch folgende kleine Rede hielt[20]:

»Meine lieben Brüder und Schwestern! Es ist mir ja ungemein[21] angenehm, dass ihr mich so ehren wollt! Ich muss euch aber gestehen[22], dass es mir noch angenehmer sein würde[23], wenn ich nicht die Fleißigste gewesen wäre[24]. Man lebt doch nicht bloß, um sich totzuschuften[25]!«

»Wozu denn?[26]«, schrien die Ameisen ihres Stammes – und sie schmissen[27] die große Rednerin schnell in die Bratpfanne – sonst hätte[28] dieses dumme Tier noch mehr geredet.

Paul Scheerbart, „Die gebratene Ameise", 1902

Arbeit mit dem Text 2

A. Die folgenden Sätze fassen den Inhalt der Fabel zusammen. Vervollständigen Sie sie.

Nach _______[a] Tagen wählen die Ameisen die fleißigste Ameise ihres Stammes aus und braten sie. Dann essen sie sie am neunten Tag auf, damit sie auch so _______[b] werden. Einmal ist es passiert, dass so eine sehr fleißige _______[c] eine kleine Rede hielt, weil sie nicht verspeist werden wollte. Die anderen Ameisen schmissen sie trotzdem in die _______[d].

B. Welche typische Eigenschaft der Ameisen dominiert in dieser Fabel? Kreuzen Sie an und unterstreichen Sie im Text alle Wörter, die damit zu tun haben.

- ☐ Gehorsam
- ☐ Disziplin
- ☐ Fleiß
- ☐ Kooperation

C. In Fabeln geht es oft um Kritik an der Gesellschaft oder an den schlechten Eigenschaften der Menschen. Die große Ehre für die fleißigste Ameise ist es zu sterben. Was will der Autor mit dieser Fabel sagen?

Nach dem Lesen

Schreiben Sie Ihre eigene Fabel. Wählen Sie 1 oder 2.

1. Nehmen Sie einen ähnlichen Titel, z. B., „Der gebratene Hamster" oder „Der gebratene Tiger", und schreiben Sie die Geschichte noch einmal. Sie können sie auch ein bisschen verändern.
2. Schreiben Sie eine andere Fabel auf, die Sie kennen oder die Sie erfinden möchten.

[17]gemeinschaftlich ...: *collectively eaten* [18]*eaters* [19]*occurred* [20]Rede ...: *gave a speech* [21]*exceedingly* [22]*confess* [23]*would* [24]gewesen ...: *had been* [25]sich ...: *to work oneself to death* [26]Wozu ...: *Why else?* [27]*flung* [28]*would have*

Einkaufen und Kochen

Grammatik 8.4

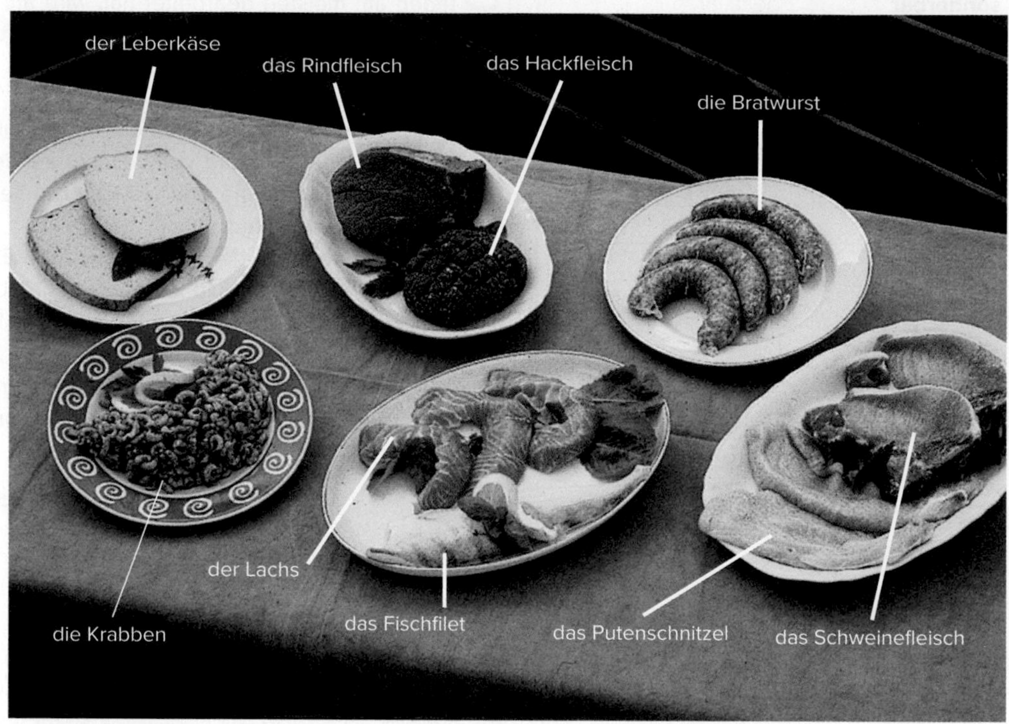

das Fleisch und der Fisch

© Dirk E. Hasenpusch

das Gemüse

© Dirk E. Hasenpusch

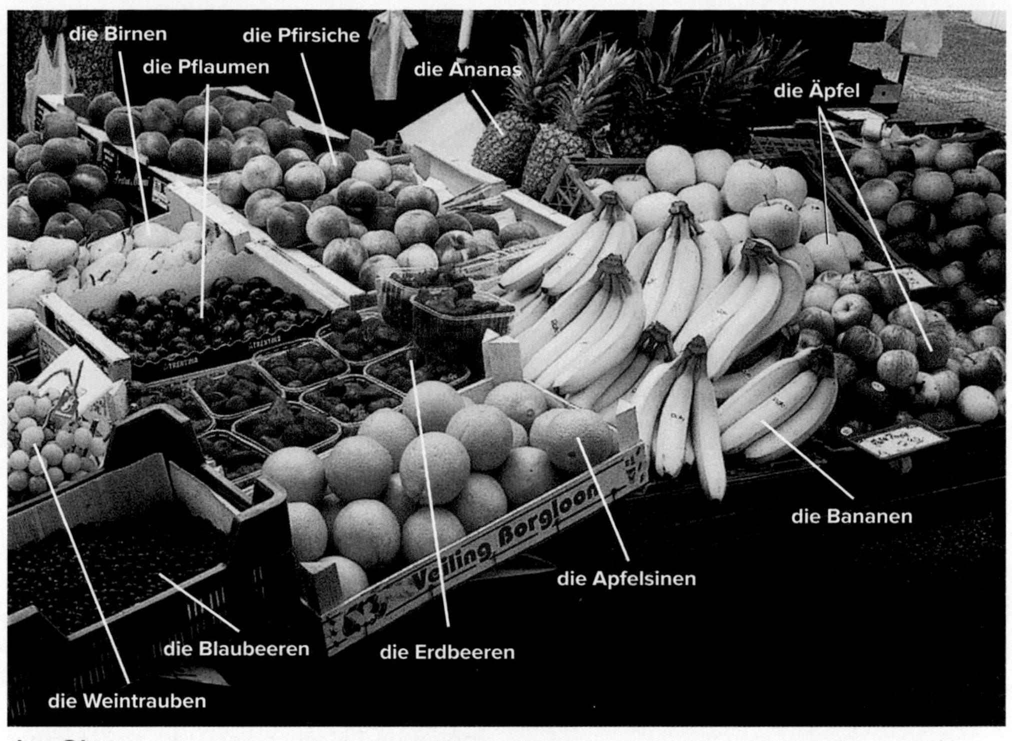

das Obst

Situation 9 Bildgeschichte: Michaels bestes Gericht

Michael kocht heute wieder sein bestes Gericht: Omelett *à la haute cuisine* ...

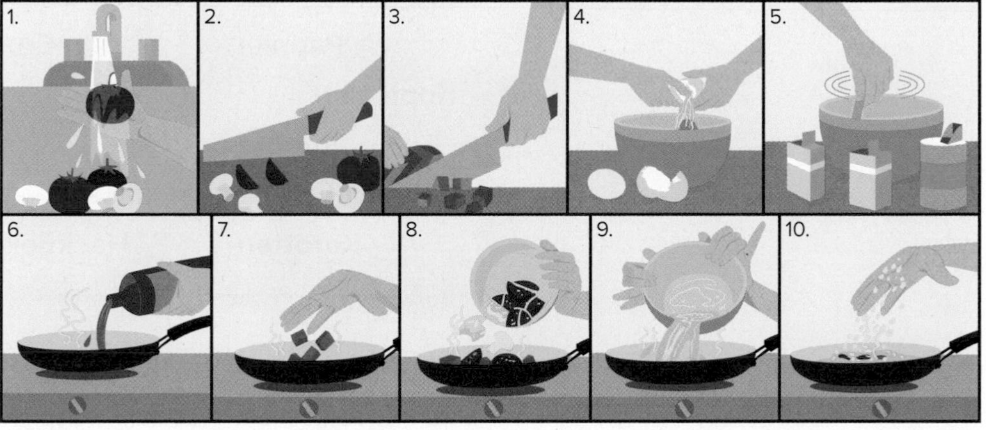

Deutsche Bioeier: Bioprodukte werden immer beliebter.
© Jochen Tack/imageBROKER/agefotostock

Situation 10 Einkaufsliste

Sie wollen heute Abend kochen. Was wollen Sie kochen? Was brauchen Sie? (Sie finden Ideen im Wortkasten.) Machen Sie für jedes Gericht eine Einkaufsliste. Denken Sie auch an Salat, Gemüse und Gewürze, an Vorspeise und Nachspeise und an Getränke.

1. ein italienisches Gericht
2. ein amerikanisches Gericht
3. ein türkisches Gericht
4. ein deutsches Gericht
5. ein französisches Gericht

Fisch **Nudeln** **Salz** **Bohnen**
Paprika **Erbsen** **Oliven**
Kopfsalat **Zwiebeln** **Pfeffer** **Knoblauch**
Tomaten **Gurken** **Pilze**
Schnitzel **Schafskäse** **Tomatensoße** **Kartoffeln**
Karotten **Hackfleisch** **Essig und Öl**

Situation 11 Interview: Einkaufen und Kochen

1. Kannst du kochen? Was zum Beispiel?
2. Kochst du oft? Wer kocht in deiner Familie?
3. Was kochst du am liebsten? Welche Zutaten braucht man dazu?
4. Kaufst du jeden Tag ein? Wenn nicht, wie oft in der Woche? An welchen Tagen? Wo kaufst du meistens ein?

Filmlektüre

Bella Martha

Vor dem Lesen

Szene aus dem Film *Bella Martha*
© AF archive/Alamy

FILMANGABEN

Titel: *Bella Martha*
Genre: Komödie/Drama
Erscheinungsjahr: 2001
Land: Deutschland
Dauer: 109 min
Regisseurin: Sandra Nettelbeck
Hauptrollen: Martina Gedeck, Maxime Foerste, Sergio Castellitto

A. Sehen Sie sich das Foto an und beantworten Sie die folgenden Fragen.

1. Wo sind die Personen?
2. Was sind sie von Beruf? Was machen sie? Wie ist Ihre Arbeit?
3. Wer, glauben Sie, ist das kleine Mädchen?

B. Lesen Sie die Wörter im Miniwörterbuch. Suchen Sie sie in der Inhaltsangabe und unterstreichen Sie sie.

Miniwörterbuch

alleinstehende	single
die **Leidenschaft**	passion
im Sinn haben	to have in mind
ertragen	to tolerate
verunglücken	to have an accident
aufnehmen	to take in
die **Umstellung**	adjustment
dickköpfige	headstrong
zurechtkommen	to get along
einstellen	to hire
ernst nehmen	to take seriously
argwöhnisch	suspicious
sich verstehen mit	to get along with
schaffen	to manage
das **Verhältnis**	relationship
letztendlich	in the end
kündigen	to quit, to resign

Inhaltsangabe

Martha Klein (Martina Gedeck) ist eine alleinstehende Frau, die nur eine Leidenschaft hat: Kochen. Als Chefköchin eines vornehmen Restaurants hat sie nichts anderes im Sinn und für niemanden Zeit. Sie ist Perfektionistin und kann es nicht ertragen, nicht die Beste zu sein. Als eines Tages ihre Schwester bei einem Autounfall tödlich verunglückt, nimmt Martha ihre achtjährige Nichte Lina (Maxime Foerste) bei sich auf. Über den Vater von Lina weiß sie nur, dass er Italiener ist und Giuseppe heißt. Um Lina zu helfen, verspricht Martha herauszufinden, wo ihr Vater lebt. Martha versucht für Lina da zu sein, doch die Umstellung ist für beide sehr schwer. Die dickköpfige Lina vermisst ihre Mutter sehr, mag die Schule nicht und isst nur wenig.

Indem Martha versucht mit dieser neuen Situation zurechtzukommen, stellt das Restaurant den charmanten Mario (Sergio Castellitto) als Koch ein. Mario, ein bisschen exzentrisch und ganz anders als Martha, hat ein besonderes Rezept für sein Leben: Er nimmt das Leben nicht so ernst. Mario bewundert die Kochkünste von Martha, doch Martha ist sehr argwöhnisch und mag Mario zuerst nicht. Mario aber versteht sich gut mit Lina, und schafft es, dass sie wieder anfängt zu essen. Dadurch wird das Verhältnis zwischen Mario und Martha langsam besser und sie verlieben sich.

Letztendlich findet Martha Linas Vater, der ein LKW-Fahrer aus Italien ist, und Lina zieht mit ihrem Vater nach Italien. Auch Martha wartet nicht mehr lange. Sie kündigt im Restaurant, fährt nach Italien, eröffnet dort ein Restaurant und heiratet Mario.

Arbeit mit dem Text

Beantworten Sie die folgenden Fragen.

1. Beschreiben Sie Martha. Wie ist sie? Was macht sie von Beruf? Was macht sie gern?
2. Wie stirbt ihre Schwester?
3. Wer ist der Vater von ihrer Nichte? Woher kommt er? Was ist er von Beruf?
4. Wie ist die Beziehung zwischen Martha und Mario am Anfang?
5. Wie endet der Film?

FILMCLIP

NOTE: For copyright reasons, the films referenced in the **FILMCLIP** feature have not been provided by the publisher. The film can be purchased as a DVD or found online at various sites such as YouTube, Amazon, or the iTunes store. The time codes mentioned below are for the North American DVD version of the film.

Szene: DVD, Szene 2, „Tragedy", 13:55–18:15 Min.

Martha ist im Restaurant und sehr beschäftigt. Das Restaurant ist voll und es ist hektisch. Plötzlich klingelt das Telefon.

Schauen Sie sich die Szene an und beantworten Sie die folgenden Fragen.

1. Wer, glaubt Martha, ist am Telefon?
   a. Mario b. ihre Schwester c. die Polizei
2. Wer ist wahrscheinlich am Telefon?
   a. Mario b. ihre Schwester c. die Polizei
3. Wie reagiert Martha auf die Nachricht?
   a. froh b. gleichgültig c. geschockt
4. Martha bricht zusammen und weint. Wo ist sie wahrscheinlich?
   a. zu Hause b. im Restaurant c. im Krankenhaus
5. Als Martha nach Hause kommt, hört sie eine Nachricht von ihrer Schwester auf dem Anrufbeantworter an. Ihre Schwester sagt, ...
   a. dass sie gleich da ist. b. dass sie erst später kommt. c. dass sie noch 100 km fahren muss.

Am nächsten Tag spricht Martha mit einem Arzt im Krankenhaus. Richtig oder falsch? Verbessern Sie die falschen Aussagen.

_____ 6. Der Arzt sagt: „Lina weiß, dass ihre Mutter tot ist."

_____ 7. Der Arzt will wissen, wo Linas Vater ist.

Nach dem Lesen

Was ist Ihre Leidenschaft im Leben? Was machen Sie gern?

Im Restaurant

Grammatik 8.5

a. —Ist hier noch frei?
—Ja, bitte schön.

b. —Was darf ich Ihnen bringen?
—Kann ich bitte die Speisekarte haben?
—Ja, gern, einen Moment, bitte.

c. —Ein Wasser, bitte.
—Ein Mineralwasser. Kommt sofort!

d. —Wir würden gern zahlen.
—Gern. Das waren zwei Wiener Schnitzel, ein Glas Wein und eine Limo ...

e. —38,80 Franken, bitte schön.
—Das stimmt so.
—Vielen Dank.
—Können Sie mir dafür eine Quittung geben?
—Selbstverständlich.

f. —Darf ich Sie noch zu einem Kaffee einladen?
—Das ist nett, aber leider muss ich mich jetzt beeilen.

Situation 12 **Was sagen Sie?**

Wählen Sie für jede Situation eine passende Aussage.

1. Sie sitzen an einem Tisch im Restaurant. Sie haben Hunger, aber noch keine Speisekarte. Sie sehen die Kellnerin und sagen: ______
2. Sie haben mit Ihren Freunden im Restaurant gegessen. Sie haben es eilig und möchten zahlen. Sie rufen den Kellner und sagen: ______
3. Ihr Essen und Trinken hat 19 Euro 20 gekostet. Sie haben der Kellnerin einen Zwanzigeuroschein gegeben. 80 Cent sind Trinkgeld. Sie sagen: ______
4. Sie essen mit Ihren Eltern in einem feinen Restaurant. Da stellen Sie fest, dass eine Fliege in der Suppe schwimmt. Sie rufen den Kellner und sagen: ______
5. Sie haben einen Sauerbraten mit Knödeln bestellt. Die Kellnerin bringt Ihnen einen Schweinebraten. Sie sagen: ______

Nein, danke.
Zahlen, bitte.
Das stimmt so.
Ja, bitte sehr.
Das kann nicht stimmen.
Ich habe doch einen Sauerbraten bestellt.
Die Speisekarte, bitte.
Morgen fliege ich in die USA.
Leider habe ich kein Geld.
Ich liebe Schweinebraten.
Herr Kellner, bitte, sehen Sie sich das mal an.

Situation 13 **Dialog: Melanie und Josef gehen aus.**

Melanie und Josef haben sich einen Tisch ausgesucht und sich hingesetzt. Der Kellner kommt an ihren Tisch.

KELLNER: Bitte schön?
MELANIE: Können wir die ________ haben?
KELLNER: Natürlich. Möchten Sie etwas trinken?
MELANIE: Für mich ein ________ bitte.
JOSEF: Und ________ ein Bier.
KELLNER: Gern.
[*etwas später*]
KELLNER: __________, was Sie essen möchten?
MELANIE: Ich möchte das Rumpsteak mit Pilzen und Kroketten.
JOSEF: Und ich hätte gern die Forelle „blau" mit Kräuterbutter, grünem Salat und Salzkartoffeln. Dazu __________ bitte.
KELLNER: Gern. Darf ich _______ auch noch etwas zu trinken bringen?
MELANIE: Nein, danke, im Moment nicht.

Situation 14 **Rollenspiel: Im Restaurant**

S1: Sie sind im Restaurant und möchten etwas zu essen und zu trinken bestellen. Wenn Sie mit dem Essen fertig sind, bezahlen Sie und geben Sie der Bedienung ein Trinkgeld.

KULTUR ... LANDESKUNDE ... INFORMATIONEN

ÖSTERREICH

- Was wissen Sie über Österreich? Wofür ist es bekannt?
- Wo liegt es? Was ist die Hauptstadt? Schauen Sie auf die Landkarte in ***Kontakte***.
- Welche berühmten Österreicherinnen und Österreicher kennen Sie?
- Kennen Sie eine Österreicherin oder einen Österreicher persönlich? Was erzählt er oder sie über sein Land?
- Waren Sie schon mal in Österreich? Erzählen Sie!

Lesen Sie den Text und suchen Sie die Antworten auf diese Fragen:

- Woher kommt der Ausdruck[1] „Felix Austria"?
- Wann wurde der Name Österreich zum ersten Mal erwähnt[2]?
- Was geschah 1804?
- Welche heutigen Länder umfasste die k. u. k. Monarchie?
- Wie lange wurde Österreich nach dem Zweiten Weltkrieg von den Alliierten verwaltet[3]?
- Wo wohnen die meisten Österreicher?
- Aus welchen Bundesländern besteht Österreich?
- Wofür ist das Reiseland Österreich vor allem bekannt?
- Welche deutschen Komponisten wirkten in Wien?
- Für welche kulinarischen Köstlichkeiten[4] ist Österreich bekannt?

Felix Austria! „Andere mögen[5] Kriege führen, du, glückliches Österreich, heirate!" Ein Motto, das ursprünglich auf die Heiratspolitik der Habsburger verwies[6], drückt das Lebensgefühl[7] eines Landes aus, das zum Großteil (ca. 60 Prozent) in den Alpen liegt und zum anderen Teil an der Donau.

Österreich wurde 976 als Ostarrîchi zum ersten Male urkundlich erwähnt. 1156 wurde es ein eigenes Herzogtum[8]. Die Habsburger, die über viele Jahrhunderte hinweg die deutschen Könige und Kaiser waren, erhoben[9] es 1278 zum Erzherzogtum. In der frühen Neuzeit (15.–17. Jahrhundert) musste sich Österreich vor allem gegen die Türken wehren[10], die zweimal Wien belagerten[11]. 1804 wurde das Kaiserreich Österreich gegründet und 1867 die kaiserlich-königliche (k. u. k.) Monarchie Österreich-Ungarn, die neben Österreich und Ungarn auch die heutigen Länder Tschechien, die Slowakei, Slowenien, Kroatien, Bosnien sowie Teile Italiens, Rumäniens, Polens und der Ukraine umfasste. Nach dem Ersten Weltkrieg wurde der Vielvölkerstaat Österreich-Ungarn zerschlagen[12] und Österreich wurde in seinen jetzigen Grenzen gegründet. Nach dem Zweiten Weltkrieg wurde Österreich zehn Jahre lang von den Alliierten verwaltet und erst 1955 entstand die jetzige Zweite Republik. Über die Zeit nach dem Zweiten Weltkrieg erzählt der britische Spielfilm *Der dritte Mann* mit Orson Welles in der Hauptrolle und mit dem immer noch bekannten Harry-Lime-Thema, das von dem Österreicher Anton Karas komponiert und auf der Zither gespielt wurde.

Kaffee, Kuchen und ein Glas Wasser
© Rainer Hackenberg/dpa/Corbis

Österreich ist ein relativ kleines Land, etwas größer als die Schweiz, und steht auf Platz 113 der Länder der Welt von seiner Fläche[13] her und auf Platz 92 von der Bevölkerungszahl. Von den 8,4 Millionen Einwohnern wohnen 2,4 Millionen in der Metropolregion Wien. Die Bundesländer Vorarlberg, Tirol, Salzburg, Kärnten und die Steiermark liegen in den Alpen und die Bundesländer Oberösterreich, Niederösterreich und Wien an der Donau. Das Burgenland liegt im Südosten an der Grenze zu Ungarn. Seit 1995 ist Österreich Mitglied der EU; seine Währung ist der Euro.

Die zentrale Lage in Europa, seine wunderschönen Berg- und Flußlandschaften und seine historischen Städte machen Österreich zu einem Reiseland *par excellence*. In Europa liegt Österreich auf Platz 2 der Länder, die durch den Tourismus besonders viel Geld verdienen. Die meisten Touristen kommen aus Deutschland. Im Winter kommen viele Leute zum Skifahren oder Snowboarden. Im Sommer gibt es viele Touristen, die wandern oder klettern. Im 18. und 19. Jahrhundert war Wien das Zentrum klassischer Musik. Die Wiener Klassik um Joseph Haydn und Wolfgang Amadeus Mozart zog den in Bonn geborenen Ludwig van Beethoven an, die Romantik um Franz Schubert und Anton Bruckner den in Hamburg geborenen Johannes Brahms. Sowohl Beethoven als auch Brahms lebten bis zu ihrem Lebensende in Wien.

Kulinarisch ist Österreich für vieles bekannt. Das Wiener Schnitzel, die Sachertorte, der Apfelstrudel, der Kaiserschmarren und die Palatschinke sind nur einige wenige Beispiele für österreichische Köstlichkeiten. Eine der schönsten Erfindungen[14] sind jedoch die Wiener Kaffeehäuser, in denen man stundenlang vor einer Kaffeeköstlichkeit sitzen und Zeitung lesen oder sich unterhalten kann.

[1]*expression* [2]*mentioned* [3]*governed* [4]*delicacies* [5]*may* [6]*referred* [7]*attitude to life* [8]*duchy* [9]*elevated* [10]*defend* [11]*laid siege to* [12]*broken apart* [13]*area* [14]*inventions*

Situation 15 Bildgeschichte: Abendessen mit Hindernissen

Situation 16 Interview

1. Gehst du oft essen? Wie oft in der Woche isst du nicht zu Hause? Wirst du heute Abend zu Hause essen?
2. Isst du oft im Studentenheim? Wirst du morgen im Studentenheim essen? Schmeckt dir das Essen da?
3. Gehst du oft in Fast-Food-Restaurants? Wirst du vielleicht noch diese Woche in so einem Restaurant essen?
4. Warst du schon mal in einem deutschen Restaurant? Wenn ja, was hast du gegessen? Wenn nein, was wirst du bestellen, wenn du mal in einem deutschen Restaurant bist?
5. In welchem Restaurant schmeckt es dir am besten? Gibt es ein Restaurant, in dem du oft isst? Wie heißt es? Was isst du da? Wirst du diese Woche noch einmal hingehen?
6. Was ist das feinste Restaurant in unserer Stadt? Wie viel muss man da für ein gutes Essen bezahlen?

Videoecke

Perspektiven

Was ist für dich *gesundes Essen*?

Lebensmittel vom Biomarkt

Aufgabe 1 Fleisch oder Gemüse?

Was ist für diese Leute *gesundes Essen*?

1. Sandra ___

2. Simone ___

3. Hend ___

4. Martin ___

5. Tina ___

6. Nadezda ___

7. Felicitas ___

8. Pascal ___

a. das Essen genießen
b. Joghurt und Vollkornbrot
c. keine Pommes
d. Lebensmittel vom Biomarkt
e. nicht zu viel Fleisch
f. Obst und Gemüse
g. ökologisches Essen
h. viel Obst und möglichst wenig Schokolade
i. wenig Fett und wenig Zucker
j. viel selber anbauen

Interviews

- Was isst du zum Frühstück?
- Was isst du zum Mittag?
- Wie oft gehst du essen?
- Wo isst du am liebsten?
- Was isst du da?
- Was isst du nicht gerne?
- Was kannst du besonders gut kochen?
- Wie macht man das?

Tanja

Susan

Aufgabe 2 Tanja oder Susan?

Sehen Sie sich das Video an und kreuzen Sie an.

	Tanja	Susan
1. Wer isst zum Frühstück etwas Süßes vom Bäcker?	☐	☐
2. Wer isst zum Frühstück einen Joghurt?	☐	☐
3. Wer isst zum Mittag normalerweise einen Salat oder eine Suppe?	☐	☐
4. Wer geht jeden Tag in der Mensa essen?	☐	☐
5. Wer isst am liebsten in der Mensa?	☐	☐
6. Wer isst am liebsten in Auerbachs Keller?	☐	☐
7. Wer isst gern Kassler mit Sauerkraut und Klößen?	☐	☐
8. Wer mag kein Rindfleisch?	☐	☐
9. Wer mag keinen Rosenkohl und keinen Spinat[15]?	☐	☐

Aufgabe 3 Nudeln mit Shrimps

Susan macht Nudeln mit Shrimps. Bringen Sie die Sätze in die richtige Reihenfolge.

_____ Am Ende kommt noch Sahne dazu.
_____ Dann gebe ich die Shrimps hinein.
_____ Dann würze ich es mit Chili, Salz und Pfeffer.
_____ Ganz zum Schluss kommt oben drauf noch Parmesan-Käse.
_____ Zuerst brate ich die Zwiebeln an.

Aufgabe 4 Interview

Interviewen Sie eine Partnerin oder einen Partner. Stellen Sie dieselben Fragen.

[15] *spinach*

Wortschatz

Frühstück	**Breakfast**
die **Wurst, -̈e**	sausage
der **Käse**	cheese
der **Schinken**	ham
der **Speck**	bacon
das **Brötchen, -**	roll
das **Ei, -er**	egg
gebratene Eier	fried eggs
gekochte Eier	boiled eggs
das **Würstchen, -**	frank(furter); hot dog

Ähnliche Wörter

die **Marmelade, -n;** der **Honig;** das **Omelett, -s**

Mittagessen und Abendessen	**Lunch and Dinner**
die **Forelle, -n**	trout
die **Krabbe, -n**	shrimp
die **Mahlzeit, -en**	meal
die **Nachspeise, -n**	dessert
die **Vorspeise, -n**	appetizer
der **Braten, -**	roast
der **Knödel, -**	dumpling
der **Pilz, -e**	mushroom
das **Brot, -e**	bread
das **belegte Brot,** die **belegten Brote**	open-face sandwich
das **Fleisch**	meat
das **Hackfleisch**	ground beef (or pork)
das **Rindfleisch**	beef
das **Schweinefleisch**	pork
die **Pommes (frites)** [frit] or [frits] (*pl.*)	French fries

Ähnliche Wörter

die **Krokette, -n;** die **Muschel, -n;** die **Nudel, -n;** der **Fisch, -e;** der **Reis;** das **Rumpsteak, -s;** das **Schnitzel, -**

Obst und Nüsse	**Fruit and Nuts**
die **Apfelsine, -n**	orange
die **Birne, -n**	pear
die **Erdbeere, -n**	strawberry
die **Weintraube, -n**	grape
die **Zitrone, -n**	lemon
der **Pfirsich, -e**	peach

Ähnliche Wörter

die **Banane, -n;** die **Nuss, -̈e;** die **Pflaume, -n**

Gemüse	**Vegetables**
die **Bohne, -n**	bean
die **Erbse, -n**	pea
die **Gurke, -n**	cucumber
saure Gurken	pickles
die **Kartoffel, -n**	potato
die **Salzkartoffeln**	boiled potatoes
die **Zwiebel, -n**	onion
der **Kohl**	cabbage
der **Blumenkohl**	cauliflower
der **Rosenkohl**	Brussels sprouts

Ähnliche Wörter

die **Karotte, -n;** die **Olive, -n;** die **Tomate, -n;** der **Salat, -e** (R); der **Kopfsalat**

Getränke	**Beverages**
der **Saft, -̈e**	juice
der **Apfelsaft**	apple juice
der **Orangensaft**	orange juice

Ähnliche Wörter

die **Milch;** der **Kakao** [kakau]; das **Mineralwasser**

Zutaten	**Ingredients**
der **Essig**	vinegar
der **Knoblauch**	garlic
der **Senf**	mustard
das **Gewürz, -e**	spice; seasoning
die **Kräuter** (*pl.*)	herbs

Ähnliche Wörter

die **Butter;** die **Kräuterbutter;** die **Soße, -n;** der **Pfeffer;** der **Zucker;** das **Öl** (R); das **Salz**

Küche und Zubereitung	**Cooking and Preparation**
bestreuen	to sprinkle
braten, brät, gebraten (R)	to fry
bräunen	to brown, fry
erhitzen	to heat
geben, gibt, gegeben (in + *akk.***)**	to put (into)
gießen, gegossen	to pour
schlagen, schlägt, geschlagen	to beat
würzen	to season

Im Restaurant	At the Restaurant
die **Bedienung**	service; waiter, waitress
die **Fliege, -n**	fly
die **Geschäftsführerin, -nen**	manager (female)
die **Kellnerin, -nen**	waitress
die **Quittung, -en**	receipt, check
die **Speisekarte, -n**	menu
die **Suppe, -n**	soup
der **Geschäftsführer, -**	manager (male)
der **Kellner, -**	waiter
der **Schein, -e**	bill, note (*of currency*)
der **Teller, -**	plate
das **Gericht, -e**	dish
das **Stück, -e**	slice; piece

Ähnliche Wörter

der **Schweizer Franken, -**; das **Eiscafé, -s**; das **Trinkgeld, -er**

Im Haushalt	In the Household
die **Dose, -n**	can
die **Gabel, -n**	fork
die **Küchenmaschine, -n**	mixer
die **Schere, -n**	scissors
die **Schnur, ¨-e**	string
die **Schüssel, -n**	bowl
die **Serviette, -n**	napkin
die **Zange, -n**	pliers, tongs
der **Dosenöffner, -**	can opener
der **Haken, -**	hook
der **Löffel, -**	spoon
der **Mülleimer, -**	garbage can
der **Nagel, ¨**	nail
der **Strom**	electricity, power
der **Wäschetrockner, -**	clothes dryer
das **Gerät, -e**	appliance
das **Handtuch, ¨-er**	hand towel
das **Messer, -**	knife
das **Paket, -e**	package
das **Streichholz, ¨-er**	match
das **Werkzeug, -e**	tool

Ähnliche Wörter

die **Teekanne, -n**; der **Flaschenöffner, -**; der **Hammer, ¨**; der **Korkenzieher, -**; der **Teekessel, -**

Sonstige Substantive	Other Nouns
die **Ehre, -n**	honor
die **Lehre, -n**	moral, teaching
der **Geist**	spirit, mind

Sonstige Verben	Other Verbs
ab·nehmen, nimmt ... ab, abgenommen	to lose weight
ab·schneiden, abgeschnitten	to cut off
auf·hängen (R)	to hang up
aus·fallen, fällt ... aus, ist ausgefallen	to go out (*power*)
sich beeilen	to hurry
berechnen (+ *dat.*)	to charge (*someone*)
sich beschweren (bei)	to complain (to)
bestellen	to order (*food*)
fest·stellen	to establish
herrschen	to reign, dominate
stimmen	to be right
das stimmt so	that's right; keep the change
trauen	to trust
über·gehen, ist übergegangen	to transfer, pass across
ziehen, gezogen	to pull
zu·bereiten, zubereitet	to prepare (*food*)

Adjektive und Adverbien	Adjectives and Adverbs
fettig	fat; greasy
frei	free, empty, available
ist hier noch frei?	is this seat available?
gebraten	roasted; broiled; fried
gemein	mean
gesund	healthy
vorsichtig	cautious(ly)
zart	tender
zugebunden	tied shut

Ähnliche Wörter

elegant, elektrisch, fein, frisch, gegrillt, gekocht, gemischt, gesalzen, holländisch, japanisch, mexikanisch, russisch, sauer

Sonstige Wörter und Ausdrücke	Other Words and Expressions
am wenigsten	the least
danach	afterward
dazu	in addition
eilig	rushed
es eilig haben	to be in a hurry
meistens	usually, mostly
normalerweise	normally
der **Schluss, ¨-e**	end
zum Schluss	in the end, finally
selbst gemacht	homemade
selten	rare(ly), seldom

Strukturen und Übungen

8.1 Adjectives: an overview

Attributive adjectives precede nouns and have endings. Predicate adjectives follow the verb **sein** and have no endings.

A. Attributive and Predicate Adjectives

Adjectives that precede nouns are called *attributive adjectives* and have endings similar to the forms of the definite article: **kalter, kaltes, kalte, kalten, kaltem.** Adjectives that follow the verb **sein** and a few other verbs are called *predicate adjectives* and do not have any endings.

VERKÄUFER:	**Heiße** Würstchen! Ich verkaufe **heiße** Würstchen!	VENDOR:	*Hot dogs! I'm selling hot dogs!*
KUNDE:	Verzeihung, sind die Würstchen auch wirklich **heiß?**	CUSTOMER:	*Excuse me, are the hot dogs really hot?*
VERKÄUFER:	Natürlich, was denken Sie denn?!	VENDOR:	*Of course, what do you think?!*

B. Attributive Adjectives with and without Preceding Article

If *no* article or article-like word (**mein, dein, dieser,** or the like) precedes the adjective, then the adjective itself has the ending of the definite article **(der, das, die).** This means that the adjective provides the information about the gender, number, and case of the noun that follows.

Ich esse gern gegrillt**en** Fisch. *I like to eat grilled fish.*	**den** Fisch = masculine accusative
Stefan isst gern frisch**es** Müsli. *Stefan likes to eat fresh cereal.*	**das** Müsli = neuter accusative

If an article or article-like word precedes the adjective but does not have an ending, the adjective—again—has the ending of the definite article. **Ein**-words (the indefinite article **ein,** the negative article **kein,** and the possessive determiners **mein, dein,** etc.) do *not* have an ending in the masculine nominative and in the neuter nominative and accusative. In these instances, as you might expect, the adjective again gives the information about the gender, number, and case of the noun that follows.

Ein groß**er** Topf steht auf dem Herd. *There is a large pot on the stove.*	**der** Topf = masculine nominative
Ich esse ein frisch**es** Brötchen. *I am eating a fresh roll.*	**das** Brötchen = neuter accusative

If an article or article-like word with an ending precedes the adjective, the adjective ends in either **-e** or **-en.** (See Sections 8.2 and 8.4.)

Ich nehme das holländisch**e** Bier.	*I'll take the Dutch beer.*
Ich nehme die deutsch**en** Äpfel.	*I'll take the German apples.*

8.2 Attributive adjectives in the nominative and accusative cases

Rules of thumb:

1. In many instances, the adjective ending is the same as the ending of the definite article.
2. *But:* after **der** (nominative masculine) and **das,** the adjective ending is **-e.***
3. *But:* after **die** (plural), the adjective ending is **-en.**

As described in Section 8.1, adjective endings vary according to the gender, number, and case of the noun they describe and according to whether this information is already indicated by an article or article-like word. In essence, however, there are only a very limited number of possibilities. Study the following chart carefully and try to come up with some easy rules of thumb that will help you remember the adjective endings.

	Masculine	Neuter	Feminine	Plural
Nominative	der kalt**e** Tee	das kalt**e** Bier	die kalt**e** Milch	die kalt**en** Getränke
	ein kalt**er** Tee	ein kalt**es** Bier	eine kalt**e** Milch	
	kalt**er** Tee	kalt**es** Bier	kalt**e** Milch	kalt**e** Getränke
Accusative	den kalt**en** Tee	das kalt**e** Bier	die kalt**e** Milch	die kalt**en** Getränke
	einen kalt**en** Tee	ein kalt**es** Bier	eine kalt**e** Milch	
	kalt**en** Tee	kalt**es** Bier	kalt**e** Milch	kalt**e** Getränke

Nouns that come from adjectives take the same endings as those adjectives.

deutsch	*German* (adjective)
der Deutsch**e**	*the German (man)*
die Deutsch**e**	*the German (woman)*
Deutsch**e**	*Germans*
die Deutsch**en**	*the Germans*
Ich kenne einen Deutsch**en.**	*I know a German (man).*
Kennst du diese Deutsch**e?**	*Do you know this German (woman)?*

Übung 1 Spezialitäten!

Jedes Land hat eine Spezialität: ein Gericht oder ein Getränk, das aus diesem Land einfach am besten schmeckt. An welche Länder denken Sie bei den folgenden Gerichten oder Getränken?

amerikanisch **dänisch** **englisch** **russisch** **deutsch** **französisch** **griechisch** **norwegisch** **italienisch** **holländisch** **japanisch** **polnisch** **kolumbianisch** **neuseeländisch** **ungarisch**

MODELL: Salami → Italienische Salami!

1. Steak (*n.*)
2. Kaviar (*m.*)
3. Oliven (*pl.*)
4. Sushi (*n.*)
5. Champagner (*m.*)
6. Wurst (*f.*)
7. Käse (*m.*)
8. Spaghetti (*pl.*)
9. Paprika (*m.*)
10. Marmelade (*f.*)
11. Kaffee (*m.*)
12. Kiwis (*pl.*)

*Remember this rule as "**der** (nominative masculine)" because, as you will learn in Section 8.4, when **der** refers to the dative feminine, the adjective ending will be **-en**.

Übung 2 Der Gourmet

Michael isst und trinkt nicht alles, sondern nur, was er für fein hält. Übernehmen Sie Michaels Rolle.

MODELL: Kognak (*m.*) / französisch →
Ich trinke nur französischen Kognak!

1. Brot (*n.*) / deutsch
2. Kaviar (*m.*) / russisch
3. Salami (*f.*) / italienisch
4. Kaffee (*m.*) / kolumbianisch
5. Kiwis (*pl.*) / neuseeländisch
6. Wein (*m.*) / französisch
7. Bier (*n.*) / belgisch
8. Muscheln (*pl.*) / spanisch
9. Marmelade (*f.*) / englisch
10. Thunfisch (*m.*) / japanisch

Übung 3 Im Geschäft

Michael hat kein Geld, aber er möchte alles kaufen. Maria muss ihn immer bremsen.

MODELL: der schicke Anzug / teuer →
MICHAEL: Ich möchte den schicken Anzug da.
MARIA: Nein, der schicke Anzug ist viel zu teuer.

1. der graue Wintermantel / schwer
2. die gelbe Hose / bunt
3. das schicke Hemd / teuer
4. die roten Socken / warm
5. der schwarze Schlafanzug / dünn
6. die grünen Schuhe / groß
7. der modische Hut / klein
8. die schwarzen Winterstiefel / leicht
9. die elegante Sonnenbrille / bunt
10. die roten Tennisschuhe / grell

Übung 4 Minidialoge

Ergänzen Sie die Adjektivendungen.

1. HERR RUF: Na, wie ist denn Ihr neu_____[a] Auto?
FRAU WAGNER: Ach, der alt_____[b] Mercedes war mir lieber.
HERR RUF: Dann hätte ich mir aber keinen neu_____[c] Wagen gekauft!
2. KELLNER: Wie schmeckt Ihnen denn der italienisch_____[a] Wein?
MICHAEL: Sehr gut. Ich bestelle gleich noch eine weiter_____[b] Flasche.
3. MICHAEL: Heute repariere ich mein kaputt_____[a] Fahrrad.
MARIA: Prima! Dann kannst du meinen blöd_____[b] Computer auch reparieren. Er ist schon wieder kaputt.
MICHAEL: Na gut, aber dann habe ich wieder kein frei_____[c] Wochenende.

8.3 Destination vs. location: *stellen/stehen, legen/liegen, setzen/sitzen, hängen/hängen*

WISSEN SIE NOCH?

Prepositions of location are usually followed by the dative case, while prepositions of destination are usually followed by the accusative case.

Review grammar 6.2.

Destination implies the accusative case; location implies the dative case.

DESTINATION	LOCATION
Verbs of action and direction used with two-way prepositions followed by the accusative	Verbs of condition and location used with two-way prepositions followed by the dative

Maria stellt eine Flasche Wein **auf den** Tisch.

Die Flasche Wein steht **auf dem** Tisch.

stellen/stehen = vertical position

Stellen and **stehen** designate vertical placement or position. They are used with people and animals, as well as with objects that have a base and can "stand" without falling over.

DESTINATION

Michael legt eine Flasche Wein **ins** Weinregal.

LOCATION

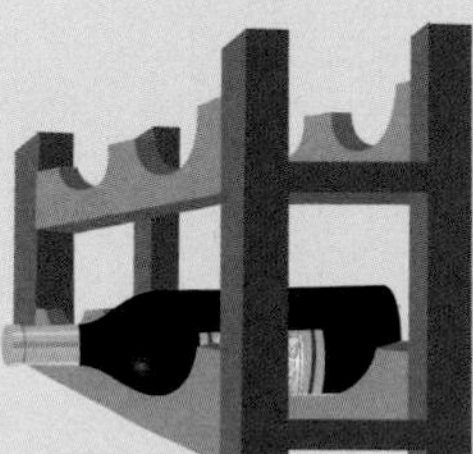

Die Flasche Wein liegt **im** Weinregal.

legen/liegen = horizontal position

Legen and **liegen** designate horizontal placement or position. They are used with people and animals, as well as with objects.

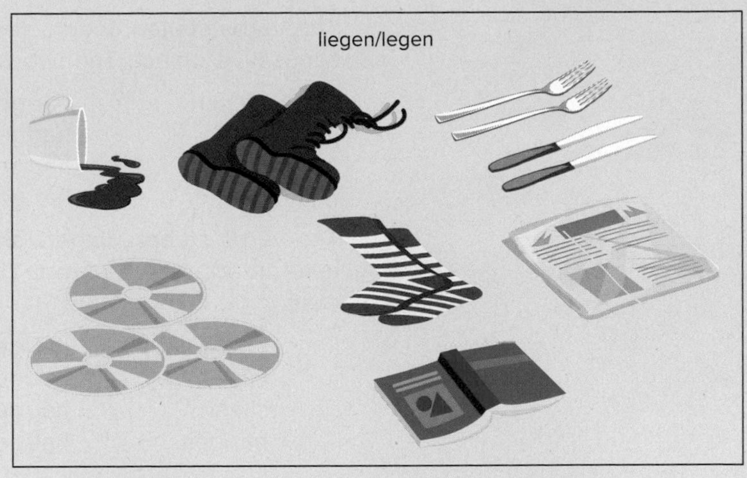

DESTINATION

Frau Wagner setzt Paula **in den** Hochstuhl.

LOCATION

Paula sitzt **im** Hochstuhl.

sitzen/setzen = sitting position (people and certain animals)

Setzen designates the act of being seated; **sitzen** the state of sitting. These verbs are used only with people and with animals that are capable of sitting.

DESTINATION

Eske hängt das Handtuch **an den** Haken.

LOCATION

Das Handtuch hängt **am** Haken.

hängen/hängen = hanging position

Hängen (gehängt) designates the act of being hung; **hängen (gehangen)** the state of hanging.

The verbs **stellen, legen, setzen,** and **hängen** are weak verbs that require an accusative object. The two-way preposition is used with the accusative case.

stellen	hat gestellt
legen	hat gelegt
setzen	hat gesetzt
hängen	hat gehängt

The verbs **stehen, liegen, sitzen,** and **hängen** are strong verbs that cannot take an accusative object. The two-way preposition is used with the dative case.

stehen	hat gestanden
liegen	hat gelegen
sitzen	hat gesessen
hängen	hat gehangen

Übung 5 Vor dem Abendessen

Beschreiben Sie die Bilder.

NÜTZLICHE WÖRTER:

legen/liegen
setzen/sitzen
stehen/stellen
der Küchenschrank
der Schrank
die Schublade
die Serviette
das Sofa
der Teller
der Tisch

MODELLE: Die Schuhe → Die Schuhe liegen auf dem Boden.

Peter → Peter stellt die Schuhe vor die Tür.

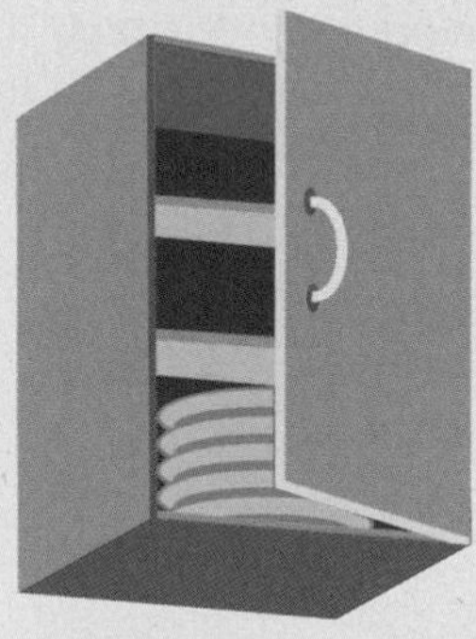

1. Die Teller ______.

2. Albert ______.

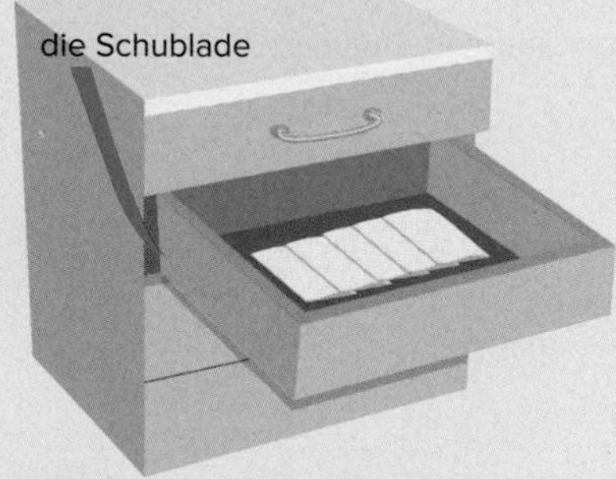

3. Die Servietten ______.

4. Monika ______.

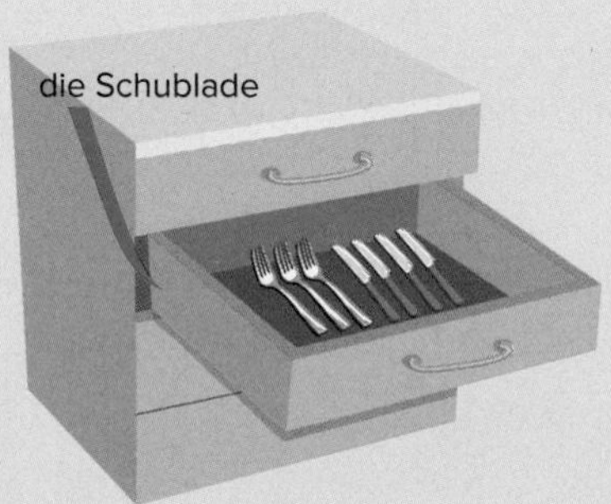

5. Messer und Gabeln ______.

6. Stefan ______.

7. Die Kerze ______.

8. Heidi ______.

9. Thomas ______.

8.4 Adjectives in the dative case

In the dative case, nouns are usually preceded by an article **(dem, der, den; einem, einer)** or an article-like word **(diesem, dieser, diesen; meinem, meiner, meinen).** When adjectives occur before such nouns they end in **-en.***

Jutta geht mit ihr**em** neu**en** Freund spazieren.	*Jutta is going for a walk with her new friend.*
Jens gießt sein**er** krank**en** Tante die Blumen.	*Jens is watering the flowers for his sick aunt.*
Ich spreche nicht mehr mit dies**en** unhöflich**en** Menschen.	*I'm not talking with these impolite people any more.*

	Masculine	Neuter	Feminine	Plural
Dative	dies**em** lieb**en** Vater	dies**em** lieb**en** Kind	dies**er** lieb**en** Mutter	dies**en** lieb**en** Eltern
	mein**em** lieb**en** Vater	mein**em** lieb**en** Kind	mein**er** lieb**en** Mutter	mein**en** lieb**en** Eltern

Übung 6 Was machen diese Leute?

Schreiben Sie Sätze.

MODELL: Jens / seine alte Tante / einen Brief schreiben →
Jens schreibt sein**er** alt**en** Tante einen Brief.

1. Jutta / ihr neuer Freund / ihre Lieblings-DVD leihen
2. Jens / der kleine Bruder von Jutta / eine Ratte verkaufen
3. Hans / nur seine besten Freunde / die Ratte zeigen
4. Jutta / ihre beste Freundin / ein Buch schenken
5. Jens / sein wütender Lehrer / eine Krawatte kaufen
6. Ernst / seine große Schwester / einen Witz erzählen
7. Jutta / die netten Leute von nebenan / Kaffee kochen
8. Ernst / das süße Baby von nebenan / einen Kuss geben

ACHTUNG!

All nouns have an **-n** in the dative plural unless their plural ends in **-s.**

Nominative: die Freunde

Dative: den Freunde**n**, *but:* den Hobbys

*Unpreceded adjectives in the dative case follow the same pattern as in the nominative and accusative case, that is, they have the ending of the definite article. For example, **mit frischem Honig** (*with fresh honey*), **mit kalter Milch** (*with cold milk*).

8.5 Talking about the future: the present and future tenses

future tense = **werden** + infinitive

You already know that **werden** is the equivalent of English *to become.*

Ich möchte Ärztin werden.	*I'd like to become a physician.*

You can also use a form of **werden** plus infinitive to talk about future events.

Wo **wirst** du morgen sein?	*Where will you be tomorrow?*
Morgen **werde** ich wahrscheinlich zu Hause sein.	*Tomorrow I will probably be at home.*

When an adverb of time is present or when it is otherwise clear that future actions or events are indicated, German speakers normally use the present tense rather than the future tense to talk about what will happen in the future.

Nächstes Jahr **fahren** wir nach Schweden.	*Next year we're going to Sweden.*
Was **machst** du, wenn du in Schweden bist?	*What are you going to do when you're in Sweden?*

Use **wohl** with the future tense to express present or future probability.

The future tense with **werden** can express present or future probability. In such cases, the sentence often includes an adverb such as **wohl** (*probably*).

Mein Freund wird jetzt **wohl** zu Hause sein.	*My friend should be home now.*
Morgen Abend werden wir **wohl** zu Hause bleiben.	*Tomorrow evening we'll probably stay home.*

Don't forget to put **werden** at the end of the dependent clause.

Ich weiß nicht, ob ich einmal heiraten **werde.**	*I don't know if I'm ever going to get married.*

Übung 7 Vorsätze

Sie wollen ein neues Leben beginnen? Schreiben Sie sechs Dinge auf, die Sie ab morgen machen werden oder nicht mehr machen werden.

MODELL: Ich werde nicht mehr so oft in Fast-Food-Restaurants gehen. Ich werde mehr Obst und Gemüse essen.

weniger/mehr Kurse belegen
weniger/mehr fernsehen
weniger oft / öfter selbst kochen
weniger/mehr arbeiten
weniger oft / öfter ins Kino gehen
weniger/mehr lernen
früher/später ins Bett gehen
weniger/mehr SMS schicken
weniger gesund / gesünder essen

Übung 8 Morgen ist Samstag

Was machen Frau Schulz und ihre Studenten morgen?

MODELL: Katrin geht morgen ins Kino.

Katrin

1. Frau Schulz

2. Heidi

3. Peter

4. Monika

5. Stefan

6. Nora

7. Albert

8. Thomas

Übung 9 Vorhersagen

Machen Sie sechs Vorhersagen, die in diesem oder im nächsten Jahr eintreffen werden.

MODELL: Dieses Jahr werden die Steelers den Superbowl gewinnen.
Nächstes Jahr werden wir einen neuen Präsidenten wählen.

mit dem Studium fertig werden
einen tollen Job bekommen
die Studiengebühren fallen/steigen
der Papst nach Indien fliegen
in eine andere Wohnung ziehen
gute Noten bekommen
weniger Steuern bezahlen
die Wimbledon-Spiele gewinnen

KAPITEL 9

Kindheit und Jugend

Kapitel 9 deals with memories and past events. You will have the opportunity to talk about your childhood and you will learn more about the tales that are an important part of childhood in the German-speaking world.

Themen

Kindheit

Jugend

Geschichten

Märchen

Kulturelles

KLI: Gebrüder Grimm

Musikszene: „Wir beide" (Juli)

KLI: 1989

Filmclip: *Nordwand* (Philipp Stölzl)

Videoecke: Schule

Lektüren

Film: *Nordwand* (Philipp Stölzl)

Märchen: *Rotkäppchen – Ein Märchen der Gebrüder Grimm*

Strukturen

9.1 The conjunction **als** with dependent-clause word order

9.2 The simple past tense of **werden,** the modal verbs, and **wissen**

9.3 Time: **als, wenn, wann**

9.4 The simple past tense of strong and weak verbs (receptive)

9.5 Sequence of events in past narration: the past perfect tense and the conjunction **nachdem** (receptive)

Walter Firle: *Märchen* (o. J.), Privatbesitz

KUNST UND KÜNSTLER

Walter Firle (1859–1929) ist ein deutscher Künstler aus Breslau (heute Polen), der als Porträt- und Genremaler bekannt ist. Er zog nach München, um zu studieren und später zu arbeiten. Seine Porträts von bayrischen Herrschern[1] und anderen Persönlichkeiten wie zum Beispiel Paul von Hindenburg sind auf alten bayrischen Briefmarken zu sehen.

Schauen Sie sich das Bild an und beantworten Sie folgende Fragen.

1. Welche Personen sehen Sie auf dem Bild?
2. Wo sind die Personen?
3. Was tragen die Kinder?
4. Welche Haarfarbe und welche Frisuren haben die Mädchen?
5. Was machen die Mädchen?
6. Welche Farben dominieren?
7. Welche Stimmung ruft das Bild hervor?

[1] *rulers*

Situationen

Kindheit

Grammatik 9.1

Jens hat seinem Onkel den Rasen gemäht.

Uli hat im Garten Äpfel gepflückt.

Richard hat mit seiner Mutter Kuchen gebacken.

Günter hat Staub gesaugt und sauber gemacht.

Willi hat seiner Oma die Blumen gegossen.

Jochen hat seinem kleinen Bruder Geschichten vorgelesen.

Situation 1 Melanies erstes Haustier

Als Melanie sechs Jahre alt war, hat sie einen Hund zum Geburtstag bekommen. Sie hat ihn Bruno genannt. Was hat sie wohl am nächsten Tag mit ihm gemacht? Ordnen Sie die Aktivitäten den Zeiten zu.

MODELL: Um sechs Uhr ist sie gemeinsam mit Bruno aufgestanden.

6.00 Uhr **6.30 Uhr** **7.00 Uhr** **10.00 Uhr** **10.15 Uhr** **14.30 Uhr** **12.00 Uhr** **15.00 Uhr** **16.00 Uhr** **19.30 Uhr**

1. Sie ist zusammen mit Bruno eingeschlafen.
2. Sie hat mit ihm gespielt.
3. Sie hat Brunos Korb[1] sauber gemacht.
4. Sie ist mit Bruno spazieren gegangen.
5. Sie ist gemeinsam mit Bruno aufgestanden.
6. Sie hat Bruno gefüttert.
7. Sie hat ihn ihren Freunden gezeigt.
8. Sie hat ihm eine Schleife[2] ins Haar gebunden.
9. Sie hat Bruno in der Badewanne gewaschen.
10. Sie hat ihm einen großen Knochen[3] gekauft.

Situation 2 Umfrage

MODELL: S1: Hast du als Kind Computerspiele gespielt?
S2: Ja.
S1: Unterschreib bitte hier.

	UNTERSCHRIFT
1. Computerspiele gespielt	______
2. viel ferngesehen	______
3. dich mit den Geschwistern gestritten	______
4. manchmal die Nachbarn geärgert	______
5. einen Hund oder eine Katze gehabt	______
6. in einer Baseballmannschaft gespielt	______
7. Ballettunterricht genommen	______
8. Fensterscheiben kaputt gemacht	______
9. viel Zeit mit Facebook verbracht	______

Situation 3 Interaktion: Als ich 12 Jahre alt war ...

Wie oft haben Sie das gemacht, als Sie 12 Jahre alt waren: **oft, manchmal, selten** oder **nie?**

1. mein Zimmer aufgeräumt
2. Kuchen gebacken
3. Liebesromane gelesen
4. Videos angeschaut
5. heimlich jemanden geliebt
6. spät aufgestanden
7. Freunde eingeladen
8. allein verreist
9. zu einem Fußballspiel gegangen

Situation 4 Interview

Als du acht Jahre alt warst ...

1. Wo hast du gewohnt? Hattest du Geschwister? Freunde? Wo hat dein Vater gearbeitet? deine Mutter? Was hast du am liebsten gegessen?
2. In welche Grundschule bist du gegangen? Wann hat die Schule angefangen? Wann hat sie aufgehört? Welchen Lehrer / Welche Lehrerin hattest du am liebsten? Welche Fächer hattest du am liebsten? Was hast du in den Pausen gespielt? Was hast du nach der Schule gemacht?
3. Hast du viel ferngesehen? Was hast du am liebsten gesehen? Hast du gern gelesen? Was? Hast du Sport getrieben? Was? Was hast du gar nicht gern gemacht?

[1]*basket* [2]*bow* [3]*bone*

KULTUR ... LANDESKUNDE ... INFORMATIONEN

GEBRÜDER GRIMM

- Was assoziieren Sie mit den Gebrüdern Grimm?
- Kennen Sie Märchen der Gebrüder Grimm? Welche?
- Haben Sie ein Lieblingsmärchen? Wie heißt es auf Deutsch?

Lesen Sie die Wörter im Miniwörterbuch. Suchen Sie sie in den Fragen und im Text und unterstreichen Sie sie.

Miniwörterbuch	
die **Sage, -n**	legend
außer	besides, in addition to
die **Wissenschaft, -en**	science, field of study

Lesen Sie den Text und suchen Sie die Antworten auf diese Fragen.

- Wann wurden die *Kinder- und Hausmärchen* der Gebrüder Grimm ins Englische übersetzt?
- Wie wurden Märchen und Sagen früher weitergegeben?
- Was haben die Gebrüder Grimm außer der Märchenkunde noch begründet?
- Wie heißt die erste Lautverschiebung auf Englisch?
- Wo studierten die Gebrüder Grimm?
- Wer waren die „Göttinger Sieben" und gegen was protestierten sie?
- Wohin gingen Jacob und Wilhelm Grimm nach ihrer Entlassung?
- Was sollte das „Deutsche Wörterbuch" dokumentieren?
- Wie viele Bände hat das Wörterbuch?
- Wie lange brauchte man, um es fertigzustellen?
- In welcher Stadt starben die Gebrüder Grimm?

Man kennt die Brüder Jacob (1785–1863) und Wilhelm Grimm (1786–1859) allgemein als Sammler und Herausgeber[4] der berühmten *Kinder- und Hausmärchen,* die 1812 auf Deutsch erschienen und bereits 1823 ins Englische übersetzt wurden. Sie sind in der ganzen westlichen Welt bekannt. Vor den Grimms wurden Märchen und Sagen vor allem mündlich erzählt. Die Brüder Grimm begründeten mit ihrem Werk die Märchenkunde[5] als Wissenschaft.

Die Grimms wurden in Hanau, in der Nähe von Frankfurt geboren. Sie sind bekannte Sprach- und Literaturwissenschaftler und werden als Begründer der modernen Germanistik angesehen[6]. Jacob Grimm zum Beispiel beschäftigte sich in seinem Werk *Die Deutsche Grammatik* mit den indogermanischen Sprachen, verglich Flexion[7], Wort- und Lautbildung[8] und den Laut- und Bedeutungswandel[9]. Er beschrieb Entwicklungsstufen[10] der Sprachen und die erste Lautverschiebung[11], die in der angelsächsischen Welt als „Grimm's Law" bekannt ist.

Eine Vorlesung bei Jacob Grimm
© FineArt/Alamy

Nach dem Jurastudium in Marburg lebten und arbeiteten die Grimms zunächst als Bibliothekare und Gelehrte in Kassel. Dann gingen sie nach Göttingen, wo Jacob und später auch Wilhelm Professoren an der Georg-August-Universität wurden.

Die Grimms betätigten sich aber auch politisch. Im Jahr 1837 beteiligten sich beide zusammen mit fünf anderen bekannten Professoren aus Göttingen am Protest gegen König Ernst August von Hannover. Dieser hatte ihrer Meinung nach eigenmächtig und unrechtmäßig die Verfassung aufgehoben[12]. „Die Göttinger Sieben" verloren ihre Professuren und mussten das Land verlassen, aber durch ihren Kampf für mehr Freiheit und Menschenrechte wurden sie in ganz Deutschland bekannt.

Jacob und Wilhelm Grimm gingen nach ihrer Entlassung aus dem Universitätsdienst zurück nach Kassel und begannen 1838 ihr Hauptwerk, das *Deutsche Wörterbuch*. Mit diesem Werk wollten die Grimms die Herkunft und den Gebrauch für jedes deutsche Wort dokumentieren. Das war in einer Zeit, in der Deutschland aus vielen Kleinstaaten bestand, auch von nationaler Bedeutung. Das *Deutsche Wörterbuch* war eine unglaublich große Aufgabe. Wilhelm arbeitete bis zu seinem Tod am Buchstaben D und Jacob Grimm schrieb die Beiträge für A, B, C und E. Er starb bei der Arbeit am Buchstaben F. Der erste Band[13] des Wörterbuches wurde 1854 veröffentlicht. Insgesamt hat das *Deutsche Wörterbuch* 32 Bände, und es dauerte 123 Jahre, bis es fertig war. Dann begann man sofort mit einer Neubearbeitung[14].

Die Brüder Grimm lebten von 1841 bis zu ihrem Tod in Berlin, lehrten anfangs dort auch an der Universität und waren Mitglieder der Akademie der Wissenschaften.

[4]*editors* [5]*study of fairy tales* [6]werden angesehen *are considered* [7]*inflection* [8]Wort- ... *morphology and phonetics* [9]*semantic change* [10]*developmental stages* [11]*sound shift* [12]eigenmächtig ... *arbitrarily and illegally suspended the constitution* [13]*volume* [14]*revised edition*

Jugend

Grammatik 9.2–9.3

1. Sybille Gretter war sehr begabt. In der Schule wusste sie immer alles.

2. Sie brauchte für die Prüfungen nicht viel zu lernen.

3. Sie konnte auch sehr gut tanzen und wollte Ballerina werden.

4. Dreimal in der Woche musste sie zum Ballettunterricht.

5. Als sie in der letzten Klasse war, hatte sie einen Freund.

6. Ihr Vater durfte nichts davon wissen, denn er war sehr streng.

7. Eines Tages hat sie ihren Freund ihren Eltern vorgestellt.

8. Aber ihr Vater mochte ihn nicht und sie mussten sich trennen.

Situation 5 Dialog: Jugendsünden

Michael Pusch geht zum zehnten Klassentreffen seiner Abiturklasse. Er trifft seinen alten Freund Alexander. Die beiden sprechen über ihre gemeinsame Schulzeit.

MICHAEL: Schön, dich mal wieder zu sehen, Alex. Was hast du eigentlich nach dem Abi ________?

ALEXANDER: Ich habe eine Tanzschule ________.

MICHAEL: Nicht schlecht. Gern und gut ________ hast du ja früher schon.

ALEXANDER: Stimmt. Erinnerst du dich an das Drama mit Frau Müller damals?

MICHAEL: Ach, als wir in ihrem Deutschunterricht laut Musik ________ und getanzt haben?

ALEXANDER: Genau. Sie war noch nicht in der Klasse, uns war langweilig und Hans hatte zufällig ein bisschen Musik dabei.

MICHAEL: Und als Frau Müller hereinkam, haben alle wild getanzt und ________. Das war ein Spaß.

ALEXANDER: Danach hat es nur leider viel Ärger mit dem Direktor ________.

MICHAEL: Richtig. Dabei hatten wir diese Sache noch nicht einmal ________.

ALEXANDER: Und als wir Herrn Riedel die Geschichtsklausuren[1] ________ oder das Auto der Französischlehrerin Frau Häuser mit Toilettenpapier ________ haben ...

MICHAEL: Es war eigentlich eine schöne Zeit auf dem Gymnasium.

ALEXANDER: Na ja. Denk doch nur an die vielen Klassenarbeiten.

Situation 6 Interview

1. Musstest du früh aufstehen, als du zur Schule gegangen bist? Wann?
2. Wann musstest du von zu Hause weggehen?
3. Musstest du zur Schule, wenn du krank warst?
4. Durftest du abends lange fernsehen oder im Internet surfen, wenn du morgens früh aufstehen musstest?
5. Konntest du zu Fuß zur Schule gehen?
6. Wolltest du manchmal lieber zu Hause bleiben? Warum?
7. Was wolltest du werden, als du ein Kind warst?
8. Durftest du abends ausgehen? Wann musstest du zu Hause sein?

Situation 7 Geständnisse

Sagen Sie, was in diesen Situationen passiert ist oder was Sie gemacht haben.

MODELL: Als ich zum ersten Mal allein verreist bin, habe ich meinen Teddy mitgenommen.

1. Als ich zum ersten Mal allein verreist bin
2. Als ich zum ersten Mal Kaffee getrunken hatte
3. Wenn ich zu spät nach Hause gekommen bin
4. Als ich mein erstes F bekommen hatte
5. Wenn ich keine Hausaufgaben gemacht habe
6. Wenn ich total verliebt war
7. Als ich zum ersten Mal verliebt war
8. Als ich einmal meinen Hausschlüssel verloren hatte
9. Wenn ich eine schlechte Note bekommen habe
10. Als ich einmal mit einem Jungen / einem Mädchen im Kino war

Situation 8 Rollenspiel: Das Klassentreffen

S1: Sie sind auf dem dritten Klassentreffen Ihrer alten High-School-Klasse. Sie unterhalten sich mit einem alten Schulfreund / einer alten Schulfreundin. Fragen Sie, was er/sie nach Abschluss der High School gemacht hat, was er/sie jetzt macht und was seine/ihre Pläne für die nächsten Jahre sind. Sprechen Sie auch über die gemeinsame Schulzeit.

[1] *history exams*

MUSIKSZENE

„Wir beide“ (2006, Deutschland) *Juli*

Biografie Juli ist in Deutschland eine der meist gespielten Bands im Radio. Sie nennen ihren Musikstil Alternativ-Pop und wollen vor allem interessante Texte schreiben. Die Band gibt es seit 2002. Die Leadsängerin Eva Briegel und die Gitarristen Simon Triebel und Jonas Pfetzing schreiben alle Lieder der Band und sie schreiben sie alle auf Deutsch. Ihre erste Single *Perfekte Welle* wurde eine Art Hymne der neuen deutschsprachigen Popmusik seit dem Jahr 2004. Die Single *„Wir beide“* stammt aus dem Jahr 2006. Sie ist eine Hymne an die Freundschaft und war 10 Wochen in den deutschen Charts.

Eva Briegel

NOTE: For copyright reasons, the songs referenced in **MUSIKSZENE** have not been provided by the publisher. The song can be found online at various sites such as YouTube, Amazon, or the iTunes store.

Vor dem Hören Haben Sie einen besten Freund / eine beste Freundin? Was macht ihn oder sie so besonders? Was erwarten Sie, dass Ihr bester Freund oder Ihre beste Freundin für Sie tut? Was würden Sie für Ihren besten Freund oder Ihre beste Freundin tun?

Nach dem Hören Was ist richtig? Korrigieren Sie die fett gedruckten Wörter.

MODELL: Du bist mir jederzeit ~~**untreu**~~.
loyal

1. Keine **findet** mich so gut wie du.
2. Immer werden wir **uns ändern.**
3. Wir beide sind jung und frei und **reich.**
4. Wir stehen **Tag für Tag** auf der guten Seite.
5. Wir **weinen** über schlechte Zeiten.
6. Deine **Probleme** sind auch meine.
7. Du schlägst dich mit meinen **Freunden.**
8. Ich vertraue **mir** mehr als **dir.**
9. Du vergisst niemals, was das **Schöne** ist.

Miniwörterbuch

die **Welle**	wave
eine Art	a kind of
untreu	unfaithful
(sich) ändern	to change (oneself)
der **Schmerz, -en**	pain
der **Feind, -e**	enemy
vertrauen	to trust

Geschichten

Grammatik 9.4

Als Willi mal allein zu Hause war ...

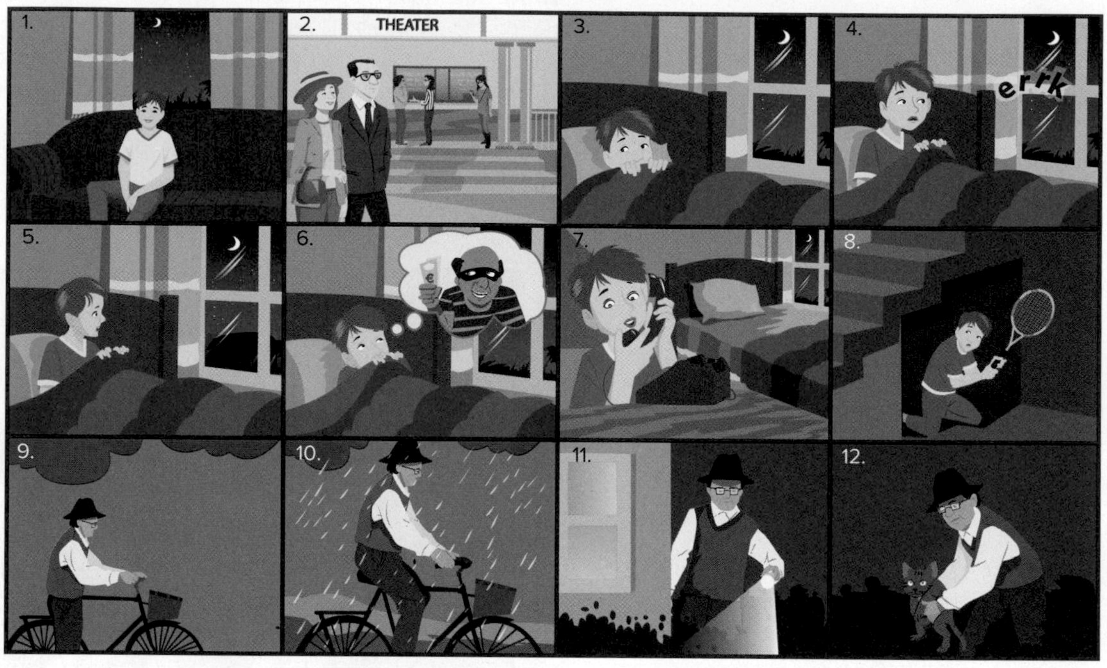

Situation 9 Informationsspiel: Was ist passiert?

MODELL: Was ist Sofie passiert? / Was ist dir passiert?
Wann ist es passiert?
Wo ist es passiert?
Warum ist es passiert?

	Sofie	Mehmet	Ernst	mein Partner / meine Partnerin
Was?		hat sein Flugzeug verpasst		
Wann?	als sie im Kino war		als er über den Zaun geklettert ist	
Wo?		in Frankfurt		
Warum?	weil ihre Jackentasche ein Loch hatte		weil der Zaun zu hoch war	

Situation 10 Und dann?

Suchen Sie für jede Situation eine logische Folge.

MODELL: Jutta konnte ihren Hausschlüssel nicht finden und kletterte durch das Fenster.

1. Ernst machte die Fensterscheibe kaputt
2. Jens reparierte sein Fahrrad
3. Richard sparte ein ganzes Jahr
4. Claire kam in Innsbruck an
5. Michael bekam ein neues Fahrrad
6. Rolf lernte sechs Jahre Englisch
7. Josef arbeitete drei Monate im Krankenhaus
8. Silvia wohnte zwei Semester allein
9. Melanie bekam ihren ersten Kuss

a. machte dann Urlaub in Spanien.
b. fuhr gleich gegen einen Baum.
c. kaufte sich ein Motorrad.
d. kaufte sich einen neuen Pulli.
e. lief weg.
f. machte eine Radtour.
g. flog dann nach Amerika.
h. sagte leise: „Oh, mein Gott!"
i. zog dann in eine WG.
j. ?

Situation 11 Bildgeschichte: Beim Zirkus

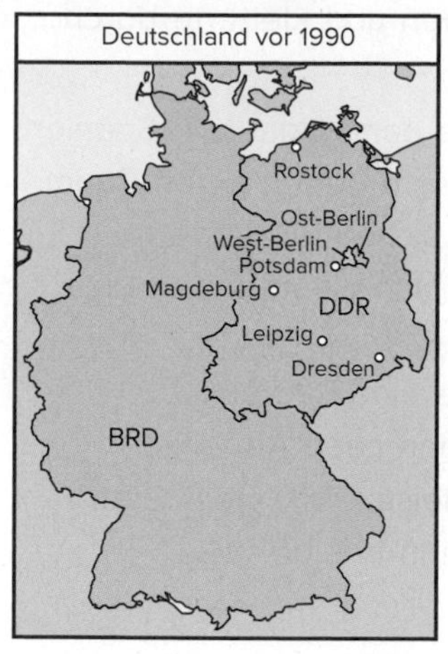

Die Mauer ist gefallen.
© Lutz Schmidt/AP Photo

1989

- Was geschah 1989?
- Wo liegt Leipzig? Wie alt ist Leipzig? Wie viele Einwohner hat Leipzig?
- Was wissen Sie sonst noch über Leipzig?

Lesen Sie den Text und suchen Sie die Antworten auf die folgenden Fragen:

1. Wie nennt man in Deutschland das Jahr 1989?
2. Wie lange war Deutschland geteilt[1]?
3. Wie hieß der östliche Teil Deutschlands? Wann wurde er gegründet[2]?
4. Was war mit West-Berlin?
5. Wie ging es der DDR in den 1980er Jahren wirtschaftlich[3]?
6. Was empörte[4] die DDR-Bürger besonders?
7. Welches sozialistische Land öffnete seine Grenzen als erstes?
8. Wo fanden die Montagsdemonstrationen statt?
9. War es gefahrlos[5], an den Montagsdemonstrationen teilzunehmen?
10. Wie viele Menschen demonstrierten am 9. Oktober?
11. Wann fiel die Berliner Mauer?
12. An welches Ereignis erinnert der deutsche Nationalfeiertag?
13. Was findet jedes Jahr am 9. Oktober in Leipzig statt?
14. Nun eine Frage an Sie persönlich: Welche historischen Ereignisse des 20. Jahrhunderts waren für Sie und Ihre Familie wichtig?

1989 ist als das Wendejahr[6] bekannt, das Jahr der friedlichen Revolution. Nach dem verlorenen 2. Weltkrieg[7] war Deutschland zuerst vier Jahre lang besetzt[8] und dann 40 Jahre lang geteilt. 1949 schlossen sich die Besatzungszonen der Westalliierten zur Bundesrepublik Deutschland (BRD) zusammen und aus der sowjetischen Besatzungszone wurde die Deutsche Demokratische Republik, die DDR. Der Kalte Krieg zwischen den USA und der Sowjetunion teilte nicht nur Deutschland, sondern Europa und die ganze Welt. Westdeutschland war eine Demokratie und Teil des kapitalistischen Westens und die DDR war eine Diktatur und Teil des sozialistischen Ostens. West-Berlin war eine westdeutsche Insel in der DDR.

1989 wurde immer deutlicher, dass das diktatorische und sozialistische System der DDR zum Scheitern verurteilt[9] war. Wirtschaftlich stand das Land vor dem Staatsbankrott, die Altstädte verfielen[10] und die Umwelt verkam[11]. Gesellschaftlich[12] empörten sich immer mehr Menschen über den Überwachungsstaat[13], der seine Bürger in Unmündigkeit[14] hielt und ihnen Vieles verbot, insbesondere den Kontakt zum Westen. Viele Leute kamen ins Gefängnis und wurden dort auch gefoltert[15]. Mindestens 200 DDR-Bürger, die versuchten, über die Grenze[16] zu fliehen[17], wurden erschossen[18].

1989 flohen immer mehr Bürger der DDR über Ungarn, das als erstes seine Grenzen öffnete, in den Westen. Am 4. September fand die erste Montagsdemonstration in Leipzig statt. Sie begann im Anschluss an die Friedensgebete[19] in der Nikolaikirche, die seit Mitte der 1980er Jahre immer montags um 17 Uhr stattfanden. Diesen Montagsdemonstrationen schlossen sich von Woche zu Woche immer mehr Menschen an[20], obwohl die Sicherheitskräfte der DDR brutal gegen die Demonstranten vorgingen[21]. Auch in anderen Städten der DDR kam es zu Demonstrationen. Am 9. Oktober demonstrierten 70.000 Menschen in Leipzig unter der Parole: Wir sind das Volk. Eine Woche später waren es bereits 120.000 und an den folgenden Montagen waren es jeweils 300.000. Am 7. November trat die gesamte DDR-Regierung[22] zurück und am 9. November wurde die Berliner Mauer geöffnet. Die Bürger der DDR waren frei.

Am 18. März 1990 gab es die ersten freien Wahlen[23] in der DDR und am 3. Oktober 1990 trat die DDR der Bundesrepublik Deutschland bei[24]. Deutschland war wiedervereinigt[25]. Der 3. Oktober wurde zum Nationalfeiertag Deutschlands. In Leipzig wird jedes Jahr am 9. Oktober ein großes Lichtfest gefeiert, in Erinnerung an die *friedliche Revolution*, bei dem ebenso wie 1989 viele Menschen mit Kerzen[26] durch die Innenstadt Leipzigs ziehen.

[1]*divided* [2]*established* [3]*economically* [4]*outraged* [5]*without danger* [6]*turning-point year* [7]*world war* [8]*occupied* [9]zum ... *doomed to collapse* [10]*were falling into disrepair* [11]*was deteriorating* [12]*In society* [13]*surveillance state* [14]*dependency* [15]*tortured* [16]*border* [17]*flee* [18]*shot dead* [19]*prayers of peace* [20]schlossen sich an *joined* [21]*acted* [22]*government* [23]*elections* [24]trat bei *joined* [25]*reunited* [26]*candles*

Filmlektüre

Nordwand

Vor dem Lesen

A. Beantworten Sie die folgenden Fragen.

1. Wer sind die Personen auf dem Poster? Wie alt sind sie? Was haben sie vor?
2. Schauen Sie sich den Hintergrund[27] an. Wie finden Sie das, was sie vorhaben?
3. Suchen Sie im Internet. Auf welchem Berg finden Sie die berühmte „Nordwand"? Wie heißt das Gebirge? In welchem Land ist dieser Berg?

© Majestic Filmverleih GmbH

B. Lesen Sie die Wörter im Miniwörterbuch. Suchen Sie sie im Text und unterstreichen Sie sie.

FILMANGABEN

Titel: Nordwand
Genre: Drama
Erscheinungsjahr: 2008
Land: Deutschland, Österreich, Schweiz
Regisseur: Phillip Stölzl
Hauptrollen: Benno Fürmann, Florian Lukas, Johanna Wokalek, Georg Friedrich, Simon Schwarz, Ulrich Tukur

Miniwörterbuch

die **Begebenheit**	event
spannend	suspenseful
das **Abenteuer**	adventure
die **Nordwand**	north wall; *here:* north face
erklettern	to climb
schaffen	to achieve
besteigen	to climb
jmdn. aufrufen etwas zu machen	to call on someone to do something
der **Bergsteiger**	mountaineer
erfahren	experienced
überzeugt	convinced
bezwingen	to conquer
vorhaben	to intend
die **Vorbereitung**	preparation
überraschend	by surprise
der **Stein**	stone
ihren Lauf nehmen	to take its course
umschlagen	to change
der **Nullpunkt**	zero point on a scale measuring degrees Celsius
toben	to rampage
die **Lawine**	avalanche
der **Drang**	quest
der **Ruhm**	fame
der **Erfolg**	success

Inhaltsangabe

Der Film basiert auf einer wahren Begebenheit und berichtet von dem spannenden und dramatischen Abenteuer einer Expedition, die 1936 stattgefunden hat. Das Ziel war es, die Eiger-Nordwand, die nicht nur extrem gefährlich sondern auch unter dem Namen „die Mordwand" bekannt war, zu erklettern. Schon viele hatten es probiert, doch bisher hatte es noch niemand geschafft, diese Wand zu besteigen. Als im Jahre 1936, kurz vor den Olympischen Spielen, die Nazis dazu aufrufen, wird es der Traum von vielen Bergsteigern aus ganz Europa. Auch die erfahrenen Kletterer Toni Kurz (Benno Fürmann) aus Berchtesgaden und Andreas Hinterstoißer (Florian Lukas) denken an nichts anderes und sind davon überzeugt, dass sie den Berg bezwingen können. Zwei ehrgeizige Österreicher, Willy (Simon Schwarz) und Edi (Georg Friedrich), haben es ebenso vor, und glauben, die ersten zu sein.

Während der Vorbereitungen am Fuß der Nordwand treffen Toni und Andi überraschend auf Luise (Johanna Wokalek), die sie schon aus ihrer frühen Kindheit kennen. Luise arbeitet jetzt als Fotoreporterin für eine Berliner Zeitung und soll über die Erstbesteigung berichten. Luise merkt bald, dass sie in Toni verliebt ist. Bald beginnt der Aufstieg und das Rennen beginnt. Zunächst läuft alles hervorragend, und beide Teams kommen schnell voran. Doch bereits am Anfang hat Willy einen Unfall. Ein Stein verletzt ihn am Kopf. Die Katastrophe nimmt ihren Lauf, als das Wetter umschlägt. Die Temperaturen fallen tief unter den Nullpunkt. Ein Schneesturm tobt. Lawinen drohen. Der Drang nach Ruhm und Erfolg führt zu einem Kampf auf Leben und Tod.

[27] *background*

Arbeit mit dem Text

Richtig oder falsch? Verbessern Sie die falschen Aussagen.

_____ 1. Der Film ist eine erfundene Geschichte.

_____ 2. Bergsteigen war Teil der Olympischen Spiele.

_____ 3. Luise lernt Toni und Andi während der Vorbereitungen am Fuß der Nordwand kennen.

_____ 4. Toni hat sich sehr früh verletzt.

_____ 5. Die Österreicher glauben, dass sie schneller sind als Toni und Andi.

_____ 6. Viele Bergsteiger haben schon versucht, die Nordwand zu besteigen.

_____ 7. Luise ist in Toni verliebt.

FILMCLIP

NOTE: For copyright reasons, the films referenced in the **FILMCLIP** feature have not been provided by the publisher. The film can be purchased as a DVD or found online at various sites such as YouTube, Amazon, or the iTunes store. The time codes mentioned below are for the North American DVD version of the film.

Szene: DVD, Kapitel 1, „Ersteigung eines Berges", 7:47–13:18 Min.

Toni und Andi verlassen die Kaserne[28] und Toni bereitet sich auf die Besteigung eines Berges vor. Die Szene zeigt, wie sie (Toni im blauen Hemd und Andi im rotkarierten Hemd) zusammen den Gipfel[29] der Zugspitze erklettern.

Sehen Sie sich die Szene an und beantworten Sie die folgenden Fragen.

1. Welche Begrüßungen kommen vor? Kreuzen Sie alle richtigen Antworten an.
   - ☐ a. Grüß Gott!
   - ☐ b. Heil Hitler!
   - ☐ c. Servus!
   - ☐ d. Berg Heil!
2. Toni und Andi haben Probleme und können nicht weiter klettern. Toni will ...
   - ☐ a. umkehren.
   - ☐ b. woanders hochklettern (unten überqueren).
   - ☐ c. direkt hochklettern.
3. Was passiert beim Hochklettern?
   - ☐ a. Andi rutscht aus[30].
   - ☐ b. Ein Haken bricht los[31].
   - ☐ c. Das Seil reißt.
4. Was machen die beiden, nachdem sie auf dem Gipfel angekommen sind? Kreuzen Sie alle richtigen Antworten an.
   - ☐ a. Toni markiert die Route in seinem Tagebuch.
   - ☐ b. Andi ruht sich aus.
   - ☐ c. Sie essen etwas.
5. Richtig oder falsch? Verbessern Sie die falschen Aussagen.
   - _____ a. Andi versucht, Toni zu überreden[32], die Eiger-Nordwand zu besteigen.
   - _____ b. Toni sagt, dass er Schiss[33] hat.

Nach dem Lesen

Beantworten Sie die folgenden Fragen.

1. Haben Sie schon einmal einen Berg bestiegen? Wie war dieses Erlebnis?
2. Hatten Sie schon einmal ein Erlebnis, das sehr aufregend oder gefährlich war? Was haben Sie gemacht?
3. Suchen Sie im Internet. Wie viele Menschen haben versucht, die Eiger-Nordwand zu besteigen? Wie viele sind dabei tödlich verunglückt?

[28]*barracks* [29]*summit* [30]rutscht ... *slips* [31]bricht ... *breaks loose* [32]*convince* [33]Angst

Märchen

Grammatik 9.4–9.5

der König und die Königin

die böse Hexe

der Frosch
(der verwunschene Prinz)

der Schatz

die gute Fee

das Schloss

der Jäger

Die böse Stiefmutter vergiftet Schneewittchen.

Der Prinz erlöst die Prinzessin.

Der Prinz tötet den Drachen.

Situation 12 Schneewittchen

Bringen Sie die Sätze in die richtige Reihenfolge.

_____ Die Königin starb bald darauf, und der König heiratete wieder.

_____ Der Prinz und Schneewittchen heirateten, aber die böse Stiefmutter musste sterben.

_____ Ein Jäger brachte Schneewittchen in den dunklen Wald.

_____ Eines Tages kam ein Königssohn. Als er Schneewittchen sah, verliebte er sich in sie und wollte sie mit nach Hause nehmen.

_____ Die böse Stiefmutter hasste Schneewittchen, weil sie so schön war.

_____ Schneewittchen blieb bei den Zwergen[1] und führte ihnen den Haushalt.

_____ Es war einmal eine Königin, die bekam eine Tochter, die so weiß war wie Schnee, so rot wie Blut und so schwarzhaarig wie Ebenholz[2].

_____ Die Stiefmutter hörte bald von ihrem Spiegel, dass Schneewittchen noch am Leben war.

_____ Schneewittchen lief durch den Wald und kam zu den sieben Zwergen.

_____ Die Zwerge weinten und legten sie in einen gläsernen Sarg.

_____ Als seine Diener[3] den Sarg wegtrugen, stolperte[4] ein Diener. Das giftige Apfelstück rutschte aus Schneewittchens Hals und sie wachte auf.

_____ Die Stiefmutter verkaufte Schneewittchen einen giftigen Apfel, Schneewittchen biss hinein und fiel tot um.

[1]*dwarves* [2]*ebony* [3]*servants* [4]*tripped*

Situation 13 Bildgeschichte: Dornröschen

Situation 14 Was ist passiert?

1. Nachdem Schneewittchen den giftigen Apfel gegessen hatte,
2. Nachdem Hänsel und Gretel durch den dunklen Wald gelaufen waren,
3. Nachdem die Prinzessin den Frosch geküsst hatte,
4. Nachdem die Müllerstochter keinen Schmuck mehr hatte,
5. Nachdem Aschenputtel alle Linsen[5] eingesammelt[6] hatte,
6. Nachdem der Wolf die Großmutter gefressen hatte,
7. Nachdem der Prinz Dornröschen geküsst hatte,
8. Nachdem Rumpelstilzchen seinen Namen gehört hatte,

a. legte er sich in ihr Bett.
b. wurde er sehr wütend.
c. wachte sie auf.
d. fiel sie tot um.
e. verwandelte er sich in einen Prinzen.
f. ging sie auf den Ball.
g. kamen sie zum Haus der Hexe.
h. versprach sie Rumpelstilzchen ihr erstes Kind.

[5]*lentils* [6]*gathered*

Situation 15 Wer weiß – gewinnt

Aus welchem Märchen ist das?

Dornröschen

Rumpelstilzchen

Aschenputtel

Der Froschkönig

Rotkäppchen

Hänsel und Gretel

Schneewittchen

1. „Knusper, knusper, knäuschen, wer knuspert an meinem Häuschen?" „Der Wind, der Wind, das himmlische Kind."

2. „Spieglein, Spieglein an der Wand, wer ist die Schönste im ganzen Land?" „Frau Königin, Ihr seid die Schönste hier, aber die junge Königin ist tausendmal schöner als Ihr."

3. „Ei, Großmutter, was hast du für große Ohren!" „Damit ich dich besser hören kann." „Ei, Großmutter, was hast du für große Augen!" „Damit ich dich besser sehen kann." „Ei, Großmutter, was hast du für ein großes Maul!" „Damit ich dich besser fressen kann."

4. „Die Königstochter soll an ihrem fünfzehnten Geburtstag in einen tiefen Schlaf fallen, der hundert Jahre dauert."

5. „Wenn ich am Tisch neben dir sitzen und von deinem Teller essen und aus deinem Becher trinken und in deinem Bett schlafen darf, dann will ich deinen goldenen Ball aus dem Brunnen heraufholen."

6. „Rucke di guh, rucke di guh,
Blut ist im Schuh:
Der Schuh ist zu klein,
die rechte Braut sitzt noch daheim."

7. „Heute back ich, morgen
brau ich, übermorgen hol'
ich der Königin ihr Kind:
ach, wie gut, dass niemand weiß,
dass ich ______ heiß!"

Situation 16 Zum Schreiben: Es war einmal ...

Schreiben Sie ein Märchen. Wählen Sie aus jeder der vier Kategorien etwas aus oder erfinden Sie etwas.

DIE GUTEN	DIE BÖSEN
eine schöne Prinzessin	eine böse Hexe
ein armer Student	eine grausame Professorin
eine tapfere Königin	ein hungriger Drache
ein treuer Diener[7]	ein böser Stiefvater
?	?

DIE AUSGANGSLAGE	DIE AUFGABE
frisst Menschen und Tiere	drei Rätsel lösen
hat lange Zeit geschlafen	mit einem Riesen kämpfen
bekommt immer nur Fs	etwas Verlorenes wiederfinden
vergiftet das Wasser	eine List erfinden
?	?

Lektüre

Vor dem Lesen

LESEHILFE

The fairy tale "Rotkäppchen" is one of many fairy tales that are well-known across cultures. Just as in English, a fairy tale in German typically contains certain formulaic expressions or phrases. English fairy tales often begin with the phrase *Once upon a time*. Look at the beginning of this fairy tale to see how German fairy tales typically begin.

Fairy tales in German are typically written and told in the simple past tense. As you read this one, you will encounter many such verb forms. Avoid the temptation to look them up upon the first reading; you will deal with them during the **Arbeit mit dem Text** activities at the end of the reading.

© Lebrecht Music and Arts Photo Library/Alamy

[7]*servant*

A. Märchenfiguren. Märchen, auch wenn man sie nicht kennt, sind vorhersagbar[8]. Im Märchen vom Rotkäppchen kommen vier wichtige Figuren vor: ein kleines Mädchen namens Rotkäppchen, ihre Großmutter, der Wolf und der Jäger. Welche Eigenschaften und Tätigkeiten sind typisch für jede Figur? Schreiben Sie neben jede Eigenschaft oder Tätigkeit, ob sie für Rotkäppchen, die Großmutter, den Wolf oder den Jäger typisch ist.

1. Er hat große Ohren und ein großes Maul.
2. Er ist listig[9].
3. Er schießt mit seinem Gewehr.
4. Er schnarcht sehr laut.
5. Er schneidet ihm den Bauch auf.
6. Er sieht nach, ob jemand was fehlt[10].
7. Er verschlingt[11] die Großmutter.
8. Er zieht ihm den Pelz ab[12].
9. Er zieht ihre Kleider an.
10. Jeder hat sie lieb.
11. Sie guckt sich gern um[13].
12. Sie hört gern die Vöglein singen.
13. Sie ist klein und süß.
14. Sie ist krank und schwach.
15. Sie pflückt gern Blumen.
16. Sie trinkt gern Wein und isst gern Kuchen.
17. Sie wohnt draußen im Wald.

B. Suchen Sie die Wörter aus dem Miniwörterbuch im Text und unterstreichen Sie sie.

Miniwörterbuch

das **Dorf**	village
zerbrechen	to break
der **Korb**	basket
zart	tender
pflücken	to pick
packen	*here:* to grab
das **Maul**	mouth (*of animals*)
schnarchen	to snore
das **Gewehr**	rifle
der **Bauch**	belly
erschrocken	startled, scared
atmen	to breathe

Rotkäppchen – Ein Märchen der Gebrüder Grimm

Es war einmal ein kleines, süßes Mädchen, das hatte jeder lieb, der sie nur ansah, am allerliebsten aber ihre Großmutter, die wusste gar nicht, was sie dem Kind alles geben sollte. Einmal schenkte sie dem Mädchen ein Käppchen aus rotem Samt, und weil es Rotkäppchen so gut stand und sie nichts anders mehr tragen wollte, hieß sie nur das Rotkäppchen. Eines Tages sprach ihre Mutter zu ihr: „Komm, Rotkäppchen, da hast du ein Stück Kuchen und eine Flasche Wein, bring das der Großmutter hinaus; sie ist krank und schwach. Dadurch wird sie zu Kräften kommen. Gehe los, bevor es heiß wird, und wenn du aus dem Dorf gehst, so geh anständig und komm nicht vom Weg ab, sonst fällst du und zerbrichst das Glas, und die Großmutter hat nichts. Und wenn du in ihr Haus kommst, vergiss nicht, guten Morgen zu sagen."

„Ich werde schon alles richtig machen", sagte Rotkäppchen zur Mutter und gab ihr die Hand darauf. Die Großmutter aber wohnte draußen im Wald, eine halbe Stunde vom Dorf. Als Rotkäppchen in den Wald kam, begegnete ihr der Wolf. Rotkäppchen aber wusste nicht, was das für ein böses Tier war, und fürchtete sich nicht vor ihm. „Guten Tag, Rotkäppchen", sprach er. „Guten Tag, Wolf." „Wo gehst du so früh hin, Rotkäppchen?" „Zur Großmutter." „Was trägst du in deinem Korb?" „Kuchen und Wein: gestern haben wir gebacken, davon soll sich die kranke und schwache Großmutter etwas stärken." „Rotkäppchen, wo wohnt deine Großmutter?" „Noch eine gute Viertelstunde weiter im Wald hinein, unter den drei großen Eichbäumen, da steht ihr Haus. Da unten sind die Nusshecken, das wirst du ja kennen", sagte Rotkäppchen.

[8]*predictable* [9]*cunning* [10]ob ... *whether anyone needs anything* [11]*devours* [12]zieht den Pelz ab *skins*
[13]guckt sich um *explores*

Der Wolf dachte bei sich: „Das junge zarte Ding, das ist ein fetter Bissen, der wird noch besser schmecken als die Alte: ich muss es listig anfangen, damit ich beide bekomme." Da ging er ein Weilchen neben Rotkäppchen her, dann sprach er: „Rotkäppchen, sieh einmal die schönen Blumen, die hier überall stehen, warum guckst du dich nicht um? Ich glaube, du hörst gar nicht, wie die Vöglein so lieblich singen? Du läufst als wärst du auf dem Weg zur Schule. Dabei ist es so lustig im Wald!"

Rotkäppchen schlug die Augen auf, und als sie sah, wie die Sonnenstrahlen durch die Bäume hin und her tanzten und alles voll schöner Blumen war, dachte sie: „Wenn ich der Großmutter einen frischen Strauß mitbringe, der wird ihr auch Freude machen. Es ist noch so früh am Tag, dass ich doch nicht zu spät komme." Rotkäppchen lief vom Weg in den Wald hinein und suchte Blumen. Und wenn sie eine Blume gepflückt hatte, meinte sie, weiter im Wald steht eine schönere, lief dorthin, und geriet immer tiefer in den Wald hinein. Der Wolf aber ging geradeswegs zum Haus der Großmutter und klopfte an die Tür. „Wer ist da?" „Rotkäppchen, ich bringe Kuchen und Wein. Mach auf!" „Die Tür ist offen, komm herein", rief die Großmutter, „ich bin zu schwach und kann nicht aufstehen." Der Wolf drückte auf die Klinke, die Tür sprang auf. Er ging, ohne ein Wort zu sagen, direkt zum Bett der Großmutter und verschluckte sie. Dann zog er ihre Kleider an, setzte ihre Haube auf, legte sich in ihr Bett und zog die Vorhänge zu.

Rotkäppchen aber suchte immer noch Blumen, und als sie so viele gesammelt hatte, dass sie keine mehr tragen konnte, fiel ihr die Großmutter wieder ein, und sie machte sich auf den Weg zu ihr. Sie wunderte sich, dass die Tür von Großmutters Haus offen stand. Als sie in das Haus trat, so kam es ihr seltsam darin vor. Rotkäppchen dachte: „Ach, du meine Güte! Warum graut es mich heute so, obwohl ich sonst so gerne bei der Großmutter bin?" Rotkäppchen rief „Guten Morgen", bekam aber keine Antwort. Darauf ging sie zum Bett und zog die Vorhänge zurück. Da lag die Großmutter und hatte die Haube tief ins Gesicht gezogen und sah so wunderlich aus. „Ei, Großmutter, was hast du für große Ohren!" „Dass ich dich besser hören kann." „Ei, Großmutter, was hast du für große Augen!" „Dass ich dich besser sehen kann." „Ei, Großmutter, was hast du für große Hände!" „Dass ich dich besser packen kann." „Aber, Großmutter, was hast du für ein entsetzlich großes Maul!" „Dass ich dich besser fressen kann." Kaum hatte der Wolf das gesagt, sprang er aus dem Bett und verschlang das arme Rotkäppchen.

Als der Wolf seine Gelüste gestillt hatte, legte er sich wieder ins Bett, schlief ein und fing an, sehr laut zu schnarchen. Da ging der Jäger an dem Haus vorbei und dachte: „Wie die alte Frau schnarcht, ich muss mal sehen, ob ihr etwas fehlt." Da trat er in das Haus, und als er zum Bett kam, sah er, dass der Wolf darin lag. „Finde ich dich hier, du alter Sünder", sagte er, „ich habe dich lange gesucht." Nun wollte er sein Gewehr anlegen, da fiel ihm ein, dass der Wolf die Großmutter vielleicht gefressen hatte und er könnte sie noch retten. Der Jäger schoss nicht, sondern nahm eine Schere und begann, dem schlafenden Wolf den Bauch aufzuschneiden. Als er ein paar Schnitte gemacht hatte, sah er das rote Käppchen leuchten, und noch ein paar Schnitte, da sprang das Mädchen heraus und rief: „Ach, wie war ich erschrocken! Es war so dunkel in dem Bauch des Wolfes!" Und dann kam die alte Großmutter auch noch lebendig heraus und konnte kaum atmen. Rotkäppchen aber holte schnell große Steine. Damit füllten sie den Leib des Wolfes, und als er aufwachte, wollte er davon springen. Aber die Steine waren so schwer, dass er zurück in das Bett fiel und nie wieder aufwachte.

Da waren alle drei vergnügt: Der Jäger zog dem Wolf den Pelz ab und ging damit nach Hause. Die Großmutter aß den Kuchen und trank den Wein, den Rotkäppchen ihr gebracht hatte, und erholte sich wieder. Rotkäppchen aber dachte: „Ich werde nie wieder allein vom Weg in den Wald laufen, wenn die Mutter es mir verboten hat."

—frei nach den Gebrüdern Grimm

Arbeit mit dem Text

A. Wer sagt das? Lesen Sie den Text und finden Sie heraus, wer die folgenden Sätze denkt oder sagt.

1. Geh anständig[14] und komm nicht vom Weg ab[15], sonst fällst du und zerbrichst das Glas.
2. Ich werde schon alles richtig machen.
3. Was trägst du in deinem Korb?
4. Ihr Haus steht unter den drei großen Eichbäumen[16].
5. Das junge, zarte Ding, das wird noch besser schmecken als die Alte.
6. Ich muss es listig anfangen, damit ich beide bekomme.
7. Du läufst als wärst du[17] auf dem Weg zur Schule.
8. Es ist noch so früh am Tag, dass ich doch nicht zu spät komme.
9. Die Tür ist offen, komm herein.
10. Ach, du meine Güte[18]! Warum graut es mich[19] heute so?
11. Was hast du für große Augen!
12. Dass ich dich besser packen kann.
13. Wie die alte Frau schnarcht, ich muss mal sehen, ob ihr etwas fehlt.
14. Finde ich dich hier, du alter Sünder[20], ich habe dich lange gesucht.
15. Ach, wie war ich erschrocken!
16. Ich werde nie wieder allein vom Weg in den Wald laufen.

B. Richtig oder falsch? Verbessern Sie die falschen Aussagen.

_______ 1. Die Mutter schenkte ihrer Tochter ein Käppchen aus rotem Samt[21].

_______ 2. Rotkäppchen versprach ihrer Mutter, dass sie alles richtig machen wird.

_______ 3. Die Großmutter wohnte draußen im Wald, eine Stunde vom Dorf.

_______ 4. Rotkäppchen hatte Angst vor dem Wolf.

_______ 5. Rotkäppchen verriet[22] dem Wolf, wo ihre Großmutter wohnt.

_______ 6. Der Wolf fraß Rotkäppchen und ging dann zum Haus ihrer Großmutter.

_______ 7. Rotkäppchen lief immer tiefer in den Wald hinein, weil sie immer mehr Blumen pflücken wollte.

_______ 8. Der Wolf machte das Licht aus und legte sich ins Bett der Großmutter.

_______ 9. Als Rotkäppchen zum Haus der Großmutter kam, war die Tür verschlossen.

_______ 10. Als der Jäger zum Haus der Großmutter kam, war alles ruhig.

_______ 11. Der Jäger schoss den Wolf in den Bauch.

_______ 12. Als die Großmutter aus dem Bauch des Wolfes herauskam, konnte sie kaum noch atmen.

_______ 13. Rotkäppchen füllte den Leib[23] des Wolfes mit großen Steinen.

_______ 14. Der Wolf lief davon und wurde[24] nie wieder gesehen.

_______ 15. Rotkäppchen wollte nie wieder etwas tun, was ihr ihre Mutter verboten hatte.

[14]*properly* [15]abkommen ... *diverge* [16]*oak trees* [17]als wärst du ... *as though you were* [18]*goodness* [19]graut ... *am I afraid* [20]du ... *you old rascal* [21]*velvet* [22]*told* [23]*body* [24]*was*

C. Suchen Sie die folgenden Wörter im Text und unterstreichen Sie sie. Ergänzen Sie die Tabelle mit Infinitiv und englischer Übersetzung.

Präteritum	Infinitiv	Englisch
ansah		
wusste		
stand		
sprach		
gab		
kam		
dachte		
ging		
schlug ... auf		
lief		
geriet ... hinein		
rief		
sprang ... auf		
zog ... an		
zog ... zu		
fiel ... ein		
trat		
kam ... vor		
bekam		
zog ... zurück		
lag		
sah ... aus		
schlief ... ein		
fing ... an		
schoss		
begann		
sprang ... heraus		
fiel		
zog ... ab		
aß		
trank		

Nach dem Lesen

Vor Gericht[25] (Alternativende). Der Wolf hat überlebt[26] und bringt Rotkäppchen vor Gericht. Sie soll ins Gefängnis[27], weil sie seinen Bauch mit Steinen gefüllt hat und ihn umbringen[28] wollte. Spielen Sie die Szene im Gericht mit verteilten Rollen. Sie brauchen einen Richter[29], der die Fragen stellt. Rotkäppchen erzählt ihre Geschichte, der Wolf erzählt seine Geschichte. Die Großmutter und der Jäger sind die Zeugen[30] und erzählen, was sie gesehen und erlebt haben. Am Ende spricht der Richter sein Urteil[31].

[25]Vor ... *In court* [26]*survived* [27]*prison* [28]*kill* [29]*judge* [30]*witnesses* [31]spricht sein Urteil ... *gives his verdict*

Videoecke

Perspektiven

Was ist gut, was ist schlecht an der Schule in Deutschland?

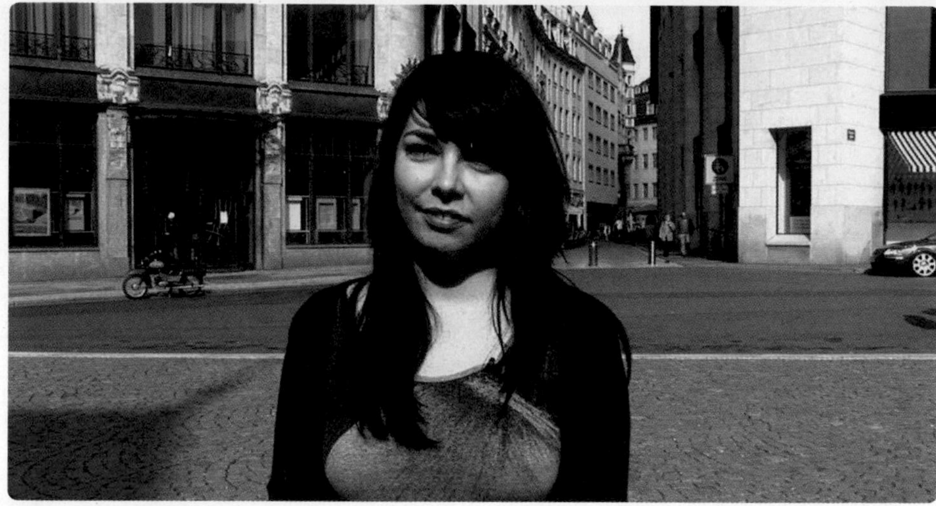

Gut finde ich, dass man nicht dafür bezahlen muss.

Aufgabe 1 Gut oder schlecht?

Was finden die Leute gut, was finden sie schlecht? Schreiben Sie *gut* oder *schlecht* neben die Aussagen.

1. Die Materialien sind kostenlos.
2. Es gibt unterschiedliche Systeme in den Bundesländern.
3. Man lernt Inhalte, die für das spätere Leben nicht so nützlich sind.
4. Jeder kann die Schule besuchen, die er will.
5. Es werden immer weniger Lehrer eingestellt.
6. Die Klassen sind groß.
7. Die Lehrer sind gut ausgebildet.
8. Es gibt keine Schuluniformen.
9. Die Schulen sind öffentlich.
10. Es gibt ein dreigliedriges Schulsystem.

Interviews

- Wann hast du Abitur gemacht?
- Hattest du gute Noten?
- In welchen Fächern warst du besonders gut?
- Was hat dir daran gefallen?
- Welchen Lehrer oder welche Lehrerin fandest du besonders gut? Warum?
- Erzähl etwas, was dieser Lehrer gemacht hat.

Carolyn

Martin

Aufgabe 2 Die Lehrergeschichte

Auf wen treffen die folgenden Aussagen zu, auf den Lehrer von Carolyn oder auf den Lehrer von Martin? Schreiben Sie **C** für Carolyn oder **M** für Martin neben die Aussagen.

1. Er war sehr engagiert.
2. Er hat auf Schulfesten immer aufgepasst.
3. Er war mit seinen Schülern auf Klassenfahrt in Amsterdam.
4. Einmal hat ihm die Musik sehr gut gefallen.
5. Er hat im Hotel ein Bier ausgegeben.
6. Er ist mit ihm/ihr ins Gespräch gekommen.
7. Er hat auf der Tanzfläche eine ganze Nacht lang getanzt.
8. Auch Lehrer sind nur Menschen.

Aufgabe 3 Abitur

Wann haben sie Abitur gemacht? Sehen Sie sich das Video an und schreiben Sie Carolyns und Martins Antworten auf.

	Carolyn	Martin
Wann hat sie/er Abitur gemacht?		
Hatte sie/er gute Noten?		
In welchen Fächern war sie/er besonders gut?		
Was hat ihr/ihm daran gefallen?		
Welche/n Lehrer/in fand sie/er besonders gut?		
Warum?		

Aufgabe 4 Interview

Interviewen Sie eine Partnerin oder einen Partner. Stellen Sie dieselben Fragen.

Wortschatz

Kindheit und Jugend	Childhood and Youth
die **Grundschule, -n**	elementary school
die **Klasse, -n**	grade (level)
die **Note, -n**	grade
der **Abschluss**	graduation
der **Unterricht**	class, instruction
das **Klassentreffen, -**	class reunion
das **Mädchen, -**	girl

Ähnliche Wörter

die **Ballerina, -s;** der **Clown, -s;** der **Spielplatz, ¨e;** der **Zirkus, -se**

Märchen	Fairy Tales
die **Braut, ¨e**	bride
die **Fee, -n**	fairy
die **Hexe, -n**	witch
die **Königin, -nen**	queen
die **List, -en**	deception, trick
die **Wissenschaft, -en**	science, field of study
der **Brunnen, -**	well; fountain
der **Drache, -n** (*wk. masc.*)	dragon
der **Jäger, -**	hunter
der **König, -e**	king
der **Riese, -n** (*wk. masc.*)	giant
der **Sarg, ¨e**	coffin
der **Schatz, ¨e**	treasure
das **Märchen, -**	fairy tale
das **Rätsel, -**	puzzle, riddle
ein Rätsel lösen	to solve a puzzle/riddle
das **Schloss, ¨er**	castle
erlösen	to rescue, free
kämpfen	to fight
klettern, ist geklettert	to climb
küssen	to kiss
sterben, stirbt, starb, ist gestorben	to die
töten	to kill
träumen	to dream
um·fallen, fällt ... um, fiel ... um, ist umgefallen	to fall over
vergiften	to poison
sich verwandeln in (+ *akk.*)	to change into
verwünschen	to curse, cast a spell on
böse	evil, mean
giftig	poisonous
gläsern	glass
grausam	cruel
heimlich	secret
tapfer	brave
tot	dead
treu	loyal, true
verwunschen	cursed; enchanted

Ähnliche Wörter

die **Prinzessin, -nen;** die **Stiefmutter, ¨;** der **Prinz, -en** (*wk. masc.*); der **Stiefvater, ¨;** das **Blut;** das **Feuer, -**

Natur und Tiere	Nature and Animals
der **Baum, ¨e**	tree
der **Frosch, ¨e**	frog
der **Schnee**	snow
das **Pferd, -e** (R)	horse
beißen, biss, gebissen	to bite
fressen, frisst, fraß, gefressen	to eat (*said of an animal*)
füttern	to feed
pflücken	to pick

Ähnliche Wörter

der **Busch, ¨e;** der **Dorn, -en;** der **Elefant, -en** (*wk. masc.*); der **Wind, -e;** der **Wolf, ¨e;** das **Schwein, -e**

Sonstige Substantive	Other Nouns
die **Direktorin, -nen**	female (school) principal, director
die **Einbrecherin, -nen**	female burglar
die **Feier, -n**	celebration, party
die **Fensterscheibe, -n**	windowpane
die **Freude, -n**	joy, pleasure
die **Mannschaft, -en**	team
die **Taschenlampe, -n**	flashlight
die **Verspätung, -en**	delay
der **Ärger**	trouble
der **Becher, -**	cup, mug
der **Direktor, -en**	male (school) principal, director
der **Einbrecher, -**	male burglar
der **Flug, ¨e**	flight
der **Gruß, ¨e**	greeting
der **Hals, ¨e**	neck; throat
der **Schatten, -**	shadow, shade
der **Schlüssel, -**	key
der **Hausschlüssel, -**	house key
der **Zaun, ¨e**	fence
das **Geräusch, -e**	sound, noise
das **Leben, -**	life
am Leben sein	to be alive
das **Loch, ¨er**	hole

Ähnliche Wörter

der **Haushalt, -e;** der **Schlaf;** das **Video, -s;** das **Werk, -e**

Sonstige Verben — Other Verbs

ändern	to change
sich erinnern (**an** + *akk.*)	to remember
eröffnen	to open
hassen	to hate
holen	to fetch, (go) get
los·fahren, fährt ... los, fuhr ... los, ist losgefahren (R)	to drive/ride off
rutschen, ist gerutscht	to slide, slip
schimpfen	to cuss; to scold
stehlen, stiehlt, stahl, gestohlen	to steal
streiten, gestritten	to argue, quarrel
übersetzen	to translate
sich unterhalten, unterhält, unterhielt, unterhalten	to converse
sich verlieben (**in** + *akk.*)	to fall in love (with)
verpassen	to miss
sich verstecken	to hide
vor·lesen, liest ... vor, las ... vor, vorgelesen	to read aloud
wachsen, wächst, wuchs, ist gewachsen	to grow
zerreißen, zerriss, zerrissen	to tear

Ähnliche Wörter

fallen, fällt, fiel, ist gefallen; planen; weg·tragen, trägt ... weg, trug ... weg, weggetragen

Adjektive und Adverbien — Adjectives and Adverbs

arm	poor
bald	soon
bald darauf	soon thereafter
begabt	gifted
daheim	at home
damals	back then
endlich	finally
hinein	in(ward)
leise	quiet(ly)
mitten	in the middle
neulich	recently
plötzlich	suddenly
streng	strict
übermorgen	the day after tomorrow
unterwegs	on the road
verboten	forbidden, prohibited
vorbei	past, over
zufällig	accidental(ly)
zurück	back

Ähnliche Wörter

hungrig, liberal

Sonstige Wörter und Ausdrücke — Other Words and Expressions

außer (+ *dat.*)	besides, in addition to
denn	for, because
gegen (+ *akk.*)	against
nachdem	after (*conj.*)
neben	next to
nichts	nothing

Strukturen und Übungen

9.1 The conjunction *als* with dependent-clause word order

The conjunction **als** (*when*) is commonly used to express that two events or circumstances happened at the same time. The **als-**clause establishes a point of reference in the past for an action or event described in the main clause.

Als ich zwölf Jahre alt war, bin ich zum ersten Mal allein verreist.	*When I was twelve years old, I traveled alone for the first time.*

WISSEN SIE NOCH?

An **als-**clause is a type of dependent clause. As in other dependent clauses, the conjugated verb appears at the end of the clause.

Review grammar 3.4 and 7.1.

When an **als-**clause introduces a sentence, it occupies the first position. Consequently, the conjugated verb in the main clause occupies the second position and the subject of the main clause is in the third position.

(1) Als ich 12 Jahre alt **war,** (2) **bin** (3) **ich** zum ersten Mal allein verreist.

Note that the conjugated verb in the **als-**clause appears at the end of that clause.

Übung 1 Meilensteine

Schreiben Sie 10–15 Sätze über Ihr Leben. Beginnen Sie jeden Satz mit **als.**

MODELL: Als ich eins war, habe ich laufen gelernt.
Als ich zwei war, habe ich sprechen gelernt.
Als ich fünf war, bin ich in die Schule gekommen.
Als ich ...

9.2 The simple past tense of *werden,* the modal verbs, and *wissen*

Use the simple past tense of **haben, sein, werden, wissen,** and the modal verbs in both writing and conversation.

The simple past tense is preferred over the perfect tense with some frequently used verbs, even in conversational German. These verbs include **haben, sein, werden,** the modal verbs, and the verb **wissen.**

Frau Gretter **war** sehr begabt.	*Mrs. Gretter was very talented.*
In der Schule **wusste** sie immer alles.	*In school she always knew everything.*
Sie **hatte** viele Freunde.	*She had many friends.*

The conjugations of **werden,** the modal verbs, and **wissen** appear below. For **haben** and **sein,** refer back to **Strukturen 7.5.** Notice that the **ich-** and **er/sie/es-**forms are the same for all these verbs.

A. The Verb **werden**

Michael **wurde** Tierpfleger.	*Michael became an animal caretaker.*
Im August **wurde** er sehr krank.	*In August he became very sick.*

werden			
ich	wurde	*wir*	wurden
du	wurdest	*ihr*	wurdet
Sie	wurden	*Sie*	wurden
er / *sie* / *es*	wurde	*sie*	wurden

B. Modal Verbs

To form the simple past tense of modal verbs, use the stem, drop any umlauts, and add **-te-** plus the appropriate ending.

können → könn → k**o**nn → konn**te** → du konnte**st**

Wir **wollten** mitkommen.	*We wanted to come along.*
Mehmet **musste** jeden Tag um sechs aufstehen.	*Mehmet had to get up at six every morning.*
Eske und Damla **durften** mit fünf Jahren noch nicht fernsehen.	*When they were five, Eske and Damla weren't yet allowed to watch TV.*

Here are the simple past-tense forms of the modal verbs.

	können	**müssen**	**dürfen**	**sollen**	**wollen**	**mögen**
ich	konnte	musste	durfte	sollte	wollte	mochte
du	konntest	musstest	durftest	solltest	wolltest	mochtest
Sie	konnten	mussten	durften	sollten	wollten	mochten
er *sie* *es*	konnte	musste	durfte	sollte	wollte	mochte
wir	konnten	mussten	durften	sollten	wollten	mochten
ihr	konntet	musstet	durftet	solltet	wolltet	mochtet
Sie	konnten	mussten	durften	sollten	wollten	mochten
sie	konnten	mussten	durften	sollten	wollten	mochten

Note the consonant change in the past tense of **mögen: mo*ch*te.**

C. The Verb **wissen**

The forms of the verb **wissen** are similar to those of the modal verbs.

Ich **wusste** nicht, dass du keine Erdbeeren magst.	*I didn't know that you don't like strawberries.*

Here are the simple past-tense forms.

wissen			
ich	wusste	*wir*	wussten
du	wusstest	*ihr*	wusstet
Sie	wussten	*Sie*	wussten
er *sie* *es*	wusste	*sie*	wussten

Übung 2 Fragen und Antworten

Hier sind die Fragen. Was sind die Antworten?

MODELL: Lydia, warum bist du nicht mit ins Kino gegangen? (nicht können)
→ Ich konnte nicht.

1. Ernst, warum bist du nicht mit zum Schwimmen gekommen? (nicht dürfen)
2. Maria, warum bist du nicht gekommen? (nicht wollen)
3. Jens, gestern war Juttas Geburtstag! (das / nicht wissen)
4. Jutta, warum hast du eine neue Frisur? (eine/wollen)
5. Jochen, warum hast du das Essen nicht gekocht? (das / nicht sollen)

Übung 3 Minidialoge

Setzen Sie Modalverben oder **wissen** ein.

1. SILVIA: Was hast du gemacht, wenn du nicht zur Schule gehen _______[a]?
   JÜRGEN: Ich habe gesagt: „Ich bin krank."
   SILVIA: Haben deine Eltern das geglaubt?
   JÜRGEN: Nein, meine Mutter _______[b] immer, was los war.
2. ERNST: Hans, warum bist du gestern nicht auf den Spielplatz gekommen?
   HANS: Ich _______[a] nicht. Ich habe eine Fünf in Mathe geschrieben und _______[b] zu Hause bleiben.
   ERNST: Schade. Wir _______[c] Fußball spielen, aber dann _______[d] wir nicht genug Spieler finden.
3. HERR RUF: Guten Tag, Frau Gretter. Tut mir leid, dass ich neulich nicht zu Ihrer kleinen Feier kommen _______[a]. Aber ich _______[b] meine alte Tante in Würzburg besuchen.
   FRAU GRETTER: Ja, wirklich schade. Ich _______[c] gar nicht, dass Sie eine Tante in Würzburg haben.
   HERR RUF: Sie zieht diese Woche nach Düsseldorf zu ihrer Tochter, und ich _______[d] sie noch einmal besuchen.

9.3 Time: *als, wenn, wann*

Als refers to a circumstance (time period) in the past or to a single event (point in time) in the past or present, but never in the future.

TIME PERIOD

Als ich 15 Jahre alt war, sind meine Eltern nach Texas gezogen.
When I was 15 years old, my parents moved to Texas.

POINT IN TIME

Als wir in Texas angekommen sind, war es sehr heiß.
When we arrived in Texas, it was very hot.

Als Veronika ins Zimmer kommt, klingelt das Telefon.
When (As) Veronika comes into the room, the phone rings.

Wenn has three distinct meanings: a conditional meaning and two temporal meanings. In conditional sentences, **wenn** means *if.* In the temporal sense, **wenn** may be used to describe events that happen or happened one or more times (*when*[*ever*]) or to describe events that will happen in the future (*when*).

CONDITION

Wenn man auf diesen Knopf drückt, öffnet sich die Tür.
If you press this button, the door will open.

REPEATED EVENTS

Wenn Herr Wagner nach Hause kam, freuten sich die Kinder.
When(ever) Mr. Wagner came home, the children were happy.

Wenn Herr Wagner nach Hause kommt, freuen sich die Kinder.
When(ever) Mr. Wagner comes home, the children are happy.

FUTURE EVENT

Wenn ich in Frankfurt ankomme, rufe ich dich an.
When I arrive in Frankfurt, I'll call you.

In the simple past, **wenn** refers to a habit or an action or event that happened repeatedly or customarily; **als** refers to a specific action or event that happened once, over a particular time period or at a particular point in time in the past.

Wenn ich nicht zur Schule gehen wollte, habe ich gesagt, dass ich krank bin.	*When(ever) I didn't want to go to school, I said that I was sick.*
Als ich mein erstes F bekommen habe, habe ich geweint.	*When I got my first F, I cried.*

Wann is an adverb of time meaning *at what time*. It is used in both direct and indirect questions.

Wann hast du deinen ersten Kuss bekommen?	*When did you get your first kiss?*
Ich weiß nicht, **wann** der Zug kommt.	*I don't know when the train is coming.*

Note that when **wann** is used in an indirect question, the conjugated verb comes at the end of the clause.

When	
Single event in past or present (*at one time*)	**als**
Circumstance in the past	
Condition (*if*)	
Repeated event in past, present, or future (*whenever*)	**wenn**
Single event in the future (*when*)	
Adverb of time (*at what time?*)	**wann**

Übung 4 Minidialoge

Wann, wenn oder **als?**

1. ERNST: _______[a] darf ich fernsehen?
   FRAU WAGNER: _______[b] du deine Hausaufgaben gemacht hast.
2. ROLF: Oma, _______[a] hast du Opa kennengelernt?
   OMA: _______[b] ich siebzehn war.
3. STEFAN: Was habt ihr gemacht, _______ ihr in München wart?
   NORA: Wir haben sehr viele Filme gesehen.
4. NESRIN: _______[a] hast du Sofie getroffen?
   WILLI: Gestern, _______[b] ich an der Uni war.
5. ALBERT: _______[a] fliegst du nach Europa?
   PETER: _______[b] ich genug Geld habe.
6. MONIKA: Du spielst sehr gut Tennis. _______[a] hast du das gelernt?
   HEIDI: _______[b] ich noch klein war.

Übung 5 Eine Mail

Wann, wenn oder **als?**

Liebe Tina, gestern Nachmittag musste ich meiner Oma mal wieder Kuchen und Wein bringen. Immer _____[a] ich mich mit meinen Freunden verabrede[1], will mein Vater irgendetwas[2] von mir. Ich war ganz schön wütend. _____[b] ich den Korb[3] zusammengepackt habe, habe ich leise geschimpft. _____[c] ich meine Oma besuche, muss ich immer ein bisschen dableiben und mich mit ihr unterhalten. Das ist langweilig und anstrengend[4], denn die Oma hört nicht mehr so gut. Außerdem wohnt sie am anderen Ende der Stadt. Auch _____[d] ich mit dem Bus fahre, dauert es mindestens zwei Stunden.

[1]*make plans* [2]*something* [3]*basket* [4]*strenuous*

_____[e] ich aus dem Haus gekommen bin, habe ich an der Ecke Billy auf seinem Moped gesehen. _____[f] ich ihn das letzte Mal gesehen habe, haben wir uns prima unterhalten.

„_____[g] kommst du mal wieder ins Jugendzentrum?" hat Billy gerufen. „Vielleicht heute Abend", habe ich geantwortet. _____[h] ich mich auf den Weg gemacht habe, hat es auch noch angefangen zu regnen. Und natürlich ... wie immer ... _____[i] es regnet, habe ich keinen Regenschirm dabei.

Liebe Grüße, deine Jutta

9.4 The simple past tense of strong and weak verbs (receptive)

In written texts, the simple past tense is frequently used instead of the perfect to refer to past events.

Jutta **fuhr** allein in Urlaub.	*Jutta went on vacation alone.*
Ihr Vater **brachte** sie zum Bahnhof.	*Her father took her to the train station.*

In the simple past tense, just as in the present tense, separable-prefix verbs are separated in independent clauses but joined in dependent clauses.

Rolf **stand** um acht Uhr **auf.** Es war selten, dass er so früh **aufstand.**	*Rolf got up at eight. It was rare that he got up so early.*

weak verbs = **-(e)te-**

A. Weak Verbs

You can recognize the simple past of weak verbs by the **-(e)te-** that is inserted between the stem and the ending.

PRESENT		SIMPLE PAST
du sagst	:	du sag**te**st
sie arbeitet	:	sie arbeit**ete**

Wir bad**ete**n, bau**te**n Sandburgen und spiel**te**n Volleyball.	*We went swimming, built sand castles, and played volleyball.*

Like modal verbs, simple past-tense forms do not have an ending in the **ich-** or the **er/sie/es-**forms: **ich sagte, er sagte.** Here are the simple past-tense forms of the verb **machen.**

machen			
ich	machte	*wir*	machten
du	machtest	*ihr*	machtet
Sie	machten	*Sie*	machten
er *sie* *es*	machte	*sie*	machten

irregular weak verbs = stem-vowel change + **-te-**

For a few weak verbs, the stem of the simple past is the same as the one used to form the past participle.

PRESENT	SIMPLE PAST	PERFECT	
bringen	brachte	hat gebracht	*to bring*
denken	dachte	hat gedacht	*to think*
kennen	kannte	hat gekannt	*to know, be acquainted with*
wissen	wusste	hat gewusst	*to know (as a fact)*

B. Strong Verbs

strong verbs = stem-vowel change

All strong verbs have a different stem in the simple past: **schwimmen/ schwamm, singen/sang, essen/aß.** Since English also has a number of verbs with irregular stems in the past (*swim/swam*, *sing/sang*, *eat/ate*), you will usually have no trouble recognizing simple past stems. You will also recognize the **ich-** and **er/sie/es-**forms of strong verbs easily, because they do not have an ending.

Through practice reading texts in the simple past, you will gradually become familiar with the various patterns of stem change that exist. Here are some common past-tense forms you are likely to encounter in your reading.* A more complete list of irregular verbs, including stem-changing verbs, can be found in **Appendix E**.

bleiben	blieb	*to stay*
essen	aß	*to eat*
fahren	fuhr	*to drive*
fliegen	flog	*to fly*
geben	gab	*to give*
gehen	ging	*to go*
lesen	las	*to read*
nehmen	nahm	*to take*
rufen	rief	*to call*
schlafen	schlief	*to sleep*
schreiben	schrieb	*to write*
sehen	sah	*to see*
sprechen	sprach	*to speak*
stehen	stand	*to stand*
tragen	trug	*to carry*
waschen	wusch	*to wash*

Der Bus **fuhr** um sieben Uhr ab.	*The bus left at seven o'clock.*
Sechs Kinder **schliefen** in einem Zimmer.	*Six children were sleeping in one room.*
Jutta **aß** frische Krabben.	*Jutta ate fresh shrimp.*

Übung 6 Die Radtour

Setzen Sie die Verben ein:

hielten, gingen, aßen, kamen, schwammen, fuhren, standen, sprangen, schliefen

Willi und Sofie wollten eine Radtour machen, aber ihre Räder waren kaputt. Sie mussten sie reparieren, bevor sie losfahren konnten. Am Morgen der Tour _____[a] sie um sechs Uhr auf, _____[b] in die Garage, wo die Räder waren und machten sich an die Arbeit. Gegen acht waren sie fertig, sie frühstückten noch und dann _____[c] sie ab. Gegen elf _____[d] sie an einen kleinen See. Sie _____[e] an und setzten sich ins Gras. Willis Mutter hatte ihnen Essen eingepackt. Sie waren hungrig und _____[f] alles auf. Sie _____[g] im See und legten sich dann in den Schatten und _____[h]. Am späten Nachmittag _____[i] sie noch mal ins Wasser und radelten dann zurück nach Hause.
Die Rückfahrt dauerte eine Stunde länger als die Hinfahrt.

* It is fairly easy to make an educated guess about the form of the infinitive when encountering new simple pasttense forms. The following vowel correspondences are the most common.
1. a < e/i, for example: gab < geben, fand < finden
2. i/ie < a/ei, for example: ritt < reiten, hielt < halten, schrieb < schreiben

Übung 7 Hänsel und Gretel

Ergänzen Sie die Verben.

brachten, fanden, gab, kamen, liefen, rannte, sahen, saß, schliefen, schloss, tötete, trug, wohnte

1. Vor einem großen Wald ________ eine arme Familie mit den beiden Kindern Hänsel und Gretel.

2. Als sie eines Tages nichts mehr zu essen hatten, ________ die Eltern die Kinder in den Wald.

3. Die Kinder ________ ein und als sie aufwachten, waren sie allein.

4. Dann ________ sie durch den Wald, bis sie an ein kleines Haus ________.

5. Durch das Fenster ________ sie eine alte Frau, die vor einem Kamin[1] ________ und strickte.

6. Als die Alte die Kinder bemerkte[2], holte sie sie herein und ________ ihnen etwas zu essen. Die Kinder ________ die Frau sehr freundlich.

7. Aber leider war sie eine böse Hexe. Sie packte[3] Hänsel, ________ ihn in einen Käfig und ________ die Tür. Er sollte dick werden, damit sie ihn essen konnte.

8. Gretel weinte und versuchte, Hänsel zu helfen. Sie ________ die Hexe und ________ mit Hänsel weg.

9.5 Sequence of events in past narration: the past perfect tense and the conjunction *nachdem* (receptive)

A. Uses of the Past Perfect Tense

The past perfect tense is used to describe past actions and events that were completed before other past actions and events.

Nachdem Luca zwei Stunden **ferngesehen hatte**, ging er ins Bett.	*After Luca had watched TV for two hours, he went to bed.*
Nachdem Jutta mit ihrer Freundin **telefoniert hatte,** machte sie ihre Hausaufgaben.	*After Jutta had talked with her friend on the phone, she did her homework.*

[1]*hearth* [2]*noticed* [3]*grabbed*

The past perfect tense is often used in the clause with **nachdem**. The simple past tense is then used in the concluding (main) clause.

The past perfect tense often occurs in a dependent clause with the conjunction **nachdem** (*after*); the verb of the main clause is in the simple past or the perfect tense.

Nachdem Jens seine erste Zigarette **geraucht hatte, wurde** ihm schlecht.	*After Jens had smoked his first cigarette, he got sick.*

A dependent clause introduced by **nachdem** usually precedes the main clause. This results in the pattern "verb, verb."

DEPENDENT CLAUSE	MAIN CLAUSE
1	2
Nachdem ich die Schule **beendet hatte,**	**machte** ich eine Lehre.

After I had finished school, I learned a trade.

The conjugated verb of the dependent clause is at the end of the dependent clause; the conjugated verb of the main clause is at the beginning of the main clause. Because the entire dependent clause holds the first position in the sentence, the verb-second rule applies here.

past perfect tense = **hatte/war** + past participle

B. Formation of the Past Perfect Tense

The past perfect tense of a verb consists of the simple past tense of the auxiliary **haben** or **sein** and the past participle of the verb.

Ich **hatte** schon **bezahlt** und wir konnten gehen.	*I had already paid, and we could go.*
Als wir ankamen, **waren** sie schon **weggegangen**.	*When we arrived, they had already left.*

Übung 8 Was ist zuerst passiert?

Bilden Sie logische Sätze mit Satzteilen aus beiden Spalten.

MODELL: Nachdem Jutta den Schlüssel verloren hatte, kletterte sie durch das Fenster.

1. Nachdem Jutta den Schlüssel verloren hatte,	a. flog er nach Australien.
2. Nachdem Ernst die Fensterscheibe eingeworfen hatte,	b. ging er ins Bett.
3. Nachdem Claire angekommen war,	c. kletterte sie durch das Fenster.
4. Nachdem Hans seine Hausaufgaben gemacht hatte,	d. lief er weg.
5. Nachdem Jens sein Fahrrad repariert hatte,	e. machte er eine Radtour.
6. Nachdem Michael die Seiltänzerin[1] gesehen hatte,	f. rief er den Großvater an.
7. Nachdem Richard ein ganzes Jahr gespart hatte,	g. rief sie Melanie an.
8. Nachdem Silvia zwei Semester allein gewohnt hatte,	h. war er ganz verliebt.
9. Nachdem Willi ein Geräusch gehört hatte,	i. zog sie in eine WG.

[1] *tightrope walker*

KAPITEL **10**

Auf Reisen

Kapitel 10 focuses on travel. You will also learn to get around in the German-speaking world by following directions and reading maps.

Themen

Reisepläne

Nach dem Weg fragen

Urlaub am Strand

Tiere

Kulturelles

KLI: Universitätsstadt Göttingen

Musikszene: „Dieser Weg“ (Xavier Naidoo)

KLI: Die deutsche Einwanderung in die USA

Filmclip: *Die fetten Jahre sind vorbei* (Hans Weingartner)

Videoecke: Urlaub

Lektüren

Gedicht: Die Stadt (Theodor Storm)

Reiseführer: Husum

Film: *Die fetten Jahre sind vorbei* (Hans Weingartner)

Strukturen

10.1 Prepositions to talk about places: **aus, bei, nach, von, zu**

10.2 Requests and instructions: the imperative (summary review)

10.3 Prepositions for giving directions: **an ... vorbei, bis zu, entlang, gegenüber von, über**

10.4 Being polite: the subjunctive form of modal verbs

10.5 Focusing on the action: the passive voice

Franz Marc: *Turm der blauen Pferde* (1913), verschollen

KUNST UND KÜNSTLER

Franz Marc (1880–1916) ist ein führender Vertreter des deutschen Expressionismus. Er studierte an der Münchner Kunstakademie und war Mitbegründer des Kunstalmanachs *Der Blaue Reiter*, für den er auch Artikel über Kunsttheorie schrieb. Im August 1914 meldete er sich freiwillig zum Kriegsdienst. Er fiel 1916 bei seinem letzten Einsatz nahe Verdun in Frankreich. Die Nazis bezeichneten sein Werk als „Entartete Kunst". Manche seiner Bilder wurden anschließend ins Ausland verkauft oder zerstört. Viele seiner Bilder sind von geometrischen und abstrakten Formen gekennzeichnet[1] und stellen oft Tiermotive, zum Beispiel Pferde, dar. Die Farben Blau, Gelb und Rot hatten für ihn eine symbolische Bedeutung. Blau steht für das Männliche, Herbe[2] und Geistige[3], Gelb für das Weibliche, Sanfte, Heitere[4] und Sinnliche[5], und Rot für das Brutale und Schwere.

Schauen Sie sich das Bild an und beantworten Sie die folgenden Fragen.

1. Welche Farben überwiegen[6] in dem Bild?
2. Beschreiben Sie die Atmosphäre des Bildes und gehen Sie dabei auf Marcs Farbensymbolik ein[7].
3. Was könnte die Körperhaltung[8] des Pferds im Vordergrund ausdrücken[9]?
4. Stellen Sie sich vor, Sie wären der Künstler oder die Künstlerin. Welche Farbe hätten Sie dem Pferd gegeben, Blau oder Gelb? Warum?

[1]gekennzeichnet von *characterized by* [2]*harsh* [3]*intellectual* [4]*cheerful* [5]*sensual* [6]*dominate* [7]eingehen auf *consider* [8]*posture* [9]*express*

Situationen

Reisepläne

Grammatik 10.1

WILLI: Wo warst du in deinem letzten Urlaub?
NESRIN: Ich war in Schweden.

WILLI: Was hast du dort gemacht?
NESRIN: Ich bin Kanu gefahren und viel gewandert.

WILLI: Bist du geflogen?
NESRIN: Nein, ich bin mit dem Auto gefahren und war die ganzen zwei Wochen dort auch mit dem Auto unterwegs.

WILLI: Wo hast du gewohnt?
NESRIN: Ich habe auf Campingplätzen gezeltet.

THOMAS: Ich will nächsten Sommer nach Australien fliegen.
PETER: Wie lange möchtest du dort bleiben?

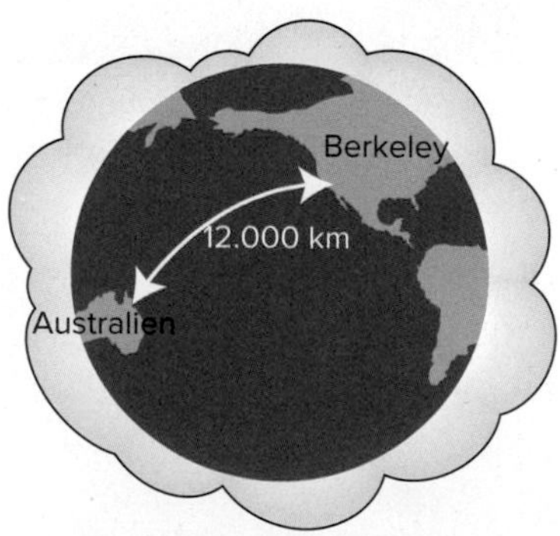

THOMAS: Vielleicht drei Wochen.
PETER: Warum willst du nach Australien? Das ist doch so weit weg und der Flug ist sehr teuer.

THOMAS: Ich möchte die vielen interessanten Tiere sehen, zum Beispiel Kängurus. Und dann will ich meine Freundin in Sydney besuchen.
PETER: Da musst du dir bestimmt ein Auto mieten.

THOMAS: Nein, ich trampe. Und wohnen werde ich bei meiner Freundin und in Jugendherbergen. Dann wird alles ein bisschen billiger.
PETER: Gute Idee. Viel Spaß in Australien.

Situation 1 Urlaub

Wer ist das, Nesrin (N) oder Thomas (T)?

_____ ist Kanu gefahren und viel gewandert.
_____ möchte Kängurus sehen.
_____ war in Schweden.
_____ hat auf Campingplätzen gezeltet.
_____ will in Jugendherbergen wohnen.
_____ war mit dem Auto unterwegs.
_____ möchte drei Wochen bleiben.
_____ will seine Freundin in Sydney besuchen.

Situation 2 Informationsspiel: Reisen

MODELL: S1: Woher kommt Sofie? S2: Aus _____.
S1: Wohin fährt sie in den Ferien? S2: Nach/In _____.
S1: Wo wohnt sie? S2: Bei _____. Was macht sie da?
S1: Sie kauft Bücher und besucht Verwandte. Wann kommt sie zurück? S2: In _____.

	Richard	Sofie	Mehmet	Peter	Jürgen	mein(e) Partner(in)
Woher?	aus Innsbruck		aus Izmir		aus Bad Harzburg	
Wohin?	nach Frankreich		nach Italien		in die Alpen	
Wo?	bei einer Gastfamilie			bei seiner Schwester		
Was?		Bücher kaufen; Verwandte besuchen		einen Vulkan besteigen		
Wann?	in drei Monaten				in zwei Wochen	

KULTUR ... LANDESKUNDE ... INFORMATIONEN

UNIVERSITÄTSSTADT GÖTTINGEN

- Welche Personen in ***Kontakte*** studieren in Göttingen?
- Welche Person wohnt in den Ferien bei ihren Eltern in Göttingen?
- Was wissen Sie schon über die Stadt?
- Welches Tier sehen Sie im Wappen der Stadt Göttingen?

Lesen Sie den Text und suchen Sie Antworten auf diese Fragen:

- Wofür ist Göttingen vor allem bekannt?
- Wer besuchte im Mittelalter die Pfalz[1] Grona?
- Wann wurde die Universität Göttingen offiziell eingeweiht[2]?
- Nach wem ist die Universität benannt?
- Wer arbeitete in der Göttinger Sternwarte[3]?
- Wer protestierte gegen König Ernst August von Hannover?
- In welchen Fächern erhielten sechs Wissenschaftler der Universität den Nobelpreis?
- Wogegen war die „Göttinger Erklärung[4]" aus dem Jahr 1957?

Die Stadt Göttingen liegt ziemlich genau in der Mitte von Deutschland und ist vor allem bekannt für ihre Universität. Die traditionsreiche „Georgia Augusta" ist die größte und älteste Universität in Niedersachsen und die Zahl der Studierenden macht 20 Prozent der Bevölkerung aus.

Das heutige Göttingen kann man bis ins 7. Jahrhundert zurückverfolgen[5]. Es gab ein Dorf Gutingi, über das man nicht viel weiß, und die Pfalz Grona, die zwei Kilometer entfernt lag. Diese Pfalz wurde im Mittelalter von Kaisern und Königen besucht. Heinrich II. und seine Frau Kunigunde liebten Grona, heute ein Stadtteil Göttingens, und Heinrich starb dort im Jahre 1024.

Später profitierte Göttingen wirtschaftlich von seiner guten Lage zwischen Lübeck und Frankfurt am Main und war zeitweise sogar Mitglied der Hanse*.

Im Jahre 1734 wurden in Göttingen die ersten Studenten unterrichtet und 1737 wurde die Universität feierlich eingeweiht. Georg II. August von Großbritannien, der auch Kurfürst von Braunschweig-Lüneburg war, gab der neuen Hochschule seinen Namen.

Neben der Universität hatte Göttingen auch eine Akademie der Wissenschaften, die 1766 von Benjamin Franklin besucht wurde, und eine Sternwarte, in der der berühmte Mathematiker und Physiker Carl Friedrich Gauß (1777–1855) arbeitete.

Bekannt wurde Göttingen im 19. Jahrhundert durch den Protest der „Göttinger Sieben" gegen König Ernst August von Hannover, der die recht freiheitliche Verfassung aufgehoben hatte. Zu den sieben Professoren, die protestierten, gehörten auch die Germanisten Jacob und Wilhelm Grimm.

Im 20. Jahrhundert erhielten sechs Wissenschaftler der Georgia Augusta den Nobelpreis in Chemie oder Physik. Aber auch in anderen akademischen Fächern studierten und lehrten bedeutende Wissenschaftler in Göttingen, z. B. die Philosophin Edith Stein und der Mediziner Robert Koch. Im Jahr 1957 veröffentlichten bedeutende Wissenschaftler, unter anderem Otto Hahn, Werner Heisenberg und Carl Friedrich von Weizsäcker aus Göttingen, die „Göttinger Erklärung" gegen die atomare Bewaffnung[6] der Bundeswehr.

Das Wahrzeichen der Stadt Göttingen ist das „Gänseliesel", eine Brunnenfigur auf dem Marktplatz, die Göttinger Doktoranden[7] küssen müssen, wenn sie ihren Doktorhut erhalten.

Wappen der Stadt Göttingen
© Tibor Bognar/Alamy

Das Gänseliesel
© imageBROKER/Alamy

*the Hanseatic League, a medieval trade organization based around cities of northern Europe

[1]*imperial residence* [2]*inaugurated* [3]*astronomical observatory* [4]*declaration* [5]*trace back* [6]*armament* [7]*doctoral graduates*

Situation 3 Dialog: Am Fahrkartenschalter

Silvia steht am Fahrkartenschalter und möchte mit dem Zug von Göttingen nach München fahren.

BAHNANGESTELLTER: Bitte schön?

SILVIA: Eine __________ nach München, bitte.

BAHNANGESTELLTER: Einfach oder hin und zurück?

SILVIA: Hin und zurück bitte, mit BahnCard __________ Klasse.

BAHNANGESTELLTER: Wann wollen Sie fahren?

SILVIA: Ich würde gern __________ in München sein.

BAHNANGESTELLTER: Wenn Sie um 8.06 Uhr fahren, sind Sie um 12.11 Uhr in München.

SILVIA: Das ist gut. Wissen Sie, wo der Zug __________?

BAHNANGESTELLTER: Aus Gleis 10.

SILVIA: Ach ja, ich würde gern mit VISA bezahlen. __________?

BAHNANGESTELLTER: Selbstverständlich. Das macht dann 115 Euro 20.

SILVIA: Bitte sehr.

Göttingen → München Hbf

530 km

Ab	Zug	Umsteigen	An	Ab	Zug	An	Verkehrstage
5.56	ICE 997	Fulda	6.49	7.00	ICE 987	10.11	01
5.56	ICE 997	Fulda	6.52	7.02	ICE 987	10.11	02
7.03	ICE 581					10.58	täglich
8.06	ICE 783					12.11	täglich
9.03	ICE 583					12.58	täglich
9.47	IC 1081	Augsburg Hbf	14.04	14.10	SE 21139	14.54	täglich
10.03	ICE 91	Nürnberg Hbf	12.26	12.30	IC 523	14.17	täglich
10.30	IC 1087	Nürnberg Hbf	13.23	13.34	IC 813	15.17	03

From DB Bahn, Germany

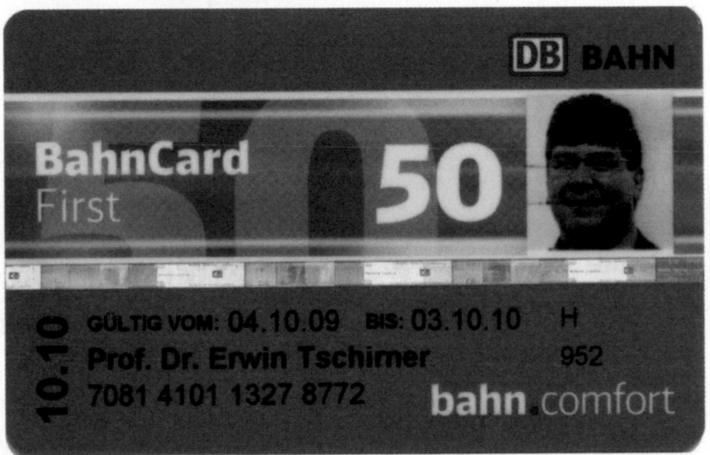

Mit der BahnCard spart man.

Situation 4 Interview

1. Wo machst du gern Urlaub?
2. Fliegst du gern? Was gefällt dir daran? Stört dich etwas beim Fliegen? Was?
3. Wie suchst du dir deine Urlaubsziele aus? Wie besorgst[8] du dir dein Ticket?
4. Wie packst du für eine Flugreise? Was nimmst du alles mit?
5. Erzähl von einer deiner letzten Reisen. Wo warst du? Wie bist du dahin gekommen? Warst du allein? Hast du jemanden kennengelernt? Was hast du am liebsten gemacht? Was war das Interessanteste, was dir passiert ist?

[8]*get*

Nach dem Weg fragen

Grammatik 10.2–10.3

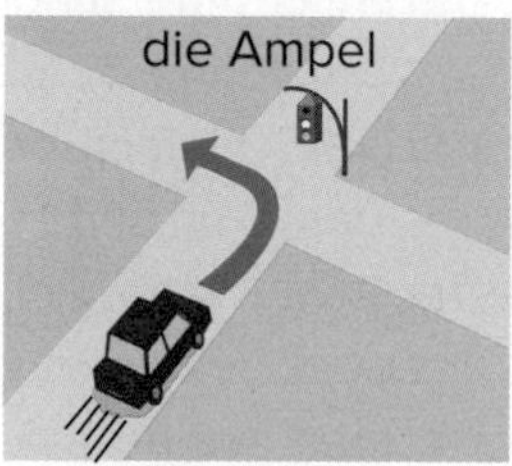

Biegen Sie an der Ampel nach links ab.

Gehen Sie über den Zebrastreifen.

Gehen Sie geradeaus, bis Sie eine Kirche sehen.

Gehen Sie an der Kirche vorbei, immer geradeaus.

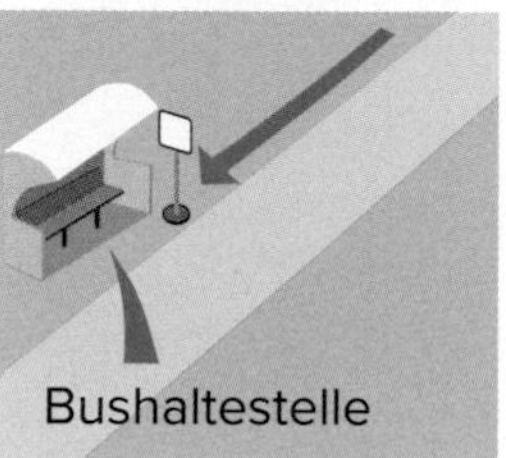

Gehen Sie die Goethe-allee entlang bis zur Bushaltestelle.

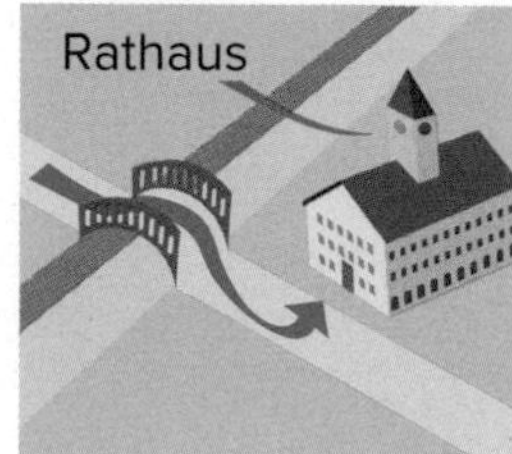

Gehen Sie über die Brücke. Auf der linken Seite ist dann das Rathaus.

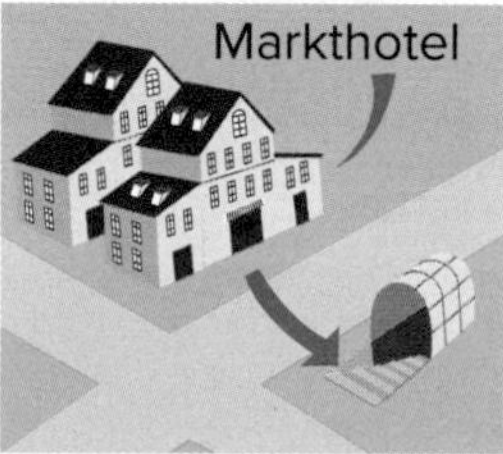

Die U-Bahnhaltestelle ist gegenüber vom Markthotel.

Gehen Sie die Treppe hinauf und dann ist es die zweite Tür links.

Situation 5 Mit dem Stadtplan° unterwegs in Regensburg

°*city street map*

Suchen Sie sich ein Ziel in Regensburg aus dem Stadtplan aus. Beschreiben Sie Ihrem Partner / Ihrer Partnerin den Weg, ohne das Ziel zu verraten[1]. Wenn er/sie dort richtig ankommt, bekommen Sie einen Punkt und es wird gewechselt.

MODELL: Also, wir sind jetzt an der Steinernen Brücke, auf dem Stadtplan oben in der Mitte. Siehst du die Steinerne Brücke? Gut. Von der Steinernen Brücke aus geh bitte nach links in die Goldene-Bären-Straße hinein und an der nächsten Straße gleich wieder rechts. Du kommst dann zum Krauterermarkt und zum Dom. Geh geradeaus über den Krauterermarkt hinüber und durch die Residenzstraße zum Neupfarrplatz. Dort gehst du bitte wieder links, die Schwarze-Bären-Straße ganz durch und über die Maximilianstraße hinüber. Noch ein paar Schritte weiter und du bist am ______.

[1]*give away*

NÜTZLICHE AUSDRÜCKE

links/rechts die (Goliath)straße entlang
links/rechts in die (Kram)gasse hinein
geradeaus über den (Krauterer)markt / über die (Kepler)straße hinüber
weiter bis zum/zur ______
an der (Steinernen Brücke) vorbei

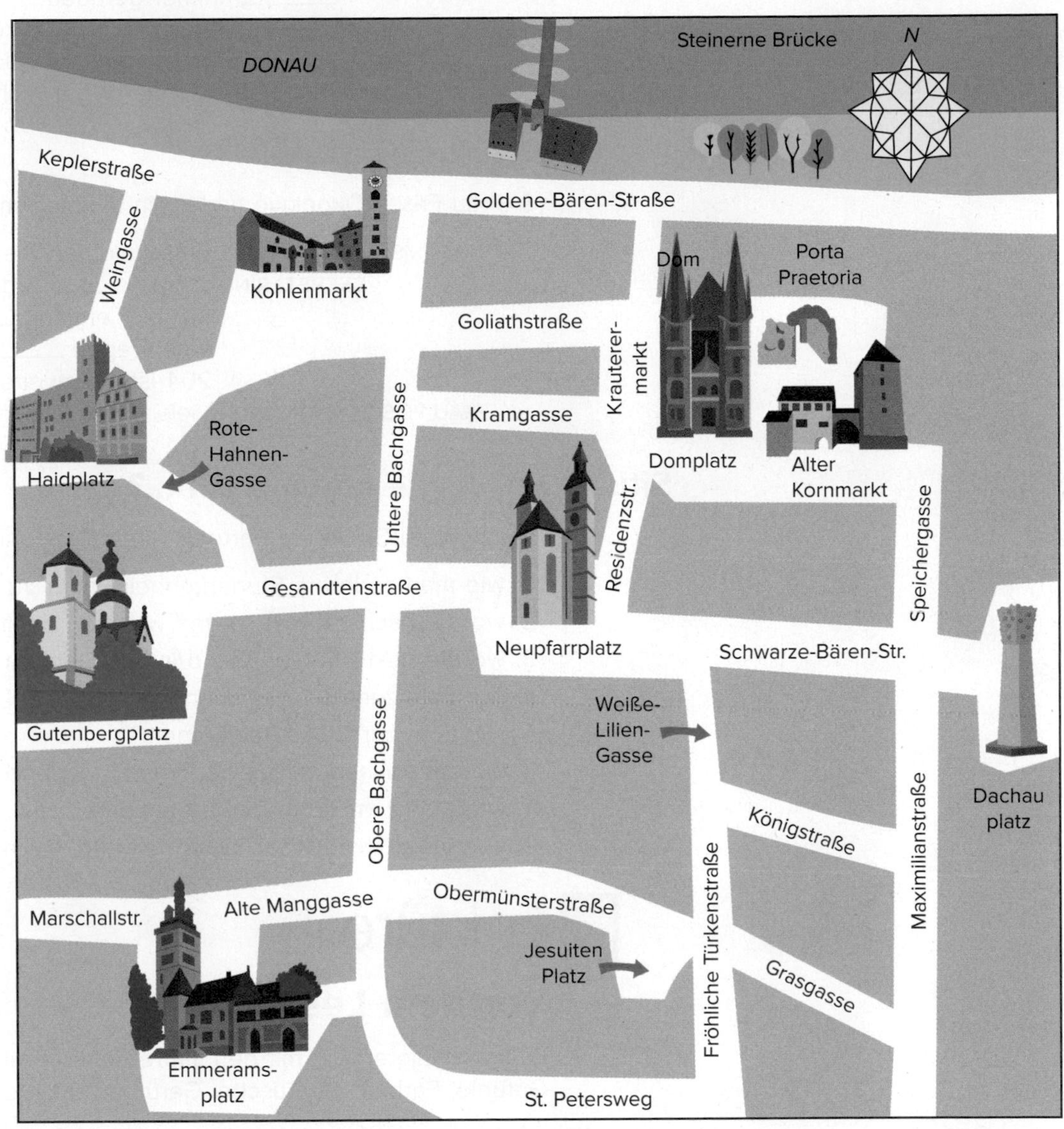

Situation 6 Dialoge

1. Jürgen ist bei Silvias Mutter zum Geburtstag eingeladen.

JÜRGEN: Wie komme ich denn zu eurem Haus?

SILVIA: Das ist ganz einfach. Wenn du ______ Bahnhofsgebäude herauskommst, siehst du rechts ______ anderen Seite der Straße einen Supermarkt. Geh ______ Straße, links ______ Supermarkt vorbei, und wenn du einfach geradeaus weitergehst, kommst du ______ Bismarckstraße. Die musst du nur ganz hinaufgehen, bis du ______ Kreisverkehr kommst. Direkt ______ anderen Seite ist unser Haus.

2. Claire und Melanie sind in Göttingen und suchen die Universitätsbibliothek.

MELANIE: Entschuldige, kannst du uns sagen, wo die Universitätsbibliothek ist?

STUDENT: Ach, da seid ihr aber ganz schön falsch. Also, geht erst die Straße mal wieder zurück ________ großen Kreuzung. ________ Kreuzung ________ und ________ Fußgängerzone ________. Immer geradeaus ________ Fußgängerzone ________ Prinzenstraße. Da rechts. ________ rechten Seite seht ihr dann die Post. Direkt ____________ Post ist die Bibliothek. Könnt ihr gar nicht verfehlen.

MELANIE UND CLAIRE: Danke.

3. Frau Frisch-Okonkwo findet ein Zimmer im Rathaus nicht.

FRAU FRISCH-OKONKWO: Entschuldigen Sie, ich suche Zimmer 204.

SEKRETÄRIN: Das ist ________ dritten Stock. Gehen Sie den Korridor entlang ________ Treppenhaus[2]. Dann eine Treppe ________ und oben links. Zimmer 204 ist die zweite Tür ________ rechten Seite.

FRAU FRISCH-OKONKWO: Vielen Dank. Da hätte ich ja lange suchen können ...

Situation 7 Wie komme ich ...?

Beschreiben Sie Ihrem Partner / Ihrer Partnerin,

1. wie man zu Ihrem Studentenheim oder zu Ihrer Wohnung kommt.
2. wo die nächste Post ist und wie man dahin kommt.
3. wo die beste Kneipe/Disko in der Stadt ist und wie man dahin kommt.
4. wie man zum Schwimmbad kommt.
5. wie man zur Bibliothek kommt.
6. wo der nächste billige Kopierladen ist und wie man dahin kommt.
7. wie man zum Büro von Ihrem Lehrer / Ihrer Lehrerin kommt.
8. wo der nächste Waschsalon[3] ist und wie man dahin kommt.

Lektüre

Vor dem Lesen 1

Was assoziieren Sie mit den Jahreszeiten Frühling und Herbst? Schreiben Sie Gefühle, Farben, Geräusche, Gerüche, Tätigkeiten und Erinnerungen auf.

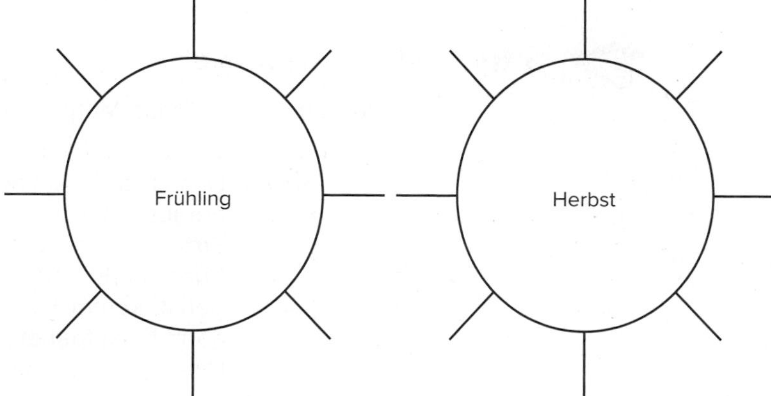

LESEHILFE

A short text like a poem usually requires intensive reading. Every single word is carefully chosen to convey the meaning and feelings one desires to express. One of the most famous poems of Theodor Storm, a well-known German poet and novelist, describes his hometown, Husum.

[2]*stairwell* [3]*laundromat*

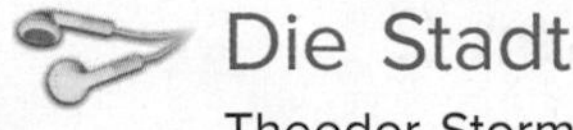

Die Stadt

Theodor Storm

Miniwörterbuch	
seitab	off to the side
brausen	to rage
eintönig	monotonously
rauschen	to rustle
ohne Unterlass	incessantly
wehen	to blow
der **Zauber**	charm
für und für	forever

Am grauen Strand, am grauen Meer
Und seitab liegt die Stadt;
Der Nebel drückt die Dächer schwer,
Und durch die Stille braust das Meer
Eintönig um die Stadt.

Es rauscht kein Wald, es schlägt im Mai
Kein Vogel ohn' Unterlaß;
Die Wandergans mit hartem Schrei
Nur fliegt in Herbstesnacht vorbei,
Am Strande weht das Gras.

Doch hängt mein ganzes Herz an dir,
Du graue Stadt am Meer;
Der Jugend Zauber für und für
Ruht lächelnd doch auf dir, auf dir,
Du graue Stadt am Meer.

Theodor Storm, „Die Stadt", 1852

Arbeit mit dem Text 1

A. Suchen Sie Beispiele aus dem Gedicht für die folgenden Kategorien: Landschaft, Wetter/Jahreszeit, Fauna und Flora, Geräusche. Schreiben Sie sie in die Tabelle.

Landschaft	Wetter/Jahreszeit	Fauna und Flora	Geräusche

B. Kontraste

1. Die ersten beiden Zeilen der zweiten Strophe und die drei weiteren bilden einen Kontrast. Welches Bild oder welche Farbe hat man bei Wald, Mai, Vögel vor Augen und woran denkt man bei Wandergans, Herbstesnacht, Strand und Gras?
2. Die dritte Strophe steht im Kontrast zu den ersten beiden. Warum? Welches Wort ist hier sehr wichtig?

C. Wie ist die Stimmung in dem Gedicht? Fröhlich, melancholisch, dramatisch? Wie erreicht der Dichter das? Denken Sie an Rhythmus, Klang[4] und Lautmalerei[5].

Nach dem Lesen 1

Sind Sie Dichter oder Dichterin? Schreiben Sie ein Gedicht über Ihre Heimatstadt, über die Natur, über die Liebe oder über sich selbst. Das Gedicht muss sich nicht reimen. Es kann auch ein modernes Gedicht sein.

[4] *sound* [5] *onomatopoeia*

LESEHILFE

This selection is taken from a travel guide published by the German automobile association ADAC. Husum, a small town of 25,000, is best known for being the birthplace of Theodor Storm.

Vor dem Lesen 2

A. Was für Informationen erwartet man in einem Reiseführer? Kreuzen Sie an.

- ☐ Museen
- ☐ Restaurants und Kneipen
- ☐ Wetter und Klima
- ☐ Attraktionen
- ☐ Rezepte
- ☐ berühmte Personen
- ☐ Unterkunft
- ☐ Stadtplan[6]
- ☐ Kultur und Feste
- ☐ Zugfahrplan
- ☐ Nachtleben
- ☐ Wörterbuch

B. Überfliegen Sie den Text „Husum" und bestimmen Sie, in welcher Reihenfolge die folgenden Informationen gegeben werden.

_____ Anziehungspunkte in Husum

_____ Informationen zu Theodor Storm, der in Husum geboren wurde

_____ Kirchen und Museen

_____ Vorschläge für einen Stadtrundgang

© Jochen Kallhardt

Husum

Husum ist die Stadt Theodor Storms. Als „Graue Stadt am Meer" hat er sie liebevoll in seinem Gedicht angeredet, das er ihr gewidmet hat. Storm wurde 1817 in Husum geboren und schuf hier einen Teil seiner Gedichte und Novellen. Husum gehörte damals zu den Herzogtümern Schleswig und Holstein und gehörte damit zu Dänemark. Von 1852 bis 1864 konnte der Dichter, der im bürgerlichen Leben als Anwalt, später als Amtsrichter tätig war, nicht in seiner Vaterstadt leben, weil er gegenüber der dänischen Herrschaft die deutsche Sache vertrat. Er starb 1888 in Hademarschen, doch liegt er in Husum begraben.

Sie können in Husum Häuser anschauen, in denen Storm gelebt, und andere, die er in seinen Novellen geschildert hat. Weitere Anziehungspunkte sind der Hafen mit den Krabbenkuttern, das Schloss mit seinen Wiesen, auf denen im Frühling Millionen von Krokussen blühen, sowie die alten Kaufmannshäuser am Markt und in der Großstraße.

Ein Rundgang beginnt am Markt an der Großstraße, führt durch die Hohle Gasse und die Wasserreihe zum Hafen, durch das Westerende und die Nordhusumer Straße zum „Ostenfelder Haus", einem Freilichtmuseum mit einem Niedersachsenhaus des 16./17. Jahrhunderts. Über den alten Friedhof und den Totengang geht man über die Neustadt zum Schloss (Sitz des Kreisarchivs) mit dem als „Cornils'sches Haus" bekannten Torhaus (1612) und durch den Schlossgang zum Markt zurück. Storms Grab auf dem Klosterkirchhof erreichen Sie vom Markt aus durch die Norderstraße.

Miniwörterbuch

gewidmet	dedicated
schaffen, schuf	to create
das **Herzogtum**	duchy
der **Amtsrichter**	district judge
die **Sache**	*here:* cause
vertreten, vertrat	to plead for
schildern	to portray
der **Anziehungspunkt**	attraction
der **Rundgang**	(walking) tour
das **Freilichtmuseum**	open-air museum
sich befinden	to be located

[6] *city street map*

Das Haus in der Wasserreihe 31, in dem der Dichter zwischen 1866 und 1880 wohnte, dient heute als Storm-Museum (täglich geöffnet von April bis Oktober). Im Nissenhaus befindet sich das Nordfriesische Museum zu den Themen Erd- und Vorgeschichte, Landschaftskunde und Kulturgeschichte (täglich geöffnet). Die Marktkirche Husums gilt als der bedeutendste klassizistische Kirchenbau Schleswig-Holsteins.

(aus: ADAC-Reiseführer Norddeutschland)

Arbeit mit dem Text 2

A. Ein Rundgang durch Husum. Zeichnen Sie den Weg, der im Reiseführer beschrieben wird, in den Stadtplan ein.

B. Storms Leben. Welche dieser Jahreszahlen und Ereignisse stehen im Text, welche nicht? Unterstreichen Sie sie im Text, wenn sie im Text stehen.

		IM TEXT?
1817	wird Theodor Storm in Husum geboren	________
1843–1852	ist er Rechtsanwalt in Husum	________
1846	erste Heirat mit Konstanze Esmarch	________
1852–1856	ist er Assessor in Potsdam	________
1852–1864	lebt er aus politischen Gründen nicht in Husum	________
1856–1864	ist er Richter in Heiligenstadt	________
1864–1867	ist er Landvogt[7] in Husum	________
1866	zweite Heirat mit Dorothea Jensen	________
1866–1880	wohnt er in der Wasserreihe 31	________
1867	wird er Amtsrichter	________
1888	stirbt er in Hademarschen	________

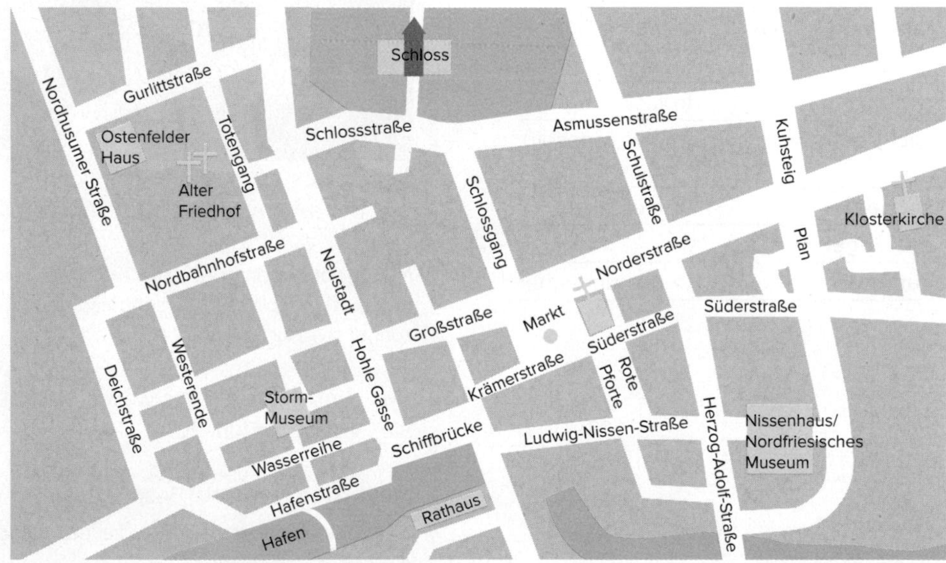

Nach dem Lesen 2

Suchen Sie im Internet mehr Informationen über Husum und über Theodor Storm und stellen Sie sie in der Klasse vor.

[7] *governor*

MUSIKSZENE

„Dieser Weg" (2005, Deutschland) *Xavier Naidoo*

Biografie Xavier Naidoo ist 1971 in Mannheim geboren. Sein Vater kam aus Sri Lanka und seine Mutter aus Südafrika. Er ist einer der erfolgreichsten deutschen Sänger, sowohl als Mitglied der Band *Söhne Mannheims* als auch als Solosänger. Naidoo ist ein bekennender Christ. Viele seiner Texte handeln von Gott und der Nächstenliebe, aber auch von der Wichtigkeit, Fremdenhass zu bekämpfen. Er engagiert sich für christliche Projekte wie *Zeichen der Zeit* und für Projekte gegen Fremdenhass wie *Rock gegen Rechts* und *Brothers Keepers*. Die Single „Dieser Weg" stammt aus dem Jahr 2005 und wurde die Hymne der deutschen Fußballnationalmannschaft während der Europameisterschaft in Deutschland 2006.

Der Sänger Xavier Naidoo

Miniwörterbuch

das **Mitglied**	member
bekennend	avowed
handeln von	deal with
die **Nächstenliebe**	charity
der **Fremdenhass**	xenophobia
bekämpfen	to combat
sich engagieren	to get involved
das **Zeichen**	sign
erreichen	to achieve
(ent-)lang gehen	to walk along
steinig	rocky
treten	to kick
(sich) aufgeben	to give (oneself) up
segnen	to bless
das **Segel**	sail
aufbrausen	to pile up higher

NOTE: For copyright reasons, the songs referenced in **MUSIKSZENE** have not been provided by the publisher. The song can be found online at various sites such as YouTube, Amazon, or the iTunes store.

Vor dem Hören Was wollen Sie in Ihrem Leben erreichen? Was ist wichtig für Sie? Wohin führt Ihr Weg?

Nach dem Hören Beantworten Sie die Fragen.

1. Wohin führt die Straße, die der Sänger entlang geht?
2. Was spielt im Sänger?
3. Wie ist *dieser* Weg?
4. Was machen manche Menschen mit jemandem?
5. Wann soll man sein Segel nicht setzen?

Urlaub am Strand

Grammatik 10.4

Situation 8 Umfrage: Urlaub am Strand

MODELL: S1: Hast du schon mal eine Sandburg gebaut?
S2: Ja.
S1: Unterschreib bitte hier.

	UNTERSCHRIFT
1. Hast du schon mal eine Sandburg gebaut?	________
2. Hast du eine Luftmatratze?	________
3. Bist du schon mal im Meer geschwommen?	________
4. Kannst du Wellen reiten?	________
5. Sammelst du gern Muscheln?	________
6. Warst du schon mal windsurfen?	________
7. Liegst du gern im Liegestuhl?	________
8. Hast du schon mal auf einer Luftmatratze gelegen?	________
9. Bekommst du leicht einen Sonnenbrand?	________
10. Benutzt du viel Sonnenmilch?	________

Situation 9 Informationsspiel: Wo wollen wir übernachten?

MODELL: Wie viel kostet _____?
Haben die Zimmer im (in der) _____ eine eigene Dusche und Toilette?
Gibt es im (in der) _____ Einzelzimmer?
Gibt es im (in der, auf dem) _____ einen Fernseher?
Ist das Frühstück im (in der, auf dem) _____ inbegriffen?
Ist die Lage von dem (von der) _____ zentral/ruhig?
Gibt es im (in der, auf dem) _____ Internet?

	das Hotel Strandpromenade	das Hotel Ostseeblick	die Jugendherberge	der Campingplatz
Preis pro Person	88,- Euro		18,50 Euro	16,- Euro
Dusche/Toilette	ja	ja		
Einzelzimmer	ja	ja	nein	
Fernseher			im Aufenthaltsraum	
Frühstück		inbegriffen		
zentrale Lage	ja	ja	im Wald	direkt am Strand
ruhige Lage	ja	ja	ja	
Internet				

Situation 10 Dialog: Auf Zimmersuche

Herr und Frau Ruf suchen ein Zimmer.

HERR RUF: Guten Tag, haben Sie noch ein Doppelzimmer mit Dusche frei?
WIRTIN: Wie lange möchten Sie denn ________?
HERR RUF: ________.
WIRTIN: Ja, da habe ich ein Zimmer ________ und Toilette.
FRAU RUF: Ist das Zimmer auch ruhig?
WIRTIN: Natürlich. Unsere Zimmer sind alle ruhig.
FRAU RUF: ________ das Zimmer denn?
WIRTIN: 54 Euro ________.
HERR RUF: Ist Frühstück dabei?
WIRTIN: Selbstverständlich ist Frühstück dabei.
FRAU RUF: Gut, wir nehmen das Zimmer.
HERR RUF: Und wann können wir ________?
WIRTIN: ____________ im Frühstückszimmer.

Situation 11 Rollenspiel: Im Hotel

S1: Sie sind im Hotel und möchten ein Zimmer mit Dusche und Toilette. Außerdem möchten Sie ein ruhiges Zimmer. Fragen Sie auch nach Preisen, Frühstück, Internet und wann Sie morgens abreisen müssen.

KULTUR ... LANDESKUNDE ... INFORMATIONEN

DIE DEUTSCHE EINWANDERUNG IN DIE USA

- Gibt es in Ihrer Nähe Orte, Städte oder Stadtteile mit deutschem Namen? Wie heißen sie? Wann wurden sie gegründet[1]?
- Gibt es in Ihrer Stadt ein Viertel[2] mit deutschen Geschäften und Restaurants?
- Welche deutschen Einwanderer[3] spielten eine wichtige Rolle in der Geschichte der USA (oder Ihres Landes)?

Lesen Sie den Text und suchen Sie die Antworten auf diese Fragen:

- Aus welchem Land kamen die meisten Einwanderer in die USA? Und die zweitmeisten?
- Wie viele Millionen US-Amerikaner sagen, dass *Deutsch* ihre Hauptabstammung[4] ist?
- Wie hieß die erste deutsche Siedlung in Nordamerika? Wo lag sie? Wann wurde sie gegründet?
- Welchen Religionen gehörten viele Deutsche an, die im 18. Jahrhundert in Pennsylvania einwanderten[5]?
- Warum hieß eine besonders große Einwanderergruppe die *Forty-Eighters*?
- Welche Region wurde der *German Belt* genannt?
- Wie viel Prozent der Einwohner San Antonios sprach im 19. Jahrhundert Deutsch?
- Welche Industrie wurde besonders von Deutschen dominiert?
- Welche Berufsgruppe und welche Religionsgruppe wanderte in den 1930er Jahren vor allem nach Amerika aus?

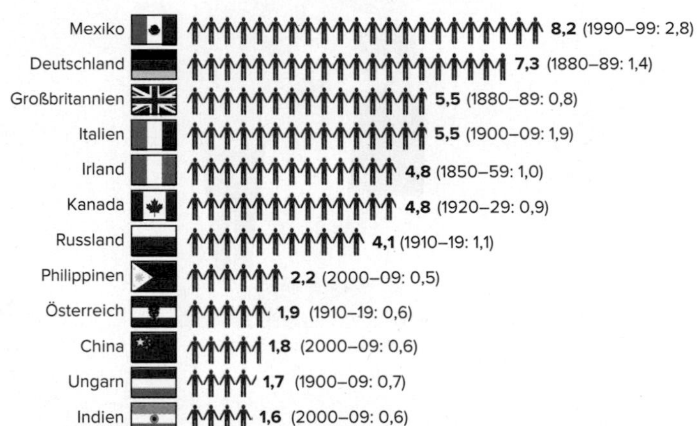

Quelle: Yearbook of Immigration Statistics, U.S. Department of Homeland Security, http://www.dhs.gov

Die USA sind eines der großen Einwanderungsländer. Jedes Jahr wandern zahlreiche Personen in die USA ein. Deutschland steht hoch auf der Liste der Einwanderungsnationen. Allein in New York City leben circa 500.000 Deutschstämmige[6]. Bei einer Volkszählung[7] im Jahr 2006 gaben 51 Millionen US-Amerikaner *Deutsch* als ihre Hauptabstammung an.

Die deutsche Einwanderung in die USA hat eine lange Tradition. 1683 wurde in Pennsylvanien die erste deutsche Siedlung[8] mit dem Namen „Germantown" gegründet. 90 Jahre später war ein Drittel der pennsylvanischen Bevölkerung[9] deutschstämmig. Sie gehörten größtenteils protestantischen Religionen an. Auf der Basis pfälzischer Dialekte entwickelten die Deutschamerikaner eine eigene Sprache – das Pennsylvania Dutch. Dieser Dialekt wird auch heute noch in einigen Teilen Pennsylvaniens gesprochen.

Im 19. Jahrhundert wanderten fast 8 Millionen deutschsprachige Menschen in den USA ein, aus allen deutschsprachigen Ländern: aus der Schweiz, aus Österreich-Ungarn und aus dem Deutschen Reich. Viele kamen nach der gescheiterten[10] deutschen Revolution von 1848. Diese Einwanderer hießen die *Forty-Eighters*. Unter ihnen befanden sich viele Intellektuelle und Bürgerrechtskämpfer[11], ebenso wie viele Juden[12], wie zum Beispiel Abraham Jacobi, der 1860 das erste Kinderkrankenhaus der USA eröffnete[13] und Emil Berliner, der Erfinder der Schallplatte. Sehr viele Deutsche ließen sich im Mittleren Westen nieder[14], insbesondere in der Region zwischen Cincinnati, St. Louis und Milwaukee, die der *German Belt* genannt wurde. 1890 waren 69% der Einwohner Milwaukees deutschstämmig, in Cincinnati waren es zu Beginn des 20. Jahrhunderts 60%. Auch in Texas wanderten sehr viele Deutsche ein. Ein Drittel der Einwohner von San Antonio sprach 1870 Deutsch.

Viele berühmte Amerikaner waren deutscher Abstammung. John Jacob Astor kam 1784 als junger Mann nach New York. Der Chemiker Karl Pfizer gründete 1849 in Brooklyn ein Pharmaunternehmen[15], das es heute noch gibt. Levi Strauss, der Erfinder der Jeans, wanderte 1853 aus Bayern ein. Ein Monopol hatten deutsche Einwanderer in der Bierindustrie: Yuengling, Anheuser-Busch, Joseph Schlitz und Coors wurden alle im 19. Jahrhundert gegründet. Auch Hamburger, Frankfurter, Bratwurst, Schnitzel, Strudel und Brezel sind deutschen Ursprungs[16].

Nach der Machtergreifung[17] der Nazis in den 1930er Jahren verließen wiederum viele Deutsche ihre Heimat, vor allem Akademiker und vor allem jüdische Akademiker. Zu den Auswanderern gehörten der Physiker Albert Einstein, die Mathematikerin Emmy Noether, die Philosophin Hannah Arendt, der Schriftsteller Thomas Mann, der Architekt Walter Gropius und die Schauspielerin Marlene Dietrich.

[1]*established* [2]*neighborhood* [3]*immigrants* [4]*primary ancestry* [5]*immigrated* [6]*people of German descent* [7]*census* [8]*settlement* [9]*populace* [10]*failed* [11]*civil rights activist* [12]*Jews* [13]*opened* [14]ließen sich nieder *settled* [15]*pharmaceutical company* [16]*origin* [17]*takeover*

Tiere

Grammatik 10.5

Juttas Ratte wird gegen Tollwut geimpft.

Ernsts Meerschweinchen wird oft gebadet.

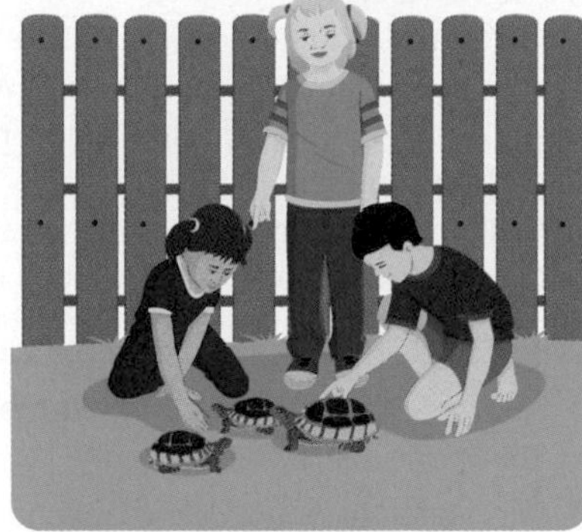

Schildkröten werden oft als Haustiere gehalten.

In der Wüste muss man aufpassen, dass man nicht von einer Schlange gebissen wird.

Gestern wurde Silvia von einer Biene gestochen.

Als Josef und Melanie gestern beim Baden waren, wurden sie von tausend Mücken gestochen.

Situation 12 Ratespiel

die Klapperschlange

die Schildkröte

die Schnecke

der Kolibri

der Gepard

die Fledermaus

1. Das größte Landsäugetier[1]: Es hat einen Rüssel[2] und zwei Stoßzähne[3] aus Elfenbein; wegen des Elfenbeins wird es oft illegal gejagt.
2. Die schnellste Katze der Welt: Sie läuft mindestens 80 Kilometer in der Stunde.
3. Das schwerste Tier: Es lebt im Wasser, aber es ist kein Fisch.
4. Das langsamste Tier: Es trägt oft ein Haus auf seinem Rücken und hat keine Beine.
5. Es sieht aus wie ein Hund, ist aber nicht so zahm.
6. Dieses Tier lebt länger als der Elefant.
7. Das ist die giftigste Schlange in Nordamerika.
8. Dieser Wasservogel hat eine Spannweite von mehr als drei Metern.
9. Dieses Tier hat die höchste Herzfrequenz, mit zirka 1.000 Schlägen pro Minute.
10. Dieses Tier hört besser als ein Delfin.

a. der Kolibri
b. der Elefant
c. die Riesenschildkröte
d. die Schnecke
e. die Fledermaus
f. der Blauwal
g. der Gepard[4]
h. die Klapperschlange
i. der Albatros
j. der Wolf

der Blauwal

[1]*land mammal* [2]*trunk* [3]*tusks* [4]*cheetah*

Situation 13 Informationsspiel: Tiere

MODELL: Welche Tiere findet _____ am tollsten?
Vor welchem Tier hat _____ am meisten Angst?
Welches Tier hätte _____ gern als Haustier?
Welches wilde Tier möchte _____ gern in freier Natur sehen?
Wenn _____ an Afrika denkt, an welche Tiere denkt er/sie?
Wenn _____ an die Wüste denkt, an welches Tier denkt er/sie dann zuerst?
Welche Vögel findet _____ am schönsten?
Welchen Fisch findet _____ am gefährlichsten?
Welchem Tier möchte _____ nicht im Wald begegnen?

	Ernst	Maria	mein(e) Partner(in)
Lieblingstier		eine Katze	
Angst	vor dem Hund von nebenan		
Haustier	eine Schlange		
wildes Tier		eine Giraffe	
Afrika	an Löwen		
Wüste		an ein Kamel	
Vögel		Eulen	
Fisch	den weißen Hai		
Wald	einem Wolf		

Zwei ältere Menschen gehen mit Hunden aufs Gassi.

Situation 14 Interview: Tiere

1. Was ist dein Lieblingstier? Warum?
2. Vor welchen Tieren fürchtest du dich?
3. Welches Tier findest du am interessantesten?
4. Welches Tier findest du am hässlichsten?
5. Welches Tier wärst du am liebsten? Warum?
6. Hast du oder hattest du ein Haustier? Was für eins? Wie heißt oder wie hieß es? Beschreib es. Erzähl eine Geschichte von ihm!
7. Findest du es wichtig, dass Kinder mit Tieren aufwachsen? Wenn ja, mit welchen? Warum?

Situation 15 Bildgeschichte: Lydias Hamster

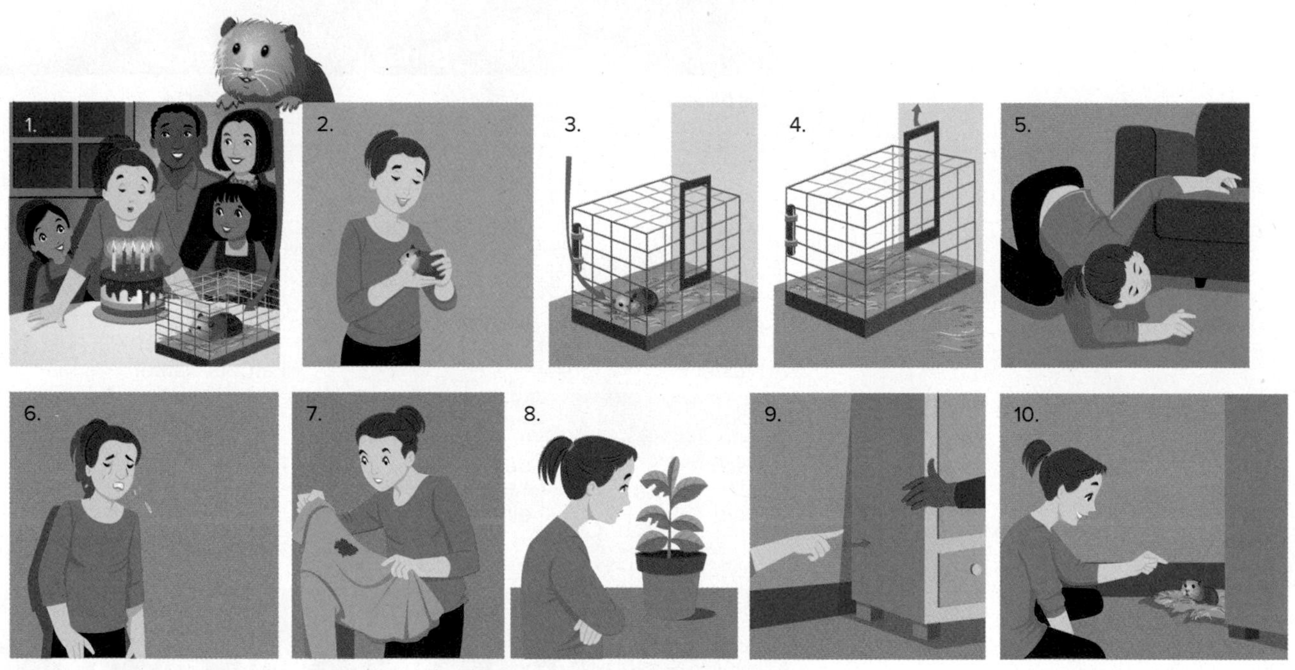

Situation 16 Tiere in Sprichwörtern

In vielen Sprachen gibt es Sprichwörter, in denen Tiere vorkommen. Welche Sprichwörter fallen Ihnen auf Englisch ein? Ordnen Sie jeder Zeichnung das passende Sprichwort (1 - 6) zu.

1. Wenn dem Esel zu wohl ist, geht er aufs Eis.
2. Einem geschenkten Gaul (= Pferd) sieht man nicht ins Maul (= Mund).
3. Kaum ist die Katze aus dem Haus, tanzen die Mäuse auf dem Tisch.
4. Den letzten beißen die Hunde.
5. In der Not[5] frisst der Teufel Fliegen.
6. Ein blindes Huhn findet auch mal ein Korn.

[5]*emergency*

__

__

__

__

__

__

Was bedeuten die Sprichwörter? Kombinieren Sie die Definitionen (a - f) mit den Sprichwörtern (1 - 6).

a. Wenn man etwas geschenkt bekommt, sollte man nicht zu kritisch damit sein.
b. Wenn man etwas nötig braucht, muss man nehmen, was da ist.
c. Wenn der Chef nicht da ist, machen die Angestellten, was sie wollen.
d. Jemandem, der sonst wenig Erfolg hat, kann auch etwas gelingen.
e. Wenn man sich nicht beeilt, ergeht es einem schlecht.
f. Leute, die zu viel Erfolg oder Glück haben, werden übermütig[6].

[6] *cocky*

Filmlektüre

Die fetten Jahre sind vorbei

Vor dem Lesen

A. Schauen Sie sich das Foto an.

1. Welche Personen sehen Sie auf dem Foto?
2. Beschreiben Sie die junge Frau.
3. Welchen der beiden jungen Männer finden Sie am sympathischsten? Warum?

FILMANGABEN

Titel: *Die fetten Jahre sind vorbei*
Genre: Drama
Land: Österreich
Erscheinungsjahr: 2004
Dauer: 127 Min.
Regisseur: Hans Weingartner
Hauptrollen: Daniel Brühl, Julia Jentsch, Stipe Erceg, Burghart Klaußner

... denn sie wissen nicht, was sie tun

B. Lesen Sie die Wörter im Miniwörterbuch. Suchen Sie sie im Text und unterstreichen Sie sie.

Miniwörterbuch

verteilt	distributed
einbrechen	to break in
verrücken	to move, to disarrange
verstecken	to hide
die **Nachricht**	message
unterzeichnen	to sign
die **Erziehungsberechtigten**	legal guardians
im Überschwang der Gefühle	in exuberance
schulden	to owe
das **Vermögen**	fortune
beschädigen	to damage
entführen	to kidnap
die **Berghütte**	mountain cabin
aufbegehren	to revolt
verraten	to betray
auf etwas verzichten	to go without
aufbrechen	to take off

Inhaltsangabe

Besitz und Geld sind auf der Welt ungerecht verteilt. Jan (Daniel Brühl) und Peter (Stipe Erceg), zwei junge Berliner, wollen diese Situation ändern und haben eine eigene Methode dafür gefunden: Sie brechen nachts in Villen reicher Leute ein. Sie stehlen nichts, sondern verrücken Möbel, hängen Bilder um und verstecken wertvolle Gegenstände im Kühlschrank oder werfen sie in den Swimmingpool. Die Nachrichten, die sie für die Hausbesitzer hinterlassen, lauten: „Die fetten Jahre sind vorbei" oder „Sie haben zu viel Geld", unterzeichnet mit „Die Erziehungsberechtigten". Ihr Ziel: Die Reichen sollen über ihren Luxus nachdenken.

Alles läuft immer nach Plan. Jan und Peter haben ihren Spaß und die Villenbewohner sind geschockt beim Anblick ihrer Häuser. Doch als sich dann Jan und Jule (Julia Jentsch), Peters Freundin, ineinander verlieben, brechen die beiden im Überschwang der Gefühle und ohne Peter in die Villa des Geschäftsmannes Justus Hardenberg (Burghart Klaußner) ein. Dem schuldet Jule ein halbes Vermögen, weil sie bei einem Unfall sein teures Auto beschädigt hat. Ein harmloser Einbruch wie die anderen wird es nicht, denn sie werden vom Hausbesitzer überrascht. Jan und Jule schlagen Hardenberg nieder, entführen ihn und bringen ihn mit Peters Hilfe in die Berghütte von Jules Onkel am Tiroler Achensee. In der Berghütte stellt sich heraus, dass Hardenberg in seinen jungen Jahren genauso gegen das etablierte Bürgertum aufbegehrte wie Jan, Peter und Jule jetzt. Seine Ideale von früher hat Hardenberg jedoch verraten.

Am Ende bringen die drei Entführer Hardenberg in seine Villa zurück. Er verzichtet auf das Geld, das ihm Jule wegen des Autounfalls schuldet. Als die Polizei wenig später die Wohnung der drei jungen Leute stürmt, sind sie schon verschwunden. Am Schluss brechen sie mit Hardenbergs Motorjacht zu neuen Taten auf.

Arbeit mit dem Text

Welche Aussagen sind falsch? Verbessern Sie die falschen Aussagen.

1. Peter und Jan brechen in Villen ein, weil sie Geld brauchen.
2. Jule ist Peters Freundin, verliebt sich aber in Jan.
3. Peter und Jan brechen in die Villa von Justus Hardenberg ein.
4. Hardenberg schuldet Jule sehr viel Geld.
5. Jan, Peter und Jule entführen Hardenberg, weil er sie beim Einbruch in seine Villa überrascht hat.
6. Die drei Entführer lassen Hardenberg in Tirol frei.
7. Jan, Peter und Jule melden sich[7] bei der Polizei.

FILMCLIP

NOTE: For copyright reasons, the films referenced in the **FILMCLIP** feature have not been provided by the publisher. The film can be purchased as a DVD or found online at various sites such as YouTube, Amazon, or the iTunes store. The time codes mentioned below are for the North American DVD version of the film.

Szene: DVD, Kapitel 1, „Start", 4:30–8:25 Min.

Jule nimmt an einer Demonstration gegen Sweatshops teil. Jan sitzt im Park und fährt dann mit der Straßenbahn nach Hause. Schauen Sie sich den Filmclip an. Bringen Sie die Sätze in die richtige Reihenfolge.

___*1*___ Jule nimmt an einer Demo gegen Sweatshops teil und erklärt zwei Mädchen, wie viel die kleinen Kinder in Entwicklungsländern arbeiten müssen.
______ Als die Polizisten und die zwei festgenommenen jungen Leute im Polizeiwagen sitzen, versuchen die Demonstranten ihn umzukippen.
______ Die Demonstranten wollen das verhindern, die Polizei greift aber brutal durch.
______ Die Polizei nimmt die zwei jungen Männer fest und bringt sie in den Polizeiwagen.
______ Ein Kontrolleur geht hinter Jan her und will ihn anpöbeln, Jan aber lässt sich das nicht gefallen.
______ Jan (Brühl) sitzt im Park und sieht einer jungen Familie zu.
______ Jan mischt sich ein und gibt dem alten Mann seine Karte.
______ Jan sitzt in der Straßenbahn und sieht, wie drei Kontrolleure einen Obdachlosen schlecht behandeln.
______ Jule geht nach Hause und bekommt eine Räumungsklage.
______ Jule hat zwar die Miete bezahlt, aber sechs Monate zu spät.
______ Jule liest die Räumungsklage und ist völlig verzweifelt.
______ Zwei Angestellte des Sportladens werfen die zwei jungen Männer hinaus.
______ Zwei junge Männer gehen in einen Sportladen und erklären, wie die Turnschuhe hergestellt werden.

Miniwörterbuch

das **Entwicklungsland**	developing country
festnehmen	to apprehend
umkippen	to topple
verhindern	to prevent
durchgreifen	to crack down
anpöbeln	to accost
sich einmischen	to intervene
der **Obdachlose**	homeless person
behandeln	to treat
die **Räumungsklage**	eviction notice
zwar ... aber	to be sure ... but
verzweifelt	desperate
herstellen	to make

Nach dem Lesen

Beantworten Sie die folgenden Fragen.

1. Glauben Sie, dass Jan, Peter und Jule mit ihren Aktionen Erfolg[8] haben? Rüttelt man mit so etwas die Gesellschaft wach[9]? Kann oder muss man die drei ernst[10] nehmen? Warum? Warum nicht?
2. Justus Hardenberg gehörte zu den sogenannten „68ern". 1968 war ein aufregendes Jahr in der alten BRD. Forschen Sie im Internet nach, was in diesem Jahr in Westdeutschland passierte und welche Rolle die Studenten gespielt haben.

[7]melden ... *turn themselves in* [8]*success* [9]rüttelt wach *shakes awake* [10]*seriously*

Videoecke

Perspektiven

Was machst du gern im Urlaub?

Ich wandere gerne und mache viel Sport.

Aufgabe 1 Wer sagt das?

Ordnen Sie die Aussagen den Personen zu.

1. Sandra ___

2. Shaimaa ___

3. Tina ___

4. Jenny ___

5. Nadezda ___

6. Pascal ___

7. Simone ___

8. Michael ___

a. Ich entspanne am liebsten am Strand.
b. Ich erhole mich gern.
c. Ich fahre am liebsten ans Meer.
d. Ich gucke mir Gebäude an und treffe Menschen, die dort leben.
e. Ich mache gerne gar nichts.
f. Ich schwimme sehr gerne.
g. Ich wandere gerne.
h. Reisen und schlafen.

Interviews

- Wohin fährst du gern in Urlaub?
- Was war dein bisher schönster Urlaub?
- Was war daran so besonders?
- Gab es mal einen Urlaub, in dem etwas schief ging?
- Hast du oder hattest du ein Haustier?
- Gibt es über dein Haustier eine lustige Geschichte?

Tina

Tabea

Aufgabe 2 Tina oder Tabea?

Sehen Sie sich das Video an und schreiben Sie Tina oder Tabea neben die Aussagen.

1. Mein schönster Urlaub war meine Reise nach Spanien.
2. Mein schönster Urlaub war am Meer.
3. Wir haben den tollen Sternenhimmel gesehen.
4. Ich habe dort drei Monate gearbeitet.
5. Mir wurde einmal die Tasche geklaut.
6. Wir haben Flusskrebse gebraten.
7. Wir haben die Hälfte des Essens vergessen.
8. Ich hatte ein Zwergkaninchen.
9. Ich hatte einen Hasen namens Milan.
10. Ich dachte immer, es wär ein Junge.

Aufgabe 3 Tinas Tasche

Tina wird die Tasche geklaut. Hören Sie sich die Geschichte an und erzählen Sie sie dann mithilfe der folgenden Notizen.

Tasche neben sich stellen **im Bus sitzen**

einen Moment unachtsam **problematisch**

gut aufpassen **Tasche weg**

Personalausweis und Portemonnaie drin

Aufgabe 4 Tinas Hase

Tina hatte einen Hasen. Hören Sie sich die Geschichte an, machen Sie sich Notizen und erzählen Sie dann die Geschichte mithilfe Ihrer Notizen.

Aufgabe 5 Interview

Interviewen Sie eine Partnerin oder einen Partner. Stellen Sie dieselben Fragen.

Wortschatz

Reisen und Tourismus	Travel and Tourism
die **Fahrt, -en** (R)	trip
die **Haltestelle, -n**	stop
die **Jugendherberge, -n**	youth hostel
die **Klasse, -n** (R)	class
erster Klasse fahren	to travel first class
die **Lage, -n**	place; position
die **Luftmatratze, -n**	air mattress
die **Möwe, -n**	seagull
die **Reisende, -n**	female traveler
die **Schiene, -n**	train track
die **Sonnenmilch**	suntan lotion
die **Welle, -n**	wave
der **Aufenthaltsraum, ¨e**	lounge, recreation room
der **Ausweis, -e**	identification card
der **Hafen, ¨**	harbor, port
der **Reisende, -n** **(ein Reisender)**	male traveler
der **Sonnenbrand, ¨e**	sunburn
der **Sonnenschirm, -e**	sunshade, beach parasol
der **Strandkorb, ¨e**	beach chair
der **Wirt, -e**	host, innkeeper; bar keeper
der **Zug, ¨e** (R)	train
das **Einzelzimmer, -**	single room
das **Gleis, -e**	(set of) train tracks
das **Kanu, -s**	canoe
Kanu fahren	to go canoeing
das **Ziel, -e**	destination

Ähnliche Wörter

die **Idee, -n;** die **Rezeption, -en;** der **Campingplatz, ¨e;** das **Doppelzimmer, -** (R); das **Fernsehzimmer, -; packen**

Den Weg beschreiben	Giving Directions
ab·biegen, bog ... ab, ist abgebogen	to turn
entlang·gehen	to go along
verfehlen	to miss, not notice
vorbei·gehen (an + *dat.*)	to go by
weiter·gehen	to keep on walking
entlang	along
gegenüber von (R)	across from
geradeaus	straight ahead
her(·kommen)	(to come) this way
heraus(·kommen)	(to come) out this way
herein(·kommen)	(to get/go) in this way
hin(·gehen)	(to go) that way
hinauf(·gehen)	(to go) up that way
hinüber(·gehen)	(to go) over that way
links (R)	left
oben	above
rechts (R)	right

In der Stadt	In the City
die **Brücke, -n**	bridge
die **Bushaltestelle, -n**	bus stop
die **Gasse, -n**	narrow street; alley
die **Kreuzung, -en**	intersection
der **Dom, -e**	cathedral
der **Kopierladen, ¨**	copy shop
der **Kreisverkehr, -e**	traffic roundabout
der **Zebrastreifen, -**	crosswalk
das **Gebäude, -**	building

Ähnliche Wörter

die **Fußgängerzone, -n;** der **Markt, ¨e**

Tiere	Animals
die **Biene, -n**	bee
die **Mücke, -n**	mosquito
die **Schildkröte, -n**	turtle
die **Schlange, -n**	snake
der **Adler, -**	eagle
der **Hai, -e**	shark
der **Löwe, -n** (*wk. masc.*)	lion
der **Papagei, -en**	parrot
der **Vogel, ¨**	bird
das **Meerschweinchen, -**	guinea pig
das **Tier, -e** (R)	animal
das **Haustier, -e**	pet

Ähnliche Wörter

die **Giraffe, -n;** die **Maus, ¨e;** die **Ratte, -n;** der **Delfin, -e;** der **Hamster, -;** der **Piranha, -s;** der **Skorpion, -e;** das **Känguru, -s;** das **Krokodil, -e;** das **Wildschwein, -e;** das **Zebra, -s**

Sonstige Substantive	Other Nouns
die **Bürgerin, -nen**	female citizen
die **Fensterbank, ¨-e**	windowsill
die **Tollwut**	rabies
der **Bürger, -**	male citizen
der **Käfig, -e**	cage
das **Elfenbein**	ivory

Ähnliche Wörter

der **Staat, -en;** das **Nest, -er; in freier Natur**

Sonstige Verben	Other Verbs
ab·reisen, ist abgereist	to depart
ein·schalten	to turn on
ein·steigen (R), **stieg ... ein, ist eingestiegen**	to board
entschuldigen	to excuse
entschuldigen Sie!	excuse me!
sich erkundigen nach	to ask about, get information about
sich fürchten vor (+ *dat.*)	to be afraid of
grüßen	to greet, say hello to
impfen gegen	to vaccinate against
sammeln	to collect
sonnenbaden gehen	to go sunbathing
stechen, sticht, stach, gestochen	to sting; to bite (*of insects*)
trampen, ist getrampt	to hitchhike

Ähnliche Wörter

antworten (+ *dat.*) (R); **windsurfen gehen**

Adjektive und Adverbien	Adjectives and Adverbs
einfach	one-way (trip)
gefährlich	dangerous
gemütlich	cozy
komisch	funny, strange
nützlich	useful
überall	everywhere
ungeduldig	impatient
zahm	tame

Ähnliche Wörter

extra, voll, zentral

Sonstige Wörter und Ausdrücke	Other Words and Expressions
an ... vorbei	by
aus	of; from; out of
außerdem (R)	besides
bei (R)	at; with; near
bis zu	as far as; up to
danach (R)	afterward
hin und zurück (R)	there and back; round-trip
inbegriffen	included
nach (R)	to (*a place*)
nach Hause (R)	(to) home
selbstverständlich	of course
vielen Dank	many thanks
von (R)	of; from
zu (R)	to (*a place*)
zu Hause (R)	at home

Strukturen und Übungen

10.1 Prepositions to talk about places: *aus, bei, nach, von, zu*

WISSEN SIE NOCH?

The prepositions **aus** (*from*), **bei** (*near, with*), **mit** (*with*), **nach** (*to*), **von** (*from*), and **zu** (*to*) are prepositions that take the dative case.

Review grammar B.6, 6.4, and 6.6.

Use the prepositions **aus** and **von** to indicate origin; **bei** to indicate a fixed location; and **nach** and **zu** to indicate destination. These five prepositions are always used with nouns and pronouns in the dative case.

Woher (kommt sie?)	Wo (ist sie?)	Wohin (geht/fährt sie?)
aus Spanien		nach Spanien
aus dem Zimmer		nach Hause
von rechts		nach links
von Erika	bei Erika	zu Erika
vom Strand		zum Strand

A. The Prepositions **aus** and **von**

aus: enclosed spaces
countries
towns
buildings

1. Use **aus** to indicate that someone or something comes from an enclosed or defined space, such as a country, a town, or a building.

Diese Fische kommen **aus der Donau.**	*These fish come from the Danube river.*
Jens kam **aus seinem Zimmer.**	*Jens came out of his room.*

Most country and city names are neuter; no article is used with these names.

Josef kommt **aus Deutschland.**

However, the article is included when the country name is masculine, feminine, or plural.

Richards Freund Ali kommt **aus dem Iran.**
Mehmets Familie kommt **aus der Türkei.**
Ich komme **aus den USA.**

von: open spaces
directions
persons

2. Use **von** to indicate that someone or something comes not from an enclosed space but from an open space, from a particular direction, or from a person.

Melanie kommt gerade **vom Markt** zurück.	*Melanie's just returning from the market.*
Das rote Auto kam **von rechts.**	*The red car came from the right.*
Michael hat es mir gesagt. Ich weiß es **von ihm.**	*Michael told me. I know it through (from) him.*

B. The Preposition **bei**

bei: place of work
residence

Use **bei** before the name of a place where someone works or a place where someone lives or is staying.

Albert arbeitet **bei McDonald's.**	*Albert works at McDonald's.*
Rolf wohnt **bei einer Gastfamilie.**	*Rolf is staying with a hostfamily.*
Treffen wir uns **bei Katrin.**	*Let's meet at Katrin's.*

ACHTUNG!

von + dem = vom
bei + dem = beim
zu + dem = zum
zu + der = zur

nach: cities
countries without articles
direction
nach Hause (idiom)

C. The Prepositions **nach** and **zu**

Use **nach** with neuter names of cities and countries (no article), to indicate direction, and in the idiom **nach Hause** ([*going*] *home*).

Wir fahren morgen **nach Salzburg.**	*We'll go to Salzburg tomorrow.*
Biegen Sie an der Ampel **nach links ab.**	*Turn left at the light.*
Gehen Sie **nach Westen.**	*Go west.*
Ich muss jetzt **nach Hause.**	*I have to go home now.*

zu: places
persons
zu Hause (idiom)

Use **zu** to indicate movement toward a place or a person, and in the idiom **zu Hause** (*at home*).

Wir fahren heute **zum Strand.**	*We'll go to the beach today.*
Wir gehen morgen **zu Tante Julia.**	*We'll go to Aunt Julia's tomorrow.*
Rolf ist nicht **zu Hause.**	*Rolf is not at home.*

Übung 1 Die Familie Ruf

Kombinieren Sie Fragen und Antworten.

1. Hier kommt Herr Ruf. Er hat seine Hausschuhe an. Woher kommt er gerade?
2. Hans hat noch seine Schultasche auf dem Rücken. Woher kommt er?
3. Frau Ruf kommt mit zwei Taschen voll Obst und Gemüse herein. Woher kommt sie?
4. Jutta kommt herein. Sie hat eine neue Frisur[1]. Woher kommt sie?
5. Gestern Abend war Jutta nicht zu Hause. Wo war sie?
6. Ihre Mutter war auch nicht zu Hause. Wo war sie?
7. Morgen geht Herr Ruf aus. Wohin geht er?
8. Hans fährt am Wochenende weg. Wohin fährt er?
9. Frau Ruf ist am Wochenende geschäftlich unterwegs. Wohin fährt sie?
10. Jutta möchte mit ihrem Freund einen Skiurlaub machen. Wohin wollen sie?

a. Aus der Schule.
b. Aus seinem Zimmer.
c. Bei ihrem Freund.
d. Bei Frau Körner.
e. Nach Innsbruck.
f. Nach Berlin.
g. Vom Friseur.
h. Vom Markt.
i. Zu Herrn Thelen, Karten spielen.
j. Zu seiner Tante.

Übung 2 Melanies Reise nach Dänemark

Beantworten Sie die Fragen. Verwenden Sie die Präpositionen **aus, bei, nach, von** oder **zu.**

MODELL: CLAIRE: Wohin bist du gefahren? (Dänemark) →
MELANIE: Nach Dänemark.

1. Wohin genau? (Kopenhagen)
2. Wohin bist du am ersten Tag gegangen? (der Strand)
3. Und deine Freundin Fatima? Wohin ist sie gegangen? (ihre Tante Sule)

[1]*hairstyle*

4. Woher kommt die Tante deiner Freundin? (die Türkei)
5. Kommt deine Freundin auch aus der Türkei? (nein / der Iran)
6. Am Strand hast du Peter getroffen, nicht? Woher ist der plötzlich gekommen? (das Wasser)
7. Sein Freund war auch dabei, nicht? Woher ist der gekommen? (der Markt)
8. Weißt du, wo die beiden übernachten wollten? (ja / uns)
9. Und wo haben sie übernachtet? (Fatimas Tante)
10. Wohin seid ihr am nächsten Morgen gefahren? (Hause)

10.2 Requests and instructions: the imperative (summary review)

WISSEN SIE NOCH?

The imperative is used to form commands, sentences in which you tell others how to act.

Review grammar 2.6.

As you have already learned, the imperative (command form) in German is used to make requests, to give instructions and directions, and to issue orders. To soften requests or to make them more polite, words such as **doch, mal,** and **bitte** are often included in imperative sentences.

Mach mal das Fenster **zu!**	*Close the window!*
Bringen Sie mir **bitte** noch einen Kaffee.	*Bring me another cup of coffee, please.*

The imperative has four forms: the familiar singular (**du**), the familiar plural (**ihr**), the polite (**Sie**), and the first-person plural (**wir**).

A. **Sie** and **wir**

In both the **Sie-** and the **wir**-forms, the verb begins the sentence and the pronoun follows.

Kontrollieren Sie bitte das Öl.	*Please check the oil.*
Gehen wir doch heute ins Kino!	*Let's go to the movies today.*

B. **ihr**

The familiar plural imperative consists of the present-tense **ihr-**form of the verb but does not include the pronoun **ihr.**

Lydia und Yamina, **kommt her** und **hört** mir **zu**!	*Lydia and Yamina, come here and listen to me.*
Sagt immer die Wahrheit!	*Always tell the truth.*

C. **du**

The familiar singular imperative consists of the present-tense **du-**form of the verb without the -(**s**)**t** ending and without the pronoun **du.**

du kommst	**Komm!**
du tanzt	**Tanz!**
du isst	**Iss!**

In written German, you will sometimes see a final **-e (komme, gehe),** but this **-e** is usually omitted in the spoken language for all verbs except those for which the present-tense **du-**form ends in **-est.**

du arbeitest	**Arbeite!**
du öffnest	**Öffne!**

Verbs that have a stem-vowel change from **-a-** to **-ä-** or **-au-** to **-äu-** do not have an umlaut in the **du-**imperative.

du fährst	**Fahr!**
du läufst	**Lauf!**

D. **sein**

The verb **sein** has irregular imperative forms.

du → **Sei** leise! ihr → **Seid** leise! Sie → **Seien Sie** leise!	*Be quiet!*	(*Paul!*) (*You two!*) (*Mrs. Smith!*)
wir → **Seien wir** leise!	*Let's be quiet!*	
Sei so gut und gib mir die Butter, Andrea.	*Be so kind and pass me the butter, Andrea.*	
Seid keine Egoisten!	*Don't be such egotists!*	

Übung 3 Hans und sein Vater

Hans und sein Vater sind zu Hause. Hans fragt seinen Vater, was er tun darf oder tun muss. Spielen Sie die Rolle seines Vaters. Sie brauchen auch einen guten Grund!

MODELL: Darf ich den Fernseher einschalten? →
Ja, schalte ihn ein. Es kommt ein guter Film.
oder Nein, schalte ihn nicht ein. Ich möchte Musik hören.

1. Muss ich jetzt Klavier üben?
2. Darf ich Jens anrufen?
3. Darf ich die Schokolade essen?
4. Darf ich das Fenster aufmachen?
5. Muss ich dir einen Kuss geben?
6. Kann ich mit dir reden?
7. Muss ich das Geschirr spülen?
8. Darf ich in den Garten gehen?
9. Darf ich morgen mit dem Fahrrad in die Schule fahren?

Übung 4 Aufforderungen!

Sie sind die erste Person in jeder Zeile. Was sagen Sie?

MODELL: Frau Wagner: Jens und Ernst / Zimmer aufräumen →
Jens und Ernst, räumt euer Zimmer auf!

1. Herr Wagner: Jens und Ernst / nicht so laut sein
2. Michael: Maria / bitte an der nächsten Ampel halten
3. Frau Wagner: Uli / an der nächsten Straße nach links abbiegen
4. Herr Ruf: Jutta / mehr Obst essen
5. Herr Siebert: Herr Pusch / nicht so schnell fahren
6. Jutta: Jens / an der Ecke auf mich warten
7. Frau Frisch-Okonkwo: Sumita und Yamina / nicht ungeduldig sein
8. Herr Thelen: Andrea und Paula / Vater von mir grüßen
9. Frau Ruf: Hans / mal schnell zu Papa laufen
10. Aydan: Eske und Damla / jeden Tag die Zeitung lesen

Übung 5 Minidialoge

helfen	machen
vergessen	
warten	sprechen

Verwenden Sie die Verben im Kasten.

1. FRAU RUF: Ich sitze jetzt schon wieder seit sechs Stunden vor dem Computer.
   HERR RUF: Du arbeitest zu viel. _____ mal eine Pause.
2. HERR SIEBERT: _____ bitte lauter, ich verstehe Sie nicht.
   MARIA: Ja, wie laut soll ich denn sprechen? Wollen Sie, dass ich schreie?
3. MICHAEL: Na, was ist? Kommen Sie nun oder kommen Sie nicht?
   FRAU KÖRNER: Ich bin ja gleich fertig. Bitte _____ doch noch einen Moment.
4. HANS: Kann ich mit euch zum Schwimmen gehen?
   JENS: Ja, komm und _____ deine Badehose nicht.
5. AYDAN: _____ mir bitte, ich kann die Koffer nicht allein tragen.
   ESKE UND DAMLA: Aber natürlich, Mama, wir helfen dir doch gern.

10.3 Prepositions for giving directions: *an ... vorbei, bis zu, entlang, gegenüber von, über*

ACCUSATIVE:
entlang (follows the noun)
über (precedes the noun)

A. **entlang** (*along*) and **über** (*over*) + Accusative

Use the prepositions **entlang** and **über** with nouns in the accusative case. Note that **entlang** follows the noun.

Fahren Sie **den Fluss entlang.**	*Drive along the river.*
Gehen Sie **über den Zebrastreifen.**	*Walk across the crosswalk.*

The preposition **über** may also be used as the equivalent of English *via*.

Der Zug fährt **über** Frankfurt und Hannover nach Hamburg.	*The train goes to Hamburg via Frankfurt and Hanover.*

DATIVE:
an ... vorbei (encloses the noun)
bis zu (precedes the noun)
gegenüber von (precedes the noun)

B. **an ... vorbei** (*past*), **bis zu** (*up to, as far as*), **gegenüber von** (*across from*) + Dative

Use **an ... vorbei, bis zu,** and **gegenüber von** with the noun in the dative case. Note that **an ... vorbei** encloses the noun.

Gehen Sie **am Supermarkt vorbei.**	*Go past the supermarket.*
Fahren Sie **bis zur Fußgängerzone** und biegen Sie links ab.	*Drive to the pedestrian zone and turn left.*
Die U-Bahnhaltestelle ist **gegenüber vom Markthotel.**	*The subway station is across from the Markthotel.*

Übung 6 Wie komme ich dahin?

Ein Ortsfremder[1] fragt Sie nach dem Weg. Antworten Sie! Nützliche Wörter:

entlang	an ... vorbei	gegenüber von
über	bis zu	

1. Wie muss ich fahren?

2. Wie muss ich gehen?

3. Wie muss ich gehen?

4. Wie muss ich fahren?

[1] *stranger*

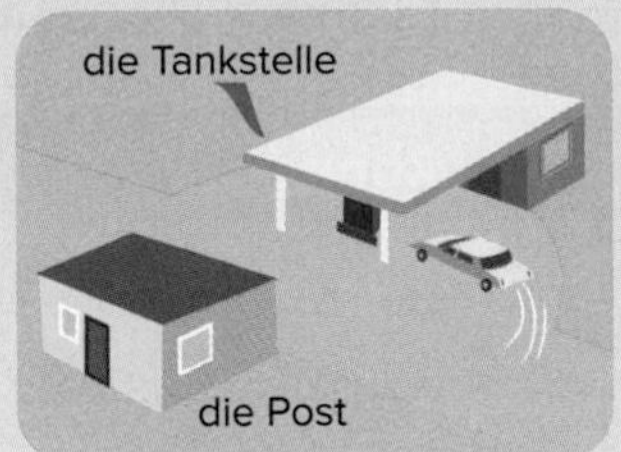

5. Wo ist die Tankstelle?

6. Wie komme ich zum Zug?

7. Immer geradeaus?

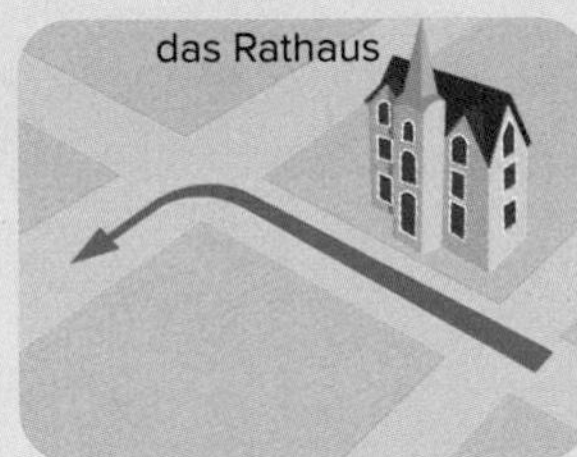

8. Vor dem Rathaus links?

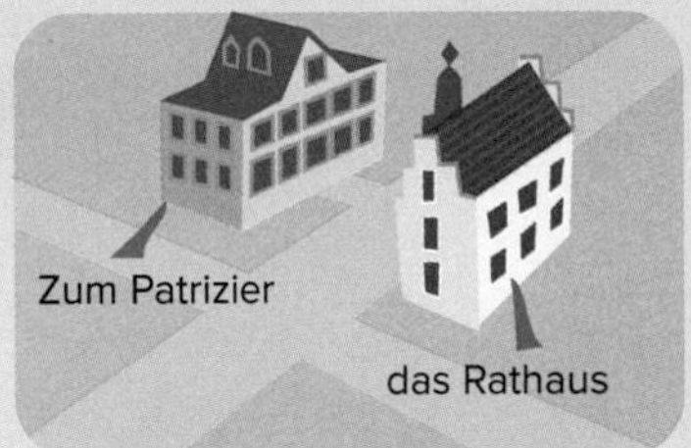

9. Das Hotel „Zum Patrizier"?

10. Wie komme ich nach Nürnberg?

10.4 Being polite: the subjunctive form of modal verbs

Use the subjunctive form of modal verbs to be more polite.

Könnten Sie mir bitte dafür eine Quittung geben?	*Could you please give me a receipt for that?*
Ich **müsste** mal telefonieren.	*I have to make a phone call.*
Dürfte ich Ihr Telefon benutzen?	*Could I use your phone?*

The subjunctive is formed from the simple past-tense stem. Add an umlaut if there is an umlaut in the infinitive.

To form the subjunctive of a modal verb, add an umlaut to the simple past form if there is also one in the infinitive. If the modal verb has no umlaut in the infinitive (**sollen** and **wollen**), the subjunctive form is the same as the simple past form.

Infinitive	Past	Subjunctive
dürfen	ich durfte	ich d**ü**rfte
können	ich konnte	ich k**ö**nnte
mögen	ich mochte	ich m**ö**chte
müssen	ich musste	ich m**ü**sste
sollen	ich sollte	ich sollte
wollen	ich wollte	ich wollte

Below are the subjunctive forms of **können** and **wollen.**

können			
ich	könnte	*wir*	könnten
du	könntest	*ihr*	könntet
Sie	könnten	*Sie*	könnten
er *sie* *es*	könnte	*sie*	könnten

wollen			
ich	wollte	*wir*	wollten
du	wolltest	*ihr*	wolltet
Sie	wollten	*Sie*	wollten
er *sie* *es*	wollte	*sie*	wollten

In modern German, **möchte,** the subjunctive form of **mögen,** has become almost a synonym of **wollen.**

—Wohin wollen Sie fliegen? *Where do you want to go (fly)?*
—Wir möchten nach Kanada fliegen. *We want / would like to fly to Canada.*

Another polite form, **hätte gern,** is often used instead of **möchte**, especially in conversational exchanges involving goods and services.

Ich hätte gern eine Cola, bitte. *I'd like a Coke, please.*
Wir hätten gern die Speisekarte, bitte. *We'd like the menu, please.*

Übung 7 Überredungskünste

Versuchen Sie, jemanden zu überreden[1], etwas anderes zu machen als das, was er/sie machen will.

MODELL: S1: Ich fahre jetzt. (bleiben)
S2: Ach, könntest du nicht bleiben?

1. Ich koche Kaffee. (Tee, Suppe, ?)
2. Ich lese jetzt. (später, morgen, ?)
3. Ich sehe jetzt fern. (etwas Klavier spielen, mit mir sprechen, ?)
4. Ich rufe meine Mutter an. (deinen Vater, deine Tante, ?)
5. Ich gehe nach Hause. (noch eine Stunde bleiben, bis morgen bleiben, ?)

MODELL: S1: Wir fahren nach Spanien. (Italien)
S2: Könnten wir nicht mal nach Italien fahren?

6. Wir übernachten im Zelt. (Hotel, Campingbus, ?)
7. Wir kochen selbst. (essen gehen, fasten, ?)
8. Wir gehen jeden Tag wandern. (schwimmen, ins Kino, ?)
9. Wir schreiben viele Briefe. (nur eine E-Mail, nur Postkarten, ?)
10. Wir sehen uns alle Museen an. (in der Sonne liegen, viel schlafen, ?)

[1]*convince*

Übung 8 Eine Autofahrt

Sie wollen mit einem Freund ausgehen und fahren in seinem Auto mit. Stellen Sie Fragen. Versuchen Sie, besonders freundlich und höflich zu sein.

MODELL: wir / jetzt nicht fahren können →
Könnten wir jetzt nicht fahren?

1. du / nicht noch tanken müssen
2. wir / nicht Jens abholen sollen
3. zwei Freunde von mir / auch mitfahren können
4. wir / nicht zuerst in die Stadt fahren sollen
5. du / nicht zur Bank wollen
6. du / etwas langsamer fahren können
7. ich / das Autoradio anmachen dürfen
8. ich / das Fenster aufmachen dürfen

10.5 Focusing on the action: the passive voice

A. Uses of the Passive Voice

The passive voice is used in German to focus on the action of the sentence itself rather than on the person or thing performing the action.

ACTIVE VOICE

Der Arzt impft die Kinder. — *The physician inoculates the children.*

PASSIVE VOICE

Die Kinder **werden geimpft.** — *The children are (being) inoculated.*

Note that the accusative (direct) object of the active sentence, **die Kinder,** becomes the nominative subject of the passive sentence.

In passive sentences, the agent of the action is often unknown or unspecified. In the following sentences, there is no mention of who performs each action.

Schildkröten werden oft als Haustiere gehalten. — *Turtles are often kept as pets.*

1088 wurde die erste Universität gegründet. — *The first university wasfounded in 1088.*

WISSEN SIE NOCH?

In addition to the passive auxiliary, **werden** can be used as a main verb meaning "to become" or as a future auxiliary with an infinitive to form the future tense.

Review grammar 5.3 and 8.5.

B. Forming the Passive Voice

passive = **werden** + past participle

The passive voice is formed with the auxiliary **werden** and the past participle of the verb. The present-tense and simple past-tense forms are the tenses you will encounter most frequently in the passive voice.

Passive Voice, Present Tense fragen			
ich	werde gefragt	*wir*	werden gefragt
du	wirst gefragt	*ihr*	werdet gefragt
Sie	werden gefragt	*Sie*	werden gefragt
er *sie* *es*	wird gefragt	*sie*	werden gefragt

Passive Voice, Past Tense fragen			
ich	wurde gefragt	*wir*	wurden gefragt
du	wurdest gefragt	*ihr*	wurdet gefragt
Sie	wurden gefragt	*Sie*	wurden gefragt
er *sie* *es*	wurde gefragt	*sie*	wurden gefragt

Passive agents are indicated by **von** + noun.

C. Expressing the Agent in the Passive Voice

In most passive sentences in German, the agent (the person or thing performing the action) is not mentioned. When the agent is expressed, the construction **von** + dative is used.

ACTIVE VOICE

Die Kinder füttern die Tiere. — *The children are feeding the animals.*

PASSIVE VOICE
AGENT: **von** + DATIVE

Die Tiere werden **von den Kindern** gefüttert. — *The animals are being fed by the children.*

Übung 9 Geschichte

Hier sind die Antworten. Was sind die Fragen?

MODELL: 1492 → Wann wurde Amerika entdeckt?

1. vor 50.000 Jahren
2. um 2500 v. Chr.[1]
3. 44 v. Chr.
4. 800 n. Chr.[2]
5. 1088
6. 1789
7. 1885
8. 1945
9. 1963
10. 1990

a. Deutschland vereinigen
b. John F. Kennedy erschießen
c. die amerikanische Verfassung unterschreiben
d. die erste Universität (Bologna) gründen
e. die Atombomben auf Hiroshima und Nagasaki werfen
f. die ersten Pyramiden bauen
g. Cäsar ermorden
h. in Kanada die transkontinentale Eisenbahn vollenden
i. Karl den Großen zum Kaiser krönen
j. Australien von den Aborigines besiedeln

[1]vor Christus [2]nach Christus

Übung 10 Der Mensch und das Tier

MODELL: die Giraffe / langsam aus ihrem Lebensraum verdrängt →
Die Giraffe wird langsam aus ihrem Lebensraum verdrängt.

1. Mäuse
2. Meerschweinchen
3. Bienen
4. Mücken
5. die Fledermaus
6. Schnecken
7. der Gepard
8. die meisten Papageien
9. Delfine
10. viele Haie

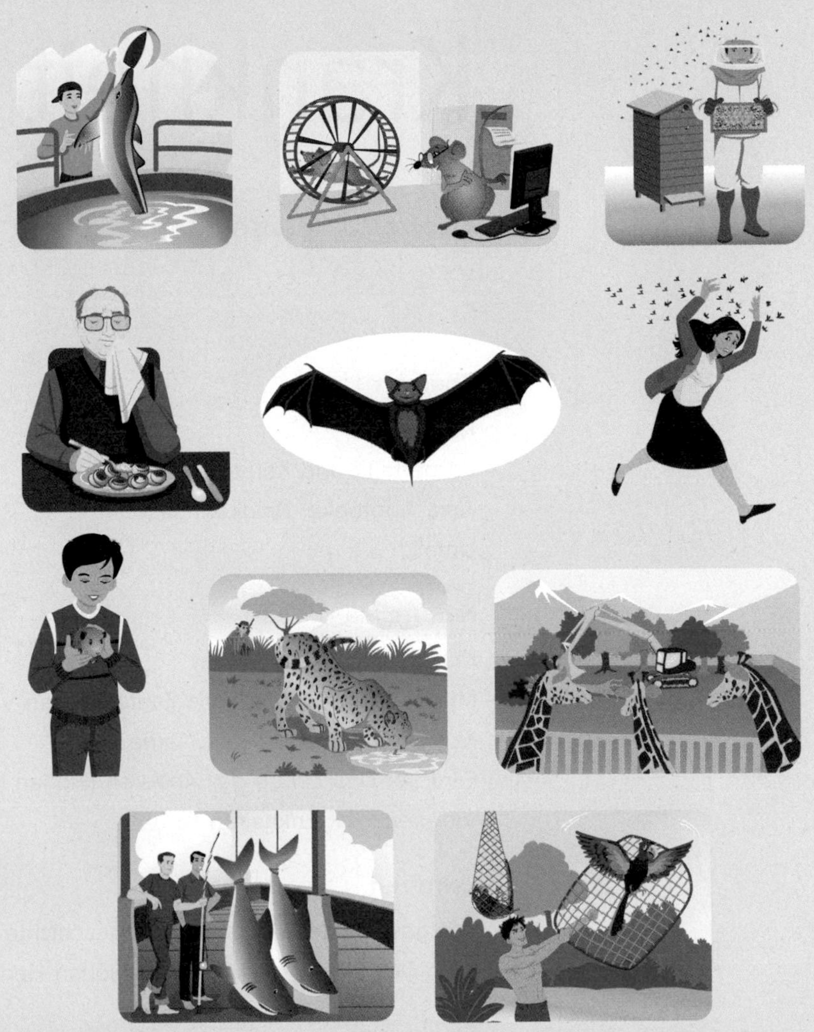

jedes Jahr gefischt
in der Wildnis gefangen
wegen ihrer Intelligenz bewundert[3]
durch Parfum und Kosmetikprodukte angelockt[4]
in vielen Labortests benutzt
oft mit Butter- und Knoblauchsoße gegessen
oft als Haustiere gehalten
wegen ihrer Honigproduktion geschätzt[5]
langsam aus ihrem Lebensraum verdrängt[6]
immer noch für seinen Pelz getötet
in vielen Kulturen mit Vampiren assoziiert

[3]*admired* [4]*attracted* [5]*valued* [6]*displaced*

KAPITEL **11**

Gesundheit und Krankheit

Kapitel 11 focuses on health and fitness. You will talk about how to stay fit and about illness and accidents.

Themen

Krankheit

Körperteile und Körperpflege

Arzt, Apotheke, Krankenhaus

Unfälle

Kulturelles

KLI: Hausmittel

Musikszene: „Danke" (Die Fantastischen Vier)

KLI: Geschichte der Psychiatrie

Filmclip: *Das Leben der Anderen* (Florian Henckel von Donnersmarck)

Videoecke: Krankheiten

Lektüren

Kurzgeschichte: Montagmorgengeschichte (Susanne Kilian)

Film: *Das Leben der Anderen* (Florian Henckel von Donnersmarck)

Strukturen

11.1 Accusative reflexive pronouns

11.2 Dative reflexive pronouns

11.3 Word order of accusative and dative objects

11.4 Indirect questions: **Wissen Sie, wo ...?**

11.5 Word order in dependent and independent clauses (summary review)

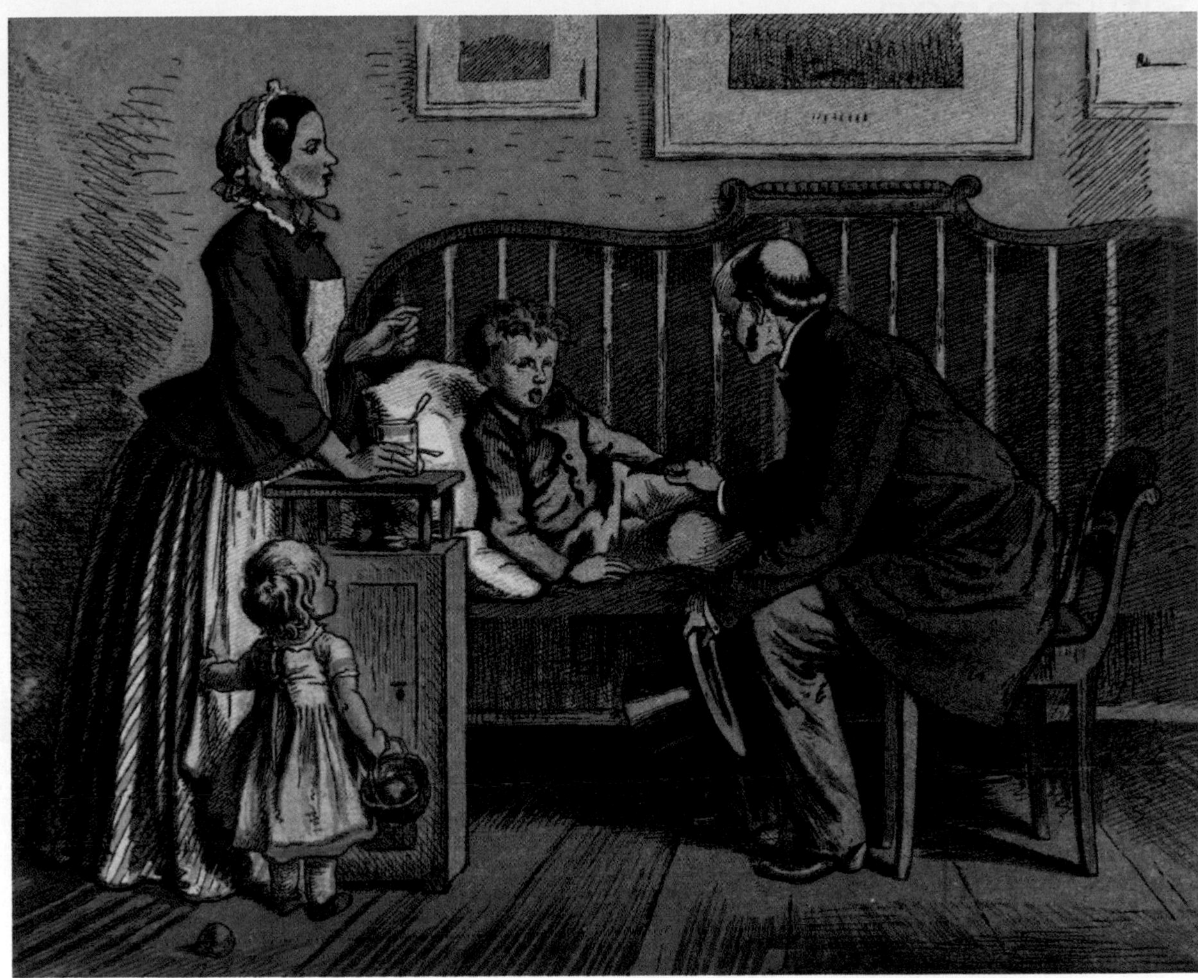

Oscar Pletsch: *Fieber* (1871), aus dem Buch *Kinderland*

KUNST UND KÜNSTLER

Oskar Pletsch (1830–1888) ist ein bekannter Maler und Kinderbuch-Illustrator aus Berlin, der an der Kunstakademie in Dresden studierte und dort auch arbeitete. Später ging er wieder nach Berlin zurück und widmete sich der Genremalerei mit Motiven aus dem bürgerlichen Alltags- und Familienleben. Seine Bilderbücher waren sehr beliebt und wurden auch in England, Frankreich, Schweden und den USA gelesen. Seine bevorzugte[1] Technik war der Holzschnitt.

Schauen Sie sich das Bild an und beantworten Sie folgende Fragen.

1. Welche Personen sehen Sie auf dem Bild?
2. Wo sind die Personen?
3. Warum liegt der kleine Junge auf dem Sofa?
4. Was ist der Mann von Beruf?
5. Was macht er mit dem Jungen?
6. Was macht der Junge und warum macht er das?
7. Welche Kleidung tragen die Menschen?
8. Was steht auf dem kleinen Schrank?
9. Was fragt die Mutter den Arzt vielleicht?
10. Welche Farben dominieren in der Illustration?
11. Was assoziieren Sie damit?

[1]*preferred*

Situationen

Krankheit

Grammatik 11.1

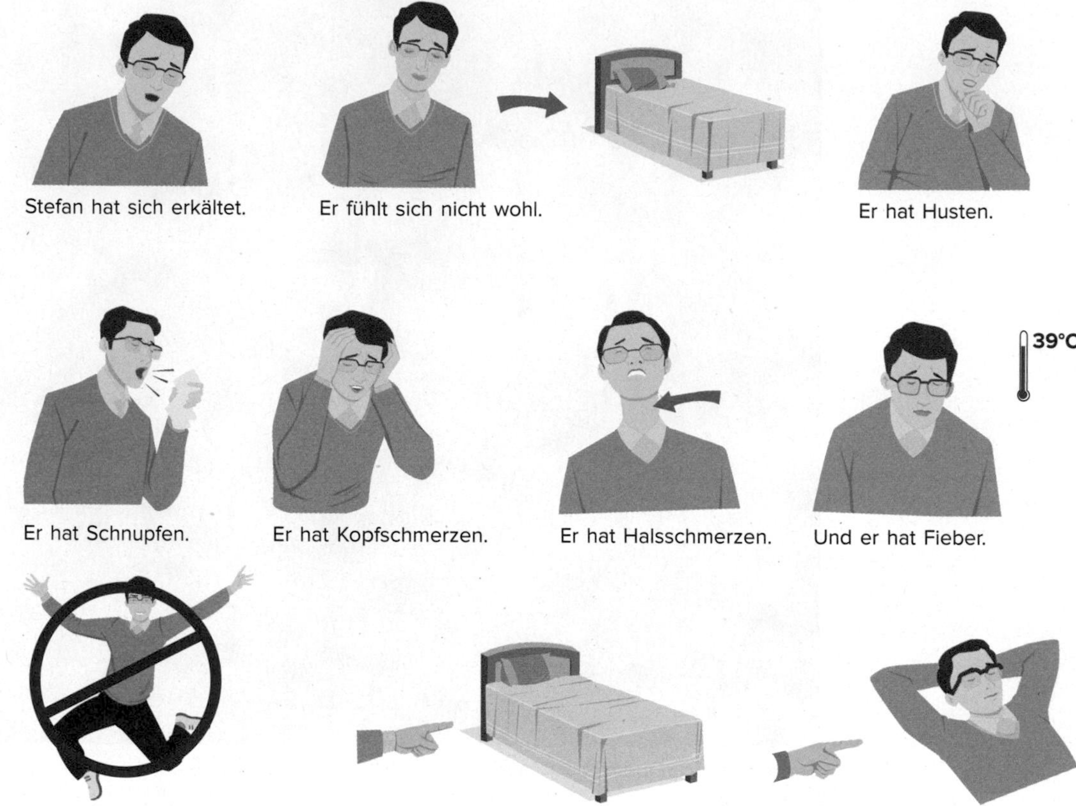

Stefan hat sich erkältet.
Er fühlt sich nicht wohl.
Er hat Husten.
Er hat Schnupfen.
Er hat Kopfschmerzen.
Er hat Halsschmerzen.
39°C
Und er hat Fieber.
Er darf sich nicht aufregen.
Er muss sich ins Bett legen.
Er muss sich ausruhen.

Situation 1 Hausmittel°

°*Home remedies*

Was machst du immer, manchmal, nie?

1. Wenn ich Fieber habe,
   a. lege ich mich ins Bett.
   b. nehme ich zwei Aspirin.
   c. gehe ich zum Arzt.
   d. rege ich mich auf.
2. Wenn ich Husten habe,
   a. nehme ich Hustensaft.
   b. trinke ich heißen Tee mit Zitrone.
   c. rauche ich eine Zigarette.
   d. lutsche ich Hustenbonbons.
3. Wenn ich mich erkältet habe,
   a. gehe ich schwimmen.
   b. ruhe ich mich aus.
   c. gehe ich in die Sauna.
   d. ärgere ich mich furchtbar.
4. Wenn ich Kopfschmerzen habe,
   a. gehe ich zum Friseur.
   b. nehme ich zwei Aspirin.
   c. bleibe ich im Bett.
   d. nehme ich ein heißes Bad.
5. Wenn ich Zahnschmerzen habe,
   a. trinke ich heißen Kaffee.
   b. gehe ich zum Zahnarzt.
   c. nehme ich Tabletten.
   d. setze ich mich aufs Sofa.

6. Wenn ich mich verletzt habe,
   a. desinfiziere ich die Wunde.
   b. falle ich in Ohnmacht.
   c. hole ich ein Pflaster.
   d. ziehe ich mich aus.
7. Wenn ich Muskelkater habe,
   a. lasse ich mich massieren.
   b. gehe ich zum Arzt.
   c. mache ich Muskeltraining.
   d. lege ich mich aufs Sofa.
8. Wenn ich mich in den Finger geschnitten habe,
   a. ärgere ich mich furchtbar.
   b. hole ich ein Pflaster.
   c. nehme ich Hustensaft.
   d. desinfiziere ich die Wunde.
9. Wenn ich einen Kater habe,
   a. gehe ich ins Krankenhaus.
   b. nehme ich zwei Aspirin.
   c. schlafe ich den ganzen Tag.
   d. gehe ich joggen.
10. Wenn ich Magenschmerzen habe,
   a. lege ich mich aufs Sofa.
   b. trinke ich Kamillentee.
   c. ziehe ich mich aus.
   d. esse ich viel Schokolade.

Situation 2 **Was tut dir weh?**

MODELL: Du warst in einem Rockkonzert. →
Ich habe Ohrenschmerzen.

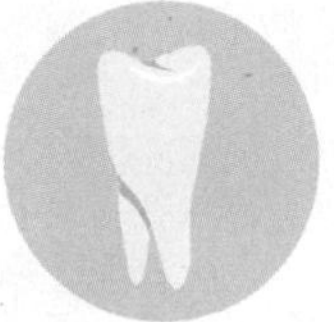
Zahnschmerzen

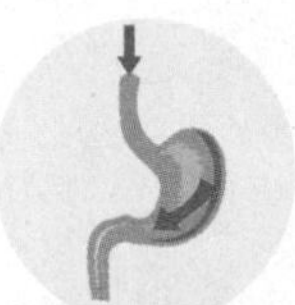
Magenschmerzen

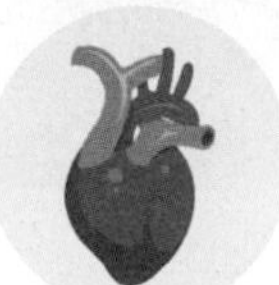
Herzschmerzen

Mir tut die Zunge weh.

AUCH
Mir tut die Nase weh.
Mir tut der Rücken weh.
Mir tun die Augen weh.
Mir tun die Füße weh.

AUCH
Kopfschmerzen
Halsschmerzen

1. Du hast den ganzen Tag in der Bibliothek gesessen und Bücher gelesen.
2. Du hast zwei große Teller Chili gegessen.
3. Jemand hat dich auf die Nase geschlagen.
4. Du bist 20 Kilometer gewandert.
5. Du hast gestern Abend zu viel Kaffee getrunken.
6. Du warst bei einem Basketballspiel und hast viel geschrien.
7. Du hast zu viele Bonbons gegessen.
8. Du hast furchtbaren Liebeskummer.
9. Du hast zwei Stunden Schnee geschaufelt[1].
10. Der Kaffee, den du getrunken hast, war zu heiß.

[1]*shoveled*

KULTUR ... LANDESKUNDE ... INFORMATIONEN

HAUSMITTEL

- Welche von diesen Hausmitteln kennen Sie? Wogegen helfen sie?
  - ☐ Eisbeutel
  - ☐ grüner Tee
  - ☐ heißer Tee mit Zitrone
  - ☐ Hühnersuppe
  - ☐ Kamillentee
  - ☐ Knoblauch
  - ☐ Salzwasser
  - ☐ warme Umschläge[2]
- Benutzen Sie Hausmittel, wenn Sie sich nicht wohl fühlen? Wenn ja, welche?

Lesen Sie die drei Zeitungstexte. Kennen Sie diese Hausmittel? Glauben Sie, dass sie wirken? Warum?

Bei Husten warmes Zuckerwasser mit Eidotter[3] vermischen. Das mildert den Hustenreiz[4]. Oder Hustenbier trinken: Einen halben Liter Bier erhitzen, mit fünf Löffeln flüssigem Honig verrühren[5] und abends trinken.

Wenn die Augen müde sind, Hände reiben[6] bis sie warm sind, sie auf die geschlossenen Augen legen und an die Farbe Schwarz denken.

Bei Fieber Zitronenscheiben auf die Schläfen[7] legen. Oder eine Kette aus Rettichscheiben[8] über Nacht um den Hals binden.

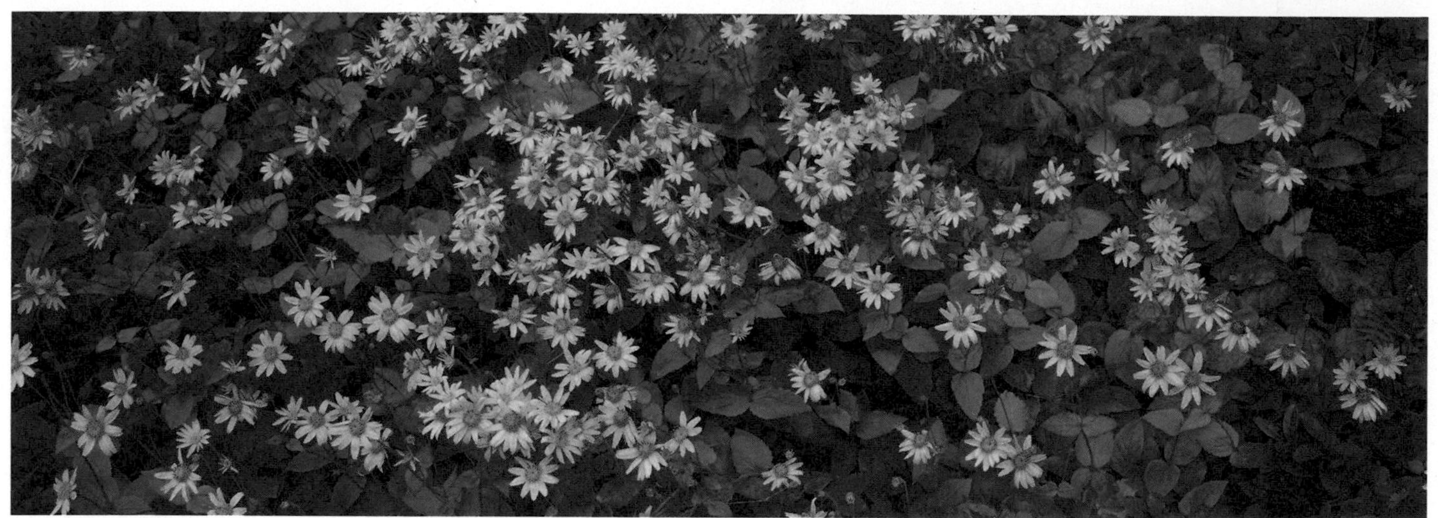

Arnika ist eine beliebte Heilpflanze.

Hausmittel stehen oftmals der Pflanzenheilkunde[9] nahe[10]. Die Arnikapflanze ist nur ein Beispiel. Lesen Sie den Text und beantworten Sie die Fragen.

- Wo wächst die Arnika?
- Wofür wird Arnika verwendet?
- In welcher Form kann man heute Arnika bekommen?

Die Arnika wächst in den Alpen. Seit jeher wird die Alpenpflanze von den Menschen in den Bergen bei Prellungen[11], Stauchungen[12] und schmerzenden Beinen verwendet. Man hat herausgefunden, dass die Arnika die Beine besonders gut durchblutet, Schmerzen lindert, Schwellungen[13] abbaut und entzündungshemmend[14] wirkt. Deshalb eignet sich Arnika bei Sportverletzungen sehr gut. Heute kann man Arnika-Salben, -Gels und -Beinsprays kaufen.

[2]*compresses* [3]*egg yolk* [4]*irritation of the throat* [5]*stir* [6]*rub* [7]*temples* [8]*radish slices* [9]*herbal medicine* [10]nahestehen *to be closely connected to* [11]*bruises* [12]*sprains* [13]*swelling* [14]*as an anti-inflammatory*

Situation 3 Umfrage

MODELL: S1: Legst du dich ins Bett, wenn du dich erkältet hast?
S2: Ja.
S1: Unterschreib bitte hier.

	UNTERSCHRIFT
1. Ruhst du dich aus, wenn du Kopfschmerzen hast?	____________
2. Ärgerst du dich, wenn du in den Ferien krank wirst?	____________
3. Legst du dich ins Bett, wenn du eine Grippe hast?	____________
4. Bist du gegen Katzen allergisch?	____________
5. Hast du einen niedrigen Blutdruck[15]?	____________
6. Freust du dich, wenn dein Lehrer / deine Lehrerin krank ist?	____________
7. Regst du dich auf, wenn du dich verletzt hast?	____________
8. Erkältest du dich oft?	____________
9. Nimmst du Tabletten, wenn du dich nicht wohl fühlst?	____________

Körperteile und Körperpflege

Grammatik 11.2–11.3

Ich wasche mich.

Ich wasche mir die Haare.

Ich trockne mich ab.

Ich trockne mir die Hände ab.

Ich kämme mir die Haare.

Ich schminke mich.

Ich rasiere mich.

Ich putze mir die Zähne.

Ich ziehe mich an.

[15]niedrigen ... *low blood pressure*

Miniwörterbuch

der **Sprechgesang**	rap
ehren	to honor
der **Verdienst**	merit
die **Todesgefahr**	danger of death
wäre gewesen	would have been
zwangsbeatmen	to resuscitate (forcefully)
begraben	to bury
der **Tod**	death
die **Schranke**	gate (at a railway crossing)
sich verheddern	to get tangled up
der **Gurt**	(seat) belt
das **Bahngleis**	railway track
die **Achterbahn**	roller coaster
sich aushängen	to get unhinged, uncoupled
anschnallen	to buckle up
sich verschlucken	to choke (on sth.)

„Danke" (2010, Deutschland) *Die Fantastischen Vier*

Biografie Die Fantastischen Vier (kurz: Fanta 4) machten den deutschsprachigen Hip-Hop populär. Sie nannten ihn den deutschen Sprechgesang. Fanta 4 besteht aus Michael Bernd Schmidt alias Smudo, Thomas Dürr, Michael Beck und Andreas Rieke und kommt aus Stuttgart. Ihren ersten Auftritt hatte die Gruppe schon 1989. 2000 nahmen sie ein MTV-Unplugged-Album auf, nach Herbert Grönemeyer damals erst die zweiten deutschen Künstler, die so geehrt wurden. 2009 erhielten sie als erste Musikgruppe den Paul-Lincke-Ring der Stadt Goslar für ihre Verdienste im deutschen Sprechgesang. Die Single „Danke" entstammt ihrem Album *Für dich immer noch Fanta Sie* aus dem Jahre 2010.

Die Fantastischen Vier
© Fabrizio Bensch/Reuters/Corbis

NOTE: For copyright reasons, the songs referenced in **MUSIKSZENE** have not been provided by the publisher. The song can be found online at various sites such as YouTube, Amazon, or the iTunes store.

Vor dem Hören Waren Sie schon mal in Todesgefahr? Was ist passiert?

Nach dem Hören

A. Beantworten Sie die Fragen zum Anfang und zum Refrain.

1. Was wäre schön gewesen?
2. Wo liegt der Sänger?
3. Was wollen die Leute mit ihm noch machen?
4. Wann nur könnte der Sänger noch einmal *danke* sagen?

B. Der erste Tod. Bringen Sie die Sätze in die richtige Reihenfolge.

_______ Er fragt sich, warum es auch hinter ihm eine Schranke gibt.
_______ Er fragt sich, wo es eine Tankstelle gibt.
_______ Er steht mit seinem Auto vor einer Schranke.
_______ Er verheddert sich in seinem Gurt.

C. Woran stirbt der Sänger in den vier Strophen? Verbinden Sie Ort und Todesart.

1. auf dem Bahngleis	**a.** Er bekommt einen elektrischen Schock.
2. im Restaurant	**b.** Der Wagen hängt sich aus und er ist nicht angeschnallt.
3. auf der Achterbahn	**c.** Er verschluckt sich an einem Fisch.
4. im Bad	**d.** Er wird von einem Zug überfahren.

Situation 4 Körperteile

MODELL: S1: Was macht man mit den Augen?
S2: Mit den Augen sieht man.

greifen sprechen denken atmen küssen kauen hören fühlen riechen gehen

1. mit den Ohren
2. mit den Händen
3. mit dem Gehirn
4. mit der Nase
5. mit der Lunge
6. mit den Zähnen
7. mit den Lippen
8. mit den Beinen
9. mit dem Mund
10. mit dem Herzen

Situation 5 Körperpflege

1. Wenn meine Haut trocken ist,
   a. creme ich sie ein.
   b. gehe ich schwimmen.
   c. gehe ich zum Arzt.
2. Wenn meine Fingernägel lang sind,
   a. bade ich mich.
   b. schneide ich sie mir.
   c. kaue ich sie ab.
3. Wenn meine Haare fettig sind,
   a. putze ich mir die Zähne.
   b. schneide ich sie mir.
   c. wasche ich sie mir.
4. Wenn ich ins Theater gehe,
   a. schminke ich mich.
   b. rasiere ich mich.
   c. schneide ich mir die Haare.
5. Wenn ich ins Bett gehe,
   a. ziehe ich mir warme Schuhe an.
   b. putze ich mir die Zähne.
   c. schneide ich mir die Fingernägel.
6. Wenn ich mich geduscht habe,
   a. ziehe ich mich aus.
   b. trockne ich mich ab.
   c. föhne ich mir die Haare.
7. Wenn ich mich erholen will,
   a. gehe ich in die Sauna.
   b. rasiere ich mir die Beine.
   c. nehme ich Tabletten.
8. Wenn es draußen kalt ist,
   a. dusche ich mich heiß.
   b. ziehe ich mir eine warme Hose an.
   c. ziehe ich mich aus.
9. Wenn ich eine Verabredung habe,
   a. schminke ich mich.
   b. wasche ich mir die Haare.
   c. esse ich viel Knoblauch.

Situation 6 Bildgeschichte: Maria hat eine Verabredung.

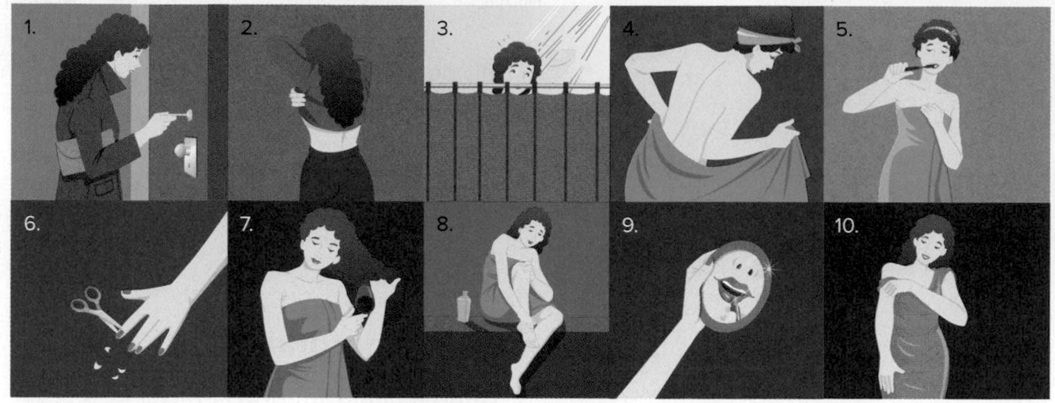

Situation 7 Interview: Körperpflege

1. (für Frauen) Schminkst du dich jeden Tag? Was machst du?
2. (für Männer) Rasierst du dich jeden Tag? Hattest du schon mal einen Bart? Was für einen (Schnurrbart, Vollbart, Spitzbart[1], Backenbart[2])? Wie war das? Wenn du einen Bart hast: Seit wann hast du einen Bart?
3. Wäschst du dir jeden Tag die Haare? Föhnst du sie dir auch? Was für Haar hast du (trockenes, fettiges, normales Haar)?
4. Putzt du dir jeden Tag die Zähne? Gehst du oft zum Zahnarzt?
5. Wie oft gehst du zum Friseur? Hattest du mal eine Dauerwelle[3]? Wie hast du ausgesehen?
6. Hast du trockene Haut? Cremst du dich oft ein?
7. Treibst du regelmäßig Sport? Was machst du? Wie oft? Gehst du manchmal in die Sauna oder ins Solarium?

Lektüre

Vor dem Lesen

A. Lesen Sie die Wörter im Miniwörterbuch. Suchen Sie sie im Text und unterstreichen Sie sie.

Miniwörterbuch	
aufpassen	to pay attention
rasen	to rush, speed
sich verabreden mit	to agree to meet with
begreifen	to comprehend
vorsichtig	carefully
die **Bahre, -n**	stretcher
der **Ranzen, -**	schoolbag
der **Bub, -en**	(*slang*) boy
rausschießen, ist rausgeschossen	to shoot out
anfertigen	to prepare
schuld sein	to be at fault

[1]*goatee* [2]*sideburns* [3]*perm*

B. Schreiben Sie mit den folgenden Stichwörtern und Ausdrücken eine kleine Geschichte: Was ist an diesem Montagmorgen passiert?

- Autofahrer
- Adresse
- Schultasche
- Bremsenquietschen[4]
- Konferenz
- neunjähriger Junge
- Unfall
- schnell laufen
- Krankenhaus
- Kreidestriche auf der Straße
- nicht aufgepasst
- Polizeirevier[5]

LESEHILFE

This text is written in the style of a police report. What narrative point of view would you expect in such a text?

C. Orientierung. Sehen Sie sich jetzt den Text an; lesen Sie zuerst nur das **fett Gedruckte** und das *kursiv Gedruckte*. Aus welchen Teilen besteht die Geschichte?

Montagmorgengeschichte

von Susanne Kilian

So stand es in der Zeitung:
»Nicht aufgepasst«

Nicht genügend aufgepasst hatte ein neunjähriger Junge, der ...

So ist es passiert:

7 Uhr 30
Herr Langen hat in Ruhe gefrühstückt. Um 8 Uhr 10 hat er eine Vertreterkonferenz. Er ist ausgeruht und gut vorbereitet. Er hat keine Eile. Sorgfältig und in Ruhe startet er seinen Wagen.

7 Uhr 42
Lothar Bernich hat um 8 Uhr Schule. Heute ist alles verquer. Nicht mal Zeit zum Frühstücken hat er. Das Brot isst er auf dem Schulweg. Das geht doch auch mal!

7 Uhr 44
Herr Langen fährt die stille Seitenstraße auf dem Weg zu seinem Büro entlang. Zum x-ten Mal. Er kennt diese Straße genau.

Lothar Bernich rast: Ihm fällt ein, dass er sich heute mit dem Martin verabredet hat. Er will mit ihm zusammen zur Schule gehen.

Er rennt aus der Tür. Er rennt zwischen den parkenden Autos einfach durch. Er rennt direkt in das Auto von Herrn Langen.

7 Uhr 46
Lothar liegt auf der Straße. Das Auto von Herrn Langen hat ihn erwischt. Herr Langen kann das nicht begreifen. Er hat das Kind nicht gesehen. Als er es sah, hat er gebremst. Das Auto stand sofort. Lothar tut alles weh. Er denkt an die Schule. An den Martin. Wieso liegt er jetzt auf der Straße? Wie ging das so schnell? Ihm tut alles weh.

7 Uhr 47
Herr Hartmann hat das Bremsenquietschen gehört. Er rennt ans Fenster. Sieht das Kind vor dem Auto auf der Straße liegen. Sofort ruft er das Unfallkommando an.

7 Uhr 49
Das Unfallkommando der Polizei hat die Arbeitersamariter verständigt. Lothar weiß nicht, was überhaupt mit ihm passiert. Leute starren ihn an. Die Sanitäter heben ihn schnell und vorsichtig hoch. An den Beinen und am Kopf. Er spürt eine weiche Bahre unter sich. „Meine Mama, wo ist bloß meine Mama ... ", jammert er.

7 Uhr 55
Lothar wird ins Krankenhaus gefahren. Die Polizei trifft an der Unfallstelle ein. Sie untersucht Lothars Ranzen. Findet seine Adresse im Ranzendeckel stehen –

[4]*squealing of brakes* [5]*police station*

sie wird per Funk zum Revier durchgegeben. Von dort wird Lothars Mutter verständigt. Herr Langen wird zum Unfall vernommen. Wo der Junge lag, wird mit Kreidestrichen eine Skizze auf die Straße gemalt. Herr Langen hat das nicht gewollt. Er ist nervös. Er zittert. Er hat den Jungen nicht zwischen den Autos hervorrennen sehen. Er gibt der Polizei seine Papiere.

Die Zeugen:

Alte Frau:
Klar. Der Mann ist doch gerast wie verrückt. Heute ist man doch auf der Straße wie Freiwild. Rasen einfach. Für Fußgänger ist kein Platz mehr. Das arme Kind, das kleine!

Mann:
Na, also der Bub ist doch zwischen den Autos nur so rausgeschossen. Den konnte der im Auto doch gar nicht sehen. Das war ganz unmöglich. Der hat überhaupt nicht aufgepasst ... hat sicher an ganz was anderes gedacht ...

Mädchen:
Ich weiß nicht. Also, ich weiß nicht ... das ging alles so schnell. Eben hab' ich noch den Jungen rennen gesehen, da lag er schon auf der Straße. Ich weiß wirklich nicht. Bremsenquietschen hab' ich gehört.

Später auf dem Polizeirevier:

Der Unfallbericht wird angefertigt.

Ist Lothar schuld, der es so eilig hatte? Weil er ein paar Sekunden nicht aufgepasst hat?

Ist Herr Langen schuld, der gar nicht wusste, was geschah, bis Lothar vor seinem Auto lag? Wer ist schuld? Feststeht: das kann jedem jeden Tag passieren.

Susanne Killian, "Montagmorgengeschichte"
in *Am Montag fängt die Woche an.* Used with permission.

Arbeit mit dem Text

A. Was ist wann passiert? Ordnen Sie die Sätze aus dem Berichtteil der richtigen Zeit zu. Achtung: Meistens gehören mehrere Sätze zu einer Zeit.

7 Uhr 30 **7 Uhr 46** **7 Uhr 49** **7 Uhr 55** **7 Uhr 42** **7 Uhr 44** **7 Uhr 47**

7 Uhr _____: Lothar rennt zwischen den parkenden Autos durch auf die Straße.

7 Uhr _____: Herr Langen ist auf dem Weg in sein Büro.

7 Uhr _____: Die Polizei kommt an die Unfallstelle.

7 Uhr _____: Die Sanitäter kommen und legen Lothar auf eine Bahre.

7 Uhr _____: Lothar rast, weil er sich mit Martin verabredet hat.

7 Uhr _____: Lothar hat keine Zeit, weil er schon um 8 Uhr da sein muss.

7 Uhr _____: Die Sanitäter bringen Lothar ins Krankenhaus.

7 Uhr _____: Lothar ist vor ein Auto gelaufen und liegt jetzt auf der Straße.

7 Uhr *30*: Herr Langen muss erst um 8 Uhr 10 zu einer Konferenz und hat keine Eile.

7 Uhr _____: Die Polizei ruft Lothars Mutter an.

7 Uhr _____: Herr Langen ist sehr nervös und schockiert, als er der Polizei seine Papiere gibt.

7 Uhr _____: Lothar weiß gar nicht, was los ist.

7 Uhr _____: Herr Hartmann ruft das Unfallkommando der Polizei an.

7 Uhr _____: Die Polizei malt mit Kreide eine Skizze auf die Straße.

B. Wer sagt was? (die alte Frau, der Mann, das Mädchen)

„Ich weiß nicht, wer schuld ist. Es ging alles viel zu schnell."
„Der Autofahrer ist schuld, weil er viel zu schnell gefahren ist."
„Der Junge ist schuld, weil er nicht aufgepasst hat."

C. Was glauben Sie? Ist der Junge schuld, der Autofahrer oder jemand anderes? Was könnte man machen, damit so was nicht passiert?

D. Lesen Sie wieder den ersten Absatz. Wobei hat der Junge nicht aufgepasst? Schreiben Sie den Relativsatz zu Ende.

Nicht genügend aufgepasst hatte ein neunjähriger Junge, der ...

Nach dem Lesen

Beschreiben Sie einen Unfall, den Sie einmal hatten oder gesehen haben. Wie ist der Unfall passiert? Welche Schäden oder Verletzungen gab es? Was haben Sie gemacht? Machen Sie sich zuerst Notizen, berichten Sie dann in der Gruppe und schreiben Sie schließlich Ihren Bericht auf.

Arzt, Apotheke, Krankenhaus

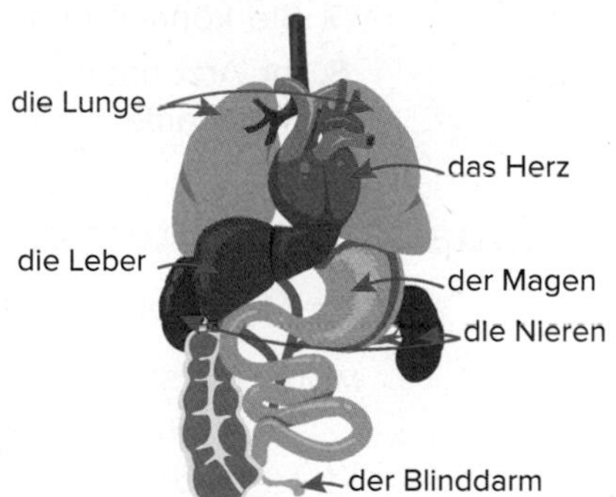

Jürgen hat sich das Bein gebrochen. Jetzt muss er einen Gips tragen.

Silvia bekommt eine Spritze.

Josef bekommt einen Verband.

Der Zahnarzt zieht Melanie einen Zahn.

Die Ärztin gibt Claire ein Rezept.

Situation 8 Medizinische Berufe

Wohin gehen Sie?

ins Krankenhaus	in die Drogerie
in die Apotheke	zum Hausarzt
zum Zahnarzt	zum Psychiater
zum Tierarzt	zum Augenarzt

1. Sie haben sich erkältet und brauchen Hustensaft.
2. Sie haben schon seit zwei Wochen eine schlimme Halsentzündung und wollen Antibiotika.
3. Ihr Freund / Ihre Freundin hat sich in den Finger geschnitten. Der Finger blutet stark.
4. Ihr Freund / Ihre Freundin hat Sie verlassen und Sie sind sehr deprimiert.
5. Ihr Goldfisch frisst schon seit mehreren Tagen nicht mehr.
6. Sie haben furchtbare Zahnschmerzen.
7. Sie können im Unterricht nicht lesen, was an der Tafel steht.
8. Ihr Arzt hat Ihnen ein Rezept ausgeschrieben und Sie wollen sich das Medikament abholen.

Situation 9 Interaktion: Ich bin krank

Ein Mitstudent / Eine Mitstudentin ist krank. Was raten Sie ihm/ihr?

MODELL: S1: Ich habe Fieber.
S2: Leg dich ins Bett.

1. Ich habe Fieber.	a. Geh zum Arzt.
2. Ich habe Kopfschmerzen.	b. Nimm Hustensaft.
3. Ich fühle mich nicht wohl.	c. Leg dich ins Bett.
4. Ich habe starken Husten.	d. Geh nach Hause.
5. Ich habe mich in den Finger geschnitten.	e. Kauf dir Kopfschmerztabletten.
6. Ich habe mich erkältet.	f. Ruh dich aus.
7. Ich habe Zahnschmerzen.	g. Nimm ein warmes Bad.
8. Ich bin allergisch gegen Katzen.	h. Zieh dich warm an.
9. Mir tun die Augen weh.	i. Verkauf deine Katze.
10. Ich habe Magenschmerzen.	j. Geh zum Zahnarzt.
	k. Kauf dir eine Brille.
	l. _________?

KULTUR ... LANDESKUNDE ... INFORMATIONEN

GESCHICHTE DER PSYCHIATRIE

- Haben Sie manchmal schlechte Laune[1]? Was machen Sie dann?
- Manche Leute essen Schokolade, wenn sie traurig oder depressiv sind. Was hilft Ihnen?
- Kennen Sie Sigmund Freud? Was assoziieren Sie mit ihm und der Psychoanalyse?

Lesen Sie den Text und suchen Sie Antworten auf diese Fragen:

- Wie versuchte man in der Antike und im Mittelalter kranke Menschen zu heilen?
- Was machte man bis zum Ende des 18. Jahrhunderts mit den „Irren"[2] und warum?
- Wo stand das erste psychiatrische Krankenhaus in Europa?
- Wer suchte schon im 18. Jahrhundert den Ursprung für Krankheiten in der Seele[3]?
- Was spielt in Sigmund Freuds Psychoanalyse eine zentrale Rolle?
- Was begründete C.G. Jung und welche bekannten Charakterisierungen für Menschen stammen von ihm?

Seelische[4] Störungen sind in der modernen Welt die häufigsten Krankheiten und nehmen immer mehr zu[5]. Schon in der Antike hat man sich mit psychischen Krankheitsbildern beschäftigt. Grundlage war das Gleichgewicht[6] der Elemente Luft, Feuer, Erde und Wasser und damit verbunden die Körpersäfte[7] Blut, gelbe Galle, schwarze Galle und Schleim[8]. Die revolutionären Theorien des römischen Arztes Galen hatten über 1.500 Jahre in Europa Gültigkeit und finden sich in vielen Aspekten der kulturellen Tradition, nicht nur in der Medizin. Wenn die vier Säfte im Gleichgewicht waren, war der Mensch gesund. Waren Menschen krank, versuchte man sie z. B. durch Ernährung oder Diäten zu kurieren. An diesen antiken Grundsatz der Medizin hielten sich noch so berühmte Persönlichkeiten wie die Äbtissin und Ärztin Hildegard von Bingen (1098–1179) im Mittelalter und später der Naturheilkundler[9] Sebastian Kneipp (1821–1897).

Das Narrenhaus, Wilhelm von Kaulbach (1834)

Bis Ende des 18. Jahrhunderts wurden die „Irren" hauptsächlich eingesperrt[10], wenn sie nicht gefoltert[11] oder verbrannt wurden, weil man glaubte, dass sie vom Teufel besessen[12] waren. Ein Fortschritt war es schon, wenn sie nicht mit Armen und Kriminellen in den sogenannten „Zuchthäusern"[13] angekettet[14] waren, sondern wie in Wien seit 1784 im „Narrenturm"[15], dem ersten psychiatrischen Krankenhaus in Europa, leben konnten.

Im 18. und 19. Jahrhundert beschäftigten sich viele europäische Wissenschaftler mit seelischen Krankheiten und suchten nach Ursachen und Behandlungsmethoden, z. B. der deutsche Chemiker und Mediziner Georg Ernst Stahl (1659–1734), der den Ursprung von Krankheiten in der Seele[16] suchte und wie später Sigmund Freud die Bedeutung des Unbewussten[17] propagierte.

Der Österreicher Sigmund Freud (1856–1939) gilt weltweit als Begründer der Psychoanalyse. Bei einem Besuch in Paris lernte er die Hypnose als Behandlung für die Hysterie kennen und entwickelte dann seine eigenen Methoden, um die Ursachen von Neurosen und Psychosen zu verstehen und zu behandeln. Die Traumdeutung[18] und das Unterbewusste[19] spielen dabei eine zentrale Rolle.

Ein anderer bekannter Wissenschaftler war der Schweizer Psychiater Carl Gustav Jung (1875–1961), der z. B. das Phänomen der „gespaltenen[20] Persönlichkeit" beschrieb und die analytische Psychologie begründete. Jung beschäftigte sich intensiv mit dem kollektiven Unbewussten, dem Archetypus und entwickelte die Begriffe vom introvertierten und extrovertierten Menschen. Er pflegte mehrere Jahre eine intensive Freundschaft mit Freud. Als Präsident der IAÄGP (Internationale Allgemeine Ärztliche Gesellschaft für Psychotherapie) versuchte er von 1933 bis 1939 in Deutschland die Psychotherapie über die Zeit des Nationalsozialismus zu retten.

[1]*mood* [2]*lunatics* [3]*mind, soul* [4]*Mental* [5]nehmen zu *increase* [6]*balance* [7]here: *bodily fluids* [8]*mucus* [9]*naturopath* [10]*locked up* [11]*tortured* [12]*possessed* [13]*penitentiary* [14]*put in chains* [15]*fools' tower* [16]*soul, mind* [17]*unconscious* [18]*dream interpretation* [19]*subconscious* [20]*split*

Situation 10 Informationsspiel: Krankheitsgeschichte

MODELL: Hat Claire sich (Hast du dir) schon mal etwas gebrochen? Was?
Ist Claire (Bist du) schon mal im Krankenhaus gewesen? Warum?
Hat Herr Thelen (Hast du) schon mal eine Spritze bekommen? Gegen was?
Erkältet sich Herr Thelen (Erkältest du dich) oft?
Ist Claire (Bist du) gegen etwas allergisch? Gegen was?
Hat man Claire (Hat man dir) schon mal einen Zahn gezogen?
Hatte Herr Thelen (Hattest du) schon mal hohes Fieber? Wie hoch?

	Claire	Herr Thelen	mein(e) Partner(in)
sich etwas brechen		das Bein	
im Krankenhaus sein		Lungenentzündung	
eine Spritze bekommen	Diphtherie		
sich oft erkälten	ja		
gegen etwas allergisch sein		Katzen	
einen Zahn gezogen haben		ja	
hohes Fieber haben	104° F		

Situation 11 Dialoge

1. Herr Thelen möchte einen Termin beim Arzt.

HERR THELEN: Guten Tag, ich hätte gern __________ für nächste Woche.
SPRECHSTUNDENHILFE: Gern, vormittags oder nachmittags?
HERR THELEN: Das ist mir eigentlich __________.
SPRECHSTUNDENHILFE: Mittwochmorgen um neun?
HERR THELEN: Ja, __________. Vielen Dank.

2. Frau Körner geht in die Apotheke.

FRAU KÖRNER: Ich habe schon seit Tagen __________. Können Sie mir etwas __________ geben?
APOTHEKERIN: Wir haben gerade etwas ganz Neues bekommen, Magenex.
FRAU KÖRNER: Hauptsache, __________.
APOTHEKERIN: Es soll sehr gut __________. Hier ist es.

3. Frau Frisch-Okonkwo ist bei ihrem Hausarzt.

HAUSARZT: Guten Tag, Frau Frisch-Okonkwo, wie geht es Ihnen?
FRAU FRISCH-OKONKWO: Ich fühle mich gar nicht wohl. __________ ... alles tut mir weh.
HAUSARZT: Das klingt nach __________. Sagen Sie mal bitte „Ah“.

Situation 12 Rollenspiel: Anruf beim Arzt

S1: Sie fühlen sich nicht wohl. Wahrscheinlich haben Sie Grippe. Rufen Sie beim Arzt an, sagen Sie, was Ihnen fehlt, und lassen Sie sich einen Termin geben. Es ist dringend, aber Sie haben einen vollen Terminkalender.

Situation 13 Interview

1. Warst du schon mal schwer krank? Wann? Was hat dir gefehlt?
2. Warst du schon mal im Krankenhaus? Wann? Warum? Wie lange? Hat man dich untersucht? Hat man dir Blut abgenommen? Hast du eine Spritze bekommen?
3. Hast du dir schon mal etwas gebrochen? Was? Hattest du einen Gips? Wie lange?
4. Hat man dich schon mal geröntgt? Wann? Warum?
5. Erkältest du dich oft? Was machst du, wenn du eine Erkältung hast?
6. Bist du gegen etwas allergisch? Gegen was?
7. Bist du schon mal in Ohnmacht gefallen? Warum?

Unfälle

Grammatik 11.4–11.5

Zwei Autos sind zusammengestoßen. Eine Frau ist schwer verletzt.

Situation 14 Ein Autounfall

Eine Polizistin spricht mit einem Zeugen über einen Unfall. Bringen Sie die Sätze in eine logische Reihenfolge.

_____ Können Sie mir sagen, wie spät es ungefähr war?
_____ Also, heute Morgen war ich auf dem Weg zur Uni.
__1__ Bitte erzählen Sie genau, was passiert ist.
_____ Ein Auto ist aus einer Einfahrt[1] gekommen.
_____ Ich glaube nicht, er hat jedenfalls nicht gebremst, bevor er auf die Straße gefahren ist.
_____ Wissen Sie, ob der Fahrer auf den Verkehr geachtet hat?
_____ Ja, ein anderes Auto kam von rechts und dann sind sie zusammengestoßen.
_____ So zwischen halb und Viertel vor neun.
_____ Was haben Sie da gesehen?
_____ Und dann?
_____ Vielen Dank für Ihre Hilfe.

Situation 15 Unfälle

Welcher Satz passt zu welchem Bild?

a.

b.

c.

d.

e.

f.

g.

h.

1. Michael und Maria waren beim Segeln, als das Boot umkippte[2].
2. Sofie schnitt gerade Tomaten, als plötzlich vor ihrem Haus ein Mann von einem Auto überfahren wurde.
3. Jutta und Hans waren auf dem Weg ins Konzert, als Jutta ausrutschte und hinfiel.
4. Jürgen saß gerade in der Bibliothek, als auf der Straße zwei Autos zusammenstießen.
5. Herr Okonkwo fuhr gerade zur Arbeit, als ihm ein Hund vors Auto lief.
6. Als Ernst mit seinen Freunden Fußball spielte, brach er sich das Bein.
7. Maria und Michael liefen Schlittschuh, als ein Kind ins Eis einbrach.
8. Rolf wollte gerade nach Hawaii fliegen, als ein Flugzeug abstürzte.

[1]*driveway* [2]*turned over*

Situation 16 **Notfälle**

Was machst du, wenn ...

1. du einen Unfall siehst?
2. der Verletzte einen Schock hat?
3. der Fahrer von dem anderen Auto flüchtet?
4. du im Fahrstuhl stecken bleibst?
5. du ausrutschst und hinfällst?
6. du dir den Arm gebrochen hast?
7. du ins Wasser fällst?
8. es im Nachbarhaus brennt?
9. du dir die Zunge verbrannt hast?

a. den Krankenwagen rufen
b. die Feuerwehr rufen
c. die Autonummer aufschreiben
d. die Polizei rufen
e. eine Decke holen und den Verletzten zudecken
f. fluchen
g. liegen bleiben und warten, dass jemand kommt
h. schwimmen
i. um Hilfe rufen
j. _______?

Situation 17 **Bildgeschichte: Paulas Unfall**

Filmlektüre

Das Leben der Anderen

Vor dem Lesen

FILMANGABEN

Titel: *Das Leben der Anderen*
Genre: Drama
Erscheinungsjahr: 2006
Land: Deutschland
Dauer: 137 min
Regisseur: Florian Henckel von Donnersmarck
Hauptrollen: Ulrich Mühe, Sebastian Koch, Thomas Thieme, Martina Gedeck

Der Lauscher auf dem Dachboden
© Sony Pictures Classics/Photofest

A. Beantworten Sie die folgenden Fragen.

1. Was macht der Mann auf dem Bild? Warum macht er das?
2. Beschreiben Sie das Gesicht des Mannes. Was hört er?
3. Was wissen Sie über die DDR und die Rolle der Stasi[3]?

B. Lesen Sie die Wörter im Miniwörterbuch. Suchen Sie sie im Text und unterstreichen Sie sie.

Inhaltsangabe

Ost-Berlin 1984. Der pflichtbewusste Stasi-Mitarbeiter Gerd Wiesler (Ulrich Mühe) soll den bekannten und angeblich regimetreuen Dramaturgen Georg Dreyman (Sebastian Koch) bespitzeln. Wiesler hat einen guten Spürsinn und glaubt, dass Dreyman nicht so treu ist, wie er tut. Kulturminister Hempf (Thomas Thieme) unterstützt die Bewachung des Theaterschriftstellers, weil er ihn aus dem Weg schaffen will, um freie Bahn bei dessen Freundin, der Schauspielerin Christa-Maria Sieland (Martina Gedeck) zu haben.

Dreymans Wohnung wird verwanzt, und auf dem Dachboden des Hauses installiert Wiesler Abhörgeräte. Wiesler, der allein in einer Neubauwohnung lebt und kein aufregendes Privatleben hat, erlebt durch die Bewachung Dreymans eine für ihn völlig neue Welt: nämlich die der Kunst, der Literatur, des freien Geistes und der Liebe. Das Leben des Dramaturgen und der Schauspielerin beeindruckt den Stasi-Mann so sehr, dass er aufhört, belastendes Material über Dreyman zu sammeln. Wieslers Berichte über den Theaterschriftsteller sind trivial. Er unternimmt auch nichts, als Dreyman nach dem Selbstmord eines befreundeten Regisseurs anonym einen Essay über die hohe Selbstmordrate in der DDR veröffentlicht. Wiesler schützt Dreyman sogar, indem er die Schreibmaschine, auf der Dreyman den Essay für den *Spiegel* geschrieben hat, aus ihrem Versteck nimmt und verschwinden lässt.

Ein Opfer gibt es dennoch: Die psychisch labile Schauspielerin Christa-Maria Sieland verrät der Stasi, dass Dreyman den Essay geschrieben hat und wo die Schreibmaschine versteckt ist. Dann flüchtet sie, läuft vor ein Auto und stirbt. Als Dreyman nach der Wende Einsicht in seine Stasi-Akten bekommt, erfährt er, dass ein Stasi-Mitarbeiter ihn geschützt hat. Seine Erinnerungen schreibt Dreyman in einem Roman nieder. Sein Buch widmet er seinem Stasi-Spitzel Wiesler unter dessen Stasi-Deckcode-Namen HGW XX/7–in Dankbarkeit.

[3]Ministerium für Staatssicherheit

Miniwörterbuch

pflichtbewusst	conscientious
regimetreu	loyal to the regime
bespitzeln	to spy on
der **Spürsinn**	perceptiveness
die **Bewachung**	guarding
jemanden aus dem Weg schaffen	to get rid of someone
verwanzt	bugged
das **Abhörgerät**	bugging device
belastendes Material	incriminating evidence
das **Versteck**	hiding place
verschwinden lassen	to make disappear
das **Opfer**	victim
verraten	to reveal
widmen	to dedicate

Arbeit mit dem Text

Welche Aussagen sind falsch? Verbessern Sie die falschen Aussagen.

1. „Das Leben der Anderen" spielt vor dem Fall der Berliner Mauer.
2. Der Dramaturg Dreyman scheint ein Fan des DDR-Regimes zu sein.
3. Gerd Wiesler arbeitet für die Polizei und den Kulturminister.
4. Wiesler hat den Auftrag, den Dramaturgen Dreyman und dessen Freundin zu überwachen.
5. Dreyman unterschreibt den Essay im *Spiegel* mit seinem Namen.
6. Wiesler meldet seinem Chef, dass sich Dreyman nicht regimetreu verhält.
7. Christa-Maria Sieland schützt Dreyman und muss deshalb sterben.
8. Nach der Wiedervereinigung schreibt Dreyman ein Buch über seine Erinnerungen.

FILMCLIP

NOTE: For copyright reasons, the films referenced in the **FILMCLIP** feature have not been provided by the publisher. The film can be purchased as a DVD or found online at various sites such as YouTube, Amazon, or the iTunes store. The time codes mentioned below are for the North American DVD version of the film.

Szene: DVD Kapitel 4, 25.06–29.08 Min.

Die Männer der Stasi verwanzen die Wohnung von Georg und Christa-Maria. Die Nachbarin Frau Meineke sieht dies zufällig und wird von den Stasi-Leuten gezwungen, niemandem etwas davon zu sagen. Sonst würde man ihrer Tochter den Medizinstudienplatz wegnehmen. Georg ist in der Zwischenzeit bei seinem Freund, Albert Jerska. Jerska war Regisseur. Vor sieben Jahren hat man ihm ein Berufsverbot erteilt, weil er das Regime kritisiert hatte. Georg versucht, seinen Freund aufzumuntern und möchte ihn zu seiner Geburtstagsfeier einladen. Das vergisst er aber. Als Georg nach Hause kommt, schmückt Christa-Maria gerade die Wohnung für die Feier.

Schauen Sie sich die Szene an und beantworten Sie die Fragen.

1. Welche Beziehung haben Georg und Christa-Maria?
2. Was soll in ihrer Wohnung stattfinden? Was macht Christa-Maria dafür?
3. Was erzählt Georg Christa-Maria über seinen Besuch bei Albert Jerska?
4. Was hält Christa-Maria von Albert?
5. Wie alt wird Georg?
6. Was schenkt ihm Christa-Maria zum Geburtstag?
7. Was sagt Georg, dass er tun kann? Stimmt das?
8. Warum braucht Georg die Hilfe der Nachbarin Frau Meineke?
9. Wie reagiert die Nachbarin auf Georg? Warum?
10. Warum verlässt Georg die Wohnung?
11. Was macht Christa-Maria, als Georg gegangen ist? Warum macht sie das vielleicht?
12. Was denken Sie? Wie geht die Beziehung zwischen Georg und Christa-Maria weiter?

Miniwörterbuch

zufällig	by chance
zwingen,gezwungen	to coerce
sonst	otherwise
das **Berufsverbot**	occupational ban
aufmuntern	to cheer up
schmücken	to decorate
die **Beziehung**	relationship
einen Schlips binden	to tie a tie
zusperren, zugesperrt	to lock

Nach dem Lesen

Georg Dreymans Tagebuch: Schreiben Sie zu einer der folgenden Situationen einen Eintrag aus Dreymans Perspektive.

a. nach der Veröffentlichung des Essays im *Spiegel*
b. nach dem Unfall von Christa-Maria Sieland
c. nach Einsicht in die eigenen Stasi-Akten nach der Wende

Videoecke

Perspektiven

Was hältst du von Tattoos oder Piercings?

Tattoos und Piercings gehören zum Zeitgeschmack.

Aufgabe 1 Tattoos und Piercings

Wer mag Tattoos? Wer mag Piercings? Wer mag weder Tattoos noch Piercings? Ordnen Sie die Personen in drei Kategorien.

1. Michael

2. Judith

3. Nadezda

4. Pascal

5. Sophie

6. Felicitas

7. Tina

8. Martin

Tattoos	Piercings	Weder/noch

Aufgabe 2 Wer sagt das?

Schreiben Sie die Namen neben die Aussagen.

1. Ich finde Tattoos und Piercings absolut hässlich.
2. Ich habe ein Tattoo.
3. Ich habe zu große Angst vor den Schmerzen.
4. Man soll möglichst natürlich aussehen.
5. Piercings sind nicht so mein Ding.
6. Tattoos und Piercings kommen irgendwann wieder aus der Mode.

Interviews

- Wann warst du das letzte Mal krank?
- Was hattest du?
- Was machst du, wenn du dich erkältet hast?
- Hattest du schon mal einen Unfall? Erzähl mal.
- Was für eine Krankenversicherung hast du?
- Wie viel kostet sie?
- Bist du mit ihr zufrieden?

Albrecht

Michael

Aufgabe 3 Albrecht oder Michael?

Sehen Sie sich das Video an und ergänzen Sie die Tabelle.

	Albrecht	Michael
Wann war er das letzte Mal krank?		
Was hatte er?		
Was macht er, wenn er sich erkältet hat?		
Was für eine Krankenversicherung hat er?		
Wie viel kostet sie?		
Ist er damit zufrieden?		

Wortschatz

Krankheit und Gesundheit	Illness and Health
die **Entzündung, -en**	infection
die **Erkältung, -en**	(head) cold
die **Gesundheit**	health
die **Grippe**	influenza, flu
die **Krankheit, -en**	illness, sickness
die **Ohnmacht**	unconsciousness
in Ohnmacht fallen	to faint
der **Husten**	cough
der **Hustensaft, ¨e**	cough syrup
der **Kater, -**	hangover
der **Liebeskummer**	lovesickness
der **Muskelkater, -**	sore muscles
der **Schmerz, -en**	pain
die **Halsschmerzen**	sore throat
die **Kopfschmerzen**	headache
die **Magenschmerzen**	stomachache
die **Ohrenschmerzen**	earache
die **Zahnschmerzen**	toothache
der **Schnupfen, -**	cold (*with a runny nose*), sniffles
das **Bonbon, -s**	drop, lozenge
sich ärgern (R)	to get angry
sich auf·regen	to get excited, get upset
sich erkälten	to catch a cold
fehlen (+ *dat.*) (R)	to be wrong with, be the matter with (*a person*)
weh·tun, tat ... weh, wehgetan	to hurt

Ähnliche Wörter

das **Fieber; (sich) fühlen; sich wohl fühlen**

Der Körper	The Body
die **Haut, ¨e**	skin
die **Niere, -n**	kidney
die **Zunge, -n**	tongue
der **Blinddarm, ¨e**	appendix
der **Magen, ¨**	stomach
der **Zahn, ¨e**	tooth
das **Gehirn, -e**	brain
atmen	to breathe
greifen, griff, gegriffen	to grab, grasp
kauen	to chew
lutschen	to suck
riechen, roch, gerochen	to smell

Ähnliche Wörter

die **Leber, -n;** die **Lippe, -n;** die **Lunge, -n;** die **Nase, -n** (R); der **Finger, -;** der **Fingernagel, ¨;** das **Haar, -e** (R); das **Herz, -en**

Apotheke und Krankenhaus	Pharmacy and Hospital
die **Apothekerin, -nen**	female pharmacist
die **Ärztin, -nen** (R)	female doctor, physician
die **Hausärztin, -nen**	family doctor
die **Zahnärztin, -nen**	dentist
die **Arztpraxis, Arztpraxen**	doctor's office
die **Psychiaterin, -nen**	female psychiatrist
die **Spritze, -n**	shot, injection
der **Apotheker, -**	male pharmacist
der **Arzt, ¨e** (R)	male doctor, physician
der **Hausarzt, ¨e**	family doctor
der **Zahnarzt, ¨e**	dentist
der **Gips**	cast (*plaster*)
der **Psychiater, -**	male psychiatrist
der **Verband, ¨e**	bandage
das **Medikament, -e**	medicine
ein **Medikament gegen**	medicine for
das **Pflaster, -**	adhesive bandage
das **Rezept, -e**	prescription
ab·nehmen, nimmt ... ab, nahm ... ab, abgenommen	to remove; to lose weight
Blut abnehmen	to take blood
röntgen	to X-ray
wirken	to work, take effect

Ähnliche Wörter

die **Diphtherie;** die **Tablette, -n;** die **Wunde, -n;** der **Schock;** der **Tetanus;** das **Blut** (R); die **Antibiotika** (*pl.*); **bluten; desinfizieren**

Unfälle	Accidents
die **Feuerwehr**	fire department
die **Verletzte, -n**	female injured person
die **Zeugin, -nen**	female witness
der **Verletzte, -n (ein Verletzter)**	male injured person
der **Zeuge, -n** (*wk. masc.*)	male witness
ab·stürzen, ist abgestürzt	to crash
aus·rutschen, ist ausgerutscht	to slip
bremsen	to brake
brennen, brannte, gebrannt	to burn
hin·fallen, fällt ... hin, fiel ... hin, ist hingefallen	to fall down
schlagen, schlägt, schlug, geschlagen (R)	to hit
stecken bleiben, blieb ... stecken, ist stecken geblieben	to get stuck

überfahren, überfährt, überfuhr, überfahren	to run over
verbrennen, verbrannte, verbrannt	to burn
sich (die Zunge) verbrennen	to burn (one's tongue)
sich verletzen	to injure oneself
zu·decken	to cover
zusammen·stoßen, stößt ... zusammen, stieß ... zusammen, ist zusammengestoßen	to crash

Ähnliche Wörter

der **Krankenwagen, -; brechen, bricht, brach, gebrochen; sich (den Arm) brechen**

Körperpflege — Personal Hygiene

die **Dauerwelle, -n**	perm
die **Seife, -n**	soap
das **Solarium, Solarien**	tanning salon
sich ab·trocknen (R)	to dry oneself off
sich an·ziehen, zog ... an, angezogen (R)	to get dressed
sich aus·ruhen (R)	to rest
sich aus·ziehen, zog... aus, ausgezogen (R)	to get undressed
(sich) duschen (R)	to shower (take a shower)
sich ein·cremen	to put lotion on
sich erholen	to recuperate
sich (die Haare) föhnen (R)	to blow dry (one's hair)
sich (die Zähne) putzen	to brush (one's teeth)
sich rasieren	to shave
sich schminken	to put makeup on
(sich) schneiden, schnitt, geschnitten (R)	to cut (oneself)

Ähnliche Wörter

die **Sauna, -s; (sich) baden** (R); **sich (die Haare) kämmen** (R); **(sich) waschen, wäscht, wusch, gewaschen** (R)

Sonstige Substantive — Other Nouns

die **Decke, -n**	blanket
die **Tüte, -n**	(paper or plastic) bag
die **Verabredung, -en**	appointment; date
der **Rat, Ratschläge**	advice
der **Termin, -e** (R)	appointment
der **Terminkalender, -**	appointment calendar
der **Unterricht** (R)	class, instruction
der **Verkehr**	traffic
das **Fahrzeug, -e**	vehicle

Ähnliche Wörter

die **Zigarette, -n;** der **Goldfisch, -e**

Sonstige Verben — Other Verbs

achten auf (+ *akk.*)	to watch out for; to pay attention to
auf·schreiben, schrieb ... auf, aufgeschrieben	to write down
beschreiben, beschrieb, beschrieben	to describe
fluchen	to curse, swear
flüchten, ist geflüchtet	to flee
sich freuen über (+ *akk.*)	to be happy about
herunter·klettern, ist heruntergeklettert	to climb down
sich hin·legen	to lie down
klingen (wie), klang, geklungen	to sound (like)
lassen, lässt, ließ, gelassen	to let
sich einen Termin geben lassen	to get an appointment
passen (R)	to fit
das passt gut	that fits well
raten, rät, riet, geraten (+ *dat.*)	to advise
rufen, rief, gerufen (R)	to call
verlassen, verlässt, verließ, verlassen	to leave; to abandon

Ähnliche Wörter

sich setzen (R)

Adjektive und Adverbien — Adjectives and Adverbs

deprimiert	depressed
fettig (R)	greasy
regelmäßig	regularly
schlimm	bad
stark	heavy, severe
trocken	dry
verletzt	injured
schwer verletzt	critically injured

Ähnliche Wörter

allergisch, medizinisch

Sonstige Wörter und Ausdrücke — Other Words and Expressions

aber (R)	but
als (R)	when (*conj.*)
bevor	before (*conj.*)
bis (R)	until (*prep., conj.*)
dagegen	*here:* for it
Haben Sie etwas dagegen?	Do you have something for it (*illness*)?
damit	so that
dass	that (*conj.*)
denn (R)	for, because

draußen	outside
gemeinsam (R)	together; common
herunter	down (*toward the speaker*)
Hilfe!	Help!
jedenfalls	in any case
mal	(*word used to soften commands*)
Komm mal vorbei!	Come on over!
nachdem (R)	after (*conj.*)
ob (R)	whether
obwohl	although
oder (R)	or
seit (R)	since, for (*prep.*)
seit mehreren Tagen	for several days
sondern (R)	on the contrary
und (R)	and
während	during
weil (R)	because
wenn (R)	if; whenever

Strukturen und Übungen

11.1 Accusative reflexive pronouns

Reflexive pronouns are generally used to express the fact that someone is doing something to or for himself or herself.

Ich lege das Baby ins Bett.	*I'm putting the baby to bed.*
Ich lege mich ins Bett.	*I'm putting myself to bed (lying down).*

Some verbs are always used with a reflexive pronoun in German, whereas their English counterparts may not be.

Ich habe mich erkältet.	*I caught a cold.*
Warum regst du dich auf?	*Why are you getting excited?*

Here are some common reflexive verbs.

sich ärgern	*to get angry*
sich aufregen	*to get excited, get upset*
sich ausruhen	*to rest*
sich erkälten	*to catch a cold*
sich freuen	*to be happy*
sich (wohl) fühlen	*to feel (well)*
sich hinlegen	*to lie down*
sich verletzen	*to get hurt*

In most instances the forms of the reflexive pronoun are the same as those of the personal object pronouns. The only reflexive form that is distinct is **sich,** which corresponds to **er, sie** (*she*), **es, sie** (*they*), and **Sie*** (*you*).

Accusative Reflexive Pronouns	
ich → mich	wir → uns
du → dich Sie → sich	ihr → euch Sie → sich
er sie → sich es	sie → sich

Ich fühle mich nicht wohl.	*I don't feel well.*
Michael hat sich verletzt.	*Michael hurt himself.*

Verbs with reflexive pronouns use the auxiliary **haben** in the perfect and past perfect tenses.

Heidi hat sich in den Finger geschnitten.	*Heidi cut her finger.*

*Even when it refers to **Sie,** the polite form of *you*, **sich** is not capitalized.

Übung 1 Minidialoge

Ergänzen Sie das Verb und das Reflexivpronomen.

sich ärgern (geärgert)
sich aufregen (aufgeregt)
sich ausruhen (ausgeruht)
sich erkälten (erkältet)
sich freuen (gefreut)
sich fühlen (gefühlt)
sich legen (gelegt)
sich schneiden (geschnitten)
sich verletzen (verletzt)

1. SILVIA: Ich _____ _____[a] gar nicht wohl.
   JÜRGEN: Warum denn?
   SILVIA: Ich glaube, ich habe _____ _____[b].
   JÜRGEN: Du Ärmste! Du musst _____ gleich ins Bett _____[c].
2. MICHAEL: Du, weißt du, dass Herr Thelen einen Herzinfarkt[1] hatte?
   MARIA: Kein Wunder, er hat _____ auch immer so furchtbar _____[a].
   MICHAEL: Na, jetzt muss er _____ erst mal ein paar Wochen _____[b].
3. FRAU RUF: Du blutest ja! Hast du _____ _____[a]?
   HERR RUF: Ja, ich habe _____ in den Finger _____[b].
4. HEIDI: Warum _____ du _____[a], Stefan?
   STEFAN: Ich habe in meiner Prüfung ein D bekommen.
   HEIDI: Du solltest _____ _____[b], dass du kein F bekommen hast.

11.2 Dative reflexive pronouns

When a clause contains another object in addition to the reflexive pronoun, then the reflexive pronoun is in the dative case; the other object, usually a thing or a part of the body, is in the accusative case.

DAT. (mir) ACC. (den Mantel) Ich ziehe mir den Mantel aus.	*I'm taking off my coat.*

Note that the accusative object (the piece of clothing or part of the body) is preceded by the definite article.

Wäschst du dir jeden Tag **die** Haare?	*Do you wash your hair every day?*
Sumita hat sich **den** Arm gebrochen.	*Sumita broke her arm.*

Only the reflexive pronouns that correspond to **ich** and **du** have different dative and accusative forms.

Reflexive Pronouns

		Accusative	*Dative*
SINGULAR	*ich* →	mich	mir
	du →	dich	dir
	Sie →	sich	
	er/sie/es →		
PLURAL	*wir* →	uns	
	ihr →	euch	
	Sie →	sich	
	sie →		

[1]*heart attack*

Übung 2 Meine Morgentoilette

In welcher Reihenfolge machen Sie das?

MODELL: Erst stehe ich auf. Dann dusche ich mich. Dann ...

sich abtrocknen	sich die Haare föhnen
sich anziehen	sich die Haare kämmen
aufstehen	sich die Haare waschen
sich duschen	sich rasieren
sich die Fingernägel putzen	sich schminken
frühstücken	sich die Zähne putzen
sich das Gesicht waschen	zur Uni gehen

Übung 3 Körperpflege

Wer macht das? Sie, Ihre Freundin, Ihr Vater ...?

1. sich jeden Morgen rasieren	ich
2. sich zu sehr schminken	meine Freundin
3. sich nicht oft genug die Haare waschen	mein Freund
4. sich nach jeder Mahlzeit die Zähne putzen	mein Vater
5. sich immer verrückt anziehen	meine Mutter
6. sich jeden Tag duschen	meine Schwester
7. sich nie kämmen	meine Oma
8. sich nie die Haare föhnen	mein Onkel
9. sich nicht gern baden	_______?
10. sich immer elegant anziehen	

11.3 Word order of accusative and dative objects

When the accusative object and the dative object are both *nouns,* then the dative object precedes the accusative object.

DAT. ACC.

Ich schenke **meiner Mutter einen Ring**. *I'm giving my mother a ring.*

When either the accusative object or the dative object is a *pronoun* and the other object is a *noun,* then the pronoun precedes the noun regardless of case.

DAT. ACC.

Ich schenke **ihr einen Ring.** *I'm giving her a ring.*

ACC. DAT.

Ich schenke **ihn meiner Mutter.** *I'm giving it to my mother.*

The dative object precedes the accusative object, unless the accusative object is a pronoun.

When the accusative object and the dative object are both *pronouns,* then the accusative object precedes the dative object.

Ich schenke **ihn ihr.** *I'm giving it to her.*

Note that English speakers use a similar word order. Remember that German speakers do *not* use a preposition to emphasize the dative object as English speakers often do (*to my mother, to her*).

Übung 4 Im Hotel

Sie sind mit Ihrem Partner / Ihrer Partnerin in einem Hotel. Sie sind gerade aufgestanden und packen Ihre gemeinsame Toilettentasche aus.

MODELL: S1: Brauchst du den Lippenstift?
S2: Ja, kannst du ihn mir geben?
oder Nein, ich brauche ihn nicht.

1. Brauchst du das Shampoo?
2. Brauchst du den Spiegel?
3. Brauchst du den Rasierapparat?
4. Brauchst du die Seife?
5. Brauchst du das Handtuch?
6. Brauchst du den Föhn?
7. Brauchst du die Creme?
8. Brauchst du das Rasierwasser?
9. Brauchst du den Kamm?

Übung 5 Gute Ratschläge!

Geben Sie Ihrem Partner / Ihrer Partnerin Rat.

MODELL: S1: Meine Hände sind schmutzig.
S2: Warum wäschst du sie dir nicht?

1. Mein Bart ist zu lang.
2. Meine Füße sind schmutzig.
3. Meine Fingernägel sind zu lang.
4. Meine Haut ist ganz trocken.
5. Meine Haare sind nass.
6. Mein Hals ist schmutzig.
7. Meine Nase läuft.
8. Meine Haare sind zu lang.
9. Mein Gesicht ist ganz trocken.
10. Meine Haare sind fettig.

eincremen **waschen**
putzen **föhnen**
schneiden **stutzen**[1]

11.4 Indirect questions: *Wissen Sie, wo ...?*

Indirect questions:
- dependent clause begins with a question word or **ob**
- conjugated verb in the dependent clause appears at the end of the clause

Indirect questions are dependent clauses that are commonly preceded by an introductory clause such as **Wissen Sie, ...** or **Ich weiß nicht, ...** Recall that the conjugated verb is in last position in a dependent clause.

Wissen Sie, **wo** das Kind gefunden **wurde?**	*Do you know where the child was found?*
Können Sie mir sagen, **wann** die Polizei **ankommt?**	*Can you tell me when the police will arrive?*

The question word of the direct question functions as a subordinating conjunction in an indirect question.

DIRECT QUESTION: **Wie** komme ich zur Apotheke?
INDIRECT QUESTION: Ich weiß nicht, **wie** ich zur Apotheke **komme.**

Use the conjunction **ob** (*whether, if*) when the corresponding direct question does not begin with a question word but with a verb.

DIRECT QUESTION: **Kommt** Michael heute Abend?
INDIRECT QUESTION: Ich weiß nicht, **ob** Michael heute Abend **kommt.**

[1]*to trim*

Übung 6 Bitte etwas freundlicher!

Verwandeln Sie die folgenden direkten Fragen in etwas höflichere indirekte Fragen. Beginnen Sie mit **Wissen Sie, ...** oder **Können Sie mir sagen, ...**

MODELL: Wo war Herr Langen um sieben Uhr fünfzehn? →
Wissen Sie, wo Herr Langen um sieben Uhr fünfzehn war?
oder Können Sie mir sagen, wo Herr Langen um sieben Uhr fünfzehn war?

1. Was ist hier passiert?
2. Hat das Kind das Auto gesehen?
3. Wer war daran schuld?
4. Warum hat Herr Langen das Kind nicht gesehen?
5. Hat Herr Langen gebremst?
6. Wann hat er gebremst?
7. Wie oft fährt Herr Langen diese Straße zur Arbeit?
8. Wie lange lag Lothar auf der Straße?
9. Wann hat die Polizei Lothars Mutter angerufen?

11.5 Word order in dependent and independent clauses (summary review)

WISSEN SIE NOCH?

Dependent clauses may be introduced by subordinating conjunctions, such as **als** (*when, as*), **wenn** (*when, whenever*), and **wann** (*when*); by relative pronouns such as **der, die,** and **das** (*who, whom, that, or which*); or by question words such as **was** (*what*), **wie** (*how*), and **warum** (*why*) in indirect questions. Main verbs in dependent clauses appear at the end of the clause.

Review grammar 3.4, 7.1, 9.1, 9.3, 9.5, and 11.4.

To connect thoughts more effectively, two or more clauses may be combined in one sentence. There are essentially two kinds of combinations:

1. Coordination: both clauses are equally important and do not depend on each other structurally.
2. Subordination: one clause depends on the other one; it does not make sense when it stands alone.

COORDINATION

Heute ist ein kalter Tag und es schneit.	*Today is a cold day, and it is snowing.*

SUBORDINATION

Gestern war es wärmer, weil die Sonne schien.	*Yesterday was warmer because the sun was shining.*

A. Coordination

These are the five most common coordinating conjunctions.

und	*and*
oder	*or*
aber	*but*
sondern	*but, on the contrary*
denn	*because*

In clauses joined with these conjunctions, the conjugated verb is in second position in both statements.

CLAUSE 1	CONJ.	CLAUSE 2
I II		I II
Ich muss noch viel lernen,	denn	ich habe morgen eine Prüfung.

I still have to study a lot, since I have a test tomorrow.

B. Subordination

Clauses joined by subordinating conjunctions follow one of two word order patterns.

1. When the sentence begins with the main clause, that clause has regular word order (verb second in statements) and the dependent clause introduced by the conjunction has dependent word order (verb last).

CLAUSE 1	CONJ.	CLAUSE 2	
I II		I	LAST
Ich muss noch viel lernen,	weil	ich morgen eine Prüfung habe.	

I still have to study a lot because I have a test tomorrow.

2. When the sentence begins with the dependent clause, the entire dependent clause is considered the first part of the main clause and occupies first position. The verb-second rule applies, then, moving the subject of the main clause after the verb.

CLAUSE 1	CLAUSE 2
I	II SUBJECT
Weil ich morgen eine Prüfung habe,	muss ich noch viel lernen.

Because I have a test tomorrow, I still have to study a lot.

Here are the most commonly used subordinating conjunctions.

als	*when*
bevor	*before*
bis	*until*
damit	*so that*
dass	*that*
nachdem	*after*
ob	*whether, if*
obwohl	*although*
während	*while*
weil	*because, since*
wenn	*if, when*

Übung 7 Opa ist im Garten

Ergänzen Sie **dass, ob, weil, damit** oder **wenn.**

1. OMA: Weißt du, _____[a] Opa schon den Rasen gemäht hat?
   ESKE: Ich weiß nur, _____[b] er schon seit zwei Stunden im Garten ist.
   OMA: _____[c] Opa schon so lange im Garten ist, liegt er bestimmt in der Sonne.
2. ESKE: Du, Opi, was machst du denn im Gras?
   OPA: Ich habe mich nur kurz hingelegt, _____[a] mich die Nachbarn nicht sehen.
   ESKE: Aber warum sollen die dich denn nicht sehen?
   OPA: _____[b] ich mich heute noch nicht rasiert habe.

Übung 8 Minidialoge

Ergänzen Sie **obwohl, als, nachdem, bevor** oder **während.**

1. HERR THELEN: Was hat denn deine Tochter gesagt, ______[a] du mit deiner neuen Frisur nach Hause gekommen bist?

   HERR SIEBERT: Zuerst gar nichts. Erst ______[b] sie ein paar Mal um mich herumgegangen war, hat sie angefangen zu lachen und gesagt: „Aber, Papi, erst fast eine Glatze und jetzt so viele Haare. Das sieht aber komisch aus!"

2. FRAU ROWOHLT: Guten Tag, Herr Okonkwo! Kommen Sie doch bitte erst zu mir, ______ Sie mit Ihrer Arbeit beginnen.

   HERR OKONKWO: Aber natürlich, Frau Direktorin.

3. JOSEF: Ja, seid ihr denn immer noch nicht fertig? Was habt ihr eigentlich die ganze Zeit gemacht?

   MELANIE: ______ du dich stundenlang geduscht hast, haben wir die ganze Wohnung aufgeräumt.

4. MARIA: Aber, Herr Wachtmeister, könnten Sie nicht mal ein Auge zudrücken? Die Ampel war doch schon fast wieder grün.

   POLIZIST: Nein, leider nicht, ______ ich es gern tun würde, meine gnädige[1] Frau. Aber Sie wissen ja, Pflicht ist Pflicht.

[1]*dear*

KAPITEL 12

Die moderne Gesellschaft

In **Kapitel 12,** you will discuss politics, social relationships and some of the issues that arise in modern multicultural societies. In addition, you will learn to talk about money matters and about German art and literature.

Themen

Politik

Multikulturelle Gesellschaft

Das liebe Geld

Kunst und Literatur

Kulturelles

KLI: Politische Parteien

Musikszene: „Cüs Junge" (Muhabbet [mit Fler])

KLI: Wie bezahlt man in Europa?

Filmclip: *Sophie Scholl – Die letzten Tage* (Marc Rothemund)

Videoecke: Medien und Finanzen

Lektüren

Anekdote: Sternzeichen (Rafik Schami)

Film: *Sophie Scholl – Die letzten Tage* (Marc Rothemund)

Strukturen

12.1 The genitive case

12.2 Expressing possibility: **würde, hätte,** and **wäre**

12.3 Causality and purpose: **weil, damit, um ... zu**

12.4 Principles of case (summary review)

Ismail Çoban: *Der Fremde in mir* (2008), Privatbesitz des Künstlers

KUNST UND KÜNSTLER

Ismail Çoban ist ein türkisch-deutscher Maler und Grafiker. Er wurde 1945 in Çorum in der Türkei geboren. Nach einer Schneiderlehre und einer Ausbildung zum Volksschullehrer besuchte er die Hochschule für angewandte Werkkunst in Istanbul. 1969 ging Çoban nach Deutschland. Seit 1971 lebt er als freischaffender Maler und Grafiker in Wuppertal.

Çobans Kunst bewegt sich zwischen der deutschen und der türkischen Kultur und ist geprägt[1] vom Wunsch nach Verständigung. Er selbst sieht sich als „Welten-Künstler" und ist nicht nur Künstler, sondern auch Mystiker, Philosoph und politischer Mensch.

2006 gründete Çoban die „Ismail-Çoban-Stiftung[2]" zur Förderung internationaler junger Künstler. Çoban will damit Kunst, die zur Verständigung zwischen Deutschen und Migranten beiträgt, fördern[3] und den Menschen ein Forum geben, die zum Teil immer noch als „Fremde" in Deutschland leben.

Schauen Sie sich das Bild an und beantworten Sie die folgenden Fragen.

1. Was für Assoziationen weckt der Titel des Bildes in Ihnen? Passt das zu dem Eindruck[4], den das Bild auf Sie macht?
2. Welche Farben dominieren in dem Bild?
3. Das Bild hat drei Teile. Was sehen Sie im linken Teil? Was könnte das symbolisieren? Was steht dem im rechten Teil gegenüber? Was sehen Sie in der Mitte?
4. Betrachten Sie die Körperteile der Menschen. Welche Körperteile sind besonders groß? Was könnte das symbolisieren?
5. Wie ist der Gesichtsausdruck der Menschen? Wie ist ihre Körperhaltung? Wie nah sind sie beieinander?

[1]*characterized* [2]*foundation* [3]*promote* [4]*impression*

Situationen

Politik

Grammatik 12.1–12.2

Die Rede der Bundeskanzlerin erhielt viel Applaus.

Die Mehrheit der Bevölkerung unterstützt den Kurs der Regierung.

Die Koalition vereinbart eine Reform des Rentensystems.

Im nächsten Jahr soll ein neues Wahlrecht eingeführt werden.

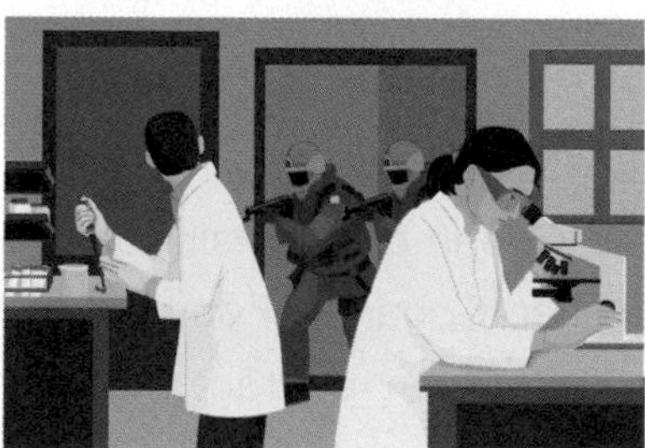

Die Freiheit der Wissenschaft ist in manchen Ländern in Gefahr.

Die Grenzen des Wachstums sind bald erreicht.

Situation 1 Wer im Kurs ...?

1. kennt eine Politikerin oder einen Politiker
2. ist schon mal zur Wahl gegangen
3. tritt für die Rechte[1] von Minderheiten ein
4. hat schon mal eine politische Rede gehalten
5. gehört einer politischen Partei an
6. ist dafür, dass die Steuern[2] gesenkt werden
7. hat schon mal eine Kandidatin unterstützt?
8. möchte nicht in der Öffentlichkeit stehen
9. hat eine Strategie gegen den Terrorismus
10. möchte einmal eine Schlüsselposition besetzen

[1]*rights* [2]*taxes*

POLITISCHE PARTEIEN

- Wie heißen die wichtigsten Parteien in Ihrem Land?
- Wann wurden sie gegründet?
- Welche Grundwerte[3] haben sie?
- Welche Partei ist zur Zeit an der Regierung?
- Wofür kämpft Ihre jetzige Regierung?

Lesen Sie den Text und suchen Sie Antworten auf diese Fragen:

- Welche Partei ist die älteste Partei Deutschlands? Wie alt ist sie? Wo wurde sie gegründet?
- Welche Partei war am längsten in der Regierung? Wie alt ist sie? Wo liegen ihre Wurzeln[4]?
- Welche Partei ist die Nachfolgepartei der Regierungspartei der DDR? Wie hieß diese Regierungspartei? Wie nannte sie sich nach der friedlichen Revolution im Jahre 1989?
- Welche Partei ist die ökologische Partei? Wie alt ist sie? Ist sie eher links oder rechts?
- Welche Partei möchte aus dem Euro austreten[5]? Wie alt ist sie? Ist sie eher links oder rechts?
- Welche Parteien sagen, ihre Grundwerte sind Freiheit, Solidarität und Gerechtigkeit[6]?
- Welche Parteien setzen sich besonders für die Frauen ein?

Die zwei größten Parteien in Deutschland sind die Christlich Demokratische Union Deutschlands (CDU) und die Sozialdemokratische Partei Deutschlands (SPD) mit jeweils[7] knapp einer halben Million Mitglieder.

Die CDU wurde 1945 gegründet und nennt sich selbst die Volkspartei der Mitte. Sie ist die Partei, die am häufigsten den Bundeskanzler[8] oder die Bundeskanzlerin gestellt[9] hat. Die CDU hat religiöse Wurzeln. In ihrem Grundsatzprogramm steht, dass ihre Grundwerte Freiheit, Solidarität und Gerechtigkeit sind. Diese Grundwerte sind christlich orientiert. Die CDU sieht Ehe und Familie als Fundament der Gesellschaft. Sie möchte die Würde[10] des Menschen vom Beginn bis zum Ende des Lebens schützen und seine natürlichen Lebensgrundlagen (Umwelt, Klima, Nahrungsmittel) bewahren. Sie fördert Eigeninitiative und sie steht für die Freiheit und Sicherheit der Bürger.

Die SPD ist die älteste deutsche Partei. Sie wurde 1863 als *Allgemeiner Deutscher Arbeiterverein* (ADAV) in Leipzig gegründet. Seit 1890 nennt sie sich Sozialdemokratische Partei Deutschlands. Ihre Grundwerte sind ebenfalls Freiheit, Gerechtigkeit und Solidarität. Die SPD steht den Gewerkschaften[11] nahe. Besonders wichtig ist für sie der Sozialstaat, der Menschen bei Krankheit, Behinderung oder Arbeitslosigkeit unterstützt. Sie möchte, dass Menschen von ihrer Arbeit leben können. Sie kämpft für Menschenrechte und Frieden und baut dabei auf Dialog und Konfliktlösung. Sie ist für eine gesetzliche[12] Frauenquote, die die Emanzipation der Frau fördern soll.

[3]*basic values* [4]*roots* [5]austreten aus *withdraw from* [6]*justice* [7]*each* [8]*Federal Chancellor* [9]here: *provided* [10]*dignity* [11]*labor unions* [12]*legally (mandated)*

Ein Wahlplakat der SPD um 1920

Die Grünen und die Linke sind kleinere Parteien. Sie sind aber wichtig für Koalitionen mit den zwei großen Parteien.

Die Grünen wurden 1980 gegründet. 1993 fusionierten[13] sie mit dem Bündnis 90, das aus der Bürgerbewegung der DDR hervorging, zum Bündnis 90/Die Grünen. Die Grundwerte der Grünen sind Ökologie, Selbstbestimmung[14], Gerechtigkeit und Demokratie. Sie wollen eine grüne Marktwirtschaft, die ökologisch und gerecht[15] ist. Sie wollen das Klima retten[16] und Armut bekämpfen. Sie sind für eine multikulturelle Gesellschaft und kämpfen für die Integration von Einwanderern. Sie wollen gleiche Rechte für Männer und Frauen und sind für die Anerkennung von gleichgeschlechtlichen[17] Lebenspartnerschaften.

Die Linke bezeichnet sich selbst als eine sozialistische Partei. Sie wurde 2007 gegründet, als die Partei des demokratischen Sozialismus (PDS) mit einer linken Partei aus den alten Bundesländern fusionierte. Die PDS war die Nachfolgepartei der Sozialistischen Einheitspartei Deutschlands (SED), die die DDR von ihrer Gründung 1949 bis zu ihrem Ende 1990 regierte. Die Linke ist vor allem in den neuen Bundesländern stark. Sie möchte, dass große Unternehmen und reiche Menschen deutlich mehr Steuern zahlen, bis zu 50%, und will Löhne[18], Renten[19] und das Kindergeld erhöhen[20]. Sie kämpft dafür, dass Menschen weniger arbeiten müssen. Sie will eine stärkere staatliche Kontrolle der Finanzmärkte, ist dagegen, dass die Bundeswehr[21] im Ausland eingesetzt wird, und möchte das Adoptionsrecht für gleichgeschlechtliche Paare[22].

Die Alternative für Deutschland (AfD) wurde erst 2013 gegründet. Sie ist eine typische Protestpartei. Ihr Hauptziel ist seit ihrer Gründung, den Euro abzuschaffen[23]. Die AfD ist eine nationalkonservative Partei mit rechtspopulistischen Zügen[24]. Sie möchte die Einwanderung begrenzen, ist dagegen, dass die Türkei in die EU aufgenommen wird, und möchte kein Freihandelsabkommen[25] mit den USA. Einwanderer, die nicht genügend Geld haben, um in Deutschland zu leben, sollen in ihr Heimatland zurückkehren.

[13]*merged* [14]*self-determination* [15]*just* [16]*save* [17]*same-sex* [18]*wages* [19]*pensions* [20]*increase* [21]*federal military* [22]*couples* [23]*do away with* [24]*tendencies* [25]*free trade agreement*

Situation 2 Die deutschen Parteien

Welche Ziele passen zu welchen Parteien? Lesen Sie die KLI auf Seite 413 und entscheiden Sie, welche Ziele zu den folgenden Parteien passen.

1. das Klima retten
2. Einkommen und Vermögen gerechter verteilen
3. für starke Gewerkschaften sorgen
4. das Adoptionsrecht für gleichgeschlechtliche Paare einführen
5. die Explosion der Mietpreise verhindern
6. aus dem EURO austreten
7. die Renten und Löhne erhöhen
8. dafür sorgen, dass Menschen von ihrer Arbeit leben können
9. kriminelle Einwanderer in ihr Heimatland zurückschicken
10. für den Tierschutz eintreten
11. die Steuern senken
12. die europäische Integration unterstützen
13. eine gesetzliche Frauenquote einführen
14. christliche Werte betonen

a. AfD (Alternative für Deutschland): nationalkonservativ
© Peter Probst/Alamy

b. CDU (Christlich-demokratische Union): konservativ
© Peter Probst/Alamy

c. Die Grünen: liberal, ökologisch
© Peter Probst/Alamy

DIE LINKE.

d. Die Linke: sozialistisch
© Peter Probst/Alamy

e. SPD (Sozialdemokratische Partei Deutschlands): sozialdemokratisch
© Peter Probst/Alamy

Der deutsche Bundestag
© dpa picture alliance/Alamy

Situation 3 Diskussion

Was ist gut (G) und was ist schlecht (S) für eine Demokratie?

1. Weniger als die Hälfte der Wahlbevölkerung geht zur Wahl.
2. Der neue Präsident bringt seine Verwandten und Freunde mit in die Regierung.
3. Die Wissenschaft wird zensiert, weil sie nicht mit dem konform ist, was in der Bibel steht.
4. Der Kampf gegen den Terrorismus schadet der persönlichen Freiheit der Bürger.
5. Jeder kann seine Meinung sagen, auch wenn es gegen die Regierung gerichtet ist.
6. Weil das Land in einer tiefen Rezession ist, steigt die Zahl der Arbeitslosen immer weiter.
7. Die Regierung kann ihre Bürger nicht vor Kriminellen schützen.
8. Proteste von Bürgerinitiativen haben Erfolg, weil die Regierung zum Kompromiss bereit ist.
9. Die Medien sind engagiert und können schreiben, was sie wollen.
10. Weil das Land sehr reich ist, wollen viele Ausländer aus ärmeren Ländern einwandern.
11. Die Finanzmärkte werden vom Staat immer stärker kontrolliert.
12. Abgeordnete werden für 10 Jahre gewählt, damit es eine stabile Regierung gibt.

Situation 4 Interview

1. Wie würdest du deine politische Einstellung einschätzen[26]? eher konservativ oder eher liberal? und die politische Einstellung deiner Eltern?
2. Kannst du dir vorstellen, eine Abgeordnete oder ein Abgeordneter zu sein? Welcher Partei? Wofür würdest du eintreten?
3. Wenn du in Deutschland wählen könntest, welche Partei würdest du wählen? Warum?
4. Hast du schon mal mit einem wichtigen Politiker oder einer Politikerin gesprochen? Wer war das? Wie ist es dazu gekommen? Worüber habt ihr gesprochen?
5. Was hältst du von der jetzigen Regierung in deinem Land? Was macht sie gut? Was macht sie schlecht?
6. Welche Gesetze sollten sich, deiner Meinung nach, in deinem Land ändern? Wähle eins und erkläre warum.
7. Welche großen politischen oder gesellschaftlichen Probleme gibt es in der Welt?
8. Bist du schon einmal mit etwas gescheitert (z. B. mit einer neuen Idee, in der Schule, in der Uni, im Freundeskreis)? Was war es? Wie ist es passiert? Was hast du danach gemacht?

[26]*assess*

Multikulturelle Gesellschaft

Grammatik 12.3

CLAIRE: Ist Deutschland eigentlich ein multikulturelles Land?
JOSEF: Ja, natürlich. Es leben hier über acht Millionen Ausländer.

RENATE: Unsere ausländischen Mitbürger bereichern Deutschland mit ihrer Kultur und ihren Traditionen.

MEHMET: Deutschland braucht in vielen Branchen ausländische Arbeitskräfte, insbesondere im MINT-Bereich.

JÜRGEN: Wie in jedem anderen Land müssen Ausländer auch in Deutschland ihre Aufenthalts-und Arbeitserlaubnis beantragen. Dazu müssen sie viele Formulare ausfüllen.

Situation 5 **Definitionen**

1. das Formular
2. die Aufenthaltserlaubnis
3. die Arbeitserlaubnis
4. der MINT-Bereich
5. das multikulturelle Land
6. etwas beantragen

a. Die braucht man, damit man in Deutschland wohnen darf.
b. Das muss man ausfüllen, um zum Beispiel eine Arbeitserlaubnis zu bekommen.
c. Die braucht man, damit man arbeiten kann.
d. Land, in dem Menschen aus verschiedenen Kulturen zusammen leben
e. Formulare ausfüllen und in einem Büro abgeben
f. Mathematik, Informatik, Naturwissenschaften, Technik

Situation 6 **Interview**

1. Sind deine Vorfahren eingewandert? Wenn ja, weißt du, wann deine Vorfahren eingewandert sind? Woher kamen sie? Welche Sprache haben sie gesprochen? Warum haben sie ihre Heimat verlassen?
2. Spricht man in deiner Familie mehr als eine Sprache? Welche? Welche Vorteile oder Nachteile hat das für dich?
3. Kennst du Einwanderer? Woher kommen sie? Sprechen sie Englisch? Warum sind sie eingewandert?

4. Weißt du, welche Formalitäten man erfüllen muss, um legal hier wohnen und arbeiten zu dürfen?
5. Welche Probleme können Einwanderer haben? Wie kann man diese Probleme lösen?

Situation 7 Diskussion: Leben in einer fremden Kultur

Welche Probleme können Ausländer haben? Was ist für die Integration von Ausländern wichtig? Arbeiten Sie in kleinen Gruppen. Schreiben Sie in jede Spalte fünf Dinge, die Sie für wichtig halten. Ordnen Sie die Dinge: das Wichtigste zuerst. Einige Ideen finden Sie im Wortkasten.

Probleme von Ausländern	für die Integration wichtig

Geld verdienen **eine Wohnung finden**
eine gute Schulbildung bekommen
Feste gemeinsam feiern **Heimweh haben**
Freunde finden **gemeinsam Sport treiben**
ein Kulturzentrum gründen **die Sprache lernen**
sich über die Kultur des anderen informieren
einen Arbeitsplatz finden **seine Religion ausüben**
_______?

Situation 8 Diskussion: Extremismus

1. Gibt es Extremisten in Ihrem Land? Wo? Was für Ziele haben sie? Was machen sie?
2. Was ist, Ihrer Meinung nach, ein typischer Extremist?
   - ☐ Frau ☐ Mann
   - ☐ jung ☐ alt
   - ☐ schlecht ausgebildet ☐ gut ausgebildet
   - ☐ arm ☐ reich
   - ☐ sympathisches Äußeres ☐ unsympathisches Äußeres
   - ☐ arbeitslos ☐ mit gutem Arbeitsplatz
   - ☐ Einzelgänger[1] ☐ nur in der Gruppe stark
3. Wodurch fallen Extremisten auf?
4. Was kann man gegen Extremismus tun?

[1] *loner*

MUSIKSZENE

„Cüs Junge" (2007, Deutschland) *Muhabbet* (*mit Fler*)

Biografie Muhabbet (Murat Ersen) ist 1984 in Köln geboren. Er singt auf Deutsch und Türkisch. Seine Musik verbindet arabeske Elemente orientalischer Popmusik mit R'n'B. Muhabbet nennt diesen Stil R'n'Besk, so auch der Titel seines zweiten Albums. 2007 spielte er auf dem Sommerfest des damaligen Bundespräsidenten, Horst Köhler, und er nahm das Lied „Deutschland" mit dem damaligen Außenminister Deutschlands, Frank-Walter Steinmeier, auf. Mit diesem Lied wollte er für ein modernes und tolerantes Deutschland werben. Der Song „Cüs Junge" entstand 2006 aus einer Zusammenarbeit mit dem deutschen Rapper Fler.

Miniwörterbuch	
verbinden	to combine
der **Außenminister**	foreign minister, secretary of state
aufnehmen	to record
werben	to promote
die **Herausforderung**	challenge
der **Wahnsinn**	madness

Muhabbet und der deutsche Außenminister nehmen den Song „Deutschland" auf.

NOTE: For copyright reasons, the songs referenced in **MUSIKSZENE** have not been provided by the publisher. The song can be found online at various sites such as YouTube, Amazon, or the iTunes store.

Vor dem Hören Sind Sie bi-kulturell oder kennen Sie jemanden, der zwei Kulturen in sich vereint? Welche Vorteile haben bi-kulturelle Menschen? Welchen Herausforderungen müssen sie sich stellen?

Nach dem Hören

1. Beantworten Sie die Fragen zum Anfang und zum Refrain.
- **a.** Für wen ist dieser Song?
- **b.** Wie ist der Klub?
- **c.** Wer sitzt an der Bar?
- **d.** Was sagt der Sänger, als er sie sieht?

2. Was sagt der Sänger NICHT über die junge Frau im Klub?
- **a.** Sie hat Klasse.
- **b.** Sie ist seine Traumfrau.
- **c.** Sie ist der pure Wahnsinn.
- **d.** Sie ist heute Abend der Star.
- **e.** Sie hat viel Geld.
- **f.** Sie füllt den Raum mit Licht.
- **g.** Sie macht alle schwach.
- **h.** Sie ist das Beste aus zwei Welten.
- **i.** Sie ist eine Mischung aus Gold und Platin.

Lektüre

Vor dem Lesen

A. Beantworten Sie die folgenden Fragen.

1. Wann haben Sie Geburtstag? Wissen Sie, um wie viel Uhr Sie geboren sind? Wie war das Wetter? Welche berühmten Persönlichkeiten sind am gleichen Tag wie Sie geboren?
2. Welches Sternzeichen sind Sie? Welche Eigenschaften hat Ihr Sternzeichen? Spielen Sternzeichen eine Rolle in Ihrem Leben?

LESEHILFE

The reading "Sternzeichen" is an anecdote (German **die Anekdote**). An anecdote is typically a very short story with a direct, humorous tone. The telling of an anecdote often leads to a climactic twist (German **die Pointe**) near the end, almost a sort of punch line. As you read this text, be on the lookout for the **Pointe.**

© Roger-Viollet/The Image Works

B. Lesen Sie die Wörter im Miniwörterbuch. Suchen Sie sie in den Aktivitäten und im Text und unterstreichen Sie sie.

Miniwörterbuch	
die **Eigenschaft**	characteristic
Krebs	Cancer
sei	*here:* was
ablehnen	to refuse
die **Schüchternheit**	shyness
meiden	to avoid
die **Neigung**	inclination
trotz	despite
der **Schriftsteller**	writer
der **Bahnhof**	train station
die **Geburt**	birth
berühmt	famous
die **Persönlichkeit**	personality, celebrity
das **Sternzeichen**	sign of the zodiac
der **Ausweis**	identification card
die **Aprikose, -n**	apricot
der **Kampf,** die **Kämpfe**	battle, fight
der **Berg, -e**	mountain
hämisch	gloatingly
die **Begegnung**	encounter

Sternzeichen

von Rafik Schami

„Typisch Krebs", sagt H., ein Bekannter aus Heidelberg. Er glaubt noch genau zu wissen, daß ich Ende Juni geboren sei. Wir haben zusammen studiert. Heute ist er erfolgreicher Astrologe. Nach fünfzehn Jahren sehen wir uns zum ersten Mal wieder. Er will mir sofort ein exaktes Horoskop erstellen. Als ich es ablehne, führt er das auf die für Krebse angeblich typische Schüchternheit zurück.

„Nicht von ungefähr geht der Krebs seitlich. Er meidet jeden Konflikt, aber wenn es darauf ankommt, ist er sehr wehrhaft", denke ich laut. H. lacht. Typisch für den Krebs sei seine Neigung zur Kunst, entgegnet er.

Entscheidend für die gesamte Ausrichtung meines Lebens ist also seiner Meinung nach mein Geburtstag gewesen. Und er spricht ständig von Achsen und Sternenkonstellationen. Er habe schon damals in Heidelberg sicher gewußt, daß ich trotz des Chemiestudiums in meinem tiefsten Innern ein Künstler sei. „Und was ist dann aus dir geworden, hm? Vielleicht ein Chemiker? Nein, ein Schriftsteller."

Araber feiern vieles, aber Geburtstage nie. Denn wenn man seinen Geburtstag genau kennt, wird man nur älter. Mir kommen die Europäer manchmal so vor, als wären sie alle am Bahnhof geboren. Sie wissen nicht bloß das Datum, sondern sogar die genaue Uhrzeit ihrer Geburt. H., mein Bekannter, weiß auch die Temperatur und das Himmelsbild jenes Tages, sogar die berühmten Persönlichkeiten, die unter demselben Sternzeichen wie er auf die Welt kamen, hat er parat.

Als er geht, rufe ich meine Mutter in Damaskus an und frage sie, wann ich geboren wurde, denn ich mißtraue meinem Ausweis. „Anfang bis Mitte April", antwortet sie. „Die Aprikosen standen in voller Blüte. Wir mußten uns aber wegen der Kämpfe in der Hauptstadt in den Bergen verstecken. Deshalb konnten wir dich erst danach in der Hauptstadt registrieren lassen, das war dann Ende Juni."

Und ich freue mich hämisch auf die nächste Begegnung mit H., dem Astrologen.

Arbeit mit dem Text

A. Personen. Es gibt drei Personen im Text: H., Rafik Schami und Rafiks Mutter. Was erfahren wir über sie? Kreuzen Sie alles an, was richtig ist.

1. H.
   - ☐ a. Er hat mit Rafik studiert.
   - ☐ b. Er ist Astrologe.
   - ☐ c. Er ist aus Heidelberg
   - ☐ d. Er hat Chemie studiert.
2. Rafik Schami
   - ☐ a. Er ist Krebs.
   - ☐ b. Er hat Astrologie studiert.
   - ☐ c. Er ist Schriftsteller.
   - ☐ d. Er hat H. 15 Jahre nicht gesehen.
3. Rafiks Mutter
   - ☐ a. Sie wohnt in Damaskus.
   - ☐ b. Sie feiert gern ihren Geburtstag.
   - ☐ c. Als Rafik geboren wurde, war sie in den Bergen.
   - ☐ d. Sie liebt Aprikosen.

B. Handlung. Die folgenden Sätze fassen die Handlung zusammen. Bringen Sie sie in die richtige Reihenfolge.

_____ Rafik lehnt ab.

_____ Die Mutter erzählt Rafik, dass er im April geboren ist.

_____ H. will für Rafik ein Horoskop erstellen.

_____ H. trifft Rafik nach 15 Jahren wieder.

_____ Als H. geht, ruft Rafik seine Mutter an.

_____ H. meint, dass Rafik schon immer ein Künstler war.

C. Inhalt. Beantworten Sie die Fragen.

1. Welche Eigenschaften hat ein Krebs?
2. Warum feiern Araber ihren Geburtstag nicht?
3. Warum, glaubt Rafik, sind Europäer am Bahnhof geboren?
4. Was weiß H. alles über seine eigene Geburt?
5. Warum steht in Rafiks Ausweis, dass er Ende Juni geboren ist?
6. Warum freut sich Rafik hämisch auf die nächste Begegnung mit seinem Bekannten?

Nach dem Lesen

Rafik Schami kommt aus Syrien. Er ist 1971 nach Deutschland ausgewandert, hat dort Chemie studiert und 1979 seinen Doktortitel bekommen. Seit 1982 ist er freier Schriftsteller und seit 2002 ist er Mitglied der Bayerischen Akademie der Künste. Die Geschichte *Sternzeichen* stammt aus dem Buch *Gesammelte Olivenkerne* von 1997. In diesem Buch schreibt Rafik Schami über seine Begegnungen mit Deutschen und Arabern. Suchen Sie im Internet Antworten auf die folgenden Fragen.

1. Wann ist Rafik Schami geboren?
2. Wo wohnt er?
3. Welche weiteren Bücher hat er geschrieben? Nennen Sie vier.
4. Welche Preise hat er bekommen? Nennen Sie zwei.
5. In wie viele Sprachen wurden seine Bücher übersetzt?
6. Warum hat er sein Buch von 1997 *Gesammelte Olivenkerne* genannt?

Das liebe Geld

Grammatik 12.4

—Ich möchte gern ein Konto eröffnen.
—Ein Spar- oder ein Girokonto?

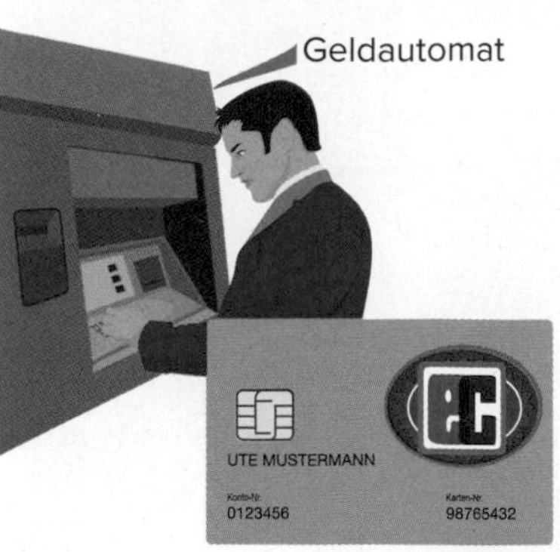

Für diesen Geldautomaten braucht man eine EC-Karte.

Wenn man Geld auf einem Sparkonto hat, bekommt man Zinsen.

Wenn man Schulden hat, muss man Zinsen zahlen.

Wenn man Geld überweisen möchte, kann man das auch per Online-Banking tun.

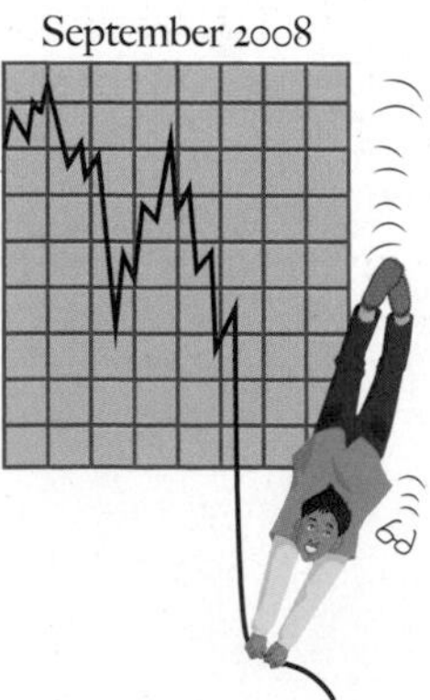

Der Börsenkrach vom September 2008 war einer der schlimmsten in der Geschichte.

Situation 9 Wer weiß – gewinnt: Geld

1. der Ort, an dem mit Aktien gehandelt wird
2. die Karte, mit der man bargeldlos[1] bezahlen kann
3. die zahlt man, wenn man Schulden hat
4. der Kurs, zu dem man ausländische Währung kaufen oder verkaufen kann
5. Automat, aus dem man Bargeld holen kann
6. die Münzen und Geldscheine einer Währung
7. das offizielle Zahlungsmittel eines Landes
8. das macht man, wenn man Rechnungen per Online-Banking bezahlt
9. das Konto für den täglichen Gebrauch
10. das macht man, wenn man bei einer Bank neu ist

a. das Bargeld
b. das Girokonto
c. der Geldautomat
d. der Wechselkurs
e. die Börse
f. die Kreditkarte
g. die Währung
h. die Zinsen
i. ein Konto eröffnen
j. Geld überweisen

Situation 10 Dialog: Auf der Bank

PETER: Guten Tag, ich möchte ein Konto ___________.

BANKANGESTELLTE: Ein Spar- oder ein Girokonto?

PETER: Ein Girokonto.

BANKANGESTELLTE: Würden Sie dann bitte dieses Formular ausfüllen?

PETER: Bekomme ich bei dem ___________ auch eine EC-Karte?

BANKANGESTELLTE: Die müssen Sie extra beantragen, aber das ist kein Problem, wenn regelmäßig auf das Konto ________ wird.

PETER: Ich bekomme ein Stipendium. Das soll auf dieses Konto überwiesen werden.

BANKANGESTELLTE: Gut. Die EC-Karte und Ihre ___________ bekommen Sie mit der Post.

PETER: Bekomme ich auf mein Guthaben auch ___________?

BANKANGESTELLTE: Nein, Zinsen gibt es nur auf Sparkonten.

PETER: Habe ich bei dem _________ einen Überziehungskredit[2]?

BANKANGESTELLTE: Ja, die Höhe richtet sich nach Ihrem Einkommen.

PETER: Kann ich meine ___________ auch übers Internet ausführen?

BANKANGESTELLTE: Natürlich. Meine Kollegin, Frau Schröder, hilft Ihnen da weiter.

PETER: Vielen Dank. Auf Wiedersehen.

BANKANGESTELLTE: Auf Wiedersehen.

Situation 11 Interview

1. Hast du ein Konto bei der Bank? Welche Konten hast du?
2. Hast du eine Kreditkarte? Wie viel kannst du damit ausgeben? Wie viel Zinsen musst du bezahlen?
3. Wie viel sparst du im Monat? Worauf sparst du? Wenn du jetzt nicht sparen kannst: Worauf würdest du sparen, wenn du Geld hättest?
4. Womit zahlst du meistens: mit Schecks, mit Kreditkarte oder mit Bargeld?
5. Wie viel Geld hast du im Monat? Wie viel Geld gibst du aus? Wofür gibst du das meiste Geld aus?
6. Hast du schon einmal einen Kredit aufgenommen? Wie hast du das gemacht?

[1] *cash-free* [2] *overdraft protection*

KULTUR ... LANDESKUNDE ... INFORMATIONEN

WIE BEZAHLT MAN IN EUROPA?

Wie ist es bei Ihnen?

- Wie bezahlen Sie meistens, wenn Sie im Supermarkt einkaufen?
- Wie bezahlen Sie Ihre Miete?
- Wie bezahlen Sie Ihre Telefonrechnung?
- Wie bezahlen Sie, wenn Sie ein Kleidungsstück kaufen? Ein Smartphone oder einen Computer? Ein Auto?
- Wie werden Sie in Ihrem Job bezahlt? In bar, mit Scheck oder Überweisung?
- In welcher Form bekommen Sie Geld von Ihren Eltern oder finanzielle Unterstützung für Ihr Studium?

In manchen Restaurants werden keine Kreditkarten akzeptiert.

Man zahlt daher meistens bar.

Wie ist es in Europa? Lesen Sie den Text und beantworten Sie die Fragen.

1. Wie heißt die Karte, die in Deutschland am häufigsten zum Einkaufen benutzt wird?
2. Was muss man für diese Karte bei der Bank haben?
3. Wie heißt die „elektronische Geldbörse", die man in Österreich benutzt?
4. Wie wird in Österreich immer noch am häufigsten bezahlt?
5. Wie bezahlt man normalerweise in Deutschland Miete und Rechnungen?
6. Was ist ein Dauerauftrag?
7. Wie viel Prozent der Deutschen nehmen am Online-Banking teil?

Auch in vielen Ländern Europas bezahlt man inzwischen nicht mehr so häufig mit Bargeld wie noch vor einigen Jahren. Für bargeldlose Transaktionen wird in Deutschland die EC-Karte am häufigsten benutzt. Man kann mit ihr im Supermarkt, beim Tanken und in den meisten Einzelhandelsgeschäften[3] bezahlen und Geld aus dem Geldautomaten bekommen. Für eine EC-Karte braucht man ein Konto bei einer Bank, das – anders als bei Kreditkarten – bei jeder Transaktion sofort belastet[4] wird. Manchmal muss man allerdings beim Einkauf außerdem noch seinen Personalausweis zeigen.

In Österreich ist die Quickcard eine beliebte Alternative zum Bargeld. Sie funktioniert wie eine elektronische Geldbörse[5]. Man muss sie „aufladen[6]" und kann dann z. B. an Parkautomaten, in Geschäften und an Tankstellen auch kleine Beträge[7] bezahlen. Die dominierende Zahlungsform in Österreich ist aber immer noch die Bargeldtransaktion. Die beliebteste Kreditkarte in Deutschland ist die Eurocard, die zu der Organisation von Mastercard (USA) gehört. Danach kommt die Visa-Karte.

Rechnungen für Telefon, Nebenkosten oder Miete bezahlt man bargeldlos mit Überweisungen vom Girokonto oder Bankeinzug[8] (der Betrag wird automatisch von der Bank des Empfängers[9] eingezogen). Damit man die monatlichen Zahlungen nicht vergisst, kann man sie per Dauerauftrag[10] überweisen lassen. Das heißt, man gibt seiner Bank einmal den Auftrag[11] und zu einem bestimmten Termin wird der Betrag automatisch überwiesen. Sehr beliebt ist auch das Online-Banking, das inzwischen von ungefähr 70% der Deutschen genutzt wird, vor allem für Überweisungen und Daueraufträge oder zum Überprüfen des Kontostandes.

[3]*retail shops* [4]*debited* [5]*wallet* [6]*charge, recharge* [7]*amounts* [8]*automatic withdrawal, i.e. electronic funds transfer* [9]*payee* [10]*standing order, i.e. recurring bill-pay* [11]*order*

Situation 12 Rollenspiel: Auf der Bank

S1: Sie haben ein Stipendium für ein Jahr an der Universität Leipzig bekommen. Sie wollen bei der Deutschen Bank ein Konto eröffnen. Fragen Sie auch nach den Zinsen, nach Online-Zugang und EC-Karte und ob Sie Ihr Konto überziehen dürfen.

Kunst und Literatur

Situation 13 Wer weiß – gewinnt: Kunst und Literatur

1. Welches Instrument gehört normalerweise nicht in ein Symphonie-Orchester?
2. Was braucht ein Bildhauer für seine Kunst?
3. Was war Theodor Storm von Beruf?
4. Von wem sind die Brandenburgischen Konzerte?
5. Was war Marlene Dietrich von Beruf?
6. Was brauchte Paul Klee für seine Kunst?
7. Wer schrieb die Tragödie *Faust*?
8. Welches Instrument spielt der Musiker David Garrett?

a. Blockflöte
b. Geige
c. Stein, Hammer und Meißel
d. Staffelei, Pinsel und Farben
e. Schriftsteller/in
f. Schauspieler/in
g. Johann Sebastian Bach
h. Johann Wolfgang von Goethe

Situation 14 Interview

1. Hörst du gern Musik? Was für Musik? Hast du einen Lieblingskomponisten oder eine Lieblingskomponistin?
2. Spielst du ein Instrument oder singst du?
3. Liest du gern? Was liest du gern: Romane, Gedichte, Dramen, Comics? Welche Schriftsteller magst du besonders? Hast du etwas von deutschen Schriftstellern gelesen?
4. Hast du schon mal etwas geschrieben? Was?
5. Welche Maler, Bildhauer oder Grafiker magst du am liebsten?
6. Malst oder zeichnest du? Welche Motive magst du am liebsten? (Berge? das Meer? eine Blumenvase?) Arbeitest du mit anderen Materialien wie Holz, Ton oder Stein?
7. Gehst du gern ins Theater? Welche Stücke gefallen dir besonders gut?
8. Hast du schon mal Theater gespielt? Welche Rollen hast du gespielt? Wie war das?

Situation 15 Faust: Die einfache Version

Eins der bekanntesten Werke der deutschen Literatur ist die Tragödie *Faust* von Goethe. Was in *Faust* geschieht, finden Sie in den folgenden Sätzen. Bringen Sie die Sätze in die richtige Reihenfolge.

TEIL 1

_____ Als Faust an einem Osternachmittag spazieren geht, sieht er einen schwarzen Pudel, der ihm nach Hause folgt.

_____ Nach ihrer Unterhaltung gehen Mephisto und Faust in eine Hexenküche. Dort zeigt ihm Mephisto einen magischen Spiegel.

__1__ Faust ist ein berühmter Wissenschaftler, der sehr unzufrieden ist, weil er nicht alles weiß.

_____ Faust spricht lange mit Mephisto und verspricht ihm seine Seele für einen Augenblick vollkommenen Glücks.

_____ In Fausts Studierzimmer verwandelt sich der Pudel in Mephisto.

_____ Im Spiegel sieht Faust eine wunderschöne Frau.

_____ Kurz danach lernt Faust Gretchen kennen und verliebt sich in sie.

TEIL 2

_____ Aber Gretchen will nicht vom Teufel gerettet werden und bittet Gott um Vergebung.

_____ Als Gretchen stirbt, hört man eine Stimme von oben, die sagt: „Sie ist gerettet."

_____ Als Gretchen vom Tod ihres Bruders hört, wird sie wahnsinnig, und als ihr Kind geboren wird, tötet sie es.

_____ Auf dem Brocken hat Faust eine Vision von Gretchen, und er und Mephisto eilen ins Gefängnis, um sie zu retten.

_____ Faust und Valentin kämpfen. Faust tötet Valentin und verlässt die Stadt.

_____ Gretchen wird ins Gefängnis geworfen und zum Tode verurteilt.

__1__ Gretchen wird schwanger. Valentin, ihr Bruder, will Faust deshalb töten.

_____ Während Gretchen im Gefängnis sitzt, steigen Faust und Mephisto in der Walpurgisnacht auf den Brocken und feiern mit den Hexen.

Faust und Mephisto feiern mit den Hexen
Yale University Art Gallery

Situation 16 Rollenspiel: An der Kinokasse

S1: Sie wollen mit vier Freunden in die „Rocky Horror Picture Show". Das Kino ist schon ziemlich ausverkauft. Sie wollen aber unbedingt mit ihren Freunden zusammensitzen und Reis werfen. Fragen Sie, wann, zu welchem Preis und wo noch fünf Plätze übrig sind.

Filmlektüre

Sophie Scholl – Die letzten Tage

Vor dem Lesen

FILMANGABEN

Titel: *Sophie Scholl — Die letzten Tage*
Genre: Drama
Erscheinungsjahr: 2005
Land: Deutschland
Dauer: 116 min.
Regisseur: Marc Rothemund
Hauptrollen: Julia Jentsch, Fabian Hinrichs, Alexander Held

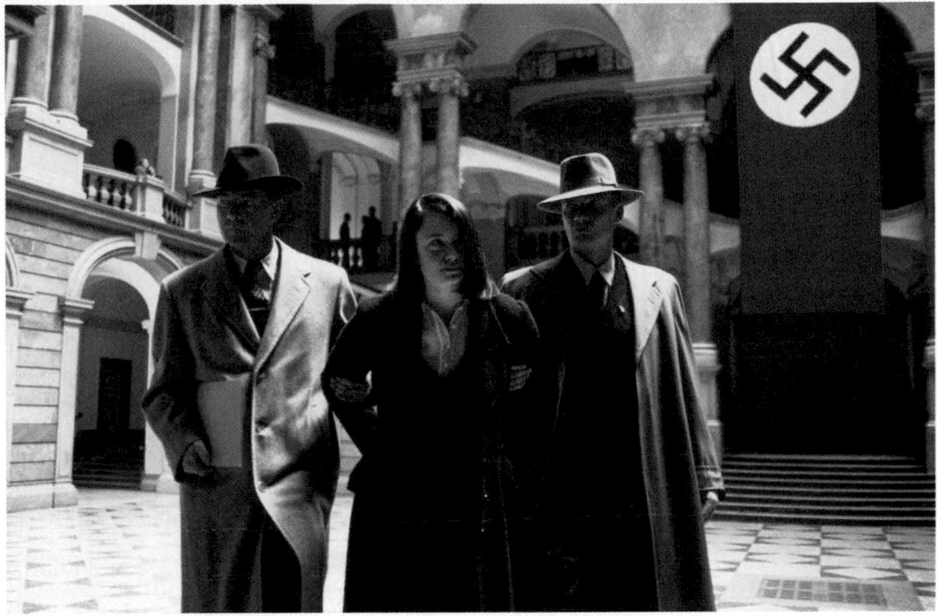

Sophie Scholl wird verhaftet.

Miniwörterbuch

das **Mitglied**	member
der **Widerstand**	resistance
das **Flugblatt**	pamphlet
verteilen	to distribute
erwischen	to catch
die **Gestapo**	*Geheime Staatspolizei* (secret police)
verhören	to interrogate
überzeugen	to convince
gestehen	to admit
der **Prozess**	(court) trial
der **Hochverrat**	high treason
anklagen	to accuse
gnadenlos	merciless
der **Gerichtssaal**	court room
verurteilen	to sentence
die **Abschlusserklärung**	final declaration
hinrichten	to kill by execution
verraten	to betray
der **Umschlag**	envelope
sich durchsetzen	to prevail
der **Hörsaaldiener**	lecture hall attendant

A. Filmplakat. Sehen Sie sich das Foto an. Welche Assoziationen weckt das Foto? An welche Zeit denken Sie? Welcher Ort könnte das sein? Was könnte die Frau gemacht haben? Wer könnten die Männer sein?

B. Lesen Sie die Wörter im Miniwörterbuch. Suchen Sie sie im Text und in den Aufgaben und unterstreichen Sie sie.

Inhaltsangabe

Sophie Scholl (Julia Jentsch) und ihr Bruder Hans (Fabian Hinrichs) sind Mitglieder der *Weißen Rose*, einer Widerstandsgruppe im Dritten Reich. Als sie ein Flugblatt an der Universität München verteilen wollen, werden sie erwischt und kommen ins Gefängnis. Dort werden sie vom Gestapobeamten Mohr (Alexander Held) verhört. Zuerst gelingt es Sophie, Mohr von ihrer Unschuld zu überzeugen. Als aber Hans gesteht, dass er die Flugblätter geschrieben hat, gesteht auch Sophie und versucht alle Schuld auf sich zu nehmen. Die Geschwister behaupten, sie hätten ganz alleine gehandelt, um die anderen Mitglieder der *Weißen Rose* zu schützen. Es kommt zum Prozess. Sophie, Hans und ein weiteres Mitglied werden wegen Hochverrat angeklagt. Der Nazirichter Freisler ist gnadenlos, aber Sophie hat keine Angst und es kommt zu großen Rededuellen zwischen den beiden im Gerichtssaal. Am Ende werden alle drei zum Tode verurteilt. Sophie sagt in ihrer Abschlusserklärung: „Wo wir heute stehen, werdet ihr bald stehen." Noch am selben Tag werden die drei hingerichtet.

Arbeit mit dem Text

Welche Aussagen sind falsch? Verbessern Sie die falschen Aussagen.

_____ 1. Die *Weiße Rose* war eine Widerstandsgruppe im Dritten Reich.
_____ 2. Sophie und Hans Scholl werden erwischt, als sie ein Flugblatt drucken.
_____ 3. Die Geschwister Scholl werden vom Gestapobeamten Freisler verhört.
_____ 4. Sophie verrät die Namen der anderen Mitglieder der *Weißen Rose*.
_____ 5. Während des Prozesses fängt Sophie Scholl an zu weinen.
_____ 6. Die Geschwister werden zum Tode verurteilt und einen Monat später hingerichtet.

FILMCLIP

NOTE: For copyright reasons, the films referenced in the **FILMCLIP** feature have not been provided by the publisher. The film can be purchased as a DVD or found online at various sites such as YouTube, Amazon, or the iTunes store. The time codes mentioned below are for the North American DVD version of the film.

Szene: DVD, Kapitel 3, 4 „Peaceful Resistance, Arrested" 9:45–16:50 Min.

Die Mitglieder der *Weißen Rose* haben ein neues Flugblatt gedruckt, aber sie haben nicht mehr genügend Umschläge, um es zu versenden. Deshalb kommen sie auf die Idee, die Flugblätter an der Uni zu verteilen. Das ist sehr gefährlich, aber Hans setzt sich durch. Am nächsten Tag gehen Sophie und Hans an die Uni.

Schauen Sie sich die Szene an und beantworten Sie die Fragen.

1. Wo sind Sophie und Hans?
2. Was machen sie?
3. Warum gehen sie noch einmal ganz nach oben?
4. Was macht Sophie, als es klingelt?
5. Was passiert, als Sophie und Hans mit den anderen Studenten die Treppe hinuntergehen?
6. Was gibt Sophie zu?
7. Was findet der Hörsaaldiener bei Hans im Büro des Universitätsdirektors?
8. Wer betritt dann das Zimmer?
9. Wie erklärt Hans dem Gestapobeamten, dass er das Flugblatt hat?

Nach dem Lesen

Dieser Film basiert auf einer wahren Begebenheit. Recherchieren Sie im Internet und suchen Sie Antworten auf die folgenden Fragen.

1. Wer waren die Mitglieder der *Weißen Rose*?
2. Von wann bis wann gab es sie?
3. Was empörte die Mitglieder der *Weißen Rose* besonders?
4. Welche Ziele verfolgten sie?
5. Woher kam der Name?
6. Was erinnert in München an die *Weiße Rose*?

Videoecke

Perspektiven

Wie informierst du dich? Wie hältst du dich auf dem Laufenden?

Ich informiere mich über das Internet.

Aufgabe 1 Internet, Radio, Zeitungen oder Fernsehen?

Wie informieren sich die Personen? Ordnen Sie die Medien den Personen zu.

1. Shaimaa ___

2. Martin ___

3. Michael ___

4. Felicitas ___

5. Sandra ___

6. Pascal ___

7. Tina ___

8. Sophie ___

1. Fernsehen
2. Internet
3. Radio
4. Zeitung

Interviews

- Was hältst du von der Ehe?
- Wie erledigst du deine Geldgeschäfte?
- Wie viel Geld steht dir pro Monat zur Verfügung?
- Wofür gibst du es aus?
- Hast du schon mal einen Kredit aufgenommen?
- Wie hast du das gemacht?

Jenny

Nadezda

Aufgabe 2 Jenny, Nadezda oder beide?

Sehen Sie sich das Video an und beantworten Sie die Fragen.

	Jenny	Nadezda	Beide
1. Wer findet Hochzeiten schön, möchte aber trotzdem nicht heiraten?	☐	☐	☐
2. Wer erledigt Geldgeschäfte per Onlinebanking?	☐	☐	☐
3. Wer hat eine Kreditkarte?	☐	☐	☐
4. Wer bekommt BaföG?	☐	☐	☐
5. Wer gibt das meiste Geld für die Miete aus?	☐	☐	☐
6. Wer hat schon mal einen Kredit aufgenommen?	☐	☐	☐

Aufgabe 3 Jenny

Wie viel Geld hat Jenny, und wofür gibt sie es aus? Ergänzen Sie die Sätze.

1. Jenny bekommt im Monat _____ Euro BAföG.
2. Sie bezahlt _____ Euro Miete.
3. _____ Euro gibt sie monatlich für Telefon und Internet aus.
4. Dazu braucht sie noch Geld für __________, __________ und Hobbys.

Aufgabe 4 Interview

Interviewen Sie eine Partnerin oder einen Partner. Stellen Sie dieselben Fragen.

Wortschatz

Politik	Politics
die **Abgeordnete, -n** (*adj. noun*)	female representative
die **Bevölkerung, -en**	population
die **Freiheit, -en**	freedom
die **Gefahr, -en**	danger
die **Gesellschaft, -en**	society
die **Grenze, -n**	limit, border
die **Lage, -n**	situation
die **Mehrheit, -en**	majority
die **Minderheit, -en**	minority
die **Öffentlichkeit, -en**	public
die **Rede, -n**	speech
die **Regierung, -en**	government
die **Wahl, -en**	election
der **Abgeordnete, -n** (*adj. noun*)	male representative
der **Frieden, -**	peace
der **Krieg, -e**	war
das **Volk, ¨-er**	people
das **Wachstum**	growth
besetzen	to occupy
einführen, eingeführt	to introduce
eintreten für (+ *akk.*), **tritt ... ein, trat ... ein, ist eingetreten**	to champion, stand up for
erhalten, erhält, erhielt, erhalten (R)	to receive
erreichen (R)	to reach
scheitern	to fail
senken	to lower
unterstützen	to support
vereinbaren, vereinbart	to agree upon
verteilen	to distribute
wählen	to elect
entscheidend	decisive
gesellschaftlich	societal

Ähnliche Wörter

die **Alternative, -n**; die **Demokratie, -n**; die **Kandidatin, -nen**; die **Koalition, -en**; die **Krise, -n**; die **Ministerin, -nen**; die **Organisation, -en**; die **Partei, -en**; die **Politik**; die **Strategie, -n**; der **Kandidat, -en** (*wk. masc.*); der **Konflikt, -e**; der **Minister, -**; der **Staat, -en** (R); der **Terrorismus**; das **System, -e**; **integrieren**; **demokratisch, national, politisch, staatlich**

Multikulturelle Gesellschaft	Multicultural Society
die **Arbeitserlaubnis, -se**	work permit
die **Arbeitskraft, ¨-e**	labor; employee
die **Aufenthaltserlaubnis, -se**	residence permit
die **Ausländerin, -nen**	female foreigner
die **Branche, -n**	sector
die **Formalität, -en**	formality
die **Türkin, -nen**	Turkish woman
der **Ausländer, -**	male foreigner
der **Bereich, -e**	sector, area
der **Einwanderer, -**	immigrant
der **Türke, -n** (*wk. masc.*)	Turkish man
der **Vorfahre, -n** (*wk. masc.*)	ancestor
auf·fallen, fällt ... auf, fiel ... auf, ist aufgefallen	to be noticeable
aus·üben	to practice
aus·wandern, ist ausgewandert	to emigrate
beantragen	to apply for
bereichern	to enrich
ein·wandern, ist eingewandert	to immigrate

Ähnliche Wörter

die **Heimat**; die **Integration**; die **Kultur, -en**; die **Million, -en**; die **Tradition, -en** (R); das **Heimatland, ¨-er**

Das liebe Geld	Beloved Money
die **Aktie, -n**	share, stock
die **Börse, -n**	stock exchange
die **Geheimzahl, -en**	secret PIN (personal identification number)
die **Höhe, -n**	height; amount (*of money*)
die **Kundin, -nen**	female customer
die **Überweisung, -en**	transfer (*of money*)
die **Währung, -en**	currency
der **Börsenkrach, ¨-e**	stock market crash
der **Gebrauch, ¨-e**	use
der **Geldautomat, -en** (*wk. masc.*)	automatic teller machine (ATM)
der **Kunde, -n** (*wk. masc.*)	male customer
der **Zugang**	access
das **Bargeld**	cash

das **Einkommen**	income
das **Formular, -e**	form
das **Girokonto, Girokonten**	checking account
das **Guthaben**	bank balance
das **Sparkonto, Sparkonten**	savings account
das **Zahlungsmittel**	means of payment
die **Zinsen** (*pl.*)	interest
ab·zahlen	to pay off
auf·nehmen, nimmt ... auf, nahm ... auf, aufgenommen	to take out (*a loan*)
aus·geben, gibt ... aus, gab ... aus, ausgegeben	to spend
aus·führen	to carry out, execute

Ähnliche Wörter

der **Geldschein, -e**

Kunst und Literatur — Art and Literature

die **Bildhauerin, -nen**	female sculptor
die **Blockflöte, -n**	recorder (type of flute)
die **Kasse, -n** (R)	cashier window
die **Ölfarbe, -n**	oil color (paint)
die **Orgel, -n**	organ
die **Querflöte, -n**	(transverse) flute
die **Seele, -n**	soul
die **Staffelei, -en**	easel
die **Stimme, -n**	voice
die **Töpferscheibe, -n**	potter's wheel
die **Wissenschaftlerin, -nen**	female scientist
der **Bildhauer, -**	male sculptor
der **Brennofen, ¨**	kiln
der **Meißel, -**	chisel
der **Pinsel, -**	paintbrush
der **Stein, -e**	stone
der **Teufel, -**	devil
der **Tod, -e**	death
der **Ton**	clay
der **Wissenschaftler, -**	male scientist
das **Holz, ¨er**	wood
das **Motiv, -e**	motif, theme
das **Schlagzeug, -e**	percussion, drums
malen	to paint
vollkommen	flawless, perfect
wahnsinnig	crazy, insane

Ähnliche Wörter

die **Figur, -en**; die **Mundharmonika, -s**; die **Trompete, -n**; der **Gott, ¨er**; das **Instrument, -e**; das **Material, -ien**; **magisch**

Sonstige Substantive — Other Nouns

die **Ehe, -n**	marriage
die **Einstellung, -en**	attitude
die **Gewalt**	violence
die **Schuld, -en**	debt; guilt

Ähnliche Wörter

der **Charakter;** der **Fanatiker, -;** der **Preis, -e** (R); der **Text, -e**

Sonstige Verben — Other Verbs

an·gehören (+ *dat.*)	to belong to (*an organization*)
auf·wachsen, wächst ... auf, wuchs ... auf, ist aufgewachsen	to grow up
bitten (um + *akk.*), **bat, gebeten**	to ask (for)
erwarten	to expect
halten von, hält, hielt, gehalten	to think of
sich informieren über (+ *akk.*)	to inform oneself about
sich kümmern um	to take care of
sich verlieben in (+ *akk.*) (R)	to fall in love with
verliebt sein	to be in love

Adjektive und Adverbien — Adjectives and Adverbs

arbeitslos	unemployed
ausgebildet	educated
ausländisch	foreign
eng	tight; narrow; small
fleißig	industrious
geduldig	patient
gleich	equal, same
lustig	fun, funny
neugierig	curious
täglich	daily
trotzdem	nonetheless
verschieden	different, various

Ähnliche Wörter

afro-deutsch, ideal, illegal

Sonstige Wörter und Ausdrücke — Other Words and Expressions

anstatt (+ *gen.*)	instead of
eher	rather
statt (+ *gen.*)	instead of
trotz (+ *gen.*)	in spite of
um ... zu	in order to
wohl	probably

Strukturen und Übungen

12.1 The genitive case

Spoken German: Possession may be indicated by **von.**

Written German: Use the genitive case to indicate possession.

WISSEN SIE NOCH?

You can show possession by using possessive determiners, such as **mein** (*my*), **dein** (*your*), and **sein** (*his/its*), or by placing an **-s** after someone's name, for example **Julias Buch.**

Review grammar B.7 and 2.4.

As you have learned, the preposition **von** followed by the dative case is commonly used in spoken German to express possession.

Das ist das Haus **von meinen Eltern.** *This is my parents' house.*

In writing, and sometimes in speech, this relationship between two noun phrases may also be expressed with the genitive case. The genitive case in German is equivalent to both the *of*-phrase and the possessive with *'s* in English.

Die Freiheit **der Wissenschaft** ist in Gefahr. — *The freedom of science is in danger.*

Die Grenzen **des Wachstums** sind erreicht. — *We have reached the limits of growth.*

The genitive is also required by certain prepositions. The most common ones are these:

(an)statt	*instead of*
trotz	*in spite of*
während	*during*
wegen	*because of*

Anstatt eines Fernsehers hätte ich mir ein neues Fahrrad gekauft. — *Instead of a TV, I would have bought myself a new bike.*

Trotz des vielen Regens ist noch nicht genügend Wasser in den Tanks. — *In spite of all the rain, there's still not enough water in the tanks.*

Während der letzten Tage bin ich nicht viel aus dem Haus gekommen. — *During the last few days I haven't gotten out of the house much.*

Wegen dieser dummen Situation kann ich jetzt nicht zur Hochzeit kommen. — *Because of this stupid situation, I can't come to the wedding now.*

English tends to use the possessive *'s* with nouns denoting people (for example, *the girl's mother*). In German, **-s** (without the apostrophe) is added only to *proper names* of people and places.

Nora**s** Vater — *Nora's father*

England**s** Rettung — *England's salvation*

A. Nouns in the Genitive

Feminine nouns and plural nouns do not add any endings in the genitive case. In the singular genitive, masculine and neuter nouns of more than one syllable add **-s** and those of one syllable add **-es: die Farbe des Vogels, die Größe des Hauses.**

Masculine	Neuter	Feminine	Plural
des Vater**s**	des Kind**es**	der Mutter	der Eltern

B. Articles and Article-like Words in the Genitive

In the genitive case, all determiners (**der**-words and **ein**-words) end in **-es** in the masculine and neuter singular, and in **-er** in the feminine singular and all plural forms.

Masculine	Neuter	Feminine	Plural
d**es** Mannes	d**es** Kindes	d**er** Frau	d**er** Eltern
ein**es** Mannes	ein**es** Kindes	ein**er** Frau	kein**er** Eltern
mein**es** Mannes	mein**es** Kindes	mein**er** Frau	mein**er** Eltern
dies**es** Mannes	dies**es** Kindes	dies**er** Frau	dies**er** Eltern

C. Adjectives in the Genitive

In the genitive, all adjectives end in **-en** when preceded by a determiner.*

Masculine and Neuter	Feminine and Plural
des arm**en** Mannes	der arm**en** Frau
des arm**en** Kindes	der arm**en** Leute

Eine mögliche Rolle des modernen Mannes ist es, zu Hause zu bleiben und auf die Kinder aufzupassen.

A possible role for a modern man is to stay home and take care of the children.

Übung 1 Minidialoge

Ergänzen Sie die richtige Form der Wörter in Klammern. Achten Sie auf die Deklination!

1. KATRIN: Ist das dein Auto?
   ALBERT: Nein, das ist das Auto _______ Bruders. (mein)
2. BEAMTER: Was ist das Alter_______ Kinder? (Ihr)
   FRAU FRISCH-OKONKWO: Sumita ist fünf, Yamina ist sechs und Lydia ist neun Jahre alt.
3. FRAU SCHULZ: Wie war die Rede ______ Bundeskanzlerin? (die)
   THOMAS: Ich fand sie gut und sie hat viel Applaus erhalten.
4. MONIKA: Möchtest du mit mir in die Berge fahren? Meine Eltern haben da ein Wochenendhaus.
   ROLF: Wo ist denn das Wochenendhaus _______ Eltern? (dein)
   MONIKA: In der Nähe von Lake Tahoe.
5. HEIDI: Kennst du den Film „M – Mörder unter uns“?
   ROLF: Ja.
   HEIDI: Wie heißt noch mal der Regisseur _______ Films? (dies-)
6. ROLF: Die neue Regierung ist kaum gewählt, schon steckt sie in der Krise.
   PETER: Ja, leider. Niemand unterstützt den Kurs des _______ Präsidenten. (neu)
7. FRAU GRETTER: Wer ist denn das?
   FRAU KÖRNER: Das ist die zweite Frau meines _______ Mannes. (erst-)
8. FRAU AUGENTHALER: 24352 – was ist denn das für eine Telefonnummer?
   RICHARD: Das ist die Telefonnummer meiner _______ Freundin. (alt)

*Unpreceded masculine and neuter adjectives also end in **-en**; unpreceded feminine and plural adjectives end in **-er**. Unpreceded adjectives, however, rarely occur in the genitive.

Übung 2 Worüber sprechen sie?

Bilden Sie Sätze.

MODELL: Albert sagt, dass sein Auto rot ist. →
Albert spricht über die Farbe seines Autos.

das Alter **die Wahl**
der Beruf
die Kleidung **die Qualität**
die Sprache
die Situation **die Länge**

1. Monika sagt, dass ihre Schwester Politikerin ist.
2. Thomas sagt, dass bald ein neuer Präsident gewählt wird.
3. Frau Schulz sagt, dass ihre Nichten fünf und acht Jahre alt sind.
4. Stefan sagt, dass sein Studium insgesamt fünf Jahre dauert.
5. Albert sagt, dass seine Großeltern nur Spanisch sprechen.
6. Nora sagt, dass ihr Freund gern Jeans und lange Pullover trägt.
7. Thomas sagt, dass das Leitungswasser in Berkeley sehr gut ist.
8. Katrin sagt, dass die neue Regierung in der Krise steckt.

Übung 3 Minidialoge

Ergänzen Sie **statt, trotz, während** oder **wegen.**

1. KATRIN: Bist du _______ des Regens spazieren gegangen?
   THOMAS: Ja, so ein bisschen Regen macht doch nichts.
2. MONIKA: Hast du gestern Flugblätter verteilt?
   HEIDI: Nein, _______ des schlechten Wetters bin ich zu Hause geblieben.
3. ALBERT: Was hast du _______ der Rede des neuen Präsidenten gemacht?
   PETER: Ich habe sie leider gar nicht gehört. Ich war so müde und habe geschlafen.
4. JÜRGEN: Ich muss _______ meiner Erkältung zur Uni.
   SILVIA: Du Ärmster, leg dich lieber ins Bett!
5. PETER: Fährst du nächste Woche weg?
   KATRIN: Ich kann doch _______ des Semesters nicht verreisen!
6. JOCHEN: Warum bist du mit dem Bus gefahren?
   JUTTA: _______ des schlechten Wetters.
7. MARIA: Na, wie war die Wahl? Hast du wieder die Grünen gewählt?
   MICHAEL: Nein, _______ der Grünen habe ich diesmal SPD gewählt.
8. KATRIN: In deinem Zimmer ist es _______ der Heizung kalt!
   STEFAN: Tut mir leid, sie funktioniert nicht richtig.

12.2 Expressing possibility: *würde, hätte,* and *wäre*

WISSEN SIE NOCH?

Würde functions like a modal verb. In sentences with modal verbs, the infinitive appears at the end of the sentence.

Review grammar 3.1 and 3.2.

Use the construction **würde** + infinitive to talk about possibilities: things you would do, if you were in that particular situation.

Stell dir vor, du würdest nach Deutschland fliegen.	*Imagine you were flying to Germany.*
Wo würdest du übernachten?	*Where would you stay for the night?*

Here are the forms of **würde,** which are the subjunctive forms of the verb **werden.**

würde = would

werden			
ich	würde	*wir*	würden
du	würdest	*ihr*	würdet
Sie	würden	*Sie*	würden
er *sie* *es*	würde	*sie*	würden

Instead of using **würde sein** and **würde haben,** German speakers prefer to say **wäre** (*would be*) and **hätte** (*would have*).

Ich glaube, dass ich eine gute Politikerin **wäre.**	*I believe I would be a good politician.*
Ich **hätte** sicher viel Zeit für meine Wähler.	*I'm sure I would have plenty of time for my voters.*

Here are the forms of **wäre** and **hätte,** which are the subjunctive forms of **sein** and **haben.**

sein			
ich	wäre	*wir*	wären
du	wärst	*ihr*	wärt
Sie	wären	*Sie*	wären
er *sie* *es*	wäre	*sie*	wären

haben			
ich	hätte	*wir*	hätten
du	hättest	*ihr*	hättet
Sie	hätten	*Sie*	hätten
er *sie* *es*	hätte	*sie*	hätten

Übung 4 Kein Problem

Was würden Sie in diesen Situationen machen? Beantworten Sie die Fragen! Was würden Sie machen, ...

1. wenn Sie sich in einen Politiker / eine Politikerin verlieben würden?
2. wenn Sie sich um Ihre Eltern kümmern müssten?
3. wenn Ihr Partner / Ihre Partnerin eine andere Partei wählen würde als Sie?
4. wenn Sie in ein anderes Land ziehen würden?
5. wenn Sie mit dem Studium aufhören müssten?

Übung 5 Was wäre, wenn ...

Schreiben Sie für jede Perspektive drei Sätze darüber, wie Ihr Leben aussehen würde. Verwenden Sie **hätte, wäre** und **würde** in Ihrer Antwort. Sie müssen nicht nur über sich selbst, sondern können auch über andere (z. B. Kinder, Eltern, Partner und Freunde) schreiben.

MODELL: Wenn ich Kinder hätte, würde ich nicht so oft ins Kino gehen. Ich hätte wahrscheinlich viel mehr Arbeit. Abends wäre ich bestimmt müder.

Was wäre, wenn ...

1. Sie (keine) Kinder hätten?
2. Sie für ein politisches Ziel kämpfen würden?
3. Sie (kein / sehr viel) Geld hätten?
4. Sie in einem anderen Land leben würden?
5. Sie ein berühmter Politiker / eine berühmte Politikerin wären?

12.3 Causality and purpose: *weil*, *damit*, *um ... zu*

weil = reason for action
damit = goal of action
um ... zu = goal of action

Use **weil** + dependent clause to express the reason for a particular action. Use **damit** or **um ... zu** to express the goal of an action.

Viele Deutsche wanderten nach Australien aus, **weil ihnen Deutschland zu eng war.**	*Many Germans emigrated to Australia because Germany was too crowded for them.*
Sie wanderten nach Australien aus, **um dort eine bessere Arbeit zu finden.**	*They emigrated to Australia in order to find better jobs there.*

WISSEN SIE NOCH?

You can show reasons for action with the conjunctions **weil** and **denn.**

Review grammar 3.4 and 11.5.

Weil and **damit** introduce a dependent clause. Recall that the conjugated verb is in last position in a dependent clause.

Albert steht auf, damit Frau Schulz sich setzen **kann.**	*Albert gets up so that Frau Schulz can sit down.*

Um ... zu clauses have no expressed subjects.

Damit and **um ... zu** both express the aim or goal of an action. But whereas **damit** introduces a dependent clause complete with subject and conjugated verb, **um ... zu** introduces a dependent infinitive without a subject and without a conjugated verb. Use **damit** when the subject of the main clause is different from the subject of the dependent clause.

Heidi macht das Fenster zu, **damit** Stefan nicht friert.
Heidi closes the window so that Stefan won't be cold.

Use **um ... zu** when the understood subject of the dependent infinitive is the same as the subject of the main clause.

Heidi macht das Fenster zu, **damit** sie nicht friert. *Heidi closes the window so that she won't be cold.*	Heidi macht das Fenster zu, **um** nicht **zu** frieren. *Heidi closes the window so as not to be cold.*

Übung 6 Erfolgsgeschichten

Was muss man tun, um Erfolg an der Universität zu haben?

MODELL: Um gute Noten zu bekommen, muss man fleißig lernen.

1. morgens munter[1] sein	a. früh ins Bett gehen
2. die Professoren kennenlernen	b. in die Sprechstunde gehen
3. die Mitstudenten kennenlernen	c. jeden Tag zum Unterricht kommen
4. am Wochenende nicht allein sein	d. Leute einladen
5. die Kurse bekommen, die man will	e. regelmäßig essen
6. in vier Jahren fertig werden	f. sich so früh wie möglich einschreiben
7. nicht verhungern	g. viel Gruppenarbeit machen
8. eine gute Note in Deutsch bekommen	h. viel lernen and wenig Feste feiern

Übung 7 Gute Gründe?

Verbinden Sie Sätze aus der ersten Gruppe mit Sätzen aus der zweiten Gruppe mit Hilfe der Konjunktionen **weil, damit, um ... zu.** Wenn Ihnen ein Grund nicht gefällt, suchen Sie einen besseren Grund.

MODELL: Ich möchte immer hier leben. Dieses Land ist das beste Land der Welt. →
Ich möchte immer hier leben, weil dieses Land das beste Land der Welt ist.

GRUPPE 1

Ich möchte immer hier leben.
Ich möchte für ein paar Jahre in Deutschland leben.
Ausländer haben oft Probleme.
Wenn ich Kinder habe, möchte ich hier leben.
Viele Ausländer kommen hierher.
Englisch sollte die einzige offizielle Sprache (der USA, Kanadas, Australiens, usw.) sein.

GRUPPE 2

Ausländer verstehen die Sprache und Kultur des Gastlandes nicht.
Ich möchte richtig gut Deutsch lernen.
Dieses Land ist das beste Land der Welt.
Hier kann man gut Geld verdienen.
Meine Kinder sollen als (Amerikaner, Kanadier, Australier, usw.) aufwachsen.
Aus der multikulturellen Bevölkerung soll eine homogene Gemeinschaft werden.

[1] *wide awake*

12.4 Principles of case (summary review)

Three main factors determine the choice of a particular case for a given noun: function, prepositions, and verbs.

A. Function

Function refers to the role a particular noun plays within a sentence: the subject, the direct object, the indirect object, or the possessive. The subject of a sentence (who or what is doing something) is in the nominative case; the direct object (the thing or person to which or to whom the action is done) is in the accusative case; and the indirect object (usually the person who benefits from the action) is in the dative case.

NOM DAT ACC
Maria schreibt ihrer Freundin einen Scheck.
Maria is writing her friend a check.

Possessives express relationships of various kinds, such as belonging to or being part of someone or something. Possessives are in the genitive case.

Der Kurs **des Euro** ist leider wieder gefallen. — *The exchange rate of the euro has unfortunately fallen again.*

B. Prepositions

Nouns or pronouns that follow prepositions are always in a case other than the nominative. You have encountered four groups of prepositions so far: those that take the accusative, those that take the dative, two-way prepositions that take either the accusative or the dative according to the meaning of the clause, and those that take the genitive.

ACCUSATIVE	DATIVE	ACCUSATIVE OR DATIVE	GENITIVE
durch	aus	an	(an)statt
für	außer	auf	trotz
gegen	bei	hinter	während
ohne	mit	in	wegen
um	nach	neben	
	seit	über	
	von	unter	
	zu	vor	
		zwischen	

Bargeld können Sie **aus dem Geldautomaten** bekommen. — *You can get cash from the ATM.*

Wegen des Feiertags bleiben die Banken geschlossen. — *Because of the holiday, the banks remain closed.*

Two-way prepositions require accusative objects when movement toward a *destination* is involved. They require dative objects when no such destination is expressed, when the focus is on the setting of the action or state (*location*).

Ich habe kein Geld **auf meinem Sparkonto.** — *I don't have any money in my savings account.*

Ich muss Geld **auf mein Sparkonto** überweisen. — *I have to transfer money to my savings account.*

C. Verbs

Certain verbs, just like prepositions, require a noun or pronoun to be in a particular case. The verbs **sein, werden, bleiben,** and **heißen** establish identity relationships between the subject and the predicate, and therefore require a predicate noun in the *nominative* case.

Thomas ist **ein fleißiger Student.**	*Thomas is a conscientious student.*

The following verbs are among those that require *dative* objects.

antworten	*to answer*
begegnen	*to meet*
fehlen	*to be missing*
gefallen	*to be to one's liking*
gehören	*to belong to*
gratulieren	*to congratulate*
helfen	*to help*
passen	*to fit*
schaden	*to be harmful (to)*
schmecken	*to taste good (to)*
stehen	*to suit, look good on* (e.g., clothing)
zuhören	*to listen to*

Die Aktien gehören **meiner Mutter.**	*The stocks belong to my mother.*
Eine schwache Wirtschaft schadet **den Aktienmärkten.**	*A weak economy hurts the stock markets.*

Most other verbs require the accusative, if they require an object at all.

Ich habe für mein Konto **keinen Überziehungskredit.**	*I don't have any overdraft protection for my account.*

Übung 8 Der Umzug

Bestimmen Sie den Kasus (**Nom, Akk, Dat** oder **Gen**) der unterstrichenen Nominalphrasen und geben Sie an, ob dieser Kasus wegen der Funktion (F), wegen der Präposition (P) oder wegen des Verbs (V) benutzt wurde.

	KASUS	GRUND
1. Meine Freundin braucht einen neuen Schrank.	*Nom*	*F*
2. Sie möchte Schriftstellerin werden.	______	______
3. Die Möbel meiner Freundin sind ultramodern.	______	______
4. Morgen kaufe ich ihr eine schöne Lampe.	______	______
5. Diesen Teppich mag sie sicher nicht.	______	______
6. Meine Tapeten gefallen ihr sicher auch nicht.	______	______
7. Setzen wir uns doch an diesen Tisch.	______	______
8. Ich habe nichts gegen Vorhänge.	______	______
9. Das Bett tragen wir am besten zusammen.	______	______
10. Der Wecker steht auf dem Regal.	______	______
11. Diese Decke gehört mir.	______	______
12. Der Umzug findet wegen schlechten Wetters nicht statt.	______	______

Übung 9 Jutta hat sich wieder verliebt!

Ergänzen Sie die richtigen Endungen. Unten finden Sie das Genus wichtiger Substantive.

die Adresse
die Augen (*pl.*)
der Brief
die Disko
die Eltern (*pl.*)
der Fernseher
das Fest
die Hausaufgaben (*pl.*)
die Hose
die Jacke
der Mann
der Name
der Park
die Schule
die Stadt
die Tür
der Weg

Jutta hat sich total verliebt. Sie sah vor einem Monat auf ein________[1] Klassenfest ein________[2] jungen Mann und jetzt denkt sie nur noch an ihn.

Er trug an jenem Abend ein________[3] Jeansjacke, unter sein________[4] Jacke ein altes Unterhemd und ein________[5] uralte Hose. Er stand die ganze Zeit neben d________[6] Tür. Seine Kleidung und sein________[7] blauen Augen gefielen ihr sehr. Er schaute oft zu ihr hin, aber sie sprach ihn nicht an, sie war zu schüchtern.

Jetzt träumt sie von ihm. Sie möchte mit ihm durch d________[8] Park gehen und in d________[9] Stadt. Vielleicht könnten sie auch mal für ein paar Tage ohne d________[10] Eltern wegfahren. Sie möchte ihm gern ein________[11] Brief schreiben, aber sie weiß sein________[12] Adresse nicht. Sie kennt nur sein________[13] Vornamen, Florian. Dies________[14] Namen wird sie nie mehr vergessen!

Morgens in d________[15] Schule denkt sie an ihn, mittags auf d________[16] Weg nach Hause, nachmittags bei d________[17] Hausaufgaben, abends vor d________[18] Fernseher oder in d________[19] Disko.

Ach, wenn sie ihn doch nur noch einmal treffen könnte! Diesmal würde sie sicher zu ihm gehen und ihn ansprechen.

APPENDIX **A**

Informationsspiele: 2. Teil

Einführung A

Situation 6 10 Fragen

Stellen Sie zehn Fragen. Für jedes „Ja" gibt es einen Punkt.

MODELL: S2: Trägt Frau Körner einen Hut?
S1: Nein. Trägt Nora einen Mantel?
S2: Nein.

	HERR SIEBERT JA ODER NEIN	FRAU KÖRNER JA ODER NEIN
einen Anzug	______	______
eine Bluse	______	______
eine Brille	______	______
ein Hemd	______	______
eine Hose	______	______
einen Hut	______	*N*
eine Jacke	______	______
eine Jeans	______	______
ein Kleid	______	______
eine Krawatte	______	______

	HERR SIEBERT JA ODER NEIN	FRAU KÖRNER JA ODER NEIN
einen Mantel	______	______
einen Pullover	______	______
einen Rock	______	______
ein Sakko	______	______
Schuhe	______	______
Sportschuhe	______	______
Stiefel	______	______
ein Stirnband	______	______
ein T-Shirt	______	______

Situation 12 **Zahlenrätsel**

Verbinden Sie die Punkte. Sagen Sie Ihrem Partner oder Ihrer Partnerin, wie er oder sie die Punkte verbinden soll. Dann sagt Ihr Partner oder Ihre Partnerin Ihnen, wie Sie die Punkte verbinden sollen. Was zeigen Ihre Bilder?

S2: Start ist Nummer 1. Geh zu 17, zu 5, zu 60, zu 23, zu 14, zu 3, zu 19, zu 7, zu 21, zu 12, zu 6, zu 33, zu 8, zu 11, zu 40, zu 25, zu 13, zu 4, zu 15, zu 35, zu 50, zu 9, und zum Schluss zu 16. Was zeigt dein Bild?

Einführung B

Situation 7 **Familie**

MODELL: S2: Wie heißt Richards Vater?
S1: Er heißt _____.
S2: Wie schreibt man das?
S1: _____. Wie alt ist er?
S2: Er ist 39 Jahre alt. Wo wohnt er?
S1: Er wohnt in _____. Wie heißt Richards Mutter?
S2: Sie heißt Maria.
S1: Wie schreibt man das?
S2: M–A–R–I–A.

		Richard	Sofie	Mehmet
Vater	*Name*			Kenan
	Alter	39		
	Wohnort		Dresden	
Mutter	*Name*	Maria		
	Alter	38	47	54
	Wohnort			Izmir
Bruder	*Name*		Erwin	
	Alter			
	Wohnort	Innsbruck	Leipzig	Istanbul
Schwester	*Name*	Elisabeth	—	Fatima
	Alter	16	—	31
	Wohnort		—	

Situation 9 Temperaturen

MODELL: S2: Wie viel Grad Fahrenheit sind 18 Grad Celsius?
S1: ______ Grad Fahrenheit.

°F	90		32		−5	
°C	32	18	0	−18	−21	−39

Kapitel 1

Situation 2 Freizeit

MODELL: S2: Wie alt ist Richard?
S1: ______.
S2: Woher kommt Rolf?
S1: Aus ______.
S2: Was macht Jürgen gern?
S1: Er ______.
S2: Wie alt bist du?
S1: ______.
S2: Woher kommst du?
S1: ______.
S2: Was machst du gern?
S1: ______.

	Alter	Wohnort	Hobby
Richard		Innsbruck	geht gern in die Berge
Rolf	20		spielt gern Tennis
Jürgen		Göttingen	
Sofie			kocht gern
Jutta	16	München	
Melanie		Regensburg	
mein Partner / meine Partnerin			

Situation 7 Juttas Stundenplan

MODELL: S1: Was hat Jutta am Montag um acht Uhr?
S2: Sie hat Latein.

Uhr	Montag	Dienstag	Mittwoch	Donnerstag	Freitag
8.00–8.45	Latein			Biologie	
8.50–9.35		Englisch	Englisch		Physik
9.35–9.50	← Pause →				
9.50–10.35			Mathematik		Religion
10.40–11.25	Geschichte	Französisch		Mathematik	
11.25–11.35	← Pause →				
11.35–12.20		Musik		Sport	
12.25–13.10	Erdkunde		Kunst		frei

Situation 12 Diese Woche

MODELL: S1: Was macht Silvia am Montag?
S2: Sie steht um 6 Uhr auf.
S1: Was machst du am Montag?
S2: Ich _______.

	Silvia Mertens	Mehmet Sengün	mein(e) Partner(in)
Montag	Sie steht um 6 Uhr auf.		
Dienstag		Er lernt eine neue Kollegin kennen.	
Mittwoch	Sie schreibt eine Prüfung.		
Donnerstag	Sie ruft ihre Eltern an.		
Freitag		Er hört um 15 Uhr mit der Arbeit auf.	
Samstag		Er räumt seine Wohnung auf.	
Sonntag		Er repariert sein Motorrad.	

Kapitel 2

Situation 3 Was machen sie morgen?

MODELL: S1: Schreibt Silvia morgen eine E-Mail?
S2: Ja.
S1: Schreibst du morgen eine E-Mail?
S2: Ja. (Nein.)

	Jürgen	Silvia	mein(e) Partner(in)
1. *schreibt/schreibst ... eine E-Mail*		+	
2. *kauft/kaufst ... ein Buch*		+	
3. *schaut/schaust ... einen Film an*	–	–	
4. *ruft/rufst ... eine Freundin an*			
5. *macht/machst ... Hausaufgaben*		+	
6. *treibt/treibst ... Sport*	+	–	
7. *besucht/besuchst ... einen Freund*			
8. *räumt/räumst ... das Zimmer auf*		–	

Situation 15 Was machen sie gern?

MODELL: S1: Was trägt Richard gern?
S2: Pullis.
S1: Was trägst du gern?
S2: _______

	Richard	Josef und Melanie	mein(e) Partner(in)
fahren		Zug	
tragen	Pullis		
essen		Pizza	
sehen		Gruselfilme	
vergessen	seine Hausaufgaben		
waschen		ihr Auto	
treffen		ihre Lehrer	
einladen		ihre Eltern	
sprechen	Italienisch		

Kapitel 3

Situation 2 Kann Katrin kochen?

MODELL: S1: Kann Peter kochen?
S2: Ja, fantastisch.
S1: Kannst du kochen?
S2: Ja, aber nicht so gut.

[+]
ausgezeichnet
fantastisch
sehr gut
gut

[0]
ganz gut

[−]
nicht so gut
nur ein bisschen
gar nicht
kein bisschen

	Katrin	Peter	mein(e) Partner(in)
kochen		fantastisch	
zeichnen	sehr gut		
tippen		ganz gut	
Witze erzählen		ganz gut	
tanzen	fantastisch		
stricken	gar nicht		
Skateboard fahren		nicht so gut	
Geige spielen		nur ein bisschen	
Schlittschuh laufen		nur ein bisschen	
ein Auto reparieren	nicht so gut		

Situation 13 Was machen sie, wenn ...?

MODELL: S1: Was macht Renate, wenn sie müde ist?
S2: Sie trinkt Kaffee.
S1: Was machst du, wenn du müde bist?
S2: Ich gehe ins Bett.

	Renate	Ernst	mein(e) Partner(in)
1. *traurig ist/bist*		weint	
2. *müde ist/bist*	trinkt Kaffee		
3. *in Eile ist/bist*	nimmt ein Taxi		
4. *wütend ist/bist*		schreit ganz laut	
5. *krank ist/bist*	geht zum Arzt		
6. *glücklich ist/bist*		lacht ganz laut	
7. *Hunger hat/hast*			
8. *Langeweile hat/hast*	liest ein Buch	ärgert seine Schwester	
9. *Durst hat/hast*		trinkt Limo	
10. *Angst hat/hast*	schließt die Tür ab		

Kapitel 4

Situation 9 Geburtstage

MODELL: S1: Wann ist Sofie geboren?
S2: Am neunten November 1995.

Person	Geburtstag
Willi	
Sofie	9. November 1995
Claire	
Melanie	3. April 1992
Nora	
Thomas	17. Januar 1998
Heidi	
mein(e) Partner(in)	
sein/ihr Vater	
seine/ihre Mutter	

Situation 15 Zum ersten Mal

MODELL: S1: Wann hat Herr Thelen seinen ersten Kuss bekommen?
S2: Als er zwölf war.

	Herr Thelen	Frau Gretter	mein(e) Partner(in)
seinen/ihren/ deinen ersten Kuss bekommen	als er 12 war		
zum ersten Mal ausgegangen		als sie 15 war	
seinen/ihren/ deinen Führerschein gemacht	mit 18		
sein/ihr/dein erstes Bier getrunken		mit 18	
seinen/ihren/ deinen ersten Preis gewonnen	mit 21		
zum ersten Mal nachts nicht nach Hause gekommen	noch nie		

Kapitel 6

Situation 8 Gestern und heute

Arbeiten Sie zu zweit und stellen Sie Fragen wie im Modell.

MODELL: S1: Früher war hier eine Reinigung. Was ist heute hier?
S2: Heute ist hier ein Schreibwarengeschäft.

FRÜHER

HEUTE

Situation 16 Haus- und Gartenarbeit

MODELL: S1: Was macht Nora am liebsten?
S2: Sie geht am liebsten einkaufen.
S1: Was hat Thomas letztes Wochenende gemacht?
S2: Er hat das Geschirr gespült.
S1: Was muss Nora diese Woche noch machen?
S2: Sie muss den Boden wischen.

S2: Was machst du am liebsten?
S1: Ich _______ am liebsten _______.

	Thomas	Nora	mein(e) Partner(in)
am liebsten		einkaufen gehen	
am wenigsten gern	das Bad putzen		
jeden Tag	nichts von alledem		
einmal in der Woche		die Wäsche waschen	
letztes Wochenende	das Geschirr spülen		
gestern	die Blumen gießen		
diese Woche		den Boden wischen	
bald mal wieder		Staub wischen	

Kapitel 7

Situation 3 Deutschlandreise

Wo liegen die folgenden Städte? Schreiben Sie die Namen der Städte auf die Landkarte.

Augsburg, Braunschweig, Bremen, Düsseldorf, Frankfurt/Oder, Halle, Kiel, Nürnberg, Rostock, Stuttgart

MODELL: S1: Wo liegt Hannover?
S2: Hannover liegt im Norden.
S1: Wo genau?
S2: Südlich von Hamburg.

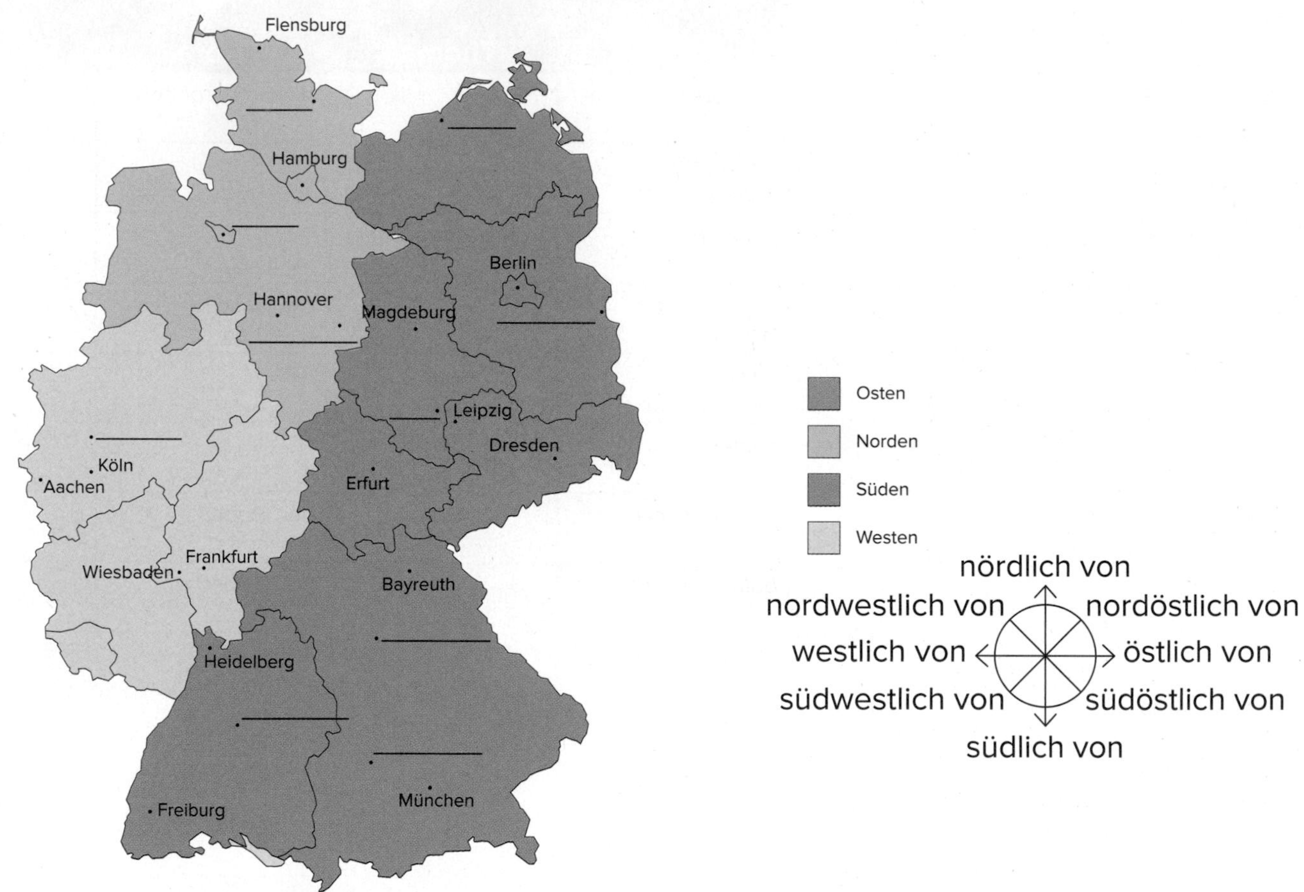

Situation 11 **Ein Auto kaufen**

S2: Sie wollen einen neueren Gebrauchtwagen[1] kaufen und lesen deshalb Anzeigen im Internet. Die Anzeigen für einen VW Golf und einen VW Touareg Hybrid sind interessant. Rufen Sie an und stellen Sie Fragen.

Sie haben auch eine Anzeige im Internet aufgegeben, weil Sie Ihren Opel Corsa und Ihren Ford Fiesta verkaufen wollen. Antworten Sie auf die Fragen der Leute.

MODELL: Guten Tag, ich rufe wegen des VW Golf an.
Wie alt ist der Wagen?
Welche Farbe hat er?
Wie ist der Kilometerstand[2]?
Wie lange hat er noch TÜV?
Wie viel Benzin braucht er?
Was kostet der Wagen?

Modell	VW Golf	VW Touareg Hybrid	Opel Corsa	Ford Fiesta
Baujahr			2010	2011
Farbe			schwarz	blaugrün
Kilometerstand			84.500 km	52.000 km
TÜV			6 Monate	fast 2 Jahre
Benzinverbrauch pro 100 km			6 Liter	6,5 Liter
Preis			5.000 Euro	4.000 Euro

Kapitel 8

Situation 2 **Mahlzeiten und Getränke**

MODELL: S2: Was isst Frau Gretter zum Frühstück?
S1: ______.

	Frau Gretter	Stefan	Andrea
zum Frühstück essen		frisches Müsli	Brot mit selbst gemachter Marmelade
zum Frühstück trinken		kalten Orangensaft	
zu Mittag essen	frisches Gemüse und Hähnchen		heiße Würstchen
zu Abend essen		italienische Spaghetti	
nach dem Sport trinken	nichts, sie treibt keinen Sport	kalten Tee mit Zitrone	
auf einem Fest trinken	deutschen Sekt		eiskalte Limonade
essen, wenn er/sie groß ausgeht		frischen Fisch mit französischer Soße	

[1]*used car* [2]*number of kilometers driven*

Kapitel 9

Situation 9 Was ist passiert?

MODELL: Was ist Mehmet passiert? / Was ist dir passiert?
Wann ist es passiert?
Wo ist es passiert?
Warum ist es passiert?

	Sofie	Mehmet	Ernst	mein(e) Partner / meine Partnerin
Was?	hat ihre Schlüssel verloren		hat seine Hose zerrissen	
Wann?		als er in die Türkei fliegen wollte		
Wo?	in Leipzig		bei seiner Tante	
Warum?		weil der Flug aus Berlin Verspätung hatte		

Kapitel 10

Situation 2 Reisen

MODELL: S2: Woher kommt Richard?
S1: Aus ______.
S2: Wohin fährt er in den Ferien?
S1: Nach/In ______.
S2: Wo wohnt er?
S1: Bei ______. Was macht er da?
S2: Er lernt Französisch. Wann kommt er zurück?
S1: In ______.

	Richard	Sofie	Mehmet	Peter	Jürgen	mein(e) Partner(in)
Woher?		aus Dresden		aus Berkeley		
Wohin?		nach Düsseldorf		nach Hawaii		
Wo?		bei ihrer Tante	bei alten Freunden		bei einem Freund	
Was?	Französisch lernen		am Strand liegen; schwimmen		Ski fahren natürlich	
Wann?		in einer Woche	in zwei Wochen	nächstes Wochenende		

Situation 9 Wo wollen wir übernachten?

MODELL: Wie viel kostet _____?
Haben die Zimmer im (in der) _____ eine eigene Dusche und Toilette?
Gibt es im (in der) _____ Einzelzimmer?
Gibt es im (in der, auf dem) _____ einen Fernseher?
Ist das Frühstück im (in der, auf dem) _____ inbegriffen?
Ist die Lage von dem (von der) _____ zentral/ruhig?
Gibt es im (in der, auf dem) _____ Internet?

	das Hotel Strandpromenade	das Hotel Ostseeblick	die Jugendherberge	der Campingplatz
Preis pro Person		52,- Euro		
Dusche/Toilette			nein	nein
Einzelzimmer				natürlich nicht
Fernseher	in jedem Zimmer	nicht in allen Zimmern		natürlich nicht
Frühstück	inbegriffen		kostet extra	nein
zentrale Lage				
ruhige Lage				ja
Internet	in jedem Zimmer	im Frühstückszimmer	ja	nein

Situation 13 Tiere

MODELL: Welche Tiere findet _____ am tollsten?
Vor welchem Tier hat _____ am meisten Angst?
Welches Tier hätte _____ gern als Haustier?
Welches wilde Tier möchte _____ gern in freier Natur sehen?
Wenn _____ an Afrika denkt, an welche Tiere denkt er/sie?
Wenn _____ an die Wüste denkt, an welches Tier denkt er/sie dann zuerst?
Welche Vögel findet _____ am schönsten?
Welchen Fisch findet _____ am gefährlichsten?
Welchem Tier möchte _____ nicht im Wald begegnen?

	Ernst	Maria	mein(e) Partner(in)
Lieblingstier	ein Krokodil		
Angst		vor Mäusen	
Haustier		einen Papagei	
wildes Tier	einen Elefanten		
Afrika		an Zebras	
Wüste	an einen Skorpion		
Vögel	Adler		
Fisch		den Piranha	
Wald		einem Wildschwein	

Kapitel 11

Situation 10 Krankheitsgeschichte

MODELL: Hat Herr Thelen sich (Hast du dir) schon mal etwas gebrochen? Was?
Ist Herr Thelen (Bist du) schon mal im Krankenhaus gewesen? Warum?
Hat Claire (Hast du) schon mal eine Spritze bekommen? Gegen was?
Erkältet sich Claire (Erkältest du dich) oft?
Ist Herr Thelen (Bist du) gegen etwas allergisch? Gegen was?
Hat man Herrn Thelen (Hat man dir) schon mal einen Zahn gezogen?
Hatte Claire (Hattest du) schon mal hohes Fieber? Wie hoch?

	Claire	Herr Thelen	mein(e) Partner(in)
sich etwas brechen	den Arm		
im Krankenhaus sein	Nierenentzündung		
eine Spritze bekommen		Tetanus	
sich oft erkälten		nein	
gegen etwas allergisch sein	Sonne		
einen Zahn gezogen haben	nein		
hohes Fieber haben		41,2° C	

APPENDIX B

Rollenspiele: 2. Teil

Einführung A

Situation 10 Begrüßen

S2: Begrüßen Sie einen Mitstudenten oder eine Mitstudentin. Schütteln Sie dem Mitstudenten oder der Mitstudentin die Hand. Sagen Sie Ihren Namen. Fragen Sie, wie alt er oder sie ist. Verabschieden Sie sich.

Einführung B

Situation 12 Herkunft

S2: Sie sind Student/Studentin an einer Universität in Deutschland. Sie lernen einen neuen Studenten / eine neue Studentin kennen. Fragen Sie, wie er/sie heißt, woher er/sie kommt, woher seine/ihre Familie kommt und welche Sprachen er/sie spricht.

Kapitel 1

Situation 15 Auf dem Auslandsamt°

study abroad office

S2: Sie arbeiten auf dem Auslandsamt der Universität. Ein Student / Eine Studentin kommt zu Ihnen und möchte ein Stipendium für Österreich.

Fragen Sie nach den persönlichen Angaben und schreiben Sie sie auf: Name, Adresse, Telefon, E-Mail-Adresse, Geburtstag, Studienfach.

Sagen Sie „Auf Wiedersehen".

Kapitel 2

Situation 8 Am Telefon

S2: Das Telefon klingelt. Ein Freund / Eine Freundin ruft an. Er/Sie lädt Sie ein. Fragen Sie: **wo, wann, um wie viel Uhr, wer kommt mit**. Sagen Sie „ja" oder „nein", und sagen Sie „tschüss".

Kapitel 3

Situation 11 In der Mensa

S2: Sie sind Student/Studentin an der Uni in Regensburg und sind in der Mensa. Jemand möchte sich an Ihren Tisch setzen. Fragen Sie, wie er/sie heißt, woher er/sie kommt und was er/sie studiert.

Kapitel 4

Situation 16 **Das Studentenleben**

S2: Sie sind Student/Studentin an einer Uni in Ihrem Land. Ein Reporter / Eine Reporterin aus Österreich fragt Sie viel und Sie antworten gern. Sie wollen aber auch wissen, was der Reporter / die Reporterin gestern alles gemacht hat: am Vormittag, am Mittag, am Nachmittag und am Abend.

Kapitel 5

Situation 12 **Bei der Berufsberatung**

S2: Sie sind Student/Studentin und gehen zur Berufsberatung, weil Sie nicht wissen, was Sie nach dem Studium machen sollen. Beantworten Sie die Fragen des Berufsberaters / der Berufsberaterin.

Kapitel 6

Situation 12 **Zimmer zu vermieten**

S2: Sie möchten ein Zimmer in Ihrem Haus vermieten. Das Zimmer ist 25 Quadratmeter groß und hat Zentralheizung. Es kostet warm 410 Euro im Monat. Es hat große Fenster und ist sehr ruhig. Das Zimmer hat keine Küche und auch kein Bad, aber der Mieter / die Mieterin darf Ihre Küche und Ihr Bad benutzen. Der Mieter / Die Mieterin darf Freunde einladen, aber sie dürfen nicht zu lange bleiben. Sie haben kleine Kinder, die früh ins Bett müssen. Fragen Sie, was der Student / die Studentin studiert, ob er/sie raucht, ob er/sie oft laute Musik hört, ob er/sie Haustiere hat und ob er/sie Möbel hat.

Kapitel 7

Situation 9 **Am Fahrkartenschalter**

S2: Sie arbeiten am Fahrkartenschalter im Bahnhof von Bremen. Ein Fahrgast möchte eine Fahrkarte nach München kaufen. Hier ist der Fahrplan. Alle Züge fahren über Hannover und Würzburg.

	Abfahrt	Ankunft	2. Kl.	1. Kl.
IC	4.25	15.40	142 Euro	237 Euro
ICE	7.15	14.05	152 Euro	249 Euro
IC	7.30	20.45	142 Euro	237 Euro

Kapitel 8

Situation 14 **Im Restaurant**

S2: Sie arbeiten als Kellner/Kellnerin in einem Restaurant. Ein Gast setzt sich an einen freien Tisch. Bedienen Sie ihn.

Kapitel 9

Situation 8 **Das Klassentreffen**

S2: Sie sind auf dem dritten Klassentreffen Ihrer alten High-School-Klasse. Sie unterhalten sich mit einem alten Schulfreund / einer alten Schulfreundin. Fragen Sie: was er/sie nach Abschluss der High School gemacht hat, was er/sie jetzt macht und was seine/ihre Pläne für die nächsten Jahre sind. Sprechen Sie auch über die gemeinsame Schulzeit.

Kapitel 10

Situation 11 **Im Hotel**

S2: Sie arbeiten an der Rezeption von einem Hotel. Alle Zimmer haben Dusche und Toilette. Manche haben auch Internet. Frühstück ist inklusive. Das Hotel ist im Moment ziemlich voll. Ein Reisender / Eine Reisende kommt herein und erkundigt sich nach Zimmern. Denken Sie zuerst darüber nach: Was für Zimmer sind noch frei? Was kosten die Zimmer? Bis wann müssen die Gäste abreisen?

Kapitel 11

Situation 12 **Anruf beim Arzt**

S2: Sie arbeiten in einer Arztpraxis. Ein Patient / Eine Patientin ruft an und möchte einen Termin. Fragen Sie, was er/sie hat und wie dringend es ist. Der Terminkalender für diesen Tag ist schon sehr voll.

Kapitel 12

Situation 12 **Auf der Bank**

S2: Sie sind Bankangestellte(r) bei der Deutschen Bank und ein Kunde / eine Kundin möchte ein Konto eröffnen. Fragen Sie, ob der Kunde / die Kundin ein Girokonto oder ein Sparkonto eröffnen möchte. Zinsen gibt es nur auf Sparkonten. Eine EC-Karte bekommt man nur, wenn man ein festes Einkommen hat. Online-Zugang ist kostenlos. Man darf das Konto nicht überziehen.

Situation 16 **An der Kinokasse**

S2: Sie arbeiten an der Kinokasse und sind gestresst, weil Sie den ganzen Tag Karten verkauft haben. Sie haben vielleicht noch zehn Karten für die „Rocky Horror Picture Show" heute Abend, alles Einzelplätze. Auch die nächsten Tage sind schon völlig ausverkauft. Jetzt freuen Sie sich auf Ihren Feierabend, weil Sie dann mit Ihren Freunden selbst in die „Rocky Horror Picture Show" gehen wollen. Sie haben sich fünf ganz tolle Plätze besorgt, in der ersten Reihe. Da kommt noch ein Kunde.

APPENDIX C

Phonetics Summary Tables

I. Phoneme-Grapheme Relationships (Overview)

Note: The ***Kontakte*** *Workbook / Lab Manual* presents the phoneme-grapheme relationship in reverse: The graphemes (letters of the alphabet) are the starting point for variations in pronunciation.

Vowels

Sound Group	Phonemes/ Sounds	Graphemes	Examples
a-sounds	[aː]	**a**	T**a**fel
		ah	Z**ah**l
		aa	H**aa**r
	[a]	**a**	H**a**llo
i-sounds	[iː]	**i**	**I**da
		ie	L**ie**be
		ih	**ih**r
		ieh	sich anz**ieh**en
	[ɪ]	**i**	St**i**ft
e-sounds	[eː]	**e**	P**e**ter
		eh	s**eh**en
		ee	T**ee**
	[ɛ]	**e**	H**e**rr
		ä	**Ä**rger
	[ɛː]	**ä**	C**ä**sar
		äh	z**äh**len
o-sounds	[oː]	**o**	H**o**se
		oh	**Oh**r
		oo	B**oo**t
	[ɔ]	**o**	K**o**pf
u-sounds	[uː]	**u**	F**u**ß
		uh	**Uh**r
	[ʊ]	**u**	M**u**nd
ö-sounds	[øː]	**ö**	h**ö**ren
		öh	fr**öh**lich
	[œ]	**ö**	**ö**ffnen
ü-sounds	[yː]	**ü**	**Ü**bung
		üh	fr**üh**
		y	T**y**p
	[ʏ]	**ü**	tsch**ü**ss
		y	**Y**psilon

Sound Group	Phonemes/ Sounds	Graphemes	Examples
reduced vowels	[ə]	**e**	b**e**ginnen
	[ɐ]	**er**	Vat**er**
	[ɐ̯]	**r**	Oh**r**
diphthongs	[aɛ̯]	**ei/ai**	Kl**ei**d/M**ai**
		ey/ay	M**ey**er, B**ay**ern
	[aɔ̯]	**au**	**Au**ge
	[ɔø̯]	**eu**	n**eu**n
		äu	H**äu**ser

Rules

1. **Long vowels** may be represented in writing by doubled vowels and by <ie>—for example, *T**ee**, B**oo**t, L**ie**be*.
2. **Long vowels** may also be represented by a vowel followed by <h>, which is not pronounced but only indicates vowel length—for example, *Za**h**l, se**h**en, frü**h***.
3. **Single vowels** are often long when they appear in an open or potentially open syllable. Such syllables end in vowels—that is, they have no following end-consonant—for example, *Ü-bung, Ho-se, hörst* (from *hö-ren*), *gut* (from *gu-te*), *Fuß* (from *Fü-ße*). This rule applies above all to verbs, nouns, and adjectives.
4. **Diphthongs** consist of two closely associated short vowels within a syllable. Diphthongs are always long vowels—for example, ***Au**ge, Kl**ei**d, n**eu**n*.
5. **Short vowels** generally precede double consonants—for example: *ö**ff**nen, Bri**ll**e, do**pp**elt*.
6. **Short vowels** may precede, though not always, a cluster of multiple consonants—for example, *Wu**rst**, Gesi**cht**, Her**bst***.

Consonants

Sound Group	Phonemes/ Sounds	Graphemes	Examples
plosives	[p]	**p**	**P**aula
		pp	do**pp**elt
		-b	gel**b**
	[b]	**b**	**B**rille
		bb	Kra**bb**e
	[t]	**t**	**T**ür
		tt	bi**tt**e
		-d	Hem**d**
		th	**Th**eorie
		dt	Sta**dt**
	[d]	**d**	re**d**en
		dd	Te**dd**y
	[k]	**k**	**K**leid
		ck	Ro**ck**
		-g	Ta**g**
	[g]	**g**	Au**g**e

Sound Group	Phonemes/ Sounds	Graphemes	Examples
fricatives	[f]	**f**	**F**rau
		ff	ö**ff**nen
		v	**V**ater
	[v]	**w**	**W**ort
		v	**V**iktor
		(q)u	beq**u**em
	[s]	**s**	Hau**s**
		ss	Profe**ss**or
		ß	hei**ß**en
	[z]	**s**	Ho**s**e
	[ʃ]	**sch**	**Sch**ule
		s(t)	**S**tiefel
		s(p)	**S**prache
	[ʒ]	**j**	**J**ournalist
		g	Eta**g**e
	[ç]	**ch** ("**ich**-sound")	Gesi**ch**t
		-ig	zwanzi**g**
	[j]	**j**	**j**a
	[x]	**ch** ("**ach**-sound")	Bau**ch**
r-sounds	[r]	**r**	**r**ot
		rr	He**rr**
		rh	**Rh**ythmus
	[ɐ̯]	**r**	Tü**r**
	[ɐ]	**er**	Vat**er**
nasals	[m]	**m**	**M**antel
		mm	ko**mm**en
	[n]	**n**	**N**ame
		nn	Ma**nn**
	[ŋ]	**ng**	spri**ng**en
		n(k)	da**n**ke
liquids	[l]	**l**	**L**ehrer
		ll	Bri**ll**e
aspirants	[h]	**h**	**H**ose
glottal stops	[ʔ]		be·antworten
affricates	[pf]	**pf**	Ko**pf**
	[ts]	**z**	**z**ählen
		tz	se**tz**en
		ts	rech**ts**
		-t(ion)	Lek**t**ion
		zz	Pi**zz**a
	[ks]	**x**	Te**x**t
		ks	lin**ks**
		gs	du sa**gs**t
		chs	se**chs**

Rules

1. Double consonants are pronounced the same as single consonants; they merely indicate that the preceding vowel is short.
2. The letter pair <ch> is pronounced as:
   - a so-called "**ach**-sound" [x] after <u, o, a, au>, for example, *su**ch**en, To**ch**ter, Spra**ch**e, au**ch**;*
   - a so-called "**ich**-sound" [ç] after all other vowels as well as after <l, n, r> and in -***ch**en*—for example, *ni**ch**t, Bü**ch**er, Tö**ch**ter, Nä**ch**te, lei**ch**t, eu**ch**, Mil**ch**, dur**ch**, man**ch**mal, Mäd**ch**en;*
   - [k] in the cluster <chs> as well as at the beginning of certain foreign words and German names—for example, *se**ch**s, **Ch**arakter, **Ch**emnitz.*
3. [ʃ] is represented:
   - by the letters <sch>: ***sch**ön, Ta**sch**e; but not in Häuschen* (*Häus-chen*);
   - by <s(t)>: ***S**traße;* <s(p)>: ***S**prache.*
4. <r> can be clearly heard pronounced as a fricative, uvular, or trilled consonant [r]:
   - at the beginning of a word or syllable: ***r**ot, hö-**r**en;*
   - after consonants and before vowels: *grün;*
   - after short vowels (when clearly enunciated): *Wo**r**t, He**rr**.*
5. <r> is pronounced as a vowel [ɐ]:
   - after long vowels: *Uhr;*
   - in the unstressed combinations ***er-**, **ver-**, **zer-**, and **-er:** **er**zählen, **Ver**käuf**er**, **zer**stören, Lehr**er**, ab**er**.*

II. German Vowels and Their Features (Overview)

There are 16 or 17 vowels (+ the vocalic pronunciation of <r>). They can be differentiated by:

- **quantity** (in their length)—they are either short or long;
- **quality** (in their tenseness)—they are either lax or tense.
  Quantity and quality are combined in German. The short vowels are lax; that is, in contrast to long vowels, they are formed with less muscular tension, less use of the lips, and less raising of the tongue. The **a**-vowels are only long and short. In addition, there is a long, open [ɛ:] as well as the reduced [ə] and [ɐ] (schwa).

  The following minimal pairs illustrate these differences:

  [a:] – [a] Herr M**ah**ler – Herr M**a**ller
  [e:] – [ɛ] Herr M**eh**ler – Herr M**e**ller
  [i:] – [ɪ] Herr M**ie**ler – Herr Miller
  [o:] – [ɔ] Herr M**oh**ler – Herr M**o**ller
  [u:] – [ʊ] Herr M**uh**ler – Herr M**u**ller
  [ø:] – [œ] Herr M**öh**ler – Herr M**ö**ller
  [y:] – [Y] Herr M**üh**ler – Herr M**ü**ller

  Quality and quantity do not play a role with the reduced vowels [ə] as in *ein**e*** or [ɐ] as in *ein**er**.*

- the raising of the tongue—either the front, middle, or back of the tongue is raised. The following minimal pairs illustrate the differences in front vowels:

  [e:] – [ɛ] Herr M**eh**ler – Herr M**e**ller
  [i:] – [ɪ] Herr M**ie**ler – Herr Miller
  [ø:] – [œ] Herr M**öh**ler – Herr M**ö**ller
  [y:] – [Y] Herr M**üh**ler – Herr M**ü**ller

The following minimal pairs illustrate the differences in mid vowels:

[aː] – [a] Herr **Mah**ler – Herr **Ma**ller
[ə] – [ɐ] ein**e** – ein**er**

The following minimal pairs illustrate the differences in back vowels:

[oː] – [ɔ] Herr **Moh**ler – Herr **Mo**ller
[uː] – [ʊ] Herr **Muh**ler – Herr **Mu**ller

- the rounding of the lips—there are rounded and unrounded vowels. The following minimal pairs illustrate the differences between rounded and unrounded vowels:

  [øː] – [eː] Herr **Möh**ler – Herr **Meh**ler
  [œː] – [ɛ] Herr **Mö**ller – Herr **Me**ller
  [yː] – [iː] Herr **Müh**ler – Herr **Mie**ler
  [Yː] – [ɪ] Herr **Mü**ller – Herr **Mi**ller

The German vowels can be systematized according to features:

	front		mid	back
long + tense	iː eː ɛː	yː øː	 aː	uː oː
short + lax	ɪ ɛ	y œ	 a	ʊ ɔː
unstressed			ə ɐ	
		rounded		rounded

III. German Consonants and Their Features (Overview)

German consonants are differentiated according to:

- point of articulation: they are formed from the lips (in the front) to the velum (in the back) at different points in the mouth (see overview table below);
- type of articulation:

  There are plosives/stops, in which the passage of air is interrupted:

  [p] as in *Li**pp**en,* [b] as in *lie**b**en,* [t] as in *re**tt**en,* [d] as in *re**d**en,* [k] as in *we**ck**en,* [g] as in *we**g**en*

  There are fricatives, in which the passage of air creates friction:

  [f] as in ***v**ier,* [v] as in ***w**ir,* [s] as in *Hau**s**,* [z] as in *Häu**s**er,* [ʃ] as in *Ta**sch**e,* [ʒ] as in *Gara**g**e,* [ç] as in *Mäd**ch**en,* [j] as in ***j**a,* [x] as in *To**ch**ter,* [r] as in *To**r**te*

  There are nasals, in which air passes through the nose:

  [m] as in ***M**ai,* [n] as in ***n**ie,* [ŋ] as in *la**ng**e*

  There are isolated consonants—the liquid [l] as in *he**ll**,* the aspirant [h] as in ***h**ier.*

- tension—there are tense consonants that are always voiceless:

  [p] as in *Li**pp**en,* [t] as in *re**tt**en,* [k] as in *we**ck**en,* [f] as in ***v**ier,* [s] as in *Hau**s**,* [ʃ] as in *Ta**sch**e,* [ç] as in *Mäd**ch**en,* [x] as in *To**ch**ter*

  There are lax consonants that are voiced after vowels and voiced consonants:

  [b] as in *lie**b**en,* [d] as in *re**d**en,* [g] as in *we**g**en,* [v] as in *be**w**egen,* [z] as in *Häu**s**er,* [ʒ] as in *Gara**g**e,* [j] as in *Ka**j**ak*

After a pause in speech (for example at the beginning of a sentence after a pause) and after voiceless consonants, these consonants are also pronounced voiceless:

[b̥] as in *mit**b**ringen,* [d̥] as in *bis **d**rei,* [g̊] as in *ins Haus **g**ehen,* [v̥] as in *auch **w**ir,* [z̥] as in *ab **s**ieben,* [ʒ̊] as in *das **J**ournal,* [j̊] as in ach ***j**a*

At the end of words and syllables, the following consonants are pronounced voiceless and tense—that is, as fortis consonants. This phenomenon is known as final devoicing:

[b → p] as in *lie**b**,* [d → t] as in *un**d**,* [g → k] as in *we**g**,* [v → f] as in *explosi**v**,* [z → s] as in *Hau**s***

The German consonants can be systematized according to their features as follows:

	front					back
PLOSIVE	p					
fortis				t		k
lenis	b			d		g
FRICATIVE						
fortis		f	s	ʃ	ç	x
lenis		v	z	ʒ	j	r
NASAL	m		n			ŋ
ISOLATED			l			h

IV. Rules for Melody and Accentuation

Melody

1. Melody falls at the end of a sentence (terminal) in:
   - statements—*Ich heiße Anna.* ↘
   - questions with question words—*Woher kommst du?* ↘
   - double questions—*Kommst du aus Bonn oder aus Berlin?* ↘
   - imperatives—*Setz dich!* ↘
2. Melody rises at the end of a sentence (interrogative) in:
   - yes-no questions—*Kommst du aus Bonn?* ↗
   - follow-up questions—*Woher kommst du?* ↘ *Aus Bonn?* ↗
   - questions posed in a friendly or curious tone of voice—*Wie heißt du?* ↗ *Was möchtest du trinken?* ↗
   - imperatives and statements made in a friendly tone of voice—*Bleib noch hier!* ↗ *Die Blumen sind für dich.* ↗
3. Melody remains neutral (doesn't change) directly before pauses in incomplete sentences (progredient)—*Peter kommt aus Bonn,* → *Anna kommt aus Berlin* → *und Ute kommt aus Wien.* ↘

Sentence Stress

1. The most important word is stressed:
   *Ich möchte ein Glas **Wein.*** (*kein Bier*)
   *Ich möchte ein **Glas** Wein.* (*keine Flasche*)
   *Ich möchte **ein** Glas Wein.* (*nicht zwei*)
2. Longer sentences are divided by pauses into accent (rhythmic) groups, in which there is always a main accent:
   *Ich möchte ein Glas **Wein,** / ein Stück **Brot,** / etwas **Käse** / und viel **Wasser.***

Word Stress

1. The stem is stressed:
   - in simple German words: ***Mo**de, **hö**ren;*
   - in words with the prefixes ***be-, ge-, er-, ver-, zer-:** be**halt**en;*
   - in verbs with inseparable prefixes and in nouns ending in *-ung* that are derived from them—for example, *wieder**ho**len → Wieder**ho**lung.*
2. The beginning of a word (prefix) is stressed:
   - in verbs with separable prefixes and in nouns derived from them—***aus**sprechen → die **Aus**sprache;*
   - in compounds with *un-* and *ur-* *—**Ur**laub, **un**genau.*
3. The principally defining word is stressed:
   - in compound nouns and adjectives—***Schlaf**zimmer, **dunkel**grün.*
4. The final syllable is stressed:
   - in German words with the suffix *-ei—Poli**zei**;*
   - in abbreviations in which each letter is pronounced separately—*AB**C***;
   - in words that end in *-ion—Explo**sion**.*

APPENDIX D

Grammar Summary Tables

I. Personal Pronouns

Nominative	Accusative	Accusative Reflexive	Dative	Dative Reflexive
ich	mich	mich	mir	mir
du	dich	dich	dir	dir
Sie	Sie	sich	Ihnen	sich
er	ihn	sich	ihm	sich
sie	sie	sich	ihr	sich
es	es	sich	ihm	sich
wir	uns	uns	uns	uns
ihr	euch	euch	euch	euch
Sie	Sie	sich	Ihnen	sich
sie	sie	sich	ihnen	sich

II. Definite Articles / Pronouns Declined Like Definite Articles

dieser/dieses/diese	*this*
mancher/manches/manche	*some, many a*
welcher/welches/welche	*which*
jeder/jedes/jede *(singular)*	*each, every*
alle *(plural)*	*all*

	Singular			Plural
	MASCULINE	**NEUTER**	**FEMININE**	
Nominative	der	das	die	die
	dieser	dieses	diese	diese
Accusative	den	das	die	die
	diesen	dieses	diese	diese
Dative	dem	dem	der	den
	diesem	diesem	dieser	diesen
Genitive	des	des	der	der
	dieses	dieses	dieser	dieser

III. Indefinite Articles / Negative Articles / Possessive Determiners

mein/meine	*my*
dein/deine	*your (familiar singular)*
Ihr/Ihre	*your (polite singular)*
sein/seine	*his, its*
ihr/ihre	*her, its*
unser/unsere	*our*
euer/eure	*your (familiar plural)*
Ihr/Ihre	*your (polite plural)*
ihr/ihre	*their*

	Singular			Plural
	MASCULINE	NEUTER	FEMININE	
Nominative	ein	ein	eine	—
	kein	kein	keine	keine
	mein	mein	meine	meine
Accusative	einen	ein	eine	—
	keinen	kein	keine	keine
	meinen	mein	meine	meine
Dative	einem	einem	einer	—
	keinem	keinem	keiner	keinen
	meinem	meinem	meiner	meinen
Genitive	eines	eines	einer	—
	keines	keines	keiner	keiner
	meines	meines	meiner	meiner

IV. Relative Pronouns

	Singular			Plural
	MASCULINE	NEUTER	FEMININE	
Nominative	der	das	die	die
Accusative	den	das	die	die
Dative	dem	dem	der	deren
Genitive	dessen	dessen	deren	deren

V. Question Pronouns

	People	Things and Concepts
Nominative	wer	was
Accusative	wen	was
Dative	wem	—
Genitive	wessen	—

VI. Attributive Adjectives

		Masculine	Neuter	Feminine	Plural
Nominative	*strong*	guter	gutes	gute	gute
	weak	gute	gute	gute	guten
Accusative	*strong*	guten	gutes	gute	gute
	weak	guten	gute	gute	guten
Dative	*strong*	gutem	gutem	guter	guten
	weak	guten	guten	guten	guten
Genitive	*strong*	guten	guten	guter	guter
	weak	guten	guten	guten	guten

Nouns declined like adjectives: Angestellte, Deutsche, Geliebte, Reisende, Verletzte, Verwandte

VII. Comparative and Superlative

A. *Regular Patterns*

schnell	schneller	am schnellsten
intelligent	intelligenter	am intelligentesten
heiß	heißer	am heißesten
teuer	teurer	am teuersten
dunkel	dunkler	am dunkelsten

B. *Umlaut Patterns*

alt	älter	am ältesten
groß	größer	am größten
jung	jünger	am jüngsten

Similarly: arm, dumm, hart, kalt, krank, kurz, lang, oft, scharf, schwach, stark, warm

C. *Irregular Patterns*

gern	lieber	am liebsten
gut	besser	am besten
hoch	höher	am höchsten
nah	näher	am nächsten
viel	mehr	am meisten

VIII. Weak Masculine Nouns

These nouns add **-(e)n** in the accusative, dative, and genitive.

A. *International nouns ending in* **-t** *denoting male persons:* Dirigent, Komponist, Patient, Polizist, Präsident, Soldat, Student, Tourist

B. *Nouns ending in* **-e** *denoting male persons or animals:* Drache, Junge, Kunde, Löwe, Neffe, Riese, Vorfahre, Zeuge

C. *The following nouns:* Elefant, Herr, Mensch, Nachbar, Name[1]

	Singular	Plural
Nominative	der Student	die Studenten
	der Junge	die Jungen
Accusative	den Studenten	die Studenten
	den Jungen	die Jungen
Dative	dem Studenten	den Studenten
	dem Jungen	den Jungen
Genitive	des Studenten	der Studenten
	des Jungen	der Jungen

IX. Prepositions

Accusative	Dative	Accusative/Dative	Genitive
durch	aus	an	(an)statt
für	außer	auf	trotz
gegen	bei	hinter	während
ohne	mit	in	wegen
um	nach	neben	
	seit	über	
	von	unter	
	zu	vor	
		zwischen	

X. Dative Verbs

antworten	*to answer*
begegnen	*to meet*
danken	*to thank*
erlauben	*to allow*
fehlen	*to be missing*
folgen	*to follow*
gefallen	*to please, be pleasing to*
gehören	*to belong to*
glauben	*to believe*
gratulieren	*to congratulate*
helfen	*to help*
leidtun	*to be sorry; to feel sorry for*
passen	*to fit*
passieren	*to happen*
raten	*to advise*
schaden	*to be harmful*
schmecken	*to taste (good)*
stehen	*to suit*
wehtun	*to hurt*
zuhören	*to listen to*

[1]*genitive:* des Namens

XI. Reflexive Verbs

sich anziehen	*to get dressed*
sich ärgern	*to get angry*
sich aufregen	*to get excited*
sich ausruhen	*to rest*
sich ausziehen	*to get undressed*
sich beeilen	*to hurry*
sich erholen	*to relax, recover*
sich erkälten	*to catch a cold*
sich erkundigen	*to ask*
sich (die Haare) föhnen	*to blow-dry (one's hair)*
sich fragen (ob)	*to wonder (if)*
sich freuen	*to be happy*
sich (wohl) fühlen	*to feel (well)*
sich fürchten	*to be afraid*
sich gewöhnen an	*to get used to*
sich hinlegen	*to lie down*
sich infizieren	*to get infected*
sich informieren	*to get information*
sich interessieren für	*to be interested in*
sich kümmern um	*to take care of*
sich rasieren	*to shave*
sich schminken	*to put on makeup*
sich setzen	*to sit down*
sich umsehen	*to look around*
sich unterhalten	*to have a conversation*
sich verletzen	*to get hurt*
sich verloben	*to get engaged*
sich vorstellen	*to imagine*

XII. Verbs + Prepositions

ACCUSATIVE

bitten um	*to ask for*
denken an	*to think about*
glauben an	*to believe in*
nachdenken über	*to think about; to ponder*
schreiben an	*to write to*
schreiben/sprechen über	*to write/talk about*
sorgen für	*to care for*
verzichten auf	*to renounce, do without*
warten auf	*to wait for*

SICH + ACCUSATIVE

sich ärgern über	*to be angry at*/about
sich erinnern an	*to remember*
sich freuen über	*to be happy about*
sich gewöhnen an	*to get used to*
sich interessieren für	*to be interested in*
sich kümmern um	*to take care of*
sich verlieben in	*to fall in love with*

DATIVE

fahren/reisen mit	*to go/travel by*
halten von	*to think of; to value*
handeln von	*to deal with*
träumen von	*to dream of*

SICH + DATIVE

sich erkundigen nach	*to ask about*
sich fürchten vor	*to be afraid of*

XIII. Inseparable Prefixes of Verbs

A. *Common*

be–	bedeuten, bekommen, bestellen, besuchen, bezahlen
er–	erfinden, erkälten, erklären, erlauben, erreichen
ver–	verbrennen, verdienen, vergessen, verlassen, verletzen

B. *Less Common*

ent–	entdecken, entscheiden, entschuldigen
ge–	gefallen, gehören, gewinnen, gewöhnen
zer–	zerreißen, zerstören

APPENDIX E

Verbs

I. Conjugation Patterns

A. *Simple tenses and principal parts*

		Present	Simple Past	Subjunctive	Aux. + Past Participle
Strong	ich	komme	kam	käme	bin gekommen
	du	kommst	kamst	kämst	bist gekommen
	er/sie/es	kommt	kam	käme	ist gekommen
	wir	kommen	kamen	kämen	sind gekommen
	ihr	kommt	kamt	kämt	seid gekommen
	sie, Sie	kommen	kamen	kämen	sind gekommen
Weak	ich	glaube	glaubte	glaubte	habe geglaubt
	du	glaubst	glaubtest	glaubtest	hast geglaubt
	er/sie/es	glaubt	glaubte	glaubte	hat geglaubt
	wir	glauben	glaubten	glaubten	haben geglaubt
	ihr	glaubt	glaubtet	glaubtet	habt geglaubt
	sie, Sie	glauben	glaubten	glaubten	haben geglaubt
Irregular Weak	ich	weiß	wusste	wüsste	habe gewusst
	du	weißt	wusstest	wüsstest	hast gewusst
	er/sie/es	weiß	wusste	wüsste	hat gewusst
	wir	wissen	wussten	wüssten	haben gewusst
	ihr	wisst	wusstet	wüsstet	habt gewusst
	sie, Sie	wissen	wussten	wüssten	haben gewusst
Modal	ich	kann	konnte	könnte	habe gekonnt
	du	kannst	konntest	könntest	hast gekonnt
	er/sie/es	kann	konnte	könnte	hat gekonnt
	wir	können	konnten	könnten	haben gekonnt
	ihr	könnt	konntet	könntet	habt gekonnt
	sie, Sie	können	konnten	könnten	haben gekonnt
haben	ich	habe	hatte	hätte	habe gehabt
	du	hast	hattest	hättest	hast gehabt
	er/sie/es	hat	hatte	hätte	hat gehabt
	wir	haben	hatten	hätten	haben gehabt
	ihr	habt	hattet	hättet	habt gehabt
	sie, Sie	haben	hatten	hätten	haben gehabt

		Present	Simple Past	Subjunctive	Aux. + Past Participle
sein	ich	bin	war	wäre	bin gewesen
	du	bist	warst	wärst	bist gewesen
	er/sie/es	ist	war	wäre	ist gewesen
	wir	sind	waren	wären	sind gewesen
	ihr	seid	wart	wärt	seid gewesen
	sie, Sie	sind	waren	wären	sind gewesen
werden	ich	werde	wurde	würde	bin geworden
	du	wirst	wurdest	würdest	bist geworden
	er/sie/es	wird	wurde	würde	ist geworden
	wir	werden	wurden	würden	sind geworden
	ihr	werdet	wurdet	würdet	seid geworden
	sie, Sie	werden	wurden	würden	sind geworden

B. *Compound tenses*

1. *Active voice*

	Perfect	Past Perfect	Future	Subjunctive
Strong	ich habe genommen	hatte genommen	werde nehmen	würde nehmen
	ich bin gefahren	war gefahren	werde fahren	würde fahren
Weak	ich habe gekauft	hatte gekauft	werde kaufen	würde kaufen
	ich bin gesegelt	war gesegelt	werde segeln	würde segeln
Irregular Weak	ich habe gewusst	hatte gewusst	werde wissen	würde wissen
Modal	ich habe gekonnt	hatte gekonnt	werde können	würde können
haben	ich habe gehabt	hatte gehabt	werde haben	würde haben
sein	ich bin gewesen	war gewesen	werde sein	würde sein
werden	ich bin geworden	war geworden	werde werden	würde werden

2. *Passive voice*

	Present	Simple Past	Perfect
Strong	es wird genommen	wurde genommen	ist genommen worden
Weak	es wird gekauft	wurde gekauft	ist gekauft worden

II. Strong and Irregular Weak Verbs

backen (backt/bäckt)	backte	hat gebacken	*to bake*
beginnen (beginnt)	begann	hat begonnen	*to begin*
beißen (beißt)	biss	hat gebissen	*to bite*
bekommen (bekommt)	bekam	hat bekommen	*to get, receive*
beschreiben (beschreibt)	beschrieb	hat beschrieben	*to describe*
besitzen (besitzt)	besaß	hat besessen	*to own, possess*
besteigen (besteigt)	bestieg	hat bestiegen	*to climb*
bitten (bittet)	bat	hat gebeten	*to ask*
bleiben (bleibt)	blieb	ist geblieben	*to stay*

braten (brät)	briet	hat gebraten	*to roast, fry*
brechen (bricht)	brach	hat gebrochen	*to break*
brennen (brennt)	brannte	hat gebrannt	*to burn*
bringen (bringt)	brachte	hat gebracht	*to bring*
denken (denkt)	dachte	hat gedacht	*to think*
dürfen (darf)	durfte	hat gedurft	*to be allowed to*
empfehlen (empfiehlt)	empfahl	hat empfohlen	*to recommend*
entscheiden (entscheidet)	entschied	hat entschieden	*to decide*
erfinden (erfindet)	erfand	hat erfunden	*to invent*
essen (isst)	aß	hat gegessen	*to eat*
fahren (fährt)	fuhr	ist gefahren	*to go, drive*
fallen (fällt)	fiel	ist gefallen	*to fall*
fangen (fängt)	fing	hat gefangen	*to catch*
finden (findet)	fand	hat gefunden	*to find*
fliegen (fliegt)	flog	ist geflogen	*to fly*
fliehen (flieht)	floh	ist geflohen	*to flee*
fließen (fließt)	floss	ist geflossen	*to flow*
fressen (frisst)	fraß	hat gefressen	*to eat*
geben (gibt)	gab	hat gegeben	*to give*
gefallen (gefällt)	gefiel	hat gefallen	*to please, be pleasing to*
gehen (geht)	ging	ist gegangen	*to go, walk*
gewinnen (gewinnt)	gewann	hat gewonnen	*to win*
gießen (gießt)	goss	hat gegossen	*to water*
haben (hat)	hatte	hat gehabt	*to have*
halten (hält)	hielt	hat gehalten	*to hold*
hängen (hängt)	hing	hat gehangen	*to hang, be suspended*
heben (hebt)	hob	hat gehoben	*to lift*
heißen (heißt)	hieß	hat geheißen	*to be called*
helfen (hilft)	half	hat geholfen	*to help*
kennen (kennt)	kannte	hat gekannt	*to know*
klingen (klingt)	klang	hat geklungen	*to sound*
kommen (kommt)	kam	ist gekommen	*to come*
können (kann)	konnte	hat gekonnt	*to be able to*
laden (lädt)	lud	hat geladen	*to load*
lassen (lässt)	ließ	hat gelassen	*to let, leave*
laufen (läuft)	lief	ist gelaufen	*to run*
leihen (leiht)	lieh	hat geliehen	*to lend, borrow*
lesen (liest)	las	hat gelesen	*to read*
liegen (liegt)	lag	hat gelegen	*to lie*
mögen (mag)	mochte	hat gemocht	*to like*
müssen (muss)	musste	hat gemusst	*to have to*
nehmen (nimmt)	nahm	hat genommen	*to take*
nennen (nennt)	nannte	hat genannt	*to name*
raten (rät)	riet	hat geraten	*to advise*
reiten (reitet)	ritt	ist geritten	*to ride*
riechen (riecht)	roch	hat gerochen	*to smell*
rufen (ruft)	rief	hat gerufen	*to call*

scheiden (scheidet)	schied	hat geschieden	*to separate*
schießen (schießt)	schoss	hat geschossen	*to shoot*
schlafen (schläft)	schlief	hat geschlafen	*to sleep*
schlagen (schlägt)	schlug	hat geschlagen	*to strike, beat*
schließen (schließt)	schloss	hat geschlossen	*to shut, close*
schneiden (schneidet)	schnitt	hat geschnitten	*to cut*
schreiben (schreibt)	schrieb	hat geschrieben	*to write*
schwimmen (schwimmt)	schwamm	ist geschwommen	*to swim*
sehen (sieht)	sah	hat gesehen	*to see*
sein (ist)	war	ist gewesen	*to be*
senden (sendet)	sandte	hat gesandt	*to send*
singen (singt)	sang	hat gesungen	*to sing*
sinken (sinkt)	sank	ist gesunken	*to sink*
sitzen (sitzt)	saß	hat gesessen	*to sit*
sprechen (spricht)	sprach	hat gesprochen	*to speak*
springen (springt)	sprang	ist gesprungen	*to spring, jump*
stehen (steht)	stand	hat gestanden	*to stand*
steigen (steigt)	stieg	ist gestiegen	*to climb*
sterben (stirbt)	starb	ist gestorben	*to die*
stoßen (stößt)	stieß	hat gestoßen	*to shove, push*
streiten (streitet)	stritt	hat gestritten	*to quarrel, fight*
tragen (trägt)	trug	hat getragen	*to wear, carry*
treffen (trifft)	traf	hat getroffen	*to meet, hit*
treiben (treibt)	trieb	hat getrieben	*to do (sports)*
trinken (trinkt)	trank	hat getrunken	*to drink*
tun (tut)	tat	hat getan	*to do*
verbrennen (verbrennt)	verbrannte	hat verbrannt	*to burn; to incinerate*
verbringen (verbringt)	verbrachte	hat verbracht	*to spend (time)*
vergessen (vergisst)	vergaß	hat vergessen	*to forget*
verlassen (verlässt)	verließ	hat verlassen	*to leave (a place)*
verlieren (verliert)	verlor	hat verloren	*to lose*
verschwinden (verschwindet)	verschwand	ist verschwunden	*to disappear*
versprechen (verspricht)	versprach	hat versprochen	*to promise*
wachsen (wächst)	wuchs	ist gewachsen	*to grow*
waschen (wäscht)	wusch	hat gewaschen	*to wash*
werden (wird)	wurde	ist geworden	*to become*
wissen (weiß)	wusste	hat gewusst	*to know*

APPENDIX F

Answers to Grammar Exercises

Einführung A

Übung 1: 1. Hören Sie zu! 2. Geben Sie mir die Hausaufgabe! 3. Öffnen Sie das Buch! 4. Schauen Sie an die Tafel! 5. Nehmen Sie einen Stift! 6. Sagen Sie „Guten Tag"! 7. Schließen Sie das Buch! 8. Schreiben Sie „Tschüss"! **Übung 2:** 1.a. heißt b. heiße c. heiße 2.a. heißen b. heiße 3.a. heiße b. heiße c. heißt **Übung 3:** 1. Sie 2. Es 3. Er 4. Sie 5. Es 6. Sie 7. Er 8. Sie 9. Sie 10. Er **Übung 4:** 1. Er ist orange. 2. Sie ist grün. 3. Es ist gelb. 4. Er ist schwarz und rot. 5. Sie sind rosa. *or* Sie sind lila. 6. Sie sind braun. 7. Sie ist weiß. **Übung 5:** 1. du 2. Sie 3. du 4. ihr 5. Sie 6. Sie 7. Sie 8. ihr

Einführung B

Übung 1: 1.a. ein b. der c. rot 2.a. ein b. der c. grün 3.a. eine b. die c. grau 4.a. eine b. die c. braun 5.a. ein b. das c. orange 6.a. eine b. die c. schwarz **Übung 2:** 1. Nein, das ist eine Lampe. 2. Nein, das ist eine Tafel. 3. Nein, das ist ein Fenster. 4. Nein, das ist ein Kind. 5. Nein, das ist ein Heft. 6. Nein, das ist eine Uhr. 7. Nein, das ist ein Tisch. 8. Nein, das ist eine Tür. **Übung 3:** 1.a. bist b. bin c. sind 2.a. ist b. sind 3.a. seid b. bin c. ist. 4.a. bin b. bin **Übung 4:** 1.a. haben b. habe 2. hast 3.a. Habt b. hat c. haben d. habe **Übung 5:** Der Mensch hat zwei Arme, zwei Augen, zwei Beine, zehn Finger, zwei Füße, zwei Hände, eine Nase, zwei Ohren, und zwei Schultern. **Übung 6:** (*Numbers will vary.*) In meinem Zimmer sind viele Bücher, vier Computer, ein Fenster, zwei Lampen, zwei Stühle, ein Tisch, eine Tür, eine Uhr, vier Wände. **Übung 7:** 1. Er ist schwarz. *oder* Er ist schwarz und weiß. *oder* Er ist schwarz und lila. *oder* Er ist schwarz und lila und weiß. 2. Es ist weiß. *oder* Es ist hellblau. *oder* Es ist grau. 3. Sie ist blau. 4. Sie ist gelb. 5. Sie sind weiß. 6. Es ist rot. 7. Er ist lila. 8. Sie sind braun. 9. Sie ist grün. 10. Er ist rosa. **Übung 8:** 1.a. kommst b. komme 2.a. kommt b. aus c. Woher d. kommen e. ich f. aus 3.a. sie b. kommen 4.a. ihr b. wir **Übung 9:** 1. Ihre 2.a. dein b. mein 3.a. mein b. mein c. Dein 4.a. Ihre b. Meine c. mein **Übung 10:** (*Answers will vary.*) 1. Ich komme aus ______. 2. Meine Mutter kommt aus ______. 3. Mein Vater kommt aus ______. 4. Meine Großeltern kommen aus ______. / Mein Großvater kommt aus ______ und meine Großmutter kommt aus ______. 5. Mein Professor / Meine Professorin kommt aus ______. 6. Ein Student aus meinem Deutschkurs heißt ______ und er kommt aus ______. 7. Eine Studentin aus meinem Deutschkurs heißt ______ und sie kommt aus ______.

Kapitel 1

Übung 1: (*Answers may vary.*) 1. Ich besuche Freunde. 2. Ihr geht ins Kino. 3. Jutta und Jens lernen Spanisch. 4. Du spielst gut Tennis. 5. Melanie studiert in Regensburg. 6. Ich lese ein Buch. 7. Wir reisen nach Deutschland. 8. Richard hört gern Musik. 9. Jürgen und Silvia kochen Spaghetti. **Übung 2:** 1. sie 2. Sie 3.a. du b. Ich 4.a. ihr b. Wir 5.a. Ich b. ihr c. Wir **Übung 3:** 1.a. (tanz)t b. (tanz)e c. (tanz)t 2.a. (geh)t b. (mach)en c. (reis)t d. (arbeit)et 3.a. (koch)en b. (mach)t c. (besuch)en 4.a. (Schreib)st b. (Chatt)est c. (mach)e **Übung 4:** (*Answers may vary slightly.*) 1. Monika und Albert spielen gern Schach. 2. Heidi arbeitet gern. 3. Stefan besucht gern Freunde. 4. Nora geht gern ins Kino. 5. Peter hört gern Musik. 6. Katrin macht gern Fotos. 7. Monika zeltet gern. 8. Albert trinkt gern Tee. **Übung 5:** 1. Frau Ruf liegt gern in der Sonne. Jutta liegt auch gern in der Sonne, aber Herr Ruf liegt nicht gern in der Sonne. 2. Jens reitet gern. Ernst reitet auch gern, aber Jutta reitet nicht gern. 3. Jens kocht gern. Jutta kocht auch gern, aber Andrea kocht nicht gern. 4. Michael und Maria spielen gern Karten. Die Rufs spielen auch gern Karten, aber die Wagners spielen nicht gern Karten. **Übung 6:** 1. Es ist halb acht. 2. Es ist elf Uhr. 3. Es ist Viertel vor fünf. 4. Es ist halb eins. 5. Es ist zehn vor sieben. 6. Es ist Viertel nach zwei. 7. Es ist fünfundzwanzig nach fünf. 8. Es ist halb elf. **Übung 7:** 1. (Rolf) nach 2. (er) vor 3. (Seine Stiefmutter) nach 4. (Rolf) vor 5. (er) vor 6. (er) vor 7. (er) vor 8. (Er) nach **Übung 8:** (*Answers will vary.*) 1. Ich studiere ______. 2. Im Moment wohne ich in ______. 3. Heute koche ich ______. 4. Manchmal trinke ich ______. 5. Ich spiele gern ______. 6. Mein Freund heißt ______. 7. Jetzt wohnt er in ______. 8. Manchmal spielen wir ______. **Übung 9:** 1. auf 2. auf 3. ein 4. an 5. aus 6. ab 7. ein 8. aus 9. auf **Übung 10:** (*Answers may vary.*) 1. Rolf kommt in San Francisco an. 2. Thomas räumt das Zimmer auf. 3. Heidi ruft Thomas an. 4. Albert füllt das Formular aus. 5. Peter holt Monika ab. 6. Peter und Monika gehen aus. 7. Frau Schulz packt die Bücher ein. 8. Stefan steht um sieben Uhr auf. **Übung 11:** 1. Wann bist du geboren? 2. Woher kommst du? 3. Wie groß bist du? 4. Studierst du? 5. Welche Fächer studierst du? 6. Wie viele Stunden arbeitest du? 7. Was machst du gern? **Übung 12:** (*Answers may vary.*) 1. Wie heißt du? 2. Kommst du aus München? 3. Woher kommst du? 4. Was studierst du? 5. Wie heißt dein Freund? 6. Wo wohnt er? 7. Spielst du Tennis? 8. Tanzt du gern? 9. Trinkst du gern Cola? 10. Trinkt Willi gern Bier?

Kapitel 2

Übung 1: Ernst kauft die Tasche, die Stühle und den Schreibtisch. Melanie kauft die Tasche, das Regal und den Schreibtisch. Jutta kauft den Pullover, die Lampe und den DVD-Spieler. Ich kaufe ... (*Answers will vary.*) **Übung 2:** (*Answers will vary. Possible answer:*) Ich habe ein Bett, Bilder, Bücher, einen Fernseher, eine Lampe, einen Laptop, einen Sessel und ein Smartphone. **Übung 3:** (*Sentences will vary.*) Heidi hat einen Teppich, aber keinen Fernseher. Sie hat eine Gitarre, aber kein Fahrrad. Sie hat einen Computer, aber keine Bilder. Sie hat ein Smartphone. Monika hat keinen Teppich, keinen Fernseher und keine Gitarre. Aber sie hat ein Fahrrad, einen Computer, Bilder und ein Smartphone. Ich habe ______. **Übung 4:** (*Answers will vary. Possible answers:*) 1. Ich möchte ein Auto und eine Sonnenbrille. 2. Mein bester Freund möchte eine Katze. 3. Meine Eltern möchten einen Laptop. 4. Meine Mitbewohnerin und ich möchten einen Fernseher. 5. Mein Nachbar in der Klasse möchte ein Motorrad. 6. Meine Professorin möchte einen Koffer. 7. Mein Bruder möchte einen Hund. **Übung 5:** Seine Haare; Seine Augen; Seine Halskette; Seine Schuhe; Seine Gitarre; Sein Zimmer; Sein Fenster; Ihre Haare; Ihre Augen; Ihre Halskette ist kurz. Ihre Schuhe sind sauber. Ihre Gitarre ist neu. Ihr Zimmer ist klein. Ihr Fenster ist groß. **Übung 6:** 1. Ihren 2. Deine 3. eure 4. Deine 5. Ihr 6. deine 7. Euren **Übung 7:** (*Answers will vary.*) **Übung 8:** 1.a. ihr b. wir 2.a. Sie b. Ich 3.a. sie b. er 4.a. du b. Ich c. ihr d. Wir **Übung 9:** a. machen b. fährt c. sieht d. Isst e. isst f. isst g. macht h. lese i. schläft j. fahren **Übung 10:** (*Answers will vary.*) 1. Wir sprechen (nicht) gern Deutsch. Sprecht ihr auch (nicht) gern Deutsch? 2. Ich lade (nicht) gern Freunde ein. Lädst du auch (nicht) gern Freunde ein? 3. Ich laufe (nicht) gern im Wald. Läufst du auch (nicht) gern im Wald? 4. Ich trage (nicht) gern Pullis. Trägst du auch (nicht) gern Pullis? 5. Wir sehen (nicht) gern fern. Seht ihr auch (nicht) gern fern? 6. Ich fahre (nicht) gern Fahrrad. Fährst du auch (nicht) gern Fahrrad? 7. Wir vergessen (nicht) gern die Hausaufgabe. Vergesst ihr auch (nicht) gern die Hausaufgabe? 8. Ich schlafe (nicht) gern. Schläfst du auch (nicht) gern? 9. Wir lesen (nicht) gern online. Lest ihr auch (nicht) gern online? **Übung 11:** 1. Schlaf nicht den ganzen Tag! 2. Lieg nicht den ganzen Tag in der Sonne! 3. Vergiss deine Hausaufgaben nicht! 4. Lies deine Bücher! 5. Sieh nicht den ganzen Tag fern! 6. Trink nicht zu viel Cola! 7. Sitz nicht den ganzen Tag am Computer! 8. Trag deine Brille! 9. Spiel nicht immer Computerspiele! 10. Treib Sport! **Übung 12:** 1. Trag heute ein T-Shirt! 2. Spiel keine laute Musik! 3. Lern den Wortschatz! 4. Ruf deine Freunde an! 5. Lauf nicht allein im Park! 6. Lieg nicht zu lange in der Sonne! 7. Räum dein Zimmer auf! 8. Iss heute Abend in einem Restaurant! 9. Steh früh auf!

Kapitel 3

Übung 1: (*Predicates and sequence will vary. Subjects and their corresponding conjugated verbs are given here.*) A. Mein Freund / Meine Freundin kann ______. Meine Eltern können ______. Ich kann / Wir können ______. Mein Bruder / Meine Schwester kann ______. Der Professor / Die Professorin kann ______. B.1. Kannst du / Könnt ihr Gedichte schreiben? 2. Kannst du / Könnt ihr Auto fahren? 3. Kannst du / Könnt ihr tippen? 4. Kannst du / Könnt ihr stricken? 5. Kannst du / Könnt ihr zeichnen? **Übung 2:** (*Answers will vary.*) 1. Heute Abend will ich ______. 2. Morgen kann ich nicht ______. 3. Mein Freund / Meine Freundin kann gut ______. 4. Am Samstag will mein Freund / meine Freundin ______. 5. Mein Freund / Meine Freundin und ich wollen ______. 6. Im Winter wollen meine Eltern / meine Freunde ______. 7. Meine Eltern / Meine Freunde können gut ______. **Übung 3:** 1. Sie darf nicht mit Jens zusammen lernen. 2. Sie darf nicht den ganzen Abend chatten. 3. Sie muss in der Klasse aufpassen und mitschreiben. 4. Sie darf nicht jeden Tag tanzen gehen. 5. Sie muss jeden Tag ihren Wortschatz lernen. 6. Sie muss amerikanische Filme im Original sehen. 7. Sie muss ihren Englischlehrer zum Abendessen einladen. 8. Sie muss für eine Woche nach London fahren. 9. Sie muss die englische Grammatik fleißig lernen. **Übung 4:** 1.a. Willst b. will c. kann d. muss 2.a. darf b. musst c. kann d. darfst e. könnt 3.a. sollst b. kann c. musst **Übung 5:** 1. dich 2.a. mich b. dich 3. uns 4. euch 5.a. dich b. dich 6.a. mich b. Sie 7. Sie **Übung 6:** 1. Ja, ich mache es gern. / Nein, ich mache es nicht gern. 2. Ja, ich kann es aufsagen. / Nein, ich kann es nicht aufsagen. 3. Ja, ich kenne ihn. / Nein, ich kenne ihn nicht. 4. Ja, ich lese sie gern. / Nein, ich lese sie nicht gern. 5. Ja, ich lerne ihn gern. / Nein, ich lerne ihn nicht gern. 6. Ja, ich kenne sie. / Nein, ich kenne sie nicht. 7. Ja, ich vergesse sie oft. / Nein, ich vergesse sie nicht oft. 8. Ja, ich mag ihn/sie. / Nein, ich mag ihn/sie nicht. **Übung 7:** 1. Nein, sie liest ihn nicht, sie schreibt ihn. 2. Nein, er isst sie nicht, er trinkt sie. 3. Nein, sie macht ihn nicht an, sie macht ihn aus. 4. Nein, er kauft es nicht, er verkauft es. 5. Nein, er zieht sie nicht aus, er zieht sie an. 6. Nein, sie trägt ihn nicht, sie kauft ihn. 7. Nein, er bestellt es nicht, er isst es. 8. Nein, er besucht ihn nicht, er ruft ihn an.

9. Nein, sie kämmt es nicht, sie wäscht es. 10. Nein, er bläst sie nicht aus, er zündet sie an. **Übung 8:** (*Answers may vary. Possible answers:*) 1. Weil ich krank bin. 2. Weil er müde ist. 3. Weil wir Hunger haben. 4. Weil sie keine Zeit hat. 5. Weil sie Langeweile hat. 6. Weil ich traurig bin. 7. Weil sie Durst haben. 8. Weil ich Angst habe. 9. Weil er glücklich ist. 10. Weil ich lernen muss. **Übung 9:** (*Answers will vary. Possible answers:*) 1. S1: Was macht Albert, wenn er müde ist? S2: Wenn Albert müde ist, geht er nach Hause. S1: Und du? S2: Wenn ich müde bin, ______. 2. S1: Was macht Maria, wenn sie glücklich ist? S2: Wenn Maria glücklich ist, trifft sie Michael. S1: Und du? S2: Wenn ich glücklich bin, ______. 3. S1: Was macht Herr Ruf, wenn er Durst hat? S2: Wenn Herr Ruf Durst hat, trinkt er eine Cola. S1: Und du? S2: Wenn ich Durst habe, ______. 4. S1: Was macht Frau Wagner, wenn sie in Eile ist? S2: Wenn Frau Wagner in Eile ist, fährt sie mit dem Taxi. S1: Und du? S2: Wenn ich in Eile bin, ______. 5. S1: Was macht Heidi, wenn sie Hunger hat? S2: Wenn Heidi Hunger hat, kauft sie einen Hamburger. S1: Und du? S2: Wenn ich Hunger habe, ______. 6. S1: Was macht Frau Schulz, wenn sie Ferien hat? S2: Wenn Frau Schulz Ferien hat, fliegt sie nach Deutschland. S1: Und du? S2: Wenn ich Ferien habe, ______. 7. S1: Was macht Hans, wenn er Angst hat? S2: Wenn Hans Angst hat, ruft er, „Mama, Mama". S1: Und du? S2: Wenn ich Angst habe, ______. 8. S1: Was macht Stefan, wenn er krank ist? S2: Wenn Stefan krank ist, geht er zum Arzt. S1: Und du? S2: Wenn ich krank bin, ______. **Übung 10:** (*Answers may vary. Possible answers:*) 1. Jürgen ist wütend, weil er immer so früh aufstehen muss. 2. Silvia ist froh, weil sie heute nicht arbeiten muss. 3. Claire ist in Eile, weil sie noch einkaufen muss. 4. Josef ist traurig, weil Melanie ihn nicht anruft. 5. Thomas geht nicht zu Fuß, weil seine Freundin ihn zur Uni mitnimmt. 6. Willi hat selten Langeweile, weil er immer fernsieht. 7. Nesrin hat Angst vor Wasser, weil sie nicht schwimmen kann. 8. Mehmet fährt in die Türkei, weil er seine Eltern besuchen will.

Kapitel 4

Übung 1: a. hat b. ist c. hat d. hat e. ist f. sind g. ist h. hat i. hat **Fragen:** 1. Yamina ist um 7 Uhr aufgestanden. 2. Sie sind zur Schule gegangen. 3. Frau Dehne ist die Lehrerin. 4. Sie hat „Herzlich Willkommen" an die Tafel geschrieben. **Übung 2:** a. haben b. sind c. haben d. sind e. sind f. haben g. haben h. sind i. haben j. sind **Fragen:** 1. Josef und Melanie sind mit dem Taxi zum Bahnhof gefahren. 2. Sie sind um 5.30 mit dem Zug abgefahren. 3. Sie haben im Speisewagen gefrühstückt. 4. Nachts sind sie in den Schlafwagen gegangen und haben schlecht geschlafen. **Übung 3:** a. aufgestanden b. geduscht c. gefrühstückt d. gegangen e. gehört f. getroffen g. getrunken h. gearbeitet i. gegessen **Übung 4:** 1. Hast du schon gefrühstückt? 2. Bist du schon geschwommen? 3. Hast du schon eine Geschichte gelesen? 4. Hast du schon Klavier gespielt? 5. Hast du schon geschlafen? 6. Hast du schon gegessen? 7. Hast du schon Geschirr gespült? 8. Hast du den Brief schon geschrieben? 9. Bist du schon ins Bett gegangen? **Übung 5:** 1. Katrin hat bis 9 Uhr im Bett gelegen. 2. Sie hat einen Rock getragen. 3. Sie hat mit Frau Schulz gesprochen. 4. Sie hat ein Referat gehalten. 5. Sie hat Freunde getroffen. 6. Sie hat gearbeitet. 7. Es hat geregnet. 8. Sie ist nach Hause gekommen. 9. Sie hat ihre Wäsche gewaschen. 10. Sie ist abends zu Hause geblieben. **Übung 6:** 1. (*Answers will vary.*) 2. (*Answers will vary.*) 3. (*Answers will vary.*) 4. (*Answers will vary.*) 5. Am ersten Januar. 6. Am ersten November. *oder* Vom 31. Oktober bis zum 2. November. 7. (*Answers will vary.*) 8. (*Answers will vary.*) 9. (*Answers will vary.*) 10. (*Answers will vary.*) **Übung 7:** a. im b. im c. — d. am e. Am f. um g. um h. Am i. im j. am **Übung 8:** (*Answers will vary.*) **Übung 9:** A: 1. R 2. F 3. R 4. R 5. R

B: Partizipien mit **ge-**:

aufgestanden	aufstehen
gehört	hören
gegangen	gehen
gekocht	kochen
gefahren	fahren
geparkt	parken
zurückgekommen	zurückkommen
gewaschen	waschen
aufgeräumt	aufräumen
gefallen	fallen
eingelaufen	einlaufen
abgebrannt	abbrennen

Partizipien ohne **ge-**:

verschlafen	verschlafen
bekommen	bekommen
bezahlt	bezahlen
zerbrochen	zerbrechen

Übung 10: a. ist ... angekommen b. hat ... begrüßt c. getrunken d. ist ... gegangen e. hat ... geschlafen f. ist ... gegangen g. haben ... gefragt h. hat ... gesprochen i. haben ... getrunken j. sind ... gegangen **Übung 11:** (*Answers will vary. Possible answers follow.*) 1. —Bist du gestern früh aufgestanden? —Ja. —Wann? —Um 6 Uhr. 2. —Hast du gestern jemanden fotografiert? —Ja. —Wen? —Jane. 3. —Hast du gestern jemanden besucht? —Ja. —Wen? —Alan. 4. —Bist du gestern ausgegangen? —Ja. —Wohin? —Ins Kino. 5. —Hast du gestern etwas bezahlt? —Ja. —Was? —Die Rechnung. 6. —Hast du gestern etwas repariert? —Ja. —Was? —Mein Auto. 7. —Hast du gestern etwas Neues probiert? —Ja. —Was? —Segeln. 8. —Hast du gestern ferngesehen? —Ja. —Wie lange? —Eine Stunde. 9. — Hast du gestern etwas nicht verstanden? —Ja. —Was? —Sophies Referat. 10. —Hast du gestern dein Zimmer aufgeräumt? —Ja. —Wann? —Um 4 Uhr. *or* Um 4 Uhr nachmittags. *or* Um 16 Uhr.

Kapitel 5

Übung 1: (*Answers will vary.*) Ich backe meiner Tante einen Kuchen. Ich erkläre meinem Partner einen Witz. Ich erzähle meiner Kusine ein Geheimnis. Ich gebe meinem Freund einen Kuss. Ich kaufe meinem Vater eine Krawatte. Ich koche meiner Mitbewohnerin Kaffee. Ich leihe meinem Bruder fünfzig Dollar. Ich schenke meiner Großmutter ein Buch. Ich schreibe meiner Mutter einen Brief. Ich verkaufe meinem Mitbewohner mein Deutschbuch. **Übung 2:** (*Answers will vary.*) Heidi erklärt ihrer Freundin die Grammatik. Peter erzählt seinem Vetter ein Geheimnis. Thomas gibt seiner Mutter ein Armband. Katrin kauft ihrem Mann einen Rucksack. Stefan kocht seinem Freund eine Suppe. Albert leiht seinen Eltern einen Regenschirm. Monika schenkt ihrer Schwester einen Bikini. Frau Schulz schreibt ihrer Tante eine Karte. Nora verkauft ihrem Professor ein Zelt. **Übung 3:** 1. Wer 2. Wen 3. Wem 4. Wen 5. Wem 6. wer **Übung 4:** 1. Was passiert am Abend? d. Es wird dunkel. 2. Was passiert, wenn man Bücher schreibt? b. Man wird bekannt. 3. Was passiert, wenn man krank wird? h. Man bekommt Fieber. 4. Was passiert im Frühling? i. Die Tage werden länger. 5. Was passiert im Herbst? c. Die Blätter werden bunt. 6. Was passiert, wenn Kinder älter werden? e. Sie werden größer. 7. Was passiert, wenn man in der Lotterie gewinnt? j. Man wird reich. 8. Was passiert, wenn man Medizin studiert? a. Man wird Arzt. 9. Was passiert am Morgen? g. Es wird hell. 10. Was passiert im Sommer? f. Es wird wärmer. **Übung 5:** 1. Vielleicht wird sie Köchin. 2. Vielleicht wird sie Apothekerin. 3. Vielleicht wird er Pilot. 4. Vielleicht wird er Lehrer. 5. Vielleicht wird sie Architektin. 6. Vielleicht wird sie Bibliothekarin. 7. Vielleicht wird er Krankenpfleger. 8. Vielleicht wird sie Dirigentin. **Übung 6:** 1. Was macht man im Kino? Man sieht einen Film. 2. Was macht man auf der Post? Man kauft Briefmarken. 3. Was macht man an der Tankstelle? Man tankt Benzin. 4. Was macht man in der Disko? Man tanzt. 5. Was macht man in der Kirche? Man betet. 6. Was macht man auf der Bank? Man wechselt Geld. 7. Was macht man im Meer? Man schwimmt. 8. Was macht man in der Bibliothek? Man liest ein Buch. 9. Was macht man im Park? Man geht spazieren. **Übung 7:** 1. Monika ist in der Kirche. 2. Albert ist im Meer. 3. Heidi ist auf der Polizei. 4. Nora ist in einem Hotel. 5. Katrin ist im Schwimmbad. 6. Thomas ist auf der Post. 7. Frau Schulz ist in der Küche. 8. Das Poster ist an der Wand. 9. Der Topf ist auf dem Herd. 10. Der Wein ist im Kühlschrank. **Übung 8:** 1. mir 2. dir 3. euch 4. Ihnen 5. uns **Übung 9:** 1. Er hat ihr einen Regenschirm geschenkt. 2. Sie hat ihm ihr Auto geliehen. 3. Er hat ihm 500 Euro geliehen. 4. Sie hat ihr ein Geheimnis erzählt. 5. Er hat ihnen eine Geschichte erzählt. 6. Sie hat ihr ihre Sonnenbrille verkauft. 7. Er hat ihnen seinen Fernseher verkauft. 8. Sie hat ihm ihr Büro gezeigt. 9. Er hat ihm seine Wohnung gezeigt. 10. Sie hat ihr eine neue Brille gekauft. 11. Er hat ihr einen Kinderwagen gekauft.

Kapitel 6

Übung 1: 1. gefällt 2. gratuliere 3. helfen 4. Schmeckt 5. passt 6. gehört 7. Fehlt 8. begegnet 9. schadet 10. zugehört **Übung 2:** (*Answers will vary.*) **Übung 3:** (*Answers may vary.*) 1. Albert ist unter der Dusche. 2. Der Spiegel hängt an der Wand. 3. Der Kühlschrank steht neben dem Fernseher. 4. Das Deutschbuch liegt im Kühlschrank. 5. Die Lampe hängt über dem Tisch. 6. Der Computer steht auf dem Schreibtisch. 7. Die Schuhe liegen auf dem Bett. 8. Die Hose liegt auf dem Tisch. 9. Das Poster von Berlin hängt über dem Fernseher. 10. Die Katze liegt unter dem Bett. **Übung 4:** (*Answers will vary.*) **Übung 5:** (*Answers will vary. Possible answers:*) 1. Ich bin heute Abend in der Bibliothek. 2. Ich bin am Nachmittag in der Mensa. 3. Ich bin um 16 Uhr bei Freunden. 4. Ich bin in der Nacht im Bett. 5. Ich bin am frühen Morgen am Frühstückstisch. 6. Ich bin am Montag in der Klasse. 7. Ich bin am 1. August im Urlaub. 8. Ich bin an Weihnachten auf einer Party. 9. Ich bin im Winter bei meinen Eltern. 10. Ich bin am Wochenende auf einer Party. **Übung 6:** 1. Er geht zum Arzt. 2. Er geht zum Fußballplatz. 3. Sie geht ins Hotel. 4. Er fährt zur Tankstelle. 5. Er geht in den Supermarkt. 6. Er geht auf die Post. 7. Sie gehen in den Wald. 8. Sie geht zu ihrem Freund. 9. Er fährt zum Flughafen. 10. Sie geht ins Theater. 11. Sie geht in die Schule. **Übung 7:** 1. aufstehst 2.a. hör b. zu c. mache d. aus 3.a. kommt b. an 4.a. zieht b. um 5. einladen 6.a. räumt b. auf 7.a. mitkommen b. mitnimmst 8.a. rufst b. an **Übung 8:** Andrea hat ferngesehen. Katrin und Peter sind ausgegangen. Heidi hat Frau Schulz angerufen. Herr Ruf hat das Geschirr abgetrocknet. Jürgen ist ausgezogen. Jutta hat ihr Abendkleid angezogen. Maria ist aus Bulgarien zurückgekommen. Herr Thelen ist aufgewacht. **Übung 9:** 1. Womit kochst du Kaffee? Mit der Kaffeemaschine. 2. Womit saugst du Staub? Mit dem Staubsauger. 3. Womit fegst du den Boden? Mit dem Besen. 4. Womit bügelst du? Mit dem Bügeleisen. 5. Womit tippst du einen Brief? Mit dem Computer. 6. Womit gießt du die Blumen im Garten? Mit dem Gartenschlauch. 7. Womit wischst du den Boden? Mit dem Putzlappen. 8. Womit gießt du die Blumen in der Wohnung? Mit der Gießkanne. **Übung 10:** 1.a. mit b. mit c. Mit d. bei 2.a. bei b. mit c. bei d. mit 3.a. mit b. mit c. bei.

Kapitel 7

Übung 1: (*Answers will vary. Possible answers:*) 1. Ich mag Leute, die laut lachen. 2. Ich mag keine Leute, die viel sprechen. 3. Ich mag eine Stadt, die Spaß macht. 4. Ich mag keine Stadt, die langweilig ist. 5. Ich mag einen Mann, der gern verreist. 6. Ich mag keinen Mann, der interessant aussieht. 7. Ich mag eine Frau, die nett ist. 8. Ich mag keine Frau, die betrunken ist. 9. Ich mag einen Urlaub, der exotisch ist. 10. Ich mag ein Auto, das schnell fährt. **Übung 2:** 1. h Europa → Wie heißt der Kontinent, der eigentlich eine Halbinsel von Asien ist? 2. j Mississippi → Wie heißt der Fluss, von dem Mark Twain erzählt? 3. c San Francisco → Wie heißt die Stadt, die an einer Bucht liegt? 4. g die Alpen → Wie heißen die Berge, in denen man sehr gut Ski fahren kann? 5. e Washington → Wie heißt der Staat in den USA, dem ein Präsident seinen Namen gegeben hat? 6. f das Tal des Todes → Wie heißt das Tal, in dem es sehr heiß ist? 7. b Ellis → Wie heißt die Insel, die man von New York sieht? 8. i der Pazifik → Wie heißt das Meer, über das man nach Hawaii fliegt? 9. d die Sahara → Wie heißt die Wüste, die man aus vielen Filmen kennt? 10. a der Große Salzsee → Wie heißt der See in Utah, auf dem man segeln kann? **Übung 3:** 1. Berlin ist größer als Zürich. 2. München ist älter als San Francisco. 3. Athen ist wärmer als Hamburg. 4. Der Mount Everest ist höher als das Matterhorn. 5. Der Mississippi ist länger als der Rhein. 6. Liechtenstein ist kleiner als die Schweiz. 7. Leipzig ist kälter als Kairo. 8. Ein Fernseher ist billiger als eine Waschmaschine. 9. Schnaps ist stärker als Bier. 10. Ein Haus auf dem Land ist schöner als ein Haus in der Stadt. (*oder* Ein Haus in der Stadt ist schöner als ein Haus auf dem Land.) 11. Zehn Euro sind mehr als zehn Cent. 12. Ein Appartement ist teurer als eine Wohnung in einem Studentenheim. 13. Ein Motorrad ist schneller als ein Fahrrad. 14. Ein Sofa ist schwerer als ein Stuhl. 15. Bier ist besser als Milch. (*oder* Milch ist besser als Bier.) **Übung 4:** 1. Heidi ist schwerer als Monika. 2. Thomas und Stefan sind am schwersten. 3. Thomas ist besser in Deutsch als Stefan. 4. Heidi ist in Deutsch am besten. 5. Heidi ist kleiner als Stefan. 6. Monika ist am kleinsten. 7. Stefan ist jünger als Thomas. 8. Stefan ist am jüngsten. 9. Thomas' Haare sind länger als Heidis. 10. Monikas Haare sind am längsten. 11. Heidis Haare sind kürzer als Monikas. 12. Stefans Haare sind am kürzesten. 13. Monika ist schlechter in Deutsch als Heidi. 14. Stefan ist in Deutsch am schlechtesten. **Übung 5:** 1. a. In Athen ist es am heißesten. 2. g. In Moskau ist es am kältesten. 3. f. Monaco ist am kleinsten. 4. e. Frankreich ist am ältesten. 5. j. Südafrika ist am jüngsten. 6. i. Der Nil ist am längsten. 7. d. Frankfurt liegt am nördlichsten. 8. h. Der Mount Everest ist am höchsten. 9. c. Deutschland ist am größten. **Übung 6:** a. darauf b. daneben c. Dazwischen d. Darin e. Davor/daneben f. darüber g. Daran h. Darunter i. dahinter **Übung 7:** 1. Mit wem gehen Sie am liebsten ins Theater? 2. Worauf freuen Sie sich am meisten? 3. Auf wen müssen Sie immer warten? 4. Über wen haben Sie sich in letzter Zeit geärgert? 5. Woran denken Sie, wenn Sie „USA" hören? 6. Womit fahren Sie zur Schule? 7. Worüber schreiben Sie nicht gern? 8. An wen haben Sie Ihren letzten Brief geschrieben? 9. Von wem halten Sie nicht viel? **Übung 8:** 1. bin 2.a. hat b. bin 3.a. habe b. bin 4. bin 5. bin 6.a. habe b. bin 7.a. habe b. ist 8.a. haben b. ist 9.a. ist/sind b. hat **Übung 9:** 1. Ich habe schon Frühstück gemacht. 2. Ich habe meine Milch schon getrunken. 3. Ich habe den Tisch schon sauber gemacht. 4. Ich bin schon zum Bäcker gelaufen. 5. Ich habe schon Brötchen mitgebracht. 6. Ich habe schon Geld mitgenommen. 7. Ich habe den Hund schon gefüttert. 8. Ich habe die Tür schon zugemacht. **Übung 10:** 1.a. Hatten b. hatte 2. Waren 3.a. wart b. hatten 4.a. Warst b. war 5. hatte 6. hattest 7.a. Warst b. war c. hatte.

Kapitel 8

Übung 1: (*Answers will vary.*) 1. Amerikanisches Steak! 2. Russischer Kaviar! 3. Griechische Oliven! 4. Japanisches Sushi! 5. Französischer Champagner! 6. Deutsche Wurst! 7. Dänischer Käse! 8. Italienische Spaghetti! 9. Ungarischer Paprika! 10. Englische Marmelade! 11. Kolumbianischer Kaffee! 12. Neuseeländische Kiwis! **Übung 2:** 1. Ich esse nur deutsches Brot. 2. Ich esse nur russischen Kaviar. 3. Ich esse nur italienische Salami. 4. Ich trinke nur kolumbianischen Kaffee. 5. Ich esse nur neuseeländische Kiwis. 6. Ich trinke nur französischen Wein. 7. Ich trinke nur belgisches Bier. 8. Ich esse nur spanische Muscheln. 9. Ich esse nur englische Marmelade. 10. Ich esse nur japanischen Thunfisch. **Übung 3:** 1. Michael: Ich möchte den grauen Wintermantel da. Maria: Nein, der graue Wintermantel ist viel zu schwer. 2. Michael: Ich möchte die gelbe Hose da. Maria: Nein, die gelbe Hose ist viel zu bunt. 3. Michael: Ich möchte das schicke Hemd da. Maria: Nein, das schicke Hemd ist viel zu teuer. 4. Michael: Ich möchte die roten Socken da. Maria: Nein, die roten Socken sind viel zu warm. 5. Michael: Ich möchte den schwarzen Schlafanzug da. Maria: Nein, der schwarze Schlafanzug ist viel zu dünn. 6. Michael: Ich möchte die grünen Schuhe da. Maria: Nein, die grünen Schuhe sind viel zu groß. 7. Michael: Ich möchte den modischen Hut da. Maria: Nein, der modische Hut ist viel zu klein. 8. Michael: Ich möchte die schwarzen Winterstiefel da. Maria: Nein, die schwarzen Winterstiefel sind viel zu leicht. 9. Michael: Ich möchte die elegante Sonnenbrille da. Maria: Nein, die elegante Sonnenbrille ist viel zu bunt. 10. Michael: Ich möchte die roten Tennisschuhe da. Maria: Nein, die roten Tennisschuhe sind viel zu grell. **Übung 4:** 1.a. Ihr neues Auto b. der alte Mercedes c. keinen neuen Wagen 2.a. der italienische Wein b. eine weitere Flasche 3.a. mein kaputtes Fahrrad b. meinen blöden Computer c. kein freies Wochenende **Übung 5:** 1. Die Teller stehen im Küchenschrank. 2. Albert stellt die Teller auf den Tisch. 3. Die Servietten liegen in der Schublade. 4. Monika legt die Servietten auf den Tisch. 5. Messer und Gabeln liegen in der Schublade. 6. Stefan legt Messer und Gabeln auf den Tisch. 7. Die Kerze steht auf dem Schrank. 8. Heidi stellt die Kerze auf den Tisch. 9. Thomas sitzt auf dem Sofa. **Übung 6:** 1. Jutta leiht ihrem neuen Freund ihre Lieblings-DVD. 2. Jens verkauft dem kleinen Bruder von Jutta eine Ratte. 3. Hans zeigt die Ratte nur seinen besten Freunden. 4. Jutta schenkt ihrer besten Freundin ein Buch. 5. Jens kauft seinem wütenden Lehrer eine Krawatte. 6. Ernst erzählt seiner großen Schwester einen Witz. 7. Jutta kocht den netten Leuten von nebenan Kaffee. 8. Ernst gibt dem süßen Baby von nebenan einen Kuss. **Übung 7:** (*Answers and sequence will vary.*) 1. Ich werde weniger fernsehen. 2. Ich werde mehr lernen. 3. Ich werde weniger oft ins Kino gehen. 4. Ich werde früher ins Bett gehen. 5. Ich werde mehr arbeiten. 6. Ich werde öfter selbst kochen. **Übung 8:** (*Answers may vary.*) 1. Frau Schulz repariert morgen das Auto. 2. Heidi fährt morgen aufs Land. 3. Peter spielt morgen Fußball. 4. Monika schreibt morgen eine E-Mail. 5. Stefan geht morgen einkaufen. 6. Nora heiratet morgen. 7. Albert geht morgen in den Supermarkt. 8. Thomas räumt morgen sein Zimmer auf. **Übung 9:** (*Answers will vary.*)

Kapitel 9

Übung 1: (*Answers will vary.*) **Übung 2:** (*Answers will vary*). 1. Ich durfte nicht. 2. Ich wollte nicht. 3. Das wusste ich nicht. 4. Ich wollte eine. 5. Ich sollte das nicht. **Übung 3:** 1.a. wolltest b. wusste 2.a. durfte b. musste c. wollten d. konnten 3.a. konnte b. musste c. wusste d. wollte **Übung 4:** 1.a. Wann b. Wenn 2.a. wann b. Als 3. als 4.a. Wann b. als 5.a. Wann b. Wenn 6.a. Wann b. Als **Übung 5:** a. wenn b. Als c. Wenn d. wenn e. Als f. Als g. Wann h. Als i. wenn **Übung 6:** a. standen b. gingen c. fuhren d. kamen e. hielten f. aßen g. schwammen h. schliefen i. sprangen **Übung 7:** 1. wohnte 2. brachten 3. schliefen 4. liefen, kamen 5. sahen, saß 6. gab, fanden 7. trug, schloss 8. tötete, rannte **Übung 8:** 1.c Nachdem Jutta den Schlüssel verloren hatte, kletterte sie durch das Fenster. 2.d Nachdem Ernst die Fensterscheibe eingeworfen hatte, lief er weg. 3.g Nachdem Claire angekommen war, rief sie Melanie an. 4.b Nachdem Hans seine Hausaufgaben gemacht hatte, ging er ins Bett. 5.e Nachdem Jens sein Fahrrad repariert hatte, machte er eine Radtour. 6.h Nachdem Michael die Seiltänzerin gesehen hatte, war er ganz verliebt. 7.a Nachdem Richard ein ganzes Jahr gespart hatte, flog er nach Australien. 8.i Nachdem Silvia zwei Semester allein gewohnt hatte, zog sie in eine WG. 9.f Nachdem Willi ein Geräusch gehört hatte, rief er den Großvater an.

Kapitel 10

Übung 1: 1. b. 2. a. 3. h. 4. g. 5. c. 6. d. 7. i. 8. j. 9. f. 10. e. **Übung 2:** 1. Nach Kopenhagen. 2. Zum Strand. 3. Zu ihrer Tante Sule. 4. Aus der Türkei. 5. Nein, aus dem Iran. 6. Aus dem Wasser. 7. Vom Markt. 8. Ja, bei uns. 9. Bei Fatimas Tante. 10. Nach Hause. **Übung 3:** (*Answers will vary.*) 1. Ja, üb jetzt Klavier. Du hast morgen Klavierstunde. (*oder* Nein, üb jetzt nicht Klavier. Wir gehen gleich aus.) 2. Ja, ruf ihn an. Er wollte mit dir sprechen. (*oder* Nein, ruf ihn nicht an. Du musst deine Hausaufgaben machen.) 3. Ja, iss sie mal. Du hast heute noch keine Süßigkeiten gegessen. (*oder* Nein, iss sie nicht. Wir essen gleich zu Abend.) 4. Ja, mach es auf. Die Luft ist hier schlecht. (*oder* Nein, mach es nicht auf. Es ist draußen zu kalt.) 5. Ja, gib mir einen Kuss. Ich fahre weg. (*oder* Nein, gib mir keinen Kuss. Du hast gerade Schokolade auf den Lippen.) 6. Ja, rede doch mal mit mir. Du hast wohl etwas zu erklären. (*oder* Nein, rede im Moment nicht mit mir. Ich bin beschäftigt.) 7. Ja, spül bitte das Geschirr. Ich bin nicht dazu gekommen. (*oder* Nein, spül das Geschirr nicht. Ich mache es nachher.) 8. Ja, geh mal in den Garten. Du brauchst die frische Luft. (*oder* Nein, geh nicht in den Garten. Es regnet.) 9. Ja, fahr mal morgen mit dem Fahrrad in die Schule. Ich kann dich mit dem Auto nicht hinbringen. (*oder* Nein, fahr morgen nicht mit dem Fahrrad in die Schule. Ich bringe dich mit dem Auto hin.) **Übung 4:** 1. Jens und Ernst, seid nicht so laut! 2. Maria, halte bitte an der nächsten Ampel! 3. Uli, bieg an der nächsten Straße nach links ab! 4. Jutta, iss mehr Obst! 5. Herr Pusch, fahren Sie nicht so schnell! 6. Jens, warte an der Ecke auf mich! 7. Sumita und Yamina, seid nicht ungeduldig! 8. Andrea und Paula, grüßt euren Vater von mir! 9. Hans, lauf mal schnell zu Papa! 10. Eske und Damla, lest jeden Tag die Zeitung! **Übung 5:** 1. Mach 2. Sprechen Sie 3. warten Sie 4. vergiss 5. Helft **Übung 6:** (*Answers may vary.*) 1. Fahren Sie den Fluss entlang. 2. Gehen Sie über die Brücke. 3. Gehen Sie an der Kirche vorbei. 4. Fahren Sie bis zum Bahnhof und dann links. 5. Die Tankstelle ist gegenüber von der Post. 6. Gehen Sie über die Schienen. 7. Ja, bis zur Bismarckstraße und dann rechts. 8. Nein, gehen Sie am Rathaus vorbei und dann links. 9. Das Hotel „Zum Patrizier" ist gegenüber vom Rathaus. 10. Fahren Sie 10 km die Straße entlang. **Übung 7:** (*Answers will vary.*) **Übung 8:** 1. Müsstest du nicht noch tanken? 2. Sollten wir nicht Jens abholen? 3. Könnten zwei Freunde von mir auch mitfahren? 4. Sollten wir nicht zuerst in die Stadt fahren? 5. Wolltest du nicht zur Bank? 6. Könntest du etwas langsamer fahren? 7. Dürfte ich das Autoradio anmachen? 8. Dürfte ich das Fenster aufmachen? **Übung 9:** 1. vor 50 000 Jahren → Wann wurde Australien von den Aborigines besiedelt? 2. um 2500 v. Chr. → Wann wurden die ersten Pyramiden gebaut? 3. 44 v. Chr. → Wann wurde Cäsar ermordet? 4. 800 n. Chr. → Wann wurde Karl der Große zum Kaiser gekrönt? 5. 1088 → Wann wurde die erste Universität (Bologna) gegründet? 6. 1789 → Wann wurde die amerikanische Verfassung unterschrieben? 7. 1885 → Wann wurde in Kanada die transkontinentale Eisenbahn vollendet? 8. 1945 → Wann wurden die Atombomben auf Hiroshima und Nagasaki geworfen? 9. 1963 → Wann wurde John F. Kennedy erschossen? 10. 1990 → Wann wurde Deutschland vereinigt? **Übung 10:** 1. Mäuse werden in vielen Labortests benutzt. 2. Meerschweinchen werden oft als Haustiere gehalten. 3. Bienen werden wegen ihrer Honigproduktion geschätzt. 4. Mücken werden durch Parfum und Kosmetikprodukte angelockt. 5. Die Fledermaus wird in vielen Kulturen mit Vampiren assoziiert. 6. Schnecken werden oft mit Butter- und Knoblauchsoße gegessen. 7. Der Gepard wird immer noch für seinen Pelz getötet. 8. Die meisten Papageien werden in der Wildnis gefangen. 9. Delfine werden wegen ihrer Intelligenz bewundert. 10. Viele Haie werden jedes Jahr gefischt.

Kapitel 11

Übung 1: 1.a. fühle mich b. mich erkältet c. dich ... legen 2.a. sich ... aufgeregt b. sich ... ausruhen 3.a. dich verletzt b. mich ... geschnitten 4.a. ärgerst ... dich b. dich freuen **Übung 2:** (*Answers will vary.*) Erst stehe ich auf. Dann dusche ich mich. Dann wasche ich mir das Gesicht. Dann wasche ich mir die Haare. Dann trockne ich mich ab. Dann putze ich mir die Fingernägel. Dann rasiere ich mich. Dann kämme ich mir die Haare. Dann ziehe ich mich an. Dann frühstücke ich. Dann putze ich mir die Zähne und gehe zur Uni. **Übung 3:** (*Answers will vary.*) 1. Ich rasiere mich jeden Morgen. 2. Meine Oma schminkt sich zu sehr. 3. Mein Freund wäscht sich nicht oft genug die Haare. 4. Mein Vater putzt sich nach jeder Mahlzeit die Zähne. 5. Mein Onkel zieht sich immer verrückt an. 6. Meine Schwester duscht sich jeden Tag. 7. Meine Freundin kämmt sich nie. 8. Mein Bruder föhnt sich nie die Haare. 9. Meine Kusine badet sich nicht gern. 10. Meine Mutter zieht sich immer elegant an. **Übung 4:** 1. Ja, kannst du es mir geben? / Nein, ich brauche es nicht. 2. Ja, kannst du ihn mir geben? / Nein, ich brauche ihn nicht. 3. Ja, kannst du ihn mir geben? / Nein, ich brauche ihn nicht. 4. Ja, kannst du sie mir geben? / Nein, ich brauche sie nicht. 5. Ja, kannst du es mir geben? / Nein, ich brauche es nicht. 6. Ja, kannst du ihn mir geben? / Nein, ich brauche ihn nicht. 7. Ja, kannst du sie mir geben? / Nein, ich brauche sie nicht. 8. Ja, kannst du es mir geben? / Nein, ich brauche es nicht. 9. Ja, kannst du ihn mir geben? / Nein, ich brauche ihn nicht. **Übung 5:** 1. Warum schneidest du ihn dir nicht? *or* Warum stutzt du ihn dir nicht? 2. Warum wäschst du sie dir nicht? 3. Warum schneidest du sie dir nicht? 4. Warum cremst du sie dir nicht ein? 5. Warum föhnst du sie dir nicht? 6. Warum wäschst du ihn dir nicht? 7. Warum putzt du sie dir nicht? 8. Warum lässt du sie dir nicht schneiden? *or* Warum schneidest du sie dir nicht? 9. Warum cremst du es dir nicht ein? 10. Warum wäschst du sie dir nicht? **Übung 6:** (*Some answers will vary.*) 1. Wissen Sie, was hier passiert ist? (*oder* Können Sie mir sagen, was hier passiert ist?) 2. Wissen Sie, ob das Kind das Auto gesehen hat? (*oder* Können Sie mir sagen, ob das Kind das Auto gesehen hat?) 3. Wissen Sie, wer daran schuld war? (*oder* Können Sie mir sagen, wer daran schuld war?) 4. Wissen Sie, warum Herr Langen das Kind nicht gesehen hat? (*oder* Können Sie mir sagen, warum Herr Langen das Kind nicht gesehen hat?) 5. Wissen Sie, ob Herr Langen gebremst hat? (*oder* Können Sie mir sagen, ob Herr Langen gebremst hat?) 6. Wissen Sie, wann er gebremst hat? (*oder* Können Sie mir sagen, wann er gebremst hat?) 7. Wissen Sie, wie oft Herr Langen diese Straße zur Arbeit fährt? (*oder* Können Sie mir sagen, wie oft Herr Langen diese Straße zur Arbeit fährt?) 8. Wissen Sie, wie lange Lothar auf der Straße lag? (*oder* Können Sie mir sagen, wie lange Lothar auf der Straße lag?) 9. Wissen Sie, wann die Polizei Lothars Mutter angerufen hat? (*oder* Können Sie mir sagen, wann die Polizei Lothars Mutter angerufen hat?) **Übung 7:** 1.a. ob b. dass c. Wenn 2.a. damit b. Weil **Übung 8:** 1.a. als b. nachdem 2. bevor 3. Während 4. obwohl

Kapitel 12

Übung 1: 1. meines 2. Ihrer 3. der 4. deiner 5. dieses 6. neuen 7. ersten 8. alten **Übung 2:** 1. Monika spricht über den Beruf ihrer Schwester. 2. Thomas spricht über die Wahl eines neuen Präsidenten. 3. Frau Schulz spricht über das Alter ihrer Nichten. 4. Stefan spricht über die Länge seines Studiums. 5. Albert spricht über die Sprache seiner Großeltern. 6. Nora spricht über die Kleidung ihres Freundes. 7. Thomas spricht über die Qualität des Leitungswassers in Berkeley. 8. Katrin spricht über die Situation der Regierung. **Übung 3:** 1. trotz 2. wegen 3. während 4. trotz 5. während 6. Wegen 7. statt 8. trotz **Übung 4:** (*Answers will vary.*) **Übung 5:** (*Answers will vary.*)

Übung 6: 1. Um morgens munter zu sein, muss man früh ins Bett gehen. 2. Um die Professoren kennenzulernen, muss man in die Sprechstunde gehen. 3. Um die Mitstudenten kennenzulernen, muss man viel Gruppenarbeit machen. 4. Um am Wochenende nicht allein zu sein, muss man Leute einladen. 5. Um die Kurse zu bekommen, die man will, muss man sich so früh wie möglich einschreiben. 6. Um in vier Jahren fertig zu werden, muss man viel lernen und wenig Feste feiern. 7. Um nicht zu verhungern, muss man regelmäßig essen. 8. Um eine gute Note in Deutsch zu bekommen, muss man jeden Tag zum Unterricht kommen. **Übung 7:** (*Answers may vary.*) 1. Ich möchte immer hier leben, weil dieses Land das beste Land der Welt ist. 2. Ich möchte für ein paar Jahre in Deutschland leben, um richtig gut Deutsch zu lernen. 3. Ausländer haben oft Probleme, weil sie die Sprache und Kultur des Gastlandes nicht verstehen. 4. Wenn ich Kinder habe, möchte ich hier leben, damit meine Kinder als (Amerikaner, Kanadier, Australier usw.) aufwachsen. 5. Viele Ausländer kommen hierher, weil man hier gut Geld verdienen kann. 6. Englisch sollte die einzige offizielle Sprache (der USA, Kanadas, Australiens usw.) sein, damit aus der multikulturellen Bevölkerung eine homogene Gemeinschaft wird. **Übung 8:** 1. Nom, F 2. Nom, V 3. Gen, F 4. Dat, F 5. Nom, F 6. Dat, V 7. Akk, P 8. Akk, P 9. Akk, F 10. Dat, P 11. Dat, V 12. Gen, P **Übung 9:** 1. em 2. en 3. e 4. er 5. e 6. er 7. e 8. en 9. ie 10. ie 11. en 12. e 13. en 14. en 15. er 16. em 17. en 18. em 19. er

Vokabeln

Deutsch-Englisch

Note to Students: The definitions in this vocabulary are based on the words as used in this text. For additional meanings, please refer to a dictionary.

Proper nouns are given only if the name is feminine or masculine or if the spelling is different from that in English. Compound words that do not appear in the chapter vocabulary lists have generally been omitted if they are easily analyzable and their constituent parts appear elsewhere in the vocabulary.

The letters or numbers in parentheses following the entries refer to the chapters in which the words occur in the chapter vocabulary lists.

Abbreviations

acc.	accusative	*gen.*	genitive	*p.p.*	past participle
adj.	adjective	*inf.*	infinitive	*prep.*	preposition
adv.	adverb	*infor.*	informal	*pron.*	pronoun
coll.	colloquial	*Interj.*	interjection	*rel. pron.*	relative pronoun
coord. conj.	coordinating conjunction	*masc.*	masculine	*sg.*	singular
dat.	dative	*n.*	noun	*s.o.*	someone
def. art.	definite article	*neut.*	neuter	*s.th.*	something
dem. pron.	demonstrative pronoun	*nom.*	nominative	*subord. conj.*	subordinating conjunction
fem.	feminine	*o.s.*	oneself	*v.*	verb
for.	formal	*pl.*	plural	*wk.*	weak masculine noun

ab (+ *dat.*) from; as of, effective

ab·bauen, abgebaut to reduce

ab·biegen (biegt ... ab), bog ... ab, ist abgebogen to turn (10)

das **Abbild, -er** likeness

ab·brennen (brennt ... ab), brannte ... ab, ist abgebrannt to burn down

der **Abend, -e** evening (1, 4); **am Abend** in the evening (4); **gestern Abend** last night (4); **guten Abend** good evening (A); **heute Abend** this evening (2); **morgen Abend** tomorrow evening; **zu Abend essen** to dine, have dinner (4)

das **Abendessen, -** dinner, supper, evening meal (1); **zum Abendessen** for dinner

das **Abendmahl, -e** dinner, supper, evening meal

abends evenings, in the evening (4)

aber (*coord. conj.*) but (A, 11)

ab·fahren (fährt ... ab), fuhr ... ab, ist abgefahren to leave, depart (4)

die **Abfahrt, -en** departure (7)

ab·geben (gibt ... ab), gab ... ab, abgegeben to hand over (to); to deliver (to)

der/die **Abgeordnete, -n (ein Abgeordneter)** representative (12)

ab·holen, abgeholt to pick (*s.o./s.th.*) up (from a place) (1)

das **Abhörgerät, -e** listening device, bug

das **Abi** (*coll.*) = das **Abitur** high school graduation exam (4)

ab·kauen, abgekaut to chew off

ab·kommen (kommt ... ab), kam ... ab, ist abgekommen: vom Weg abkommen to leave the path, go off course

die **Abkürzung, -en** abbreviation

ab·lehnen, abgelehnt to reject

die **Ablehnung, -en** rejection

ab·lenken, abgelenkt to divert; to change / get off the subject

ab·nehmen (nimmt ... ab), nahm ... ab, abgenommen to remove (11); to lose weight (8, 11); **Blut abnehmen** to take blood (11)

ab·räumen, abgeräumt to clear (3); to remove; **den Tisch abräumen** to clear the table (3)

ab·rechnen, abgerechnet to tally up; to settle an account

ab·reisen, ist abgereist to depart (10)

der **Absatz, ¨-e** paragraph

ab·saugen, abgesaugt to vacuum

ab·schaffen, abgeschafft to abolish, repeal

ab·schließen (schließt ... ab), schloss ... ab, abgeschlossen to lock (up)

abschließend in conclusion

der **Abschluss, ¨-e** completion; graduation (9)

die **Abschlusserklärung, -en** closing statement

ab·schneiden (schneidet ... ab), schnitt ... ab, abgeschnitten to cut off (8)

absolut absolute(ly)

die **Abstammung, -en** descent

abstrakt abstract(ly)

ab·stürzen, ist abgestürzt to crash (11)

der **Abt, ¨-e** / die **Äbtissin, -nen** abbot/abbess

ab·trocknen, abgetrocknet to dry (*dishes*) (6); **sich abtrocknen** to dry oneself off (11)

ab·waschen (wäscht ... ab), wusch ... ab, abgewaschen to wash (dishes)

ab·wischen, abgewischt to wipe clean (6)

ab·zahlen, abgezahlt to pay off (12)

ab·ziehen (zieht ... ab), zog ... ab, abgezogen to pull off; to withdraw (*troops*)

ach oh; **ach so** I see

die **Achse, -n** axis

acht eight (A)

acht- eighth (4)

achten (auf + *acc.*), **geachtet** to watch out (for); to pay attention (to) (11)

die **Achterbahn, -en** roller coaster

achtundzwanzig twenty-eight (A)

die **Achtung** attention (7)

achtzehn eighteen (A)

achtzig eighty (A)

adaptieren, adaptiert to adapt

das **Adjektiv, -e** adjective

der **Adler, -** eagle (10)

das **Adoptionsrecht, -e** right to adopt children

die **Adresse, -n** address (1)

der **Adventskalender, -** Advent calendar

das **Adverb, -ien** adverb

der **Affe, -n** (*wk.*) monkey; ape

(das) **Afrika** Africa (B)

afro-deutsch Afro-German (*adj.*) (12)

aggressiv aggressive(ly)

ägyptisch Egyptian (*adj.*)

ähnlich similar(ly)

die **Ahnung, -en** idea; suspicion

die **Akademie, -n** academy

der **Akademiker, -** / die **Akademikerin, -nen** academic (*person*)

akademisch academic(ally)

das **Akkordeon, -s** accordion

das **Akronym, -e** acronym

die **Akte, -n** (document) file

die **Aktie, -n** share, stock (12)

die **Aktion, -en** action

aktiv active(ly)

die **Aktivität, -en** activity

aktuell current(ly); present-day

akzeptieren, akzeptiert to accept

der **Albatros, -se** albatross

die **Albernheit, -en** foolishness

das **Album, Alben** album

der **Alkohol** alcohol

all all; **alle** (*pl.*) everybody; **alle zwei Jahre** every two years; **nichts von alledem** none of this; **vor allem** above all

die **Allee, -n** avenue

allein(e) alone; by oneself

alleinstehend single

allerdings however (6); of course

allergisch (gegen + *acc.***)** allergic (to) (11)

allerliebst- most favorite

alles everything (2); **alles Mögliche** everything possible (2)

allgemein general(ly)

die **Alliierten** (*pl.*) the Allies

der **Alltag, -e** daily routine (4)

alltäglich everyday, daily

die **Alltagssprache** everyday language

die **Alpen** (*pl.*) the Alps (7)

das **Alphabet, -e** alphabet (3)

als (*after comparative*) than; (*subord. conj.*) as; when (5, 11); **als ich acht Jahre alt war** when I was eight years old (5); **als ob** as if; as though; **als was?** as what? (5); **anders als** different from

also well; so; thus (4)

alt (älter, ältest-) old (A)

der **Altbau, -ten** *building built before the end of World War II*

das **Alter, -** age (1)

alternativ alternative(ly)

die **Alternative, -n** alternative (12)

der **Altgeselle, -n** (*wk.*) senior journeyman

die **Altstadt, ¨-e** old part of town

am = **an dem** at/on the

die **Ameise, -n** ant

(das) **Amerika** America, the USA (B)

der **Amerikaner, -** / die **Amerikanerin, -nen** American (*person*) (B)

amerikanisch American (*adj.*)

die **Ampel, -n** traffic light

der **Amtsrichter, -** / die **Amtsrichterin, -nen** local or district court judge

an (+ *acc./dat.*) at; on; to; in (2, 4); **am Abend** in the evening (4); **am Leben sein** to be alive (9); **am liebsten** (*like to do s.th.*) best (7); **am Samstag** on Saturday (2); **am Schalter** at the ticket booth (5); **am Telefon** on the phone (2); **am wenigsten** the least (8); **am Wochenende** over the weekend (1); **an der Tankstelle** at the gas station (5); **an** (+ *dat.*) **... vorbei** by (10); **an welchem Tag?** on what day? (4); **ans Meer** to the sea (2); **das Bild an die Wand hängen** to hang the picture on the wall (3)

analysieren, analysiert to analyze

analytisch analytical(ly)

die **Ananas, -** *or* **-se** pineapple

an·bauen, angebaut to grow, cultivate

an·bieten (bietet ... an), bot ... an, angeboten to offer

der **Anblick, -e** sight

an·braten (brät ... an), briet ... an, angebraten to brown, fry

an·bringen (bringt ... an), brachte ... an, angebracht to put up; to display

das **Andenken, -** souvenir

ander- other; different; **anders** different(ly); **(et)was anderes** something else; **jemand anderes** someone else; **unter anderem** among other things

(sich) **ändern, geändert** to change (9)

androgyn androgynous(ly)

die **Anerkennung, -en** recognition, acknowledgment; appreciation

der **Anfang, ¨-e** beginning

an·fangen (fängt ... an), fing ... an, angefangen to begin (4)

anfangs at first, initially

an·fassen, angefasst to touch

an·fertigen, angefertigt to make, prepare

an·führen, angeführt to lead

die **Angabe, -n** information; (*pl.*) particulars (1)

an·geben (gibt ... an), gab ... an, angegeben to state

angeblich alleged(ly)

das **Angebot, -e** offer; offering

an·gehören (+ *dat.*), **angehört** to belong to (*an organization*) (12)

die **Angeln** (*pl.*) Angles (*Germanic tribe*)

angelsächsisch Anglo-Saxon (*adj.*)

angenehm pleasant(ly) (6)

angespannt tense

der/die **Angestellte, -n (ein Angestellter)** employee; clerk (7)

angewandt applied

die **Angst, ¨-e** fear (3); **Angst einjagen** (+ *dat.*) to scare; **Angst haben (vor** + *dat.***)** to be afraid (of) (3)

sich (*dat.*) **an·gucken, angeguckt** (*coll.*) to look at

an·halten (hält ... an), hielt ... an, angehalten to stop (7)

sich (*acc.*) **an·hören, angehört** to sound; **das hört sich toll an** that sounds great (4)

sich (*dat.*) **an·hören, angehört** to listen to

an·ketten, angekettet to chain up

an·klagen, angeklagt to accuse

an·kommen (kommt ... an), kam ... an, ist angekommen to arrive (1)

an·kreuzen, angekreuzt to mark with an X

die **Ankunft, ¨-e** arrival (7)

an·legen, angelegt to put on; to aim

an·locken, angelockt to attract

an·machen, angemacht to turn on, switch on (3)

die **Anmeldung, -en** registration

an·nehmen (nimmt ... an), nahm ... an, angenommen to accept; to take; to adopt

die **Annonce, -n** advertisement

anonym anonymous(ly)

an·passen, angepasst to adapt, conform

an·pöbeln, angepöbelt (*coll.*) to abuse

an·reden, angeredet to speak to; to address

der **Anruf, -e** phone call

an·rufen (ruft ... an), rief ... an, angerufen to call up (*on the telephone*) (1)

ans = **an das** to/on the

ansässig resident (*adj.*)

(sich) (*dat.*) **an·schauen, angeschaut** to look at (2); to watch

sich **an·schließen** (+ *dat.*) **(schließt ... an), schloss ... an, angeschlossen** to join; to follow

anschließend subsequent(ly)

der **Anschluss, ¨-e** connection

an·schnallen, angeschnallt to strap in

(sich) (*dat.*) **an·sehen (sieht ... an), sah ... an, angesehen** to look at; to watch (3); to regard

an·sprechen (spricht ... an), sprach ... an, angesprochen to speak to (*s.o.*)

anständig respectable, respectably

an·starren, angestarrt to stare at

anstatt (+ *gen.*) instead of (12)

anstrengend strenuous; tiring

der **Anteil, -e** share

antiautoritär anti-authoritarian

das **Antibiotikum, Antibiotika** antibiotic (11)

antik antique; classical

die **Antike** classical antiquity

die **Antwort, -en** answer (A)

antworten (+ *dat.*), **geantwortet** to answer (*s.o.*) (4, 10); **auf eine Frage antworten** to answer a question

der **Anwalt, ⸚e** / die **Anwältin, -nen** lawyer (5)

an·werben (wirbt ... an), warb ... an, angeworben to recruit

die **Anwerbung, -en** recruitment

die **Anzahl** number

die **Anzeige, -n** ad (6)

an·ziehen (zieht ... an), zog ... an, angezogen to put on (*clothes*); to attract (3); **sich anziehen** to get dressed (11)

der **Anziehungspunkt, -e** attraction

der **Anzug, ⸚e** suit (A)

an·zünden, angezündet to light (3); to set on fire

der **Apfel, ⸚** apple

der **Apfelsaft** apple juice (8)

die **Apfelschorle, -n** mixture of apple juice and mineral water

die **Apfelsine, -n** orange (8)

die **Apotheke, -n** pharmacy (6)

der **Apotheker, -** / die **Apothekerin, -nen** pharmacist (11)

das **Appartement, -s** apartment

der **Applaus** applause

die **Aprikose, -n** apricot

der **April** April (B)

der **Araber, -** / die **Araberin, -nen** Arab

die **Arabeske, -n** arabesque

(das) **Arabisch** Arabic (*language*) (B)

die **Arbeit, -en** work (1); **sich an die Arbeit machen** to get down to work; **von der Arbeit** from work (3); **zur Arbeit gehen** to go to work (1)

arbeiten, gearbeitet to work (1); **Arbeiten Sie mit einem Partner.** Work with a partner. (A)

der **Arbeiter, -** / die **Arbeiterin, -nen** worker (5)

der **Arbeitersamariter, -** emergency aid worker

der **Arbeitnehmer, -** / die **Arbeitnehmerin, -nen** employee

das **Arbeitsbuch, ⸚er** workbook (3)

die **Arbeitserlaubnis, -se** work permit (12)

die **Arbeitskraft, ⸚e** labor; employee (12)

arbeitslos unemployed (12)

die **Arbeitslosigkeit** unemployment

die **Arbeitsteilung** division of labor

der **Archetypus, Archetypen** archetype

der **Architekt, -en** (*wk.*) / die **Architektin, -nen** architect (5)

die **Architektur, -en** architecture

der **Ärger** trouble (9); annoyance

ärgern, geärgert to tease, annoy (1, 3); **sich ärgern (über** + *acc.***)** to get angry (about) (11)

argwöhnisch suspicious(ly)

arm (ärmer, ärmst-) poor (9)

der **Arm, -e** arm (B); **jemanden auf den Arm nehmen** to tease someone; to pull someone's leg; **sich den Arm brechen** to break one's arm (11)

das **Armband, ⸚er** bracelet (2)

die **Armbanduhr, -en** (wrist)watch (A)

die **Armut** poverty

die **Arnika** arnica

die **Art, -en** kind, type (2)

der **Artikel, -** article

der **Arzt, ⸚e** / die **Ärztin, -nen** doctor; physician (3, 5, 11); **zum Arzt** to the doctor (3)

ärztlich medical(ly)

die **Arztpraxis, Arztpraxen** doctor's office (11)

die **Asche, -n** ash(es)

(das) **Aschenputtel** Cinderella

(das) **Asien** Asia (B)

der **Aspekt, -e** aspect

die **Asphaltschindel, -n** asphalt shingle

das **Aspirin** aspirin

der **Assessor, -en** / die **Assessorin, -nen** assistant judge

die **Assoziation, -en** association

assoziieren (mit + *dat.***), assoziiert** to associate (with)

der **Astrologe, -n** (*wk.*) / die **Astrologin, -nen** astrologer

die **Astrologie** astrology

das **Atelier, -s** studio

(das) **Athen** Athens

atmen, geatmet to breathe (11)

die **Atmosphäre, -n** atmosphere

atomar atomic

die **Atombombe, -n** atomic bomb

die **Attraktion, -en** attraction

attraktiv attractive(ly) (6)

au oh

auch also; too; as well (A); **auch wenn** (*subord. conj.*) even if

auf (+ *dat./acc.*) on; upon; on top of; onto; to; at; **auf dem Bahnhof** at the train station (5); **auf dem Land** in the country (*rural*) (6); **auf der Uni(versität)** at the university (1, 5); **auf Deutsch** in German; **auf eine Party gehen** to go to a party (1); **auf jeden Fall** by all means (4); **auf Reisen sein** to be on a trip (7); **auf Wiederhören** good-bye (*on the telephone*) (6); **auf Wiedersehen** good-bye (A)

auf·begehren, aufbegehrt to rebel

auf·brausen, ist aufgebraust to surge up

auf·brechen (bricht ... auf), brach ... auf, ist aufgebrochen to set out; to start off

der **Aufenthalt, -e** stay, sojourn (5)

die **Aufenthaltserlaubnis, -se** residence permit (12)

die **Aufenthaltsgenehmigung, -en** residence permit

der **Aufenthaltsraum, ⸚e** lounge, recreation room (10)

auf·fallen (fällt ... auf), fiel ... auf, ist aufgefallen to be noticeable (12)

die **Aufforderung, -en** request; instruction

die **Aufgabe, -n** assignment (4); task; homework; job

auf·geben (gibt ... auf), gab ... auf, aufgegeben to give up (1); to hand in; to assign

aufgrund (+ *gen.*) on the basis of

auf·hängen, aufgehängt to hang up (2)

auf·heben (hebt ... auf), hob ... auf, aufgehoben to pick up; to abolish, repeal

auf·hören (mit + *dat.***), aufgehört** to stop (*doing s.th.*) (1); to be over

die **Aufklärung** the Enlightenment

auf·laden (lädt ... auf), lud ... auf, aufgeladen to load; to charge

auf·leben, ist aufgelebt to revive

auf·machen, aufgemacht to open (3)

aufmerksam attentive(ly)

die **Aufmerksamkeit** attention, attentiveness

auf·muntern, aufgemuntert to cheer up

die **Aufnahmeprüfung, -en** entrance examination

auf·nehmen (nimmt ... auf), nahm ... auf, aufgenommen to pick up; to record; to take in; to take out (*a loan*) (12); **einen Kredit aufnehmen** to take out a loan

auf·passen (auf + *acc.***), aufgepasst** to pay attention (to) (3); to watch out (for)

auf·räumen, aufgeräumt to clean (up) (1); to tidy up

sich **auf·regen, aufgeregt** to get excited; to get upset (11)

aufregend exciting

die **Aufregung** excitement; agitation

sich **auf·richten, aufgerichtet** to stand up; to get back up

auf·rufen (zu + *inf.***) (ruft ... auf), rief ... auf, aufgerufen** to call on (*to do s.th*)

aufs = auf das on/onto/to the

auf·sagen, aufgesagt to recite

auf·saugen, aufgesaugt to vacuum

auf·schlagen (schlägt ... auf), schlug ... auf, aufgeschlagen to open

auf·schneiden (schneidet ... auf), schnitt ... auf, aufgeschnitten to chop; to cut open

der **Aufschnitt** cold cuts

auf·schreiben (schreibt ... auf), schrieb ... auf, aufgeschrieben to write down (11)

der **Aufschwung** upswing

auf·setzen, aufgesetzt to put on

auf·springen (springt ... auf), sprang ... auf, ist aufgesprungen to spring open

auf·stehen (steht ... auf), stand ... auf, ist aufgestanden to get up (1); to rise; to stand up; **mit dem linken Fuß aufstehen** to get up on the wrong side of bed (4); **stehen Sie auf** get up, stand up (A)

die **Aufstiegschance, -n** chance of promotion

der **Auftrag, ⸚e** instruction; task; order

der **Auftritt, -e** appearance

auf·wachen, ist aufgewacht to wake up (2, 4)

auf·wachsen (wächst ... auf), wuchs ... auf, ist aufgewachsen to grow up (12)

auf·wischen, aufgewischt to mop (up)

der **Aufzug, ⸚e** elevator (6)

das **Auge, -n** eye (B)

der **Augenarzt, ⸚e** / die **Augenärztin, -nen** eye doctor

der **Augenblick, -e** moment

die **Augenfarbe, -n** color of eyes (1)

der **August** August (B)

aus (+ *dat.*) from; of; out of (10); **aus Seide** of/from silk (2)

die **Ausbildung, -en** education; training

aus·blasen (bläst ... aus), blies ... aus, ausgeblasen to blow out

der **Ausblick, -e** view (6)

aus·breiten, ausgebreitet to spread out

der **Ausdruck, ⸚e** expression

aus·drücken, ausgedrückt to express

auseinander apart

aus·fallen (fällt ... aus), fiel ... aus, ist ausgefallen to fall out; to fail; to go out (*power*) (8)

das **Ausflugsziel, -e** destination of an excursion

aus·führen, ausgeführt to carry out, execute (12)

ausführlich thorough(ly) (5)

aus·füllen, ausgefüllt to fill out (1)

die **Ausgangslage, -n** starting position; initial situation

der **Ausgangspunkt, -e** starting point

aus·geben (gibt ... aus), gab ... aus, ausgegeben to spend (*money*) (12)

ausgebildet educated (12)

aus·gehen (geht ... aus), ging ... aus, ist ausgegangen to go out (1); to end, turn out (7); **es ist gut ausgegangen** it ended well (7)

ausgezeichnet excellent(ly) (3)

sich **aus·hängen, ausgehängt** to become unfastened; to get uncoupled

das **Aushängeschild, -er** advertising sign

aus·koppeln, ausgekoppelt to uncouple

das **Ausland** foreign countries (6); **im Ausland** abroad (6)

der **Ausländer, -** / die **Ausländerin, -nen** foreigner (12)

ausländisch foreign (12)

das **Auslandsamt, ⸚er** center for study abroad

aus·leeren, ausgeleert to empty (3)

aus·machen, ausgemacht to turn off (3)

aus·packen, ausgepackt to unpack

aus·probieren, ausprobiert to try out

aus·rauben, ausgeraubt to rob (completely)

ausreichend sufficient(ly)

die **Ausrichtung, -en** orientation; organization

der **Ausruf, -e** cry

sich **aus·ruhen, ausgeruht** to rest (11)

aus·rutschen, ist ausgerutscht to slip (11)

die **Aussage, -n** statement

aus·schlafen (schläft ... aus), schlief ... aus, ausgeschlafen to get enough sleep

aus·schließen (schließt ... aus), schloss ... aus, ausgeschlossen to exclude

der **Ausschnitt, -e** excerpt

aus·schreiben (schreibt ... aus), schrieb ... aus, ausgeschrieben to write out

aus·sehen (sieht ... aus), sah ... aus, ausgesehen to look (2); to appear; **Es sieht gut aus.** It looks good. (2)

das **Aussehen** appearance

der **Außenminister, -** / die **Außenministerin, -nen** foreign minister

die **Außenwelt** outside world

außer (+ *dat.*) besides, in addition to (9); except

außerdem besides (3, 10)

das **Äußere (ein Äußeres)** outward appearance

außergewöhnlich unusual(ly)

außerhalb (+ *gen.*) outside of

außerordentlich extraordinary, extraordinarily

aus·steigen (steigt ... aus), stieg ... aus, ist ausgestiegen to get out, get off

aus·stellen, ausgestellt to exhibit

die **Ausstellung, -en** exhibition

aus·strecken, ausgestreckt to stretch out

aus·suchen, ausgesucht to choose; to pick out

der **Austauschstudent, -en** (*wk.*) / die **Austauschstudentin, -nen** exchange student

aus·tragen (trägt ... aus), trug ... aus, ausgetragen to deliver (5); **Zeitungen austragen** to deliver newspapers (5)

(das) **Australien** Australia (B)

der **Australier, -** / die **Australierin, -nen** Australian (*person*) (B)

aus·treten aus (+ *dat.*) **(tritt ... aus), trat ... aus, ist ausgetreten** to leave, resign from

aus·üben, ausgeübt to practice (12)

ausverkauft sold out (5)

aus·wandern, ist ausgewandert to emigrate (12)

der **Ausweis, -e** identification card (10)

aus·ziehen (zieht ... aus), zog ... aus, ausgezogen to take off (*clothes*) (3); **sich ausziehen** to get undressed (11)

das **Auto, -s** car (A); **Auto fahren** to drive (*a car*)

die **Autobahn, -en** freeway (7)

der **Automat, -en** (*wk.*) vending machine

automatisch automatic(ally)

die **Autonummer, -n** license plate number

der **Autor, -en** / die **Autorin, -nen** author

das **Autoradio, -s** car radio (7)

das **Baby, -s** baby (7)

der **Bachelor, -s** bachelor's degree

backen (backt/bäckt), backte, gebacken to bake (5)

der **Backenbart, ⸚e** sideburns

der **Bäcker, -** / die **Bäckerin, -nen** baker

die **Bäckerei, -en** bakery (5); **in der Bäckerei** at the bakery (5)

der **Backofen, ⸚** oven (5)

das **Bad, ⸚er** bathroom; bath (6)

die **Badehose, -n** swim(ming) trunks (5)

der **Bademantel, ⸚** bathrobe (2)

der **Bademeister, -** / die **Bademeisterin, -nen** swimming-pool attendant (5)

baden, gebadet to bathe (3, 11); to swim; **sich baden** to bathe (*o.s.*) (11)

baden-württembergisch of Baden-Württemberg (*German state*)

die **Badewanne, -n** bathtub (6)

das **BAföG** = das **Bundesausbildungsförderungsgesetz** *financial aid for students from the German government*

das **Baguette, -s** baguette

die **Bahn, -en** path, way; railroad (7)

der/die **Bahnangestellte, -n (ein Bahnangestellter)** train agent; railway employee

die **Bahncard, -s** *discount card for rail travel in Germany*

die **Bahnfahrt, -en** train trip (7)

der **Bahnhof, ⸚e** train station (4, 5); **auf dem Bahnhof** at the train station (5)

die **Bahre, -n** stretcher

bald soon (9); **bald darauf** soon thereafter (9); **bis bald** so long; see you soon (A)

der **Balkon, -e** balcony (6)

der **Ball, ⸚e** ball (A, 1)

die **Ballerina, -s** ballerina (9)

der **Ballettunterricht** ballet class

die **Banane, -n** banana (8)

der **Band, ⸚e** volume

die **Band, -s** band, music group

der **Bandscheibenvorfall, ⸚e** slipped disc

die **Bank, ⸚e** bench

die **Bank, -en** bank (5); **auf der Bank** at the bank (5); **bei einer Bank** at a bank (6)

der/die **Bankangestellte, -n (ein Bankangestellter)** bank employee (5)
der **Bankeinzug, ¨-e** automatic withdrawal; electronic transfer of funds
bar (in) cash
die **Bar, -s** bar
der **Bär, -en** (*wk.*) bear
das **Bargeld** cash (12)
bargeldlos cash-free
die **Bar-Mizwa, -s** bar mitzvah (*coming-of-age ceremony for Jewish boys*)
das/der **Barock** (*n.*) baroque
der **Bart, ¨-e** beard (B)
der **Baseball, ¨-e** baseball
die **Baseballmannschaft, -en** baseball team
(das) **Basel** Basel
basieren (auf + *dat.***), basiert** to be based (on)
die **Basis, Basen** basis
das **Baskenland** Basque country
der **Basketball, ¨-e** basketball (2)
der **Bass, ¨-e** bass (guitar)
die **Bat-Mizwa, -s** bat mitzvah (*coming-of-age ceremony for Jewish girls*)
der **Bauarbeiter, -** / die **Bauarbeiterin, -nen** construction worker (5)
der **Bauch, ¨-e** belly, stomach (B)
bauen, gebaut to build
das **Bauernbrot, -e** (loaf of) farmer's bread
das **Bauernhaus, ¨-er** farmhouse (6)
das **Baujahr, -e** year of construction
der **Baum, ¨-e** tree (9)
bayerisch Bavarian (*adj.*)
(das) **Bayern** Bavaria
bayrisch Bavarian (*adj.*)
der **Beamer, -** data projector (B)
der **Beamte, -n (ein Beamter)** / die **Beamtin, -nen** civil servant; official
beantragen, beantragt to apply for (12)
beantworten, beantwortet to answer (7)
der **Becher, -** cup; mug (9); glass
bedeuten, bedeutet to mean
bedeutend important, significant(ly)
die **Bedeutung, -en** meaning (6); significance
der **Bedeutungswandel** semantic change
bedienen, bedient to serve
die **Bedienung** service; waiter, waitress (8)
das **Bedürfnis, -se** need
sich **beeilen, beeilt** to hurry (8)
beeindrucken, beeindruckt to impress
beenden, beendet to end
sich **befinden (befindet), befand, befunden** to be located; to be situated
befragen, befragt to interview; to interrogate
befreien, befreit to set free
die **Befreiung** liberation
befreundet (mit) (*adj.*) friends (with)
befriedigend satisfactory, satisfactorily
begabt gifted (9)
die **Begebenheit, -en** event, occurrence
begegnen (+ *dat.*), **ist begegnet** to meet (6); to encounter
die **Begegnung, -en** meeting, encounter
begeistert (*p.p. of* **begeistern**) thrilled; enthusiastic
der **Beginn** beginning
beginnen (beginnt), begann, begonnen to begin, start (1)
begleiten, begleitet to accompany
begraben (begräbt), begrub, begraben to bury
begreifen (begreift), begriff, begriffen to understand
begrenzen, begrenzt to limit, restrict
der **Begriff, -e** concept
begründen, begründet to found
der **Begründer, -** / die **Begründerin, -nen** founder
begrüßen, begrüßt to greet
die **Begrüßung, -en** greeting
behalten (behält), behielt, behalten to keep, retain
behandeln, behandelt to handle, treat, deal with
die **Behandlung, -en** treatment
behaupten, behauptet to maintain, assert
beherrschen, beherrscht to have a command of
behindert (*p.p. of* **behindern**) handicapped
die **Behörde, -n** public authority
bei (+ *dat.*) with; at; near (2, 6, 10); during; upon; among; **bei deinen Eltern** with your parents, at your parents' (6); **bei einer Bank** at a bank (6); **bei Monika** at Monika's (2); **bei Regen** in rainy weather (7)
beide both
beieinander together
beim = **bei dem** at/with/near the
das **Bein, -e** leg (B)
beisammen together
das **Beispiel, -e** example (3); **zum Beispiel (z. B.)** for example (3)
beißen (beißt), biss, gebissen to bite (9)
der **Beitrag, ¨-e** contribution
bei·tragen (+ *dat.*) **(trägt ... bei), trug ... bei, beigetragen** to contribute
bei·treten (+ *dat.*) **(tritt ... bei), trat ... bei, ist beigetreten** to join
bekämpfen, bekämpft to fight against
bekannt well-known
der/die **Bekannte, -n (ein Bekannter)** acquaintance
bekennend admitted, avowed
bekommen (bekommt), bekam, bekommen to get; to receive (3)
belagern, belagert to besiege
belasten, belastet to load; to debit; **belastendes Material** incriminating evidence
beleben, belebt to liven up
belegen, belegt to cover; to take (*a course*) (4); **das belegte Brot** open-faced sandwich (8)
belehren, belehrt to teach, instruct
(das) **Belgien** Belgium (B)
belgisch Belgian (*adj.*)
beliebt popular (3)
bemerken, bemerkt to notice
der/die **Benachteiligte, -n (ein Benachteiligter)** disadvantaged person
benennen (nach + *dat.***) (benennt), benannte, benannt** to name (after)
benutzen, benutzt to use (7)
das **Benzin** gasoline (6)
der **Benzinverbrauch** gasoline consumption
berechnen (+ *dat.*), **berechnet** to charge (8)
der **Bereich, -e** sector, area (12)
bereichern, bereichert to enrich (12)
bereit ready; prepared
bereits already; just
der **Berg, -e** mountain (1); **in den Bergen wandern** to hike in the mountains (1); **in die Berge gehen** to go to the mountains (1)
bergauf uphill
die **Berghütte, -n** mountain cabin
der **Bergsteiger, -** / die **Bergsteigerin, -nen** mountaineer
der **Bericht, -e** report
berichten, berichtet to report
Berliner (*adj.*) (of) Berlin; die **Berliner Mauer** the Berlin Wall; die **Berliner Weiße** *light, fizzy beer served with raspberry syrup*
der **Berliner, -** / die **Berlinerin, -nen** person from Berlin
(das) **Bern** Bern(e)
der **Beruf, -e** profession; career (1, 5); **Was sind Sie von Beruf?** What's your profession? (1)
der **Berufsberater, -** / die **Berufsberaterin, -nen** career counselor (5)
die **Berufsberatung, -en** job counseling
das **Berufsleben** career, professional life
berufstätig working; employed
das **Berufsverbot, -e** *prohibition from practicing a particular profession*
beruhen (auf + *dat.***), beruht** to be based (on)
berühmt famous (7)
berühren, berührt to touch
die **Besatzungszone, -n** occupation zone
beschädigen, beschädigt to damage
beschaffen, beschafft to get; to obtain
sich **beschäftigen (mit** + *dat.***), beschäftigt** to occupy oneself (with); **beschäftigt** busy
die **Beschäftigung, -en** activity
der **Bescheid, -e** information; **Bescheid wissen** to know; to have an idea
beschließen (beschließt), beschloss, beschlossen to resolve; to decide
beschreiben (beschreibt), beschrieb, beschrieben to describe (11); **den Weg beschreiben** to give directions
die **Beschreibung, -en** description (B)

sich **beschweren (bei** + *dat.*), **beschwert** to complain (to) (8)

beschwören (beschwört), beschwor, beschworen to swear

der **Besen, -** broom (6)

besetzen, besetzt to occupy (12); **besetzt** occupied, taken

besichtigen, besichtigt to visit, sightsee (7)

besiedeln, besiedelt to settle

besiegen, besiegt to conquer

der **Besitz** possessions (2)

besitzen (besitzt), besaß, besessen to possess

besonder- special, particular

besonders particularly (3)

(sich) (*dat.*) **besorgen, besorgt** to get (*o.s.*); to buy (*o.s.*)

bespitzeln, bespitzelt to spy on

besser better (2)

(sich) **bessern, gebessert** to improve

best- best

der **Bestandteil, -e** part, component

das **Besteck** silverware, cutlery (5)

bestehen (besteht), bestand, bestanden to exist; to last; to pass (*a test*); (**aus** + *dat.*) to consist (of)

besteigen (besteigt), bestieg, bestiegen to climb (7)

bestellen, bestellt to order (*food*) (8)

bestimmen, bestimmt to determine

bestimmt definite(ly); certain(ly) (3)

bestreuen, bestreut to sprinkle (8)

der **Besuch, -e** visit (3); **zu Besuch kommen** to visit (3)

besuchen, besucht to visit (1); to attend (*school*)

der **Besucher, -** / die **Besucherin, -nen** visitor

sich **betätigen, betätigt** to occupy oneself

sich **beteiligen (an** + *dat.*), **beteiligt** to participate (in)

beten, gebetet to pray

der **Beton** concrete

betonen, betont to emphasize

betrachten, betrachtet to look at

der **Betrag, ¨-e** amount (*of money*)

betragen (beträgt), betrug, betragen to amount to

betreffen (betrifft), betraf, betroffen to concern, deal with (6); to affect; **betroffen** upset; affected

betreten (betritt), betrat, betreten to enter

betreuen, betreut to take care of; to look after

der **Betrieb, -e** workplace, operation (7)

betrunken drunk(en), drunkenly

das **Bett, -en** bed (1, 6); **ins Bett gehen** to go to bed (1)

sich **beugen, gebeugt** to bend down

die **Beule, -n** bump, bulge

die **Bevölkerung, -en** population (12)

bevor (*subord. conj.*) before (11)

die **Bewachung** watch(ing); observation

die **Bewaffnung, -en** arming

bewahren, bewahrt to protect, preserve

sich **bewegen, bewegt** to move

die **Bewegung, -en** movement

sich **bewerben (um** + *acc.*) **(bewirbt), bewarb, beworben** to apply (for)

bewerten, bewertet to rate

bewirken, bewirkt to cause; to bring about

die **Bewirtung, -en** service

der **Bewohner, -** / die **Bewohnerin, -nen** occupant; inhabitant

bewundern, bewundert to admire

bewusstlos unconscious(ly)

bezahlen, bezahlt to pay (for) (4)

bezeichnen (als), bezeichnet to describe (as)

die **Beziehung, -en** relationship

beziehungsweise or; and . . . respectively

bezwingen (bezwingt), bezwang, bezwungen to defeat

die **Bibel, -n** Bible

die **Bibliothek, -en** library (4)

der **Bibliothekar, -e** / die **Bibliothekarin, -nen** librarian (5)

die **Biene, -n** bee (10)

das **Bier, -e** beer (2)

bieten (bietet), bot, geboten to offer

der **Bikini, -s** bikini (5)

bi-kulturell bicultural(ly)

das **Bild, -er** picture (2); image; **das Bild an die Wand hängen** to hang the picture on the wall (3); **Was zeigen Ihre Bilder?** What do your pictures show? (A)

bilden, gebildet to form; **die bildenden Künste** (*pl.*) the plastic arts

der **Bilderrahmen, -** picture frame

der **Bildhauer, -** / die **Bildhauerin, -nen** sculptor (12)

das **Bildnis, -se** portrait

billig cheap(ly), inexpensive(ly) (2)

binden (an + *acc.*) **(bindet), band, gebunden** to tie (to)

das **Bioei, -er** organic egg

die **Biografie, -n** biography

die **Biologie** biology (1)

der **Biomarkt, ¨-e** organic produce market

das **Bioprodukt, -e** organic product

die **Birne, -n** pear (8)

bis (*prep. + acc.; subord. conj.*) until (2, 4, 11); **bis acht Uhr** until eight o'clock (2); **bis bald** so long; see you soon (A); **bis um vier Uhr (früh)** until four o'clock (in the morning) (4); **bis zu** as far as; up to (10)

bisher thus far; up to now

bisschen: ein bisschen a little (bit) (3); **kein bisschen** not at all (3)

der **Bissen, -** mouthful

bitte please (A); **Bitte schön?** Yes, please? May I help you? (7); **Bitte schön/sehr.** There you go. **Unterschreib bitte hier.** Sign here, please. (A)

bitten (um + *acc.*) **(bittet), bat, gebeten** to ask (for) (12)

blass pale(ly)

das **Blatt, ¨-er** leaf; sheet (*of paper*)

blau blue (A, B); **der Blaue Reiter** *a group of artists in Munich (1911–14)*

die **Blaubeere, -n** blueberry

der **Blauwal, -e** blue whale

das **Blechdach, ¨-er** tin roof

bleiben (bleibt), blieb, ist geblieben to stay, remain; **sitzen bleiben (bleibt ... sitzen), blieb ... sitzen, ist sitzen geblieben** to remain seated; to be held back a grade; **stecken bleiben (bleibt ... stecken), blieb ... stecken, ist stecken geblieben** to get stuck (11)

bleichen, gebleicht to bleach

der **Bleistift, -e** pencil (A, B)

der **Blick, -e** look; glance; view

blind blind(ly)

der **Blinddarm, ¨-e** appendix (11)

die **Blockflöte, -n** recorder (*type of flute*) (12)

blöd stupid(ly)

blond blond(e) (B)

bloß mere(ly); only; just

blühen, geblüht to bloom

die **Blume, -n** flower (3); **die Blumen gießen** to water the flowers (3)

der **Blumenkohl** cauliflower (8)

die **Bluse, -n** blouse (A)

das **Blut** blood (9, 11); **Blut abnehmen** to take blood (11)

der **Blutdruck** blood pressure; **niedrigen/hohen Blutdruck haben** to have low/high blood pressure

die **Blüte, -n** bloom

bluten, geblutet to bleed (11)

das **Blütenblatt, ¨-er** petal

der **Boden, ¨-** floor (B)

die **Bohne, -n** bean (8)

das **Bonbon, -s** drop, lozenge (11)

das **Boot, -e** boat (2)

die **Börse, -n** stock exchange (12)

der **Börsenkrach, ¨-e** stock market crash (12)

böse evil, mean (9)

(das) **Bosnien** Bosnia

der **Boss, -e** boss

die **Boutique, -n** boutique (6)

bowlen, gebowlt to bowl

die **Bowlingbahn, -en** bowling alley

die **Box, -en** stereo speaker

boxen, geboxt to box (1)

die **Branche, -n** sector (12)

die **Brandenburgischen Konzerte** (*pl.*) the Brandenburg Concertos

braten (brät), briet, gebraten to grill, fry (2, 8)

der **Braten, -** roast (8)

die **Bratpfanne, -n** frying pan

die **Bratwurst, ¨-e** (fried) sausage

der **Brauch, ¨-e** custom

brauchen, gebraucht to need; to use (1)

brauen, gebraut to brew

braun brown (A)

bräunen, gebräunt to brown, fry (8)

(das) **Braunschweig** Braunschweig, Brunswick

brausen, gebraust to roar; to rage

die **Braut, ¨-e** bride (9)

die **BRD** = die **Bundesrepublik Deutschland** Federal Republic of Germany

brechen (bricht), brach, gebrochen to break (11); **sich den Arm brechen** to break one's arm (11)

der **Brei, -e** mush, purée

breit broad, wide

die **Bremse, -n** brake (7)

bremsen, gebremst to brake (11)

das **Bremsenquietschen** squealing of brakes

brennen (brennt), brannte, gebrannt to burn (11)

der **Brennofen, ¨** kiln (12)

(das) **Breslau** Wrocław (*city in Poland*)

das **Brett, -er** board; **das schwarze Brett** bulletin board

die **Brezel, -n** pretzel

der **Brief, -e** letter, epistle (1)

die **Briefmarke, -n** (postage) stamp (5)

die **Brille, -n** (eye)glasses (A)

bringen (bringt), brachte, gebracht to bring (2)

britisch British

der **Brocken** *highest mountain in the Harz range*

das **Brot, -e** (loaf of) bread (8); **das belegte Brot** open-face sandwich (8)

das **Brötchen, -** (bread) roll (8)

das **Brotsortiment, -e** assortment of breads

die **Brücke, -n** bridge (10)

der **Bruder, ¨** brother (B)

die **Bruderschaft, -en** fraternity

brüllen, gebrüllt to roar

der **Brunnen, -** well; fountain (9)

brutal brutal(ly)

das **Bruttogehalt, ¨-er** gross salary

der **Bub, -en** (*wk.*) boy

das **Buch, ¨-er** book (A, B, 2)

der **Bucheinband, ¨-e** book cover

buchen, gebucht to book (7)

der **Bücherwurm, ¨-er** bookworm

der **Buchstabe, -n** letter (*of the alphabet*)

die **Bucht, -en** bay (7)

das **Bügeleisen, -** iron (6)

bügeln, gebügelt to iron (6)

die **Bulette, -n** rissole, meatball, hamburger patty

(das) **Bulgarien** Bulgaria (B)

das **Bundesausbildungsförderungsgesetz (BAföG)** *financial aid for students from the German government*

der **Bundeskanzler, -** / die **Bundeskanzlerin, -nen** (federal) chancellor

das **Bundesland, ¨-er** (German or Austrian) state

der **Bundespräsident, -en** (*wk.*) / die **Bundespräsidentin, -nen** (federal) president

die **Bundesrepublik** federal republic; **die Bundesrepublik Deutschland** Federal Republic of Germany

die **Bundeswehr** (German) armed forces

das **Bündnis, -se** alliance

bunt colorful(ly)

das **Burgenland** *Austrian state*

der **Bürger, -** / die **Bürgerin, -nen** citizen (10)

die **Bürgerinitiative, -n** citizens' action group

bürgerlich bourgeois, middle-class

der **Bürgerrechtskämpfer, -** / die **Bürgerrechtskämpferin, -nen** campaigner for civil rights

das **Bürgertum** bourgeoisie, middle class

das **Büro, -s** office (5); **im Büro** at the office (5)

die **Bürste, -n** brush

der **Bus, -se** bus (2, 7)

der **Busch, ¨-e** bush (9)

die **Bushaltestelle, -n** bus stop (10)

die **Butter** butter (8)

ca. = **circa/zirka** circa

das **Café, -s** café; **im Café** at the café (4)

die **Cafeteria, -s** cafeteria

das **Camping** camping (10)

der **Campingplatz, ¨-e** campsite (10)

der **Cartoon, -s** cartoon

(der) **Cäsar** Caesar

das **Casino, -s** casino

die **CD, -s** CD, compact disc (A, 3)

der **CD-Spieler, -** CD player (2)

die **CDU** = die **Christlich-Demokratische Union** Christian Democratic Party

Celsius Celsius, centigrade (B); **18 Grad Celsius** 18 degrees Celsius (B)

der **Cent, -** cent (*one hundredth of a euro*)

der **Champagner, -** champagne

das **Chaos** chaos (5)

der **Charakter, -e** character; personality (12)

die **Charakterisierung, -en** characterization

der **Chat, -s** (online) chat

chatten, gechattet to chat (*online*) (1)

checken, gecheckt to check

der **Chef, -s** / die **Chefin, -nen** boss; director

die **Chemie** chemistry (1)

der **Chemiker, -** / die **Chemikerin, -nen** chemist

der **Chili, -s** chili

chillen, gechillt to relax, hang (1)

(das) **China** China (B)

der **Chinese, -n** (*wk.*) / die **Chinesin, -nen** Chinese (*person*)

chinesisch Chinese (*adj.*)

(das) **Chinesisch** Chinese (*language*) (B)

die **Chipkarte, -n** chip card, smart card

cholerisch irascible, irascibly

der **Chor, ¨-e** choir; chorus

der **Christ, -en** (*wk.*) / die **Christin, -nen** Christian (*person*)

der **Christkindlmarkt, ¨-e** Christmas market

christlich Christian (*adj.*)

die **Christlich-Demokratische Union (CDU)** Christian Democratic Party

(der) **Christus** Christ; **Christi Geburt** the birth of Christ

circa = **zirka** circa

der **Clip, -s** (video) clip

der **Clown, -s** clown (9)

die **Cola, -s** cola

die **Collagetechnik, -en** collage technique

das **College, -s** college

der **Comic, -s** comic strip; comic book

der **Computer, -** computer (2)

cool cool(ly); fabulous(ly)

die **Creme, -s** cream

das **Croissant, -s** croissant

ct = der **Cent, -** cent (*one hundredth of a euro*)

da (*adv.*) there (2); then; (*subord. conj.*) as, since

dabei in that connection; while doing so; (along) with it; **dabei sein** to be present; **Ist ein/eine ... dabei?** Does it come with a . . . ? (6)

dabei·haben (hat ... dabei), hatte ... dabei, dabeigehabt to have (*s.th.*) with/on (*s.o.*)

da·bleiben (bleibt ... da), blieb ... da, ist dageblieben to stay, remain (there)

das **Dach, ¨-er** roof (6)

der **Dachboden, ¨** attic

der **Dada(ismus)** Dada(ism)

dadurch through it/them

dafür for it/them; for that reason; on behalf of it

dagegen against it/them; **Haben Sie etwas dagegen?** Do you have something for it (*illness*)? (11)

daheim at home (9)

daher therefore; **das kommt daher ...** the reason for that is . . .

dahin (to) there

damalig (*adj.*) back then, at that time

damals (*adv.*) back then, at that time (9)

(das) **Damaskus** Damascus

die **Damenmalschule, -n** painting school for ladies

damit (*adv.*) with it/them; (*subord. conj.*) so that (11)

danach after it/them; afterward (8, 10)
(das) **Dänemark** Denmark (B)
dänisch Danish (*adj.*)
(das) **Dänisch** Danish (*language*)
der **Dank** thanks; **vielen Dank** many thanks (10)
die **Dankbarkeit** gratitude
danke thank you (A)
danken (+ *dat.*), **gedankt** to thank
dann then (A)
daran at/on/to it/them
darauf after/for/on it/them; afterward, then; **bald darauf** soon thereafter (9)
darauffolgend following, subsequent
daraufhin following that, thereupon
darin in it/them (6)
dar·stellen, dargestellt to represent, depict
darüber over/above/about it/them
das (*def. art., neut. nom./acc.*) the; (*dem. pron., neut. nom./acc.*) this/that; (*rel. pron., neut. nom./acc.*) which, who(m); **Das ist ...** This/That is . . . (B); **Das ist es ja!** That's just it! (4); **Das macht nichts.** That doesn't matter. (1); **Das sind ...** These/Those are . . . (B)
dass (*subord. conj.*) that (11)
die **Daten** (*pl.*) data; **persönliche Daten** biographical information (1)
die **Datenverarbeitung** data processing
das **Datum, Daten** date (4); **Welches Datum ist heute?** What is today's date? (4)
die **Dauer, -** duration
der **Dauerauftrag, ¨-e** standing order
dauern, gedauert to last (4)
die **Dauerwelle, -n** perm
der **Daumen, -** thumb
davon of/from/about it/them
davon·fliegen (fliegt ... davon), flog ... davon, ist davongeflogen to fly away
davon·laufen (läuft ... davon), lief ... davon, ist davongelaufen to run away
dazu to it/them; in addition (8)
dazu·kommen (kommt ... dazu), kam ... dazu, ist dazugekommen to turn up, arrive
dazu·schreiben (schreibt ... dazu), schrieb ... dazu, dazugeschrieben to add in writing
die **DDR** = die **Deutsche Demokratische Republik** German Democratic Republic (*former East Germany*)
der **Deckcode-Name, -n** (*wk.*) code name
die **Decke, -n** ceiling (B); blanket (11)
der **Deckel, -** lid; top
decken, gedeckt to cover; set (5); **den Tisch decken** to set the table (5)
die **Definition, -en** definition
deftig good and solid
dein(e) (*infor. sg.*) your (B, 2)
die **Deklination, -en** declension
der **Delfin, -e** dolphin (10)
dem (*def. art., masc./neut. dat.*) the; (*dem. pron., masc./neut. dat.*) this/that; (*rel. pron., masc./neut. dat.*) which, whom
die **Demo, -s** = die **Demonstration, -en** (*coll.*) demonstration; rally
die **Demokratie, -n** democracy (12)
demokratisch democratic(ally) (12)
der **Demonstrant, -en** (*wk.*) / die **Demonstrantin, -nen** demonstrator
die **Demonstration, -en** demonstration; rally
demonstrieren, demonstriert to demonstrate
den (*def. art., masc. acc., pl. dat.*) the; (*dem. pron., masc. acc.*) this/that; (*rel. pron., masc. acc.*) which, whom
denen (*dem. pron., pl. dat.*) these/those; (*rel. pron., pl. dat.*) which, whom
denken (denkt), dachte, gedacht to think (4); **denken an** (+ *acc.*) to think of (4); **denken über** (+ *acc.*) to think about
denn (*coord. conj.*) for, because (9, 11); *particle used in questions*: **Wo willst du denn hin?** Where are you going? (A)
dennoch nevertheless
die **Deportation, -en** deportation
deportieren, deportiert to deport
depressiv depressive
deprimiert depressed (11)
der (*def. art., masc. nom., fem. dat./gen., pl. gen.*) the; (*dem. pron., masc. nom., fem. dat.*) this/that; (*rel. pron., masc. nom., fem. dat.*) which, who(m)
deren (*dem. pron., fem. gen., pl. gen.*) of this/that/these/those; (*rel. pron., fem. gen., pl. gen.*) of which, whose
derselbe, dasselbe, dieselbe(n) the same
des (*def. art., masc./neut. gen.*) (of) the
deshalb therefore; that's why (4)
desinfizieren, desinfiziert to disinfect (11)
dessen (*dem. pron., masc./neut. gen.*) of this/that; (*rel. pron., masc./neut. gen.*) of which, whose
deutlich clear(ly); distinct(ly)
deutsch German (*adj.*)
(das) **Deutsch** German (*language*) (B); **auf Deutsch** in German
der/die **Deutsche, -n (ein Deutscher)** German (*person*) (B); **Ich bin Deutsche/r.** I am German. (B)
die **Deutsche Demokratische Republik (DDR)** German Democratic Republic (*former East Germany*)
die **Deutschkenntnisse** (*pl.*) knowledge of German (*language*)
der **Deutschkurs, -e** German (*language*) course; German class (A)
(das) **Deutschland** Germany (B); **die Bundesrepublik Deutschland** Federal Republic of Germany
deutschlandweit throughout Germany
deutschschweizerisch German-Swiss (*adj.*)
deutschsprachig German-speaking
deutschstämmig of German origin
der **Dezember** December (B)
der **Dialekt, -e** dialect
der **Dialog, -e** dialogue
die **Diät, -en** diet
dich (*infor. sg. acc.*) you (2)
der **Dichter, -** / die **Dichterin, -nen** poet
dick large; fat (2)
dickköpfig headstrong
die (*def. art., fem. nom./acc., pl. nom./acc.*) the; (*dem. pron., fem. nom./acc., pl. nom./acc.*) this/that/these/those; (*rel. pron., fem. nom./acc., pl. nom./acc.*) which, who(m)
dienen (als), gedient to serve (as)
der **Diener, -** / die **Dienerin, -nen** servant
der **Dienst, -e** service; work
der **Dienstag, -e** Tuesday (1)
der **Dienstwagen, -** company car
dieser, dies(es), diese this, these; that, those (2, 4)
diesmal this time
die **Digitalkamera, -s** digital camera
diktatorisch dictatorial(ly)
die **Diktatur, -en** dictatorship
das **Ding, -e** thing (2)
die **Diphtherie** diphtheria (11)
das **Diplom, -e** degree; diploma
dir (*infor. sg. dat.*) you
direkt direct(ly); right
der **Direktor, -en** / die **Direktorin, -nen** (school) principal; director (9)
der **Dirigent, -en** (*wk.*) / die **Dirigentin, -nen** (orchestra) conductor (5)
die **Disko, -s** disco (3)
die **Diskussion, -en** discussion
diskutieren, diskutiert to discuss (4)
die **Dissertation, -en** dissertation
die **Disziplin, -en** discipline
die **DM** = die **D-Mark (Deutsche Mark)** German mark (*former monetary unit*)
doch however; nevertheless; yet; **doch!** yes (on the contrary)! (4)
der **Doktor, -en** / die **Doktorin, -nen** doctor
der **Doktorand, -en** (*wk.*) / die **Doktorandin, -nen** doctoral student
dokumentieren, dokumentiert to document
doll (*coll.*) neat, great
der **Dollar, -** dollar
der **Dom, -e** cathedral (10)
dominant dominant(ly)
dominieren, dominiert to dominate
die **Donau** Danube (River)
der **Donnerstag, -e** Thursday (1)
doppelt double
das **Doppelzimmer, -** double room (5, 10)

das **Dorf, ¨-er** village
der **Dorn, -en** thorn (9)
(das) **Dornröschen** Sleeping Beauty, Briar Rose
dort there (7)
dorthin there, thither, to a specific place
die **Dose, -n** can (8)
der **Dosenöffner, -** can opener (8)
der **Dozent, -en** (*wk.*) / die **Dozentin, -nen** lecturer
Dr. = **Doktor** (*as a title*)
der **Drache, -n** (*wk.*) dragon (9)
das **Drama, Dramen** drama
dramatisch dramatic(ally)
der **Dramaturg, -en** (*wk.*) / die **Dramaturgin, -nen** dramaturge, theatrical adviser
dran = **daran** at/on/to it/them; **Du bist dran.** (*coll.*) It's your turn.
der **Drang** drive, urge; quest
drauf = **darauf** after/for/on it/them
draußen outside (11)
drei three (A)
dreigliedrig divided into three parts
dreimal three times (3)
dreißig thirty (A)
dreißigst- thirtieth
dreiundzwanzig twenty-three (A)
dreizehn thirteen (A)
dreizehnt- thirteenth (4)
dressieren, dressiert to train
drin = **darin** in it/them (6)
dringend urgent(ly) (2)
dritt- third (4); **das Dritte Reich** the Third Reich (Nazi Germany)
das **Drittel, -** third
droben up there
die **Droge, -n** drug
die **Drogerie, -n** drugstore (6)
drucken, gedruckt to print
drücken, gedrückt to press
der **Drudenfuß, ¨-e** pentagram
drum = **darum** therefore
der **Dschungel, -** jungle
du (*infor. sg. nom.*) you
dumm (dümmer, dümmst-) dumb, stupid(ly) (6)
dümmlich simple-minded(ly)
dunkel dark (5)
dünn thin
durch (+ *acc.*) through (7); by means of
durchbluten, durchblutet to supply with blood
durcheinander in confusion
durch·geben (gibt ... durch), gab ... durch, durchgegeben to announce
durch·greifen (greift ... durch), griff ... durch, durchgegriffen to take action
durch·kreuzen, durchgekreuzt to thwart
durch·rennen (rennt ... durch), rannte ... durch, ist durchgerannt to run through
durchs = **durch das** through the
durch·schneiden (schneidet ... durch), schnitt ... durch, durchgeschnitten to cut through
der **Durchschnitt** average; **im Durchschnitt** on average
durchschnittlich (on) average
(sich) **durch·setzen, durchgesetzt** to assert (o.s.)
dürfen (darf), durfte, gedurft to be permitted (to), may (3); **nicht dürfen** must not
dürr withered; scrawny
der **Durst** thirst (3); **Durst haben** to be thirsty (3)
die **Dusche, -n** shower (5)
(sich) **duschen, geduscht** to (take a) shower (1, 11)
die **DVD, -s** DVD
der **DVD-Spieler, -** DVD player (2, 3)
die **Dynamik** dynamic(s)
eben simply, just; just now
ebenfalls also, likewise
das **Ebenholz** ebony
ebenso likewise; just as
das **E-Book, -s** e-book
echt real(ly) (2)
die **EC-Karte, -n** = die **Eurocheque-Karte, -n** Eurocheque card (*debit card*)
die **Ecke, -n** corner (5); **(gleich) um die Ecke** (right) around the corner (5, 6)
eckig angular
die **EDV** = die **elektronische Datenverarbeitung** electronic data processing
egal equal(ly), same (6); **Das ist mir egal.** It doesn't matter to me. (6)
der **Egoist, -en** (*wk.*) / die **Egoistin, -nen** egoist
die **Ehe, -n** marriage (12)
die **Eheleute** (*pl.*) married couple
ehemalig former
eher rather (12); more
der **Ehering, -e** wedding ring
der **Ehevertrag, ¨-e** prenuptial agreement
die **Ehre, -n** honor (8)
ehren, geehrt to honor
ehrgeizig ambitious(ly)
ei (*interj.*) oh
das **Ei, -er** egg (8); **gebratene Eier** (*pl.*) fried eggs (8); **gekochte Eier** (*pl.*) boiled eggs (8)
der **Eichbaum, ¨-e** oak tree
der/das **Eidotter** egg yolk
die **Eifersucht** jealousy
eifersüchtig jealous(ly) (3)
eigen own (3, 6)
eigenmächtig on one's own authority, unauthorized
die **Eigenschaft, -en** trait, characteristic
eigensinnig stubborn(ly)
eigentlich actual(ly) (3)
eigenverantwortlich autonomous(ly)
sich **eignen, geeignet** to be suitable
die **Eile** hurry (3); **in Eile sein** to be in a hurry (3)
eilen, geeilt to hurry
eilig rushed (8); **es eilig haben** to be in a hurry (8)
ein, eine a(n); one (A); **ein bisschen** a little (bit) (3); some; **ein paar** a few (2)
einander one another, each other (3)
die **Einbahnstraße, -n** one-way street (7)
ein·biegen (biegt ... ein), bog ... ein, ist eingebogen to turn
ein·brechen (in + *acc.*) **(bricht ... ein), brach ... ein, ist eingebrochen** to break in(to); to break through; **ins Eis einbrechen** to go through the ice
der **Einbrecher, -** / die **Einbrecherin, -nen** burglar (9)
der **Einbruch, ¨-e** burglary; break-in
(sich) **ein·cremen, eingecremt** to put lotion on (*o.s.*) (11)
der **Eindruck, ¨-e** impression (5)
einer, eine, ein(e)s one (*pron.*)
einfach simple, simply (2); one-way (*trip*) (10)
die **Einfachheit** simplicity
die **Einfahrt, -en** driveway
ein·fallen (+ *dat.*) **(fällt ... ein), fiel ... ein, ist eingefallen** to occur (*to s.o.*)
das **Einfamilienhaus, ¨-er** single-family home (6)
ein·führen, eingeführt to introduce (12)
die **Einführung, -en** introduction (A)
der **Eingang, ¨-e** entrance
ein·geben (gibt ... ein), gab ... ein, eingegeben to give
ein·gehen (geht ... ein), ging ... ein, ist eingegangen to arrive; **darauf eingehen** to get into something
sich **ein·gewöhnen (in** + *acc.*), **eingewöhnt** to get accustomed (to)
ein·gravieren, eingraviert to engrave
die **Einheit** unity
einige some; several; a few
ein·jagen, eingejagt: jemandem Angst einjagen to scare someone
der **Einkauf, ¨-e** purchase
ein·kaufen, eingekauft to shop (for) (1); **einkaufen gehen** to go shopping (1, 5)
das **Einkommen, -** income (12)
ein·laden (lädt ... ein), lud ... ein, eingeladen to invite (2)
die **Einladung, -en** invitation (2)
sich **ein·lassen mit** (+ *dat.*) **(lässt ... ein), ließ ... ein, eingelassen** to get involved with
ein·laufen (läuft ... ein), lief ... ein, ist eingelaufen to shrink
einmal once (4); for once; **Es war einmal ...** Once upon a time there was . . . ; **noch einmal** one more time; **Warst du schon einmal ...?** Were you ever . . . ? (4)

sich **ein·mischen, eingemischt** to interfere
ein·packen, eingepackt to pack up (1)
eins one (*cardinal number*) (A)
ein·sammeln, eingesammelt to gather, collect
der **Einsatz, ⸚e** deployment
ein·schalten, eingeschaltet to turn on (10)
ein·schätzen, eingeschätzt to judge, assess
ein·schlafen (schläft ... ein), schlief ... ein, ist eingeschlafen to fall asleep (7)
sich **ein·schreiben (schreibt ... ein), schrieb ... ein, eingeschrieben** to register, enroll
ein·sehen (sieht ... ein), sah ... ein, eingesehen to see, realize
ein·setzen, eingesetzt to use, bring into action
die **Einsicht, -en** view, look
ein·sperren, eingesperrt to lock up
ein·steigen (steigt ... ein), stieg ... ein, ist eingestiegen to board (10); to get in/on
ein·stellen, eingestellt to hire; to employ
die **Einstellung, -en** attitude (12)
(sich) **ein·teilen, eingeteilt** to divide up; to organize
eintönig monotonous(ly)
ein·tragen (trägt ... ein), trug ... ein, eingetragen to enter (*into a list or ledger*)
ein·treffen (trifft ... ein), traf ... ein, ist eingetroffen to arrive
ein·treten für (+ *acc.***) (tritt ... ein), trat ... ein, ist eingetreten** to champion, stand up for (12)
die **Eintrittskarte, -n** admissions ticket
einundzwanzig twenty-one (A)
der **Einwanderer, -** / die **Einwanderin, -nen** immigrant (12)
ein·wandern, ist eingewandert to immigrate (12)
die **Einwanderung** immigration
ein·weihen, eingeweiht to open, dedicate
ein·werfen (wirft ... ein), warf ... ein, eingeworfen to break, smash (*a window*)
der **Einwohner, -** / die **Einwohnerin, -nen** inhabitant, resident
ein·zahlen, eingezahlt to pay in; to deposit
der **Einzelgänger, -** / die **Einzelgängerin, -nen** loner
das **Einzelhandelsgeschäft, -e** retail shop, retail store
der **Einzelplatz, ⸚e** single seat
das **Einzelzimmer, -** single room (10)
ein·ziehen (zieht ... ein), zog ... ein, hat eingezogen to collect; to withdraw
ein·ziehen (in + *acc.***) (zieht ... ein), zog ... ein, ist eingezogen** to move in(to)
einzig only; single; sole
das **Eis** ice; ice cream (2); **ins Eis einbrechen** to go through the ice
der **Eisbeutel, -** ice pack
das **Eiscafé, -s** ice cream parlor (8)
die **Eisenbahn, -en** railroad
das **Eisenwalzwerk, -e** iron mill, steel press
eiskalt ice-cold
der **Eistee** iced tea
der **Elan** zest, vigor
der **Elefant, -en** (*wk.*) elephant (9)
elegant elegant(ly) (8)
elektrisch electric(ally) (8)
der **Elektro** electro, electronic dance music
elektronisch electronic(ally); **die elektronische Datenverarbeitung (EDV)** electronic data processing
der **Elektrotechniker, -** / die **Elektrotechnikerin, -nen** electrician; electronics technician
das **Element, -e** element
elf eleven (A)
das **Elfenbein** ivory (10)
elft- eleventh (4)
die **Eltern** (*pl.*) parents (B)
der **Elternteil, -e** parent
die **E-Mail, -s** e-mail (1, 2)
die **Emanzipation, -en** emancipation
der **Emmentaler** Emmenthaler (cheese)
die **Emotion, -en** emotion
empfangen (empfängt), empfing, empfangen to receive
der **Empfänger, -** / die **Empfängerin, -nen** recipient; payee
empören, empört to outrage; **sich empören** to become outraged
das **Ende, -n** end
enden, geendet to end
endgültig final; conclusive(ly)
endlich finally (9)
das **Endspiel, -e** final (game)
die **Endung, -en** ending
eng tight, narrow, small (12); closely
das **Engagement, -s** commitment, involvement
sich **engagieren (für** + *acc.***), engagiert** to commit oneself (to); **engagiert** (*adj.*) committed, involved
(das) **England** England (B)
der **Engländer, -** / die **Engländerin, -nen** English (*person*) (B)
englisch English (*adj.*)
(das) **Englisch** English (*language*) (B)
der **Enkel, -** / die **Enkelin, -nen** grandson/granddaughter (5)
entartet (*adj.*) degenerate
entdecken, entdeckt to discover (4)
sich **entfernen (von** + *dat.***), entfernt** to go away (from); **entfernt** (*adj.*) distant, away
sich **entfremden (von** + *dat.***), entfremdet** to become estranged (from)
entführen, entführt to kidnap
der **Entführer, -** / die **Entführerin, -nen** kidnapper
entgegnen, entgegnet to reply
entlang along (10)
entlang·fahren (fährt ... entlang), fuhr ... entlang, ist entlanggefahren to drive along
entlang·gehen (geht ... entlang), ging ... entlang, ist entlanggegangen to go along (10)
entlassen (entlässt), entließ, entlassen to release
die **Entlassung, -en** release, discharge
(sich) **entscheiden (entscheidet), entschied, entschieden** to decide (4)
entscheidend decisive(ly) (12)
entschlossen (*p.p. of* **entschließen**) resolute(ly)
entschuldigen, entschuldigt to excuse (10); **Entschuldigen Sie!** Excuse me! (10)
die **Entschuldigung, -en** excuse; **Entschuldigung!** Excuse me! (3)
entsetzlich terrible, terribly
entsetzt (*p.p. of* **entsetzen**) horrified
entspannen, entspannt to relax
entstammen (+ *dat.*), **ist entstammt** to come from
entstehen (aus + *dat.***) (entsteht), entstand, ist entstanden** to originate (from)
entweder ... oder either . . . or
entwerfen (entwirft), entwarf, entworfen to design
(sich) **entwickeln, entwickelt** to develop
die **Entwicklungsland, ⸚er** developing country
die **Entwicklungsstufe, -n** stage of development
die **Entzündung, -en** infection (11); inflammation
entzündungshemmend anti-inflammatory
die **Epoche, -n** epoch, era, period
er (*pron., masc. nom.*) he, it
erarbeiten, erarbeitet to work on/out
erblicken, erblickt to see
die **Erbse, -n** pea (8)
die **Erdbeere, -n** strawberry (8)
die **Erde, -n** earth; ground; soil, dirt
die **Erdgeschichte** history of the earth
das **Erdgeschoss, -e** first floor, ground floor
die **Erdkunde** earth science; geography
das **Ereignis, -se** event
erfahren (erfährt), erfuhr, erfahren to find out, learn; to experience; to discover
die **Erfahrung, -en** experience
erfinden (erfindet), erfand, erfunden to invent (4)
der **Erfinder, -** / die **Erfinderin, -nen** inventor
die **Erfindung, -en** invention
der **Erfolg, -e** success; **Erfolg haben** to be successful
erfolgreich successful(ly)
erfüllen, erfüllt to fulfill
die **Erfüllung: in Erfüllung gehen** to become true
ergänzen, ergänzt to complete, fill in the blanks (4)
sich **ergeben (aus** + *dat.***) (ergibt), ergab, ergeben** to arise (from)

das **Ergebnis, -se** result

ergehen (+ *dat.*) **(ergeht), erging, ist ergangen** to go (well or badly) (*for a person*)

erhalten (erhält), erhielt, erhalten to receive (5, 12)

erheben (erhebt), erhob, erhoben to raise

erhitzen, erhitzt to heat (8)

erhöhen, erhöht to raise, increase

sich **erholen, erholt** to recuperate (11)

erinnern (an + *acc.***), erinnert** to remind (*of s.o./s.th.*); to commemorate (*s.o./s.th.*); **sich erinnern (an** + *acc.***), erinnert** to remember (*s.o./s.th.*) (9)

die **Erinnerung, -en** memory, remembrance (4)

sich **erkälten, erkältet** to catch a cold (11)

die **Erkältung, -en** (head) cold (11)

erkennen (an + *dat.***) (erkennt), erkannte, erkannt** to recognize (by)

erklären, erklärt to explain (5)

die **Erklärung, -en** explanation; declaration

erklettern, erklettert to climb

sich **erkundigen (nach** + *dat.***), erkundigt** to ask (about), get information (about) (10)

erlauben, erlaubt to permit (7)

die **Erlaubnis, -se** permission

erleben, erlebt to experience (3)

das **Erlebnis, -se** experience (7)

erledigen, erledigt to take care of; to handle; to settle

erleiden (erleidet), erlitt, erlitten to suffer

erlösen, erlöst to rescue, free (9)

ermitteln, ermittelt to investigate

ermöglichen, ermöglicht to make possible

ermorden, ermordet to murder

die **Ernährung** nutrition

ernst serious(ly); **ernst nehmen** to take seriously

erobern, erobert to conquer

eröffnen, eröffnet to open (9); **ein Konto eröffnen** to open a bank account (5)

erreichbar reachable

erreichen, erreicht to reach (5, 12); to achieve

erscheinen (erscheint), erschien, ist erschienen to appear

das **Erscheinungsjahr, -e** year of publication

erschießen (erschießt), erschoss, erschossen to shoot dead

erschrocken frightened

erst (*adv.*) first; not until (4); **erst mal** for now; **erst um vier Uhr** not until four o'clock (4)

erst- first (*ordinal number*) (4); **am ersten Oktober** on the first of October (4); **der erste Oktober** the first of October (4); **erster Klasse fahren** to travel first class (10); **im ersten Stock** on the second floor (6); **zum ersten Mal** for the first time (4)

das **Erstaunen** astonishment

erstellen, erstellt to draw up

ersticken, ist erstickt to suffocate

erstmal *old spelling of* **erst mal** for now

das **Erststudium, Erststudien** undergraduate study

erteilen, erteilt to give

ertragen (erträgt), ertrug, ertragen to tolerate

ertränken, ertränkt to drown (*s.o./s.th.*)

erwachen, ist erwacht to wake up

erwachsen grown-up

erwarten, erwartet to expect (12)

die **Erwartung, -en** expectation

erwischen, erwischt to catch

erzählen, erzählt to tell (*a story, joke*) (3, 5); **Witze erzählen** to tell jokes (3)

das **Erzherzogtum, ¨-er** archduchy

erziehen (erzieht), erzog, erzogen to raise, bring up

der/die **Erziehungsberechtigte, -n (ein Erziehungsberechtigter)** parent or legal guardian

es (*pron., neut. nom./acc.*) it (B); **Gibt es ...?** Is there . . . ? / Are there . . . ? (A)

der **Esel, -** donkey

der/das **Essay, -s** essay

essen (isst), aß, gegessen to eat (2, 4); **zu Abend essen** to dine, have dinner (4)

das **Essen** food

der **Essig** vinegar (8)

die **Essiggurke, -n** pickle

das **Esszimmer, -** dining room (6)

(das) **Estland** Estonia

etablieren, etabliert to establish

etwa approximately

etwas something (2, 4, 5); anything (5); somewhat; a little; **etwas anderes** something else; **etwas Interessantes/Neues** something interesting/new (4); **Haben Sie etwas dagegen?** Do you have something for it (*illness*)? (11); **Sonst noch etwas?** Anything else? (5)

die **EU** = die **Europäische Union** European Union

euch (*infor. pl. pron., acc./dat.*) you; yourselves

euer, eu(e)re (*infor. pl.*) your

die **Eule, -n** owl

der **Euro, -** euro (*European monetary unit*) (7)

die **Eurocard** *European credit card*

(das) **Europa** Europe (B)

der **Europäer, -** / die **Europäerin, -nen** European (*person*)

europäisch European (*adj.*)

die **Europäische Union (EU)** European Union

die **Euroscheckkarte, -n** Eurocheque Card (*debit card*)

der **Euroschein, -e** *banknote in euros*; **der Zwanzigeuroschein, -e** twenty-euro note

die **Eurozone** *countries of the European Union in which the euro is the unit of currency*

evozieren, evoziert to evoke

ewig eternal(ly)

exakt exact(ly)

existieren, existiert to exist

exotisch exotic(ally) (7)

die **Explosion, -en** explosion

der **Exportartikel, -** export article

der **Expressionismus** expressionism

extra extra (10)

extrem extreme(ly)

der **Extremismus** extremism

der **Extremist, -en** (*wk.*) / die **Extremistin, -nen** extremist

extrovertiert extroverted

die **Fabel, -n** fable

die **Fabrik, -en** factory (5); **in der Fabrik** in the factory (5)

das **Fach, ¨-er** academic subject (1)

die **Fachhochschule, -n** university of applied arts and sciences

der **Fachleistungskurs, -e** extension course

die **Fähigkeit, -en** ability, capability

fahren (fährt), fuhr, ist/hat gefahren to drive; to ride (2); **Auto fahren** to drive a car; **erster Klasse fahren** to travel first class (10); **ins Schwimmbad fahren** to drive/go to the swimming pool (1); **Kanu fahren** to go canoeing (10); **Rad fahren** to ride a bicycle (7); **Ski fahren** to ski (3)

Fahrenheit Fahrenheit (B)

der **Fahrer, -** / die **Fahrerin, -nen** driver (7)

der **Fahrgast, ¨-e** passenger

die **Fahrkarte, -n** ticket (4)

der **Fahrkartenschalter, -** ticket window (7)

der **Fahrplan, ¨-e** timetable, schedule

das **Fahrrad, ¨-er** bicycle (2, 7); **Fahrrad fahren** to ride a bicycle

der **Fahrradhelm, -e** bicycle helmet (5)

der **Fahrstuhl, ¨-e** elevator

die **Fahrt, -en** trip (7, 10)

das **Fahrzeug, -e** vehicle (7, 11)

der **Faktor, -en** factor

der **Fall, ¨-e** fall, collapse; case; **auf jeden Fall** by all means (4)

fallen (fällt), fiel, ist gefallen to fall (9); **in Ohnmacht fallen** to faint (11); **schwer fallen** (+ *dat.*) to seem/feel difficult (*to s.o.*)

falls (*subord. conj.*) if; in case

falsch wrong(ly) (2); false(ly)

fälschen, gefälscht to fake

die **Familie, -n** family (B)

das **Familienfest, -e** family celebration

der **Familienname, -n** (*wk.*) family name, surname (A, 1)

familienversichert covered by family health insurance

der **Fan, -s** fan, enthusiast

der **Fanatiker, -** / die **Fanatikerin, -nen** fanatic (12)

fangen (fängt), fing, gefangen to catch

die **Fantasie, -n** imagination

fantastisch fantastic(ally)

die **Farbe, -n** color (A, 1); **Welche Farbe hat ...?** What color is . . . ? (A)

die **Farbensymbolik** color symbolism

der **Farbfilm** color film

fassen, gefasst to grab, grasp

fassungslos stunned; bewildered

fast almost (5)

fasten, gefastet to fast

das **Fast Food** fast food

faul lazy, lazily

die **Fauna** fauna; animal life

der **Februar** February (B)

die **Fee, -n** fairy (9)

fegen, gefegt to sweep (5)

fehlen (+ *dat.*), **gefehlt** to lack; to be missing (6); to be wrong with, be the matter with (*a person*) (11)

die **Feier, -n** celebration, party (9)

feierlich ceremonial, with ceremony

feiern, gefeiert to celebrate (4, 5)

der **Feiertag, -e** holiday (4)

fein fine(ly) (8)

der **Feind, -e** / die **Feindin, -nen** enemy

das **Feld, -er** field (7)

das **Fenster, -** window (B); **unter dem Fenster** under the window (5)

die **Fensterbank, ¨-e** windowsill (10)

die **Fensterscheibe, -n** windowpane (9)

die **Ferien** (*pl.*) vacation (1)

das **Ferienhaus, ¨-er** vacation house (4)

die **Fernreise, -n** long-distance trip

fern·sehen (sieht ... fern), sah ... fern, ferngesehen to watch TV (1)

das **Fernsehen** television

der **Fernseher, -** TV set (2)

die **Fernsehsendung, -en** TV show

das **Fernsehzimmer, -** TV room (10)

fertig ready; finished (3)

fertig·stellen, fertiggestellt to complete

fest steady; fixed

das **Fest, -e** celebration (4)

fest·nehmen (nimmt ... fest), nahm ... fest, festgenommen to arrest

fest·stehen (steht ... fest), stand ... fest, festgestanden to stand fast

fest·stellen, festgestellt to establish (8); to detect; to realize

die **Fete, -n** (*coll.*) party

fett fat; bold; **fett gedruckt** in bold print, boldface; **fette Jahre** good times, years of plenty

das **Fett, -e** fat

fettig fat(ty), greasy (8, 11)

feucht humid (B)

das **Feuer, -** fire (9)

die **Feuerwehr** fire department (11)

das **Fieber** fever (11)

fies nasty, nastily

die **Figur, -en** figure (12); character

der **Film, -e** film (2)

die **Finanzen** (*pl.*) finances

der **Finanzmarkt, ¨-e** financial market

finanziell financial(ly)

finden (findet), fand, gefunden to find (2); **Wie findest du das?** How do you like that?

der **Finger, -** finger (11)

der **Fingernagel, ¨** fingernail (11)

(das) **Finnland** Finland (B)

die **Firma, Firmen** company, firm (3)

der **Fisch, -e** fish (8)

fischen, gefischt to fish

das **Fischfilet, -s** fish fillet

flach flat

der **Flachbildschirm, -e** flat-screen (monitor) (2)

die **Fläche, -n** surface (7); area

das **Fladenbrot, -e** unleavened bread

die **Flasche, -n** bottle (5)

der **Flaschenöffner, -** bottle opener (8)

die **Fledermaus, ¨-e** bat

das **Fleisch** meat (8)

das **Fleischchüechli** (*Swiss*) rissole, meatball, hamburger patty

der **Fleiß** diligence

fleißig industrious(ly) (12); diligent(ly)

die **Flexion, -en** (grammatical) inflection

die **Fliege, -n** fly (8)

fliegen (fliegt), flog, ist/hat geflogen to fly (1)

fliehen (flieht), floh, ist geflohen to flee

fließen (fließt), floss, ist geflossen to flow (7)

flippig (*coll.*) funky; stylish

der **Flohmarkt, ¨-e** flea market (2)

die **Flora** flora; plant life

der **Fluch, ¨-e** curse

fluchen, geflucht to curse, swear (11)

die **Flucht** flight; **auf der Flucht** on the run

flüchten, ist geflüchtet to flee (11)

der **Flug, ¨-e** flight (9)

das **Flugblatt, ¨-er** pamphlet

der **Flughafen, ¨** airport (6)

das **Flugzeug, -e** airplane (7)

der **Fluss, ¨-e** river (7)

flüssig (*adj.*) liquid, fluid

flüstern, geflüstert to whisper

der **Föhn, -e** blow-dryer, hair-dryer

föhnen, geföhnt to blow dry (2); **die Haare föhnen** to blow dry hair (2); **sich (die Haare) föhnen** to blow dry (one's hair) (11)

die **Folge, -n** consequence, result; sequence

folgen (+ *dat.*), **ist gefolgt** to follow

folgend following

die **Folklore** folk music

foltern, gefoltert to torture

fördern, gefördert to promote, support

die **Forderung, -en** demand

die **Förderung** promotion, support

die **Forelle, -n** trout (8)

die **Form, -en** form

formal formal(ly)

die **Formalität, -en** formality (12)

das **Formular, -e** form (12)

fort away

fort·rennen (rennt ... fort), rannte ... fort, ist fortgerannt to run away

der **Fortschritt, -e** progress

(sich) **fort·setzen, fortgesetzt** to continue

die **Fortsetzung, -en** continuation

das **Forum, Foren/Fora** forum

das **Foto, -s** photo (1); **Fotos machen** to take photos

die **Fotografie** photography

fotografieren, fotografiert to take pictures (4)

die **Fotomontage** photomontage

die **Frage, -n** question (A); **eine Frage stellen** to ask a question (5)

fragen, gefragt to ask; (**nach** + *dat.*) to inquire (about); **nach dem Weg fragen** to ask for directions

der **Franken, -** (**der Schweizer Franken)** (Swiss) franc (8)

die **Frankfurter, -** frankfurter (sausage)

(das) **Frankreich** France (B)

der **Franzose, -n** (*wk.*) / die **Französin, -nen** French (*person*) (B)

französisch French (*adj.*)

(das) **Französisch** French (*language*) (B)

die **Frau, -en** woman; Mrs., Ms.; wife (A, B)

die **Frauensache, -n** woman's job, woman's concern

frech impudent(ly)

frei free(ly); empty, available (8); **in freier Natur** out in the open (country) (10); **Ist hier noch frei?** Is this seat available? (8)

der **Freigang** work-release day pass

frei·haben (hat ... frei), hatte ... frei, freigehabt to have free; to have time off

das **Freihandelsabkommen, -** free trade agreement

die **Freiheit, -en** freedom (12)

freiheitlich liberal(ly)

frei·lassen (lässt ... frei), ließ ... frei, freigelassen to set free

das **Freilichtmuseum, Freilichtmuseen** open-air museum

freischaffend freelance

der **Freitag, -e** Friday (1)

freitags on Friday(s)

das **Freiwild** fair game

freiwillig voluntary; optional; voluntarily, willingly

die **Freizeit** leisure time (1)

fremd foreign

der/die **Fremde, -n (ein Fremder)** foreigner

der **Fremdenhass** xenophobia

die **Fremdsprache, -n** foreign language (1)

fressen (frisst), fraß, gefressen to eat (*said of an animal*) (9)

die **Freude, -n** joy; pleasure (9); **vor Freude** for/with joy

sich **freuen, gefreut (über** + *acc.*) to be happy (about) (11); **(auf** + *acc.*) to look forward (to)

der **Freund, -e** / die **Freundin, -nen** friend; boyfriend/girlfriend (A)

freundlich friendly (B)

die **Freundschaft, -en** friendship

der **Frieden, -** peace (12)

der **Friedhof, ⁻̈e** cemetery

friedlich peaceful(ly)

frieren (friert), fror, gefroren to freeze

frisch fresh(ly) (8)

der **Friseur, -e** / die **Friseurin, -nen** hairdresser (5)

die **Frisur, -en** hairstyle

froh happy; cheerful

fröhlich happy; cheerful(ly)

der **Frosch, ⁻̈e** frog (9)

„Der Froschkönig" "The Frog Prince" (*fairy tale*)

die **Frucht, ⁻̈e** fruit

früh early (1); in the morning (4); **bis um vier Uhr früh** until four in the morning (4); **früher** former(ly); **morgen früh** tomorrow morning

der **Frühjahrsputz** spring cleaning (6)

der **Frühling, -e** spring (B); **im Frühling** in the spring (B)

das **Frühstück, -e** breakfast (2); **zum Frühstück** for breakfast

frühstücken, gefrühstückt to eat breakfast (1)

das **Frühstückszimmer, -** breakfast room/nook

der **Fuchs, ⁻̈e** fox

(sich) **fühlen, gefühlt** to feel (3, 11); to touch; **Ich fühle mich ...** I feel . . . (3); **sich wohl fühlen** to feel well (11); **Wie fühlst du dich?** How do you feel? (3)

führen, geführt to lead; **den Haushalt führen** (+ *dat.*) to keep house (*for s.o.*); **Krieg führen** to wage war

der **Führerschein, -e** driver's license (4)

die **Führungsposition, -en** leadership position

füllen, gefüllt to fill

das **Fundament, -e** foundation, basis

fünf five (A)

die **Fünf: eine Fünf** poor (*school grade*)

fünft- fifth (4)

fünfundzwanzig twenty-five (A)

fünfzehn fifteen (A)

fünfzehnt- fifteenth

fünfzig fifty (A)

der **Funk** radio

die **Funktion, -en** function

funktionieren, funktioniert to work, function

für (+ *acc.*) for (2); in favor of; **was für** what kind of (3)

furchtbar terrible, terribly (4)

sich **fürchten (vor** + *dat.*), **gefürchtet** to be afraid (of) (10)

fürs = **für das** for the

fusionieren, fusioniert to merge

der **Fuß, ⁻̈e** foot (B); **mit dem linken Fuß aufstehen** to get up on the wrong side of bed (4); **zu Fuß** on foot (3)

der **Fußball, ⁻̈e** soccer ball; soccer (A, 1)

der **Fußballplatz, ⁻̈e** soccer field

der **Fußgänger, -** pedestrian (7)

der **Fußgängerweg, -e** sidewalk

die **Fußgängerzone, -n** pedestrian mall (10)

das **Futter** feed; fodder

füttern, gefüttert to feed (9)

die **Gabel, -n** fork (8)

die **Galle, -n** gall

der **Gang, ⁻̈e** gear (7)

die **Gang, -s** gang

das **Gänseliesel** *famous fountain in Göttingen*

ganz whole (2); complete(ly); quite (2); rather; **den ganzen Tag** all day long (1); **die ganze Nacht** all night long (3); **ganz gut** quite good; **ganz schön viel** quite a bit (3)

gar: gar kein(e) no . . . at all; **gar nicht** not at all, not a bit (3); **gar nichts** nothing at all

die **Garage, -n** garage (6)

der **Garten, ⁻̈** garden (6); yard; **im Garten** in the garden (4)

der **Gartenschlauch, ⁻̈e** garden hose (6)

die **Gasse, -n** narrow street; alley (10)

der **Gast, ⁻̈e** guest; patron, customer

der **Gastarbeiter, -** / die **Gastarbeiterin, -nen** foreign worker

das **Gästehaus, ⁻̈er** bed and breakfast (inn) (10)

die **Gastfamilie, -n** host family

die **Gastfreundschaft** hospitality

das **Gastland, ⁻̈er** host country

die **Gaststätte, -n** restaurant (5); **in der Gaststätte** at the restaurant (5)

der **Gaul, ⁻̈e** horse

das **Gebäude, -** building (10)

geben (gibt), gab, gegeben to give (6); **(in** + *acc.*) to put (into) (8); **eine Party geben** to throw a party; **Es gibt ...** There is/are . . . (6); **geben Sie mir** give me (A); **Gibt es ...?** Is there . . . ? / Are there . . . ? (A, 6); **sich einen Termin geben lassen** to get an appointment (11)

das **Gebet, -e** prayer

das **Gebirge, -** mountains, mountain range (7)

geboren born (1); **Wann sind Sie geboren?** When were you born? (1)

gebraten (*p.p. of* **braten**) roasted; broiled; fried (8); **gebratene Eier** (*pl.*) fried eggs (8)

der **Gebrauch, ⁻̈e** use (12)

gebrauchen, gebraucht to use

der **Gebrauchtwagen, -** used car

die **Gebrüder** (*pl.*) brothers

die **Gebühr, -en** fee

die **Geburt, -en** birth

der **Geburtstag, -e** birthday (1, 2); **zum Geburtstag** for someone's birthday (2)

die **Geburtstagskarte, -n** birthday card (2)

der **Gedanke, -n** (*wk.*) thought; **sich Gedanken machen (über** + *acc.*) to think (about)

gedenken (+ *dat.*) **(gedenkt), gedachte, gedacht** to remember

das **Gedicht, -e** poem (3)

geduldig patient(ly) (12)

die **Gefahr, -en** danger (12)

gefährlich dangerous(ly) (10)

gefahrlos safe(ly)

gefallen (+ *dat.*) **(gefällt), gefiel, gefallen** to be to one's liking; to please (6); **es gefällt mir** I like it (6); **sich (etwas) gefallen lassen** (*coll.*) to put up with (*s.th.*)

der **Gefallen, -** favor

die **Gefangenschaft** captivity

das **Gefängnis, -se** prison; jail (6)

das **Gefrierfach, ⁻̈er** freezer compartment

die **Gefriertruhe, -n** freezer

das **Gefühl, -e** feeling (3)

gegen (+ *acc.*) against (9); around; **ein Medikament gegen** medicine for (11)

der **Gegenstand, ⁻̈e** object

gegenüber (+ *dat.*) opposite; across (6); **(von** + *dat.*) across from (10); **gleich gegenüber** right across the way (6)

gegenüber·stehen (+ *dat.*) **(steht ... gegenüber), stand ... gegenüber, gegenübergestanden** to stand opposite (*s.o./s.th.*)

der **Gegner, -** / die **Gegnerin, -nen** opponent

gegrillt (*p.p. of* **grillen**) grilled; broiled; barbecued (8)

geheim secret(ly); die **Geheime Staatspolizei (Gestapo)** Secret State Police (*in Nazi Germany*)

das **Geheimnis, -se** secret (5)

die **Geheimniskrämerei, -en** secret-mongering

geheimnisvoll mysterious(ly)

die **Geheimzahl, -en** secret PIN (personal identification number) (12)

gehen (geht), ging, ist gegangen to go; to walk (A); **auf eine Party gehen** to go to a party (1); **einkaufen gehen** to go shopping (1, 5); **ich gehe lieber ...** I'd rather go . . . (2); **in die Berge gehen** to go to the mountains (1); **in Erfüllung gehen** to come true; **ins Bett gehen** to go to bed (1); **nach Hause gehen** to go home (1); **schief gehen** to go wrong; **Wie geht es dir?** (*infor.*) / **Wie geht es Ihnen?** (*for.*) How are you? **zur Uni gehen** to go to the university (1)

das **Gehirn, -e** brain (11)
gehören (+ *dat.*), **gehört** to belong to (*s.o.*) (6); **gehören zu** (+ *dat.*) to belong (*to s.th.*)
gehörlos deaf
gehorsam obedient(ly)
der **Gehorsam** obedience
die **Geige, -n** violin (3)
der **Geist** spirit, mind (8)
geistig mental(ly); intellectual(ly)
der **Geizhals, ⸚e** skinflint
gekocht (*p.p. of* **kochen**) cooked; boiled (8); **gekochte Eier** (*pl.*) boiled eggs (8)
das **Gel, -s** gel
gelb yellow (A)
das **Geld** money (2)
der **Geldautomat, -en** (*wk.*) automatic teller machine (ATM) (12)
die **Geldbörse, -n** purse; wallet
der **Geldbote, -n** (*wk.*) / die **Geldbotin, -nen** money runner
das **Geldgeschäft, -e** financial transaction
der **Geldschein, -e** note, bill (*of currency*) (12)
der/die **Gelehrte, -n (ein Gelehrter)** scholar
der/die **Geliebte, -n (ein Geliebter)** beloved friend, love (3)
gelingen (gelingt), gelang, ist gelungen to succeed
gelten (als) (gilt), galt, gegolten to be valid (as); to be regarded (as)
das **Gelüst, -e** craving
das **Gemälde, -** painting
gemein mean(ly) (8)
gemeinsam in common, together (6, 11)
die **Gemeinschaft, -en** community
gemeinschaftlich together
das **Gemisch, -e** mixture
gemischt (*p.p. of* **mischen**) mixed (8)
das **Gemüse, -** vegetable (8)
gemütlich cozy (10)
genau exact(ly) (B)
genauso just as
die **Generation, -en** generation
genießen (genießt), genoss, genossen to enjoy
der **Genosse, -n** (*wk.*) / die **Genossin, -nen** comrade
das **Genre, -s** genre
genug enough (3)
genügend sufficient(ly)
das **Genus, Genera** gender
die **Geografie** geography (B, 1)
geometrisch geometric(ally)
der **Gepard, -e** cheetah
gerade right now; just (at the moment); straight; upright; **gerade stellen** to straighten (3)
geradeaus straight ahead (10)
geradewegs straight; directly
das **Gerät, -e** appliance (8)
das **Geräusch, -e** sound, noise (9)
gerecht just(ly), fair(ly)
die **Gerechtigkeit** justice
das **Gericht, -e** court(house) (5); dish (8); **auf dem Gericht** at the courthouse (5)
der **Gerichtssaal, -säle** courtroom
gering low
der **Germane, -n** (*wk.*) / die **Germanin, -nen** ancient German, Teuton
germanisch Germanic (*adj.*)
der **Germanist, -en** (*wk.*) / die **Germanistin, -nen** Germanist, German scholar
die **Germanistik** (*sg.*) German studies
gern(e) gladly (5); willingly; with pleasure; (*with verb*) to like to; **ich habe ... gern** I like (*s.o./s.th.*); **ich hätte gern** I would like (to have) (*s.th.*) (5); **Trägst du gern ...?** Do you like to wear . . . ? (A); **Wir singen gern.** We like to sing. (1)
der **Geruch, ⸚e** smell
gesalzen salted (8)
gesamt whole; combined
die **Gesamtschule, -n** comprehensive school
das **Geschäft, -e** store (2)
geschäftlich (*relating to*) business
die **Geschäftsfrau, -en** businesswoman
der **Geschäftsführer, -** / die **Geschäftsführerin, -nen** manager (8)
der **Geschäftsmann, Geschäftsleute** businessman
geschehen (geschieht), geschah, ist geschehen to happen; to occur
das **Geschenk, -e** present (2)
der **Geschenkgutschein, -e** gift certificate (2)
die **Geschichte, -n** history (1); story
die **Geschichtsklausur, -en** history test
das **Geschirr** (*sg.*) dishes (4, 5); **Geschirr spülen** to wash the dishes (4)
der **Geschirrspüler, -** dishwasher (5)
die **Geschirrspülmaschine, -n** dishwasher
die **Geschlechterrolle, -n** gender role
geschlechtertypisch gender-biased, typical for a particular sex
der **Geschmack, ⸚er** taste
die **Geschwister** (*pl.*) brother(s) and sister(s), siblings (B)
die **Geselligkeit** sociability; conviviality
die **Gesellschaft, -en** society (12)
gesellschaftlich social(ly); societal(ly) (12)
das **Gesetz, -e** law
gesetzlich legal(ly); statutory
das **Gesicht, -er** face (B)
der **Gesichtsausdruck, ⸚e** facial expression
das **Gespräch, -e** conversation
gestalten, gestaltet to form; to create
die **Gestaltung, -en** design
das **Geständnis, -se** confession
die **Gestapo** = die **Geheime Staatspolizei** Secret State Police (*in Nazi Germany*)
die **Geste, -n** gesture
gestehen (gesteht), gestand, gestanden to confess
gestern yesterday (4); **gestern Abend** last night (4)
gestresst (*p.p. of* **stressen**) (*coll.*) stressed out; under stress
das **Gesuch, -e** request; application
gesund (gesünder, gesündest-) healthy (8)
die **Gesundheit** health (11)
das **Getöse** racket, din
das **Getränk, -e** beverage (8)
das **Getreide** grain
getrennt separate(ly); separate checks (5)
die **Gewalt** violence (12); force
das **Gewehr, -e** rifle
die **Gewerkschaft, -en** labor union
gewinnen (gewinnt), gewann, gewonnen to win (4); **in der Lotterie gewinnen** to win the lottery (5)
das **Gewitter, -** storm; thunderstorm
gewöhnlich ordinary, ordinarily
das **Gewürz, -e** spice; seasoning (8)
gießen (gießt), goss, gegossen to pour (8); to water (3); **die Blumen gießen** to water the flowers (3)
die **Gießkanne, -n** watering can
giftig poisonous (9)
der **Gipfel, -** mountaintop (7)
der **Gips** cast (*plaster*) (11)
die **Giraffe, -n** giraffe (10)
das **Girokonto, Girokonten** checking account (12)
die **Gitarre, -n** guitar (1)
der **Gitarrenverstärker, -** guitar amplifier
der **Gitarrist, -en** (*wk.*) / die **Gitarristin, -nen** guitarist
das **Glas, ⸚er** glass (5)
gläsern (*adj.*) (made of) glass (9)
die **Glatze, -n** bald head
glauben (an + *acc.*)**, geglaubt** to believe (in) (2)
gleich (*adj.*) equal, same (12); (*adv.*) right away, immediately; right, directly (6); **gleich gegenüber** right across the way (6); **gleich um die Ecke** right around the corner (6)
gleichgeschlechtlich same-sex
das **Gleichgewicht** balance
das **Gleis, -e** (set of) train tracks (10)
der **Gletscher, -** glacier (7)
glitzern, geglitzert to twinkle
das **Glück** luck; happiness (3); **Glück haben** to have luck, be lucky; **Viel Glück!** Lots of luck! Good luck! (3)
glücklich happy, happily (B)
gnadenlos merciless(ly)

gnädig gracious, kind, dear; **gnädige Frau** *very formal way of addressing a woman*

das **Gold** gold

golden gold(en)

der **Goldfisch, -e** goldfish (11)

das **Golf** golf (1)

der **Gott, ¨-er** god; God (12); **grüß Gott** good afternoon; hello (*for.; southern Germany, Austria*) (A)

Göttinger (*adj.*) (of) Göttingen

der **Gourmet, -s** gourmet

der **Gouverneur, -e** governor

das **Grab, ¨-er** grave, tomb

graben (gräbt), grub, gegraben to dig

der **Grad, -e** degree; **18 Grad Celsius** 18 degrees Celsius (B)

die **Grafik, -en** drawing; graphic(s)

der **Grafiker, -** / die **Grafikerin, -nen** graphic designer

die **Grammatik, -en** grammar (A)

das **Gras, ¨-er** grass

gratulieren (+ *dat.*), **gratuliert** to congratulate (6)

grau gray (A)

grauen, gegraut: es graut mir/mich I dread

grausam cruel(ly) (9)

greifen (greift), griff, gegriffen to grab, grasp (11)

grell gaudy, shrill; cool, neat (2)

die **Grenze, -n** limit, border (12)

(das) **Griechenland** Greece (B)

griechisch Greek (*adj.*)

der **Grill, -s** grill, barbecue (2)

grinsen, gegrinst to grin

die **Grippe, -n** influenza, flu (11)

groß (größer, größt-) large, big; tall (B); great; in a big way; **ziemlich groß** pretty big (2)

großartig magnificent(ly)

(das) **Großbritannien** Great Britain (B)

die **Größe, -n** size; height (1)

die **Großeltern** (*pl.*) grandparents (B)

die **Großmutter, ¨-** grandmother (B)

größtenteils for the most part

der **Großvater, ¨-** grandfather (B)

grüezi hi (*Switzerland*) (A)

grün green (A); **Die Grünen** (*pl.*) The Greens (*political party*)

der **Grund, ¨-e** reason; basis; **Grund-** basic (*prefixed to nouns*); **im Grunde** in principle; basically

gründen, gegründet to found

die **Grundlage, -n** basis, foundation; principle

der **Grundsatz, ¨-e** principle

grundsätzlich in principle; fundamental(ly)

die **Grundschule, -n** elementary school (9)

das **Grundstück, -e** property, lot (*land*)

die **Gründung, -en** foundation, establishment

das **Grundwasser** ground water

die **Gruppe, -n** group

der **Gruselfilm, -e** horror film (2)

der **Gruß, ¨-e** greeting (9)

grüßen, gegrüßt to greet; to say hello to (10); **grüß dich** hello (*infor.; southern Germany, Austria*); **grüß Gott** good afternoon; hello (*for.; southern Germany, Austria*) (A)

die **Grütze, -n** groats; **rote Grütze** red fruit pudding

gucken, geguckt (*coll.*) to look (at); to watch

die **Gültigkeit** validity

der **Gummibaum, ¨-e** rubber tree

die **Gurke, -n** cucumber (8); **saure Gurken** (*pl.*) pickles (8)

der **Gurt, -e** strap

der **Gürtel, -** belt (2)

gut good; well; **Das passt gut.** That fits well. (11); **Das steht / Die stehen dir gut!** That looks / Those look good on you! (2); **Es ist gut ausgegangen.** It ended well. (7); **Es sieht gut aus.** It looks good. (2); **ganz gut** very good; quite well; **gut gekleidet sein** to be well dressed (2); **guten Abend** good evening (A); **guten Morgen** good morning (A); **guten Tag** good afternoon; hello (*for.*) (A); **mach's gut** take care (*infor.*) (A)

die **Güte** goodness; **Du meine Güte!** (*coll.*) My goodness!

das **Guthaben, -** bank balance (12)

das **Gymnasium, Gymnasien** high school, college preparatory school (6)

das **Haar, -e** hair (A, B, 11); **die Haare föhnen** to blow dry hair (2); **Haare schneiden** to cut hair (3); **sich die Haare föhnen** to blow dry one's hair (11); **sich die Haare kämmen** to comb one's hair (11)

die **Haarfarbe, -n** color of hair (1)

die **Haarmode, -n** hairstyle

der **Haarschnitt, -e** haircut (2)

der **Haarstreifen, -** strip of hair

der **Haartrockner, -** hair dryer (2)

haben (hat), hatte, gehabt to have (A); **Angst haben (vor** + *dat.***)** to be afraid (of) (3); **es eilig haben** to be in a hurry (8); **Haben Sie etwas dagegen?** Do you have something for it (*illness*)? (11); **Hunger haben** to be hungry (3); **ich habe ... gern** I like (*s.o./s.th.*); **ich hätte gern** I would like to (have) (*s.th.*) (5); **Interesse haben an** (+ *dat.*) to be interested in (5); **Lust haben** to feel like (*doing s.th.*) (3); **recht haben** to be right (2); **Welche Farbe hat ...?** What color is . . . ? (A)

die **Habgier** greed

der **Habsburger, -** Habsburg

das **Hackfleisch** ground beef (or pork) (8)

der **Hafen, ¨-** harbor, port (10)

das **Hähnchen, -** (grilled) chicken

der **Hai, -e** shark (10)

der **Haken, -** hook (8)

halb half; **um halb drei** at two thirty (1)

die **Halbinsel, -n** peninsula (7)

die **Hälfte, -n** half

hallo hi (*infor.*) (A)

der **Hals, ¨-e** neck; throat (9)

die **Halsentzündung, -en** inflammation of the throat

die **Halskette, -n** necklace (2)

die **Halsschmerzen** (*pl.*) sore throat (11)

das **Halstuch, ¨-er** bandanna

halten (hält), hielt, gehalten to hold (4); to keep; to stop (7); **ein Referat halten** to give a paper / oral report (4); **halten für** (+ *acc.*) to consider; to think of as; **halten von** (+ *dat.*) to think of (12); **sich halten an** (+ *acc.*) to keep to, follow

die **Haltestelle, -n** stop (10)

die **Haltung, -en** posture

der **Hamburger, -** hamburger

hämisch malicious(ly)

der **Hammer, ¨-** hammer (8)

der **Hamster, -** hamster (10)

die **Hand, ¨-e** hand (B); **die Hand schütteln** to shake hands (A)

handeln, gehandelt to act; **handeln von** (+ *dat.*) to be about

die **Handlung, -en** action; plot

der **Handschuh, -e** glove (2)

das **Handtuch, ¨-er** hand towel (8)

handwerklich handy

das **Handy, -s** cellular phone (2)

hängen (hängt), hing, gehangen to hang, be in a hanging position (3)

hängen, gehängt to hang (up), put in a hanging position (3); **das Bild an die Wand hängen** to hang the picture on the wall (3)

(das) **Hannover** Hanover

die **Hanse** Hanseatic League

harmlos harmless(ly)

hart (härter, härtest-) hard

hartnäckig obstinate(ly), stubborn(ly)

der **Hartz IV-Typ, -en** (*wk.*) (*coll.*) person who collects unemployment benefits

der **Hase, -n** (*wk.*) hare

hassen, gehasst to hate (9)

hässlich ugly (2)

die **Haube, -n** bonnet; cap

häufig often, frequent(ly); common(ly)

Haupt- main (*prefixed to nouns*)

die **Hauptabstammung, -en** main line of descent

hauptsächlich main(ly), principal(ly)

die **Hauptstadt, ¨-e** capital city (B)

das **Haus, ¨-er** house (1, 2, 6); home (2); **nach Hause gehen** to go home (1, 10); **zu Hause sein** to be at home (A, 1, 10)

der **Hausarrest, -e** house arrest

der **Hausarzt, ¨-e** / die **Hausärztin, -nen** family doctor (11) die **Hausaufgabe, -n** homework (assignment) (A)

der **Hausbesitzer, -** / die **Hausbesitzerin, -nen** homeowner

das **Häuschen, -** small house, cottage

die **Hausfrau, -en** housewife, (*female*) homemaker

der **Haushalt, -e** household; housekeeping (9)

häuslich domestic

der **Hausmann, ¨-er** (*male*) homemaker

der **Hausmeister, -** / die **Hausmeisterin, -nen** custodian (5)

das **Hausmittel, -** home remedy

die **Hausnummer, -n** house number (1)

der **Hausschlüssel, -** house key (9)

der **Hausschuh, -e** slipper

das **Haustier, -e** pet (10)

die **Haut, ¨-e** skin (11)

heben (hebt), hob, gehoben to lift, raise

die **Hefe, -n** yeast

das **Heft, -e** notebook (B)

das **Heidebrot, -e** *type of rye-wheat bread from northwestern Germany*

heilen, geheilt to heal (5)

die **Heilpflanze, -n** medicinal plant

die **Heimat, -en** home, hometown, homeland (12)

das **Heimatland, ¨-er** homeland

die **Heimatstadt, ¨-e** hometown (6)

heimlich secret(ly) (9)

das **Heimweh** homesickness (3); **Heimweh haben** to be homesick (3)

die **Heirat, -en** marriage

heiraten, geheiratet to marry (3, 5)

heiß hot (B)

heißen (heißt), hieß, geheißen to be called, to be named (A); **Ich heiße ...** My name is . . . (A); **Wie heißen Sie?** (*for.*) / **Wie heißt du?** (*infor.*) What's your name? (A)

heiter cheerful(ly)

die **Heizung, -en** heating

der **Held, -en** (*wk.*) / die **Heldin, -nen** hero/heroine

helfen (+ *dat.*) **(hilft), half, geholfen** to help (6)

hell light (6); bright(ly)

das **Hemd, -en** shirt (A)

her (to) here, hither; this way (10); **hin und her** to and fro; back and forth; **von** (+ *dat.*) **... her** as far as . . . is concerned

herab·kommen (kommt ... herab), kam ... herab, ist herabgekommen to come down

herauf·holen, heraufgeholt to bring up, retrieve

heraus out this way (10); (**aus** + *dat.* **heraus**) out (of)

heraus·bringen (bringt ... heraus), brachte ... heraus, herausgebracht to bring out; to utter, say

heraus·finden (findet ... heraus), fand ... heraus, herausgefunden to find out

die **Herausforderung, -en** challenge

der **Herausgeber, -** / die **Herausgeberin, -nen** publisher; editor

heraus·kommen (kommt ... heraus), kam ... heraus, ist herausgekommen to come out this way (10)

heraus·springen (springt ... heraus), sprang ... heraus, ist herausgesprungen to jump out

sich **heraus·stellen, herausgestellt** to turn out

heraus·suchen, herausgesucht to pick out

heraus·ziehen (zieht ... heraus), zog ... heraus, herausgezogen to pull out

herb sharp; harsh; bitter

herbei·schleppen, herbeigeschleppt to drag (*s.th.*) over

die **Herbergseltern** (*pl.*) wardens of a youth hostel

der **Herbst, -e** fall, autumn (B)

der **Herd, -e** stove (5)

herein in this way (10)

herein·holen, hereingeholt to bring in

herein·kommen (kommt ... herein), kam ... herein, ist hereingekommen to get/go in this way (10)

her·gehen (geht ... her), ging ... her, ist hergegangen to go along

her·kommen (kommt ... her), kam ... her, ist hergekommen to come this way (10)

die **Herkunft, ¨-e** origin (B); nationality

der **Herr, -en** (*wk.*) gentleman; Mr. (A); master

die **Herrschaft, -en** rule; dominion

herrschen, geherrscht to reign, dominate (8)

der **Herrscher, -** / die **Herrscherin, -nen** ruler

her·schauen, hergeschaut (*coll.*) to look this way

her·stellen, hergestellt to produce

die **Herstellung** production, manufacture

herum around, round about; **um** (+ *acc.*) **... herum** around

herum·gehen (um + *acc.*) **(geht ... herum), ging ... herum, ist herumgegangen** to go around (*s.th.*)

herum·schwirren, ist herumgeschwirrt to buzz around

herum·tragen (trägt ... herum), trug ... herum, herumgetragen to carry around

herum·trampeln, hat/ist herumgetrampelt to stomp around

herunter down (*toward the speaker*) (11)

herunter·klettern, ist heruntergeklettert to climb down (11)

herunter·kommen (kommt ... herunter), kam ... herunter, ist heruntergekommen to come down

herunter·laden (lädt ... herunter), lud ... herunter, heruntergeladen to download

herunter·steigen (steigt ... herunter), stieg ... herunter, ist heruntergestiegen to climb down

herunter·werfen (wirft ... herunter), warf ... herunter, heruntergeworfen to throw down

hervor·gehen (geht ... hervor), ging ... hervor, ist hervorgegangen to emerge

hervor·rennen (rennt ... hervor), rannte ... hervor, ist hervorgerannt to run out in front

hervor·rufen (ruft ... hervor), rief ... hervor, hervorgerufen to evoke, call forth (1)

das **Herz, -en** heart (11)

der **Herzanfall, ¨-e** heart attack

die **Herzfrequenz, -en** heart rate

der **Herzinfarkt, -e** heart attack

herzlich hearty, heartily

das **Herzogtum, ¨-er** duchy

die **Herzschmerzen** (*pl.*) heartache

heute today (B); **heute Abend** this evening (2); **heute früh** this morning; **heute Morgen** this morning; **Welcher Tag ist heute?** What day is today? (1); **Welches Datum ist heute?** What is today's date? (4)

heutig (*adj.*) of today; present-day

die **Hexe, -n** witch (9)

hier here (A); **Ist hier noch frei?** Is this seat available? (8)

die **Hierarchie, -n** hierarchy

hierher (to) here, hither

die **Hilfe, -n** help; **Hilfe!** Help! (11)

der **Himmel, -** sky; heaven(s)

himmlisch heavenly

hin (to) there, thither; that way (10); **die Hin- und Rückfahrt** round-trip (7); **hin und her** to and fro; back and forth; **hin und wieder** now and then; **hin und zurück** there and back; round-trip (5, 10); **Wo willst du denn hin?** Where are you going? (A)

hinauf up that way (10)

hinauf·gehen (geht ... hinauf), ging ... hinauf, ist hinaufgegangen to go up that way (10)

hinauf·steigen (steigt ... hinauf), stieg ... hinauf, ist hinaufgestiegen to climb up

hinaus·bringen (bringt ... hinaus), brachte ... hinaus, hinausgebracht to bring out

hinaus·gehen (geht ... hinaus), ging ... hinaus, ist hinausgegangen to go out

hinaus·werfen (wirft ... hinaus), warf ... hinaus, hinausgeworfen to throw out

hin·bringen (bringt ... hin), brachte ... hin, hingebracht to take (*s.o/s.th. somewhere*)

das **Hindernis, -se** obstacle

hinein in(ward) (9); (**in** + *acc.*) into

hinein·beißen (beißt ... hinein), biss ... hinein, hineingebissen to bite in

hinein·biegen (biegt ... hinein), bog ... hinein, ist hineingebogen to turn

hinein·geben (gibt ... hinein), gab ... hinein, hineingegeben to put in

hinein·geraten (in + *acc.*) **(gerät ... hinein), geriet ... hinein, ist hineingeraten** to get (into)

hinein·laufen (in + *acc.*) **(läuft ... hinein), lief ... hinein, ist hineingelaufen** to run (into)

hinein·mischen, hineingemischt to mix in

sich **hinein·trauen, hineingetraut** to dare to go inside

hin·fahren (fährt ... hin), fuhr ... hin, ist hingefahren to go/drive (that way)

die **Hinfahrt, -en** journey there; outbound journey; **die Hin- und Rückfahrt** round-trip (7)

hin·fallen (fällt ... hin), fiel ... hin, ist hingefallen to fall down (11)

hin·gehen (geht ... hin), ging ... hin, ist hingegangen to go that way (10)

sich **hin·legen, hingelegt** to lie down (11)

hin·richten, hingerichtet to execute

hin·schauen, hingeschaut to look

sich **hin·setzen, hingesetzt** to sit down

sich **hin·stellen, hingestellt** to stand; to position oneself

hinten in the back

hinter (*prep.* + *dat./acc.*) behind; (*adj.*) back

hintereinander in a row (3)

der **Hintergrund, ¨-e** background

der **Hinterhof, ¨-e** courtyard

hinterher afterwards

hinterlassen (hinterlässt), hinterließ, hinterlassen to leave behind

hinüber over that way (10)

hinüber·gehen (geht ... hinüber), ging ... hinüber, ist hinübergegangen to go over that way (10)

hinüber·rufen (ruft ... hinüber), rief ... hinüber, hinübergerufen to call over

hinüber·ziehen (zieht ... hinüber), zog ... hinüber, hinübergezogen to pull over/across

hinunter·gehen (geht ... hinunter), ging ... hinunter, ist hinuntergegangen to go/walk down

hinunter·tauchen, ist hinuntergetaucht to dive down

hinweg: über viele Jahrhunderte hinweg for many centuries

hin·weisen (auf + *acc.*) **(weist ... hin), wies ... hin, hingewiesen** to point (to)

hinzu·fügen, hinzugefügt to add

der **Hirschbraten, -** roast venison

historisch historical(ly)

der **Hit, -s** (*coll.*) hit

die **Hitze** heat

das **Hobby, -s** hobby (1)

hoch (höher, höchst-) high(ly) (7); **einen hohen Lebensstandard haben** to have a high standard of living

das **Hochhaus, ¨-er** high-rise building (6)

hoch·heben (hebt ... hoch), hob ... hoch, hochgehoben to lift up

hochqualifiziert highly qualified

der **Hochschulabschluss, ¨-e** college/university degree

die **Hochschule, -n** college, university

der **Höchstsatz, ¨-e** maximum rate

der **Hochstuhl, ¨-e** high chair

der **Hochverrat** high treason

die **Hochzeit, -en** wedding

hoffen, gehofft to hope (3)

die **Hoffnung, -en** hope (2)

höflich polite(ly)

die **Höhe, -n** height; amount (*of money*) (12)

der **Höhepunkt, -e** highlight (7)

die **Höhle, -n** cave

holen, geholt to fetch, (go) get (9)

(das) **Holland** Holland (B)

holländisch Dutch (*adj.*) (8)

das **Holz, ¨-er** wood (12)

die **Holzschindel, -n** wooden shingle

homogen homogeneous

der **Honig** honey (8)

hoppla oops

hören, gehört to hear; to listen (to) (1)

das **Horoskop, -e** horoscope

der **Hörsaaldiener, -** / die **Hörsaaldienerin, -nen** lecture hall custodian

die **Hose, -n** pants, trousers (A)

das **Hotel, -s** hotel (2, 5); **im Hotel** at the hotel (5)

hübsch pretty (A, 2)

der **Hügel, -** hill (7)

das **Huhn, ¨-er** chicken

die **Hühnersuppe, -n** chicken soup

humorvoll humorous(ly)

der **Hund, -e** dog (2)

das **Hundefutter** dog food

die **Hunderasse, -n** breed of dog

hundert hundred (A)

hundertst- hundredth (4)

der **Hunger** hunger (3); **Hunger haben** to be hungry (3)

hungrig hungry, hungrily (9)

die **Hupe, -n** horn (7)

hupen, gehupt to honk (7)

husten, gehustet to cough

der **Husten, -** cough (11)

das **Hustenbier** warm beer with honey

das **Hustenbonbon, -s** cough drop

der **Hustenreiz** tickling in the throat; need to cough

der **Hustensaft, ¨-e** cough syrup (11)

der **Hut, ¨-e** hat (A)

der **Hybrid, -e** hybrid (car)

die **Hymne, -n** hymn; anthem

die **Hypnose, -n** hypnosis

die **Hysterie, -n** hysteria

die **IAÄGP** = die **Internationale Allgemeine Ärztliche Gesellschaft für Psychotherapie** International General Medical Society for Psychotherapy

ich I

ideal ideal(ly) (12)

die **Idee, -n** idea (10)

identifizieren, identifiziert to identify

die **Identität, -en** identity

das **Idol, -e** idol

das **Iglu, -s** igloo

ihm (*dat.*) him, it

ihn (*acc.*) him, it (2)

ihnen (*dat.*) them

Ihnen (*for. dat.*) you

ihr (*dat. sg.*) her; (*infor. nom. pl.*) you

ihr(e) her, its (1, 2); their (2)

Ihr(e) (*for.*) your (B, 2)

illegal illegal(ly) (12)

illusionslos without illusions

im = **in dem** in the; see **in**

immer always (3); **immer mehr** more and more; **immer noch** still

die **Immobilien** (*pl.*) real estate

impfen (gegen + *acc.*), **geimpft** to vaccinate (against) (10)

das **Importland, ¨-er** importer, country that imports

impressionistisch impressionistic(ally)

das **Impressum, Impressen** imprint

in (+ *dat./acc.*) in; into; at (A, 4); **im Ausland** abroad (6); **im Büro** at the office (5); **im ersten Stock** on the second floor (6); **im Frühling** in the spring (B); **im Internet surfen** to surf the Internet (1); **im Januar** in January (B); **im Moment** at the moment; right now (1); **in den Bergen wandern** to hike in the mountains (1); **in der Nähe** in the vicinity (6); **in der Schule** at school (5); **in der Woche** during the week (1); **in die Berge gehen** to go to the mountains (1); **in Eile sein** to be in a hurry (3); **ins Bett gehen** to go to bed (1); **ins Schwimmbad fahren** to drive/go to the swimming pool (1)

inbegriffen included (10)

indem (*subord. conj.*) while; as

indirekt indirect(ly)

indoeuropäisch Indo-European (*adj.*)

indogermanisch Indo-European (*adj.*)

die **Industrie, -n** industry

ineinander in one another; **sich ineinander verlieben** to fall in love with each other

der **Infinitiv, -e** infinitive

die **Info, -s** (*coll.*) info, information

die **Informatik** computer science (1)

die **Information, -en** information (4)

(sich) **informieren (über** + *acc.*), **informiert** to inform (*o.s.*) (about) (12)

der **Inhalt, -e** contents

die **Initiative, -n** initiative

inklusive (inkl.) included (*utilities*) (6)

die **Innenstadt, ¨-e** downtown (6)

das **Innere (ein Inneres)** inside

ins = **in das** in(to) the

insbesondere especially

die **Inschrift, -en** inscription
die **Insel, -n** island (7)
insgesamt altogether
installieren, installiert to install
das **Institut, -e** institute (7)
die **Instruktion, -en** instruction
das **Instrument, -e** instrument (12)
die **Integration, -en** integration (12)
integrieren, integriert to integrate (12)
der/die **Intellektuelle, -n (ein Intellektueller)** intellectual
intelligent intelligent(ly) (7)
die **Intelligenz, -en** intelligence
intensiv intensive(ly)
interessant interesting (7); **etwas Interessantes** something interesting (4)
das **Interesse, -n** interest (5); **Interesse haben an** (+ *dat.*) to be interested in (5)
interessieren, interessiert to interest (5); **sich interessieren für** (+ *acc.*) to be interested in (5)
international international(ly)
das **Internet** Internet; **im Internet surfen** to surf the Internet (1)
das **Interview, -s** interview (4)
interviewen, interviewt to interview
introvertiert introverted
inzwischen in the meantime, meanwhile
das **iPad, -s** iPad
der **iPod, -s** iPod
der **Iran** Iran
irgendetwas something; anything
irgendwann sometime; anytime
(das) **Irland** Ireland (B)
ironisch ironic(ally)
islamisch Islamic
(das) **Italien** Italy (B)
italienisch Italian (*adj.*)
(das) **Italienisch** Italian (*language*) (B)
ja yes; indeed (4); **Das ist es ja!** That's just it! (4); **wenn ja** if so
die **Jacke, -n** jacket (A)
die **Jackentasche, -n** jacket pocket
jagen, gejagt to hunt
der **Jäger, -** / die **Jägerin, -nen** hunter (9)
das **Jahr, -e** year (B); **im Jahr(e) ...** in the year . . . ; **seit zwei Jahren** for (the last) two years (4)
der **Jahrestag, -e** anniversary
die **Jahreszahl, -en** date (year)
die **Jahreszeit, -en** season (B)
das **Jahrhundert, -e** century
-jährig -year-old (*adj.*)
jährlich annual(ly)
das **Jahrzehnt, -e** decade (4)
jammern, gejammert to wail, moan
der **Januar** January (B); **im Januar** in January (B)
japanisch Japanese (*adj.*) (8)
je ever; each; **je nach Betrag** depending on the amount
je (*interj.*): **Oh je!** Oh dear!
die **Jeans** (*pl.*) jeans (2)
die **Jeansjacke, -n** denim jacket
jedenfalls in any case (11)
jeder, jedes, jede each; every (3, 5); **auf jeden Fall** by all means (4); **jede Woche** every week (3); **jeden Tag** every day (1)
jederzeit at any time
jedoch however
jeher: seit jeher always; since time immemorial
jemand someone, somebody (3)
jener, jenes, jene (*dem. pron.*) that, those
jetzig present, current
jetzt now (3)
jeweilig particular
jeweils each time; each; every
der **Job, -s** job
das **Joch, -e** yoke
joggen, ist gejoggt to jog
der **Joghurt** yogurt
(der) **Jom Kippur** Yom Kippur (*Jewish holiday*)
journalistisch journalistic(ally)
der **Jude, -n** (*wk.*) / die **Jüdin, -nen** Jewish man/woman
jüdisch Jewish
die **Jugend** youth (9); young people
die **Jugendherberge, -n** youth hostel (10)
der/die **Jugendliche, -n (ein Jugendlicher)** young person
der **Jugendschutz** protection of young people
das **Jugendschutzgesetz, -e** law for the protection of minors
die **Jugendsünde, -n** youthful folly
(das) **Jugoslawien** Yugoslavia
der **Juli** July (B)
jung (jünger, jüngst-) young (B)
der **Junge, -n** (*wk.*) boy
der **Jünger, -** / die **Jüngerin, -nen** disciple
der **Juni** June (B)
das **Jurastudium** law study
der **Jux, -e** joke; prank
der **Kaffee** coffee (1)
der **Kaffeefilter, -** coffee filter
die **Kaffeemaschine, -n** coffee machine (5)
die **Kaffeemühle, -n** coffee grinder
der **Käfig, -e** cage (10)
kahl bald
(das) **Kairo** Cairo
der **Kaiser, -** / die **Kaiserin, -nen** emperor/ empress
kaiserlich imperial
das **Kaiserreich, -e** empire
der **Kaiserschmarren** *pancake pieces sprinkled with powdered sugar and served with fruit sauce*
der **Kakao** cocoa; hot chocolate (8)
kalorienarm low in calories
kalorienbewusst calorie-conscious
kalt (kälter, kältest-) cold(ly) (B)
das **Kamel, -e** camel
die **Kamera, -s** camera (2)
der **Kamillentee** chamomile tea
der **Kamin, -e** hearth, fireplace
der **Kamm, ¨-e** comb
kämmen, gekämmt to comb (3); **sich (die Haare) kämmen** to comb one's hair (11)
der **Kampf, ¨-e** battle; struggle
kämpfen, gekämpft to fight (9)
(das) **Kanada** Canada (B)
der **Kanadier, -** / die **Kanadierin, -nen** Canadian (*person*) (B)
der **Kandidat, -en** (*wk.*) / die **Kandidatin, -nen** candidate (12)
das **Känguru, -s** kangaroo (10)
das **Kaninchen, -** rabbit
der **Kanton, -e** canton (*division of Switzerland*)
das **Kanu, -s** canoe (10); **Kanu fahren** to go canoeing (10)
der **Kapitalismus** capitalism
kapitalistisch capitalistic
das **Kapitel, -** chapter (A)
das **Käppchen, -** little cap; little hood
kaputt broken (A)
kaputt·machen, kaputtgemacht to break; to ruin
Karl der Große Charlemagne
das **Karnickel, -** rabbit (*dialectal*)
(das) **Kärnten** Carinthia
die **Karotte, -n** carrot (8)
die **Karriere, -n** career
die **Karte, -n** card; ticket; map (1, 2)
die **Kartoffel, -n** potato (8)
der **Kartoffelbrei** mashed potatoes
der **Kartoffelchip, -s** potato chips
der **Käse, -** cheese (8)
die **Kasse, -n** ticket booth (5); cashier window (12); **an der Kasse** at the ticket booth (5)
das **Kassler, -s** *salted and smoked pork*
die **Kastanie, -n** chestnut
der **Kasus, -** (grammatical) case
die **Kategorie, -n** category
der **Kater, -** tomcat; hangover (11)
die **Katze, -n** cat (2)
der **Katzenliebhaber, -** / die **Katzenliebhaberin, -nen** cat lover
kauen, gekaut to chew (11)
kaufen, gekauft to buy (1)
der **Käufer, -** / die **Käuferin, -nen** buyer; customer

das **Kaufhaus, ¨-er** department store (5); **im Kaufhaus** at the department store (5)

(das) **Kaufland** *department store chain*

das **Kaufmannshaus, ¨-er** merchant's house

kaum hardly

die **Kaution, -en** security deposit

der **Kaviar, -e** caviar

kein(e) no; none (2); **gar kein(e)** no . . . at all; **kein bisschen** not at all (3); **kein Wunder** no wonder (4)

der **Keller, -** basement, cellar (4, 6)

der **Kellner, -** / die **Kellnerin, -nen** waiter/ waitress (8)

kennen (kennt), kannte, gekannt to know, be acquainted with (B)

kennen·lernen, kennengelernt to meet, get acquainted with (1)

die **Kenntnisse** (*pl.*) skills; knowledge about a field (5)

kennzeichnen, gekennzeichnet to label; to characterize

der **Kern, -e** seed, pit

die **Kerze, -n** candle (3)

die **Kette, -n** chain

der **Kilometer, -** kilometer (2)

der **Kilometerstand** mileage

das **Kind, -er** child (B)

der **Kindergarten, ¨-** kindergarten (6)

das **Kindergeld** child benefit/allowance

der **Kinderreim, -e** nursery rhyme

der **Kinderwagen, -** baby carriage (7)

die **Kindheit** childhood (9)

das **Kino, -s** movie theater, cinema (1); **ins Kino gehen** to go to the movies (1)

die **Kinokarte, -n** movie ticket

die **Kirche, -n** church (5); **in der Kirche** at church (5)

der **Kirchenbau, -ten** church building

das **Kissen, -** cushion, pillow

die **Kiwi, -s** kiwi (fruit)

Kl. = die **Klasse, -n** class

die **Klammer, -n** bracket; parenthesis

die **Klamotten** (*pl., coll.*) clothes

der **Klang, ¨-e** sound; tone

die **Klapperschlange, -n** rattlesnake

klar clear(ly); **Klar!** Of course! (2)

die **Klarinette, -n** clarinet

klasse (*coll.*) great

die **Klasse, -n** class (5, 10); grade, level (9); **erster Klasse fahren** to travel first class (10)

die **Klassenarbeit, -en** (written) class test

der **Klassenkamerad, -en** (*wk.*) / die **Klassenkameradin, -nen** classmate

der **Klassenlehrer, -** / die **Klassenlehrerin, -nen** homeroom teacher

das **Klassentreffen, -** class reunion (9)

die **Klassik** classical period

klassisch classical

klassizistisch classical

klauen, geklaut (*coll.*) to steal

das **Klavier, -e** piano (2)

die **Klavierstunde, -n** piano lesson

kleben, geklebt to stick, adhere; to be sticky

das **Kleid, -er** dress (A); (*pl.*) clothes

(sich) **kleiden, gekleidet** to clothe (*o.s.*); **gut gekleidet sein** to be well dressed (2)

der **Kleiderschrank, ¨-e** clothes closet, wardrobe (6)

die **Kleidung** clothes (A, 2)

klein small, little; short (B)

klettern, ist geklettert to climb (9)

das **Klima, -s** climate

klingeln, geklingelt to ring (2)

klingen (wie) (klingt), klang, geklungen to sound (like) (11); (**nach** + *dat.*) to sound (like)

die **Klinik, -en** clinic; hospital

die **Klinke, -n** door handle

klopfen, geklopft to knock

der **Kloß, ¨-e** dumpling

das **Kloster, ¨-** cloister; monastery; convent

der **Klub, -s** club; nightclub

km = der **Kilometer, -** kilometer

knacken, geknackt to crack; (*coll.*) to break into

knapp meager; scarce(ly); just, barely (4)

die **Kneipe, -n** bar, tavern (1, 4)

das **Knie, -** knee

der **Knoblauch** garlic (8)

der **Knochen, -** bone

der **Knödel, -** dumpling (8)

der **Knopf, ¨-e** button

knuddeln, geknuddelt to hug

knuspern (an + *dat.*), **geknuspert** to nibble noisily (at)

die **Koalition, -en** coalition (12)

der **Koch, ¨-e** / die **Köchin, -nen** cook, chef (5)

kochen, gekocht to cook (1); to boil

der **Koffer, -** suitcase (3)

der **Kofferraum, ¨-e** trunk (7)

der **Kognak, -s** cognac

der **Kohl** cabbage (8)

der **Kolibri, -s** hummingbird

der **Kollege, -n** (*wk.*) / die **Kollegin, -nen** colleague, co-worker

kollektiv collective(ly)

(das) **Köln** Cologne

kolumbianisch Colombian (*adj.*)

das **Koma, -s** coma

die **Kombination, -en** combination

kombinieren, kombiniert to combine (3)

der **Komfort** comfort

komisch funny, strange (10)

kommen (kommt), kam, ist gekommen to come (B); **kommen aus** (+ *dat.*) to come from (*a place*) (B); **auf andere Gedanken kommen** to keep one's mind off something; **das kommt daher ...** the reason for that is . . . ; **Woher kommst du?** Where do you come from? (*infor.*); **zu Besuch kommen** to visit (3)

der **Kommentar, -e** commentary

kommentieren, kommentiert to comment on

der **Kommilitone, -n** (*wk.*) / die **Kommilitonin, -nen** fellow student

die **Kommode, -n** dresser (6); chest of drawers

die **Kommunikation, -en** communication

die **Komödie, -n** comedy

komponieren, komponiert to compose

der **Komponist, -en** (*wk.*) / die **Komponistin, -nen** composer

das **Kompositum, Komposita** compound noun

der **Kompromiss, -e** compromise

die **Konferenz, -en** conference

die **Konfession, -en** religious denomination, church

der **Konflikt, -e** conflict (12)

konform in agreement

der **König, -e** / die **Königin, -nen** king/queen (9)

königlich royal

die **Konjunktion, -en** conjunction

können (kann), konnte, gekonnt to be able (to), can (1); may (3)

konservativ conservative(ly) (B)

das **Konservatorium, Konservatorien** conservatory

der **Konsonant, -en** (*wk.*) consonant

die **Konstellation, -en** constellation

der **Kontakt, -e** contact

der **Kontinent, -e** continent

das **Konto, Konten** bank account (5); **ein Konto eröffnen** to open a bank account (5)

der **Kontostand, ¨-e** balance; account status

der **Kontrast, -e** contrast

die **Kontrolle, -n** control; scrutiny

der **Kontrolleur, -e** / die **Kontrolleurin, -nen** police inspector

kontrollieren, kontrolliert to check; to control; **das Öl kontrollieren** to check the oil (5)

kontrovers controversial(ly)

sich **konzentrieren (auf** + *acc.*), **konzentriert** to concentrate (on)

der **Konzern, -e** group (of companies)

das **Konzert, -e** concert (1); concerto; **die Brandenburgischen Konzerte** (*pl.*) the Brandenburg Concertos; **ins Konzert gehen** to go to a concert (1)

die **Konzertkarte, -n** concert ticket (5)

der **Konzertsaal, -säle** concert hall

die **Kooperation, -en** cooperation

(das) **Kopenhagen** Copenhagen

der **Kopf, ¨-e** head (B)

die **Kopfbedeckung, -en** headgear

das **Kopfkissen, -** pillow

der **Kopfsalat, -e** lettuce (8)

die **Kopfschmerzen** (*pl.*) headache (11)
die **Kopfschmerztablette, -n** headache tablet
der **Kopierladen, ¨** copy shop (10)
der **Korb, ¨e** basket
der **Korkenzieher, -** corkscrew (8)
das **Korn, ¨er** grain; corn
der **Körper, -** body (B)
körperlich physical(ly)
die **Körperpflege** personal hygiene
der **Körpersaft, ¨e** bodily fluid
der **Korridor, -e** corridor, hall
korrigieren, korrigiert to correct (4)
das **Kosmetikprodukt, -e** cosmetic product
der **Kosmonaut, -en** (*wk.*) / die **Kosmonautin, -nen** cosmonaut
kosten, gekostet to cost (2, 6)
kostenlos free of charge
die **Köstlichkeit, -en** delicacy
das **Kostüm, -e** costume
die **Krabbe, -n** shrimp (8)
der **Krabbenkutter, -** shrimp boat
die **Kraft, ¨e** power
kräftig powerful(ly); strong(ly)
die **Krähe, -n** crow
krank sick (3)
das **Krankenhaus, ¨er** hospital (3, 5); **im Krankenhaus** in the hospital (5)
der **Krankenpfleger, -** / die **Krankenpflegerin, -nen** nurse (5)
krankenversichert covered by health insurance
die **Krankenversicherung, -en** health insurance
der **Krankenwagen, -** ambulance (11)
die **Krankheit, -en** illness, sickness (11)
krass (*coll.*) awesome, intense, incredible
das **Kraut, ¨er** herb (8)
die **Kräuterbutter** herb butter (8)
die **Krawatte, -n** tie, necktie (A)
kreativ creative(ly)
die **Kreativität** creativity
der **Krebs, -e** crab; Cancer (*astrological sign*)
der **Kredit, -e** credit; loan; **einen Kredit aufnehmen** to take out a loan
die **Kreide, -n** chalk
der **Kreidestrich, -e** chalk line
der **Kreis, -e** circle; (administrative) district
das **Kreisarchiv, -e** district archives
kreischen, gekreischt to screech
der **Kreisverkehr, -e** traffic roundabout (10)
die **Kreuzung, -en** intersection (10)
der **Krieg, -e** war (12); **Krieg führen** to wage war
der **Kriegsdienst** military service
die **Kriegsgefangenschaft** captivity (as a prisoner of war)
der **Krimi, -s** crime thriller (*book, film, etc.*)
die **Kriminalität** crime
kriminell criminal(ly)
der/die **Kriminelle, -n (ein Krimineller)** criminal
die **Krise, -n** crisis (12)
die **Kritik, -en** criticism, critique
kritisch critical(ly)
kritisieren, kritisiert to criticize
(das) **Kroatien** Croatia
die **Krokette, -n** croquette (8)
das **Krokodil, -e** crocodile (10)
der **Krokus, -se** crocus
die **Krone, -n** crown; top (*of a tree*)
krönen, gekrönt to crown
krumm crooked(ly)
die **Küche, -n** kitchen (5); cooking (8); cuisine
der **Kuchen, -** cake (5)
die **Küchenarbeit, -en** kitchen work
die **Küchenlampe, -n** kitchen lamp (5)
die **Küchenmaschine, -n** mixer (8)
der **Küchentisch, -e** kitchen table (5)
die **Küchenuhr, -en** kitchen clock (5)
die **Küchenwaage, -n** kitchen scale (5)
der **Kugelschreiber, -** ballpoint pen (4)
kühl cool(ly) (B)
der **Kühlschrank, ¨e** refrigerator (5)
k. u. k. = **kaiserlich und königlich** imperial and royal (*pertaining to the dual monarchy of Austria-Hungary*)
kulinarisch culinary
die **Kultur, -en** culture (12)
kulturell cultural(ly)
der **Kulturminister, -** / die **Kulturministerin, -nen** minister for the arts
der **Kummer** sorrow; grief; trouble
sich **kümmern (um** + *acc.*)**, gekümmert** to take care (of) (12); to pay attention (to)
der **Kunde, -n** (*wk.*) / die **Kundin, -nen** customer (12)
kündigen, gekündigt to quit, resign
die **Kunst, ¨e** art (1)
der **Kunstalmanach, -e** art yearbook
die **Kunstgeschichte** art history (1)
die **Kunstgewerbeschule, -n** school of arts and crafts
der **Künstler, -** / die **Künstlerin, -nen** artist
künstlerisch artistic(ally)
künstlich artificial(ly); **sich künstlich auf·regen, aufgeregt** (*coll.*) to get excited/upset about nothing
die **Kunsttheorie, -n** art theory
der **Kurfürst, -en** (*wk.*) elector (*in the Holy Roman Empire*)
kurieren, kuriert to cure
der **Kurs, -e** (*academic*) course, class (A, 1); exchange rate
die **Kursfahrt, -en** cruise; boat trip
kursiv italic; **kursiv gedruckt** printed in italics
die **Kurve, -n** curve (7)
kurz (kürzer, kürzest-) short(ly) (A, B); brief(ly)
die **Kurzgeschichte, -n** short story
kurzsichtig nearsighted, myopic
die **Kusine, -n** (female) cousin (B)
der **Kuss, ¨e** kiss (4)
küssen, geküsst to kiss (9)
die **Küste, -n** coast (7)
labil unstable
das **Labor, -s** laboratory
lächeln, gelächelt to smile
lachen, gelacht to laugh (3); **vor Lachen** from laughing (so hard)
der **Lachs, -e** salmon
der **Lack, -e** varnish, lacquer
der **Laden, ¨** store, shop
die **Lage, -n** place; position (10); situation (12); location
die **Lampe, -n** lamp (B)
das **Land, ¨er** land, country; state; country (*rural*) (6); **auf dem Land** in the country (6)
die **Landkarte, -n** map (7)
das **Landsäugetier, -e** land mammal
die **Landschaft, -en** landscape; scenery; region
die **Landschaftskunde** study of the region
die **Landsleute** (*pl.*) compatriots
der **Landvogt, ¨e** governor (*of an imperial province*)
lang (länger, längst-) long (A, B)
lange (*adv.*) a long time; **lange nicht gesehen** haven't seen (you / each other) for a long time (1)
die **Länge, -n** length
die **Langeweile** boredom (3); **Langeweile haben** to be bored (3)
lang·gehen (geht ... lang), ging ... lang, ist langgegangen (*coll.*) to go along
langsam slow(ly)
sich **langweilen, gelangweilt** to be bored
langweilig boring (2)
der **Laptop, -s** laptop (computer) (B, 2)
der **Lärm** noise
lassen (lässt), ließ, gelassen to let (11); to leave alone; to have something done; **sich einen Termin geben lassen** to get an appointment (11)
der **Lastwagen, -** truck (7)
(das) **Latein** Latin (*language*) (1)
die **Laterne, -n** lamp; lantern
der **Lauf, ¨e** course; **im Laufe der Zeit** in the course of time; **seinen Lauf nehmen** to take its course
laufen (läuft), lief, ist gelaufen to go; to run (A, 2); **im Wald laufen** to run in the woods (2); **Schlittschuh laufen** to go ice-skating (3)
laufend current; **sich auf dem Laufenden halten** to keep oneself up-to-date
die **Laune, -n** mood
die **Lausitz** Lusatia (*region on the German-Polish border*)
laut loud(ly)

der **Laut, -e** sound
die **Lautbildung** articulation, formation of sounds
lauten, gelautet to read, go, run (*of text, an utterance, words*)
die **Lautmalerei, -en** onomatopoeia
der **Lautsprecher, -** loudspeaker (2)
die **Lautstärke** volume
die **Lautverschiebung, -en** sound-shift
der **Lautwandel** sound change
die **Lawine, -n** avalanche
das **Layout, -s** layout
der **Leadsänger, -** / die **Leadsängerin, -nen** lead singer
leben, gelebt to live (3)
das **Leben, -** life (9); **am Leben sein** to be alive (9)
lebendig alive
das **Lebensgefühl** awareness of life
das **Lebensmittel, -** food; groceries
das **Lebensmittelgeschäft, -e** grocery store
der **Lebensraum, ¨-e** living space; habitat
der **Lebensstandard, -s** standard of living
die **Leber, -n** liver (11)
der **Leberkäse** *loaf made of minced liver, eggs, and spices*
leblos lifeless
lecker delicious
das **Leder, -** leather (2)
ledig unmarried (1)
leer empty (5)
legal legal(ly)
legen, gelegt to lay, put, place (*in a horizontal position*); **sich legen** to lie down
die **Legende, -n** legend
die **Lehre, -n** apprenticeship (5); moral, teaching (8)
lehren, gelehrt to teach
der **Lehrer, -** / die **Lehrerin, -nen** teacher, instructor (A, 1)
der **Lehrjunge, -n** (*wk.*) (*young male*) apprentice
die **Lehrkraft, ¨-e** teacher(s)
das **Lehrmädchen, -** (*young female*) apprentice
der **Leib, -er** body; belly
leicht easy, easily; light (6)
das **Leid** suffering
leiden (an + *dat.*) **(leidet), litt, gelitten** to suffer (from)
die **Leidenschaft, -en** passion
leider unfortunately (B)
leid·tun (tut ... leid), tat ... leid, leidgetan: to be sorry; to feel sorry for; **tut mir leid** I'm sorry (4, 5)
leihen (leiht), lieh, geliehen to lend (5)
leise quiet(ly) (9); soft(ly)
die **Leistung, -en** achievement, accomplishment
leiten, geleitet to lead; to be head of
die **Leiter, -n** ladder

das **Leitungswasser** tap water
die **Lektüre, -n** reading material
das **Lenkrad, ¨-er** steering wheel (7)
lernen, gelernt to learn; to study (1)
lesen (liest), las, gelesen to read (A, 1); **Zeitung lesen** to read the newspaper (1)
(das) **Lettland** Latvia
letzt- last (4); **das letzte Mal** the last time (4); **letzten Montag** last Monday (4); **letzten Sommer** last summer (4); **letztendlich** ultimately, in the end; **letztes Wochenende** last weekend (4); **letzte Woche** last week (4)
leuchten, geleuchtet to shine
die **Leute** (*pl.*) people (7)
liberal liberal(ly) (9)
das **Licht, -er** light (3)
der **Lichtblick, -e** bright spot
der **Lichthof, ¨-e** light-well, atrium
lieb dear (7); beloved; sweet, lovable; **am liebsten** like (*to do s.th.*) best (7); **lieb haben** to love; to be fond of
die **Liebe, -n** love
lieben, geliebt to love (3)
lieber rather (2); **ich gehe lieber ...** I'd rather go . . . (2)
der **Liebeskummer** lovesickness (11)
der **Liebesroman, -e** romance novel
liebevoll loving(ly)
lieblich charming(ly)
Lieblings- favorite (A)
das **Lieblingsfach, ¨-er** favorite subject (5)
die **Lieblingsfarbe, -n** favorite color (A)
der **Lieblingsname, -n** (*wk.*) favorite name (A)
(das) **Liechtenstein** Liechtenstein (B)
das **Lied, -er** song
der **Liedermacher, -** / die **Liedermacherin, -nen** singer-songwriter
liegen (liegt), lag, gelegen to lie, be (in a horizontal position) (1); to recline; to be situated; **in der Sonne liegen** to lie in the sun (1); **liegen bleiben (bleibt ... liegen), blieb ... liegen, ist liegen geblieben** to remain lying down
der **Liegestuhl, ¨-e** deck chair (4)
lila purple (A)
die **Limo, -s** = die **Limonade, -n** soft drink; lemonade
lindern, gelindert to relieve, soothe
die **Linguistik** linguistics (1)
die **Linie, -n** line
link- (*adj.*), **links** (*adv.*) left; on/to the left (4, 10); **Die Linke** The Left (*political party*); **mit dem linken Fuß aufstehen** to get up on the wrong side of bed (4); **nach links** (to the) left
die **Linse, -n** lentil
die **Lippe, -n** lip (11)
der **Lippenstift, -e** lipstick
die **List, -en** deception, trick (9)

die **Liste, -n** list (5)
listen, gelistet to list
listig cunning(ly)
(das) **Litauen** Lithuania
der **Liter, -** liter (7)
die **Literatur, -en** literature (1)
das **Loch, ¨-er** hole (9)
der **Löffel, -** spoon (8)
logisch logical(ly)
der **Lohn, ¨-e** pay; wages, salary
die **Lokomotive, -n** locomotive (7)
los loose; away; **Was ist los?** What's happening? What's the matter?
lösen, gelöst to solve; **ein Rätsel lösen** to solve a puzzle/riddle (9); **sich lösen** to free oneself
los·fahren (fährt ... los), fuhr ... los, ist losgefahren to drive/ride off (4, 9)
los·gehen (geht ... los), ging ... los, ist losgegangen to set off; to get started
los·rennen (rennt ... los), rannte ... los, ist losgerannt to run off, start running
die **Lösung, -en** solution (1)
der **Lösungsvorschlag, ¨-e** suggested solution
die **Lotterie, -n** lottery (5); **in der Lotterie gewinnen** to win the lottery (5)
der **Löwe, -n** (*wk.*) lion (10)
loyal loyal(ly)
die **Luft, ¨-e** air (7)
die **Luftmatratze, -n** air mattress (10)
lügen, log, gelogen to lie, tell a falsehood
die **Lunge, -n** lung (11)
die **Lungenentzündung** pneumonia
die **Lust, ¨-e** desire (3); **Lust haben** to feel like (*doing s.th.*) (3)
lustig fun, funny (12); cheerful, jolly
lutschen, gelutscht to suck (11)
(das) **Luxemburg** Luxembourg
der **Luxus** luxury
machen, gemacht to make; to do; **Das macht nichts.** That doesn't matter. (1); **mach's gut** take care (*infor.*) (A); **sauber machen** to clean (3); **selbst gemacht** homemade (8); **sich an die Arbeit machen** to get down to work; **Spaß machen** to be fun; **Urlaub machen** to take a vacation
die **Machtergreifung, -en** seizure of power
das **Mädchen, -** girl (9)
die **Mafia, -s** Mafia
der **Magen, ¨** stomach (11)
die **Magen-Darm-Grippe** gastrointestinal flu
die **Magenschmerzen** (*pl.*) stomachache (11)
die **Magersucht** anorexia
magersüchtig anorexic
die **Magie** magic
magisch magical(ly) (12)
der **Magister, -** master's degree

mähen, gemäht to mow (5)

mahlen (mahlt), mahlte, gemahlen to grind

die **Mahlzeit, -en** meal (8)

(das) **Mähren** Moravia

der **Mai** May (B)

die **Mail, -s** e-mail (1)

mailen, gemailt to send e-mail

der **Main** Main (*river*)

mal once; (*word used to soften commands*) (11); **Komm mal vorbei!** Come on over! (11); **nicht mal** not even

das **Mal, -e** time (3, 4); **das letzte Mal** the last time (4); **das nächste Mal** the next time (3); **ein paar Mal** a few times; **mit einem Mal** all of a sudden; **zum ersten Mal** for the first time (4)

malen, gemalt to paint (12)

der **Maler, -** / die **Malerin, -nen** painter

die **Malerei, -en** painting

die **Mama, -s** mama, mom

die **Mami, -s** mommy

man one (*pron.*); people, they; **Wie schreibt man das?** How do you spell that? (A)

manch- some

manchmal sometimes (B)

mangelhaft poor, deficient, unsatisfactory

der **Mann, ¨er** man; husband (A, B)

männlich masculine; male

die **Mannschaft, -en** team (9)

der **Mantel, ¨** coat; overcoat (A)

das **Märchen, -** fairy tale (9)

märchenhaft as in a fairy tale

die **Märchenkunde** study of fairy tales

der **Markt, ¨e** market (10)

die **Marktkirche, -n** church on the market square

der **Marktplatz, ¨e** marketplace; market square (6)

die **Marktwirtschaft, -en** market economy

die **Marmelade, -n** jam; marmelade (8)

der **März** March (B)

der **Maschinenbau** mechanical engineering (1)

das **Massaker, -** massacre

der **Massenmord, -e** mass murder

massieren, massiert to massage

das **Masterstudium, -studien** course of study for a master's degree

das **Material, -ien** material, substance (12)

die **Mathe** math

die **Mathematik** mathematics (1)

der **Mathematiker, -** / die **Mathematikerin, -nen** mathematician

das **Matterhorn** *mountain in Switzerland*

die **Mauer, -n** wall; **die Berliner Mauer** the Berlin Wall

das **Maul, ¨er** mouth (of an animal)

die **Maus, ¨e** mouse (10)

die **Medien** (*pl.*) media

das **Medikament, -e** medicine (11); **ein Medikament gegen** (+ *acc.*) medicine for (11)

die **Medizin** medicine

der **Mediziner, -** / die **Medizinerin, -nen** doctor, physician

medizinisch medical(ly) (11)

das **Meer, -e** sea (1, 7); **ans Meer** to the sea (2); **im Meer schwimmen** to swim in the sea (1)

der **Meerrettich** horseradish

das **Meerschweinchen, -** guinea pig (10)

mehr more (7); **immer mehr** more and more; **nicht mehr** no longer; **nie mehr** never again

mehrere (*pl.*) several; **seit mehreren Tagen** for several days (11)

das **Mehrfamilienhaus, ¨er** house with several apartments

die **Mehrheit, -en** majority (12)

mehrmals several times (5)

die **Mehrzimmerwohnung, -en** multi-bedroom apartment

meiden (meidet), mied, gemieden to avoid

die **Meile, -n** mile

der **Meilenstein, -e** milestone

mein(e) my (A, 2)

meinen, gemeint to mean; to think

die **Meinung, -en** opinion; **Ihrer Meinung nach** (*for.*) in your opinion

der **Meißel, -** chisel (12)

meist most(ly) (3); **am meisten** mostly; the most; **die meisten** most (of)

meistens usually; mostly (8)

der **Meister, -** / die **Meisterin, -nen** master

die **Meisterschaft, -en** championship

melancholisch melancholy

(sich) **melden, gemeldet** to report

die **Mengenlehre** set theory

die **Mensa, Mensen** student cafeteria (2)

der **Mensch, -en** (*wk.*) person (2); human being; **Mensch!** (*coll.*) Man! Oh boy! (2)

menschengerecht suitable for humans

das **Menschenrecht, -e** human right

menschlich human

der **Mercedes** *make of car*

die **Messe, -n** trade fair

das **Messer, -** knife (8)

der **Meter, -** meter

die **Methode, -n** method

die **Metropolregion, -en** metropolitan area

die **Metzgerei, -en** butcher shop (6)

der **Mexikaner, -** / die **Mexikanerin, -nen** Mexican (*person*) (B)

mexikanisch Mexican (*adj.*) (8)

(das) **Mexiko** Mexico

mich (*acc.*) me

mies (*coll.*) crummy

die **Miete, -n** rent (6); **zur Miete** for rent

mieten, gemietet to rent (6)

der **Mieter, -** / die **Mieterin, -nen** renter (6)

der **Mietpreis, -e** rent, rental charge

das **Mietrecht** tenancy law, rental law

der **Migrant, -en** (*wk.*) / die **Migrantin, -nen** immigrant; emigrant

der **Migrationshintergrund** immigrant background

die **Mikrowelle, -n** microwave (oven)

die **Milch** milk (8)

mildern, gemildert to relieve; to soothe

die **Million, -en** million (12)

Min. = die **Minute, -n** minute

die **Minderheit, -en** minority (12)

mindestens at least (7)

das **Mineralwasser** mineral water (8)

der **Minister, -** / die **Ministerin, -nen** (government) minister (12)

das **Miniwörterbuch, ¨er** mini-dictionary

der **MINT-Bereich (= Mathematik, Informatik, Naturwissenschaft, Technik)** STEM fields (science, technology, engineering, mathematics)

die **Minute, -n** minute

mir (*dat.*) me; **mit mir** with me (3)

das **Mischbrot, -e** bread made from rye and wheat

die **Mischung, -en** mixture

die **Misshandlung, -en** mistreatment

misstrauen (+ *dat.*), **misstraut** to mistrust

mit (+ *dat.*) with (A); **mit dem linken Fuß aufstehen** to get up on the wrong side of bed (4); **mit mir** with me (3)

der **Mitarbeiter, -** / die **Mitarbeiterin, -nen** co-worker; collaborator

der **Mitbegründer, -** / die **Mitbegründerin, -nen** cofounder

der **Mitbewohner, -** / die **Mitbewohnerin, -nen** roommate; housemate (2)

mit·bringen (bringt ... mit), brachte ... mit, mitgebracht to bring along (3)

der **Mitbürger, -** / die **Mitbürgerin, -nen** fellow citizen

miteinander with each other, together (1, 3)

mit·fahren (fährt ... mit), fuhr ... mit, ist mitgefahren to ride/travel along

die **Mitfahrzentrale, -n** ride-share agency

das **Mitglied, -er** member

mithilfe (+ *gen.*) with the aid of

mit·kommen (kommt ... mit), kam ... mit, ist mitgekommen to come along

mit·machen, mitgemacht to participate; to join in

mit·nehmen (nimmt ... mit), nahm ... mit, mitgenommen to take along (3)

mit·schreiben (schreibt ... mit), schrieb ... mit, mitgeschrieben to write along (at the same time)

der **Mitschüler, -** / die **Mitschülerin, -nen** schoolmate, fellow pupil

mit·spielen, mitgespielt to play along, join in the game der **Mitstudent, -en** (*wk.*) / die **Mitstudentin, -nen** fellow student (A)

der **Mittag, -e** midday, noon (3); **zu Mittag essen** to eat lunch

das **Mittagessen, -** midday meal, lunch (3); **zum Mittagessen** for lunch (3)

mittags at noon (2)

die **Mitte** middle, center; in the middle of

das **Mittelalter** Middle Ages

mittelalterlich medieval

(das) **Mitteleuropa** Central Europe

der **Mittelfinger, -** middle finger

das **Mittelmeer** Mediterranean Sea (B)

mitten in the middle (9); **mitten in der Nacht** in the middle of the night

die **Mitternacht** midnight; **um Mitternacht** at midnight

mittler- (*adj.*) middle

der **Mittwoch, -e** Wednesday (1)

mit·versorgen, mitversorgt to be equally responsible for taking care of

die **Möbel** (*pl.*) furniture (6)

das **Möbelstück, -e** piece of furniture

möbliert furnished (6)

das **Modalverb, -en** modal verb

die **Mode, -n** fashion

das **Modell, -e** model, example

modern modern, in a modern fashion (6)

der **Modeschnickschnack** fashionable frills

der **Modezeichner, -** / die **Modezeichnerin, -nen** fashion designer

modisch fashionable, fashionably (2)

mögen (mag), mochte, gemocht to like (to); to care for (1, 3); **möchte** would like (to) (2, 3)

möglich possible; **alles Mögliche** everything possible (2)

möglicherweise possibly

die **Möglichkeit, -en** possibility (5)

möglichst (+ *adv.*) as . . . as possible (6)

der **Moment, -e** moment (1); **im Moment** at the moment; right now (1)

die **Monarchie, -n** monarchy

der **Monat, -e** month (B)

monatlich monthly

das **Monopol, -e** monopoly

der **Montag, -e** Monday (1); **letzten Montag** last Monday (4)

montags on Monday(s)

das **Moped, -s** moped

der **Mörder, -** / die **Mörderin, -nen** murderer

morgen tomorrow (2); **morgen Abend** tomorrow evening; **morgen früh** tomorrow morning

der **Morgen, -** morning; **am Morgen** in the morning; **guten Morgen** good morning (A); **heute Morgen** this morning

das **Morgengrauen** dawn, daybreak

morgens in the morning(s)

die **Morgentoilette** morning grooming routine

(das) **Moskau** Moscow

das **Motiv, -e** motif, theme (12)

die **Motorjacht, -en** motor yacht

das **Motorrad, ¨-er** motorcycle (1, 7); **Motorrad fahren** to ride a motorcycle (1)

das **Motto, -s** motto, slogan

die **Möwe, -n** seagull (10)

der **MP3-Player, -** MP3 player

der **MP3-Spieler, -** MP3 player (2, 5)

die **Mücke, -n** mosquito (10)

müde tired (3)

die **Mühle, -n** mill

der **Müll** trash; garbage (6)

der **Mülleimer, -** garbage can (8)

der **Müllermeister, -** / die **Müllermeisterin, -nen** master miller

die **Müllerstochter, ¨-** miller's daughter

multikulturell multicultural(ly)

(das) **München** Munich

Münchner (*adj.*) (of) Munich

der Mund, ¨-er mouth (B)

die **Mundharmonika, -s** harmonica (12)

mündlich oral(ly); verbal(ly)

munter cheerful(ly); lively; wide awake

die **Münze, -n** coin

die **Murmel, -n** marble

die **Muschel, -n** mussel (8); seashell

das **Museum, Museen** museum (1); **ins Museum gehen** to go to a museum (1)

das **Musical, -s** musical (*stage play*)

die **Musik, -en** music (1)

der **Musiker, -** / die **Musikerin, -nen** musician

der **Muskelkater, -** sore muscles (11)

das **Muskeltraining** muscle exercise

das **Müsli, -s** granola

müssen (muss), musste, gemusst to have to, must (3); **nicht müssen** not to have to, not to need to

der **Mut** courage (1)

mutig brave(ly)

die **Mutter, ¨-** mother (B)

die **Muttersprache, -n** mother tongue, native language

der **Muttertag** Mother's Day

die **Mutti, -s** mom, mommy

die **Mütze, -n** cap (5)

mysteriös mysterious(ly)

der **Mystiker, -** / die **Mystikerin, -nen** mystic

na (*interj.*) well (3); so; **na, gut** well, okay; **na ja** all right; **na, klar** of course

der **Nabel, -** navel

nach (+ *dat.*) after; past; according to; toward; to (*a place*) (3, 10); **je nach Betrag** depending on the amount; **nach dem Weg fragen** to ask for directions; **nach Hause gehen** to go home (1, 10); **nach links** (to the) left; **nach oben** upwards; **um zwanzig nach fünf** at twenty after/past five (1)

der **Nachbar, -n** (*wk.*) / die **Nachbarin, -nen** neighbor (4)

das **Nachbarhaus, ¨-er** house next door

nachdem (*subord. conj.*) after (9, 11)

nach·denken (über + *acc.*) **(denkt ... nach), dachte ... nach, nachgedacht** to think (about); to consider (7)

nacheinander one after the other

die **Nachfolgepartei, -en** successor party

nach·forschen, nachgeforscht to investigate

nach·gehen (+ *dat.*) **(geht ... nach), ging ... nach, ist nachgegangen** to follow

nachher afterward

die **Nachhilfe** tutoring (3)

nachlässig lax; careless(ly)

der **Nachmieter, -** / die **Nachmieterin, -nen** subletter

der **Nachmittag, -e** afternoon (4); **am Nachmittag** in the afternoon; **heute Nachmittag** this afternoon

nachmittags afternoons, in the afternoon (4)

die **Nachricht, -en** report; message; (*pl.*) news (7)

nach·sehen (sieht ... nach), sah ... nach, nachgesehen to check; to go and see

die **Nachspeise, -n** dessert (8)

nächst- next; nearest; **das nächste Mal** the next time (3)

die **Nächstenliebe** charity, brotherly love

die **Nacht, ¨-e** night (3); **die ganze Nacht** all night long (3); **mitten in der Nacht** in the middle of the night

der **Nachteil, -e** disadvantage (7)

das **Nachthemd, -en** nightshirt (2)

nachts nights, at night (4)

der **Nachttisch, -e** nightstand, bedside table (6)

der **Nacken, -** neck

der **Nagel, ¨-** nail (8)

nah (näher, nächst-) close, nearby (7)

nahe (+ *dat.*) near, close to

die **Nähe** closeness, proximity; vicinity (6); **in der Nähe** in the vicinity (6)

sich **nähern, genähert** to approach

das **Nahrungsmittel, -** food

der **Name, -n** (*wk.*) name (A, 1)

namens by the name of; called

nämlich namely; actually

die **Narbe, -n** scar

die **Nase, -n** nose (B, 11)

nass wet (3)

die **Nation, -en** nation

national national(ly) (12)

die **Nationalität, -en** nationality

der **Nationalpark, -s** national park (2)

nationalsozialistisch (*adj.*) National Socialist, Nazi

der **Nationalsozialismus** National Socialism, Nazism

nativ native; natural

die **Natur, -en** nature (9); disposition, temperament; **in freier Natur** out in the open (country) (10)

naturalistisch naturalistic(ally)

der **Naturheilkundler, -** / die **Naturheilkundlerin, -nen** naturopath

natürlich natural(ly) (2); of course

der **Naturschutz** nature conservation

der **Nazi, -s** Nazi

der **Nebel, -** fog, mist

neben (+ *dat./acc.*) next to (9); in addition to (3)

nebenan next door (5); **von nebenan** from next door (5)

die **Nebenkosten** (*pl.*) extra costs (*e.g., utilities*) (6)

das **Nebenzimmer, -** next room, adjacent room

der **Neffe, -n** (*wk.*) nephew (B)

negativ negative(ly)

nehmen (nimmt), nahm, genommen to take (A); **jemanden auf den Arm nehmen** to tease someone; to pull someone's leg

der **Neid** envy, jealousy

die **Neigung, -en** inclination; tendency

nein no (A)

nennen (nennt), nannte, genannt to name; to call; **sich nennen** to be called

nervös nervous(ly) (1)

das **Nest, -er** nest (10)

nett nice(ly) (3)

das **Netz, -e** net

das **Netzwerk, -e** network

neu new(ly) (A); **etwas Neues** something new (4)

der **Neubau, -ten** *building completed after 1 Dec. 1949*

die **Neubearbeitung, -en** new version, revision

die **Neugier** curiosity, inquisitiveness

neugierig curious(ly) (12)

neulich recently (9)

neun nine (A)

neunt- ninth (4)

neunundzwanzig twenty-nine (A)

neunzehn nineteen (A)

neunzehnt- nineteenth

neunzig ninety (A)

die **Neurose, -n** neurosis

(das) **Neuseeland** New Zealand (B)

neuseeländisch of/from New Zealand

die **Neustadt, ¨-e** new part of town

die **Neuverfilmung, -en** remake (*film*)

die **Neuzeit** modern era

nicht not (A); **gar nicht** not at all, not a bit (3); **lange nicht gesehen** haven't seen (you / each other) for a long time (1); **nicht mehr** no longer; **nicht (wahr)?** isn't that right?; **noch nicht** not yet

die **Nichte, -n** niece (B)

nichts nothing (9); **Das macht nichts.** That doesn't matter. (1); **gar nichts** nothing at all

nie never (2); **nie mehr** never again; **noch nie** never (before)

die **Niederlande** (*pl.*) the Netherlands (B)

niederländisch Dutch

sich **nieder·lassen (lässt ... nieder), ließ ... nieder, niedergelassen** to settle

(das) **Niederösterreich** Lower Austria

(das) **Niedersachsen** Lower Saxony

nieder·schlagen (schlägt ... nieder), schlug ... nieder, niedergeschlagen to knock down

nieder·schreiben (schreibt ... nieder), schrieb ... nieder, niedergeschrieben to write down

niedrig low

niemals never

niemand no one, nobody (2)

die **Niere, -n** kidney (11)

die **Nierenentzündung** kidney infection

das **Nikotin** nicotine

der **Nil** Nile (*river*)

das **Niveau, -s** level

der **Nobelpreis, -e** Nobel Prize

noch even, still (B); yet; else; in addition; **immer noch** still; **Ist hier noch frei?** Is this seat available? (8); **noch ein(e)** another, an additional (one); **noch einmal** one more time; **noch nicht** not yet; **noch nie** never (before); **sonst noch** in addition; else; **Sonst noch etwas?** Anything else? (5)

die **Nominalphrase, -n** noun phrase

nord- north

(das) **Nordamerika** North America (B)

(das) **Nordbayern** Northern Bavaria

norddeutsch North German (*adj.*)

der **Norden** north

nordfriesisch North Frisian (*adj.*)

nordgermanisch North Germanic (*adj.*)

nördlich (von + *dat.*) north (of) (7)

nordöstlich (von + *dat.*) northeast (of)

die **Nordsee** North Sea (B)

die **Nordwand, ¨-e** north wall; north face (*of a mountain*)

nordwestlich (von + *dat.*) northwest (of)

die **Norm, -en** norm

normal normal(ly) (5)

normalerweise normally (8)

(das) **Norwegen** Norway (B)

(das) **Norwegisch** Norwegian (*language*)

die **Not, ¨-e** need; hardship; trouble

die **Note, -n** grade, mark (*in school*) (9)

das **Notebook, -s** notebook (computer)

der **Notfall, ¨-e** emergency

nötig necessary; **nötig brauchen** to need urgently

die **Notiz, -en** note

die **Novelle, -n** novella

der **November** November (B)

die **Nudel, -n** noodle (8)

null zero (A)

der **Nullpunkt** freezing point, zero degrees Celsius (= 32 degrees Fahrenheit)

die **Nummer, -n** number (1)

das **Nummernschild, -er** license plate (7)

nun now; well

nur only (3)

(das) **Nürnberg** Nuremberg

die **Nuss, ¨-e** nut (8)

die **Nusshecke, -n** nut thicket

nutzen, genutzt to use

nützen, genützt to do some good; to be of use

nützlich useful(ly) (10)

ob (*subord. conj.*) if, whether (6, 11)

der/die **Obdachlose, -n (ein Obdachloser)** homeless person

oben above (10); on top; upstairs; **nach oben** upwards

der **Oberarm, -e** upper arm

(das) **Oberösterreich** Upper Austria

die **Oberschule, -n** secondary school

das **Objekt, -e** object

das **Obst** fruit (8)

obwohl (*subord. conj.*) although (11)

oder (*coord. conj.*) or (A, 11)

die **Odyssee, -n** odyssey

der **Ofen, ¨** oven

offen open(ly)

öffentlich public(ly) (7)

die **Öffentlichkeit, -en** public (12)

offiziell official(ly)

öffnen, geöffnet to open (A)

oft (öfter, öftest) often (A)

öfters now and then, once in a while

oftmals often

oh je (*interj.*) oh dear

ohne (+ *acc.*) without

die **Ohnmacht, -en** unconsciousness (11); **in Ohnmacht fallen** to faint (11)

das **Ohr, -en** ear (B)

die **Ohrenschmerzen** (*pl.*) earache (11)

der **Ohrring, -e** earring (A, 2)

okay (*coll.*) okay

die **Ökologie** ecology

ökologisch ecological(ly)

der **Oktober** October (B); **am ersten Oktober** on the first of October (4); **der erste Oktober** the first of October (4)

das **Oktoberfest, -e** Octoberfest (*annual beer festival in Munich*) (7)

das **Öl** oil (5, 8); **das Öl kontrollieren** to check the oil (5)

die **Ölfarbe, -n** oil color (*paint*) (12)

die **Olive, -n** olive (8)

die **Oma, -s** grandma

das **Omelett, -s** omelet (8)

der **Onkel, -** uncle (B)
online online
der **Onlinezugang, ¨-e** online access
der **Opa, -s** grandpa
das **Opfer, -** sacrifice; victim
das **Opferfest** Festival of the Sacrifice (*Eid al-Adha*)
der **Opi, -s** grandpa
orange orange (*color*) (A)
die **Orange, -n** orange
der **Orangensaft** orange juice (8)
die **Ordinalzahl, -en** ordinal number
ordnen, geordnet to arrange, put in order
die **Organisation, -en** organization (12)
organisch organic(ally)
die **Orgel, -n** organ (*musical instrument*) (12)
orientalisch oriental(ly)
der **Orientexpress** Orient Express (*train*)
orientieren, orientiert to orient
die **Orientierung, -en** orientation
das **Original, -e** original
das **Originaldrehbuch, ¨-er** original screenplay
der **Ort, -e** place (1, 5); town
der/die **Ortsfremde, -n (ein Ortsfremder)** stranger, nonresident
die **Oskar-Nominierung, -en** Oscar (Academy Award) nomination
ost- east
(das) **Ostdeutschland** (*former*) East Germany
der **Osten** east
das **Ostern, -** Easter
(das) **Österreich** Austria (B)
der **Österreicher, -** / die **Österreicherin, -nen** Austrian (*person*) (B)
österreichisch Austrian (*adj.*)
östlich (von + *dat.*) east (of) (7)
die **Ostsee** Baltic Sea (B)
paar: ein paar a few (2); a couple of; **ein paar Mal** a few times
das **Paar, -e** couple; pair (of)
packen, gepackt to pack (10)
der **Pädagoge, -n** (*wk.*) / die **Pädagogin, -nen** teacher; educational theorist
pädagogisch educational(ly)
das **Paket, -e** package (8)
die **Palatschinke, -n** *pancake with sweet filling*
der **Papa, -s** daddy, dad
der **Papagei, -en** parrot (10)
der **Papi, -s** daddy
das **Papier, -e** paper (B)
der **Papierkorb, ¨-e** wastebasket (3)
das **Papiertuch, ¨-er** paper towel (5)
der **Paprika** paprika
die **Paprika, -s** bell pepper
der **Papst, ¨-e** pope
parallel parallel
parat ready

das **Parfüm, -e** perfume (5)
der **Park, -s** park (1); **im Park spazieren gehen** to go for a walk in the park (1)
der **Parkautomat, -en** (*wk.*) parking meter
parken, geparkt to park (7)
die **Parole, -n** slogan
die **Partei, -en** (political) party (12)
das **Partizip, -ien** participle
der **Partner, -** / die **Partnerin, -nen** partner; **Arbeiten Sie mit einem Partner.** Work with a partner. (A)
die **Partnerschaft, -en** partnership
die **Party, -s** party (1, 2); **auf eine Party gehen** to go to a party (1)
der **Pass, ¨-e** passport (7)
passen, gepasst (+ *dat.*) to fit (6, 11); to suit; to match, go with (2); **(zu** + *dat.*) to go (with), fit in (with); **Das passt gut.** That fits well. (11)
passend fitting; proper
passieren, passiert to happen (4)
das **Passwort, ¨-er** password (7)
der **Patient, -en** (*wk.*) / die **Patientin, -nen** patient (5)
der **Patrizier, -** patrician
die **Pause, -n** recess, break (1); **Pause machen** to take a break
der **Pazifik** Pacific Ocean
das **Pech** pitch; bad luck; **Pech haben** to be unlucky
(das) **Peking** Beijing
der **Pelz, -e** fur
pendeln, ist gependelt to commute
das **Penizillin** penicillin
(das) **Pennsylvanien** Pennsylvania
pennsylvanisch Pennsylvanian (*adj.*)
per per, by means of
perfekt perfect(ly)
die **Person, -en** person, individual (1)
der **Personalausweis, -e** (personal) ID card (1)
persönlich personal(ly); **persönliche Daten** biographical information (1)
die **Persönlichkeit, -en** personality
die **Perspektive, -n** perspective
die **Pfalz, -en** palace, Palatinate; **die Pfalz Grona** *medieval royal palace formerly on the site of present-day Göttingen*
pfälzisch of/from the Palatinate
die **Pfanne, -n** (frying) pan (5)
der **Pfeffer, -** (black) pepper (8)
die **Pfeife, -n** pipe
das **Pfeifenwölkchen, -** little cloud of pipe smoke
das **Pferd, -e** horse (2, 9)
der **Pfifferling, -e** chanterelle (*type of mushroom*)
der **Pfirsich, -e** peach (8)
die **Pflanze, -n** plant (3, 6)
die **Pflanzenheilkunde** herbal medicine
das **Pflaster, -** adhesive bandage (11)

die **Pflaume, -n** plum (8)
pflegen, gepflegt to attend to; to nurse (5); to nurture
die **Pflicht, -en** duty; requirement; obligation (3)
pflichtbewusst conscientious(ly)
der **Pflichtunterricht** required instruction
pflücken, gepflückt to pick (9)
das **Phänomen, -e** phenomenon
die **Phantasie, -n** fantasy
das **Pharmaunternehmen, -** pharmaceutical company
der **Philosoph, -en** (*wk.*) / die **Philosophin, -nen** philosopher
die **Physik** physics (1)
der **Physiker, -** / die **Physikerin, -nen** physicist
das **Picknick, -s** picnic (4)
das **Piercing, -s** piercing (2)
der **Pilot, -en** (*wk.*) / die **Pilotin, -nen** pilot (5)
der **Pilz, -e** mushroom (8)
die **Pinnwand, ¨-e** bulletin board
der **Pinsel, -** paintbrush (12)
der **Pionier, -e** pioneer; Pioneer (*member of an East German youth organization*)
der **Piranha, -s** piranha (10)
die **Pistole, -n** pistol
die **Pizza, Pizzen** pizza (2)
das **Plakat, -e** poster; placard
der **Plan, ¨-e** plan (3)
planen, geplant to plan (9)
das **Platin** platinum
der **Platz, ¨-e** place; seat; room, space; plaza, square (3); **Platz nehmen** to take a seat
die **Playliste, -n** playlist
plötzlich sudden(ly) (9)
plus plus
(das) **Polen** Poland (B)
die **Politik** politics (12)
der **Politiker, -** / die **Politikerin, -nen** politician
politisch political(ly) (12)
die **Polizei** police; police station (5); **auf der Polizei** at the police station (5)
der **Polizist, -en** (*wk.*) / die **Polizistin, -nen** police officer (5)
die **Pommes (frites)** (*pl.*) French fries (8)
die **Popmusik** pop music
populär popular(ly)
das **Portal, -e** portal, gateway
das **Portemonnaie, -s** wallet
das **Porträt, -s** portrait
(das) **Portugal** Portugal (B)
(das) **Portugiesisch** Portuguese (*language*) (B)
(das) **Posen** Poznan (*city in Poland*)
positiv positive(ly)
der **Possessivartikel, -** possessive determiner
die **Post, -** mail; post office (5); **auf der Post** at the post office (5)
das **Poster, -** poster (6)

die **Postkarte, -n** postcard (2)
potentiell potential(ly)
das **Präfix, -e** prefix
prägen, geprägt to impress; to shape
pragmatisch pragmatic(ally)
praktisch practical(ly) (5)
die **Präposition, -en** preposition
präsentieren, präsentiert to present
der **Präsident, -en** (*wk.*) / die **Präsidentin, -nen** president (5)
der **Preis, -e** price (7, 12); prize (4)
preisgünstig at a favorable price; inexpensive(ly)
die **Prellung, -en** bruise
pressen, gepresst to press, squeeze
das **Prestige** prestige (5)
der **Priester, -** / die **Priesterin, -nen** priest/ priestess (5)
prima great (6)
der **Prinz, -en** (*wk.*) / die **Prinzessin, -nen** prince/princess (9)
privat private(ly)
pro per (2)
die **Probe, -n** test; rehearsal
probieren, probiert to try; to taste (3)
das **Problem, -e** problem
problematisch problematic
die **Produktion, -en** production
der **Produzent, -en** (*wk.*) / die **Produzentin, -nen** producer
der **Professor, -en** / die **Professorin, -nen** professor (A, B)
die **Professur, -en** professorship
der **Profikoch, ¨-e** / die **Profiköchin, -nen** professional cook, chef
profitieren, profitiert to profit
das **Programm, -e** program
das **Projekt, -e** project
die **Proportion, -en** proportion
der **Protest, -e** protest
protestieren, protestiert to protest
protestantisch Protestant (*adj.*)
provisionsfrei without commission
das **Prozent, -e** percent, percentage (4)
der **Prozentsatz, ¨-e** percentage
prozentual by percentage
die **Prozentzahl, -en** percentage
der **Prozess, -e** trial
die **Prüfung, -en** test, exam (1)
die **Prüfungskommission, -en** examination committee
die **Prügel** (*pl.*) beating(s)
prügeln, geprügelt to beat
der **Psychiater, -** / die **Psychiaterin, -nen** psychiatrist (11)
die **Psychiatrie** psychiatry
psychiatrisch psychiatric
psychisch psychological(ly), mental(ly)
die **Psychoanalyse** psychoanalysis
die **Psychologie** psychology
die **Psychose, -n** psychosis
die **Psychotherapie** psychotherapy
der **Pudel, -** poodle
der **Pulli, -s** = der **Pullover, -** pullover; sweater (2)
der **Pullover, -** pullover; sweater
das **Pult, -e** desk
der **Punkt, -e** point (A); dot
punkten, gepunktet to score points
pünktlich punctual(ly); on time (4)
die **Pünktlichkeit** punctuality
pur pure
das **Putenschnitzel, -** turkey cutlet
putzen, geputzt to clean (6); **sich (die Zähne) putzen** to brush (one's teeth) (11)
der **Putzlappen, -** cloth, rag (*for cleaning*) (6)
die **Pyramide, -n** pyramid
qm = der **Quadratmeter, -** square meter (m2) (6)
das **Quadrat, -e** square
der **Quadratmeter, -** square meter (m^2) (6)
quälen, gequält to torment
die **Qualität, -en** quality
der **Quatsch** nonsense
die **Quelle, -n** source
die **Querflöte, -n** (transverse) flute (12)
die **Quickcard** *Austrian debit card*
die **Quittung, -en** receipt, check (8)
die **Quote, -n** proportion; rate; figures
das **Rad, ¨-er** wheel (7); bicycle; **Rad fahren (fährt ... Rad), fuhr ... Rad, ist Rad gefahren** to ride a bicycle (7)
radeln, ist geradelt to ride a bicycle
der **Radfahrer, -** / die **Radfahrerin, -nen** bicyclist
das **Radieschen, -** radish
das **Radio, -s** radio (2)
das **Radium** radium
die **Radtour, -en** bicycle tour (9)
die **Rakete, -n** rocket (7)
das **Ranking, -s** ranking
der **Ranzen, -** schoolbag; knapsack; satchel
der **Rapper, -** / die **Rapperin, -nen** rapper, rap singer
rasen, ist gerast to race, rush
der **Rasen, -** lawn (5)
der **Rasenmäher, -** lawnmower (6)
der **Rasierapparat, -e** shaver, (electric) razor
sich **rasieren, rasiert** to shave (11)
die **Rasierklinge, -n** razor blade
das **Rasierwasser** aftershave lotion
der **Rat** (*pl.* **Ratschläge**) advice (5, 11)
die **Rate, -n** rate
raten (rät), riet, geraten to guess; (+ *dat.*) to advise (*s.o.*) (11)
das **Ratespiel, -e** guessing game; quiz
das **Rathaus, ¨-er** town/city hall (1, 6); **auf dem Rathaus** at the town hall (1)
der **Ratschlag, ¨-e** (piece of) advice (5, 11)
das **Rätsel, -** puzzle, riddle (9); **ein Rätsel lösen** to solve a puzzle/riddle (9)
die **Ratte, -n** rat (10)
rauchen, geraucht to smoke (3)
der **Raum, ¨-e** room; space; area
die **Räumungsklage, -n** eviction notice
raus = **heraus** out
rauschen, gerauscht to rustle
raus·schießen (schießt ... raus), schoss ... raus, ist rausgeschossen to dart out
reagieren, reagiert to react
der **Realismus** realism
recherchieren, recherchiert to investigate
rechnen, gerechnet to do arithmetic
die **Rechnung, -en** bill; check (*in restaurant*) (4)
recht (*adv.*) really
recht- (*adj.*); **rechts** (*adv.*) right; on/to the right (7, 10)
das **Recht, -e** right; law
recht haben (hat ... recht), hatte ... recht, recht gehabt to be right (2)
rechtlich legal(ly)
der **Rechtsanwalt, ¨-e** / die **Rechtsanwältin, -nen** lawyer
das **Rechtschreiben** spelling
rechtspopulistisch right-wing populist (*adj.*)
die **Rede, -n** speech (12)
das **Rededuell, -e** duel of words
reden, geredet to speak, talk
der **Redner, -** / die **Rednerin, -nen** orator
das **Referat, -e** report (3); (term) paper; **ein Referat halten** to give a paper / oral report (4)
die **Reform, -en** reform
der **Reformator, -en** / die **Reformatorin, -nen** reformer
der **Refrain, -s** refrain
das **Regal, -e** bookshelf, bookcase (2); rack
regelmäßig regular(ly) (11)
regeln, geregelt to regulate
der **Regen, -** rain (7); **bei Regen** in rainy weather (7)
sich **regen, geregt** to move, stir
der **Regenschirm, -e** umbrella (5)
regieren, regiert to rule
die **Regierung, -en** government (12)
die **Regierungspartei, -en** ruling party
das **Regime, -** regime
regimetreu loyal to the regime
die **Region, -en** region
regional regional(ly)
der **Regisseur, -e** / die **Regisseurin, -nen** stage/ film director

registrieren, registriert to register; **sich registrieren lassen** to get registered
regnen, geregnet to rain; **es regnet** it is raining; it rains (B)
reiben (reibt), rieb, gerieben to rub
reich rich(ly)
das **Reich, -e** empire; kingdom; realm; **das Dritte Reich** the Third Reich (Nazi Germany)
der **Reifen, -** tire (7)
die **Reifenpanne, -n** flat tire (7)
die **Reihe, -n** row
die **Reihenfolge, -n** order, sequence (2, 4)
das **Reihenhaus, ¨-er** row house, town house (6)
sich **reimen, gereimt** to rhyme
rein = herein in
rein·gehen (geht ... rein), ging ... rein, ist reingegangen to go inside (1)
die **Reinigung, -en** dry cleaner's (6)
der **Reis** rice (8)
die **Reise, -n** trip, journey (7); **auf Reisen sein** to be on a trip (7)
das **Reisebüro, -s** travel agency (6)
das **Reiseerlebnis, -se** travel experience (7)
reisefertig ready to leave
der **Reiseführer, -** travel guidebook (5)
das **Reiseland, ¨-er** tourist country
reisen, ist gereist to travel (1)
der/die **Reisende, -n (ein Reisender)** traveler (10)
der **Reisepass, ¨-e** passport (1)
der **Reiseplan, ¨-e** travel plan; itinerary
das **Reiseziel, -e** destination
reißen (reißt), riss, gerissen to tear, rip
reiten (reitet), ritt, ist geritten to ride (on horseback) (1); **Wellen reiten** to ride the waves, surf
der **Reiter, -** / die **Reiterin, -nen** (horseback) rider; **der Blaue Reiter** *a group of artists in Munich (1911–14)*
relativ relative(ly)
der **Relativsatz, ¨-e** relative clause
die **Religion, -en** religion (1)
die **Religionsmündigkeit** religious coming-of-age
religiös religious(ly) (B)
die **Renaissance** Renaissance
rennen (rennt), rannte, ist gerannt to run (7)
die **Rente, -n** pension
die **Reparatur, -en** repair
reparieren, repariert to repair (1)
der **Reporter, -** / die **Reporterin, -nen** reporter (4)
repräsentativ representative(ly)
die **Republik, -en** republic
republikanisch Republican (*adj.*)
reservieren, reserviert to reserve (7)
die **Residenz, -en** (royal) residence
die **Resonanz, -en** resonance

der **Rest, -e** rest, remainder
das **Restaurant, -s** restaurant (2)
das **Resultat, -e** result
retten, gerettet to save; to rescue
die **Rettichscheibe, -n** radish slice
die **Rettung** rescue; salvation
das **Revier, -e** station
die **Revolution, -en** revolution
revolutionär revolutionary
das **Rezept, -e** recipe; prescription (11)
die **Rezeption, -en** reception desk (10)
die **Rezession, -en** recession
der **Rhein** Rhine (*river*)
der **Rhythmus, Rhythmen** rhythm
richten, gerichtet to direct; to turn; **sich richten (nach** + *dat.***)** to depend (on); to comply (with)
der **Richter, -** / die **Richterin, -nen** judge (5)
richtig right(ly), correct(ly) (2)
die **Richtung, -en** direction (7)
riechen (riecht), roch, gerochen to smell (11)
der **Riese, -n** (*wk.*) / die **Riesin, -nen** giant (9)
riesig gigantic; tremendous(ly)
das **Rindfleisch** beef (8)
der **Ring, -e** ring (2)
der **Rock, ¨-e** skirt (A); (*sg. only*) rock music
das **Rockkonzert, -e** rock concert
der **Roggen** rye
roh raw
die **Rolle, -n** role; part (4)
die **Rollenverteilung, -en** assignment of roles
der **Roman, -e** novel (5)
die **Romantik** Romantic period/movement
römisch Roman (*adj.*)
röntgen, geröntgt to X-ray (11)
rosa pink (A)
die **Rose, -n** rose; **die Weiße Rose** the White Rose (*name of an anti-Nazi resistance group*)
der **Rosenkohl** Brussels sprouts (8)
rostig rusty
rot red (A); **rote Grütze** red fruit pudding
(das) **Rotkäppchen** Little Red Riding Hood
das **Roulette** roulette
der **Rücken, -** back (B)
die **Rückfahrt, -en** return journey; **die Hin- und Rückfahrt** round-trip (7)
rückgängig machen, gemacht to reverse
der **Rucksack, ¨-e** backpack (2)
die **Rückseite, -n** back (side); reverse (*of a coin*)
rufen (ruft), rief, gerufen to call, shout (7, 11)
die **Ruhe** silence; peace
ruhen, geruht to rest
ruhig quiet(ly), calm(ly) (B)
der **Ruhm** fame
sich **rühren, gerührt** to move, stir

(das) **Rumänien** Romania (B)
(das) **Rumpelstilzchen** Rumpelstiltskin
das **Rumpsteak, -s** rump steak (8)
rund round
der **Rundgang, ¨-e** walking tour
runter·bringen (bringt ... runter), brachte ... runter, runtergebracht = herunter·bringen to bring down
runter·kriegen, runtergekriegt (*coll.*) to get down
der **Rüssel, -** trunk (*of an elephant*)
russisch Russian (*adj.*) (8)
(das) **Russisch** Russian (*language*) (B)
(das) **Russland** Russia (B)
rustikal country-style
die **Rute, -n** switch, rod
rutschen, ist gerutscht to slide, slip (9)
rütteln, gerüttelt to shake
die **Sache, -n** thing (2); cause
die **Sachertorte, -n** *type of chocolate cake*
der **Sachse, -n** (*wk.*) Saxon
(das) **Sachsen** Saxony
sächsisch Saxon (*adj.*)
der **Sachtext, -e** non-fiction text
der **Saft, ¨-e** juice (8)
die **Sage, -n** legend, saga
sagen, gesagt to say; to tell (A, 5)
die **Sahara** Sahara (Desert)
die **Sahne** cream
das **Sakko, -s** sports jacket (A)
die **Salami, -** salami
der **Salat, -e** salad (8)
die **Salatschüssel, -n** salad (mixing) bowl (5)
die **Salbe, -n** ointment
das **Salz** salt (8)
salzig salty (7)
die **Salzkartoffeln** (*pl.*) boiled potatoes (8)
sammeln, gesammelt to collect (10); to gather
der **Sammler, -** / die **Sammlerin, -nen** collector
der **Samstag, -e** Saturday (1); **am Samstag** on Saturday (2)
samstags on Saturday(s)
der **Samt** velvet
der **Sand** sand (7)
die **Sandale, -n** sandal (2)
die **Sandburg, -en** sandcastle (4)
sanft soft(ly); gentle, gently
der **Sänger, -** / die **Sängerin, -nen** singer
der **Sanitäter, -** / die **Sanitäterin, -nen** paramedic
der **Sarg, ¨-e** coffin (9)
der **Satz, ¨-e** sentence (3)
die **Satzklammer, -n** sentence bracket
der **Satzteil, -e** part of sentence, clause
sauber clean (B); **sauber machen** to clean (3)
sauer sour (8); angry, angrily (5); **saure Gurken** (*pl.*) pickles (8)

der **Sauerbraten, -** sauerbraten (*marinated beef roast*)
das **Sauerkraut** sauerkraut, pickled cabbage
saugen, gesaugt to vacuum; **Staub saugen** to vacuum (6)
die **Sauna, -s** sauna (11)
das **Schach** chess (1)
schade! too bad! (6)
schaden (+ *dat.*), **geschadet** to be harmful to (6)
der **Schaden, ¨** damage
schaffen (schafft), schuf, geschaffen to create
schaffen, geschafft to manage; to achieve; **jemanden aus dem Weg schaffen** to get someone out of the way
der **Schafskäse** feta cheese
der **Schal, -s** scarf (2)
die **Schallplatte, -n** (phonograph) record
der **Schallplattenspieler, -** record player (2)
der **Schalter, -** ticket booth (5); **am Schalter** at the ticket booth (5)
der **Schatten, -** shadow; shade (9)
der **Schatz, ¨e** treasure (9); darling
schätzen, geschätzt to value
die **Schätzung, -en** estimate
schauen (an/auf + *acc.*), **geschaut** to look (at) (A)
schaufeln, geschaufelt to shovel
der **Schauspieler, -** / die **Schauspielerin, -nen** actor/actress
der **Scheck, -s** check
die **Scheibe, -n** slice; windowpane (7)
der **Scheibenwischer, -** windshield wiper (7)
der **Scheiblettenkäse** processed cheese slices
die **Scheidung, -en** divorce
der **Schein, -e** bill, note (*of currency*) (8)
scheinen (scheint), schien, geschienen to shine; to seem, appear
scheitern, gescheitert to fail (12)
schenken, geschenkt to give (as a present) (5)
die **Schere, -n** scissors (8)
der **Scherz, -e** joke
schick chic, stylish(ly), smart(ly) (2)
schicken, geschickt to send (2)
schief crooked
schief·gehen (geht ... schief), ging ... schief, ist schiefgegangen (*coll.*) to go wrong
die **Schiene, -n** train track (10)
schießen (schießt), schoss, geschossen to shoot
das **Schild, -er** sign (7)
schildern, geschildert to depict
die **Schildkröte, -n** turtle (10); tortoise
schimpfen, geschimpft to cuss; to scold (9)
der **Schinken, -** ham (8)
der **Schlaf** sleep (9)
der **Schlafanzug, ¨e** pajamas
die **Schläfe, -n** temple
schlafen (schläft), schlief, geschlafen to sleep (2); **lange schlafen** to sleep late
der **Schlafsack, ¨e** sleeping bag (2)
der **Schlafwagen, -** sleeping car
das **Schlafzimmer, -** bedroom (6)
der **Schlag, ¨e** (heart)beat
schlagen (schlägt), schlug, geschlagen to beat (8); to strike, hit (11)
der **Schlager, -** pop song
der **Schlagertext, -e** pop lyrics
das **Schlagzeug, -e** percussion, drums (12)
die **Schlange, -n** snake (10)
schlank slender, slim (B)
schlau clever(ly), cunning(ly)
das **Schlauchboot, -e** inflatable dinghy
schlecht bad(ly) (2)
die **Schleife, -n** bow, ribbon
der **Schleim, -e** phlegm, mucus
(das) **Schlesien** Silesia
schleudern, geschleudert to hurl
schließen (schließt), schloss, geschlossen to close, shut (A)
schließlich finally (7); after all
schlimm bad (11)
der **Schlittschuh, -e** ice skate (3); **Schlittschuh laufen** to go ice-skating (3)
das **Schloss, ¨er** castle (9)
der **Schlossgang, ¨e** castle passageway
der **Schluss, ¨e** end (8); conclusion; **zum Schluss** in the end, finally (8); in conclusion
der **Schlüssel, -** key (9)
die **Schlüsselposition, -en** key position
schmal narrow; thin
schmecken (+ *dat.*), **geschmeckt** to taste good (to) (6)
schmeißen (schmeißt), schmiss, geschmissen to fling, hurl
der **Schmerz, -en** pain (11)
sich **schminken, geschminkt** to put makeup on (11)
der **Schmuck** jewelry (2)
schmücken, geschmückt to decorate
schmutzig dirty (A)
der **Schnaps, ¨e** spirit; schnapps
schnarchen, geschnarcht to snore
die **Schnecke, -n** snail
der **Schnee** snow (9)
(das) **Schneewittchen** Snow White
schneiden (schneidet), schnitt, geschnitten to cut (3, 11); **Haare schneiden** to cut hair (3); **sich schneiden** to cut oneself (11)
die **Schneiderlehre** tailoring apprenticeship
schneien, geschneit to snow; **es schneit** it is snowing; it snows (B)
schnell quick(ly), fast (7)
der **Schnitt, -e** cut, incision; **im Schnitt** on average
das **Schnitzel, -** (veal/beef/pork) cutlet (8)
der **Schnupfen, -** cold (*with a runny nose*), sniffles (11)
die **Schnur, ¨e** string (8)
der **Schnurrbart, ¨e** mustache (A)
der **Schock, -s** shock (11)
schocken, geschockt (*coll.*) to shock
schockieren, schockiert to shock
die **Schokolade, -n** chocolate
schon already (2, 4); indeed; **schon wieder** once again (3); **Warst du schon einmal ...?** Were you ever . . . ? (4)
schön pretty, beautiful; nice (B); **Bitte schön.** There you go. **Bitte schön?** Yes please? May I help you? (7); **ganz schön viel** quite a bit (3)
das **Schönheitsideal, -e** ideal of beauty
der **Schrank, ¨e** wardrobe cabinet, cupboard (2, 6)
die **Schranke, -n** barrier
der **Schrei, -e** cry; shout; scream
schreiben (schreibt), schrieb, geschrieben to write; to spell (A, 1); **eine SMS schreiben** to write a text message (1); **schreiben an** (+ *acc.*) to write to; **schreiben über** (+ *acc.*) to write about; **Wie schreibt man das?** How do you spell that? (A)
die **Schreibmaschine, -n** typewriter
der **Schreibtisch, -e** desk (2)
das **Schreibwarengeschäft, -e** stationery store (6)
schreien (schreit), schrie, geschrien to scream, yell (3)
die **Schrift, -en** script; writing
der **Schriftsteller, -** / die **Schriftstellerin, -nen** writer (5)
der **Schritt, -e** step
die **Schublade, -n** drawer (5)
schüchtern shy(ly) (B)
die **Schüchternheit** shyness
der **Schuh, -e** shoe (A)
das **Schuhgeschäft, -e** shoe store (6)
die **Schulbildung** education, schooling
schuld: schuld sein (an + *dat.*) to be at fault (for)
die **Schuld, -en** debt; fault; guilt (12)
schulden, geschuldet to owe
die **Schule, -n** school (1, 5); **in der Schule** at school (5)
der **Schüler, -** / die **Schülerin, -nen** student; pupil (1)
der **Schulhof, ¨e** schoolyard, playground
der **Schulleiter, -** / die **Schulleiterin, -nen** principal, headmaster
die **Schultasche, -n** book bag
die **Schulter, -n** shoulder (B)
die **Schuluniform, -en** school uniform
der **Schulweg, -e** way to school
die **Schulzeit** school days
die **Schüssel, -n** bowl (8)

schütteln, geschüttelt to shake; **die Hand schütteln** to shake hands (A)

schützen, geschützt to protect

schwach (schwächer, schwächst-) weak(ly)

schwanger pregnant

schwarz (schwärzer, schwärzest-) black (A); **das schwarze Brett** bulletin board

das **Schwarzbier, -e** *very dark beer*

schwarzhaarig black-haired

Schwarzwälder (*adj.*) (of the) Black Forest

schweben, geschwebt to float

(das) **Schweden** Sweden (B)

schwedisch Swedish (*adj.*)

(das) **Schwedisch** Swedish (*language*) (B)

schweigen (schweigt), schwieg, geschwiegen to become silent; to be silent, say nothing

das **Schweigen** silence

das **Schwein, -e** pig (9)

der **Schweinebraten, -** pork roast

das **Schweinefleisch** pork (8)

der **Schweinestall, ¨-e** pigpen

die **Schweiz** Switzerland (B)

Schweizer Swiss (*adj.*); **der Schweizer Franken, -** Swiss franc (8)

der **Schweizer, -** / die **Schweizerin, -nen** Swiss (*person*) (B)

die **Schwellung, -en** swelling

schwer heavy, heavily; hard; difficult (3); **schwer verletzt** critically injured (11)

die **Schwester, -n** sister (B)

schwierig difficult (2)

die **Schwierigkeit, -en** difficulty

das **Schwimmbad, ¨-er** swimming pool (1, 5); **im Schwimmbad** at the swimming pool (5); **ins Schwimmbad fahren/gehen** to drive/go to the swimming pool (1)

schwimmen (schwimmt), schwamm, ist/hat geschwommen to swim (7); **im Meer schwimmen** to swim in the sea (1); **schwimmen gehen** to go swimming (1)

sich **schwingen (schwingt), schwang, geschwungen** to swing oneself

schwitzen, geschwitzt to sweat, perspire

sechs six (A)

sechst- sixth (4)

sechsundzwanzig twenty-six (A)

sechzehn sixteen (A)

sechzig sixty (A)

der **See, -n** lake (7)

die **Seele, -n** soul (12)

seelisch mental(ly), psychological(ly)

das **Segel, -** sail

segeln, ist/hat gesegelt to sail (1)

segnen, gesegnet to bless

sehen (sieht), sah, gesehen to see (2); **lange nicht gesehen** haven't seen (you / each other) for a long time (1)

die **Sehnsucht** longing

sehr very (B); **Bitte sehr.** There you go.

die **Seide, -n** silk (2); **aus Seide** of/from silk (2)

die **Seife, -n** soap (11)

der **Seiltänzer, -** / die **Seiltänzerin, -nen** tightrope walker

sein (ist), war, ist gewesen to be (A, 4)

sein(e) his, its (1, 2)

seit (*prep.*) since; for (4, 11); **seit mehreren Tagen** for several days (11); **seit zwei Jahren** for two years (4)

seitab off to the side

die **Seite, -n** side; page (6)

die **Seitenstraße, -n** side street

seitlich sideways

der **Sekretär, -e** / die **Sekretärin, -nen** secretary (5)

der **Sekt, -e** sparkling wine

die **Sekunde, -n** second (1)

selber, selbes, selbe same

selbst even (2); oneself (2); myself, yourself, himself, herself, itself; ourselves, yourselves, themselves; by (one)self; **selbst gemacht** homemade (8)

die **Selbstbestimmung** self-determination

der **Selbstmord, -e** suicide

selbstvergessen oblivious to one's surroundings

selbstverständlich of course (10)

selten rare(ly), seldom (8)

seltsam strange(ly)

das **Semester, -** semester (1)

die **Semesterferien** (*pl.*) semester break

das **Seminar, -e** seminar

der **Seminarraum, ¨-e** classroom (B)

die **Semmel, -n** (bread) roll

die **Sendung, -en** broadcast

der **Senf** mustard (8)

senken, gesenkt to lower (12)

der **September** September (B)

die **Serviette, -n** napkin (8)

servus hello; good-bye (*infor.; southern Germany, Austria*) (A)

der **Sessel, -** armchair (2, 6)

setzen, gesetzt to put, place, set (*in a sitting position*) (7); **sich setzen** to sit down (A, 11)

das **Shampoo, -s** shampoo

sich oneself, himself, herself, itself, yourself; themselves, yourselves

sicher sure(ly) (1); of course; safe(ly)

die **Sicherheit** safety

der **Sicherheitsgurt, -e** seat belt (7)

die **Sicherheitskraft, ¨-e** security officer

sicherlich certainly (3)

sichtbar visible, visibly

sie (*pron., fem. nom./acc.*) she, her, it; (*nom./acc. pl.*) they, them

Sie (*for. sg./pl.*) you

sieben seven (A)

siebenundzwanzig twenty-seven (A)

siebt- seventh (4)

siebzehn seventeen (A)

siebzig seventy (A)

die **Siedlung, -en** settlement

der **Sieg, -e** victory

signalisieren, signalisiert to signal; to indicate

silbern silver (*adj.*), silvery

die **Silvesternacht, ¨-e** night of New Year's Eve

simsen, gesimst (*coll.*) to text (1)

singen (singt), sang, gesungen to sing (1)

die **Single, -s** single (record)

sinken (sinkt), sank, ist gesunken to sink

der **Sinn, -e** sense; **im Sinn haben** to have in mind

sinnlich sensual(ly)

die **Sitte, -n** custom

die **Situation, -en** situation

der **Sitz, -e** seat (7)

sitzen (sitzt), saß, gesessen to sit, be in a seated position (2, 4); **sitzen bleiben** to remain seated

(das) **Skandinavien** Scandinavia

das **Skateboard, -s** skateboard (3); **Skateboard fahren** to skateboard (3)

der **Ski, -er** ski (3); **Ski fahren** to ski (3)

die **Skihütte, -n** ski lodge (6)

die **Skizze, -n** sketch

der **Sklave, -n** (*wk.*) / die **Sklavin, -nen** slave

der **Skorpion, -e** scorpion (10)

skypen, geskypt to Skype

die **Slowakei** Slovakia (B)

(das) **Slowenien** Slovenia (B)

das **Smartphone, -s** smartphone (2)

die **SMS** SMS, text message; **eine SMS schreiben** to write a text message (1)

das **Snowboard, -s** snowboard (1); **Snowboard fahren** to snowboard

das **Snowboarden** snowboarding

so so; such; that way (A); **das stimmt so** that's right; keep the change (8); **so viel** so much; **so was** something like that; some such thing

sobald (*subord. conj.*) as soon as

die **Socke, -n** sock (2)

das **Sofa, -s** sofa, couch (6)

sofort immediately (3)

sogar even

sogenannt so-called

der **Sohn, ¨-e** son (B)

das **Solarium, Solarien** tanning salon (11)

solcher, solches, solche such

die **Solidarität** solidarity

sollen (soll), sollte, gesollt to be supposed to (3)

die **Solokarriere, -n** solo career

der **Solosänger, -** / die **Solosängerin, -nen** solo artist (singer) der **Sommer, -** summer (B); **letzten Sommer** last summer (4)

sonderbar strange(ly)

sondern but, rather, on the contrary (A, 11)

die **Sonderschulpädagogik** special education

der **Song, -s** song

das **Songbuch, ⁻̈er** songbook (2)

der **Songwriter, -** / die **Songwriterin, -nen** songwriter

die **Sonne, -n** sun; **in der Sonne liegen** to lie in the sun (1)

das **Sonnenbaden** sunbathing

sonnenbaden gehen (geht ... sonnenbaden), ging ... sonnenbaden, ist sonnenbaden gegangen to go sunbathing (10)

der **Sonnenbrand, ⁻̈e** sunburn (10)

die **Sonnenbrille, -n** sunglasses (1, 2)

die **Sonnenmilch** suntan lotion (10)

der **Sonnenschirm, -e** sunshade; beach parasol (10)

der **Sonnenstrahl, -en** ray of sunlight

sonnig sunny (B)

der **Sonntag, -e** Sunday (1)

sonst otherwise (2, 5); **sonst noch** in addition; else; **Sonst noch etwas?** Anything else? (5)

sonstig other

sorgen für (+ *acc.*), **gesorgt** to take care of

sorgfältig careful(ly)

die **Sorte, -n** sort, type, kind

sortieren, sortiert to sort

die **Soße, -n** sauce (8); (salad) dressing

das **Souvenir, -s** souvenir (7)

sowie as well as

sowjetisch Soviet (*adj.*)

die **Sowjetunion** Soviet Union

sowohl als/wie as well as

sozial social(ly)

sozialdemokratisch Social Democratic

der **Sozialismus** socialism

der **Sozialist, -en** (*wk.*) / die **Sozialistin, -nen** socialist (*person*)

sozialistisch socialist (*adj.*)

die **Sozialkunde** social studies (1)

der **Sozialstaat, -en** welfare state

die **Soziologie** sociology (1)

die **Spaghetti** (*pl.*) spaghetti

die **Spalte, -n** column

(das) **Spanien** Spain (B)

spanisch Spanish (*adj.*)

(das) **Spanisch** Spanish (*language*) (B)

spannend suspenseful

die **Spannweite, -n** wingspan

sparen, gespart to save (*money*) (5); **sparen auf** + (*acc.*) to save up for

das **Sparkonto, Sparkonten** savings account (12)

der **Spaß, ⁻̈e** fun; joke **Spaß haben** to have fun; **Spaß machen** to be fun; **viel Spaß** have fun (A)

spät late (1); **später** later (1); **Wie spät ist es?** What time is it? (1)

die **Spätzle** (*pl.*) spaetzle (*kind of noodles*)

spazieren gehen (geht ... spazieren), ging ... spazieren, ist spazieren gegangen to go for a walk (1); **im Park spazieren gehen** to go for a walk in the park (1)

die **SPD** = die **Sozialdemokratische Partei Deutschlands** Social Democratic Party of Germany

der **Speck** bacon (8)

speichern, gespeichert to store

die **Speisekarte, -n** menu (8)

der **Speisesaal, -säle** dining hall

der **Speisewagen, -** dining car

spekulieren, spekuliert to speculate

spenden, gespendet to donate

der **Sperrmüll** bulky waste

die **Spezialität, -en** specialty

speziell special; especially

der **Spiegel, -** mirror (6); *title of a German news magazine*

das **Spieglein, -** (*diminutive form of* **der Spiegel**) little mirror

das **Spiel, -e** game; match

spielen, gespielt to play (1)

der **Spieler, -** / die **Spielerin, -nen** player

der **Spielfilm, -e** theatrical feature film

der **Spielplatz, ⁻̈e** playground (9)

der **Spinat** spinach

spitz pointed

der **Spitzbart, ⁻̈e** goatee

der **Spitzel, -** informer

der **Spitzname, -n** (*wk.*) nickname

splittern, gesplittert to splinter

der **Sport** sport(s); physical education (1); **Sport treiben** to do sports (2)

die **Sporthose, -n** tights, sports pants (2)

sportlich athletic(ally) (B)

der **Sportplatz, ⁻̈e** sports field; playing field

der **Sportschuh, -e** athletic shoe (A)

die **Sporttasche, -n** athletic bag

die **Sportverletzung, -en** sports injury

die **Sprache, -n** language (B)

die **Sprachfamilie, -n** language family

das **Sprachlabor, -s** language laboratory (4)

der **Sprachwissenschaftler, -** / die **Sprachwissenschaftlerin, -nen** linguist

sprachlos speechless(ly)

der/das **Spray, -s** spray

sprechen (spricht), sprach, gesprochen to speak, talk (B); **sprechen über** (+ *acc.*) to talk about

der **Sprechgesang** spoken song

die **Sprechsituation, -en** conversational situation (A)

die **Sprechstunde, -n** office hour (3)

die **Sprechstundenhilfe** (doctor's) receptionist

sprengen, gesprengt to water, sprinkle

das **Sprichwort, ⁻̈er** proverb, saying

springen (springt), sprang, ist gesprungen to jump (A)

die **Spritze, -n** shot, injection (11)

sprühen, gesprüht to spray

das **Spülbecken, -** sink (5)

spülen, gespült to wash; to rinse (4); **Geschirr spülen** to wash the dishes (4)

spüren, gespürt to feel

der **Spürsinn** intuition

der **Staat, -en** state (10, 12); nation

staatlich state, government (*adj.*) (12)

die **Staatsangehörigkeit, -en** nationality, citizenship

der **Staatsbankrott, -e** government bankruptcy

die **Staatsbürgerschaft, -en** citizenship

das **Staatsbürgerschaftsrecht, -e** citizenship law

der **Staatschef, -s** / die **Staatschefin, -nen** head of state

das **Staatsexamen, -** *final university examination*

die **Staatspolizei** state police; **die Geheime Staatspolizei (Gestapo)** Secret State Police (*in Nazi Germany*)

der **Staatsrat, ⁻̈e** / die **Staatsrätin, -nen** state councilor

die **Staatssicherheit** State Security (*former East German secret police*)

stabil sturdy, sturdily; solid(ly)

die **Stadt, ⁻̈e** town, city (2, 6)

das **Stadtbild, -er** townscape, cityscape

der **Stadtplan, ⁻̈e** city street map

der **Stadtrand, ⁻̈er** city limits (6)

die **Stadtrundfahrt, -en** tour of the city (7)

der **Stadtteil, -e** district, neighborhood (6)

das **Stadtviertel, -** quarter, district, neighborhood (6)

die **Staffelei, -en** easel (12)

der **Stamm, ⁻̈e** tribe

stammen (aus/von + *dat.*), **gestammt** to come (from), originate (in)

der **Stammgast, ⁻̈e** regular customer

die **Stammkneipe, -n** usual bar/pub

ständig constant(ly)

der **Star, -s** star, celebrity

stark (stärker, stärkst-) strong(ly); heavy, heavily; severe(ly) (11); (*coll.*) great

der **Start, -s** start

starten, ist gestartet to start; to take off

die **Stasi** (*coll.*) = die **Staatssicherheit** State Security (*former East German secret police*)

statt (+ *gen.*) instead of (12)

statt·finden (findet ... statt), fand ... statt, stattgefunden to take place (5)

die **Statue, -n** statue

der **Staub** dust; **Staub saugen** to vacuum (6); **Staub wischen** to (wipe) dust der **Staubsauger, -** vacuum cleaner (6)

die **Stauchung, -en** compression

das **Staunen** amazement

das **Steak, -s** steak

stechen (sticht), stach, gestochen to prick; to sting; to bite (*of insects*) (10)

stecken, gesteckt to stick; to put; to be; **stecken bleiben (bleibt ... stecken), blieb ... stecken, ist stecken geblieben** to get stuck (11)

stehen (steht), stand, gestanden to stand (*be in a vertical position*) (2, 6); to be (situated); (+ *dat.*) to suit (6); **stehen für** (+ *acc.*) to stand (for); **Das steht / Die stehen dir gut!** That looks / Those look good on you! (2)

stehlen (stiehlt), stahl, gestohlen to steal (9)

die **Steiermark** Styria (*Austrian state*)

steigen (steigt), stieg, ist gestiegen to climb; to ascend; to increase

der **Stein, -e** stone (12)

steinern (*adj.*) (made of) stone

steinig stony, rocky

stellen, gestellt to stand up, put, place (upright) (3, 5); **eine Frage stellen** to ask a question (5); **gerade stellen** to straighten (3)

sterben (stirbt), starb, ist gestorben to die (9)

die **Stereoanlage, -n** stereo system

der **Stern, -e** star

das **Sternchen, -** asterisk

die **Sternwarte, -n** observatory

das **Sternzeichen, -** astrological sign, sign of the zodiac

die **Steuer, -n** tax

das **Stichwort, ¨-er** keyword

sticken, gestickt to do embroidery

der **Stiefel, -** boot (A)

die **Stiefmutter, ¨-** stepmother (9)

der **Stiefsohn, ¨-e** stepson

die **Stieftochter, ¨-** stepdaughter

der **Stiefvater, ¨-** stepfather (9)

der **Stift, -e** pen (A, B)

der **Stil, -e** style

still quiet(ly), silent(ly)

die **Stille** quiet; silence

stillen, gestillt to still, stop

das **Stillleben, -** still life

die **Stimme, -n** voice (12)

stimmen, gestimmt to be right (8); **das stimmt so** that's right; keep the change (8); **Stimmt!** That's right!

die **Stimmung, -en** mood; atmosphere

das **Stipendium, Stipendien** scholarship (1)

die **Stirn, -en** forehead

das **Stirnband, ¨-er** headband (A)

stöbern, gestöbert (*coll.*) to browse, rummage about

der **Stock, ¨-e** stick; walking stick

der **Stock** (*pl.* **Stockwerke**) floor, story (6); **im ersten Stock** on the second floor (6)

das **Stockwerk, -e** floor, story (6)

stolpern, ist gestolpert to trip, stumble

der **Stopp, -s** stop

stören, gestört to disturb (3)

die **Störung, -en** disturbance

der **Stoßzahn, ¨-e** tusk

stottern, gestottert to stutter

der **Strafzettel, -** (parking or speeding) ticket

der **Strand, ¨-e** shore, beach (4, 7)

der **Strandkorb, ¨-e** beach chair (10)

die **Strandpromenade, -n** (beach) promenade

die **Straße, -n** street, road (6)

die **Straßenbahn, -en** streetcar (7)

der **Straßenrand, ¨-er** roadside

die **Strategie, -n** strategy (12)

der **Strauß, ¨-e** bouquet

das **Streichholz, ¨-er** match (8)

(sich) **streiten (streitet), stritt, gestritten** to argue, quarrel (9)

streng strict(ly) (9)

stricken, gestrickt to knit (3)

der **Strom, ¨-e** current; electricity, power (8)

die **Strophe, -n** strophe; verse

der **Strudel, -** strudel (*pastry*)

die **Strumpfhose, -n** tights; pantyhose

das **Stück, -e** piece; slice (8)

der **Student, -en** (*wk.*) / die **Studentin, -nen** student (A, B)

das **Studentenheim, -e** dorm (2, 6)

die **Studentenkneipe, -n** student pub (1)

das **Studentenleben** student life

das **Studentenwerk, -e** student union

die **Studie, -n** study

das **Studienfach, ¨-er** academic subject

der **Studiengang, ¨-e** course of study

die **Studiengebühr, -en** registration fee, tuition

studieren, studiert to study; to attend a university/college (1)

der/die **Studierende, -n (ein Studierender)** student

das **Studium, Studien** university studies (1); course of studies (3)

der **Stuhl, ¨-e** chair (B, 2)

die **Stunde, -n** hour (1, 2)

stundenlang for hours

der **Stundenplan, ¨-e** schedule (1)

stürmen, gestürmt to storm

stürzen, ist gestürzt to fall

das **Substantiv, -e** noun

die **Suchanzeige, -n** housing-wanted ad

die **Suche, -n** search

suchen, gesucht to look for (1)

die **Suchmaschine, -n** search engine

süd- south

(das) **Südafrika** South Africa

(das) **Südamerika** South America (B)

süddeutsch Southern German (*adj.*)

der **Süden** south

südlich (von + *dat.*) south (of) (7)

der **Südosten** southeast

südöstlich (von + *dat.*) southeast (of)

südwestlich (von + *dat.*) southwest (of)

der **Sünder, -** / die **Sünderin, -nen** sinner

super super

der **Superbowl** Super Bowl

der **Supermarkt, ¨-e** supermarket (5, 6); **im Supermarkt** at the supermarket (5)

die **Suppe, -n** soup (8)

das **Surfbrett, -er** surfboard (2)

surfen, gesurft to surf, go surfing; **im Internet surfen** to surf the Internet (1)

das **Sushi** sushi

süß sweet(ly) (2); **voll süß** totally sweet (2)

die **Süßigkeit, -en** sweet, candy

der **Swimmingpool, -s** swimming pool

das **Symbol, -e** symbol

symbolisch symbolic(ally)

symbolisieren, symbolisiert to symbolize

der **Sympathisant, -en** (*wk.*) / die **Sympathisantin, -nen** sympathizer

sympathisch congenial(ly), appealing(ly)

das **Symphonieorchester, -** symphony orchestra

(das) **Syrien** Syria

das **System, -e** system (12)

die **Szene, -n** scene

das **Szenepublikum** trendy following, in-crowd

die **Tabelle, -n** table; list

die **Tablette, -n** tablet, pill (11)

die **Tafel, -n** blackboard; whiteboard (A, B)

der **Tag, -e** day (1); **an welchem Tag?** on what day? (4); **den ganzen Tag** all day long (1); **eines Tages** one day; **guten Tag** good afternoon; hello (*for.*) (A); **jeden Tag** every day (1); **seit mehreren Tagen** for several days (11); **Welcher Tag ist heute?** What day is today? (1)

das **Tagebuch, ¨-er** diary (4)

der **Tagesablauf, ¨-e** daily routine (1); course of (one's) day

die **Tageszeitung, -en** daily newspaper (5)

täglich daily (12)

das **Tal, ¨-er** valley (7)

das **Talent, -e** talent (3)

der **Tank, -s** (fuel) tank (7)

tanken, getankt to fill up (with gas)

die **Tankstelle, -n** gas station (5); **an der Tankstelle** at the gas station (5)

die **Tante, -n** aunt (B)

tanzen, getanzt to dance (1)

die **Tanzfläche, -n** dance floor

die **Tanzschule, -n** dancing school

das **Tanzstudio, -s** dance studio

die **Tapete, -n** wallpaper

tapfer brave(ly) (9)
die **Tasche, -n** (hand)bag; purse; pocket (1)
die **Taschenlampe, -n** flashlight (9)
das **Taschentuch, ¨-er** handkerchief (3)
die **Tasse, -n** cup (2, 5)
die **Tat, -en** act; deed
tätig active
die **Tätigkeit, -en** activity (6)
tätowieren, tätowiert to tattoo
der **Tätowierer, -** / die **Tätowiererin, -nen** tattoo artist
das **Tattoo, -s** tattoo
die **Taube, -n** pigeon; dove
der/die **Taubstumme, -n (ein Taubstummer)** person who is hearing-impaired and cannot speak
tauchen, hat/ist getaucht to dive (3)
die **Taufe, -n** baptism, christening (4)
taugen, getaugt: nichts taugen to be no good
tausend thousand
tausendmal a thousand times
das **Taxi, -s** taxi (3, 7)
der **Taxifahrer, -** / die **Taxifahrerin, -nen** taxi driver (5)
die **Technik** technology; engineering
das/der **Techno** techno (music)
der **Teddy, -s** = der **Teddybär, -en** (*wk.*) teddy bear (A)
der **Tee, -s** tea (4)
die **Teekanne, -n** teapot (8)
der **Teekessel, -** tea kettle (8)
der **Teenager, -** teenager
der **Teig, -e** dough
der **Teil, -e** part (7); **zum Teil** partly
teilen, geteilt to divide; to share
teil·nehmen (an + *dat.*) **(nimmt ... teil), nahm ... teil, teilgenommen** to participate (*in s.th.*)
das **Telefon, -e** telephone (1, 2); **am Telefon** on the phone (2)
das **Telefonat, -e** telephone call
telefonieren, telefoniert to telephone, talk on the phone (4)
die **Telefonkarte, -n** telephone card (2)
die **Telefonnummer, -n** telephone number (1)
die **Telefonzelle, -n** telephone booth
der **Teller, -** plate (8)
die **Temperatur, -en** temperature
das **Tennis** tennis (1)
der **Teppich, -e** carpet (2); rug
der **Termin, -e** appointment (5, 11); **sich einen Termin geben lassen** to get an appointment (11)
der **Terminkalender, -** appointment calendar (11)
die **Terrasse, -n** terrace, deck (6)
der **Terrorismus** terrorism (12)
der **Test, -s** test
der **Tetanus** tetanus (11)
teuer expensive(ly) (2)
der **Teufel, -** devil (12)
der **Text, -e** text (12)
das **Theater, -** theater (6)
das **Thema, Themen** theme, topic, subject (4)
der **Theologe, -n** (*wk.*) / die **Theologin, -nen** theologian
die **Theorie, -n** theory
der **Therapeut, -en** (*wk.*) / die **Therapeutin, -nen** therapist
der **Thunfisch, -e** tuna
(das) **Thüringen** Thuringia
das **Ticket, -s** ticket
tief deep(ly) (7)
das **Tier, -e** animal (3, 7, 10)
der **Tierarzt, ¨-e** / die **Tierärztin, -nen** veterinarian
der **Tierschutz** animal protection
der **Tiger, -** tiger
tippen, getippt to type (3, 6)
(das) **Tirol** Tyrol (*Austrian state*)
Tiroler (*adj.*) Tyrolean
der **Tisch, -e** table (B); **den Tisch abräumen** to clear the table (3); **den Tisch decken** to set the table (5)
das **Tischtennis** table tennis (3)
der **Titel, -** title
der **Toaster, -** toaster
toben, getobt to rampage
die **Tochter, ¨** daughter (B)
der **Tod, -e** death (12)
die **Todesart, -en** way to die
die **Todesgefahr, -en** mortal danger
die **Toilette, -n** toilet (6)
das **Toilettenpapier** toilet paper
die **Toilettentasche, -n** cosmetic bag
tolerant tolerant(ly) (7)
toll (*coll.*) neat, great (2); **das hört sich toll an** that sounds great (4)
die **Tollwut** rabies (10)
die **Tomate, -n** tomato (8)
der **Ton, -e** clay (12)
der **Ton, ¨-e** tone; musical note
der **Topf, ¨-e** pot, pan (5)
die **Töpferscheibe, -n** potter's wheel (12)
der **Topflappen, -** potholder (5)
das **Topjahrzehnt, -e** top decade
das **Torhaus, ¨-er** gatehouse
tot dead (9)
total total(ly) (4)
töten, getötet to kill (9)
der **Totengang, ¨-e** path of the dead
der **Totenkopf, ¨-e** skull; death's head
sich **tot·schuften, totgeschuftet** (*coll.*) to work oneself to death
die **Tour, -en** tour; trip
das **Tourette-Syndrom** Tourette syndrome
der **Tourismus** tourism (10)
der **Tourist, -en** (*wk.*) / die **Touristin, -nen** tourist
die **Tradition, -en** tradition (4, 12)
traditionell traditional(ly)
traditionsreich rich in tradition
tragen (trägt), trug, getragen to carry; to wear (A); **Trägst du gern ...?** Do you like to wear . . . ? (A)
die **Tragikomödie, -n** tragicomedy
die **Tragödie, -n** tragedy
der **Trailer, -** trailer
trampeln, hat/ist getrampelt to stomp
trampen, ist getrampt to hitchhike (10)
der **Tramper, -** / die **Tramperin, -nen** hitchhiker
die **Transaktion, -en** transaction
transkontinental transcontinental(ly)
transportieren, transportiert to transport (7)
das **Transportmittel, -** means of transportation (7)
trauen (+ *dat.*), **getraut** to trust (8)
die **Trauer** sorrow (2)
der **Traum, ¨-e** dream
die **Traumdeutung** interpretation of dreams
träumen (von + *dat.*), **geträumt** to dream (of/about) (9)
traurig sad(ly) (B)
(sich) **treffen (trifft), traf, getroffen** to meet (2); **Treffen wir uns ...** Let's meet . . . (2)
treiben (treibt), trieb, getrieben to drive; to carry out, do; **Sport treiben** to do sports (2)
trennbar separable
(sich) **trennen, getrennt** to separate (7); to break up (*people*); to divide
die **Treppe, -n** stairway (6)
das **Treppenhaus, ¨-er** stairwell
treten (tritt), trat, ist getreten to step
treu loyal(ly); true (9)
die **Treue** loyalty, fidelity
(das) **Triest** Trieste (*city in Italy*)
trinken (trinkt), trank, getrunken to drink (1)
das **Trinkgeld, -er** tip (8)
trivial trivial(ly); trite(ly)
trocken dry (11)
die **Trompete, -n** trumpet (12)
der **Trost** consolation
trotz (+ *gen.*) in spite of (12)
trotzdem in spite of that; nonetheless (12)
die **Truppen** (*pl.*) troops
(das) **Tschechien** Czech Republic (B)
(die) **Tschechische Republik** Czech Republic
tschüss bye (*infor.*) (A)
das **T-Shirt, -s** T-shirt (2)
tun (tut), tat, getan to do (1)
die **Tür, -en** door (A)
der **Türke, -n** (*wk.*) / die **Türkin, -nen** Turkish man/woman (12)
die **Türkei** Turkey (B)

türkisch Turkish (*adj.*)
(das) **Türkisch** Turkish (*language*) (B)
der **Turnschuh, -e** gym shoe
die **Türschwelle, -n** threshold
die **Tüte, -n** (paper or plastic) bag (11)
der **TÜV** = der **Technische Überwachungsverein** Technical Control Board (*German agency that checks vehicular safety*)
twittern, getwittert to use Twitter, tweet
der **Typ, -en** (*coll.*) character, person, guy
typisch typical(ly)
u. a. = **unter anderem** among others
die **U-Bahn, -en** = die **Untergrundbahn, -en** subway (7)
übel bad, nasty; **übel sein** (+ *dat.*) to feel sick, **mir ist übel** I feel sick
üben, geübt to practice; to exercise
über (+ *dat./acc.*) over (4); above; about; across; **übers Wochenende** over the weekend (4)
überall everywhere (10)
überfahren (überfährt), überfuhr, überfahren to run over (11)
überfallen (überfällt), überfiel, überfallen to hold up (*bank/store*)
überfliegen (überfliegt), überflog, überflogen to skim
über·gehen (geht ... über), ging ... über, ist übergegangen to transfer, pass across (8)
überhaupt anyway; at all
überleben, überlebt to survive
überlegen, überlegt to consider, think about
übermorgen the day after tomorrow (9)
übermütig in high spirits, cocky
übernachten, übernachtet to stay overnight (6)
die **Übernachtung, -en** overnight stay
übernehmen (übernimmt), übernahm, übernommen to take on (7); to take over, adopt
überprüfen, überprüft to check, inspect
überqueren, überquert to cross, go across
überraschen, überrascht to surprise
überreden, überredet to convince, persuade
die **Überredungskunst, ¨-e** powers of persuasion
überrollen, überrollt to overrun
übers = **über das** over/about the; see **über**
der **Überschwang** exuberance
übersetzen, übersetzt to translate (9)
überwachen, überwacht to monitor
die **Überwachung, -en** surveillance
überweisen (überweist), überwies, überwiesen to transfer (*money*)
die **Überweisung, -en** transfer (*of money*) (12)
überwiegen (überwiegt), überwog, überwogen to predominate
überzeugen, überzeugt to convince
überziehen (überzieht), überzog, überzogen to overdraw
der **Überziehungskredit, -e** overdraft protection
üblich usual, customary
übrig remaining, left over
die **Übung, -en** exercise (A)
die **UdSSR** = die **Union der Sozialistischen Sowjetrepubliken** USSR, Soviet Union
die **Uhr, -en** clock (B); watch; (*sg. only*) o'clock; **bis acht Uhr** until eight o'clock (2); **bis um vier Uhr (früh)** until four o'clock (in the morning) (4); **erst um vier Uhr** not until four o'clock (4); **um sechs Uhr** at six o'clock (1); **um sieben Uhr zwanzig** at seven twenty (1); **Um wie viel Uhr ...?** At what time . . . ? (1); **Wie viel Uhr ist es?** What time is it? (1)
die **Uhrzeit, -en** time
die **Ukraine** Ukraine
um around; about; at; for; **(gleich) um die Ecke** (right) around the corner (5, 6); **um halb drei** at two thirty (1); **um sechs (Uhr)** at six o'clock (1); **um Viertel vor vier** at a quarter to four (1); **Um wie viel Uhr ...?** At what time . . . ? (1); **um zwanzig nach fünf** at twenty after/past five (1)
um ... zu (+ *inf.*) in order to (12)
um·bringen (bringt ... um), brachte ... um, umgebracht to kill
um·fallen (fällt ... um), fiel ... um, ist umgefallen to fall over (9)
umfassen, umfasst to embrace; to include
die **Umfrage, -n** survey (4)
der **Umgang** contact
umgeben (umgibt), umgab, umgeben to surround, enclose
die **Umgebung, -en** surrounding area, environs
um·gehen (mit + *dat.*) **(geht ... um), ging ... um, umgegangen** to treat, handle
sich **um·gucken, umgeguckt** (*coll.*) to look around
um·hängen, umgehängt to hang somewhere else
um·kippen, ist/hat umgekippt to turn over; to knock over
die **Umkleidekabine, -n** dressing room
ums = **um das** around/about/at/for the
der **Umsatz, ¨-e** sales, returns
der **Umschlag, ¨-e** cover; envelope; **warmer Umschlag** warm compress, poultice
um·schlagen (schlägt ... um), schlug ... um, ist/hat umgeschlagen to change
der **Umstand, ¨-e** circumstance
die **Umstellung, -en** adjustment
um·steigen (steigt ... um), stieg ... um, ist umgestiegen to change (*trains etc.*) (7)
der **Umweg, -e** circuitous route, detour
die **Umwelt** environment
die **Umweltkunde** environmental studies
der **Umweltschutz** environmental protection
um·werfen (wirft ... um), warf ... um, umgeworfen to knock over/down
um·ziehen (zieht ... um), zog ... um, ist umgezogen to move (*to another residence*); (sich) **umziehen, hat umgezogen** to change clothes
der **Umzug, ¨-e** move, relocation
der **Umzugsservice, -s** moving service
unachtsam inattentive(ly)
unangenehm unpleasant(ly)
unbedingt without fail; absolute(ly)
das **Unbewusste** (*declined as adj.*) unconscious
und (*coord. conj.*) and (A, 11); **und so weiter (usw.)** and so forth
unerwartet unexpected(ly)
der **Unfall, ¨-e** accident (4, 7)
das **Unfallkommando, -s** accident response unit
die **Unfallstelle, -n** scene of an accident
(das) **Ungarn** Hungary (B)
ungeduldig impatient(ly) (10)
ungefähr approximate(ly) (7)
ungeheuer enormous(ly); terrible, terribly
ungemein exceptional(ly)
ungenügend inadequate(ly); unsatisfactory, unsatisfactorily
ungerecht unjust(ly), unfair(ly)
ungewöhnlich unusual(ly)
ungezogen naughty, naughtily; badly behaved
unglaublich incredible, incredibly (5)
unhöflich impolite(ly)
die **Uni, -s** (*coll.*) = die **Universität, -en** university (B, 1); **auf der Uni sein** to be at the university (1); **zur Uni gehen** to go to the university (1, 2)
die **Union, -en** union; die **Europäische Union (EU)** European Union
die **Universität, -en** university (B, 1, 5); **auf der Universität** at the university (5)
unklug unwise(ly)
unmöglich impossible, impossibly
die **Unmündigkeit** dependence
unrechtmäßig illegal(ly)
uns (*acc./dat.*) us; ourselves (1)
unsaniert unrestored, unrenovated
die **Unschuld** innocence
unser(e) our (2)
unsympathisch uncongenial(ly); disagreeable, disagreeably; unpleasant(ly)
unten (*adv.*) below; down; downstairs
unter (+ *dat./acc.*) under, underneath (5); below, beneath; among (6); (*adj.*) lower; **unter anderem** among other things; **unter dem Fenster** under the window (5)
das **Unterbewusste** (*declined as adj.*) subconscious
die **Untergrundbahn, -en** = die **U-Bahn, -en** subway
sich **unterhalten (unterhält), unterhielt, unterhalten** to converse (9)

die **Unterhaltung, -en** conversation; entertainment

das **Unterhemd, -en** undershirt (2)

die **Unterhose, -n** underpants (2)

die **Unterkunft, ¨-e** lodging

der **Unterlass: ohne Unterlass** incessantly

unternehmen (unternimmt), unternahm, unternommen to undertake

das **Unternehmen** undertaking; enterprise; company

der **Unterricht** class, instruction (9, 11)

unterrichten, unterrichtet to teach, instruct (5)

der **Unterschied, -e** difference

unterschiedlich different; various(ly)

unterschreiben (unterschreibt), unterschrieb, unterschrieben to sign (1); **Unterschreib bitte hier.** Sign here please. (A)

die **Unterschrift, -en** signature (1)

unterstreichen (unterstreicht), unterstrich, unterstrichen to underline

unterstützen, unterstützt to support (12)

die **Unterstützung** support

untersuchen, untersucht to investigate; to examine (5)

unterwegs underway; on the road (9)

unterzeichnen, unterzeichnet to sign

untreu disloyal; unfaithful

das **Unverständnis** lack of understanding

unvorhergesehen unforeseen; unexpected(ly)

unwichtig unimportant

unzufrieden dissatisfied

uralt very old, ancient

der **Uranus** Uranus

urkundlich erwähnt mentioned in a document

der **Urlaub, -e** vacation (4, 5); **Urlaub machen** to take a vacation

der **Urlauber, -** / die **Urlauberin, -nen** vacationer

die **Ursache, -n** cause

der **Ursprung, ¨-e** origin

ursprünglich original(ly)

das **Urteil, -e** verdict

die **USA** (*pl.*) USA (B)

der **US-Amerikaner, -** / die **US-Amerikanerin, -nen** American (from the USA) (*person*)

US-amerikanisch American (from the USA) (*adj.*)

usw. = **und so weiter** and so forth

die **Utopie, -n** utopia

der **Valentinstag** Valentine's Day

der **Vampir, -e** vampire

die **Vase, -n** vase

der **Vater, ¨** father (B)

die **Vaterstadt, ¨-e** hometown

der **Vati, -s** dad, daddy

der **Vegetarier, -** / die **Vegetarierin, -nen** vegetarian (*person*)

sich **verabreden (mit** + *dat.*), **verabredet** to make a date (with), make an appointment (with)

die **Verabredung, -en** appointment; date (11)

sich **verabschieden, verabschiedet** to say good-bye, take leave

(sich) **verändern, verändert** to change (6)

die **Veränderung, -en** change

verarbeiten zu (+ *dat.*), **verarbeitet** to make into

das **Verb, -en** verb

der **Verband, ¨-e** bandage (11)

verbessern, verbessert to improve; to correct

verbieten (verbietet), verbot, verboten to forbid

verbinden (verbindet), verband, verbunden to connect (A); to combine

verboten (*p.p. of* **verbieten**) forbidden, prohibited (9)

verbrennen (verbrennt), verbrannte, verbrannt to burn (11); **sich (die Zunge) verbrennen** to burn (one's tongue) (11)

verbringen (verbringt), verbrachte, verbracht to spend (*time*) (3)

verdienen, verdient to earn (4)

der **Verdienst, -e** earnings

verdrängen, verdrängt to drive out, displace

verdutzt taken aback

der **Verein, -e** society, association

vereinbaren, vereinbart to agree upon (12)

vereinen, vereint to unite

vereinigen, vereinigt to unite

verfallen (verfällt), verfiel, ist verfallen to decline; to deteriorate

die **Verfassung, -en** constitution; **körperliche und geistige Verfassung** physical and mental state

verfehlen, verfehlt to miss, not notice (10)

verfolgen, verfolgt to persecute

die **Verfügung, -en** order; **zur Verfügung** at one's disposal

die **Vergebung** forgiveness

vergehen (vergeht), verging, ist vergangen to pass, go by (*time*)

vergessen (vergisst), vergaß, vergessen to forget (2)

vergiften, vergiftet to poison (9)

der **Vergleich, -e** comparison

vergleichbar comparable

vergleichen (vergleicht), verglich, verglichen to compare (7)

das **Vergnügen** pleasure (2); entertainment

vergnügt cheerful(ly); happy, happily

verhaften, verhaftet to arrest

sich **verhalten (verhält), verhielt, verhalten** to behave, act

das **Verhältnis, -se** relationship

verharren, verharrt to remain

sich **verheddern (in** + *dat.*), **verheddert** to get tangled up (in)

verheimlichen, verheimlicht to conceal, keep secret

sich **verheiraten (mit** + *dat.*), **verheiratet** to get married (to)

verheiratet married (1)

verhindern, verhindert to prevent

verhören, verhört to interrogate

verhungern, ist verhungert to starve (to death)

verkaufen, verkauft to sell (2, 5); **zu verkaufen** for sale

der **Verkäufer, -** / die **Verkäuferin, -nen** salesperson (5)

der **Verkehr** traffic (7, 11)

das **Verkehrsmittel, -** means of transportation

das **Verkehrsschild, -er** traffic sign (7)

verkommen (verkommt), verkam, ist verkommen to degenerate, go bad

verlassen (verlässt), verließ, verlassen to leave; to abandon (11)

sich **verletzen, verletzt** to injure oneself (11)

verletzt injured (11); **schwer verletzt** critically injured (11)

der/die **Verletzte, -n (ein Verletzter)** injured person (11)

die **Verletzung, -en** injury

sich **verlieben (in** + *acc.*), **verliebt** to fall in love (with) (9, 12)

verliebt (sein) (to be) in love (4, 12)

verlieren (verliert), verlor, verloren to lose (7)

sich **verloben (mit** + *dat.*), **verlobt** to get engaged (to); **verlobt sein** to be engaged

vermieten, vermietet to rent out (6)

der **Vermieter, -** / die **Vermieterin, -nen** landlord/landlady (6)

vermischen, vermischt to mix

vermissen, vermisst to miss

das **Vermögen, -** fortune

vernehmen (vernimmt), vernahm, vernommen to question

veröffentlichen, veröffentlicht to publish

die **Veröffentlichung, -en** publication

verpassen, verpasst to miss (9)

verquer: heute ist alles verquer everything is going wrong today

verraten (verrät), verriet, verraten to betray; to disclose, give away (*a secret*)

verreisen, ist verreist to go on a trip (3)

verrücken, verrückt to move, shift

verrückt crazy, crazily

verrühren, verrührt to stir together

verschieden different(ly); various(ly) (12)

verschlafen (verschläft), verschlief, verschlafen to sleep in, oversleep

verschlingen (verschlingt), verschlang, verschlungen to devour, swallow up

verschlossen reserved; taciturn

verschlucken, verschluckt to swallow; **sich verschlucken (an** + *dat.*), **verschluckt** to choke (on)

verschollen lost; missing

verschütten, verschüttet to spill

verschwinden (verschwindet), verschwand, ist verschwunden to disappear

versenden (versendet), versandte/versendete, versandt/versendet to send

die **Versetzung, -en** promotion (*to next grade in school*)

die **Versicherung, -en** insurance (5)

die **Version, -en** version

die **Verspätung, -en** lateness; delay (9)

verspeisen, verspeist to consume

versprechen (verspricht), versprach, versprochen to promise (7)

verständigen, verständigt to notify, inform

die **Verständigung** communication

das **Verständnis** understanding

verstauen, verstaut to stow (7)

das **Versteck, -e** hiding place

(sich) **verstecken, versteckt** to hide (9)

verstehen (versteht), verstand, verstanden to understand (4); **sich verstehen mit jemandem** to understand someone

versuchen, versucht to try, attempt (4)

verteilen, verteilt to distribute (12)

die **Verteilung** distribution

der **Vertrag, ¨-e** contract

vertrauen (+ *dat.*), **vertraut** to trust

vertreiben (vertreibt), vertrieb, vertrieben to drive (*s.o./s.th.*) out/away, expel

vertreten (vertritt), vertrat, vertreten to represent; to plead for

der **Vertreter, -** / die **Vertreterin, -nen** representative

die **Vertretung, -en** delegation

verunglücken, verunglückt to have an accident

verurteilen, verurteilt to sentence; to condemn

vervollständigen, vervollständigt to complete

verwalten, verwaltet to administer

die **Verwaltung, -en** administration

verwandeln, verwandelt to convert, transform; **sich verwandeln (in** + *acc.***)** to change (into) (9)

verwandt (mit + *dat.***)** related (to)

der/die **Verwandte, -n (ein Verwandter)** relative (2)

verwanzen, verwanzt to bug, plant listening devices

verweisen (auf + *acc.***) (verweist), verwies, verwiesen** to refer (to)

verwenden, verwendet to use

verwunschen cursed, enchanted (9)

verwünschen, verwünscht to curse, cast a spell on (9)

verzaubert (*p.p. of* **verzaubern**) bewitched

die **Verzeihung** forgiveness

verzichten (auf + *acc.***), verzichtet** to do without, renounce (*s.th.*)

verzweifelt desperate(ly); despairing(ly)

der **Vetter, -n** (male) cousin (B)

das **Video, -s** video (9)

das **Videospiel, -e** video game (5)

viel (*sg.*) a lot (of), much (A); **viele** (*pl.*) many (A); **ganz schön viel** quite a bit (3); **Um wie viel Uhr ...?** At what time . . . ? (1); **vielen Dank** many thanks (10); **Viel Glück!** Lots of luck! Good luck! (3); **viel Spaß** have fun (A); **Wie viel ...?** How much . . . ?; **Wie viele ...?** How many . . . ? (A); **Wie viel Uhr ist es?** What time is it? (1)

vielleicht perhaps (2); maybe

der **Vielvölkerstaat, -en** multinational state

vier four (A)

viert- fourth (4)

das **Viertel, -** quarter; **um Viertel vor vier** at a quarter to four (1)

vierundzwanzig twenty-four (A)

vierzehn fourteen (A)

vierzig forty (A)

vierzigst- fortieth

die **Villa, Villen** villa (6)

violett violet

die **Visa-Karte, -n** Visa card

die **Vision, -en** vision

der **Vogel, ¨** bird (10)

das **Vöglein, -** little bird

der **Vokal, -e** vowel

das **Volk, ¨-er** people (12)

die **Volksmusik** folk music

die **Volkspartei, -en** people's party

der **Volksschullehrer, -** / die **Volksschullehrerin, -nen** primary school teacher

die **Volkszählung, -en** census

voll full; full of; fully (10); **voll süß** totally sweet (2)

vollenden, vollendet to complete, finish

der **Volleyball, ¨-e** volleyball (1)

völlig fully, completely

vollkommen perfect(ly); flawless(ly) (12); complete(ly)

das **Vollkornbrot, -e** whole grain bread

vollständig complete(ly)

voll·tanken, vollgetankt to fill up (with gas) (5)

vom = **von dem** of/from/by the

von (+ *dat.*) of; from (A, 10); by; **von der Arbeit** from work (3); **von nebenan** from next door (5); **Was sind Sie von Beruf?** What's your profession? (1)

vor (+ *dat./acc.*) before; in front of; ago (4); because of; **um Viertel vor vier** at a quarter to four (1); **vor allem** above all; **vor Lachen** from laughing (so hard); **vor zwei Tagen** two days ago (4)

die **Voraussetzung, -en** prerequisite

voraussichtlich expected; probably

vorbei past, over (9); **an** (+ *dat.*) **... vorbei** by (10)

vorbei·fahren (fährt ... vorbei), fuhr ... vorbei, ist vorbeigefahren to go by

vorbei·fliegen (fliegt ... vorbei), flog ... vorbei, ist vorbeigeflogen to fly by

vorbei·gehen (an + *dat.***) (geht ... vorbei), ging ... vorbei, ist vorbeigegangen** to go by (10)

vorbei·kommen (kommt ... vorbei), kam ... vorbei, ist vorbeigekommen to come by; to visit (3); **Komm mal vorbei!** Come on over! (11)

vorbei·schieben (schiebt ... vorbei), schob ... vorbei, vorbeigeschoben to push past

(sich) **vor·bereiten, vorbereitet** to prepare (4)

die **Vorbereitung, -en** preparation

das **Vorbild, -er** role model, idol

vorder- (*adj.*) front

der **Vordergrund** foreground

die **Vorderseite, -n** front (side); obverse (*of a coin*)

der **Vorfahre, -n** (*wk.*) ancestor (12)

die **Vorfahrt** right-of-way

vor·gehen (gegen + *acc.***) (geht ... vor), ging ... vor, ist vorgegangen** to take action (against)

die **Vorgeschichte** prehistory

vorgestern the day before yesterday

vor·haben (hat ... vor), hatte ... vor, vorgehabt to plan, intend

der **Vorhang, ¨-e** drapery, curtain (6)

vor·heizen, vorgeheizt to preheat

vorhersagbar predictable

die **Vorhersage, -n** prediction

vor·kommen (kommt ... vor), kam ... vor, ist vorgekommen to occur; (+ *dat.*) to seem (to *s.o.*)

vor·lesen (liest ... vor), las ... vor, vorgelesen to read aloud (9)

die **Vorlesung, -en** lecture (4)

der **Vormittag, -e** late morning (4)

vormittags in the morning(s)

der **Vorname, -n** (*wk.*) first name, given name (A, 1)

vorne at the front; **von vorne** from the beginning

der **Vorreiter, -** / die **Vorreiterin, -nen** pioneer

vors = **vor das** in front of the

der **Vorschlag, ¨-e** suggestion (5)

vorsichtig cautious(ly) (8)

die **Vorspeise, -n** appetizer (8)

vor·spielen, vorgespielt to perform

(sich) **vor·stellen, vorgestellt** to introduce (*o.s.*); to present (*o.s.*) (6); **sich** (*dat.*) **etwas vorstellen** to imagine something (6)

der **Vorteil, -e** advantage (7)

der **Vortrag, ¨-e** talk; lecture; presentation

der **Vulkan, -e** volcano

wach·rütteln, wachgerüttelt to rouse out of apathy

wachsen (wächst), wuchs, ist gewachsen to grow (9)

das **Wachstum** growth (12)

der **Wachtmeister, -** / die **Wachtmeisterin, -nen** (police) constable

wagen, gewagt to dare; to risk

der **Wagen, -** car (7)

der **Waggon, -s** train car (7)

die **Wahl, -en** choice; election (12)

wählen, gewählt to choose, select; to vote (for); to elect (12)

der **Wähler, -** / die **Wählerin, -nen** voter

wahlfrei optional

der **Wahlpflichtunterricht** compulsory class

der **Wahnsinn** insanity, madness

wahnsinnig crazy, crazily; insane(ly) (12)

wahr true (3); **nicht wahr?** isn't it so?

während (+ *gen.*) during (11); (*subord. conj.*) while

die **Wahrheit, -en** truth

wahrscheinlich probable, probably (1)

das **Wahrzeichen, -** symbol

die **Währung, -en** currency (12)

der **Wald, ¨er** forest, woods (2, 7); **im Wald laufen** to run in the woods (2)

die **Walpurgisnacht** Walpurgis Night (*the witches' sabbath, April 30*)

der **Walzer, -** waltz (3)

die **Wand, ¨e** wall (B); **das Bild an die Wand hängen** to hang the picture on the wall (3)

die **Wandergans, ¨e** migratory goose

wandern, ist gewandert to hike (1); **in den Bergen wandern** to hike in the mountains (1)

der **Wanderschuh, -e** hiking shoe/boot (2)

die **Wanderung, -en** hike (7)

die **Wange, -n** cheek

wann when (B, 1); **Wann sind Sie geboren?** When were you born? (1)

der **Wannsee** *a lake in Berlin*

das **Wappen, -** coat of arms

warm (wärmer, wärmst-) warm(ly) (B); (*of room/ apartment*) heated, heat included (6)

warnen, gewarnt to warn (7)

warten (auf + *acc.*)**, gewartet** to wait (for) (7); **ein Auto warten** to do maintenance on a car

das **Wartezimmer, -** waiting room

warum why (3)

was what (B); **was für** (+ *acc.*) what kind of (3); **Was sind Sie von Beruf?** What's your profession? (1); **Was zeigen Ihre Bilder?** What do your pictures show? (A)

das **Waschbecken, -** (wash)basin (6)

die **Wäsche** laundry (4)

(sich) **waschen (wäscht), wusch, gewaschen** to wash (*o.s.*) (2, 11)

der **Wäschetrockner, -** clothes dryer (8)

die **Waschmaschine, -n** washing machine (6)

der **Waschraum, ¨e** laundry room

der **Waschsalon, -s** laundromat

das **Wasser** water

der **Wasserhahn, ¨e** faucet (5)

der **Wasservogel, ¨** waterfowl

der **Wechselkurs, -e** exchange rate

wechseln, gewechselt to change; **Geld wechseln** to exchange money

wecken, geweckt to wake (*s.o.*) up (7)

der **Wecker, -** alarm clock (2)

weder ... noch neither . . . nor

weg away; **Wie weit weg?** How far away? (6)

der **Weg, -e** way; road; path; **den Weg beschreiben** to give directions; **nach dem Weg fragen** to ask for directions; **sich auf den Weg machen** to go on one's way, set off

weg·bringen (bringt ... weg), brachte ... weg, weggebracht to take out; to take away (5)

wegen (+ *gen.*) on account of; about (6)

weg·fahren (fährt ... weg), fuhr ... weg, ist weggefahren to drive off, leave

weg·führen, weggeführt to lead away

weg·gehen (geht ... weg), ging ... weg, ist weggegangen to go away, leave (4)

weg·laufen (läuft ... weg), lief ... weg, ist weggelaufen to run away

weg·nehmen (nimmt ... weg), nahm ... weg, weggenommen to take away

weg·stellen, weggestellt to put away (5)

weg·tragen (trägt ... weg), trug ... weg, weggetragen to carry away (9)

weg·trampen, ist weggetrampt to hitchhike away

weg·ziehen (zieht ... weg), zog ... weg, ist weggezogen to move away

wehen, geweht to blow

sich **wehren, gewehrt** to defend oneself

wehrhaft able to defend oneself

weh·tun (tut ... weh), tat ... weh, wehgetan to hurt (11)

weiblich female; feminine(ly)

weich soft(ly)

die **Weihe, -n** harrier (*type of hawk*)

(das) **Weihnachten** Christmas (4)

das **Weihnachtsgeschenk, -e** Christmas present (5)

weil (*subord. conj.*) because (3, 11)

das **Weilchen, -** little while

die **Weile, -n** while

Weimarer (*adj.*) (of) Weimar

der **Wein, -e** wine (7)

weinen, geweint to cry (3)

die **Weintraube, -n** grape (8)

weiß white (A); **die Weiße Rose** the White Rose (*name of an anti-Nazi resistance group*)

die **Weiße:** die **Berliner Weiße** *light, fizzy beer served with raspberry syrup*

weit far (6); **Wie weit weg?** How far away? (6)

weiter (*adj.*) additional; (*adv.*) farther; further; **und so weiter (usw.)** and so forth

die **Weiterbildung** continuing education

weiter·entwickeln, weiterentwickelt to develop further

weiter·gehen (geht ... weiter), ging ... weiter, ist weitergegangen to keep on walking (10); to go on, continue

weiter·helfen (hilft ... weiter), half ... weiter, weitergeholfen to help further

weiterhin still; in addition

weiter·leben, weitergelebt to go on living

das **Weizenmehl** wheat flour

welch- which, what (B); **an welchem Tag?** on what day? (4); **Welche Farbe hat ...?** What color is . . . ? (A); **Welcher Tag ist heute?** What day is today? (1); **Welches Datum ist heute?** What is today's date? (4)

die **Welle, -n** wave (10); well, shaft

die **Welt, -en** world (7); **alle Welt** (*coll.*) the whole world; everybody

das **Weltkulturerbe** world cultural heritage

weltweit worldwide; all over the world

wem whom (*dat.*) (5)

wen whom (*acc.*) (4)

die **Wende, -n** change

wenig (*sg.*) little; **am wenigsten** the least (8); **wenige** (*pl.*) few; **wenigstens** at least (1)

wenn (*subord. conj.*) if; when(ever) (2, 11); **wenn ja** if so

wer who (A, B)

werben (wirbt), warb, geworben to advertise

das **Werbeplakat, -e** advertising sign

die **Werbung, -en** advertisement

werden (wird), wurde, ist geworden to become (5)

werfen (wirft), warf, geworfen to throw (3)

das **Werk, -e** work, product (9)

die **Werkkunst, ¨e** applied art

die **Werkstatt, ¨en** workshop; repair shop, garage (5)

das **Werkzeug, -e** tool (8)

wert worth

der **Wert, -e** value

wertvoll valuable, expensive (2)

weshalb why

wessen whose

west- west

(das) **Westdeutschland** (*former*) West Germany

der **Westen** west

westgermanisch West Germanic (*adj.*)

westlich (von + *dat.*) west (of) (7)

das **Wetter** weather (B)

das **Wettrennen, -** race

die **WG, -s** = die **Wohngemeinschaft, -en** shared housing (6)

wichtig important (2)

die **Wichtigkeit** importance

wider·spiegeln, widergespiegelt to reflect

der **Widerstand, ¨e** resistance

widmen, gewidmet to dedicate

wie how (B); **Um wie viel Uhr ...?** At what time . . . ? (1); **Wie fühlst du dich?** How do you feel? (3); **Wie heißen Sie?** (*for.*) / **Wie heißt du?** (*infor.*) What's your name? (A); **Wie schreibt man das?** How do you spell that? (A); **Wie spät ist es?** What time is it? (1); **Wie viel ...?** How much . . . ?; **Wie viele ...?** How many . . . ? (A); **Wie viel Uhr ist es?** What time is it? (1); **Wie weit weg?** How far away? (6)

wieder again (3); **hin und wieder** now and then; **schon wieder** once again (3)
wiederholen, wiederholt to repeat (6)
das **Wiederhören: auf Wiederhören** good-bye (*on the phone*) (6)
wiederkehrend recurring
das **Wiedersehen: auf Wiedersehen** good-bye (A)
wiederum again
die **Wiedervereinigung, -en** reunification
(das) **Wien** Vienna
Wiener Viennese (*adj.*); **das Wiener Schnitzel** breaded veal cutlet
die **Wiese, -n** meadow, pasture (7)
wieso why
wild wild(ly)
die **Wildnis, -se** wilderness
das **Wildschwein, -e** wild boar (10)
willkommen welcome
der **Wind, -e** wind (9)
windig windy (B)
windsurfen gehen (geht ... windsurfen), ging ... windsurfen, ist windsurfen gegangen to go windsurfing (10)
der **Winter, -** winter (B)
wir we
wirken, gewirkt to work, take effect (11)
wirklich real(ly) (B)
die **Wirklichkeit, -en** reality
der **Wirt, -e** / die **Wirtin, -nen** host/hostess; innkeeper; barkeeper (10)
die **Wirtschaft, -en** economy; economics (1)
wirtschaftlich economic(ally)
wirtschaftsfreundlich pro-business (*adj.*)
wischen, gewischt to wipe (7); to mop (6); **Staub wischen** to (wipe) dust
wissen (weiß), wusste, gewusst to know (*as a fact*) (2); **Bescheid wissen** to know; to have an idea
die **Wissenschaft, -en** science, field of study (9)
der **Wissenschaftler, -** / die **Wissenschaftlerin, -nen** scientist (12)
der **Witz, -e** joke (3); **Witze erzählen** to tell jokes (3)
wo where (B); **Wo willst du denn hin?** Where are you going? (A)
wobei with/at/during what
die **Woche, -n** week (1); **in der Woche** during the week (1); **jede Woche** every week (3); **letzte Woche** last week (4)
das **Wochenende, -n** weekend (1); **am Wochenende** over the weekend (1); **letztes Wochenende** last weekend (4)
das **Wochenendhaus, ¨er** weekend cabin/cottage
der **Wochenplan, ¨e** weekly schedule
wodurch through what
wofür for what
wogegen against what
woher from where (B); whence
wohin where to (3); whither
wohl probably (12); well; **sich wohl fühlen** to feel well (11)
wohlbekannt well-known
wohlhabend prosperous
der **Wohlstand** prosperity
der **Wohnblock, -s** *or* **¨e** residential block, apartment complex
wohnen (in + *dat.***), gewohnt** to live (in) (B)
die **Wohngemeinschaft, -en** = die **WG, -s** shared housing (6)
das **Wohnheim, -e** dorm
der **Wohnkomfort** comfortable living
die **Wohnmöglichkeit, -en** living arrangements
der **Wohnort, -e** place of residence (1)
der/die **Wohnraumbietende, -n (ein Wohnraumbietender)** person offering housing
der/die **Wohnraumsuchende, -n (ein Wohnraumsuchender)** person looking for housing
die **Wohnung, -en** apartment (1, 2)
die **Wohnungsbörse, -n** apartment brokerage
die **Wohnungssuche, -n** search for an apartment; **auf Wohnungssuche** looking for a room or apartment
das **Wohnviertel, -** residential district
der **Wohnwagen, -** mobile home
das **Wohnzimmer, -** living room (6)
der **Wolf, ¨e** wolf (9)
die **Wolke, -n** cloud
wollen (will), wollte, gewollt to want (to); to intend (to); to plan (to) (3); **Wo willst du denn hin?** Where are you going? (A)
womit with what, by what means
woran at/on/of what
worauf on/for what
das **Wort, ¨er/-e** word; **Worte** words (*connected discourse*); **Wörter** words (*individual vocabulary items*) (A)
das **Wörterbuch, ¨er** dictionary (2)
der **Wortkasten, ¨** word box
der **Wortschatz, ¨e** vocabulary (A)
worüber about what
wozu to/for what
die **Wunde, -n** wound (11)
das **Wunder, -** miracle, wonder (4); **kein Wunder** no wonder (4)
wunderlich strange(ly)
sich **wundern, gewundert** to be surprised
wunderschön exceedingly beautiful(ly) (6)
der **Wunsch, ¨e** wish
sich (*dat.*) **wünschen, gewünscht** to wish for
der **Wunschzettel, -** wish list (*of things one would like to have*)
die **Würde** dignity
die **Wurst, ¨e** sausage (8); cold cuts
das **Würstchen, -** sausage; frank(furter); hot dog (8)
die **Wurzel, -n** root
würzen, gewürzt to season (8)
die **Wüste, -n** desert (7)
wütend angry, angrily (3)
x-t-: zum x-ten Mal (*coll.*) for the umpteenth time
die **Zahl, -en** number (A); figure
zahlen, gezahlt to pay (for) (5); **Miete zahlen** to pay rent; **Zahlen, bitte.** The check, please.
zählen, gezählt to count (A); **(zu +** *dat.***)** to be among
das **Zahlenrätsel, -** number puzzle
zahlreich numerous
die **Zahlung, -en** payment
das **Zahlungsmittel, -** means of payment (12); **offizielles Zahlungsmittel** legal tender
zahm tame(ly) (10)
der **Zahn, ¨e** tooth (11); **sich die Zähne putzen** to brush one's teeth (11)
der **Zahnarzt, ¨e** / die **Zahnärztin, -nen** dentist (11)
die **Zahnschmerzen** (*pl.*) toothache (11)
die **Zange, -n** pliers; tongs (8)
zart tender(ly) (8)
der **Zauber, -** magic; charm
der **Zauberer, -** wizard
der **Zaun, ¨e** fence (9)
z. B. = **zum Beispiel** for example (3)
das **Zebra, -s** zebra (10)
der **Zebrastreifen, -** crosswalk (10)
zehn ten (A)
zehnt- tenth (4)
das **Zeichen, -** sign
zeichnen, gezeichnet to draw (3, 5)
die **Zeichnung, -en** drawing
der **Zeigefinger, -** index finger
zeigen, gezeigt to show; **sich zeigen** to appear; **Was zeigen Ihre Bilder?** What do your pictures show? (A)
die **Zeile, -n** line
die **Zeit, -en** time (4); **in letzter Zeit** recently; **lange Zeit** (for) a long time; **zu dieser Zeit** at this time; **zur Zeit** at present
der **Zeitgeschmack, ¨er** contemporary taste
die **Zeitschrift, -en** magazine
die **Zeitung, -en** newspaper (2); **Zeitungen austragen** to deliver newspapers (5); **Zeitung lesen** to read the newspaper (1)
zeitweise occasionally
das **Zelt, -e** tent (2)
zelten, gezeltet to camp (1)
zensieren, zensiert to censor
der **Zentimeter, -** centimeter
zentral central(ly) (10)
die **Zentralheizung** central heating (6)
das **Zentrum, Zentren** center
der **Zeppelin, -e** zeppelin, dirigible
zerbeulen, zerbeult to dent up

zerbrechen (zerbricht), zerbrach, hat/ist zerbrochen to break into pieces

zerreißen (zerreißt), zerriss, zerrissen to tear (to pieces) (9)

zerschlagen (zerschlägt), zerschlug, zerschlagen to smash to bits

zerstören, zerstört to destroy

der **Zeuge, -n** (*wk.*) / die **Zeugin, -nen** witness (11)

das **Zeugnis, -se** report card

der **Ziegel, -** clay tile

ziehen (zieht), zog, ist gezogen to move (2); (*p.p. with* **haben**) to pull (8)

das **Ziel, -e** goal; destination (10)

zielen, gezielt to aim

ziellos aimless(ly)

ziemlich rather (2); **ziemlich groß** pretty big (2)

die **Zigarette, -n** cigarette (11)

die **Zigarre, -n** cigar

das **Zimmer, -** room (1, 2)

die **Zimmersuche, -n** search for a room (*to rent*)

die **Zinsen** (*pl.*) interest (12)

zirka circa, about, approximately

der **Zirkus, -se** circus (9)

die **Zither, -n** zither

die **Zitrone, -n** lemon (8)

zittern, gezittert to tremble

der **Zivilisationskritiker, -** / die **Zivilisationskritikerin, -nen** critic of civilization

der **Zoo, -s** zoo

zu (*adj.*) closed; (*adv.*) too (2); **zu viel** too much

zu (+ *dat.*) to (*a place*) (2, 10); for (*an occasion*) (2); for the purpose of; **bis zu** as far as; up to (10); **um ... zu** (+ *inf.*) in order to (12); **zu Abend essen** to dine, have dinner (4); **zu Besuch kommen** to visit (3); **zu Fuß** on foot (3); **zu Hause sein** to be at home (A, 1, 10); **zum Arzt** to the doctor (3); **zum Beispiel (z. B.)** for example (3); **zum ersten Mal** for the first time (4); **zum Geburtstag** for someone's birthday (2); **zum Mittagessen** for lunch (3); **zum Schluss** in the end, finally (8); **zum Teil** partly; **zur Uni gehen** to go to the university (1, 2)

zu·bereiten, zubereitet to prepare (*food*) (8)

die **Zubereitung, -en** preparation (8)

zu·binden (bindet ... zu), band ... zu, zugebunden to tie shut

der **Zucker** sugar (8)

zu·decken, zugedeckt to cover (*with a blanket*) (11)

zu·drücken, zugedrückt to squeeze shut; **ein Auge zudrücken** to look the other way

zuerst first (4, 7)

der **Zufall, ¨-e** coincidence (1)

zufällig accidental(ly) (9)

zufolge (+ *dat.*) according to

zufrieden satisfied

der **Zug, ¨-e** train (7, 10); characteristic

der **Zugang, ¨-e** access (12)

zugebunden (*p.p. of* **zubinden**) tied shut (8)

die **Zugfahrkarte, -n** train ticket

die **Zugspitze** *mountain on the German-Austrian border*

zu·hören (+ *dat.*), **zugehört** to listen (to) (6); **hören Sie zu** listen (A)

die **Zukunft, ¨-e** future

zukünftig future (*adj.*)

zulässig permissible

zum = **zu dem** to/for the; see **zu**

zu·machen, zugemacht to close (3)

zumindest at least

zunächst at first

der **Zuname, -n** (*wk.*) surname, last name

zu·nehmen (nimmt ... zu), nahm ... zu, zugenommen to increase

die **Zunge, -n** tongue (11); **sich die Zunge verbrennen** to burn one's tongue (11)

zu·ordnen (+ *dat.*), **zugeordnet** to classify (as)

zur = **zu der** to/for the; see **zu**

zurecht·kommen (mit + *dat.*) **(kommt ... zurecht), kam ... zurecht, ist zurechtgekommen** to get along (with)

(das) **Zürich** Zurich

zurück back (9); **hin und zurück** there and back; round-trip (5, 10)

zurück·bekommen (bekommt ... zurück), bekam ... zurück, zurückbekommen to get back

zurück·bringen (bringt ... zurück), brachte ... zurück, zurückgebracht to bring back

zurück·führen (auf + *acc.*), **zurückgeführt** to lead back (to); to be attributable (to)

zurück·geben (gibt ... zurück), gab ... zurück, zurückgegeben to give back, return; to reply

zurück·gehen (geht ... zurück), ging ... zurück, ist zurückgegangen to go back

zurück·kehren, ist zurückgekehrt to come back, return

zurück·kommen (kommt ... zurück), kam ... zurück, ist zurückgekommen to come back, return (6)

zurück·rufen (ruft ... zurück), rief ... zurück, zurückgerufen to call back

zurück·schicken, zurückgeschickt to send back

zurück·treten (tritt ... zurück), trat ... zurück, ist zurückgetreten to step back; to step down

zurück·verfolgen, zurückverfolgt to trace back

zurück·ziehen (zieht ... zurück), zog ... zurück, zurückgezogen to pull back, draw back; **sich zurückziehen** to withdraw

zusammen together (2)

die **Zusammenarbeit, -en** collaboration

zusammen·brechen (bricht ... zusammen), brach ... zusammen, zusammengebrochen to collapse, break down

der **Zusammenbruch, ¨-e** breakdown; collapse

zusammen·fassen, zusammengefasst to summarize

die **Zusammenfassung, -en** summary

zusammen·gehören, zusammengehört to belong together

zusammen·packen, zusammengepackt to pack up

sich **zusammen·schließen (schließt ... zusammen), schloss ... zusammen, zusammengeschlossen** to join together

sich **zusammen·setzen (aus** + *dat.*), **zusammengesetzt** to be composed (of)

zusammen·sitzen (mit + *dat.*) **(sitzt ... zusammen), saß ... zusammen, zusammengesessen** to sit together (with)

zusammen·stoßen (stößt ... zusammen), stieß ... zusammen, ist zusammengestoßen to crash (11)

zusammen·treffen (mit + *dat.*) **(trifft ... zusammen), traf ... zusammen, ist zusammengetroffen** to meet

zusätzlich additional, in addition

zu·schauen, zugeschaut to watch

zu·schlagen (schlägt ... zu), schlug ... zu, zugeschlagen to slam shut

zu·schnüren, zugeschnürt to tie up; to constrict

zu·sehen (+ *dat.*) **(sieht ... zu), sah ... zu, zugesehen** to watch

zu·sperren, zugesperrt to lock

der **Zustand, ¨-e** condition

die **Zutat, -en** ingredient (8)

(sich) **zu·wenden** (+ *dat.*) **(wendet ... zu), wandte ... zu, zugewandt** to turn toward

zu·winken (+ *dat.*), **zugewinkt/zugewunken** to wave to

zu·ziehen (zieht ... zu), zog ... zu, zugezogen to pull shut; to draw (*curtains*)

zwangsbeatmen, zwangsbeatmet to administer artificial respiration by force

zwangsneurotisch obsessive-compulsive

zwanzig twenty (A)

der **Zwanzigeuroschein, -e** twenty-euro note

zwanzigst- twentieth (4)

zwar to be sure

zwei two (A)

der **Zweifel, -** doubt

zweimal twice (5)

zweit: zu zweit arbeiten/leben to work/live together (*two people*)

zweit- second (4)

zweitmeist- second-most

zweiundzwanzig twenty-two (A)

der **Zwerg, -e** dwarf

die **Zwiebel, -n** onion (8)

zwingen (zwingt), zwang, gezwungen to force

zwischen (+ *dat./acc.*) between (7); among

die **Zwischenmiete, -n: zur Zwischenmiete** for sublet

die **Zwischenzeit** interim; **in der Zwischenzeit** in the meantime

zwölf twelve (A)

zwölft- twelfth (4)

(das) **Zypern** Cyprus

Vokabeln

Englisch-Deutsch

This list contains the words from the chapter vocabulary lists.

to abandon **verlassen (verlässt), verließ, verlassen** (11)

able: to be able (to) **können (kann), konnte, gekonnt** (1, 3)

about **wegen** (+ *gen.*) (6)

above (*adv.*) **oben** (10); (*prep.*) **über** (4)

abroad **im Ausland** (6)

academic subject **das Fach, ¨er** (1)

access **der Zugang** (12)

accident **der Unfall, ¨e** (4, 7)

accidental(ly) **zufällig** (9)

account: bank account **das Konto, Konten** (5); checking account **das Girokonto, Girokonten** (12); on account of **wegen** (+ *gen.*) (6); to open a bank account **ein Konto eröffnen** (5); savings account **das Sparkonto, Sparkonten** (12)

acquainted: to get acquainted with **kennen·lernen, kennengelernt** (1)

across **gegenüber** (+ *dat.*) (6); across from **gegenüber von** (+ *dat.*) (10); to pass across **über·gehen (geht ... über), ging ... über, ist übergegangen** (8); right across the way **gleich gegenüber** (6)

activity **die Tätigkeit, -en** (6)

actually **eigentlich** (3)

ad **die Anzeige, -n** (6)

addition: in addition **dazu** (8); in addition to **neben** (+ *dat./acc.*) (3); **außer** (+ *dat.*) (9)

address **die Adresse, -n** (1)

adhesive bandage **das Pflaster, -** (11)

advantage **der Vorteil, -e** (7)

advice **der Rat, Ratschläge** (5, 11)

to advise **raten** (+ *dat.*) **(rät), riet, geraten** (11)

afraid: to be afraid **Angst haben** (3); to be afraid of **sich fürchten vor** (+ *dat.*), **gefürchtet** (10)

Africa **(das) Afrika** (B)

Afro-German (*adj.*) **afro-deutsch** (12)

after (*prep.*) **nach** (+ *dat.*) (3); (*subord. conj.*) **nachdem** (9, 11); at twenty after five **um zwanzig nach fünf** (1); the day after tomorrow **übermorgen** (9)

afternoon **der Nachmittag, -e** (4); afternoons, in the afternoon **nachmittags** (4); good afternoon (*for.*) **guten Tag** (A); good afternoon (*for.; southern Germany, Austria*) **grüß Gott** (A)

afterward **danach** (8, 10)

again **wieder** (3); once again **schon wieder** (3)

against **gegen** (+ *acc.*) (9)

age **das Alter** (1)

agency: travel agency **das Reisebüro, -s** (6)

ago **vor** (+ *dat.*) (4); two days ago **vor zwei Tagen** (4)

to agree upon **vereinbaren, vereinbart** (12)

ahead: straight ahead **geradeaus** (10)

air **die Luft** (7); air mattress **die Luftmatratze, -n** (10)

airplane **das Flugzeug, -e** (7)

airport **der Flughafen, ¨** (6)

alarm clock **der Wecker, -** (2)

alive: to be alive **am Leben sein** (9)

all: all day long **den ganzen Tag** (1); all night long **die ganze Nacht** (3); by all means **auf jeden Fall** (4); not at all **kein bisschen** (3)

allergic **allergisch** (11)

alley **die Gasse, -n** (10)

almost **fast** (5)

along **entlang** (10); to bring along **mit·bringen (bringt ... mit), brachte ... mit, mitgebracht** (3); to go along **entlang·gehen (geht ... entlang), ging ... entlang, ist entlanggegangen** (10); to take along **mit·nehmen (nimmt ... mit), nahm ... mit, mitgenommen** (3)

aloud: to read aloud **vor·lesen (liest ... vor), las ... vor, vorgelesen** (9)

alphabet **das Alphabet, -e** (3)

the Alps **die Alpen** (*pl.*) (7)

already **schon** (2, 4)

also **auch** (A)

alternative **die Alternative, -n** (12)

although (*subord. conj.*) **obwohl** (11)

always **immer** (3)

ambulance **der Krankenwagen, -** (11)

America **(das) Amerika** (B); North America **(das) Nordamerika** (B); South America **(das) Südamerika** (B)

American (*person*) **der Amerikaner, - / die Amerikanerin, -nen** (B)

among **unter** (+ *dat./acc.*) (6)

amount (*of money*) **die Höhe, -n** (12)

ancestor **der Vorfahre, -n** (*wk.*) (12)

and (*coord. conj.*) **und** (A, 11)

angry **wütend** (3); **sauer** (5); to get angry **sich ärgern, geärgert** (11)

animal **das Tier, -e** (3, 7, 10)

to annoy **ärgern, geärgert** (1, 3)

another: one another **einander** (3)

answer **die Antwort, -en** (A); to answer **antworten** (+ *dat.*), **geantwortet** (4, 10); **beantworten, beantwortet** (7)

antibiotics **die Antibiotika** (*pl.*) (11)

any: in any case **jedenfalls** (11)

anything **etwas** (5); Anything else? **Sonst noch etwas?** (5)

apartment **die Wohnung, -en** (1, 2)

appendix **der Blinddarm, ¨e** (11)

appetizer **die Vorspeise, -n** (8)

apple juice **der Apfelsaft** (8)

appliance **das Gerät, -e** (8)

to apply for **beantragen, beantragt** (12)

appointment **der Termin, -e** (5, 11); **die Verabredung, -en** (11); appointment calendar **der Terminkalender, -** (11); to get an appointment **sich einen Termin geben lassen** (11)

apprenticeship **die Lehre, -n** (5)

approximately **ungefähr** (7)

April **der April** (B)

Arabic (*language*) **(das) Arabisch** (B)

architect **der Architekt, -en** (*wk.*) **/ die Architektin, -nen** (5)

area **der Bereich, -e** (12)

to argue **streiten (streitet), stritt, gestritten** (9)

arm **der Arm, -e** (B); to break one's arm **sich den Arm brechen** (11)

armchair **der Sessel, -** (2, 6)

army: German army **die Bundeswehr** (5); in the German army **bei der Bundeswehr** (5)

around the corner **um die Ecke** (5); right around the corner **gleich um die Ecke** (6)

arrival **die Ankunft, ¨e** (7)

to arrive **an·kommen (kommt ... an), kam ... an, ist angekommen** (1)

art **die Kunst** (1); art history **die Kunstgeschichte** (1)

as **als** (5); as . . . as possible **möglichst** (+ *adv.*) (6); as far as **bis zu** (+ *dat.*) (10); as well **auch** (A); as what? **als was?** (5)

Asia **(das) Asien** (B)

to ask (for) **bitten (um** + *acc.*) **(bittet), bat, gebeten** (12); to ask about **sich erkundigen nach** (+ *dat.*), **erkundigt** (10); to ask a question **eine Frage stellen** (5) asleep: to fall asleep **ein·schlafen (schläft ... ein), schlief ... ein, ist eingeschlafen** (7)

assignment **die Aufgabe, -n** (4)

at **an** (+ *dat.*) (2); **bei** (+ *dat.*) (2, 6, 10); **in** (+ *dat./acc.*) (4); at a bank **bei einer Bank** (6); at home **zu Hause** (A, 1, 10); **daheim** (9); at least **wenigstens** (1), **mindestens** (7); at Monika's **bei Monika** (2); at night **nachts** (4); at noon **mittags** (2); at school **in der Schule** (5); at six o'clock **um sechs (Uhr)** (1); at the cafe **im Café** (4); at the courthouse **auf dem Gericht** (5); at the department store **im Kaufhaus** (5); at the gas station **an der Tankstelle** (5); at the moment **im Moment** (1); at the ticket booth **an der Kasse** (5); **am Schalter** (5); at the university **auf der Universität** (5); at two thirty **um halb drei** (1); At what time . . . ? **Um wie viel Uhr ...?** (1); at your parents' **bei deinen Eltern** (6)

athletic **sportlich** (B); athletic shoe **der Sportschuh, -e** (A)

ATM (automatic teller machine) **der Geldautomat, -en** (*wk.*) (12)

to attempt **versuchen, versucht** (4)

to attend to **pflegen, gepflegt** (5)

attendant: swimming-pool attendant **der Bademeister, - / die Bademeisterin, -nen** (5)

attention **die Achtung** (7); to pay attention **auf·passen, aufgepasst** (3); to pay attention to **achten auf** (+ *acc.*), **geachtet** (11)

attitude **die Einstellung, -en** (12)

to attract **an·ziehen (zieht ... an), zog ... an, angezogen** (3)

attractive **attraktiv** (6)

August **der August** (B)

aunt **die Tante, -n** (B)

Australia **(das) Australien** (B)

Australian (*person*) **der Australier, - / die Australierin, -nen** (B)

Austria **(das) Österreich** (B)

Austrian (*person*) **der Österreicher, - / die Österreicherin, -nen** (B)

automatic teller machine (ATM) **der Geldautomat, -en** (*wk.*) (12)

autumn **der Herbst, -e** (B)

available **frei** (8); Is this seat available? **Ist hier noch frei?** (8)

away: to carry away **weg·tragen (trägt ... weg), trug ... weg, weggetragen** (9); How far away? **Wie weit weg?** (6); to put away **weg·stellen, weggestellt** (5); right away **gleich** (6); to take away **weg·bringen (bringt ... weg), brachte ... weg, weggebracht** (5)

baby **das Baby, -s** (7); baby carriage **der Kinderwagen, -** (7)

back (*adv.*) **zurück** (9); back then **damals** (9); to come back **zurück·kommen (kommt ... zurück), kam ... zurück, ist zurückgekommen** (6); there and back **hin und zurück** (10)

back (*n.*) **der Rücken, -** (B)

backpack **der Rucksack, ¨-e** (2)

bacon **der Speck** (8)

bad **schlecht** (2); **schlimm** (11); too bad! **schade!** (6)

bag **die Tasche, -n** (1); (*paper or plastic*) **die Tüte, -n** (11); sleeping bag **der Schlafsack, ¨-e** (2)

to bake **backen (backt/bäckt), backte, gebacken** (5)

bakery **die Bäckerei, -en** (5); at the bakery **in der Bäckerei** (5)

balance: bank balance **das Guthaben** (12)

balcony **der Balkon, -e** (6)

ball **der Ball, ¨-e** (A, 1); soccer ball **der Fußball, ¨-e** (A, 1)

ballerina **die Ballerina, -s** (9)

ballpoint pen **der Kugelschreiber, -** (4)

Baltic Sea **die Ostsee** (B)

banana **die Banane, -n** (8)

bandage **der Verband, ¨-e** (11); adhesive bandage **das Pflaster, -** (11)

bank **die Bank, -en** (5); at a bank **bei einer Bank** (6); at the bank **auf der Bank** (5); bank account **das Konto, Konten** (5); bank balance **das Guthaben** (12); bank employee **der/die Bankangestellte, -n (ein Bankangestellter)** (5); to open a bank account **ein Konto eröffnen** (5)

baptism **die Taufe, -n** (4)

bar **die Kneipe, -n** (1, 4)

barbecue grill **der Grill, -s** (2)

barbecued **gegrillt** (8)

barely **knapp** (4)

barkeeper **der Wirt, -e / die Wirtin, -nen** (10)

basement **der Keller, -** (4, 6)

basin **das Waschbecken, -** (6)

basketball **der Basketball, ¨-e** (2)

bath, bathroom **das Bad, ¨-er** (6)

to bathe **baden, gebadet** (3); **(sich) baden, gebadet** (11)

bathrobe **der Bademantel, ¨-** (2)

bathtub **die Badewanne, -n** (6)

bay **die Bucht, -en** (7)

to be **sein (ist), war, ist gewesen** (A, 4); to be in a seated position **sitzen, saß, gesessen** (2)

beach **der Strand, ¨-e** (4, 7); beach chair **der Strandkorb, ¨-e** (10); beach parasol **der Sonnenschirm, -e** (10)

bean **die Bohne, -n** (8)

bear: teddy bear **der Teddybär, -en** (*wk.*) (A)

beard **der Bart, ¨-e** (B)

to beat **schlagen (schlägt), schlug, geschlagen** (8)

beautiful **schön** (B); exceedingly beautiful **wunderschön** (6)

because (*subord. conj.*) **weil** (3, 11); (*coord. conj.*) **denn** (9, 11)

to become **werden (wird), wurde, ist geworden** (5)

bed **das Bett, -en** (1, 6); bed-and-breakfast (inn) **das Gästehaus, ¨-er** (10); to get up on the wrong side of bed **mit dem linken Fuß auf·stehen (steht ... auf), stand ... auf, aufgestanden** (4); to go to bed **ins Bett gehen** (1)

bedroom **das Schlafzimmer, -** (6)

bedside table **der Nachttisch, -e** (6)

bee **die Biene, -n** (10)

beef **das Rindfleisch** (8); ground beef **das Hackfleisch** (8)

beer **das Bier, -e** (2)

before (*subord. conj.*) **bevor** (11)

to begin **beginnen (beginnt), begann, begonnen** (1); **an·fangen (fängt ... an), fing ... an, angefangen** (4)

Belgium **(das) Belgien** (B)

to believe **glauben, geglaubt** (2)

belly **der Bauch, ¨-e** (B)

to belong to **gehören** (+ *dat.*), **gehört** (6); to belong to (*an organization*) **an·gehören** (+ *dat.*), **angehört** (12)

beloved female friend **die Geliebte, -n** (3)

below **unter** (+ *dat./acc.*) (6)

belt **der Gürtel, -** (2); seat belt **der Sicherheitsgurt, -e** (7)

beneath **unter** (+ *dat./acc.*) (6)

beside **neben** (+ *dat./acc.*) (3)

besides (*adv.*) **außerdem** (3, 10); (*prep.*) **außer** (+ *dat.*) (9)

best: like (*to do*) best **am liebsten** (7)

better **besser** (2)

between **zwischen** (+ *dat./acc.*) (7)

beverage **das Getränk, -e** (8)

bicycle **das Fahrrad, ¨-er** (2, 7); bicycle helmet **der Fahrradhelm, -e** (5); bicycle tour **die Radtour, -en** (9); to ride a bicycle **Rad fahren (fährt ... Rad), fuhr ... Rad, ist Rad gefahren** (7)

big **groß** (B); pretty big **ziemlich groß** (2)

bikini **der Bikini, -s** (5)

bill **die Rechnung, -en** (4); (*of currency*) **der Schein, -e** (8); **der Geldschein, -e** (12)

biographical information **persönliche Angaben** (*pl.*) (1)

biology **die Biologie** (1)

bird **der Vogel, ¨-** (10)

birthday **der Geburtstag, -e** (1, 2); birthday card **die Geburtstagskarte, -n** (2); for someone's birthday **zum Geburtstag** (2)

bit: a little bit **ein bisschen** (3); not a bit **gar nicht** (3); quite a bit **ganz schön viel** (3)

to bite **beißen (beißt), biss, gebissen** (9); (*of insects*) **stechen (sticht), stach, gestochen** (10)

black **schwarz** (A)

blackboard **die Tafel, -n** (A, B)

blanket **die Decke, -n** (11)

to bleed **bluten, geblutet** (11)

blond **blond** (B)

blood **das Blut** (9, 11); to take blood **Blut ab·nehmen** (11)

blouse **die Bluse, -n** (A)

to blow dry (one's hair) **sich (die Haare) föhnen, geföhnt** (2, 11)

blue **blau** (A, B)

boar: wild boar **das Wildschwein, -e** (10)

to board **ein·steigen (steigt ... ein), stieg ... ein, ist eingestiegen** (10)

boat **das Boot, -e** (2)

body **der Körper, -** (B)

boiled **gekocht** (8); boiled eggs **gekochte Eier** (*pl.*) (8); boiled potatoes **die Salzkartoffeln** (*pl.*) (8)

to book **buchen, gebucht** (7)

book **das Buch, ¨er** (A, B, 2)

bookcase, bookshelf **das Regal, -e** (2)

boot **der Stiefel, -** (A)

booth: ticket booth **die Kasse, -n** (5), **der Schalter, -** (5); at the ticket booth **an der Kasse** (5), **am Schalter** (5)

border **die Grenze, -n** (12)

bored: to be bored **Langeweile haben** (3)

boredom **die Langeweile** (3)

boring **langweilig** (2)

born **geboren** (1); When were you born? **Wann sind Sie geboren?** (1)

bottle **die Flasche, -n** (5); bottle opener **der Flaschenöffner, -** (8)

boutique **die Boutique, -n** (6)

bowl **die Schüssel, -n** (8); salad (mixing) bowl **die Salatschüssel, -n** (5)

to box **boxen, geboxt** (1)

boy: Oh boy! (*coll.*) **Mensch!** (2)

boyfriend **der Freund, -e** (A)

bracelet **das Armband, ¨er** (2)

brain **das Gehirn, -e** (11)

to brake **bremsen, gebremst** (11)

brake **die Bremse, -n** (7)

brave **tapfer** (9)

bread **das Brot, -e** (8)

to break **brechen (bricht), brach, gebrochen** (11); to break one's arm **sich den Arm brechen** (11)

break **die Pause, -n** (1)

breakfast **das Frühstück, -e** (2); to eat breakfast **frühstücken, gefrühstückt** (1)

to breathe **atmen, geatmet** (11)

bride **die Braut, ¨e** (9)

bridge **die Brücke, -n** (10)

to bring **bringen (bringt), brachte, gebracht** (2); to bring along **mit·bringen (bringt ... mit), brachte ... mit, mitgebracht** (3)

broiled **gebraten** (8); **gegrillt** (8)

broken **kaputt** (A)

broom **der Besen, -n** (6)

brother **der Bruder, ¨** (B)

brown **braun** (A); to brown **bräunen, gebräunt** (8)

to brush (one's teeth) **sich (die Zähne) putzen** (11)

Brussels sprouts **der Rosenkohl** (8)

building **das Gebäude, -** (10); high-rise building **das Hochhaus, ¨er** (6)

Bulgaria **(das) Bulgarien** (B)

burglar **der Einbrecher, - / die Einbrecherin, -nen** (9)

to burn **brennen (brennt), brannte, gebrannt** (11); **verbrennen (verbrennt), verbrannte, verbrannt** (11); to burn (one's tongue) **sich (die Zunge) verbrennen** (11)

bus **der Bus, -se** (2, 7); bus stop **die Bushaltestelle, -n** (10)

bush **der Busch, ¨e** (9)

but (*coord. conj.*) **aber** (A, 11); but (rather / on the contrary) **sondern** (A)

butcher shop **die Metzgerei, -en** (6)

butter **die Butter** (8); herb butter **die Kräuterbutter** (8)

to buy **kaufen, gekauft** (1)

by **an ... vorbei** (10); by all means **auf jeden Fall** (4)

bye (*infor.*) **tschüss** (A)

cabbage **der Kohl** (8)

cabinet: wardrobe cabinet **der Schrank, ¨e** (2, 6)

café **das Café, -s** (4); at the café **im Café** (4)

cafeteria: student cafeteria **die Mensa, Mensen** (2)

cage **der Käfig, -e** (10)

cake **der Kuchen, -** (5)

calendar: appointment calendar **der Terminkalender, -** (11)

to call **rufen (ruft), rief, gerufen** (7, 11); to call forth **hervor·rufen (ruft ... hervor), rief ... hervor, hervorgerufen** (1); to call up **an·rufen (ruft ... an), rief ... an, angerufen** (1)

called: to be called **heißen (heißt), hieß, geheißen** (A)

calm **ruhig** (B)

camera **die Kamera, -s** (2)

to camp **zelten, gezeltet** (1)

camping **das Camping** (10)

campsite **der Campingplatz, ¨e** (10)

can (*v.*) **können (kann), konnte, gekonnt** (1, 3)

can (*n.*) **die Dose, -n** (8); can opener **der Dosenöffner, -** (8); garbage can **der Mülleimer, -** (8)

Canada **(das) Kanada** (B)

Canadian (*person*) **der Kanadier, - / die Kanadierin, -nen** (B)

candidate **der Kandidat, -en** (*wk.*) **/ die Kandidatin, -nen** (12)

candle **die Kerze, -n** (3)

canoe **das Kanu, -s** (10)

canoeing: to go canoeing **Kanu fahren** (10)

cap **die Mütze, -n** (5)

capital city **die Hauptstadt, ¨e** (B)

car **das Auto, -s** (A); **der Wagen, -** (7); car radio **das Autoradio, -s** (7); train car **der Waggon, -s** (7)

card **die Karte, -n** (1, 2); birthday card **die Geburtstagskarte, -n** (2); identification card **der Ausweis, -e** (10); personal ID card **der Personalausweis, -e** (1); telephone card **die Telefonkarte, -n** (2)

to care for **mögen (mag), mochte, gemocht** (3)

care: take care (*infor.*) **mach's gut** (A); to take care of **sich kümmern um** (+ *acc.*), **gekümmert** (12)

career **der Beruf, -e** (5); career counselor **der Berufsberater, - / die Berufsberaterin, -nen** (5); practical career training **praktische Ausbildung** (5)

carpet **der Teppich, -e** (2)

carriage: baby carriage **der Kinderwagen, -** (7)

carrot **die Karotte, -n** (8)

to carry away **weg·tragen (trägt ... weg), trug ... weg, weggetragen** (9); to carry out **aus·führen, ausgeführt** (12)

case: in any case **jedenfalls** (11)

cash **das Bargeld** (12)

cashier window **die Kasse, -n** (12)

cast (*plaster*) **der Gips** (11)

to cast a spell on **verwünschen, verwünscht** (9)

castle **das Schloss, ¨er** (9)

cat **die Katze, -n** (2)

to catch a cold **sich erkälten, erkältet** (11)

cathedral **der Dom, -e** (10)

cauliflower **der Blumenkohl** (8)

cautious(ly) **vorsichtig** (8)

CD **die CD, -s** (A, 3); CD player **der CD-Spieler, -** (2)

ceiling **die Decke, -n** (B)

to celebrate **feiern, gefeiert** (4, 5)

celebration **das Fest, -e** (4); **die Feier, -n** (9)

cellar **der Keller, -** (4, 6)

cellular phone **das Handy, -s** (2)

Celsius **Celsius** (B)

central **zentral** (10); central heating **die Zentralheizung** (6)

certainly **bestimmt** (3); **sicherlich** (3)

certificate: gift certificate **der Geschenkgutschein, -e** (2)

chair **der Stuhl, ¨e** (B, 2); beach chair **der Strandkorb, ¨e** (10); deck chair **der Liegestuhl, ¨e** (4)

to champion **ein·treten für** (+ *acc.*) **(tritt ... ein), trat ... ein, ist eingetreten** (12)

to change **verändern, verändert** (6); **ändern, geändert** (9); to change (*trains*) **um·steigen (steigt ... um), stieg ... um, ist umgestiegen** (7); to change into **sich verwandeln in** (+ *acc.*), **verwandelt** (9)

change: keep the change **das stimmt so** (8)

chaos **das Chaos** (5)

chapter **das Kapitel, -** (A)

to charge **berechnen** (+ *dat.*), **berechnet** (8)

to chat **chatten, gechattet** (1)

cheap **billig** (2)

to check the oil **das Öl kontrollieren** (5)

check (*in restaurant*) **die Rechnung, -en** (4); **die Quittung, -en** (8); separate checks **getrennt** (5)

checking account **das Girokonto, Girokonten** (12)

cheese **der Käse** (8)

chemistry **die Chemie** (1)

chess **das Schach** (1)

to chew **kauen, gekaut** (11)

chic **schick** (2)

child **das Kind, -er** (B)

childhood **die Kindheit** (9)

China **(das) China** (B)

Chinese (*language*) **(das) Chinesisch** (B)

chisel **der Meißel, -** (12)

chocolate: hot chocolate **der Kakao** (8)

christening **die Taufe, -n** (4)

Christmas **(das) Weihnachten** (4); Christmas present **das Weihnachtsgeschenk, -e** (5)

church **die Kirche, -n** (5); at church **in der Kirche** (5)

cigarette **die Zigarette, -n** (11)

cinema **das Kino, -s** (1)

circus **der Zirkus, -se** (9)

citizen **der Bürger, - / die Bürgerin, -nen** (10)

city **die Stadt, ¨-e** (2, 6); capital city **die Hauptstadt, ¨-e** (B); city limits **der Stadtrand, ¨-er** (6); tour of the city **die Stadtrundfahrt, -en** (7)

class **der Kurs, -e** (A, 1); **die Klasse, -n** (5, 10); **der Unterricht** (9, 11); class reunion **das Klassentreffen, -** (9); German class **der Deutschkurs, -e** (A); to travel first class **erster Klasse fahren** (10)

classroom **der Seminarraum, ¨-e** (B)

clay **der Ton** (12)

to clean **sauber machen, sauber gemacht** (3); **putzen, geputzt** (6); to clean (up) **auf·räumen, aufgeräumt** (1); to wipe clean **ab·wischen, abgewischt** (6)

clean (*adj.*) **sauber** (B);

cleaner: dry cleaner's **die Reinigung, -en** (6); vacuum cleaner **der Staubsauger, -** (6)

cleaning: spring cleaning **der Frühjahrsputz** (6)

to clear **ab·räumen, abgeräumt** (3); to clear the table **den Tisch ab·räumen** (3)

clerk **der/die Angestellte, -n (ein Angestellter)** (7)

to climb **besteigen (besteigt), bestieg, bestiegen** (7); **klettern, ist geklettert** (9); to climb down **herunter·klettern, ist heruntergeklettert** (11)

clock **die Uhr, -en** (B); alarm clock **der Wecker, -** (2); kitchen clock **die Küchenuhr, -en** (5)

to close **schließen (schließt), schloss, geschlossen** (A); **zu·machen, zugemacht** (3)

close **nah** (7)

closet **der Schrank, ¨-e** (6); clothes closet **der Kleiderschrank, ¨-e** (6)

cloth (*for cleaning*) **der Putzlappen, -** (6)

clothes **die Kleidung** (A, 2); clothes closet **der Kleiderschrank, ¨-e** (6); clothes dryer **der Wäschetrockner, -** (8)

clown **der Clown, -s** (9)

coalition **die Koalition, -en** (12)

coast **die Küste, -n** (7)

coat **der Mantel, ¨** (A)

cocoa **der Kakao** (8)

coffee **der Kaffee** (1); coffee machine **die Kaffeemaschine, -n** (5)

coffin **der Sarg, ¨-e** (9)

coincidence **der Zufall, ¨-e** (1)

cold (*adj.*) **kalt** (B)

cold (*n.*) (*head cold*) **die Erkältung, -en** (11); to catch a cold **sich erkälten, erkältet** (11); cold (*with a runny nose*) **der Schnupfen, -** (11)

to collect **sammeln, gesammelt** (10)

college preparatory school **das Gymnasium, Gymnasien** (6)

color **die Farbe, -n** (A, 1); color of eyes **die Augenfarbe, -n** (1); color of hair **die Haarfarbe, -n** (1); favorite color **die Lieblingsfarbe, -n** (A); oil color (*paint*) **die Ölfarbe, -n** (12); What color is . . . ? **Welche Farbe hat ...?** (A)

to comb **kämmen, gekämmt** (3); to comb (one's hair) **sich (die Haare) kämmen, gekämmt** (11)

to combine **kombinieren, kombiniert** (3)

to come (from) **kommen (aus** + *dat.***) (kommt), kam, ist gekommen** (B); to come back **zurück·kommen (kommt ... zurück), kam ... zurück, ist zurückgekommen** (6); to come by **vorbei·kommen (kommt ... vorbei), kam ... vorbei, ist vorbeigekommen** (3); Come on over! **Komm mal vorbei!** (11); to come out this way **heraus·kommen (kommt ... heraus), kam ... heraus, ist herausgekommen** (10); to come this way **her·kommen (kommt ... her), kam ... her, ist hergekommen** (10); Does it come with a . . . ? **Ist ein/eine ... dabei?** (6)

common, in common **gemeinsam** (6, 11)

company **die Firma, Firmen** (3)

to compare **vergleichen (vergleicht), verglich, verglichen** (7)

to complain (to) **sich beschweren (bei** + *dat.***), beschwert** (8)

to complete **ergänzen, ergänzt** (4)

computer **der Computer, -** (2); computer science **die Informatik** (1)

to concern **betreffen (betrifft), betraf, betroffen** (6)

concert **das Konzert, -e** (1); concert ticket **die Konzertkarte, -n** (5); to go to a concert **ins Konzert gehen** (1)

conductor: orchestra conductor **der Dirigent, -en** (*wk.*) **/ die Dirigentin, -nen** (5)

conflict **der Konflikt, -e** (12)

to congratulate **gratulieren** (+ dat.), **gratuliert** (6)

to connect **verbinden (verbindet), verband, verbunden** (A)

conservative **konservativ** (B)

to consider **nach·denken (über** + *acc.***) (denkt ... nach), dachte ... nach, nachgedacht** (7)

construction worker **der Bauarbeiter, - / die Bauarbeiterin, -nen** (5)

contrary: but (rather / on the contrary) **sondern** (A, 11); on the contrary! **doch!** (4)

conversational situation **die Sprechsituation, -en** (A)

to converse **sich unterhalten (unterhält), unterhielt, unterhalten** (9)

cook **der Koch, ¨-e / die Köchin, -nen** (5);

to cook **kochen, gekocht** (1)

cooked **gekocht** (8)

cooking **die Küche** (8)

cool **kühl** (B); (*coll.*) **grell** (2)

copy shop **der Kopierladen, ¨** (10)

corkscrew **der Korkenzieher, -** (8)

corner **die Ecke, -n** (5); (right) around the corner **um die Ecke** (5, 6)

to correct **korrigieren, korrigiert** (4)

correct **richtig** (2)

to cost **kosten, gekostet** (2, 6)

cost: extra costs (*e.g., utilities*) **die Nebenkosten** (*pl.*) (6)

couch **das Sofa, -s** (6)

cough **der Husten** (11); cough syrup **der Hustensaft, ¨-e** (11)

counselor: career counselor **der Berufsberater, - / die Berufsberaterin, -nen** (5)

to count **zählen, gezählt** (A)

country **das Land, ¨-er** (6); foreign countries **das Ausland** (6); in the country (*rural*) **auf dem Land** (6)

courage **der Mut** (1)

course **der Kurs, -e** (A, 1); course of studies **das Studium, Studien** (3); German course **der Deutschkurs, -e** (A); of course **selbstverständlich** (10); Of course! **Klar!** (2)

courthouse **das Gericht, -e** (5); at the courthouse **auf dem Gericht** (5)

cousin: female cousin **die Kusine, -n** (B); male cousin **der Vetter, -n** (B)

to cover **decken, gedeckt** (5); **zu·decken, zugedeckt** (11)

cozy **gemütlich** (10)

to crash (*airplane*) **ab·stürzen, ist abgestürzt** (11); (*cars*) **zusammen·stoßen (stößt ... zusammen), stieß ...zusammen, ist zusammengestoßen** (11)

crash: stock market crash **der Börsenkrach, ¨-e** (12)

crazy **verrückt** (B); **wahnsinnig** (12)

crisis **die Krise, -n** (12)

critically injured **schwer verletzt** (11)

crocodile **das Krokodil, -e** (10)

croquette **die Krokette, -n** (8)

crosswalk **der Zebrastreifen, -** (10)

cruel **grausam** (9)

to cry **weinen, geweint** (3)

cucumber **die Gurke, -n** (8)

culture **die Kultur, -en** (12)

cup **die Tasse, -n** (2, 5); **der Becher, -** (9)

cupboard **der Schrank, ¨-e** (2, 6)

curious **neugierig** (12)

currency **die Währung, -en** (12)

to curse **verwünschen, verwünscht** (9); **fluchen, geflucht** (11)

cursed **verwunschen** (9)

curtain **der Vorhang, ¨-e** (6)

curve **die Kurve, -n** (7)

to cuss **schimpfen, geschimpft** (9)

custodian **der Hausmeister, - / die Hausmeisterin, -nen** (5)

customer **der Kunde, -n** (*wk.*) **/ die Kundin, -nen** (12)

to cut **schneiden (schneidet), schnitt, geschnitten** (3); to cut (oneself) **(sich) schneiden** (11); to cut hair **Haare schneiden** (3); to cut off **ab·schneiden (schneidet ... ab), schnitt ... ab, abgeschnitten** (8)

cutlery **das Besteck** (5)

cutlet **das Schnitzel, -** (8)

Czech Republic **(das) Tschechien** (B)

daily **täglich** (12); daily newspaper **die Tageszeitung, -en** (5); daily routine **der Tagesablauf, ¨-e** (1); **der Alltag** (4)

to dance **tanzen, getanzt** (1)

danger **die Gefahr, -en** (12)

dangerous **gefährlich** (10)

dark **dunkel** (5)

data: data projector **der Beamer, -** (B)

date **das Datum, Daten** (4); (*appointment*) **die Verabredung, -en** (11); What is today's date? **Welches Datum ist heute?** (4)

daughter **die Tochter, ¨-** (B)

day **der Tag, -e** (1); all day long **den ganzen Tag** (1); day after tomorrow **übermorgen** (9); every day **jeden Tag** (1); for several days **seit mehreren Tagen** (11); on what day? **an welchem Tag?** (4); two days ago **vor zwei Tagen** (4); What day is today? **Welcher Tag ist heute?** (1)

dead **tot** (9)

to deal with **betreffen (betrifft), betraf, betroffen** (6)

dear **lieb** (7)

death **der Tod, -e** (12)

debt **die Schuld, -en** (12)

decade **das Jahrzehnt, -e** (4)

December **der Dezember** (B)

deception **die List, -en** (9)

to decide **entscheiden (entscheidet), entschied, entschieden** (4)

decisive **entscheidend** (12)

deck chair **der Liegestuhl, ¨-e** (4)

deep **tief** (7)

definitely **bestimmt** (3)

degree **der Grad, -e** (B)

delay **die Verspätung, -en** (9)

to deliver **aus·tragen (trägt ... aus), trug ... aus, ausgetragen** (5); to deliver newspapers **Zeitungen aus·tragen** (5)

democracy **die Demokratie, -n** (12)

democratic(ally) **demokratisch** (12)

Denmark **(das) Dänemark** (B)

dentist **der Zahnarzt, ¨-e / die Zahnärztin, -nen** (11)

to depart **ab·fahren (fährt ... ab), fuhr ... ab, ist abgefahren** (4); **ab·reisen, ist abgereist** (10)

department: department store **das Kaufhaus, ¨-er** (5); at the department store **im Kaufhaus** (5); fire department **die Feuerwehr** (11)

departure **die Abfahrt, -en** (7)

depressed **deprimiert** (11)

to describe **beschreiben (beschreibt), beschrieb, beschrieben** (11)

description **die Beschreibung, -en** (B)

desert **die Wüste, -n** (7)

desire **die Lust** (3)

desk **der Schreibtisch, -e** (2)

dessert **die Nachspeise, -n** (8)

destination **das Ziel, -e** (10)

devil **der Teufel, -** (12)

diary **das Tagebuch, ¨-er** (4)

dictionary **das Wörterbuch, ¨-er** (2)

to die **sterben (stirbt), starb, ist gestorben** (9)

different **verschieden** (12)

difficult **schwierig** (2); **schwer** (3)

to dine **zu Abend essen** (4)

dining room **das Esszimmer, -** (6)

dinner: to have dinner **zu Abend essen** (4)

diphtheria **die Diphtherie** (11)

direction **die Richtung, -en** (7)

directly **gleich** (6)

director **der Direktor, -en / die Direktorin, -nen** (9)

dirty **schmutzig** (A)

disadvantage **der Nachteil, -e** (7)

disco **die Disko, -s** (3)

to discover **entdecken, entdeckt** (4)

to discuss **diskutieren, diskutiert** (4)

dish **das Gericht, -e** (8); dishes **das Geschirr** (4, 5); to wash the dishes **Geschirr spülen, gespült** (4)

dishwasher **der Geschirrspüler, -** (5)

to disinfect **desinfizieren, desinfiziert** (11)

to distribute **verteilen, verteilt** (12)

district **der Stadtteil, -e** (6); **das Stadtviertel, -** (6)

to disturb **stören, gestört** (3)

to dive **tauchen, hat/ist getaucht** (3)

to do **tun (tut), tat, getan** (1); to do sports **Sport treiben (treibt ... Sport), trieb ... Sport, Sport getrieben** (2)

doctor **der Arzt, ¨-e / die Ärztin, -nen** (5, 11); doctor's office **die Arztpraxis, Arztpraxen** (11); family doctor **der Hausarzt, ¨-e / die Hausärztin, -nen** (11); to the doctor **zum Arzt** (3)

dog **der Hund, -e** (2); hot dog **das Würstchen, -** (8)

dolphin **der Delfin, -e** (10)

to dominate **herrschen, geherrscht** (8)

door **die Tür, -en** (A); (from) next door **(von) nebenan** (5)

dorm **das Studentenheim, -e** (2, 6)

double room **das Doppelzimmer, -** (5, 10)

down (*toward the speaker*) **herunter** (11); to climb down **herunter·klettern, ist heruntergeklettert** (11); to fall down **hin·fallen (fällt ... hin), fiel ... hin, ist hingefallen** (11); to lie down **sich hin·legen, hingelegt** (11); to sit down **sich setzen, gesetzt** (11); to write down **auf·schreiben (schreibt ... auf), schrieb ... auf, aufgeschrieben** (11)

downtown **die Heimatstadt, ¨-e** (6)

dragon **der Drache, -n** (*wk.*) (9)

drapery **der Vorhang, ¨-e** (6)

to draw **zeichnen, gezeichnet** (3, 5)

drawer **die Schublade, -n** (5)

to dream **träumen, geträumt** (9)

dress **das Kleid, -er** (A)

dressed: to be well dressed **gut gekleidet sein** (2); to get dressed **sich an·ziehen (zieht ... an), zog ... an, angezogen** (11)

dresser **die Kommode, -n** (6)

dressing, salad dressing **die Soße, -n** (8)

to drink **trinken (trinkt), trank, getrunken** (1)

to drive **fahren (fährt), fuhr, ist/hat gefahren** (2); to drive off **los·fahren (fährt ... los), fuhr ... los, ist losgefahren** (4, 9)

driver **der Fahrer, - / die Fahrerin, -nen** (7); driver's license **der Führerschein, -e** (4); taxi driver **der Taxifahrer, - / die Taxifahrerin, -nen** (5)

drop **das Bonbon, -s** (11)

drugstore **die Drogerie, -n** (6)

drums **das Schlagzeug, -e** (12)

to dry (*dishes*) **ab·trocknen, abgetrocknet** (6); to blow dry (one's hair) **sich (die Haare) föhnen, geföhnt** (2, 11); to dry oneself off **sich ab·trocknen** (11)

dry **trocken** (11); dry cleaner's **die Reinigung, -en** (6)

dryer: clothes dryer **der Wäschetrockner, -** (8); hair dryer **der Haartrockner, -** (2)

dumb **dumm** (6)

dumpling **der Knödel, -** (8)

during **während** (+ *gen.*) (11); during the week **in der Woche** (1)

Dutch (*adj.*) **holländisch** (8)

duty **die Pflicht, -en** (3)

DVD player **der DVD-Spieler, -** (2, 3)

each **jeder, jedes, jede** (1, 3, 5); each other **einander** (3); with each other **miteinander** (1, 3)

eagle **der Adler, -** (10)

ear **das Ohr, -en** (B)

earache **die Ohrenschmerzen** (*pl.*) (11)

early **früh** (1)

to earn **verdienen, verdient** (4)

earring **der Ohrring, -e** (A, 2)

easel **die Staffelei, -en** (12)

east (of) **östlich** (**von** + *dat.*) (7)

easy **leicht** (6)

to eat **essen (isst), aß, gegessen** (2, 4); (*said of an animal*) **fressen (frisst), fraß, gefressen** (9); to eat breakfast **frühstücken, gefrühstückt** (1); to eat dinner **zu Abend essen** (4)

economics **die Wirtschaft** (1)

educated **ausgebildet** (12)

education **die Schulbildung** (5)

effect: to take effect **wirken, gewirkt** (11)

egg **das Ei, -er** (8); boiled eggs **gekochte Eier** (*pl.*) (8); fried eggs **gebratene Eier** (*pl.*) (8)

eight **acht** (A)

eighteen **achtzehn** (A)

eighth **acht-** (4)

eighty **achtzig** (A)

to elect **wählen, gewählt** (12)

election **die Wahl, -en** (12)

electric(al) **elektrisch** (8)

electricity **der Strom** (8)

elegant **elegant** (8)

elementary school **die Grundschule, -n** (9)

elephant **der Elefant, -en** (*wk.*) (9)

elevator **der Aufzug, ¨-e** (6)

eleven **elf** (A)

eleventh **elft-** (4)

else: Anything else? **Sonst noch etwas?** (5)

e-mail **die E-Mail, -s** (1, 2), **die Mail, -s** (1)

to emigrate **aus·wandern, ist ausgewandert** (12)

employee **die Arbeitskraft, ¨-e** (12); bank employee **der/die Bankangestellte, -n (ein Bankangestellter)** (5)

to empty **aus·leeren, ausgeleert** (3)

empty (*adj.*) **leer** (5); **frei** (8)

enchanted **verwunschen** (9)

to end **aus·gehen (geht ... aus), ging ... aus, ist ausgegangen** (7); it ended well **es ist gut ausgegangen** (7)

end **der Schluss, ¨-e** (8); in the end **zum Schluss** (8)

engineering: mechanical engineering **der Maschinenbau** (1)

England **(das) England** (B)

English (*language*) **(das) Englisch** (B); (*person*) **der Engländer, - / die Engländerin, -nen** (B)

enough **genug** (3)

to enrich **bereichern, bereichert** (12)

equal **egal** (6); **gleich** (12)

to establish **fest·stellen, festgestellt** (8)

euro **der Euro, -** (7)

Europe **(das) Europa** (B)

even **noch** (B); **selbst** (2)

evening **der Abend, -e** (1, 4); evening meal **das Abendessen, -** (1); evenings **abends** (4); good evening **guten Abend** (A); in the evening **abends** (4), **am Abend** (4); this evening **heute Abend** (2)

ever: Were you ever . . . ? **Warst du schon einmal ...?** (4)

every **jeder, jede, jedes** (1, 3); every day **jeden Tag** (1); every week **jede Woche** (3)

everything **alles** (2); everything possible **alles Mögliche** (2)

everywhere **überall** (10)

evil **böse** (9)

to evoke **hervor·rufen (ruft ... hervor), rief ... hervor, hervorgerufen** (1)

exactly **genau** (B)

exam: high school graduation exam **das Abitur** (4)

to examine **untersuchen, untersucht** (5)

example **das Beispiel, -e** (3); for example **zum Beispiel (z. B.)** (3)

exceedingly beautiful **wunderschön** (6)

excellent **ausgezeichnet** (3)

exchange: stock exchange **die Börse, -n** (12)

excited: to get excited **sich auf·regen, aufgeregt** (11)

to excuse **entschuldigen, entschuldigt** (10); Excuse me! **Entschuldigung!** (3), **Entschuldigen Sie!** (10)

to execute **aus·führen, ausgeführt** (12)

exercise **die Übung, -en** (A)

exotic **exotisch** (7)

to expect **erwarten, erwartet** (12)

expensive **teuer** (2); **wertvoll** (2)

to experience **erleben, erlebt** (3); experience **das Erlebnis, -se** (7); travel experience **das Reiseerlebnis, -se** (7)

to explain **erklären, erklärt** (5)

extra **extra** (10); extra costs (*e.g., utilities*) **die Nebenkosten** (*pl.*) (6)

eye **das Auge, -n** (B); color of eyes **die Augenfarbe, -n** (1)

face **das Gesicht, -er** (B)

factory **die Fabrik, -en** (5); in the factory **in der Fabrik** (5)

Fahrenheit **Fahrenheit** (B)

to fail **scheitern, gescheitert** (12)

to faint **in Ohnmacht fallen** (11)

fairy **die Fee, -n** (9); fairy tale **das Märchen, -** (9)

to fall **fallen (fällt), fiel, ist gefallen** (9); to fall asleep **ein·schlafen (schläft ... ein), schlief ... ein, ist eingeschlafen** (7); to fall down **hin·fallen (fällt ... hin), fiel ... hin, ist hingefallen** (11); to fall in love (with) **sich verlieben (in** + *acc.***), verliebt** (9, 12); to fall over **um·fallen (fällt ... um), fiel ... um, ist umgefallen** (9)

fall (*autumn*) **der Herbst, -e** (B)

family **die Familie, -n** (B); family doctor **der Hausarzt, ¨-e / die Hausärztin, -nen** (11); family name **der Familienname, -n** (*wk.*) (A, 1)

famous **berühmt** (7)

fanatic **der Fanatiker, -** (12)

far **weit** (6); as far as **bis zu** (+ *dat.*) (10); How far away? **Wie weit weg?** (6)

farmhouse **das Bauernhaus, ¨-er** (6)

fashionable **modisch** (2)

fast **schnell** (7)

fat **dick** (2); **fettig** (8)

father **der Vater, ¨** (B)

faucet **der Wasserhahn, ¨-e** (5)

favorite **Lieblings-** (A); favorite color **die Lieblingsfarbe, -n** (A); favorite name **der Lieblingsname, -n** (*wk.*) (A); favorite subject **das Lieblingsfach, ¨-er** (5)

fear **die Angst, ¨-e** (3)

February **der Februar** (B)

to feed **füttern, gefüttert** (9)

to feel **(sich) fühlen, gefühlt** (3, 11); to feel like (*doing s.th.*) **Lust haben** (3); to feel well **sich wohl fühlen** (11); How do you feel? **Wie fühlst du dich?** (3); I feel . . . **Ich fühle mich ...** (3)

feeling **das Gefühl, -e** (3)

fellow student **der Mitstudent, -en** (*wk.*) **/ die Mitstudentin, -nen** (A)

fence **der Zaun, ¨-e** (9)

to fetch **holen, geholt** (9)

fever **das Fieber** (11)

few: a few **ein paar** (2)

field **das Feld, -er** (7); field of study **die Wissenschaft, -en** (9)

fifteen **fünfzehn** (A)

fifth **fünft-** (4)

fifty **fünfzig** (A)

to fight **kämpfen, gekämpft** (9)

figure **die Figur, -en** (12)

to fill in the blanks **ergänzen, ergänzt** (4); to fill out **aus·füllen, ausgefüllt** (1); to fill up (with gas) **voll·tanken, vollgetankt** (5)

film **der Film, -e** (2); horror film **der Gruselfilm, -e** (2)

finally **schließlich** (7); **zum Schluss** (8); **endlich** (9)

to find **finden (findet), fand, gefunden** (2)

fine **fein** (8)

finger **der Finger, -** (11)

fingernail **der Fingernagel, ¨** (11)

finished **fertig** (3)

Finland **(das) Finnland** (B)

fire **das Feuer, -** (9); fire department **die Feuerwehr** (11)

firm **die Firma, Firmen** (3)

first (*adj.*) **erst-** (4); (*adv.*) **zuerst** (4, 7); first floor **das Erdgeschoss, -e** (6); first name **der Vorname, -n** (*wk.*) (A, 1); the first of October **der erste Oktober** (4); for the first time **zum ersten Mal** (4); on the first of October **am ersten Oktober** (4); to travel first class **erster Klasse fahren** (10)

fish **der Fisch, -e** (8)

to fit **passen** (+ *dat.*), **gepasst** (6, 11); That fits well. **Das passt gut.** (11)

five **fünf** (A)

flashlight **die Taschenlampe, -n** (9)

flat-screen (monitor) **der Flachbildschirm, -e** (2)

flat tire **die Reifenpanne, -n** (7)

flawless **vollkommen** (12)

flea market **der Flohmarkt, ¨-e** (2)

to flee **flüchten, ist geflüchtet** (11)

flight **der Flug, ¨-e** (9)

to float **schwimmen (schwimmt), schwamm, ist geschwommen** (7)

floor **der Boden, ¨-** (B); (*story*) **der Stock, Stockwerke** (6); first floor **das Erdgeschoss, -e** (6); on the second floor **im ersten Stock** (6)

to flow **fließen (fließt), floss, ist geflossen** (7)

flower **die Blume, -n** (3); to water the flowers **die Blumen gießen (gießt), goss, gegossen** (3)

flu **die Grippe** (11)

flute: transverse flute **die Querflöte, -n** (12)

to fly **fliegen (fliegt), flog, ist/hat geflogen** (1)

fly **die Fliege, -n** (8)

foot **der Fuß, ¨-e** (B); on foot **zu Fuß** (3)

for (*prep.*) **für** (+ *acc.*) (2); **seit** (+ *dat.*) (4, 11); **zu** (+ *dat.*) (2); (*coord. conj.*) **denn** (9, 11); Do you have something for it (*illness*)? **Haben Sie etwas dagegen?** (11); for example **zum Beispiel (z. B.)** (3); for lunch **zum Mittagessen** (3); for several days **seit mehreren Tagen** (11); for someone's birthday **zum Geburtstag** (2); for the first time **zum ersten Mal** (4); for two years **seit zwei Jahren** (4); medicine for **ein Medikament gegen** (+ *acc.*) (11)

forbidden **verboten** (9)

foreign **ausländisch** (12); foreign countries **das Ausland** (6); foreign language **die Fremdsprache, -n** (1)

foreigner **der Ausländer, - / die Ausländerin, -nen** (12)

forest **der Wald, ¨-er** (2, 7)

to forget **vergessen (vergisst), vergaß, vergessen** (2)

fork **die Gabel, -n** (8)

form **das Formular, -e** (12)

formality **die Formalität, -en** (12)

forth: to call forth **hervor·rufen (ruft ... hervor), rief ... hervor, hervorgerufen** (1)

forty **vierzig** (A)

fountain **der Brunnen, -** (9)

four **vier** (A)

fourteen **vierzehn** (A)

fourth **viert-** (4)

franc (Swiss) **der Schweizer Franken, -** (8)

France **(das) Frankreich** (B)

frank(furter) **das Würstchen, -** (8)

free **frei** (8)

to free **erlösen, erlöst** (9)

freedom **die Freiheit, -en** (12)

freeway **die Autobahn, -en** (7)

French (*language*) **(das) Französisch** (B); (*person*) **der Franzose, -n** (*wk.*) **/ die Französin, -nen** (B); French fries **die Pommes (frites)** (*pl.*) (8)

fresh **frisch** (8)

Friday **der Freitag** (1)

fried **gebraten** (8); fried eggs **gebratene Eier** (*pl.*) (8)

friend **der Freund, -e / die Freundin, -nen** (A); beloved female friend **die Geliebte, -n** (3)

friendly **freundlich** (B)

fries: French fries **die Pommes (frites)** (*pl.*) (8)

frog **der Frosch, ¨-e** (9)

from **von** (+ *dat.*) (A, 10); **aus** (+ *dat.*) (10); from next door **von nebenan** (5); from silk **aus Seide** (2); from where **woher** (B); from work **von der Arbeit** (3)

fruit **das Obst** (8)

to fry **braten (brät), briet, gebraten** (2, 8); **bräunen, gebräunt** (8)

frying pan **die Pfanne, -n** (5)

fuel tank **der Tank, -s** (7)

full(y) **voll** (10)

fun **lustig** (12); have fun **viel Spaß** (A)

funny **komisch** (10); **lustig** (12)

furnished **möbliert** (6)

furniture **die Möbel** (*pl.*) (6)

game: video game **das Videospiel, -e** (5)

garage **die Werkstatt, ¨-en** (5); **die Garage, -n** (6)

garbage **der Müll** (6); garbage can **der Mülleimer, -** (8)

garden **der Garten, ¨-** (6); garden hose **der Gartenschlauch, ¨-e** (6); in the garden **im Garten** (4)

garlic **der Knoblauch** (8)

gas station **die Tankstelle, -n** (5); at the gas station **an der Tankstelle** (5)

gasoline **das Benzin** (6)

gaudy **grell** (2)

gear **der Gang, ¨-e** (7)

gentleman **der Herr, -en** (*wk.*) (A)

geography **die Geografie** (B, 1)

German (*language*) **(das) Deutsch** (B); (*person*) **der/die Deutsche, -n (ein Deutscher)** (B); German army **die Bundeswehr** (5); German class/course **der Deutschkurs, -e** (A); I am German. **Ich bin Deutsche/r.** (B); in the German army **bei der Bundeswehr** (5)

Germany **(das) Deutschland** (B)

to get **bekommen (bekommt), bekam, bekommen** (3); **holen, geholt** (9); to get acquainted with **kennen·lernen, kennengelernt** (1); to get an appointment **sich einen Termin geben lassen** (11); to get angry **sich ärgern, geärgert** (11); to get dressed **sich an·ziehen (zieht ... an), zog ... an, angezogen** (11); to get excited **sich auf·regen, aufgeregt** (11); to get information about **sich erkundigen nach** (+ *dat.*), **erkundigt** (10); to get in this way **herein·kommen (kommt ... herein), kam ... herein, ist hereingekommen** (10); to get stuck **stecken bleiben (bleibt ... stecken), blieb ... stecken, ist stecken geblieben** (11); to get undressed **sich aus·ziehen (zieht ... aus), zog ... aus, ausgezogen** (11); to get up **auf·stehen (steht ... auf), stand ... auf, ist aufgestanden** (A, 1); to get up on the wrong side of bed **mit dem linken Fuß auf·stehen** (4); to get upset **sich auf·regen, aufgeregt** (11)

giant **der Riese, -n** (*wk.*) (9)

gift certificate **der Geschenkgutschein, -e** (2)

gifted **begabt** (9)

giraffe **die Giraffe, -n** (10)

girl **das Mädchen, -** (9)

girlfriend **die Freundin, -nen** (A)

to give **geben (gibt), gab, gegeben** (A, 6); (*as a present*) **schenken, geschenkt** (5); to give a paper / oral report **ein Referat halten (hält), hielt, gehalten** (4); give me **geben Sie mir** (A); to give up **auf·geben (gibt ... auf), gab ... auf, aufgegeben** (1)

given name **der Vorname, -n** (*wk.*) (A, 1)

glacier **der Gletscher, -** (7)

gladly **gern** (1, 5)

glass (*n.*) **das Glas, ¨-er** (5); (*adj.*) **gläsern** (9)

glasses (*pair of eyeglasses*) **die Brille, -n** (A)

glove **der Handschuh, -e** (2)

to go **gehen (geht), ging, ist gegangen** (A); **laufen (läuft), lief, ist gelaufen** (A); to go along **entlang·gehen (geht ... entlang), ging ... entlang, ist entlanggegangen** (10); to go away **weg·gehen (geht ... weg), ging ... weg, ist weggegangen** (4); to go by **vorbei·fahren (fährt ... vorbei), fuhr ... vorbei, ist vorbeigefahren** (10); to go canoeing **Kanu fahren** (10); to go for a walk **spazieren gehen (geht ... spazieren), ging ... spazieren, ist spazieren gegangen** (1); to go get **holen, geholt** (9); to go home **nach Hause gehen** (1); to go ice-skating **Schlittschuh laufen** (3); to go in this way **herein·kommen (kommt ... herein), kam ... herein, ist hereingekommen** (10); to go inside **rein·gehen (geht ... rein), ging ... rein, ist reingegangen** (1); to go on a trip **verreisen, ist verreist** (3); to go out **aus·gehen (geht ... aus), ging ... aus, ist ausgegangen** (1); to go out (*power*)

aus·fallen (fällt ... aus), fiel ... aus, ausgefallen (8); to go over that way **hinüber·gehen (geht ... hinüber), ging ... hinüber, ist hinübergegangen** (10); to go shopping **ein·kaufen gehen (geht ... einkaufen), ging ... einkaufen, ist einkaufen gegangen** (1, 5); to go that way **hin·gehen (geht ... hin), ging ... hin, ist hingegangen** (10); to go to a party **auf eine Party gehen** (1); to go to bed **ins Bett gehen** (1); to go to the mountains **in die Berge gehen** (1); to go to the movies **ins Kino gehen** (1); to go to the swimming pool **ins Schwimmbad fahren** (1); to go to the university **zur Uni gehen** (1); to go to work **zur Arbeit gehen** (1); to go up that way **hinauf·gehen (geht ... hinauf), ging ... hinauf, ist hinaufgegangen** (10); to go with **passen** (+ *dat.*), **gepasst** (2); I'd rather go . . . **Ich gehe lieber ...** (2); Where are you going? **Wo willst du denn hin?** (A)

god, God **der Gott, ⸚er** (12)

goldfish **der Goldfisch, -e** (11)

golf **das Golf** (1)

good: good afternoon (*for.*) **guten Tag** (A); (*for.; southern Germany, Austria*) **grüß Gott** (A); good evening **guten Abend** (A); Good luck! **Viel Glück!** (3); good morning **guten Morgen** (A); It looks good. **Es sieht gut aus.** (2); to taste good to **schmecken** (+ *dat.*), **geschmeckt** (6); That looks / Those look good on you! **Das steht / Die stehen dir gut!** (2)

good-bye **auf Wiedersehen** (A); (*infor.; southern Germany, Austria*) **servus** (A); (*on the phone*) **auf Wiederhören** (6)

government **die Regierung, -en** (12), **der Staat, -en** (12); (*adj.*) **staatlich** (12)

to grab **greifen (greift), griff, gegriffen** (11)

grade (*level*) **die Klasse, -n** (9); (*mark*) **die Note, -n** (9)

graduation **der Abschluss** (9); high school graduation exam **das Abitur** (4)

grammar **die Grammatik, -en** (A)

granddaughter **die Enkelin, -nen** (5)

grandfather **der Großvater, ¨** (B)

grandmother **die Großmutter, ¨** (B)

grandparents **die Großeltern** (*pl.*) (B)

grandson **der Enkel, -** (5)

grape **die Weintraube, -n** (8)

to grasp **greifen (greift), griff, gegriffen** (11)

gray **grau** (A)

greasy **fettig** (8, 11)

great (*coll.*) **toll** (2); great! **prima!** (6); That sounds great. **Das hört sich toll an.** (4)

Great Britain **(das) Großbritannien** (B)

Greece **(das) Griechenland** (B)

green **grün** (A)

to greet **grüßen, gegrüßt** (10)

greeting **der Gruß, ⸚e** (9)

to grill **braten (brät), briet, gebraten** (2)

grill **der Grill, -s** (2)

grilled **gegrillt** (8)

ground beef (or pork) **das Hackfleisch** (8)

to grow **wachsen (wächst), wuchs, ist gewachsen** (9); to grow up **auf·wachsen (wächst ... auf), wuchs ... auf, ist aufgewachsen** (12)

growth **das Wachstum** (12)

guidebook: travel guidebook **der Reiseführer, -** (5)

guilt **die Schuld, -en** (12)

guinea pig **das Meerschweinchen, -** (10)

guitar **die Gitarre, -n** (1)

hair **das Haar, -e** (B, 11); to blow dry (one's) hair **(sich) die Haare föhnen** (2, 11); color of hair **die Haarfarbe** (1); to comb one's hair **sich die Haare kämmen** (11); to cut hair **Haare schneiden (schneidet), schnitt, geschnitten** (3); hair dryer **der Haartrockner, -** (2); with the short/long hair **mit dem kurzen/langen Haar** (A)

haircut **der Haarschnitt, -e** (2)

hairdresser **der Friseur, -e / die Friseurin, -nen** (5)

hall: town hall **das Rathaus, ⸚er** (1, 6); at the town hall **auf dem Rathaus** (1)

ham **der Schinken** (8)

hammer **der Hammer, ¨** (8)

hamster **der Hamster, -** (10)

hand **die Hand, ⸚e** (B); hand towel **das Handtuch, ⸚er** (8); to shake hands **die Hand schütteln** (A)

handkerchief **das Taschentuch, ⸚er** (3)

to hang (*be in a hanging position*) **hängen (hängt), hing, gehangen** (3); (*coll.*) **chillen, gechillt** (1); to hang (up) **hängen, gehängt** (3); **auf·hängen, aufgehängt** (2); to hang the picture on the wall **das Bild an die Wand hängen** (3)

hangover **der Kater, -** (11)

to happen **passieren, ist passiert** (4)

happiness **das Glück** (3)

happy **glücklich** (B); to be happy about **sich freuen über** (+ *acc.*), **gefreut** (11)

harbor **der Hafen, ¨** (10)

hard **schwer** (3)

harmful: to be harmful to **schaden** (+ *dat.*), **geschadet** (6)

harmonica **die Mundharmonika, -s** (12)

hat **der Hut, ⸚e** (A)

to hate **hassen, gehasst** (9)

to have **haben (hat), hatte, gehabt** (A); have fun **viel Spaß** (A); to have to **müssen (muss), musste, gemusst** (3); haven't seen (you / each other) for a long time **lange nicht gesehen** (1)

head **der Kopf, ⸚e** (B); head cold **die Erkältung, -en** (11)

headache **die Kopfschmerzen** (*pl.*) (11)

headband **das Stirnband, ⸚er** (A)

to heal **heilen, geheilt** (5)

health **die Gesundheit** (11)

healthy **gesund** (8)

to hear **hören, gehört** (1)

heart **das Herz, -en** (11)

to heat **erhitzen, erhitzt** (8)

heated, heat included **warm** (6)

heating: central heating **die Zentralheizung** (6)

heavy **schwer** (3); **stark** (11)

height **die Größe, -n** (1); **die Höhe, -n** (12)

hello (*for.*) **guten Tag** (A); (*for.; southern Germany, Austria*) **grüß Gott** (A); (*infor.; southern Germany, Austria*) **servus** (A); to say hello to **grüßen, gegrüßt** (10)

helmet: bicycle helmet **der Fahrradhelm, -e** (5)

to help **helfen** (+ *dat.*) **(hilft), half, geholfen** (6); Help! **Hilfe!** (11); May I help you? **Bitte schön?** (7)

her **ihr(e)** (1, 2)

herb butter **die Kräuterbutter** (8)

herbs **die Kräuter** (*pl.*) (8)

here **hier** (A)

hi (*infor.*) **hallo** (A); (*Switzerland*) **grüezi** (A)

to hide **sich verstecken, versteckt** (9)

high **hoch** (7); high-rise building **das Hochhaus, ⸚er** (6); high school **das Gymnasium, Gymnasien** (6); high school graduation exam **das Abitur** (4)

highlight **der Höhepunkt, -e** (7)

hike **die Wanderung, -en** (7); to hike **wandern, ist gewandert** (1); to hike in the mountains **in den Bergen wandern** (1)

hiking shoe **der Wanderschuh, -e** (2)

hill **der Hügel, -** (7)

him (*acc.*) **ihn** (2)

his **sein(e)** (1, 2)

history **die Geschichte** (1); history: art history **die Kunstgeschichte** (1)

to hit **schlagen (schlägt), schlug, geschlagen** (11)

to hitchhike **trampen, ist getrampt** (10)

hobby **das Hobby, -s** (1)

to hold **halten (hält), hielt, gehalten** (4)

hole **das Loch, ⸚er** (9)

holiday **der Feiertag, -e** (4)

Holland **(das) Holland** (B)

home **das Haus, ⸚er** (2); **die Heimat, -en** (12); at home **zu Hause** (A, 1, 10); **daheim** (9); to go home **nach Hause gehen** (1, 10); single-family home **das Einfamilienhaus, ⸚er** (6)

homeland **die Heimat, -en** (12)

homemade **selbst gemacht** (8)

homesick: to be homesick **Heimweh haben** (3)

homesickness **das Heimweh** (3)

hometown **die Heimatstadt, ⸚e** (6); **die Heimat, -en** (12)

homework **die Hausaufgabe, -n** (A)

honey **der Honig** (8)

to honk **hupen, gehupt** (7)

honor **die Ehre, -n** (8)

hook **der Haken, -** (8)

to hope **hoffen, gehofft** (3)

hope **die Hoffnung, -en** (2)

horn **die Hupe, -n** (7)

horror film **der Gruselfilm, -e** (2)

horse **das Pferd, -e** (2, 9)

hose: garden hose **der Gartenschlauch, ¨e** (6)

hospital **das Krankenhaus, ¨er** (3, 5); in the hospital **im Krankenhaus** (5)

host **der Wirt, -e / die Wirtin, -nen** (10)

hostel: youth hostel **die Jugendherberge, -n** (10)

hot **heiß** (B); hot chocolate **der Kakao** (8); hot dog **das Würstchen, -** (8)

hotel **das Hotel, -s** (2, 5); at the hotel **im Hotel** (5)

hour **die Stunde, -n** (1, 2); office hour **die Sprechstunde, -n** (3)

house **das Haus, ¨er** (1, 2, 6); house key **der Hausschlüssel, -** (9); house number **die Hausnummer, -n** (1); row house, town house **das Reihenhaus, ¨er** (6); vacation house **das Ferienhaus, ¨er** (4)

household **der Haushalt, -e** (9)

housekeeping **der Haushalt, -e** (9)

housemate **der Mitbewohner, - / die Mitbewohnerin, -nen** (2)

housing: shared housing **die WG, -s (die Wohngemeinschaft, -en)** (6)

how **wie** (B); How do you feel? **Wie fühlst du dich?** (3); How do you spell that? **Wie schreibt man das?** (A); How far away? **Wie weit weg?** (6); how many . . . ? **wie viele ...?** (A)

however **allerdings** (6)

humid **feucht** (B)

hundred **hundert** (A)

hundredth **hundertst-** (4)

Hungary **(das) Ungarn** (B)

hunger **der Hunger** (3)

hungry **hungrig** (9); to be hungry **Hunger haben** (3)

hunter **der Jäger, - / die Jägerin, -nen** (9)

to hurry **sich beeilen, beeilt** (8)

hurry **die Eile** (3); to be in a hurry **in Eile sein** (3); **es eilig haben** (8); to hurt **weh·tun (tut ... weh), tat ... weh, wehgetan** (11)

husband **der Mann, ¨er** (B)

ice **das Eis** (2); ice cream parlor **das Eiscafé, -s** (8); ice skate **der Schlittschuh, -e** (3); to go ice-skating **Schlittschuh laufen** (3)

idea **die Idee, -n** (10)

ideal **ideal** (12)

identification card **der Personalausweis, -e** (1); **der Ausweis, -e** (10)

if (*subord. conj.*) **wenn** (2, 11); **ob** (6)

illegal **illegal** (12)

illness **die Krankheit, -en** (11)

to imagine something **sich etwas vor·stellen, vorgestellt** (6)

immediately **sofort** (3)

immigrant **der Einwanderer, -** (12)

to immigrate **ein·wandern, ist eingewandert** (12)

impatient **ungeduldig** (10)

important **wichtig** (2)

impression **der Eindruck, ¨e** (5)

in **in** (+ *dat./acc.*) (A, 4); **an** (+ *dat./acc.*) (4) in addition **dazu** (8); in addition to **neben** (+ *dat./acc.*) (3), **außer** (+ *dat.*) (9); in any case **jedenfalls** (11); in a row **hintereinander** (3); in common **gemeinsam** (6); in it **drin/darin** (6); in January **im Januar** (B); in love **verliebt** (4); in order to **um ... zu** (12); in rainy weather **bei Regen** (7); in spite of **trotz** (+ *gen.*) (12); in the afternoon **nachmittags** (4); in the country (*rural*) **auf dem Land** (6); in the end **zum Schluss** (8); in the evening **am Abend** (4), **abends** (4); in the garden **im Garten** (4); in the German army **bei der Bundeswehr** (5); in the middle **mitten** (9); in the morning **früh** (4); in the spring **im Frühling** (B); in the vicinity **in der Nähe** (6); in this way **herein** (10); in(ward) **hinein** (9)

included **inbegriffen** (10); (*utilities*) **inklusive** (6); heat included **warm** (6)

income **das Einkommen** (12)

incredible **unglaublich** (5)

indeed **ja** (4)

industrious **fleißig** (12)

inexpensive **billig** (2)

infection **die Entzündung, -en** (11)

influenza **die Grippe** (11)

to inform oneself about **sich informieren über** (+ *acc.*), **informiert** (12)

information **die Information, -en** (4); biographical information **persönliche Angaben** (*pl.*) (1); to get information about **sich erkundigen nach** (+ *dat.*), **erkundigt** (10)

ingredient **die Zutat, -en** (8)

injection **die Spritze, -n** (11)

to injure oneself **sich verletzen, verletzt** (11)

injured **verletzt** (11); critically injured **schwer verletzt** (11); injured person **der/die Verletzte, -n (ein Verletzter)** (11)

inn (bed-and-breakfast) **das Gästehaus, ¨er** (10)

innkeeper **der Wirt, -e / die Wirtin, -nen** (10)

insane **wahnsinnig** (12)

inside: to go inside **rein·gehen (geht ... rein), ging ... rein, ist reingegangen** (1)

instead of **anstatt** (+ *gen.*) (12); **statt** (+ *gen.*) (12)

institute **das Institut, -e** (7)

to instruct **unterrichten, unterrichtet** (5)

instruction **der Unterricht** (9, 11)

instructor **der Lehrer, - / die Lehrerin, -nen** (A, 1)

instrument **das Instrument, -e** (12)

insurance **die Versicherung, -en** (5)

to integrate **integrieren, integriert** (12)

integration **die Integration** (12)

intelligent **intelligent** (7)

to intend (to) **wollen (will), wollte, gewollt** (3)

to interest **interessieren, interessiert** (5); to be interested in **Interesse haben an** (+ *dat.*) (5); **sich interessieren für** (+ *acc.*) (5)

interest **das Interesse, -n** (5); (*money*) **die Zinsen** (*pl.*) (12)

interesting **interessant** (7); something interesting **etwas Interessantes** (4)

Internet: to surf the Internet **im Internet surfen** (1)

intersection **die Kreuzung, -en** (10)

interview **das Interview, -s** (4)

into **in** (+ *acc.*) (A)

to introduce **vor·stellen, vorgestellt** (6); **ein·führen, eingeführt** (12)

introduction **die Einführung, -en** (A)

to invent **erfinden (erfindet), erfand, erfunden** (4)

to investigate **untersuchen, untersucht** (5)

invitation **die Einladung, -en** (2)

to invite **ein·laden (lädt ... ein), lud ... ein, eingeladen** (2)

Ireland **(das) Irland** (B)

iron **das Bügeleisen, -** (6); to iron **bügeln, gebügelt** (6)

island **die Insel, -n** (7)

it **es** (B)

Italian (*language*) **(das) Italienisch** (B)

Italy **(das) Italien** (B)

its (*fem.*) **ihr(e)** (2); (*masc./neut.*) **sein(e)** (2)

ivory **das Elfenbein** (10)

jacket **die Jacke, -n** (A); sports jacket **das Sakko, -s** (A)

jail **das Gefängnis, -se** (6)

jam **die Marmelade, -n** (8)

January **der Januar** (B); in January **im Januar** (B)

Japanese (*adj.*) **japanisch** (8)

jealous **eifersüchtig** (3)

jeans **die Jeans** (*pl.*) (2)

jewelry **der Schmuck** (2)

joke **der Witz, -e** (3); to tell jokes **Witze erzählen** (3)

journey **die Reise, -n** (7)

joy **die Freude, -n** (9)

judge **der Richter, - / die Richterin, -nen** (5)

juice **der Saft, ¨e** (8); apple juice **der Apfelsaft** (8); orange juice **der Orangensaft** (8)

July **der Juli** (B)

to jump **springen (springt), sprang, ist gesprungen** (A)

June **der Juni** (B)

just **knapp** (4); That's just it! **Das ist es ja!** (4)

kangaroo **das Känguru, -s** (10)

to keep: keep the change **das stimmt so** (8); to keep on walking **weiter·gehen (geht ... weiter), ging ... weiter, ist weitergegangen** (10)

kettle: tea kettle **der Teekessel, -** (8)

key **der Schlüssel, -** (9); house key **der Hausschlüssel, -** (9)

kidney **die Niere, -n** (11)

to kill **töten, getötet** (9)

kiln **der Brennofen, ¨** (12)

kilometer **der Kilometer, -** (2)

kind **die Art, -en** (2); what kind of **was für** (+ *acc.*) (3)

kindergarten **der Kindergarten, ¨** (6)

king **der König, -e** (9)

to kiss **küssen, geküsst** (9)

kiss **der Kuss, ¨e** (4)

kitchen **die Küche, -n** (5); kitchen clock **die Küchenuhr, -en** (5); kitchen lamp **die Küchenlampe, -n** (5); kitchen scale **die Küchenwaage, -n** (5); kitchen table **der Küchentisch, -e** (5)

knife **das Messer, -** (8)

to knit **stricken, gestrickt** (3)

to know **kennen (kennt), kannte, gekannt** (B); **wissen (weiß), wusste, gewusst** (2)

knowledge about a field **die Kenntnisse** (*pl.*) (5)

labor **die Arbeitskraft, ¨e** (12)

laboratory: language laboratory **das Sprachlabor, -s** (4)

lake **der See, -n** (7)

lamp **die Lampe, -n** (B); kitchen lamp **die Küchenlampe, -n** (5)

landlord/landlady **der Vermieter, - / die Vermieterin, -nen** (6)

language **die Sprache, -n** (B); foreign language **die Fremdsprache, -n** (1); language laboratory **das Sprachlabor, -s** (4)

laptop (computer) **der Laptop, -s** (B, 2)

large **dick** (2)

last **letzt-** (4); last Monday **letzten Montag** (4); last night **gestern Abend** (4); last summer **letzten Sommer** (4); last week **letzte Woche** (4); last weekend **letztes Wochenende** (4); the last time **das letzte Mal** (4)

to last **dauern, gedauert** (4)

late(r) **spät(er)** (1); late morning **der Vormittag, -e** (4)

Latin (*language*) **das Latein** (1)

to laugh **lachen, gelacht** (3)

laundry **die Wäsche** (4)

lawn **der Rasen** (5); lawn mower **der Rasenmäher, -** (6)

to learn **lernen, gelernt** (1)

least: at least **wenigstens** (1), **mindestens** (7); the least **am wenigsten** (8)

leather **das Leder** (2)

to leave **verlassen (verlässt), verließ, verlassen** (11)

lecture **die Vorlesung, -en** (4)

left **links** (4, 10)

leg **das Bein, -e** (B)

leisure time **die Freizeit** (1)

lemon **die Zitrone, -n** (8)

to lend **leihen (leiht), lieh, geliehen** (5)

to let **lassen (lässt), ließ, gelassen** (11); Let's meet . . . **Treffen wir uns ...** (2)

letter **der Brief, -e** (1)

lettuce **der Kopfsalat** (8)

liberal **liberal** (9)

librarian **der Bibliothekar, -e / die Bibliothekarin, -nen** (5)

library **die Bibliothek, -en** (4)

license: driver's license **der Führerschein, -e** (4); license plate **das Nummernschild, -er** (7)

to lie **liegen (liegt), lag, gelegen** (1); to lie down **sich hin·legen, hingelegt** (11); to lie in the sun **in der Sonne liegen** (1)

Liechtenstein **(das) Liechtenstein** (B)

life **das Leben, -** (9)

to light **an·zünden, angezündet** (3)

light (*adj., color*) **hell** (6); (*adj., weight*) **leicht** (6)

light (*n.*) **das Licht, -er** (3)

to like **mögen (mag), mochte, gemocht** (1, 3); to be to one's liking **gefallen** (+ *dat.*) **(gefällt), gefiel, gefallen** (6); Do you like to wear . . . ? **Trägst du gern ...?** (A); I like it. **Es gefällt mir.** (6); I would like **ich hätte gern** (5); like (*to do*) best **am liebsten** (7); We like to sing. **Wir singen gern.** (1); would like (to) **möchte** (2, 3)

limit **die Grenze, -n** (12); city limits **der Stadtrand, ¨er** (6)

linguistics **die Linguistik** (1)

lion **der Löwe, -n** (*wk.*) (10)

lip **die Lippe, -n** (11)

list **die Liste, -n** (5)

to listen **zu·hören, zugehört** (A); to listen (to) **hören, gehört** (1); to listen to **zu·hören** (+ *dat.*), **zugehört** (6)

liter **der Liter, -** (7)

literature **die Literatur** (1)

little: a little bit **ein bisschen** (3)

to live **leben, gelebt** (3); to live (in) **wohnen (in** + *dat.*)**, gewohnt** (B)

liver **die Leber, -n** (11)

living room **das Wohnzimmer, -** (6)

locomotive **die Lokomotive, -n** (7)

lodge: ski lodge **die Skihütte, -n** (6)

long **lang** (B); all day long **den ganzen Tag** (1); all night long **die ganze Nacht** (3); haven't seen (you /each other) for a long time **lange nicht gesehen** (1); so long **bis bald** (A); with the long hair **mit dem langen Haar** (A)

to look **schauen, geschaut** (A); **aus·sehen (sieht ... aus), sah ... aus, ausgesehen** (2); It looks good. **Es sieht gut aus.** (2); to look at **an·schauen, angeschaut** (2); **an·sehen (sieht ... an), sah ... an, angesehen** (3); to look for **suchen, gesucht** (1); That looks / Those look good on you! **Das steht / Die stehen dir gut!** (2)

to lose **verlieren (verliert), verlor, verloren** (7); to lose weight **ab·nehmen (nimmt ... ab), nahm ... ab, abgenommen** (8, 11)

lot: a lot **viel** (A); Lots of luck! **Viel Glück!** (3)

lotion: suntan lotion **die Sonnenmilch** (10); to put lotion on **sich ein·cremen, eingecremt** (11)

lottery **die Lotterie, -n** (5); to win the lottery **in der Lotterie gewinnen** (5)

loudspeaker **der Lautsprecher, -** (2)

lounge **der Aufenthaltsraum, ¨e** (10)

to love **lieben, geliebt** (3); to be in love **verliebt sein** (4, 12); to fall in love (with) **sich verlieben (in** + *acc.*)**, verliebt** (9, 12); love (*beloved female friend*) **die Geliebte, -n** (3)

lovesickness **der Liebeskummer** (11)

to lower **senken, gesenkt** (12)

loyal **treu** (9)

lozenge **das Bonbon, -s** (11)

luck **das Glück** (3); Good luck! Lots of luck! **Viel Glück!** (3)

lunch **das Mittagessen, -** (3); for lunch **zum Mittagessen** (3)

lung **die Lunge, -n** (11)

machine: automatic teller machine (ATM) **der Geldautomat, -en** (*wk.*) (12); coffee machine **die Kaffeemaschine, -n** (5); washing machine **die Waschmaschine, -n** (6)

magical **magisch** (12)

mail: e-mail **die Mail, -s** (1)

majority **die Mehrheit, -en** (12)

makeup: to put makeup on **sich schminken, geschminkt** (11)

mall: pedestrian mall **die Fußgängerzone, -n** (10)

man **der Mann, ¨er** (A, B); Man! (*coll.*) **Mensch!** (2)

manager **der Geschäftsführer, - / die Geschäftsführerin, -nen** (8)

mansion **die Villa, Villen** (6)

many **viele** (A); how many . . . ? **wie viele ...?** (A); many thanks **vielen Dank** (10)

map **die Landkarte, -n** (7)

March **der März** (B)

market **der Markt, ¨e** (10); flea market **der Flohmarkt, ¨e** (2); marketplace, market square **der Marktplatz, ¨e** (6); stock market crash **der Börsenkrach, ¨e** (12)

marmalade **die Marmelade, -n** (8)

marriage **die Ehe, -n** (12)

married **verheiratet** (1)

to marry **heiraten, geheiratet** (3, 5)

to match **passen** (+ *dat.*), **gepasst** (2)

match **das Streichholz, ¨er** (8)

material **das Material, -ien** (12)

mathematics **die Mathematik** (1)

matter: It doesn't matter to me. **Das ist mir egal.** (6); That doesn't matter. **Das macht nichts.** (1); to be the matter with (*a person*) **fehlen** (+ *dat.*), **gefehlt** (11)

mattress: air mattress **die Luftmatratze, -n** (10)

May **der Mai** (B)

may (*v.*) **dürfen (darf), durfte, gedurft** (3); **können (kann), konnte, gekonnt** (3); May I help you? **Bitte schön?** (7)

meadow **die Wiese, -n** (7)

meal **die Mahlzeit, -en** (8); evening meal **das Abendessen, -** (1); midday meal **das Mittagessen, -** (3)

mean **gemein** (8); **böse** (9)

meaning **die Bedeutung, -en** (6)

means: by all means **auf jeden Fall** (4); means of payment **das Zahlungsmittel, -** (12); means of transportation **das Transportmittel, -** (7)

meat **das Fleisch** (8)

mechanical engineering **der Maschinenbau** (1)

medical **medizinisch** (11)

medicine **das Medikament, -e** (11); medicine for **ein Medikament gegen** (+ *acc.*) (11)

Mediterranean Sea **das Mittelmeer** (B)

to meet **treffen (trifft), traf, getroffen** (2); **begegnen** (+ *dat.*), **begegnet** (6); Let's meet . . . **Treffen wir uns ...** (2)

memory **die Erinnerung, -en** (4)

menu **die Speisekarte, -n** (8)

meter: square meter (m^2) **der Quadratmeter (qm), -** (6)

Mexican (*adj.*) **mexikanisch** (8); Mexican (*person*) **der Mexikaner, - / die Mexikanerin, -nen** (B)

midday **der Mittag, -e** (3); midday meal **das Mittagessen, -** (3)

mileage **der Kilometerstand** (7)

milk **die Milch** (8)

million **die Million, -en** (12)

mind **der Geist** (8)

mineral water **das Mineralwasser** (8)

minister (*government*) **der Minister, - / die Ministerin, -nen** (12)

minority **die Minderheit, -en** (12)

miracle **das Wunder, -** (4)

mirror **der Spiegel, -** (6)

to miss **verpassen, verpasst** (9); **verfehlen, verfehlt** (10); to be missing **fehlen** (+ *dat.*), **gefehlt** (6)

mixed **gemischt** (8)

mixer **die Küchenmaschine, -n** (8)

modern **modern** (6)

moment **der Moment, -e** (1); at the moment **im Moment** (1)

Monday **der Montag** (1); last Monday **letzten Montag** (4)

money **das Geld** (2)

monitor: flat-screen monitor **der Flachbildschirm, -e** (2)

month **der Monat, -e** (B)

to mop **wischen, gewischt** (6)

moral **die Lehre, -n** (8)

more **mehr** (7)

morning: good morning **guten Morgen** (A); in the morning **früh** (4); late morning **der Vormittag, -e** (4); until four in the morning **bis um vier Uhr früh** (4)

mosquito **die Mücke, -n** (10)

most **meist** (3); mostly **meist** (3); **meistens** (8)

mother **die Mutter, ¨** (B)

motif **das Motiv, -e** (12)

motorcycle **das Motorrad, ¨er** (1, 7); to ride a motorcycle **Motorrad fahren** (1)

mountain **der Berg, -e** (1); to go to the mountains **in die Berge gehen** (1); to hike in the mountains **in den Bergen wandern** (1); mountain range **das Gebirge, -** (7)

mountaintop **der Gipfel, -** (7)

mouse **die Maus, ¨e** (10)

mouth **der Mund, ¨er** (B)

to move **ziehen (zieht), zog, ist gezogen** (2)

movie: to go to the movies **ins Kino gehen** (1); movie theater **das Kino, -s** (1)

to mow **mähen, gemäht** (5)

mower: lawn mower **der Rasenmäher, -** (6)

MP3 player **der MP3-Spieler, -** (2, 5)

Mr. **der Herr, -en** (*wk.*) (A)

Mrs.; Ms. **die Frau, -en** (A)

much **viel** (A)

mug **der Becher, -** (9)

muscle: sore muscles **der Muskelkater, -** (11)

museum **das Museum, Museen** (1); to go to the museum **ins Museum gehen** (1)

mushroom **der Pilz, -e** (8)

music **die Musik** (1)

mussel **die Muschel, -n** (8)

must **müssen (muss), musste, gemusst** (3)

mustache **der Schnurrbart, ¨e** (A)

mustard **der Senf** (8)

my **mein(e)** (A, 2)

nail **der Nagel, ¨** (8)

name **der Name, -n** (*wk.*) (A, 1); family name **der Familienname, -n** (*wk.*) (A, 1); favorite name **der Lieblingsname, -n** (*wk.*) (A); first/given name **der Vorname, -n** (*wk.*) (A, 1); What's your name? **Wie heißen Sie?** (*for.*) **/ Wie heißt du?** (*infor.*) (A)

named: to be named **heißen (heißt), hieß, geheißen** (A)

napkin **die Serviette, -n** (8)

narrow **eng** (12); narrow street **die Gasse, -n** (10)

national(ly) **national** (12); national park **der Nationalpark, -s** (2)

naturally **natürlich** (2)

nature **die Natur** (9)

near **bei** (+ *dat.*) (10)

nearby **nah** (7)

neat (*coll.*) **grell** (2); **toll** (2)

neck **der Hals, ¨e** (9)

necklace **die Halskette, -n** (2)

to need **brauchen, gebraucht** (1)

neighbor **der Nachbar, -n** (*wk.*) **/ die Nachbarin, -nen** (4)

neighborhood **der Stadtteil, -e** (6); **das Stadtviertel, -** (6)

nephew **der Neffe, -n** (*wk.*) (B)

nervous **nervös** (1)

nest **das Nest, -er** (10)

the Netherlands **die Niederlande** (*pl.*) (B)

never **nie** (2)

new **neu** (A); something new **etwas Neues** (4)

news **die Nachrichten** (*pl.*) (7)

newspaper **die Zeitung, -en** (2); daily newspaper **die Tageszeitung, -en** (5); to deliver newspapers **Zeitungen aus·tragen** (5); to read the newspaper **Zeitung lesen** (1)

New Zealand **(das) Neuseeland** (B)

next: (from) next door **(von) nebenan** (5); next to **neben** (+ *dat./acc.*) (9); **the next time** das nächste Mal (3)

nice **nett** (3); (*weather*) **schön** (B)

niece **die Nichte, -n** (B)

night **die Nacht, ¨e** (3); all night long **die ganze Nacht** (3); last night **gestern Abend** (4); nights, at night **nachts** (4)

nightshirt **das Nachthemd, -en** (2)

nightstand **der Nachttisch, -e** (6)

nine **neun** (A)

nineteen **neunzehn** (A)

ninety **neunzig** (A)

ninth **neunt-** (4)

no **nein** (A); **kein(e)** (2); no one **niemand** (2); no wonder **kein Wunder** (4)

nobody **niemand** (2)

noise **das Geräusch, -e** (9)

none **kein(e)** (2)

nonetheless **trotzdem** (12)

noodle **die Nudel, -n** (8)

noon **der Mittag, -e** (3); at noon **mittags** (2)

normal **normal** (5); normally **normalerweise** (8)

north (of) **nördlich (von** + *dat.*) (7)

North America **(das) Nordamerika** (B)

North Sea **die Nordsee** (B)

Norway **(das) Norwegen** (B)

nose **die Nase, -n** (B, 11)

not **nicht** (A); not a bit **gar nicht** (3); not at all **kein bisschen** (3); not until (four o'clock) **erst (um vier Uhr)** (4)

note (*of currency*) **der Schein, -e** (8); **der Geldschein, -e** (12)

notebook **das Heft, -e** (B)

nothing **nichts** (9)

notice: not to notice **verfehlen, verfehlt** (10)

noticeable: to be noticeable **auf·fallen (fällt ... auf), fiel ... auf, ist aufgefallen** (12)

novel **der Roman, -e** (5)

November **der November** (B)

now **jetzt** (3)

number **die Zahl, -en** (A); **die Nummer, -n** (1); house number **die Hausnummer, -n** (1); secret PIN (personal identification number) **die Geheimzahl, -en** (12); telephone number **die Telefonnummer, -n** (1)

nurse **der Krankenpfleger, - / die Krankenpflegerin, -nen** (5); to nurse **pflegen, gepflegt** (5)

nut **die Nuss, ¨-e** (8)

obligation **die Pflicht, -en** (3)

to occupy **besetzen, besetzt** (12)

o'clock: at six o'clock **um sechs (Uhr)** (1); until four o'clock **bis um vier Uhr** (4)

October **der Oktober** (B)

Octoberfest (*annual beer festival in Munich*) **das Oktoberfest, -e** (7)

of **von** (+ *dat.*) (A, 10); **aus** (+ *dat.*) (10); Of course! **Klar!** (2); **selbstverständlich** (10); of silk **aus Seide** (2)

office **das Büro, -s** (5); at the office **im Büro** (5); at the post office **auf der Post** (5); doctor's office **die Arztpraxis, Arztpraxen** (11); office hour **die Sprechstunde, -n** (3); post office **die Post** (5)

officer: police officer **der Polizist, -en** (*wk.*) **/ die Polizistin, -nen** (5)

often **oft** (A)

Oh boy! (*coll.*) **Mensch!** (2)

oil **das Öl** (5, 8); to check the oil **das Öl kontrollieren, kontrolliert** (5); oil color (*paint*) **die Ölfarbe, -n** (12)

old **alt** (A)

olive **die Olive, -n** (8)

omelet **das Omelett, -s** (8)

on **an** (+ *dat./acc.*) (2, 4); Come on over! **Komm mal vorbei!** (11); on account of **wegen** (+ *gen.*) (6); on foot **zu Fuß** (3); on Saturday **am Samstag** (2); on the contrary **sondern** (11); on the contrary! **doch!** (4); on the first of October **am ersten Oktober** (4); on the phone **am Telefon** (2); on the road **unterwegs** (9); on the second floor **im ersten Stock** (6); on time **pünktlich** (4); on what day? **an welchem Tag?** (4)

once **einmal** (4); once again **schon wieder** (3)

one **eins** (A); one another **einander** (3); one-way (*trip*) **einfach** (10); one-way street **die Einbahnstraße, -n** (7)

oneself **selbst** (2)

onion **die Zwiebel, -n** (8)

only **nur** (3)

to open **öffnen, geöffnet** (A); **auf·machen, aufgemacht** (3); **eröffnen, eröffnet** (9); to open a bank account **ein Konto eröffnen** (5)

open: out in the open (country) **in freier Natur** (10); open-face sandwich **das belegte Brot, die belegten Brote** (8)

opener: bottle opener **der Flaschenöffner, -** (8); can opener **der Dosenöffner, -** (8)

operation **der Betrieb, -e** (7)

opposite **gegenüber** (+ *dat.*) (6)

or (*coord. conj.*) **oder** (A, 11)

oral: to give an oral report **ein Referat halten** (4)

orange (*adj.*) **orange** (A)

orange (*n.*) **die Apfelsine, -n** (8); orange juice **der Orangensaft** (8)

orchestra conductor **der Dirigent, -en** (*wk.*) **/ die Dirigentin, -nen** (5)

order **die Reihenfolge, -n** (2); in order to **um ... zu** (12); to order (*food*) **bestellen, bestellt** (8)

organ **die Orgel, -n** (12)

organization **die Organisation, -en** (12)

origin **die Herkunft, ¨-e** (B)

other: each other **einander** (3); with each other **miteinander** (1, 3)

otherwise **sonst** (2, 5)

our **unser(e)** (2)

ourselves **uns** (1)

out (of) **aus** (+ *dat.*) (10); out in the open (country) **in freier Natur** (10); out this way **heraus** (10); to turn out **aus·gehen (geht ... aus), ging ... aus, ist ausgegangen** (7)

outside **draußen** (11)

oven **der Backofen, ¨** (5)

over (*prep.*) **über** (+ *dat./acc.*) (4); (*adv.*) **vorbei** (9); Come on over! **Komm mal vorbei!** (11); over that way **hinüber** (10); over the weekend **am Wochenende** (1), **übers Wochenende** (4); to run over **überfahren (überfährt), überfuhr, überfahren** (11)

overcoat **der Mantel, ¨** (A)

overnight: to stay overnight **übernachten, übernachtet** (6)

own **eigen** (3, 6)

to pack **packen, gepackt** (10); to pack up **ein·packen, eingepackt** (1)

package **das Paket, -e** (8)

page **die Seite, -n** (6)

pain **der Schmerz, -en** (11)

to paint **malen, gemalt** (12)

paintbrush **der Pinsel, -** (12)

pan **der Topf, ¨-e** (5); **die Pfanne, -n** (5)

pants **die Hose, -n** (A); sports pants **die Sporthose, -n** (2)

paper **das Papier, -e** (B); to give a paper / oral report **ein Referat halten** (4); paper towel **das Papiertuch, ¨-er** (5)

parasol: beach parasol **der Sonnenschirm, -e** (10)

parents **die Eltern** (*pl.*) (B); with your parents, at your parents' **bei deinen Eltern** (6)

to park **parken, geparkt** (7)

park **der Park, -s** (1); to go for a walk in the park **im Park spazieren gehen** (1); national park **der Nationalpark, -s** (2)

parlor: ice cream parlor **das Eiscafé, -s** (8)

parrot **der Papagei, -en** (10)

part **der Teil, -e** (7)

particularly **besonders** (3)

particulars **die Angaben** (*pl.*) (1)

partner: work with a partner **arbeiten Sie mit einem Partner** (A)

party **die Party, -s** (1, 2); **die Feier, -n** (9); (*political*) **die Partei, -en** (12); to go to a party **auf eine Party gehen** (1)

to pass across **über·gehen (geht ... über), ging ... über, ist übergegangen** (8)

passport **der Reisepass, ¨-e** (1); **der Pass, ¨-e** (7)

password **das Passwort, ¨-er** (7)

past (*adv.*) **vorbei** (9); (*prep.*) at twenty past five **um zwanzig nach fünf** (1)

pasture **die Wiese, -n** (7)

patient (*adj.*) **geduldig** (12)

patient (*n.*) **der Patient, -en** (*wk.*) **/ die Patientin, -nen** (5)

to pay **zahlen, gezahlt** (5); to pay (for) **bezahlen, bezahlt** (4); to pay attention **auf·passen, aufgepasst** (3); to pay attention to **achten auf** (+ *acc.*), **geachtet** (11); to pay off **ab·zahlen, abgezahlt** (12)

payment: means of payment **das Zahlungsmittel, -** (12)

pea **die Erbse, -n** (8)

peace **der Frieden, -** (12)

peach **der Pfirsich, -e** (8)

pear **die Birne, -n** (8)

pedestrian **der Fußgänger, -** (7); pedestrian mall **die Fußgängerzone, -n** (10)

pen **der Stift, -e** (A, B); ballpoint pen **der Kugelschreiber, -** (4)

pencil **der Bleistift, -e** (A, B)

peninsula **die Halbinsel, -n** (7)

people **die Leute** (*pl.*) (7); **das Volk, ¨-er** (12)

pepper (black) **der Pfeffer** (8)

per **pro** (2); per cent **das Prozent, -e** (4)

percussion **das Schlagzeug, -e** (12)

perfect **vollkommen** (12)

perfume **das Parfüm, -e** (5)

perhaps **vielleicht** (2)

to permit **erlauben, erlaubt** (7)

permit: residence permit **die Aufenthaltserlaubnis, -se** (12); work permit **die Arbeitserlaubnis, -se** (12)

permitted: to be permitted (to) **dürfen (darf), durfte, gedurft** (3)

person **die Person, -en** (1); **der Mensch, -en** (*wk.*) (2)

personal ID card **der Personalausweis, -e** (1); secret PIN (personal identification number) **die Geheimzahl, -en** (12)

pet **das Haustier, -e** (10)

pharmacist **der Apotheker, - / die Apothekerin, -nen** (11)

pharmacy **die Apotheke, -n** (6)

phone **das Telefon, -e** (1, 2); cellular phone **das Handy, -s** (2); on the phone **am Telefon** (2); phone number **die Telefonnummer, -n** (1); to talk on the phone **telefonieren, telefoniert** (4) photo **das Foto, -s** (1)

to photograph **fotografieren, fotografiert** (4)

physician **der Arzt, ¨-e / die Ärztin, -nen** (3, 5, 11)

physics **die Physik** (1)

piano **das Klavier, -e** (2)

to pick **pflücken, gepflückt** (9); to pick (*s.o.*) up (from a place) **ab·holen, abgeholt** (1)

pickles **saure Gurken** (8)

picnic **das Picknick, -s** (4)

picture **das Bild, -er** (2); to hang the picture on the wall **das Bild an die Wand hängen** (3); What do your pictures show? **Was zeigen Ihre Bilder?** (A)

piece **das Stück, -e** (8)

piercing **das Piercing, -s** (2)

pig **das Schwein, -e** (9); guinea pig **das Meerschweinchen, -** (10)

pilot **der Pilot, -en** (*wk.*) **/ die Pilotin, -nen** (5)

PIN: secret PIN (personal identification number) **die Geheimzahl, -en** (12)

pink **rosa** (A)

piranha **der Piranha, -s** (10)

pizza **die Pizza, Pizzen** (2)

to place (*in an upright position*) **stellen, gestellt** (3, 5); (*in a sitting position*) **setzen, gesetzt** (7)

place **der Ort, -e** (1, 5); **der Platz, ¨-e** (3); **die Lage, -n** (10); marketplace **der Marktplatz, ¨-e** (6); to take place **statt·finden (findet ... statt), fand ... statt, stattgefunden** (5)

to plan **planen, geplant** (9); to plan (to) **wollen (will), wollte, gewollt** (3)

plan **der Plan, ¨-e** (3)

plant **die Pflanze, -n** (3, 6)

plate **der Teller, -** (8); license plate **das Nummernschild, -er** (7)

to play **spielen, gespielt** (1)

player: CD player **der CD-Spieler, -** (2); DVD player **der DVD-Spieler, -** (2, 3); MP3 player **der MP3-Spieler, -** (2, 5); record player **der Schallplattenspieler, -** (2)

playground **der Spielplatz, ¨-e** (9)

pleasant **angenehm** (6)

to please **gefallen** (+ *dat.*) **(gefällt), gefiel, gefallen** (6)

please **bitte** (A); Sign here, please. **Unterschreib bitte hier** (A); Yes please? **Bitte schön?** (7)

pleasure **das Vergnügen** (2); **die Freude, -n** (9); with pleasure **gern** (1)

pliers **die Zange, -n** (8)

plum **die Pflaume, -n** (8)

pocket **die Tasche, -n** (1)

poem **das Gedicht, -e** (3)

point **der Punkt, -e** (A)

to poison **vergiften, vergiftet** (9)

poisonous **giftig** (9)

Poland **(das) Polen** (B)

police: police officer **der Polizist, -en** (*wk.*) **/ die Polizistin, -nen** (5); police station **die Polizei** (5); at the police station **auf der Polizei** (5)

political(ly) **politisch** (12)

politics **die Politik** (12)

pool: swimming pool **das Schwimmbad, ¨-er** (1, 5); at the swimming pool **im Schwimmbad** (5); to go to the swimming pool **ins Schwimmbad fahren** (1)

poor **arm** (9)

popular **beliebt** (3)

population **die Bevölkerung, -en** (12)

pork **das Schweinefleisch** (8); ground pork (or beef) **das Hackfleisch** (8)

port **der Hafen, ¨** (10)

Portugal **(das) Portugal** (B)

Portuguese (*language*) **(das) Portugiesisch** (B)

position **die Lage, -n** (10); to be in a seated position **sitzen (sitzt), saß, gesessen** (2)

possessions **der Besitz** (2)

possibility **die Möglichkeit, -en** (5)

possible: as . . . as possible **möglichst** (+ *adv.*) (6); everything possible **alles Mögliche** (2)

post office **die Post** (5); at the post office **auf der Post** (5)

postcard **die Postkarte, -n** (2)

poster **das Poster, -** (6)

pot **der Topf, ¨-e** (5)

potato **die Kartoffel, -n** (8); boiled potatoes **die Salzkartoffeln** (*pl.*) (8)

potholder **der Topflappen, -** (5)

potter's wheel **die Töpferscheibe, -n** (12)

to pour **gießen (gießt), goss, gegossen** (8)

power **der Strom** (8)

practical (career) training **praktische Ausbildung** (5)

practical(ly) **praktisch** (5)

to practice **aus·üben, ausgeübt** (12)

preparation **die Zubereitung, -en** (8)

to prepare **vor·bereiten, vorbereitet** (4); (*food*) **zu·bereiten, zubereitet** (8)

prescription **das Rezept, -e** (11)

to present **vor·stellen, vorgestellt** (6)

present **das Geschenk, -e** (2); Christmas present **das Weihnachtsgeschenk, -e** (5)

president **der Präsident, -en** (*wk.*) **/ die Präsidentin, -nen** (5)

prestige **das Prestige** (5)

pretty **hübsch** (A, 2); **schön** (B); pretty big **ziemlich groß** (2)

price **der Preis, -e** (7, 12)

priest **der Priester, - / die Priesterin, -nen** (5)

prince **der Prinz, -en** (*wk.*) (9)

princess **die Prinzessin, -nen** (9)

principal **der Direktor, -en / die Direktorin, -nen** (9)

prison **das Gefängnis, -se** (6)

prize **der Preis, -e** (4)

probably **wahrscheinlich** (1); **wohl** (12)

profession **der Beruf, -e** (1, 5); What's your profession? **Was sind Sie von Beruf?** (1)

professor **der Professor, -en / die Professorin, -nen** (A, B)

prohibited **verboten** (9)

projector: data projector **der Beamer, -** (B)

to promise **versprechen (verspricht), versprach, versprochen** (7)

psychiatrist **der Psychiater, - / die Psychiaterin, -nen** (11)

pub: student pub **die Studentenkneipe, -n** (1)

public (*adj.*) **öffentlich** (7)

public (*n.*) **die Öffentlichkeit, -en** (12)

to pull **ziehen (zieht), zog, gezogen** (8)

pullover **der Pullover, - (der Pulli, -s)** (2)

punctual **pünktlich** (4)

pupil **der Schüler, - / die Schülerin, -nen** (1)

purple **lila** (A)

purse **die Tasche, -n** (1)

to put (*in a sitting position*) **setzen, gesetzt** (7); (*in an upright position*) **stellen, gestellt** (3, 5); to put (into) **geben (in** + *acc.*) **(gibt), gab, gegeben** (8); to put away **weg·stellen, weggestellt** (5); to put lotion on **sich ein·cremen, eingecremt** (11); to put makeup on **sich schminken, geschminkt** (11); to put on (*clothes*) **an·ziehen (zieht ... an), zog ... an, angezogen** (3)

puzzle **das Rätsel, -** (9); to solve a puzzle **ein Rätsel lösen** (9)

to quarrel **streiten (streitet), stritt, gestritten** (9)

quarter: at a quarter to four **um Viertel vor vier** (1)

queen **die Königin, -nen** (9)

question **die Frage, -n** (A); to ask a question **eine Frage stellen** (5)

quick **schnell** (7)

quiet(ly) **ruhig** (B); **leise** (9)

quite **ganz** (2); quite a bit **ganz schön viel** (3)

rabies **die Tollwut** (10)

radio **das Radio, -s** (2); car radio **das Autoradio, -s** (7)

rag (*for cleaning*) **der Putzlappen, -** (6)

railroad **die Bahn, -en** (7)

rain **der Regen** (7); to rain **regnen, geregnet** (B)

rainy: in rainy weather **bei Regen** (7)

range: mountain range **das Gebirge, -** (7)

rare(ly) **selten** (8)

rat **die Ratte, -n** (10)

rather **ziemlich** (2); **lieber** (2); **eher** (12); but (rather / on the contrary) **sondern** (A); I'd rather go . . . **Ich gehe lieber ...** (2)

to reach **erreichen, erreicht** (5, 12)

to read **lesen (liest), las, gelesen** (A, 1); to read aloud **vor·lesen (liest ... vor), las ... vor, vorgelesen** (9); to read the newspaper **Zeitung lesen** (1) ready **fertig** (3)

real(ly) **echt** (2); really **wirklich** (B)

receipt **die Quittung, -en** (8)

to receive **bekommen (bekommt), bekam, bekommen** (3); **erhalten (erhält), erhielt, erhalten** (5, 12)

recently **neulich** (9)

reception (desk) **die Rezeption, -en** (10)

recess **die Pause, -n** (1)

record player **der Schallplattenspieler, -** (2)

recorder (*type of flute*) **die Blockflöte, -n** (12)

recreation room **der Aufenthaltsraum, ¨-e** (10)

to recuperate **sich erholen, erholt** (11)

red **rot** (A)

refrigerator **der Kühlschrank, ¨-e** (5)

regularly **regelmäßig** (11)

to reign **herrschen, geherrscht** (8)

relatives **die Verwandten** (*pl.*) (2)

to relax (*coll.*) **chillen, gechillt** (1)

religion **die Religion** (1)

religious **religiös** (B)

to remain **bleiben (bleibt), blieb, ist geblieben** (3)

to remember **sich erinnern (an** + *acc.*)**, erinnert** (9)

remembrance **die Erinnerung, -en** (4)

to remove **ab·nehmen (nimmt ... ab), nahm ... ab, abgenommen** (11)

rent **die Miete, -n** (6); to rent **mieten, gemietet** (6); to rent out **vermieten, vermietet** (6)

renter **der Mieter, - / die Mieterin, -nen** (6)

to repair **reparieren, repariert** (1)

repair shop **die Werkstatt, ¨-en** (5)

to repeat **wiederholen, wiederholt** (6)

report **das Referat, -e** (3); to give a paper / oral report **ein Referat halten** (4)

reporter **der Reporter, - / die Reporterin, -nen** (4)

representative **der/die Abgeordnete, -n (ein Abgeordneter)** (12)

requirement **die Pflicht, -en** (3)

to rescue **erlösen, erlöst** (9)

to reserve **reservieren, reserviert** (7)

residence **der Wohnort, -e** (1); residence permit **die Aufenthaltserlaubnis, -se** (12)

to rest **sich aus·ruhen, ausgeruht** (11)

restaurant **das Restaurant, -s** (2); **die Gaststätte, -n** (5); at the restaurant **in der Gaststätte** (5)

reunion: class reunion **das Klassentreffen, -** (9)

rice **der Reis** (8)

riddle **das Rätsel, -** (9); to solve a riddle **ein Rätsel lösen** (9)

to ride **fahren (fährt), fuhr, ist/hat gefahren** (2); (*on horseback*) **reiten (reitet), ritt, ist geritten** (1); to ride a bicycle **Rad fahren (fährt ... Rad), fuhr ... Rad, ist Rad gefahren** (7); to ride a motorcycle **Motorrad fahren** (1); to ride off **los·fahren (fährt ... los), fuhr ... los, ist losgefahren** (9)

right (*adj.*) **richtig** (2); (*adv.*) **rechts** (10); to be right (*of a person*) **recht haben (hat ... recht), hatte ... recht, recht gehabt** (2); to be right, correct **stimmen, gestimmt** (8); right (away) **gleich** (6); right across the way **gleich gegenüber** (6); right around the corner **gleich um die Ecke** (6); that's right **das stimmt so** (8); to the right **rechts** (7)

to ring **klingeln, geklingelt** (2)

ring **der Ring, -e** (2)

to rinse **spülen, gespült** (4)

river **der Fluss, ¨-e** (7)

road **die Straße, -n** (6); on the road **unterwegs** (9)

roast **der Braten, -** (8)

roasted **gebraten** (8)

rocket **die Rakete, -n** (7)

role **die Rolle, -n** (4)

roll **das Brötchen, -** (8)

Romania **(das) Rumänien** (B)

roof **das Dach, ¨-er** (6)

room **das Zimmer, -** (1, 2); dining room **das Esszimmer, -** (6); double room **das Doppelzimmer, -** (5, 10); living room **das Wohnzimmer, -** (6); recreation room **der Aufenthaltsraum, ¨-e** (10); single room **das Einzelzimmer, -** (10); TV room **das Fernsehzimmer, -** (10)

roommate **der Mitbewohner, - / die Mitbewohnerin, -nen** (2)

roundabout: traffic roundabout **der Kreisverkehr** (10)

round-trip **hin und zurück** (5, 10); **die Hin- und Rückfahrt** (7)

routine: daily routine **der Tagesablauf, ¨-e** (1); **der Alltag** (4)

row: in a row **hintereinander** (3); row house **das Reihenhaus, ¨-er** (6)

rump steak **das Rumpsteak, -s** (8)

to run **laufen (läuft), lief, ist gelaufen** (A, 2); **rennen, ist gerannt** (7); to run in the woods **im Wald laufen** (2); to run over **überfahren (überfährt), überfuhr, überfahren** (11)

rushed **eilig** (8)

Russia **(das) Russland** (B)

Russian (*adj.*) **russisch** (8); (*language*) **(das) Russisch** (B)

sad **traurig** (B)

to sail **segeln, gesegelt** (1)

salad **der Salat, -e** (8); salad (mixing) bowl **die Salatschüssel, -n** (5); salad dressing **die Soße, -n** (8)

salesperson **der Verkäufer, - / die Verkäuferin, -nen** (5)

salon: tanning salon **das Solarium, Solarien** (11)

salt **das Salz** (8)

salted **gesalzen** (8)

salty **salzig** (7)

same **egal** (6); **gleich** (12)

sand **der Sand** (7)

sandal **die Sandale, -n** (2)

sandcastle **die Sandburg, -en** (4)

sandwich: open-face sandwich **das belegte Brot, die belegten Brote** (8)

Saturday **der Samstag** (1)

Saturday: on Saturday **am Samstag** (2)

sauce **die Soße, -n** (8)

sauna **die Sauna, -s** (11)

sausage **die Wurst, ¨-e** (8)

to save **sparen, gespart** (5)

savings account **das Sparkonto, Sparkonten** (12)

to say **sagen, gesagt** (A, 5); to say hello to **grüßen, gegrüßt** (10)

scale: kitchen scale **die Küchenwaage, -n** (5)

scarf **der Schal, -s** (2)

schedule **der Stundenplan, ¨-e** (1)

scholarship **das Stipendium, Stipendien** (1)

school **die Schule, -n** (1, 3, 5); at school **in der Schule** (5); elementary school **die Grundschule, -n** (9); high school, college preparatory school **das Gymnasium, Gymnasien** (6); high school graduation exam **das Abitur** (4); school principal **der Direktor, -en / die Direktorin, -nen** (9)

schooling **die Schulbildung** (5)

science **die Wissenschaft, -en** (9); computer science **die Informatik** (1)

scientist **der Wissenschaftler, - / die Wissenschaftlerin, -nen** (12)

scissors **die Schere, -n** (8)

to scold **schimpfen, geschimpft** (9)

scorpion **der Skorpion, -e** (10)

to scream **schreien (schreit), schrie, geschrien** (3)

screen: flat-screen (monitor) **der Flachbildschirm, -e** (2)

sculptor **der Bildhauer, - / die Bildhauerin, -nen** (12)

sea **das Meer, -e** (1, 7); to swim in the sea **im Meer schwimmen** (1); to the sea **ans Meer** (2)

seagull **die Möwe, -n** (10)

to season **würzen, gewürzt** (8)

season **die Jahreszeit, -en** (B)

seasoning **das Gewürz, -e** (8)

seat **der Sitz, -e** (7); Is this seat available? **Ist hier noch frei?** (8); seat belt **der Sicherheitsgurt, -e** (7)

seated: to be in a seated position **sitzen (sitzt), saß, gesessen** (2)

second (*adj.*) **zweit-** (4); on the second floor **im ersten Stock** (6)

second (*n.*) **die Sekunde, -n** (1)

secret (*adj.*) **heimlich** (9); secret PIN (personal identification number) **die Geheimzahl, -en** (12)

secret (*n.*) **das Geheimnis, -se** (5)

secretary **der Sekretär, -e / die Sekretärin, -nen** (5)

sector **die Branche, -n** (12); **der Bereich, -e** (12)
to see **sehen (sieht), sah, gesehen** (2); haven't seen (you / each other) for a long time **lange nicht gesehen** (1); see you soon **bis bald** (A)
seldom **selten** (8)
to sell **verkaufen, verkauft** (2, 5)
semester **das Semester, -** (1)
to send **schicken, geschickt** (2)
sentence **der Satz, ¨-e** (3)
to separate **trennen, getrennt** (7)
separately, separate checks **getrennt** (5)
September **der September** (B)
sequence **die Reihenfolge, -n** (2, 4)
service **die Bedienung** (8)
to set **decken, gedeckt** (5); **setzen, gesetzt** (7); to set the table **den Tisch decken** (5)
seven **sieben** (A)
seventeen **siebzehn** (A)
seventh **siebt-** (4)
seventy **siebzig** (A)
several: for several days **seit mehreren Tagen** (11); several times **mehrmals** (5)
severe **stark** (11)
shade, shadow **der Schatten, -** (9)
to shake hands **die Hand schütteln, geschüttelt** (A)
share **die Aktie, -n** (12)
shared housing **die WG, -s (die Wohngemeinschaft, -en)** (6)
shark **der Hai, -e** (10)
to shave **sich rasieren, rasiert** (11)
shirt **das Hemd, -en** (A); T-shirt **das T-Shirt, -s** (2)
shock **der Schock, -s** (11)
shoe **der Schuh, -e** (A); athletic shoe **der Sportschuh, -e** (A); hiking shoe **der Wanderschuh, -e** (2); shoe store **das Schuhgeschäft, -e** (6)
to shop (for) **ein·kaufen, eingekauft** (1)
shop: butcher shop **die Metzgerei, -en** (6); copy shop **der Kopierladen, ¨-** (10); repair shop **die Werkstatt, ¨-en** (5)
shopping: to go shopping **ein·kaufen gehen (geht ... einkaufen), ging ... einkaufen, ist einkaufen gegangen** (1, 5)
shore **der Strand, ¨-e** (7)
short **kurz** (B); **klein** (B); with the short hair **mit dem kurzen Haar** (A)
shot **die Spritze, -n** (11)
shoulder **die Schulter, -n** (B)
to shout **rufen (ruft), rief, gerufen** (7)
to show: What do your pictures show? **Was zeigen Ihre Bilder?** (A)
shower **die Dusche, -n** (5); to shower **(sich) duschen, geduscht** (1, 11)
shrill **grell** (2)
shrimp **die Krabbe, -n** (8)
to shut **schließen (schließt), schloss, geschlossen** (A); tied shut **zugebunden** (8)
shy **schüchtern** (B)
siblings **die Geschwister** (*pl.*) (B)
sick **krank** (3)
sickness **die Krankheit, -en** (11)
side **die Seite, -n** (6)
to sightsee **besichtigen, besichtigt** (7)
to sign **unterschreiben (unterschreibt), unterschrieb, unterschrieben** (1); Sign here, please. **Unterschreib bitte hier.** (A)
sign **das Schild, -er** (7); traffic sign **das Verkehrsschild, -er** (7)
signature **die Unterschrift, -en** (1)
silk **die Seide, -n** (2); of/from silk **aus Seide** (2)
silverware **das Besteck** (5)
simple, simply **einfach** (2)
since **seit** (+ *dat.*) (4, 11)
to sing **singen (singt), sang, gesungen** (1); We like to sing. **Wir singen gern.** (1)
single-family home **das Einfamilienhaus, ¨-er** (6); single room **das Einzelzimmer, -** (10)
sink **das Spülbecken, -** (5)
sister **die Schwester, -n** (B)
to sit **sitzen (sitzt), saß, gesessen** (2, 4); to sit down **sich setzen, gesetzt** (A, 11)
situation **die Lage, -n** (12); conversational situation **die Sprechsituation, -en** (A)
six **sechs** (A)
sixteen **sechzehn** (A)
sixth **sechst-** (4)
sixty **sechzig** (A)
skateboard **das Skateboard, -s** (3); to skateboard **Skateboard fahren (fährt ... Skateboard), fuhr ... Skateboard, ist Skateboard gefahren** (3)
ski **der Ski, -er** (3); to ski **Ski fahren (fährt ... Ski), fuhr ... Ski, ist Ski gefahren** (3); ski lodge **die Skihütte, -n** (6)
skills **die Kenntnisse** (*pl.*) (5)
skin **die Haut, ¨-e** (11)
skirt **der Rock, ¨-e** (A)
sleep **der Schlaf** (9); to sleep **schlafen (schläft), schlief, geschlafen** (2)
sleeping bag **der Schlafsack, ¨-e** (2)
slender **schlank** (B)
slice **das Stück, -e** (8)
to slide **rutschen, ist gerutscht** (9)
slim **schlank** (B)
to slip **rutschen, ist gerutscht** (9); **aus·rutschen, ist ausgerutscht** (11)
Slovakia **die Slowakei** (B)
Slovenia **(das) Slowenien** (B)
small **klein** (B); **eng** (12)
smartphone **das Smartphone, -s** (2)
to smell **riechen (riecht), roch, gerochen** (11)
to smoke **rauchen, geraucht** (3)
snake **die Schlange, -n** (10)
sniffles **der Schnupfen, -** (11)
snow **der Schnee** (9); to snow **schneien, geschneit** (B)
snowboard **das Snowboard, -s** (1)
so **so** (A); **also** (4)
so long **bis bald** (A)
so that (*subord. conj.*) **damit** (11)
soap **die Seife, -n** (11)
soccer (ball) **der Fußball, ¨-e** (A, 1)
social studies **die Sozialkunde** (1)
societal **gesellschaftlich** (12)
society **die Gesellschaft, -en** (12)
sociology **die Soziologie** (1)
sock **die Socke, -n** (2)
sofa **das Sofa, -s** (6)
sojourn **der Aufenthalt, -e** (5)
sold out **ausverkauft** (5)
solution **die Lösung, -en** (1)
to solve a puzzle/riddle **ein Rätsel lösen, gelöst** (9)
somebody, someone **jemand** (3)
something **etwas** (2, 4, 5); Do you have something for it (*illness*)? **Haben Sie etwas dagegen?** (11); something interesting/new **etwas Interessantes/Neues** (4)
sometimes **manchmal** (B)
son **der Sohn, ¨-e** (B)
songbook **das Songbuch, ¨-er** (2)
soon **bald** (9); see you soon **bis bald** (A); soon thereafter **bald darauf** (9)
sore muscles **der Muskelkater, -** (11); sore throat **die Halsschmerzen** (*pl.*) (11)
sorrow **die Trauer** (2)
sorry: to be sorry **leid·tun** (+ *dat.*) **(tut ... leid), tat ... leid, leidgetan** (5); I'm sorry. **Tut mir leid.** (4, 5)
soul **die Seele, -n** (12)
to sound (like) **klingen (wie) (klingt), klang, geklungen** (11); That sounds great. **Das hört sich toll an.** (4)
sound **das Geräusch, -e** (9)
soup **die Suppe, -n** (8)
sour **sauer** (8)
south (of) **südlich (von** + *dat.*) (7)
South America **(das) Südamerika** (B)
souvenir **das Souvenir, -s** (7)
Spain **(das) Spanien** (B)
Spanish (*language*) **(das) Spanisch** (B)
to speak **sprechen (spricht), sprach, gesprochen** (B)
specialized training **die Ausbildung** (5)
speech **die Rede, -n** (12)
to spell **schreiben (schreibt), schrieb, geschrieben** (A); How do you spell that? **Wie schreibt man das?** (A)
spell: to cast a spell on **verwünschen, verwünscht** (9)
to spend (*money*) **aus·geben (gibt ... aus), gab ... aus, ausgegeben** (12); (*time*) **verbringen (verbringt), verbrachte, verbracht** (3)

spice **das Gewürz, -e** (8)

spirit **der Geist** (8)

spite: in spite of **trotz** (+ *gen.*) (12)

spoon **der Löffel, -** (8)

sports **der Sport** (1); to do sports **Sport treiben (treibt ... Sport), trieb ... Sport, Sport getrieben** (2); sports jacket **das Sakko, -s** (A); sports pants **die Sporthose, -n** (2)

spring **der Frühling, -e** (B); in the spring **im Frühling** (B); spring cleaning **der Frühjahrsputz** (6)

to sprinkle **bestreuen, bestreut** (8)

sprout: Brussels sprouts **der Rosenkohl** (8)

square: market square **der Marktplatz, ¨-e** (6); square meter (m^2) **der Quadratmeter (qm), -** (6)

stairway **die Treppe, -n** (6)

stamp **die Briefmarke, -n** (5)

to stand **stehen (steht), stand, gestanden** (2, 6); to stand up **auf·stehen (steht ... auf), stand ... auf, ist aufgestanden** (A); to stand up for **ein·treten für** (+ *acc.*) **(tritt ... ein), trat ... ein, ist eingetreten** (12)

state **der Staat, -en** (10, 12); (*adj.*) **staatlich** (12)

station: gas station **die Tankstelle, -n** (5); at the gas station **an der Tankstelle** (5); police station **die Polizei** (5); at the police station **auf der Polizei** (5); train station **der Bahnhof, ¨-e** (4, 5); at the train station **auf dem Bahnhof** (5)

stationery store **das Schreibwarengeschäft, -e** (6)

to stay **bleiben (bleibt), blieb, ist geblieben** (3); to stay overnight **übernachten, übernachtet** (6)

stay **der Aufenthalt, -e** (5)

steak: rump steak **das Rumpsteak, -s** (8)

to steal **stehlen (stiehlt), stahl, gestohlen** (9)

steering wheel **das Lenkrad, ¨-er** (7)

stepfather **der Stiefvater, ¨** (9)

stepmother **die Stiefmutter, ¨** (9)

still **noch** (B)

to sting **stechen (sticht), stach, gestochen** (10)

stock **die Aktie, -n** (12); stock exchange **die Börse, -n** (12); stock market crash **der Börsenkrach, ¨-e** (12)

stomach **der Bauch, ¨-e** (B); **der Magen, ¨** (11)

stomachache **die Magenschmerzen** (*pl.*) (11)

stone **der Stein, -e** (12)

to stop **an·halten (hält ... an), hielt ... an, angehalten** (7); **halten (hält), hielt, gehalten** (7); to stop (*doing s.th.*) **auf·hören (mit** + *dat.*)**, aufgehört** (1)

stop **die Haltestelle, -n** (10); bus stop **die Bushaltestelle, -n** (10)

store **das Geschäft, -e** (2); department store **das Kaufhaus, ¨-er** (5); at the department store **im Kaufhaus** (5); shoe store **das Schuhgeschäft, -e** (6); stationery store **das Schreibwarengeschäft, -e** (6)

story **der Stock, Stockwerke** (6)

stove **der Herd, -e** (5)

to stow **verstauen, verstaut** (7)

straight ahead **geradeaus** (10)

to straighten **gerade stellen, gerade gestellt** (3)

strange **komisch** (10)

strategy **die Strategie, -n** (12)

strawberry **die Erdbeere, -n** (8)

street **die Straße, -n** (6); narrow street **die Gasse, -n** (10); one-way street **die Einbahnstraße, -n** (7)

streetcar **die Straßenbahn, -en** (7)

strict **streng** (9)

string **die Schnur, ¨-e** (8)

stuck: to get stuck **stecken bleiben (bleibt ... stecken), blieb ... stecken, ist stecken geblieben** (11)

student **der Student, -en** (*wk.*) **/ die Studentin, -nen** (A, B); fellow student **der Mitstudent, -en** (*wk.*) **/ die Mitstudentin, -nen** (A); student cafeteria **die Mensa, Mensen** (2); student pub **die Studentenkneipe, -n** (1)

to study (*at a university/college*) **studieren, studiert** (1); to study (*for a test*) **lernen, gelernt** (1)

study: course of studies, university studies **das Studium, Studien** (3); field of study **die Wissenschaft, -en** (9); social studies **die Sozialkunde** (1)

stupid **dumm** (6)

stylish **schick** (2)

subject **das Thema, Themen** (4); academic subject **das Fach, ¨-er** (1); favorite subject **das Lieblingsfach, ¨-er** (5)

subway **die U-Bahn, -en** (7)

to suck **lutschen, gelutscht** (11)

suddenly **plötzlich** (9)

sugar **der Zucker** (8)

suggestion **der Vorschlag, ¨-e** (5)

to suit **stehen** (+ *dat.*) **(steht), stand, gestanden** (6)

suit **der Anzug, ¨-e** (A)

suitcase **der Koffer, -** (3)

summer **der Sommer, -** (B); last summer **letzten Sommer** (4)

sun: to lie in the sun **in der Sonne liegen** (1)

sunbathing: to go sunbathing **sonnenbaden gehen** (10)

sunburn **der Sonnenbrand, ¨-e** (10)

Sunday **der Sonntag** (1)

sunglasses **die Sonnenbrille, -n** (1, 2)

sunny **sonnig** (B)

sunshade **der Sonnenschirm, -e** (10)

suntan lotion **die Sonnenmilch** (10)

supermarket **der Supermarkt, ¨-e** (5, 6); at the supermarket **im Supermarkt** (5)

supper **das Abendessen, -** (1)

to support **unterstützen, unterstützt** (12)

supposed: to be supposed to **sollen (soll), sollte, gesollt** (3)

sure **sicher** (1)

to surf the Internet **im Internet surfen, gesurft** (1)

surface **die Fläche, -n** (7)

surfboard **das Surfbrett, -er** (2)

surname **der Familienname, -n** (*wk.*) (A, 1)

survey **die Umfrage, -n** (4)

to swear **fluchen, geflucht** (11)

Sweden **(das) Schweden** (B)

Swedish (*language*) **(das) Schwedisch** (B)

to sweep **fegen, gefegt** (5)

sweet **süß** (2); totally sweet **voll süß** (2)

to swim **schwimmen (schwimmt), schwamm, ist geschwommen** (7); to go swimming **schwimmen gehen (geht ... schwimmen), ging ... schwimmen, ist schwimmen gegangen** (1); to swim in the sea **im Meer schwimmen** (1)

swimming pool **das Schwimmbad, ¨-er** (1, 5); at the swimming pool **im Schwimmbad** (5); to go to the swimming pool **ins Schwimmbad fahren** (1); swimming pool attendant **der Bademeister, - / die Bademeisterin, -nen** (5)

swim(ming) trunks **die Badehose, -n** (5)

Swiss (*person*) **der Schweizer, - / die Schweizerin, -nen** (B); Swiss franc **der Schweizer Franken, -** (8)

to switch on **an·machen, angemacht** (3)

Switzerland **die Schweiz** (B)

syrup: cough syrup **der Hustensaft, ¨-e** (11)

system **das System, -e** (12)

table **der Tisch, -e** (B); bedside table **der Nachttisch, -e** (6); to clear the table **den Tisch ab·räumen** (3); kitchen table **der Küchentisch, -e** (5); to set the table **den Tisch decken** (5); table tennis **das Tischtennis** (3)

tablet **die Tablette, -n** (11)

to take **nehmen (nimmt), nahm, genommen** (A); to take (*a course*) **belegen, belegt** (4); to take along **mit·nehmen (nimmt ... mit), nahm ... mit, mitgenommen** (3); to take a shower **(sich) duschen, geduscht** (11); to take away **weg·bringen (bringt ... weg), brachte ... weg, weggebracht** (5); to take blood **Blut ab·nehmen (nimmt ... ab), nahm ... ab, abgenommen** (11); take care (*infor.*) **mach's gut** (A); to take care of **sich kümmern um** (+ *acc.*)**, gekümmert** (12); to take effect **wirken, gewirkt** (11); to take off (*clothes*) **aus·ziehen (zieht ... aus), zog ... aus, ausgezogen** (3); to take on **übernehmen (übernimmt), übernahm, übernommen** (7); to take out (*a loan*) **auf·nehmen (nimmt ... auf), nahm ... auf, aufgenommen** (12); to take place **statt·finden (findet ... statt), fand ... statt, stattgefunden** (5)

tale: fairy tale **das Märchen, -** (9)

talent **das Talent, -e** (3)

to talk on the phone **telefonieren, telefoniert** (4)

tall **groß** (B)

tame **zahm** (10)

tank: fuel tank **der Tank, -s** (7)
tanning salon **das Solarium, Solarien** (11)
to taste **probieren, probiert** (3); to taste good to **schmecken** (+ *dat.*), **geschmeckt** (6)
tavern **die Kneipe, -n** (1, 4)
taxi **das Taxi, -s** (3, 7); taxi driver **der Taxifahrer, - / die Taxifahrerin, -nen** (5)
tea **der Tee** (4); tea kettle **der Teekessel, -** (8)
to teach **unterrichten, unterrichtet** (5)
teacher **der Lehrer, - / die Lehrerin, -nen** (A, 1)
teaching **die Lehre, -n** (8)
team **die Mannschaft, -en** (9)
teapot **die Teekanne, -n** (8)
to tear **zerreißen (zerreißt), zerriss, zerrissen** (9)
to tease **ärgern, geärgert** (1, 3)
teddy bear **der Teddybär, -en** (*wk.*) (A)
to telephone **telefonieren, telefoniert** (4)
telephone **das Telefon, -e** (1, 2); telephone card **die Telefonkarte, -n** (2); telephone number **die Telefonnummer, -n** (1)
to tell **erzählen, erzählt** (3, 5); **sagen, gesagt** (5); to tell jokes **Witze erzählen** (3)
teller: automatic teller machine (ATM) **der Geldautomat, -en** (*wk.*) (12)
ten **zehn** (A)
tender **zart** (8)
tennis **das Tennis** (1); table tennis **das Tischtennis** (3)
tent **das Zelt, -e** (2)
tenth **zehnt-** (4)
terrace **die Terrasse, -n** (6)
terrible **furchtbar** (4)
terrorism **der Terrorismus** (12)
test **die Prüfung, -en** (1)
tetanus **der Tetanus** (11)
to text **simsen, gesimst** (1)
text **der Text, -e** (12); to write a text message **eine SMS schreiben** (1)
thank you **danke** (A)
thanks: many thanks **vielen Dank** (10)
that (*dem. pron.*) **dieser, dies(es), diese** (4); over that way **hinüber** (10); That doesn't matter. **Das macht nichts.** (1); That is . . . **Das ist ...** (B); That's just it! **Das ist es ja!** (4); that's why **deshalb** (4); that way **hin** (10); up that way **hinauf** (10)
that (*subord. conj.*) **dass** (11); so that (*subord. conj.*) **damit** (11)
theater **das Theater, -** (6); movie theater **das Kino, -s** (1)
their **ihr(e)** (2)
theme **das Thema, Themen** (4); **das Motiv, -e** (12)
then **dann** (A); back then **damals** (9)
there **da** (2); **dort** (7); Is/Are there . . . ? **Gibt es ...?** (A, 6); there and back **hin und zurück** (10); There is/are . . . **Es gibt ...** (6)
thereafter: soon thereafter **bald darauf** (9)
therefore **deshalb** (4)
these **diese** (2, 4); These are . . . **Das sind ...** (B)
thing **das Ding, -e** (2); **die Sache, -n** (2)
to think (about) **nach·denken (über** + *acc.*) **(denkt ... nach), dachte ... nach, nachgedacht** (7); to think (of/about) **denken (an** + *acc.*) **(denkt), dachte, gedacht** (4); to think of **halten von** (+ *dat.*) **(hält), hielt, gehalten** (12)
third **dritt-** (4)
thirst **der Durst** (3)
thirsty: to be thirsty **Durst haben** (3)
thirteen **dreizehn** (A)
thirteenth **dreizehnt-** (4)
thirty **dreißig** (A)
this **dieser, dies(es), diese** (2, 4); in this way **herein** (10); out this way **heraus** (10); this evening **heute Abend** (2); This is . . . **Das ist ...** (B); this way **her** (10)
thorn **der Dorn, -en** (9)
thorough **ausführlich** (5)
those **diese** (4); Those are . . . **Das sind ...** (B)
three **drei** (A); three times **dreimal** (3)
throat **der Hals, ¨-e** (9); sore throat **die Halsschmerzen** (*pl.*) (11)
through **durch** (+ *acc.*) (7)
to throw **werfen (wirft), warf, geworfen** (3)
Thursday **der Donnerstag** (1)
thus **also** (4)
ticket **die Karte, -n** (2); **die Fahrkarte, -n** (4); concert ticket **die Konzertkarte, -n** (5); ticket booth **die Kasse, -n** (5), **der Schalter, -** (5); at the ticket booth **an der Kasse** (5), **am Schalter** (5); ticket window **der Fahrkartenschalter, -** (7)
tie **die Krawatte, -n** (A)
tied shut **zugebunden** (8)
tight **eng** (12)
tights **die Sporthose, -n** (2)
time **die Zeit, -en** (4); **das Mal, -e** (3, 4); At what time . . . ? **Um wie viel Uhr ...?** (1); for the first time **zum ersten Mal** (4); haven't seen (you / each other) for a long time **lange nicht gesehen** (1); leisure time **die Freizeit** (1); on time **pünktlich** (4); several times **mehrmals** (5); the last time **das letzte Mal** (4); the next time **das nächste Mal** (3); three times **dreimal** (3); What time is it? **Wie spät ist es?** (1), **Wie viel Uhr ist es?** (1)
tip **das Trinkgeld, -er** (8)
tire **der Reifen, -** (7); flat tire **die Reifenpanne, -n** (7)
tired **müde** (3)
to **an** (+ *acc.*) (2); **zu** (+ *dat.*) (2, 10); **nach** (+ *dat.*) (3, 10); at a quarter to four **um Viertel vor vier** (1); to the doctor **zum Arzt** (3); to the right **rechts** (7); to the sea **ans Meer** (2); to the university **zur Uni** (2); up to **bis zu** (+ *dat.*) (10)
today **heute** (B); What day is today? **Welcher Tag ist heute?** (1); What is today's date? **Welches Datum ist heute?** (4)
together **miteinander** (1); **zusammen** (2); **gemeinsam** (6, 11)
toilet **die Toilette, -n** (6)
tolerant **tolerant** (7)
tomato **die Tomate, -n** (8)
tomorrow **morgen** (2); the day after tomorrow **übermorgen** (9)
tongs **die Zange, -n** (8)
tongue **die Zunge, -n** (11)
to burn one's tongue **sich die Zunge verbrennen** (11)
too **auch** (A); **zu** (2); too bad! **schade!** (6)
tool **das Werkzeug, -e** (8)
tooth **der Zahn, ¨-e** (11)
to brush one's teeth **sich die Zähne putzen** (11)
toothache **die Zahnschmerzen** (*pl.*) (11)
topic **das Thema, Themen** (4)
total(ly) **total** (4); totally sweet **voll süß** (2)
tour: bicycle tour **die Radtour, -en** (9); tour of the city **die Stadtrundfahrt, -en** (7)
tourism **der Tourismus** (10)
towel: hand towel **das Handtuch, ¨-er** (8); paper towel **das Papiertuch, ¨-er** (5)
town **die Stadt, ¨-e** (6); town hall **das Rathaus, ¨-er** (1, 6); at the town hall **auf dem Rathaus** (1); town house **das Reihenhaus, ¨-er** (6)
track: train track **die Schiene, -n** (10); (set of) train tracks **das Gleis, -e** (10)
tradition **die Tradition, -en** (4, 12)
traffic **der Verkehr** (7, 11); traffic roundabout **der Kreisverkehr** (10); traffic sign **das Verkehrsschild, -er** (7)
train **der Zug, ¨-e** (7, 10); train car **der Waggon, -s** (7); train station **der Bahnhof, ¨-e** (4, 5); at the train station **auf dem Bahnhof** (5); train track **die Schiene, -n** (10); (set of) train tracks **das Gleis, -e** (10); train trip **die Bahnfahrt, -en** (7)
training: practical (career) training **praktische Ausbildung** (5); specialized training **die Ausbildung** (5)
to transfer **über·gehen (geht ... über), ging ... über, ist übergegangen** (8)
transfer (*of money*) **die Überweisung, -en** (12)
to translate **übersetzen, übersetzt** (9)
to transport **transportieren, transportiert** (7)
transportation: means of transportation **das Transportmittel, -** (7)
transverse flute **die Querflöte, -n** (12)
trash **der Müll** (6)
to travel **reisen, ist gereist** (1); to travel first class **erster Klasse fahren** (10)
travel: travel agency **das Reisebüro, -s** (6); travel experience **das Reiseerlebnis, -se** (7); travel guidebook **der Reiseführer, -** (5)
traveler **der/die Reisende, -n (ein Reisender)** (10)
treasure **der Schatz, ¨-e** (9)
tree **der Baum, ¨-e** (9)
trick **die List, -en** (9)

trip **die Reise, -n** (7); **die Fahrt, -en** (7, 10); to be on a trip **auf Reisen sein** (7); to go on a trip **verreisen, ist verreist** (3); round-trip **die Hin- und Rückfahrt** (7); train trip **die Bahnfahrt, -en** (7)

trouble **der Ärger** (9)

trout **die Forelle, -n** (8)

truck **der Lastwagen, -** (7)

true **wahr** (3); **treu** (9)

trumpet **die Trompete, -n** (12)

trunk **der Kofferraum, ¨-e** (7)

trunks: swim(ming) trunks **die Badehose, -n** (5)

to trust **trauen** (+ *dat.*), **getraut** (8)

to try **probieren, probiert** (3); **versuchen, versucht** (4)

T-shirt **das T-Shirt, -s** (2)

Tuesday **der Dienstag** (1)

Turkey **die Türkei** (B)

Turkish (*language*) **(das) Türkisch** (B); (*person*) **der Türke, -n** (*wk.*) **/ die Türkin, -nen** (12)

to turn **ab·biegen (biegt ... ab), bog ... ab, ist abgebogen** (10); to turn off **aus·machen, ausgemacht** (3); to turn on **an·machen, angemacht** (3); **ein·schalten, eingeschaltet** (10); to turn out **aus·gehen (geht ... aus), ging ... aus, ist ausgegangen** (7)

turtle **die Schildkröte, -n** (10)

tutoring **die Nachhilfe** (3)

TV room **das Fernsehzimmer, -** (10); TV set **der Fernseher, -** (2); to watch TV **fern·sehen (sieht ... fern), sah ... fern, ferngesehen** (1)

twelfth **zwölft-** (4)

twelve **zwölf** (A)

twentieth **zwanzigst-** (4)

twenty **zwanzig** (A)

twenty-one **einundzwanzig** (A)

twice **zweimal** (5)

two **zwei** (A)

to type **tippen, getippt** (3, 6)

type **die Art, -en** (2)

ugly **hässlich** (2)

umbrella **der Regenschirm, -e** (5)

uncle **der Onkel, -** (B)

unconsciousness **die Ohnmacht** (11)

under, underneath **unter** (+ *dat./acc.*) (5); under the window **unter dem Fenster** (5)

underpants **die Unterhose, -n** (2)

undershirt **das Unterhemd, -en** (2)

to understand **verstehen (versteht), verstand, verstanden** (4)

undressed: to get undressed **sich aus·ziehen (zieht ... aus), zog ... aus, ausgezogen** (11)

unemployed **arbeitslos** (12)

unfortunately **leider** (B)

university **die Universität, -en** (B, 1, 5); (*coll.*) **die Uni, -s** (B, 1); at the university **auf der Universität** (5); to be at the university **auf der Uni sein** (1); (to go) to the university **zur Uni (gehen)** (1, 2); university studies **das Studium, Studien** (1)

unmarried **ledig** (1)

until (*prep.*) **bis** (+ *acc.*) (2, 4, 11); (*subord. conj.*) **bis** (11); not until (four o'clock) **erst (um vier Uhr)** (4); until eight o'clock **bis acht Uhr** (2); until four in the morning **bis um vier Uhr früh** (4)

up: to give up **auf·geben (gibt ... auf), gab ... auf, aufgegeben** (1); to hang up **auf·hängen, aufgehängt** (2); to stand up for **ein·treten für** (+ *acc.*) **(tritt ... ein), trat ... ein, ist eingetreten** (12); up to **bis zu** (+ *dat.*) (10); up that way **hinauf** (10); to wake up **auf·wachen, ist aufgewacht** (2, 4)

upset: to get upset **sich auf·regen, aufgeregt** (11)

urgent(ly) **dringend** (2)

us (*acc./dat.*) **uns** (1)

USA **die USA** (*pl.*) (B)

to use **brauchen, gebraucht** (1); **benutzen, benutzt** (7)

use **der Gebrauch, ¨-e** (12)

useful **nützlich** (10)

usually **meistens** (8)

vacation **die Ferien** (*pl.*) (1); **der Urlaub, -e** (4, 5); vacation house **das Ferienhaus, ¨-er** (4)

to vaccinate against **impfen gegen** (+ *acc.*), **geimpft** (10)

to vacuum **Staub saugen, Staub gesaugt** (6)

vacuum cleaner **der Staubsauger, -** (6)

valley **das Tal, ¨-er** (7)

valuable **wertvoll** (2)

various **verschieden** (12)

vegetable **das Gemüse, -** (8)

vehicle **das Fahrzeug, -e** (7, 11)

very **sehr** (B)

vicinity **die Nähe** (6); in the vicinity **in der Nähe** (6)

video **das Video, -s** (9); video game **das Videospiel, -e** (5)

view **der Ausblick, -e** (6)

vinegar **der Essig** (8)

violence **die Gewalt** (12)

violin **die Geige, -n** (3)

to visit **besuchen, besucht** (1); **zu Besuch kommen** (3); **vorbei·kommen (kommt ... vorbei), kam ... vorbei, ist vorbeigekommen** (3); **besichtigen, besichtigt** (7)

visit **der Besuch, -e** (3)

vocabulary **der Wortschatz, ¨-e** (A)

voice **die Stimme, -n** (12)

volleyball **der Volleyball, ¨-e** (1)

to wait **warten, gewartet** (7)

waiter/waitress **der Kellner, - / die Kellnerin, -nen** (8); **die Bedienung** (8)

to wake up **auf·wachen, ist aufgewacht** (2, 4); to wake (*s.o.*) (up) **wecken, geweckt** (7)

to walk **gehen (geht), ging, ist gegangen** (A); to go for a walk **spazieren gehen (geht ... spazieren), ging ... spazieren, ist spazieren gegangen** (1); to go for a walk in the park **im Park spazieren gehen** (1); to keep on walking **weiter·gehen (geht ... weiter), ging ... weiter, ist weitergegangen** (10)

wall **die Wand, ¨-e** (B); to hang the picture on the wall **das Bild an die Wand hängen** (3)

waltz **der Walzer, -** (3)

to want (to) **wollen (will), wollte, gewollt** (3)

war **der Krieg, -e** (12)

wardrobe cabinet **der Schrank, ¨-e** (2); **der Kleiderschrank, ¨-e** (6)

warm **warm** (B)

to warn **warnen, gewarnt** (7)

to wash **waschen (wäscht), wusch, gewaschen** (2); **spülen, gespült** (4); to wash (oneself) **(sich) waschen (wäscht), wusch, gewaschen** (11); to wash the dishes **Geschirr spülen, gespült** (4)

washbasin **das Waschbecken, -** (6)

washing machine **die Waschmaschine, -n** (6)

wastebasket **der Papierkorb, ¨-e** (3)

to watch **an·sehen (sieht ... an), sah ... an, angesehen** (3); to watch out for **achten auf** (+ *acc.*), **geachtet** (11); to watch TV **fern·sehen (sieht ... fern), sah ... fern, ferngesehen** (1)

watch **die Armbanduhr, -en** (A)

to water **gießen (gießt), goss, gegossen** (3); to water the flowers **die Blumen gießen** (3)

water: mineral water **das Mineralwasser** (8)

wave **die Welle, -n** (10)

way: in this way **herein** (10); one-way street **die Einbahnstraße, -n** (7); out this way **heraus** (10); over that way **hinüber** (10); right across the way **gleich gegenüber** (6); that way **hin** (10); this way **her** (10); up that way **hinauf** (10)

to wear **tragen (trägt), trug, getragen** (A); Do you like to wear . . . ? **Trägst du gern ...?** (A)

weather **das Wetter** (B); in rainy weather **bei Regen** (7)

Wednesday **der Mittwoch** (1)

week **die Woche, -n** (1); during the week **in der Woche** (1); every week **jede Woche** (3); last week **letzte Woche** (4)

weekend **das Wochenende, -n** (1); last weekend **letztes Wochenende** (4); over the weekend **am Wochenende** (1), **übers Wochenende** (4)

weight: to lose weight **ab·nehmen (nimmt ... ab), nahm ... ab, abgenommen** (8, 11)

well (*adv.*): as well **auch** (A); to be well dressed **gut gekleidet sein** (2); to feel well **sich wohl fühlen, gefühlt** (11); it ended well **es ist gut ausgegangen** (7); That fits well. **Das passt gut.** (11)

well (*interj.*) **na** (3); **also** (4)

well (*n.*) **der Brunnen, -** (9)

west (of) **westlich (von** + *dat.*) (7)

wet **nass** (3)

what **was** (B); At what time . . . ? **Um wie viel Uhr ...?** (1); on what day? **an welchem Tag?** (4); What color is . . . ? **Welche Farbe hat ...?** (A); What day is today? **Welcher Tag ist heute?** (1); What do your pictures show? **Was zeigen Ihre Bilder?** (A); What is today's date? **Welches Datum ist heute?** (4); what kind of **was für** (+ *acc.*) (3); What's your name? **Wie heißen Sie?** (*for.*) **/ Wie heißt du?** (*infor.*) (A); What's your profession? **Was sind Sie von Beruf?** (1); What time is it? **Wie spät ist es?** (1), **Wie viel Uhr ist es?** (1)

wheel **das Rad, ¨-er** (7); potter's wheel **die Töpferscheibe, -n** (12); steering wheel **das Lenkrad, ¨-er** (7)

when **wann** (B, 1); (*subord. conj.*) **als** (5, 11); when(ever) (*subord. conj.*) **wenn** (2, 11); when I was eight years old **als ich acht Jahre alt war** (5); When were you born? **Wann sind Sie geboren?** (1)

whenever (*subord. conj.*) **wenn** (11)

where **wo** (B); from where **woher** (B); Where are you going? **Wo willst du denn hin?** (A); where to **wohin** (3)

whether (*subord. conj.*) **ob** (6, 11)

which **welch-** (B)

white **weiß** (A)

whiteboard **die Tafel, -n** (A, B)

who **wer** (A, B)

whole **ganz** (2)

whom (*acc.*) **wen** (4); (*dat.*) **wem** (5)

why **warum** (3); that's why **deshalb** (4)

wife **die Frau, -en** (B)

wild boar **das Wildschwein, -e** (10)

to win **gewinnen (gewinnt), gewann, gewonnen** (4); to win the lottery **in der Lotterie gewinnen** (5)

wind **der Wind, -e** (9)

window **das Fenster, -** (B); cashier window **die Kasse, -n** (12); ticket window **der Fahrkartenschalter, -** (7); under the window **unter dem Fenster** (5)

windowpane **die Scheibe, -n** (7); **die Fensterscheibe, -n** (9)

windowsill **die Fensterbank, ¨-e** (10)

windshield wiper **der Scheibenwischer, -** (7)

windsurfing: to go windsurfing **windsurfen gehen** (10)

windy **windig** (B)

wine **der Wein, -e** (7)

winter **der Winter, -** (B)

to wipe **wischen, gewischt** (7); to wipe clean **ab·wischen, abgewischt** (6)

wiper: windshield wiper **der Scheibenwischer, -** (7)

witch **die Hexe, -n** (9)

with **mit** (+ *dat.*) (A); **bei** (+ *dat.*) (2, 6, 10); Does it come with a . . . ? **Ist ein/eine ... dabei?** (6); to go with **passen** (+ *dat.*), **gepasst** (2); with each other **miteinander** (1, 3); with me **mit mir** (3); with pleasure **gern** (1); with the short/long hair **mit dem kurzen/langen Haar** (A); with your parents **bei deinen Eltern** (6)

witness **der Zeuge, -n** (*wk.*) **/ die Zeugin, -nen** (11)

wolf **der Wolf, ¨-e** (9)

woman **die Frau, -en** (A, B)

wonder **das Wunder, -** (4); no wonder **kein Wunder** (4)

wood **das Holz, ¨-er** (12)

woods **der Wald, ¨-er** (2, 7); to run in the woods **im Wald laufen** (2)

word **das Wort, ¨-er** (A)

to work **arbeiten, gearbeitet** (1); (*take effect*) **wirken, gewirkt** (11); work with a partner **arbeiten Sie mit einem Partner** (A)

work **die Arbeit, -en** (1); **das Werk, -e** (9); from work **von der Arbeit** (3); to go to work **zur Arbeit gehen** (1); work permit **die Arbeitserlaubnis, -se** (12)

workbook **das Arbeitsbuch, ¨-er** (3)

worker **der Arbeiter, - / die Arbeiterin, -nen** (5); construction worker **der Bauarbeiter, - / die Bauarbeiterin, -nen** (5)

workplace **der Betrieb, -e** (7)

world **die Welt, -en** (7)

would like (to) **möchte** (2, 3)

wound **die Wunde, -n** (11)

to write **schreiben (schreibt), schrieb, geschrieben** (A, 1); to write a text message **eine SMS schreiben** (1); to write down **auf·schreiben (schreibt ... auf), schrieb ... auf, aufgeschrieben** (11)

writer **der Schriftsteller, - / die Schriftstellerin, -nen** (5)

wrong **falsch** (2); to be wrong with (*a person*) **fehlen** (+ *dat.*), **gefehlt** (11); to get up on the wrong side of bed **mit dem linken Fuß auf·stehen (steht ... auf), stand ... auf, aufgestanden** (4)

to X-ray **röntgen, geröntgt** (11)

year **das Jahr, -e** (B); for two years **seit zwei Jahren** (4)

to yell **schreien (schreit), schrie, geschrien** (3)

yellow **gelb** (A)

yes (on the contrary)! **doch!** (4); Yes please? **Bitte schön?** (7)

yesterday **gestern** (4)

you (*infor. sg. acc.*) **dich** (2)

young **jung** (B)

your (*for.*) **Ihr(e)** (B, 2); (*infor. sg.*) **dein(e)** (B, 2); (*infor. pl.*) **euer, eure** (2)

youth **die Jugend** (9); youth hostel **die Jugendherberge, -n** (10)

zebra **das Zebra, -s** (10)

zero **null** (A)

Index

This index is divided into three subsections: Culture, Grammar, and Vocabulary. Reading, film, and music titles are included in the Culture section, as are artists' names.

Culture

Grammar

Vocabulary